multiply	by	to obtain
in/ft	1/0.012	mm/m
in^2	6.4516	cm^2
in^3	1/1728	ft^3
J	9.4778×10^{-4}	Btu
J	6.2415×10^{18}	eV
J	0.73756	ft-lbf
J	1.0	N·m
J/s	1.0	W
kg	2.2046	lbm
kg/m^3	0.06243	lbm/ft^3
kip	4.4480	kN
kip	1000.0	lbf
kip	4448.0	N
kip/ft	14.594	kN/m
kip/ft^2	47.880	kPa
kJ	0.94778	Btu
kJ	737.56	ft-lbf
kJ/kg	0.42992	Btu/lbm
kJ/kg·K	0.23885	Btu/lbm-°R
km	3280.8	ft
km	0.62138	mi
km/h	0.62138	mi/hr
kN	0.2248	kips
kN·m	0.73757	ft-kips
kN/m	0.06852	kips/ft
kPa	9.8692×10^{-3}	atm
kPa	0.14504	lbf/in^2
kPa	1000.0	Pa
kPa	0.02089	kips/ft^2
kPa	0.01	bar
ksi	6.8948×10^6	Pa
ksi	6894.8	kPa
kW	737.56	ft-lbf/sec
kW	44,250	ft-lbf/min
kW	1.3410	hp
kW	3413.0	Btu/hr
kW	0.9483	Btu/s
kW-hr	3413.0	Btu
kW-hr	3.60×10^6	J
L	1/102,790	ac-in
L	1000.0	cm^3
L	0.03531	ft^3
L	0.26417	gal
L	61.024	in^3
L	0.0010	m^3
L/s	2.1189	ft^3/min
L/s	15.850	gal/min
lbf	0.001	kips
lbf	4.4482	N
lbf/ft^2	0.01414	in Hg
lbf/ft^2	0.19234	in water
lbf/ft^2	0.00694	lbf/in^2
lbf/ft^2	47.880	Pa
lbf/ft^2	5.0×10^{-4}	tons/ft^2
lbf/in^2	0.06805	atm
lbf/in^2	144.0	lbf/ft^2
lbf/in^2	2.308	ft water
lbf/in^2	27.70	in water
lbf/in^2	2.0370	in Hg
lbf/in^2	6894.8	Pa
lbf/in^2	0.00050	tons/in^2
lbf/in^2	0.0720	tons/ft^2
lbm	7000.0	grains
lbm	453.59	g
lbm	0.45359	kg
lbm	4.5359×10^5	mg
lbm	5.0×10^{-4}	tons (mass)
lbm (of water)	0.12	gal (of water)
lbm/ac-ft-day	0.02296	lbm/1000 ft^3-day
lbm/ft^3	0.016018	g/cm^3
lbm/ft^3	16.018	kg/m^3
lbm/1000 ft^3-day	43.560	lbm/ac-ft-day
lbm/1000 ft^3-day	133.68	lbm/MG-day
lbm/MG	0.0070	grains/gal
lbm/MG	0.11983	mg/L
lbm/MG-day	0.00748	lbm/1000 ft^3-day
leagues	4428.0	m
m	1.0×10^{10}	angstroms

multiply		
m		
m		
m		
m		
m/s	196.85	ft/min
m^2	2.4711×10^{-4}	ac
m^2	10.764	ft^2
m^2	1/10,000	hectare
mi^2-in	53.3	ac-ft
mi^2-in/day	26.89	ft^3/s
m^3	8.1071×10^{-4}	ac-ft
m^3/m·d	80.5196	gal/day-ft
m^3/m^2·d	24.542	gal/day-ft^2
Meinzer unit	1.0	gal/day-ft^2
mg	2.2046×10^{-6}	lbm
mg/L	1.0	ppm
mg/L	0.05842	grains/gal
mg/L	8.3454	lbm/MG
MG	1.0×10^6	gal
MG/ac-day	22.968	gal/ft^2-day
MGD	1.5472	ft^3/sec
MGD	1×10^6	gal/day
MGD/ac (mgad)	22.957	gal/day-ft^2
mi	5280.0	ft
mi	80.0	chains
mi	1.6093	km
mi (statute)	0.86839	miles (nautical)
mi	320.0	rods
mi/hr	1.4667	ft/sec
mi^2	640.0	acres
micron	1.0×10^{-6}	m
micron	0.001	mm
mil (angular)	0.05625	degrees
mil (angular)	3.375	min
min (angular)	0.29630	mils
min (angular)	2.90888×10^{-4}	radians
min (time, mean solar)	60	s
mm	1/25.4	in
mm	1000.0	microns
mm/m	0.012	in/ft
MPa	1.0×10^6	Pa
N	0.22481	lbf
N	1.0×10^5	dynes
N·m	0.73756	ft-lbf
N·m	8.8511	in-lbf
N·m	1.0	J
N/m^2	1.0	Pa
oz	28.353	g
Pa	0.001	kPa
Pa	1.4504×10^{-7}	ksi
Pa	1.4504×10^{-4}	lbf/in^2
Pa	0.02089	lbf/ft^2
Pa	1.0×10^{-6}	MPa
Pa	1.0	N/m^2
ppm	0.05842	grains/gal
radian	180/π	degrees (angular)
radian	3437.7	min (angular)
rod	0.250	chain
rod	16.50	ft
rod	1/320	mi
s (time)	1/86,400	day (mean solar)
s (time)	1.1605×10^{-5}	day (sidereal)
s (time)	1/60	min
therm	1.0×10^5	Btu
ton (force)	2000.0	lbf
ton (mass)	2000.0	lbm
ton/ft^2	2000.0	lbf/ft^2
ton/ft^2	13.889	lbf/in^2
W	3.413	Btu/hr
W	0.73756	ft-lbf/sec
W	1.3410×10^{-3}	hp
W	1.0	J/s
yd	1/22	chain
yd	0.91440	m
yd^3	27	ft^3
yd^3	201.97	gal

Environmental Engineering Reference Manual

for the PE Exam

Third Edition

Michael R. Lindeburg, PE

The Power to Pass®
www.ppi2pass.com

Professional Publications, Inc. • Belmont, California

Benefit by Registering This Book with PPI

- Get book updates and corrections.
- Hear the latest exam news.
- Obtain exclusive exam tips and strategies.
- Receive special discounts.

Register your book at **ppi2pass.com/register**.

Report Errors and View Corrections for This Book

PPI is grateful to every reader who notifies us of a possible error. Your feedback allows us to improve the quality and accuracy of our products. You can report errata and view corrections at **ppi2pass.com/errata**.

ENVIRONMENTAL ENGINEERING REFERENCE MANUAL FOR THE PE EXAM

Third Edition

Current printing of this edition: 1

Printing History

edition number	printing number	update
2	5	Minor revisions.
2	6	Minor revisions.
3	1	New edition. Copyright update.

Printed in the United States of America.

PPI
1250 Fifth Avenue
Belmont, CA 94002
(650) 593-9119
ppi2pass.com

ISBN: 978-1-59126-475-0

Library of Congress Control Number: 2014951240

Topics

Support Material

Background and Support

Flow of Fluids

Water Treatment

Ventilation

Combustion

Solid Waste

Health, Safety, Welfare

Systems, Mgmt., Professional

Where do I find practice problems to test what I've learned in this Reference Manual?

The *Environmental Engineering Reference Manual* provides a knowledge base that will prepare you for the Environmental PE exam. But there's no better way to exercise your skills than to practice solving problems. To simplify your preparation, please consider *Practice Problems for the Environmental Engineering PE Exam: A Companion to the Environmental Engineering Reference Manual.* This publication provides you with more than 625 practice problems, each with a complete, step-by-step solution.

Practice Problems for the Environmental Engineering PE Exam may be obtained from PPI at **ppi2pass.com** or from your favorite retailer.

Table of Contents

Appendices
Table of Contents

Preface to the Third Edition

The label "environmentalist" has always encompassed people from all walks of life. All engineers have the ability to affect the environment, from the materials and processes we implicitly design into our creations, through the residuals left behind in their manufacture, to the resultant products of their use. So, if it doesn't already, the title "environmental engineer" should apply to all engineers.

For that reason, the breadth of *Environmental Engineering Reference Manual* is wide. The intended audience for this book includes not only those who already consider themselves environmental engineers, but also civil engineers needing a general, all-inclusive reference book.

I know from firsthand experience that the easiest type of engineering book to write is a collection of solved problems. Although I admit that I have authored my share of problem collections, this book is not one of them. It is a convenient collection of methods and data covering familiar but, perhaps, faintly remembered (or, impossible-to-remember) subjects.

Many people will use it initially for environmental PE exam preparation. But the broader intent of this book is to provide you with a resource for your daily work long after you have taken the exam. Therefore, it has sections that will be useful to practicing environmental engineers (and in some cases, to civil and mechanical engineers as well), and to prospective engineering licensees in those branches.

New editions of this book are published periodically to include new material, bring content in line with current practice, or improve consistency and readability. This third edition is being published for all of those reasons; it is also the first time the *Environmental Engineering Reference Manual* will be produced using PPI's integrated book development and maintenance system. That fact is of significance only to PPI. However, it is strategically too important to go unmentioned. And, I'll just leave it at that.

This third edition also includes revised material in the following chapters: Protection of Wetlands, Air Quality, Toxicology, Industrial Hygiene, Health and Safety, Biological Effects of Radiation, Shielding, and Emergency Management. All content in these chapters was reviewed and, in many cases, updated to reflect current environmental practices and standards. Additionally, I have used this new edition as an opportunity to make thousands of improvements, most too small to be noticed except through a side-by-side comparison. These improvements include updates to terminology, descriptions, and references; clarifications and rewordings of explanations; additional typical data; updated chapter nomenclature; improved consistency between chapters; and the inevitable correction of author's errata.

While writing this book, I also tried to adhere to a number of conventions, including the following.

- Each topic is presented in its own stand-alone section. In most cases, each chapter within a section builds on previous chapters, and each successive section builds on the information in previous sections.

- As much as possible, the chapters use common nomenclature.

- The nomenclature used in the various source documents has been adopted.

- The most current references available at the time of writing were used.

- Some environmental subjects are traditionally evaluated in customary U.S. (English, British, etc.) units, while others use a mix of metric systems. In this book, SI units, equations, data, and examples are presented in parallel with their customary U.S. counterparts.

- Answers are given to the example problems.

- The contents have been extensively indexed. Most entries have been entered in several different ways.

- The sources of figures and tables have been rigorously cited, even when the original sources are in the public domain.

- All appendices are located at the end of the book.

- Many of the appendices—including long listings of original regulations and huge reference sets of data—are abbreviated in this book. In such cases, the entire resource is available for downloading at **ppi2pass.com/ENVRM/originalsources.html**.

Safety and the environment are subject to ongoing and ever-changing legislation. Because of this, many of the tables in this book are given titles such as "Typical data," "Representative values," or "Sample values." These "wiggle" titles should alert you to the need for additional research when compliance is your intention. To assist you with the additional research, I have included many footnotes and web addresses pointing to the original research or official regulatory sources.

The choice of subjects and data to be included in this book, though greatly influenced by hundreds of engineers, was ultimately mine, as was the choice of degree and depth of coverage of these subjects. After all these years, I think I have a pretty good feel for the academic needs of a working engineer. But time marches on. You can help me keep this book relevant by doing what thousands of engineers have already done: Stay in touch. The next edition will be the result of your input.

Thanks!

Michael R. Lindeburg, PE

Acknowledgments

Every new edition is a lot of work, for someone. Sometimes, as when new content is needed, the work of bringing out a new edition falls to the author; sometimes the work falls to the publisher's editorial and production staff, as when existing content is reformatted and reorganized; and sometimes the work falls to one or more subject matter experts, as when specialized knowledge is required to integrate changes in standards, codes, and federal legislation. Sometimes, one player shoulders a disproportionate share of the load; sometimes one player gets off easier than another.

I know enough about environmental engineering to know how little I know. So, for the most critical chapters, I asked several environmental experts from industry and academia to fill in holes in my outline. Their contributions ranged from small sections to entire chapters. When their contributions were submitted to me, I edited every sentence and checked every calculation. However, the genius of the chapters remains theirs. This is how 9 of the 58 chapters of the first edition of this book were created. The following experts were invaluable in helping me prepare the first edition manuscript in an accurate and timely manner.

> Jeffrey S. Mueller, MS, PE, CHP: Biological Effects of Radiation, Shielding (see Chap. 48 and Chap. 49)
> R. Wane Schneiter, PhD, PE, DEE: Air Quality/Pollution (see Chap. 38) plus technical review of Protection of Wetlands (see Chap. 45)
> James R. Sheetz, PE, DEE: Risk Analysis, Emergency Response, Protection of the Wetlands, Toxicology, Industrial Hygiene, Health and Safety (see Chap. 11 [part], Chap. 28, Chap. 45, Chap. 46, Chap. 47, and Chap. 51)

Now, I'd like to introduce you to the "I couldn't have done it without you" crew, the talented team at PPI.

Editing and Production: Tom Bergstrom, David Chu, Nicole Evans, Hilary Flood, Kate Hayes, Tyler Hayes, Julia Lopez, Scott Marley, Ellen Nordman, Heather Turbeville, and Ian A. Walker

Management: Sarah Hubbard, director of product development and implementation; Cathy Schrott, production services manager; and Jenny Lindeburg King, Chelsea Logan, Magnolia Molcan, and Julia White, editorial project managers

This edition incorporates the comments, questions, suggestions, and errata submitted by many people who have used the previous editions for their own preparations. As an author, I am humbled to know that these individuals have read the previous editions in such detail as to notice typos, illogic, and other errata, and that they subsequently took the time to share their observations with me. Their suggestions have been incorporated into this edition, and their attention to detail will benefit you and all future readers. The following is a partial list (in alphabetical order) of some of those who have improved this book through their comments.

> Kevin Berry, Anna Buseman-Williams, Stephen Ertman, David Graiver, Samuel Haffey, Alia Johnson, Richard Mestan, Catherine Regan, Coralynn Revis, Il-Won Shin, Alison Skwarski, Evan Winchester, and Janine Yieh

This edition shares a common developmental heritage with its previous editions, and I have not forgotten those that submitted errata for them. For this edition, though, there isn't a single contributor that I intentionally excluded. Still, I could have slipped up and forgotten to mention you. I hope you'll let me know if you should have been credited, but were inadvertently left out. I'd appreciate the opportunity to list your name in the next printing of this edition.

Near the end of the acknowledgments, after mentioning a lot of people who contributed to the book and, therefore, could be blamed for a variety of types of errors, it is common for an author to say something like, "I take responsibility for all of the errors you find in this book." Or, "All of the mistakes are mine." This is certainly true, given the process of publishing, since the author sees and approves the final version before his/her book goes to the printer. You would think that after more than 35 years of writing, I would have figured out how to write something without making a mistake and how to proofread without missing those blunders that are so obvious to readers. However, such perfection continues to elude me. So, yes, the finger points straight at me.

All I can say instead is that I'll do my best to respond to any suggestions and errata that you report through PPI's website, **ppi2pass.com/errata**. I'd love to see your name in the acknowledgments for the next edition.

Thank you, everyone!

Michael R. Lindeburg, PE

Codes Used to Prepare This Book

The *Environmental Engineering Reference Manual* is based on the codes, standards, and regulations listed later in the section entitled "What the Well-Heeled Environmental Engineer Would Take to the PE Exam." The versions and editions (i.e., dates) that I used were the most current available. However, as with engineering practice itself, the PE examination is not always based on the currently available standards, as adoptions by state and local agencies often lag issurance by several years.

PPI lists on its website the dates of the codes, standards, and regulations on which NCEES has announced the exams are based. However, these postings do not contain any announcements specifically about the environmental exam. The conclusion you and I must reach from such an omission is that the exam is not sensitive to changes in standards, regulations, or announcements in the Federal Register.

Introduction

PART 1: HOW YOU CAN USE THIS BOOK

QUICKSTART

If you never read the material at the front of your books anyway, and if you're in a hurry to begin and only want to read one paragraph, here it is:

> Most chapters in this book are independent. Start with any one and look through it. Use the index extensively. Decide if you are going to work problems in that topic. If so, solve as many problems in that topic as time allows. Don't stop studying until the exam. Start right now! Quickly! Good luck.

However, if you want to begin a thorough review, you should probably try to find out everything there is to know about the PE exam. The rest of this Introduction is for you.

IF YOU ARE A PRACTICING ENGINEER

If you are a practicing engineer and have obtained this book as a general reference handbook, it will probably sit in your bookcase until you have a specific need.

However, if you are preparing for the PE examination in environmental engineering, the following suggestions may help.

- Find out the current edition of this book. You might be reading this book long after it was published. Newer editions mean that older editions are no longer appropriate for the current exam. **Newer editions mean that the codes, standards, and regulations on which the exam is based are not represented in the older edition, that the exam body of knowledge has changed, and/or the exam format and policies have changed.** New editions are published for a reason, and it's not reasonable for you to expect the older edition to serve your needs when it is out of date.

- Be reasonable in what you expect from this book. Much like any textbook, this book is a compilation of material designed to help you learn certain subjects—in this case, subjects on the exam. This book does not contain "everything" that you need to know to pass the exam. You will need to assemble a library of other references. This book is not a substitute for the experience, general knowledge, and judgment that you are expected to demonstrate on the exam. This book will help you learn subjects. It won't help you pass the exam if you go into the exam unprepared or unqualified.

- Become intimately familiar with this book. This means knowing the order of the chapters, the approximate locations of important figures and tables, what appendices are available, and so on.

- Use the subject title tabs along the side of each page.

- Use Table 1 of this Introduction to learn which subjects in this book are not specific exam subjects. Some chapters in this book are supportive and do not cover specific exam topics. However, these chapters provide background and support for the other chapters.

- Some engineers read every page in a chapter. Some merely skim through a chapter and its appendices. In either case, you must familiarize yourself with the subjects before starting to solve practice problems.

- Identify and obtain a set of 10–30 solved practice problems for each of the exam subjects. I have written an accompanying book, *Practice Problems for the Environmental Engineering PE Exam*, for this purpose. Other resources include the *Environmental Engineering Practice PE Exams* and books in the *Six-Minute* series all published by PPI. You may use problem sets from your old textbooks, college notes, or review course if they are more convenient. Regardless of the books you use, you should know that you will encounter two types of practice problems. Some problems look like examination problems. They are short and have multiple-choice answers. This type of problem is good for familiarizing yourself with the exam format. However, it is not very effective for exposing you to the integration of multiple concepts in problem solving, for familiarizing you with this book, and for making sure you have seen all of the "gotchas" that are possible in a subject. To address those requirements, you'll need some longer problems. *Practice Problems for the Environmental Engineering PE Exam* contains both types of problems.

- Most of the problems in *Practice Problems for the Environmental Engineering PE Exam* are presented in both customary U.S. (English) and SI units. Initially, work through the problems in U.S. units. If you have time at the end of your review, start over and solve all of the problems in SI units.

- Set a reasonable limit on the time you spend on each subject. It isn't necessary to solve an infinite number of practice problems. The number of practice problems you attempt will depend on how much time you have and how skilled you are in the subject.

- If it isn't already your habit, practice carrying units along in all calculations. Many errors are caused by, and many incorrect exam answer options are based on, common mistakes with units. Pounds don't work in $F = ma$; ft^3/sec doesn't cancel gpm. When working in customary U.S. (English) units, you will find equations in this book in which the quantity g/g_c appears. For calculations at standard gravity, the numerical value of this fraction is 1.00. Therefore, it is necessary to incorporate this quantity only in calculations with a nonstandard gravity or when you are being meticulous with units.

- Use the solutions to your practice problems to check your work. If your answer isn't correct, figure out why.

- To minimize time spent in searching for often-used formulas and data, prepare a one-page summary of all the important formulas and information in each subject area. You can then use these summaries during the examination instead of searching in this book.

- Use the index extensively. Every significant term, law, theorem, and concept has been indexed in every conceivable way—backward and forward—using fuzzy logic synonyms in anticipation of frantic exam searches. If you don't recognize a term used, look for it in the index. Many engineers bring a separate copy of the index with them to the exam.

- Some subjects appear in more than one chapter. Use the index liberally to learn all there is to know about a particular subject.

IF YOU ARE AN INSTRUCTOR

If you are teaching a review course for the PE examination without the benefit of recent, firsthand exam experience, you can use the material in this book as a guide to prepare your lectures. You may want to incorporate additional material into your lectures to better meet the needs and expectations of your class.

I have always tried to overprepare my students. For that reason, the homework problems (i.e., example problems in this book and practice problems in the companion *Practice Problems for the Environmental Engineering PE Exam* book) are often more difficult and more varied than actual examination questions. Also, you will appreciate the fact that it is more efficient to cover several procedural steps in one problem than to ask simple "one-liners" or definition questions. That is the reason that the example and homework problems are often harder and longer than actual exam problems.

To do all the homework for some chapters requires approximately 15 to 20 hours. If you are covering one or more chapters per week, that's a lot of homework per week. "Capacity assignment" is the goal in my review courses. If you assign 20 hours of homework and a student is able to put in only 10 hours that week, that student will have worked to his or her capacity. After the PE examination, that student will honestly say that he or she could not have prepared any more than he or she did in your course. For that reason, you have to assign homework on the basis of what is required to become proficient in the subjects of your lecture. You must resist assigning only the homework that you think can be completed in an arbitrary number of hours.

Homework assignments in my review courses are not individually graded. Instead, students are permitted to make use of existing solutions to learn procedures and techniques to the problems in their homework set, such as those in the companion *Practice Problems for the Environmental Engineering PE Exam* book, which contains solutions to all practice problems. However, each student must turn in a completed set of problems for credit each week. Though I don't correct the homework problems, I address comments or questions emailed to me, posted on the course forum, or written on the assignments.

I believe that students should start preparing for the PE exam at least six months before the examination date. However, most wait until three or four months before getting serious. Because of that, I have found that a 13- or 14-week format works well for a live PE review course. It's a little rushed, but the course is over before everyone gets bored with my jokes. Each week, there is a three-hour meeting, which includes lecture and a short break. Table 1 outlines a course format that might work for you. If you can add more course time, your students will appreciate it. Another lecture covering water resources or environmental engineering would be wonderful. However, I don't think you can cover the full breadth of material in much less time or in many fewer weeks.

I have tried to order the subjects in a logical, progressive manner, keeping my eye on "playing the high-probability subjects." I cover the subjects that everyone can learn (e.g., fluids) early in the course, and I leave the subjects that only daily practitioners should attempt to the end.

Lecture coverage of some examination subjects is necessarily brief; other subjects are not covered at all. These omissions are intentional; they are not the result of scheduling omissions. Why? First, time is not on our side in a review course. Second, some subjects rarely contribute to the examination. Third, some subjects are not well-received by the students. For example, I have found that very few people try to become proficient in protection of wetlands if they don't already work in those areas. Some environmental engineers have a civil engineering bent; others have a mechanical engineering bent. Most civil engineers are comfortable with water

Table 1 *Typical PE Exam Review Course Format*

meeting	subject covered	chapters
1	Introduction to the Exam, Applicable Federal Regulations, Statistical Analysis, Risk Assessment, and Hazard Anaysis	11
2	SI Units, Moles, Measures of Concentration, Units of Regulatory Limits	2, 22
3	Fluids, Conduit Flow	14–17
4	Pumps, Wells, and Groundwater	18, 21
5	Hydrology; Culvert Design and Analysis	19–20
6	Water Supply	23–24
7	Wastewater	25–27
8	Ideal Gases, Airflow Calculations, Ventilation, and Psychrometrics	29, 31–32
9	Combustion, Incineration, and Air Quality	37–38
10	Solid Waste, Hazardous Waste, and Landfills	39
11	Remediation Techniques	40–42
12	Biology, Bacteriology, and Toxicology	43–45
13	Hygiene, Health, and Safety	46–47
14	Radiation, Shielding, and Sound	48–50

supply and wastewater; most mechanical engineers are comfortable with combustion and air pollution. Unless you have six months in which to teach your PE review, your students' time can be better spent covering other subjects.

All the skipped chapters and any related practice problems are presented as floating assignments to be made up in the students' "free time."

I strongly believe in exposing my students to a realistic sample examination, but I no longer administer an in-class mock exam. Since the review course usually ends only a few days before the real PE examination, I hesitate to make students sit for several hours in the late evening to take a "final exam." Rather, I distribute and assign a take-home sample exam at the first meeting of the review course.

If the practice test is to be used as an indication of preparedness, caution your students not to even look at the sample exam prior to taking it. Looking at the sample examination, or otherwise using it to direct their review, will produce unwarranted specialization in subjects contained in the sample examination.

There are many ways to organize a PE review course, depending on your available time, budget, intended audience, facilities, and enthusiasm. However, all good course formats have the same result: The students struggle with the workload during the course, and then they breeze through the examination after the course.

PART 2: EVERYTHING YOU EVER WANTED TO KNOW ABOUT THE PE EXAM

WHAT IS THE FORMAT OF THE PE EXAM?

The NCEES PE examination in environmental engineering consists of two four-hour sessions separated by a one-hour lunch period. Both four-hour sessions contain 50 questions in multiple-choice (i.e., "objective") format. You must answer all questions in each session correctly to receive full credit. There are no optional questions. You must be approved by your state licensing board before you can register for the exam using the "MyNCEES" system on the NCEES website.

WHAT SUBJECTS ARE ON THE PE EXAM?

NCEES has published a description of subjects on the examination. Irrespective of the published examination structure, the exact number of questions that will appear in each subject area cannot be predicted reliably.

There is no guarantee that any single subject will occur in any quantity. One of the reasons for this is that some of the questions span several disciplines. You might consider a pump selection question to come from the subject of fluids, while NCEES might categorize it as sanitary engineering.

Most examinees find the list to be formidable in appearance. The percentage breakdowns in Table 2 are according to NCEES, but these percentages are approximate. NCEES adds,

> The examination is developed with questions that require a variety of approaches and methodologies including design, analysis, application, and operations. Some questions may require knowledge of engineering economics. These areas are examples of the kinds of knowledge that will be tested but are not exclusive or exhaustive categories.

WHAT IS THE TYPICAL QUESTION FORMAT?

Almost all of the questions are standalone—that is, they are completely independent. However, NCEES allows that some sets of questions may start with a statement of a "situation" that will apply to (typically) two to five following questions. Such grouped questions are increasingly rare, however.

Table 2 Environmental PE Exam Specifications

Water   32%
  basic principles, 4%
  wastewater, 10%
  stormwater, 4%
  potable water, 10%
  water resources, 4%

Solid Waste   18%
  basic principles, 3%
  municipal and industrial solid waste, 6%
  hazardous waste, 6%
  medical, radioactive, and other waste, 3%

Air   23%
  basic principles, 7%
  pollution control, 16%

Site Assessment and Remediation   12%
  basic principles, 5%
  site assessment, 3%
  remediation, 4%

Environmental Health and Safety   8%
  industrial hygiene, health, and safety
  security, emergency preparedness, and
    incident response procedures
  exposure assessments
  radiation protection/health physics
  vector control, sanitation, and biohazards
  noise pollution
  indoor air quality
  codes, standards, regulations, and guidelines

Associated Engineering Principles   7%
  mathematics and statistics
  economics and project management
  sustainable design
  mass and energy balance

Each of the questions will have four answer options, labeled "A," "B," "C," and "D." If the answer options are numerical, they will be displayed in increasing value. One of the answer options is correct (or, will be "most nearly correct," as described in the following section). The remaining answer options are incorrect and may consist of one or more "logical distractors," the term used by NCEES to designate incorrect options that look correct.

NCEES intends the questions to be unrelated. Questions are independent or start with new given data. A mistake on one of the questions shouldn't cause you to get a subsequent question wrong. However, considerable time may be required to repeat previous calculations with a new set of given data.

HOW MUCH "LOOK-UP" IS REQUIRED ON THE EXAM?

Since the questions are multiple choice in design, all required data will appear in the situation statement. Since the examination would be unfair if it was possible to arrive at an incorrect answer after making valid

assumptions or using plausible data, you will not generally be required to come up with numerical data that might affect your success on the problem. There will also be superfluous information in the majority of questions.

WHAT DOES "MOST NEARLY" REALLY MEAN?

One of the more disquieting aspects of these questions is that the available answer choices are seldom exact. Answer choices generally have only two or three significant digits. Exam questions ask, "Which answer choice is most nearly the correct value?" or they instruct you to complete the sentence, "The value is approximately ..." A lot of self-confidence is required to move on to the next question when you don't find an exact match for the answer you calculated, or if you have had to split the difference because no available answer choice is close.

NCEES describes it like this:

> Many of the questions on NCEES exams require calculations to arrive at a numerical answer. Depending on the method of calculation used, it is very possible that examinees working correctly will arrive at a range of answers. The phrase "most nearly" is used to accommodate answers that have been derived correctly but that may be slightly different from the correct answer choice given on the exam. You should use good engineering judgment when selecting your choice of answer. For example, if the question asks you to calculate an electrical current or determine the load on a beam, you should literally select the answer option that is most nearly what you calculated, regardless of whether it is more or less than your calculated value. However, if the question asks you to select a fuse or circuit breaker to protect against a calculated current or to size a beam to carry a load, you should select an answer option that will safely carry the current or load. Typically, this requires selecting a value that is closest to but larger than the current or load.

The difference is significant. Suppose you were asked to calculate "most nearly" the volumetric pure water flow required to dilute a contaminated stream to an acceptable concentration. Suppose, also, that you calculated 823 gpm. If the answer choices were (A) 600 gpm, (B) 800 gpm, (C) 1000 gpm, and (D) 1200 gpm, you would go with answer choice (B), because it is most nearly what you calculated. If, however, you were asked to select a pump or pipe with the same rated capacities, you would have to go with choice (C) because an 800 gpm pump wouldn't be sufficient. Got it?

HOW MUCH MATHEMATICS IS NEEDED FOR THE EXAM?

There are no pure mathematics questions (algebra, geometry, trigonometry, etc.) on the exam. However, you will need to apply your knowledge of these subjects to the exam questions.

Generally, only simple algebra, trigonometry, and geometry are needed on the PE exam. You will need to use the trigonometric, logarithm, square root, exponentiation, and similar buttons on your calculator. There is no need to use any other method for these functions.

Except for simple quadratic equations, you will probably not need to find the roots of higher-order equations. For second-order (quadratic) equations, the exam does not care if you find roots by factoring, completing the square, using the quadratic equation, or using your calculator's root finder. Occasionally, it will be convenient to use the equation-solving capability of your calculator. However, other solution methods will always exist.

There is essentially no use of calculus on the exam. Rarely, you may need to take a simple derivative to find a maximum or minimum of some simple algebraic function. Even rarer is the need to integrate to find an average.

There is essentially no need to solve differential equations. Questions involving radioactive decay and fluid mixing have appeared from time to time. However, these applications are extremely rare, have usually been first-order, and could usually be handled without having to solve differential equations.

Basic statistical analysis of observed data may be necessary. Statistical calculations are generally limited to finding means, medians, standard deviations, variances, percentiles, and confidence limits. Since the problems are multiple choice, you won't have to draw a histogram, although you might have to interpret one. Usually, the only population distribution you need to be familiar with is the normal curve. Probability, reliability, hypothesis testing, and statistical quality control are not explicit exam subjects, though their concepts may appear peripherally in some problems. You will not have to use linear or nonlinear regression and other curve fitting techniques to correlate data.

The PE exam is concerned with numerical answers, not with proofs or derivations. You will not be asked to prove or derive formulas, use deductive reasoning, or validate theorems, corollaries, or lemmas.

Inasmuch as first assumptions can significantly affect the rate of convergence, problems requiring trial-and-error solutions are unlikely. Rarely, a calculation may require an iterative solution method. Generally, there is no need to complete more than two iterations. You will not need to program your calculator to obtain an "exact" answer, nor will you generally need to use complex numerical methods.

HOW ABOUT ENGINEERING ECONOMICS?

For most of the early years of engineering licensing, questions on engineering economics appeared frequently on the examinations. This is no longer the case. However, in its outline of exam subjects, NCEES notes: "Some questions may require knowledge of engineering economics." What this means is that engineering economics might appear in several questions on the exam, or the subject might be totally absent. While the degree of engineering economics knowledge may have decreased somewhat, the basic economic concepts (e.g., time value of money, present worth, non-annual compounding, comparison of alternatives, etc.) are still valid test subjects.

If engineering economics is incorporated into other questions, its "disguise" may be totally transparent. For example, you might need to compare the economics of buying and operating two blowers for remediation of a hydrocarbon spill—blowers whose annual costs must be calculated from airflow rates and heads. Also, you may need to use engineering economics concepts and tables in problems that don't even mention "dollars" (e.g., when you need to predict future water demand).

WHAT ABOUT PROFESSIONALISM AND ETHICS?

For many decades, NCEES has considered adding professionalism and ethics questions to the PE exam. However, these subjects are not part of the test outline, and there has yet to be an ethics question on the PE exam.

IS THE EXAM TRICKY?

Other than providing superfluous data, the PE exam is not a "tricky exam." The exam does not overtly try to get you to fail. Examinees manage to fail on a regular basis with perfectly straightforward questions. The exam questions are difficult in their own right. NCEES does not need to provide misleading or conflicting statements. However, you will find that commonly made mistakes are represented in the available answer choices. Thus, the alternative answers (known as distractors) will be logical.

Questions are generally practical, dealing with common and plausible situations that you might experience in your job. You will not be asked to design a structure for reduced gravity on the moon, to design a mud-brick road, to analyze the effects of a nuclear bomb blast on a structure, or to use bamboo for tension reinforcement.

DOES NCEES WRITE EXAM QUESTIONS AROUND THIS BOOK?

Only NCEES knows what NCEES uses to write its exam questions. However, it is irrelevant, because this book is not intended to (1) be everything you need to pass the exam, (2) expose exam secrets or exam questions, or (3) help you pass when you don't deserve to pass. NCEES knows about this book, but worrying about NCEES writing exam questions based on information that is or is not in this book means you are placing too much dependency on this book. This book, for example, will provide instruction in certain principles. Expecting that you will not need to learn anything else is unrealistic. This book presents many facts, definitions, and numerical values. Expecting that you will not need to know other facts, definitions, and numerical values is unrealistic. What NCEES uses to write exam questions won't have any effect on what you need to do to prepare for the exam.

WHAT MAKES THE QUESTIONS DIFFICULT?

Some questions are difficult because the pertinent theory is not obvious. Many reduction-of-experimental-data questions are of this variety. There is only one acceptable procedure—it is heuristic—and nothing else is acceptable. For example, if you don't know the standard procedure for calculating the Most Probable Number (MPN) in a coliform test, you are out of luck.

Some questions are difficult because the data needed are hard to find. Some data just aren't available unless you happen to have brought the right reference book. Many exposure-related questions can be this way. There is no way to answer an exposure question without having the latest regulation.

Some questions are difficult because they defy visualization. If you cannot visualize the situation—if you cannot get an intuitive feeling about what is going on—you probably cannot analyze it. Some open channel flow and psychrometric processes fit this description.

Some questions are difficult because the computational burden is high and they just take a long time to solve. Pipe networks analyzed with the Hardy-Cross method fall into this category.

Some questions are difficult because the terminology is obscure, and you just don't know what the terms mean. This can happen in almost any subject.

DOES THE PE EXAM USE SI UNITS?

The PE exam in environmental engineering requires working in both customary U.S. units (also known as "English units," "inch-pound units," and "British units") and a variety of other metric systems, including SI. Questions use the units that correspond to commonly accepted industry standards. Some questions, such as those covering fluid and air-handling subjects, primarily use units of pounds, feet, seconds, gallons, and degrees Fahrenheit. Metric units are used in chemical-related subjects, including electrical power (watts) and water-supply and wastewater (mg/L) questions. Either system can be used for fluids, although the use of metric units is still rare.

Unlike this book, the exam does not differentiate between lbf and lbm (pounds-force and pounds-mass). Similarly, the exam does not follow this book's practice of meticulously separating the concepts of mass and weight; density and specific weight; and gravity, g, and the gravitational constant, g_c.

WHY DOES NCEES REUSE SOME QUESTIONS?

NCEES reuses some of the more reliable questions from each exam. The percentage of repeat questions isn't high—no more than 25% of the exam. NCEES repeats questions in order to equate the performance of one group of examinees with the performance of an earlier group. The repeated questions are known as *equaters*, and together, they are known as the *equating subtest*.

Occasionally, a new question appears on the exam that very few of the examinees do well on. Usually, the reason for this is that the subject is too obscure or the question is too difficult. Also, there have been cases where a low percentage of the examinees get the answer correct because the question was inadvertently stated in a poor or confusing manner. Questions that everyone gets correct are also considered defective.

NCEES tracks the usage and "success" of each of the exam questions. "Rogue" questions are not repeated without modification. This is one of the reasons historical analysis of question types shouldn't be used as the basis of your review.

DOES NCEES USE THE EXAM TO PRE-TEST FUTURE QUESTIONS?

NCEES does not use the PE exam to "pre-test" or qualify future questions. (It does use this procedure on the FE exam, however.) All of the questions you work will contribute toward your final score.

ARE THE EXAMPLE PROBLEMS IN THIS BOOK REPRESENTATIVE OF THE EXAM?

The example problems in this book are intended to be instructional and informative. They were written to illustrate how their respective concepts can be implemented. Example problems are not intended to represent exam problems or provide guidance on what you should study.

ARE THE PRACTICE PROBLEMS REPRESENTATIVE OF THE EXAM?

The practice problems in the companion *Practice Problems for the Environmental Engineering PE Exam* book were chosen to cover the most likely exam subjects. Some of the practice problems are multiple choice, and some require free-format solutions. However, they are generally more comprehensive and complex than actual exam problems, regardless of their formats. Some of the practice problems are marked "*Time limit: one hour.*" Compared to the four-and-a-half-minute problems on the environmental PE exam, such one-hour problems are considerably more time-consuming.

Practice problems in the companion book were selected to complement subjects in the *Environmental Engineering Reference Manual.* Over the editions of both books, the practice problems have developed into a comprehensive review of the most important environmental engineering subjects covered on the exam.

All of the practice problems are original. Since NCEES does not release old exams, none of the practice problems are actual exam problems.

WHAT REFERENCE MATERIAL IS PERMITTED IN THE EXAM?

The PE examination is an open-book exam. Most states do not have any limits on the numbers and types of books you can use. Personal notes in a three-ring binder and other semipermanent covers can usually be used.

Some states use a "shake test" to eliminate loose papers from binders. Make sure that nothing escapes from your binders when they are inverted and shaken.

The references you bring into the examination room in the morning do not have to be the same as the references you use in the afternoon. However, you cannot share books with other examinees during the exam.

A few states do not permit collections of solved problems such as *Schaum's Outline Series*, sample exams, and solutions manuals. A few states maintain a formal list of banned books.

Strictly speaking, loose paper and scratch pads are not permitted in the examination. Certain types of pre-printed graphs (e.g., psychrometric charts) and logarithmically scaled graph papers (which are almost never needed) should be three-hole punched and brought in a three-ring binder. An exception to this restriction may be made for laminated and oversize charts, graphs, and tables that are commonly needed for particular types of questions. However, there probably aren't any such items for the environmental PE exam.

HOW MANY BOOKS SHOULD YOU BRING?

You shouldn't need many books in the examination. The trouble is, you can't know in advance which ones you will need. That's the reason why many examinees show up with boxes and boxes of books. Since this book is not a substitute for your own experience and knowledge, without a doubt, there are things that you will need that are not in this book. But there are not so many that you need to bring your company's entire library. The examination is very fast-paced. You will not have time to use books with which you are not thoroughly familiar. The exam doesn't require you to know obscure solution methods or to use difficult-to-find data. You won't need articles printed in an industry magazine; you won't need doctoral theses or industry proceedings; and, you won't need to know about recent industry events.

So, it really is unnecessary to bring a large quantity of books with you. Essential books are identified in Table 3 in this Introduction, and you should be able to decide which support you need for the areas in which you intend to work. This book and five to ten other references of your choice should be sufficient for most of the questions you answer.[1]

MAY TABS BE PLACED ON PAGES?

It is common to tab pages in your books in an effort to reduce the time required to locate useful sections. Inasmuch as some states consider Post-it® notes to be "loose paper," your tabs should be of the more permanent variety. Although you can purchase tabs with gummed attachment points, it is also possible simply to use transparent tape to securely attach the Post-its you have already placed in your books.

CAN YOU WRITE AND MARK IN YOUR BOOKS?

During your preparation, you may write anything you want, anywhere in your books, including this one. You can use pencil, pen, or highlighter in order to further your understanding of the content. However, during the exam, you must avoid the appearance of taking notes about the exam. This means that you should only write on the scratch paper that is provided. During the exam, other than drawing a line across a wide table of numbers, or using your pencil to follow a line on a graph, you should not write in your books.

WHAT ABOUT CALCULATORS?

The exam requires use of a scientific calculator. However, it may not be obvious that you should bring a spare calculator with you to the examination. It is always unfortunate when an examinee is not able to finish because his or her calculator was dropped or stolen or stopped working for some unknown reason.

To protect the integrity of its exams, NCEES has banned communicating and text-editing calculators from the exam site. NCEES provides a list of calculator models acceptable for use during the exam. Calculators not included in the list are not permitted. Check the current list of permissible devices at the PPI website (**ppi2pass.com/calculators**). Contact your state board to determine if nomographs and specialty slide rules are permitted.

The exam has not been optimized for any particular brand or type of calculator. In fact, for most calculations,

---

[1]For decades, this Introduction has recommended that you bring an engineering/scientific dictionary with you, but this recommendation is no longer valid. Printed engineering/scientific dictionaries appear to be things of the past. Those that still exist are not very good, and none are targeted enough to be helpful.

xxiiENVIRONMENTAL ENGINEERING REFERENCE MANUAL

a $15 scientific calculator will produce results as satisfactory as those from a $200 calculator. There are definite benefits to having built-in statistical functions, graphing, unit-conversion, and equation-solving capabilities. However, these benefits are not so great as to give anyone an unfair advantage.

It is essential that a calculator used for the environmental PE examination have the following functions.

- trigonometric and inverse trigonometric functions
- hyperbolic and inverse hyperbolic functions
- π
- $\sqrt{x}$ and x^2
- both common and natural logarithms
- y^x and e^x

For maximum speed and utility, your calculator should also have or be programmed for the following functions.

- interpolation
- extracting roots of quadratic and higher-order equations
- calculating factors for economic analysis questions

You may not share calculators with other examinees. Be sure to take your calculator with you whenever you leave the examination room for any length of time.

Laptop, palmtop, and tablet computers (including the iPad®), and electronic readers (e.g., Nook® and Kindle™), are not permitted in the examination. Their use has been considered, but no states actually permit them. However, considering the nature of the exam questions, it is very unlikely that these devices would provide any advantage.

ARE CELL PHONES PERMITTED?

You may not possess or use a walkie-talkie, cell phone, or other communications or text-messaging device during the exam, regardless of whether it is on. You won't be frisked upon entrance to the exam, but should a proctor discover that you are in possession of a communication device, you should expect to be politely excluded from the remainder of the examination.

HOW YOU SHOULD GUESS

There is no deduction for incorrect answers, so guessing is encouraged. However, since NCEES produces defensible licensing exams, there is no pattern to the placement of correct responses. Since the quantitative responses are sequenced according to increasing values, the placement of a correct answer among other

numerical distractors is a function of the distractors, not of some statistical normalizing routine. Therefore, it is irrelevant whether you choose all "A," "B," "C," or "D" when you get into guessing mode during the last minute or two of the exam period.

The proper way to guess is as an engineer. You should use your knowledge of the subject to eliminate illogical answer choices. Illogical answer choices are those that violate good engineering principles, that are outside normal operating ranges, or that require extraordinary assumptions. Of course, this requires you to have some basic understanding of the subject in the first place. Otherwise, it's back to random guessing. That's the reason that the minimum passing score is higher than 25%.

You won't get any points using the "test-taking skills" that helped you in college—the skills that helped with tests prepared by amateurs. You won't be able to eliminate any [verb] answer choices from "Which [noun] ..." questions. You won't find problems with options of the "more than 50" and "less than 50" variety. You won't find one answer choice among the four that has a different number of significant digits, or has a verb in a different tense, or has some singular/plural discrepancy with the stem. The distractors will always match the stem, and they will be logical.

HOW IS THE EXAM GRADED AND SCORED?

The maximum number of points you can earn on the environmental engineering PE exam is 80. The minimum number of points for passing (referred to by NCEES as the *cut score*) varies from exam to exam. The cut score is determined through a rational procedure, without the benefit of knowing examinees' performance on the exam. That is, the exam is not graded on a curve. The cut score is selected based on what you are expected to know, not based on passing a certain percentage of engineers.

Each of the questions is worth one point. Grading is straightforward, since a computer grades your score sheet. Either you get the question right or you don't. If you mark two or more answers for the same problem, no credit is given for the problem.

You will receive the results of your examination from your state board (not NCEES) by mail. Eight to ten weeks will pass before NCEES releases the results to the state boards. However, the state boards take varying amounts of additional time before notifying examinees. You should allow three to four months for notification.

Your score may or may not be revealed to you, depending on your state's procedure. Even if the score is reported to you, it may have been scaled or normalized to 100%. It may be difficult to determine whether the reported score is out of 80 points or is out of 100%.

PPI • www.ppi2pass.com

Table 3 Most Useful Environmental References (alphabetical order by title)

Air Pollution Control: A Design Approach, C.D. Cooper and F.C. Alley
Air Pollution: Its Origin and Control, Kenneth Wark, C.F., Warner, and W.T. Davis
CERCLA Regulations, EPA
Chemical Fate and Transport in the Environment, Harold F. Hemond and Elizabeth J. Fechner-Levy
Chemistry for Environmental Engineering and Science, C.N. Sawyer, P.O. McCarty, and G.F. Parkin
Control Technologies for Hazardous Air Pollutants, EPA
Definitions, Conversions, and Calculations for Occupational Safety and Health Professionals, Edward W. Finucane, ed.
Elements of Bioenvironmental Engineering, Anthony Gaudy and Elizabeth Gaudy
Engineering Field Reference Manual, American Industrial Hygeine Association
Environmental Engineers' Handbook, David H.F. Liu et al., eds.
Environmental Engineering: A Design Approach, A.P. Sincero and G.A. Sincero
Environmental Engineering and Sanitation, J.A. Salvato
Fundamentals of Air Quality Systems, Kenneth E. Noll
Fundamentals of Industrial Hygiene, National Safety Council
Hazardous Waste Management, M.D. LeGrega et al.
Hazardous Waste Site Remediation, Domenic Grasso
HAZWOPER Regulations, OSHA
Health and Environmental Risk Analysis: Fundamentals with Applications, J.F. Louvar and B.D. Louvar
Industrial Ventilation: A Manual of Recommended Practice, ACGIH
Integrated Design and Operation of Water Treatment Facilities, S. Kawamura
Integrated Solid Waste Management: Engineering Principles and Management Issues, George Tchobanoglous et al.
Keeping Buildings Healthy: How to Monitor and Prevent Indoor Environmental Problems, James T. O'Reilly et al.
NIOSH Pocket Guide to Chemical Hazards, NIOSH
NPDES Regulations, EPA
The Occupational Environment: Its Evaluation, Control, and Management, Salvatore R. DiNardi, ed.
Occupational Safety and Health Guidance Manual for Hazardous Waste Site Activities, OSHA
OSHA Technical Manual (OTM), OSHA
Principles of Radiation Protection: A Textbook of Health Physics, K.Z. Morgan and J.E. Turner
Principles of Radiological Health, Earnest F. Gloyna and J.O. Ledbetter
Radiation Protection: A Guide for Scientists, Regulators, and Physicians, Jacob Shapiro
RCRA Regulations, EPA
Recommended Standard for Wastewater Facilities ("Ten States' Standards"), Health Education Services, Health Resources, Inc.
SARA Regulations, EPA
Semiconductor Industrial Hygiene Handbook: Monitoring, Ventilation, Equipment, and Ergonomics, Michael E. Williams et al.
Sources and Control of Air Pollution, Robert J. Heinsohn and Robert L. Kabel
Standard Handbook of Environmental Engineering, Robert A. Corbitt, ed.
Standard Methods for the Examination of Water and Wastewater, a joint publication of the American Public Health Association, the American Water Works Association, and the Water Pollution Control Federation
Urban Hydrology for Small Watersheds (Techical Release TR-55), United States Department of Agriculture, Natural Resources Conservation Service (previously the Soil Conservation Service)
Wastewater Engineering: Treatment, Disposal, and Reuse (Metcalf & Eddy), George Tchobanoglous and Franklin L. Burton
Water Chemistry, Vernon L. Snoeyink and David Jenkins
Water Supply and Pollution Control, Warner Viessman
Water Quality and Treatment: A Handbook of Community Water Supplies, American Water Works Association
Water Quality Management, P.A. Krenkel and Vladimir Novotny
Water Resources Engineering, George Tchobanoglous et al.
Water Treatment Principles and Design, James M. Montgomery

HOW IS THE CUT SCORE ESTABLISHED?

The raw cut score may be established by NCEES before or after the exam is administered. Final adjustments may be made following the exam date.

NCEES uses a process known as the modified *Angoff procedure* to establish the cut score. This procedure starts with a small group (the cut score panel) of professional engineers and educators selected by NCEES. Each individual in the group reviews each problem and makes an estimate of its difficulty. Specifically, each individual estimates the number of minimally qualified engineers out of a hundred examinees who should know the correct answer to the problem. (This is equivalent to predicting the percentage of minimally qualified engineers who will answer correctly.)

Next, the panel assembles, and the estimates for each problem are openly compared and discussed. Eventually, a consensus value is obtained for each. When the panel has established a consensus value for every problem, the values are summed and divided by 100 to establish the cut score.

Various minor adjustments can be made to account for examinee population (as characterized by the average performance on any equater questions) and any flawed problems. Rarely, security breaches result in compromised problems or examinations. How equater questions, examination flaws, and security issues affect examinee performance is not released by NCEES to the public.

WHAT IS THE HISTORICAL PASSING RATE?

Since the environmental PE exam became a no-choice exam with multiple-choice questions, the passing rate has stayed around 70%. The passing rate for repeat examinees is lower, but complete statistics are not available. The overall passing rate used to vary considerably from exam to exam. The existing format has apparently reduced the variability in the passing rate considerably.

CHEATING AND EXAM SUBVERSION

There aren't very many ways to cheat on an open-book test. The proctors are well trained in spotting the few ways that do exist. It goes without saying that you should not talk to other examinees in the room, nor should you pass notes back and forth. You should not write anything into your books or take notes on the contents of the exam. You shouldn't use your cell phone. The number of people who are released to use the restroom may be limited to prevent discussions.

NCEES regularly reuses good problems that have appeared on previous exams. Therefore, examination integrity is a serious issue with NCEES, which goes to great lengths to make sure nobody copies the questions. You may not keep your exam booklet or scratch paper, enter text of questions into your calculator, or copy problems into your own material.

NCEES has become increasingly unforgiving about loss of its intellectual property. NCEES routinely prosecutes violators and seeks financial redress for loss of its examination problems, as well as invalidating any engineering license you may have earned by taking one of its examinations while engaging in prohibited activities. Your state board may impose additional restrictions on your right to retake any examination if you are convicted of such activities. In addition to tracking down the sources of any examination problem compilations that it becomes aware of, NCEES is also aggressive in pursuing and prosecuting examinees who disclose the contents of the exam in internet forum and "chat" environments. Your constitutional rights to free speech and expression will not protect you from civil prosecution for violating the nondisclosure agreement that NCEES requires you to sign before taking the examination. If you wish to participate in a dialog about a particular exam subject, you must do so in such a manner that does not violate the essence of your nondisclosure agreement. This requires decoupling your discussion from the examination and reframing the question to avoid any examination particulars.

The proctors are concerned about exam subversion, which generally means activity that might invalidate the examination or the examination process. The most common form of exam subversion involves trying to copy exam problems for future use. However, in their zeal to enforce and protect, proctors have shown unforgiving intolerance of otherwise minor infractions such as using your own pencil, using a calculator not on the approved list, possessing a cell phone, or continuing to write for even an instant after "pencils down" is called. For such infractions, you should expect to have the results of your examination invalidated, and all of your pleas and arguments in favor of forgiveness to be ignored. Even worse, since you will summarily be considered to have cheated, your state board will most likely prohibit you from retaking the exam for a number of examination cycles. There is no mercy built into the NCEES and state board procedures.

PART 3: HOW TO PREPARE FOR AND PASS THE PE EXAM IN ENVIRONMENTAL ENGINEERING

WHAT SHOULD YOU STUDY?

The exam covers many diverse subjects. Strictly speaking, you don't have to study every subject on the exam in order to pass. However, the more subjects you study, the more you'll improve your chances of passing. You should decide early in the preparation process which subjects you are going to study. The strategy you select will depend on your background. Following are the four most common strategies.

A broad approach is the key to success for examinees who have recently completed their academic studies. This strategy is to review the fundamentals in a broad range of undergraduate subjects (which means studying all or most of the chapters in this book). The examination includes enough fundamentals problems to make this strategy worthwhile. Overall, it's the best approach.

Engineers who have little time for preparation tend to concentrate on the subject areas in which they hope to find the most problems. By studying the list of examination subjects, some have been able to focus on those subjects that will give them the highest probability of finding enough problems that they can answer. This strategy works as long as the examination cooperates and has enough of the types of questions they need. Too often, though, examinees who pick and choose subjects to review can't find enough problems to complete the exam.

Engineers who have been away from classroom work for a long time tend to concentrate on the subjects in which they have had extensive experience, in the hope that the exam will feature lots of problems in those subjects. This method is seldom successful.

Some engineers plan on modeling their solutions from similar problems they have found in textbooks, collections of solutions, and old exams. These engineers often spend a lot of time compiling and indexing the example and sample problem types in all of their books. This is not a legitimate preparation method, and it is almost never successful.

DO YOU NEED A CLASSROOM REVIEW COURSE?

Approximately 60% of first-time PE examinees take an instructor-led review course of some form. Live classroom and internet courses of various types, as well as previously recorded lessons, are available for some or all of the exam topics. Live courses and instructor-moderated internet courses provide several significant advantages over self-directed study, some of which may apply to you.

- A course structures and paces your review. It ensures that you keep going forward without getting bogged down in one subject.

- A course focuses you on a limited amount of material. Without a course, you might not know which subjects to study.

- A course provides you with the questions you need to solve. You won't have to spend time looking for them.

- A course spoon-feeds you the material. You may not need to read the book!

- The course instructor can answer your questions when you are stuck.

You probably already know if any of these advantages apply to you. A review course will be less valuable if you are thorough, self-motivated, and highly disciplined.

HOW LONG SHOULD YOU STUDY?

We've all heard stories of the person who didn't crack a book until the week before the exam and still passed it with flying colors. Yes, these people really exist. However, I'm not one of them, and you probably aren't either. In fact, after having taught thousands of engineers in my own classes, I'm convinced that these people are as rare as the ones who have taken the exam five times and still can't pass it.

A thorough review takes approximately 300 hours. Most of this time is spent solving problems. Some of it may be spent in class; some is spent at home. Some examinees spread this time over a year. Others try to cram it all

into two months. Most classroom review courses last for three or four months. The best time to start studying will depend on how much time you can spend per week.

WHAT THE WELL-HEELED ENVIRONMENTAL ENGINEER SHOULD BEGIN ACCUMULATING

There are many references and resources that you should begin to assemble for review and for use in the examination.

It is unlikely that you could pass the PE exam without accumulating other books and resources. There certainly isn't much margin for error if you show up with only one book. True, references aren't needed to answer some fluids, hydrology, and combustion questions. However, there are many questions that require access to current regulations and that you cannot answer without the proper references. You would have to be truly lucky to go in "bare," find the right mix of questions, and pass.

Few examinees are able to accumulate all of the references needed to support the exam's entire body of knowledge. The accumulation process is too expensive and time-consuming, and the sources are too diverse. Like purchasing an insurance policy, what you end up with will be more a function of your budget than of your needs. In some cases, one book will satisfy several needs.

The list in Table 3 was compiled from surveys conducted of examinees in four administrations of the environmental engineering PE exam. The books and other items listed are regularly cited by examinees as being particularly useful to them. This listing only includes the major "named" books that have become standard references in the industry. These books are in addition to any textbooks or resources that you might choose to bring.

ADDITIONAL REVIEW MATERIAL

In addition to this manual and its accompanying *Practice Problems* book, PPI can provide you with the following references and study aids.

- *Environmental Engineering Solved Problems*, R. Wane Schneiter, PhD, PE. PPI

- *Environmental Engineering Practice PE Exams*, R. Wane Schneiter, PhD, PE. PPI

- *Engineering Unit Conversions*, Michael R. Lindeburg, PE. PPI

DON'T FORGET THE DOWNLOADS

Many of the tables and appendices in this book are representative abridgments with just enough data to (a) do the practice problems in the companion book and (b) give you a false sense of security. You can download or link to additional data, explanations, and references by visiting PPI's website, **ppi2pass.com/ENVRM/ originalsources.html**.

WHAT YOU WON'T NEED

Generally, people bring too many things to the examination. One general rule is that you shouldn't bring books that you have not looked at during your review. If you didn't need a book while doing the problems in this book or its companion *Practice Problems* book, you won't need it during the exam.

There are some other things that you won't need.

- Books on basic and introductory subjects: You won't need books that cover trigonometry, geometry, or calculus.

- Books that cover background engineering subjects that appear on the exam, such as fluids, thermodynamics, and chemistry: The exam is more concerned with the applications of these bodies of knowledge than with the bodies of knowledge themselves.

- Books on non-exam subjects: Such subjects as materials science, statics, dynamics, mechanics of materials, drafting, history, the English language, geography, and philosophy are not part of the exam.

- Books on mathematical analysis, numerical analysis, or extensive mathematics tabulations

- Extensive collections of properties: You will not be expected to know the properties and characteristics of chemical compounds, obscure or exotic alloys, uncommon liquids and gases, or biological organisms. Most characteristics affecting performance are provided as part of the question statement.

- Obscure books and materials: Books that are in foreign languages, doctoral theses, and papers presented at technical societies won't be needed during the exam.

- Old textbooks or obsolete, rare, and ancient books: NCEES exam committees are aware of which textbooks are in use. Material that is available only in out-of-print publications and old editions won't be used.

- Handbooks in other disciplines: You probably won't need a mechanical, electrical, or industrial engineering handbook.

- The *Handbook of Chemistry and Physics*

- Computer science books: You won't need to bring books covering computer logic, programming, algorithms, program design, or subroutines for BASIC, FORTRAN, C, Pascal, HTML, Java, or any other language.

- Crafts- and trades-oriented books: The exam does not expect you to have detailed knowledge of trades

or manufacturing operations (e.g., carpentry, plumbing, electrical wiring, roofing, sheetrocking, foundry, metal turning, sheet-metal forming, or designing jigs and fixtures).

- Manufacturer's literature and catalogs: No part of the exam requires you to be familiar with products that are proprietary to any manufacturer.

- U.S. government publications: With the exceptions of the publications mentioned and referenced in this book, no government publications are required in the PE exam.

- The text of federal acts, policies, and treaties such as the *Clean Air Act, Clean Water Act, Resource Recovery and Conservation Act, Oil Pollution Act, Atomic Energy Act*, and *Nuclear Waste Policy Act.*

- Your state's laws: The PE exam is a national exam. Nothing unique to your state will appear on it. (However, federal legislation affecting engineers, particularly in environmental areas, is fair game.)

- Local or state building codes

SHOULD YOU LOOK FOR OLD EXAMS?

The traditional approach to preparing for standardized tests includes working sample tests. However, NCEES does not release old tests or questions after they are used. Therefore, there are no official questions or tests available from legitimate sources. NCEES publishes booklets of sample questions and solutions to illustrate the format of the exam. However, these questions have been compiled from various previous exams, and the resulting publication is not a true "old exam." Furthermore, NCEES sometimes constructs its sample questions books from questions that have been pulled from active use for various reasons, including poor performance. Such marginal questions, while accurately reflecting the format of the examination, are not always representative of actual exam subjects.

WHAT SHOULD YOU MEMORIZE?

You get lucky here, because it isn't necessary to actually memorize anything. The exam is open-book, so you can look up any procedure, formula, or piece of information that you need. You can speed up your problem-solving response time significantly if you don't have to look up the conversion from gal/min to ft^3/sec, the definition of the sine of an angle, and the chemical formula for carbon dioxide, but you don't even have to memorize these kinds of things. As you work practice problems in the companion *Practice Problems for the Environmental Engineering PE Exam* book, you will automatically memorize the things that you come across more than a few times.

DO YOU NEED A REVIEW SCHEDULE?

It is important that you develop and adhere to a review outline and schedule. Once you have decided which subjects you are going to study, you can allocate the available time to those subjects in a manner that makes sense to you. If you are not taking a classroom review course (where the order of preparation is determined by the lectures), you should make an outline of subjects for self-study to use for scheduling your preparation. A fill-in-the-dates schedule is provided in Table 4 at the end of this Introduction.

A SIMPLE PLANNING SUGGESTION

Designate some location (a drawer, a corner, a cardboard box, or even a paper shopping bag left on the floor) as your "exam catch-all." Use your catch-all during the months before the exam when you have revelations about things you should bring with you. For example, you might realize that the plastic ruler marked off in tenths of an inch that is normally kept in the kitchen junk drawer can help you with some soil pressure questions. Or, you might decide that a certain book is particularly valuable. Or, that it would be nice to have dental floss after lunch. Or, that large rubber bands and clips are useful for holding books open.

It isn't actually necessary to put these treasured items in the catch-all during your preparation. You can, of course, if it's convenient. But if these items will have other functions during the time before the exam, at least write yourself a note and put the note into the catch-all. When you go to pack your exam kit a few days before the exam, you can transfer some items immediately, and the notes will be your reminders for the other items that are back in the kitchen drawer.

HOW YOU CAN MAKE YOUR REVIEW REALISTIC

In the exam, you must be able to quickly recall solution procedures, formulas, and important data. You must remain sharp for eight hours or more. When you played a sport back in school, your coach tried to put you in game-related situations. Preparing for the PE exam isn't much different from preparing for a big game. Some part of your preparation should be realistic and representative of the examination environment.

There are several things you can do to make your review more representative. For example, if you gather most of your review resources (i.e., books) in advance and try to use them exclusively during your review, you will become more familiar with them. (Of course, you can also add to or change your references if you find inadequacies.)

Learning to use your time wisely is one of the most important lessons you can learn during your review.

You will undoubtedly encounter questions that end up taking much longer than you expected. In some instances, you will cause your own delays by spending too much time looking through books for things you need (or just by looking for the books themselves!). Other times, the questions will entail too much work. Learn to recognize these situations so that you can make an intelligent decision about skipping such questions in the exam.

WHAT TO DO A FEW DAYS BEFORE THE EXAM

There are a few things you should do a week or so before the examination. You should arrange for child-care and transportation. Since the examination does not always start or end at the designated time, make sure that your childcare and transportation arrangements are flexible.

Check PPI's website for last-minute updates and errata to any PPI books you might have and are bringing to the exam.

Prepare a separate copy of this book's index. You may download a copy of the index at **ppi2pass.com/ envrmindex**.

If it's convenient, visit the exam location in order to find the building, parking areas, examination room, and restrooms. If it's not convenient, you may find driving directions and/or site maps on the web.

Take the battery cover off your calculator and check to make sure you are bringing the correct size replacement batteries. Some calculators require a different kind of battery for their "permanent" memories. Put the cover back on and secure it with a piece of masking tape. Write your name on the tape to identify your calculator.

If your spare calculator is not the same as your primary calculator, spend a few minutes familiarizing yourself with how it works. In particular, you should verify that your spare calculator is functional.

PREPARE YOUR CAR

[ ] Gather snow chains, shovel, and tarp to kneel on while installing chains.
[ ] Check tire pressures.
[ ] Check your spare tire.
[ ] Check for tire installation tools.
[ ] Verify that you have the vehicle manual.
[ ] Check fluid levels (oil, gas, water, brake fluid, transmission fluid, window-washing solution).
[ ] Fill up with gas.
[ ] Check battery and charge if necessary.
[ ] Know something about your fuse system (where they are, how to replace them, etc.).

[ ] Assemble all required maps.
[ ] Fix anything that might slow you down (missing wiper blades, etc.).
[ ] Check your taillights.
[ ] Affix the recently arrived DMV license sticker.
[ ] Fix anything that might get you pulled over on the way to the exam (burned-out taillight or headlight, broken lenses, bald tires, missing license plate, noisy muffler).
[ ] Treat the inside windows with anti-fog solution.
[ ] Put a roll of paper towels in the back seat.
[ ] Gather exact change for any bridge tolls or toll roads.
[ ] Find your electronic toll tag (FasTrak®/E-Z Pass®, etc.).
[ ] Put $20 in your glove box.
[ ] Check for current registration and proof of insurance.
[ ] Locate a spare door and ignition key.
[ ] Find your AAA or other roadside-assistance cards and phone numbers.
[ ] Plan out alternate routes.

PREPARE YOUR EXAM KITS

Second in importance to your scholastic preparation is the preparation of your two examination kits. The first kit consists of a bag, box (plastic milk crates hold up better than cardboard in the rain), or wheeled travel suitcase containing items to be brought with you into the examination room.

[ ] letter admitting you to the examination
[ ] photographic identification (e.g., driver's license)
[ ] this book
[ ] a separate, bound copy of this book's index
[ ] other textbooks and reference books
[ ] regular dictionary
[ ] review course notes in a three-ring binder
[ ] cardboard boxes or plastic milk crates to use as bookcases
[ ] primary calculator
[ ] spare calculator
[ ] instruction booklets for your calculators
[ ] extra calculator batteries
[ ] straightedge and rulers
[ ] protractor
[ ] scissors
[ ] stapler
[ ] transparent tape
[ ] magnifying glass
[ ] small (jeweler's) screwdriver for fixing your glasses or for removing batteries from your calculator
[ ] unobtrusive (quiet) snacks or candies, already unwrapped
[ ] two small plastic bottles of water
[ ] travel pack of tissue (keep in your pocket)
[ ] handkerchief

[ ] headache remedy
[ ] personal medication
[ ] $5.00 in assorted coinage
[ ] spare contact lenses and wetting solution
[ ] backup reading glasses
[ ] eye drops
[ ] light, comfortable sweater
[ ] loose shoes or slippers
[ ] cushion for your chair
[ ] earplugs
[ ] wristwatch with alarm
[ ] several large trash bags ("raincoats" for your boxes of books)
[ ] roll of paper towels
[ ] wire coat hanger (to hang up your jacket or to get back into your car in an emergency)
[ ] extra set of car keys on a string around your neck

The second kit consists of the following items and should be left in a separate bag or box in your car in case it is needed.

[ ] copy of your application
[ ] proof of delivery
[ ] light lunch
[ ] beverage in thermos or cans
[ ] sunglasses
[ ] extra pair of prescription glasses
[ ] raincoat, boots, gloves, hat, and umbrella
[ ] street map of the examination area
[ ] parking permit
[ ] battery-powered desk lamp
[ ] your cell phone
[ ] length of rope

The following items cannot be used during the examination and should be left at home.

[ ] personal pencils and erasers (NCEES distributes mechanical pencils at the exam.)
[ ] fountain pens
[ ] radio, CD player, or iPod™
[ ] battery charger
[ ] extension cords
[ ] scratch paper
[ ] note pads
[ ] drafting compass
[ ] circular ("wheel") slide rules

PREPARE FOR THE WORST

All of the occurrences listed in this section have happened to examinees. Granted, you cannot prepare for every eventuality. But, even though each of these occurrences taken individually is a low-probability event, taken together, they are worth considering in advance.

- Imagine getting a flat tire, getting stuck in traffic, or running out of gas on the way to the exam.

- Imagine rain and snow as you are carrying your cardboard boxes of books into the exam room.

- Imagine arriving late. Can you get into the exam without having to make two trips from your car?

- Imagine having to park two blocks from the exam site. How are you going to get everything to the exam room? Can you actually carry everything that far? Could you use a furniture dolly, a supermarket basket, or perhaps a helpmate?

- Imagine a Star Trek convention, square-dancing contest, construction, or auction in the next room.

- Imagine a site without any heat, with poor lighting, or with sunlight streaming directly into your eyes.

- Imagine a hard folding chair and a table with one short leg.

- Imagine a site next to an airport with frequent take-offs, or next to a construction site with a pile driver, or next to the NHRA's Drag Racing Championship.

- Imagine a seat where someone nearby chews gum with an open mouth; taps his pencil or drums her fingers; or wheezes, coughs, and sneezes for eight hours.

- Imagine the distraction of someone crying or of proctors evicting yelling and screaming examinees who have been found cheating.

- Imagine the tragedy of another examinee's serious medical emergency.

- Imagine a delay of an hour while they find someone to unlock the building, turn on the heat, or wait for the head proctor to bring instructions.

- Imagine a power outage occurring sometime during the exam.

- Imagine a proctor who (a) tells you that one of your favorite books can't be used in the exam, (b) accuses you of cheating, or (c) calls "time up" without giving you any warning.

- Imagine not being able to get your lunch out of your car or find a restaurant.

- Imagine getting sick or nervous in the exam.

- Imagine someone stealing your calculator during lunch.

WHAT TO DO THE DAY BEFORE THE EXAM

Take the day before the examination off from work to relax. Do not cram the last night. A good night's sleep is the best way to start the examination. If you live a considerable distance from the examination site, consider getting a hotel room in which to spend the night.

Practice setting up your examination work environment. Carry your boxes to the kitchen table. Arrange your "bookcases" and supplies. Decide what stays on the floor in boxes and what gets an "honored position" on the tabletop.

Use your checklist to make sure you have everything. Make sure your exam kits are packed and ready to go. Wrap your boxes in plastic bags in case it's raining when you carry them from the car to the exam room.

Calculate your wake-up time and set the alarms on two bedroom clocks. Select and lay out your clothing items. (Dress in layers.) Select and lay out your breakfast items.

If it's going to be hot on exam day, put your (plastic) bottles of water in the freezer.

Make sure you have gas in your car and money in your wallet.

WHAT TO DO THE DAY OF THE EXAM

Turn off the quarterly and hourly alerts on your wristwatch. Leave your pager or cell phone at home. If you must bring them, change them to silent mode. Bring or buy a morning newspaper.

You should arrive at least 30 minutes before the examination starts. This will allow time for finding a convenient parking place, bringing your materials to the examination room, making room and seating changes, and calming down. Be prepared, though, to find that the examination room is not open or ready at the designated time.

Once you have arranged the materials around you on your table, take out your morning newspaper and look cool. (Only nervous people work crossword puzzles.)

WHAT TO DO DURING THE EXAM

All of the procedures typically associated with timed, proctored, computer-graded assessment tests will be in effect when you take the PE examination.

The proctors will distribute the examination booklets and answer sheets if they are not already on your tables. However, you should not open the booklets until instructed to do so. You may read the information on the front and back covers, and you should write your name in any appropriate blank spaces.

Listen carefully to everything the proctors say. Do not ask your proctors any engineering questions. Even if they are knowledgeable in engineering, they will not be permitted to answer your questions.

Answers to questions are recorded on an answer sheet contained in the test booklet. The proctors will guide you through the process of putting your name and other biographical information on this sheet when the time comes, which will take approximately 15 minutes. You will be given the full four hours to answer questions.

Time to initialize the answer sheet is not part of your four hours.

The common suggestions to "completely fill the bubbles and erase completely" apply here. NCEES provides each examinee with a mechanical pencil with HB lead. Use of ballpoint pens and felt-tip markers is prohibited for several reasons.

If you finish the exam early and there are still more than 30 minutes remaining, you will be permitted to leave the room. If you finish less than 30 minutes before the end of the exam, you may be required to remain until the end. This is done to be considerate of the people who are still working.

Be prepared to stop working immediately when the proctors call "pencils down" or "time is up." Continuing to work for even a few seconds will completely invalidate your examination.

When you leave, you must return your exam booklet. You may not keep the exam booklet for later review.

If there are any questions that you think were flawed, in error, or unsolvable, ask a proctor for a "reporting form" on which you can submit your comments. Follow your proctor's advice in preparing this document.

WHAT ABOUT EATING AND DRINKING IN THE EXAM ROOM?

The official rule is probably the same in every state: no eating or drinking in the exam. That makes sense, for a number of reasons. Some exam sites don't want (or don't permit) stains and messes. Others don't want crumbs to attract ants and rodents. Your table partners don't want spills or smells. Nobody wants the distractions. Your proctors can't give you a new exam booklet when the first one is ruined with coffee.

How this rule is administered varies from site to site and from proctor to proctor. Some proctors enforce the letter of law, threatening to evict you from the exam room when they see you chewing gum. Others may permit you to have bottled water, as long as you store the bottles on the floor where any spills will not harm what's on the table. No one is going to let you crack peanuts while you work on the exam, but I can't see anyone complaining about a hard candy melting away in your mouth. You'll just have to find out when you get there.

HOW TO SOLVE MULTIPLE-CHOICE QUESTIONS

When you begin each session of the exam, observe the following suggestions:

- Use only the pencil provided.
- Do not spend an inordinate amount of time on any single question. If you have not answered a question in a reasonable amount of time, make a note of it and move on.

- Set your wristwatch alarm for five minutes before the end of each four-hour session, and use that remaining time to guess at all of the remaining questions. Odds are that you will be successful with about 25% of your guesses, and these points will more than make up for the few points that you might earn by working during the last five minutes.

- Make mental notes about any question for which you cannot find a correct response, that appears to have two correct responses, or that you believe has some technical flaw. Errors in the exam are rare, but they do occur. Such errors are usually discovered during the scoring process and discounted from the examination, so it is not necessary to tell your proctor, but be sure to mark the one best answer before moving on.

- Make sure all of your responses on the answer sheet are dark and completely fill the bubbles.

SOLVE QUESTIONS CAREFULLY

Many points are lost to carelessness. Keep the following items in mind when you are solving practice problems. Hopefully, these suggestions will be automatic in the exam.

[ ] Did you recheck your mathematical equations?
[ ] Do the units cancel out in your calculations?
[ ] Did you convert between radius and diameter?
[ ] Did you convert between feet and inches?
[ ] Did you convert from gage to absolute pressures?
[ ] Did you convert between pounds and kips, or between kPa and Pa?
[ ] Did you recheck all data obtained from other sources, tables, and figures? (In finding the friction factor, did you enter the Moody diagram at the correct Reynolds number?)

SHOULD YOU TALK TO OTHER EXAMINEES AFTER THE EXAM?

The jury is out on this question. People react quite differently to the examination experience. Some people are energized. Most are exhausted. Some people need to unwind by talking with other examinees, describing every detail of their experience, and dissecting every examination question. Other people need lots of quiet space, and prefer to just get into a hot tub to soak and sulk. Most engineers, apparently, are in this latter category.

Since everyone who took the exam has seen it, you will not be violating your "oath of silence" if you talk about the details with other examinees immediately after the exam. It's difficult not to ask how someone else approached a question that had you completely stumped. However, keep in mind that it is very disquieting to think you answered a question correctly, only to have someone tell you where you went wrong.

To ensure you do not violate the nondisclosure agreement you signed before taking the exam, make sure you do not discuss any exam particulars with people who have not also taken the exam.

AFTER THE EXAM

Yes, there is something to do after the exam. Most people return home, throw their exam "kits" into the corner, and collapse. A week later, when they can bear to think about the experience again, they start integrating their exam kits back into their normal lives. The calculators go back into the desk, the books go back on the shelves, the $5.00 in change goes back into the piggy bank, and all of the miscellaneous stuff you brought with you to the exam is put back wherever it came from.

Here's what I suggest you do as soon as you get home, before you collapse.

[ ] Thank your spouse and children for helping you during your preparation.
[ ] Take any paperwork you received on exam day out of your pocket, purse, or wallet. Put this inside your *Environmental Engineering Reference Manual*.
[ ] Reflect on any statements regarding exam secrecy to which you signed your agreement in the exam.
[ ] Call your employer and tell him/her that you need to take a mental health day off on Monday.

A few days later, when you can face the world again, do the following.

[ ] Make notes about anything you would do differently if you had to take the exam over again.
[ ] Consolidate all of your application paperwork, correspondence to/from your state, and any paperwork that you received on exam day.
[ ] If you took a live review course, call or email the instructor (or write a note) to say "Thanks."
[ ] Visit the Engineering Exam Forum part of PPI's website and see what other people are saying about the exam you took.
[ ] Return any books you borrowed.
[ ] Write thank-you notes to all of the people who wrote letters of recommendation or reference for you.
[ ] Find and read the chapter in this book that covers ethics. There were no ethics questions on your PE exam, but it doesn't make any difference. Ethical behavior is expected of a PE in any case. Spend a few minutes reflecting on how your performance (obligations, attitude, presentation, behavior, appearance, etc.) might be about to change once you are licensed. Consider how you are going to be a role model for others around you.
[ ] Put all of your review books, binders, and notes someplace where they will be out of sight.

FINALLY

By the time you've "undone" all of your preparations, you might have thought of a few things that could help future examinees. If you have any sage comments about how to prepare, any suggestions about what to do in or bring to the exam, any comments on how to improve this book, or any funny anecdotes about your experience, I hope you will share these with me. By this time, you'll be the "expert," and I'll be your biggest fan.

AND THEN, THERE'S THE WAIT ...

Waiting for the exam results is its own form of mental torture.

Yes, I know the exam is 100% multiple-choice, and grading should be almost instantaneous. But, you are going to wait, nevertheless. There are many reasons for the delay.

Although the actual machine grading "only takes seconds," consider the following facts: (a) NCEES prepares multiple exams for each administration, in case one becomes unusable (i.e., is inappropriately released) before the exam date. (b) Since the actual version of the exam used is not known until after it is finally given, the cut-score determination occurs after the exam date.

I wouldn't be surprised to hear that NCEES receives dozens, if not hundreds, of claims from well-meaning examinees who were 100% certain that the exams they took were seriously flawed to some degree—that there wasn't a correct answer for such-and-such question—that there were two answers for such-and-such question—or even, perhaps, that such-and-such question was missing from their exam booklet altogether. Each of these claims must be considered as a potential adjustment to the cut-score.

Then, the exams must actually be graded. Since grading nearly 50,000 exams (counting all the FE and PE exams) requires specialized equipment, software, and training not normally possessed by the average employee, as well as time to do the work (also not normally possessed by the average employee), grading is invariably outsourced.

Outsourced grading cannot begin until all of the states have returned their score sheets to NCEES and NCEES has sorted, separated, organized, and consolidated the score sheets into whatever "secret sauce sequence" is best. During grading, some of the score sheets "pop out" with any number of abnormalities that demand manual scoring.

After the individual exams are scored, the results are analyzed in a variety of ways. Some of the analysis looks at passing rates by such delineators as degree, major, university, site, and state. Part of the analysis looks for similarities between physically adjacent examinees (to look for cheating). Part of the analysis looks for exam sites that have statistically abnormal group performance. And, some of the analysis looks for exam questions that have a disproportionate fraction of successful or unsuccessful examinees. Anyway, you get the idea: It's not merely putting your exam sheet in an electronic reader. All of these steps have to be completed for 100% of the examinees before any results can go out.

Once NCEES has graded your test and notified your state, when you hear about it depends on when the work is done by your state. Some states have to approve the results at a board meeting; others prepare the certificates before sending out notifications. Some states are more computerized than others. Some states have 50 examinees, while others have 10,000. Some states are shut down by blizzards and hurricanes; others are administratively challenged—understaffed, inadequately trained, or over budget.

There is no pattern to the public release of results. None. The exam results are not released to all states simultaneously. (The states with the fewest examinees often receive their results soonest.) They are not released by discipline. They are not released alphabetically by state or examinee name. The people who failed are not notified first (or last). Your coworker might receive his or her notification today, and you might be waiting another three weeks for yours.

Some states post the names of the successful examinees, or unsuccessful examinees, or both on their official state websites before the results go out. Others update their websites after the results go out. Some states don't list much of anything on their websites.

Remember, too, that the size or thickness of the envelope you receive from your state does not mean anything. Some states send a big congratulations package and certificate. Others send a big package with a new application to repeat the exam. Some states send a postcard. Some send a one-page letter. Some states send you an invoice for your license fees. (Ahh, what a welcome bill!) You just have to open it to find out.

AND WHEN YOU PASS . . .

[ ] Celebrate.
[ ] Notify the people who wrote letters of recommendation or reference for you.
[ ] Read "FAQs about What Happens After You Pass the Exam" at **ppi2pass.com/postexam**.
[ ] Ask your employer for a raise.
[ ] Tell the folks at PPI (who have been rootin' for you all along) the good news.

Table 4 Schedule for Self-Study

chapter number	subject	date to start	date to finish
1	Systems of Units		
2	Properties of Areas		
3	Algebra		
4	Linear Algebra		
5	Vectors		
6	Trigonometry		
7	Analytic Geometry		
8	Differential Calculus		
9	Integral Calculus		
10	Differential Equations		
11	Probability, Statistics, and Risk Analysis		
12	Numerical Analysis		
13	Energy, Work, and Power		
14	Fluid Properties		
15	Fluid Statics		
16	Fluid Flow Parameters		
17	Fluid Dynamics		
18	Hydraulic Machines		
19	Open Channel Flow		
20	Meteorology, Climatology, and Hydrology		
21	Groundwater		
22	Inorganic Chemistry		
23	Water Supply Quality and Testing		
24	Water Supply Treatment and Distribution		
25	Wastewater Quantity and Quality		
26	Wastewater Treatment: Equipment and Processes		
27	Activated Sludge and Sludge Processing		
28	Protection of Wetlands		
29	Thermodynamic Properties		
30	Changes in Thermodynamic Properties		
31	HVAC: Psychrometrics		
32	HVAC: Ventilation		
33	HVAC: Heating Load		
34	HVAC: Cooling Load		
35	Air Conditioning Equipment and Controls		
36	Fans, Ductwork, and Terminal Devices		
37	Fuels and Combustion		
38	Air Quality		

(continued)

Table 4 Schedule for Self-Study (continued)

chapter number	subject	date to start	date to finish
39	Municipal Solid Waste		
40	Environmental Pollutants		
41	Disposition of Hazardous Materials		
42	Environmental Remediation		
43	Organic Chemistry		
44	Biology and Bacteriology		
45	Toxicology		
46	Industrial Hygiene		
47	Health and Safety		
48	Biological Effects of Radiation		
49	Shielding		
50	Illumination and Sound		
51	Emergency Management		
52	Electrical Systems and Equipment		
53	Instrumentation and Digital Numbering		
54	Engineering Economic Analysis		
55	Project Management, Budgeting, and Scheduling		
56	Professional Services, Contracts, and Engineering Law		
57	Engineering Ethics		
58	Engineering Licensing in the United States		

Topic I: Background and Support

Chapter

1 Systems of Units

1. INTRODUCTION

The purpose of this chapter is to eliminate some of the confusion regarding the many units available for each engineering variable. In particular, an effort has been made to clarify the use of the so-called English systems, which for years have used the *pound* unit both for force and mass—a practice that has resulted in confusion even for those familiar with it.

2. COMMON UNITS OF MASS

The choice of a mass unit is the major factor in determining which system of units will be used in solving a problem. It is obvious that one will not easily end up with a force in pounds if the rest of the problem is stated in meters and kilograms. Actually, the choice of a mass unit determines more than whether a conversion factor will be necessary to convert from one system to another (e.g., between SI and English units). An inappropriate choice of a mass unit may actually require a conversion factor *within* the system of units.

The common units of mass are the gram, pound, kilogram, and slug.[1] There is nothing mysterious about these units. All represent different quantities of matter, as Fig. 1.1 illustrates. In particular, note that the pound and slug do not represent the same quantity of matter.[2]

Figure 1.1 Common Units of Mass

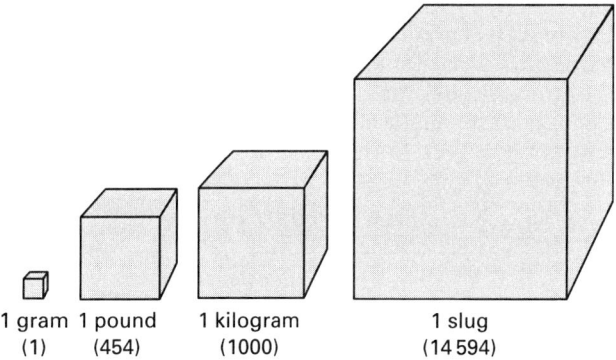

1 gram 1 pound   1 kilogram   1 slug
(1)   (454)   (1000)   (14 594)

3. MASS AND WEIGHT

In SI, *kilograms* are used for mass and *newtons* for weight (force). The units are different, and there is no confusion between the variables. However, for years the term *pound* has been used for both mass and weight. This usage has obscured the distinction between the two: mass is a constant property of an object; weight varies with the gravitational field. Even the conventional use of the abbreviations *lbm* and *lbf* (to distinguish between pounds-mass and pounds-force) has not helped eliminate the confusion.

It is true that an object with a mass of one pound will have an earthly weight of one pound, but this is true only on the earth. The weight of the same object will be much less on the moon. Therefore, care must be taken when working with mass and force in the same problem.

The relationship that converts mass to weight is familiar to every engineering student.

$$W = mg \qquad\qquad 1.1$$

---

[1]Normally, one does not distinguish between a unit and a multiple of that unit, as is done here with the gram and the kilogram. However, these two units actually are bases for different consistent systems.
[2]A slug is approximately equal to 32.1740 pounds-mass.

Equation 1.1 illustrates that an object's weight will depend on the local acceleration of gravity as well as the object's mass. The mass will be constant, but gravity will depend on location. Mass and weight are not the same.

4. ACCELERATION OF GRAVITY

Gravitational acceleration on the earth's surface is usually taken as 32.2 ft/sec^2 or 9.81 m/s^2. These values are rounded from the more precise values of 32.1740 ft/sec^2 and 9.806 65 m/s^2. However, the need for greater accuracy must be evaluated on a problem-by-problem basis. Usually, three significant digits are adequate, since gravitational acceleration is not constant anyway but is affected by location (primarily latitude and altitude) and major geographical features.

The term *standard gravity*, g_0, is derived from the acceleration at essentially any point at sea level and approximately 45° N latitude. If additional accuracy is needed, the gravitational acceleration can be calculated from Eq. 1.2. This equation neglects the effects of large land and water masses. ϕ is the latitude in degrees.

$$g_{\text{surface}} = g'\Big(1 + (5.3024 \times 10^{-3})\sin^2 \phi$$
$$- (5.8 \times 10^{-6})\sin^2 2\phi\Big)$$
$$g' = 32.08769 \text{ ft/sec}^2$$
$$= 9.780\,327 \text{ m/s}^2 \qquad 1.2$$

If the effects of the earth's rotation are neglected, the gravitational acceleration at an altitude h above the earth's surface is given by Eq. 1.3. r_e is the earth's radius.

$$g_h = g_{\text{surface}}\left(\frac{r_e}{r_e + h}\right)^2$$
$$r_e = 3959 \text{ mi}$$
$$= 6.378\,1 \times 10^6 \text{ m} \qquad 1.3$$

5. CONSISTENT SYSTEMS OF UNITS

A set of units used in a calculation is said to be *consistent* if no conversion factors are needed.[3] For example, a moment is calculated as the product of a force and a lever arm length.

$$M = Fr \qquad 1.4$$

A calculation using Eq. 1.4 would be consistent if M was in newton-meters, F was in newtons, and r was in meters. The calculation would be inconsistent if M was in ft-kips, F was in kips, and r was in inches (because a conversion factor of 1/12 would be required).

The concept of a consistent calculation can be extended to a system of units. A *consistent system of units* is one in which no conversion factors are needed for any calculation. For example, Newton's second law of motion can be written without conversion factors. Newton's second law simply states that the force required to accelerate an object is proportional to the acceleration of the object. The constant of proportionality is the object's mass.

$$F = ma \qquad 1.5$$

Equation 1.5 is $F = ma$, not $F = Wa/g$ or $F = ma/g_c$. Equation 1.5 is consistent: it requires no conversion factors. This means that in a consistent system where conversion factors are not used, once the units of m and a have been selected, the units of F are fixed. This has the effect of establishing units of work and energy, power, fluid properties, and so on.

The decision to work with a consistent set of units is desirable but not necessary. Problems in fluid flow and thermodynamics are routinely solved in the United States with inconsistent units. This causes no more of a problem than working with inches and feet when calculating a moment. It is necessary only to use the proper conversion factors.

6. THE ENGLISH ENGINEERING SYSTEM

Through common and widespread use, pounds-mass (lbm) and pounds-force (lbf) have become the standard units for mass and force in the *English Engineering System*. (The English Engineering System is used in this book.)

There are subjects in the United States in which the use of pounds for mass is firmly entrenched. For example, most thermodynamics, fluid flow, and heat transfer problems have traditionally been solved using the units of lbm/ft^3 for density, Btu/lbm for enthalpy, and Btu/lbm-°F for specific heat. Unfortunately, some equations contain both lbm-related and lbf-related variables, as does the steady flow conservation of energy equation, which combines enthalpy in Btu/lbm with pressure in lbf/ft^2.

The units of pounds-mass and pounds-force are as different as the units of gallons and feet, and they cannot be canceled. A mass conversion factor, g_c, is needed to make the equations containing lbf and lbm dimensionally consistent. This factor is known as the *gravitational constant* and has a value of 32.1740 lbm-ft/lbf-sec^2. The numerical value is the same as the standard acceleration of gravity, but g_c is not the local gravitational acceleration, g.[4] g_c is a conversion constant, just as 12.0 is the conversion factor between feet and inches.

The English Engineering System is an inconsistent system as defined according to Newton's second law.

---

[3]The terms *homogeneous* and *coherent* are also used to describe a consistent set of units.

[4]It is acceptable (and recommended) that g_c be rounded to the same number of significant digits as g. Therefore, a value of 32.2 for g_c would typically be used.

$F = ma$ cannot be written if lbf, lbm, and ft/sec^2 are the units used. The g_c term must be included.

$$F \text{ in lbf} = \frac{(m \text{ in lbm})\left(a \text{ in } \dfrac{\text{ft}}{\text{sec}^2}\right)}{g_c \text{ in } \dfrac{\text{lbm-ft}}{\text{lbf-sec}^2}} \qquad 1.6$$

In Eq. 1.6, g_c does more than "fix the units." Since g_c has a numerical value of 32.1740, it actually changes the calculation numerically. A force of 1.0 pound will not accelerate a 1.0-pound mass at the rate of 1.0 ft/sec^2.

In the English Engineering System, work and energy are typically measured in ft-lbf (mechanical systems) or in British thermal units, Btu (thermal and fluid systems). One Btu is equal to 778.17 ft-lbf.

Example 1.1

Calculate the weight in lbf of a 1.00 lbm object in a gravitational field of 27.5 ft/sec^2.

Solution

From Eq. 1.6,

$$F = \frac{ma}{g_c} = \frac{(1.00 \text{ lbm})\left(27.5 \dfrac{\text{ft}}{\text{sec}^2}\right)}{32.2 \dfrac{\text{lbm-ft}}{\text{lbf-sec}^2}}$$

$$= 0.854 \text{ lbf}$$

7. OTHER FORMULAS AFFECTED BY INCONSISTENCY

It is not a significant burden to include g_c in a calculation, but it may be difficult to remember when g_c should be used. Knowing when to include the gravitational constant can be learned through repeated exposure to the formulas in which it is needed, but it is safer to carry the units along in every calculation.

The following is a representative (but not exhaustive) listing of formulas that require the g_c term. In all cases, it is assumed that the standard English Engineering System units will be used.

- kinetic energy

$$E = \frac{m\text{v}^2}{2g_c} \quad (\text{in ft-lbf}) \qquad 1.7$$

- potential energy

$$E = \frac{mgz}{g_c} \quad (\text{in ft-lbf}) \qquad 1.8$$

- pressure at a depth

$$p = \frac{\rho gh}{g_c} \quad (\text{in lbf/ft}^2) \qquad 1.9$$

Example 1.2

A rocket that has a mass of 4000 lbm travels at 27,000 ft/sec. What is its kinetic energy in ft-lbf?

Solution

From Eq. 1.7,

$$E_k = \frac{m\text{v}^2}{2g_c} = \frac{(4000 \text{ lbm})\left(27{,}000 \dfrac{\text{ft}}{\text{sec}}\right)^2}{(2)\left(32.2 \dfrac{\text{lbm-ft}}{\text{lbf-sec}^2}\right)}$$

$$= 4.53 \times 10^{10} \text{ ft-lbf}$$

8. WEIGHT AND WEIGHT DENSITY

Weight, W, is a force exerted on an object due to its placement in a gravitational field. If a consistent set of units is used, Eq. 1.1 can be used to calculate the weight of a mass. In the English Engineering System, however, Eq. 1.10 must be used.

$$W = \frac{mg}{g_c} \qquad 1.10$$

Both sides of Eq. 1.10 can be divided by the volume of an object to derive the *weight density*, γ, of the object. Equation 1.11 illustrates that the weight density (in units of lbf/ft^3) can also be calculated by multiplying the mass density (in units of lbm/ft^3) by g/g_c. Since g and g_c usually have the same numerical values, the only effect of Eq. 1.12 is to change the units of density.

$$\frac{W}{V} = \left(\frac{m}{V}\right)\left(\frac{g}{g_c}\right) \qquad 1.11$$

$$\gamma = \frac{W}{V} = \left(\frac{m}{V}\right)\left(\frac{g}{g_c}\right) = \frac{\rho g}{g_c} \qquad 1.12$$

Weight does not occupy volume. Only mass has volume. The concept of weight density has evolved to simplify certain calculations, particularly fluid calculations. For example, pressure at a depth is calculated from Eq. 1.13. (Compare this with Eq. 1.9.)

$$p = \gamma h \qquad 1.13$$

9. THE ENGLISH GRAVITATIONAL SYSTEM

Not all English systems are inconsistent. Pounds can still be used as the unit of force as long as pounds are not used as the unit of mass. Such is the case with the consistent *English Gravitational System*.

If acceleration is given in ft/sec^2, the units of mass for a consistent system of units can be determined from Newton's second law. The combination of units in Eq. 1.14 is known as a *slug*. g_c is not needed at all since this system is consistent. It would be needed only to convert slugs to another mass unit.

$$\text{units of } m = \frac{\text{units of } F}{\text{units of } a} = \frac{\text{lbf}}{\frac{\text{ft}}{\text{sec}^2}} = \frac{\text{lbf-sec}^2}{\text{ft}} \qquad \textit{1.14}$$

Slugs and pounds-mass are not the same, as Fig. 1.1 illustrates. However, both are units for the same quantity: mass. Equation 1.15 will convert between slugs and pounds-mass.

$$\text{no. of slugs} = \frac{\text{no. of lbm}}{g_c} \qquad \textit{1.15}$$

The number of slugs is not derived by dividing the number of pounds-mass by the local gravity. g_c is used regardless of the local gravity. The conversion between feet and inches is not dependent on local gravity; neither is the conversion between slugs and pounds-mass.

Since the English Gravitational System is consistent, Eq. 1.16 can be used to calculate weight. The local gravitational acceleration is used.

$$W \text{ in lbf} = (m \text{ in slugs})\left(g \text{ in } \frac{\text{ft}}{\text{sec}^2}\right) \qquad \textit{1.16}$$

10. THE ABSOLUTE ENGLISH SYSTEM

The obscure *Absolute English System* takes the approach that mass must have units of pounds-mass (lbm) and the units of force can be derived from Newton's second law. The units for F cannot be simplified any more than they are in Eq. 1.17. This particular combination of units is known as a *poundal*.[5] A poundal is not the same as a pound.

$$\begin{aligned} \text{units of } F &= (\text{units of } m)(\text{units of } a) \\ &= (\text{lbm})\left(\frac{\text{ft}}{\text{sec}^2}\right) \\ &= \frac{\text{lbm-ft}}{\text{sec}^2} \qquad \textit{1.17} \end{aligned}$$

Poundals have not seen widespread use in the United States. The English Gravitational System (using slugs for mass) has greatly eclipsed the Absolute English System in popularity. Both are consistent systems, but there seems to be little need for poundals in modern engineering. Figure 1.2 shows the poundal in comparison to other common units of force.

11. METRIC SYSTEMS OF UNITS

Strictly speaking, a *metric system* is any system of units that is based on meters or parts of meters. This broad definition includes *mks systems* (based on meters,

[5]A poundal is equal to 0.03108 pounds-force.

Figure 1.2 *Common Force Units*

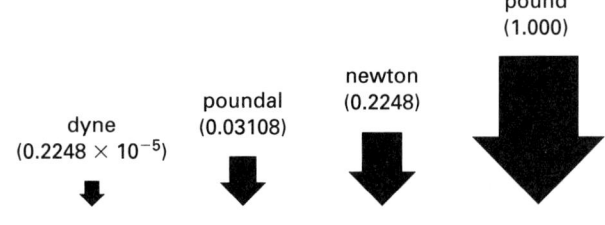

kilograms, and seconds) as well as *cgs systems* (based on centimeters, grams, and seconds).

Metric systems avoid the pounds-mass versus pounds-force ambiguity in two ways. First, a unit of weight is not established at all. All quantities of matter are specified as mass. Second, force and mass units do not share a common name.

The term *metric system* is not explicit enough to define which units are to be used for any given variable. For example, within the cgs system there is variation in how certain electrical and magnetic quantities are represented (resulting in the ESU and EMU systems). Also, within the mks system, it is common practice in some industries to use kilocalories as the unit of thermal energy, while the SI unit for thermal energy is the joule. This shows a lack of uniformity even within the metricated engineering community.[6]

The "metric" parts of this book use SI, which is the most developed and codified of the so-called metric systems.[7] There will be occasional variances with local engineering custom, but it is difficult to anticipate such variances within a book that must itself be consistent.

12. THE cgs SYSTEM

The cgs system is used widely by chemists and physicists. It is named for the three primary units used to construct its derived variables: the centimeter, the gram, and the second.

When Newton's second law is written in the cgs system, the following combination of units results.

$$\begin{aligned} \text{units of force} &= (m \text{ in g})\left(a \text{ in } \frac{\text{cm}}{\text{s}^2}\right) \\ &= \text{g·cm/s}^2 \qquad \textit{1.18} \end{aligned}$$

[6]In the "field test" of the metric system conducted over the past 200 years, other conventions are to use kilograms-force (kgf) instead of newtons and kgf/cm² for pressure (instead of pascals).
[7]SI units are an outgrowth of the *General Conference of Weights and Measures*, an international treaty organization that established the *Système International d'Unités* (*International System of Units*) in 1960. The United States subscribed to this treaty in 1975.

This combination of units for force is known as a *dyne*. Energy variables in the cgs system have units of dyne·cm or, equivalently, g·cm^2/s^2. This combination is known as an *erg*. There is no uniformly accepted unit of power in the cgs system, although calories per second is frequently used.

The fundamental volume unit in the cgs system is the cubic centimeter (cc). Since this is the same volume as one thousandth of a liter, units of milliliters (mL) are also used.

13. SI UNITS (THE mks SYSTEM)

SI units comprise an *mks system* (so named because it uses the meter, kilogram, and second as base units). All other units are derived from the base units, which are completely listed in Table 1.1. This system is fully consistent, and there is only one recognized unit for each physical quantity (variable).

Table 1.1 SI Base Units

quantity	name	symbol
length	meter	m
mass	kilogram	kg
time	second	s
electric current	ampere	A
temperature	kelvin	K
amount of substance	mole	mol
luminous intensity	candela	cd

Two types of units are used: base units and derived units. The *base units* are dependent only on accepted standards or reproducible phenomena. The *derived units* (Table 1.2 and Table 1.3) are made up of combinations of base units. Prior to 1995, radians and steradians were classified as *supplementary units*.

In addition, there is a set of non-SI units that may be used. This concession is primarily due to the significance and widespread acceptance of these units. Use of the non-SI units listed in Table 1.4 will usually create an inconsistent expression requiring conversion factors.

The SI unit of force can be derived from Newton's second law. This combination of units for force is known as a *newton*.

$$\text{units of force} = (m \text{ in kg})\left(a \text{ in } \frac{\text{m}}{\text{s}^2}\right)$$
$$= \text{kg·m/s}^2 \qquad \qquad 1.19$$

Energy variables in SI units have units of N·m or, equivalently, kg·m^2/s^2. Both of these combinations are known as a *joule*. The units of power are joules per second, equivalent to a *watt*.

Table 1.2 Some SI Derived Units with Special Names

quantity	name	symbol	expressed in terms of other units
frequency	hertz	Hz	1/s
force	newton	N	kg·m/s^2
pressure, stress	pascal	Pa	N/m^2
energy, work, quantity of heat	joule	J	N·m
power, radiant flux	watt	W	J/s
quantity of electricity, electric charge	coulomb	C	
electric potential, potential difference, electromotive force	volt	V	W/A
electric capacitance	farad	F	C/V
electric resistance	ohm	Ω	V/A
electric conductance	siemens	S	A/V
magnetic flux	weber	Wb	V·s
magnetic flux density	tesla	T	Wb/m^2
inductance	henry	H	Wb/A
luminous flux	lumen	lm	
illuminance	lux	lx	lm/m^2
plane angle	radian	rad	–
solid angle	steradian	sr	–

Example 1.3

A 10 kg block hangs from a cable. What is the tension in the cable? (Standard gravity equals 9.81 m/s^2.)

Solution

$$F = mg$$
$$= (10 \text{ kg})\left(9.81 \frac{\text{m}}{\text{s}^2}\right)$$
$$= 98.1 \text{ kg·m/s}^2 \quad (98.1 \text{ N})$$

Example 1.4

A 10 kg block is raised vertically 3 m. What is the change in potential energy?

Solution

$$\Delta E_p = mg\Delta h$$
$$= (10 \text{ kg})\left(9.81 \frac{\text{m}}{\text{s}^2}\right)(3 \text{ m})$$
$$= 294 \text{ kg·m}^2/\text{s}^2 \quad (294 \text{ J})$$

14. RULES FOR USING SI UNITS

In addition to having standardized units, the set of SI units also has rigid syntax rules for writing the units and combinations of units. Each unit is abbreviated with a

Table 1.3 *Some SI Derived Units*

quantity	description	symbol
area	square meter	m^2
volume	cubic meter	m^3
speed—linear	meter per second	m/s
speed—angular	radian per second	rad/s
acceleration—linear	meter per second squared	m/s^2
acceleration—angular	radian per second squared	rad/s^2
density, mass density	kilogram per cubic meter	kg/m^3
concentration (of amount of substance)	mole per cubic meter	mol/m^3
specific volume	cubic meter per kilogram	m^3/kg
luminance	candela per square meter	cd/m^2
absolute viscosity	pascal second	Pa·s
kinematic viscosity	square meters per second	m^2/s
moment of force	newton meter	N·m
surface tension	newton per meter	N/m
heat flux density, irradiance	watt per square meter	W/m^2
heat capacity, entropy	joule per kelvin	J/K
specific heat capacity, specific entropy	joule per kilogram kelvin	J/kg·K
specific energy	joule per kilogram	J/kg
thermal conductivity	watt per meter kelvin	W/m·K
energy density	joule per cubic meter	J/m^3
electric field strength	volt per meter	V/m
electric charge density	coulomb per cubic meter	C/m^3
surface density of charge, flux density	coulomb per square meter	C/m^2
permittivity	farad per meter	F/m
current density	ampere per square meter	A/m^2
magnetic field strength	ampere per meter	A/m
permeability	henry per meter	H/m
molar energy	joule per mole	J/mol
molar entropy, molar heat capacity	joule per mole kelvin	J/mol·K
radiant intensity	watt per steradian	W/sr

Table 1.4 *Acceptable Non-SI Units*

quantity	unit name	symbol or abbreviation	relationship to SI unit
area	hectare	ha	1 ha = 10 000 m^2
energy	kilowatt-hour	kW·h	1 kW·h = 3.6 MJ
mass	metric ton[a]	t	1 t = 1000 kg
plane angle	degree (of arc)	°	1° = 0.017 453 rad
speed of rotation	revolution per minute	r/min	1 r/min = $2\pi/60$ rad/s
temperature interval	degree Celsius	°C	1°C = 1K
time	minute	min	1 min = 60 s
	hour	h	1 h = 3600 s
	day (mean solar)	d	1 d = 86 400 s
	year (calendar)	a	1 a = 31 536 000 s
velocity	kilometer per hour	km/h	1 km/h = 0.278 m/s
volume	liter[b]	L	1 L = 0.001 m^3

[a]The international name for metric ton is *tonne*. The metric ton is equal to the *megagram* (Mg).
[b]The international symbol for liter is the lowercase l, which can be easily confused with the numeral 1. Several English-speaking countries have adopted the script ℓ or uppercase L (as does this book) as a symbol for liter in order to avoid any misinterpretation.

specific symbol. The following rules for writing and combining these symbols should be adhered to.

- The expressions for derived units in symbolic form are obtained by using the mathematical signs of multiplication and division. For example, units of velocity are m/s. Units of torque are N·m (not N-m or Nm).

- Scaling of most units is done in multiples of 1000.

- The symbols are always printed in roman type, regardless of the type used in the rest of the text. The only exception to this is in the use of the symbol for liter, where the use of the lower case "el" (l) may be confused with the numeral one (1). In this case, "liter" should be written out in full, or the script ℓ or L used. (L is used in this book.)

- Symbols are not pluralized: 45 kg (not 45 kgs).

- A period after a symbol is not used, except when the symbol occurs at the end of a sentence.

- When symbols consist of letters, there must always be a full space between the quantity and the symbols: 45 kg (not 45kg). However, for planar angle designations, no space is left: 42°12′45″ (not 42° 12′ 45″).

- All symbols are written in lowercase, except when the unit is derived from a proper name: m for meter, s for second, A for ampere, Wb for weber, N for newton, W for watt.

- Prefixes are printed without spacing between the prefix and the unit symbol (e.g., km is the symbol for kilometer). Table 1.5 lists common prefixes, their symbols, and their values.

- In text, when no number is involved, the unit should be spelled out. Example: Carpet is sold by the square meter, not by the m^2.

- Where a decimal fraction of a unit is used, a zero should always be placed before the decimal marker: 0.45 kg (not .45 kg). This practice draws attention to the decimal marker and helps avoid errors of scale.

- A practice in some countries is to use a comma as a decimal marker, while the practice in North America, the United Kingdom, and some other countries is to use a period as the decimal marker. Furthermore, in some countries that use the decimal comma, a period is frequently used to divide long numbers into groups of three. Because of these differing practices, spaces must be used instead of commas to separate long lines of digits into easily readable blocks of three digits with respect to the decimal marker: 32 453.246 072 5. A space (half-space preferred) is optional with a four-digit number: 1 234 or 1234.

- The word *ton* has multiple meanings. In the United States and Canada, the *short ton* of 2000 lbm (907.18 kg; 8896.44 N) is used. Previously, for commerce within the United Kingdom, a *long ton* of 2240 lbm (1016.05 kg) was used. A *metric ton* (or, *tonne*) is 1000 kg (10^6 Mg; 2205 lbm). In air conditioning industries, a *ton of refrigeration* is equivalent to a cooling rate of 200 Btu/min (12,000 Btu/hr). For explosives, a *ton of explosive power* is approximately the energy given off by 2000 lbm of TNT, standardized by international convention as 10^9 cal (1 Gcal; 4.184 GJ) for both "ton" and "tonne" designations. Various definitions are used in shipping, where *freight ton* (*measurement ton* or MTON) refers to volume, usually 40 ft^3. Many other specialty definitions are also in use.

Table 1.5 *SI Prefixes**

prefix	symbol	value
exa	E	10^{18}
peta	P	10^{15}
tera	T	10^{12}
giga	G	10^{9}
mega	M	10^{6}
kilo	k	10^{3}
hecto	h	10^{2}
deka (or deca)	da	10^{1}
deci	d	10^{-1}
centi	c	10^{-2}
milli	m	10^{-3}
micro	μ	10^{-6}
nano	n	10^{-9}
pico	p	10^{-12}
femto	f	10^{-15}
atto	a	10^{-18}

*There is no "B" (billion) prefix. In fact, the word "billion" means 10^9 in the United States but 10^{12} in most other countries. This unfortunate ambiguity is handled by avoiding the use of the term "billion."

15. UNCOMMON UNITS ENCOUNTERED IN THE UNITED STATES

Table 1.6 lists some units and their abbreviations that may be encountered in the United States. These units are found in specific industries and engineering specialty areas. In some cases, they may be archaic.

Table 1.6 *Uncommon Units Encountered in the United States (by abbreviation)*

abbreviation or symbol	meaning
℧	reciprocal of ohms (SI siemens)
□′	square feet
⌗	square feet
barn	area of a uranium nucleus (10^{-28} square meters)
BeV	billions (10^9) of electron-volts (SI GeV)
csf	hundreds of square feet
kBh	thousands of Btus per hour (ambiguous)
MBh	thousands of Btus per hour (ambiguous)
MeV	millions (10^6) of electron volts
mho	reciprocal of ohms (SI siemens)
MMBtuh	millions (10^6) of Btus per hour
square	100 square feet (roofing)

16. PRIMARY DIMENSIONS

Regardless of the system of units chosen, each variable representing a physical quantity will have the same *primary dimensions*. For example, velocity may be expressed in miles per hour (mph) or meters per second (m/s), but both units have dimensions of length per unit time. Length and time are two of the primary dimensions, as neither can be broken down into more basic dimensions. The concept of primary dimensions is useful when converting little-used variables between different systems of units, as well as in correlating experimental results (i.e., dimensional analysis).

There are three different sets of primary dimensions in use.[8] In the *MLθT system*, the primary dimensions are mass (M), length (L), time (θ), and temperature (T). All symbols are uppercase. In order to avoid confusion between time and temperature, the Greek letter theta is used for time.[9]

All other physical quantities can be derived from these primary dimensions.[10] For example, work in SI units has units of N·m. Since a newton is a kg·m/s^2, the primary dimensions of work are ML^2/θ^2. The primary dimensions for many important engineering variables are shown in Table 1.7. If it is more convenient to stay with traditional English units, it may be more desirable to work in the *FMLθTQ* system (sometimes called the *engineering dimensional system*). This system adds the primary dimensions of force (F) and heat (Q). Work (ft-lbf in the English system) has the primary dimensions of FL.

---

[8]One of these, the *FLθT* system, is not discussed here.

[9]This is the most common usage. There is a lack of consistency in the engineering world about the symbols for the primary dimensions in dimensional analysis. Some writers use t for time instead of θ. Some use H for heat instead of Q. And, in the worst mix-up of all, some have reversed the use of T and θ.

[10]A *primary dimension* is the same as a *base unit* in the SI set of units. The SI units add several other base units, as shown in Table 1.1, to deal with variables that are difficult to derive in terms of the four primary base units.

Table 1.7 Dimensions of Common Variables

variable (common symbol)	dimensional system		
	$ML\theta T$	$FL\theta T$	$FMLT\theta Q$
mass (m)	M	$F\theta^2/L$	M
force (F)	ML/θ^2	F	F
length (L)	L	L	L
time (θ or t)	θ	θ	θ
temperature (T)	T	T	T
work (W)	ML^2/θ^2	FL	FL
heat (Q)	ML^2/θ^2	FL	Q
acceleration (a)	L/θ^2	L/θ^2	L/θ^2
frequency (n or f)	$1/\theta$	$1/\theta$	$1/\theta$
area (A)	L^2	L^2	L^2
coefficient of thermal expansion (β)	$1/T$	$1/T$	$1/T$
density (ρ)	M/L^3	$F\theta^2/L^4$	M/L^3
dimensional constant (g_c)	1.0	1.0	$ML/\theta^2 F$
specific heat at constant pressure (c_p); at constant volume (c_v)	$L^2/\theta^2 T$	$L^2/\theta^2 T$	Q/MT
heat transfer coefficient (h); overall (U)	$M/\theta^3 T$	$F/\theta LT$	$Q/\theta L^2 T$
power (P)	ML^2/θ^3	FL/θ	FL/θ
heat flow rate ($\dot{Q}$)	ML^2/θ^3	FL/θ	Q/θ
kinematic viscosity (ν)	L^2/θ	L^2/θ	L^2/θ
mass flow rate ($\dot{m}$)	M/θ	$F\theta/L$	M/θ
mechanical equivalent of heat (J)	–	–	FL/Q
pressure (p)	$M/L\theta^2$	F/L^2	F/L^2
surface tension (σ)	M/θ^2	F/L	F/L
angular velocity (ω)	$1/\theta$	$1/\theta$	$1/\theta$
volumetric flow rate ($\dot{m}/\rho = \dot{V}$)	L^3/θ	L^3/θ	L^3/θ
conductivity (k)	$ML/\theta^3 T$	$F/\theta T$	$Q/L\theta T$
thermal diffusivity (α)	L^2/θ	L^2/θ	L^2/θ
velocity (v)	L/θ	L/θ	L/θ
viscosity, absolute (μ)	$M/L\theta$	$F\theta/L^2$	$F\theta/L^2$
volume (V)	L^3	L^3	L^3

(Compare this with the primary dimensions for work in the $ML\theta T$ system.) Thermodynamic variables are similarly simplified.

Dimensional analysis will be more conveniently carried out when one of the four-dimension systems ($ML\theta T$ or $FL\theta T$) is used. Whether the $ML\theta T$, $FL\theta T$, or $FML\theta TQ$ system is used depends on what is being derived and who will be using it, and whether or not a consistent set of variables is desired. Conversion constants such as g_c and J will almost certainly be required if the $ML\theta T$ system is used to generate variables for use in the English systems. It is also much more convenient to use the $FML\theta TQ$ system when working in the fields of thermodynamics, fluid flow, heat transfer, and so on.

17. DIMENSIONLESS GROUPS

A *dimensionless group* is derived as a ratio of two forces or other quantities. Considerable use of dimensionless groups is made in certain subjects, notably fluid mechanics and heat transfer. For example, the Reynolds number, Mach number, and Froude number are used to distinguish between distinctly different flow regimes in pipe flow, compressible flow, and open channel flow, respectively.

Table 1.8 contains information about the most common dimensionless groups used in fluid mechanics and heat transfer.

18. LINEAL AND BOARD FOOT MEASUREMENTS

The term *lineal* is often mistaken as a typographical error for *linear*. Although "lineal" has its own specific meaning very different from "linear," the two are often used interchangeably by engineers.[11] The adjective *lineal* is often encountered in the building trade (e.g., 12 lineal feet of lumber), where the term is used to distinguish it from board feet measurement.

A *board foot* (abbreviated bd-ft) is not a measure of length. Rather, it is a measure of volume used with lumber. Specifically, a board foot is equal to 144 in^3 (2.36×10^{-3} m^3). The name is derived from the volume of a board 1 foot square and 1 inch thick. In that sense, it is parallel in concept to the acre-foot. Since lumber cost is directly related to lumber weight and volume, the board foot unit is used in determining the overall lumber cost.

19. AREAL MEASUREMENTS

Areal is an adjective that means "per unit area." For example, "areal bolt density" is the number of bolts per unit area. In computer work, the *areal density* of a magnetic disk is the number of bits per square inch (square centimeter).

20. DIMENSIONAL ANALYSIS

Dimensional analysis is a means of obtaining an equation that describes some phenomenon without understanding the mechanism of the phenomenon. The most serious limitation is the need to know beforehand which variables influence the phenomenon. Once these are known or assumed, dimensional analysis can be applied by a routine procedure.

The first step is to select a system of primary dimensions. (See Sec. 1.16.) Usually the $ML\theta T$ system is used, although this choice may require the use of g_c and J in the final results.

---

[11]*Lineal* is best used when discussing a line of succession (e.g., a lineal descendant of a particular person). *Linear* is best used when discussing length (e.g., a linear dimension of a room).

Table 1.8 *Common Dimensionless Groups*

name	symbol	formula	interpretation
Biot number	Bi	$\dfrac{hL}{k_s}$	$\dfrac{\text{surface conductance}}{\text{internal conduction of solid}}$
Cauchy number	Ca	$\dfrac{\text{v}^2}{\dfrac{B_s}{\rho}} = \dfrac{\text{v}^2}{a^2}$	$\dfrac{\text{inertia force}}{\text{compressive force}} = \text{Mach number}^2$
Eckert number	Ec	$\dfrac{\text{v}^2}{2c_p \Delta T}$	$\dfrac{\text{temperature rise due to energy conversion}}{\text{temperature difference}}$
Eötvös number	Eo	$\dfrac{\rho g L^2}{\sigma}$	$\dfrac{\text{buoyancy}}{\text{surface tension}}$
Euler number	Eu	$\dfrac{\Delta p}{\rho \text{v}^2}$	$\dfrac{\text{pressure force}}{\text{inertia force}}$
Fourier number	Fo	$\dfrac{kt}{\rho c_p L^2} = \dfrac{\alpha t}{L^2}$	$\dfrac{\text{rate of conduction of heat}}{\text{rate of storage of energy}}$
Froude number*	Fr	$\dfrac{\text{v}^2}{gL}$	$\dfrac{\text{inertia force}}{\text{gravity force}}$
Graetz number*	Gz	$\left(\dfrac{D}{L}\right)\left(\dfrac{\text{v}\rho c_p D}{k}\right)$	$\dfrac{(\text{Re})(\text{Pr})}{L/D}$ $\dfrac{\text{heat transfer by convection in entrance region}}{\text{heat transfer by conduction}}$
Grashof number*	Gr	$\dfrac{g\beta \Delta T L^3}{\nu^2}$	$\dfrac{\text{buoyancy force}}{\text{viscous force}}$
Knudsen number	Kn	$\dfrac{\lambda}{L}$	$\dfrac{\text{mean free path of molecules}}{\text{characteristic length of object}}$
Lewis number*	Le	$\dfrac{\alpha}{D_c}$	$\dfrac{\text{thermal diffusivity}}{\text{molecular diffusivity}}$
Mach number	M	$\dfrac{\text{v}}{a}$	$\dfrac{\text{macroscopic velocity}}{\text{speed of sound}}$
Nusselt number	Nu	$\dfrac{hL}{k}$	$\dfrac{\text{temperature gradient at wall}}{\text{overall temperature difference}}$
Péclet number	Pé	$\dfrac{\text{v}\rho c_p D}{k}$	$\dfrac{(\text{Re})(\text{Pr})}{\dfrac{\text{heat transfer by convection}}{\text{heat transfer by conduction}}}$
Prandtl number	Pr	$\dfrac{\mu c_p}{k} = \dfrac{\nu}{\alpha}$	$\dfrac{\text{diffusion of momentum}}{\text{diffusion of heat}}$
Reynolds number	Re	$\dfrac{\rho \text{v} L}{\mu} = \dfrac{\text{v}L}{\nu}$	$\dfrac{\text{inertia force}}{\text{viscous force}}$
Schmidt number	Sc	$\dfrac{\mu}{\rho D_c} = \dfrac{\nu}{D_c}$	$\dfrac{\text{diffusion of momentum}}{\text{diffusion of mass}}$
Sherwood number*	Sh	$\dfrac{k_D L}{D_c}$	$\dfrac{\text{mass diffusivity}}{\text{molecular diffusivity}}$
Stanton number	St	$\dfrac{h}{\text{v}\rho c_p} = \dfrac{h}{c_p G}$	$\dfrac{\text{heat transfer at wall}}{\text{energy transported by stream}}$
Stokes number	Sk	$\dfrac{\Delta p L}{\mu \text{v}}$	$\dfrac{\text{pressure force}}{\text{viscous force}}$
Strouhal number*	Sl	$\dfrac{L}{t\text{v}} = \dfrac{L\omega}{\text{v}}$	$\dfrac{\text{frequency of vibration}}{\text{characteristic frequency}}$
Weber number	We	$\dfrac{\rho \text{v}^2 L}{\sigma}$	$\dfrac{\text{inertia force}}{\text{surface tension force}}$

*Multiple definitions exist, most often the square or square root of the formula shown.

The second step is to write a functional relationship between the dependent variable and the independent variable, x_i.

$$y = f(x_1, x_2, \ldots, x_m) \qquad 1.20$$

This function can be expressed as an exponentiated series. The C_1, a_i, b_i, ..., z_i in Eq. 1.21 are unknown constants.

$$y = C_1 x_1^{a_1} x_2^{b_1} x_3^{c_1} \cdots x_m^{z_1} + C_2 x_1^{a_2} x_2^{b_2} x_3^{c_2} \cdots x_m^{z_2} + \cdots$$

$$1.21$$

The key to solving Eq. 1.21 is that each term on the right-hand side must have the same dimensions as y. Simultaneous equations are used to determine some of the a_i, b_i, c_i, and z_i. Experimental data are required to determine the C_i and remaining exponents. In most analyses, it is assumed that the $C_i = 0$ for $i \geq 2$.

Since this method requires working with m different variables and n different independent dimensional quantities (such as M, L, θ, and T), an easier method is desirable. One simplification is to combine the m variables into dimensionless groups called *pi-groups*. (See Table 1.7.)

If these dimensionless groups are represented by π_1, π_2, π_3, ..., π_k, the equation expressing the relationship between the variables is given by the *Buckingham π-theorem*.

$$f(\pi_1, \pi_2, \pi_3, \ldots, \pi_k) = 0 \qquad 1.22$$

$$k = m - n \qquad 1.23$$

The dimensionless pi-groups are usually found from the m variables according to an intuitive process.

Example 1.5

A solid sphere rolls down a submerged incline. Find an equation for the velocity, v.

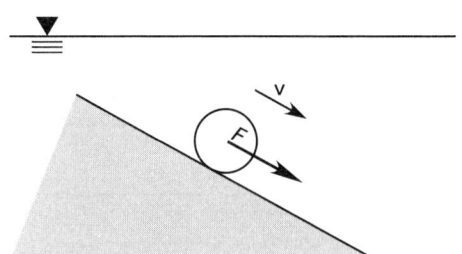

Solution

Assume that the velocity depends on the force, F, due to gravity, the diameter of the sphere, D, the density of the fluid, ρ, and the viscosity of the fluid, μ.

$$v = f(F, D, \rho, \mu) = CF^a D^b \rho^c \mu^d$$

This equation can be written in terms of the primary dimensions of the variables.

$$\frac{L}{\theta} = C\left(\frac{ML}{\theta^2}\right)^a L^b \left(\frac{M}{L^3}\right)^c \left(\frac{M}{L\theta}\right)^d$$

Since L on the left-hand side has an implied exponent of one, a necessary condition is

$$1 = a + b - 3c - d \quad (L)$$

Similarly, the other necessary conditions are

$$-1 = -2a - d \quad (\theta)$$
$$0 = a + c + d \quad (M)$$

Solving simultaneously yields

$$b = -1$$
$$c = a - 1$$
$$d = 1 - 2a$$
$$v = CF^a D^{-1} \rho^{a-1} \mu^{1-2a}$$
$$= C\left(\frac{\mu}{D\rho}\right)\left(\frac{F\rho}{\mu^2}\right)^a$$

C and a must be determined experimentally.

2 Properties of Areas

Nomenclature

A	area	units^2
b	base distance	units
c	distance to extreme fiber	units
d	separation distance	units
h	height distance	units
I	moment of inertia	units^4
J	polar moment of inertia	units^4
L	length	units
P	product of inertia	units^4
Q	first moment of the area	units^3
r	radius	units
r	radius of gyration	units
S	section modulus	units^3
u	distance in the u-direction	units
v	distance in the v-direction	units
V	volume	units^3
x	distance in the x-direction	units
y	distance in the y-direction	units

Symbols

θ	angle	deg

Subscripts

c	centroidal

1. CENTROID OF AN AREA

The *centroid of an area* is analogous to the center of gravity of a homogeneous body.[1] The centroid is often described as the point at which a thin homogeneous plate would balance. This definition, however, combines the definitions of centroid and center of gravity and

---

[1]The analogy has been simplified. A three-dimensional body also has a centroid. The centroid and center of gravity will coincide when the body is homogeneous.

implies that gravity is required to identify the centroid, which is not true.

The location of the centroid of an area bounded by the x- and y-axes and the mathematical function $y = f(x)$ can be found by the *integration method* and by using Eq. 2.1 through Eq. 2.4. The centroidal location depends only on the geometry of the area and is identified by the coordinates (x_c, y_c). Some references place a bar over the coordinates of the centroid to indicate an average point, such as $(\overline{x}, \overline{y})$.

$$x_c = \frac{\int x \, dA}{A} \qquad 2.1$$

$$y_c = \frac{\int y \, dA}{A} \qquad 2.2$$

$$A = \int f(x) \, dx \qquad 2.3$$

$$dA = f(x)dx = g(y)dy \qquad 2.4$$

The locations of the centroids of *basic shapes*, such as triangles and rectangles, are well known. The most common basic shapes have been included in App. 2.A. There should be no need to derive centroidal locations for these shapes by the integration method.

The centroid of a complex area can be found from Eq. 2.5 and Eq. 2.6 if the area can be divided into the basic shapes in App. 2.A. This process is simplified when all or most of the subareas adjoin the reference axis. Example 2.1 illustrates this method.

$$x_c = \frac{\sum_i A_i x_{c,i}}{\sum_i A_i} \qquad 2.5$$

$$y_c = \frac{\sum_i A_i y_{c,i}}{\sum_i A_i} \qquad 2.6$$

Example 2.1

An area is bounded by the x- and y-axes, the line $x = 2$, and the function $y = e^{2x}$. Find the x-component of the centroid.

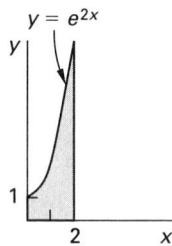

Solution

First, use Eq. 2.3 to find the area.

$$A = \int f(x)\,dx = \int_{x=0}^{x=2} e^{2x}\,dx$$

$$= \tfrac{1}{2}e^{2x}\Big|_0^2 = 27.3 - 0.5$$

$$= 26.8 \text{ units}^2$$

Since y is a function of x, dA must be expressed in terms of x. From Eq. 2.4,

$$dA = f(x)\,dx = e^{2x}\,dx$$

Finally, use Eq. 2.1 to find x_c.

$$x_c = \frac{\int x\,dA}{A} = \frac{1}{26.8}\int_{x=0}^{x=2} xe^{2x}\,dx$$

$$= \left(\frac{1}{26.8}\right)\left(\tfrac{1}{2}xe^{2x} - \tfrac{1}{4}e^{2x}\Big|_0^2\right)$$

$$= 1.54 \text{ units}$$

Example 2.2

Find the y-coordinate of the centroid of the area shown.

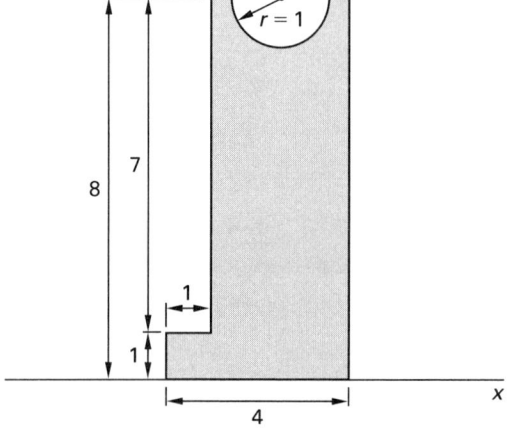

Solution

The x-axis is the reference axis. The area is divided into basic shapes of a 1×1 square, a 3×8 rectangle, and a half-circle of radius 1. (The area could also be divided into 1×4 and 3×7 rectangles and the half-circle, but then the 3×7 rectangle would not adjoin the x-axis.)

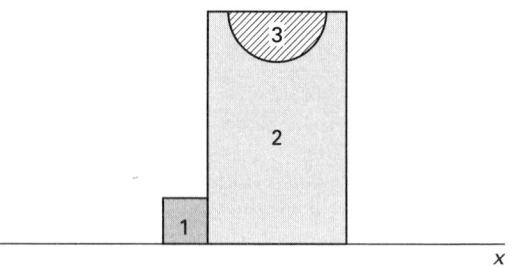

First, calculate the areas of the basic shapes. Notice that the half-circle area is negative since it represents a cutout.

$$A_1 = (1.0)(1.0) = 1.0 \text{ units}^2$$

$$A_2 = (3.0)(8.0) = 24.0 \text{ units}^2$$

$$A_3 = -\tfrac{1}{2}\pi r^2 = -\tfrac{1}{2}\pi(1.0)^2 = -1.57 \text{ units}^2$$

Next, find the y-components of the centroids of the basic shapes. Most are found by inspection, but App. 2.A can be used for the half-circle. The centroidal location for the half-circle is positive.

$$y_{c,1} = 0.5 \text{ units}$$

$$y_{c,2} = 4.0 \text{ units}$$

$$y_{c,3} = 8.0 - 0.424 = 7.576 \text{ units}$$

Finally, use Eq. 2.6.

$$y_c = \frac{\sum A_i y_{c,i}}{\sum A_i}$$

$$= \frac{(1.0)(0.5) + (24.0)(4.0) + (-1.57)(7.576)}{1.0 + 24.0 - 1.57}$$

$$= 3.61 \text{ units}$$

2. FIRST MOMENT OF THE AREA

The quantity $\int x\,dA$ is known as the *first moment of the area* or *first area moment* with respect to the y-axis. Similarly, $\int y\,dA$ is known as the first moment of the area with respect to the x-axis. By rearranging Eq. 2.1 and Eq. 2.2, the first moment of the area can be calculated from the area and centroidal distance.

$$Q_y = \int x\,dA = x_c A \qquad 2.7$$

$$Q_x = \int y\,dA = y_c A \qquad 2.8$$

In basic engineering, the two primary applications of the first moment concept are to determine centroidal locations and shear stress distributions. In the latter application, the first moment of the area is known as the *statical moment*.

3. CENTROID OF A LINE

The location of the *centroid of a line* can be defined by Eq. 2.9 and Eq. 2.10, which are analogous to the equations used for centroids of areas.

$$x_c = \frac{\int x \, dL}{L} \qquad 2.9$$

$$y_c = \frac{\int y \, dL}{L} \qquad 2.10$$

Since equations of lines are typically in the form $y = f(x)$, dL must be expressed in terms of x or y.

$$dL = \left(\sqrt{\left(\frac{dy}{dx}\right)^2 + 1} \right) dx \qquad 2.11$$

$$dL = \left(\sqrt{\left(\frac{dx}{dy}\right)^2 + 1} \right) dy \qquad 2.12$$

4. THEOREMS OF PAPPUS-GULDINUS

The *Theorems of Pappus-Guldinus* define the surface and volume of revolution (i.e., the surface area and volume generated by revolving a curve around a fixed axis).

- *Theorem I:* The area of a surface of revolution is equal to the product of the length of the generating curve and the distance traveled by the centroid of the curve while the surface is being generated. (See Fig. 2.1.)

Figure 2.1 *Surface of Revolution*

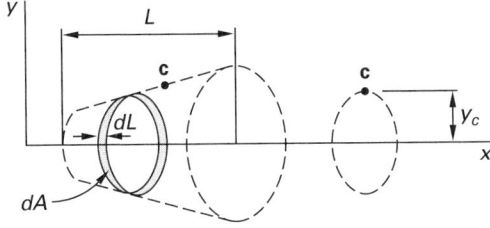

When a part, dL, of a line, L, is revolved about the x-axis, a differential ring having surface area dA is generated.

$$dA = 2\pi y \, dL \qquad 2.13$$

$$A = \int dA = 2\pi \int y \, dL = 2\pi y_c L \qquad 2.14$$

- *Theorem II:* The volume of a surface of revolution is equal to the generating area times the distance traveled by the centroid of the area in generating the volume. (See Fig. 2.2.)

Figure 2.2 *Volume of Revolution*

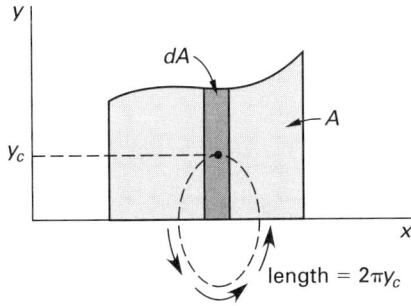

When a differential plane area, dA, is revolved about the x-axis and does not intersect the y-axis, it generates a ring of volume dV.

$$dV = \pi y^2 \, dx = \pi y \, dA \qquad 2.15$$

$$V = 2\pi y_c A \qquad 2.16$$

5. MOMENT OF INERTIA OF AN AREA

The *moment of inertia*, I, of an area is needed in mechanics of materials problems. It is convenient to think of the moment of inertia of a beam's cross-sectional area as a measure of the beam's ability to resist bending. Given equal loads, a beam with a small moment of inertia will bend more than a beam with a large moment of inertia.

Since the moment of inertia represents a resistance to bending, it is always positive. Since a beam can be unsymmetrical (e.g., a rectangular beam) and can be stronger in one direction than another, the moment of inertia depends on orientation. Therefore, a reference axis or direction must be specified.

The moment of inertia taken with respect to one of the axes in the rectangular coordinate system is sometimes referred to as the *rectangular moment of inertia*.

The symbol I_x is used to represent a moment of inertia with respect to the x-axis. Similarly, I_y is the moment of inertia with respect to the y-axis. I_x and I_y do not normally combine and are not components of some resultant moment of inertia.

Any axis can be chosen as the reference axis, and the value of the moment of inertia will depend on the reference selected. The moment of inertia taken with respect to an axis passing through the area's centroid is known as the *centroidal moment of inertia, I_{cx}* or *I_{cy}*. The centroidal moment of inertia is the smallest possible moment of inertia for the shape.

The *integration method* can be used to calculate the moment of inertia of a function that is bounded by the *x*- and *y*-axes and a curve $y = f(x)$. From Eq. 2.17 and Eq. 2.18, it is apparent why the moment of inertia is also known as the *second moment of the area* or *second area moment*.

$$I_x = \int y^2 \, dA \qquad 2.17$$

$$I_y = \int x^2 \, dA \qquad 2.18$$

$$dA = f(x) \, dx = g(y) \, dy \qquad 2.19$$

The moments of inertia of the *basic shapes* are well known and are listed in App. 2.A.

Example 2.3

What is the centroidal moment of inertia with respect to the *x*-axis of a rectangle 5.0 units wide and 8.0 units tall?

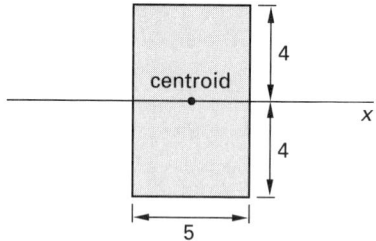

Solution

Since the centroidal moment of inertia is needed, the reference line passes through the centroid. So, from App. 2.A, the centroidal moment of inertia is

$$I_{cx} = \frac{bh^3}{12} = \frac{(5)(8)^3}{12} = 213.3 \text{ units}^4$$

Example 2.4

What is the moment of inertia with respect to the *y*-axis of the area bounded by the *y*-axis, the line $y = 8.0$, and the parabola $y^2 = 8x$?

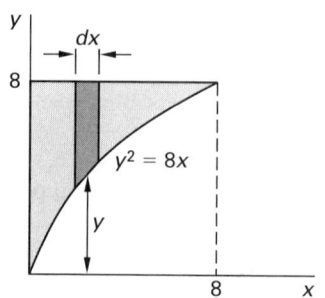

Solution

This problem is more complex than it first appears, since the area is above the curve, bounded not by $y = 0$ but by $y = 8$. In particular, dA must be determined correctly.

$$y = \sqrt{8x}$$

Use Eq. 2.4.

$$dA = \left(8 - f(x)\right) dx = (8 - y) \, dx$$
$$= (8 - \sqrt{8x}) \, dx$$

Equation 2.18 is used to calculate the moment of inertia with respect to the *y*-axis.

$$I_y = \int x^2 \, dA = \int_0^8 x^2 (8 - \sqrt{8x}) \, dx$$
$$= \frac{8}{3} x^3 - \left(\frac{4\sqrt{2}}{7}\right) x^{7/2} \Big|_0^8$$
$$= 195.0 \text{ units}^4$$

6. PARALLEL AXIS THEOREM

If the moment of inertia is known with respect to one axis, the moment of inertia with respect to another parallel axis can be calculated from the *parallel axis theorem*, also known as the *transfer axis theorem*. In Eq. 2.20, *d* is the distance between the centroidal axis and the second, parallel axis.

$$I_{\text{parallel axis}} = I_c + Ad^2 \qquad 2.20$$

The second term in Eq. 2.20 is often much larger than the first term. Areas close to the centroidal axis do not affect the moment of inertia significantly. This principle is exploited by structural steel shapes (such as is shown in Fig. 2.3) that derive bending resistance from *flanges* located away from the centroidal axis. The *web* does not contribute significantly to the moment of inertia.

Figure 2.3 *Structural Steel W-Shape*

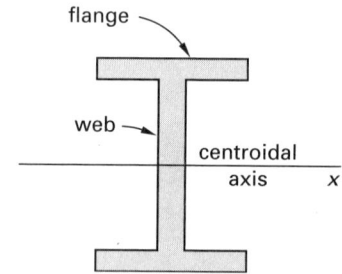

Example 2.5

Find the moment of inertia about the x-axis for the inverted-T area shown.

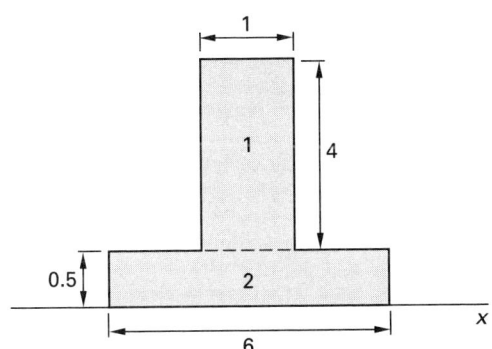

Solution

The area is divided into two basic shapes: 1 and 2. From App. 2.A, the moment of inertia of basic shape 2 with respect to the x-axis is

$$I_{x,2} = \frac{bh^3}{3} = \frac{(6.0)(0.5)^3}{3} = 0.25 \text{ units}^4$$

The moment of inertia of basic shape 1 about its own centroid is

$$I_{cx,1} = \frac{bh^3}{12} = \frac{(1)(4)^3}{12} = 5.33 \text{ units}^4$$

The x-axis is located 2.5 units from the centroid of basic shape 1. From the parallel axis theorem, Eq. 2.20, the moment of inertia of basic shape 1 about the x-axis is

$$I_{x,1} = I_{cx,1} + A_1 d_1^2 = 5.33 + (4)(2.5)^2$$
$$= 30.33 \text{ units}^4$$

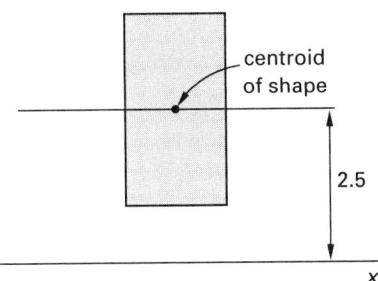

The total moment of inertia of the T-area is

$$I_x = I_{x,1} + I_{x,2} = 30.33 \text{ units}^4 + 0.25 \text{ units}^4$$
$$= 30.58 \text{ units}^4$$

Example 2.6

Find the moment of inertia about the horizontal centroidal axis for the inverted-T area shown in Ex. 2.5.

Solution

The first step is to find the location of the centroid. The areas and centroidal locations (with respect to the x-axis) of the two basic shapes are

$$A_1 = (4.0)(1.0) = 4.0 \text{ units}^2$$

$$A_2 = (0.5)(6.0) = 3.0 \text{ units}^2$$

$$y_{c,1} = 2.5 \text{ units}$$

$$y_{c,2} = 0.25 \text{ units}$$

From Eq. 2.6, the composite centroid is located at

$$y_c = \frac{A_1 y_{c,1} + A_2 y_{c,2}}{A_1 + A_2} = \frac{(4.0)(2.5) + (3.0)(0.25)}{4.0 + 3.0}$$
$$= 1.536 \text{ units}$$

The distances between the centroids of the basic shapes and the composite shape are

$$d_1 = 2.5 - 1.536 = 0.964 \text{ units}$$

$$d_2 = 1.536 - 0.25 = 1.286 \text{ units}$$

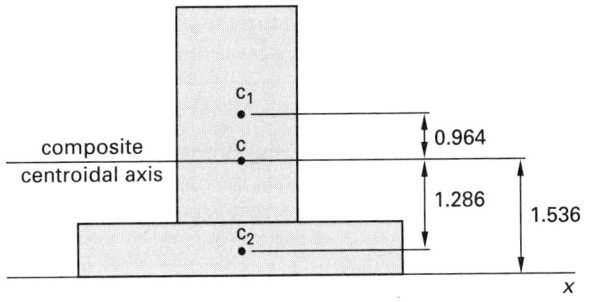

The centroidal moments of inertia of the basic shapes with respect to an axis parallel to the x-axis are

$$I_{cx,1} = \frac{bh^3}{12} = \frac{(1)(4)^3}{12} = 5.33 \text{ units}^4$$

$$I_{cx,2} = \frac{bh^3}{12} = \frac{(6)(0.5)^3}{12} = 0.0625 \text{ units}^4$$

Using Eq. 2.20, the centroidal moment of inertia of the inverted-T area is

$$I_{cx} = I_{cx,1} + A_1 d_1^2 + I_{cx,2} + A_2 d_2^2$$
$$= 5.33 + (4.0)(0.964)^2 + 0.0625 + (3.0)(1.286)^2$$
$$= 14.07 \text{ units}^4$$

7. POLAR MOMENT OF INERTIA

The *polar moment of inertia*, J, is required in torsional shear stress calculations.[2] It can be thought of as a measure of an area's resistance to torsion (twisting). The definition of a polar moment of inertia of a two-dimensional area requires three dimensions because the reference axis for a polar moment of inertia of a plane area is perpendicular to the plane area.

The polar moment of inertia is derived from Eq. 2.21.

$$J = \int (x^2 + y^2) dA \qquad 2.21$$

It is often easier to use the *perpendicular axis theorem* to quickly calculate the polar moment of inertia.

- *perpendicular axis theorem:* The polar moment of inertia of a plane area about an axis normal to the plane is equal to the sum of the moments of inertia about any two mutually perpendicular axes lying in the plane and passing through the given axis.

$$J = I_x + I_y \qquad 2.22$$

Since the two perpendicular axes can be chosen arbitrarily, it is most convenient to use the centroidal moments of inertia.

$$J = I_{cx} + I_{cy} \qquad 2.23$$

Example 2.7

What is the centroidal polar moment of inertia of a circular area of radius r?

Solution

From App. 2.A, the centroidal moment of inertia of a circle with respect to the x-axis is

$$I_{cx} = \frac{\pi r^4}{4}$$

Since the area is symmetrical, I_{cy} and I_{cx} are the same. From Eq. 2.23,

$$J_c = I_{cx} + I_{cy} = \frac{\pi r^4}{4} + \frac{\pi r^4}{4} = \frac{\pi r^4}{2}$$

8. RADIUS OF GYRATION

Every nontrivial area has a centroidal moment of inertia. Usually, some portions of the area are close to the centroidal axis and other portions are farther away. The *transverse radius of gyration*, or just *radius of gyration*, r, is an imaginary distance from the centroidal axis at which the entire area can be assumed to exist without affecting the moment of inertia. Despite the name

---

[2]The symbols I_z and I_{xy} are also encountered. However, the symbol J is more common.

"radius," the radius of gyration is not limited to circular shapes or to polar axes. This concept is illustrated in Fig. 2.4.

Figure 2.4 *Radius of Gyration*

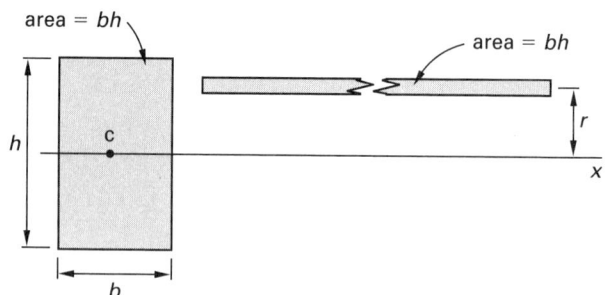

The method of calculating the radius of gyration is based on the parallel axis theorem. If all of the area is located a distance, r, from the original centroidal axis, there will be no I_c term in Eq. 2.20. Only the Ad^2 term will contribute to the moment of inertia.

$$I = r^2 A \qquad 2.24$$

$$r = \sqrt{\frac{I}{A}} \qquad 2.25$$

The concept of *least radius of gyration* comes up frequently in column design problems. (The column will tend to buckle about an axis that produces the smallest radius of gyration.) Usually, finding the least radius of gyration for symmetrical section will mean solving Eq. 2.25 twice: once with I_x to find r_x and once with I_y to find r_y. The smallest value of r is the least radius of gyration.

The analogous quantity in the polar system is

$$r = \sqrt{\frac{J}{A}} \qquad 2.26$$

Just as the polar moment of inertia, J, can be calculated from the two rectangular moments of inertia, the polar radius of gyration can be calculated from the two rectangular radii of gyration.

$$r^2 = r_x^2 + r_y^2 \qquad 2.27$$

Example 2.8

What is the radius of gyration of the rectangular shape in Ex. 2.3?

Solution

The area of the rectangle is

$$A = bh = (5)(8) = 40 \text{ units}^2$$

From Eq. 2.25, the radius of gyration is

$$r_x = \sqrt{\frac{I_x}{A}} = \sqrt{\frac{213.3 \text{ units}^4}{40 \text{ units}^2}} = 2.31 \text{ units}$$

2.31 units is the distance from the centroidal x-axis that an infinitely long strip with an area of 40 square units would have to be located in order to have a moment of inertia of 213.3 units[4].

9. PRODUCT OF INERTIA

The *product of inertia*, P_{xy}, of a two-dimensional area is useful when relating properties of areas evaluated with respect to different axes. It is found by multiplying each differential element of area by its x- and y-coordinate and then summing over the entire area. (See Fig. 2.5.)

$$P_{xy} = \int xy \, dA \qquad 2.28$$

Figure 2.5 *Calculating the Product of Inertia*

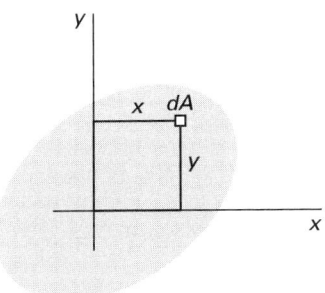

The product of inertia is zero when either axis is an axis of symmetry. Since the axes can be chosen arbitrarily, the area may be in one of the negative quadrants, and the product of inertia may be negative.

The parallel axis theorem for products of inertia, as shown in Fig. 2.6, is given by Eq. 2.29. (Both axes are allowed to move to new positions.) x'_c and y'_c are the coordinates of the centroid in the new coordinate system.

$$P_{x'y'} = P_{c,xy} + x'_c y'_c A \qquad 2.29$$

10. SECTION MODULUS

In the analysis of beams, the outer compressive (or tensile) surface is known as the *extreme fiber*. The distance, c, from the centroidal axis of the beam cross section to the extreme fiber is the distance to the extreme fiber. The *section modulus*, S, combines the centroidal moment of inertia and the distance to the extreme fiber.

$$S = \frac{I_c}{c} \qquad 2.30$$

Figure 2.6 *Parallel Axis Theorem for Products of Inertia*

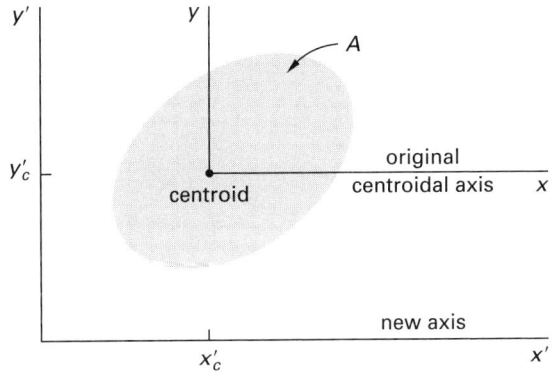

11. ROTATION OF AXES

Figure 2.7 shows rotation of the x-y axes through an angle, θ, into a new set of u-v axes, without rotating the area. If the moments and product of inertia of the area are known with respect to the old x-y axes, the new properties can be calculated from Eq. 2.31 through Eq. 2.33.

$$I_u = I_x \cos^2\theta - 2P_{xy}\sin\theta\cos\theta + I_y\sin^2\theta$$
$$= \tfrac{1}{2}(I_x + I_y) + \tfrac{1}{2}(I_x - I_y)\cos 2\theta$$
$$- P_{xy}\sin 2\theta \qquad 2.31$$

$$I_v = I_x\sin^2\theta + 2P_{xy}\sin\theta\cos\theta + I_y\cos^2\theta$$
$$= \tfrac{1}{2}(I_x + I_y) - \tfrac{1}{2}(I_x - I_y)\cos 2\theta$$
$$+ P_{xy}\sin 2\theta \qquad 2.32$$

$$P_{uv} = I_x\sin\theta\cos\theta + P_{xy}(\cos^2\theta - \sin^2\theta)$$
$$- I_y\sin\theta\cos\theta$$
$$= \tfrac{1}{2}(I_x - I_y)\sin 2\theta + P_{xy}\cos 2\theta \qquad 2.33$$

Since the polar moment of inertia about a fixed axis perpendicular to any two orthogonal axes in the plane is constant, the polar moment of inertia is unchanged by the rotation.

$$J_{xy} = I_x + I_y = I_u + I_v = J_{uv} \qquad 2.34$$

Figure 2.7 *Rotation of Axes*

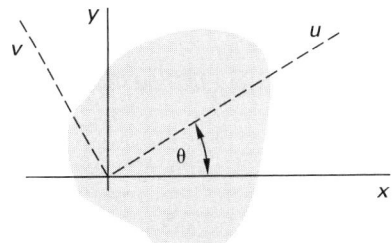

Example 2.9

What is the centroidal area moment of inertia of a 6×6 square that is rotated $45°$ from its "flat" orientation?

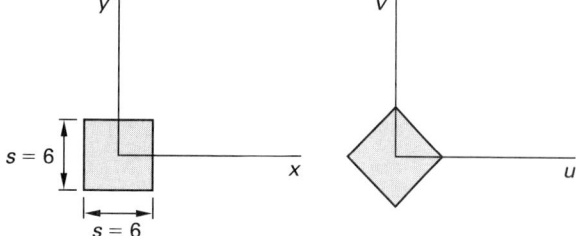

Solution

From App. 2.A, the centroidal moments of inertia with respect to the x- and y-axes are

$$I_x = I_y = \frac{s^4}{12} = \frac{(6)^4}{12} = 108 \text{ units}^4$$

Since the centroidal x- and y-axes are axes of symmetry, the product of inertia is zero.

Use Eq. 2.31.

$$\begin{aligned} I_u &= I_x \cos^2\theta - 2P_{xy}\sin\theta\cos\theta + I_y\sin^2\theta \\ &= (108)\cos^2 45° - 0 + (108)\sin^2 45° \\ &= 108 \text{ units}^4 \end{aligned}$$

The centroidal moment of inertia of a square is the same regardless of rotation angle.

12. PRINCIPAL AXES FOR AREA PROPERTIES

Referring to Fig. 2.7, there is one angle, θ, that will maximize the moment of inertia, I_u. This angle can be found from calculus by setting $dI_u/d\theta = 0$. The resulting equation defines two angles, one that maximizes I_u and one that minimizes I_u.

$$\tan 2\theta = \frac{-2P_{xy}}{I_x - I_y} \qquad 2.35$$

The two angles that satisfy Eq. 2.35 are $90°$ apart. The set of u-v axes defined by Eq. 2.35 are known as *principal axes*. The moments of inertia about the principal axes are defined by Eq. 2.36 and are known as the *principal moments of inertia*.

$$I_{\max,\min} = \tfrac{1}{2}(I_x + I_y) \pm \sqrt{\tfrac{1}{4}(I_x - I_y)^2 + P_{xy}^2} \qquad 2.36$$

13. MOHR'S CIRCLE FOR AREA PROPERTIES

Once I_x, I_y, and P_{xy} are known, *Mohr's circle* can be drawn to graphically determine the moments of inertia about the principal axes. The procedure for drawing Mohr's circle is given as follows.

step 1: Determine I_x, I_y, and P_{xy} for the existing set of axes.

step 2: Draw a set of I-P_{xy} axes.

step 3: Plot the center of the circle, point **c**, by calculating distance c along the I-axis. (See Fig. 2.8.)

$$c = \tfrac{1}{2}(I_x + I_y) \qquad 2.37$$

Figure 2.8 *Mohr's Circle*

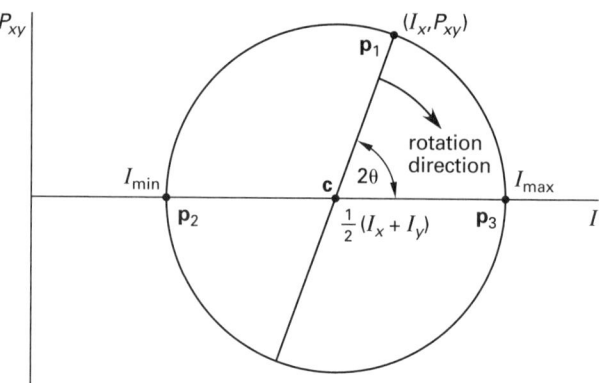

step 4: Plot the point $\mathbf{p}_1 = (I_x, P_x)$.

step 5: Draw a line from point $\mathbf{p}_1$ through center **c** and extend it an equal distance below the I-axis. This is the diameter of the circle.

step 6: Using the center **c** and point $\mathbf{p}_1$, draw the circle. An alternate method of constructing the circle is to draw a circle of radius r.

$$r = \sqrt{\tfrac{1}{4}(I_x - I_y)^2 + P_{xy}^2} \qquad 2.38$$

step 7: Point $\mathbf{p}_2$ defines $I_{\min}$. Point $\mathbf{p}_3$ defines $I_{\max}$.

step 8: Determine the angle θ as half of the angle 2θ on the circle. This angle corresponds to $I_{\max}$. (The axis giving the minimum moment of inertia is perpendicular to the maximum axis.) The sense of this angle and the sense of the rotation are the same. That is, the direction that the diameter would have to be turned in order to coincide with the $I_{\max}$-axis has the same sense as the rotation of the x-y axes needed to form the principal u-v axes. (See Fig. 2.9.)

Figure 2.9 *Principal Axes from Mohr's Circle*

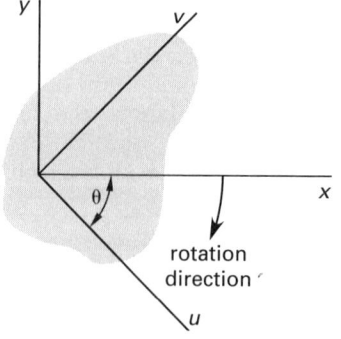

3 Algebra

1. INTRODUCTION

Engineers working in design and analysis encounter mathematical problems on a daily basis. Although algebra and simple trigonometry are often sufficient for routine calculations, there are many instances when certain advanced subjects are needed. This chapter and the following, in addition to supporting the calculations used in other chapters, consolidate the mathematical concepts most often needed by engineers.

2. SYMBOLS USED IN THIS BOOK

Many symbols, letters, and Greek characters are used to represent variables in the formulas used throughout this book. These symbols and characters are defined in the nomenclature section of each chapter. However, some of the other symbols in this book are listed in Table 3.2.

3. GREEK ALPHABET

Table 3.1 lists the Greek alphabet.

Table 3.1 The Greek Alphabet

A	α	alpha	N	ν	nu
B	β	beta	Ξ	ξ	xi
Γ	γ	gamma	O	o	omicron
Δ	δ	delta	Π	π	pi
E	ϵ	epsilon	P	ρ	rho
Z	ζ	zeta	Σ	σ	sigma
H	η	eta	T	τ	tau
Θ	θ	theta	Υ	υ	upsilon
I	ι	iota	Φ	ϕ	phi
K	κ	kappa	X	χ	chi
Λ	λ	lambda	Ψ	ψ	psi
M	μ	mu	Ω	ω	omega

4. TYPES OF NUMBERS

The *numbering system* consists of three types of numbers: real, imaginary, and complex. *Real numbers*, in turn, consist of rational numbers and irrational numbers. *Rational real numbers* are numbers that can be written as the ratio of two integers (e.g., 4, $^2/_5$, and $^1/_3$).[1] *Irrational real numbers* are nonterminating, nonrepeating numbers that cannot be expressed as the ratio of two integers (e.g., π and $\sqrt{2}$). Real numbers can be positive or negative.

Imaginary numbers are square roots of negative numbers. The symbols i and j are both used to represent the square root of -1.[2] For example, $\sqrt{-5} = \sqrt{5}\sqrt{-1} = \sqrt{5}i$. *Complex numbers* consist of combinations of real and imaginary numbers (e.g., $3 - 7i$).

5. SIGNIFICANT DIGITS

The significant digits in a number include the leftmost, nonzero digits to the rightmost digit written. Final answers from computations should be rounded off to the number of decimal places justified by the data. The answer can be no more accurate than the least accurate number in the data. Of course, rounding should be done on final calculation results only. It should not be done on interim results.

There are two ways that significant digits can affect calculations. For the operations of multiplication and division, the final answer is rounded to the number of

---

[1] Notice that 0.3333333... is a nonterminating number, but as it can be expressed as a ratio of two integers (i.e., 1/3), it is a rational number.
[2] The symbol j is used to represent the square root of -1 in electrical calculations to avoid confusion with the current variable, i.

Table 3.2 *Symbols Used in This Book*

symbol	name	use	example
$\sum$	sigma	series summation	$\sum\limits_{i=1}^{3} x_i = x_1 + x_2 + x_3$
π	pi	3.1415926...	$P = \pi D$
e	base of natural logs	2.71828...	
Π	pi	series multiplication	$\prod\limits_{i=1}^{3} x_i = x_1 x_2 x_3$
Δ	delta	change in quantity	$\Delta h = h_2 - h_1$
$\overline{}$	over bar	average value	$\overline{x}$
$\dot{}$	over dot	per unit time	$\dot{m} = $ mass flowing per second
!	factorial[a]		$x! = x(x-1)(x-2)\cdots(2)(1)$
$\vert\;\vert$	absolute value[b]		$\vert -3 \vert = +3$
$\sim$	similarity		$\Delta \text{ABC} \sim \Delta \text{DEF}$
$\approx$	approximately equal to		$x \approx 1.5$
$\cong$	congruency		$\text{ST} \cong \text{UV}$
$\propto$	proportional to		$x \propto y$
$\equiv$	equivalent to		$a + bi \equiv re^{i\theta}$
∞	infinity		$x \rightarrow \infty$
log	base-10 logarithm		$\log 5.74$
ln	natural logarithm		$\ln 5.74$
exp	exponential power		$\exp(x) \equiv e^x$
rms	root-mean-square		$V_{\text{rms}} = \sqrt{\dfrac{1}{n}\sum\limits_{i=1}^{n} V_i^2}$
$\angle$	phasor or angle		$\angle 53°$
$\in$	element of	set/membership	$Y = \sum\limits_{j\in\{0,1,2,...,n\}} y_j$

[a]*Zero factorial* (0!) is frequently encountered in the form of $(n-n)!$ when calculating permutations and combinations. Zero factorial is defined as 1.
[b]The notation abs(x) is also used to indicate the absolute value.

significant digits in the least significant multiplicand, divisor, or dividend. So, $2.0 \times 13.2 = 26$ since the first multiplicand (2.0) has two significant digits only.

For the operations of addition and subtraction, the final answer is rounded to the position of the least significant digit in the addenda, minuend, or subtrahend. So, $2.0 + 13.2 = 15.2$ because both addenda are significant to the tenth's position; but $2 + 13.4 = 15$ since the 2 is significant only in the ones' position.

The multiplication rule should not be used for addition or subtraction, as this can result in strange answers. For example, it would be incorrect to round $1700 + 0.1$ to 2000 simply because 0.1 has only one significant digit. Table 3.3 gives examples of significant digits.

Table 3.3 *Examples of Significant Digits*

number as written	number of significant digits	implied range
341	3	340.5–341.5
34.1	3	34.05–34.15
0.00341	3	0.003405–0.003415
341×10^7	3	340.5×10^7–341.5×10^7
3.41×10^{-2}	3	3.405×10^{-2}–3.415×10^{-2}
3410	3	3405–3415
3410[*]	4	3409.5–3410.5
341.0	4	340.95–341.05

[*]It is permitted to write "3410." to distinguish the number from its 3-significant-digit form, although this is rarely done.

6. EQUATIONS

An *equation* is a mathematical statement of equality, such as $5 = 3 + 2$. *Algebraic equations* are written in terms of *variables*. In the equation $y = x^2 + 3$, the value of variable y depends on the value of variable x. Therefore, y is the *dependent variable* and x is the *independent variable*. The dependency of y on x is clearer when the equation is written in *functional form*: $y = f(x)$.

A *parametric equation* uses one or more independent variables (*parameters*) to describe a function.[3] For example, the parameter θ can be used to write the parametric equations of a unit circle.

$$x = \cos\theta \qquad \textit{3.1}$$

$$y = \sin\theta \qquad \textit{3.2}$$

A unit circle can also be described by a *nonparametric equation*.[4]

$$x^2 + y^2 = 1 \qquad \textit{3.3}$$

[3]As used in this section, there is no difference between a parameter and an independent variable. However, the term *parameter* is also used as a descriptive measurement that determines or characterizes the form, size, or content of a function. For example, the radius is a parameter of a circle, and mean and variance are parameters of a probability distribution. Once these parameters are specified, the function is completely defined.

[4]Since only the coordinate variables are used, this equation is also said to be in *Cartesian equation form*.

7. FUNDAMENTAL ALGEBRAIC LAWS

Algebra provides the rules that allow complex mathematical relationships to be expanded or condensed. Algebraic laws may be applied to complex numbers, variables, and real numbers. The general rules for changing the form of a mathematical relationship are given as follows.

- commutative law for addition:
$$A + B = B + A \qquad 3.4$$

- commutative law for multiplication:
$$AB = BA \qquad 3.5$$

- associative law for addition:
$$A + (B + C) = (A + B) + C \qquad 3.6$$

- associative law for multiplication:
$$A(BC) = (AB)C \qquad 3.7$$

- distributive law:
$$A(B + C) = AB + AC \qquad 3.8$$

8. POLYNOMIALS

A *polynomial* is a rational expression—usually the sum of several variable terms known as *monomials*—that does not involve division. The *degree of the polynomial* is the highest power to which a variable in the expression is raised. The following *standard polynomial forms* are useful when trying to find the roots of an equation.

$$(a + b)(a - b) = a^2 - b^2 \qquad 3.9$$

$$(a \pm b)^2 = a^2 \pm 2ab + b^2 \qquad 3.10$$

$$(a \pm b)^3 = a^3 \pm 3a^2b + 3ab^2 \pm b^3 \qquad 3.11$$

$$(a^3 \pm b^3) = (a \pm b)(a^2 \mp ab + b^2) \qquad 3.12$$

$$(a^n - b^n) = (a - b)\left(\begin{matrix} a^{n-1} + a^{n-2}b + a^{n-3}b^2 \\ + \cdots + b^{n-1} \end{matrix}\right)$$

[n is any positive integer] $\qquad 3.13$

$$(a^n + b^n) = (a + b)\left(\begin{matrix} a^{n-1} - a^{n-2}b + a^{n-3}b^2 \\ - \cdots + b^{n-1} \end{matrix}\right)$$

[n is any positive odd integer] $\qquad 3.14$

The *binomial theorem* defines a polynomial of the form $(a + b)^n$.

$$(a + b)^n = \underset{[i=0]}{a^n} + \underset{[i=1]}{na^{n-1}b} + \underset{[i=2]}{C_2 a^{n-2}b^2} + \cdots$$

$$+ C_i a^{n-i}b^i + \cdots + nab^{n-1} + b^n \qquad 3.15$$

$$C_i = \frac{n!}{i!(n-i)!} \qquad [i = 0, 1, 2, \ldots, n] \qquad 3.16$$

The coefficients of the expansion can be determined quickly from *Pascal's triangle*—each entry is the sum of the two entries directly above it. (See Fig. 3.1.)

Figure 3.1 *Pascal's Triangle*

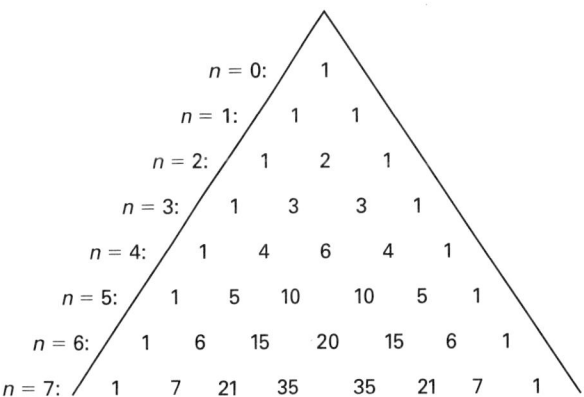

The values $r_1, r_2, \ldots, r_n$ of the independent variable x that satisfy a polynomial equation $f(x) = 0$ are known as *roots* or *zeros* of the polynomial. A polynomial of degree n with real coefficients will have at most n real roots, although they need not all be distinctly different.

9. ROOTS OF QUADRATIC EQUATIONS

A *quadratic equation* is an equation of the general form $ax^2 + bx + c = 0$ [$a \neq 0$]. The *roots*, x_1 and x_2, of the equation are the two values of x that satisfy it.

$$x_1, x_2 = \frac{-b \pm \sqrt{b^2 - 4ac}}{2a} \qquad 3.17$$

$$x_1 + x_2 = -\frac{b}{a} \qquad 3.18$$

$$x_1 x_2 = \frac{c}{a} \qquad 3.19$$

The types of roots of the equation can be determined from the *discriminant* (i.e., the quantity under the radical in Eq. 3.17).

- If $b^2 - 4ac > 0$, the roots are real and unequal.

- If $b^2 - 4ac = 0$, the roots are real and equal. This is known as a *double root*.

- If $b^2 - 4ac < 0$, the roots are complex and unequal.

10. ROOTS OF GENERAL POLYNOMIALS

It is difficult to find roots of cubic and higher-degree polynomials because few general techniques exist. *Cardano's formula (method)* uses closed-form equations to laboriously calculate roots for general cubic (3rd degree) polynomials. Compromise methods are used when solutions are needed on the spot.

- *inspection and trial and error:* Finding roots by inspection is equivalent to making reasonable guesses about the roots and substituting into the polynomial.

- *graphing:* If the value of a polynomial $f(x)$ is calculated and plotted for different values of x, an approximate value of a root can be determined as the value of x at which the plot crosses the x-axis.

- *factoring:* If at least one root (say, $x = r$) of a polynomial $f(x)$ is known, the quantity $x - r$ can be factored out of $f(x)$ by long division. The resulting quotient will be lower by one degree, and the remaining roots may be easier to determine. This method is particularly applicable if the polynomial is in one of the standard forms presented in Sec. 3.8.

- *special cases:* Certain polynomial forms can be simplified by substitution or solved by standard formulas if they are recognized as being special cases. (The standard solution to the quadratic equation is such a special case.) For example, $ax^4 + bx^2 + c = 0$ can be reduced to a polynomial of degree 2 if the substitution $u = x^2$ is made.

- *numerical methods:* If an approximate value of a root is known, numerical methods (bisection method, Newton's method, etc.) can be used to refine the value. The more efficient techniques are too complex to be performed by hand.

11. EXTRANEOUS ROOTS

With simple equalities, it may appear possible to derive roots by basic algebraic manipulations.[5] However, multiplying each side of an equality by a power of a variable may introduce *extraneous roots.* Such roots do not satisfy the original equation even though they are derived according to the rules of algebra. Checking a calculated root is always a good idea, but is particularly necessary if the equation has been multiplied by one of its own variables.

Example 3.1

Use algebraic operations to determine a value that satisfies the following equation. Determine if the value is a valid or extraneous root.

$$\sqrt{x-2} = \sqrt{x} + 2$$

Solution

Square both sides.
$$x - 2 = x + 4\sqrt{x} + 4$$

[5]In this sentence, *equality* means a combination of two expressions containing an equal sign. Any two expressions can be linked in this manner, even those that are not actually equal. For example, the expressions for two nonintersecting ellipses can be equated even though there is no intersection point. Finding extraneous roots is more likely when the underlying equality is false to begin with.

Subtract x from each side, and combine the constants.
$$4\sqrt{x} = -6$$

Solve for x.
$$x = \left(\frac{-6}{4}\right)^2 = \frac{9}{4}$$

Substitute $x = 9/4$ into the original equation.
$$\sqrt{\frac{9}{4} - 2} = \sqrt{\frac{9}{4}} + 2$$
$$\frac{1}{2} = \frac{7}{2}$$

Since the equality is not established, $x = 9/4$ is an extraneous root.

12. DESCARTES' RULE OF SIGNS

Descartes' rule of signs determines the maximum number of positive (and negative) real roots that a polynomial will have by counting the number of sign reversals (i.e., changes in sign from one term to the next) in the polynomial. The polynomial $f(x) = 0$ must have real coefficients and must be arranged in terms of descending powers of x.

- The number of positive roots of the polynomial equation $f(x) = 0$ will not exceed the number of sign reversals.

- The difference between the number of sign reversals and the number of positive roots is an even number.

- The number of negative roots of the polynomial equation $f(x) = 0$ will not exceed the number of sign reversals in the polynomial $f(-x)$.

- The difference between the number of sign reversals in $f(-x)$ and the number of negative roots is an even number.

Example 3.2

Determine the possible numbers of positive and negative roots that satisfy the following polynomial equation.

$$4x^5 - 5x^4 + 3x^3 - 8x^2 - 2x + 3 = 0$$

Solution

There are four sign reversals, so up to four positive roots exist. To keep the difference between the number of positive roots and the number of sign reversals an even number, the number of positive real roots is limited to zero, two, and four.

Substituting $-x$ for x in the polynomial results in
$$-4x^5 - 5x^4 - 3x^3 - 8x^2 + 2x + 3 = 0$$

There is only one sign reversal, so the number of negative roots cannot exceed one. There must be exactly one negative real root in order to keep the difference to an even number (zero in this case).

13. RULES FOR EXPONENTS AND RADICALS

In the expression $b^n = a$, b is known as the *base* and n is the *exponent* or *power*. In Eq. 3.20 through Eq. 3.33, a, b, m, and n are any real numbers with limitations listed.

$$b^0 = 1 \quad [b \neq 0] \qquad 3.20$$

$$b^1 = b \qquad 3.21$$

$$b^{-n} = \frac{1}{b^n} = \left(\frac{1}{b}\right)^n \quad [b \neq 0] \qquad 3.22$$

$$\left(\frac{a}{b}\right)^n = \frac{a^n}{b^n} \quad [b \neq 0] \qquad 3.23$$

$$(ab)^n = a^n b^n \qquad 3.24$$

$$b^{m/n} = \sqrt[n]{b^m} = \left(\sqrt[n]{b}\right)^m \qquad 3.25$$

$$\left(b^n\right)^m = b^{nm} \qquad 3.26$$

$$b^m b^n = b^{m+n} \qquad 3.27$$

$$\frac{b^m}{b^n} = b^{m-n} \quad [b \neq 0] \qquad 3.28$$

$$\sqrt[n]{b} = b^{1/n} \qquad 3.29$$

$$\left(\sqrt[n]{b}\right)^n = \left(b^{1/n}\right)^n = b \qquad 3.30$$

$$\sqrt[n]{ab} = \sqrt[n]{a}\sqrt[n]{b} = a^{1/n}b^{1/n}$$
$$= (ab)^{1/n} \qquad 3.31$$

$$\sqrt[n]{\frac{a}{b}} = \frac{\sqrt[n]{a}}{\sqrt[n]{b}} = \left(\frac{a}{b}\right)^{1/n} \quad [b \neq 0] \qquad 3.32$$

$$\sqrt[m]{\sqrt[n]{b}} = \sqrt[mn]{b} = b^{1/mn} \qquad 3.33$$

14. LOGARITHMS

Logarithms can be considered to be exponents. For example, the exponent n in the expression $b^n = a$ is the logarithm of a to the base b. Therefore, the two expressions $\log_b a = n$ and $b^n = a$ are equivalent.

The base for *common logs* is 10. Usually, "log" will be written when common logs are desired, although "$\log_{10}$" appears occasionally. The base for *natural (Napierian) logs* is 2.71828..., a number which is given the symbol e. When natural logs are desired, usually "ln" will be written, although "$\log_e$" is also used.

Most logarithms will contain an integer part (the *characteristic*) and a decimal part (the *mantissa*). The common and natural logarithms of any number less than one are negative. If the number is greater than one, its common and natural logarithms are positive. Although the logarithm may be negative, the mantissa is always positive. For negative logarithms, the characteristic is found by expressing the logarithm as the sum of a negative characteristic and a positive mantissa.

For common logarithms of numbers greater than one, the characteristics will be positive and equal to one less than the number of digits in front of the decimal. If the number is less than one, the characteristic will be negative and equal to one more than the number of zeros immediately following the decimal point.

If a negative logarithm is to be used in a calculation, it must first be converted to *operational form* by adding the characteristic and mantissa. The operational form should be used in all calculations and is the form displayed by scientific calculators.

The logarithm of a negative number is a complex number.

Example 3.3

Use logarithm tables to determine the operational form of $\log_{10} 0.05$.

Solution

Since the number is less than one and there is one leading zero, the characteristic is found by observation to be -2. From a book of logarithm tables, the mantissa of 5.0 is 0.699. Two ways of expressing the combination of mantissa and characteristic are used.

method 1: $\overline{2}.699$

method 2: $8.699 - 10$

The operational form of this logarithm is $-2 + 0.699 = -1.301$.

15. LOGARITHM IDENTITIES

Prior to the widespread availability of calculating devices, logarithm identities were used to solve complex calculations by reducing the solution method to table look-up, addition, and subtraction. Logarithm identities are still useful in simplifying expressions containing exponentials and other logarithms. In Eq. 3.34 through Eq. 3.45, $a \neq 1$, $b \neq 1$, $x > 0$, and $y > 0$.

$$\log_b b = 1 \qquad 3.34$$

$$\log_b 1 = 0 \qquad 3.35$$

$$\log_b b^n = n \qquad 3.36$$

$$\log x^a = a \log x \qquad 3.37$$

$$\log \sqrt[n]{x} = \log x^{1/n} = \frac{\log x}{n} \qquad 3.38$$

$$b^{n \log_b x} = x^n = \text{antilog}(n \log_b x) \qquad 3.39$$

$$b^{\log_b x/n} = x^{1/n} \qquad 3.40$$

$$\log xy = \log x + \log y \qquad 3.41$$

$$\log \frac{x}{y} = \log x - \log y \qquad 3.42$$

$$\log_a x = \log_b x \log_a b \qquad 3.43$$

$$\ln x = \ln 10 \log_{10} x \approx 2.3026 \log_{10} x \qquad 3.44$$

$$\log_{10} x = \log_{10} \ln x \, e \approx 0.4343 \ln x \qquad 3.45$$

Example 3.4

The surviving fraction, f, of a radioactive isotope is given by $f = e^{-0.005t}$. For what value of t will the surviving percentage be 7%?

Solution

$$f = 0.07 = e^{-0.005t}$$

Take the natural log of both sides.

$$\ln 0.07 = \ln e^{-0.005t}$$

From Eq. 3.36, $\ln e^x = x$. Therefore,

$$-2.66 = -0.005t$$

$$t = 532$$

16. PARTIAL FRACTIONS

The method of *partial fractions* is used to transform a proper polynomial fraction of two polynomials into a sum of simpler expressions, a procedure known as *resolution*.[6,7] The technique can be considered to be the act of "unadding" a sum to obtain all of the addends.

Suppose $H(x)$ is a proper polynomial fraction of the form $P(x)/Q(x)$. The object of the resolution is to determine the partial fractions u_1/v_1, u_2/v_2, and so on, such that

$$H(x) = \frac{P(x)}{Q(x)} = \frac{u_1}{v_1} + \frac{u_2}{v_2} + \frac{u_3}{v_3} + \cdots \qquad 3.46$$

The form of the denominator polynomial $Q(x)$ will be the main factor in determining the form of the partial fractions. The task of finding the u_i and v_i is simplified by categorizing the possible forms of $Q(x)$.

case 1: $Q(x)$ factors into n different linear terms.

$$Q(x) = (x - a_1)(x - a_2) \cdots (x - a_n) \qquad 3.47$$

Then,

$$H(x) = \sum_{i=1}^{n} \frac{A_i}{x - a_i} \qquad 3.48$$

case 2: $Q(x)$ factors into n identical linear terms.

$$Q(x) = (x - a)(x - a) \cdots (x - a) \qquad 3.49$$

Then,

$$H(x) = \sum_{i=1}^{n} \frac{A_i}{(x - a)^i} \qquad 3.50$$

case 3: $Q(x)$ factors into n different quadratic terms, $x^2 + p_i x + q_i$.

Then,

$$H(x) = \sum_{i=1}^{n} \frac{A_i x + B_i}{x^2 + p_i x + q_i} \qquad 3.51$$

case 4: $Q(x)$ factors into n identical quadratic terms, $x^2 + px + q$.

Then,

$$H(x) = \sum_{i=1}^{n} \frac{A_i x + B_i}{(x^2 + px + q)^i} \qquad 3.52$$

Once the general forms of the partial fractions have been determined from inspection, the *method of undetermined coefficients* is used. The partial fractions are all cross multiplied to obtain $Q(x)$ as the denominator, and the coefficients are found by equating $P(x)$ and the cross-multiplied numerator.

Example 3.5

Resolve $H(x)$ into partial fractions.

$$H(x) = \frac{x^2 + 2x + 3}{x^4 + x^3 + 2x^2}$$

Solution

Here, $Q(x) = x^4 + x^3 + 2x^2$ factors into $x^2(x^2 + x + 2)$. This is a combination of cases 2 and 3.

$$H(x) = \frac{A_1}{x} + \frac{A_2}{x^2} + \frac{A_3 + A_4 x}{x^2 + x + 2}$$

Cross multiplying to obtain a common denominator yields

$$\frac{(A_1 + A_4)x^3 + (A_1 + A_2 + A_3)x^2 + (2A_1 + A_2)x + 2A_2}{x^4 + x^3 + 2x^2}$$

Since the original numerator is known, the following simultaneous equations result.

$$A_1 + A_4 = 0$$
$$A_1 + A_2 + A_3 = 1$$
$$2A_1 + A_2 = 2$$
$$2A_2 = 3$$

The solutions are $A_1 = 0.25$; $A_2 = 1.50$; $A_3 = -0.75$; and $A_4 = -0.25$.

$$H(x) = \frac{1}{4x} + \frac{3}{2x^2} - \frac{x + 3}{4(x^2 + x + 2)}$$

---

[6]To be a *proper polynomial fraction*, the degree of the numerator must be less than the degree of the denominator. If the polynomial fraction is improper, the denominator can be divided into the numerator to obtain whole and fractional polynomials. The method of partial fractions can then be used to reduce the fractional polynomial.
[7]This technique is particularly useful for calculating integrals and inverse Laplace transforms in subsequent chapters.

17. SIMULTANEOUS LINEAR EQUATIONS

A *linear equation* with n variables is a polynomial of degree 1 describing a geometric shape in n-space. A *homogeneous linear equation* is one that has no constant term, and a *nonhomogeneous linear equation* has a constant term.

A solution to a set of simultaneous linear equations represents the intersection point of the geometric shapes in n-space. For example, if the equations are limited to two variables (e.g., $y = 4x - 5$), they describe straight lines. The solution to two simultaneous linear equations in 2-space is the point where the two lines intersect. The set of the two equations is said to be a *consistent system* when there is such an intersection.[8]

Simultaneous equations do not always have unique solutions, and some have none at all. In addition to crossing in 2-space, lines can be parallel or they can be the same line expressed in a different equation format (i.e., dependent equations). In some cases, parallelism and dependency can be determined by inspection. In most cases, however, matrix and other advanced methods must be used to determine whether a solution exists. A set of linear equations with no simultaneous solution is known as an *inconsistent system*.

Several methods exist for solving linear equations simultaneously by hand.[9]

- *graphing:* The equations are plotted and the intersection point is read from the graph. This method is possible only with two-dimensional problems.

- *substitution:* An equation is rearranged so that one variable is expressed as a combination of the other variables. The expression is then substituted into the remaining equations wherever the selected variable appears.

- *reduction:* All terms in the equations are multiplied by constants chosen to eliminate one or more variables when the equations are added or subtracted. The remaining sum can then be solved for the other variables. This method is also known as *eliminating the unknowns*.

- *Cramer's rule:* This is a procedure in linear algebra that calculates determinants of the original coefficient matrix $\mathbf{A}$ and of the n matrices resulting from the systematic replacement of column $\mathbf{A}$ by the constant matrix $\mathbf{B}$.

Example 3.6

Solve the following set of linear equations by (a) substitution and (b) reduction.

$$2x + 3y = 12 \quad \text{[Eq. I]}$$
$$3x + 4y = 8 \quad \text{[Eq. II]}$$

Solution

(a) From Eq. I, solve for variable x.

$$x = 6 - 1.5y \quad \text{[Eq. III]}$$

Substitute $6 - 1.5y$ into Eq. II wherever x appears.

$$(3)(6 - 1.5y) + 4y = 8$$
$$18 - 4.5y + 4y = 8$$
$$y = 20$$

Substitute 20 for y in Eq. III.

$$x = 6 - (1.5)(20) = -24$$

The solution $(-24, 20)$ should be checked to verify that it satisfies both original equations.

(b) Eliminate variable x by multiplying Eq. I by 3 and Eq. II by 2.

$$3 \times \text{Eq. I:} \ 6x + 9y = 36 \quad \text{[Eq. I}'\text{]}$$
$$2 \times \text{Eq. II:} \ 6x + 8y = 16 \quad \text{[Eq. II}'\text{]}$$

Subtract Eq. II$'$ from Eq. I$'$.

$$y = 20 \quad \text{[Eq. I}' - \text{Eq. II}'\text{]}$$

Substitute $y = 20$ into Eq. I$'$.

$$6x + (9)(20) = 36$$
$$x = -24$$

The solution $(-24, 20)$ should be checked to verify that it satisfies both original equations.

18. COMPLEX NUMBERS

A *complex number*, $\mathbf{Z}$, is a combination of real and imaginary numbers. When expressed as a sum (e.g., $a + bi$), the complex number is said to be in *rectangular* or *trigonometric form*. The complex number can be plotted on the real-imaginary coordinate system known as the *complex plane*, as illustrated in Fig. 3.2.

Figure 3.2 *A Complex Number in the Complex Plane*

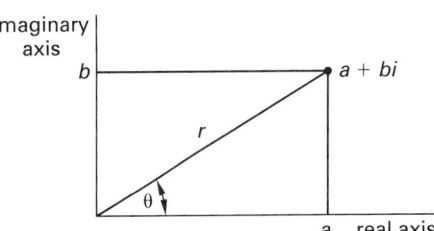

---

[8]A homogeneous system always has at least one solution: the *trivial solution*, in which all variables have a value of zero.
[9]Other methods exist, but they require a computer.

The complex number $\mathbf{Z} = a + bi$ can also be expressed in *exponential form*.[10] The quantity r is known as the *modulus* of $\mathbf{Z}$; θ is the *argument*.

$$a + bi \equiv re^{i\theta} \qquad 3.53$$

$$r = \text{mod } \mathbf{Z} = \sqrt{a^2 + b^2} \qquad 3.54$$

$$\theta = \text{arg } \mathbf{Z} = \arctan \frac{b}{a} \qquad 3.55$$

Similarly, the *phasor form* (also known as the *polar form*) is

$$\mathbf{Z} = r\angle\theta \qquad 3.56$$

The *rectangular form* can be determined from r and θ.

$$a = r\cos\theta \qquad 3.57$$

$$b = r\sin\theta \qquad 3.58$$

$$\mathbf{Z} = a + bi = r\cos\theta + ir\sin\theta$$
$$= r(\cos\theta + i\sin\theta) \qquad 3.59$$

The *cis form* is a shorthand method of writing a complex number in rectangular (trigonometric) form.

$$a + bi = r(\cos\theta + i\sin\theta) = r\text{ cis }\theta \qquad 3.60$$

Euler's equation, as shown in Eq. 3.61, expresses the equality of complex numbers in exponential and trigonometric form.

$$e^{i\theta} = \cos\theta + i\sin\theta \qquad 3.61$$

Related expressions are

$$e^{-i\theta} = \cos\theta - i\sin\theta \qquad 3.62$$

$$\cos\theta = \frac{e^{i\theta} + e^{-i\theta}}{2} \qquad 3.63$$

$$\sin\theta = \frac{e^{i\theta} - e^{-i\theta}}{2i} \qquad 3.64$$

Example 3.7

What is the exponential form of the complex number $\mathbf{Z} = 3 + 4i$?

Solution

$$r = \sqrt{a^2 + b^2} = \sqrt{3^2 + 4^2} = \sqrt{25}$$
$$= 5$$

$$\theta = \arctan \frac{b}{a} = \arctan \frac{4}{3} = 0.927 \text{ rad}$$

$$\mathbf{Z} = re^{i\theta} = 5e^{i(0.927)}$$

19. OPERATIONS ON COMPLEX NUMBERS

Most algebraic operations (addition, multiplication, exponentiation, etc.) work with complex numbers, but notable exceptions are the inequality operators. The concept of one complex number being less than or greater than another complex number is meaningless.

---

[10]The terms *polar form*, *phasor form*, and *exponential form* are all used somewhat interchangeably.

When adding two complex numbers, real parts are added to real parts, and imaginary parts are added to imaginary parts.

$$(a_1 + ib_1) + (a_2 + ib_2) = (a_1 + a_2) + i(b_1 + b_2) \qquad 3.65$$

$$(a_1 + ib_1) - (a_2 + ib_2) = (a_1 - a_2) + i(b_1 - b_2) \qquad 3.66$$

Multiplication of two complex numbers in rectangular form is accomplished by the use of the algebraic distributive law and the definition $i^2 = -1$.

Division of complex numbers in rectangular form requires use of the *complex conjugate*. The complex conjugate of the complex number $(a + bi)$ is $(a - bi)$. By multiplying the numerator and the denominator by the complex conjugate, the denominator will be converted to the real number $a^2 + b^2$. This technique is known as *rationalizing* the denominator and is illustrated in Ex. 3.8(c).

Multiplication and division are often more convenient when the complex numbers are in exponential or phasor forms, as Eq. 3.67 and Eq. 3.68 show.

$$(r_1 e^{i\theta_1})(r_2 e^{i\theta_2}) = r_1 r_2 e^{i(\theta_1 + \theta_2)} \qquad 3.67$$

$$\frac{r_1 e^{i\theta_1}}{r_2 e^{i\theta_2}} = \left(\frac{r_1}{r_2}\right) e^{i(\theta_1 - \theta_2)} \qquad 3.68$$

Taking powers and roots of complex numbers requires *de Moivre's theorem*, Eq. 3.69 and Eq. 3.70.

$$\mathbf{Z}^n = (re^{i\theta})^n = r^n e^{in\theta} \qquad 3.69$$

$$\sqrt[n]{\mathbf{Z}} = (re^{i\theta})^{1/n} = \sqrt[n]{r}\, e^{i(\theta + k360°/n)}$$
$$[k = 0, 1, 2, \ldots, n-1] \qquad 3.70$$

Example 3.8

Perform the following complex arithmetic.

(a) $(3 + 4i) + (2 + i)$

(b) $(7 + 2i)(5 - 3i)$

(c) $\dfrac{2 + 3i}{4 - 5i}$

Solution

(a)
$$(3 + 4i) + (2 + i) = (3 + 2) + (4 + 1)i$$
$$= 5 + 5i$$

(b)
$$(7 + 2i)(5 - 3i) = (7)(5) - (7)(3i) + (2i)(5)$$
$$- (2i)(3i)$$
$$= 35 - 21i + 10i - 6i^2$$
$$= 35 - 21i + 10i - (6)(-1)$$
$$= 41 - 11i$$

(c) Multiply the numerator and denominator by the complex conjugate of the denominator.

$$\frac{2+3i}{4-5i}=\frac{(2+3i)(4+5i)}{(4-5i)(4+5i)}=\frac{-7+22i}{(4)^2+(5)^2}$$

$$=\frac{-7}{41}+i\frac{22}{41}$$

20. LIMITS

A *limit* (*limiting value*) is the value a function approaches when an independent variable approaches a target value. For example, suppose the value of $y=x^2$ is desired as x approaches 5. This could be written as

$$\lim_{x\to 5}x^2 \qquad\qquad 3.71$$

The power of limit theory is wasted on simple calculations such as this but is appreciated when the function is undefined at the target value. The object of limit theory is to determine the limit without having to evaluate the function at the target. The general case of a limit evaluated as x approaches the target value a is written as

$$\lim_{x\to a}f(x) \qquad\qquad 3.72$$

It is not necessary for the actual value $f(a)$ to exist for the limit to be calculated. The function $f(x)$ may be undefined at point a. However, it is necessary that $f(x)$ be defined on both sides of point a for the limit to exist. If $f(x)$ is undefined on one side, or if $f(x)$ is discontinuous at $x=a$ (as in Fig. 3.3(c) and Fig. 3.3(d)), the limit does not exist.

Figure 3.3 *Existence of Limits*

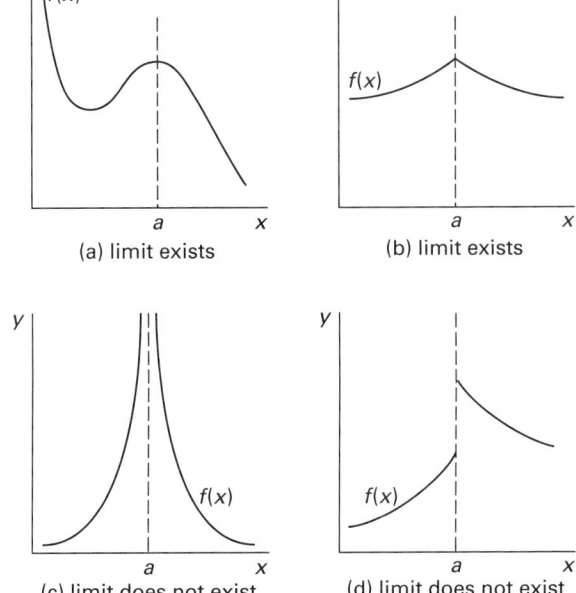

(a) limit exists

(b) limit exists

(c) limit does not exist

(d) limit does not exist

The following theorems can be used to simplify expressions when calculating limits.

$$\lim_{x\to a}x=a \qquad\qquad 3.73$$

$$\lim_{x\to a}(mx+b)=ma+b \qquad\qquad 3.74$$

$$\lim_{x\to a}b=b \qquad\qquad 3.75$$

$$\lim_{x\to a}\big(kF(x)\big)=k\lim_{x\to a}F(x) \qquad\qquad 3.76$$

$$\lim_{x\to a}\left(F_1(x)\left\{\begin{array}{c}+\\-\\\times\\\div\end{array}\right\}F_2(x)\right)$$

$$=\lim_{x\to a}\big(F_1(x)\big)\left\{\begin{array}{c}+\\-\\\times\\\div\end{array}\right\}\lim_{x\to a}\big(F_2(x)\big) \qquad 3.77$$

The following identities can be used to simplify limits of trigonometric expressions.

$$\lim_{x\to 0}\sin x=0 \qquad\qquad 3.78$$

$$\lim_{x\to 0}\left(\frac{\sin x}{x}\right)=1 \qquad\qquad 3.79$$

$$\lim_{x\to 0}\cos x=1 \qquad\qquad 3.80$$

The following standard methods (tricks) can be used to determine limits.

- If the limit is taken to infinity, all terms can be divided by the largest power of x in the expression. This will leave at least one constant. Any quantity divided by a power of x vanishes as x approaches infinity.

- If the expression is a quotient of two expressions, any common factors should be eliminated from the numerator and denominator.

- *L'Hôpital's rule*, Eq. 3.81, should be used when the numerator and denominator of the expression both approach zero or both approach infinity.[11] $P^k(x)$ and $Q^k(x)$ are the kth derivatives of the functions $P(x)$ and $Q(x)$, respectively. (L'Hôpital's rule can be applied repeatedly as required.)

$$\lim_{x\to a}\left(\frac{P(x)}{Q(x)}\right)=\lim_{x\to a}\left(\frac{P^k(x)}{Q^k(x)}\right) \qquad 3.81$$

---

[11]L'Hôpital's rule should not be used when only the denominator approaches zero. In that case, the limit approaches infinity regardless of the numerator.

Example 3.9

Evaluate the following limits.

(a) $\lim\limits_{x\to3}\left(\dfrac{x^3-27}{x^2-9}\right)$

(b) $\lim\limits_{x\to\infty}\left(\dfrac{3x-2}{4x+3}\right)$

(c) $\lim\limits_{x\to2}\left(\dfrac{x^2+x-6}{x^2-3x+2}\right)$

Solution

(a) Factor the numerator and denominator. (L'Hôpital's rule can also be used.)

$$\lim\limits_{x\to3}\left(\dfrac{x^3-27}{x^2-9}\right)=\lim\limits_{x\to3}\left(\dfrac{(x-3)(x^2+3x+9)}{(x-3)(x+3)}\right)$$

$$=\lim\limits_{x\to3}\left(\dfrac{x^2+3x+9}{x+3}\right)$$

$$=\dfrac{(3)^2+(3)(3)+9}{3+3}$$

$$=9/2$$

(b) Divide through by the largest power of x. (L'Hôpital's rule can also be used.)

$$\lim\limits_{x\to\infty}\left(\dfrac{3x-2}{4x+3}\right)=\lim\limits_{x\to\infty}\left(\dfrac{3-\dfrac{2}{x}}{4+\dfrac{3}{x}}\right)$$

$$=\dfrac{3-\dfrac{2}{\infty}}{4+\dfrac{3}{\infty}}=\dfrac{3-0}{4+0}$$

$$=3/4$$

(c) Use L'Hôpital's rule. (Factoring can also be used.) Take the first derivative of the numerator and denominator.

$$\lim\limits_{x\to2}\left(\dfrac{x^2+x-6}{x^2-3x+2}\right)=\lim\limits_{x\to2}\left(\dfrac{2x+1}{2x-3}\right)$$

$$=\dfrac{(2)(2)+1}{(2)(2)-3}=\dfrac{5}{1}$$

$$=5$$

21. SEQUENCES AND PROGRESSIONS

A *sequence*, $\{A\}$, is an ordered *progression* of numbers, a_i, such as $1, 4, 9, 16, 25, \ldots$ The *terms* in a sequence can be all positive, all negative, or of alternating signs. a_n is known as the *general term* of the sequence.

$$\{A\}=\{a_1,a_2,a_3,\ldots,a_n\} \qquad 3.82$$

A sequence is said to *diverge* (i.e., be *divergent*) if the terms approach infinity or if the terms fail to approach any finite value, and it is said to *converge* (i.e., be *convergent*) if the terms approach any finite value (including zero). That is, the sequence converges if the limit defined by Eq. 3.83 exists.

$$\lim\limits_{n\to\infty}a_n\begin{cases}\text{converges if the limit is finite}\\\text{diverges if the limit is infinite}\\\text{or does not exist}\end{cases}\qquad 3.83$$

The main task associated with a sequence is determining the next (or the general) term. If several terms of a sequence are known, the next (unknown) term must usually be found by intuitively determining the pattern of the sequence. In some cases, though, the method of *Rth-order differences* can be used to determine the next term. This method consists of subtracting each term from the following term to obtain a set of differences. If the differences are not all equal, the next order of differences can be calculated.

Example 3.10

What is the general term of the sequence $\{A\}$?

$$\{A\}=\left\{3,\ \dfrac{9}{2},\ \dfrac{27}{6},\ \dfrac{81}{24},\ldots\right\}$$

Solution

The solution is purely intuitive. The numerator is recognized as a power series based on the number 3. The denominator is recognized as the factorial sequence. The general term is

$$a_n=\dfrac{3^n}{n!}$$

Example 3.11

Find the sixth term in the sequence $\{A\}=\{7, 16, 29, 46, 67, a_6\}$.

Solution

The sixth term is not intuitively obvious, so the method of Rth-order differences is tried. The pattern is not obvious from the first order differences, but the second order differences are all 4.

$$\delta_5-21=4$$

$$\delta_5=25$$

$$a_6-67=\delta_5=25$$

$$a_6=92$$

Example 3.12

Does the sequence with general term e^n/n converge or diverge?

Solution

See if the limit exists.

$$\lim_{n \to \infty} \left(\frac{e^n}{n} \right) = \frac{\infty}{\infty}$$

Since ∞/∞ is inconclusive, apply L'Hôpital's rule. Take the derivative of both the numerator and the denominator with respect to n.

$$\lim_{n \to \infty} \left(\frac{e^n}{1} \right) = \frac{\infty}{1} = \infty$$

The sequence diverges.

22. STANDARD SEQUENCES

There are four standard sequences: the geometric, arithmetic, harmonic, and p-sequence.

- *geometric sequence:* The geometric sequence converges for $-1 < r \leq 1$ and diverges otherwise. a is known as the *first term*; r is known as the *common ratio*.

$$a_n = ar^{n-1} \quad \begin{bmatrix} a \text{ is a constant} \\ n = 1, 2, 3, \ldots, \infty \end{bmatrix} \qquad 3.84$$

  example: $\{1, 2, 4, 8, 16, 32\}$ $(a = 1, \ r = 2)$

- *arithmetic sequence:* The arithmetic sequence always diverges.

$$a_n = a + (n-1)d \quad \begin{bmatrix} a \text{ and } d \text{ are constants} \\ n = 1, 2, 3, \ldots, \infty \end{bmatrix} \qquad 3.85$$

  example: $\{2, 7, 12, 17, 22, 27\}$ $(a = 2, \ d = 5)$

- *harmonic sequence:* The harmonic sequence always converges.

$$a_n = \frac{1}{n} \quad [n = 1, 2, 3, \ldots, \infty] \qquad 3.86$$

  example: $\{1, 1/2, 1/3, 1/4, 1/5, 1/6\}$

- *p-sequence:* The p-sequence converges if $p \geq 0$ and diverges if $p < 0$. (This is different from the p-series whose convergence depends on the sum of its terms.)

$$a_n = \frac{1}{n^p} \quad [n = 1, 2, 3, \ldots, \infty] \qquad 3.87$$

  example: $\{1, 1/4, 1/9, 1/16, 1/25, 1/36\}$ $(p = 2)$

23. APPLICATION: GROWTH RATES

Models of *population growth* commonly assume arithmetic or geometric growth rates over limited periods of time.[12] *Arithmetic growth rate*, also called *constant growth rate* and *linear growth rate*, is appropriate when a population increase involves limited resources or occurs at specific intervals. Means of subsistence, such as areas of farmable land by generation, and budgets and enrollments by year, for example, are commonly assumed to increase arithmetically with time. Arithmetic growth is equivalent to simple interest compounding. Given a starting population P_0 that increases every period by a constant *growth rate amount*, R, the population after t periods is

$$P_t = P_0 + tR \quad \text{[arithmetic]} \qquad 3.88$$

The *average annual growth rate* is conventionally defined as

$$r_{\text{ave},\%} = \frac{P_t - P_0}{tP_0} \times 100\% = \frac{R}{P_0} \times 100\% \qquad 3.89$$

Geometric growth (*Malthusian growth*) is appropriate when resources to support growth are infinite. The population changes by a fixed *growth rate fraction*, r, each period. Geometric growth is equivalent to discrete period interest compounding, and $(F/P, r\%, t)$ economic interest factors can be used. The population at time t (i.e., after t periods) is

$$P_t = P_0(1 + r)^t \equiv P_0(P/F, r\%, t) \quad \text{[geometric]} \qquad 3.90$$

Geometric growth can be expressed in terms of a time constant, τ_b, associated with a specific base, b. Commonly, only three bases are used. For $b = 2$, τ is the *doubling time*, T. For $b = 1/2$, τ is the *half-life*, $t_{1/2}$. For $b = e$, τ is the *e-folding time*, or just *time constant*, τ. Base-e growth is known as *exponential growth* or *instantaneous growth*. It is appropriate for continuous growth (not discrete time periods) and is equivalent to continuous interest compounding.

$$P_t = P_0(1 + r)^t = P_0 e^{t/\tau} = P_0(2)^{t/T}$$
$$= P_0 \left(\tfrac{1}{2}\right)^{t/t_{1/2}} \quad \text{[geometric]} \qquad 3.91$$

Taking the logarithm of Eq. 3.90 results in a *log-linear form*. Log-linear functions graph as straight lines.

$$\log P_t = \log P_0 + t \log(1 + r) \qquad 3.92$$

24. SERIES

A *series* is the sum of terms in a sequence. There are two types of series. A *finite series* has a finite number of terms, and an *infinite series* has an infinite number of

---

[12]Another population growth model is the logistic (S-shaped) curve.

terms.[13] The main tasks associated with series are determining the sum of the terms and whether the series converges. A series is said to *converge* (be *convergent*) if the sum, S_n, of its term exists.[14] A finite series is always convergent.

The performance of a series based on standard sequences (defined in Sec. 3.22) is well known.

- *geometric series:*

$$S_n = \sum_{i=1}^{n} ar^{i-1} = \frac{a(1-r^n)}{1-r} \quad \text{[finite series]} \qquad 3.93$$

$$S_n = \sum_{i=1}^{\infty} ar^{i-1} = \frac{a}{1-r} \quad \begin{bmatrix} \text{infinite series} \\ -1 < r < 1 \end{bmatrix} \qquad 3.94$$

- *arithmetic series:* The infinite series diverges unless $a = d = 0$.

$$S_n = \sum_{i=1}^{n} \big(a + (i-1)d\big)$$

$$= \frac{n\big(2a + (n-1)d\big)}{2} \quad \text{[finite series]} \qquad 3.95$$

- *harmonic series:* The infinite series diverges.

- *p-series:* The infinite series diverges if $p \le 1$. The infinite series converges if $p > 1$. (This is different from the *p*-sequence whose convergence depends only on the last term.)

25. TESTS FOR SERIES CONVERGENCE

It is obvious that all finite series (i.e., series having a finite number of terms) converge. That is, the sum, S_n, defined by Eq. 3.96 exists.

$$S_n = \sum_{i=1}^{n} a_i \qquad 3.96$$

Convergence of an infinite series can be determined by taking the limit of the sum. If the limit exists, the series converges; otherwise, it diverges.

$$\lim_{n \to \infty} S_n = \lim_{n \to \infty} \sum_{i=1}^{n} a_i \qquad 3.97$$

In most cases, the expression for the general term a_n will be known, but there will be no simple expression for the sum S_n. Therefore, Eq. 3.97 cannot be used to determine convergence. It is helpful, but not conclusive, to look at the limit of the general term. If the limit, as defined in Eq. 3.98, is nonzero, the series

diverges. If the limit equals zero, the series may either converge or diverge. Additional testing is needed in that case.

$$\lim_{n \to \infty} a_n \begin{cases} = 0 & \text{inconclusive} \\ \neq 0 & \text{diverges} \end{cases} \qquad 3.98$$

Two tests can be used independently or after Eq. 3.98 has proven inconclusive: the ratio and comparison tests. The *ratio test* calculates the limit of the ratio of two consecutive terms.

$$\lim_{n \to \infty} \frac{a_{n+1}}{a_n} \begin{cases} < 1 & \text{converges} \\ = 1 & \text{inconclusive} \\ > 1 & \text{diverges} \end{cases} \qquad 3.99$$

The *comparison test* is an indirect method of determining convergence of an unknown series. It compares a standard series (geometric and *p*-series are commonly used) against the unknown series. If all terms in a positive standard series are smaller than the terms in the unknown series and the standard series diverges, the unknown series must also diverge. Similarly, if all terms in the standard series are larger than the terms in the unknown series and the standard series converges, then the unknown series also converges.

In mathematical terms, if A and B are both series of positive terms such that $a_n < b_n$ for all values of n, then (a) B diverges if A diverges, and (b) A converges if B converges.

Example 3.13

Does the infinite series A converge or diverge?

$$A = 3 + \frac{9}{2} + \frac{27}{6} + \frac{81}{24} + \cdots$$

Solution

The general term was found in Ex. 3.10 to be

$$a_n = \frac{3^n}{n!}$$

Since limits of factorials are not easily determined, use the ratio test.

$$\lim_{n \to \infty} \left(\frac{a_{n+1}}{a_n} \right) = \lim_{n \to \infty} \left(\frac{\dfrac{3^{n+1}}{(n+1)!}}{\dfrac{3^n}{n!}} \right) = \lim_{n \to \infty} \left(\frac{3}{n+1} \right) = \frac{3}{\infty}$$

$$= 0$$

Since the limit is less than 1, the infinite series converges.

---

[13]The term *infinite series* does not imply the sum is infinite.
[14]This is different from the definition of convergence for a sequence where only the last term was evaluated.

Example 3.14

Does the infinite series A converge or diverge?

$$A = 2 + \frac{3}{4} + \frac{4}{9} + \frac{5}{16} + \cdots$$

Solution

By observation, the general term is

$$a_n = \frac{1 + n}{n^2}$$

The general term can be expanded by partial fractions to

$$a_n = \frac{1}{n} + \frac{1}{n^2}$$

However, $1/n$ is the harmonic series. Since the harmonic series is divergent and this series is larger than the harmonic series (by the term $1/n^2$), this series also diverges.

26. SERIES OF ALTERNATING SIGNS[15]

Some series contain both positive and negative terms. The ratio and comparison tests can both be used to determine if a series with alternating signs converges. If a series containing all positive terms converges, then the same series with some negative terms also converges. Therefore, the all-positive series should be tested for convergence. If the all-positive series converges, the original series is said to be *absolutely convergent*. (If the all-positive series diverges and the original series converges, the original series is said to be *conditionally convergent*.)

Alternatively, the ratio test can be used with the absolute value of the ratio. The same criteria apply.

$$\lim_{n \to \infty} \left| \frac{a_{n+1}}{a_n} \right| \begin{cases} < 1 & \text{converges} \\ = 1 & \text{inconclusive} \\ > 1 & \text{diverges} \end{cases} \qquad 3.100$$

---

[15]This terminology is commonly used even though it is not necessary that the signs strictly alternate.

4 Linear Algebra

1. MATRICES

A *matrix* is an ordered set of *entries* (*elements*) arranged rectangularly and set off by brackets.[1] The entries can be variables or numbers. A matrix by itself has no particular value—it is merely a convenient method of representing a set of numbers.

The size of a matrix is given by the number of rows and columns, and the nomenclature $m \times n$ is used for a matrix with m rows and n columns. For a *square matrix*, the number of rows and columns will be the same, a quantity known as the *order* of the matrix.

Bold uppercase letters are used to represent matrices, while lowercase letters represent the entries. For example, a_{23} would be the entry in the second row and third column of matrix $\mathbf{A}$.

$$\mathbf{A} = \begin{bmatrix} a_{11} & a_{12} & a_{13} \\ a_{21} & a_{22} & a_{23} \\ a_{31} & a_{32} & a_{33} \end{bmatrix}$$

A *submatrix* is the matrix that remains when selected rows or columns are removed from the original matrix.[2] For example, for matrix $\mathbf{A}$, the submatrix remaining

after the second row and second column have been removed is

$$\begin{bmatrix} a_{11} & a_{13} \\ a_{31} & a_{33} \end{bmatrix}$$

An *augmented matrix* results when the original matrix is extended by repeating one or more of its rows or columns or by adding rows and columns from another matrix. For example, for the matrix $\mathbf{A}$, the augmented matrix created by repeating the first and second columns is

$$\begin{bmatrix} a_{11} & a_{12} & a_{13} & | & a_{11} & a_{12} \\ a_{21} & a_{22} & a_{23} & | & a_{21} & a_{22} \\ a_{31} & a_{32} & a_{33} & | & a_{31} & a_{32} \end{bmatrix}$$

2. SPECIAL TYPES OF MATRICES

Certain types of matrices are given special designations.

- *cofactor matrix:* the matrix formed when every entry is replaced by the cofactor (see Sec. 4.4) of that entry

- *column matrix:* a matrix with only one column

- *complex matrix:* a matrix with complex number entries

- *diagonal matrix:* a square matrix with all zero entries except for the a_{ij} for which $i = j$

- *echelon matrix:* a matrix in which the number of zeros preceding the first nonzero entry of a row increases row by row until only zero rows remain. A *row-reduced echelon matrix* is an echelon matrix in which the first nonzero entry in each row is a 1 and all other entries in the columns are zero.

- *identity matrix:* a diagonal (square) matrix with all nonzero entries equal to 1, usually designated as $\mathbf{I}$, having the property that $\mathbf{AI} = \mathbf{IA} = \mathbf{A}$

- *null matrix:* the same as a zero matrix

- *row matrix:* a matrix with only one row

- *scalar matrix:*[3] a diagonal (square) matrix with all diagonal entries equal to some scalar k

---

[1]The term *array* is synonymous with *matrix*, although the former is more likely to be used in computer applications.
[2]By definition, a matrix is a submatrix of itself.

[3]Although the term *complex matrix* means a matrix with complex entries, the term *scalar matrix* means more than a matrix with scalar entries.

- *singular matrix:* a matrix whose determinant is zero (see Sec. 4.10)

- *skew symmetric matrix:* a square matrix whose transpose (see Sec. 4.9) is equal to the negative of itself (i.e., $\mathbf{A} = -\mathbf{A}^t$)

- *square matrix:* a matrix with the same number of rows and columns (i.e., $m = n$)

- *symmetric(al) matrix:* a square matrix whose transpose is equal to itself (i.e., $\mathbf{A}^t = \mathbf{A}$), which occurs only when $a_{ij} = a_{ji}$

- *triangular matrix:* a square matrix with zeros in all positions above or below the diagonal

- *unit matrix:* the same as the identity matrix

- *zero matrix:* a matrix with all zero entries

Figure 4.1 shows examples of special matrices.

Figure 4.1 *Examples of Special Matrices*

$$\begin{bmatrix} 9 & 0 & 0 & 0 \\ 0 & -6 & 0 & 0 \\ 0 & 0 & 1 & 0 \\ 0 & 0 & 0 & 5 \end{bmatrix}$$
(a) diagonal

$$\begin{bmatrix} 2 & 18 & 2 & 18 \\ 0 & 0 & 1 & 9 \\ 0 & 0 & 0 & 9 \\ 0 & 0 & 0 & 0 \end{bmatrix}$$
(b) echelon

$$\begin{bmatrix} 1 & 9 & 0 & 0 \\ 0 & 0 & 1 & 0 \\ 0 & 0 & 0 & 1 \\ 0 & 0 & 0 & 0 \end{bmatrix}$$
(c) row-reduced echelon

$$\begin{bmatrix} 1 & 0 & 0 & 0 \\ 0 & 1 & 0 & 0 \\ 0 & 0 & 1 & 0 \\ 0 & 0 & 0 & 1 \end{bmatrix}$$
(d) identity

$$\begin{bmatrix} 3 & 0 & 0 & 0 \\ 0 & 3 & 0 & 0 \\ 0 & 0 & 3 & 0 \\ 0 & 0 & 0 & 3 \end{bmatrix}$$
(e) scalar

$$\begin{bmatrix} 2 & 0 & 0 & 0 \\ 7 & 6 & 0 & 0 \\ 9 & 1 & 1 & 0 \\ 8 & 0 & 4 & 5 \end{bmatrix}$$
(f) triangular

3. ROW EQUIVALENT MATRICES

A matrix $\mathbf{B}$ is said to be *row equivalent* to a matrix $\mathbf{A}$ if it is obtained by a finite sequence of *elementary row operations* on $\mathbf{A}$:

- interchanging the ith and jth rows

- multiplying the ith row by a nonzero scalar

- replacing the ith row by the sum of the original ith row and k times the jth row

However, two matrices that are row equivalent as defined do not necessarily have the same determinants. (See Sec. 4.5.)

Gauss-Jordan elimination is the process of using these elementary row operations to row-reduce a matrix to echelon or row-reduced echelon forms, as illustrated in Ex. 4.8. When a matrix has been converted to a row-reduced echelon matrix, it is said to be in *row canonical form*. The phrases *row-reduced echelon form* and *row canonical form* are synonymous.

4. MINORS AND COFACTORS

Minors and cofactors are determinants of submatrices associated with particular entries in the original square matrix. The *minor* of entry a_{ij} is the determinant of a submatrix resulting from the elimination of the single row i and the single column j. For example, the minor corresponding to entry a_{12} in a 3×3 matrix $\mathbf{A}$ is the determinant of the matrix created by eliminating row 1 and column 2.

$$\text{minor of } a_{12} = \begin{vmatrix} a_{21} & a_{23} \\ a_{31} & a_{33} \end{vmatrix} \qquad 4.1$$

The *cofactor* of entry a_{ij} is the minor of a_{ij} multiplied by either $+1$ or -1, depending on the position of the entry. (That is, the cofactor either exactly equals the minor or it differs only in sign.) The sign is determined according to the following positional matrix.[4]

$$\begin{bmatrix} +1 & -1 & +1 & \cdots \\ -1 & +1 & -1 & \cdots \\ +1 & -1 & +1 & \cdots \\ \vdots & \vdots & \vdots & \end{bmatrix}$$

For example, the cofactor of entry a_{12} in matrix $\mathbf{A}$ (described in Sec. 4.4) is

$$\text{cofactor of } a_{12} = -\begin{vmatrix} a_{21} & a_{23} \\ a_{31} & a_{33} \end{vmatrix} \qquad 4.2$$

Example 4.1

What is the cofactor corresponding to the -3 entry in the following matrix?

$$\mathbf{A} = \begin{bmatrix} 2 & 9 & 1 \\ -3 & 4 & 0 \\ 7 & 5 & 9 \end{bmatrix}$$

Solution

The minor's submatrix is created by eliminating the row and column of the -3 entry.

$$\mathbf{M} = \begin{bmatrix} 9 & 1 \\ 5 & 9 \end{bmatrix}$$

---

[4]The sign of the cofactor a_{ij} is positive if $(i+j)$ is even and is negative if $(i+j)$ is odd.

The minor is the determinant of **M**.

$$|\mathbf{M}| = (9)(9) - (5)(1) = 76$$

The sign corresponding to the -3 position is negative. Therefore, the cofactor is -76.

5. DETERMINANTS

A *determinant* is a scalar calculated from a square matrix. The determinant of matrix **A** can be represented as $D\{\mathbf{A}\}$, $\text{Det}(\mathbf{A})$, $\Delta\mathbf{A}$, or $|\mathbf{A}|$.[5] The following rules can be used to simplify the calculation of determinants.

- If **A** has a row or column of zeros, the determinant is zero.

- If **A** has two identical rows or columns, the determinant is zero.

- If **B** is obtained from **A** by adding a multiple of a row (column) to another row (column) in **A**, then $|\mathbf{B}| = |\mathbf{A}|$.

- If **A** is triangular, the determinant is equal to the product of the diagonal entries.

- If **B** is obtained from **A** by multiplying one row or column in **A** by a scalar k, then $|\mathbf{B}| = k|\mathbf{A}|$.

- If **B** is obtained from the $n \times n$ matrix **A** by multiplying by the scalar matrix k, then $|\mathbf{kA}| = k^n|\mathbf{A}|$.

- If **B** is obtained from **A** by switching two rows or columns in **A**, then $|\mathbf{B}| = -|\mathbf{A}|$.

Calculation of determinants is laborious for all but the smallest or simplest of matrices. For a 2×2 matrix, the formula used to calculate the determinant is easy to remember.

$$\mathbf{A} = \begin{bmatrix} a & b \\ c & d \end{bmatrix}$$

$$|\mathbf{A}| = \begin{vmatrix} a & b \\ c & d \end{vmatrix} = ad - bc \qquad 4.3$$

Two methods are commonly used for calculating the determinant of 3×3 matrices by hand. The first uses an augmented matrix constructed from the original matrix and the first two columns (as shown in Sec. 4.1).[6] The determinant is calculated as the sum of the products in the left-to-right downward diagonals less the sum of the products in the left-to-right upward diagonals.

---

[5]The vertical bars should not be confused with the square brackets used to set off a matrix, nor with absolute value.

[6]It is not actually necessary to construct the augmented matrix, but doing so helps avoid errors.

$$\mathbf{A} = \begin{bmatrix} a & b & c \\ d & e & f \\ g & h & i \end{bmatrix}$$

augmented **A** =

$$4.4$$

$$|\mathbf{A}| = aei + bfg + cdh - gec - hfa - idb \qquad 4.5$$

The second method of calculating the determinant is somewhat slower than the first for a 3×3 matrix but illustrates the method that must be used to calculate determinants of 4×4 and larger matrices. This method is known as *expansion by cofactors*. One row (column) is selected as the base row (column). The selection is arbitrary, but the number of calculations required to obtain the determinant can be minimized by choosing the row (column) with the most zeros. The determinant is equal to the sum of the products of the entries in the base row (column) and their corresponding cofactors.

$$\mathbf{A} = \begin{bmatrix} a & b & c \\ d & e & f \\ g & h & i \end{bmatrix}$$

$$|\mathbf{A}| = a\begin{vmatrix} e & f \\ h & i \end{vmatrix} - d\begin{vmatrix} b & c \\ h & i \end{vmatrix} + g\begin{vmatrix} b & c \\ e & f \end{vmatrix} \qquad 4.6$$

Example 4.2

Calculate the determinant of matrix **A** (a) by cofactor expansion, and (b) by the augmented matrix method.

$$\mathbf{A} = \begin{bmatrix} 2 & 3 & -4 \\ 3 & -1 & -2 \\ 4 & -7 & -6 \end{bmatrix}$$

Solution

(a) Since there are no zero entries, it does not matter which row or column is chosen as the base. Choose the first column as the base.

$$|\mathbf{A}| = 2\begin{vmatrix} -1 & -2 \\ -7 & -6 \end{vmatrix} - 3\begin{vmatrix} 3 & -4 \\ -7 & -6 \end{vmatrix} + 4\begin{vmatrix} 3 & -4 \\ -1 & -2 \end{vmatrix}$$

$$= (2)(6 - 14) - (3)(-18 - 28) + (4)(-6 - 4)$$

$$= 82$$

(b)

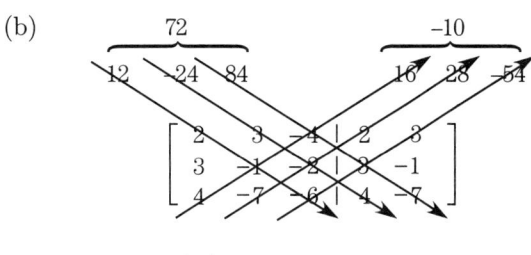

$$|\mathbf{A}| = 72 - (-10) = 82$$

6. MATRIX ALGEBRA[7]

Matrix algebra differs somewhat from standard algebra.

- *equality:* Two matrices, **A** and **B**, are equal only if they have the same numbers of rows and columns *and* if all corresponding entries are equal.

- *inequality:* The $>$ and $<$ operators are not used in matrix algebra.

- *commutative law of addition:*

$$\mathbf{A} + \mathbf{B} = \mathbf{B} + \mathbf{A} \qquad 4.7$$

- *associative law of addition:*

$$\mathbf{A} + (\mathbf{B} + \mathbf{C}) = (\mathbf{A} + \mathbf{B}) + \mathbf{C} \qquad 4.8$$

- *associative law of multiplication:*

$$(\mathbf{AB})\mathbf{C} = \mathbf{A}(\mathbf{BC}) \qquad 4.9$$

- *left distributive law:*

$$\mathbf{A}(\mathbf{B} + \mathbf{C}) = \mathbf{AB} + \mathbf{AC} \qquad 4.10$$

- *right distributive law:*

$$(\mathbf{B} + \mathbf{C})\mathbf{A} = \mathbf{BA} + \mathbf{CA} \qquad 4.11$$

- *scalar multiplication:*

$$k(\mathbf{AB}) = (k\mathbf{A})\mathbf{B} = \mathbf{A}(k\mathbf{B}) \qquad 4.12$$

Except for trivial and special cases, matrix multiplication is not commutative. That is,

$$\mathbf{AB} \neq \mathbf{BA}$$

7. MATRIX ADDITION AND SUBTRACTION

Addition and subtraction of two matrices is possible only if both matrices have the same numbers of rows and columns (i.e., order). They are accomplished by adding or subtracting the corresponding entries of the two matrices.

---

[7]Since matrices are used to simplify the presentation and solution of sets of linear equations, matrix algebra is also known as *linear algebra*.

8. MATRIX MULTIPLICATION AND DIVISION

A matrix can be multiplied by a scalar, an operation known as *scalar multiplication*, in which case all entries of the matrix are multiplied by that scalar. For example, for the 2×2 matrix **A**,

$$k\mathbf{A} = \begin{bmatrix} ka_{11} & ka_{12} \\ ka_{21} & ka_{22} \end{bmatrix}$$

A matrix can be multiplied by another matrix, but only if the left-hand matrix has the same number of columns as the right-hand matrix has rows. *Matrix multiplication* occurs by multiplying the elements in each left-hand matrix row by the entries in each right-hand matrix column, adding the products, and placing the sum at the intersection point of the participating row and column.

Matrix division can only be accomplished by multiplying by the inverse of the denominator matrix. There is no specific division operation in matrix algebra.

Example 4.3

Determine the product matrix **C**.

$$\mathbf{C} = \begin{bmatrix} 1 & 4 & 3 \\ 5 & 2 & 6 \end{bmatrix} \begin{bmatrix} 7 & 12 \\ 11 & 8 \\ 9 & 10 \end{bmatrix}$$

Solution

The left-hand matrix has three columns, and the right-hand matrix has three rows. Therefore, the two matrices can be multiplied.

The first row of the left-hand matrix and the first column of the right-hand matrix are worked with first. The corresponding entries are multiplied, and the products are summed.

$$c_{11} = (1)(7) + (4)(11) + (3)(9) = 78$$

The intersection of the top row and left column is the entry in the upper left-hand corner of the matrix **C**.

The remaining entries are calculated similarly.

$$c_{12} = (1)(12) + (4)(8) + (3)(10) = 74$$
$$c_{21} = (5)(7) + (2)(11) + (6)(9) = 111$$
$$c_{22} = (5)(12) + (2)(8) + (6)(10) = 136$$

The product matrix is

$$\mathbf{C} = \begin{bmatrix} 78 & 74 \\ 111 & 136 \end{bmatrix}$$

9. TRANSPOSE

The *transpose*, $\mathbf{A}^t$, of an $m \times n$ matrix $\mathbf{A}$ is an $n \times m$ matrix constructed by taking the ith row and making it the ith column. The diagonal is unchanged. For example,

$$\mathbf{A} = \begin{bmatrix} 1 & 6 & 9 \\ 2 & 3 & 4 \\ 7 & 1 & 5 \end{bmatrix}$$

$$\mathbf{A}^t = \begin{bmatrix} 1 & 2 & 7 \\ 6 & 3 & 1 \\ 9 & 4 & 5 \end{bmatrix}$$

Transpose operations have the following characteristics.

$$(\mathbf{A}^t)^t = \mathbf{A} \qquad 4.13$$

$$(k\mathbf{A})^t = k(\mathbf{A}^t) \qquad 4.14$$

$$\mathbf{I}^t = \mathbf{I} \qquad 4.15$$

$$(\mathbf{AB})^t = \mathbf{B}^t \mathbf{A}^t \qquad 4.16$$

$$(\mathbf{A} + \mathbf{B})^t = \mathbf{A}^t + \mathbf{B}^t \qquad 4.17$$

$$|\mathbf{A}^t| = |\mathbf{A}| \qquad 4.18$$

10. SINGULARITY AND RANK

A *singular matrix* is one whose determinant is zero. Similarly, a *nonsingular* matrix is one whose determinant is nonzero.

The *rank* of a matrix is the maximum number of linearly independent row or column vectors.[8] A matrix has rank r if it has at least one nonsingular square submatrix of order r but has no nonsingular square submatrix of order more than r. While the submatrix must be square (in order to calculate the determinant), the original matrix need not be.

The rank of an $m \times n$ matrix will be, at most, the smaller of m and n. The rank of a null matrix is zero. The ranks of a matrix and its transpose are the same. If a matrix is in echelon form, the rank will be equal to the number of rows containing at least one nonzero entry. For a 3×3 matrix, the rank can either be 3 (if it is nonsingular), 2 (if any one of its 2×2 submatrices is nonsingular), 1 (if it and all 2×2 submatrices are singular), or 0 (if it is null).

The determination of rank is laborious if done by hand. Either the matrix is reduced to echelon form by using elementary row operations, or exhaustive enumeration is used to create the submatrices and many determinants are calculated. If a matrix has more rows than columns and row-reduction is used, the work required to put the matrix in echelon form can be reduced by working with the transpose of the original matrix.

---

[8]The *row rank* and *column rank* are the same.

Example 4.4

What is the rank of matrix $\mathbf{A}$?

$$\mathbf{A} = \begin{bmatrix} 1 & -2 & -1 \\ -3 & 3 & 0 \\ 2 & 2 & 4 \end{bmatrix}$$

Solution

Matrix $\mathbf{A}$ is singular because $|\mathbf{A}| = 0$. However, there is at least one 2×2 nonsingular submatrix:

$$\begin{vmatrix} 1 & -2 \\ -3 & 3 \end{vmatrix} = (1)(3) - (-3)(-2) = -3$$

Therefore, the rank is 2.

Example 4.5

Determine the rank of matrix $\mathbf{A}$ by reducing it to echelon form.

$$\mathbf{A} = \begin{bmatrix} 7 & 4 & 9 & 1 \\ 0 & 2 & -5 & 3 \\ 0 & 4 & -10 & 6 \end{bmatrix}$$

Solution

By inspection, the matrix can be row-reduced by subtracting two times the second row from the third row. The matrix cannot be further reduced. Since there are two nonzero rows, the rank is 2.

$$\begin{bmatrix} 7 & 4 & 9 & 1 \\ 0 & 2 & -5 & 3 \\ 0 & 0 & 0 & 0 \end{bmatrix}$$

11. CLASSICAL ADJOINT

The *classical adjoint* is the transpose of the cofactor matrix. (See Sec. 4.2.) The resulting matrix can be designated as $\mathbf{A}_{adj}$, $adj\{\mathbf{A}\}$, or $\mathbf{A}^{adj}$.

Example 4.6

What is the classical adjoint of matrix $\mathbf{A}$?

$$\mathbf{A} = \begin{bmatrix} 2 & 3 & -4 \\ 0 & -4 & 2 \\ 1 & -1 & 5 \end{bmatrix}$$

Solution

The matrix of cofactors is determined to be

$$\begin{bmatrix} -18 & 2 & 4 \\ -11 & 14 & 5 \\ -10 & -4 & -8 \end{bmatrix}$$

The transpose of the matrix of cofactors is

$$\mathbf{A}_{adj} = \begin{bmatrix} -18 & -11 & -10 \\ 2 & 14 & -4 \\ 4 & 5 & -8 \end{bmatrix}$$

12. INVERSE

The product of a matrix $\mathbf{A}$ and its inverse, $\mathbf{A}^{-1}$, is the identity matrix, $\mathbf{I}$. Only square matrices have inverses, but not all square matrices are invertible. A matrix has an inverse if and only if it is nonsingular (i.e., its determinant is nonzero).

$$\mathbf{A}\mathbf{A}^{-1} = \mathbf{A}^{-1}\mathbf{A} = \mathbf{I} \qquad 4.19$$

$$(\mathbf{A}\mathbf{B})^{-1} = \mathbf{B}^{-1}\mathbf{A}^{-1} \qquad 4.20$$

The inverse of a 2 × 2 matrix is most easily determined by formula.

$$\mathbf{A} = \begin{bmatrix} a & b \\ c & d \end{bmatrix}$$

$$\mathbf{A}^{-1} = \frac{\begin{bmatrix} d & -b \\ -c & a \end{bmatrix}}{|\mathbf{A}|} \qquad 4.21$$

For any matrix, the inverse is determined by dividing every entry in the classical adjoint by the determinant of the original matrix.

$$\mathbf{A}^{-1} = \frac{\mathbf{A}_{adj}}{|\mathbf{A}|} \qquad 4.22$$

Example 4.7

What is the inverse of matrix $\mathbf{A}$?

$$\mathbf{A} = \begin{bmatrix} 4 & 5 \\ 2 & 3 \end{bmatrix}$$

Solution

The determinant is calculated as

$$|\mathbf{A}| = (4)(3) - (2)(5) = 2$$

Using Eq. 4.22, the inverse is

$$\mathbf{A}^{-1} = \frac{\begin{bmatrix} 3 & -5 \\ -2 & 4 \end{bmatrix}}{2} = \begin{bmatrix} \frac{3}{2} & -\frac{5}{2} \\ -1 & 2 \end{bmatrix}$$

Check.

$$\mathbf{A}\mathbf{A}^{-1} = \begin{bmatrix} 4 & 5 \\ 2 & 3 \end{bmatrix}\begin{bmatrix} \frac{3}{2} & -\frac{5}{2} \\ -1 & 2 \end{bmatrix} = \begin{bmatrix} 6-5 & -10+10 \\ 3-3 & -5+6 \end{bmatrix}$$

$$= \begin{bmatrix} 1 & 0 \\ 0 & 1 \end{bmatrix}$$

$$= \mathbf{I} \qquad [\text{OK}]$$

13. WRITING SIMULTANEOUS LINEAR EQUATIONS IN MATRIX FORM

Matrices are used to simplify the presentation and solution of sets of simultaneous linear equations. For example, the following three methods of presenting simultaneous linear equations are equivalent:

$$a_{11}x_1 + a_{12}x_2 = b_1$$
$$a_{21}x_1 + a_{22}x_2 = b_2$$

$$\begin{bmatrix} a_{11} & a_{12} \\ a_{21} & a_{22} \end{bmatrix}\begin{bmatrix} x_1 \\ x_2 \end{bmatrix} = \begin{bmatrix} b_1 \\ b_2 \end{bmatrix}$$

$$\mathbf{A}\mathbf{X} = \mathbf{B}$$

In the second and third representations, $\mathbf{A}$ is known as the *coefficient matrix*, $\mathbf{X}$ as the *variable matrix*, and $\mathbf{B}$ as the *constant matrix*.

Not all systems of simultaneous equations have solutions, and those that do may not have unique solutions. The existence of a solution can be determined by calculating the determinant of the coefficient matrix. Solution-existence rules are summarized in Table 4.1.

- If the system of linear equations is homogeneous (i.e., $\mathbf{B}$ is a zero matrix) and $|\mathbf{A}|$ is zero, there are an infinite number of solutions.

- If the system is homogeneous and $|\mathbf{A}|$ is nonzero, only the trivial solution exists.

- If the system of linear equations is nonhomogeneous (i.e., $\mathbf{B}$ is not a zero matrix) and $|\mathbf{A}|$ is nonzero, there is a unique solution to the set of simultaneous equations.

- If $|\mathbf{A}|$ is zero, a nonhomogeneous system of simultaneous equations may still have a solution. The requirement is that the determinants of all substitutional matrices (see Sec. 4.14) are zero, in which case there will be an infinite number of solutions. Otherwise, no solution exists.

Table 4.1 *Solution-Existence Rules for Simultaneous Equations*

	$\mathbf{B} = 0$	$\mathbf{B} \neq 0$
$|\mathbf{A}| = 0$	infinite number of solutions (linearly dependent equations)	either an infinite number of solutions or no solution at all
$|\mathbf{A}| \neq 0$	trivial solution only ($x_i = 0$)	unique nonzero solution

14. SOLVING SIMULTANEOUS LINEAR EQUATIONS

Gauss-Jordan elimination can be used to obtain the solution to a set of simultaneous linear equations. The coefficient matrix is augmented by the constant matrix. Then, elementary row operations are used to reduce the coefficient matrix to canonical form. All of the operations performed on the coefficient matrix are performed on the constant matrix. The variable values that satisfy the simultaneous equations will be the entries in the constant matrix when the coefficient matrix is in canonical form.

Determinants are used to calculate the solution to linear simultaneous equations through a procedure known as *Cramer's rule.*

The procedure is to calculate determinants of the original coefficient matrix $\mathbf{A}$ and of the n matrices resulting from the systematic replacement of a column in $\mathbf{A}$ by the constant matrix $\mathbf{B}$. For a system of three equations in three unknowns, there are three substitutional matrices, $\mathbf{A}_1$, $\mathbf{A}_2$, and $\mathbf{A}_3$, as well as the original coefficient matrix, for a total of four matrices whose determinants must be calculated.

The values of the unknowns that simultaneously satisfy all of the linear equations are

$$x_1 = \frac{|\mathbf{A}_1|}{|\mathbf{A}|} \qquad \qquad 4.23$$

$$x_2 = \frac{|\mathbf{A}_2|}{|\mathbf{A}|} \qquad \qquad 4.24$$

$$x_3 = \frac{|\mathbf{A}_3|}{|\mathbf{A}|} \qquad \qquad 4.25$$

Example 4.8

Use Gauss-Jordan elimination to solve the following system of simultaneous equations.

$$2x + 3y - 4z = 1$$
$$3x - y - 2z = 4$$
$$4x - 7y - 6z = -7$$

Solution

The augmented matrix is created by appending the constant matrix to the coefficient matrix.

$$\begin{bmatrix} 2 & 3 & -4 & | & 1 \\ 3 & -1 & -2 & | & 4 \\ 4 & -7 & -6 & | & -7 \end{bmatrix}$$

Elementary row operations are used to reduce the coefficient matrix to canonical form. For example, two times the first row is subtracted from the third row. This step obtains the 0 needed in the a_{31} position.

$$\begin{bmatrix} 2 & 3 & -4 & | & 1 \\ 3 & -1 & -2 & | & 4 \\ 0 & -13 & 2 & | & -9 \end{bmatrix}$$

This process continues until the following form is obtained.

$$\begin{bmatrix} 1 & 0 & 0 & | & 3 \\ 0 & 1 & 0 & | & 1 \\ 0 & 0 & 1 & | & 2 \end{bmatrix}$$

$x = 3$, $y = 1$, and $z = 2$ satisfy this system of equations.

Example 4.9

Use Cramer's rule to solve the following system of simultaneous equations.

$$2x + 3y - 4z = 1$$
$$3x - y - 2z = 4$$
$$4x - 7y - 6z = -7$$

Solution

The determinant of the coefficient matrix is

$$|\mathbf{A}| = \begin{vmatrix} 2 & 3 & -4 \\ 3 & -1 & -2 \\ 4 & -7 & -6 \end{vmatrix} = 82$$

The determinants of the substitutional matrices are

$$|\mathbf{A}_1| = \begin{vmatrix} 1 & 3 & -4 \\ 4 & -1 & -2 \\ -7 & -7 & -6 \end{vmatrix} = 246$$

$$|\mathbf{A}_2| = \begin{vmatrix} 2 & 1 & -4 \\ 3 & 4 & -2 \\ 4 & -7 & -6 \end{vmatrix} = 82$$

$$|\mathbf{A}_3| = \begin{vmatrix} 2 & 3 & 1 \\ 3 & -1 & 4 \\ 4 & -7 & -7 \end{vmatrix} = 164$$

The values of x, y, and z that will satisfy the linear equations are

$$x = \frac{246}{82} = 3$$

$$y = \frac{82}{82} = 1$$

$$z = \frac{164}{82} = 2$$

15. EIGENVALUES AND EIGENVECTORS

Eigenvalues and eigenvectors (also known as *characteristic values* and *characteristic vectors*) of a square matrix $\mathbf{A}$ are the scalars k and matrices $\mathbf{X}$ such that

$$\mathbf{A}\mathbf{X} = k\mathbf{X} \qquad 4.26$$

The scalar k is an eigenvalue of $\mathbf{A}$ if and only if the matrix $(k\mathbf{I} - \mathbf{A})$ is singular; that is, if $|k\mathbf{I} - \mathbf{A}| = 0$. This equation is called the *characteristic equation* of the matrix $\mathbf{A}$. When expanded, the determinant is called the *characteristic polynomial*. The method of using the characteristic polynomial to find eigenvalues and eigenvectors is illustrated in Ex. 4.10.

If all of the eigenvalues are unique (i.e., nonrepeating), then Eq. 4.27 is valid.

$$[k\mathbf{I} - \mathbf{A}]\mathbf{X} = 0 \qquad 4.27$$

Example 4.10

Find the eigenvalues and nonzero eigenvectors of the matrix $\mathbf{A}$.

$$\mathbf{A} = \begin{bmatrix} 2 & 4 \\ 6 & 4 \end{bmatrix}$$

Solution

$$k\mathbf{I} - \mathbf{A} = \begin{bmatrix} k & 0 \\ 0 & k \end{bmatrix} - \begin{bmatrix} 2 & 4 \\ 6 & 4 \end{bmatrix} = \begin{bmatrix} k-2 & -4 \\ -6 & k-4 \end{bmatrix}$$

The characteristic polynomial is found by setting the determinant $|k\mathbf{I} - \mathbf{A}|$ equal to zero.

$$(k-2)(k-4) - (-6)(-4) = 0$$
$$k^2 - 6k - 16 = (k-8)(k+2) = 0$$

The roots of the characteristic polynomial are $k = +8$ and $k = -2$. These are the eigenvalues of $\mathbf{A}$.

Substituting $k = 8$,

$$k\mathbf{I} - \mathbf{A} = \begin{bmatrix} 8-2 & -4 \\ -6 & 8-4 \end{bmatrix} = \begin{bmatrix} 6 & -4 \\ -6 & 4 \end{bmatrix}$$

The resulting system can be interpreted as the linear equation $6x_1 - 4x_2 = 0$. The values of x that satisfy this equation define the eigenvector. An eigenvector $\mathbf{X}$ associated with the eigenvalue $+8$ is

$$\mathbf{X} = \begin{bmatrix} x_1 \\ x_2 \end{bmatrix} = \begin{bmatrix} 4 \\ 6 \end{bmatrix}$$

All other eigenvectors for this eigenvalue are multiples of $\mathbf{X}$. Normally $\mathbf{X}$ is reduced to smallest integers.

$$\mathbf{X} = \begin{bmatrix} 2 \\ 3 \end{bmatrix}$$

Similarly, the eigenvector associated with the eigenvalue -2 is

$$\mathbf{X} = \begin{bmatrix} x_1 \\ x_2 \end{bmatrix} = \begin{bmatrix} +4 \\ -4 \end{bmatrix}$$

Reducing this to smallest integers gives

$$\mathbf{X} = \begin{bmatrix} +1 \\ -1 \end{bmatrix}$$

5 Vectors

1. INTRODUCTION

Some characteristics can be described by scalars, vectors, or tensors. A *scalar* has only magnitude. Knowing its value is sufficient to define the characteristic. Mass, enthalpy, density, and speed are examples of scalar characteristics.

Force, momentum, displacement, and velocity are examples of vector characteristics. A *vector* is a directed straight line with a specific magnitude and that is specified completely by its direction (consisting of the vector's *angular orientation* and its *sense*) and magnitude. A vector's *point of application* (*terminal point*) is not needed to define the vector.[1] Two vectors with the same direction and magnitude are said to be *equal vectors* even though their *lines of action* may be different.[2]

A vector can be designated by a boldface variable (as in this book) or as a combination of the variable and some other symbol. For example, the notations $\mathbf{V}$, $\overline{V}$, $\hat{V}$, $\vec{V}$, and $\underline{V}$ are used by different authorities to represent vectors. In this book, the magnitude of a vector can be designated by either $|\mathbf{V}|$ or V (italic but not bold).

Stress, dielectric constant, and magnetic susceptibility are examples of tensor characteristics. A *tensor* has magnitude in a specific direction, but the direction is not unique. Tensors are frequently associated with *anisotropic materials* that have different properties in different directions. A tensor in three-dimensional space is defined

by nine components, compared with the three that are required to define vectors. Those components are written in matrix form. Stress, σ, at a point, for example, would be defined by the following tensor matrix.

$$\sigma \equiv \begin{bmatrix} \sigma_{xx} & \sigma_{xy} & \sigma_{xz} \\ \sigma_{yx} & \sigma_{yy} & \sigma_{yz} \\ \sigma_{zx} & \sigma_{zy} & \sigma_{zz} \end{bmatrix}$$

2. VECTORS IN *n*-SPACE

In some cases, a vector, $\mathbf{V}$, will be designated by its two endpoints in *n*-dimensional vector space. The usual vector space is three-dimensional force-space. Usually, one of the points will be the origin, in which case the vector is said to be "based at the origin," "origin-based," or "zero-based."[3] If one of the endpoints is the origin, specifying a terminal point P would represent a force directed from the origin to point P.

If a coordinate system is superimposed on the vector space, a vector can be specified in terms of the *n* coordinates of its two endpoints. The magnitude of the vector $\mathbf{V}$ is the distance in vector space between the two points, as given by Eq. 5.1. Similarly, the direction is defined by the angle the vector makes with one of the axes. Figure 5.1 illustrates a vector in two dimensions.

$$|\mathbf{V}| = \sqrt{(x_2 - x_1)^2 + (y_2 - y_1)^2} \qquad 5.1$$

$$\phi = \arctan \frac{y_2 - y_1}{x_2 - x_1} \qquad 5.2$$

Figure 5.1 *Vector in Two-Dimensional Space*

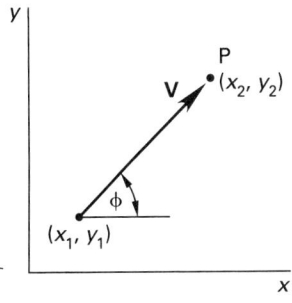

---

[1] A vector that is constrained to act at or through a certain point is a *bound vector* (*fixed vector*). A *sliding vector* (*transmissible vector*) can be applied anywhere along its line of action. A *free vector* is not constrained and can be applied at any point in space.

[2] A distinction is sometimes made between equal vectors and equivalent vectors. *Equivalent vectors* produce the same effect but are not necessarily equal.

[3] Any vector directed from P_1 to P_2 can be transformed into a zero-based vector by subtracting the coordinates of point P_1 from the coordinates of terminal point P_2. The transformed vector will be equivalent to the original vector.

The *components* of a vector are the projections of the vector on the coordinate axes. (For a zero-based vector, the components and the coordinates of the endpoint are the same.) Simple trigonometric principles are used to resolve a vector into its components. A vector reconstructed from its components is known as a *resultant vector*.

$$V_x = |\mathbf{V}|\cos\phi_x \qquad 5.3$$

$$V_y = |\mathbf{V}|\cos\phi_y \qquad 5.4$$

$$V_z = |\mathbf{V}|\cos\phi_z \qquad 5.5$$

$$|\mathbf{V}| = \sqrt{V_x^2 + V_y^2 + V_z^2} \qquad 5.6$$

In Eq. 5.3 through Eq. 5.5, ϕ_x, ϕ_y, and ϕ_z are the *direction angles*—the angles between the vector and the x-, y-, and z-axes, respectively. Figure 5.2 shows the location of direction angles. The cosines of these angles are known as *direction cosines*. The sum of the squares of the direction cosines is equal to 1.

$$\cos^2\phi_x + \cos^2\phi_y + \cos^2\phi_z = 1 \qquad 5.7$$

Figure 5.2 *Direction Angles of a Vector*

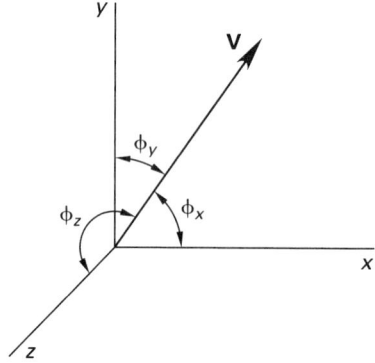

3. UNIT VECTORS

Unit vectors are vectors with unit magnitudes (i.e., magnitudes of 1). They are represented in the same notation as other vectors. (Unit vectors in this book are written in boldface type.) Although they can have any direction, the standard unit vectors (the *Cartesian unit vectors* **i**, **j**, and **k**) have the directions of the x-, y-, and z-coordinate axes and constitute the *Cartesian triad*, as illustrated in Fig. 5.3.

A vector **V** can be written in terms of unit vectors and its components.

$$\mathbf{V} = |\mathbf{V}|\mathbf{a} = V_x\mathbf{i} + V_y\mathbf{j} + V_z\mathbf{k} \qquad 5.8$$

Figure 5.3 *Cartesian Unit Vectors*

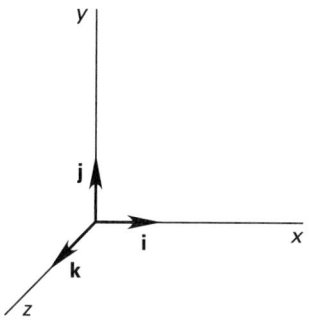

The unit vector, **a**, has the same direction as the vector **V** but has a length of 1. This unit vector is calculated by dividing the original vector, **V**, by its magnitude, $|\mathbf{V}|$.

$$\mathbf{a} = \frac{\mathbf{V}}{|\mathbf{V}|} = \frac{V_x\mathbf{i} + V_y\mathbf{j} + V_z\mathbf{k}}{\sqrt{V_x^2 + V_y^2 + V_z^2}} \qquad 5.9$$

4. VECTOR REPRESENTATION

The most common method of representing a vector is by writing it in *rectangular form*—a vector sum of its orthogonal components. In rectangular form, each of the orthogonal components has the same units as the resultant vector.

$$\mathbf{A} \equiv A_x\mathbf{i} + A_y\mathbf{j} + A_z\mathbf{k} \quad \text{[three dimensions]}$$

However, the vector is also completely defined by its magnitude and associated angle. These two quantities can be written together in *phasor form*, sometimes referred to as *polar form*.

$$\mathbf{A} \equiv |\mathbf{A}|\angle\phi = A\angle\phi$$

5. CONVERSION BETWEEN SYSTEMS

The choice of the **ijk** triad may be convenient but is arbitrary. A vector can be expressed in terms of any other set of orthogonal unit vectors, **uvw**.

$$\mathbf{V} = V_x\mathbf{i} + V_y\mathbf{j} + V_z\mathbf{k} = V_x'\mathbf{u} + V_y'\mathbf{v} + V_z'\mathbf{w} \qquad 5.10$$

The two representations are related.

$$V_x' = \mathbf{V}\cdot\mathbf{u} = (\mathbf{i}\cdot\mathbf{u})V_x + (\mathbf{j}\cdot\mathbf{u})V_y + (\mathbf{k}\cdot\mathbf{u})V_z$$
$$5.11$$

$$V_y' = \mathbf{V}\cdot\mathbf{v} = (\mathbf{i}\cdot\mathbf{v})V_x + (\mathbf{j}\cdot\mathbf{v})V_y + (\mathbf{k}\cdot\mathbf{v})V_z$$
$$5.12$$

$$V_z' = \mathbf{V}\cdot\mathbf{w} = (\mathbf{i}\cdot\mathbf{w})V_x + (\mathbf{j}\cdot\mathbf{w})V_y + (\mathbf{k}\cdot\mathbf{w})V_z$$
$$5.13$$

Equation 5.11 through Eq. 5.13 can be expressed in matrix form. The dot products are known as the *coefficients of transformation*, and the matrix containing them is the *transformation matrix*.

$$\begin{pmatrix} V'_x \\ V'_y \\ V'_z \end{pmatrix} = \begin{pmatrix} \mathbf{i}\cdot\mathbf{u} & \mathbf{j}\cdot\mathbf{u} & \mathbf{k}\cdot\mathbf{u} \\ \mathbf{i}\cdot\mathbf{v} & \mathbf{j}\cdot\mathbf{v} & \mathbf{k}\cdot\mathbf{v} \\ \mathbf{i}\cdot\mathbf{w} & \mathbf{j}\cdot\mathbf{w} & \mathbf{k}\cdot\mathbf{w} \end{pmatrix} \begin{pmatrix} V_x \\ V_y \\ V_z \end{pmatrix} \qquad 5.14$$

6. VECTOR ADDITION

Addition of two vectors by the *polygon method* is accomplished by placing the tail of the second vector at the head (tip) of the first. The sum (i.e., the *resultant vector*) is a vector extending from the tail of the first vector to the head of the second, as shown in Fig. 5.4. Alternatively, the two vectors can be considered as the two sides of a parallelogram, while the sum represents the diagonal. This is known as addition by the *parallelogram method*.

Figure 5.4 Addition of Two Vectors

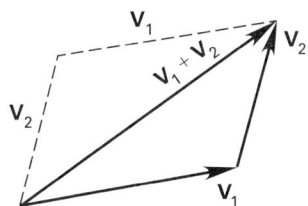

The components of the resultant vector are the sums of the components of the added vectors (that is, $V_{1x} + V_{2x}$, $V_{1y} + V_{2y}$, $V_{1z} + V_{2z}$).

Vector addition is both commutative and associative.

$$\mathbf{V}_1 + \mathbf{V}_2 = \mathbf{V}_2 + \mathbf{V}_1 \qquad 5.15$$

$$\mathbf{V}_1 + (\mathbf{V}_2 + \mathbf{V}_3) = (\mathbf{V}_1 + \mathbf{V}_2) + \mathbf{V}_3 \qquad 5.16$$

7. MULTIPLICATION BY A SCALAR

A vector, $\mathbf{V}$, can be multiplied by a scalar, c. If the original vector is represented by its components, each of the components is multiplied by c.

$$c\mathbf{V} = c|\mathbf{V}|\mathbf{a} = cV_x\mathbf{i} + cV_y\mathbf{j} + cV_z\mathbf{k} \qquad 5.17$$

Scalar multiplication is distributive.

$$c(\mathbf{V}_1 + \mathbf{V}_2) = c\mathbf{V}_1 + c\mathbf{V}_2 \qquad 5.18$$

8. VECTOR DOT PRODUCT

The *dot product (scalar product)*, $\mathbf{V}_1 \cdot \mathbf{V}_2$, of two vectors is a scalar that is proportional to the length of the projection of the first vector onto the second vector, as illustrated in Fig. 5.5.[4]

Figure 5.5 Vector Dot Product

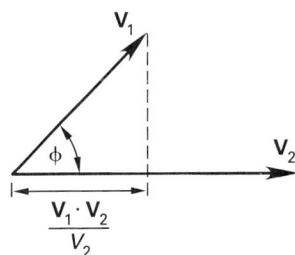

The dot product is commutative and distributive.

$$\mathbf{V}_1 \cdot \mathbf{V}_2 = \mathbf{V}_2 \cdot \mathbf{V}_1 \qquad 5.19$$

$$\mathbf{V}_1 \cdot (\mathbf{V}_2 + \mathbf{V}_3) = \mathbf{V}_1 \cdot \mathbf{V}_2 + \mathbf{V}_1 \cdot \mathbf{V}_3 \qquad 5.20$$

The dot product can be calculated in two ways, as Eq. 5.21 indicates. ϕ is the acute angle between the two vectors and is less than or equal to 180°.

$$\mathbf{V}_1 \cdot \mathbf{V}_2 = |\mathbf{V}_1|\,|\mathbf{V}_2|\cos\phi$$
$$= V_{1x}V_{2x} + V_{1y}V_{2y} + V_{1z}V_{2z} \qquad 5.21$$

When Eq. 5.21 is solved for the angle between the two vectors, ϕ, it is known as the *Cauchy-Schwartz theorem*.

$$\cos\phi = \frac{V_{1x}V_{2x} + V_{1y}V_{2y} + V_{1z}V_{2z}}{|\mathbf{V}_1|\,|\mathbf{V}_2|} \qquad 5.22$$

The dot product can be used to determine whether a vector is a unit vector and to show that two vectors are orthogonal (perpendicular). For any unit vector, $\mathbf{u}$,

$$\mathbf{u}\cdot\mathbf{u} = 1 \qquad 5.23$$

For two non-null orthogonal vectors,

$$\mathbf{V}_1 \cdot \mathbf{V}_2 = 0 \qquad 5.24$$

Equation 5.23 and Eq. 5.24 can be extended to the Cartesian unit vectors.

$$\mathbf{i}\cdot\mathbf{i} = 1 \qquad 5.25$$
$$\mathbf{j}\cdot\mathbf{j} = 1 \qquad 5.26$$
$$\mathbf{k}\cdot\mathbf{k} = 1 \qquad 5.27$$
$$\mathbf{i}\cdot\mathbf{j} = 0 \qquad 5.28$$
$$\mathbf{i}\cdot\mathbf{k} = 0 \qquad 5.29$$
$$\mathbf{j}\cdot\mathbf{k} = 0 \qquad 5.30$$

---

[4]The dot product is also written in parentheses without a dot; that is, $(\mathbf{V}_1\mathbf{V}_2)$.

Example 5.1

What is the angle between the zero-based vectors $V_1 = (-\sqrt{3}, 1)$ and $V_2 = (2\sqrt{3}, 2)$ in an x-y coordinate system?

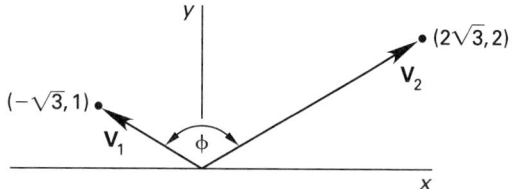

Solution

From Eq. 5.22,

$$\cos \phi = \frac{V_{1x} V_{2x} + V_{1y} V_{2y}}{|V_1| \, |V_2|} = \frac{V_{1x} V_{2x} + V_{1y} V_{2y}}{\sqrt{V_{1x}^2 + V_{1y}^2} \sqrt{V_{2x}^2 + V_{2y}^2}}$$

$$= \frac{(-\sqrt{3})(2\sqrt{3}) + (1)(2)}{\sqrt{(-\sqrt{3})^2 + (1)^2} \sqrt{(2\sqrt{3})^2 + (2)^2}}$$

$$= \frac{-4}{8} \quad \left(-\tfrac{1}{2}\right)$$

$$\phi = \arccos\left(-\tfrac{1}{2}\right) = 120°$$

9. VECTOR CROSS PRODUCT

The *cross product (vector product)*, $V_1 \times V_2$, of two vectors is a vector that is orthogonal (perpendicular) to the plane of the two vectors.[5] The unit vector representation of the cross product can be calculated as a third-order determinant. Figure 5.6 illustrates the vector cross product.

$$V_1 \times V_2 = \begin{vmatrix} i & V_{1x} & V_{2x} \\ j & V_{1y} & V_{2y} \\ k & V_{1z} & V_{2z} \end{vmatrix} \qquad 5.31$$

Figure 5.6 *Vector Cross Product*

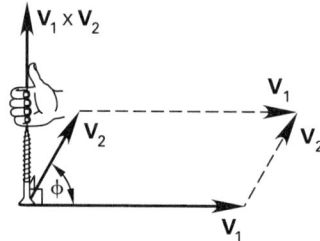

---

[5]The cross product is also written in square brackets without a cross, that is, $[V_1 V_2]$.

The direction of the cross-product vector corresponds to the direction a right-hand screw would progress if vectors V_1 and V_2 are placed tail-to-tail in the plane they define and V_1 is rotated into V_2. The direction can also be found from the *right-hand rule*.

The magnitude of the cross product can be determined from Eq. 5.32, in which ϕ is the angle between the two vectors and is limited to 180°. The magnitude corresponds to the area of a parallelogram that has V_1 and V_2 as two of its sides.

$$|V_1 \times V_2| = |V_1||V_2| \sin \phi \qquad 5.32$$

Vector cross multiplication is distributive but not commutative.

$$V_1 \times V_2 = -(V_2 \times V_1) \qquad 5.33$$

$$c(V_1 \times V_2) = (cV_1) \times V_2 = V_1 \times (cV_2) \qquad 5.34$$

$$V_1 \times (V_2 + V_3) = V_1 \times V_2 + V_1 \times V_3 \qquad 5.35$$

If the two vectors are parallel, their cross product will be zero.

$$i \times i = j \times j = k \times k = 0 \qquad 5.36$$

Equation 5.31 and Eq. 5.33 can be extended to the unit vectors.

$$i \times j = -j \times i = k \qquad 5.37$$

$$j \times k = -k \times j = i \qquad 5.38$$

$$k \times i = -i \times k = j \qquad 5.39$$

Example 5.2

Find a unit vector orthogonal to $V_1 = i - j + 2k$ and $V_2 = 3j - k$.

Solution

The cross product is a vector orthogonal to V_1 and V_2.

$$V_1 \times V_2 = \begin{vmatrix} i & 1 & 0 \\ j & -1 & 3 \\ k & 2 & -1 \end{vmatrix}$$

$$= -5i + j + 3k$$

Check to see whether this is a unit vector.

$$|V_1 \times V_2| = \sqrt{(-5)^2 + (1)^2 + (3)^2} = \sqrt{35}$$

Since its length is $\sqrt{35}$, the vector must be divided by $\sqrt{35}$ to obtain a unit vector.

$$a = \frac{-5i + j + 3k}{\sqrt{35}}$$

10. MIXED TRIPLE PRODUCT

The *mixed triple product* (*triple scalar product* or just *triple product*) of three vectors is a scalar quantity representing the volume of a parallelepiped with the three vectors making up the sides. It is calculated as a determinant. Since Eq. 5.40 can be negative, the absolute value must be used to obtain the volume in that case. Figure 5.7 shows a mixed triple product.

$$\mathbf{V}_1 \cdot (\mathbf{V}_2 \times \mathbf{V}_3) = \begin{vmatrix} V_{1x} & V_{1y} & V_{1z} \\ V_{2x} & V_{2y} & V_{2z} \\ V_{3x} & V_{3y} & V_{3z} \end{vmatrix} \qquad 5.40$$

Figure 5.7 *Vector Mixed Triple Product*

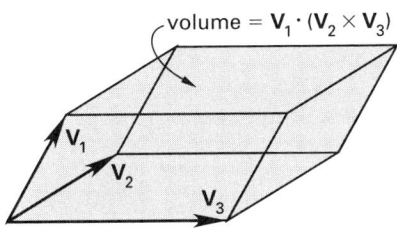

The mixed triple product has the property of *circular permutation*, as defined by Eq. 5.41.

$$\mathbf{V}_1 \cdot (\mathbf{V}_2 \times \mathbf{V}_3) = (\mathbf{V}_1 \times \mathbf{V}_2) \cdot \mathbf{V}_3 \qquad 5.41$$

11. VECTOR TRIPLE PRODUCT

The *vector triple product* (also known as the *vector triple cross product*, *BAC-CAB identity* (spoken as "BAC minus CAB"), and *Lagrange's formula*) is a vector defined by Eq. 5.42.[6] The cross products in the parentheses on the right-hand side are scalars, so the vector triple product is the scaled difference between $\mathbf{V}_2$ and $\mathbf{V}_3$. Geometrically, the resultant vector is perpendicular to $\mathbf{V}_1$ and is in the plane defined by $\mathbf{V}_2$ and $\mathbf{V}_3$.

$$\mathbf{V}_1 \times (\mathbf{V}_2 \times \mathbf{V}_3) = (\mathbf{V}_1 \cdot \mathbf{V}_3)\mathbf{V}_2 - (\mathbf{V}_1 \cdot \mathbf{V}_2)\mathbf{V}_3 \qquad 5.42$$

12. VECTOR FUNCTIONS

A vector can be a function of another parameter. For example, a vector $\mathbf{V}$ is a function of variable t when its V_x, V_y, and V_z components are functions of t. For example,

$$\mathbf{V}(t) = (2t - 3)\mathbf{i} + (t^2 + 1)\mathbf{j} + (-7t + 5)\mathbf{k}$$

When the functions of t are differentiated (or integrated) with respect to t, the vector itself is differentiated (integrated).[7]

$$\frac{d\mathbf{V}(t)}{dt} = \frac{dV_x}{dt}\mathbf{i} + \frac{dV_y}{dt}\mathbf{j} + \frac{dV_z}{dt}\mathbf{k} \qquad 5.43$$

Similarly, the integral of the vector is

$$\int \mathbf{V}(t)\, dt = \mathbf{i} \int V_x\, dt + \mathbf{j} \int V_y\, dt + \mathbf{k} \int V_z\, dt \qquad 5.44$$

---

[6]This concept is usually encountered only in electrodynamic field theory.
[7]This is particularly valuable when converting among position, velocity, and acceleration vectors.

6 Trigonometry

1. DEGREES AND RADIANS

Degrees and *radians* are two units for measuring angles. One complete circle is divided into 360 degrees (written 360°) or 2π radians (abbreviated *rad*).[1] The conversions between degrees and radians are

multiply	by	to obtain
radians	$\dfrac{180}{\pi}$	degrees
degrees	$\dfrac{\pi}{180}$	radians

The number of radians in an angle, θ, corresponds to twice the area within a circular sector with arc length θ and a radius of one, as shown in Fig. 6.1. Alternatively,

Figure 6.1 *Radians and Area of Unit Circle*

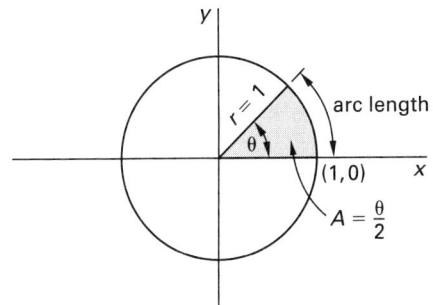

the area of a sector with central angle θ radians is $\theta/2$ for a *unit circle* (i.e., a circle with a radius of one unit).

2. PLANE ANGLES

A *plane angle* (usually referred to as just an *angle*) consists of two intersecting lines and an intersection point known as the *vertex*. The angle can be referred to by a capital letter representing the vertex (e.g., B in Fig. 6.2), a letter representing the angular measure (e.g., B or β), or by three capital letters, where the middle letter is the vertex and the other two letters are two points on different lines, and either the symbol $\angle$ or $\sphericalangle$ (e.g., $\sphericalangle$ ABC).

Figure 6.2 *Angle*

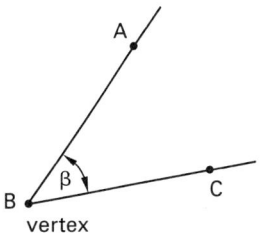

The angle between two intersecting lines generally is understood to be the smaller angle created.[2] Angles have been classified as follows.

- *acute angle:* an angle less than 90° ($\pi/2$ rad)
- *obtuse angle:* an angle more than 90° ($\pi/2$ rad) but less than 180° (π rad)
- *reflex angle:* an angle more than 180° (π rad) but less than 360° (2π rad)
- *related angle:* an angle that differs from another by some multiple of 90° ($\pi/2$ rad)
- *right angle:* an angle equal to 90° ($\pi/2$ rad)
- *straight angle:* an angle equal to 180° (π rad); that is, a straight line

Complementary angles are two angles whose sum is 90° ($\pi/2$ rad). *Supplementary angles* are two angles whose sum is 180° (π rad). *Adjacent angles* share a common vertex and one (the interior) side. Adjacent angles are

---

[1] The abbreviation *rad* is also used to represent *radiation absorbed dose*, a measure of radiation exposure.

[2] In books on geometry, the term *ray* is used instead of *line*.

supplementary if, and only if, their exterior sides form a straight line.

Vertical angles are the two angles with a common vertex and with sides made up by two intersecting straight lines, as shown in Fig. 6.3. Vertical angles are equal.

Angle of elevation and *angle of depression* are surveying terms referring to the angle above and below the horizontal plane of the observer, respectively.

Figure 6.3 *Vertical Angles*

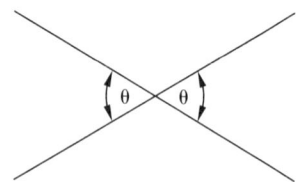

3. TRIANGLES

A *triangle* is a three-sided closed polygon with three angles whose sum is 180° (π rad). Triangles are identified by their vertices and the symbol Δ (e.g., ΔABC in Fig. 6.4). A side is designated by its two endpoints (e.g., AB in Fig. 6.4) or by a lowercase letter corresponding to the capital letter of the opposite vertex (e.g., c).

In *similar triangles*, the corresponding angles are equal and the corresponding sides are in proportion. (Since there are only two independent angles in a triangle, showing that two angles of one triangle are equal to two angles of the other triangle is sufficient to show similarity.) The symbol for similarity is $\sim$. In Fig. 6.4, ΔABC $\sim$ ΔDEF (i.e., ΔABC is similar to ΔDEF).

Figure 6.4 *Similar Triangles*

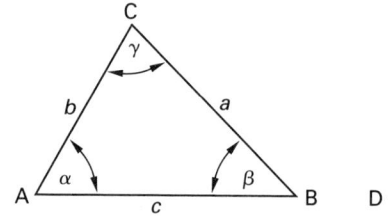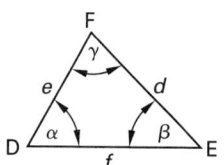

4. RIGHT TRIANGLES

A *right triangle* is a triangle in which one of the angles is 90° ($\pi/2$ rad). The remaining two angles are complementary. If one of the acute angles is chosen as the reference, the sides forming the right angle are known as the *adjacent side*, x, and the *opposite side*, y. The longest side is known as the *hypotenuse*, r. The *Pythagorean theorem* relates the lengths of these sides.

$$x^2 + y^2 = r^2 \qquad 6.1$$

In certain cases, the lengths of unknown sides of right triangles can be determined by inspection.[3] This occurs when the lengths of the sides are in the ratios of 3:4:5, 1:1:$\sqrt{2}$, 1:$\sqrt{3}$:2, or 5:12:13. Figure 6.5 illustrates a 3:4:5 triangle.

Figure 6.5 *3:4:5 Right Triangle*

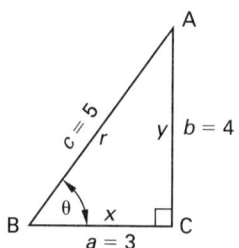

5. CIRCULAR TRANSCENDENTAL FUNCTIONS

The *circular transcendental functions* (usually referred to as the *transcendental functions*, *trigonometric functions*, or *functions of an angle*) are calculated from the sides of a right triangle. Equation 6.2 through Eq. 6.7 refer to Fig. 6.5.

$$\text{sine: } \sin\theta = \frac{y}{r} = \frac{\text{opposite}}{\text{hypotenuse}} \qquad 6.2$$

$$\text{cosine: } \cos\theta = \frac{x}{r} = \frac{\text{adjacent}}{\text{hypotenuse}} \qquad 6.3$$

$$\text{tangent: } \tan\theta = \frac{y}{x} = \frac{\text{opposite}}{\text{adjacent}} \qquad 6.4$$

$$\text{cotangent: } \cot\theta = \frac{x}{y} = \frac{\text{adjacent}}{\text{opposite}} \qquad 6.5$$

$$\text{secant: } \sec\theta = \frac{r}{x} = \frac{\text{hypotenuse}}{\text{adjacent}} \qquad 6.6$$

$$\text{cosecant: } \csc\theta = \frac{r}{y} = \frac{\text{hypotenuse}}{\text{opposite}} \qquad 6.7$$

Three of the transcendental functions are reciprocals of the others. However, while the tangent and cotangent functions are reciprocals of each other, the sine and cosine functions are not.

$$\cot\theta = \frac{1}{\tan\theta} \qquad 6.8$$

$$\sec\theta = \frac{1}{\cos\theta} \qquad 6.9$$

$$\csc\theta = \frac{1}{\sin\theta} \qquad 6.10$$

The trigonometric functions correspond to the lengths of various line segments in a right triangle with a unit

---

[3]These cases are almost always contrived examples. There is nothing intrinsic in nature to cause the formation of triangles with these proportions.

hypotenuse. Figure 6.6 shows such a triangle inscribed in a unit circle.

Figure 6.6 Trigonometric Functions in a Unit Circle

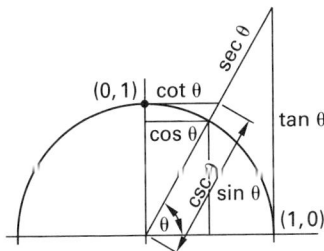

6. SMALL ANGLE APPROXIMATIONS

When an angle is very small, the hypotenuse and adjacent sides are essentially equal in length, and certain approximations can be made. (The angle θ must be expressed in radians in Eq. 6.11 and Eq. 6.12.)

$$\sin \theta \approx \tan \theta \approx \theta \big|_{\theta < 10° \ (0.175 \text{ rad})} \qquad 6.11$$

$$\cos \theta \approx 1 \big|_{\theta < 5° \ (0.0873 \text{ rad})} \qquad 6.12$$

7. GRAPHS OF THE FUNCTIONS

Figure 6.7 illustrates the periodicity of the sine, cosine, and tangent functions.[4]

Figure 6.7 Graphs of Sine, Cosine, and Tangent Functions

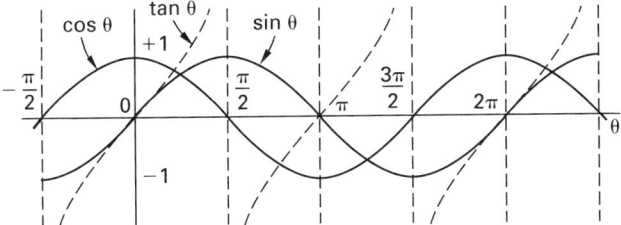

8. SIGNS OF THE FUNCTIONS

Table 6.1 shows how the sine, cosine, and tangent functions vary in sign with different values of θ. All three functions are positive for angles $0° \leq \theta \leq 90°$ ($0 \leq \theta \leq \pi/2$ rad), but only the sine is positive for angles $90° < \theta \leq 180°$ ($\pi/2$ rad $< \theta \leq \pi$ rad). The concept of quadrants is used to summarize the signs of the functions: angles up to $90°$ ($\pi/2$ rad) are in quadrant I, angles between $90°$ and $180°$ ($\pi/2$ and π rad) are in quadrant II, and so on.

9. FUNCTIONS OF RELATED ANGLES

Figure 6.7 shows that the sine, cosine, and tangent curves are symmetrical with respect to the horizontal

Table 6.1 Signs of the Functions by Quadrant

| | quadrant | | | |
function	I	II	III	IV
sine	+	+	−	−
cosine	+	−	−	+
tangent	+	−	+	−

axis. Furthermore, portions of the curves are symmetrical with respect to a vertical axis. The values of the sine and cosine functions repeat every $360°$ (2π rad), and the absolute values repeat every $180°$ (π rad). This can be written as

$$\sin(\theta + 180°) = -\sin \theta \qquad 6.13$$

Similarly, the tangent function will repeat every $180°$ (π rad), and its absolute value will repeat every $90°$ ($\pi/2$ rad).

Table 6.2 summarizes the functions of the related angles.

Table 6.2 Functions of Related Angles

$f(\theta)$	$-\theta$	$90° - \theta$	$90° + \theta$	$180° - \theta$	$180° + \theta$
sin	$-\sin \theta$	$\cos \theta$	$\cos \theta$	$\sin \theta$	$-\sin \theta$
cos	$\cos \theta$	$\sin \theta$	$-\sin \theta$	$-\cos \theta$	$-\cos \theta$
tan	$-\tan \theta$	$\cot \theta$	$-\cot \theta$	$-\tan \theta$	$\tan \theta$

10. TRIGONOMETRIC IDENTITIES

There are many relationships between trigonometric functions. For example, Eq. 6.14 through Eq. 6.16 are well known.

$$\sin^2 \theta + \cos^2 \theta = 1 \qquad 6.14$$

$$1 + \tan^2 \theta = \sec^2 \theta \qquad 6.15$$

$$1 + \cot^2 \theta = \csc^2 \theta \qquad 6.16$$

Other relatively common identities are listed as follows.[5]

- *double-angle formulas*

$$\sin 2\theta = 2 \sin \theta \cos \theta = \frac{2 \tan \theta}{1 + \tan^2 \theta} \qquad 6.17$$

$$\cos 2\theta = \cos^2 \theta - \sin^2 \theta = 1 - 2 \sin^2 \theta$$
$$= 2\cos^2 \theta - 1 = \frac{1 - \tan^2 \theta}{1 + \tan^2 \theta} \qquad 6.18$$

$$\tan 2\theta = \frac{2 \tan \theta}{1 - \tan^2 \theta} \qquad 6.19$$

$$\cot 2\theta = \frac{\cot^2 \theta - 1}{2 \cot \theta} \qquad 6.20$$

---

[4]The remaining functions, being reciprocals of these three functions, are also periodic.

[5]It is an idiosyncrasy of the subject that these formulas are conventionally referred to as *formulas*, not *identities*.

- *two-angle formulas*

$$\sin(\theta \pm \phi) = \sin\theta\cos\phi \pm \cos\theta\sin\phi \qquad 6.21$$

$$\cos(\theta \pm \phi) = \cos\theta\cos\phi \mp \sin\theta\sin\phi \qquad 6.22$$

$$\tan(\theta \pm \phi) = \frac{\tan\theta \pm \tan\phi}{1 \mp \tan\theta\tan\phi} \qquad 6.23$$

$$\cot(\theta \pm \phi) = \frac{\cot\phi\cot\theta \mp 1}{\cot\phi \pm \cot\theta} \qquad 6.24$$

- *half-angle formulas* $(\theta < 180°)$

$$\sin\frac{\theta}{2} = \sqrt{\frac{1 - \cos\theta}{2}} \qquad 6.25$$

$$\cos\frac{\theta}{2} = \sqrt{\frac{1 + \cos\theta}{2}} \qquad 6.26$$

$$\tan\frac{\theta}{2} = \sqrt{\frac{1 - \cos\theta}{1 + \cos\theta}} = \frac{\sin\theta}{1 + \cos\theta} = \frac{1 - \cos\theta}{\sin\theta} \qquad 6.27$$

- *miscellaneous formulas* $(\theta < 90°)$

$$\sin\theta = 2\sin\frac{\theta}{2}\cos\frac{\theta}{2} \qquad 6.28$$

$$\sin\theta = \sqrt{\frac{1 - \cos 2\theta}{2}} \qquad 6.29$$

$$\cos\theta = \cos^2\frac{\theta}{2} - \sin^2\frac{\theta}{2} \qquad 6.30$$

$$\cos\theta = \sqrt{\frac{1 + \cos 2\theta}{2}} \qquad 6.31$$

$$\tan\theta = \frac{2\tan\frac{\theta}{2}}{1 - \tan^2\frac{\theta}{2}} = \frac{2\sin\frac{\theta}{2}\cos\frac{\theta}{2}}{\cos^2\frac{\theta}{2} - \sin^2\frac{\theta}{2}} \qquad 6.32$$

$$\tan\theta = \sqrt{\frac{1 - \cos 2\theta}{1 + \cos 2\theta}} = \frac{\sin 2\theta}{1 + \cos 2\theta} = \frac{1 - \cos 2\theta}{\sin 2\theta} \qquad 6.33$$

$$\cot\theta = \frac{\cot^2\frac{\theta}{2} - 1}{2\cot\frac{\theta}{2}}$$

$$= \frac{\cos^2\frac{\theta}{2} - \sin^2\frac{\theta}{2}}{2\sin\frac{\theta}{2}\cos\frac{\theta}{2}} \qquad 6.34$$

$$\cot\theta = \sqrt{\frac{1 + \cos 2\theta}{1 - \cos 2\theta}} = \frac{1 + \cos 2\theta}{\sin 2\theta} = \frac{\sin 2\theta}{1 - \cos 2\theta} \qquad 6.35$$

11. INVERSE TRIGONOMETRIC FUNCTIONS

Finding an angle from a known trigonometric function is a common operation known as an *inverse trigonometric operation*. The inverse function can be designated by adding "inverse," "arc," or the superscript -1 to the name of the function. For example,

$$\text{inverse } \sin 0.5 \equiv \arcsin 0.5 \equiv \sin^{-1} 0.5 = 30°$$

12. ARCHAIC FUNCTIONS

A few archaic functions are occasionally encountered, primarily in navigation and surveying.[6] The *versine* (*versed sine* or *flipped sine*) is

$$\text{vers } \theta = 1 - \cos\theta = 2\sin^2\frac{\theta}{2} \qquad 6.36$$

The *coversed sine* (*versed cosine*) is

$$\text{covers } \theta = 1 - \sin\theta \qquad 6.37$$

The *haversine* is one-half of the versine.

$$\text{havers } \theta = \frac{1 - \cos\theta}{2} = \sin^2\frac{\theta}{2} \qquad 6.38$$

The *exsecant* is

$$\text{exsec } \theta = \sec\theta - 1 \qquad 6.39$$

13. HYPERBOLIC TRANSCENDENTAL FUNCTIONS

Hyperbolic transcendental functions (normally referred to as *hyperbolic functions*) are specific equations containing combinations of the terms e^{θ} and $e^{-\theta}$. These combinations appear regularly in certain types of problems (e.g., analysis of cables and heat transfer through fins) and are given specific names and symbols to simplify presentation.[7]

- *hyperbolic sine*

$$\sinh\theta = \frac{e^{\theta} - e^{-\theta}}{2} \qquad 6.40$$

- *hyperbolic cosine*

$$\cosh\theta = \frac{e^{\theta} + e^{-\theta}}{2} \qquad 6.41$$

---

[6]These functions were useful in eras prior to the widespread use of calculators, because logarithms of their functions eliminated the need for calculating squares and square roots.

[7]The hyperbolic sine and cosine functions are pronounced (by some) as "sinch" and "cosh," respectively.

- *hyperbolic tangent*

$$\tanh\theta = \frac{e^\theta - e^{-\theta}}{e^\theta + e^{-\theta}} = \frac{\sinh\theta}{\cosh\theta} \qquad 6.42$$

- *hyperbolic cotangent*

$$\coth\theta = \frac{e^\theta + e^{-\theta}}{e^\theta - e^{-\theta}} = \frac{\cosh\theta}{\sinh\theta} \qquad 6.43$$

- *hyperbolic secant*

$$\text{sech}\,\theta = \frac{2}{e^\theta + e^{-\theta}} = \frac{1}{\cosh\theta} \qquad 6.44$$

- *hyperbolic cosecant*

$$\text{csch}\,\theta = \frac{2}{e^\theta - e^{-\theta}} = \frac{1}{\sinh\theta} \qquad 6.45$$

Hyperbolic functions cannot be related to a right triangle, but they are related to a rectangular (equilateral) hyperbola, as shown in Fig. 6.8. For a *unit hyperbola* ($a^2 = 1$), the shaded area has a value of $\theta/2$ and is sometimes given the units of *hyperbolic radians*.

$$\sinh\theta = \frac{y}{a} \qquad 6.46$$

$$\cosh\theta = \frac{x}{a} \qquad 6.47$$

$$\tanh\theta = \frac{y}{x} \qquad 6.48$$

$$\coth\theta = \frac{x}{y} \qquad 6.49$$

$$\text{sech}\,\theta = \frac{a}{x} \qquad 6.50$$

$$\text{csch}\,\theta = \frac{a}{y} \qquad 6.51$$

Figure 6.8 *Equilateral Hyperbola and Hyperbolic Functions*

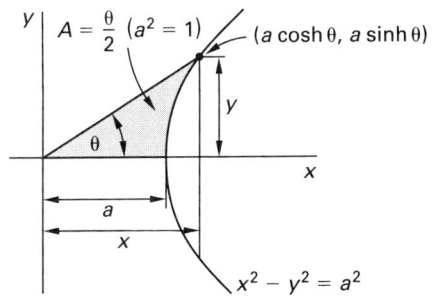

14. HYPERBOLIC IDENTITIES

The hyperbolic identities are different from the standard trigonometric identities. Some of the most important identities are presented as follows.

$$\cosh^2\theta - \sinh^2\theta = 1 \qquad 6.52$$

$$1 - \tanh^2\theta = \text{sech}^2\theta \qquad 6.53$$

$$1 - \coth^2\theta = -\text{csch}^2\theta \qquad 6.54$$

$$\cosh\theta + \sinh\theta = e^\theta \qquad 6.55$$

$$\cosh\theta - \sinh\theta = e^{-\theta} \qquad 6.56$$

$$\sinh(\theta \pm \phi) = \sinh\theta\cosh\phi \pm \cosh\theta\sinh\phi \qquad 6.57$$

$$\cosh(\theta \pm \phi) = \cosh\theta\cosh\phi \pm \sinh\theta\sinh\phi \qquad 6.58$$

$$\tanh(\theta \pm \phi) = \frac{\tanh\theta \pm \tanh\phi}{1 \pm \tanh\theta\tanh\phi} \qquad 6.59$$

15. GENERAL TRIANGLES

A *general triangle* (also known as an *oblique triangle*) is one that is not specifically a right triangle, as shown in Fig. 6.9. Equation 6.60 calculates the area of a general triangle.

$$\text{area} = \tfrac{1}{2}ab\sin C = \tfrac{1}{2}bc\sin A = \tfrac{1}{2}ca\sin B \qquad 6.60$$

Figure 6.9 *General Triangle*

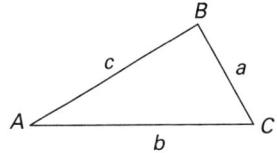

The *law of sines*, as shown in Eq. 6.61, relates the sides and the sines of the angles.

$$\frac{\sin A}{a} = \frac{\sin B}{b} = \frac{\sin C}{c} \qquad 6.61$$

The *law of cosines* relates the cosine of an angle to an opposite side. (Equation 6.62 can be extended to the two remaining sides.)

$$a^2 = b^2 + c^2 - 2bc\cos A \qquad 6.62$$

The *law of tangents* relates the sum and difference of two sides. (Equation 6.63 can be extended to the two remaining sides.)

$$\frac{a-b}{a+b} = \frac{\tan\left(\dfrac{A-B}{2}\right)}{\tan\left(\dfrac{A+B}{2}\right)} \qquad 6.63$$

16. SPHERICAL TRIGONOMETRY

A *spherical triangle* is a triangle that has been drawn on the surface of a sphere, as shown in Fig. 6.10. The *trihedral angle* $O-ABC$ is formed when the vertices A, B, and C are joined to the center of the sphere. The *face angles* (BOC, COA, and AOB in Fig. 6.10) are used to measure the sides (a, b, and c in Fig. 6.10). The *vertex angles* are A, B, and C. Angles are used to measure both vertex angles and sides.

Figure 6.10 *Spherical Triangle ABC*

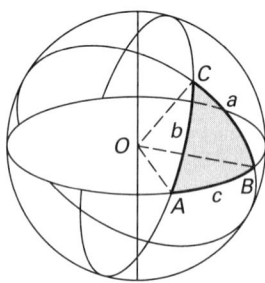

The following rules are valid for spherical triangles for which each side and angle is less than 180°.

- The sum of the three vertex angles is greater than 180° and less than 540°.

$$180° < A + B + C < 540° \qquad 6.64$$

- The sum of any two sides is greater than the third side.

- The sum of the three sides is less than 360°.

$$0° < a + b + c < 360° \qquad 6.65$$

- If the two sides are equal, the corresponding angles opposite are equal, and the converse is also true.

- If two sides are unequal, the corresponding angles opposite are unequal. The greater angle is opposite the greater side.

The *spherical excess*, ϵ, is the amount by which the sum of the vertex angles exceeds 180°. The *spherical defect*, d, is the amount by which the sum of the sides differs from 360°.

$$\epsilon = A + B + C - 180° \qquad 6.66$$
$$d = 360° - (a + b + c) \qquad 6.67$$

There are many trigonometric identities that define the relationships between angles in a spherical triangle. Some of the more common identities are presented as follows.

- *law of sines*

$$\frac{\sin A}{\sin a} = \frac{\sin B}{\sin b} = \frac{\sin C}{\sin c} \qquad 6.68$$

- *first law of cosines*

$$\cos a = \cos b \cos c + \sin b \sin c \cos A \qquad 6.69$$

- *second law of cosines*

$$\cos A = -\cos B \cos C + \sin B \sin C \cos a \qquad 6.70$$

17. SOLID ANGLES

A *solid angle*, ω, is a measure of the angle subtended at the vertex of a cone, as shown in Fig. 6.11. The solid angle has units of *steradians* (abbreviated *sr*). A steradian is the solid angle subtended at the center of a unit sphere (i.e., a sphere with a radius of one) by a unit area on its surface. Since the surface area of a sphere of radius r is r^2 times the surface area of a unit sphere, the solid angle is equal to the area cut out by the cone divided by r^2.

$$\omega = \frac{\text{surface area}}{r^2} \qquad 6.71$$

Figure 6.11 *Solid Angle*

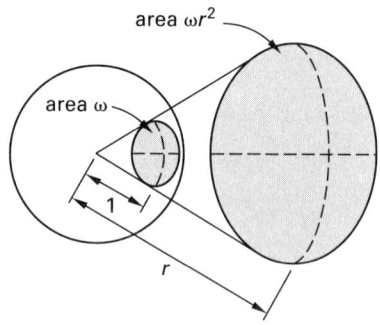

7 Analytic Geometry

1. MENSURATION OF REGULAR SHAPES

The dimensions, perimeter, area, and other geometric properties constitute the *mensuration* (i.e., the measurements) of a geometric shape. Appendix 7.A and App. 7.B contain formulas and tables used to calculate these properties.

Example 7.1

In the study of open channel fluid flow, the hydraulic radius is defined as the ratio of flow area to wetted perimeter. What is the hydraulic radius of a 6 in inside diameter pipe filled to a depth of 2 in?

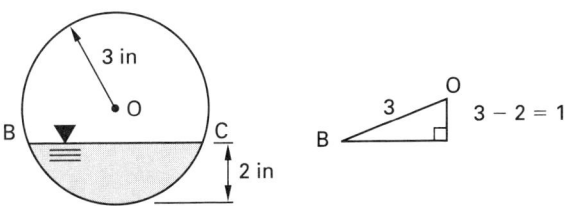

Solution

Points O, B, and C constitute a circular segment and are used to find the central angle of the circular segment.

$$\tfrac{1}{2}\angle \text{BOC} = \arccos\tfrac{1}{3} = 70.53°$$

$$\phi = \angle \text{BOC} = \frac{(2)(70.53°)(2\pi)}{360°} = 2.462 \text{ rad}$$

From App. 7.A, the area in flow and arc length are

$$A = \tfrac{1}{2}r^2(\phi - \sin\phi)$$
$$= \left(\tfrac{1}{2}\right)(3 \text{ in})^2\big(2.462 \text{ rad} - \sin(2.462 \text{ rad})\big)$$
$$= 8.251 \text{ in}^2$$
$$s = r\phi = (3 \text{ in})(2.462 \text{ rad})$$
$$= 7.386 \text{ in}$$

The hydraulic radius is

$$r_h = \frac{A}{s} = \frac{8.251 \text{ in}^2}{7.386 \text{ in}}$$
$$= 1.12 \text{ in}$$

2. AREAS WITH IRREGULAR BOUNDARIES

Areas of sections with irregular boundaries (such as creek banks) cannot be determined precisely, and approximation methods must be used. If the irregular side can be divided into a series of cells of equal width, either the trapezoidal rule or Simpson's rule can be used. Figure 7.1 shows an example of an irregular area.

Figure 7.1 Irregular Areas

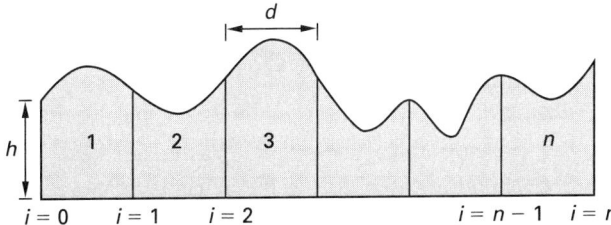

If the irregular side of each cell is fairly straight, the *trapezoidal rule* is appropriate.

$$A = \frac{d}{2}\left(h_0 + h_n + 2\sum_{i=1}^{n-1} h_i\right) \qquad 7.1$$

If the irregular side of each cell is curved (parabolic), *Simpson's rule* should be used. (n must be even to use Simpson's rule.)

$$A = \frac{d}{3}\left(h_0 + h_n + 4\sum_{\substack{i \text{ odd} \\ i=1}}^{n-1} h_i + 2\sum_{\substack{i \text{ even} \\ i=2}}^{n-2} h_i\right) \qquad 7.2$$

3. GEOMETRIC DEFINITIONS

The following terms are used in this book to describe the relationship or orientation of one geometric figure to another. Figure 7.2 illustrates some of the following geometric definitions.

- *abscissa:* the horizontal coordinate, typically designated as x in a rectangular coordinate system

- *asymptote:* a straight line that is approached but not intersected by a curved line

- *asymptotic:* approaching the slope of another line; attaining the slope of another line in the limit

- *center:* a point equidistant from all other points

- *collinear:* falling on the same line

- *concave:* curved inward (in the direction indicated)[1]

- *convex:* curved outward (in the direction indicated)

- *convex hull:* a closed figure whose surface is convex everywhere

- *coplanar:* falling on the same plane

- *inflection point:* a point where the second derivative changes sign or the curve changes from concave to convex; also known as a *point of contraflexure*

- *locus of points:* a set or collection of points having some common property and being so infinitely close together as to be indistinguishable from a line

- *node:* a point on a line from which other lines enter or leave

- *normal:* rotated 90°; being at right angles

- *ordinate:* the vertical coordinate, typically designated as y in a rectangular coordinate system

- *orthogonal:* rotated 90°; being at right angles

---
[1]This is easily remembered since one must go inside to explore a cave.

- *saddle point:* a point on a three-dimensional surface where all adjacent points are higher than the saddle point in one direction (the direction of the saddle) and are lower than the saddle point in an orthogonal direction (the direction of the sides)

- *tangent point:* having equal slopes at a common point

Figure 7.2 *Geometric Definitions*

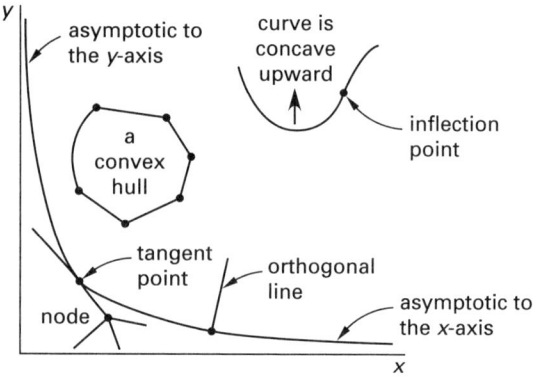

4. CONCAVE CURVES

Concavity is a term that is applied to curved lines. A *concave up curve* is one whose function's first derivative increases continuously from negative to positive values. Straight lines drawn tangent to concave up curves are all below the curve. The graph of such a function may be thought of as being able to "hold water."

The first derivative of a *concave down curve* decreases continuously from positive to negative. A graph of a concave down function may be thought of as "spilling water." (See Fig. 7.3.)

Figure 7.3 *Concave Curves*

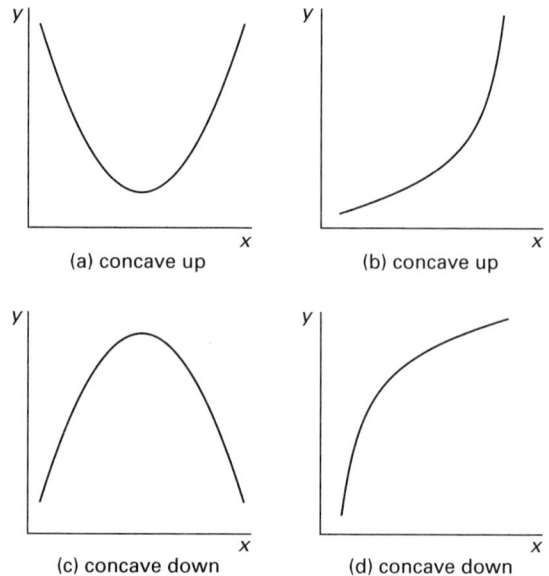

5. CONVEX REGIONS

Convexity is a term that is applied to sets and regions.[2] It plays an important role in many mathematics subjects. A set or multidimensional region is *convex* if it contains the line segment joining any two of its points; that is, if a straight line is drawn connecting any two points in a convex region, that line will lie entirely within the region. For example, the interior of a parabola is a convex region, as is a solid sphere. The *void* or *null region* (i.e., an empty set of points), single points, and straight lines are convex sets. A convex region bounded by separate, connected line segments is known as a *convex hull*. (See Fig. 7.4.)

Figure 7.4 *Convexity*

  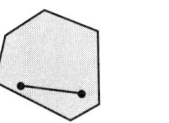 

convex region   convex hull   nonconvex region

Within a convex region, a local maximum is also the global maximum. Similarly, a local minimum is also the global minimum. The intersection of two convex regions is also convex.

6. CONGRUENCY

Congruence in geometric figures is analogous to *equality* in algebraic expressions. Congruent line segments are segments that have the same length. Congruent angles have the same angular measure. Congruent triangles have the same vertex angles and side lengths.

In general, *congruency*, indicated by the symbol $\cong$, means that there is one-to-one correspondence between all points on two objects. This correspondence is defined by the *mapping function* or *isometry*, which can be a translation, rotation, or reflection. Since the identity function is a valid mapping function, every geometric shape is congruent to itself.

Two congruent objects can be in different spaces. For example, a triangular area in three-dimensional space can be mapped into a triangle in two-dimensional space.

7. COORDINATE SYSTEMS

The manner in which a geometric figure is described depends on the coordinate system that is used. The three-dimensional system (also known as the *rectangular coordinate system* and *Cartesian coordinate system*)

with its x-, y-, and z-coordinates is the most commonly used in engineering. Table 7.1 summarizes the components needed to specify a point in the various coordinate systems. Figure 7.5 illustrates the use of and conversion between the coordinate systems.

Table 7.1 *Components of Coordinate Systems*

name	dimensions	components
rectangular	2	x, y
rectangular	3	x, y, z
polar	2	r, θ
cylindrical	3	r, θ, z
spherical	3	r, θ, ϕ

Figure 7.5 *Different Coordinate Systems*

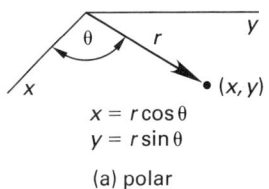

$$x = r\cos\theta$$
$$y = r\sin\theta$$

(a) polar

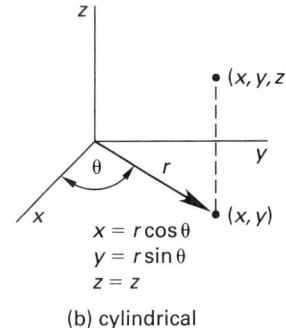

$$x = r\cos\theta$$
$$y = r\sin\theta$$
$$z = z$$

(b) cylindrical

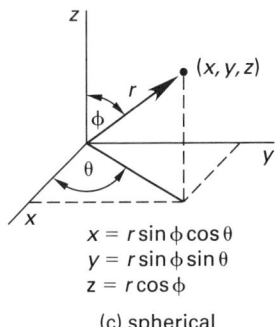

$$x = r\sin\phi\cos\theta$$
$$y = r\sin\phi\sin\theta$$
$$z = r\cos\phi$$

(c) spherical

8. CURVES

A *curve* (commonly called a *line*) is a function over a finite or infinite range of the independent variable. When a curve is drawn in two- or three-dimensional space, it is known as a *graph of the curve*. It may or may not be possible to describe the curve mathematically. The *degree of a curve* is the highest exponent in

---

[2]It is tempting to define regions that fail the convexity test as being "concave." However, it is more proper to define such regions as "nonconvex." In any case, it is important to recognize that convexity depends on the reference point: An observer within a sphere will see the spherical boundary as convex; an observer outside the sphere may see the boundary as nonconvex.

the function. For example, Eq. 7.3 is a fourth-degree curve.

$$f(x) = 2x^4 + 7x^3 + 6x^2 + 3x + 9 = 0 \qquad 7.3$$

An *ordinary cycloid* ("wheel line") is a curve traced out by a point on the rim of a wheel that rolls without slipping. (See Fig. 7.6.) Cycloids that start with the tracing point down (i.e., on the x-axis) are described in parametric form by Eq. 7.4 and Eq. 7.5 and in rectangular form by Eq. 7.6. In Eq. 7.4 through Eq. 7.6, using the minus sign results in a bottom (downward) cusp at the origin; using the plus sign results in a vertex (trough) at the origin and a top (upward) cusp.

$$x = r(\theta \pm \sin\theta) \qquad 7.4$$

$$y = r(1 \pm \cos\theta) \qquad 7.5$$

$$x = r\arccos\frac{r - y}{r} \pm \sqrt{2ry - y^2} \qquad 7.6$$

Figure 7.6 Cycloid (cusp at origin shown)

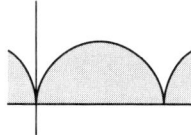

An *epicycloid* is a curve generated by a point on the rim of a wheel that rolls on the outside of a circle. A *hypocycloid* is a curve generated by a point on the rim of a wheel that rolls on the inside of a circle. The equation of a hypocycloid of four cusps is

$$x^{2/3} + y^{2/3} = r^{2/3} \quad \text{[4 cusps]} \qquad 7.7$$

9. SYMMETRY OF CURVES

Two points, P and Q, are symmetrical with respect to a line if the line is a perpendicular bisector of the line segment PQ. If the graph of a curve is unchanged when y is replaced with $-y$, the curve is symmetrical with respect to the x-axis. If the curve is unchanged when x is replaced with $-x$, the curve is symmetrical with respect to the y-axis.

Repeating waveforms can be symmetrical with respect to the y-axis. A curve $f(x)$ is said to have *even symmetry* if $f(x) = f(-x)$. (Alternatively, $f(x)$ is said to be a *symmetrical function*.) With even symmetry, the function to the left of $x = 0$ is a reflection of the function to the right of $x = 0$. (In effect, the y-axis is a mirror.) The cosine curve is an example of a curve with even symmetry.

A curve is said to have *odd symmetry* if $f(x) = -f(-x)$. (Alternatively, $f(x)$ is said to be an *asymmetrical function*.[3]) The sine curve is an example of a curve with odd symmetry.

---

[3]Although they have the same meaning, the semantics of "odd symmetry" and "asymmetrical function" seem contradictory.

A curve is said to have *rotational symmetry (half-wave symmetry)* if $f(x) = -f(x + \pi)$.[4] Curves of this type are identical except for a sign reversal on alternate half-cycles. (See Fig. 7.7.)

Figure 7.7 Waveform Symmetry

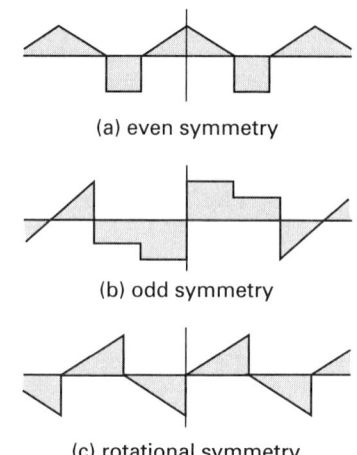

(a) even symmetry

(b) odd symmetry

(c) rotational symmetry

Table 7.2 describes the type of function resulting from the combination of two functions.

Table 7.2 Combinations of Functions

	operation			
	$+$	$-$	$\times$	$\div$
$f_1(x)$ even, $f_2(x)$ even	even	even	even	even
$f_1(x)$ odd, $f_2(x)$ odd	odd	odd	even	even
$f_1(x)$ even, $f_2(x)$ odd	neither	neither	odd	odd

10. STRAIGHT LINES

Figure 7.8 illustrates a straight line in two-dimensional space. The *slope* of the line is m, the *y-intercept* is b, and the *x-intercept* is a. The equation of the line can be represented in several forms. The procedure for finding the equation depends on the form chosen to represent the line. In general, the procedure involves substituting one or more known points on the line into the equation in order to determine the coefficients.

Figure 7.8 Straight Line

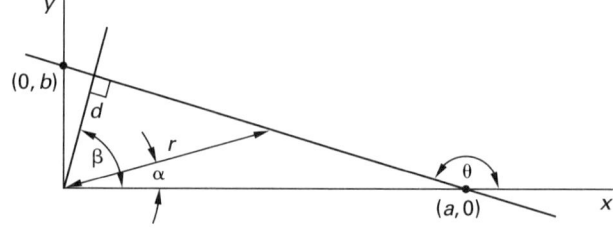

---

[4]The symbol π represents half of a full cycle of the waveform, not the value 3.141....

- *general form*

$$Ax + By + C = 0 \qquad \text{7.8}$$

$$A = -mB \qquad \text{7.9}$$

$$B = \frac{-C}{b} \qquad \text{7.10}$$

$$C = -aA = -bB \qquad \text{7.11}$$

- *slope-intercept form*

$$y = mx + b \qquad \text{7.12}$$

$$m = \frac{-A}{B} = \tan\theta = \frac{y_2 - y_1}{x_2 - x_1} \qquad \text{7.13}$$

$$b = \frac{-C}{B} \qquad \text{7.14}$$

$$a = \frac{-C}{A} \qquad \text{7.15}$$

- *point-slope form*

$$y - y_1 = m(x - x_1) \qquad \text{7.16}$$

- *intercept form*

$$\frac{x}{a} + \frac{y}{b} = 1 \qquad \text{7.17}$$

- *two-point form*

$$\frac{y - y_1}{x - x_1} = \frac{y_2 - y_1}{x_2 - x_1} \qquad \text{7.18}$$

- *normal form*

$$x\cos\beta + y\sin\beta - d = 0 \qquad \text{7.19}$$

(d and β are constants; x and y are variables.)

- *polar form*

$$r = \frac{d}{\cos(\beta - \alpha)} \qquad \text{7.20}$$

(d and β are constants; r and α are variables.)

11. DIRECTION NUMBERS, ANGLES, AND COSINES

Given a directed line or vector, $\mathbf{R}$, from $(x_1,\ y_1,\ z_1)$ to $(x_2,\ y_2,\ z_2)$, the *direction numbers* are

$$L = x_2 - x_1 \qquad \text{7.21}$$

$$M = y_2 - y_1 \qquad \text{7.22}$$

$$N = z_2 - z_1 \qquad \text{7.23}$$

The distance between two points is

$$d = \sqrt{L^2 + M^2 + N^2} \qquad \text{7.24}$$

The *direction cosines* are

$$\cos\alpha = \frac{L}{d} \qquad \text{7.25}$$

$$\cos\beta = \frac{M}{d} \qquad \text{7.26}$$

$$\cos\gamma = \frac{N}{d} \qquad \text{7.27}$$

The sum of the squares of direction cosines is equal to 1.

$$\cos^2\alpha + \cos^2\beta + \cos^2\gamma = 1 \qquad \text{7.28}$$

The *direction angles* are the angles between the axes and the lines. They are found from the inverse functions of the direction cosines.

$$\alpha = \arccos\frac{L}{d} \qquad \text{7.29}$$

$$\beta = \arccos\frac{M}{d} \qquad \text{7.30}$$

$$\gamma = \arccos\frac{N}{d} \qquad \text{7.31}$$

The direction cosines can be used to write the equation of the straight line in terms of the unit vectors. The line $\mathbf{R}$ would be defined as

$$\mathbf{R} = d(\mathbf{i}\cos\alpha + \mathbf{j}\cos\beta + \mathbf{k}\cos\gamma) \qquad \text{7.32}$$

Similarly, the line may be written in terms of its direction numbers.

$$\mathbf{R} = L\mathbf{i} + M\mathbf{j} + N\mathbf{k} \qquad \text{7.33}$$

Example 7.2

A directed line, $\mathbf{R}$, passes through the points $(4, 7, 9)$ and $(0, 1, 6)$. Write the equation of the line in terms of its (a) direction numbers and (b) direction cosines.

Solution

(a) The direction numbers are

$$L = 4 - 0 = 4$$

$$M = 7 - 1 = 6$$

$$N = 9 - 6 = 3$$

Using Eq. 7.33,

$$\mathbf{R} = 4\mathbf{i} + 6\mathbf{j} + 3\mathbf{k}$$

(b) The distance between the two points is

$$d = \sqrt{(4)^2 + (6)^2 + (3)^2} = 7.81$$

The line in terms of its direction cosines is

$$\mathbf{R} = \frac{4\mathbf{i} + 6\mathbf{j} + 3\mathbf{k}}{7.81}$$

$$= 0.512\mathbf{i} + 0.768\mathbf{j} + 0.384\mathbf{k}$$

12. INTERSECTION OF TWO LINES

The intersection of two lines is a point. The location of the intersection point can be determined by setting the two equations equal and solving them in terms of a common variable. Alternatively, Eq. 7.34 and Eq. 7.35 can be used to calculate the coordinates of the intersection point.

$$x = \frac{B_2 C_1 - B_1 C_2}{A_2 B_1 - A_1 B_2} \qquad 7.34$$

$$y = \frac{A_1 C_2 - A_2 C_1}{A_2 B_1 - A_1 B_2} \qquad 7.35$$

13. PLANES

A *plane* in three-dimensional space, as shown in Fig. 7.9, is completely determined by one of the following:

- three noncollinear points

- two nonparallel vectors $\mathbf{V}_1$ and $\mathbf{V}_2$ and their intersection point P_0

- a point P_0 and a vector, $\mathbf{N}$, normal to the plane (i.e., the *normal vector*)

Figure 7.9 *Plane in Three-Dimensional Space*

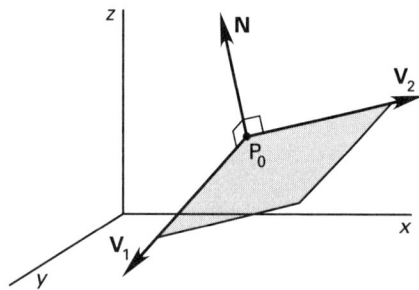

The plane can be specified mathematically in one of two ways: in rectangular form or as a parametric equation. The general form is

$$A(x - x_0) + B(y - y_0) + C(z - z_0) = 0 \qquad 7.36$$

x_0, y_0, and z_0 are the coordinates of the intersection point of any two vectors in the plane. The coefficients

A, B, and C are the same as the coefficients of the normal vector, $\mathbf{N}$.

$$\mathbf{N} = \mathbf{V}_1 \times \mathbf{V}_2 = A\mathbf{i} + B\mathbf{j} + C\mathbf{k} \qquad 7.37$$

Equation 7.36 can be simplified as follows.

$$Ax + By + Cz + D = 0 \qquad 7.38$$

$$D = -(Ax_0 + By_0 + Cz_0) \qquad 7.39$$

The following procedure can be used to determine the equation of a plane from three noncollinear points, P_1, P_2, and P_3, or from a normal vector and a single point.

step 1: (If the normal vector is known, go to step 3.) Determine the equations of the vectors $\mathbf{V}_1$ and $\mathbf{V}_2$ from two pairs of the points. For example, determine $\mathbf{V}_1$ from points P_1 and P_2, and determine $\mathbf{V}_2$ from P_1 and P_3. Express the vectors in the form $A\mathbf{i} + B\mathbf{j} + C\mathbf{k}$.

$$\mathbf{V}_1 = (x_2 - x_1)\mathbf{i} + (y_2 - y_1)\mathbf{j}$$
$$+ (z_2 - z_1)\mathbf{k} \qquad 7.40$$
$$\mathbf{V}_2 = (x_3 - x_1)\mathbf{i} + (y_3 - y_1)\mathbf{j}$$
$$+ (z_3 - z_1)\mathbf{k} \qquad 7.41$$

step 2: Find the normal vector, $\mathbf{N}$, as the cross product of the two vectors.

$$\mathbf{N} = \mathbf{V}_1 \times \mathbf{V}_2$$
$$= \begin{vmatrix} \mathbf{i} & (x_2 - x_1) & (x_3 - x_1) \\ \mathbf{j} & (y_2 - y_1) & (y_3 - y_1) \\ \mathbf{k} & (z_2 - z_1) & (z_3 - z_1) \end{vmatrix} \qquad 7.42$$

step 3: Write the general equation of the plane in rectangular form (see Eq. 7.36) using the coefficients A, B, and C from the normal vector and any one of the three points as P_0.

The parametric equations of a plane also can be written as a linear combination of the components of two vectors in the plane. Referring to Fig. 7.9, the two known vectors are

$$\mathbf{V}_1 = V_{1x}\mathbf{i} + V_{1y}\mathbf{j} + V_{1z}\mathbf{k} \qquad 7.43$$

$$\mathbf{V}_2 = V_{2x}\mathbf{i} + V_{2y}\mathbf{j} + V_{2z}\mathbf{k} \qquad 7.44$$

If s and t are scalars, the coordinates of each point in the plane can be written as Eq. 7.45 through Eq. 7.47. These are the parametric equations of the plane.

$$x = x_0 + sV_{1x} + tV_{2x} \qquad 7.45$$

$$y = y_0 + sV_{1y} + tV_{2y} \qquad 7.46$$

$$z = z_0 + sV_{1z} + tV_{2z} \qquad 7.47$$

Example 7.3

A plane is defined by the following points.

$$P_1 = (2, 1, -4)$$
$$P_2 = (4, -2, -3)$$
$$P_3 = (2, 3, -8)$$

Determine the equation of the plane in (a) general form and (b) parametric form.

Solution

(a) Use the first two points to find a vector, $\mathbf{V}_1$.

$$\mathbf{V}_1 = (x_2 - x_1)\mathbf{i} + (y_2 - y_1)\mathbf{j} + (z_2 - z_1)\mathbf{k}$$
$$= (4 - 2)\mathbf{i} + (-2 - 1)\mathbf{j} + \left(-3 - (-4)\right)\mathbf{k}$$
$$= 2\mathbf{i} - 3\mathbf{j} + 1\mathbf{k}$$

Similarly, use the first and third points to find $\mathbf{V}_2$.

$$\mathbf{V}_2 = (x_3 - x_1)\mathbf{i} + (y_3 - y_1)\mathbf{j} + (z_3 - z_1)\mathbf{k}$$
$$= (2 - 2)\mathbf{i} + (3 - 1)\mathbf{j} + \left(-8 - (-4)\right)\mathbf{k}$$
$$= 0\mathbf{i} + 2\mathbf{j} - 4\mathbf{k}$$

From Eq. 7.42, determine the normal vector as a determinant.

$$\mathbf{N} = \begin{vmatrix} \mathbf{i} & 2 & 0 \\ \mathbf{j} & -3 & 2 \\ \mathbf{k} & 1 & -4 \end{vmatrix}$$

Expand the determinant across the top row.

$$\mathbf{N} = \mathbf{i}(12 - 2) - 2(-4\mathbf{j} - 2\mathbf{k})$$
$$= 10\mathbf{i} + 8\mathbf{j} + 4\mathbf{k}$$

The rectangular form of the equation of the plane uses the same constants as in the normal vector. Use the first point and write the equation of the plane in the form of Eq. 7.36.

$$(10)(x - 2) + (8)(y - 1) + (4)(z + 4) = 0$$

The three constant terms can be combined by using Eq. 7.39.

$$D = -\left((10)(2) + (8)(1) + (4)(-4)\right) = -12$$

The equation of the plane is

$$10x + 8y + 4z - 12 = 0$$

(b) The parametric equations based on the first point and for any values of s and t are

$$x = 2 + 2s + 0t$$
$$y = 1 - 3s + 2t$$
$$z = -4 + 1s - 4t$$

The scalars s and t are not unique. Two of the three coordinates can also be chosen as the parameters. Dividing the rectangular form of the plane's equation by 4 to isolate z results in an alternate set of parametric equations.

$$x = x$$
$$y = y$$
$$z = 3 - 2.5x - 2y$$

14. DISTANCES BETWEEN GEOMETRIC FIGURES

The smallest distance, d, between various geometric figures is given by the following equations.

- between two points in (x, y, z) format:

$$d = \sqrt{(x_2 - x_1)^2 + (y_2 - y_1)^2 + (z_2 - z_1)^2} \qquad 7.48$$

- between a point (x_0, y_0) and a line $Ax + By + C = 0$:

$$d = \frac{|Ax_0 + By_0 + C|}{\sqrt{A^2 + B^2}} \qquad 7.49$$

- between a point (x_0, y_0, z_0) and a plane $Ax + By + Cz + D = 0$:

$$d = \frac{|Ax_0 + By_0 + Cz_0 + D|}{\sqrt{A^2 + B^2 + C^2}} \qquad 7.50$$

- between two parallel lines $Ax + By + C = 0$:

$$d = \left| \frac{|C_2|}{\sqrt{A_2^2 + B_2^2}} - \frac{|C_1|}{\sqrt{A_1^2 + B_1^2}} \right| \qquad 7.51$$

Example 7.4

What is the minimum distance between the line $y = 2x + 3$ and the origin $(0, 0)$?

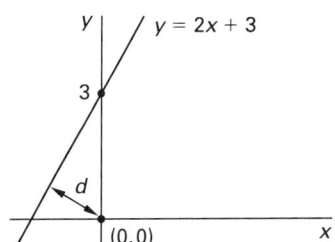

Solution

Put the equation in general form.

$$Ax + By + C = 2x - y + 3 = 0$$

Use Eq. 7.49 with $(x, y) = (0,0)$.

$$d = \frac{|Ax + By + C|}{\sqrt{A^2 + B^2}} = \frac{(2)(0) + (-1)(0) + 3}{\sqrt{(2)^2 + (-1)^2}}$$

$$= \frac{3}{\sqrt{5}}$$

15. ANGLES BETWEEN GEOMETRIC FIGURES

The angle, ϕ, between various geometric figures is given by the following equations.

- between two lines in $Ax + By + C = 0$, $y = mx + b$, or direction angle formats:

$$\phi = \arctan\left(\frac{A_1 B_2 - A_2 B_1}{A_1 A_2 + B_1 B_2}\right) \qquad 7.52$$

$$\phi = \arctan\left(\frac{m_2 - m_1}{1 + m_1 m_2}\right) \qquad 7.53$$

$$\phi = |\arctan m_1 - \arctan m_2| \qquad 7.54$$

$$\phi = \arccos\left(\frac{L_1 L_2 + M_1 M_2 + N_1 N_2}{d_1 d_2}\right) \qquad 7.55$$

$$\phi = \arccos\left(\begin{array}{c}\cos\alpha_1\cos\alpha_2 + \cos\beta_1\cos\beta_2 \\ + \cos\gamma_1\cos\gamma_2\end{array}\right) \qquad 7.56$$

If the lines are parallel, then $\phi = 0$.

$$\frac{A_1}{A_2} = \frac{B_1}{B_2} \qquad 7.57$$

$$m_1 = m_2 \qquad 7.58$$

$$\alpha_1 = \alpha_2;\ \beta_1 = \beta_2;\ \gamma_1 = \gamma_2 \qquad 7.59$$

If the lines are perpendicular, then $\phi = 90°$.

$$A_1 A_2 = -B_1 B_2 \qquad 7.60$$

$$m_1 = \frac{-1}{m_2} \qquad 7.61$$

$$\alpha_1 + \alpha_2 = \beta_1 + \beta_2 = \gamma_1 + \gamma_2 = 90° \qquad 7.62$$

- between two planes in $A\mathbf{i} + B\mathbf{j} + C\mathbf{k} = 0$ format, the coefficients A, B, and C are the same as the coefficients for the normal vector. (See Eq. 7.37.) ϕ is equal to the angle between the two normal vectors.

$$\cos\phi = \frac{|A_1 A_2 + B_1 B_2 + C_1 C_2|}{\sqrt{A_1^2 + B_1^2 + C_1^2}\sqrt{A_2^2 + B_2^2 + C_2^2}} \qquad 7.63$$

Example 7.5

Use Eq. 7.52, Eq. 7.53, and Eq. 7.54 to find the angle between the lines.

$$y = -0.577x + 2$$
$$y = +0.577x - 5$$

Solution

Write both equations in general form.

$$-0.577x - y + 2 = 0$$
$$0.577x - y - 5 = 0$$

(a) From Eq. 7.52,

$$\phi = \arctan\left(\frac{A_1 B_2 - A_2 B_1}{A_1 A_2 + B_1 B_2}\right)$$

$$= \arctan\left(\frac{(-0.577)(-1) - (0.577)(-1)}{(-0.577)(0.577) + (-1)(-1)}\right)$$

$$= 60°$$

(b) Use Eq. 7.53.

$$\phi = \arctan\left(\frac{m_2 - m_1}{1 + m_1 m_2}\right)$$

$$= \arctan\left(\frac{0.577 - (-0.577)}{1 + (0.577)(-0.577)}\right)$$

$$= 60°$$

(c) Use Eq. 7.54.

$$\phi = |\arctan m_1 - \arctan m_2|$$

$$= |\arctan(-0.577) - \arctan(0.577)|$$

$$= |-30° - 30°|$$

$$= 60°$$

16. CONIC SECTIONS

A *conic section* is any one of several curves produced by passing a plane through a cone as shown in Fig. 7.10. If α is the angle between the vertical axis and the cutting plane and β is the cone generating angle, Eq. 7.64 gives the *eccentricity*, ϵ, of the conic section. Values of the eccentricity are given in Fig. 7.10.

$$\epsilon = \frac{\cos\alpha}{\cos\beta} \qquad 7.64$$

Figure 7.10 *Conic Sections Produced by Cutting Planes*

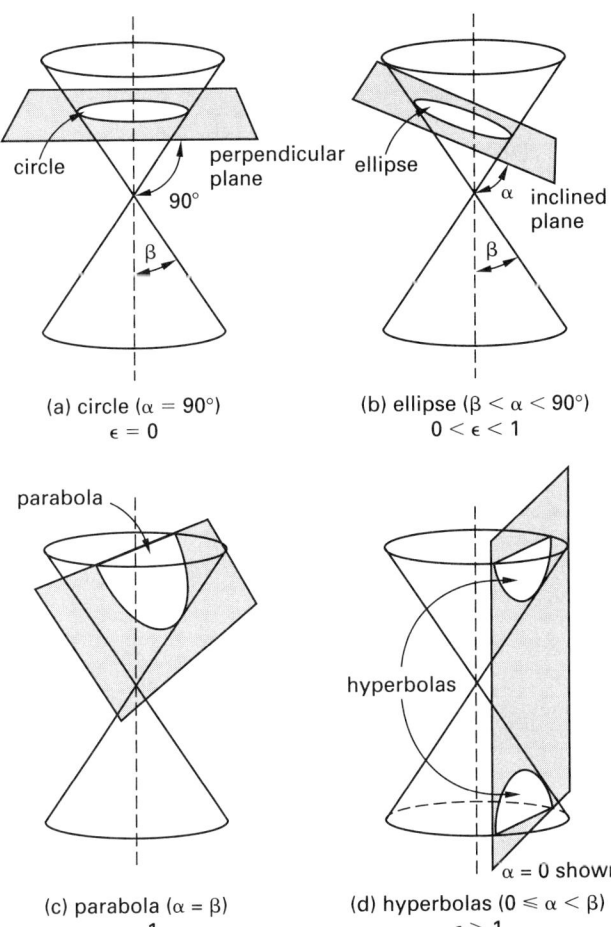

(a) circle ($\alpha = 90°$)
$\epsilon = 0$

(b) ellipse ($\beta < \alpha < 90°$)
$0 < \epsilon < 1$

(c) parabola ($\alpha = \beta$)
$\epsilon = 1$

(d) hyperbolas ($0 \leq \alpha < \beta$)
$\epsilon > 1$

Figure 7.11 *Determining Conic Sections from Quadratic Equations*

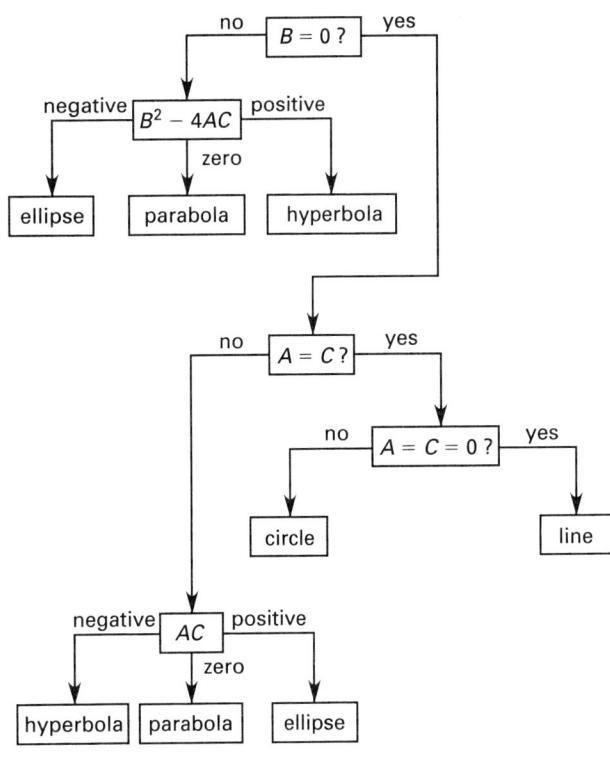

All conic sections are described by second-degree polynomials (i.e., *quadratic equations*) of the following form.[5]

$$Ax^2 + Bxy + Cy^2 + Dx + Ey + F = 0 \qquad 7.65$$

This is the *general form*, which allows the figure axes to be at any angle relative to the coordinate axes. The *standard forms* presented in the following sections pertain to figures whose axes coincide with the coordinate axes, thereby eliminating certain terms of the general equation.

Figure 7.11 can be used to determine which conic section is described by the quadratic function. The quantity $B^2 - 4AC$ is known as the *discriminant*. Figure 7.11 determines only the type of conic section; it does not determine whether the conic section is degenerate (e.g., a circle with a negative radius).

---

[5]One or more straight lines are produced when the cutting plane passes through the cone's vertex. Straight lines can be considered to be quadratic functions without second-degree terms.

Example 7.6

What geometric figures are described by the following equations?

(a) $4y^2 - 12y + 16x + 41 = 0$

(b) $x^2 - 10xy + y^2 + x + y + 1 = 0$

(c) $x^2 + 4y^2 + 2x - 8y + 1 = 0$

(d) $x^2 + y^2 - 6x + 8y + 20 = 0$

Solution

(a) Referring to Fig. 7.11, $B = 0$ since there is no xy term, $A = 0$ since there is no x^2 term, and $AC = (0)(4) = 0$. This is a parabola.

(b) $B \neq 0$; $B^2 - 4AC = (-10)^2 - (4)(1)(1) = +96$. This is a hyperbola.

(c) $B = 0$; $A \neq C$; $AC = (1)(4) = +4$. This is an ellipse.

(d) $B = 0$; $A = C$; $A = C = 1 \ (\neq 0)$. This is a circle.

17. CIRCLE

The general form of the equation of a circle, as illustrated in Fig. 7.12, is

$$Ax^2 + Ay^2 + Dx + Ey + F = 0 \qquad 7.66$$

Figure 7.12 *Circle*

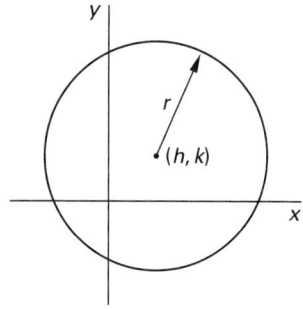

The *center-radius form* of the equation of a circle with radius r and center at (h, k) is

$$(x - h)^2 + (y - k)^2 = r^2 \qquad 7.67$$

The two forms can be converted by use of Eq. 7.68 through Eq. 7.70.

$$h = \frac{-D}{2A} \qquad 7.68$$

$$k = \frac{-E}{2A} \qquad 7.69$$

$$r^2 = \frac{D^2 + E^2 - 4AF}{4A^2} \qquad 7.70$$

If the right-hand side of Eq. 7.70 is positive, the figure is a circle. If it is zero, the circle shrinks to a point. If the right-hand side is negative, the figure is imaginary. A *degenerate circle* is one in which the right-hand side is less than or equal to zero.

18. PARABOLA

A *parabola* is the locus of points equidistant from the *focus* (point F in Fig. 7.13) and a line called the *directrix*. A parabola is symmetric with respect to its *parabolic axis*. The line normal to the parabolic axis and passing through the focus is known as the *latus rectum*. The eccentricity of a parabola is 1.

Figure 7.13 *Parabola*

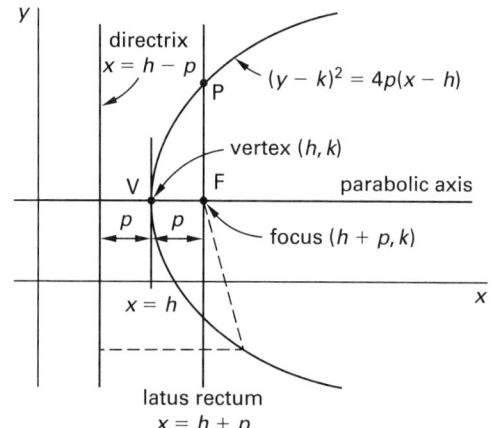

There are two common types of parabolas in the Cartesian plane—those that open right and left, and those that open up and down. Equation 7.65 is the general form of the equation of a parabola. With Eq. 7.71, the parabola points horizontally to the right if $CD > 0$ and to the left if $CD < 0$. With Eq. 7.72, the parabola points vertically up if $AE > 0$ and down if $AE < 0$.

$$Cy^2 + Dx + Ey + F = 0 \Big|_{\substack{C, D \neq 0 \\ \text{opens horizontally}}} \qquad 7.71$$

$$Ax^2 + Dx + Ey + F = 0 \Big|_{\substack{A, E \neq 0 \\ \text{opens vertically}}} \qquad 7.72$$

The *standard form* of the equation of a parabola with vertex at (h, k), focus at $(h + p, k)$, and directrix at $x = h - p$, and that opens to the right or left is given by Eq. 7.73. The parabola opens to the right (points to the left) if $p > 0$ and opens to the left (points to the right) if $p < 0$.

$$(y - k)^2 = 4p(x - h)\Big|_{\text{opens horizontally}} \qquad 7.73$$

$$y^2 = 4px\Big|_{\substack{\text{vertex at origin} \\ h = k = 0}} \qquad 7.74$$

The *standard form* of the equation of a parabola with vertex at (h, k), focus at $(h, k + p)$, and directrix at $y = k - p$, and that opens up or down is given by Eq. 7.75. The parabola opens up (points down) if $p > 0$ and opens down (points up) if $p < 0$.

$$(x - h)^2 = 4p(y - k)\Big|_{\text{opens vertically}} \qquad 7.75$$

$$x^2 = 4py\Big|_{\text{vertex at origin}} \qquad 7.76$$

The general and vertex forms of the equations can be reconciled with Eq. 7.77 through Eq. 7.79. Whether the first or second forms of these equations are used depends on whether the parabola opens horizontally or vertically (i.e., whether $A = 0$ or $C = 0$), respectively.

$$h = \begin{cases} \dfrac{E^2 - 4CF}{4CD} & \text{[opens horizontally]} \\ \dfrac{-D}{2A} & \text{[opens vertically]} \end{cases} \qquad 7.77$$

$$k = \begin{cases} \dfrac{-E}{2C} & \text{[opens horizontally]} \\ \dfrac{D^2 - 4AF}{4AE} & \text{[opens vertically]} \end{cases} \qquad 7.78$$

$$p = \begin{cases} \dfrac{-D}{4C} & \text{[opens horizontally]} \\ \dfrac{-E}{4A} & \text{[opens vertically]} \end{cases} \qquad 7.79$$

19. ELLIPSE

An *ellipse* has two foci separated along its *major axis* by a distance $2c$ as shown in Fig. 7.14. The line perpendicular to the major axis passing through the center of the ellipse is the *minor axis*. The two lines passing through the foci perpendicular to the major axis are the *latera recta*. The distance between the two vertices is $2a$. The ellipse is the locus of those points whose distances from the two foci add up to $2a$. For each point P on the ellipse,

$$F_1P + PF_2 = 2a \qquad \text{7.80}$$

Figure 7.14 *Ellipse*

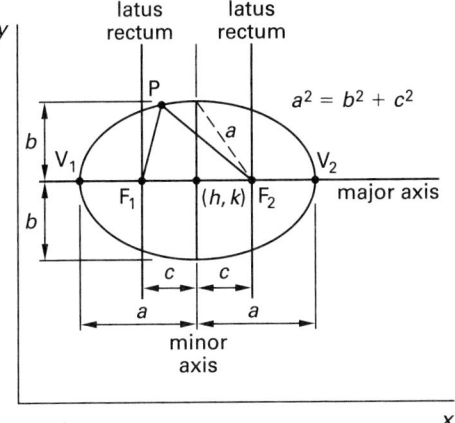

Equation 7.81 is the standard equation used for an ellipse with axes parallel to the coordinate axes, while Eq. 7.65 is the general form. F is not independent of A, C, D, and E for the ellipse.

$$Ax^2 + Cy^2 + Dx + Ey + F = 0\Big|_{\substack{AC > 0 \\ A \neq C}} \qquad \text{7.81}$$

Equation 7.82 gives the standard form of the equation of an ellipse centered at (h, k). Distances a and b are known as the *semimajor distance* and *semiminor distance*, respectively.

$$\frac{(x-h)^2}{a^2} + \frac{(y-k)^2}{b^2} = 1 \qquad \text{7.82}$$

The distance between the two foci is $2c$.

$$2c = 2\sqrt{a^2 - b^2} \qquad \text{7.83}$$

The *aspect ratio* of the ellipse is

$$\text{aspect ratio} = \frac{a}{b} \qquad \text{7.84}$$

The *eccentricity*, ϵ, of the ellipse is always less than 1. If the eccentricity is zero, the figure is a circle (another form of a *degenerative ellipse*).

$$\epsilon = \frac{\sqrt{a^2 - b^2}}{a} < 1 \qquad \text{7.85}$$

The standard and center forms of the equations of an ellipse can be reconciled by using Eq. 7.86 through Eq. 7.89.

$$h = \frac{-D}{2A} \qquad \text{7.86}$$

$$k = \frac{-E}{2C} \qquad \text{7.87}$$

$$a = \sqrt{C} \qquad \text{7.88}$$

$$b = \sqrt{A} \qquad \text{7.89}$$

20. HYPERBOLA

A *hyperbola* has two foci separated along its *transverse axis* (*major axis*) by a distance $2c$, as shown in Fig. 7.15. The line perpendicular to the transverse axis and midway between the foci is the *conjugate axis* (*minor axis*). The distance between the two vertices is $2a$. If a line is drawn parallel to the conjugate axis through either vertex, the distance between the points where it intersects the asymptotes is $2b$. The hyperbola is the locus of those points whose distances from the two foci differ by $2a$. For each point P on the hyperbola,

$$F_2P - PF_1 = 2a \qquad \text{7.90}$$

Equation 7.91 is the standard equation of a hyperbola. Coefficients A and C have opposite signs.

$$Ax^2 + Cy^2 + Dx + Ey + F = 0\big|_{AC < 0} \qquad \text{7.91}$$

Figure 7.15 *Hyperbola*

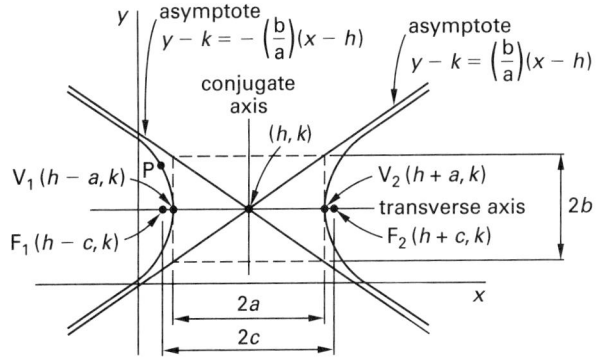

Background and Support

Equation 7.92 gives the standard form of the equation of a hyperbola centered at (h, k) and opening to the left and right.

$$\frac{(x - h)^2}{a^2} - \frac{(y - k)^2}{b^2} = 1 \bigg|_{\text{opens horizontally}} \qquad 7.92$$

Equation 7.93 gives the standard form of the equation of a hyperbola that is centered at (h, k) and is opening up and down.

$$\frac{(y - k)^2}{a^2} - \frac{(x - h)^2}{b^2} = 1 \bigg|_{\text{opens vertically}} \qquad 7.93$$

The distance between the two foci is $2c$.

$$2c = 2\sqrt{a^2 + b^2} \qquad 7.94$$

The *eccentricity*, ϵ, of the hyperbola is calculated from Eq. 7.95 and is always greater than 1.

$$\epsilon = \frac{c}{a} = \frac{\sqrt{a^2 + b^2}}{a} > 1 \qquad 7.95$$

The hyperbola is asymptotic to the lines given by Eq. 7.96 and Eq. 7.97.

$$y = \pm\frac{b}{a}(x - h) + k \bigg|_{\text{opens horizontally}} \qquad 7.96$$

$$y = \pm\frac{a}{b}(x - h) + k \bigg|_{\text{opens vertically}} \qquad 7.97$$

For a *rectangular (equilateral) hyperbola*, the asymptotes are perpendicular, $a = b$, $c = \sqrt{2}a$, and the eccentricity is $\epsilon = \sqrt{2}$. If the hyperbola is centered at the origin (i.e., $h = k = 0$), then the equations are $x^2 - y^2 = a^2$ (opens horizontally) and $y^2 - x^2 = a^2$ (opens vertically).

If the asymptotes are the x- and y-axes, the equation of the hyperbola is simply

$$xy = \pm\frac{a^2}{2} \qquad 7.98$$

The general and center forms of the equations of a hyperbola can be reconciled by using Eq. 7.99 through Eq. 7.103. Whether the hyperbola opens left and right or up and down depends on whether M/A or M/C is positive, respectively, where M is defined by Eq. 7.99.

$$M = \frac{D^2}{4A} + \frac{E^2}{4C} - F \qquad 7.99$$

$$h = \frac{-D}{2A} \qquad 7.100$$

$$k = \frac{-E}{2C} \qquad 7.101$$

$$a = \begin{cases} \sqrt{-C} & \text{[opens horizontally]} \\ \sqrt{-A} & \text{[opens vertically]} \end{cases} \qquad 7.102$$

$$b = \begin{cases} \sqrt{A} & \text{[opens horizontally]} \\ \sqrt{C} & \text{[opens vertically]} \end{cases} \qquad 7.103$$

21. SPHERE

Equation 7.104 is the general equation of a sphere. The coefficient A cannot be zero.

$$Ax^2 + Ay^2 + Az^2 + Bx + Cy + Dz + E = 0 \qquad 7.104$$

Equation 7.105 gives the standard form of the equation of a sphere centered at (h, k, l) with radius r.

$$(x - h)^2 + (y - k)^2 + (z - l)^2 = r^2 \qquad 7.105$$

The general and center forms of the equations of a sphere can be reconciled by using Eq. 7.106 through Eq. 7.109.

$$h = \frac{-B}{2A} \qquad 7.106$$

$$k = \frac{-C}{2A} \qquad 7.107$$

$$l = \frac{-D}{2A} \qquad 7.108$$

$$r = \sqrt{\frac{B^2 + C^2 + D^2}{4A^2} - \frac{E}{A}} \qquad 7.109$$

22. HELIX

A *helix* is a curve generated by a point moving on, around, and along a cylinder such that the distance the point moves parallel to the cylindrical axis is proportional to the angle of rotation about that axis. (See Fig. 7.16.) For a cylinder of radius r, Eq. 7.110 through Eq. 7.112 define the three-dimensional positions of points along the helix. The quantity $2\pi k$ is the *pitch* of the helix.

$$x = r\cos\theta \qquad 7.110$$

$$y = r\sin\theta \qquad 7.111$$

$$z = k\theta \qquad 7.112$$

Figure 7.16 *Helix*

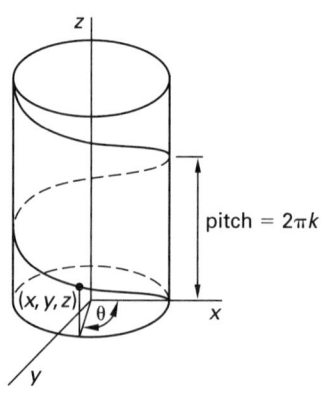

8 Differential Calculus

1. Derivative of a Function 8-1
2. Elementary Derivative Operations 8-1
3. Critical Points 8-2
4. Derivatives of Parametric Equations 8-3
5. Partial Differentiation 8-4
6. Implicit Differentiation 8-4
7. Tangent Plane Function 8-5
8. Gradient Vector 8-5
9. Directional Derivative 8-6
10. Normal Line Vector 8-6
11. Divergence of a Vector Field 8-7
12. Curl of a Vector Field 8-7
13. Taylor's Formula 8-8
14. Maclaurin Power Approximations 8-8

1. DERIVATIVE OF A FUNCTION

In most cases, it is possible to transform a continuous function, $f(x_1, x_2, x_3, \ldots)$, of one or more independent variables into a derivative function.[1] In simple cases, the *derivative* can be interpreted as the slope (tangent or rate of change) of the curve described by the original function. Since the slope of the curve depends on x, the derivative function will also depend on x. The derivative, $f'(x)$, of a function $f(x)$ is defined mathematically by Eq. 8.1. However, limit theory is seldom needed to actually calculate derivatives.

$$f'(x) = \lim_{\Delta x \to 0} \frac{\Delta f(x)}{\Delta x} \qquad 8.1$$

The derivative of a function $f(x)$, also known as the *first derivative*, is written in various ways, including

$$f'(x), \frac{df(x)}{dx}, \frac{df}{dx}, \mathbf{D}f(x), \mathbf{D}_x f(x), \dot{f}(x), sf(s)$$

A *second derivative* may exist if the derivative operation is performed on the first derivative—that is, a derivative is taken of a derivative function. This is written as

$$f''(x), \frac{d^2 f(x)}{dx^2}, \frac{d^2 f}{dx^2}, \mathbf{D}^2 f(x), \mathbf{D}_x^2 f(x), \ddot{f}(x), s^2 f(s)$$

---

[1]A function, $f(x)$, of one independent variable, x, is used in this section to simplify the discussion. Although the derivative is taken with respect to x, the independent variable can be anything.

Newton's notation (e.g., $\dot{m}$ and $\ddot{x}$) is generally only used with functions of time.

A *regular* (*analytic* or *holomorphic*) *function* possesses a derivative. A point at which a function's derivative is undefined is called a *singular point*, as Fig. 8.1 illustrates.

Figure 8.1 *Derivatives and Singular Points*

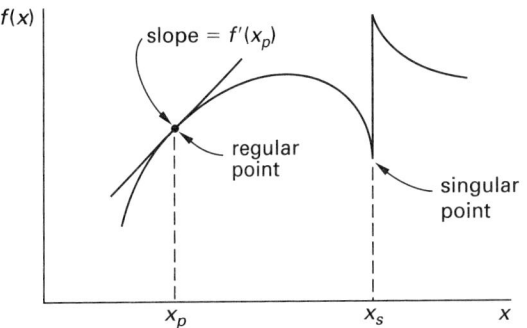

2. ELEMENTARY DERIVATIVE OPERATIONS

Equation 8.2 through Eq. 8.5 summarize the elementary derivative operations on polynomials and exponentials. Equation 8.2 and Eq. 8.3 are particularly useful. (a, n, and k represent constants. $f(x)$ and $g(x)$ are functions of x.)

$$\mathbf{D}k = 0 \qquad 8.2$$

$$\mathbf{D}x^n = nx^{n-1} \qquad 8.3$$

$$\mathbf{D}\ln x = \frac{1}{x} \qquad 8.4$$

$$\mathbf{D}e^{ax} = ae^{ax} \qquad 8.5$$

Equation 8.6 through Eq. 8.17 summarize the elementary derivative operations on transcendental (trigonometric) functions.

$$\mathbf{D}\sin x = \cos x \qquad 8.6$$

$$\mathbf{D}\cos x = -\sin x \qquad 8.7$$

$$\mathbf{D}\tan x = \sec^2 x \qquad 8.8$$

$$\mathbf{D}\cot x = -\csc^2 x \qquad 8.9$$

$$\mathbf{D}\sec x = \sec x \tan x \qquad 8.10$$

$$\mathbf{D}\csc x = -\csc x \cot x \qquad 8.11$$

$$\mathbf{D} \arcsin x = \frac{1}{\sqrt{1-x^2}} \qquad \text{8.12}$$

$$\mathbf{D} \arccos x = -\mathbf{D} \arcsin x \qquad \text{8.13}$$

$$\mathbf{D} \arctan x = \frac{1}{1+x^2} \qquad \text{8.14}$$

$$\mathbf{D} \operatorname{arccot} x = -\mathbf{D} \arctan x \qquad \text{8.15}$$

$$\mathbf{D} \operatorname{arcsec} x = \frac{1}{x\sqrt{x^2-1}} \qquad \text{8.16}$$

$$\mathbf{D} \operatorname{arccsc} x = -\mathbf{D} \operatorname{arcsec} x \qquad \text{8.17}$$

Equation 8.18 through Eq. 8.23 summarize the elementary derivative operations on hyperbolic transcendental functions. Derivatives of hyperbolic functions are not completely analogous to those of the regular transcendental functions.

$$\mathbf{D} \sinh x = \cosh x \qquad \text{8.18}$$

$$\mathbf{D} \cosh x = \sinh x \qquad \text{8.19}$$

$$\mathbf{D} \tanh x = \operatorname{sech}^2 x \qquad \text{8.20}$$

$$\mathbf{D} \coth x = -\operatorname{csch}^2 x \qquad \text{8.21}$$

$$\mathbf{D} \operatorname{sech} x = -\operatorname{sech} x \tanh x \qquad \text{8.22}$$

$$\mathbf{D} \operatorname{csch} x = -\operatorname{csch} x \coth x \qquad \text{8.23}$$

Equation 8.24 through Eq. 8.29 summarize the elementary derivative operations on functions and combinations of functions.

$$\mathbf{D} k f(x) = k \mathbf{D} f(x) \qquad \text{8.24}$$

$$\mathbf{D}\big(f(x) \pm g(x)\big) = \mathbf{D} f(x) \pm \mathbf{D} g(x) \qquad \text{8.25}$$

$$\mathbf{D}\big(f(x) \cdot g(x)\big) = f(x)\mathbf{D} g(x) + g(x)\mathbf{D} f(x) \qquad \text{8.26}$$

$$\mathbf{D}\left(\frac{f(x)}{g(x)}\right) = \frac{g(x)\mathbf{D} f(x) - f(x)\mathbf{D} g(x)}{\big(g(x)\big)^2} \qquad \text{8.27}$$

$$\mathbf{D}\big(f(x)\big)^n = n\big(f(x)\big)^{n-1}\mathbf{D} f(x) \qquad \text{8.28}$$

$$\mathbf{D} f\big(g(x)\big) = \mathbf{D}_g f(g)\mathbf{D}_x g(x) \qquad \text{8.29}$$

Example 8.1

What is the slope at $x = 3$ of the curve $f(x) = x^3 - 2x$?

Solution

The derivative function found from Eq. 8.3 determines the slope.

$$f'(x) = 3x^2 - 2$$

The slope at $x = 3$ is

$$f'(3) = (3)(3)^2 - 2 = 25$$

Example 8.2

What are the derivatives of the following functions?

(a) $f(x) = 5\sqrt[3]{x^5}$

(b) $f(x) = \sin x \cos^2 x$

(c) $f(x) = \ln(\cos e^x)$

Solution

(a) Using Eq. 8.3 and Eq. 8.24,

$$\begin{aligned}
f'(x) &= 5\mathbf{D}\sqrt[3]{x^5} = 5\mathbf{D}(x^5)^{1/3} \\
&= (5)\big(\tfrac{1}{3}\big)(x^5)^{-2/3}\mathbf{D} x^5 \\
&= (5)\big(\tfrac{1}{3}\big)(x^5)^{-2/3}(5)(x^4) \\
&= 25x^{2/3}/3
\end{aligned}$$

(b) Using Eq. 8.26,

$$\begin{aligned}
f'(x) &= \sin x \mathbf{D}\cos^2 x + \cos^2 x \mathbf{D}\sin x \\
&= (\sin x)(2\cos x)(\mathbf{D}\cos x) + \cos^2 x \cos x \\
&= (\sin x)(2\cos x)(-\sin x) + \cos^2 x \cos x \\
&= -2\sin^2 x \cos x + \cos^3 x
\end{aligned}$$

(c) Using Eq. 8.29,

$$\begin{aligned}
f'(x) &= \left(\frac{1}{\cos e^x}\right)\mathbf{D}\cos e^x \\
&= \left(\frac{1}{\cos e^x}\right)(-\sin e^x)\mathbf{D} e^x \\
&= \left(\frac{-\sin e^x}{\cos e^x}\right)e^x \\
&= -e^x \tan e^x
\end{aligned}$$

3. CRITICAL POINTS

Derivatives are used to locate the local *critical points* of functions of one variable—that is, *extreme points* (also known as *maximum* and *minimum* points) as well as the *inflection points* (*points of contraflexure*). The plurals *extrema*, *maxima*, and *minima* are used without the word "points." These points are illustrated in Fig. 8.2. There is usually an inflection point between two adjacent local extrema.

The first derivative is calculated to determine the locations of possible critical points. The second derivative is calculated to determine whether a particular point is a local maximum, minimum, or inflection point, according to the following conditions. With this method, no distinction is made between local and global extrema. Therefore, the extrema should be compared with the function values at the endpoints of the interval, as

Figure 8.2 *Extreme and Inflection Points*

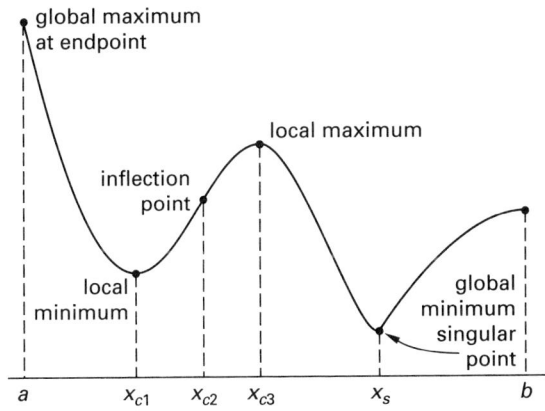

illustrated in Ex. 8.3.[2] Generally, $f'(x) \neq 0$ at an inflection point.

$$f'(x_c) = 0 \text{ at any extreme point, } x_c \qquad 8.30$$

$$f''(x_c) < 0 \text{ at a maximum point} \qquad 8.31$$

$$f''(x_c) > 0 \text{ at a minimum point} \qquad 8.32$$

$$f''(x_c) = 0 \text{ at an inflection point} \qquad 8.33$$

Example 8.3

Find the global extrema of the function $f(x)$ on the interval $[-2, +2]$.

$$f(x) = x^3 + x^2 - x + 1$$

Solution

The first derivative is

$$f'(x) = 3x^2 + 2x - 1$$

Since the first derivative is zero at extreme points, set $f'(x)$ equal to zero and solve for the roots of the quadratic equation.

$$3x^2 + 2x - 1 = (3x - 1)(x + 1) = 0$$

The roots are $x_1 = {}^1\!/_3$, $x_2 = -1$. These are the locations of the two local extrema.

---

[2]It is also necessary to check the values of the function at singular points (i.e., points where the derivative does not exist).

The second derivative is

$$f''(x) = 6x + 2$$

Substituting x_1 and x_2 into $f''(x)$,

$$f''(x_1) = (6)\left(\tfrac{1}{3}\right) + 2 = 4$$

$$f''(x_2) = (6)(-1) + 2 = -4$$

Therefore, x_1 is a local minimum point (because $f''(x_1)$ is positive), and x_2 is a local maximum point (because $f''(x_2)$ is negative). The inflection point between these two extrema is found by setting $f''(x)$ equal to zero.

$$f''(x) = 6x + 2 = 0 \text{ or } x = -\tfrac{1}{3}$$

Since the question asked for the global extreme points, it is necessary to compare the values of $f(x)$ at the local extrema with the values at the endpoints.

$$f(-2) = -1$$

$$f(-1) = 2$$

$$f\left(\tfrac{1}{3}\right) = 22/27$$

$$f(2) = 11$$

Therefore, the actual global extrema are at the endpoints.

4. DERIVATIVES OF PARAMETRIC EQUATIONS

The derivative of a function $f(x_1, x_2, \ldots, x_n)$ can be calculated from the derivatives of the parametric equations $f_1(s), f_2(s), \ldots, f_n(s)$. The derivative will be expressed in terms of the parameter, s, unless the derivatives of the parametric equations can be expressed explicitly in terms of the independent variables.

Example 8.4

A circle is expressed parametrically by the equations

$$x = 5 \cos \theta$$

$$y = 5 \sin \theta$$

Express the derivative dy/dx (a) as a function of the parameter θ and (b) as a function of x and y.

Solution

(a) Taking the derivative of each parametric equation with respect to θ,

$$\frac{dx}{d\theta} = -5 \sin \theta$$

$$\frac{dy}{d\theta} = 5 \cos \theta$$

Then,

$$\frac{dy}{dx} = \frac{\dfrac{dy}{d\theta}}{\dfrac{dx}{d\theta}} = \frac{5 \cos \theta}{-5 \sin \theta} = -\cot \theta$$

(b) The derivatives of the parametric equations are closely related to the original parametric equations.

$$\frac{dx}{d\theta} = -5 \sin \theta = -y$$

$$\frac{dy}{d\theta} = 5 \cos \theta = x$$

$$\frac{dy}{dx} = \frac{\dfrac{dy}{d\theta}}{\dfrac{dx}{d\theta}} = \frac{-x}{y}$$

5. PARTIAL DIFFERENTIATION

Derivatives can be taken with respect to only one independent variable at a time. For example, $f'(x)$ is the derivative of $f(x)$ and is taken with respect to the independent variable x. If a function, $f(x_1, x_2, x_3, \ldots)$, has more than one independent variable, a *partial derivative* can be found, but only with respect to one of the independent variables. All other variables are treated as constants. Symbols for a partial derivative of f taken with respect to variable x are $\partial f/\partial x$ and $f_x(x, y)$.

The geometric interpretation of a partial derivative $\partial f/\partial x$ is the slope of a line tangent to the surface (a sphere, ellipsoid, etc.) described by the function when all variables except x are held constant. In three-dimensional space with a function described by $z = f(x, y)$, the partial derivative $\partial f/\partial x$ (equivalent to $\partial z/\partial x$) is the slope of the line tangent to the surface in a plane of constant y. Similarly, the partial derivative $\partial f/\partial y$ (equivalent to $\partial z/\partial y$) is the slope of the line tangent to the surface in a plane of constant x.

Example 8.5

What is the partial derivative $\partial z/\partial x$ of the following function?

$$z = 3x^2 - 6y^2 + xy + 5y - 9$$

Solution

The partial derivative with respect to x is found by considering all variables other than x to be constants.

$$\frac{\partial z}{\partial x} = 6x - 0 + y + 0 - 0 = 6x + y$$

Example 8.6

A surface has the equation $x^2 + y^2 + z^2 - 9 = 0$. What is the slope of a line that lies in a plane of constant y and is tangent to the surface at $(x, y, z) = (1, 2, 2)$?[3]

Solution

Solve for the dependent variable. Then, consider variable y to be a constant.

$$z = \sqrt{9 - x^2 - y^2}$$

$$\frac{\partial z}{\partial x} = \frac{\partial (9 - x^2 - y^2)^{1/2}}{\partial x}$$

$$= \left(\tfrac{1}{2}\right)(9 - x^2 - y^2)^{-1/2}\left(\frac{\partial (9 - x^2 - y^2)}{\partial x}\right)$$

$$= \left(\tfrac{1}{2}\right)(9 - x^2 - y^2)^{-1/2}(-2x)$$

$$= \frac{-x}{\sqrt{9 - x^2 - y^2}}$$

At the point $(1, 2, 2)$, $x = 1$ and $y = 2$.

$$\left.\frac{\partial z}{\partial x}\right|_{(1,2,2)} = \frac{-1}{\sqrt{9 - (1)^2 - (2)^2}} = -\frac{1}{2}$$

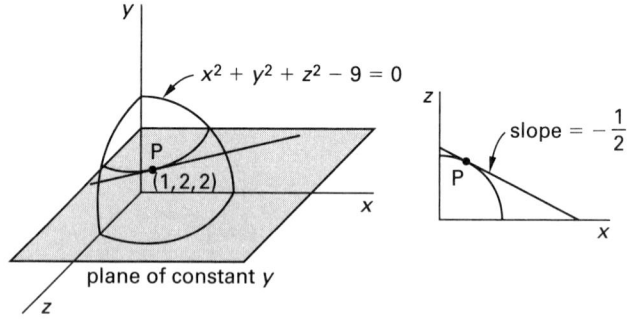

6. IMPLICIT DIFFERENTIATION

When a relationship between n variables cannot be manipulated to yield an explicit function of $n - 1$ independent variables, that relationship implicitly defines the nth variable. Finding the derivative of the implicit variable with respect to any other independent variable is known as *implicit differentiation*.

---

[3]Although only implied, it is required that the point actually be on the surface (i.e., it must satisfy the equation $f(x, y, z) = 0$).

An implicit derivative is the quotient of two partial derivatives. The two partial derivatives are chosen so that dividing one by the other eliminates a common differential. For example, if z cannot be explicitly extracted from $f(x,y,z) = 0$, the partial derivatives $\partial z/\partial x$ and $\partial z/\partial y$ can still be found as follows.

$$\frac{\partial z}{\partial x} = \frac{\frac{\partial f}{\partial x}}{\frac{\partial f}{\partial z}} \qquad 8.34$$

$$\frac{\partial z}{\partial y} = \frac{\frac{\partial f}{\partial y}}{\frac{\partial f}{\partial z}} \qquad 8.35$$

Example 8.7

Find the derivative dy/dx of

$$f(x,y) = x^2 + xy + y^3$$

Solution

Implicit differentiation is required because x cannot be extracted from $f(x,y)$.

$$\frac{\partial f}{\partial x} = 2x + y$$

$$\frac{\partial f}{\partial y} = x + 3y^2$$

$$\frac{dy}{dx} = \frac{-\frac{\partial f}{\partial x}}{\frac{\partial f}{\partial y}} = \frac{-(2x+y)}{x+3y^2}$$

Example 8.8

Solve Ex. 8.6 using implicit differentiation.

Solution

$$f(x,y,z) = x^2 + y^2 + z^2 - 9 = 0$$

$$\frac{\partial f}{\partial x} = 2x$$

$$\frac{\partial f}{\partial z} = 2z$$

$$\frac{\partial z}{\partial x} = \frac{-\frac{\partial f}{\partial x}}{\frac{\partial f}{\partial z}} = \frac{-2x}{2z} = -\frac{x}{z}$$

At the point $(1,2,2)$,

$$\frac{\partial z}{\partial x} = -\frac{1}{2}$$

7. TANGENT PLANE FUNCTION

Partial derivatives can be used to find the equation of a plane tangent to a three-dimensional surface defined by $f(x,y,z) = 0$ at some point, P_0.

$$T(x_0, y_0, z_0) = (x - x_0)\frac{\partial f(x,y,z)}{\partial x}\Big|_{P_0}$$
$$+ (y - y_0)\frac{\partial f(x,y,z)}{\partial y}\Big|_{P_0}$$
$$+ (z - z_0)\frac{\partial f(x,y,z)}{\partial z}\Big|_{P_0}$$
$$= 0 \qquad 8.36$$

The coefficients of x, y, and z are the same as the coefficients of $\mathbf{i}$, $\mathbf{j}$, and $\mathbf{k}$ of the normal vector at point P_0. (See Sec. 8.10.)

Example 8.9

What is the equation of the plane that is tangent to the surface defined by $f(x,y,z) = 4x^2 + y^2 - 16z = 0$ at the point $(2,4,2)$?

Solution

Calculate the partial derivatives and substitute the coordinates of the point.

$$\frac{\partial f(x,y,z)}{\partial x}\Big|_{P_0} = 8x|_{(2,4,2)} = (8)(2) = 16$$

$$\frac{\partial f(x,y,z)}{\partial y}\Big|_{P_0} = 2y|_{(2,4,2)} = (2)(4) = 8$$

$$\frac{\partial f(x,y,z)}{\partial z}\Big|_{P_0} = -16|_{(2,4,2)} = -16$$

$$T(2,4,2) = (16)(x-2) + (8)(y-4) - (16)(z-2)$$
$$= 16x + 8y - 16z - 32$$

Substitute into Eq. 8.36, and divide both sides by 8.

$$2x + y - 2z - 4 = 0$$

8. GRADIENT VECTOR

The slope of a function is the change in one variable with respect to a distance in a chosen direction. Usually, the direction is parallel to a coordinate axis. However, the maximum slope at a point on a surface may not be in a direction parallel to one of the coordinate axes.

The *gradient vector function* $\nabla f(x, y, z)$ (pronounced "del f") gives the maximum rate of change of the function $f(x, y, z)$.

$$\nabla f(x, y, z) = \frac{\partial f(x, y, z)}{\partial x}\mathbf{i} + \frac{\partial f(x, y, z)}{\partial y}\mathbf{j}$$
$$+ \frac{\partial f(x, y, z)}{\partial z}\mathbf{k} \qquad \text{8.37}$$

Example 8.10

A two-dimensional function is defined by

$$f(x, y) = 2x^2 - y^2 + 3x - y$$

(a) What is the gradient vector for this function? (b) What is the direction of the line passing through the point $(1, -2)$ that has a maximum slope? (c) What is the maximum slope at the point $(1, -2)$?

Solution

(a) It is necessary to calculate two partial derivatives in order to use Eq. 8.37.

$$\frac{\partial f(x, y)}{\partial x} = 4x + 3$$

$$\frac{\partial f(x, y)}{\partial y} = -2y - 1$$

$$\nabla f(x, y) = (4x + 3)\mathbf{i} + (-2y - 1)\mathbf{j}$$

(b) Find the direction of the line passing through $(1, -2)$ with maximum slope by inserting $x = 1$ and $y = -2$ into the gradient vector function.

$$\mathbf{V} = \Big((4)(1) + 3\Big)\mathbf{i} + \Big((-2)(-2) - 1\Big)\mathbf{j}$$

$$= 7\mathbf{i} + 3\mathbf{j}$$

(c) The magnitude of the slope is

$$|\mathbf{V}| = \sqrt{(7)^2 + (3)^2} = 7.62$$

9. DIRECTIONAL DERIVATIVE

Unlike the gradient vector (covered in Sec. 8.8), which calculates the maximum rate of change of a function, the *directional derivative*, indicated by $\nabla_u f(x, y, z)$, $D_u f(x, y, z)$, or $f'_u(x, y, z)$, gives the rate of change in the direction of a given vector, $\mathbf{u}$ or $\mathbf{U}$. The subscript u implies that the direction vector is a unit vector, but it

does not need to be, as only the direction cosines are calculated from it.

$$\nabla_u f(x, y, z) = \left(\frac{\partial f(x, y, z)}{\partial x}\right)\cos\alpha$$
$$+ \left(\frac{\partial f(x, y, z)}{\partial y}\right)\cos\beta$$
$$+ \left(\frac{\partial f(x, y, z)}{\partial z}\right)\cos\gamma \qquad \text{8.38}$$

$$\mathbf{U} = U_x\mathbf{i} + U_y\mathbf{j} + U_z\mathbf{k} \qquad \text{8.39}$$

$$\cos\alpha = \frac{U_x}{|\mathbf{U}|} = \frac{U_x}{\sqrt{U_x^2 + U_y^2 + U_z^2}} \qquad \text{8.40}$$

$$\cos\beta = \frac{U_y}{|\mathbf{U}|} \qquad \text{8.41}$$

$$\cos\gamma = \frac{U_z}{|\mathbf{U}|} \qquad \text{8.42}$$

Example 8.11

What is the rate of change of $f(x, y) = 3x^2 + xy - 2y^2$ at the point $(1, -2)$ in the direction $4\mathbf{i} + 3\mathbf{j}$?

Solution

The direction cosines are given by Eq. 8.40 and Eq. 8.41.

$$\cos\alpha = \frac{U_x}{|\mathbf{U}|} = \frac{4}{\sqrt{(4)^2 + (3)^2}} = \frac{4}{5}$$

$$\cos\beta = \frac{U_y}{|\mathbf{U}|} = \frac{3}{5}$$

The partial derivatives are

$$\frac{\partial f(x, y)}{\partial x} = 6x + y$$

$$\frac{\partial f(x, y)}{\partial y} = x - 4y$$

The directional derivative is given by Eq. 8.38.

$$\nabla_u f(x, y) = \left(\frac{4}{5}\right)(6x + y) + \left(\frac{3}{5}\right)(x - 4y)$$

Substituting the given values of $x = 1$ and $y = -2$,

$$\nabla_u f(1, -2) = \left(\frac{4}{5}\right)\Big((6)(1) - 2\Big) + \left(\frac{3}{5}\right)\Big(1 - (4)(-2)\Big)$$

$$= \frac{43}{5} \quad (8.6)$$

10. NORMAL LINE VECTOR

Partial derivatives can be used to find the vector normal to a three-dimensional surface defined by $f(x, y, z) = 0$

at some point P_0. The coefficients of **i**, **j**, and **k** are the same as the coefficients of x, y, and z calculated for the equation of the tangent plane at point P_0. (See Sec. 8.7.)

$$\mathbf{N} = \left.\frac{\partial f(x,y,z)}{\partial x}\right|_{P_0} \mathbf{i} + \left.\frac{\partial f(x,y,z)}{\partial y}\right|_{P_0} \mathbf{j}$$
$$+ \left.\frac{\partial f(x,y,z)}{\partial z}\right|_{P_0} \mathbf{k} \qquad 8.43$$

Example 8.12

What is the vector normal to the surface of $f(x,y,z) = 4x^2 + y^2 - 16z = 0$ at the point $(2,4,2)$?

Solution

The equation of the tangent plane at this point was calculated in Ex. 8.9 to be

$$T(2,4,2) = 2x + y - 2z - 4 = 0$$

A vector that is normal to the tangent plane through this point is

$$\mathbf{N} = 2\mathbf{i} + \mathbf{j} - 2\mathbf{k}$$

11. DIVERGENCE OF A VECTOR FIELD

The *divergence*, div **F**, of a vector field $\mathbf{F}(x,y,z)$ is a scalar function defined by Eq. 8.45 and Eq. 8.46.[4] The divergence of **F** can be interpreted as the *accumulation* of flux (i.e., a flowing substance) in a small region (i.e., at a point). One of the uses of the divergence is to determine whether flow (represented in direction and magnitude by **F**) is compressible. Flow is incompressible if div $\mathbf{F} = 0$, since the substance is not accumulating.

$$\mathbf{F} = P(x,y,z)\mathbf{i} + Q(x,y,z)\mathbf{j} + R(x,y,z)\mathbf{k} \qquad 8.44$$

$$\text{div } \mathbf{F} = \frac{\partial P}{\partial x} + \frac{\partial Q}{\partial y} + \frac{\partial R}{\partial z} \qquad 8.45$$

It may be easier to calculate divergence from Eq. 8.46.

$$\text{div } \mathbf{F} = \nabla \cdot \mathbf{F} \qquad 8.46$$

The vector del operator, ∇, is defined as

$$\nabla = \frac{\partial}{\partial x}\mathbf{i} + \frac{\partial}{\partial y}\mathbf{j} + \frac{\partial}{\partial z}\mathbf{k} \qquad 8.47$$

If there is no divergence, then the dot product calculated in Eq. 8.46 is zero.

---

[4] A bold letter, **F**, is used to indicate that the vector is a function of x, y, and z.

Example 8.13

Calculate the divergence of the following vector function.

$$\mathbf{F}(x,y,z) = xz\mathbf{i} + e^x y\mathbf{j} + 7x^3 y\mathbf{k}$$

Solution

From Eq. 8.45,

$$\text{div } \mathbf{F} = \frac{\partial}{\partial x}xz + \frac{\partial}{\partial y}e^x y + \frac{\partial}{\partial z}7x^3 y = z + e^x + 0$$
$$= z + e^x$$

12. CURL OF A VECTOR FIELD

The *curl*, curl **F**, of a vector field $\mathbf{F}(x,y,z)$ is a vector field defined by Eq. 8.49 and Eq. 8.50. The curl **F** can be interpreted as the *vorticity* per unit area of flux (i.e., a flowing substance) in a small region (i.e., at a point). One of the uses of the curl is to determine whether flow (represented in direction and magnitude by **F**) is rotational. Flow is irrotational if curl $\mathbf{F} = 0$.

$$\mathbf{F} = P(x,y,z)\mathbf{i} + Q(x,y,z)\mathbf{j} + R(x,y,z)\mathbf{k} \qquad 8.48$$

$$\text{curl } \mathbf{F} = \left(\frac{\partial R}{\partial y} - \frac{\partial Q}{\partial z}\right)\mathbf{i} + \left(\frac{\partial P}{\partial z} - \frac{\partial R}{\partial x}\right)\mathbf{j}$$
$$+ \left(\frac{\partial Q}{\partial x} - \frac{\partial P}{\partial y}\right)\mathbf{k} \qquad 8.49$$

It may be easier to calculate the curl from Eq. 8.50. (The vector del operator, ∇, was defined in Eq. 8.47.)

$$\text{curl } \mathbf{F} = \nabla \times \mathbf{F}$$
$$= \begin{vmatrix} \mathbf{i} & \mathbf{j} & \mathbf{k} \\ \dfrac{\partial}{\partial x} & \dfrac{\partial}{\partial y} & \dfrac{\partial}{\partial z} \\ P(x,y,z) & Q(x,y,z) & R(x,y,z) \end{vmatrix} \qquad 8.50$$

If the velocity vector is **V**, then the vorticity is

$$\omega = \nabla \times \mathbf{V} = \omega_x \mathbf{i} + \omega_y \mathbf{j} + \omega_z \mathbf{k} \qquad 8.51$$

The *circulation* is the line integral of the velocity, **V**, along a closed curve.

$$\Gamma = \oint V \cdot ds = \oint \omega \cdot dA \qquad 8.52$$

Example 8.14

Calculate the curl of the following vector function.

$$\mathbf{F}(x,y,z) = 3x^2\mathbf{i} + 7e^x y\mathbf{j}$$

Solution

Using Eq. 8.50,

$$\text{curl } \mathbf{F} = \begin{vmatrix} \mathbf{i} & \mathbf{j} & \mathbf{k} \\ \dfrac{\partial}{\partial x} & \dfrac{\partial}{\partial y} & \dfrac{\partial}{\partial z} \\ 3x^2 & 7e^x y & 0 \end{vmatrix}$$

Expand the determinant across the top row.

$$\mathbf{i}\left(\frac{\partial}{\partial y}(0) - \frac{\partial}{\partial z}(7e^x y)\right) - \mathbf{j}\left(\frac{\partial}{\partial x}(0) - \frac{\partial}{\partial z}(3x^2)\right)$$

$$+ \mathbf{k}\left(\frac{\partial}{\partial x}(7e^x y) - \frac{\partial}{\partial y}(3x^2)\right)$$

$$= \mathbf{i}(0-0) - \mathbf{j}(0-0) + \mathbf{k}(7e^x y - 0)$$

$$= 7e^x y\, \mathbf{k}$$

13. TAYLOR'S FORMULA

Taylor's formula (*series*) can be used to expand a function around a point (i.e., approximate the function at one point based on the function's value at another point). The approximation consists of a series, each term composed of a derivative of the original function and a polynomial. Using Taylor's formula requires that the original function be continuous in the interval $[a, b]$ and have the required number of derivatives. To expand a function, $f(x)$, around a point, a, in order to obtain $f(b)$, Taylor's formula is

$$f(b) = f(a) + \frac{f'(a)}{1!}(b-a) + \frac{f''(a)}{2!}(b-a)^2$$

$$+ \cdots + \frac{f^n(a)}{n!}(b-a)^n + R_n(b) \qquad 8.53$$

In Eq. 8.53, the expression f^n designates the nth derivative of the function $f(x)$. To be a useful approximation, two requirements must be met: (1) point a must be relatively close to point b, and (2) the function and its derivatives must be known or easy to calculate. The last term, $R_n(b)$, is the uncalculated remainder after n derivatives. It is the difference between the exact and approximate values. By using enough terms, the remainder can be made arbitrarily small. That is, $R_n(b)$ approaches zero as n approaches infinity.

It can be shown that the remainder term can be calculated from Eq. 8.54, where c is some number in the interval $[a, b]$. With certain functions, the constant c can be completely determined. In most cases, however, it is possible only to calculate an upper bound on the remainder from Eq. 8.55. M_n is the maximum (positive) value of $f^{n+1}(x)$ on the interval $[a, b]$.

$$R_n(b) = \frac{f^{n+1}(c)}{(n+1)!}(b-a)^{n+1} \qquad 8.54$$

$$|R_n(b)| \le M_n \frac{|(b-a)^{n+1}|}{(n+1)!} \qquad 8.55$$

14. MACLAURIN POWER APPROXIMATIONS

If $a = 0$ in the Taylor series, Eq. 8.53 is known as the *Maclaurin series*. The Maclaurin series can be used to approximate functions at some value of x between 0 and 1. The following common approximations may be referred to as Maclaurin series, Taylor series, or power series approximations.

$$\sin x \approx x - \frac{x^3}{3!} + \frac{x^5}{5!} - \frac{x^7}{7!} + \cdots + (-1)^n \frac{x^{2n+1}}{(2n+1)!} \qquad 8.56$$

$$\cos x \approx 1 - \frac{x^2}{2!} + \frac{x^4}{4!} - \frac{x^6}{6!} + \cdots + (-1)^n \frac{x^{2n}}{(2n)!} \qquad 8.57$$

$$\sinh x \approx x + \frac{x^3}{3!} + \frac{x^5}{5!} + \frac{x^7}{7!} + \cdots + \frac{x^{2n+1}}{(2n+1)!} \qquad 8.58$$

$$\cosh x \approx 1 + \frac{x^2}{2!} + \frac{x^4}{4!} + \frac{x^6}{6!} + \cdots + \frac{x^{2n}}{(2n)!} \qquad 8.59$$

$$e^x \approx 1 + x + \frac{x^2}{2!} + \frac{x^3}{3!} + \cdots + \frac{x^n}{n!} \qquad 8.60$$

$$\ln(1+x) \approx x - \frac{x^2}{2} + \frac{x^3}{3} - \frac{x^4}{4} + \cdots + (-1)^{n+1}\frac{x^n}{n} \qquad 8.61$$

$$\frac{1}{1-x} \approx 1 + x + x^2 + x^3 + \cdots + x^n \qquad 8.62$$

Integral Calculus

1. INTEGRATION

Integration is the inverse operation of differentiation. For that reason, *indefinite integrals* are sometimes referred to as *antiderivatives*.[1] Although expressions can be functions of several variables, integrals can only be taken with respect to one variable at a time. The *differential term* (dx in Eq. 9.1) indicates that variable. In Eq. 9.1, the function $f'(x)$ is the *integrand*, and x is the variable of integration.

$$\int f'(x)\,dx = f(x) + C \qquad 9.1$$

While most of a function, $f(x)$, can be "recovered" through integration of its derivative, $f'(x)$, a constant term will be lost. This is because the derivative of a constant term vanishes (i.e., is zero), leaving nothing to recover from. A *constant of integration*, C, is added to the integral to recognize the possibility of such a constant term.

2. ELEMENTARY OPERATIONS

Equation 9.2 through Eq. 9.8 summarize the elementary integration operations on polynomials and exponentials.[2]

[1]The difference between an indefinite and definite integral (covered in Sec. 9.7) is simple: An *indefinite integral* is a function, while a *definite integral* is a number.
[2]More extensive listings, known as *tables of integrals*, are widely available. (See App. 9.A.)

Equation 9.2 and Eq. 9.3 are particularly useful. (C and k represent constants. $f(x)$ and $g(x)$ are functions of x.)

$$\int k\,dx = kx + C \qquad 9.2$$

$$\int x^m\,dx = \frac{x^{m+1}}{m+1} + C \quad [m \neq -1] \qquad 9.3$$

$$\int \frac{1}{x}\,dx = \ln|x| + C \qquad 9.4$$

$$\int e^{kx}\,dx = \frac{e^{kx}}{k} + C \qquad 9.5$$

$$\int xe^{kx}\,dx = \frac{e^{kx}(kx-1)}{k^2} + C \qquad 9.6$$

$$\int k^{ax}\,dx = \frac{k^{ax}}{a\ln k} + C \qquad 9.7$$

$$\int \ln x\,dx = x\ln x - x + C \qquad 9.8$$

Equation 9.9 through Eq. 9.20 summarize the elementary integration operations on transcendental functions.

$$\int \sin x\,dx = -\cos x + C \qquad 9.9$$

$$\int \cos x\,dx = \sin x + C \qquad 9.10$$

$$\int \tan x\,dx = \ln|\sec x| + C$$
$$= -\ln|\cos x| + C \qquad 9.11$$

$$\int \cot x\,dx = \ln|\sin x| + C \qquad 9.12$$

$$\int \sec x\,dx = \ln|\sec x + \tan x| + C$$
$$= \ln\left|\tan\left(\frac{x}{2}+\frac{\pi}{4}\right)\right| + C \qquad 9.13$$

$$\int \csc x\,dx = \ln|\csc x - \cot x| + C$$
$$= \ln\left|\tan\frac{x}{2}\right| + C \qquad 9.14$$

$$\int \frac{dx}{k^2+x^2} = \frac{1}{k}\arctan\frac{x}{k} + C \qquad 9.15$$

$$\int \frac{dx}{\sqrt{k^2-x^2}} = \arcsin\frac{x}{k} + C \quad [k^2 > x^2] \qquad 9.16$$

$$\int \frac{dx}{x\sqrt{x^2 - k^2}} = \frac{1}{k}\operatorname{arcsec}\frac{x}{k} + C \quad [x^2 > k^2] \qquad 9.17$$

$$\int \sin^2 x\, dx = \tfrac{1}{2}x - \tfrac{1}{4}\sin 2x + C \qquad 9.18$$

$$\int \cos^2 x\, dx = \tfrac{1}{2}x - \tfrac{1}{4}\sin 2x + C \qquad 9.19$$

$$\int \tan^2 x\, dx = \tan x - x + C \qquad 9.20$$

Equation 9.21 through Eq. 9.26 summarize the elementary integration operations on hyperbolic transcendental functions. Integrals of hyperbolic functions are not completely analogous to those of the regular transcendental functions.

$$\int \sinh x\, dx = \cosh x + C \qquad 9.21$$

$$\int \cosh x\, dx = \sinh x + C \qquad 9.22$$

$$\int \tanh x\, dx = \ln|\cosh x| + C \qquad 9.23$$

$$\int \coth x\, dx = \ln|\sinh x| + C \qquad 9.24$$

$$\int \operatorname{sech} x\, dx = \arctan(\sinh x) + C \qquad 9.25$$

$$\int \operatorname{csch} x\, dx = \ln\left|\tanh\frac{x}{2}\right| + C \qquad 9.26$$

Equation 9.27 through Eq. 9.30 summarize the elementary integration operations on functions and combinations of functions.

$$\int kf(x)\, dx = k\int f(x)\, dx \qquad 9.27$$

$$\int \big(f(x) + g(x)\big)\, dx = \int f(x)\, dx + \int g(x)\, dx \qquad 9.28$$

$$\int \frac{f'(x)}{f(x)}\, dx = \ln|f(x)| + C \qquad 9.29$$

$$\int f(x)\, dg(x) = f(x)\int dg(x) - \int g(x)\, df(x) + C$$
$$= f(x)g(x) - \int g(x)\, df(x) + C \qquad 9.30$$

Example 9.1

Find the integral with respect to x of

$$3x^2 + \tfrac{1}{3}x - 7$$

Solution

This is a polynomial function, and Eq. 9.3 can be applied to each of the three terms.

$$\int \big(3x^2 + \tfrac{1}{3}x - 7\big)\, dx = x^3 + \tfrac{1}{6}x^2 - 7x + C$$

3. INTEGRATION BY PARTS

Equation 9.30, repeated here, is known as *integration by parts*. $f(x)$ and $g(x)$ are functions. The use of this method is illustrated by Ex. 9.2.

$$\int f(x)\, dg(x) = f(x)g(x) - \int g(x)\, df(x) + C \qquad 9.31$$

Example 9.2

Find the following integral.

$$\int x^2 e^x\, dx$$

Solution

$x^2 e^x$ is factored into two parts so that integration by parts can be used.

$$f(x) = x^2$$
$$dg(x) = e^x\, dx$$
$$df(x) = 2x\, dx$$

$$g(x) = \int dg(x) = \int e^x\, dx = e^x$$

From Eq. 9.31, disregarding the constant of integration (which cannot be evaluated),

$$\int f(x)\, dg(x) = f(x)g(x) - \int g(x)\, df(x)$$

$$\int x^2 e^x\, dx = x^2 e^x - \int e^x(2x)\, dx$$

The second term is also factored into two parts, and integration by parts is used again. This time,

$$f(x) = x$$
$$dg(x) = e^x\, dx$$
$$df(x) = dx$$

$$g(x) = \int dg(x) = \int e^x\, dx = e^x$$

From Eq. 9.31,

$$\int 2xe^x\,dx = 2\int xe^x\,dx = 2\left(xe^x - \int e^x\,dx\right)$$
$$= 2(xe^x - e^x)$$

Then, the complete integral is

$$\int x^2 e^x\,dx = x^2 e^x - 2(xe^x - e^x) + C$$
$$= e^x(x^2 - 2x + 2) + C$$

4. SEPARATION OF TERMS

Equation 9.28 shows that the integral of a sum of terms is equal to a sum of integrals. This technique is known as *separation of terms*. In many cases, terms are easily separated. In other cases, the technique of *partial fractions* can be used to obtain individual terms. These techniques are illustrated by Ex. 9.3 and Ex. 9.4.

Example 9.3

Find the following integral.

$$\int \frac{(2x^2 + 3)^2}{x}\,dx$$

Solution

$$\int \frac{(2x^2 + 3)^2}{x}\,dx = \int \frac{4x^4 + 12x^2 + 9}{x}\,dx$$
$$= \int \left(4x^3 + 12x + \frac{9}{x}\right)dx$$
$$= x^4 + 6x^2 + 9\ln|x| + C$$

Example 9.4

Find the following integral.

$$\int \frac{3x + 2}{3x - 2}\,dx$$

Solution

The integrand is larger than 1, so use long division to simplify it.

$$
\begin{array}{r}
1 \text{ rem } 4 \quad \left(1 + \dfrac{4}{3x-2}\right) \\
3x - 2\,\overline{\big)\,3x + 2} \\
\underline{3x - 2} \\
4 \text{ remainder}
\end{array}
$$

$$\int \frac{3x + 2}{3x - 2}\,dx = \int \left(1 + \frac{4}{3x - 2}\right)dx$$
$$= \int dx + \int \frac{4}{3x - 2}\,dx$$
$$= x + \tfrac{4}{3}\ln|(3x - 2)| + C$$

5. DOUBLE AND HIGHER-ORDER INTEGRALS

A function can be successively integrated. (This is analogous to successive differentiation.) A function that is integrated twice is known as a *double integral*; if integrated three times, it is a *triple integral*, and so on. Double and triple integrals are used to calculate areas and volumes, respectively.

The successive integrations do not need to be with respect to the same variable. Variables not included in the integration are treated as constants.

There are several notations used for a multiple integral, particularly when the product of length differentials represents a differential area or volume. A double integral (i.e., two successive integrations) can be represented by one of the following notations.

$$\iint f(x, y)\,dx\,dy, \quad \int_{R^2} f(x, y)\,dx\,dy,$$
$$\text{or } \iint_{R^2} f(x, y)\,dA$$

A triple integral can be represented by one of the following notations.

$$\iiint f(x, y, z)\,dx\,dy\,dz, \quad \int_{R^3} f(x, y, z)\,dx\,dy\,dz,$$
$$\text{or } \iiint_{R^3} f(x, y, z)\,dV$$

Example 9.5

Find the following double integral.

$$\iint (x^2 + y^3 x)\,dx\,dy$$

Solution

$$\int (x^2 + y^3 x)\,dx = \tfrac{1}{3}x^3 + \tfrac{1}{2}y^3 x^2 + C_1$$
$$\int \left(\tfrac{1}{3}x^3 + \tfrac{1}{2}y^3 x^2 + C_1\right)dy = \tfrac{1}{3}yx^3 + \tfrac{1}{8}y^4 x^2 + C_1 y + C_2$$

So,

$$\iint (x^2 + y^3 x)\, dx\, dy = \tfrac{1}{3}yx^3 + \tfrac{1}{8}y^4 x^2 + C_1 y + C_2$$

6. INITIAL VALUES

The constant of integration, C, can be found only if the value of the function $f(x)$ is known for some value of x_0. The value $f(x_0)$ is known as an *initial value* or *initial condition*. To completely define a function, as many initial values, $f(x_0)$, $f'(x_1)$, $f''(x_2)$, and so on, as there are integrations are needed. x_0, x_1, x_2, and so on, can be, but do not have to be, the same.

Example 9.6

It is known that $f(x) = 4$ when $x = 2$ (i.e., the initial value is $f(2) = 4$). Find the original function.

$$\int (3x^3 - 7x)\, dx$$

Solution

The function is

$$f(x) = \int (3x^3 - 7x)\, dx = \tfrac{3}{4}x^4 - \tfrac{7}{2}x^2 + C$$

Substituting the initial value determines C.

$$4 = \left(\tfrac{3}{4}\right)(2)^4 - \left(\tfrac{7}{2}\right)(2)^2 + C$$
$$4 = 12 - 14 + C$$
$$C = 6$$

The function is

$$f(x) = \tfrac{3}{4}x^4 - \tfrac{7}{2}x^2 + 6$$

7. DEFINITE INTEGRALS

A *definite integral* is restricted to a specific range of the independent variable. (Unrestricted integrals of the types shown in all preceding examples are known as *indefinite integrals*.) A definite integral restricted to the region bounded by *lower* and *upper limits* (also known as *bounds*), x_1 and x_2, is written as

$$\int_{x_1}^{x_2} f(x)\, dx$$

Equation 9.32 indicates how definite integrals are evaluated. It is known as the *fundamental theorem of calculus*.

$$\int_{x_1}^{x_2} f'(x)\, dx = f(x)\Big|_{x_1}^{x_2} = f(x_2) - f(x_1) \qquad 9.32$$

A common use of a definite integral is the calculation of work performed by a force, F, that moves an object from position x_1 to x_2. The force can be constant or a function of x.

$$W = \int_{x_1}^{x_2} F\, dx \qquad 9.33$$

Example 9.7

Evaluate the following definite integral.

$$\int_{\pi/4}^{\pi/3} \sin x\, dx$$

Solution

From Eq. 9.32,

$$\int_{\pi/4}^{\pi/3} \sin x\, dx = -\cos x\Big|_{\pi/4}^{\pi/3} = -\cos\frac{\pi}{3} - \left(-\cos\frac{\pi}{4}\right)$$
$$= -0.5 - (-0.707)$$
$$= 0.207$$

8. AVERAGE VALUE

The average value of a function $f(x)$ that is integrable over the interval $[a, b]$ is

$$\text{average value} = \frac{1}{b-a}\int_a^b f(x)\, dx \qquad 9.34$$

9. AREA

Equation 9.35 calculates the area, A, bounded by $x = a$, $x = b$, $f_1(x)$ above, and $f_2(x)$ below. ($f_2(x) = 0$ if the area is bounded by the x-axis.) This is illustrated in Fig. 9.1.

$$A = \int_a^b \big(f_1(x) - f_2(x)\big)\, dx \qquad 9.35$$

Figure 9.1 *Area Between Two Curves*

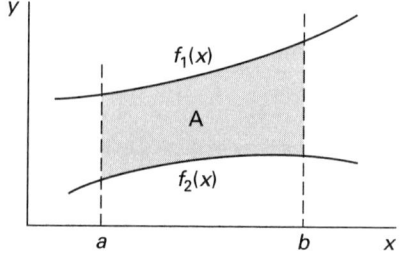

Example 9.8

Find the area between the x-axis and the parabola $y = x^2$ in the interval $[0, 4]$.

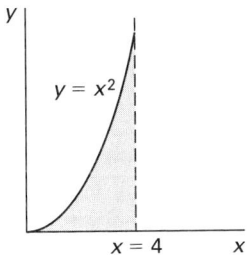

Solution

Referring to Eq. 9.35,

$$f_1(x) = x^2$$
$$f_2(x) = 0$$

$$A = \int_a^b \left(f_1(x) - f_2(x)\right) dx = \int_0^4 x^2\,dx$$
$$= \frac{x^3}{3}\Big|_0^4$$
$$= 64/3$$

10. ARC LENGTH

Equation 9.36 gives the length of a curve defined by $f(x)$ whose derivative exists in the interval $[a, b]$.

$$\text{length} = \int_a^b \sqrt{1 + \left(f'(x)\right)^2}\,dx \qquad 9.36$$

11. PAPPUS' THEOREMS[3]

The first and second theorems of Pappus are:[4]

- *first theorem:* Given a curve, C, that does not intersect the y-axis, the area of the *surface of revolution* generated by revolving C around the y-axis is equal to the product of the length of the curve and the circumference of the circle traced by the centroid of curve C.

$$A = \text{length} \times \text{circumference}$$
$$= \text{length} \times 2\pi \times \text{radius} \qquad 9.37$$

- *second theorem:* Given a plane region, R, that does not intersect the y-axis, the *volume of revolution* generated by revolving R around the y-axis is equal

to the product of the area and the circumference of the circle traced by the centroid of area R.

$$V = \text{area} \times \text{circumference}$$
$$= \text{area} \times 2\pi \times \text{radius} \qquad 9.38$$

12. SURFACE OF REVOLUTION

The surface area obtained by rotating $f(x)$ about the x-axis is

$$A = 2\pi \int_{x=a}^{x=b} f(x)\sqrt{1 + \left(f'(x)\right)^2}\,dx \qquad 9.39$$

The surface area obtained by rotating $f(y)$ about the y-axis is

$$A = 2\pi \int_{y=c}^{y=d} f(y)\sqrt{1 + \left(f'(y)\right)^2}\,dy \qquad 9.40$$

Example 9.9

The curve $f(x) = \frac{1}{2}x$ over the region $x = [0, 4]$ is rotated about the x-axis. What is the surface of revolution?

Solution

The surface of revolution is

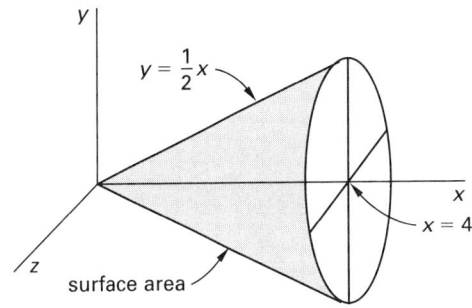

Since $f(x) = \frac{1}{2}x$, $f'(x) = 1/2$. From Eq. 9.39, the area is

$$A = 2\pi \int_{x=a}^{x=b} f(x)\sqrt{1 + \left(f'(x)\right)^2}\,dx$$
$$= 2\pi \int_0^4 \tfrac{1}{2}x\sqrt{1 + \left(\tfrac{1}{2}\right)^2}\,dx$$
$$= \frac{\sqrt{5}}{2}\pi \int_0^4 x\,dx$$
$$= \frac{\sqrt{5}}{2}\pi \frac{x^2}{2}\Big|_0^4$$
$$= \frac{\sqrt{5}}{2}\pi \left(\frac{(4)^2 - (0)^2}{2}\right)$$
$$= 4\sqrt{5}\pi$$

---

[3]This section is an introduction to surfaces and volumes of revolution but does not involve integration.
[4]Some authorities call the first theorem the second, and vice versa.

13. VOLUME OF REVOLUTION

The volume obtained by rotating $f(x)$ about the x-axis is given by Eq. 9.41. $f^2(x)$ is the square of the function, not the second derivative. Equation 9.41 is known as the *method of discs*.

$$V = \pi \int_{x=a}^{x=b} f^2(x)\,dx \qquad 9.41$$

The volume obtained by rotating $f(x)$ about the y-axis can be found from Eq. 9.41 (i.e., using the method of discs) by rewriting the limits and equation in terms of y, or alternatively, the *method of shells* can be used, resulting in the second form of Eq. 9.42.

$$V = \pi \int_{y=c}^{y=d} f^2(y)\,dy$$
$$= 2\pi \int_{x=a}^{x=b} xf(x)\,dx \qquad 9.42$$

Example 9.10

The curve $f(x) = x^2$ over the region $x = [0, 4]$ is rotated about the x-axis. What is the volume of revolution?

Solution

The volume of revolution is

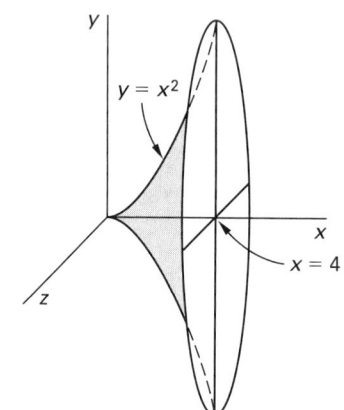

$$V = \pi \int_a^b f^2(x)\,dx = \pi \int_0^4 (x^2)^2\,dx = \pi \frac{x^5}{5}\Big|_0^4$$
$$= \pi\left(\frac{1024}{5} - 0\right)$$
$$= 204.8\pi$$

14. MOMENTS OF A FUNCTION

The *first moment of a function* is a concept used in finding centroids and centers of gravity. Equation 9.43 and Eq. 9.44 are for one- and two-dimensional problems,

respectively. It is the exponent of x (1 in this case) that gives the moment its name.

$$\text{first moment} = \int xf(x)\,dx \qquad 9.43$$

$$\text{first moment} = \iint xf(x, y)\,dx\,dy \qquad 9.44$$

The *second moment of a function* is a concept used in finding moments of inertia with respect to an axis. Equation 9.45 and Eq. 9.46 are for two- and three-dimensional problems, respectively. Second moments with respect to other axes are analogous.

$$(\text{second moment})_x = \iint y^2 f(x, y)\,dy\,dx \qquad 9.45$$

$$(\text{second moment})_x = \iiint (y^2 + z^2) f(x, y, z)\,dy\,dz\,dx \qquad 9.46$$

15. FOURIER SERIES

Any periodic waveform can be written as the sum of an infinite number of sinusoidal terms, known as *harmonic terms* (i.e., an infinite series). Such a sum of terms is known as a *Fourier series*, and the process of finding the terms is *Fourier analysis*. (Extracting the original waveform from the series is known as *Fourier inversion*.) Since most series converge rapidly, it is possible to obtain a good approximation to the original waveform with a limited number of sinusoidal terms.

Fourier's theorem is Eq. 9.47.[5] The object of a Fourier analysis is to determine the coefficients a_n and b_n. The constant a_0 can often be determined by inspection since it is the average value of the waveform.

$$f(t) = a_0 + a_1 \cos \omega t + a_2 \cos 2\omega t + \cdots$$
$$+ b_1 \sin \omega t + b_2 \sin 2\omega t + \cdots \qquad 9.47$$

ω is the *natural (fundamental) frequency* of the waveform. It depends on the actual waveform period, T.

$$\omega = \frac{2\pi}{T} \qquad 9.48$$

To simplify the analysis, the time domain can be normalized to the radian scale. The normalized scale is obtained by dividing all frequencies by ω. Then, the Fourier series becomes

$$f(t) = a_0 + a_1 \cos t + a_2 \cos 2t + \cdots$$
$$+ b_1 \sin t + b_2 \sin 2t + \cdots \qquad 9.49$$

---

[5]The independent variable used in this section is t, since Fourier analysis is most frequently used in the time domain.

The coefficients a_n and b_n are found from the following relationships.

$$a_0 = \frac{1}{2\pi}\int_0^{2\pi} f(t)\,dt$$

$$= \frac{1}{T}\int_0^T f(t)\,dt \qquad 9.50$$

$$a_n = \frac{1}{\pi}\int_0^{2\pi} f(t)\cos nt\,dt$$

$$= \frac{2}{T}\int_0^T f(t)\cos nt\,dt \quad [n \ge 1] \qquad 9.51$$

$$b_n = \frac{1}{\pi}\int_0^{2\pi} f(t)\sin nt\,dt$$

$$= \frac{2}{T}\int_0^T f(t)\sin nt\,dt \quad [n \ge 1] \qquad 9.52$$

While Eq. 9.51 and Eq. 9.52 are always valid, the work of integrating and finding a_n and b_n can be greatly simplified if the waveform is recognized as being symmetrical. Table 9.1 summarizes the simplifications.

Example 9.11

Find the first four terms of a Fourier series that approximates the repetitive step function illustrated.

$$f(t) = \left\{ \begin{array}{ll} 1 & 0 < t < \pi \\ 0 & \pi < t < 2\pi \end{array} \right\}$$

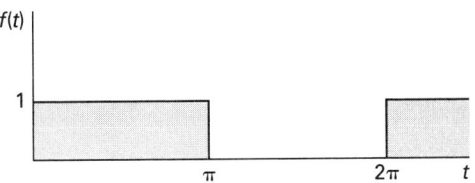

Solution

From Eq. 9.50,

$$a_0 = \frac{1}{2\pi}\int_0^{\pi}(1)\,dt + \frac{1}{2\pi}\int_{\pi}^{2\pi}(0)\,dt = \frac{1}{2}$$

This value of $1/2$ corresponds to the average value of $f(t)$. It could have been found by observation.

$$a_1 = \frac{1}{\pi}\int_0^{\pi}(1)\cos t\,dt + \frac{1}{\pi}\int_{\pi}^{2\pi}(0)\cos t\,dt$$

$$= \frac{1}{\pi}\sin t\Big|_0^{\pi} + 0$$

$$= 0$$

In general,

$$a_n = \frac{1}{\pi}\frac{\sin nt}{n}\Big|_0^{\pi} = 0$$

Table 9.1 *Fourier Analysis Simplifications for Symmetrical Waveforms*

	even symmetry $f(-t) = f(t)$	odd symmetry $f(-t) = -f(t)$
full-wave symmetry* $f(t + 2\pi) = f(t)$ $\lvert A_2\rvert = \lvert A_1\rvert$ $\lvert A_{\text{total}}\rvert = \lvert A_1\rvert$ *any repeating waveform	$b_n = 0$ [all n] $a_n = \frac{1}{\pi}\int_0^{2\pi} f(t)\cos nt\,dt$ [all n]	$a_0 = 0$ $a_n = 0$ [all n] $b_n = \frac{1}{\pi}\int_0^{2\pi} f(t)\sin nt\,dt$ [all n]
half-wave symmetry* $f(t + \pi) = -f(t)$ $\lvert A_2\rvert = \lvert A_1\rvert$ $\lvert A_{\text{total}}\rvert = 2\lvert A_1\rvert$ *same as rotational symmetry	$a_n = 0$ [even n] $b_n = 0$ [all n] $a_n = \frac{2}{\pi}\int_0^{\pi} f(t)\cos nt\,dt$ [odd n]	$a_0 = 0$ $a_n = 0$ [all n] $b_n = 0$ [even n] $b_n = \frac{2}{\pi}\int_0^{\pi} f(t)\sin nt\,dt$ [odd n]
quarter-wave symmetry $f(t + \pi) = -f(t)$ $\lvert A_2\rvert = \lvert A_1\rvert$ $\lvert A_{\text{total}}\rvert = 4\lvert A_1\rvert$	$a_0 = 0$ $a_n = 0$ [even n] $b_n = 0$ [all n] $a_n = \frac{4}{\pi}\int_0^{\frac{\pi}{2}} f(t)\cos nt\,dt$ [odd n]	$a_0 = 0$ $a_n = 0$ [all n] $b_n = 0$ [even n] $b_n = \frac{4}{\pi}\int_0^{\frac{\pi}{2}} f(t)\sin nt\,dt$ [odd n]

$$b_1 = \frac{1}{\pi}\int_0^{\pi}(1)\sin t\,dt + \frac{1}{\pi}\int_{\pi}^{2\pi}(0)\sin t\,dt$$

$$= \frac{1}{\pi} - \cos t\Big|_0^{\pi}$$

$$= 2/\pi$$

In general,

$$b_n = \frac{1}{\pi}\frac{-\cos nt}{n}\Big|_0^{\pi} = \left\{ \begin{array}{ll} 0 & \text{for } n \text{ even} \\ \frac{2}{\pi n} & \text{for } n \text{ odd} \end{array} \right.$$

The series is

$$f(t) = \frac{1}{2} + \frac{2}{\pi}(\sin t + \tfrac{1}{3}\sin 3t + \tfrac{1}{5}\sin 5t + \cdots)$$

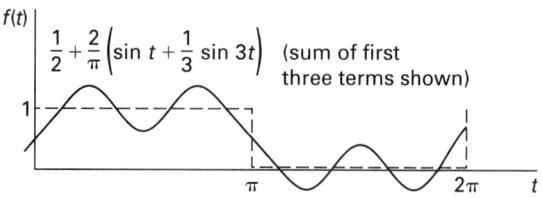

16. FAST FOURIER TRANSFORMS

Many mathematical operations are needed to implement a true Fourier transform. While the terms of a Fourier series might be slowly derived by integration, a faster method is needed to analyze real-time data. The *fast Fourier transform* (FFT) is a computer algorithm implemented in *spectrum analyzers* (*signal analyzers* or *FFT analyzers*) and replaces integration and multiplication operations with table look-ups and additions.[6]

Since the complexity of the transform is reduced, the transformation occurs more quickly, enabling efficient analysis of waveforms with little or no periodicity.[7]

Using a spectrum analyzer requires choosing the frequency band (e.g., 0–20 kHz) to be monitored. (This step automatically selects the sampling period. The lower the frequencies sampled, the longer the sampling period.) If they are not fixed by the analyzer, the numbers of time-dependent input variable samples (e.g., 1024) and frequency-dependent output variable values (e.g., 400) are chosen.[8] There are half as many frequency lines as data points because each line contains two pieces of information—real (amplitude) and imaginary (phase). The *resolution* of the resulting frequency analysis is

$$\text{resolution} = \frac{\text{frequency bandwidth}}{\text{no. of output variable values}} \qquad 9.53$$

17. INTEGRAL FUNCTIONS

Integrals that cannot be evaluated as finite combinations of elementary functions are called *integral functions*. These functions are evaluated by series expansion. Some of the more common functions are listed as follows.[9]

- *integral sine function*

$$\text{Si}(x) = \int_0^x \frac{\sin x}{x}\,dx$$
$$= x - \frac{x^3}{3\cdot 3!} + \frac{x^5}{5\cdot 5!} - \frac{x^7}{7\cdot 7!} + \cdots \qquad 9.54$$

- *integral cosine function*

$$\text{Ci}(x) = \int_{-\infty}^x \frac{\cos x}{x}\,dx = -\int_x^\infty \frac{\cos x}{x}\,dx$$
$$= C_E + \ln x - \frac{x^2}{2\cdot 2!} + \frac{x^4}{4\cdot 4!} - \cdots \qquad 9.55$$

- *integral exponential function*

$$\text{Ei}(x) = \int_{-\infty}^x \frac{e^x}{x}\,dx = -\int_{-x}^\infty \frac{e^{-x}}{x}\,dx$$
$$= C_E + \ln x + x + \frac{x^2}{2\cdot 2!} + \frac{x^3}{3\cdot 3!} + \cdots \qquad 9.56$$

C_E in Eq. 9.55 and Eq. 9.56 is *Euler's constant*.

$$C_E = \int_{+\infty}^0 e^{-x}\ln x\,dx$$
$$= \lim_{m\to\infty}\left(1 + \tfrac{1}{2} + \tfrac{1}{3} + \cdots + \frac{1}{m} - \ln m\right)$$
$$= 0.577215665$$

- *error function*

$$\text{erf}(x) = \frac{2}{\sqrt{\pi}}\int_0^x e^{-x^2}\,dx$$
$$= \left(\frac{2}{\sqrt{\pi}}\right)\left(\frac{x}{1\cdot 0!} - \frac{x^3}{3\cdot 1!} + \frac{x^5}{5\cdot 2!} - \frac{x^7}{7\cdot 3!} + \cdots\right)$$
$$\qquad 9.57$$

---

[6]*Spectrum analysis*, also known as *frequency analysis*, *signature analysis*, and *time-series analysis*, develops a relationship (usually graphical) between some property (e.g., amplitude or phase shift) and frequency.
[7]Hours and days of manual computations are compressed into milliseconds.
[8]Two samples per time-dependent cycle (at the maximum frequency) is the lower theoretical limit for sampling, but the practical minimum rate is approximately 2.5 samples per cycle. This will ensure that *alias components* (i.e., low-level frequency signals) do not show up in the frequency band of interest.

[9]Other integral functions include the Fresnel integral, gamma function, and elliptic integral.

10 Differential Equations

1. TYPES OF DIFFERENTIAL EQUATIONS

A *differential equation* is a mathematical expression combining a function (e.g., $y = f(x)$) and one or more of its derivatives. The *order* of a differential equation is the highest derivative in it. *First-order differential equations* contain only first derivatives of the function, *second-order differential equations* contain second derivatives (and may contain first derivatives as well), and so on.

A *linear differential equation* can be written as a sum of products of multipliers of the function and its derivatives. If the multipliers are scalars, the differential equation is said to have *constant coefficients*. If the function or one of its derivatives is raised to some power (other than one) or is embedded in another function (e.g., y embedded in $\sin y$ or e^y), the equation is said to be *nonlinear*.

Each term of a *homogeneous differential equation* contains either the function (y) or one of its derivatives; that is, the sum of derivative terms is equal to zero. In a *nonhomogeneous differential equation*, the sum of derivative terms is equal to a nonzero *forcing function* of the independent variable (e.g., $g(x)$). In order to solve a nonhomogeneous equation, it is often necessary to solve the homogeneous equation first. The homogeneous equation corresponding to a nonhomogeneous equation is known as a *reduced equation* or *complementary equation*.

The following examples illustrate the types of differential equations.

$y' - 7y = 0$	homogeneous, first-order linear, with constant coefficients
$y'' - 2y' + 8y = \sin 2x$	nonhomogeneous, second-order linear, with constant coefficients
$y'' - (x^2 - 1)y^2 = \sin 4x$	nonhomogeneous, second-order, nonlinear

An *auxiliary equation* (also called the *characteristic equation*) can be written for a homogeneous linear differential equation with constant coefficients, regardless of order. This auxiliary equation is simply the polynomial formed by replacing all derivatives with variables raised to the power of their respective derivatives.

The purpose of solving a differential equation is to derive an expression for the function in terms of the independent variable. The expression does not need to be explicit in the function, but there can be no derivatives in the expression. Since, in the simplest cases, solving a differential equation is equivalent to finding an indefinite integral, it is not surprising that *constants of integration* must be evaluated from knowledge of how the system behaves. Additional data are known as *initial values*, and any problem that includes them is known as an *initial value problem*.[1]

Most differential equations require lengthy solutions and are not efficiently solved by hand. However, several types are fairly simple and are presented in this chapter.

---

[1]The term *initial* implies that time is the independent variable. While this may explain the origin of the term, initial value problems are not limited to the time domain. A *boundary value problem* is similar, except that the data come from different points. For example, additional data in the form $y(x_0)$ and $y'(x_0)$ or $y(x_0)$ and $y'(x_1)$ that need to be simultaneously satisfied constitute an initial value problem. Data of the form $y(x_0)$ and $y(x_1)$ constitute a boundary value problem. Until solved, it is difficult to know whether a boundary value problem has zero, one, or more than one solution.

Example 10.1

Write the complementary differential equation for the following nonhomogeneous differential equation.

$$y'' + 6y' + 9y = e^{-14x} \sin 5x$$

Solution

The complementary equation is found by eliminating the forcing function, $e^{-14x} \sin 5x$.

$$y'' + 6y' + 9y = 0$$

Example 10.2

Write the auxiliary equation to the following differential equation.

$$y'' + 4y' + y = 0$$

Solution

Replacing each derivative with a polynomial term whose degree equals the original order, the auxiliary equation is

$$r^2 + 4r + 1 = 0$$

2. HOMOGENEOUS, FIRST-ORDER LINEAR DIFFERENTIAL EQUATIONS WITH CONSTANT COEFFICIENTS

A homogeneous, first-order linear differential equation with constant coefficients will have the general form of Eq. 10.1.

$$y' + ky = 0 \qquad\qquad 10.1$$

The auxiliary equation is $r + k = 0$ and it has a root of $r = -k$. Equation 10.2 is the solution.

$$y = Ae^{rx} = Ae^{-kx} \qquad\qquad 10.2$$

If the initial condition is known to be $y(0) = y_0$, the solution is

$$y = y_0 e^{-kx} \qquad\qquad 10.3$$

3. FIRST-ORDER LINEAR DIFFERENTIAL EQUATIONS

A first-order linear differential equation has the general form of Eq. 10.4. $p(x)$ and $g(x)$ can be constants or any function of x (but not of y). However, if $p(x)$ is a constant and $g(x)$ is zero, it is easier to solve the equation as shown in Sec. 10.2.

$$y' + p(x)y = g(x) \qquad\qquad 10.4$$

The *integrating factor* (which is usually a function) to this differential equation is

$$u(x) = \exp\left(\int p(x)\,dx \right) \qquad\qquad 10.5$$

The closed-form solution to Eq. 10.4 is

$$y = \frac{1}{u(x)}\left(\int u(x)g(x)\,dx + C \right) \qquad\qquad 10.6$$

For the special case where $p(x)$ and $g(x)$ are both constants, Eq. 10.4 becomes

$$y' + ay = b \qquad\qquad 10.7$$

If the initial condition is $y(0) = y_0$, then the solution to Eq. 10.7 is

$$y = \frac{b}{a}(1 - e^{-ax}) + y_0 e^{-ax} \qquad\qquad 10.8$$

Example 10.3

Find a solution to the following differential equation.

$$y' - y = 2xe^{2x} \qquad y(0) = 1$$

Solution

This is a first-order linear equation with $p(x) = -1$ and $g(x) = 2xe^{2x}$. The integrating factor is

$$u(x) = \exp\left(\int p(x)\,dx \right) = \exp\left(\int -1\,dx \right) = e^{-x}$$

The solution is given by Eq. 10.6.

$$\begin{aligned}
y &= \frac{1}{u(x)}\left(\int u(x)g(x)\,dx + C \right) \\
&= \frac{1}{e^{-x}}\left(\int e^{-x} 2xe^{2x}\,dx + C \right) \\
&= e^x \left(2 \int xe^x\,dx + C \right) \\
&= e^x (2xe^x - 2e^x + C) \\
&= e^x (2e^x(x - 1) + C)
\end{aligned}$$

From the initial condition,

$$y(0) = 1$$
$$e^0\big((2)(e^0)(0 - 1) + C \big) = 1$$
$$1\big((2)(1)(-1) + C \big) = 1$$

Therefore, $C = 3$. The complete solution is

$$y = e^x\big(2e^x(x - 1) + 3 \big)$$

4. FIRST-ORDER SEPARABLE DIFFERENTIAL EQUATIONS

First-order separable differential equations can be placed in the form of Eq. 10.9. For clarity and convenience, y' is written as dy/dx.

$$m(x) + n(y)\frac{dy}{dx} = 0 \qquad 10.9$$

Equation 10.9 can be placed in the form of Eq. 10.10, both sides of which are easily integrated. An initial value will establish the constant of integration.

$$m(x)\,dx = -n(y)\,dy \qquad 10.10$$

5. FIRST-ORDER EXACT DIFFERENTIAL EQUATIONS

A *first-order exact differential equation* has the form

$$f_x(x,y) + f_y(x,y)y' = 0 \qquad 10.11$$

$f_x(x,y)$ is the exact derivative of $f(x,y)$ with respect to x, and $f_y(x,y)$ is the exact derivative of $f(x,y)$ with respect to y. The solution is

$$f(x,y) - C = 0 \qquad 10.12$$

6. HOMOGENEOUS, SECOND-ORDER LINEAR DIFFERENTIAL EQUATIONS WITH CONSTANT COEFFICIENTS

Homogeneous second-order linear differential equations with constant coefficients have the form of Eq. 10.13. They are most easily solved by finding the two roots of the auxiliary equation, Eq. 10.14.

$$y'' + k_1 y' + k_2 y = 0 \qquad 10.13$$
$$r^2 + k_1 r + k_2 = 0 \qquad 10.14$$

There are three cases. If the two roots of Eq. 10.14 are real and different, the solution is

$$y = A_1 e^{r_1 x} + A_2 e^{r_2 x} \qquad 10.15$$

If the two roots are real and the same, the solution is

$$y = A_1 e^{rx} + A_2 x e^{rx} \qquad 10.16$$
$$r = \frac{-k_1}{2} \qquad 10.17$$

If the two roots are imaginary, they will be of the form $(\alpha + i\omega)$ and $(\alpha - i\omega)$, and the solution is

$$y = A_1 e^{\alpha x}\cos\omega x + A_2 e^{\alpha x}\sin\omega x \qquad 10.18$$

In all three cases, A_1 and A_2 must be found from the two initial conditions.

Example 10.4

Solve the following differential equation.

$$y'' + 6y' + 9y = 0$$
$$y(0) = 0 \qquad y'(0) = 1$$

Solution

The auxiliary equation is

$$r^2 + 6r + 9 = 0$$
$$(r+3)(r+3) = 0$$

The roots to the auxiliary equation are $r_1 = r_2 = -3$. Therefore, the solution has the form of Eq. 10.16.

$$y = A_1 e^{-3x} + A_2 x e^{-3x}$$

The first initial condition is

$$y(0) = 0$$
$$A_1 e^0 + A_2(0)e^0 = 0$$
$$A_1 + 0 = 0$$
$$A_1 = 0$$

To use the second initial condition, the derivative of the equation is needed. Making use of the known fact that $A_1 = 0$,

$$y' = \frac{d}{dx}A_2 x e^{-3x} = -3A_2 x e^{-3x} + A_2 e^{-3x}$$

Using the second initial condition,

$$y'(0) = 1$$
$$-3A_2(0)e^0 + A_2 e^0 = 1$$
$$0 + A_2 = 1$$
$$A_2 = 1$$

The solution is

$$y = x e^{-3x}$$

7. NONHOMOGENEOUS DIFFERENTIAL EQUATIONS

A nonhomogeneous equation has the form of Eq. 10.19. $f(x)$ is known as the *forcing function*.

$$y'' + p(x)y' + q(x)y = f(x) \qquad 10.19$$

The solution to Eq. 10.19 is the sum of two equations. The *complementary solution*, y_c, solves the complementary (i.e., homogeneous) problem. The *particular solution*, y_p, is any specific solution to the nonhomogeneous Eq. 10.19 that is known or can be found. Initial values are used to evaluate any unknown coefficients in the complementary solution *after* y_c and y_p have been combined. (The particular solution will not have any unknown coefficients.)

$$y = y_c + y_p \qquad 10.20$$

Two methods are available for finding a particular solution. The *method of undetermined coefficients*, as presented here, can be used only when $p(x)$ and $q(x)$ are constant coefficients and $f(x)$ takes on one of the forms in Table 10.1.

The particular solution can be read from Table 10.1 if the forcing function is of one of the forms given. Of course, the coefficients A_i and B_i are not known—these are the *undetermined coefficients*. The exponent s is the smallest nonnegative number (and will be 0, 1, or 2), which ensures that no term in the particular solution, y_p, is also a solution to the complementary equation, y_c. s must be determined prior to proceeding with the solution procedure.

Once y_p (including s) is known, it is differentiated to obtain y_p' and y_p'', and all three functions are substituted into the original nonhomogeneous equation. The resulting equation is rearranged to match the forcing function, $f(x)$, and the unknown coefficients are determined, usually by solving simultaneous equations.

If the forcing function, $f(x)$, is more complex than the forms shown in Table 10.1, or if either $p(x)$ or $q(x)$ is a function of x, the method of *variation of parameters* should be used. This complex and time-consuming method is not covered in this book.

Table 10.1 *Particular Solutions**

form of $f(x)$	form of y_p
$P_n(x) = a_0 x^n + a_1 x^{n-1}$ $+ \cdots + a_n$	$x^s \left(\begin{array}{c} A_0 x^n + A_1 x^{n-1} \\ + \cdots + A_n \end{array} \right)$
$P_n(x) e^{\alpha x}$	$x^s \left(\begin{array}{c} A_0 x^n + A_1 x^{n-1} \\ + \cdots + A_n \end{array} \right) e^{\alpha x}$
$P_n(x) e^{\alpha x} \left\{ \begin{array}{c} \sin \omega x \\ \cos \omega x \end{array} \right\}$	$x^s \left(\begin{array}{c} \left(\begin{array}{c} A_0 x^n + A_1 x^{n-1} \\ + \cdots + A_n \end{array} \right) \\ \times (e^{\alpha x} \cos \omega x) \\ + \left(\begin{array}{c} B_0 x^n + B_1 x^{n-1} \\ + \cdots + B_n \end{array} \right) \\ \times (e^{\alpha x} \sin \omega x) \end{array} \right)$

* $P_n(x)$ is a polynomial of degree n.

Example 10.5

Solve the following nonhomogeneous differential equation.

$$y'' + 2y' + y = e^x \cos x$$

Solution

step 1: Find the solution to the complementary (homogeneous) differential equation.

$$y'' + 2y' + y = 0$$

Since this is a differential equation with constant coefficients, write the auxiliary equation.

$$r^2 + 2r + 1 = 0$$

The auxiliary equation factors in $(r + 1)^2 = 0$ with two identical roots at $r = -1$. Therefore, the solution to the homogeneous differential equation is

$$y_c(x) = C_1 e^{-x} + C_2 x e^{-x}$$

step 2: Use Table 10.1 to determine the form of a particular solution. Since the forcing function has the form $P_n(x) e^{\alpha x} \cos \omega x$ with $P_n(x) = 1$ (equivalent to $n = 0$), $\alpha = 1$, and $\omega = 1$, the particular solution has the form

$$y_p(x) = x^s (A e^x \cos x + B e^x \sin x)$$

step 3: Determine the value of s. Check to see if any of the terms in $y_p(x)$ will themselves solve the homogeneous equation. Try $A e^x \cos x$ first.

$$\frac{d}{dx}(A e^x \cos x) = A e^x \cos x - A e^x \sin x$$

$$\frac{d^2}{dx^2}(A e^x \cos x) = -2A e^x \sin x$$

Substitute these quantities into the homogeneous equation.

$$y'' + 2y' + y = 0$$
$$-2A e^x \sin x + 2A e^x \cos x$$
$$- 2A e^x \sin x + A e^x \cos x = 0$$
$$3A e^x \cos x - 4A e^x \sin x = 0$$

Disregarding the trivial (i.e., $A = 0$) solution, $A e^x \cos x$ does not solve the homogeneous equation.

Next, try $B e^x \sin x$.

$$\frac{d}{dx}(B e^x \sin x) = B e^x \cos x + B e^x \sin x$$

$$\frac{d^2}{dx^2}(B e^x \sin x) = 2B e^x \cos x$$

Substitute these quantities into the homogeneous equation.

$$y'' + 2y' + y = 0$$

$$2Be^x \cos x + 2Be^x \cos x$$
$$+ 2Be^x \sin x + Be^x \sin x = 0$$

$$3Be^x \sin x + 4Be^x \cos x = 0$$

Disregarding the trivial $(B = 0)$ case, $Be^x \sin x$ does not solve the homogeneous equation.

Since none of the terms in $y_p(x)$ solve the homogeneous equation, $s = 0$, and a particular solution has the form

$$y_p(x) = Ae^x \cos x + Be^x \sin x$$

step 4: Use the method of unknown coefficients to determine A and B in the particular solution. Drawing on the previous steps, substitute the quantities derived from the particular solution into the nonhomogeneous equation.

$$y'' + 2y' + y = e^x \cos x$$

$$-2Ae^x \sin x + 2Be^x \cos x$$
$$+ 2Ae^x \cos x - 2Ae^x \sin x$$
$$+ 2Be^x \cos x + 2Be^x \sin x$$
$$+ Ae^x \cos x + Be^x \sin x = e^x \cos x$$

Combining terms,

$$(-4A + 3B)e^x \sin x + (3A + 4B)e^x \cos x$$
$$= e^x \cos x$$

Equating the coefficients of like terms on either side of the equal sign results in the following simultaneous equations.

$$-4A + 3B = 0$$

$$3A + 4B = 1$$

The solution to these equations is

$$A = \frac{3}{25}$$

$$B = \frac{4}{25}$$

A particular solution is

$$y_p(x) = \tfrac{3}{25} e^x \cos x + \tfrac{4}{25} e^x \sin x$$

step 5: Write the general solution.

$$y(x) = y_c(x) + y_p(x)$$
$$= C_1 e^{-x} + C_2 x e^{-x} + \tfrac{3}{25} e^x \cos x$$
$$+ \tfrac{4}{25} e^x \sin x$$

The values of C_1 and C_2 would be determined at this time if initial conditions were known.

8. NAMED DIFFERENTIAL EQUATIONS

Some differential equations with specific forms are named after the individuals who developed solution techniques for them.

- *Bessel equation of order ν*

$$x^2 y'' + xy' + (x^2 - \nu^2)y = 0 \qquad \text{10.21}$$

- *Cauchy equation*

$$a_0 x^n \frac{d^n y}{dx^n} + a_1 x^{n-1} \frac{d^{n-1}y}{dx^{n-1}} + \cdots$$
$$+ a_{n-1} x \frac{dy}{dx} + a_n y = f(x) \qquad \text{10.22}$$

- *Euler's equation*

$$x^2 y'' + \alpha xy' + \beta y = 0 \qquad \text{10.23}$$

- *Gauss' hypergeometric equation*

$$x(1-x)y'' + \big(c - (a+b+1)x\big)y' - aby = 0 \qquad \text{10.24}$$

- *Legendre equation of order λ*

$$(1 - x^2)y'' - 2xy' + \lambda(\lambda + 1)y = 0 \quad [-1 < x < 1]$$
$$\text{10.25}$$

9. LAPLACE TRANSFORMS

Traditional methods of solving nonhomogeneous differential equations by hand are usually difficult and/or time consuming. *Laplace transforms* can be used to reduce many solution procedures to simple algebra.

Every mathematical function, $f(t)$, for which Eq. 10.26 exists has a Laplace transform, written as $\mathcal{L}(f)$ or $F(s)$. The transform is written in the s-domain, regardless of the independent variable in the original function.[2] (The variable s is equivalent to a derivative operator, although it may be handled in the equations as a simple

---

[2]It is traditional to write the original function as a function of the independent variable t rather than x. However, Laplace transforms are not limited to functions of time.

variable.) Equation 10.26 converts a function into a Laplace transform.

$$\mathcal{L}\big(f(t)\big) = F(s) = \int_0^\infty e^{-st} f(t)\, dt \qquad 10.26$$

Equation 10.26 is not often needed because tables of transforms are readily available. (Appendix 10.A contains some of the most common transforms.)

Extracting a function from its transform is the *inverse Laplace transform* operation. Although other methods exist, this operation is almost always done by finding the transform in a set of tables.[3]

$$f(t) = \mathcal{L}^{-1}\big(F(s)\big) \qquad 10.27$$

Example 10.6

Find the Laplace transform of the following function.

$$f(t) = e^{at} \qquad [s > a]$$

Solution

Applying Eq. 10.26,

$$\mathcal{L}(e^{at}) = \int_0^\infty e^{-st} e^{at}\, dt = \int_0^\infty e^{-(s-a)t}\, dt$$

$$= -\frac{e^{-(s-a)t}}{s-a}\bigg|_0^\infty$$

$$= \frac{1}{s-a} \qquad [s > a]$$

10. STEP AND IMPULSE FUNCTIONS

Many forcing functions are sinusoidal or exponential in nature; others, however, can only be represented by a step or impulse function. A *unit step function*, u_t, is a function describing the disturbance of magnitude 1 that is not present before time t but is suddenly there after time t. A step of magnitude 5 at time $t = 3$ would be represented as $5u_3$. (The notation $5u(t-3)$ is used in some books.)

The *unit impulse function*, δ_t, is a function describing a disturbance of magnitude 1 that is applied and removed so quickly as to be instantaneous. An impulse of magnitude 5 at time 3 would be represented by $5\delta_3$. (The notation $5\delta(t-3)$ is used in some books.)

Example 10.7

What is the notation for a forcing function of magnitude 6 that is applied at $t = 2$ and that is completely removed at $t = 7$?

Solution

The notation is $f(t) = 6(u_2 - u_7)$.

---

[3]Other methods include integration in the complex plane, convolution, and simplification by partial fractions.

Example 10.8

Find the Laplace transform of u_0, a unit step at $t = 0$.

$$f(t) = 0 \text{ for } t < 0$$

$$f(t) = 1 \text{ for } t \geq 0$$

Solution

Since the Laplace transform is an integral that starts at $t = 0$, the value of $f(t)$ prior to $t = 0$ is irrelevant.

$$\mathcal{L}(u_0) = \int_0^\infty e^{-st}(1)\, dt = -\frac{e^{-st}}{s}\bigg|_0^\infty$$

$$= 0 - \frac{-1}{s}$$

$$= \frac{1}{s}$$

11. ALGEBRA OF LAPLACE TRANSFORMS

Equations containing Laplace transforms can be simplified by applying the following principles.

- *linearity theorem* (c is a constant.)

$$\mathcal{L}\big(cf(t)\big) = c\mathcal{L}\big(f(t)\big) = cF(s) \qquad 10.28$$

- *superposition theorem* ($f(t)$ and $g(t)$ are different functions.)

$$\mathcal{L}\big(f(t) \pm g(t)\big) = \mathcal{L}\big(f(t)\big) \pm \mathcal{L}\big(g(t)\big)$$
$$= F(s) \pm G(s) \qquad 10.29$$

- *time-shifting theorem* (*delay theorem*)

$$\mathcal{L}\big(f(t-b)u_b\big) = e^{-bs}F(s) \qquad 10.30$$

- *Laplace transform of a derivative*

$$\mathcal{L}\big(f^n(t)\big) = -f^{n-1}(0) - sf^{n-2}(0) - \cdots$$
$$- s^{n-1}f(0) + s^n F(s) \qquad 10.31$$

- *other properties*

$$\mathcal{L}\left(\int_0^t f(u)\, du\right) = \frac{1}{s}F(s) \qquad 10.32$$

$$\mathcal{L}\big(tf(t)\big) = -\frac{dF}{ds} \qquad 10.33$$

$$\mathcal{L}\left(\frac{1}{t}f(t)\right) = \int_0^\infty F(u)\, du \qquad 10.34$$

12. CONVOLUTION INTEGRAL

A complex Laplace transform, $F(s)$, will often be recognized as the product of two other transforms, $F_1(s)$ and

$F_2(s)$, whose corresponding functions $f_1(t)$ and $f_2(t)$ are known. Unfortunately, Laplace transforms cannot be computed with ordinary multiplication. That is, $f(t) \neq f_1(t)f_2(t)$ even though $F(s) = F_1(s)F_2(s)$.

However, it is possible to extract $f(t)$ from its *convolution*, $h(t)$, as calculated from either of the *convolution integrals* in Eq. 10.35. This process is demonstrated in Ex. 10.9. χ is a dummy variable.

$$f(t) = \mathcal{L}^{-1}\big(F_1(s)F_2(s)\big)$$
$$= \int_0^t f_1(t-\chi)f_2(\chi)\,d\chi$$
$$= \int_0^t f_1(\chi)f_2(t-\chi)\,d\chi \qquad \text{10.35}$$

Example 10.9

Use the convolution integral to find the inverse transform of

$$F(s) = \frac{3}{s^2(s^2+9)}$$

Solution

$F(s)$ can be factored as

$$F_1(s)F_2(s) = \left(\frac{1}{s^2}\right)\left(\frac{3}{s^2+9}\right)$$

As the inverse transforms of $F_1(s)$ and $F_2(s)$ are $f_1(t) = t$ and $f_2(t) = \sin 3t$, respectively, the convolution integral from Eq. 10.35 is

$$f(t) = \int_0^t (t-\chi)\sin 3\chi\,d\chi$$
$$= \int_0^t (t\sin 3\chi - \chi\sin 3\chi)\,d\chi$$
$$= t\int_0^t \sin 3\chi\,d\chi - \int_0^t \chi\sin 3\chi\,d\chi$$

Expand using integration by parts.

$$f(t) = -\tfrac{1}{3}t\cos 3\chi + \tfrac{1}{3}\chi\cos 3\chi - \tfrac{1}{9}\sin 3\chi\Big|_0^t$$
$$= \frac{3t - \sin 3t}{9}$$

13. USING LAPLACE TRANSFORMS

Any nonhomogeneous linear differential equation with constant coefficients can be solved with the following procedure, which reduces the solution to simple algebra. A complete table of transforms simplifies or eliminates step 5.

step 1: Put the differential equation in standard form (i.e., isolate the y'' term).

$$y'' + k_1 y' + k_2 y = f(t) \qquad \text{10.36}$$

step 2: Take the Laplace transform of both sides. Use the linearity and superposition theorems. (See Eq. 10.28 and Eq. 10.29.)

$$\mathcal{L}(y'') + k_1\mathcal{L}(y') + k_2\mathcal{L}(y) = \mathcal{L}\big(f(t)\big) \qquad \text{10.37}$$

step 3: Use Eq. 10.38 and Eq. 10.39 to expand the equation. (These are specific forms of Eq. 10.31.) Use a table to evaluate the transform of the forcing function.

$$\mathcal{L}(y'') = s^2\mathcal{L}(y) - sy(0) - y'(0) \qquad \text{10.38}$$
$$\mathcal{L}(y') = s\mathcal{L}(y) - y(0) \qquad \text{10.39}$$

step 4: Use algebra to solve for $\mathcal{L}(y)$.

step 5: If needed, use partial fractions to simplify the expression for $\mathcal{L}(y)$.

step 6: Take the inverse transform to find $y(t)$.

$$y(t) = \mathcal{L}^{-1}\big(\mathcal{L}(y)\big) \qquad \text{10.40}$$

Example 10.10

Find $y(t)$ for the following differential equation.

$$y'' + 2y' + 2y = \cos t$$
$$y(0) = 1 \qquad y'(0) = 0$$

Solution

step 1: The equation is already in standard form.

step 2: $\mathcal{L}(y'') + 2\mathcal{L}(y') + 2\mathcal{L}(y) = \mathcal{L}(\cos t)$

step 3: Use Eq. 10.38 and Eq. 10.39. Use App. 10.A to find the transform of $\cos t$.

$$s^2\mathcal{L}(y) - sy(0) - y'(0) + 2s\mathcal{L}(y) - 2y(0) + 2\mathcal{L}(y)$$
$$= \frac{s}{s^2+1}$$

But, $y(0) = 1$ and $y'(0) = 0$.

$$s^2\mathcal{L}(y) - s + 2s\mathcal{L}(y) - 2 + 2\mathcal{L}(y) = \frac{s}{s^2+1}$$

step 4: Combine terms and solve for $\mathcal{L}(y)$.

$$\mathcal{L}(y)(s^2 + 2s + 2) - s - 2 = \frac{s}{s^2+1}$$

$$\mathcal{L}(y) = \frac{\dfrac{s}{s^2+1} + s + 2}{s^2 + 2s + 2}$$
$$= \frac{s^3 + 2s^2 + 2s + 2}{(s^2+1)(s^2+2s+2)}$$

step 5: Expand the expression for $\mathcal{L}(y)$ by partial fractions.

$$
\begin{aligned}
\mathcal{L}(y) &= \frac{s^3 + 2s^2 + 2s + 2}{(s^2+1)(s^2+2s+2)} \\
&= \frac{A_1 s + B_1}{s^2+1} + \frac{A_2 s + B_2}{s^2+2s+2} \\
&= \frac{\begin{aligned}s^3(A_1 + A_2) + s^2(2A_1 + B_1 + B_2) \\ + s(2A_1 + 2B_1 + A_2) + (2B_1 + B_2)\end{aligned}}{(s^2+1)(s^2+2s+2)}
\end{aligned}
$$

The following simultaneous equations result.

$$
\begin{aligned}
A_1 &+ A_2 & & & &= 1 \\
2A_1 & & &+ B_1 &+ B_2 &= 2 \\
2A_1 &+ A_2 &+ 2B_1 & & &= 2 \\
& & & 2B_1 &+ B_2 &= 2
\end{aligned}
$$

These equations have the solutions $A_1 = 1/5$, $A_2 = 4/5$, $B_1 = 2/5$, and $B_2 = 6/5$.

step 6: Refer to App. 10.A and take the inverse transforms. The numerator of the second term is rewritten from $(4s+6)$ to $((4s+4)+2)$.

$$
\begin{aligned}
y &= \mathcal{L}^{-1}\Big(\mathcal{L}(y)\Big) \\
&= \mathcal{L}^{-1}\left(\frac{\left(\frac{1}{5}\right)(s+2)}{s^2+1} + \frac{\left(\frac{1}{5}\right)(4s+6)}{s^2+2s+2}\right) \\
&= \left(\tfrac{1}{5}\right)\left(\begin{array}{l} \mathcal{L}^{-1}\left(\dfrac{s}{s^2+1}\right) + 2\mathcal{L}^{-1}\left(\dfrac{1}{s^2+1}\right) \\[2ex] + 4\mathcal{L}^{-1}\left(\dfrac{s-(-1)}{\big(s-(-1)\big)^2+1}\right) \\[2ex] + 2\mathcal{L}^{-1}\left(\dfrac{1}{\big(s-(-1)\big)^2+1}\right) \end{array}\right) \\
&= \left(\tfrac{1}{5}\right)(\cos t + 2\sin t + 4e^{-t}\cos t + 2e^{-t}\sin t)
\end{aligned}
$$

14. THIRD- AND HIGHER-ORDER LINEAR DIFFERENTIAL EQUATIONS WITH CONSTANT COEFFICIENTS

The solutions of third- and higher-order linear differential equations with constant coefficients are extensions of the solutions for second-order equations of this type. Specifically, if an equation is homogeneous, the auxiliary equation is written and its roots are found. If the equation is nonhomogeneous, Laplace transforms can be used to simplify the solution.

Consider the following homogeneous differential equation with constant coefficients.

$$y^n + k_1 y^{n-1} + \cdots + k_{n-1}y' + k_n y = 0 \qquad 10.41$$

The auxiliary equation to Eq. 10.41 is

$$r^n + k_1 r^{n-1} + \cdots + k_{n-1}r + k_n = 0 \qquad 10.42$$

For each real and distinct root r, the solution contains the term

$$y = Ae^{rx} \qquad 10.43$$

For each real root r that repeats m times, the solution contains the term

$$y = (A_1 + A_2 x + A_3 x^2 + \cdots + A_m x^{m-1})e^{rx} \qquad 10.44$$

For each pair of complex roots of the form $r = \alpha \pm i\omega$ the solution contains the terms

$$y = e^{\alpha x}(A_1 \sin \omega x + A_2 \cos \omega x) \qquad 10.45$$

15. APPLICATION: ENGINEERING SYSTEMS

There is a wide variety of engineering systems (mechanical, electrical, fluid flow, heat transfer, and so on) whose behavior is described by linear differential equations with constant coefficients.

16. APPLICATION: MIXING

A typical mixing problem involves a liquid-filled tank. The liquid may initially be pure or contain some solute. Liquid (either pure or as a solution) enters the tank at a known rate. A drain may be present to remove thoroughly mixed liquid. The concentration of the solution (or, equivalently, the amount of solute in the tank) at some given time is generally unknown. (See Fig. 10.1.)

If $m(t)$ is the mass of solute in the tank at time t, the rate of solute change will be $m'(t)$. If the solute is being added at the rate of $a(t)$ and being removed at the rate of $r(t)$, the rate of change is

$$
\begin{aligned}
m'(t) &= \text{rate of addition} - \text{rate of removal} \\
&= a(t) - r(t) \qquad 10.46
\end{aligned}
$$

The rate of solute addition $a(t)$ must be known and, in fact, may be constant or zero. However, $r(t)$ depends on the concentration, $c(t)$, of the mixture and volumetric flow rates at time t. If $o(t)$ is the volumetric flow rate out of the tank, then

$$r(t) = c(t)o(t) \qquad 10.47$$

Figure 10.1 *Fluid Mixture Problem*

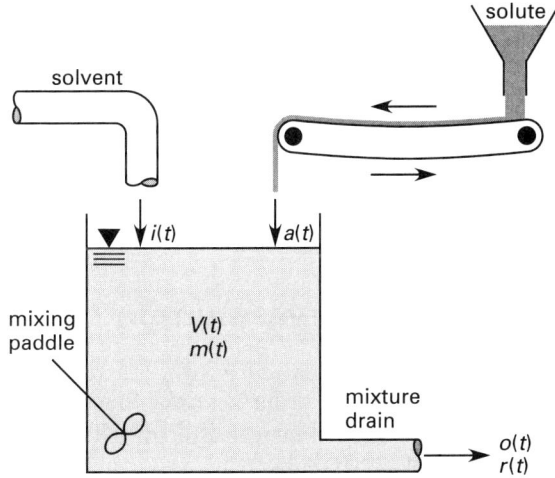

However, the concentration depends on the mass of solute in the tank at time t. Recognizing that the volume, $V(t)$, of the liquid in the tank may be changing with time,

$$c(t) = \frac{m(t)}{V(t)} \qquad 10.48$$

The differential equation describing this problem is

$$m'(t) = a(t) - \frac{m(t)o(t)}{V(t)} \qquad 10.49$$

Example 10.11

A tank contains 100 gal of pure water at the beginning of an experiment. Pure water flows into the tank at a rate of 1 gal/min. Brine containing $\frac{1}{4}$ lbm of salt per gallon enters the tank from a second source at a rate of 1 gal/min. A perfectly mixed solution drains from the tank at a rate of 2 gal/min. How much salt is in the tank 8 min after the experiment begins?

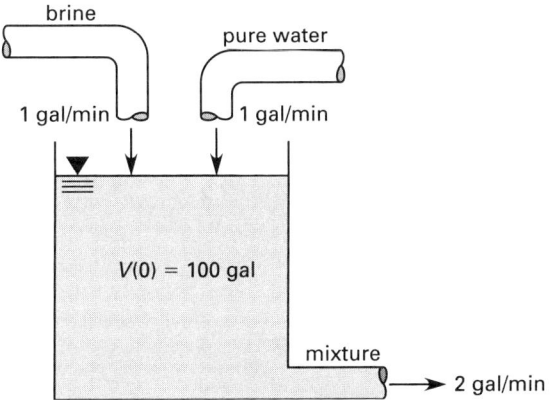

Solution

Let $m(t)$ represent the mass of salt in the tank at time t. 0.25 lbm of salt enters the tank per minute (that is, $a(t) = 0.25$ lbm/min). The salt removal rate depends on the concentration in the tank. That is,

$$r(t) = o(t)c(t) = \left(2 \ \frac{\text{gal}}{\text{min}}\right)\left(\frac{m(t)}{100 \ \text{gal}}\right)$$

$$= \left(0.02 \ \frac{1}{\text{min}}\right)m(t)$$

From Eq. 10.46, the rate of change of salt in the tank is

$$m'(t) = a(t) - r(t)$$

$$= 0.25 \ \frac{\text{lbm}}{\text{min}} - \left(0.02 \ \frac{1}{\text{min}}\right)m(t)$$

$$m'(t) + \left(0.02 \ \frac{1}{\text{min}}\right)m(t) = 0.25 \ \text{lbm/min}$$

This is a first-order linear differential equation of the form of Eq. 10.7. Since the initial condition is $m(0) = 0$, the solution is

$$m(t) = \left(\frac{0.25 \ \frac{\text{lbm}}{\text{min}}}{0.02 \ \frac{1}{\text{min}}}\right)\left(1 - e^{-\left(0.02 \frac{1}{\text{min}}\right)t}\right)$$

$$= (12.5 \ \text{lbm})\left(1 - e^{-\left(0.02 \frac{1}{\text{min}}\right)t}\right)$$

At $t = 8$,

$$m(t) = (12.5 \ \text{lbm})\left(1 - e^{-\left(0.02 \frac{1}{\text{min}}\right)(8 \ \text{min})}\right)$$

$$= (12.5 \ \text{lbm})(1 - 0.852)$$

$$= 1.85 \ \text{lbm}$$

17. APPLICATION: EXPONENTIAL GROWTH AND DECAY

Equation 10.50 describes the behavior of a substance (e.g., radioactive and irradiated molecules) whose quantity, $m(t)$, changes at a rate proportional to the quantity present. The constant of proportionality, k, will be negative for decay (e.g., *radioactive decay*) and positive for growth (e.g., compound interest).

$$m'(t) = km(t) \qquad 10.50$$

$$m'(t) - km(t) = 0 \qquad 10.51$$

If the initial quantity of substance is $m(0) = m_0$, then Eq. 10.51 has the solution

$$m(t) = m_0 e^{kt} \qquad 10.52$$

If $m(t)$ is known for some time t, the constant of proportionality is

$$k = \frac{1}{t} \ln\left(\frac{m(t)}{m_0}\right) \qquad 10.53$$

For the case of a decay, the *half-life*, $t_{1/2}$, is the time at which only half of the substance remains. The relationship between k and $t_{1/2}$ is

$$kt_{1/2} = \ln\tfrac{1}{2} = -0.693 \qquad 10.54$$

18. APPLICATION: EPIDEMICS

During an epidemic in a population of n people, the density of sick (contaminated, contagious, affected, etc.) individuals is $\rho_s(t) = s(t)/n$, where $s(t)$ is the number of sick individuals at a given time, t. Similarly, the density of well (uncontaminated, unaffected, susceptible, etc.) individuals is $\rho_w(t) = w(t)/n$, where $w(t)$ is the number of well individuals. Assuming there is no quarantine, the population size is constant, individuals move about freely, and sickness does not limit the activities of individuals, the rate of contagion, $\rho_s'(t)$, will be $k\rho_s(t)\rho_w(t)$, where k is a proportionality constant.

$$\rho_s'(t) = k\rho_s(t)\rho_w(t) = k\rho_s(t)\big(1 - \rho_s(t)\big) \qquad 10.55$$

This is a separable differential equation that has the solution

$$\rho_s(t) = \frac{\rho_s(0)}{\rho_s(0) + \big(1 - \rho_s(0)\big)e^{-kt}} \qquad 10.56$$

19. APPLICATION: SURFACE TEMPERATURE

Newton's law of cooling states that the surface temperature, T, of a cooling object changes at a rate proportional to the difference between the surface and ambient temperatures. The constant k is a positive number.

$$T'(t) = -k\big(T(t) - T_{\text{ambient}}\big) \quad [k > 0] \qquad 10.57$$

$$T'(t) + kT(t) - kT_{\text{ambient}} = 0 \quad [k > 0] \qquad 10.58$$

This first-order linear differential equation with constant coefficients has the following solution (from Eq. 10.8).

$$T(t) = T_{\text{ambient}} + \big(T(0) - T_{\text{ambient}}\big)e^{-kt} \qquad 10.59$$

If the temperature is known at some time t, the constant k can be found from Eq. 10.60.

$$k = \frac{-1}{t} \ln\left(\frac{T(t) - T_{\text{ambient}}}{T(0) - T_{\text{ambient}}}\right) \qquad 10.60$$

20. APPLICATION: EVAPORATION

The mass of liquid evaporated from a liquid surface is proportional to the exposed surface area. Since quantity, mass, and remaining volume are all proportional, the differential equation is

$$\frac{dV}{dt} = -kA \qquad 10.61$$

For a spherical drop of radius r, Eq. 10.61 reduces to

$$\frac{dr}{dt} = -k \qquad 10.62$$

$$r(t) = r(0) - kt \qquad 10.63$$

For a cube with sides of length s, Eq. 10.61 reduces to

$$\frac{ds}{dt} = -2k \qquad 10.64$$

$$s(t) = s(0) - 2kt \qquad 10.65$$

11 Probability, Statistical Analysis of Data, and Risk Analysis

Nomenclature

ABS	dust absorption rate, decimal fraction	–
AT	averaging time (same as ED for noncarcinogens, 70 yr for carcinogens)	d
BW	body mass	kg
C	concentration	mg/L water or mg/m^3 air
CDI	chronic daily intake (also known as intake)	mg/kg·d
CR	contact rate	L/d water or m^3/d air
ED	exposure duration	yr
EF	exposure frequency	d/yr
HI	hazard index, decimal ratio	–
R	lifetime risk (decimal probability) of excess cancer	–
RfD	chronic reference dose (also known as acceptable daily intake) (for noncarcinogens)	mg/kg·d
RR	dust retention rate, decimal fraction	–
SF	slope factor (for carcinogens)	(mg/kg·d)$^{-1}$
TS	toxicity score	kg·d/L

Subscripts

a/w	air or water exposure
c	carcinogens
dust	dust exposure
ep	measured at exposure point
n	noncarcinogens
max	maximum
w	well

1. SET THEORY

A *set* (usually designated by a capital letter) is a population or collection of individual items known as *elements* or *members*. The *null set*, Ø, is empty (i.e., contains no members). If A and B are two sets, A is a *subset* of B if every member in A is also in B. A is a *proper subset* of B if B consists of more than the elements in A. These relationships are denoted as follows.

$$A \subseteq B \quad [\text{subset}]$$

$$A \subset B \quad [\text{proper subset}]$$

The *universal set*, U, is one from which other sets draw their members. If A is a subset of U, then A' (also designated as A^{-1}, $\tilde{A}$, $-A$, and $\overline{A}$) is the *complement* of A and consists of all elements in U that are not in A. This is illustrated by the *Venn diagram* in Fig. 11.1(a).

The *union of two sets*, denoted by $A \cup B$ and shown in Fig. 11.1(b), is the set of all elements that are either in A or B or both. The *intersection of two sets*, denoted by $A \cap B$ and shown in Fig. 11.1(c), is the set of all elements that belong to both A and B. If $A \cap B = Ø$, A and B are said to be *disjoint sets*.

Figure 11.1 *Venn Diagrams*

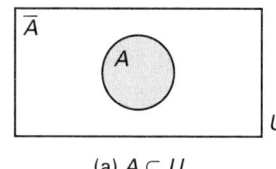

(a) $A \subset U$

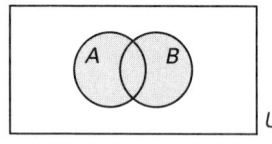

(b) $A \cup B$

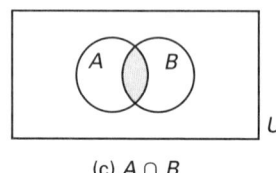

(c) $A \cap B$

If A, B, and C are subsets of the universal set, the following laws apply.

- *identity laws*

$$A \cup \emptyset = A \qquad 11.1$$
$$A \cup U = U \qquad 11.2$$
$$A \cap \emptyset = \emptyset \qquad 11.3$$
$$A \cap U = A \qquad 11.4$$

- *idempotent laws*

$$A \cup A = A \qquad 11.5$$
$$A \cap A = A \qquad 11.6$$

- *complement laws*

$$A \cup A' = U \qquad 11.7$$
$$(A')' = A \qquad 11.8$$
$$A \cap A' = \emptyset \qquad 11.9$$
$$U' = \emptyset \qquad 11.10$$

- *commutative laws*

$$A \cup B = B \cup A \qquad 11.11$$
$$A \cap B = B \cap A \qquad 11.12$$

- *associative laws*

$$(A \cup B) \cup C = A \cup (B \cup C) \qquad 11.13$$
$$(A \cap B) \cap C = A \cap (B \cap C) \qquad 11.14$$

- *distributive laws*

$$A \cup (B \cap C) = (A \cup B) \cap (A \cup C) \qquad 11.15$$
$$A \cap (B \cup C) = (A \cap B) \cup (A \cap C) \qquad 11.16$$

- *de Morgan's laws*

$$(A \cup B)' = A' \cap B' \qquad 11.17$$
$$(A \cap B)' = A' \cup B' \qquad 11.18$$

2. COMBINATIONS OF ELEMENTS

There are a finite number of ways in which n elements can be combined into distinctly different groups of r items. For example, suppose a farmer has a hen, a rooster, a duck, and a cage that holds only two birds. The possible *combinations* of three birds taken two at a time are (hen, rooster), (hen, duck), and (rooster, duck). The birds in the cage will not remain stationary, and the combination (rooster, hen) is not distinctly different from (hen, rooster). That is, the groups are not *order conscious*.

The number of combinations of n items taken r at a time is written $C(n, r)$, C_r^n, $_nC_r$, or $\binom{n}{r}$ (pronounced "n choose r") and given by Eq. 11.19. It is sometimes referred to as the *binomial coefficient*.

$$\binom{n}{r} = C(n, r) = \frac{n!}{(n-r)!r!} \quad \text{[for } r \leq n\text{]} \qquad 11.19$$

Example 11.1

Six people are on a sinking yacht. There are four life jackets. How many combinations of survivors are there?

Solution

The groups are not order conscious. From Eq. 11.19,

$$C(6,4) = \frac{n!}{(n-r)!r!} = \frac{6!}{(6-4)!4!} = \frac{6 \cdot 5 \cdot 4 \cdot 3 \cdot 2 \cdot 1}{(2 \cdot 1)(4 \cdot 3 \cdot 2 \cdot 1)}$$
$$= 15$$

3. PERMUTATIONS

An order-conscious subset of r items taken from a set of n items is the *permutation* $P(n, r)$, also written P_r^n and $_nP_r$. The permutation is order conscious because the arrangement of two items (say a_i and b_i) as $a_i b_i$ is different from the arrangement $b_i a_i$. The number of permutations is

$$P(n, r) = \frac{n!}{(n-r)!} \quad \text{[for } r \leq n\text{]} \qquad 11.20$$

If groups of the entire set of n items are being enumerated, the number of permutations of n items taken n at a time is

$$P(n, n) = \frac{n!}{(n-n)!} = \frac{n!}{0!} = n! \qquad 11.21$$

A *ring permutation* is a special case of n items taken n at a time. There is no identifiable beginning or end, and the number of permutations is divided by n.

$$P_{\text{ring}}(n, n) = \frac{P(n, n)}{n} = (n-1)! \qquad 11.22$$

Example 11.2

A pianist knows four pieces but will have enough stage time to play only three of them. Pieces played in a different order constitute a different program. How many different programs can be arranged?

Solution

The groups are order conscious. From Eq. 11.20,

$$P(4, 3) = \frac{n!}{(n-r)!} = \frac{4!}{(4-3)!} = \frac{4 \cdot 3 \cdot 2 \cdot 1}{1} = 24$$

Example 11.3

Seven diplomats from different countries enter a circular room. The only furnishings are seven chairs arranged around a circular table. How many ways are there of arranging the diplomats?

Solution

All seven diplomats must be seated, so the groups are permutations of seven objects taken seven at a time. Since there is no head chair, the groups are ring permutations. From Eq. 11.22,

$$P_{\text{ring}}(7, 7) = (7 - 1)! = 6 \cdot 5 \cdot 4 \cdot 3 \cdot 2 \cdot 1$$
$$= 720$$

4. PROBABILITY THEORY

The act of conducting an experiment (trial) or taking a measurement is known as *sampling*. *Probability theory* determines the relative likelihood that a particular event will occur. An *event, e,* is one of the possible outcomes of the *trial*. Taken together, all of the possible events constitute a finite *sample space, $E = [e_1, e_2, \ldots, e_n]$*. The trial is drawn from the *population* or *universe*. Populations can be finite or infinite in size.

Events can be numerical or nonnumerical, discrete or continuous, and dependent or independent. An example of a nonnumerical event is getting tails on a coin toss. The number from a roll of a die is a discrete numerical event. The measured diameter of a bolt produced from an automatic screw machine is a numerical event. Since the diameter can (within reasonable limits) take on any value, its measured value is a continuous numerical event.

An event is *independent* if its outcome is unaffected by previous outcomes (i.e., previous runs of the experiment) and *dependent* otherwise. Whether or not an event is independent depends on the population size and how the sampling is conducted. Sampling (a trial) from an infinite population is implicitly independent. When the population is finite, *sampling with replacement* produces independent events, while *sampling without replacement* changes the population and produces dependent events.

The terms *success* and *failure* are loosely used in probability theory to designate obtaining and not obtaining, respectively, the tested-for condition. "Failure" is not the same as a *null event* (i.e., one that has a zero probability of occurrence).

The *probability* of event e_1 occurring is designated as $p\{e_1\}$ and is calculated as the ratio of the total number of ways the event can occur to the total number of outcomes in the sample space.

Example 11.4

There are 380 students in a rural school—200 girls and 180 boys. One student is chosen at random and is checked for gender and height. (a) Define and categorize the population. (b) Define and categorize the sample space. (c) Define the trials. (d) Define and categorize the events. (e) In determining the probability that the student chosen is a boy, define success and failure. (f) What is the probability that the student is a boy?

Solution

(a) The population consists of 380 students and is finite.

(b) In determining the gender of the student, the sample space consists of the two outcomes $E = $ [girl, boy]. This sample space is nonnumerical and discrete. In determining the height, the sample space consists of a range of values and is numerical and continuous.

(c) The trial is the actual sampling (i.e., the determination of gender and height).

(d) The events are the outcomes of the trials (i.e., the gender and height of the student). These events are independent if each student returns to the population prior to the random selection of the next student; otherwise, the events are dependent.

(e) The event is a success if the student is a boy and is a failure otherwise.

(f) From the definition of probability,

$$p\{\text{boy}\} = \frac{\text{no. of boys}}{\text{no. of students}} = \frac{180}{380} = \frac{9}{19}$$
$$= 0.47$$

5. JOINT PROBABILITY

Joint probability rules specify the probability of a combination of events. If n mutually exclusive events from the set E have probabilities $p\{e_i\}$, the probability of any one of these events occurring in a given trial is the sum of the individual probabilities. The events in Eq. 11.23 come from a single sample space and are linked by the word *or*.

$$p\{e_1 \text{ or } e_2 \text{ or} \cdots \text{or } e_k\} = p\{e_1\} + p\{e_2\} + \cdots + p\{e_k\}$$

$$11.23$$

When given two independent sets of events, E and G, Eq. 11.24 will give the probability that events e_i and g_i will both occur. The events in Eq. 11.24 are independent and are linked by the word *and*.

$$p\{e_i \text{ and } g_i\} = p\{e_i\}p\{g_i\} \qquad 11.24$$

When given two independent sets of events, E and G, Eq. 11.25 will give the probability that either event e_i or g_i will occur. The events in Eq. 11.25 are mutually exclusive and are linked by the word *or*.

$$p\{e_i \text{ or } g_i\} = p\{e_i\} + p\{g_i\} - p\{e_i\}p\{g_i\} \qquad 11.25$$

Example 11.5

A bowl contains five white balls, two red balls, and three green balls. What is the probability of getting either a white ball or a red ball in one draw from the bowl?

Solution

Since the two possible events are mutually exclusive and come from the same sample space, Eq. 11.23 can be used.

$$p\{\text{white or red}\} = p\{\text{white}\} + p\{\text{red}\} = \frac{5}{10} + \frac{2}{10}$$

$$= 7/10$$

Example 11.6

One bowl contains five white balls, two red balls, and three green balls. Another bowl contains three yellow balls and seven black balls. What is the probability of getting a red ball from the first bowl and a yellow ball from the second bowl in one draw from each bowl?

Solution

Equation 11.24 can be used because the two events are independent.

$$p\{\text{red and yellow}\} = p\{\text{red}\}p\{\text{yellow}\}$$

$$= \left(\frac{2}{10}\right)\left(\frac{3}{10}\right)$$

$$= 6/100$$

6. COMPLEMENTARY PROBABILITIES

The probability of an event occurring is equal to one minus the probability of the event not occurring. This is known as *complementary probability*.

$$p\{e_i\} = 1 - p\{\text{not } e_i\} \qquad 11.26$$

Equation 11.26 can be used to simplify some probability calculations. Specifically, calculation of the probability of numerical events being "greater than" or "less than" or quantities being "at least" a certain number can often be simplified by calculating the probability of the complementary event.

Example 11.7

A fair coin is tossed five times.[1] What is the probability of getting at least one tail?

Solution

The probability of getting at least one tail in five tosses could be calculated as

$$p\{\text{at least 1 tail}\} = p\{1 \text{ tail}\} + p\{2 \text{ tails}\}$$

$$+ p\{3 \text{ tails}\} + p\{4 \text{ tails}\}$$

$$+ p\{5 \text{ tails}\}$$

However, it is easier to calculate the complementary probability of getting no tails (i.e., getting all heads).

From Eq. 11.25 and Eq. 11.26 (for calculating the probability of getting no tails in five successive tosses),

$$p\{\text{at least 1 tail}\} = 1 - p\{0 \text{ tails}\}$$

$$= 1 - (0.5)^5$$

$$= 0.96875$$

7. CONDITIONAL PROBABILITY

Given two dependent sets of events, E and G, the probability that event e_k will occur given the fact that the dependent event g has already occurred is written as $p\{e_k|g\}$ and given by *Bayes' theorem*, Eq. 11.27.

$$p\{e_k|g\} = \frac{p\{e_k \text{ and } g\}}{p\{g\}} = \frac{p\{g|e_k\}p\{e_k\}}{\sum\limits_{i=1}^{n} p\{g|e_i\}p\{e_i\}} \qquad 11.27$$

8. PROBABILITY DENSITY FUNCTIONS

A *density function* is a nonnegative function whose integral taken over the entire range of the independent variable is unity. A *probability density function* is a mathematical formula that gives the probability of a

---

[1]It makes no difference whether one coin is tossed five times or five coins are each tossed once.

discrete numerical event occurring. A *discrete numerical event* is an occurrence that can be described (usually) by an integer. For example, 27 cars passing through a bridge toll booth in an hour is a discrete numerical event. Figure 11.2 shows a graph of a typical probability density function.

Figure 11.2 *Probability Density Function*

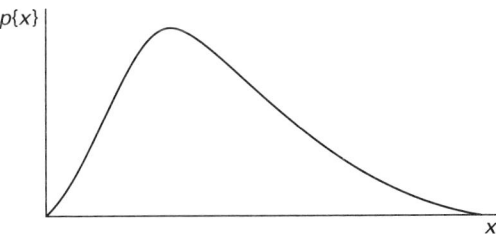

A probability density function, $f(x)$, gives the probability that discrete event x will occur. That is, $p\{x\} = f(x)$. Important discrete probability density functions are the binomial, hypergeometric, and Poisson distributions.

9. BINOMIAL DISTRIBUTION

The *binomial probability density function* (*binomial distribution*) is used when all outcomes can be categorized as either successes or failures. The probability of success in a single trial is $\hat{p}$, and the probability of failure is the complement, $\hat{q} = 1 - \hat{p}$. The population is assumed to be infinite in size so that sampling does not change the values of $\hat{p}$ and $\hat{q}$. (The binomial distribution can also be used with finite populations when sampling with replacement.)

Equation 11.28 gives the probability of x successes in n independent *successive trials*. The quantity $\binom{n}{x}$ is the *binomial coefficient*, identical to the number of combinations of n items taken x at a time.

$$p\{x\} = f(x) = \binom{n}{x}\hat{p}^x\hat{q}^{n-x} \qquad 11.28$$

$$\binom{n}{x} = \frac{n!}{(n-x)!x!} \qquad 11.29$$

Equation 11.28 is a discrete distribution, taking on values only for discrete integer values up to n. The mean, μ, and variance, σ^2 (see Sec. 11.23 and Sec. 11.24), of this distribution are

$$\mu = n\hat{p} \qquad 11.30$$

$$\sigma^2 = n\hat{p}\hat{q} \qquad 11.31$$

Example 11.8

Five percent of a large batch of high-strength steel bolts purchased for bridge construction are defective. (a) If seven bolts are randomly sampled, what is the

probability that exactly three will be defective? (b) What is the probability that two or more bolts will be defective?

Solution

(a) The bolts are either defective or not, so the binomial distribution can be used.

$$\hat{p} = 0.05 \quad [\text{success} = \text{defective}]$$

$$\hat{q} = 1 - 0.05 = 0.95 \quad [\text{failure} = \text{not defective}]$$

From Eq. 11.28,

$$p\{3\} = f(3) = \binom{n}{x}\hat{p}^x\hat{q}^{n-x} = \binom{7}{3}(0.05)^3(0.95)^{7-3}$$

$$= \left(\frac{7\cdot6\cdot5\cdot4\cdot3\cdot2\cdot1}{4\cdot3\cdot2\cdot1\cdot3\cdot2\cdot1}\right)(0.05)^3(0.95)^4$$

$$= 0.00356$$

(b) The probability that two or more bolts will be defective could be calculated as

$$p\{x \geq 2\} = p\{2\} + p\{3\} + p\{4\} + p\{5\} + p\{6\} + p\{7\}$$

This method would require six probability calculations. It is easier to use the complement of the desired probability.

$$p\{x \geq 2\} = 1 - p\{x \leq 1\} = 1 - \Big(p\{0\} + p\{1\}\Big)$$

$$p\{0\} = \binom{n}{x}\hat{p}^x\hat{q}^{n-x} = \binom{7}{0}(0.05)^0(0.95)^7 = (0.95)^7$$

$$p\{1\} = \binom{n}{x}\hat{p}^x\hat{q}^{n-x} = \binom{7}{1}(0.05)^1(0.95)^6$$

$$= (7)(0.05)(0.95)^6$$

$$p\{x \geq 2\} = 1 - \Big((0.95)^7 + (7)(0.05)(0.95)^6\Big)$$

$$= 1 - (0.6983 + 0.2573)$$

$$= 0.0444$$

10. HYPERGEOMETRIC DISTRIBUTION

Probabilities associated with sampling from a finite population without replacement are calculated from the *hypergeometric distribution*. If a population of finite size M contains K items with a given characteristic (e.g., red color, defective construction), then the probability of finding x items with that characteristic in a sample of n items is

$$p\{x\} = f(x) = \frac{\binom{K}{x}\binom{M-K}{n-x}}{\binom{M}{n}} \qquad [\text{for } x \leq n] \qquad 11.32$$

11. MULTIPLE HYPERGEOMETRIC DISTRIBUTION

Sampling without replacement from finite populations containing several different types of items is handled by the *multiple hypergeometric distribution*. If a population of finite size M contains K_i items of type i (such that $\Sigma K_i = M$), the probability of finding x_1 items of type 1, x_2 items of type 2, and so on, in a sample size of n (such that $\Sigma x_i = n$) is

$$p\{x_1, x_2, x_3, \ldots\} = \frac{\binom{K_1}{x_1}\binom{K_2}{x_2}\binom{K_3}{x_3}\cdots}{\binom{M}{n}} \qquad 11.33$$

12. POISSON DISTRIBUTION

Certain discrete events occur relatively infrequently but at a relatively regular rate. The probability of such an event occurring is given by the *Poisson distribution*. Suppose an event occurs, on the average, λ times per period. The probability that the event will occur x times per period is

$$p\{x\} = f(x) = \frac{e^{-\lambda}\lambda^x}{x!} \qquad [\lambda > 0] \qquad 11.34$$

λ is both the mean and the variance of the Poisson distribution.

$$\mu = \lambda \qquad 11.35$$

$$\sigma^2 = \lambda \qquad 11.36$$

Example 11.9

The number of customers arriving at a hamburger stand in the next period is a Poisson distribution having a mean of eight. What is the probability that exactly six customers will arrive in the next period?

Solution

$\lambda = 8$, and $x = 6$. From Eq. 11.34,

$$p\{6\} = \frac{e^{-\lambda}\lambda^x}{x!} = \frac{e^{-8}(8)^6}{6!} = 0.122$$

13. CONTINUOUS DISTRIBUTION FUNCTIONS

Most numerical events are *continuously distributed* and are not constrained to discrete or integer values. For example, the resistance of a 10% 1 Ω resistor may be any value between 0.9 Ω and 1.1 Ω. The probability of an exact numerical event is zero for continuously distributed variables. That is, there is no chance that a numerical event will be *exactly* x.[2] It is possible to determine only the probability that a numerical event will be less than x, greater than x, or between the values of x_1 and x_2, but not exactly equal to x.

Since an expression, $f(x)$, for a probability density function cannot always be written, it is more common to specify the *continuous distribution function*, $F(x_0)$, which gives the probability of numerical event x_0 or less occurring, as illustrated in Fig. 11.3.

$$p\{X < x_0\} = F(x_0) = \int_0^{x_0} f(x)\,dx \qquad 11.37$$

$$f(x) = \frac{dF(x)}{dx} \qquad 11.38$$

Figure 11.3 Continuous Distribution Function

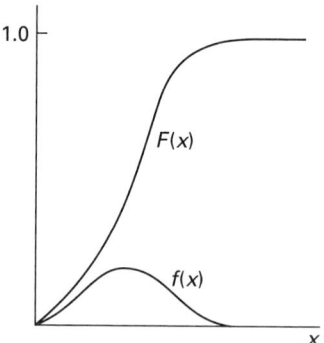

14. EXPONENTIAL DISTRIBUTION

The continuous *exponential distribution* is given by its probability density and continuous distribution functions.

$$f(x) = \lambda e^{-\lambda x} \qquad 11.39$$

$$p\{X < x\} = F(x) = 1 - e^{-\lambda x} \qquad 11.40$$

The mean and variance of the exponential distribution are

$$\mu = \frac{1}{\lambda} \qquad 11.41$$

$$\sigma^2 = \frac{1}{\lambda^2} \qquad 11.42$$

---

[2]It is important to understand the rationale behind this statement. Since the variable can take on any value and has an infinite number of significant digits, we can infinitely continue to increase the precision of the value. For example, the probability is zero that a resistance will be exactly 1 Ω because the resistance is really 1.03 or 1.0260008 or 1.02600080005, and so on.

15. NORMAL DISTRIBUTION

The *normal distribution (Gaussian distribution)* is a symmetrical distribution commonly referred to as the *bell-shaped curve*, which represents the distribution of outcomes of many experiments, processes, and phenomena. (See Fig. 11.4.) The probability density and continuous distribution functions for the normal distribution with mean μ and variance σ^2 are

$$f(x) = \frac{e^{-\frac{1}{2}\left(\frac{x-\mu}{\sigma}\right)^2}}{\sigma\sqrt{2\pi}} \quad [-\infty < x < +\infty] \qquad 11.43$$

$$p\{\mu < X < x_0\} = F(x_0)$$

$$= \frac{1}{\sigma\sqrt{2\pi}} \int_0^{x_0} e^{-\frac{1}{2}\left(\frac{x-\mu}{\sigma}\right)^2} dx \qquad 11.44$$

Figure 11.4 *Normal Distribution*

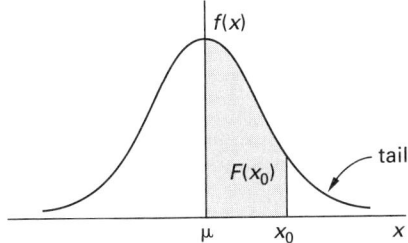

Since $f(x)$ is difficult to integrate, Eq. 11.43 is seldom used directly, and a *standard normal table* is used instead. (See App. 11.A.) The standard normal table is based on a normal distribution with a mean of zero and a standard deviation of 1. Since the range of values from an experiment or phenomenon will not generally correspond to the standard normal table, a value, x_0, must be converted to a *standard normal value*, z. In Eq. 11.45, μ and σ are the mean and standard deviation, respectively, of the distribution from which x_0 comes. For all practical purposes, all normal distributions are completely bounded by $\mu \pm 3\sigma$.

$$z = \frac{x_0 - \mu}{\sigma} \qquad 11.45$$

Numbers in the standard normal table, as given by App. 11.A, are the probabilities of the normalized x being between zero and z and represent the areas under the curve up to point z. When x is less than μ, z will be negative. However, the curve is symmetrical, so the table value corresponding to positive z can be used. The probability of x being greater than z is the complement of the table value. The curve area past point z is known as the *tail of the curve*.

Example 11.10

The mass, m, of a particular hand-laid fiberglass (Fiberglas™) part is normally distributed with a mean of 66 kg and a standard deviation of 5 kg. (a) What percent of the parts will have a mass less than 72 kg? (b) What percent of the parts will have a mass in excess of 72 kg? (c) What percent of the parts will have a mass between 61 kg and 72 kg?

Solution

(a) The 72 kg value must be normalized, so use Eq. 11.45. The standard normal variable is

$$z = \frac{x - \mu}{\sigma} = \frac{72 \text{ kg} - 66 \text{ kg}}{5 \text{ kg}} = 1.2$$

Reading from App. 11.A, the area under the normal curve is 0.3849. This represents the probability of the mass, m, being between 66 kg and 72 kg (i.e., z being between 0 and 1.2). However, the probability of the mass being less than 66 kg is also needed. Since the curve is symmetrical, this probability is 0.5. Therefore,

$$p\{m < 72 \text{ kg}\} = p\{z < 1.2\} = 0.5 + 0.3849 = 0.8849$$

(b) The probability of the mass exceeding 72 kg is the area under the tail past point z.

$$p\{m > 72 \text{ kg}\} = p\{z > 1.2\} = 0.5 - 0.3849 = 0.1151$$

(c) The standard normal variable corresponding to $m = 61$ kg is

$$z = \frac{x - \mu}{\sigma} = \frac{61 \text{ kg} - 66 \text{ kg}}{5 \text{ kg}} = -1$$

Since the two masses are on opposite sides of the mean, the probability will have to be determined in two parts.

$$p\{61 < m < 72\} = p\{61 < m < 66\} + p\{66 < m < 72\}$$
$$= p\{-1 < z < 0\} + p\{0 < z < 1.2\}$$
$$= 0.3413 + 0.3849$$
$$= 0.7262$$

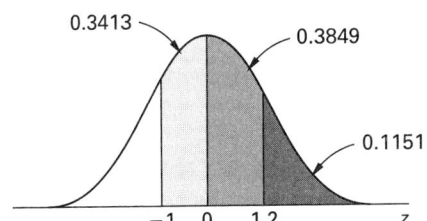

16. STUDENT'S *t*-DISTRIBUTION

In many cases requiring statistical analysis, including setting of confidence limits and hypothesis testing, the true population standard deviation, σ, and mean, μ, are not known. In such cases, the sample standard deviation, s, is used as an estimate for the population standard deviation, and the sample mean, $\bar{x}$, is used to estimate the population mean, μ. To account for the additional uncertainty of not knowing the population parameters exactly, a *t-distribution* is used rather than the normal distribution. The *t*-distribution essentially relaxes or expands the confidence intervals to account for these additional uncertainties.

The *t*-distribution is actually a family of distributions. They are similar in shape (symmetrical and bell-shaped) to the normal distribution, although wider, and flatter in the tails. *t*-distributions are more likely to result in (or, accept) values located farther from the mean than the normal distribution. The specific shape is dependent on the sample size, *n*. The smaller the sample size, the wider and flatter the distribution tails. The shape of the distribution approaches the standard normal curve as the sample size increases. Generally, the two distributions have the same shape for $n > 50$.

An important parameter needed to define a *t*-distribution is the *degrees of freedom*, df. (The symbol, ν, Greek nu, is also used.) The degrees of freedom is usually 1 less than the sample size.

$$\text{df} = n - 1 \qquad 11.46$$

Student's t-distribution is a standardized *t*-distribution, centered on zero, just like the standard normal variable, z. With a normal distribution, the population standard deviation would be multiplied by $z = 1.96$ to get a 95% confidence interval. When using the *t*-distribution, the sample standard deviation, s, is multiplied by a number, t, coming from the *t*-distribution.

$$t = \frac{\overline{x} - \mu}{\dfrac{s}{\sqrt{n}}} \qquad 11.47$$

That number is designated as $t_{\alpha,\text{df}}$ or $t_{\alpha,n-1}$ for a one-tail test/confidence interval, and would be designated as $t_{\alpha/2,\text{df}}$ or $t_{\alpha/2,n-1}$ for a two-tail test/confidence interval. For example, for a two-tail confidence interval with confidence level $C = 1 - \alpha$, the upper and lower confidence limits are

$$\text{UCL, LCL} = \overline{x} \pm t_{\alpha/2,n-1} \frac{s}{\sqrt{n}} \qquad 11.48$$

17. CHI-SQUARED DISTRIBUTION

The *chi-squared* (*chi square*, χ^2) *distribution* is a distribution of the sum of squared standard normal deviates, z_i. It has numerous useful applications. The distribution's only parameter, its *degrees of freedom*, ν or df, is the number of standard normal deviates being summed. Chi-squared distributions are positively skewed, but the skewness decreases and the distribution approaches a normal distribution as degrees of freedom increases. Appendix 11.B tabulates the distribution. The mean and variance of a chi-squared distribution are related to its degrees of freedom.

$$\mu = \nu \qquad 11.49$$
$$\sigma^2 = 2\nu \qquad 11.50$$

After taking n samples from a standard normal distribution, the *chi-squared statistic* is defined as

$$\chi^2 = \frac{(n-1)s^2}{\sigma^2} \qquad 11.51$$

The sum of chi-squared variables is itself a chi-squared variable. For example, after taking three measurements from a standard normal distribution, squaring each term, and adding all three squared terms, the sum is a chi-squared variable. A chi-squared distribution with three degrees of freedom can be used to determine the probability of that summation exceeding another number. This characteristic makes the chi-squared distribution useful in determining whether or not a population's variance has shifted, or whether two populations have the same variance. The distribution is also extremely useful in categorical hypothesis testing.

18. LOG-NORMAL DISTRIBUTION

With a *log-normal* (*lognormal* or *Galton's*) *distribution*, the logarithm of the independent variable, $\ln(x)$ or $\log(x)$, is normally distributed, not x. Log-normal distributions are rare in engineering; they are more common in social, political, financial, biological, and environmental applications.[3] The log-normal distribution may be applicable whenever a normal-appearing distribution is skewed to the left, or when the independent variable is a function (e.g., product) of multiple positive independent variables. Depending on its parameters (i.e., skewness), a log-normal distribution can take on different shapes, as shown in Fig. 11.5.

Figure 11.5 *Log-Normal Probability Density Function*

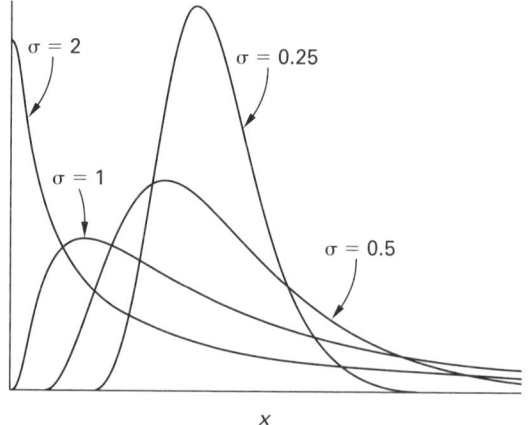

The symbols μ and σ are used to represent the *parameter values* of the transformed (logarithmitized) distribution; μ is the *location parameter* (*log mean*), the mean of the transformed values; and σ is the *scale parameter*

---

[3]Log-normal is a popular reliability distribution. In electrical engineering, some transmission losses are modeled as log-normal. Some geologic mineral concentrations follow the log-normal pattern.

(*log standard deviation*), whose unbiased estimator is the sample standard deviation, s, of the transformed values. These parameter values are used to calculate the standard normal variable, z, in order to determine event probabilities (i.e., areas under the normal curve). The *expected value*, $E\{x\}$, and standard deviation, σ_x, of the non-transformed distribution, x, are

$$E\{x\} = e^{\mu + \frac{1}{2}\sigma^2} \qquad 11.52$$

$$\sigma_x = \sqrt{\left(e^{\sigma^2} - 1\right)e^{2\mu + \sigma^2}} \qquad 11.53$$

$$z = \frac{x_0 - \mu}{\sigma} \qquad 11.54$$

19. ERROR FUNCTION

The *error function*, $\mathrm{erf}(x)$, and its complement, the *complementary error function*, $\mathrm{erfc}(x)$, are defined by Eq. 11.55 and Eq. 11.56. The functions can be used to determine the probable error of a measurement, but they also appear in many engineering formulas. Values are seldom calculated from Eq. 11.55 or Eq. 11.56. Rather, approximation and tabulations (e.g., App. 11.D) are used.

$$\mathrm{erf}(x_0) = \frac{2}{\sqrt{\pi}} \int_0^{x_0} e^{-x^2}\, dx \qquad 11.55$$

$$\mathrm{erfc}(x_0) = 1 - \mathrm{erf}(x_0) \qquad 11.56$$

The error function can be used to calculate areas under the normal curve. Combining Eq. 11.44, Eq. 11.45, and Eq. 11.55,

$$\frac{1}{\sqrt{2\pi}} \int_0^z e^{-u^2/2}\, du = \frac{1}{2}\mathrm{erf}\left(\frac{z}{\sqrt{2}}\right) \qquad 11.57$$

The error function has the following properties.

$$\mathrm{erf}(0) = 0$$
$$\mathrm{erf}(+\infty) = 1$$
$$\mathrm{erf}(-\infty) = -1$$
$$\mathrm{erf}(-x_0) = -\mathrm{erf}(x_0)$$

20. APPLICATION: RELIABILITY

Introduction

Reliability, $R\{t\}$, is the probability that an item will continue to operate satisfactorily up to time t. The *bathtub distribution*, Fig. 11.6, is often used to model the probability of failure of an item (or, the number of failures from a large population of items) as a function of time. Items initially fail at a high rate, a phenomenon known as *infant mortality*. For the majority of the operating time, known as the *steady-state operation*, the failure rate is constant (i.e., is due to random causes). After

a long period of time, the items begin to deteriorate and the failure rate increases. (No mathematical distribution describes all three of these phases simultaneously.)

Figure 11.6 *Bathtub Reliability Curve*

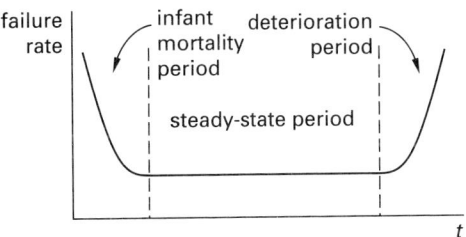

The *hazard function*, $z\{t\}$, represents the *conditional probability of failure*—the probability of failure in the next time interval, given that no failure has occurred thus far.[4]

$$z\{t\} = \frac{f(t)}{R(t)} = \frac{\dfrac{dF(t)}{dt}}{1 - F(t)} \qquad 11.58$$

A *proof test* is a comprehensive validation of an item that checks 100% of all failure mechanisms. It tests for faults and degradation so that the item can be certified as being in "as new" condition. The *proof test interval* is the time after initial installation at which an item must be either proof tested or replaced. If the policy is to replace rather than test an item, the item's lifetime is equal to the proof test interval and is known as *mission time*.

Exponential Reliability

Steady-state reliability is often described by the *negative exponential distribution*. This assumption is appropriate whenever an item fails only by random causes and does not experience deterioration during its life. The parameter λ is related to the *mean time to failure* (MTTF) of the item.[5]

$$R\{t\} = e^{-\lambda t} = e^{-t/\mathrm{MTTF}} \qquad 11.59$$

$$\lambda = \frac{1}{\mathrm{MTTF}} \qquad 11.60$$

Equation 11.59 and the exponential continuous distribution function, Eq. 11.40, are complementary.

$$R\{t\} = 1 - F(t) = 1 - \left(1 - e^{-\lambda t}\right) = e^{-\lambda t} \qquad 11.61$$

---

[4]The symbol $z\{t\}$ is traditionally used for the hazard function and is not related to the standard normal variable.

[5]The term "mean time *between* failures" is improper. However, the term *mean time before failure* (MTBF) is acceptable.

The hazard function for the negative exponential distribution is

$$z\{t\} = \lambda \qquad 11.62$$

Therefore, the hazard function for exponential reliability is constant and does not depend on t (i.e., on the age of the item). In other words, the expected future life of an item is independent of the previous history (length of operation). This lack of memory is consistent with the assumption that only random causes contribute to failure during steady-state operations. And since random causes are unlikely discrete events, their probability of occurrence can be represented by a Poisson distribution with mean λ. That is, the probability of having x failures in any given period is

$$p\{x\} = \frac{e^{-\lambda}\lambda^x}{x!} \qquad 11.63$$

Serial System Reliability

In the analysis of system reliability, the binary variable X_i is defined as 1 if item i operates satisfactorily and 0 if otherwise. Similarly, the binary variable Φ is 1 only if the entire system operates satisfactorily. Therefore, Φ will depend on a *performance function* containing the X_i.

A *serial system* is one for which all items must operate correctly for the system to operate. Each item has its own reliability, R_i. For a serial system of n items, the performance function is

$$\Phi = X_1 X_2 X_3 \cdots X_n = \min(X_i) \qquad 11.64$$

The probability of a serial system operating correctly is

$$p\{\Phi = 1\} = R_{\text{serial system}} = R_1 R_2 R_3 \cdots R_n \qquad 11.65$$

Parallel System Reliability

A *parallel system* with n items will fail only if all n items fail. Such a system is said to be *redundant* to the nth degree. Using redundancy, a highly reliable system can be produced from components with relatively low individual reliabilities.

The performance function of a redundant system is

$$\Phi = 1 - (1 - X_1)(1 - X_2)(1 - X_3)\cdots(1 - X_n)$$
$$= \max(X_i) \qquad 11.66$$

The reliability of the parallel system is

$$R = p\{\Phi = 1\}$$
$$= 1 - (1 - R_1)(1 - R_2)(1 - R_3)\cdots(1 - R_n) \qquad 11.67$$

With a fully redundant, parallel k-out-of-n system of n independent, identical items, any k of which maintain system functionality, the ratio of redundant MTTF to single-item MTTF $(1/\lambda)$ is given by Table 11.1.

Table 11.1 MTTF Multipliers for k-out-of-n Systems

| | \multicolumn{5}{c}{n} |
k	1	2	3	4	5
1	1	3/2	11/6	25/12	137/60
2		1/2	5/6	13/12	77/60
3			1/3	7/12	47/60
4				1/4	9/20
5					1/5

Example 11.11

The reliability of an item is exponentially distributed with mean time to failure (MTTF) of 1000 hr. What is the probability that the item will not have failed before 1200 hr of operation?

Solution

The probability of not having failed before time t is the reliability. From Eq. 11.60 and Eq. 11.61,

$$\lambda = \frac{1}{\text{MTTF}} = \frac{1}{1000 \text{ hr}} = 0.001 \text{ hr}^{-1}$$

$$R\{1200\} = e^{-\lambda t} = e^{(-0.001 \text{ hr}^{-1})(1200 \text{ hr})} = 0.3$$

Example 11.12

What are the reliabilities of the following systems?

(a)

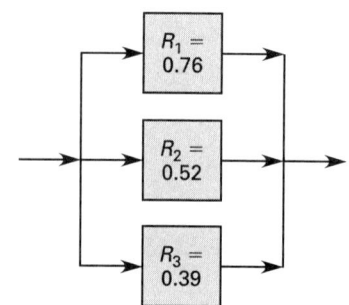

(b)

Solution

(a) This is a serial system. From Eq. 11.65,

$$R = R_1 R_2 R_3 R_4 = (0.93)(0.98)(0.91)(0.87)$$
$$= 0.72$$

(b) This is a parallel system. From Eq. 11.67,

$$R = 1 - (1 - R_1)(1 - R_2)(1 - R_3)$$
$$= 1 - (1 - 0.76)(1 - 0.52)(1 - 0.39)$$
$$= 0.93$$

21. ANALYSIS OF EXPERIMENTAL DATA

Experiments can take on many forms. An experiment might consist of measuring the mass of one cubic foot of concrete or measuring the speed of a car on a roadway. Generally, such experiments are performed more than once to increase the precision and accuracy of the results.

Both systematic and random variations in the process being measured will cause the observations to vary, and the experiment would not be expected to yield the same result each time it was performed. Eventually, a collection of experimental outcomes (observations) will be available for analysis.

The *frequency distribution* is a systematic method for ordering the observations from small to large, according to some convenient numerical characteristic. The *step interval* should be chosen so that the data are presented in a meaningful manner. If there are too many intervals, many of them will have zero frequencies; if there are too few intervals, the frequency distribution will have little value. Generally, 10 to 15 intervals are used.

Once the frequency distribution is complete, it can be represented graphically as a *histogram*. The procedure in drawing a histogram is to mark off the interval limits (also known as *class limits*) on a number line and then draw contiguous bars with lengths that are proportional to the frequencies in the intervals and that are centered on the midpoints of their respective intervals. The continuous nature of the data can be depicted by a *frequency polygon*. The number or percentage of observations that occur up to and including some value can be shown in a *cumulative frequency table*.

Example 11.13

The number of cars that travel through an intersection between 12 noon and 1 p.m. is measured for 30 consecutive working days. The results of the 30 observations are

79, 66, 72, 70, 68, 66, 68, 76, 73, 71, 74, 70, 71, 69, 67, 74, 70, 68, 69, 64, 75, 70, 68, 69, 64, 69, 62, 63, 63, 61

(a) What are the frequency and cumulative distributions? (Use a distribution interval of two cars per hour.) (b) Draw the histogram. (Use a cell size of two cars per hour.) (c) Draw the frequency polygon. (d) Graph the cumulative frequency distribution.

Solution

(a) Tabulate the frequency, cumulative frequency, and cumulative percent distributions.

cars per hour	frequency	cumulative frequency	cumulative percent
60–61	1	1	3
62–63	3	4	13
64–65	2	6	20
66–67	3	9	30
68–69	8	17	57
70–71	6	23	77
72–73	2	25	83
74–75	3	28	93
76–77	1	29	97
78–79	1	30	100

(b) Draw the histogram.

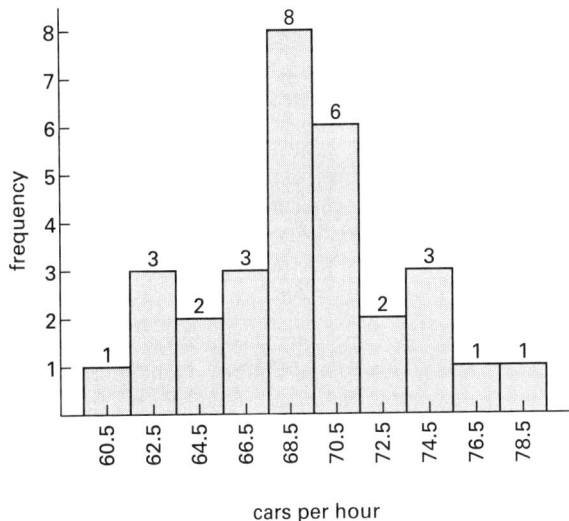

(c) Draw the frequency polygon.

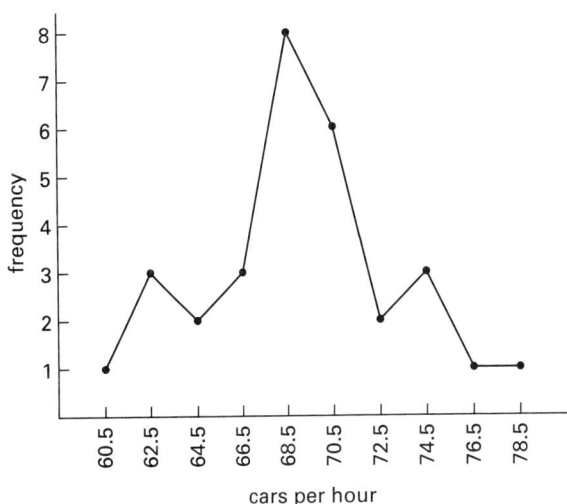

(d) Graph the cumulative frequency distribution.

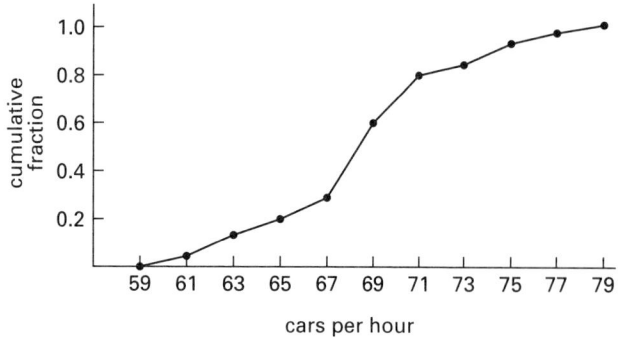

22. MEASURES OF EXPERIMENTAL ADEQUACY

An experiment is said to be *accurate* if it is unaffected by experimental error. In this case, *error* is not synonymous with *mistake*, but rather includes all variations not within the experimenter's control.

For example, suppose a gun is aimed at a point on a target and five shots are fired. The mean distance from the point of impact to the sight in point is a measure of the alignment accuracy between the barrel and sights. The difference between the actual value and the experimental value is known as *bias*.

Precision is not synonymous with accuracy. Precision is concerned with the repeatability of the experimental results. If an experiment is repeated with identical results, the experiment is said to be precise.

The average distance of each impact from the centroid of the impact group is a measure of the precision of the experiment. It is possible to have a highly precise experiment with a large bias.

Most of the techniques applied to experiments in order to improve the accuracy (i.e., reduce bias) of the experimental results (e.g., repeating the experiment, refining the experimental methods, or reducing variability) actually increase the precision.

Sometimes the word *reliability* is used with regard to the precision of an experiment. In this case, a "reliable estimate" is used in the same sense as a "precise estimate."

Stability and *insensitivity* are synonymous terms. A stable experiment will be insensitive to minor changes in the experimental parameters. For example, suppose the centroid of a bullet group is 2.1 in from the target point at 65°F and 2.3 in away at 80°F. The sensitivity of the experiment to temperature change would be

$$\text{sensitivity} = \frac{\Delta x}{\Delta T} = \frac{2.3 \text{ in} - 2.1 \text{ in}}{80°F - 65°F} = 0.0133 \text{ in/}°F$$

23. MEASURES OF CENTRAL TENDENCY

It is often unnecessary to present the experimental data in their entirety, either in tabular or graphical form. In such cases, the data and distribution can be represented by various parameters. One type of parameter is a measure of *central tendency*. Mode, median, and mean are measures of central tendency.

The *mode* is the observed value that occurs most frequently. The mode may vary greatly between series of observations. Therefore, its main use is as a quick measure of the central value since little or no computation is required to find it. Beyond this, the usefulness of the mode is limited.

The *median* is the point in the distribution that partitions the total set of observations into two parts containing equal numbers of observations. It is not influenced by the extremity of scores on either side of the distribution. The median is found by counting up (from either end of the frequency distribution) until half of the observations have been accounted for.

For even numbers of observations, the median is estimated as some value (i.e., the average) between the two center observations.

Similar in concept to the median are *percentiles* (*percentile ranks*), *quartiles*, and *deciles*. The median could also have been called the *50th percentile* observation. Similarly, the 80th percentile would be the observed value (e.g., the number of cars per hour) for which the cumulative frequency was 80%. The quartile and decile points on the distribution divide the observations or distribution into segments of 25% and 10%, respectively.

The *arithmetic mean* is the arithmetic average of the observations. The sample mean, $\bar{x}$, can be used as an *unbiased estimator* of the population mean, μ. The *mean* may be found without ordering the data (as was necessary to find the mode and median). The mean can be found from the following formula.

$$\bar{x} = \left(\frac{1}{n}\right)(x_1 + x_2 + \cdots + x_n) = \frac{\sum x_i}{n} \qquad 11.68$$

The *geometric mean* is used occasionally when it is necessary to average ratios. The geometric mean is calculated as

$$\text{geometric mean} = \sqrt[n]{x_1 x_2 x_3 \cdots x_n} \quad [x_i > 0] \qquad 11.69$$

The *harmonic mean* is defined as

$$\text{harmonic mean} = \frac{n}{\dfrac{1}{x_1} + \dfrac{1}{x_2} + \cdots + \dfrac{1}{x_n}} \qquad 11.70$$

The *root-mean-squared (rms) value* of a series of observations is defined as

$$x_{\text{rms}} = \sqrt{\frac{\sum x_i^2}{n}} \qquad 11.71$$

The *ratio of exceedance* of a single measurement, x_i, from the mean (or, any other value), μ, is

$$\text{ER} = \left| \frac{x_i - \mu}{\mu} \right| \qquad 11.72$$

The exceedance of a single measurement from the mean can be expressed in standard deviations. For a normal distribution, this exceedance is usually given the variable z and is referred to as the *standard normal value, variable, variate,* or *deviate*.

$$z = \left| \frac{x_i - \mu}{\sigma} \right| \qquad 11.73$$

Example 11.14

Find the mode, median, and arithmetic mean of the distribution represented by the data given in Ex. 11.13.

Solution

First, resequence the observations in increasing order.
61, 62, 63, 63, 64, 64, 66, 66, 67, 68, 68, 68, 68, 69, 69, 69, 69, 70, 70, 70, 70, 71, 71, 72, 73, 74, 74, 75, 76, 79

The mode is the interval 68–69, since this interval has the highest frequency. If 68.5 is taken as the interval center, then 68.5 would be the mode.

The 15th and 16th observations are both 69, so the median is

$$\frac{69 + 69}{2} = 69$$

The mean can be found from the raw data or from the grouped data using the interval center as the assumed observation value. Using the raw data,

$$\overline{x} = \frac{\sum x_i}{n} = \frac{2069}{30} = 68.97$$

24. MEASURES OF DISPERSION

The simplest statistical parameter that describes the variability in observed data is the *range*. The range is found by subtracting the smallest value from the largest. Since the range is influenced by extreme (low probability) observations, its use as a measure of variability is limited.

The *population standard deviation* is a better estimate of variability because it considers every observation.

That is, in Eq. 11.74, N is the total population size, not the sample size, n.

$$\sigma = \sqrt{\frac{\sum (x_i - \mu)^2}{N}} = \sqrt{\frac{\sum x_i^2}{N} - \mu^2} \qquad 11.74$$

The standard deviation of a sample (particularly a small sample) is a biased (i.e., not a good) estimator of the population standard deviation. An *unbiased estimator* of the population standard deviation is the *sample standard deviation*, s, also known as the *standard error* of a sample.[6]

$$s = \sqrt{\frac{\sum (x_i - \overline{x})^2}{n - 1}} = \sqrt{\frac{\sum x_i^2 - \frac{\left(\sum x_i\right)^2}{n}}{n - 1}} \qquad 11.75$$

If the sample standard deviation, s, is known, the standard deviation of the sample, σ_{sample}, can be calculated.

$$\sigma_{\text{sample}} = s\sqrt{\frac{n - 1}{n}} \qquad 11.76$$

The *variance* is the square of the standard deviation. Since there are two standard deviations, there are two variances. The *variance of the sample* is σ^2, and the *sample variance* is s^2.

The *relative dispersion* is defined as a measure of dispersion divided by a measure of central tendency. The *coefficient of variation* is a relative dispersion calculated from the sample standard deviation and the mean.

$$\text{coefficient of variation} = \frac{s}{\overline{x}} \qquad 11.77$$

Example 11.15

For the data given in Ex. 11.13, calculate (a) the sample range, (b) the standard deviation of the sample, (c) an unbiased estimator of the population standard deviation, (d) the variance of the sample, and (e) the sample variance.

Solution

$$\sum x_i = 2069$$

$$\left(\sum x_i\right)^2 = (2069)^2 = 4{,}280{,}761$$

$$\sum x_i^2 = 143{,}225$$

---

[6]There is a subtle yet significant difference between *standard deviation of the sample*, σ (obtained from Eq. 11.74 for a finite sample drawn from a larger population), and the *sample standard deviation*, s (obtained from Eq. 11.75). While σ can be calculated, it has no significance or use as an estimator. It is true that the difference between σ and s approaches zero when the sample size, n, is large, but this convergence does nothing to legitimize the use of σ as an estimator of the true standard deviation. (Some people say "large" is 30, others say 50 or 100.)

$$n = 30$$

$$\bar{x} = \frac{2069}{30} = 68.967$$

(a) The sample range is

$$R = x_{max} - x_{min} = 79 - 61 = 18$$

(b) Normally, the standard deviation of the sample would not be calculated. In this case, it was specifically requested. From Eq. 11.74, using n for N and $\bar{x}$ for μ,

$$\sigma = \sqrt{\frac{\sum x_i^2}{n} - (\bar{x})^2} = \sqrt{\frac{143{,}225}{30} - \left(\frac{2069}{30}\right)^2}$$
$$= 4.215$$

(c) From Eq. 11.75,

$$s = \sqrt{\frac{\sum x_i^2 - \frac{\left(\sum x_i\right)^2}{n}}{n-1}}$$
$$= \sqrt{\frac{143{,}225 - \frac{4{,}280{,}761}{30}}{29}}$$
$$= 4.287$$

(d) The variance of the sample is

$$\sigma^2 = (4.215)^2 = 17.77$$

(e) The sample variance is

$$s^2 = (4.287)^2 = 18.38$$

25. SKEWNESS

Skewness is a measure of a distribution's lack of symmetry. Distributions that are pushed to the left have negative skewness, while distributions pushed to the right have positive skewness. Various formulas are used to calculate skewness. *Pearson's skewness*, sk, is a simple normalized difference between the mean and mode (see Eq. 11.78). Since the mode is poorly represented when sampling from a distribution, the difference is estimated as three times the deviation from the mean. *Fisher's skewness*, γ_1, and more modern methods are based on the *third moment about the mean* (see Eq. 11.79). Normal distributions have zero skewness, although zero skewness is not a sufficient requirement for determining normally distributed data. Square

root, log, and reciprocal transformations of a variable can reduce skewness.

$$\text{sk} = \frac{\mu - \text{mode}}{\sigma} \approx \frac{3(\bar{x} - \text{median})}{s} \qquad \textit{11.78}$$

$$\gamma_1 = \frac{\sum_{i=1}^{N}(x_i - \mu)^3}{N\sigma^3} \approx \frac{n\sum_{i=1}^{n}(x_i - \bar{x})^3}{(n-1)(n-2)s^3} \qquad \textit{11.79}$$

26. KURTOSIS

While skewness refers to the symmetry of the distribution about the mean, *kurtosis* refers to the contribution of the tails to the distribution, or alternatively, how flat the peak is. A *mesokurtic distribution* ($\beta_2 = 3$) is statistically normal. Compared to a normal curve, a *leptokurtic distribution* ($\beta_2 < 3$) is "fat in the tails," longer-tailed, and more sharp-peaked, while a *platykurtic distribution* ($\beta_2 > 3$) is "thin in the tails," shorter-tailed, and more flat-peaked. Different methods are used to calculate kurtosis. The most common, *Fisher's kurtosis*, β_2, is defined as the *fourth standardized moment (fourth moment of the mean)*.

$$\beta_2 = \frac{\sum_{i=1}^{N}(x_i - \mu)^4}{N\sigma^4} \qquad \textit{11.80}$$

Some statistical analyses, including those in Microsoft Excel, calculate a related statistic, *excess kurtosis (kurtosis excess* or *Pearson's kurtosis)*, by subtracting 3 from the kurtosis. A normal distribution has zero excess kurtosis; a peaked distribution has positive excess kurtosis; and a flat distribution has negative excess kurtosis.

$$\gamma_2 = \beta_2 - 3$$

$$\approx \frac{n(n+1)\sum_{i=1}^{n}(x_i - \bar{x})^4}{(n-1)(n-2)(n-3)s^4} - \frac{3(n-1)^2}{(n-2)(n-3)} \qquad \textit{11.81}$$

27. CENTRAL LIMIT THEOREM

Measuring a sample of n items from a population with mean μ and standard deviation σ is the general concept of an experiment. The sample mean, $\bar{x}$, is one of the parameters that can be derived from the experiment. This experiment can be repeated k times, yielding a set of averages $(\bar{x}_1, \bar{x}_2, \ldots, \bar{x}_k)$. The k numbers in the set themselves represent samples from distributions of averages. The average of averages, $\bar{\bar{x}}$, and sample standard deviation of averages, $s_{\bar{x}}$ (known as the *standard error of the mean*), can be calculated.

The *central limit theorem* characterizes the distribution of the sample averages. The theorem can be stated in several ways, but the essential elements are the following points.

1. The averages, $\bar{x}_i$, are normally distributed variables, even if the original data from which they are calculated are not normally distributed.

2. The grand average, $\bar{\bar{x}}$ (i.e., the average of the averages), approaches and is an unbiased estimator of μ.

$$\mu \approx \bar{\bar{x}} \qquad 11.82$$

The standard deviation of the original distribution, σ, is much larger than the standard error of the mean.

$$\sigma \approx \sqrt{n} s_{\bar{x}} \qquad 11.83$$

28. CONFIDENCE LEVEL

The results of experiments are seldom correct 100% of the time. Recognizing this, researchers accept a certain probability of being wrong. In order to minimize this probability, experiments are repeated several times. The number of repetitions depends on the desired level of confidence in the results.

If the results have a 5% probability of being wrong, the *confidence level*, C, is 95% that the results are correct, in which case the results are said to be *significant*. If the results have only a 1% probability of being wrong, the confidence level is 99%, and the results are said to be *highly significant*. Other confidence levels (90%, 99.5%, etc.) are used as appropriate.

The complement of the confidence level is α, referred to as *alpha*. α is the *significance level* and may be given as a decimal value or percentage. Alpha is also known as *alpha risk* and *producer risk*, as well as the probability of a type I error. A *type I error*, also known as a *false positive error*, occurs when the null hypothesis is incorrectly rejected, and an action occurs that is not actually required. For a random sample of manufactured products, the null hypothesis would be that the distribution of sample measurements is not different from the historical distribution of those measurements. If the null hypothesis is rejected, all of the products (not just the sample) will be rejected, and the producer will have to absorb the expense. It is not uncommon to use 5% as the producer risk in noncritical business processes.

$$\alpha = 100\% - C \qquad 11.84$$

β (*beta* or *beta risk*) is the *consumer risk*, the probability of a type II error. A *type II error* occurs when the null hypothesis (e.g., "everything is fine") is incorrectly accepted, and no action occurs when action is actually required. If a batch of defective products is accepted, the products are distributed to consumers who then suffer the consequences. Generally, smaller values of α coincide

with larger values of β because requiring overwhelming evidence to reject the null increases the chances of a type II error. β can be minimized while holding α constant by increasing sample sizes. The *power of the test* is the probability of rejecting the null hypothesis when it is false.

$$\text{power of the test} = 1 - \beta \qquad 11.85$$

29. NULL AND ALTERNATIVE HYPOTHESES

All statistical conclusions involve constructing two mutually exclusive hypotheses, termed the *null hypothesis* (written as H_0) and *alternative hypothesis* (written as H_1). Together, the hypotheses describe all possible outcomes of a statistical analysis. The purpose of the analysis is to determine which hypothesis to accept and which to reject.

Usually, when an improvement is made to a program, treatment, or process, the change is expected to make a difference. The null hypothesis is so named because it refers to a case of "no difference" or "no effect." Typical null hypotheses are:

> H_0: There has been no change in the process.
> H_0: The two distributions are the same.
> H_0: The change has had no effect.
> H_0: The process is in control and has not changed.
> H_0: Everything is fine.

The alternative hypothesis is that there has been an effect, and there is a difference. The null and alternative hypotheses are mutually exclusive.

30. APPLICATION: CONFIDENCE LIMITS

As a consequence of the central limit theorem, sample means of n items taken from a normal distribution with mean μ and standard deviation σ will be normally distributed with mean μ and variance σ^2/n. The probability that any given average, $\bar{x}$, exceeds some value, L, is

$$p\{\bar{x} > L\} = p\left\{ z > \left| \frac{L - \mu}{\frac{\sigma}{\sqrt{n}}} \right| \right\} \qquad 11.86$$

L is the *confidence limit* for the confidence level $1 - p\{\bar{x} > L\}$ (normally expressed as a percent). Values of z are read directly from the standard normal table. As an example, $z = 1.645$ for a 95% confidence level since only 5% of the curve is above that value of z in the upper tail. This is known as a *one-tail confidence limit* because all of the probability is given to one side of the variation. Similar values are given in Table 11.2.

Table 11.2 *Values of z for Various Confidence Levels*

confidence level, C	one-tail limit, z	two-tail limit, z
90%	1.28	1.645
95%	1.645	1.96
97.5%	1.96	2.17
99%	2.33	2.575
99.5%	2.575	2.81
99.75%	2.81	3.00

With *two-tail confidence limits*, the probability is split between the two sides of variation. There will be upper and lower confidence limits, UCL and LCL, respectively.

$$p\{\text{LCL} < \bar{x} < \text{UCL}\} = p\left\{\left|\frac{\text{LCL} - \mu}{\frac{\sigma}{\sqrt{n}}}\right| < z < \left|\frac{\text{UCL} - \mu}{\frac{\sigma}{\sqrt{n}}}\right|\right\}$$

$$11.87$$

31. APPLICATION: BASIC HYPOTHESIS TESTING

A *hypothesis test* is a procedure that answers the question, "Did these data come from [a particular type of] distribution?" There are many types of tests, depending on the distribution and parameter being evaluated. The simplest hypothesis test determines whether an average value obtained from n repetitions of an experiment could have come from a population with known mean, μ, and standard deviation, σ. A practical application of this question is whether a manufacturing process has changed from what it used to be or should be. Of course, the answer (i.e., "yes" or "no") cannot be given with absolute certainty—there will be a confidence level associated with the answer.

The following procedure is used to determine whether the average of n measurements can be assumed (with a given confidence level) to have come from a known population.

step 1: Assume random sampling from a normal population.

step 2: Choose the desired confidence level, C.

step 3: Decide on a one-tail or two-tail test. If the hypothesis being tested is that the average has or has not *increased* or *decreased*, choose a one-tail test. If the hypothesis being tested is that the average has or has not *changed*, choose a two-tail test.

step 4: Use Table 11.2 or the standard normal table to determine the z-value corresponding to the confidence level and number of tails.

step 5: Calculate the actual standard normal variable, z'.

$$z' = \left|\frac{\bar{x} - \mu}{\frac{\sigma}{\sqrt{n}}}\right|$$

$$11.88$$

step 6: If $z' \geq z$, the average can be assumed (with confidence level C) to have come from a different distribution.

Example 11.16

When it is operating properly, a cement plant has a daily production rate that is normally distributed with a mean of 880 tons/day and a standard deviation of 21 tons/day. During an analysis period, the output is measured on 50 consecutive days, and the mean output is found to be 871 tons/day. With a 95% confidence level, determine whether the plant is operating properly.

Solution

step 1: Given.

step 2: $C = 95\%$ is given.

step 3: Since a specific direction in the variation is not given (i.e., the example does not ask whether the average has decreased), use a two-tail hypothesis test.

step 4: The population mean and standard deviation are known. The standard normal distribution may be used. From Table 11.2, $z = 1.96$.

step 5: From Eq. 11.88,

$$z' = \left|\frac{\bar{x} - \mu}{\frac{\sigma}{\sqrt{n}}}\right| = \left|\frac{871 - 880}{\frac{21}{\sqrt{50}}}\right| = 3.03$$

Since $3.03 > 1.96$, the distributions are not the same. There is at least a 95% probability that the plant is not operating correctly.

32. APPLICATION: STATISTICAL PROCESS CONTROL

All manufacturing processes contain variation due to random and nonrandom causes. Random variation cannot be eliminated. *Statistical process control* (SPC) is the act of monitoring and adjusting the performance of a process to detect and eliminate nonrandom variation.

Statistical process control is based on taking regular (hourly, daily, etc.) samples of n items and calculating the mean, $\bar{x}$, and range, R, of the sample. To simplify the calculations, the range is used as a measure of the dispersion. These two parameters are graphed on their respective x-bar and *R-control charts*, as shown in Fig. 11.7.[7] Confidence limits are drawn at $\pm 3\sigma/\sqrt{n}$. From a statistical standpoint, the control chart tests a hypothesis each time a point is plotted. When a point falls outside these limits, there is a 99.75% probability that the process is out of control. Until a point exceeds the control limits, no action is taken.[8]

Figure 11.7 *Typical Statistical Process Control Charts*

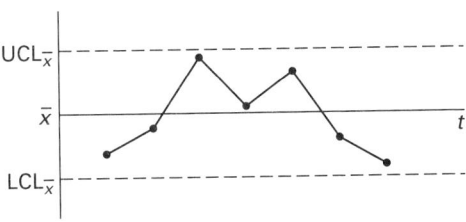

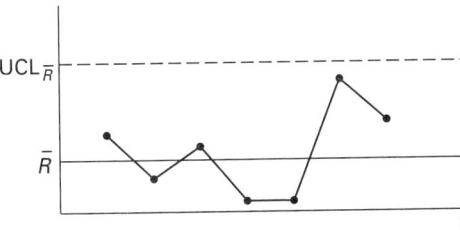

33. LINEAR REGRESSION

If it is necessary to draw a straight line $(y = mx + b)$ through n data points $(x_1, y_1), (x_2, y_2), \ldots, (x_n, y_n)$, the following method based on the *method of least squares* can be used.

step 1: Calculate the following nine quantities.

$$\sum x_i \quad \sum x_i^2 \quad \left(\sum x_i\right)^2 \quad \bar{x} = \frac{\sum x_i}{n} \quad \sum x_i y_i$$

$$\sum y_i \quad \sum y_i^2 \quad \left(\sum y_i\right)^2 \quad \bar{y} = \frac{\sum y_i}{n}$$

step 2: Calculate the slope, m, of the line.

$$m = \frac{n\sum x_i y_i - \sum x_i \sum y_i}{n\sum x_i^2 - \left(\sum x_i\right)^2} \qquad 11.89$$

step 3: Calculate the y-intercept, b.

$$b = \bar{y} - m\bar{x} \qquad 11.90$$

---

[7]Other charts (e.g., the *sigma chart*, *p-chart*, and *c-chart*) are less common but are used as required.
[8]Other indications that a correction may be required are seven measurements on one side of the average and seven consecutively increasing measurements. Rules such as these detect shifts and trends.

step 4: To determine the goodness of fit, calculate the *correlation coefficient, r.*

$$r = \frac{n\sum x_i y_i - \sum x_i \sum y_i}{\sqrt{\left(n\sum x_i^2 - \left(\sum x_i\right)^2\right)\left(n\sum y_i^2 - \left(\sum y_i\right)^2\right)}}$$

$$11.91$$

If m is positive, r will be positive; if m is negative, r will be negative. As a general rule, if the absolute value of r exceeds 0.85, the fit is good; otherwise, the fit is poor. r equals 1.0 if the fit is a perfect straight line.

A low value of r does not eliminate the possibility of a nonlinear relationship existing between x and y. It is possible that the data describe a parabolic, logarithmic, or other nonlinear relationship. (Usually this will be apparent if the data are graphed.) It may be necessary to convert one or both variables to new variables by taking squares, square roots, cubes, or logarithms, to name a few of the possibilities, in order to obtain a linear relationship. The apparent shape of the line through the data will give a clue to the type of variable transformation that is required. The curves in Fig. 11.8 may be used as guides to some of the simpler variable transformations.

Figure 11.8 *Nonlinear Data Curves*

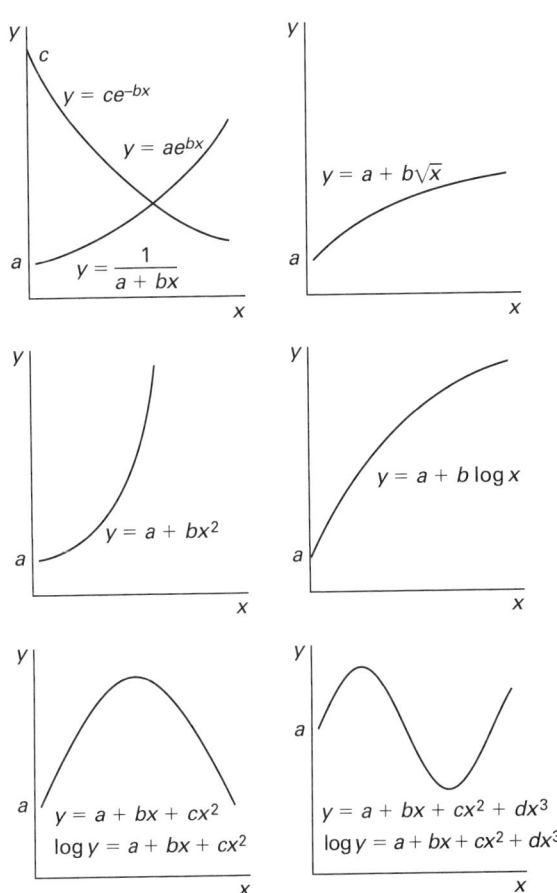

Figure 11.9 illustrates several common problems encountered in trying to fit and evaluate curves from experimental data. Figure 11.9(a) shows a graph of clustered data with several extreme points. There will be moderate correlation due to the weighting of the extreme points, although there is little actual correlation at low values of the variables. The extreme data should be excluded, or the range should be extended by obtaining more data.

Figure 11.9 *Common Regression Difficulties*

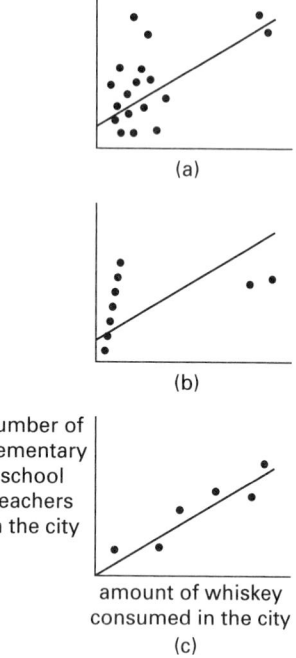

(a)

(b)

number of
elementary
school
teachers
in the city

amount of whiskey
consumed in the city

(c)

Figure 11.9(b) shows that good correlation exists in general, but extreme points are missed, and the overall correlation is moderate. If the results within the small linear range can be used, the extreme points should be excluded. Otherwise, additional data points are needed, and curvilinear relationships should be investigated.

Figure 11.9(c) illustrates the problem of drawing conclusions of cause and effect. There may be a predictable relationship between variables, but that does not imply a cause and effect relationship. In the case shown, both variables are functions of a third variable, the city population. But there is no direct relationship between the plotted variables.

Example 11.17

An experiment is performed in which the dependent variable, y, is measured against the independent variable, x. The results are as follows.

x	y
1.2	0.602
4.7	5.107
8.3	6.984
20.9	10.031

(a) What is the least squares straight line equation that best represents this data? (b) What is the correlation coefficient?

Solution

(a) Calculate the following quantities.

$$\sum x_i = 35.1$$
$$\sum y_i = 22.72$$
$$\sum x_i^2 = 529.23$$
$$\sum y_i^2 = 175.84$$
$$\left(\sum x_i\right)^2 = 1232.01$$
$$\left(\sum y_i\right)^2 = 516.38$$
$$\overline{x} = 8.775$$
$$\overline{y} = 5.681$$
$$\sum x_i y_i = 292.34$$
$$n = 4$$

From Eq. 11.89, the slope is

$$m = \frac{n\sum x_i y_i - \sum x_i \sum y_i}{n\sum x_i^2 - \left(\sum x_i\right)^2} = \frac{(4)(292.34) - (35.1)(22.72)}{(4)(529.23) - (35.1)^2}$$
$$= 0.42$$

From Eq. 11.90, the y-intercept is

$$b = \overline{y} - m\overline{x} = 5.681 - (0.42)(8.775)$$
$$= 2.0$$

The equation of the line is

$$y = 0.42x + 2.0$$

(b) From Eq. 11.91, the correlation coefficient is

$$r = \frac{n\sum x_i y_i - \sum x_i \sum y_i}{\sqrt{\left(n\sum x_i^2 - \left(\sum x_i\right)^2\right)\left(n\sum y_i^2 - \left(\sum y_i\right)^2\right)}}$$
$$= \frac{(4)(292.34) - (35.1)(22.72)}{\sqrt{\begin{array}{c}((4)(529.23) - 1232.01) \\ \times ((4)(175.84) - 516.38)\end{array}}}$$
$$= 0.914$$

Example 11.18

Repeat Ex. 11.17 assuming the relationship between the variables is nonlinear.

Solution

The first step is to graph the data. Since the graph has the appearance of the fourth case in Fig. 11.8, it can be assumed that the relationship between the variables has the form of $y = a + b \log x$. Therefore, the variable change $z = \log x$ is made, resulting in the following set of data.

z	y
0.0792	0.602
0.672	5.107
0.919	6.984
1.32	10.031

If the regression analysis is performed on this set of data, the resulting equation and correlation coefficient are

$$y = 7.599z + 0.000247$$

$$r = 0.999$$

This is a very good fit. The relationship between the variable x and y is approximately

$$y = 7.599 \log x + 0.000247$$

34. QUANTITATIVE RISK ANALYSIS

Introduction

The study of risk involves two aspects, risk analysis and risk management, that are related but distinctly different. *Risk analysis*, also known as *risk assessment*, is a *technical* evaluation of the overall probability of an undesirable consequence of exposure to a hazard. *Risk management* is a *political* (or management) decision regarding the selection of actions to be taken in response to exposure to a risk. While an environmental engineer may make the key decisions involved in risk analysis, upper-level managers, politicians, or attorneys may be the key decision-makers in risk management decisions, which typically involve choosing between social, economic, and technical alternatives or impacts. This section will cover only risk analysis.

Risk analysis can be applied to many aspects of the human-built and natural environments. Our everyday lives involve numerous risk analysis evaluations that we take for granted; often we are unaware of the mental processes involved. Without much effort, we quickly analyze whether we can make it through an intersection, or whether another cigarette or doughnut is an acceptable risk to our health. These risks are known as *assumed risks* because we "assume," or take, them and accept the consequences, even if we are not fully aware of what those consequences are.

Assumed risks are collectively known as the *background risk*, meaning the sum of the risks to which we are exposed excluding the risk of additional activities being evaluated. *Incremental risk* is the risk associated with the additional activities being evaluated, and *total risk* is the sum of the background risk and the incremental risk. Incremental risk is also known as *imposed risk* because the risk associated with the activity may be involuntarily imposed on the exposed population.

Environmental engineers readily understand that judgmental factors are involved in making risk management decisions. Policy and remediation alternatives must be weighed with social, economic, and political factors to come up with a solution. It is important to realize also that there is always a level of uncertainty in the risk analysis process. However, to the extent possible, the risk analysis process attempts to use facts to evaluate the environmental and human health effects of exposure to hazardous materials and situations.

Quantitative risk analysis determines a quantitative, numerical estimate of risk associated with an activity. (This is contrasted with a *qualitative analysis* whereby the potential risk may be identified but its magnitude is not estimated.) Quantitative risk analysis is typically described as a four-step process:

step 1: hazard identification

step 2: toxicity assessment

step 3: exposure assessment

step 4: risk characterization

The risk analysis process involves the application of scientific knowledge and engineering principles in the areas of toxicology, chemistry, and fate and transport modeling. The four-step process is applied by the EPA and state regulatory agencies to the remediation of hazardous waste sites in order to determine appropriate levels of cleanup and to select alternative remedies.

Hazard Identification

Hazard identification involves identifying substances and determining that they are indeed health hazards. A hazard is not the same as a risk. *Hazard* refers to the capability of the substance to cause an adverse effect on humans, aquatic life, or any other organism. The hazard posed by the substance depends on its toxicity, mobility, and persistence, and on the measures taken to prevent its release to the environment (e.g., noncorrosive containers, spill containment). By contrast, *risk* is the probability of the adverse effect occurring. Exposure, or potential exposure, to the hazard is required for a risk to be present.

Although an environmental engineer will most commonly deal with hazard identification and risk analysis in situations involving the remediation of hazardous waste sites, another important situation occurs in the design and operation of industrial processes that involve hazardous substances. Process hazard analysis should be a part of the design of industrial processes involving hazardous substances. The Occupational Safety and Health Administration (OSHA) has identified the procedure for performing a process hazard analysis. (See App. 11.B.)

For a typical risk analysis of chemicals encountered, the first step is to identify the chemicals that are present at the site and the media that are affected. Samples are collected to determine the contaminant levels in the air, groundwater, surface water, soils, and sediments associated with the site. The concentration and distribution of the chemicals present in each sample are determined. Moreover, the ways by which the chemicals can move off the site to expose the receptor population need to be known in order to perform the subsequent fate and transport analysis.

A large number of chemicals may be identified at a site, so an initial screening is necessary to reduce the number of chemicals that must be evaluated to a manageable number (perhaps 10 to 20). The objective of the initial screening is to identify the chemicals that will account for 99% of the risk for both carcinogens and noncarcinogens. "99% of the risk" does not mean 99% of the chemicals; only a few chemicals might account for this level of risk.

The initial screening procedure involves six steps.

step 1: Sort the chemicals by medium, and then into carcinogens and noncarcinogens.

step 2: Calculate the mean and maximum concentrations for each chemical.

step 3: Determine the slope factors (SF) for carcinogens and the reference doses (RfD) for noncarcinogens. (See Table 11.3.)

step 4: Determine the toxicity scores for each chemical by medium.

step 5: Rank the chemicals by toxicity scores for each medium.

step 6: Select the chemicals comprising 99% of the total toxicity score for each medium.

Beyond initial screening based on toxicity score, the chemicals present should also be analyzed for persistence, mobility, and other factors that may increase or decrease the hazard. Chemicals that are highly persistent and readily mobile by the pathways present at the site should be retained for further analysis.

In step 4, the toxicity scores are determined in the initial screening as given in Eq. 11.92 and Eq. 11.93. Strictly, *toxicity score*, TS, contains units, though it is generally reported unitless.

For noncarcinogens,

$$TS_n = \frac{C_{max}}{RfD} \qquad 11.92$$

For carcinogens,

$$TS_c = C_{max}(SF) \qquad 11.93$$

Toxicity Assessment

The purpose of toxicity assessment is to quantify the hazard of the substances. Although identified as a separate step of risk analysis, in practice at least part of the toxicity assessment must be done concurrently with hazard identification because the reference doses (noncarcinogens) and the slope factors (carcinogens) are used in the initial screening.

EPA Toxicity Information

In most hazardous waste site remediation projects, EPA toxicity indices are used. The toxicity data are available in the Integrated Risk Information Systems (IRIS) database. Where data are missing or ambiguous, the regulatory agency (state or EPA) will usually provide toxicity indices.

The toxicity indices are different for noncarcinogens and carcinogens. Noncarcinogens are assigned thresholds below which the toxicant dose does not result in an observable adverse health effect. The threshold, known as the *no observable adverse effect level* (NOAEL), is used to calculate a *reference dose* (RfD). The RfD is calculated by dividing the NOAEL by uncertainty factors of several orders of magnitude. The IRIS database contains oral and inhalation RfDs for more than 500 chemicals.

The RfD is the lifetime dose that a healthy person could be exposed to daily (*chronic daily intake*, CDI) without an appreciable risk. The RfD is expressed in mg of chemical per kg body mass per day (mg/kg·d). The *hazard index* is used to analyze the risk for noncarcinogens. It is the ratio of the actual dose to the RfD, which if greater than 1.0 represents the possibility of an adverse health effect from the given level of exposure.

Carcinogens are believed not to have thresholds, and any exposure is believed to cause a *response* (mutation of the human DNA). Since there is no level of carcinogens that can be considered safe for continued human exposure, the EPA method extrapolates the dose-response curve from high doses to very low exposures in order to specify what are known as *slope factors* (SFs). (Slope factors are also known as *potency factors* and *carcinogen potency factors*.) The SF is the probability of cancer produced by lifetime exposure to 1.0 mg/kg·d of the carcinogen. The SF is selected at the 95% upper confidence limit of the dose-response curve. The SF is the inverse of dose and is expressed as $(mg/kg·d)^{-1}$. The risk (probability) is the product of the SF and the dose.

Extracts from the EPA's IRIS database are given in Table 11.3. The EPA's carcinogenicity classification system is given in Table 11.4.

Table 11.3 Typical Toxicological Data from IRIS Database

chemical	class	RfD (oral) $(mg/kg \cdot d)$	RfD (inhalation) $(mg/kg \cdot d)$	SF (oral) $(mg/kg \cdot d)^{-1}$	SF (inhalation) $(mg/kg \cdot d)^{-1}$
benzene	A	–	–	0.0290	0.0290
1,2 dichloroethane	B2	–	–	0.091	0.091
1,2 dibromoethane (EDB)	B2	–	–	85	0.77
1,3 dichloropropene	B2	0.0003	0.00572	no data	no data
acrylamide	B2	0.0002	–	4.50	4.50
bromodichloromethane	B2	0.002	no data	0.062	no data
chloroform	B2	0.01	–	0.0061	0.081
dieldrin	B2	0.00005	no data	16	16
chlorobenzene	D	0.02	pending	no data	no data
endrin	D	0.0003	no data	no data	no data
nitrobenzene	D	0.0005	–	no data	no data
pentachlorobenzene	D	0.0008	no data	no data	no data
pyrene	D	0.003	no data	no data	no data
toluene	D	0.2	1.40	–	–

Source: Environmental Protection Agency

Table 11.4 EPA Carcinogenicity Classification System

group	description
A	human carcinogen
B1 or B2	probable human carcinogen: B1 indicates that human data are available. B2 indicates sufficient evidence in animals but inadequate or no evidence in humans.
C	possible human carcinogen
D	not classifiable as to human carcinogenicity
E	evidence of noncarcinogenicity for humans

Source: Environmental Protection Agency

Example 11.19

The project engineer for cleanup of a hazardous waste site in eastern Oregon has obtained data on chemicals present at the site and their potential pathways of exposure to residents in the area. She needs to perform an initial screening to determine the chemicals or surrogate chemicals that will be used in the exposure assessment. The project engineer notes that she will need to develop a slope factor for 1,3 dichloropropene for the oral route since no data is given in the IRIS database. She reviews available toxicity studies and, with the help of a toxicologist, determines that a slope factor of 8×10^{-3} $(mg/kg \cdot d)^{-1}$ would be appropriate. Which of the chemicals noted below should she select for subsequent analysis for the groundwater route?

chemical	groundwater mean (mg/L)	groundwater max (mg/L)	air mean (mg/m^3)	air max (mg/m^3)
carcinogens				
1,2 dichloro-ethane	0.12	0.92	61	103
1,2 dibromo-ethane (EDB)	0.034	0.096	4.3	7.9
1,3 dichloro-propene	9.0×10^{-4}	9.9×10^{-4}	5.8	9.9
dieldrin	0.4	0.44	ND	ND

noncarcinogens				
chlorobenzene	60	73	0.33	4.6
endrin	6.1	8.1	0.041	0.093
pyrene	9	48	ND	ND
toluene	0.8	4.3	0.71	0.98

Solution

Steps 1 and 2 of the screening procedure appear in the problem statement.

step 3: Determine the slope factors (SF) for carcinogens and the reference doses (RfD) for noncarcinogens.

The reference doses and slope factors are determined from the EPA IRIS database, extracts of which are given in Table 11.3. The problem statement gives a slope factor of 8×10^{-3} $(mg/kg \cdot d)^{-1}$.

step 4: Determine the toxicity scores for each chemical by medium.

step 5: Rank the chemicals by toxicity scores by medium.

In this case, only the groundwater route is being considered.

For noncarcinogens, $TS_n = C_{max}/RfD$.

For carcinogens, $TS_c = C_{max}(SF)$.

The calculations are shown in the following table for steps 4 and 5.

chemical	noncarcinogens C_{max} (mg/L)	RfD, oral $(mg/kg \cdot d)$	TS_n $(kg \cdot d/L)$	rank
endrin	8.1	3×10^{-4}	27 000	1
pyrene	48	3×10^{-3}	16 000	2
dieldrin	0.44	5×10^{-5}	8800	3
chlorobenzene	73	0.02	3650	4
toluene	4.3	0.2	21.5	5
1,3 dichloro-propene	9.9×10^{-4}	3×10^{-4}	3.3	6
		total	55 474.8	

chemical	carcinogens C_{max} (mg/L)	SF, oral $(mg/kg \cdot d)^{-1}$	TS_c	rank
1,2 dibromo-ethane (EDB)	0.096	85	8.16	1
dieldrin	0.44	16	7.04	2
1,2 dichloro-ethane	0.92	9.1×10^{-2}	0.08372	3
1,3 dichloro-propene	9.9×10^{-4}	8×10^{-3}	7.92×10^{-6}	4
		total	15.2837	

Note that dieldrin and 1,3 dichloropropene show both carcinogenic and noncarcinogenic effects, so they must be analyzed for both effects.

step 6: Select the chemicals comprising 99% of the total toxicity score by medium.

The chemicals ranked 1 through 4 comprise 99% of the total toxicity score for noncarcinogens.

The correct answer for noncarcinogens is endrin, pyrene, dieldrin, and chlorobenzene.

The chemicals ranked 1 and 2 comprise 99% of the total toxicity score for carcinogens.

The correct answer for carcinogens is 1,2 dibromoethane (EDB) and dieldrin.

Exposure Assessment

Exposure assessment is the process of tracking a substance from its source to the potential receptors. This is by no means a trivial undertaking and can quickly become very complicated due to the multiplicities of chemicals, release mechanisms, exposure pathways, fate and transport mechanisms, and receptors, among other things. The general approach is to formulate conservative but realistic scenarios of the mechanisms and processes involved, and then to quantitatively evaluate the intake (exposure) of the receptor. Factors to consider for hazardous waste sites often include workers at industrial sites or other nearby activities, trespassers, residential use, recreational use, and construction of remediation elements. The necessary components of the scenarios are described below.

Sources

The potential sources of chemical releases must first be identified. In some cases, this may be very obvious, such as a pile of drums. In other instances, more evaluation is required, such as to identify buried wastes, to determine if a lagoon has an impermeable liner, or to delineate contaminated soil areas. The concentrations and total masses of the chemicals in these sources must be determined or estimated.

Release Mechanisms

The release mechanisms must be identified. Some may be obvious, such as visible leakage from corroded drums. Others will require more investigation to determine, such as volatilization from open pools of chemicals, leakage from the bottom of a lagoon, or spillage from tanks or ponds. Contaminated soil particles can be windblown from the site, and waste may be burned, releasing chemicals with the combustion products. Leachate may be generated from precipitation onto waste piles or burial areas. Storm water may wash contaminated soil from a site.

Transport

All of the relevant *transport mechanisms* associated with the release mechanisms must be identified in the scenario. Transport mechanisms may include air transport of volatilized chemicals, groundwater transport of leakage, surface water transport of leachate and contaminated storm water, and many others.

Transfer

Transfer means the exchange of contaminants from one transport mechanism to another. For example, leachate may flow into surface water that subsequently percolates into the soil and reaches groundwater. Fugitive dust might be blown from the site and settle in surface water. Chemicals may volatilize from surface water or groundwater, or be adsorbed onto the soils or sediments through which they pass. Plants that are used as food for humans or animals can take up chemicals. Precipitation may cause chemicals in soil to dissolve and be transported in runoff. Chemicals may be washed out of the air by precipitation, and airborne particles may settle onto lawns and gardens.

Transformation

Transformation occurs when a chemical is transformed into different substances that may be more or less toxic. While many organic compounds biodegrade to less toxic constituents, others (such as vinyl chloride) may change to more potent products. Photochemical degradation and oxidation by atmospheric ozone are additional examples of transformation.

Exposure Point

The *exposure point* is the location where the receptor and the chemical make contact or have the potential to do so. Exposure points include the original site, a water supply well for a community or residence, planting areas with soil contaminated from airborne particles, and surface water where contaminated fish are caught, to name a few. Identification of the receptor population needs to be done concurrently with identification of the exposure point.

Receptor Population

The *receptor population* is the group of humans that could potentially be exposed at the exposure points. Both existing and future populations need to be identified. Receptors at high risk, such as the elderly and children, need to be identified. The remediation site workers must also be included in the receptor population.

Exposure Point Concentrations

Major effort may be required to calculate the concentrations of the selected chemicals at the exposure points. All pathways and contaminants must be evaluated for present and future conditions as hypothesized in the scenarios. The support of an experienced modeler is essential in performing fate and transport modeling.

Receptor Doses

The last part of an exposure assessment is to calculate the receptor doses. This involves determining the quantity of the medium to which the receptor is exposed and multiplying it by the exposure point concentration. If the receptor is exposed to the same contaminant from more than one pathway, the total dose for that contaminant is the sum of the dose by each pathway. Although *chronic exposure* is the usual case, *acute exposure*, such as from a spill to surface water, will need to be evaluated also.

The *receptor dose* is the total mass of contaminant taken into the body divided by the body mass of the receptor and averaged over the days on which the risk factor is based. For noncarcinogens, the averaging time is the same as the exposure period, while for carcinogens the averaging time is 70 yr to correspond to lifetime risk.

The receptor dose through air or water exposure can be calculated from Eq. 11.94.

$$\text{CDI}_{a/w} = \frac{C(\text{CR})(\text{EF})(\text{ED})}{(\text{BW})(\text{AT})} \qquad 11.94$$

Equation 11.94 also applies to dust exposure. For inhalation of fugitive dust, the intake must be modified to account for the decimal fractions of the *retention rate*, RR, (most dust is trapped and removed by the respiratory system or is exhaled) and the *absorption rate*, AR (only a fraction of the contaminants in dust that remains in the lungs reaches the bloodstream). Equation 11.95 can be used to calculate the receptor dose for exposure to dust.

$$\text{CDI}_{\text{dust}} = \frac{C(\text{CR})(\text{EF})(\text{ED})(\text{RR})(\text{ABS})}{(\text{BW})(\text{AT})} \qquad 11.95$$

In some cases, dermal (direct skin) contact and exposures may need to be evaluated.

Example 11.20

For the situation given in Ex. 11.19, the project engineer and modeler have developed five scenarios for the release, transport, transfer, transformation, and exposure point concentrations for potential receptors. They have determined the exposure point concentrations for two of the selected contaminants, as follows.

exposure point	receptor	pathway	chemical	concentration
residential water supply well	adult	groundwater, oral	dieldrin chlorobenzene	0.36 mg/L 36 mg/L
air in residence	adult	volatilization of groundwater, inhalation	dieldrin chlorobenzene	ND* 2.4 mg/m^3

*No data available.

The engineer and modeler estimate that natural attenuation will reduce the concentrations of the indicated chemicals to nondetectable levels after 20 years and estimate the exposure duration to be 350 days per year. Standard contact rate (CR) factors are 2 L/d adult water consumption and 20 m^3/d adult air consumption. What are the CDIs for each toxicant?

Solution

For water supply well exposure,

CR = 2 L/d adult water consumption

EF = 350 d/yr

ED = 20 yr

BW = 70 kg

AT = 70 yr for carcinogens

AT = 20 yr for noncarcinogens

For dieldrin, the well exposure is

$$\text{CDI}_w = \frac{C(\text{CR})(\text{EF})(\text{ED})}{(\text{BW})(\text{AT})}$$

$$= \frac{\left(0.36 \; \frac{\text{mg}}{\text{L}}\right)\left(2 \; \frac{\text{L}}{\text{d}}\right)\left(350 \; \frac{\text{d}}{\text{yr}}\right)(20 \; \text{yr})}{(70 \; \text{kg})(70 \; \text{yr})\left(365 \; \frac{\text{d}}{\text{yr}}\right)}$$

$$= 2.818 \times 10^{-3} \; \text{mg/kg·d} \quad (0.003 \; \text{mg/kg·d})$$

For chlorobenzene, the well exposure is

$$CDI_w = \frac{C(CR)(EF)(ED)}{(BW)(AT)}$$

$$= \frac{\left(36 \ \frac{mg}{L}\right)\left(2 \ \frac{L}{d}\right)\left(350 \ \frac{d}{yr}\right)(20 \ yr)}{(70 \ kg)(20 \ yr)\left(350 \ \frac{d}{yr}\right)}$$

$$= 1.029 \ mg/kg \cdot d$$

For air exposure,

$CR = 20 \ m^3/d$ adult air consumption

$EF = 350 \ d/yr$

$ED = 20 \ yr$

$BW = 70 \ kg$

$AT = 70 \ yr$ for carcinogens

$AT = 20 \ yr$ for noncarcinogens

From the table, for dieldrin,

$$CDI_a = ND$$

For chlorobenzene, the air exposure is

$$CDI_a = \frac{C(CR)(EF)(ED)}{(BW)(AT)}$$

$$= \frac{\left(2.4 \ \frac{mg}{m^3}\right)\left(20 \ \frac{m^3}{d}\right)\left(350 \ \frac{d}{yr}\right)(20 \ yr)}{(70 \ kg)(20 \ yr)\left(350 \ \frac{d}{yr}\right)}$$

$$= 0.686 \ mg/kg \cdot d$$

The CDIs are

	CDI (mg/kg·d)	
chemical	water well	air
dieldrin	0.003	ND
chlorobenzene	1.029	0.686

Risk Characterization

The final step in the risk analysis process is to quantify the total noncarcinogenic and carcinogenic risks by combining the estimated exposures with the calculated potencies.

For noncarcinogens, the risk is characterized by the *hazard index*, which is the ratio of the dose to the reference dose. An acceptable exposure occurs when the total exposures by all pathways and exposure routes for each chemical and for the sum of the hazard indices for all chemicals are less than 1.0.

$$HI = \frac{CDI}{RfD} \qquad 11.96$$

For carcinogens, the risks for each exposure scenario are summed for each receptor group by exposure route. The receptor groups should be differentiated between the group experiencing average exposure and the group experiencing maximum exposure. Both the mean and maximum exposure point concentrations should be used to characterize the range of risk. The *carcinogen risk* is calculated as the product of the chronic daily intake and the slope factor.

$$R_c = (CDI)(SF) \qquad 11.97$$

Example 11.21

The project engineer is now ready to characterize the risks for the situation given in Ex. 11.19. She is evaluating one of the scenarios for exposure of an adult in a residence near the site. The pathways, exposure routes, exposure points, chemicals, reference doses, and slope factors are given in *Table for Example 11.21*. Characterize the risk to this receptor.

Solution

For noncarcinogens, use Eq. 11.96.

$$HI = \frac{CDI}{RfD}$$

chemical	CDI	RfD	HI
chlorobenzene			
oral	1.034	0.02	51.7
inhalation	0.731	0.002	365.5
pyrene			
oral	0.041	0.003	13.7
inhalation	0.077	0.0003	256.7
total HI			
oral			65.4
inhalation			622.2

The risk due to noncarcinogens is unacceptable because the hazard index is much greater than 1.0 for individual and combined exposures.

For carcinogens, use Eq. 11.97.

$$R_c = (CDI)(SF)$$

chemical	CDI	SF	R_c
dieldrin			
oral	0.012	16	0.192
inhalation	0.003	16	0.048

The risk due to carcinogens is also unacceptable because the risk for dieldrin is outside the range of the normally acceptable EPA risk of 10^{-4} to 10^{-6}.

Table for Example 11.21

pathway	exposure point	exposure route	chemical	CDI (mg/kg·d)	RfD (mg/kg·d)	SF $((\text{mg/kg·d})^{-1})$
groundwater	water well	oral	chlorobenzene	1.028		
			pyrene	0.036		
			dieldrin	0.003		
groundwater	volatilization at tap	inhalation	chlorobenzene	0.686		
			pyrene	0.044		
			dieldrin	ND		
fugitive dust	air at residence	inhalation	chlorobenzene	0.045		
			pyrene	0.033		
			dieldrin	0.003		
surface water	garden at residence	oral	chlorobenzene	0.006		
			pyrene	0.005		
			dieldrin	0.009		
total		oral	chlorobenzene	1.034	0.02	–
			pyrene	0.041	0.003	–
			dieldrin	0.012	–	16
total		inhalation	chlorobenzene	0.731	0.002	–
			pyrene	0.077	0.0003	–
			dieldrin	0.003	–	16

12 Numerical Analysis

1. NUMERICAL METHODS

Although the roots of second-degree polynomials are easily found by a variety of methods (by factoring, completing the square, or using the quadratic equation), easy methods of solving cubic and higher-order equations exist only for specialized cases. However, cubic and higher-order equations occur frequently in engineering, and they are difficult to factor. Trial and error solutions, including graphing, are usually satisfactory for finding only the general region in which the root occurs.

Numerical analysis is a general subject that covers, among other things, iterative methods for evaluating roots to equations. The most efficient numerical methods are too complex to present and, in any case, work by hand. However, some of the simpler methods are presented here. Except in critical problems that must be solved in real time, a few extra calculator or computer iterations will make no difference.[1]

2. FINDING ROOTS: BISECTION METHOD

The *bisection method* is an iterative method that "brackets" ("straddles") an interval containing the *root* or *zero* of a particular equation.[2] The size of the interval is halved after each iteration. As the method's name suggests, the best estimate of the root after any iteration is the midpoint of the interval. The maximum error is half the interval length. The procedure continues until the size of the maximum error is "acceptable."[3]

The disadvantages of the bisection method are (a) the slowness in converging to the root, (b) the need to know

---

[1]Most advanced hand-held calculators have "root finder" functions that use numerical methods to iteratively solve equations.

[2]The equation does not have to be a pure polynomial. The bisection method requires only that the equation be defined and determinable at all points in the interval.

[3]The bisection method is not a closed method. Unless the root actually falls on the midpoint of one iteration's interval, the method continues indefinitely. Eventually, the magnitude of the maximum error is small enough not to matter.

the interval containing the root before starting, and (c) the inability to determine the existence of or find other real roots in the starting interval.

The bisection method starts with two values of the independent variable, $x = L_0$ and $x = R_0$, which straddle a root. Since the function passes through zero at a root, $f(L_0)$ and $f(R_0)$ will almost always have opposite signs. The following algorithm describes the remainder of the bisection method.

Let n be the iteration number. Then, for $n = 0, 1, 2, \ldots$, perform the following steps until sufficient accuracy is attained.

step 1: Set $m = \frac{1}{2}(L_n + R_n)$.

step 2: Calculate $f(m)$.

step 3: If $f(L_n)f(m) \leq 0$, set $L_{n+1} = L_n$ and $R_{n+1} = m$. Otherwise, set $L_{n+1} = m$ and $R_{n+1} = R_n$.

step 4: $f(x)$ has at least one root in the interval $[L_{n+1}, R_{n+1}]$. The estimated value of that root, x^*, is

$$x^* \approx \tfrac{1}{2}(L_{n+1} + R_{n+1})$$

The maximum error is $\frac{1}{2}(R_{n+1} - L_{n+1})$.

Example 12.1

Use two iterations of the bisection method to find a root of

$$f(x) = x^3 - 2x - 7$$

Solution

The first step is to find L_0 and R_0, which are the values of x that straddle a root and have opposite signs. A table can be made and values of $f(x)$ calculated for random values of x.

x	-2	-1	0	$+1$	$+2$	$+3$
$f(x)$	-11	-6	-7	-8	-3	$+14$

Since $f(x)$ changes sign between $x = 2$ and $x = 3$, $L_0 = 2$ and $R_0 = 3$.

First iteration, $n = 0$:

$$m = \tfrac{1}{2}(L_n + R_n) = \left(\tfrac{1}{2}\right)(2 + 3) = 2.5$$

$$f(2.5) = x^3 - 2x - 7 = (2.5)^3 - (2)(2.5) - 7 = 3.625$$

Since $f(2.5)$ is positive, a root must exist in the interval $[2, 2.5]$. Therefore, $L_1 = 2$ and $R_1 = 2.5$. Or, using step 3, $f(2)f(2.5) = (-3)(3.625) = -10.875 \leq 0$. So, $L_1 = 2$, and $R_1 = 2.5$. At this point, the best estimate of the root is

$$x^* \approx \left(\tfrac{1}{2}\right)(2 + 2.5) = 2.25$$

The maximum error is $\left(\tfrac{1}{2}\right)(2.5 - 2) = 0.25$.

Second iteration, $n = 1$:

$$m = \left(\tfrac{1}{2}\right)(2 + 2.5) = 2.25$$

$$f(2.25) = (2.25)^3 - (2)(2.25) - 7 = -0.1094$$

Since $f(2.25)$ is negative, a root must exist in the interval $[2.25, 2.5]$. Or, using step 3, $f(2)f(2.25) = (-3)(-0.1096) = 0.3288 > 0$. Therefore, $L_2 = 2.25$, and $R_2 = 2.5$. The best estimate of the root is

$$x^* \approx \tfrac{1}{2}(L_{n+1} + R_{n+1}) \approx \left(\tfrac{1}{2}\right)(2.25 + 2.5) = 2.375$$

The maximum error is $\left(\tfrac{1}{2}\right)(2.5 - 2.25) = 0.125$.

3. FINDING ROOTS: NEWTON'S METHOD

Many other methods have been developed to overcome one or more of the disadvantages of the bisection method. These methods have their own disadvantages.[4]

Newton's method is a particular form of *fixed-point iteration*. In this sense, "fixed point" is often used as a synonym for "root" or "zero." Fixed-point iterations get their name from functions with the characteristic property $x = g(x)$ such that the limit of $g(x)$ is the fixed point (i.e., is the root).

All fixed-point techniques require a starting point. Preferably, the starting point will be close to the actual root.[5] And, while Newton's method converges quickly, it requires the function to be continuously differentiable.

Newton's method algorithm is simple. At each iteration ($n = 0, 1, 2$, etc.), Eq. 12.1 estimates the root. The maximum error is determined by looking at how much the estimate changes after each iteration. If the change between the previous and current estimates (representing the magnitude of error in the estimate) is too large, the current estimate is used as the independent variable for the subsequent iteration.[6]

$$x_{n+1} = g(x_n) = x_n - \frac{f(x_n)}{f'(x_n)} \qquad \textit{12.1}$$

---

[4]The *regula falsi (false position) method* converges faster than the bisection method but is unable to specify a small interval containing the root. The *secant method* is prone to round-off errors and gives no indication of the remaining distance to the root.
[5]Theoretically, the only penalty for choosing a starting point too far away from the root will be a slower convergence to the root.
[6]Actually, the theory defining the maximum error is more definite than this. For example, for a large enough value of n, the error decreases approximately linearly. Therefore, the consecutive values of x_n converge linearly to the root as well.

Example 12.2

Solve Ex. 12.1 using two iterations of Newton's method. Use $x_0 = 2$.

Solution

The function and its first derivative are

$$f(x) = x^3 - 2x - 7$$

$$f'(x) = 3x^2 - 2$$

First iteration, $n = 0$:

$$x_0 = 2$$

$$f(x_0) = f(2) = (2)^3 - (2)(2) - 7 = -3$$

$$f'(x_0) = f'(2) = (3)(2)^2 - 2 = 10$$

$$x_1 = x_0 - \frac{f(x_0)}{f'(x_0)} = 2 - \frac{-3}{10} = 2.3$$

Second iteration, $n = 1$:

$$x_1 = 2.3$$

$$f(x_1) = (2.3)^3 - (2)(2.3) - 7 = 0.567$$

$$f'(x_1) = (3)(2.3)^2 - 2 = 13.87$$

$$x_2 = x_1 - \frac{f(x_1)}{f'(x_1)} = 2.3 - \frac{0.567}{13.87} = 2.259$$

4. NONLINEAR INTERPOLATION: LAGRANGIAN INTERPOLATING POLYNOMIAL

Interpolating between two points of known data is common in engineering. Primarily due to its simplicity and speed, straight-line interpolation is used most often. Even if more than two points on the curve are explicitly known, they are not used. Since straight-line interpolation ignores all but two of the points on the curve, it ignores the effects of curvature.

A more powerful technique that accounts for the curvature is the *Lagrangian interpolating polynomial*. This method uses an nth degree parabola (polynomial) as the interpolating curve.[7] This method requires that $f(x)$ be continuous and real-valued on the interval $[x_0, x_n]$ and that $n + 1$ values of $f(x)$ are known corresponding to $x_0, x_1, x_2, \ldots, x_n$.

---

[7]The Lagrangian interpolating polynomial reduces to straight-line interpolation if only two points are used.

The procedure for calculating $f(x)$ at some intermediate point, x, starts by calculating the Lagrangian interpolating polynomial for each known point.

$$L_k(x^*) = \prod_{\substack{i=0 \\ i \neq k}}^{n} \frac{x^* - x_i}{x_k - x_i} \qquad 12.2$$

The value of $f(x)$ at x^* is calculated from Eq. 12.3.

$$f(x^*) = \sum_{k=0}^{n} f(x_k) L_k(x^*) \qquad 12.3$$

The Lagrangian interpolating polynomial has two primary disadvantages. The first is that a large number of additions and multiplications are needed.[8] The second is that the method does not indicate how many interpolating points should be used. Other interpolating methods have been developed that overcome these disadvantages.[9]

Example 12.3

A real-valued function has the following values.

$$f(1) = 3.5709$$

$$f(4) = 3.5727$$

$$f(6) = 3.5751$$

Use the Lagrangian interpolating polynomial to estimate the value of the function at 3.5.

Solution

The procedure for applying Eq. 12.2, the Lagrangian interpolating polynomial, is illustrated in tabular form. Notice that the term corresponding to $i = k$ is omitted from the product.

$k = 0$: $i = 0$ $i = 1$ $i = 2$

$$L_0(3.5) = \left(\frac{3.5 - 1}{1 - 1}\right)\left(\frac{3.5 - 4}{1 - 4}\right)\left(\frac{3.5 - 6}{1 - 6}\right)$$

$$= 0.08333$$

$k = 1$:

$$L_1(3.5) = \left(\frac{3.5 - 1}{4 - 1}\right)\left(\frac{3.5 - 4}{4 - 4}\right)\left(\frac{3.5 - 6}{4 - 6}\right)$$

$$= 1.04167$$

$k = 2$:

$$L_2(3.5) = \left(\frac{3.5 - 1}{6 - 1}\right)\left(\frac{3.5 - 4}{6 - 4}\right)\left(\frac{3.5 - 6}{6 - 6}\right)$$

$$= -0.12500$$

---

[8]As with the numerical methods for finding roots previously discussed, the number of calculations probably will not be an issue if the work is performed by a calculator or computer.
[9]Other common methods for performing interpolation include the *Newton form* and *divided difference table*.

Equation 12.3 is used to calculate the estimate.

$$f(3.5) = (3.5709)(0.08333) + (3.5727)(1.04167)$$

$$+ (3.5751)(-0.12500)$$

$$= 3.57225$$

5. NONLINEAR INTERPOLATION: NEWTON'S INTERPOLATING POLYNOMIAL

Newton's form of the interpolating polynomial is more efficient than the Lagrangian method of interpolating between known points.[10] Given $n + 1$ known points for $f(x)$, the *Newton form of the interpolating polynomial* is

$$f(x^*) = \sum_{i=0}^{n} \left(f[x_0, x_1, \ldots, x_i] \prod_{j=0}^{i-1} (x^* - x_j) \right) \qquad 12.4$$

$f[x_0, x_1, \ldots, x_i]$ is known as the ith *divided difference*.

$$f[x_0, x_1, \ldots, x_i] = \sum_{k=0}^{i} \left(\frac{f(x_k)}{(x_k - x_0) \cdots (x_k - x_{k-1})} \right. $$
$$\left. \times (x_k - x_{k+1}) \cdots (x_k - x_i) \right) $$
$$12.5$$

It is necessary to define the following two terms.

$$f[x_0] = f(x_0) \qquad 12.6$$

$$\prod (x^* - x_j) = 1 \quad [i = 0] \qquad 12.7$$

Example 12.4

Repeat Ex. 12.3 using Newton's form of the interpolating polynomial.

Solution

Since there are $n + 1 = 3$ data points, $n = 2$. Evaluate the terms for $i = 0$, 1, and 2.

$i = 0$:

$$f[x_0] \prod_{j=0}^{-1} (x^* - x_j) = f[x_0](1) = f(x_0)$$

---

[10]In this case, "efficiency" relates to the ease in adding new known points without having to repeat all previous calculations.

$i = 1$:

$$f[x_0, x_1] \prod_{j=0}^{0} (x^* - x_j) = f[x_0, x_1](x^* - x_0)$$

$$f[x_0, x_1] = \frac{f(x_0)}{x_0 - x_1} + \frac{f(x_1)}{x_1 - x_0}$$

$i = 2$:

$$f[x_0, x_1, x_2] \prod_{j=0}^{1} (x^* - x_j) = f[x_0, x_1, x_2](x^* - x_0)(x^* - x_1)$$

$$f[x_0, x_1, x_2] = \frac{f(x_0)}{(x_0 - x_1)(x_0 - x_2)}$$

$$+ \frac{f(x_1)}{(x_1 - x_0)(x_1 - x_2)}$$

$$+ \frac{f(x_2)}{(x_2 - x_0)(x_2 - x_1)}$$

Use Eq. 12.4. Substitute known values.

$$f(3.5) = \sum_{i=0}^{n} \left(f[x_0, x_1, \ldots, x_i] \prod_{j=0}^{i-1} (x^* - x_j) \right)$$

$$= 3.5709 + \left(\frac{3.5709}{1 - 4} + \frac{3.5727}{4 - 1} \right)(3.5 - 1)$$

$$+ \left(\frac{3.5709}{(1 - 4)(1 - 6)} + \frac{3.5727}{(4 - 1)(4 - 6)} \right.$$

$$\left. + \frac{3.5751}{(6 - 1)(6 - 4)} \right)$$

$$\times (3.5 - 1)(3.5 - 4)$$

$$= 3.57225$$

This answer is the same as that determined in Ex. 12.3.

13

Energy, Work, and Power

Nomenclature

c	specific heat	Btu/lbm-°F	J/kg·°C
C	molar specific heat	Btu/lbmol-°F	J/kmol·°C
E	energy	ft-lbf	J
F	force	lbf	N
g	gravitational acceleration, 32.2 (9.81)	ft/sec^2	m/s^2
g_c	gravitational constant, 32.2	lbm-ft/lbf-sec^2	n.a.
h	height	ft	m
I	mass moment of inertia	lbm-ft^2	kg·m^2
J	Joule's constant, 778.17	ft-lbf/Btu	n.a.
k	spring constant	lbf/ft	N/m
m	mass	lbm	kg
MW	molecular weight	lbm/lbmol	kg/kmol
p	pressure	lbf/ft^2	Pa
P	power	ft-lbf/sec	W
q	heat	Btu/lbm	J/kg
Q	heat	Btu	J
r	radius	ft	m
s	distance	ft	m
t	time	sec	s
T	temperature	°F	°C
T	torque	ft-lbf	N·m
u	specific energy	ft-lbf/lbm	J/kg
U	internal energy	Btu	J
v	velocity	ft/sec	m/s
W	work	ft-lbf	J

Symbols

δ	deflection	ft	m
η	efficiency	–	–
θ	angular position	rad	rad
ρ	mass density	lbm/ft^3	kg/m^3
υ	specific volume	ft^3/lbm	m^3/kg
ϕ	angle	deg	deg
ω	angular velocity	rad/sec	rad/s

Subscripts

f	frictional
p	constant pressure
v	constant volume

1. ENERGY OF A MASS

The *energy* of a mass represents the capacity of the mass to do work. Such energy can be stored and released. There are many forms that it can take, including mechanical, thermal, electrical, and magnetic energies. Energy is a positive, scalar quantity (although the change in energy can be either positive or negative).

The total energy of a body can be calculated from its mass, m, and its *specific energy*, u (i.e., the energy per unit mass).[1]

$$E = mu \qquad 13.1$$

Typical units of mechanical energy are foot-pounds and joules. (A joule is equivalent to the units of N·m and kg·m^2/s^2.) In traditional English-unit countries, the *British thermal unit* (Btu) is used for thermal energy, whereas the kilocalorie (kcal) is still used in some applications in SI countries. *Joule's constant*, or the *Joule equivalent* (778.17 ft-lbf/Btu, usually shortened to 778, three significant digits), is used to convert between English mechanical and thermal energy units.

$$\text{energy in Btu} = \frac{\text{energy in ft-lbf}}{J} \qquad 13.2$$

Two other units of large amounts of energy are the therm and the quad. A *therm* is 10^5 Btu (1.055×10^8 J). A *quad* is equal to a quadrillion (10^{15}) Btu. This is 1.055×10^{18} J, or roughly the energy contained in 200 million barrels of oil.

2. LAW OF CONSERVATION OF ENERGY

The *law of conservation of energy* says that energy cannot be created or destroyed. However, energy can be converted into different forms. Therefore, the sum of all energy forms is constant.

$$\sum E = \text{constant} \qquad 13.3$$

---

[1]The use of symbols E and u for energy is not consistent in the engineering field.

3. WORK

Work, W, is the act of changing the energy of a particle, body, or system. For a mechanical system, *external work* is work done by an external force, whereas *internal work* is done by an internal force. Work is a signed, scalar quantity. Typical units are inch-pounds, foot-pounds, and joules. Mechanical work is seldom expressed in British thermal units or kilocalories.

For a mechanical system, work is positive when a force acts in the direction of motion and helps a body move from one location to another. Work is negative when a force acts to oppose motion. (Friction, for example, always opposes the direction of motion and can do only negative work.) The work done on a body by more than one force can be found by superposition.

From a thermodynamic standpoint, work is positive if a particle or body does work on its surroundings. Work is negative if the surroundings do work on the object. (For example, blowing up a balloon represents negative work to the balloon.) This is consistent with the conservation of energy, since the sum of negative work and the positive energy increase is zero (i.e., no net energy change in the system).[2]

The work performed by a variable force and torque are calculated as the dot products of Eq. 13.4 and Eq. 13.5.

$$W_{\text{variable force}} = \int \mathbf{F} \cdot d\mathbf{s} \quad \text{[linear systems]} \qquad 13.4$$

$$W_{\text{variable torque}} = \int \mathbf{T} \cdot d\theta \quad \text{[rotational systems]} \qquad 13.5$$

The work done by a force or torque of constant magnitude is

$$W_{\text{constant force}} = \mathbf{F} \cdot \mathbf{s} = Fs \cos \phi \quad \text{[linear systems]}$$
$$13.6$$

$$W_{\text{constant torque}} = \mathbf{T} \cdot \theta$$
$$= Fr\theta \cos \phi \quad \text{[rotational systems]} \qquad 13.7$$

The nonvector forms, Eq. 13.6 and Eq. 13.7, illustrate that only the component of force or torque in the direction of motion contributes to work. (See Fig. 13.1.)

Figure 13.1 *Work of a Constant Force*

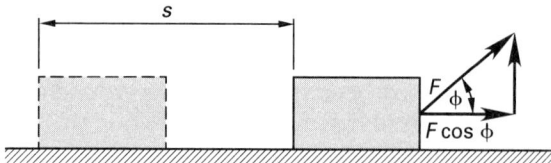

Common applications of the work done by a constant force are frictional work and gravitational work. The work to move an object a distance s against a frictional force of F_f is

$$W_{\text{friction}} = F_f s \qquad 13.8$$

The work done against gravity when a mass m changes in elevation from h_1 to h_2 is

$$W_{\text{gravity}} = mg(h_2 - h_1) \qquad \text{[SI]} \quad 13.9(a)$$

$$W_{\text{gravity}} = \frac{mg(h_2 - h_1)}{g_c} \qquad \text{[U.S.]} \quad 13.9(b)$$

The work done by or on a *linear spring* whose length or deflection changes from δ_1 to δ_2 is given by Eq. 13.10.[3] It does not make any difference whether the spring is a compression spring or an extension spring.

$$W_{\text{spring}} = \tfrac{1}{2}k(\delta_2^2 - \delta_1^2) \qquad 13.10$$

Example 13.1

A lawn mower engine is started by pulling a cord wrapped around a sheave. The sheave radius is 8.0 cm. The cord is wrapped around the sheave two times. If a constant tension of 90 N is maintained in the cord during starting, what work is done?

Solution

The starting torque on the engine is

$$T = Fr = (90 \text{ N}) \left(\frac{8 \text{ cm}}{100 \frac{\text{cm}}{\text{m}}} \right) = 7.2 \text{ N·m}$$

The cord wraps around the sheave $(2)(2\pi) = 12.6$ rad. From Eq. 13.7, the work done by a constant torque is

$$W = T\theta = (7.2 \text{ N·m})(12.6 \text{ rad})$$
$$= 90.7 \text{ J}$$

Example 13.2

A 200 lbm crate is pushed 25 ft at constant velocity across a warehouse floor. There is a frictional force of 60 lbf between the crate and floor. What work is done by the frictional force on the crate?

Solution

From Eq. 13.8,

$$W_{\text{friction}} = F_f s = (60 \text{ lbf})(25 \text{ ft})$$
$$= 1500 \text{ ft-lbf}$$

---

[2]This is just a partial statement of the *first law of thermodynamics*.

[3]A *linear spring* is one for which the linear relationship $F = kx$ is valid.

4. POTENTIAL ENERGY OF A MASS

Potential energy (gravitational energy) is a form of mechanical energy possessed by a body due to its relative position in a gravitational field. Potential energy is lost when the elevation of a body decreases. The lost potential energy usually is converted to kinetic energy or heat.

$$E_{\text{potential}} = mgh \qquad \text{[SI]} \qquad 13.11(a)$$

$$E_{\text{potential}} = \frac{mgh}{g_c} \qquad \text{[U.S.]} \qquad 13.11(b)$$

In the absence of friction and other nonconservative forces, the change in potential energy of a body is equal to the work required to change the elevation of the body.

$$W = \Delta E_{\text{potential}} \qquad 13.12$$

5. KINETIC ENERGY OF A MASS

Kinetic energy is a form of mechanical energy associated with a moving or rotating body. The kinetic energy of a body moving with instantaneous linear velocity, v, is

$$E_{\text{kinetic}} = \tfrac{1}{2}mv^2 \qquad \text{[SI]} \qquad 13.13(a)$$

$$E_{\text{kinetic}} = \frac{mv^2}{2g_c} \qquad \text{[U.S.]} \qquad 13.13(b)$$

A body can also have rotational kinetic energy.

$$E_{\text{rotational}} = \tfrac{1}{2}I\omega^2 \qquad \text{[SI]} \qquad 13.14(a)$$

$$E_{\text{rotational}} = \frac{I\omega^2}{2g_c} \qquad \text{[U.S.]} \qquad 13.14(b)$$

According to the *work-energy principle* (see Sec. 13.9), the kinetic energy is equal to the work necessary to initially accelerate a stationary body or to bring a moving body to rest.

$$W = \Delta E_{\text{kinetic}} \qquad 13.15$$

Example 13.3

A solid disk flywheel ($I = 200$ kg·m^2) is rotating with a speed of 900 rpm. What is its rotational kinetic energy?

Solution

The angular rotational velocity is

$$\omega = \frac{\left(900\ \frac{\text{rev}}{\text{min}}\right)\left(2\pi\ \frac{\text{rad}}{\text{rev}}\right)}{60\ \frac{\text{s}}{\text{min}}} = 94.25\ \text{rad/s}$$

From Eq. 13.14, the rotational kinetic energy is

$$E = \tfrac{1}{2}I\omega^2 = (\tfrac{1}{2})(200\ \text{kg·m}^2)\left(94.25\ \frac{\text{rad}}{\text{s}}\right)^2$$
$$= 888 \times 10^3\ \text{J} \quad (888\ \text{kJ})$$

6. SPRING ENERGY

A spring is an energy storage device because the spring has the ability to perform work. In a perfect spring, the amount of energy stored is equal to the work required to compress the spring initially. The stored spring energy does not depend on the mass of the spring. Given a spring with spring constant (stiffness) k, the *spring energy* is

$$E_{\text{spring}} = \tfrac{1}{2}k\delta^2 \qquad 13.16$$

Example 13.4

A body of mass m falls from height h onto a massless, simply supported beam. The mass adheres to the beam. If the beam has a lateral stiffness k, what will be the deflection, δ, of the beam?

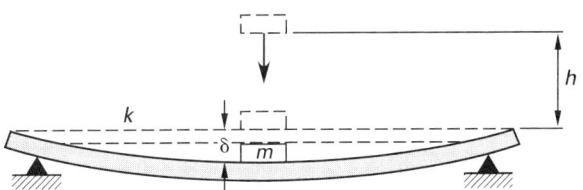

Solution

The initial energy of the system consists of only the potential energy of the body. Using consistent units, the change in potential energy is

$$E = mg(h + \delta) \quad \text{[consistent units]}$$

All of this energy is stored in the spring. Therefore, from Eq. 13.16,

$$\tfrac{1}{2}k\delta^2 = mg(h + \delta)$$

Solving for the deflection,

$$\delta = \frac{mg \pm \sqrt{mg(2hk + mg)}}{k}$$

7. PRESSURE ENERGY OF A MASS

Since work is done in increasing the pressure of a system (e.g., work is done in blowing up a balloon), mechanical energy can be stored in pressure form. This is known as *pressure energy, static energy, flow energy, flow work,* and p-V work (*energy*). For a system of pressurized mass m, the pressure energy is

$$E_{\text{flow}} = \frac{mp}{\rho} = mpv \quad [v = \text{specific volume}] \qquad 13.17$$

8. INTERNAL ENERGY OF A MASS

The property of internal energy is encountered primarily in thermodynamics problems. Typical units are British thermal units, joules, and kilocalories. The total internal energy, usually given the symbol U, of a body increases when the body's temperature increases.[4] In the absence of any work done on or by the body, the change in internal energy is equal to the heat flow, Q, into the body. Q is positive if the heat flow is into the body and negative otherwise.

$$U_2 - U_1 = Q \qquad 13.18$$

An increase in internal energy is needed to cause a rise in temperature. Different substances differ in the quantity of heat needed to produce a given temperature increase. The heat, Q, required to change the temperature of a mass, m, by an amount, ΔT, is called the *specific heat* (*heat capacity*) of the substance, c.

$$c = \frac{Q}{m\Delta T} \qquad 13.19$$

Because specific heats of solids and liquids are slightly temperature dependent, the mean specific heats are used when evaluating processes covering a large temperature range.

The lowercase c implies that the units are Btu/lbm-°F or J/kg·°C. Typical values of specific heat are given in Table 13.1. The *molar specific heat*, designated by the symbol C, has units of Btu/lbmol-°F or J/kmol·°C.

$$C = (\text{MW}) \times c \qquad 13.20$$

For gases, the specific heat depends on the type of process during which the heat exchange occurs. Specific heats for constant-volume and constant-pressure processes are designated by c_v and c_p, respectively.

$$Q = mc_v\Delta T \quad [\text{constant-volume process}] \qquad 13.21$$

$$Q = mc_p\Delta T \quad [\text{constant-pressure process}] \qquad 13.22$$

Values of c_p and c_v for solids and liquids are essentially the same. However, the designation c_p is often encountered for solids and liquids. Table 13.1 gives values of c_p for selected liquids and solids.

---

[4]The *thermal energy*, represented by the body's enthalpy, is the sum of internal and pressure energies.

Table 13.1 *Approximate Specific Heats of Selected Liquids and Solids**

substance	c_p	
	Btu/lbm-°F	kJ/kg·°C
aluminum, pure	0.23	0.96
aluminum, 2024-T4	0.2	0.84
ammonia	1.16	4.86
asbestos	0.20	0.84
benzene	0.41	1.72
brass, red	0.093	0.39
bronze	0.082	0.34
concrete	0.21	0.88
copper, pure	0.094	0.39
Freon-12	0.24	1.00
gasoline	0.53	2.20
glass	0.18	0.75
gold, pure	0.031	0.13
ice	0.49	2.05
iron, pure	0.11	0.46
iron, cast (4% C)	0.10	0.42
lead, pure	0.031	0.13
magnesium, pure	0.24	1.00
mercury	0.033	0.14
oil, light hydrocarbon	0.5	2.09
silver, pure	0.06	0.25
steel, 1010	0.10	0.42
steel, stainless 301	0.11	0.46
tin, pure	0.055	0.23
titanium, pure	0.13	0.54
tungsten, pure	0.032	0.13
water	1.0	4.19
wood (typical)	0.6	2.50
zinc, pure	0.088	0.37

(Multiply Btu/lbm-°F by 4.1868 to obtain kJ/kg·°C.)
*Values in Btu/lbm-°F are the same as cal/g·°C. Values in kJ/kg·°C are the same as kJ/kg·K.

9. WORK-ENERGY PRINCIPLE

Since energy can neither be created nor destroyed, external work performed on a conservative system goes into changing the system's total energy. This is known as the *work-energy principle* (or *principle of work and energy*).

$$W = \Delta E = E_2 - E_1 \qquad 13.23$$

Generally, the term *work-energy principle* is limited to use with mechanical energy problems (i.e., conversion of work into kinetic or potential energies). When energy is limited to kinetic energy, the work-energy principle is a direct consequence of Newton's second law but is valid for only inertial reference systems.

By directly relating forces, displacements, and velocities, the work-energy principle introduces some simplifications into many mechanical problems.

- It is not necessary to calculate or know the acceleration of a body to calculate the work performed on it.

- Forces that do not contribute to work (e.g., are normal to the direction of motion) are irrelevant.

- Only scalar quantities are involved.

- It is not necessary to individually analyze the particles or component parts in a complex system.

Example 13.5

A 4000 kg elevator starts from rest, accelerates uniformly to a constant speed of 2.0 m/s, and then decelerates uniformly to a stop 20 m above its initial position. Neglecting friction and other losses, what work was done on the elevator?

Solution

By the work-energy principle, the work done on the elevator is equal to the change in the elevator's energy. Since the initial and final kinetic energies are zero, the only mechanical energy change is the potential energy change.

Taking the initial elevation of the elevator as the reference (i.e., $h_1 = 0$), and combining Eq. 13.9(a) and Eq. 13.23,

$$W = E_{2,\text{potential}} - E_{1,\text{potential}} = mg(h_2 - h_1)$$

$$= (4000 \text{ kg})\left(9.81 \ \frac{\text{m}}{\text{s}^2}\right)(20 \text{ m})$$

$$= 785 \times 10^3 \text{ J} \quad (785 \text{ kJ})$$

10. CONVERSION BETWEEN ENERGY FORMS

Conversion of one form of energy into another does not violate the conservation of energy law. However, most problems involving conversion of energy are really just special cases of the work-energy principle. For example, consider a falling body that is acted upon by a gravitational force. The conversion of potential energy into kinetic energy can be interpreted as equating the work done by the constant gravitational force to the change in kinetic energy.

In general terms, *Joule's law* states that one energy form can be converted without loss into another. There are two specific formulations of Joule's law. As related to electricity, $P = I^2 R = V^2/R$ is the common formulation of Joule's law. As related to thermodynamics and ideal gases, Joule's law states that "the change in internal energy of an ideal gas is a function of the temperature change, not of the volume." This latter form can also be stated more formally as "at constant temperature, the internal energy of a gas approaches a finite value that is independent of the volume as the pressure goes to zero."

Example 13.6

A 2 lbm projectile is launched straight up with an initial velocity of 700 ft/sec. Neglecting air friction, calculate the (a) kinetic energy immediately after launch, (b) kinetic energy at maximum height, (c) potential energy at maximum height, (d) total energy at an elevation where the velocity has dropped to 300 ft/sec, and (e) maximum height attained.

Solution

(a) From Eq. 13.13, the kinetic energy is

$$E_{\text{kinetic}} = \frac{mv^2}{2g_c} = \frac{(2 \text{ lbm})\left(700 \ \frac{\text{ft}}{\text{sec}}\right)^2}{(2)\left(32.2 \ \frac{\text{lbm-ft}}{\text{lbf-sec}^2}\right)}$$

$$= 15{,}217 \text{ ft-lbf}$$

(b) The velocity is zero at the maximum height. Therefore, the kinetic energy is zero.

(c) At the maximum height, all of the kinetic energy has been converted into potential energy. Therefore, the potential energy is 15,217 ft-lbf.

(d) Although some of the kinetic energy has been transformed into potential energy, the total energy is still 15,217 ft-lbf.

(e) Since all of the kinetic energy has been converted into potential energy, the maximum height can be found from Eq. 13.11.

$$E_{\text{potential}} = \frac{mgh}{g_c}$$

$$15{,}217 \text{ ft-lbf} = \frac{(2 \text{ lbm})\left(32.2 \ \frac{\text{ft}}{\text{sec}^2}\right)h}{32.2 \ \frac{\text{lbm-ft}}{\text{lbf-sec}^2}}$$

$$h = 7609 \text{ ft}$$

Example 13.7

A 4500 kg ore car rolls down an incline and passes point A traveling at 1.2 m/s. The ore car is stopped by a spring bumper that compresses 0.6 m. A constant friction force of 220 N acts on the ore car at all times. What is the spring constant?

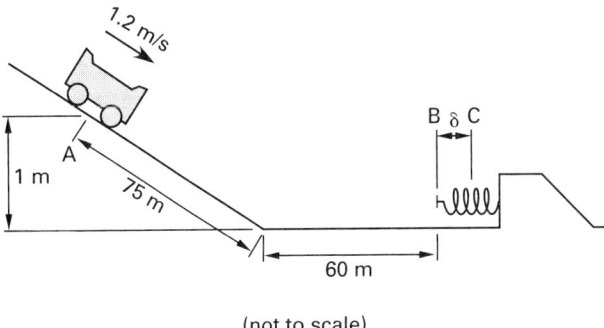

(not to scale)

Solution

The car's total energy at point A is the sum of the kinetic and potential energies.

$$E_{\text{total,A}} = E_{\text{kinetic}} + E_{\text{potential}} = \tfrac{1}{2}m\text{v}^2 + mgh$$

$$= \left(\tfrac{1}{2}\right)(4500 \text{ kg})\left(1.2 \ \tfrac{\text{m}}{\text{s}}\right)^2$$

$$+ (4500 \text{ kg})\left(9.81 \ \tfrac{\text{m}}{\text{s}^2}\right)(1 \text{ m})$$

$$= 47\,385 \text{ J}$$

At point B, the potential energy has been converted into additional kinetic energy. However, except for friction, the total energy is the same as at point A. Since the frictional force does negative work, the total energy remaining at point B is

$$E_{\text{total,B}} = E_{\text{total,A}} - W_{\text{friction}}$$

$$= 47\,385 \text{ J} - (220 \text{ N})(75 \text{ m} + 60 \text{ m})$$

$$= 17\,685 \text{ J}$$

At point C, the maximum compression point, the remaining energy has gone into compressing the spring a distance $\delta = 0.6$ m and performing a small amount of frictional work.

$$E_{\text{total,B}} = E_{\text{total,C}} = W_{\text{spring}} + W_{\text{friction}}$$

$$= \tfrac{1}{2}k\delta^2 + F_f\delta$$

$$17\,685 = \tfrac{1}{2}k(0.6 \text{ m})^2 + (220 \text{ N})(0.6 \text{ m})$$

The spring constant can be determined directly.

$$k = 97\,520 \text{ N/m} \quad (97.5 \text{ kN/m})$$

11. POWER

Power is the amount of work done per unit time. It is a scalar quantity. (Although power is calculated from two vectors, the vector dot-product operation is seldom needed.) Typical basic units of power are ft-lbf/sec and watts (J/s), although *horsepower* is widely used. Table 13.2 can be used to convert units of power.

$$P = \frac{W}{\Delta t} \qquad 13.24$$

For a body acted upon by a force or torque, the instantaneous power can be calculated from the velocity.

$$P = F\text{v} \quad \text{[linear systems]} \qquad 13.25$$

$$P = T\omega \quad \text{[rotational systems]} \qquad 13.26$$

For a fluid flowing at a rate of $\dot{m}$, the unit of time is already incorporated into the flow rate (e.g., lbm/sec). If the fluid experiences a specific energy change of Δu, the power generated or dissipated will be

$$P = \dot{m}\Delta u \qquad 13.27$$

Table 13.2 *Useful Power Conversion Formulas*

1 hp	= 550 ft-lbf/sec
	= 33,000 ft-lbf/min
	= 0.7457 kW
	= 0.7068 Btu/sec
1 kW	= 737.6 ft-lbf/sec
	= 44,250 ft-lbf/min
	= 1.341 hp
	= 0.9483 Btu/sec
1 Btu/sec	= 778.17 ft-lbf/sec
	= 46,680 ft-lbf/min
	= 1.415 hp

Example 13.8

When traveling at 100 km/h, a car supplies a constant horizontal force of 50 N to the hitch of a trailer. What tractive power (in horsepower) is required for the trailer alone?

Solution

From Eq. 13.25, the power being generated is

$$P = F\text{v} = \frac{(50 \text{ N})\left(100 \ \tfrac{\text{km}}{\text{h}}\right)\left(1000 \ \tfrac{\text{m}}{\text{km}}\right)}{\left(60 \ \tfrac{\text{s}}{\text{min}}\right)\left(60 \ \tfrac{\text{min}}{\text{h}}\right)\left(1000 \ \tfrac{\text{W}}{\text{kW}}\right)}$$

$$= 1.389 \text{ kW}$$

Using a conversion from Table 13.2, the horsepower is

$$P = \left(1.341 \ \frac{\text{hp}}{\text{kW}}\right)(1.389 \text{ kW}) = 1.86 \text{ hp}$$

12. EFFICIENCY

For energy-using systems (such as cars, electrical motors, elevators, etc.), the *energy-use efficiency*, η, of a system is the ratio of an ideal property to an actual property. The property used is commonly work, power, or, for thermodynamics problems, heat. When the rate of work is constant, either work or power can be used to calculate the efficiency. Otherwise, power should be used. Except in rare instances, the numerator and denominator of the ratio must have the same units.[5]

$$\eta = \frac{P_{\text{ideal}}}{P_{\text{actual}}} \quad [P_{\text{actual}} \geq P_{\text{ideal}}] \qquad 13.28$$

For energy-producing systems (such as electrical generators, prime movers, and hydroelectric plants), the *energy-production efficiency* is

$$\eta = \frac{P_{\text{actual}}}{P_{\text{ideal}}} \quad [P_{\text{ideal}} \geq P_{\text{actual}}] \qquad 13.29$$

The efficiency of an *ideal machine* is 1.0 (100%). However, all *real machines* have efficiencies of less than 1.0.

---

[5]The *energy-efficiency ratio* used to evaluate refrigerators, air conditioners, and heat pumps, for example, has units of Btu per watt-hour (Btu/W-hr).

Topic II: Water Resources #1: Flow of Fluids

Flow of Fluids

Chapter

14. Fluid Properties

15. Fluid Statics

16. Fluid Flow Parameters

17. Fluid Dynamics

18. Hydraulic Machines

19. Open Channel Flow

20. Meteorology, Climatology, and Hydrology

21. Groundwater

14 Fluid Properties

Nomenclature

a	speed of sound	ft/sec	m/s
A	area	ft^2	m^2
d	diameter	ft	m
E	bulk modulus	lbf/ft^2	Pa
F	force	lbf	N
g	gravitational acceleration, 32.2 (9.81)	ft/sec^2	m/s^2
G	gravimetric (mass) fraction	–	–
g_c	gravitational constant, 32.2	lbm-ft/lbf-sec^2	n.a.
h	height	ft	m
k	ratio of specific heats	–	–
L	length	ft	m
m	mass	lbm	kg
M	molar concentration	lbmol/ft^3	kmol/m^3
M	Mach number	–	–
MW	molecular weight	lbm/lbmol	kg/kmol
n	number of moles	–	–
p	pressure	lbf/ft^2	Pa
r	radius	ft	m
R	specific gas constant	ft-lbf/lbm-°R	J/kg·K
R^*	universal gas constant, 1545.35 (8314.47)	ft-lbf/lbmol-°R	J/kmol·K
SG	specific gravity	–	–
T	absolute temperature	°R	K
v	velocity	ft/sec	m/s
V	volume	ft^3	m^3
VBI	viscosity blending index	–	–
x	mole fraction	–	–
y	distance	ft	m
Z	compressibility factor	–	–

Symbols

β	compressibility	ft^2/lbf	Pa^{-1}
β	contact angle	deg	deg
γ	specific weight	lbf/ft^3	n.a.
μ	absolute viscosity	lbf-sec/ft^2	Pa·s
ν	kinematic viscosity	ft^2/sec	m^2/s
π	osmotic pressure	lbf/ft^2	Pa
ρ	density	lbm/ft^3	kg/m^3
σ	surface tension	lbf/ft	N/m
τ	shear stress	lbf/ft^2	Pa
v	specific volume	ft^3/lbm	m^3/kg

Subscripts

0	zero velocity (wall face)
c	critical
p	constant pressure
s	constant entropy
T	constant temperature
v	vapor

1. CHARACTERISTICS OF A FLUID

Liquids and gases can both be categorized as fluids, although this chapter is primarily concerned with incompressible liquids. There are certain characteristics shared by all fluids, and these characteristics can be used, if necessary, to distinguish between liquids and gases.[1]

- *Compressibility:* Liquids are only slightly compressible and are assumed to be incompressible for most purposes. Gases are highly compressible.

- *Shear resistance:* Liquids and gases cannot support shear, and they deform continuously to minimize applied shear forces.

- *Shape and volume:* As a consequence of their inability to support shear forces, liquids and gases take on the shapes of their containers. Only liquids have free surfaces. Liquids have fixed volumes, regardless of their container volumes, and these volumes are not significantly affected by temperature and pressure. Unlike liquids, gases take on the volumes of their containers. If allowed to do so, gas densities will change as temperature and pressure are varied.

[1]The differences between liquids and gases become smaller as temperature and pressure are increased. Gas and liquid properties become the same at the critical temperature and pressure.

- *Resistance to motion:* Due to viscosity, liquids resist instantaneous changes in velocity, but the resistance stops when liquid motion stops. Gases have very low viscosities.

- *Molecular spacing:* Molecules in liquids are relatively close together and are held together with strong forces of attraction. Liquid molecules have low kinetic energy. The distance each liquid molecule travels between collisions is small. In gases, the molecules are relatively far apart and the attractive forces are weak. Kinetic energy of the molecules is high. Gas molecules travel larger distances between collisions.

- *Pressure:* The pressure at a point in a fluid is the same in all directions. Pressure exerted by a fluid on a solid surface (e.g., container wall) is always normal to that surface.

2. TYPES OF FLUIDS

For computational convenience, fluids are generally divided into two categories: ideal fluids and real fluids. (See Fig. 14.1.) *Ideal fluids* are assumed to have no viscosity (and therefore, no resistance to shear), be incompressible, and have uniform velocity distributions when flowing. In an ideal fluid, there is no friction between moving layers of fluid, and there are no eddy currents or turbulence.

Figure 14.1 Types of Fluids

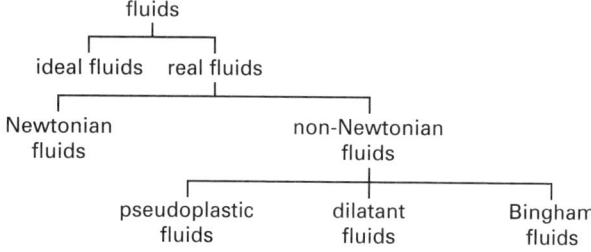

Real fluids exhibit finite viscosities and nonuniform velocity distributions, are compressible, and experience friction and turbulence in flow. Real fluids are further divided into *Newtonian fluids* and *non-Newtonian fluids*, depending on their viscous behavior. The differences between Newtonian and the various types of non-Newtonian fluids are described in Sec. 14.9.

For convenience, most fluid problems assume real fluids with Newtonian characteristics. This is an appropriate assumption for water, air, gases, steam, and other simple fluids (alcohol, gasoline, acid solutions, etc.). However, slurries, pastes, gels, suspensions, and polymer/electrolytesolutions may not behave according to simple fluid relationships.

3. FLUID PRESSURE AND VACUUM

In the English system, fluid pressure is measured in pounds per square inch (lbf/in^2 or psi) and pounds per square foot (lbf/ft^2 or psf), although tons (2000 pounds) per square foot (tsf) are occasionally used. In SI units, pressure is measured in pascals (Pa). Because a pascal is very small, kilopascals (kPa) and megapascals (MPa) are usually used. Other units of pressure include bars, millibars, atmospheres, inches and feet of water, torrs, and millimeters, centimeters, and inches of mercury. (See Fig. 14.2.)

Figure 14.2 Relative Sizes of Pressure Units

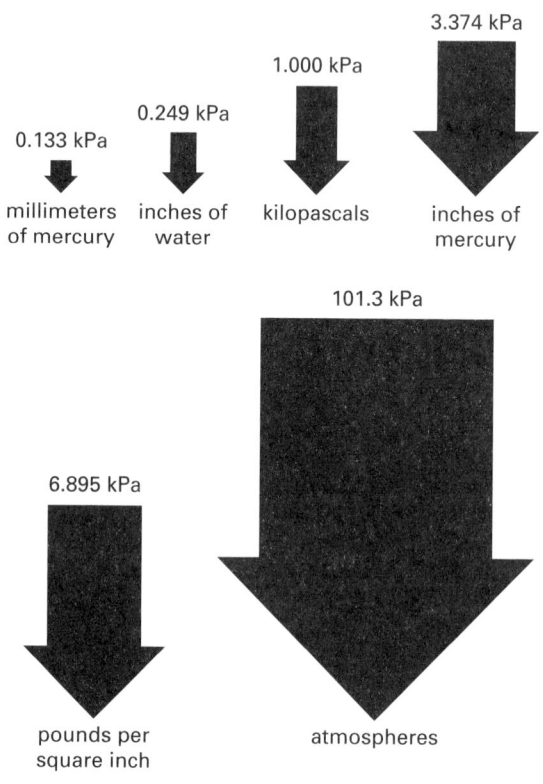

Fluid pressures are measured with respect to two pressure references: zero pressure and atmospheric pressure. Pressures measured with respect to a true zero pressure reference are known as *absolute pressures*. Pressures measured with respect to atmospheric pressure are known as *gage pressures.*[2] Most pressure gauges read the excess of the test pressure over atmospheric pressure (i.e., the gage pressure). To distinguish between these two pressure measurements, the letters "a" and "g" are traditionally added to the the unit symbols in the English unit system (e.g., 14.7 psia and 4015 psfg). For SI units, the actual words "gage" and "absolute" can be added to the measurement (e.g., 25.1 kPa absolute). Alternatively, the pressure is assumed to be absolute unless the "g" is used (e.g., 15 kPag or 98 barg).

---

[2]The spelling *gage* persists even though pressures are measured with *gauges*. In some countries, the term *meter pressure* is used instead of gage pressure.

Absolute and gage pressures are related by Eq. 14.1. It should be mentioned that $p_{atmospheric}$ in Eq. 14.1 is the actual atmospheric pressure existing when the gage measurement is taken. It is not standard atmospheric pressure, unless that pressure is implicitly or explicitly applicable. Also, since a barometer measures atmospheric pressure, *barometric pressure* is synonymous with atmospheric pressure. Table 14.1 lists standard atmospheric pressure in various units.

$$p_{absolute} = p_{gage} + p_{atmospheric} \qquad 14.1$$

Table 14.1 Standard Atmospheric Pressure

1.000 atm	(atmosphere)
14.696 psia	(pounds per square inch absolute)
2116.2 psfa	(pounds per square foot absolute)
407.1 in wg	(inches of water, inches water gage)
33.93 ft wg	(feet of water, feet water gage)
29.921 in Hg	(inches of mercury)
760.0 mm Hg	(millimeters of mercury)
760.0 torr	
1.013 bars	
1013 millibars	
1.013×10^5 Pa	(pascals)
101.3 kPa	(kilopascals)

A *vacuum* measurement is implicitly a pressure below atmospheric (i.e., a negative gage pressure). It must be assumed that any measured quantity given as a vacuum is a quantity to be subtracted from the atmospheric pressure. For example, when a condenser is operating with a vacuum of 4.0 in Hg (4 in of mercury), the absolute pressure is 29.92 in Hg − 4.0 in Hg = 25.92 in Hg. Vacuums are generally stated as positive numbers.

$$p_{absolute} = p_{atmospheric} - p_{vacuum} \qquad 14.2$$

A difference in two pressures may be reported with units of *psid* (i.e., a *differential* in psi).

4. DENSITY

The *density*, ρ, of a fluid is its mass per unit volume.[3] In SI units, density is measured in kg/m^3. In a consistent English system, density would be measured in $slugs/ft^3$, even though fluid density is exclusively reported in lbm/ft^3.

The density of a fluid in a liquid form is usually given, known in advance, or easily obtained from tables in any one of a number of sources. (See Table 14.2.) Most English fluid data are reported on a per pound basis,

---

[3]Mass is an absolute property of a substance. Weight is not absolute, since it depends on the local gravity. The equations using γ that result (such as Bernoulli's equation) cannot be used with SI data, since the equations are not consistent. Thus, engineers end up with two different equations for the same thing.

and the data included in this book follow that tradition. To convert pounds to slugs, divide by g_c.

$$\rho_{slugs} = \frac{\rho_{lbm}}{g_c} \qquad 14.3$$

The density of an ideal gas can be found from the specific gas constant and the ideal gas law.

$$\rho = \frac{p}{RT} \qquad 14.4$$

Table 14.2 Approximate Densities of Common Fluids

fluid	lbm/ft^3	kg/m^3
air (STP)	0.0807	1.29
air (70°F, 1 atm)	0.075	1.20
alcohol	49.3	790
ammonia	38	602
gasoline	44.9	720
glycerin	78.8	1260
mercury	848	13 600
water	62.4	1000

(Multiply lbm/ft^3 by 16.01 to obtain kg/m^3.)

Example 14.1

The density of water is typically taken as 62.4 lbm/ft^3 for engineering problems where greater accuracy is not required. What is the value in (a) $slugs/ft^3$ and (b) kg/m^3?

Solution

(a) Equation 14.3 can be used to calculate the slug-density of water.

$$\rho = \frac{\rho_{lbm}}{g_c} = \frac{62.4 \frac{lbm}{ft^3}}{32.2 \frac{lbm\text{-}ft}{lbf\text{-}sec^2}} = 1.94 \text{ lbf-sec}^2/\text{ft-ft}^3$$

$$= 1.94 \text{ slugs/ft}^3$$

(b) The conversion between lbm/ft^3 and kg/m^3 is approximately 16.0, derived as follows.

$$\rho = \left(62.4 \ \frac{lbm}{ft^3}\right)\left(\frac{35.31 \frac{ft^3}{m^3}}{2.205 \frac{lbm}{kg}}\right)$$

$$= \left(62.4 \ \frac{lbm}{ft^3}\right)\left(16.01 \ \frac{kg\text{-}ft^3}{m^3\text{-}lbm}\right)$$

$$= 999 \text{ kg/m}^3$$

In SI problems, it is common to take the density of water as 1000 kg/m^3.

5. SPECIFIC VOLUME

Specific volume, v, is the volume occupied by a unit mass of fluid.[4] Since specific volume is the reciprocal of density, typical units will be ft^3/lbm, $ft^3/lbmol$, or m^3/kg.[5]

$$v = \frac{1}{\rho} \qquad 14.5$$

6. SPECIFIC GRAVITY

Specific gravity (SG) is a dimensionless ratio of a fluid's density to some standard reference density.[6] For liquids and solids, the reference is the density of pure water. There is some variation in this reference density, however, since the temperature at which the water density is evaluated is not standardized. Temperatures of $39.2°F$ ($4°C$), $60°F$ ($16.5°C$), and $70°F$ ($21.1°C$) have been used.[7]

Fortunately, the density of water is the same to three significant digits over the normal ambient temperature range: 62.4 lbm/ft^3 or 1000 kg/m^3. However, to be precise, the temperature of both the fluid and water should be specified (e.g., "...the specific gravity of the $20°C$ fluid is 1.05 referred to $4°C$ water...").

$$SG_{liquid} = \frac{\rho_{liquid}}{\rho_{water}} \qquad 14.6$$

Since the SI density of water is very nearly 1.000 g/cm^3 (1000 kg/m^3), the numerical values of density in g/cm^3 and specific gravity are the same. Such is not the case with English units.

The standard reference used to calculate the specific gravity of gases is the density of air. Since the density of a gas depends on temperature and pressure, both must be specified for the gas and air (i.e., two temperatures and two pressures must be specified). While STP (standard temperature and pressure) conditions are commonly specified, they are not universal.[8] Table 14.3 lists several common sets of standard conditions.

$$SG_{gas} = \frac{\rho_{gas}}{\rho_{air}} \qquad 14.7$$

Table 14.3 Commonly Quoted Values of Standard Temperature and Pressure

system	temperature	pressure
SI	273.15K	101.325 kPa
scientific	0.0°C	760 mm Hg
U.S. engineering	32°F	14.696 psia
natural gas industry (U.S.)	60°F	14.65, 14.73, or 15.025 psia
natural gas industry (Canada)	60°F	14.696 psia

If it is known or implied that the temperature and pressure of the air and gas are the same, the specific gravity of the gas will be equal to the ratio of molecular weights and the inverse ratio of specific gas constants. The density of air evaluated at STP is listed in Table 14.2. At $70°F$ ($21.1°C$) and 1.0 atm, the density is approximately 0.075 lbm/ft^3 (1.20 kg/m^3).

$$SG_{gas} = \frac{MW_{gas}}{MW_{air}} = \frac{MW_{gas}}{29.0} = \frac{R_{air}}{R_{gas}}$$
$$= \frac{53.3\ \frac{ft\text{-}lbf}{lbm\text{-}°R}}{R_{gas}} \qquad 14.8$$

Specific gravities of petroleum liquids and aqueous solutions (of acid, antifreeze, salts, etc.) can be determined by use of a *hydrometer*. (See Fig. 14.3.) In its simplest form, a hydrometer is constructed as a graduated scale weighted at one end so it will float vertically. The height at which the hydrometer floats depends on the density of the fluid, and the graduated scale can be calibrated directly in specific gravity.[9]

Figure 14.3 Hydrometer

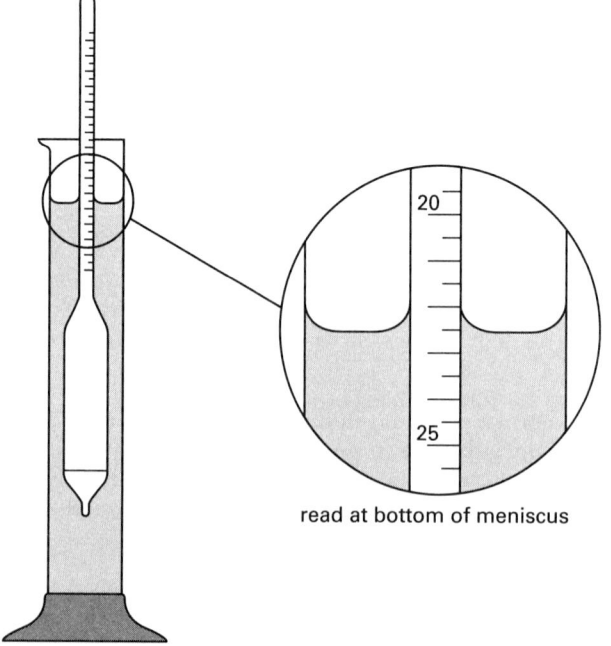

read at bottom of meniscus

---

[4]Care must be taken to distinguish between the symbol upsilon, v, used for specific volume, and italic roman "vee," v, used for velocity in many engineering textbooks.

[5]Units of $ft^3/slug$ are also possible, but this combination of units is almost never encountered.

[6]The symbols S.G., sp.gr., S, and G are also used. In fact, petroleum engineers in the United States use γ, a symbol that civil engineers use for specific weight. There is no standard engineering symbol for specific gravity.

[7]Density of liquids is sufficiently independent of pressure to make consideration of pressure in specific gravity calculations unnecessary.

[8]The abbreviation "SC" (standard conditions) is interchangeable with "STP."

[9]This is a direct result of the buoyancy principle of Archimedes.

There are two standardized hydrometer scales (i.e., methods for calibrating the hydrometer stem).[10] Both state specific gravity in degrees, although temperature is not being measured. The *American Petroleum Institute* (API) scale (°API) may be used with all liquids, not only with oils or other hydrocarbons. For the specific gravity value, a standard reference temperature of 60°F (15.6°C) is implied for both the liquid and the water.

$$°\text{API} = \frac{141.5}{\text{SG}} - 131.5 \qquad 14.9$$

$$\text{SG} = \frac{141.5}{°\text{API} + 131.5} \qquad 14.10$$

The *Baumé scale* (°Be) is used in the wine, honey, and acid industries. It is somewhat confusing because there are actually two Baumé scales—one for liquids heavier than water and another for liquids lighter than water. (There is also a discontinuity in the scales at SG = 1.00.) As with the API scale, the specific gravity value assumes 60°F (15.6°C) is the standard temperature for both scales.

$$\text{SG} = \frac{140.0}{130.0 + °\text{Be}} \quad [\text{SG} < 1.00] \qquad 14.11$$

$$\text{SG} = \frac{145.0}{145.0 - °\text{Be}} \quad [\text{SG} > 1.00] \qquad 14.12$$

Example 14.2

Determine the specific gravity of carbon dioxide gas (molecular weight = 44) at 66°C (150°F) and 138 kPa (20 psia) using STP air as a reference. The specific gas constant of carbon dioxide is 35.1 ft-lbf/lbm-°R.

SI Solution

Since the specific gas constant for carbon dioxide was not given in SI units, it must be calculated from the universal gas constant.

$$R = \frac{R^*}{\text{MW}} = \frac{8314.47 \dfrac{\text{J}}{\text{kmol·K}}}{44 \dfrac{\text{kg}}{\text{kmol}}} = 189 \text{ J/kg·K}$$

$$\rho = \frac{p}{RT} = \frac{1.38 \times 10^5 \text{ Pa}}{\left(189 \dfrac{\text{J}}{\text{kg·K}}\right)(66°\text{C} + 273°)}$$

$$= 2.15 \text{ kg/m}^3$$

From Table 14.2, the density of STP air is 1.29 kg/m³. From Eq. 14.7, the specific gravity of carbon dioxide at the conditions given is

$$\text{SG} = \frac{\rho_{\text{gas}}}{\rho_{\text{air}}} = \frac{2.15 \dfrac{\text{kg}}{\text{m}^3}}{1.29 \dfrac{\text{kg}}{\text{m}^3}} = 1.67$$

Customary U.S. Solution

Since the conditions of the carbon dioxide and air are different, Eq. 14.8 cannot be used. Therefore, it is necessary to calculate the density of the carbon dioxide from Eq. 14.4. The density is

$$\rho = \frac{p}{RT} = \frac{\left(20 \dfrac{\text{lbf}}{\text{in}^2}\right)\left(12 \dfrac{\text{in}}{\text{ft}}\right)^2}{\left(35.1 \dfrac{\text{ft-lbf}}{\text{lbm-°R}}\right)(150°\text{F} + 460°)}$$

$$= 0.135 \text{ lbm/ft}^3$$

From Table 14.2, the density of STP air is 0.0807 lbm/ft³. From Eq. 14.7, the specific gravity of carbon dioxide at the conditions given is

$$\text{SG} = \frac{\rho_{\text{gas}}}{\rho_{\text{air}}} = \frac{0.135 \dfrac{\text{lbm}}{\text{ft}^3}}{0.0807 \dfrac{\text{lbm}}{\text{ft}^3}} = 1.67$$

7. SPECIFIC WEIGHT

Specific weight (unit weight), γ, is the weight of fluid per unit volume. The use of specific weight is most often encountered in civil engineering projects in the United States, where it is commonly called "density." The usual units of specific weight are lbf/ft³.[11] Specific weight is not an absolute property of a fluid, since it depends not only on the fluid, but on the local gravitational field as well.

$$\gamma = g\rho \qquad [\text{SI}] \qquad 14.13(a)$$

$$\gamma = \rho \times \frac{g}{g_c} \qquad [\text{U.S.}] \qquad 14.13(b)$$

If the gravitational acceleration is 32.2 ft/sec², as it is almost everywhere on the earth, the specific weight in lbf/ft³ will be numerically equal to the density in lbm/ft³. This is illustrated in Ex. 14.3.

Example 14.3

What is the sea level ($g = 32.2$ ft/sec²) specific weight (in lbf/ft³) of liquids with densities of (a) 1.95 slug/ft³, and (b) 58.3 lbm/ft³?

---

[10]In addition to °Be and °API mentioned in this chapter, the *Twaddell scale* (°Tw) is used in chemical processing, the *Brix* and *Balling scales* are used in the sugar industry, and the *Salometer scale* is used to measure salt (NaCl and CaCl₂) solutions.

[11]Notice that the units are lbf/ft³, not lbm/ft³. Pound-mass (lbm) is a mass unit, not a weight (force) unit.

Solution

(a) Equation 14.13(a) can be used with any consistent set of units, including densities involving slugs.

$$\gamma = g\rho = \left(32.2 \ \frac{ft}{sec^2}\right)\left(1.95 \ \frac{slug}{ft^3}\right)$$
$$= \left(32.2 \ \frac{ft}{sec^2}\right)\left(1.95 \ \frac{lbf\text{-}sec^2}{ft\text{-}ft^3}\right)$$
$$= 62.8 \ lbf/ft^3$$

(b) From Eq. 14.13(b),

$$\gamma = \rho \times \frac{g}{g_c} = \left(58.3 \ \frac{lbm}{ft^3}\right)\left(\frac{32.2 \ \frac{ft}{sec^2}}{32.2 \ \frac{lbm\text{-}ft}{lbf\text{-}sec^2}}\right)$$

$$= 58.3 \ lbf/ft^3$$

8. MOLE FRACTION

Mole fraction is an important parameter in many practical engineering problems, particularly in chemistry and chemical engineering. The composition of a fluid consisting of two or more distinctly different substances, A, B, C, and so on, can be described by the mole fractions, x_A, x_B, x_C, and so on, of each substance. (There are also other methods of specifying the composition.) The mole fraction of component A is the number of moles of that component, n_A, divided by the total number of moles in the combined fluid mixture.

$$x_A = \frac{n_A}{n_A + n_B + n_C + \cdots} \qquad 14.14$$

Mole fraction is a number between 0 and 1. *Mole percent* is the mole fraction multiplied by 100%, expressed in percent.

9. VISCOSITY

The *viscosity* of a fluid is a measure of that fluid's resistance to flow when acted upon by an external force such as a pressure differential or gravity. Some fluids, such as heavy oils, jellies, and syrups, are very viscous. Other fluids, such as water, lighter hydrocarbons, and gases, are not as viscous.

The more viscous the fluid, the more time will be required for the fluid to leak out of a container. *Saybolt Seconds Universal* (SSU) and *Saybolt Seconds Furol* (SSF) are scales of such viscosity measurement based on the smaller and larger orifices, respectively. Seconds can be converted (empirically) to viscosity in other units. The following relations are approximate conversions between SSU and stokes.

- For SSU < 100 sec,
$$\nu_{stokes} = 0.00226(SSU) - \frac{1.95}{SSU} \qquad 14.15$$

- For SSU > 100 sec,
$$\nu_{stokes} = 0.00220(SSU) - \frac{1.35}{SSU} \qquad 14.16$$

Most common liquids will flow more easily when their temperatures are raised. However, the behavior of a fluid when temperature, pressure, or stress is varied will depend on the type of fluid. The different types of fluids can be determined with a *sliding plate viscometer test*.[12]

Consider two plates of area A separated by a fluid with thickness y_0, as shown in Fig. 14.4. The bottom plate is fixed, and the top plate is kept in motion at a constant velocity, v_0, by a force, F.

Figure 14.4 *Sliding Plate Viscometer*

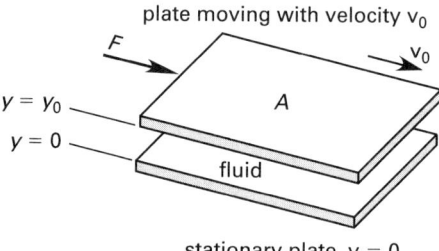

Experiments with water and most common fluids have shown that the force, F, required to maintain the velocity, v_0, is proportional to the velocity and the area and is inversely proportional to the separation of the plates. That is,

$$\frac{F}{A} \propto \frac{dv}{dy} \qquad 14.17$$

The constant of proportionality needed to make Eq. 14.17 an equality is the *absolute viscosity*, μ, also known as the *coefficient of viscosity*.[13] The reciprocal of absolute viscosity, $1/\mu$, is known as the *fluidity*.

$$\frac{F}{A} = \mu \frac{dv}{dy} \qquad 14.18$$

F/A is the *fluid shear stress*, τ. The quantity dv/dy (v_0/y_0) is known by various names, including *rate of strain*, *shear rate*, *velocity gradient*, and *rate of shear*

---

[12]This test is conceptually simple but is not always practical, since the liquid leaks out between the plates. In research work with liquids, it is common to determine viscosity with a *concentric cylinder viscometer*, also known as a *cup-and-bob viscometer*. Viscosities of perfect gases can be predicted by the kinetic theory of gases. Viscosity can also be measured by a *Saybolt viscometer*, which is essentially a container that allows a given quantity of fluid to leak out through one of two different-sized orifices.

[13]Another name for absolute viscosity is *dynamic viscosity*. The name *absolute viscosity* is preferred, if for no other reason than to avoid confusion with *kinematic viscosity*.

formation. Equation 14.18 is known as *Newton's law of viscosity*, from which *Newtonian fluids* get their name. Sometimes Eq. 14.19 is written with a minus sign to compare viscous behavior with other behavior. However, the direction of positive shear stress is arbitrary. Equation 14.19 is simply the equation of a straight line.

$$\tau = \mu \frac{dv}{dy} \qquad \textit{14.19}$$

Not all fluids are Newtonian (although most common fluids are), and Eq. 14.19 is not universally applicable. Figure 14.5 (known as a *rheogram*) illustrates how differences in fluid shear stress behavior (at constant temperature and pressure) can be used to define Bingham, pseudoplastic, and dilatant fluids, as well as Newtonian fluids.

Gases, water, alcohol, and benzene are examples of Newtonian fluids. In fact, all liquids with a simple chemical formula are Newtonian. Also, most solutions of simple compounds, such as sugar and salt, are Newtonian. For a more viscous fluid, the straight line will be closer to the τ axis (i.e., the slope will be higher). (See Fig. 14.5.) For low-viscosity fluids, the straight line will be closer to the dv/dy axis (i.e., the slope will be lower).

Figure 14.5 *Shear Stress Behavior for Different Types of Fluids*

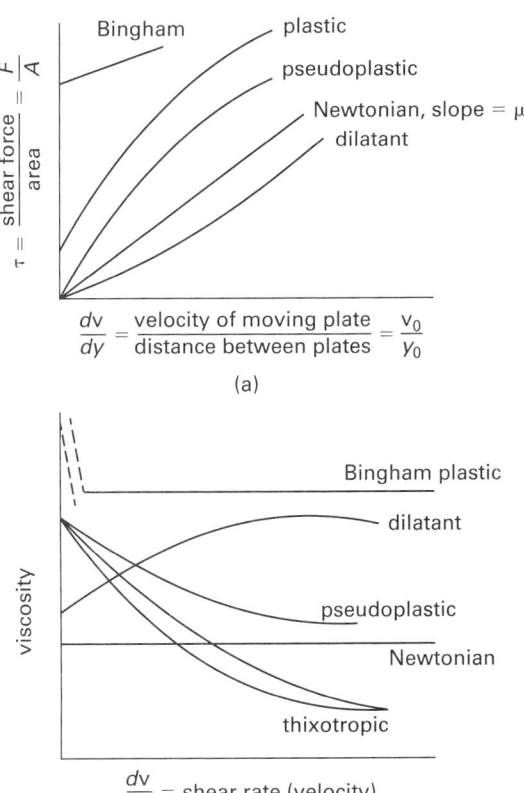

(a)

(b)

Pseudoplastic fluids (muds, motor oils, polymer solutions, natural gums, and most slurries) exhibit viscosities that decrease with an increasing velocity gradient. Such fluids present no serious pumping problems.

Plastic materials, such as tomato catsup, behave similarly to pseudoplastic fluids once movement begins; that is, their viscosities decrease with agitation. However, a finite force must be applied before any fluid movement occurs.

Bingham fluids (Bingham plastics), typified by toothpaste, jellies, bread dough, and some slurries, are capable of indefinitely resisting a small shear stress but move easily when the stress becomes large—that is, Bingham fluids become pseudoplastic when the stress increases.

Dilatant fluids are rare but include clay slurries, various starches, some paints, milk chocolate with nuts, and other candy compounds. They exhibit viscosities that increase with increasing agitation (i.e., with increasing velocity gradients), but they return rapidly to their normal viscosity after the agitation ceases. Pump selection is critical for dilatant fluids because these fluids can become almost solid if the shear rate is high enough.

Viscosity can also change with time (all other conditions being constant). If viscosity decreases with time during agitation, the fluid is said to be a *thixotropic fluid*. If viscosity increases (usually up to a finite value) with time during agitation, the fluid is a *rheopectic fluid*. Viscosity does not change in time-independent fluids. *Colloidal materials*, such as gelatinous compounds, lotions, shampoos, and low-temperature solutions of soaps in water and oil, behave like *thixotropic liquids*—their viscosities decrease as the agitation continues. However, viscosity does not return to its original state after the agitation ceases.

Molecular cohesion is the dominating cause of viscosity in liquids. As the temperature of a liquid increases, these cohesive forces decrease, resulting in a decrease in viscosity.

In gases, the dominant cause of viscosity is random collisions between gas molecules. This molecular agitation increases with increases in temperature. Therefore, viscosity in gases increases with temperature.

Although viscosity of liquids increases slightly with pressure, the increase is insignificant over moderate pressure ranges. Therefore, the absolute viscosity of both gases and liquids is usually considered to be essentially independent of pressure.[14]

The units of absolute viscosity, as derived from Eq. 14.19, are lbf-sec/ft^2. Such units are actually used in the English engineering system.[15] Absolute viscosity is measured in pascal-seconds (Pa·s) in SI units. Another common unit used throughout the world is the *poise* (abbreviated P), equal to a dyne·s/cm^2. These dimensions are the same primary dimensions as in the English system, $F\theta/L^2$ or $M/L\theta$, and are functionally the same as a g/cm·s. Since the poise is a large unit, the *centipoise* (abbreviated cP) scale is generally used. The viscosity of pure water at room temperature is approximately 1 cP.

---

[14]This is not true for kinematic viscosity, however.

[15]Units of lbm/ft-sec are also used for absolute viscosity in the English system. These units are obtained by multiplying lbf-sec/ft^2 units by g_c.

Example 14.4

A liquid ($\mu = 5.2 \times 10^{-5}$ lbf-sec/ft^2) is flowing in a rectangular duct. The equation of the symmetrical facial velocity distribution (in ft/sec) is approximately $v = 3y^{0.7}$ ft/sec, where y is measured in inches from the wall. (a) What is the velocity gradient at $y = 3.0$ in from the duct wall? (b) What is the shear stress in the fluid at that point?

Solution

(a) The velocity is not a linear function of y, so dv/dy must be calculated as a derivative.

$$\frac{dv}{dy} = \frac{d}{dy} 3y^{0.7} = (3)(0.7y^{-0.3}) = 2.1y^{-0.3}$$

At $y = 3$ in,

$$\frac{dv}{dy} = (2.1)(3)^{-0.3} = 1.51 \text{ ft/sec-in}$$

(b) From Eq. 14.19, the shear stress is

$$\tau = \mu \frac{dv}{dy}$$

$$= \left(5.2 \times 10^{-5} \frac{\text{lbf-sec}}{\text{ft}^2}\right)\left(1.51 \frac{\text{ft}}{\text{sec-in}}\right)\left(12 \frac{\text{in}}{\text{ft}}\right)$$

$$= 9.42 \times 10^{-4} \text{ lbf/ft}^2$$

10. KINEMATIC VISCOSITY

Another quantity with the name *viscosity* is the ratio of absolute viscosity to mass density. This combination of variables, known as *kinematic viscosity*, ν, appears sufficiently often in fluids and other problems as to warrant its own symbol and name. Typical units are ft^2/sec and cm^2/s (the *stoke*, St). It is also common to give kinematic viscosity in *centistokes*, cSt. The SI units of kinematic viscosity are m^2/s.

$$\nu = \frac{\mu}{\rho} \qquad \text{[SI]} \quad 14.20(a)$$

$$\nu = \frac{\mu g_c}{\rho} = \frac{\mu g}{\gamma} \qquad \text{[U.S.]} \quad 14.20(b)$$

It is essential that consistent units be used with Eq. 14.20. The following sets of units are consistent.

$$\text{ft}^2/\text{sec} = \frac{\text{lbf-sec/ft}^2}{\text{slugs/ft}^3}$$

$$\text{m}^2/\text{s} = \frac{\text{Pa·s}}{\text{kg/m}^3}$$

$$\text{St (stoke)} = \frac{\text{P (poise)}}{\text{g/cm}^3}$$

$$\text{cSt (centistokes)} = \frac{\text{cP (centipoise)}}{\text{g/cm}^3}$$

Unlike absolute viscosity, kinematic viscosity is greatly dependent on both temperature and pressure, since these variables affect the density of the fluid. Referring to Eq. 14.20, even if absolute viscosity is independent of temperature or pressure, the change in density will change the kinematic viscosity.

11. VISCOSITY CONVERSIONS

The most common units of absolute and kinematic viscosity are listed in Table 14.4.

Table 14.4 Common Viscosity Units

	absolute, μ	kinematic, ν
English	lbf-sec/ft^2 (slug/ft-sec)	ft^2/sec
conventional metric	dyne·s/cm^2 (poise)	cm^2/s (stoke)
SI	Pa·s (N·s/m^2)	m^2/s

Table 14.5 contains conversions between the various viscosity units.

Example 14.5

Water at 60°F has a specific gravity of 0.999 and a kinematic viscosity of 1.12 cSt. What is the absolute viscosity in lbf-sec/ft^2?

Solution

The density of a liquid expressed in g/cm^3 is numerically equal to its specific gravity.

$$\rho = 0.999 \text{ g/cm}^3$$

Table 14.5 Viscosity Conversions*

multiply	by	to obtain
absolute viscosity, μ		
dyne·s/cm²	0.10	Pa·s
lbf-sec/ft²	478.8	P
lbf-sec/ft²	47,880	cP
lbf-sec/ft²	47.88	Pa·s
slug/ft-sec	47.88	Pa·s
lbm/ft-sec	1.488	Pa·s
cP	1.0197×10^{-4}	kgf·s/m²
cP	2.0885×10^{-5}	lbf-sec/ft²
cP	0.001	Pa·s
Pa·s	0.020885	lbf-sec/ft²
Pa·s	1000	cP
reyn	144	lbf-sec/ft²
reyn	1.0	lbf-sec/in²
kinematic viscosity, ν		
ft²/sec	92,903	cSt
ft²/sec	0.092903	m²/s
m²/s	10.7639	ft²/sec
m²/s	1×10^6	cSt
cSt	1×10^{-6}	m²/s
cSt	1.0764×10^{-5}	ft²/sec
absolute viscosity to kinematic viscosity		
cP	$1/\rho$ in g/cm³	cSt
cP	$6.7195 \times 10^{-4}/\rho$ in lbm/ft³	ft²/sec
lbf-sec/ft²	$32.174 /\rho$ in lbm/ft³	ft²/sec
kgf·s/m²	$9.807 /\rho$ in kg/m³	m²/s
Pa·s	$1000 /\rho$ in g/cm³	cSt
kinematic viscosity to absolute viscosity		
cSt	ρ in g/cm³	cP
cSt	$0.001 \times \rho$ in g/cm³	Pa·s
cSt	$1.6 \times 10^{-5} \times \rho$ in lbm/ft³	Pa·s
m²/s	$0.10197 \times \rho$ in kg/m³	kgf·s/m²
m²/s	$1000 \times \rho$ in g/cm³	Pa·s
m²/s	ρ in kg/m³	Pa·s
ft²/sec	$0.031081 \times \rho$ in lbm/ft³	lbf-sec/ft²
ft²/sec	$1488.2 \times \rho$ in lbm/ft³	cP

*cP: centipoise; cSt: centistoke; kgf: kilogram-force; P: poise

The centistoke (cSt) is a measure of kinematic viscosity. Kinematic viscosity is converted first to the absolute viscosity units of centipoise. From Table 14.5,

$$\mu_{cP} = \nu_{cSt}\rho_{g/cm^3}$$
$$= (1.12 \text{ cSt})\left(0.999 \ \frac{g}{cm^3}\right)$$
$$= 1.119 \text{ cP}$$

Next, centipoise is converted to lbf-sec/ft².

$$\mu_{lbf-sec/ft^2} = \mu_{cP}(2.0885 \times 10^{-5})$$
$$= (1.119 \text{ cP})(2.0885 \times 10^{-5})$$
$$= 2.34 \times 10^{-5} \text{ lbf-sec/ft}^2$$

12. VISCOSITY GRADE

The ISO *viscosity grade* (VG) as specified in ISO 3448, is commonly used to classify oils. (See Table 14.6.) Viscosity at 104°F (40°C), the approximate temperature of machinery, in centistokes (same as mm²/s), is used as the index. Each subsequent viscosity grade within the classification has approximately a 50% higher viscosity, whereas the minimum and maximum values of each grade range ±10% from the midpoint.

Table 14.6 ISO Viscosity Grade

ISO 3448 viscosity grade	kinematic viscosity at 40°C (cSt)		
	minimum	midpoint	maximum
ISO VG 2	1.98	2.2	2.42
ISO VG 3	2.88	3.2	3.52
ISO VG 5	4.14	4.6	5.06
ISO VG 7	6.12	6.8	7.48
ISO VG 10	9.0	10	11.0
ISO VG 15	13.5	15	16.5
ISO VG 22	19.8	22	24.2
ISO VG 32	28.8	32	35.2
ISO VG 46	41.4	46	50.6
ISO VG 68	61.2	68	74.8
ISO VG 100	90	100	110
ISO VG 150	135	150	165
ISO VG 220	198	220	242
ISO VG 320	288	320	352
ISO VG 460	414	460	506
ISO VG 680	612	680	748
ISO VG 1000	900	1000	1100
ISO VG 1500	1350	1500	1650

13. VISCOSITY INDEX

Viscosity index (VI) is a measure of a fluid's viscosity sensitivity to changes in temperature. It has traditionally been applied to crude and refined oils through use of a 100-point scale.[16] The viscosity is measured at two temperatures: 100°F and 210°F (38°C and 99°C). These viscosities are converted into a viscosity index in accordance with standard ASTM D2270. (See App. 14.L.)

14. VAPOR PRESSURE

Molecular activity in a liquid will allow some of the molecules to escape the liquid surface. Strictly speaking, a small portion of the liquid vaporizes. Molecules of the vapor also condense back into the liquid. The vaporization and condensation at constant temperature are equilibrium processes. The equilibrium pressure exerted by these free molecules is known as the *vapor pressure* or *saturation pressure*. (Vapor pressure does not include the pressure of other substances in the mixture.) Typical values of vapor pressure are given in Table 14.7.

[16]Use of the *viscosity index* has been adopted by other parts of the chemical process industry (CPI), including in the manufacture of solvents, polymers, and other synthetics. The 100-point scale may be exceeded (on both ends) for these uses. Refer to standard ASTM D2270 for calculating extreme values of the viscosity index.

Table 14.7 Typical Vapor Pressures

fluid	lbf/ft², 68°F	kPa, 20°C
mercury	0.00362	0.000173
turpentine	1.115	0.0534
water	48.9	2.34
ethyl alcohol	122.4	5.86
ether	1231	58.9
butane	4550	218
Freon-12	12,200	584
propane	17,900	855
ammonia	18,550	888

(Multiply lbf/ft² by 0.04788 to obtain kPa.)

Some liquids, such as propane, butane, ammonia, and Freon, have significant vapor pressures at normal temperatures. Liquids near their boiling points or that vaporize easily are said to be *volatile liquids*.[17] Other liquids, such as mercury, have insignificant vapor pressures at normal temperatures. Liquids with low vapor pressures are used in accurate barometers.

The tendency toward vaporization is dependent on the temperature of the liquid. *Boiling* occurs when the liquid temperature is increased to the point that the vapor pressure is equal to the local ambient pressure. Therefore, a liquid's boiling temperature depends on the local ambient pressure as well as on the liquid's tendency to vaporize.

Vapor pressure is usually considered to be a nonlinear function of temperature only. It is possible to derive correlations between vapor pressure and temperature, and such correlations usually involve a logarithmic transformation of vapor pressure.[18] Vapor pressure can also be graphed against temperature in a (logarithmic) *Cox chart* (see App. 14.H) when values are needed over larger temperature extremes. Although there is also some variation with external pressure, the external pressure effect is negligible under normal conditions.

15. OSMOTIC PRESSURE

Osmosis is a special case of diffusion in which molecules of the *solvent* move under pressure from one fluid to another (i.e., from the *solvent* to the *solution*) in one direction only, usually through a *semipermeable membrane*.[19] Osmosis continues until sufficient solvent has passed through the membrane to make the activity (or solvent pressure) of the solution equal to that of the solvent.[20] The pressure at equilibrium is known as the *osmotic pressure*, π.

---

[17]Because a liquid that vaporizes easily has an aroma, the term *aromatic liquid* is also occasionally used.

[18]The *Clausius-Clapeyron equation* and *Antoine equation* are two such logarithmic correlations of vapor pressure with temperature.

[19]A semipermeable membrane will be impermeable to the solute but permeable for the solvent.

[20]Two solutions in equilibrium (i.e., whose activities are equal) are said to be in *isopiestic equilibrium*.

Figure 14.6 illustrates an *osmotic pressure apparatus*. The fluid column can be interpreted as the result of an osmotic pressure that has developed through diffusion into the solution. The fluid column will continue to increase in height until equilibrium is reached. Alternatively, the fluid column can be adjusted so that the solution pressure just equals the osmotic pressure that would develop otherwise, in order to prevent the flow of solvent. For the arrangement in Fig. 14.6, the osmotic pressure can be calculated from the difference in fluid level heights, h.

$$\pi = \rho g h \qquad \text{[SI]} \qquad 14.21(a)$$

$$\pi = \frac{\rho g h}{g_c} \qquad \text{[U.S.]} \qquad 14.21(b)$$

Figure 14.6 Osmotic Pressure Apparatus

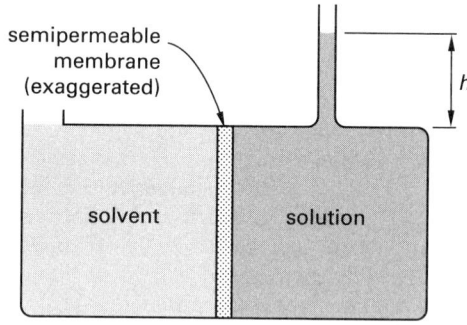

In dilute solutions, osmotic pressure follows the ideal gas law. The solute acts like a gas in exerting pressure against the membrane. The solvent exerts no pressure since it can pass through. In Eq. 14.22, M is the molarity (concentration). The value of the *universal gas constant*, R^*, depends on the units used. Common values include 1545.35 ft-lbf/lbmol-°R, 8314.47 J/kmol·K, and 0.08206 atm·L/mol·K. Consistent units must be used.

$$\pi = MR^*T \qquad 14.22$$

Example 14.6

An aqueous solution is in isopiestic equilibrium with a 0.1 molarity sucrose solution at 22°C. What is the osmotic pressure?

Solution

Referring to Eq. 14.22,

$$M = 0.1 \text{ mol/L of solution}$$

$$T = 22°C + 273° = 295\text{K}$$

$$\pi = MR^*T = \left(0.1 \frac{\text{mol}}{\text{L}}\right)\left(0.08206 \frac{\text{atm·L}}{\text{mol·K}}\right)(295\text{K})$$

$$= 2.42 \text{ atm}$$

16. SURFACE TENSION

The membrane or "skin" that seems to form on the free surface of a fluid is due to the intermolecular cohesive forces and is known as *surface tension*, σ. Surface tension is the reason that insects are able to walk and a needle is able to float on water. Surface tension also causes bubbles and droplets to take on a spherical shape, since any other shape would have more surface area per unit volume.

Data on the surface tension of liquids is important in determining the performance of heat-, mass-, and momentum-transfer equipment, including heat transfer devices.[21] Surface tension data is needed to calculate the nucleate boiling point (i.e., the initiation of boiling) of liquids in a pool (using the *Rohsenow equation*) and the maximum heat flux of boiling liquids in a pool (using the *Zuber equation*).

Surface tension can be interpreted as the tension between two points a unit distance apart on the surface or as the amount of work required to form a new unit of surface area in an apparatus similar to that shown in Fig. 14.7. Typical units of surface tension are lbf/ft (ft-lbf/ft^2) dyne/cm, and N/m.

Figure 14.7 Wire Frame for Stretching a Film

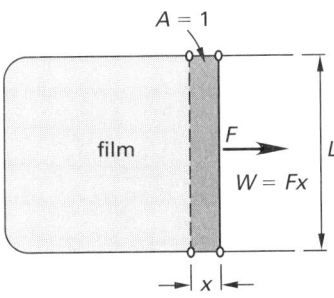

The apparatus shown in Fig. 14.7 consists of a wire frame with a sliding side that has been dipped in a liquid to form a film. Surface tension is determined by measuring the force necessary to keep the sliding side stationary against the surface tension pull of the film.[22] (The film does not act like a spring, since the force, F, does not increase as the film is stretched.) Since the film has two surfaces (i.e., two surface tensions), the surface tension is

$$\sigma = \frac{F}{2L} \qquad \text{14.23}$$

Alternatively, surface tension can also be determined by measuring the force required to pull a wire ring out

of the liquid, as shown in Fig. 14.8.[23] Since the ring's inner and outer sides are in contact with the liquid, the wetted perimeter is twice the circumference. The surface tension is

$$\sigma = \frac{F}{4\pi r} \qquad \text{14.24}$$

Figure 14.8 Du Nouy Ring Surface Tension Apparatus

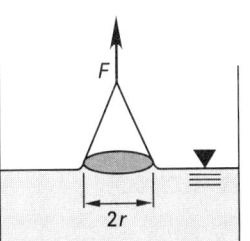

Surface tension depends slightly on the gas in contact with the free surface. Surface tension values are usually quoted for air contact. Typical values of surface tension are listed in Table 14.8.

Table 14.8 Approximate Values of Surface Tension (air contact)

fluid	lbf/ft, 68°F	N/m, 20°C
n-octane	0.00149	0.0217
ethyl alcohol	0.00156	0.0227
acetone	0.00162	0.0236
kerosene	0.00178	0.0260
carbon tetrachloride	0.00185	0.0270
turpentine	0.00186	0.0271
toluene	0.00195	0.0285
benzene	0.00198	0.0289
olive oil	0.0023	0.034
glycerin	0.00432	0.0631
water	0.00499	0.0728
mercury	0.0356	0.519

(Multiply lbf/ft by 14.59 to obtain N/m.)
(Multiply dyne/cm by 0.001 to obtain N/m.)

At temperatures below freezing, the substance will be a solid, so surface tension is a moot point. As the temperature of a liquid is raised, the surface tension decreases because the cohesive forces decrease. Surface tension is zero at a substance's critical temperature. If a substance's critical temperature is known, the *Othmer correlation*, Eq. 14.25, can be used to determine the surface tension at one temperature from the surface tension at another temperature.

$$\sigma_2 = \sigma_1 \left(\frac{T_c - T_2}{T_c - T_1} \right)^{11/9} \qquad \text{14.25}$$

Surface tension is the reason that the pressure on the inside of bubbles and droplets is greater than on the

---

[21]Surface tension plays a role in processes involving dispersion, emulsion, flocculation, foaming, and solubilization. It is not surprising that surface tension data are particularly important in determining the performance of equipment in the chemical process industry (CPI), such as distillation columns, packed towers, wetted-wall columns, strippers, and phase-separation equipment.

[22]The force includes the weight of the sliding side wire if the frame is oriented vertically, with gravity acting on the sliding side wire to stretch the film.

[23]This apparatus is known as a *Du Nouy torsion balance*. The ring is made of platinum with a diameter of 4.00 cm.

FLUID PROPERTIES **14-11**

PPI • w w w . p p i 2 p a s s . c o m

outside. Equation 14.26 gives the relationship between the surface tension in a hollow bubble surrounded by a gas and the difference between the inside and outside pressures. For a spherical droplet or a bubble in a liquid, where in both cases there is only one surface in tension, the surface tension is twice as large. (r is the radius of the bubble or droplet.)

$$\sigma_{\text{bubble}} = \frac{r(p_{\text{inside}} - p_{\text{outside}})}{4} \qquad 14.26$$

$$\sigma_{\text{droplet}} = \frac{r(p_{\text{inside}} - p_{\text{outside}})}{2} \qquad 14.27$$

17. CAPILLARY ACTION

Capillary action (*capillarity*) is the name given to the behavior of a liquid in a thin-bore tube. Capillary action is caused by surface tension between the liquid and a vertical solid surface.[24] In the case of liquid water in a glass tube, the adhesive forces between the liquid molecules and the surface are greater than (i.e., dominate) the cohesive forces between the water molecules themselves.[25] The adhesive forces cause the water to attach itself to and climb a solid vertical surface. It can be said that the water "reaches up and tries to wet as much of the interior surface as it can." In so doing, the water rises above the general water surface level. The surface is *hydrophilic* (*lyophilic*). This is illustrated in Fig. 14.9.

Figure 14.9 *Capillarity of Liquids*

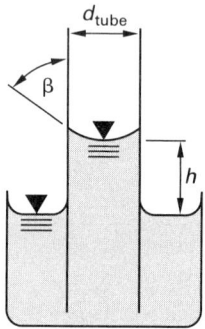
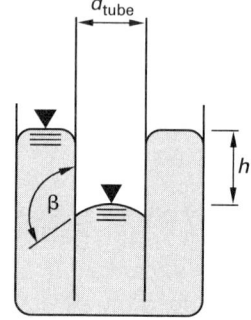

(a) adhesive force dominates   (b) cohesive force dominates

Figure 14.9 also illustrates that the same surface tension forces that keep a droplet spherical are at work on the surface of the liquid in the tube. The curved liquid surface, known as the *meniscus*, can be considered to be an incomplete droplet. If the inside diameter of the tube is less than approximately 0.1 in (2.5 mm), the meniscus is essentially hemispherical, and $r_{\text{meniscus}} = r_{\text{tube}}$.

---

[24]In fact, observing the rise of liquid in a capillary tube is another method of determining the surface tension of a liquid.
[25]*Adhesion* is the attractive force between molecules of different substances. *Cohesion* is the attractive force between molecules of the same substance.

For a few other liquids, such as mercury, the molecules have a strong affinity for each other (i.e., the cohesive forces dominate). The liquid avoids contact with the tube surface. The surface is *hydrophobic* (*lyophobic*). In such liquids, the meniscus in the tube will be below the general surface level.

The *angle of contact*, β, is an indication of whether adhesive or cohesive forces dominate. For contact angles less than 90°, adhesive forces dominate. For contact angles greater than 90°, cohesive forces dominate.

Equation 14.28 can be used to predict the capillary rise in a small-bore tube. Surface tension and contact angles can be obtained from Table 14.8 and Table 14.9, respectively.

$$h = \frac{4\sigma \cos \beta}{\rho d_{\text{tube}} g} \qquad \text{[SI]} \quad 14.28(a)$$

$$h = \frac{4\sigma \cos \beta}{\rho d_{\text{tube}}} \times \frac{g_c}{g} \qquad \text{[U.S.]} \quad 14.28(b)$$

$$\sigma = \frac{h \rho d_{\text{tube}} g}{4 \cos \beta} \qquad \text{[SI]} \quad 14.29(a)$$

$$\sigma = \frac{h \rho d_{\text{tube}}}{4 \cos \beta} \times \frac{g}{g_c} \qquad \text{[U.S.]} \quad 14.29(b)$$

$$r_{\text{meniscus}} = \frac{d_{\text{tube}}}{2 \cos \beta} \qquad 14.30$$

Table 14.9 *Contact Angles, β*

materials	angle
mercury–glass	140°
water–paraffin	107°
water–silver	90°
silicone oil–glass	20°
kerosene–glass	26°
glycerin–glass	19°
water–glass	0°
ethyl alcohol–glass	0°

If it is assumed that the meniscus is hemispherical, then $r_{\text{meniscus}} = r_{\text{tube}}$, $\beta = 0°$, and $\cos \beta = 1.0$, and the above equations can be simplified. (Such an assumption can only be made when the diameter of the capillary tube is less than 0.1 in.)

Example 14.7

To what height will 68°F (20°C) ethyl alcohol rise in a 0.005 in (0.127 mm) internal diameter glass capillary tube? The alcohol's density is 49 lbm/ft³ (790 kg/m³).

SI Solution

From Table 14.8 and Table 14.9, respectively, the surface tension and contact angle are

$$\sigma = 0.0227 \text{ N/m}$$

$$\beta = 0°$$

From Eq. 14.28, the height is

$$h = \frac{4\sigma\cos\beta}{\rho d_{\text{tube}}g} = \frac{(4)\left(0.0227\,\frac{\text{N}}{\text{m}}\right)(1.0)\left(1000\,\frac{\text{mm}}{\text{m}}\right)}{\left(790\,\frac{\text{kg}}{\text{m}^3}\right)(0.127\text{ mm})\left(9.81\,\frac{\text{m}}{\text{s}^2}\right)}$$

$$= 0.0923 \text{ m}$$

Customary U.S. Solution

From Table 14.8 and Table 14.9, respectively, the surface tension and contact angle are

$$\sigma = 0.00156 \text{ lbf/ft}$$

$$\beta = 0°$$

From Eq. 14.28, the height is

$$h = \frac{4\sigma\cos\beta g_c}{\rho d_{\text{tube}}g}$$

$$= \frac{(4)\left(0.00156\,\frac{\text{lbf}}{\text{ft}}\right)(1.0)\left(32.2\,\frac{\text{lbm-ft}}{\text{lbf-sec}^2}\right)\left(12\,\frac{\text{in}}{\text{ft}}\right)}{\left(49\,\frac{\text{lbm}}{\text{ft}^3}\right)(0.005\text{ in})\left(32.2\,\frac{\text{ft}}{\text{sec}^2}\right)}$$

$$= 0.306 \text{ ft}$$

18. COMPRESSIBILITY[26]

Compressibility (also known as the *coefficient of compressibility*), β, is the fractional change in the volume of a fluid per unit change in pressure in a constant-temperature process.[27] Typical units are in^2/lbf, ft^2/lbf, 1/atm, and 1/kPa. (See Table 14.10.) It is the reciprocal of the bulk modulus, a quantity that is more commonly tabulated than compressibility. Equation 14.31 is written with a negative sign to show that volume decreases as pressure increases.

$$\beta = \frac{-\dfrac{\Delta V}{V_0}}{\Delta p} = \frac{1}{E} \qquad\qquad 14.31$$

---

[26]Compressibility should not be confused with the *thermal coefficient of expansion*, $(1/V_0)(\partial V/\partial T)_p$, which is the fractional change in volume per unit temperature change in a constant-pressure process (with units of 1/°F or 1/°C), or the dimensionless *compressibility factor*, Z, which is used with the ideal gas law.
[27]Other symbols used for compressibility are c, C, and K.

Table 14.10 Approximate Compressibilities of Common Liquids at 1 atm

liquid	temperature	β (in^2/lbf)	β (1/atm)
mercury	32°F	0.027×10^{-5}	0.39×10^{-5}
glycerin	60°F	0.16×10^{-5}	2.4×10^{-5}
water	60°F	0.33×10^{-5}	4.9×10^{-5}
ethyl alcohol	32°F	0.68×10^{-5}	10×10^{-5}
chloroform	32°F	0.68×10^{-5}	10×10^{-5}
gasoline	60°F	1.0×10^{-5}	15×10^{-5}
hydrogen	20K	11×10^{-5}	160×10^{-5}
helium	2.1K	48×10^{-5}	700×10^{-5}

(Multiply 1/psi by 14.696 to obtain 1/atm.)
(Multiply in^2/lbf by 0.145 to obtain 1/kPa.)

Compressibility can also be written in terms of partial derivatives.

$$\beta = \left(\frac{-1}{V_0}\right)\left(\frac{\partial V}{\partial p}\right)_T = \left(\frac{1}{\rho_0}\right)\left(\frac{\partial \rho}{\partial p}\right)_T \qquad 14.32$$

Compressibility changes only slightly with temperature. The small compressibility of liquids is typically considered to be insignificant, giving rise to the common understanding that liquids are incompressible.

The density of a compressible fluid depends on the fluid's pressure. For small changes in pressure, Eq. 14.33 can be used to calculate the density at one pressure from the density at another pressure.

$$\rho_2 \approx \rho_1\left(1 + \beta(p_2 - p_1)\right) \qquad 14.33$$

Gases, of course, are easily compressed. The compressibility of an ideal gas depends on its pressure, p, its ratio of specific heats, k, and the nature of the process.[28] Depending on the process, the compressibility may be known as *isothermal compressibility* or (*adiabatic*) *isentropic compressibility*. Of course, compressibility is zero for constant-volume processes and is infinite (or undefined) for constant-pressure processes.

$$\beta_T = \frac{1}{p} \quad \text{[isothermal ideal gas processes]} \qquad 14.34$$

$$\beta_s = \frac{1}{kp} \quad \text{[adiabatic ideal gas processes]} \qquad 14.35$$

Example 14.8

Water at 68°F (20°C) and 1 atm has a density of 62.3 lbm/ft^3 (997 kg/m^3). What is the new density if the pressure is isothermally increased from 14.7 lbf/in^2 to 400 lbf/in^2 (100 kPa to 2760 kPa)? The bulk modulus has a constant value of 320,000 lbf/in^2 (2.2×10^6 kPa).

---

[28]For air, $k = 1.4$.

SI Solution

Compressibility is the reciprocal of the bulk modulus.

$$\beta = \frac{1}{E} = \frac{1}{2.2 \times 10^6 \text{ kPa}} = 4.55 \times 10^{-7} \text{ 1/kPa}$$

From Eq. 14.33,

$$\begin{aligned}\rho_2 &= \rho_1 \Big(1 + \beta(p_2 - p_1)\Big) \\ &= \left(997 \ \frac{\text{kg}}{\text{m}^3}\right)\left(\begin{aligned}&1 + \left(4.55 \times 10^{-7} \ \frac{1}{\text{kPa}}\right) \\ &\times (2760 \text{ kPa} - 100 \text{ kPa})\end{aligned}\right) \\ &= 998.2 \text{ kg/m}^3\end{aligned}$$

Customary U.S. Solution

Compressibility is the reciprocal of the bulk modulus.

$$\beta = \frac{1}{E} = \frac{1}{320,000 \ \frac{\text{lbf}}{\text{in}^2}} = 0.3125 \times 10^{-5} \text{ in}^2/\text{lbf}$$

From Eq. 14.33,

$$\begin{aligned}\rho_2 &= \rho_1 \Big(1 + \beta(p_2 - p_1)\Big) \\ &= \left(62.3 \ \frac{\text{lbm}}{\text{ft}^3}\right)\left(\begin{aligned}&1 + \left(0.3125 \times 10^{-5} \ \frac{\text{in}^2}{\text{lbf}}\right) \\ &\times \left(400 \ \frac{\text{lbf}}{\text{in}^2} - 14.7 \ \frac{\text{lbf}}{\text{in}^2}\right)\end{aligned}\right) \\ &= 62.38 \text{ lbm/ft}^3\end{aligned}$$

19. BULK MODULUS

The *bulk modulus*, E, of a fluid is analogous to the modulus of elasticity of a solid.[29] Typical units are lbf/in^2, atm, and kPa. The term Δp in Eq. 14.36 represents an increase in stress. The term $\Delta V/V_0$ is a *volumetric strain*. Analogous to Hooke's law describing elastic formation, the *bulk modulus* of a fluid (liquid or gas) is given by Eq. 14.36.

$$E = \frac{\text{stress}}{\text{strain}} = \frac{-\Delta p}{\frac{\Delta V}{V_0}} \qquad 14.36(a)$$

$$E = -V_0 \left(\frac{\partial p}{\partial V}\right)_T \qquad 14.36(b)$$

The term *secant bulk modulus* is associated with Eq. 14.36(a) (the average slope), while the terms *tangent bulk modulus* and *point bulk modulus* are associated with Eq. 14.36(b) (the instantaneous slope).

---
[29]To distinguish it from the modulus of elasticity, the bulk modulus is represented by the symbol B when dealing with solids.

The bulk modulus is the reciprocal of compressibility.

$$E = \frac{1}{\beta} \qquad 14.37$$

The bulk modulus changes only slightly with temperature. The bulk modulus of water is usually taken as $300{,}000 \text{ lbf/in}^2$ (2.1×10^6 kPa) unless greater accuracy is required, in which case Table 14.11 or App. 14.A can be used.

Table 14.11 *Approximate Bulk Modulus of Water*

pressure (lbf/in^2)	32°F	68°F	120°F	200°F	300°F
	(thousands of lbf/in^2)				
15	292	320	332	308	–
1500	300	330	340	319	218
4500	317	348	362	338	271
15,000	380	410	420	405	350

(Multiply lbf/in^2 by 6.8948 to obtain kPa.)

Reprinted with permission from Victor L. Streeter, *Handbook of Fluid Dynamics*, © 1961, by McGraw-Hill Book Company.

20. SPEED OF SOUND

The *speed of sound* (*acoustic velocity* or *sonic velocity*), a, in a fluid is a function of its bulk modulus (or, equivalently, of its compressibility).[30] Equation 14.38 gives the speed of sound through a liquid.

$$a = \sqrt{\frac{E}{\rho}} = \sqrt{\frac{1}{\beta\rho}} \qquad \text{[SI]} \quad 14.38(a)$$

$$a = \sqrt{\frac{Eg_c}{\rho}} = \sqrt{\frac{g_c}{\beta\rho}} \qquad \text{[U.S.]} \quad 14.38(b)$$

Equation 14.39 gives the speed of sound in an ideal gas. The temperature, T, must be in degrees absolute (i.e., °R or K). For air, the ratio of specific heats is $k = 1.4$, and the molecular weight is 28.967. The universal gas constant is $R^* = 1545.35 \text{ ft-lbf/lbmol-°R}$ (8314.47 J/kmol·K).

$$\begin{aligned}a &= \sqrt{\frac{E}{\rho}} = \sqrt{\frac{kp}{\rho}} \\ &= \sqrt{kRT} = \sqrt{\frac{kR^*T}{\text{MW}}} \qquad \text{[SI]} \quad 14.39(a)\end{aligned}$$

$$\begin{aligned}a &= \sqrt{\frac{Eg_c}{\rho}} = \sqrt{\frac{kg_cp}{\rho}} \\ &= \sqrt{kg_cRT} = \sqrt{\frac{kg_cR^*T}{\text{MW}}} \qquad \text{[U.S.]} \quad 14.39(b)\end{aligned}$$

Since k and R are constant for an ideal gas, the speed of sound is a function of temperature only. Equation 14.40

---
[30]The symbol c is also used for the speed of sound.

can be used to calculate the new speed of sound when temperature is varied.

$$\frac{a_1}{a_2} = \sqrt{\frac{T_1}{T_2}} \qquad 14.40$$

The *Mach number*, M, of an object is the ratio of the object's speed to the speed of sound in the medium through which it is traveling. (See Table 14.12.)

$$M = \frac{v}{a} \qquad 14.41$$

Table 14.12 Approximate Speeds of Sound (at one atmospheric pressure)

	speed of sound	
material	(ft/sec)	(m/s)
air	1130 at 70°F	330 at 0°C
aluminum	16,400	4990
carbon dioxide	870 at 70°F	260 at 0°C
hydrogen	3310 at 70°F	1260 at 0°C
steel	16,900	5150
water	4880 at 70°F	1490 at 20°C

The term *subsonic travel* implies $M < 1$.[31] Similarly, *supersonic travel* implies $M > 1$, but usually $M < 5$. Travel above $M = 5$ is known as *hypersonic travel*. Travel in the transition region between subsonic and supersonic (i.e., $0.8 < M < 1.2$) is known as *transonic travel*. A *sonic boom* (a shock-wave phenomenon) occurs when an object travels at supersonic speed.

Example 14.9

What is the speed of sound in 150°F (66°C) water? The density is 61.2 lbm/ft³ (980 kg/m³), and the bulk modulus is 328,000 lbf/in² (2.26×10^6 kPa).

SI Solution

From Eq. 14.38,

$$a = \sqrt{\frac{E}{\rho}} = \sqrt{\frac{(2.26 \times 10^6 \text{ kPa})\left(1000 \frac{\text{Pa}}{\text{kPa}}\right)}{980 \frac{\text{kg}}{\text{m}^3}}}$$

$$= 1519 \text{ m/s}$$

Customary U.S. Solution

From Eq. 14.38,

$$a = \sqrt{\frac{Eg_c}{\rho}} = \sqrt{\frac{\left(328{,}000 \frac{\text{lbf}}{\text{in}^2}\right)\left(12 \frac{\text{in}}{\text{ft}}\right)^2\left(32.2 \frac{\text{lbm-ft}}{\text{lbf-sec}^2}\right)}{61.2 \frac{\text{lbm}}{\text{ft}^3}}}$$

$$= 4985 \text{ ft/sec}$$

---

[31]In the language of compressible fluid flow, this is known as the *subsonic flow regime*.

Example 14.10

What is the speed of sound in 150°F (66°C) air at standard atmospheric pressure?

SI Solution

The specific gas constant, R, for air is

$$R = \frac{R^*}{MW} = \frac{8314.47 \frac{\text{J}}{\text{kmol·K}}}{28.967 \frac{\text{kg}}{\text{kmol}}} = 287.03 \text{ J/kg·K}$$

The absolute temperature is

$$T = 66°C + 273° = 339\text{K}$$

From Eq. 14.39(a),

$$a = \sqrt{kRT} = \sqrt{(1.4)\left(287.03 \frac{\text{J}}{\text{kg·K}}\right)(339\text{K})}$$

$$= 369 \text{ m/s}$$

Customary U.S. Solution

The specific gas constant, R, for air is

$$R = \frac{R^*}{MW} = \frac{1545.35 \frac{\text{ft-lbf}}{\text{lbmol-°R}}}{28.967 \frac{\text{lbm}}{\text{lbmol}}}$$

$$= 53.35 \text{ ft-lbf/lbm-°R}$$

The absolute temperature is

$$T = 150°F + 460° = 610°\text{R}$$

From Eq. 14.39(b),

$$a = \sqrt{kg_c RT}$$

$$= \sqrt{(1.4)\left(32.2 \frac{\text{lbm-ft}}{\text{lbf-sec}^2}\right)\left(53.35 \frac{\text{ft-lbf}}{\text{lbm-°R}}\right)(610°\text{R})}$$

$$= 1211 \text{ ft/sec}$$

21. PROPERTIES OF MIXTURES OF NONREACTING LIQUIDS

There are very few convenient ways of predicting the properties of nonreacting, nonvolatile organic and aqueous solutions (acids, brines, alcohol mixtures, coolants, etc.) from the individual properties of the components.

Volumes of two nonreacting organic liquids (e.g., acetone and chloroform) in a mixture are essentially additive. The volume change upon mixing will seldom be more than a few tenths of a percent. The volume change in aqueous solutions is often slightly greater, but is still limited to a few percent (e.g., 3% for some solutions of methanol and water). Therefore, the specific gravity (density, specific weight, etc.) can be considered to be a volumetric weighting of the individual specific gravities.

A rough estimate of the absolute viscosity of a mixture of two or more liquids having different viscosities (at the same temperature) can be found from the mole fractions, x_i, of the components.

$$\mu_{\text{mixture}} \approx \frac{1}{\sum_i \dfrac{x_i}{\mu_i}} = \frac{\sum_i \dfrac{m_i}{\text{MW}_i}}{\sum_i \dfrac{m_i}{(\text{MW}_i)\mu_i}} \qquad 14.42$$

A three-step procedure for calculating a more reliable estimate of the viscosity of a mixture of two or more liquids starts by using the *Refutas equation* to calculate a linearized *viscosity blending index*, VBI, (or, *viscosity blending number*, VBN) for each component.

$$\text{VBI}_i = 10.975 + 14.534 \times \ln\big(\ln(\nu_{i,\text{cSt}} + 0.8)\big) \qquad 14.43$$

The second step calculates the average VBI of the mixture from the gravimetrically weighted component VBIs.

$$\text{VBI}_{\text{mixture}} = \sum_i G_i \times \text{VBI}_i \qquad 14.44$$

The final step extracts the mixture viscosity from the mixture VBI by inverting the Refutas equation.

$$\nu_{\text{mixture,cSt}} = \exp\left(\exp\left(\frac{\text{VBI}_{\text{mixture}} - 10.975}{14.534}\right)\right) - 0.8 \qquad 14.45$$

Components of mixtures of hydrocarbons may be referred to as pseudocomponents.[32] *Raoult's law* can be used to calculate the vapor pressure above a liquid mixture from the vapor pressures of the liquid components. (See App. 14.H.)

$$p_{v,\text{mixture}} = \sum_i x_i p_{v,i} \qquad 14.46$$

---

[32]A *pseudocomponent* represents a mixture of components and has the thermodynamic behavior of a mixture, rather than that of a single component. The term can refer to the performance of a blend of known components, or to the performance of a component of a known mixture.

15 Fluid Statics

Nomenclature

a	acceleration	ft/sec^2	m/s^2
A	area	ft^2	m^2
b	base length	ft	m
d	diameter	ft	m
e	eccentricity	ft	m
F	force	lbf	N
FS	factor of safety	–	–
g	gravitational acceleration, 32.2 (9.81)	ft/sec^2	m/s^2
g_c	gravitational constant, 32.2	lbm-ft/lbf-sec^2	n.a.
h	height	ft	m
I	moment of inertia	ft^4	m^4
k	radius of gyration	ft	m
k	ratio of specific heats	–	–
L	length	ft	m
m	mass	lbm	kg
M	mechanical advantage	–	–
M	moment	ft-lbf	N·m
n	polytropic exponent	–	–
N	normal force	lbf	N
p	pressure	lbf/ft^2	Pa
r	radius	ft	m
R	resultant force	lbf	N
R	specific gas constant	ft-lbf/lbm-°R	J/kg·K
SG	specific gravity	–	–
T	temperature	°R	K
v	velocity	ft/sec	m/s
V	volume	ft^3	m^3
w	width	ft	m
W	weight	lbf	n.a.
x	distance	ft	m
x	fraction	–	–
y	distance	ft	m

Symbols

γ	specific weight	lbf/ft^3	n.a.
η	efficiency	–	–
θ	angle	deg	deg
μ	coefficient of friction	–	–
ρ	density	lbm/ft^3	kg/m^3
υ	specific volume	ft^3/lbm	m^3/kg
ω	angular velocity	rad/sec	rad/s

Subscripts

a	atmospheric
b	buoyant
bg	between CB and CG
c	centroidal
f	frictional
F	force
l	lever or longitudinal
m	manometer fluid, mercury, or metacentric
p	plunger
r	ram
R	resultant
t	tank
v	vapor or vertical
w	water

1. PRESSURE-MEASURING DEVICES

There are many devices for measuring and indicating fluid pressure. Some devices measure gage pressure; others measure absolute pressure. The effects of nonstandard atmospheric pressure and nonstandard gravitational acceleration must be determined, particularly for devices relying on columns of liquid to indicate pressure. Table 15.1 lists the common types of devices and the ranges of pressure appropriate for each.

Table 15.1 *Common Pressure-Measuring Devices*

device	approximate range (in atm)
water manometer	0–0.1
mercury barometer	0–1
mercury manometer	0.001–1
metallic diaphragm	0.01–200
transducer	0.001–15,000
Bourdon pressure gauge	1–3000
Bourdon vacuum gauge	0.1–1

The *Bourdon pressure gauge* is the most common pressure-indicating device. (See Fig. 15.1.) This mechanical device consists of a C-shaped or helical hollow tube that tends to straighten out (i.e., unwind) when the tube is subjected to an internal pressure. The gauge is referred to as a *C-Bourdon gauge* because of the shape of the hollow tube. The degree to which the coiled tube unwinds depends on the difference between the internal and external pressures. A Bourdon gauge directly indicates *gage pressure*. Extreme accuracy is generally not a characteristic of Bourdon gauges.

Figure 15.1 *C-Bourdon Pressure Gauge*

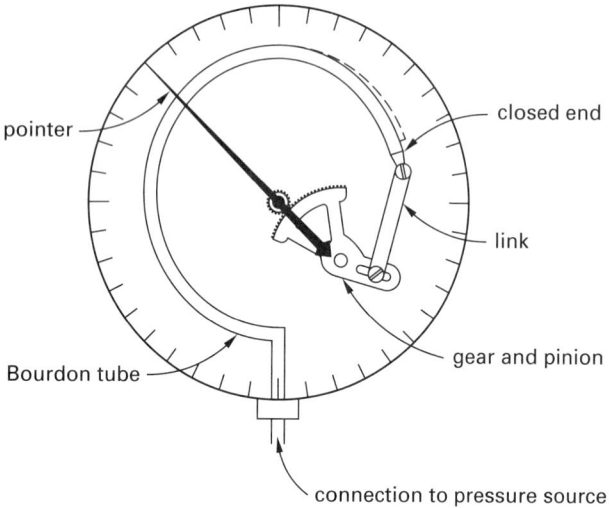

In non-SI installations, gauges are always calibrated in psi (or psid). Vacuum pressure is usually calibrated in inches of mercury. SI gauges are marked in kilopascals (kPa) or bars, although kg/cm^2 may be used in older gauges. Negative numbers are used to indicate vacuum. The gauge dial will be clearly marked if other units are indicated.

The *barometer* is a common device for measuring the absolute pressure of the atmosphere.[1] It is constructed

by filling a long tube open at one end with mercury (or alcohol, or some other liquid) and inverting the tube so that the open end is below the level of a mercury-filled container. If the vapor pressure of the mercury in the tube is neglected, the fluid column will be supported only by the atmospheric pressure transmitted through the container fluid at the lower, open end.

Strain gauges, diaphragm gauges, quartz-crystal transducers, and other devices using the *piezoelectric effect* are also used to measure stress and pressure, particularly when pressure fluctuates quickly (e.g., as in a rocket combustion chamber). With these devices, calibration is required to interpret pressure from voltage generation or changes in resistance, capacitance, or inductance. These devices are generally unaffected by atmospheric pressure or gravitational acceleration.

Manometers (*U-tube manometers*) can also be used to indicate small pressure differences, and for this purpose they provide great accuracy. (Manometers are not suitable for measuring pressures much larger than 10 lbf/in^2 (70 kPa), however.) A difference in manometer fluid surface heights is converted into a pressure difference. If one end of a manometer is open to the atmosphere, the manometer indicates gage pressure. It is theoretically possible, but impractical, to have a manometer indicate absolute pressure, since one end of the manometer would have to be exposed to a perfect vacuum.

A *static pressure tube* (*piezometer tube*) is a variation of the manometer. (See Fig. 15.2.) It is a simple method of determining the static pressure in a pipe or other vessel, regardless of fluid motion in the pipe. A vertical transparent tube is connected to a hole in the pipe wall.[2] (None of the tube projects into the pipe.) The static pressure will force the contents of the pipe up into the tube. The height of the contents will be an indication of gage pressure in the pipe.

Figure 15.2 *Static Pressure Tube*

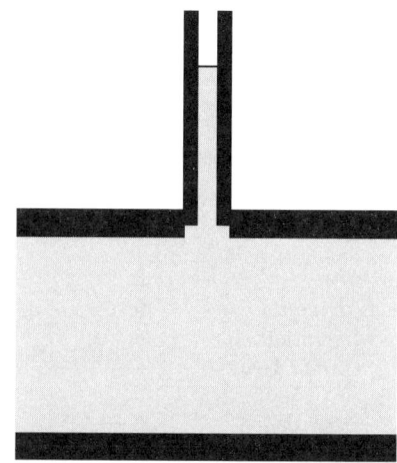

---

[1] A barometer can be used to measure the pressure inside any vessel. However, the barometer must be completely enclosed in the vessel, which may not be possible. Also, it is difficult to read a barometer enclosed within a tank.

[2] Where greater accuracy is required, multiple holes may be drilled around the circumference of the pipe and connected through a manifold (*piezometer ring*) to the pressure-measuring device.

The device used to measure the pressure should not be confused with the method used to obtain exposure to the pressure. For example, a static pressure *tap* in a pipe is merely a hole in the pipe wall. A Bourdon gauge, manometer, or transducer can then be used with the tap to indicate pressure.

Tap holes are generally $^{1}/_{8}$–$^{1}/_{4}$ in (3–6 mm) in diameter, drilled at right angles to the wall, and smooth and flush with the pipe wall. No part of the gauge or connection projects into the pipe. The tap holes should be at least 5 to 10 pipe diameters downstream from any source of turbulence (e.g., a bend, fitting, or valve).

2. MANOMETERS

Figure 15.3 illustrates a simple U-tube manometer used to measure the difference in pressure between two vessels. When both ends of the manometer are connected to pressure sources, the name *differential manometer* is used. If one end of the manometer is open to the atmosphere, the name *open manometer* is used.[3] The open manometer implicitly measures gage pressures.

Figure 15.3 Simple U-Tube Manometer

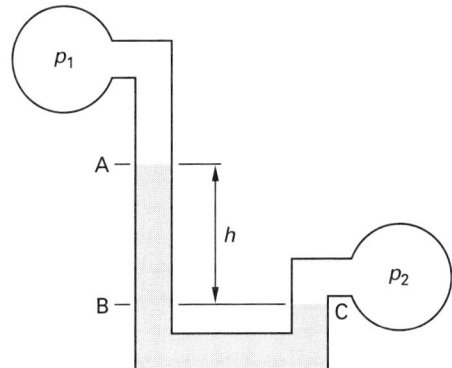

Since the pressure at point B in Fig. 15.3 is the same as at point C, the pressure differential produces the vertical fluid column of height h. In Eq. 15.2, A is the area of the tube. In the absence of any capillary action, the inside diameters of the manometer tubes are irrelevant.

$$F_{\text{net}} = F_{\text{C}} - F_{\text{A}} = \text{weight of fluid column AB} \qquad 15.1$$

$$(p_2 - p_1)A = \rho_m g h A \qquad 15.2$$

$$p_2 - p_1 = \rho_m g h \qquad \text{[SI]} \quad 15.3(a)$$

$$p_2 - p_1 = \rho_m h \times \frac{g}{g_c} = \gamma_m h \qquad \text{[U.S.]} \quad 15.3(b)$$

[3]If one of the manometer legs is inclined, the term *inclined manometer* or *draft gauge* is used. Although only the vertical distance between the manometer fluid surfaces should be used to calculate the pressure difference, with small pressure differences it may be more accurate to read the inclined distance (which is larger than the vertical distance) and compute the vertical distance from the angle of inclination.

The quantity g/g_c has a value of 1.0 lbf/lbm in almost all cases, so γ_m is numerically equal to ρ_m, with units of lbf/ft^3.

Equation 15.3(a) and Eq. 15.3(b) assume that the manometer fluid height is small, or that only low-density gases fill the tubes above the manometer fluid. If a high-density fluid (such as water) is present above the measuring fluid, or if the columns h_1 or h_2 are very long, corrections will be necessary. (See Fig. 15.4.)

Figure 15.4 Manometer Requiring Corrections

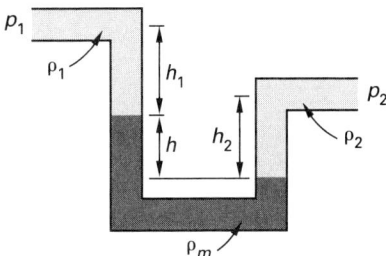

Fluid column h_2 "sits on top" of the manometer fluid, forcing the manometer fluid to the left. This increase must be subtracted out. Similarly, the column h_1 restricts the movement of the manometer fluid. The observed measurement must be increased to correct for this restriction.

$$p_2 - p_1 = g(\rho_m h + \rho_1 h_1 - \rho_2 h_2) \qquad \text{[SI]} \quad 15.4(a)$$

$$p_2 - p_1 = (\rho_m h + \rho_1 h_1 - \rho_2 h_2) \times \frac{g}{g_c}$$
$$= \gamma_m h + \gamma_1 h_1 - \gamma_2 h_2 \qquad \text{[U.S.]} \quad 15.4(b)$$

When a manometer is used to measure the pressure difference across an orifice or other fitting where the same liquid exists in both manometer sides (shown in Fig. 15.5), it is not necessary to correct the manometer reading for all of the liquid present above the manometer fluid. This is because parts of the correction for both sides of the manometer are the same. Therefore, the distance y in Fig. 15.5 is an irrelevant distance.

Figure 15.5 Irrelevant Distance, y

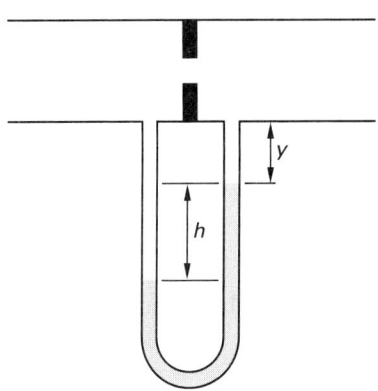

Manometer tubes are generally large enough in diameter to avoid significant capillary effects. Corrections for capillarity are seldom necessary.

Example 15.1

The pressure at the bottom of a water tank ($\rho = 62.4$ lbm/ft^3; $\rho = 998$ kg/m^3) is measured with a mercury manometer located below the tank bottom, as shown. (The density of mercury is 848 lbm/ft^3; 13 575 kg/m^3.) What is the gage pressure at the bottom of the water tank?

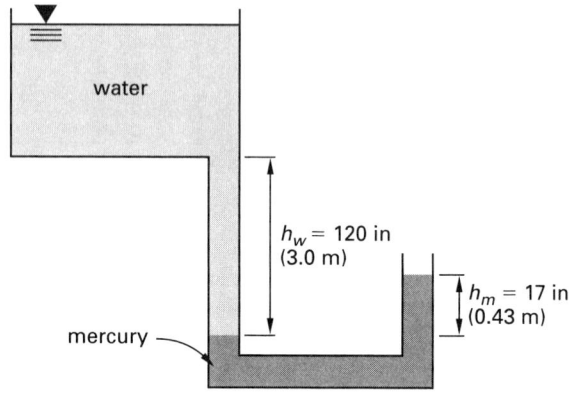

SI Solution

From Eq. 15.4(a),

$$\Delta p = g(\rho_m h_m - \rho_w h_w)$$

$$= \left(9.81 \ \frac{\text{m}}{\text{s}^2}\right) \left(\begin{array}{l} \left(13\,575 \ \frac{\text{kg}}{\text{m}^3}\right)(0.43 \ \text{m}) \\ - \left(998 \ \frac{\text{kg}}{\text{m}^3}\right)(3.0 \ \text{m}) \end{array} \right)$$

$$= 27\,892 \ \text{Pa} \quad (27.9 \ \text{kPa gage})$$

Customary U.S. Solution

From Eq. 15.4(b),

$$\Delta p = (\rho_m h_m - \rho_w h_w) \times \frac{g}{g_c}$$

$$= \frac{\left(848 \ \frac{\text{lbm}}{\text{ft}^3}\right)(17 \ \text{in}) - \left(62.4 \ \frac{\text{lbm}}{\text{ft}^3}\right)(120 \ \text{in})}{\left(12 \ \frac{\text{in}}{\text{ft}}\right)^3}$$

$$\times \left(\frac{32.2 \ \frac{\text{ft}}{\text{sec}^2}}{32.2 \ \frac{\text{lbm-ft}}{\text{lbf-sec}^2}} \right)$$

$$= 4.01 \ \text{lbf/in}^2 \quad (4.01 \ \text{psig})$$

3. HYDROSTATIC PRESSURE

Hydrostatic pressure is the pressure a fluid exerts on an immersed object or container walls.[4] Pressure is equal to the force per unit area of surface.

$$p = \frac{F}{A} \qquad \qquad 15.5$$

Hydrostatic pressure in a stationary, incompressible fluid behaves according to the following characteristics.

- Pressure is a function of vertical depth (and density) only. The pressure will be the same at two points with identical depths.

- Pressure varies linearly with (vertical and inclined) depth.

- Pressure is independent of an object's area and size and the weight (mass) of water above the object. Figure 15.6 illustrates the *hydrostatic paradox*. The pressures at depth h are the same in all four columns because pressure depends on depth, not volume.

Figure 15.6 *Hydrostatic Paradox*

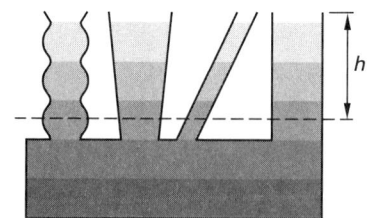

- Pressure at a point has the same magnitude in all directions (*Pascal's law*). Therefore, pressure is a scalar quantity.

- Pressure is always normal to a surface, regardless of the surface's shape or orientation. (This is a result of the fluid's inability to support shear stress.)

- The resultant of the pressure distribution acts through the *center of pressure*.

- The center of pressure rarely coincides with the average depth.

4. FLUID HEIGHT EQUIVALENT TO PRESSURE

Pressure varies linearly with depth. The relationship between pressure and depth (i.e., the *hydrostatic head*) for an incompressible fluid is given by Eq. 15.6.

$$p = \rho g h \qquad \qquad \text{[SI]} \quad 15.6(a)$$

$$p = \frac{\rho g h}{g_c} = \gamma h \qquad \text{[U.S.]} \quad 15.6(b)$$

---

[4]The term *hydrostatic* is used with all fluids, not only with water.

Since ρ and g are constants, Eq. 15.6 shows that p and h are linearly related. Knowing one determines the other.[5] For example, the height of a fluid column needed to produce a pressure is

$$h = \frac{p}{\rho g} \qquad \text{[SI]} \quad 15.7(a)$$

$$h = \frac{pg_c}{\rho g} = \frac{p}{\gamma} \qquad \text{[U.S.]} \quad 15.7(b)$$

Table 15.2 lists six common fluid height equivalents.[6]

Table 15.2 *Approximate Fluid Height Equivalents at 68°F (20°C)*

liquid	height equivalents	
water	0.0361 psi/in	27.70 in/psi
water	62.4 psf/ft	0.01603 ft/psf
water	9.81 kPa/m	0.1019 m/kPa
water	0.4329 psi/ft	2.31 ft/psi
mercury	0.491 psi/in	2.036 in/psi
mercury	133.3 kPa/m	0.00750 m/kPa

A barometer is a device that measures atmospheric pressure by the height of a fluid column. If the vapor pressure of the barometer liquid is neglected, the atmospheric pressure will be given by Eq. 15.8.

$$p_a = \rho g h \qquad \text{[SI]} \quad 15.8(a)$$

$$p_a = \frac{\rho g h}{g_c} = \gamma h \qquad \text{[U.S.]} \quad 15.8(b)$$

If the vapor pressure of the barometer liquid is significant (as it would be with alcohol or water), the vapor pressure effectively reduces the height of the fluid column, as Eq. 15.9 illustrates.

$$p_a - p_v = \rho g h \qquad \text{[SI]} \quad 15.9(a)$$

$$p_a - p_v = \frac{\rho g h}{g_c} = \gamma h \qquad \text{[U.S.]} \quad 15.9(b)$$

Example 15.2

A vacuum pump is used to drain a flooded mine shaft of 68°F (20°C) water.[7] The vapor pressure of water at this temperature is 0.34 lbf/in² (2.34 kPa). The pump is incapable of lifting the water higher than 400 in (10.16 m). What is the atmospheric pressure?

---

[5]In fact, pressure and height of a fluid column can be used interchangeably. The height of a fluid column is known as *head*. For example: "The fan developed a static head of 3 in of water," or "The pressure head at the base of the water tank was 8 m." When the term "head" is used, it is essential to specify the fluid.
[6]Of course, these values are recognized to be the approximate specific weights of the liquids.
[7]A reciprocating or other direct-displacement pump would be a better choice to drain a mine.

SI Solution

From Eq. 15.9,

$$p_a = p_v + \rho g h$$

$$= 2.34 \text{ kPa} + \frac{\left(998 \frac{\text{kg}}{\text{m}^3}\right)\left(9.81 \frac{\text{m}}{\text{s}^2}\right)(10.16 \text{ m})}{1000 \frac{\text{Pa}}{\text{kPa}}}$$

$$= 101.8 \text{ kPa}$$

(*Alternate SI solution, using Table 15.2*)

$$p_a = p_v + \rho g h = 2.34 \text{ kPa} + \left(9.81 \frac{\text{kPa}}{\text{m}}\right)(10.16 \text{ m})$$

$$= 102 \text{ kPa}$$

Customary U.S. Solution

From Table 15.2, the height equivalent of water is approximately 0.0361 psi/in. The unit psi/in is the same as lbf/in³, the units of γ. From Eq. 15.9, the atmospheric pressure is

$$p_a = p_v + \rho g h = p_v + \gamma h$$

$$= 0.34 \frac{\text{lbf}}{\text{in}^2} + \left(0.0361 \frac{\text{lbf}}{\text{in}^3}\right)(400 \text{ in})$$

$$= 14.78 \text{ lbf/in}^2 \quad (14.78 \text{ psia})$$

5. MULTIFLUID BAROMETERS

It is theoretically possible to fill a barometer tube with several different immiscible fluids.[8] Upon inversion, the fluids will separate, leaving the most dense fluid at the bottom and the least dense fluid at the top. All of the fluids will contribute, by superposition, to the balance between the external atmospheric pressure and the weight of the fluid column.

$$p_a - p_v = g\sum \rho_i h_i \qquad \text{[SI]} \quad 15.10(a)$$

$$p_a - p_v = \frac{g}{g_c}\sum \rho_i h_i = \sum \gamma_i h_i \qquad \text{[U.S.]} \quad 15.10(b)$$

The pressure at any intermediate point within the fluid column is found by starting at a location where the pressure is known, and then adding or subtracting $\rho g h$ terms to get to the point where the pressure is needed. Usually, the known pressure will be the atmospheric pressure located in the barometer barrel at the level (elevation) of the fluid outside of the barometer.

---

[8]In practice, barometers are never constructed this way. This theory is more applicable to a category of problems dealing with up-ended containers, as illustrated in Ex. 15.3.

Example 15.3

Neglecting vapor pressure, what is the pressure of the air at point E in the container shown? The atmospheric pressure is 1.0 atm (14.7 lbf/in^2, 101 300 Pa).

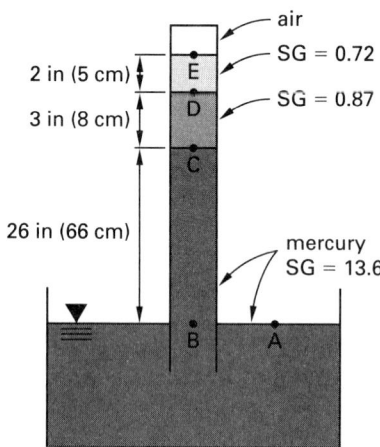

SI Solution

$$p_E = p_a - g\rho_w \sum (SG_i)h_i$$

$$= 101\,300 \text{ Pa} - \left(9.81 \ \frac{\text{m}}{\text{s}^2}\right)\left(1000 \ \frac{\text{kg}}{\text{m}^3}\right)$$

$$\times \begin{pmatrix} (0.66 \text{ m})(13.6) + (0.08 \text{ m})(0.87) \\ + (0.05 \text{ m})(0.72) \end{pmatrix}$$

$$= 12\,210 \text{ Pa} \quad (12.2 \text{ kPa})$$

Customary U.S. Solution

The pressure at point B is the same as the pressure at point A—1.0 atm. The density of mercury is $13.6 \times 0.0361 \text{ lbm/in}^3 = 0.491 \text{ lbm/in}^3$, and the specific weight is 0.491 lbf/in^3. Therefore, the pressure at point C is

$$p_C = 14.7 \ \frac{\text{lbf}}{\text{in}^2} - (26 \text{ in})\left(0.491 \ \frac{\frac{\text{lbf}}{\text{in}^2}}{\text{in}}\right)$$

$$= 1.93 \text{ lbf/in}^2 \quad (1.93 \text{ psia})$$

Similarly, the pressure at point E (and anywhere within the captive air space) is

$$p_E = 14.7 \ \frac{\text{lbf}}{\text{in}^2} - (26 \text{ in})\left(0.491 \ \frac{\text{lbf}}{\text{in}^3}\right)$$

$$- (3 \text{ in})(0.87)\left(0.0361 \ \frac{\text{lbf}}{\text{in}^3}\right)$$

$$- (2 \text{ in})(0.72)\left(0.0361 \ \frac{\text{lbf}}{\text{in}^3}\right)$$

$$= 1.79 \text{ lbf/in}^2 \quad (1.79 \text{ psia})$$

6. PRESSURE ON A HORIZONTAL PLANE SURFACE

The pressure on a horizontal plane surface is uniform over the surface because the depth of the fluid is uniform. (See Fig. 15.7.) The resultant of the pressure distribution acts through the center of pressure of the surface, which corresponds to the centroid of the surface.

Figure 15.7 *Hydrostatic Pressure on a Horizontal Plane Surface*

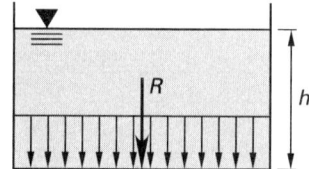

The uniform pressure at depth h is given by Eq. 15.11.[9]

$$p = \rho g h \quad \text{[SI]} \quad 15.11(a)$$

$$p = \frac{\rho g h}{g_c} = \gamma h \quad \text{[U.S.]} \quad 15.11(b)$$

The total vertical force on the horizontal plane of area A is given by Eq. 15.12.

$$R = pA \quad 15.12$$

It is tempting, but not always correct, to calculate the vertical force on a submerged surface as the weight of the fluid above it. Such an approach works only when there is no change in the cross-sectional area of the fluid above the surface. This is a direct result of the *hydrostatic paradox*. (See Sec. 15.3.) Figure 15.8 illustrates two containers with the same pressure distribution (force) on their bottom surfaces.

Figure 15.8 *Two Containers with the Same Pressure Distribution*

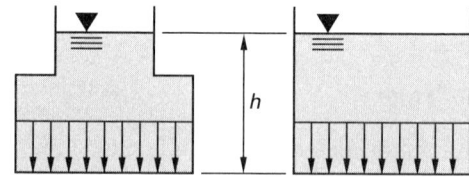

7. PRESSURE ON A RECTANGULAR VERTICAL PLANE SURFACE

The pressure on a vertical rectangular plane surface increases linearly with depth. The pressure distribution will be triangular, as in Fig. 15.9(a), if the plane surface extends to the surface; otherwise, the distribution will be trapezoidal, as in Fig. 15.9(b).

---

[9]The phrase *pressure at a depth* is universally understood to mean the *gage pressure*, as given by Eq. 15.11.

Figure 15.9 Hydrostatic Pressure on a Vertical Plane Surface

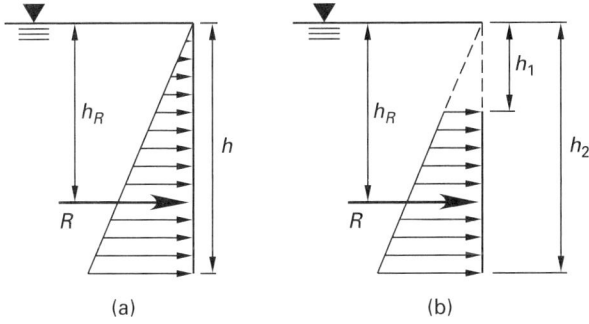

(a)　　　　　　　(b)

The resultant force is calculated from the *average pressure*.

$$\bar{p} = \tfrac{1}{2}(p_1 + p_2) \qquad\qquad 15.13$$

$$\bar{p} = \tfrac{1}{2}\rho g(h_1 + h_2) \qquad [SI] \quad 15.14(a)$$

$$\bar{p} = \frac{\tfrac{1}{2}\rho g(h_1 + h_2)}{g_c} = \tfrac{1}{2}\gamma(h_1 + h_2) \quad [U.S.] \quad 15.14(b)$$

$$R = \bar{p}A \qquad\qquad 15.15$$

Although the resultant is calculated from the average depth, it does not act at the average depth. The resultant of the pressure distribution passes through the centroid of the pressure distribution. For the triangular distribution of Fig. 15.9(a), the resultant is located at a depth of $h_R = \tfrac{2}{3}h$. For the more general case of Fig. 15.9(b), the resultant is located from Eq. 15.16.

$$h_R = \tfrac{2}{3}\left(h_1 + h_2 - \frac{h_1 h_2}{h_1 + h_2}\right) \qquad 15.16$$

8. PRESSURE ON A RECTANGULAR INCLINED PLANE SURFACE

The average pressure and resultant force on an inclined rectangular plane surface are calculated in a similar fashion as that for the vertical plane surface. (See Fig. 15.10.) The pressure varies linearly with depth. The resultant is

Figure 15.10 Hydrostatic Pressure on an Inclined Rectangular Plane Surface

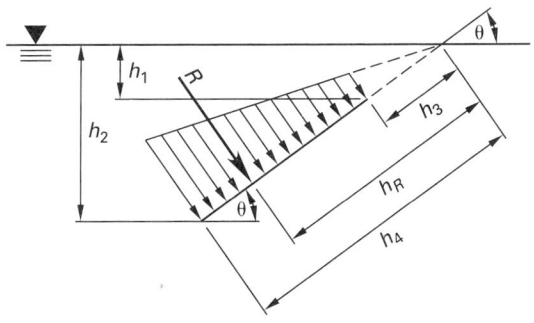

calculated from the average pressure, which, in turn, depends on the average depth.

The average pressure and resultant are found using Eq. 15.17 through Eq. 15.19.

$$\bar{p} = \tfrac{1}{2}(p_1 + p_2) \qquad\qquad 15.17$$

$$\begin{aligned}\bar{p} &= \tfrac{1}{2}\rho g(h_1 + h_2) \\ &= \tfrac{1}{2}\rho g(h_3 + h_4)\sin\theta \end{aligned} \qquad [SI] \quad 15.18(a)$$

$$\begin{aligned}\bar{p} &= \frac{\tfrac{1}{2}\rho g(h_1 + h_2)}{g_c} = \frac{\tfrac{1}{2}\rho g(h_3 + h_4)\sin\theta}{g_c} \\ &= \tfrac{1}{2}\gamma(h_1 + h_2) = \tfrac{1}{2}\gamma(h_3 + h_4)\sin\theta \end{aligned} \quad [U.S.] \quad 15.18(b)$$

$$R = \bar{p}A \qquad\qquad 15.19$$

As with the vertical plane surface, the resultant acts at the centroid of the pressure distribution, not at the average depth. Equation 15.16 is rewritten in terms of inclined depths.[10]

$$\begin{aligned}h_R &= \left(\frac{\tfrac{2}{3}}{\sin\theta}\right)\left(h_1 + h_2 - \frac{h_1 h_2}{h_1 + h_2}\right) \\ &= \tfrac{2}{3}\left(h_3 + h_4 - \frac{h_3 h_4}{h_3 + h_4}\right) \end{aligned} \qquad 15.20$$

Example 15.4

The tank shown is filled with water ($\rho = 62.4 \text{ lbm/ft}^3$; $\rho = 1000 \text{ kg/m}^3$). (a) What is the total resultant force on a 1 ft (1 m) width of the inclined portion of the wall?[11] (b) At what depth (vertical distance) is the resultant force on the inclined portion of the wall located?

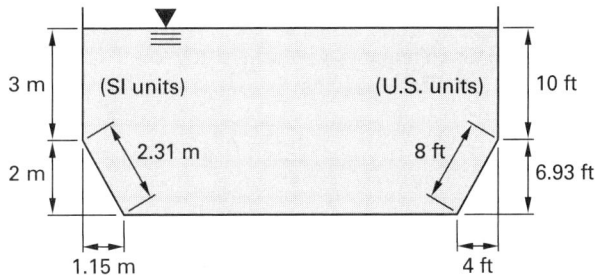

SI Solution

(a) The depth of the tank bottom is

$$h_2 = 3 \text{ m} + 2 \text{ m} = 5 \text{ m}$$

---

[10] h_R is an inclined distance. If a vertical distance is wanted, it must usually be calculated from h_R and $\sin\theta$. Equation 15.20 can be derived simply by dividing Eq. 15.16 by $\sin\theta$.

[11] Since the width of the tank (the distance into and out of the illustration) is unknown, it is common to calculate the pressure or force per unit width of tank wall. This is the same as calculating the pressure on a 1 ft (1 m) wide section of wall.

Flow of Fluids

From Eq. 15.18, the average gage pressure on the inclined section is

$$\overline{p} = \left(\tfrac{1}{2}\right)\left(1000 \ \frac{\text{kg}}{\text{m}^3}\right)\left(9.81 \ \frac{\text{m}}{\text{s}^2}\right)(3 \ \text{m} + 5 \ \text{m})$$

$$= 39\,240 \ \text{Pa} \quad \text{(gage)}$$

The total resultant force on a 1 m section of the inclined portion of the wall is

$$R = \overline{p}A = (39\,240 \ \text{Pa})(2.31 \ \text{m})(1 \ \text{m})$$

$$= 90\,644 \ \text{N} \quad (90.6 \ \text{kN})$$

(b) θ must be known to determine h_R.

$$\theta = \arctan \frac{2 \ \text{m}}{1.15 \ \text{m}} = 60°$$

From Eq. 15.20, the location of the resultant can be calculated once h_3 and h_4 are known.

$$h_3 = \frac{3 \ \text{m}}{\sin 60°} = 3.464 \ \text{m}$$

$$h_4 = \frac{5 \ \text{m}}{\sin 60°} = 5.774 \ \text{m}$$

$$h_R = \left(\tfrac{2}{3}\right)\left(3.464 \ \text{m} + 5.774 \ \text{m} - \frac{(3.464 \ \text{m})(5.774 \ \text{m})}{3.464 \ \text{m} + 5.774 \ \text{m}}\right)$$

$$= 4.715 \ \text{m} \quad [\text{inclined}]$$

The vertical depth at which the resultant force on the inclined portion of the wall acts is

$$h = h_R \sin \theta = (4.715 \ \text{m})\sin 60°$$

$$= 4.08 \ \text{m} \quad [\text{vertical}]$$

Customary U.S. Solution

(a) The water density is given in traditional U.S. mass units. The specific weight, γ, is

$$\gamma = \frac{\rho g}{g_c} = \frac{\left(62.4 \ \frac{\text{lbm}}{\text{ft}^3}\right)\left(32.2 \ \frac{\text{ft}}{\text{sec}^2}\right)}{32.2 \ \frac{\text{lbm-ft}}{\text{lbf-sec}^2}}$$

$$= 62.4 \ \text{lbf/ft}^3$$

The depth of the tank bottom is

$$h_2 = 10 \ \text{ft} + 6.93 \ \text{ft} = 16.93 \ \text{ft}$$

From Eq. 15.18, the average gage pressure on the inclined section is

$$\overline{p} = \left(\tfrac{1}{2}\right)\left(62.4 \ \frac{\text{lbf}}{\text{ft}^3}\right)(10 \ \text{ft} + 16.93 \ \text{ft})$$

$$= 840.2 \ \text{lbf/ft}^2 \quad \text{(gage)}$$

The total resultant force on a 1 ft section of the inclined portion of the wall is

$$R = \overline{p}A = \left(840.2 \ \frac{\text{lbf}}{\text{ft}^2}\right)(8 \ \text{ft})(1 \ \text{ft}) = 6722 \ \text{lbf}$$

(b) θ must be known to determine h_R.

$$\theta = \arctan \frac{6.93 \ \text{ft}}{4 \ \text{ft}} = 60°$$

From Eq. 15.20, the location of the resultant can be calculated once h_3 and h_4 are known.

$$h_3 = \frac{10 \ \text{ft}}{\sin 60°} = 11.55 \ \text{ft}$$

$$h_4 = \frac{16.93 \ \text{ft}}{\sin 60°} = 19.55 \ \text{ft}$$

From Eq. 15.20,

$$h_R = \left(\tfrac{2}{3}\right)\left(11.55 \ \text{ft} + 19.55 \ \text{ft} - \frac{(11.55 \ \text{ft})(19.55 \ \text{ft})}{11.55 \ \text{ft} + 19.55 \ \text{ft}}\right)$$

$$= 15.89 \ \text{ft} \quad [\text{inclined}]$$

The vertical depth at which the resultant force on the inclined portion of the wall acts is

$$h = h_R \sin \theta = (15.89 \ \text{ft})\sin 60°$$

$$= 13.76 \ \text{ft} \quad [\text{vertical}]$$

9. PRESSURE ON A GENERAL PLANE SURFACE

Figure 15.11 illustrates a nonrectangular plane surface that may or may not extend to the liquid surface and that may or may not be inclined, as was shown in Fig. 15.10. As with other regular surfaces, the resultant force depends on the average pressure and acts through the *center of pressure* (CP). The average pressure is calculated from the depth of the surface's centroid (center of gravity, CG).

$$\overline{p} = \rho g h_c \sin \theta \qquad \qquad \text{[SI]} \quad \textit{15.21(a)}$$

$$\overline{p} = \frac{\rho g h_c \sin \theta}{g_c} = \gamma h_c \sin \theta \qquad \text{[U.S.]} \quad \textit{15.21(b)}$$

$$R = \overline{p}A \qquad \qquad \textit{15.22}$$

Figure 15.11 *General Plane Surface*

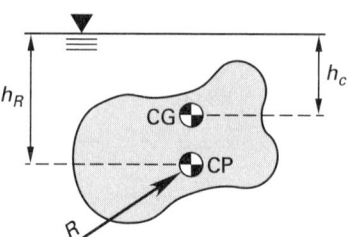

The resultant force acts at depth h_R normal to the plane surface. I_c in Eq. 15.23 is the centroidal area moment of inertia about an axis parallel to the surface. Both h_c and h_R are measured parallel to the plane surface. That is, if the plane surface is inclined, h_c and h_R are inclined distances.

$$h_R = h_c + \frac{I_c}{Ah_c} \qquad \textit{15.23}$$

Example 15.5

The top edge of a vertical circular observation window is located 4.0 ft (1.25 m) below the surface of the water. The window is 1.0 ft (0.3 m) in diameter. The water's density is 62.4 lbm/ft^3 (1000 kg/m^3). Neglect the salinity of the water. (a) What is the resultant force on the window? (b) At what depth does the resultant force act?

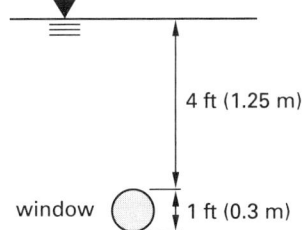

SI Solution

(a) The radius of the window is

$$r = \frac{0.3 \text{ m}}{2} = 0.15 \text{ m}$$

The depth at which the centroid of the circular window is located is

$$h_c = 1.25 \text{ m} + 0.15 \text{ m} = 1.4 \text{ m}$$

The area of the circular window is

$$A = \pi r^2 = \pi (0.15 \text{ m})^2 = 0.0707 \text{ m}^2$$

The average pressure is

$$\bar{p} = \rho g h_c = \left(1000 \; \frac{\text{kg}}{\text{m}^3}\right)\left(9.81 \; \frac{\text{m}}{\text{s}^2}\right)(1.4 \text{ m})$$

$$= 13\,734 \text{ Pa} \quad \text{(gage)}$$

The resultant is calculated from Eq. 15.19.

$$R = \bar{p}A = (13\,734 \text{ Pa})(0.0707 \text{ m}^2) = 971 \text{ N}$$

(b) The centroidal area moment of inertia of a circle is

$$I_c = \frac{\pi}{4} r^4 = \left(\frac{\pi}{4}\right)(0.15 \text{ m})^4 = 3.976 \times 10^{-4} \text{ m}^4$$

From Eq. 15.23, the depth at which the resultant force acts is

$$h_R = h_c + \frac{I_c}{Ah_c} = 1.4 \text{ m} + \frac{3.976 \times 10^{-4} \text{ m}^4}{(0.0707 \text{ m}^2)(1.4 \text{ m})}$$

$$= 1.404 \text{ m}$$

Customary U.S. Solution

(a) The radius of the window is

$$r = \frac{1 \text{ ft}}{2} = 0.5 \text{ ft}$$

The depth at which the centroid of the circular window is located is

$$h_c = 4.0 \text{ ft} + 0.5 \text{ ft} = 4.5 \text{ ft}$$

The area of the circular window is

$$A = \pi r^2 = \pi (0.5 \text{ ft})^2 = 0.7854 \text{ ft}^2$$

The average pressure is

$$\bar{p} = \gamma h_c = \left(62.4 \; \frac{\text{lbf}}{\text{ft}^3}\right)(4.5 \text{ ft})$$

$$= 280.8 \text{ lbf/ft}^2 \quad \text{(psfg)}$$

The resultant is calculated from Eq. 15.19.

$$R = \bar{p}A = \left(280.8 \; \frac{\text{lbf}}{\text{ft}^2}\right)(0.7854 \text{ ft}^2) = 220.5 \text{ lbf}$$

(b) The centroidal area moment of inertia of a circle is

$$I_c = \frac{\pi}{4} r^4 = \left(\frac{\pi}{4}\right)(0.5 \text{ ft})^4 = 0.049 \text{ ft}^4$$

From Eq. 15.23, the depth at which the resultant force acts is

$$h_R = h_c + \frac{I_c}{Ah_c} = 4.5 \text{ ft} + \frac{0.049 \text{ ft}^4}{(0.7854 \text{ ft}^2)(4.5 \text{ ft})}$$

$$= 4.514 \text{ ft}$$

10. SPECIAL CASES: VERTICAL SURFACES

Several simple wall shapes and configurations recur frequently. Figure 15.12 indicates the depths, h_R, of their hydrostatic pressure resultants (*centers of pressure*). In all cases, the surfaces are vertical and extend to the liquid's surface.

Figure 15.12 *Centers of Pressure for Common Configurations*

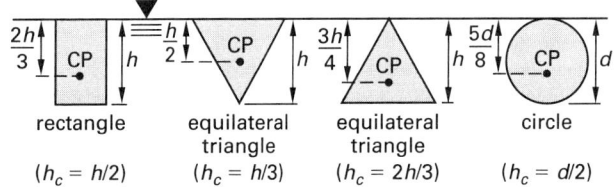

11. FORCES ON CURVED AND COMPOUND SURFACES

Figure 15.13 illustrates a curved surface cross section, BA. The resultant force acting on such a curved surface is not difficult to determine, although the x- and y-components of the resultant usually must be calculated first. The magnitude and direction of the resultant are found by conventional methods.

$$R = \sqrt{R_x^2 + R_y^2} \qquad 15.24$$

$$\theta = \arctan \frac{R_y}{R_x} \qquad 15.25$$

Figure 15.13 *Pressure Distributions on a Curved Surface*

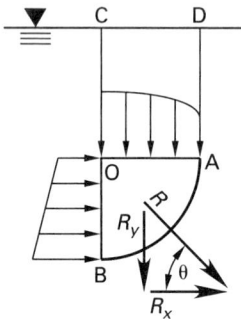

The horizontal component of the resultant hydrostatic force is found in the same manner as for a vertical plane surface.

The fact that the surface is curved does not affect the calculation of the horizontal force. In Fig. 15.13, the horizontal pressure distribution on curved surface BA is the same as the horizontal pressure distribution on imaginary projected surface BO.

The vertical component of force on the curved surface is most easily calculated as the weight of the liquid above it.[12] In Fig. 15.13, the vertical component of force on the curved surface BA is the weight of liquid within the area ABCD, with a vertical line of action passing through the centroid of the area ABCD.

Figure 15.14 illustrates a curved surface with no liquid above it. However, it is not difficult to show that the resultant force acting upward on the curved surface HG is equal in magnitude (and opposite in direction) to the force that would be acting downward due to the missing area EFGH. Such an imaginary area used to calculate hydrostatic pressure is known as an *equivalent area*.

---

[12]Calculating the vertical force component in this manner is not in conflict with the hydrostatic paradox as long as the cross-sectional area of liquid above the curved surface does not decrease between the curved surface and the liquid's free surface. If there is a change in the cross-sectional area, the vertical component of force is equal to the weight of fluid in an unchanged cross-sectional area (i.e., the equivalent area).

Figure 15.14 *Equivalent Area*

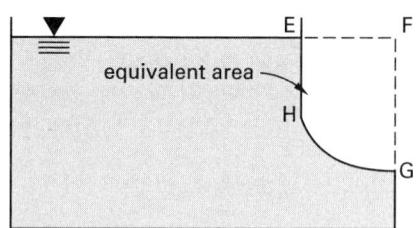

Example 15.6

What is the total resultant force on a 1 ft section of the entire wall in Ex. 15.4?

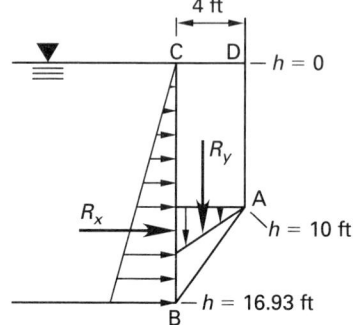

Solution

The average depth is

$$\bar{h} = \left(\tfrac{1}{2}\right)(0 + 16.93 \text{ ft}) = 8.465 \text{ ft}$$

The average pressure and horizontal component of the resultant on a 1 ft section of wall are

$$\bar{p} = \gamma \bar{h} = \left(62.4 \ \frac{\text{lbf}}{\text{ft}^3}\right)(8.465 \text{ ft})$$

$$= 528.2 \text{ lbf/ft}^2 \quad (528.2 \text{ psfg})$$

$$R_x = \bar{p}A = \left(528.2 \ \frac{\text{lbf}}{\text{ft}^2}\right)(16.93 \text{ ft})(1 \text{ ft})$$

$$= 8942 \text{ lbf}$$

The volume of a 1 ft section of area ABCD is

$$V_{\text{ABCD}} = (1 \text{ ft})\left((4 \text{ ft})(10 \text{ ft}) + \left(\tfrac{1}{2}\right)(4 \text{ ft})(6.93 \text{ ft})\right)$$

$$= 53.86 \text{ ft}^3$$

The vertical component is

$$R_y = \gamma V = \left(62.4 \ \frac{\text{lbf}}{\text{ft}^3}\right)(53.86 \text{ ft}^3) = 3361 \text{ lbf}$$

The total resultant force is

$$R = \sqrt{(8942 \text{ lbf})^2 + (3361 \text{ lbf})^2}$$

$$= 9553 \text{ lbf}$$

12. TORQUE ON A GATE

When an openable gate or door is submerged in such a manner as to have unequal depths of liquid on its two sides, or when there is no liquid present on one side, the hydrostatic pressure will act to either open or close the door. If the gate does not move, this pressure is resisted, usually by a latching mechanism on the gate itself.[13] The magnitude of the resisting latch force can be determined from the *hydrostatic torque* (*hydrostatic moment*) acting on the gate. (See Fig. 15.15.) The moment is almost always taken with respect to the gate hinges.

Figure 15.15 *Torque on a Hinge (Gate)*

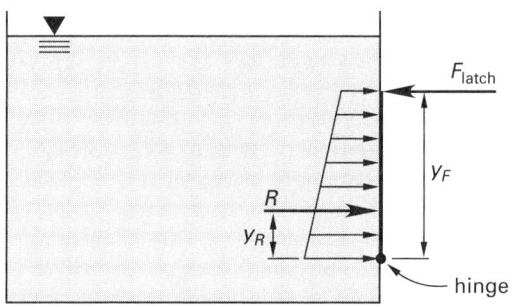

The applied moment is calculated as the product of the resultant force on the gate and the distance from the hinge to the resultant on the gate. This applied moment is balanced by the resisting moment, calculated as the latch force times the separation of the latch and hinge.

$$M_{\text{applied}} = M_{\text{resisting}} \qquad 15.26$$

$$Ry_R = F_{\text{latch}} y_F \qquad 15.27$$

13. HYDROSTATIC FORCES ON A DAM

The concepts presented in the preceding sections are applicable to dams. That is, the horizontal force on the dam face can be found as in Ex. 15.6, regardless of inclination or curvature of the dam face. The vertical force on the dam face is calculated as the weight of the water above the dam face. Of course, the vertical force is zero if the dam face is vertical.

Figure 15.16 illustrates a typical dam, defining its *heel*, *toe*, and *crest*. x_{CG}, the horizontal distance from the toe to the dam's center of gravity, is not shown.

There are several stability considerations for gravity dams.[14] Most notably, the dam must not tip over or slide away due to the hydrostatic pressure. Furthermore, the pressure distribution within the soil under the dam is not uniform, and soil loading must not be excessive.

---

[13]Any contribution to resisting force from stiff hinges or other sources of friction is typically neglected.

[14]A *gravity dam* is one that is held in place and orientation by its own mass (weight) and the friction between its base and the ground.

Figure 15.16 *Dam*

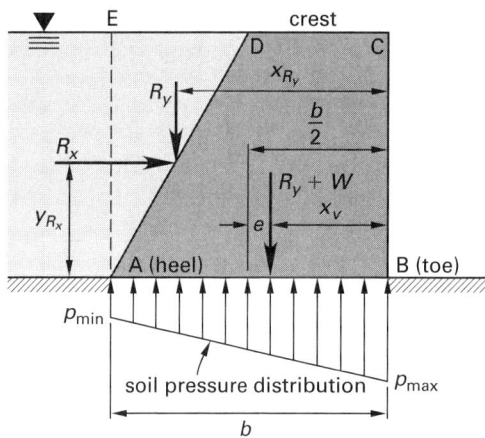

The *overturning moment* is a measure of the horizontal pressure's tendency to tip the dam over, pivoting it about the toe of the dam (point B in Fig. 15.16). (Usually, moments are calculated with respect to the pivot point.) The overturning moment is calculated as the product of the horizontal component of hydrostatic pressure (i.e., the x-component of the resultant) and the vertical distance between the toe and the line of action of the force.

$$M_{\text{overturning}} = R_x y_{R_x} \qquad 15.28$$

In most configurations, the overturning is resisted jointly by moments from the dam's own weight, W, and the vertical component of the resultant, R_y (i.e., the weight of area EAD in Fig. 15.16).[15]

$$M_{\text{resisting}} = R_y x_{R_y} + W x_{\text{CG}} \qquad 15.29$$

The *factor of safety against overturning* is

$$(\text{FS})_{\text{overturning}} = \frac{M_{\text{resisting}}}{M_{\text{overturning}}} \qquad 15.30$$

In addition to causing the dam to tip over, the horizontal component of hydrostatic force will also cause the dam to tend to slide along the ground. This tendency is resisted by the frictional force between the dam bottom and soil. The frictional force, F_f, is calculated as the product of the *normal force*, N, and the *coefficient of static friction*, μ.

$$F_f = \mu_{\text{static}} N = \mu_{\text{static}}(W + R_y) \qquad 15.31$$

The *factor of safety against sliding* is

$$(\text{FS})_{\text{sliding}} = \frac{F_f}{R_x} \qquad 15.32$$

The soil pressure distribution beneath the dam is usually assumed to vary linearly from a minimum to a

---

[15]The density of concrete or masonry with steel reinforcing is usually taken to be approximately 150 lbm/ft³ (2400 kg/m³).

maximum value. (The minimum value must be greater than zero, since soil cannot be in a state of tension. The maximum pressure should not exceed the allowable soil pressure.) Equation 15.33 predicts the minimum and maximum soil pressures.

$$p_{\max}, p_{\min} = \left(\frac{R_y + W}{b}\right)\left(1 \pm \frac{6e}{b}\right) \quad \text{[per unit width]}$$

$$15.33$$

The *eccentricity*, e, in Eq. 15.33 is the distance between the mid-length of the dam and the line of action of the total vertical force, $W + R_y$. (The eccentricity must be less than $b/6$ for the entire base to be in compression.) Distances x_v and x_{CG} are different.

$$e = \frac{b}{2} - x_v \qquad 15.34$$

$$x_v = \frac{M_{\text{resisting}} - M_{\text{overturning}}}{R_y + W} \qquad 15.35$$

14. PRESSURE DUE TO SEVERAL IMMISCIBLE LIQUIDS

Figure 15.17 illustrates the nonuniform pressure distribution due to two immiscible liquids (e.g., oil on top and water below).

Figure 15.17 *Pressure Distribution from Two Immiscible Liquids*

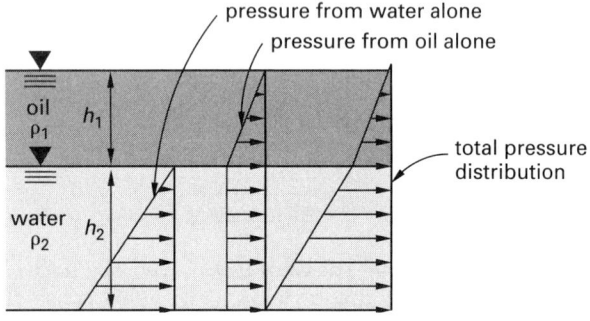

The pressure due to the upper liquid (oil), once calculated, serves as a *surcharge* to the liquid below (water). The pressure at the tank bottom is given by Eq. 15.36. (The principle can be extended to three or more immiscible liquids as well.)

$$p_{\text{bottom}} = \rho_1 g h_1 + \rho_2 g h_2 \qquad \text{[SI]} \quad 15.36(a)$$

$$\begin{aligned} p_{\text{bottom}} &= \frac{\rho_1 g h_1}{g_c} + \frac{\rho_2 g h_2}{g_c} \\ &= \gamma_1 h_1 + \gamma_2 h_2 \end{aligned} \qquad \text{[U.S.]} \quad 15.36(b)$$

15. PRESSURE FROM COMPRESSIBLE FLUIDS

Fluid density, thus far, has been assumed to be independent of pressure. In reality, even "incompressible" liquids are slightly compressible. Sometimes, the effect of this compressibility cannot be neglected.

The familiar $p = \rho g h$ equation is a special case of Eq. 15.37. (It is assumed that $h_2 > h_1$. The minus sign in Eq. 15.37 indicates that pressure decreases as elevation (height) increases.)

$$\int_{p_1}^{p_2} \frac{dp}{\rho g} = -(h_2 - h_1) \qquad \text{[SI]} \quad 15.37(a)$$

$$\int_{p_1}^{p_2} \frac{g_c \, dp}{\rho g} = -(h_2 - h_1) \qquad \text{[U.S.]} \quad 15.37(b)$$

If the fluid is a perfect gas, and if compression is an isothermal (i.e., constant temperature) process, then the relationship between pressure and density is given by Eq. 15.38. The isothermal assumption is appropriate, for example, for the earth's *stratosphere* (i.e., above 35,000 ft (11 000 m)), where the temperature is assumed to be constant at approximately $-67°\text{F}$ ($-55°\text{C}$).

$$pv = \frac{p}{\rho} = RT = \text{constant} \qquad 15.38$$

In the isothermal case, Eq. 15.38 can be rewritten as Eq. 15.39. (For air, $R = 53.35$ ft-lbf/lbm-°R; $R = 287.03$ J/kg·K.) Of course, the temperature, T, must be in degrees absolute (i.e., in °R or K). Equation 15.39 is known as the *barometric height relationship*[16] because knowledge of atmospheric temperature and the pressures at two points is sufficient to determine the elevation difference between the two points.

$$h_2 - h_1 = \frac{RT}{g} \ln \frac{p_1}{p_2} \qquad \text{[SI]} \quad 15.39(a)$$

$$h_2 - h_1 = \frac{g_c RT}{g} \ln \frac{p_1}{p_2} \qquad \text{[U.S.]} \quad 15.39(b)$$

The pressure at an elevation (height) h_2 in a layer of perfect gas that has been isothermally compressed is given by Eq. 15.40.

$$p_2 = p_1 e^{g(h_1 - h_2)/RT} \qquad \text{[SI]} \quad 15.40(a)$$

$$p_2 = p_1 e^{g(h_1 - h_2)/g_c RT} \qquad \text{[U.S.]} \quad 15.40(b)$$

---

[16]This is equivalent to the work done in an isothermal compression process. The elevation (height) difference, $h_2 - h_1$ (with units of feet), can be interpreted as the work done per unit mass during compression (with units of ft-lbf/lbm).

If the fluid is a perfect gas, and if compression is an *adiabatic process*, the relationship between pressure and density is given by Eq. 15.41,[17] where k is the *ratio of specific heats*, a property of the gas. ($k = 1.4$ for air, hydrogen, oxygen, and carbon monoxide, among others.)

$$pv^k = p\left(\frac{1}{\rho}\right)^k = \text{constant} \qquad 15.41$$

The following three equations apply to adiabatic compression of an ideal gas.

$$h_2 - h_1 = \left(\frac{k}{k-1}\right)\left(\frac{RT_1}{g}\right)\left(1 - \left(\frac{p_2}{p_1}\right)^{(k-1)/k}\right)$$
$$\text{[SI]} \quad 15.42(a)$$

$$h_2 - h_1 = \left(\frac{k}{k-1}\right)\left(\frac{g_c}{g}\right)RT_1\left(1 - \left(\frac{p_2}{p_1}\right)^{(k-1)/k}\right)$$
$$\text{[U.S.]} \quad 15.42(b)$$

$$p_2 = p_1\left(1 - \left(\frac{k-1}{k}\right)\left(\frac{g}{RT_1}\right)(h_2 - h_1)\right)^{k/(k-1)}$$
$$\text{[SI]} \quad 15.43(a)$$

$$p_2 = p_1\left(1 - \left(\frac{k-1}{k}\right)\left(\frac{g}{g_c}\right)\left(\frac{h_2 - h_1}{RT_1}\right)\right)^{k/(k-1)}$$
$$\text{[U.S.]} \quad 15.43(b)$$

$$T_2 = T_1\left(1 - \left(\frac{k-1}{k}\right)\left(\frac{g}{RT_1}\right)(h_2 - h_1)\right) \quad \text{[SI]} \quad 15.44(a)$$

$$T_2 = T_1\left(1 - \left(\frac{k-1}{k}\right)\left(\frac{g}{g_c}\right)\left(\frac{h_2 - h_1}{RT_1}\right)\right) \quad \text{[U.S.]} \quad 15.44(b)$$

The three adiabatic compression equations can be used for the more general *polytropic compression* case simply by substituting the *polytropic exponent*, n, for k.[18] Unlike the ratio of specific heats, the polytropic exponent is a function of the process, not of the gas. The polytropic compression assumption is appropriate for the earth's *troposphere*.[19] Assuming a linear decrease in temperature along with an altitude of $-0.00356°F/\text{ft}$ ($-0.00649°C/\text{m}$), the polytropic exponent is $n = 1.235$.

---

[17]There is no heat or energy transfer to or from the ideal gas in an adiabatic process. However, this is not the same as an isothermal process.
[18]Actually, polytropic compression is the general process. Isothermal compression is a special case ($n = 1$) of the polytropic process, as is adiabatic compression ($n = k$).
[19]The *troposphere* is the part of the earth's atmosphere we live in and where most atmospheric disturbances occur. The *stratosphere*, starting at approximately 35,000 ft (11 000 m), is cold, clear, dry, and still. Between the troposphere and the stratosphere is the *tropopause*, a transition layer that contains most of the atmosphere's dust and moisture. Temperature actually increases with altitude in the stratosphere and decreases with altitude in the troposphere, but is constant in the tropopause.

Example 15.7

The air pressure and temperature at sea level are 1.0 standard atmosphere and 68°F (20°C), respectively. Assume polytropic compression with $n = 1.235$. What is the pressure at an altitude of 5000 ft (1525 m)?

SI Solution

The absolute temperature of the air is $20°C + 273° = 293K$. From Eq. 15.43 (substituting $k = n = 1.235$ for polytropic compression), the pressure at 1525 m altitude is

$$p_2 = (1.0 \text{ atm})$$
$$\times \left(1 - \left(\frac{1.235 - 1}{1.235}\right)\left(9.81 \frac{\text{m}}{\text{s}^2}\right) \right.$$
$$\left. \times \left(\frac{1525 \text{ m}}{\left(287.03 \frac{\text{J}}{\text{kg·K}}\right)(293K)}\right) \right)^{1.235/(1.235-1)}$$
$$= 0.834 \text{ atm}$$

Customary U.S. Solution

The absolute temperature of the air is $68°F + 460° = 528°R$. From Eq. 15.43 (substituting $k = n = 1.235$ for polytropic compression), the pressure at an altitude of 5000 ft is

$$p_2 = (1.0 \text{ atm})$$
$$\times \left(1 - \left(\frac{1.235 - 1}{1.235}\right)\left(\frac{32.2 \frac{\text{ft}}{\text{sec}^2}}{32.2 \frac{\text{lbm-ft}}{\text{lbf-sec}^2}}\right) \right.$$
$$\left. \times \left(\frac{5000 \text{ ft}}{\left(53.35 \frac{\text{ft-lbf}}{\text{lbm-°R}}\right)(528°R)}\right) \right)^{1.235/(1.235-1)}$$
$$= 0.835 \text{ atm}$$

16. EXTERNALLY PRESSURIZED LIQUIDS

If the gas above a liquid in a closed tank is pressurized to a gage pressure of p_t, this pressure will add to the hydrostatic pressure anywhere in the fluid. The pressure at the tank bottom illustrated in Fig. 15.18 is given by Eq. 15.45.

$$p_{\text{bottom}} = p_t + \rho gh \qquad \text{[SI]} \quad 15.45(a)$$

$$p_{\text{bottom}} = p_t + \frac{\rho gh}{g_c} = p_t + \gamma h \qquad \text{[U.S.]} \quad 15.45(b)$$

Figure 15.18 *Externally Pressurized Liquid*

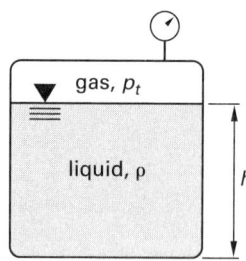

17. HYDRAULIC RAM

A *hydraulic ram* (*hydraulic jack, hydraulic press, fluid press*, etc.) is illustrated in Fig. 15.19. This is a force-multiplying device. A force, F_p, is applied to the *plunger*, and a useful force, F_r, appears at the *ram*. Even though the pressure in the hydraulic fluid is the same on the ram and plunger, the forces on them will be proportional to their respective cross-sectional areas.

Figure 15.19 *Hydraulic Ram*

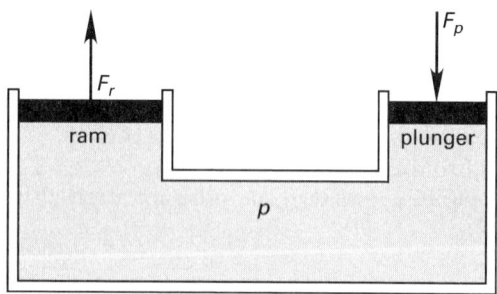

Since the pressure, p, is the same on both the plunger and ram, Eq. 15.46 can be solved for it.

$$F = pA = p\pi\left(\frac{d^2}{4}\right) \qquad 15.46$$

$$p_p = p_r \qquad 15.47$$

$$\frac{F_p}{A_p} = \frac{F_r}{A_r} \qquad 15.48$$

$$\frac{F_p}{d_p^2} = \frac{F_r}{d_r^2} \qquad 15.49$$

A small, manually actuated hydraulic ram will have a lever handle to increase the *mechanical advantage* of the ram from 1.0 to M, as illustrated in Fig. 15.20. In most cases, the pivot and connection mechanism will not be frictionless, and some of the applied force will be used to overcome the friction. This friction loss is accounted for by a *lever efficiency* or *lever effectiveness*, η.

$$M = \frac{L_1}{L_2} \qquad 15.50$$

$$F_p = \eta M F_l \qquad 15.51$$

Figure 15.20 *Hydraulic Ram with Mechanical Advantage*

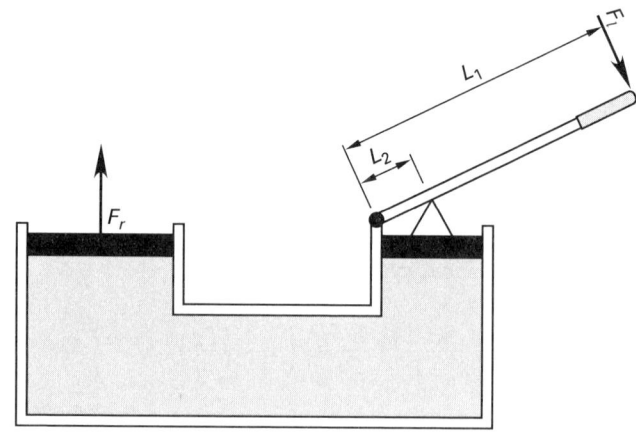

$$\frac{\eta M F_l}{A_p} = \frac{F_r}{A_r} \qquad 15.52$$

18. BUOYANCY

The *buoyant force* is an upward force that acts on all objects that are partially or completely submerged in a fluid. The fluid can be a liquid, as in the case of a ship floating at sea, or the fluid can be a gas, as in a balloon floating in the atmosphere.

There is a buoyant force on all submerged objects, not just those that are stationary or ascending. There will be, for example, a buoyant force on a rock sitting at the bottom of a pond. There will also be a buoyant force on a rock sitting exposed on the ground, since the rock is "submerged" in air. For partially submerged objects floating in liquids, such as icebergs, a buoyant force due to displaced air also exists, although it may be insignificant.

Buoyant force always acts to counteract an object's weight (i.e., buoyancy acts against gravity). The magnitude of the buoyant force is predicted from *Archimedes' principle* (*the buoyancy theorem*): The buoyant force on a submerged object is equal to the weight of the displaced fluid.[20] An equivalent statement of Archimedes' principle is: A floating object displaces liquid equal in weight to its own weight.

$$F_{\text{buoyant}} = \rho g V_{\text{displaced}} \qquad \text{[SI]} \quad 15.53(a)$$

$$F_{\text{buoyant}} = \frac{\rho g V_{\text{displaced}}}{g_c} = \gamma V_{\text{displaced}} \qquad \text{[U.S.]} \quad 15.53(b)$$

In the case of stationary (i.e., not moving vertically) objects, the buoyant force and object weight are in equilibrium. If the forces are not in equilibrium, the

---

[20]The volume term in Eq. 15.53 is the total volume of the object only in the case of complete submergence.

object will rise or fall until equilibrium is reached. That is, the object will sink until its remaining weight is supported by the bottom, or it will rise until the weight of displaced liquid is reduced by breaking the surface.[21]

The specific gravity (SG) of an object submerged in water can be determined from its dry and submerged weights. Neglecting the buoyancy of any surrounding gases,

$$\text{SG} = \frac{W_{\text{dry}}}{W_{\text{dry}} - W_{\text{submerged}}} \qquad 15.54$$

Figure 15.21 illustrates an object floating partially exposed in a liquid. Neglecting the insignificant buoyant force from the displaced air (or other gas), the fractions, x, of volume exposed and submerged are easily determined.

$$x_{\text{submerged}} = \frac{\rho_{\text{object}}}{\rho_{\text{liquid}}} = \frac{(\text{SG})_{\text{object}}}{(\text{SG})_{\text{liquid}}} \qquad 15.55$$

$$x_{\text{exposed}} = 1 - x_{\text{submerged}} \qquad 15.56$$

Figure 15.21 Partially Submerged Object

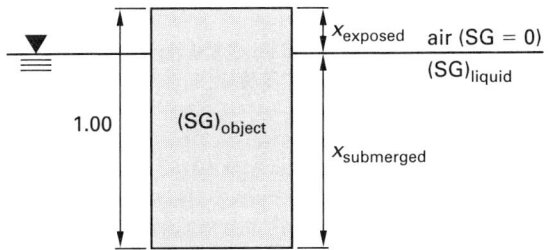

Figure 15.22 illustrates a somewhat more complicated situation—that of an object floating at the interface between two liquids of different densities. The fractions of immersion in each liquid are given by the following equations.

$$x_1 = \frac{(\text{SG})_2 - (\text{SG})_{\text{object}}}{(\text{SG})_2 - (\text{SG})_1} \qquad 15.57$$

$$x_2 = 1 - x_1 = \frac{(\text{SG})_{\text{object}} - (\text{SG})_1}{(\text{SG})_2 - (\text{SG})_1} \qquad 15.58$$

A more general case of a floating object is shown in Fig. 15.23. Situations of this type are easily evaluated by equating the object's weight with the sum of the buoyant forces.

---

[21]An object can also stop rising or falling due to a change in the fluid's density. The buoyant force will increase with increasing depth in the ocean due to an increase in density at great depths. The buoyant force will decrease with increasing altitude in the atmosphere due to a decrease in density at great heights.

Figure 15.22 Object Floating in Two Liquids

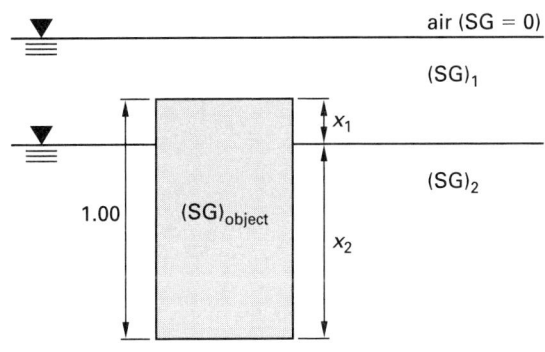

Figure 15.23 General Two-Liquid Buoyancy Problem

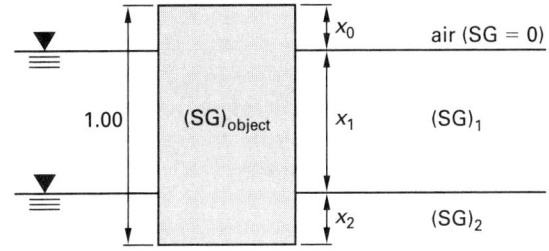

In the case of Fig. 15.23 (with two liquids), the following relationships apply, where x_0 is the fraction, if any, extending into the air above.

$$(\text{SG})_{\text{object}} = x_1(\text{SG})_1 + x_2(\text{SG})_2 \qquad 15.59$$

$$(\text{SG})_{\text{object}} = (1 - x_0 - x_2)(\text{SG})_1 + x_2(\text{SG})_2 \qquad 15.60$$

$$(\text{SG})_{\text{object}} = x_1(\text{SG})_1 + (1 - x_0 - x_1)(\text{SG})_2 \qquad 15.61$$

Example 15.8

An empty polyethylene telemetry balloon and payload have a mass of 500 lbm (225 kg). The balloon is filled with helium when the atmospheric conditions are 60°F (15.6°C) and 14.8 psia (102 kPa). The specific gas constant of helium is 2079 J/kg·K (386.3 ft-lbf/lbm-°R). What minimum volume of helium is required for lift-off from a sea-level platform?

SI Solution

$$\rho_{\text{air}} = \frac{p}{RT} = \frac{1.02 \times 10^5 \text{ Pa}}{\left(287.03 \dfrac{\text{J}}{\text{kg·K}}\right)(15.6°\text{C} + 273°)}$$

$$= 1.231 \text{ kg/m}^3$$

$$\rho_{\text{helium}} = \frac{1.02 \times 10^5 \text{ Pa}}{\left(2079 \dfrac{\text{J}}{\text{kg·K}}\right)(288.6\text{K})} = 0.17 \text{ kg/m}^3$$

Flow of Fluids

$$m = 225 \text{ kg} + \left(0.17 \frac{\text{kg}}{\text{m}^3}\right) V_{\text{He}}$$

$$m_b = \left(1.231 \frac{\text{kg}}{\text{m}^3}\right) V_{\text{He}}$$

$$225 \text{ kg} + \left(0.17 \frac{\text{kg}}{\text{m}^3}\right) V_{\text{He}} = \left(1.231 \frac{\text{kg}}{\text{m}^3}\right) V_{\text{He}}$$

$$V_{\text{He}} = 212.1 \text{ m}^3$$

Customary U.S. Solution

The gas densities are

$$\rho_{\text{air}} = \frac{p}{RT} = \frac{\left(14.8 \frac{\text{lbf}}{\text{in}^2}\right)\left(12 \frac{\text{in}}{\text{ft}}\right)^2}{\left(53.35 \frac{\text{ft-lbf}}{\text{lbm-}°\text{R}}\right)(60°\text{F} + 460°)}$$

$$= 0.07682 \text{ lbm/ft}^3$$

$$\gamma_{\text{air}} = \rho \times \frac{g}{g_c} = 0.07682 \text{ lbf/ft}^3$$

$$\rho_{\text{helium}} = \frac{\left(14.8 \frac{\text{lbf}}{\text{in}^2}\right)\left(12 \frac{\text{in}}{\text{ft}}\right)^2}{\left(386.3 \frac{\text{ft-lbf}}{\text{lbm-}°\text{R}}\right)(520°\text{R})}$$

$$= 0.01061 \text{ lbm/ft}^3$$

$$\gamma_{\text{helium}} = 0.01061 \text{ lbf/ft}^3$$

The total weight of the balloon, payload, and helium is

$$W = 500 \text{ lbf} + \left(0.01061 \frac{\text{lbf}}{\text{ft}^3}\right) V_{\text{He}}$$

The buoyant force is the weight of the displaced air. Neglecting the payload volume, the displaced air volume is the same as the helium volume.

$$F_b = \left(0.07682 \frac{\text{lbf}}{\text{ft}^3}\right) V_{\text{He}}$$

At lift-off, the weight of the balloon is just equal to the buoyant force.

$$W = F_b$$

$$500 \text{ lbf} + \left(0.01061 \frac{\text{lbf}}{\text{ft}^3}\right) V_{\text{He}} = \left(0.07682 \frac{\text{lbf}}{\text{ft}^3}\right) V_{\text{He}}$$

$$V_{\text{He}} = 7552 \text{ ft}^3$$

19. BUOYANCY OF SUBMERGED PIPELINES

Whenever possible, submerged pipelines for river crossings should be completely buried at a level below river scour. This will reduce or eliminate loads and movement due to flutter, scour and fill, drag, collisions, and buoyancy. Submerged pipelines should cross at right angles to the river. For maximum flexibility and ductility, pipelines should be made of thick-walled mild steel.

Submerged pipelines should be weighted to achieve a minimum of 20% negative buoyancy (i.e., an average density of 1.2 times the environment, approximately 72 lbm/ft^3 or 1200 kg/m^3). Metal or concrete clamps can be used for this purpose, as well as concrete coatings. Thick steel clamps have the advantage of a smaller lateral exposed area (resulting in less drag from river flow), while brittle concrete coatings are sensitive to pipeline flutter and temperature fluctuations.

Due to the critical nature of many pipelines and the difficulty in accessing submerged portions for repair, it is common to provide a parallel auxiliary line. The auxiliary and main lines are provided with crossover and mainline valves, respectively, on high ground at both sides of the river to permit either or both lines to be used.

20. INTACT STABILITY: STABILITY OF FLOATING OBJECTS

A stationary object is said to be in *static equilibrium*. However, an object in static equilibrium is not necessarily stable. For example, a coin balanced on edge is in static equilibrium, but it will not return to the balanced position if it is disturbed. An object is said to be *stable* (i.e., in *stable equilibrium*) if it tends to return to the equilibrium position when slightly displaced.

Stability of floating and submerged objects is known as *intact stability*.[22] There are two forces acting on a stationary floating object: the buoyant force and the object's weight. The buoyant force acts upward through the centroid of the displaced volume. This centroid is known as the *center of buoyancy*. The gravitational force on the object (i.e., the object's weight) acts downward through the object's center of gravity.

For a totally submerged object (as in the balloon and submarine shown in Fig. 15.24) to be stable, the center of buoyancy must be above the center of gravity. The object will be stable because a righting moment will be created if the object tips over, since the center of buoyancy will move outward from the center of gravity.

The stability criterion is different for partially submerged objects (e.g., surface ships). If the vessel shown in

---

[22]The subject of intact stability, being a part of naval architecture curriculum, is not covered extensively in most fluids books. However, it is covered extensively in basic ship design and naval architecture books.

Figure 15.24 *Stability of a Submerged Object*

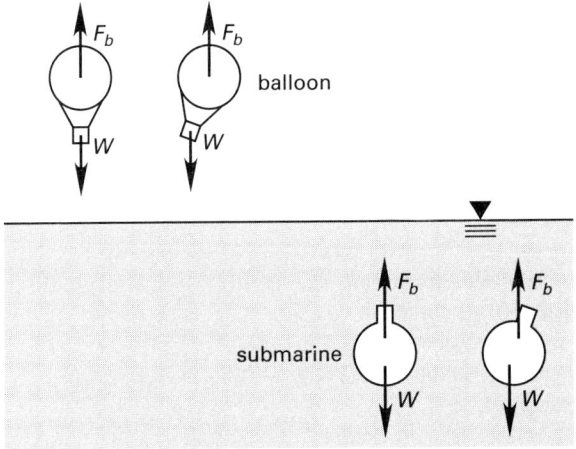

Fig. 15.25 heels (i.e., lists or rolls), the location of the center of gravity of the object does not change.[23] However, the center of buoyancy shifts to the centroid of the new submerged section 123. The centers of buoyancy and gravity are no longer in line.

Figure 15.25 *Stability of a Partially Submerged Floating Object*

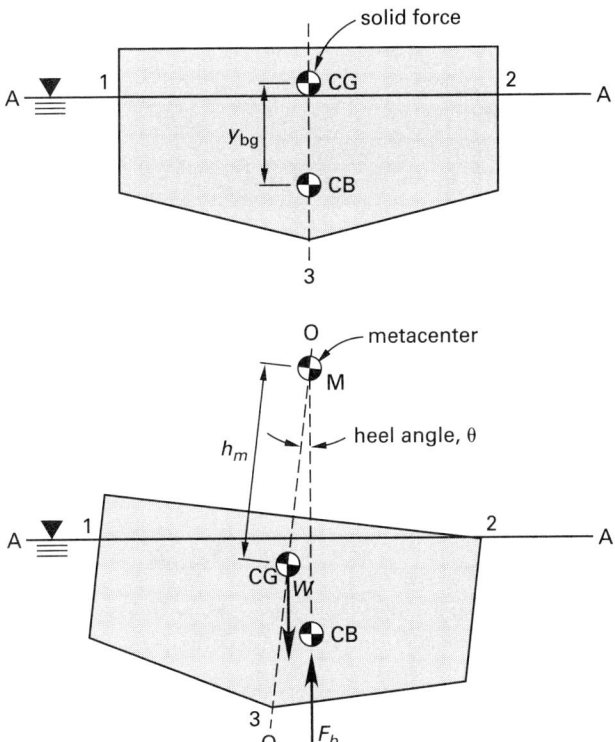

This righting couple exists when the extension of the buoyant force, F_b, intersects line O–O above the center of gravity at M, the *metacenter*. For partially submerged objects to be stable, the metacenter must be above the center of gravity. If M lies below the center

---

[23]The verbs *roll*, *list*, and *heel* are synonymous.

of gravity, an overturning couple will exist. The distance between the center of gravity and the metacenter is called the *metacentric height*, and it is reasonably constant for heel angles less than 10°. Also, for angles less than 10°, the center of buoyancy follows a locus for which the metacenter is the instantaneous center.

The metacentric height is one of the most important and basic parameters in ship design. It determines the ship's ability to remain upright as well as the ship's roll and pitch characteristics.

"Acceptable" minimum values of the metacentric height have been established from experience, and these depend on the ship type and class. For example, many submarines are required to have a metacentric height of 1 ft (0.3 m) when surfaced. This will increase to approximately 3.5 ft (1.2 m) for some of the largest surface ships. If an acceptable metacentric height is not achieved initially, the center of gravity must be lowered or the keel depth increased. The beam width can also be increased slightly to increase the waterplane moment of inertia.

For a surface vessel rolling through an angle less than approximately 10°, the distance between the vertical center of gravity and the metacenter can be found from Eq. 15.62. Variable I is the centroidal area moment of inertia of the original waterline (free surface) cross section about a longitudinal (fore and aft) waterline axis; V is the displaced volume.

If the distance, y_{bg}, separating the centers of buoyancy and gravity is known, Eq. 15.62 can be solved for the metacentric height. y_{bg} is positive when the center of gravity is above the center of buoyancy. This is the normal case. Otherwise, y_{bg} is negative.

$$y_{bg} + h_m = \frac{I}{V} \qquad 15.62$$

The *righting moment* (also known as the *restoring moment*) is the stabilizing moment exerted when the ship rolls. Values of the righting moment are typically specified with units of foot-tons (MN·m).

$$M_{\text{righting}} = h_m \gamma_w V_{\text{displaced}} \sin \theta \qquad 15.63$$

The transverse (roll) and longitudinal (pitch) *periods* also depend on the metacentric height. The roll characteristics are found from the differential equation formed by equating the righting moment to the product of the ship's transverse mass moment of inertia and the angular acceleration. Larger metacentric heights result in lower roll periods. If k is the radius of gyration about the roll axis, the roll period is

$$T_{\text{roll}} = \frac{2\pi k}{\sqrt{g h_m}} \qquad 15.64$$

The roll and pitch periods must be adjusted for the appropriate level of crew and passenger comfort. A "beamy" ice-breaking ship will have a metacentric height much larger than normally required for intact stability, resulting in a short, nauseating roll period. The designer of a passenger ship, however, would have to decrease the intact stability (i.e., decrease the metacentric height) in order to achieve an acceptable ride characteristic. This requires a moderate metacentric height that is less than approximately 6% of the beam length.

Example 15.9

A 600,000 lbm (280 000 kg) rectangular barge has external dimensions of 24 ft width, 98 ft length, and 12 ft height (7 m × 30 m × 3.6 m). It floats in seawater (γ_w = 64.0 lbf/ft^3; ρ_w = 1024 kg/m^3). The center of gravity is 7.8 ft (2.4 m) from the top of the barge as loaded. Find (a) the location of the center of buoyancy when the barge is floating on an even keel, and (b) the approximate location of the metacenter when the barge experiences a 5° heel.

SI Solution

(a) Refer to the following diagram. Let dimension y represent the depth of the submerged barge.

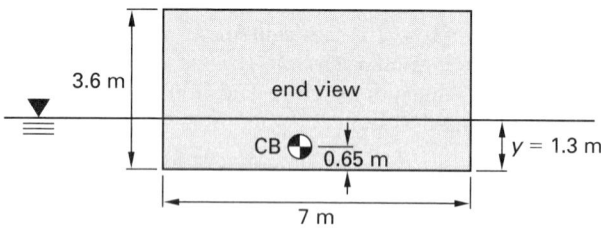

From Archimedes' principle, the buoyant force equals the weight of the barge. This, in turn, equals the weight of the displaced seawater.

$$F_b = W = V\rho_w g$$

$$(280\,000 \text{ kg})\left(9.81 \frac{\text{m}}{\text{s}^2}\right) = y\big((7 \text{ m})(30 \text{ m})\big)\left(1024 \frac{\text{kg}}{\text{m}^3}\right)$$

$$\times \left(9.81 \frac{\text{m}}{\text{s}^2}\right)$$

$$y = 1.3 \text{ m}$$

The center of buoyancy is located at the centroid of the submerged cross section. When floating on an even keel, the submerged cross section is rectangular with a height of 1.3 m. The height of the center of buoyancy above the keel is

$$\frac{1.3 \text{ m}}{2} = 0.65 \text{ m}$$

(b) While the location of the new center of buoyancy can be determined, the location of the metacenter does not change significantly for small angles of heel. For approximate calculations, the angle of heel is not significant.

The area moment of inertia of the longitudinal waterline cross section is

$$I = \frac{Lw^3}{12} = \frac{(30 \text{ m})(7 \text{ m})^3}{12} = 858 \text{ m}^4$$

The submerged volume is

$$V = (1.3 \text{ m})(7 \text{ m})(30 \text{ m}) = 273 \text{ m}^3$$

The distance between the center of gravity and the center of buoyancy is

$$y_{bg} = 3.6 \text{ m} - 2.4 \text{ m} - 0.65 \text{ m} = 0.55 \text{ m}$$

The metacentric height measured above the center of gravity is

$$h_m = \frac{I}{V} - y_{bg} = \frac{858 \text{ m}^4}{273 \text{ m}^3} - 0.55 \text{ m} = 2.6 \text{ m}$$

Customary U.S. Solution

(a) Refer to the following diagram. Let dimension y represent the depth of the submerged barge.

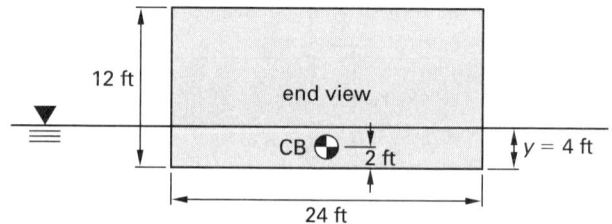

From Archimedes' principle, the buoyant force equals the weight of the barge. This, in turn, equals the weight of the displaced seawater.

$$F_b = W = V\gamma_w$$

$$600,000 \text{ lbf} = y\big((24 \text{ ft})(98 \text{ ft})\big)\left(64 \frac{\text{lbf}}{\text{ft}^3}\right)$$

$$y = 4 \text{ ft}$$

The center of buoyancy is located at the centroid of the submerged cross section. When floating on an even keel, the submerged cross section is rectangular with a height of 4 ft. The height of the center of buoyancy above the keel is

$$\frac{4 \text{ ft}}{2} = 2 \text{ ft}$$

(b) While the location of the new center of buoyancy can be determined, the location of the metacenter does not change significantly for small angles of heel. Therefore, for approximate calculations, the angle of heel is not significant.

The area moment of inertia of the longitudinal waterline cross section is

$$I = \frac{Lw^3}{12} = \frac{(98 \text{ ft})(24 \text{ ft})^3}{12} = 112{,}900 \text{ ft}^4$$

The submerged volume is

$$V = (4 \text{ ft})(24 \text{ ft})(98 \text{ ft}) = 9408 \text{ ft}^3$$

The distance between the center of gravity and the center of buoyancy is

$$y_{\text{bg}} = 12 \text{ ft} - 7.8 \text{ ft} - 2 \text{ ft} = 2.2 \text{ ft}$$

The metacentric height measured above the center of gravity is

$$\begin{aligned} h_m &= \frac{I}{V} - y_{\text{bg}} = \frac{112{,}900 \text{ ft}^4}{9408 \text{ ft}^3} - 2.2 \text{ ft} \\ &= 9.8 \text{ ft} \end{aligned}$$

21. FLUID MASSES UNDER EXTERNAL ACCELERATION

If a fluid mass is subjected to an external acceleration (moved sideways, rotated, etc.), an additional force will be introduced. This force will change the equilibrium position of the fluid surface as well as the hydrostatic pressure distribution.

Figure 15.26 illustrates a liquid mass subjected to constant accelerations in the vertical and/or horizontal directions. (a_y is negative if the acceleration is downward.) The surface is inclined at the angle predicted by Eq. 15.65. The planes of equal hydrostatic pressure beneath the surface are also inclined at the same angle.[24]

$$\theta = \arctan \frac{a_x}{a_y + g} \qquad \qquad 15.65$$

$$p = \rho(g + a_y)h \qquad \qquad \text{[SI]} \quad 15.66(a)$$

$$p = \frac{\rho(g + a_y)h}{g_c} = \gamma h\left(1 + \frac{a_y}{g}\right) \qquad \text{[U.S.]} \quad 15.66(b)$$

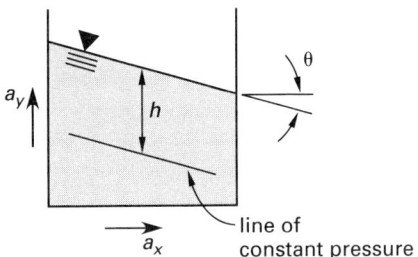

Figure 15.26 *Fluid Mass Under Constant Linear Acceleration*

Figure 15.27 illustrates a fluid mass rotating about a vertical axis at constant angular velocity, ω, in rad/sec.[25] The resulting surface is parabolic in shape. The elevation of the fluid surface at point A at distance r from the axis of rotation is given by Eq. 15.68. The distance h in Fig. 15.27 is measured from the lowest fluid elevation during rotation. h is not measured from the original elevation of the stationary fluid.

$$\theta = \arctan \frac{\omega^2 r}{g} \qquad \qquad 15.67$$

$$h = \frac{(\omega r)^2}{2g} = \frac{\text{v}^2}{2g} \qquad \qquad 15.68$$

Figure 15.27 *Rotating Fluid Mass*

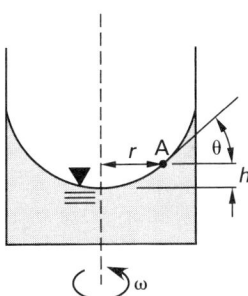

---

[24]Once the orientation of the surface is known, the pressure distribution can be determined without considering the acceleration. The hydrostatic pressure at a point depends only on the height of the liquid above that point. The acceleration affects that height but does not change the $p = \rho g h$ relationship.

[25]Even though the rotational speed is not increasing, the fluid mass experiences a constant *centripetal acceleration* radially outward from the axis of rotation.

16 Fluid Flow Parameters

Nomenclature

A	area	ft^2	m^2
C	correction factor	–	–
C	Hazen-Williams coefficient	–	–
d	depth	ft	m
D	diameter	ft	m
D-load	ASTM C497 load rating	lbf/ft^2	N/m^2
E	joint efficiency	–	–
E	specific energy	ft-lbf/lbm	J/kg
F	force	lbf	N
g	gravitational acceleration, 32.2 (9.81)	ft/sec^2	m/s^2
g_c	gravitational constant, 32.2	lbm-ft/ lbf-sec^2	n.a.
G	mass flow rate per unit area	lbm/ft^2-sec	kg/m^2·s
h	height or head	ft	m
L	length	ft	m
n	Manning's roughness coefficient	–	–
p	pressure	lbf/ft^2	Pa
r	radius	ft	m
Re	Reynolds number	–	–
s	wetted perimeter	ft	m
S	allowable stress	lbf/ft^2	Pa
v	velocity	ft/sec	m/s
$\dot{V}$	volumetric flow rate	ft^3/sec	m^3/s
y	distance	ft	m
z	elevation	ft	m

Symbols

α	angle	rad	rad
γ	specific weight	lbf/ft^3	n.a.
θ	time	sec	s
μ	absolute viscosity	lbf-sec/ft^2	Pa·s
ν	kinematic viscosity	ft^2/sec	m^2/s
ρ	density	lbm/ft^3	kg/m^3
ϕ	angle	rad	rad

Subscripts

ave	average
e	equivalent
h	hydraulic
i	impact or inner
max	maximum
o	outer
p	pressure
r	radius
s	static
t	total
v	velocity
z	potential

1. INTRODUCTION TO FLUID ENERGY UNITS

Several important fluids and thermodynamics equations, such as Bernoulli's equation and the steady-flow energy equation, are special applications of the *conservation of energy* concept. However, it is not always obvious how some formulations of these equations can be termed "energy." For example, elevation, z, with units of feet, is often called *gravitational energy*.

Fluid energy is expressed per unit mass, as indicated by the name, *specific energy*. Units of fluid specific energy are commonly ft-lbf/lbm and J/kg.[1]

2. KINETIC ENERGY

Energy is needed to accelerate a stationary body. Therefore, a moving mass of fluid possesses more energy than an identical, stationary mass. The energy difference is the *kinetic energy* of the fluid.[2] If the kinetic energy is

---

[1]The ft-lbf/lbm unit may be thought of as just "ft," although lbf and lbm do not really cancel out. The combination of variables $E \times (g_c/g)$ is required in order for the units to resolve into "ft."

[2]The terms *velocity energy* and *dynamic energy* are used less often.

evaluated per unit mass, the term *specific kinetic energy* is used. Equation 16.1 gives the specific kinetic energy corresponding to a fluid flow with average velocity, v.

$$E_v = \frac{v^2}{2} \qquad \text{[SI]} \quad 16.1(a)$$

$$E_v = \frac{v^2}{2g_c} \qquad \text{[U.S.]} \quad 16.1(b)$$

3. POTENTIAL ENERGY

Work is performed in elevating a body. Therefore, a mass of fluid at a high elevation will have more energy than an identical mass of fluid at a lower elevation. The energy difference is the *potential energy* of the fluid.[3] Like kinetic energy, potential energy is usually expressed per unit mass. Equation 16.2 gives the potential energy of fluid at an elevation z.[4]

$$E_z = zg \qquad \text{[SI]} \quad 16.2(a)$$

$$E_z = \frac{zg}{g_c} \qquad \text{[U.S.]} \quad 16.2(b)$$

The units of potential energy in a consistent system are again m^2/s^2 or ft^2/sec^2. The units in a traditional English system are ft-lbf/lbm.

z is the elevation of the fluid. The reference point (i.e., zero elevation point) is entirely arbitrary and can be chosen for convenience. This is because potential energy always appears in a difference equation (i.e., ΔE_z), and the reference point cancels out.

4. PRESSURE ENERGY

Work is performed and energy is added when a substance is compressed. Therefore, a mass of fluid at a high pressure will have more energy than an identical mass of fluid at a lower pressure. The energy difference is the *pressure energy* of the fluid.[5] Pressure energy is usually found in equations along with kinetic and potential energies and is expressed as energy per unit mass. Equation 16.3 gives the pressure energy of fluid at pressure p.

$$E_p = \frac{p}{\rho} \qquad 16.3$$

5. BERNOULLI EQUATION

The *Bernoulli equation* is an ideal energy conservation equation based on several reasonable assumptions. The equation assumes the following.

- The fluid is incompressible.
- There is no fluid friction.
- Changes in thermal energy are negligible.[6]

The Bernoulli equation states that the *total energy* of a fluid flowing without friction losses in a pipe is constant.[7] The total energy possessed by the fluid is the sum of its pressure, kinetic, and potential energies. Drawing on Eq. 16.1, Eq. 16.2, and Eq. 16.3, the Bernoulli equation is written as

$$E_t = E_p + E_v + E_z \qquad 16.4$$

$$E_t = \frac{p}{\rho} + \frac{v^2}{2} + zg \qquad \text{[SI]} \quad 16.5(a)$$

$$E_t = \frac{p}{\rho} + \frac{v^2}{2g_c} + \frac{zg}{g_c} \qquad \text{[U.S.]} \quad 16.5(b)$$

Equation 16.5 is valid for both laminar and turbulent flows. Since the original research by Bernoulli assumed laminar flow, when Eq. 16.11 is used for turbulent flow, it may be referred to as the "steady-flow energy equation" instead of the "Bernoulli equation." It can also be used for gases and vapors if the incompressibility assumption is valid.[8]

The quantities known as *total head*, h_t, and *total pressure*, p_t, can be calculated from total energy.

$$h_t = \frac{E_t}{g} \qquad \text{[SI]} \quad 16.6(a)$$

$$h_t = E_t \times \frac{g_c}{g} \qquad \text{[U.S.]} \quad 16.6(b)$$

$$p_t = \rho g h_t \qquad \text{[SI]} \quad 16.7(a)$$

$$p_t = \rho h_t \times \frac{g}{g_c} \qquad \text{[U.S.]} \quad 16.7(b)$$

Example 16.1

A pipe draws water from the bottom of a reservoir and discharges it freely at point C, 100 ft (30 m) below the surface. The flow is frictionless. (a) What is the total specific energy at an elevation 50 ft (15 m) below the water surface (i.e., point B)? (b) What is the velocity at point C?

---

[3]The term *gravitational energy* is also used.
[4]Since $g = g_c$ (numerically), it is tempting to write $E_z = z$. In fact, many engineers in the United States do just that.
[5]The terms *static energy* and *flow energy* are also used. The name *flow energy* results from the need to push (pressurize) a fluid to get it to flow through a pipe. However, flow energy and kinetic energy are not the same.

[6]In thermodynamics, the fluid flow is said to be *adiabatic*.
[7]Strictly speaking, this is the *total specific energy*, since the energy is per unit mass. However, the word "specific," being understood, is seldom used. Of course, "the total energy of the system" means something else and requires knowing the fluid mass in the system.
[8]A gas or vapor can be considered to be incompressible as long as its pressure does not change by more than 10% between the entrance and exit, and its velocity is less than Mach 0.3 everywhere.

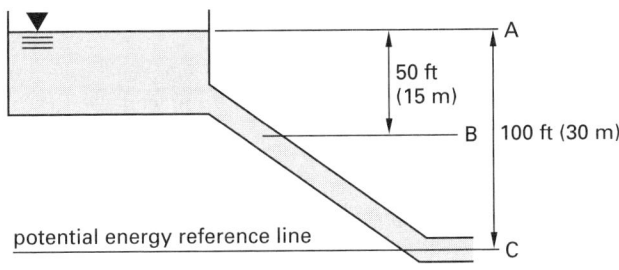

potential energy reference line

SI Solution

(a) At point A, the velocity and gage pressure are both zero. Therefore, the total energy consists only of potential energy. Choose point C as the reference ($z = 0$) elevation.

$$E_A = z_A g = (30\text{ m})\left(9.81\,\frac{\text{m}}{\text{s}^2}\right)$$
$$= 294.3\text{ m}^2/\text{s}^2 \quad (\text{J/kg})$$

At point B, the fluid is moving and possesses kinetic energy. The fluid is also under hydrostatic pressure and possesses pressure energy. These energy forms have come at the expense of potential energy. (This is a direct result of the Bernoulli equation.) Also, the flow is frictionless. Therefore, there is no net change in the total energy between points A and B.

$$E_B = E_A = 294.3\text{ m}^2/\text{s}^2 \quad (\text{J/kg})$$

(b) At point C, the gage pressure and pressure energy are again zero, since the discharge is at atmospheric pressure. The potential energy is zero, since $z = 0$. The total energy of the system has been converted to kinetic energy. From Eq. 16.5,

$$E_t = 294.3\,\frac{\text{m}^2}{\text{s}^2} = 0 + \frac{\text{v}^2}{2} + 0$$
$$\text{v} = 24.3\text{ m/s}$$

Customary U.S. Solution

(a) $$E_A = \frac{z_A g}{g_c} = \frac{(100\text{ ft})\left(32.2\,\frac{\text{ft}}{\text{sec}^2}\right)}{32.2\,\frac{\text{lbm-ft}}{\text{lbf-sec}^2}}$$
$$= 100\text{ ft-lbf/lbm}$$
$$E_t = E_B = E_A = 100\text{ ft-lbf/lbm}$$

(b) $$E_t = E_C = 100\,\frac{\text{ft-lbf}}{\text{lbm}} = 0 + \frac{\text{v}^2}{2g_c} + 0$$
$$\text{v}^2 = 2g_c E_t = (2)\left(32.2\,\frac{\text{lbm-ft}}{\text{lbf-sec}^2}\right)\left(100\,\frac{\text{ft-lbf}}{\text{lbm}}\right)$$
$$= 6440\text{ ft}^2/\text{sec}^2$$
$$\text{v} = \sqrt{6440\,\frac{\text{ft}^2}{\text{sec}^2}} = 80.2\text{ ft/sec}$$

Example 16.2

Water (62.4 lbm/ft^3; 1000 kg/m^3) is pumped up a hillside into a reservoir. The pump discharges water with a velocity of 6 ft/sec (2 m/s) and a pressure of 150 psig (1000 kPa). Disregarding friction, what is the maximum elevation (above the centerline of the pump's discharge) of the reservoir's water surface?

SI Solution

At the centerline of the pump's discharge, the potential energy is zero. The pressure and velocity energies are

$$E_p = \frac{p}{\rho} = \frac{(1000\text{ kPa})\left(1000\,\frac{\text{Pa}}{\text{kPa}}\right)}{1000\,\frac{\text{kg}}{\text{m}^3}} = 1000\text{ J/kg}$$

$$E_v = \frac{\text{v}^2}{2} = \frac{\left(2\,\frac{\text{m}}{\text{s}}\right)^2}{2} = 2\text{ J/kg}$$

The total energy at the pump's discharge is

$$E_{t,1} = E_p + E_v = 1000\,\frac{\text{J}}{\text{kg}} + 2\,\frac{\text{J}}{\text{kg}}$$
$$= 1002\text{ J/kg}$$

Since the flow is frictionless, the same energy is possessed by the water at the reservoir's surface. Since the velocity and gage pressure at the surface are zero, all of the available energy has been converted to potential energy.

$$E_{t,2} = E_{t,1}$$
$$z_2 g = 1002\text{ J/kg}$$
$$z_2 = \frac{E_{t,2}}{g} = \frac{1002\,\frac{\text{J}}{\text{kg}}}{9.81\,\frac{\text{m}}{\text{s}^2}} = 102.1\text{ m}$$

The volumetric flow rate of the water is not relevant since the water velocity was known. Similarly, the pipe size is not needed.

Customary U.S. Solution

$$E_p = \frac{p}{\rho} = \frac{\left(150\,\frac{\text{lbf}}{\text{in}^2}\right)\left(12\,\frac{\text{in}}{\text{ft}}\right)^2}{62.4\,\frac{\text{lbm}}{\text{ft}^3}}$$
$$= 346.15\text{ ft-lbf/lbm}$$

$$E_v = \frac{\text{v}^2}{2g_c} = \frac{\left(6\,\frac{\text{ft}}{\text{sec}}\right)^2}{(2)\left(32.2\,\frac{\text{lbm-ft}}{\text{lbf-sec}^2}\right)}$$
$$= 0.56\text{ ft-lbf/lbm}$$

$$E_{t,1} = E_p + E_v = 346.15 \; \frac{\text{ft-lbf}}{\text{lbm}} + 0.56 \; \frac{\text{ft-lbf}}{\text{lbm}}$$

$$= 346.71 \; \text{ft-lbf/lbm}$$

$$E_{t,2} = E_{t,1}$$

$$\frac{z_2 g}{g_c} = 346.71 \; \text{ft-lbf/lbm}$$

$$z_2 = \frac{E_{t,2} g_c}{g} = \frac{\left(346.71 \; \frac{\text{ft-lbf}}{\text{lbm}}\right)\left(32.2 \; \frac{\text{lbm-ft}}{\text{lbf-sec}^2}\right)}{32.2 \; \frac{\text{ft}}{\text{sec}^2}}$$

$$= 346.71 \; \text{ft}$$

6. IMPACT ENERGY

Impact energy, E_i, (also known as *stagnation energy* and *total energy*), is the sum of the kinetic and pressure energy terms.[9] Equation 16.8 is applicable to liquids and gases flowing with velocities less than approximately Mach 0.3.

$$E_i = E_p + E_v \qquad \qquad 16.8$$

$$E_i = \frac{p}{\rho} + \frac{\text{v}^2}{2} \qquad \text{[SI]} \quad 16.9(a)$$

$$E_i = \frac{p}{\rho} + \frac{\text{v}^2}{2g_c} \qquad \text{[U.S.]} \quad 16.9(b)$$

Impact head, h_i, is calculated from the impact energy in a manner analogous to Eq. 16.6. Impact head represents the height the liquid will rise in a piezometer-pitot tube when the liquid has been brought to rest (i.e., stagnated) in an adiabatic manner. Such a case is illustrated in Fig. 16.1. If a gas or high-velocity, high-pressure liquid is flowing, it will be necessary to use a mercury manometer or pressure gauge to measure stagnation head.

Figure 16.1 *Pitot Tube-Piezometer Apparatus*

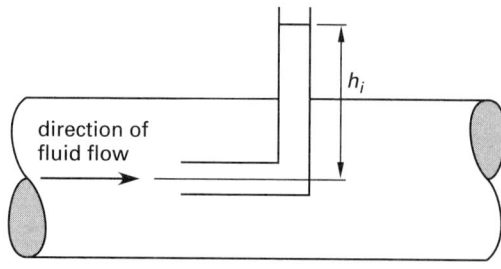

---

[9]It is confusing to label Eq. 16.8 *total* when the gravitational energy term has been omitted. However, the reference point for gravitational energy is arbitrary, and in this application the reference coincides with the centerline of the fluid flow. In truth, the effective pressure developed in a fluid which has been brought to rest adiabatically does not depend on the elevation or altitude of the fluid. This situation is seldom ambiguous. The application will determine which definition of total head or total energy is intended.

7. PITOT TUBE

A *pitot tube* (also known as an *impact tube* or *stagnation tube*) is simply a hollow tube that is placed longitudinally in the direction of fluid flow, allowing the flow to enter one end at the fluid's *velocity of approach*. (See Fig. 16.2.) It is used to measure velocity of flow and finds uses in both subsonic and supersonic applications.

Figure 16.2 *Pitot Tube*

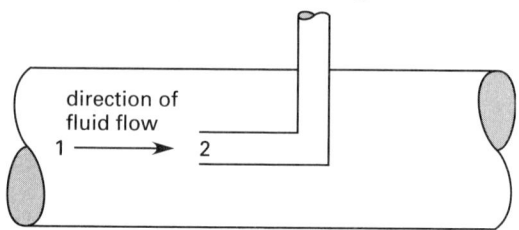

When the fluid enters the pitot tube, it is forced to come to a stop (at the *stagnation point*), and the velocity energy is transformed into pressure energy. If the fluid is a low-velocity gas, the stagnation is assumed to occur without compression heating of the gas. If there is no friction (the common assumption), the process is said to be adiabatic.

Bernoulli's equation can be used to predict the static pressure at the stagnation point. Since the velocity of the fluid within the pitot tube is zero, the upstream velocity can be calculated if the static and stagnation pressures are known.

$$\frac{p_1}{\rho} + \frac{\text{v}_1^2}{2} = \frac{p_2}{\rho} \qquad \qquad 16.10$$

$$\text{v}_1 = \sqrt{\frac{2(p_2 - p_1)}{\rho}} \qquad \text{[SI]} \quad 16.11(a)$$

$$\text{v}_1 = \sqrt{\frac{2g_c(p_2 - p_1)}{\rho}} \qquad \text{[U.S.]} \quad 16.11(b)$$

In reality, both friction and heating occur, and the fluid may be compressible. These errors are taken care of by a correction factor known as the *impact factor*, C_i, which is applied to the derived velocity. C_i is usually very close to 1.00 (e.g., 0.99 or 0.995).

$$\text{v}_{\text{actual}} = C_i \text{v}_{\text{indicated}}$$

Since accurate measurements of fluid velocity are dependent on one-dimensional fluid flow, it is essential that any obstructions or pipe bends be more than 10 pipe diameters upstream from the pitot tube.

Example 16.3

The static pressure of air (0.075 lbm/ft^3; 1.20 kg/m^3) flowing in a pipe is measured by a precision gauge to be 10.00 psig (68.95 kPa). A pitot tube-manometer indicates 20.6 in (0.523 m) of mercury. The density of mercury is 0.491 lbm/in^3 ($13\,600$ kg/m^3). Losses are insignificant. What is the velocity of the air in the pipe?

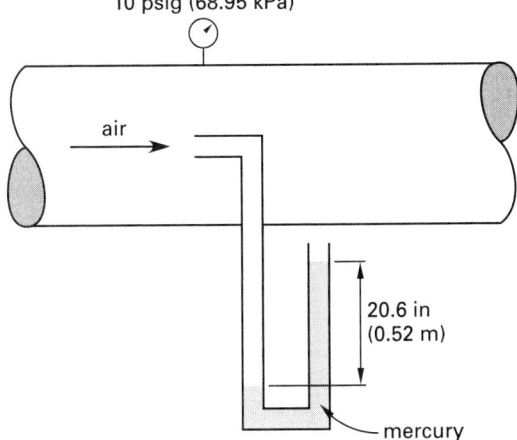

10 psig (68.95 kPa)

air

20.6 in
(0.52 m)

mercury

SI Solution

The impact pressure is

$$p_i = \rho g h$$

$$= \frac{\left(13\,600 \ \frac{\text{kg}}{\text{m}^3}\right)\left(9.81 \ \frac{\text{m}}{\text{s}^2}\right)(0.523 \ \text{m})}{1000 \ \frac{\text{Pa}}{\text{kPa}}}$$

$$= 69.78 \ \text{kPa}$$

From Eq. 16.11, the velocity is

$$\text{v} = \sqrt{\frac{2(p_i - p_s)}{\rho}}$$

$$= \sqrt{\frac{(2)(69.78 \ \text{kPa} - 68.95 \ \text{kPa})\left(1000 \ \frac{\text{Pa}}{\text{kPa}}\right)}{1.20 \ \frac{\text{kg}}{\text{m}^3}}}$$

$$= 37.2 \ \text{m/s}$$

Customary U.S. Solution

The impact pressure is

$$p_i = \frac{\rho g h}{g_c} = \frac{\left(0.491 \ \frac{\text{lbm}}{\text{in}^3}\right)\left(32.2 \ \frac{\text{ft}}{\text{sec}^2}\right)(20.6 \ \text{in})}{32.2 \ \frac{\text{lbm-ft}}{\text{lbf-sec}^2}}$$

$$= 10.11 \ \text{lbf/in}^2 \quad (10.11 \ \text{psig})$$

Since impact pressure is the sum of the static and kinetic (velocity) pressures, the kinetic pressure is

$$p_\text{v} = p_i - p_s$$

$$= 10.11 \ \text{psig} - 10.00 \ \text{psig}$$

$$= 0.11 \ \text{psi}$$

From Eq. 16.11, the velocity is

$$\text{v} = \sqrt{\frac{2g_c(p_i - p_s)}{\rho}}$$

$$= \sqrt{\frac{(2)\left(32.2 \ \frac{\text{lbm-ft}}{\text{lbf-sec}^2}\right)\left(0.11 \ \frac{\text{lbf}}{\text{in}^2}\right)\left(12 \ \frac{\text{in}}{\text{ft}}\right)^2}{0.075 \ \frac{\text{lbm}}{\text{ft}^3}}}$$

$$= 116.6 \ \text{ft/sec}$$

8. HYDRAULIC RADIUS

The *hydraulic radius* is defined as the area in flow divided by the *wetted perimeter*.[10] (The hydraulic radius is not the same as the radius of a pipe.) The area in flow is the cross-sectional area of the fluid flowing. When a fluid is flowing under pressure in a pipe (i.e., *pressure flow* in a *pressure conduit*), the area in flow will be the internal area of the pipe. However, the fluid may not completely fill the pipe and may flow simply because of a sloped surface (i.e., *gravity flow* or *open channel flow*).

The wetted perimeter is the length of the line representing the interface between the fluid and the pipe or channel. It does not include the *free surface* length (i.e., the interface between fluid and atmosphere).

$$r_h = \frac{\text{area in flow}}{\text{wetted perimeter}} = \frac{A}{s} \qquad \textit{16.12}$$

Consider a circular pipe flowing completely full. The area in flow is πr^2. The wetted perimeter is the entire circumference, $2\pi r$. The hydraulic radius is

$$r_{h,\text{pipe}} = \frac{\pi r^2}{2\pi r} = \frac{r}{2} = \frac{D}{4} \qquad \textit{16.13}$$

The hydraulic radius of a pipe flowing half full is also $r/2$, since the flow area and wetted perimeter are both halved. However, it is time-consuming to calculate the hydraulic radius for pipe flow at any intermediate depth, due to the difficulty in evaluating the flow area and wetted perimeter. Appendix 16.A greatly simplifies such calculations.

---

[10]The hydraulic radius can also be calculated as one-fourth of the hydraulic diameter of the pipe or channel, as will be subsequently shown. That is, $r_h = \frac{1}{4}D_h$.

Example 16.4

A pipe (internal diameter = 6) carries water with a depth of 2 flowing under the influence of gravity. (a) Calculate the hydraulic radius analytically. (b) Verify the result by using App. 16.A.

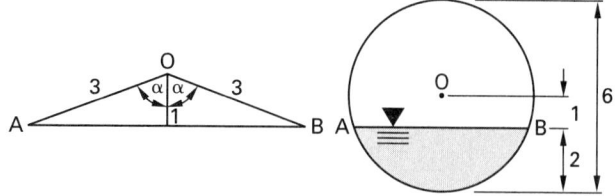

Solution

(a) Use App. 7.A. The equations for a circular segment must be used. The radius is $6/2 = 3$.

Points A, O, and B are used to find the central angle of the circular segment.

$$\phi = 2\alpha = 2\arccos\tfrac{1}{3} = (2)(70.53°)$$
$$= 141.06°$$

ϕ must be expressed in radians.

$$\phi = 2\pi\left(\frac{141.06°}{360°}\right) = 2.46 \text{ rad}$$

The area of the circular segment (i.e., the area in flow) is

$$A = \tfrac{1}{2}r^2(\phi - \sin\phi) \quad [\phi \text{ in radians}]$$
$$= \left(\tfrac{1}{2}\right)(3)^2\left(2.46 \text{ rad} - \sin(2.46 \text{ rad})\right)$$
$$= 8.235$$

The arc length (i.e., the wetted perimeter) is

$$s = r\phi = (3)(2.46 \text{ rad}) = 7.38$$

The hydraulic radius is

$$r_h = \frac{A}{s} = \frac{8.235}{7.38} = 1.12$$

(b) The ratio d/D is needed to use App. 16.A.

$$\frac{d}{D} = \frac{2}{6} = 0.333$$

From App. 16.A,

$$\frac{r_h}{D} \approx 0.186$$

$$r_h = (0.186)(6) = 1.12$$

9. HYDRAULIC DIAMETER

Many fluid, thermodynamic, and heat transfer processes are dependent on the physical length of an object. This controlling variable is generally known as the *characteristic dimension*. The characteristic

dimension in evaluating fluid flow is the *hydraulic diameter* (also known as the *equivalent hydraulic diameter*).[11] The hydraulic diameter for a full-flowing pipe is simply its inside diameter. The hydraulic diameters of other cross sections in flow are given in Table 16.1. If the hydraulic radius is known, it can be used to calculate the hydraulic diameter.

$$D_h = 4r_h \qquad 16.14$$

Table 16.1 Hydraulic Diameters for Common Conduit Shapes

conduit cross section	D_h
flowing full	
circle	D
annulus (outer diameter D_o, inner diameter D_i)	$D_o - D_i$
square (side L)	L
rectangle (sides L_1 and L_2)	$\dfrac{2L_1 L_2}{L_1 + L_2}$
flowing partially full	
half-filled circle (diameter D)	D
rectangle (h deep, L wide)	$\dfrac{4hL}{L + 2h}$
wide, shallow stream (h deep)	$4h$
triangle, vertex down (h deep, L broad, s side)	$\dfrac{hL}{s}$
trapezoid (h deep, a wide at top, b wide at bottom, s side)	$\dfrac{2h(a + b)}{b + 2s}$

Example 16.5

Determine the hydraulic diameter and hydraulic radius for the open trapezoidal channel shown.

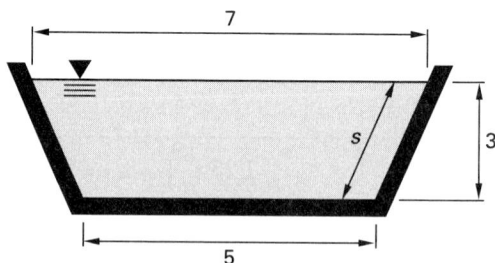

---

[11]The engineering community is very inconsistent, but the three terms *hydraulic depth*, *hydraulic diameter*, and *equivalent diameter* do not have the same meanings. Hydraulic depth (flow area divided by exposed surface width) is a characteristic length used in Froude number and other open channel flow calculations. Hydraulic diameter (four times the area in flow divided by the wetted surface) is a characteristic length used in Reynolds number and friction loss calculations. Equivalent diameter $(1.3(ab)^{0.625}/(a + b)^{0.25})$ is the diameter of a round duct or pipe that will have the same friction loss per unit length as a noncircular duct. Unfortunately, these terms are often used interchangeably.

Solution

The batter of the inclined walls is $(7-5)/2$ walls $= 1$.

$$s = \sqrt{(3)^2 + (1)^2} = 3.16$$

Using Table 16.1,

$$D_h = \frac{2h(a+b)}{b+2s} = \frac{(2)(3)(7+5)}{5+(2)(3.16)} = 6.36$$

From Eq. 16.14,

$$r_h = \frac{D_h}{4} = \frac{6.36}{4} = 1.59$$

10. REYNOLDS NUMBER

The *Reynolds number*, Re, is a dimensionless number interpreted as the ratio of inertial forces to viscous forces in the fluid.[12]

$$\text{Re} = \frac{\text{inertial forces}}{\text{viscous forces}} \qquad 16.15$$

The inertial forces are proportional to the flow diameter, velocity, and fluid density. (Increasing these variables will increase the momentum of the fluid in flow.) The viscous force is represented by the fluid's absolute viscosity, μ. Thus, the Reynolds number is calculated as

$$\text{Re} = \frac{D_h \text{v} \rho}{\mu} \qquad \text{[SI]} \quad 16.16(a)$$

$$\text{Re} = \frac{D_h \text{v} \rho}{g_c \mu} \qquad \text{[U.S.]} \quad 16.16(b)$$

Since μ/ρ is defined as the *kinematic viscosity*, ν, Eq. 16.16 can be simplified.

$$\text{Re} = \frac{D_h \text{v}}{\nu} \qquad 16.17$$

Occasionally, the *mass flow rate per unit area*, $G = \rho \text{v}$, will be known. This variable expresses the quantity of fluid flowing in kg/m^2·s or lbm/ft^2-sec.

$$G = \frac{\dot{m}}{A} \qquad 16.18$$

$$\text{Re} = \frac{D_h G}{\mu} \qquad \text{[SI]} \quad 16.19(a)$$

$$\text{Re} = \frac{D_h G}{g_c \mu} \qquad \text{[U.S.]} \quad 16.19(b)$$

---

[12]Engineering authors are not in agreement about the symbol for the Reynolds number. In addition to Re (used in this book), engineers commonly use **Re**, R, $\Re$, N_{Re}, and N_R.

11. LAMINAR FLOW

Laminar flow gets its name from the word *laminae* (layers). If all of the fluid particles move in paths parallel to the overall flow direction (i.e., in layers), the flow is said to be *laminar*. (The terms *viscous flow* and *streamline flow* are also used.) This occurs in pipeline flow when the Reynolds number is less than (approximately) 2100. Laminar flow is typical when the flow channel is small, the velocity is low, and the fluid is viscous. Viscous forces are dominant in laminar flow.

In laminar flow, a stream of dye inserted in the flow will continue from the source in a continuous, unbroken line with very little mixing of the dye and surrounding liquid. The fluid particle paths coincide with imaginary *streamlines*. (Streamlines and velocity vectors are always tangent to each other.) A "bundle" of these streamlines (i.e., a *streamtube*) constitutes a complete fluid flow.

12. TURBULENT FLOW

A fluid is said to be in *turbulent flow* if the Reynolds number is greater than (approximately) 4000. Turbulent flow is characterized by a three-dimensional movement of the fluid particles superimposed on the overall direction of motion. A stream of dye injected into a turbulent flow will quickly disperse and uniformly mix with the surrounding flow. Inertial forces dominate in turbulent flow. At very high Reynolds numbers, the flow is said to be *fully turbulent*.

13. CRITICAL FLOW

The flow is said to be in a *critical zone* or *transition region* when the Reynolds number is between 2100 and 4000. These numbers are known as the lower and upper *critical Reynolds numbers* for fluid flow, respectively. (Critical Reynolds numbers for other processes are different.) It is difficult to design for the transition region, since fluid behavior is not consistent and few processes operate in the critical zone. In the event a critical zone design is required, the conservative assumption of turbulent flow will result in the greatest value of friction loss.

14. FLUID VELOCITY DISTRIBUTION IN PIPES

With laminar flow, the viscosity makes some fluid particles adhere to the pipe wall. The closer a particle is to the pipe wall, the greater the tendency will be for the fluid to adhere to the pipe wall. The following statements characterize laminar flow.

- The velocity distribution is parabolic.
- The velocity is zero at the pipe wall.

• The velocity is maximum at the center and equal to twice the average velocity.

$$v_{ave} = \frac{\dot{V}}{A} = \frac{v_{max}}{2} \quad \text{[laminar]} \qquad 16.20$$

With turbulent flow, there is generally no distinction made between the velocities of particles near the pipe wall and particles at the pipe centerline.[13] All of the fluid particles are assumed to have the same velocity. This velocity is known as the *average* or *bulk velocity*. It can be calculated from the volume flowing.

$$v_{ave} = \frac{\dot{V}}{A} \quad \text{[turbulent]} \qquad 16.21$$

Laminar and turbulent velocity distributions are shown in Fig. 16.3. In actuality, no flow is completely turbulent, and there is a difference between the *centerline velocity* and the *average velocity*. The error decreases as the Reynolds number increases. The ratio v_{ave}/v_{max} starts at approximately 0.75 for Re = 4000 and increases to approximately 0.86 at Re = 10^6. Most problems ignore the difference between v_{ave} and v_{max}, but care should be taken when a centerline measurement (as from a pitot tube) is used to evaluate the average velocity.

Figure 16.3 *Laminar and Turbulent Velocity Distributions*

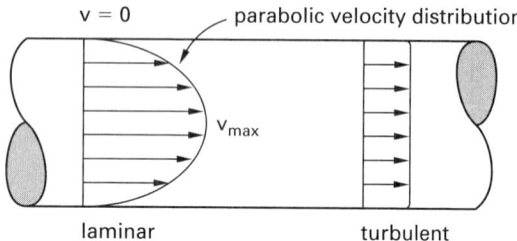

The ratio of the average velocity to maximum velocity is known as the *pipe coefficient* or *pipe factor*. Considering all the other coefficients used in pipe flow, these names are somewhat vague and ambiguous. Therefore, they are not in widespread use.

For turbulent flow (Re $\approx 10^5$) in a smooth, circular pipe of radius r_o, the velocity at a radial distance r from the centerline is given by the $1/7$-*power law*.

$$v_r = v_{max}\left(\frac{r_o - r}{r_o}\right)^{1/7} \quad \text{[turbulent flow]} \qquad 16.22$$

The fluid's *velocity profile*, given by Eq. 16.22, is valid in smooth pipes up to a Reynolds number of approximately 100,000. Above that, up to a Reynolds number of approximately 400,000, an exponent of $1/8$ fits experimental data better. For rough pipes, the exponent is larger (e.g., $1/5$).

---

[13]This disregards the *boundary layer*, a thin layer near the pipe wall, where the velocity goes from zero to v_{ave}.

Equation 16.22 can be integrated to determine the average velocity.

$$v_{ave} = \left(\frac{49}{60}\right)v_{max} = 0.817 v_{max} \quad \text{[turbulent flow]} \qquad 16.23$$

When the flow is laminar, the velocity profile within a pipe will be parabolic and of the form of Eq. 16.24. (Equation 16.22 is for turbulent flow and does not describe a parabolic velocity profile.) The velocity at a radial distance r from the centerline in a pipe of radius r_o is

$$v_r = v_{max}\left(\frac{r_o^2 - r^2}{r_o^2}\right) \quad \text{[laminar flow]} \qquad 16.24$$

When the velocity profile is parabolic, the flow rate and pressure drop can easily be determined. The average velocity is half of the maximum velocity given in the velocity profile equation.

$$v_{ave} = \tfrac{1}{2}v_{max} \quad \text{[laminar flow]} \qquad 16.25$$

The average velocity is used to determine the flow quantity and friction loss. The friction loss is determined by traditional means.

$$\dot{V} = A v_{ave} \qquad 16.26$$

The kinetic energy of laminar flow can be found by integrating the velocity profile equation, resulting in Eq. 16.27.

$$E_v = v_{ave}^2 \quad \text{[laminar flow]} \qquad \text{[SI]} \quad 16.27(a)$$

$$E_v = \frac{v_{ave}^2}{g_c} \quad \text{[laminar flow]} \qquad \text{[U.S.]} \quad 16.27(b)$$

15. ENERGY GRADE LINE

The *energy grade line* (EGL) is a graph of the total energy (total specific energy) along a length of pipe.[14] In a frictionless pipe without pumps or turbines, the total specific energy is constant, and the EGL will be horizontal. (This is a restatement of the Bernoulli equation.)

$$\text{elevation of EGL} = h_p + h_v + h_z \qquad 16.28$$

The *hydraulic grade line* (HGL) is the graph of the sum of the pressure and gravitational heads, plotted as a position along the pipeline. Since the pressure head can increase at the expense of the velocity head, the HGL can increase in elevation if the flow area is increased.

$$\text{elevation of HGL} = h_p + h_z \qquad 16.29$$

---

[14]The term *energy line* (EL) is also used.

The difference between the EGL and the HGL is the velocity head, h_v, of the fluid.

$$h_v = \text{elevation of EGL} - \text{elevation of HGL} \qquad 16.30$$

The following rules apply to these grade lines in a frictionless environment, in a pipe flowing full (i.e., under pressure), without pumps or turbines. (See Fig. 16.4.)

- The EGL is always horizontal.

- The HGL is always equal to or below the EGL.

- For still ($v = 0$) fluid at a free surface, EGL = HGL (i.e., the EGL coincides with the fluid surface in a reservoir).

- If flow velocity is constant (i.e., flow in a constant-area pipe), the HGL will be horizontal and parallel to the EGL, regardless of pipe orientation or elevation.

- When the flow area decreases, the HGL decreases.

- When the flow area increases, the HGL increases.

- In a free jet (i.e., a stream of water from a hose), the HGL coincides with the jet elevation, following a parabolic path.

Figure 16.4 *Energy and Hydraulic Grade Lines Without Friction*

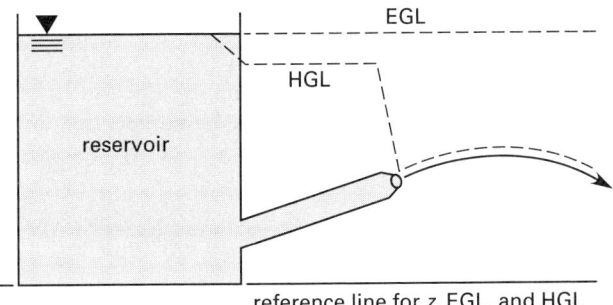

reference line for z, EGL, and HGL

16. SPECIFIC ENERGY

Specific energy is a term that is used primarily with open channel flow. It is the total energy with respect to the channel bottom, consisting of pressure and velocity energy contributions only.

$$E_{\text{specific}} = E_p + E_v \qquad 16.31$$

Since the channel bottom is chosen as the reference elevation ($z = 0$) for gravitational energy, there is no contribution by gravitational energy to specific energy.

$$E_{\text{specific}} = \frac{p}{\rho} + \frac{v^2}{2} \qquad \text{[SI]} \qquad 16.32(a)$$

$$E_{\text{specific}} = \frac{p}{\rho} + \frac{v^2}{2g_c} \qquad \text{[U.S.]} \qquad 16.32(b)$$

However, since p is the hydrostatic pressure at the channel bottom due to a fluid depth, the p/ρ term can be interpreted as the depth of the fluid, d.

$$E_{\text{specific}} = d + \frac{v^2}{2g} \qquad \text{[open channel]} \qquad 16.33$$

Specific energy is constant when the flow depth and width are constant (i.e., *uniform flow*). A change in channel width will cause a change in flow depth, and since width is not part of the equation for specific energy, there will be a corresponding change in specific energy. There are other ways that specific energy can decrease, also.[15]

17. PIPE MATERIALS AND SIZES

Many materials are used for pipes. The material used depends on the application. Water supply distribution, wastewater collection, and air conditioning refrigerant lines place different demands on pipe material performance. Pipe materials are chosen on the basis of tensile strength to withstand internal pressures, compressive strength to withstand external loads from backfill and traffic, smoothness, corrosion resistance, chemical inertness, cost, and other factors.

The following are characteristics of the major types of legacy and new installation commercial pipe materials that are in use.

- *asbestos cement:* immune to electrolysis and corrosion, light in weight but weak structurally; environmentally limited

- *concrete:* durable, watertight, low maintenance, smooth interior

- *copper and brass:* used primarily for water, condensate, and refrigerant lines; in some cases, easily bent by hand, good thermal conductivity

- *ductile cast iron:* long lived, strong, impervious, heavy, scour resistant, but costly

- *plastic* (PVC, CPVC, HDPE, and ABS):[16] chemically inert, resistant to corrosion, very smooth, light-weight, low cost

- *steel:* high strength, ductile, resistant to shock, very smooth interior, but susceptible to corrosion

- *vitrified clay:* resistant to corrosion, acids (e.g., hydrogen sulfide from septic sewage), scour, and erosion

The required wall thickness of a pipe is proportional to the pressure the pipe must carry. However, not all pipes operate at high pressures. Therefore, pipes and tubing

---

[15]Specific energy changes dramatically in a *hydraulic jump* or *hydraulic drop*.
[16]PVC: polyvinyl chloride; CPVC: chlorinated polyvinyl chloride; HDPE: high-density polyethylene; ABS: acrylonitrile-butadiene-styrene.

may be available in different wall thicknesses (*schedules*, *series*, or *types*). Steel pipe, for example, is available in schedules 40, 80, and others.[17]

For initial estimates, the approximate schedule of steel pipe can be calculated from Eq. 16.34. p is the operating pressure in psig; S is the allowable stress in the pipe material; and, E is the *joint efficiency*, also known as the *joint quality factor* (typically 1.00 for seamless pipe, 0.85 for electric resistance-welded pipe, 0.80 for electric fusion-welded pipe, and 0.60 for furnace butt-welded pipe). For seamless carbon steel (A53) pipe used below 650°F (340°C), the allowable stress is approximately between 12,000 psi and 15,000 psi. So, with butt-welded joints, a value of 6500 psi is often used for the product SE.

$$\text{schedule} \approx \frac{1000p}{SE} \qquad 16.34$$

Steel pipe is available in black (i.e., plain *black pipe*) and galvanized (inside, outside, or both) varieties. Steel pipe is manufactured in plain-carbon and stainless varieties. AISI 316 stainless steel pipe is particularly corrosion resistant.

The actual dimensions of some pipes (concrete, clay, some cast iron, etc.) coincide with their *nominal dimensions*. For example, a 12 in concrete pipe has an inside diameter of 12 in, and no further refinement is needed. However, some pipes and tubing (e.g., steel pipe, copper and brass tubing, and some cast iron) are called out by a nominal diameter that has nothing to do with the internal diameter of the pipe. For example, a 16 in schedule-40 steel pipe has an actual inside diameter of 15 in. In some cases, the nominal size does not coincide with the external diameter, either.

PVC (polyvinyl chloride) pipe is used extensively as water and sewer pipe due to its combination of strength, ductility, and corrosion resistance. Manufactured lengths range approximately 10–13 ft (3–3.9 m) for sewer pipe and 20 ft (6 m) for water pipe, with integral gasketed joints or solvent-weld bells. Infiltration is very low (less than 50 gal/in-mile-day), even in the wettest environments. The low Manning's roughness constant (0.009 typical) allows PVC sewer pipe to be used with flatter grades or smaller diameters. PVC pipe is resistant to corrosive soils and sewerage gases and is generally resistant to abrasion from pipe-cleaning tools.

Truss pipe is a double-walled PVC or ABS pipe with radial or zigzag (diagonal) reinforcing ribs between the thin walls and with lightweight concrete filling all voids. It is used primarily for underground sewer service because

of its smooth interior surface ($n = 0.009$), infiltration-resistant, impervious exterior, and resistance to bending.

Prestressed concrete pipe (PSC) or *reinforced concrete pipe* (RCP) consists of a concrete core that is compressed by a circumferential wrap of high-tensile strength wire and covered with an exterior mortar coating. *Prestressed concrete cylinder pipe* (PCCP), or *lined concrete pipe* (LCP), is constructed with a concrete core, a thin steel cylinder, prestressing wires, and an exterior mortar coating. The concrete core is the primary structural, load-bearing component. The prestressing wires induce a uniform compressive stress in the core that offsets tensile stresses in the pipe. If present, the steel cylinder acts as a water barrier between concrete layers. The mortar coating protects the prestressing wires from physical damage and external corrosion. PSC and PCCP are ideal for large-diameter pressurized service and are used primarily for water supply systems. PSC is commonly available in diameters from 16 in to 60 in (400 mm to 1500 mm), while PCCP diameters as large as 144 in (3600 mm) are available.

It is essential that tables of pipe sizes, such as App. 16.B, be used when working problems involving steel and copper pipes, since there is no other way to obtain the inside diameters of such pipes.[18]

18. MANUFACTURED PIPE STANDARDS

There are many different standards governing pipe diameters and wall thicknesses. A pipe's nominal outside diameter is rarely sufficient to determine the internal dimensions of the pipe. A manufacturing specification and class or category are usually needed to completely specify pipe dimensions.

Cast-iron pressure pipe was formerly produced to ANSI/AWWA C106/A21.6 standards but is now obsolete. Modern cast-iron soil (i.e., sanitary) pipe is produced according to ASTM A74. Ductile iron (CI/DI) pipe is produced to ANSI/AWWA C150/A21.50 and C151/A21.51 standards. Gasketed PVC sewer pipe up to 15 in inside diameter is produced to ASTM D3034 standards. Gasketed sewer PVC pipe from 18 in to 48 in is produced to ASTM F679 standards. PVC pressure pipe for water distribution is manufactured to ANSI/AWWA C900 standards. Reinforced concrete pipe (RCP) for culvert, storm drain, and sewer applications is manufactured to ASTM/AASHTO C76/M170 standards.

19. PIPE CLASS

The term *pipe class* has several valid meanings that can be distinguished by designation and context. For plastic and metallic (i.e., steel, cast and ductile iron, cast bronze, and wrought copper) pipe and fittings, pressure class

---

[17]Other schedules of steel pipe, such as 30, 60, 120, and so on, also exist, but in limited sizes, as App. 16.B indicates. Schedule-40 pipe roughly corresponds to the standard weight (S) designation used in the past. Schedule-80 roughly corresponds to the extra-strong (X) designation. There is no uniform replacement designation for double-extra-strong (XX) pipe.

[18]It is a characteristic of standard steel pipes that the schedule number does not affect the outside diameter of the pipe. An 8 in schedule-40 pipe has the same exterior dimensions as an 8 in schedule-80 pipe. However, the interior flow area will be less for the schedule-80 pipe.

designations such as 25, 150, and 300 refer to the pressure ratings and dimensional design systems as defined in the appropriate ASME, ANSI, AWWA, and other standards. Generally, the class corresponds roughly to a maximum operating pressure category in psig.

ASME *Code* pipe classes 1, 2, and 3 refer to the maximum stress categories allowed in pipes, piping systems, components, and supports, as defined in the ASME *Boiler and Pressure Vessel Code* (BPVC), Sec. III, Div. 1, "Rules for Construction of Nuclear Facility Components."

For precast concrete pipe manufactured according to ASTM C76, the pipe classes 1, 2, 3, 4, and 5 correspond to the minimum vertical loading (D-load) capacity as determined in a *three-edge bearing test* (ASTM C497), according to Table 16.2.[19] Each pipe has two D-load ratings: the pressure that induces a crack 0.01 in (0.25 mm) wide at least 1 ft (100 mm) long ($D_{0.01}$), and the pressure that results in structural collapse (D_{ult}).[20]

$$\text{D-load} = \frac{F_{lbf}}{D_{ft}L_{ft}} \quad \text{[ASTM C76 pipe]} \qquad 16.35$$

Table 16.2 *ASTM C76 Concrete Pipe D-Load Equivalent Pipe Class**

ASTM C76 pipe class	$D_{0.01}$-load rating (lbf/ft^2)	D_{ult}-load rating (lbf/ft^2)
class 1 (I)	800	1200
class 2 (II)	1000	1500
class 3 (III)	1350	2000
class 4 (IV)	2000	3000
class 5 (V)	3000	3750

(Multiply lbf/ft^2 by 0.04788 to obtain kPa.)
*As defined in ASTM C76 and AASHTO M170.

20. PAINTS, COATINGS, AND LININGS

Various materials have been used to protect steel and ductile iron pipes against rust and other forms of corrosion. *Red primer* is a shop-applied, rust-inhibiting primer applied to prevent short-term rust prior to shipment and the application of subsequent coatings. *Asphaltic coating* ("tar" coating) is applied to the exterior of underground pipes. *Bituminous coating* refers to a similar coating made from tar pitch. Both asphaltic and bituminous coatings should be completely removed or sealed with a synthetic resin prior to the pipe being finish-coated, since their oils may bleed through otherwise.

Though bituminous materials (i.e., asphaltic materials) continue to be cost effective, epoxy-based products are

now extensively used. Epoxy products are delivered as a two-part formulation (a polyamide resin and liquid chemical hardener) that is mixed together prior to application. *Coal tar epoxy*, also referred to as *epoxy coal tar*, a generic name, sees frequent use in pipes exposed to high humidity, seawater, other salt solutions, and crude oil. Though suitable for coating steel penstocks of hydroelectric installations, coal tar epoxy is generally not suitable for potable water delivery systems. Though it is self-priming, appropriate surface preparation is required for adequate adhesion. Coal tar epoxy has a density range of 1.9–2.3 lbm/gal (230–280 g/L).

21. CORRUGATED METAL PIPE

Corrugated metal pipe (CMP, also known as *corrugated steel pipe*) is frequently used for culverts. Pipe is made from corrugated sheets of galvanized steel that are rolled and riveted together along a longitudinal seam. Aluminized steel may also be used in certain ranges of soil pH. Standard round pipe diameters range from 8 in to 96 in (200 mm to 2450 mm). Metric dimensions of standard diameters are usually rounded to the nearest 25 mm or 50 mm (e.g., a 42 in culvert would be specified as a 1050 mm culvert, not 1067 mm).

Larger and noncircular culverts can be created out of curved steel plate. Standard section lengths are 10–20 ft (3–6 m). Though most corrugations are transverse (i.e., annular), helical corrugations are also used. Metal gauges of 8, 10, 12, 14, and 16 are commonly used, depending on the depth of burial.

The most common corrugated steel pipe has transverse corrugations that are $1/2$ in (13 mm) deep and $2^2/_3$ in (68 mm) from crest to crest. These are referred to as "$2^1/_2$ inch" or "68×13" corrugations. For larger culverts, corrugations with a 2 in (25 mm) depth and 3 in, 5 in, or 6 in (76 mm, 125 mm, or 152 mm) pitches are used. Plate-based products using 6 in $\times$ 2 in (152 mm $\times$ 51 mm) corrugations are known as *structural plate corrugated steel pipe* (SPCSP) and *multiplate* after the trade-named product "Multi-Plate™."

The flow area for circular culverts is based on the nominal culvert diameter, regardless of the gage of the plate metal used to construct the pipe. Flow area is calculated to (at most) three significant digits.

A Hazen-Williams coefficient, C, of 60 is typically used with all sizes of corrugated pipe. Values of C and Manning's coefficient, n, for corrugated pipe are generally not affected by age. *Design Charts for Open Channel Flow* (U.S. Department of Transportation, 1979) recommends a Manning constant of $n = 0.024$ for all cases. The U.S. Department of the Interior recommends the following values.

For standard ($2^2/_3$ in $\times$ $1/2$ in or 68 mm $\times$ 13 mm) corrugated pipe with the diameter ranges given: 12 in (457 mm), 0.027; 24 in (610 mm), 0.025; 36–48 in

---

[19]The D-load rating for ASTM C14 concrete pipe is D-load = F/L in lbf/ft.

[20]This D-load rating for concrete pipe is analogous to the *pipe stiffness* (PS) rating (also known as *ring stiffness*) for PVC and other plastic pipe, although the units are different. The PS rating is stated in lbf/in^2 (pounds per inch of pipe length per inch of pipe diameter).

(914–1219 mm), 0.024; 60–84 in (1524–2134 mm), 0.023; 96 in (2438 mm), 0.022.

For (6 in × 2 in or 152 mm × 51 mm) multiplate construction with the diameter ranges given: 5–6 ft (1.5–1.8 m), 0.034; 7–8 ft (2.1–2.4 m), 0.033; 9–11 ft (2.7–3.3 m), 0.032; 12–13 ft (3.6–3.9 m), 0.031; 14–15 ft (4.2–4.5 m), 0.030; 16–18 ft (4.8–5.4 m), 0.029; 19–20 ft (5.8–6.0 m), 0.028; 21–22 ft (6.3–6.6 m), 0.027.

If the inside of the corrugated pipe has been asphalted completely smooth 360° circumferentially, Manning's n ranges from 0.009 to 0.011. For culverts with 40% asphalted inverts, $n = 0.019$. For other percentages of paved invert, the resulting value is proportional to the percentage and the values normally corresponding to that diameter pipe. For field-bolted corrugated metal pipe arches, $n = 0.025$.

It is also possible to calculate the Darcy friction loss if the corrugation depth, 0.5 in (13 mm) for standard corrugations and 2.0 in (51 mm) for multiplate, is taken as the specific roughness.

22. TYPES OF VALVES

Valves used for *shutoff service* (e.g., gate, plug, ball, and some butterfly valves) are used fully open or fully closed. *Gate valves* offer minimum resistance to flow. They are used in clean fluid and slurry services when valve operation is infrequent. Many turns of the handwheels are required to raise or lower their gates. *Plug valves* provide for tight shutoff. A 90° turn of their handles is sufficient to rotate the plugs fully open or closed. *Eccentric plug valves*, in which the plug rotates out of the fluid path when open, are among the most common wastewater valves. *Plug cock valves* have a hollow passageway in their plugs through which fluid can flow. Both eccentric plug valves and plug cock valves are referred to as "plug valves." *Ball valves* offer an unobstructed flow path and tight shutoff. They are often used with slurries and viscous fluids, as well as with cryogenic fluids. A 90° turn of their handles rotates the balls fully open or closed. *Butterfly valves* (when specially designed with appropriate seats) can be used for shutoff operation. They are particularly applicable to large flows of low-pressure (vacuum up to 200 psig (1.4 MPa)) gases or liquids, although high-performance butterfly valves can operate as high as 600 psig (4.1 MPa). Their straight-through, open-disk design results in minimal solids build-up and low pressure drops.

Other valve types (e.g., globe, needle, Y-, angle, and some butterfly valves) are more suitable for *throttling service*. *Globe valves* provide positive shutoff and precise metering on clean fluids. However, since the seat is parallel to the direction of flow and the fluid makes two right-angle turns, there is substantial resistance and pressure drop through them, as well as relatively fast erosion of the seat. Globe valves are intended for frequent operation. *Needle valves* are similar to globe valves, except that the plug is a tapered, needlelike cone.

Needle valves provide accurate metering of small flows of clean fluids. Needle valves are applicable to cryogenic fluids. *Y-valves* are similar to globe valves in operation, but their seats are inclined to the direction of flow, offering more of a straight-through passage and unobstructed flow than the globe valve. *Angle valves* are essentially globe valves where the fluid makes a 90° turn. They can be used for throttling and shut-off of clean or viscous fluids and slurries. *Butterfly valves* are often used for throttling services with the same limitations and benefits as those listed for shutoff use.

Other valves are of the *check (nonreverse-flow, anti-reversal)* variety. These react automatically to changes in pressure to prevent reversals of flow. Special check valves can also prevent excess flow. Figure 16.5 illustrates *swing, lift,* and *angle lift check valves,* and Table 16.3 gives typical characteristics of common valve types.

23. SAFETY AND PRESSURE RELIEF VALVES

A *pressure relief device* typically incorporates a disk or needle that is held in place by spring force, and that is lifted off its seat (i.e., opens) when the static pressure at the valve inlet exceeds the opening pressure. A *relief valve* is a pressure relief device used primarily for liquid service. It has a gradual lift that is approximately proportional (though not necessarily linear) to the increase in pressure over opening pressure. A *safety valve* is used for compressible steam, air, and gas services. Its performance is characterized by rapid opening ("pop action"). In a *low-lift safety valve*, the discharge area depends on the disc position and is limited by the lift amount. In a *full-lift (high-lift) safety valve*, the discharge area is not determined by the position of the disc. Since discs in low-lift valves have lift distances of only approximately 1/24th of the bore diameter, the discharge capacities tend to be much lower than those of high-lift valves.

A *safety relief valve* is a dual-function valve that can be used for either a liquid or a compressible fluid. Its operation is characterized by rapid opening (for gases) and by slower opening in proportion to the increase in inlet pressure over the opening pressure (for liquids), depending on the application.

24. AIR RELEASE AND VACUUM VALVES

Small amounts of air in fluid flows tend to accumulate at high points in the line. These air pockets reduce flow capacities, increase pumping power, and contribute to pressure surges and water hammer. *Air release valves* should be installed at high points in the line, particularly with line diameters greater than 12 in (3000 mm).[21] They are float operated and open against the line's internal pressure to release accumulated air and gases. Air release valves are essential when lines are being filled. Valves for

[21]Smaller diameter lines may develop flow velocities sufficiently high (e.g., greater than 5 ft/sec (1.5 m/s)) to flush gas accumulations.

Figure 16.5 Types of Valves

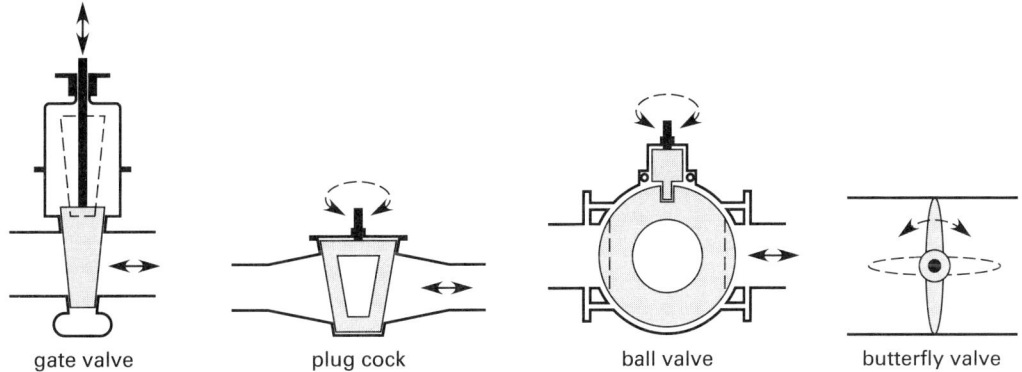

gate valve   plug cock   ball valve   butterfly valve

(a) valves for shut-off service

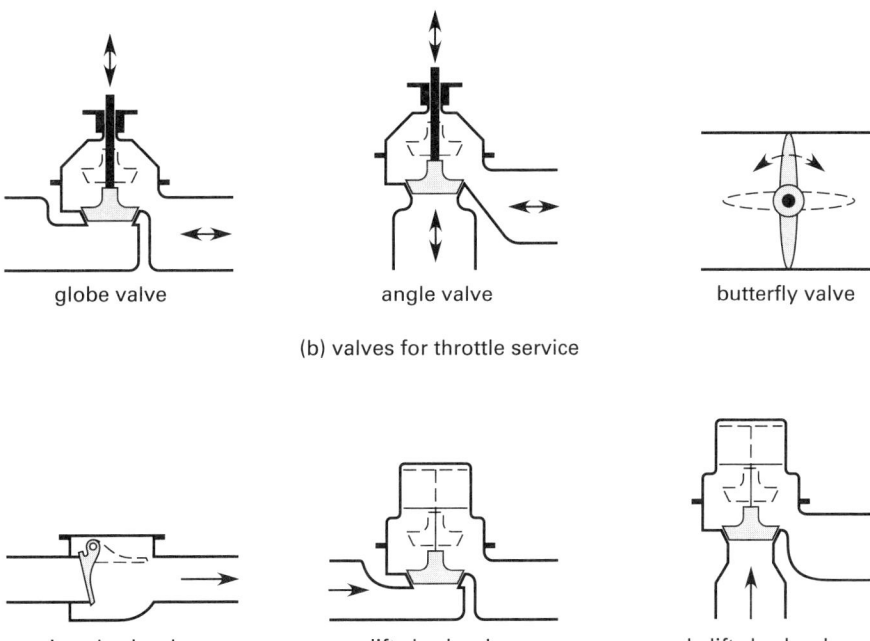

globe valve   angle valve   butterfly valve

(b) valves for throttle service

swing check valve   lift check valve   angle lift check valve

(c) valves for antireversal service

Table 16.3 Typical Characteristics of Common Valve Types

valve type	fluid condition	switching frequency	pressure drop (fully open)	typical control response	typical maximum pressure (atm)	typical maximum temperature (°C)
ball	clean	low	low	very poor	160	300
butterfly	clean	low	low	poor	200	400
diaphragm* (not shown)	clean to slurried	very high	low to medium	very good	16	150
gate	clean	low	low	very poor	50	400
globe	clean	high	medium to high	very good	80	300
plug	clean	low	low	very poor	160	300

(Multiply atm by 101.33 to obtain kPa.)

*Diaphragm valves use a flexible diaphragm to block the flow path. The diaphragm may be manually or pneumatically actuated. Such valves are suitable for both throttling and shut-off service.

Flow of Fluids

sewage lines prevent buildups of sewage gases (e.g., hydrogen sulfide) and are similar in functionality, though they contain features to prevent clogging and fouling.

Vacuum valves are similar in design, except that they permit air to enter a line when the line is drained, preventing pipeline collapse. *Combination air valves* (CAV), also known as *double orifice air valves*, combine the functions of both air release valves and vacuum valves. Vacuum valves should not be used in drinking water supply distribution systems due to the potential for contamination.

If unrestricted, air moving through a valve orifice will reach a maximum velocity of approximately 300 ft/sec (91 m/s) at 6.7 lbf/in^2 (46 kPa) differential pressure. It is good practice to limit the flow to 10 ft/sec (3 m/s) (occurring at about 1 lbf/in^2 (7 kPa)), and this can be accomplished with slow-closing units.

Siphon air valves (*make-or-break valves*) are a type of air vacuum valve that includes a paddle which hangs down in the flow. As long as the flow exists, the valve remains closed. When the flow stops or reverses, the valve opens. In a typical application, a pump is used to initiate siphon flow, and once flow is started, the pump is removed.

25. STEAM TRAPS

Steam traps are installed in steam lines. Since the heat content of condensate is significantly less than that of the steam, condensate is not as useful in downstream applications. Traps collect and automatically release condensed water while permitting vapor to pass through. They may also release accumulated noncondensing gases (e.g., air).

Inverted bucket steam traps are simple, mechanically robust, and reliable. They are best for applications with water hammer. *Thermostatic steam traps* operate with either a bimetallic or a balanced pressure design to detect the difference in temperature between live steam and condensate or air. They are suitable for cold start-up conditions, but less desirable with variable loadings. The *float steam trap* is the best choice for variable loadings, but it is less robust due to its mechanical complexity and is less resistant to water hammer. A *thermodynamic disc steam trap* uses flash steam as the stream passes through the trap and is both simple and robust. It is suitable for intermittent, not continuous, operation only.

17 Fluid Dynamics

Nomenclature

a	speed of sound	ft/sec	m/s
A	area	ft^2	m^2
B	magnetic field strength	T	T
c_P	Poisson's effect coefficient	–	–
C	coefficient	–	–
C	Hazen-Williams coefficient	–	–
d	diameter	in	cm
D	diameter	ft	m
E	bulk modulus	lbf/ft^2	Pa
E	specific energy	ft-lbf/lbm	J/kg
f	Darcy friction factor	–	–
f	fraction split	–	–
F	force	lbf	N
Fr	Froude number	–	–
g	gravitational acceleration, 32.2 (9.81)	ft/sec^2	m/s^2
g_c	gravitational constant, 32.2	lbm-ft/lbf-sec^2	n.a.
G	mass flow rate per unit area	lbm/ft^2-sec	kg/m^2·s
h	height or head	ft	m
I	impulse	lbf-sec	N·s
k	magmeter instrument constant	–	–
k	ratio of specific heats	–	–
K	minor loss coefficient	–	–
L	length	ft	m
m	mass	lbm	kg
$\dot{m}$	mass flow rate	lbm/sec	kg/s
MW	molecular weight	lbm/lbmol	kg/kmol

n	Manning roughness constant	–	–
n	flow rate exponent	–	–
n	rotational speed	rev/min	rev/min
p	pressure	lbf/ft^2	Pa
P	momentum	lbm-ft/sec	kg·m/s
P	power	ft-lbf/sec	W
P	wetted perimeter	ft	m
Q	flow rate	gal/min	n.a.
r	radius	ft	m
R	resultant force	lbf	N
R^*	universal gas constant, 1545.35 (8314.47)	ft-lbf/lbmol-°R	J/kmol·K
Re	Reynolds number	–	–
SG	specific gravity	–	–
t	thickness	ft	m
t	time	sec	s
T	absolute temperature	°R	K
u	x-component of velocity	ft/sec	m/s
v	y-component of velocity	ft/sec	m/s
v	velocity	ft/sec	m/s
V	induced voltage	V	V
V	volume	ft^3	m^3
$\dot{V}$	volumetric flow rate	ft^3/sec	m^3/s
W	work	ft-lbf	J
We	Weber number	–	–
WHP	water horsepower	hp	n.a.
x	x-coordinate of position	ft	m
y	y-coordinate of position	ft	m
Y	expansion factor	–	–
z	elevation	ft	m

Symbols

β	diameter ratio	–	–
γ	specific weight	lbf/ft^3	N/m^3
Γ	circulation	ft^2/sec	m^2/s
δ	flow rate correction	ft^3/sec	m^3/s
ϵ	specific roughness	ft	m
η	efficiency	–	–
η	non-Newtonian viscosity	lbf-sec/ft^2	Pa·s
θ	angle	deg	deg
μ	absolute viscosity	lbf-sec/ft^2	Pa·s
ν	kinematic viscosity	ft^2/sec	m^2/s
ν	Poisson's ratio	–	–
ρ	density	lbm/ft^3	kg/m^3
σ	surface tension	lbf/ft	N/m
τ	shear stress	lbf/ft^2	Pa
υ	specific volume	ft^3/lbm	m^3/kg
ϕ	angle	deg	deg
Φ	stream potential	–	–
ψ	sphericity	–	–
Ψ	stream function	–	–
ω	angular velocity	rad/sec	rad/s

Subscripts

0	critical (yield)
a	assumed
A	added (by pump)
b	blade or buoyant
c	contraction
d	discharge
D	drag
e	equivalent
E	extracted (by turbine)
f	flow or friction
h	hydraulic
i	inside
I	instrument
L	lift
m	manometer fluid, minor, or model
o	orifice or outside
p	pressure or prototype
r	ratio
s	static
t	tank, theoretical, or total
v	valve
v	velocity
va	velocity of approach
z	potential

1. HYDRAULICS AND HYDRODYNAMICS

This chapter covers fluid moving through pipes, measurements with venturis and orifices, and other motion-related topics such as model theory, lift and drag, and pumps. In a strict interpretation, any fluid-related phenomenon that is not hydro*statics* should be hydro*dynamics*. However, tradition has separated the study of moving fluids into the fields of hydraulics and hydrodynamics.

In a general sense, *hydraulics* is the study of the practical laws of fluid flow and resistance in pipes and open channels. Hydraulic formulas are often developed from experimentation, empirical factors, and curve fitting, without an attempt to justify why the fluid behaves the way it does.

On the other hand, *hydrodynamics* is the study of fluid behavior based on theoretical considerations. Hydrodynamicists start with Newton's laws of motion and try to develop models of fluid behavior. Models developed in this manner are complicated greatly by the inclusion of viscous friction and compressibility. Therefore, hydrodynamic models assume a perfect fluid with constant density and zero viscosity. The conclusions reached by hydrodynamicists can differ greatly from those reached by hydraulicians.[1]

2. CONSERVATION OF MASS

Fluid mass is always conserved in fluid systems, regardless of the pipeline complexity, orientation of the flow, and fluid. This single concept is often sufficient to solve simple fluid problems.

$$\dot{m}_1 = \dot{m}_2 \qquad 17.1$$

---

[1]Perhaps the most disparate conclusion is *D'Alembert's paradox*. In 1744, D'Alembert derived theoretical results "proving" that there is no resistance to bodies moving through an ideal (non-viscous) fluid.

When applied to fluid flow, the conservation of mass law is known as the *continuity equation.*

$$\rho_1 A_1 v_1 = \rho_2 A_2 v_2 \qquad 17.2$$

If the fluid is incompressible, then $\rho_1 = \rho_2$.

$$A_1 v_1 = A_2 v_2 \qquad 17.3$$

$$\dot{V}_1 = \dot{V}_2 \qquad 17.4$$

Various units and symbols are used for *volumetric flow rate.* (Though this book uses $\dot{V}$, the symbol Q is commonly used when the flow rate is expressed in gallons.) MGD (millions of gallons per day) and MGPCD (millions of gallons per capita day) are units commonly used in municipal water works problems. MMSCFD (millions of standard cubic feet per day) may be used to express gas flows.

Calculation of flow rates is often complicated by the interdependence between flow rate and friction loss. Each affects the other, so many pipe flow problems must be solved iteratively. Usually, a reasonable friction factor is assumed and used to calculate an initial flow rate. The flow rate establishes the flow velocity, from which a revised friction factor can be determined.

3. TYPICAL VELOCITIES IN PIPES

Fluid friction in pipes is kept at acceptable levels by maintaining reasonable fluid velocities. Table 17.1 lists typical maximum fluid velocities. Higher velocities may be observed in practice, but only with a corresponding increase in friction and pumping power.

4. STREAM POTENTIAL AND STREAM FUNCTION

An application of hydrodynamic theory is the derivation of the stream function from stream potential. The *stream potential function (velocity potential function)*, Φ, is the algebraic sum of the component velocity potential functions.[2]

$$\Phi = \Phi_x(x, y) + \Phi_y(x, y) \qquad 17.5$$

The velocity component of the resultant in the x-direction is

$$u = \frac{\partial \Phi}{\partial x} \qquad 17.6$$

The velocity component of the resultant in the y-direction is

$$v = \frac{\partial \Phi}{\partial y} \qquad 17.7$$

---

[2]The two-dimensional derivation of the stream function can be extended to three dimensions, if necessary. The stream function can also be expressed in the cylindrical coordinate system.

Table 17.1 Typical Full-Pipe Bulk Fluid Velocities

fluid and application	velocity	
	ft/sec	m/s
water: city service	2–10	0.6–2.1
3 in diameter	4	1.2
6 in diameter	5	1.5
12 in diameter	9	2.7
water: boiler feed	8–15	2.4–4.5
water: pump suction	4	1.2
water: pump discharge	4–8.5	1.2–2.5
water, sewage: partially filled sewer	2.5 (min)	0.75 (min)
brine, water: chillers and coolers	6–8 typ (3–10)	1.8–2.4 typ (0.9–3)
air: compressor suction	75–200	23–60
air: compressor discharge	100–250	30–75
air: HVAC forced air	15–25	5–8
natural gas: overland pipeline	< 150 (60 typ)	< 45 (18 typ)
steam, saturated: heating	65–100	20–30
steam, saturated: miscellaneous	100–200	30–60
50–100 psia	< 150	< 45
150–400 psia	< 130	< 39
400–600 psia	< 100	< 30
steam, superheated: turbine feed	160–250	50–75
hydraulic fluid: fluid power	7–15	2.1–4.6
liquid sodium ($T > 525°C$): heat transfer	10 typ (0.3–40)	3 typ (0.1–12)
ammonia: compressor suction	85 (max)	25 (max)
ammonia: compressor discharge	100 (max)	30 (max)
oil, crude: overland pipeline	4–12	1.2–3.6
oil, lubrication: pump suction	< 2	< 0.6
oil, lubrication: pump discharge	3–7	0.9–2.1

(Multiply ft/sec by 0.3048 to obtain m/s.)

The total derivative of the stream potential function is

$$d\Phi = \frac{\partial \Phi}{\partial x}\,dx + \frac{\partial \Phi}{\partial y}\,dy = u\,dx + v\,dy \qquad 17.8$$

An *equipotential line* is a line along which the function Φ is constant (i.e., $d\Phi = 0$). The slope of the equipotential line is derived from Eq. 17.8.

$$\left.\frac{dy}{dx}\right|_{\text{equipotential}} = -\frac{u}{v} \qquad 17.9$$

For flow through a porous, permeable medium, pressure will be constant along equipotential lines (i.e., along lines of constant Φ). (See Fig. 17.1.) However, for an ideal, nonviscous fluid flowing in a frictionless environment, Φ has no physical significance.

Figure 17.1 *Equipotential Lines and Streamlines*

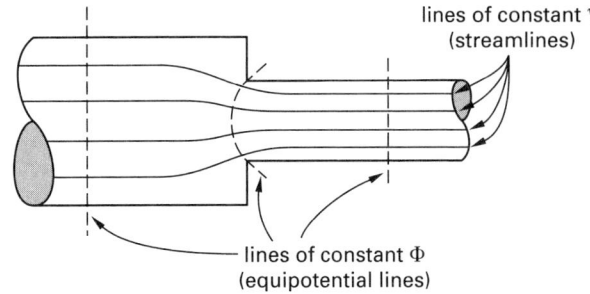

The *stream function* (*Lagrange stream function*), $\Psi(x, y)$, defines the direction of flow at a point.

$$u = \frac{\partial \Psi}{\partial y} \qquad 17.10$$

$$v = -\frac{\partial \Psi}{\partial x} \qquad 17.11$$

The stream function can also be written in total derivative form.

$$d\Psi = \frac{\partial \Psi}{\partial x}\, dx + \frac{\partial \Psi}{\partial y}\, dy$$

$$= -v\, dx + u\, dy \qquad 17.12$$

The stream function, $\Psi(x, y)$, satisfies Eq. 17.12. For a given streamline, $d\Psi = 0$, and each streamline is a line representing a constant value of Ψ. A streamline is perpendicular to an equipotential line.

$$\left.\frac{dy}{dx}\right|_{\text{streamline}} = -\frac{v}{u} \qquad 17.13$$

Example 17.1

The stream potential function for water flowing through a particular valve is $\Phi = 3xy - 2y$. What is the stream function, Ψ?

Solution

First, work with Φ to obtain u and v.

$$u = \frac{\partial \Phi}{\partial x} = \frac{\partial(3xy - 2y)}{\partial x} = 3y$$

$$v = \frac{\partial \Phi}{\partial y} = 3x - 2$$

u and v are also related to the stream function, Ψ. From Eq. 17.10,

$$u = \frac{\partial \Psi}{\partial y}$$

$$\partial\Psi = u\, \partial y$$

$$\Psi = \int 3y\, dy = \tfrac{3}{2}y^2 + \text{some function of } x + C_1$$

Similarly, from Eq. 17.11,

$$v = -\frac{\partial \Psi}{\partial x}$$

$$\partial\Psi = -v\, \partial x$$

$$\Psi = -\int (3x - 2)\, dx$$

$$= 2x - \tfrac{3}{2}x^2 + \text{some function of } y + C_2$$

The entire stream function is found by superposition of these two results.

$$\Psi = \tfrac{3}{2}y^2 + 2x - \tfrac{3}{2}x^2 + C$$

5. HEAD LOSS DUE TO FRICTION

The original Bernoulli equation was based on an assumption of frictionless flow. In actual practice, friction occurs during fluid flow. This friction "robs" the fluid of energy, so that the fluid at the end of a pipe section has less energy than it does at the beginning.[3]

$$E_1 > E_2 \qquad 17.14$$

Most formulas for calculating friction loss use the symbol h_f to represent the *head loss due to friction*.[4] This loss is added into the original Bernoulli equation to restore the equality. Of course, the units of h_f must be the same as the units for the other terms in the Bernoulli equation. (See Eq. 17.23.) If the Bernoulli equation is written in terms of energy, the units will be ft-lbf/lbm or J/kg.

$$E_1 = E_2 + E_f \qquad 17.15$$

Consider the constant-diameter, horizontal pipe in Fig. 17.2. An incompressible fluid is flowing at a steady rate. Since the elevation of the pipe does not change, the potential energy is constant. Since the pipe has a constant area, the kinetic energy (velocity) is constant. Therefore, the friction energy loss must show up as a decrease in pressure energy. Since the fluid is incompressible, this can only occur if the pressure decreases in the direction of flow.

Figure 17.2 *Pressure Drop in a Pipe*

1		2
v_1		$v_2 = v_1$
z_1		$z_2 = z_1$
ρ_1		$\rho_2 = \rho_1$
p_1		$p_2 = p_1 - \Delta p_f$

---

[3]The friction generates minute amounts of heat. The heat is lost to the surroundings.

[4]Other names and symbols for this friction loss are *friction head loss* (h_L), *lost work* (LW), *friction heating* ($\mathcal{F}$), *skin friction loss* (F_f), and *pressure drop due to friction* (Δp_f). All terms and symbols essentially mean the same thing, although the units may be different.

6. RELATIVE ROUGHNESS

It is intuitive that pipes with rough inside surfaces will experience greater friction losses than smooth pipes.[5] *Specific roughness*, ϵ, is a parameter that measures the average size of imperfections inside the pipe. Table 17.2 lists values of ϵ for common pipe materials. (Also, see App. 17.A.)

Table 17.2 *Values of Specific Roughness for Common Pipe Materials*

| | ϵ | |
material	ft	m
plastic (PVC, ABS)	0.000005	1.5×10^{-6}
copper and brass	0.000005	1.5×10^{-6}
steel	0.0002	6.0×10^{-5}
plain cast iron	0.0008	2.4×10^{-4}
concrete	0.004	1.2×10^{-3}

(Multiply ft by 0.3048 to obtain m.)

However, an imperfection the size of a sand grain will have much more effect in a small-diameter hydraulic line than in a large-diameter sewer. Therefore, the *relative roughness*, ϵ/D, is a better indicator of pipe roughness. Both ϵ and D have units of length (e.g., feet or meters), and the relative roughness is dimensionless.

7. FRICTION FACTOR

The *Darcy friction factor*, f, is one of the parameters used to calculate friction loss.[6] The friction factor is not constant but decreases as the Reynolds number (fluid velocity) increases, up to a certain point known as *fully turbulent flow* (or *rough-pipe flow*). Once the flow is fully turbulent, the friction factor remains constant and depends only on the relative roughness and not on the Reynolds number. (See Fig. 17.3.) For very smooth pipes, fully turbulent flow is achieved only at very high Reynolds numbers.

The friction factor is not dependent on the material of the pipe but is affected by the roughness. For example, for a given Reynolds number, the friction factor will be the same for any smooth pipe material (glass, plastic, smooth brass and copper, etc.).

The friction factor is determined from the relative roughness, ϵ/D, and the Reynolds number, Re, by various methods. These methods include explicit and implicit equations, the Moody diagram, and tables. The values

[5]Surprisingly, this intuitive statement is valid only for turbulent flow. The roughness does not (ideally) affect the friction loss for laminar flow.
[6]There are actually two friction factors: the Darcy friction factor and the *Fanning friction factor*, f_{Fanning}, also known as the *skin friction coefficient* and *wall shear stress factor*. Both factors are in widespread use, sharing the same symbol, f. Civil and (most) mechanical engineers use the Darcy friction factor. The Fanning friction factor is encountered more often by chemical engineers. One can be derived from the other: $f_{\text{Darcy}} = 4f_{\text{Fanning}}$.

Figure 17.3 *Friction Factor as a Function of Reynolds Number*

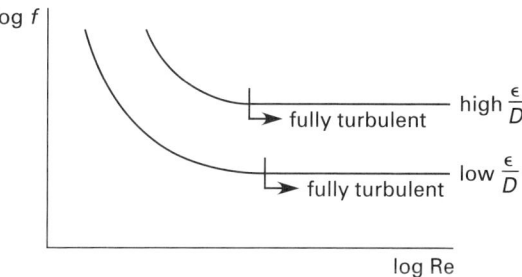

obtained are based on experimentation, primarily the work of J. Nikuradse in the early 1930s.

When a moving fluid initially encounters a parallel surface (as when a moving gas encounters a flat plate or when a fluid first enters the mouth of a pipe), the flow will generally not be turbulent, even for very rough surfaces. The flow will be laminar for a certain *critical distance* before becoming turbulent.

Friction Factors for Laminar Flow

The easiest method of obtaining the friction factor for laminar flow (Re < 2100) is to calculate it. Equation 17.16 illustrates that roughness is not a factor in determining the frictional loss in ideal laminar flow.

$$f = \frac{64}{\text{Re}} \quad \text{[circular pipe]} \qquad \textit{17.16}$$

Table 17.3 gives friction factors for laminar flow in various cross sections.

Friction Factors for Turbulent Flow: by Formula

One of the earliest attempts to predict the friction factor for turbulent flow in smooth pipes resulted in the *Blasius equation* (claimed "valid" for 3000 < Re < 100,000).

$$f = \frac{0.316}{\text{Re}^{0.25}} \qquad \textit{17.17}$$

The *Nikuradse equation* can also be used to determine the friction factor for smooth pipes (i.e., when $\epsilon/D = 0$). Unfortunately, this equation is implicit in f and must be solved iteratively.

$$\frac{1}{\sqrt{f}} = 2.0 \log_{10}(\text{Re}\sqrt{f}) - 0.80 \qquad \textit{17.18}$$

The *Karman-Nikuradse equation* predicts the fully turbulent friction factor (i.e., when Re is very large).

$$\frac{1}{\sqrt{f}} = 1.74 - 2 \log_{10} \frac{2\epsilon}{D} \qquad \textit{17.19}$$

The most widely known method of calculating the friction factor for any pipe roughness and Reynolds number

Flow of Fluids

Table 17.3 Friction Factors for Laminar Flow in Various Cross Sections*

tube geometry	D_h (full)	c/d or θ	friction factor, f
circle	D	–	64.00/Re
rectangle	$\dfrac{2cd}{c+d}$	1	56.92/Re
		2	62.20/Re
		3	68.36/Re
		4	72.92/Re
		6	78.80/Re
		8	82.32/Re
		∞	96.00/Re
ellipse	$\dfrac{cd}{\sqrt{\frac{1}{2}(c^2+d^2)}}$	1	64.00/Re
		2	67.28/Re
		4	72.96/Re
		8	76.60/Re
		16	78.16/Re
isosceles triangle	$\dfrac{d\sin\theta}{2(1+\sin\frac{\theta}{2})}$	10°	50.80/Re
		30°	52.28/Re
		60°	53.32/Re
		90°	52.60/Re
		120°	50.96/Re

*Re $= \mathrm{v_{bulk}} D_h/\nu$, and $D_h = 4A/P$.

is another implicit formula, the *Colebrook equation*. Most other equations are variations of this equation. (Notice that the relative roughness, ϵ/D, is used to calculate f.)

$$\frac{1}{\sqrt{f}} = -2\log_{10}\left(\frac{\frac{\epsilon}{D}}{3.7} + \frac{2.51}{\mathrm{Re}\sqrt{f}}\right) \qquad 17.20$$

A suitable approximation would appear to be the *Swamee-Jain equation*, which claims to have less than 1% error (as measured against the Colebrook equation) for relative roughnesses between 0.000001 and 0.01, and for Reynolds numbers between 5000 and 100,000,000.[7] Even with a 1% error, this equation produces more accurate results than can be read from the Moody friction factor chart.

$$f = \frac{0.25}{\left(\log_{10}\left(\dfrac{\frac{\epsilon}{D}}{3.7} + \dfrac{5.74}{\mathrm{Re}^{0.9}}\right)\right)^2} \qquad 17.21$$

---

[7]*ASCE Hydraulic Division Journal*, Vol. 102, May 1976, p. 657. This is not the only explicit approximation to the Colebrook equation in existence.

Friction Factors for Turbulent Flow: by Moody Chart

The *Moody friction factor chart*, shown in Fig. 17.4, presents the friction factor graphically as a function of Reynolds number and relative roughness. There are different lines for selected discrete values of relative roughness. Due to the complexity of this graph, it is easy to mislocate the Reynolds number or use the wrong curve. Nevertheless, the Moody chart remains the most common method of obtaining the friction factor.

Friction Factors for Turbulent Flow: by Table

Appendix 17.B (based on the Colebrook equation), or a similar table, will usually be the most convenient method of obtaining friction factors for turbulent flow.

Example 17.2

Determine the friction factor for a Reynolds number of Re = 400,000 and a relative roughness of $\epsilon/D = 0.004$ using (a) the Moody diagram, (b) Appendix 17.B, and (c) the Swamee-Jain approximation. (d) Check the table value of f with the Colebrook equation.

Solution

(a) From Fig. 17.4, the friction factor is approximately 0.028.

(b) Appendix 17.B lists the friction factor as 0.0287.

(c) From Eq. 17.21,

$$f = \frac{0.25}{\left(\log_{10}\left(\dfrac{\frac{\epsilon}{D}}{3.7} + \dfrac{5.74}{\mathrm{Re}^{0.9}}\right)\right)^2}$$

$$= \frac{0.25}{\left(\log_{10}\left(\dfrac{0.004}{3.7} + \dfrac{5.74}{(400,000)^{0.9}}\right)\right)^2}$$

$$= 0.0288$$

(d) From Eq. 17.20,

$$\frac{1}{\sqrt{f}} = -2\log_{10}\left(\frac{\frac{\epsilon}{D}}{3.7} + \frac{2.51}{\mathrm{Re}\sqrt{f}}\right)$$

$$\frac{1}{\sqrt{0.0287}} = -2\log_{10}\left(\frac{0.004}{3.7} + \frac{2.51}{400,000\sqrt{0.0287}}\right)$$

$$5.903 = 5.903$$

8. ENERGY LOSS DUE TO FRICTION: LAMINAR FLOW

Two methods are available for calculating the frictional energy loss for fluids experiencing laminar flow. The most common is the *Darcy equation* (which is also

Figure 17.4 *Moody Friction Factor Chart*

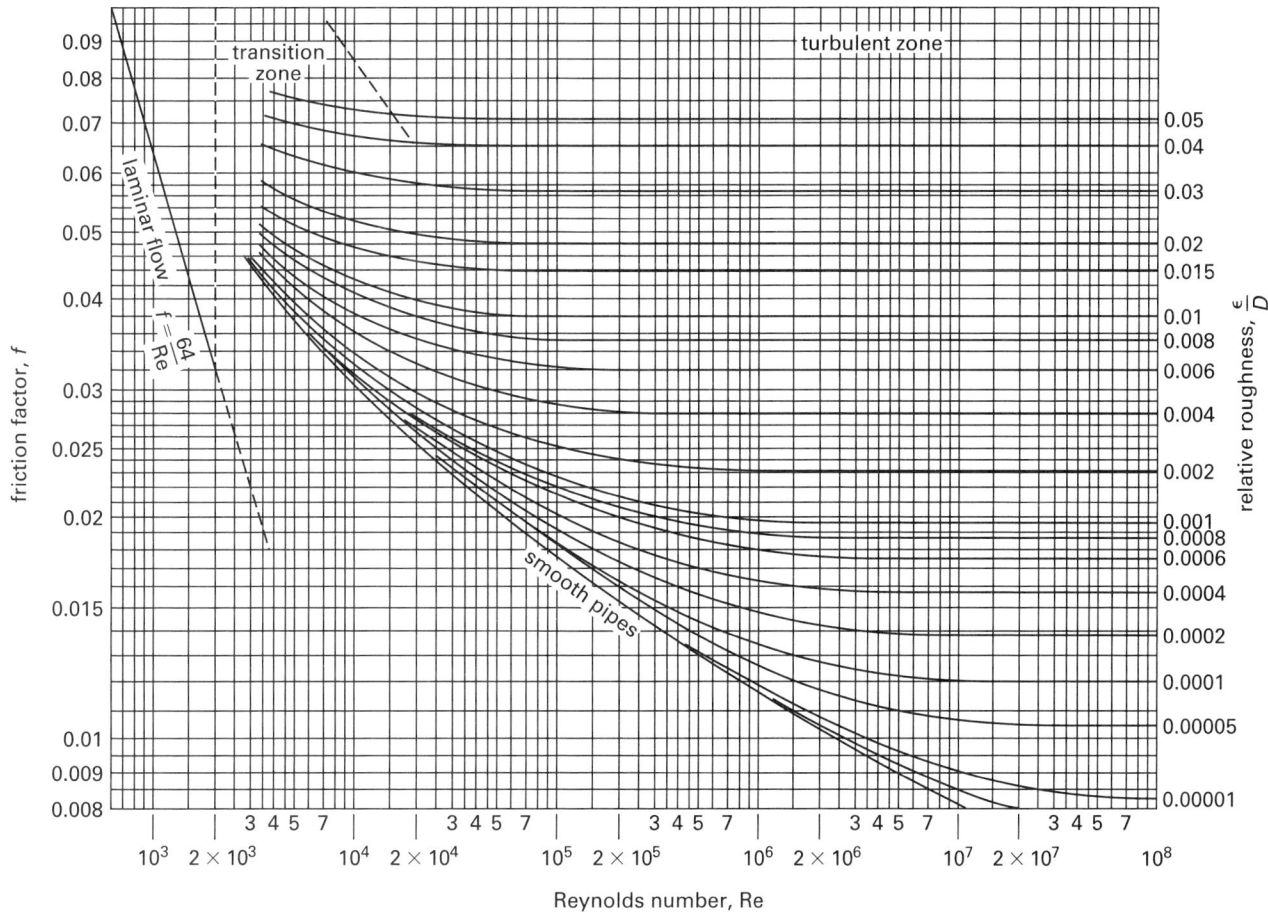

known as the *Weisbach equation* or the *Darcy-Weisbach equation*), which can be used for both laminar and turbulent flow.[8] One of the advantages of using the Darcy equation is that the assumption of laminar flow does not need to be confirmed if f is known.

$$h_f = \frac{fLv^2}{2Dg} \qquad 17.22$$

$$E_f = h_f g = \frac{fLv^2}{2D} \qquad \text{[SI]} \quad 17.23(a)$$

$$E_f = h_f \times \frac{g}{g_c} = \frac{fLv^2}{2Dg_c} \qquad \text{[U.S.]} \quad 17.23(b)$$

If the flow is truly laminar and the fluid is flowing in a circular pipe, then the *Hagen-Poiseuille equation* can be used.

$$E_f = \frac{32\mu v L}{D^2 \rho} = \frac{128\mu \dot{V} L}{\pi D^4 \rho} \qquad 17.24$$

---

[8]The difference is that the friction factor can be derived by hydrodynamics: $f = 64/\text{Re}$. For turbulent flow, f is empirical.

If necessary, h_f can be converted to an actual pressure drop in lbf/ft^2 or Pa by multiplying by the fluid density.

$$\Delta p = h_f \times \rho g \qquad \text{[SI]} \quad 17.25(a)$$

$$\Delta p = h_f \times \rho \left(\frac{g}{g_c}\right) = h_f \gamma \qquad \text{[U.S.]} \quad 17.25(b)$$

Values of the Darcy friction factor, f, are often quoted for new, clean pipe. The friction head losses and pumping power requirements calculated from these values are minimum values. Depending on the nature of the service, scale and impurity buildup within pipes may decrease the pipe diameters over time. Since the frictional loss is proportional to the fifth power of the diameter, such diameter decreases can produce dramatic increases in the friction loss.

$$\frac{h_{f,\text{scaled}}}{h_{f,\text{new}}} = \left(\frac{D_{\text{new}}}{D_{\text{scaled}}}\right)^5 \qquad 17.26$$

Equation 17.26 accounts only for the decrease in diameter. Any increase in roughness (i.e., friction factor) will produce a proportional increase in friction loss.

Because the "new, clean" condition is transitory in most applications, an uprating factor of 10–30% is often applied to either the friction factor, f, or the head loss, h_f. Of course, even larger increases should be considered when extreme fouling is expected.

Another approach eliminates the need to estimate the scaled pipe diameter. This simplistic approach multiplies the initial friction loss by a factor based on the age of the pipe. For example, for schedule-40 pipe between 4 in and 10 in (10 cm and 25 cm) in diameter, the multipliers of 1.4, 2.2, and 5.0 have been proposed for pipe ages of 5, 10, and 20 years, respectively. For larger pipes, the corresponding multipliers are 1.3, 1.6, and 2.0. Obviously, use of these values should be based on a clear understanding of the method's limitations.

9. ENERGY LOSS DUE TO FRICTION: TURBULENT FLOW

The *Darcy equation* is used almost exclusively to calculate the head loss due to friction for turbulent flow.

$$h_f = \frac{fLv^2}{2Dg} \qquad 17.27$$

The head loss can be converted to pressure drop.

$$\Delta p = h_f \times \rho g \qquad \text{[SI]} \quad 17.28(a)$$

$$\Delta p = h_f \times \rho \left(\frac{g}{g_c}\right) = h_f \gamma \qquad \text{[U.S.]} \quad 17.28(b)$$

In problems where the pipe size is unknown, it will be impossible to obtain an accurate initial value of the friction factor, f (since f depends on velocity). In such problems, an iterative solution will be necessary.

Civil engineers commonly use the *Hazen-Williams equation* to calculate head loss. This method requires knowing the Hazen-Williams *roughness coefficient*, C, values of which are widely tabulated.[9] (See App. 17.A.) The advantage of using this equation is that C does not depend on the Reynolds number. The Hazen-Williams equation is empirical and is not dimensionally homogeneous. It is taken as a matter of faith that the units of h_f are feet.

$$h_{f,\text{ft}} = \frac{3.022 v_{\text{ft/sec}}^{1.85} L_{\text{ft}}}{C^{1.85} D_{\text{ft}}^{1.17}} = \frac{10.44 L_{\text{ft}} Q_{\text{gpm}}^{1.85}}{C^{1.85} d_{\text{in}}^{4.87}} \qquad \text{[U.S.]} \quad 17.29$$

---

[9]An approximate value of $C = 140$ is often chosen for initial calculations for new water pipe. $C = 100$ is more appropriate for water pipe that has been in service for some time. For sludge, C values are 20–40% lower than the equivalent water pipe values.

The Hazen-Williams equation should be used only for turbulent flow. It gives good results for liquids that have kinematic viscosities around 1.2×10^{-5} ft^2/sec $(1.1 \times 10^{-6}$ m^2/s), which corresponds to the viscosity of 60°F (16°C) water. At extremely high and low temperatures, the Hazen-Williams equation can be 20% or more in error for water.

Example 17.3

50°F water is pumped through 1000 ft of 4 in, schedule-40 welded steel pipe at the rate of 300 gpm. What friction loss (in ft-lbf/lbm) is predicted by the Darcy equation?

Solution

The fluid viscosity, pipe dimensions, and other parameters can be found from the appendices.

From App. 14.A, $\nu = 1.41 \times 10^{-5}$ ft^2/sec.

From App. 17.A, $\epsilon = 0.0002$ ft.

From App. 16.B,

$$D = 0.3355 \text{ ft}$$

$$A = 0.0884 \text{ ft}^2$$

The flow quantity is converted from gallons per minute to cubic feet per second.

$$\dot{V} = \frac{300 \dfrac{\text{gal}}{\text{min}}}{\left(7.4805 \dfrac{\text{gal}}{\text{ft}^3}\right)\left(60 \dfrac{\text{sec}}{\text{min}}\right)} = 0.6684 \text{ ft}^3/\text{sec}$$

The velocity is

$$v = \frac{\dot{V}}{A} = \frac{0.6684 \dfrac{\text{ft}^3}{\text{sec}}}{0.0884 \text{ ft}^2} = 7.56 \text{ ft/sec}$$

The Reynolds number is

$$\text{Re} = \frac{Dv}{\nu} = \frac{(0.3355 \text{ ft})\left(7.56 \dfrac{\text{ft}}{\text{sec}}\right)}{1.41 \times 10^{-5} \dfrac{\text{ft}^2}{\text{sec}}}$$

$$= 1.8 \times 10^5$$

The relative roughness is

$$\frac{\epsilon}{D} = \frac{0.0002 \text{ ft}}{0.3355 \text{ ft}} = 0.0006$$

From the friction factor table, App. 17.B (or the Moody friction factor chart), $f = 0.0195$. Equation 17.23(b) is used to calculate the friction loss.

$$E_f = h_f \times \frac{g}{g_c} = \frac{fL\mathrm{v}^2}{2Dg_c}$$

$$= \frac{(0.0195)(1000 \text{ ft})\left(7.56 \, \frac{\text{ft}}{\text{sec}}\right)^2}{(2)(0.3355 \text{ ft})\left(32.2 \, \frac{\text{lbm-ft}}{\text{lbf-sec}^2}\right)}$$

$$= 51.6 \text{ ft-lbf/lbm}$$

Example 17.4

Calculate the head loss due to friction for the pipe in Ex. 17.3 using the Hazen-Williams formula. Assume $C = 100$.

Solution

Substituting the parameters derived in Ex. 17.3 into Eq. 17.29,

$$h_f = \frac{3.022\mathrm{v}_{\text{ft/sec}}^{1.85} L_{\text{ft}}}{C^{1.85} D_{\text{ft}}^{1.17}} = \frac{(3.022)\left(7.56 \, \frac{\text{ft}}{\text{sec}}\right)^{1.85}(1000 \text{ ft})}{(100)^{1.85}(0.3355 \text{ ft})^{1.17}}$$

$$= 91.3 \text{ ft}$$

Alternatively, the given data can be substituted directly into Eq. 17.29.

$$h_f = \frac{10.44 L_{\text{ft}} Q_{\text{gpm}}^{1.85}}{C^{1.85} d_{\text{in}}^{4.87}} = \frac{(10.44)(1000 \text{ ft})(300 \text{ gpm})^{1.85}}{(100)^{1.85}(4.026 \text{ in})^{4.87}}$$

$$= 90.4 \text{ ft}$$

10. FRICTION LOSS FOR WATER FLOW IN STEEL PIPES

Since water's specific volume is essentially constant within the normal temperature range, tables and charts can be used to determine water velocity. Friction loss and velocity for water flowing through steel pipe (as well as for other liquids and other pipe materials) in table and chart form are widely available. (Appendix 17.C is an example of such a table.) Tables and charts almost always give the friction loss per 100 ft or 10 m of pipe. The pressure drop is proportional to the length, so the value read can be scaled for other pipe lengths. Flow velocity is independent of pipe length.

These tables and charts are unable to compensate for the effects of fluid temperature and different pipe roughness. Unfortunately, the assumptions made in developing the tables and charts are seldom listed. Another disadvantage is that the values can be read to only a few significant figures. Friction loss data should be considered accurate to only ±20%. Alternatively, a 20% safety margin should be established in choosing pumps and motors.

11. FRICTION LOSS IN NONCIRCULAR DUCTS

The frictional energy loss by a fluid flowing in a rectangular, annular, or other noncircular duct can be calculated from the Darcy equation by using the *hydraulic diameter*, D_h, in place of the diameter, D.[10] The friction factor, f, is determined in any of the conventional manners.

12. FRICTION LOSS FOR OTHER LIQUIDS, STEAM, AND GASES

The Darcy equation can be used to calculate the frictional energy loss for all incompressible liquids, not just for water. Alcohol, gasoline, fuel oil, and refrigerants, for example, are all handled well, since the effect of viscosity is considered in determining the friction factor, f.[11]

In fact, the Darcy equation is commonly used with noncondensing vapors and compressed gases, such as air, nitrogen, and steam.[12] In such cases, reasonable accuracy will be achieved as long as the fluid is not moving too fast (i.e., less than Mach 0.3) and is incompressible. The fluid is assumed to be incompressible if the pressure (or density) change along the section of interest is less than 10% of the starting pressure.

If possible, it is preferred to base all calculations on the average properties of the fluid as determined at the midpoint of a pipe.[13] Specifically, the fluid velocity would normally be calculated as

$$\mathrm{v} = \frac{\dot{m}}{\rho_{\text{ave}} A} \qquad 17.30$$

However, the average density of a gas depends on the average pressure, which is unknown at the start of a problem. The solution is to write the Reynolds number and Darcy equation in terms of the constant mass flow rate per unit area, G, instead of velocity, v, which varies.

$$G = \mathrm{v}_{\text{ave}} \rho_{\text{ave}} \qquad 17.31$$

$$\mathrm{Re} = \frac{DG}{\mu} \qquad \text{[SI]} \quad 17.32(a)$$

$$\mathrm{Re} = \frac{DG}{g_c \mu} \qquad \text{[U.S.]} \quad 17.32(b)$$

---

[10]Although it is used for both, this approach is better suited for turbulent flow than for laminar flow. Also, the accuracy of this method decreases as the flow area becomes more noncircular. The friction drop in long, narrow slit passageways is poorly predicted, for example. However, there is no other convenient method of predicting friction drop. Experimentation should be used with a particular flow geometry if extreme accuracy is required.
[11]Since viscosity is not an explicit factor in the formula, it should be obvious that the Hazen-Williams equation is primarily used for water.
[12]Use of the Darcy equation is limited only by the availability of the viscosity data needed to calculate the Reynolds number.
[13]Of course, the entrance (or exit) conditions can be used if great accuracy is not needed.

$$\Delta p_f = p_1 - p_2 = \rho_{\text{ave}} h_f g = \frac{fLG^2}{2D\rho_{\text{ave}}} \quad \text{[SI]} \quad 17.33(a)$$

$$\Delta p_f = p_1 - p_2 = \gamma_{\text{ave}} h_f = \rho_{\text{ave}} h_f \times \frac{g}{g_c}$$

$$= \frac{fLG^2}{2D\rho_{\text{ave}} g_c} \quad \text{[U.S.]} \quad 17.33(b)$$

Assuming a perfect gas with a molecular weight of MW, the ideal gas law can be used to calculate ρ_{ave} from the absolute temperature, T, and $p_{\text{ave}} = (p_1 + p_2)/2$.

$$p_1^2 - p_2^2 = \frac{fLG^2 R^* T}{D(\text{MW})} \quad \text{[SI]} \quad 17.34(a)$$

$$p_1^2 - p_2^2 = \frac{fLG^2 R^* T}{Dg_c(\text{MW})} \quad \text{[U.S.]} \quad 17.34(b)$$

To summarize, use the following guidelines when working with compressible gases or vapors flowing in a pipe or duct. (a) If the pressure drop, based on the entrance pressure, is less than 10%, the fluid can be assumed to be incompressible, and the gas properties can be evaluated at any point known along the pipe. (b) If the pressure drop is between 10% and 40%, use of the midpoint properties will yield reasonably accurate friction losses. (c) If the pressure drop is greater than 40%, the pipe can be divided into shorter sections and the losses calculated for each section, or exact calculations based on compressible flow theory must be made.

Calculating a friction loss for steam flow using the Darcy equation can be frustrating if steam viscosity data are unavailable. Generally, the steam viscosities listed in compilations of heat transfer data are sufficiently accurate. Various empirical methods are also in use. For example, the *Babcock formula*, given by Eq. 17.35, for pressure drop when steam with a specific volume of v flows in a pipe of diameter d is

$$\Delta p_{\text{psi}} = 0.470 \left(\frac{d_{\text{in}} + 3.6}{d_{\text{in}}^6} \right) \dot{m}_{\text{lbm/sec}}^2 L_{\text{ft}} v_{\text{ft}^3/\text{lbm}} \quad 17.35$$

Use of empirical formulas is not limited to steam. Theoretical formulas (e.g., the *complete isothermal flow equation*) and specialized empirical formulas (e.g., the *Weymouth, Panhandle,* and *Spitzglass formulas*) have been developed, particularly by the gas pipeline industry. Each of these provides reasonable accuracy within their operating limits. However, none should be used without knowing the assumptions and operational limitations that were used in their derivations.

Example 17.5

0.0011 kg/s of 25°C nitrogen gas flows isothermally through a 175 m section of smooth tubing (inside diameter = 0.012 m). The viscosity of the nitrogen is

1.8×10^{-5} Pa·s. The pressure of the nitrogen is 200 kPa originally. At what pressure is the nitrogen delivered?

SI Solution

The flow area of the pipe is

$$A = \frac{\pi D^2}{4} = \frac{\pi(0.012 \text{ m})^2}{4} = 1.131 \times 10^{-4} \text{ m}^2$$

The mass flow rate per unit area is

$$G = \frac{\dot{m}}{A} = \frac{0.0011 \frac{\text{kg}}{\text{s}}}{1.131 \times 10^{-4} \text{ m}^2} = 9.73 \text{ kg/m}^2\text{·s}$$

The Reynolds number is

$$\text{Re} = \frac{DG}{\mu} = \frac{(0.012 \text{ m})\left(9.73 \frac{\text{kg}}{\text{m}^2\text{·s}}\right)}{1.8 \times 10^{-5} \text{ Pa·s}} = 6487$$

The flow is turbulent, and the pipe is said to be smooth. Therefore, the friction factor is interpolated (from App. 17.B) as 0.0347.

Since two atoms of nitrogen form a molecule of nitrogen gas, the molecular weight of nitrogen is twice the atomic weight, or 28.0 kg/kmol. The temperature must be in degrees absolute: $T = 25°C + 273° = 298K$. The universal gas constant is 8314.47 J/kmol·K.

From Eq. 17.34, the final pressure is

$$p_2^2 = p_1^2 - \frac{fLG^2 R^* T}{D(\text{MW})}$$

$$= (200\,000 \text{ Pa})^2 - \frac{(0.0347)(175 \text{ m})\left(9.73 \frac{\text{kg}}{\text{m}^2\text{·s}}\right)^2 \times \left(8314.47 \frac{\text{J}}{\text{kmol·K}}\right)(298K)}{(0.012 \text{ m})\left(28 \frac{\text{kg}}{\text{kmol}}\right)}$$

$$= 3.576 \times 10^{10} \text{ Pa}^2$$

$$p_2 = \sqrt{3.576 \times 10^{10} \text{ Pa}^2} = 1.89 \times 10^5 \text{ Pa} \quad (189 \text{ kPa})$$

The percentage drop in pressure should not be more than 10%.

$$\frac{200 \text{ kPa} - 189 \text{ kPa}}{200 \text{ kPa}} = 0.055 \quad (5.5\%) \quad [\text{OK}]$$

Example 17.6

Superheated steam at 140 psi and 500°F enters a 200 ft long steel pipe (Darcy friction factor of 0.02) with an internal diameter of 3.826 in. The pipe is insulated so that there is no heat loss. (a) Use the Babcock formula to determine the maximum velocity and mass flow rate such that the steam does not experience more than a 10% drop in pressure. (b) Verify the velocity by calculating the pressure drop with the Darcy equation.

Solution

(a) From superheated steam tables (see App. 29.C), the interpolated specific volume of the steam is

$$v = 5.588 \ \frac{\text{ft}^3}{\text{lbm}} - \left(\frac{140 \text{ psia} - 100 \text{ psia}}{150 \text{ psia} - 100 \text{ psia}} \right)$$

$$\times \left(5.588 \ \frac{\text{ft}^3}{\text{lbm}} - 3.680 \ \frac{\text{ft}^3}{\text{lbm}} \right)$$

$$= 4.062 \text{ ft}^3/\text{lbm}$$

The maximum pressure drop is 10% of 140 psi or 14 psi.

From Eq. 17.35,

$$\Delta p_{\text{psi}} = 0.470 \left(\frac{d_{\text{in}} + 3.6}{d_{\text{in}}^6} \right) \dot{m}^2_{\text{lbm/sec}} L_{\text{ft}} v_{\text{ft}^3/\text{lbm}}$$

$$14 \text{ psi} = (0.470) \left(\frac{3.826 \text{ in} + 3.6}{(3.826 \text{ in})^6} \right) \dot{m}^2$$

$$\times (200 \text{ ft}) \left(4.062 \ \frac{\text{ft}^3}{\text{lbm}} \right)$$

$$\dot{m} = 3.935 \text{ lbm/sec}$$

The steam velocity is

$$\text{v} = \frac{\dot{V}}{A} = \frac{\dot{m}}{\rho A} = \frac{\dot{m} v}{A}$$

$$= \frac{\left(3.935 \ \frac{\text{lbm}}{\text{sec}} \right) \left(4.062 \ \frac{\text{ft}^3}{\text{lbm}} \right)}{\left(\frac{\pi}{4} \right) \left(\frac{3.826 \text{ in}}{12 \ \frac{\text{in}}{\text{ft}}} \right)^2}$$

$$= 200.2 \text{ ft/sec}$$

(b) The steam friction head is

$$h_f = \frac{f L \text{v}^2}{2 D g} = \frac{(0.02)(200 \text{ ft}) \left(200.2 \ \frac{\text{ft}}{\text{sec}} \right)^2}{(2) \left(\frac{3.826 \text{ in}}{12 \ \frac{\text{in}}{\text{ft}}} \right) \left(32.2 \ \frac{\text{ft}}{\text{sec}^2} \right)}$$

$$= 7808 \text{ ft of steam}$$

From Eq. 17.33,

$$\Delta p = \rho h_f \times \frac{g}{g_c} = \frac{h_f}{v} \times \frac{g}{g_c}$$

$$= \frac{7808 \text{ ft}}{\left(4.062 \ \frac{\text{ft}^3}{\text{lbm}} \right) \left(12 \ \frac{\text{in}}{\text{ft}} \right)^2} \times \left(\frac{32.2 \ \frac{\text{ft}}{\text{sec}^2}}{32.2 \ \frac{\text{lbm-ft}}{\text{lbf-sec}^2}} \right)$$

$$= 13.3 \text{ lbf/in}^2 \quad (13.3 \text{ psi})$$

13. EFFECT OF VISCOSITY ON HEAD LOSS

Friction loss in a pipe is affected by the fluid viscosity. For both laminar and turbulent flow, viscosity is considered when the Reynolds number is calculated. When viscosities substantially increase without a corresponding decrease in flow rate, two things usually happen: (a) the friction loss greatly increases, and (b) the flow becomes laminar.

It is sometimes necessary to estimate head loss for a new fluid viscosity based on head loss at an old fluid viscosity. The estimation procedure used depends on the flow regimes for the new and old fluids.

For laminar flow, the friction factor is directly proportional to the viscosity. If the flow is laminar for both fluids, the ratio of new-to-old head losses will be equal to the ratio of new-to-old viscosities. Therefore, if a flow is already known to be laminar at one viscosity and the fluid viscosity increases, a simple ratio will define the new friction loss.

If both flows are fully turbulent, the friction factor will not change. If flow is fully turbulent and the viscosity decreases, the Reynolds number will increase. Theoretically, this will have no effect on the friction loss.

There are no analytical ways of estimating the change in friction loss when the flow regime changes between laminar and turbulent or between semiturbulent and fully turbulent. Various graphical methods are used, particularly by the pump industry, for calculating power requirements.

14. FRICTION LOSS WITH SLURRIES AND NON-NEWTONIAN FLUIDS

A *slurry* is a mixture of a liquid (usually water) and a solid (e.g., coal, paper pulp, foodstuffs). The liquid is generally used as the transport mechanism (i.e., the *carrier*) for the solid.

Friction loss calculations for slurries vary in sophistication depending on what information is available. In many cases, only the slurry's specific gravity is known. In that case, use is made of the fact that friction loss can be reasonably predicted by multiplying the friction loss based on the pure carrier (e.g., water) by the specific gravity of the slurry.

Another approach is possible if the density and viscosity in the operating range are known. The traditional Darcy equation (see Eq. 17.27) and Reynolds number can be used for thin slurries as long as the flow velocity is high enough to keep solids from settling. (Settling is more of a concern for laminar flow. With turbulent flow, the direction of velocity components fluctuates, assisting the solids to remain in suspension.)

The most analytical approach to slurries or other non-Newtonian fluids requires laboratory-derived rheological data. *Non-Newtonian viscosity* (η, in Pa·s) is fitted to data of the shear rate (dv/dy, in s^{-1}) according to two common models: the power-law model and the Bingham-plastic model. These two models are applicable to both laminar and turbulent flow, although each has its advantages and disadvantages.

The *power-law model* has two empirical constants, m and n, that must be determined.

$$\eta = m\left(\frac{dv}{dy}\right)^{n-1} \qquad 17.36$$

The *Bingham-plastic model* also requires finding two empirical constants: the *yield* (or *critical*) *stress*, τ_0 (in units of Pa) below which the fluid is immobile, and the *Bingham-plastic limiting viscosity*, μ_∞ (in units of Pa·s).

$$\eta = \frac{\tau_0}{\dfrac{dv}{dy}} + \mu_\infty \qquad 17.37$$

Once m and n (or τ_0 and μ_∞) have been determined, the friction factor is determined from one of various models (e.g., Buckingham-Reiner, Dodge-Metzner, Metzner-Reed, Hanks-Ricks, Darby, or Hanks-Dadia). Specialized texts and articles cover these models in greater detail. The friction loss is calculated from the traditional Darcy equation.

15. MINOR LOSSES

In addition to the frictional energy lost due to viscous effects, friction losses also result from fittings in the line, changes in direction, and changes in flow area. These losses are known as *minor losses* or *local losses*, since they are usually much smaller in magnitude than the pipe wall frictional loss.[14] Two methods are used to calculate minor losses: equivalent lengths and loss coefficients.

With the *method of equivalent lengths*, each fitting or other flow variation is assumed to produce friction equal to the pipe wall friction from an *equivalent length* of pipe. For example, a two inch globe valve may produce

the same amount of friction as 54 feet (its equivalent length) of two inch pipe. The equivalent lengths for all minor losses are added to the pipe length term, L, in the Darcy equation. The method of equivalent lengths can be used with all liquids, but it is usually limited to turbulent flow by the unavailability of laminar equivalent lengths, which are significantly larger than turbulent equivalent lengths.

$$L_t = L + \sum L_e \qquad 17.38$$

Equivalent lengths are simple to use, but the method depends on having a table of equivalent length values. The actual value for a fitting will depend on the fitting manufacturer, as well as the fitting material (e.g., brass, cast iron, or steel) and the method of attachment (e.g., weld, thread, or flange).[15] Because of these many variations, it may be necessary to use a "generic table" of equivalent lengths during the initial design stages. (See App. 17.D.)

An alternative method of calculating the minor loss for a fitting is to use the *method of loss coefficients*. Each fitting has a *loss coefficient*, K, associated with it, which, when multiplied by the kinetic energy, gives the loss. (See Table 17.4.) Therefore, a loss coefficient is the minor loss expressed in fractions (or multiples) of the velocity head.

$$h_m = K h_v \qquad 17.39$$

The loss coefficient for any minor loss can be calculated if the equivalent length is known. However, there is no advantage to using one method over the other, other than convention and for consistency in calculations.

$$K = \frac{f L_e}{D} \qquad 17.40$$

Exact friction loss coefficients for bends, fittings, and valves are unique to each manufacturer. Furthermore, except for contractions, enlargements, exits, and entrances, the coefficients decrease fairly significantly (according to the fourth power of the diameter ratio) with increases in valve size. Therefore, a single K value is seldom applicable to an entire family of valves. Nevertheless, generic tables and charts have been developed. These compilations can be used for initial estimates as long as the general nature of the data is recognized.

Loss coefficients for specific fittings and valves must be known in order to be used. They cannot be derived theoretically. However, the loss coefficients for certain changes in flow area can be calculated from the following equations.[16]

---

[14]Example and practice problems often include the instruction to "Ignore minor losses." In some industries, valves are considered to be "components," not fittings. In such cases, instructions to "Ignore minor losses in fittings" would be ambiguous, since minor losses in valves would be included in the calculations. However, this interpretation is rare in examples and practice problems.

---

[15]In the language of pipe fittings, a *threaded fitting* is known as a *screwed fitting*, even though no screws are used.
[16]No attempt is made to imply great accuracy with these equations. Correlation between actual and theoretical losses is fair.

Table 17.4 Typical Loss Coefficients[a]

device	K
angle valve	5
bend, close return	2.2
butterfly valve,[b] 2–8 in	$45f_t$
butterfly valve, 10–14 in	$35f_t$
butterfly valve, 16–24 in	$25f_t$
check valve, swing, fully open	2.3
corrugated bends	1.3–1.6 times value for smooth bend
standard 90° elbow	0.9
long radius 90° elbow	0.6
45° elbow	0.42
gate valve, fully open	0.19
gate valve, $1/4$ closed	1.15
gate valve, $1/2$ closed	5.6
gate valve, $3/4$ closed	24
globe valve	10
meter disk or wobble	3.4–10
meter, rotary (star or cog-wheel piston)	10
meter, reciprocating piston	15
meter, turbine wheel (double flow)	5–7.5
tee, standard	1.8

[a]The actual loss coefficient will usually depend on the size of the valve. Average values are given.
[b]Loss coefficients for butterfly valves are calculated from the friction factors for the pipes with complete turbulent flow.

- *sudden enlargements:* (D_1 is the smaller of the two diameters)

$$K = \left(1 - \left(\frac{D_1}{D_2}\right)^2\right)^2 \qquad 17.41$$

- *sudden contractions:* (D_1 is the smaller of the two diameters)

$$K = \frac{1}{2}\left(1 - \left(\frac{D_1}{D_2}\right)^2\right) \qquad 17.42$$

- *pipe exit:* (projecting exit, sharp-edged, or rounded)

$$K = 1.0 \qquad 17.43$$

- *pipe entrance:*

  reentrant: $K = 0.78$
  sharp-edged: $K = 0.50$
  rounded:

bend radius D	K
0.02	0.28
0.04	0.24
0.06	0.15
0.10	0.09
0.15	0.04

Flow of Fluids

- *tapered diameter changes:*

$$\beta = \frac{\text{small diameter}}{\text{large diameter}} = \frac{D_1}{D_2}$$

$$\phi = \text{wall-to-horizontal angle}$$

enlargement, $\phi \le 22°$:

$$K = 2.6 \sin\phi(1 - \beta^2)^2 \qquad 17.44$$

enlargement, $\phi > 22°$:

$$K = (1 - \beta^2)^2 \qquad 17.45$$

contraction, $\phi \le 22°$:

$$K = 0.8 \sin\phi(1 - \beta^2) \qquad 17.46$$

contraction, $\phi > 22°$:

$$K = 0.5\sqrt{\sin\phi}(1 - \beta^2) \qquad 17.47$$

Example 17.7

Determine the total equivalent length of the piping system shown. The pipeline contains one gate valve, five regular 90° elbows, one tee (flow through the run), and 228 ft of straight pipe. All fittings are 1 in screwed steel pipe. Disregard entrance and exit losses.

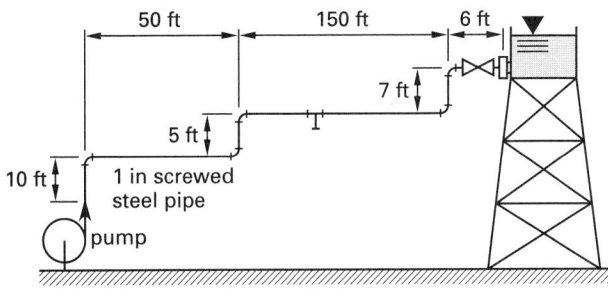

(not to scale)

Solution

From App. 17.D, the individual and total equivalent lengths are

1	gate valve	1 × 0.84 ft	= 0.84 ft
5	regular elbows	5 × 5.2 ft	= 26.00 ft
1	tee run	1 × 3.2 ft	= 3.20 ft
	straight pipe		= 228.00 ft
		total L_t	= 258.04 ft

16. VALVE FLOW COEFFICIENTS

Valve flow capacities depend on the geometry of the inside of the valve. The *flow coefficient*, C_v, for a valve (particularly a control valve) relates the flow quantity (in gallons per minute) of a fluid with specific gravity to the pressure drop (in pounds per square inch). (The flow coefficient for a valve is not the same as the coefficient of flow for an orifice or venturi meter.) As Eq. 17.48 shows, the flow coefficient is not dimensionally homogeneous.

Metricated countries use a similar concept with a different symbol, K_v, (not the same as the loss coefficient, K) to distinguish the valve flow coefficient from customary U.S. units. K_v is defined[17] as the flow rate in cubic meters per hour of water at a temperature of 16°C with a pressure drop across the valve of 1 bar. To further distinguish it from its U.S. counterpart, K_v may also be referred to as a *flow factor*. C_v and K_v are linearly related by Eq. 17.51.

$$\dot{V}_{\text{m}^3/\text{h}} = K_v \sqrt{\frac{\Delta p_{\text{bars}}}{\text{SG}}} \qquad \text{[SI]} \quad \textit{17.48(a)}$$

$$Q_{\text{gpm}} = C_v \sqrt{\frac{\Delta p_{\text{psi}}}{\text{SG}}} \qquad \text{[U.S.]} \quad \textit{17.48(b)}$$

$$K_v = 0.86 C_v \qquad \textit{17.49}$$

When selecting a control valve for a particular application, the value of C_v is first calculated. Depending on the application and installation, C_v may be further modified by dividing by *piping geometry* and *Reynolds number factors*. (These additional procedures are often specified by the valve manufacturer.) Then, a valve with the required value of C_v is selected.

Although the flow coefficient concept is generally limited to control valves, its use can be extended to all fittings and valves. The relationship between C_v and the loss coefficient, K, is

$$C_v = \frac{29.9 d_{\text{in}}^2}{\sqrt{K}} \qquad \text{[U.S.]} \quad \textit{17.50}$$

17. SHEAR STRESS IN CIRCULAR PIPES

Shear stress in fluid always acts to oppose the motion of the fluid. (That is the reason the term *pipe friction* is used.) Shear stress for a fluid in laminar flow can be calculated from the basic definition of absolute viscosity.

$$\tau = \mu \frac{d\text{v}}{dy} \qquad \textit{17.51}$$

In the case of the flow in a circular pipe, dr can be substituted for dy in the expression for *shear rate (velocity gradient)*, $d\text{v}/dy$.

$$\tau = \mu \frac{d\text{v}}{dr} \qquad \textit{17.52}$$

Equation 17.53 calculates the shear stress between fluid layers a distance r from the pipe centerline in terms of the pressure drop across a length L of the pipe.[18] Equation 17.53 is valid for both laminar and turbulent flows.

$$\tau = \frac{(p_1 - p_2)r}{2L} \qquad \left[r \leq \frac{D}{2}\right] \quad \textit{17.53}$$

The quantity $(p_1 - p_2)$ can be calculated from the Darcy equation. (See Eq. 17.27.) If v is the average flow velocity, the shear stress at the wall (where $r = D/2$) is

$$\tau_{\text{wall}} = \frac{f \rho \text{v}^2}{8} \qquad \text{[SI]} \quad \textit{17.54(a)}$$

$$\tau_{\text{wall}} = \frac{f \rho \text{v}^2}{8 g_c} \qquad \text{[U.S.]} \quad \textit{17.54(b)}$$

Equation 17.53 can be rearranged to give the relationship between the pressure gradient along the flow path and the shear stress at the wall.

$$\frac{dp}{dL} = \frac{4\tau_{\text{wall}}}{D} \qquad \textit{17.55}$$

Equation 17.54 can be combined with the Hagen-Poiseuille equation (given in Eq. 17.24) if the flow is laminar. (v in Eq. 17.56 is the average velocity of fluid flow.)

$$\tau = \frac{16 \mu \text{v} r}{D^2} \qquad \left[\text{laminar; } r \leq \frac{D}{2}\right] \quad \textit{17.56}$$

At the pipe wall, $r = D/2$, and the shear stress is maximum. (See Fig. 17.5.)

$$\tau_{\text{wall}} = \frac{8 \mu \text{v}}{D} \qquad \textit{17.57}$$

Figure 17.5 *Shear Stress Distribution in a Circular Pipe*

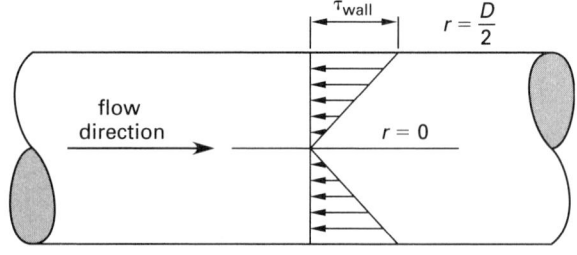

---

[17]Several definitions of both C_v and K_v are in use. A definition of C_v based on Imperial gallons is used in Great Britain. Definitions of K_v based on pressure drops in kilograms-force and volumes in liters per minute are in use. Other differences in the definition include the applicable temperature, which may be given as 5–30°C or 5–40°C instead of 16°C.

[18]In highly turbulent flow, shear stress is not caused by viscous effects but rather by momentum effects. Equation 17.53 is derived from a shell momentum balance. Such an analysis requires the concept of *momentum flux*. In a circular pipe with laminar flow, momentum flux is maximum at the pipe wall, zero at the flow centerline, and varies linearly in between.

Flow of Fluids

18. INTRODUCTION TO PUMPS AND TURBINES

A *pump* adds energy to the fluid flowing through it. (See Fig. 17.6.) The amount of energy that a pump puts into the fluid stream can be determined by the difference between the total energy on either side of the pump. The specific energy added (a positive number) on a per-unit mass basis (i.e., ft-lbf/lbm or J/kg) is given by Eq. 17.58. In most situations, a pump will add primarily pressure energy.

$$E_A = E_{t,2} - E_{t,1} \qquad 17.58$$

Figure 17.6 *Pump and Turbine Representation*

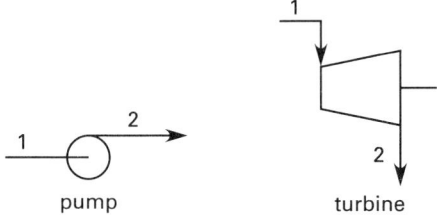

The *head added* by a pump is

$$h_A = \frac{E_A}{g} \qquad \text{[SI]} \qquad 17.59(a)$$

$$h_A = \frac{E_A g_c}{g} \qquad \text{[U.S.]} \qquad 17.59(b)$$

The specific energy added by a pump can also be calculated from the input power if the mass flow rate is known. The input power to the pump will be the output power of the electric motor or engine driving the pump.

$$E_A = \frac{\left(1000 \; \frac{\text{W}}{\text{kW}}\right) P_{\text{kW,input}} \eta_{\text{pump}}}{\dot{m}} \qquad \text{[SI]} \qquad 17.60(a)$$

$$E_A = \frac{\left(550 \; \frac{\text{ft-lbf}}{\text{sec-hp}}\right) P_{\text{hp,input}} \eta_{\text{pump}}}{\dot{m}} \qquad \text{[U.S.]} \qquad 17.60(b)$$

The *water horsepower* (WHP, also known as the *hydraulic horsepower* and *theoretical horsepower*) is the amount of power actually entering the fluid.

$$\text{WHP} = P_{\text{hp,input}} \eta_{\text{pump}} \qquad 17.61$$

A *turbine* extracts energy from the fluid flowing through it. As with a pump, the energy extraction can be obtained by determining the total energy on both sides of the turbine and taking the difference. The energy extracted (a positive number) on a per-unit mass basis is given by Eq. 17.62.

$$E_E = E_{t,1} - E_{t,2} \qquad 17.62$$

19. EXTENDED BERNOULLI EQUATION

The original Bernoulli equation assumes frictionless flow and does not consider the effects of pumps and turbines. When friction is present and when there are minor losses such as fittings and other energy-related devices in a pipeline, the energy balance is affected. The *extended Bernoulli equation* takes these additional factors into account.

$$(E_p + E_v + E_z)_1 + E_A$$
$$= (E_p + E_v + E_z)_2 + E_E + E_f + E_m \qquad 17.63$$

$$\frac{p_1}{\rho} + \frac{v_1^2}{2} + z_1 g + E_A$$
$$= \frac{p_2}{\rho} + \frac{v_2^2}{2} + z_2 g + E_E + E_f + E_m \qquad \text{[SI]} \qquad 17.64(a)$$

$$\frac{p_1}{\rho} + \frac{v_1^2}{2g_c} + \frac{z_1 g}{g_c} + E_A$$
$$= \frac{p_2}{\rho} + \frac{v_2^2}{2g_c} + \frac{z_2 g}{g_c} + E_E + E_f + E_m \qquad \text{[U.S.]} \qquad 17.64(b)$$

As defined, E_A, E_E, and E_f are all positive terms. None of the terms in Eq. 17.63 and Eq. 17.64 are negative.

The concepts of sources and sinks can be used to decide whether the friction, pump, and turbine terms appear on the left or right side of the Bernoulli equation. An *energy source* puts energy into the system. The incoming fluid and a pump contribute energy to the system. An *energy sink* removes energy from the system. The leaving fluid, friction, and a turbine remove energy from the system. In an energy balance, all energy must be accounted for, and the energy sources just equal the energy sinks.

$$\sum E_{\text{sources}} = \sum E_{\text{sinks}} \qquad 17.65$$

Therefore, the energy added by a pump always appears on the entrance side of the Bernoulli equation. Similarly, the frictional energy loss always appears on the discharge side.

20. ENERGY AND HYDRAULIC GRADE LINES WITH FRICTION

The *energy grade line* (EGL, also known as *total energy line*) is a graph of the total energy versus position in a pipeline. Since a pitot tube measures total (stagnation) energy, EGL will always coincide with the elevation of a pitot-piezometer fluid column. When friction is present, the EGL will always slope down, in the direction of flow. Figure 17.7 illustrates the EGL for a complex pipe network. The difference between $EGL_{frictionless}$ and $EGL_{with friction}$ is the energy loss due to friction.

Figure 17.7 Energy and Hydraulic Grade Lines

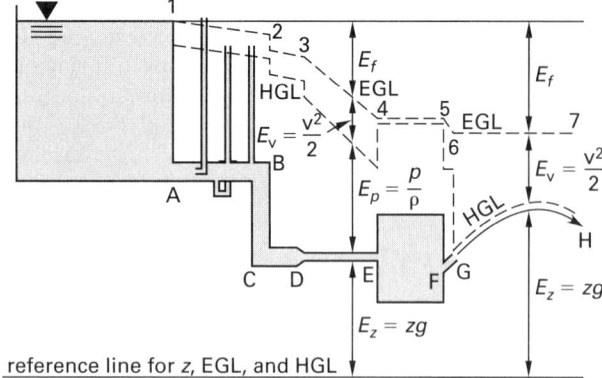

The EGL line in Fig. 17.7 is discontinuous at point 2, since the friction in pipe section B–C cannot be portrayed without disturbing the spatial correlation of points in the figure. Since the friction loss is proportional to v^2, the slope is steeper when the fluid velocity increases (i.e., when the pipe decreases in flow area), as it does in section D–E. Disregarding air friction, the EGL becomes horizontal at point 6 when the fluid becomes a free jet.

The *hydraulic grade line* (HGL) is a graph of the sum of pressure and potential energies versus position in the pipeline. (That is, the EGL and HGL differ by the kinetic energy.) The HGL will always coincide with the height of the fluid column in a static piezometer tube. The reference point for elevation is arbitrary, and the pressure energy is usually referenced to atmospheric pressure. Therefore, the pressure energy, E_p, for a free jet will be zero, and the HGL will consist only of the potential energy, as shown in section G–H.

The easiest way to draw the energy and hydraulic grade lines is to start with the EGL. The EGL can be drawn simply by recognizing that the rate of divergence from the horizontal $EGL_{frictionless}$ line is proportional to v^2. Then, since EGL and HGL differ by the velocity head, the HGL can be drawn parallel to the EGL when the pipe diameter is constant. The larger the pipe diameter, the closer the two lines will be.

The EGL for a pump will increase in elevation by E_A across the pump. (The actual energy "path" taken by the fluid is unknown, and a dotted line is used to indicate a lack of knowledge about what really happens in the pump.) The placement of the HGL for a pump will depend on whether the pump increases the fluid velocity and elevation, as well as the fluid pressure. In most cases, only the pressure will be increased. Figure 17.8 illustrates the HGL for the case of a pressure increase only.

The EGL and HGL for minor losses (fittings, contractions, expansions, etc.) are shown in Fig. 17.9.

Figure 17.8 EGL and HGL for a Pump

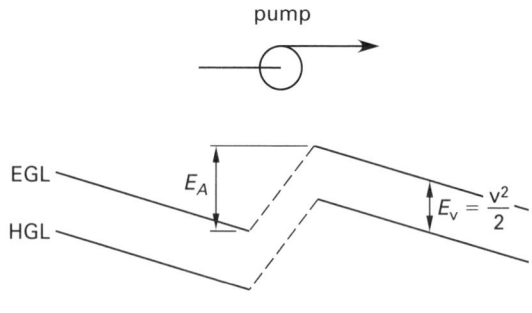

Figure 17.9 EGL and HGL for Minor Losses

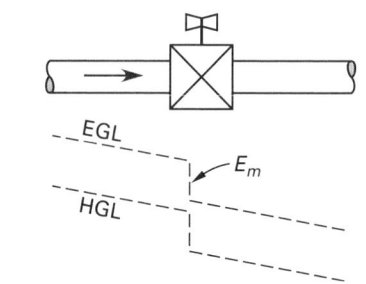

(a) valve, fitting, or obstruction

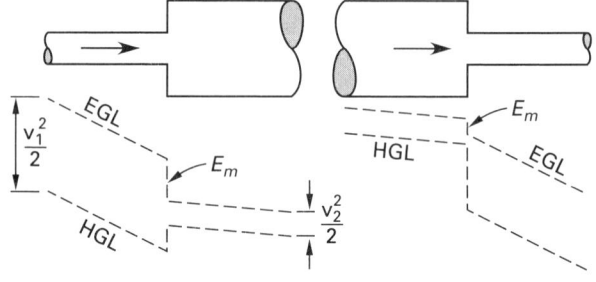

(b) sudden enlargement   (c) sudden contraction

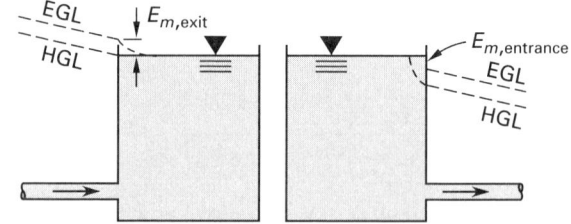

(d) transition to reservoir   (e) transition to pipeline

21. DISCHARGE FROM TANKS

The velocity of a jet issuing from an orifice in a tank can be determined by comparing the total energies at the free fluid surface and the jet itself. (See Fig. 17.10.) At the fluid surface, $p_1 = 0$ (atmospheric) and $v_1 = 0$. (v_1 is

Figure 17.10 Discharge from a Tank

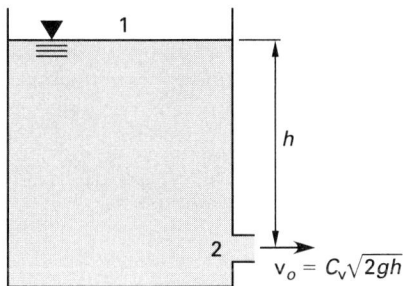

$v_o = C_v\sqrt{2gh}$

known as the *velocity of approach*.) The only energy the fluid has is potential energy. At the jet, $p_2 = 0$. All of the potential energy difference ($z_1 - z_2$) has been converted to kinetic energy. The theoretical velocity of the jet can be derived from the Bernoulli equation. Equation 17.66 is known as the equation for *Torricelli's speed of efflux*.

$$v_t = \sqrt{2gh} \qquad\qquad 17.66$$

$$h = z_1 - z_2 \qquad\qquad 17.67$$

The actual jet velocity is affected by the orifice geometry. The *coefficient of velocity*, C_v, is an empirical factor that accounts for the friction and turbulence at the orifice. Typical values of C_v are given in Table 17.5.

$$v_o = C_v\sqrt{2gh} \qquad\qquad 17.68$$

$$C_v = \frac{\text{actual velocity}}{\text{theoretical velocity}} = \frac{v_o}{v_t} \qquad 17.69$$

The specific energy loss due to turbulence and friction at the orifice is calculated as a multiple of the jet's kinetic energy.

$$E_f = \left(\frac{1}{C_v^2} - 1\right)\frac{v_o^2}{2} = \left(1 - C_v^2\right)gh \qquad \text{[SI]} \quad 17.70(a)$$

$$E_f = \left(\frac{1}{C_v^2} - 1\right)\frac{v_o^2}{2g_c} = \left(1 - C_v^2\right)h \times \frac{g}{g_c} \quad \text{[U.S.]} \quad 17.70(b)$$

The total head producing discharge (*effective head*) is the difference in elevations that would produce the same velocity from a frictionless orifice.

$$h_{\text{effective}} = C_v^2 h \qquad\qquad 17.71$$

The orifice guides quiescent water from the tank into the jet geometry. Unless the orifice is very smooth and the transition is gradual, momentum effects will continue to cause the jet to contract after it has passed through. The velocity calculated from Eq. 17.68 is usually assumed to be the velocity at the *vena contracta*, the section of smallest cross-sectional area. (See Fig. 17.11.)

Figure 17.11 Vena Contracta of a Fluid Jet

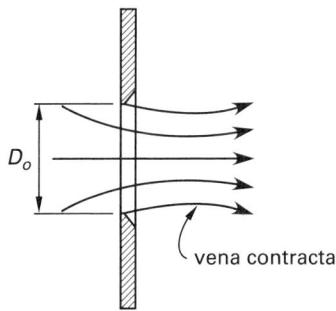

vena contracta

Table 17.5 Approximate Orifice Coefficients for Turbulent Water

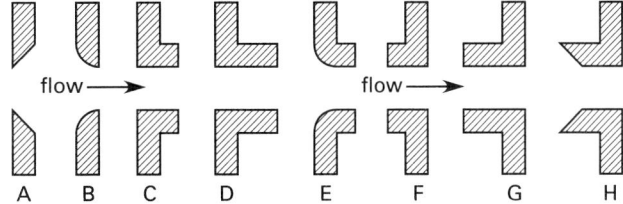

flow →   flow →

A   B   C   D   E   F   G   H

illustration	description	C_d	C_c	C_v
A	sharp-edged	0.62	0.63	0.98
B	round-edged	0.98	1.00	0.98
C	short tube* (fluid separates from walls)	0.61	1.00	0.61
D	sharp tube (no separation)	0.82	1.00	0.82
E	short tube with rounded entrance	0.97	0.99	0.98
F	reentrant tube, length less than one-half of pipe diameter	0.54	0.55	0.99
G	reentrant tube, length 2 to 3 pipe diameters	0.72	1.00	0.72
H	Borda	0.51	0.52	0.98
(none)	smooth, well-tapered nozzle	0.98	0.99	0.99

*A short tube has a length less than approximately 3 diameters.

For a thin plate or sharp-edged orifice, the vena contracta is often assumed to be located approximately one half an orifice diameter past the orifice, although the actual distance can vary from $0.3D_o$ to $0.8D_o$. The area of the vena contracta can be calculated from the orifice area and the *coefficient of contraction, C_c*. For water flowing with a high Reynolds number through a small sharp-edged orifice, the contracted area is approximately 61–63% of the orifice area.

$$A_{\text{vena contracta}} = C_c A_o \qquad 17.72$$

$$C_c = \frac{A_{\text{vena contracta}}}{A_o} \qquad 17.73$$

The theoretical discharge rate from a tank is $\dot{V}_t = A_o\sqrt{2gh}$. However, this relationship needs to be corrected for friction and contraction by multiplying by C_v and C_c. The *coefficient of discharge, C_d*, is the product of the coefficients of velocity and contraction.

$$\dot{V} = C_c v_o A_o = C_d v_t A_o = C_d A_o \sqrt{2gh} \qquad 17.74$$

$$C_d = C_v C_c = \frac{\dot{V}}{\dot{V}_t} \qquad 17.75$$

22. DISCHARGE FROM PRESSURIZED TANKS

If the gas or vapor above the liquid in a tank is at gage pressure p, and the discharge is to atmospheric pressure, the head causing discharge will be

$$h = z_1 - z_2 + \frac{p}{\rho g} \qquad \text{[SI]} \quad 17.76(a)$$

$$h = z_1 - z_2 + \frac{p}{\rho} \times \frac{g_c}{g} = z_1 - z_2 + \frac{p}{\gamma} \qquad \text{[U.S.]} \quad 17.76(b)$$

The discharge velocity can be calculated from Eq. 17.68 using the increased discharge head. (See Fig. 17.12.)

Figure 17.12 *Discharge from a Pressurized Tank*

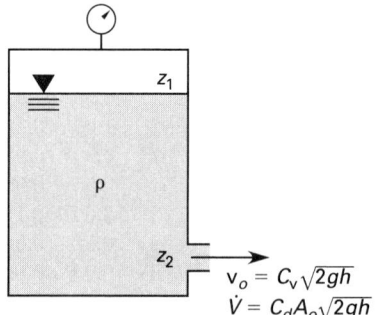

23. COORDINATES OF A FLUID STREAM

Fluid discharged from an orifice in a tank gets its initial velocity from the conversion of potential energy. After discharge, no additional energy conversion occurs, and all subsequent velocity changes are due to external forces. (See Fig. 17.13.)

Figure 17.13 *Coordinates of a Fluid Stream*

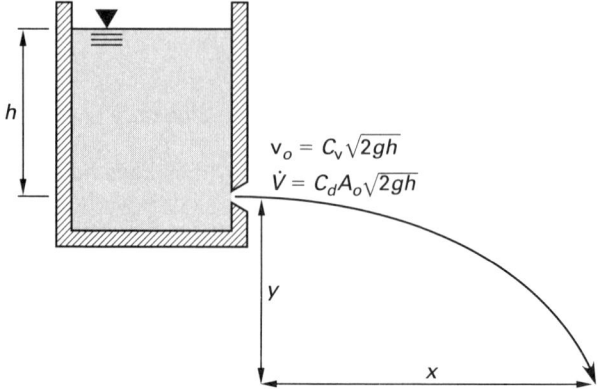

In the absence of air friction (drag), there are no retarding or accelerating forces in the x-direction on the fluid stream. The x-component of velocity is constant. Projectile motion equations can be used to predict the path of the fluid stream.

$$v_x = v_o \qquad \text{[horizontal discharge]} \qquad 17.77$$

$$x = v_o t = v_o \sqrt{\frac{2y}{g}} = 2C_v \sqrt{hy} \qquad 17.78$$

After discharge, the fluid stream is acted upon by a constant gravitational acceleration. The y-component of velocity is zero at discharge but increases linearly with time.

$$v_y = gt \qquad 17.79$$

$$y = \frac{gt^2}{2} = \frac{gx^2}{2v_o^2} = \frac{x^2}{4hC_v^2} \qquad 17.80$$

24. DISCHARGE FROM LARGE ORIFICES

When an orifice diameter is large compared with the discharge head, the jet velocity at the top edge of the orifice will be less than the velocity at the bottom edge. Since the velocity is related to the square root of the head, the distance used to calculate the effective jet velocity should be measured from the fluid surface to a point above the centerline of the orifice.

This correction is generally neglected, however, since it is small for heads of more than twice the orifice diameter. Furthermore, if an orifice is intended to work regularly with small heads, the orifice should be calibrated in place. The discrepancy can then be absorbed into the discharge coefficient, C_d.

25. TIME TO EMPTY A TANK

If the fluid in an open or vented tank is not replenished at the rate of discharge, the static head forcing discharge through the orifice will decrease with time. If the tank has a varying cross section, A_t, Eq. 17.81 specifies the basic relationship between the change in elevation and elapsed time. (The negative sign indicates that z decreases as t increases.)

$$\dot{V}\,dt = -A_t\,dz \qquad \textit{17.81}$$

If A_t can be expressed as a function of h, Eq. 17.82 can be used to determine the time to lower the fluid elevation from z_1 to z_2.

$$t = \int_{z_1}^{z_2} \frac{-A_t\,dz}{C_d A_o \sqrt{2gz}} \qquad \textit{17.82}$$

For a tank with a constant cross-sectional area, A_t, the time required to lower the fluid elevation is

$$t = \frac{2A_t(\sqrt{z_1} - \sqrt{z_2})}{C_d A_o \sqrt{2g}} \qquad \textit{17.83}$$

If a tank is replenished at a rate of $\dot{V}_{\text{in}}$, Eq. 17.84 can be used to calculate the discharge time. If the tank is replenished at a rate greater than the discharge rate, t in Eq. 17.84 will represent the time to raise the fluid level from z_1 to z_2.

$$t = \int_{z_1}^{z_2} \frac{A_t\,dz}{C_d A_o \sqrt{2gz} - \dot{V}_{\text{in}}} \qquad \textit{17.84}$$

If the tank is not open or vented but is pressurized, the elevation terms, z_1 and z_2, in Eq. 17.83 must be replaced by the total head terms, h_1 and h_2, that include the effects of pressurization. (See Eq. 17.86.)

Example 17.8

A vertical, cylindrical tank 15 ft in diameter discharges 150°F water $(\rho = 61.20 \text{ lbm/ft}^3)$ through a sharp-edged, 1 in diameter orifice $(C_d = 0.62)$ in the tank bottom. The original water depth is 12 ft. The tank is continually pressurized to 50 psig. What is the time to empty the tank?

Solution

The area of the orifice is

$$A_o = \frac{\pi D^2}{4} = \frac{\pi (1 \text{ in})^2}{(4)\left(12 \ \frac{\text{in}}{\text{ft}}\right)^2} = 0.00545 \text{ ft}^2$$

The tank area is constant with respect to depth.

$$A_t = \frac{\pi D^2}{4} = \frac{\pi (15 \text{ ft})^2}{4} = 176.7 \text{ ft}^2$$

The total initial head includes the effect of the pressurization. Use Eq. 17.76.

$$h_1 = z_1 - z_2 + \frac{p}{\rho} \times \frac{g_c}{g}$$

$$= 12 \text{ ft} + \frac{\left(50 \ \frac{\text{lbf}}{\text{in}^2}\right)\left(12 \ \frac{\text{in}}{\text{ft}}\right)^2}{61.2 \ \frac{\text{lbm}}{\text{ft}^3}} \times \frac{32.2 \ \frac{\text{lbm-ft}}{\text{lbf-sec}^2}}{32.2 \ \frac{\text{ft}}{\text{sec}^2}}$$

$$= 12 \text{ ft} + 117.6 \text{ ft}$$

$$= 129.6 \text{ ft}$$

When the fluid has reached the level of the orifice, the fluid potential head will be zero, but the pressurization will remain.

$$h_2 = 117.6 \text{ ft}$$

The time to empty the tank is given by Eq. 17.83.

$$t = \frac{2A_t(\sqrt{z_1} - \sqrt{z_2})}{C_d A_o \sqrt{2g}}$$

$$= \frac{(2)(176.7 \text{ ft}^2)(\sqrt{129.6 \text{ ft}} - \sqrt{117.6 \text{ ft}})}{(0.62)(0.00545 \text{ ft}^2)\sqrt{(2)\left(32.2 \ \frac{\text{ft}}{\text{sec}^2}\right)}}$$

$$= 7036 \text{ sec}$$

26. PRESSURE CULVERTS

A *culvert* is a water path (usually a large diameter pipe) used to channel water around or through an obstructing feature. (See Fig. 17.14.) In most instances, a culvert is used to restore a natural water path obstructed by a manufactured feature. For example, when a road is built across (perpendicular to) a natural drainage, a culvert is used to channel water under the road.

Figure 17.14 *Simple Pipe Culvert*

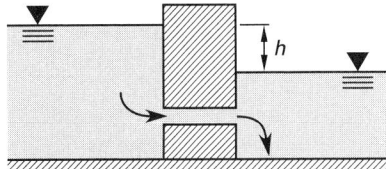

Because culverts often operate only partially full and with low heads, Torricelli's equation does not apply. Therefore, most culvert designs are empirical. However, if the entrance and exit of a culvert are both submerged, the culvert will flow full, and the discharge will be independent of the barrel slope. Equation 17.85 can be used to calculate the discharge.

$$\dot{V} = C_d A_v = C_d A \sqrt{2gh_{\text{effective}}} \qquad 17.85$$

If the culvert is long (more than 60 ft or 20 m), or if the entrance is not gradual, the available energy will be divided between friction and velocity heads. The effective head used in Eq. 17.85 should be

$$h_{\text{effective}} = h - h_{f,\text{barrel}} - h_{m,\text{entrance}} \qquad 17.86$$

The friction loss in the barrel can be found in the usual manner, from either the Darcy equation or the Hazen-Williams equation. The entrance loss is calculated using the standard method of loss coefficients. Representative values of the loss coefficient, K, are given in Table 17.6. Since the fluid velocity is not initially known but is needed to find the friction factor, a trial-and-error solution will be necessary.

Table 17.6 Representative Loss Coefficients for Culvert Entrances

entrance	K
smooth and gradual transition	0.08
flush vee or bell shape	0.10
projecting vee or bell shape	0.15
flush, square-edged	0.50
projecting, square-edged	0.90

27. SIPHONS

A *siphon* is a bent or curved tube that carries fluid from a fluid surface at a high elevation to another fluid surface at a lower elevation. Normally, it would not seem difficult to have a fluid flow to a lower elevation. However, the fluid seems to flow "uphill" in a portion of a siphon. Figure 17.15 illustrates a siphon.

Figure 17.15 Siphon

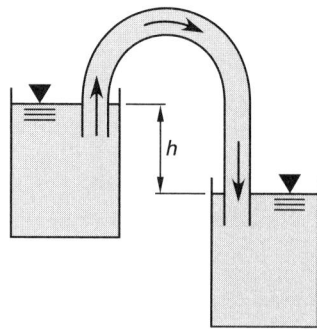

Starting a siphon requires the tube to be completely filled with liquid. Then, since the fluid weight is greater in the longer arm than in the shorter arm, the fluid in the longer arm "falls" out of the siphon, "pulling" more liquid into the shorter arm and over the bend.

Operation of a siphon is essentially independent of atmospheric pressure. The theoretical discharge is the same as predicted by the Torricelli equation. A correction for discharge is necessary, but little data is available on typical values of C_d. Therefore, siphons should be tested and calibrated in place.

$$\dot{V} = C_d A_v = C_d A \sqrt{2gh} \qquad 17.87$$

28. SERIES PIPE SYSTEMS

A system of pipes in series consists of two or more lengths of different-diameter pipes connected end-to-end. In the case of the series pipe from a reservoir discharging to the atmosphere shown in Fig. 17.16, the available head will be split between the velocity head and the friction loss.

$$h = h_v + h_f \qquad 17.88$$

Figure 17.16 Series Pipe System

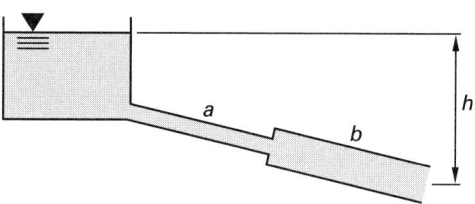

If the flow rate or velocity in any part of the system is known, the friction loss can easily be found as the sum of the friction losses in the individual sections. The solution is somewhat more simple than it first appears to be, since the velocity of all sections can be written in terms of only one velocity.

$$h_{f,t} = h_{f,a} + h_{f,b} \qquad 17.89$$

$$A_a v_a = A_b v_b \qquad 17.90$$

If neither the velocity nor the flow quantity is known, a trial-and-error solution will be required, since a friction factor must be known to calculate h_f. A good starting point is to assume fully turbulent flow.

When velocity and flow rate are both unknown, the following procedure using the Darcy friction factor can be used.[19]

step 1: Calculate the relative roughness, ϵ/D, for each section. Use the Moody diagram to determine f_a and f_b for fully turbulent flow (i.e., the horizontal portion of the curve).

---

[19]If Hazen-Williams constants are given for the pipe sections, the procedure for finding the unknown velocities is similar, although considerably more difficult since v^2 and $v^{1.85}$ cannot be combined. A first approximation, however, can be obtained by replacing $v^{1.85}$ in the Hazen-Williams equation for friction loss. A trial and error method can then be used to find velocity.

step 2: Write all of the velocities in terms of one unknown velocity.

$$\dot{V}_a = \dot{V}_b \qquad 17.91$$

$$v_b = \left(\frac{A_a}{A_b}\right) v_a \qquad 17.92$$

step 3: Write the total friction loss in terms of the unknown velocity.

$$h_{f,t} = \frac{f_a L_a v_a^2}{2 D_a g} + \left(\frac{f_b L_b}{2 D_b g}\right)\left(\frac{A_a}{A_b}\right)^2 v_a^2$$

$$= \left(\frac{v_a^2}{2g}\right)\left(\frac{f_a L_a}{D_a} + \left(\frac{f_b L_b}{D_b}\right)\left(\frac{A_a}{A_b}\right)^2\right) \qquad 17.93$$

step 4: Solve for the unknown velocity using the Bernoulli equation between the free reservoir surface ($p = 0$, $v = 0$, $z = h$) and the discharge point ($p = 0$, if free discharge; $z = 0$). Include pipe friction, but disregard minor losses for convenience.

$$h = \frac{v_b^2}{2g} + h_{f,t} = \left(\frac{v_a^2}{2g}\right)\left(\left(\frac{A_a}{A_b}\right)^2\left(1 + \frac{f_b L_b}{D_b}\right) + \frac{f_a L_a}{D_a}\right)$$

$$17.94$$

step 5: Using the value of v_a, calculate v_b. Calculate the Reynolds number, and check the values of f_a and f_b from step 4. Repeat steps 3 and 4 if necessary.

29. PARALLEL PIPE SYSTEMS

Adding a second pipe in parallel with a first is a standard method of increasing the capacity of a line. A *pipe loop* is a set of two pipes placed in parallel, both originating and terminating at the same junction. The two pipes are referred to as *branches* or *legs*. (See Fig. 17.17.)

Figure 17.17 *Parallel Pipe System*

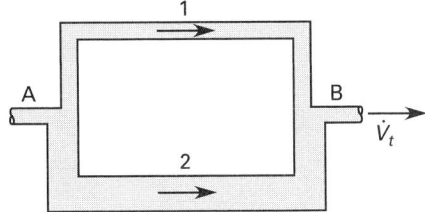

There are three principles that govern the distribution of flow between the two branches.

- The flow divides in such a manner as to make the head loss in each branch the same.

$$h_{f,1} = h_{f,2} \qquad 17.95$$

- The head loss between the junctions A and B is the same as the head loss in branches 1 and 2.

$$h_{f,A\text{-}B} = h_{f,1} = h_{f,2} \qquad 17.96$$

- The total flow rate is the sum of the flow rates in the two branches.

$$\dot{V}_t = \dot{V}_1 + \dot{V}_2 \qquad 17.97$$

If the pipe diameters are known, Eq. 17.95 and Eq. 17.97 can be solved simultaneously for the branch velocities. For first estimates, it is common to neglect minor losses, the velocity head, and the variation in the friction factor, f, with velocity.

If the system has only two parallel branches, the unknown branch flows can be determined by solving Eq. 17.98 and Eq. 17.100 simultaneously.

$$\frac{f_1 L_1 v_1^2}{2 D_1 g} = \frac{f_2 L_2 v_2^2}{2 D_2 g} \qquad 17.98$$

$$\dot{V}_1 + \dot{V}_2 = \dot{V}_t \qquad 17.99$$

$$\frac{\pi}{4}(D_1^2 v_1 + D_2^2 v_2) = \dot{V}_t \qquad 17.100$$

However, if the system has three or more parallel branches, it is easier to use the following iterative procedure. This procedure can be used for problems (a) where the flow rate is unknown but the pressure drop between the two junctions is known, or (b) where the total flow rate is known but the pressure drop and velocity are both unknown. In both cases, the solution iteratively determines the friction coefficients, f.

step 1: Solve the friction head loss, h_f, expression (either Darcy or Hazen-Williams) for velocity in each branch. If the pressure drop is known, first convert it to friction head loss.

$$v = \sqrt{\frac{2 D g h_f}{f L}} \quad \text{[Darcy]} \qquad 17.101$$

$$v = \frac{0.355 \, C D^{0.63} h_f^{0.54}}{L^{0.54}} \quad \text{[Hazen-Williams; SI]}$$

$$17.102(a)$$

$$v = \frac{0.550 \, C D^{0.63} h_f^{0.54}}{L^{0.54}} \quad \text{[Hazen-Williams; U.S.]}$$

$$17.102(b)$$

step 2: Solve for the flow rate in each branch. If they are unknown, friction factors, f, must be assumed for each branch. The fully turbulent assumption provides a good initial estimate. (The value of k' will be different for each branch.)

$$\dot{V} = Av = A\sqrt{\frac{2Dgh_f}{fL}}$$

$$= k'\sqrt{h_f} \quad \text{[Darcy]} \qquad \textit{17.103}$$

step 3: Write the expression for the conservation of flow. Calculate the friction head loss from the total flow rate. For example, for a three-branch system,

$$\dot{V}_t = \dot{V}_1 + \dot{V}_2 + \dot{V}_3$$

$$= (k'_1 + k'_2 + k'_3)\sqrt{h_f} \qquad \textit{17.104}$$

step 4: Check the assumed values of the friction factor. Repeat as necessary.

Example 17.9

3 ft^3/sec of water enter the parallel pipe network shown at junction A. All pipes are schedule-40 steel with the nominal sizes shown. Minor losses are insignificant. What is the total friction head loss between junctions A and B?

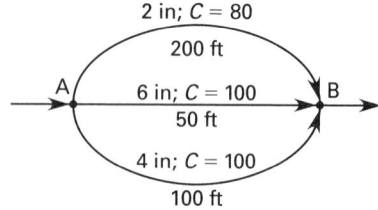

Solution

step 1: Collect the pipe dimensions from App. 16.B.

	2 in	4 in	6 in
flow area	0.0233 ft^2	0.0884 ft^2	0.2006 ft^2
diameter	0.1723 ft	0.3355 ft	0.5054 ft

Follow the procedure given in Sec. 17.29.

$$h_f = \frac{3.022\, v^{1.85} L}{C^{1.85} D^{1.165}}$$

$$v = \frac{0.550 C D^{0.63} h_f^{0.54}}{L^{0.54}}$$

The velocity (expressed in ft/sec) in the 2 in diameter pipe branch is

$$v_{2\,in} = \frac{(0.550)(80)(0.1723 \text{ ft})^{0.63} h_f^{0.54}}{(200 \text{ ft})^{0.54}}$$

$$= 0.831 h_f^{0.54}$$

The friction head loss is the same in all parallel branches. The velocities in the other two branches are

$$v_{6\,in} = 4.327 h_f^{0.54}$$

$$v_{4\,in} = 2.299 h_f^{0.54}$$

step 2: The flow rates are

$$\dot{V} = Av$$

$$\dot{V}_{2\,in} = (0.0233 \text{ ft}^2)0.831 h_f^{0.54}$$

$$= 0.0194 h_f^{0.54}$$

$$\dot{V}_{6\,in} = (0.2006 \text{ ft}^2)4.327 h_f^{0.54}$$

$$= 0.8680 h_f^{0.54}$$

$$\dot{V}_{4\,in} = (0.0884 \text{ ft}^2)2.299 h_f^{0.54}$$

$$= 0.2032 h_f^{0.54}$$

step 3: The total flow rate is

$$\dot{V}_t = \dot{V}_{2\,in} + \dot{V}_{6\,in} + \dot{V}_{4\,in}$$

$$3\,\frac{\text{ft}^3}{\text{sec}} = 0.0194 h_f^{0.54} + 0.8680 h_f^{0.54} + 0.2032 h_f^{0.54}$$

$$= (0.0194 + 0.8680 + 0.2032) h_f^{0.54}$$

$$h_f = 6.5 \text{ ft}$$

30. MULTIPLE RESERVOIR SYSTEMS

In the *three-reservoir problem*, there are many possible choices for the unknown quantity (pipe length, diameter, head, flow rate, etc.). In all but the simplest cases, the solution technique is by trial and error based on conservation of mass and energy. (See Fig. 17.18.)

Figure 17.18 *Three-Reservoir System*

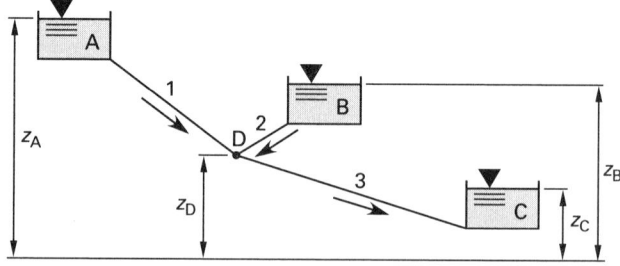

For simplification, velocity heads and minor losses are usually insignificant and can be neglected. However, the presence of a pump in any of the lines must be included in the solution procedure. This is most easily done by adding the pump head to the elevation of the reservoir feeding the pump. If the pump head is not known or depends on the flow rate, it must be determined iteratively as part of the solution procedure.

Case 1: Given all lengths, diameters, and elevations, find all flow rates.

Although an analytical solution method is possible, this type of problem is easily solved iteratively. The following procedure makes an initial estimate of a flow rate and uses it to calculate the pressure at the junction, p_D. Since this method may not converge if the initial estimate of $\dot{V}_1$ is significantly in error, it is helpful to use other information (e.g., normal pipe velocities; see Sec. 17.3) to obtain the initial estimate. An alternate procedure is simply to make several estimates of p_D and calculate the corresponding values of flow rate.

step 1: Assume a reasonable value for $\dot{V}_1$. Calculate the corresponding friction loss, $h_{f,1}$. Use the Bernoulli equation to find the corresponding value of p_D. Disregard minor losses and velocity head.

$$v_1 = \frac{\dot{V}_1}{A_1} \qquad 17.105$$

$$z_A = z_D + \frac{p_D}{\gamma} + h_{f,1} \qquad 17.106$$

step 2: Use the value of p_D to calculate $h_{f,2}$. Use the friction loss to determine v_2. Use v_2 to determine $\dot{V}_2$. If flow is out of reservoir B, $h_{f,2}$ should be added. If $z_D + (p_D/\gamma) > z_B$, flow will be into reservoir B. In this case, $h_{f,2}$ should be subtracted.

$$z_B = z_D + \frac{p_D}{\gamma} \pm h_{f,2} \qquad 17.107$$

$$\dot{V}_2 = v_2 A_2 \qquad 17.108$$

step 3: Similarly, use the value of p_D to calculate $h_{f,3}$. Use the friction loss to determine v_3. Use v_3 to determine $\dot{V}_3$.

$$z_C = z_D + \frac{p_D}{\gamma} - h_{f,3} \qquad 17.109$$

$$\dot{V}_3 = v_3 A_3 \qquad 17.110$$

step 4: Check if $\dot{V}_1 \pm \dot{V}_2 = \dot{V}_3$. If it does not, repeat steps 1 through 4. After the second iteration, plot $\dot{V}_1 \pm \dot{V}_2 - \dot{V}_3$ versus $\dot{V}_1$. Interpolate or extrapolate the value of $\dot{V}_1$ that makes the difference zero.

Case 2: Given $\dot{V}_1$ and all lengths, diameters, and elevations except z_C, find z_C.

step 1: Calculate v_1.

$$v_1 = \frac{\dot{V}_1}{A_1} \qquad 17.111$$

step 2: Calculate the corresponding friction loss, $h_{f,1}$. Use the Bernoulli equation to find the corresponding value of p_D. Disregard minor losses and velocity head.

$$z_A = z_D + \frac{p_D}{\gamma} + h_{f,1} \qquad 17.112$$

step 3: Use the value p_D to calculate $h_{f,2}$. Use the friction loss to determine v_2. Use v_2 to determine $\dot{V}_2 A_2$. If flow is out of reservoir B, $h_{f,2}$ should be added. If $z_D + (p_D/\gamma) > z_B$, flow will be into reservoir B. In this case, $h_{f,2}$ should be subtracted.

$$z_B = z_D + \frac{p_D}{\gamma} \pm h_{f,2} \qquad 17.113$$

$$\dot{V}_2 = v_2 A_2 \qquad 17.114$$

step 4: $$\dot{V}_3 = \dot{V}_1 \pm \dot{V}_2 \qquad 17.115$$

step 5: $$v_3 = \frac{\dot{V}_3}{A_3} \qquad 17.116$$

step 6: Calculate $h_{f,3}$.

step 7: $$z_C = z_D + \frac{p_D}{\gamma} - h_{f,3} \qquad 17.117$$

Case 3: Given $\dot{V}_1$, all lengths, all elevations, and all diameters except D_3, find D_3.

step 1: Repeat step 1 from case 2.

step 2: Repeat step 2 from case 2.

step 3: Repeat step 3 from case 2.

step 4: Repeat step 4 from case 2.

step 5: Calculate $h_{f,3}$ from

$$z_C = z_D + \frac{p_D}{\gamma} - h_{f,3} \qquad 17.118$$

step 6: Calculate D_3 from $h_{f,3}$.

Case 4: Given all lengths, diameters, and elevations except z_D, find all flow rates.

step 1: Calculate the head loss between each reservoir and junction D. Combine as many terms as possible into constant k'.

$$\dot{V} = Av = A\sqrt{\frac{2Dgh_f}{fL}}$$
$$= k'\sqrt{h_f} \quad \text{[Darcy]} \qquad 17.119$$

$$\dot{V} = Av = \frac{A(0.550)CD^{0.63}h_f^{0.54}}{L^{0.54}}$$
$$= k'h_f^{0.54} \quad \text{[Hazen-Williams; U.S.]} \qquad 17.120$$

step 2: Assume that the flow direction in all three pipes is toward junction D. Write the conservation equation for junction D.

$$\dot{V}_{D,t} = \dot{V}_1 + \dot{V}_2 + \dot{V}_3 = 0 \qquad 17.121$$

$$k_1' \sqrt{h_{f,1}} + k_2' \sqrt{h_{f,2}} + k_3' \sqrt{h_{f,3}} = 0 \quad \text{[Darcy]} \qquad 17.122$$

$$k_1' h_{f,1}^{0.54} + k_2' h_{f,2}^{0.54} + k_3' h_{f,3}^{0.54} = 0 \quad \text{[Hazen-Williams]}$$
$$17.123$$

step 3: Write the Bernoulli equation between each reservoir and junction D. Since $p_A = p_B = p_C = 0$, and $v_A = v_B = v_C = 0$, the friction loss in branch 1 is

$$h_{f,1} = z_A - z_D - \frac{p_D}{\gamma} \qquad 17.124$$

However, z_D and p_D can be combined since they are related constants in any particular situation. Define the correction, δ_D, as

$$\delta_D = z_D + \frac{p_D}{\gamma} \qquad 17.125$$

Then, the friction head losses in the branches are

$$h_{f,1} = z_A - \delta_D \qquad 17.126$$
$$h_{f,2} = z_B - \delta_D \qquad 17.127$$
$$h_{f,3} = z_C - \delta_D \qquad 17.128$$

step 4: Assume a value for δ_D. Calculate the corresponding h_f values. Use Eq. 17.119 to find $\dot{V}_1$, $\dot{V}_2$, and $\dot{V}_3$. Calculate the corresponding $\dot{V}_t$ value. Repeat until $\dot{V}_t$ converges to zero. It is not necessary to calculate p_D or z_D once all of the flow rates are known.

31. PIPE NETWORKS

Network flows in a *multiloop system* cannot be determined by any closed-form equation. (See Fig. 17.19.) Most real-world problems involving multiloop systems are analyzed iteratively on a computer. Computer programs are based on the *Hardy Cross method*, which can also be performed manually when there are only a few loops. In this method, flows in all of the branches are first assumed, and adjustments are made in consecutive iterations to the assumed flow.

Figure 17.19 *Multiloop System*

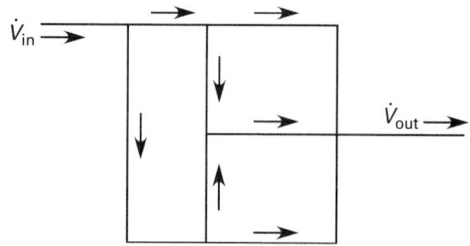

The Hardy Cross method is based on the following principles.

Principle 1: Conservation—The flows entering a junction equal the flows leaving the junction.

Principle 2: The algebraic sum of head losses around any closed loop is zero.

The friction head loss has the form $h_f = K' \dot{V}^n$, with h_f having units of feet. (k' used in Eq. 17.103 is equal to $\sqrt{1/K'}$.) The Darcy friction factor, f, is usually assumed to be the same in all parts of the network. For a Darcy head loss, the exponent is $n = 2$. For a Hazen-Williams loss, $n = 1.85$.

- For $\dot{V}$ in ft^3/sec, L in feet, and D in feet, the friction coefficient is

$$K' = \frac{0.02517 f L}{D^5} \quad \text{[Darcy]} \qquad 17.129$$

$$K' = \frac{4.727 L}{D^{4.87} C^{1.85}} \quad \text{[Hazen-Williams]} \qquad 17.130$$

- For $\dot{V}$ in gal/min, L in feet, and d in inches, the friction coefficient is

$$K' = \frac{0.03109 f L}{d^5} \quad \text{[Darcy]} \qquad 17.131$$

$$K' = \frac{10.44 L}{d^{4.87} C^{1.85}} \quad \text{[Hazen-Williams]} \qquad 17.132$$

- For $\dot{V}$ in gal/min, L in feet, and D in feet, the friction coefficient is

$$K' = \frac{1.251 \times 10^{-7} f L}{D^5} \quad \text{[Darcy]} \qquad 17.133$$

$$K' = \frac{5.862 \times 10^{-5} L}{D^{4.87} C^{1.85}} \quad \text{[Hazen-Williams]} \qquad 17.134$$

- For $\dot{V}$ in MGD (millions of gallons per day), L in feet, and D in feet, the friction coefficient is

$$K' = \frac{0.06026 f L}{D^5} \quad \text{[Darcy]} \qquad 17.135$$

$$K' = \frac{10.59 L}{D^{4.87} C^{1.85}} \quad \text{[Hazen-Williams]} \qquad 17.136$$

If $\dot{V}_a$ is the assumed flow in a pipe, the true value, $\dot{V}$, can be calculated from the difference (correction), δ.

$$\dot{V} = \dot{V}_a + \delta \qquad 17.137$$

The friction loss term for the assumed value and its correction can be expanded as a series. Since the correction is small, higher order terms can be omitted.

$$h_f = K'(\dot{V}_a + \delta)^n$$
$$\approx K' \dot{V}_a^n + n K' \delta \dot{V}_a^{n-1} \qquad 17.138$$

From Principle 2, the sum of the friction drops is zero around a loop. The correction, δ, is the same for all pipes in the loop and can be taken out of the summation. Since the loop closes on itself, all elevations can be omitted.

$$\sum h_f = \sum K' \dot{V}_a^n + n\delta \sum K' \dot{V}_a^{n-1} = 0 \qquad 17.139$$

This equation can be solved for δ.

$$\delta = \frac{-\sum K' \dot{V}_a^n}{n \sum \left| K' \dot{V}_a^{n-1} \right|}$$

$$= -\frac{\sum h_f}{n \sum \left| \dfrac{h_f}{\dot{V}_a} \right|} \qquad 17.140$$

The Hardy Cross procedure is as follows.

step 1: Determine the value of n. For a Darcy head loss, the exponent is $n = 2$. For a Hazen-Williams loss, $n = 1.85$.

step 2: Arbitrarily select a positive direction (e.g., clockwise).

step 3: Label all branches and junctions in the network.

step 4: Separate the network into independent loops such that each branch is included in at least one loop.

step 5: Calculate K' for each branch in the network.

step 6: Assume consistent and reasonable flow rates and directions for each branch in the network.

step 7: Calculate the correction, δ, for each independent loop. (The numerator is the sum of head losses around the loop, taking signs into consideration.) It is not necessary for the loop to be at the same elevation everywhere. Disregard elevations. Since the loop closes on itself, all elevations can be omitted.

step 8: Apply the correction, δ, to each branch in the loop. The correction must be applied in the same sense to each branch in the loop. If clockwise has been taken as the positive direction, then δ is added to clockwise flows and subtracted from counterclockwise flows.

step 9: Repeat steps 7 and 8 until the correction is sufficiently small.

Example 17.10

A two-loop pipe network is shown. All junctions are at the same elevation. The Darcy friction factor is 0.02 for all pipes in the network. For convenience, use the nominal pipe sizes shown in the figure. Determine the flows in all branches.

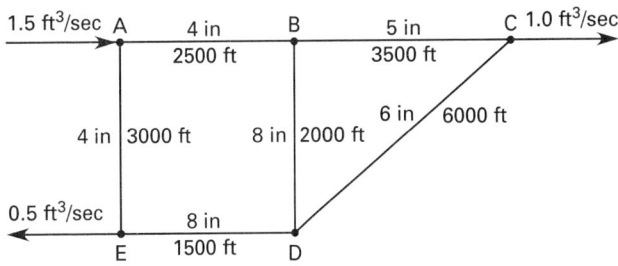

Solution

step 1: The Darcy friction factor is given, so $n = 2$.

step 2: Select clockwise as the positive direction.

step 3: Use the junction letters in the illustration.

step 4: Two independent loops are needed. Work with loops ABDE and BCD. (Loop ABCDE could also be used but would be more complex than loop BCD.)

step 5: Work with branch AB.

$$D_{\text{AB}} = \frac{4 \text{ in}}{12 \, \dfrac{\text{in}}{\text{ft}}} = 0.3333 \text{ ft}$$

Use Eq. 17.129.

$$K'_{\text{AB}} = \frac{0.02517 fL}{D^5} = \frac{(0.02517)(0.02)(2500 \text{ ft})}{(0.3333 \text{ ft})^5}$$

$$= 306.0$$

Similarly,

$$K'_{\text{BC}} = 140.5$$
$$K'_{\text{DC}} = 96.8$$
$$K'_{\text{BD}} = 7.7$$
$$K'_{\text{ED}} = 5.7$$
$$K'_{\text{AE}} = 367.4$$

step 6: Assume the direction and flow rates shown.

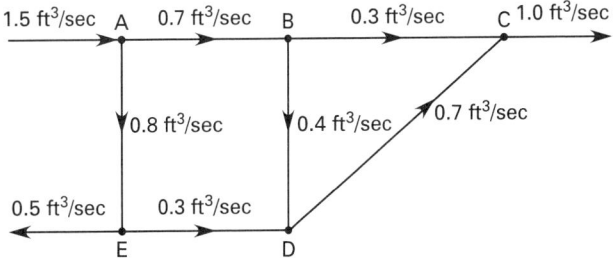

Flow of Fluids

step 7: Use Eq. 17.140.

$$\delta = \frac{-\sum K' \dot{V}_a^n}{n \sum \left| K' \dot{V}_a^{n-1} \right|}$$

$$\delta_{\text{ABDE}} = \frac{-\left(\begin{array}{c} (306.0)\left(0.7 \; \frac{\text{ft}^3}{\text{sec}}\right)^2 + (7.7)\left(0.4 \; \frac{\text{ft}^3}{\text{sec}}\right)^2 \\ - (5.7)\left(0.3 \; \frac{\text{ft}^3}{\text{sec}}\right)^2 - (367.4)\left(0.8 \; \frac{\text{ft}^3}{\text{sec}}\right)^2 \end{array} \right)}{(2)\left(\begin{array}{c} (306.0)\left(0.7 \; \frac{\text{ft}^3}{\text{sec}}\right) + (7.7)\left(0.4 \; \frac{\text{ft}^3}{\text{sec}}\right) \\ + (5.7)\left(0.3 \; \frac{\text{ft}^3}{\text{sec}}\right) + (367.4)\left(0.8 \; \frac{\text{ft}^3}{\text{sec}}\right) \end{array} \right)}$$

$$= 0.08 \; \text{ft}^3/\text{sec}$$

$$\delta_{\text{BCD}} = \frac{-\left(\begin{array}{c} (140.5)\left(0.3 \; \frac{\text{ft}^3}{\text{sec}}\right)^2 - (96.8)\left(0.7 \; \frac{\text{ft}^3}{\text{sec}}\right)^2 \\ - (7.7)\left(0.4 \; \frac{\text{ft}^3}{\text{sec}}\right)^2 \end{array} \right)}{(2)\left(\begin{array}{c} (140.5)\left(0.3 \; \frac{\text{ft}^3}{\text{sec}}\right) + (96.8)\left(0.7 \; \frac{\text{ft}^3}{\text{sec}}\right) \\ + (7.7)\left(0.4 \; \frac{\text{ft}^3}{\text{sec}}\right) \end{array} \right)}$$

$$= 0.16 \; \text{ft}^3/\text{sec}$$

step 8: The corrected flows are

$$\dot{V}_{\text{AB}} = 0.7 \; \frac{\text{ft}^3}{\text{sec}} + 0.08 \; \frac{\text{ft}^3}{\text{sec}} = 0.78 \; \text{ft}^3/\text{sec}$$

$$\dot{V}_{\text{BC}} = 0.3 \; \frac{\text{ft}^3}{\text{sec}} + 0.16 \; \frac{\text{ft}^3}{\text{sec}} = 0.46 \; \text{ft}^3/\text{sec}$$

$$\dot{V}_{\text{DC}} = 0.7 \; \frac{\text{ft}^3}{\text{sec}} - 0.16 \; \frac{\text{ft}^3}{\text{sec}} = 0.54 \; \text{ft}^3/\text{sec}$$

$$\dot{V}_{\text{BD}} = 0.4 \; \frac{\text{ft}^3}{\text{sec}} + 0.08 \; \frac{\text{ft}^3}{\text{sec}} - 0.16 \; \frac{\text{ft}^3}{\text{sec}}$$
$$= 0.32 \; \text{ft}^3/\text{sec}$$

$$\dot{V}_{\text{ED}} = 0.3 \; \frac{\text{ft}^3}{\text{sec}} - 0.08 \; \frac{\text{ft}^3}{\text{sec}} = 0.22 \; \text{ft}^3/\text{sec}$$

$$\dot{V}_{\text{AE}} = 0.8 \; \frac{\text{ft}^3}{\text{sec}} - 0.08 \; \frac{\text{ft}^3}{\text{sec}} = 0.72 \; \text{ft}^3/\text{sec}$$

32. FLOW MEASURING DEVICES

A device that measures flow can be calibrated to indicate either velocity or volumetric flow rate. There are many methods available to obtain the flow rate. Some are indirect, requiring the use of transducers and solid-state electronics, and others can be evaluated using the Bernoulli equation. Some are more appropriate for one variety of fluid than others, and some are limited to specific ranges of temperature and pressure.

Table 17.7 categorizes a few common flow measurement methods. Many other methods and variations thereof exist, particularly for specialized industries. Some of the methods listed are so basic that only a passing mention will be made of them. Others, particularly those that can be analyzed with energy and mass conservation laws, will be covered in greater detail in subsequent sections.

Table 17.7 Flow Measuring Devices

I   direct (primary) measurements
      positive-displacement meters
      volume tanks
      weight and mass scales
II indirect (secondary) measurements
      obstruction meters
          − flow nozzles
          − orifice plate meters
          − variable-area meters
          − venturi meters
      velocity probes
          − direction sensing probes
          − pitot-static meters
          − pitot tubes
          − static pressure probes
      miscellaneous methods
          − hot-wire meters
          − magnetic flow meters
          − mass flow meters
          − sonic flow meters
          − turbine and propeller meters

The utility meters used to measure gas and water usage are examples of *displacement meters*. Such devices are cyclical, fixed-volume devices with counters to record the numbers of cycles. Displacement devices are generally unpowered, drawing on only the pressure energy to overcome mechanical friction. Most configurations for positive-displacement pumps (e.g., reciprocating piston, helical screw, and nutating disk) have also been converted to measurement devices.

The venturi nozzle, orifice plate, and flow nozzle are examples of *obstruction meters*. These devices rely on a decrease in static pressure to measure the flow velocity. One disadvantage of these devices is that the pressure drop is proportional to the square of the velocity, limiting the range over which any particular device can be used.

Flow of Fluids

An obstruction meter that somewhat overcomes the velocity range limitation is the *variable-area meter*, also known as a *rotameter*, illustrated in Fig. 17.20.[20] This device consists of a float (which is actually more dense than the fluid) and a transparent sight tube. With proper design, the effects of fluid density and viscosity can be minimized. The sight glass can be directly calibrated in volumetric flow rate, or the height of the float above the zero position can be used in a volumetric calculation.

Figure 17.20 Variable-Area Rotameter

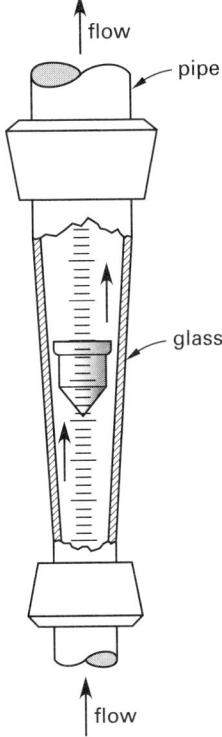

It is necessary to be able to measure static pressures in order to use obstruction meters and pitot-static tubes. In some cases, a *static pressure probe* is used. Figure 17.21 illustrates a simplified static pressure probe. In practice, such probes are sensitive to burrs and irregularities in the tap openings, orientation to the flow (i.e., *yaw*), and interaction with the pipe walls and other probes. A *direction-sensing probe* overcomes some of these problems.

A weather station *anemometer* used to measure wind velocity is an example of a simple *turbine meter*. Similar devices are used to measure the speed of a stream or river, in which case the name *current meter* may be used. Turbine meters are further divided into cup-type meters and propeller-type meters, depending on the orientation of the turbine axis relative to the flow direction. (The turbine axis and flow direction are

Figure 17.21 Simple Static Pressure Probe

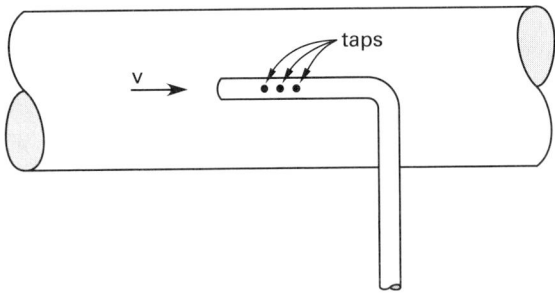

parallel for propeller-type meters; they are perpendicular for cup-type meters.) Since the wheel motion is proportional to the flow velocity, the velocity is determined by counting the number of revolutions made by the wheel per unit time.

More sophisticated turbine flowmeters use a reluctance-type pickup coil to detect wheel motion. The permeability of a magnetic circuit changes each time a wheel blade passes the pole of a permanent magnet in the meter body. This change is detected to indicate velocity or flow rate.

A *hot-wire anemometer* measures velocity by determining the cooling effect of fluid (usually a gas) flowing over an electrically heated tungsten, platinum, or nickel wire. Cooling is primarily by convection; radiation and conduction are neglected. Circuitry can be used either to keep the current constant (in which case, the changing resistance or voltage is measured) or to keep the temperature constant (in which case, the changing current is measured). Additional circuitry can be used to compensate for thermal lag if the velocity changes rapidly.

Modern *magnetic flowmeters* (*magmeter*, *electromagnetic flowmeter*) measure fluid velocity by detecting a voltage (potential difference, electromotive force) that is generated in response to the fluid passing through a magnetic field applied by the meter. The magnitude of the induced voltage is predicted by *Faraday's law of induction*, which states that the induced voltage is equal to the negative of the time rate of change of a magnetic field. The voltage is not affected by changes in fluid viscosity, temperature, or density. From Faraday's law, the induced voltage is proportional to the fluid velocity. For a magmeter with a dimensionless instrument constant, k, and a magnetic field strength, B, the induced voltage will be

$$V = kB\mathrm{v}D_{\mathrm{pipe}} \qquad 17.141$$

Since no parts of the meter extend into the flow, magmeters are ideal for corrosive fluids and slurries of large particles. Normally, the fluid has to be at least slightly electrically conductive, making magmeters ideal for measuring liquid metal flow. It may be necessary to dope electrically neutral fluids with precise quantities of conductive ions in order to obtain measurements by this method. Most magmeters have integral instrumentation and are direct-reading.

[20]The rotameter has its own disadvantages, however. It must be installed vertically; the fluid cannot be opaque; and it is more difficult to manufacture for use with high-temperature, high-pressure fluids.

In an *ultrasonic flowmeter*, two electric or magnetic transducers are placed a short distance apart on the outside of the pipe. One transducer serves as a transmitter of ultrasonic waves; the other transducer is a receiver. As an ultrasonic wave travels from the transmitter to the receiver, its velocity will be increased (or decreased) by the relative motion of the fluid. The phase shift between the fluid-carried waves and the waves passing through a stationary medium can be measured and converted to fluid velocity.

33. PITOT-STATIC GAUGE

Measurements from pitot tubes are used to determine total (stagnation) energy. Piezometer tubes and wall taps are used to measure static pressure energy. The difference between the total and static energies is the kinetic energy of the flow. Figure 17.22 illustrates a comparative method of directly measuring the velocity head for an incompressible fluid.

$$\frac{v^2}{2} = \frac{p_t - p_s}{\rho} = hg \qquad \text{[SI]} \quad 17.142(a)$$

$$\frac{v^2}{2g_c} = \frac{p_t - p_s}{\rho} = h \times \frac{g}{g_c} \qquad \text{[U.S.]} \quad 17.142(b)$$

$$v = \sqrt{2gh} \qquad\qquad 17.143$$

The pitot tube and static pressure tap shown in Fig. 17.22 can be combined into a *pitot-static gauge*. (See Fig. 17.23.) In a pitot-static gauge, one end of the manometer is acted upon by the static pressure (also referred to as the *transverse pressure*). The other end of the manometer experiences the total (stagnation or impact) pressure. The difference in elevations of the manometer fluid columns is the velocity head. This distance must be corrected if the density of the flowing fluid is significant.

$$\frac{v^2}{2} = \frac{p_t - p_s}{\rho} = \frac{h(\rho_m - \rho)g}{\rho} \qquad \text{[SI]} \quad 17.144(a)$$

$$\frac{v^2}{2g_c} = \frac{p_t - p_s}{\rho} = \frac{h(\rho_m - \rho)}{\rho} \times \frac{g}{g_c} \qquad \text{[U.S.]} \quad 17.144(b)$$

$$v = \sqrt{\frac{2gh(\rho_m - \rho)}{\rho}} \qquad\qquad 17.145$$

Another correction, which is seldom made, is to multiply the velocity calculated from Eq. 17.145 by C_I, the *coefficient of the instrument*. Since the flow past the pitot-static tube is slightly faster than the free-fluid velocity,

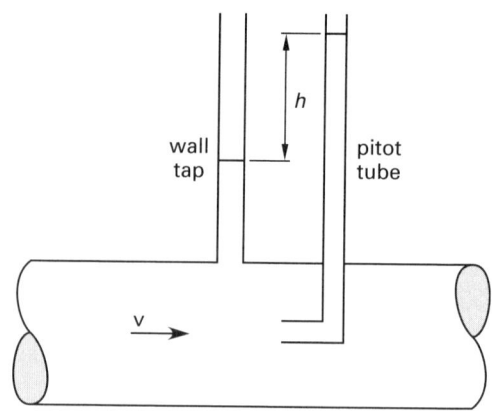

Figure 17.22 *Comparative Velocity Head Measurement*

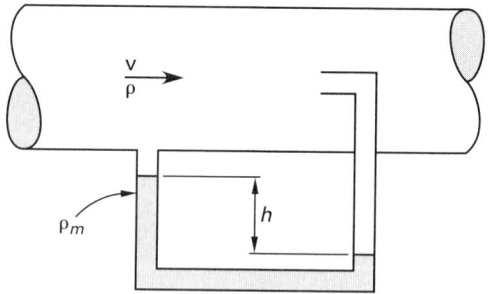

Figure 17.23 *Pitot-Static Gauge*

the static pressure measured will be slightly lower than the true value. This makes the indicated velocity slightly higher than the true value. C_I, a number close to but less than 1.0, corrects for this.

Pitot tube measurements are sensitive to the condition of the opening and errors in installation alignment. The *yaw angle* (i.e., the acute angle between the pitot tube axis and the flow streamline) should be zero.

A pitot-static tube indicates the velocity at only one point in a pipe. If the flow is laminar, and if the pitot-static tube is in the center of the pipe, v_{max} will be determined. The average velocity, however, will be only half the maximum value.

$$v = v_{max} = v_{ave} \quad \text{[turbulent]} \qquad 17.146$$

$$v = v_{max} = 2v_{ave} \quad \text{[laminar]} \qquad 17.147$$

Example 17.11

Water (62.4 lbm/ft^3; $\rho = 1000 \text{ kg/m}^3$) is flowing through a pipe. A pitot-static gauge with a mercury manometer registers 3 in (0.076 m) of mercury. What is the velocity of the water in the pipe?

SI Solution

The density of mercury is $\rho = 13\,580 \text{ kg/m}^3$. The velocity can be calculated directly from Eq. 17.145.

$$\text{v} = \sqrt{\frac{2gh(\rho_m - \rho)}{\rho}}$$

$$= \sqrt{\frac{(2)\left(9.81 \dfrac{\text{m}}{\text{s}^2}\right)(0.076 \text{ m})}{1000 \dfrac{\text{kg}}{\text{m}^3}} \times \left(13\,580 \dfrac{\text{kg}}{\text{m}^3} - 1000 \dfrac{\text{kg}}{\text{m}^3}\right)}$$

$$= 4.33 \text{ m/s}$$

Customary U.S. Solution

The density of mercury is $\rho = 848.6 \text{ lbm/ft}^3$.

From Eq. 17.145,

$$\text{v} = \sqrt{\frac{2gh(\rho_m - \rho)}{\rho}}$$

$$= \sqrt{\frac{(2)\left(32.2 \dfrac{\text{ft}}{\text{sec}^2}\right)(3 \text{ in}) \times \left(848.6 \dfrac{\text{lbm}}{\text{ft}^3} - 62.4 \dfrac{\text{lbm}}{\text{ft}^3}\right)}{\left(62.4 \dfrac{\text{lbm}}{\text{ft}^3}\right)\left(12 \dfrac{\text{in}}{\text{ft}}\right)}}$$

$$= 14.24 \text{ ft/sec}$$

34. VENTURI METER

Figure 17.24 illustrates a simple *venturi*. (Sometimes the venturi is called a *converging-diverging nozzle*.) This flow-measuring device can be inserted directly into a pipeline. Since the diameter changes are gradual, there is very little friction loss. Static pressure measurements are taken at the throat and upstream of the diameter change. These measurements are traditionally made by a manometer.

Figure 17.24 *Venturi Meter*

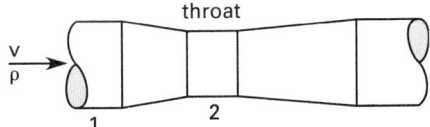

The analysis of *venturi meter* performance is relatively simple. The traditional derivation of upstream velocity starts by assuming a horizontal orientation and frictionless, incompressible, and turbulent flow. Then, the Bernoulli equation is written for points 1 and 2. Equation 17.148 shows that the static pressure decreases as the velocity increases. This is known as the *venturi effect*.

$$\frac{\text{v}_1^2}{2} + \frac{p_1}{\rho} = \frac{\text{v}_2^2}{2} + \frac{p_2}{\rho} \qquad \text{[SI]} \quad 17.148(a)$$

$$\frac{\text{v}_1^2}{2g_c} + \frac{p_1}{\rho} = \frac{\text{v}_2^2}{2g_c} + \frac{p_2}{\rho} \qquad \text{[U.S.]} \quad 17.148(b)$$

The two velocities are related by the continuity equation.

$$A_1 \text{v}_1 = A_2 \text{v}_2 \qquad\qquad 17.149$$

Combining Eq. 17.148 and Eq. 17.149 and eliminating the unknown v_1 produces an expression for the throat velocity. A *coefficient of velocity* is used to account for the small effect of friction. (C_v is very close to 1.0, usually 0.98 or 0.99.)

$$\text{v}_2 = C_\text{v}\text{v}_{2,\text{ideal}}$$

$$= \left(\frac{C_\text{v}}{\sqrt{1 - \left(\dfrac{A_2}{A_1}\right)^2}}\right)\sqrt{\frac{2(p_1 - p_2)}{\rho}} \qquad \text{[SI]} \quad 17.150(a)$$

$$\text{v}_2 = C_\text{v}\text{v}_{2,\text{ideal}}$$

$$= \left(\frac{C_\text{v}}{\sqrt{1 - \left(\dfrac{A_2}{A_1}\right)^2}}\right)\sqrt{\frac{2g_c(p_1 - p_2)}{\rho}} \quad \text{[U.S.]} \quad 17.150(b)$$

The *velocity of approach factor*, F_va, also known as the *meter constant*, is the reciprocal of the first term of the denominator of the first term of Eq. 17.150. The *beta ratio* can be incorporated into the formula for F_va.

$$\beta = \frac{D_2}{D_1} \qquad\qquad 17.151$$

$$F_\text{va} = \frac{1}{\sqrt{1 - \left(\dfrac{A_2}{A_1}\right)^2}} = \frac{1}{\sqrt{1 - \beta^4}} \qquad 17.152$$

If a manometer is used to measure the pressure difference directly, Eq. 17.150 can be rewritten in terms of the manometer fluid reading. (See Fig. 17.25.)

$$v_2 = C_v v_{2,\text{ideal}} = \left(\frac{C_v}{\sqrt{1-\beta^4}}\right)\sqrt{\frac{2g(\rho_m - \rho)h}{\rho}}$$

$$= C_v F_{va}\sqrt{\frac{2g(\rho_m - \rho)h}{\rho}} \qquad 17.153$$

Figure 17.25 *Venturi Meter with Manometer*

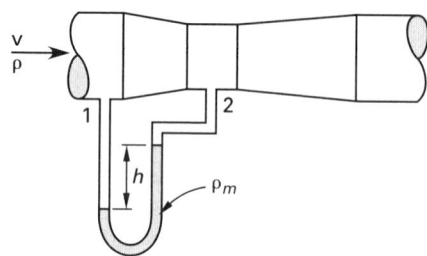

The flow rate through a venturi meter can be calculated from the throat area. There is an insignificant amount of contraction of the flow as it passes through the throat, and the *coefficient of contraction* is seldom encountered in venturi meter work. The *coefficient of discharge* ($C_d = C_c C_v$) is essentially the same as the coefficient of velocity. Values of C_d range from slightly less than 0.90 to over 0.99, depending on the Reynolds number. C_d is seldom less than 0.95 for turbulent flow. (See Fig. 17.26.)

$$\dot{V} = C_d A_2 v_{2,\text{ideal}} \qquad 17.154$$

Figure 17.26 *Typical Venturi Meter Discharge Coefficients (long radius venturi meter)*

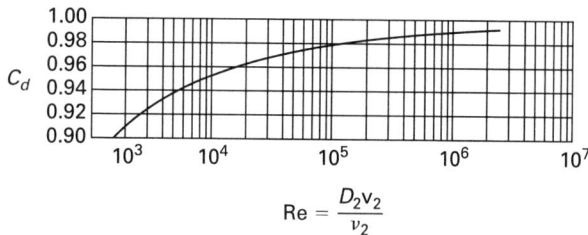

The product $C_d F_{va}$ is known as the *coefficient of flow* or *flow coefficient*, not to be confused with the coefficient of discharge.[21] This factor is used for convenience, since it combines the losses with the meter constant.

$$C_f = C_d F_{va} = \frac{C_d}{\sqrt{1-\beta^4}} \qquad 17.155$$

---

[21]Some writers use the symbol K for the flow coefficient.

$$\dot{V} = C_f A_2 \sqrt{\frac{2g(\rho_m - \rho)h}{\rho}} \qquad 17.156$$

35. ORIFICE METER

The *orifice meter* (or *orifice plate*) is used more frequently than the venturi meter to measure flow rates in small pipes. It consists of a thin or sharp-edged plate with a central, round hole through which the fluid flows. Such a plate is easily clamped between two flanges in an existing pipeline. (See Fig. 17.27.)

Figure 17.27 *Orifice Meter with Differential Manometer*

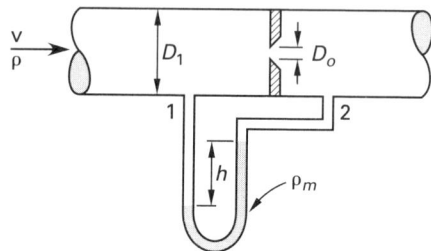

While (for small pipes) the orifice meter may consist of a thin plate without significant thickness, various types of bevels and rounded edges are also used with thicker plates. There is no significant difference in the analysis procedure between "flat plate," "sharp-edged," or "square-edged" orifice meters. Any effect that the orifice edges have is accounted for in the discharge and flow coefficient correlations. Similarly, the direction of the bevel will affect the coefficients but not the analysis method.

As with the venturi meter, pressure taps are used to obtain the static pressure upstream of the orifice plate and at the *vena contracta* (i.e., at the point of minimum pressure).[22] A differential manometer connected to the two taps conveniently indicates the difference in static pressures.

Although the orifice meter is simpler and less expensive than a venturi meter, its discharge coefficient is much less than that of a venturi meter. C_d usually ranges from 0.55 to 0.75, with values of 0.60 and 0.61 often being quoted. (The coefficient of contraction has a large effect,

---

[22]Calibration of the orifice meter is sensitive to tap placement. Upstream taps are placed between one-half and two pipe diameters upstream from the orifice. (An upstream distance of one pipe diameter is often quoted and used.) There are three tap-placement options: flange, vena contracta, and standardized. Flange taps are used with prefabricated orifice meters that are inserted (by flange bolting) in pipes. If the location of the vena contracta is known, a tap can be placed there. However, the location of the vena contracta depends on the diameter ratio $\beta = D_o/D$ and varies from approximately 0.4 to 0.7 pipe diameters downstream. Due to the difficulty of locating the vena contracta, the standardized $1D$-$1/2D$ configuration is often used. The upstream tap is one diameter before the orifice; the downstream tap is one-half diameter after the orifice. Since approaching flow should be stable and uniform, care must be taken not to install the orifice meter less than approximately five diameters after a bend or elbow.

since $C_d = C_v C_c$.) Also, its pressure recovery is poor (i.e., there is a permanent pressure reduction), and it is susceptible to inaccuracies from wear and abrasion.[23]

The derivation of the governing equations for an orifice meter is similar to that of the venturi meter. (The obvious falsity of assuming frictionless flow through the orifice is corrected by the coefficient of discharge.) The major difference is that the coefficient of contraction is taken into consideration in writing the mass continuity equation, since the pressure is measured at the vena contracta, not the orifice.

$$A_2 = C_c A_o \qquad 17.157$$

$$v_o = \left(\frac{C_d}{\sqrt{1 - \left(\frac{C_c A_o}{A_1}\right)^2}}\right)\sqrt{\frac{2(p_1 - p_2)}{\rho}} \qquad \text{[SI]} \quad 17.158(a)$$

$$v_o = \left(\frac{C_d}{\sqrt{1 - \left(\frac{C_c A_o}{A_1}\right)^2}}\right)\sqrt{\frac{2g_c(p_1 - p_2)}{\rho}} \qquad \text{[U.S.]} \quad 17.158(b)$$

If a manometer is used to indicate the differential pressure $p_1 - p_2$, the velocity at the vena contracta can be calculated from Eq. 17.159.

$$v_o = \left(\frac{C_d}{\sqrt{1 - \left(\frac{C_c A_o}{A_1}\right)^2}}\right)\sqrt{\frac{2g(\rho_m - \rho)h}{\rho}} \qquad 17.159$$

The *velocity of approach factor*, F_{va}, for an orifice meter is defined differently than for a venturi meter, since it takes into consideration the contraction of the flow. However, the velocity of approach factor is still combined with the coefficient of discharge into the flow coefficient, C_f. Figure 17.28 illustrates how the flow coefficient varies with the area ratio and the Reynolds number.

[23]The actual loss varies from 40% to 90% of the differential pressure. The loss depends on the diameter ratio $\beta = D_o/D_1$, and is not particularly sensitive to the Reynolds number for turbulent flow. For $\beta = 0.5$, the loss is 73% of the measured pressure difference, $p_1 - p_2$. This decreases to approximately 56% of the pressure difference when $\beta = 0.65$ and to 38%, when $\beta = 0.8$. For any diameter ratio, the pressure drop coefficient, K, in multiples of the orifice velocity head is $K = (1 - \beta^2)/C_f^2$.

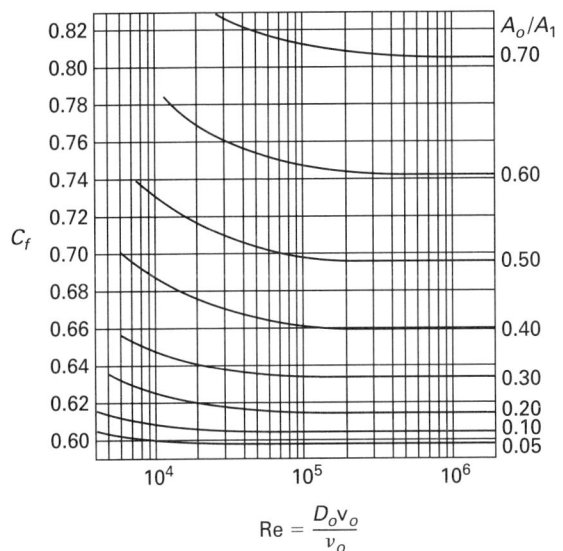

Figure 17.28 *Typical Flow Coefficients for Orifice Plates*

$$\text{Re} = \frac{D_o v_o}{v_o}$$

$$F_{va} = \frac{1}{\sqrt{1 - \left(\frac{C_c A_o}{A_1}\right)^2}} \qquad 17.160$$

$$C_f = C_d F_{va} \qquad 17.161$$

The flow rate through an orifice meter is given by Eq. 17.162.

$$\dot{V} = C_f A_o \sqrt{\frac{2g(\rho_m - \rho)h}{\rho}} = C_f A_o \sqrt{\frac{2(p_1 - p_2)}{\rho}} \qquad \text{[SI]} \quad 17.162(a)$$

$$\dot{V} = C_f A_o \sqrt{\frac{2g(\rho_m - \rho)h}{\rho}} = C_f A_o \sqrt{\frac{2g_c(p_1 - p_2)}{\rho}} \qquad \text{[U.S.]} \quad 17.162(b)$$

Example 17.12

150°F water ($\rho = 61.2$ lbm/ft^3) flows in an 8 in schedule-40 steel pipe at the rate of 2.23 ft^3/sec. A sharp-edged orifice with a 7 in diameter hole is placed in the line. A mercury differential manometer is used to record the pressure difference. If the orifice has a flow coefficient, C_f, of 0.62, what deflection in inches of mercury is observed? (Mercury has a density of 848.6 lbm/ft^3.)

Solution

The orifice area is

$$A_o = \frac{\pi D_o^2}{4} = \frac{\pi\left(\frac{7 \text{ in}}{12 \frac{\text{in}}{\text{ft}}}\right)^2}{4} = 0.2673 \text{ ft}^2$$

Equation 17.162 is solved for h.

$$h = \frac{\dot{V}^2 \rho}{2g C_f^2 A_o^2 (\rho_m - \rho)}$$

$$= \frac{\left(2.23 \ \frac{\text{ft}^3}{\text{sec}}\right)^2 \left(61.2 \ \frac{\text{lbm}}{\text{ft}^3}\right)\left(12 \ \frac{\text{in}}{\text{ft}}\right)}{(2)\left(32.2 \ \frac{\text{ft}}{\text{sec}^2}\right)(0.62)^2}$$

$$\times (0.2673 \ \text{ft}^2)^2 \left(848.6 \ \frac{\text{lbm}}{\text{ft}^3} - 61.2 \ \frac{\text{lbm}}{\text{ft}^3}\right)$$

$$= 2.62 \ \text{in}$$

36. FLOW NOZZLE

A typical flow nozzle is illustrated in Fig. 17.29. This device consists only of a converging section. It is somewhat between an orifice plate and a venturi meter in performance, possessing some of the advantages and disadvantages of each. The venturi performance equations can be used for the flow nozzle.

Figure 17.29 Flow Nozzle

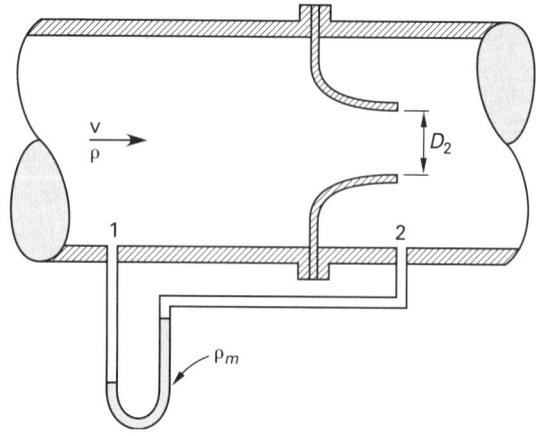

The geometry of the nozzle entrance is chosen to prevent separation of the fluid from the wall. The converging portion and the subsequent parallel section keep the coefficients of velocity and contraction close to 1.0. However, the absence of a diffuser section disrupts the orderly return of fluid to its original condition. The permanent pressure drop is more similar to that of the orifice meter than the venturi meter.

Since the nozzle geometry greatly affects the nozzle's performance, values of C_d and C_f have been established for only a limited number of specific proprietary nozzles.[24]

---

[24]Some of the proprietary nozzles for which detailed performance data exist are the ASME long-radius nozzle (low-β and high-β series) and the International Standards Association (ISA) nozzle (German standard nozzle).

37. FLOW MEASUREMENTS OF COMPRESSIBLE FLUIDS

Volume measurements of compressible fluids (i.e., gases) are not very meaningful. The volume of a gas will depend on its temperature and pressure. For that reason, flow quantities of gases discharged should be stated as mass flow rates.

$$\dot{m} = \rho_2 A_2 \text{v}_2 \qquad 17.163$$

Equation 17.163 requires that the velocity and area be measured at the same point. More importantly, the density must be measured at that point as well. However, it is common practice in flow measurement work to use the density of the upstream fluid at position 1. (This is not the stagnation density.)

The significant error introduced by this simplification is corrected by the use of an *expansion factor*, Y. For venturi meters and flow nozzles, values of the expansion factor are generally calculated theoretical values. Values of Y are determined experimentally for orifice plates. (See Fig. 17.30.)

$$\dot{m} = Y \rho_1 A_2 \text{v}_2 \qquad 17.164$$

Figure 17.30 Approximate Expansion Factors, k = 1.4 (air)

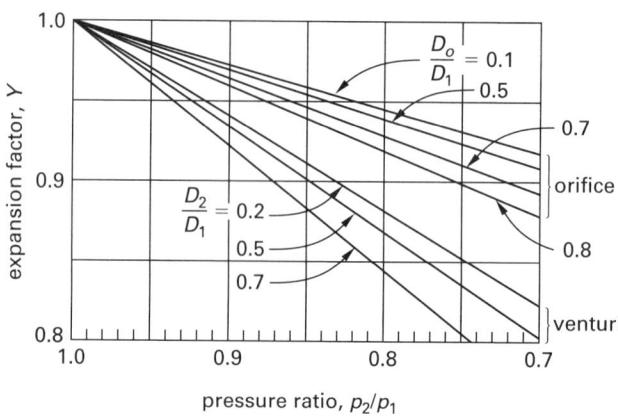

Derivation of the theoretical formula for the expansion factor for venturi meters and flow nozzles is based on thermodynamic principles and an assumption of adiabatic flow.

$$Y = \sqrt{\frac{(1 - \beta^4)\left(\left(\frac{p_2}{p_1}\right)^{2/k} - \left(\frac{p_2}{p_1}\right)^{(k+1)/k}\right)}{\left(\frac{k-1}{k}\right)\left(1 - \frac{p_2}{p_1}\right)\left(1 - \beta^4 \left(\frac{p_2}{p_1}\right)^{2/k}\right)}} \qquad 17.165$$

Once the expansion factor is known, it can be used with the standard flow rate, $\dot{V}$, equations for venturi meters, flow nozzles, and orifice meters. For example, for a venturi meter, the mass flow rate would be calculated from Eq. 17.166.

$$\dot{m} = Y\dot{m}_{\text{ideal}} = \left(\frac{Y\,C_d A_2}{\sqrt{1-\beta^4}}\right)\sqrt{2\rho_1(p_1 - p_2)}$$

$$\text{[SI]} \quad 17.166(a)$$

$$\dot{m} = Y\dot{m}_{\text{ideal}} = \left(\frac{Y\,C_d A_2}{\sqrt{1-\beta^4}}\right)\sqrt{2g_c\rho_1(p_1 - p_2)}$$

$$\text{[U.S.]} \quad 17.166(b)$$

38. IMPULSE-MOMENTUM PRINCIPLE

(The convention of this section is to make F and x positive when they are directed toward the right. F and y are positive when directed upward. Also, the fluid is assumed to flow horizontally from left to right, and it has no initial y-component of velocity.)

The *momentum*, $\mathbf{P}$ (also known as *linear momentum* to distinguish it from *angular momentum*, which is not considered here), of a moving object is a vector quantity defined as the product of the object's mass and velocity.[25]

$$\mathbf{P} = m\mathbf{v} \qquad \text{[SI]} \quad 17.167(a)$$

$$\mathbf{P} = \frac{m\mathbf{v}}{g_c} \qquad \text{[U.S.]} \quad 17.167(b)$$

The *impulse*, $\mathbf{I}$, of a constant force is calculated as the product of the force and the length of time the force is applied.

$$\mathbf{I} = \mathbf{F}\Delta t \qquad 17.168$$

The *impulse-momentum principle* (*law of conservation of momentum*) states that the impulse applied to a body is equal to the change in momentum. Equation 17.169 is one way of stating Newton's second law of motion.

$$\mathbf{I} = \Delta \mathbf{P} \qquad 17.169$$

$$F\Delta t = m\Delta\text{v} = m(\text{v}_2 - \text{v}_1) \qquad \text{[SI]} \quad 17.170(a)$$

$$F\Delta t = \frac{m\Delta\text{v}}{g_c} = \frac{m(\text{v}_2 - \text{v}_1)}{g_c} \qquad \text{[U.S.]} \quad 17.170(b)$$

---

[25]The symbol B is also used for momentum. In many texts, however, momentum is given no symbol at all.

For fluid flow, there is a mass flow rate, $\dot{m}$, but no mass per se. Since $\dot{m} = m/\Delta t$, the impulse-momentum equation can be rewritten as follows.

$$F = \dot{m}\Delta\text{v} \qquad \text{[SI]} \quad 17.171(a)$$

$$F = \frac{\dot{m}\Delta\text{v}}{g_c} \qquad \text{[U.S.]} \quad 17.171(b)$$

Equation 17.171 calculates the constant force required to accelerate or retard a fluid stream. This would occur when fluid enters a reduced or enlarged flow area. If the flow area decreases, for example, the fluid will be accelerated by a wall force up to the new velocity. Ultimately, this force must be resisted by the pipe supports.

As Eq. 17.171 illustrates, fluid momentum is not always conserved, since it is generated by the external force, F. Examples of external forces are gravity (considered zero for horizontal pipes), gage pressure, friction, and turning forces from walls and vanes. Only if these external forces are absent is fluid momentum conserved.

Since force is a vector, it can be resolved into its x- and y-components of force.

$$F_x = \dot{m}\Delta\text{v}_x \qquad \text{[SI]} \quad 17.172(a)$$

$$F_x = \frac{\dot{m}\Delta\text{v}_x}{g_c} \qquad \text{[U.S.]} \quad 17.172(b)$$

$$F_y = \dot{m}\Delta\text{v}_y \qquad \text{[SI]} \quad 17.173(a)$$

$$F_y = \frac{\dot{m}\Delta\text{v}_y}{g_c} \qquad \text{[U.S.]} \quad 17.173(b)$$

If the flow is initially at velocity v but is directed through an angle θ with respect to the original direction, the x- and y-components of velocity can be calculated from Eq. 17.174 and Eq. 17.175.

$$\Delta\text{v}_x = \text{v}(\cos\theta - 1) \qquad 17.174$$

$$\Delta\text{v}_y = \text{v}\sin\theta \qquad 17.175$$

Since F and v are vector quantities and Δt and m are scalars, F must have the same direction as $\text{v}_2 - \text{v}_1$. (See Fig. 17.31.) This provides an intuitive method of determining the direction in which the force acts. Essentially, one needs to ask, "In which direction must the force act in order to push the fluid stream into its new direction?" (The force, F, is the force on the fluid. The force on the pipe walls or pipe supports has the same magnitude but is opposite in direction.)

If a jet is completely stopped by a flat plate placed perpendicular to its flow, then $\theta = 90°$ and $\Delta\text{v}_x = -\text{v}$. If a jet is turned around so that it ends up returning to where it originated, then $\theta = 180°$ and $\Delta\text{v}_x = -2\text{v}$. A positive Δv indicates an increase in velocity. A negative Δv indicates a decrease in velocity.

Flow of Fluids

Flow of Fluids

Figure 17.31 Force on a Confined Fluid Stream

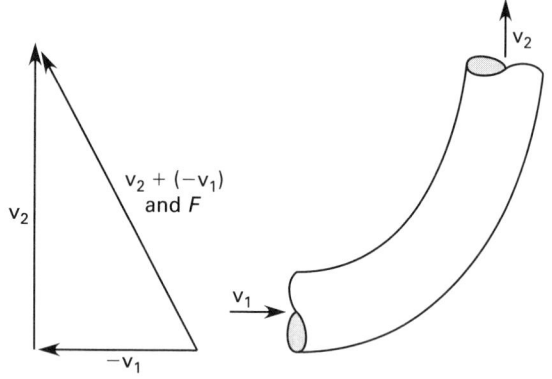

39. JET PROPULSION

A basic application of the impulse-momentum principle is the analysis of jet propulsion. Air enters a jet engine and is mixed with fuel. The air and fuel mixture is compressed and ignited, and the exhaust products leave the engine at a greater velocity than was possessed by the original air. The change in momentum of the air produces a force on the engine. (See Fig. 17.32.)

Figure 17.32 Jet Engine

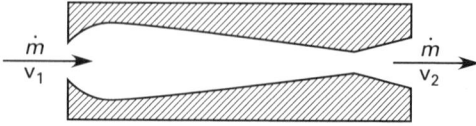

The governing equation for a jet engine is Eq. 17.176. The mass of the jet fuel is small compared with the air mass, and the fuel mass is commonly disregarded.

$$F_x = \dot{m}(v_2 - v_1) \qquad 17.176$$

$$F_x = \dot{V}_2\rho_2 v_2 - \dot{V}_1\rho_1 v_1 \qquad \text{[SI]} \quad 17.177(a)$$

$$F_x = \frac{\dot{V}_2\rho_2 v_2 - \dot{V}_1\rho_1 v_1}{g_c} \qquad \text{[U.S.]} \quad 17.177(b)$$

40. OPEN JET ON A VERTICAL FLAT PLATE

Figure 17.33 illustrates an open jet on a vertical flat plate. The fluid approaches the plate with no vertical component of velocity; it leaves the plate with no horizontal component of velocity. (This is another way of saying there is no splash-back.) Therefore, all of the velocity in the x-direction is canceled. (The minus sign in Eq. 17.178 indicates that the force is opposite the initial velocity direction.)

$$\Delta v = -v \qquad 17.178$$

Figure 17.33 Jet on a Vertical Plate

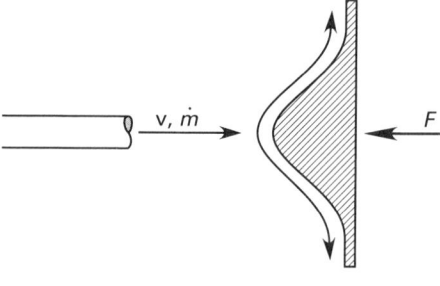

$$F_x = -\dot{m}v \qquad \text{[SI]} \quad 17.179(a)$$

$$F_x = \frac{-\dot{m}v}{g_c} \qquad \text{[U.S.]} \quad 17.179(b)$$

Since the flow is divided, half going up and half going down, the net velocity change in the y-direction is zero. There is no force in the y-direction on the fluid.

41. OPEN JET ON A HORIZONTAL FLAT PLATE

If a jet of fluid is directed upward, its velocity will decrease due to the effect of gravity. The force exerted on the fluid by the plate will depend on the fluid velocity at the plate surface, v_y, not the original jet velocity, v_o. All of this velocity is canceled. Since the flow divides evenly in both horizontal directions ($\Delta v_x = 0$), there is no force component in the x-direction. (See Fig. 17.34.)

$$v_y = \sqrt{v_o^2 - 2gh} \qquad 17.180$$

$$\Delta v_y = -\sqrt{v_o^2 - 2gh} \qquad 17.181$$

$$F_y = -\dot{m}\sqrt{v_o^2 - 2gh} \qquad \text{[SI]} \quad 17.182(a)$$

$$F_y = \frac{-\dot{m}\sqrt{v_o^2 - 2gh}}{g_c} \qquad \text{[U.S.]} \quad 17.182(b)$$

Figure 17.34 Open Jet on a Horizontal Plate

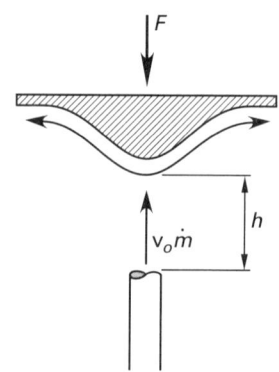

42. OPEN JET ON AN INCLINED PLATE

An open jet will be diverted both up and down (but not laterally) by a stationary, inclined plate, as shown in Fig. 17.35. In the absence of friction, the velocity in each diverted flow will be v, the same as in the approaching jet. The fractions f_1 and f_2 of the jet that are diverted up and down can be found from Eq. 17.183 through Eq. 17.186.

$$f_1 = \frac{1 + \cos\theta}{2} \qquad 17.183$$

$$f_2 = \frac{1 - \cos\theta}{2} \qquad 17.184$$

$$f_1 - f_2 = \cos\theta \qquad 17.185$$

$$f_1 + f_2 = 1.0 \qquad 17.186$$

Figure 17.35 Open Jet on an Inclined Plate

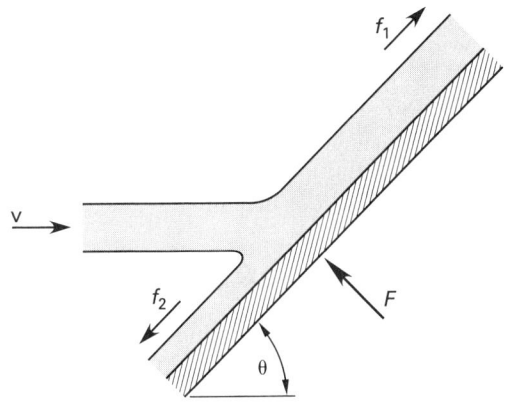

If the flow along the plate is frictionless, there will be no force component parallel to the plate. The force perpendicular to the plate is given by Eq. 17.187.

$$F = \dot{m}\text{v}\sin\theta \qquad \text{[SI]} \quad 17.187(a)$$

$$F = \frac{\dot{m}\text{v}\sin\theta}{g_c} \qquad \text{[U.S.]} \quad 17.187(b)$$

43. OPEN JET ON A SINGLE STATIONARY BLADE

Figure 17.36 illustrates a fluid jet being turned through an angle θ by a stationary blade (also called a *vane*). It is common to assume that $|\text{v}_2| = |\text{v}_1|$, although this will not be strictly true if friction between the blade and fluid is considered. Since the fluid is both retarded (in the x-direction) and accelerated (in the y-direction), there will be two components of force on the fluid.

$$\Delta\text{v}_x = \text{v}_2\cos\theta - \text{v}_1 \qquad 17.188$$

$$\Delta\text{v}_y = \text{v}_2\sin\theta \qquad 17.189$$

$$F_x = \dot{m}(\text{v}_2\cos\theta - \text{v}_1) \qquad \text{[SI]} \quad 17.190(a)$$

$$F_x = \frac{\dot{m}(\text{v}_2\cos\theta - \text{v}_1)}{g_c} \qquad \text{[U.S.]} \quad 17.190(b)$$

Figure 17.36 Open Jet on a Stationary Blade

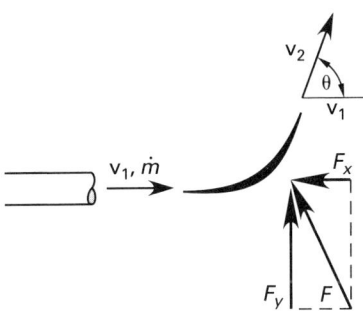

$$F_y = \dot{m}\text{v}_2\sin\theta \qquad \text{[SI]} \quad 17.191(a)$$

$$F_y = \frac{\dot{m}\text{v}_2\sin\theta}{g_c} \qquad \text{[U.S.]} \quad 17.191(b)$$

$$F = \sqrt{F_x^2 + F_y^2} \qquad 17.192$$

44. OPEN JET ON A SINGLE MOVING BLADE

If a blade is moving away at velocity v_b from the source of the fluid jet, only the *relative velocity difference* between the jet and blade produces a momentum change. Furthermore, not all of the fluid jet overtakes the moving blade. The equations used for the single stationary blade can be used by substituting $(\text{v} - \text{v}_b)$ for v and by using the effective mass flow rate, $\dot{m}_{\text{eff}}$. (See Fig. 17.37.)

$$\Delta\text{v}_x = (\text{v} - \text{v}_b)(\cos\theta - 1) \qquad 17.193$$

$$\Delta\text{v}_y = (\text{v} - \text{v}_b)\sin\theta \qquad 17.194$$

$$\dot{m}_{\text{eff}} = \left(\frac{\text{v} - \text{v}_b}{\text{v}}\right)\dot{m} \qquad 17.195$$

$$F_x = \dot{m}_{\text{eff}}(\text{v} - \text{v}_b)(\cos\theta - 1) \qquad \text{[SI]} \quad 17.196(a)$$

$$F_x = \frac{\dot{m}_{\text{eff}}(\text{v} - \text{v}_b)(\cos\theta - 1)}{g_c} \qquad \text{[U.S.]} \quad 17.196(b)$$

$$F_y = \dot{m}_{\text{eff}}(\text{v} - \text{v}_b)\sin\theta \qquad \text{[SI]} \quad 17.197(a)$$

$$F_y = \frac{\dot{m}_{\text{eff}}(\text{v} - \text{v}_b)\sin\theta}{g_c} \qquad \text{[U.S.]} \quad 17.197(b)$$

Figure 17.37 Open Jet on a Moving Blade

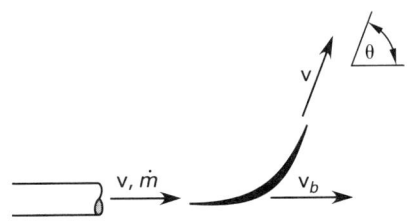

Flow of Fluids

45. OPEN JET ON A MULTIPLE-BLADED WHEEL

An *impulse turbine* consists of a series of blades (buckets or vanes) mounted around a wheel. (See Fig. 17.38.) The tangential velocity of the blades is approximately parallel to the jet. The effective mass flow rate, $\dot{m}_{eff}$, used in calculating the reaction force is the full discharge rate, since when one blade moves away from the jet, other blades will have moved into position. All of the fluid discharged is captured by the blades. Equation 17.196 and Eq. 17.197 are applicable if the total flow rate is used. The tangential blade velocity is

$$v_b = \frac{2\pi r n_{rpm}}{60} = \omega r \qquad 17.198$$

Figure 17.38 *Impulse Turbine*

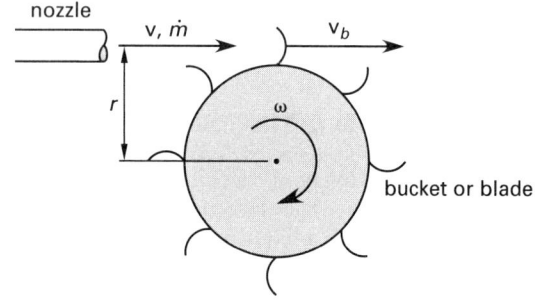

46. IMPULSE TURBINE POWER

The total power potential of a fluid jet can be calculated from the kinetic energy of the jet and the mass flow rate.[26] (This neglects the pressure energy, which is small by comparison.)

$$P_{jet} = \frac{\dot{m}v^2}{2} \qquad \text{[SI]} \quad 17.199(a)$$

$$P_{jet} = \frac{\dot{m}v^2}{2g_c} \qquad \text{[U.S.]} \quad 17.199(b)$$

The power transferred from a fluid jet to the blades of a turbine is calculated from the x-component of force on the blades. The y-component of force does no work.

$$P = F_x v_b \qquad 17.200$$

$$P = \dot{m}v_b(v - v_b)(1 - \cos\theta) \qquad \text{[SI]} \quad 17.201(a)$$

$$P = \frac{\dot{m}v_b(v - v_b)(1 - \cos\theta)}{g_c} \qquad \text{[U.S.]} \quad 17.201(b)$$

---

[26]The full jet discharge is used in this section. If only a single blade is involved, the effective mass flow rate, $\dot{m}_{eff}$, must be used.

The maximum theoretical blade velocity is the velocity of the jet: $v_b = v$. This is known as the *runaway speed* and can only be achieved when the turbine is unloaded. If Eq. 17.201 is maximized with respect to v_b, the maximum power will occur when the blade is traveling at half of the jet velocity: $v_b = v/2$. The power (force) is also affected by the deflection angle of the blade. Power is maximized when $\theta = 180°$. Figure 17.39 illustrates the relationship between power and the variables θ and v_b.

Figure 17.39 *Turbine Power*

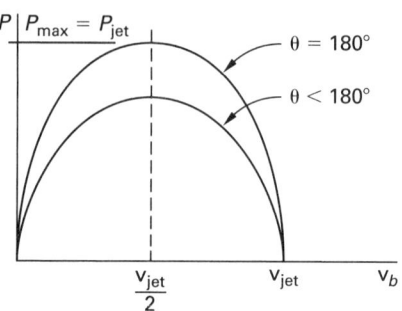

Putting $\theta = 180°$ and $v_b = v/2$ into Eq. 17.201 results in $P_{max} = \dot{m}v^2/2$, which is the same as P_{jet} in Eq. 17.199. If the machine is 100% efficient, 100% of the jet power can be transferred to the machine.

47. CONFINED STREAMS IN PIPE BENDS

As presented in Sec. 17.38, momentum can also be changed by pressure forces. Such is the case when fluid enters a pipe fitting or bend. (See Fig. 17.40.) Since the fluid is confined, the forces due to static pressure must be included in the analysis. (The effects of gravity and friction are neglected.)

$$F_x = p_2 A_2 \cos\theta - p_1 A_1 + \dot{m}(v_2 \cos\theta - v_1) \qquad \text{[SI]} \quad 17.202(a)$$

$$F_x = p_2 A_2 \cos\theta - p_1 A_1 + \frac{\dot{m}(v_2 \cos\theta - v_1)}{g_c} \qquad \text{[U.S.]} \quad 17.202(b)$$

$$F_y = (p_2 A_2 + \dot{m}v_2)\sin\theta \qquad \text{[SI]} \quad 17.203(a)$$

$$F_y = \left(p_2 A_2 + \frac{\dot{m}v_2}{g_c}\right)\sin\theta \qquad \text{[U.S.]} \quad 17.203(b)$$

Figure 17.40 *Pipe Bend*

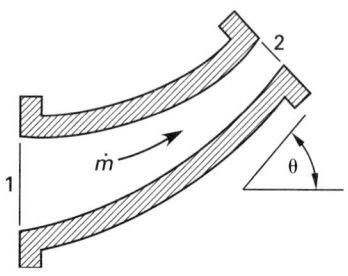

Example 17.13

60°F water ($\rho = 62.4$ lbm/ft^3) at 40 psig enters a 12 in × 8 in reducing elbow at 8 ft/sec and is turned through an angle of 30°. Water leaves 26 in higher in elevation. (a) What is the resultant force exerted on the water by the elbow? (b) What other forces should be considered in the design of supports for the fitting?

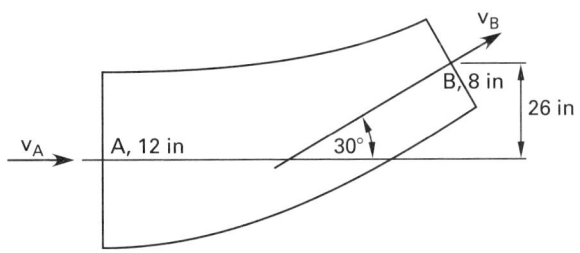

Solution

(a) The velocity and pressure at point B are both needed. The velocity is easily calculated from the continuity equation.

$$A_A = \frac{\pi D_A^2}{4} = \frac{\pi\left(\dfrac{12\text{ in}}{12\,\dfrac{\text{in}}{\text{ft}}}\right)^2}{4}$$

$$= 0.7854\text{ ft}^2$$

$$A_B = \frac{\pi D_B^2}{4} = \frac{\pi\left(\dfrac{8\text{ in}}{12\,\dfrac{\text{in}}{\text{ft}}}\right)^2}{4}$$

$$= 0.3491\text{ ft}^2$$

$$v_B = \frac{v_A A_A}{A_B}$$

$$= \left(8\,\frac{\text{ft}}{\text{sec}}\right)\left(\frac{0.7854\text{ ft}^2}{0.3491\text{ ft}^2}\right)$$

$$= 18\text{ ft/sec}$$

$$p_A = \left(40\,\frac{\text{lbf}}{\text{in}^2}\right)\left(12\,\frac{\text{in}}{\text{ft}}\right)^2$$

$$= 5760\text{ lbf/ft}^2$$

The Bernoulli equation is used to calculate p_B. (Gage pressures are used for this calculation. Absolute pressures

could also be used, but the addition of p_{atm}/ρ to both sides of the Bernoulli equation would not affect p_B.)

$$\frac{p_A}{\rho} + \frac{v_A^2}{2g_c} = \frac{p_B}{\rho} + \frac{v_B^2}{2g_c} + \frac{zg}{g_c}$$

$$\frac{5760\,\dfrac{\text{lbf}}{\text{ft}^2}}{62.4\,\dfrac{\text{lbm}}{\text{ft}^3}} + \frac{\left(8\,\dfrac{\text{ft}}{\text{sec}}\right)^2}{(2)\left(32.2\,\dfrac{\text{lbm-ft}}{\text{lbf-sec}^2}\right)}$$

$$= \frac{p_B}{62.4\,\dfrac{\text{lbm}}{\text{ft}^3}} + \frac{\left(18\,\dfrac{\text{ft}}{\text{sec}}\right)^2}{(2)\left(32.2\,\dfrac{\text{lbm-ft}}{\text{lbf-sec}^2}\right)}$$

$$+ \frac{26\text{ in}}{12\,\dfrac{\text{in}}{\text{ft}}} \times \frac{32.2\,\dfrac{\text{ft}}{\text{sec}^2}}{32.2\,\dfrac{\text{lbm-ft}}{\text{lbf-sec}^2}}$$

$$p_B = 5373\text{ lbf/ft}^2$$

The mass flow rate is

$$\dot{m} = \dot{V}\rho = v A \rho = \left(8\,\frac{\text{ft}}{\text{sec}}\right)(0.7854\text{ ft}^2)\left(62.4\,\frac{\text{lbm}}{\text{ft}^3}\right)$$

$$= 392.1\text{ lbm/sec}$$

From Eq. 17.202,

$$F_x = p_2 A_2 \cos\theta - p_1 A_1 + \frac{\dot{m}(v_2\cos\theta - v_1)}{g_c}$$

$$= \left(5373\,\frac{\text{lbf}}{\text{ft}^2}\right)(0.3491\text{ ft}^2)(\cos 30°)$$

$$- \left(5760\,\frac{\text{lbf}}{\text{ft}^2}\right)(0.7854\text{ ft}^2)$$

$$+ \frac{\left(392.1\,\dfrac{\text{lbm}}{\text{sec}}\right)\left(\left(18\,\dfrac{\text{ft}}{\text{sec}}\right)(\cos 30°) - 8\,\dfrac{\text{ft}}{\text{sec}}\right)}{32.2\,\dfrac{\text{lbm-ft}}{\text{lbf-sec}^2}}$$

$$= -2807\text{ lbf}$$

From Eq. 17.203,

$$F_y = \left(p_2 A_2 + \frac{\dot{m}v_2}{g_c}\right)\sin\theta$$

$$= \left(\frac{\left(5373\,\dfrac{\text{lbf}}{\text{ft}^2}\right)(0.3491\text{ ft}^2)}{+ \dfrac{\left(392.1\,\dfrac{\text{lbm}}{\text{sec}}\right)\left(18\,\dfrac{\text{ft}}{\text{sec}}\right)}{32.2\,\dfrac{\text{lbm-ft}}{\text{lbf-sec}^2}}}\right)\sin 30°$$

$$= 1047\text{ lbf}$$

The resultant force on the water is

$$R = \sqrt{F_x^2 + F_y^2} = \sqrt{(-2807 \text{ lbf})^2 + (1047 \text{ lbf})^2}$$
$$= 2996 \text{ lbf}$$

(b) In addition to counteracting the resultant force, R, the support should be designed to carry the weight of the elbow and the water in it. Also, the support must carry a part of the pipe and water weight tributary to the elbow.

48. WATER HAMMER

Water hammer in a long pipe is an increase in fluid pressure caused by a sudden velocity decrease. (See Fig. 17.41.) The sudden velocity decrease will usually be caused by a valve closing. Analysis of the water hammer phenomenon can take two approaches, depending on whether or not the pipe material is assumed to be elastic.

Figure 17.41 Water Hammer

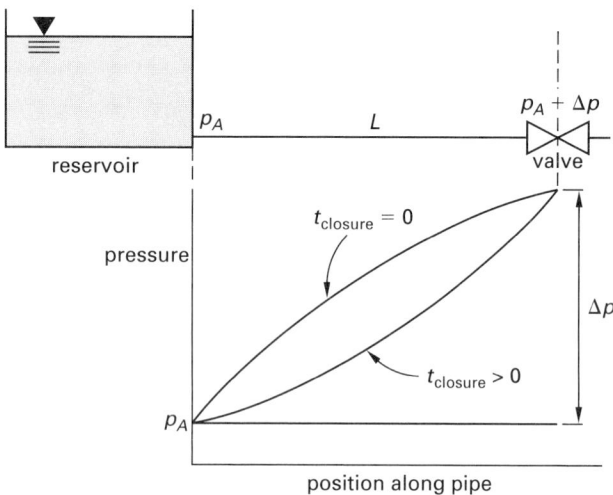

The *water hammer wave speed*, a, is given by Eq. 17.204. The first term on the right-hand side represents the effect of fluid compressibility, and the second term represents the effect of pipe elasticity.

$$\frac{1}{a^2} = \frac{d\rho}{dp} + \frac{\rho}{A}\frac{dA}{dp} \qquad 17.204$$

For a compressible fluid within an inelastic (rigid) pipe, $dA/dp = 0$, and the wave speed is simply the speed of sound in the fluid.

$$a = \sqrt{\frac{dp}{d\rho}} = \sqrt{\frac{E}{\rho}} \quad \begin{bmatrix} \text{inelastic pipe;} \\ \text{consistent units} \end{bmatrix} \qquad 17.205$$

If the pipe material is assumed to be inelastic (such as cast iron, ductile iron, and concrete), the time required

for the water hammer shock wave to travel from the suddenly closed valve to a point of interest depends only on the velocity of sound in the fluid, a, and the distance, L, between the two points. This is also the time required to bring all of the fluid in the pipe to rest.

$$t = \frac{L}{a} \qquad 17.206$$

When the water hammer shock wave reaches the original source of water, the pressure wave will dissipate. A rarefaction wave (at the pressure of the water source) will return at velocity a to the valve. The time for the compression shock wave to travel to the source and the rarefaction wave to return to the valve is given by Eq. 17.207. This is also the length of time that the pressure is constant at the valve.

$$t = \frac{2L}{a} \qquad 17.207$$

The fluid pressure increase resulting from the shock wave is calculated by equating the kinetic energy change of the fluid with the average pressure during the compression process. The pressure increase is independent of the length of pipe. If the velocity is decreased by an amount Δv instantaneously, the increase in pressure will be

$$\Delta p = \rho a \Delta v \qquad [\text{SI}] \quad 17.208(a)$$

$$\Delta p = \frac{\rho a \Delta v}{g_c} \qquad [\text{U.S.}] \quad 17.208(b)$$

It is interesting that the pressure increase at the valve depends on Δv but not on the actual length of time it takes to close the valve, as long as the valve is closed when the wave returns to it. Therefore, there is no difference in pressure buildups at the valve for an "instantaneous closure," "rapid closure," or "sudden closure."[27] It is only necessary for the closure to occur rapidly.

Water hammer does not necessarily occur just because a return shockwave increases the pressure. When the velocity is low—less than 7 ft/sec (2.1 m/s), with 5 ft/sec (1.5 m/s) recommended—there isn't enough force generated to create water hammer. The cost of using larger pipe, fittings, and valves is the disadvantage of keeping velocity low.

Having a very long pipe is equivalent to assuming an instantaneous closure. When the pipe is long, the time for the shock wave to travel round-trip is much longer than the time to close the valve. The valve will be closed when the rarefaction wave returns to the valve.

If the pipe is short, it will be difficult to close the valve before the rarefaction wave returns to the valve. With a short pipe, the pressure buildup will be less than is predicted by Eq. 17.208. (Having a short pipe is equivalent to the case of "slow closure.") The actual pressure

---

[27]The pressure elsewhere along the pipe, however, will be lower for slow closures than for instantaneous closures.

history is complex, and no simple method exists for calculating the pressure buildup in short pipes.

Installing a *surge tank, accumulator, slow-closing valve* (e.g., a gate valve), or *pressure-relief valve* in the line will protect against water hammer damage. (See Fig. 17.42.) The surge tank (or *surge chamber*) is an open tank or reservoir. Since the water is unconfined, large pressure buildups do not occur. An accumulator is a closed tank that is partially filled with air. Since the air is much more compressible than the water, it will be compressed by the water hammer shock wave. The energy of the shock wave is dissipated when the air is compressed.

Figure 17.42 *Water Hammer Protective Devices*

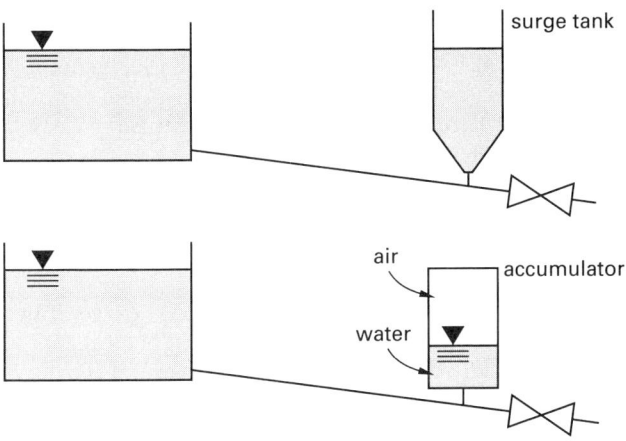

For an incompressible fluid within an elastic pipe, $d\rho/dp$ in Eq. 17.205 is zero. In that case, the wave speed is

$$a = \sqrt{\frac{A}{\rho}\frac{dp}{dA}} \quad \begin{bmatrix} \text{elastic pipe;} \\ \text{consistent units} \end{bmatrix} \qquad 17.209$$

For long elastic pipes (the common assumption for steel and plastic), the degree of pipe anchoring and longitudinal stress propagation affect the wave velocity. Longitudinal stresses resulting from circumferential stresses (i.e., Poisson's effect) must be considered by including a Poisson's effect factor, c_P. In Eq. 17.210, for pipes anchored at their upstream ends only, $c_P = 1 - 0.5\nu_{\text{pipe}}$; for pipes anchored throughout, $c_P = 1 - \nu_{\text{pipe}}^2$; for anchored pipes with expansion joints throughout, $c_P = 1$ (i.e., the commonly used *Korteweg formula*). Pipe sections joined with gasketed integral bell ends and gasketed ends satisfy the requirement for expansion joints. Equation 17.210 shows that the wave velocity can be reduced by increasing the pipe diameter.

$$a = \sqrt{\frac{\dfrac{E_{\text{fluid}}t_{\text{pipe}}E_{\text{pipe}}}{t_{\text{pipe}}E_{\text{pipe}} + c_P D_{\text{pipe}}E_{\text{fluid}}}}{\rho}} \qquad 17.210$$

At room temperature, the modulus of elasticity of ductile steel is approximately 2.9×10^7 lbf/in^2 (200 GPa); for ductile cast iron, 2.2–2.5×10^7 lbf/in^2 (150–170 GPa); for PVC, 3.5–4.1×10^5 lbf/in^2 (2.4–2.8 GPa); for ABS, 3.2–3.5×10^5 lbf/in^2 (2.2–2.4 GPa).

Example 17.14

Water ($\rho = 1000$ kg/m^3, $E = 2 \times 10^9$ Pa), is flowing at 4 m/s through a long length of 4 in schedule-40 steel pipe ($D_i = 0.102$ m, $t = 0.00602$ m, $E = 2 \times 10^{11}$ Pa) with expansion joints when a valve suddenly closes completely. What is the theoretical increase in pressure?

Solution

For pipes with expansion joints, $c_P = 1$. From Eq. 17.210, the modulus of elasticity to be used in calculating the speed of sound is

$$
\begin{aligned}
a &= \sqrt{\frac{\dfrac{E_{\text{fluid}}t_{\text{pipe}}E_{\text{pipe}}}{t_{\text{pipe}}E_{\text{pipe}} + c_P D_{\text{pipe}}E_{\text{fluid}}}}{\rho}} \\[2mm]
&= \sqrt{\frac{\dfrac{(2 \times 10^9 \text{ Pa})(0.00602 \text{ m})(2 \times 10^{11} \text{ Pa})}{\begin{array}{c}(0.00602 \text{ m})(2 \times 10^{11} \text{ Pa}) \\ + (1)(0.102 \text{ m})(2 \times 10^9 \text{ Pa})\end{array}}}{1000 \ \dfrac{\text{kg}}{\text{m}^3}}} \\[2mm]
&= 1307.75 \text{ m/s}
\end{aligned}
$$

From Eq. 17.208, the pressure increase is

$$
\begin{aligned}
\Delta p &= \rho a \Delta v = \left(1000 \ \frac{\text{kg}}{\text{m}^3}\right)\left(1307.75 \ \frac{\text{m}}{\text{s}}\right)\left(4 \ \frac{\text{m}}{\text{s}}\right) \\
&= 5.23 \times 10^6 \text{ Pa}
\end{aligned}
$$

49. LIFT

Lift is an upward force that is exerted on an object (flat plate, airfoil, rotating cylinder, etc.) as the object passes through a fluid. Lift combines with drag to form the resultant force on the object, as shown in Fig. 17.43.

Figure 17.43 *Lift and Drag on an Airfoil*

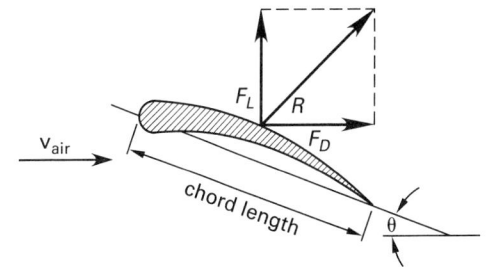

Flow of Fluids

The generation of lift from air flowing over an airfoil is predicted by Bernoulli's equation. Air molecules must travel a longer distance over the top surface of the airfoil than over the lower surface, and, therefore, they travel faster over the top surface. Since the total energy of the air is constant, the increase in kinetic energy comes at the expense of pressure energy. The static pressure on the top of the airfoil is reduced, and a net upward force is produced.

Within practical limits, the lift produced can be increased at lower speeds by increasing the curvature of the wing. This increased curvature is achieved by the use of *flaps*. (See Fig. 17.44.) When a plane is traveling slowly (e.g., during takeoff or landing), its flaps are extended to create the lift needed.

Figure 17.44 *Use of Flaps in an Airfoil*

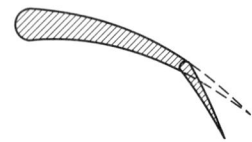

The lift produced can be calculated from Eq. 17.211, whose use is not limited to airfoils.

$$F_L = \frac{C_L A \rho \mathrm{v}^2}{2} \qquad \text{[SI]} \quad \textit{17.211(a)}$$

$$F_L = \frac{C_L A \rho \mathrm{v}^2}{2 g_c} \qquad \text{[U.S.]} \quad \textit{17.211(b)}$$

The dimensions of an airfoil or wing are frequently given in terms of chord length and aspect ratio. The *chord length* (or just *chord*) is the front-to-back dimension of the airfoil. The *aspect ratio* is the ratio of the *span* (wing length) to chord length. The area, A, in Eq. 17.211 is the airfoil's area projected onto the plane of the chord.

$$A = \text{chord length} \times \text{span} \quad \left[\begin{array}{c}\text{rectangular}\\\text{airfoil}\end{array}\right] \quad \textit{17.212}$$

The dimensionless *coefficient of lift*, C_L, modifies the effectiveness of the airfoil. The coefficient of lift depends on the shape of the airfoil, the Reynolds number, and the angle of attack. The theoretical coefficient of lift for a thin flat plate in two-dimensional flow at a low angle of attack, θ, is given by Eq. 17.213. Actual airfoils are able to achieve only 80–90% of this theoretical value.

$$C_L = 2\pi \sin \theta \quad \text{[flat plate]} \quad \textit{17.213}$$

The coefficient of lift for an airfoil cannot be increased without limit merely by increasing θ. Eventually, the *stall angle* is reached, at which point the coefficient of lift decreases dramatically. (See Fig. 17.45.)

Figure 17.45 *Typical Plot of Lift Coefficient*

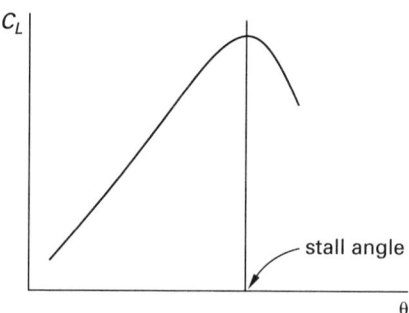

50. CIRCULATION

A theoretical concept for calculating the lift generated by an object (an airfoil, propeller, turbine blade, etc.) is *circulation*. Circulation, Γ, is defined by Eq. 17.214.[28] Its units are length²/time.

$$\Gamma = \oint \mathrm{v} \cos \theta \, dL \qquad \textit{17.214}$$

Figure 17.46 illustrates an arbitrary closed curve drawn around a point (or body) in steady flow. The tangential components of velocity, v, at all points around the curve are $\mathrm{v}\cos\theta$. It is a fundamental theorem that circulation has the same value for every closed curve that can be drawn around a body.

Figure 17.46 *Circulation Around a Point*

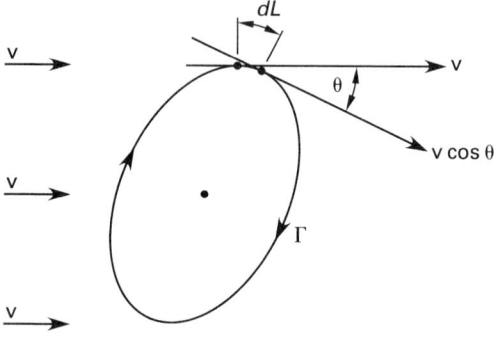

Lift on a body traveling with relative velocity v through a fluid of density ρ can be calculated from the circulation by using Eq. 17.215.[29]

$$F_L = \rho \mathrm{v} \Gamma \times \text{chord length} \qquad \textit{17.215}$$

---

[28]Equation 17.214 is analogous to the calculation of work being done by a constant force moving around a curve. If the force makes an angle of θ with the direction of motion, the work done as the force moves a distance dl around a curve is $W = \oint F \cos\theta \, dl$. In calculating circulation, velocity takes the place of force.

[29]U is the traditional symbol of velocity in circulation studies.

There is no actual circulation of air "around" an airfoil, but this mathematical concept can be used, nevertheless. However, since the flow of air "around" an airfoil is not symmetrical in path or velocity, experimental determination of C_L is favored over theoretical calculations of circulation.

51. LIFT FROM ROTATING CYLINDERS

When a cylinder is placed transversely to a relative airflow of velocity v_∞, the velocity at a point on the surface of the cylinder is $2v_\infty \sin\theta$. Since the flow is symmetrical, however, no lift is produced. (See Fig. 17.47.)

Figure 17.47 Flow Over a Cylinder

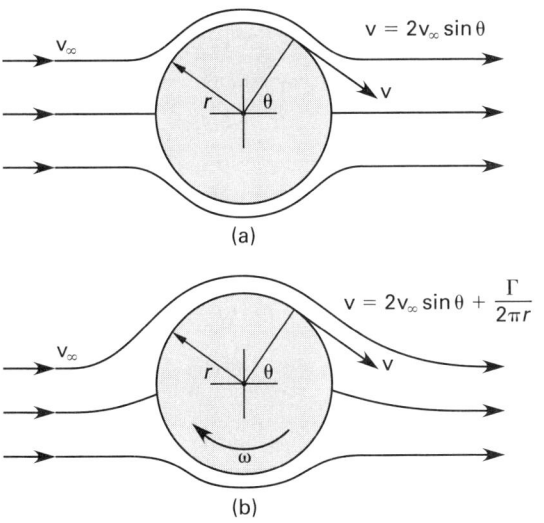

(a)

(b)

If the cylinder with radius r and length L rotates at ω rad/sec while moving with a relative velocity v_∞ through the air, the *Kutta-Joukowsky result* (*theorem*) can be used to calculate the lift per unit length of cylinder.[30] This is known as the *Magnus effect*.

$$\frac{F_L}{L} = \rho v_\infty \Gamma \qquad \text{[SI]} \quad 17.216(a)$$

$$\frac{F_L}{L} = \frac{\rho v_\infty \Gamma}{g_c} \qquad \text{[U.S.]} \quad 17.216(b)$$

$$\Gamma = 2\pi r^2 \omega \qquad 17.217$$

Equation 17.216 assumes that there is no slip (i.e., that the air drawn around the cylinder by rotation moves at ω), and in that ideal case, the maximum coefficient of lift is 4π. Practical rotating devices, however, seldom achieve a coefficient of lift in excess of 9 or 10, and even then, the power expenditure to keep the cylinder rotating is excessive.

---

[30]A similar analysis can be used to explain why a pitched baseball curves. The rotation of the ball produces a force that changes the path of the ball as it travels.

52. DRAG

Drag is a frictional force that acts parallel but opposite to the direction of motion. The total drag force is made up of *skin friction* and *pressure drag* (also known as *form drag*). These components, in turn, can be subdivided and categorized into *wake drag*, *induced drag*, and *profile drag*. (See Fig. 17.48.)

Figure 17.48 Components of Total Drag

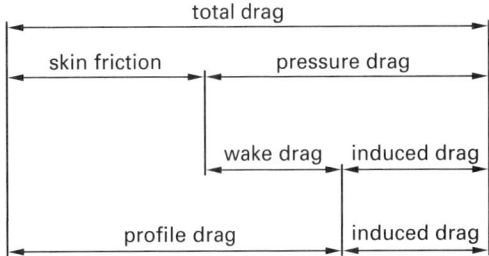

Most aeronautical engineering books contain descriptions of these drag terms. However, the difference between the situations where either skin friction drag or pressure drag predominates is illustrated in Fig. 17.49.

Figure 17.49 Extreme Cases of Pressure Drag and Skin Friction

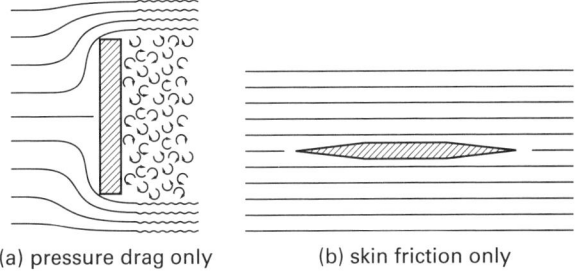

(a) pressure drag only   (b) skin friction only

Total drag is most easily calculated from the dimensionless *drag coefficient*, C_D. The drag coefficient depends only on the Reynolds number.

$$F_D = \frac{C_D A \rho v^2}{2} \qquad \text{[SI]} \quad 17.218(a)$$

$$F_D = \frac{C_D A \rho v^2}{2g_c} \qquad \text{[U.S.]} \quad 17.218(b)$$

In most cases, the area, A, in Eq. 17.218 is the projected area (i.e., the *frontal area*) normal to the stream. This is appropriate for spheres, cylinders, and automobiles. In a few cases (e.g., for airfoils and flat plates) as determined by how C_D was derived, the area is a projection of the object onto a plane parallel to the stream.

Typical drag coefficients for production cars vary from approximately 0.25 to approximately 0.70, with most modern cars being nearer the lower end. By comparison, other low-speed drag coefficients are approximately 0.05 (aircraft wing), 0.10 (sphere in turbulent flow), and 1.2 (flat plate). (See Table 17.8.)

Table 17.8 *Typical Ranges of Vehicle Drag Coefficients*

vehicle	low	medium	high
experimental race	0.17	0.21	0.23
sports	0.27	0.31	0.38
performance	0.32	0.34	0.38
60s muscle car	0.38	0.44	0.50
sedan	0.34	0.39	0.50
motorcycle	0.50	0.90	1.00
truck	0.60	0.90	1.00
tractor-trailer	0.60	0.77	1.20

Aero horsepower is a term used by automobile manufacturers to designate the power required to move a car horizontally at 50 mi/hr (80.5 km/h) against the drag force. Aero horsepower varies from approximately 7 hp (5.2 kW) for a streamlined subcompact car (approximate drag coefficient 0.35) to approximately 100 hp (75 kW) for a box-shaped truck (approximate drag coefficient 0.9).

In aerodynamic studies performed in *wind tunnels*, accuracy in lift and drag measurement is specified in *drag counts*. One drag count is equal to a C_D (or, C_L) of 0.0001. Drag counts are commonly used when describing the effect of some change to a fuselage or wing geometry. For example, an increase in C_D from 0.0450 to 0.0467 would be reported at an increase of 17 counts. Some of the best wind tunnels can measure C_D down to 1 count; most can only get to about 5 counts. Outside of the wind tunnel, an accuracy of 1 count is the target in computational fluid dynamics.

53. DRAG ON SPHERES AND DISKS

The drag coefficient varies linearly with the Reynolds number for laminar flow around a sphere or disk. In laminar flow, the drag is almost entirely due to skin friction. For Reynolds numbers below approximately 0.4, experiments have shown that the drag coefficient can be calculated from Eq. 17.219.[31] In calculating the Reynolds number, the sphere or disk diameter should be used as the characteristic dimension.

$$C_D = \frac{24}{\text{Re}} \qquad 17.219$$

Substituting this value of C_D into Eq. 17.218 results in *Stokes' law* (see Eq. 17.220), which is applicable to laminar slow motion (ascent or descent) of spherical particles and bubbles through a fluid. Stokes' law is based on the assumptions that (a) flow is laminar, (b) Newton's law of

viscosity is valid, and (c) all higher-order velocity terms (v^2, etc.) are negligible.

$$F_D = 6\pi\mu v D \qquad 17.220$$

The drag coefficients for disks and spheres operating outside the region covered by Stokes' law have been determined experimentally. In the turbulent region, pressure drag is predominant. Figure 17.50 can be used to obtain approximate values for C_D.

Figure 17.50 *Drag Coefficients for Spheres and Circular Flat Disks*

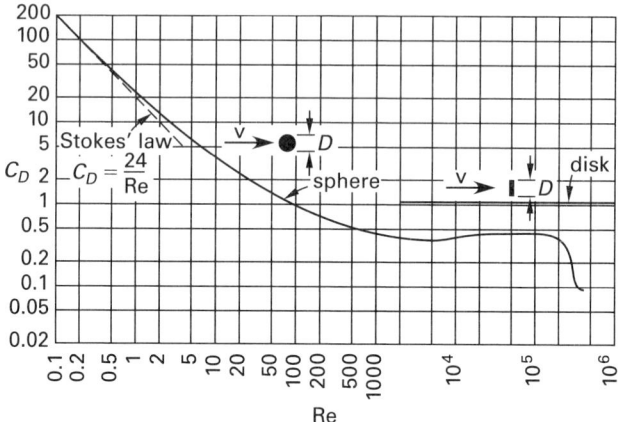

Figure 17.50 shows that there is a dramatic drop in the drag coefficient around Re = 10^5. The explanation for this is that the point of separation of the boundary layer shifts, decreasing the width of the wake. (See Fig. 17.51.) Since the drag force is primarily pressure drag at higher Reynolds numbers, a reduction in the wake reduces the pressure drag. Therefore, anything that can be done to a sphere (scuffing or wetting a baseball, dimpling a golf ball, etc.) to induce a smaller wake will reduce the drag. There can be no shift in the boundary layer separation point for a thin disk, since the disk has no depth in the direction of flow. Therefore, the drag coefficient for a thin disk remains the same at all turbulent Reynolds numbers.

Figure 17.51 *Turbulent Flow Around a Sphere at Various Reynolds Numbers*

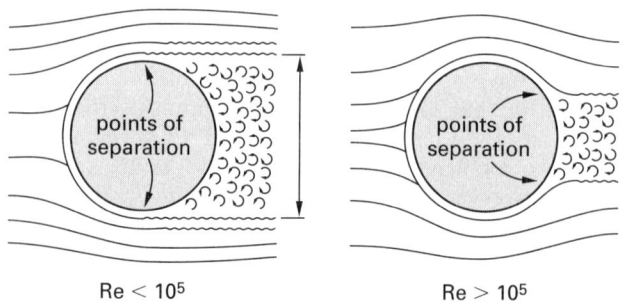

Re < 10^5 Re > 10^5

---

[31]Some sources report that the region in which Stokes' law applies extends to Re = 1.0.

54. TERMINAL VELOCITY

The velocity of an object falling through a fluid will continue to increase until the drag force equals the net downward force (i.e., the weight less the buoyant force). The maximum velocity attained is known as the *terminal velocity* (*settling velocity*).

$$F_D = mg - F_b \quad \begin{bmatrix} \text{at terminal} \\ \text{velocity} \end{bmatrix} \quad \text{[SI]} \quad 17.221(a)$$

$$F_D = \frac{mg}{g_c} - F_b \quad \begin{bmatrix} \text{at terminal} \\ \text{velocity} \end{bmatrix} \quad \text{[U.S.]} \quad 17.221(b)$$

If the drag coefficient is known, the terminal velocity can be calculated from Eq. 17.222. For small, heavy objects falling in air, the buoyant force (represented by the ρ_{fluid} term) can be neglected.[32]

$$v_{\text{terminal}} = \sqrt{\frac{2(mg - F_b)}{C_D A \rho_{\text{fluid}}}} = \sqrt{\frac{2 V g (\rho_{\text{object}} - \rho_{\text{fluid}})}{C_D A \rho_{\text{fluid}}}}$$

$$17.222$$

For a sphere of diameter D, the terminal velocity is

$$v_{\text{terminal,sphere}} = \sqrt{\frac{4 D g (\rho_{\text{sphere}} - \rho_{\text{fluid}})}{3 C_D \rho_{\text{fluid}}}} \quad 17.223$$

If the spherical particle is very small, Stokes' law may apply. In that case, the terminal velocity can be calculated from Eq. 17.224.

$$v_{\text{terminal}} = \frac{D^2 g (\rho_{\text{sphere}} - \rho_{\text{fluid}})}{18 \mu} \quad \text{[laminar]}$$

$$\text{[SI]} \quad 17.224(a)$$

$$v_{\text{terminal}} = \frac{D^2 (\rho_{\text{sphere}} - \rho_{\text{fluid}})}{18 \mu} \times \frac{g}{g_c} \quad \text{[laminar]}$$

$$\text{[U.S.]} \quad 17.224(b)$$

55. NONSPHERICAL PARTICLES

Few particles are actually spherical. One method of overcoming the complexity of dealing with the flow of real particles is to correlate performance with *sphericity*, ψ. (See Fig. 17.52.) For a particle and a sphere with the same volume, the sphericity is defined by Eq. 17.225 as the ratio of surface areas. Sphericity is always less than or equal to 1.0. The *equivalent diameter* is the diameter of a sphere having the same volume as the particle.

$$\psi = \frac{A_{\text{sphere}}}{A_{\text{particle}}} \quad 17.225$$

---

[32] A skydiver's terminal velocity can vary from approximately 120 mi/hr (54 m/s) in horizontal configuration to 200 mi/hr (90 m/s) in vertical configuration.

Figure 17.52 Drag Coefficients for Nonspherical Particles

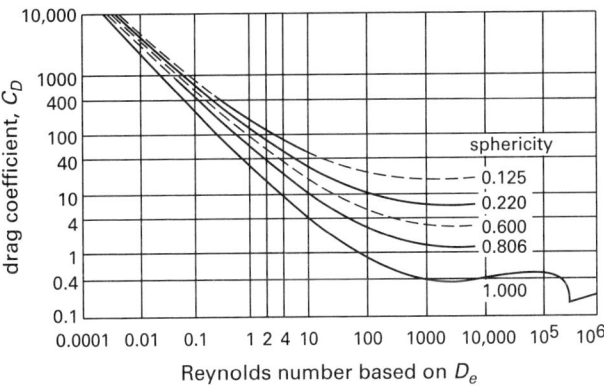

Used with permission from *Unit Operations* by George Granger Brown, et al., John Wiley & Sons, Inc., © 1950.

56. FLOW AROUND A CYLINDER

The characteristic drag coefficient plot for cylinders placed normal to the fluid flow is similar to the plot for spheres. The plot shown in Fig. 17.53 is for infinitely long cylinders, since there is additional wake drag at the cylinder ends. In calculating the Reynolds number, the cylinder diameter should be used as the characteristic dimension.

Figure 17.53 Drag Coefficient for Cylinders

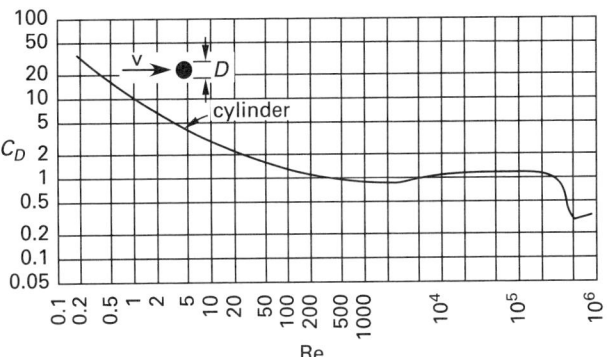

Example 17.15

A 50 ft (15 m) high flagpole is manufactured from a uniformly smooth cylinder 10 in (25 cm) in diameter. The surrounding air is at 40°F (0°C) and 14.6 psia (100 kPa). What is the total drag force on the flagpole in a 30 mi/hr (50 km/h) gust? (Neglect variations in the wind speed with height above the ground.)

SI Solution

At 0°C, the absolute viscosity of air is

$$\mu_{0°C} = 1.709 \times 10^{-5} \text{ Pa·s}$$

The density of the air is

$$\rho = \frac{p}{RT} = \frac{(100 \text{ kPa})\left(1000 \frac{\text{Pa}}{\text{kPa}}\right)}{\left(287.03 \frac{\text{J}}{\text{kg·K}}\right)(0°\text{C} + 273°)}$$

$$= 1.276 \text{ kg/m}^3$$

The kinematic viscosity is

$$\nu = \frac{\mu}{\rho} = \frac{1.709 \times 10^{-5} \text{ Pa·s}}{1.276 \frac{\text{kg}}{\text{m}^3}}$$

$$= 1.339 \times 10^{-5} \text{ m}^2/\text{s}$$

The wind speed is

$$v = \frac{\left(50 \frac{\text{km}}{\text{h}}\right)\left(1000 \frac{\text{m}}{\text{km}}\right)}{3600 \frac{\text{s}}{\text{h}}} = 13.89 \text{ m/s}$$

The characteristic dimension of a cylinder is its diameter. The Reynolds number is

$$\text{Re} = \frac{Lv}{\nu} = \frac{Dv}{\nu} = \frac{\left(\frac{25 \text{ cm}}{100 \frac{\text{cm}}{\text{m}}}\right)\left(13.89 \frac{\text{m}}{\text{s}}\right)}{1.339 \times 10^{-5} \frac{\text{m}^2}{\text{s}}}$$

$$= 2.59 \times 10^5$$

From Fig. 17.53, the drag coefficient is approximately 1.2. The frontal area of the flagpole is

$$A = DL = \frac{(25 \text{ cm})(15 \text{ m})}{100 \frac{\text{cm}}{\text{m}}}$$

$$= 3.75 \text{ m}^2$$

From Eq. 17.218(a), the drag on the flagpole is

$$F_D = \frac{C_D A \rho v^2}{2}$$

$$= \frac{(1.2)(3.75 \text{ m}^2)\left(1.276 \frac{\text{kg}}{\text{m}^3}\right)\left(13.89 \frac{\text{m}}{\text{s}}\right)^2}{2}$$

$$= 554 \text{ N}$$

Customary U.S. Solution

At 40°F, the absolute viscosity of air is

$$\mu_{40°\text{F}} = 3.62 \times 10^{-7} \text{ lbf-sec/ft}^2$$

The density of the air is

$$\rho = \frac{p}{RT} = \frac{\left(14.6 \frac{\text{lbf}}{\text{in}^2}\right)\left(12 \frac{\text{in}}{\text{ft}}\right)^2}{\left(53.35 \frac{\text{ft-lbf}}{\text{lbm-}°\text{R}}\right)(40°\text{F} + 460°)}$$

$$= 0.0788 \text{ lbm/ft}^3$$

The kinematic viscosity is

$$\nu = \frac{\mu g_c}{\rho} = \frac{\left(3.62 \times 10^{-7} \frac{\text{lbf-sec}}{\text{ft}^2}\right)\left(32.2 \frac{\text{lbm-ft}}{\text{lbf-sec}^2}\right)}{0.0788 \frac{\text{lbm}}{\text{ft}^3}}$$

$$= 1.479 \times 10^{-4} \text{ ft}^2/\text{sec}$$

The wind speed is

$$v = \frac{\left(30 \frac{\text{mi}}{\text{hr}}\right)\left(5280 \frac{\text{ft}}{\text{mi}}\right)}{3600 \frac{\text{sec}}{\text{hr}}} = 44 \text{ ft/sec}$$

The characteristic dimension of a cylinder is its diameter. The Reynolds number is

$$\text{Re} = \frac{Lv}{\nu} = \frac{Dv}{\nu} = \frac{\left(\frac{10 \text{ in}}{12 \frac{\text{in}}{\text{ft}}}\right)\left(44 \frac{\text{ft}}{\text{sec}}\right)}{1.479 \times 10^{-4} \frac{\text{ft}^2}{\text{sec}}}$$

$$= 2.48 \times 10^5$$

From Fig. 17.53, the drag coefficient is approximately 1.2. The frontal area of the flagpole is

$$A = DL = \frac{(10 \text{ in})(50 \text{ ft})}{12 \frac{\text{in}}{\text{ft}}} = 41.67 \text{ ft}^2$$

From Eq. 17.218(b), the drag on the flagpole is

$$F_D = \frac{C_D A \rho v^2}{2g_c}$$

$$= \frac{(1.2)(41.67 \text{ ft}^2)\left(0.0788 \frac{\text{lbm}}{\text{ft}^3}\right)\left(44 \frac{\text{ft}}{\text{sec}}\right)^2}{(2)\left(32.2 \frac{\text{lbm-ft}}{\text{lbf-sec}^2}\right)}$$

$$= 118.5 \text{ lbf}$$

57. FLOW OVER A PARALLEL FLAT PLATE

The drag experienced by a flat plate oriented parallel to the direction of flow is almost totally skin friction drag. (See Fig. 17.54.) Prandtl's *boundary layer theory* can be used to evaluate the frictional effects. Such an analysis predicts that the shape of the boundary layer profile is a function of the Reynolds number. The characteristic dimension used in calculating the Reynolds number is the chord length, L (i.e., the dimension of the plate parallel to the flow).

Figure 17.54 *Flow Over a Parallel Flat Plate (one side)*

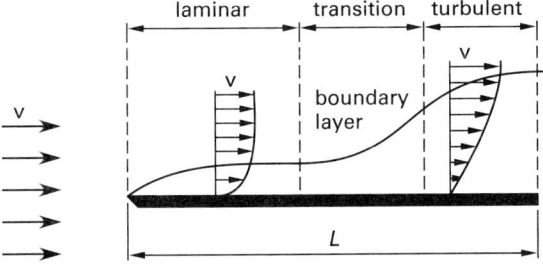

When skin friction predominates, it is common to use the symbol C_f (i.e., the *skin friction coefficient*) for the drag coefficient. For laminar flow over a smooth, flat plate, the drag coefficient based on the boundary layer theory is given by Eq. 17.226, which is known as the *Blasius solution*.[33] The critical Reynolds number for laminar flow over a flat plate is often reported to be 530,000. However, the transition region between laminar flow and turbulent flow actually occupies the Reynolds number range of 100,000–1,000,000.

$$C_f = \frac{1.328}{\sqrt{\text{Re}}} \quad \text{[laminar flow]} \qquad 17.226$$

Prandtl reported that the skin friction coefficient for turbulent flow is

$$C_f = \frac{0.455}{(\log \text{Re})^{2.58}} \quad \text{[turbulent flow]} \qquad 17.227$$

The drag force is calculated from Eq. 17.228. The factor 2 appears because there is friction on two sides of the flat plate. The area, A, is

$$F_D = 2\left(\frac{C_f A \rho \text{v}^2}{2}\right) = C_f A \rho \text{v}^2 \quad \text{[SI]} \quad 17.228(a)$$

$$F_D = 2\left(\frac{C_f A \rho \text{v}^2}{2g_c}\right) = \frac{C_f A \rho \text{v}^2}{g_c} \quad \text{[U.S.]} \quad 17.228(b)$$

58. SIMILARITY

Similarity considerations between a *model* (subscript m) and a full-sized object (subscript p, for *prototype*) imply that the model can be used to predict the performance of the prototype. Such a model is said to be *mechanically similar* to the prototype.

Complete *mechanical similarity* requires both geometric and dynamic similarity.[34] *Geometric similarity* means that the model is true to scale in length, area, and volume. The *model scale* (*length ratio*) is defined as

$$L_r = \frac{\text{size of model}}{\text{size of prototype}} \qquad 17.229$$

The area and volume ratios are based on the model scale.

$$\frac{A_m}{A_p} = L_r^2 \qquad 17.230$$

$$\frac{V_m}{V_p} = L_r^3 \qquad 17.231$$

Dynamic similarity means that the ratios of all types of forces are equal for the model and the prototype. These forces result from inertia, gravity, viscosity, elasticity (i.e., fluid compressibility), surface tension, and pressure.

The number of possible ratios of forces is large. For example, the ratios of viscosity/inertia, inertia/gravity, and inertia/surface tension are only three of the ratios of forces that must match for every corresponding point on the model and prototype. Fortunately, some force ratios can be neglected because the forces are negligible or are self-canceling.

In some cases, the geometric scale may be deliberately distorted. For example, with scale models of harbors and rivers, the water depth might be only a fraction of an inch if the scale is maintained precisely. Not only will the surface tension be excessive, but the shifting of the harbor or river bed may not be properly observed. Therefore, the vertical scale is chosen to be different from the horizontal scale. Such models are known as *distorted models*. Experience is needed to interpret observations made of distorted models.

The following sections deal with the most common similarity problems, but not all. For example, similarity of steady laminar flow in a horizontal pipe requires the Stokes number, similarity of high-speed (near-sonic) aircraft requires the Mach number, and similarity of capillary rise in a tube requires the *Eötvös number* ($g\rho L^2/\sigma$).

59. VISCOUS AND INERTIAL FORCES DOMINATE

For completely submerged objects, such as the items listed in Table 17.9, surface tension is negligible. The fluid can be assumed to be incompressible for low velocity. Gravity does not change the path of the fluid particles significantly during the passage of the object.

Only inertial, viscous, and pressure forces are significant. Because they are the only forces that are acting, these three forces are in equilibrium. Since they are in equilibrium, knowing any two forces will define the third

---

[33]Other correlations substitute the coefficient 1.44, the original value calculated by Prandtl, for 1.328 in Eq. 17.226. The Blasius solution is considered to be more accurate.

[34]Complete mechanical similarity also requires kinematic and thermal similarity, which are not discussed in this book.

Table 17.9 *Applications of Reynolds Number Similarity*

aircraft (subsonic)
airfoils (subsonic)
closed-pipe flow (turbulent)
drainage through tank orifices
fans
flow meters
open channel flow (without wave action)
pumps
submarines
torpedoes
turbines

force. This third force is dependent and can be omitted from the similarity analysis. For submerged objects, pressure is traditionally chosen as the dependent force.

The dimensionless ratio of the inertial forces to the viscous forces is the Reynolds number. Equating the model's and prototype's Reynolds numbers will ensure similarity.[35]

$$\text{Re}_m = \text{Re}_p \qquad 17.232$$

$$\frac{L_m \text{v}_m}{\nu_m} = \frac{L_p \text{v}_p}{\nu_p} \qquad 17.233$$

If the model is tested in the same fluid and at the same temperature in which the prototype is expected to operate, setting the Reynolds numbers equal is equivalent to setting $L_m \text{v}_m = L_p \text{v}_p$.

Example 17.16

A $^1/_{30}$ size scale model of a helicopter fuselage is tested in a wind tunnel at 120 mi/hr (190 km/h). The conditions in the wind tunnel are 50 psia and 100°F (350 kPa and 50°C). What is the corresponding speed of a prototype traveling in 14 psia and 40°F still air (100 kPa, 0°C)?

SI Solution

The absolute viscosity of air at atmospheric pressure is

$$\mu_{p,0°C} = 1.709 \times 10^{-5} \text{ Pa·s}$$

$$\mu_{m,50°C} = 1.951 \times 10^{-5} \text{ Pa·s}$$

The densities of air at the two conditions are

$$\rho_p = \frac{p}{RT} = \frac{(100 \text{ kPa})\left(1000 \frac{\text{Pa}}{\text{kPa}}\right)}{\left(287.03 \frac{\text{J}}{\text{kg·K}}\right)(0°C + 273°)}$$

$$= 1.276 \text{ kg/m}^3$$

$$\rho_m = \frac{p}{RT} = \frac{(350 \text{ kPa})\left(1000 \frac{\text{Pa}}{\text{kPa}}\right)}{\left(287.03 \frac{\text{J}}{\text{kg·K}}\right)(50°C + 273°)}$$

$$= 3.775 \text{ kg/m}^3$$

The kinematic viscosities are

$$\nu_p = \frac{\mu}{\rho} = \frac{1.709 \times 10^{-5} \text{ Pa·s}}{1.276 \frac{\text{kg}}{\text{m}^3}}$$

$$= 1.339 \times 10^{-5} \text{ m}^2/\text{s}$$

$$\nu_m = \frac{\mu}{\rho} = \frac{1.951 \times 10^{-5} \text{ Pa·s}}{3.775 \frac{\text{kg}}{\text{m}^3}}$$

$$= 5.168 \times 10^{-6} \text{ m}^2/\text{s}$$

From Eq. 17.233,

$$\text{v}_p = \text{v}_m \left(\frac{L_m}{L_p}\right)\left(\frac{\nu_p}{\nu_m}\right)$$

$$= \frac{\left(190 \frac{\text{km}}{\text{h}}\right)\left(\frac{1}{30}\right)\left(1.339 \times 10^{-5} \frac{\text{m}^2}{\text{s}}\right)}{5.168 \times 10^{-6} \frac{\text{m}^2}{\text{s}}}$$

$$= 16.4 \text{ km/h}$$

Customary U.S. Solution

Since surface tension and gravitational forces on the air particles are insignificant and since the flow velocities are low, viscous and inertial forces dominate. The Reynolds numbers of the model and prototype are equated.

The kinematic viscosity of air must be evaluated at the respective temperatures and pressures. Absolute viscosity is essentially independent of pressure.

$$\mu_{p,40°F} = 3.62 \times 10^{-7} \text{ lbf-sec/ft}^2$$

$$\mu_{m,100°F} = 3.96 \times 10^{-7} \text{ lbf-sec/ft}^2$$

The densities of air at the two conditions are

$$\rho_p = \frac{p}{RT} = \frac{\left(14 \frac{\text{lbf}}{\text{in}^2}\right)\left(12 \frac{\text{in}}{\text{ft}}\right)^2}{\left(53.35 \frac{\text{ft-lbf}}{\text{lbm-°R}}\right)(40°F + 460°)}$$

$$= 0.0756 \text{ lbm/ft}^3$$

$$\rho_m = \frac{p}{RT} = \frac{\left(50 \frac{\text{lbf}}{\text{in}^2}\right)\left(12 \frac{\text{in}}{\text{ft}}\right)^2}{\left(53.35 \frac{\text{ft-lbf}}{\text{lbm-°R}}\right)(100°F + 460°)}$$

$$= 0.241 \text{ lbm/ft}^2$$

---

[35]An implied assumption is that the drag coefficients are the same for the model and the prototype. In the case of pipe flow, it is assumed that flow will be in the turbulent region with the same relative roughness.

The kinematic viscosities are

$$\nu_p = \frac{\mu g_c}{\rho} = \frac{\left(3.62 \times 10^{-7}\ \frac{\text{lbf-sec}}{\text{ft}^2}\right) g_c}{0.0756\ \frac{\text{lbm}}{\text{ft}^3}}$$

$$= (4.79 \times 10^{-6}\ \text{ft}^2/\text{sec}) g_c$$

$$\nu_m = \frac{\mu g_c}{\rho} = \frac{\left(3.96 \times 10^{-7}\ \frac{\text{lbf-sec}}{\text{ft}^3}\right) g_c}{0.241\ \frac{\text{lbm}}{\text{ft}^3}}$$

$$= (1.64 \times 10^{-6}\ \text{ft}^2/\text{sec}) g_c$$

From Eq. 17.233,

$$\text{v}_p = \text{v}_m \left(\frac{L_m}{L_p}\right)\left(\frac{\nu_p}{\nu_m}\right)$$

$$= \frac{\left(120\ \frac{\text{mi}}{\text{hr}}\right)\left(\frac{1}{30}\right)\left(4.79 \times 10^{-6}\ \frac{\text{ft}^2}{\text{sec}}\right) g_c}{\left(1.64 \times 10^{-6}\ \frac{\text{ft}^2}{\text{sec}}\right) g_c}$$

$$= 11.7\ \text{mi/hr}$$

60. INERTIAL AND GRAVITATIONAL FORCES DOMINATE

Table 17.10 lists the cases when elasticity and surface tension forces can be neglected but gravitational forces cannot. Omitting these two forces from the similarity calculations leaves pressure, inertia, viscosity, and gravity forces, which are in equilibrium. Pressure is chosen as the dependent force and is omitted from the analysis.

There are only two possible combinations of the remaining three forces. The ratio of inertial to viscous forces is the Reynolds number. The ratio of the inertial forces to the gravitational forces is the *Froude number*, Fr.[36] The Froude number is used when gravitational forces are significant, such as in wave motion produced by a ship or seaplane hull.

$$\text{Fr} = \frac{\text{v}^2}{Lg} \qquad 17.234$$

Similarity is ensured when Eq. 17.235 and Eq. 17.236 are satisfied.

$$\text{Re}_m = \text{Re}_p \qquad 17.235$$

$$\text{Fr}_m = \text{Fr}_p \qquad 17.236$$

Table 17.10 Applications of Froude Number Similarity

bow waves from ships
flow over spillways
flow over weirs
motion of a fluid jet
open channel flow with varying surface levels
oscillatory wave action
seaplane hulls
surface ships
surface wave action
surge and flood waves

As an alternative, Eq. 17.235 and Eq. 17.236 can be solved simultaneously. This results in the following requirement for similarity, which indicates that it is necessary to test the model in a manufactured liquid with a specific viscosity.

$$\frac{\nu_m}{\nu_p} = \left(\frac{L_m}{L_p}\right)^{3/2} = L_r^{3/2} \qquad 17.237$$

Sometimes it is not possible to satisfy Eq. 17.235 and Eq. 17.236. This occurs when a model fluid viscosity is called for that is not available. If only one of the equations is satisfied, the model is said to be *partially similar*. In such a case, corrections based on other factors are used.

Another problem with trying to achieve similarity in open channel flow problems is the need to scale surface drag. It can be shown that the ratio of Manning's roughness constants is given by Eq. 17.238. In some cases, it may not be possible to create a surface smooth enough to satisfy this requirement.

$$n_r = L_r^{1/6} \qquad 17.238$$

61. SURFACE TENSION FORCE DOMINATES

Table 17.11 lists some of the cases where surface tension is the predominant force. Such cases can be handled by equating the Weber numbers, We, of the model and prototype.[37] The *Weber number* is the ratio of inertial force to surface tension.

$$\text{We} = \frac{\text{v}^2 L \rho}{\sigma} \qquad 17.239$$

$$\text{We}_m = \text{We}_p \qquad 17.240$$

Table 17.11 Applications of Weber Number Similarity

air entrainment
bubbles
droplets
waves

---

[36]There are two definitions of the Froude number. Dimensional analysis determines the Froude number as Eq. 17.234 (v²/Lg), a form that is used in model similitude. However, in open channel flow studies performed by civil engineers, the Froude number is taken as the square root of Eq. 17.234. Whether the derived form or its square root is used can sometimes be determined from the application. If the Froude number is squared (e.g., as in $dE/dx = 1 - \text{Fr}^2$), the square root form is probably needed. In similarity problems, it doesn't make any difference which definition is used.

[37]There are two definitions of the Weber number. The alternate definition is the square root of Eq. 17.239. In similarity problems, it does not make any difference which definition of the Weber number is used.

18 Hydraulic Machines

Nomenclature

BHP	brake horsepower	hp	n.a.
C	coefficient	–	–
C	factor	–	–
D	diameter	ft	m
E	specific energy	ft-lbf/lbm	J/kg
f	Darcy friction factor	–	–
f	frequency	Hz	Hz
FHP	friction horsepower	hp	n.a.
g	gravitational acceleration, 32.2 (9.81)	ft/sec^2	m/s^2
g_c	gravitational constant, 32.2	lbm-ft/ lbf-sec^2	n.a.
h	height or head	ft	m
K	compressibility factor	–	–
L	length	ft	m
$\dot{m}$	mass flow rate	lbm/sec	kg/s
n	dimensionless exponent	–	–
n	rotational speed	rev/min	rev/min
NPSHA	net positive suction head available	ft	m
NPSHR	net positive suction head required	ft	m
p	pressure	lbf/ft^2	Pa
P	power	ft-lbf/sec	W
Q	volumetric flow rate	gal/min	L/s
r	radius	ft	m
SA	suction specific speed available	rev/min	rev/min
SF	service factor	–	–
SG	specific gravity	–	–
t	time	sec	s
T	torque	ft-lbf	N·m
v	velocity	ft/sec	m/s
$\dot{V}$	volumetric flow rate	ft^3/sec	m^3/s
W	work	ft-lbf	kW·h
WHP	water horsepower	hp	n.a.
WkW	water kilowatts	n.a.	kW
z	elevation	ft	m

Symbols

γ	specific weight	lbf/ft^3	n.a.
η	efficiency	–	–
θ	angle	deg	deg
ν	kinematic viscosity	ft^2/sec	m^2/s
ρ	density	lbm/ft^3	kg/m^3
σ	cavitation number	–	–
ω	angular velocity	rad/sec	rad/s

Subscripts

ac	acceleration
atm	atmospheric
A	added (by pump)
b	blade
cr	critical
d	discharge
f	friction
j	jet
m	motor
n	nozzle
p	pressure or pump
s	specific or suction
ss	suction specific
t	tangential or total
th	theoretical
v	velocity
v	volumetric
vp	vapor pressure
z	potential

1. HYDRAULIC MACHINES

Pumps and turbines are the two basic types of hydraulic machines discussed in this chapter. Pumps convert mechanical energy into fluid energy, increasing the energy possessed by the fluid. Turbines convert fluid energy into mechanical energy, extracting energy from the fluid.

2. TYPES OF PUMPS

Pumps can be classified according to how energy is transferred to the fluid: intermittently or continuously.

The most common types of *positive displacement pumps* (*PD pumps*) are *reciprocating action pumps* (which use pistons, plungers, diaphragms, or bellows) and *rotary action pumps* (which use vanes, screws, lobes, or progressing cavities). Such pumps discharge a fixed volume for each stroke or revolution. Energy is added intermittently to the fluid.

Kinetic pumps transform fluid kinetic energy into fluid static pressure energy. The pump imparts the kinetic energy; the pump mechanism or housing is constructed in a manner that causes the transformation. *Jet pumps* and *ejector pumps* fall into the kinetic pump category, but centrifugal pumps are the primary examples.

In the operation of a *centrifugal pump*, liquid flowing into the *suction side* (the *inlet*) is captured by the *impeller* and thrown to the outside of the pump casing. Within the casing, the velocity imparted to the fluid by the impeller is converted into pressure energy. The fluid leaves the pump through the *discharge line* (the *exit*). It is a characteristic of most centrifugal pumps that the fluid is turned approximately 90° from the original flow direction. (See Table 18.1.)

Table 18.1 *Generalized Characteristics of Positive Displacement and Kinetic Pumps*

characteristic	positive displacement pumps	kinetic pumps
flow rate	low	high
pressure rise per stage	high	low
constant quantity over operating range	flow rate	pressure rise
self-priming	yes	no
discharge stream	pulsing	steady
works with high viscosity fluids	yes	no

3. RECIPROCATING POSITIVE DISPLACEMENT PUMPS

Reciprocating positive displacement (PD) pumps can be used with all fluids, and are useful with viscous fluids and slurries (up to about 8000 SSU), when the fluid is sensitive to shear, and when a high discharge pressure is required.[1] By entrapping a volume of fluid in the cylinder, reciprocating pumps provide a fixed-displacement volume per cycle. They are self-priming and inherently leak-free. Within the pressure limits of the line and pressure relief valve and the current capacity of the motor circuit, reciprocating pumps can provide an infinite discharge pressure.[2]

There are three main types of reciprocating pumps: power, direct-acting, and diaphragm. A *power pump* is a *cylinder-operated pump*. It can be single-acting or double-acting. A *single-acting pump* discharges liquid (or takes suction) only on one side of the piston, and there is only one transfer operation per crankshaft revolution. A *double-acting pump* discharges from both sides, and there are two transfers per revolution of the crank.

Traditional reciprocating pumps with pistons and rods can be either single-acting or double-acting and are suitable up to approximately 2000 psi (14 MPa). *Plunger pumps* are only single-acting and are suitable up to approximately 10,000 psi (70 MPa).

Simplex pumps have one cylinder, *duplex pumps* have two cylinders, *triplex pumps* have three cylinders, and so forth. *Direct-acting pumps* (sometimes referred to as *steam pumps*) are always double-acting. They use steam, unburned fuel gas, or compressed air as a motive fluid.

PD pumps are limited by both their NPSHR characteristics, acceleration head, and (for rotary pumps) slip.[3] Because the flow is unsteady, a certain amount of energy, the *acceleration head*, h_{ac}, is required to accelerate the fluid flow each stroke or cycle. If the acceleration head is too large, the NPSHR requirements may not be attainable. Acceleration head can be reduced by increasing the pipe diameter, shortening the suction piping, decreasing the pump speed, or placing a *pulsation damper* (*stabilizer*) in the suction line.[4]

Generally, friction losses with pulsating flows are calculated based on the maximum velocity attained by the fluid. Since this is difficult to determine, the maximum velocity can be approximated by multiplying the average velocity (calculated from the rated capacity) by the factors in Table 18.2.

When the suction line is "short," the acceleration head can be calculated from the length of the suction line, the average velocity in the line, and the rotational speed.[5]

---

[1]For viscosities of Saybolt seconds universal (SSU) greater than 240, multiply SSU viscosity by 0.216 to get viscosity in centistokes.

[2]For this reason, a relief valve should be included in every installation of positive displacement pumps. Rotary pumps typically have integral relief valves, but external relief valves are often installed to provide easier adjusting, cleaning, and inspection.

[3]Manufacturers of PD pumps prefer the term *net positive inlet pressure* (NPIP) to NPSH. NPIPA corresponds to NPSHA; NPIPR corresponds to NPSHR. Pressure and head are related by $p = \gamma h$.

[4]Pulsation dampers are not needed with rotary-action PD pumps, as the discharge is essentially constant.

[5]With a properly designed pulsation damper, the effective length of the suction line is reduced to approximately 10 pipe diameters.

Table 18.2 Typical v_{max}/v_{ave} Velocity Ratios[a,b]

pump type	single-acting	double-acting
simplex	3.2	2.0
duplex	1.6	1.3
triplex	1.1	1.1
quadriplex	1.1	1.1
quintuplex and up	1.05	1.05

[a]Without stabilization. With properly sized stabilizers, use 1.05–1.1 for all cases.
[b]Multiply the values by 1.3 for metering pumps where lost fluid motion is relied on for capacity control.

In Eq. 18.1, C and K are dimensionless factors. K represents the relative compressibility of the liquid. (Typical values are 1.4 for hot water; 1.5 for amine, glycol, and cold water; and 2.5 for hot oil.) Values of C are given in Table 18.3.

$$h_{ac} = \frac{C}{K}\left(\frac{L_{suction}v_{ave}n}{g}\right) \qquad 18.1$$

Table 18.3 Typical Acceleration Head C-Values[*]

pump type	single-acting	double-acting
simplex	0.4	0.2
duplex	0.2	0.115
triplex	0.066	0.066
quadriplex	0.040	0.040
quintuplex and up	0.028	0.028

[*]Typical values for common connecting rod lengths and crank radii.

4. ROTARY PUMPS

Rotary pumps are *positive displacement* (PD) pumps that move fluid by means of screws, progressing cavities, gears, lobes, or vanes turning within a fixed casing (the *stator*). Rotary pumps are useful for high viscosities (up to 4×10^6 SSU for screw pumps). The rotation creates a cavity of fixed volume near the pump input; atmospheric or external pressure forces the fluid into that cavity. Near the outlet, the cavity is collapsed, forcing the fluid out. Figure 18.1 illustrates the external circumferential piston rotary pump.

Discharge from rotary pumps is relatively smooth. Acceleration head is negligible. Pulsation dampers and suction stabilizers are not required.

Slip in rotary pumps is the amount (sometimes expressed as a percentage) of each rotational fluid volume that "leaks" back to the suction line on each revolution. Slip reduces pump capacity. It is a function of clearance, differential pressure, and viscosity. Slip is proportional to the third power of the clearance between the rotating element and the casing. Slip decreases with increases in viscosity; it increases linearly with increases in differential pressure. Slip is not affected by rotational speed. The *volumetric efficiency* is defined by Eq. 18.2.

Figure 18.1 External Circumferential Piston Rotary Pump

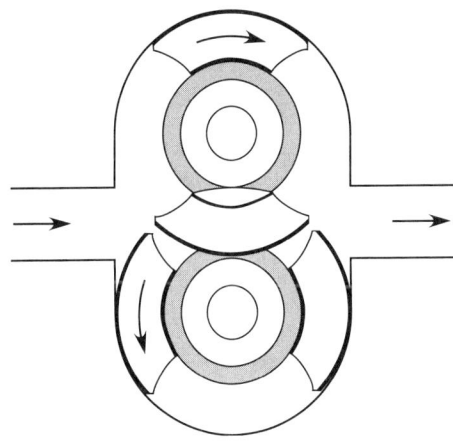

Figure 18.2 illustrates the relationship between flow rate, speed, slip, and differential pressure.

$$\eta_v = \frac{Q_{actual}}{Q_{ideal}} = \frac{Q_{ideal} - Q_{slip}}{Q_{ideal}} \qquad 18.2$$

Figure 18.2 Slip in Rotary Pumps

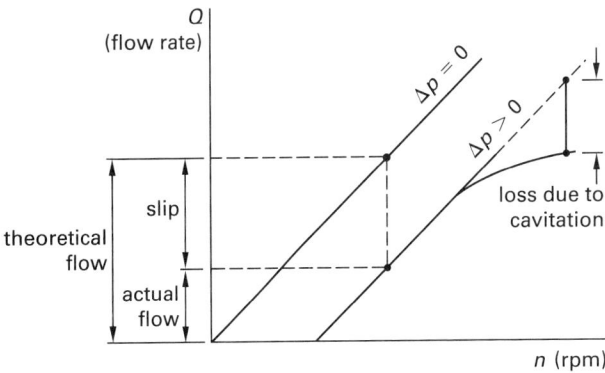

Except for screw pumps, rotary pumps are generally not used for handling abrasive fluids or materials with suspended solids.

5. DIAPHRAGM PUMPS

Hydraulically operated *diaphragm pumps* have a diaphragm that completely separates the pumped fluid from the rest of the pump. A reciprocating plunger pressurizes and moves a hydraulic fluid that, in turn, flexes the diaphragm. Single-ball check valves in the suction and discharge lines determine the direction of flow during both phases of the diaphragm action.

Metering is a common application of diaphragm pumps. Diaphragm metering pumps have no packing and are essentially leakproof. This makes them ideal when fugitive emissions are undesirable. Diaphragm pumps are suitable for pumping a wide range of materials, from liquefied gases to coal slurries, though the upper viscosity limit is approximately 3500 SSU. Within the limits of their reactivities, hazardous and reactive materials can also be handled.

Diaphragm pumps are limited by capacity, suction pressure, and discharge pressure and temperature. Because of their construction and size, most diaphragm pumps are limited to discharge pressures of 5000 psi (35 MPa) or less, and most high-capacity pumps are limited to 2000 psi (14 MPa). Suction pressures are similarly limited to 5000 psi (35 MPa). A pressure range of 3–9 psi (20–60 kPa) is often quoted as the minimum liquid-side pressure for metering applications.

The discharge is inherently pulsating, and dampers or stabilizers are often used. (The acceleration head term is required when calculating NPSHR.) The discharge can be smoothed out somewhat by using two or three (i.e., duplex or triplex) plungers.

Diaphragms are commonly manufactured from stainless steel (type 316) and polytetrafluoroethylene (PTFE) or other elastomers. PTFE diaphragms are suitable in the range of $-50°F$ to $300°F$ ($-45°C$ to $150°C$) while metal diaphragms (and some ketone resin diaphragms) are used up to approximately $400°F$ ($200°C$) with life expectancy being reduced at higher temperatures. Although most diaphragm pumps usually operate below 200 spm (strokes per minute), diaphragm life will be improved by limiting the maximum speed to 100 spm.

6. CENTRIFUGAL PUMPS

Centrifugal pumps and their impellers can be classified according to the way energy is imparted to the fluid. Each category of pump is suitable for a different application and (specific) speed range. (See Table 18.9.) Figure 18.3 illustrates a typical centrifugal pump and its schematic symbol.

Radial-flow impellers impart energy primarily by centrifugal force. Fluid enters the impeller at the hub and flows radially to the outside of the casing. Radial-flow pumps are suitable for adding high pressure at low fluid flow rates. *Axial-flow impellers* impart energy to the fluid by acting as compressors. Fluid enters and exits along the axis of rotation. Axial-flow pumps are suitable for adding low pressures at high fluid flow rates.[6]

Radial-flow pumps can be designed for either single- or double-suction operation. In a *single-suction pump*, fluid enters from only one side of the impeller. In a *double-suction pump*, fluid enters from both sides of the impeller.[7] (That is, the impeller is two-sided.) Operation is

---

[6]There is a third category of centrifugal pumps known as *mixed flow pumps*. Mixed flow pumps have operational characteristics between those of radial flow and axial flow pumps.

[7]The double-suction pump can handle a greater fluid flow rate than a single-suction pump with the same specific speed. Also, the double-suction pump will have a lower NPSHR.

similar to having two single-suction pumps in parallel. (See Fig. 18.4.)

Figure 18.3 *Centrifugal Pump and Symbol*

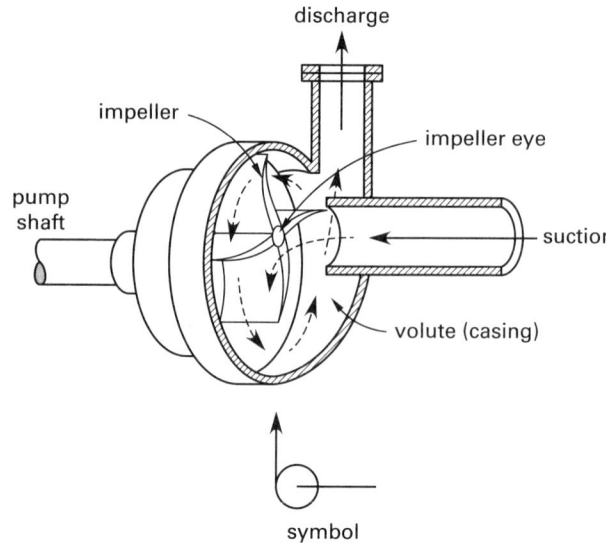

Figure 18.4 *Radial- and Axial-Flow Impellers*

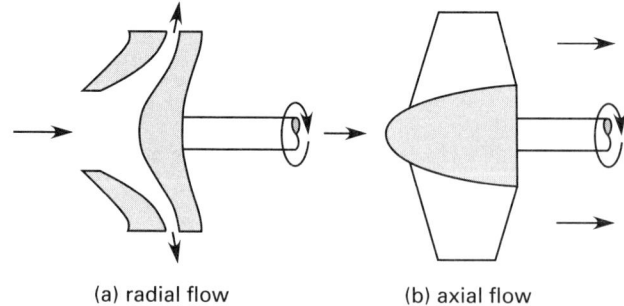

(a) radial flow   (b) axial flow

A *multiple-stage pump* consists of two or more impellers within a single casing. The discharge of one stage feeds the input of the next stage, and operation is similar to having several pumps in series. In this manner, higher heads are achieved than would be possible with a single impeller.

The *pitch* (*circular blade pitch*) is the impeller's circumference divided by the number of impeller vanes. The impeller *tip speed* is calculated from the impeller diameter and rotational speed. The impeller "tip speed" is actually the tangential velocity at the periphery. Tip speed is typically somewhat less than 1000 ft/sec (300 m/s).

$$v_{\text{tip}} = \frac{\pi D n}{60 \frac{\text{sec}}{\text{min}}} = \frac{D\omega}{2} \qquad 18.3$$

7. SEWAGE PUMPS

The primary consideration in choosing a pump for sewage and large solids is resistance to clogging. Centrifugal pumps should always be the single-suction type with nonclog, open impellers. (Double-suction pumps are prone to clogging because rags catch and wrap around the shaft extending through the impeller eye.) Clogging can be further minimized by limiting the number of impeller blades to two or three, providing for large passageways, and using a bar screen ahead of the pump.

Though made of heavy construction, nonclog pumps are constructed for ease of cleaning and repair. Horizontal pumps usually have a split casing, half of which can be removed for maintenance. A hand-sized cleanout opening may also be built into the casing. A sewage pump should normally be used with a grit chamber for prolonged bearing life.

The solids-handling capacity of a pump may be specified in terms of the largest sphere that can pass through it without clogging, usually about 80% of the inlet diameter. For example, a wastewater pump with a 6 in (150 mm) inlet should be able to pass a 4 in (100 mm) sphere. The pump must be capable of handling spheres with diameters slightly larger than the bar screen spacing.

Figure 18.5 shows a simplified wastewater pump installation. Not shown are instrumentation and water level measurement devices, baffles, lighting, drains for the dry well, electrical power, pump lubrication equipment, and access ports. (Totally submerged pumps do not require dry wells. However, such pumps without dry wells are more difficult to access, service, and repair.)

The multiplicity and redundancy of pumping equipment is not apparent from Fig. 18.5. The number of pumps used in a wastewater installation largely depends on the expected demand, pump capacity, and design criteria for backup operation. It is good practice to install pumps in sets of two, with a third backup pump being available for each set of pumps that performs the same function. The number of pumps and their capacities should be able to handle the peak flow when one pump in the set is out of service.

8. SLUDGE PUMPS AND GRAVITY FLOW

Centrifugal and reciprocating pumps are extensively used for pumping sludge. Progressive cavity screw impeller pumps are also used.

As further described in Sec. 18.28, the pumping power is proportional to the specific gravity. Accordingly, pumping power for dilute and well-digested sludges is typically only 10–25% higher than for water. However, most sludges are non-Newtonian fluids, often flow in a laminar mode, and have characteristics that may change with the season. Also, sludge characteristics change greatly during the pumping cycle; engineering judgment and rules of thumb are often important in choosing sludge pumps. For example, a general rule is to choose sludge pumps capable of developing at least 50–100% excess head.

One method of determining the required pumping power is to multiply the power required for pumping pure water by a service factor. Historical data is the best method of selecting this factor. Choice of initial values is a matter of judgment. Guidelines are listed in Table 18.4.

Table 18.4 *Pumping Power Multiplicative Factors*

solids concentration	digested sludge	untreated, primary, and concentrated sludge
0%	1.0	1.0
2%	1.2	1.4
4%	1.3	2.5
6%	1.7	4.1
8%	2.2	7.0
10%	3.0	10.0

Derived from *Wastewater Engineering: Treatment, Disposal, Reuse,* 3rd ed., by Metcalf & Eddy, et al., © 1991, with permission from The McGraw-Hill Companies, Inc.

Generally, sludge will thin out during a pumping cycle. The most dense sludge components will be pumped first, with more watery sludge appearing at the end of the pumping cycle. With a constant power input, the reduction in load at the end of pumping cycles may cause centrifugal pumps to operate far from the desired operating point and to experience overload failures. The operating point should be evaluated with high-, medium-, and low-density sludges.

To avoid cavitation, sludge pumps should always be under a positive suction head of at least 4 ft (1.2 m), and suction lifts should be avoided. The minimum diameters of suction and discharge lines for pumped sludge are typically 6 in (150 mm) and 4 in (100 mm), respectively.

Not all sludge is moved by pump action. Some installations rely on gravity flow to move sludge. The minimum diameter of sludge gravity transfer lines is typically 8 in (200 mm), and the recommended minimum slope is 3%.

To avoid clogging due to settling, sludge velocity should be above the transition from laminar to turbulent flow,

Figure 18.5 *Typical Wastewater Pump Installation (greatly simplified)*

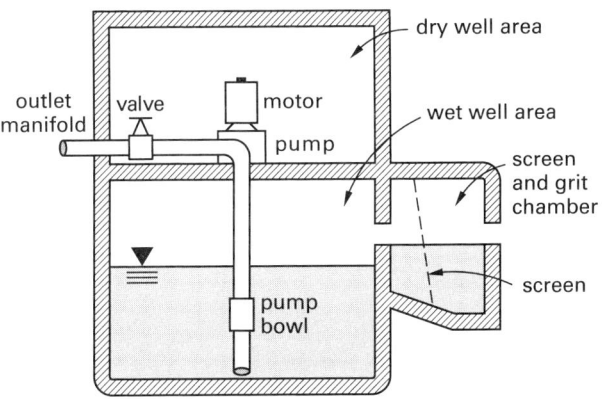

known as the *critical velocity*. The critical velocity for most sludges is approximately 3.5 ft/sec (1.1 m/s). Velocity ranges of 5–8 ft/sec (1.5–2.4 m/s) are commonly quoted and are adequate.

9. TERMINOLOGY OF HYDRAULIC MACHINES AND PIPE NETWORKS

A pump has an inlet (designated the *suction*) and an outlet (designated the *discharge*). The subscripts s and d refer to the inlet and outlet of the pump, not of the pipeline.

All of the terms that are discussed in this section are *head* terms and, as such, have units of length. When working with hydraulic machines, it is common to hear such phrases as "a pressure head of 50 feet" and "a static discharge head of 15 meters." The term *head* is often substituted for pressure or pressure drop. Any head term (*pressure head, atmospheric head, vapor pressure head*, etc.) can be calculated from pressure by using Eq. 18.4.[8]

$$h = \frac{p}{g\rho} \qquad \text{[SI]} \quad 18.4(a)$$

$$h = \frac{p}{\rho} \times \frac{g_c}{g} = \frac{p}{\gamma} \qquad \text{[U.S.]} \quad 18.4(b)$$

Some of the terms used in the description of pipe networks appear to be similar (e.g., suction head and total suction head). The following general rules will help to clarify the meanings.

rule 1: The word *suction* or *discharge* limits the quantity to the suction line or discharge line, respectively. The absence of either word implies that both the suction and discharge lines are included. Example: discharge head.

rule 2: The word *static* means that static head only is included (not velocity head, friction head, etc.). Example: static suction head.

rule 3: The word *total* means that static head, velocity head, and friction head are all included. (Total does not mean the combination of suction and discharge.) Example: total suction head.

The following terms are commonly encountered.

- *friction head, h_f:* The head required to overcome resistance to flow in the pipes, fittings, valves, entrances, and exits.

$$h_f = \frac{fLv^2}{2Dg} \qquad 18.5$$

- *velocity head, h_v:* The specific kinetic energy of the fluid. Also known as *dynamic head.*[9]

$$h_v = \frac{v^2}{2g} \qquad 18.6$$

- *static suction head, $h_{z(s)}$:* The vertical distance above the centerline of the pump inlet to the free level of the fluid source. If the free level of the fluid is below the pump inlet, $h_{z(s)}$ will be negative and is known as *static suction lift.* (See Fig. 18.6.)

- *static discharge head, $h_{z(d)}$:* The vertical distance above the centerline of the pump inlet to the point of free discharge or surface level of the discharge tank. (See Fig. 18.7.)

The ambiguous term *effective head* is not commonly used when discussing hydraulic machines, but when used, the term most closely means *net head* (i.e., starting head less losses). Consider a hydroelectric turbine that is fed by water with a static head of H. After frictional and other losses, the net head available to the turbine will be less than H. The turbine output will coincide with an ideal turbine being acted upon by the net or effective head. Similarly, the actual increase in pressure across a pump will be the effective head added (i.e., the head net of internal losses and geometric effects).

Figure 18.6 Static Suction Lift

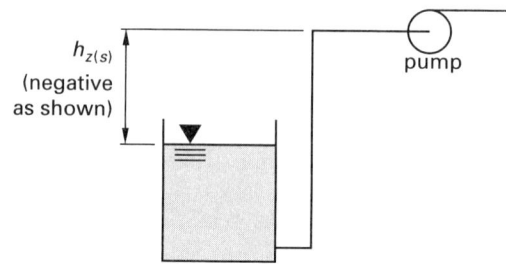

Figure 18.7 Static Discharge Head

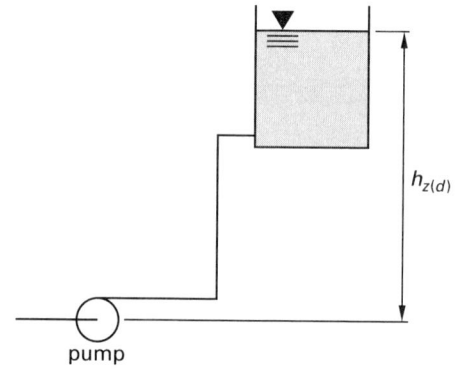

---

[8]Equation 18.4 can be used to define *pressure head, atmospheric head*, and *vapor pressure head*, whose meanings and derivations should be obvious.

[9]The term *dynamic* is not as consistently applied as are the terms described in the rules. In particular, it is not clear whether dynamic head includes friction head.

Example 18.1

Write the symbolic equations for the following terms:
(a) the total suction head, (b) the total discharge head, and (c) the total head added.

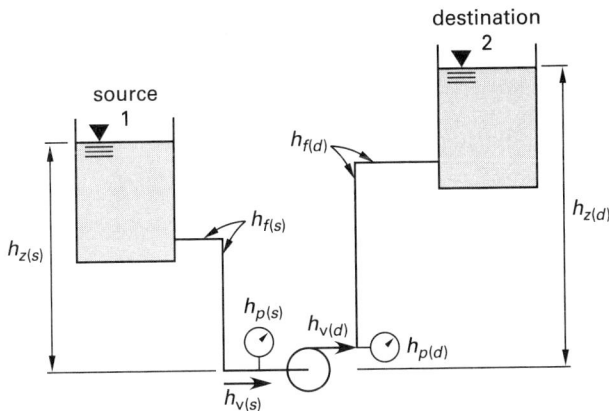

Solution

(a) The *total suction head* at the pump inlet is the sum of static (pressure) head and velocity head at the pump suction.

$$h_{t(s)} = h_{p(s)} + h_{v(s)}$$

Total suction head can also be calculated from the conditions existing at the source (1), in which case suction line friction would also be considered. (With an open reservoir, $h_{p(1)}$ will be zero if gage pressures are used, and $h_{v(1)}$ will be zero if the source is large.)

$$h_{t(s)} = h_{p(1)} + h_{z(s)} + h_{v(1)} - h_{f(s)}$$

(b) The *total discharge head* at the pump outlet is the sum of the static (pressure) and velocity heads at the pump outlet. Friction head is zero since the fluid has not yet traveled through any length of pipe when it is discharged.

$$h_{t(d)} = h_{p(d)} + h_{v(d)}$$

The total discharge head can also be evaluated at the destination (2) if the friction head, $h_{f(d)}$, between the discharge and the destination is known. (With an open reservoir, $h_{p(2)}$ will be zero if gage pressures are used, and $h_{v(2)}$ will be zero if the destination is large.)

$$h_{t(d)} = h_{p(2)} + h_{z(d)} + h_{v(2)} + h_{f(d)}$$

(c) The *total head added* by the pump is the total discharge head less the total suction head.

$$h_t = h_A = h_{t(d)} - h_{t(s)}$$

Assuming suction from and discharge to reservoirs exposed to the atmosphere and assuming negligible reservoir velocities,

$$h_t = h_A \approx h_{z(d)} - h_{z(s)} + h_{f(d)} + h_{f(s)}$$

10. PUMPING POWER

The energy (head) added by a pump can be determined from the difference in total energy on either side of the pump. Writing the Bernoulli equation for the discharge and suction conditions produces Eq. 18.7, an equation for the *total dynamic head*, often abbreviated TDH.

$$h_A = h_t = h_{t(d)} - h_{t(s)} \qquad 18.7$$

$$h_A = \frac{p_d - p_s}{\rho g} + \frac{v_d^2 - v_s^2}{2g} + z_d - z_s \qquad \text{[SI]} \qquad 18.8(a)$$

$$h_A = \frac{(p_d - p_s)g_c}{\rho g} + \frac{v_d^2 - v_s^2}{2g} + z_d - z_s \qquad \text{[U.S.]} \qquad 18.8(b)$$

In most applications, the change in velocity and potential heads is either zero or small in comparison to the increase in pressure head. Equation 18.8 then reduces to Eq. 18.9.

$$h_A = \frac{p_d - p_s}{\rho g} \qquad \text{[SI]} \qquad 18.9(a)$$

$$h_A = \frac{p_d - p_s}{\rho} \times \frac{g_c}{g} \qquad \text{[U.S.]} \qquad 18.9(b)$$

It is important to recognize that the variables in Eq. 18.8 and Eq. 18.9 refer to the conditions at the pump's immediate inlet and discharge, not to the distant ends of the suction and discharge lines. However, the total dynamic head added by a pump can be calculated in another way. For example, for a pump raising water from one open reservoir to another, the total dynamic head would consider the total elevation rise, the velocity head (often negligible), and the friction losses in the suction and discharge lines.

The head added by the pump can also be calculated from the impeller and fluid speeds. Equation 18.10 is useful for radial- and mixed-flow pumps for which the incoming fluid has little or no rotational velocity component (i.e., up to a specific speed of approximately 2000 U.S. or 40 SI). In Eq. 18.10, $v_{impeller}$ is the tangential impeller velocity at the radius being considered, and v_{fluid} is the average tangential velocity imparted to the fluid by the impeller. The impeller efficiency, $\eta_{impeller}$, is typically 0.85–0.95. This is much higher than the total pump efficiency (see Sec. 18.11) because it does not include mechanical and fluid friction losses.

$$h_A = \frac{\eta_{impeller} v_{impeller} v_{fluid}}{g} \qquad 18.10$$

Head added can be thought of as the energy added per unit mass. The total pumping power depends on the head added, h_A, and the mass flow rate. For example, the product $\dot{m} h_A$ has the units of foot-pounds per second (in customary U.S. units), which can be easily converted to horsepower. Pump output power is known as *hydraulic power* or *water power*. Hydraulic power is the net power actually transferred to the fluid.

Horsepower is a common unit of power, which results in the terms *hydraulic horsepower* and *water horsepower*, WHP, being used to designate the power that is transferred into the fluid. Various relationships for finding the hydraulic horsepower are given in Table 18.5.

The unit of power in SI units is the watt (kilowatt). Table 18.6 can be used to determine *hydraulic kilowatts*, WkW.

Table 18.5 Hydraulic Horsepower Equations[a]

	Q (gal/min)	$\dot{m}$ (lbm/sec)	$\dot{V}$ (ft^3/sec)
h_A in feet	$\dfrac{h_A Q(SG)}{3956}$	$\dfrac{h_A \dot{m}}{550} \times \dfrac{g}{g_c}$	$\dfrac{h_A \dot{V}(SG)}{8.814}$
Δp in psi[b]	$\dfrac{\Delta p Q}{1714}$	$\dfrac{\Delta p \dot{m}}{(238.3)(SG)} \times \dfrac{g}{g_c}$	$\dfrac{\Delta p \dot{V}}{3.819}$
Δp in psf[b]	$\dfrac{\Delta p Q}{2.468 \times 10^5}$	$\dfrac{\Delta p \dot{m}}{(34{,}320)(SG)} \times \dfrac{g}{g_c}$	$\dfrac{\Delta p \dot{V}}{550}$
W in $\dfrac{\text{ft-lbf}}{\text{lbm}}$	$\dfrac{W Q(SG)}{3956}$	$\dfrac{W \dot{m}}{550}$	$\dfrac{W \dot{V}(SG)}{8.814}$

(Multiply horsepower by 0.7457 to obtain kilowatts.)
[a]based on $\rho_{\text{water}} = 62.4$ lbm/ft^3 and $g = 32.2$ ft/sec^2
[b]Velocity head changes must be included in Δp.

Table 18.6 Hydraulic Kilowatt Equations[a]

	Q (L/s)	$\dot{m}$ (kg/s)	$\dot{V}$ (m^3/s)
h_A in meters	$\dfrac{9.81 h_A Q(SG)}{1000}$	$\dfrac{9.81 h_A \dot{m}}{1000}$	$9.81 h_A \dot{V}(SG)$
Δp in kPa[b]	$\dfrac{\Delta p Q}{1000}$	$\dfrac{\Delta p \dot{m}}{1000(SG)}$	$\Delta p \dot{V}$
W in $\dfrac{\text{J}}{\text{kg}}$[b]	$\dfrac{W Q(SG)}{1000}$	$\dfrac{W \dot{m}}{1000}$	$W \dot{V}(SG)$

(Multiply kilowatts by 1.341 to obtain horsepower.)
[a]based on $\rho_{\text{water}} = 1000$ kg/m^3 and $g = 9.81$ m/s^2
[b]Velocity head changes must be included in Δp.

Example 18.2

A pump adds 550 ft of pressure head to 100 lbm/sec of water ($\rho = 62.4$ lbm/ft^3 or 1000 kg/m^3). (a) Complete the following table of performance data. (b) What is the hydraulic power in horsepower and kilowatts?

item	customary U.S.	SI
$\dot{m}$	100 lbm/sec	___kg/s
h	550 ft	___m
Δp	___lbf/ft^2	___kPa
$\dot{V}$	___ft^3/sec	___m^3/s
W	___ft-lbf/lbm	___J/kg
P	___hp	___kW

Solution

(a) Work initially with the customary U.S. data.

$$\Delta p = \rho h \times \frac{g}{g_c} = \left(62.4 \ \frac{\text{lbm}}{\text{ft}^3}\right)(550 \text{ ft}) \times \frac{g}{g_c}$$
$$= 34{,}320 \text{ lbf/ft}^2$$

$$\dot{V} = \frac{\dot{m}}{\rho} = \frac{100 \ \frac{\text{lbm}}{\text{sec}}}{62.4 \ \frac{\text{lbm}}{\text{ft}^3}} = 1.603 \text{ ft}^3/\text{sec}$$

$$W = h \times \frac{g}{g_c} = 550 \text{ ft} \times \frac{g}{g_c}$$
$$= 550 \text{ ft-lbf/lbm}$$

Convert to SI units.

$$\dot{m} = \frac{100 \ \frac{\text{lbm}}{\text{sec}}}{2.201 \ \frac{\text{lbm}}{\text{kg}}} = 45.43 \text{ kg/s}$$

$$h = \frac{550 \text{ ft}}{3.281 \ \frac{\text{ft}}{\text{m}}} = 167.6 \text{ m}$$

$$\Delta p = \left(34{,}320 \ \frac{\text{lbf}}{\text{ft}^2}\right)\left(\frac{1}{\left(12 \ \frac{\text{in}}{\text{ft}}\right)^2}\right)\left(6.895 \ \frac{\text{kPa}}{\frac{\text{lbf}}{\text{in}^2}}\right)$$
$$= 1643 \text{ kPa}$$

$$\dot{V} = \left(1.603 \ \frac{\text{ft}^3}{\text{sec}}\right)\left(0.0283 \ \frac{\text{m}^3}{\text{ft}^3}\right) = 0.0454 \text{ m}^3/\text{s}$$

$$W = \left(550 \ \frac{\text{ft-lbf}}{\text{lbm}}\right)\left(1.356 \ \frac{\text{J}}{\text{ft-lbf}}\right)\left(2.201 \ \frac{\text{lbm}}{\text{kg}}\right)$$
$$= 1642 \text{ J/kg}$$

(b) From Table 18.5, the hydraulic horsepower is

$$\text{WHP} = \frac{h_A \dot{m}}{550} \times \frac{g}{g_c}$$
$$= \frac{(550 \text{ ft})\left(100 \ \frac{\text{lbm}}{\text{sec}}\right)}{550 \ \frac{\text{ft-lbf}}{\text{hp-sec}}} \times \frac{g}{g_c}$$
$$= 100 \text{ hp}$$

From Table 18.6, the power is

$$\text{WkW} = \frac{\Delta p \dot{m}}{(1000)(\text{SG})} = \frac{(1643 \text{ kPa})\left(45.43 \; \frac{\text{kg}}{\text{s}}\right)}{\left(1000 \; \frac{\text{W}}{\text{kW}}\right)(1.0)}$$

$$= 74.6 \text{ kW}$$

11. PUMPING EFFICIENCY

Hydraulic power is the net energy actually transferred to the fluid per unit time. The input power delivered by the motor to the pump is known as the *brake pump power*. The term *brake horsepower*, BHP, is commonly used to designate both the quantity and its units. Due to frictional losses between the fluid and the pump and mechanical losses in the pump itself, the brake pump power will be greater than the hydraulic power.

The ratio of hydraulic power to brake pump power is the pump efficiency, η_p. Figure 18.8 gives typical pump efficiencies as a function of the pump's specific speed. The difference between the brake and hydraulic powers is known as the *friction power* (or *friction horsepower*, FHP).

$$\text{BHP} = \frac{\text{WHP}}{\eta_p} \qquad 18.11$$

$$\text{FHP} = \text{BHP} - \text{WHP} \qquad 18.12$$

Figure 18.8 *Average Pump Efficiency Versus Specific Speed*

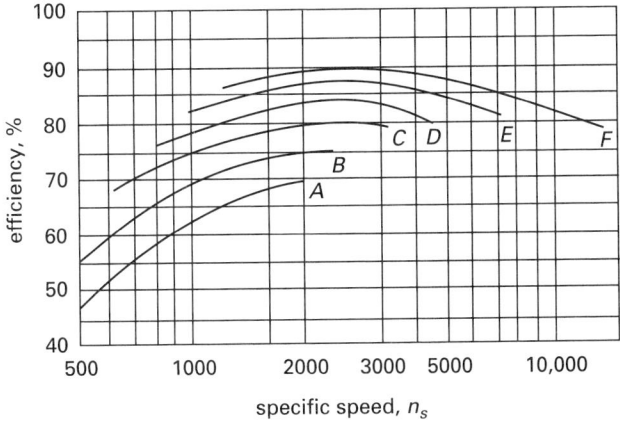

curve A: 100 gal/min
curve B: 200 gal/min
curve C: 500 gal/min
curve D: 1000 gal/min
curve E: 3000 gal/min
curve F: 10,000 gal/min

Pumping efficiency is not constant for any specific pump; rather, it depends on the operating point and the speed of the pump. (See Sec. 18.23.) A pump's characteristic efficiency curves will be published by its manufacturer.

With pump characteristic curves given by the manufacturer, the efficiency is not determined from the intersection of the system curve and the efficiency curve. (See Sec. 18.22.) Rather, the efficiency is a function of only the flow rate. Therefore, the operating efficiency is read from the efficiency curve directly above or below the operating point. (See Sec. 18.23 and Fig. 18.9.)

Figure 18.9 *Typical Centrifugal Pump Efficiency Curves*

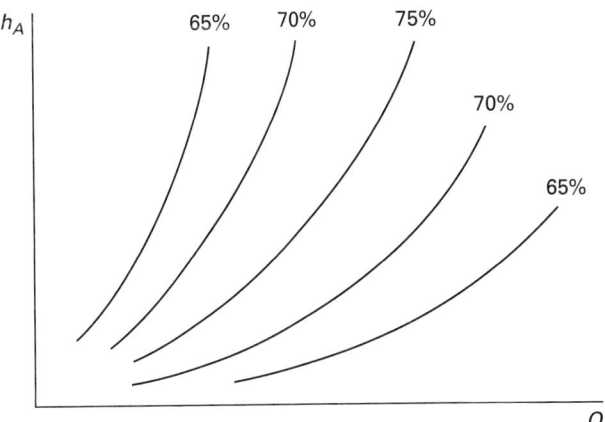

Efficiency curves published by a manufacturer will not include such losses in the suction elbow, discharge diffuser, couplings, bearing frame, seals, or pillow blocks. Up to 15% of the motor horsepower may be lost to these factors. Upon request, the manufacturer may provide the pump's installed *wire-to-water efficiency* (i.e., the fraction of the electrical power drawn that is converted to hydraulic power).

The pump must be driven by an engine or motor.[10] The power delivered to the motor is greater than the power delivered to the pump, as accounted for by the motor efficiency, η_m. If the pump motor is electrical, its input power requirements will be stated in kilowatts.[11]

$$P_{m,\text{hp}} = \frac{\text{BHP}}{\eta_m} \qquad 18.13$$

$$P_{m,\text{kW}} = \frac{0.7457 \text{BHP}}{\eta_m} \qquad 18.14$$

The *overall efficiency* of the pump installation is the product of the pump and motor efficiencies.

$$\eta = \eta_p \eta_m = \frac{\text{WHP}}{P_{m,\text{hp}}} \qquad 18.15$$

12. COST OF ELECTRICITY

The power utilization of pump motors is usually measured in kilowatts. The kilowatt usage represents the rate that energy is transferred by the pump motor. The total amount of work, W, done by the pump motor is

---

[10] The source of power is sometimes called the *prime mover*.
[11] A *watt* is a joule per second.

found by multiplying the rate of energy usage (i.e., the delivered power, P), by the length of time the pump is in operation.

$$W = Pt \qquad 18.16$$

Although the units horsepower-hours are occasionally encountered, it is more common to measure electrical work in *kilowatt-hours* (kW-hr). Accordingly, the cost of electrical energy is stated per kW-hr (e.g., $0.10 per kW-hr).

$$\text{cost} = \frac{W_{\text{kW-hr}}(\text{cost per kW-hr})}{\eta_m} \qquad 18.17$$

13. STANDARD MOTOR SIZES AND SPEEDS

An effort should be made to specify standard motor sizes when selecting the source of pumping power. Table 18.7 lists NEMA (National Electrical Manufacturers Association) standard motor sizes by horsepower *nameplate rating*.[12] Other motor sizes may also be available by special order.

Table 18.7 NEMA Standard Motor Sizes (brake horsepower)

$\frac{1}{8}$,*	$\frac{1}{6}$,*	$\frac{1}{4}$,*	$\frac{1}{3}$,*	$\frac{1}{2}$,*	$\frac{3}{4}$*		
1,	1.5,	2,	3,	5,	7.5	10,	15
20,	25,	30,	40,	50,	60,	75,	100
125,	150,	200,	250,	300,	350,	400,	450
500,	600,	700,	800,	900,	1000,	1250,	1500
1750,	2000,	2250,	2500,	2750,	3000,	3500,	4000
4500,	5000,	6000,	7000,	8000			

*fractional horsepower series

For industrial grade motors, the rated (nameplate) power is the power at the output shaft that the motor can produce continuously while operating at the nameplate ambient conditions and without exceeding a particular temperature rise dependent on its wiring type. However, if a particular motor is housed in a larger, open frame that has better cooling, it will be capable of producing more power than the nameplate rating. NEMA defines the *service factor*, SF, as the amount of continual overload capacity designed into a motor without reducing its useful life (provided voltage, frequency, and ambient temperature remain normal). Common motor service factors are 1.0, 1.15, and 1.25. When a motor develops more than the nameplate power, efficiency, power factor, and operating temperature will be affected adversely.

$$SF = \frac{P_{\text{max continuous}}}{P_{\text{rated}}} \qquad 18.18$$

[12]The nameplate rating gets its name from the information stamped on the motor's identification plate. Besides the horsepower rating, other nameplate data used to classify the motor are the service class, voltage, full-load current, speed, number of phases, frequency of the current, and maximum ambient temperature (or, for older motors, the motor temperature rise).

Larger horsepower motors are usually three-phase induction motors. The *synchronous speed*, n in rpm, of such motors is the speed of the rotating field, which depends on the number of poles per stator phase and the frequency, f. The number of poles must be an even number. The frequency is typically 60 Hz, as in the United States, or 50 Hz, as in European countries. Table 18.8 lists common synchronous speeds.

$$n = \frac{120f}{\text{no. of poles}} \qquad 18.19$$

Table 18.8 Common Synchronous Speeds

number of poles	n (rpm) 60 Hz	50 Hz
2	3600	3000
4	1800	1500
6	1200	1000
8	900	750
10	720	600
12	600	500
14	514	428
18	400	333
24	300	250
48	150	125

Induction motors do not run at their synchronous speeds when loaded. Rather, they run at slightly less than synchronous speed. The deviation is known as the *slip*. Slip is typically around 4% and is seldom greater than 10% at full load for motors in the 1–75 hp range.

slip (in rpm)

$$= \text{synchronous speed} - \text{actual speed} \qquad 18.20$$

slip (in percent)

$$= \frac{\text{synchronous speed} - \text{actual speed}}{\text{synchronous speed}} \times 100\% \qquad 18.21$$

Induction motors may also be specified in terms of their kVA (kilovolt-amp) ratings. The kVA rating is not the same as the power in kilowatts, although one can be derived from the other if the motor's *power factor* is known. Such power factors typically range from 0.8 to 0.9, depending on the installation and motor size.

$$\text{kVA rating} = \frac{\text{motor power in kW}}{\text{power factor}} \qquad 18.22$$

Example 18.3

A pump driven by an electrical motor moves 25 gal/min of water from reservoir A to reservoir B, lifting the water a total of 245 ft. The efficiencies of the pump and motor are 64% and 84%, respectively. Electricity costs $0.08/kW-hr. Neglect velocity head, friction, and minor losses. (a) What minimum size motor is required? (b) How much does it cost to operate the pump for 6 hr?

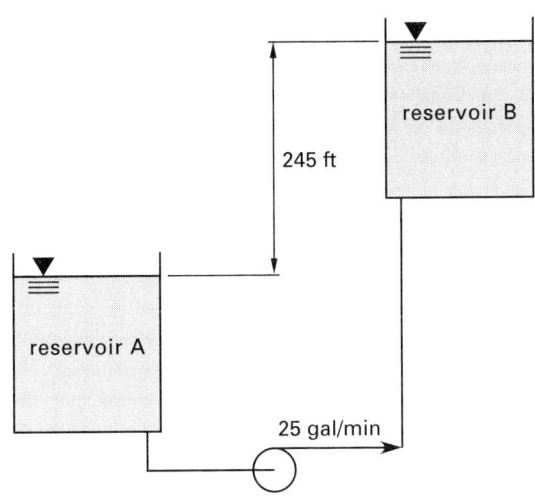

Solution

(a) The head added is 245 ft. Using Table 18.5 and incorporating the pump efficiency, the motor power required is

$$P = \frac{h_A Q(\text{SG})}{3956\eta_p} = \frac{(245 \text{ ft})\left(25 \dfrac{\text{gal}}{\text{min}}\right)(1.0)}{\left(3956 \dfrac{\text{ft-gal}}{\text{hp-min}}\right)(0.64)}$$

$$= 2.42 \text{ hp}$$

From Table 18.7, select a 3 hp motor.

(b) The developed power, not the motor's rated power, is used. From Eq. 18.16 and Eq. 18.17,

$$\text{cost} = (\text{cost per kW-hr})\left(\frac{Pt}{\eta_m}\right)$$

$$= \left(0.08 \dfrac{\$}{\text{kW-hr}}\right)\left(0.7457 \dfrac{\text{kW}}{\text{hp}}\right)\left(\dfrac{(2.42 \text{ hp})(6 \text{ hr})}{0.84}\right)$$

$$= \$1.03$$

14. PUMP SHAFT LOADING

The torque on a pump or motor shaft can be calculated from the brake power and rotational speed.

$$T_{\text{in-lbf}} = \frac{63{,}025 P_{\text{hp}}}{n} \qquad \textit{18.23}$$

$$T_{\text{ft-lbf}} = \frac{5252 P_{\text{hp}}}{n} \qquad \textit{18.24}$$

$$T_{\text{N·m}} = \frac{9549 P_{\text{kW}}}{n} \qquad \textit{18.25}$$

The actual (developed) torque can be calculated from the change in momentum of the fluid flow. For radial impellers, the fluid enters through the eye and is turned

90°. The direction change is related to the increase in momentum and the shaft torque. When fluid is introduced axially through the eye of the impeller, the tangential velocity at the inlet (eye), $\text{v}_{t(s)}$, is zero.[13]

$$T_{\text{actual}} = \dot{m}\left(\text{v}_{t(d)} r_{\text{impeller}} - \text{v}_{t(s)} r_{\text{eye}}\right) \qquad \text{[SI]} \quad \textit{18.26(a)}$$

$$T_{\text{actual}} = \frac{\dot{m}}{g_c}\left(\text{v}_{t(d)} r_{\text{impeller}} - \text{v}_{t(s)} r_{\text{eye}}\right) \qquad \text{[U.S.]} \quad \textit{18.26(b)}$$

Centrifugal pumps can be driven directly from a motor, or a speed changer can be used. Rotary pumps generally require a speed reduction. *Gear motors* have integral speed reducers. V-belt drives are widely used because of their initial low cost, although timing belts and chains can be used in some applications.

When a belt or chain is used, the pump's and motor's maximum overhung loads must be checked. This is particularly important for high-power, low-speed applications (such as rotary pumps). *Overhung load* is the side load (force) put on shafts and bearings. The overhung load is calculated from Eq. 18.27. The empirical factor K is 1.0 for chain drives, 1.25 for timing belts, and 1.5 for V-belts.

$$\text{overhung load} = \frac{2KT}{D_{\text{sheave}}} \qquad \textit{18.27}$$

If a direct drive cannot be used and the overhung load is excessive, the installation can be modified to incorporate a jack shaft or outboard bearing.

Example 18.4

A centrifugal pump delivers 275 lbm/sec (125 kg/s) of water while turning at 850 rpm. The impeller has straight radial vanes and an outside diameter of 10 in (25.4 cm). Water enters the impeller through the eye. The driving motor delivers 30 hp (22 kW). What are the (a) theoretical torque, (b) pump efficiency, and (c) total dynamic head?

SI Solution

(a) From Eq. 18.3, the impeller's tangential velocity is

$$\text{v}_t = \frac{\pi D n}{60 \dfrac{\text{s}}{\text{min}}} = \frac{\pi (25.4 \text{ cm})\left(850 \dfrac{\text{rev}}{\text{min}}\right)}{\left(60 \dfrac{\text{s}}{\text{min}}\right)\left(100 \dfrac{\text{cm}}{\text{m}}\right)}$$

$$= 11.3 \text{ m/s}$$

---

[13]The tangential component of fluid velocity is sometimes referred to as the *velocity of whirl*.

Since water enters axially, the incoming water has no tangential component. From Eq. 18.26, the developed torque is

$$T = \dot{m}v_{t(d)}r_{impeller} = \frac{\left(125\ \frac{kg}{s}\right)\left(11.3\ \frac{m}{s}\right)\left(\frac{25.4\ cm}{2}\right)}{100\ \frac{cm}{m}}$$

$$= 179.4\ \text{N·m}$$

(b) From Eq. 18.25, the developed power is

$$P_{kW} = \frac{nT_{N·m}}{9549} = \frac{\left(850\ \frac{rev}{min}\right)(179.4\ N·m)}{9549\ \frac{N·m}{kW·min}}$$

$$= 15.97\ \text{kW}$$

The pump efficiency is

$$\eta_p = \frac{P_{developed}}{P_{input}} = \frac{15.97\ kW}{22\ kW}$$

$$= 0.726 \quad (72.6\%)$$

(c) From Table 18.6, the total dynamic head is

$$h_A = \frac{P}{\dot{m}g} = \frac{(15.97\ kW)\left(1000\ \frac{W}{kW}\right)}{\left(125\ \frac{kg}{s}\right)\left(9.81\ \frac{m}{s^2}\right)}$$

$$= 13.0\ \text{m}$$

Customary U.S. Solution

(a) From Eq. 18.3, the impeller's tangential velocity is

$$v_t = \frac{\pi Dn}{60\ \frac{sec}{min}} = \frac{\pi(10\ in)\left(850\ \frac{rev}{min}\right)}{\left(60\ \frac{sec}{min}\right)\left(12\ \frac{in}{ft}\right)}$$

$$= 37.09\ \text{ft/sec}$$

Since water enters axially, the incoming water has no tangential component. From Eq. 18.26, the developed torque is

$$T = \frac{\dot{m}v_{t(d)}r_{impeller}}{g_c} = \frac{\left(275\ \frac{lbm}{sec}\right)\left(37.08\ \frac{ft}{sec}\right)\left(\frac{10\ in}{2}\right)}{\left(32.2\ \frac{lbm-ft}{lbf-sec^2}\right)\left(12\ \frac{in}{ft}\right)}$$

$$= 131.9\ \text{ft-lbf}$$

(b) From Eq. 18.24, the developed power is

$$P_{hp} = \frac{T_{ft-lbf}n}{5252} = \frac{(131.9\ ft-lbf)\left(850\ \frac{rev}{min}\right)}{5252\ \frac{ft-lbf}{hp-min}}$$

$$= 21.35\ \text{hp}$$

The pump efficiency is

$$\eta_p = \frac{P_{developed}}{P_{input}} = \frac{21.35\ hp}{30\ hp}$$

$$= 0.712 \quad (71.2\%)$$

(c) From Table 18.5, the total dynamic head is

$$h_A = \frac{\left(550\ \frac{ft-lbf}{hp-sec}\right)P_{hp}}{\dot{m}} \times \frac{g_c}{g}$$

$$= \frac{\left(550\ \frac{ft-lbf}{hp-sec}\right)(21.35\ hp)}{275\ \frac{lbm}{sec}} \times \frac{32.2\ \frac{lbm-ft}{lbf-sec^2}}{32.2\ \frac{ft}{sec^2}}$$

$$= 42.7\ \text{ft}$$

15. SPECIFIC SPEED

The capacity and efficiency of a centrifugal pump are partially governed by the impeller design. For a desired flow rate and added head, there will be one optimum impeller design. The quantitative index used to optimize the impeller design is known as *specific speed*, n_s, also known as *impeller specific speed*. Table 18.9 lists the impeller designs that are appropriate for different specific speeds.[14]

Table 18.9 *Specific Speed versus Impeller Design*

impeller type	approximate range of specific speed (rpm)	
	customary U.S. units	SI units
radial vane	500 to 1000	10 to 20
Francis (mixed) vane	2000 to 3000	40 to 60
mixed flow	4000 to 7000	80 to 140
axial flow	9000 and above	180 and above

(Divide customary U.S. specific speed by 51.64 to obtain SI specific speed.)

---

[14]Specific speed is useful for more than just selecting an impeller type. Maximum suction lift, pump efficiency, and net positive suction head required (NPSHR) can be correlated with specific speed.

Highest heads per stage are developed at low specific speeds. However, for best efficiency, specific speed should be greater than 650 (13 in SI units). If the specific speed for a given set of conditions drops below 650 (13), a multiple-stage pump should be selected.[15]

Specific speed is a function of a pump's capacity, head, and rotational speed at peak efficiency, as shown in Eq. 18.28. For a given pump and impeller configuration, the specific speed remains essentially constant over a range of flow rates and heads. (Q or $\dot{V}$ in Eq. 18.28 is half of the full flow rate for double-suction pumps.)

$$n_s = \frac{n\sqrt{\dot{V}}}{h_A^{0.75}} \qquad \text{[SI]} \qquad 18.28(a)$$

$$n_s = \frac{n\sqrt{Q}}{h_A^{0.75}} \qquad \text{[U.S.]} \qquad 18.28(b)$$

A common definition of specific speed is the speed (in rpm) at which a *homologous pump* would have to turn in order to deliver one gallon per minute at one foot total added head.[16] This definition is implicit to Eq. 18.28 but is not very useful otherwise. While specific speed is not dimensionless, the units are meaningless. Specific speed may be assigned units of rpm, but most often it is expressed simply as a pure number.

The numerical range of acceptable performance for each impeller type is redefined when SI units are used. The SI specific speed is obtained by dividing the customary U.S. specific speed by 51.64.

Specific speed can be used to determine the type of impeller needed. Once a pump is selected, its specific speed and Eq. 18.28 can be used to determine other operational parameters (e.g., maximum rotational speed). Specific speed can be used with Fig. 18.8 to obtain an approximate pump efficiency.

Example 18.5

A centrifugal pump powered by a direct-drive induction motor is needed to discharge 150 gal/min against a 300 ft total head when turning at the fully loaded speed of 3500 rpm. What type of pump should be selected?

Solution

From Eq. 18.28, the specific speed is

$$n_s = \frac{n\sqrt{Q}}{h_A^{0.75}} = \frac{\left(3500\ \frac{\text{rev}}{\text{min}}\right)\sqrt{150\ \frac{\text{gal}}{\text{min}}}}{(300\ \text{ft})^{0.75}} = 595$$

From Table 18.9, the pump should be a radial vane type. However, pumps achieve their highest efficiencies when specific speed exceeds 650. (See Fig. 18.8.) To increase the specific speed, the rotational speed can be increased, or the total added head can be decreased. Since the pump is direct-driven and 3600 rpm is the maximum speed for induction motors (see Table 18.8), the total added head should be divided evenly between two stages, or two pumps should be used in series.

In a two-stage system, the specific speed would be

$$n_s = \frac{\left(3500\ \frac{\text{rev}}{\text{min}}\right)\sqrt{150\ \frac{\text{gal}}{\text{min}}}}{\left(\frac{300\ \text{ft}}{2}\right)^{0.75}} = 1000$$

This is satisfactory for a radial vane pump.

Example 18.6

An induction motor turning at 1200 rpm is to be selected to drive a single-stage, single-suction centrifugal water pump through a direct drive. The total dynamic head added by the pump is 26 ft. The flow rate is 900 gal/min. What minimum size motor should be selected?

Solution

The specific speed is

$$n_s = \frac{n\sqrt{Q}}{h_A^{0.75}} = \frac{\left(1200\ \frac{\text{rev}}{\text{min}}\right)\sqrt{900\ \frac{\text{gal}}{\text{min}}}}{(26\ \text{ft})^{0.75}} = 3127$$

From Fig. 18.8, the pump efficiency will be approximately 82%.

From Table 18.5, the minimum motor horsepower is

$$P_{\text{hp}} = \frac{h_A\,Q(\text{SG})}{3956\eta_p} = \frac{(26\ \text{ft})\left(900\ \frac{\text{gal}}{\text{min}}\right)(1.0)}{\left(3956\ \frac{\text{ft-gal}}{\text{hp-min}}\right)(0.82)} = 7.2\ \text{hp}$$

From Table 18.7, select a 7.5 hp or larger motor.

Example 18.7

A single-stage pump driven by a 3600 rpm motor is currently delivering 150 gal/min. The total dynamic head is 430 ft. What would be the approximate increase in efficiency per stage if the single-stage pump is replaced by a double-stage pump?

---

[15]*Partial emission, forced vortex centrifugal pumps* allow operation down to specific speeds of 150 (3 in SI). Such pumps have been used for low-flow, high-head applications, such as high-pressure petrochemical cracking processes.

[16]*Homologous pumps* are geometrically similar. This means that each pump is a scaled up or down version of the others. Such pumps are said to belong to a *homologous family*.

Solution

The specific speed is

$$n_s = \frac{n\sqrt{Q}}{h_A^{0.75}} = \frac{\left(3600 \ \frac{\text{rev}}{\text{min}}\right)\sqrt{150 \ \frac{\text{gal}}{\text{min}}}}{(430 \ \text{ft})^{0.75}} = 467$$

From Fig. 18.8, the approximate efficiency is 45%.

In a two-stage pump, each stage adds half of the head. The specific speed per stage would be

$$n_s = \frac{n\sqrt{Q}}{h_A^{0.75}} = \frac{\left(3600 \ \frac{\text{rev}}{\text{min}}\right)\sqrt{150 \ \frac{\text{gal}}{\text{min}}}}{\left(\frac{430 \ \text{ft}}{2}\right)^{0.75}} = 785$$

From Fig. 18.8, the efficiency for this configuration is approximately 60%.

The increase in stage efficiency is 60% − 45% = 15%. Whether or not the cost of multistaging is worthwhile in this low-volume application would have to be determined.

16. CAVITATION

Cavitation is a spontaneous vaporization of the fluid inside the pump, resulting in a degradation of pump performance. Wherever the fluid pressure is less than the vapor pressure, small pockets of vapor will form. These pockets usually form only within the pump itself, although cavitation slightly upstream within the suction line is also possible. As the vapor pockets reach the surface of the impeller, the local high fluid pressure collapses them. Noise, vibration, impeller pitting, and structural damage to the pump casing are manifestations of cavitation.

Cavitation can be caused by any of the following conditions.

- discharge head far below the pump head at peak efficiency
- high suction lift or low suction head
- excessive pump speed
- high liquid temperature (i.e., high vapor pressure)

17. NET POSITIVE SUCTION HEAD

The occurrence of cavitation is predictable. Cavitation will occur when the net pressure in the fluid drops below the vapor pressure. This criterion is commonly stated in terms of head: Cavitation occurs when the available head is less than the required head for satisfactory operation. (See Eq. 18.34.)

The minimum fluid energy required at the pump inlet for satisfactory operation (i.e., the required head) is

known as the *net positive suction head required*, NPSHR.[17] NPSHR is a function of the pump and will be given by the pump manufacturer as part of the pump performance data.[18] NPSHR is dependent on the flow rate. However, if NPSHR is known for one flow rate, it can be determined for another flow rate from Eq. 18.34.

$$\frac{\text{NPSHR}_2}{\text{NPSHR}_1} = \left(\frac{Q_2}{Q_1}\right)^2 \qquad 18.29$$

Net positive suction head available, NPSHA, is the actual total fluid energy at the pump inlet. There are two different methods for calculating NPSHA, both of which are correct and will yield identical answers. Equation 18.30 is based on the conditions at the fluid surface at the top of an open fluid source (e.g., tank or reservoir). There is a potential energy term but no kinetic energy term. Equation 18.31 is based on the conditions at the immediate entrance (suction, subscript s) to the pump. At that point, some of the potential head has been converted to velocity head. Frictional losses are implicitly part of the reduced pressure head, as is the atmospheric pressure head. Since the pressure head, $h_{p(s)}$, is absolute, it includes the atmospheric pressure head, and the effect of higher altitudes is explicit in Eq. 18.30 and implicit in Eq. 18.31. If the source was pressurized instead of being open to the atmosphere, the pressure head would replace h_{atm} in Eq. 18.30 but would be implicit in $h_{p(s)}$ in Eq. 18.31.

$$\text{NPSHA} = h_{\text{atm}} + h_{z(s)} - h_{f(s)} - h_{\text{vp}} \qquad 18.30$$

$$\text{NPSHA} = h_{p(s)} + h_{v(s)} - h_{\text{vp}} \qquad 18.31$$

The net positive suction head available (NPSHA) for most positive displacement pumps includes a term for acceleration head.[19]

$$\text{NPSHA} = h_{\text{atm}} + h_{z(s)} - h_{f(s)} - h_{\text{vp}} - h_{\text{ac}} \qquad 18.32$$

$$\text{NPSHA} = h_{p(s)} + h_{v(s)} - h_{\text{vp}} - h_{\text{ac}} \qquad 18.33$$

If NPSHA is less than NPSHR, the fluid will cavitate. The criterion for cavitation is given by Eq. 18.34. (In practice, it is desirable to have a safety margin.)

$$\text{NPSHA} < \text{NPSHR} \quad \begin{bmatrix} \text{criterion for} \\ \text{cavitation} \end{bmatrix} \qquad 18.34$$

---

[17]If NPSHR (a head term) is multiplied by the fluid specific weight, it is known as the *net inlet pressure required*, NIPR. Similarly, NPSHA can be converted to NIPA.

[18]It is also possible to calculate NPSHR from other information, such as suction specific speed. However, this still depends on information provided by the manufacturer.

[19]The friction loss and the acceleration terms are both maximum values, but they do not occur in phase. Combining them is conservative.

Example 18.8

2.0 ft³/sec (56 L/s) of 60°F (16°C) water are pumped from an elevated feed tank to an open reservoir through 6 in (15.2 cm), schedule-40 steel pipe, as shown. The friction loss for the piping and fittings in the suction line is 2.6 ft (0.9 m). The friction loss for the piping and fittings in the discharge line is 13 ft (4.3 m). The atmospheric pressure is 14.7 psia (101 kPa). What is the NPSHA?

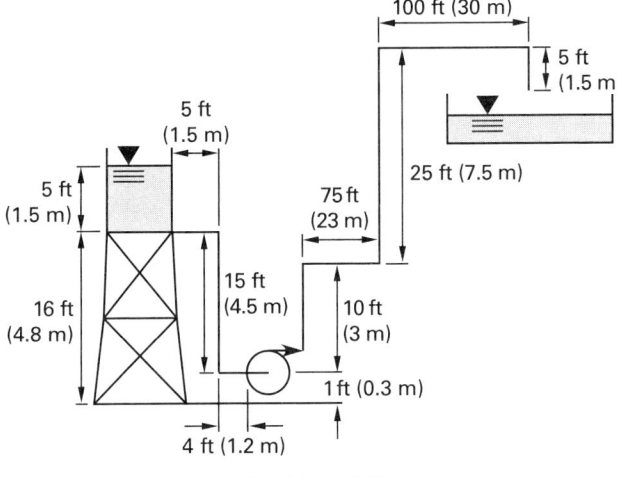

(not to scale)

SI Solution

The density of water is approximately 1000 kg/m³. The atmospheric head is

$$h_{\text{atm}} = \frac{p}{\rho g} = \frac{(101 \text{ kPa})\left(1000 \frac{\text{Pa}}{\text{kPa}}\right)}{\left(1000 \frac{\text{kg}}{\text{m}^3}\right)\left(9.81 \frac{\text{m}}{\text{s}^2}\right)}$$

$$= 10.3 \text{ m}$$

For 16°C water, the vapor pressure is approximately 0.01818 bars. The vapor pressure head is

$$h_{\text{vp}} = \frac{p}{\rho g} = \frac{(0.01818 \text{ bar})\left(1 \times 10^5 \frac{\text{Pa}}{\text{bar}}\right)}{\left(1000 \frac{\text{kg}}{\text{m}^3}\right)\left(9.81 \frac{\text{m}}{\text{s}^2}\right)}$$

$$= 0.2 \text{ m}$$

From Eq. 18.30, the NPSHA is

$$\text{NPSHA} = h_{\text{atm}} + h_{z(s)} - h_{f(s)} - h_{\text{vp}}$$

$$= 10.3 \text{ m} + 1.5 \text{ m} + 4.8 \text{ m} - 0.3 \text{ m}$$

$$- 0.9 \text{ m} - 0.2 \text{ m}$$

$$= 15.2 \text{ m}$$

Customary U.S. Solution

The specific weight of water is approximately 62.4 lbf/ft³. The atmospheric head is

$$h_{\text{atm}} = \frac{p}{\gamma} = \frac{\left(14.7 \frac{\text{lbf}}{\text{in}^2}\right)\left(12 \frac{\text{in}}{\text{ft}}\right)^2}{62.4 \frac{\text{lbf}}{\text{ft}^3}}$$

$$= 33.9 \text{ ft}$$

For 60°F water, the vapor pressure head is 0.59 ft. Use 0.6 ft.

From Eq. 18.30, the NPSHA is

$$\text{NPSHA} = h_{\text{atm}} + h_{z(s)} - h_{f(s)} - h_{\text{vp}}$$

$$= 33.9 \text{ ft} + 5 \text{ ft} + 16 \text{ ft} - 1 \text{ ft} - 2.6 \text{ ft} - 0.6 \text{ ft}$$

$$= 50.7 \text{ ft}$$

18. PREVENTING CAVITATION

Cavitation is prevented by increasing NPSHA or decreasing NPSHR. NPSHA can be increased by

- increasing the height of the fluid source
- lowering the pump
- reducing friction and minor losses by shortening the suction line or using a larger pipe size
- reducing the temperature of the fluid at the pump entrance
- pressurizing the fluid supply tank
- reducing the flow rate or velocity (i.e., reducing the pump speed)

NPSHR can be reduced by

- placing a throttling valve or restriction in the discharge line[20]
- using an oversized pump
- using a double-suction pump
- using an impeller with a larger eye
- using an inducer

High NPSHR applications, such as boiler feed pumps needing 150–250 ft (50–80 m), should use one or more booster pumps in front of each high-NPSHR pump. Such booster pumps are typically single-stage, double-suction pumps running at low speed. Their NPSHR can be 25 ft (8 m) or less.

Throttling the input line to a pump and venting or evacuating the receiving tank both increase cavitation. Throttling the input line increases the friction head and

---

[20]This will increase the total head, h_A, added by the pump, thereby reducing the pump's output and driving the pump's operating point into a region of lower NPSHR.

decreases NPSHA. Evacuating the receiving tank increases the flow rate, increasing NPSHR while simultaneously increasing the friction head and reducing NPSHA.

19. CAVITATION COEFFICIENT

The *cavitation coefficient* (or *cavitation number*), σ, is a dimensionless number that can be used in modeling and extrapolating experimental results. The actual cavitation coefficient is compared with the *critical cavitation number* obtained experimentally. If the actual cavitation number is less than the critical cavitation number, cavitation will occur. Absolute pressure must be used for the fluid pressure term, p.

$$\sigma < \sigma_{\mathrm{cr}} \quad \text{[criterion for cavitation]} \qquad \textit{18.35}$$

$$\sigma = \frac{2(p - p_{\mathrm{vp}})}{\rho v^2} = \frac{\mathrm{NPSHA}}{h_A} \qquad \text{[SI]} \quad \textit{18.36(a)}$$

$$\sigma = \frac{2g_c(p - p_{\mathrm{vp}})}{\rho v^2} = \frac{\mathrm{NPSHA}}{h_A} \qquad \text{[U.S.]} \quad \textit{18.36(b)}$$

The two forms of Eq. 18.36 yield slightly different results. The first form is essentially the ratio of the net pressure available for collapsing a vapor bubble to the velocity pressure creating the vapor. It is useful in model experiments. The second form is applicable to tests of production model pumps.

20. SUCTION SPECIFIC SPEED

The formula for *suction specific speed*, n_{ss}, can be derived by substituting NPSHR for total head in the expression for specific speed. Q and $\dot{V}$ are halved for double-suction pumps.

$$n_{\mathrm{ss}} = \frac{n\sqrt{\dot{V}}}{(\mathrm{NPSHR\ in\ m})^{0.75}} \qquad \text{[SI]} \quad \textit{18.37(a)}$$

$$n_{\mathrm{ss}} = \frac{n\sqrt{Q}}{(\mathrm{NPSHR\ in\ ft})^{0.75}} \qquad \text{[U.S.]} \quad \textit{18.37(b)}$$

Suction specific speed is an index of the suction characteristics of the impeller. Ideally, it should be approximately 8500 (165 in SI) for both single- and double-suction pumps. This assumes the pump is operating at or near its point of optimum efficiency.

Suction specific speed can be used to determine the maximum recommended operating speed by substituting 8500 (165 in SI) for n_{ss} in Eq. 18.37 and solving for n.

If the suction specific speed is known, it can be used to determine the NPSHR. If the pump is known to be operating at or near its optimum efficiency, an approximate NPSHR value can be found by substituting 8500 (165 in SI) for n_{ss} in Eq. 18.37 and solving for NPSHR.

Suction specific speed available, SA, is obtained when NPSHA is substituted for total head in the expression for specific speed. The suction specific speed available must be less than the suction specific speed required to prevent cavitation.[21]

21. PUMP PERFORMANCE CURVES

For a given impeller diameter and constant speed, the head added will decrease as the flow rate increases. This is shown graphically on the *pump performance curve* (*pump curve*) supplied by the pump manufacturer. Other operating characteristics (e.g., power requirement, NPSHR, and efficiency) also vary with flow rate, and these are usually plotted on a common graph, as shown in Fig. 18.10.[22] Manufacturers' pump curves show performance over a limited number of calibration speeds. If an operating point is outside the range of published curves, the affinity laws can be used to estimate the speed at which the pump gives the required performance.

Figure 18.10 *Pump Performance Curves*

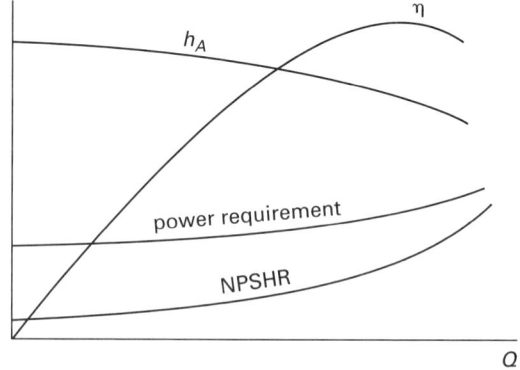

On the pump curve, the *shutoff point* (also known as *churn*) corresponds to a closed discharge valve (i.e., zero flow); the *rated point* is where the pump operates with rated 100% of capacity and head; the *overload point* corresponds to 65% of the rated head.

Figure 18.10 is for a pump with a fixed impeller diameter and rotational speed. The characteristics of a pump operated over a range of speeds or for different impeller diameters are illustrated in Fig. 18.11.

22. SYSTEM CURVES

A *system curve* (or *system performance curve*) is a plot of the static and friction energy losses experienced by the fluid for different flow rates. Unlike the pump curve,

---

[21]Since speed and flow rate are constants, this is another way of saying NPSHA must equal or exceed NPSHR.

[22]The term *pump curve* is commonly used to designate the h_A versus Q characteristics, whereas *pump characteristics curve* refers to all of the pump data.

Figure 18.11 *Centrifugal Pump Characteristics Curves*

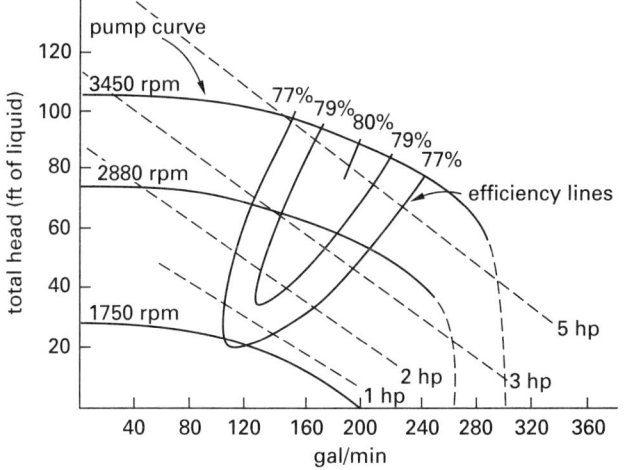

(a) variable speed

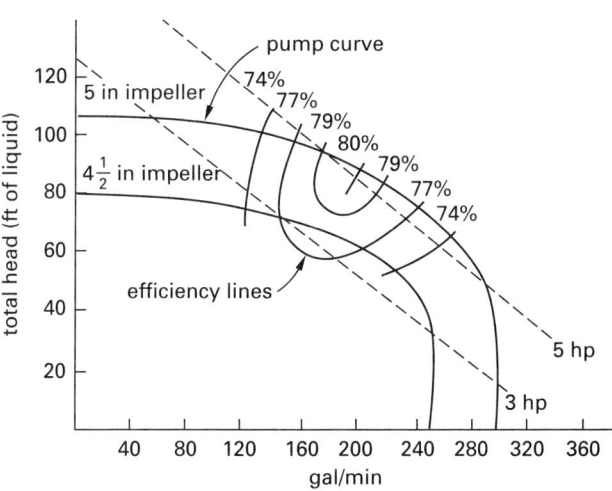

(b) variable impeller diameter

which depends only on the pump, the system curve depends only on the configuration of the suction and discharge lines. (The following equations assume equal pressures at the fluid source and destination surfaces, which is the case for pumping from one atmospheric reservoir to another. The velocity head is insignificant and is disregarded.)

$$h_A = h_z + h_f \qquad 18.38$$

$$h_z = h_{z(d)} - h_{z(s)} \qquad 18.39$$

$$h_f = h_{f(s)} + h_{f(d)} \qquad 18.40$$

If the fluid reservoirs are large, or if the fluid reservoir levels are continually replenished, the net static suction head $(h_{z(d)} - h_{z(s)})$ will be constant for all flow rates. The friction loss, h_f, varies with v^2 (and, therefore, with Q^2) in the Darcy friction formula. This makes it easy to

find friction losses for other flow rates (subscript 2) once one friction loss (subscript 1) is known.[23]

$$\frac{h_{f,1}}{h_{f,2}} = \left(\frac{Q_1}{Q_2}\right)^2 \qquad 18.41$$

Figure 18.12 illustrates a system curve following Eq. 18.38 with a positive added head (i.e., a fluid source below the fluid destination). The system curve is shifted upward, intercepting the vertical axis at some positive value of h_A.

Figure 18.12 *System Curve*

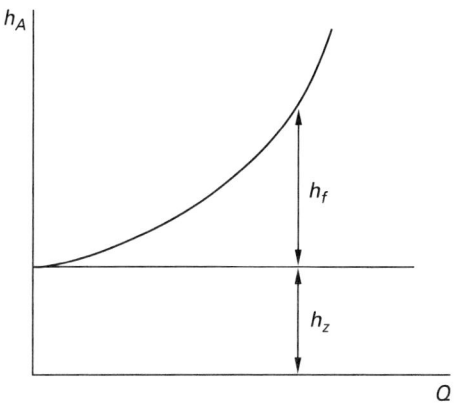

23. OPERATING POINT

The intersection of the pump curve and the system curve determines the *operating point*, as shown in Fig. 18.13. The operating point defines the system head and system flow rate.

Figure 18.13 *Extreme Operating Points*

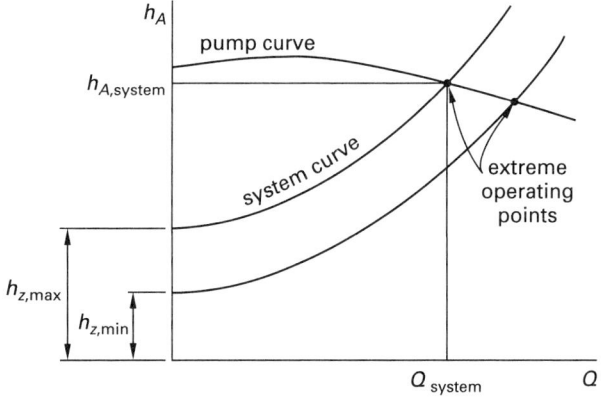

---

[23]Equation 18.41 implicitly assumes that the friction factor, f, is constant. This may be true over a limited range of flow rates, but it is not true over large ranges. Nevertheless, Eq. 18.41 is often used to quickly construct preliminary versions of the system curve.

When selecting a pump, the system curve is plotted on manufacturers' pump curves for different speeds and/or impeller diameters (i.e., Fig. 18.11). There will be several possible operating points corresponding to the various pump curves shown. Generally, the design operating point should be close to the highest pump efficiency. This, in turn, will determine speed and impeller diameter.

In some systems, the static head varies as the source reservoir is drained or as the destination reservoir fills. The system head is then defined by a pair of matching system friction curves intersecting the pump curve. The two intersection points are called the *extreme operating points*—the maximum and minimum capacity requirements.

After a pump is installed, it may be desired to change the operating point. This can be done without replacing the pump by placing a throttling valve in the discharge line. The operating point can then be moved along the pump curve by partially opening or closing the valve, as is illustrated in Fig. 18.14. (A throttling valve should never be placed in the suction line since that would reduce NPSHA.)

Figure 18.14 Throttling the Discharge

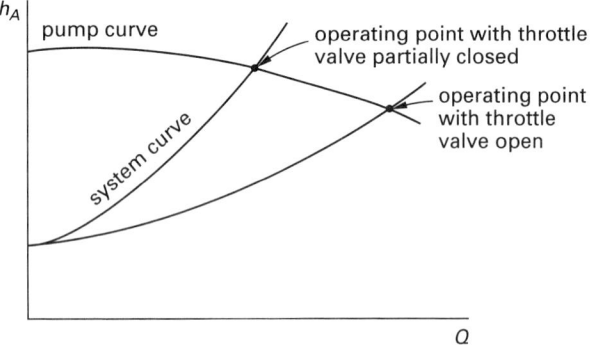

24. PUMPS IN PARALLEL

Parallel operation is obtained by having two pumps discharging into a common header. This type of connection is advantageous when the system demand varies greatly or when high reliability is required. A single pump providing total flow would have to operate far from its optimum efficiency at one point or another. With two pumps in parallel, one can be shut down during low demand. This allows the remaining pump to operate close to its optimum efficiency point.

Figure 18.15 illustrates that parallel operation increases the capacity of the system while maintaining the same total head.

The performance curve for a set of pumps in parallel can be plotted by adding the capacities of the two pumps at various heads. A second pump will operate only when its discharge head is greater than the discharge head of the

Figure 18.15 Pumps Operating in Parallel

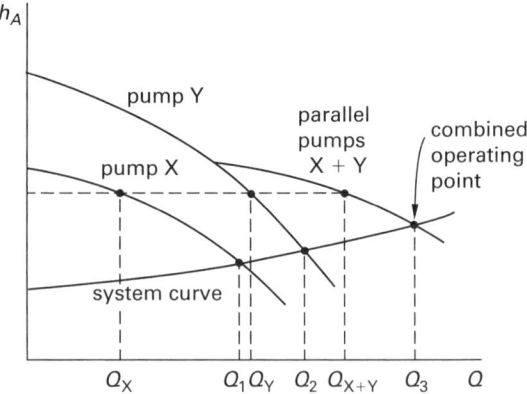

pump already running. Capacity does not increase at heads above the maximum head of the smaller pump.

When the parallel performance curve is plotted with the system head curve, the operating point is the intersection of the system curve with the X + Y curve. With pump X operating alone, the capacity is given by Q_1. When pump Y is added, the capacity increases to Q_3 with a slight increase in total head.

25. PUMPS IN SERIES

Series operation is achieved by having one pump discharge into the suction of the next. This arrangement is used primarily to increase the discharge head, although a small increase in capacity also results. (See Fig. 18.16.)

The performance curve for a set of pumps in series can be plotted by adding the heads of the two pumps at various capacities.

Figure 18.16 Pumps Operating in Series

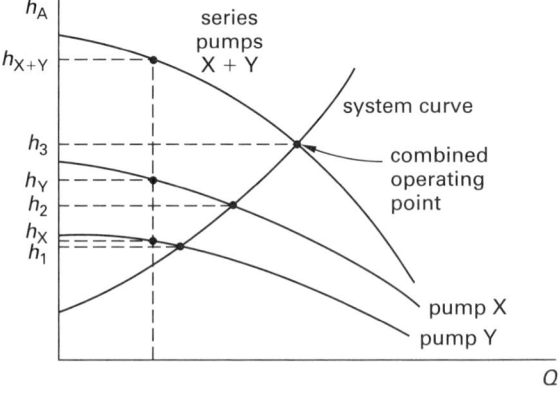

26. AFFINITY LAWS

Most parameters (impeller diameter, speed, and flow rate) determining a specific pump's performance can be modified. If the impeller diameter is held constant and the speed is varied, the following ratios are maintained with no change in efficiency.[24]

$$\frac{Q_2}{Q_1} = \frac{n_2}{n_1} \qquad 18.42$$

$$\frac{h_2}{h_1} = \left(\frac{n_2}{n_1}\right)^2 = \left(\frac{Q_2}{Q_1}\right)^2 \qquad 18.43$$

$$\frac{P_2}{P_1} = \left(\frac{n_2}{n_1}\right)^3 = \left(\frac{Q_2}{Q_1}\right)^3 \qquad 18.44$$

If the speed is held constant and the impeller size is reduced (i.e., the impeller is trimmed), while keeping the pump body, volute, shaft diameter, and suction and discharge openings the same, the following ratios may be used.[25]

$$\frac{Q_2}{Q_1} = \frac{D_2}{D_1} \qquad 18.45$$

$$\frac{h_2}{h_1} = \left(\frac{D_2}{D_1}\right)^2 \qquad 18.46$$

$$\frac{P_2}{P_1} = \left(\frac{D_2}{D_1}\right)^3 \qquad 18.47$$

The affinity laws are based on the assumption that the efficiency stays the same. In reality, larger pumps are somewhat more efficient than smaller pumps, and extrapolations to greatly different sizes should be avoided. Equation 18.48 can be used to estimate the efficiency of a differently sized pump. The dimensionless exponent, n, varies from 0 to approximately 0.26, with 0.2 being a typical value.

$$\frac{1 - \eta_{\text{smaller}}}{1 - \eta_{\text{larger}}} = \left(\frac{D_{\text{larger}}}{D_{\text{smaller}}}\right)^n \qquad 18.48$$

Example 18.9

A pump operating at 1770 rpm delivers 500 gal/min against a total head of 200 ft. Changes in the piping system have increased the total head to 375 ft. At what speed should this pump be operated to achieve this new head at the same efficiency?

Solution

From Eq. 18.43,

$$n_2 = n_1 \sqrt{\frac{h_2}{h_1}} = \left(1770 \ \frac{\text{rev}}{\text{min}}\right) \sqrt{\frac{375 \ \text{ft}}{200 \ \text{ft}}}$$

$$= 2424 \ \text{rpm}$$

---

[24]See Sec. 18.27 if the entire pump is scaled to a different size.

[25]One might ask, "How is it possible to change a pump's impeller diameter?" In practice, a different impeller may be available from the manufacturer, but more often the impeller is taken out and shaved down on a lathe. Equation 18.45, Eq. 18.46, and Eq. 18.47 are limited in use to radial flow machines, and with reduced accuracy, to mixed-flow impellers. Changing the impeller diameter significantly impacts other design relationships, and the accuracy of performance prediction decreases if the diameter is changed much more than 20%.

Example 18.10

A pump is required to pump 500 gal/min against a total dynamic head of 425 ft. The hydraulic system has no static head change. Only the 1750 rpm performance curve is known for the pump. At what speed must the pump be turned to achieve the desired performance with no change in efficiency or impeller size?

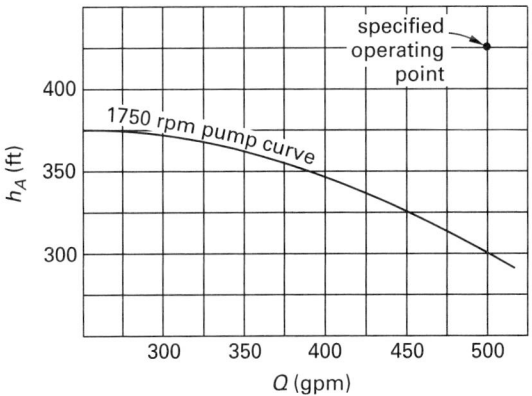

Solution

A flow of 500 gal/min with a head of 425 ft does not correspond to any point on the 1750 rpm curve.

From Eq. 18.43, the quantity h/Q^2 is constant.

$$\frac{h}{Q^2} = \frac{425 \ \text{ft}}{\left(500 \ \frac{\text{gal}}{\text{min}}\right)^2} = 1.7 \times 10^{-3} \ \text{ft-min}^2/\text{gal}^2$$

In order to use the affinity laws, the operating point on the 1750 rpm curve must be determined. Random values of Q are chosen and the corresponding values of h are determined such that the ratio h/Q^2 is unchanged.

Q	h
475	383
450	344
425	307
400	272

These points are plotted as the system curve. The intersection of the system and 1750 rpm pump curve at 440 gal/min defines the operating point at 1750 rpm. From Eq. 18.42, the required pump speed is

$$n_2 = \frac{n_1 Q_2}{Q_1} = \frac{\left(1750 \ \frac{\text{rev}}{\text{min}}\right)\left(500 \ \frac{\text{gal}}{\text{min}}\right)}{440 \ \frac{\text{gal}}{\text{min}}}$$

$$= 1989 \ \text{rpm}$$

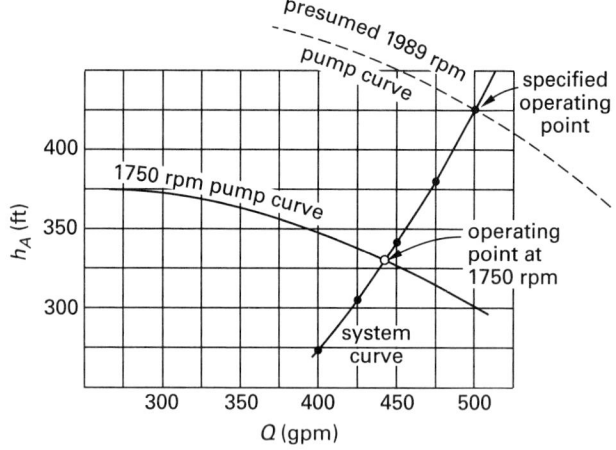

27. PUMP SIMILARITY

The performance of one pump can be used to predict the performance of a *dynamically similar (homologous)* pump. This can be done by using Eq. 18.49 through Eq. 18.54.

$$\frac{n_1 D_1}{\sqrt{h_1}} = \frac{n_2 D_2}{\sqrt{h_2}} \qquad 18.49$$

$$\frac{Q_1}{D_1^2 \sqrt{h_1}} = \frac{Q_2}{D_2^2 \sqrt{h_2}} \qquad 18.50$$

$$\frac{P_1}{\rho_1 D_1^2 h_1^{1.5}} = \frac{P_2}{\rho_2 D_2^2 h_2^{1.5}} \qquad 18.51$$

$$\frac{Q_1}{n_1 D_1^3} = \frac{Q_2}{n_2 D_2^3} \qquad 18.52$$

$$\frac{P_1}{\rho_1 n_1^3 D_1^5} = \frac{P_2}{\rho_2 n_2^3 D_2^5} \qquad 18.53$$

$$\frac{n_1 \sqrt{Q_1}}{h_1^{0.75}} = \frac{n_2 \sqrt{Q_2}}{h_2^{0.75}} \qquad 18.54$$

These *similarity laws* (also known as *scaling laws*) assume that both pumps

- operate in the turbulent region
- have the same pump efficiency
- operate at the same percentage of wide-open flow

Similar pumps also will have the same specific speed and cavitation number.

As with the affinity laws, these relationships assume that the efficiencies of the larger and smaller pumps are the same. In reality, larger pumps will be more efficient than smaller pumps. Therefore, extrapolations to much larger or much smaller sizes should be avoided.

Example 18.11

A 6 in pump operating at 1770 rpm discharges 1500 gal/min of cold water (SG = 1.0) against an 80 ft head at 85% efficiency. A homologous 8 in pump

operating at 1170 rpm is being considered as a replacement. (a) What total head and capacity can be expected from the new pump? (b) What would be the new motor horsepower requirement?

Solution

(a) From Eq. 18.49,

$$h_2 = \left(\frac{D_2 n_2}{D_1 n_1}\right)^2 h_1 = \left(\frac{(8 \text{ in})\left(1170 \frac{\text{rev}}{\text{min}}\right)}{(6 \text{ in})\left(1770 \frac{\text{rev}}{\text{min}}\right)}\right)^2 (80 \text{ ft})$$

$$= 62.14 \text{ ft}$$

From Eq. 18.52,

$$Q_2 = \left(\frac{n_2 D_2^3}{n_1 D_1^3}\right) Q_1 = \left(\frac{\left(1170 \frac{\text{rev}}{\text{min}}\right)(8 \text{ in})^3}{\left(1770 \frac{\text{rev}}{\text{min}}\right)(6 \text{ in})^3}\right)\left(1500 \frac{\text{gal}}{\text{min}}\right)$$

$$= 2350 \text{ gal/min}$$

(b) From Table 18.5, the hydraulic horsepower is

$$\text{WHP}_2 = \frac{h_2 Q_2 (\text{SG})}{3956} = \frac{(62.14 \text{ ft})\left(2350 \frac{\text{gal}}{\text{min}}\right)(1.0)}{3956 \frac{\text{ft-gal}}{\text{hp-min}}}$$

$$= 36.91 \text{ hp}$$

From Eq. 18.48,

$$\eta_{\text{larger}} = 1 - \frac{1 - \eta_{\text{smaller}}}{\left(\frac{D_{\text{larger}}}{D_{\text{smaller}}}\right)^n} = 1 - \frac{1 - 0.85}{\left(\frac{8 \text{ in}}{6 \text{ in}}\right)^{0.2}} = 0.858$$

$$\text{BHP}_2 = \frac{\text{WHP}_2}{\eta_p} = \frac{36.92 \text{ hp}}{0.858} = 43.0 \text{ hp}$$

28. PUMPING LIQUIDS OTHER THAN COLD WATER

Many liquid pump parameters are determined from tests with cold, clear water at 85°F (29°C). The following guidelines can be used when pumping water at other temperatures or when pumping other liquids.

- Head developed is independent of the liquid's specific gravity. Pump performance curves from tests with water can be used with other Newtonian fluids (e.g., gasoline, alcohol, and aqueous solutions) having similar viscosities.

- The hydraulic horsepower depends on the specific gravity of the liquid. If the pump characteristic curve is used to find the operating point, multiply the horsepower reading by the specific gravity. Table 18.5 and Table 18.6 incorporate the specific gravity term in the calculation of hydraulic power where required.

- Efficiency is not affected by changes in temperature that cause only the specific gravity to change.

- Efficiency is nominally affected by changes in temperature that cause the viscosity to change. Equation 18.55 is an approximate relationship suggested by the Hydraulics Institute when extrapolating the efficiency (in decimal form) from cold water to hot water. n is an experimental exponent established by the pump manufacturer, generally in the range of 0.05–0.1.

$$\eta_{\text{hot}} = 1 - (1 - \eta_{\text{cold}}) \left(\frac{\nu_{\text{hot}}}{\nu_{\text{cold}}} \right)^n \qquad 18.55$$

- NPSHA depends significantly on liquid temperature.

- NPSHR is not significantly affected by variations in the liquid temperature.

- When hydrocarbons are pumped, the NPSHR determined from cold water can usually be reduced. This reduction is apparently due to the slow vapor release of complex organic liquids. If the hydrocarbon's vapor pressure at the pumping temperature is known, Fig. 18.17 will give the percentage of the cold-water NPSHR.

- Pumping many fluids requires expertise that goes far beyond simply extrapolating parameters in proportion to the fluid's specific gravity. Such special cases include pumping liquids containing abrasives, liquids that solidify, highly corrosive liquids, liquids with vapor or gas, highly viscous fluids, paper stock, and hazardous fluids.

- Head, flow rate, and efficiency are all reduced when pumping highly viscous non-Newtonian fluids. No exact method exists for determining the reduction factors, other than actual tests of an installation using both fluids. Some sources have published charts of correction factors based on tests over limited viscosity and size ranges.[26]

Figure 18.17 Hydrocarbon NPSHR Correction Factor

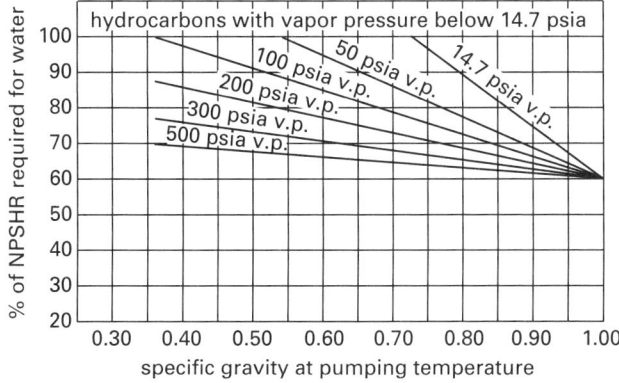

---

[26]A chart published by the Hydraulics Institute is widely distributed.

Example 18.12

A centrifugal pump has an NIPR (NPSHR) of 12 psi based on cold water. 10°F liquid isobutane has a specific gravity of 0.60 and a vapor pressure of 15 psia. What NPSHR should be used with 10°F liquid isobutane?

Solution

From Fig. 18.17, the intersection of a specific gravity of 0.60 and 15 psia is above the horizontal 100% line. The full NIPR of 12 psi should be used.

29. TURBINE SPECIFIC SPEED

Like centrifugal pumps, turbines are classified according to the manner in which the impeller extracts energy from the fluid flow. This is measured by the turbine-specific speed equation, which is different from the equation used to calculate specific speed for pumps.

$$n_s = \frac{n \sqrt{P_{\text{kW}}}}{h_t^{1.25}} \qquad \text{[SI]} \qquad 18.56(a)$$

$$n_s = \frac{n \sqrt{P_{\text{hp}}}}{h_t^{1.25}} \qquad \text{[U.S.]} \qquad 18.56(b)$$

30. IMPULSE TURBINES

An *impulse turbine* consists of a rotating shaft (called a *turbine runner*) on which buckets or blades are mounted. (This is commonly called a *Pelton wheel*.[27]) A jet of water (or other fluid) hits the buckets and causes the turbine to rotate. The kinetic energy of the jet is converted into rotational kinetic energy. The jet is essentially at atmospheric pressure. (See Fig. 18.18.)

Impulse turbines are generally employed where the available head is very high, above 800–1600 ft (250–500 m). (There is no exact value for the limiting head, hence the range. What is important is that impulse turbines are *high-head turbines*.)

The total available head in this installation is h_t, but not all of this energy can be extracted. Some of the energy is lost to friction in the penstock. Minor losses also occur, but these small losses are usually disregarded. In the penstock, immediately before entering the nozzle, the remaining head is divided between the pressure head and the velocity head.[28]

$$h' = h_t - h_f = (h_p + h_v)_{\text{penstock}} \qquad 18.57$$

---

[27]In a Pelton wheel turbine, the spoon-shaped buckets are divided into two halves, with a ridge between the halves. Half of the water is thrown to each side of the bucket. A Pelton wheel is known as a *tangential turbine* (*tangential wheel*) because the centerline of the jet is directed at the periphery of the wheel.
[28]Care must be taken to distinguish between the conditions existing in the penstock, the nozzle throat, and the jet itself. The velocity in the nozzle throat and jet will be the same, but this is different from the penstock velocity. Similarly, the pressure in the jet is zero, although it is nonzero in the penstock.

Figure 18.18 *Impulse Turbine Installation*

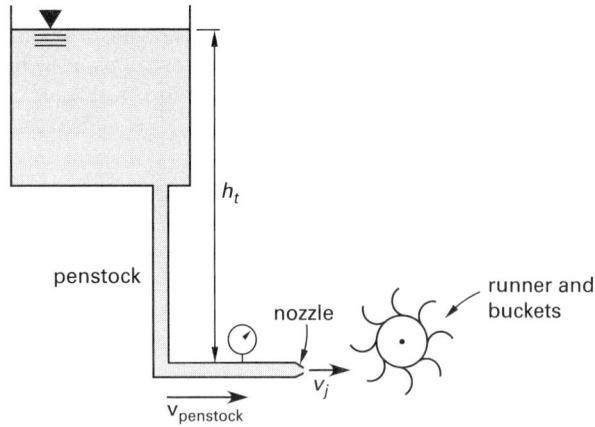

Another loss, h_n, occurs in the nozzle itself. The head remaining to turn the turbine is

$$h'' = h' - h_n$$
$$= h_t - h_f - h_n \qquad 18.58$$

h_f is calculated from either the Darcy or the Hazen-Williams equation. The nozzle loss is calculated from the *nozzle coefficient* (*coefficient of velocity*), C_v.

$$h_n = h'(1 - C_v^2) \qquad 18.59$$

The total head at the nozzle exit is converted into velocity head according to the Torricelli equation.

$$v_j = \sqrt{2gh''} = C_v\sqrt{2gh'} \qquad 18.60$$

In order to maximize power output, buckets are usually designed to partially reverse the direction of the water jet flow. The forces on the turbine buckets can be found from the impulse-momentum equations. If the water is turned through an angle θ and the wheel's *tangential velocity* is v_b, the energy transmitted by each unit mass of water to the turbine runner is[29]

$$E = v_b(v_j - v_b)(1 - \cos\theta) \qquad \text{[SI]} \quad 18.61(a)$$

$$E = \left(\frac{v_b(v_j - v_b)}{g_c}\right)(1 - \cos\theta) \qquad \text{[U.S.]} \quad 18.61(b)$$

$$v_b = \frac{2\pi n r}{60 \frac{\text{sec}}{\text{min}}} = \omega r \qquad 18.62$$

The theoretical *turbine power* is found by multiplying Eq. 18.61 by the mass flow rate. The actual power will be less than the theoretical output. Efficiencies are in the range of 80–90%, with the higher efficiencies being associated with turbines having two or more jets per runner. For a mass flow rate in kg/s, the theoretical

---

[29]$\theta = 180°$ would be ideal. However, the actual angle is limited to approximately 165° to keep the deflected jet out of the way of the incoming jet.

power (in kilowatts) will be as shown in Eq. 18.63(a). For a mass flow rate in lbm/sec, the theoretical horse-power will be as shown in Eq. 18.63(b).

$$P_{th} = \frac{\dot{m}E}{1000 \frac{\text{W}}{\text{kW}}} \qquad \text{[SI]} \quad 18.63(a)$$

$$P_{th} = \frac{\dot{m}E}{550 \frac{\text{ft-lbf}}{\text{hp-sec}}} \qquad \text{[U.S.]} \quad 18.63(b)$$

Example 18.13

A Pelton wheel impulse turbine develops 100 hp (brake) while turning at 500 rpm. The water is supplied from a penstock with an internal area of 0.3474 ft². The water subsequently enters a nozzle with a reduced flow area. The total head is 200 ft before nozzle loss. The turbine efficiency is 80%, and the nozzle coefficient, C_v, is 0.95. Disregard penstock friction losses. What are the (a) flow rate (in ft³/sec), (b) area of the jet, and (c) pressure head in the penstock just before the nozzle?

Solution

(a) Use Eq. 18.60, with $h' = 200$ ft, to find the jet velocity.

$$v_j = C_v\sqrt{2gh'} = 0.95\sqrt{(2)\left(32.2 \frac{\text{ft}}{\text{sec}^2}\right)(200 \text{ ft})}$$
$$= 107.8 \text{ ft/sec}$$

From Eq. 18.59, the nozzle loss is

$$h_n = h'(1 - C_v^2) = (200 \text{ ft})\left(1 - (0.95)^2\right)$$
$$= 19.5 \text{ ft}$$

From Table 18.5, the flow rate is

$$\dot{V} = \frac{8.814P}{h_A(\text{SG})\eta} = \frac{\left(8.814 \frac{\text{ft}^4}{\text{hp-sec}}\right)(100 \text{ hp})}{(200 \text{ ft} - 19.5 \text{ ft})(1)(0.8)}$$
$$= 6.104 \text{ ft}^3/\text{sec}$$

(b) The jet area is

$$A_j = \frac{\dot{V}}{v_j} = \frac{6.104 \frac{\text{ft}^3}{\text{sec}}}{107.8 \frac{\text{ft}}{\text{sec}}} = 0.0566 \text{ ft}^2$$

(c) The velocity in the penstock is

$$v_{penstock} = \frac{\dot{V}}{A} = \frac{6.104 \frac{\text{ft}^3}{\text{sec}}}{0.3474 \text{ ft}^2}$$
$$= 17.57 \text{ ft/sec}$$

The pressure head in the penstock is

$$h_p = h' - h_v = h' - \frac{v^2}{2g}$$

$$= 200 \text{ ft} - \frac{\left(17.57 \frac{\text{ft}}{\text{sec}}\right)^2}{(2)\left(32.2 \frac{\text{ft}}{\text{sec}^2}\right)}$$

$$= 195.2 \text{ ft}$$

31. REACTION TURBINES

Reaction turbines (also known as *Francis turbines* or *radial-flow turbines*) are essentially centrifugal pumps operating in reverse. (See Fig. 18.19.) They are used when the total available head is small, typically below 600–800 ft (183–244 m). However, their energy conversion efficiency is higher than that of impulse turbines, typically in the 85–95% range.

In a reaction turbine, water enters the turbine housing with a pressure greater than atmospheric pressure. The water completely surrounds the turbine runner (impeller) and continues through the draft tube. There is no vacuum or air pocket between the turbine and the tailwater.

All of the power, affinity, and similarity relationships used with centrifugal pumps can be used with reaction turbines.

Figure 18.19 Reaction Turbine

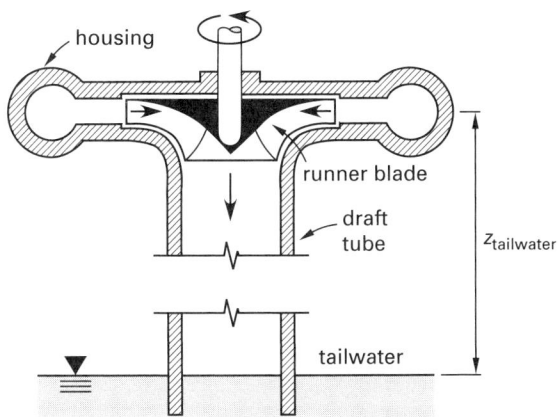

Example 18.14

A reaction turbine with a draft tube develops 500 hp (brake) when 50 ft^3/sec water flow through it. Water enters the turbine at 20 ft/sec with a 100 ft pressure head. The elevation of the turbine above the tailwater level is 10 ft. Disregarding friction, what are the (a) total available head and (b) turbine efficiency?

Solution

(a) The available head is the difference between the forebay and tailwater elevations. The tailwater depression is known, but the height of the forebay above the turbine is not known. However, at the turbine entrance, this unknown potential energy has been converted to pressure and velocity head. The total available head (exclusive of friction) is

$$h_t = z_{\text{forebay}} - z_{\text{tailwater}} = h_p + h_v - z_{\text{tailwater}}$$

$$= 100 \text{ ft} + \frac{\left(20 \frac{\text{ft}}{\text{sec}}\right)^2}{(2)\left(32.2 \frac{\text{ft}}{\text{sec}^2}\right)} - (-10 \text{ ft})$$

$$= 116.2 \text{ ft}$$

(b) From Table 18.5, the theoretical hydraulic horsepower is

$$P_{\text{th}} = \frac{h_A \dot{V}(\text{SG})}{8.814} = \frac{(116.2 \text{ ft})\left(50 \frac{\text{ft}^3}{\text{sec}}\right)(1.0)}{8.814 \frac{\text{ft}^4}{\text{hp-sec}}}$$

$$= 659.2 \text{ hp}$$

The efficiency of the turbine is

$$\eta = \frac{P_{\text{brake}}}{P_{\text{th}}} = \frac{500 \text{ hp}}{659.2 \text{ hp}} = 0.758 \quad (75.8\%)$$

32. TYPES OF REACTION TURBINES

Each of the three types of turbines is associated with a range of specific speeds.

- *Axial-flow reaction turbines* (also known as *propeller turbines*) are used for low heads, high rotational speeds, and large flow rates. (See Fig. 18.20.) These propeller turbines operate with specific speeds in the 70–260 range (266–988 in SI). Their best efficiencies, however, are produced with specific speeds between 120 and 160 (460 and 610 in SI).

- For *mixed-flow reaction turbines*, the specific speed varies from 10 to 90 (38 to 342 in SI). Best efficiencies are found in the 40 to 60 (150 to 230 in SI) range with heads below 600 ft to 800 ft (180 m to 240 m).

- *Radial-flow reaction turbines* have the lowest flow rates and specific speeds but are used when heads are high. These turbines have specific speeds between 1 and 20 (3.8 and 76 in SI).

33. HYDROELECTRIC GENERATING PLANTS

In a typical hydroelectric generating plant using reaction turbines, the turbine is generally housed in a *powerhouse*, with water conducted to the turbine through the *penstock* piping. Water originates in a reservoir, dam, or

Figure 18.20 *Axial-Flow Turbine*

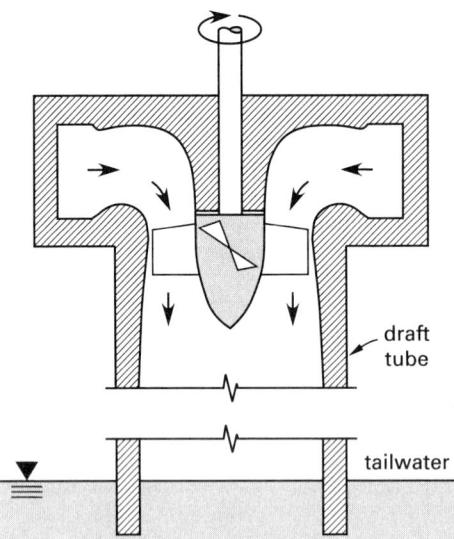

Figure 18.21 *Typical Hydroelectric Plant*

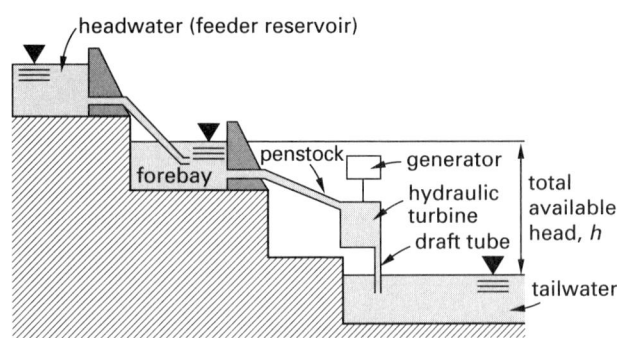

forebay (in the instance where the reservoir is a long distance from the turbine). (See Fig. 18.21.)

After the water passes through the turbine, it is discharged through the draft tube to the receiving reservoir, known as the *tailwater*. The difference between the elevations of the turbine and the surface of the tailwater is the *tailwater depression*. The *draft tube* is used to keep the turbine up to 15 ft (5 m) above the tailwater surface,

while still being able to extract the total available head. If a draft tube is not employed, water may be returned to the tailwater by way of a channel known as the *tail race*. The turbine, draft tube, and all related parts comprise what is known as the *setting*.

When a forebay is not part of the generating plant's design, it will be desirable to provide a *surge chamber* in order to relieve the effects of rapid changes in flow rate. In the case of a sudden power demand, the surge chamber would provide an immediate source of water, without waiting for a contribution from the feeder reservoir. Similarly, in the case of a sudden decrease in discharge through the turbine, the excess water would surge back into the surge chamber.

19
Open Channel Flow

Nomenclature

a	velocity-head coefficient	–	–
A	area	ft^2	m^2
b	weir or channel width	ft	m
C	Hazen-Williams coefficient	$ft^{1/2}/sec$	$m^{1/2}/s$
d	depth of flow	ft	m
d	diameter	in	m
D	diameter	ft	m
E	specific energy	ft	m
f	friction factor	–	–
Fr	Froude number	–	–
g	gravitational acceleration, 32.2 (9.81)	ft/sec^2	m/s^2
g_c	gravitational constant, 32.2	$lbm\text{-}ft/lbf\text{-}sec^2$	n.a.
h	head	ft	m
H	total hydraulic head	ft	m
k	minor loss coefficient	–	–
K	conveyance	ft^3/sec	m^3/s
K'	modified conveyance	–	–
L	channel length	ft	m
m	cotangent of side slope angle	–	–
$\dot{m}$	mass flow rate	lbm/sec	kg/s
n	Manning roughness coefficient	–	–
N	number of end contractions	–	–
p	pressure	lbf/ft^2	Pa
P	wetted perimeter	ft	m
q	flow per unit width	$ft^3/sec\text{-}ft$	$m^3/s\cdot m$
Q	flow quantity	ft^3/sec	m^3/s
R	hydraulic radius	ft	m
S	slope of energy line (energy gradient)	–	–
S_0	channel slope	–	–
T	width at surface	ft	m
v	velocity	ft/sec	m/s
w	channel width	ft	m
x	distance	ft	m
y	distance	ft	m
Y	weir height	ft	m
z	height above datum	ft	m

Symbols

α	velocity-head coefficient	–	–
γ	specific weight	lbf/ft^3	n.a.
ρ	density	lbm/ft^3	kg/m^3
θ	angle	deg	deg

Subscripts

b	brink
c	critical or composite
d	discharge
e	entrance or equivalent
f	friction
h	hydraulic
n	normal
o	channel or culvert barrel
s	spillway
t	total
w	weir

1. INTRODUCTION

An *open channel* is a fluid passageway that allows part of the fluid to be exposed to the atmosphere. This type of channel includes natural waterways, canals, culverts, flumes, and pipes flowing under the influence of gravity (as opposed to pressure conduits, which always flow full). A *reach* is a straight section of open channel with uniform shape, depth, slope, and flow quantity.

There are difficulties in evaluating open channel flow. The unlimited geometric cross sections and variations in roughness have contributed to a relatively small number of scientific observations upon which to estimate the required coefficients and exponents. Therefore, the analysis of open channel flow is more empirical and less exact than that of pressure conduit flow. This lack of precision, however, is more than offset by the percentage error in runoff calculations that generally precede the channel calculations.

Flow can be categorized on the basis of the channel material, for example, concrete or metal pipe or earth material. Except for a short discussion of erodible canals in Sec. 19.36, this chapter assumes the channel is non-erodible.

2. TYPES OF FLOW

Flow in open channels is almost always turbulent; laminar flow will occur only in very shallow channels or at very low fluid velocities. However, within the turbulent category are many somewhat confusing categories of flow. Flow can be a function of time and location. If the flow quantity (volume per unit of time across an area in flow) is invariant, it is said to be *steady flow*. (Flow that varies with time, such as stream flow during a storm, known as *varied flow*, is not covered in this chapter.) If the flow cross section does not depend on the location along the channel, it is said to be *uniform flow*. Steady flow can also be *nonuniform*, as in the case of a river with a varying cross section or on a steep slope. Furthermore, uniform channel construction does not ensure uniform flow, as will be seen in the case of hydraulic jumps.

Table 19.1 summarizes some of the more common categories and names of steady open channel flow. All of the subcategories are based on variations in depth and flow area with respect to location along the channel.

3. MINIMUM VELOCITIES

The minimum permissible velocity in a sewer or other nonerodible channel is the lowest that prevents sedimentation and plant growth. Velocity ranges of 2 ft/sec to 3 ft/sec (0.6 m/s to 0.9 m/s) keep all but the heaviest silts in suspension. 2.5 ft/sec (0.75 m/s) is considered the minimum to prevent plant growth.

Table 19.1 *Categories of Steady Open Channel Flow*

subcritical flow (tranquil flow)
    uniform flow
        normal flow
    nonuniform flow
        accelerating flow
        decelerating flow (retarded flow)
critical flow
supercritical flow (rapid flow, shooting flow)
    uniform flow
        normal flow
    nonuniform flow
        accelerating flow
        decelerating flow

4. VELOCITY DISTRIBUTION

Due to the adhesion between the wetted surface of the channel and the water, the velocity will not be uniform across the area in flow. The velocity term used in this chapter is the *mean velocity*. The mean velocity, when multiplied by the flow area, gives the flow quantity.

$$Q = A\mathrm{v} \qquad\qquad 19.1$$

The location of the mean velocity depends on the distribution of velocities in the waterway, which is generally quite complex. The procedure for measuring the velocity of a channel (called *stream gauging*) involves measuring the average channel velocity at multiple locations and depths across the channel width. These sub-average velocities are averaged to give a grand average (mean) flow velocity. (See Fig. 19.1.)

Figure 19.1 *Velocity Distribution in an Open Channel*

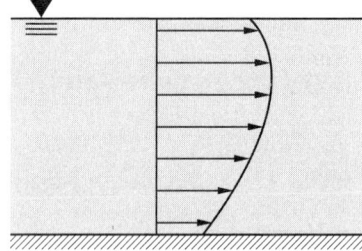

5. PARAMETERS USED IN OPEN CHANNEL FLOW

The *hydraulic radius* is the ratio of the area in flow to the wetted perimeter.[1]

$$R = \frac{A}{P} \qquad\qquad 19.2$$

---

[1]The hydraulic radius is also referred to as the *hydraulic mean depth*. However, this name is easily confused with "mean depth" and "hydraulic depth," both of which have different meanings. Therefore, the term "hydraulic mean depth" is not used in this chapter.

Flow of Fluids

Table 19.2 *Hydraulic Parameters of Basic Channel Sections*

section	area, A	wetted perimeter, P	hydraulic radius, R
rectangle	dw	$2d + w$	$\dfrac{dw}{w + 2d}$
trapezoid	$\left(b + \dfrac{d}{\tan\theta}\right)d$	$b + 2\left(\dfrac{d}{\sin\theta}\right)$	$\dfrac{bd\sin\theta + d^2\cos\theta}{b\sin\theta + 2d}$
triangle	$\dfrac{d^2}{\tan\theta}$	$\dfrac{2d}{\sin\theta}$	$\dfrac{d\cos\theta}{2}$
circle circle	$\frac{1}{8}(\theta - \sin\theta)D^2$ [θ in radians]	$\frac{1}{2}\theta D$ [θ in radians]	$\frac{1}{4}\left(1 - \dfrac{\sin\theta}{\theta}\right)D$ [θ in radians]

For a circular channel flowing either full or half-full, the hydraulic radius is one-fourth of the *hydraulic diameter*, $D_h/4$. The hydraulic radii of other channel shapes are easily calculated from the basic definition. Table 19.2 summarizes parameters for the basic shapes. For very wide channels such as rivers, the hydraulic radius is approximately equal to the depth.

The *hydraulic depth* is the ratio of the area in flow to the width of the channel at the fluid surface.[2]

$$D_h = \frac{A}{T} \qquad 19.3$$

The uniform flow *section factor* represents a frequently occurring variable group. The section factor is often evaluated against depth of flow when working with discharge from irregular cross sections.

$$\text{section factor} = AR^{2/3} \quad \text{[general uniform flow]} \qquad 19.4$$

$$\text{section factor} = A\sqrt{D_h} \quad \text{[critical flow only]} \qquad 19.5$$

The *slope*, S, is the gradient of the energy line. In general, the slope can be calculated from the Bernoulli

[2]For a rectangular channel, $D_h = d$.

equation as the energy loss per unit length of channel. For small slopes typical of almost all natural waterways, the channel length and horizontal run are essentially identical.

$$S = \frac{dE}{dL} \qquad 19.6$$

If the flow is uniform, the slope of the energy line will parallel the water surface and channel bottom, and the *energy gradient* will equal the *geometric slope*, S_0.

$$S_0 = \frac{\Delta z}{L} \approx S \quad \text{[uniform flow]} \qquad 19.7$$

Any open channel performance equation can be written using the geometric slope, S_0, instead of the hydraulic slope, S, but only under the condition of uniform flow.

In most problems, the geometric slope is a function of the terrain and is known. However, it may be necessary to calculate the slope that results in some other specific parameter. The slope that produces flow at some normal depth, d, is called the *normal slope*. The slope that produces flow at some critical depth, d_c, is called the *critical slope*. Both are determined by solving the Manning equation for slope.

6. GOVERNING EQUATIONS FOR UNIFORM FLOW

Since water is incompressible, the continuity equation is

$$A_1 v_1 = A_2 v_2 \qquad 19.8$$

The most common equation used to calculate the flow velocity in open channels is the 1768 *Chezy equation*.[3]

$$v = C\sqrt{RS} \qquad 19.9$$

Various methods for evaluating the *Chezy coefficient, C,* or "Chezy's C," have been proposed.[4] If the channel is small and very smooth, Chezy's own formula can be used. The friction factor, f, is dependent on the Reynolds number and can be found in the usual manner from the Moody diagram.

$$C = \sqrt{\frac{8g}{f}} \qquad 19.10$$

If the channel is large and the flow is fully turbulent, the friction loss will not depend so much on the Reynolds number as on the channel roughness. The 1888 *Manning formula* is frequently used to evaluate the constant C.[5] The value of C depends only on the channel roughness and geometry. (The conversion constant 1.49 in Eq. 19.11(b) is reported as 1.486 by some authorities. 1.486 is the correct SI-to-English conversion, but it is doubtful whether this equation warrants four significant digits.)

$$C = \left(\frac{1}{n}\right) R^{1/6} \qquad \text{[SI]} \quad 19.11(a)$$

$$C = \left(\frac{1.49}{n}\right) R^{1/6} \qquad \text{[U.S.]} \quad 19.11(b)$$

n is the *Manning roughness coefficient (Manning constant)*. Typical values of Manning's n are given in App. 19.A. Judgment is needed in selecting values since tabulated values often differ by as much as 30%. More important to recognize for sewer work is the layer of slime that often coats the sewer walls. Since the slime characteristics can change with location in the sewer, there can be variations in Manning's roughness coefficient along the sewer length.

---

[3]Pronounced "Shay'-zee." This equation does not appear to be dimensionally consistent. However, the coefficient C is not a pure number. Rather, it has units of length$^{1/2}$/time (i.e., acceleration$^{1/2}$).

[4]Other methods of evaluating C include the *Kutter equation* (also known as the *G.K. formula*) and the *Bazin formula*. These methods are interesting from a historical viewpoint, but both have been replaced by the Manning equation.

[5]This equation was originally proposed in 1868 by Gaukler and again in 1881 by Hagen, both working independently. For some reason, the Frenchman Flamant attributed the equation to an Irishman, R. Manning. In Europe and many other places, the Manning equation may be known as the *Strickler equation*.

Combining Eq. 19.9 and Eq. 19.11 produces the *Manning equation*, also known as the *Chezy-Manning equation*.

$$v = \left(\frac{1}{n}\right) R^{2/3}\sqrt{S} \qquad \text{[SI]} \quad 19.12(a)$$

$$v = \left(\frac{1.49}{n}\right) R^{2/3}\sqrt{S} \qquad \text{[U.S.]} \quad 19.12(b)$$

All of the coefficients and constants in the Manning equation may be combined into the *conveyance, K*.

$$Q = vA = \left(\frac{1}{n}\right) A R^{2/3}\sqrt{S}$$
$$= K\sqrt{S} \qquad \text{[SI]} \quad 19.13(a)$$

$$Q = vA = \left(\frac{1.49}{n}\right) A R^{2/3}\sqrt{S}$$
$$= K\sqrt{S} \qquad \text{[U.S.]} \quad 19.13(b)$$

Example 19.1

A rectangular channel on a 0.002 slope is constructed of finished concrete. The channel is 8 ft (2.4 m) wide. Water flows at a depth of 5 ft (1.5 m). What is the flow rate?

SI Solution

The hydraulic radius is

$$R = \frac{A}{P} = \frac{(2.4 \text{ m})(1.5 \text{ m})}{1.5 \text{ m} + 2.4 \text{ m} + 1.5 \text{ m}}$$
$$= 0.667 \text{ m}$$

From App. 19.A, the roughness coefficient for finished concrete is 0.012. The Manning coefficient is determined from Eq. 19.11(a).

$$C = \left(\frac{1}{n}\right) R^{1/6} = \left(\frac{1}{0.012}\right)(0.667 \text{ m})^{1/6}$$
$$= 77.9$$

The flow rate is

$$Q = vA = C\sqrt{RS}\,A$$
$$= \left(77.9 \; \frac{\sqrt{\text{m}}}{\text{s}}\right) \sqrt{(0.667 \text{ m})(0.002)}\,(1.5 \text{ m})(2.4 \text{ m})$$
$$= 10.2 \text{ m}^3/\text{s}$$

Customary U.S. Solution

The hydraulic radius is

$$R = \frac{A}{P} = \frac{(8 \text{ ft})(5 \text{ ft})}{5 \text{ ft} + 8 \text{ ft} + 5 \text{ ft}}$$
$$= 2.22 \text{ ft}$$

From App. 19.A, the roughness coefficient for finished concrete is 0.012. The Manning coefficient is determined from Eq. 19.11(b).

$$C = \left(\frac{1.49}{n}\right) R^{1/6}$$
$$= \left(\frac{1.49}{0.012}\right)(2.22 \text{ ft})^{1/6}$$
$$= 141.8$$

The flow rate is

$$Q = \text{v}A = C\sqrt{RS}A$$
$$= \left(141.8 \; \frac{\sqrt{\text{ft}}}{\text{sec}}\right)\sqrt{(2.22 \text{ ft})(0.002)}(8 \text{ ft})(5 \text{ ft})$$
$$= 377.9 \text{ ft}^3/\text{sec}$$

7. VARIATIONS IN THE MANNING CONSTANT

The value of n also depends on the depth of flow, leading to a value (n_{full}) specifically intended for use with full flow. (It is seldom clear from tabulations, such as App. 19.A, whether the values are for full flow or general use.) The variation in n can be taken into consideration using *Camp's correction*, shown in App. 19.C. However, this degree of sophistication cannot be incorporated into an analysis problem unless a specific value of n is known for a specific depth of flow.

For most calculations, however, n is assumed to be constant. The accuracy of other parameters used in open-flow calculations often does not warrant considering the variation of n with depth, and the choice to use a constant or varying n-value is left to the engineer.

If it is desired to acknowledge variations in n with respect to depth, it is expedient to use tables or graphs of hydraulic elements prepared for that purpose. Table 19.3 lists such hydraulic elements under the assumption that n varies. Appendix 19.C can be used for both varying and constant n.

Table 19.3 Circular Channel Ratios (varying n)

$\dfrac{d}{D}$	$\dfrac{Q}{Q_{\text{full}}}$	$\dfrac{\text{v}}{\text{v}_{\text{full}}}$
0.1	0.02	0.31
0.2	0.07	0.48
0.3	0.14	0.61
0.4	0.26	0.71
0.5	0.41	0.80
0.6	0.56	0.88
0.7	0.72	0.95
0.8	0.87	1.01
0.9	0.99	1.04
0.95	1.02	1.03
1.00	1.00	1.00

Example 19.2

2.5 ft^3/sec (0.07 m^3/s) of water flow in a 20 in (0.5 m) diameter sewer line ($n = 0.015$, $S = 0.001$). The Manning coefficient, n, varies with depth. Flow is uniform and steady. What are the velocity and depth?

SI Solution

The hydraulic radius is

$$R = \frac{D}{4} = \frac{0.5 \text{ m}}{4} = 0.125 \text{ m}$$

From Eq. 19.12(a),

$$\text{v}_{\text{full}} = \left(\frac{1}{n}\right) R^{2/3}\sqrt{S}$$
$$= \left(\frac{1}{0.015}\right)(0.125 \text{ m})^{2/3}\sqrt{0.001}$$
$$= 0.53 \text{ m/s}$$

If the pipe was flowing full, it would carry Q_{full}.

$$Q_{\text{full}} = \text{v}_{\text{full}} A$$
$$= \left(0.53 \; \frac{\text{m}}{\text{s}}\right)\left(\frac{\pi}{4}\right)(0.5 \text{ m})^2$$
$$= 0.10 \text{ m}^3/\text{s}$$

$$\frac{Q}{Q_{\text{full}}} = \frac{0.07 \; \frac{\text{m}^3}{\text{s}}}{0.10 \; \frac{\text{m}^3}{\text{s}}} = 0.7$$

From App. 19.C, $d/D = 0.68$, and $\text{v}/\text{v}_{\text{full}} = 0.94$.

$$\text{v} = (0.94)\left(0.53 \; \frac{\text{m}}{\text{s}}\right) = 0.50 \text{ m/s}$$
$$d = (0.68)(0.5 \text{ m}) = 0.34 \text{ m}$$

Customary U.S. Solution

The hydraulic radius is

$$R = \frac{D}{4} = \frac{\dfrac{20 \text{ in}}{12 \; \frac{\text{in}}{\text{ft}}}}{4} = 0.417 \text{ ft}$$

From Eq. 19.12(b),

$$\text{v}_{\text{full}} = \left(\frac{1.49}{n}\right) R^{2/3}\sqrt{S}$$
$$= \left(\frac{1.49}{0.015}\right)(0.417 \text{ ft})^{2/3}\sqrt{0.001}$$
$$= 1.75 \text{ ft/sec}$$

If the pipe was flowing full, it would carry Q_{full}.

$$Q_{\text{full}} = v_{\text{full}} A$$

$$= \left(1.75 \ \frac{\text{ft}}{\text{sec}}\right)\left(\frac{\pi}{4}\right)\left(\frac{20 \ \text{in}}{12 \ \frac{\text{in}}{\text{ft}}}\right)^2$$

$$= 3.82 \ \text{ft}^3/\text{sec}$$

$$\frac{Q}{Q_{\text{full}}} = \frac{2.5 \ \dfrac{\text{ft}^3}{\text{sec}}}{3.82 \ \dfrac{\text{ft}^3}{\text{sec}}} = 0.65$$

From App. 19.C, $d/D = 0.66$, and $v/v_{\text{full}} = 0.92$.

$$v = (0.92)\left(1.75 \ \frac{\text{ft}}{\text{sec}}\right) = 1.61 \ \text{ft/sec}$$

$$d = (0.66)(20 \ \text{in}) = 13.2 \ \text{in}$$

8. HAZEN-WILLIAMS VELOCITY

The empirical Hazen-Williams open channel velocity equation was developed in the early 1920s. It is still occasionally used in the United States for sizing gravity sewers. It is applicable to water flows at reasonably high Reynolds numbers and is based on sound dimensional analysis. However, the constants and exponents were developed experimentally.

The equation uses the Hazen-Williams coefficient, C, to characterize the roughness of the channel. Since the equation is used only for water within "normal" ambient conditions, the effects of temperature, pressure, and viscosity are disregarded. The primary advantage of this approach is that the coefficient, C, depends only on the roughness, not on the fluid characteristics. This is also the method's main disadvantage, since professional judgment is required in choosing the value of C.

$$v = 0.85 C R^{0.63} S_0^{0.54} \qquad \text{[SI]} \qquad \textbf{\textit{19.14(a)}}$$

$$v = 1.318 C R^{0.63} S_0^{0.54} \qquad \text{[U.S.]} \qquad \textbf{\textit{19.14(b)}}$$

9. NORMAL DEPTH

When the depth of flow is constant along the length of the channel (i.e., the depth is neither increasing nor decreasing), the flow is said to be *uniform*. The depth of flow in that case is known as the *normal depth, d_n*. If the normal depth is known, it can be compared with the actual depth of flow to determine if the flow is uniform.[6]

The difficulty with which the normal depth is calculated depends on the cross section of the channel. If the width is very large compared to the depth, the flow cross section will essentially be rectangular and the Manning

---

[6]Normal depth is a term that applies only to uniform flow. The two alternate depths that can occur in nonuniform flow are not normal depths.

equation can be used. (Equation 19.15 assumes that the hydraulic radius equals the normal depth.)

$$d_n = \left(\frac{nQ}{w\sqrt{S}}\right)^{3/5} \qquad [w \gg d_n] \qquad \text{[SI]} \qquad \textbf{\textit{19.15(a)}}$$

$$d_n = 0.788\left(\frac{nQ}{w\sqrt{S}}\right)^{3/5} \qquad [w \gg d_n] \qquad \text{[U.S.]} \qquad \textbf{\textit{19.15(b)}}$$

Normal depth in circular channels can be calculated directly only under limited conditions. If the circular channel is flowing full, the normal depth is the inside pipe diameter.

$$D = d_n = 1.548\left(\frac{nQ}{\sqrt{S}}\right)^{3/8} \qquad \text{[full]} \qquad \text{[SI]} \qquad \textbf{\textit{19.16(a)}}$$

$$D = d_n = 1.335\left(\frac{nQ}{\sqrt{S}}\right)^{3/8} \qquad \text{[full]} \qquad \text{[U.S.]} \qquad \textbf{\textit{19.16(b)}}$$

If a circular channel is flowing half full, the normal depth is half of the inside pipe diameter.

$$D = 2d_n = 2.008\left(\frac{nQ}{\sqrt{S}}\right)^{3/8} \qquad \text{[half full]} \qquad \text{[SI]} \qquad \textbf{\textit{19.17(a)}}$$

$$D = 2d_n = 1.731\left(\frac{nQ}{\sqrt{S}}\right)^{3/8} \qquad \text{[half full]} \qquad \text{[U.S.]} \qquad \textbf{\textit{19.17(b)}}$$

For other cases of uniform flow (trapezoidal, triangular, etc.), it is more difficult to determine normal depth. Various researchers have prepared tables and figures to assist in the calculations. For example, Table 19.3 is derived from App. 19.C and can be used for circular channels flowing other than full or half full.

In the absence of tables or figures, trial-and-error solutions are required. The appropriate expressions for the flow area and hydraulic radius are used in the Manning equation. Trial values are used in conjunction with graphical techniques, linear interpolation, or extrapolation to determine the normal depth. The Manning equation is solved for flow rate with various assumed values of d_n. The calculated value is compared to the actual known flow quantity, and the normal depth is approached iteratively.

For a rectangular channel whose width is small compared to the depth, the hydraulic radius and area in flow are

$$R = \frac{wd_n}{w + 2d_n} \qquad\qquad \textbf{\textit{19.18}}$$

$$A = wd_n \qquad\qquad \textbf{\textit{19.19}}$$

$$Q = \left(\frac{1}{n}\right)wd_n\left(\frac{wd_n}{w + 2d_n}\right)^{2/3}\sqrt{S} \qquad \text{[rectangular]}$$

$$\text{[SI]} \qquad \textbf{\textit{19.20(a)}}$$

$$Q = \left(\frac{1.49}{n}\right)wd_n\left(\frac{wd_n}{w + 2d_n}\right)^{2/3}\sqrt{S} \quad \text{[rectangular]}$$

[U.S.] *19.20(b)*

For a trapezoidal channel with exposed surface width w, base width b, side length s, and normal depth of flow d_n, the hydraulic radius and area in flow are

$$R = \frac{d_n(b + w)}{2(b + 2s)} \quad \text{[trapezoidal]}$$ *19.21*

$$A = \frac{d_n(w + b)}{2} \quad \text{[trapezoidal]}$$ *19.22*

For a symmetrical triangular channel with exposed surface width w, side slope 1:z (vertical:horizontal), and normal depth of flow d_n, the hydraulic radius and area in flow are

$$R = \frac{zd_n}{2\sqrt{1 + z^2}} \quad \text{[triangular]}$$ *19.23*

$$A = zd_n^2 \quad \text{[triangular]}$$ *19.24*

10. ENERGY AND FRICTION RELATIONSHIPS

Bernoulli's equation is an expression for the conservation of energy along a fluid streamline. The Bernoulli equation can also be written for two points along the bottom of an open channel.

$$\frac{p_1}{\rho g} + \frac{v_1^2}{2g} + z_1 = \frac{p_2}{\rho g} + \frac{v_2^2}{2g} + z_2 + h_f \quad \text{[SI]}$$ *19.25(a)*

$$\frac{p_1}{\gamma} + \frac{v_1^2}{2g} + z_1 = \frac{p_2}{\gamma} + \frac{v_2^2}{2g} + z_2 + h_f \quad \text{[U.S.]}$$ *19.25(b)*

However, $p/\rho g = p/\gamma = d$.

$$d_1 + \frac{v_1^2}{2g} + z_1 = d_2 + \frac{v_2^2}{2g} + z_2 + h_f$$ *19.26*

And, since $d_1 = d_2$ and $v_1 = v_2$ for uniform flow at the bottom of a channel,

$$h_f = z_1 - z_2$$ *19.27*

$$S_0 = \frac{z_1 - z_2}{L}$$ *19.28*

The channel slope, S_0, and the hydraulic energy gradient, S, are numerically the same for uniform flow. Therefore, the total friction loss along a channel is

$$h_f = LS$$ *19.29*

Combining Eq. 19.29 with the Manning equation (see Eq. 19.12) results in a method for calculating friction loss.

$$h_f = \frac{Ln^2v^2}{R^{4/3}} \quad \text{[SI]}$$ *19.30(a)*

$$h_f = \frac{Ln^2v^2}{2.208R^{4/3}} \quad \text{[U.S.]}$$ *19.30(b)*

Example 19.3

The velocities upstream and downstream, v_1 and v_2, of a 12 ft (4.0 m) wide sluice gate are both unknown. The upstream and downstream depths are 6 ft (2.0 m) and 2 ft (0.6 m), respectively. Flow is uniform and steady. What is the downstream velocity, v_2?

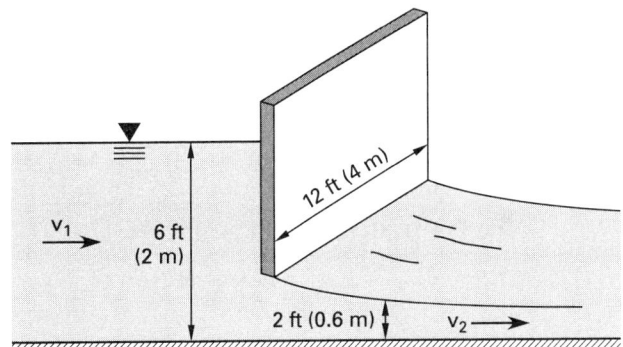

SI Solution

Since the channel bottom is essentially level on either side of the gate, $z_1 = z_2$. Bernoulli's equation reduces to

$$d_1 + \frac{v_1^2}{2g} = d_2 + \frac{v_2^2}{2g}$$

$$2 \text{ m} + \frac{v_1^2}{2g} = 0.6 \text{ m} + \frac{v_2^2}{2g}$$

v_1 and v_2 are related by continuity.

$$Q_1 = Q_2$$

$$A_1v_1 = A_2v_2$$

$$(2 \text{ m})(4 \text{ m})v_1 = (0.6 \text{ m})(4 \text{ m})v_2$$

$$v_1 = 0.3v_2$$

Substituting the expression for v_1 into the Bernoulli equation gives

$$2 \text{ m} + \frac{(0.3v_2)^2}{(2)\left(9.81 \frac{\text{m}}{\text{s}^2}\right)} = 0.6 \text{ m} + \frac{v_2^2}{(2)\left(9.81 \frac{\text{m}}{\text{s}^2}\right)}$$

$$2 \text{ m} + 0.004587v_2^2 = 0.6 \text{ m} + 0.050968v_2^2$$

$$v_2 = 5.5 \text{ m/s}$$

Customary U.S. Solution

Since the channel bottom is essentially level on either side of the gate, $z_1 = z_2$. Bernoulli's equation reduces to

$$d_1 + \frac{v_1^2}{2g} = d_2 + \frac{v_2^2}{2g}$$

$$6 \text{ ft} + \frac{v_1^2}{2g} = 2 \text{ ft} + \frac{v_2^2}{2g}$$

v_1 and v_2 are related by continuity.

$$Q_1 = Q_2$$
$$A_1 v_1 = A_2 v_2$$
$$(6 \text{ ft})(12 \text{ ft})v_1 = (2 \text{ ft})(12 \text{ ft})v_2$$
$$v_1 = \frac{v_2}{3}$$

Substituting the expression for v_1 into the Bernoulli equation gives

$$6 \text{ ft} + \frac{v_2^2}{(3)^2(2)\left(32.2 \frac{\text{ft}}{\text{sec}^2}\right)} = 2 \text{ ft} + \frac{v_2^2}{(2)\left(32.2 \frac{\text{ft}}{\text{sec}^2}\right)}$$

$$6 \text{ ft} + 0.00173 v_2^2 = 2 \text{ ft} + 0.0155 v_2^2$$

$$v_2 = 17.0 \text{ ft/sec}$$

Example 19.4

In Ex. 19.1, the open channel experiencing normal flow had the following characteristics: $S = 0.002$, $n = 0.012$, $v = 9.447$ ft/sec (2.9 m/s), and $R = 2.22$ ft (0.68 m). What is the energy loss per 1000 ft (100 m)?

SI Solution

There are two methods for finding the energy loss. From Eq. 19.29,

$$h_f = LS = (100 \text{ m})(0.002)$$
$$= 0.2 \text{ m}$$

From Eq. 19.30(a),

$$h_f = \frac{Ln^2 v^2}{R^{4/3}}$$

$$= \frac{(100 \text{ m})(0.012)^2 \left(2.9 \frac{\text{m}}{\text{s}}\right)^2}{(0.68 \text{ m})^{4/3}}$$

$$= 0.2 \text{ m}$$

Customary U.S. Solution

There are two methods for finding the energy loss. From Eq. 19.29,

$$h_f = LS = (1000 \text{ ft})(0.002)$$
$$= 2 \text{ ft}$$

From Eq. 19.30(b),

$$h_f = \frac{Ln^2 v^2}{2.208 R^{4/3}}$$

$$= \frac{(1000 \text{ ft})(0.012)^2 \left(9.447 \frac{\text{ft}}{\text{sec}}\right)^2}{(2.208)(2.22 \text{ ft})^{4/3}}$$

$$= 2 \text{ ft}$$

11. SIZING TRAPEZOIDAL AND RECTANGULAR CHANNELS

Trapezoidal and rectangular cross sections are commonly used for artificial surface channels. The flow through a trapezoidal channel is easily determined from the Manning equation when the cross section is known. However, when the cross section or uniform depth is unknown, a trial-and-error solution is required. (See Fig. 19.2.)

Figure 19.2 Trapezoidal Cross Section

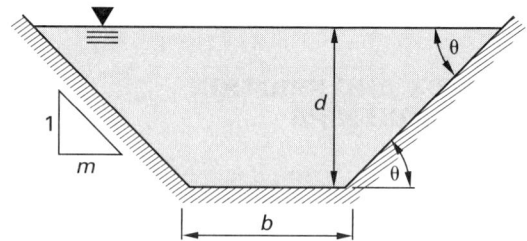

For such problems involving rectangular and trapezoidal channels, it is common to calculate and plot the *conveyance*, K (or alternatively, the product Kn), against depth. For trapezoidal sections, it is particularly convenient to write the uniform flow, Q, in terms of a modified conveyance, K'. b is the base width of the channel, d is the depth of flow, and m is the cotangent of the side slope angle. m and the ratio d/b are treated as independent variables. Values of K' are tabulated in App. 19.F.

$$Q = \frac{K' b^{8/3} \sqrt{S_0}}{n} \qquad 19.31$$

$$K' = \left(\frac{\left(1 + m\left(\frac{d}{b}\right)\right)^{5/3}}{\left(1 + 2\left(\frac{d}{b}\right)\sqrt{1+m^2}\right)^{2/3}}\right)\left(\frac{d}{b}\right)^{5/3} \qquad \text{[SI]} \quad 19.32(a)$$

$$K' = \left(\frac{1.49\left(1 + m\left(\frac{d}{b}\right)\right)^{5/3}}{\left(1 + 2\left(\frac{d}{b}\right)\sqrt{1+m^2}\right)^{2/3}}\right)\left(\frac{d}{b}\right)^{5/3} \qquad \text{[U.S.]} \quad 19.32(b)$$

$$m = \cot\theta \qquad 19.33$$

For any fixed value of m, enough values of K' are calculated over a reasonable range of the d/b ratio ($0.05 < d/b < 0.5$) to define a curve. Given specific values of Q, n, S_0, and b, the value of K' can be calculated from the expression for Q. The graph is used to determine the ratio d/b, giving the depth of uniform flow, d, since b is known.

When the ratio of d/b is very small (less than 0.02), it is satisfactory to consider the trapezoidal channel as a wide rectangular channel with area $A = bd$.

12. MOST EFFICIENT CROSS SECTION

The most efficient open channel cross section will maximize the flow for a given Manning coefficient, slope, and flow area. Accordingly, the Manning equation requires that the hydraulic radius be maximum. For a given flow area, the wetted perimeter will be minimum.

Semicircular cross sections have the smallest wetted perimeter; therefore, the cross section with the highest efficiency is the semicircle. Although such a shape can be constructed with concrete, it cannot be used with earth channels.

The most efficient cross section is also generally assumed to minimize construction cost. This is true only in the most simplified cases, however, since the labor and material costs of excavation and formwork must be considered. Rectangular and trapezoidal channels are much easier to form than semicircular channels. So in this sense the "least efficient" (i.e., most expensive) cross section (i.e., semicircular) is also the "most efficient." (See Fig. 19.3.)

The most efficient rectangle is one having depth equal to one-half of the width (i.e., is one-half of a square).

$$d = \frac{w}{2} \quad \text{[most efficient rectangle]} \qquad 19.34$$

$$A = dw = \frac{w^2}{2} = 2d^2 \qquad 19.35$$

$$P = d + w + d = 2w = 4d \qquad 19.36$$

$$R = \frac{w}{4} = \frac{d}{2} \qquad 19.37$$

The most efficient trapezoidal channel is always one in which the flow depth is twice the hydraulic radius. If the side slope is adjustable, the sides of the most efficient trapezoid should be inclined at 60° from the horizontal. Since the surface width will be equal to twice the sloping side length, the most efficient trapezoidal channel will be half of a regular hexagon (i.e., three adjacent equilateral triangles of side length $2d/\sqrt{3}$). If the side slope

is any other angle, only the $d = 2R$ criterion is applicable.

$$d = 2R \quad \text{[most efficient trapezoid]} \qquad 19.38$$

$$b = \frac{2d}{\sqrt{3}} \qquad 19.39$$

$$A = \sqrt{3}d^2 \qquad 19.40$$

$$P = 3b = 2\sqrt{3}d \quad \text{[most efficient trapezoid]} \qquad 19.41$$

$$R = \frac{d}{2} \qquad 19.42$$

A semicircle with its center at the middle of the water surface can always be inscribed in a cross section with maximum efficiency.

Figure 19.3 *Circles Inscribed in Efficient Channels*

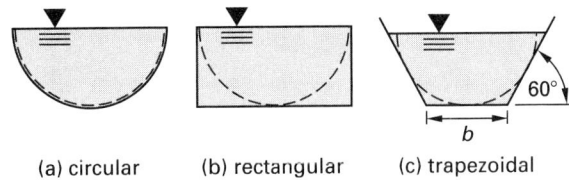

  (a) circular   (b) rectangular   (c) trapezoidal

Example 19.5

A rubble masonry open channel is being designed to carry 500 ft³/sec (14 m³/s) of water on a 0.0001 slope. Using $n = 0.017$, find the most efficient dimensions for a rectangular channel.

SI Solution

Let the depth and width be d and w, respectively. For an efficient rectangle, $d = w/2$.

$$A = dw = \left(\frac{w}{2}\right)w = \frac{w^2}{2}$$

$$P = d + w + d = \frac{w}{2} + w + \frac{w}{2} = 2w$$

$$R = \frac{A}{P} = \frac{\frac{w^2}{2}}{2w} = \frac{w}{4}$$

Using Eq. 19.13(a),

$$Q = \left(\frac{1}{n}\right)AR^{2/3}\sqrt{S}$$

$$14 \ \frac{\text{m}^3}{\text{s}} = \left(\frac{1}{0.017}\right)\left(\frac{w^2}{2}\right)\left(\frac{w}{4}\right)^{2/3}\sqrt{0.0001}$$

$$14 \ \frac{\text{m}^3}{\text{s}} = 0.1167w^{8/3}$$

$$w = 6.02 \text{ m}$$

$$d = \frac{w}{2} = \frac{6.02 \text{ m}}{2} = 3.01 \text{ m}$$

Customary U.S. Solution

Let the depth and width be d and w, respectively. For an efficient rectangle, $d = w/2$.

$$A = dw = \left(\frac{w}{2}\right)w = \frac{w^2}{2}$$

$$P = d + w + d = \frac{w}{2} + w + \frac{w}{2} = 2w$$

$$R = \frac{A}{P} = \frac{\frac{w^2}{2}}{2w} = \frac{w}{4}$$

Using Eq. 19.13(b),

$$Q = \left(\frac{1.49}{n}\right) A R^{2/3} \sqrt{S}$$

$$500 \ \frac{\text{ft}^3}{\text{sec}} = \left(\frac{1.49}{0.017}\right)\left(\frac{w^2}{2}\right)\left(\frac{w}{4}\right)^{2/3}\sqrt{0.0001}$$

$$500 \ \frac{\text{ft}^3}{\text{sec}} = 0.1739 w^{8/3}$$

$$w = 19.82 \ \text{ft}$$

$$d = \frac{w}{2} = \frac{19.82 \ \text{ft}}{2} = 9.91 \ \text{ft}$$

13. ANALYSIS OF NATURAL WATERCOURSES

Natural watercourses do not have uniform paths or cross sections. This complicates their analysis considerably. Frequently, analyzing the flow from a river is a matter of making the most logical assumptions. Many evaluations can be solved with a reasonable amount of error.

As was seen in Eq. 19.30, the friction loss (and hence the hydraulic gradient) depends on the square of the roughness coefficient. Therefore, an attempt must be made to evaluate the roughness constant as accurately as possible. If the channel consists of a river with flood plains (as in Fig. 19.4), it should be treated as parallel channels. The flow from each subdivision can be calculated independently and the separate values added to obtain the total flow. (The common interface between adjacent subdivisions is not included in the wetted perimeter.) Alternatively, a composite value of the roughness coefficient, n_c, can be approximated from the *Horton-Einstein equation* using the individual values of n and the corresponding wetted perimeters. Equation 19.43 assumes the flow velocities and lengths are the same in all cross sections.

$$n_c = \left(\frac{\sum P_i n_i^{3/2}}{\sum P_i}\right)^{2/3} \qquad 19.43$$

If the channel is divided (as in Fig. 19.5) by an island into two channels, some combination of flows will usually be known. For example, if the total flow, Q, is known, Q_1 and Q_2 may be unknown. If the slope is

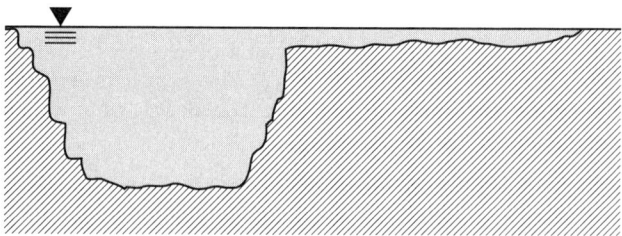

Figure 19.4 *River with Flood Plain*

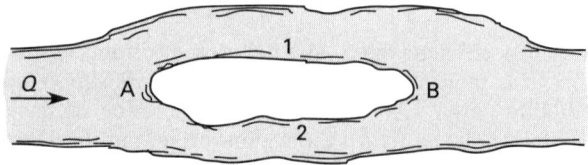

Figure 19.5 *Divided Channel*

known, Q_1 and Q_2 may be known. Iterative trial-and-error solutions are often required.

Since the elevation drop $(z_A - z_B)$ between points A and B is the same regardless of flow path,

$$S_1 = \frac{z_A - z_B}{L_1} \qquad 19.44$$

$$S_2 = \frac{z_A - z_B}{L_2} \qquad 19.45$$

Once the slopes are known, initial estimates Q_1 and Q_2 can be calculated from Eq. 19.13. The sum of Q_1 and Q_2 will probably not be the same as the given flow quantity, Q. In that case, Q should be prorated according to the ratios of Q_1 and Q_2 to $Q_1 + Q_2$.

If the lengths L_1 and L_2 are the same or almost so, the Manning equation may be solved for the slope by writing Eq. 19.46.

$$Q = Q_1 + Q_2$$
$$= \left(\left(\frac{A_1}{n_1}\right)R_1^{2/3} + \left(\frac{A_2}{n_2}\right)R_2^{2/3}\right)\sqrt{S} \qquad \text{[SI]} \quad 19.46(a)$$

$$Q = Q_1 + Q_2$$
$$= 1.49\left(\left(\frac{A_1}{n_1}\right)R_1^{2/3} + \left(\frac{A_2}{n_2}\right)R_2^{2/3}\right)\sqrt{S} \quad \text{[U.S.]} \quad 19.46(b)$$

Equation 19.46 yields only a rough estimate of the flow quantity, as the geometry and roughness of a natural channel changes considerably along its course.

14. FLOW MEASUREMENT WITH WEIRS

A *weir* is an obstruction in an open channel over which flow occurs. Although a dam spillway is a specific type of weir, most weirs are intended specifically for flow measurement.

Measurement weirs consist of a vertical flat plate with sharp edges. Because of their construction, they are called *sharp-crested weirs*. Sharp-crested weirs are most frequently rectangular, consisting of a straight, horizontal crest. However, weirs may also have trapezoidal and triangular openings.

For any given width of weir opening (referred to as the *weir length*), the discharge will be a function of the head over the weir. The head (or sometimes surface elevation) can be determined by a standard *staff gauge* mounted adjacent to the weir.

The full channel flow usually goes over the weir. However, it is also possible to divert a small portion of the total flow through a measurement channel. The full channel flow rate can be extrapolated from a knowledge of the split fractions.

If a rectangular weir is constructed with an opening width less than the channel width, the falling liquid sheet (called the *nappe*) decreases in width as it falls. Because of this *contraction* of the nappe, these weirs are known as *contracted weirs*, although it is the nappe that is actually contracted, not the weir. If the opening of the weir extends the full channel width, the weir is known as a *suppressed weir*, since the contractions are suppressed. (See Fig. 19.6.)

Figure 19.6 *Contracted and Suppressed Weirs*

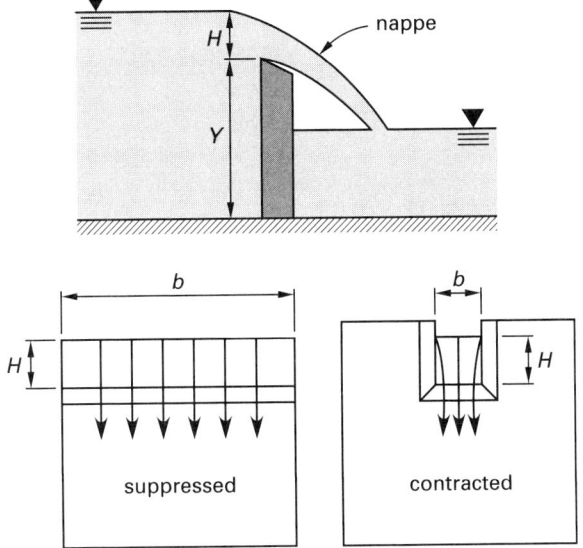

The derivation of an expression for the quantity flowing over a weir is dependent on many simplifying assumptions. The basic weir equation (see Eq. 19.47 or Eq. 19.48) is, therefore, an approximate result requiring correction by experimental coefficients.

If it is assumed that the contractions are suppressed, upstream velocity is uniform, flow is laminar over the crest, nappe pressure is zero, the nappe is fully ventilated, and viscosity, turbulence, and surface tension effects are negligible, then the following equation may be derived from the Bernoulli equation.

$$Q = \tfrac{2}{3} b \sqrt{2g} \left(\left(H + \frac{v_1^2}{2g} \right)^{3/2} - \left(\frac{v_1^2}{2g} \right)^{3/2} \right) \qquad 19.47$$

If the velocity of approach, v_1, is negligible, then

$$Q = \tfrac{2}{3} b \sqrt{2g} H^{3/2} \qquad 19.48$$

Equation 19.48 must be corrected for all of the assumptions made, primarily for a nonuniform velocity distribution. This is done by introducing an empirical discharge coefficient, C_1. Equation 19.49 is known as the *Francis weir equation*.

$$Q = \tfrac{2}{3} C_1 b \sqrt{2g} H^{3/2} \qquad 19.49$$

Many investigations have been done to evaluate C_1 analytically. Perhaps the most widely known is the coefficient formula developed by *Rehbock*.[7]

$$C_1 = \left(0.6035 + 0.0813 \left(\frac{H}{Y} \right) + \frac{0.000295}{Y} \right)$$
$$\times \left(1 + \frac{0.00361}{H} \right)^{3/2} \quad \text{[U.S. only]} \qquad 19.50$$

$$C_1 \approx 0.602 + 0.083 \left(\frac{H}{Y} \right) \quad \text{[U.S. and SI]} \qquad 19.51$$

When $H/Y < 0.2$, C_1 approaches 0.61–0.62. In most cases, a value in this range is adequate. Other constants (i.e., $\tfrac{2}{3}$ and $\sqrt{2g}$) can be taken out of Eq. 19.49. In that case,

$$Q \approx 1.84 b h^{3/2} \quad \text{[SI]} \qquad 19.52(a)$$

$$Q \approx 3.33 b h^{3/2} \quad \text{[U.S.]} \qquad 19.52(b)$$

If the contractions are not suppressed (i.e., one or both sides do not extend to the channel sides), then the actual width, b, should be replaced with the *effective width*. In Eq. 19.53, N is 1 if one side is contracted and N is 2 if there are two end contractions.

$$b_{\text{effective}} = b_{\text{actual}} - 0.1NH \qquad 19.53$$

A *submerged rectangular weir* requires a more complex analysis because of the difficulty in measuring H and because the discharge depends on both the upstream

---

[7]There is much variation in how different investigators calculate the discharge coefficient, C_1. For ratios of H/b less than 5, $C_1 = 0.622$ gives a reasonable value. With the questionable accuracy of some of the other variables used in open channel flow problems, the pursuit of greater accuracy is of dubious value.

and downstream depths. (See Fig. 19.7.) The following equation, however, may be used with little difficulty.

$$Q_{\text{submerged}} = Q_{\text{free flow}} \left(1 - \left(\frac{H_{\text{downstream}}}{H_{\text{upstream}}} \right)^{3/2} \right)^{0.385}$$

$$19.54$$

Equation 19.54 is used by first finding the flow rate, Q, from Eq. 19.49 and then correcting it with the bracketed quantity.

Figure 19.7 Submerged Weir

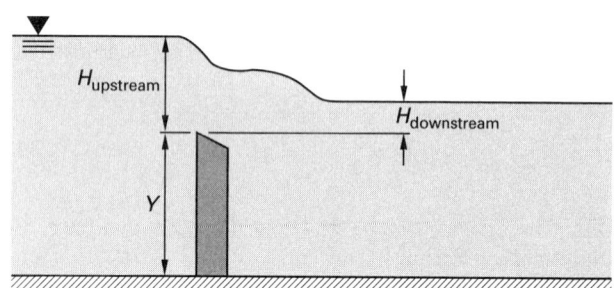

Example 19.6

The crest of a sharp-crested, rectangular weir with two contractions is 2.5 ft (1.0 m) high above the channel bottom. The crest is 4 ft (1.6 m) long. A 4 in (100 mm) head exists over the weir. What is the velocity of approach?

SI Solution

$$H = \frac{100 \text{ mm}}{1000 \frac{\text{mm}}{\text{m}}} = 0.1 \text{ m}$$

The number of contractions, N, is 2. From Eq. 19.53, the effective width is

$$b_{\text{effective}} = b_{\text{actual}} - 0.1NH = 1.6 \text{ m} - (0.1)(2)(0.1 \text{ m})$$
$$= 1.58 \text{ m}$$

From Eq. 19.51, the Rehbock coefficient is

$$C_1 \approx 0.602 + 0.083 \left(\frac{H}{Y} \right)$$
$$= 0.602 + (0.083) \left(\frac{0.1}{1} \right)$$
$$= 0.61$$

From Eq. 19.49, the flow is

$$Q = \tfrac{2}{3} C_1 b \sqrt{2g} H^{3/2}$$
$$= \left(\tfrac{2}{3} \right)(0.61)(1.58 \text{ m}) \sqrt{(2) \left(9.81 \frac{\text{m}}{\text{s}^2} \right)} (0.10 \text{ m})^{3/2}$$
$$= 0.090 \text{ m}^3/\text{s}$$

$$\text{v} = \frac{Q}{A} = \frac{0.090 \frac{\text{m}^3}{\text{s}}}{(1.6 \text{ m})(1.0 \text{ m} + 0.1 \text{ m})}$$
$$= 0.051 \text{ m/s}$$

Customary U.S. Solution

$$H = \frac{4 \text{ in}}{12 \frac{\text{in}}{\text{ft}}} = 0.333 \text{ ft}$$

The number of contractions, N, is 2. From Eq. 19.53, the effective width is

$$b_{\text{effective}} = b_{\text{actual}} - 0.1NH = 4 \text{ ft} - (0.1)(2)(0.333 \text{ ft})$$
$$= 3.93 \text{ ft}$$

From Eq. 19.50, the Rehbock coefficient is

$$C_1 = \left(0.6035 + 0.0813 \left(\frac{H}{Y} \right) + \frac{0.000295}{Y} \right)$$
$$\times \left(1 + \frac{0.00361}{H} \right)^{3/2}$$
$$= \left(0.6035 + (0.0813) \left(\frac{0.333 \text{ ft}}{2.5 \text{ ft}} \right) + \frac{0.000295}{2.5 \text{ ft}} \right)$$
$$\times \left(1 + \frac{0.00361}{0.333 \text{ ft}} \right)^{3/2}$$
$$= 0.624$$

From Eq. 19.49, the flow is

$$Q = \tfrac{2}{3} C_1 b \sqrt{2g} H^{3/2}$$
$$= \left(\tfrac{2}{3} \right)(0.624)(3.93 \text{ ft}) \sqrt{(2) \left(32.2 \frac{\text{ft}}{\text{sec}^2} \right)} (0.333 \text{ ft})^{3/2}$$
$$= 2.52 \text{ ft}^3/\text{sec}$$

$$\text{v} = \frac{Q}{A} = \frac{2.52 \frac{\text{ft}^3}{\text{sec}}}{(4 \text{ ft})(2.5 \text{ ft} + 0.333 \text{ ft})}$$
$$= 0.222 \text{ ft/sec}$$

15. TRIANGULAR WEIRS

Triangular weirs (*V-notch weirs*) should be used when small flow rates are to be measured. The flow coefficient over a triangular weir depends on the notch angle, θ, but generally varies from 0.58 to 0.61. For a 90° weir, $C_2 \approx 0.593$. (See Fig. 19.8.)

$$Q = C_2\left(\frac{8}{15}\tan\frac{\theta}{2}\right)\sqrt{2g}H^{5/2} \qquad 19.55$$

$$Q \approx 1.4H^{2.5} \quad [90° \text{ weir}] \qquad [\text{SI}] \qquad 19.56(a)$$

$$Q \approx 2.5H^{2.5} \quad [90° \text{ weir}] \qquad [\text{U.S.}] \qquad 19.56(b)$$

Figure 19.8 Triangular Weir

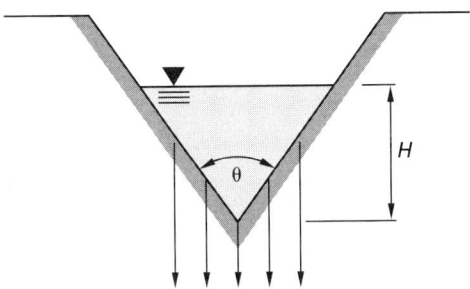

16. TRAPEZOIDAL WEIRS

A *trapezoidal weir* is essentially a rectangular weir with a triangular weir on either side. (See Fig. 19.9.) If the angle of the sides from the vertical is approximately 14° (i.e., 4 vertical and 1 horizontal), the weir is known as a *Cipoletti weir*. The discharge from the triangular ends of a Cipoletti weir approximately make up for the contractions that would reduce the flow over a rectangular weir. Therefore, no correction is theoretically necessary. This is not completely accurate, and for this reason, Cipoletti weirs are not used where great accuracy is required. The discharge is

$$Q = \tfrac{2}{3}C_d b\sqrt{2g}H^{3/2} \qquad 19.57$$

The average value of the discharge coefficient is 0.63. The discharge from a Cipoletti weir is found by using Eq. 19.58.

$$Q = 1.86bH^{3/2} \qquad [\text{SI}] \qquad 19.58(a)$$

$$Q = 3.367bH^{3/2} \qquad [\text{U.S.}] \qquad 19.58(b)$$

17. BROAD-CRESTED WEIRS AND SPILLWAYS

Most weirs used for flow measurement are sharp-crested. However, the flow over spillways, broad-crested weirs, and similar features can be calculated using Eq. 19.49 even though flow measurement is not the primary

Figure 19.9 Trapezoidal Weir

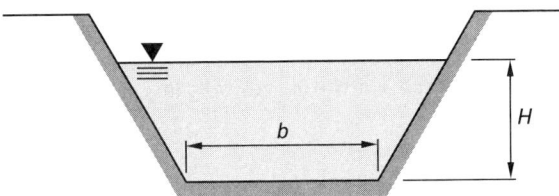

function of the feature. (A weir is broad-crested if the weir thickness is greater than half of the head, H.)

A *dam's spillway* (*overflow spillway*) is designed for a capacity based on the dam's inflow hydrograph, turbine capacity, and storage capacity. Spillways frequently have a cross section known as an *ogee*, which closely approximates the underside of a nappe from a sharp-crested weir. This cross section minimizes the cavitation that is likely to occur if the water surface breaks contact with the spillway due to upstream heads that are higher than designed for.[8]

Discharge from an overflow spillway is derived in the same manner as for a weir. Equation 19.59 can be used for broad-crested weirs ($C_1 = 0.5$–0.57) and ogee spillways ($C_1 = 0.60$–0.75).

$$Q = \tfrac{2}{3}C_1 b\sqrt{2g}H^{3/2} \qquad 19.59$$

The *Horton equation* (see Eq. 19.60) for broad-crested weirs combines all of the coefficients into a spillway (weir) coefficient and adds the velocity of approach to the upstream head. The *Horton coefficient*, C_s, is specific to the Horton equation. (C_s and C_1 differ by a factor of about 5 and cannot easily be mistaken for each other.)

$$Q = C_s b\left(H + \frac{v^2}{2g}\right)^{3/2} \qquad 19.60$$

If the velocity of approach is insignificant, the discharge is found using Eq. 19.61.

$$Q = C_s bH^{3/2} \qquad 19.61$$

C_s is a *spillway coefficient*, which varies from about 3.3 ft$^{1/2}$/sec to 3.98 ft$^{1/2}$/sec (1.8 m$^{1/2}$/s to 2.2 m$^{1/2}$/s) for ogee spillways. 3.97 ft$^{1/2}$/sec (2.2 m$^{1/2}$/s) is frequently used for first approximations. For broad-crested weirs, C_s varies between 2.63 ft$^{1/2}$/sec and 3.33 ft$^{1/2}$/sec (1.45 m$^{1/2}$/s and 1.84 m$^{1/2}$/s). (Use 3.33 ft$^{1/2}$/sec (1.84 m$^{1/2}$/sec) for initial estimates.) C_s increases as the upstream design head above the spillway top, H, increases, and the larger values apply to the higher heads.

Broad-crested weirs and spillways can be calibrated to obtain greater accuracy in predicting flow rates.

---

[8]Cavitation and separation will not normally occur as long as the actual head, H, is less than twice the design value. The shape of the ogee spillway will be a function of the design head.

Scour protection is usually needed at the toe of a spill-way to protect the area exposed to a hydraulic jump. This protection usually takes the form of an extended horizontal or sloping apron. Other measures, however, are needed if the tailwater exhibits large variations in depth.

18. PROPORTIONAL WEIRS

The *proportional weir (Sutro weir)* is used in water level control because it demonstrates a linear relationship between Q and H. Figure 19.10 illustrates a proportional weir whose sides are hyperbolic in shape.

$$Q = C_d K \left(\frac{\pi}{2}\right) \sqrt{2gH} \qquad 19.62$$

$$K = 2x\sqrt{y} \qquad 19.63$$

Figure 19.10 Proportional Weir

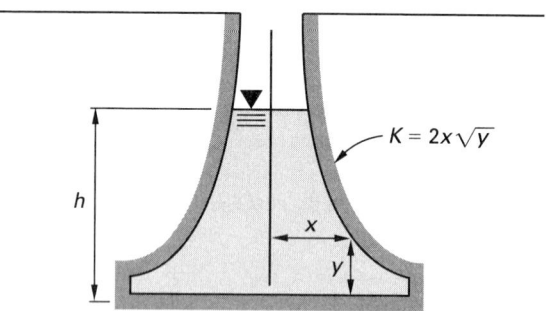

19. FLOW MEASUREMENT WITH PARSHALL FLUMES

The Parshall flume is widely used for measuring open channel wastewater flows. It performs well when head losses must be kept to a minimum and when there are high amounts of suspended solids. (See Fig. 19.11.)

The Parshall flume is constructed with a converging upstream section, a throat, and a diverging downstream section. The walls of the flume are vertical, but the floor of the throat section drops. The length, width, and height of the flume are essentially predefined by the anticipated flow rate.[9]

The throat geometry in a Parshall flume has been designed to force the occurrence of critical flow (described in Sec. 19.25) at that point. Following the critical section is a short length of supercritical flow followed by a hydraulic jump. (See Sec. 19.33.) This design eliminates any dead water region where debris and silt can accumulate (as are common with flat-topped weirs).

The discharge relationship for a Parshall flume is given for submergence ratios of H_b/H_a up to 0.7. Above 0.7,

---

[9]This chapter does not attempt to design the Parshall flume, only to predict flow rates through its use.

Figure 19.11 Parshall Flume

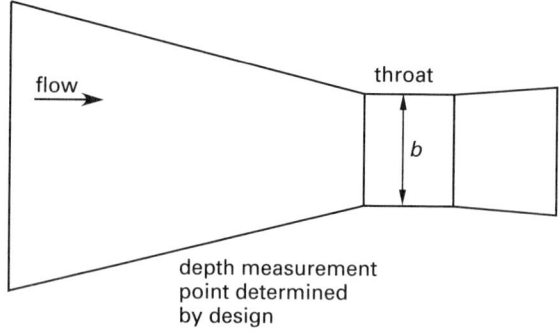

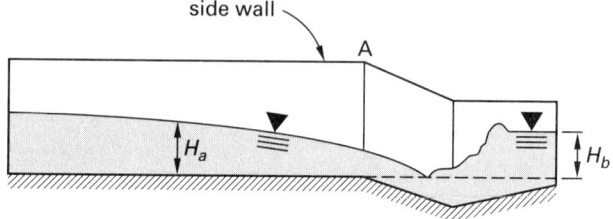

the true discharge is less than predicted by Eq. 19.64. Values of K are given in Table 19.4, although using a value of 4.0 is accurate for most purposes.

$$Q = KbH_a^n \qquad 19.64$$

$$n = 1.522b^{0.026} \qquad 19.65$$

Above a certain tailwater height, the Parshall flume no longer operates in the *free-flow mode*. Rather, it operates in a *submerged mode*. A very high tailwater reduces the flow rate through the flume. Equation 19.64 predicts the flow rate with reasonable accuracy, however, even for 50–80% submergence (calculated as H_b/H_a). For large submergence, the tailwater height must be known and a different analysis method must be used.

Table 19.4 Parshall Flume K-Values

b (ft (m))	K
0.25 (0.076)	3.97
0.50 (0.15)	4.12
0.75 (0.229)	4.09
1.0 (0.305)	4.00
1.5 (0.46)	4.00
2.0 (0.61)	4.00
3.0 (0.92)	4.00
4.0 (1.22)	4.00

(Multiply ft by 0.3048 to obtain m.)

20. UNIFORM AND NONUNIFORM STEADY FLOW

Steady flow is constant-volume flow. However, the flow may be uniform or nonuniform (varied) in depth. There may be significant variations over long and short distances without any change in the flow rate.

Figure 19.12 illustrates the three definitions of "slope" existing for open channel flow. These three slopes are the slope of the channel bottom, the slope of the water surface, and the slope of the energy gradient line.

Figure 19.12 *Slopes Used in Open Channel Flow*

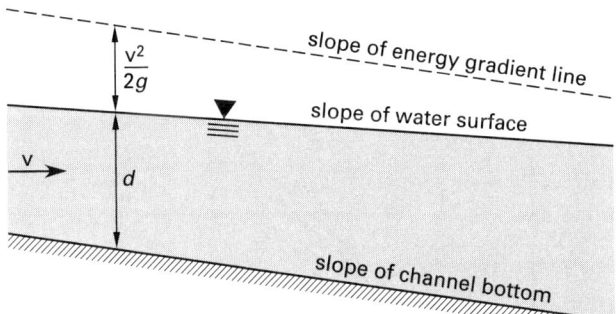

Under conditions of uniform flow, all of these three slopes are equal since the flow quantity and flow depth are constant along the length of flow.[10] With nonuniform flow, however, the flow velocity and depth vary along the length of channel and the three slopes are not necessarily equal.

If water is introduced down a path with a steep slope (as after flowing over a spillway), the effect of gravity will cause the velocity to increase. As the velocity increases, the depth decreases in accordance with the continuity of flow equation. The downward velocity is opposed by friction. Because the gravitational force is constant but friction varies with the square of velocity, these two forces eventually become equal. When equal, the velocity stops increasing, the depth stops decreasing, and the flow becomes uniform. Until they become equal, however, the flow is nonuniform (varied).

21. SPECIFIC ENERGY

The total head possessed by a fluid is given by the Bernoulli equation.

$$E_t = \frac{p}{\rho g} + \frac{v^2}{2g} + z \qquad \text{[SI]} \qquad 19.66(a)$$

$$E_t = \frac{p}{\gamma} + \frac{v^2}{2g} + z \qquad \text{[U.S.]} \qquad 19.66(b)$$

Specific energy, E, is defined as the total head with respect to the channel bottom. In this case, $z = 0$ and $p/\gamma = d$.

$$E = d + \frac{v^2}{2g} \qquad 19.67$$

[10]As a simplification, this chapter deals only with channels of constant width. If the width is varied, changes in flow depth may not coincide with changes in flow quantity.

Equation 19.67 is not meant to imply that the potential energy is an unimportant factor in open channel flow problems. The concept of specific energy is used for convenience only, and it should be clear that the Bernoulli equation is still the valid energy conservation equation.

In uniform flow, total head also decreases due to the frictional effects, but specific energy is constant. In nonuniform flow, total head also decreases, but specific energy may increase or decrease.

Since $v = Q/A$, Eq. 19.67 can be written as

$$E = d + \frac{Q^2}{2gA^2} \qquad \text{[general case]} \qquad 19.68$$

For a rectangular channel, the velocity can be written in terms of the width and flow depth.

$$v = \frac{Q}{A} = \frac{Q}{wd} \qquad 19.69$$

The specific energy equation for a rectangular channel is given by Eq. 19.70 and shown in Fig. 19.13.

$$E = d + \frac{Q^2}{2g(wd)^2} \qquad \text{[rectangular]} \qquad 19.70$$

Specific energy can be used to differentiate between flow regimes. Figure 19.13 illustrates how specific energy is affected by depth, and accordingly, how specific energy relates to critical depth (described in Sec. 19.25) and the Froude number (described in Sec. 19.27).

Figure 19.13 *Specific Energy Diagram*

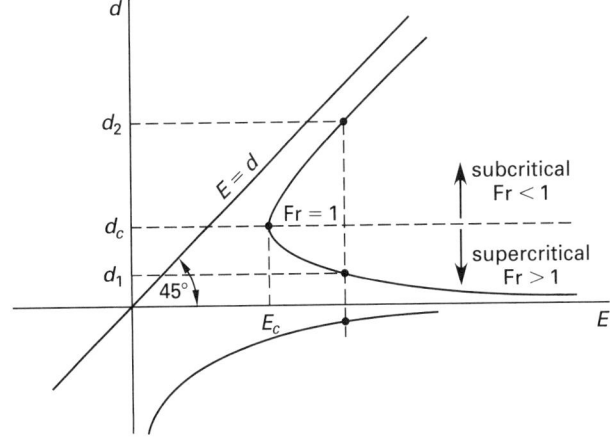

22. SPECIFIC FORCE

The *specific force* of a general channel section is the total force per unit weight acting on the water. Equivalently, specific force is the total force that a submerged object would experience. In Eq. 19.71, $\overline{d}$ is the distance from the free surface to the centroid of the flowing area cross section, A.

$$\frac{F}{g\rho} = \frac{Q^2}{gA} + \overline{d}A \qquad \text{[SI]} \qquad 19.71(a)$$

$$\frac{F}{\gamma} = \frac{Q^2}{gA} + \overline{d}A \qquad \text{[U.S.]} \qquad 19.71(b)$$

The first term in Eq. 19.71 represents the momentum flow through the channel per unit time and per unit mass of water. The second term is the pressure force per unit mass of water. Graphs of specific force and specific energy are similar in appearance and predict equivalent results for the critical and alternate depths. (See Sec. 19.24.)

23. CHANNEL TRANSITIONS

Sudden changes in channel width or bottom elevation are known as *channel transitions*. (Contractions in width are not covered in this chapter.) For sudden vertical steps in channel bottom, the Bernoulli equation, written in terms of the specific energy, is used to predict the flow behavior (i.e., the depth).

$$E_1 + z_1 = E_2 + z_2 \qquad 19.72$$

$$E_1 - E_2 = z_2 - z_1 \qquad 19.73$$

The maximum possible change in bottom elevation without affecting the energy equality occurs when the depth of flow over the step is equal to the critical depth (d_2 equals d_c). (See Sec. 19.25.)

24. ALTERNATE DEPTHS

Since the area depends on the depth, fixing the channel shape and slope and assuming a depth will determine the flow rate, Q, as well as the specific energy. Since Eq. 19.70 is a cubic equation, there are three values of depth of flow, d, that will satisfy it. One of them is negative, as Fig. 19.13 shows. Since depth cannot be negative, that value can be discarded. The two remaining values are known as *alternate depths*.

For a given flow rate, the two alternate depths have the same energy. One represents a high velocity with low depth; the other represents a low velocity with high depth. The former is called *supercritical (rapid) flow*; the latter is called *subcritical (tranquil) flow*.

The Bernoulli equation cannot predict which of the two alternate depths will occur for any given flow quantity. The concept of *accessibility* is required to evaluate the two depths. Specifically, the upper and lower limbs of the energy curve are not accessible from each other unless there is a local restriction in the flow.

Energy curves can be drawn for different flow quantities, as shown in Fig. 19.14 for flow quantities Q_A and Q_B. Suppose that flow is initially at point 1. Since the flow is on the upper limb, the flow is initially subcritical. If there is a step up in the channel bottom, Eq. 19.73 predicts that the specific energy will decrease.

Figure 19.14 *Specific Energy Curve Families*

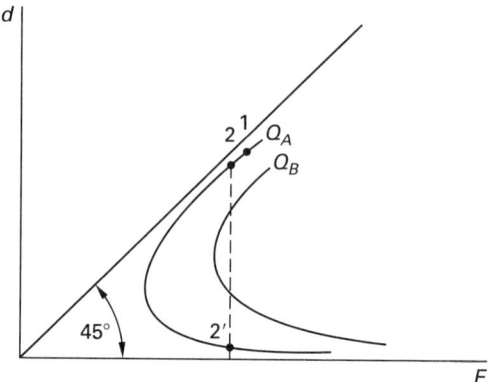

However, the flow cannot arrive at point $2'$ without the flow quantity changing (i.e., going through a specific energy curve for a different flow quantity). Therefore, point $2'$ is not accessible from point 1 without going through point 2 first.[11]

If the flow is well up on the top limb of the specific energy curve (as it is in Ex. 19.7), the water level will drop only slightly. Since the upper limb is asymptotic to a 45° diagonal line, any change in specific energy will result in almost the same change in depth.[12] Therefore, the surface level will remain almost the same.

$$\Delta d \approx \Delta E \qquad \text{[fully subcritical]} \qquad 19.74$$

However, if the initial point on the limb is close to the critical point (i.e., the nose of the curve), then a small change in the specific energy (such as might be caused by a small variation in the channel floor) will cause a large change in depth. That is why severe turbulence commonly occurs near points of critical flow.

Example 19.7

4 ft/sec (1.2 m/s) of water flow in a 7 ft (2.1 m) wide, 6 ft (1.8 m) deep open channel. The flow encounters a 1.0 ft (0.3 m) step in the channel bottom. What is the depth of flow above the step?

---

[11]Actually, specific energy curves are typically plotted for flow per unit width, $q = Q/w$. If that is the case, a jump from one limb to the other could take place if the width were allowed to change as well as depth.
[12]A rise in the channel bottom does not always produce a drop in the water surface. Only if the flow is initially subcritical will the water surface drop upon encountering a step. The water surface will rise if the flow is initially supercritical.

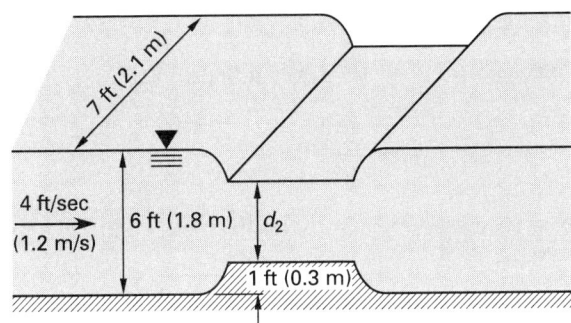

SI Solution

The initial specific energy is found from Eq. 19.67.

$$E_1 = d + \frac{v^2}{2g}$$
$$= 1.8 \text{ m} + \frac{\left(1.2 \frac{\text{m}}{\text{s}}\right)^2}{(2)\left(9.81 \frac{\text{m}}{\text{s}^2}\right)}$$
$$= 1.87 \text{ m}$$

From Eq. 19.73, the specific energy above the step is

$$E_2 = E_1 + z_1 - z_2$$
$$= 1.87 \text{ m} + 0 - 0.3 \text{ m}$$
$$= 1.57 \text{ m}$$

The quantity flowing is

$$Q = Av$$
$$= (2.1 \text{ m})(1.8 \text{ m})\left(1.2 \frac{\text{m}}{\text{s}}\right)$$
$$= 4.54 \text{ m}^3/\text{s}$$

Substituting Q into Eq. 19.70 gives

$$E = d + \frac{Q^2}{2g(wd)^2}$$
$$1.57 \text{ m} = d_2 + \frac{\left(4.54 \frac{\text{m}^3}{\text{s}}\right)^2}{(2)\left(9.81 \frac{\text{m}}{\text{s}^2}\right)(2.1 \text{ m})^2 d_2^2}$$

By trial and error or a calculator's equation solver, the alternate depths are $d_2 = 0.46$ m, 1.46 m.

Since the 0.46 m depth is not accessible from the initial depth of 1.8 m, the depth over the step is 1.5 m. The drop in the water level is

$$1.8 \text{ m} - (1.46 \text{ m} + 0.3 \text{ m}) = 0.04 \text{ m}$$

Customary U.S. Solution

The initial specific energy is found from Eq. 19.67.

$$E_1 = d + \frac{v^2}{2g}$$
$$= 6 \text{ ft} + \frac{\left(4 \frac{\text{ft}}{\text{sec}}\right)^2}{(2)\left(32.2 \frac{\text{ft}}{\text{sec}^2}\right)}$$
$$= 6.25 \text{ ft}$$

From Eq. 19.73, the specific energy over the step is

$$E_2 = E_1 + z_1 - z_2 = 6.25 \text{ ft} + 0 - 1 \text{ ft}$$
$$= 5.25 \text{ ft}$$

The quantity flowing is

$$Q = Av = (7 \text{ ft})(6 \text{ ft})\left(4 \frac{\text{ft}}{\text{sec}}\right)$$
$$= 168 \text{ ft}^3/\text{sec}$$

Substituting Q into Eq. 19.70 gives

$$E = d + \frac{Q^2}{2g(wd)^2}$$
$$5.25 \text{ ft} = d_2 + \frac{\left(168 \frac{\text{ft}^3}{\text{sec}}\right)^2}{(2)\left(32.2 \frac{\text{ft}}{\text{sec}^2}\right)(7 \text{ ft})^2 d_2^2}$$

By trial and error or a calculator's equation solver, the alternate depths are $d_2 = 1.6$ ft, 4.9 ft.

Since the 1.6 ft depth is not accessible from the initial depth of 6 ft, the depth over the step is 4.9 ft. The drop in the water level is

$$6 \text{ ft} - (4.9 \text{ ft} + 1 \text{ ft}) = 0.1 \text{ ft}$$

25. CRITICAL FLOW AND CRITICAL DEPTH IN RECTANGULAR CHANNELS

There is one depth, known as the *critical depth*, that minimizes the energy of flow. (The depth is not minimized, however.) The critical depth for a given flow depends on the shape of the channel.

For a rectangular channel, if Eq. 19.70 is differentiated with respect to depth in order to minimize the specific energy, Eq. 19.75 results.

$$d_c^3 = \frac{Q^2}{gw^2} \quad \text{[rectangular]} \qquad 19.75$$

Geometrical and analytical methods can be used to correlate the critical depth and the minimum specific energy.

$$d_c = \tfrac{2}{3}E_c \qquad 19.76$$

For a rectangular channel, $Q = d_c w v_c$. Substituting this into Eq. 19.75 produces an equation for the *critical velocity*.

$$v_c = \sqrt{gd_c} \qquad 19.77$$

The expression for critical velocity also coincides with the expression for the velocity of a low-amplitude *surface wave (surge wave)* moving in a liquid of depth d_c. Since surface disturbances are transmitted as ripples upstream (and downstream) at velocity v_c, it is apparent that a surge wave will be stationary in a channel moving at the critical velocity. Such motionless waves are known as *standing waves*.

If the flow velocity is less than the surge wave velocity (for the actual depth), then a ripple can make its way upstream. If the flow velocity exceeds the surge wave velocity, the ripple will be swept downstream.

Example 19.8

500 ft^3/sec (14 m^3/s) of water flow in a 20 ft (6 m) wide rectangular channel. What are the (a) critical depth and (b) critical velocity?

SI Solution

(a) From Eq. 19.75, the critical depth is

$$d_c^3 = \frac{Q^2}{gw^2}$$

$$= \frac{\left(14\ \frac{m^3}{s}\right)^2}{\left(9.81\ \frac{m}{s^2}\right)(6\ m)^2}$$

$$d_c = 0.822\ m$$

(b) From Eq. 19.77, the critical velocity is

$$v_c = \sqrt{gd_c}$$

$$= \sqrt{\left(9.81\ \frac{m}{s^2}\right)(0.822\ m)}$$

$$= 2.84\ m/s$$

Customary U.S. Solution

(a) From Eq. 19.75, the critical depth is

$$d_c^3 = \frac{Q^2}{gw^2}$$

$$= \frac{\left(500\ \frac{ft^3}{sec}\right)^2}{\left(32.2\ \frac{ft}{sec^2}\right)(20\ ft)^2}$$

$$d_c = 2.687\ ft$$

(b) From Eq. 19.77, the critical velocity is

$$v_c = \sqrt{gd_c} = \sqrt{\left(32.2\ \frac{ft}{sec^2}\right)(2.687\ ft)}$$

$$= 9.30\ ft/sec$$

26. CRITICAL FLOW AND CRITICAL DEPTH IN NONRECTANGULAR CHANNELS

For nonrectangular shapes (including trapezoidal channels), the critical depth can be found by trial and error from the following equation in which T is the surface width. To use Eq. 19.78, assume trial values of the critical depth, use them to calculate dependent quantities in the equation, and then verify the equality.

$$\frac{Q^2}{g} = \frac{A^3}{T} \qquad \text{[nonrectangular]} \qquad 19.78$$

Equation 19.78 is particularly difficult to use with circular channels. Appendix 19.D is a convenient method of determining critical depth in circular channels.

27. FROUDE NUMBER

The dimensionless *Froude number*, Fr, is a convenient index of the flow regime. It can be used to determine whether the flow is subcritical or supercritical. L is the *characteristic length*, also referred to as the *characteristic (length) scale*, hydraulic depth, mean hydraulic depth, and others, depending on the channel configuration. d is the depth corresponding to velocity v. For circular channels flowing half full, $L = \pi D/8$. For a rectangular channel, $L = d$. For trapezoidal and semicircular channels, and in general, L is the area in flow divided by the top width, T.

$$Fr = \frac{v}{\sqrt{gL}} \qquad 19.79$$

When the Froude number is less than one, the flow is subcritical (i.e., the depth of flow is greater than the critical depth) and the velocity is less than the critical velocity.

For convenience, the Froude number can be written in terms of the flow rate per average unit width.

$$\text{Fr} = \frac{\frac{Q}{b}}{\sqrt{gd^3}} \quad \text{[rectangular]} \qquad \textit{19.80}$$

$$\text{Fr} = \frac{\frac{Q}{b_{\text{ave}}}}{\sqrt{g\left(\frac{A}{b_{\text{ave}}}\right)^3}} \quad \text{[nonrectangular]} \qquad \textit{19.81}$$

When the Froude number is greater than one, the flow is supercritical. The depth is less than critical depth, and the flow velocity is greater than the critical velocity.

When the Froude number is equal to one, the flow is critical.[13]

The Froude number has another form. Dimensional analysis determines it to be v^2/gL, a form that is also used in analyzing similarity of models. Whether the derived form or the square root form is used can sometimes be determined by observing the form of the intended application. If the Froude number is squared (as it is in Eq. 19.82), then the square root form is probably intended. For open channel flow, the Froude number is always the square root of the derived form.

28. PREDICTING OPEN CHANNEL FLOW BEHAVIOR

Upon encountering a variation in the channel bottom, the behavior of an open channel flow is dependent on whether the flow is initially subcritical or supercritical. Open channel flow is governed by Eq. 19.82 in which the Froude number is the primary independent variable.

$$\frac{dd}{dx}(1 - \text{Fr}^2) + \frac{dz}{dx} = 0 \qquad \textit{19.82}$$

The quantity dd/dx is the slope of the surface (i.e., it is the derivative of the depth with respect to the channel length). The quantity dz/dx is the slope of the channel bottom.

For an upward step, $dz/dx > 0$. If the flow is initially subcritical (i.e., $\text{Fr} < 1$), then Eq. 19.82 requires that $dd/dx < 0$, a drop in depth.

This logic can be repeated for other combinations of the terms. Table 19.5 lists the various behaviors of open channel flow surface levels based on Eq. 19.82.

Table 19.5 *Surface Level Change Behavior*

initial flow	step up	step down
subcritical	surface drops	surface rises
supercritical	surface rises	surface drops

---

[13]The similarity of the Froude number to the Mach number used to classify gas flows is more than coincidental. Both bodies of knowledge employ parallel concepts.

If $dz/dx = 0$ (i.e., a horizontal slope), then either the depth must be constant or the Froude number must be unity. The former case is obvious. The latter case predicts critical flow. Such critical flow actually occurs where the slope is horizontal over broad-crested weirs and at the top of a rounded spillway. Since broad-crested weirs and spillways produce critical flow, they represent a class of controls on flow.

29. OCCURRENCES OF CRITICAL FLOW

The critical depth not only minimizes the energy of flow, but also maximizes the quantity flowing for a given cross section and slope. Critical flow is generally quite turbulent because of the large changes in energy that occur with small changes in elevation and depth. Critical depth flow is often characterized by successive water surface undulations over a very short stretch of channel. (See Fig. 19.15.)

For any given discharge and cross section, there is a unique slope that will produce and maintain flow at critical depth. Once d_c is known, this critical slope can be found from the Manning equation. In all of the instances of critical depth, Eq. 19.77 can be used to calculate the actual velocity.

Figure 19.15 *Occurrence of Critical Depth*

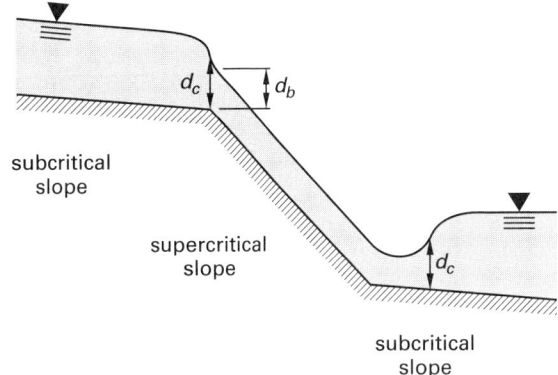

Critical depth occurs at free outfall from a channel of mild slope. The occurrence is at the point of curvature inversion, which is just upstream from the brink. (See Fig. 19.16.) For mild slopes, the *brink depth* is approximately

$$d_b = 0.715 d_c \qquad \textit{19.83}$$

Figure 19.16 *Free Outfall*

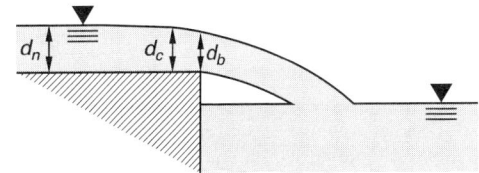

Flow of Fluids

Critical flow can occur across a broad-crested weir, as shown in Fig. 19.17.[14] With no obstruction to hold the water, it falls from the normal depth to the critical depth, but it can fall no more than that because there is no source to increase the specific energy (to increase the velocity). This is not a contradiction of the previous free outfall case where the brink depth is less than the critical depth. The flow curvatures in free outfall are a result of the constant gravitational acceleration.

Figure 19.17 Broad-Crested Weir

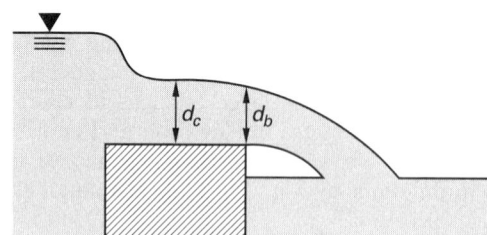

Critical depth can also occur when a channel bottom has been raised sufficiently to choke the flow. A raised channel bottom is essentially a broad-crested weir. (See Fig. 19.18.)

Figure 19.18 Raised Channel Bottom with Choked Flow

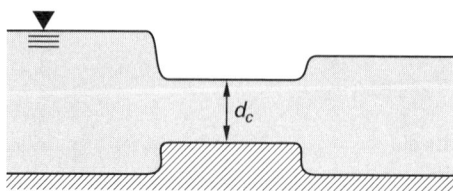

Example 19.9

At a particular point in an open rectangular channel ($n = 0.013$, $S = 0.002$, and $w = 10$ ft (3 m)), the flow is 250 ft³/sec (7 m³/s) and the depth is 4.2 ft (1.3 m).

(a) Is the flow tranquil, critical, or rapid?

(b) What is the normal depth?

(c) If the channel ends in a free outfall, what is the brink depth?

SI Solution

(a) From Eq. 19.75, the critical depth is

$$d_c = \left(\frac{Q^2}{gw^2}\right)^{1/3}$$

$$= \left(\frac{\left(7 \; \frac{\text{m}^3}{\text{s}}\right)^2}{(3 \text{ m})^2 \left(9.81 \; \frac{\text{m}}{\text{s}^2}\right)}\right)^{1/3}$$

$$= 0.82 \text{ m}$$

Since the actual depth exceeds the critical depth, the flow is tranquil.

(b) From Eq. 19.13,

$$Q = \left(\frac{1}{n}\right) A R^{2/3} \sqrt{S}$$

$$R = \frac{A}{P} = \frac{d_n(3 \text{ m})}{2d_n + 3 \text{ m}}$$

Substitute the expression for R into Eq. 19.13 and solve for d_n.

$$7 \; \frac{\text{m}^3}{\text{s}} = \left(\frac{1}{0.013}\right) d_n(3 \text{ m}) \left(\frac{d_n(3 \text{ m})}{2d_n + 3 \text{ m}}\right)^{2/3} \sqrt{0.002}$$

By trial and error or a calculator's equation solver, $d_n = 0.97$ m. Since the actual and normal depths are different, the flow is nonuniform.

(c) From Eq. 19.83, the brink depth is

$$d_b = 0.715 d_c = (0.715)(0.82 \text{ m})$$

$$= 0.59 \text{ m}$$

Customary U.S. Solution

(a) From Eq. 19.75, the critical depth is

$$d_c = \left(\frac{Q^2}{gw^2}\right)^{1/3}$$

$$= \left(\frac{\left(250 \; \frac{\text{ft}^3}{\text{sec}}\right)^2}{\left(32.2 \; \frac{\text{ft}}{\text{sec}^2}\right)(10 \text{ ft})^2}\right)^{1/3}$$

$$= 2.69 \text{ ft}$$

Since the actual depth exceeds the critical depth, the flow is tranquil.

---

[14]Figure 19.17 is an example of a *hydraulic drop*, the opposite of a hydraulic jump. A hydraulic drop can be recognized by the sudden decrease in depth over a short length of channel.

(b) From Eq. 19.13,

$$Q = \left(\frac{1.49}{n}\right) A R^{2/3} \sqrt{S}$$

$$R = \frac{A}{P} = \frac{d_n(10 \text{ ft})}{2d_n + 10 \text{ ft}}$$

Substitute the expression for R into Eq. 19.13 and solve for d_n.

$$250 \ \frac{\text{ft}^3}{\text{sec}} = \left(\frac{1.49}{0.013}\right) d_n(10 \text{ ft}) \left(\frac{d_n(10 \text{ ft})}{2d_n + 10 \text{ ft}}\right)^{2/3} \sqrt{0.002}$$

By trial and error or a calculator's equation solver, $d_n = 3.1$ ft. Since the actual and normal depths are different, the flow is nonuniform.

(c) From Eq. 19.83, the brink depth is

$$d_b = 0.715 d_c$$
$$= (0.715)(2.69 \text{ ft})$$
$$= 1.92 \text{ ft}$$

30. CONTROLS ON FLOW

In general, any feature that affects depth and discharge rates is known as a *control on flow*. Controls may consist of constructed control structures (weirs, gates, sluices, etc.), forced flow through critical depth (as in a free outfall), sudden changes of slope (which forces a hydraulic jump or hydraulic drop to the new normal depth), or free flow between reservoirs of different surface elevations. A downstream control may also be an upstream control, as Fig. 19.19 shows.

If flow is subcritical, then a disturbance downstream will be able to affect the upstream conditions. Since the flow velocity is less than the critical velocity, a ripple will be able to propagate upstream to signal a change in the downstream conditions. Any object downstream that affects the flow rate, velocity, or depth upstream is known as a *downstream control*.

If a flow is supercritical, then a downstream obstruction will have no effect upstream, since disturbances cannot propagate upstream faster than the flow velocity. The only effect on supercritical flow is from an upstream obstruction. Such an obstruction is said to be an *upstream control*.

Figure 19.19 Control on Flow

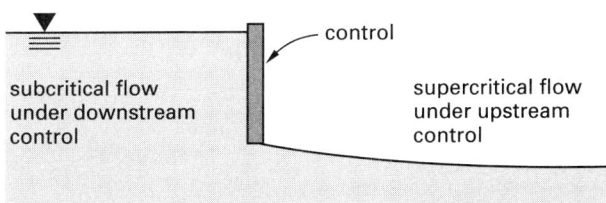

31. FLOW CHOKING

A channel feature that causes critical flow to occur is known as a *choke*, and the corresponding flow past the feature and downstream is known as *choked flow*.

In the case of vertical transitions (i.e., upward or downward steps in the channel bottom), choked flow will occur when the step size is equal to the difference between the upstream specific energy and the critical flow energy.

$$\Delta z = E_1 - E_c \quad \text{[choked flow]} \qquad 19.84$$

In the case of a rectangular channel, combining Eq. 19.68 and Eq. 19.77 the maximum variation in channel bottom will be

$$\Delta z = E_1 - \left(d_c + \frac{\text{v}_c^2}{2g}\right)$$
$$= E_1 - \tfrac{3}{2}d_c \qquad 19.85$$

The flow downstream from a choke point can be subcritical or supercritical, depending on the downstream conditions. If there is a downstream control, such as a sluice gate, the flow downstream will be subcritical. If there is additional gravitational acceleration (as with flow down the side of a dam spillway), then the flow will be supercritical.

32. VARIED FLOW

Accelerated flow occurs in any channel where the actual slope exceeds the friction loss per foot.

$$S_0 > \frac{h_f}{L} \qquad 19.86$$

Retarded flow occurs when the actual slope is less than the unit friction loss.

$$S_0 < \frac{h_f}{L} \qquad 19.87$$

In sections AB and CD of Fig. 19.20, the slopes are less than the energy gradient, so the flows are retarded. In section BC, the slope is greater than the energy gradient, so the velocity increases (i.e., the flow is accelerated). If section BC were long enough, the friction loss

Figure 19.20 Varied Flow

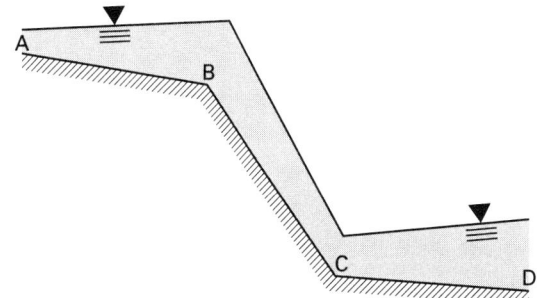

would eventually become equal to the accelerating energy and the flow would become uniform.

The distance between points 1 and 2 with two known depths in accelerated or retarded flow can be determined from the average velocity. Equation 19.88 and Eq. 19.89 assume that the friction losses are the same for varied flow as for uniform flow.

$$S_{ave} = \left(\frac{n v_{ave}}{R_{ave}^{2/3}}\right)^2 \qquad \text{[SI]} \quad 19.88(a)$$

$$S_{ave} = \left(\frac{n v_{ave}}{1.49 R_{ave}^{2/3}}\right)^2 \qquad \text{[U.S.]} \quad 19.88(b)$$

$$v_{ave} = \tfrac{1}{2}(v_1 + v_2) \qquad 19.89$$

S is the slope of the energy gradient from Eq. 19.88, not the channel slope S_0. The usual method of finding the *depth profile* is to start at a point in the channel where d_2 and v_2 are known. Then, assume a depth d_1, find v_1 and S, and solve for L. Repeat as needed.

$$L = \frac{\left(d_1 + \dfrac{v_1^2}{2g}\right) - \left(d_2 + \dfrac{v_2^2}{2g}\right)}{S - S_0}$$

$$= \frac{E_1 - E_2}{S - S_0} \qquad 19.90$$

In Eq. 19.89 and Eq. 19.90, d_1 is always the smaller of the two depths.

Example 19.10

How far from the point described in Ex. 19.9 will the depth be 4 ft (1.2 m)?

SI Solution

The difference between 1.3 m and 1.2 m is small, so a one-step calculation will probably be sufficient.

$$d_1 = 1.2 \text{ m}$$

$$v_1 = \frac{Q}{A} = \frac{7 \ \frac{m^3}{s}}{(1.2 \text{ m})(3 \text{ m})}$$

$$= 1.94 \text{ m/s}$$

$$E_1 = d_1 + \frac{v_1^2}{2g} = 1.2 \text{ m} + \frac{\left(1.94 \ \frac{m}{s}\right)^2}{(2)\left(9.81 \ \frac{m}{s^2}\right)}$$

$$= 1.39 \text{ m}$$

$$R_1 = \frac{A_1}{P_1} = \frac{(1.2 \text{ m})(3 \text{ m})}{1.2 \text{ m} + 3 \text{ m} + 1.2 \text{ m}}$$

$$= 0.67 \text{ m}$$

$$d_2 = 1.3 \text{ m}$$

$$v_2 = \frac{7 \ \frac{m^3}{s}}{(1.3 \text{ m})(3 \text{ m})}$$

$$= 1.79 \text{ m/s}$$

$$E_2 = d_2 + \frac{v_2^2}{2g} = 1.3 \text{ m} + \frac{\left(1.79 \ \frac{m}{s}\right)^2}{(2)\left(9.81 \ \frac{m}{s^2}\right)}$$

$$= 1.46 \text{ m}$$

$$R_2 = \frac{A_2}{P_2} = \frac{(1.3 \text{ m})(3 \text{ m})}{1.3 \text{ m} + 3 \text{ m} + 1.3 \text{ m}}$$

$$= 0.70 \text{ m}$$

$$v_{ave} = \tfrac{1}{2}(v_1 + v_2) = \left(\tfrac{1}{2}\right)\left(1.94 \ \frac{m}{s} + 1.79 \ \frac{m}{s}\right)$$

$$= 1.865 \text{ m/s}$$

$$R_{ave} = \tfrac{1}{2}(R_1 + R_2) = \left(\tfrac{1}{2}\right)(0.70 \text{ m} + 0.67 \text{ m})$$

$$= 0.685 \text{ m}$$

From Eq. 19.88,

$$S = \left(\frac{n v_{ave}}{R_{ave}^{2/3}}\right)^2$$

$$= \left(\frac{(0.013)\left(1.865 \ \frac{m}{s}\right)}{(0.685 \text{ m})^{2/3}}\right)^2$$

$$= 0.000973$$

From Eq. 19.90,

$$L = \frac{E_1 - E_2}{S - S_0} = \frac{1.39 \text{ m} - 1.46 \text{ m}}{0.000973 - 0.002}$$

$$= 68.2 \text{ m}$$

Customary U.S. Solution

The difference between 4 ft and 4.2 ft is small, so a one-step calculation will probably be sufficient.

$$d_1 = 4 \text{ ft}$$

$$v_1 = \frac{Q}{A} = \frac{250 \ \frac{ft^3}{sec}}{(4 \text{ ft})(10 \text{ ft})}$$

$$= 6.25 \text{ ft/sec}$$

$$E_1 = d_1 + \frac{v_1^2}{2g}$$

$$= 4 \text{ ft} + \frac{\left(6.25 \; \frac{\text{ft}}{\text{sec}}\right)^2}{(2)\left(32.2 \; \frac{\text{ft}}{\text{sec}^2}\right)}$$

$$= 4.607 \text{ ft}$$

$$R_1 = \frac{A_1}{P_1} = \frac{(4 \text{ ft})(10 \text{ ft})}{4 \text{ ft} + 10 \text{ ft} + 4 \text{ ft}}$$

$$= 2.22 \text{ ft}$$

$$d_2 = 4.2 \text{ ft}$$

$$v_2 = \frac{250 \; \frac{\text{ft}^3}{\text{sec}}}{(4.2 \text{ ft})(10 \text{ ft})}$$

$$= 5.95 \text{ ft/sec}$$

$$E_2 = d_2 + \frac{v_2^2}{2g} = 4.2 \text{ ft} + \frac{\left(5.95 \; \frac{\text{ft}}{\text{sec}}\right)^2}{(2)\left(32.2 \; \frac{\text{ft}}{\text{sec}^2}\right)}$$

$$= 4.75 \text{ ft}$$

$$R_2 = \frac{A_2}{P_2} = \frac{(4.2 \text{ ft})(10 \text{ ft})}{4.2 \text{ ft} + 10 \text{ ft} + 4.2 \text{ ft}}$$

$$= 2.28 \text{ ft}$$

$$v_{\text{ave}} = \tfrac{1}{2}(v_1 + v_2)$$

$$= \left(\tfrac{1}{2}\right)\left(6.25 \; \frac{\text{ft}}{\text{sec}} + 5.95 \; \frac{\text{ft}}{\text{sec}}\right)$$

$$= 6.1 \text{ ft/sec}$$

$$R_{\text{ave}} = \tfrac{1}{2}(R_1 + R_2)$$

$$= \left(\tfrac{1}{2}\right)(2.22 \text{ ft} + 2.28 \text{ ft})$$

$$= 2.25 \text{ ft}$$

From Eq. 19.88,

$$S = \left(\frac{n v_{\text{ave}}}{1.49 R_{\text{ave}}^{2/3}}\right)^2$$

$$= \left(\frac{(0.013)\left(6.1 \; \frac{\text{ft}}{\text{sec}}\right)}{(1.49)(2.25 \text{ ft})^{2/3}}\right)^2$$

$$= 0.000961$$

From Eq. 19.90,

$$L = \frac{E_1 - E_2}{S - S_0} = \frac{4.607 \text{ ft} - 4.75 \text{ ft}}{0.000961 - 0.002}$$

$$= 138 \text{ ft}$$

33. HYDRAULIC JUMP

If water is introduced at high (supercritical) velocity to a section of slow-moving (subcritical) flow (as shown in Fig. 19.21), the velocity will be reduced rapidly over a short length of channel. The abrupt rise in the water surface is known as a *hydraulic jump*. The increase in depth is always from below the critical depth to above the critical depth.[15] The depths on either side of the hydraulic jump are known as *conjugate depths*. The conjugate depths and the relationship between them are as follows.

$$d_1 = -\tfrac{1}{2}d_2 + \sqrt{\frac{2v_2^2 d_2}{g} + \frac{d_2^2}{4}} \quad \begin{bmatrix} \text{rectangular} \\ \text{channels} \end{bmatrix} \qquad 19.91$$

$$d_2 = -\tfrac{1}{2}d_1 + \sqrt{\frac{2v_1^2 d_1}{g} + \frac{d_1^2}{4}} \quad \begin{bmatrix} \text{rectangular} \\ \text{channels} \end{bmatrix} \qquad 19.92$$

$$\frac{d_2}{d_1} = \tfrac{1}{2}\left(\sqrt{1 + 8(\text{Fr}_1)^2} - 1\right) \quad \begin{bmatrix} \text{rectangular} \\ \text{channels} \end{bmatrix} \qquad 19.93(a)$$

$$\frac{d_1}{d_2} = \tfrac{1}{2}\left(\sqrt{1 + 8(\text{Fr}_2)^2} - 1\right) \quad \begin{bmatrix} \text{rectangular} \\ \text{channels} \end{bmatrix} \qquad 19.93(b)$$

Figure 19.21 *Conjugate Depths*

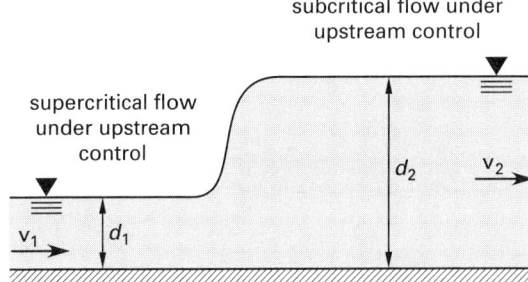

If the depths d_1 and d_2 are known, then the upstream velocity can be found from Eq. 19.94.

$$v_1^2 = \left(\frac{g d_2}{2 d_1}\right)(d_1 + d_2) \quad \begin{bmatrix} \text{rectangular} \\ \text{channels} \end{bmatrix} \qquad 19.94$$

Conjugate depths are not the same as alternate depths. Alternate depths are derived from the conservation of energy equation (i.e., a variation of the Bernoulli equation). Conjugate depths (calculated in Eq. 19.91 through Eq. 19.93) are derived from a conservation of momentum equation. Conjugate depths are calculated only when there has been an abrupt energy loss such as occurs in a hydraulic jump or drop.

[15]This provides a way of determining if a hydraulic jump can occur in a channel. If the original depth is above the critical depth, the flow is already subcritical. Therefore, a hydraulic jump cannot form. Only a hydraulic drop could occur.

Hydraulic jumps have practical applications in the design of stilling basins. Stilling basins are designed to intentionally reduce energy of flow through hydraulic jumps. In the case of a concrete apron at the bottom of a dam spillway, the apron friction is usually low, and the water velocity will decrease only gradually. However, supercritical velocities can be reduced to much slower velocities by having the flow cross a series of baffles on the channel bottom.

The specific energy lost in the jump is the energy lost per pound of water flowing.

$$\Delta E = \left(d_1 + \frac{v_1^2}{2g}\right) - \left(d_2 + \frac{v_2^2}{2g}\right) \approx \frac{(d_2 - d_1)^3}{4 d_1 d_2} \qquad 19.95$$

Evaluation of hydraulic jumps in stilling basins starts by determining the depth at the toe. The depth of flow at the toe of a spillway is found from an energy balance. Neglecting friction, the total energy at the toe equals the total upstream energy before the spillway. The total upstream energy before the spillway is

$$E_{\text{upstream}} = E_{\text{toe}} \qquad 19.96$$

$$y_{\text{crest}} + H + \frac{v^2}{2g} = d_{\text{toe}} + \frac{v_{\text{toe}}^2}{2g} \qquad 19.97$$

The upstream velocity, v, is the velocity before the spillway (which is essentially zero), not the velocity over the brink. If the brink depth is known, the velocity over the brink can be used with the continuity equation to calculate the upstream velocity, but the velocity over the brink should not be used with H to determine total energy since $v_b \neq v$. (See Fig. 19.22.)

Figure 19.22 *Total Energy Upstream of a Spillway*

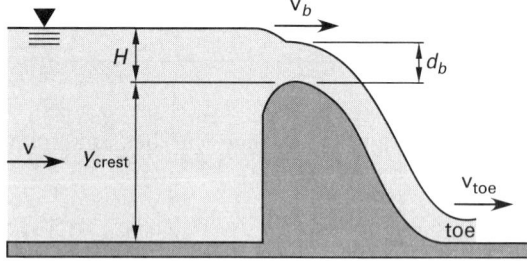

If water is drained quickly from the apron so that the tailwater depth is small or zero, no hydraulic jump will form. This is because the tailwater depth is already less than the critical depth.

A hydraulic jump will form along the apron at the bottom of the spillway when the actual tailwater depth equals the conjugate depth d_2 corresponding to the depth at the toe. That is, the jump is located at the toe when $d_2 = d_{\text{tailwater}}$, where d_2 and $d_1 = d_{\text{toe}}$ are conjugate depths. The tailwater and toe depths are implicitly the conjugate depths. This is shown in

Fig. 19.23(a) and is the proper condition for energy dissipation in a stilling basin.

When the actual tailwater depth is less than the conjugate depth d_2 corresponding to d_{toe}, but still greater than the critical depth, flow will continue along the apron until the depth increases to conjugate depth d_1 corresponding to the actual tailwater depth. This is shown in Fig. 19.23(b). A hydraulic jump will form at that point to increase the depth to the tailwater depth. (Another way of saying this is that the hydraulic jump moves downstream from the toe.) This is an undesirable condition, since the location of the jump is often inadequately protected from scour.

If the tailwater depth is greater than the conjugate depth corresponding to the depth at the toe, as in Fig. 19.23(c), the hydraulic jump may occur up on the spillway, or it may be completely submerged (i.e., it will not occur at all).

Figure 19.23 *Hydraulic Jump to Reach Tailwater Level*

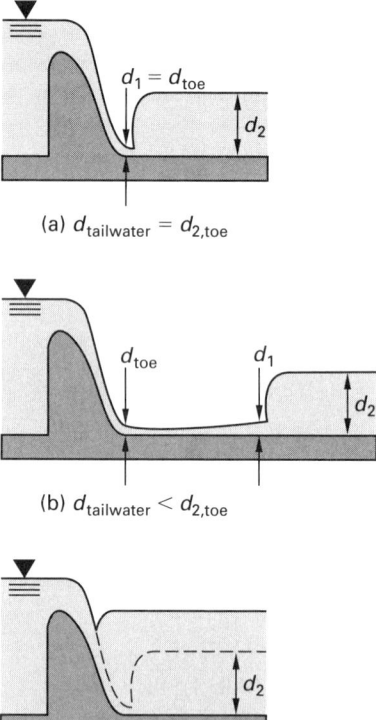

(a) $d_{\text{tailwater}} = d_{2,\text{toe}}$

(b) $d_{\text{tailwater}} < d_{2,\text{toe}}$

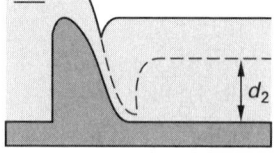

(c) $d_{\text{tailwater}} > d_{2,\text{toe}}$

Example 19.11

A hydraulic jump is produced at a point in a 10 ft (3 m) wide channel where the depth is 1 ft (0.3 m). The flow rate is 200 ft³/sec (5.7 m³/s). (a) What is the depth after the jump? (b) What is the total power dissipated?

SI Solution

(a) From Eq. 19.69,

$$v_1 = \frac{Q}{A} = \frac{5.7 \ \frac{m^3}{s}}{(3 \ m)(0.3 \ m)}$$
$$= 6.33 \ m/s$$

From Eq. 19.92,

$$d_2 = -\tfrac{1}{2}d_1 + \sqrt{\frac{2v_1^2 d_1}{g} + \frac{d_1^2}{4}}$$
$$= -\left(\tfrac{1}{2}\right)(0.3 \ m)$$
$$+ \sqrt{\frac{(2)\left(6.33 \ \frac{m}{s}\right)^2 (0.3 \ m)}{9.81 \ \frac{m}{s^2}} + \frac{(0.3 \ m)^2}{4}}$$
$$= 1.42 \ m$$

(b) The mass flow rate is

$$\dot{m} = \left(5.7 \ \frac{m^3}{s}\right)\left(1000 \ \frac{kg}{m^3}\right)$$
$$= 5700 \ kg/s$$

The velocity after the jump is

$$v_2 = \frac{Q}{A_2} = \frac{5.7 \ \frac{m^3}{s}}{(3 \ m)(1.42 \ m)}$$
$$= 1.33 \ m/s$$

From Eq. 19.95, the change in specific energy is

$$\Delta E = \left(d_1 + \frac{v_1^2}{2g}\right) - \left(d_2 + \frac{v_2^2}{2g}\right)$$
$$= \left(0.3 \ m + \frac{\left(6.33 \ \frac{m}{s}\right)^2}{(2)\left(9.81 \ \frac{m}{s^2}\right)}\right)$$
$$- \left(1.423 \ m + \frac{\left(1.33 \ \frac{m}{s}\right)^2}{(2)\left(9.81 \ \frac{m}{s^2}\right)}\right)$$
$$= 0.83 \ m$$

The total power dissipated is

$$P = \dot{m}g\Delta E$$
$$= \left(5700 \ \frac{kg}{s}\right)\left(9.81 \ \frac{m}{s^2}\right)\left(\frac{0.83 \ m}{1000 \ \frac{W}{kW}}\right)$$
$$= 46.4 \ kW$$

Customary U.S. Solution

(a) From Eq. 19.69,

$$v_1 = \frac{Q}{A} = \frac{200 \ \frac{ft^3}{sec}}{(10 \ ft)(1 \ ft)}$$
$$= 20 \ ft/sec$$

From Eq. 19.92,

$$d_2 = -\tfrac{1}{2}d_1 + \sqrt{\frac{2v_1^2 d_1}{g} + \frac{d_1^2}{4}}$$
$$= -\left(\tfrac{1}{2}\right)(1 \ ft) + \sqrt{\frac{(2)\left(20 \ \frac{ft}{sec}\right)^2 (1 \ ft)}{32.2 \ \frac{ft}{sec^2}} + \frac{(1 \ ft)^2}{4}}$$
$$= 4.51 \ ft$$

(b) The mass flow rate is

$$\dot{m} = \left(200 \ \frac{ft^3}{sec}\right)\left(62.4 \ \frac{lbm}{ft^3}\right)$$
$$= 12{,}480 \ lbm/sec$$

The velocity after the jump is

$$v_2 = \frac{Q}{A_2} = \frac{200 \ \frac{ft^3}{sec}}{(10 \ ft)(4.51 \ ft)}$$
$$= 4.43 \ ft/sec$$

From Eq. 19.95, the change in specific energy is

$$\Delta E = \left(d_1 + \frac{v_1^2}{2g}\right) - \left(d_2 + \frac{v_2^2}{2g}\right)$$
$$= \left(1 \ ft + \frac{\left(20 \ \frac{ft}{sec}\right)^2}{(2)\left(32.2 \ \frac{ft}{sec^2}\right)}\right)$$
$$- \left(4.51 \ ft + \frac{\left(4.43 \ \frac{ft}{sec}\right)^2}{(2)\left(32.2 \ \frac{ft}{sec^2}\right)}\right)$$
$$= 2.4 \ ft$$

The total power dissipated is

$$P = \frac{\dot{m}g\Delta E}{g_c}$$
$$= \frac{\left(12{,}480 \ \frac{lbm}{sec}\right)\left(32.2 \ \frac{ft}{sec^2}\right)(2.4 \ ft)}{\left(32.2 \ \frac{lbm\text{-}ft}{lbf\text{-}sec^2}\right)\left(550 \ \frac{ft\text{-}lbf}{hp\text{-}sec}\right)}$$
$$= 54.5 \ hp$$

Flow of Fluids

34. LENGTH OF HYDRAULIC JUMP

For practical stilling basin design, it is helpful to have an estimate of the length of the hydraulic jump. Lengths of hydraulic jumps are difficult to measure because of the difficulty in defining the endpoints of the jumps. However, the length of the jump, L, varies within the limits of $5 < L/d_2 < 6.5$, in which d_2 is the conjugate depth after the jump. Where greater accuracy is warranted, Table 19.6 can be used. This table correlates the length of the jump to the upstream Froude number.

Table 19.6 Approximate Lengths of Hydraulic Jumps

Fr_1	L/d_2
3	5.25
4	5.8
5	6.0
6	6.1
7	6.15
8	6.15

35. HYDRAULIC DROP

A *hydraulic drop* is the reverse of a hydraulic jump. If water is introduced at low (subcritical) velocity to a section of fast-moving (supercritical) flow, the velocity will be increased rapidly over a short length of channel. The abrupt drop in the water surface is known as a hydraulic drop. The decrease in depth is always from above the critical depth to below the critical depth.

Water flowing over a spillway and down a long, steep chute typically experiences a hydraulic drop, with critical depth occurring just before the brink. This is illustrated in Fig. 19.17 and Fig. 19.22.

The depths on either side of the hydraulic drop are the *conjugate depths*, which are determined from Eq. 19.91 and Eq. 19.92. The equations for calculating specific energy and power changes are the same for hydraulic jumps and drops.

36. ERODIBLE CHANNELS

Given an appropriate value of the Manning coefficient, the analysis of channels constructed of erodible materials is similar to that for concrete or pipe channels.

However, for design problems, maximum velocities and permissible side slopes must also be considered. The present state of knowledge is not sufficiently sophisticated to allow for precise designs. The usual uniform flow equations are insufficient because the stability of erodible channels is dependent on the properties of the channel material rather than on the hydraulics of flow. Two methods of design exist: (a) the tractive force method and (b) the simpler maximum permissible velocity method. Maximum velocities that should be used with erodible channels are given in Table 19.7.

Table 19.7 Suggested Maximum Velocities

soil type or lining (earth; no vegetation)	clear water	water carrying fine silts	water carrying sand and gravel
fine sand (noncolloidal)	1.5	2.5	1.5
sandy loam (noncolloidal)	1.7	2.5	2.0
silt loam (noncolloidal)	2.0	3.0	2.0
ordinary firm loam	2.5	3.5	2.2
volcanic ash	2.5	3.5	2.0
fine gravel	2.5	5.0	3.7
stiff clay (very colloidal)	3.7	5.0	3.0
graded, loam to cobbles (noncolloidal)	3.7	5.0	5.0
graded, silt to cobbles (colloidal)	4.0	5.5	5.0
alluvial silts (noncolloidal)	2.0	3.5	2.0
alluvial silts (colloidal)	3.7	5.0	3.0
coarse gravel (noncolloidal)	4.0	6.0	6.5
cobbles and shingles	5.0	5.5	6.5
shales and hard pans	6.0	6.0	5.0

(Multiply ft/sec by 0.3048 to obtain m/s.)

Source: Special Committee on Irrigation Research, ASCE, 1926.

The sides of the channel should not have a slope exceeding the natural angle of repose for the material used. Although there are other factors that determine the maximum permissible side slope, Table 19.8 lists some guidelines.

Table 19.8 Recommended Side Slopes

type of channel	side slope (horizontal:vertical)
firm rock	vertical to $\frac{1}{4}$:1
concrete-lined stiff clay	$\frac{1}{2}$:1
fissured rock	$\frac{1}{2}$:1
firm earth with stone lining	1:1
firm earth, large channels	1:1
firm earth, small channels	$1\frac{1}{2}$:1
loose, sandy earth	2:1
sandy, porous loam	3:1

37. CULVERTS[16]

A *culvert* is a pipe that carries water under or through some feature (usually a road or highway) that would otherwise block the flow of water. For example, highways are often built at right angles to ravines draining hillsides and other watersheds. Culverts under the

---

[16]The methods of culvert flow analysis in this chapter are based on Bodhaine, G.L., 1968, *Measurement of Peak Discharge by Indirect Methods*, U.S. Geological Survey, *Techniques of Water Resources Investigations*, book 3, chapter A3.

highway keep the construction fill from blocking the natural runoff.

Culverts are classified according to which of their ends controls the discharge capacity: inlet control or outlet control. If water can flow through and out of the culvert faster than it can enter, the culvert is under *inlet control*. If water can flow into the culvert faster than it can flow through and out, the culvert is under *outlet control*. Culverts under inlet control will always flow partially full. Culverts under outlet control can flow either partially full or full.

The culvert length is one of the most important factors in determining whether the culvert flows full. A culvert may be known as "hydraulically long" if it runs full and "hydraulically short" if it does not.[17]

All culvert design theory is closely dependent on energy conservation. However, due to the numerous variables involved, no single formula or procedure can be used to design a culvert. Culvert design is often an empirical, trial-and-error process. Figure 19.24 illustrates some of the important variables that affect culvert performance.

A culvert can operate with its entrance partially or totally submerged. Similarly, the exit can be partially or totally submerged, or it can have free outfall. The upstream head, h, is the water surface level above the lowest part of the culvert barrel, known as the *invert*.[18]

In Fig. 19.24, the three lowermost surface level profiles are of the type that would be produced with inlet control. Such a situation can occur if the culvert is short and the slope is steep. Flow at the entrance is critical as the water falls over the brink. Since critical flow occurs, the flow is choked and the inlet controls the flow rate. Downstream variations cannot be transmitted past the critical section.

Figure 19.24 *Flow Profiles in Culvert Design*

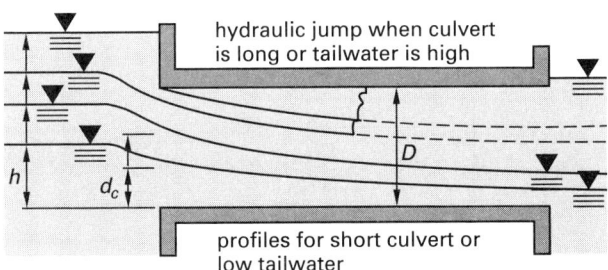

If the tailwater covers the culvert exit completely (i.e., a submerged exit), the culvert will be full at that point, even though the inlet control forces the culvert to be only partially full at the inlet. The transition from partially full to totally full occurs in a hydraulic jump, the location of which depends on the flow resistance and

water levels. If the flow resistance is very high, or if the headwater and tailwater levels are high enough, the jump will occur close to or at the entrance.

If the flow in a culvert is full for its entire length, then the flow is under outlet control. The discharge will be a function of the differences in tailwater and headwater levels, as well as the flow resistance along the barrel length.

38. DETERMINING TYPE OF CULVERT FLOW

For convenience, culvert flow is classified into six different types on the basis of the type of control, the steepness of the barrel, the relative tailwater and headwater heights, and in some cases, the relationship between critical depth and culvert size. These parameters are quantified through the use of the ratios in Table 19.9.[19] The six types are illustrated in Fig. 19.25. Identification of the type of flow beyond the guidelines in Table 19.9 requires a trial-and-error procedure.

In the following cases, several variables appear repeatedly. C_d is the discharge coefficient, a function of the barrel inlet geometry. Orifice data can be used to approximate the discharge coefficient when specific information is unavailable. v_1 is the average velocity of the water approaching the culvert entrance and is often insignificant. The velocity-head coefficient, α, also called the *Coriolis coefficient*, accounts for a nonuniform distribution of velocities over the channel section. However, it represents only a second-order correction and is normally neglected (i.e., assumed equal to 1.0). d_c is the critical depth, which may not correspond to the actual depth of flow. (It must be calculated from the flow conditions.) h_f is the friction loss in the identified section. For culverts flowing full, the friction loss can be found in the usual manner developed for pipe flow: from the Darcy formula and the Moody friction factor chart. For partial flow, the Manning equation and its variations (e.g., Eq. 19.30) can also be used. The Manning equation is particularly useful since it eliminates the need for trial-and-error solutions. The friction head loss between sections 2 and 3, for example, can be calculated from Eq. 19.98.

$$h_{f,2\text{-}3} = \frac{LQ^2}{K_2 K_3} \qquad 19.98$$

$$K = \left(\frac{1}{n}\right) R^{2/3} A \qquad [\text{SI}] \quad 19.99(a)$$

$$K = \left(\frac{1.49}{n}\right) R^{2/3} A \qquad [\text{U.S.}] \quad 19.99(b)$$

---

[17]Proper design of culvert entrances can reduce the importance of length on culvert filling.

[18]The highest part of the culvert barrel is known as the *soffit* or *crown*.

[19]The six cases presented here do not exhaust the various possibilities for entrance and exit control. Culvert design is complicated by this multiplicity of possible flows. Since only the easiest problems can be immediately categorized as one of the six cases, each situation needs to be carefully evaluated.

Figure 19.25 Culvert Flow Classifications

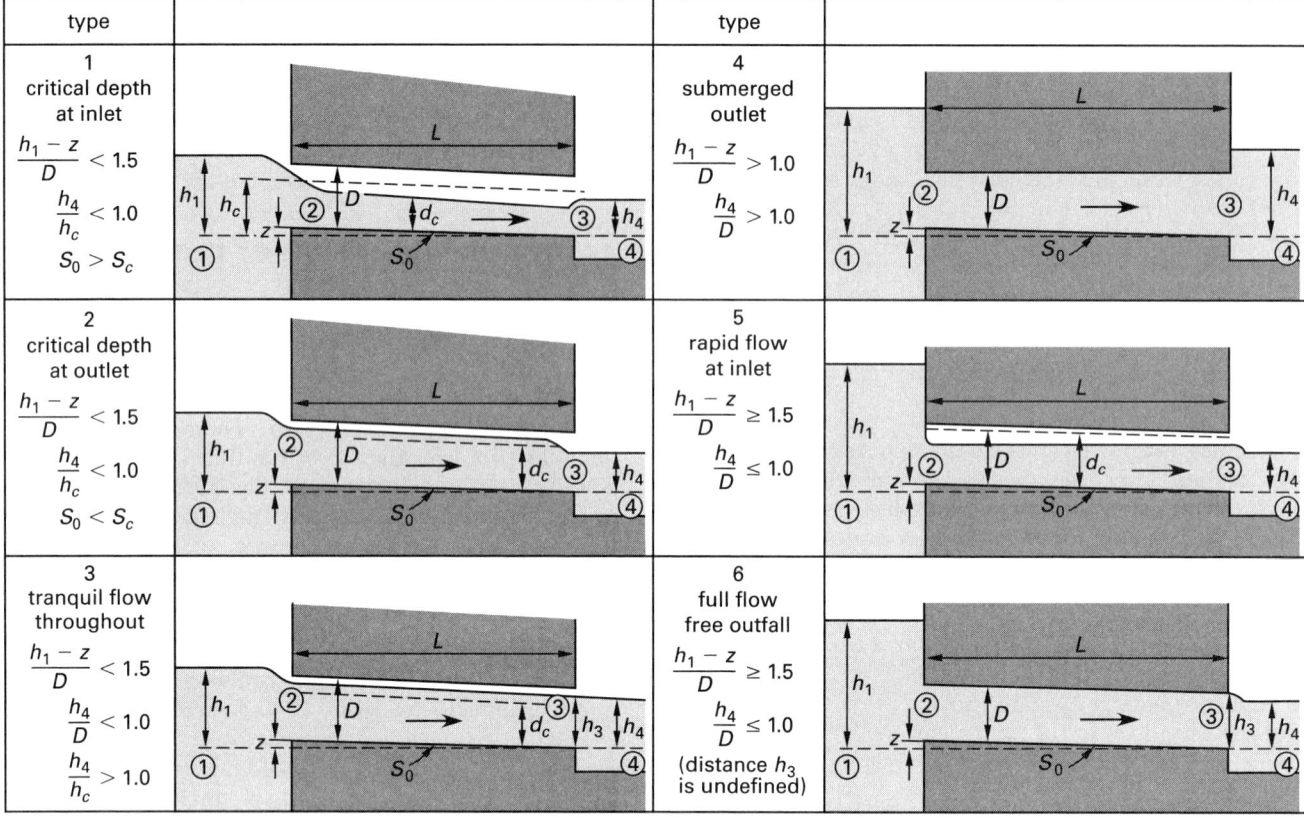

Table 19.9 Culvert Flow Classification Parameters

flow type	$\dfrac{h_1 - z}{D}$	$\dfrac{h_4}{h_c}$	$\dfrac{h_4}{D}$	culvert slope	barrel flow	location of control	kind of control
1	< 1.5	< 1.0	≤ 1.0	steep	partial	inlet	critical depth
2	< 1.5	< 1.0	≤ 1.0	mild	partial	outlet	critical depth
3	< 1.5	> 1.0	≤ 1.0	mild	partial	outlet	backwater
4	> 1.0		≥ 1.0	any	full	outlet	backwater
5	≥ 1.5		≤ 1.0	any	partial	inlet	entrance geometry
6	≥ 1.5		≤ 1.0	any	full	outlet	entrance and barrel geometry

The total hydraulic head available, H, (equal to $h_1 - h_4$ in Fig. 19.25) is divided between the velocity head in the culvert, the entrance loss from Table 19.10 (if considered), and the friction.

$$H = \frac{v^2}{2g} + k_e\left(\frac{v^2}{2g}\right) + \frac{v^2 n^2 L}{R^{4/3}} \quad \text{[SI]} \quad 19.100(a)$$

$$H = \frac{v^2}{2g} + k_e\left(\frac{v^2}{2g}\right) + \frac{v^2 n^2 L}{2.21 R^{4/3}} \quad \text{[U.S.]} \quad 19.100(b)$$

Equation 19.100 can be solved directly for the velocity. Equation 19.101 is valid for culverts of any shape.

$$v = \sqrt{\frac{H}{\frac{1+k_e}{2g} + \frac{n^2 L}{R^{4/3}}}} \quad \text{[SI]} \quad 19.101(a)$$

$$v = \sqrt{\frac{H}{\frac{1+k_e}{2g} + \frac{n^2 L}{2.21 R^{4/3}}}} \quad \text{[U.S.]} \quad 19.101(b)$$

Table 19.10 Minor Entrance Loss Coefficients

k_e	condition of entrance
0.08	smooth, tapered
0.10	flush concrete groove
0.10	flush concrete bell
0.15	projecting concrete groove
0.15	projecting concrete bell
0.50	flush, square-edged
0.90	projecting, square-edged

A. Type-1 Flow

Water passes through the critical depth near the culvert entrance, and the culvert flows partially full. The slope of the culvert barrel is greater than the critical slope, and the tailwater elevation is less than the elevation of the water surface at the control section.

The discharge is

$$Q = C_d A_c \sqrt{2g\left(h_1 - z + \frac{\alpha \mathrm{v}_1^2}{2g} - d_c - h_{f,1\text{-}2}\right)} \qquad 19.102$$

The area, A, used in the discharge equation is not the culvert area since the culvert does not flow full. A_c is the area in flow at the critical section.

B. Type-2 Flow

As in type-1 flow, flow passes through the critical depth at the culvert outlet, and the barrel flows partially full. The slope of the culvert is less than critical, and the tailwater elevation does not exceed the elevation of the water surface at the control section.

$$Q = C_d A_c \sqrt{2g\left(h_1 + \frac{\alpha \mathrm{v}_1^2}{2g} - d_c - h_{f,1\text{-}2} - h_{f,2\text{-}3}\right)}$$
$$19.103$$

The area, A, used in the discharge equation is not the culvert area since the culvert does not flow full. A_c is the area in flow at the critical section.

C. Type-3 Flow

When backwater is the controlling factor in culvert flow, the critical depth cannot occur. The upstream water-surface elevation for a given discharge is a function of the height of the tailwater. For type-3 flow, flow is subcritical for the entire length of the culvert, with the flow being partial. The outlet is not submerged, but the tailwater elevation does exceed the elevation of critical depth at the terminal section.

$$Q = C_d A_3 \sqrt{2g\left(h_1 + \frac{\alpha \mathrm{v}_1^2}{2g} - h_3 - h_{f,1\text{-}2} - h_{f,2\text{-}3}\right)}$$
$$19.104$$

The area, A, used in the discharge equation is not the culvert area since the culvert does not flow full. A_3 is the area in flow at numbered section 3 (i.e., the exit).

D. Type-4 Flow

As in type-3 flow, the backwater elevation is the controlling factor in this case. Critical depth cannot occur, and the upstream water surface elevation for a given discharge is a function of the tailwater elevation.

Discharge is independent of barrel slope. The culvert is submerged at both the headwater and the tailwater. No differentiation between low head and high head is made for this case. If the velocity head at section 1 (the entrance), the entrance friction loss, and the exit friction loss are neglected, the discharge can be calculated. A_o is the culvert area.

$$Q = C_d A_o \sqrt{2g\left(\frac{h_1 - h_4}{1 + \dfrac{29 C_d^2 n^2 L}{R^{4/3}}}\right)} \qquad 19.105$$

The complicated term in the denominator corrects for friction. For rough estimates and for culverts less than 50 ft (15 m) long, the friction loss can be ignored.

$$Q = C_d A_o \sqrt{2g(h_1 - h_4)} \qquad 19.106$$

E. Type-5 Flow

Partially full flow under a high head is classified as type-5 flow. The flow pattern is similar to the flow downstream from a sluice gate, with rapid flow near the entrance. Usually, type-5 flow requires a relatively square entrance that causes contraction of the flow area to less than the culvert area. In addition, the barrel length, roughness, and bed slope must be sufficient to keep the velocity high throughout the culvert.

It is difficult to distinguish in advance between type-5 and type-6 flow. Within a range of the important parameters, either flow can occur.[20] A_o is the culvert area.

$$Q = C_d A_o \sqrt{2g(h_1 - z)} \qquad 19.107$$

F. Type-6 Flow

Type-6 flow, like type-5 flow, is considered a high-head flow. The culvert is full under pressure with free outfall. The discharge is

$$Q = C_d A_o \sqrt{2g(h_1 - h_3 - h_{f,2\text{-}3})} \qquad 19.108$$

Equation 19.108 is inconvenient because h_3 (the true piezometric head at the outfall) is difficult to evaluate without special graphical aids. The actual hydraulic head driving the culvert flow is a function of the Froude number. For conservative first approximations, h_3 can be taken as the barrel diameter. This will give the minimum hydraulic head. In reality, h_3 varies from somewhat less than half the barrel diameter to the full diameter.

---

[20]If the water surface ever touches the top of the culvert, the passage of air in the culvert will be prevented and the culvert will flow full everywhere. This is type-6 flow.

If h_3 is taken as the barrel diameter, the total hydraulic head ($H = h_1 - h_3$) will be split between the velocity head and friction. In that case, Eq. 19.101 can be used to calculate the velocity. The discharge is easily calculated from Eq. 19.109.[21] A_o is the culvert area.

$$Q = A_o v \qquad\qquad 19.109$$

Example 19.12

Size a square culvert with an entrance fluid level 5 ft above the barrel top and a free exit to operate with the following characteristics.

$$\text{slope} = 0.01$$

$$\text{length} = 250 \text{ ft}$$

$$\text{capacity} = 45 \text{ ft}^3/\text{sec}$$

$$n = 0.013$$

Solution

Since the h_1 dimension is measured from the culvert invert, it is difficult to classify the type of flow at this point. However, either type 5 or type 6 is likely since the head is high.

step 1: Assume a trial culvert size. Select a square opening with 1 ft sides.

step 2: Calculate the flow assuming case 5 (entrance control). The entrance will act like an orifice.

$$A_o = (1 \text{ ft})(1 \text{ ft}) = 1 \text{ ft}^2$$

$$\begin{aligned} H &= h_1 - z \\ &= \big(5 \text{ ft} + 1 \text{ ft} + (0.01)(250 \text{ ft})\big) \\ &\quad - (0.01)(250 \text{ ft}) \\ &= 6 \text{ ft} \end{aligned}$$

C_d is approximately 0.62 for square-edged openings with separation from the wall. From Eq. 19.107,

$$\begin{aligned} Q &= C_d A_o \sqrt{2g(h_1 - z)} \\ &= (0.62)(1 \text{ ft}^2)\sqrt{(2)\left(32.2 \ \frac{\text{ft}}{\text{sec}^2}\right)(6 \text{ ft})} \\ &= 12.2 \text{ ft}^3/\text{sec} \end{aligned}$$

Since this size has insufficient capacity, try a larger culvert. Choose a square opening with 2 ft sides.

$$H = h_1 - z = 5 \text{ ft} + 2 \text{ ft} = 7 \text{ ft}$$

$$\begin{aligned} Q &= C_d A_o \sqrt{2g(h_1 - z)} \\ &= (0.62)(4 \text{ ft}^2)\sqrt{(2)\left(32.2 \ \frac{\text{ft}}{\text{sec}^2}\right)(7 \text{ ft})} \\ &= 52.7 \text{ ft}^3/\text{sec} \end{aligned}$$

step 3: Begin checking the entrance control assumption by calculating the maximum hydraulic radius. The upper surface of the culvert is not wetted because the flow is entrance controlled. The hydraulic radius is maximum at the entrance.

$$\begin{aligned} R &= \frac{A_o}{P} \\ &= \frac{4 \text{ ft}^2}{2 \text{ ft} + 2 \text{ ft} + 2 \text{ ft}} \\ &= 0.667 \text{ ft} \end{aligned}$$

step 4: Calculate the velocity using the Manning equation for open channel flow. Since the hydraulic radius is maximum, the velocity will also be maximum.

$$\begin{aligned} v &= \left(\frac{1.49}{n}\right) R^{2/3} \sqrt{S} \\ &= \left(\frac{1.49}{0.013}\right)(0.667 \text{ ft})^{2/3}\sqrt{0.01} \\ &= 8.75 \text{ ft/sec} \end{aligned}$$

step 5: Calculate the normal depth, d_n.

$$\begin{aligned} d_n &= \frac{Q}{vw} = \frac{45 \ \dfrac{\text{ft}^3}{\text{sec}}}{\left(8.75 \ \dfrac{\text{ft}}{\text{sec}}\right)(2 \text{ ft})} \\ &= 2.57 \text{ ft} \end{aligned}$$

Since the normal depth is greater than the culvert size, the culvert will flow full under pressure. (It was not necessary to calculate the critical depth since the flow is implicitly subcritical.) The entrance control assumption was, therefore, not valid for this size culvert.[22] At this point, two things can be done: A larger culvert can be chosen if entrance control is desired, or the solution can continue by

---

[21]Equation 19.109 does not include the discharge coefficient. Velocity, v, when calculated from Eq. 19.101, is implicitly the velocity in the barrel.

[22]If the normal depth had been less than the barrel diameter, it would still be necessary to determine the critical depth of flow. If the normal depth was less than the critical depth, the entrance control assumption would have been valid.

checking to see if the culvert has the required capacity as a pressure conduit.

step 6: Check the capacity as a pressure conduit. H is the total available head.

$$H = h_1 - h_3$$
$$= \left(5 \text{ ft} + 2 \text{ ft} + (0.01)(250 \text{ ft})\right) - 2 \text{ ft}$$
$$= 7.5 \text{ ft}$$

step 7: Since the pipe is flowing full, the hydraulic radius is

$$R = \frac{A}{P} = \frac{4 \text{ ft}^2}{8 \text{ ft}} = 0.5 \text{ ft}$$

step 8: Equation 19.101 can be used to calculate the flow velocity. Since the culvert has a square-edged entrance, a loss coefficient of $k_e = 0.5$ is used. However, this does not greatly affect the velocity.

$$v = \sqrt{\frac{H}{\frac{1+k_e}{2g} + \frac{n^2 L}{2.21 R^{4/3}}}}$$
$$= \sqrt{\frac{7.5 \text{ ft}}{\frac{1+0.5}{(2)\left(32.2 \frac{\text{ft}}{\text{sec}}\right)} + \frac{(0.013)^2(250 \text{ ft})}{(2.21)(0.5 \text{ ft})^{4/3}}}}$$
$$= 10.24 \text{ ft/sec}$$

step 9: Check the capacity.

$$Q = vA_o = \left(10.24 \frac{\text{ft}}{\text{sec}}\right)(4 \text{ ft}^2)$$
$$= 40.96 \text{ ft}^3/\text{sec}$$

The culvert size is not acceptable since its discharge under the maximum head does not have a capacity of 45 ft³/sec.

step 10: Repeat from step 2, trying a larger-size culvert. With a 2.5 ft side, the following values are obtained.

$$A_o = (2.5 \text{ ft})(2.5 \text{ ft}) = 6.25 \text{ ft}^2$$
$$H = 5 \text{ ft} + 2.5 \text{ ft} = 7.5 \text{ ft}$$

$$Q = (0.62)(6.25 \text{ ft}^2)$$
$$\times \sqrt{(2)\left(32.2 \frac{\text{ft}}{\text{sec}^2}\right)(7.5 \text{ ft})}$$
$$= 85.2 \text{ ft}^3/\text{sec}$$
$$R = \frac{6.25 \text{ ft}^2}{7.5 \text{ ft}} = 0.833 \text{ ft}$$
$$v = \left(\frac{1.49}{0.013}\right)(0.833 \text{ ft})^{2/3}\sqrt{0.01}$$
$$= 10.15 \text{ ft/sec}$$
$$d_n = \frac{Q}{vw} = \frac{45 \frac{\text{ft}^3}{\text{sec}}}{\left(10.15 \frac{\text{ft}}{\text{sec}}\right)(2.5 \text{ ft})}$$
$$= 1.77 \text{ ft}$$

step 11: Calculate the critical depth. For rectangular channels, Eq. 19.75 can be used.

$$d_c = \left(\frac{Q^2}{gw^2}\right)^{1/3}$$
$$= \left(\frac{\left(45 \frac{\text{ft}^3}{\text{sec}}\right)^2}{\left(32.2 \frac{\text{ft}}{\text{sec}^2}\right)(2.5 \text{ ft})^2}\right)^{1/3}$$
$$= 2.16 \text{ ft}$$

Since the normal depth is less than the critical depth, the flow is supercritical. The entrance control assumption was correct for the culvert. The culvert has sufficient capacity to carry 45 ft³/sec.

39. CULVERT DESIGN

Designing a culvert is somewhat easier than culvert analysis because of common restrictions placed on designers and the flexibility to change almost everything else. For example, culverts may be required to (a) never be more than 50% full (deep), (b) always be under inlet control, or (c) always operate with some minimum head (above the centerline or crown). In the absence of any specific guidelines, a culvert may be designed using the following procedure.

step 1: Determine the required flow rate.

step 2: Determine all water surface elevations, lengths, and other geometric characteristics.

step 3: Determine the material to be used for the culvert and its roughness.

step 4: Assume type 1 flow (inlet control).

step 5: Select a trial diameter.

step 6: Assume a reasonable slope.

step 7: Position the culvert entrance such that the ratio of headwater depth (inlet to water surface) to culvert diameter is 1:2 to 1:2.5.

step 8: Calculate the flow. Repeat step 5 through step 7 until the capacity is adequate.

step 9: Determine the location of the outlet. Check for outlet control. If the culvert is outlet controlled, repeat step 5 through step 7 using a different flow model.

step 10: Calculate the discharge velocity. Specify rip-rap, concrete, or other protection to prevent erosion at the outlet.

20 Meteorology, Climatology, and Hydrology

Nomenclature

a	storm constant	–	–
A	area	ft^2	m^2
A_d	drainage area	ac	km^2
b	storm constant	min	min
c	storm constant	–	–
C	constant	–	–
C	rational runoff coefficient	–	–
CN	curve number	–	–
d	distance between stations	mi	km
E	evaporation	in/day	cm/d
F	storm constant	–	–
F	frequency of occurrence	1/yr	1/yr
F	infiltration	in	cm
H	elevation difference	ft	m
I	rainfall intensity	in/hr	cm/h
I_a	initial abstraction	in	cm
Imp	imperviousness	%	%
K	coefficient	–	–
K, K'	storm constant	in-min/hr	cm·min/h
K_p	pan coefficient	–	–
L	length	ft	m
M	order number	–	–
n	Manning roughness coefficient	–	–
n	number of years	yr	yr
n_y	number of years of streamflow data	yr	yr
N	normal annual precipitation	in	cm
P	precipitation	in	cm
q	runoff	ft^3/mi^2-in	m^3/km^2·cm
Q	flow rate	ft^3/sec	m^3/s
S	storage capacity	in	cm
$S_{decimal}$	slope	ft/ft	m/m
$S_{percent}$	slope	%	%
t	time	min	min
t_c	time to concentration	min	min
t_p	time from start of storm to peak runoff	hr	h
t_R	rainstorm duration	hr	h
v	flow velocity	ft/sec	m/s
V	volume	ft^3	m^3
W	width of unit hydrograph	min	min

Symbols

ϕ	watershed conveyance factor	–	–

Subscripts

ave	average
b	base
c	concentration
d	drainage
n	period n
o	overland
p	peak, pond, or pan
R	rain (storm) or reservoir
t	time
u	unit
x	missing station

1. HYDROLOGIC CYCLE

The *hydrologic cycle* is the full "life cycle" of water. The cycle begins with *precipitation*, which encompasses all of the hydrometeoric forms, including rain, snow, sleet, and hail from a storm. Precipitation can (a) fall on vegetation and structures and evaporate back into the atmosphere, (b) be absorbed into the ground and either make its way to the water table or be absorbed by plants after which it evapotranspires back into the atmosphere, or (c) travel as surface water to a depression, watershed, or creek from which it either evaporates back into the atmosphere, infiltrates into the ground water system, or flows off in streams and rivers to an ocean or lakes. The cycle is completed when lake and ocean water evaporates into the atmosphere.

The *water balance equation* (*water budget equation*) is the application of conservation to the hydrologic cycle.

total precipitation = net change in surface water removed

+ net change in ground water removed

+ evapotranspiration

+ interception evaporization

+ net increase in surface water storage

+ net increase in ground water storage

$$P = Q + E + \Delta S \qquad 20.1$$

The total amount of water that is intercepted (and subsequently evaporates) and absorbed into ground water before runoff begins is known as the *initial abstraction*. Even after runoff begins, the soil continues to absorb some infiltrated water. Initial abstraction and infiltration do not contribute to surface runoff. Equation 20.1 can be restated as Eq. 20.2.

total precipitation = initial abstraction + infiltration

+ surface runoff   20.2

2. STORM CHARACTERISTICS

Storm rainfall characteristics include the duration, total volume, intensity, and areal distribution of a storm. Storms are also characterized by their recurrence intervals. (See Sec. 20.6.)

The duration of storms is measured in hours and days. The volume of rainfall is simply the total quantity of precipitation dropping on the watershed. Average rainfall intensity is the volume divided by the duration of the storm. Average rainfall can be considered to be generated by an equivalent theoretical storm that drops the same volume of water uniformly and constantly over the entire watershed area.

A *storm hyetograph* is the instantaneous rainfall intensity measured as a function of time, as shown in Fig. 20.1(a). Hyetographs are usually bar graphs showing constant rainfall intensities over short periods of time. Hyetograph data can be reformulated as a *cumulative rainfall curve*,

Figure 20.1 *Storm Hyetograph and Cumulative Rainfall Curves*

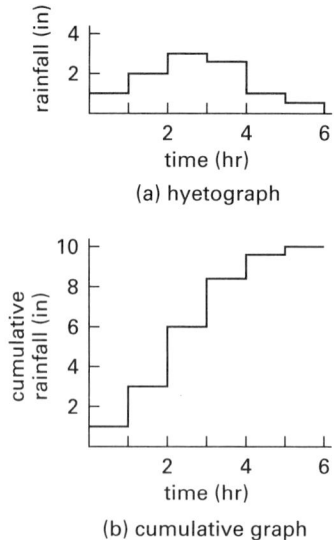

(a) hyetograph

(b) cumulative graph

as shown in Fig. 20.1(b). Cumulative rainfall curves are also known as *rainfall mass curves*.

3. PRECIPITATION DATA

Precipitation data on rainfall can be collected in a number of ways, but use of an open precipitation rain gauge is quite common. This type of gauge measures only the volume of rain collected between readings, usually 24 hr.

The *average precipitation* over a specific area can be found from station data in several ways.

Method 1: If the stations are uniformly distributed over a flat site, their precipitations can be averaged. This also requires that the individual precipitation records not vary too much from the mean.

Method 2: The *Thiessen method* calculates the average weighting station measurements by the area of the assumed watershed for each station. These assumed watershed areas are found by drawing dotted lines between all stations and bisecting these dotted lines with solid lines (which are extended outward until they connect with other solid lines). The solid lines will form a polygon whose area is the assumed watershed area.

Method 3: The *isohyetal method* requires plotting lines of constant precipitation (*isohyets*) and weighting the isohyet values by the areas enclosed by the lines. This method is the most accurate of the three methods, as long as there are enough observations to permit drawing of isohyets. Station data are used to draw isohyets, but they are not used in the calculation of average rainfall.

4. ESTIMATING UNKNOWN PRECIPITATION

If a precipitation measurement at a location is unknown, it may still be possible to estimate the value by one of the following procedures.

Method 1: Choose three stations close to and evenly spaced around the location with missing data. If the normal annual precipitations at the three sites do not vary more than 10% from the missing station's normal annual precipitation, the rainfall can be estimated as the arithmetic mean of the three neighboring stations' precipitations for the period in question.

Method 2: If the precipitation difference between locations is more than 10%, the *normal-ratio method* can be used. In Eq. 20.3, P_x is the precipitation at the missing station; N_x is the long-term normal precipitation at the missing station; P_A, P_B, and P_C are the precipitations at known stations; and N_A, N_B, and N_C are the long-term normal precipitations at the known stations.

$$P_x = \frac{1}{3}\left(\left(\frac{N_x}{N_A}\right)P_A + \left(\frac{N_x}{N_B}\right)P_B + \left(\frac{N_x}{N_C}\right)P_C\right) \qquad 20.3$$

Method 3: Use data from stations in the four nearest quadrants (north, south, east, and west of the unknown station) and weight the data with the inverse squares of the distance between the stations. In Eq. 20.4, P_x, P_A, P_B, P_C, and P_D are defined as in Method 2, and d_{A-x}, d_{B-x}, d_{C-x}, and d_{D-x} are the distances between stations A and x, B and x, and so on, respectively.

$$P_x = \frac{\dfrac{P_A}{d_{A-x}^2} + \dfrac{P_B}{d_{B-x}^2} + \dfrac{P_C}{d_{C-x}^2} + \dfrac{P_D}{d_{D-x}^2}}{\dfrac{1}{d_{A-x}^2} + \dfrac{1}{d_{B-x}^2} + \dfrac{1}{d_{C-x}^2} + \dfrac{1}{d_{D-x}^2}} \qquad 20.4$$

5. TIME OF CONCENTRATION

Time of concentration, t_c, is defined as the time of travel from the hydraulically most remote (timewise) point in the watershed to the watershed outlet or other design point. For points (e.g., manholes) along storm drains being fed from a watershed, time of concentration is taken as the largest combination of overland flow time (sheet flow), swale or ditch flow (shallow concentrated flow), and storm drain, culvert, or channel time. It is unusual for time of concentration to be less than 0.1 hr (6 min) when using the Natural Resources Conservation Service (NRCS, previously known as the Soil Conservation Service (SCS)) method, or less than 10 min when using the rational method.

$$t_c = t_{\text{sheet}} + t_{\text{shallow}} + t_{\text{channel}} \qquad 20.5$$

The NRCS specifies using the *Manning kinematic equation* (Overton and Meadows 1976 formulation) for calculating *sheet flow* travel time over distances less than 300 ft (100 m). In Eq. 20.6, n is the Manning roughness coefficient for sheet flow, as given in Table 20.1. P_2 is the 2 yr, 24 hr rainfall in inches, and S is the slope of the hydraulic grade line in ft/ft.

$$t_{\text{sheet flow}} = \frac{0.007(nL_o)^{0.8}}{\sqrt{P_2}\, S_{\text{decimal}}^{0.4}} \qquad 20.6$$

Table 20.1 Manning Roughness Coefficient for Sheet Flow

surface	n
smooth surfaces (concrete, asphalt, gravel, or bare soil)	0.011
fallow (no residue cover)	0.05
cultivated soils	
residue cover $\leq 20\%$	0.06
residue cover $> 20\%$	0.17
grasses	
short prairie grass	0.15
dense grass[a]	0.24
Bermuda grass	0.41
range, natural	0.13
woods[b]	
light underbrush	0.40
dense underbrush	0.80

[a]This includes species such as weeping lovegrass, bluegrass, buffalo grass, blue grama grass, and native grass mixtures.
[b]When selecting a value of n, consider the cover to a height of about 0.1 ft (3 cm). This is the only part of the plant that will obstruct sheet flow.

Reprinted from *Urban Hydrology for Small Watersheds*, Technical Release TR-55, United States Department of Agriculture, Natural Resources Conservation Service, Table 3-1, after Engman (1986).

After about 300 ft (100 m), the flow usually becomes a shallow concentrated flow (swale, ditch flow). Travel time is calculated as L/v. Velocity can be found from the Manning equation if the flow geometry is well defined, but must be determined from other correlations, such as those specified by the NRCS in Eq. 20.7 and Eq. 20.8.

$$v_{\text{shallow,ft/sec}} = 16.1345\sqrt{S_{\text{decimal}}} \quad [\text{unpaved}] \qquad 20.7$$

$$v_{\text{shallow,ft/sec}} = 20.3282\sqrt{S_{\text{decimal}}} \quad [\text{paved}] \qquad 20.8$$

Storm drain (channel) time is found by dividing the storm drain length by the actual or an assumed channel velocity. Storm drain velocity is found from either the Manning or the Hazen-Williams equation. Since size and velocity are related, an iterative trial-and-error solution is generally required.[1]

There are a variety of methods available for estimating time of concentration. Early methods include Kirpich (1940), California Culverts Practice (1942), Hathaway (1945), and Izzard (1946). More recent methods include those from the Federal Aviation Administration, or FAA (1970), the kinematic wave formulas of Morgali (1965) and Aron (1973), the NRCS lag equation (1975), and NRCS average velocity charts (1975). Estimates of time of concentration from these methods can vary by as much as 100%. These differences carry over into estimates of peak flow, hence the need to carefully determine the validity of any method used.

---

[1]If the pipe or channel size is known, the velocity can be found from $Q = Av$. If the pipe size is not known, the area will have to be estimated. In that case, one might as well estimate velocity instead. 5 ft/sec (1.5 m/s) is a reasonable flow velocity for open channel flow. The minimum velocity for a *self-cleansing pipe* is 2 ft/sec (0.6 m/s).

Flow of Fluids

The distance L_o in the various equations that follow is the longest distance to the collection point, as shown in Fig. 20.2.

Figure 20.2 *Overland Flow Distances*

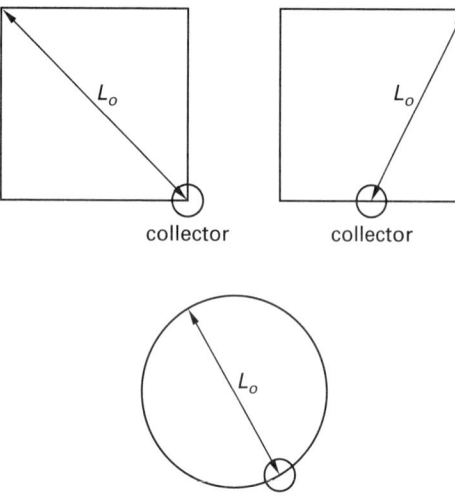

For irregularly shaped drainage areas, it may be necessary to evaluate several alternative overland flow distances. For example, Fig. 20.3 shows a drainage area with a long tongue. Although the tongue area contributes little to the drainage area, it does lengthen the overland flow time. Depending on the intensity-duration-frequency curve, the longer overland flow time (resulting in a lower rainfall intensity) may offset the increase in area due to the tongue. Therefore, two runoffs need to be compared, one ignoring and the other including the tongue.

Figure 20.3 *Irregular Drainage Area*

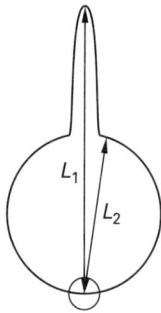

The *FAA formula*, Eq. 20.9, was developed from airfield drainage data collected by the Army Corps of Engineers. However, it has been widely used for urbanized areas. C is the rational method runoff coefficient. The slope S, in Eq. 20.9 is in percent. L_o is in ft.

$$t_{c,\min} = \frac{1.8(1.1 - C)\sqrt{L_{o,\text{ft}}}}{S_{\text{percent}}^{1/3}} \qquad 20.9$$

The *kinematic wave formula* is particularly accurate for uniform planar homogenous areas (e.g., paved areas such as parking lots and streets). It requires iteration, since both intensity and time of concentration are unknown. In Eq. 20.10, L_o is in ft, n is the *Manning overland roughness coefficient (retardance roughness coefficient)*, I is the intensity in in/hr, and S_{decimal} is the slope in ft/ft. Recommended values of n for this application are different than for open channel flow and are: smooth impervious surfaces, 0.011–0.014; smooth bare-packed soil, free of stones, 0.05; poor grass, moderately bare surface, 0.10; pasture or average grass cover, 0.20; and dense grass or forest, 0.40. Equation 20.10 is solved iteratively since the intensity depends on the time to concentration.

$$t_{c,\min} = \frac{0.94 L_{o,\text{ft}}^{0.6} n^{0.6}}{I_{\text{in/hr}}^{0.4} S_{\text{decimal,ft/ft}}^{0.3}} \qquad 20.10$$

The NRCS *lag equation* was developed from observations of agricultural watersheds where overland flow paths are poorly defined and channel flow is absent. However, it has been adapted to small urban watersheds under 2000 ac. The equation performs reasonably well for areas that are completely paved, as well. Correction factors are used to account for channel improvement and impervious areas. L_o is in ft, CN is the NRCS runoff curve number, S_{in} is the potential maximum retention in the watershed after runoff begins in inches, and S_{percent} is the average slope in percent. The factor 1.67 converts the watershed lag time to the time of concentration. Since the formula overestimates time for mixed areas, different adjustment factors have been proposed. The NRCS lag equation performs poorly when channel flow is a significant part of the time of concentration.

$$t_{c,\min} = 1.67 t_{\text{watershed lag time,min}}$$

$$= \frac{\left(60 \, \dfrac{\min}{\text{hr}}\right) L_{o,\text{ft}}^{0.8} (S_{\text{in}} + 1)^{0.7}}{1900 \sqrt{S_{\text{percent}}}}$$

$$= \frac{\left(60 \, \dfrac{\min}{\text{hr}}\right) L_{o,\text{ft}}^{0.8} \left(\dfrac{1000}{\text{CN}} - 9\right)^{0.7}}{1900 \sqrt{S_{\text{percent}}}} \qquad 20.11$$

If the velocity of runoff water is known, the time of concentration can be easily determined. The NRCS has published charts of average velocity as functions of watercourse slope and surface cover. (See Fig. 20.4.) The charts are best suited for flow paths of at least several hundred feet (at least 70 m). The time of concentration is easily determined from these charts as

$$t_c = \frac{\sum L_{o,\text{ft}}}{v \left(60 \, \dfrac{\sec}{\min}\right)} \qquad 20.12$$

6. RAINFALL INTENSITY

Effective design of a surface feature depends on its geographical location and required degree of protection.

Figure 20.4 *NRCS Average Velocity Chart for Overland Flow Travel Time*

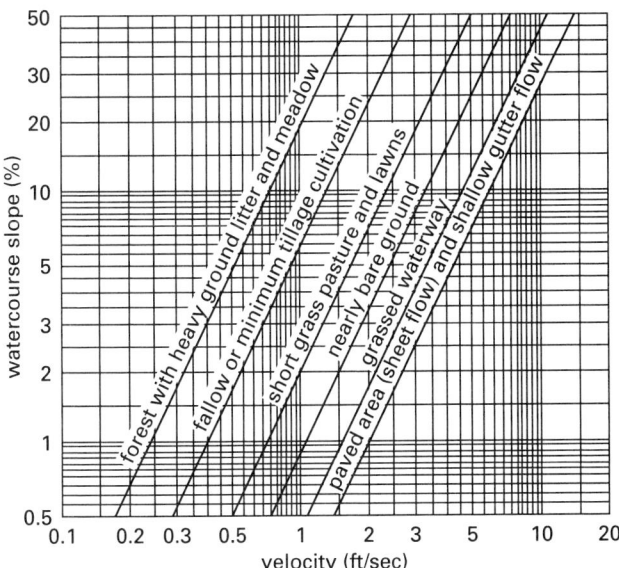

Reprinted from NRCS TR-55-1975. TR-55-1986 contains a similar graph for shallow-concentrated flow over paved and unpaved surfaces, but it does not contain this particular graph.

Once the location of the feature is known, the *design storm* (or *design flood*) must be determined based on some probability of recurrence.

Rainfall intensity is the amount of precipitation per hour. The instantaneous intensity changes throughout the storm. However, it may be averaged over short time intervals or over the entire storm duration. Average intensity will be low for most storms, but it can be high for some. These high-intensity storms can be expected infrequently, say, every 20, 50, or 100 years. The average number of years between storms of a given intensity is known as the *frequency of occurrence* (*recurrence interval*, *return interval*, or *storm frequency*).

In general, the design storm may be specified by its recurrence interval (e.g., "100-year storm"), its annual probability of occurrence (e.g., "1% storm"), or a nickname (e.g., "century storm"). A 1% storm is a storm that would be exceeded in severity only once every hundred years on the average.

The average intensity of a storm over a time period t can be calculated from Eq. 20.13 (and similar correlations). (When using the rational method described in Sec. 20.15, t is the time of concentration.) In the United States, it is understood that the units of intensity calculated using Eq. 20.13 will be in in/hr.

$$I = \frac{K' F^a}{(t + b)^c} \qquad 20.13$$

K', F, a, b, and c are constants that depend on the conditions, recurrence interval, and location of a storm. For many reasons, these constants may be unavailable.

The *Steel formula* is a simplification of Eq. 20.13. Steel formula rainfall regions are shown in Fig. 20.5.

$$I = \frac{K}{t_c + b} \qquad 20.14$$

Figure 20.5 *Steel Formula Rainfall Regions*

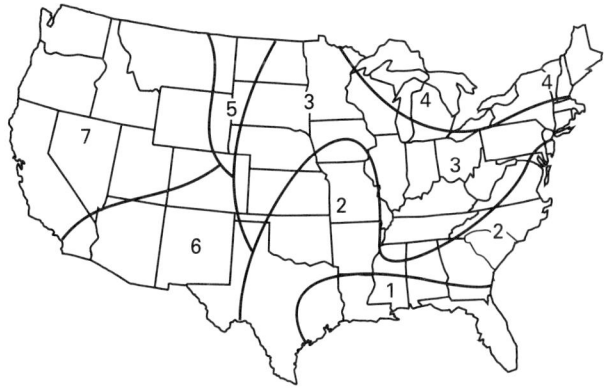

Values of the constants K and b in Eq. 20.14 are not difficult to obtain once the intensity-duration-frequency curve is established. Although a logarithmic transformation could be used to convert the data to straight-line form, an easier method exists. This method starts by taking the reciprocal of Eq. 20.14 and converting the equation to a straight line.

$$\begin{aligned} \frac{1}{I} &= \frac{t_c + b}{K} \\ &= \frac{t_c}{K} + \frac{b}{K} \\ &= C_1 t_c + C_2 \qquad 20.15 \end{aligned}$$

Once C_1 and C_2 have been found, K and b can be calculated.

$$K = \frac{1}{C_1} \qquad 20.16$$

$$b = \frac{C_2}{C_1} \qquad 20.17$$

Published values of K and b can be obtained from compilations, but these values are suitable only for very rough estimates. Table 20.2 is typical of some of this general data.

The total rainfall can be calculated from the average intensity and duration.

$$P = It \qquad 20.18$$

Rainfall data can be compiled into *intensity-duration-frequency curves* (*IDF curves*) similar to those shown in Fig. 20.6.

Table 20.2 *Steel Formula Coefficients (for intensities of in/hr)*

frequency in years	coefficients	region 1	2	3	4	5	6	7
2	K	206	140	106	70	70	68	32
	b	30	21	17	13	16	14	11
4	K	247	190	131	97	81	75	48
	b	29	25	19	16	13	12	12
10	K	300	230	170	111	111	122	60
	b	36	29	23	16	17	23	13
25	K	327	260	230	170	130	155	67
	b	33	32	30	27	17	26	10
50	K	315	350	250	187	187	160	65
	b	28	38	27	24	25	21	8
100	K	367	375	290	220	240	210	77
	b	33	36	31	28	29	26	10

(Multiply in/hr by 2.54 to obtain cm/h.)

Figure 20.6 *Typical Intensity-Duration-Frequency Curves*

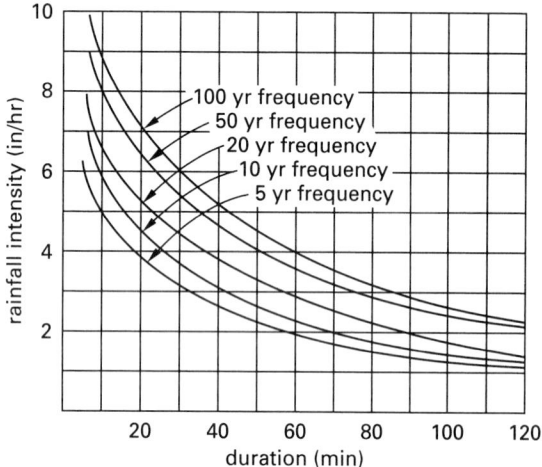

Example 20.1

A storm has an intensity given by

$$I = \frac{100}{t_c + 10}$$

15 min are required for runoff from the farthest corner of a 5 ac watershed to reach a discharge culvert. What is the design intensity?

Solution

The intensity is

$$I = \frac{100}{t_c + 10} = \frac{100 \frac{\text{in-min}}{\text{hr}}}{15 \text{ min} + 10 \text{ min}}$$
$$= 4 \text{ in/hr}$$

7. FLOODS

A *flood* occurs when more water arrives than can be drained away. When a watercourse (i.e., a creek or river) is too small to contain the flow, the water overflows the banks.

The flooding may be categorized as nuisance, damaging, or devastating. *Nuisance floods* result in inconveniences such as wet feet, tire spray, and soggy lawns. *Damaging floods* soak flooring, carpeting, and first-floor furniture. *Devastating floods* wash buildings, vehicles, and livestock downstream.

Although rain causes flooding, large storms do not always cause floods. The size of a flood depends not only on the amount of rainfall, but also on the conditions within the watershed before and during the storm. Runoff will occur only when the rain falls on a very wet watershed that is unable to absorb additional water, or when a very large amount of rain falls on a dry watershed faster than it can be absorbed.

Specific terms are sometimes used to designate the degree of protection required. For example, the *probable maximum flood* (PMF) is a hypothetical flood that can be expected to occur as a result of the most severe combination of critical meteorologic and hydrologic conditions possible within a region.

Designing for the *probable maximum precipitation* (PMP) or probable maximum flood is very conservative and usually uneconomical since the recurrence interval for these events exceeds 100 years and may even approach 1000 years. Designing for 100 year floods and floods with even lower recurrence intervals is more common. (100 year floods are not necessarily caused by 100 year storms.)

The *design flood* or *design basis flood* (DBF) depends on the site. It is the flood that is adopted as the basis for design of a particular project. The DBF is usually determined from economic considerations, or it is specified as part of the contract document.

The *standard flood* or *standard project flood* (SPF) is a flood that can be selected from the most severe combinations of meteorological and hydrological conditions reasonably characteristic of the region, excluding extremely rare combinations of events. SPF volumes are commonly 40–60% of the PMF volumes.

The probability that a flooding event in any given year will equal a design basis flood with a *recurrence interval frequency* (*return interval*) of F is

$$p\{F \text{ event in one year}\} = \frac{1}{F} \qquad 20.19$$

The probability of an F event occurring in n years is

$$p\{F \text{ event in } n \text{ years}\} = 1 - \left(1 - \frac{1}{F}\right)^n \qquad 20.20$$

Planning for a 1% flood has proven to be a good compromise between not doing enough and spending too much. Although the 1% flood is a common choice for the design basis flood, shorter recurrence intervals are often used, particularly in low-value areas such as cropland. For example, a 5 year value can be used in residential areas, a 10 year value in business sections, and a 15 year value for high-value districts where flooding will result in more extensive damage. The ultimate choice of recurrence interval, however, must be made on the basis of economic considerations and trade-offs.

Example 20.2

A wastewater treatment plant has been designed to be in use for 40 years. What is the probability that a 1% flood will occur within the useful lifetime of the plant?

Solution

Use Eq. 20.20.

$$p\{F \text{ event in } n \text{ years}\} = 1 - \left(1 - \frac{1}{F}\right)^n$$

$$p\{100 \text{ yr flood in 40 years}\} = 1 - \left(1 - \frac{1}{100}\right)^{40}$$

$$= 0.33 \quad (33\%)$$

8. TOTAL SURFACE RUNOFF FROM STREAM HYDROGRAPH

After a rain, runoff and groundwater increases stream flow. A plot of the stream discharge versus time is known as a *hydrograph*. Hydrograph periods may be very short (e.g., hours) or very long (e.g., days, weeks, or months). A typical hydrograph is shown in Fig. 20.7. The *time base* is the length of time that the stream flow exceeds the original *base flow*. The flow rate increases on the *rising limb* (*concentration curve*) and decreases on the *falling limb* (*recession curve*).

Figure 20.7 Stream Hydrograph

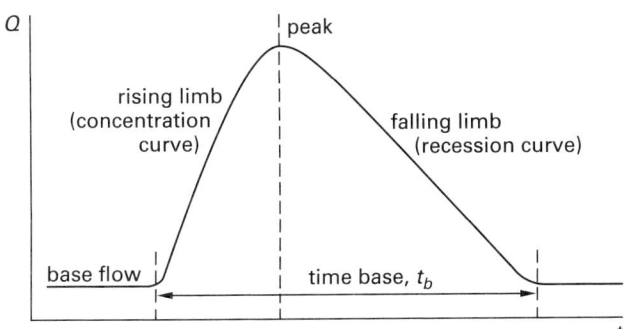

9. HYDROGRAPH SEPARATION

The stream discharge consists of both surface runoff and subsurface groundwater flows. A procedure known as *hydrograph separation* or *hydrograph analysis* is used to separate runoff (*surface flow*, *net flow*, or *overland flow*) and groundwater (*subsurface flow*, *base flow*).[2]

There are several methods of separating groundwater from runoff. Most of the methods are somewhat arbitrary. Three methods that are easily carried out manually are presented here.

Method 1: In the *straight-line method*, a horizontal line is drawn from the start of the rising limb to the falling limb. All of the flow under the horizontal line is considered base flow. This assumption is not theoretically accurate, but the error can be small. This method is illustrated in Fig. 20.8.

Figure 20.8 Straight-Line Method

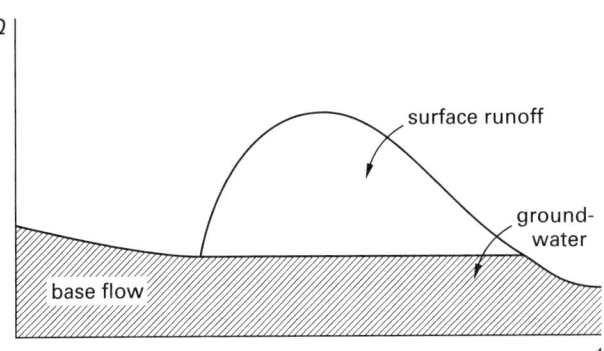

Method 2: In the *fixed-base method*, shown in Fig. 20.9, the base flow existing before the storm is projected graphically down to a point directly under the peak of the hydrograph. Then, a straight line is used to connect the projection to the falling limb. The duration of the recession limb is determined by inspection, or it can be calculated from correlations with the drainage area.

Figure 20.9 Fixed-Base Method

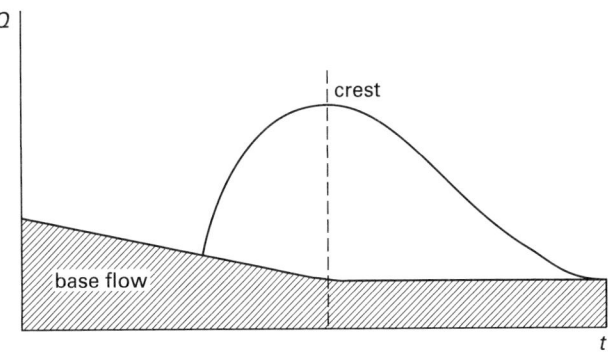

---

[2]The total rain dropped by a storm is the *gross rain*. The rain that actually appears as immediate runoff can be called *surface runoff, overland flow, surface flow,* and *net rain*. The water that is absorbed by the soil and that does not contribute to the surface runoff can be called *base flow, groundwater, infiltration,* and *dry weather flow.*

Method 3: The *variable-slope method*, as shown in Fig. 20.10, recognizes that the shape of the base flow curve before the storm will probably match the shape of the base flow curve after the storm. The groundwater curve after the storm is projected back under the hydrograph to a point under the inflection point of the falling limb. The separation line under the rising limb is drawn arbitrarily.

Figure 20.10 *Variable-Slope Method*

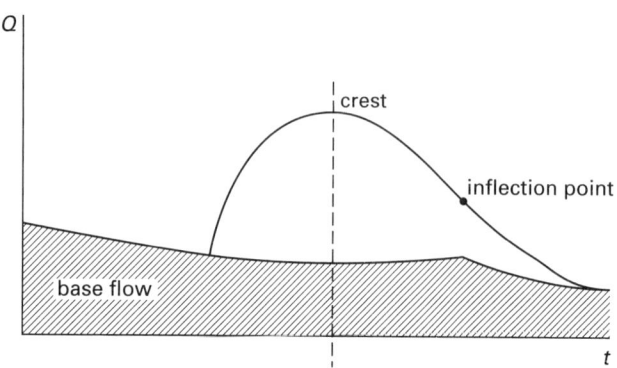

Once the base flow is separated out, the hydrograph of surface runoff will have the approximate appearance of Fig. 20.11.

Figure 20.11 *Overland Flow Hydrograph*

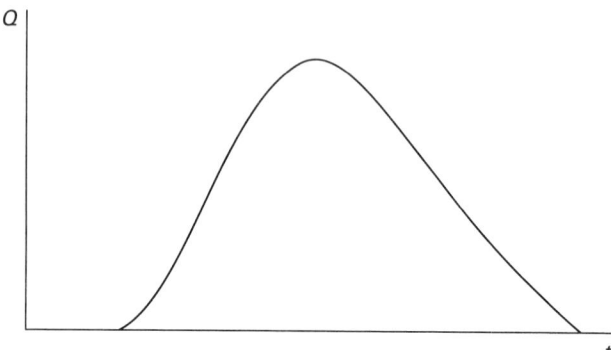

10. UNIT HYDROGRAPH

Once the overland flow hydrograph for a watershed has been developed, the total runoff (i.e., "excess rainfall") volume, V, from the storm can be found as the area under the curve. Although this can be found by integration, planimetry, or computer methods, it is often sufficiently accurate to approximate the hydrograph with a histogram and to sum the areas of the rectangles. (See Fig. 20.12.)

Since the area of the watershed is known, the average depth, P_{ave}, of the excess precipitation can be calculated. (Consistent units must be used in Eq. 20.21.)

$$V = A_d P_{ave,excess} \qquad 20.21$$

Figure 20.12 *Hydrograph Histogram*

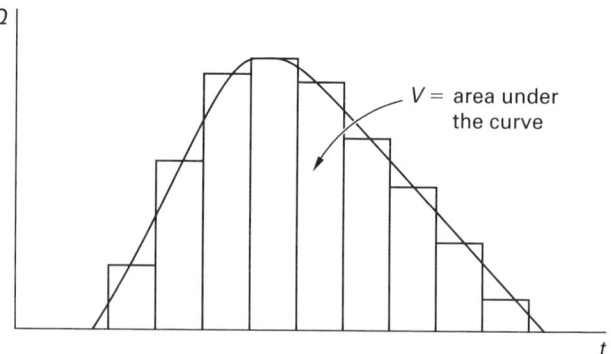

A *unit hydrograph* (UH) is developed by dividing every point on the overland flow hydrograph by the average excess precipitation, $P_{ave,excess}$. This is a hydrograph of a storm dropping 1 in (1 cm) of excess precipitation (runoff) evenly on the entire watershed. Units of the unit hydrograph are in/in (cm/cm). Figure 20.13 shows how a unit hydrograph compares to its surface runoff hydrograph.

Figure 20.13 *Unit Hydrograph*

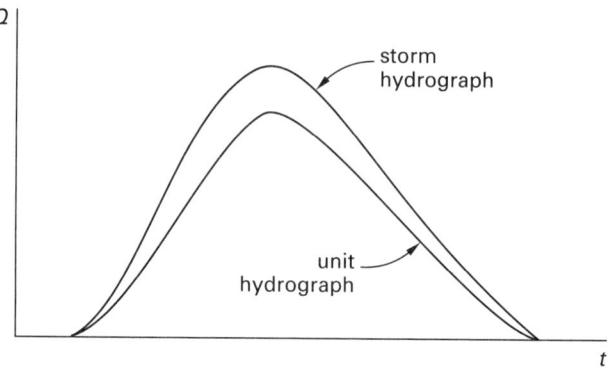

Once a unit hydrograph has been developed from historical data of a particular storm volume, it can be used for other storm volumes. Such application is based on several assumptions: (a) All storms in the watershed have the same duration, (b) the time base is constant for all storms, (c) the shape of the rainfall curve is the same for all storms, and (d) only the total amount of rainfall varies from storm to storm.

The hydrograph of a storm producing more or less than 1 in (1 cm) of rain is found by multiplying all ordinates on the unit hydrograph by the total precipitation of the storm.

A unit hydrograph can be used to predict the runoff for storms that have durations somewhat different than the storms used to develop the unit hydrograph. Generally, storm durations differing by up to ±25% are considered to be equivalent.

Example 20.3

After a 2 hr storm, a station downstream from a 45 mi^2 (115 km^2) drainage watershed records a peak discharge of 9400 ft^3/sec (250 m^3/s) and a total runoff of 3300 ac-ft(4×10^6 m^3). (a) What is the unit hydrograph peak discharge? (b) What would be the peak runoff and design flood volume if a 2 hr storm dropped 2.5 in (6 cm) net precipitation?

SI Solution

(a) Use Eq. 20.21 to find the average precipitation for the drainage watershed.

$$P_{ave} = \frac{V}{A_d}$$

$$= \frac{4 \times 10^6 \text{ m}^3}{(115 \text{ km}^2)\left(1000 \frac{\text{m}}{\text{km}}\right)^2}$$

$$= 0.0348 \text{ m} \quad (3.5 \text{ cm})$$

The ordinates for the unit hydrograph are found by dividing every point on the 3.5 cm hydrograph by 3.5 cm. Therefore, the peak discharge for the unit hydrograph is

$$Q_{p,unit} = \frac{250 \frac{\text{m}^3}{\text{s}}}{3.5 \text{ cm}}$$

$$= 71.4 \text{ m}^3/\text{s·cm}$$

(b) The hydrograph for a storm that is producing more than 1 cm of rain is found by multiplying the ordinates of the unit hydrograph by the total precipitation. For a 6 cm storm, the peak discharge is

$$Q_p = \left(71.4 \frac{\text{m}^3}{\text{s·cm}}\right)(6 \text{ cm})$$

$$= 428 \text{ m}^3/\text{s}$$

To find the design flood volume, first use Eq. 20.21 to find the unit hydrograph total volume.

$$V = A_d P$$

$$= (115 \text{ km}^2)\left(1000 \frac{\text{m}}{\text{km}}\right)^2\left(\frac{1 \text{ m}}{100 \text{ cm}}\right)\left(1 \frac{\text{cm}}{\text{cm}}\right)$$

$$= 1.15 \times 10^6 \text{ m}^3/\text{cm}$$

The design flood volume for a 6 cm storm is

$$V = (6 \text{ cm})\left(1.15 \times 10^6 \frac{\text{m}^3}{\text{cm}}\right)$$

$$= 6.9 \times 10^6 \text{ m}^3$$

Customary U.S. Solution

(a) Use Eq. 20.21 to find the average precipitation for the drainage watershed.

$$P_{ave} = \frac{V}{A_d}$$

$$= \frac{(3300 \text{ ac-ft})\left(43{,}560 \frac{\text{ft}^2}{\text{ac}}\right)\left(12 \frac{\text{in}}{\text{ft}}\right)}{(45 \text{ mi}^2)\left(5280 \frac{\text{ft}}{\text{mi}}\right)^2}$$

$$= 1.375 \text{ in}$$

The ordinates for the unit hydrograph are found by dividing every point on the 1.375 in hydrograph by 1.375 in. Therefore, the peak discharge for the unit hydrograph is

$$Q_{p,unit} = \frac{9400 \frac{\text{ft}^3}{\text{sec}}}{1.375 \text{ in}}$$

$$= 6836 \text{ ft}^3/\text{sec-in}$$

(b) The hydrograph for a storm that is producing more than 1 in of rain is found by multiplying the ordinates of the unit hydrograph by the total precipitation. For a 2.5 in storm, the peak discharge is

$$Q_p = \left(6836 \frac{\text{ft}^3}{\text{sec-in}}\right)(2.5 \text{ in})$$

$$= 17{,}090 \text{ ft}^3/\text{sec}$$

To find the design flood volume, first use Eq. 20.21 to find the unit hydrograph total volume.

$$V = A_d P$$

$$= (45 \text{ mi}^2)\left(\frac{640 \frac{\text{ac}}{\text{mi}^2}}{12 \frac{\text{in}}{\text{ft}}}\right)\left(1 \frac{\text{in}}{\text{in}}\right)$$

$$= 2400 \text{ ac-ft/in}$$

The design flood volume for a 2.5 in storm is

$$V = (2.5 \text{ in})\left(2400 \frac{\text{ac-ft}}{\text{in}}\right) = 6000 \text{ ac-ft}$$

Example 20.4

A 6 hr storm rains on a 25 mi^2 (65 km^2) drainage watershed. Records from a stream gaging station draining the watershed are shown. (a) Construct the unit hydrograph for the 6 hr storm. (b) Find the runoff rate at $t = 15$ hr from a two-storm system if the first storm drops 2 in (5 cm) starting at $t = 0$ and the second storm drops 5 in (12 cm) starting at $t = 12$ hr.

t (hr)	Q (ft³/sec)	Q (m³/s)
0	0	0
3	400	10
6	1300	35
9	2500	70
12	1700	50
15	1200	35
18	800	20
21	600	15
24	400	10
27	300	10
30	200	5
33	100	3
36	0	0
totals	9500	263

SI Solution

(a) Plot the stream gaging data.

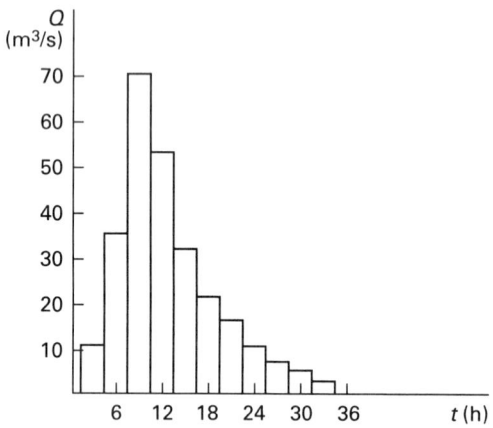

The total runoff is the total area under the curve, found by multiplying the average runoff by the rainfall interval. This is equivalent to summing the areas of each rectangle of the histogram. Each of the histogram bars is 3 h "wide."

$$V = \left(263 \; \frac{m^3}{s}\right)(3 \; h)\left(3600 \; \frac{s}{h}\right)$$
$$= 2.84 \times 10^6 \; m^3$$

The watershed drainage area is

$$A_d = (65 \; km^2)\left(1000 \; \frac{m}{km}\right)^2$$
$$= 65 \times 10^6 \; m^2$$

The average precipitation is calculated from Eq. 20.21.

$$P = \frac{V}{A_d}$$
$$= \frac{(2.84 \times 10^6 \; m^3)\left(100 \; \frac{cm}{m}\right)}{65 \times 10^6 \; m^2}$$
$$= 4.37 \; cm$$

The unit hydrograph has the same shape as the actual hydrograph with all ordinates reduced by a factor of 4.37.

(b) To find the flow at 15 h, add the contributions from each storm. For the 5 cm storm, the contribution is the 15 h runoff multiplied by its scaling factors; for the 12 cm storm, the contribution is the 15 h − 12 h = 3 h runoff multiplied by its scaling factors.

$$Q = \frac{(5 \; cm)\left(35 \; \frac{m^3}{s}\right)}{4.37 \; cm} + \frac{(12 \; cm)\left(10 \; \frac{m^3}{s}\right)}{4.37 \; cm}$$
$$= 67.5 \; m^3/s$$

Customary U.S. Solution

(a) Plot the stream gaging data.

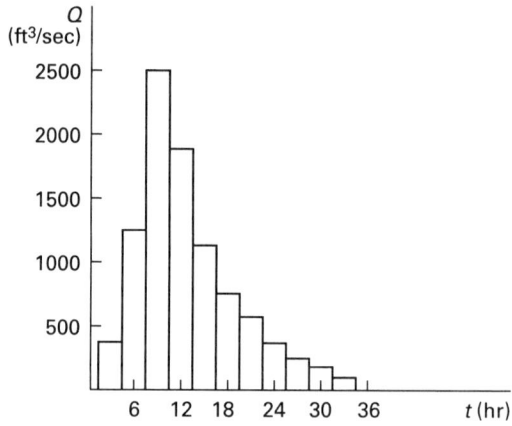

The total runoff is the area under the curve, found by multiplying the average runoff by the rainfall interval. This is equivalent to summing the areas of each rectangle of the histogram. Each of the histogram bars is 3 hr "wide."

$$V = \left(9500 \; \frac{ft^3}{sec}\right)(3 \; hr)\left(3600 \; \frac{sec}{hr}\right)$$
$$= 1.026 \times 10^8 \; ft^3$$

The watershed drainage area is

$$A_d = (25 \; mi^2)\left(5280 \; \frac{ft}{mi}\right)^2$$
$$= 6.97 \times 10^8 \; ft^2$$

The average precipitation is calculated from Eq. 20.21.

$$P = \frac{V}{A_d}$$

$$= \frac{(1.026 \times 10^8 \text{ ft}^3)\left(12 \frac{\text{in}}{\text{ft}}\right)}{6.97 \times 10^8 \text{ ft}^2}$$

$$= 1.766 \text{ in}$$

The unit hydrograph has the same shape as the actual hydrograph with all ordinates reduced by a factor of 1.766.

(b) To find the flow at $t = 15$ hr, add the contributions from each storm. For the 2 in storm, the contribution is the 15 hr runoff multiplied by its scaling factors; for the 5 in storm, the contribution is the 15 hr $-$ 12 hr $=$ 3 hr runoff multiplied by its scaling factors.

$$Q = \frac{(2 \text{ in})\left(1200 \frac{\text{ft}^3}{\text{sec}}\right)}{1.766 \text{ in}} + \frac{(5 \text{ in})\left(400 \frac{\text{ft}^3}{\text{sec}}\right)}{1.766 \text{ in}}$$

$$= 2492 \text{ ft}^3/\text{sec}$$

11. NRCS SYNTHETIC UNIT HYDROGRAPH

If a watershed is ungauged such that no historical records are available to produce a unit hydrograph, the *synthetic hydrograph* can still be reasonably approximated. The process of developing a synthetic hydrograph is known as *hydrograph synthesis.*

Pioneering work was done in 1938 by Snyder, who based his analysis on Appalachian highland watersheds with areas between 10 mi^2 and 10,000 mi^2. Snyder's work has been largely replaced by more sophisticated analyses, including the NRCS methods.

The NRCS developed a synthetic unit hydrograph based on the *curve number*, CN. The method was originally intended for use with rural watersheds up to 2000 ac, but it appears to be applicable for urban conditions up to 4000–5000 ac.

In order to draw the NRCS synthetic unit hydrograph, it is necessary to calculate the time to peak flow, t_p, and the peak discharge, Q_p. Provisions for calculating both of these parameters are included in the method.

$$t_p = 0.5t_R + t_1 \qquad \textit{20.22}$$

t_R in Eq. 20.22 is the storm duration (i.e., of the rainfall). t_1 in Eq. 20.22 is the *lag time* (i.e., the time from the centroid of the rainfall distribution to the peak discharge). Lag time can be determined from correlations with geographical region and drainage area or calculated from Eq. 20.23. Although Eq. 20.23 was developed for natural watersheds, limited studies of urban watersheds indicate that it does not change significantly for urbanized watersheds. S in Eq. 20.23 is the

soil water storage capacity in inches, computed as a function of the curve number. (See Eq. 20.43.)

$$t_{1,\text{hr}} = \frac{L_{o,\text{ft}}^{0.8}(S+1)^{0.7}}{1900\sqrt{S_{\text{percent}}}} \qquad \textit{20.23}$$

The peak runoff is calculated as

$$Q_p = \frac{0.756 A_{d,\text{ac}}}{t_p} \qquad \textit{20.24}$$

$$Q_p = \frac{484 A_{d,\text{mi}^2}}{t_p} \qquad \textit{20.25}$$

Q_p and t_p only contribute one point to the construction of the unit hydrograph. To construct the remainder, Table 20.3 must be used. Using time as the independent variable, selections of time (different from t_p) are arbitrarily made, and the ratio t/t_p is calculated. The curve is then used to obtain the ratio of Q_t/Q_p.

Table 20.3 NRCS Dimensionless Unit Hydrograph and Mass Curve Ratios

time ratios (t/t_p)	discharge ratios (Q/Q_p)	cumulative mass curve fraction
0.0	0.000	0.000
0.1	0.030	0.001
0.2	0.100	0.006
0.3	0.190	0.012
0.4	0.310	0.035
0.5	0.470	0.065
0.6	0.660	0.107
0.7	0.820	0.163
0.8	0.930	0.228
0.9	0.990	0.300
1.0	1.000	0.375
1.1	0.990	0.450
1.2	0.930	0.522
1.3	0.860	0.589
1.4	0.780	0.650
1.5	0.680	0.700
1.6	0.560	0.751
1.7	0.460	0.790
1.8	0.390	0.822
1.9	0.330	0.849
2.0	0.280	0.871
2.2	0.207	0.908
2.4	0.147	0.934
2.6	0.107	0.953
2.8	0.077	0.967
3.0	0.055	0.977
3.2	0.040	0.984
3.4	0.029	0.989
3.6	0.021	0.993
3.8	0.015	0.995
4.0	0.011	0.997
4.5	0.005	0.999
5.0	0.000	1.000

Source: National Engineering Handbook, Part 630, Hydrology, NRCS, 1972.

12. NRCS SYNTHETIC UNIT TRIANGULAR HYDROGRAPH

The NRCS unit triangular hydrograph is shown in Fig. 20.14. It is found from the peak runoff, the time to peak, and the duration of runoff. Peak runoff, Q_p, is found from Eq. 20.24 or Eq. 20.25. Time to peak and duration are correlated with the time of concentration. These are generalizations that apply to specific storm and watershed types.

Figure 20.14 NRCS Synthetic Unit Triangular Hydrograph

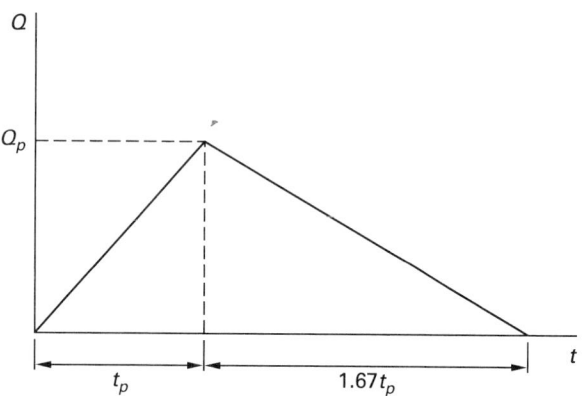

Equation 20.26 can be used to estimate the time to peak.

$$t_p = 0.5t_R + t_1 \qquad \textit{20.26}$$

$$t_1 \approx 0.6t_c \qquad \textit{20.27}$$

Alternatively, the time to peak has been roughly correlated to the time of concentration.

$$t_p = 0.67t_c \qquad \textit{20.28}$$

The total duration of the unit hydrograph is the sum of time to peak and length of recession limb, assumed to be $1.67t_p$.

$$t_b = t_p + t_{\text{recession}} = t_p + 1.67t_p = 2.67t_p \qquad \textit{20.29}$$

13. ESPEY SYNTHETIC UNIT HYDROGRAPH

The *Espey method* calculates the time to peak (t_p, in min), peak discharge (Q_p, in ft^3/sec), total hydrograph base (t_b, in min), and the hydrograph widths at 50% and 75% of the peak discharge rates (W_{50} and W_{75}, in min). These values depend on the watershed area (A, in mi^2), main channel flow path length (L, in ft), slope (S, in ft/ft), roughness, and percent imperviousness (Imp). ϕ is a dimensionless watershed conveyance factor ($0.6 < \phi < 1.3$) that depends on the percent imperviousness and weighted

Figure 20.15 Espey Watershed Conveyance Factor

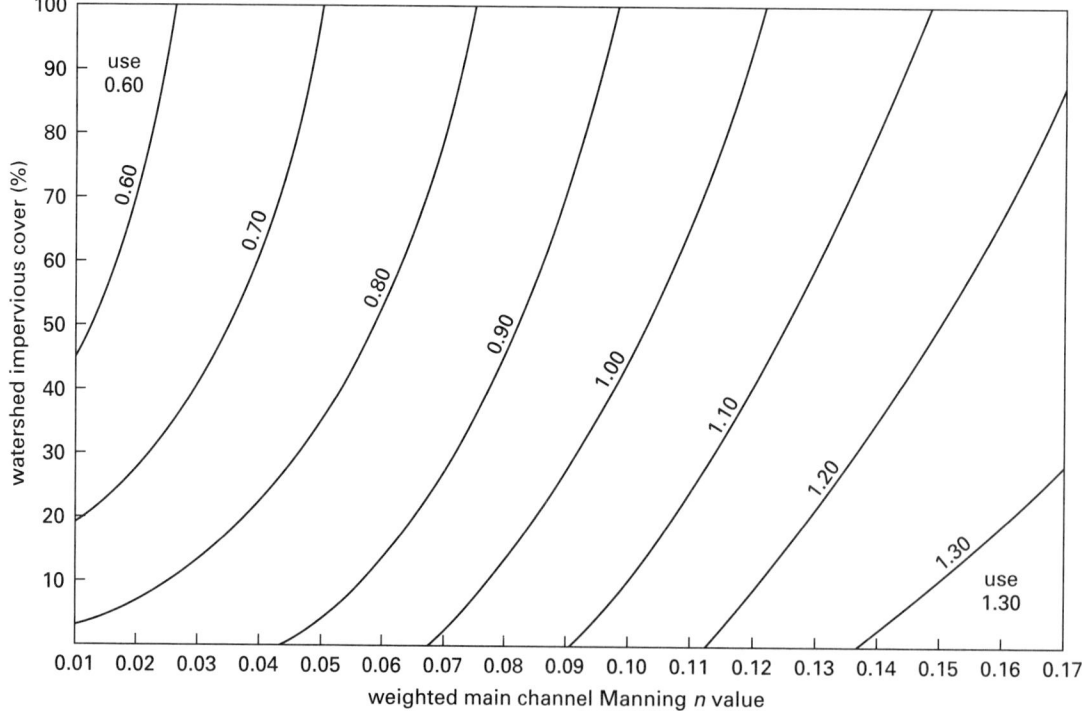

Reprinted from *Recommended Hydrologic Procedures for Computing Urban Runoff from Small Watersheds in Pennsylvania*, Commonwealth of Pennsylvania, Fig. 6-3, 1982, Department of Environmental Resources.

main channel Manning roughness coefficient, n. ϕ is found graphically from Fig. 20.15.

$$t_p = 3.1 L^{0.23} S_{\text{decimal}}^{-0.25} (\text{Imp}_{\text{percent}})^{-0.18} \phi^{1.57} \qquad 20.30$$

$$Q_p = (31.62 \times 10^3) A^{0.96} t_p^{-1.07} \qquad 20.31$$

$$t_b = (125.89 \times 10^3) A Q_p^{-0.95} \qquad 20.32$$

$$W_{50} = (16.22 \times 10^3) A^{0.93} Q_p^{-0.92} \qquad 20.33$$

$$W_{75} = (3.24 \times 10^3) A^{0.79} Q_p^{-0.78} \qquad 20.34$$

To use this method, the geometric slope, S, used in Eq. 20.30 is specifically calculated from Eq. 20.35. H is the difference in elevation of points A and B. Point A is the channel bottom a distance $0.2L$ downstream from the upstream watershed boundary. Point B is the channel bottom at the downstream watershed boundary.

$$S_{\text{decimal}} = \frac{H}{0.8L} \qquad 20.35$$

The unit hydrograph is drawn by manually "fitting" a smooth curve over the seven computed points. The widths of W_{50} and W_{75} are allocated in a 1:2 ratio to the rising and falling hydrograph limbs, respectively. After the curve is drawn, it is adjusted to be a unit hydrograph. The resulting curve is sometimes referred to as an *Espey 10-minute unit hydrograph*. (See Fig. 20.16.)

Figure 20.16 *Espey Synthetic Hydrograph*

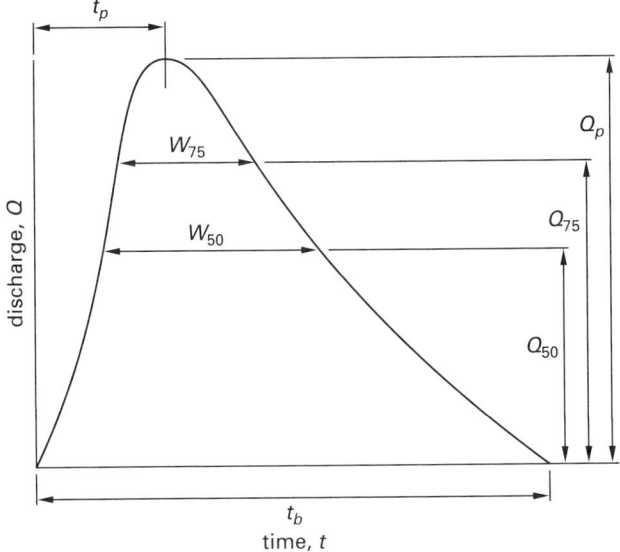

14. HYDROGRAPH SYNTHESIS

If a storm's duration is not the same as, or close to, the hydrograph base length, the unit hydrograph cannot be used to predict runoff. For example, the runoff from a six hour storm cannot be predicted from a unit hydrograph derived from a two hour storm. However, the technique of hydrograph synthesis can be used to construct the hydrograph of the longer storm from the unit hydrograph of a shorter storm.

A. Lagging Storm Method

If a unit hydrograph for a storm of duration t_R is available, the *lagging storm method* can be used to construct the hydrograph of a storm whose duration is a whole multiple of t_R. (See Fig. 20.17.) For example, a six hour storm hydrograph can be constructed from a two hour unit hydrograph.

Let the whole multiple number be n. To construct the longer hydrograph, draw n unit hydrographs, each separated by time t_R. Then add the ordinates to obtain a hydrograph for an nt_R duration storm. Since the total rainfall from this new hydrograph is n inches (having been constructed from n unit hydrographs), the curve will have to be reduced (i.e., divided) by n everywhere to produce a unit hydrograph.

Figure 20.17 *Lagging Storm Method*

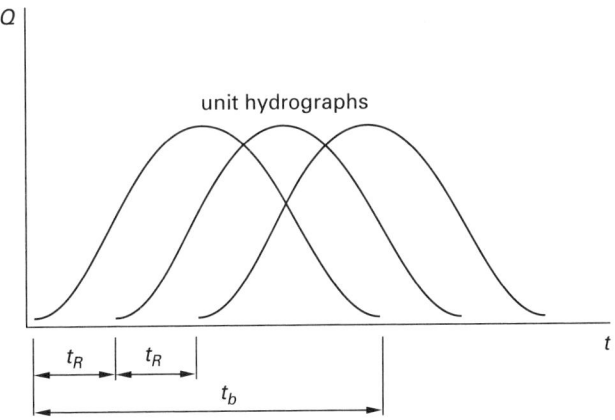

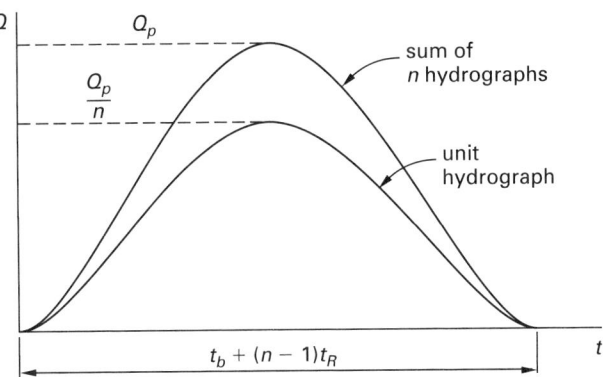

B. S-Curve Method

The *S-curve method* can be used to construct hydrographs from unit hydrographs with longer or shorter durations, even when the storm durations are not multiples. This method begins by adding the ordinates of many unit hydrographs, each lagging the other by time t_R, the duration of the storm that produced the unit hydrograph. After a sufficient number of lagging unit hydrographs have been added together, the accumulation will level off and remain constant. At that point, the lagging can be stopped. The resulting accumulation is known as an S-curve. (See Fig. 20.18.)

Figure 20.18 *Constructing the S-Curve*

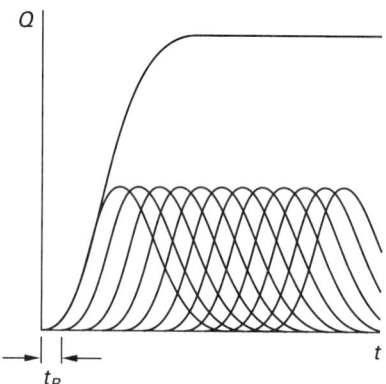

If two S-curves are drawn, one lagging the other by time t'_R, the area between the two curves represents a hydrograph area for a storm of duration t'_R. (See Fig. 20.19.) The differences between the two curves can be plotted and scaled to a unit hydrograph by multiplying by the ratio of t_R/t'_R.

Figure 20.19 *Using S-Curves to Construct a t' Hydrograph*

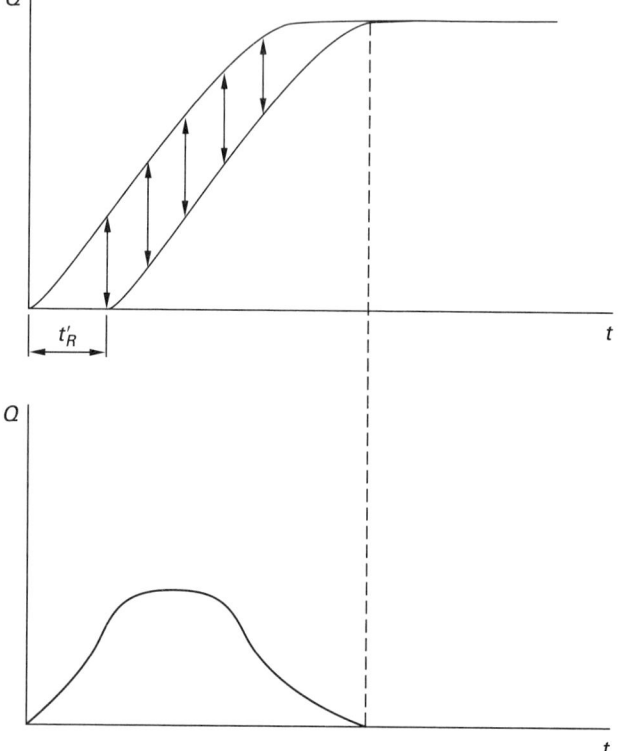

15. PEAK RUNOFF FROM THE RATIONAL METHOD

Although total runoff volume is required for reservoir and dam design, the instantaneous peak runoff is needed to size culverts and storm drains.

The *rational formula* ("method," "equation," etc.), shown in Eq. 20.36, for peak discharge has been in widespread use in the United States since the early 1900s.[3] It is applicable to small areas (i.e., less than several hundred acres or so), but is seldom used for areas greater than 1–2 mi². The intensity used in Eq. 20.36 depends on the time of concentration and the degree of protection desired (i.e., the recurrence interval).[4]

$$Q_p = CIA_d \qquad \textit{20.36}$$

Since A_d is in acres, Q_p is in ac-in/hr. However, Q_p is taken as ft³/sec since the conversion factor between these two units is 1.008.

Typical values of C coefficients are found in App. 20.A. If more than one area contributes to the runoff, the coefficient is weighted by the areas.

Accurate values of the C coefficient depend not only on the surface cover and soil type, but also on the recurrence interval, antecedent moisture content, rainfall intensity, drainage area, slope, and fraction of imperviousness. These factors have been investigated and quantified by Rossmiller (1981), who correlated these effects with the NRCS curve number. The *Schaake, Geyer, and Knapp (1967) equation* developed at Johns Hopkins University was intended for use in urban areas, correlating the impervious fraction and slope.

$$C = 0.14 + 0.65(\text{Imp}_{\text{decimal}}) + 0.05 S_{\text{percent}} \qquad \textit{20.37}$$

The rational method assumes that rainfall occurs at a constant rate. If this is true, then the peak runoff will occur when the entire drainage area is contributing to surface runoff, which will occur at t_c. Other assumptions include (a) the recurrence interval of the peak flow is the same as for the design storm, (b) the runoff coefficient is constant, and (c) the rainfall is spatially uniform over the drainage area.

Example 20.5

Two adjacent fields, as shown, contribute runoff to a collector whose capacity is to be determined. The storm

---

[3]In Great Britain, the rational equation is known as the *Lloyd-Davies equation.*
[4]When using intensity-duration-frequency curves to size storm sewers, culverts, and other channels, it is assumed that the frequencies and probabilities of flood damage and storms are identical. This is not generally true, but the assumption is usually made anyway.

intensity after 25 min is 3.9 in/hr. (a) Calculate the time to concentration. (b) Use the rational method to calculate the peak flow.

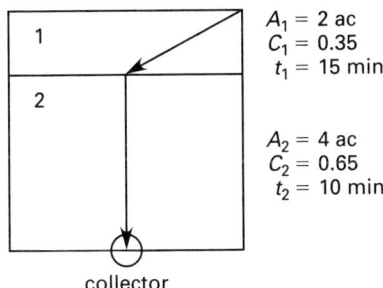

collector

Solution

(a) The overland flow time is given for both areas. The time for water from the farthest corner to reach the collector is

$$t_c = 15 \text{ min} + 10 \text{ min} = 25 \text{ min}$$

(b) The runoff coefficients are given for each area. Since the pipe carrying the total runoff needs to be sized, the coefficients are weighted by their respective contributing areas.

$$C = \frac{(2 \text{ ac})(0.35) + (4 \text{ ac})(0.65)}{2 \text{ ac} + 4 \text{ ac}}$$
$$= 0.55$$

The intensity after 25 min was given as 3.9 in/hr.

The total area is 4 ac + 2 ac = 6 ac.

The peak flow is found from Eq. 20.36.

$$Q_p = CIA_d = (0.55)\left(3.9 \ \frac{\text{in}}{\text{hr}}\right)(6 \text{ ac})$$
$$= 12.9 \text{ ac-in/hr} \quad (12.9 \text{ ft}^3/\text{sec})$$

Example 20.6

Three watersheds contribute runoff to a storm drain. The watersheds have the following characteristics.

watershed	area (ac)	overland flow time (min)	C
A	10	20	0.3
B	2	5	0.7
C	15	25	0.4

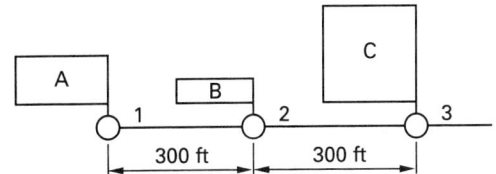

The instantaneous rainfall intensity (in/hr) for the area and specified storm frequency is

$$I = \frac{115}{t_c + 15}$$

The collector inlets are 300 ft apart. The pipe slope is 0.009. The Manning roughness coefficient is 0.015. Assume the flow velocity is 5 ft/sec in all sections. The storm duration is long enough to permit all three areas to contribute to the combined flow through section 3. (a) What should be the pipe size in section 2? (b) What is the maximum flow through section 3?

Solution

The maximum flow occurs when the overland flow from both areas A and B have reached the second manhole. The time of concentration for area A is the time of concentration to the first manhole plus the travel time in pipe section 1.

$$t_c = 20 \text{ min} + \frac{300 \text{ ft}}{\left(5 \ \frac{\text{ft}}{\text{sec}}\right)\left(60 \ \frac{\text{sec}}{\text{min}}\right)}$$
$$= 21 \text{ min}$$

The time of concentration for area B is 5 min. The pipe in section 2 should be sized for flow at 21 min.

$$I = \frac{115}{t_c + 15} = \frac{115 \ \frac{\text{in-hr}}{\text{hr}}}{21 \text{ min} + 15 \text{ min}}$$
$$= 3.19 \text{ in/hr}$$

Use the sum of the CA values in Eq. 20.36.

$$\sum CA = (0.3)(10 \text{ ac}) + (0.7)(2 \text{ ac})$$
$$= 4.4 \text{ ac}$$

Use the rational formula, Eq. 20.36.

$$Q = \sum CA_d I = (4.4 \text{ ac})\left(3.19 \ \frac{\text{in}}{\text{hr}}\right)$$
$$= 14.0 \text{ ac-in/hr} \quad (14.0 \text{ ft}^3/\text{sec})$$

This value of Q should be used to design section 2 of the pipe. From Eq. 19.16 and assuming the pipe is flowing full,

$$D = 1.335 \left(\frac{nQ}{\sqrt{S_{\text{decimal}}}}\right)^{3/8}$$

$$= (1.335)\left(\frac{(0.015)\left(14.0 \ \frac{\text{ft}^3}{\text{sec}}\right)}{\sqrt{0.009}}\right)^{3/8}$$

$$= 1.798 \text{ ft} \quad [\text{round to } 2.0 \text{ ft}]$$

The maximum flow through section 3 occurs when the overland flow from plots A, B, and C reach the third manhole.

For plot A,

$$t_c = 20 \text{ min} + 1 \text{ min} + 1 \text{ min} = 22 \text{ min}$$

For plot B,

$$t_c = 5 \text{ min} + 1 \text{ min} = 6 \text{ min}$$

For plot C,

$$t_c = 25 \text{ min}$$

The maximum runoff will occur 25 min after the start of the storm.

$$I = \frac{115}{t_c + 15} = \frac{115 \ \frac{\text{in-min}}{\text{hr}}}{25 \text{ min} + 15 \text{ min}}$$

$$= 2.875 \text{ in/hr}$$

The sum of the CA values is

$$\sum CA = (0.3)(10 \text{ ac}) + (0.7)(2 \text{ ac}) + (0.4)(15 \text{ ac})$$

$$= 10.4 \text{ ac}$$

$$Q = \sum CA_d I = (10.4 \text{ ac})\left(2.875 \ \frac{\text{in}}{\text{hr}}\right)$$

$$= 29.9 \text{ ac-in/hr} \quad (29.9 \text{ ft}^3/\text{sec})$$

16. NRCS CURVE NUMBER

Several methods of calculating total and peak runoff have been developed over the years by the U.S. Natural Resources Conservation Service. These methods have generally been well correlated with actual experience, and the NRCS methods have become dominant in the United States.

The NRCS methods classify the land use and soil type by a single parameter called the *curve number*, CN. This method can be used for any size homogeneous watershed with a known percentage of imperviousness. If the watershed varies in soil type or in cover, it generally

should be divided into regions to be analyzed separately. A composite curve number can be calculated by weighting the curve number for each region by its area. Alternatively, the runoffs from each region can be calculated separately and added.

The NRCS method of using precipitation records and an assumed distribution of rainfall to construct a synthetic storm is based on several assumptions. First, a type II storm is assumed. Type I storms, which drop most of their precipitation early, are applicable to Hawaii, Alaska, and the coastal side of the Sierra Nevada and Cascade mountains in California, Oregon, and Washington. Type II distributions are typical of the rest of the United States, Puerto Rico, and the Virgin Islands.

This method assumes that initial abstraction (depression storage, evaporation, and interception losses) is equal to 20% of the storage capacity.

$$I_a = 0.2S \qquad\qquad 20.38$$

For there to be any runoff at all, the gross rain must equal or exceed the initial abstraction.

$$P \geq I_a \qquad\qquad 20.39$$

The storage capacity must be great enough to absorb the initial abstraction plus the infiltration.

$$S \geq I_a + F \qquad\qquad 20.40$$

The following steps constitute the NRCS method.

step 1: Classify the soil into a *hydrologic soil group* (HSG) according to its infiltration rate. Soil is classified into HSG A (low runoff potential) through D (high runoff potential).

Group A: High infiltration rates (> 0.30 in/hr (0.76 cm/h)) even if thoroughly saturated; chiefly deep sands and gravels with good drainage and high moisture transmission. In urbanized areas, this category includes sand, loamy sand, and sandy loam.

Group B: Moderate infiltration rates if thoroughly wetted (0.15–0.30 in/hr (0.38–0.76 cm/h)), moderate rates of moisture transmission, and consisting chiefly of coarse to moderately fine textures. In urbanized areas, this category includes silty loam and loam.

Group C: Slow infiltration rates if thoroughly wetted (0.05–0.15 in/hr (0.13–0.38 cm/h)), and slow moisture transmission; soils having moderately fine to fine textures or that impede the downward movement of water. In urbanized areas, this category includes sandy clay loam.

Group D: Very slow infiltration rates (less than 0.05 in/hr (0.13 cm/h)) if thoroughly wetted, very slow water transmission, and consisting primarily of clay soils with high potential for swelling; soils with permanent high water tables;

or soils with an impervious layer near the surface. In urbanized areas, this category includes clay loam, silty clay loam, sandy clay, silty clay, and clay.

(Note that as a result of urbanization, the underlying soil may be disturbed or covered by a new layer. The original classification will no longer be applicable, and the "urbanized" soil HSGs are applicable.)

step 2: Determine the preexisting soil conditions. The soil condition is classified into *antecedent runoff conditions* (ARC) I through III.[5] Generally, "average" conditions (ARC II) are assumed.

ARC I: Dry soils, prior to or after plowing or cultivation, or after periods without rain.

ARC II: Typical conditions existing before maximum annual flood.

ARC III: Saturated soil due to heavy rainfall (or light rainfall with freezing temperatures) during 5 days prior to storm.

step 3: Classify *cover type* and hydrologic condition of the soil-cover complex. For pasture, range, row crops, arid, and semi-arid lands, the NRCS method includes additional tables to classify the cover and hydrologic conditions. In order to use these tables, it is necessary to characterize the surface coverage. The condition is "good" if it is lightly grazed or has plant cover over 75% or more of its area. The condition is "fair" if plant coverage is 50–75% or not heavily grazed. The condition is "poor" if the area is heavily grazed, has no mulch, or has plant cover over less than 50% of the area.

step 4: Use Table 20.4 or Table 20.5 to determine the *curve number*, CN, corresponding to the soil classification for ARC II.

step 5: If the soil is ARC I or ARC III, convert the curve number from step 4 by using Eq. 20.41 or Eq. 20.42 and rounding up.

$$\mathrm{CN_I} = \frac{4.2\,\mathrm{CN_{II}}}{10 - 0.058\,\mathrm{CN_{II}}} \qquad 20.41$$

$$\mathrm{CN_{III}} = \frac{23\,\mathrm{CN_{II}}}{10 + 0.13\,\mathrm{CN_{II}}} \qquad 20.42$$

step 6: If any significant fraction of the watershed is impervious (i.e., CN = 98 for pavement), or if the watershed consists of areas with different curve numbers, calculate the composite curve number by weighting by the runoff areas (same as weighting by the impervious and pervious fractions). If the watershed's impervious fraction is different from the value implicit in step 3's classification, or if the impervious area is not connected directly to a storm drainage system, then the NRCS method includes direct and graphical adjustments.

step 7: Estimate the time of concentration of the watershed.[6] (See Sec. 20.5.)

step 8: Determine the *gross (total) rainfall*, *P*, from the storm. (See Eq. 20.18.) To do this, it is necessary to assume the storm length and recurrence interval. It is a characteristic of the NRCS methods to use a 24 hr storm. Maps from the U.S. Weather Bureau can be used to read gross point rainfalls for storms with frequencies from 1 to 100 years.[7]

step 9: Multiply the gross rain point value from step 8 by a factor from Fig. 20.20 to make the gross rain representative of larger areas. This is the *areal rain*.

step 10: The NRCS method assumes that infiltration follows an exponential decay curve with time. Storage capacity of the soil (i.e., the potential maximum retention after runoff begins), *S*, is calculated from the curve number by using Eq. 20.43.

$$S_{\mathrm{in}} = \frac{1000}{\mathrm{CN}} - 10 \qquad 20.43$$

step 11: Calculate the total runoff (net rain, precipitation excess, etc.), *Q*, in inches from the areal rain. Equation 20.44 subtracts losses from interception, storm period evaporation, depression storage, and infiltration from the *gross rain* to obtain the *net rain*. (Equation 20.44 can be derived from the water balance equation, Eq. 20.2.)

$$Q_{\mathrm{in}} = \frac{(P_g - I_a)^2}{P_g - I_a + S} = \frac{(P_g - 0.2S)^2}{P_g + 0.8S} \qquad 20.44$$

17. NRCS GRAPHICAL PEAK DISCHARGE METHOD

Two NRCS methods are available for calculating the peak discharge. When a full hydrograph is not needed, the so-called graphical method can be used. If a hydrograph is needed, the tabular method can be used.[8] The graphical method is applicable when (a) CN > 40, (b) 0.1 hr < t_c < 10 hr, (c) the watershed is relatively homogeneous or uniformly mixed, (d) all streams have the same time of concentration, and (e) there is no interim storage along the stream path.

---

[6]The NRCS method uses T_c, not t_c, as the symbol for time of concentration.

[7]*Rainfall Frequency Atlas of the United States for Durations from 30 Minutes to 24 Hours and Return Periods from 1 to 100 Years* (1961), U.S. Weather Bureau, Technical Paper 40.

[8]The graphical and tabular methods both rely on graphs and tables that are contained in TR-55. These graphs and tables are not included in this chapter. The NRCS tabular method is not described in this book.

Flow of Fluids

Table 20.4 *Runoff Curve Numbers of Urban Areas (ARC II)*

cover description		curve numbers for hydrologic soil			
cover type and hydrologic condition	average percent impervious area	group A	group B	group C	group D
fully developed urban areas (vegetation established)					
open space (lawns, parks, golf courses, cemeteries, etc.)					
poor condition (grass cover < 50%)		68	79	86	89
fair condition (grass cover 50–75%)		49	69	79	84
good condition (grass cover > 75%)		39	61	74	80
impervious areas					
paved parking lots, roofs, driveways, etc., (excluding right-of-way)		98	98	98	98
streets and roads					
paved; curbs and storm sewers (excluding right-of-way)		98	98	98	98
paved; open ditches (including right-of-way)		83	89	92	93
gravel (including right-of-way)		76	85	89	91
dirt (including right-of-way)		72	82	87	89
western desert urban areas					
natural desert landscaping (pervious areas only)		63	77	85	88
artificial desert landscaping (impervious weed barrier, desert shrub with 1–2 in sand or gravel mulch and basin borders)		96	96	96	96
urban districts					
commercial and business	85	89	92	94	95
industrial	72	81	88	91	93
residential districts by average lot size					
$\frac{1}{8}$ acre or less (townhouses)	65	77	85	90	92
$\frac{1}{4}$ acre	38	61	75	83	87
$\frac{1}{3}$ acre	30	57	72	81	86
$\frac{1}{2}$ acre	25	54	70	80	85
1 acre	20	51	68	79	84
2 acres	12	46	65	77	82
developing urban areas					
newly graded areas (pervious areas only, no vegetation)		77	86	91	94

Reprinted from *Urban Hydrology for Small Watersheds*, Technical Release TR-55, United States Department of Agriculture, Natural Resources Conservation Service, Table 2-2a, 1986.

Figure 20.20 *Point to Areal Rain Conversion Factors*

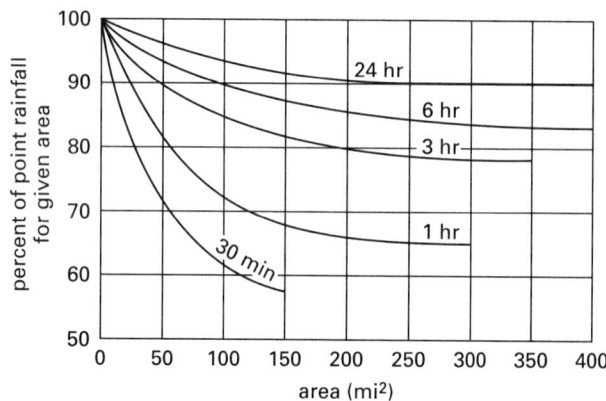

Equation 20.45 is the NRCS peak discharge equation. Q_p is the peak discharge (ft³/sec), q_u is the unit peak discharge (cubic feet per square mile per inch of runoff, csm/in), A_{mi^2} is the drainage area in mi², Q_{inches} is the runoff (in), and F_p is a pond and swamp adjustment factor.

$$Q_p = q_u A_{\text{mi}^2} Q_{\text{in}} F_p \qquad 20.45$$

The graphical method begins with obtaining the design for the 24 hr rainfall, P, for the area. Total runoff, Q, and the curve number, CN, are determined by the same method as in Sec. 20.16. Then, the initial abstraction, I_a, is obtained from Table 20.6, and the ratio I_a/P is calculated. Next, NRCS curves are entered with t_c and the I_a/P ratio to determine the runoff in cubic feet per square mile per inch of runoff (csm/in). The appropriate pond and swamp adjustment factor is selected from the

Table 20.5 Runoff Curve Numbers for Cultivated Agricultural Lands (ARC II)

cover description			curve numbers for hydrologic soil			
cover type	treatment[a]	hydrologic condition[b]	group A	group B	group C	group D
fallow	bare soil	–	77	86	91	94
	crop residue cover (CR)	poor	76	85	90	93
		good	74	83	88	90
row crops	straight row (SR)	poor	72	81	88	91
		good	67	78	85	89
	SR + CR	poor	71	80	87	90
		good	64	75	82	85
	contoured (C)	poor	70	79	84	88
		good	65	75	82	86
	C + CR	poor	69	78	83	87
		good	64	74	81	85
	contoured and terraced (C&T)	poor	66	74	80	82
		good	62	71	78	81
	C&T + CR	poor	65	73	79	81
		good	61	70	77	80
small grain	SR	poor	65	76	84	88
		good	63	75	83	87
	SR + CR	poor	64	75	83	86
		good	60	72	80	84
	C	poor	63	74	82	85
		good	61	73	81	84
	C + CR	poor	62	73	81	84
		good	60	72	80	83
	C&T	poor	61	72	79	82
		good	59	70	78	81
	C&T + CR	poor	60	71	78	81
		good	58	69	77	80
close-seeded or broadcast legumes or rotation meadow	SR	poor	66	77	85	89
		good	58	72	81	85
	C	poor	64	75	83	85
		good	55	69	78	83
	C&T	poor	63	73	80	83
		good	51	67	76	80

[a]*Crop residue cover* applies only if residue is on at least 5% of the surface throughout the year.
[b]Hydrologic condition is based on a combination of factors that affect infiltration and runoff, including (a) density and canopy of vegetative areas, (b) amount of year-round cover, (c) amount of grass or close-seeded legumes in rotations, (d) percent of residue cover on the land surface (good ≥ 20%), and (e) degree of surface roughness.
 Poor: Factors impair infiltration and tend to increase runoff.
 Good: Factors encourage average and better-than-average infiltration and tend to decrease runoff.

Reprinted from *Urban Hydrology for Small Watersheds*, Technical Release TR-55, United States Department of Agriculture, Natural Resources Conservation Service, Table 2-b, 1986.

NRCS literature. ($F_p = 1$ if there are no ponds or swamps.)

18. RESERVOIR SIZING: MODIFIED RATIONAL METHOD

An effective method of preventing flooding is to store surface runoff temporarily. After the storm is over, the stored water can be gradually released. An *impounding*

reservoir (*retention watershed* or *detention watershed*) is a watershed used to store excess flow from a stream or river. The stored water is released when the stream flow drops below the minimum level that is needed to meet water demand. The *impoundment depth* is the design depth. Finding the impoundment depth is equivalent to finding the design storage capacity of the reservoir.

The purpose of a *reservoir sizing* (*reservoir yield*) analysis is to determine the proper size of a reservoir or dam,

Table 20.6 *Initial Abstraction versus Curve Number*

curve number	I_a (in)	curve number	I_a (in)
40	3.000	70	0.857
41	2.878	71	0.817
42	2.762	72	0.778
43	2.651	73	0.740
44	2.545	74	0.703
45	2.444	75	0.667
46	2.348	76	0.632
47	2.255	77	0.597
48	2.167	78	0.564
49	2.082	79	0.532
50	2.000	80	0.500
51	1.922	81	0.469
52	1.846	82	0.439
53	1.774	83	0.410
54	1.704	84	0.381
55	1.636	85	0.353
56	1.571	86	0.326
57	1.509	87	0.299
58	1.448	88	0.273
59	1.390	89	0.247
60	1.333	90	0.222
61	1.279	91	0.198
62	1.226	92	0.174
63	1.175	93	0.151
64	1.125	94	0.128
65	1.077	95	0.105
66	1.030	96	0.083
67	0.985	97	0.062
68	0.941	98	0.041
69	0.899		

(Multiply in by 2.54 to obtain cm.)

Reprinted from *Urban Hydrology for Small Watersheds*, Technical Release TR-55, United States Department of Agriculture, Natural Resources Conservation Service, Table 4-1, 1986.

or to evaluate the ability of an existing reservoir to meet water demands.

The volume of a reservoir needed to hold streamflow from a storm is simply the total area of the hydrograph. Similarly, when comparing two storms, the incremental volume needed is simply the difference in the areas of their two hydrographs.

Poertner's 1974 *modified rational method*, as illustrated in Fig. 20.21, can be used to design detention storage facilities for small areas (up to 20 ac). A trapezoidal hydrograph is constructed with the peak flow calculated from the rational equation. The total hydrograph base (i.e., the total duration of surface runoff) is the storm duration plus the time to concentration. The durations of the rising and falling limbs are both taken as t_c.

To size a detention watershed using this method, a trapezoidal hydrograph is drawn for several storm durations greater than $2t_c$ (e.g., 10, 15, 20, and 30 min). For each of the storm durations, the total rainfall is calculated as the hydrograph area. The difference between

Figure 20.21 *Modified Rational Method Hydrograph*

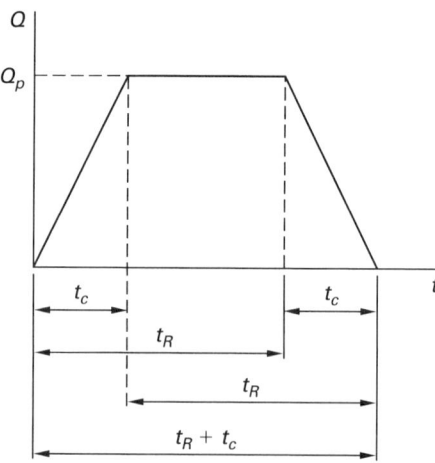

the total rainfall and the total released water (calculated as the product of the allowable release rate and the storm duration) is the required storage capacity.

Detention watershed sizing for areas larger than 20 ac should be accomplished using full hydrograph methods in combination with reservoir routing.

19. RESERVOIR SIZING: NONSEQUENTIAL DROUGHT METHOD

The *nonsequential drought method* is somewhat complex, but it has the advantage of giving an estimate of the required reservoir size, rather than merely evaluating a trial size. In the absence of synthetic drought information, it is first necessary to develop intensity-duration-frequency curves from stream flow records.

step 1: Choose a duration. Usually, the first duration used will be 7 days, although choosing 15 days will not introduce too much error.

step 2: Search the stream flow records to find the smallest flow during the duration chosen. (The first time through, for example, find the smallest discharge totaled over any 7 days.) The days do not have to be sequential.

step 3: Continue searching the discharge records to find the next smallest discharge over the number of days in the period. Continue searching and finding the smallest discharges (which gradually increase) until all of the days in the record have been used up. Do not use the same day more than once.

step 4: Give the values of smallest discharge order numbers; that is, give $M = 1$ to the smallest discharge, $M = 2$ to the next smallest, etc.

Flow of Fluids

step 5: For each observation, calculate the recurrence interval as

$$F = \frac{n_y}{M} \qquad 20.46$$

n_y is the number of years of stream flow data that was searched to find the smallest discharges.

step 6: Plot the points as discharge on the y-axis versus F in years on the x-axis. Draw a reasonably continuous curve through the points.

step 7: Return to step 1 for the next duration. Repeat for all of the following durations: 7, 15, 30, 60, 120, 183, and 365 days.

A synthetic drought can be constructed for any recurrence interval. For example, in Fig. 20.22, a 5 year drought is being planned for, so the discharges V_7, $V_{15}, V_{30}, \ldots, V_{365}$ are read from the appropriate curves for $F = 5$ yr.

Figure 20.22 *Sample Family of Synthetic Inflow Curves*

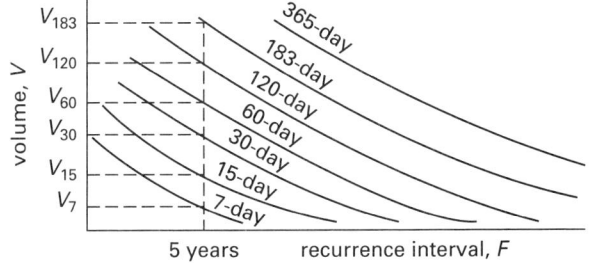

The next step is to plot the reservoir *mass diagram* (also known as a *Rippl diagram*). This is a plot of the cumulative net volume—a simultaneous plot of the cumulative demand (known as *draft*) and cumulative inflow. The mass diagram is used to graphically determine the reservoir storage requirements (i.e., size).

As long as the slopes of the cumulative demand and inflow lines are equal, the water reserve in the reservoir will not change. When the slope of the inflow is less than the slope of the demand, the inflow cannot by itself satisfy the community's water needs, and the reservoir is drawn down to make up the difference. A peak followed by a trough is, therefore, a drought condition.

If the reservoir is to be sized so that the community will not run dry during a drought, the required capacity is the maximum separation between two parallel lines (pseudo-demand lines with slopes equal to the demand rate) drawn tangent to a peak and a subsequent trough. If the mass diagram covers enough time so that multiple droughts are present, the largest separation between peaks and subsequent troughs represents the capacity.

In order for the reservoir to supply enough water during a drought condition, the reservoir must be full prior to

the start of the drought. This fact is not represented when the mass diagram is drawn, hence the need to draw a pseudo-demand line parallel to the peak.

After a drought equal to the capacity of the reservoir, the reservoir will again be empty. At the trough, however, the reservoir begins to fill up again. When the cumulative excess exceeds the reservoir capacity, the reservoir will have to "spill" (i.e., release) water. This occurs when the cumulative inflow line crosses the prior peak's pseudo-demand line, as shown in Fig. 20.23.

A *flood-control dam* is built to keep water in and must be sized so that water is not spilled. The mass diagram can still be used, but the maximum separation between troughs and subsequent peaks (not peaks followed by troughs) is the required capacity.

Figure 20.23 *Reservoir Mass Diagram (Rippl Diagram)*

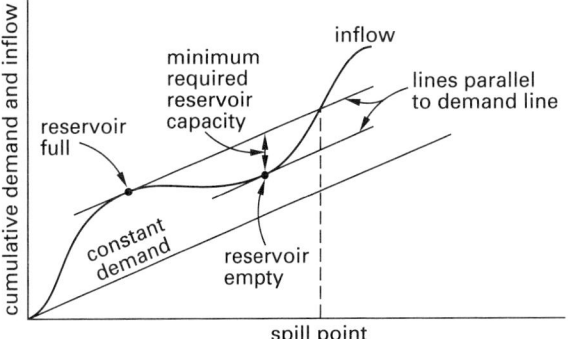

20. RESERVOIR SIZING: RESERVOIR ROUTING

Reservoir routing is the process by which the outflow hydrograph (i.e., the outflow over time) of a reservoir is determined from the inflow hydrograph (i.e., the inflow over time), the initial storage, and other characteristics of the reservoir. The simplest method is to keep track of increments in inflow, storage, and outflow period by period in a tabular simulation. This is the basis of the *storage indication method*, which is basically a bookkeeping process. The validity of this method is dependent on choosing time increments that are as small as possible.

step 1: Determine the starting storage volume, V_n. If the starting volume is zero or considerably different from the average steady-state storage, a large number of iterations will be required before the simulation reaches its steady-state results. A convergence criterion should be determined before the simulation begins.

step 2: For the next iteration, determine the inflow, discharge, evaporation, and seepage. The starting storage volume for the next iteration is found by solving Eq. 20.47.

$$V_{n+1} = V_n + (\text{inflow})_n - (\text{discharge})_n$$
$$- (\text{seepage})_n - (\text{evaporation})_n \qquad 20.47$$

Repeat step 2 as many times as necessary.

Loss due to seepage is generally very small compared to inflow and discharge, and it is often neglected. Reservoir *evaporation* can be estimated from analytical relationships or by evaluating data from *evaporation pans*. Pan data is extended to reservoir evaporation by the pan coefficient formula. In Eq. 20.48, the summation is taken over the number of days in the simulation period. Units of evaporation are typically in/day. The *pan coefficient*, K_p, is typically 0.7–0.8.

$$E_R = \sum K_p E_p \qquad 20.48$$

Inflow can be taken from actual past history. However, since it is unlikely that history will repeat exactly, the next method is preferred.

21. RESERVOIR ROUTING: STOCHASTIC SIMULATION

The *stochastic simulation* method of reservoir routing is the same as tabular simulation except for the method of determining the inflow. This description uses a *Monte Carlo simulation* technique that is dependent on enough historical data to establish a cumulative inflow distribution. A Monte Carlo simulation is suitable if long periods are to be simulated. If short periods are to be simulated, the simulation should be performed several times and the results averaged.

step 1: Tabulate or otherwise determine a frequency distribution of inflow quantities.

step 2: Form a cumulative distribution of inflow quantities.

step 3: Multiply the cumulative x-axis (which runs from 0 to 1) by 100 or 1000, depending on the accuracy needed.

step 4: Generate random numbers between 0 and 100 (or 0 and 1000). Use of a random number table, such as App. 20.B, is adequate for hand simulation.

step 5: Locate the inflow quantity corresponding to the random number from the cumulative distribution x-axis.

Example 20.7

A well-monitored stream has been observed for 50 years and has the following frequency distribution of total annual discharges.

discharge (units)	frequency (yr)	fraction of time
0 to 0.5	5	0.10
0.5 to 1.0	21	0.42
1.0 to 1.5	17	0.34
1.5 to 2.0	7	0.14

It is proposed that a dam be placed across the stream to create a reservoir with a capacity of 1.8 units. The reservoir is to support a town that will draw 1.2 units per year. Use stochastic simulation to simulate 10 yr of reservoir operation, assuming it is initially filled with 1.5 units.

Solution

step 1: The frequency distribution is given.

steps 2 and 3: Make a table to form the cumulative distribution with multiplied frequencies.

discharge	cumulative frequency	cumulative frequency × 100
0 to 0.5	0.10	10
0.5 to 1.0	0.52	52
1.0 to 1.5	0.86	86
1.5 to 2.0	1.00	100

step 4: From App. 20.B, choose 10 two-digit numbers. The starting point within the table is arbitrary, but the numbers must come sequentially from a row or column. Use the first row for this simulation.

78, 46, 68, 33, 26, 96, 58, 98, 87, 27

step 5: For the first year, the random number is 78. Since 78 is greater than 52 but less than 86, the inflow is in the 1.0–1.5 unit range. The midpoint of this range is taken as the inflow, which would be 1.25. The reservoir volume after the first year would be $1.5 + 1.25 - 1.20 = 1.55$. The remaining years can be similarly simulated.

year	starting volume	+ inflow	− usage	= ending volume	+ spill
1	1.5	1.25	1.2	1.55	
2	1.55	0.75	1.2	1.1	
3	1.1	1.25	1.2	1.15	
4	1.15	0.75	1.2	0.7	
5	0.7	0.75	1.2	0.25	
6	0.25	1.75	1.2	0.8	
7	0.8	1.25	1.2	0.85	
8	0.85	1.75	1.2	1.4	
9	1.4	1.75	1.2	1.95	0.15
10	1.8	0.75	1.2	1.35	

No shortages are experienced; one spill is required.

22. STORMWATER/WATERSHED MANAGEMENT MODELING

Flood routing and *channel routing* are terms used to describe the passage of a flood or runoff wave through a complex system. Due to flow times, detention, and processing, the wave front appears at different points in a system at different times. The ability to simulate flood, channel, and reservoir routing is particularly useful when evaluating competing features (e.g., treatment, routing, storage options).

The interaction of stormwater features (e.g., watersheds, sewers and storm drains, detention watersheds, treatment facilities) during and after a storm is complex. In addition to continuity considerations, the interaction is affected by hydraulic considerations (e.g., the characteristics of the flow paths) and topography (e.g., the elevation changes from point to point).

Many computer programs have been developed to predict the performance of such complex systems. These simulation programs vary considerably in complexity, degree of hydraulic detail, length of simulation interval, and duration of simulation study. The Environmental Protection Agency (EPA) Stormwater Management Model is a well-known micro-scale model, particularly well-suited to areas with a high impervious fraction. The Army Corps of Engineers' STORM model is a macro-scale model. The NRCS TR-20 program is consistent with other NRCS methodology.

23. FLOOD CONTROL CHANNEL DESIGN

For many years, the traditional method of flood control was *channelization*, converting a natural stream to a uniform channel cross section. However, this method does not always work as intended and may fail at critical moments. Some of the reasons for reduced capacities are sedimentation, increased flow resistance, and inadequate maintenance.

The design of artificial channels is often based on clearwater hydraulics without sediment. However, the capacity of such channels is greatly affected by *sedimentation*. Silting and sedimentation can double or triple the Manning roughness coefficient. Every large flood carries appreciable amounts of bed load, significantly increasing the composite channel roughness. Also, when the channel is unlined, scour and erosion can significantly change the channel cross-sectional area.

To minimize cost and right-of-way requirements, shallow supercritical flow is often intended when the channel is designed. However, supercritical flow can occur only with low bed and side roughness. When the roughness increases during a flood, the flow shifts back to deeper, slower-moving subcritical flows.

Debris carried downstream by floodwaters can catch on bridge pilings and culverts, obstructing the flow. Actual flood profiles can resemble a staircase consisting of a series of backwater pools behind obstructed bridges, rather than a uniformly sloping surface. In such situations, the most effective flood control method may be replacement of bridges or improved emergency maintenance procedures to remove debris.

Vegetation and other debris that collects during dry periods in the flow channel reduces the flow capacity. Maintenance of channels to eliminate the vegetation is often haphazard.

Flood control channel design programs frequently emphasize "creek protection" and the long-term management of natural channels. Elements of such programs include (a) excavation of a low-flow channel within the natural channel, (b) periodic intervention to keep the channel clear, (c) establishment of a wide flood plain terrace along the channel banks, incorporating wetlands, vegetation, and public-access paths, and (d) planting riparian vegetation and trees along the terrace to slow bank erosion and provide a continuous corridor for wildlife.

Flow of Fluids

21 Groundwater

Nomenclature

A	area	ft^2	m^2
A	availability	–	–
b	aquifer width	ft	m
C	concentration	ppm	ppm
C	constant	various	various
D	diameter (grain size)	ft	m
e	void ratio	–	–
f	infiltration	in/hr	cm/h
F	cumulative infiltration	in	cm
FS	factor of safety	–	–
g	acceleration of gravity, 32.2 (9.81)	ft/sec^2	m/s^2
g_c	gravitational constant, 32.2	$lbm\text{-}ft/lbf\text{-}sec^2$	n.a.
H	total hydraulic head	ft	m
i	hydraulic gradient	ft/ft	m/m
I	rainfall intensity	in/hr	cm/h
j	equipotential index	–	–
k	infiltration decay constant	1/hr	1/h
k	intrinsic permeability	ft^2	m^2
K	hydraulic conductivity	$ft^3/day\text{-}ft^2$	$m^3/d\cdot m^2$
L	length	ft	m
m	mass	lbm	kg
n	porosity	–	–
N	neutral stress coefficient	–	–
N	quantity (number of)	–	–
p	pressure	lbf/ft^2	Pa
q	specific discharge	$ft^3/ft^2\text{-}sec$	$m^3/m^2\cdot s$
Q	flow quantity	ft^3/sec	m^3/s
r	radial distance from well	ft	m
Re	Reynolds number	–	–

s	drawdown	ft	m
S	storage constant	–	–
S_r	specific retention	–	–
S_y	specific yield	–	–
t	time	sec	s
T	transmissivity	$ft^3/day\text{-}ft$	$m^3/d\cdot m$
u	well function argument	–	–
U	uplift force	lbf	N
v	flow velocity	ft/sec	m/s
V	volume	ft^3	m^3
w	moisture content	–	–
$W(u)$	well function	–	–
y	aquifer thickness after drawdown	ft	m
Y	original aquifer phreatic zone thickness	ft	m
z	distance below datum	–	–

Symbols

γ	specific weight	lbf/ft^3	n.a.
μ	absolute viscosity	$lbf\text{-}sec/ft^2$	Pa·s
ν	kinematic viscosity	ft^2/sec	m^2/s
ρ	density	lbm/ft^3	kg/m^3

Subscripts

0	initial
c	equilibrium
e	effective
f	flow
o	at well
p	equipotential
r	radius
s	solid
t	time or total
u	uniformity or uplift
v	void
w	water

1. AQUIFERS

Underground water, also known as *subsurface water*, is contained in saturated geological formations known as *aquifers*. Aquifers are divided into two zones by the water table surface. The *vadose zone* is above the elevation of the water table. Pores in the vadose zone may be either saturated, partially saturated, or empty. The *phreatic zone* is below the elevation of the water table. Pores are always saturated in the phreatic zone.

An aquifer whose water surface is at atmospheric pressure and that can rise or fall with changes in volume is a *free aquifer*, also known as an *unconfined aquifer*. If a well is drilled into an unconfined aquifer, the water level in the well will correspond to the water table. Such a well is known as a *gravity well*.

An aquifer that is bounded on all extents is known as a *confined aquifer*. The water in confined aquifers may be under pressure. If a well is drilled into such an aquifer, the water in the well will rise to a height corresponding to the hydrostatic pressure. The *piezometric height* of the rise is

$$H = \frac{p}{\rho g} \qquad \text{[SI]} \quad 21.1(a)$$

$$H = \frac{p}{\gamma} = \frac{p}{\rho} \times \frac{g_c}{g} \qquad \text{[U.S.]} \quad 21.1(b)$$

If the confining pressure is high enough, the water will be expelled from the surface, and the source is known as an *artesian well*.

2. AQUIFER CHARACTERISTICS

Soil moisture content (*water content*), w, can be determined by oven drying a sample of soil and measuring the change in mass.[1] The water content is the ratio of the mass of water to the mass of solids, expressed as a percentage. The water content can also be determined with a *tensiometer*, which measures the vapor pressure of the moisture in the soil.

$$w = \frac{m_w}{m_s} = \frac{m_t - m_s}{m_s} \qquad 21.2$$

The *porosity*, n, of the aquifer is the percentage of void volume to total volume.[2]

$$n = \frac{V_v}{V_t} = \frac{V_t - V_s}{V_t} \qquad 21.3$$

The *void ratio*, e, is

$$e = \frac{V_v}{V_s} = \frac{V_t - V_s}{V_s} \qquad 21.4$$

Void ratio and porosity are related.

$$e = \frac{n}{1-n} \qquad 21.5$$

Some pores and voids are dead ends or are too small to contribute to seepage. Only the *effective porosity*, n_e, 95–98% of the total porosity, contributes to groundwater flow.

The *hydraulic gradient*, i, is the change in hydraulic head over a particular distance. The hydraulic head at a point is determined as the piezometric head at observation wells.

$$i = \frac{\Delta H}{L} \qquad 21.6$$

3. PERMEABILITY

The flow of a liquid through a permeable medium is affected by both the fluid and the medium. The effects of the medium (independent of the fluid properties) are characterized by the *intrinsic permeability* (*specific permeability*), k. Intrinsic permeability has dimensions of length squared. (See Table 21.1.) The *darcy* has been widely accepted as the unit of intrinsic permeability. One darcy is 0.987×10^{-8} cm^2.

For studies involving the flow of water through an aquifer, effects of intrinsic permeability and the water are combined into the *hydraulic conductivity*, also known as the *coefficient of permeability* or simply the *permeability*, K. Hydraulic conductivity can be determined from a number of water-related tests.[3] It has units of volume per unit area per unit time, which is equivalent to length divided by time (i.e., units of velocity). Volume may be expressed as cubic feet and cubic meters, or gallons and liters.

$$K = \frac{kg\rho}{\mu} \qquad \text{[SI]} \quad 21.7(a)$$

$$K = \frac{k\gamma}{\mu} \qquad \text{[U.S.]} \quad 21.7(b)$$

For many years in the United States, hydraulic conductivity was specified in *Meinzer units* (gallons per day per square foot). To avoid confusion related to multiple definitions and ambiguities in these definitions, hydraulic conductivity is now often specified in units of ft/day (m/d).

The coefficient of permeability is proportional to the square of the mean particle diameter.

$$K = C D_{\text{mean}}^2 \qquad 21.8$$

Hazen's empirical formula can be used to calculate an approximate coefficient of permeability for clean, uniform sands. D_{10} is the *effective* size in mm (i.e., the size for which 10% of the distribution is finer).

$$K_{\text{cm/s}} \approx C D_{10,\text{mm}}^2 \quad [0.1 \text{ mm} \leq D_{10,\text{mm}} \leq 3.0 \text{ mm}] \qquad 21.9$$

The coefficient C is 0.4–0.8 for very fine sand (poorly sorted) or fine sand with appreciable fines; 0.8–1.2 for medium sand (well sorted) or coarse sand (poorly sorted); and 1.2–1.5 for coarse sand (well sorted and clean).

---

[1]It is common in civil engineering to use the term "weight" in place of mass. For example, the *water content* would be defined as the ratio of the weight of water to the weight of solids, expressed as a percentage.
[2]The symbol θ is sometimes used for porosity.

---

[3]Permeability can be determined from constant-head permeability tests (sands), falling-head permeability tests (fine sands and silts), consolidation tests (clays), and field tests of wells (in situ gravels and sands).

Table 21.1 *Typical Permeabilities*

	k (cm^2)	k (m^2)	k (darcys)	K (cm/s)	K (gal/day-ft^2)
gravel	10^{-5} to 10^{-3}	10^{-1} to 10	10^3 to 10^5	0.5 to 50	10^4 to 10^6
gravelly sand	10^{-5}	10^{-1}	10^3	0.5	10^4
clean sand	10^{-6}	10^{-2}	10^2	0.05	10^3
sandstone	10^{-8}	10^{-3}	10	0.005	10^2
dense shale or limestone	10^{-9}	10^{-5}	10^{-1}	0.00005	1
granite or quartzite	10^{-11}	10^{-7}	10^{-3}	0.0000005	10^{-2}
clay	10^{-11}	10^{-7}	10^{-3}	0.0000005	10^{-2}

(Multiply gal/day-ft^2 by 0.1337 to obtain ft^3/day-ft^2.)
(Multiply darcys by 0.987×10^{-8} to obtain cm^2.)
(Multiply darcys by 0.987×10^{-12} to obtain m^2.)
(Multiply cm^2 by 10^{-4} to obtain m^2.)
(Multiply ft^2 by 9.4135×10^{10} to obtain darcys.)
(Multiply m^2 by 10^4 to obtaim cm^2.)
(Multiply gal/day-ft^2 by 4.716×10^{-5} to obtain cm/s.)

4. DARCY'S LAW

Movement of groundwater through an aquifer is given by *Darcy's law*, Eq. 21.10.[4] The hydraulic gradient may be specified in either ft/ft (m/m) or ft/mi (m/km), depending on the units of area used.

$$Q = -KiA_{\text{gross}} = -v_e A_{\text{gross}} \qquad 21.10$$

The *specific discharge* is the same as the *effective velocity*, v_e.

$$q = \frac{Q}{A_{\text{gross}}} = Ki = v_e \qquad 21.11$$

Darcy's law is applicable only when the Reynolds number is less than 1. Significant deviations have been noted when the Reynolds number is even as high as 2. In Eq. 21.12, D_{mean} is the mean grain diameter.

$$\text{Re} = \frac{\rho q D_{\text{mean}}}{\mu} = \frac{q D_{\text{mean}}}{\nu} \qquad 21.12$$

5. TRANSMISSIVITY

Transmissivity (also known as the *coefficient of transmissivity*) is an index of the rate of groundwater movement. The transmissivity of flow from a saturated aquifer of thickness Y and width b is given by Eq. 21.13. The thickness, Y, of a confined aquifer is the difference in elevations

of the bottom and top of the saturated formation. For permeable soil, the thickness, Y, is the difference in elevations of the impermeable bottom and the water table.

$$T = KY \qquad 21.13$$

Combining Eq. 21.10 and Eq. 21.13 gives

$$Q = bTi \qquad 21.14$$

6. SPECIFIC YIELD, RETENTION, AND CAPACITY

The dimensionless *storage constant* (*storage coefficient*), S, of a confined aquifer is the change in aquifer water volume per unit surface area of the aquifer per unit change in head. That is, the storage constant is the amount of water that is removed from a column of the aquifer 1 ft^2 (1 m^2) in plan area when the water table drops 1 ft (1 m). For unconfined aquifers, the storage coefficient is virtually the same as the specific yield. Various methods have been proposed for calculating the storage coefficient directly from properties of the rock and water. It can also be determined from unsteady flow analysis of wells. (See Sec. 21.12.)

The *specific yield*, S_y, is the water yielded when water-bearing material drains by gravity. It is the volume of water removed per unit area when a drawdown of one length unit is experienced. A time period may be given for different values of specific yield.

$$S_y = \frac{V_{\text{yielded}}}{V_{\text{total}}} \qquad 21.15$$

---

[4]The negative sign in Darcy's law accounts for the fact that flow is in the direction of decreasing head. That is, the hydraulic gradient is negative in the direction of flow. The negative sign is omitted in the remainder of this chapter, or appropriate equations are rearranged to be positive.

The *specific retention* is the volume of water that, after being saturated, will remain in the aquifer against the pull of gravity.

$$S_r = \frac{V_{\text{retained}}}{V_{\text{total}}} = n - S_y \qquad 21.16$$

The *specific capacity* of an aquifer is the discharge rate divided by the drawdown.

$$\text{specific capacity} = \frac{Q}{s} \qquad 21.17$$

7. DISCHARGE VELOCITY AND SEEPAGE VELOCITY

The *pore velocity* (*linear velocity, flow front velocity,* and *seepage velocity*) is given by Eq. 21.18.[5]

$$v_{\text{pore}} = \frac{Q}{A_{\text{net}}} = \frac{Q}{nA_{\text{gross}}} = \frac{Q}{nbY} = \frac{Ki}{n} \qquad 21.18$$

The gross cross-sectional area in flow depends on the aquifer dimensions. However, water can only flow through voids (pores) in the aquifer. If the gross cross-sectional area is known, it will be necessary to multiply by the porosity to reduce the area in flow. (The hydraulic conductivity is not affected.)

The *effective velocity* (also known as the *apparent velocity, Darcy velocity, Darcian velocity, Darcy flux, discharge velocity, specific discharge, superficial velocity, face velocity,* and *approach velocity*) through a porous medium is the velocity of flow averaged over the gross aquifer cross-sectional area.

$$v_e = n v_{\text{pore}} = \frac{Q}{A_{\text{gross}}} = \frac{Q}{bY} = Ki \qquad 21.19$$

Contaminants introduced into an aquifer will migrate from place to place relative to the surface at the pore velocity given by Eq. 21.18. The effective velocity given in Eq. 21.19 should not be used to determine the overland time taken and distance moved by a contaminant.

8. FLOW DIRECTION

Flow direction will be from an area with a high piezometric head (as determined from observation wells) to an area of low piezometric head. Piezometric head is assumed to vary linearly between points of known head. Similarly, all points along a line joining two points with the same piezometric head can be considered to have the same head.

---

[5]Terms used to describe pore and effective velocities are not used consistently.

9. WELLS

Water in aquifers can be extracted from *gravity wells.* However, *monitor wells* may also be used to monitor the quality and quantity of water in an aquifer. *Relief wells* are used to dewater soil.

Wells may be dug, bored, driven, jetted, or drilled in a number of ways, depending on the aquifer material and the depth of the well. Wells deeper than 100 ft (30 m) are usually drilled. After construction, the well is *developed,* which includes the operations of removing any fine sand and mud. Production is *stimulated* by increasing the production rate. The fractures in the rock surrounding the well are increased in size by injecting high-pressure water or using similar operations.

Well equipment is *sterilized* by use of chlorine or other disinfectants. The strength of any chlorine solution used to disinfect well equipment should not be less than 100 ppm by weight (i.e., 100 kg of chlorine per 10^6 kg of water). Calcium hypochlorite (which contains 65% available chlorine) and sodium hypochlorite (which contains 12.5% available chlorine) are commonly used for this purpose. The mass of any chlorine-supplying compound with fractional availability A required to produce V gallons of disinfectant with concentration C is

$$m_{\text{lbm}} = (8.33 \times 10^{-6})\, V_{\text{gal}} \left(\frac{C_{\text{ppm}}}{A_{\text{decimal}}} \right) \qquad 21.20$$

Figure 21.1 illustrates a typical water-supply well. Water is removed from the well through the *riser pipe* (*eductor pipe*). Water enters the well through a perforated or slotted casing known as the *screen.* The required *open area* depends on the flow rate and is limited by the maximum permissible entrance velocity that will not lift grains larger than a certain size. Table 21.2 recommends maximum flow velocities as functions of the grain diameter. A safety factor of 1.5–2.0 is also used to account for the fact that parts of the screen may become blocked. As a general rule, the openings should also be smaller than D_{50} (i.e., smaller than 50% of the screen material particles).

Screens do not necessarily extend the entire length of the well. The required screen length can be determined from the total amount of open area required and the open area per unit length of casing. For confined aquifers, screens are usually installed in the middle 70–80% of the well. For unconfined aquifers, screens are usually installed in the lower 30–40% of the well.

It is desirable to use screen openings as large as possible to reduce entrance friction losses. Larger openings can be tolerated if the well is surrounded by a *gravel pack* to prevent fine material from entering the well. Gravel packs are generally required in soils where the D_{90} size (i.e., the sieve size retaining 90% of the soil) is less than 0.01 in (0.25 mm) and when the well goes through layers of sand and clay.

Flow of Fluids

Figure 21.1 *Typical Gravity Well*

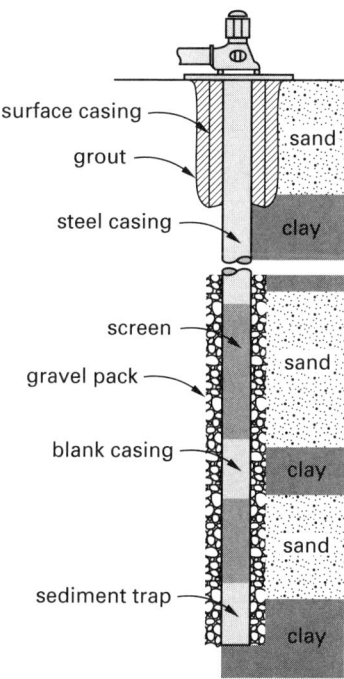

surface casing
grout
steel casing
sand
clay
screen
gravel pack
sand
blank casing
clay
sand
sediment trap
clay

Table 21.2 *Lifting Velocity of Water**

	maximum water velocity	
grain diameter (mm)	(ft/sec)	(m/s)
0 to 0.25	0.10	0.030
0.25 to 0.50	0.22	0.066
0.50 to 1.00	0.33	0.10
1.00 to 2.00	0.56	0.17
2.00 to 4.00	2.60	0.78

(Multiply ft/sec by 0.3 to obtain m/s.)
*spherical particles with specific gravity of 2.6

10. DESIGN OF GRAVEL SCREENS AND POROUS FILTERS

Gravel and other porous materials may be used as filters as long as their voids are smaller than the particles to be excluded. Actually, only the largest 15% of the particles need to be filtered out, since the agglomeration of these particles will themselves create even smaller openings, and so on.

In specifying the opening sizes for screens, perforated pipe, and fabric filters, the following criteria are in widespread use as the basis for filter design in sandy gravels. D_{15}, the interpolated grain size that passes 15%, is known as the *permeability protection limit*. D_{85} is known as the *piping predicting limit*.

$$D_{\text{opening,filter}} \leq D_{85,\text{soil}} \quad \text{[screen filters]} \quad 21.21$$

$$[\text{filtering criterion}]^6 \quad D_{15,\text{filter}} \leq 5D_{85,\text{soil}} \quad \text{[filter beds]}$$
$$21.22$$

$$[\text{permeability criterion}]^7 \quad D_{15,\text{filter}} \leq 5D_{15,\text{soil}} \quad \text{[filter beds]}$$
$$21.23$$

As stated in Sec. 21.9, the screen openings should also be smaller than D_{50} (i.e., smaller than 50% of the screen material particles).

The *coefficient of uniformity*, C_u, is used to determine whether the filter particles are properly graded. Only uniform materials ($C_u < 2.5$) and well-graded materials ($2.5 < C < 6$) are suitable for use as filters.

$$C_u = \frac{D_{60}}{D_{10}} \quad 21.24$$

11. WELL DRAWDOWN IN AQUIFERS

An aquifer with a well is shown in Fig. 21.2. Once pumping begins, the water table will be lowered in the vicinity of the well. The resulting water table surface is referred to as a *cone of depression*. The decrease in water level at some distance r from the well is known as the *drawdown*, s_r. The drawdown at the well is denoted as s_o.

Figure 21.2 *Well Drawdown in an Unconfined Aquifer*

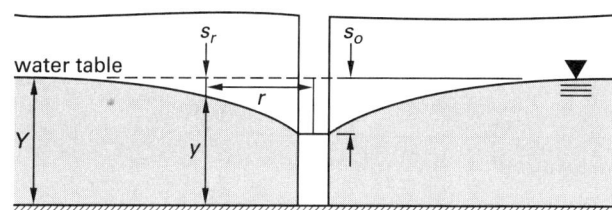

water table

If the drawdown is small with respect to the aquifer phreatic zone thickness, Y, and the well completely penetrates the aquifer, the equilibrium (steady-state) well discharge is given by the *Dupuit equation*, Eq. 21.25. Equation 21.25 can only be used to determine the equilibrium flow rate a "long time" after pumping has begun.

$$Q = \frac{\pi K(y_1^2 - y_2^2)}{\ln \frac{r_1}{r_2}} \quad 21.25$$

In Eq. 21.25, y_1 and y_2 are the aquifer depths at radial distances r_1 and r_2, respectively, from the well. y_1 can also be taken as the original aquifer depth, Y, if r_1 is the well's *radius of influence*, the distance at which the well has no effect on the water table level.

---

[6]Some authorities replace "5" with "9" for uniform soils.
[7]Some authorities replace "5" with "4."

In very thick unconfined aquifers where the drawdown is negligible compared to the aquifer thickness, or in confined aquifers where there is no cone of depression at all, $y_1 + y_2$ is essentially equal to $2Y$. Then, since $y_2 - y_1 = s_1 - s_2$, $y_2^2 - y_1^2 = (y_2 - y_1)(y_2 + y_1) = 2Y(s_1 - s_2)$. Using this equality and Eq. 21.13, Eq. 21.25 can be written as

$$Q = \frac{2\pi T(s_2 - s_1)}{\ln \frac{r_1}{r_2}} \qquad 21.26$$

For an artesian well fed by a confined aquifer of thickness Y, the discharge is given by the *Thiem equation*.

$$Q = \frac{2\pi KY(y_1 - y_2)}{\ln \frac{r_1}{r_2}} \qquad 21.27$$

The rate (or amount in some period) of water that can be extracted without experiencing some undesirable result is known as the *safe yield*. Undesirable results include deteriorations in the water quality, large pump lifts, and infringement on the water rights of others. Extractions in excess of the safe yield are known as *overdrafts*.

Example 21.1

A 9 in (25 cm) diameter well is pumped at the rate of 50 gal/min (0.2 m³/min). The aquifer is 100 ft (30 m) thick. After some time, the well sides cave in and are replaced with an 8 in (20 cm) diameter tube. The drawdown is 6 ft (2 m). The water table recovers its original thickness 2500 ft (750 m) from the well. What will be the steady flow from the new well?

SI Solution

$$r_2 = \frac{25 \text{ cm}}{2} = 12.5 \text{ cm} \quad (0.125 \text{ m})$$

$$y_2 = 30 \text{ m} - 2 \text{ m} = 28 \text{ m}$$

Rearrange Eq. 21.25 to determine hydraulic conductivity from the given information.

$$K = \frac{Q \ln \frac{r_1}{r_2}}{\pi(y_1^2 - y_2^2)}$$
$$= \frac{\left(0.2 \frac{\text{m}^3}{\text{min}}\right) \ln \frac{750 \text{ m}}{0.125 \text{ m}}}{\pi\left((30 \text{ m})^2 - (28 \text{ m})^2\right)}$$
$$= 0.004774 \text{ m/min} \quad (0.007957 \text{ cm/s})$$

For the relined well,

$$r_2 = \frac{20 \text{ cm}}{2} = 10 \text{ cm} \quad (0.10 \text{ m})$$

Use Eq. 21.25 to solve for the new flow rate.

$$Q = \frac{\pi K(y_1^2 - y_2^2)}{\ln \frac{r_1}{r_2}}$$
$$= \pi\left(0.004774 \frac{\text{m}}{\text{min}}\right)\left(\frac{(30 \text{ m})^2 - (28 \text{ m})^2}{\ln \frac{750 \text{ m}}{0.10 \text{ m}}}\right)$$
$$= 0.195 \text{ m}^3/\text{min}$$

Customary U.S. Solution

$$r_2 = \frac{9 \text{ in}}{(2)\left(12 \frac{\text{in}}{\text{ft}}\right)} = 0.375 \text{ ft}$$

$$y_2 = 100 \text{ ft} - 6 \text{ ft} = 94 \text{ ft}$$

$$Q = \left(50 \frac{\text{gal}}{\text{min}}\right)\left(0.002228 \frac{\text{ft}^3\text{-min}}{\text{sec-gal}}\right)$$
$$= 0.1114 \text{ ft}^3/\text{sec}$$

Rearrange Eq. 21.25 to find the hydraulic conductivity.

$$K = \frac{Q \ln \frac{r_1}{r_2}}{\pi(y_1^2 - y_2^2)}$$
$$= \frac{\left(50 \frac{\text{gal}}{\text{min}}\right)\left(1440 \frac{\text{min}}{\text{day}}\right) \ln \frac{2500 \text{ ft}}{0.375 \text{ ft}}}{\pi\left((100 \text{ ft})^2 - (94 \text{ ft})^2\right)}$$
$$= 173.4 \text{ gal/day-ft}^2$$

For the relined well,

$$r_2 = \frac{8 \text{ in}}{(2)\left(12 \frac{\text{in}}{\text{ft}}\right)} = 0.333 \text{ ft}$$

Use Eq. 21.25 to solve for the new flow rate.

$$Q = \frac{\pi K(y_1^2 - y_2^2)}{\ln \frac{r_1}{r_2}}$$
$$= \frac{\pi\left(173.4 \frac{\text{gal}}{\text{day-ft}^2}\right)\left((100 \text{ ft})^2 - (94 \text{ ft})^2\right)}{\ln \frac{2500 \text{ ft}}{0.333 \text{ ft}}}$$
$$= 71{,}057 \text{ gal/day}$$

12. UNSTEADY FLOW

When pumping first begins, the removed water also comes from the aquifer above the equilibrium cone of depression. Therefore, Eq. 21.26 cannot be used, and a nonequilibrium analysis is required.

Nonequilibrium solutions to well problems have been formulated in terms of dimensionless numbers. For small drawdowns compared with the initial thickness of the aquifer, the *Theis equation* for the drawdown at a distance r from the well and after pumping for time t is

$$s_{r,t} = \left(\frac{Q}{4\pi K Y}\right) W(u)$$

$$= \left(\frac{Q}{4\pi T}\right) W(u) \quad \begin{bmatrix} \text{consistent} \\ \text{units} \end{bmatrix} \qquad 21.28$$

$W(u)$ is a dimensionless *well function*. (See Table 21.3.) Though it is possible to obtain $W(u)$ from u, extracting u from $W(u)$ is more difficult. The relationship between u and $W(u)$ is often given in tabular or graphical forms. In Eq. 21.30, S is the aquifer *storage constant*.

$$W(u) = -0.577216 - \ln u + u - \frac{u^2}{(2)(2!)}$$

$$+ \frac{u^3}{(3)(3!)} - \frac{u^4}{(4)(4!)} + \cdots \qquad 21.29$$

$$u = \frac{r^2 S}{4 K Y t} = \frac{r^2 S}{4 T t} \quad \begin{bmatrix} \text{consistent} \\ \text{units} \end{bmatrix} \qquad 21.30$$

Accordingly, for any two different times in the pumping cycle,

$$s_1 - s_2 = y_2 - y_1 = \left(\frac{Q}{4\pi K Y}\right)\left(W(u_1) - W(u_2)\right) \qquad 21.31$$

If $u < 0.01$, then *Jacob's equation* can be used.

$$s_{r,t} = \left(\frac{Q}{4\pi T}\right) \ln \frac{2.25 T t}{r^2 S} \qquad 21.32$$

13. PUMPING POWER

Various types of pumps are used in wells. Problems with excessive suction lift are avoided by the use of submersible pumps.

Pumping power can be determined from hydraulic (water) power equations. The total head is the sum of static lift, velocity head, drawdown, pipe friction, and minor entrance losses from the casing, strainer, and screen. The Hazen-Williams equation is commonly used with a coefficient of $C = 100$ to determine the pipe friction.

14. FLOW NETS

Groundwater seepage is from locations of high hydraulic head to locations of lower hydraulic head. Relatively complex two-dimensional problems may be evaluated using a graphical technique that shows the decrease in hydraulic head along the flow path. The resulting graphic representation of pressure and flow path is called a *flow net*.

The flow net concept as discussed here is limited to cases where the flow is steady, two-dimensional, incompressible, and through a homogeneous medium, and where the liquid has a constant viscosity. This is the ideal case of groundwater seepage.

Flow nets are constructed from streamlines and equipotential lines. *Streamlines* (*flow lines*) show the path taken by the seepage. *Equipotential lines* are contour lines of constant driving (differential) hydraulic head. (This head does not include static head, which varies with depth.)

The object of a graphical flow net solution is to construct a network of flow paths (outlined by the streamlines) and equal pressure drops (bordered by equipotential lines).

Table 21.3 *Well Function W(u) for Various Values of u*

u	1.0	2.0	3.0	4.0	5.0	6.0	7.0	8.0	9.0
$\times 1$	0.219	0.049	0.013	0.0038	0.0011	0.00036	0.00012	0.000038	0.000012
$\times 10^{-1}$	1.82	1.22	0.91	0.70	0.56	0.45	0.37	0.31	0.26
$\times 10^{-2}$	4.04	3.35	2.96	2.68	2.47	2.30	2.15	2.03	1.92
$\times 10^{-3}$	6.33	5.64	5.23	4.95	4.73	4.54	4.39	4.26	4.14
$\times 10^{-4}$	8.63	7.94	7.53	7.25	7.02	6.84	6.69	6.55	6.44
$\times 10^{-5}$	10.94	10.24	9.84	9.55	9.33	9.14	8.99	8.86	8.74
$\times 10^{-6}$	13.24	12.55	12.14	11.85	11.63	11.45	11.29	11.16	11.04
$\times 10^{-7}$	15.54	14.85	14.44	14.15	13.93	13.75	13.60	13.46	13.34
$\times 10^{-8}$	17.84	17.15	16.74	16.46	16.23	16.05	15.90	15.76	15.65
$\times 10^{-9}$	20.15	19.45	19.05	18.76	18.54	18.35	18.20	18.07	17.95
$\times 10^{-10}$	22.45	21.76	21.35	21.06	20.84	20.66	20.50	20.37	20.25
$\times 10^{-11}$	24.75	24.06	23.65	23.36	23.14	22.96	22.81	22.67	22.55
$\times 10^{-12}$	27.05	26.36	25.96	25.67	25.44	25.26	25.11	24.97	24.86
$\times 10^{-13}$	29.36	28.66	28.26	27.97	27.75	27.56	27.41	27.28	27.16
$\times 10^{-14}$	31.66	30.97	30.56	30.27	30.05	29.87	29.71	29.58	29.46
$\times 10^{-15}$	33.96	33.27	32.86	32.58	32.35	32.17	32.02	31.88	31.76

Reprinted from L. K. Wenzel, "Methods for Determining Permeability of Water Bearing Materials with Special Reference to Discharging Well Methods," U.S. Geological Survey, 1942, Water-Supply Paper 887.

No fluid flows across streamlines, and a constant amount of fluid flows between any two streamlines.

Flow nets are constructed according to the following rules.

Rule 21.1: Streamlines enter and leave pervious surfaces perpendicular to those surfaces.

Rule 21.2: Streamlines approach the line of seepage (above which there is no hydrostatic pressure) asymptotically to (i.e., parallel but gradually approaching) that surface.

Rule 21.3: Streamlines are parallel to but cannot touch impervious surfaces that are streamlines.

Rule 21.4: Streamlines are parallel to the flow direction.

Rule 21.5: Equipotential lines are drawn perpendicular to streamlines such that the resulting cells are approximately square and the intersections are 90° angles. Theoretically, it should be possible to draw a perfect circle within each cell that touches all four boundaries, even though the cell is not actually square.

Rule 21.6: Equipotential lines enter and leave impervious surfaces perpendicular to those surfaces.

Many flow nets with differing degrees of detail can be drawn, and all will be more or less correct. Generally, three to five streamlines are sufficient for initial graphical evaluations. The size of the cells is determined by the number of intersecting streamlines and equipotential lines. As long as the rules are followed, the ratio of stream flow channels to equipotential drops will be approximately constant regardless of whether the grid is coarse or fine.

Figure 21.3 shows flow nets for several common cases. A careful study of the flow nets will help to clarify the rules and conventions previously listed.

15. SEEPAGE FROM FLOW NETS

Once a flow net is drawn, it can be used to calculate the seepage. First, the number of flow channels, N_f, between the streamlines is counted. Then, the number of equipotential drops, N_p, between equipotential lines is counted. The total hydraulic head, H, is determined as a function of the water surface levels.

$$Q = KH\left(\frac{N_f}{N_p}\right) \quad \text{[per unit width]} \qquad 21.33$$

$$H = H_1 - H_2 \qquad 21.34$$

Figure 21.3 *Typical Flow Nets*

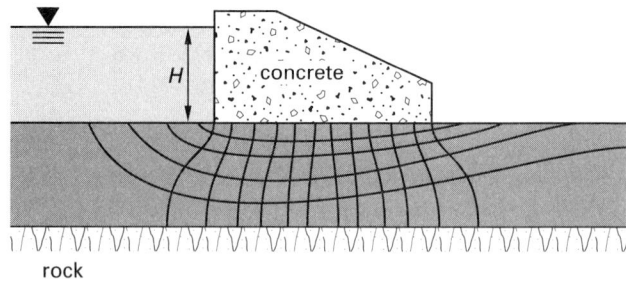
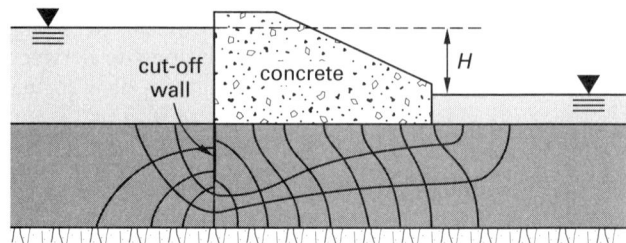
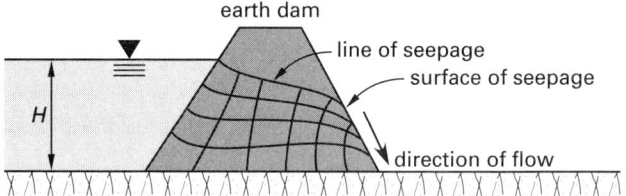
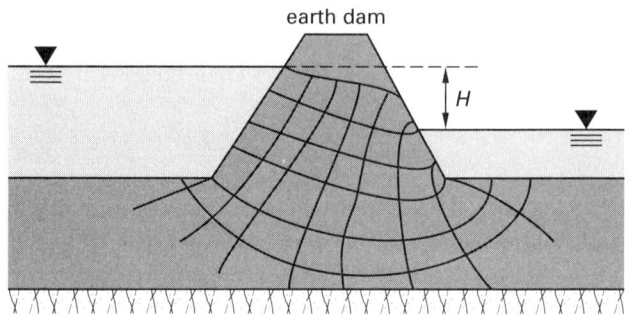
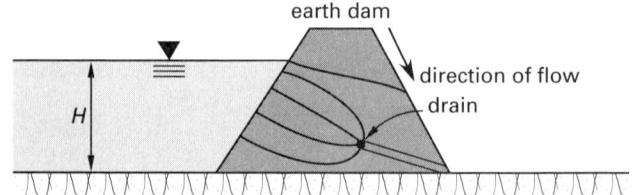

16. HYDROSTATIC PRESSURE ALONG FLOW PATH

The hydrostatic pressure at equipotential drop j (counting the last line on the downstream side as zero) in a flow net with a total of N_p equipotential drops is given

Flow of Fluids

by Eq. 21.35. If the point being investigated is along the bottom of a dam or other structure, the hydrostatic pressure is referred to as *uplift pressure*. Since uplift due to velocity is negligible, all uplift is due to the neutral pore pressure. Therefore, *neutral pressure* is synonymous with uplift pressure.

$$p_u = \left(\frac{j}{N_p}\right) H g \rho_w \qquad \text{[SI]} \qquad 21.35(a)$$

$$p_u = \left(\frac{j}{N_p}\right) H \gamma_w \qquad \text{[U.S.]} \qquad 21.35(b)$$

If the point being investigated is located a distance z below the datum, then the pressure at the point is given by Eq. 21.36. If the point being investigated is above the datum, z is negative.

$$p_u = \left(\left(\frac{j}{N_p}\right) H + z\right) g \rho_w \qquad \text{[SI]} \qquad 21.36(a)$$

$$p_u = \left(\left(\frac{j}{N_p}\right) H + z\right) \gamma_w \qquad \text{[U.S.]} \qquad 21.36(b)$$

The actual *uplift force* on a surface of area A is given by Eq. 21.37. N is the *neutral stress coefficient*, defined as the fraction of the surface that is exposed to the hydrostatic pressure. For soil, $N \approx 1$. However, for fractured rock, concrete, porous limestone, or sandstone, it varies from 0.5–1.0. For impervious materials such as marble or granite, it can be as low as 0.1.

$$U = N p_u A \qquad 21.37$$

As Table 21.1 indicates, some clay layers are impervious enough to prevent the passing of water. However, the hydrostatic uplift pressure below a clay layer can still be transmitted through the clay layer. The transfer geometry for thick layers is complex, making estimates of the amount of uplift force difficult. Upward forces on an impervious layer due to hydrostatic pressure below is known as *heave*. A factor of safety, FS, against heave of at least 1.5 is desirable.

$$(\text{FS})_{\text{heave}} = \frac{\text{downward pressure}}{\text{uplift pressure}} \qquad 21.38$$

17. INFILTRATION

Aquifers can be recharged (refilled) in a number of natural and artificial ways. The *Horton equation* gives a lower bound on the *infiltration capacity* (in inches (centimeters) of rainfall per unit time), which is the capacity of an aquifer to absorb water from a saturated source above as a function of time. When the rainfall supply exceeds the infiltration capacity, infiltration decreases exponentially over time. In Eq. 21.39, f_c is the final or

equilibrium capacity, f_0 is the initial infiltration capacity, and f_t is the infiltration capacity at time t. k is an infiltration decay constant with dimensions of 1/hr determined from a *double-ring infiltration test*.

$$f_t = f_c + (f_0 - f_c)e^{-kt} \qquad 21.39$$

The cumulative infiltration will not correspond to the increase in water table elevation due to the effects of porosity. However, the increase in water table elevation can be determined from a knowledge of the aquifer properties.

Infiltration rates for intermediate, silty soils are approximately 50% of those for sand. Clay infiltration rates are approximately 10% of those for sand. The actual infiltration rate at any time is equal (in the Horton model) to the smaller of the rainfall intensity, $i(t)$, and the instantaneous capacity.

$$f = i \quad [i \leq f_t] \qquad 21.40$$

$$f = f_t \quad [i > f_t] \qquad 21.41$$

The cumulative infiltration over time is

$$F_t = f_c t + \left(\frac{f_0 - f_c}{k}\right)\left(1 - e^{-kt}\right) \qquad 21.42$$

The average infiltration rate is found by dividing Eq. 21.42 by the duration of infiltration.

$$\bar{f} = \frac{F_t}{t_{\text{infiltration}}} \qquad 21.43$$

The depth of water stored in the layer, S, is equal to the total depth of infiltration, F_t, minus the depth of water that has leaked out the bottom of the layer. Except in very permeable soils (such as sand or gravel), water leaks out the bottom of the layer slowly. In that case, the water storage is

$$S = F_t - f_c t \qquad 21.44$$

The maximum depth of water that the layer can store is

$$S_{\max} = \frac{f_0 - f_c}{k} \qquad 21.45$$

Horton's equation assumes that the rate of precipitation, I, is greater than the infiltration rate, f_{in}, throughout the storm period. The overland flow, Q_{overland}, experienced when the rainfall rate is greater than the infiltration rate is

$$Q_{\text{overland}} = I - f_{\text{in}} \qquad 21.46$$

Flow of Fluids

The rate at which water is stored in the layer is equal to the infiltration rate minus the leakage outflow rate through the bottom of the layer.

$$\frac{dS}{dt} = f_{\text{in}} - f_{\text{out}}$$ 21.47

Equation 21.47 is easily solved for constant precipitation rates greater than the infiltration rate. If the precipitation rate is ever lower than the infiltration rate, the layer will lose water to lower levels, and Horton's theory must be modified. Rather than Eq. 21.47, finite difference techniques are used to calculate the infiltration rate and storage incrementally. The finite difference infiltration equation is given by Eq. 21.48. Of course, the infiltration can never exceed the rate of precipitation, no matter what the capacity.

$$S(t + \Delta t) = S(t) + \Delta t(f_{\text{in}} - f_c)$$ 21.48

Example 21.2

A soil has the following characteristics

initial infiltration capacity (f_0), 6.8 cm/h

equilibrium infiltration capacity (f_c), 0.3 cm/h

infiltration decay constant (k), 0.75 1/h

Calculate the (a) infiltration rate and (b) the total depth of infiltrated water after 3 hours of hard rain. (c) What is the maximum depth of water that the layer can store?

Solution

(a) Use Eq. 21.39 to calculate the infiltration rate.

$$f_t = f_c + (f_0 - f_c)e^{-kt}$$
$$= 0.3 \ \frac{\text{cm}}{\text{h}} + \left(6.8 \ \frac{\text{cm}}{\text{h}} - 0.3 \ \frac{\text{cm}}{\text{h}}\right) e^{-\left(0.75 \frac{1}{\text{h}}\right)(3 \text{ h})}$$
$$= 0.985 \text{ cm/h}$$

(b) Use Eq. 21.42 to calculate the total depth of infiltrated water.

$$F(t) = f_c t + \left(\frac{f_0 - f_c}{k}\right)\left(1 - e^{-kt}\right)$$
$$= \left(0.3 \ \frac{\text{cm}}{\text{h}}\right)(3 \text{ h})$$
$$+ \left(\frac{6.8 \ \frac{\text{cm}}{\text{h}} - 0.3 \ \frac{\text{cm}}{\text{h}}}{0.75 \ \frac{1}{\text{h}}}\right)\left(1 - e^{-\left(0.75 \frac{1}{\text{h}}\right)(3 \text{ h})}\right)$$
$$= 8.65 \text{ cm}$$

(c) The maximum storage is calculated from Eq. 21.45.

$$S_{\text{max}} = \frac{f_0 - f_c}{k} = \frac{6.8 \ \frac{\text{cm}}{\text{h}} - 0.3 \ \frac{\text{cm}}{\text{h}}}{0.75 \ \frac{1}{\text{h}}}$$
$$= 8.67 \text{ cm}$$

Topic III: Water Resources #2: Water and Wastewater Treatment

Water Treatment

22 Inorganic Chemistry

Nomenclature

A	atomic weight	lbm/lbmol	kg/kmol
C	concentration	n.a.	mg/L
EW	equivalent weight	lbm/lbmol	kg/kmol
F	formality	n.a.	FW/L
FW	formula weight	lbm/lbmol	kg/kmol
H	enthalpy	Btu/lbmol	kcal/mol
H	Henry's law constant	atm	atm
k	reaction rate constant	1/min	1/min
K	equilibrium constant	–	–
m	mass	lbm	kg
m	molality	n.a.	mol/1000 g
M	molarity	n.a.	mol/L
MW	molecular weight	lbm/lbmol	kg/kmol
n	number of moles	–	–
N	normality	n.a.	GEW/L
N_A	Avogadro's number	n.a.	1/mol
p	pressure	lbf/ft^2	Pa
R^*	universal gas constant, 1545.35 (8314.47)	ft-lbf/lbmol-°R	J/kmol·K
T	temperature	°R	K
v	rate of reaction	n.a.	mol/L·s
V	volume	ft^3	m^3
x	gravimetric fraction	–	–
x	mole fraction	–	–
x	relative abundance	–	–
X	fraction ionized	–	–
Z	atomic number	–	–

Subscripts

a	acid
A	Avogadro
b	base
eq	equilibrium
f	formation
r	reaction
sp	solubility product
t	total

1. ATOMIC STRUCTURE

An *element* is a substance that cannot be decomposed into simpler substances during ordinary chemical reactions.[1] An *atom* is the smallest subdivision of an element that can take part in a chemical reaction. A *molecule* is the smallest subdivision of an element or compound that can exist in a natural state.

The atomic nucleus consists of neutrons and protons, known as *nucleons*. The masses of neutrons and protons are essentially the same—one *atomic mass unit*, amu.

---

[1]Atoms of an element can be decomposed into subatomic particles in nuclear reactions.

One amu is exactly $^1/_{12}$ of the mass of an atom of carbon-12, approximately equal to 1.66×10^{-27} kg.[2] The *relative atomic weight* or *atomic weight*, A, of an atom is approximately equal to the number of protons and neutrons in the nucleus.[3] The *atomic number*, Z, of an atom is equal to the number of protons in the nucleus.

The atomic number and atomic weight of an element E are written in symbolic form as $_Z E^A$, E_Z^A, or $_Z^A E$. For example, carbon is the sixth element; radioactive carbon has an atomic mass of 14. Therefore, the symbol for carbon-14 is C_6^{14}. Since the atomic number is superfluous if the chemical symbol is given, the atomic number can be omitted (e.g., C^{14}).

2. ISOTOPES

Although an element can have only a single atomic number, atoms of that element can have different atomic weights. Many elements possess *isotopes*. The nuclei of isotopes differ from one another only in the number of neutrons. Isotopes behave the same way chemically.[4] Therefore, isotope separation must be done physically (e.g., by centrifugation or gaseous diffusion) rather than chemically.

Hydrogen has three isotopes. H_1^1 is *normal hydrogen* with a single proton nucleus. H_1^2 is known as *deuterium* (*heavy hydrogen*), with a nucleus of a proton and neutron. (This nucleus is known as a *deuteron*.) Finally, H_1^3 (*tritium*) has two neutrons in the nucleus. While normal hydrogen and deuterium are stable, tritium is radioactive. Many elements have more than one stable isotope. Tin, for example, has ten.

The *relative abundance*, x_i, of an isotope i is equal to the fraction of that isotope in a naturally occurring sample of the element. The *chemical atomic weight* is the weighted average of the isotope weights.

$$A_{\text{average}} = x_1 A_1 + x_2 A_2 + \cdots \qquad 22.1$$

3. PERIODIC TABLE

The *periodic table*, as shown in Table 22.1, is organized around the *periodic law:* The properties of the elements depend on the atomic structure and vary with the atomic number in a systematic way. Elements are arranged in order of increasing atomic numbers from left to right. Adjacent elements in horizontal rows differ

decidedly in both physical and chemical properties. However, elements in the same column have similar properties. Graduations in properties, both physical and chemical, are most pronounced in the *periods* (i.e., the horizontal rows).

The vertical columns are known as *groups*, numbered in Roman numerals. Elements in a group are called *cogeners*. Each vertical group except 0 and VIII has A and B subgroups (*families*). The elements of a family resemble each other more than they resemble elements in the other family of the same group. Graduations in properties are definite but less pronounced in vertical families. The trend in any family is toward more *metallic properties* as the atomic weight increases.

Metals (elements at the left end of the periodic chart) have low electron affinities and electronegativities, are reducing agents, form positive ions, and have positive oxidation numbers. They have high electrical conductivities, luster, generally high melting points, ductility, and malleability.

Nonmetals (elements at the right end of the periodic chart) have high electron affinities and electronegativities, are oxidizing agents, form negative ions, and have negative oxidation numbers. They are poor electrical conductors, have little or no luster, and form brittle solids. Of the common nonmetals, fluorine has the highest electronic affinity and electronegativity, with oxygen having the next highest values.

The *metalloids* (e.g., boron, silicon, germanium, arsenic, antimony, tellurium, and polonium) have characteristics of both metals and nonmetals. Electrically, they are semiconductors.

The electron-attracting power of an atom is called its *electronegativity*. Metals have low electronegativities. Group VIIA elements (fluorine, chlorine, etc.) are most strongly electronegative. The alkali metals (Group IA) are the most weakly electronegative. Generally, the most *electronegative elements* are those at the right ends of the periods. Elements with low electronegativities are found at the beginning (i.e., left end) of the periods. Electronegativity decreases as you go down a group.

Elements in the periodic table are often categorized into the following groups.

- *actinides:* same as actinons
- *actinons:* elements 90–103[5]
- *alkali metals:* group IA
- *alkaline earth metals:* group IIA
- *halogens:* group VIIA
- *heavy metals:* metals near the center of the chart
- *inner transition elements:* same as transition metals

---

[2]Until 1961, the atomic mass unit was defined as $^1/_{16}$ of the mass of one atom of oxygen-16.

[3]The term *weight* is used even though all chemical calculations involve mass. The atomic weight of an atom includes the mass of the electrons. Published *chemical atomic weights* of elements are averages of all the atomic weights of stable isotopes, taking into consideration the relative abundances of the isotopes.

[4]There are slight differences, known as *isotope effects*, in the chemical behavior of isotopes. These effects usually influence only the rate of reaction, not the kind of reaction.

[5]The *actinons* resemble element 89, *actinium*. Therefore, element 89 is sometimes included as an actinon.

Table 22.1 Periodic Table of the Elements (referred to Carbon-12)

The Periodic Table of Elements (Long Form)

The number of electrons in filled shells is shown in the column at the extreme left; the remaining electrons for each element are shown immediately below the symbol for each element. Atomic numbers are enclosed in brackets. Atomic weights (rounded, based on carbon-12) are shown above the symbols. Atomic weight values in parentheses are those of the isotopes of longest half-life for certain radioactive elements whose atomic weights cannot be precisely quoted without knowledge of origin of the element.

metals — transition metals — nonmetals

periods	IA	IIA	IIIB	IVB	VB	VIB	VIIB	VIII	VIII	VIII	IB	IIB	IIIA	IVA	VA	VIA	VIIA	0
1 / 0	1.00794 H[1] 1																	4.00260 He[2] 2
2 / 2	6.941 Li[3] 1	9.01218 Be[4] 2											10.811 B[5] 3	12.0107 C[6] 4	14.0067 N[7] 5	15.9994 O[8] 6	18.9984 F[9] 7	20.1797 Ne[10] 8
3 / 2,8	22.9898 Na[11] 1	24.3050 Mg[12] 2											26.9815 Al[13] 3	28.0855 Si[14] 4	30.9738 P[15] 5	32.065 S[16] 6	35.453 Cl[17] 7	39.948 Ar[18] 8
4 / 2,8	39.0983 K[19] 8,1	40.078 Ca[20] 8,2	44.9559 Sc[21] 9,2	47.867 Ti[22] 10,2	50.9415 V[23] 11,2	51.9961 Cr[24] 13,1	54.9380 Mn[25] 13,2	55.845 Fe[26] 14,2	58.9332 Co[27] 15,2	58.6934 Ni[28] 16,2	63.546 Cu[29] 18,1	65.38 Zn[30] 18,2	69.723 Ga[31] 18,3	72.64 Ge[32] 18,4	74.9216 As[33] 18,5	78.96 Se[34] 18,6	79.904 Br[35] 18,7	83.798 Kr[36] 18,8
5 / 2,8,18	85.4678 Rb[37] 8,1	87.62 Sr[38] 8,2	88.9059 Y[39] 9,2	91.224 Zr[40] 10,2	92.9064 Nb[41] 12,1	95.96 Mo[42] 13,1	(98) Tc[43] 14,1	101.07 Ru[44] 15,1	102.906 Rh[45] 16,1	106.42 Pd[46] 18	107.868 Ag[47] 18,1	112.411 Cd[48] 18,2	114.818 In[49] 18,3	118.710 Sn[50] 18,4	121.760 Sb[51] 18,5	127.60 Te[52] 18,6	126.904 I[53] 18,7	131.293 Xe[54] 18,8
6 / 2,8,18	132.905 Cs[55] 18,8,1	137.327 Ba[56] 18,8,2	* (57-71)	178.49 Hf[72] 32,10,2	180.948 Ta[73] 32,11,2	183.84 W[74] 32,12,2	186.207 Re[75] 32,13,2	190.23 Os[76] 32,14,2	192.217 Ir[77] 32,15,2	195.084 Pt[78] 32,17,1	196.967 Au[79] 32,18,1	200.59 Hg[80] 32,18,2	204.383 Tl[81] 32,18,3	207.2 Pb[82] 32,18,4	208.980 Bi[83] 32,18,5	(209) Po[84] 32,18,6	(210) At[85] 32,18,7	(222) Rn[86] 32,18,8
7 / 2,8,18,32	(223) Fr[87] 18,8,1	(226) Ra[88] 18,8,2	† (89-103)	(265) Rf[104] 32,10,2	(268) Db[105] 32,11,2	(271) Sg[106] 32,12,2	(272) Bh[107] 32,13,2	(270) Hs[108] 32,14,2	(276) Mt[109] 32,15,2	(281) Ds[110] 32,17,1	(280) Rg[111] 32,18,1	(285) Cn[112] 32,18,2						

*lathanide series:
138.905 La[57] 18,9,2 | 140.116 Ce[58] 20,8,2 | 140.908 Pr[59] 21,8,2 | 144.242 Nd[60] 22,8,2 | (145) Pm[61] 23,8,2 | 150.36 Sm[62] 24,8,2 | 151.964 Eu[63] 25,8,2 | 157.25 Gd[64] 25,9,2 | 158.925 Tb[65] 27,8,2 | 162.500 Dy[66] 28,8,2 | 164.930 Ho[67] 29,8,2 | 167.259 Er[68] 30,8,2 | 168.934 Tm[69] 31,8,2 | 173.054 Yb[70] 32,8,2 | 174.967 Lu[71] 32,9,2

†actinide series:
(227) Ac[89] 18,9,2 | 232.038 Th[90] 18,10,2 | 231.036 Pa[91] 20,9,2 | 238.029 U[92] 21,9,2 | (237) Np[93] 23,8,2 | (244) Pu[94] 24,8,2 | (243) Am[95] 25,8,2 | (247) Cm[96] 25,9,2 | (247) Bk[97] 26,9,2 | (251) Cf[98] 28,8,2 | (252) Es[99] 29,8,2 | (257) Fm[100] 30,8,2 | (258) Md[101] 31,8,2 | (259) No[102] 32,8,2 | (262) Lr[103] 32,9,2

Water Treatment

- *lanthanides:* same as lanthanons
- *lanthanons:* elements 58–71[6]
- *light metals:* elements in the first two groups
- *metals:* everything except the nonmetals
- *metalloids:* elements along the dark line in the chart separating metals and nonmetals
- *noble gases:* group 0
- *nonmetals:* elements 2, 5–10, 14–18, 33–36, 52–54, 85, and 86
- *rare earths:* same as lanthanons
- *transition elements:* same as transition metals
- *transition metals:* all B families and group VIII B[7]

4. OXIDATION NUMBER

The *oxidation number* (*oxidation state*) is an electrical charge assigned by a set of prescribed rules. It is actually the charge assuming all bonding is ionic. The sum of the oxidation numbers equals the net charge. For monoatomic ions, the oxidation number is equal to the charge. The oxidation numbers of some common ions and radicals are given in Table 22.2.

In covalent compounds, all of the bonding electrons are assigned to the ion with the greater electronegativity. For example, non-metals are more electronegative than metals. Carbon is more electronegative than hydrogen.

For atoms in a free-state molecule, the oxidation number is zero. Hydrogen gas is a diatomic molecule, H_2. Therefore, the oxidation number of the hydrogen molecule, H_2, is zero. The same is true for the atoms in O_2, N_2, Cl_2, and so on. Also, the sum of all the oxidation numbers of atoms in a neutral molecule is zero.

Fluorine is the most electronegative element, and it has an oxidation number of -1. Oxygen is second only to fluorine in electronegativity. Usually, the oxidation number of oxygen is -2, except in peroxides, where it is -1, and when combined with fluorine, where it is $+2$. Hydrogen is usually $+1$, except in hydrides, where it is -1.

For a charged *radical* (a group of atoms that combine as a single unit), the net oxidation number is equal to the charge on the radical, known as the *charge number*.

Example 22.1

What are the oxidation numbers of all the elements in the chlorate ($ClO_3{}^{-1}$) and permanganate ($MnO_4{}^{-1}$) ions?

---

[6]The *lanthanons* resemble element 57, *lanthanum*. Therefore, element 57 is sometimes included as a lanthanon.

[7]The *transition metals* are elements whose electrons occupy the d sublevel. They can have various oxidation numbers, including +2, +3, +4, +6, and +7.

Table 22.2 *Oxidation Numbers of Selected Atoms and Charge Numbers of Radicals*

name	symbol	oxidation or charge number
acetate	$C_2H_3O_2$	-1
aluminum	Al	$+3$
ammonium	NH_4	$+1$
barium	Ba	$+2$
borate	BO_3	-3
boron	B	$+3$
bromine	Br	-1
calcium	Ca	$+2$
carbon	C	$+4, -4$
carbonate	CO_3	-2
chlorate	ClO_3	-1
chlorine	Cl	-1
chlorite	ClO_2	-1
chromate	CrO_4	-2
chromium	Cr	$+2, +3, +6$
copper	Cu	$+1, +2$
cyanide	CN	-1
dichromate	Cr_2O_7	-2
fluorine	F	-1
gold	Au	$+1, +3$
hydrogen	H	$+1$ (-1 in hydrides)
hydroxide	OH	-1
hypochlorite	ClO	-1
iron	Fe	$+2, +3$
lead	Pb	$+2, +4$
lithium	Li	$+1$
magnesium	Mg	$+2$
mercury	Hg	$+1, +2$
nickel	Ni	$+2, +3$
nitrate	NO_3	-1
nitrite	NO_2	-1
nitrogen	N	$-3, +1, +2, +3, +4, +5$
oxygen	O	-2 (-1 in peroxides)
perchlorate	ClO_4	-1
permanganate	MnO_4	-1
phosphate	PO_4	-3
phosphorus	P	$-3, +3, +5$
potassium	K	$+1$
silicon	Si	$+4, -4$
silver	Ag	$+1$
sodium	Na	$+1$
sulfate	SO_4	-2
sulfite	SO_3	-2
sulfur	S	$-2, +4, +6$
tin	Sn	$+2, +4$
zinc	Zn	$+2$

Solution

For the chlorate ion, the oxygen is more electronegative than the chlorine. (Only fluorine is more electronegative than oxygen.) Therefore, the oxidation number of oxygen is -2. In order for the net oxidation number to be -1, the chlorine must have an oxidation number of $+5$.

For the permanganate ion, the oxygen is more electronegative than the manganese. Therefore, the oxidation number of oxygen is −2. For the net oxidation number to be −1, the manganese must have an oxidation number of +7.

5. FORMATION OF COMPOUNDS

Compounds form according to the *law of definite (constant) proportions:* A pure compound is always composed of the same elements combined in a definite proportion by mass. For example, common table salt is always NaCl. It is not sometimes NaCl and other times Na_2Cl or $NaCl_3$ (which do not exist, in any case).

Furthermore, compounds form according to the *law of (simple) multiple proportions:* When two elements combine to form more than one compound, the masses of one element that combine with the same mass of the other are in the ratios of small integers.

In order to evaluate whether a compound formula is valid, it is necessary to know the *oxidation numbers* of the interacting atoms. Although some atoms have more than one possible oxidation number, most do not.

The sum of the oxidation numbers must be zero if a neutral compound is to form. For example, H_2O is a valid compound since the two hydrogen atoms have a total positive oxidation number of $2 \times 1 = +2$. The oxygen ion has an oxidation number of −2. These oxidation numbers sum to zero.

On the other hand, $NaCO_3$ is not a valid compound formula. The sodium (Na) ion has an oxidation number of +1. However, the carbonate radical has a charge number of −2. The correct sodium carbonate molecule is Na_2CO_3.

6. NAMING COMPOUNDS

Combinations of elements are known as *compounds*. *Binary compounds* contain two elements; *ternary (tertiary) compounds* contain three elements. A *chemical formula* is a representation of the relative numbers of each element in the compound. For example, the formula $CaCl_2$ shows that there are one calcium atom and two chlorine atoms in one molecule of calcium chloride.

Generally, the numbers of atoms are reduced to their lowest terms. However, there are exceptions. For example, acetylene is C_2H_2, and hydrogen peroxide is H_2O_2.

For binary compounds with a metallic element, the positive metallic element is listed first. The chemical name ends in the suffix "-ide." For example, NaCl is sodium chloride. If the metal has two oxidation states, the suffix "-ous" is used for the lower state, and "-ic" is used for the higher state. Alternatively, the element name can be used with the oxidation number written in Roman numerals. For example,

$FeCl_2$: ferrous chloride, or iron (II) chloride
$FeCl_3$: ferric chloride, or iron (III) chloride

For binary compounds formed between two nonmetals, the more positive element is listed first. The number of atoms of each element is specified by the prefixes "di-" (2), "tri-" (3), "tetra-" (4), and "penta-" (5), and so on. For example,

N_2O_5: dinitrogen pentoxide

Binary acids start with the prefix "hydro-," list the name of the nonmetallic element, and end with the suffix "-ic." For example,

HCl: hydrochloric acid

Ternary compounds generally consist of an element and a radical. The positive part is listed first in the formula. *Ternary acids* (also known as *oxyacids*) usually contain hydrogen, a nonmetal, and oxygen, and can be grouped into families with different numbers of oxygen atoms. The most common acid in a family (i.e., the root acid) has the name of the nonmetal and the suffix "-ic." The acid with one more oxygen atom than the root is given the prefix "per-" and the suffix "-ic." The acid containing one less oxygen atom than the root is given the ending "-ous." The acid containing two less oxygen atoms than the root is given the prefix "hypo-" and the suffix "-ous."

For example,

HClO: hypochlorous acid
$HClO_2$: chlorous acid
$HClO_3$: chloric acid (the root)
$HClO_4$: perchloric acid

7. MOLES AND AVOGADRO'S LAW

The *mole* is a measure of the quantity of an element or compound. Specifically, a mole of an element will have a mass equal to the element's atomic (or molecular) weight. The three main types of moles are based on mass being measured in grams, kilograms, and pounds. Obviously, a gram-based mole of carbon (12.0 grams) is not the same quantity as a pound-based mole of carbon (12.0 pounds). Although "mol" is understood in SI countries to mean a gram-mole, the term *mole* is ambiguous, and the units mol (gmol), kmol (kgmol), or lbmol must be specified.[8]

One gram-mole of any substance has the same number of particles (atoms, molecules, ions, electrons, etc.), 6.022×10^{23}, *Avogadro's number*, N_A. A pound-mole contains approximately 454 times the number of particles in a gram-mole.

Avogadro's law (hypothesis) holds that equal volumes of all gases at the same temperature and pressure contain equal numbers of gas molecules. Specifically, at standard scientific conditions (1.0 atm and 0°C), one

---

[8]There are also variations on the presentation of these units, such as g mol, gmole, g-mole, kmole, kg-mol, lb-mole, pound-mole, and p-mole. In most cases, the intent is clear.

gram-mole of any gas contains 6.022×10^{23} molecules and occupies 22.4 L. A pound-mole occupies 454 times that volume, 359 ft^3.

"Molar" is used as an adjective when describing properties of a mole. For example, a *molar volume* is the volume of a mole.

Example 22.2

How many electrons are in 0.01 g of gold? ($A = 196.97$; $Z = 79$.)

Solution

The number of gram-moles of gold present is

$$n = \frac{m}{MW} = \frac{0.01 \text{ g}}{196.97 \frac{\text{g}}{\text{mol}}} = 5.077 \times 10^{-5} \text{ mol}$$

The number of gold nuclei is

$$N = nN_A = \left(5.077 \times 10^{-5} \text{ mol}\right)\left(6.022 \times 10^{23} \frac{\text{nuclei}}{\text{mol}}\right)$$
$$= 3.057 \times 10^{19} \text{ nuclei}$$

Since the atomic number is 79, there are 79 protons and 79 electrons in each gold atom. The number of electrons is

$$N_{\text{electrons}} = (3.057 \times 10^{19})(79) = 2.42 \times 10^{21}$$

8. FORMULA AND MOLECULAR WEIGHTS

The *formula weight*, FW, of a molecule (compound) is the sum of the atomic weights of all elements in the molecule. The *molecular weight*, MW, is generally the same as the formula weight. The units of molecular weight are g/mol, kg/kmol, or lbm/lbmol. However, units are sometimes omitted because weights are relative. For example,

$$CaCO_3: FW = MW = 40.1 + 12 + 3 \times 16 = 100.1$$

An *ultimate analysis* (which determines how much of each element is present in a compound) will not necessarily determine the molecular formula. It will determine only the formula weight based on the relative proportions of each element. Therefore, except for hydrated molecules and other linked structures, the molecular weight will be an integer multiple of the formula weight.

For example, an ultimate analysis of hydrogen peroxide (H_2O_2) will show that the compound has one oxygen atom for each hydrogen atom. In this case, the formula would be assumed to be HO and the formula weight would be approximately 17, although the actual molecular weight is 34.

For *hydrated molecules* (e.g., $FeSO_4 \cdot 7H_2O$), the mass of the *water of hydration* (also known as the *water of crystallization*) is included in the formula and in the molecular weight.

9. EQUIVALENT WEIGHT

The *equivalent weight*, EW (i.e., an *equivalent*), is the amount of substance (in grams) that supplies one gram-mole (i.e., 6.022×10^{23}) of reacting units. For acid-base reactions, an acid equivalent supplies one gram-mole of H$^+$ ions. A base equivalent supplies one gram-mole of OH$^-$ ions. In oxidation-reduction reactions, an equivalent of a substance gains or loses a gram-mole of electrons. Similarly, in electrolysis reactions an equivalent weight is the weight of substance that either receives or donates one gram-mole of electrons at an electrode.

The equivalent weight can be calculated as the molecular weight divided by the change in oxidation number experienced by a compound in a chemical reaction. A compound can have several equivalent weights.

$$EW = \frac{MW}{\Delta \text{ oxidation number}} \qquad 22.2$$

Example 22.3

What are the equivalent weights of the following compounds?

(a) Al in the reaction

$$Al^{+++} + 3e^- \rightarrow Al$$

(b) H_2SO_4 in the reaction

$$H_2SO_4 + H_2O \rightarrow 2H^+ + SO_4^{-2} + H_2O$$

(c) NaOH in the reaction

$$NaOH + H_2O \rightarrow Na^+ + OH^- + H_2O$$

Solution

(a) The atomic weight of aluminum is approximately 27. Since the change in the oxidation number is 3, the equivalent weight is $27/3 = 9$.

(b) The molecular weight of sulfuric acid is approximately 98. Since the acid changes from a neutral molecule to ions with two charges each, the equivalent weight is $98/2 = 49$.

(c) Sodium hydroxide has a molecular weight of approximately 40. The originally neutral molecule goes to a singly charged state. Therefore, the equivalent weight is $40/1 = 40$.

10. GRAVIMETRIC FRACTION

The *gravimetric fraction*, x_i, of an element i in a compound is the fraction by weight of that element in the compound. The gravimetric fraction is found from an *ultimate analysis* (also known as a *gravimetric analysis*) of the compound.

$$x_i = \frac{m_i}{m_1 + m_2 + \cdots + m_i + \cdots + m_n} = \frac{m_i}{m_t} \qquad 22.3$$

The *percentage composition* is the gravimetric fraction converted to percentage.

$$\% \text{ composition} = x_i \times 100\% \qquad 22.4$$

If the gravimetric fractions are known for all elements in a compound, the *combining weights* of each element can be calculated. (The term *weight* is used even though mass is the traditional unit of measurement.)

$$m_i = x_i m_t \qquad 22.5$$

11. EMPIRICAL FORMULA DEVELOPMENT

It is relatively simple to determine the *empirical formula* of a compound from the atomic and combining weights of elements in the compound. The empirical formula gives the relative number of atoms (i.e., the formula weight is calculated from the empirical formula).

step 1: Divide the gravimetric fractions (or percentage compositions) by the atomic weight of each respective element.

step 2: Determine the smallest ratio from step 1.

step 3: Divide all of the ratios from step 1 by the smallest ratio.

step 4: Write the chemical formula using the results from step 3 as the numbers of atoms. Multiply through as required to obtain all integer numbers of atoms.

Example 22.4

A clear liquid is analyzed, and the following gravimetric percentage compositions are recorded: carbon, 37.5%; hydrogen, 12.5%; oxygen, 50%. What is the chemical formula for the liquid?

Solution

step 1: Divide the percentage compositions by the atomic weights.

$$\text{C:} \quad \frac{37.5}{12} = 3.125$$

$$\text{H:} \quad \frac{12.5}{1} = 12.5$$

$$\text{O:} \quad \frac{50}{16} = 3.125$$

step 2: The smallest ratio is 3.125.

step 3: Divide all ratios by 3.125.

$$\text{C:} \quad \frac{3.125}{3.125} = 1$$

$$\text{H:} \quad \frac{12.5}{3.125} = 4$$

$$\text{O:} \quad \frac{3.125}{3.125} = 1$$

step 4: The empirical formula is CH_4O.

If it had been known that the liquid behaved as though it contained a hydroxyl (OH) radical, the formula would have been written as CH_3OH. This is recognized as methyl alcohol.

12. CHEMICAL REACTIONS

During chemical reactions, bonds between atoms are broken and new bonds are usually formed. The starting substances are known as *reactants*; the ending substances are known as *products*. In a chemical reaction, reactants are either converted to simpler products or synthesized into more complex compounds. There are four common types of reactions.

- *direct combination* (or *synthesis*): This is the simplest type of reaction where two elements or compounds combine directly to form a compound.

$$2H_2 + O_2 \rightarrow 2H_2O$$

$$SO_2 + H_2O \rightarrow H_2SO_3$$

- *decomposition* (or *analysis*): Bonds within a compound are disrupted by heat or other energy to produce simpler compounds or elements.

$$2HgO \rightarrow 2Hg + O_2$$

$$H_2CO_3 \rightarrow H_2O + CO_2$$

- *single displacement* (or *replacement*[9]): This type of reaction has one element and one compound as reactants.

$$2Na + 2H_2O \rightarrow 2NaOH + H_2$$

$$2KI + Cl_2 \rightarrow 2KCl + I_2$$

- *double displacement* (or *replacement*): These are reactions with two compounds as reactants and two compounds as products.

$$AgNO_3 + NaCl \rightarrow AgCl + NaNO_3$$

$$H_2SO_4 + ZnS \rightarrow H_2S + ZnSO_4$$

13. BALANCING CHEMICAL EQUATIONS

The coefficients in front of element and compound symbols in chemical reaction equations are the numbers of molecules or moles taking part in the reaction.

---

[9]Another name for replacement is *metathesis*.

(For gaseous reactants and products, the coefficients also represent the numbers of volumes. This is a direct result of Avogadro's hypothesis that equal numbers of molecules in the gas phase occupy equal volumes under the same conditions.)[10]

Since atoms cannot be changed in a normal chemical reaction (i.e., mass is conserved), the numbers of each element must match on both sides of the equation. When the numbers of each element match, the equation is said to be "balanced." The total atomic weights on both sides of the equation will be equal when the equation is balanced.

Balancing simple chemical equations is largely a matter of deductive trial and error. More complex reactions require use of oxidation numbers.

Example 22.5

Balance the following reaction equation.

$$Al + H_2SO_4 \rightarrow Al_2(SO_4)_3 + H_2$$

Solution

As written, the reaction is not balanced. For example, there is one aluminum on the left, but there are two on the right. The starting element in the balancing procedure is chosen somewhat arbitrarily.

step 1: Since there are two aluminums on the right, multiply Al by 2.

$$2Al + H_2SO_4 \rightarrow Al_2(SO_4)_3 + H_2$$

step 2: Since there are three sulfate radicals (SO_4) on the right, multiply H_2SO_4 by 3.

$$2Al + 3H_2SO_4 \rightarrow Al_2(SO_4)_3 + H_2$$

step 3: Now there are six hydrogens on the left, so multiply H_2 by 3 to balance the equation.

$$2Al + 3H_2SO_4 \rightarrow Al_2(SO_4)_3 + 3H_2$$

14. STOICHIOMETRIC REACTIONS

Stoichiometry is the study of the proportions in which elements and compounds react and are formed. A *stoichiometric reaction* (also known as a *perfect reaction* or an *ideal reaction*) is one in which just the right amounts of reactants are present. After the reaction stops, there are no unused reactants.

---

[10]When water is part of the reaction, the interpretation that the coefficients are volumes is valid only if the reaction takes place at a high enough temperature to vaporize the water.

Stoichiometric problems are known as *weight and proportion problems* because their solutions use simple ratios to determine the masses of reactants required to produce given masses of products, or vice versa. The procedure for solving these problems is essentially the same regardless of the reaction.

step 1: Write and balance the chemical equation.

step 2: Determine the atomic (molecular) weight of each element (compound) in the equation.

step 3: Multiply the atomic (molecular) weights by their respective coefficients and write the products under the formulas.

step 4: Write the given mass data under the weights determined in step 3.

step 5: Fill in the missing information by calculating simple ratios.

Example 22.6

Caustic soda (NaOH) is made from sodium carbonate (Na_2CO_3) and slaked lime ($Ca(OH)_2$) according to the given reaction. How many kilograms of caustic soda can be made from 2000 kg of sodium carbonate?

Solution

	Na_2CO_3	+ $Ca(OH)_2$	$\rightarrow$ 2NaOH	+ $CaCO_3$
molecular weights	106	74	2×40	100
given data	2000 kg		m kg	

The simple ratio used is

$$\frac{NaOH}{Na_2CO_3} = \frac{80}{106} = \frac{m}{2000 \text{ kg}}$$

Solving for the unknown mass, $m = 1509$ kg.

15. NONSTOICHIOMETRIC REACTIONS

In many cases, it is not realistic to assume a stoichiometric reaction because an excess of one or more reactants is necessary to assure that all of the remaining reactants take part in the reaction. Combustion is an example where the stoichiometric assumption is, more often than not, invalid. Excess air is generally needed to ensure that all of the fuel is burned.

With nonstoichiometric reactions, the reactant that is used up first is called the *limiting reactant*. The amount of product will be dependent on (limited by) the limiting reactant.

The *theoretical yield* or *ideal yield* of a product is the maximum mass of product per unit mass of limiting reactant that can be obtained from a given reaction if the reaction goes to completion. The *percentage yield* is a measure of the efficiency of the actual reaction.

$$\text{percentage yield} = \frac{\text{actual yield} \times 100\%}{\text{theoretical yield}} \qquad 22.6$$

16. SOLUTIONS OF GASES IN LIQUIDS

When a liquid is exposed to a gas, a small amount of the gas will dissolve in the liquid. Diffusion alone is sufficient for this to occur; bubbling or collecting the gas over a liquid is not necessary. Given enough time, at equilibrium, the concentration of the gas will reach a maximum known as the *saturation concentration*. Due to the large amount of liquid compared to the small amount of dissolved gas, a liquid exposed to multiple gases will eventually become saturated by all of the gases; the presence of one gas does not affect the solubility of another gas.

The characteristics of a solution of one or more gases in a liquid is predicted by *Henry's law*. In one formulation specifically applicable to liquids exposed to mixtures of gases, Henry's law states that, at equilibrium, the partial pressure, p_i, of a gas in a mixture will be proportional to the gas mole fraction, x_i, of that dissolved gas in solution. In Eq. 22.7, Henry's law constant, H, (see Table 22.3) has units of pressure, typically reported in the literature in atmospheres (same as atm/mole fraction).

$$p_i = H_i x_i \qquad 22.7$$

Table 22.3 *Approximate Values of Henry's Law Constant (solutions of gases in water)*

gas	Henry's law constant, H (atm) (Multiply all values by 10^3.)	
	20°C	30°C
CO	53.6	62.0
CO_2	1.42	1.86
H_2S	48.3	60.9
N_2	80.4	92.4
NO	26.4	31.0
O_2	40.1	47.5
SO_2	0.014	0.016

Adapted from *Scrubber Systems Operational Review* (APTI Course SI:412C), Second Edition, *Lesson 11: Design Review of Absorbers Used for Gaseous Pollutants*, 1998, North Carolina State University for the U.S. Environmental Protection Agency.

Since, for mixtures of ideal gases, the mole fraction, x_i, volumetric fraction, B_i, and partial pressure fraction, p_i/p_{total}, all have the same numerical values, these measures can all be integrated into Henry's law.

Henry's law is stated in several incompatible formulations, and the corresponding equations and Henry's law constants are compatible only with their own formulations. Also, a variety of variables are used for Henry's

law constant, including H, k_H, K_H, and so on. The context and units of the Henry's law constant must be used to determine Henry's law.

equation	typical units of Henry's law constant	Henry's law statement (at equilibrium)
$p_i = H_i x_i$	H: pressure (atm)	Partial pressure is proportional to mole fraction.
$p_i = k_{H(p/C),i} C_i$	$k_{H(p/C)}$: pressure divided by concentration (atm·L/mol; atm·L/mg)	Partial pressure is proportional to concentration.[*]
$p_i = \dfrac{C_i}{k_{H(C/p),i}}$	$k_{H(C/p)}$: concentration divided by pressure (mol/atm·L; mg/atm·L)	
$C_{i,gas} = \alpha_i C_{i,liquid}$	α: dimensionless (L_{gas}/L_{liquid})	Concentration in the gas mixture is proportional to concentration in the liquid solution.

[*]The statements of Henry's law are the same. However, the values of Henry's law constants are inverses.

The dimensionless form of the Henry's law constant is also known as the *absorption coefficient*, *coefficient of absorption*, and *solubility coefficient*. It represents the volume of a gas at a specific temperature and pressure that can be dissolved in a unit volume of liquid. Typical units are L/L (dimensionless). Approximate values for gases in water at 1 atm and 20°C are: CO, 0.023; CO_2, 0.88; He, 0.009; H_2, 0.017; H_2S, 2.62; N_2, 0.015; NH_3, 710; O_2, 0.028.

The amount of gas dissolved in a liquid varies with the temperature of the liquid and the concentration of dissolved salts in the liquid. Generally, the solubility of gases in liquids decreases with increasing temperature. Appendix 22.D lists the saturation values of dissolved oxygen in water at various temperatures and for various amounts of chloride ion (also referred to as *salinity*).

Example 22.7

At 20°C and 1 atm, 1 L of water can absorb 0.043 g of oxygen and 0.017 g of nitrogen. Atmospheric air is 20.9% oxygen by volume, and the remainder is assumed to be nitrogen. What masses of oxygen and nitrogen will be absorbed by 1 L of water exposed to 20°C air at 1 atm?

Solution

Since partial pressure is volumetrically weighted,

$$m_{\text{oxygen}} = (0.209)\left(0.043 \ \frac{\text{g}}{\text{L}}\right)$$

$$= 0.009 \ \text{g/L}$$

$$m_{\text{nitrogen}} = (1.000 - 0.209)\left(0.017 \ \frac{\text{g}}{\text{L}}\right)$$

$$= 0.0134 \ \text{g/L}$$

Example 22.8

At an elevation of 4000 ft, the barometric pressure is 660 mm Hg. What is the dissolved oxygen concentration of 18°C water with a 800 mg/L chloride concentration at that elevation?

Solution

From App. 22.D, oxygen's saturation concentration for 18°C water corrected for a 800 mg/L chloride concentration is

$$C_s = 9.54 \ \frac{\text{mg}}{\text{L}} - \left(\frac{800 \ \frac{\text{mg}}{\text{L}}}{100 \ \frac{\text{mg}}{\text{L}}}\right)\left(0.009 \ \frac{\text{mg}}{\text{L}}\right)$$

$$= 9.44 \ \text{mg/L}$$

Use the appendix footnote to correct for the barometric pressure.

$$C'_s = \left(9.44 \ \frac{\text{mg}}{\text{L}}\right)\left(\frac{660 \ \text{mm} - 16 \ \text{mm}}{760 \ \text{mm} - 16 \ \text{mm}}\right)$$

$$= 8.17 \ \text{mg/L}$$

17. PROPERTIES OF SOLUTIONS

There are very few convenient ways of predicting the properties of nonreacting, nonvolatile organic and aqueous solutions (acids, brines, alcohol mixtures, coolants, etc.) from the individual properties of the components.

Volumes of two nonreacting organic liquids (e.g., acetone and chloroform) in a mixture are essentially additive. The volume change upon mixing will seldom be more than a few tenths of a percent. The volume change in aqueous solutions is often slightly greater, but is still limited to a few percent (e.g., 3% for some solutions of methanol and water). Therefore, the specific gravity (density, specific weight, etc.) can be considered to be a volumetric weighting of the individual specific gravities.

Most other fluid properties of aqueous solutions, such as viscosity, compressibility, surface tension, and vapor pressure, must be measured.

18. SOLUTIONS OF SOLIDS IN LIQUIDS

When a solid is added to a liquid, the solid is known as the *solute*, and the liquid is known as the *solvent*.[11] If the dispersion of the solute throughout the solvent is at the molecular level, the mixture is known as a *solution*. If the solute particles are larger than molecules, the mixture is known as a *suspension*.[12]

In some solutions, the solvent and solute molecules bond loosely together. This loose bonding is known as *solvation*. If water is the solvent, the bonding process is also known as *aquation* or *hydration*.

The solubility of most solids in liquid solvents usually increases with increasing temperature. Pressure has very little effect on the solubility of solids in liquids.

When the solvent has absorbed as much solute as it can, it is a *saturated solution*.[13] Adding more solute to an already saturated solution will cause the excess solute to settle to the bottom of the container, a process known as *precipitation*. Other changes (in temperature, concentration, etc.) can be made to cause precipitation from saturated and unsaturated solutions. Precipitation in a chemical reaction is indicated by a downward arrow (i.e., "↓"). For example, the precipitation of silver chloride from an aqueous solution of silver nitrate ($AgNO_3$) and potassium chloride (KCl) would be written:

$$AgNO_3(aq) + KCl(aq) \rightarrow AgCl(s)\downarrow + KNO_3(aq)$$

19. UNITS OF CONCENTRATION

Several units of concentration are commonly used to express solution strengths.

 F— formality: The number of gram formula weights (i.e., molecular weights in grams) per liter of solution.

 m— molality: The number of gram-moles of solute per 1000 grams of solvent. A "molal" solution contains 1 gram-mole per 1000 grams of solvent.

 M— molarity: The number of gram-moles of solute per liter of solution. A "molar" (i.e., 1 M) solution contains 1 gram-mole per liter of solution. Molarity is related to normality as shown in Eq. 22.8.

$$N = M \times \Delta \, \text{oxidation number} \qquad \textit{22.8}$$

---

[11]The term *solvent* is often associated with volatile liquids, but the term is more general than that. (A *volatile liquid* evaporates rapidly and readily at normal temperatures.) Water is the solvent in aqueous solutions.

[12]An *emulsion* is not a mixture of a solid in a liquid. It is a mixture of two immiscible liquids.

[13]Under certain circumstances, a *supersaturated solution* can exist for a limited amount of time.

N— *normality:* The number of gram equivalent weights of solute per liter of solution. A solution is "normal" (i.e., 1 N) if there is exactly one gram equivalent weight per liter of solution.

x— *mole fraction:* The number of moles of solute divided by the number of moles of solvent and all solutes.

meq/L— *milligram equivalent weights of solute per liter of solution:* calculated by multiplying normality by 1000 or dividing concentration in mg/L by equivalent weight.

mg/L— *milligrams per liter:* The number of milligrams of solute per liter of solution. Same as ppm for solutions of water.

ppm— *parts per million:* The number of pounds (or grams) of solute per million pounds (or grams) of solution. Same as mg/L for solutions of water.

For compounds whose molecules do not dissociate in solution (e.g., table sugar), there is no difference between molarity and formality. There is a difference, however, for compounds that dissociate into ions (e.g., table salt). Consider a solution derived from 1 gmol of magnesium nitrate $Mg(NO_3)_2$ in enough water to bring the volume to 1 L. The formality is 1 F (i.e., the solution is 1 formal). However, 3 moles of ions will be produced: 1 mole of Mg^{++} ions and 2 moles of NO_3^- ions. Therefore, molarity is 1 M for the magnesium ion and 2 M for the nitrate ion.

The use of formality avoids the ambiguity in specifying concentrations for ionic solutions. Also, the use of formality avoids the problem of determining a molecular weight when there are no discernible molecules (e.g., as in a crystalline solid such as NaCl). Unfortunately, the distinction between molarity and formality is not always made, and molarity is used as if it were formality.

Example 22.9

A solution is made by dissolving 0.353 g of $Al_2(SO_4)_3$ in 730 g of water. Assuming 100% ionization, what is the concentration expressed as normality, molarity, and mg/L?

Solution

The molecular weight of $Al_2(SO_4)_3$ is

$$MW = (2)\left(26.98 \ \frac{g}{mol}\right) + (3)\left(32.06 \ \frac{g}{mol}\right)$$
$$+ (4)(3)\left(16 \ \frac{g}{mol}\right)$$
$$= 342.14 \ g/mol$$

Either the aluminum or sulfate ion can be used to determine the net charge transfer (i.e., the oxidation number). Since each aluminum ion has a charge of 3, and since there are two aluminum ions in the molecule, the oxidation number is $(3)(2) = 6$.

The equivalent weight is

$$EW = \frac{MW}{oxidation \ number} = \frac{342.14 \ \frac{g}{mol}}{6}$$
$$= 57.02 \ g/mol$$

The number of gram equivalent weights used is

$$\frac{0.353 \ g}{57.02 \ \frac{g}{mol}} = 6.19 \times 10^{-3} \ GEW$$

The volume of solution (same as the solvent volume if the small amount of solute is neglected) is 0.73 L.

The normality is

$$N = \frac{6.19 \times 10^{-3} \ GEW}{0.73 \ L} = 8.48 \times 10^{-3}$$

The number of moles of solute used is

$$\frac{0.353 \ g}{342.14 \ \frac{g}{mol}} = 1.03 \times 10^{-3} \ mol$$

The molarity is

$$M = \frac{1.03 \times 10^{-3} \ mol}{0.73 \ L} = 1.41 \times 10^{-3} \ mol/L$$

The concentration is

$$C = \frac{m}{V} = \frac{(0.353 \ g)\left(1000 \ \frac{mg}{g}\right)}{0.73 \ L} = 483.6 \ mg/L$$

20. pH AND pOH

A standard measure of the strength of an acid or base is the number of hydrogen or hydroxide ions in a liter of solution. Since these are very small numbers, a logarithmic scale is used.

$$pH = -\log_{10}[H^+] = \log_{10}\frac{1}{[H^+]} \qquad \text{22.9}$$

$$pOH = -\log_{10}[OH^-] = \log_{10}\frac{1}{[OH^-]} \qquad \text{22.10}$$

The quantities $[H^+]$ and $[OH^-]$ in square brackets are the *ionic concentrations* in moles of ions per liter. The number of moles can be calculated from Avogadro's law by dividing the actual number of ions per liter by 6.022×10^{23}. Alternatively, for a partially ionized compound in a solution of known molarity, M, the ionic concentration is

$$[ion] = XM \qquad \text{22.11}$$

A *neutral solution* has a pH of 7. Solutions with a pH below 7 are acidic; the smaller the pH, the more acidic the solution. Solutions with a pH above 7 are basic.

The relationship between pH and pOH is

$$pH + pOH = 14 \qquad 22.12$$

Example 22.10

A 4.2% ionized 0.01M ammonia solution is prepared from ammonium hydroxide (NH_4OH). Calculate the pH, pOH, and concentrations of $[H^+]$ and $[OH^-]$.

Solution

From Eq. 22.11,

$$[OH^-] = XM = (0.042)(0.01)$$
$$= 4.2 \times 10^{-4} \text{ mol/L}$$

From Eq. 22.10,

$$pOH = -\log[OH^-] = -\log(4.2 \times 10^{-4})$$
$$= 3.38$$

From Eq. 22.12,

$$pH = 14 - pOH = 14 - 3.38$$
$$= 10.62$$

The $[H^+]$ ionic concentration can be extracted from the definition of pH.

$$[H^+] = 10^{-pH} = 10^{-10.62}$$
$$= 2.4 \times 10^{-11} \text{ mol/L}$$

21. BUFFERS

A *buffer solution* resists changes in acidity and maintains a relatively constant pH when a small amount of an acid or base is added to it. Buffers are usually combinations of weak acids and their salts. A buffer is most effective when the acid and salt concentrations are equal.

22. NEUTRALIZATION

Acids and bases neutralize each other to form water.

$$H^+ + OH^- \rightarrow H_2O$$

Assuming 100% ionization of the solute, the volumes, V, required for complete neutralization can be calculated from the normalities, N, or the molarities, M.

$$V_b N_b = V_a N_a \qquad 22.13$$

$$V_b M_b \Delta_{b,\text{charge}} = V_a M_a \Delta_{a,\text{charge}} \qquad 22.14$$

23. REVERSIBLE REACTIONS

Reversible reactions are capable of going in either direction and do so to varying degrees (depending on the concentrations and temperature) simultaneously. These reactions are characterized by the simultaneous presence of all reactants and all products. For example, the chemical equation for the exothermic formation of ammonia from nitrogen and hydrogen is

$$N_2 + 3H_2 \rightleftharpoons 2NH_3 \quad (\Delta H = -92.4 \text{ kJ})$$

At *chemical equilibrium*, reactants and products are both present. Concentrations of the reactants and products do not change after equilibrium is reached.

24. LE CHATELIER'S PRINCIPLE

Le Châtelier's principle predicts the direction in which a reversible reaction initially at equilibrium will go when some condition (e.g., temperature, pressure, concentration) is "stressed" (i.e., changed). The principle says that when an equilibrium state is stressed by a change, a new equilibrium is formed that reduces that stress.

Consider the formation of ammonia from nitrogen and hydrogen. (See Sec. 22.23.) When the reaction proceeds in the forward direction, energy in the form of heat is released and the temperature increases. If the reaction proceeds in the reverse direction, heat is absorbed and the temperature decreases. If the system is stressed by increasing the temperature, the reaction will proceed in the reverse direction because that direction absorbs heat and reduces the temperature.

For reactions that involve gases, the reaction equation coefficients can be interpreted as volumes. In the nitrogen-hydrogen reaction, four volumes combine to form two volumes. If the equilibrium system is stressed by increasing the pressure, then the forward reaction will occur because this direction reduces the volume and pressure.[14]

If the concentration of any participating substance is increased, the reaction proceeds in a direction away from the substance with the increase in concentration. (For example, an increase in the concentration of the reactants shifts the equilibrium to the right, increasing the amount of products formed.)

The *common ion effect* is a special case of Le Châtelier's principle. If a salt containing a common ion is added to a solution of a weak acid, almost all of the salt will dissociate, adding large quantities of the common ion to the solution. Ionization of the acid will be greatly suppressed, a consequence of the need to have an unchanged equilibrium constant.

---

[14]The exception to this rule is the addition of an inert or nonparticipating gas to a gaseous equilibrium system. Although there is an increase in total pressure, the position of the equilibrium is not affected.

25. IRREVERSIBLE REACTION KINETICS

The rate at which a compound is formed or used up in an irreversible (one-way) reaction is known as the *rate of reaction, speed of reaction, reaction velocity*, and so on. The rate, v, is the change in concentration per unit time, usually measured in mol/L·s.

$$v = \frac{\text{change in concentration}}{\text{time}} \qquad 22.15$$

According to the *law of mass action*, the rate of reaction varies with the concentrations of the reactants and products. Specifically, the rate is proportional to the molar concentrations (i.e., the molarities). The rate of the formation or conversion of substance A is represented in various forms, such as r_A, dA/dt, and $d[A]/dt$, where the variable A or [A] can represent either the mass or the concentration of substance A. Substance A can be either a pure element or a compound.

The rate of reaction is generally not affected by pressure, but it does depend on five other factors.

- *type of substances in the reaction:* Some substances are more reactive than others.

- *exposed surface area:* The rate of reaction is proportional to the amount of contact between the reactants.

- *concentrations:* The rate of reaction increases with increases in concentration.

- *temperature:* The rate of reaction approximately doubles with every 10°C increase in temperature.

- *catalysts:* If a catalyst is present, the rate of reaction increases. However, the equilibrium point is not changed. (A catalyst is a substance that increases the reaction rate without being consumed in the reaction.)

26. ORDER OF THE REACTION

The *order of the reaction* is the total number of reacting molecules in or before the slowest step in the process.[15] The order must be determined experimentally. However, for an irreversible elementary reaction, the order is usually assumed from the stoichiometric reaction equation as the sum of the combining coefficients for the reactants.[16,17] For example, for the reaction

---

[15]This definition is valid for elementary reactions. For complex reactions, the order is an empirical number that need not be an integer.

[16]The overall order of the reaction is the sum of the orders with respect to the individual reactants. For example, in the reaction $2NO + O_2 \rightarrow 2NO_2$, the reaction is second order with respect to NO, first order with respect to O_2, and third order overall.

[17]In practice, the order of the reaction must be known, given, or determined experimentally. It is not always equal to the sum of the combining coefficients for the reactants. For example, in the reaction $H_2 + I_2 \rightarrow 2HI$, the overall order of the reaction is indeed 2, as expected. However, in the reaction $H_2 + Br_2 \rightarrow 2HBr$, the overall order is found experimentally to be 3/2, even though the two reactions have the same stoichiometry, and despite the similarities of iodine and bromine.

$mA + nB \rightarrow pC$, the overall order of the forward reaction is assumed to be $m + n$.

Many reactions (e.g., dissolving metals in acid or the evaporation of condensed materials) have *zero-order reaction rates*. These reactions do not depend on the concentrations or temperature at all, but rather, are affected by other factors such as the availability of reactive surfaces or the absorption of radiation. The formation (conversion) rate of a compound in a zero-order reaction is constant. That is, $dA/dt = -k_0$. k_0 is known as the *reaction rate constant*. (The subscript "0" refers to the zero-order.) Since the concentration (amount) of the substance decreases with time, dA/dt is negative. Since the negative sign is explicit in rate equations, the reaction rate constant is generally reported as a positive number.

Once a reaction rate equation is known, it can be integrated to obtain an expression for the concentration (mass) of the substance at various times. The time for half of the substance to be formed (or converted) is the *half-life*, $t_{1/2}$. Table 22.4 contains reaction rate and half-life equations for various types of low-order reactions.

Example 22.11

Nitrogen pentoxide decomposes according to the following first-order reaction.

$$N_2O_5 \rightarrow 2NO_2 + \tfrac{1}{2}O_2$$

At a particular temperature, the decomposition of nitrogen pentoxide is 85% complete after 11 min. What is the reaction rate constant?

Solution

The reaction is given as first order. Use the integrated reaction rate equation from Table 22.4. Since the decomposition reaction is 85% complete, the surviving fraction is 15% (0.15).

$$\ln \frac{[A]}{[A]_0} = k_1 t$$

$$\ln(0.15) = k_1(11 \text{ min})$$

$$k_1 = -0.172 \text{ 1/min} \quad (0.172 \text{ 1/min})$$

(The rate constant is reported as a positive number.)

27. REVERSIBLE REACTION KINETICS

Consider the following reversible reaction.

$$aA + bB \rightleftharpoons cC + dD \qquad 22.16$$

In Eq. 22.17 and Eq. 22.18, the *reaction rate constants* are k_{forward} and k_{reverse}. The order of the forward reaction is $a + b$; the order of the reverse reaction is $c + d$.

$$v_{\text{forward}} = k_{\text{forward}}[A]^a[B]^b \qquad 22.17$$

Table 22.4 Reaction Rates and Half-Life Equations

reaction	order	rate equation	integrated forms
$A \to B$	zero	$\dfrac{d[A]}{dt} = -k_0$	$[A] = [A]_0 - k_0 t$ $t_{1/2} = \dfrac{[A]_0}{2k_0}$
$A \to B$	first	$\dfrac{d[A]}{dt} = -k_1[A]$	$\ln\dfrac{[A]}{[A]_0} = -k_1 t$ $t_{1/2} = \dfrac{1}{k_1}\ln 2$
$A + A \to P$	second, type I	$\dfrac{d[A]}{dt} = -k_2[A]^2$	$\dfrac{1}{[A]} - \dfrac{1}{[A]_0} = k_2 t$ $t_{1/2} = \dfrac{1}{k_2[A]_0}$
$aA + bB \to P$	second, type II	$\dfrac{d[A]}{dt} = -k_2[A][B]$	$\ln\dfrac{[A]_0 - [B]}{[B]_0 - \left(\frac{b}{a}\right)[X]} = \ln\dfrac{[A]}{[B]}$ $= \left(\dfrac{b[A]_0 - a[B]_0}{a}\right)k_2 t + \ln\dfrac{[A]_0}{[B]_0}$ $t_{1/2} = \left(\dfrac{a}{k_2(b[A]_0 - a[B]_0)}\right)\ln\left(\dfrac{a[B]_0}{2a[B]_0 - b[A]_0}\right)$

$$v_{\text{reverse}} = k_{\text{reverse}}[C]^c[D]^d \qquad 22.18$$

At equilibrium, the forward and reverse speeds of reaction are equal.

$$v_{\text{forward}} = v_{\text{reverse}}|_{\text{equilibrium}} \qquad 22.19$$

28. EQUILIBRIUM CONSTANT

For reversible reactions, the *equilibrium constant*, K, is proportional to the ratio of the reverse rate of reaction to the forward rate of reaction.[18] Except for catalysis, the equilibrium constant depends on the same factors affecting the reaction rate. For the complex reversible reaction given by Eq. 22.16, the equilibrium constant is given by the *law of mass action*.

$$K = \dfrac{[C]^c[D]^d}{[A]^a[B]^b} = \dfrac{k_{\text{forward}}}{k_{\text{reverse}}} \qquad 22.20$$

If any of the reactants or products are in pure solid or pure liquid phases, their concentrations are omitted from the calculation of the equilibrium constant. For example, in weak aqueous solutions, the concentration of water, H_2O, is very large and essentially constant; therefore, that concentration is omitted.

For gaseous reactants and products, the concentrations (i.e., the numbers of atoms) will be proportional to the

partial pressures. Therefore, an equilibrium constant can be calculated directly from the partial pressures and is given the symbol K_p. For example, for the formation of ammonia gas from nitrogen and hydrogen, the equilibrium constant is

$$K_p = \dfrac{[p_{NH_3}]^2}{[p_{N_2}][p_{H_2}]^3} \qquad 22.21$$

K and K_p are not numerically the same, but they are related by Eq. 22.22. Δn is the number of moles of products minus the number of moles of reactants.

$$K_p = K(R^* T)^{\Delta n} \qquad 22.22$$

Example 22.12

A particularly weak solution of acetic acid ($HC_2H_3O_2$) in water has the ionic concentrations (in mol/L) given. What is the equilibrium constant?

$$HC_2H_3O_2 + H_2O \rightleftharpoons H_3O^+ + C_2H_3O_2^-$$
$$[HC_2H_3O_2] = 0.09866$$
$$[H_2O] = 55.5555$$
$$[H_3O^+] = 0.00134$$
$$[C_2H_3O_2^-] = 0.00134$$

---

[18]The symbols K_c (in molarity units) and K_{eq} are occasionally used for the equilibrium constant.

Solution

The concentration of the water molecules is not included in the calculation of the equilibrium or ionization constant. Therefore, the equilibrium constant is

$$K = K_a = \frac{[H_3O^+][C_2H_3O_2^-]}{[HC_2H_3O_2]} = \frac{(0.00134)(0.00134)}{0.09866}$$

$$= 1.82 \times 10^{-5}$$

29. IONIZATION CONSTANT

The equilibrium constant for a weak solution is essentially constant and is known as the *ionization constant* (also known as a *dissociation constant*). (See Table 22.5 and App. 22.G.) For weak acids, the symbol K_a and name *acid constant* are used. For weak bases, the symbol K_b and the name *base constant* are used. For example, for the ionization of hydrocyanic acid,

$$HCN \rightleftharpoons H^+ + CN^-$$

$$K_a = \frac{[H^+][CN^-]}{[HCN]}$$

Pure water is itself a very weak electrolyte and ionizes only slightly.

$$2H_2O \rightleftharpoons H_3O^+ + OH^- \qquad 22.23$$

At equilibrium, the ionic concentrations are equal.

$$[H_3O^+] = 10^{-7}$$

$$[OH^-] = 10^{-7}$$

From Eq. 22.20, the ionization constant (*ion product*) for pure water is

$$K_w = K_{a,water} = [H_3O^+][OH^-] = (10^{-7})(10^{-7})$$

$$= 10^{-14} \qquad 22.24$$

If the molarity, M, and *fraction of ionization*, X, are known, the ionization constant can be calculated from Eq. 22.25.

$$K_{ionization} = \frac{MX^2}{1-X} \quad [K_a \text{ or } K_b] \qquad 22.25$$

The reciprocal of the ionization constant is the *stability constant* (*overall stability constant*), also known as the *formation constant*. Stability constants are used to describe complex ions that dissociate readily.

Example 22.13

A 0.1 molar (0.1M) acetic acid solution is 1.34% ionized. Find the (a) hydrogen ion concentration, (b) acetate ion concentration, (c) un-ionized acid concentration, and (d) ionization constant.

Solution

(a) From Eq. 22.11, the hydrogen (hydronium) ion concentration is

$$[H_3O^+] = XM = (0.0134)(0.1)$$

$$= 0.00134 \text{ mol/L}$$

(b) Since every hydronium ion has a corresponding acetate ion, the acetate and hydronium ion concentrations are the same.

$$[C_2H_3O_2^-] = [H_3O^+] = 0.00134 \text{ mol/L}$$

(c) The concentration of un-ionized acid can be derived from Eq. 22.11.

$$[HC_2H_3O_2] = (1-X)M = (1-0.0134)(0.1)$$

$$= 0.09866 \text{ mol/L}$$

(d) The ionization constant is calculated from Eq. 22.25.

$$K_a = \frac{MX^2}{1-X} = \frac{(0.1)(0.0134)^2}{1-0.0134}$$

$$= 1.82 \times 10^{-5}$$

Table 22.5 *Approximate Ionization Constants of Common Water Supply Chemicals*

substance	0°C	5°C	10°C	15°C	20°C	25°C
Ca(OH)$_2$						3.74×10^{-3}
HClO	2.0×10^{-8}	2.3×10^{-8}	2.6×10^{-8}	3.0×10^{-8}	3.3×10^{-8}	3.7×10^{-8}
HC$_2$H$_3$O$_2$	1.67×10^{-5}	1.70×10^{-5}	1.73×10^{-5}	1.75×10^{-5}	1.75×10^{-5}	1.75×10^{-5}
HBrO					$\approx 2 \times 10^{-9}$	
H$_2$CO$_3$ (K_1)	2.6×10^{-7}	3.04×10^{-7}	3.44×10^{-7}	3.81×10^{-7}	4.16×10^{-7}	4.45×10^{-7}
HClO$_2$					$\approx 1.1 \times 10^{-2}$	
NH$_3$	1.37×10^{-5}	1.48×10^{-5}	1.57×10^{-5}	1.65×10^{-5}	1.71×10^{-5}	1.77×10^{-5}
NH$_4$OH						1.79×10^{-5}
water*	14.9435	14.7338	14.5346	14.3463	14.1669	13.9965

$^*-\log_{10}K$ given

Water Treatment

Example 22.14

The ionization constant for acetic acid is 1.82×10^{-5}. What is the hydrogen ion concentration for a 0.2 M solution?

Solution

From Eq. 22.25,

$$K_a = \frac{MX^2}{1-X}$$
$$1.82 \times 10^{-5} = \frac{0.2X^2}{1-X}$$

Since acetic acid is a weak acid, X is known to be small. Therefore, the computational effort can be reduced by assuming that $1 - X \approx 1$.

$$1.82 \times 10^{-5} = 0.2X^2$$
$$X = 9.54 \times 10^{-3}$$

From Eq. 22.11, the concentration of the hydrogen ion is

$$[H_3O^+] = XM = (9.54 \times 10^{-3})(0.2)$$
$$= 1.9 \times 10^{-3} \text{ mol/L}$$

Example 22.15

The ionization constant for acetic acid ($HC_2H_3O_2$) is 1.82×10^{-5}. What is the hydrogen ion concentration of a solution with 0.1 mol of 80% ionized ammonium acetate ($NH_4C_2H_3O_2$) in one liter of 0.1M acetic acid?

Solution

The acetate ion ($C_2H_3O_2^-$) is a common ion, since it is supplied by both the acetic acid and the ammonium acetate. Both sources contribute to the ionic concentration. However, the ammonium acetate's contribution dominates. Since the acid dissociates into an equal number of hydrogen and acetate ions,

$$[C_2H_3O_2^-]_{total} = [C_2H_3O_2^-]_{acid}$$
$$+ [C_2H_3O_2^-]_{ammonium\,acetate}$$
$$= [H_3O^+] + (0.8)(0.1)$$
$$\approx (0.8)(0.1) = 0.08$$

As a result of the common ion effect and Le Châtelier's law, the acid's dissociation is essentially suppressed by the addition of the ammonium acetate. The concentration of un-ionized acid is

$$[HC_2H_3O_2] = 0.1 - [H_3O^+]$$
$$\approx 0.1$$

The ionization constant is unaffected by the number of sources of the acetate ion.

$$K_a = \frac{[H_3O^+][C_2H_3O_2^-]}{[HC_2H_3O_2]}$$
$$1.82 \times 10^{-5} = \frac{[H_3O^+](0.08)}{0.1}$$
$$[H_3O^+] = 2.3 \times 10^{-5} \text{ mol/L}$$

30. IONIZATION CONSTANTS FOR POLYPROTIC ACIDS

A *polyprotic acid* has as many ionization constants as it has acidic hydrogen atoms. For oxyacids (see Sec. 22.6), each successive ionization constant is approximately 10^5 times smaller than the preceding one. For example, phosphoric acid (H_3PO_4) has three ionization constants.

$$K_1 = 7.1 \times 10^{-3} \quad (H_3PO_4)$$
$$K_2 = 6.3 \times 10^{-8} \quad (H_2PO_4^-)$$
$$K_3 = 4.4 \times 10^{-13} \quad (HPO_4^{-2})$$

31. SOLUBILITY PRODUCT

When an ionic solid is dissolved in a solvent, it dissociates. For example, consider the ionization of silver chloride in water.

$$AgCl(s) \rightleftharpoons Ag^+(aq) + Cl^-(aq)$$

If the equilibrium constant is calculated, the terms for pure solids and liquids (in this case, [AgCl] and [H₂O]) are omitted. Therefore, the *solubility product*, K_{sp}, consists only of the ionic concentrations. As with the general case of ionization constants, the solubility product for slightly soluble solutes is essentially constant at a standard value.

$$K_{sp} = [Ag^+][Cl^-] \qquad 22.26$$

When the product of terms exceeds the standard value of the solubility product, solute will precipitate out until the product of the remaining ion concentrations attains the standard value. If the product is less than the standard value, the solution is not saturated.

The solubility products of nonhydrolyzing compounds are relatively easy to calculate. (Example 22.16 illustrates a method.) Such is the case for chromates (CrO_4^{-2}), halides (F^-, Cl^-, Br^-, I^-), sulfates (SO_4^{-2}), and iodates (IO_3^-). However, compounds that hydrolyze (i.e., combine with water molecules) must be treated differently. The method used in Ex. 22.16 cannot be used for hydrolyzing compounds. Appendix 22.F lists some common solubility products.

Example 22.16

At a particular temperature, it takes 0.038 grams of lead sulfate ($PbSO_4$, molecular weight $= 303.25$) per liter of water to prepare a saturated solution. What is the solubility product of lead sulfate if all of the lead sulfate ionizes?

Solution

Sulfates are not one of the hydrolyzing ions. Therefore, the solubility product can be calculated from the concentrations.

Since one liter of water has a mass of 1 kg, the number of moles of lead sulfate dissolved per saturated liter of solution is

$$n = \frac{m}{\text{MW}} = \frac{0.038 \text{ g}}{303.25 \, \frac{\text{g}}{\text{mol}}} = 1.25 \times 10^{-4} \text{ mol}$$

Lead sulfate ionizes according to the following reaction.

$$PbSO_4(s) \rightleftharpoons Pb^{+2}(aq) + SO_4^-(aq) \quad \text{[in water]}$$

Since all of the lead sulfate ionizes, the number of moles of each ion is the same as the number of moles of lead sulfate. Therefore,

$$K_{sp} = [Pb^{+2}][SO_4^{-2}] = (1.25 \times 10^{-4})(1.25 \times 10^{-4})$$
$$= 1.56 \times 10^{-8}$$

32. ENTHALPY OF FORMATION

Enthalpy, H, is the useful energy that a substance possesses by virtue of its temperature, pressure, and phase.[19] The *enthalpy of formation (heat of formation)*, ΔH_f, of a compound is the energy absorbed during the formation of 1 gmol of the compound from the elements in their free, standard states.[20] The enthalpy of formation is assigned a value of zero for elements in their free states at 25°C and 1 atm. This is the so-called *standard state* for enthalpies of formation.

Table 22.6 contains enthalpies of formation for some common elements and compounds. The enthalpy of formation depends on the temperature and phase of the compound. A standard temperature of 25°C is used in most tables of enthalpies of formation.[21] Compounds are solid (s) unless indicated to be gaseous (g) or liquid (l). Some aqueous (aq) values are also encountered.

---

[19]The older term *heat* is rarely encountered today.
[20]The symbol H is used to denote molar enthalpies. The symbol h is used for specific enthalpies (i.e., enthalpy per kilogram or per pound).
[21]It is possible to correct the enthalpies of formation to account for other reaction temperatures.

Table 22.6 Standard Enthalpies of Formation (at 25°C)

element/compound	ΔH_f (kcal/mol)
Al (s)	0.00
Al_2O_3 (s)	−399.09
C (graphite)	0.00
C (diamond)	0.45
C (g)	171.70
CO (g)	−26.42
CO_2 (g)	−94.05
CH_4 (g)	−17.90
C_2H_2 (g)	54.19
C_2H_4 (g)	12.50
C_2H_6 (g)	−20.24
CCl_4 (g)	−25.5
$CHCl_4$ (g)	−24
CH_2Cl_2 (g)	−21
CH_3Cl (g)	−19.6
CS_2 (g)	27.55
COS (g)	−32.80
$(CH_3)_2S$ (g)	−8.98
CH_3OH (g)	−48.08
C_2H_5OH (g)	−56.63
$(CH_3)_2O$ (g)	−44.3
C_3H_6 (g)	9.0
C_6H_{12} (g)	−29.98
C_6H_{10} (g)	−1.39
C_6H_6 (g)	19.82
Fe (s)	0.00
Fe (g)	99.5
Fe_2O_3 (s)	−196.8
Fe_3O_4 (s)	−267.8
H_2 (g)	0.00
H_2O (g)	−57.80
H_2O (l)	−68.32
H_2O_2 (g)	−31.83
H_2S (g)	−4.82
N_2 (g)	0.00
NO (g)	21.60
NO_2 (g)	8.09
NO_3 (g)	13
NH_3 (g)	−11.04
O_2 (g)	0.00
O_3 (g)	34.0
S (g)	0.00
SO_2 (g)	−70.96
SO_3 (g)	−94.45

(Multiply kcal/mol by 4.184 to obtain kJ/mol.)
(Multiply kcal/mol by 1800/MW to obtain Btu/lbm.)

33. ENTHALPY OF REACTION

The *enthalpy of reaction (heat of reaction)*, ΔH_r, is the energy absorbed during a chemical reaction under constant volume conditions. It is found by summing the enthalpies of formation of all products and subtracting the sum of enthalpies of formation of all reactants. This is essentially a restatement of the energy conservation principle and is known as *Hess' law of energy summation*.

$$\Delta H_r = \sum_{\text{products}} \Delta H_f - \sum_{\text{reactants}} \Delta H_f \qquad \textit{22.27}$$

Reactions that give off energy (i.e., have negative enthalpies of reaction) are known as *exothermic reactions*. Many (but not all) exothermic reactions begin spontaneously. On the other hand, *endothermic reactions* absorb energy and require thermal or electrical energy to begin.

Example 22.17

Using enthalpies of formation, calculate the heat of stoichiometric combustion (standardized to 25°C) of gaseous methane (CH_4) and oxygen.

Solution

The balanced chemical equation for the stoichiometric combustion of methane is

$$CH_4 + 2O_2 \rightarrow 2H_2O + CO_2$$

The enthalpy of formation of oxygen gas (its free-state configuration) is zero. Using enthalpies of formation from Table 22.6 in Eq. 22.27, the enthalpy of reaction per mole of methane is

$$\Delta H_r = 2\Delta H_{f,H_2O} + \Delta H_{f,CO_2} - \Delta H_{f,CH_4} - 2\Delta H_{f,O_2}$$
$$= (2)\left(-57.80 \ \frac{\text{kcal}}{\text{mol}}\right) + \left(-94.05 \ \frac{\text{kcal}}{\text{mol}}\right)$$
$$- \left(-17.90 \ \frac{\text{kcal}}{\text{mol}}\right) - (2)(0)$$
$$= -191.75 \ \text{kcal/mol } CH_4 \quad [\text{exothermic}]$$

Using the footnote to Table 22.6, this value can be converted to Btu/lbm. The molecular weight of methane is

$$MW_{CH_4} = 12 + (4)(1) = 16$$

$$\text{higher heating value} = \frac{\left(191.75 \ \dfrac{\text{kcal}}{\text{mol}}\right)\left(1800 \ \dfrac{\text{Btu-mol}}{\text{lbm-kcal}}\right)}{16}$$
$$= 21{,}572 \ \text{Btu/lbm}$$

34. GALVANIC ACTION

Galvanic action (*galvanic corrosion* or *two-metal corrosion*) results from the difference in oxidation potentials of metallic ions. The greater the difference in oxidation potentials, the greater will be the galvanic action. If two metals with different oxidation potentials are placed in an electrolytic medium (e.g., seawater), a galvanic cell will be created. The metal with the higher potential (i.e., the more "active" metal) will act as an anode and will corrode. The metal with the lower potential (the more

"noble" metal), being the cathode, will be unchanged. In one extreme type of intergranular corrosion known as *exfoliation*, open endgrains separate into layers.

Metals are often classified according to their positions in the *galvanic series* listed in Table 22.7. As would be expected, the metals in this series are in approximately the same order as their half-cell potentials. However, alloys and proprietary metals are also included in the series.

Table 22.7 Galvanic Series in Seawater (top to bottom anodic (sacrificial, active) to cathodic (noble, passive))

magnesium
zinc
Alclad 3S
cadmium
2024 aluminum alloy
low-carbon steel
cast iron
stainless steels (active)
no. 410
no. 430
no. 404
no. 316
Hastelloy A
lead
lead-tin alloys
tin
nickel
brass (copper-zinc)
copper
bronze (copper-tin)
90/10 copper-nickel
70/30 copper-nickel
Inconel
silver solder
silver
stainless steels (passive)
Monel metal
Hastelloy C
titanium
graphite
gold

Precautionary measures can be taken to inhibit or reduce galvanic action when use of dissimilar metals is unavoidable.

- Use dissimilar metals that are close neighbors in the galvanic series.

- Use *sacrificial anodes*. In marine saltwater applications, sacrificial zinc plates can be used.

- Use protective coatings, oxides, platings, or inert spacers to reduce the access of corrosive environments to the metals.[22]

---

[22]While cadmium, nickel, chromium, and zinc are often used as protective deposits on steel, porosities in the surfaces can act as small galvanic cells, resulting in invisible subsurface corrosion.

35. CORROSION

Corrosion is an undesirable degradation of a material resulting from a chemical or physical reaction with the environment. Conditions within the crystalline structure can accentuate or retard corrosion. Corrosion rates are reported in units of mils per year (mpy) and micrometers per year (μm/y). (A *mil* is a thousandth of an inch.)

Uniform rusting of steel and oxidation of aluminum over entire exposed surfaces are examples of *uniform attack corrosion*. Uniform attack is usually prevented by the use of paint, plating, and other protective coatings.

Some metals are particularly sensitive to *intergranular corrosion*, IGC—selective or localized attack at metal-grain boundaries. For example, the Cr_2O_3 oxide film on stainless steel contains numerous imperfections at grain boundaries, and these boundaries can be attacked and enlarged by chlorides.

Intergranular corrosion may occur after a metal has been heated, in which case it may be known as *weld decay*. In the case of type 304 austenitic stainless steels, heating to 930–1300°F (500–700°C) in a welding process causes chromium carbides to precipitate out, reducing the corrosion resistance.[23] Reheating to 1830–2010°F (1000–1100°C) followed by rapid cooling will redissolve the chromium carbides and restore corrosion resistance.

Pitting is a localized perforation on the surface. It can occur even where there is little or no other visible damage. Chlorides and other halogens in the presence of water (e.g., HF and HCl) foster pitting in passive alloys, especially in stainless steels and aluminum alloys.

Concentration-cell corrosion (also known as *crevice corrosion* and *intergranular attack*, IGA) occurs when a metal is in contact with different electrolyte concentrations. It usually occurs in crevices, between two assembled parts, under riveted joints, or where there are scale and surface deposits that create stagnant areas in a corrosive medium.

Erosion corrosion is the deterioration of metals buffeted by the entrained solids in a corrosive medium.

Selective leaching is the dealloying process in which one of the alloy ingredients is removed from the solid solution. This occurs because the lost ingredient has a lower corrosion resistance than the remaining ingredient. *Dezincification* is the classic case where zinc is selectively destroyed in brass. Other examples include the

dealloying of nickel from copper-nickel alloys, iron from steel, and aluminum from copper-aluminum alloys.

Hydrogen damage occurs when hydrogen gas diffuses through and decarburizes steel (i.e., reacts with carbon to form methane). *Hydrogen embrittlement* (also known as *caustic embrittlement*) is hydrogen damage from hydrogen produced by caustic corrosion.

When subjected to sustained surface tensile stresses (including low residual stresses from manufacturing) in corrosive environments, certain metals exhibit catastrophic *stress corrosion cracking*, SCC. When the stresses are cyclic, *corrosion fatigue* failures can occur well below normal fatigue lives.

Stress corrosion occurs because the more highly stressed grains (at the crack tip) are slightly more anodic than neighboring grains with lower stresses. Although intergranular cracking (at grain boundaries) is more common, corrosion cracking may be *intergranular* (between the grains), *transgranular* (through the grains), or a combination of the two, depending on the alloy. Cracks propagate, often with extensive branching, until failure occurs.

The precautionary measures that can be taken to inhibit or eliminate stress corrosion are as follows.

- Avoid using metals that are susceptible to stress corrosion in corrosive environments. These include austenitic stainless steels without heat treatment in seawater; certain tempers of the aluminum alloys 2124, 2219, 7049, and 7075 in seawater; and copper alloys exposed to ammonia.

- Protect open-grain surfaces from the environment. For example, press-fitted parts in drilled holes can be assembled with wet zinc chromate paste. Also, weldable aluminum can be "buttered" with pure aluminum rod.

- Stress-relieve by annealing heat treatment after welding or cold working.

Fretting corrosion occurs when two highly loaded members have a common surface (known as a *faying surface*) at which rubbing and sliding take place. The phenomenon is a combination of wear and chemical corrosion. Metals that depend on a film of surface oxide for protection, such as aluminum and stainless steel, are especially susceptible.

Fretting corrosion can be reduced by the following methods.

- Lubricate the faying surfaces.

- Seal the faying surfaces.

- Reduce vibration and movement.

Cavitation is the formation and sudden collapse of minute bubbles of vapor in liquids. It is caused by a

---

[23]*Austenitic stainless steels* are the 300 series. They consist of chromium nickel alloys with up to 8% nickel. They are not hardenable by heat treatment, are nonmagnetic, and offer the greatest resistance to corrosion. *Martensitic stainless steels* are hardenable and magnetic. *Ferritic stainless steels* are magnetic and not hardenable.

Water Treatment

combination of reduced pressure and increased velocity in the fluid. In effect, very small amounts of the fluid vaporize (i.e., boil) and almost immediately condense. The repeated collapse of the bubbles hammers and work-hardens the surface.

When the surface work-hardens, it becomes brittle. Small amounts of the surface flake away, and the surface becomes pitted. This is known as *cavitation corrosion*. Eventually, the entire piece may work-harden and become brittle, leading to structural failure.

36. WATER SUPPLY CHEMISTRY

Most municipal water supply composition data are not given in units of molarity, normality, molality, and so on. Rather, the most common measure of solution strength is the *$CaCO_3$ equivalent*. With this method, substances are reported in milligrams per liter (mg/L, same as parts per million, ppm) "as $CaCO_3$," even when $CaCO_3$ is unrelated to the substance or reaction that produced the substance.

Actual gravimetric amounts of a substance can be converted to amounts as $CaCO_3$ by use of the conversion factors in App. 22.C. These factors are easily derived from stoichiometric principles.

The reason for converting all substance quantities to amounts as $CaCO_3$ is to simplify calculations for non-technical personnel. Equal $CaCO_3$ amounts constitute stoichiometric reaction quantities. For example, 100 mg/L as $CaCO_3$ of sodium ion (Na^+) will react with 100 mg/L as $CaCO_3$ of chloride ion (Cl^-) to produce 100 mg/L as $CaCO_3$ of salt (NaCl), even though the gravimetric quantities differ and $CaCO_3$ is not part of the reaction.

Example 22.18

Lime is added to water to remove carbon dioxide gas.

$$CO_2 + Ca(OH)_2 \rightarrow CaCO_3\downarrow + H_2O$$

If water contains 5 mg/L of CO_2, how much lime is required for its removal?

Solution

From App. 22.C, the factor that converts CO_2 as substance to CO_2 as $CaCO_3$ is 2.27.

$$CO_2 \text{ as } CaCO_3 \text{ equivalent} = (2.27)\left(5 \frac{mg}{L}\right)$$
$$= 11.35 \text{ mg/L as } CaCO_3$$

Therefore, the $CaCO_3$ equivalent of lime required will also be 11.35 mg/L.

From App. 22.C again, the factor that converts lime as $CaCO_3$ to lime as substance is (1/1.35).

$$Ca(OH)_2 \text{ substance} = \frac{11.35 \frac{mg}{L}}{1.35}$$
$$= 8.41 \text{ mg/L as substance}$$

This problem could also have been solved stoichiometrically.

37. ACIDITY AND ALKALINITY IN MUNICIPAL WATER SUPPLIES

Acidity is a measure of acids in solutions. Acidity in surface water (e.g., lakes and streams) is caused by formation of *carbonic acid* (H_2CO_3) from carbon dioxide in the air.[24] Acidity in water is typically given in terms of the $CaCO_3$ equivalent.

$$CO_2 + H_2O \rightarrow H_2CO_3 \qquad \text{22.28}$$

$$H_2CO_3 + H_2O \rightarrow HCO_3^- + H_3O^+ \quad [pH > 4.5] \qquad \text{22.29}$$

$$HCO_3^- + H_2O \rightarrow CO_3^{--} + H_3O^+ \quad [pH > 8.3] \qquad \text{22.30}$$

Alkalinity is a measure of the amount of negative (basic) ions in the water. Specifically, OH^-, CO_3^{--}, and HCO_3^- all contribute to alkalinity.[25] The measure of alkalinity is the sum of concentrations of each of the substances measured as $CaCO_3$.

Alkalinity and acidity of a titrated sample is determined from color changes in indicators added to the titrant.

Example 22.19

Water from a city well is analyzed and is found to contain 20 mg/L as substance of HCO_3^- and 40 mg/L as substance of CO_3^{--}. What is the alkalinity of this water as $CaCO_3$?

Solution

From App. 22.C, the factors converting HCO_3^- and CO_3^{--} ions to $CaCO_3$ equivalents are 0.82 and 1.67, respectively.

$$\text{alkalinity} = (0.82)\left(20 \frac{mg}{L}\right) + (1.67)\left(40 \frac{mg}{L}\right)$$
$$= 83.2 \text{ mg/L as } CaCO_3$$

---

[24]Carbonic acid is very aggressive and must be neutralized to eliminate the cause of water pipe corrosion. If the pH of water is greater than 4.5, carbonic acid ionizes to form bicarbonate. (See Eq. 22.29.) If the pH is greater than 8.3, carbonate ions form that cause water hardness by combining with calcium. (See Eq. 22.30.)

[25]Other ions, such as NO_3^-, also contribute to alkalinity, but their presence is rare. If detected, they should be included in the calculation of alkalinity.

38. WATER HARDNESS

Water hardness is caused by multivalent (doubly charged, triply charged, etc., but not singly charged) positive metallic ions such as calcium, magnesium, iron, and manganese. (Iron and manganese are not as common, however.) Hardness reacts with soap to reduce its cleansing effectiveness and to form scum on the water surface and a ring around the bathtub.

Water containing bicarbonate (HCO_3^-) ions can be heated to precipitate carbonate molecules.[26] This hardness is known as *temporary hardness* or *carbonate hardness*.[27]

$$Ca^{++} + 2HCO_3^- + heat \rightarrow CaCO_3\downarrow + CO_2 + H_2O$$
$$22.31$$

$$Mg^{++} + 2HCO_3^- + heat \rightarrow MgCO_3\downarrow + CO_2 + H_2O$$
$$22.32$$

Remaining hardness due to sulfates, chlorides, and nitrates is known as *permanent hardness* or *noncarbonate hardness* because it cannot be removed by heating. The amount of permanent hardness can be determined numerically by causing precipitation, drying, and then weighing the precipitate.

$$Ca^{++} + SO_4^{--} + Na_2CO_3 \rightarrow 2Na^+ + SO_4^{--} + CaCO_3\downarrow$$
$$22.33$$

$$Mg^{++} + 2Cl^- + 2NaOH \rightarrow 2Na^+ + 2Cl^- + Mg(OH)_2\downarrow$$
$$22.34$$

Total hardness is the sum of temporary and permanent hardnesses, both expressed in mg/L as $CaCO_3$.

39. COMPARISON OF ALKALINITY AND HARDNESS

Hardness measures the presence of positive, multivalent ions in the water supply. Alkalinity measures the presence of negative (basic) ions such as hydrates, carbonates, and bicarbonates. Since positive and negative ions coexist, an alkaline water can also be hard.

If certain assumptions are made, it is possible to draw conclusions about the water composition from the hardness and alkalinity. For example, if the effects of Fe^{+2} and OH^- are neglected, the following rules apply. (All concentrations are measured as $CaCO_3$.)

- *hardness = alkalinity*: There is no noncarbonate hardness. There are no SO_4^-, Cl^-, or NO_3^- ions present.

- *hardness > alkalinity*: Noncarbonate hardness is present.

- *hardness < alkalinity*: All hardness is carbonate hardness. The extra HCO_3^- comes from other sources (e.g., $NaHCO_3$).

Titration with indicator solutions is used to determine the alkalinity. The *phenolphthalein alkalinity* (or "P reading" in mg/L as $CaCO_3$) measures hydrate alkalinity and half of the carbonate alkalinity. The *methyl orange alkalinity* (or "M reading" in mg/L as $CaCO_3$) measures the total alkalinity (including the phenolphthalein alkalinity). Table 22.8 can be used to interpret these tests.[28]

Table 22.8 *Interpretation of Alkalinity Tests*

case	hydrate as $CaCO_3$	carbonate as $CaCO_3$	bicarbonate as $CaCO_3$
$P = 0$	0	0	M
$0 < P < \dfrac{M}{2}$	0	2P	M − 2P
$P = \dfrac{M}{2}$	0	2P	0
$\dfrac{M}{2} < P < M$	2P − M	2(M − P)	0
$P = M$	M	0	0

40. WATER SOFTENING WITH LIME

Water softening can be accomplished with lime and soda ash to precipitate calcium and magnesium ions from the solution. Lime treatment has the added benefits of disinfection, iron removal, and clarification. Practical limits of *precipitation softening* are 30 mg/L of $CaCO_3$ and 10 mg/L of $Mg(OH)_2$ (as $CaCO_3$) because of intrinsic solubilities. Water treated by this method usually leaves the softening apparatus with a hardness between 50 mg/L and 80 mg/L as $CaCO_3$.

41. WATER SOFTENING BY ION EXCHANGE

In the *ion exchange process* (also known as *zeolite process* or *base exchange method*), water is passed through a filter bed of exchange material. This exchange material is known as *zeolite*. Ions in the insoluble exchange material are displaced by ions in the water.

The processed water will have a zero hardness. However, if there is no need for water with zero hardness (as in municipal water supply systems), some water can be bypassed around the unit.

There are three types of ion exchange materials. *Greensand (glauconite)* is a natural substance that is mined and treated with manganese dioxide. *Siliceous-gel zeolite* is an artificial solid used in small volume deionizer

---

[26]Hard water forms scale when heated. This scale, if it forms in pipes, eventually restricts water flow. Even in small quantities, the scale insulates boiler tubes. Therefore, water used in steam-producing equipment must be essentially hardness-free.

[27]The hardness is known as *carbonate* hardness even though it is caused by *bicarbonate* ions, not carbonate ions.

[28]The titration may be affected by the presence of silica and phosphates, which also contribute to alkalinity. The effect is small, but the titration may not be a completely accurate measure of carbonates and bicarbonates.

columns. *Polystyrene resins* are also synthetic and dominate the softening field.

The earliest synthetic zeolites were gelular *ion exchange resins* using a three-dimensional copolymer (e.g., styrene-divinyl benzene). Porosity through the continuous-phase gel was near zero, and dry contact surface areas of 500 ft^2/lbm (0.1 m^2/g) or less were common.

Macroreticular synthetic resins are discontinuous, three-dimensional copolymer beads in a rigid-sponge type formation. Each bead is made up of thousands of microspheres of the gel resin. Porosity is increased, and dry contact surface areas are approximately 270,000 ft^2/lbm to 320,000 ft^2/lbm (55 m^2/g to 65 m^2/g).

During operation, the calcium and magnesium ions are removed according to the following reaction in which R represents the zeolite anion. The resulting sodium compounds are soluble.

$$\begin{Bmatrix} Ca \\ Mg \end{Bmatrix} \begin{Bmatrix} (HCO_3)_2 \\ SO_4 \\ Cl_2 \end{Bmatrix} + Na_2R$$

$$\rightarrow Na_2 \begin{Bmatrix} (HCO_3)_2 \\ SO_4 \\ Cl_2 \end{Bmatrix} + \begin{Bmatrix} Ca \\ Mg \end{Bmatrix} R \qquad 22.35$$

Typical saturation capacities of synthetic resins are 1.0 meq/mL to 1.5 meq/mL for anion exchange resins and 1.7 meq/mL to 1.9 meq/mL for cation exchange resins. However, working capacities are more realistic measures. Working capacities are approximately 10 kilograins/ft^3 to 15 kilograins/ft^3 (23 kg/m^3 to 35 kg/m^3) before regeneration.

Flow rates through the bed are typically 1 gpm/ft^3 to 6 gpm/ft^3 (2 L/s·m^3 to 13 L/s·m^3) of resin volume. The flow rate in terms of gpm/ft^2 (L/s·m^2) across the exposed surface will depend on the geometry of the bed, but values of 3 gpm/ft^2 to 15 gpm/ft^2 (2 L/s·m^2 to 10 L/s·m^2) are typical.[29]

Example 22.20

A municipal plant processes water with a total initial hardness of 200 mg/L. The designed discharge hardness is 50 mg/L. If an ion exchange unit is used, what is the bypass factor?

Solution

The water passing through the ion exchange unit is reduced to zero hardness. If x is the water fraction bypassed around the zeolite bed,

$$(1-x)\left(0 \; \frac{mg}{L}\right) + x\left(200 \; \frac{mg}{L}\right) = 50 \; mg/L$$
$$x = 0.25$$

42. REGENERATION OF ION EXCHANGE RESINS

Ion exchange material has a finite capacity for ion removal. When the zeolite is saturated or has reached some other prescribed limit, it must be regenerated (rejuvenated).

Standard ion exchange units are regenerated when the alkalinity of their effluent increases to the *set point*. Most condensate polishing units that also collect crud are operated to a *pressure-drop endpoint*. The pressure drop through the ion exchange unit is primarily dependent on the amount of crud collected. When the pressure drop reaches a set point, the resin is regenerated.

Regeneration of synthetic ion exchange resins is accomplished by passing a *regenerating solution* over/through the resin. Although regeneration can occur in the ion exchange unit itself, external regeneration is common. This involves removing the bed contents hydraulically, backwashing to separate the components (for mixed beds), regenerating the bed components separately, washing, then recombining and transferring the bed components back into service.

Common regeneration compounds are NaCl (for water hardness removal units), H_2SO_4 (for cation exchange resins), and NaOH (for anion exchange resins). The amount of regeneration solution depends on the resin's degree of saturation. A rule of thumb is to expect to use 6 lbm to 10 lbm of regeneration compound per cubic foot of resin (100 kg to 160 kg per cubic meter). Alternatively, dosage of the regeneration compound may be specified in terms of hardness removed (e.g., 0.4 lbm of salt per 1000 grains of hardness removed). These rates are applicable to deionization plants for boiler make-up water. For condensate polishing, saturation levels of 10 lbm/ft^3 to 25 lbm/ft^3 (160 kg/m^3 to 400 kg/m^3) are used.

43. CHARACTERISTICS OF BOILER FEEDWATER

Water that is converted to steam in a boiler is known as *feedwater*. Water that is added to replace losses is known as *make-up water*. The purity of the feedwater returned to the boiler (after condensing) depends on the purity of the make-up water, since impurities continually build up. *Blowdown* is the intentional periodic

---

[29]Much higher values, up to 15 gpm/ft^3 to 20 gpm/ft^3 or 40 gpm/ft^2 to 50 gpm/ft^2 (33 L/s·m^3 to 45 L/s·m^3 or 27 L/s·m^2 to 34 L/s·m^2), may occur in certain types of units and at certain times (e.g., start-up and leak conditions).

release of some of the feedwater in order to remove chemicals whose concentrations have built up over time.

Water impurities cause *scaling* (which reduces fluid flow and heat transfer rates) and corrosion (which reduces strength). Deposits from calcium, magnesium, and silica compounds are particularly troublesome.[30] As feedwater impurity increases, deposits from copper, iron, and nickel oxides (corrosion products from pipe and equipment) become more problematic.

Water can also contain dissolved oxygen, nitrogen, and carbon dioxide. The nitrogen is inert and does not need to be considered. However, both the oxygen and carbon dioxide need to be removed. High-temperature dissolved oxygen readily attacks pipe and boiler metal. Most of the oxygen in make-up water is removed by heating the water. Although the solubility of oxygen in water decreases with temperature, water at high pressures can hold large amounts of oxygen. Hydrazine (N_2H_4) and sodium sulfite (Na_2SO_3) are used for *oxygen scavenging*.[31] Because sodium sulfite forms sodium sulfate at high pressures, only hydrazine should be used above 1500 psi to 1800 psi (10.3 MPa to 12.4 MPa).

Carbon dioxide combines with water to form *carbonic acid*. In power plants, this is more likely found in condenser return lines than elsewhere. Acidity of the condensate is reduced by *neutralizing amines* such as morpholine, cyclohexylamine, ethanolamine (the tetra-sodium salt of ethylenediamine tetra-acetic acid, also known as Na_4EDTA), and diethylaminoethanol.

Most types of corrosion are greatly reduced when the pH is within the range of 9 to 10. Below this range, corrosion and deposits of sulfates, carbonates, and silicates become a major problem. For this reason, boilers are often shut down when the pH drops below 8.

Filming amines (*polar amines*) such as octadecylamine do not neutralize acidity or raise pH. They form a non-wettable layer which protects surfaces from corrosive compounds. Filming and neutralizing amines are often used in conjunction, though they are injected at different locations in the system.

Other purity requirements depend on the boiler equipment, and in particular, the steam pressure. Also, specific locations within the system can tolerate different concentrations.[32] The following guidelines apply to boilers operating in the 900 psig to 2500 psig (6.2 MPa to 17.3 MPa) range.[33]

- All water feedwater entering the boiler should be free from dissolved oxygen, carbon dioxide, suspended solids, and hardness.

- pH of water entering the boiler should be in the range of 8.5 to 9.0. pH of water in the boiler should be in the range of 10.8 to 11.5.

- Silica in boiler water should be limited to 5 ppm (as $CaCO_3$) at 900 psig (6.2 MPa), with a gradual reduction to 1 ppm (as $CaCO_3$) at 2500 psig (17.3 MPa).

44. MONITORING OF BOILER FEEDWATER QUALITY

Water chemistry of boiler feedwater is monitored through continuous *inline sampling* and periodic *grab sampling*. Water impurities are expressed in milligrams per liter (mg/L), parts per million (ppm), parts per billion (ppb), unit equivalents per million (epm), and milliequivalents per milliliter (meq/mL).[34,35] ppm and ppb can be "as $CaCO_3$" equivalents or "as substance." Electrical conductivity (a measure of total dissolved solids) is measured in microsiemens[36] (μS) or ppm.

Water quality is maintained by monitoring pH, electrical resistivity silica, and hardness (calcium, magnesium, and bicarbonates), the primary corrosion species (sulfates, chlorides, and sodium), dissolved gases (primarily oxygen), and the concentrations of buffering chemicals (e.g., phosphate, hydrazine, or AVT).

Continuous monitoring of ionic concentrations is complicated by a process known as *hideout,* in which a chemical species (e.g., *crevice salts*) disappears by precipitation or adsorption during low-flow and high heat transfer. Upon cooling (as during a shut-down), the hideout chemicals reappear.

45. PRODUCTION AND REGENERATION OF BOILER FEEDWATER

The processes used to treat raw water for use in power-generating plants depend on the incoming water quality. Depending on need, filters may be used to remove solids, softeners or exchange units may remove permanent (sulfate) hardness, and activated carbon may remove organics.

---

[30]Silica is most troubling because silicate deposits cannot be removed by chemical means. They must be removed mechanically.

[31]8 ppm of sodium nitrate react with 1 ppm of oxygen. To achieve a complete reaction, an excess of 2 ppm to 3 ppm of sodium nitrate is required. Hydrazine reacts on a one-to-one basis and is commonly available as a 35% solution. Therefore, approximately 3 ppm of solution is required per ppm of oxygen.

[32]For example, silica in the boiler feedwater may be limited to 1 ppm to 5 ppm, but the silica content in steam should be limited to 0.01 ppm to 0.03 ppm.

[33]Tolerable concentrations at cold start-up can be as much as 5 to 100 times higher.

[34]Milligrams per liter (mg/L) is essentially the same as parts per million (ppm). The older units of grains per gallon are still occasionally encountered. 1 *grain* equals 1/7000th of a pound. Multiply grains per gallon (gpg) by 17.1 to get ppm. Unit equivalents per million are derived by dividing the concentration in ppm by the equivalent weight. Equivalent weight is the molecular weight divided by the valence or oxidation number.

[35]Use the following relationship to convert meq/mL to ppm.

$$ppm = (1000)\left(\frac{meq}{mL}\right)(\text{equivalent weight})$$
$$= \frac{(1000)\left(\frac{meq}{mL}\right)(\text{formula weight})}{valence}$$

[36]Microsiemens were previously known as "micromhos," where "ohms" is spelled backward to indicate its inverse.

Bicarbonate hardness is converted to sludge when the water is boiled, is removed during blow-down, and does not need to be chemically removed. However, a strong *dispersant (sludge conditioner)* must be added to prevent scale formation.[37] Alternatively, the calcium salts can be converted to sludges of phosphate salts by adding phosphates. Prior to the 1970s, *caustic phosphate treatments* using sodium phosphate, SP, in the form of Na_3PO_4, $NaHPO_4$, and NaH_2PO_4, was the most common method used to treat low-pressure boiler feedwater with high solids. Sodium phosphate converts impurities and buffers pH.

However, magnesium phosphate is a sticky sludge, and using phosphates may cause more problems than not using them. Also, with higher pressures and temperatures, pipe wall *wasting* becomes problematic.[38] Therefore, many modern plants now use an *all-volatile treatment*, AVT, also known as a *zero-solids treatment*. Early AVTs used ammonia (NH_3) for pH control, but ammonia causes *denting* in systems that operate without blowdown.[39]

Chelant AVTs do not add any solid chemicals to the boiler. They work by keeping calcium and magnesium in solution. The compounds are removed through continuous blowdown.[40] Chelant treatments can be used in boilers up to approximately 1500 psi (10.3 MPa) but require high-purity (low-solids) feedwater.[41]

Subsequent advances in corrosion protection have been based on use of stainless steel or titanium-tubed condensers, condensate polishing, and even better AVT formulations. Supplementing ammonia-AVTs are volatile amines, chelants, and boric acid treatments. Boric acid effectively combats denting, intergranular attack, and intergranular stress corrosion cracking.

Water that is processed through an ion-exchange unit is known as *deionized water*. Deionized water is "hungry" for minerals and picks up contamination as it passes through the power plant. When the operating pressure is 1500 psig (10.3 MPa) or higher, condensed steam cannot be reused without additional demineralization.[42] *Demineralization* is not generally needed for lower-pressure units, but some *condensate polishing* is still necessary.

In low-pressure plants, condensate polishing may consist of passing the water through a filter (of sand, anthracite, or pre-coat cellulose), a strong-acid cation unit, and a mixed-bed unit. The filter removes gross particles (referred to as *crud*), the strong-acid unit removes the dissolved iron, and the mixed-bed unit polishes the condensate.[43]

In modern high-pressure plants, the filtration step is often omitted, and crud removal occurs in the strong-acid unit. Furthermore, if the mixed-bed unit is made large enough or the cation component is increased, the strong-acid unit can also be omitted. A mixed-bed unit operating by itself is known as a *naked mixed-bed unit*. Demineralization of condensed steam is commonly accomplished with naked mixed beds.

---

[37]Typical sludge conditioner dispersants are natural organics (lignins, tannins, and starches) and synthetics (sodium polyacrylate, sodium polymethacrylate, sulfonated polystyrene, and maleic anhydride).

[38]*Wasting* is a term used in the power-generating industry to describe the process of general pipe wall thinning.

[39]*Denting* is the constricting of the intersection between tubes and support plates in boilers due to a buildup of corrosion.

[40]AVTs work in a completely different way than phosphates. It is not necessary to use phosphates and AVTs simultaneously.

[41]If the feedwater becomes contaminated, a supplementary phosphate treatment will be required.

[42]The term *deionization* refers to the process that produces makeup water. *Demineralization* refers to the process used to prepare condensed steam for reuse.

[43]*Crud* is primarily iron-corrosion products ranging from dissolved to particulate matter.

23 Water Supply Quality and Testing

Nomenclature

A	acidity	mg/L CaCO$_3$
Ca	calcium	mg/L CaCO$_3$
H	total hardness	mg/L CaCO$_3$
L	free lime	mg/L CaCO$_3$
M	alkalinity	mg/L CaCO$_3$
N	normality	gEW/L
O	hydroxides	mg/L CaCO$_3$
P	phenolphthalein alkalinity	mg/L CaCO$_3$
S	sulfate	mg/L CaCO$_3$
V	volume	mL

1. CATIONS AND ANIONS IN NEUTRAL SOLUTIONS

Equivalency concepts provide a useful check on the accuracy of water analyses. For the water to be electrically neutral, the sum of anion equivalents must equal the sum of cation equivalents.

Concentrations of dissolved compounds in water are usually expressed in mg/L, not equivalents. (See Sec. 22.18.) However, anionic and cationic substances can be converted to their equivalent concentrations in milliequivalents per liter (meq/L) by dividing their concentrations in mg/L by their equivalent weights. Appendix 22.B is useful for this purpose.

Since water is an excellent solvent, it will contain the ions of the inorganic compounds to which it is exposed. A chemical analysis listing these ions does not explicitly determine the compounds from which the ions originated. Several graphical methods, such as bar graphs and Piper (Hill) trilinear diagrams, can be used for this purpose. The bar graph, also known as a *milliequivalent per liter bar chart*, is constructed by listing the cations (positive ions) in the sequence of calcium (Ca^{++}), magnesium (Mg^{++}), sodium (Na^+), and potassium (K^+), and pairing them with the anions (negative ions) in the sequence of carbonate (CO_3^{--}), bicarbonate (HCO_3^-), sulfate (SO_4^{--}), and chloride (Cl^-). A bar chart, such as the one shown in Ex. 23.1, can be used to deduce the hypothetical combinations of positive and negative ions that would have resulted in the given water analysis.

Example 23.1

A water analysis reveals the following ionic components in solution: Ca^{++}, 29.0 mg/L; Mg^{++}, 16.4 mg/L; Na^+, 23.0 mg/L; K^+, 17.5 mg/L; HCO_3^-, 171 mg/L; SO_4^{--}, 36.0 mg/L; Cl^-, 24.0 mg/L. Verify that the analysis is reasonably accurate. Draw the milliequivalent per liter bar chart.

Solution

Use the equivalent weights in App. 22.B to complete the following table.

compound	concentration (mg/L)	equivalent weight (g/mol)	equivalent (meq/L)
cations			
Ca^{++}	29.0	20.0	1.45
Mg^{++}	16.4	12.2	1.34
Na^+	23.0	23.0	1.00
K^+	17.5	39.1	0.44
		total	4.23
anions			
HCO_3^-	171.0	61.0	2.80
SO_4^{--}	36.0	48.0	0.75
Cl^-	24.0	35.5	0.68
		total	4.23

The sums of the cation equivalents and anion equivalents are equal. The analysis is presumed to be reasonably accurate.

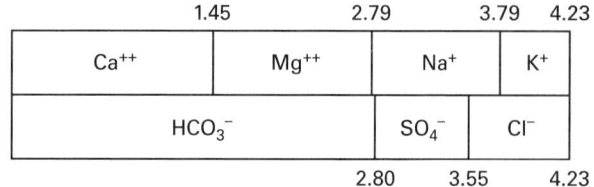

2. ACIDITY

Acidity is a measure of acids in solution. Acidity in surface water is caused by formation of carbonic acid (H_2CO_3) from carbon dioxide in the air. (See Eq. 23.1.) Carbonic acid is aggressive and must be neutralized to eliminate a cause of water pipe corrosion. If the pH of water is greater than 4.5, carbonic acid ionizes to form bicarbonate. (See Eq. 23.2.) If the pH is greater than 8.3, carbonate ions form. (See Eq. 23.3.)

$$CO_2 + H_2O \rightarrow H_2CO_3 \qquad 23.1$$

$$H_2CO_3 + H_2O \rightarrow HCO_3^- + H_3O^+ \; (pH > 4.5) \qquad 23.2$$

$$HCO_3^- + H_2O \rightarrow CO_3^{--} + H_3O^+ \; (pH > 8.3) \qquad 23.3$$

Measurement of acidity is done by titration with a standard basic measuring solution. Acidity, A, in water is typically given in terms of the $CaCO_3$ equivalent that would neutralize the acid. The constant 50 000 used in Eq. 23.4 is the product of the equivalent weight of $CaCO_3$ (50 g) and 1000 mg/g.

$$A_{mg/L\,as\,CaCO_3} = \frac{V_{titrant,mL} N_{titrant}(50\,000)}{V_{sample,mL}} \qquad 23.4$$

The standard titration procedure for determining acidity measures the amount of titrant needed to raise the pH to 3.7 plus the amount needed to raise the pH to 8.3. The total of these two is used in Eq. 23.4. Most water samples, unless grossly polluted with industrial wastes, will exist at a pH greater than 3.7, so the titration will be a one-step process.

3. ALKALINITY

Alkalinity is a measure of the ability of a water to neutralize acids (i.e., to absorb hydrogen ions without significant pH change). The principal alkaline ions are OH^-, CO_3^{--}, and HCO_3^-. Other radicals, such as NO_3^-, also contribute to alkalinity, but their presence is rare. The measure of alkalinity is the sum of concentrations of each of the substances measured as equivalent $CaCO_3$.

The standard titration method for determining alkalinity measures the amount of acidic titrant needed to lower the pH to 8.3 plus the amount needed to lower the pH to 4.5. Therefore, alkalinity, M, is the sum of all titratable base concentrations down to a pH of 4.5. The

constant 50 000 in Eq. 23.5 is the product of the equivalent weight of $CaCO_3$ (50 g) and 1000 mg/g.

$$M_{mg/L\,as\,CaCO_3} = \frac{V_{titrant,mL} N_{titrant}(50\,000)}{V_{sample,mL}} \qquad 23.5$$

Example 23.2

Water from a city well is analyzed and found to contain 20 mg/L as substance of HCO_3^- and 40 mg/L as substance of CO_3^{--}. What is the alkalinity of the water expressed as $CaCO_3$?

Solution

From App. 22.B, the equivalent weight of HCO_3^- is 61 g/mol, the equivalent weight of CO_3^{--} is 30 g/mol, and the equivalent weight of $CaCO_3$ is 50 g/mol.

$$M_{mg/L\,of\,CaCO_3} = \left(20\,\frac{mg}{L}\right)\left(\frac{50\,\frac{g}{mol}}{61\,\frac{g}{mol}}\right)$$
$$+ \left(40\,\frac{mg}{L}\right)\left(\frac{50\,\frac{g}{mol}}{30\,\frac{g}{mol}}\right)$$
$$= 83.1\ mg/L\ as\ CaCO_3$$

4. INDICATOR SOLUTIONS

End points for acidity and alkalinity titrations are determined by color changes in indicator dyes that are pH sensitive. Several commonly used indicators are listed in Table 23.1.

Table 23.1 Indicator Solutions Commonly Used in Water Chemistry

indicator	titration	end point pH	color change
bromophenol blue	acidity	3.7	yellow to blue
phenolphthalein	acidity	8.3	colorless to red-violet
phenolphthalein	alkalinity	8.3	red-violet to colorless
mixed bromocresol/ green-methyl red	alkalinity	4.5	grayish to orange-red

Depending on pH, alkaline samples can contain hydroxide alone, hydroxide and carbonate, carbonate alone, carbonate and bicarbonate, or bicarbonate alone. Samples containing hydroxide or hydroxide and carbonate have a high pH, usually greater than 10. If the titration is complete at the phenolphthalein end point (i.e., the mixed bromocresol green-methyl red indicator does not change color), the alkalinity is hydroxide alone. Samples containing carbonate and bicarbonate alkalinity have a pH greater than 8.3, and the titration to the phenolphthalein end point represents stoichiometrically one-half of the carbonate alkalinity. If the volume of phenolphthalein titrant equals the volume of titrant needed to reach the mixed bromocresol green-methyl red end point, all of the alkalinity is in the form of carbonate.

If abbreviations of P for the measured phenolphthalein alkalinity and M for the total alkalinity are used, the following relationships define the possible alkalinity states of the sample. In sequence, the relationships can be used to determine the actual state of a sample.

(state I)	hydroxide $= P = M$
(state II)	hydroxide $= 2P - M$ and carbonate $= 2(M - P)$
(state III)	carbonate $= 2P = M$
(state IV)	carbonate $= 2P$ and bicarbonate $= M - 2P$
(state V)	bicarbonate $= M$

Example 23.3

A 100 mL sample is titrated for alkalinity by using 0.02 N sulfuric acid solution. To reach the phenolphthalein end point requires 3.0 mL of the acid solution, and an additional 12.0 mL is added to reach the mixed bromocresol green-methyl red end point. Calculate the (a) phenolphthalein alkalinity and (b) total alkalinity. (c) What are the ionic forms of alkalinity present?

Solution

(a) Use Eq. 23.5.

$$P = \frac{V_{\text{titrant,mL}} N_{\text{titrant}}(50\,000)}{V_{\text{sample,mL}}}$$

$$= \frac{(3.0 \text{ mL})\left(0.02 \dfrac{\text{gEW}}{\text{L}}\right)\left(50\,000 \dfrac{\text{mg}}{\text{gEW}}\right)}{100 \text{ mL}}$$

$$= 30 \text{ mg/L (as CaCO}_3)$$

(b) Use Eq. 23.5.

$$M = \frac{V_{\text{titrant,mL}} N_{\text{titrant}}(50\,000)}{V_{\text{sample,mL}}}$$

$$= \frac{(3.0 \text{ mL} + 12.0 \text{ mL})}{100 \text{ mL}} \times \left(0.02 \dfrac{\text{gEW}}{\text{L}}\right)\left(50\,000 \dfrac{\text{mg}}{\text{gEW}}\right)$$

$$= 150 \text{ mg/L (as CaCO}_3)$$

(c) Test the state relationships in sequence.

(I) hydroxide $= P = M$:
$$30 \text{ mg/L} \neq 150 \text{ mg/L}$$
$$\text{(invalid: } P \neq M)$$

(II) hydroxide $= 2P - M$:
$$(2)\left(30 \dfrac{\text{mg}}{\text{L}}\right) - 150 \dfrac{\text{mg}}{\text{L}}$$
$$= -90 \text{ mg/L}$$
$$\text{(invalid: negative value)}$$

(III) carbonate $= 2P = M$:
$$(2)\left(30 \dfrac{\text{mg}}{\text{L}}\right) \neq 150 \text{ mg/L}$$
$$\text{(invalid: } 2P \neq M)$$

(IV) carbonate $= 2P$:
$$(2)\left(30 \dfrac{\text{mg}}{\text{L}}\right) = 60 \text{ mg/L}$$
bicarbonate $= M - 2P$
$$150 \dfrac{\text{mg}}{\text{L}} - (2)\left(30 \dfrac{\text{mg}}{\text{L}}\right) = 90 \text{ mg/L}$$

The process ends at (IV) since a valid answer is obtained. It is not necessary to check relationship (V), as (IV) gives consistent results. The alkalinity is composed of 60 mg/L as $CaCO_3$ of carbonate and 90 mg/L as $CaCO_3$ of bicarbonate.

5. HARDNESS

Hardness in natural water is caused by the presence of polyvalent (but not singly charged) metallic cations. Principal cations causing hardness in water and the major anions associated with them are presented in Table 23.2. Because the most prevalent of these species are the divalent cations of calcium and magnesium, total hardness is typically defined as the sum of the concentration of these two elements and is expressed in terms of milligrams per liter as $CaCO_3$. (Hardness is occasionally expressed in units of *grains per gallon*, where 7000 grains are equal to a pound.)

Table 23.2 Principal Cations and Anions Indicating Hardness

cations	anions
Ca^{++}	HCO_3^-
Mg^{++}	SO_4^{--}
Sr^{++}	Cl^-
Fe^{++}	NO_3^-
Mn^{++}	SiO_3^{--}

Carbonate hardness is caused by cations from the dissolution of calcium or magnesium carbonate and bicarbonate in the water. Carbonate hardness is hardness that is chemically equivalent to alkalinity, where most of the alkalinity in natural water is caused by the bicarbonate and carbonate ions.

Noncarbonate hardness is caused by cations from calcium (i.e., calcium hardness) and magnesium (i.e., magnesium hardness) compounds of sulfate, chloride, or silicate that are dissolved in the water. Noncarbonate hardness is equal to the total hardness minus the carbonate hardness.

Hardness can be classified as shown in Table 23.3. Although high values of hardness do not present a health risk, they have an impact on the aesthetic acceptability of water for domestic use. (Hardness reacts with soap to reduce its cleansing effectiveness and to form scum on the water surface.) Where feasible, carbonate hardness in potable water should be reduced to the 25–40 mg/L range and total hardness reduced to the 50–75 mg/L range.

Table 23.3 *Relationship of Hardness Concentration to Classification*

hardness (mg/L as $CaCO_3$)	classification
0 to 60	soft
61 to 120	moderately hard
121 to 180	hard
181 to 350	very hard
> 350	saline; brackish

Water containing bicarbonate (HCO_3^-) can be heated to precipitate carbonate (CO_3^{--}) as a *scale*. Water used in steam-producing equipment (e.g., boilers) must be essentially hardness-free to avoid deposit of scale.

$$Ca^{++} + 2HCO_3^- + heat \rightarrow CaCO_3 \downarrow + CO_2 + H_2O$$
23.6

$$Mg^{++} + 2HCO_3^- + heat \rightarrow MgCO_3 \downarrow$$
$$+ CO_2 + H_2O$$
23.7

Noncarbonate hardness, also called *permanent hardness*, cannot be removed by heating. It can be removed by precipitation softening processes (typically the lime-soda ash process) or by ion exchange processes using resins selective for ions causing hardness.

Hardness is measured in the laboratory by titrating the sample using a standardized solution of ethylenediaminetetraacetic acid (EDTA) and an indicator dye such as Eriochrome Black T. The sample is titrated at a pH of approximately 10 until the dye color changes from red to blue. The standardized solution of EDTA is usually prepared such that 1 mL of EDTA is equivalent to 1 mg/L of hardness, but Eq. 23.8 can be used to determine hardness with any strength EDTA solution.

$$H_{mg/L\,as\,CaCO_3} =$$
$$\frac{V_{titrant,mL}(CaCO_3 \text{ equivalent of EDTA})(1000)}{V_{sample,mL}}$$
23.8

Example 23.4

A water sample contains sodium (Na^+, 15 mg/L), magnesium (Mg^{++}, 70 mg/L), and calcium (Ca^{++}, 40 mg/L). What is the hardness?

Solution

Sodium is singly charged, so it does not contribute to hardness. The necessary approximate equivalent weights are found in App. 22.B.

    Mg: 12.2 g/mol
    Ca: 20.0 g/mol
    $CaCO_3$: 50 g/mol

The equivalent hardness is

$$H = \left(70 \; \frac{mg}{L}\right) \left(\frac{50 \; \frac{g}{mol}}{12.2 \; \frac{g}{mol}}\right) + \left(40 \; \frac{mg}{L}\right)\left(\frac{50 \; \frac{g}{mol}}{20 \; \frac{g}{mol}}\right)$$
$$= 387 \text{ mg/L (as } CaCO_3)$$

Example 23.5

A 75 mL water sample was titrated using 8.1 mL of an EDTA solution formulated such that 1 mL of EDTA is equivalent to 0.8 mg/L of $CaCO_3$. What is the hardness of the sample, measured in terms of mg/L of $CaCO_3$?

Solution

Use Eq. 23.8.

$$H = \frac{V_{titrant,mL}(CaCO_3 \text{ equivalent of EDTA})(1000)}{V_{sample,mL}}$$
$$= \frac{(8.1 \text{ mL})\left(0.8 \; \frac{gEW}{L}\right)\left(1000 \; \frac{mg}{gEW}\right)}{75 \text{ mL}}$$
$$= 86.4 \text{ mg/L (as } CaCO_3)$$

6. HARDNESS AND ALKALINITY

Hardness is caused by multi-positive ions. Alkalinity is caused by negative ions. Both positive and negative ions are present simultaneously. Therefore, an alkaline water can also be hard.

With some assumptions and minimal information about the water composition, it is possible to determine the ions in the water from the hardness and alkalinity. For example, Fe^{++} is an unlikely ion in most water supplies, and it is often neglected. Figure 23.1 can be used to quickly deduce the compounds in the water from hardness and alkalinity.

If hardness and alkalinity (both as $CaCO_3$) are the same and there are no monovalent cations, then there are no SO_4^{--}, Cl^-, or NO_3^- ions present. That is, there is no noncarbonate (permanent) hardness. If hardness is greater than the alkalinity, however, then noncarbonate hardness is present, and the carbonate (temporary) hardness is equal to the alkalinity. If hardness is less than the alkalinity, then all hardness is carbonate hardness, and the extra HCO_3^- comes from other sources (such as $NaHCO_3$).

Figure 23.1 Hardness and Alkalinity[a,b]

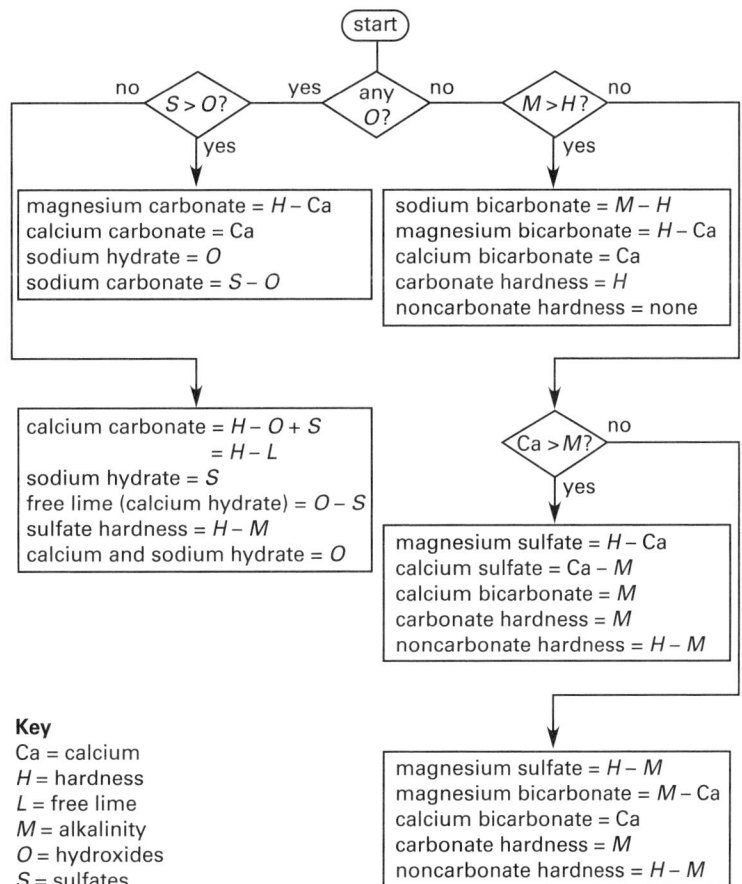

Key
Ca = calcium
H = hardness
L = free lime
M = alkalinity
O = hydroxides
S = sulfates

[a]All concentrations are expressed as $CaCO_3$.
[b]Not for use when other ionic species are present in significant quantities.

Example 23.6

A water sample analysis results in the following: alkalinity, 220 mg/L; hardness, 180 mg/L; Ca^{++}, 140 mg/L; OH^-, insignificant. All concentrations are expressed as $CaCO_3$. (a) What is the noncarbonate hardness? (b) What is the Mg^{++} content in mg/L as substance?

Solution

(a) Use Fig. 23.1 with $M > H$ (alkalinity greater than hardness).

$$[NaHCO_3] = M - H = 220 \ \frac{mg}{L} - 180 \ \frac{mg}{L}$$
$$= 40 \ mg/L \ (as \ CaCO_3)$$

$$[Mg(HCO_3)_2] = H - Ca = 180 \ \frac{mg}{L} - 140 \ \frac{mg}{L}$$
$$= 40 \ mg/L \ (as \ CaCO_3)$$

$$[Ca(HCO_3)_2] = Ca = 140 \ mg/L \ (as \ CaCO_3)$$

$$carbonate \ hardness = H = 180 \ mg/L \ (as \ CaCO_3)$$

$$noncarbonate \ hardness = 0$$

(b) The Mg^{++} ion content as $CaCO_3$ is equal to the $Mg(HCO_3)_2$ content as $CaCO_3$. Use App. 22.B to convert $CaCO_3$ equivalents to amounts as substance. To convert Mg^{++} as $CaCO_3$ to Mg^{++} as substance, use a factor of 4.1.

$$Mg^{++} = \frac{40 \ \frac{mg}{L}}{4.1} = 9.8 \ mg/L \ (as \ substance)$$

7. NATIONAL PRIMARY DRINKING WATER STANDARDS

Following passage of the Safe Drinking Water Act in the United States, the Environmental Protection Agency (EPA) established minimum primary drinking water standards. These standards set limits on the amount of various substances in drinking water. Every public water supply serving at least 15 service connections or 25 or more people must ensure that its water meets these minimum standards.

Accordingly, the EPA has established the National Primary Drinking Water Standards, outlined in App. 24.A, and the National Secondary Drinking Water Standards. The primary standards establish *maximum contaminant levels* (MCL) and *maximum contaminant level goals* (MCLG) for materials that are known or suspected health hazards. The MCL is the enforceable level that the water supplier must not exceed, while the MCLG is an unenforceable health goal equal to the maximum level of a contaminant that is not expected to cause any adverse health effects over a lifetime of exposure.

8. NATIONAL SECONDARY DRINKING WATER STANDARDS

The national secondary drinking water standards, outlined in Table 23.4, are not designed to protect public health. Instead, they are intended to protect "public welfare" by providing helpful guidelines regarding the taste, odor, color, and other aesthetic aspects of drinking water.

Table 23.4 National Secondary Drinking Water Standards (Code of Federal Regulations (CFR) Title 40, Ch. I, Part 143)

contaminant	suggested levels	effects
aluminum	0.05–0.2 mg/L	discoloration of water
chloride	250 mg/L	salty taste and pipe corrosion
color	15 color units	visible tint
copper	1.0 mg/L	metallic taste and staining
corrosivity	noncorrosive	taste, staining, and corrosion
fluoride	2.0 mg/L	dental fluorosis
foaming agents	0.5 mg/L	froth, odor, and bitter taste
iron	0.3 mg/L	taste, staining, and sediment
manganese	0.05 mg/L	taste and staining
odor	3 TON[*]	"rotten egg," musty, and chemical odor
pH	6.5–8.5	low pH—metallic taste and corrosion; high pH—slippery feel, soda taste, and deposits
silver	0.1 mg/L	discoloration of skin and graying of eyes
sulfate	250 mg/L	salty taste and laxative effect
total dissolved solids (TDS)	500 mg/L	taste, corrosivity, and soap interference
zinc	5 mg/L	metallic taste

[*]threshold odor number

9. IRON

Even at low concentrations, iron is objectionable because it stains porcelain bathroom fixtures, causes a brown color in laundered clothing, and can be tasted. Typically, iron is a problem in groundwater pumped from anaerobic aquifers in contact with iron compounds. Soluble ferrous ions can be formed under these conditions, which, when exposed to atmospheric air at the surface or to dissolved oxygen in the water system, are oxidized to the insoluble ferric state, causing the color and staining problems mentioned. Iron determinations are made through colorimetric (i.e., wet titration) analysis.

10. MANGANESE

Manganese ions are similar in formation, effect, and measurement to iron ions.

11. FLUORIDE

Natural fluoride is found in groundwaters as a result of dissolution from geologic formations. Surface waters generally contain much smaller concentrations of fluoride. An absence or low concentration of ingested fluoride causes the formation of tooth enamel less resistant to decay, resulting in a high incidence of dental cavities in children's teeth. Excessive concentration of fluoride causes *fluorosis*, a brownish discoloration of dental enamel. The MCL of 4.0 mg/L established by the EPA is to prevent unsightly fluorosis.

Communities with water supplies deficient in natural fluoride may chemically add fluoride during the treatment process. Since water consumption is influenced by climate, the recommended optimum concentrations listed in Table 23.5 are based on the annual average of the maximum air temperatures based on a minimum of five years of records.

Table 23.5 Recommended Optimum Concentrations of Fluoride in Drinking Water

average air temperature range (°F (°C))	recommended optimum concentration (mg/L)
53.7 (12.0) and below	1.2
53.8 to 58.3 (12.1–14.6)	1.1
58.4 to 63.8 (14.7–17.6)	1.0
63.9 to 70.6 (17.7–21.4)	0.9
70.7 to 79.2 (21.5–26.2)	0.8
79.3 to 90.5 (26.3–32.5)	0.7

Compounds commonly used as fluoride sources in water treatment are listed in Table 23.6.

Table 23.6 Fluoridation Chemicals

compound	formula	percentage F^- ion (%)
sodium fluoride	NaF	45
sodium silicofluoride	Na_2SiF_6	61
hydrofluosilicic acid	H_2SiF_6	79
ammonium silicofluoride[*]	$(NH_4)_2SiF_6$	64

[*]used in conjunction with chlorine disinfection where it is desired to maintain a chloramine residual in the distribution system

Example 23.7

A liquid feeder adds a 4.0% saturated sodium fluoride solution to a water supply, increasing the fluoride concentration from the natural fluoride level of 0.4 mg/L to 1.0 mg/L. The commercial NaF powder used to prepare the NaF solution contains 45% fluoride by weight. (a) How many pounds (kilograms) of NaF are required per million gallons (liters) treated? (b) What volume of 4% NaF solution is used per million gallons (liters)?

SI Solution

$$m_{\text{NaF}} = \frac{\left(1.0 \ \frac{\text{mg}}{\text{L}} - 0.4 \ \frac{\text{mg}}{\text{L}}\right)\left(10^6 \ \frac{\text{L}}{\text{ML}}\right)}{(0.45)\left(10^6 \ \frac{\text{mg}}{\text{kg}}\right)}$$

$$= 1.33 \ \text{kg/ML} \quad \text{(kg per million liters)}$$

(b) One liter of water has a mass of one kilogram (10^6 mg). Therefore, a 4% NaF solution contains 40 000 mg NaF per liter. The concentration needs to be increased from 0.4 mg/L to 1.0 mg/L.

$$\left(40\,000 \ \frac{\text{mg}}{\text{L}}\right)(0.45) = 18\,000 \ \text{mg/L}$$

$$V = \frac{\left(1.0 \times 10^6 \ \frac{\text{L}}{\text{ML}}\right) \times \left(1.0 \ \frac{\text{mg}}{\text{L}} - 0.4 \ \frac{\text{mg}}{\text{L}}\right)}{18\,000 \ \frac{\text{mg}}{\text{L}}}$$

$$= 33.3 \ \text{L/ML}$$

Customary U.S. Solution

$$m_{\text{NaF}} = \left(\frac{1.0 \ \frac{\text{mg}}{\text{L}} - 0.4 \ \frac{\text{mg}}{\text{L}}}{0.45}\right)\left(8.345 \ \frac{\text{lbm-L}}{\text{mg-MG}}\right)$$

$$= 11.1 \ \text{lbm/MG} \quad \text{(lbm per million gallons)}$$

(b) A 4% NaF solution contains 40,000 mg NaF per liter. The concentration needs to be increased from 0.4 mg/L to 1.0 mg/L.

$$\left(40,000 \ \frac{\text{mg}}{\text{L}}\right)(0.45) = 18,000 \ \text{mg/L}$$

$$V = \frac{\left(1.0 \times 10^6 \ \frac{\text{gal}}{\text{MG}}\right) \times \left(1.0 \ \frac{\text{mg}}{\text{L}} - 0.4 \ \frac{\text{mg}}{\text{L}}\right)}{18,000 \ \frac{\text{mg}}{\text{L}}}$$

$$= 33.3 \ \text{gal/MG}$$

12. PHOSPHORUS

Phosphate content is more of a concern in wastewater treatment than in supply water, although phosphorus can enter water supplies in large amounts from runoff. Excessive phosphate discharge contributes to aquatic plant (phytoplankton, algae, and macrophytes) growth and subsequent *eutrophication*. (Eutrophication is an "over-fertilization" of receiving waters.)

Phosphorus is of considerable interest in the management of lakes and reservoirs because phosphorus is a nutrient that has a major effect on aquatic plant growth. Algae normally have a phosphorus content of 0.1–1% of dry weight. The molar N:P ratio for ideal algae growth is 16:1.

Phosphorus exists in several forms in aquatic environments. Soluble phosphorus occurs as *orthophosphate*, as condensed *polyphosphates* (from detergents), and as various organic species. Orthophosphates ($H_2PO_4^-$, HPO_4^{--}, and PO_4^{---}) and polyphosphates (such as $Na_3(PO_3)_6$) result from the use of synthetic detergents (*syndets*). A sizable fraction of the soluble phosphorus is in the organic form, originating from the decay or excretion of nucleic acids and algal storage products. *Particulate phosphorus* occurs in the organic form as a part of living organisms and detritus as well as in the inorganic form of minerals such as apatite.

Phosphorus is normally measured by colorimetric or digestion methods. The results are reported in terms of mg/L of phosphorus (e.g., "mg/L of P" or "mg/L of total P"), although the tests actually measure the concentration of orthophosphate. The concentration of a particular compound is found by multiplying the concentration as P by the molecular weight of the compound and dividing by the atomic weight of phosphorus (30.97).

$$[\text{X}] = \frac{[\text{P}](\text{MW}_\text{X})}{30.97} \qquad 23.9$$

A substantial amount of the phosphorus that enters lakes is probably not available to aquatic plants. Bioavailable compounds include orthophosphates, polyphosphates, most soluble organic phosphorus, and a portion of the particulate fraction. Studies have indicated that bioavailable phosphorus generally does not exceed 60% of the total phosphorus.

In aquatic systems, phosphorus does not enter into any redox reactions, nor are any common species volatile. Therefore, lakes retain a significant portion of the entering phosphorus. The main mechanism for retaining phosphorus is simple sedimentation of particles containing the phosphorus. Particulate phosphorus can originate from the watershed (*allochthonous material*) or can be formed within the lake (*autochthonous material*).

A large fraction of the phosphorus that enters a lake is recycled, and much of the recycling occurs at the sediment-water interface. Recycling of phosphorus is linked to iron and manganese recycling. Soluble phosphorus in the water column is removed by adsorption onto iron and manganese hydroxides, which precipitate under aerobic conditions. However, when the *hypolimnion* (i.e., the lower part of the lake that is essentially stagnant) becomes anaerobic, the iron (or manganese) is reduced, freeing up phosphorus. This is consistent with a fairly general observation that phosphorus release rates are nearly an order of magnitude higher under anaerobic conditions than under aerobic conditions.

Factors that control phosphorus recycling rates are not well understood, although it is clear that oxygen status, phosphorus speciation, temperature, and pH are important variables. Phosphorus release from sediments is usually considered to be constant (usually less than 1 mg/m^2-day under aerobic conditions).

In addition to direct regeneration from sediments, *macrophytes* (large aquatic plants) often play a significant role in phosphorus recycling. Macrophytes with highly developed root systems derive most of their phosphorus from the sediments. Regeneration to the water column can occur by excretion or through decay, effectively "pumping" phosphorus from the sediments. The internal loading generated by recycling is particularly important in shallow, eutrophic lakes. In several cases where phosphorus inputs have been reduced to control algal blooms, regeneration of phosphorus from phosphorus-rich sediments has slowed the rate of recovery.

13. NITROGEN

Compounds containing nitrogen are not abundant in virgin surface waters. However, nitrogen can reach large concentrations in ground waters that have been contaminated with barnyard runoff or that have percolated through heavily fertilized fields. Sources of surface water contamination include agricultural runoff and discharge from sewage treatment facilities.

Of greatest interest, in order of decreasing oxidation state, are nitrates (NO_3^-), nitrites (NO_2^-), ammonia (NH_3), and organic nitrogen. These three compounds are reported as *total nitrogen*, TN, with units of "mg/L of N" or "mg/L of total N." The concentration of a particular compound is found by multiplying the concentration as N by the molecular weight of the compound and dividing by the atomic weight of nitrogen (14.01).

$$[X] = \frac{[N](MW_X)}{14.01} \qquad 23.10$$

Excessive amounts of nitrate in water can contribute to the illness in infants known as *methemoglobinemia* ("blue baby" syndrome). As with phosphorus, nitrogen stimulates aquatic plant growth.

Un-ionized ammonia is a colorless gas at standard temperature and pressure. A pungent odor is detectable at levels above 50 mg/L. Ammonia is very soluble in water at low pH.

Ammonia levels in zero-salinity surface water increase with increasing pH and temperature (see Table 23.7). At low pH and temperature, ammonia combines with water to produce ammonium (NH_4^+) and hydroxide (OH^-) ions. The ammonium ion is nontoxic to aquatic life and not of great concern. The un-ionized ammonia (NH_3), however, can easily cross cell membranes and have a toxic effect on a wide variety of fish. The EPA has established the criteria for fresh and saltwater fish that depend on temperature, pH, species, and averaging period.

Table 23.7 *Percentage of Total Ammonia Present in Toxic, Un-Ionized Form*

temp (°F (°C))	pH								
	6.0	6.5	7.0	7.5	8.0	8.5	9.0	9.5	10.0
41 (5)	0.013	0.040	0.12	0.39	1.2	3.8	11	28	56
50 (10)	0.019	0.059	0.19	0.59	1.8	5.6	16	37	65
59 (15)	0.027	0.087	0.27	0.86	2.7	8.0	21	46	73
68 (20)	0.040	0.13	0.40	1.2	3.8	11	28	56	80
77 (25)	0.057	0.18	0.57	1.8	5.4	15	36	64	85
86 (30)	0.080	0.25	0.80	2.5	7.5	20	45	72	89

Ammonia is usually measured by a distillation and titration technique or with an ammonia-selective electrode. The results are reported as ammonia nitrogen. Nitrites are measured by a colorimetric method. The results are reported as nitrite nitrogen. Nitrates are measured by ultraviolet spectrophotometry, selective electrode, or reduction methods. The results are reported as nitrate nitrogen. Organic nitrogen is determined by a digestion process that identifies organic and ammonia nitrogen combined. Organic nitrogen is found by subtracting the ammonia nitrogen value from the digestion results.

14. COLOR

Color in water is caused by substances in solution, known as *true color*, and by substances in suspension, mostly organics, known as *apparent* or *organic color*. Iron, copper, manganese, and industrial wastes all can cause color. Color is aesthetically undesirable, and it stains fabrics and porcelain bathroom fixtures.

Water color is determined by comparison with standard platinum/cobalt solutions or by spectrophotometric methods. The standard color scales range from 0 (clear) to 70. Water samples with more intense color can be evaluated using a dilution technique.

15. TURBIDITY

Turbidity is a measure of the light-transmitting properties of water and is comprised of suspended and colloidal material. Turbidity is expressed in *nephelometric turbidity units* (NTU). Viruses and bacteria become attached to these particles, where they can be protected from the bactericidal and viricidal effects of chlorine, ozone, and other disinfecting agents. The organic material included in turbidity has also been identified as a potential precursor to carcinogenic disinfection by-products.

Turbidity in excess of 5 NTU is noticeable by visual observation. Turbidity in a typical clear lake is approximately 25 NTU, and muddy water exceeds 100 NTU. Turbidity is measured using an electronic instrument called a nephelometer, which detects light scattered by the particles when a focused light beam is shown through the sample.

16. SOLIDS

Solids present in a sample of water can be classified in several ways.

- *total solids* (TS): Total solids are the material residue left in the vessel after the evaporation of the sample at 103–105°C. Total solids include total suspended solids and total dissolved solids.

- *total suspended solids* (TSS): The material retained on a standard glass-fiber filter disk is defined as the suspended solids in a sample. The filter is weighed before filtration, dried at 103–105°C, and weighed again. The gain in weight is the amount of suspended solids. Suspended solids can also be categorized into *volatile suspended solids* (VSS) and *fixed suspended solids* (FSS).

- *total dissolved solids* (TDS): These solids are in solution and pass through the pores of the standard glass-fiber filter. Dissolved solids are determined by passing the sample through a filter, collecting the filtrate in a weighed drying dish, and evaporating the liquid at 180°C. The gain in weight represents the dissolved solids. Dissolved solids can be categorized into *volatile dissolved solids* (VDS) and *fixed dissolved solids* (FDS).

- *total volatile solids* (TVS): The residue from one of the previous determinations is ignited to constant weight in an electric muffle furnace at 550°C. The loss in weight during the ignition process represents the volatile solids.

- *total fixed solids* (TFS): The weight of solids that remain after the ignition used to determine volatile solids represents the fixed solids.

- *settleable solids*: The volume (mL/L) of settleable solids is measured by allowing a sample to settle for one hour in a graduated conical container (*Imhoff cone*).

The following relationships exist.

$$TS = TSS + TDS = TVS + TFS \qquad 23.11$$

$$TSS = VSS + FSS \qquad 23.12$$

$$TDS = VDS + FDS \qquad 23.13$$

$$TVS = VSS + VDS \qquad 23.14$$

$$TFS = FSS + FDS \qquad 23.15$$

Waters with high concentrations of suspended solids are classified as *turbid waters*. Waters with high concentrations of dissolved solids often can be tasted. Therefore, a limit of 500 mg/L has been established in the National Secondary Drinking Water Standards for dissolved solids.

17. CHLORINE AND CHLORAMINES

Chlorine is the most common disinfectant used in water treatment. It is a strong oxidizer that deactivates microorganisms. Its oxidizing capability also makes it useful in removing soluble iron and manganese ions.

Chlorine gas in water forms *hydrochloric* and *hypochlorous acids*. At a pH greater than 9, hypochlorous acid dissociates to hydrogen and hypochlorite ions, as shown in Eq. 23.16.

$$Cl_2 + H_2O \underset{pH<4}{\overset{pH>4}{\rightleftharpoons}} HCl + HOCl \underset{pH<9}{\overset{pH>9}{\rightleftharpoons}} H^+ + OCl^- $$
$$23.16$$

Free chlorine, hypochlorous acid, and hypochlorite ions left in water after treatment are known as *free chlorine residuals*. Hypochlorous acid reacts with ammonia (if it is present) to form *chloramines*. Chloramines are known as *combined residuals*. Chloramines are more stable than free residuals, but their disinfecting ability is less.

$$HOCl + NH_3 \rightarrow H_2O + NH_2Cl \text{ (monochloramine)}$$
$$23.17$$

$$HOCl + NH_2Cl \rightarrow H_2O + NHCl_2 \text{ (dichloramine)}$$
$$23.18$$

$$HOCl + NHCl_2 \rightarrow H_2O + NCl_3 \text{ (trichloramine)}$$
$$23.19$$

Free and combined residual chlorine can be determined by color comparison, by titration, and with chlorine-sensitive electrodes. Color comparison is the most common field method.

18. HALOGENATED COMPOUNDS

Halogenated compounds have become a subject of concern in the treatment of water due to their potential as carcinogens. The use of chlorine as a disinfectant in water treatment generates these compounds by reacting with organic substances in the water. (For this reason there are circumstances under which chlorine may not be the most appropriate disinfectant.)

The organic substances, called *precursors*, are not in themselves harmful, but the chlorinated end products, known by the term *disinfection by-products* (DBPs) raise serious health concerns. Typical precursors are decay by-products such as humic and fulvic acids. Several of the DBPs contain bromine, which is found in low concentrations in most surface waters and can also occur as an impurity in commercial chlorine gas.

The DBPs can take a variety of forms depending on the precursors present, the concentration of free chlorine, the contact time, the pH, and the temperature. The most common ones are the trihalomethanes (THMs), haloacetic acids (HAAs), dihaloacetonitriles (DHANs), and various trichlorophenol isomers.

19. TRIHALOMETHANES

Only four trihalomethane (THM) compounds are normally found in chlorinated waters.

$CHCl_3$	trichloromethane (chloroform)
$CHBrCl_2$	bromodichloromethane
$CHBr_2Cl$	dibromochloromethane
$CHBr_3$	tribromomethane (bromoform)

Trihalomethanes are regulated by the EPA under the National Primary Drinking Water Standards. Standards for total *haloacetic acids* (referred to as "HAA5" in consideration of the five compounds identified) have also been established.

20. HALOACETIC ACIDS

The haloacetic acids (HAAs) exist in tri-, di-, and mono-forms, abbreviated as THAAs, DHAAs, and MHAAs. All of the haloacetic acids are toxic and are suspected or proven carcinogens. Maximum concentration limits for HAAs are listed in App. 24.A.

$C_2HCl_3O_2$	trichloroacetic acid	(TCAA)
$C_2HBrCl_2O_2$	bromodichloroacetic acid	(BDCAA)
$C_2HBr_2ClO_2$	dibromochloroacetic acid	(DBCAA)
$C_2HBr_3O_2$	tribromoacetic acid	(TBAA)
$C_2H_2Cl_2O_2$	dichloroacetic acid	(DCAA)
$C_2H_2BrClO_2$	bromochloroacetic acid	(BCAA)
$C_2H_2Br_2O_2$	dibromoacetic acid	(DBAA)
$C_2H_3ClO_2$	monochloroacetic acid	(MCAA)
$C_2H_3BrO_2$	monobromoacetic acid	(MBAA)

21. DIHALOACETONITRILES

The dihaloacetonitriles (DHANs) are formed when acetonitrile (methyl cyanide: C_2H_3N (structurally H_3CCN)) is exposed to chlorine. All are toxic and suspected carcinogens. Maximum concentration limits have not been established by the EPA for DHANs.

C_2HCl_2N	dichloroacetonitrile	(DCAN)
$C_2HBrClN$	bromochloroacetonitrile	(BCAN)
C_2HBr_2N	dibromoacetonitrile	(DBAN)

22. TRICHLOROPHENOL

Trichlorophenol can exist in six isomeric forms, with varying potential toxicities. As with all halogenated organics, they are potential carcinogens.

2,3,4-trichlorophenol	
2,3,5-trichlorophenol	
2,3,6-trichlorophenol	
2,4,5-trichlorophenol	irritant
2,4,6-trichlorophenol	fungicide, bactericide
3,4,5-trichlorophenol	

23. AVOIDANCE OF DISINFECTION BY-PRODUCTS IN DRINKING WATER

Reduction of DBPs can best be achieved by avoiding their production in the first place. The best strategy dictates using source water with few or no organic precursors.

Often source water choices are limited, necessitating tailoring treatment processes to produce the desired result. This entails removing the precursors prior to the application of chlorine, applying chlorine at certain points in the treatment process that minimize production of DBPs, using disinfectants that do not produce significant DBPs, or a combination of these techniques.

Removal of precursors is achieved by preventing growth of vegetative material (algae, plankton, etc.) in the source water and by collecting source water at various depths to avoid concentrations of precursors. Oxidizers such as potassium permanganate and chlorine dioxide can often reduce the concentration of the precursors without forming the DBPs. Use of activated carbon, pH-adjustment processes or dechlorination can reduce the impact of chlorination.

Chlorine application should be delayed if possible until after the flocculation, coagulation, settling, and filtration processes have been completed. In this manner, turbidity and common precursors will be reduced. If it is necessary to chlorinate early to facilitate treatment processes, chlorination can be followed by dechlorination to reduce contact time. Granular activated carbon has been used to some extent to remove DBPs after they form, but the carbon needs frequent regeneration.

Alternative disinfectants include ozone, chloramines, chlorine dioxide, iodine, bromine, potassium permanganate, hydrogen peroxide, and ultraviolet radiation. Ozone and chloramines, singularly or together, are often used for control of THMs. However, ozone creates other DBPs, including aldehydes, hydrogen peroxide, carboxylic acids, ketones, and phenols. When ozone is used as the primary disinfectant, a secondary disinfectant such as chlorine or chloramine must be used to provide an active residual that can be measured within the distribution system.

24. SALINITY IN IRRIGATION WATER

Crop yield is affected greatly by *salinity*. Plants can only transpire pure water. Therefore, the primary effect of high salinity on crop productivity is the inability of plants to compete with ions in the soil solution for water. The higher the salinity, the less water is available to plants, even though the soil is saturated. The specific sensitivity to salinity depends greatly on the crop, type of soil, and salinity depth.

Salinity can be characterized in several ways, including ionic concentrations and electrical conductivity. When salinity is characterized by relative portions of various ions, the deleterious effects are referred to as *salinity hazard*, *sodium hazard*, *boron hazard*, and *bicarbonate hazard*.

The *soluble sodium percent (percent sodium content)*, SSP, is defined by Eq. 23.20, where the concentrations are expressed in meq/L. Water quality for irrigation is roughly classified based on the SSP according to: $< 20\%$, excellent; 20–40%, good; 20–60% permissible; 60–80% doubtful; and $> 80\%$, unsuitable.

$$\text{SSP} = \frac{[\text{Na}^+] \times 100\%}{[\text{Ca}^{++}] + [\text{Mg}^{++}] + [\text{Na}^+] + [\text{K}^+]} \quad 23.20$$

In waters with high concentrations of bicarbonate ions, calcium and magnesium may precipitate out as the water in the soil becomes more concentrated. As a result, the relative proportion of sodium in the form of sodium carbonate increases. The *relative proportion of sodium carbonate*, RSC, is defined by Eq. 23.21, where all of the concentrations are expressed as meq/L. The U.S. Department of Agriculture categorizes water according to its RSC as good (< 1.5 meq/L), doubtful (1.5–2.5 meq/L), and unsuitable (> 2.5 meq/L).

$$\text{RSC} = \left([\text{HCO}_3^-] + [\text{CO}_3^{--}]\right) - \left([\text{Ca}^{++}] + [\text{Mg}^{++}]\right) \quad 23.21$$

Percent sodium and RSC are based on chemical species, but not on reactions with the soil itself. A better measure of the sodium hazard for irrigation is the *sodium adsorption ratio*, SAR. SAR is an indicator of the amount of sodium in the water relative to calcium and magnesium. Sodium hazard is low with SAR < 9. For SAR 10–17, hazard is medium, and soil amendments (e.g., gypsum) and leaching may be required. Water with SAR 18–25 has a high sodium hazard and cannot be used continuously. An SAR in excess of 25 represents unsuitable water.

$$\text{SAR} = \frac{[\text{Na}^+]}{\sqrt{\dfrac{[\text{Ca}^{++}] + [\text{Mg}^{++}]}{2}}} \quad 23.22$$

Salinity, like total dissolved solids, is also commonly categorized by electrical conductivity (EC) and reported in dS/m.[1,2] Water is classified according to EC as excellent (< 0.25 dS/m), good (0.25–0.75 dS/m), permissible (0.75–2.0 dS/m), doubtful (2.0–3.0 dS/m), and unsuitable (> 3.0 dS/m).

[1]The subscripts w, iw, or e may be used (e.g., EC_w) to indicate water, irrigation water, or saturated soil extract.
[2]Decisiemens per meter, dS/m, is numerically the same as mS/cm. The siemens, S, is equivalent to the obsolete mho unit.

24 Water Supply Treatment and Distribution

Nomenclature

A	surface area	ft^2	m^2
AADF	average annual daily flow	gal/day	L/d
b	width, length, distance	ft	m
BFF	basic fire flow	gal/min	L/s
C	coefficient	–	–
C	concentration	mg/L	mg/L
C	construction factor	–	–
D	diameter	ft	m
D	dose	mg/L	mg/L
E	modulus of elasticity	lbf/ft^2	Pa
f	frequency	Hz	Hz
f	removal fraction	–	–
F	construction class coefficient	–	–
F	feed rate	lbm/day	kg/d
F	fire fighting construction factor	–	–
F	force	lbf	N
g	acceleration of gravity, 32.2 (9.81)	ft/sec^2	m/s^2
g_c	gravitational constant, 32.2	lbm-ft/lbf-sec^2	n.a.
G	availability	–	–
G	gravimetric fraction	–	–
G	mixing velocity gradient	1/sec	1/s
h	head (depth), height	ft	m
I	moment of inertia	ft^4	m^4
K	mixing rate constant	1/sec	1/s
K_a	ionization constant	–	–
L	length	ft	m
m	mass	lbm	kg
M	alkalinity	mg/L	mg/L
M	demand multiplier	–	–
M	dose (miscellaneous)	mg/L	mg/L
MPN	most probable number	–	–
n	number of waters of hydration	–	–
n	rotational speed	rev/sec	rev/s
N	dimensionless number	–	–
NFF	needed fire flow	gal/min	L/s
O	occupancy combustible factor	–	–
P	communication factor	–	–
P	population	thousands	thousands
P	power	hp	kW
P	purity	–	–

Q	flow rate	ft^3/sec or MGD	m^3/s or L/s
Re	Reynolds number	–	–
SG	specific gravity	–	–
t	time	sec	s
T	temperature	°F	°C
TDS	total dissolved solids	mg/L	mg/L
TON	threshold odor number	–	–
TSS	total suspended solids	mg/L	mg/L
v	velocity	ft/sec	m/s
v*	overflow rate	gal/day-ft^2	L/d·m^2
V	volume	ft^3	m^3
x	mass fraction remaining	–	–
X	exposure factor	–	–
y	vertical distance	ft	m

Symbols

γ	specific weight	lbf/ft^3	N/m^3
η	collection efficiency	–	–
μ	absolute viscosity	lbf-sec/ft^2	Pa·s
ν	kinematic viscosity	ft^2/sec	m^2/s
ρ	density	lbm/ft^3	kg/m^3

Subscripts

c	critical
d	detention
D	drag
eq	equilibrium
f	flow-through or fraction
i	in
L	per unit length
o	out
P	power
Q	flow
s	settling
sat	saturation
sc	settling column
sp	solubility product
v	velocity

1. WATER TREATMENT PLANT LOCATION

The location chosen for a water treatment plant is influenced by many factors. The most common factors include availability of resources such as (a) local water, (b) power, and (c) sewerage services; economic factors such as (d) land cost and (e) annual taxes; and environmental factors such as (f) traffic and (g) other concerns that would be identified on an environmental impact report.

It is imperative to locate water treatment plants (a) above the flood plain and (b) where a 15–20 ft (4.5–6 m) elevation difference exists. This latter requirement will eliminate the need to pump the water between processes. Traditional treatment requires a total head of approximately 15 ft (4.5 m), whereas advanced processes such as granular activated charcoal and ozonation increase the total head required to approximately 20 ft (6 m).

The average operational life of equipment and facilities is determined by economics. However, it is unlikely that a lifetime of less than 50 years would be economically viable. Equipment lifetimes of 25–30 years are typical.

2. PROCESS INTEGRATION

The processes and sequences used in a water treatment plant depend on the characteristics of the incoming water. However, some sequences are more appropriate than others due to economic and hydraulic considerations. *Conventional filtration*, also referred to as *complete filtration*, is a term used to describe the traditional sequence of adding coagulation chemicals, flash mixing, coagulation-flocculation, sedimentation, and subsequent filtration. Coagulants, chlorine (or an alternative disinfectant), fluoride, and other chemicals are added at various points along the path, as indicated by Fig. 24.1. Conventional filtration is still the best choice when incoming water has high color, turbidity, or other impurities.

Direct filtration refers to a modern sequence of adding coagulation chemicals, flash mixing, minimal flocculation, and subsequent filtration. In direct filtration, the physical chemical reactions of flocculation occur to some extent, but special flocculation and sedimentation facilities are eliminated. This reduces the amount of sludge that has to be treated and disposed of. Direct filtration is applicable when the incoming water is of high initial quality.

In-line filtration refers to another modern sequence that starts with adding coagulation chemicals at the filter inlet pipe. Mixing occurs during the turbulent flow toward a filter, which is commonly of the pressure-filter variety. As with direct filtration, flocculation and sedimentation facilities are not used.

Table 24.1 provides guidelines based on incoming water characteristics for choosing processes required to achieve satisfactory quality.

3. PRETREATMENT

Preliminary treatment is a general term that usually includes all processes prior to the first flocculation operation (i.e., pretreatment, screening, presedimentation, microstraining, aeration, and chlorination). Flow measurement is usually considered to be part of the pretreatment sequence.

4. SCREENING

Screens are used to protect pumps and mixing equipment from large objects. The degree of screening required will depend on the nature of solids expected. Screens can be either manually or automatically cleaned.

Water Treatment

Figure 24.1 *Chemical Application Points*

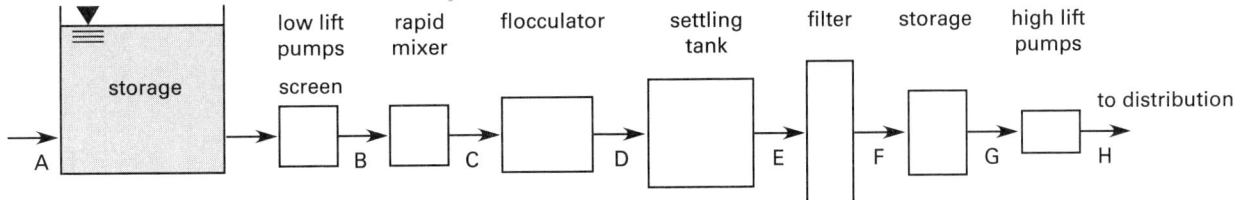

typical flow diagram of water treatment plant

category of chemicals	possible points of application							
	A	B	C	D	E	F	G	H
algicide	X				X			
disinfectant		X	X		X	X	X	X
activated carbon		X	X	X	X			
coagulants		X	X					
coagulation aids		X	X		X			
alkali								
for flocculation			X					
for corrosion control						X		
for softening			X					
acidifier			X		X			
fluoride					X			
cupric-chloramine					X			
dechlorinating agent					X			X

Note: With solids contact reactors, point C is the same as point D.

5. MICROSTRAINING

Microstrainers are effective at removing 50–95% of the algae in incoming water. Microstrainers are constructed from woven stainless steel fabric mounted on a hollow drum that rotates at 4–7 rpm. Flow is usually radially outward through the drum. The accumulated filter cake is removed by backwashing.

6. ALGAE PRETREATMENT

Biological growth in water from impounding reservoirs, lakes, storage reservoirs, and settling basins can be prevented or eliminated with an *algicide* such as copper sulfate. Such growth can produce unwanted taste and odors, clog fine-mesh filters, and contribute to the buildup of slime.

Typical dosages of copper sulfate vary from 0.54 lbm/ac-ft to 5.4 lbm/ac-ft (0.2 mg/L to 2.0 mg/L), with the lower dose usually being adequate for waters that are soft or have alkalinities less than 50 mg/L as $CaCO_3$. (For aqueous solutions, units of mg/L and ppm are equivalent.) Since it is toxic, copper sulfate should not be used without considering and monitoring the effects on aquatic life (e.g., fish).

7. PRECHLORINATION

A prechlorination process was traditionally employed in most water treatment plants. Many plants have now eliminated all or part of the prechlorination due to the formation of THMs. In so doing, benefits such as algae control have been eliminated. Alternative disinfection chemicals (e.g., ozone and potassium permanganate) can be used. Otherwise, coagulant doses must be increased.

8. PRESEDIMENTATION

The purpose of presedimentation is to remove easily settled sand and grit. This can be accomplished by using pure sedimentation basins, sand and grit chambers, and various passive cyclone degritters. *Trash racks* may be integrated into sedimentation basins to remove leaves and other floating debris.

9. FLOW MEASUREMENT

Flow measurement is incorporated into the treatment process whenever the water is conditioned enough to be compatible with the measurement equipment. Flow measurement devices should not be exposed to scour from grit or highly corrosive chemicals (e.g., chlorine).

Flow measurement often takes place in a Parshall flume. Chemicals may be added in the flume to take advantage of the turbulent mixing that occurs at that point. All of the traditional fluid measurement devices are also applicable, including venturi meters, orifice plates, propeller and turbine meters, and modern passive devices such as magnetic and ultrasonic flowmeters.

Table 24.1 *Treatment Methods*

incoming water quality		pretreatment				treatment					special treatments			
constituents	concentration (mg/L)	screening	prechlorination	plain settling	aeration	lime softening	coagulation and sedimentation	rapid sand filtration	slow sand filtration	postchlorination	superchlorination[a] or chloramoniation	active carbon	special chemical treatment	salt water conversion[b]
coliform monthly average (MPN[e]/100 mL)	0–20									E				
	20–100			O			O	O	O	E				
	100–5000		E				E	E	O	E				
	5000		E	O[c]			E	E		E	O			
suspended solids	0–100	O							O					
	100–200	O					E	E						
	200	O		O[d]			E	E						
color, (mg/L)	20–70						O	O				O		
	70						E	E				O		
tastes and odors	noticeable		O		O					O	O	E		
CaCO₃, (mg/L)	200					E	E	E					E	
pH	6.5–8.5 (normal)													
iron and manganese, (mg/L)	≤0.3		O	O										
	0.3–1.0				O		E	E	O					
	1.0		E		E		E	E	O				O	
chloride, (mg/L)	0–250													
	250–500													O
	500													E
phenolic compounds, (mg/L)	0–0.005						O	O			O	O		
	0.005						E	E			O	E	O	
toxic chemicals							E	E				E	O	
less critical chemicals							O	O				O	O	

Note: E = essential, O = optional
[a]Superchlorination shall be followed by dechlorination.
[b]As an alternative, dilute with low chloride water.
[c]Double settling shall be provided for coliform exceeding 20,000 MPN/100 mL.
[d]For extremely muddy water, presedimentation by plain settling may be provided.
[e]MPN = most probable number

10. AERATION

Aeration is used to reduce taste- and odor-causing compounds, to lower the concentration of dissolved gases (e.g., hydrogen sulfide), to increase dissolved CO_2 (i.e., recarbonation) or decrease CO_2, to reduce iron and manganese, and to increase dissolved oxygen.

Various types of aerators are used. The best transfer efficiencies are achieved when the air-water contact area is large, the air is changed rapidly, and the aeration period is long. *Force draft air injection* is common. The release depth varies from 10 ft to 25 ft (3 m to 7.5 m). The ideal compression power required to aerate water with a simple air injector depends on the air flow rate, Q, and head (i.e., which must be greater than the release depth), h, at the point where the air is injected. Motor-compression-distribution efficiencies are typically around 75%.

$$P_{kW} = \frac{Q_{L/s} h_m}{100} \qquad \text{[SI]} \quad 24.1(a)$$

$$P_{hp} = \frac{Q_{cfm} h_{ft}}{528} \qquad \text{[U.S.]} \quad 24.1(b)$$

Diffused air systems with compressed air at 5–10 psig (35–70 kPa) are the most efficient methods of aerating water. The air injection volume is 0.2–0.3 ft³/gal. However, the equipment required to produce and deliver compressed air is more complex than with simple injectors. The *transfer efficiency* of a diffused air system varies with depth and bubble size. With injection depths of 5–10 ft (1.5–3.0 m), if coarse bubbles are produced, only 4–8% of the available oxygen will be transferred to the water. With medium-sized bubbles, the efficiency can be 6–15%, and it can approach 10–30% with fine bubble systems.

11. SEDIMENTATION PHYSICS

Water containing suspended sediment can be held in a *plain sedimentation tank* (basin) that allows the particles to settle out.[1] Settling velocity and settling time for sediment depends on the water temperature (i.e., viscosity), particle size, and particle specific gravity. (The specific gravity of sand is usually taken as 2.65.) Typical settling velocities are as follows: gravel, 1 m/s; coarse sand, 0.1 m/s; fine sand, 0.01 m/s; and silt, 0.0001 m/s. Bacteria and colloidal particles are generally considered to be nonsettleable during the detention periods available in water treatment facilities.

Settlement time can be calculated from the settling velocity and the depth of the tank. If it is necessary to determine the settling velocity of a particle with diameter D, the following procedure can be used.[2]

step 1: Assume a settling velocity, v_s.

step 2: Calculate the settling Reynolds number, Re.

$$\text{Re} = \frac{v_s D}{\nu} \qquad 24.2$$

step 3: (a) If Re < 1, use *Stokes' law.*

$$
\begin{aligned}
v_{s,\text{m/s}} &= \frac{(\rho_{\text{particle}} - \rho_{\text{water}})D_{\text{m}}^2 g}{18\mu} \\
&= \frac{(\text{SG}_{\text{particle}} - 1)D_{\text{m}}^2 g}{18\nu} \qquad \text{[SI]} \quad 24.3(a)
\end{aligned}
$$

$$
\begin{aligned}
v_{s,\text{ft/sec}} &= \frac{(\rho_{\text{particle}} - \rho_{\text{water}})D_{\text{ft}}^2 g}{18\mu g_c} \\
&= \frac{(\text{SG}_{\text{particle}} - 1)D_{\text{ft}}^2 g}{18\nu} \qquad \text{[U.S.]} \quad 24.3(b)
\end{aligned}
$$

(b) If 1 < Re < 2000, use Fig. 24.2, which gives theoretical settling velocities in 68°F (20°C) water for spherical particles with specific gravities of 1.05, 1.2, and 2.65. Actual settling velocities will be much less than shown because particles are not actually spherical.

(c) If Re > 2000, use *Newton's first law of motion* and balance the weight of the particle against the buoyant and drag forces. A spherical shape is often assumed in determining the drag coefficient, C_D. In that case, for laminar descent, use $C_D = 24/\text{Re}$.

$$v_s = \sqrt{\frac{4gD(\text{SG}_{\text{particle}} - 1)}{3C_D}} \qquad 24.4$$

---

[1] The term "plain" refers to the fact that no chemicals are used as coagulants.
[2] This calculation procedure is appropriate for the *Type I settling* that describes sand, sediment, and grit particle performance. It is not appropriate for *Type II settling*, which describes floc and other particles that grow as they settle.

Figure 24.2 *Settling Velocities, 68°F (20°C)*

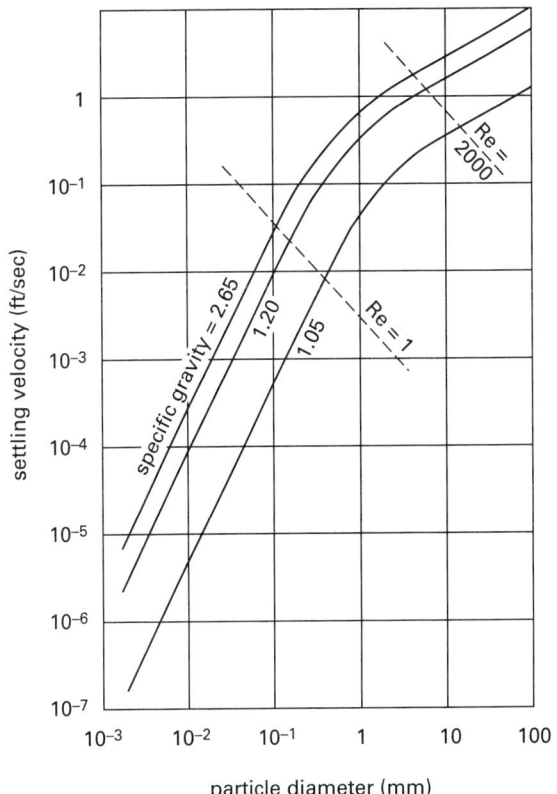

12. SEDIMENTATION TANKS

Sedimentation tanks are usually concrete, rectangular or circular in plan, and are equipped with scrapers or raking arms to periodically remove accumulated sediment. (See Fig. 24.3.) Steel should be used only for small or temporary installations. Where steel parts are unavoidable, as in the case of some rotor parts, adequate corrosion resistance is necessary.

Figure 24.3 *Sedimentation Basin*

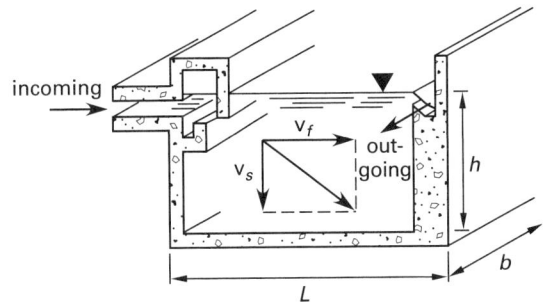

Water flows through the tank at the average *flow-through velocity*, v_f. The flow-through velocity can be found by injecting a colored dye into the tank. It should not exceed 1 ft/min (0.5 cm/sec). The time that water spends in the tank depends on the flow-through velocity and the tank length, L, typically 100–200 ft (30–60 m).

Water Treatment

The minimum settling time depends on the tank depth, h, typically 6–15 ft (1.8–4.5 m).

$$t_{\text{settling}} = \frac{h}{v_s} \qquad 24.5$$

The time that water remains in the basin is known as the *detention time* (*retention time, detention period*, etc.). Typical detention times range from 2 hr to 6 hr, although periods from 1 hr to 12 hr are used depending on the size of particles. All particles will be removed whose t_{settling} is less than t_d.

$$t_d = \frac{V_{\text{tank}}}{Q} = \frac{Ah}{Q} \qquad 24.6$$

Rectangular basins are preferred. Rectangular basins should be constructed with aspect ratios greater than 3:1, and preferably greater than 4:1. The bottom should be sloped toward the drain at no less than 1%. Multiple inlet ports along the entire inlet wall should be used. If there are fewer than four inlet ports, an inlet baffle should be provided.

Square or circular basins are appropriate only when space is limited. The slope toward the drain should be greater, typically 8%. A baffled center inlet should be provided. For radial-flow basins, the diameter is on the order of 100 ft (30 m).

Basin efficiency can approach 80% for fine sediments. Virtually all of the coarse particles are removed. Theoretically, all particles with settling velocities greater than the *overflow rate*, v^*, also known as the *surface loading* or *critical velocity*, will be removed. In Eq. 24.7, b is the tank width, typically 30–40 ft (9–12 m). At least two basins should be constructed (both normally operating) so that one can be out of service for cleaning during low-volume periods without interrupting plant operation. Therefore, Q_{filter} in Eq. 24.7 should be calculated by dividing the total plant flow by 2 or more. The overflow rate is typically 600–1000 gpd/ft^2 (24–40 kL/d·m^2) for rectangular tanks. For square and circular basins, the range is approximately 500–750 gpd/ft^2 (20–30 kL/d·m^2).

$$v^* = \frac{Q_{\text{filter}}}{A_{\text{surface}}} = \frac{Q_{\text{filter}}}{bL} \qquad 24.7$$

In some modern designs, sedimentation has been enhanced by the installation of vertically inclined *laminar tubes* or inclined plates (*lamella plates*) at the tank bottom. These passive tubes, which are typically about 2 in (50 mm) in diameter, allow the particles to fall only a short distance in more turbulent water before entering the tube in which the flow is laminar. Incoming solids settle to the lower surface of the tube, slide downward, exit the tube, and settle to the sedimentation basin floor.

Weir loading is the daily flow rate divided by the total effluent weir length. Weir loading is commonly specified as 15,000–20,000 gal/day-ft (190–250 kL/d·m), but certainly less than 50,000 gal/day-ft (630 kL/d·m).

The accumulated sediment is referred to as *sludge*. It is removed either periodically or on a continual basis, when it has reached a concentration of 25 mg/L or is organic. Various methods of removing the sludge are used, including scrapers and pumps. The linear velocity of sludge scrapers should be 15 ft/min (7.5 cm/sec) or higher.

13. SEDIMENTATION REMOVAL EFFICIENCY: UNMIXED BASINS

Sedimentation efficiency, also known as *removal fraction* and *collection efficiency*, refers to the gravimetric fraction (i.e., fraction by weight) of suspended particles that is removed by sedimentation. Since the suspended particles vary in size, shape, and specific gravity, the distribution of particles is characterized by a distribution of settling velocities, v_s. The distribution is determined, almost always, by a *laboratory settling column test*. Turbid water is placed in a 3–8 ft (1–2.4 m) long vertical column and allowed to settle, with the settled mass being measured over time.

The settling test produces multiple discrete cumulative mass fraction settled, $1 - x$, versus settling time, t_s, data pairs. These are transformed to mass fraction remaining, x, versus settling velocity, $v_s = h_{\text{sc}}/t_s$ data pairs, where h_{sc} is the height of the settling column. The distribution data can be presented in several ways, including settling test results and cumulative curves of fraction remaining or fraction removed, as shown in Fig. 24.4. The *fraction remaining curve* (shown in Fig. 24.4(c)) plots the mass fraction, x, remaining in suspension and is also known as the *mass fraction curve*. The curve shape is the same as the original settling test results. This curve is usually drawn visually through the settling test data points, although more sophisticated curve fitting methods can be used. The curve may also be drawn simply as linear segments between the data points.

Consider a *quiet sedimentation basin* where water enters on one side and is withdrawn on the other side, without any mixing. Some particles have such a high settling velocity that they will settle no matter where they are introduced into the settling basin. Other particles have such a low settling velocity that they will never settle out during the hydraulic retention time of the basin. In between, some particles will settle to various degrees, depending on their initial depths in the settling basin. This leads to two parameters being used to describe the removal fraction: the removal fraction of particles that are completely removed and the total fraction of all particles removed.[3]

Whether or not a particle settles depends on its settling velocity, the time available to settle, and its initial location. The time available to settle depends on the horizontal component of flow velocity and on the horizontal distance between the settling basin's points of flow entry and removal.

---

[3]That is, the "fraction completely collected" is different from the "total fraction collected."

Figure 24.4 *Settling Test Data Representations*

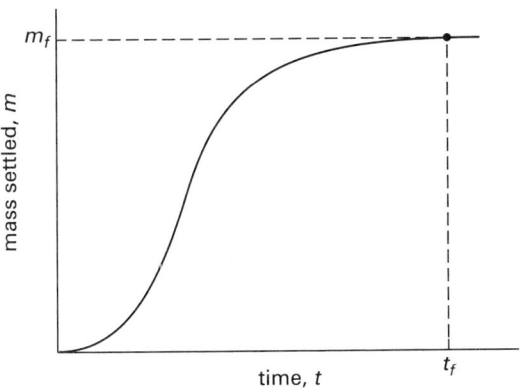

(a) settling test results (mass settled vs. time)

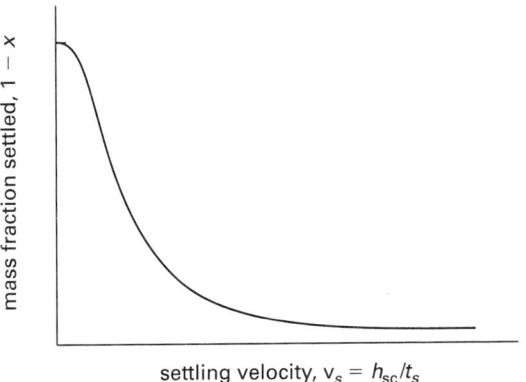

(b) fraction settled vs. settling velocity

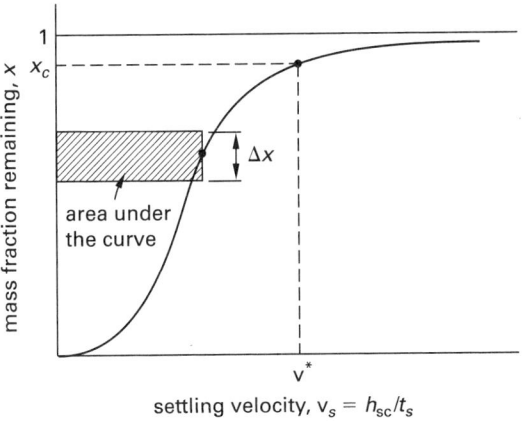

(c) fraction remaining vs. settling velocity

Due to the continuous outflow, the tank contents experience a horizontal flow-through velocity, v_f, between the entry and removal points. If a particle settles with speed v_s, its vertical fall, y, over the flow path length (generally, the length of the tank), L, is

$$y = v_s t = \frac{v_s L}{v_f} \qquad 24.8$$

The vertical fall distance, y, may be less than or greater than the basin depth, h. If $y \geq h$, the particle will reach

the bottom before reaching the end of the basin and will be collected. If $y < h$, the particle may or may not settle, depending on the depth at which it starts. If it starts settling close to the bottom, it will settle; otherwise, it will not fall enough and will escape with the outflow.

The *critical settling velocity*, v_c, is numerically the same as the overflow rate, v^*, and corresponds to the limiting velocity for particles experiencing total removal. Therefore, the fraction of particles experiencing total removal (a collection efficiency of 100%) can be read directly from the mass fraction remaining curve as $1 - x_c$, where x_c is the mass fraction remaining corresponding to $v_s = v_c = v^*$. If the mass fraction curve has not been plotted (or, has been plotted only roughly, as is usually the case), the value of x_c can be determined from the settling test data points by linear interpolation.

The total fraction of all particles removed, f, includes the fraction of particles with settling velocities greater than the critical velocity plus some fraction of the particles with settling velocities less than the critical velocity. For particles with settling velocities less than the critical velocity, the collection efficiency in both rectangular and circular sedimentation basins is

$$\eta = \frac{v_s}{v^*} \quad [v_s \leq v^*] \qquad 24.9$$

The total removal fraction (total collection efficiency) is

$$f = 1 - x_c + \int_{x=0}^{x=x_c} \frac{v_s}{v^*} \, dx$$

$$\approx 1 - x_c + \frac{1}{v^*} \sum_{x=0}^{x=x_c} v_s \Delta x \qquad 24.10$$

The area beneath the mass fraction curve corresponds to the integral term in Eq. 24.10.[4] Since it will almost never be possible to describe the mass fraction curve mathematically, integrating the area of the curve must be done manually by dividing the area under the curve into rectangles (i.e., replacing the curve with a histogram), as is represented by the second form of Eq. 24.10. The rectangle widths, Δx, may vary and are chosen for convenience. The values of v_s in the second form of Eq. 24.10 correspond to the midpoints of the rectangle widths.

14. SEDIMENTATION EFFICIENCY: MIXED BASINS

Quiet (unmixed) settling basins have the highest removal fractions, but mixing is incorporated into some designs for various reasons. Two types of ideal basins that can be mathematically evaluated are basins with pure transverse mixing and basins with thorough mixing. With transverse mixing, a particle may be moved

---

[4]The variable of integration, x, is the fraction remaining and is plotted vertically on a mass fraction curve. x is not the horizontal axis. The "area under the curve" is approximated by horizontal (not vertical) histogram bars.

vertically or side to side in either direction, randomly, but it will not move forward or backwards. In other words, the particle is constrained to a transverse plane perpendicular to the overall travel path between entry and discharge points. The collection efficiency as a function of settling velocity for a specific particle is given by Eq. 24.11. While a quiet settling basin has a 100% removal efficiency for a particle with $v_s = v^*$, a basin with transverse mixing has a 63.2% efficiency for the same particle.

$$\eta = 1 - e^{-v_s/v^*} \quad \text{[transverse mixing]} \qquad 24.11$$

For a basin with thorough mixing, particles may move three dimensionally—vertically, horizontally, and longitudinally. The collection efficiency is given by Eq. 24.12. When $v_s = v^*$, the efficiency is 50%.

$$\eta = \frac{v_s}{v_s + v^*} \quad \text{[thorough mixing]} \qquad 24.12$$

As a function of v_s (or as a function of the dimensionless ratio v_s/v^*), the removal efficiencies for transverse- and thoroughly-mixed basins both approach 100% asymptotically. Since large values of v_s/v^* correspond to large basins, any basin removal efficiency can be achieved by making the basin large enough. However, the smallest basin size will be achieved when the basin is unmixed.

15. COAGULANTS

Various chemicals can be added to remove fine solids. There are two main categories of coagulating chemicals: hydrolyzing metal ions (based on either aluminum or iron) and ionic polymers. Since the chemicals work by agglomerating particles in the water to form floc, they are known as *coagulants*. *Floc* is the precipitate that forms when the coagulant allows the colloidal particles to agglomerate.

Common *hydrolyzing metal ion* coagulants are aluminum sulfate ($Al_2(SO_4)_3 \cdot nH_2O$, commonly referred to as "alum"), ferrous sulfate ($FeSO_4 \cdot 7H_2O$, sometimes referred to as "copperas"), and chlorinated copperas (a mixture of ferrous sulfate and ferric chloride). n is the number of waters of hydration, approximately 14.3.

Alum is, by far, the most common compound used in water treatment.[5] Alum provides the positive charges needed to attract and neutralize negative colloidal particles. It reacts with alkalinity in the water to form gelatinous aluminum hydroxide ($Al(OH)_3$) in the proportion of 1 mg/L alum:$^1/_2$ mg/L alkalinity. The aluminum hydroxide forms the nucleus for floc agglomeration.

For alum coagulation to be effective, the following requirements must be met: (a) Enough alum must be used to neutralize all of the incoming negative particles. (b) Enough alkalinity must be present to permit as complete as possible a conversion of the aluminum

sulfate to aluminum hydroxide. (c) The pH must be maintained within the effective range.

Alum dosage is generally 5–50 mg/L, depending on the turbidity. Alum floc is effective within a wide pH range of 5.5–8.0. However, the hydrolysis of the aluminum ion depends on pH and other factors in complex ways, and stoichiometric relationships rarely tell the entire story. Although a pH of 6–7 is a typical operating range for alum coagulation, depending on the contaminant, the pH can range as high as 9. Alum removal of chromium, for example, is effective within a pH range of 7–9.

Alum reacts with natural alkalinity in the water according to the following reaction.

$$Al_2(SO_4)_3 \cdot nH_2O + 3Ca(HCO_3)_2$$
$$\rightarrow 2Al(OH)_3\downarrow + 3CaSO_4 + 6CO_2 + nH_2O \qquad 24.13$$

Since alum is naturally acidic, if the water is not sufficiently alkaline, lime (CaO) and soda ash ("caustic soda," Na_2CO_3) can be added for preliminary pH adjustment.[6] If there is inadequate alkalinity and lime is added, the reaction is

$$Al_2(SO_4)_3 \cdot nH_2O + 3Ca(OH)_2$$
$$\rightarrow 2Al(OH)_3\downarrow + 3CaSO_4 + nH_2O \qquad 24.14$$

Alum reacts with soda ash according to

$$Al_2(SO_4)_3 \cdot nH_2O + 3Na_2CO_3 + 3H_2O$$
$$\rightarrow 2Al(OH)_3\downarrow + 3Na_2SO_4 + 3CO_2 + nH_2O \qquad 24.15$$

Ferrous sulfate ($FeSO_4 \cdot 7H_2O$) reacts with slaked lime ($Ca(OH)_2$) to flocculate ferric hydroxide ($Fe(OH)_3$). This is an effective method of clarifying turbid water at higher pH.

Ferric sulfate ($Fe_2(SO_4)_3$) reacts with natural alkalinity and lime. It can also be used for color removal at low pH. At high pH, it is useful for iron and manganese removal, as well as a coagulant with precipitation softening.

If alkalinity is high, it may be necessary to use hydrochloric or sulfuric acid for pH control rather than adding lime. The iron salts are effective above a pH of 7, so they may be advantageous in alkaline waters.[7]

Polymers are chains or groups of repeating identical molecules (*mers*) with many available active adsorption sites. Their molecular weights range from several hundred to several million. Polymers can be positively charged (cationic polymers), negatively charged (anionic polymers), or neutral (nonionic polymers). The charge can vary with pH. *Organic polymers* (*polyelectrolytes*),

---

[5]Lime is the second most-used compound.

[6]The cost of soda ash used for pH adjustment is about three times that of lime. Also, soda ash leaves sodium ions in solution, an increasingly undesirable contaminant for people with hypertension.

[7]Optimum pH for ferric floc is approximately 7.5. However, iron salts are active over the pH range of 3.8–10.

Water Treatment

such as starches and polysaccharides, and *synthetic polymers*, such as polyacrylamides, are used.

Polymers are useful in specialized situations, such as when particular metallic ions are to be removed. For example, conventional alum will not remove positive iron ions. However, an anionic polymer will combine with the iron. Polymers are effective in narrow ranges of turbidity and alkalinity. In some cases, it may be necessary to artificially seed the water with clay or alum (to produce floc).

16. FLOCCULATION ADDITIVES

Flocculation additives improve the coagulation efficiency by changing the floc size. Additives include *weighting agents* (e.g., bentonite clays), *adsorbents* (e.g., powdered activated carbon), and *oxidants* (chlorine). Polymers are also used in conjunction with metallic ion coagulants.

17. DOSES OF COAGULANTS AND OTHER COMPOUNDS

The amount of a substance added to a water supply is known as the *dose*. Purity, P, and *availability*, G, both affect dose. A substance's purity decreases when it is mixed with another substance. A substance's availability decreases when some of the substance doesn't dissolve or react chemically.

For example, *lime* is a term that describes various calcium compounds. Lime is typically quarried from naturally occurring limestone ($CaCO_3$). Limestone contains silica and other non-calcium compounds, which reduces the as-delivered lime purity. *Quicklime* (CaO) is produced by heating (i.e., burning) $CaCO_3$ in a lime kiln. Accordingly, quicklime may contain some calcium in the form of calcium carbonate rather than calcium oxide. The *total lime* (i.e., total calcium) content includes both calcium compounds, while the *available lime* only includes the CaO form.

Feed rate, F, of a compound with purity P and fractional availability G can be calculated from the *dose equation*, Eq. 24.16. Doses in mg/L and ppm for aqueous solutions are the same. Each 1% by weight of concentration is equivalent to 10,000 mg/L of solution. Doses may be given in terms of grains per gallon. There are 7000 grains per pound.

$$F_{kg/d} = \frac{D_{mg/L} Q_{ML/d}}{PG} \quad \text{[SI]} \quad 24.16(a)$$

$$F_{lbm/day} = \frac{D_{mg/L} Q_{MGD} \left(8.345 \frac{lbm\text{-}L}{mg\text{-}MG}\right)}{PG} \quad \text{[U.S.]} \quad 24.16(b)$$

Example 24.1

Incoming water contains 2.5 mg/L as a substance of natural alkalinity (HCO_3^-). The flow rate is 2.5 MGD (9.5 ML/day). (a) What feed rate is required if the alum (purity 48%) dose is 7 mg/L? (b) How much lime (purity 85%) in the form of $Ca(OH)_2$ is required to react completely with the alum?

SI Solution

(a) The dose is specifically in terms of mg/L of alum. Therefore, Eq. 24.16 can be used.

$$F_{kg/d} = \frac{D_{mg/L} Q_{ML/d}}{PG} = \frac{\left(7 \frac{mg}{L}\right)\left(9.5 \frac{ML}{d}\right)}{(0.48)(1.0)}$$
$$= 138.5 \text{ kg/d}$$

(b) From Eq. 24.13, each alum molecule reacts with six ions of alkalinity.

$$Al_2(SO_4)_3 + 6HCO_3^- \rightarrow$$

| MW: | 342 | (6)(61) |
| mg/L: | X | 2.5 |

By simple proportion, the amount of alum used to counteract the natural alkalinity is

$$X = \frac{\left(2.5 \frac{mg}{L}\right)(342)}{(6)(61)} = 2.34 \text{ mg/L}$$

The alum remaining is

$$7 \frac{mg}{L} - 2.34 \frac{mg}{L} = 4.66 \text{ mg/L}$$

From Eq. 24.14, one alum molecule reacts with three slaked lime molecules.

$$Al_2(SO_4)_3 + 3Ca(OH)_2 \rightarrow$$

| MW: | 342 | (3)(74) |
| mg/L: | 4.66 | X |

By simple proportion, the amount of lime needed is

$$X = \frac{\left(4.66 \frac{mg}{L}\right)(3)(74)}{342} = 3.02 \text{ mg/L}$$

Using Eq. 24.16,

$$F_{kg/d} = \frac{D_{mg/L} Q_{ML/d}}{PG}$$
$$= \frac{\left(3.02 \frac{mg}{L}\right)\left(9.5 \frac{ML}{d}\right)}{(0.85)(1.0)}$$
$$= 33.8 \text{ kg/d}$$

Customary U.S. Solution

(a) The dose is specifically in terms of mg/L of alum. Therefore, Eq. 24.16 can be used.

$$F_{\text{lbm/day}} = \frac{D_{\text{mg/L}} Q_{\text{MGD}} \left(8.345 \ \frac{\text{lbm-L}}{\text{mg-MG}}\right)}{PG}$$

$$= \frac{\left(7 \ \frac{\text{mg}}{\text{L}}\right)(2.5 \ \text{MGD}) \left(8.345 \ \frac{\text{lbm-L}}{\text{mg-MG}}\right)}{(0.48)(1.0)}$$

$$= 304.2 \ \text{lbm/day}$$

(b) From Eq. 24.13, each alum molecule reacts with six ions of alkalinity.

$$\text{Al}_2(\text{SO}_4)_3 \quad + \quad 6\text{HCO}_3^- \quad \rightarrow$$

MW: 342   (6)(61)

mg/L: X 2.5

By simple proportion, the amount of alum used to counteract the natural alkalinity is

$$X = \frac{\left(2.5 \ \frac{\text{mg}}{\text{L}}\right)(342)}{(6)(61)} = 2.34 \ \text{mg/L}$$

The alum remaining is

$$7 \ \frac{\text{mg}}{\text{L}} - 2.34 \ \frac{\text{mg}}{\text{L}} = 4.66 \ \text{mg/L}$$

From Eq. 24.14, one alum molecule reacts with three lime molecules.

$$\text{Al}_2(\text{SO}_4)_3 \quad + \quad 3\text{Ca(OH)}_2 \quad \rightarrow$$

MW: 342   (3)(74)

mg/L: 4.66 X

By simple proportion, the amount of lime needed is

$$X = \frac{\left(4.66 \ \frac{\text{mg}}{\text{L}}\right)(3)(74)}{342} = 3.02 \ \text{mg/L}$$

Using Eq. 24.16,

$$F_{\text{lbm/day}} = \frac{D_{\text{mg/L}} Q_{\text{MGD}} \left(8.345 \ \frac{\text{lbm-L}}{\text{mg-MG}}\right)}{PG}$$

$$= \frac{\left(3.02 \ \frac{\text{mg}}{\text{L}}\right)(2.5 \ \text{MGD}) \left(8.345 \ \frac{\text{lbm-L}}{\text{mg-MG}}\right)}{(0.85)(1.0)}$$

$$= 74.1 \ \text{lbm/day}$$

18. MIXERS AND MIXING KINETICS

Coagulants and other water treatment chemicals are added in *mixers*. If the mixer adds a coagulant for the removal of colloidal sediment, a downstream location (i.e., a tank or basin) with a reduced velocity gradient may be known as a *flocculator*.

There are two basic models: plug flow mixing and complete mixing. The *complete mixing* model is appropriate when the chemical is distributed throughout by impellers or paddles. If the basin volume is small, so that time for mixing is low, the tank is known as a *flash mixer*, *rapid mixer*, or *quick mixer*. The volume of flash mixers is seldom greater than 300 ft^3 (8 m^3), and flash mixer detention time is usually 30–60 sec. Flash mixing kinetics are described by the complete mixing model.

Flash mixers are usually concrete tanks, square in horizontal cross section, and fitted with vertical shaft impellers. The size of a mixing basin can be determined from various combinations of dimensions that satisfy the volume-flow rate relationship.

$$V = tQ \qquad \qquad 24.17$$

The detention time required for complete mixing in a tank of volume V depends on the *mixing rate constant*, K, and the incoming and outgoing concentrations.

$$t_{\text{complete}} = \frac{V}{Q} = \frac{1}{K}\left(\frac{C_i}{C_o} - 1\right) \qquad 24.18$$

The *plug flow mixing* model is appropriate when the water flows through a long narrow chamber, the chemical is added at the entrance, and there is no mechanical agitation. All of the molecules remain in the plug flow mixer for the same amount of time as they flow through. For any mixer, the maximum chemical conversion will occur with plug flow, since all of the molecules have the maximum opportunity to react. The detention time in a plug flow mixer of length L is

$$t_{\text{plug flow}} = \frac{V}{Q} = \frac{L}{\text{v}_f}$$

$$= \frac{1}{K} \ln \frac{C_i}{C_o} \qquad 24.19$$

19. MIXING PHYSICS

The drag force on a paddle is given by the standard fluid drag force equation. For flat plates, the coefficient of drag, C_D, is approximately 1.8.

$$F_D = \frac{C_D A \rho \text{v}_{\text{mixing}}^2}{2} \qquad \text{[SI]} \quad 24.20(a)$$

$$F_D = \frac{C_D A \rho \text{v}_{\text{mixing}}^2}{2g_c} = \frac{C_D A \gamma \text{v}_{\text{mixing}}^2}{2g} \qquad \text{[U.S.]} \quad 24.20(b)$$

The power required is calculated from the drag force and the mixing velocity. The average *mixing velocity*, v_{mixing}, also known as the *relative paddle velocity*, is the difference in paddle and average water velocities. The mixing velocity is approximately 0.7–0.8 times the tip speed.

$$v_{paddle,ft/sec} = \frac{2\pi R n_{rpm}}{60 \; \frac{sec}{min}} \qquad 24.21$$

$$v_{mixing} = v_{paddle} - v_{water} \qquad 24.22$$

$$P_{kW} = \frac{F_D v_{mixing}}{1000 \; \frac{W}{kW}}$$

$$= \frac{C_D A \rho v_{mixing}^3}{2\left(1000 \; \frac{W}{kW}\right)} \qquad \text{[SI]} \quad 24.23(a)$$

$$P_{hp} = \frac{F_D v_{mixing}}{550 \; \frac{ft\text{-}lbf}{hp\text{-}sec}} = \frac{C_D A \rho v_{mixing}^3}{2 g_c \left(550 \; \frac{ft\text{-}lbf}{hp\text{-}sec}\right)}$$

$$= \frac{C_D A \gamma v_{mixing}^3}{2 g \left(550 \; \frac{ft\text{-}lbf}{hp\text{-}sec}\right)} \qquad \text{[U.S.]} \quad 24.23(b)$$

For slow-moving paddle mixers, the *velocity gradient*, G, varies from 20 sec^{-1} to 75 sec^{-1} for a 15 min to 30 min mixing period. Typical units in Eq. 24.24 are ft-lbf/sec for power (multiply hp by 550 to obtain ft-lbf/sec), lbf-sec/ft^2 for μ, and ft^3 for volume. In the SI system, power in kW is multiplied by 1000 to obtain W, viscosity is in Pa·s, and volume is in m^3. (Multiply viscosity in cP by 0.001 to obtain Pa·s.)

$$G = \sqrt{\frac{P}{\mu V_{tank}}} \qquad 24.24$$

Equation 24.24 can also be used for rapid mixers, in which case the mean velocity gradient is much higher: approximately 500–1000 sec^{-1} for 10–30 sec mixing period, or 3000–5000 sec^{-1} for a 0.5–1.0 sec mixing period in an in-line blender configuration.

Equation 24.24 can be rearranged to calculate the power requirement. Power is typically 0.5–1.5 hp/MGD for rapid mixers.

$$P = \mu G^2 V_{tank} \qquad 24.25$$

The dimensionless product, Gt_d, of the velocity gradient and detention time is known as the *mixing opportunity parameter*. Typical values range from 10^4 to 10^5.

$$Gt_d = \frac{V_{tank}}{Q} \sqrt{\frac{P}{\mu V_{tank}}}$$

$$= \frac{1}{Q} \sqrt{\frac{P V_{tank}}{\mu}} \qquad 24.26$$

20. IMPELLER CHARACTERISTICS

Mixing equipment uses rotating impellers on rotating shafts. The blades of *radial-flow impellers* (paddle-type impellers, turbine impellers, etc.) are parallel to the drive shaft. *Axial-flow impellers* (propellers, pitched-blade impellers, etc.) have blades inclined with respect to the drive shaft. (See Fig. 24.5.) Axial-flow impellers are better at keeping materials (e.g., water softening chemicals) in suspension.

Figure 24.5 *Typical Axial Flow Mixing Impellers*

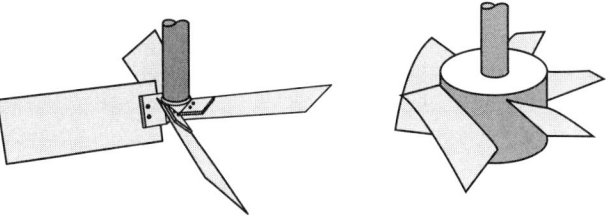

The *mixing Reynolds number* depends on the impeller diameter, D, suspension specific gravity, SG, liquid viscosity, μ, and rotational speed, n, in rev/sec. Flow is in transition for $10 < \text{Re} < 10{,}000$, is laminar below 10, and is turbulent above 10,000.

$$\text{Re} = \frac{D^2 n \rho}{\mu} \qquad \text{[SI]} \quad 24.27(a)$$

$$\text{Re} = \frac{D^2 n \rho}{g_c \mu} \qquad \text{[U.S.]} \quad 24.27(b)$$

The dimensionless *power number*, N_P, of the impeller is defined implicitly by the power required to drive the impeller. For any given level of suspension, Eq. 24.28 through Eq. 24.31 apply to geometrically similar impellers and turbulent flow.

$$P = N_P n^3 D^5 \rho \qquad \text{[SI]} \quad 24.28(a)$$

$$P = \frac{N_P n^3 D^5 \rho}{g_c} \qquad \text{[U.S.]} \quad 24.28(b)$$

$$P = \rho g Q h_v \qquad \text{[SI]} \quad 24.29(a)$$

$$P = \frac{\rho g Q h_v}{g_c} \qquad \text{[U.S.]} \quad 24.29(b)$$

The impeller's dimensionless *flow number*, N_Q, is defined implicitly by the *flow rate equation*.

$$Q = N_Q n D^3 \qquad 24.30$$

The velocity head can be calculated from a combination of the power and flow numbers.

$$h_v = \frac{N_P n^2 D^2}{N_Q g} \qquad 24.31$$

Vibration near the critical speed can be a major problem with modern, high-efficiency, high-speed impellers. Mixing speed should be well below (i.e., less than 80% of) the first critical speed of the shaft. Other important design factors include tip speed and shaft bending moment.

The critical speed of a shaft corresponds to its natural harmonic frequency of vibration when loaded as a flexing beam. If the harmonic frequency, f_c, is in hertz, the shaft's critical rotational speed (in rev/min) will be $60f_c$. The critical speed depends on many factors, including the masses and locations of paddles and other shaft loads, shaft material, length, and end restraints. It is not important whether the shaft is vertical or horizontal, and the damping effects of the mixing fluid and any stabilizing devices are disregarded in calculations. For a lightweight (i.e., massless) mixing impeller that is on a suspended (cantilever) shaft of length L, has a uniform cross section, a mass per unit length of m_L, and is supported by a fixed upper bearing, the first critical speed, f_c, in hertz, will be the same as that of a cantilever beam and can be approximated from Eq. 24.32. Other configurations, including mixed shaft materials, stepped or varying shaft diameters, two or more support bearings, substantial impeller mass, and multiple impellers require different analysis methods.

$$f_c = 0.56 \sqrt{\frac{EI}{L^4 m_L}} \qquad \text{[SI]} \quad 24.32(a)$$

$$f_c = 0.56 \sqrt{\frac{EI g_c}{L^4 m_L}} \qquad \text{[U.S.]} \quad 24.32(b)$$

21. FLOCCULATION

After flash mixing, the floc is allowed to form during a 20–60 min period of gentle mixing. Flocculation is enhanced by the gentle agitation, but floc disintegrates with violent agitation. During this period, the flow-through velocity should be limited to 0.5–1.5 ft/min (0.25–0.75 cm/s). The peripheral speed of mixing paddles should vary approximately from 0.5 ft/sec (0.15 m/s) for fragile, cold-water floc to 3.0 ft/sec (0.9 m/s) for warm-water floc.

Many modern designs make use of *tapered flocculation*, also known as *tapered energy*, a process in which the amount (severity) of flocculation gradually decreases as the treated water progresses through the flocculation basin.

Flocculation is followed by sedimentation for two to eight hours (four hours typical) in a low-velocity portion of the basin. (The flocculation time is determined from settling column data.) A good settling process will remove 90% of the settleable solids. Poor design, usually resulting in some form of *short-circuiting* of the flow path, will reduce the effective time in which particles have to settle.

22. FLOCCULATOR-CLARIFIERS

A *flocculator-clarifier* combines mixing, flocculation, and sedimentation into a single tank. Such units are called *solid contact units* and *upflow tanks*. They are generally round in construction, with mixing and flocculation taking place near the central hub and sedimentation occurring at the periphery. Flocculator-clarifiers are most suitable when combined with softening, since the precipitated solids help seed the floc.

Typical operational characteristics of flocculator-clarifiers are given in Table 24.2.

Table 24.2 *Characteristics of Flocculator-Clarifiers*

typical flocculation and mixing time	20–60 min
minimum detention time	1.5–2.0 hr
maximum weir loading	10 gpm/ft (2 L/s·m)
upflow rate	0.8–1.7 gpm/ft^2; 1.0 gpm/ft^2 typical (0.54–1.2 L/s·m^2; 0.68 L/s·m^2 typical)
maximum sludge formation rate	5% of water flow

(Multiply gpm/ft by 0.207 to obtain L/s·m.)
(Multiply gpm/ft^2 by 0.679 to obtain L/s·m^2.)

23. SLUDGE QUANTITIES

Sludge is the watery waste that carries off the settled floc and the water softening precipitates. The sludge volume produced is given by Eq. 24.33, in which G is the gravimetric fraction of solids. The gravimetric solids fraction for coagulation sludge is generally less than 0.02. For water softening sludge, it is on the order of 0.10. In Eq. 24.33, m_{sludge} can be either a specific quantity or a rate of production per unit time.

$$V_{sludge} = \frac{m_{sludge}}{\rho_{water} (\text{SG})_{sludge} G} \qquad 24.33$$

The most accurate way to calculate the mass of sludge is to extrapolate from jar or pilot test data. There is no absolute correlation between the mass generated and other water quality measurements. However, a few generalizations are possible. (a) Each unit mass (lbm, mg/L, etc.) of alum produces 0.46 unit mass of floc.[8] (b) 100% of the reduction in suspended solids (expressed as substance) shows up as floc. (c) 100% of any supplemental flocculation aids is recovered in the sludge.

Suspended solids in water may be reported in turbidity units. There is no easy way to calculate total suspended solids (TSS) in mg/L from turbidity in NTU. The ratio of TSS to NTU normally varies from 1.0–2.0, and can be as high as 10. A value of 1 or 1.5 is generally appropriate.

The rate of dry sludge production from coagulation processes can be estimated from Eq. 24.34. M is an extra factor that accounts for the use of any miscellaneous inorganic additives such as clay.

$$m_{sludge,kg/d} =$$
$$\frac{\left(86\,400\ \frac{s}{d}\right) Q_{m^3/s}\left(1000\ \frac{L}{m^3}\right)}{10^6\ \frac{mg}{kg}}$$
$$\qquad \times (0.46 D_{alum,mg/L} + \Delta TSS_{mg/L} + M_{mg/L})$$

[SI] *24.34(a)*

$$m_{sludge,lbm/day} =$$
$$\left(8.345\ \frac{lbm\text{-}L}{mg\text{-}MG}\right) Q_{MGD}$$
$$\qquad \times (0.46 D_{alum,mg/L} + \Delta TSS_{mg/L} + M_{mg/L})$$

[U.S.] *24.34(b)*

After a sludge dewatering process from sludge solids constant G_1 to G_2, the resulting volume will be

$$V_{2,sludge} \approx V_{1,sludge} \frac{G_1}{G_2} \qquad 24.35$$

24. FILTRATION

Nonsettling floc, algae, suspended precipitates from softening, and metallic ions (iron and manganese) are removed by filtering. *Sand filters* (and in particular, rapid sand filters) are commonly used for this purpose. Sand filters are beds of gravel, sand, and other granulated materials.[9]

Although filter box heights are on the order of 10 ft (3 m) to provide for expansion and freeboard during backwashing, the almost-universal specification used before 1960 for a single-media filter was 24–30 in (610–760 mm) of sand or ground coal. Filters are usually square or nearly square in plan, and they operate with a hydraulic head of 1–8 ft (0.3–2.4 m).

Although most plants have six or more filters, there should be at least three filters so that two can be operational when one is being cleaned. Cleaning of multiple filters should be staged.

Filter operation is enhanced when the top layer of sand is slightly more coarse than the deeper sand. During backwashing, however, the finest sand rises to the top. Various *dual-layer* and *multi-layer* (also known as *dual-media* and *multi-media*) filter designs using layers of coal ("anthrafilt") and (more recently) granular activated carbon, alone or in conjunction with sand, overcome this problem. Since coal has a lower specific gravity than sand, media particle size is larger, as is pore space.

Historically, the *loading rate* (i.e., *flow rate*) for *rapid sand filters* was set at 2–3 gpm/ft² (1.4–2.0 L/s·m²). A range of 4–6 gpm/ft² (2.7–4.1 L/s·m²) is a reasonable minimum rate for dual-media filters. Multi-media filters operate at 5–10 gpm/ft² (3.4–6.8 L/s·m²) and above.[10] The filter loading rate is

$$\text{loading rate} = \frac{Q}{A} \qquad 24.36$$

Filters discharge into a storage reservoir known as a *clearwell*. The *hydraulic head* (the distance between the water surfaces in the filter and clearwell) is usually 9–12 ft (2.7–3.6 m). This allows for a substantial decrease in available head prior to backwashing. Clearwell storage volume is 30–60% of the daily filter output, with a minimum capacity of 12 hr of the maximum daily demand so that demand can be satisfied from the clearwell while the filter is being cleaned or serviced.

25. FILTER BACKWASHING

The most common type of service needed by filters is backwashing, which is needed when the pores between the filter particles clog up. Typically, this occurs after 1 to 3 days of operation, when the head loss reaches 6–8 ft (1.6–2.4 m). There are two parameters that can trigger backwashing: head loss and turbidity. Head loss increases almost linearly with time, while turbidity remains constant for several days before suddenly increasing. The point of sudden increase is known as *breakthrough*.[11] Since head loss is more easily monitored than turbidity, it is desired to have head loss trigger the backwashing cycle.

Backwashing with filtered water pumped back through the filter from the bottom to the top expands the sand layer 30–50%, which dislodges trapped material. Backwashing for 3–5 min at 8–15 gpm/ft² (5.4–10 L/s·m²) is

[8]0.456 is the stoichiometric ratio of floc (Al(OH)₃) to alum (Al₂(SO₄)₃·14.3H₂O) in Eq. 24.13 when the water of hydration (14.3H₂O) is included. 0.26 is the ratio if the water of hydration is not included.
[9]Most early sand filter beds were designed when a turbidity level of 5 NTU was acceptable. With the U.S. federal MCL at 1 NTU, some states at 0.5 NTU, and planning "on the horizon" for 0.2 NTU, these early conventional filters are clearly inadequate.

[10]Testing has demonstrated that deep-bed, uniformly graded anthracite filters can operate at 10–15 gpm/ft² (7–10 L/s·m²), although this requires high-efficiency preozonation and microflocculation.
[11]Technically, *breakthrough* is the point at which the turbidity rises above the MCL permitted. With low MCLs, this occurs very soon after the beginning of filter performance degradation.

a typical specification.[12] The head loss is reduced to approximately 1 ft (0.3 m) after washing.

Experience has shown that supplementary agitation of the filter media is necessary to prevent "caking" and "mudballs" in almost all installations. Prior to backwashing, the filter material may be expanded by an *air prewash* volume of 1–8 (2–5 typical) times the sand filter volume per minute for 2–10 min (3–5 min typical). Alternatively, turbulence in the filter material may be encouraged during backwashing with an *air wash* or with rotating hydraulic surface jets.

During backwashing, the water in the filter housing will rise at a rate of 1–3 ft/min (0.5–1.5 cm/s). This rise should not exceed the settling velocity of the smallest particle that is to be retained in the filter. The wash water, which is collected in troughs for disposal, constitutes approximately 1–5% of the total processed water. The total water used is approximately 75–100 gal/ft^2 (3–4 kL/m^2). The actual amount of backwash water is

$$V = A_{\text{filter}}(\text{rate of rise})t_{\text{backwash}} \qquad 24.37$$

The temperature of the water used in backwashing is important since the viscosity changes. (The effect of temperature on water density is negligible.) 40°F (4°C) water is more viscous than 70°F (21°C) water. Media particles, therefore, may be expanded to the same extent using lower upflow rates at the lower backwash temperature.

26. OTHER FILTRATION METHODS

Pressure (sand) filters for water supply treatment operate similarly to rapid sand filters except that incoming water is typically pressurized 25–75 psig (170–520 kPa gage). Single media filter rates are 2–10 gpm/ft^2, with 4–5 gpm/ft^2 being typical (1.4–14 L/s·m^2 (2.7–3.4 L/s·m^2 typical)), with dual media filters running at 1.5 to 2.0 times these rates. Pressure filters are not used in large installations.

Ultrafilters are membranes that act as sieves to retain turbidity, microorganisms, and large organic molecules that are THM precursors, while allowing water, salts, and small molecules to pass through. Ultrafiltration is effective in removing particles ranging in size of 0.001–10 μm.[13] A pressure of 15–75 psig (100–500 kPa) is required to drive the water through the membrane.

Biofilm filtration (biofilm process) uses microorganisms to remove selected contaminants (e.g., aromatics and other hydrocarbons). Operation of biofilters is similar to trickling filters used in wastewater processing. Sand filter facilities are relatively easy to modify—sand is replaced with gravel in the 4–14 mm size range, application rates are decreased, and exposure to chlorine from incoming and backwash water is eliminated.

Slow sand filters are primarily of historical interest, though there are some similarities with modern biomethods used to remediate toxic spills. Slow sand filters operate similarly to rapid sand filters except that the exposed surface (loading) area is much larger and the flow rate is much lower (0.05–0.1 gpm/ft^2; 0.03–0.07 L/s·m^2). Slow sand filters are limited to low turbidity applications not requiring chemical treatment and where large rural areas are available to spread out the facilities. Slow sand filters can be operated as biofilm processes if a layer of biological slime is allowed to form on the top of the filter medium.[14] Slow filter cleaning usually involves removing a few inches of sand.

27. ADSORPTION

Many dissolved organic molecules and inorganic ions can be removed by adsorption processes. *Adsorption* is not a straining process, but occurs when contaminants are trapped on the surface or interior of the adsorption particles.

Granular activated carbon (GAC) is considered to be the best available technology for removal of THMs and synthetic organic chemicals from water.[15] GAC is also useful in removing compounds that contribute to taste, color, and odor. Activated carbon can be used in the form of powder (which must be subsequently removed) or granules. GAC can be integrated into the design of sand filters.[16]

Eventually, the GAC becomes saturated and must be removed and reactivated. In such dual-media filters, the reactivation interval for GAC is approximately one to two years.

28. FLUORIDATION

Fluoridation can occur any time after filtering and can involve solid compounds or liquid solutions. Small utilities may manufacturer their own liquid solution on-site from sodium silicofluoride (Na_2SiF_6, typically 22–30% purity) or sodium fluoride (NaF, typically 90–98% purity) for use with a volumetric metering system. Larger utilities use gravimetric dry feeders with sodium silicofluoride (Na_2SiF_6, typically 98–99% purity) or solution feeders with fluorsilic acid (H_2SiF_6).

Assuming 100% ionization, the application rate, F (in pounds per day), of a compound with fluoride gravimetric fraction, G, and a fractional purity, P, needed to obtain a final concentration, C, of fluoride is

$$F_{\text{kg/d}} = \frac{C_{\text{mg/L}} Q_{\text{L/d}}}{PG} \qquad [\text{SI}] \qquad 24.38(a)$$

---

[14]The biological slime is called *schmutzdecke*. The slime forms a physical barrier that traps particles.
[15]Enhanced coagulation is also a "best available" technology for THM removal that may be economically more attractive.
[16]For example, an 80 in (2032 mm) GAC layer may be placed on top of a 40 in (1016 mm) sand layer.

---

[12]While the maximum may never be used, a maximum backwash rate of 20 gpm/ft^2 (14 L/s·m^2) should be provided for.
[13]A μm is the same as a micron.

$$F_{\text{lbm/day}} = \frac{C_{\text{mg/L}} Q_{\text{MGD}} \left(8.345 \ \dfrac{\text{lbm-L}}{\text{mg-MG}} \right)}{PG}$$

[U.S.]   24.38(b)

29. IRON AND MANGANESE REMOVAL

Several methods can be used to remove iron and manganese. Most involve aeration with chemical oxidation since manganese is not easily removed by aeration alone. These processes are described in Table 24.3.

30. TASTE AND ODOR CONTROL

A number of different processes affect taste and odor. Some are more effective and appropriate than others. *Microstraining*, using a 35 μm or finer metal cloth, can be used to reduce the number of algae and other organisms in the water, since these are sources of subsequent tastes and odors. Microstraining does not remove dissolved or colloidal organic material, however.

Activated carbon removes more tastes and odors, as well as a wide variety of chemical contaminants.

Aeration can be used when dissolved oxygen is low or when hydrogen sulfide is present. Aeration has little effect on most other tastes and odors.

Chlorination disinfects and reduces odors caused by organic matter and industrial wastes. Normally, the dosage required will be several times greater than those for ordinary disinfection, and the term *superchlorination* is used to describe applying enough chlorine to maintain an excessively large residual. Subsequent dechlorination will be required to remove the excess chlorine. Similar results will be obtained with chlorine dioxide and other disinfection products.

A quantitative odor ranking is the *threshold odor number* (TON), which is determined by adding increasing amounts of odor-free dilution water to a sample until the combined sample is virtually odor free.

$$\text{TON} = \frac{V_{\text{raw sample}} + V_{\text{dilution water}}}{V_{\text{raw sample}}}$$

24.39

A TON of 3 or less is ideal. Untreated river water usually has a TON between 6 and 24. Treated water normally has a TON between 3 and 6. At TON of 5 and above, customers will begin to notice the taste and odor of their water.

31. PRECIPITATION SOFTENING

Precipitation softening using the *lime-soda ash process* adds lime (CaO), also known as *quicklime*, and soda ash (Na_2CO_3) to remove calcium and magnesium from hard

water.[17] Granular quicklime is available with a minimum purity of 90%, and soda ash is available with a 98% purity.

Lime forms *slaked lime* (also known as *hydrated lime*), $Ca(OH)_2$, in an exothermic reaction when added to feed water. The slaked lime is delivered to the water supply as a *milk of lime* suspension.

$$CaO + H_2O \rightarrow Ca(OH)_2 + \text{heat} \qquad 24.40$$

Slaked lime reacts first with any carbon dioxide dissolved in the water, as in Eq. 24.41. No softening occurs, but the carbon dioxide demand must be satisfied before any reactions involving calcium or magnesium can occur.

carbon dioxide removal:

$$CO_2 + Ca(OH)_2 \rightarrow CaCO_3 \downarrow + H_2O \qquad 24.41$$

Lime next reacts with any carbonate hardness, precipitating calcium carbonate and magnesium hydroxide, as shown in Eq. 24.42 and Eq. 24.43. The removal of carbonate hardness caused by magnesium (characterized by magnesium bicarbonate in Eq. 24.43) requires two molecules of calcium hydroxide to precipitate calcium carbonate and magnesium hydroxide.

calcium carbonate hardness removal:

$$Ca(HCO_3)_2 + Ca(OH)_2 \rightarrow 2CaCO_3 \downarrow + 2H_2O$$

24.42

magnesium carbonate hardness removal:

$$Mg(HCO_3)_2 + 2Ca(OH)_2 \rightarrow$$
$$2CaCO_3 \downarrow + Mg(OH)_2 \downarrow + 2H_2O \qquad 24.43$$

To remove noncarbonate hardness (characterized by sulfate ions in Eq. 24.44 and Eq. 24.45), it is necessary to add soda ash and more lime. The sodium sulfate that remains in solution does not contribute to hardness, for sodium is a single-valent ion (i.e., Na^+).

magnesium noncarbonate hardness removal:

$$MgSO_4 + Ca(OH)_2 \rightarrow$$
$$Mg(OH)_2 \downarrow + CaSO_4 \qquad 24.44$$

calcium noncarbonate hardness removal:

$$CaSO_4 + Na_2CO_3 \rightarrow CaCO_3 \downarrow + Na_2SO_4 \qquad 24.45$$

The calcium ion can be effectively reduced by the lime addition shown in the previous equations, raising the pH of the water to approximately 10.3. (Thus, the lime added removes itself.) Precipitation of the magnesium ion, however, requires a higher pH and the presence of excess lime in the amount of approximately 35 mg/L of CaO or 50 mg/L of $Ca(OH)_2$ above the stoichiometric

---

[17]Soda ash does not always need to be used. When lime alone is used, the process may be referred to as *lime softening*.

Table 24.3 *Processes for Iron and Manganese Removal*

processes	iron and/or manganese removed	pH required	remarks
aeration, settling, and filtration	ferrous bicarbonate	7.5	provide aeration unless incoming water
	ferrous sulfate	8.0	contains adequate dissolved oxygen
	manganous bicarbonate	10.3	
	manganous sulfate	10.0	
aeration, free residual chlorination, settling, and filtration	ferrous bicarbonate	5.0	provide aeration unless incoming water
	manganous bicarbonate	9.0	contains adequate dissolved oxygen
aeration, lime softening, settling, and filtration	ferrous bicarbonate	8.5–9.6	
	manganous bicarbonate		
aeration, coagulation, lime softening, settling, and filtration	colloidal or organic iron	8.5–9.6	require lime, and alum or iron coagulant
	colloidal or organic manganese	10.0	
ion exchange	ferrous bicarbonate	≈ 6.5	water must be devoid of oxygen
	manganous bicarbonate		iron and manganese in raw water not to exceed 2.0 mg/L
			consult manufacturers for type of ion exchange resin to be used

requirements. The practical limits of precipitation softening are 30–40 mg/L of $CaCO_3$ and 10 mg/L of $Mg(OH)_2$, both as $CaCO_3$.

After softening, the water must be recarbonated to lower its pH and to reduce its scale-forming potential. This is accomplished by bubbling carbon dioxide gas through the water.

$$Ca(OH)_2 + CO_2 \rightarrow CaCO_3\downarrow + H_2O \qquad 24.46$$

The treatment process could be designed such that all of these reactions take place sequentially. That is, first slaked lime would be added (see Eq. 24.42 and Eq. 24.43), then the water would be recarbonated (see Eq. 24.46), then excess lime would be added to raise the pH, followed by soda ash treatment and recarbonation. In fact, essentially such a sequence is used in a *double-stage process:* two chemical application points and two recarbonation points.

A *split process* can be used to reduce the amount of lime that is neutralized in recarbonation. The excess lime needed to raise the pH is added prior to the first flocculator/clarifier stage. The soda ash is added to the first stage effluent, prior to the second flocculator/clarifier stage. Recarbonation, if used, is applied to the effluent of the second stage. A portion of the flow is bypassed (i.e., is not softened) and is later recombined with softened water to obtain the desired hardness.

Example 24.2

Water contains 130 mg/L of calcium bicarbonate $(Ca(HCO_3)_2)$ as $CaCO_3$. How much slaked lime $(Ca(OH)_2)$ is required to remove the hardness?

Solution

The hardness is given as a $CaCO_3$ equivalent. Therefore, the amount of slaked lime required is implicitly the

same: 130 mg/L as $CaCO_3$. Convert the quantity to an "as substance" measurement. The conversion factor for $Ca(OH)_2$ is 1.35.

$$Ca(OH)_2: \frac{130 \dfrac{mg}{L}}{1.35} = 96.3 \text{ mg/L as substance}$$

Example 24.3

Water is received with the following characteristics.

total hardness	250 mg/L as $CaCO_3$
alkalinity	150 mg/L as $CaCO_3$
carbon dioxide	5 mg/L as substance

The water is to be treated with precipitation softening and recarbonation. Lime (90% pure) and soda ash (98% pure) are available.

Using 50 mg/L of $Ca(OH)_2$ as substance to raise the pH for magnesium precipitation, what stoichiometric amounts (as substance) of (a) slaked lime, (b) soda ash, and (c) carbon dioxide are required to reduce the hardness of the water to zero?

Solution

(a) First, convert the carbon dioxide concentration to a $CaCO_3$ equivalent. The factor is 2.27.

$$CO_2: \left(5 \dfrac{mg}{L}\right)(2.27) = 11.35 \text{ mg/L as } CaCO_3$$

Since the alkalinity is less than the hardness and no hydroxides or calcium are reported, it is concluded that the carbonate hardness is equal to the alkalinity, and the noncarbonate hardness is equal to the difference in total hardness and alkalinity.

Since the alkalinity is reported as a $CaCO_3$ equivalent, the first-state treatment to remove carbon dioxide and carbonate hardness requires lime in the amount of

$$Ca(OH)_2: 11.35 \frac{mg}{L} + 150 \frac{mg}{L} = 161.35 \text{ mg/L as } CaCO_3$$

50 mg/L of $Ca(OH)_2$ are added to raise the pH so that the magnesium (noncarbonate) hardness can be removed. The factor that converts $Ca(OH)_2$ to $CaCO_3$ is 1.35.

$$Ca(OH)_2: 161.35 \frac{mg}{L} + (1.35)\left(50 \frac{mg}{L}\right)$$
$$= 228.85 \text{ mg/L as } CaCO_3$$

Use the 90% fractional purity to convert to quantity as substance.

$$Ca(OH)_2: \frac{228.85 \frac{mg}{L}}{(1.35)(0.9)} = 188.4 \text{ mg/L as substance}$$

(b) The noncarbonate hardness is

$$250 \frac{mg}{L} - 150 \frac{mg}{L} = 100 \text{ mg/L as } CaCO_3$$

The soda ash requirement is

$$Na_2CO_3: \frac{100 \frac{mg}{L}}{(0.94)(0.98)} = 108.6 \text{ mg/L as substance}$$

(c) The recarbonation required to lower the pH depends on the amount of excess lime added.

$$CO_2: \frac{(1.35)\left(50 \frac{mg}{L}\right)}{2.27} = 29.74 \text{ mg/L as substance}$$

32. ADVANTAGES AND DISADVANTAGES OF PRECIPITATION SOFTENING

Precipitation softening is relatively inexpensive for large quantities of water. Both alkalinity and total solids are reduced. The high pH and lime help disinfect the water.

However, the process produces large quantities of sludge that constitute a disposal problem. The intrinsic solubility of some of the compounds means that complete softening cannot be achieved. Flow rates and chemical feed rates must be closely monitored.

33. WATER SOFTENING BY ION EXCHANGE

In the *ion exchange process* (also known as the *zeolite process* and the *base exchange method*), water is passed through a filter bed of exchange material. Ions in the insoluble exchange material are displaced by ions in the water. The processed water leaves with zero hardness. However, since there is no need for water with zero hardness, some of the water is typically bypassed around the process.

If dissolved solids or sodium concentration of the water are issues, then ion exchange may not be suitable.

There are several types of ion exchange materials. *Green sand* (*glauconite*) is a natural substance that is mined and treated with manganese dioxide. Green sand is not used commercially.

Synthetic zeolites have historically been the workhorses of water softening and demineralization. Porosity of continuous-phase *gelular resins* is low, and dry contact surface areas of 500 ft^2/lbm (0.1 m^2/g) or less is common. This makes gel-based zeolites suitable only for small volumes.

Synthetic *macroporous resins* (*macroreticular resins*) are suitable for use in large-volume water processing systems, when chemical resistance is required, and when specific ions are to be removed. These are discontinuous, three-dimensional copolymer beads in a rigid-sponge type formation.[18] Each bead is made up of thousands of microspheres of the resin. Porosity is much higher than with gelular resins, and dry contact specific surface areas are approximately 270,000–320,000 ft^2/lbm (55–65 m^2/g).

There are four primary resin families: strong acid, strong base, weak acid, and weak base. Each family has different resistances to fouling by organic and inorganic chemicals, stabilities, and lifetimes. For example, strong acid resins can last for 20 years. Strong base resins, on the other hand, may have lifetimes of only three years. Each family can be used in either gelular or macroporous forms.

During operation, calcium and magnesium ions in hard water are removed according to the following reaction in which R is the zeolite anion. The resulting sodium compounds are soluble.

$$\begin{Bmatrix} Ca \\ Mg \end{Bmatrix} \begin{Bmatrix} (HCO_3)_2 \\ SO_4 \\ Cl_2 \end{Bmatrix} + Na_2R$$
$$\rightarrow Na_2 \begin{Bmatrix} (HCO_3)_2 \\ SO_4 \\ Cl_2 \end{Bmatrix} + \begin{Bmatrix} Ca \\ Mg \end{Bmatrix} R \qquad 24.47$$

The typical saturation capacity ranges of synthetic resins are 1.0–1.5 meq/mL for anion exchange resins and 1.7–1.9 meq/mL for cation exchange resins. However, working capacities are more realistic measures than saturation capacities. The specific working capacity for zeolites is approximately 3–11 kilograins/ft^3 (6.9–25 kg/m^3) of hardness before regeneration.[19] For synthetic resins, the specific working capacity is

---

[18]The differences in structure between gel polymers and macroporous polymers occur during the polymerization step. Either can be obtained from the same zeolite.

[19]There are 7000 grains in a pound. 1000 grains of hardness (i.e., $^1/_7$ of a pound) is known as a *kilograin*. It should not be confused with a kilogram. Other conversions are 1 grain = 64.8 mg, and 1 grain/gal = 17.12 mg/L.

approximately 10–15 kilograins/ft³ (23–35 kg/m³) of hardness.

The volume of water that can be softened per cycle (between regenerations) is

$$V_{\text{water}} = \frac{(\text{specific working capacity}) \, V_{\text{exchange material}}}{\text{hardness}}$$

24.48

Flow rates are typically 1–6 gpm/ft³ (2–13 L/s·m³) of resin volume. The flow rate across the exposed surface will depend on the geometry of the bed, but values of 3–15 gpm/ft² (2–10 L/s·m²) are typical.

Example 24.4

A municipal plant processes water with a total initial hardness of 200 mg/L. The desired discharge hardness is 50 mg/L. If an ion exchange process is used, what is the bypass factor?

Solution

The water passing through the ion exchange unit is reduced to zero hardness. If x is the water fraction bypassed around the zeolite bed,

$$(1 - x)(0) + x\left(200 \; \frac{\text{mg}}{\text{L}}\right) = 50 \; \frac{\text{mg}}{\text{L}}$$
$$x = 0.25$$

34. REGENERATION OF ION EXCHANGE RESINS

Ion exchange material has a finite capacity for ion removal. When the zeolite approaches saturation, it is regenerated (rejuvenated). Standard ion exchange units are regenerated when the alkalinity of their effluent increases to the *set point*. Most units that collect *crud* are operated to a pressure-drop endpoint.[20] The pressure drop through the ion exchange unit is primarily dependent on the amount of crud collected.

Regeneration of synthetic ion exchange resins is accomplished by passing a *regenerating solution* over/through the resin. Although regeneration can occur in the ion exchange unit itself, external regeneration is becoming common. This involves removing the bed contents hydraulically, backwashing to separate the components (for mixed beds), regenerating the bed components separately, washing, and then recombining and transferring the bed components back into service. For complete regeneration, a contact time of 30–45 min may be required.

Common regeneration compounds are NaCl (for water hardness removal units), H_2SO_4 (for cation exchange resins), and NaOH (for anion exchange resins). The amount of regeneration solution depends on the resin's

---

[20]Since water has been filtered before reaching the zeolite bed, *crud* consists primarily of iron-corrosion products ranging from dissolved to particulate matter.

degree of saturation. A rule of thumb is to expect to use 5–25 lbm of regeneration compound per cubic foot of resin (80–400 kg/m³). Alternatively, dosage of the regeneration compound may be specified in terms of hardness removed (e.g., 0.4 lbm of salt per 1000 grains of hardness removed).

The salt requirement per regeneration cycle is

$$m_{\text{salt}} = (\text{specific working capacity}) \, V_{\text{exchange material}}$$
$$\times (\text{salt requirement})$$

24.49

35. STABILIZATION AND SCALING POTENTIAL

Water that is stable will not gain or lose ions as it passes through the distribution system. Deposits in pipes from dissolved compounds are known as *scale*. Although various types of scale are possible, deposits of calcium carbonate ($CaCO_3$, calcite) are the most prevalent. *Stabilization treatment* is used to eliminate or reduce the potential for scaling in a pipe after the treated water is distributed. Various factors influence the stability of water, including temperature, dissolved oxygen, dissolved solids, pH, and alkalinity. Increased temperature greatly increases the likelihood of scaling.

Stability is a general term used to describe a water's tendency to cause scaling. Four indices are commonly used to quantify the stability of water: the Langelier stability index, Ryznar stability index, Puckorius scaling index, and aggressive index. The Langelier and Ryznar indices are widely used in municipal water works, while the Puckorius index is encountered in installations with cooling water (e.g., cooling towers). Each index has its proponents, strengths and weaknesses, and imprecisions. Conclusions drawn from them are not always in agreement.

The common *Langelier stability index* (LSI) (also known as the *Langelier saturation index*) is significantly positive (e.g., greater than 1.5) when a water is supersaturated and will continue to deposit $CaCO_3$ scale in the pipes downstream of the treatment plant. When LSI is significantly negative (e.g., less than −1.5), the water has little scaling potential and may actually dissolve existing scale encountered.

$$\text{LSI} = \text{pH} - \text{pH}_{\text{sat}}$$

24.50

The *Ryznar stability index* (RSI) arranges the same parameters as LSI differently, but avoids the use of negative numbers. Water is considered to be essentially neutral when $6.5 < \text{RSI} < 7$. When $\text{RSI} > 8$, the water has little scaling potential and will dissolve scale deposits already present. When $\text{RSI} < 6.5$, the water has scaling potential.

$$\text{RSI} = 2\text{pH}_{\text{sat}} - \text{pH} = \text{pH} - 2(\text{LSI})$$

24.51

The *Puckorius scaling index* (PSI) (also known as the *practical scaling index*) does not use the actual pH in its calculation. To account for the buffering effects of other ions, it uses an equilibrium pH. Like RSI, PSI is always positive, and it has the same index interpretation ranges as RSI.

$$PSI = 2pH_{sat} - pH_{eq} \quad \text{24.52}$$

$$pH_{eq} = 4.54 + 1.465 \log_{10}[M] \quad \text{24.53}$$

Equation 24.51, Eq. 24.52, and Eq. 24.53 require knowing the pH when the water is saturated with $CaCO_3$. This value depends on the ionic concentrations in a complex manner, and it is calculated or estimated in a variety of ways. The saturation pH can be calculated from Eq. 24.54, where K_{sp} is the solubility product constant for $CaCO_3$, and K_a is the ionization constant for H_2CO_3.[21,22] M is the alkalinity (often reported as the HCO_3^- concentration) in mg/L as $CaCO_3$. Ca^{++} is the calcium ion content in mg/L, also as $CaCO_3$.

$$pH_{sat} = (pK_a - pK_{sp}) + pCa + pM$$
$$= -\log_{10}\left(\frac{K_a[Ca^{++}][M]}{K_{sp}}\right) \quad \text{24.54}$$

$$M = [HCO_3^-] + 2[CO_3^{--}] + [OH^-] \quad \text{24.55}$$

Since the solubility product constants are rarely known and depend in a complex manner on the total ion strength (total dissolved solids) and temperature, if the water chemistry is fully known, the correlation of Eq. 24.56 can be used.[23] Hard water is typically two-thirds calcium hardness and one-third magnesium. However, if only total hardness is known (i.e., calcium hardness is not known) the worst case of calcium hardness equaling total hardness can be assumed. Total dissolved solids (TDS) can be estimated from water conductivity using Table 24.4.

$$pH_{sat} = 9.3 + A + B - C - D \quad \text{24.56}$$

$$A = \frac{\log_{10}[TDS] - 1}{10} \quad \text{24.57}$$

$$B = -13.12 \log_{10}(T_{\circ C} + 273^\circ) + 34.55 \quad \text{24.58}$$

$$C = \log_{10}[Ca^{++}] - 0.4 \quad \text{24.59}$$

$$D = \log_{10}[M] \quad \text{24.60}$$

The *aggressive index* (AI) was originally developed to monitor water in asbestos pipe, but can be encountered in other environments. It is calculated from Eq. 24.61, but disregards the effects of dissolved solids and water

---

[21]pK = $-\log K$; pCa = $-\log[Ca^{++}]$; pM = $-\log[M]$
[22]Equation 24.54 is an approximation to the stoichiometric reaction equation because it assumes the activities of the H^+, CO_3^-, and HCO_3^- ions present in the water are all 1. This is essentially, but not exactly, true.
[23]As reported in *Industrial Water Treatment, Operation and Maintenance*, UFC 3-240-13FN, Appendix B. Department of Defense (Unified Facilities), 2005.

Table 24.4 *Total Dissolved Solids vs. Conductivity*

conductivity (S/cm)	total dissolved solids (mg/L as $CaCO_3$)
1	0.42
10.6	4.2
21.2	8.5
42.4	17.0
63.7	25.5
84.8	34.0
106.0	42.5
127.3	51.0
148.5	59.5
169.6	68.0
190.8	76.5
212.0	85.0
410.0	170.0
610.0	255.0
812.0	340.0
1008.0	425.0

temperature. Therefore, it is less accurate than the other indices.

$$AI = pH + \log_{10}[M] + \log_{10}[Ca^{++}] \quad \text{24.61}$$

LSI, RSI, PSI, and AI predict tendencies only for calcium carbonate scaling. They do not estimate scaling potential for calcium phosphate, calcium sulfate, silica, or magnesium silicate scale. They are generally only accurate with untreated water. Although water treated with phosphonates and acrylates to dissolve or stabilize calcium carbonate can be evaluated, the scaling potential of water treated with crystal modifiers (e.g., poly-maleates, sulfonated styrene/maleic anhydride, and terpolymers) should not be evaluated with these indices.

The primary stabilization processes are pH and alkalinity adjustment using lime or soda ash (if the pH is low), and with carbon dioxide, sulfuric acid, or hydrochloric acid if the pH is too high. Scaling can also be prevented within the treatment plan by using protective pipe linings and coatings. Various inhibitors and sequestering agents can also be used.[24] If lime is used for pH adjustment, care must be exercised that excess lime does not form clinker particles that are carried out into the water distribution system.

Some waters (including cooling waters in some industries) are always corrosive because of the presence of dissolved oxygen, carbon dioxide, and various solids, regardless of whether calcium carbonate is present. $CaCO_3$ scaling indices were never intended to be used as indices or corrosion potential of mild steel pipes, but that has occurred. LSI, RSI, and PSI are blind to the presence of chloride and sulfate content, as well as to the nature of the pipe and fitting materials and to the galvanic potential differences. For that reason, stability

---

[24]A *sequestering agent* (*chelant, chelator, chelating agent*) is a compound that works by sequestering positive metallic ions in a complex negative ion.

indices should not be used for corrosion prediction. However, the LSI, RSI, and PSI scaling potential continuums continue to be inappropriately divided by some authorities into scaling and corrosion regions.

36. DISINFECTION

Chlorination is commonly used for disinfection. Chlorine can be added as a gas or as a liquid. If it is added to the water as a gas, it is stored as a liquid, which vaporizes around $-31°F$ ($-35°C$). Liquid chlorine is the primary form used since it is less expensive than calcium hypochlorite solid ($Ca(OCl)_2$) and sodium hypochlorite (NaOCl).

Chlorine is corrosive and toxic. Special safety and handling procedures must be followed with its use.

37. CHLORINATION CHEMISTRY

When chlorine gas dissolves in water, it forms hydrochloric acid (HCl) and hypochlorous acid (HOCl).

$$Cl_2 + H_2O \rightarrow HCl + HOCl \qquad 24.62$$

$$HCl \rightarrow H^+ + Cl^- \qquad 24.63$$

$$HOCl \rightleftharpoons H^+ + OCl^- \qquad 24.64$$

The fraction of HOCl ionized into H^+ and OCl^- depends on the pH and can be determined from Fig. 24.6.

Figure 24.6 Ionized HOCl Fraction vs. pH

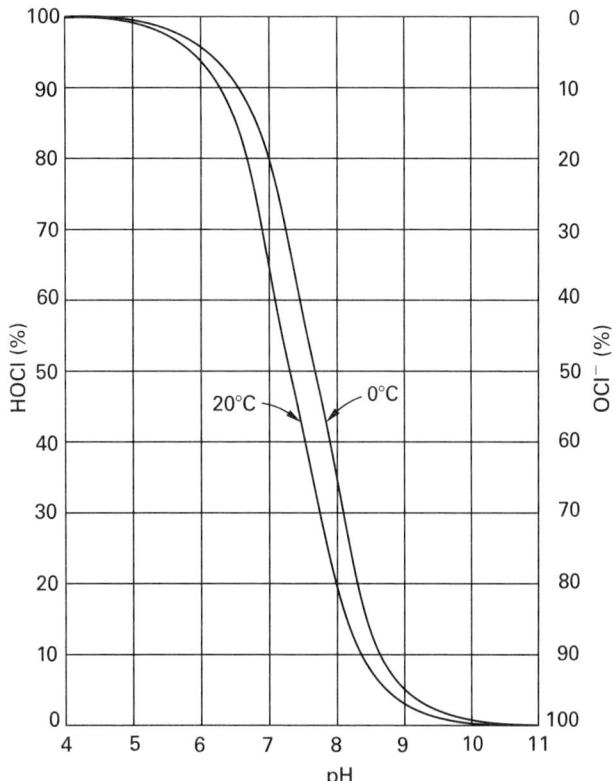

When calcium hypochlorite solid is added to water, the ionization reaction is

$$Ca(OCl)_2 \rightarrow Ca^{++} + 2OCl^- \qquad 24.65$$

38. CHLORINE DOSE

Figure 24.7 illustrates a breakpoint chlorination curve. Basically, the concept of *breakpoint chlorination* is to continue adding chlorine until the desired quantity of free residuals appears. This cannot occur until after the demand for combined residuals has been satisfied.

Figure 24.7 Breakpoint Chlorination Curve

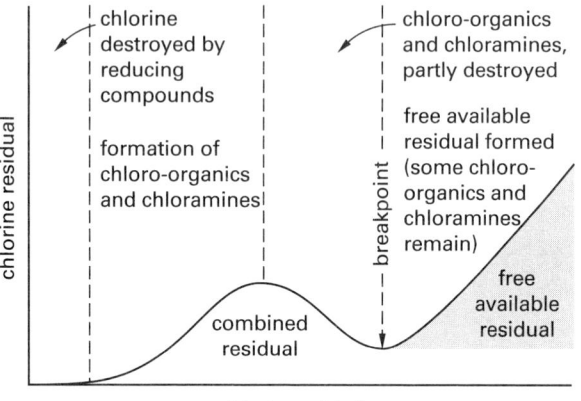

The amount of chlorine necessary for disinfection varies with the organic and inorganic material present in the water, the pH, the temperature, and the contact time. Thirty minutes of chlorine contact time is generally sufficient to deactivate giardia cysts. Satisfactory results can generally be obtained if a free chlorine residual (consisting of hypochlorous acid and hypochlorite ions) of 0.2–0.5 mg/L can be maintained throughout the distribution system. Combined residuals, if use is approved by health authorities, should be 1.0–2.0 mg/L at distant points in the distribution system.

Example 24.5

The flow rate through a treatment plant is 2 MGD (7.5 ML/day). The hypochlorite ion (OCl^-) dose is 20 mg/L as substance. The purity of calcium hypochlorite, $Ca(OCl)_2$, is 97.5% as delivered. How many pounds of calcium hypochlorite are needed to treat the water?

SI Solution

The approximate molecular weights for the components of calcium hypochlorite are

Ca:		40.1
O:	(2)(16) =	32.0
Cl:	(2)(35.5) =	71.0
	total	143.1

The fraction of available chlorine in the form of the hypochlorite ion is

$$G = \frac{32.0 + 71.0}{143.1} = 0.720$$

Use the standard dose equation with adjustments for both purity and availability.

$$
\begin{aligned}
F_{\mathrm{kg/d}} &= \frac{D_{\mathrm{mg/L}} Q_{\mathrm{ML/d}}}{PG} \\
&= \frac{\left(20\ \dfrac{\mathrm{mg}}{\mathrm{L}}\right)\left(7.5\ \dfrac{\mathrm{ML}}{\mathrm{d}}\right)}{(0.975)(0.720)} \\
&= 213.7\ \mathrm{kg/d}
\end{aligned}
$$

Customary U.S. Solution

The approximate molecular weights for the components of calcium hypochlorite are

Ca:		40.1
O:	(2)(16) =	32.0
Cl:	(2)(35.5) =	71.0
	total	143.1

The fraction of available chlorine in the form of the hypochlorite ion is

$$\frac{32.0 + 71.0}{143.1} = 0.720$$

Use the standard dose equation with adjustments for both purity and availability.

$$
\begin{aligned}
F_{\mathrm{lbm/day}} &= \frac{D_{\mathrm{mg/L}} Q_{\mathrm{MGD}}\left(8.345\ \dfrac{\mathrm{lbm\text{-}L}}{\mathrm{mg\text{-}MG}}\right)}{PG} \\
&= \frac{\left(20\ \dfrac{\mathrm{mg}}{\mathrm{L}}\right)(2\ \mathrm{MGD})\left(8.345\ \dfrac{\mathrm{lbm\text{-}L}}{\mathrm{mg\text{-}MG}}\right)}{(0.975)(0.720)} \\
&= 475.5\ \mathrm{lbm/day}
\end{aligned}
$$

39. ADVANCED OXIDATION PROCESSES

Alternatives to chlorination have become necessary since THMs were traced to the chlorination process. These alternatives are categorized as *advanced oxidation processes* (AOPs).

Both bromine and iodine have properties similar to chlorine and can be used for disinfection. They are seldom used because they are relatively costly and produce their own disinfection by-products.

Chlorine dioxide (ClO_2) is manufactured at the water treatment plant from chlorine and sodium chlorite. Its ionization by-products (chlorite and chlorate) and high cost have limited its use.

Ozone is used extensively throughout the world, though not in large quantities in the United States. Ozone is a more powerful disinfectant than chlorine. Ozone can be used alone or (in several developing technologies) in conjunction with hydrogen peroxide. Ozone is generated on-site by running high voltage electricity through dry air or pure oxygen. Gases (oxygen and unused ozone) developing during ozonation must be collected and destroyed in an ozone-destruct unit to ensure that no ozone escapes into the atmosphere.

Ultraviolet radiation is effective in disinfecting shallow (e.g., less than 10 cm) bodies of water. Its primary disadvantages are cost and the absence of any residual disinfection for downstream protection.

40. CHLORAMINATION

Before leaving the plant, treated water may be *chloraminated* (i.e., treated with both ammonia and chlorine to form chloramines). This step ensures lasting disinfection by providing a residual level of chloramines, protecting the water against bacteria regrowth as it travels through the distribution system.

41. DECHLORINATION

In addition to its routine disinfection use, chlorine can enter water supply systems during operation and maintenance activities such as disinfection of mains, testing of hydrants, and routine flushing of distribution systems. Such flows are often discharged to wastewater sewer systems. Chlorine reacts with some organic compounds to produce THMs and is toxic to aquatic life. For these reasons, most water supply and wastewater facilities dechlorinate their discharges for regulatory compliance.

Excess chlorine can be removed with a reducing agent, referred to as a *dechlor*. Table 24.5 lists common dechlor chemicals. Dechlors reduce all forms of active chlorine (chlorine gas, hypochlorites, and chloramines) to chloride. Aeration also reduces chlorine content, as does contact with granular activated charcoal (GAC). Carbon absorption processes are more costly than chemical treatments and are used only when total dechlorination is required.

Sulfur dioxide is a toxic gas and is most frequently used by large wastewater. Sodium bisulfite, sodium sulfite, and sodium thiosulfate are most frequently used by smaller water utilities. Sodium bisulfite is low in cost and has a high rate of dechlorination. Sodium sulfite tablets are easy to store and handle. Sodium thiosulfate is less hazardous and depletes oxygen less than sodium bisulfite and sodium sulfite. Ascorbic acid and sodium ascorbate are used because they do not impact DO concentrations. Monitoring is required to prevent the

Table 24.5 *Dechlorinating Chemicals Used in Water Supply and Wastewater Facilities*

name	formula	stoichiometric dose[*] (mg/mg Cl_2)
ascorbic acid	$C_6H_8O_6$	2.48
calcium thiosulfate	CaS_2O_3	1.19
hydrogen peroxide	H_2O_2	0.488
sodium ascorbate	$C_6H_7NaO_6$	2.78
sodium bisulfite	$NaHSO_3$	1.46
sodium metabisulfite	$Na_2S_2O_5$	1.34
sodium sulfite	Na_2SO_3	1.78
sodium thiosulfate	$Na_2S_2O_3$	0.556
sulfur dioxide	SO_2	0.903

[*]Doses are approximate and depend on pH and assumed reaction chemistry. Theoretical values may be used for initial approximations and equipment sizing. Under the best conditions, 10% excess is required.

over-application of chemicals that may deplete the dissolved oxygen concentration or alter the pH of receiving streams.

42. DEMINERALIZATION AND DESALINATION

Demineralization and desalination (*salt water conversion*) are required when only brackish water supplies are available.[25] These processes are carried out in distillation, electrodialysis, ion exchange, and membrane processes.

Distillation is a process whereby the raw water is vaporized, leaving the salt and minerals behind. The water vapor is reclaimed by condensation. Distillation cannot be used to economically provide large quantities of water.

Reverse osmosis is the least costly and most attractive membrane demineralization process. A thin membrane separates two solutions of different concentrations. Pore size is smaller (0.0001–0.001 μm) than with ultrafilter membranes (see Sec. 24.26), as salt ions are not permitted to pass through. Typical large-scale osmosis units operate at 150–500 psi (1.0–5.2 MPa).[26]

Nanofiltration is similar to ultrafiltration and reverse osmosis, with pore size (0.001 μm) and operating pressure (75–250 psig; 0.5–1.7 MPa) intermediate between the two. Nanofilters are commonly referred to as *softening membranes*.

In *electrodialysis*, positive and negative ions flow through selective membranes under the influence of an induced electrical current. Unlike pressure-driven filtration processes, however, the ions (not the water molecules) pass

through the membrane. The ions removed from the water form a concentrate stream that is discarded.

The *ion exchange* process is an excellent solution to demineralization and desalination. (See Sec. 24.33.)

43. WATER DEMAND

Normal water demand is specified in gallons per capita day (gpcd) or liters per capita day (Lpcd)—the average number of gallons (liters) used by each person each day. This is referred to as *average annual daily flow* (AADF) if the average is taken over a period of a year. Residential (i.e., domestic), commercial, industrial, and public uses all contribute to normal water demand, as do waste and unavoidable loss.[27]

As Table 24.6 illustrates, an AADF of 165 gpcd (625 Lpcd) is a typical minimum for planning purposes. If large industries are present (e.g., canning, steel making, automobile production, and electronics), then their special demand requirements must be added.

Table 24.6 *Annual Average Water Requirements[a]*

use	demand gpcd	Lpcd[b]
residential	75–130	284–490
commercial and industrial	70–100	265–380
public	10–20	38–80
loss and waste	10–20	38–80
totals	165–270	625–1030

(Multiply gpcd by 3.79 to obtain Lpcd.)
[a]exclusive of fire fighting requirements
[b]liters per capita-day

Water demand varies with the time of day and season. Each community will have its own demand distribution curve. Table 24.7 gives typical multipliers, M, that might be used to estimate instantaneous demand from the average daily flow.

$$Q_{instantaneous} = M(AADF) \qquad 24.66$$

Table 24.7 *Typical Demand Multipliers (to be used with average annual daily flow)*

period of usage	multiplier, M
maximum daily	1.5–1.8
maximum hourly	2.0–3.0
early morning	0.25–0.40
noon	1.5–2.0
winter	0.80
summer	1.30

[25]In the United States, Florida, Arizona, and Texas are leaders in desalinization installations. Potable water is routinely made from sea water in the Middle East.
[26]Membrane processes operate in a *fixed flux condition*. In order to keep the yield constant over time, the pressure must be constantly increased in order to compensate for the effects of fouling and compaction. Fouling by inorganic substances and biofilms is the biggest problem with membranes.

[27]"Public" use includes washing streets, flushing water and sewer mains, flushing fire hydrants, filling public fountains, and fighting fires.

Per capita demand must be multiplied by the population to obtain the total system demand. Since a population can be expected to change in number, a supply system must be designed to handle the water demand for a reasonable amount of time into the future. Several methods can be used for estimating future demand, including mathematical, comparative, and correlative methods. Various assumptions can be made for mathematical predictions, including uniform growth rate (same as straight-line extrapolation) and constant percentage growth rate. (In the case of decreasing population, the "growth" rate would be negative.)

Economic aspects will dictate the number of years of future capacity that are installed. Excess capacities of 25% for large systems and 50% for small systems are typically built into the system initially.

44. FIRE FIGHTING DEMAND

The amount of water that should be budgeted for municipal fire protection depends on the degree of protection desired, construction types, and the size of the population to be protected. Municipal capacity decisions are almost always made on the basis of economics driven by the insurance industry. The Insurance Services Office (ISO) rates municipalities according to multiple criteria it publishes in its Fire Suppression Rating Schedule.[28]

One of the criteria is the ability of a municipality to provide what the ISO refers to as the *needed fire flow*, NFF, at various locations around the community. The NFF methodology has been determined from measurement of historical events (i.e., past fires) over many years. The ISO awards credit points dependent on how well actual capacities (availability) matches the NFFs at these locations. Adequacy may be limited by the water works, main capacity, and hydrant distribution.

The ISO also calculates a community's *basic fire flow*, BFF, which is the fifth highest needed fire flow (not to exceed 3500 gpm (13 250 L/m) among the locations listed in the ISO batch report. BFF is used to determine the numbers of engine companies, crews, and number, nature, and size of the fire-fighting apparatus that the community needs.

Calculating the NFF is highly codified, requiring specific knowledge of construction types, floor areas, and usage (occupancy). NFF is calculated in gallons per minute and is rounded up or down to the nearest 250 gpm (16 L/s) if less than 2500 gpm (160 L/s), and up or down to the nearest 500 gpm (32 L/s) if greater than 2500 gpm (160 L/s). The minimum NFF value is 250 gpm (16 L/s), corresponding to the discharge from a single standard 1.125 in (29 mm) diameter smooth fire nozzle,

and the maximum value is 12,000 gpm (760 L/s). Calculated values are increased by 500 gpm (32 L/s) if the building has a wood shingle roof covering that can contribute to the spread of a fire. Reductions in NFF are given for structures having sprinkler systems designed in accordance with National Fire Protection Association (NFPA) standards.

As a convenience, the NFF for 1- and 2-family dwellings not exceeding two stories in height can be read directly from Table 24.8, without the need of calculations.

Table 24.8 *Needed Fire Flow for 1- and 2-Family Dwellings*

distance between buildings (ft)	needed fire flow (gpm)
≤ 10	1500
11–30	1000
31–100	750
> 100	500

(Multiply ft by 0.3048 to obtain m.)
(Multiply gpm by 0.06309 to obtain L/s.)

Source: *Guide for Determination of Needed Fire Flow*, Chap. 7, Insurance Services Office

For other than 1- and 2-family dwellings, NFF must be calculated based on construction type, occupancy, area, and other factors. Equation 24.67 is the ISO formula for an individual, nonsprinklered building. Some of the factors are evaluated for each exposure (side) of the building. ISO provides guidance and examples of application for selecting values for the factors, but specialized knowledge and expertise is still needed to use Eq. 24.67 accurately, particularly for buildings with mixed construction classes and occupancies.

$$NFF = CO(1 + X + P) \qquad 24.67$$

C is a factor dependent on the type of construction. In Eq. 24.68, F is a *construction class coefficient*. $F = 1.5$ for wood frame construction (class 1), 1.0 for joisted masonry (class 2), 0.8 for noncombustible construction (class 3) including masonry (class 4), and 0.6 for fire-resistive construction (classes 5 and 6). A is the effective area in square feet, determined as the area of the largest floor in the building plus 50% of all the other floors for classes 1, 2, 3, and 4. Other modifications apply to classes 5 and 6. Special rules apply to finding areas of multi-story, fire-resistant buildings with sprinkler systems. C is rounded up or down to the nearest 250 gpm (16 L/s) (prior to the rounding of NFF), may not be less than 500 gpm (32 L/s), is limited to a maximum of 8000 gpm (500 L/s) for classes 1 and 2, is limited to 6000 gpm (380 L/s) for classes 3, 4, 5, and 6, and for any 1-story building of any class.

$$C = 18F\sqrt{A} \qquad 24.68$$

The *occupancy combustible factor*, O, accounts for the combustibility of a building's contents and is 1.25 for

---

[28]Other methods exist and are in use around the world. The ISO method works well for large diameter hoses preferred in the United States and needed to achieve the high flow rates necessary with fires involving the wood-framed buildings that predominate in the United States.

rapid burning (class C-5), 1.15 for free-burning (class C-4), 1.0 for combustible (class C-3), 0.85 for limited-combustible (class C-2), and 0.75 for noncombustible (class C-1). The exact definitions and examples of these classes are specified by the ISO.

The *exposure factor*, X, reflects the influence of an adjoining or connected building less than 100 ft (30 m) from a wall of the subject building. Its value is read from an ISO table as a function of separation, lateral exposure length, and construction class. Values vary from essentially 0 to 0.25.

The *communication factor*, P, reflects any connections (passageways) between adjacent structures. Like the exposure factor, it is read from ISO tables as a function of the length and nature of the passageway and closures. Values vary from 0 to 0.35.

The exposure and communication factors are summed for all exposures (sides) of a building. However, the total of $X + P$ over all exposures need not exceed 0.6.

45. OTHER FORMULAS FOR FIRE FIGHTING NEEDS

Although modern methods calculate fire fighting demand based on the area and nature of construction to be protected, the earliest methods of predicting fire-fighting demand were based on population.[29] The 1911 *American Insurance Association equation* (also referred to as the *National Board of Fire Underwriters equation* after AIA's predecessor) for cities with populations less than 200,000 was used widely up until the 1980s. P in Eq. 24.69 is in thousands of people.

$$Q_{gpm} = 1020\sqrt{P}(1 - 0.01\sqrt{P}) \qquad \textit{24.69}$$

Other authorities who have developed alternative methods include Thomas (1959), Iowa State University (1967), Illinois Institutes of Technology Research Institute (1968), and Ontario Building Code (Office of the Fire Marshal, 1995).

46. DURATION OF FIRE FIGHTING FLOW

A municipality must continue to provide water to its domestic, commercial, and industrial customers while meeting its fire fighting needs. In the ISO's *Fire Suppression Rating Schedule*, in order for a municipality to get full credit, it is required that a fire system must be operable with the potable water system operating at the maximum 24 hour average daily rate, plus the fire demand at a minimum of 20 psig (140 kPa) for a minimum time depending on the needed fire flow: 2 hours, NFF < 3000 gpm (190 L/s); 3 hours, 3000 gpm (190 L/s) < NFF < 3500 gpm (220 L/s); and 4 hours, NFF > 3500 gpm (220 L/s).

Another approach, published[30] by the American Water Works Association, is to require the fire fighting flow to be maintained for a duration equal to $Q_{gpm}/1000$ (rounded up to the next integer) in hours.

47. FIRE HYDRANTS

Other factors considered by the ISO are the type, size, capacity, installation, and inspection program of fire hydrants. The actual separation of hydrants can be specified in building codes, local ordinances, and other published standards. However, in order to be considered for ISO credits, only hydrants within 1000 ft (300 m) of a building are considered, and this distance often is used for maximum hydrant spacing (so that no building is more than 500 ft (150 m) distant from a hydrant). Additional credit is given for hydrants within 300 ft (90 m), and this distance is often selected for hydrant spacing in residential areas. Hydrants are ordinarily located near street corners where use from four directions is possible.

48. STORAGE AND DISTRIBUTION

Water is stored to provide water pressure, equalize pumping rates, equalize supply and demand over periods of high consumption, provide surge relief, and furnish water during fires and other emergencies when power is disrupted. Storage may also serve as part of the treatment process, either by providing increased detention time or by blending water supplies to obtain a desired concentration.

Several methods are used to distribute water depending on terrain, economics, and other local conditions. *Gravity distribution* is used when a lake or reservoir is located significantly higher in elevation than the population.

Distribution from *pumped storage* is the most common option when gravity distribution cannot be used. Excess water is pumped during periods of low hydraulic and electrical demands (usually at night) into elevated storage. During periods of high consumption, water is drawn from the storage. With pumped storage, pumps are able to operate at a uniform rate and near their rated capacity most of the time.

Using pumps without storage to force water directly into the mains is the least desirable option. Without storage, pumps and motors will not always be able to run in their most efficient ranges since they must operate during low, average, and peak flows. In a power outage, all water supply will be lost unless a backup power source comes online quickly or water can be obtained by gravity flow.

Water is commonly stored in surface and elevated tanks. The elevation of the water surface in the tank directly determines the distribution pressure. This elevation is

---

[29]Freeman's formula (1892) and Kuchling's formula (1897) were among the earliest.

[30]*Distribution System Requirements for Fire Protection* (M31), American Water Works Association, 1992.

controlled by an *altitude valve* that operates on the differential in pressure between the height of the water and an adjustable spring-loaded pilot on the valve. Altitude valves are installed at ground level and, when properly adjusted, can maintain the water levels to within 4 in (102 mm). Tanks must be vented to the atmosphere. Otherwise, a rapid withdrawal of water will create a vacuum that could easily cause the tank to collapse inward.

The preferred location of an elevated tank is on the opposite side of the high-consumption district from the pumping station. During periods of high water use, the district will be fed from both sides, reducing the loss of head in the mains below what would occur without elevated storage.

Equalizing the pumping rate during the day ordinarily requires storage of at least 15–20% of the maximum daily use. Storage for fires and emergencies is more difficult to determine. Fire storage is essentially dictated by building ordinances and insurance (i.e., economic benefits to the public). Private on-site storage volume for large industries may be dictated by local codes and ordinances.

49. WATER PIPE

Several types of pipe are used in water distribution depending on the flow rate, installed location (i.e., above or below ground), depth of installation, and surface surcharge. Pipes must have adequate strength to withstand external loads from backfill, traffic, and earth movement. They must have high burst strength to withstand internal pressure, a smooth interior surface to reduce friction, corrosion resistance, and tight joints to minimize loss. Once all other requirements have been satisfied, the choice of pipe material can be made on the basis of economics.

Asbestos-cement pipe was used extensively in the past. It has a smooth inner surface and is immune to galvanic corrosion. However, it has low flexural strength. *Concrete pipe* is durable, watertight, has a smooth interior, and requires little maintenance. It is manufactured in plain and reinforced varieties. *Cast-iron pipe* (ductile and gray varieties) is strong, offers long life, and is impervious. However, it has a high initial cost and is heavy (i.e., difficult to transport and install). It may need to be coated on the exterior and interior to resist corrosion of various types. *Steel pipe* offers a smooth interior, high strength, and high ductility, but it is susceptible to corrosion inside and out. The exterior and interior may both need to be coated for protection. *Plastic pipe* (ABS and PVC) has a smooth interior and is chemically inert, corrosion resistant, and easily transported and installed. However, it has low strength.

50. SERVICE PRESSURE

For ordinary domestic use, the minimum water pressure at the tap should be 25–40 psig (170–280 kPa). A minimum of 60 psig (400 kPa) at the fire hydrant is usually adequate, since that allows for up to a 20 psig (140 kPa) pressure drop in fire hoses. 75 psig (500 kPa) and higher is common in commercial and industrial districts. *Pressure regulators* can be installed if delivery pressure is too high.

51. WATER MANAGEMENT

Unaccounted-for water is the potable water that is produced at the water treatment plant but is not accounted for in billing. It includes known unmetered uses (such as fire fighting and hydrant flushing) and all unknown uses (e.g., from leaks, broken meters, theft, and illegal connections).

Unaccounted-for water can be controlled with proper attention to meter selection, master metering, leak detection, quality control (installation and maintenance of meters), control of system pressures, and accurate data collection.

Since delivery is under pressure, infiltration of groundwater into pipe joints is not normally an issue with distribution.

Water Treatment

25 Wastewater Quantity and Quality

Nomenclature

BOD	biochemical oxygen demand	mg/L	mg/L
C	concentration	mg/L	mg/L
COD	chemical oxygen demand	mg/L	mg/L
d	depth of flow	ft	m
D	oxygen deficit	mg/L	mg/L
D	pipe diameter	ft	m
DO	dissolved oxygen	mg/L	mg/L
DO^*	dissolved oxygen of the seed	mg/L	mg/L
k	maximum rate of substrate utilization per unit mass	L/day-mg	L/d·mg
K	endogenous decay coefficient	day^{-1}	d^{-1}
K	rate constant	day^{-1}	d^{-1}
K_s	half-velocity coefficient	mg/L	mg/L
n	slope	–	–
P	population	–	–
Q	flow quantity	gal/day	L/d
r	rate of change in oxygen content	mg/L-day	mg/L·d
S	concentration of growth-limiting nutrient	mg/L	mg/L
S	slope	ft/ft	m/m
t	time	sec	s
T	temperature	°F	°C
v	velocity	ft/sec	m/s
V	volume	mL	mL
x	distance	ft	m
X	microorganism concentration	mg/L	mg/L
Y	maximum yield coefficient	mg/L	mg/L

Symbols

θ	temperature constant	–	–
μ	specific growth rate coefficient	day^{-1}	d^{-1}

Subscripts

0	initial
5	five-day
c	critical
d	decay or deoxygenation
e	equivalent
f	final
g	growth
i	initial
m	maximum
r	reoxygenation or river
sat	saturated
su	substrate utilization
t	at time t
T	at temperature T
u	ultimate carbonaceous
w	wastewater

1. DOMESTIC WASTEWATER

Domestic wastewater refers to *sanitary wastewater* discharged from residences, commercial buildings, and institutions. Other examples of domestic wastewater sources are mobile homes, hotels, schools, offices, factories, and other commercial enterprises, excluding manufacturing facilities.

The volume of domestic wastewater varies from 50–250 gallons per capita day (gpcd) (190–950 Lpcd) depending on sewer uses. A more common range for domestic wastewater flow is 100–120 gpcd (380–450 Lpcd), which assumes that residential dwellings have major water-using appliances such as dishwashers and washing machines.

Water Treatment

2. INDUSTRIAL WASTEWATER

Small manufacturing facilities within municipal limits ordinarily discharge *industrial wastewater* into the city sewer system only after pretreatment. In *joint processing* of wastewater, the municipality accepts responsibility for final treatment and disposal. The manufacturing plant may be required to pretreat the wastewater and equalize flow by holding it in a basin for stabilization prior to discharge to the sewer.

Uncontaminated streams (e.g., cooling water), in many cases, can be discharged into sewers directly. However, pretreatment at the industrial site is required for wastewaters having strengths and/or characteristics significantly different from sanitary wastewater.

To minimize the impact on the sewage treatment plant, consideration is given to modifications in industrial processes, segregation of wastes, flow equalization, and waste strength reduction. Modern industrial/manufacturing processes require segregation of separate waste streams for individual pretreatment, controlled mixing, and/or separate disposal. Process changes, equipment modifications, by-product recovery, and in-plant wastewater reuse can result in cost savings for both water supply and wastewater treatment.

Toxic waste streams are not generally accepted into the municipal treatment plant at all. Toxic substances require appropriate pretreatment prior to disposal by other means.

3. MUNICIPAL WASTEWATER

Municipal wastewater is the general name given to the liquid collected in sanitary sewers and routed to municipal sewage treatment plants. Many older cities have *combined sewer systems* where storm water and sanitary wastewaters are collected in the same lines. The combined flows are conveyed to the treatment plant for processing during dry weather. During wet weather, when the combined flow exceeds the plant's treatment capacity, the excess flow often bypasses the plant and is discharged directly into the watercourse.

4. WASTEWATER QUANTITY

Approximately 70–80% of a community's domestic and industrial water supply returns as wastewater. This water is discharged into the sewer systems, which may or may not also function as storm drains. Therefore, the nature of the return system must be known before sizing can occur.

Infiltration, due to cracks and poor joints in old or broken lines, can increase the sewer flow significantly. Infiltration per mile (kilometer) per in (mm) of pipe diameter is limited by some municipal codes to 500 gpd/mi-in (46 Lpd/km·mm). Modern piping materials and joints easily reduce the infiltration to 200 gpd/in-mi (18 Lpd/km·mm) and below. Infiltration can also be

roughly estimated as 3–5% of the peak hourly domestic rate or as 10% of the average rate.

Inflow is another contributor to the flow in sewers. Inflow is water discharged into a sewer system from such sources as roof down spouts, yard and area drains, parking area catch basins, curb inlets, and holes in manhole covers.

Sanitary sewer sizing is commonly based on an assumed average of 100–125 gpcd (380–474 Lpcd). There will be variations in the flow over time, although the variations are not as pronounced as they are for water supply. Hourly variations are the most significant. The flow rate pattern is essentially the same from day to day. Weekend flow patterns are not significantly different from weekday flow patterns. Seasonal variation depends on the location, local industries, and infiltration.

Table 25.1 lists typical *peaking factors* (i.e., peak multipliers) for treatment plant influent volume. Due to storage in ponds, clarifiers, and sedimentation basins, these multipliers are not applicable throughout all processes in the treatment plant.

Table 25.1 *Typical Variations in Wastewater Flows (based on average annual daily flow)*

flow description	typical time	peaking factor
daily average	–	1.0
daily peak	10–12 a.m.	2.25
daily minimum	4–5 a.m.	0.4
seasonal average	May, June	1.0
seasonal peak	late summer	1.25
seasonal minimum	late winter	0.9

Recommended Standards for Sewage Works (*Ten States' Standards*, abbreviated TSS) specifies that new sanitary sewer systems should be designed to have an average flow of 100 gpcd (380 Lpcd or 0.38 m³/d), which includes an allowance for normal infiltration [TSS Sec. 33.94]. However, the sewer pipe must be sized to carry the peak flow as a gravity flow. In the absence of any studies or other justifiable methods, the ratio of peak hourly flow to average flow should be estimated from the following relationship, in which P is the population served in thousands of people at a particular point in the network.

$$\frac{Q_{\text{peak}}}{Q_{\text{ave}}} = \frac{18 + \sqrt{P}}{4 + \sqrt{P}} \qquad 25.1$$

The peaking factor can also be estimated by using *Harmon's peaking factor* equation.

$$\frac{Q_{\text{peak}}}{Q_{\text{ave}}} = \frac{1 + 14}{4 + \sqrt{P}} \geq 2.5 \qquad 25.2$$

Collectors (i.e., *collector sewers, trunks,* or *mains*) are pipes that collect wastewater from individual sources and carry it to interceptors. (See Fig. 25.1.) Collectors must be designed to handle the maximum hourly flow, including domestic and infiltration, as well as additional discharge from industrial plants nearby. Peak flows of 400 gpcd (1500 Lpcd) for laterals and submains flowing full and 250 gpcd (950 Lpcd) for main, trunk, and outfall sewers can be assumed for design purposes, making the peaking factors approximately 4.0 and 2.5 for submains and mains, respectively. Both of these generalizations include generous allowances for infiltration. The lower flow rates take into consideration the averaging effect of larger contributing populations and the damping (storage) effect of larger distances from the source.

Interceptors are major sewer lines receiving wastewater from collector sewers and carrying it to a treatment plant or to another interceptor.

Figure 25.1 *Classification of Sewer Lines*

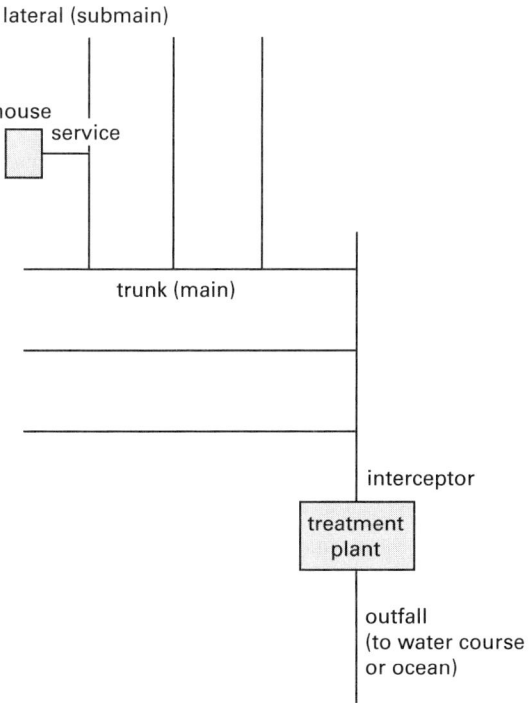

5. TREATMENT PLANT LOADING

Ideally, the quantity and organic strength of wastewater should be based on actual measurements taken throughout the year in order to account for variations that result from seasonal, climatic changes and other factors.

The *hydraulic loading* for over 75% of the sewage treatment plants in the United States is 1 MGD or less. Plants with flows less than 1 MGD are categorized as *minors*. Treatment plants handling 1 MGD or more, or serving equivalent service populations of 10,000, are categorized as *majors*.

The *organic loading* of treatment units is expressed in terms of pounds (kilograms) of biochemical oxygen

demand (BOD) per day or pounds (kilograms) of solids per day. In communities where a substantial portion of household kitchen wastes is discharged to the sewer system through garbage disposals, the average organic matter contributed by each person each day to domestic wastewater is approximately 0.24 lbm (110 g) of suspended solids and approximately 0.17–0.20 lbm (77–90 g) of BOD.

Since the average BOD of domestic waste is typically taken as 0.2 pounds per capita day (90 g per capita day), the *population equivalent* of any wastewater source, including industrial sources, can be calculated.

$$P_{e,1000s} = \frac{BOD_{mg/L}\, Q_{ML/d}}{90\, \dfrac{g}{person \cdot d}} \qquad [\text{SI}] \qquad 25.3(a)$$

$$P_{e,1000s} = \frac{BOD_{mg/L}\, Q_{gal/day} \times \left(8.345\, \dfrac{lbm\text{-}L}{MG\text{-}mg} \right)}{\left(10^6\, \dfrac{gal}{MG} \right)(1000\ persons) \times \left(0.20\, \dfrac{lbm}{person\text{-}day} \right)} \qquad [\text{U.S.}] \qquad 25.3(b)$$

6. SEWER PIPE MATERIALS

In the past, sewer pipes were constructed from clay, concrete, asbestos-cement, steel, and cast iron. Due to cost, modern large-diameter sewer lines are now almost always concrete, and new small-diameter lines are generally plastic.

Concrete pipe is used for gravity sewers and pressure mains. Circular pipe is used in most applications, although special shapes (e.g., arch, egg, elliptical) can be used. Concrete pipe in diameters up to 24 in (610 mm) is available in standard 3 ft and 4 ft (0.9 m and 1.2 m) lengths and is usually not reinforced. *Reinforced concrete pipe* (RCP) in diameters ranging from 12 in to 144 in (305–3660 mm) is available in lengths from 4 ft to 12 ft (1.2–3.6 m).

Concrete pipe is used for large diameter (16 in (406 mm) or larger) trunk and interceptor sewers. In some geographical regions, concrete pipe is used for smaller domestic sewers. However, concrete domestic lines should be selected only where stale or septic sewage is not anticipated. They should not be used with corrosive wastes or in corrosive soils.

Cast-iron pipe is particularly suited to installations where scour, high velocity waste, and high external loads are anticipated. It can be used for domestic connections, although it is more expensive than plastic. Special linings, coatings, wrappings, or encasements are required for corrosive wastes and soils.

Polyvinyl chloride (PVC) and acrylonitrile-butadiene-styrene (ABS) are two *plastic pipe* compositions that can be used for normal domestic sewage and industrial wastewater lines. They have excellent resistance to corrosive soils. However, special attention and care must be given to trench loading and pipe bedding.

ABS plastic can also be combined with concrete reinforcement for collector lines for use with corrosive domestic sewage and industrial waste. Such pipe is known as *truss pipe* due to its construction. The plastic is extruded with inner and outer web-connected pipe walls. The annular voids and the inner and outer walls are filled with lightweight concrete.

For pressure lines (i.e., *force mains*), welded steel pipe with an epoxy liner and cement-lined and coated-steel pipe are also used.

Vitrified clay pipe is resistant to acids, alkalies, hydrogen sulfide (septic sewage), erosion, and scour. Two strengths of clay pipe are available. The standard strength is suitable for pipes less than 12 in (300 mm) in diameter, D, for any depth of cover if the "$4D/3 + 8$ in trench width" ($4D/3 + 200$ mm) rule is observed. Double-strength pipe is recommended for large pipe that is deeply trenched. Clay is seldom used for diameters greater than 36 in (910 mm).

Asbestos-cement pipe has been used in the past for both gravity and pressure sewers carrying non-septic and non-corrosive waste through non-corrosive soils. The lightweight and longer laying lengths are inherent advantages of asbestos-cement pipe. However, such pipes have fallen from favor due to the asbestos content.

7. GRAVITY AND FORCE COLLECTION SYSTEMS

Wastewater collection systems are made up of a network of discharge and flow lines, drains, inlets, valve works, and connections for transporting domestic and industrial wastewater flows to treatment facilities.

Flow through gradually sloping *gravity sewers* is the most desirable means of moving sewage since it does not require pumping energy. In some instances, though, it may be necessary to use pressurized *force mains* to carry sewage uphill or over long, flat distances.

Alternative sewers are used in some remote housing developments for domestic sewage where neither the conventional sanitary sewer nor the septic tank/leach field is acceptable. Alternative sewers use pumps to force raw or communited sewage through small-diameter plastic lines to a more distant communal treatment or collection system.

There are four categories of alternative sewer systems. In the *G-P system*, a grinder-pump pushes chopped-up but untreated wastewater through a small-diameter plastic pipe. The *STEP system* (septic-tank-effluent pumping) uses a septic tank at each house, but there is

no leach field. The clear overflow goes to a pump that pushes the effluent through a small-diameter plastic sewer line operating under low pressure. The *vacuum sewer system* uses a vacuum valve at each house that periodically charges a slug of wastewater into a vacuum sewer line. The vacuum is created by a central pumping station. The *small-diameter gravity sewer* (SDG) also uses a small-diameter plastic pipe but relies on gravity instead of pumps. The pipe carries the effluent from a septic tank at each house. Costs in all four systems can be reduced greatly by having two or more houses share septic tanks, pumps, and vacuum valves.

8. SEWER VELOCITIES

The minimum design velocity actually depends on the particulate matter size. However, 2 ft/sec (0.6 m/s) is commonly quoted as the minimum *self-cleansing velocity*, although 1.5 ft/sec (0.45 m/s) may be acceptable if the line is occasionally flushed out by peak flows [TSS Sec. 33.42]. Table 25.2 gives minimum flow velocities based on fluid type.

Table 25.2 *Minimum Flow Velocities*

fluid	minimum velocity to keep particles in suspension (ft/sec)	(m/s)	minimum resuspension velocity (ft/sec)	(m/s)
raw sewage	2.5	(0.75)	3.5	(1.1)
grit tank effluent	2	(0.6)	2.5	(0.75)
primary settling tank effluent	1.5	(0.45)	2	(0.6)
mixed liquor	1.5	(0.45)	2	(0.6)
trickling filter effluent	1.5	(0.45)	2	(0.6)
secondary settling tank effluent	0.5	(0.15)	1	(0.3)

(Multiply ft/sec by 0.3 to obtain m/s.)

Table 25.3 lists the approximate minimum slope needed to achieve a 2 ft/sec (0.6 m/s) flow. Slopes slightly less than those listed may be permitted (with justification) in lines where design average flow provides a depth of flow greater than 30% of the pipe diameter. Velocity greater than 10–15 ft/sec (3–4.5 m/s) requires special provisions to protect the pipe and manholes against erosion and displacement by shock hydraulic loadings.

9. SEWER SIZING

The Manning equation is traditionally used to size gravity sewers. Depth of flow at the design flow rate is usually less than 70–80% of the pipe diameter. (See Table 25.2.) A pipe should be sized to be able to carry the peak flow (including normal infiltration) at a depth of approximately 70% of its diameter.

In general, sewers in the collection system (including laterals, interceptors, trunks, and mains) should have diameters of at least 8 in (203 mm). Building service connections can be as small as 4 in (101 mm).

Table 25.3 Minimum Slopes and Capacities for Sewers[a]

sewer diameter (in)	(mm)	minimum change in elevation[b] (ft/100 ft or m/100 m)	full flow discharge at minimum slope[c,d] (ft³/sec)
8	(200)	0.40	0.771
9	(230)	0.33	0.996
10	(250)	0.28	1.17
12	(300)	0.22	1.61
14	(360)	0.17	2.23
15	(380)	0.15	2.52
16	(410)	0.14	2.90
18	(460)	0.12	3.67
21	(530)	0.10	5.05
24	(610)	0.08	6.45
27	(690)	0.067	8.08
30	(760)	0.058	9.96
36	(910)	0.046	14.4

(Multiply in by 25.4 to obtain mm.)
(Multiply ft/100 ft by 1 to obtain m/100 m.)
(Multiply ft³/sec by 448.8 to obtain gal/min.)
(Multiply ft³/sec by 28.32 to obtain L/s.)
(Multiply gal/min by 0.0631 to obtain L/s.)
[a]to achieve a velocity of 2 ft/sec (0.6 m/s) when flowing full
[b]as specified in Sec. 24.31 of *Recommended Standards for Sewage Works* (RSSW), also known as *"Ten States' Standards"* (TSS), published by the Health Education Service, Inc. [TSS Sec. 33.41]
[c]$n = 0.013$ assumed
[d]For any diameter in inches and $n = 0.013$, calculate the full flow as
$$Q = 0.0472 d_{\text{in}}^{8/3} \sqrt{S}$$

10. STREET INLETS

Street inlets are required at all low points where ponding could occur, and they should be placed no more than 600 ft (180 m) apart; a limit of 300 ft (90 m) is preferred. A common practice is to install three inlets in a sag vertical curve—one at the lowest point and one on each side with an elevation of 0.2 ft (60 mm) above the center inlet.

Depth of gutter flow is found using Manning's equation. Inlet capacities of street inlets have traditionally been calculated from semi-empirical formulas.

Grate-type street inlets (known as *gutter inlets*) less than 0.4 ft (120 mm) deep have approximate capacities given by Eq. 25.4. Equation 25.4 should also be used for combined curb-grate inlets. All dimensions are in feet. For gutter inlets, the bars should ideally be parallel to the flow and be at least 1.5 ft (450 mm) long. Gutter inlets are more efficient than curb inlets, but clogging is a problem. Depressing the grating level below the street level increases the capacity.

$$Q_{\text{ft}^3/\text{sec}} = 3.0(\text{grate perimeter length})$$
$$\times (\text{inlet flow depth})^{3/2} \quad \textit{25.4}$$

The capacity of a *curb inlet* (where there is an opening in the vertical plane of the gutter rather than in the

horizontal plane) is given by Eq. 25.5. All dimensions are in feet. Not all curb inlets have inlet depressions. A typical curb inlet depression is 0.4 ft (130 mm).

$$Q_{\text{ft}^3/\text{sec}} = 0.7(\text{curb opening length})$$
$$\times (\text{inlet flow depth} + \text{curb inlet depression})^{3/2} \quad \textit{25.5}$$

11. MANHOLES

Manholes along sewer lines should be provided at sewer line intersections and at changes in elevation, direction, size, diameter, and slope. If a sewer line is too small for a person to enter, manholes should be placed every 400 ft (120 m) to allow for cleaning. Recommended maximum spacings are 400 ft (120 m) for pipes with diameters less than 18 in (460 mm), 500 ft (150 m) for 18–48 in (460–1220 mm) pipes, and 600–700 ft (180–210 m) for larger pipes.

12. SULFIDE ATTACK

Wastewater flowing through sewers often turns septic and releases *hydrogen sulfide* gas, H_2S. The common *Thiobacillus* sulfur bacterium converts the hydrogen sulfide to sulfuric acid. The acid attacks the crowns of concrete pipes that are not flowing full.

$$2H_2S + O_2 \rightarrow 2S + 2H_2O \quad \textit{25.6}$$

$$2S + 3O_2 + 2H_2O \rightarrow 2H_2SO_4 \quad \textit{25.7}$$

In warm climates, hydrogen sulfide can be generated in sanitary sewers placed on flat grades. Hydrogen sulfide also occurs in sewers supporting food processing industries. Sulfide generation is aggravated by the use of home garbage disposals and, to a lesser extent, by recycling. Interest in nutrition and fresh foods has also added to food wastes.

Sulfide attack in partially full sewers can be prevented by maintaining the flow rate above 5 ft/sec (1.5 m/s), raising the pH above 10.4, using biocides, precipitating the sulfides, or a combination thereof. Not all methods may be possible or economical, however. In sewers, the least expensive method generally involves intermittent (e.g., biweekly during the critical months) *shock treatments* of sodium hydroxide and ferrous chloride. The sodium hydroxide is a sterilization treatment that works by raising the pH, while the ferrous chloride precipitates the sulfides.

Sulfide attack is generally not an issue in force mains because sewage is continually moving and always makes contact with all of the pipe wall.

Water Treatment

13. WASTEWATER CHARACTERISTICS

Not all sewage flows are the same. Some sewages are stronger than others. Table 25.4 lists typical values for strong and weak domestic sewages.

Table 25.4 *Strong and Weak Domestic Sewages**

constituent	strong	weak
solids, total	1200	350
dissolved, total	850	250
fixed	525	145
volatile	325	105
suspended, total	350	100
fixed	75	30
volatile	275	70
settleable solids (mL/L)	20	5
biochemical oxygen demand, five-day, 20°C	300	100
total organic carbon	300	100
chemical oxygen demand	1000	250
nitrogen (total as N)	85	20
organic	35	8
free ammonia	50	12
nitrites	0	0
nitrates	0	0
phosphorus (total as P)	20	6
organic	5	2
inorganic	15	4
chlorides	100	30
alkalinity (as $CaCO_3$)	200	50
grease	150	50

*All concentrations are in mg/L unless otherwise noted.

14. SOLIDS

Solids in wastewater are categorized in the same manner as in water supplies. *Total solids* consist of *suspended* and *dissolved solids*. Generally, total solids constitute only a small amount of the incoming flow—less than 1/10% by mass. Therefore, wastewater fluid transport properties are essentially those of water. Figure 25.2 illustrates the solids categorization, along with typical percentages. Each category can be further divided into organic and inorganic groups.

The amount of *volatile solids* can be used as a measure of the organic pollutants capable of affecting the oxygen content. As in water supply testing, volatile solids are measured by igniting filtered solids and measuring the decrease in mass.

Refractory solids (*refractory organics*) are solids that are difficult to remove by common wastewater treatment processes.

15. MICROBIAL GROWTH

For the large numbers and mixed cultures of microorganisms found in waste treatment systems, it is more convenient to measure biomass than numbers of organisms. This is accomplished by measuring suspended or

Figure 25.2 *Wastewater Solids**

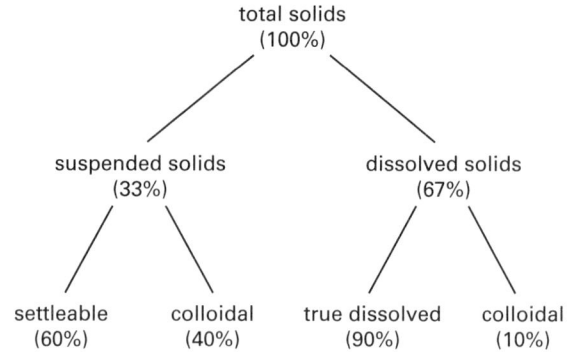

*typical percentages given

volatile suspended solids. An expression that depicts the rate of conversion of food into biomass for wastewater treatment was developed by Monod.

Equation 25.8 calculates the rate of growth, r_g, of a bacterial culture as a function of the rate of concentration of limiting food in solution. The concentration, X, of microorganisms, with dimensions of mass per unit volume (e.g., mg/L) is given by Eq. 25.8. μ is a specific growth rate factor with dimensions of time^{-1}.

$$r_g = \frac{dX}{dt} = \mu X \qquad 25.8$$

When one essential nutrient (referred to as a *substrate*) is present in a limited amount, the specific growth rate will increase up to a maximum value, as shown in Fig. 25.3 and as given by *Monod's equation*, Eq. 25.9. μ_m is the *maximum specific growth rate coefficient* with dimensions of time^{-1}. S is the concentration of the growth-limiting nutrient with dimensions of mass/unit volume. K_s is the *half-velocity coefficient*, the nutrient concentration at one half of the maximum growth rate, with dimensions of mass per unit volume.

$$\mu = \mu_m \left(\frac{S}{K_s + S} \right) \qquad 25.9$$

Combining Eq. 25.8 and Eq. 25.9, the rate of growth is

$$r_g = \frac{\mu_m X S}{K_s + S} \qquad 25.10$$

The rate of substrate (nutrient) utilization is easily calculated if the *maximum yield coefficient*, Y, defined as the ratio of the mass of cells formed to the mass of nutrient consumed (with dimensions of mass/mass) is known.

$$r_{su} = \frac{dS}{dt} = \frac{-\mu_m X S}{Y(K_s + S)} \qquad 25.11$$

Figure 25.3 *Bacterial Growth Rate with Limited Nutrient*

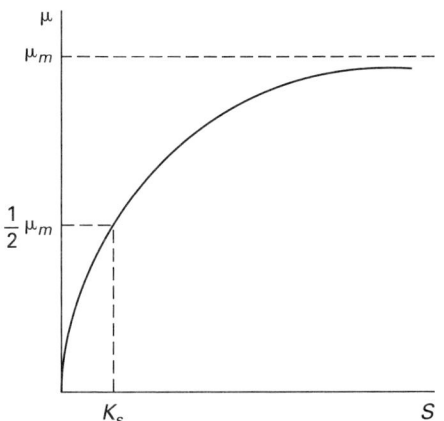

The maximum rate of substrate utilization per unit mass of microorganisms is

$$k = \frac{\mu_m}{Y} \qquad 25.12$$

Combining Eq. 25.11 and Eq. 25.12,

$$\frac{dS}{dt} = \frac{-kXS}{K_s + S} \qquad 25.13$$

The Monod equation assumes that all microorganisms are the same age. The actual distribution of ages and other factors (death and predation of cells) decreases the rate of cell growth. The decrease is referred to as *endogenous decay* and is accounted for with an *endogenous decay coefficient*, k_d, with dimensions of time^{-1}. The rate of endogenous decay is

$$r_d = -k_d X \qquad 25.14$$

The *net rate of cell growth* and *net specific growth rate* are

$$
\begin{aligned}
r'_g &= \frac{\mu_m X S}{K_s + S} - k_d X \\
&= -Y r_{su} - k_d X
\end{aligned}
\qquad 25.15
$$

$$\mu' = \mu_m \left(\frac{S}{K_s + S} \right) - k_d \qquad 25.16$$

16. DISSOLVED OXYGEN IN WASTEWATER

A body of water exposed to the atmosphere will normally become saturated with oxygen. If the dissolved oxygen content, DO, of water is less than the saturated values, there will be good reason to believe that the water is organically polluted.

The difference between the saturated and actual dissolved oxygen concentration is known as the *oxygen deficit*, D.

$$D = \mathrm{DO}_{\mathrm{sat}} - \mathrm{DO} \qquad 25.17$$

17. REOXYGENATION

The oxygen deficit can be reduced (i.e., the dissolved oxygen concentration can be increased) only by aerating the water. This occurs naturally in free-running water (e.g., in rivers). The rate of reaeration is

$$r_r = K_r(\mathrm{DO}_{\mathrm{sat}} - \mathrm{DO}) \qquad 25.18$$

K_r is the *reoxygenation (reaeration) rate constant*, which depends on the type of flow and temperature and with common units of day^{-1}. (K_r may be written as K_R and K_2 by some authorities.) Typical values of K_r (base-10) are: small stagnant ponds, 0.05–0.10; sluggish streams, 0.10–0.15; large lakes, 0.10–0.15; large streams, 0.15–0.30; swiftly flowing rivers and streams, 0.3–0.5; whitewater and waterfalls, 0.5 and above.

An exponential model is used to predict the oxygen deficit, D, as a function of time during a pure reoxygenation process. Reoxygenation constants are given for use with a base-10 and natural logarithmic base.

$$D_t = D_0 10^{-K_r t} \quad \text{[base-10]} \qquad 25.19$$

$$D_t = D_0 e^{-K'_r t} \quad \text{[base-}e\text{]} \qquad 25.20$$

The base-e constant K'_r may be written as $K_{r,\text{base-}e}$ by some authorities. The constants K_r and K'_r are different but related.

$$K'_r = 2.303 K_r \qquad 25.21$$

K'_r can be approximated from the *O'Connor and Dobbins formula* for moderate to deep natural streams with low to moderate velocities. Both v and d are average values.

$$K'_{r,20^\circ\mathrm{C}} \approx \frac{3.93\sqrt{\mathrm{v}_{\mathrm{m/s}}}}{d_{\mathrm{m}}^{1.5}} \quad \begin{bmatrix} 0.3 \text{ m} < d < 9.14 \text{ m} \\ 0.15 \text{ m/s} < \mathrm{v} < 0.49 \text{ m/s} \end{bmatrix}$$
$$\text{[SI]} \quad 25.22(a)$$

$$K'_{r,68^\circ\mathrm{F}} \approx \frac{12.9\sqrt{\mathrm{v}_{\mathrm{ft/sec}}}}{d_{\mathrm{ft}}^{1.5}} \quad \begin{bmatrix} 1 \text{ ft} < d < 30 \text{ ft} \\ 0.5 \text{ ft/sec} < \mathrm{v} < 1.6 \text{ ft} \end{bmatrix}$$
$$\text{[U.S.]} \quad 25.22(b)$$

Water Treatment

An empirical formula that has been proposed for faster moving water is the *Churchill formula*.

$$K'_{r,20°C} \approx \frac{5.049 v_{m/s}^{0.969}}{d_m^{1.67}} \left[\begin{array}{c} 0.61 \text{ m} < d < 3.35 \text{ m} \\ 0.55 \text{ m/s} < v < 1.52 \text{ m/s} \end{array} \right]$$

[SI]   *25.23(a)*

$$K'_{r,68°F} \approx \frac{11.61 v_{ft/sec}^{0.969}}{d_{ft}^{1.67}} \left[\begin{array}{c} 2 \text{ ft} < d < 11 \text{ ft} \\ 1.8 \text{ ft/sec} < v < 5 \text{ ft/sec} \end{array} \right]$$

[U.S.]   *25.23(b)*

The variation in K'_r with temperature is given approximately by Eq. 25.24. The temperature constant 1.024 has been reported as 1.016 by some authorities.

$$K'_{r,T} = K'_{r,20°C}(1.024)^{T-20°C} \qquad 25.24$$

Equation 25.24 is actually a special case of Eq. 25.25, which gives the relationship between values of K'_r between any two temperatures.

$$K'_{r,T_1} = K'_{r,T_2} \theta_r^{T_1-T_2} \qquad 25.25$$

18. DEOXYGENATION

The rate of deoxygenation at time t is

$$r_{d,t} = -K_d \text{DO} \qquad 25.26$$

The *deoxygenation rate constant*, K_d, for treatment plant effluent is approximately 0.05–0.10 day^{-1} and is typically taken as 0.1 day^{-1} (base-10 values). (K_d may be represented as K and K_1 by some authorities.) For highly polluted shallow streams, K_d can be as high as 0.25 day^{-1}. For raw sewage, it is approximately 0.15–0.30 day^{-1}. K'_d is the base-e version of K_d.

$$K'_d = 2.303 K_d \qquad 25.27$$

K_d for other temperatures can be found from Eq. 25.28. The *temperature variation constant*, θ_d, is often quoted in literature as 1.047. However, this value should not be used with temperatures below 68°F (20°C). Additional research suggests that θ_d varies from 1.135 for temperatures between 39°F and 68°F (4°C and 20°C) up to 1.056 for temperatures between 68°F and 86°F (20°C and 30°C).

$$K'_{d,T} = K'_{d,20°C} \theta_d^{T-20°C} \qquad 25.28$$

Equation 25.28 is actually a special case of Eq. 25.29, which gives the relationship between values of K'_d between any two temperatures.

$$K'_{d,T_1} = K'_{d,T_2} \theta_d^{T_1-T_2} \qquad 25.29$$

19. TESTS OF WASTEWATER CHARACTERISTICS

The most common wastewater analyses used to determine characteristics of a municipal wastewater determine biochemical oxygen demand (BOD) and suspended solids (SS). BOD and flow data are basic requirements for the operation of biological treatment units. The concentration of suspended solids relative to BOD indicates the degree that organic matter is removable by primary settling. Additionally, temperature, pH, chemical oxygen demand (COD), alkalinity, color, grease, and the quantity of heavy metals are necessary to characterize municipal wastewater characteristics.

20. BIOCHEMICAL OXYGEN DEMAND

When oxidizing organic material in water, biological organisms also remove oxygen from the water. This is typically considered to occur through oxidation of organic material to CO_2 and H_2O by microorganisms at the molecular level and is referred to as *biochemical oxygen demand* (BOD). Therefore, oxygen use is an indication of the organic waste content.

Typical values of BOD for various industrial wastewaters are given in Table 25.5. A BOD of 100 mg/L is considered to be a weak wastewater; a BOD of 200–250 mg/L is considered to be a medium strength wastewater; and a BOD above 300 mg/L is considered to be a strong wastewater.

In the past, the BOD has been determined in a lab using a traditional standardized BOD testing procedure. Since this procedure requires five days of incubating, inline measurement systems have been developed to provide

Table 25.5 Typical BOD and COD of Industrial Wastewaters

industry/ type of waste	BOD	COD
canning		
corn	19.5 lbm/ton corn	
tomatoes	8.4 lbm/ton tomatoes	
dairy milk		
processing	1150 lbm/ton raw milk 1000 mg/L	1900 mg/L
beer brewing	1.2 lbm/barrel beer	
commercial		
laundry	1250 lbm/1000 lbm dry 700 mg/L	2400 mg/L
slaughterhouse	7.7 lbm/animal	2100 mg/L
(meat packing)	1400 mg/L	
papermill	121 lbm/ton pulp	
synthetic textile	1500 mg/L	3300 mg/L
chlorophenolic		
manufacturing	4300 mg/L	5400 mg/L
milk bottling	230 mg/L	420 mg/L
cheese production	3200 mg/L	5600 mg/L
candy production	1600 mg/L	3000 mg/L

essentially instantaneous and continuous BOD information at wastewater plants.

The standardized BOD test consists of adding a measured amount of wastewater (which supplies the organic material) to a measured amount of dilution water (which reduces toxicity and supplies dissolved oxygen) and then incubating the mixture at a specific temperature. The standard procedure calls for a five-day incubation period at 68°F (20°C), though other temperatures are used. The measured BOD is designated as BOD_5 or BOD-5. The BOD of a biologically active sample at the end of the incubation period is given by Eq. 25.30.

$$BOD_5 = \frac{DO_i - DO_f}{\dfrac{V_{sample}}{V_{sample} + V_{dilution}}} \qquad 25.30$$

If more than one identical sample is prepared, the increase in BOD over time can be determined, rather than just the final BOD. Figure 25.4 illustrates a typical plot of BOD versus time. The BOD at any time t is known as the *BOD exertion*.

Figure 25.4 *BOD Exertion*

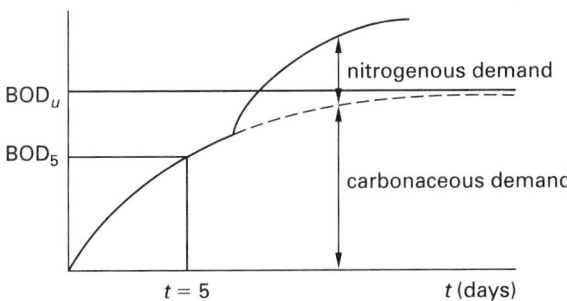

BOD exertion can be found from Eq. 25.31. The ultimate BOD (BOD_u) is the total oxygen used by *carbonaceous bacteria* if the test is run for a long period of time, usually taken as 20 days. (The symbols L, BOD_L, and BOD_{ult} are also widely used to represent the ultimate BOD.) Use the proper form of Eq. 25.31 depending on the base of the deoxygenation rate constant.

$$BOD_t = BOD_u(1 - 10^{-K_d t})$$
$$= BOD_u(1 - e^{-K'_d t}) \qquad 25.31$$

The ultimate BOD (BOD_u) usually cannot be found from long-term studies due to the effect of nitrogen-consuming bacteria in the sample. However, if K_d is 0.1 day^{-1}, the ultimate BOD can be approximated from Eq. 25.32 derived from Eq. 25.31.

$$BOD_u \approx 1.463 BOD_5 \qquad 25.32$$

The approximate variation in first-stage BOD (either BOD_5 or BOD_u) municipal wastewater with temperature is given by Eq. 25.33.[1]

$$BOD_{T°C} = BOD_{20°C}(0.02\,T_{°C} + 0.6) \qquad 25.33$$

The deviation from the expected exponential growth curve in Fig. 25.4 is due to nitrification and is considered to be *nitrogenous demand*. *Nitrification* is the use of oxygen by autotrophic bacteria. Autotrophic bacteria oxidize ammonia to nitrites and nitrates.

The number of autotrophic bacteria in a wastewater sample is initially small. Generally, 6 to 10 days are required for the autotrophic population to become sufficiently large to affect a BOD test. Therefore, the standard BOD test is terminated at five days, before the autotrophic contribution to BOD becomes significant.

Until recently, BOD has not played a significant role in the daily operations decisions of treatment plants because of the length of time it took to obtain BOD data. Other measurements, such as COD, TOC, turbidity, suspended solids, respiration rates, or flow have been used.

Example 25.1

Ten 5 mL samples of wastewater are placed in 300 mL BOD bottles and diluted to full volume. Half of the bottles are titrated immediately, and the average initial concentration of dissolved oxygen is 7.9 mg/L. The remaining bottles are incubated for five days, after which the average dissolved oxygen is determined to be 4.5 mg/L. The deoxygenation rate constant is known to be 0.13 day^{-1} (base-10). What are the (a) standard BOD and (b) ultimate carbonaceous BOD?

Solution

(a) Use Eq. 25.30.

$$\begin{aligned}
BOD_5 &= \frac{DO_i - DO_f}{\dfrac{V_{sample}}{V_{sample} + V_{dilution}}} \\[2mm]
&= \frac{7.9\,\dfrac{mg}{L} - 4.5\,\dfrac{mg}{L}}{\dfrac{5\ mL}{300\ mL}} \\[2mm]
&= 204\ mg/L
\end{aligned}$$

(b) From Eq. 25.31, the ultimate BOD is

$$\begin{aligned}
BOD_u &= \frac{BOD_t}{1 - 10^{-K_d t}} \\[2mm]
&= \frac{204\,\dfrac{mg}{L}}{1 - 10^{(-0.13\ day^{-1})(5\ days)}} \\[2mm]
&= 263\ mg/L
\end{aligned}$$

---

[1]As reported in *Water-Resources Engineering*, 4th ed., Ray K. Linsley, Joseph B. Franzini, David Freyberg, and George Tchobanoglous, McGraw-Hill Book Co., 1992, New York.

Water Treatment

21. SEEDED BOD

Industrial wastewater may lack sufficient microorganisms to metabolize the organic matter. In this situation, the standard BOD test will not accurately determine the amount of organic matter unless seed organisms are added.

The BOD in seeded samples is found by measuring dissolved oxygen in the seeded sample after 15 min (DO_i) and after five days (DO_f), as well as by measuring the dissolved oxygen of the seed material itself after 15 min (DO_i^*) and after five days (DO_f^*). In Eq. 25.34, x is the ratio of the volume of seed added to the sample to the volume of seed used to find DO^*.

$$\text{BOD} = \frac{DO_i - DO_f - x(DO_i^* - DO_f^*)}{\dfrac{V_{\text{sample}}}{V_{\text{sample}} + V_{\text{dilution}}}} \qquad 25.34$$

22. DILUTION PURIFICATION

Dilution purification (also known as *self-purification*) refers to the discharge of untreated or partially treated sewage into a large body of water such as a river. After the treatment plant effluent has mixed with the river water, the river experiences both deoxygenation (from the biological activity of the waste) and reaeration (from the agitation of the flow). If the body is large and is adequately oxygenated, the sewage's BOD may be satisfied without putrefaction.

When two streams merge, the characteristics of the streams are blended in the combined flow. If values of the characteristics of interest are known for the upstream flows, the blended concentration can be found as a weighted average. Equation 25.35 can be used to calculate the final temperature, dissolved oxygen, BOD, or suspended solids content immediately after two flows are mixed. (The subscript a is also used in literature to represent the condition immediately after mixing.)

$$C_f = \frac{C_1 Q_1 + C_2 Q_2}{Q_1 + Q_2} \qquad 25.35$$

Example 25.2

6 MGD of wastewater with a dissolved oxygen concentration of 0.9 mg/L is discharged into a 50°F river flowing at 40 ft³/sec. The river is initially saturated with oxygen. What is the oxygen content of the river immediately after mixing?

Solution

Both flow rates must have the same units. Convert MGD to ft³/sec.

$$Q_w = \frac{(6 \text{ MGD})\left(10^6 \dfrac{\text{gal}}{\text{MG}}\right)}{\left(7.48 \dfrac{\text{gal}}{\text{ft}^3}\right)\left(24 \dfrac{\text{hr}}{\text{day}}\right)\left(60 \dfrac{\text{min}}{\text{hr}}\right)\left(60 \dfrac{\text{sec}}{\text{min}}\right)}$$

$$= 9.28 \text{ ft}^3/\text{sec}$$

From App. 22.C, the saturated oxygen content at 50°F (10°C) is 11.3 mg/L. From Eq. 25.35,

$$C_f = \frac{C_w Q_w + C_r Q_r}{Q_w + Q_r}$$

$$= \frac{\left(0.9 \dfrac{\text{mg}}{\text{L}}\right)\left(9.28 \dfrac{\text{ft}^3}{\text{sec}}\right) + \left(11.3 \dfrac{\text{mg}}{\text{L}}\right)\left(40 \dfrac{\text{ft}^3}{\text{sec}}\right)}{9.28 \dfrac{\text{ft}^3}{\text{sec}} + 40 \dfrac{\text{ft}^3}{\text{sec}}}$$

$$= 9.34 \text{ mg/L}$$

23. RESPONSE TO DILUTION PURIFICATION

The oxygen deficit is the difference between actual and saturated oxygen concentrations, expressed in mg/L. Since reoxygenation and deoxygenation of a polluted river occur simultaneously, the oxygen deficit will increase if the reoxygenation rate is less than the deoxygenation rate. If the oxygen content goes to zero, anaerobic decomposition and putrefaction will occur.

The minimum dissolved oxygen concentration that will protect aquatic life in the river is the *dissolved oxygen standard* for the river. 4–6 mg/L is the generally accepted range of dissolved oxygen required to support fish populations. 5 mg/L is adequate as is evidenced in high-altitude trout lakes. However, 6 mg/L is preferable, particularly for large fish populations.

The oxygen deficit at time t is given by the *Streeter-Phelps equation*. In Eq. 25.36, BOD_u is the ultimate carbonaceous BOD of the river immediately after mixing. D_0 is the dissolved oxygen deficit immediately after mixing, which is often given the symbol D_a. (If the constants K_r and K_d are base-e constants, the base-10 exponentiation in Eq. 25.34 should be replaced with the base-e exponentiation, e^{-Kt}.)

$$D_t = \left(\frac{K_d \text{BOD}_u}{K_r - K_d}\right)\left(10^{-K_d t} - 10^{-K_r t}\right) + D_0(10^{-K_r t})$$

$$25.36$$

A graph of dissolved oxygen versus time (or distance downstream) is known as a *dissolved oxygen sag curve* and is shown in Fig. 25.5. The location at which the lowest dissolved oxygen concentration occurs is the

Figure 25.5 *Oxygen Sag Curve*

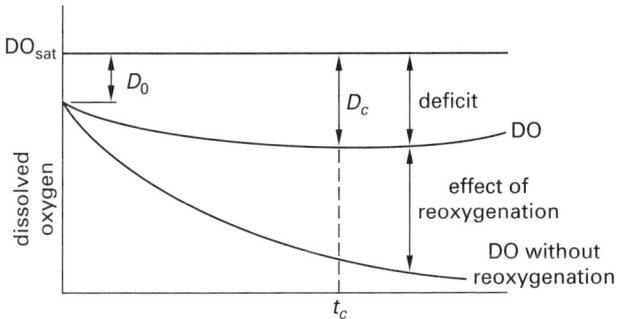

critical point. The location of the critical point is found from the river velocity and flow time to that point.

$$x_c = \mathrm{v}t_c \qquad 25.37$$

The time to the critical point is given by Eq. 25.38. The ratio K_r/K_d in Eq. 25.38 is known as the *self-purification constant*. (If the constants K_r and K_d are base-e constants, the base-10 logarithm in Eq. 25.38 should be replaced with the base-e logarithm, ln.)

$$t_c = \left(\frac{1}{K_r - K_d}\right)$$
$$\times \log_{10}\left(\left(\frac{K_d\mathrm{BOD}_u - K_rD_0 + K_dD_0}{K_d\mathrm{BOD}_u}\right)\left(\frac{K_r}{K_d}\right)\right)$$

$$25.38$$

The critical oxygen deficit can be found by substituting t_c from Eq. 25.38 into Eq. 25.36. However, it is more expedient to use Eq. 25.39.

$$D_c = \left(\frac{K_d\mathrm{BOD}_u}{K_r}\right)10^{-K_dt_c} \qquad 25.39$$

Example 25.3

A treatment plant effluent has the following characteristics.

quantity	15 ft³/sec
BOD$_{5,20°C}$	45 mg/L
DO	2.9 mg/L
temperature	24°C
$K_{d,20°C}$	0.1 day⁻¹ (base-10)
temperature variation constant for K_d, θ_d	1.047
temperature variation constant for K_r, θ_r	1.016

The outfall is located in a river carrying 120 ft³/sec of water with the following characteristics.

quantity	120 ft³/sec
velocity	0.55 ft/sec
average depth	4 ft
BOD$_{5,20°C}$	4 mg/L
DO	8.3 mg/L
temperature	16°C

When the treatment plant effluent and river are mixed, the mixture temperature persists for a long distance downstream.

(a) Determine the distance downstream where the oxygen level is at a minimum. (b) Determine if the river can support fish life at that point.

Solution

(a) Find the river conditions immediately after mixing. Use Eq. 25.35 three times.

$$C_f = \frac{C_1Q_1 + C_2Q_2}{Q_1 + Q_2}$$

$$\mathrm{BOD}_{5,20°C} = \frac{\left(15\,\frac{\mathrm{ft}^3}{\mathrm{sec}}\right)\left(45\,\frac{\mathrm{mg}}{\mathrm{L}}\right) + \left(120\,\frac{\mathrm{ft}^3}{\mathrm{sec}}\right)\left(4\,\frac{\mathrm{mg}}{\mathrm{L}}\right)}{15\,\frac{\mathrm{ft}^3}{\mathrm{sec}} + 120\,\frac{\mathrm{ft}^3}{\mathrm{sec}}}$$
$$= 8.56\ \mathrm{mg/L}$$

$$\mathrm{DO} = \frac{\left(15\,\frac{\mathrm{ft}^3}{\mathrm{sec}}\right)\left(2.9\,\frac{\mathrm{mg}}{\mathrm{L}}\right) + \left(120\,\frac{\mathrm{ft}^3}{\mathrm{sec}}\right)\left(8.3\,\frac{\mathrm{mg}}{\mathrm{L}}\right)}{15\,\frac{\mathrm{ft}^3}{\mathrm{sec}} + 120\,\frac{\mathrm{ft}^3}{\mathrm{sec}}}$$
$$= 7.7\ \mathrm{mg/L}$$

$$T = \frac{\left(15\,\frac{\mathrm{ft}^3}{\mathrm{sec}}\right)(24°C) + \left(120\,\frac{\mathrm{ft}^3}{\mathrm{sec}}\right)(16°C)}{15\,\frac{\mathrm{ft}^3}{\mathrm{sec}} + 120\,\frac{\mathrm{ft}^3}{\mathrm{sec}}}$$
$$= 16.89°C$$

Calculate the approximate rate constants. Although the temperature is low, θ_d is specified in the problem statement. From Eq. 25.28,

$$K_{d,T} = K_{d,20°C}(1.047)^{T-20°C}$$
$$K_{d,16.89°C} = (0.1\,\mathrm{day}^{-1})(1.047)^{16.89°C-20°C}$$
$$= 0.0867\,\mathrm{day}^{-1}$$

Since $\mathrm{v} = 0.55$ ft/sec, use Eq. 25.21 and Eq. 25.22 to calculate the reoxygenation constant.

$$K_{r,20°C} = \frac{K'_{r,20°C}}{2.303} = \frac{12.9\sqrt{\mathrm{v}}}{2.303d^{1.5}}$$
$$= \frac{12.9\sqrt{0.55\,\frac{\mathrm{ft}}{\mathrm{sec}}}}{(2.303)(4\,\mathrm{ft})^{1.5}}$$
$$= 0.519\,\mathrm{day}^{-1}$$

From Eq. 25.24, and using the given value of θ,

$$K_{r,T} = K_{r,20°C}(1.016)^{T-20°C}$$

$$K_{r,16.89°C} = (0.519 \text{ day}^{-1})(1.016)^{16.89°C-20°C}$$

$$= 0.494 \text{ day}^{-1}$$

Estimate BOD_u from Eq. 25.31.

$$BOD_{u,20°C} = \frac{BOD_{5,20°C}}{1 - 10^{-K_d t}}$$

$$= \frac{8.56 \frac{mg}{L}}{1 - 10^{(-0.1 \text{ day}^{-1})(5 \text{ days})}}$$

$$= 12.52 \text{ mg/L}$$

From Eq. 25.33,

$$BOD_{T°C} = BOD_{20°C}(0.02 \, T°C + 0.6)$$

$$BOD_{u,16.89°C} = \left(12.52 \frac{mg}{L}\right)\left((0.02)(16.89°C) + 0.6\right)$$

$$= 11.74 \text{ mg/L}$$

Calculate the original oxygen deficit. By interpolation from App. 22.C, the dissolved oxygen concentration at 16.89°C is approximately 9.76 mg/L.

$$D_0 = 9.76 \frac{mg}{L} - 7.7 \frac{mg}{L}$$

$$= 2.06 \text{ mg/L}$$

Calculate t_c from Eq. 25.38.

$$t_c = \left(\frac{1}{K_r - K_d}\right)$$

$$\times \log_{10}\left(\left(\frac{K_d BOD_u - K_r D_0 + K_d D_0}{K_d BOD_u}\right)\left(\frac{K_r}{K_d}\right)\right)$$

$$= \left(\frac{1}{0.494 \text{ day}^{-1} - 0.0867 \text{ day}^{-1}}\right)$$

$$\times \log_{10}\left(\left(\frac{\begin{array}{c}(0.0867 \text{ day}^{-1})\left(11.74 \frac{mg}{L}\right) \\ - (0.494 \text{ day}^{-1})\left(2.06 \frac{mg}{L}\right) \\ + (0.0867 \text{ day}^{-1})\left(2.06 \frac{mg}{L}\right)\end{array}}{(0.0867 \text{ day}^{-1})\left(11.74 \frac{mg}{L}\right)}\right) \times \left(\frac{0.494 \text{ day}^{-1}}{0.0867 \text{ day}^{-1}}\right)\right)$$

$$= 0.00107 \text{ day}$$

The distance downstream is

$$x_c = v t_c$$

$$= \frac{\left(0.55 \frac{ft}{sec}\right)(0.00107 \text{ day})\left(86{,}400 \frac{sec}{day}\right)}{5280 \frac{ft}{mi}}$$

$$= 0.0096 \text{ mi} \quad \text{[essentially, at the outfall]}$$

(b) The critical oxygen deficit is found from Eq. 25.39.

$$D_c = \left(\frac{K_d BOD_u}{K_r}\right)10^{-K_d t_c}$$

$$= \left(\frac{(0.0867 \text{ day}^{-1})\left(11.74 \frac{mg}{L}\right)}{0.494 \text{ day}^{-1}}\right)$$

$$\times \left(10^{(-0.0867 \text{ day}^{-1})(0.00107 \text{ day})}\right)$$

$$= 2.06 \text{ mg/L}$$

The saturated oxygen content at the critical point with a water temperature of 16.89°C is 9.76 mg/mL. The actual oxygen content is

$$DO = DO_{sat} - D_c = 9.76 \frac{mg}{L} - 2.06 \frac{mg}{L}$$

$$= 7.70 \text{ mg/L}$$

Since this is greater than 4–6 mg/L, fish life is supported.

24. CHEMICAL OXYGEN DEMAND

Chemical oxygen demand (COD) is a measure of oxygen removed by both biological organisms feeding on organic material, as well as oxidizable inorganic compounds (unlike biochemical oxygen demand, which is a measure of oxygen removed only by biological organisms). Therefore, COD is a good measure of total effluent strength (organic and inorganic contaminants).

COD testing is required where there is industrial chemical pollution. In such environments, the organisms necessary to metabolize organic compounds may not exist. Furthermore, the toxicity of the wastewater may make the standard BOD test impossible to carry out. Standard COD test results are usually available in a matter of hours.

If toxicity is low, BOD and COD test results can be correlated. The BOD/COD ratio typically varies from 0.4–0.8. This is a wide range but, for any given treatment plant and waste type, the correlation is essentially constant. The correlation can, however, vary along the treatment path.

25. RELATIVE STABILITY

The *relative stability* test is easier to perform than the BOD test, although it is less accurate. The relative stability of an effluent is defined as the percentage of initial BOD that has been satisfied. The test consists of taking a sample of effluent and adding a small amount of methylene blue dye. The mixture is then incubated, usually at 20°C. When all oxygen has been removed from the water, anaerobic bacteria start to remove the dye. The amount of time it takes for the color to start degrading is known as the *stabilization time* or *decoloration time*. The relative stability can be found from the stabilization time by using Table 25.6.

Table 25.6 Relative Stability[a]

stabilization time (day)	relative stability[b] (%)
$^1/_2$	11
1	21
$1^1/_2$	30
2	37
$2^1/_2$	44
3	50
4	60
5	68
6	75
7	80
8	84
9	87
10	90
11	92
12	94
13	95
14	96
16	97
18	98
20	99

[a]incubation at 20°C
[b]calculated as $(100\%)(1 - 0.794^t)$

26. CHLORINE DEMAND AND DOSE

Chlorination destroys bacteria, hydrogen sulfide, and other compounds and substances. The *chlorine demand* (*chlorine dose*) must be determined by careful monitoring of coliform counts and free residuals since there are several ways that chlorine can be used up without producing significant disinfection. Only after uncombined (free) chlorine starts showing up is it assumed that all chemical reactions and disinfection are complete.

Chlorine demand is the amount of chlorine (or its chloramine or hypochlorite equivalent) required to leave the desired residual (usually 0.5 mg/L) 15 min after mixing. Fifteen minutes is the recommended mixing and holding time prior to discharge, since during this time nearly all pathogenic bacteria in the water will have been killed.

Typical doses for wastewater effluent depend on the application point and are widely variable, though doses rarely exceed 30 mg/L. For example, chlorine may be applied at 5–25 mg/L prior to primary sedimentation, 2–6 mg/L after sand filtration, and 3–15 mg/L after trickle filtration.

27. BREAKPOINT CHLORINATION

Because of their reactivities, chlorine is initially used up in the neutralization of hydrogen sulfide and the rare ferrous and manganous (Fe^{+2} and Mn^{+2}) ions. For example, hydrogen sulfide is oxidized according to Eq. 25.40. The resulting HCl, $FeCl_2$, and $MnCl_2$ ions do not contribute to disinfection. They are known as *unavailable combined residuals*.

$$H_2S + 4H_2O + 4Cl_2 \rightarrow H_2SO_4 + 8HCl \qquad 25.40$$

Ammonia nitrogen combines with chlorine to form the family of *chloramines*. Depending on the water pH, monochloramines (NH_2Cl), dichloramines ($NHCl_2$), or trichloramines (nitrogen trichloride, NCl_3) may form. Chloramines have long-term disinfection capabilities and are therefore known as *available combined residuals*. Equation 25.41 is a typical chloramine formation reaction.

$$NH_4^+ + HOCl \rightleftharpoons NH_2Cl + H_2O + H^+ \qquad 25.41$$

Continued addition of chlorine after chloramines begin forming changes the pH and allows chloramine destruction to begin. Chloramines are converted to nitrogen gas (N_2) and nitrous oxide (N_2O). Equation 25.42 is a typical chloramine destruction reaction.

$$2NH_2Cl + HOCl \rightleftharpoons N_2 + 3HCl + H_2O \qquad 25.42$$

The destruction of chloramines continues, with the repeated application of chlorine, until no ammonia remains in the water. The point at which all ammonia has been removed is known as the *breakpoint*.

In the *breakpoint chlorination* method, additional chlorine is added after the breakpoint in order to obtain free chlorine residuals. The free residuals have a high disinfection capacity. Typical free residuals are free chlorine (Cl_2), hypochlorous acid (HOCl), and hypochlorite ions. Equation 25.43 and Eq. 25.44 illustrate the formation of these free residuals.

$$Cl_2 + H_2O \rightarrow HCl + HOCl \qquad 25.43$$
$$HOCl \rightarrow H^+ + ClO^- \qquad 25.44$$

There are several undesirable characteristics of breakpoint chlorination. First, it may not be economical to use breakpoint chlorination unless the ammonia nitrogen has been reduced. Second, free chlorine residuals produce trihalomethanes. Third, if free residuals are prohibited to prevent trihalomethanes, the water may need to be dechlorinated using sulfur dioxide gas or

Water Treatment

sodium bisulfate. (Where small concentrations of free residuals are permitted, dechlorination may be needed only during the dry months. During winter storm months, the chlorine residuals may be adequately diluted with rainwater.)

28. NITROGEN

Nitrogen occurs in water in organic, ammonia, nitrate, nitrite, and dissolved gaseous forms. Nitrogen in municipal wastewater results from human excreta, ground garbage, and industrial wastes (primarily food processing). Bacterial decomposition and the hydrolysis of urea produces ammonia, NH_3. Ammonia in water forms the ammonium ion NH_4^+, also known as *ammonia nitrogen*. Ammonia nitrogen must be removed from the waste effluent stream due to potential exertion of oxygen demand. The total of organic and ammonia nitrogen is known as *total Kjeldahl nitrogen*, or TKN. *Total nitrogen*, TN, includes TKN plus inorganic nitrates and nitrites.

Nitrification (the oxidation of ammonia nitrogen) occurs as follows.

$$NH_4^+ + 2O_2 \rightarrow NO_3^- + H_2O + 2H^+ \qquad 25.45$$

Denitrification, the reduction of nitrate nitrogen to nitrogen gas by facultative heterotrophic bacteria, occurs as follows.

$$2NO_3^- + \text{organic matter} \rightarrow N_2 + CO_2 + H_2O \qquad 25.46$$

Ammonia nitrogen can be removed chemically from water without first converting it to nitrate form by raising the pH level. This converts the ammonium ion back into ammonia, which can then be stripped from the water by passing large quantities of air through the water. Lime is added to provide the hydroxide for the reaction. The *ammonia stripping* reaction (a physical-chemical process) is

$$NH_4^+ + OH^- \rightarrow NH_3 + H_2O \qquad 25.47$$

Ammonia stripping has no effect on nitrate, which is commonly removed in an activated sludge process with a short cell detention time.

29. ORGANIC COMPOUNDS IN WASTEWATER

Biodegradable organic matter in municipal wastewater is classified into three major categories: proteins, carbohydrates, and greases (fats).

Proteins are long strings of amino acids containing carbon, hydrogen, oxygen, nitrogen, and phosphorus.

Carbohydrates consist of sugar units containing the elements of carbon, hydrogen, and oxygen. They are identified by the presence of a saccharide ring. The rings range from simple monosaccharides to polysaccharides (long chain sugars categorized as either readily degradable starches found in potatoes, rice, corn, and other edible plants, or as cellulose found in wood and similar plant tissues).

Greases are a variety of biochemical substances that have the common property of being soluble to varying degrees in organic solvents (acetone, ether, ethanol, and hexane) while being only sparingly soluble in water. The low solubility of grease in water causes problems in pipes and tanks where it accumulates, reduces contact areas during various filtering processes, and produces a sludge that is difficult to dispose of.

Of the organic matter in wastewater, 60–80% is readily available for biodegradation. Degradation of greases by microorganisms occurs at a very slow rate. A simple fat is a triglyceride composed of a glycerol unit with short-chain or long-chain fatty acids attached.

The majority of carbohydrates, fats, and proteins in wastewater are in the form of large molecules that cannot penetrate the cell membrane of microorganisms. Bacteria, in order to metabolize high molecular-weight substances, must be capable of breaking down the large molecules into diffusible fractions for assimilation into the cell.

Several organic compounds, such as cellulose, long-chain saturated hydrocarbons, and other complex compounds, although available as a bacterial substrate, are considered nonbiodegradable because of the time and environmental limitations of biological wastewater treatment systems.

Volatile organic chemicals (VOCs) are released in large quantities in industrial, commercial, agricultural, and household activities. The adverse health effects of VOCs include cancer and chronic effects on the liver, kidney, and nervous system. *Synthetic organic chemicals* (SOCs) and *inorganic chemicals* (IOCs) are used in agricultural and industrial processes as pesticides, insecticides, and herbicides.

Organic petroleum derivatives, detergents, pesticides, and other synthetic, organic compounds are particularly resistant to biodegradation in wastewater treatment plants; some are toxic and inhibit the activity of microorganisms in biological treatment processes.

30. HEAVY METALS IN WASTEWATER

Metals are classified by four categories: *Dissolved metals* are those constituents of an unacidified sample that pass through a 0.45 μm membrane filter. *Suspended metals* are those constituents of an unacidified sample retained by a 0.45 μm membrane filter. *Total metals* are the concentration determined on an unfiltered sample after vigorous digestion or the sum of both dissolved and suspended fractions. *Acid-extractable metals* remain in solution after treatment of an unfiltered sample with hot dilute mineral acid.

The metal concentration in the wastewater is calculated from Eq. 25.48. The sample size is selected so that the product of the sample size (in mL) and the metal concentration (C, in mg/L) is approximately 1000.

$$C = \frac{C_{\text{digestate}} V_{\text{digested solution}}}{V_{\text{sample}}} \quad 25.48$$

Some metals are biologically essential; others are toxic and adversely affect wastewater treatment systems and receiving waters.

31. WASTEWATER COMPOSITING

Proper sampling techniques are essential for accurate evaluation of wastewater flows. Samples should be well mixed, representative of the wastewater flow stream in composition, taken at regular time intervals, and stored properly until analysis can be performed. *Compositing* is the sampling procedure that accomplishes these goals.

Samples are collected (i.e., "grabbed," hence the name *grab sample*) at regular time intervals (e.g., every hour on the hour), stored in a refrigerator or ice chest, and then integrated to formulate the desired combination for a particular test. Flow rates are measured at each sampling to determine the wastewater flow pattern.

The total volume of the composite sample desired depends on the kinds and number of laboratory tests to be performed.

32. WASTEWATER STANDARDS

Treatment plants and industrial complexes must meet all applicable wastewater quality standards for surface waters, drinking waters, air, and effluents of various types. For example, standards for discharges from secondary treatment plants are given in terms of *maximum contaminant levels* (MCLs) for five-day BOD, suspended solids, coliform count, and pH. Some chemicals, compounds, and microbial varieties are regulated; others are unregulated but monitored. MCLs for regulated and monitored chemicals, compounds, and microbes are subject to ongoing legislation and change.

Water Treatment

26 Wastewater Treatment: Equipment and Processes

Nomenclature

A	area	ft^2	m^2
BOD	biochemical oxygen demand	mg/L	mg/L
d_p	particle diameter	ft	m
D	diameter	ft	m
f	Darcy friction factor	–	–
F	effective number of passes	–	–
g	acceleration of gravity, 32.2 (9.81)	ft/sec^2	m/s^2
k	Camp formula constant	–	–
k	removal rate constant	day^{-1}	d^{-1}
K	Velz constant	1/ft	1/m
L	length	ft	m
L_{BOD}	BOD loading	$lbm/1000\ ft^3$-day	$kg/m^3 \cdot d$
L_H	hydraulic loading	gal/day-ft^2	$m^3/d \cdot m^2$
m	exponent	–	–
n	exponent	–	–
Q	air flow quantity	ft^3/min	L/s
Q	water flow quantity	gal/day	m^3/d
R	recirculation ratio	–	–
S	BOD	mg/L	mg/L
S_a	specific surface area	ft^2/ft^3	m^2/m^3
SG	specific gravity	–	–
t	time	days	d
T	temperature	°F	°C
v	velocity	ft/sec	m/s
v^*	overflow rate	gal/day-ft^2	$m^3/d \cdot m^2$
V	volume	ft^3	m^3
w	empirical weighting factor	–	–
Z	depth (of filter or basin)	ft	m

Symbols

η	BOD removal fraction	–	–
θ	temperature constant	–	–
ρ	density	lbm/ft^3	kg/m^3

Subscripts

a	area
d	detention
e	effluent
i	in (influent)
o	original (entry)
p	particle
r	recirculation
T	at temperature T
w	wastewater

1. INDUSTRIAL WASTEWATER TREATMENT

The *National Pollution Discharge Elimination System* (NPDES) places strict controls on the discharge of industrial wastewaters into municipal sewers. Any industrial wastes that would harm subsequent municipal treatment facilities or that would upset subsequent biological processes need to be pretreated. Manufacturing plants may also be required to equalize wastewaters by holding them in basins for stabilization prior to their discharge to the sewer. Table 26.1 lists typical limitations on industrial wastewaters.

Table 26.1 *Typical Industrial Wastewater Effluent Limitations*

characteristic	concentration[a] (mg/L)
COD	300–2000
BOD	100–300
oil and grease or TPH[b]	15–55
total suspended solids	15–45
pH	6.0–9.0
temperature	less than 40°C
color	2 color units
NH_3/NO_3	1.0–10
phosphates	0.2
heavy metals	0.1–5.0
surfactants	0.5–1.0 (total)
sulfides	0.01–0.1
phenol	0.1–1.0
toxic organics	1.0 total
cyanide	0.1

[a]maximum permitted at discharge
[b]total petroleum hydrocarbons

2. CESSPOOLS

A *cesspool* is a covered pit into which domestic (i.e., household) sewage is discharged. Cesspools for temporary storage can be constructed as watertight enclosures. However, most are *leaching cesspools* that allow seepage of liquid into the soil. Cesspools are rarely used today. They are acceptable for disposal for only very small volumes (e.g., from a few families).

3. SEPTIC TANKS

A *septic tank* is a simply constructed tank that holds domestic sewage while sedimentation and digestion occur. (See Fig. 26.1.) Typical detention times are 8–24 hr. Only 30–50% of the suspended solids are digested in a septic tank. The remaining solids settle, eventually clogging the tank, and must be removed. Semi-clarified effluent percolates into the surrounding soil through lateral lines placed at the *flow line* that lead into an underground leach field. In the past, clay drainage tiles were used. The terms "tile field" and "tile bed" are still encountered even though perforated plastic pipe is now widely used.

Most septic tanks are built for use by one to three families. Larger communal tanks can be constructed for small groups. (They should be designed to hold 12–24 hr of flow plus stored sludge.) A general rule is to allow at least 30 gal (0.1 m³) of storage per person served by the tank. Typical design parameters of domestic septic tanks are given in Table 26.2.

Proper design of the percolation field is the key to successful operation. Soils studies are essential to ensure adequate absorption into the soil.

Figure 26.1 *Septic Tank*

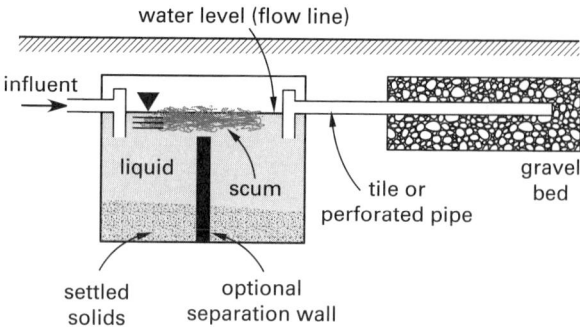

Table 26.2 *Typical Characteristics of Domestic Septic Tanks*

minimum capacity below flow line	300–500 gal (1.1–1.9 m³), plus 30 gal (0.1 m³) for each person served over 5
plan aspect ratio	1:2
minimum depth below flow line	3–4 ft (0.9–1.2 m)
minimum freeboard above flow line	1 ft (0.3 m)
tank burial depth	1–2 ft (0.3–0.6 m)
drainage field tile length	30 ft (9 m) per person
maximum drainage field tile length	60 ft (18 m)
minimum tile depth	1.5–2.5 ft (0.45–0.75 m)
lateral line spacing	6 ft (1.8 m)
gravel bed	0.33 ft (0.1 m) above lateral, 1–3 ft (0.3–1 m) below
soil layer below tile bed	10 ft (3 m)

(Multiply gal by 0.003785 to obtain m³.)
(Multiply ft by 0.3048 to obtain m.)

4. DISPOSAL OF SEPTAGE

Septage is the water and solid material pumped periodically from septic tanks, cesspools, or privies. In the United States, disposal of septage is controlled by federal sludge disposal regulations, and most local and state governments have their own regulations as well. While surface and land spreading were acceptable disposal methods in the past, septage haulers are now turning to municipal treatment plants for disposal. (Table 26.3 compares septage characteristics to those of municipal sewage. See also Table 25.4.) Since septage is many times more concentrated than sewage, it can cause *shock loads* in preliminary treatment processes. Typical problems are plugged screens and aerator inlets, reduced efficiency in grit chambers and aeration basins, and increased odors. The impact on subsequent processes is less pronounced due to the effect of dilution.

Table 26.3 *Comparison of Typical Septage and Municipal Sewage[a]*

characteristic	septage	sewage[b]
BOD	7000	220
COD	15,000	500
total solids	40,000	720
total volatile solids	25,000	365
total suspended solids	15,000	220
volatile suspended solids	10,000	165
TKN	700	40
NH_3 as N	150	25
alkalinity	1000	100
grease	8000	100
pH	6.0	n.a.

[a]all values except pH in mg/L
[b]medium strength

5. WASTEWATER TREATMENT PLANTS

For traditional *wastewater treatment plants* (WWTP), *preliminary treatment* of the wastewater stream is essentially a mechanical process intended to remove large objects, rags, and wood. Heavy solids and excessive oils and grease are also eliminated. Damage to pumps and other equipment would occur without preliminary treatment.

Odor control through chlorination or ozonation, freshening of septic waste by aeration, and flow equalization in holding basins can also be loosely categorized as preliminary processes.

After preliminary treatment, there are three "levels" of wastewater treatment: primary, secondary, and tertiary. *Primary treatment* is a mechanical (settling) process used to remove oil and most (i.e., approximately 50%) of the settleable solids. With domestic wastewater, a 25–35% reduction in BOD is also achieved, but BOD reduction is not the goal of primary treatment.

In the United States, secondary treatment is mandatory for all publicly owned wastewater treatment plants. *Secondary treatment* involves biological treatment in trickling filters, rotating contactors, biological beds, and activated sludge processes. Processing typically reduces the suspended solids and BOD content by more than 85%, volatile solids by 50%, total nitrogen by about 25%, and phosphorus by 20%.

Tertiary treatment (also known as *advanced wastewater treatment*, AWT) is targeted at specific pollutants or wastewater characteristics that have passed through previous processes in concentrations that are not allowed in the discharge. *Suspended solids* are removed by microstrainers or polishing filter beds. *Phosphorus* is removed by chemical precipitation. Aluminum and iron coagulants, as well as lime, are effective in removing phosphates. *Ammonia* can be removed by air stripping, biological denitrification, breakpoint chlorination, anion exchange, and algae ponds. Ions from *inorganic salts* can be removed by electrodialysis and reverse osmosis. The so-called *trace*

organics or *refractory substances, dissolved organic solids* that are resistant to biological processes, can be removed by filtering through carbon or ozonation.

6. WASTEWATER PLANT SITING CONSIDERATIONS

Wastewater plants should be located as far as possible from inhabited areas. A minimum distance of 1000 ft (300 m) for uncovered plants and lagoons is desired. Uncovered plants should be located downwind when a definite wind direction prevails. Soil conditions need to be evaluated, as does the proximity of the water table. Elevation in relationship to the need for sewage pumping (and for dikes around the site) is relevant.

The plant must be protected against flooding. One-hundred-year storms are often chosen as the design flood when designing dikes and similar facilities. Distance to the outfall and possible effluent pumping need to be considered.

Table 26.4 lists the approximate acreage for sizing wastewater treatment plants. Estimates of population expansion should provide for future capacity.

Table 26.4 *Treatment Plant Acreage Requirements*

type of treatment	surface area required (ac/MGD)
physical-chemical plants	1.5
activated sludge plants	2
trickling filter plants	3
aerated lagoons	16
stabilization basins	20

7. PUMPS USED IN WASTEWATER PLANTS

Wastewater treatment plants should be gravity-fed wherever possible. The influent of most plants is pumped to the starting elevation, and wastewater flows through subsequent processes by gravity thereafter. However, there are still many instances when pumping is required. Table 26.5 lists pump types by application.

At least two identical pumps should be present at every location, each capable of handling the entire peak flow. Three or more pumps are suggested for flows greater than 1 MGD, and peak flow should be handled when one of the pumps is being serviced.

8. FLOW EQUALIZATION

Equalization tanks or ponds are used to smooth out variations in flow that would otherwise overload wastewater processes. Graphical or tabular techniques similar to those used in reservoir sizing can be used to size equalization ponds. (See Fig. 26.2.) In practice, up to 25% excess capacity is added as a safety factor.

In a pure flow equalization process, there is no settling. Mechanical aerators provide the turbulence necessary to keep the solids in suspension while providing oxygen to

Table 26.5 *Pumps Used in Wastewater Plants*

flow	flow rate (gal/min (L/min))	pump type
raw sewage	<50 (190)	pneumatic ejector
	50–200 (190–760)	submersible or end-suction, nonclog centrifugal
	>200 (760)	end-suction, nonclog centrifugal
settled sewage	<500 (1900)	end-suction, nonclog centrifugal
	>500 (1900)	vertical axial or mixed-flow centrifugal
sludge, primary, thickened, or digested	–	plunger
sludge, secondary	–	end-suction, nonclog centrifugal
scum	–	plunger or recessed impeller
grit	–	recessed impeller, centrifugal, pneumatic ejector, or conveyor rake

(Multiply gal/min by 3.785 to obtain L/min.)

prevent putrefaction. For typical municipal wastewater, air is provided at the rate of 1.25–2.0 ft^3/min-1000 gal (0.01–0.015 m^3/min·m^3). Power requirements are approximately 0.02–0.04 hp/1000 gal (4–8 W/m^3).

Figure 26.2 *Equalization Volume: Mass Diagram Method*

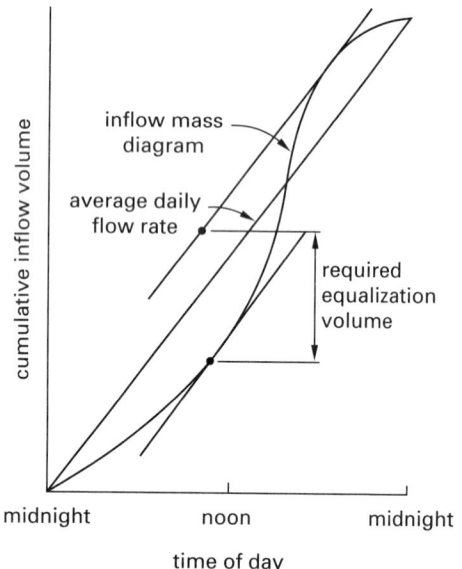

9. STABILIZATION PONDS

The term *stabilization pond* (*oxidation pond* or *stabilization lagoon*) refers to a pond used to treat organic waste by biological and physical processes. (See Table 26.6 for typical characteristics.) Aquatic plants, weeds, algae, and microorganisms stabilize the organic matter. The algae give off oxygen that is used by microorganisms to digest the organic matter. The microorganisms give off carbon dioxide, ammonia, and phosphates that the algae use. Even in modern times, such ponds may be necessary in remote areas (e.g., national parks and campgrounds). To keep toxic substances from leaching into the ground, ponds should not be used without strictly enforced industrial pretreatment requirements.

There are several types of stabilization ponds. *Aerobic ponds* are shallow ponds, less than 4 ft (1.2 m) in depth, where dissolved oxygen is maintained throughout the entire depth, mainly by action of photosynthesis. *Facultative ponds* have an anaerobic lower zone, a facultative middle zone, and an aerobic upper zone. The upper zone is maintained in an aerobic condition by photosynthesis and, in some cases, mechanical aeration at the surface. *Anaerobic ponds* are so deep and receive such a high organic loading that anaerobic conditions prevail throughout the entire pond depth. *Maturation ponds* (*tertiary ponds* or *polishing ponds*) are used for polishing effluent from secondary biological processes. Dissolved oxygen is furnished through photosynthesis and mechanical aeration. *Aerated lagoons* are oxygenated through the action of surface or diffused air aeration. They are often used with activated sludge processes. (See also Sec. 26.11.)

10. FACULTATIVE PONDS

Facultative ponds are the most common pond type selected for small communities. Approximately 25% of the municipal wastewater treatment plants in this country use ponds, and about 90% of these are located in communities of 5000 people or fewer. Long retention times and large volumes easily handle large fluctuations in wastewater flow and strength with no significant effect on effluent quality. Also, capital, operating, and maintenance costs are less than for other biological systems that provide equivalent treatment.

In a facultative pond, raw wastewater enters at the center of the pond. Suspended solids contained in the wastewater settle to the pond bottom where an anaerobic layer develops. A facultative zone develops just above the anaerobic zone. Molecular oxygen is not available in the region at all times. Generally, the zone is aerobic during the daylight hours and anaerobic during the hours of darkness.

An aerobic zone with molecular oxygen present at all times exists above the facultative zone. Some oxygen is

Table 26.6 *Typical Characteristics of Non-Aerated Stabilization Ponds*

characteristic	aerobic algae-growth maximizing (high rate)	aerobic oxygen-transfer maximizing (low rate)	facultative	facultative with surface agitation	anaerobic
size (cell)	0.5–2.5 ac (0.25–1 ha)	<10 ac (<4 ha)	2.5–10 ac (1–4 ha)	2.5–10 ac (1–4 ha)	0.5–2.5 ac (0.25–1 ha)
depth	1.0–1.5 ft (0.3–0.45 m)	3.0–5.0 ft (1–1.5 m)	3.0–7.0 ft (1–2 m)	3.0–8.5 ft (1–2.5 m)	8.0–15 ft (2.5–5 m)
BOD$_5$ loading	75–150 lbm/ac-day (80–160 kg/ha·d)	25–100 lbm/ac-day (40–120 kg/ha·d)	12–70 lbm/ac-day (15–80 kg/ha·d)	45–175 lbm/ac-day (50–200 kg/ha·d)	175–450 lbm/ac-day (200–500 kg/ha·d)
BOD$_5$ conversion	80–90%	80–90%	80–90%	80–90%	50–85%
detention time	4–6 days	10–40 days	7–30 days	7–20 days	20–50 days
temperature	40–85°F (5–30°C)	32–85°F (0–30°C)	32–120°F (0–50°C)	32–120°F (0–50°C)	40–120°F (5–50°C)
algal concentration	100–250 mg/L	40–100 mg/L	20–80 mg/L	5–20 mg/L	0–5 mg/L
suspended solids in effluent	150–300 mg/L	80–140 mg/L	40–100 mg/L	40–60 mg/L	80–160 mg/L
cell arrangement	series	parallel or series	parallel or series	parallel or series	series
minimum dike width			8 ft (2.5 m)		
maximum dike wall slope			1:3 (vertical:horizontal)		
minimum dike wall slope			1:4 (vertical:horizontal)		
minimum freeboard			3 ft (1 m)		

(Multiply lbm/ac-day by 1.12 to obtain kg/ha·d.)
(Multiply ft by 0.3048 to obtain m.)
(Multiply ac by 0.4 to obtain ha.)
(Multiply mg/L by 1.0 to obtain g/m^3.)

supplied from diffusion across the pond surface, but the majority is supplied through algal photosynthesis.

General guidelines are used to design facultative ponds. Ponds may be round, square, or rectangular. Usually, there are three cells, piped to permit operation in series or in parallel. Two of the three cells should be identical, each capable of handling half of the peak design flow. The third cell should have a minimum volume of one-third of the peak design flow.

11. AERATED LAGOONS

An *aerated lagoon* is a stabilization pond that is mechanically aerated. Such lagoons are typically deeper and have shorter detention times than nonaerated ponds. In warm climates and with floating aerators, one acre can support several hundred pounds (a hundred kilograms) of BOD per day.

The basis for the design of aerated lagoons is typically the organic loading and/or detention time. The detention time will depend on the desired BOD removal

fraction, η, and the reaction constant, k. Other factors that must be considered in the design process are solids removal requirements, oxygen requirements, temperature effects, and energy for mixing.

The BOD reduction can be calculated from the overall, first-order BOD *removal rate constant*, k_1. Typical values of $k_{1,\text{base-10}}$ (as specified by TSS Sec. 93.33) are 0.12 day^{-1} at 68°F (20°C) and 0.06 day^{-1} at 34°F (1°C).

$$t_d = \frac{V}{Q} = \frac{\eta}{k_{1,\text{base-}e}(1-\eta)} \qquad 26.1$$

$$k_{1,\text{base-}e} = 2.3 k_{1,\text{base-10}} \qquad 26.2$$

Aeration should maintain a minimum oxygen content of 2 mg/L at all times. Design depth is larger than for nonaerated lagoons—10–15 ft (3–4.5 m). Common design characteristics of mechanically aerated lagoons are shown in Table 26.7. Other characteristics are similar to those listed in Table 26.6.

Table 26.7 *Typical Characteristics of Aerated Lagoons*

aspect ratio	less than 3:1
depth	10–15 ft (3.0–4.5 m)
detention time	4–10 days
BOD loading	20–400 lbm/day-ac;
	200 lbm/day-ac typical
	(22–440 kg/ha·d;
	220 kg/ha·d typical)
operating temperature	0–38°C (21°C optimum)
typical effluent BOD	20–70 mg/L
oxygen required	0.7–1.4 times BOD removed
	(mass basis)

(Multiply ft by 0.3048 to obtain m.)
(Multiply lbm/day-ac by 1.12 to obtain kg/ha·d.)

12. RACKS AND SCREENS

Trash racks or *coarse screens* with openings 2 in (51 mm) or larger should precede pumps to prevent clogging. *Medium screens* ($^1/_2$–$1^1/_2$ in (13–38 mm) openings) and *fine screens* ($^1/_{16}$–$^1/_8$ in (1.6–3 mm)) are also used to relieve the load on grit chambers and sedimentation basins. Fine screens are rare except when used with selected industrial waste processing plants. Screens in all but the smallest municipal plants are cleaned by automatic scraping arms. A minimum of two screen units is advisable.

Screen capacities and head losses are specified by the manufacturer. Although the flow velocity must be sufficient to maintain sediment in suspension, the approach velocity should be limited to 3 ft/sec (0.9 m/s) to prevent debris from being forced through the screen.

13. GRIT CHAMBERS

Abrasive *grit* can erode pumps, clog pipes, and accumulate in excessive volumes. In a *grit chamber* (also known as a *grit clarifier* or *detritus tank*), the wastewater is slowed, allowing the grit to settle out but allowing the organic matter to continue through. (See Table 26.8 for typical characteristics.) Grit can be manually or mechanically removed with buckets or screw conveyors. A minimum of two units is needed.

Horizontal flow grit chambers are designed to keep the flow velocity as close to 1 ft/sec (0.3 m/s) as possible. If an analytical design based on settling velocity is required, the *scouring velocity* should not be exceeded. Scouring is the dislodging of particles that have already settled. Scouring velocity is not the same as settling velocity.

Scouring will be prevented if the horizontal velocity is kept below that predicted by Eq. 26.3, the *Camp formula*. SG_p is the specific gravity of the particle, typically taken as 2.65 for sand. d_p is the particle diameter. k is a dimensionless constant with typical values of 0.04 for sand and 0.06 or more for sticky, interlocking matter.

Table 26.8 *Typical Characteristics of Grit Chambers*

grit size	0.008 in (0.2 mm) and larger
grit specific gravity	2.65
grit arrival/removal rate	0.5–5 ft³/MG
	(4–40 × 10⁻⁶ m³/m³)
depth of chamber	4–10 ft (1.2–3 m)
length	40–100 ft (12–30 m)
width	varies (not critical)
detention time	90–180 sec
horizontal velocity	0.75–1.25 ft/sec
	(0.23–0.38 m/s)

(Multiply in by 25.4 to obtain mm.)
(Multiply ft³/MG by 7.48 × 10⁻⁶ to obtain m³/m³.)
(Multiply ft by 0.3048 to obtain m.)
(Multiply ft/sec by 0.3048 to obtain m/s.)

The dimensionless Darcy friction factor, f, is approximately 0.02–0.03. Any consistent set of units can be used with Eq. 26.3.

$$\mathrm{v} = \sqrt{8k\left(\frac{gd_p}{f}\right)(\mathrm{SG}_p - 1)} \qquad 26.3$$

14. AERATED GRIT CHAMBERS

An *aerated grit chamber* is a bottom-hoppered tank, as shown in Fig. 26.3, with a short detention time. Diffused aeration from one side of the tank rolls the water and keeps the organics in suspension while the grit drops into the hopper. The water spirals or rolls through the tank. Influent enters through the side, and degritted wastewater leaves over the outlet weir. A minimum of two units is needed.

Figure 26.3 *Aerated Grit Chamber*

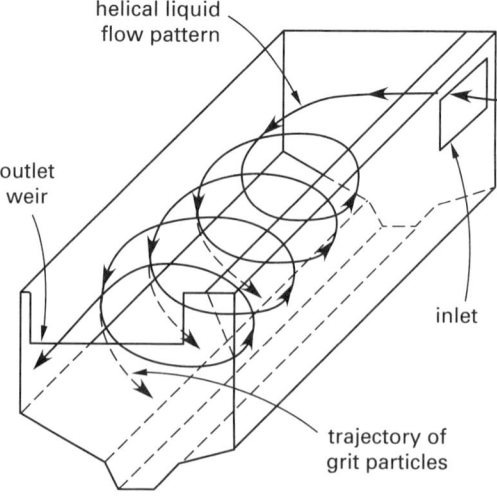

Solids are removed by pump, screw conveyer, bucket elevator, or gravity flow. However, the grit will have a significant organic content. A grit washer or cyclone separator can be used to clean the grit. Common design characteristics of aerated grit chambers are shown in Table 26.9.

Table 26.9 *Typical Characteristics of Aerated Grit Chambers*

detention time	2–5 min at peak flow; 3 min typical
air supply shallow tanks	1.5–5 cfm/ft length; 3 typical (0.13–0.45 m^3/min·m; 0.27 typical)
deep tanks	3–8 cfm/ft length; 5 typical (0.27–0.7 m^3/min·m; 0.45 typical)
grit and scum quantity	0.5–25 ft^3/MG; 2 typical (4–190 × 10^{-6} m^3/m^3; 15 typical)
length:width ratio	2.5:1–5:1 (3:1–4:1 typical)
depth	6–15 ft (1.8–4.5 m)
length	20–60 ft (6–18 m)
width	7–20 ft (2.1–6 m)

(Multiply cfm/ft by 0.0929 to obtain m^3/min·m.)
(Multiply ft^3/MG by 7.48 × 10^{-6} to obtain m^3/m^3.)
(Multiply ft by 0.3048 to obtain m.)

15. SKIMMING TANKS

If the sewage has more than 50 mg/L of floating grease or oil, a basin 8–10 ft (2.4–3 m) providing 5–15 min of detention time will allow the grease to rise as *floatables* (*scum*) to the surface. An aerating device will help coagulate and float grease to the surface. 40–80 psig (280–550 kPa) air should be provided at the approximate rate of 0.01–0.1 ft^3/gal (0.07–0.7 m^3/m^3) of influent. A small fraction (e.g., 30%) of the influent may be recycled in some cases. Grease rise rates are typically 0.2–1 ft/min (0.06–0.5 m/min), depending on degree of dispersal, amount of colloidal fines, and aeration.

Scum is mechanically removed by skimming troughs. Scum may not generally be disposed of by landfilling. It is processed with other solid wastes in anaerobic digesters. *Scum grinding* may be needed to reduce the scum to a small enough size for thorough digestion. In some cases, scum may be incinerated, although this practice may be affected by air-quality regulations.

16. SHREDDERS

Shredders (also called *comminutors*) cut waste solids to approximately $^1/_4$ in (6 mm) in size, reducing the amount of screenings that must be disposed of. Shreddings stay with the flow for later settling.

17. PLAIN SEDIMENTATION BASINS/ CLARIFIERS

Plain sedimentation basins/clarifiers (i.e., basins in which no chemicals are added to encourage clarification)

are similar in concept and design to those used to treat water supplies. Typical design characteristics for wastewater treatment sedimentation basins are listed in Table 26.10. Since the bottom slopes slightly, the depth varies with location. Therefore, the *side water depth* is usually quoted. The *surface loading* (*surface loading rate*, *overflow rate*, or *settling rate*), v*, along with sludge storage volume, is the primary design parameter.

$$v^* = \frac{Q}{A} \qquad 26.4$$

The *detention time* (*mean residence time* or *retention period*) is

$$t_d = \frac{V}{Q} \qquad 26.5$$

The *weir loading* (*weir loading rate*) is

$$\text{weir loading} = \frac{Q}{L} \qquad 26.6$$

Table 26.10 *Typical Characteristics of Clarifiers**

BOD reduction	20–40% (25–35% typical)
total suspended solids reduction	35–65%
bacteria reduction	50–60%
organic content of settled solids	50–75%
specific gravity of settled solids	1.2 or less
minimum settling velocity	4 ft/hr (1.2 m/hr) typical
plan shape	rectangular (or circular)
basin depth (side water depth)	6–15 ft; 10–12 ft typical (1.8–4.5 m; 3–3.6 m typical)
basin width	10–50 ft (3–15 m)
plan aspect ratio	3:1 to 5:1
basin diameter (circular only)	50–150 ft; 100 ft typical (15–45 m; 30 m typical)
minimum freeboard	1.5 ft (0.45 m)
minimum hopper wall angle	60°
detention time	1.5–2.5 hr
flow-through velocity	18 ft/hr (1.5 mm/s)
minimum flow-through time	30% of detention time
weir loading	10,000–20,000 gal/day-ft (125–250 m^3/d·m)
surface loading	400–2000 gal/day-ft^2; 800–1200 gal/day-ft^2 typical (16–80 m^3/d·m^2; 32–50 m^3/d·m^2 typical)
bottom slope to hopper	8%
inlet	baffled to prevent turbulence
scum removal	mechanical or manual

(Multiply ft/hr by 0.3048 to obtain m/h.)
(Multiply ft by 0.3048 to obtain m.)
(Multiply ft/hr by 0.0847 to obtain mm/s.)
(Multiply gal/day-ft by 0.0124 to obtain m^3/d·m.)
(Multiply gal/day-ft^2 by 0.0407 to obtain m^3/d·m^2.)
*See Sec. 26.26 and Sec. 26.27 for intermediate and final clarifiers, respectively.

18. CHEMICAL SEDIMENTATION BASINS/ CLARIFIERS

Chemical flocculation (*clarification* or *coagulation*) operations in chemical sedimentation basins are similar to those encountered in the treatment of water supplies except that the coagulant doses are greater. Chemical precipitation may be used when plain sedimentation is insufficient, or occasionally when the stream into which the outfall discharges is running low, or when there is a large increase in sewage flow. As with water treatment, the five coagulants used most often are (a) aluminum sulfate, $Al_2(SO_4)_3$; (b) ferric chloride, $FeCl_3$; (c) ferric sulfate, $Fe_2(SO_4)_3$; (d) ferrous sulfate, $FeSO_4$; and (e) chlorinated copperas. Lime and sulfuric acid may be used to adjust the pH for proper coagulation.

Example 26.1

A primary clarifier receives 1.4 MGD of domestic waste. The clarifier has a peripheral weir, is 50 ft in diameter, and is filled to a depth of 7 ft. Determine if the clarifier is operating within typical performance ranges.

Solution

The perimeter length is

$$L = \pi D = \pi(50 \text{ ft})$$
$$= 157 \text{ ft}$$

The surface area is

$$A = \frac{\pi}{4} D^2 = \left(\frac{\pi}{4}\right)(50 \text{ ft})^2$$
$$= 1963 \text{ ft}^2$$

The volume is

$$V = AZ = (1963 \text{ ft}^2)(7 \text{ ft})$$
$$= 13{,}741 \text{ ft}^3$$

The surface loading should be 400–2000 gal/day-ft^2.

$$v^* = \frac{Q}{A} = \frac{1.4 \times 10^6 \frac{\text{gal}}{\text{day}}}{1963 \text{ ft}^2}$$
$$= 713 \text{ gal/day-ft}^2 \quad [\text{OK}]$$

The detention time should be 1.5–2.5 hr.

$$t = \frac{V}{Q} = \frac{(13{,}741 \text{ ft}^3)\left(24 \frac{\text{hr}}{\text{day}}\right)}{\left(1.4 \times 10^6 \frac{\text{gal}}{\text{day}}\right)\left(0.1337 \frac{\text{ft}^3}{\text{gal}}\right)}$$
$$= 1.76 \text{ hr} \quad [\text{OK}]$$

The weir loading should be 10,000–20,000 gal/day-ft.

$$\frac{Q}{L} = \frac{1.4 \times 10^6 \frac{\text{gal}}{\text{day}}}{157 \text{ ft}}$$
$$= 8917 \text{ gal/day-ft} \quad [\text{low, but probably OK}]$$

19. TRICKLING FILTERS

Trickling filters (also known as *biological beds* and *fixed media filters*) consist of beds of rounded river rocks with approximate diameters of 2–5 in (50–125 mm), wooden slats, or modern synthetic media. Wastewater from primary sedimentation processing is sprayed intermittently over the bed. The biological and microbial slime growth attached to the bed purifies the wastewater as it trickles down. The water is introduced into the filter by rotating arms that move by virtue of spray reaction (reaction-type) or motors (motor-type). The clarified water is collected by an underdrain system.

The distribution rate is sometimes given by an *SK rating*, where SK is the water depth in mm deposited per pass of the distributor. Though there is strong evidence that rotational speeds of 1–2 rev/hr (high SK) produce significant operational improvement, traditional distribution arms revolve at 1–5 rev/min (low SK).

On the average, one acre of *low-rate filter* (also referred to as *standard-rate filter*) is needed for each 20,000 people served. Trickling filters can remove 70–90% of the suspended solids, 65–85% of the BOD, and 70–95% of the bacteria. Although low-rate filters have rocks to a depth of 6 ft (1.8 m), most of the reduction occurs in the first few feet of bed, and organisms in the lower part of the bed may be in a near-starvation condition.

Due to the low concentration of carbonaceous material in the water near the bottom of the filter, nitrogenous bacteria produce a highly nitrified effluent from low-rate filters. With low-rate filters, the bed will periodically slough off (unload) parts of its slime coating. Therefore, sedimentation after filtering is necessary. *Filter flies* are a major problem with low-rate filters, since fly larvae are provided with an undisturbed environment in which to breed.

Since there are limits to the heights of trickling filters, longer contact times can be achieved by returning some of the collected filter water back to the filter. This is known as *recirculation* or *recycling*. Recirculation is also used to keep the filter medium from drying out and to smooth out fluctuations in the hydraulic loading.

High-rate filters are used in most facilities. The higher hydraulic loading flushes the bed and inhibits excess biological growth. High-rate stone filters may be only 3–6 ft (0.9–1.8 m) deep. The high rate is possible because much of the filter discharge is recirculated. With the high flow rates, fly larvae are washed out, minimizing the filter fly problem. Since the biofilm is less thick and provided with carbon-based nutrients at a high rate, the effluent is nitrified only when the filter experiences low loading.

Super high-rate filters (*oxidation towers*) using synthetic media may be up to 40 ft (12 m) tall. High-rate and super high-rate trickling filters may be used as *roughing filters*, receiving wastewater at high hydraulic or organic loading and providing intermediate treatment or the first step of a multistage biological treatment process.

Figure 26.4 *Trickling Filter Process*

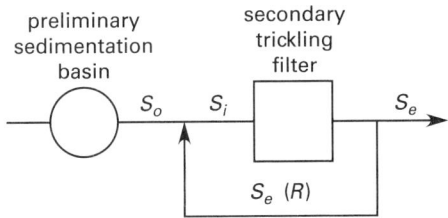

A BOD balance at the mixing point of a filter with recirculation results in Eq. 26.7, in which S_i is the BOD applied to the filter by the diluted influent, S_o is the BOD of the effluent from the primary settling tank (or the plant influent BOD if there is no primary sedimentation), and S_e is the BOD of the trickling filter effluent.

$$S_o + RS_e = (1 + R)S_i \qquad 26.7$$

$$S_i = \frac{S_o + RS_e}{1 + R} \qquad 26.8$$

BOD is reduced significantly in a trickling filter. Standard-rate filters produce an 80–85% reduction, and high-rate filters remove 65–80% of BOD, less because of reduced contact area and time. The *removal fraction* (*removal efficiency*) of a single-stage trickling filter is

$$\eta = \frac{S_{\text{removed}}}{S_o} = \frac{S_o - S_e}{S_o} \qquad 26.9$$

By definition, the *recirculation ratio*, R, is zero for standard-rate filters but can be as high as 4:1 for high-rate filters.

$$R = \frac{Q_r}{Q_w} \qquad 26.10$$

Filters may be classified as high-rate based on their hydraulic loading, organic loading, or both. The *hydraulic loading* of a trickling filter is the total water flow divided by the plan area. Typical values of hydraulic loading are 25–100 gal/day-ft^2 (1–4 m^3/d·m^2) for standard filters and 250–1000 gal/day-ft^2 (10–40 m^3/d·m^2) or higher for high-rate filters.

$$L_H = \frac{Q_w + Q_r}{A} = \frac{Q_w(1 + R)}{A} \qquad 26.11$$

The *BOD loading* (*organic loading* or *surface loading*) is calculated without considering recirculated flow. BOD loading for the filter/clarifier combination is essentially the BOD of the incoming wastewater divided by the filter volume. BOD loading is usually given in lbm per 1000 ft^3 per day (hence the 1000 term in Eq. 26.12.) Typical values are 5–25 lbm/1000 ft^3-day (0.08–0.4 kg/m^3·day) for low-rate filters and 25–110 lbm/1000 ft^3-day (0.4–1.8 kg/m^3·d) for high-rate filters.

$$L_{\text{BOD,kg/m}^3\cdot\text{d}} = \frac{Q_{w,\text{m}^3/\text{d}}S_{\text{mg/L}}}{\left(1000 \ \frac{\text{mg·m}^3}{\text{kg·L}}\right)V_{\text{m}^3}} \qquad \text{[SI]} \quad 26.12(a)$$

$$L_{\text{BOD,lbm/1000 ft}^3\text{-day}} = \frac{Q_{w,\text{MGD}}S_{\text{mg/L}}\left(8.345 \ \frac{\text{lbm-L}}{\text{MG-mg}}\right)}{V_{\text{ft}^3}} \times \left(1000 \ \frac{\text{ft}^3}{1000 \ \text{ft}^3}\right)$$

$$\text{[U.S.]} \quad 26.12(b)$$

The *specific surface area* of the filter is the total surface area of the exposed filter medium divided by the total volume of the filter.

Example 26.2

A single-stage trickling filter plant processes 1.4 MGD of raw domestic waste with a BOD of 170 mg/L. The trickling filter is 90 ft in diameter and has river rock media to a depth of 7 ft. The recirculation rate is 50%, and the filter is classified as high-rate. Water passes through clarification operations both before and after the trickling filter operation. The effluent leaves with a BOD of 45 mg/L. Determine if the trickling filter has been sized properly.

Solution

The filter area is

$$A = \frac{\pi}{4}D^2 = \left(\frac{\pi}{4}\right)(90 \ \text{ft})^2 = 6362 \ \text{ft}^2$$

The rock volume is

$$V = AZ = (6362 \ \text{ft}^2)(7 \ \text{ft}) = 44{,}534 \ \text{ft}^3$$

The hydraulic load for a high-rate filter should be 250–1000 gal/day-ft^2.

$$L_H = \frac{Q_w(1 + R)}{A}$$

$$= \frac{\left(1.4 \times 10^6 \ \frac{\text{gal}}{\text{day}}\right)(1 + 0.5)}{6362 \ \text{ft}^2}$$

$$= 330 \ \text{gal/day-ft}^2 \quad \text{[OK]}$$

From Table 26.10, the primary clarification process will remove approximately 30% of the BOD. The remaining BOD is

$$S_o = (1 - 0.3)\left(170 \ \frac{mg}{L}\right)$$
$$= 119 \ mg/L$$

The BOD loading should be approximately 25–110 lbm/ 1000 ft³-day.

$$L_{BOD} = \frac{Q_{w,MGD} S_{mg/L}\left(8.345 \ \frac{lbm\text{-}L}{MG\text{-}mg}\right)(1000)}{V_{ft^3}}$$

$$= \frac{\left(1.4 \ \frac{MG}{day}\right)\left(119 \ \frac{mg}{L}\right)}{44{,}534 \ ft^3}$$
$$\times \left(8.345 \ \frac{lbm\text{-}L}{MG\text{-}mg}\right)(1000)$$

$$= 31.2 \ lbm/day\text{-}1000 \ ft^3 \quad [OK]$$

20. TWO-STAGE TRICKLING FILTERS

If a higher BOD or solids removal fraction is needed, then two filters can be connected in series with an optional intermediate settling tank to form a *two-stage filter* system. The efficiency of the second-stage filter is considerably less than that of the first-stage filter because much of the biological food has been removed from the flow.

21. NATIONAL RESEARCH COUNCIL EQUATION

In 1946, the National Research Council (NRC) studied sewage treatment facilities at military installations. The wastewater at these facilities was stronger than typical municipal wastewater. Not surprisingly, the NRC concluded that the organic loading had a greater effect on removal efficiency than did the hydraulic loading.

If it is assumed that the biological layer and hydraulic loading are uniform, the water is at 20°C, and the filter is single-stage rock followed by a settling tank, then the *NRC equation*, Eq. 26.13, can be used to calculate the BOD removal fraction of the single-stage filter/clarifier combination. Inasmuch as installations with high BOD and low hydraulic loads were used as the basis of the studies, BOD removal efficiencies in typical municipal facilities are higher than predicted by Eq. 26.13. The value of L_{BOD} excludes recirculation returned directly from the filter outlet, which is accounted for in the value of F.

$$\eta = \frac{1}{1 + 0.0561\sqrt{\frac{L_{BOD,lbm/day}}{V_{1000s\ ft^3}F}}} \qquad 26.13$$

It is important to recognize the BOD loading, L_{BOD}, in Eq. 26.13, has units of lbm/day, not lbm/day-1000 ft³. The constant 0.0085 is often encountered in place of 0.0561 in the literature, as in Eq. 26.14. However, this value is for use with filter media volumes, V, expressed in ac-ft, not in thousands of ft³.

$$\eta = \frac{1}{1 + 0.0085\sqrt{\frac{L_{BOD,lbm/day}}{V_{ac\text{-}ft}F}}} \qquad 26.14$$

There are a number of ways to recirculate water from the output of the trickling filters back to the filter. Water can be brought back to a wet well, to the primary settling tank, to the filter itself, or to a combination of the three. Variations in performance are not significant as long as sludge is not recirculated. Equation 26.13 and Eq. 26.14 can be used with any of the recirculation schemes.

F is the *effective number of passes* of the organic material through a filter. R is the ratio of filter discharge returned to the inlet to the raw influent. In Eq. 26.15, w is a weighting factor typically assigned a value of 0.1.

$$F = \frac{1 + R}{(1 + wR)^2} \qquad 26.15$$

It is time-consuming to extract the organic loading, L_{BOD}, from Eq. 26.13 given η and R. Figure 26.5 can be used for this purpose.

Figure 26.5 *Trickling Filter Performance**

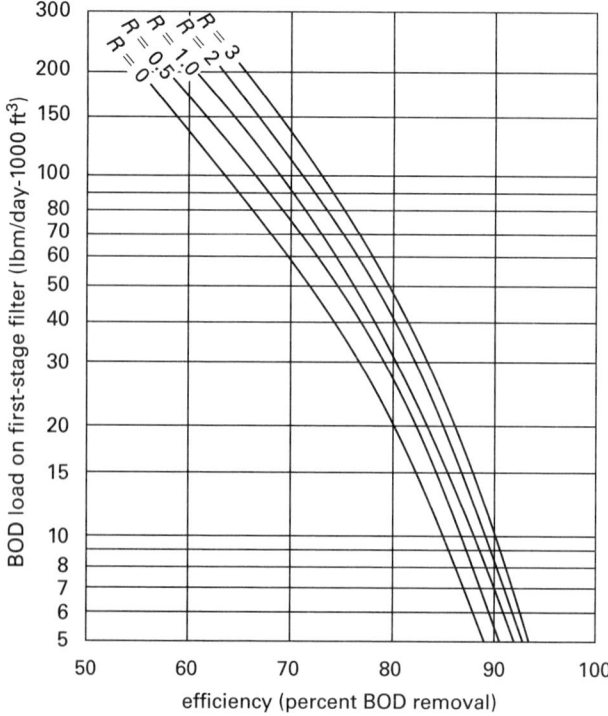

*NRC model, single-stage

Based on the NRC model, the removal fraction for a two-stage filter with an intermediate clarifier is

$$\eta_2 = \frac{1}{1 + \left(\dfrac{0.0561}{1 - \eta_1}\right)\sqrt{\dfrac{L_{\text{BOD}}}{F}}} \qquad 26.16$$

In recent years, the NRC equations have fallen from favor for several reasons. These reasons include applicability only to rock media, inability to correct adequately for temperature variations, inapplicability to industrial waste, and empirical basis.

22. VELZ EQUATION

Equation 26.17, known as the *Velz equation*, was the first semi-theoretical analysis of trickling filter performance versus depth of media. It is useful in predicting the BOD removal fraction for any generic trickling filter, including those with synthetic media. The original Velz used the term "total removable fraction of BOD," though this is understood to mean the maximum fraction of removable BOD removed, generally S_i.

Values of the *Velz decay rate*, K, an empirical rate constant, are highly dependent on the installation and operating characteristics, particularly the hydraulic loading, which is not included in the formula. Subsequent to Velz' work, the industry has developed a large database of applicable values for numerous application scenarios.

$$\frac{S_e}{S_i} = e^{-KZ} \qquad \left[\text{alternatively } \frac{S_e}{S_i} = 10^{-KZ}\right] \qquad 26.17$$

The variation in temperature was assumed to be

$$K_T = K_{20^{\circ}\text{C}}(1.047)^{T-20^{\circ}\text{C}} \qquad 26.18$$

23. MODERN FORMULATIONS

Various researchers have built upon the *Velz equation* and developed correlations of the form of Eq. 26.19 for various situations and types of filters. (In Eq. 26.19, S_a is the *specific surface area* (colonization surface area per unit volume), not the BOD.) Each researcher used different nomenclature and assumptions. For example, the inclusion of the effects of dilution by recirculation is far from universal. The specific form of the equation, values of constants and exponents, temperature coefficients, limitations, and assumptions are needed before such a correlation can be reliably used.

$$\frac{S_e}{S_i} = \exp\left\{-KZS_a^m\left(\frac{A}{Q}\right)^n\right\} \qquad 26.19$$

As an example of the difficulty in finding one model that predicts all trickling filter performance, consider the BOD removal efficiency as predicted by the *Schulze*

correlation (1960) for rock media and the *Germain correlation* (1965) for synthetic media. Though both correlations have the same form, both measured filter depths in feet, and both expressed hydraulic loadings in gal/min-ft^2, the Schulze correlation included recirculation while the Germain correlation did not. (See Eq. 26.20.) The treatability constant, k, was 0.51–0.76 day^{-1} for the Schulze correlation and 0.088 day^{-1} for the Germain correlation. The *media factor* exponent, n, was 0.67 for the Schulze correlation (rock media) and 0.5 for the Germain correlation (plastic media).

$$\frac{S_e}{S_i} = \exp\left\{\frac{-kZ}{L_H^n}\right\} \qquad 26.20$$

24. ROTATING BIOLOGICAL CONTACTORS

Rotating biological contactors, RBCs (also known as *rotating biological reactors*), consist of large-diameter plastic disks, partially immersed in wastewater, on which biofilm is allowed to grow. (See Fig. 26.6.) The disks are mounted on shafts that turn slowly. The rotation progressively wets the disks, alternately exposing the biofilm to organic material in the wastewater and to oxygen in the air. The biofilm population, since it is well oxygenated, efficiently removes organic solids from the wastewater. RBCs are primarily used for carbonaceous BOD removal, although they can also be used for nitrification or a combination of both.

Figure 26.6 *Rotating Biological Contactor*

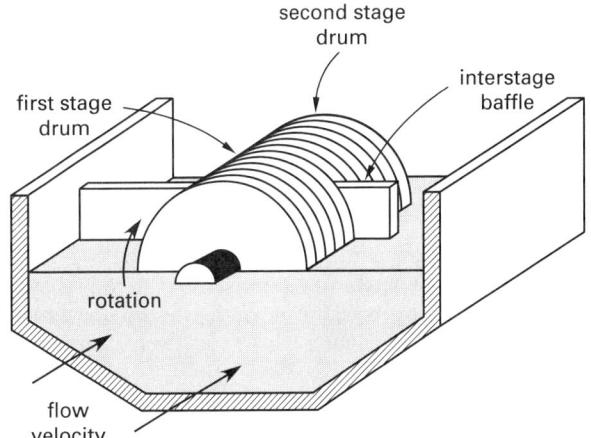

The primary design criterion is hydraulic loading, not organic (BOD) loading. For a specific hydraulic loading, the BOD removal efficiency will be essentially constant, regardless of variations in BOD. Other design criteria are listed in Table 26.11.

RBC operation is more efficient when several stages are used. Recirculation is not common with RBC processes. The process can be placed in series or in parallel with existing trickling filter or activated sludge processes.

Table 26.11 *Typical Characteristics of Rotating Biological Contactors*

number of stages	2–4
disk diameter	10–12 ft (3–3.6 m)
immersion, percentage of area	40%
hydraulic loading secondary treatment	2–4 gal/day-ft^2 (0.08–0.16 m^3/d·m^2)
tertiary treatment (nitrification)	0.75–2 gal/day-ft^2 (0.03–0.08 m^3/d·m^2)
optimum peripheral rotational speed	1 ft/sec (0.3 m/s)
tank volume	0.12 gal/ft^2 (0.0049 m^3/m^2) of biomass area
operating temperature	13–32°C
BOD removal fraction	70–80%

25. SAND FILTERS

For small populations, a *slow sand filter* (*intermittent sand filter*) can be used. Because of the lower flow rate, the filter area per person is higher than in the case of a trickling filter. Roughly one acre is needed for each 1000 people. The filter is constructed as a sand bed 2–3 ft (0.6–0.9 m) deep over a 6–12 in (150–300 mm) gravel bed. The filter is alternately exposed to water from a settling tank and to air (hence the term intermittent). Straining and aerobic decomposition clean the water. Application rates are usually 2–2.5 gal/day-ft^2 (0.08–0.1 m^3/d·m^2). Up to 95% of the BOD can be satisfied in an intermittent sand filter. The filter is cleaned by removing the top layer of clogged sand.

If the water is applied continuously as a final process following secondary treatment, the filter is known as a *polishing filter* or *rapid sand filter*. The water rate of a polishing filter is typically 2–8 gal/min-ft^2 (0.08–0.32 m^3/min·m^2) but may be as high as 10 gal/min-ft^2 (0.4 m^3/min·m^2). Although the designs are similar to those used in water supply treatment, coarser media are used since the turbidity requirements are less stringent. Backwashing is required more frequently and is more aggressive than with water supply treatment.

26. INTERMEDIATE CLARIFIERS

Sedimentation tanks located between trickling filter stages or between a filter and subsequent aeration are known as *intermediate clarifiers* or *secondary clarifiers*. Typical characteristics of intermediate clarifiers used with trickling filter processes are: maximum overflow rate, 1500 gal/day-ft^2 (61 m^3/d·m^2); water depth, 10–13 ft (3–4 m); and maximum weir loading, 10,000 gal/day-ft (125 m^3/d·m) for plants processing 1 MGD or less and 15,000 gal/day-ft (185 m^3/d·m) for plants processing over 1 MGD.

27. FINAL CLARIFIERS

Sedimentation following secondary treatment occurs in *final clarifiers*. The purpose of final clarification is to collect sloughed-off material from trickling filter processes or to collect sludge and return it for activated sludge processes, not to reduce BOD. The depth is approximately 10–12 ft (3.0–3.7 m), the average overflow rate is 500–600 gal/day-ft^2 (20–24 m^3/d·m^2), and the maximum overflow rate is approximately 1100 gal/day-ft^2 (45 m^3/d·m^2). The maximum weir loading is the same as for intermediate clarifiers, but the lower rates are preferred. For settling following extended aeration, the overflow rate and loading should be reduced approximately 50%.

28. PHOSPHORUS REMOVAL: PRECIPITATION

Phosphorus concentrations of 5–15 mg/L (as P) are experienced in untreated wastewater, most of which originates from synthetic detergents and human waste. Approximately 10% of the total phosphorus is insoluble and can be removed in primary settling. The amount that is removed by absorption in conventional biological processes is small. The remaining phosphorus is soluble and must be removed by converting it into an insoluble precipitate.

Soluble phosphorus is removed by precipitation and settling. Aluminum sulfate, ferric chloride ($FeCl_3$), and lime may be used depending on the nature of the phosphorus radical. Aluminum sulfate is more desirable since lime reacts with hardness and forms large quantities of additional precipitates. (Hardness removal is not as important as in water supply treatment.) However, the process requires about 10 lbm (10 kg) of aluminum sulfate for each pound (kilogram) of phosphorus removed. The process also produces a chemical sludge that is difficult to dewater, handle, and dispose of.

$$Al_2(SO_4)_3 + 2PO_4 \rightleftharpoons 2AlPO_4 + 3SO_4 \qquad 26.21$$

$$FeCl_3 + PO_4 \rightleftharpoons FePO_4 + 3Cl \qquad 26.22$$

Due to the many other possible reactions the compounds can participate in, the dosage should be determined from testing. The stoichiometric chemical reactions describe how the phosphorus is removed, but they do not accurately predict the quantities of coagulants needed.

29. AMMONIA REMOVAL: AIR STRIPPING

Ammonia may be removed by either biological processing or air stripping. In the biological *nitrification and denitrification process*, ammonia is first aerobically converted to nitrite and then to nitrate (nitrification) by bacteria. Then, the nitrates are converted to nitrogen gas, which escapes (denitrification).

In the *air-stripping* (*ammonia-stripping*) method, lime is added to water to increase its pH to about 10. This causes the ammonium ions, NH_4^+, to change to dissolved ammonia gas, NH_3. The water is passed through a packed tower into which air is blown at high rates. The air strips the ammonia gas out of the water. Recarbonation follows to remove the excess lime.

30. CARBON ADSORPTION

Adsorption uses high surface-area activated carbon to remove organic contaminants. Adsorption can use *granular activated carbon* (GAC) in column or fluidized-bed reactors or *powdered activated carbon* (PAC) in complete-mix reactors. Activated carbon is relatively nonspecific, and it will remove a wide variety of refractory organics as well as some inorganic contaminants. It should generally be considered for organic contaminants that are nonpolar, have low solubility, or have high molecular weights.

The most common problems associated with columns are breakthrough, excessive headloss due to plugging, and premature exhaustion. *Breakthrough* occurs when the carbon becomes saturated with the target compound and can hold no more. *Plugging* occurs when biological growth blocks the spaces between carbon particles. *Premature exhaustion* occurs when large and high molecular-weight molecules block the internal pores in the carbon particles. These latter two problems can be prevented by locating the carbon columns downstream of filtration.

31. CHLORINATION

Chlorination to disinfect and deodorize is one of the final steps prior to discharge. Vacuum-type feeders are used predominantly with chlorine gas. Chlorine under vacuum is combined with wastewater to produce a chlorine solution. A *flow-pacing* chlorinator will reduce the chlorine solution feed rate when the wastewater flow decreases (e.g., at night).

The size of the *contact tank* varies, depending on economics and other factors. An average design detention time is 30 min at average flow, some of which can occur in the plant outfall after the contact basin. Contact tanks are baffled to prevent short-circuiting that would otherwise reduce chlorination time and effectiveness.

Alternatives to disinfection by chlorine include sodium hypochlorite, ozone, ultraviolet light, bromine (as bromine chloride), chlorine dioxide, and hydrogen peroxide. All alternatives have one or more disadvantages when compared to chlorine gas.

32. DECHLORINATION

Toxicity, by-products, and strict limits on *total residual oxidants* (TROs) now make dechlorination mandatory at many installations. Sulfur dioxide (SO_2) and sodium thiosulfite ($Na_2S_2SO_3$) are the primary compounds used as dechlorinators today. Other compounds seeing limited use are sodium metabisulfate ($Na_2S_2O_5$), sodium bisulfate ($NaHSO_3$), sodium sulfite (Na_2SO_3), hydrogen peroxide (H_2O_2), and granular activated carbon.

Though reaeration was at one time thought to replace oxygen in water depleted by sulfur dioxide, this is now considered to be unnecessary unless required to meet effluent discharge requirements.

33. EFFLUENT DISPOSAL

Organic material and bacteria present in wastewater are generally not removed in their entireties, though the removal efficiency is high (e.g., better than 95% for some processes). Therefore, effluent must be discharged to large bodies of water where the remaining contaminants can be substantially diluted. Discharge to flowing surface water and oceans is the most desirable. Discharge to lakes and reservoirs should be avoided.

In some areas, *combined sewer overflow* (CSO) outfalls still channel wastewater into waterways during heavy storms when treatment plants are overworked. Such pollution can be prevented by the installation of large retention basins (*diversion chambers*), miles of tunnels, reservoirs to capture overflows for later controlled release to treatment plants, screening devices to separate solids from wastewater, swirl concentrators to capture solids for treatment while permitting the clearer portion to be chlorinated and released to waterways, and more innovative vortex solids separators.

34. WASTEWATER RECLAMATION

Treated wastewater can be used for irrigation, firefighting, road maintenance, or flushing. Public access to areas where reclaimed water is disposed of through spray-irrigation (e.g., in sod farms, fodder crops, and pasture lands) should be restricted. When the reclaimed water is to be used for irrigation of public areas, the water quality standards regarding suspended solids, fecal coliforms, and viruses should be strictly controlled.

Although it has been demonstrated that wastewater can be made potable at great expense, such practice has not gained widespread favor.

Water Treatment

27 Activated Sludge and Sludge Processing

Water Treatment

Nomenclature

A	area	ft^2	m^2
BOD	biochemical oxygen demand	mg/L	mg/L
c_p	specific heat	Btu/lbm-°F	J/kg·°C
C	cost of electricity	\$/kW-hr	\$/kW·h
D	oxygen deficit	mg/L	mg/L
DO	dissolved oxygen	mg/L	mg/L
E	efficiency of waste utilization	–	–
f	BOD_5/BOD_u ratio	–	–
F	food arrival rate	mg/day	mg/d
F:M	food to microorganism ratio	day^{-1}	d^{-1}
g	acceleration of gravity, 32.2 (9.81)	ft/sec^2	m/s^2
G	fraction of solids that are biodegradable	–	–
G	number of gravities	–	–
h	film coefficient	Btu/ hr-ft^2-°F	W/m^2·K
J	Joule's constant, 778	ft-lbf/Btu	n.a.
k	ratio of specific heats	–	–
k	thermal conductivity	Btu-ft/ hr-ft^2-°F	W/m·K
k_d	microorganism endogenous decay rate	day^{-1}	d^{-1}
K	cell yield constant (removal efficiency)	lbm/lbm	mg/mg
K_s	half-velocity coefficient	mg/L	mg/L
K_t	oxygen transfer coefficient	1/hr	1/h
L	BOD loading	lbm/ day-1000 ft^3	kg/d
L	thickness	ft	m
LHV	lower heating value	Btu/ft^3	kJ/m^3
m	mass	lbm	kg
$\dot{m}$	mass flow rate	lbm/day	kg/d
M	mass of microorganisms	lbm	kg
MLSS	total mixed liquor suspended solids	mg/L	mg/L
n	rotational speed	rev/min	rev/min
p	pressure	lbf/in^2	Pa
P	power	hp	kW
P_x	mass of sludge wasted	lbm/day	kg/d
q	heat transfer	Btu	kJ
Q	flow rate	ft^3/day	m^3/d
r	distance from center of rotation	ft	m
r	rate	mg/ft^3-day	mg/m^3·d
R	recycle ratio	–	–
R_{air}	specific gas constant	ft-lbf/lbm-°R	J/kg·K
s	gravimetric fractional solids content	–	–
S	growth-limited substrate concentration	mg/L	mg/L
SG	specific gravity	–	–
SS	suspended solids	mg/L	mg/L
SVI	sludge volume index	mL/g	mL/g
t	time	day	d
T	absolute temperature	°R	K
TSS	total suspended solids	mg/L	mg/L
U	overall coefficient of heat transfer	Btu/ hr-ft^2-°F	W/m^2·K
U	specific substrate utilization	day^{-1}	d^{-1}
v*	overflow rate	gal/day-ft^2	m^3/d·m^2
V	volume	ft^3	m^3
$\dot{V}$	volumetric flow rate	ft^3/day	m^3/d
X	mixed liquor volatile suspended solids	mg/L	mg/L
Y	yield coefficient	lbm/lbm	mg/mg

Symbols

β	oxygen saturation coefficient	–	–
η	efficiency	–	–
θ	residence/detention time	day	d
μ_m	maximum specific growth rate	day^{-1}	d^{-1}
ρ	density	lbm/ft^3	kg/m^3
ω	rotational speed	rad/sec	rad/s

Subscripts

5	5-day
a	aeration tank
c	cell or compressor
e	effluent
i	influent diluted with recycle flow
o	influent
obs	observed
r	recirculation
s	settling tank
su	substrate utilization
t	transfer
u	ultimate
w	wasted

1. SLUDGE

Sludge is the mixture of water, organic and inorganic solids, and treatment chemicals that accumulates in settling tanks. The term is also used to refer to the dried residue (screenings, grit, filter cake, and drying bed scraping) from separation and drying processes, although the term *biosolids* is becoming more common in this regard. (The term *residuals* is also used, though this term more commonly refers to sludge from water treatment plants.)

2. ACTIVATED SLUDGE PROCESS

The *activated sludge process* is a secondary biological wastewater treatment technique in which a mixture of wastewater and sludge solids is aerated. (See Fig. 27.1.) The sludge mixture produced during this oxidation process contains an extremely high concentration of aerobic bacteria, most of which are near starvation. This condition makes the sludge an ideal medium for the destruction of any organic material in the mixture. Since the bacteria are voraciously active, the sludge is called *activated sludge.*

The well-aerated mixture of wastewater and sludge, known as *mixed liquor*, flows from the aeration tank to a secondary clarifier where the sludge solids settle out. Most of the settled sludge solids are returned to the aeration tank in order to maintain the high population of bacteria needed for rapid breakdown of the organic material. However, because more sludge is produced than is needed, some of the return sludge is diverted ("wasted") for subsequent treatment and disposal. This wasted sludge is referred to as *waste activated sludge*, WAS. The volume of sludge returned to the aeration basin is typically 20–30% of the wastewater flow. The liquid fraction removed from the secondary clarifier weir is chlorinated and discharged.

Though diffused aeration and mechanical aeration are the most common methods of oxygenating the mixed liquor, various methods of staging the aeration are used, each having its own characteristic ranges of operating parameters. (See Table 27.1 for typical characteristics of activated sludge plants, Table 27.2 for additional information, and Table 27.3 for *Ten States' Standards*.)

In a traditional activated sludge plant using conventional aeration, the wastewater is typically aerated for 6–8 hours in long, rectangular aeration basins. Sufficient air, about eight volumes for each volume of wastewater treated, is provided to keep the sludge in suspension. The air is injected near the bottom of the aeration tank through a system of diffusers.

3. AERATION STAGING METHODS

Small wastewater quantities can be treated with *extended aeration*. (See Fig. 27.2 for available aeration methods.) This method uses mechanical floating or fixed subsurface aerators to oxygenate the mixed liquor for 24–36 hours in a large lagoon. There is no primary clarification, and there is generally no sludge wasting process. Sludge is allowed to accumulate at the bottom

Figure 27.1 *Typical Activated Sludge Plant*

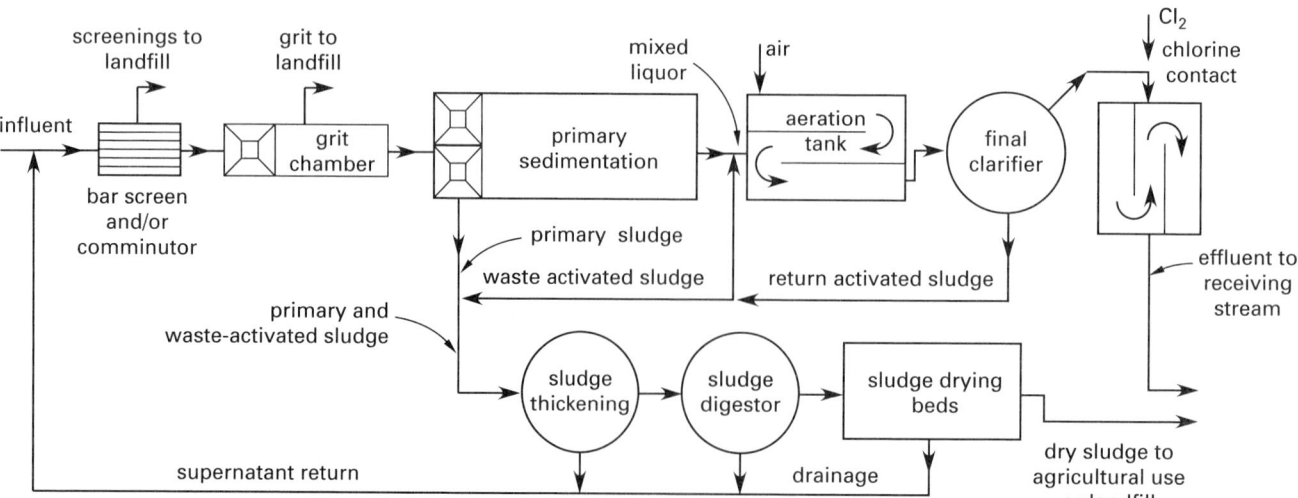

Figure 27.2 *Methods of Aeration*

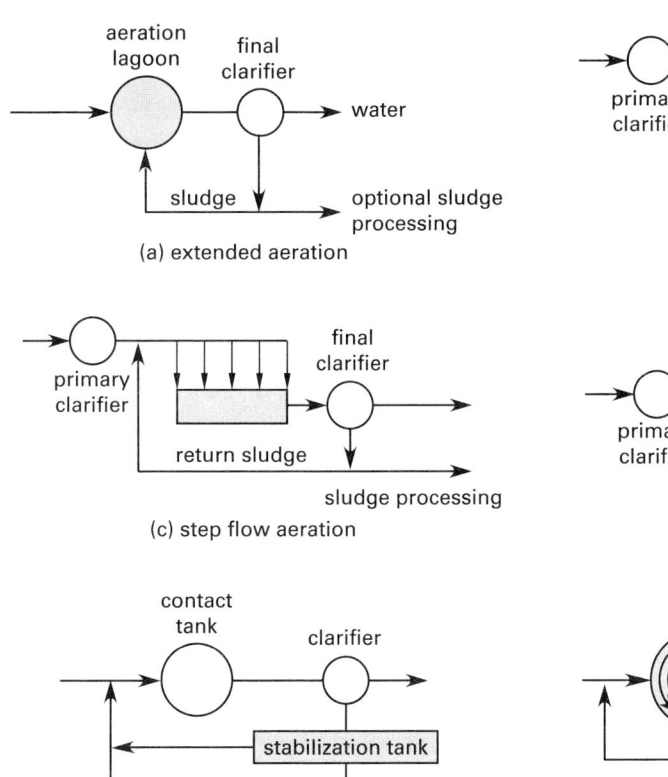

(a) extended aeration

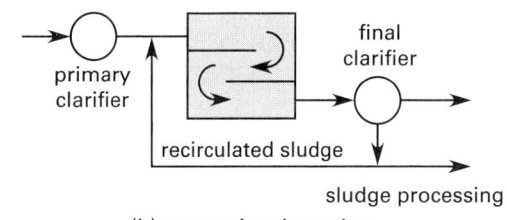

(b) conventional aeration

(c) step flow aeration

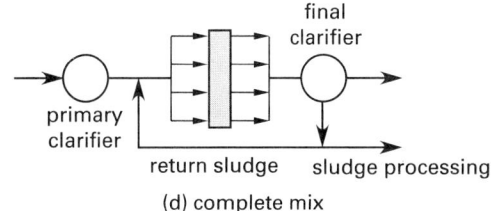

(d) complete mix

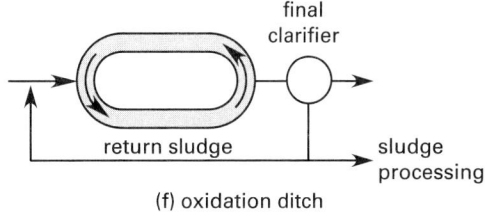

(e) contact stabilation

(f) oxidation ditch

of the lagoon for several months. Then, the system is shut down and the lagoon is pumped out. Sedimentation basins are sized very small, with low overflow rates of 200–600 gal/day-ft^2 (8.1–24 m^3/d·m^2) and long retention times.

An *oxidation ditch* is a type of extended aeration system, configured in a continuous long and narrow oval. The basin contents circulate continuously to maintain mixing and aeration. Mixing occurs by brush aerators or by a combination of low-head pumping and diffused aeration.

In *conventional aeration*, the influent is taken from a primary clarifier and then aerated. The amount of aeration is usually decreased as the wastewater travels along the aeration path since the BOD also decreases along the route. This is known as *tapered aeration*.

With *step-flow aeration*, aeration is constant along the length of the aeration path, but influent is introduced at various points along the path.

With *complete-mix aeration*, wastewater is added uniformly and the mixed liquor is removed uniformly over the length of the tank. The mixture is essentially uniform in composition throughout.

Units for the *contact stabilization* (*biosorption*) process are typically factory-built and brought to the site for installation, although permanent facilities can be built for the same process. Prebuilt units are compact but not

Table 27.1 *Typical Characteristics of Conventional Activated Sludge Plants*

BOD reduction	90–95%
effluent BOD	5–30 mg/L
effluent COD	15–90 mg/L
effluent suspended solids	5–30 mg/L
F:M	0.2–0.5
aeration chamber volume	5000 ft^3 (140 m^3) max.
aeration chamber depth	10–15 ft (3–4.5 m)
aeration chamber width	20 ft (6 m)
length:width ratio	5:1 or greater
aeration air rate	0.5–2 ft^3 air/gal (3.6–14 m^3/m^3) raw wastewater
minimum dissolved oxygen	1 mg/L
MLSS	1000–4000 mg/L
sludge volume index	50–150
settling basin depth	15 ft (4.5 m)
settling basin detention time	4–8 hr
basin overflow rate	400–2000 gal/day-ft^2; 1000 gal/day-ft^2 typical (16–81 m^3/d·m^2; 40 m^3/d·m^2 typical)
settling basin weir loading	10,000 gal/day-ft (130 m^3/d·m)
fractional sludge recycle	0.20–0.30
frequency of sludge recycle	hourly

(Multiply ft^3 by 0.0283 to obtain m^3.)
(Multiply ft by 0.3048 to obtain m.)
(Multiply ft^3/gal by 7.48 to obtain m^3/m^3.)
(Multiply gal/day-ft^2 by 0.0407 to obtain m^3/d·m^2.)
(Multiply gal/day-ft by 0.0124 to obtain m^3/d·m.)

as economical or efficient as larger plants running on the same process. The aeration tank is called a *contact tank.* The *stabilization tank* takes the sludge from the clarifier and aerates it. With this process, colloidal solids are absorbed in the activated sludge during the 30–90 min of aeration in the contact tank. Then, the sludge is removed by clarification and the return sludge is aerated for 3–6 more hours in the stabilization tank. Less time and space is required for this process because the sludge stabilization is done while the sludge is still concentrated. This method is very efficient in handling colloidal wastes.

The *high-rate aeration* method uses mechanical mixing along with aeration to decrease the aeration period and increase the BOD load per unit volume.

The *high purity oxygen aeration* method requires the use of bottled or manufactured oxygen that is introduced into closed/covered aerating tanks. Mechanical mixers are needed to take full advantage of the oxygen supply because little excess oxygen is provided. Retention times in aeration basins are longer than for other aerobic systems, producing higher concentrations of MLSS and better stabilization of sludge that is ultimately wasted. This method is applicable to high-strength sewage and industrial wastewater.

Sequencing batch reactors (SBRs) operate in a fill-and-drain sequence. The reactor is filled with influent, contents are aerated, and sludge is allowed to settle. Effluent is drawn from the basin along with some of the settled sludge, and new effluent is added to repeat the process. The "return sludge" is the sludge that remains in the basin.

4. FINAL CLARIFIERS

Table 27.4 lists typical operating characteristics for clarifiers in activated sludge processes. Sludge should be removed rapidly from the entire bottom of the clarifier.

5. SLUDGE PARAMETERS

The bacteria and other suspended material in the mixed liquor is known as *mixed liquor suspended solids* (MLSS) and is measured in mg/L. Suspended solids are further divided into fixed solids and volatile solids. *Fixed solids* (also referred to as *nonvolatile solids*) are those inert solids that are left behind after being fired in a furnace. *Volatile solids* are essentially those carbonaceous solids that are consumed in the furnace. The volatile solids are considered to be the measure of solids capable of being digested. Approximately 60–75% of sludge solids are volatile. The volatile material in the mixed liquor is known as the *mixed liquor volatile suspended solids* (MLVSS).

Table 27.2 Representative Operating Conditions for Aeration[a]

type of aeration	plant flow rate (MGD)	mean cell residence time, θ_c (d)	oxygen required (lbm/lbm BOD removed)	waste sludge (lbm/lbm BOD removed)	total plant BOD load (lbm/day)	aerator BOD load, L_{BOD} (lbm/day-1000 ft³)	F:M (lbm/lbm-day)	MLSS (mg/L)	R (%)	η_{BOD} (%)
conventional	0–0.5 0.5–1.5 1.5 up	7.5 7.5–6.0 6.0	0.8–1.1	0.4–0.6	0–1000 1000–3000 3000 up	30 30–40 40	0.2–0.5	1500–3000	30	90–95
contact stabilization	0–0.5 0.5–1.5 1.5 up	3.0[b] 3.0–2.0[b] 2.0–1.5[b]	0.8–1.1 0.4–0.6 0.4–0.6	0.4–0.6	0–1000 1000–3000 3000 up	30 30–50 50	0.2–0.5	1000–3000[b]	100	85–90
extended	0–0.5 0.5–1.5 1.5 up	24 20 16	1.4–1.6	0.15–0.3	all	10.0 12.5 15.0	0.05–0.1	3000–6000	100	85–95
high rate	0–0.5 0.5–1.5 1.5 up	4.0 3.0 2.0	0.7–0.9	0.5–0.7	2000 up	100	1.0 or less	4000–10,000	100	80–85
step aeration	0–0.5 0.5–1.5 1.5 up	7.5 7.5–5.0 5.0			0–1000 1000–3000 3000 up	30 30–50 50	0.2–0.5	2000–3500	50	85–95
high purity oxygen		3.0–1.0				100–200	0.6–1.5	6000–8000	50	90–95
oxidation ditch		36–12				5–30				

(Multiply MGD by 3785.4 to obtain m³/d.)
(Multiply lbm/day by 0.4536 to obtain kg/d.)
(Multiply lbm/day-1000 ft³ by 0.016 to obtain kg/d·m³.)
[a]compiled from a variety of sources
[b]in contact unit only

Table 27.3 *Ten States' Standards for Activated Sludge Processes*

process	aeration tank organic loading—lbm BOD_5/day per 1000 ft^3 (kg/d·m^3)	F:M lbm BOD_5/day per lbm MLVSS	MLSS^a (mg/L)
conventional step aeration complete mix	40 (0.64)	0.2–0.5	1000–3000
contact stabilization	50^b (0.80)	0.2–0.6	1000–3000
extended aeration oxidation ditch	15 (0.24)	0.05–0.1	3000–5000

(Multiply lbm/day-1000 ft^3 by 0.016 to obtain kg/d·m^3.)
(Multiply lbm/day-lbm by 1.00 to obtain kg/d·kg.)
^{a}MLSS values are dependent upon the surface area provided for sedimentation and the rate of sludge return as well as the aeration process.
^{b}Total aeration capacity; includes both contact and reaeration capacities. Normally the contact zone equals 30–35% of the total aeration capacity.

Reprinted from *Recommended Standards for Sewage Works*, 2004, Sec. 92.31, Great Lakes-Upper Mississippi River Board of State Sanitary Engineers, published by Health Education Service, Albany, NY.

The organic material in the incoming wastewater constitutes "food" for the activated organisms. The food arrival rate is given by Eq. 27.1. S_o is usually taken as the incoming BOD_5, although COD is used in rare situations.

$$F = S_o Q_o \qquad 27.1$$

The mass of microorganisms, M, is determined from the *volatile suspended solids concentration*, X, in the aeration tank.

$$M = V_a X \qquad 27.2$$

The *food-to-microorganism ratio*, (F:M) is displayed in Eq. 27.3. For conventional aeration, typical values are 0.20–0.50 lbm/lbm-day (0.20–0.50 kg/kg·d), though values between 0.05 and 1.0 have been reported. θ in Eq. 27.3 is the *hydraulic detention time*.

$$\text{F:M} = \frac{S_{o,\text{mg/L}} Q_{o,\text{MGD}}}{V_{a,\text{MG}} X_{\text{mg/L}}} = \frac{S_{o,\text{mg/L}}}{\theta_{\text{days}} X_{\text{mg/L}}} \qquad 27.3$$

Table 27.4 *Characteristics of Final Clarifiers for Activated Sludge Processes*

type of aeration	design flow (MGD)	minimum detention time (hr)	maximum overflow rate (gpd/ft^2)
conventional, high rate and step	< 0.5	3.0	600
	0.5–1.5	2.5	700
	> 1.5	2.0	800
contact stabilization	< 0.5	3.6	500
	0.5–1.5	3.0	600
	>1.5	2.5	700
extended aeration	< 0.05	4.0	300
	0.05–0.15	3.6	300
	>0.15	3.0	600

(Multiply MGD by 0.0438 to obtain m^3/s.)
(Multiply gal/day-ft^2 by 0.0407 m^3/d·m^2.)

Some authorities include the entire suspended solids content (MLSS), not just the volatile portion (X), when calculating the food-to-microorganism ratio, although this interpretation is less common.

$$\text{F:M} = \frac{S_{o,\text{mg/L}} Q_{o,\text{MGD}}}{V_{a,\text{MG}} \text{MLSS}_{\text{mg/L}}} \qquad 27.4$$

The liquid fraction of the wastewater passes through an activated sludge process in a matter of hours. However, the sludge solids are recycled continuously and have an average stay much longer in duration. There are two measures of *sludge age*: the *mean cell residence time* (also known as the *age of the suspended solids* and *solids residence time*), essentially the age of the microorganisms, and *age of the BOD*, essentially the age of the food. For conventional aeration, typical values of the mean cell residence time, θ_c, are 6 to 15 days for high-quality effluent and sludge.

$$\theta_c = \frac{V_a X}{Q_e X_e + Q_w X_w} \qquad 27.5$$

The *BOD sludge age* (*age of the BOD*), θ_{BOD}, is the reciprocal of the food-to-microorganism ratio.

$$\theta_{\text{BOD}} = \frac{1}{\text{F:M}} = \frac{V_{a,\text{m}^3} X_{\text{mg/L}}}{S_{o,\text{mg/L}} Q_{o,\text{m}^3/\text{d}}} \qquad \text{[SI]} \quad 27.6(a)$$

$$\theta_{\text{BOD}} = \frac{1}{\text{F:M}} = \frac{V_{a,\text{MG}} X_{\text{mg/L}}}{S_{o,\text{mg/L}} Q_{o,\text{MGD}}} \qquad \text{[U.S.]} \quad 27.6(b)$$

The *sludge volume index*, SVI, is a measure of the sludge's settleability. SVI can be used to determine the tendency toward *sludge bulking*. (See Sec. 27.14.) SVI is determined by taking 1 L of mixed liquor and measuring the volume of settled solids after 30 minutes. SVI is the volume in mL occupied by 1 g of settled volatile and nonvolatile suspended solids.

$$\text{SVI}_{\text{mg/L}} = \frac{\left(1000 \, \dfrac{\text{mg}}{\text{g}}\right) V_{\text{settled,mL/L}}}{\text{MLSS}_{\text{mg/L}}} \qquad \textit{27.7}$$

A similar control parameter is the *settled sludge volume*, SSV, which is the volume of sludge per liter sample after settling for 30 or 60 minutes.

$$\text{SSV} = \frac{V_{\text{settled sludge,mL}} \left(1000 \, \dfrac{\text{mL}}{\text{L}}\right)}{V_{\text{sample,mL}}} \qquad \textit{27.8}$$

The concentration of total suspended solids, TSS, in the recirculated sludge can be found from Eq. 27.9. This equation is useful in calculating the solids concentration for any sludge that is wasted from the return sludge line. (Suspended sludge solids wasted include both fixed and volatile portions.)

$$\text{TSS}_{\text{mg/L}} = \frac{\left(1000 \, \dfrac{\text{mg}}{\text{g}}\right)\left(1000 \, \dfrac{\text{mL}}{\text{L}}\right)}{\text{SVI}_{\text{mL/g}}} \qquad \textit{27.9}$$

6. SOLUBLE BOD ESCAPING TREATMENT

S is a variable used to indicate the growth-limiting substrate in solution. Specifically, S (without a subscript) is defined as the soluble BOD_5 escaping treatment (i.e., the effluent BOD_5 leaving the activated sludge process). This may need to be calculated from the effluent suspended solids and the fraction, G, of the suspended solids that is ultimately biodegradable.

$$\begin{aligned} \text{BOD}_e &= \text{BOD}_{\text{escaping treatment}} \\ &\quad + \text{BOD}_{\text{effluent suspended solids}} \\ &= S + S_e \\ &= S + 1.42 f G X_e \end{aligned} \qquad \textit{27.10}$$

Care must be taken to distinguish between the standard five-day BOD and the ultimate BOD. The ratio of these two parameters, typically approximately 70%, is

$$f = \frac{\text{BOD}_5}{\text{BOD}_u} \qquad \textit{27.11}$$

In some cases, the terms in the product $1.42fG$ in Eq. 27.10 are combined, and S_e is known directly as a percentage of X_e. However, this format is less common.

7. PROCESS EFFICIENCY

The *treatment efficiency* (*BOD removal efficiency*, *removal fraction*, etc.), typically 90–95% for conventional aeration, is given by Eq. 27.12. S_o is the BOD as received from the primary settling tank. If the overall plant efficiency is wanted, the effluent S is taken as the total BOD_5 of the effluent. Depending on convention, efficiency can also be based on only the soluble BOD escaping treatment (i.e., S from Eq. 27.10).

$$\eta_{\text{BOD}} = \frac{S_o - S}{S_o} \qquad \textit{27.12}$$

8. PLUG FLOW AND STIRRED TANK MODELS

Conventional aeration and complete-mix aeration represent two extremes for continuous-flow reactions. Ideal conventional aeration is described by a *plug-flow reaction* (PFR, also known as *tubular flow*) kinetic model. (The term *reactor* is commonly used to describe the aeration tank.) In a plug-flow reactor, material progresses along the reaction path, changing in composition as it goes. Mixing is considered to be ideal in the lateral plane but totally absent in the longitudinal direction. That is, the properties (in this case, BOD and suspended solids) change along the path, as though only a "plug" of material was progressing forward.

A complete-mix aerator, however, is a type of *continuous-flow stirred tank reactor* (CSTR or CFSTR). In a CSTR model, the material is assumed to be homogeneous in properties throughout as a result of continuous perfect mixing. It is assumed that incoming material is immediately diluted to the tank concentration, and the outflow properties are the same as the tank properties.

Modifications of the basic kinetic models are made to account for return/recycle and wastage. Many, but not all, of the same equations can be used to describe the performance of both the PFR and CSTR models. (See Fig. 27.3.)

Until the late 1950s, nearly all activated sludge aeration tanks were long and narrow with length-to-width ratios greater than 5, and performance was assumed to be predicted by PFR equations. However, extensive testing has shown that substantial longitudinal mixing actually occurs, and many conventional systems are actually CSTRs. In fact, the long basin configuration gives an oxidation ditch the attributes of a PFR while maintaining CSTR characteristics.

The kinetic model for the PFR is complex, but it can be simplified under certain conditions. If the ratio of θ_c/θ is greater than 5, the concentration of microorganisms in the influent to the reactor (aeration tank)

Figure 27.3 Kinetic Model Process Variables

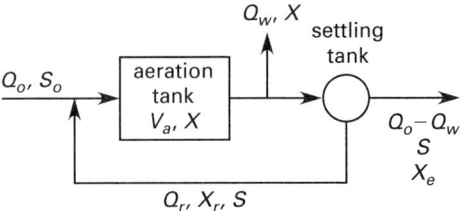

(a) plug flow reactor with recycle

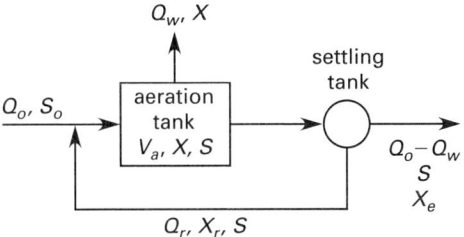

(b) continuously stirred tank reactor
with recycle

will be approximately the same as in the effluent from the reactor.

$$\frac{1}{\theta_c} = \frac{\mu_m(S_o - S)}{S_o - S + (1 + R)K_s \ln \dfrac{S_i}{S}} - k_d \quad \text{[PFR only]}$$

$$27.13$$

The *maximum specific growth rate*, μ_m, is calculated from the first-order reaction constant, k.

$$\mu_m = kY \qquad 27.14$$

The influent concentration after dilution with recycle flow is

$$S_i = \frac{S_o + RS}{1 + R} \quad \text{[PFR only]} \qquad 27.15$$

The *rate of substrate utilization* as the waste passes through the aerating reactor is

$$r_{su} = \frac{-\mu_m S X}{Y(K_s + S)} \quad \text{[PFR only]} \qquad 27.16$$

For the CSTR, the *specific substrate utilization* (*specific utilization*), U, is the food-to-microorganism ratio multiplied by the fractional process efficiency.

$$U = \eta(\text{F:M}) = \frac{-r_{su}}{X}$$

$$= \frac{S_o - S}{\theta X} \qquad 27.17$$

For the CSTR, the equivalent relationship between mean cell residence time, food-to-microorganism ratio, and specific utilization rate, U, is given by Eq. 27.18.

$$\frac{1}{\theta_c} = Y(\text{F:M})\eta - k_d$$

$$= -Y\left(\frac{r_{su}}{X}\right) - k_d$$

$$= YU - k_d \quad \text{[CSTR only]} \qquad 27.18$$

Y and k_d are the kinetic coefficients describing the bacterial growth process. Y is the *yield coefficient*, the ratio of the mass of cells formed to the mass of substrate (food) consumed, in mg/mg. k_d is the *endogenous decay coefficient* (rate) of the microorganisms in day^{-1}. For a specified waste, biological community, and set of environmental conditions, Y and k_d are fixed. Equation 27.18 is a linear equation of the general form $y = mx + b$. Figure 27.4 illustrates how these coefficients can be determined graphically from bench-scale testing.

For a continuously stirred tank reactor, the kinetic model predicts the effluent substrate concentration.

$$S = \frac{K_s(1 + k_d\theta_c)}{\theta_c(\mu_m - k_d) - 1} \quad \text{[CSTR only]} \qquad 27.19$$

The *hydraulic retention time*, θ, for the reactor is

$$\theta = \frac{V_a}{Q_o} \quad \text{[PFR and CSTR]} \qquad 27.20$$

The *hydraulic retention time*, θ_s, for the system is

$$\theta_s = \frac{V_a + V_s}{Q_o} \quad \text{[PFR and CSTR]} \qquad 27.21$$

Figure 27.4 Kinetic Growth Constants*

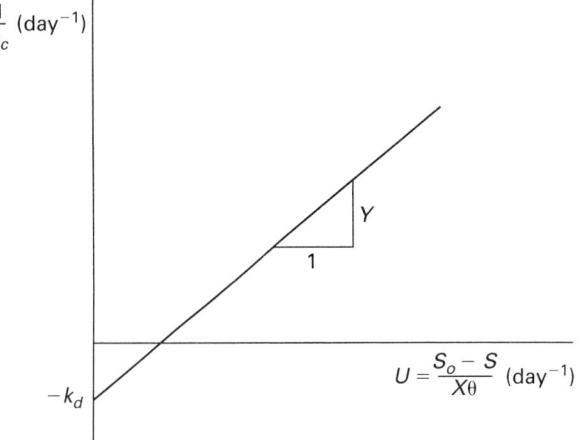

*As determined from bench-scale activated sludge studies using a continuous flow stirred tank reactor without recycle

The *reactor volume*, V_a, is

$$V_a = \theta Q_o$$
$$= \frac{\theta_c Q_o Y (S_o - S)}{X(1 + k_d \theta_c)} \quad \text{[PFR and CSTR]} \qquad 27.22$$

The average concentration of microorganisms, X, in the reactor is

$$X = \frac{\left(\dfrac{\theta_c}{\theta}\right) Y (S_o - S)}{1 + k_d \theta_c} \quad \text{[PFR and CSTR]} \qquad 27.23$$

The *observed yield* is

$$Y_{\text{obs}} = \frac{Y}{1 + k_d \theta_c} \quad \text{[PFR and CSTR]} \qquad 27.24$$

The volatile portion of the mass of activated sludge that must be wasted each day is given by Eq. 27.25. The total (fixed and volatile) mass of sludge wasted each day must be calculated from the ratio of MLVSS and MLSS.

$$P_{x,\text{kg/d}} = \frac{Q_{w,\text{m}^3/\text{d}} X_{r,\text{mg/L}}}{1000 \; \dfrac{\text{g}}{\text{kg}}}$$

$$= \frac{Y_{\text{obs,mg/mg}} Q_{o,\text{m}^3/\text{d}} (S_o - S)_{\text{mg/L}}}{1000 \; \dfrac{\text{g}}{\text{kg}}}$$

$$\text{[SI; PFR and CSTR]} \qquad 27.25(a)$$

$$P_{x,\text{lbm/day}} = Q_{w,\text{MGD}} X_{r,\text{mg/L}} \left(8.345 \; \frac{\text{lbm-L}}{\text{MG-mg}}\right)$$

$$= Y_{\text{obs,lbm/lbm}} Q_{o,\text{MGD}} (S_o - S)_{\text{mg/L}}$$

$$\times \left(8.345 \; \frac{\text{lbm-L}}{\text{MG-mg}}\right)$$

$$\text{[U.S.; PFR and CSTR]} \qquad 27.25(b)$$

The sludge that must be wasted is reduced by the sludge that escapes in the clarifier effluent, which is ideally zero. The actual sludge solids escaping can be calculated as

$$m_{e,\text{solids}} = Q_e X_e \quad \text{[PFR and CSTR]} \qquad 27.26$$

The *mean cell residence time* given in Eq. 27.5 can be calculated from Eq. 27.27. The sludge wasting rate, Q_w, can also be found from this equation. In most activated sludge processes, waste sludge is drawn from the sludge recycle line. Therefore, Q_w is the *cell wastage rate* from the recycle line, and X_r is the microorganism concentration in the return sludge line.

$$\theta_c = \frac{V_a X}{Q_w X_r + Q_e X_e}$$
$$= \frac{V_a X}{Q_w X_r + (Q_o - Q_w) X_e} \quad \text{[PFR and CSTR]} \qquad 27.27$$

Since the average mass of cells in the effluent, X_e, is very small, Eq. 27.28 gives the mean cell residence time for sludge wastage from the return line. Centrifuge testing can be used to determine the ratio of X/X_r. (There is another reason for omitting X_e from the calculation: It may be too small to measure conveniently.)

$$\theta_c \approx \frac{V_a X}{Q_w X_r} \quad \text{[PFR and CSTR]} \qquad 27.28$$

For wastage directly from the aeration basin, $X_r = X$. In this case, Eq. 27.29 shows that the process can be controlled by wasting a flow equal to the volume of the aeration tank divided by the mean cell residence time.

$$\theta_c \approx \frac{V_a}{Q_w} \quad \text{[CSTR]} \qquad 27.29$$

9. ATMOSPHERIC AIR

Atmospheric air is a mixture of oxygen, nitrogen, and small amounts of carbon dioxide, water vapor, argon, and other inert gases. If all constituents except oxygen are grouped with the nitrogen, the air composition is as given in Table 27.5. It is necessary to supply by weight $1/0.2315 = 4.32$ masses of air to obtain one mass of oxygen. The average molecular weight of air is 28.9.

Table 27.5 Composition of Air[a]

component	fraction by mass	fraction by volume
oxygen	0.2315	0.209
nitrogen	0.7685	0.791
ratio of nitrogen to oxygen	3.320	3.773[b]
ratio of air to oxygen	4.320	4.773

[a] Inert gases and CO_2 are included in N_2.
[b] This value is also reported by various sources as 3.76, 3.78, and 3.784.

10. AERATION TANKS

The *aeration period* is the same as the *hydraulic detention time* and is calculated without considering recirculation.

$$\theta = \frac{V_a}{Q_o} \qquad 27.30$$

The organic (BOD) *volumetric loading* rate (with units of kg $BOD_5/\text{d·m}^3$) for the aeration tank is given by Eq. 27.31. In the United States, volumetric loading is often specified in lbm/day-1000 ft³.

$$L_{\text{BOD}} = \frac{S_o Q_o}{V_a \left(1000 \; \dfrac{\text{mg·m}^3}{\text{kg·L}}\right)} \quad \text{[SI]} \qquad 27.31(a)$$

$$L_{\text{BOD}} = \frac{S_{o,\text{mg/L}} Q_{o,\text{MGD}} \left(8.345 \dfrac{\text{lbm-L}}{\text{MG-mg}} \right)(1000)}{V_{a,\text{ft}^3}}$$

[U.S.]   27.31(b)

The *rate of oxygen transfer* from the air to the mixed liquor during aeration is given by Eq. 27.32. K_t is a macroscopic transfer coefficient that depends on the equipment and characteristics of the mixed liquor and that has typical dimensions of 1/time. The dissolved oxygen deficit, D in mg/L, is given by Eq. 27.33. β is the mixed liquor's *oxygen saturation coefficient*, approximately 0.8 to 0.9. β corrects for the fact that the mixed liquor does not absorb as much oxygen as pure water does. The minimum dissolved oxygen content is approximately 0.5 mg/L, below which the processing would be limited by oxygen. However, a lower limit of 2 mg/L is typically specified.

$$\dot{m}_{\text{oxygen}} = K_t D \qquad\qquad 27.32$$

$$D = \beta \text{DO}_{\text{saturated water}} - \text{DO}_{\text{mixed liquor}} \qquad 27.33$$

The *oxygen demand* is given by Eq. 27.34. The factor 1.42 is the theoretical gravimetric ratio of oxygen required for carbonaceous organic material based on an ideal stoichiometric reaction. Equation 27.34 neglects nitrogenous demand, which can also be significant.

$$\dot{m}_{\text{oxygen,kg/d}} = \frac{Q_{o,\text{m}^3/\text{d}}(S_o - S)_{\text{mg/L}}}{f \left(1000 \dfrac{\text{mg}\cdot\text{m}^3}{\text{kg}\cdot\text{L}} \right)} - 1.42 P_{x,\text{kg/d}}$$

[SI]   27.34(a)

$$\dot{m}_{\text{oxygen,lbm/day}} = \frac{Q_{o,\text{MGD}}(S_o - S)_{\text{mg/L}} \left(8.345 \dfrac{\text{lbm-L}}{\text{MG-mg}} \right)}{f}$$
$$- 1.42 P_{x,\text{lbm/day}}$$

[U.S.]   27.34(b)

Air is approximately 23.2% oxygen by mass. Considering a *transfer efficiency* of η_{transfer} (typically less than 10%), the air requirement is calculated from the oxygen demand.

$$\dot{m}_{\text{air}} = \frac{\dot{m}_{\text{oxygen}}}{0.232 \eta_{\text{transfer}}} \qquad\qquad 27.35$$

The volume of air required is given by Eq. 27.36. The density of air is approximately 0.075 lbm/ft^3 (1.2 kg/m^3). Typical volumes for conventional aeration are 500–900 ft^3/lbm BOD$_5$ (30–55 m^3/kg BOD$_5$) for F:M ratios greater than 0.3, and 1200–1800 ft^3/lbm BOD$_5$ (75–115 m^3/kg BOD$_5$) for F:M ratios less than 0.3 day^{-1}. Air flows in SCFM (*standard cubic feet per minute*) are based on a temperature of 70°F (21°C) and pressure of 14.7 psia (101 kPa).

$$\dot{V}_{\text{air}} = \frac{\dot{m}_{\text{air}}}{\rho_{\text{air}}} \quad \text{[PFR and CSTR]} \qquad 27.36$$

Aeration equipment should be designed with an excess-capacity safety factor of 1.5–2.0. *Ten States' Standards* requires 200% of calculated capacity for air diffusion systems, which is assumed to be 1500 ft^3/lbm BOD$_5$ (94 m^3/kg BOD$_5$) [TSS Sec. 92.332].

The air requirement per unit volume (m^3/m^3 or ft^3/MG, calculated using consistent units) is V_{air}/Q_o. The air requirement in cubic meters per kilogram of soluble BOD$_5$ removed is

$$V_{\text{air,m}^3/\text{kg BOD}} = \frac{\dot{V}_{\text{air,m}^3/\text{d}} \left(1000 \dfrac{\text{mg}\cdot\text{m}^3}{\text{kg}\cdot\text{L}} \right)}{Q_{o,\text{m}^3/\text{d}}(S_o - S)_{\text{mg/L}}}$$

[SI]   27.37(a)

$$V_{\text{air,ft}^3/\text{lbm BOD}} = \frac{\dot{V}_{\text{air,ft}^3/\text{day}}}{Q_{o,\text{MGD}}(S_o - S)_{\text{mg/L}} \left(8.345 \dfrac{\text{lbm-L}}{\text{MG-mg}} \right)}$$

[U.S.]   27.37(b)

Example 27.1

Mixed liquor ($\beta = 0.9$, DO = 3 mg/L) is aerated at 20°C. The transfer coefficient, K_t, is 2.7 h^{-1}. What is the rate of oxygen transfer?

Solution

At 20°C, the saturated oxygen content of pure water is 9.2 mg/L. The oxygen deficit is found from Eq. 27.33.

$$D = \beta \text{DO}_{\text{saturated water}} - \text{DO}_{\text{mixed liquor}}$$
$$= (0.9)\left(9.17 \frac{\text{mg}}{\text{L}} \right) - 3 \frac{\text{mg}}{\text{L}}$$
$$= 5.25 \text{ mg/L}$$

The rate of oxygen transfer is found from Eq. 27.32.

$$\dot{m} = K_t D = (2.7 \text{ h}^{-1})\left(5.25 \frac{\text{mg}}{\text{L}} \right)$$
$$= 14.18 \text{ mg/L}\cdot\text{h}$$

11. AERATION POWER AND COST

The work required to compress a unit mass of air from atmospheric pressure to the discharge pressure depends on the nature of the compression. Various assumptions can be made, but generally an *isentropic compression* is assumed. Inefficiencies (thermodynamic and mechanical) are combined into a single *compressor efficiency*, η_c, which is essentially an *isentropic efficiency* for the compression process.

Water Treatment

The ideal power, P, to compress a volumetric flow rate, $\dot{V}$, or mass flow rate, $\dot{m}$, of air from pressure p_1 to pressure p_2 in an isentropic steady-flow compression process is given by Eq. 27.38. k is the ratio of specific heats, which has an approximate value of 1.4. c_p is the specific heat at constant pressure, which has a value of 0.24 Btu/lbm-°R (1005 J/kg·K). R_{air} is the specific gas constant, 53.3 ft-lbf/lbm-°R (287 J/kg·K). In Eq. 27.38, temperatures must be absolute. Volumetric and mass flow rates are per second.

$$P_{\text{ideal,kW}} = -\left(\frac{k p_1 \dot{V}_1}{(k-1)\left(1000\ \frac{W}{kW}\right)}\right)\left(1-\left(\frac{p_2}{p_1}\right)^{(k-1)/k}\right)$$

$$= -\left(\frac{k \dot{m} R_{air} T_1}{(k-1)\left(1000\ \frac{W}{kW}\right)}\right)\left(1-\left(\frac{p_2}{p_1}\right)^{(k-1)/k}\right)$$

$$= -\left(\frac{c_p \dot{m} T_1}{1000\ \frac{W}{kW}}\right)\left(1-\left(\frac{p_2}{p_1}\right)^{(k-1)/1}\right)$$

$$[\text{SI}] \quad 27.38(a)$$

$$P_{\text{ideal,hp}} = -\left(\frac{k p_1 \dot{V}_1}{(k-1)\left(550\ \frac{ft\text{-}lbf}{hp\text{-}sec}\right)}\right)\left(1-\left(\frac{p_2}{p_1}\right)^{(k-1)/k}\right)$$

$$= -\left(\frac{k \dot{m} R_{air} T_1}{(k-1)\left(550\ \frac{ft\text{-}lbf}{hp\text{-}sec}\right)}\right)\left(1-\left(\frac{p_2}{p_1}\right)^{(k-1)/k}\right)$$

$$= -\left(\frac{c_p J \dot{m} T_1}{550\ \frac{ft\text{-}lbf}{hp\text{-}sec}}\right)\left(1-\left(\frac{p_2}{p_1}\right)^{(k-1)/k}\right)$$

$$[\text{U.S.}] \quad 27.38(b)$$

The actual compression power is

$$P_{\text{actual}} = \frac{P_{\text{ideal}}}{\eta_c} \qquad 27.39$$

The cost of running the compressor for a duration, t, is

$$\text{total cost} = C_{\text{kW-hr}} P_{\text{actual}} t \qquad 27.40$$

12. RECYCLE RATIO AND RECIRCULATION RATE

For any given wastewater and treatment facility, the influent flow rate, BOD, and tank volume cannot be changed. Of all the variables appearing in Eq. 27.3, only

the suspended solids can be controlled. Therefore, the primary process control variable is the amount of organic material in the wastewater. This is controlled by the sludge return rate. Equation 27.41 gives the sludge *recycle ratio*, R, which is typically 0.20–0.30. (Some authorities use the symbol α to represent the recycle ratio.)

$$R = \frac{Q_r}{Q_o} \qquad 27.41$$

The actual *recirculation rate* can be found by writing a suspended solids mass balance around the inlet to the reactor and solving for the recirculation rate, Q_r, noting $X_o = 0$.

$$X_o Q_o + X_r Q_r = X(Q_o + Q_r) \qquad 27.42$$

$$\frac{Q_r}{Q_o + Q_r} = \frac{X}{X_r} \qquad 27.43$$

The theoretical required return rate for the current MLSS content can be calculated from the *sludge volume index* (SVI; see Sec. 27.5). Figure 27.5 shows the theoretical relationship between the SVI test and the return rate. Equation 27.44 assumes that the settling tank responds identically to the graduated 1000 mL cylinder used in the SVI test. Equation 27.43 and Eq. 27.44 are analogous.

$$\frac{Q_r}{Q_o + Q_r} = \frac{V_{\text{settled,mL/L}}}{1000\ \frac{mL}{L}} \qquad 27.44$$

$$R = \frac{V_{\text{settled,mL/L}}}{1000\ \frac{mL}{L} - V_{\text{settled,mL/L}}}$$

$$= \frac{1}{\dfrac{10^6}{(\text{SVI}_{\text{mL/g}})(\text{MLSS}_{\text{mg/L}})} - 1} \qquad 27.45$$

Equation 27.45 is based on settling and represents a theoretical value, but it does not necessarily correspond to the actual recirculation rate. Greater-than-necessary recirculation returns clarified liquid unnecessarily, and less recirculation leaves settled solids in the clarifier.

Figure 27.5 *Theoretical Relationship Between SVI Test and Return Rate*

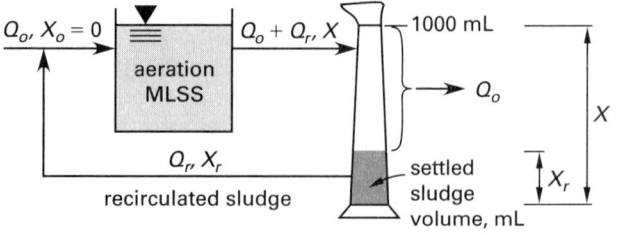

Example 27.2

Two 1 L samples of mixed liquor are taken from an aerating lagoon. After settling for 30 min in a graduated cylinder, 250 mL of solids have settled out in the first sample. The total suspended solids concentration in the second sample is found to be 2300 mg/L. (a) What is the sludge volume index? (b) What is the theoretical required sludge recycle rate?

Solution

(a) Use Eq. 27.7.

$$\text{SVI} = \frac{\left(1000 \, \frac{\text{mg}}{\text{g}}\right) V_{\text{settled,mL/L}}}{\text{MLSS}_{\text{mg/L}}}$$

$$= \frac{\left(1000 \, \frac{\text{mg}}{\text{g}}\right)\left(250 \, \frac{\text{mL}}{\text{L}}\right)}{2300 \, \frac{\text{mg}}{\text{L}}}$$

$$= 109 \, \text{mL/g}$$

(b) The theoretical required recycle rate is given by Eq. 27.45.

$$R = \frac{V_{\text{settled,mL/L}}}{1000 \, \frac{\text{mL}}{\text{L}} - V_{\text{settled,mL/L}}}$$

$$= \frac{250 \, \frac{\text{mL}}{\text{L}}}{1000 \, \frac{\text{mL}}{\text{L}} - 250 \, \frac{\text{mL}}{\text{L}}}$$

$$= 0.33$$

Example 27.3

An activated sludge plant receives 4.0 MGD of wastewater with a BOD of 200 mg/L. The primary clarifier removes 30% of the BOD. The aeration-settle-waste cycle takes 6 hr. The oxygen transfer efficiency is 6%, and one pound of oxygen is required for each pound of BOD oxidized. The food-to-microorganism ratio based on MLSS is 0.33. The SVI is 100. 30% of the MLSS is wasted each day from the aeration tank. The MLVSS/MLSS ratio is 0.7. The surface settling rate of the secondary clarifier is 800 gal/day-ft². The overflow rate in the secondary clarifier is 800 gal/day-ft². The final effluent has a BOD of 10 mg/L.

Calculate the (a) secondary BOD removal fraction, (b) aeration tank volume, (c) MLSS, (d) total suspended solids content in the recirculated sludge, (e) theoretical recirculation rate, (f) mass of MLSS wasted each day, (g) clarifier surface area, and (h) volumetric air requirements.

Solution

(a) Wastewater leaves primary settling with a BOD of $(1 - 0.3)(200 \, \text{mg/L}) = 140 \, \text{mg/L}$. The efficiency of the activated sludge processing is

$$\eta_{\text{BOD}} = \frac{140 \, \frac{\text{mg}}{\text{L}} - 10 \, \frac{\text{mg}}{\text{L}}}{140 \, \frac{\text{mg}}{\text{L}}}$$

$$= 0.929 \quad (92.9\%)$$

(b) The aeration tank volume is

$$V_a = Q_o t = \frac{(4 \, \text{MGD})\left(10^6 \, \frac{\text{gal}}{\text{MG}}\right)(6 \, \text{hr})}{\left(24 \, \frac{\text{hr}}{\text{day}}\right)\left(7.48 \, \frac{\text{gal}}{\text{ft}^3}\right)}$$

$$= 1.34 \times 10^5 \, \text{ft}^3$$

(c) The total MLSS can be determined from the food-to-microorganism ratio given. Use Eq. 27.4.

$$\text{MLSS} = \frac{S_o Q_o}{(\text{F:M}) V_a}$$

$$= \frac{\left(140 \, \frac{\text{mg}}{\text{L}}\right)(4 \, \text{MGD})\left(10^6 \, \frac{\text{gal}}{\text{MG}}\right)}{\left(0.33 \, \frac{\text{lbm}}{\text{lbm-day}}\right)(1.34 \times 10^5 \, \text{ft}^3)\left(7.48 \, \frac{\text{gal}}{\text{ft}^3}\right)}$$

$$= 1693 \, \text{mg/L}$$

(d) The suspended solids content in the recirculated sludge is given by Eq. 27.9.

$$\text{TSS}_{\text{mg/L}} = \frac{\left(1000 \, \frac{\text{mg}}{\text{g}}\right)\left(1000 \, \frac{\text{mL}}{\text{L}}\right)}{\text{SVI}_{\text{mL/g}}}$$

$$= \frac{\left(1000 \, \frac{\text{mg}}{\text{g}}\right)\left(1000 \, \frac{\text{mL}}{\text{L}}\right)}{100 \, \frac{\text{mL}}{\text{g}}}$$

$$= 10{,}000 \, \text{mg/L}$$

(e) The recirculation ratio can be calculated from the SVI test. However, the settled volume must first be calculated.

Use Eq. 27.7.

$$V_{\text{settled,mL/L}} = \frac{\text{MLSS}_{\text{mg/L}}\text{SVI}_{\text{mL/g}}}{1000 \frac{\text{mg}}{\text{g}}}$$

$$= \frac{\left(1693 \frac{\text{mg}}{\text{L}}\right)\left(100 \frac{\text{mL}}{\text{g}}\right)}{1000 \frac{\text{mg}}{\text{g}}}$$

$$= 169.3 \text{ mL/L}$$

Use Eq. 27.44.

$$R = \frac{V_{\text{settled,mL/L}}}{1000 \frac{\text{mL}}{\text{L}} - V_{\text{settled,mL/L}}}$$

$$= \frac{169.3 \frac{\text{mL}}{\text{L}}}{1000 \frac{\text{mL}}{\text{L}} - 169.3 \frac{\text{mL}}{\text{L}}}$$

$$= 0.204$$

The recirculation rate is given by Eq. 27.41.

$$Q_r = RQ_o = (0.204)(4 \text{ MGD}) = 0.816 \text{ MGD}$$

(f) The sludge wasted is

$$(0.30)(4 \text{ MGD})\left(1693 \frac{\text{mg}}{\text{L}}\right)\left(8.345 \frac{\text{lbm-L}}{\text{mg-MG}}\right)$$

$$= 16{,}954 \text{ lbm/day}$$

(g) The recirculated flow is removed by the bottom drain, so it does not influence the surface area. The clarifier surface area is

$$A = \frac{Q_o}{\text{v}^*}$$

$$= \frac{(4 \text{ MGD})\left(10^6 \frac{\text{gal}}{\text{MG}}\right)}{800 \frac{\text{gal}}{\text{day-ft}^2}}$$

$$= 5000 \text{ ft}^2$$

(h) From part (a), the fraction of BOD reduction in the activated sludge process aeration (exclusive of the wasted sludge) is 0.929.

Since one pound of oxygen is required for each pound of BOD removed, the required oxygen mass is

$$m_{\text{oxygen}} = (0.929)(4 \text{ MGD})\left(140 \frac{\text{mg}}{\text{L}}\right)\left(8.345 \frac{\text{lbm-L}}{\text{mg-MG}}\right)$$

$$= 4341 \text{ lbm/day}$$

The density of air is approximately 0.075 lbm/ft^3, and air is 20.9% oxygen by volume. Considering the efficiency of the oxygenation process, the air required is

$$\dot{V}_{\text{air}} = \frac{4341 \frac{\text{lbm}}{\text{day}}}{(0.209)(0.06)\left(0.075 \frac{\text{lbm}}{\text{ft}^3}\right)}$$

$$= 4.62 \times 10^6 \text{ ft}^3/\text{day}$$

13. SLUDGE WASTING

The activated sludge process is a biological process. Influent brings in BOD ("biomass" or organic food), and the return activated sludge (RAS) brings in microorganisms (bacteria) to digest the BOD. The food-to-microorganism ratio (F:M) is balanced by adjusting the aeration and how much sludge is recirculated and how much is wasted. As the microorganisms grow and are mixed with air, they clump together (flocculate) and are readily settled as sludge in the secondary clarifier. More sludge is produced in the secondary clarifier than is needed, and without wasting, the sludge would continue to build up.

Sludge wasting is the main control of the effluent quality and microorganism population size. The wasting rate affects the mixed liquor suspended solids (MLSS) concentration and the mean cell residence time. Excess sludge may be wasted either from the sludge return line or directly from the aeration tank as mixed liquor. Wasting from the aeration tank is preferred as the sludge concentration is fairly steady in that case. The wasted volume is normally 40% to 60% of the wastewater flow. When wasting is properly adjusted, the MLSS level will remain steady. With an excessive F:M ratio, sludge production will increase, and eventually, sludge will appear in the clarifier effluent.

14. OPERATIONAL DIFFICULTIES

Sludge bulking refers to a condition in which the sludge does not settle out. Since the solids do not settle, they leave the sedimentation tank and cause problems in subsequent processes. The sludge volume index can often (but not always) be used as a measure of settling characteristics. If the SVI is less than 100, the settling process is probably operating satisfactorily. If SVI is greater than 150, the sludge is bulking. Remedies include addition of lime, chlorination, additional aeration, and a reduction in MLSS.

Some sludge may float back to the surface after settling, a condition known as *rising sludge*. Rising sludge occurs when nitrogen gas is produced from denitrification of the nitrates and nitrites. Remedies include increasing the return sludge rate and decreasing the mean cell residence time. Increasing the speed of the sludge scraper mechanism to dislodge nitrogen bubbles may also help.

It is generally held that after 30 min of settling, the sludge should settle to between 20% and 70% of its original volume. If it occupies more than 70%, there are too many solids in the aeration basin, and more sludge should be wasted. If the sludge occupies less than 20%, less sludge should be wasted.

Sludge washout (*solids washout*) can occur during a period of peak flow or even with excess recirculation. Washout is the loss of solids from the sludge blanket in the settling tank. This often happens with insufficient wasting, such that solids build up (to more than 70% in a settling test), hinder settling in the clarifier, and cannot be contained.

15. QUANTITIES OF SLUDGE

The procedure for determining the volume and mass of sludge from activated sludge processing is essentially the same as for sludge from sedimentation and any other processes. Table 27.6 gives characteristics of sludge. The primary variables, other than mass and volume, are density and/or specific gravity.

$$m_{\text{wet}} = V\rho_{\text{sludge}} = V(\text{SG}_{\text{sludge}})\rho_{\text{water}} \qquad 27.46$$

Raw sludge is approximately 95–99% water, and the specific gravity of sludge is only slightly greater than 1.0. The actual specific gravity of sludge can be calculated from the fractional solids content, s, and specific gravity of the sludge solids, which is approximately 2.5. (If the sludge specific gravity is already known, Eq. 27.47 can be solved for the specific gravity of the solids.) In Eq. 27.47, $1 - s$ is the fraction of the sludge that is water.

$$\frac{1}{\text{SG}_{\text{sludge}}} = \frac{1-s}{1} + \frac{s}{\text{SG}_{\text{solids}}}$$

$$= \frac{1 - s_{\text{fixed}} - s_{\text{volatile}}}{1} + \frac{s_{\text{fixed}}}{\text{SG}_{\text{fixed solids}}}$$

$$+ \frac{s_{\text{volatile}}}{\text{SG}_{\text{volatile solids}}} \qquad 27.47$$

The volume of sludge can be estimated from its dried mass. Since the sludge specific gravity is approximately 1.0, the sludge volume is

$$V_{\text{sludge,wet}} = \frac{m_{\text{dried}}}{s\rho_{\text{sludge}}} \approx \frac{m_{\text{dried}}}{s\rho_{\text{water}}} \qquad 27.48$$

The dried mass of sludge solids from primary settling basins is easily determined from the decrease in solids. The decrease in suspended solids, ΔSS, due to primary settling is approximately 50% of the total incoming suspended solids.

$$m_{\text{dried,kg/d}} = \frac{(\Delta\text{SS})_{\text{mg/L}} Q_{o,\text{m}^3/\text{d}}}{1000 \ \frac{\text{mg}\cdot\text{m}^3}{\text{kg}\cdot\text{L}}} \qquad \text{[SI]} \qquad 27.49(a)$$

$$m_{\text{dried,lbm/day}} = (\Delta\text{SS})_{\text{mg/L}} Q_{o,\text{MGD}} \left(8.345 \ \frac{\text{lbm-L}}{\text{MG-mg}} \right)$$
$$\text{[U.S.]} \qquad 27.49(b)$$

The dried mass of solids for biological filters and secondary aeration (e.g., activated sludge) can be estimated on a macroscopic basis as a fraction of the change in BOD. In Eq. 27.50, K is the *cell yield*, also known as the *removal efficiency*, the fraction of the total influent BOD that ultimately appears as excess (i.e., settled) biological solids. The cell yield can be estimated from the food-to-microorganism ratio from Fig. 27.6. (The cell yield, K, and the *yield coefficient*, also known as the *sludge yield* or *biomass yield*, Y_{obs}, are not the same, because K is based on total incoming BOD and $Y_{\text{obs,lbm/day}}$ is based on consumed BOD. The term "cell yield" is often used for both Y_{obs} and K.)

$$m_{\text{dried,kg/d}} = \frac{K S_{o,\text{mg/L}} Q_{o,\text{m}^3/\text{d}}}{1000 \ \frac{\text{mg}\cdot\text{m}^3}{\text{kg}\cdot\text{L}}}$$

$$= \frac{Y_{\text{obs}}(S_o - S)_{\text{mg/L}} Q_{o,\text{m}^3/\text{d}}}{1000 \ \frac{\text{mg}\cdot\text{m}^3}{\text{kg}\cdot\text{L}}} \qquad \text{[SI]} \qquad 27.50(a)$$

$$m_{\text{dried,lbm/day}} = K S_{o,\text{mg/L}} Q_{o,\text{MGD}} \left(8.345 \ \frac{\text{lbm-L}}{\text{MG-mg}} \right)$$

$$= Y_{\text{obs}}(S_o - S)_{\text{mg/L}} Q_{o,\text{MGD}}$$

$$\times \left(8.345 \ \frac{\text{lbm-L}}{\text{MG-mg}} \right)$$
$$\text{[U.S.]} \qquad 27.50(b)$$

Table 27.6 *Characteristics of Sludge*

origin of sludge	fraction solids, s	dry mass (lbm/day-person)
primary settling tank	0.06–0.08	0.12
trickling filter	0.04–0.06	0.04
mixed primary settling and trickling filter	0.05	0.16
conventional activated sludge	0.005–0.01	0.07
mixed primary and conventional activated sludge	0.02–0.03	0.19
high-rate activated sludge	0.025–0.05	0.06
mixed primary settling and high-rate activated sludge	0.05	0.18
extended aeration activated sludge	0.02	0.02
filter backwashing water	0.01–0.1	n.a.
softening sludge	0.03–0.15	n.a.

(Multiply lbm/day-person by 0.45 to obtain kg/d·person.)

Water Treatment

Example 27.4

A trickling filter plant processes 4 MGD of domestic wastewater with 190 mg/L BOD and 230 mg/L suspended solids. The primary sedimentation tank removes 50% of the suspended solids and 30% of the BOD. The sludge yield for the trickling filter is 0.25. What is the total sludge volume from the primary sedimentation tank and trickling filter if the combined solids content of the sludge is 5%?

Solution

The mass of solids removed in primary settling is given by Eq. 27.49.

$$m_{\text{dried}} = (\Delta SS)_{\text{mg/L}} Q_{o,\text{MGD}} \left(8.345 \frac{\text{lbm-L}}{\text{MG-mg}}\right)$$

$$= (0.5)\left(230 \frac{\text{mg}}{\text{L}}\right)(4 \text{ MGD})\left(8.345 \frac{\text{lbm-L}}{\text{MG-mg}}\right)$$

$$= 3839 \text{ lbm/day}$$

The mass of solids removed from the trickling filter's clarifier is calculated from Eq. 27.50.

$$m_{\text{dried}} = K\Delta S_{\text{mg/L}} Q_{o,\text{MGD}} \left(8.345 \frac{\text{lbm-L}}{\text{MG-mg}}\right)$$

$$= (0.25)(1 - 0.3)\left(190 \frac{\text{mg}}{\text{L}}\right)(4 \text{ MGD})$$

$$\times \left(8.345 \frac{\text{lbm-L}}{\text{MG-mg}}\right)$$

$$= 1110 \text{ lbm/day}$$

The volume is calculated from Eq. 27.48.

$$V = \frac{m_{\text{dried}}}{s\rho_{\text{water}}}$$

$$= \frac{3839 \frac{\text{lbm}}{\text{day}} + 1110 \frac{\text{lbm}}{\text{day}}}{(0.05)\left(62.4 \frac{\text{lbm}}{\text{ft}^3}\right)}$$

$$= 1586 \text{ ft}^3/\text{day}$$

16. SLUDGE THICKENING

Waste-activated sludge (WAS) has a typical solids content of 0.5–1.0%. *Thickening* of sludge is used to reduce the volume of sludge prior to digestion or dewatering. Thickening is accomplished by decreasing the liquid fraction $(1 - s)$, thus increasing the solids fraction, s. Equation 27.48 shows that the volume of wet sludge is inversely proportional to its solids content.

For dewatering, thickening to at least 4% solids (i.e., 96% moisture) is required; for digestion, thickening to at least 5% solids is required. Depending on the nature of the sludge, polymers may be used with all of the thickening methods. Other chemicals (e.g., lime for

Figure 27.6 Approximate Food:Microorganism Ratio Versus Cell Yield,* K

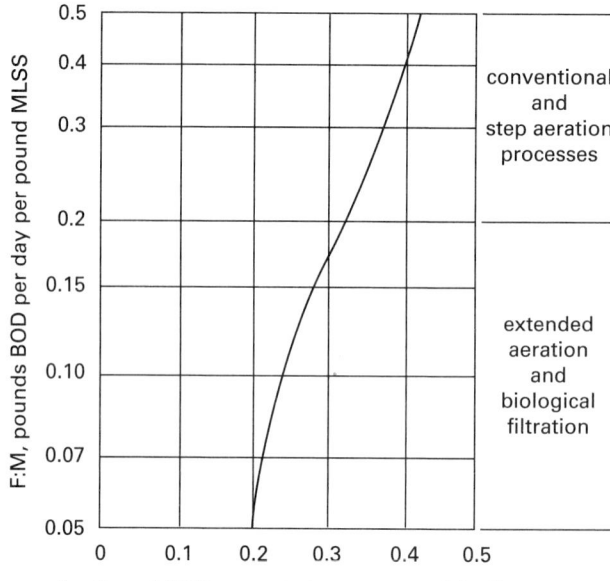

*calculated assuming an effluent BOD of approximately 30 mg/L

Hammer, Mark J., *Water and Wastewater Technology*, 3rd Edition, © 1996. Printed and electronically reproduced by permission of Pearson Education, Inc., Upper Saddle River, NJ.

stabilization through pH control and potassium permanganate to react with sulfides) can also be used. However, because of cost and increased disposal problems, the trend is toward the use of fewer inorganic chemicals with new thickening or dewatering applications unless special conditions apply.

Gravity thickening occurs in circular sedimentation tanks similar to primary and secondary clarifiers. The settling process is categorized into four zones: the *clarification zone*, containing relatively clear *supernatant*; the *hindered settling zone*, where the solids move downward at essentially a constant rate; the *transition zone*, characterized by a decrease in solids settling rate; and the *compression zone*, where motion is essentially zero.

In *batch gravity thickening*, the tank is filled with thin sludge and allowed to stand. Supernatant is decanted, and the tank is topped off with more thin sludge. The operation is repeated continually or a number of times before the underflow sludge is removed. A heavy-duty (deep-truss) scraper mechanism pushes the settled solids into a hopper in the tank bottom, and the clarified effluent is removed in a peripheral weir. A doubling of solids content is usually possible with gravity thickening. Table 27.7 contains typical characteristics of gravity thickening tanks.

With *dissolved air flotation* (DAF) *thickening* (DAFT), fine air bubbles are released into the sludge as it enters the DAF tank. The solids particles adhere to the air bubbles and float to the surface where they are skimmed away as scum. The scum has a solids content of

Table 27.7 Typical Characteristics of Gravity Thickening Tanks

shape	circular
minimum number of tanks[*]	2
overflow rate	600–800 gal/day-ft^2 (24–33 m^3/d·m^2)
maximum dry solids loading	
primary sludge	22 lbm/day-ft^2 (107 kg/d·m^2)
primary and trickling filter sludge	15 lbm/day-ft^2 (73 kg/d·m^2)
primary and modified aeration activated sludge	12 lbm/day-ft^2 (59 kg/d·m^2)
primary and conventional aeration activated sludge	8 lbm/day-ft^2 (39 mg/d·m^2)
waste-activated sludge	4 lbm/day-ft^2 (20 kg/d·m^2)
minimum detention time	6 hr
minimum sidewater depth	10 ft (3 m)
minimum freeboard	1.5 ft (0.45 m)

(Multiply gal/day-ft^2 by 0.0407 to obtain m^3/d·m^2.)
(Multiply lbm/day-ft^2 by 4.882 to obtain kg/d·m^2.)
(Multiply ft by 0.3048 to obtain m.)
[*]unless alternative methods of thickening are available

approximately 4%. Up to 85% of the total solids may be recovered in this manner, and chemical flocculants (e.g., polymers) can increase the recovery to 97%–99%.

With *gravity belt thickening* (GBT), sludge is spread over a drainage belt. Multiple stationary plows in the path of the moving belt may be used to split, turn, and recombine the sludge so that water does not pool on top of the sludge. After thickening, the sludge cake is removed from the belt by a doctor blade.

Gravity thickening is usually best for sludges from primary and secondary settling tanks, while dissolved air flotation and centrifugal thickening are better suited for activated sludge. The current trend is toward using gravity thickening for primary sludge and flotation thickening for activated sludge, then blending the thickened sludge for further processing.

Not all of the solids in sludge are captured in thickening processes. Some sludge solids escape with the liquid portion. Table 27.8 lists typical solids capture fractions.

17. SLUDGE STABILIZATION

Raw sludge is too bulky, odorous, and putrescible to be dewatered easily or disposed of by land spreading. Such sludge can be "stabilized" through chemical addition or by digestion prior to dewatering and landfilling. (Sludge that is incinerated does not need to be stabilized.) Stabilization converts the sludge to a stable, inert form that is essentially free from odors and pathogens.

Sludge can be stabilized chemically by adding lime dust to raise the pH. A rule of thumb is that adding enough lime to raise the pH to 12 for at least 2 hr will inhibit or kill bacteria and pathogens in the sludge. Kiln and cement dust is also used in this manner.

Alternatively, ammonia compounds can be added to the sludge prior to land spreading. This has the additional advantage of producing a sludge with high plant nutrient value.

18. AEROBIC DIGESTION

Aerobic digestion occurs in an open holding tank digester and is preferable for stabilized primary and combined primary-secondary sludges. Up to 70% (typically 40–50%) of the volatile solids can be removed in an aerobic digester. Mechanical aerators are used in a manner similar to aerated lagoons. Construction details of aerobic digesters are similar to those of aerated lagoons.

19. ANAEROBIC DIGESTION

Anaerobic digestion occurs in the absence of oxygen. Anaerobic digestion is more complex and more easily upset than aerobic digestion. However, it has a lower operating cost. Table 27.9 gives typical characteristics of aerobic digesters.

Three types of anaerobic bacteria are involved. The first type converts organic compounds into simple fatty or amino acids. The second group of *acid-formed bacteria* converts these compounds into simple organic acids, such as acetic acid. The third group of *acid-splitting bacteria* converts the organic acids to methane, carbon dioxide, and some hydrogen sulfide. The third phase takes the longest time and sets the rate and loadings. The pH should be 6.7–7.8, and temperature should be in the mesophilic range of 85–100°F (29–38°C) for the methane-producing bacteria to be effective. Sufficient alkalinity must be added to buffer acid production.

Table 27.8 Typical Thickening Solids Fraction and Capture Efficiencies

operation	solids capture range (%)	solids content range (%)
gravity thickeners		
primary sludge only	85–92	4–10
primary and waste-activated	80–90	2–6
flotation thickeners		
with chemicals	90–98	3–6
without chemicals	80–95	3–6
centrifuge thickeners		
with chemicals	90–98	4–8
without chemicals	80–90	3–6
vacuum filtration with chemicals	90–98	15–30
belt filter press with chemicals	85–98	15–30
filter press with chemicals	90–98	20–50
centrifuge dewatering		
with chemicals	85–98	10–35
without chemicals	55–90	10–30

Table 27.9 *Typical Characteristics of Aerobic Digesters*

minimum number of units	2
size	2–3 ft³/person
	(0.05–0.08 m³/person)
aeration period	
activated sludge only	10–15 days
activated sludge and	12–18 days
primary settling sludge	
mixture of primary sludge	15–20 days
and activated or trickling	
filter sludge	
volatile solids loading	0.1–0.3 lbm/day-ft³
	(1.6–4.8 kg/d·m³)
volatile solids reduction	40–50%
sludge age	
primary sludge	25–30 days
activated sludge	15–20 days
oxygen required (primary	1.6–1.9 lbm O₂/lbm BOD
sludge)	removed
	(1.6–1.9 kg O₂/kg BOD
	removed)
minimum dissolved oxygen	1–2 mg/L
level	
mixing energy required by	0.75–1.5 hp/1000 ft³
mechanical aerators	(0.02–0.04 kW/m³)
maximum depth	15 ft (4.5 m)

(Multiply ft³ by 0.0283 to obtain m³.)
(Multiply lbm/day-ft³ by 16.0 to obtain kg/d·m³.)
(Multiply lbm/lbm by 1.0 to obtain kg/kg.)
(Multiply hp/1000 ft³ by 0.0263 to obtain kW/m³.)
(Multiply ft by 0.3048 to obtain m.)

A simple single-stage sludge digestion tank consists of an inlet pipe, outlet pipes for removing both the digested sludge and the clear supernatant liquid, a dome, an outlet pipe for collecting and removing the digester gas, and a series of heating coils for circulating hot water. (See Fig. 27.7.)

In a single-stage, floating-cover digester, sludge is brought into the tank at the top. The contents of the digester stratify into four layers: scum on top, clear supernatant, a layer of actively digesting sludge, and a bottom layer of concentrated sludge. Some of the contents may be withdrawn, heated, and returned in order to maintain a proper digestion temperature.

Supernatant is removed along the periphery of the digester and returned to the input of the processing plant. Digested sludge is removed from the bottom and dewatered prior to disposal. Digester gas is removed from the gas dome. Heat from burning the methane can be used to warm the sludge that is withdrawn or to warm raw sludge prior to entry.

A reasonable loading and well-mixed digester can produce a well-stabilized sludge in 15 days. With poorer mixing, 30–60 days may be required.

A single-stage digester performs the functions of digestion, gravity thickening, and storage in one tank. In a

Figure 27.7 *Simple Anaerobic Digester*

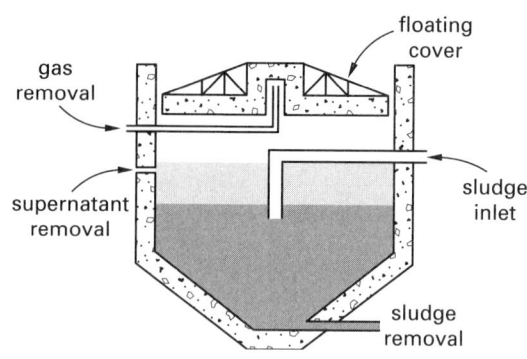

two-stage process, two digesters in series are used. Heating and mechanical mixing occur in the first digester. Since the sludge is continually mixed, it will not settle. Settling and further digestion occur in the unheated second tank.

Germany began experimenting with large spherical and egg-shaped digesters in the 1950s. These and similar designs are experiencing popularity in the United States. Volumes are on the order of 100,000–400,000 ft³ (3000–12,000 m³), and digesters can be constructed from either prestressed concrete or steel. Compared with conventional cylindrical digesters, the egg-shaped units minimize problems associated with the accumulation of solids in the lower corners and collection of foam and scum on the upper surfaces. The shape also promotes better mixing that enhances digestion. However, variations in the shape greatly affect grit collection and required mixing power. Mixing can be accomplished with uncombined gas mixing, mechanical mixing using an impeller and draft-tube, and pumped recirculation. Typical volatile suspended solids loading rates are 0.062–0.17 lbm VSS/ft³-day (1–2.8 kg VSS/m³·d) with a 15–25 day retention time. Operational problems include grit accumulation, gas leakage, and foaming.

20. METHANE PRODUCTION

Wastes that contain biologically degradable organics and are free of substances that are toxic to microorganisms (e.g., acids) will be biodegraded. Methane gas is a product of anaerobic degradation. Under anaerobic conditions, degradation of organic wastes yields methane. Inorganic wastes are not candidates for biological treatment and do not produce methane.

The theoretical volume of methane produced from sludge digestion is 5.61 ft³ (0.35 m³) per pound (per kilogram) of BOD converted. Wasted sludge does not contribute to methane production, and the theoretical amount is reduced by various inefficiencies. Therefore, the actual volume of methane produced is predicted by Eq. 27.51. The total volume of digester gas is larger than the methane volume alone since methane constitutes about ⅔ of the total volume. E is the efficiency of waste utilization, typically 0.6–0.9. S_o is the ultimate BOD of the influent. ES_o can be replaced with $S_o - S$, as the two terms are equivalent.

$$V_{\text{methane,m}^3/\text{d}} = \left(0.35\ \frac{\text{m}^3}{\text{kg}}\right)\left(\frac{\dfrac{ES_{o,\text{mg/L}}Q_{\text{m}^3/\text{d}}}{1000\ \dfrac{\text{g}}{\text{kg}}}}{-1.42P_{x,\text{kg/d}}}\right)$$

$$[\text{SI}] \quad 27.51(a)$$

$$V_{\text{methane,ft}^3/\text{day}} = \left(5.61\ \frac{\text{ft}^3}{\text{lbm}}\right)\left(\begin{array}{c} ES_{o,\text{mg/L}}Q_{\text{MGD}} \\ \times \left(8.345\ \dfrac{\text{lbm-L}}{\text{MG-mg}}\right) \\ -1.42P_{x,\text{lbm/day}} \end{array}\right)$$

$$[\text{U.S.}] \quad 27.51(b)$$

The heating value of digester gas listed in Table 27.10 is approximately 600 Btu/ft^3 (22 MJ/m^3). This value is appropriate for digester gas with the composition given: 65% methane and 35% carbon dioxide by volume. Carbon dioxide is not combustible and does not contribute to the heating effect. The actual heating value will depend on the volumetric fraction of methane. The temperature and pressure also affect the gas volume. At 60°F (15.6°C) and 1 atm, the pure methane has a lower heating value, LHV, of 913 Btu/ft^3 (34.0 MJ/m^3). The higher heating value of 1013 Btu/ft^3 (37.7 MJ/m^3) should not normally be used, as the water vapor produced during combustion is not allowed to condense out.

The total heating energy is

$$q = (\text{LHV})V_{\text{fuel}} \qquad 27.52$$

21. HEAT TRANSFER AND LOSS

In an adiabatic reactor, energy is required only to heat the influent sludge to the reactor temperature. In a nonadiabatic reactor, energy will also be required to replace energy lost to the environment through heat transfer from the exposed surfaces.

The energy required to increase the temperature of sludge from T_1 to T_2 is given by Eq. 27.53. The specific heat, c_p, of sludge is assumed to be the same as for water, 1.0 Btu/lbm-°F (4190 J/kg·°C).

$$\begin{aligned} q &= m_{\text{sludge}}c_p(T_2 - T_1) \\ &= V_{\text{sludge}}\rho_{\text{sludge}}c_p(T_2 - T_1) \qquad 27.53 \end{aligned}$$

The energy lost from a heated digester can be calculated in several ways. If the surface temperature is known, the convective heat lost from a heated surface to the surrounding air depends on the convective outside *film coefficient*, h. The film coefficient depends on the surface temperature, air temperature, orientation, and prevailing wind. Heat losses from digester sides,

top, and other surfaces must be calculated separately and combined.

$$q = hA_{\text{surface}}(T_{\text{surface}} - T_{\text{air}}) \qquad 27.54$$

If the temperature of the digester contents is known, it can be used with the *overall coefficient of heat transfer*, U (a combination of inside and outside film coefficients), as well as the conductive resistance of the digester walls to calculate the heat loss.

$$q = UA_{\text{surface}}(T_{\text{contents}} - T_{\text{air}}) \qquad 27.55$$

If the temperatures of the inner and outer surfaces of the digester walls are known, the energy loss can be calculated as a conductive loss. k is the average thermal conductivity of the digester wall, and L is the wall thickness.

$$q = \frac{kA(T_{\text{inner}} - T_{\text{outer}})}{L} \qquad 27.56$$

Table 27.10 *Typical Characteristics of Single-Stage Heated Anaerobic Digesters*

minimum number of units	2
size	
trickling filter sludge	3–4 ft^3/person (0.08–0.11 m^3/person)
primary and secondary sludge	6 ft^3 (0.16 m^3/person)
volatile solids loading	0.13–0.2 lbm/day-ft^3 (2.2–3.3 kg/d·m^3)
temperature	85–100°F; 95–98°F optimum (29–38°C; 35–36.7°C optimum)
pH	6.7–7.8; 6.9–7.2 optimum
gas production	7–10 ft^3/lbm volatile solids (0.42–0.60 m^3/ kg volatile solids) 0.5–1.0 ft^3/day-person (0.014–0.027 m^3/ d·person)
gas composition	65% methane; 35% carbon dioxide
gas heating value	500–700; 600 Btu/ft^3 typical (19–26 MJ/m^3; 22 MJ/m^3)
sludge final moisture content	90–95%
sludge detention time	30–90 days (conventional) 15–25 days (high rate)
depth	20–45 ft (6–14 m)
minimum freeboard	2 ft (0.6 m)

(Multiply ft^3 by 0.0283 to obtain m^3.)
(Multiply lbm/day-ft^3 by 16.0 to obtain kg/d·m^3.)
(Multiply ft^3/lbm by 0.0624 to obtain m^3/kg.)
(Multiply Btu/ft^3 by 0.0372 to obtain MJ/m^3.)
(Multiply ft by 0.3048 to obtain m.)

Water Treatment

22. SLUDGE DEWATERING

Once sludge has been thickened, it can be digested or dewatered prior to disposal. At water contents of 75%, sludge can be handled with shovel and spade, and is known as *sludge cake*. A 50% moisture content represents the general lower limit for conventional drying methods. Several methods of dewatering are available, including vacuum filtration, pressure filtration, centrifugation, sand and gravel drying beds, and lagooning.

Generally, vacuum filtration and centrifugation are used with undigested sludge. A common form of vacuum filtration occurs in a *vacuum drum filter*. A hollow drum covered with filter cloth is partially immersed in a vat of sludge. The drum turns at about 1 rpm. Suction is applied from within the drum to attract sludge to the filter surface and to extract moisture. The sludge cake is dewatered to about 75–80% moisture content (solids content of 20–25%) and is then scraped off the belt by a blade or delaminated by a sharp bend in the belt material. This degree of dewatering is sufficient for sanitary landfill. A higher solids content of 30–33% is needed, however, for direct incineration unless external fuel is provided. Chemical flocculants (e.g., polymers or ferric chloride, with or without lime) may be used to collect finer particles on the filter drum. Performance is measured in terms of the amount of dry solids removed per unit area. A typical performance is 3.5 lbm/hr-ft^2 (17 kg/h·m^2).

Belt filter presses accept a continuous feed of conditioned sludge and use a combination of gravity drainage and mechanically applied pressure. Sludge is applied to a continuous-loop woven belt (cotton, wool, nylon, polyester, woven stainless steel, etc.) where the sludge begins to drain by gravity. As the belt continues on, the sludge is compressed under the action of one or more rollers. Belt presses typically have belt widths of 80–140 in (2.0–3.5 m) and are loaded at the rate of 60–450 lbm/ft-hr (90–680 kg/m·h). Moisture content of the cake is typically 70–80%.

Pressure filtration is usually chosen only for sludges with poor dewaterability (such as waste-activated sludge) and where it is desired to dewater to a solids content higher than 30%. *Recessed plate filter presses* operate by forcing sludge at pressures of up to 120 psi (830 kPa) into cavities of porous filter cloth. The filtrate passes through the cloth, leaving the filter cake behind. When the filter becomes filled with cake, it is taken offline and opened. The filter cake is either manually or automatically removed.

Filter press sizing is based on a continuity of mass. In Eq. 27.57, s is the solids content as a decimal fraction in the filter cake.

$$V_{\text{press}}\rho_{\text{filter cake}}s_{\text{filter cake}}$$
$$= V_{\text{sludge, per cycle}}s_{\text{sludge}}\rho_{\text{sludge}}$$
$$= V_{\text{sludge, per cycle}}s_{\text{sludge}}\rho_{\text{water}}\text{SG}_{\text{sludge}} \quad \textit{27.57}$$

Centrifuges can operate continuously and reduce water content to about 70%, but the effluent has a high percentage of suspended solids. The types of centrifuges used include solid bowl (both cocurrent and countercurrent varieties) and imperforate basket decanter centrifuges.

With *solid bowl centrifuges*, sludge is continuously fed into a rotating bowl where it is separated into a dilute stream (the *centrate*) and a dense cake consisting of approximately 70–80% water. Since the centrate contains fine, low-density solids, it is returned to the wastewater treatment system. Equation 27.58 gives the number of gravities, G, acting on a rotating particle. Rotational speed, n, is typically 1500–2000 rpm. A polymer may be added to increase the capture efficiency (same as reducing the solids content of the centrate) and, accordingly, increase the feed rate.

$$G = \frac{\omega^2 r}{g} = \left(\frac{2\pi n}{60}\right)^2\left(\frac{r}{g}\right) \quad \textit{27.58}$$

Sand drying beds are preferable when digested sludge is to be disposed of in a landfill since sand beds produce sludge cake with a low moisture content. Drying beds are enclosed by low concrete walls. A broken stone/gravel base that is 8–20 in (20–50 cm) thick is covered by 4–8 in (10–20 cm) of sand. Drying occurs primarily through drainage, not evaporation. Water that seeps through the sand is removed by a system of drainage pipes. Typically, sludge is added to a depth of about 8–12 in (20–30 cm) and allowed to dry for 10–30 days.

There is little theory-based design, and drying beds are sized by rules of thumb. The drying area required is approximately 1–1.5 ft^2 (0.09–1.4 m^2) per person for primary digested sludge; 1.25–1.75 ft^2 (0.11–0.16 m^2) per person for primary and trickling filter digested sludge; 1.75–2.5 ft^2 (0.16–0.23 m^2) per person for primary and waste-activated digested sludge; and 2–2.5 ft^2 (0.18–0.23 m^2) per person for primary and chemically precipitated digested sludge. This area can be reduced by approximately 50% if transparent covers are placed over the beds, creating a "greenhouse effect" and preventing the spread of odors.

Other types of drying beds include paved beds (both drainage and decanting types), vacuum-assisted beds, and artificial media (e.g., stainless steel wire and high-density polyurethane panels) beds.

Sludge dewatering by freezing and thawing in exposed beds is effective during some months in parts of the country.

Example 27.5

A belt filter press handles an average flow of 19,000 gal/day of thickened sludge containing 2.5% solids by weight. Operation is 8 hr/day and 5 days/wk. The belt filter press loading is specified as 500 lbm/m-hr (lbm/hr per meter of belt width). The dewatered sludge is to have 25% solids by weight, and the suspended solids concentration in the filtrate is 900 mg/L. The belt is continuously washed on its return path at a rate of 24 gal/m-min. The specific gravities of the sludge feed, dewatered cake, and filtrate are 1.02, 1.07, and 1.01, respectively. (a) What size belt is required? (b) What is the volume of filtrate rejected? (c) What is the solids capture fraction?

Solution

(a) The daily masses of wet and dry influent sludge are

$$m_{\text{wet}} = V\rho_{\text{sludge}} = V\rho_{\text{water}}SG_{\text{sludge}}$$

$$= \left(19{,}000 \ \frac{\text{gal}}{\text{day}}\right) \left(\frac{\left(62.4 \ \frac{\text{lbm}}{\text{ft}^3}\right)(1.02)}{7.48 \ \frac{\text{gal}}{\text{ft}^3}}\right)$$

$$= 1.617 \times 10^5 \ \text{lbm/day}$$

$$m_{\text{dry}} = (\text{fraction solids})m_{\text{wet}}$$

$$= (0.025)\left(1.617 \times 10^5 \ \frac{\text{lbm}}{\text{day}}\right)$$

$$= 4042 \ \text{lbm/day}$$

Since the influent is received continuously but the belt filter press operates only 5 days a week for 8 hours per day, the average hourly processing rate is

$$\frac{\left(4042 \ \frac{\text{lbm}}{\text{day}}\right)\left(\frac{7 \ \text{days}}{5 \ \text{days}}\right)}{8 \ \frac{\text{hr}}{\text{day}}} = 707 \ \text{lbm/hr}$$

The belt size (width) required is

$$w = \frac{707 \ \frac{\text{lbm}}{\text{hr}}}{500 \ \frac{\text{lbm}}{\text{m-hr}}}$$

$$= 1.41 \ \text{m}$$

Use one 1.5 m belt filter press. A second standby unit may also be specified.

(b) The mass of solids received each day is

$$\left(707 \ \frac{\text{lbm}}{\text{hr}}\right)\left(8 \ \frac{\text{hr}}{\text{day}}\right) = 5656 \ \text{lbm/day}$$

The fractional solids in the filtrate is

$$s_{\text{filtrate}} = \frac{900 \ \frac{\text{mg}}{\text{L}}}{\left(1000 \ \frac{\text{mg}}{\text{g}}\right)\left(1000 \ \frac{\text{g}}{\text{L}}\right)} = 0.0009$$

The solids balance is

$$\text{influent solids} = \text{sludge cake solids} + \text{filtrate solids}$$

$$5656 \ \frac{\text{lbm}}{\text{day}} = \left(\frac{Q_{\text{cake,gal/day}}}{7.48 \ \frac{\text{gal}}{\text{ft}^3}}\right)\left(62.4 \ \frac{\text{lbm}}{\text{ft}^3}\right)(1.07)(0.25)$$

$$+ \left(\frac{Q_{\text{filtrate,gal/day}}}{7.48 \ \frac{\text{gal}}{\text{ft}^3}}\right)\left(62.4 \ \frac{\text{lbm}}{\text{ft}^3}\right)$$

$$\times \ (1.01)(0.0009)$$

The total liquid flow rate is the sum of the sludge flow rate and the washwater flow rate.

$$Q_{\text{total}} = Q_{\text{cake}} + Q_{\text{filtrate}}$$

$$= \left(19{,}000 \ \frac{\text{gal}}{\text{day}}\right)\left(\frac{7 \ \text{days}}{5 \ \text{days}}\right)$$

$$+ \left(24 \ \frac{\text{gal}}{\text{m-min}}\right)(1.5 \ \text{m})\left(60 \ \frac{\text{min}}{\text{hr}}\right)\left(8 \ \frac{\text{hr}}{\text{day}}\right)$$

$$= 43{,}880 \ \text{gal/day}$$

Solving the solids and liquid flow rate equations simultaneously,

$$Q_{\text{cake}} = 2396 \ \text{gal/day}$$

$$Q_{\text{filtrate}} = 41{,}484 \ \text{gal/day}$$

(c) The solids in the filtrate are

$$m_{\text{filtrate}} = \left(\frac{41{,}484 \ \frac{\text{gal}}{\text{day}}}{7.48 \ \frac{\text{gal}}{\text{ft}^3}}\right)\left(62.4 \ \frac{\text{lbm}}{\text{ft}^3}\right)(1.01)(0.0009)$$

$$= 315 \ \text{lbm/day}$$

The solids capture fraction is

$$\frac{\text{influent solids} - \text{filtrate solids}}{\text{influent solids}} = \frac{5656 \ \frac{\text{lbm}}{\text{day}} - 315 \ \frac{\text{lbm}}{\text{day}}}{5656 \ \frac{\text{lbm}}{\text{day}}}$$

$$= 0.944$$

23. SLUDGE DISPOSAL

Liquid sludge and sludge cake disposal options are limited to landfills, incineration, nonagricultural land application (e.g., strip-mine reclamation, agricultural land application, commercial applications (known as *distribution and marketing* and *beneficial reuse*) such as composting, special monofills, and surface impoundments), and various state-of-the-art methods. Different environmental contaminants are monitored with each disposal method, making the options difficult to evaluate.

If a satisfactory site is available, sludge, sludge cake, screenings, and scum can be discarded in municipal *landfills*. Approximately 40% of the sludge produced in the United States is disposed of in this manner. This has traditionally been the cheapest disposal method, even when transportation costs are included.

The sludge composition and its potential effects on landfill leachate quality must be considered. Sludge with minute concentrations of monitored substances (e.g., heavy metals such as cadmium, chromium, copper, lead, nickel, and zinc) may be classified as hazardous waste, requiring disposal in hazardous waste landfills.

Sludge-only *monofills* and other surface impoundments may be selected in place of municipal landfills. Monofills are typically open pits protected by liners, slurry cutoff walls, and levees (as protection against floods). Monofills may include a pug mill to combine sludge with soil from the site.

Approximately 25% of the sludge produced in the United States is disposed of in a *beneficial reuse program* such as surface spreading (i.e., "land application") and/or composting. Isolated cropland, pasture, and forest land (i.e., "silviculture") are perfect for surface spreading. However, sludge applied to the surface may need to be harrowed into the soil soon after spreading to prevent water pollution from sludge runoff and possible exposure to or spread of bacteria. Alternatively, the sludge can be injected 12–18 in (300–460 mm) below grade directly from distribution equipment. The actual application rate depends on sludge strength (primarily nitrogen content), the condition of the receiving soil, the crop or plants using the applied nutrients, and the spreading technology.

By itself, filter cake is not highly useful as a fertilizer since its nutrient content is low. However, it is a good filler and soil conditioner for more nutrient-rich fertilizers. Other than as filler, filter cake has few practical applications.

Where sludges have low metals contents, *composting* (in static piles or vessels) can be used. *Compost* is a dry, odor-free soil conditioner with applications in turf grass, landscaping, land reclamation, and other horticultural industries. These markets can help to defray the costs of the program. Control of odors, cold weather operations, and concern about airborne pathogens are problems with *static pile composting*. Expensive air-quality scrubbers add to the cost of processing. With *in-vessel composting*, dewatered raw sludge is mixed with a bulking material, placed in an enclosed tank, and mixed with air in the tank.

Incineration is the most expensive treatment, though it results in almost total destruction of volatile solids, is odor free, and is independent of the weather. Approximately 20% of sludge is disposed of in this manner. A solids content of 30–33% is needed for direct incineration without the need to provide external fuel (i.e., so that the sludge burns itself). Air quality controls, equipment, and the regulatory permitting needed to meet strict standards may make incineration prohibitively expensive in some areas. The incineration ash produced represents a disposal problem of its own.

State-of-the-art options include using sludge solids in bricks, tiles, and other building materials; below-grade injection; subsurface injection wells; multiple-effect (Carver-Greenfield) evaporation; combined incineration with solid waste in "mass burn" units; and fluidized-bed gasification. The first significant installation attempting to use sludge solids as powdered fuel for power generation was the 25 MW Hyperion Energy Recovery System at the Hyperion treatment plant in Los Angeles. This combined-cycle cogeneration plant unsuccessfully attempted to operate using the Carver-Greenfield process and a fluidized-bed gasifier to convert sludge solids to fuel.

Ocean dumping, once a popular option for coastal areas, has been regulated out of existence and is no longer an option for new or existing facilities.

28 Protection of Wetlands

Nomenclature

A	area	m^2
d	depth	m
ET	evapotranspiration	mm/y, m^3/s
G	groundwater flow	mm/y, m^3/s
P	precipitation	mm/y, m^3/s
Q	flow rate	m^3/s
r	renewal rate	s^{-1}
S	surface flow	mm/y, m^3/s
t	time	s
T	tidal inflow	mm/y, m^3/s
V	volume	m^3

Subscripts

ave	average
i	in
n	net
o	out
t	total

1. INTRODUCTION

Engineers have learned much about wetlands and their contributions to the natural environment. They have recognized that in addition to their other functions and values, wetlands have a great capacity to remove pollutants from water; provide habitat for wildfowl, wildlife, and aquatic life; attenuate and absorb flood waters; and provide a buffer for upland coastal areas.

2. WETLANDS IDENTIFICATION

Definition of Wetlands

The three main characteristics of wetlands are:

1. Wetlands have water at the surface or in the root zone.

2. Wetlands have unique soil conditions that usually differ from adjacent uplands.

3. Wetlands support vegetation that is adapted to wet conditions and that cannot adapt to an absence of flooding.

In 1979, the U.S. Fish and Wildlife Service (USFWS) published a scientific definition of wetlands that remains in widespread use. It defines wetlands as

> . . . lands transitional between terrestrial and aquatic systems where the water table is usually at or near the surface or the land is covered by shallow water . . . Wetlands must have one or more of the following three attributes: (1) at least periodically, the land supports predominately hydrophytes, (2) the substrate is predominately undrained hydric soil, and (3) the substrate is nonsoil and is saturated with water or covered by shallow water at some time during the growing season each year.

The USFWS definition is widely used throughout the United States to classify and inventory wetlands. However, it differs from the official regulatory definition used by the U.S. Army Corps of Engineers for issuing dredge-fill permits under Section 404 of the Clean Water Act.

> The term *wetlands* means those areas that are inundated or saturated by surface or ground water at a frequency and duration sufficient to support, and that under normal circumstances do support, a prevalence of vegetation typically adapted for life in saturated soil conditions. Wetlands generally include swamps, marshes, bogs, and similar areas. (33 CFR 328.3(b); 1999)

Jurisdictional wetlands are regulated under Section 404 of the Clean Water Act or the Swampbuster provisions of the Food Security Act. Federal manuals specify criteria for classifying wetlands as jurisdictional wetlands. The three criteria included in these manuals are the presence of (1) wetlands hydrology, (2) wetlands soils, and (3) hydrophytic vegetation.

Wetland Types

In 1979, the USFWS developed a classification system for wetlands. The major categories are coastal wetland ecosystems and inland wetland ecosystems.

Coastal Wetlands

Coastal wetlands are influenced by alternating tides. The three basic types of *coastal wetlands* are tidal salt marshes, tidal freshwater marshes, and mangrove wetlands.

Tidal salt marshes are identified as "estuarine intertidal emergent, haline." *Estuarine waters* are coastal areas where seawater mixes with freshwater, mainly from rivers. *Intertidal* means the substrate is both exposed and flooded by tides. Haline refers to the presence of ocean salt. Grasses and rushes dominate tidal salt marshes. Tidal salt marsh ecosystems have adapted to the stresses of salinity, periodic inundation, and extremes in temperature.

Tidal freshwater marshes are identified as "estuarine intertidal emergent, fresh." Tidal freshwater marshes are farther inland from tidal salt marshes and are predominately in fresh water, but still are influenced by tides. Tidal freshwater marshes are dominated by a variety of grasses and annual and perennial broad-leaved aquatic plants. These marshes lack the salinity stress of salt marshes and are very productive.

Mangrove wetlands are identified as "estuarine intertidal forested and shrub, haline." In subtropical and tropical areas, tidal salt marshes may evolve into mangrove swamps. Mangrove swamps require protection from the open ocean and occur in a wide range of areas influenced by salinity and tides.

Inland Wetlands

The four basic types of *inland wetlands* are inland freshwater marshes, northern peatlands, southern deepwater swamps, and riparian wetlands.

Freshwater marshes are identified as "palustrine emergent." They are characterized by the presence of (1) emergent, soft-stemmed aquatic plants and grasses and sedges; (2) shallow water (meaning the water depth in the deepest part of the basin is less than 2 m at low tide); and (3) shallow peat deposits. They occur in tidal areas where salinity due to ocean-derived salts is below 0.5% (5000 mg/L).

The term "freshwater marsh" was developed by the USFWS to group together the vegetated wetlands traditionally called marsh, swamp, bog, fen, and prairie. These types of wetlands are common in North America and occur in isolated basins, as fringes around lakes, and in slow-moving streams and rivers.

Northern peatlands are identified as "palustrine moss-lichen." Northern peatlands include the deep peat deposits in the north temperate areas of North America. *Peat* refers to partially carbonized vegetable matter, primarily from decomposition of moss and lichen. It has value as fuel and fertilizer.

Southern deepwater swamps are identified as "palustrine forested and scrub-shrub." *Shrub* refers to a woody plant usually less than 6 m tall that generally exhibits erect, spreading stems and has a bushy appearance. Southern deepwater swamps are freshwater woody wetlands with standing water during most or all of the growing season. They occur primarily in the southeastern United States and are normally dominated by cypress and gum/tupelo.

Riparian forested wetlands are identified as "palustrine forested and scrub-shrub." These wetlands are dry for portions of the growing season and are periodically flooded by adjacent streams and rivers. In the southeast, they are referred to as "bottomland hardwood forests." Riparian wetlands are more productive than adjacent uplands due to the inflow of nutrients during seasonal flooding.

3. WETLANDS HYDROLOGY

Hydrology is the most important factor in determining the composition, richness, and productivity of wetlands. Changes in hydrology can affect chemical and physical properties of the wetlands system, such as nutrient availability, degree of substrate anoxia, soil salinity, sedimentation, and pH. Incoming waters are the principal source of nutrients; water outflows remove decomposition products and wastes. Changes in hydrology can result in rapid and considerable changes in wetlands vegetation and animal species, as well as in productivity of wetlands.

Hydroperiod

A key characteristic of a wetland is its *hydroperiod*, which is the seasonal pattern of the rise and fall of its surface and subsurface water levels. The hydroperiod distinctly characterizes each type of wetland. If the hydroperiod is reasonably consistent from year to year, the wetland will be stable, and its productivity and value will be predictable.

The hydroperiod can be determined from the change in volume over a particular time period, a calculation known as a *water balance*, also known as a *water budget*. A water balance records all inflows and outflows of water during the particular time period.

Flood duration is the amount of time that a wetland has standing water. (The term *flood duration* does not apply to wetlands that are permanently flooded.) *Flood frequency* is the average number of times that a wetland is flooded over a given period (i.e., one year). Both flood duration and flood frequency are used to characterize periodically flooded wetlands.

Hydroperiods are qualitatively described by the U.S. Fish and Wildlife Service in Table 28.1.

Illustrative hydroperiods for several types of wetlands are shown in Fig. 28.1. A cypress dome (see Fig. 28.1(a)) may have water above the land surface during the wet summer season as well as during the dry periods in the late summer and early spring. A coastal marsh (see Fig. 28.1(b)) may have a hydroperiod composed of semi-diurnal flooding and dewatering superimposed on semi-monthly spring and ebb tides. A bottomland forest (see Fig. 28.1(c)) may have rapid changes due to seasonal flooding from local precipitation and thawing conditions. A tropical floodplain (see Fig. 28.1(d)) may exhibit large seasonal changes in hydroperiod due to the flooding of rivers.

Table 28.1 Qualitative Descriptions of Hydroperiods of Wetlands (U.S. Fish and Wildlife Service)

hydroperiod	water regime	characteristic
tidal	subtidal	land surface is permanently flooded by tidal water
tidal	irregularly exposed	land surface is exposed by tides less often than daily
tidal	regularly flooded	tidal water alternately floods and exposes the land surface at least once daily
tidal	irregularly flooded	tidal water floods the land surface less often than daily
nontidal	permanently flooded	water covers the land surface throughout the year in all years
nontidal	intermittently exposed	surface water is present throughout the year except in years of extreme drought
nontidal	semipermanently flooded	surface water persists throughout the growing season in most years, or water table is very near the land surface
nontidal	seasonally flooded	surface water is present for extended periods especially early in the growing season, but is absent by the end of the season in most years; when surface water is absent, water table is near the land surface
nontidal	saturated	land is saturated to the surface for extended periods during the growing season, but surface water is seldom present
nontidal	temporarily flooded	surface water is present for brief periods during the growing season, but the water table is usually below the soil surface for most of the season
nontidal	intermittently flooded	land surface is usually exposed, but surface water is present for variable periods without detectable seasonal periodicity; weeks, months, or years may occur between inundation events
nontidal	artificially flooded	the amount and duration of flooding is controlled by pumps, siphons, dikes, or dams

Figure 28.1 Illustrative Hydroperiods

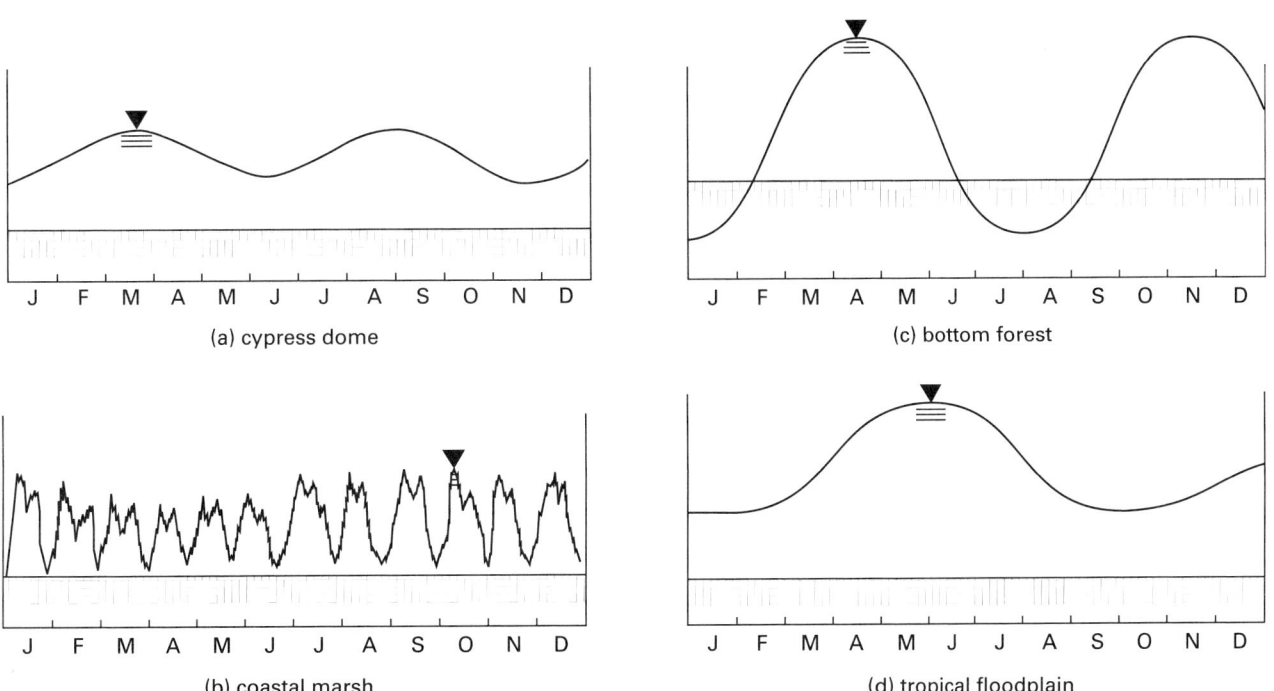

(a) cypress dome

(c) bottom forest

(b) coastal marsh

(d) tropical floodplain

The hydroperiod for a particular site can vary from year to year as well as seasonally and monthly or daily. Water depths can vary over periods of decades or just a few years. The seasonal fluctuations of water levels occurring with riverine wetlands is known as *pulsing*. Pulsing nourishes the wetland by providing nutrients and removing waste products. Pulse-fed wetlands are often the most productive and provide valuable support for adjacent ecosystems. A characteristic common to most wetlands is a seasonally fluctuating water level.

Water Budget

The *water budget*, or *water balance*, is calculated from the sum of the net precipitation, P_n, surface water inflows, S_i, and outflows, S_o, groundwater inflows, G_i, and outflows, G_o, evapotranspiration, ET, and tidal inflow, $+T$, or outflow, $-T$.

$$\frac{\Delta V}{\Delta t} = P_n + S_i + G_i - \text{ET} - S_o - G_o \pm T \qquad 28.1$$

The terms in Eq. 28.1 can have units of either depth per unit of time for a specific area (mm/y) or volume per unit of time applicable to the whole area (m^3/d).

The *average depth* at a particular time is the volume of the water stored divided by the area of the wetland.

$$d_{\text{ave}} = \frac{V}{A} \qquad 28.2$$

Over an annual cycle, the net change in storage of water in wetlands may be close to zero. Precipitation will vary with the climate of the area, and may have distinct wet and dry seasons. The surface inflows and outflows usually vary seasonally with precipitation, runoff, and flooding. Groundwater inflows and outflows vary less than surface inflows and outflows and may be absent altogether. Evapotranspiration is typically seasonal, with high rates in the summer and low rates in the winter. For wetlands that are tidally influenced, flooding will vary with elevation and will exhibit one or two tidal cycles daily.

Example 28.1

A hypothetical water budget for the Okefenokee Swamp in Georgia is given.

season	precipitation (mm)	surface + groundwater inflow (mm)	ET (mm)	surface outflow (mm)	groundwater outflow (mm)
spring	4000	1300	1300	3200	100
summer	2000	800	4000	1000	100
fall	2000	600	3000	800	100
winter	5100	1200	1000	2300	100
annual total	13 100	3900	9300	7300	400

The Okefenokee Swamp is 1770 km^2 in total area. Assume the average water depth is 8 m at the start of the spring season, precipitation is net, and the swamp is not affected by tides. What will be the average water depth at the end of spring?

Solution

For the spring season,

$$P_n = P = 4000 \text{ mm} \quad (4 \text{ m})$$
$$S_i + G_i = 1300 \text{ mm} \quad (1.30 \text{ m})$$
$$\text{ET} = 1300 \text{ mm} \quad (1.30 \text{ m})$$
$$S_o = 3200 \text{ mm} \quad (3.20 \text{ m})$$
$$G_o = 100 \text{ mm} \quad (0.10 \text{ m})$$
$$T = 0 \text{ mm}$$

The average water volume at the beginning of the spring is calculated from Eq. 28.2.

$$d_{\text{ave},1} = \frac{V_1}{A}$$
$$V_1 = A d_{\text{ave},1} = \frac{(1770 \text{ km}^2)(8 \text{ m})}{1000 \frac{\text{m}}{\text{km}}}$$
$$= 14.2 \text{ km}^3$$

The water balance is given by Eq. 28.1.

$$\frac{\Delta V}{\Delta t} = P_n + S_i + G_i - \text{ET} - S_o - G_o + T$$

For the spring season, the change in swamp water volume is

$$\Delta V = \frac{(4 \text{ m} + 1.30 \text{ m} - 1.30 \text{ m} - 3.20 \text{ m} - 0.10 \text{ m})}{1000 \frac{\text{m}}{\text{km}}} \times (1770 \text{ km}^2)$$
$$= 1.24 \text{ km}^3$$
$$V_2 = V_1 + \Delta V = 14.2 \text{ km}^3 + 1.24 \text{ km}^3$$
$$= 15.4 \text{ km}^3$$

The average water depth at the end of the spring season is given by Eq. 28.2.

$$d_{\text{ave},2} = \frac{V_2}{A} = \frac{(15.4 \text{ km}^3)\left(1000 \frac{\text{m}}{\text{km}}\right)}{1770 \text{ km}^2}$$
$$= 8.70 \text{ m}$$

Alternatively, since the change in the water depth per unit area for the spring season is 700 mm (0.70 m), the final average water depth is

$$d_{\text{ave},2} = d_{\text{ave},1} + \Delta d = 8 \text{ m} + 0.70 \text{ m}$$
$$= 8.70 \text{ m}$$

Renewal Rate

The *renewal rate*, r, is often used to describe wetlands hydrology. This is the reciprocal of the *retention time* (*residence time*).

$$r = \frac{Q_t}{V} \qquad 28.3$$

Because parts of a wetland may be poorly mixed, the actual retention time may be much shorter than that calculated from average values.

Importance of Hydrology

Hydrology largely determines the richness of the vegetation and other species present. Flowing conditions and pulsing hydroperiods usually result in productive wetlands, whereas stagnant conditions tend to depress productivity. Hydrology controls the accumulation of organic material in wetlands through its effect on primary productivity, decomposition, and removal of organic matter.

Wetlands that flood frequently exhibit greater vegetative richness than wetlands experiencing long flooding durations. Flooding rivers influence primary productivity by transporting nutrients into wetlands and removing waste materials. Wetlands that experience prolonged periods of flooding or draining usually have low primary productivity.

Nutrients are brought into wetlands by precipitation, river flooding, tides, and surface water and groundwater inflows. Nutrients are exported from wetlands primarily through outflow of surface water and groundwater. Nutrients are cycled within wetlands by biological uptake and decomposition, both of which occur at a high rate in wetlands with flowing water or pulsing hydroperiods.

4. BIOCHEMISTRY OF WETLANDS

The biochemistry of wetlands is complex and involves chemical transformations as well as chemical transport into and out of the wetlands. Transformations of nitrogen, phosphorus, sulfur, iron, manganese, and carbon affect the availability of these minerals. The transformations take place, in part, in the anaerobic conditions that occur during submergence of the soil. Microorganisms, primarily anaerobes, are responsible for the transformations. Conditions toxic to plant and animal life may result from the transformations.

Chemical transport is affected by the hydrology. Wetlands that receive primarily precipitation are usually nutrient poor, whereas flooding from surface water and other sources usually contributes more nutrients necessary to support higher levels of primary productivity.

Wetlands often absorb nitrogen and phosphorus, but they may also export nutrients and organic materials to surrounding ecosystems. Wetlands can be high- or low-nutrient systems. Sediment storage of nutrients and the role of vegetation in recycling nutrients are important biochemical features of wetlands.

A wetland is considered an *open biochemical system* when there is substantial movement of nutrients and materials across the ecosystem boundary. A *closed biochemical system* is one with little movement across its boundary.

A wetland can be considered a source, sink, or transformer of nutrients. A *source* is a net exporter of a particular nutrient (nitrogen, phosphorus, etc.), and a *sink* is a net importer. If a nutrient is only transformed in physical or chemical form, the wetland is a transformer for that nutrient. The wetland type and hydrologic conditions, and the duration the wetland has been subjected to chemical loading all determine which function the wetland may perform. For example, a wetland may become saturated with a particular chemical and cease to function as a sink, thereby becoming a source. Chemicals may be retained on a seasonal basis, such as during the growing season when there is considerable plant and microorganism activity. Wetlands may be hydrologically linked to adjacent ecosystems whereby they obtain nutrients from flooding of rivers and, in turn, export chemicals to downstream systems. Although wetlands can retain sediments and chemicals from anthropogenic activity (see Sec. 28.5), their capacity to do so is limited.

Not all wetlands are highly productive; some are low-nutrient systems with low primary productivity. Wetlands that are highly productive are *eutrophic* (rich in mineral and organic nutrients); those that are of low productivity are *oligotrophic* (lacking in nutrients). Eutrophic wetlands obtain nutrients from inflows of surface and ground water, are considered open systems, can be either a source or sink of nutrients, normally export detritus (accumulated material), and have high primary productivities. Oligotrophic wetlands obtain nutrients primarily from precipitation, are considered closed systems, are not sources or sinks of nutrients, do not export detritus, and have low primary productivities.

5. ANTHROPOGENIC EFFECTS

Wetlands have been subjected to a wide range of *anthropogenic* (i.e., caused by mankind) *effects*. Some wetlands have undergone complete destruction through conversion to agricultural, residential, industrial, or commericial use. More subtle hydrologic changes can affect the hydroperiod, thereby changing the type, composition, and richness of vegetation and animal species. Examples of anthropogenic activities that may affect wetlands are listed in Table 28.2.

6. PROTECTION OF WETLANDS

Wetlands protection became an issue in the latter part of the twentieth century. Previously, significant efforts were directed at draining wetlands to provide land for agricultural, commercial, or residential development; transportation corridors; and mosquito control.

Table 28.2 *Effect of Anthropogenic Activity*

anthropogenic activity	environmental effect	wetlands effect
land clearing	erosion	increased sedimentation; increased biochemical oxygen demand (BOD) load
stream channelization	hydrologic change; increased stream bed cutting	change in flooding frequency and hydroperiod; change in nutrient input; decrease in water level; changes in salinity
dam construction	hydrologic change; dam becomes a nutrient trap	change in flooding and hydroperiod; decreased sedimentation; reduced nutrient input
pollution discharges	increasing pollution load to water bodies	increased nutrient cycling in wetlands; increased load of toxics and organic wastes to wetlands; degraded water quality
burning	removal of vegetation	disturbance of plant and animal habitat; alteration of nutrient cycling; alteration of hydrology
flood control; stabilization of water level	alteration of hydroperiod; reduced pulsing; impeded natural drainage	change in dominant vegetation; reduction in nutrient input; accumulation of detritus
driving off-road vehicles	increased human activity	disturbance of wildlife habitat; increased pollution loads
conversion	drainage, dredging, filling	complete loss of wetlands
timber harvest	removal of forests	change in dominant vegetation; disturbance of wildlife habitat; increased erosion; degraded water quality
extraction	mining of peat and minerals; drilling oil and gas wells; pipeline construction	disturbance of wildlife and aquatic habitat; discharge of pollutants; hydrologic changes

The legal underpinning of wetlands protection is Section 404 of the Clean Water Act, which requires a permit from the U.S. Army Corps of Engineers (COE) for dredging and filling activities in the waters of the United States. This law is supplemented by federal executive orders, a "no net loss" policy, and a variety of federal and state programs. Many states have laws that supplement the federal laws. Often, the federal COE permit is issued as a joint permit with state regulatory agencies. There are also a variety of international programs, agencies, and private organizations that deal with wetlands as habitat for migratory waterfowl.

Today, strong laws exist to protect wetlands. However, wetlands alteration continues through dredging, hydrologic modification, peat mining, and water quality degradation.

Wetlands Conversion

The greatest impact on wetlands has been from conversion to agricultural, commercial, industrial, and residential uses. Wetlands are converted for these purposes by draining, dredging, and filling activities. The best wetlands protection measure is simply to avoid conversion projects.

Changes in Hydrology

Hydrologic changes have significant effects on wetlands, essentially changing their fundamental character and values. Although levees and canals provide flood protection, they often result in isolation and drainage of wetlands and prevent the seasonal flooding that is essential for bringing nutrients to wetlands. Navigation canals and highways and construction of oil and gas wells, mining sites, and pipelines have also changed wetlands hydrology.

These changes deprive wetlands of water quantity and the pulsing action associated with flooding, produce greater fluctuation of water levels, reduce sediment and nutrient supply to wetlands, and result in water stagnation and saltwater intrusion into wetlands areas that were previously freshwater. Any protective measures should minimize or eliminate this collateral damage.

Extraction

Extraction activities include peat surface mining, phosphate surface mining, coal surface mining, and water withdrawal. Peat mining destroys wetlands because the peat deposits essentially form the wetlands. Surface mining for phosphate and coal also affects wetlands by removing or altering the land surface that defines the wetlands. Pumping of groundwater for water supplies or agricultural use can result in accelerated subsidence rates, thereby lowering the elevations of wetlands.

Reclamation of phosphate mines by restoring or developing wetlands has been successful in some areas. Using wetlands for reclamation of surface coal mines has not been common.

Water Quality Degradation

Wetlands can be affected by degradation of the quality of water entering the wetlands. Although the capacity of wetlands to remove pollutants is fairly well understood, less is known about how the degraded water quality can change wetlands productivity and habitat value.

Nitrogen and phosphorus can be pollutants or nutrients, depending on the loading and the composition of the vegetation. Both nitrogen and phosphorus are nutrients that are transported into wetlands with sediments and through flooding. However, excess nitrogen and phosphorus may enter wetlands from agricultural runoff and other sources. The resulting eutrophication can change the character of the wetlands and affect aquatic and wildlife species. A wide variety of metals, oils, and organic toxics can cause drastic changes to vegetation and aquatic species. For example, acid mine drainage is low in pH and high in iron and sulfur content. Such toxics can harm plant and animal species.

Protection against water quality degradation involves interception, containment, and treatment or other control of pollutants before they enter the wetlands.

Protective Measures for Design and Construction

There is a wide variety of measures that can be taken during design and construction that will help to protect wetlands.

- Agency input: Regulatory agencies should be involved in the early stages of planning and design. Conceptual plans and issue papers should be submitted early in the process to minimize the need for redesign. Some redesign will be inevitable if the project is complex. Issues and the actions taken to address them should be documented.

- Wetlands delineation: Wetlands must be accurately delineated at the start of a project.

- Hydrology: The hydrology of the wetlands and adjacent or affected areas must be thoroughly understood and documented. The effects of the project on the hydrology must be determined. Flooding and peak runoff must be determined for existing and post-project conditions.

- Slope stability: The structural stability of preconstruction and post-project landforms must be determined and documented.

- Erosion control: The erosion potential of the existing areas affected by the project must be determined, and the anticipated effects of the project on erosion must be identified. Erosion control measures should be identified, designed, and included in contract documents. Erosion control measures should not be left solely up to the contractor. Furthermore, any maintenance required for erosion control measures should be explicitly stated in the contract documents.

- Temporary access and work area: To minimize disturbance and turbidity, the use of temporary pile-supported construction trestles should be evaluated. Sheet steel piling, wood decking, geotextile fabric, and clean granular material should also be considered.

- Construction equipment: The use of work barges, specialized tracked trenching vehicles, wide low-pressure tires, and clamshell dredges with watertight buckets should be evaluated.

- Buffer zones: Generous buffer zones should be provided to minimize impact on the wetlands and to provide time and space to react to problems encountered during construction.

- Scheduling: Construction should be scheduled to minimize impact on wetlands habitat, aquatic life, and wildlife. Construction should be planned and designed to occur during dormant vegetation periods and when aquatic life and wildlife are at the least critical stage in their life cycles.

- Sequencing: Construction sequencing should be planned and designed to minimize periods of exposure and impact between construction activities. Trenches and excavations should not be left open for significant periods. Projects should be completed as quickly as possible to minimize impact.

- Restoration: Qualified wetlands biologists should be included in the project design and construction to advise on the best practices and to monitor and document conditions prior to, during, and after construction. Documenting restoration effectiveness should be a key responsibility of the biologist. This person should be employed by the project owner and be independent of the contractor. The contractor may also be required to have a similar expert on staff to advise the contractor directly.

- Monitoring: Monitoring requirements and standards should be clearly indicated in the contract documents. Although the contractor should be required to perform environmental monitoring of construction activities, the project owner should provide an independent person to perform monitoring to ensure compliance with regulations and permit requirements (since the project owner will ultimately be responsible for compliance).

- Public information: Regular public information activities should be conducted to keep the community and other interested parties informed of the project status. Opportunities should be provided to identify and react to concerns at an early stage. A proactive approach to dealing with public concerns will generally work best.

- Compliance problems: If permit compliance problems develop, prompt and thorough action should be taken to determine the root cause and appropriate corrective action. Any problems identified should be reported to the regulatory agency along with actions taken or proposed to cure the problem. Problems should be monitored to ensure that the corrective action is effective, and actions should be taken to incorporate prevention in ongoing operations. The public should be informed if the problem is serious enough to cause public concern.

7. LEGAL PROTECTIONS

There are a variety of laws and policies administered by disparate federal agencies. This results in varied interpretations and necessitates coordination between the agencies. However, the regulatory program of the U.S. Army Corps of Engineers has developed into what is probably the most comprehensive and respected wetlands protection program.

8. RIVERS AND HARBORS ACT–1899

In 1899, Congress enacted the Rivers and Harbors Act, sometimes called the Refuse Act. Section 10 of the Act prohibited anyone from creating any obstruction to the navigable capacity of waters of the United States or building any construction on a river or harbor without a permit from the Secretary of War, upon recommendation of the Chief of Engineers. Section 13 prohibited the discharge of any refuse into navigable waters or their tributaries (other than liquid flowing from streets and sewers) or the placing of any material where it might be washed into navigable waters or their tributaries, unless the Secretary of War permitted it upon the Chief of Engineers' finding that anchorage or navigation would not be injured by such activities. These sections are still applicable to construction and other activities affecting wetlands.

The Section 13 permit process was replaced by the 1972 Clean Water Act permitting requirements, but the law remains in effect. Section 10 of the 1899 Act survives the Clean Water Act and supplements the Corps' Section 404 authority. For most practical purposes, Section 10 merges with Section 404, but Section 404 affects only "discharges," while Section 10 affects the creation of any obstruction, whether by discharge or not. On the other hand, Section 404 affects any discharge, whereas Section 10 only affects something that in some way obstructs navigable capacity. Sections 10 and 13 can only be enforced by the federal government.

Historically, Section 13 is the basis of two major components of the 1972 Clean Water Act. First, it is the basis of Section 402 of the Clean Water Act and the National Pollutant Discharge Elimination System (NPDES) requirements, which, like Section 13, adopts a no-discharge-without-permit stance. Section 13 is also the indirect authority for Section 404, the Corps' program regulating discharges of dredged and fill material.

9. CLEAN WATER ACT–1972, 1982

The Clean Water Act (CWA), originally passed in 1972 as the Federal Water Pollution Control Act, was amended in 1982 as the Clean Water Act. A key part of the Act relative to wetlands is Section 404.

The Corps Regulatory Program

Section 404 was a compromise between those who wanted to leave the discharge of dredged and fill material in the Corps' hands and those who wanted to turn it over to the EPA, as was done with all other discharges in Section 402. The EPA writes the regulations that determine whether, or to what extent, the Corps issues a Section 404 permit. These are called the 404(b)(1) Guidelines. The EPA gets veto power over permits issued by the Corps. The EPA has joint enforcement authority with the Corps for violations of Section 404. A Memorandum of Understanding between the EPA and the Corps identifies which areas each organization will exercise primary authority over.

Section 404(a) authorizes the Secretary of the Army to issue permits, after notice and opportunity for public hearings, "for the discharge of dredged or fill material into the navigable waters at specified disposal sites." Dredged and fill material are both pollutants as defined in the CWA, which requires issuance of a permit in order to make the discharge. Dredged material is excavated from the bottom area of waters of the United States. Fill material, which can come from anywhere, is material whose primary purpose is to replace an aquatic area with dry land or to change the bottom elevation of a waterway.

"Discharge into the navigable waters" means adding material to the navigable waters. For example, using a bulldozer or other mechanical equipment to move the earth around for the purpose of filling a waterway is defined as a discharge.

The CWA defines navigable waters as "waters of the United States, including the territorial seas." This definition was intended to extend the permit authority under the CWA as broadly as constitutionally possible. In other words, waters covered do not have to be navigable in any way or sense. Wetlands are included as navigable waters.

In 1987, the Corps issued a Guidance Manual defining wetlands. The EPA, however, which is responsible for overall policy under Section 404, was not satisfied with the Corps' definition. Moreover, the Fish and Wildlife Service and the Department of Agriculture's Soil Conservation Service (SCS) (now the Natural Resources Conservation Service (NRCS)) had yet a different definition.

The SCS used its definition for enforcing the "Swampbuster" provisions of the 1985 Food Security Act, which deny certain benefits to farmers who convert wetlands for agriculture.

In 1989, the four agencies reached a common agreement in a document titled Federal Manual for Identifying and Delineating Jurisdictional Wetlands. This was perceived to greatly increase the Corps' authority. Developers and farmers in particular were upset. In August 1991, the four agencies, as part of a White House policy initiative on wetlands, proposed revisions to the 1989 manual. Environmentalists, in turn, criticized this proposal because of its substantial cutback on jurisdictional wetlands. The revisions were field-tested during the fall of 1991, and substantial complaints about the proposal's practical application arose from personnel using it. The result was that Congress required the Corps to use the 1987 manual until a new manual was adopted after notice and public comment.

Wetlands hydrology, hydric soil, and hydrophytic vegetation characterize wetlands. The 1989 manual indicated that under natural, undisturbed conditions, wetlands generally possess these three characteristics, but not all three must be demonstrated. If soils and hydrology were present, the manual indicated that vegetation could be assumed. It also indicated that if certain vegetation were present, soils could be assumed. Finally, it indicated that if the area were disturbed (someone had altered its natural state), hydrology could be assumed from vegetation or soils. The primary test to evaluate saturation was whether the area's water table rose to within 450 cm of the surface for seven consecutive days during the growing season.

The 1991 manual indicated that all three characteristics had to be established independently. It indicated that there had to be standing water at the surface for 21 consecutive days during the growing season.

The 1987 manual required all three characteristics, but permitted the use of field indicators to demonstrate one or more of the characteristics. The manual indicated that the water table must rise to within 30 cm of the surface for at least 5% of the growing season. Because all characteristics must be present under the 1987 manual, if the permit is sought outside the growing season, it may be difficult to tell what the saturation will be during the growing season, unless the presence of hydrophytic vegetation and hydric soils can be used to suggest the proper hydrological condition.

For the Swampbuster provisions of the Food Security Act, the Soil Conservation Service (SCS) of the Department of Agriculture decided whether wetlands could or could not be filled. After 1989, the SCS used the 1989

manual for making its determination. This meant that wetlands could be subject to determination by two different agencies, using different criteria for certain periods of the year. This resulted in confusion for the regulated community of farmers. In 1993, as part of the Clinton Wetlands Plan, responsibility for delineating wetlands for Section 404 in farm areas was transferred from the Corps to the SCS. (Since then, the name of the Soil Conservation Service has been changed to the Natural Resource Conservation Service.) This pleased farmers and upset environmentalists.

In addition, the Clinton Wetlands Plan called upon the Corps to establish a certification program to develop private sector experts. Individuals are certified as qualified *wetlands delineators*. The Corps can rely upon the delineations of these individuals rather than having to do it all itself. The Corps generally specifies that jurisdictional delineations are good for three years, at which time they are reconsidered.

Exceptions

Although Section 404 prohibits the discharge of dredged or fill material into the waters of the United States without a permit issued by the Corps, there are statutory exceptions to this rule. The EPA is responsible for defining these exceptions. These exceptions include:

- normal farming, silviculture (timber harvesting), and ranching activities

- maintenance, including emergency reconstruction of dikes, dams, bridges, etc.

- constructing farm or stock ponds or irrigation ditches

- constructing farm or forest roads where they are built to best management practices

The first exception, for "normal farming," does not include discharges of dredged or fill material to create farmland or pastureland. Moreover, the Corps took the position that "normal" farming meant farming in the same places as historical farming, so creation of farming in new areas did not qualify as "normal" farming. When the Swampbuster provisions of the Food Security Act of 1985 were being implemented by the NRCS (SCS), the NRCS (SCS) defined certain lands that had been drained or filled for agricultural purposes before 1985 as Prior Converted Cropland. In 1990, the Corps explained that Prior Converted Cropland was not to be considered wetlands subject to Section 404, thus distinguishing Farmed Wetlands from Prior Converted Cropland.

Farmed Wetlands were those areas subject to the "normal farming" exception to Section 404's permit requirements but which continued to have wetlands characteristics. As a result, if Farmed Wetlands were to be converted to another purpose, such as development, a Section 404 permit would be required.

Water Treatment

Because Prior Converted Croplands were lands whose wetlands characteristics had essentially been destroyed, those lands may simply no longer be wetlands at all. Under the NRCS (SCS) regulations, Prior Converted Croplands can be abandoned under certain circumstances. Under the Corps/EPA regulations, if that occurs and the area still has wetlands characteristics, the area will revert to Section 404 jurisdiction.

A second exception to the requirement to obtain a permit from the Corps to fill wetlands is where states are allowed to take over Section 404 regulatory authority for certain waters within their states. States cannot regulate waters that are currently used or are reasonably adaptable to use for interstate or foreign commerce, including wetlands adjacent to those waters. The EPA is responsible for determining whether a state's program is sufficient to take over Section 404 authority and for overseeing the adequacy of the state's implementation of Section 404.

While not constituting an exception to the requirement for a permit, Section 404(e) authorizes the Corps to issue general permits, as opposed to individual permits, if the Secretary determines that the activities are similar in nature, and that they will cause only minimal adverse environmental effects when considered separately or cumulatively. These permits are good for five years.

Permits

There are three types of general permits: state program, regional, and nationwide. The District Engineer can issue state program permits where there is a state program that requires a prior permit that emulates Section 404. The applicant gives plans to the District Engineer, who circulates it to interested federal agencies. If no one objects and the state permits the use, then the state program general permit is issued. Regional permits are issued by Corps District Engineers. The Chief of Engineers issues *nationwide permits* (NWPs) as rules that are adopted after notice and comment.

States may, in effect, veto or condition NWPs in their jurisdiction either by denying or conditioning a Section 401 certification, or by denying or conditioning a Coastal Zone Management Act consistency determination (for coastal regulated matters). Under Section 401 of the Clean Water Act, any applicant for a federal license or permit for any activity that may result in a discharge to waters of the United States must provide certification from the state that any such discharge will meet applicable requirements of the Clean Water Act. Accordingly, when the Corps proposes a nationwide (or regional) permit, a state may say that it does not meet state requirements under the Clean Water Act, thereby "denying" the certification within that state. An applicant who wants to use a nationwide or regional permit in a state that has denied a Section 401 certification can go to the state for an individual Section 401 certification.

The Corps has developed several standard NWPs that include criteria to determine whether the permit can be issued.

- Maintenance, NWP 3: The structure or fill must not be put to different uses.

- Utility Line Activities, NWP 12: There must be no change in the preconstruction contours.

- Bank Stabilization, NWP 13: The activity must be less than 152 m in length and must not exceed 2.5 m^3 per meter of length.

- Linear Transportation Projects, NWP 14: The width of fill must be limited to the minimum necessary, fill in waters of the United States must be limited to 1350 m^2 in area, and fill in wetlands must not exceed 60 m in length. Individual Corps districts have the authority to impose more stringent thresholds.

- Minor Discharges, NWP 18: The discharge must not exceed 19 m^3, and there must be no more than 405 m^2 of loss in wetlands area.

- Minor Dredging, NWP 19: There must be no more than 19 m^3 dredged below the high water mark.

- Boat Ramps, NWP 36: Discharge of fill must not exceed 38 m^3, and the boat ramp cannot exceed 6 m in width.

Section 404 Permits

Section 404(b)(1) of the Clean Water Act requires the Corps to exercise its permit authority through the application of guidelines developed by the EPA.

The Guidelines set the substantive standards for permit decisions (although they are supplemented by a public interest review). The intent of the Guidelines is to protect aquatic ecosystems and wetlands by allowing permits only for the least environmentally damaging practicable alternative (LEDPA) that will accomplish the basic project purpose. Thus, the issuance of Section 404 permits generally is prohibited if there are practicable alternatives to the discharge. In order to identify the least damaging alternative, the guidelines require a rigorous examination of alternatives that satisfy this basic project purpose.

The Guidelines establish a presumption that practicable alternatives are available for non-water dependent projects unless the applicant clearly demonstrates otherwise.

Even if no alternative is available, the Guidelines require permit denial where the project will result in significant degradation of the aquatic resource. For any filling that is allowed, the Guidelines impose stringent mitigation requirements in an attempt to replicate the functions and values that were served by the wetlands damaged.

At the heart of the Guidelines is the so-called *practicable alternatives test:*

> No discharge may be permitted where there is a practicable alternative to the proposed discharge which would have less adverse impact on the aquatic ecosystem, so long as the alternative doesn't have other significant adverse environmental consequences.

A practicable alternative is one that fulfills the basic project purpose and is available and capable of being done, taking cost, existing technology, and logistics into account.

Mitigation

The Guidelines require "appropriate and practicable" mitigation as a precondition to discharge. These rules are supplemented by the Mitigation Memorandum of Agreement (MOA) entered into by the EPA and the Department of the Army in February of 1990. Key principles include the following:

1. Sequencing—The MOA notes that, in evaluating potential mitigation measures:

> The Corps... first makes a determination that potential impacts have been avoided to the maximum extent practicable; remaining unavoidable impacts will then be mitigated to the extent appropriate and practicable by requiring steps to minimize impacts and, finally, compensate for aquatic resource values.

> This hierarchy (first avoidance, then minimization, and, as a last resort, compensation) precludes the consideration of mitigation as a means of "buying down" adverse effects in the practicable alternatives analysis.

2. Compensatory Mitigation—In those situations where compensatory mitigation is appropriate, the MOA establishes the following guidelines:

(a) The goal is one-to-one replacement of functional values, not necessarily area-for-area replacement. The EPA frequently looks for two-to-one area-for-area replacement.

(b) In-kind, on-site compensation is preferred.

(c) The restoration and enhancement of existing wetlands is preferred over the creation of new wetlands.

(d) Mitigation requirements are to be incorporated into permits together with ongoing monitoring requirements designed to ensure that the mitigation is accomplishing what it was designed to accomplish. The permit should contain "reopeners" allowing for the imposition of further conditions, including further mitigation if the monitoring demonstrates inadequate results.

Superimposed over the Guidelines, the Corps' regulations subject all discharge applications to a public interest review. All factors that may be relevant to the proposal may be considered, including economics, aesthetics, general environmental concerns, historic features, floodplain values, and recreation. The Corps' regulations create a loose presumption that projects satisfying the Section 404(b)(1) guidelines are in the public interest. Thus, the burden is on the Corps to determine that the project is not in the public interest.

Environmental Impact Evaluation

Section 102(2)(c) of the National Environmental Policy Act requires an Environmental Impact Statement (EIS) whenever a major federal action may significantly affect the environment. Most Section 404 permits require only an environmental assessment (EA). EAs are required unless the project qualifies for a nationwide permit, is subject to a categorical exclusion, or already has been deemed to require an EIS. The Corps can take mitigation into account in determining whether an EIS is required, but it must be enforceable. Any decisions made regarding the need to supplement an EIS are to be upheld unless they are found to be arbitrary and capricious. The Corps is entitled to weigh the probative value of any additional information that is submitted.

Topic IV: Air Resources #1: Ventilation

Ventilation

29 Thermodynamic Properties of Substances

Nomenclature

a	van der Waals factor	atm-ft^6/lbmol	Pa·m^6/kmol
AW	atomic weight	lbm/lbmol	kg/kmol
b	van der Waals factor	ft^3/lbmol	m^3/kmol
B	volumetric fraction	–	–
c	specific heat	Btu/lbm-°F	kJ/kg·°C
C	molar specific heat	Btu/lbmol-°F	kJ/kmol·°C
E	energy	Btu	kJ
G	gravimetric fraction	–	–
h	enthalpy	Btu/lbm	kJ/kg
H	molar enthalpy	Btu/lbmol	kJ/kmol
J	Joule's constant, 778.17	ft-lbf/Btu	n.a.
k	ratio of specific heats	–	–
m	mass	lbm	kg
MW	molecular weight	lbm/lbmol	kg/kmol
n	number of moles	–	–
N	number of molecules	–	–
N_A	Avogadro's number, 6.022×10^{23}	molecules/ lbmol	molecules/ kmol
p	absolute pressure	lbf/ft^2	Pa
q	heat	Btu/lbm	kJ/kg
Q	molar heat	Btu/lbmol	kJ/kmol
Q	total heat	Btu	kJ
R	specific gas constant	ft-lbf/lbm-°R	kJ/kg·K
R^*	universal gas constant	ft-lbf/lbmol-°R	kJ/kmol·K
s	entropy	Btu/lbm-°R	kJ/kg·K
S	molar entropy	Btu/lbmol-°R	kJ/kmol·K
T	absolute temperature	°R	K
T	temperature	°F	°C
u	internal energy	Btu/lbm	kJ/kg
U	molar internal energy	Btu/lbmol	kJ/kmol
v	velocity	ft/sec	m/s
V	molar specific volume	ft^3/lbmol	m^3/kmol
V	volume	ft^3	m^3
x	mole fraction	–	–
x	quality	–	–
Z	compressibility factor	–	–

Symbols

α	isentropic compressibility	1/°R	1/K
β	isobaric compressibility	1/°R	1/K
κ	Boltzmann constant	n.a.	kJ/molecule·K
κ	isothermal compressibility	1/°R	1/K
ρ	density	lbm/ft^3	kg/m^3
v	specific volume	ft^3/lbm	m^3/kg
ϕ	entropy function	Btu/lbm-°R	kJ/kg·K

Subscripts

A	Avogadro
c	critical
f	fluid (liquid)
fg	liquid-to-gas (vaporization)

Ventilation

g gas (vapor)
i ice
k kinetic
l latent
m mean
o outside (environment)
p constant pressure or most probable
r ratio or reduced
rms   root-mean-square
s sensible or solid
sat   saturated
t total
v constant volume

1. PHASES OF A PURE SUBSTANCE

Thermodynamics is the study of a substance's energy-related properties. The properties of a substance and the procedures used to determine those properties depend on the state and the phase of the substance. The *thermodynamic state* of a substance is defined by two or more independent thermodynamic properties. For example, the temperature and pressure of a substance are two properties commonly used to define the state of a superheated vapor.

The common *phases* of a substance are solid, liquid, and gas. However, because substances behave according to different rules, it is convenient to categorize them into more than these three phases.[1]

solid: A solid does not take on the shape or volume of its container.

subcooled liquid: If a liquid is not saturated (i.e., the liquid is not at its boiling point), it is said to be subcooled. Water at 1 atm and room temperature is subcooled, as the addition of a small amount of heat will not cause vaporization.

saturated liquid: A saturated liquid has absorbed as much heat energy as it can without vaporizing. Liquid water at standard atmospheric pressure and 212°F (100°C) is an example of a saturated liquid.

liquid-vapor mixture: A liquid and vapor of the same substance can coexist at the same temperature and pressure. This is called a two-phase, liquid-vapor mixture.

saturated vapor: A vapor (e.g., steam at standard atmospheric pressure and 212°F (100°C)) that is on the verge of condensing is said to be saturated.

superheated vapor: A superheated vapor is one that has absorbed more heat than is needed merely to vaporize it. A superheated vapor will not condense when small amounts of heat are removed.

ideal gas: A gas is a highly superheated vapor. If the gas behaves according to the ideal gas law, $pV = R^*T$, it is called an ideal gas.

---

[1]Plasma, *cryogenic fluids* (*cryogens*) that boil at temperatures less than approximately 200°R (110K), and solids near absolute zero are not discussed in this chapter.

real gas: A real gas does not behave according to the ideal gas laws.

gas mixtures: Most gases mix together freely. Two or more pure gases together constitute a gas mixture.

vapor/gas mixtures: Atmospheric air is an example of a mixture of several gases and water vapor.

These phases and subphases can be illustrated with a pure substance in the piston/cylinder arrangement shown in Fig. 29.1. The pressure in this system is constant and is determined by the weight of the piston, which moves freely to permit volume changes.

Figure 29.1 Phase Changes at Constant Pressure

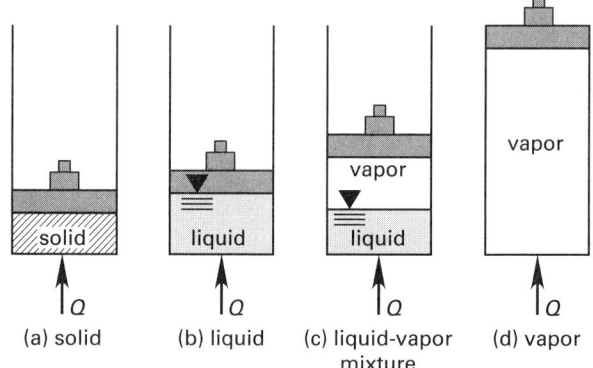

(a) solid   (b) liquid   (c) liquid-vapor mixture   (d) vapor

In Fig. 29.1(a) the volume is minimum. This is usually the solid phase. (Water is an exception. Solid ice has a lower density than liquid water.) The temperature will rise as heat, Q, is added to the solid. This increase in temperature is accompanied by a small increase in volume. The temperature increases until the melting point is reached.

The solid will begin to melt as heat is added to it at the melting point. The temperature will not increase until all of the solid has been turned into liquid. Until the ice is completely melted, solid ice and liquid water will coexist. (The term *slush ice* describes a mixture of water and small chunks of solid ice at the freezing point.) The liquid phase, with its small increase in volume, is illustrated by Fig. 29.1(b).

If the subcooled liquid continues to receive heat, its temperature will rise. This temperature increase continues until evaporation is imminent. The liquid at this point is said to be saturated. Any increase in heat energy will cause a portion of the liquid to vaporize. This is shown in Fig. 29.1(c), in which a liquid-vapor mixture exists.

As with melting, evaporation occurs at constant temperature and pressure but with a very large increase in volume. The temperature cannot increase until the last drop of liquid has been evaporated, at which point the vapor is said to be saturated. This is shown in Fig. 29.1(d).

Ventilation

Additional heat will result in high-temperature super-heated vapor. This vapor may or may not behave according to the ideal gas laws.

2. DETERMINING PHASE

It is possible to develop a three-dimensional surface that predicts the substance's phase based on the properties of pressure, temperature, and specific volume. Such an *equilibrium solid* is illustrated in Fig. 29.2.[2]

Figure 29.2 Equilibrium Solid

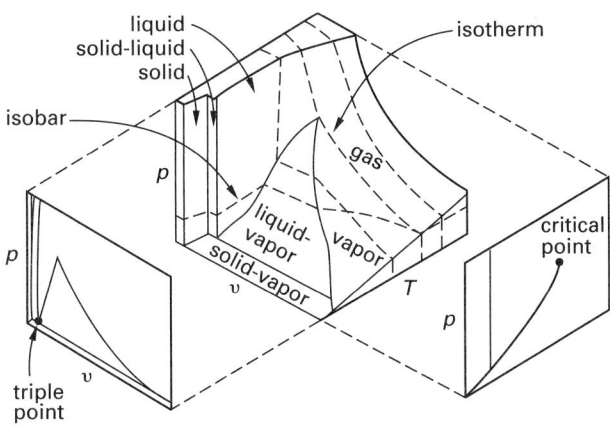

If one property is held constant during a process, a two-dimensional projection of the equilibrium solid can be used. This projection is known as an *equilibrium diagram* or a *phase diagram*, of which Fig. 29.3 is an example.

Figure 29.3 Constant Temperature Phase Diagram

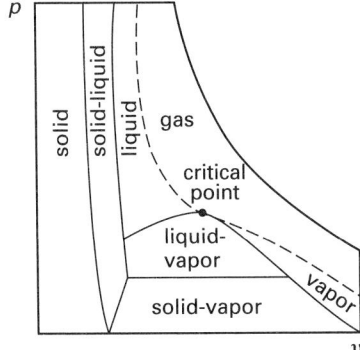

---

[2]Figure 29.2 is applicable for most substances, excluding water. In Fig. 29.2, the specific volume "steps in" (i.e., decreases) when the liquid turns to a solid as the temperature drops below freezing. However, water is unique in that it expands upon freezing. Thus, the pVT diagram for water shows a "step out" instead of a "step in" upon freezing. The remainder of the pVT diagram for water is the same as Fig. 29.2.

The most important part of a phase diagram is limited to the liquid-vapor region. A general phase diagram showing this region and the bell-shaped dividing line (known as the *vapor dome*) is shown in Fig. 29.4.

Figure 29.4 Vapor Dome with Isobars

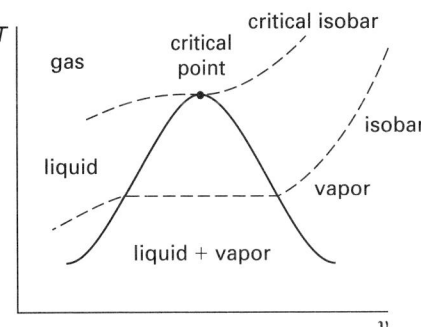

The vapor dome region can be drawn with many variables for the axes. For example, either temperature or pressure can be used for the vertical axis. Energy, specific volume, or entropy can be chosen for the horizontal axis. The principles presented here apply to all combinations.

The left-hand part of the vapor dome separates the liquid phase from the liquid-vapor phase. This part of the line is known as the *saturated liquid line*. Similarly, the right-hand part of the line separates the liquid-vapor phase from the vapor phase. This line is called the *saturated vapor line*.

Lines of constant pressure (*isobars*) can be superimposed on the vapor dome. Each isobar is horizontal as it passes through the two-phase region, indicating that both temperature and pressure remain unchanged as a liquid vaporizes.

Notice that there is no real dividing line between liquid and vapor at the top of the vapor dome. Far above the vapor dome, there is no distinction between liquids and gases, as their properties are identical. The phase is assumed to be a gas.

The implied dividing line between liquid and gas is the isobar that passes through the topmost part of the vapor dome. This is known as the *critical isobar*. The highest point of the vapor dome is known as the *critical point*. This critical isobar also provides a way to distinguish between a vapor and a gas. A substance below the critical isobar (but to the right of the vapor dome) is a vapor. Above the critical isobar, it is a gas.

Figure 29.5 illustrates a vapor dome for which pressure, p, has been chosen as the vertical axis, and enthalpy, h, has been chosen as the horizontal axis. The shape of the dome is essentially the same, but the lines of constant temperature (*isotherms*) have a different direction than isobars.

Figure 29.5 also illustrates the subscripting convention used to identify points on the saturation line. The subscript f (fluid) is used to indicate a saturated liquid. The subscript g (gas) is used to indicate a saturated vapor.[3] The subscript fg is used to indicate the difference in saturation properties and is used with vaporization properties.

Figure 29.5 *Vapor Dome with Isotherms*

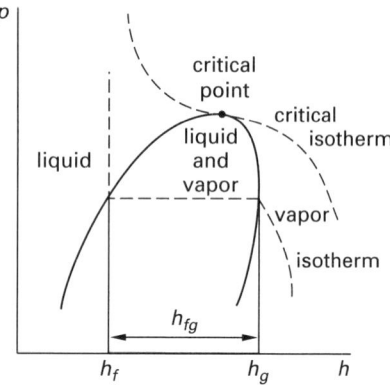

The vapor dome is a good tool for illustration, but it cannot be used to determine a substance's phase. Such a determination must be made based on the substance's pressure and temperature.

For example, consider water at a pressure of 1 atm. Its boiling temperature (the *saturation temperature*) at this pressure is 212°F (100°C). If the water has a lower temperature, for example, 85°F (29°C), the water will be liquid (i.e., will be subcooled). On the other hand, if the water's temperature (at 1 atm) is 270°F (132°C), the water must be in vapor form (i.e., must be superheated).

This example is valid only for water at atmospheric pressure. Water at other pressures will have other boiling temperatures. (The lower the pressure, the lower the boiling temperature.) However, the rules given here follow directly from the example.

Rule 1: A substance is a subcooled liquid if its temperature is less than the saturation temperature corresponding to its pressure.

Rule 2: A substance is in the liquid-vapor region if its temperature is equal to the saturation temperature corresponding to its pressure.

Rule 3: A substance is a superheated vapor if its temperature is greater than the saturation temperature corresponding to its pressure.

The rules that follow can be stated using pressure as the determining variable.

Rule 4: A substance is a subcooled liquid if its pressure is greater than the saturation pressure corresponding to its temperature.

Rule 5: A substance is in the liquid-vapor region if its pressure is equal to the saturation pressure corresponding to its temperature.

Rule 6: A substance is a superheated vapor if its pressure is less than the saturation pressure corresponding to its temperature.

3. PROPERTIES OF A SUBSTANCE

The thermodynamic *state* or condition of a substance is determined by its properties. *Intensive properties* are independent of the amount of substance present. Temperature, pressure, and stress are examples of intensive properties. *Extensive properties* are dependent on the amount of substance present. Examples of extensive properties are volume, strain, charge, and mass.

In this chapter, and in most books on thermodynamics, both lowercase and uppercase forms of the same characters are used to represent property variables. The two forms are used to distinguish between the units of mass. For example, lowercase h represents *specific enthalpy* (usually just called "enthalpy") in units of Btu/lbm or kJ/kg. Uppercase H is used to represent the *molar enthalpy* in units of Btu/lbmol or kJ/kmol.

4. MASS: *m*

The mass of a substance is a measure of its quantity. Mass is independent of location and gravitational field strength. In thermodynamics, the customary U.S. and SI units of mass, m, are pound-mass (lbm) and kilogram (kg), respectively.

5. TEMPERATURE: *T*

Temperature is a thermodynamic property of a substance that depends on internal energy content. Heat energy entering a substance will increase the temperature of that substance. Normally, heat energy will flow only from a hot object to a cold object. If two objects are in *thermal equilibrium* (i.e., are at the same temperature), no heat will flow between them.

If two systems are in thermal equilibrium, they must be at the same temperature. If both systems are in equilibrium with a third, then all three are at the same temperature. This concept is known as the *Zeroth Law of Thermodynamics*.

The scales most commonly used for measuring temperature are the Fahrenheit and Celsius scales.[4] The relationship between these two scales is

$$T_{°F} = 32° + \tfrac{9}{5} T_{°C} \qquad 29.1$$

---

[3]Although this book makes it a rule never to call a vapor a gas, this convention is not adhered to in the field of thermodynamics. The subscript g is standard for a saturated vapor.

[4]The term *centigrade* was replaced by the term *Celsius* in 1948.

The *absolute temperature scale* defines temperature independently of the properties of any particular substance. This is unlike the Celsius and Fahrenheit scales, which are based on the freezing point of water. The absolute temperature scale should be used for all calculations.

In the customary U.S. system, the absolute scale is the *Rankine scale*.[5]

$$T_{°R} = T_{°F} + 459.67° \qquad 29.2$$

$$\Delta T_{°R} = \Delta T_{°F} \qquad 29.3$$

The absolute temperature scale in the SI system is the *Kelvin scale*.[6]

$$T_K = T_{°C} + 273.15° \qquad 29.4$$

$$\Delta T_K = \Delta T_{°C} \qquad 29.5$$

The relationships between temperature differences in the customary U.S. and SI systems are independent of the freezing point of water. (See Table 29.1.)

$$\Delta T_{°C} = \tfrac{5}{9}\Delta T_{°F} \qquad 29.6$$

$$\Delta T_K = \tfrac{5}{9}\Delta T_{°R} \qquad 29.7$$

Table 29.1 *Temperature Scales*

	Kelvin	Celsius	Rankine	Fahrenheit
normal boiling point of water	373.15K	100.00°C	671.67°R	212.00°F
triple point of water (see Sec. 14)	273.16K	0.01°C	491.69°R	32.02°F
	273.15K	0.00°C	491.67°R	32.00°F ice point
absolute zero	0K	−273.15°C	0°R	−459.67°F

6. PRESSURE: *p*

Customary U.S. pressure units are pounds per square inch (psi). Standard SI pressure units are kPa or MPa, although bars are also used in tabulations of thermodynamic data.

7. DENSITY: ρ

Customary U.S. density units in tabulations of thermodynamic data are pounds per cubic foot (lbm/ft³). Standard SI density units are kilograms per cubic meter (kg/m³). Density is the reciprocal of specific volume.

$$\rho = \frac{1}{v} \qquad 29.8$$

[5]Normally, three significant temperature digits (i.e., 460°) are sufficient.
[6]Normally, three significant temperature digits (i.e., 273°) are sufficient.

8. SPECIFIC VOLUME: v AND *V*

Specific volume, v, is the volume occupied by one unit mass of a substance. Customary U.S. units in tabulations of thermodynamic data are cubic feet per pound (ft³/lbm). Standard SI specific volume units are cubic meters per kilogram (m³/kg). *Molar specific volume, V*, with units of ft³/lbmol (m³/kmol), is the volume of a mole of the substance, but is seldom encountered. Specific volume is the reciprocal of density.

$$v = \frac{1}{\rho} \qquad 29.9$$

$$V = \text{MW} \times v \qquad 29.10$$

9. INTERNAL ENERGY: *u* AND *U*

Internal energy includes all of the potential and kinetic energies of the atoms or molecules in a substance. Energies in the translational, rotational, and vibrational modes are included. Since this movement increases as the temperature increases, internal energy is a function of temperature. It does not depend on the process or path taken to reach a particular temperature.

In the United States, the *British thermal unit*, Btu, is used to measure all forms of thermodynamic energy. (One Btu is approximately the energy given off by burning one wooden match.) Standard units of *specific internal energy, u*, are Btu/lbm and kJ/kg. The units of *molar internal energy, U*, are Btu/lbmol and kJ/kmol. Equation 29.11 gives the relationship between the specific and molar quantities.

$$U = \text{MW} \times u \qquad 29.11$$

10. ENTHALPY: *h* AND *H*

Enthalpy (also known at various times in history as *total heat* and *heat content*) represents the total useful energy of a substance. Useful energy consists of two parts—the internal energy, *u*, and the *flow energy* (also known as *flow work* and *p-V work*), *pV*. Therefore, enthalpy has the same units as internal energy.

$$h = u + pv \qquad 29.12$$

$$H = U + pV \qquad 29.13$$

$$H = \text{MW} \times h \qquad 29.14$$

Enthalpy is defined as useful energy because, ideally, all of it can be used to perform useful tasks. It takes energy to increase the temperature of a substance. If that internal energy is recovered, it can be used to heat something else (e.g., to vaporize water in a boiler). Also, it takes energy to increase pressure and volume (as in blowing up a balloon). If pressure and volume are decreased, useful energy is given up.

The customary U.S. units of Eq. 29.12 and Eq. 29.13 are not consistent, since flow work (as written) has units of ft-lbf/lbm, not Btu/lbm. (There is also a consistency problem if pressure is defined in lbf/ft^2 and given in lbf/in^2.) Strictly, Eq. 29.12 should be written as

$$h = u + \frac{pv}{J} \qquad \text{[U.S.]} \quad 29.15$$

The conversion factor, J, in Eq. 29.15 is known as *Joule's constant*. It has a value of 778.17 ft-lbf/Btu. (In SI units, Joule's constant has a value of 1.0 N·m/J and is unnecessary.) As in Eq. 29.12 and Eq. 29.13, Joule's constant is often omitted from the statement of generic thermodynamic equations, but it is always needed with customary U.S. units for dimensional consistency.

11. ENTROPY: s AND S

Absolute entropy is a measure of the energy that is no longer available to perform useful work within the current environment. Other definitions used (the "disorder of the system," the "randomness of the system," etc.) are frequently quoted. Although these alternate definitions cannot be used in calculations, they are consistent with the *third law of thermodynamics* (also known as the *Nernst theorem*). This law states that the absolute entropy of a perfect crystalline solid in thermodynamic equilibrium is (approaches) zero when the temperature is (approaches) absolute zero.[7] Equation 29.16 expresses the third law mathematically.

$$\lim_{T \to 0K} s = 0 \qquad 29.16$$

An increase in entropy is known as *entropy production*. The total absolute entropy in a system is equal to the summation of all absolute entropy productions that have occurred over the life of the system.

$$s = \sum \Delta s \qquad 29.17$$

For an isothermal heat transfer process taking place at a constant temperature T_o, the entropy production depends on the amount of energy transfer.

$$\Delta s = \frac{q}{T_o} \qquad 29.18$$

$$\Delta S = \frac{Q}{T_o} \qquad 29.19$$

$$\Delta S = \text{MW} \times \Delta s \qquad 29.20$$

For heat transfer processes that occur over a varying temperature, the entropy production must be found by integration.

$$\Delta s = \int ds = \int \frac{dq}{T} \qquad 29.21$$

From Eq. 29.18 and Eq. 29.21, it is apparent that the units of specific absolute entropy are Btu/lbm-°R and kJ/kg·K. For molar absolute entropy, the units are Btu/lbmol-°R and kJ/kmol·K.

Unlike absolute entropy, *standardized entropy*, usually just called "entropy," is not referenced to absolute zero conditions but is measured with respect to some other convenient thermodynamic state. For water, the reference condition is the liquid phase at the triple point. (See Sec. 29.14.)

Example 29.1

Three planets of identical size and mass are oriented in space such that radiant energy transfers can occur. The average temperatures of planets A, B, and C are 530°R, 520°R, and 510°R (294K, 289K, and 283K), respectively. All three planets are massive enough that small energy losses or gains can be considered to be isothermal processes (i.e., they will not change the average temperature).

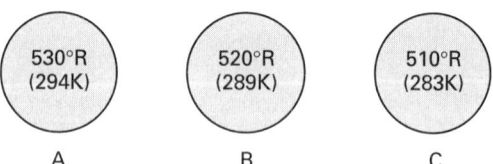

(a) Can a radiation energy transfer occur spontaneously from planet B to planet C? (b) What are the entropy productions for planets B and C if an energy transfer of 1000 Btu/lbm (2330 kJ/kg) occurs by radiation? (c) What is the overall entropy change as a result of the energy transfer in (b)? (d) Does entropy always increase? (e) Can planet B ever be returned to its original condition?

Solution

(a) A radiation transfer can occur spontaneously because planet B is hotter than planet C. Energy will flow spontaneously from a hot object to a cold object.

(b) In SI units, the entropy productions are

$$\Delta s_{\text{B}} = \frac{q}{T_o} = \frac{-2330 \, \frac{\text{kJ}}{\text{kg}}}{289\text{K}} = -8.062 \text{ kJ/kg·K}$$

$$\Delta s_{\text{C}} = \frac{2330 \, \frac{\text{kJ}}{\text{kg}}}{283\text{K}} = 8.233 \text{ kJ/kg·K}$$

---

[7]A molecule with zero entropy exists in only one quantum state. The energy state is known precisely, without *uncertainty*.

In customary U.S. units, the entropy productions are

$$\Delta s_{\mathrm{B}} = \frac{q}{T_o} = \frac{-1000 \; \dfrac{\text{Btu}}{\text{lbm}}}{520^\circ \text{R}} = -1.923 \; \text{Btu/lbm-}^\circ\text{R}$$

$$\Delta s_{\mathrm{C}} = \frac{1000 \; \dfrac{\text{Btu}}{\text{lbm}}}{510^\circ \text{R}} = 1.961 \; \text{Btu/lbm-}^\circ\text{R}$$

(c) The entropy change is not the same for the two planets. Entropy is not conserved in an energy transfer process. The overall entropy production is

$$\Delta s = \Delta s_{\mathrm{B}} + \Delta s_{\mathrm{C}} = -8.062 \; \frac{\text{kJ}}{\text{kg·K}} + 8.233 \; \frac{\text{kJ}}{\text{kg·K}}$$

$$= 0.171 \; \text{kJ/kg·K}$$

In customary U.S. units,

$$\Delta s = \Delta s_{\mathrm{B}} + \Delta s_{\mathrm{C}} = -1.923 \; \frac{\text{Btu}}{\text{lbm-}^\circ\text{R}} + 1.961 \; \frac{\text{Btu}}{\text{lbm-}^\circ\text{R}}$$

$$= 0.038 \; \text{Btu/lbm-}^\circ\text{R}$$

(d) Local entropy can decrease, as shown by planet B's negative entropy production. However, overall entropy always increases when the total universe is considered, as shown in part (c).

(e) Planet B can be brought back to its original condition if 1000 Btu/lbm (2330 kJ/kg) of energy is transferred from planet A to planet B. (Heat will not flow spontaneously from planet C to planet B, as heat will not flow spontaneously from a cold object to a hot object.[8])

12. SPECIFIC HEAT: c AND C

An increase in internal energy is needed to cause a rise in temperature. Different substances differ in the quantity of heat needed to produce a given temperature increase. The ratio of the energy (heat), Q, added to or removed from a body, to the body's mass, m, and temperature change, ΔT, is known as the *specific heat (specific heat capacity)*, c, of the body.

Because specific heats of solids and liquids are slightly temperature dependent, the mean specific heats are used when evaluating processes covering a large temperature range.

$$Q = mc\Delta T \qquad\qquad 29.22$$

$$c = \frac{Q}{m\Delta T} \qquad\qquad 29.23$$

The lowercase c implies that the units are Btu/lbm-°F or J/kg·°C. Typical values of specific heat are given in

---

[8]This is one way of stating the *Second Law of Thermodynamics*.

Table 29.2. The *molar specific heat*, designated by the symbol C, has units of Btu/lbmol-°F or J/kmol·°C.

$$C = \text{MW} \times c \qquad\qquad 29.24$$

Table 29.2 *Approximate Specific Heats of Selected Liquids and Solids**

substance	c_p	
	Btu/lbm-°F or Btu/lbm-°R	kJ/kg·°C or kJ/kg·K
aluminum, pure	0.23	0.96
aluminum, 2024-T4	0.2	0.84
ammonia	1.16	4.86
asbestos	0.20	0.84
benzene	0.41	1.72
brass, red	0.093	0.39
bronze	0.082	0.34
concrete	0.21	0.88
copper, pure	0.094	0.39
Freon-12	0.24	1.00
gasoline	0.53	2.20
glass	0.18	0.75
gold, pure	0.031	0.13
ice	0.49	2.05
iron, pure	0.11	0.46
iron, cast (4% C)	0.10	0.42
lead, pure	0.031	0.13
magnesium, pure	0.24	1.00
mercury	0.033	0.14
oil, light hydrocarbon	0.5	2.09
silver, pure	0.06	0.25
steel, 1010	0.10	0.42
steel, stainless 301	0.11	0.46
tin, pure	0.055	0.23
titanium, pure	0.13	0.54
tungsten, pure	0.032	0.13
water	1.0	4.19
wood (typical)	0.6	2.50
zinc, pure	0.088	0.37

(Multiply Btu/lbm-°F by 4.1868 to obtain kJ/kg·°C or kJ/kg·K.)
*Values in cal/g·°C are the same as Btu/lbm-°F.

For gases, the specific heat depends on the type of process during which the heat exchange occurs. Specific heats for constant-volume and constant-pressure processes are designated by c_v and c_p, respectively.

$$Q = mc_v\Delta T \quad \left[\begin{array}{l}\text{perfect gas} \\ \text{constant-volume process}\end{array}\right] \qquad 29.25$$

$$Q = mc_p\Delta T \quad \left[\begin{array}{l}\text{perfect gas} \\ \text{constant-pressure process}\end{array}\right] \qquad 29.26$$

Approximate values of c_p and c_v for common gases are given in Table 29.7. c_v and c_p for solids and liquids are essentially the same. However, the designation c_p is often encountered for solids and liquids.

The law of *Dulong and Petit* predicts the approximate molar specific heat (in cal/mol·°C) at high temperatures from the atomic weight.[9] This law is valid for solid elements having atomic weights greater than 40 and for most metallic elements. It is not valid at room temperature for carbon, silicon, phosphorus, and sulfur. 6.3 cal/mol·°C is known as the *Dulong and Petit value*.

$$c \times \text{AW} \approx 6.3 \pm 0.1 \qquad \textit{29.27}$$

Example 29.2

Compare the value of specific heat of pure iron calculated from Dulong and Petit's law with the value from Table 29.2.

Solution

The atomic weight of iron is 55.8. From Eq. 29.27,

$$c = \frac{6.3}{\text{AW}} = \frac{6.3 \; \dfrac{\text{cal}}{\text{mol·°C}}}{55.8 \; \dfrac{\text{g}}{\text{mol}}} = 0.11 \text{ cal/g·°C}$$

This is the same value as is given in Table 29.2.

13. RATIO OF SPECIFIC HEATS: *k*

For gases, the *ratio of specific heats*, k, is defined by Eq. 29.28. Typical values are given in Table 29.7.

$$k = \frac{c_p}{c_v} \qquad \textit{29.28}$$

14. TRIPLE POINT PROPERTIES

The *triple point* of a substance is a unique state at which solid, liquid, and gaseous phases can coexist. Table 29.3 lists values of the triple point for several common substances.

15. CRITICAL PROPERTIES

If the temperature and pressure of a liquid are increased, a state will eventually be reached at which the liquid and gas phases are indistinguishable. This state is known as the *critical point*, and the properties (generally temperature, pressure, and specific volume) at that point are known as the *critical properties*. At the critical point, the heat of vaporization, h_{fg}, becomes zero. Above the critical temperature, the substance will be a gas no matter how high the pressure. Critical properties are listed in Table 29.4.

---

[9]Dulong and Petit's law becomes valid at different temperatures for different substances, and a more specific definition of "high temperature" is impossible. For lead, the law is valid at 200K. For copper, it is not valid until above 400K.

Table 29.3 Approximate Triple Points

substance	pressure atm	temperature °R	K
ammonia	0.060	352	196
argon	0.676	151	84
carbon dioxide	5.10	390	217
helium	0.0508	4	2
hydrogen	0.0676	26	14
nitrogen	0.127	14	7.8
oxygen	0.00265	99	55
water	0.00592	492.02	273.34

Table 29.4 Approximate Critical Properties

substance	pressure atm	temperature °R	K
air	37.2	235.8	131.0
ammonia	111.5	730.1	405.6
argon	48.0	272.2	151.2
carbon dioxide	72.9	547.8	304.3
carbon monoxide	34.6	242.2	134.6
chlorine	75.9	751.0	417.2
ethane	48.8	549.8	305.4
ethylene	50.7	509.5	283.1
helium	2.3	10.0	5.56
hydrogen	12.8	60.5	33.6
mercury	180.0	2109.0	1171.7
methane	45.8	343.9	191.1
neon	25.7	79.0	43.9
nitrogen	33.5	227.2	126.2
oxygen	49.7	278.1	154.5
propane	42.0	666.3	370.2
sulfur dioxide	77.6	775.0	430.6
water vapor	218.2	1165.4	647.4
xenon	58.2	521.9	289.9

16. LATENT HEATS

The total energy (*total heat*, Q_t) entering a substance is the sum of the energy that changes the phase of the substance (*latent heat*, Q_l) and energy that changes the temperature of the substance (*sensible heat*, Q_s). During a phase change (solid to liquid, liquid to vapor, etc.), energy will be transferred to or from the substance without a change in temperature.[10]

$$Q_t = Q_s + Q_l \qquad \textit{29.29}$$

Examples of latent energies are the *latent heat of fusion* (i.e., change from solid to liquid), h_{sl}, *latent heat of vaporization*, h_{fg}, and *latent heat of sublimation* (i.e., direct change from solid to vapor without becoming

---

[10]Changes in crystalline form are also latent changes.

liquid), h_{ig}.[11,12] The energy required for these latent changes to occur in water is given in Table 29.5.

Table 29.5 Latent Heats for Water at One Atmosphere

effect	Btu/lbm	kJ/kg	cal/g
fusion	143.4	333.5	79.7
vaporization	970.1	2256.5	539.0
sublimation	1220	2838	677.8

Example 29.3

How much energy is required to convert 1.0 lbm (0.45 kg) of water that is originally at 75°F (24°C) and 1 atm to vapor at 212°F (100°C) and 1 atm?

SI Solution

The sensible heat required to raise the temperature of the water from 24°C to 100°C is given by Eq. 29.22.

$$Q_s = mc(T_2 - T_1)$$
$$= (0.45 \text{ kg}) \left(4.190 \ \frac{\text{kJ}}{\text{kg·°C}} \right) (100°C - 24°C)$$
$$= 143.3 \text{ kJ}$$

From Table 29.5, the latent heat required to vaporize the water is

$$Q_l = mh_{fg} = (0.45 \text{ kg}) \left(2256.5 \ \frac{\text{kJ}}{\text{kg}} \right) = 1015.4 \text{ kJ}$$

The total heat required is

$$Q_t = Q_s + Q_l = 143.3 \text{ kJ} + 1015.4 \text{ kJ}$$
$$= 1158.7 \text{ kJ}$$

Customary U.S. Solution

The sensible heat required to raise the temperature of the water from 75°F to 212°F is given by Eq. 29.22.

$$Q_s = mc(T_2 - T_1)$$
$$= (1 \text{ lbm}) \left(1.0 \ \frac{\text{Btu}}{\text{lbm-°F}} \right) (212°F - 75°F)$$
$$= 137.0 \text{ Btu}$$

From Table 29.5, the latent heat required to vaporize the water is

$$Q_l = mh_{fg} = (1 \text{ lbm}) \left(970.1 \ \frac{\text{Btu}}{\text{lbm}} \right) = 970.1 \text{ Btu}$$

The total heat required is

$$Q_t = Q_s + Q_l = 137.0 \text{ Btu} + 970.1 \text{ Btu}$$
$$= 1107.1 \text{ Btu}$$

17. QUALITY: *x*

Within the vapor dome, water is at its saturation pressure and temperature. When saturated, water can simultaneously exist in liquid and vapor phases in any proportion between 0 and 1. The *quality* is the fraction by weight of the total mass that is vapor.

$$x = \frac{m_{\text{vapor}}}{m_{\text{vapor}} + m_{\text{liquid}}} \qquad 29.30$$

18. GIBBS FUNCTION: *g* AND *G*

The *Gibbs function* is defined for a pure substance by Eq. 29.31 through Eq. 29.33.

$$g = h - Ts = u + pv - Ts \qquad 29.31$$
$$G = H - TS = U + pV - TS \qquad 29.32$$
$$G = \text{MW} \times g \qquad 29.33$$

The Gibbs function is used in investigating latent changes and chemical reactions. For a constant-temperature, constant-pressure nonflow process approaching equilibrium, the Gibbs function approaches its minimum value.

$$(dG)_{T,p} < 0 \qquad 29.34$$

Once the minimum value is obtained, the process will stop, and the Gibbs function will be constant.

$$(dG)_{T,p} = 0 \big|_{\text{equilibrium}} \qquad 29.35$$

Like enthalpy of formation, the Gibbs function, G^0, has been tabulated at the standard reference conditions of 25°C (77°F) and one atmosphere. A chemical reaction can occur spontaneously only if the change in Gibbs function is negative (i.e., the Gibbs function for the products is less than the Gibbs function for the reactants).

$$\sum_{\text{products}} nG^0 < \sum_{\text{reactants}} nG^0 \qquad 29.36$$

19. HELMHOLTZ FUNCTION: *a* AND *A*

The *Helmholtz function* is defined for a pure substance by Eq. 29.37 through Eq. 29.39.[13]

$$a = u - Ts = h - pv - Ts \qquad 29.37$$

---

[11]The subscript *s* (for "solid") is sometimes used in place of *i* (for "ice").
[12]*Sublimation* can only occur below the triple point, where it is too cold for the liquid phase to exist at all.

[13]In older references, the symbol *F* was commonly used for the Helmholtz function.

$$A = U - TS = H - pV - TS \qquad 29.38$$

$$A = \text{MW} \times a \qquad 29.39$$

Like the Gibbs function, the Helmholtz function is used in investigating equilibrium conditions. For a constant-temperature, constant-volume nonflow process approaching equilibrium, the Helmholtz function approaches its minimum value.

$$(dA)_{T,V} < 0 \qquad 29.40$$

Once the minimum value is obtained, the process will stop, and the Helmholtz function will be constant.

$$(dA)_{T,V} = 0\big|_{\text{equilibrium}} \qquad 29.41$$

20. FREE ENERGY

The Helmholtz function has, in the past, also been known as the *free energy* of the system because its change in a reversible isothermal process equals the energy that can be "freed" and converted to mechanical work. Unfortunately, the same term has also been used for the Gibbs function under analogous conditions. For example, the difference in standard Gibbs functions of reactants and products has often been called the "free energy difference."

Since there is a great possibility for confusion, it is better to refer to the Gibbs and Helmholtz functions by their actual names.

21. FUGACITY AND ACTIVITY: *f*

The *fugacity*, *f*, of a substance is a modified pressure that accounts for nonideal behavior. It has the same units as pressure. It is commonly used in conjunction with a generalized fugacity chart. Fugacity is defined by the *fugacity function* (see Eq. 29.42) and is related to the Gibbs function.

$$\lim_{p \to 0} f = p \qquad 29.42$$

$$(dG)_T = ZR^*T d(\ln p)_T = ZR^*T d(\ln f)_T \qquad 29.43$$

The ratio of fugacity at actual conditions to the fugacity at some reference state is known as the *activity*.

22. COMPRESSIBILITY: β, κ, AND α

Three different compressibilities are distinguished. The *isobaric compressibility*, β, is defined as

$$\beta = \left(\frac{1}{v}\right)\left(\frac{\partial v}{\partial T}\right)_p \qquad 29.44$$

The *isothermal compressibility*, κ, is

$$\kappa = \left(-\frac{1}{v}\right)\left(\frac{\partial v}{\partial p}\right)_T \qquad 29.45$$

The *isentropic compressibility*, α, is

$$\alpha = \left(-\frac{1}{v}\right)\left(\frac{\partial v}{\partial p}\right)_s \qquad 29.46$$

23. USING THE MOLLIER DIAGRAM

The *Mollier diagram (enthalpy-entropy diagram)* is a graph of enthalpy versus entropy for steam. (See App. 29.E.) It is particularly suitable for determining property changes between the superheated vapor and the liquid-vapor regions. For this reason, the Mollier diagram covers only a limited region.

The Mollier diagram plots the enthalpy for a unit mass of steam as the ordinate and plots the entropy as the abscissa. Lines of constant pressure (isobars) slope upward from left to right. Below the saturation line, curves of *constant moisture content* (the complement of quality) slope down from left to right. Above the saturation line are lines of constant temperature and lines of constant superheat.

Example 29.4

Find the following properties using the Mollier diagram: (a) enthalpy and entropy of steam at 700 psia and 1000°F, (b) enthalpy of steam at 1 psia and 80% quality, and (c) final temperature of steam throttled from 700 psia and 1000°F to 450 psia.

Solution

(a) Reading directly from the Mollier diagram (see App. 29.E), $h = 1510$ Btu/lbm, and $s = 1.7$ Btu/lbm-°R.

(b) 80% quality is the same as 20% moisture. Reading at the intersection of 20% moisture and 1 psia, $h = 900$ Btu/lbm.

(c) By definition, a throttling process does not change the enthalpy. This process is represented by a horizontal line to the right on the Mollier diagram. Starting at the intersection of 700 psia and 1000°F and moving horizontally to the right until 450 psia is reached defines the endpoint of the process. The final temperature is interpolated as approximately 990°F.

24. USING SATURATION TABLES

The information presented graphically on an enthalpy-entropy diagram can be obtained with greater accuracy from *saturation tables*, also known as *property tables* and (in the case of water) *steam tables*. These tables represent extensive tabulations of data for liquid and vapor phases of a substance.

Saturation tables contain values of enthalpy, h, entropy, s, internal energy, u, and specific volume, v. Within the vapor dome, these properties are functions of temperature. Appendix 29.A and App. 29.N (for water) are organized in this manner.

However, as shown in Fig. 29.4, there is a unique pressure associated with each temperature (i.e., there is only one horizontal isobar for each temperature). Since the pressure does not vary in even increments when temperature is changed, the second column of App. 29.A and App. 29.N varies irregularly. A second property table, App. 29.B and App. 29.O, is set up so that the pressure increments are uniform.

In App. 29.A and App. 29.N, the first column after temperature gives the corresponding saturation pressure. The next two columns give specific volume. The first of these gives the specific volume of the saturated liquid, v_f; the second column gives the specific volume of the saturated vapor, v_g.

The relationship between v_f, v_{fg}, and v_g is given by Eq. 29.47.

$$v_g = v_f + v_{fg} \qquad 29.47$$

The subsequent columns list the same data for enthalpy and entropy.

$$h_g = h_f + h_{fg} \qquad 29.48$$

$$s_g = s_f + s_{fg} \qquad 29.49$$

In App. 29.B and App. 29.O, the first column after the pressure gives the corresponding saturation temperature. The next two columns give specific volume in a manner similar to that of App. 29.A and App. 29.N.

25. USING SUPERHEAT TABLES

In the superheated region, pressure and temperature are independent properties. Therefore, for each pressure, a large number of temperatures is possible. Appendix 29.C and App. 29.P are superheated steam tables that give the properties of specific volume, enthalpy, and entropy for various combinations of temperature and pressure. The *degrees of superheat* (e.g., "100°F of superheat") represents the difference between actual and saturation temperatures.

26. USING COMPRESSED LIQUID TABLES

A liquid whose pressure is greater than the pressure corresponding to its saturation pressure is known as a *compressed liquid* or *subcooled liquid*. (See Sec. 29.30.) Liquids are only slightly compressible. For most thermodynamic problems, changes in properties for a liquid are negligible. In problems where the exact values are needed, tables such as App. 29.D and App. 29.Q of compressed liquid properties are available. Such tables

may present the properties directly or may give the properties at the saturation state and the corrections to those properties for various pressures.

Example 29.5

A device compresses 1.0 lbm of 300°F (150°C) saturated water to 1000 psia (7.5 MPa). What is the final specific volume?

SI Solution

$$p = \frac{(7.5 \text{ MPa})\left(10^6 \frac{\text{Pa}}{\text{MPa}}\right)}{10^5 \frac{\text{Pa}}{\text{bar}}} = 75 \text{ bars}$$

From App. 29.Q, the specific volume of the water at 75 bars and 150°C is read directly as 1085.8 cm³/kg.

Customary U.S. Solution

From App. 29.D, the specific volume of the water at 1000 psia and 300°F is read directly as 0.01738 ft³/lbm.

27. USING GAS TABLES

Gas tables are essentially superheat tables for which an assumption has been made about the pressure. For example, App. 29.F and App. 29.S are gas tables for air at low pressures. "Low pressure" means less than several hundred psi pressure. However, reasonably good results can be expected even if pressures are higher.

Gas tables are indexed by temperature. That is, implicit in their use is the assumption that properties are functions of temperature only. Gas tables are not arranged in the same way as other property tables.

The *volume ratio*, v_r, and the *pressure ratio*, p_r, columns can be used when gases take part in an isentropic process. These terms should not be confused with the reduced variables introduced in Sec. 29.42. These two columns are not the specific volume or pressure. They are ratios with arbitrary references that make analysis of isentropic processes easier. Their use is illustrated in Ex. 29.7 and is based on Eq. 29.50 and Eq. 29.51.

$$\frac{v_{r,1}}{v_{r,2}} = \frac{V_1}{V_2} \quad [\Delta s = 0] \qquad 29.50$$

$$\frac{p_{r,1}}{p_{r,2}} = \frac{p_1}{p_2} \quad [\Delta s = 0] \qquad 29.51$$

Entropy is not listed at all in App. 29.F and App. 29.S. Instead, a column of *entropy function*, ϕ, with the same units as entropy is given. The entropy function is not the same as specific entropy, but it can be used to calculate the change in entropy as the gas goes through a process. This entropy change is calculated with Eq. 29.52. Units of the gas constant must match those of s and ϕ.

$$s_2 - s_1 = \phi_2 - \phi_1 - R \ln\left(\frac{p_2}{p_1}\right) \qquad 29.52$$

Example 29.6

What is the enthalpy of air at 100°F (310K) and 50 psia (345 kPa)?

SI Solution

Since the pressure is low (less than 20 atm), App. 29.S can be used. From the $T = 310$K line, the enthalpy is read directly as 310.24 kJ/kg.

Customary U.S. Solution

Since the pressure is low (less than 300 psia), App. 29.F can be used. 100°F is the same as 560°R. From the $T = 560$°R line, the enthalpy is read directly as 133.86 Btu/lbm.

Example 29.7

Air is originally at 60°F and 14.7 psia. It is compressed isentropically to 86.5 psia. What are its new temperature and enthalpy?

Solution

From App. 29.F for $T = 60°F + 460° = 520°R$, $p_{r,1} = 1.2147$. Using Eq. 29.51,

$$p_{r,2} = \frac{p_{r,1}p_2}{p_1} = \frac{(1.2147)(86.5 \text{ psia})}{14.7 \text{ psia}} = 7.148$$

Searching the p_r column of App. 29.F results in $T = 860°R$ and $h = 206.46$ Btu/lbm.

28. USING PRESSURE-ENTHALPY CHARTS

For convenience (and by tradition), properties of refrigerants are typically shown graphically in a *pressure-enthalpy (p-h) diagram*. Figure 29.6 shows a skeleton *p-h* diagram. The vapor dome and saturation lines from Fig. 29.5 are recognizable. Lines of constant specific volume, constant entropy, and constant temperature are added.

Isotherms are horizontal within the vapor dome, corresponding to the constant saturation pressure. In the subcooled-liquid region to the left of the vapor dome, isotherms are essentially vertical, since the temperature (not the pressure) of a subcooled liquid determines enthalpy. In the superheated region, the isotherms gradually approach vertical as the gaseous refrigerant becomes more like an ideal gas.

The enthalpy of superheated refrigerant is easily determined from the pressure and temperature. Isentropic compression follows a line of constant entropy.

Figure 29.6 *Pressure-Enthalpy Diagram*

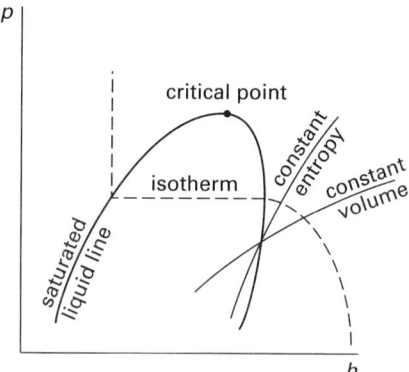

Throttling follows a line of constant enthalpy (i.e., a vertical line).

Enthalpies (plotted on the horizontal axis) may be scaled to an arbitrary "zero point." Comparing charts in traditional units and SI units can be perplexing. For most calculations, including refrigeration calculations, only the difference between two values is important. The absolute value at any particular point is seldom a useful value.

29. PROPERTIES OF SOLIDS

There are few mathematical relationships that predict the thermodynamic properties of solids. Properties such as temperature, specific heat, and density, usually are known. If the properties are not known, they must be found from tables.

The reference point for properties of solids is usually absolute zero temperature. That is, properties such as enthalpy and entropy are defined to be zero at 0°R (0K). This is an arbitrary convention. The choice of reference point does not affect the *change* in properties between two temperatures.

30. PROPERTIES OF SUBCOOLED LIQUIDS

A subcooled liquid is at a temperature less than the saturation temperature corresponding to its pressure. Unless the pressure of the liquid is very high, the various thermodynamic properties can be considered to be functions of only the liquid's temperature. In addition to compressed liquid tables (see Sec. 29.26), saturation tables (see Sec. 29.24) may be used to determine properties.

Example 29.8

What is the enthalpy of water at 30 psia (2.0 bars, 0.2 MPa) and 240°F (110°C, 383K)?

SI Solution

The logic is the same as presented in the customary U.S. solution.

$$p = 0.2 \text{ MPa} \quad (2 \text{ bars})$$

From App. 29.O,

$$T_{\text{sat},0.2\,\text{MPa}} = 120.2°C$$

$$T_{\text{actual}} < T_{\text{sat}} \quad [\text{liquid phase}]$$

From App. 29.N, for 110°C,

$$h_{f,110°C} = 461.42 \text{ kJ/kg}$$

Customary U.S. Solution

Although the substance is water, its phase is unknown. It could be liquid, vapor, or a combination of the two. From App. 29.B, the saturation (boiling) temperature for 30 psia water is approximately 250°F. Since the actual water temperature is less than the saturation temperature, the water is liquid.

As a liquid, the properties are essentially functions of temperature only. From App. 29.A for 240°F, $h = 208.5$ Btu/lbm.

31. PROPERTIES OF SATURATED LIQUIDS

Either the temperature or the pressure of a saturated liquid must be known in order to identify its thermodynamic state. One determines the other, since there is a one-to-one relationship between saturation pressure and saturation temperature. The first two columns of App. 29.A and App. 29.B (for SI, App. 29.N and App. 29.O) can be used to determine saturation temperatures and pressures.

Since the saturation tables are set up specifically for saturated substances, enthalpy, entropy, internal energy, and specific volume can be read directly from the h_f, s_f, u_f, and v_f columns, respectively. Density can be calculated as the reciprocal of the specific volume. The liquid's vapor pressure is the same as the saturation pressure listed in the table.

32. PROPERTIES OF LIQUID-VAPOR MIXTURES

When the thermodynamic state of a substance is within the vapor dome, there is a one-to-one correspondence between the saturation temperature and saturation pressure. One determines the other. The thermodynamic state is uniquely defined by any two independent properties (temperature and quality, pressure and enthalpy, entropy and quality, etc.).

If the quality of a liquid-vapor mixture is known, it can be used to calculate all of the primary thermodynamic properties. If a thermodynamic property has a value between the saturated liquid and saturated vapor values (i.e., h is between h_f and h_g), any of Eq. 29.53 through Eq. 29.56 can be solved for the quality.

$$h = h_f + xh_{fg} \qquad 29.53$$

$$s = s_f + xs_{fg} \qquad 29.54$$

$$u = u_f + xu_{fg} \qquad 29.55$$

$$v = v_f + xv_{fg} \qquad 29.56$$

Example 29.9

What is the final enthalpy of superheated steam that is expanded isentropically (i.e., with no change in entropy) from 100 psia (700 kPa) and 500°F (250°C) to 3 psia (20 kPa)?

SI Solution

From App. 29.P, the entropy, s_1, of the superheated steam at 700 kPa and 250°C is 7.1070 kJ/kg·K. Since the expansion is isentropic, this is also the final entropy, s_2.

From App. 29.O, for 20 kPa (0.20 bars) vapor, the entropy of a saturated liquid, s_f, is 0.8320 kJ/kg·K; and the entropy of a saturated vapor, s_g, is 7.9072 kJ/kg·K. Since $s_2 < s_g$, the expanded steam is in the liquid-vapor region. The quality of the mixture is given by Eq. 29.54. (There is no s_{fg} column in this appendix.)

$$x = \frac{s - s_f}{s_{fg}} = \frac{s - s_f}{s_g - s_f}$$

$$= \frac{7.1070 \frac{\text{kJ}}{\text{kg·K}} - 0.8320 \frac{\text{kJ}}{\text{kg·K}}}{7.9072 \frac{\text{kJ}}{\text{kg·K}} - 0.8320 \frac{\text{kJ}}{\text{kg·K}}}$$

$$= 0.8869$$

From App. 29.O, the enthalpy of saturated liquid, h_f, at 20 kPa is 251.42 kJ/kg. The heat of vaporization, h_{fg}, is 2357.5 kJ/kg. The final enthalpy is given by Eq. 29.53.

$$h = h_f + xh_{fg} = 251.42 \frac{\text{kJ}}{\text{kg}} + (0.8869)\left(2357.5 \frac{\text{kJ}}{\text{kg}}\right)$$

$$= 2342.3 \text{ kJ/kg}$$

Customary U.S. Solution

From App. 29.C, the entropy, s_1, of the superheated steam at 100 psia and 500°F is 1.7089 Btu/lbm-°R. Since the expansion is isentropic, this is also the final entropy, s_2.

From App. 29.B, for 3 psia vapor, the entropy of a saturated liquid, s_f, is 0.2009 Btu/lbm-°R; the entropy of vaporization, s_{fg}, is 1.6849 Btu/lbm-°R; and the entropy of a saturated vapor, s_g, is 1.8858 Btu/lbm-°R.

Since $s_2 < s_g$, the expanded steam is in the liquid-vapor region. The quality of the mixture is given by Eq. 29.54.

$$x = \frac{s - s_f}{s_{fg}} = \frac{1.7089 \; \frac{Btu}{lbm\text{-}°R} - 0.2009 \; \frac{Btu}{lbm\text{-}°R}}{1.6849 \; \frac{Btu}{lbm\text{-}°R}}$$

$$= 0.8950$$

From App. 29.B, the enthalpy of saturated liquid, h_f, at 3 psia is 109.4 Btu/lbm. The heat of vaporization, h_{fg}, is 1012.8 Btu/lbm. The final enthalpy is given by Eq. 29.53.

$$h = h_f + x h_{fg} = 109.4 \; \frac{Btu}{lbm} + (0.8950)\left(1012.8 \; \frac{Btu}{lbm}\right)$$

$$= 1015.9 \; Btu/lbm$$

Example 29.10

What is the enthalpy of 200°F (90°C, 363K) steam with a quality of 90%?

SI Solution

Using App. 29.N and Eq. 29.53,

$$h = h_f + x h_{fg} = 377.04 \; \frac{kJ}{kg} + (0.9)\left(2282.5 \; \frac{kJ}{kg}\right)$$

$$= 2431.3 \; kJ/kg$$

Customary U.S. Solution

Using App. 29.A and Eq. 29.53,

$$h = h_f + x h_{fg} = 168.13 \; \frac{Btu}{lbm} + (0.9)\left(977.6 \; \frac{Btu}{lbm}\right)$$

$$= 1048.0 \; Btu/lbm$$

33. PROPERTIES OF SATURATED VAPORS

Properties of saturated vapors can be read directly from the saturation tables. The vapor's pressure or temperature can be used to define its thermodynamic state. Enthalpy, entropy, internal energy, and specific volume can be read directly as h_g, s_g, u_g, and v_g, respectively. Saturated vapors are sometimes redundantly described as "dry" saturated vapors (e.g., "dry saturated steam") to emphasize that the quality, x, is 1.0 and that none of the substance is in the liquid state.

34. PROPERTIES OF SUPERHEATED VAPORS

Unless a vapor is highly superheated, its properties should be found from a superheat table, such as App. 29.C and App. 29.P (for water vapor). Since the temperature and pressure are independent for a superheated vapor, both must be known in order to define the thermodynamic state.

If the vapor's temperature and pressure do not correspond to the superheat table entries, single or double interpolation will be required. Such interpolation can be avoided by using more complete tables, but where required, double linear interpolation is standard practice.

Example 29.11

What is the enthalpy of water at 300 psia (20 bars, 2.0 MPa) and 900°F (500°C, 773K)?

SI Solution

The enthalpy can be read directly from App. 29.P as 3468.2 kJ/kg.

Customary U.S. Solution

It may not be obvious what phase the water is in. From App. 29.B, the saturation (boiling) temperature for 300 psia water is 417.35°F. Since the actual water temperature is higher than the saturation temperature, the water exists as a superheated vapor.

From App. 29.C, the enthalpy can be read directly as 1473.9 Btu/lbm.

35. EQUATION OF STATE FOR IDEAL GASES

An *equation of state* is a relationship that predicts the state (a property such as pressure, temperature, volume, etc.) from a set of two other independent properties.

Avogadro's law states that equal volumes of different gases at the same temperature and pressure contain equal numbers of molecules. For one mole of any gas, Avogadro's law can be stated as the *equation of state for ideal gases*. (Temperature, T, in Eq. 29.57 must be absolute.)

$$\frac{pV}{T} = R^* \qquad 29.57$$

In Eq. 29.57, R^* is known as the *universal gas constant*. It is "universal" (within a consistent system of units) because the same value can be used with any gas. Its value depends on the units used for pressure, temperature, and volume, as well as on the units of mass. (See Table 29.6.)

The ideal gas equation of state can be modified for more than one mole of gas. If there are n moles,

$$pV = nR^*T \qquad 29.58$$

The number of moles can be calculated from the substance's mass and molecular weight.

$$n = \frac{m}{MW} \qquad 29.59$$

Table 29.6 *Values of the Universal Gas Constant, R^**

units in SI and other metric systems

8.3143 kJ/kmol·K
8314.3 J/kmol·K
0.08206 atm·L/mol·K
1.986 cal/mol·K
8.314 J/mol·K
82.06 atm·cm^3/mol·K
0.08206 atm·m^3/kmol·K
8314.3 kg·m^2/s^2 kmol·K
8314.3 m^3·Pa/kmol·K
8.314 × 10^7 erg/mol·K

units in English systems

1545.33 ft-lbf/lbmol-°R
1.986 Btu/lbmol-°R
0.7302 atm-ft^3/lbmol-°R
10.73 ft^3-lbf/in^2-lbmol-°R

Equation 29.58 and Eq. 29.59 can be combined. R is the *specific gas constant*. It is specific because it is valid only for a gas with a molecular weight of MW.

$$pV = \frac{mR^*T}{\text{MW}} = m\left(\frac{R^*}{\text{MW}}\right)T = mRT \qquad 29.60$$

$$R = \frac{R^*}{\text{MW}} \qquad 29.61$$

Approximate values of the specific gas constant and the molecular weights of several common gases are given in Table 29.7.

Example 29.12

What mass of nitrogen is contained in a 2000 ft^3 (57 m^3) tank if the pressure and temperature are 1 atm and 70°F (21°C), respectively?

SI Solution

First, convert to absolute temperature.

$$T = 21°C + 273° = 294K$$

From Table 29.7, $R = 297$ J/kg·K. From Eq. 29.60,

$$m = \frac{pV}{RT} = \frac{(1\text{ atm})\left(1.013 \times 10^5 \, \frac{\text{Pa}}{\text{atm}}\right)(57\text{ m}^3)}{\left(297 \, \frac{\text{J}}{\text{kg·K}}\right)(294K)}$$

$$= 66.1\text{ kg}$$

Customary U.S. Solution

First, convert to absolute temperature.

$$T = 70°F + 460° = 530°R$$

From Table 29.7, $R = 55.16$ ft-lbf/lbm-°R. From Eq. 29.60,

$$m = \frac{pV}{RT}$$

$$= \frac{(1\text{ atm})\left(14.7 \, \frac{\text{lbf}}{\text{in}^2\text{-atm}}\right)\left(12 \, \frac{\text{in}}{\text{ft}}\right)^2(2000\text{ ft}^3)}{\left(55.16 \, \frac{\text{ft-lbf}}{\text{lbm-°R}}\right)(530°R)}$$

$$= 144.8\text{ lbm}$$

Example 29.13

A 25 ft^3 (0.71 m^3) tank contains 10 lbm (4.5 kg) of an ideal gas. The gas has a molecular weight of 44 and is at 70°F (21°C). What is the pressure of the gas?

SI Solution

From Eq. 29.61, the specific gas constant is

$$R = \frac{R^*}{\text{MW}} = \frac{8314.3 \, \frac{\text{J}}{\text{kmol·K}}}{44 \, \frac{\text{kg}}{\text{kmol}}}$$

$$= 189\text{ J/kg·K}$$

The absolute temperature is $T = 21°C + 273° = 294K$. From Eq. 29.60, the pressure is

$$p = \frac{mRT}{V} = \frac{(4.5\text{ kg})\left(189 \, \frac{\text{J}}{\text{kg·K}}\right)(294K)}{(0.71\text{ m}^3)\left(1000 \, \frac{\text{Pa}}{\text{kPa}}\right)}$$

$$= 352.2\text{ kPa}$$

Customary U.S. Solution

From Eq. 29.61, the specific gas constant is

$$R = \frac{R^*}{\text{MW}} = \frac{1545.33 \, \frac{\text{ft-lbf}}{\text{lbmol-°R}}}{44 \, \frac{\text{lbm}}{\text{lbmol}}}$$

$$= 35.12\text{ ft-lbf/lbm-°R}$$

The absolute temperature is $T = 70°F + 460° = 530°R$. From Eq. 29.60, the pressure is

$$p = \frac{mRT}{V} = \frac{(10\text{ lbm})\left(35.12 \, \frac{\text{ft-lbf}}{\text{lbm-°R}}\right)(530°R)}{25\text{ ft}^3}$$

$$= 7445\text{ lbf/ft}^2$$

Ventilation

Table 29.7 *Approximate Properties of Selected Gases*

gas	symbol	temperature (°F)	MW	customary U.S. units R ft-lbf / lbm-°R	c_p Btu / lbm-°R	c_v Btu / lbm-°R	SI units R J / kg·K	c_p J / kg·K	c_v J / kg·K	k
acetylene	C_2H_2	68	26.038	59.35	0.350	0.274	319.32	1465	1146	1.279
air		100	28.967	53.35	0.240	0.171	287.03	1005	718	1.400
ammonia	NH_3	68	17.032	90.73	0.523	0.406	488.16	2190	1702	1.287
argon	Ar	68	39.944	38.69	0.124	0.074	208.15	519	311	1.669
butane (-n)	C_4H_{10}	68	58.124	26.59	0.395	0.361	143.04	1654	1511	1.095
carbon dioxide	CO_2	100	44.011	35.11	0.207	0.162	188.92	867	678	1.279
carbon monoxide	CO	100	28.011	55.17	0.249	0.178	296.82	1043	746	1.398
chlorine	Cl_2	100	70.910	21.79	0.115	0.087	117.25	481	364	1.322
ethane	C_2H_6	68	30.070	51.39	0.386	0.320	276.50	1616	1340	1.206
ethylene	C_2H_4	68	28.054	55.08	0.400	0.329	296.37	1675	1378	1.215
Freon (R-12)*	CCl_2F_2	200	120.925	12.78	0.159	0.143	68.76	666	597	1.115
helium	He	100	4.003	386.04	1.240	0.744	2077.03	5192	3115	1.667
hydrogen	H_2	100	2.016	766.53	3.420	2.435	4124.18	14319	10195	1.405
hydrogen sulfide	H_2S	68	34.082	45.34	0.243	0.185	243.95	1017	773	1.315
krypton	Kr		83.800	18.44	0.059	0.035	99.22	247	148	1.671
methane	CH_4	68	16.043	96.32	0.593	0.469	518.25	2483	1965	1.264
neon	Ne	68	20.183	76.57	0.248	0.150	411.94	1038	626	1.658
nitrogen	N_2	100	28.016	55.16	0.249	0.178	296.77	1043	746	1.398
nitric oxide	NO	68	30.008	51.50	0.231	0.165	277.07	967	690	1.402
nitrous oxide	NO_2	68	44.01	35.11	0.221	0.176	188.92	925	736	1.257
octane vapor	C_8H_{18}		114.232	13.53	0.407	0.390	72.78	1704	1631	1.045
oxygen	O_2	100	32.000	48.29	0.220	0.158	259.82	921	661	1.393
propane	C_3H_8	68	44.097	35.04	0.393	0.348	188.55	1645	1457	1.129
sulfur dioxide	SO_2	100	64.066	24.12	0.149	0.118	129.78	624	494	1.263
water vapor*	H_2O	212	18.016	85.78	0.445	0.335	461.50	1863	1402	1.329
xenon	Xe		131.300	11.77	0.038	0.023	63.32	159	96	1.661

(Multiply Btu/lbm-°F by 4186.8 to obtain J/kg·K.)
(Multiply ft-lbf/lbm-°R by 5.3803 to obtain J/kg·K.)
*Values for steam and Freon are approximate and should be used only for low pressures and high temperatures.

36. PROPERTIES OF IDEAL GASES

A gas can be considered to behave ideally if its pressure is very low and the temperature is much higher than its critical temperature. (Otherwise, the substance is in vapor form.) Under these conditions, the molecule size is insignificant compared with the distance between molecules, and molecules do not come into contact. By definition, an *ideal gas* behaves according to the various ideal gas laws.

Values of h, u, and v for gases usually are read from gas tables. Since it contains a pv term, enthalpy can be related to the equation of state. Depending on the units chosen, a conversion factor may be needed.

$$h = u + pv = u + RT \quad\quad 29.62$$

Furthermore, density is the reciprocal of specific volume, v. Therefore, the density of an ideal gas can be

derived from the equation of state by setting $m = 1$ and solving for the reciprocal of volume.

$$\rho = \frac{1}{v} = \frac{p}{RT} \quad\quad 29.63$$

37. SPECIFIC HEATS OF IDEAL GASES

The specific heats of an ideal gas can be calculated from its specific gas constant. Depending on the units chosen, a conversion factor may or may not be needed in Eq. 29.64 through Eq. 29.67. By definition, a *perfect gas* is an ideal gas whose specific heats are constant.

$$c_p - c_v = R \quad\quad 29.64$$
$$C_p - C_v = R^* \quad\quad 29.65$$
$$c_p = \frac{Rk}{k-1} \quad\quad 29.66$$
$$C_p = \frac{R^*k}{k-1} \quad\quad 29.67$$

38. KINETIC GAS THEORY

The *kinetic gas theory* predicts the velocity distribution of gas molecules as a function of temperature. This theory makes the following assumptions.

- Gas molecules do not attract one another.

- The volume of the gas molecules is negligible compared with the volume of the gas.

- The molecules behave like hard spheres.

- The container volume is large enough that interactions with the wall are not a predominant activity of the molecules.

Equation 29.68 is the *Maxwell-Boltzmann distribution* (see Fig. 29.7) of gas molecule velocities. The variable κ is the *Boltzmann constant*, which has a value of 1.3807×10^{-23} J/molecule·K. (These units are the same as kg·m^2/s^2·molecule·K.) Kinetic gas theory calculations are traditionally performed in SI units. English units can be used, however, if all constants and units are consistent.

$$\frac{dN}{dv} = \left(\frac{4N}{\sqrt{\pi}}\right)v^2 \left(\frac{m}{2\kappa T}\right)^{3/2} e^{-mv^2/2\kappa T} \qquad 29.68$$

$$\kappa = \frac{R^*}{N_A} \qquad 29.69$$

Figure 29.7 *Maxwell-Boltzmann Velocity Distribution*

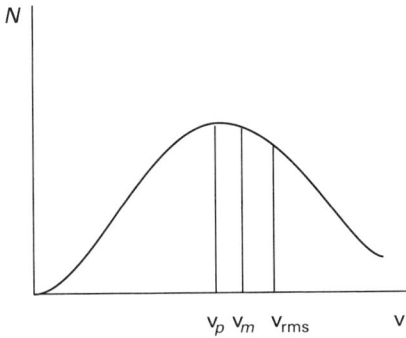

The *most probable speed* of a molecule with mass m is

$$v_p = \sqrt{\frac{2\kappa T}{m}} \qquad 29.70$$

The *mean speed* of a molecule with mass m is

$$v_m = 2\sqrt{\frac{2\kappa T}{\pi m}} \qquad 29.71$$

The *root-mean-square speed* of a molecule with mass m is

$$v_{\text{rms}} = \sqrt{\frac{3\kappa T}{m}} \qquad 29.72$$

The three velocities are illustrated in Fig. 29.7 and are related.

$$\frac{v_m}{v_p} = 1.128 \qquad 29.73$$

$$\frac{v_{\text{rms}}}{v_p} = 1.225 \qquad 29.74$$

Temperature has a molecular interpretation derived from the kinetic gas theory. It can be shown that the root-mean-square velocity is related to the absolute temperature. That is, absolute temperature is proportional to the square of the rms velocity.

$$T = \left(\frac{m}{3\kappa}\right)v_{\text{rms}}^2 \qquad 29.75$$

Since $\frac{1}{2}mv^2$ is the definition of kinetic energy, the mean translational kinetic energy of the molecule is proportional to the mean absolute temperature.

$$E_k = \frac{1}{2}mv_{\text{rms}}^2 = \frac{3}{2}\kappa T \qquad 29.76$$

The pressure of a gas can be calculated as the total change in momentum of all the gas molecules bouncing off the walls of the container.

$$p = \frac{\rho v_{\text{rms}}^2}{3} = \frac{Nmv_{\text{rms}}^2}{3V} \qquad 29.77$$

A small-enough particle suspended in a fluid will exhibit small random movements due to the statistical (i.e., random) collisions of fluid molecules on the particle's surface. Such motion is known as *Brownian movement*.

Example 29.14

What are the kinetic energy and root-mean-square velocity of 275K argon molecules (MW = 39.9)?

Solution

From Eq. 29.76,

$$E_k = \frac{3}{2}\kappa T = (1.5)\left(1.3807 \times 10^{-23} \ \frac{\text{J}}{\text{molecule·K}}\right)(275\text{K})$$
$$= 5.70 \times 10^{-21} \ \text{J/molecule}$$

The molecular mass of argon is its mass per mole divided by the number of molecules in a mole.

$$m = \frac{\text{MW}}{N_A} = \frac{39.9 \ \frac{\text{kg}}{\text{kmol}}}{\left(6.022 \times 10^{23} \ \frac{\text{molecules}}{\text{mol}}\right)\left(1000 \ \frac{\text{mol}}{\text{kmol}}\right)}$$
$$= 6.63 \times 10^{-26} \ \text{kg/molecule}$$

From Eq. 29.72, the rms velocity is

$$v_{rms} = \sqrt{\frac{3\kappa T}{m}}$$

$$= \sqrt{\frac{(3)\left(1.3807 \times 10^{-23} \, \frac{J}{molecule \cdot K}\right)(275K)}{6.63 \times 10^{-26} \, \frac{kg}{molecule}}}$$

$$= 414.5 \text{ m/s}$$

39. GRAVIMETRIC, VOLUMETRIC, AND MOLE FRACTIONS

The *gravimetric fraction*, G_A (also known as the *mass fraction* and *weight fraction*), of a component A in a mixture of components A, B, C, and so on, is the ratio of the component's mass to the total mixture mass.

$$G_A = \frac{m_A}{m} = \frac{m_A}{m_A + m_B + m_C} \qquad 29.78$$

The *volumetric fraction*, B_A, of a component A is the ratio of the component's partial volume to the overall mixture volume.

$$B_A = \frac{V_A}{V} = \frac{V_A}{V_A + V_B + V_C} \qquad 29.79$$

It is possible to convert between gravimetric and volumetric fractions.

$$G_A = \frac{B_A(MW)_A}{B_A(MW)_A + B_B(MW)_B + B_C(MW)_C} \qquad 29.80$$

$$B_A = \frac{\dfrac{G_A}{(MW)_A}}{\dfrac{G_A}{(MW)_A} + \dfrac{G_B}{(MW)_B} + \dfrac{G_C}{(MW)_C}} \qquad 29.81$$

The *mole fraction*, x_A, of a component A is the ratio of the number of moles of substance A to the total number of moles of all substances.

$$x_A = \frac{n_A}{n} = \frac{n_A}{n_A + n_B + n_C} \qquad 29.82$$

For nonreacting mixtures of ideal gases, the mole fraction and volumetric fraction (and partial pressure ratio) are the same.

$$x_A = B_A \Big|_{\text{ideal gases}} \qquad 29.83$$

40. PARTIAL PRESSURE AND PARTIAL VOLUME OF GAS MIXTURES

A *gas mixture* consists of an aggregation of molecules of each gas component, the molecules of any single component being distributed uniformly and moving as if they alone occupied the space. The *partial volume*, V_A, of a gas A in a mixture of nonreacting gases A, B, C, and so on, is the volume that gas A alone would occupy at the temperature and pressure of the mixture. (See Fig. 29.8.)

Figure 29.8 Mixture of Ideal Gases

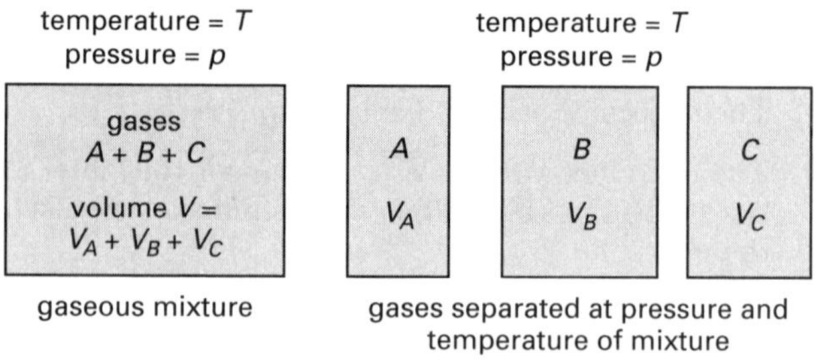

gaseous mixture / gases separated at pressure and temperature of mixture

The partial volume can be calculated from the volumetric fraction and total volume.

$$V_A = B_A V \qquad 29.84$$

Amagat's law (also known as *Amagat-Leduc's rule*) states that the total volume of a mixture of nonreacting gases is equal to the sum of the partial volumes.

$$V = V_A + V_B + V_C \qquad 29.85$$

The *partial pressure*, p_A, of gas A in a mixture of nonreacting gases A, B, C, and so on, is the pressure gas A alone would exert in the total volume at the temperature of the mixture.

$$p_A = \frac{m_A R_A T}{V} = \frac{n_A R^* T}{V} \qquad 29.86$$

The partial pressure can also be calculated from the mole fraction and the total pressure. However, for ideal gases, the partial pressure ratio, mole fraction, and volumetric fraction are the same.

$$\frac{p_A}{p} = x_A = B_A \quad \text{[ideal gas]} \qquad 29.87$$

If the average specific gas constant, $\overline{R}$, for the gas mixture is known, it can be used with the gravimetric fraction to calculate the partial pressure.

$$p_A = G_A \frac{R_A p}{\overline{R}} \qquad 29.88$$

According to *Dalton's law of partial pressures*, the *total pressure* of a gas mixture is the sum of the partial pressures.

$$p = p_A + p_B + p_C \qquad 29.89$$

41. PROPERTIES OF NONREACTING IDEAL GAS MIXTURES

A mixture's average molecular weight and density are volumetrically weighted averages of its components' values.

$$MW = \sum B_i(MW)_i \qquad 29.90$$

$$\rho = \sum B_i\rho_i \qquad 29.91$$

On the other hand, a mixture's internal energy, enthalpy, specific heats, and specific gas constant are equal to the sum of the values of its individual components (i.e., the mixture average is gravimetrically weighted).

$$u = \frac{\sum m_i u_i}{m} = \sum_i G_i u_i \qquad 29.92$$

$$h = \frac{\sum m_i h_i}{m} = \sum_i G_i h_i \qquad 29.93$$

$$c_v = \frac{\sum m_i c_{v,i}}{m} = \sum_i G_i c_{v,i} \qquad 29.94$$

$$c_p = \frac{\sum m_i c_{p,i}}{m} = \sum_i G_i c_{p,i} \qquad 29.95$$

$$R = \frac{\sum m_i R_i}{m} = \sum_i G_i R_i \qquad 29.96$$

If the mixing is reversible and adiabatic, the entropy will also be equal to the sum of the individual entropies. However, each individual entropy, s_i, must be evaluated at the temperature and volume of the mixture and at the individual partial pressure, p_i.

$$s = \frac{\sum m_i s_i}{m} = \sum_i G_i s_i \qquad 29.97$$

Equation 29.92, Eq. 29.93, and Eq. 29.97 are mathematical formulations of *Gibbs theorem* (also known as *Gibbs rule*). This theorem states that the total property (e.g., U, H, or S) of a mixture of nonreacting ideal gases is the sum of the properties that the individual gases would have if each occupied the total mixture volume alone at the same temperature.

Table 29.8 summarizes composite gas properties. Notice that the ratio of specific heats, k, is not a composite gas property. It is most expedient to find the composite ratio of specific heats from the gravimetrically weighted specific heats.

Table 29.8 Summary of Composite Ideal Gas Properties

gravimetrically (mass) weighted	volumetrically (mole fraction) weighted
u	U
h	H
c_p	C_p
c_v	C_v
R	MW
s	S
	ρ

Example 29.15

0.14 lbm (0.064 kg) of octane vapor (MW = 114) is mixed with 2.0 lbm (0.91 kg) of air (MW = 29.0) in the manifold of an engine. The total pressure in the manifold is 12.5 psia (86.1 kPa), and the temperature is 520°R (290K). Assume octane behaves ideally. (a) What is the total volume of this mixture? (b) What is the partial pressure of the air in the mixture?

SI Solution

(a) The number of moles of octane and air are

$$n_{octane} = \frac{m}{MW} = \frac{0.064 \text{ kg}}{114 \frac{\text{kg}}{\text{kmol}}} = 5.61 \times 10^{-4} \text{ kmol}$$

$$n_{air} = \frac{0.91 \text{ kg}}{29 \frac{\text{kg}}{\text{kmol}}} = 0.0314 \text{ kmol}$$

From Eq. 29.58, the total volume of any gas is

$$V = \frac{nR^*T}{p}$$

$$= (5.61 \times 10^{-4} \text{ kmol} + 0.0314 \text{ kmol})$$

$$\times \left(\frac{\left(8314.3 \frac{\text{J}}{\text{kmol·K}}\right)(290K)}{86.1 \times 10^3 \text{ Pa}}\right)$$

$$= 0.895 \text{ m}^3$$

(b) The mole fraction of the air is

$$x_{air} = \frac{n_{air}}{n_{total}} = \frac{0.0314 \text{ kmol}}{5.61 \times 10^{-4} \text{ kmol} + 0.0314 \text{ kmol}}$$

$$= 0.982$$

The partial pressure of air is

$$p_{air} = x_{air}p = (0.982)(86.1 \text{ kPa}) = 84.6 \text{ kPa}$$

Customary U.S. Solution

(a) The number of moles of octane and air are

$$n_{octane} = \frac{m}{MW} = \frac{0.14 \text{ lbm}}{114 \frac{\text{lbm}}{\text{lbmol}}} = 0.001228 \text{ lbmol}$$

$$n_{air} = \frac{2.0 \text{ lbm}}{29.0 \frac{\text{lbm}}{\text{lbmol}}} = 0.068966 \text{ lbmol}$$

From Eq. 29.58, the total volume of any gas is

$$V = \frac{nR^*T}{p}$$

$$= (0.001228 \text{ lbmol} + 0.068966 \text{ lbmol})$$

$$\times \left(\frac{\left(1545.33 \frac{\text{ft-lbf}}{\text{lbmol-}^\circ\text{R}}\right)(520^\circ\text{R})}{\left(12.5 \frac{\text{lbf}}{\text{in}^2}\right)\left(12 \frac{\text{in}}{\text{ft}}\right)^2} \right)$$

$$= 31.34 \text{ ft}^3$$

(b) The mole fraction of the air is

$$x_{air} = \frac{n_{air}}{n} = \frac{0.068966 \text{ lbmol}}{0.001228 \text{ lbmol} + 0.068966 \text{ lbmol}}$$

$$= 0.9825$$

The partial air pressure is

$$p_{air} = x_{air}p = (0.9825)\left(12.5 \frac{\text{lbf}}{\text{in}^2}\right)$$

$$= 12.3 \text{ psia}$$

42. EQUATION OF STATE FOR REAL GASES

Real gases do not meet the basic assumptions defining an ideal gas. Specifically, the molecules of a real gas occupy a volume that is not negligible in comparison with the total volume of the gas. (This is especially true for gases at low temperatures.) Furthermore, real gases are subject to *van der Waals' forces*, which are attractive forces between gas molecules.

There are two methods of accounting for real gas behavior. The first method is to modify the ideal gas equation of state with various empirical correction factors. Since the modifications are empirical, the resulting equations of state are known as *correlations*. One well-known correlation is *van der Waals' equation of state*.

$$\left(p + \frac{a}{V^2}\right)(V - b) = nR^*T \qquad 29.98$$

The van der Waals corrections usually need to be made only when a gas is below its critical temperature. For an ideal gas, the a and b terms are zero. When the spacing

between molecules is close, as it would be at low temperatures, the molecules attract each other and reduce the pressure exerted by the gas. The pressure is then corrected by the a/V^2 term. b is a constant that accounts for the molecular volume in a dense state.

Other correlations of this type include the *Clausius, Bertholet, Dieterici,* and *Beattie-Bridgeman equations of state.* However, the most accurate empirical correlation is the *virial equation of state*, which has the form shown in Eq. 29.99 and Eq. 29.100. The constants B, C, D, and so on, are called the *virial coefficients.*

$$pV = nR^*T\left(1 + \frac{B}{V} + \frac{C}{V^2} + \frac{D}{V^3} + \cdots\right) \qquad 29.99$$

$$pV = nR^*T(1 + B'p + C'p^2 + D'p^3 + \cdots) \qquad 29.100$$

The *principle (law) of corresponding states* provides a second method of correcting for real gas behavior. This law states that the *reduced properties* of all gases are identical (i.e., all gases behave similarly when their reduced variables are used). Specifically, there is one property, the *compressibility factor*, Z, that has the same value for all gases when evaluated at the same values of the *reduced variables*.[14] (The reduced variables are not the same as the ratios defined in Sec. 29.27.)

$$Z = f(T_r, p_r, v_r) \qquad 29.101$$

$$T_r = \frac{T}{T_c} \qquad 29.102$$

$$p_r = \frac{p}{p_c} \qquad 29.103$$

$$v_r = \frac{v}{v_c} \qquad 29.104$$

Compressibility factors are almost always read from *generalized compressibility charts* such as App. 29.Z. The compressibility factor can then be used to correct the ideal gas equation of state.

$$pv = ZRT \qquad 29.105$$

$$pV = mZRT \qquad 29.106$$

Example 29.16

What is the specific volume of carbon dioxide at 2680 psia (182 atm) and 300°F (150°C)?

SI Solution

The absolute temperature is

$$T = 150°C + 273° = 423K$$

---

[14]Some engineering disciplines (e.g., petroleum engineering) use the symbol z for the compressibility factor.

From Table 29.4, the critical temperature and pressure of carbon dioxide are 304.3K and 72.9 atm, respectively. The reduced variables are

$$T_r = \frac{T}{T_c} = \frac{423\text{K}}{304.3\text{K}} = 1.39$$

$$p_r = \frac{p}{p_c} = \frac{182 \text{ atm}}{72.9 \text{ atm}} = 2.5$$

From App. 29.Z, Z is read as 0.75. R is read from Table 29.7. Solving Eq. 29.105 for v,

$$v = \frac{ZRT}{p} = \frac{(0.75)\left(189 \dfrac{\text{J}}{\text{kg}\cdot\text{K}}\right)(423\text{K})}{(182 \text{ atm})\left(101.3 \times 10^3 \dfrac{\text{Pa}}{\text{atm}}\right)}$$

$$= 3.25 \times 10^{-3} \text{ m}^3/\text{kg}$$

Customary U.S. Solution

The absolute temperature is

$$T = 300°\text{F} + 460° = 760°\text{R}$$

From Table 29.4, the critical temperature and pressure of carbon dioxide are 547.8°R and 72.9 atm, respectively. The reduced variables are

$$T_r = \frac{T}{T_c} = \frac{760°\text{R}}{547.8°\text{R}} = 1.39$$

$$p_r = \frac{p}{p_c} = \frac{2680 \text{ psia}}{(72.9 \text{ atm})\left(14.7 \dfrac{\text{psia}}{\text{atm}}\right)} = 2.5$$

From App. 29.Z, Z is read as 0.75. R is read from Table 29.7. Solving Eq. 29.105 for v,

$$v = \frac{ZRT}{p} = \frac{(0.75)\left(35.11 \dfrac{\text{ft-lbf}}{\text{lbm-°R}}\right)(760°\text{R})}{\left(2680 \dfrac{\text{lbf}}{\text{in}^2}\right)\left(12 \dfrac{\text{in}}{\text{ft}}\right)^2}$$

$$= 0.0519 \text{ ft}^3/\text{lbm}$$

43. SPECIFIC HEATS OF REAL GASES

By definition, the specific heat of an ideal (perfect) gas is independent of temperature. However, the specific heat of a real gas varies with temperature and (slightly) with pressure. There are several ways to find the specific heat of a gas at different temperatures: tables, graphs, and correlations.[15]

Equation 29.107 is a typical temperature correlation. The variation with pressure, being small at low pressures, is disregarded. The coefficients A, B, C, and D will depend on the units, the temperature range over which the correlation is to be used, and the desired accuracy. Typical coefficients are given in Table 29.9.

$$c_p = A + BT + CT^2 + \frac{D}{\sqrt{T}} \qquad\qquad \textit{29.107}$$

$$c_v = c_p - R \qquad\qquad \text{[SI]} \quad \textit{29.108(a)}$$

$$c_v = c_p - \frac{R}{J} \qquad\qquad \text{[U.S.]} \quad \textit{29.108(b)}$$

---

[15]It is also possible to calculate the specific heat over a small temperature range if the enthalpies are known (e.g., from an air table) from $c_p = \Delta h/\Delta T$. However, if the enthalpies are known, it is unlikely that the specific heat will be needed.

Table 29.9 Correlation Coefficients for Calculating Specific Heat (Btu/lbm-°R or Btu/lbm-°F)

gas	°R	K	A	B	C	D
air	400 to 1200	220 to 670	0.2405	-1.186×10^{-5}	20.1×10^{-9}	0
	1200 to 4000	670 to 2220	0.2459	3.22×10^{-5}	-3.74×10^{-9}	-0.833
CH_4	400 to 1000	220 to 560	0.453	0.62×10^{-5}	268.8×10^{-9}	0
	1000 to 4000	560 to 2220	1.152	32.58×10^{-5}	-41.29×10^{-9}	-22.42
CO	400 to 1200	220 to 670	0.2534	-2.35×10^{-5}	26.88×10^{-9}	0
	1200 to 4000	670 to 2220	0.2763	3.04×10^{-5}	-3.89×10^{-9}	-1.5
CO_2	400 to 4000	220 to 2220	0.328	3.2×10^{-5}	-4.4×10^{-9}	-3.33
H_2	400 to 1000	220 to 560	2.853	145×10^{-5}	-883×10^{-9}	0
	1000 to 2500	560 to 1390	3.447	-4.7×10^{-5}	70.3×10^{-9}	0
	2500 to 4000	1390 to 2220	2.841	45×10^{-5}	-31.2×10^{-9}	0
H_2O	400 to 1800	220 to 1000	0.4267	2.425×10^{-5}	23.85×10^{-9}	0
	1800 to 4000	1000 to 2220	0.3275	14.67×10^{-5}	-13.59×10^{-9}	0
N_2	400 to 1200	220 to 670	0.2510	-1.63×10^{-5}	20.4×10^{-9}	0
	1200 to 4000	670 to 2220	0.2192	4.38×10^{-5}	-5.14×10^{-9}	-0.124
O_2	400 to 1200	220 to 670	0.213	0.188×10^{-5}	20.3×10^{-9}	0
	1200 to 4000	670 to 2220	0.340	-0.36×10^{-5}	0.616×10^{-9}	-3.19

(Multiply values in this table by 4.1868 to obtain coefficients for specific heats in kJ/kg·K or kJ/kg·°C.)

Adapted from a table published in *Engineering Thermodynamics*, C. O. Mackay, W. N. Barnard, and F. O. Ellenwood (John Wiley, New York, 1957).

30 Changes in Thermodynamic Properties

Nomenclature

c	specific heat	Btu/lbm-°F	kJ/kg·K
C	molar specific heat	Btu/lbmol-°F	kJ/kmol·K
E	energy	Btu	kJ
g	acceleration of gravity, 32.2 (9.81)	ft/sec^2	m/s^2
g_c	gravitational constant, 32.2	lbm-ft/lbf-sec^2	n.a.
h	enthalpy	Btu/lbm	kJ/kg
H	molar enthalpy	Btu/lbmol	kJ/kmol
J	Joule's constant, 778.17	ft-lbf/Btu	n.a.
k	ratio of specific heats	–	–
m	mass	lbm	kg
n	number of moles	–	–
n	polytropic exponent	–	–
p	pressure	lbf/ft^2	kPa
P	power	Btu/sec	kW
q	heat per unit mass	Btu/lbm	kJ/kg
Q	molar heat	Btu/lbmol	kJ/kmol
Q	total heat	Btu	kJ
R	specific gas constant	ft-lbf/lbm-°R	kJ/kg·K
R^*	universal gas constant	ft-lbf/lbmol-°R	kJ/kmol·K
s	entropy	Btu/lbm-°F	kJ/kg·K
S	molar entropy	Btu/lbmol-°F	kJ/kmol·K
T	temperature	°F	°C
T	absolute temperature	°R	K
u	specific internal energy	Btu/lbm	kJ/kg
U	molar internal energy	Btu/lbmol	kJ/kmol
v	velocity	ft/sec	m/s
V	molar specific volume	ft^3/lbmol	m^3/kmol
V	volume	ft^3	m^3
w	specific work	ft-lbf/lbm	kJ/kg
W	total work	ft-lbf	kJ
z	elevation	ft	m

Symbols

β	isobaric compressibility	1/°R	1/K
η	efficiency	–	–
μ_J	Joule-Thomson coefficient	°F-ft^2/lbf	K/Pa
v	specific volume	ft^3/lbm	m^3/kg
Φ	availability function	Btu/lbm	kJ/kg

Subscripts

H	hot (high-temperature)
k	kinetic
L	cold (low-temperature)
n	polytropic
o	outside (environment)
p	constant pressure or potential
t	total
th	thermal
v	constant volume
w	water

1. SYSTEMS

A *thermodynamic system* is defined as the matter enclosed within an arbitrary but precisely defined *control volume*. Everything external to the system is defined as the *surroundings*, *environment*, or *universe*. The environment and system are separated by the *system boundaries*. The surface of the control volume is known as the *control surface*. The control surface can be real (e.g., piston and cylinder walls) or imaginary.

If mass flows through the system across system boundaries, the system is an *open system*. Pumps, heat exchangers, and jet engines are examples of open systems. An important type of open system is the *steady-flow open system* in which matter enters and exits at the same rate. Pumps, turbines, heat exchangers, and boilers are all steady-flow open systems.

If no mass crosses the system boundaries, the system is said to be a *closed system*. The matter in a closed system may be referred to as a *control mass*. Closed systems can have variable volumes. The gas compressed by a piston in a cylinder is an example of a closed system with a variable control volume.

In most cases, energy in the form of heat, work, or electrical energy can enter or exit any open or closed system. Systems closed to both matter and energy transfer are known as *isolated systems*.

2. TYPES OF PROCESSES

Changes in thermodynamic properties of a system often depend on the type of process experienced. This is particularly true of gaseous systems. Several common types of processes are listed as follows, along with their heat, energy, and work relationships.[1]

- *adiabatic process*—a process in which no heat or other energy crosses the system boundary.[2] Adiabatic processes include isentropic and throttling processes.

$$Q = 0 \qquad 30.1$$
$$\Delta U = -W \qquad 30.2$$

- *constant pressure process*—also known as an *isobaric process*

$$\Delta p = 0 \qquad 30.3$$
$$Q = \Delta H \qquad 30.4$$

- *constant temperature process*—also known as an *isothermal process*

$$\Delta T = 0 \qquad 30.5$$
$$Q = W \qquad 30.6$$

- *constant volume process*—also known as an *isochoric* or *isometric process*

$$\Delta V = 0 \qquad 30.7$$
$$Q = \Delta U \qquad 30.8$$
$$W = 0 \qquad 30.9$$

- *isenthalpic process*—a process in which the enthalpy of the system substance does not change

$$\Delta H = 0 \qquad 30.10$$
$$\Delta T = 0 \quad \text{[ideal gas]} \qquad 30.11$$

- *isentropic process*—an adiabatic process in which there is no change in system entropy (i.e., is reversible)

$$\Delta S = 0 \qquad 30.12$$
$$Q = 0 \qquad 30.13$$

- *throttling process*—an adiabatic process in which there is no change in system enthalpy (i.e., the process is isenthalpic), but for which there is a significant pressure drop[3]

$$\Delta H = 0 \qquad 30.14$$
$$p_2 < p_1 \qquad 30.15$$

A system that is in equilibrium at the start and finish of a process may or may not be in equilibrium during the process. A *quasistatic process* (*quasiequilibrium process*) is one that can be divided into a series of infinitesimal deviations (steps) from equilibrium. During each step, the property changes are small, and all intensive properties are uniform throughout the system. The interim equilibrium at each step is known as a *quasiequilibrium*.

Transport processes are generally more complex than can be analyzed by the simple thermodynamic relationships in this chapter. In a *transport process*, there is a transfer of some quantity. Drying, distillation, and evaporation are examples of heat transfer processes. Fluid flow, mixing, and sedimentation are examples of momentum transfer. Distillation, absorption, extraction, and leaching are examples of mass transfer processes.

3. POLYTROPIC PROCESSES

A *polytropic process* is one that obeys Eq. 30.16, the *polytropic equation of state*. Gases always constitute the system in polytropic processes.

$$p_1 V_1^n = p_2 V_2^n \qquad 30.16$$

n is the *polytropic exponent*, a property of the equipment, not of the gas. For efficient air compressors, n is typically between 1.25 and 1.30.

Depending on the value of the polytropic exponent, the polytropic equation of state can also be used for other processes.

$n = 0$ [constant pressure process]
$n = 1$ [constant temperature process]
$n = k$ [isentropic process]
$n = \infty$ [constant volume process]

The *polytropic specific heat*, c_n, is defined as

$$c_n = \frac{n-k}{n-1} \times c_v \qquad 30.17$$
$$Q = mc_n \Delta T \qquad 30.18$$

---

[1]The heat, energy, and work relationships are easily derived from the *first law of thermodynamics*.
[2]An adiabatic process is not the same as a constant temperature process, however.

[3]The classic throttling process is the expansion of a gas in a pipe through a porous plug that offers significant resistance to flow.

4. THROTTLING PROCESSES

For ideal gases, throttling processes are constant temperature processes. Some real gases at normal temperatures, however, decrease in temperature when throttled.[4] Others increase in temperature. Whether or not an increase or a decrease in temperature occurs depends on the temperature and pressure at which the throttling occurs.

For any given set of initial conditions, there is one temperature at which no temperature change occurs when a real gas is throttled. This is called the *inversion temperature* or *inversion point*. The inversion temperature is dependent on the initial gas conditions. (See Table 30.1.)

Table 30.1 Approximate Maximum Experimental Inversion Temperatures

substance	°R	K
air	1085	603
argon	1301	723
carbon dioxide	≈ 2700	≈ 1500
helium	≈ 72	≈ 40
hydrogen	364	202
nitrogen	1118	621

The *Joule-Thomson coefficient*, μ_J, (also known as the *Joule-Kelvin coefficient*), is defined as the ratio of the change in temperature to the change in pressure when a real gas is throttled. The Joule-Thomson coefficient is zero for an ideal gas.

$$\mu_J = \left(\frac{\partial T}{\partial p}\right)_h \qquad 30.19$$

The Joule-Thomson coefficient can be calculated from the specific heat, c_p, and other properties.

$$\mu_J = \left(\frac{1}{c_p}\right)\left(T\left(\frac{\partial V}{\partial T}\right)_p - v\right) = \left(\frac{v}{c_p}\right)(T\beta - 1)$$
$$= \left(\frac{-1}{c_p}\right)\left(\frac{\partial h}{\partial p}\right)_T \qquad 30.20$$

It is convenient to plot lines of constant enthalpy (*isenthalpic curves*) on a *T-p* diagram. The various curves in Fig. 30.1 represent different sets of starting conditions (p_1 and T_1). The slope of a curve represents the Joule-Thomson coefficient, which can be zero, positive, or negative at that point. The points where the slope is zero are known as *inversion points*. The locus of inversion points is known as the *inversion curve*.

States to the right of an inversion point will have negative Joule-Thomson coefficients. States to the left of an inversion point will have positive Joule-Thomson coefficients. Throttling totally within the region bounded by the temperature axis and the inversion curve will always result in a decrease in temperature. Similarly, throttling

---

[4]This tendency can be used to liquefy gases and vapors (e.g., refrigerants) by passing them through an expansion valve.

Figure 30.1 Inversion Curve

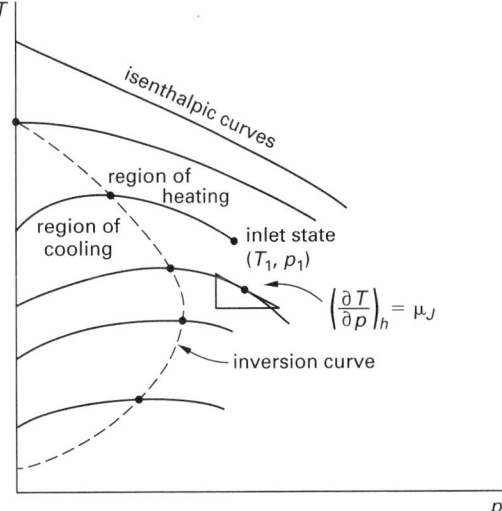

totally outside of the inversion curve will always produce an increase in temperature. Throttling across the inversion curve may produce a higher or lower final temperature, depending on initial and final conditions.

5. REVERSIBLE PROCESSES

Processes can be categorized as reversible and irreversible. A *reversible process* is one that is performed in such a way that at the conclusion of the process, *both* the system and the local surroundings can be restored to their initial states. A process that does not meet these requirements is an *irreversible process*. A reversible adiabatic process is implicitly isentropic.

Although all real-world processes result in an overall increase in entropy, it is possible to conceptualize processes that have zero entropy change. For a reversible process,

$$\Delta s = 0 \Big|_{\text{reversible}} \qquad 30.21$$

Processes that contain friction are never reversible. Other processes that are irreversible are listed as follows.

- stirring a viscous fluid
- slowing down a moving fluid
- unrestrained expansion of gas
- throttling
- changes of phase (freezing, condensation, etc.)
- chemical reaction
- diffusion
- current flow through electrical resistance
- electrical polarization
- magnetization with hysteresis
- releasing a stretched spring

• inelastic deformation

• heat conduction

A system that has experienced an irreversible process can still be returned to its original state. An example of this is the water in a closed boiler-turbine installation. The entropy increases when the water is vaporized in the boiler (an irreversible process). The entropy decreases when the steam is condensed in the condenser. When the water returns to the boiler, its entropy is increased to its original value. However, the environment cannot be returned to its original condition; hence the cycle overall is irreversible.

6. FINDING WORK AND HEAT GRAPHICALLY

A process between two thermodynamic states can be represented graphically. The line representing the locus of quasiequilibrium states between the initial and final states is known as the *path* of the process.

It is sometimes convenient to see what happens to the pressure and volume of a system by plotting the path on a p-V diagram. In addition, the work done by or on the system can be determined from the graph. This is possible because the integral calculating p-V represents area under a curve in the p-V plane.

$$W = \int_{V_1}^{V_2} p \, dV \qquad \qquad 30.22$$

Similarly, the amount of heat absorbed or released from a system can be determined as the area under the path on the T-s diagram.

$$Q = \int_{s_1}^{s_2} T \, ds \qquad \qquad 30.23$$

The variables p, V, T, and s are *point functions* because their values are independent of the path taken to arrive at the thermodynamic state. Work and heat (W and Q), however, are *path functions* because they depend on the path taken. (See Fig. 30.2.)

Figure 30.2 *Process Work and Heat*

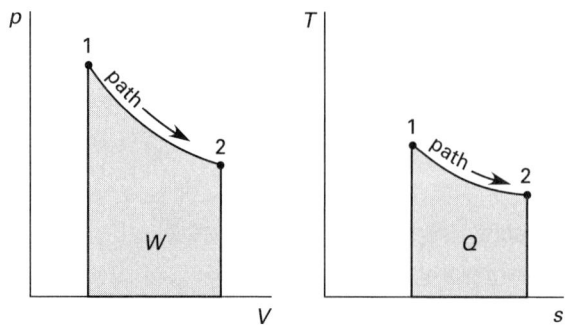

7. SIGN CONVENTION

A definite sign convention is used in calculating work, heat, and other property changes in systems.[5] This sign convention takes the system (not the environment) as the reference. (For example, a net heat gain would mean the system gained energy and the environment lost energy.)

• Heat, Q, is positive if heat flows into the system.

• Work, W, is positive if the system does work on the surroundings.

• Changes in enthalpy, entropy, and internal energy (ΔH, ΔS, and ΔU) are positive if these properties increase within the system.

8. FIRST LAW OF THERMODYNAMICS FOR CLOSED SYSTEMS

There is a basic principle that underlies all property changes as a system experiences a process: all energy must be accounted for. Energy that enters a system must either leave the system or be stored in some manner. Energy cannot be created or destroyed. These statements are the primary manifestations of the *first law of thermodynamics*: the work done in an adiabatic process depends solely on the system's endpoint conditions, not on the nature of the process.

For closed systems, the first law can be written in differential form.[6] Since Q and W are not properties, infinitesimal changes are designated as δ.

$$\delta Q = dU + \delta W \qquad \qquad 30.24$$

Most simple thermodynamic problems can be solved without resorting to differential calculus. The first law can then be written in finite terms.

$$Q = \Delta U + W \qquad \qquad 30.25$$

Equation 30.25 says that the heat, Q, entering a closed system can either increase the temperature (i.e., increase U) or be used to perform work (increase W) on the surroundings. In a nonadiabatic closed system, heat energy entering the system also can leak to the surroundings. However, in Eq. 30.25, the Q term is understood to be the net heat entering the system, exclusive of the loss.

Q, U, and W must all have the same units. This is less of a problem with SI units than with English units. For example, if Q and U are in Btu/lbm and W is in

---

[5]This sign convention is automatically followed if the formulas in this chapter are used.
[6]This formulation of the conservation of energy implicitly assumes that kinetic and potential energies are negligible.

ft-lbf/lbm, Joule's constant must be incorporated into the first law:

$$Q = \Delta U + \frac{W}{J} \qquad \text{[U.S.]} \qquad 30.26$$

In accordance with the standard sign convention given in Sec. 30.7, Q will be negative if the net heat exchange to the system is a loss. ΔU will be negative if the internal energy of the system decreases. W will be negative if the surroundings do work on the system (e.g., a piston compressing gas in a cylinder).

9. THERMAL EQUILIBRIUM

A simple application of the first law is the calculation of a thermal equilibrium point for a nonreacting system. *Thermal equilibrium* is reached when all parts of the system are at the same temperature.

The drive toward thermal equilibrium occurs spontaneously, without the addition of external work, whenever masses (two liquids, a solid and a liquid, etc.) with different temperatures are combined. Therefore, the work term, W, in the first law is zero.

The energy comes from the heat given off by the cooling mass (i.e., mass A). Energy is stored by increasing the temperature of the warmed mass (mass B). These changes are equal, but opposite, since the net change is zero.

$$Q_{\text{net}} = Q_{\text{in},A} - Q_{\text{out},B} = 0 \qquad 30.27$$

The form of the equation for Q depends on the phases of the substances and the nature of the system. If the substances are either solid or liquid, Eq. 30.28 can be used. (For use with open systems, the m in Eq. 30.28 should be replaced with $\dot{m}$.)

$$m_A c_{p,A}(T_{1,A} - T_{2,A}) = m_B c_{p,B}(T_{2,B} - T_{1,B}) \qquad 30.28$$

Example 30.1

A block of steel is removed from a furnace and quenched in an insulated aluminum tank filled with water. The water and aluminum tank are initially in equilibrium at 75°F (24°C), and the final equilibrium temperature after quenching is 100°F (38°C). What is the initial temperature of the steel?

steel block mass:
   m_{st} 2.0 lbm   (0.9 kg)
steel specific heat:
   $c_{p,\text{st}}$ 0.11 Btu/lbm-°F   (0.460 kJ/kg·K)
aluminum tank mass:
   m_{al} 5.0 lbm   (2.25 kg)
aluminum specific heat:
   $c_{p,\text{al}}$ 0.21 Btu/lbm-°F   (0.880 kJ/kg·K)
water mass:
   m_w 12.0 lbm   (5.4 kg)
water specific heat:
   $c_{p,w}$ 1.0 Btu/lbm-°F   (4.190 kJ/kg·K)

SI Solution

The heat lost by the steel is equal to the heat gained by the tank and water.

$$m_{\text{st}} c_{p,\text{st}}(T_{1,\text{st}} - T_{\text{eq}})$$
$$= (m_{\text{al}} c_{p,\text{al}} + m_w c_{p,w})(T_{\text{eq}} - T_{1,w})$$

$$(0.9 \text{ kg})\left(0.460 \ \frac{\text{kJ}}{\text{kg·K}}\right)(T_{1,\text{st}} - 38°\text{C})$$

$$= \left(\begin{array}{c} (2.25 \text{ kg})\left(0.880 \ \dfrac{\text{kJ}}{\text{kg·K}}\right) \\ + (5.4 \text{ kg})\left(4.190 \ \dfrac{\text{kJ}}{\text{kg·K}}\right) \end{array} \right)(38°\text{C} - 24°\text{C})$$

$$T_{1,\text{st}} = 870°\text{C}$$

Customary U.S. Solution

Proceeding as in the SI solution,

$$(2.0 \text{ lbm})\left(0.11 \ \frac{\text{Btu}}{\text{lbm-°F}}\right)(T_{1,\text{st}} - 100°\text{F})$$

$$= \left(\begin{array}{c} (5.0 \text{ lbm})\left(0.21 \ \dfrac{\text{Btu}}{\text{lbm-°F}}\right) \\ + (12.0 \text{ lbm})\left(1 \ \dfrac{\text{Btu}}{\text{lbm-°F}}\right) \end{array} \right)(100°\text{F} - 75°\text{F})$$

$$T_{1,\text{st}} = 1583°\text{F}$$

10. FIRST LAW OF THERMODYNAMICS FOR OPEN SYSTEMS

The first law of thermodynamics can also be written for open systems, but more terms are required to account for the many energy forms. The first law formulation is essentially the Bernoulli energy conservation equation extended to nonadiabatic processes.

If the mass flow rate is constant, the system is a *steady-flow system*, and the first law is known as the *steady-flow energy equation*, SFEE, Eq. 30.29. It is customary to express the SFEE in terms of energy per unit mass or per mole.

$$Q = \Delta U + \Delta E_p + \Delta E_k + W_{\text{flow}} + W_{\text{shaft}} \qquad 30.29$$

Some of the terms in Eq. 30.29 are illustrated in Fig. 30.3. Q is the net heat flow into or out of the system, inclusive of any losses. It can be supplied from furnace flame, electrical heating, nuclear reaction, or other sources. If the system is adiabatic, Q is zero.

ΔE_p and ΔE_k are the fluid's potential and kinetic energy changes. Generally, these terms are insignificant compared to the thermal energy transfers.

W_{shaft} is the *shaft work*—work that the steady-flow device does on the surroundings. Its name is derived from the output shaft that serves to transmit energy out of the system. For example, turbines and internal

Figure 30.3 Steady Flow Device

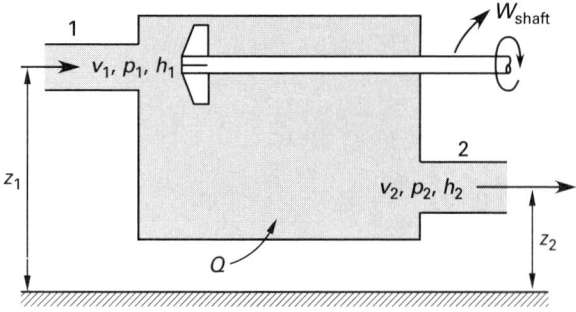

combustion engines have output shafts. W_{shaft} can be negative, as in the case of a pump or compressor.

W_{flow} is the p-V (*flow energy, flow work*, etc.). There is a pressure, p_2, at the exit of the steady-flow device in Fig. 30.3. This exit pressure opposes the entrance of the fluid. Therefore, the flow work term represents the work required to cause the flow into the system against the exit pressure. The flow work can be calculated from Eq. 30.30. (Consistent units must be used.)

$$w_{flow} = p_2 v_2 - p_1 v_1 \qquad 30.30$$

ΔU is the change in the system's internal energy. Since the combination of internal energy and flow work constitutes enthalpy, it will seldom be necessary to work with either internal energy or flow work.

$$h = u + pv \qquad 30.31$$

$$\Delta h = \Delta u + w_{flow} \qquad 30.32$$

Equation 30.33 is a useful formulation of the SFEE. Units of power can be obtained by multiplying both sides by $\dot{m}$.

$$q = h_2 - h_1 + \frac{v_2^2 - v_1^2}{2} + (z_2 - z_1)g + w_{shaft} \qquad \text{[SI]}$$
$$30.33(a)$$

$$q = h_2 - h_1 + \frac{v_2^2 - v_1^2}{2g_c J} + \frac{(z_2 - z_1)g}{g_c J} + \frac{w_{shaft}}{J}$$
$$\text{[U.S.]} \quad 30.33(b)$$

Example 30.2

4 lbm/sec (1.8 kg/s) of steam enter a turbine with a velocity of 65 ft/sec (20 m/s) and an enthalpy of 1350 Btu/lbm (3140 kJ/kg). The steam enters the condenser after being expanded to 1075 Btu/lbm (2500 kJ/kg) at 125 ft/sec (38 m/s). There is a total heat loss from the turbine casing of 50 Btu/sec (53 kJ/s). Potential energy changes are insignificant. What power is generated at the turbine shaft?

SI Solution

Equation 30.33 can be solved for the shaft work.

$$P_{shaft} = \dot{m}w_{shaft} = \dot{Q}_t + \dot{m}\left(h_1 - h_2 + \frac{v_1^2 - v_2^2}{2} \right)$$

$$= -53\,000 \ \frac{J}{s}$$

$$+ \left(1.8 \ \frac{kg}{s} \right) \left(\begin{array}{c} 3140 \times 10^3 \ \frac{J}{kg} - 2500 \times 10^3 \ \frac{J}{kg} \\ + \dfrac{\left(20 \ \frac{m}{s} \right)^2 - \left(38 \ \frac{m}{s} \right)^2}{2} \end{array} \right)$$

$$= 1.098 \times 10^6 \ W \quad (1.1 \ MW)$$

Customary U.S. Solution

Proceeding as in the SI solution,

$$P_{shaft} = \dot{m}w_{shaft} = \dot{Q}_t + \dot{m}\left(h_1 - h_2 + \frac{v_1^2 - v_2^2}{2g_c J} \right)$$

$$- -50 \ \frac{Btu}{sec}$$

$$+ \left(4.0 \ \frac{lbm}{sec} \right) \left(\begin{array}{c} 1350 \ \frac{Btu}{lbm} - 1075 \ \frac{Btu}{lbm} \\ + \dfrac{\left(65 \ \frac{ft}{sec} \right)^2 - \left(125 \ \frac{ft}{sec} \right)^2}{(2)\left(32.2 \ \frac{lbm\text{-}ft}{lbf\text{-}sec^2} \right)\left(778 \ \frac{ft\text{-}lbf}{Btu} \right)} \end{array} \right)$$

$$= 1049 \ Btu/sec$$

11. BASIC THERMODYNAMIC RELATIONS

The following relations for a simple compressible substance (though not necessarily an ideal gas) can be derived from the first law of thermodynamics and other basic definitions.[7]

$$du = T\,ds - p\,dv \qquad 30.34$$

$$dh = T\,ds + v\,dp \qquad 30.35$$

12. MAXWELL RELATIONS

Equation 30.38 through Eq. 30.41 are exact differentials with the same form.

$$dz = M\,dx + N\,dy \qquad 30.36$$

---

[7]It is impractical to list every relationship for property changes. Most formulas can be written in several forms. For example, all equations can be written either for a unit mass (i.e., per lbm or kg) or for a mole (i.e., per lbmol or kmol). For example, Eq. 30.34 on a molar basis would be written as $dU = T\,dS - p\,dV$.

For exact differentials of this form, it can be shown that the following relationship is valid.

$$\left(\frac{\partial M}{\partial y}\right)_x = \left(\frac{\partial N}{\partial x}\right)_y \qquad 30.37$$

It is possible to relate many of the thermodynamic properties of simple compressible substances. Four of these relationships are collectively known as *Maxwell relations*.

$$\left(\frac{\partial T}{\partial v}\right)_s = -\left(\frac{\partial p}{\partial s}\right)_v \qquad 30.38$$

$$\left(\frac{\partial T}{\partial p}\right)_s = \left(\frac{\partial v}{\partial s}\right)_p \qquad 30.39$$

$$\left(\frac{\partial s}{\partial v}\right)_T = \left(\frac{\partial p}{\partial T}\right)_v \qquad 30.40$$

$$\left(\frac{\partial v}{\partial T}\right)_p = -\left(\frac{\partial s}{\partial p}\right)_T \qquad 30.41$$

The Maxwell relations are useful in defining macroscopic properties (e.g., temperature) in terms of microscopic considerations and in designing experiments to measure thermodynamic properties. In particular, the relations show that pressure, temperature, and specific volume can be determined experimentally, but entropy must be derived from the other properties.

$$v = \left(\frac{\partial g}{\partial p}\right)_T = \left(\frac{\partial h}{\partial p}\right)_s \qquad 30.42$$

$$p = -\left(\frac{\partial a}{\partial v}\right)_T = -\left(\frac{\partial u}{\partial v}\right)_s \qquad 30.43$$

$$T = \left(\frac{\partial h}{\partial s}\right)_p = \left(\frac{\partial u}{\partial s}\right)_v \qquad 30.44$$

$$s = -\left(\frac{\partial a}{\partial T}\right)_v = -\left(\frac{\partial g}{\partial T}\right)_p \qquad 30.45$$

13. PROPERTY CHANGES IN IDEAL GASES

For real solids, liquids, and vapors, changes in most properties can be determined only by subtracting the initial property value from the final property value. For simple compressible substances (i.e., ideal gases), however, many changes in properties can be found directly, without knowing the initial and final property values.

Some of the relations for determining property changes do not depend on the type of process. For example, a general relationship that applies to any ideal gas experiencing any process is easily derived from the equation of state.

$$\frac{p_1 V_1}{T_1} = \frac{p_2 V_2}{T_2} \qquad 30.46$$

When temperature is held constant, Eq. 30.46 reduces to *Boyle's law.*

$$p_1 V_1 = p_2 V_2 \qquad 30.47$$

When pressure is held constant, Eq. 30.46 reduces to *Charles' law.*

$$\frac{V_1}{T_1} = \frac{V_2}{T_2} \qquad 30.48$$

When volume is held constant, Eq. 30.46 reduces to *Gay-Lussac's law.*

$$\frac{p_1}{T_1} = \frac{p_2}{T_2} \qquad 30.49$$

Similarly, the changes in enthalpy, internal energy, and entropy are independent of the process. For perfect gases (i.e., ideal gases with constant specific heats),

$$\Delta h = c_p \Delta T \quad \text{[perfect gas]} \qquad 30.50$$

$$\Delta u = c_v \Delta T \quad \text{[perfect gas]} \qquad 30.51$$

$$\Delta s = c_p \ln\left(\frac{T_2}{T_1}\right) - R\ln\left(\frac{p_2}{p_1}\right)$$
$$= c_v \ln\left(\frac{T_2}{T_1}\right) + R\ln\left(\frac{v_2}{v_1}\right) \quad \text{[perfect gas]} \qquad 30.52$$

Enthalpy and internal energy are *point functions*, so Eq. 30.50 and Eq. 30.51 are valid for all processes, even though the specific heats at constant pressure and volume are part of the calculation. These equations should not be confused with the heat transfer relations, which have a similar form. Heat, Q, is a path function and depends on the path taken.

$$q = c_p \Delta T\Big|_p \quad \text{[perfect gas]} \qquad 30.53$$

$$q = c_v \Delta T\Big|_v \quad \text{[perfect gas]} \qquad 30.54$$

Relationships between the properties follow for ideal gases experiencing specific processes. (The standard thermodynamic sign convention is automatically followed.) All relations can be written in slightly different forms, including per-unit mass and per-mole bases. For compactness, only the per-unit mass equations are listed. However, all equations can be converted to a molar basis by substituting H for h, V for v, R^* for R, C_p for c_p, and so on. Other forms can be derived by substituting the equation of state where appropriate.[8]

---
[8]For example, RT can be substituted anywhere pv appears.

Constant Pressure, Closed Systems

$$p_2 = p_1 \qquad 30.55$$

$$T_2 = T_1\left(\frac{v_2}{v_1}\right) \qquad 30.56$$

$$v_2 = v_1\left(\frac{T_2}{T_1}\right) \qquad 30.57$$

$$q = h_2 - h_1 \qquad 30.58$$

$$= c_p(T_2 - T_1) \qquad 30.59$$

$$= c_v(T_2 - T_1) + p(v_2 - v_1) \qquad 30.60$$

$$u_2 - u_1 = c_v(T_2 - T_1) \qquad 30.61$$

$$= \frac{c_v p(v_2 - v_1)}{R} \qquad 30.62$$

$$= \frac{p(v_2 - v_1)}{k - 1} \qquad 30.63$$

$$w = p(v_2 - v_1) \qquad 30.64$$

$$= R(T_2 - T_1) \qquad 30.65$$

$$s_2 - s_1 = c_p \ln\left(\frac{T_2}{T_1}\right) \qquad 30.66$$

$$= c_p \ln\left(\frac{v_2}{v_1}\right) \qquad 30.67$$

$$h_2 - h_1 = q \qquad 30.68$$

$$= c_p(T_2 - T_1) \qquad 30.69$$

$$= \frac{kp(v_2 - v_1)}{k - 1} \qquad 30.70$$

Constant Volume, Closed Systems

$$p_2 = p_1\left(\frac{T_2}{T_1}\right) \qquad 30.71$$

$$T_2 = T_1\left(\frac{p_2}{p_1}\right) \qquad 30.72$$

$$v_2 = v_1 \qquad 30.73$$

$$q = u_2 - u_1 \qquad 30.74$$

$$= c_v(T_2 - T_1) \qquad 30.75$$

$$u_2 - u_1 = q \qquad 30.76$$

$$= c_v(T_2 - T_1) \qquad 30.77$$

$$= \frac{c_v v(p_2 - p_1)}{R} \qquad 30.78$$

$$= \frac{v(p_2 - p_1)}{k - 1} \qquad 30.79$$

$$w = 0 \qquad 30.80$$

$$s_2 - s_1 = c_v \ln\left(\frac{T_2}{T_1}\right) \qquad 30.81$$

$$= c_v \ln\left(\frac{p_2}{p_1}\right) \qquad 30.82$$

$$h_2 - h_1 = c_p(T_2 - T_1) \qquad 30.83$$

$$= \frac{kv(p_2 - p_1)}{k - 1} \qquad 30.84$$

Constant Temperature, Closed Systems

$$p_2 = p_1\left(\frac{v_1}{v_2}\right) \qquad 30.85$$

$$T_2 = T_1 \qquad 30.86$$

$$v_2 = v_1\left(\frac{p_1}{p_2}\right) \qquad 30.87$$

$$q = w \qquad 30.88$$

$$= T(s_2 - s_1) \qquad 30.89$$

$$= p_1 v_1 \ln\left(\frac{v_2}{v_1}\right) \qquad 30.90$$

$$= p_1 v_1 \ln\left(\frac{p_1}{p_2}\right) \qquad 30.91$$

$$= RT \ln\left(\frac{v_2}{v_1}\right) \qquad 30.92$$

$$= RT \ln\left(\frac{p_1}{p_2}\right) \qquad 30.93$$

$$u_2 - u_1 = 0 \qquad 30.94$$

$$w = q \qquad 30.95$$

$$= p_1 v_1 \ln\left(\frac{v_1}{v_2}\right) \qquad 30.96$$

$$= p_1 v_1 \ln\left(\frac{p_1}{p_2}\right) \qquad 30.97$$

$$= RT \ln\left(\frac{v_2}{v_1}\right) \qquad 30.98$$

$$= RT \ln\left(\frac{p_1}{p_2}\right) \qquad 30.99$$

$$s_2 - s_1 = \frac{q}{T} \qquad 30.100$$

$$= R \ln\left(\frac{v_2}{v_1}\right) \qquad 30.101$$

$$= R \ln\left(\frac{p_1}{p_2}\right) \qquad 30.102$$

$$h_2 - h_1 = 0 \qquad 30.103$$

Isentropic, Closed Systems
(Reversible Adiabatic)

$$p_2 = p_1 \left(\frac{v_1}{v_2}\right)^k \qquad \text{30.104}$$

$$= p_1 \left(\frac{T_2}{T_1}\right)^{\frac{k}{k-1}} \qquad \text{30.105}$$

$$T_2 = T_1 \left(\frac{v_1}{v_2}\right)^{k-1} \qquad \text{30.106}$$

$$= T_1 \left(\frac{p_2}{p_1}\right)^{\frac{k-1}{k}} \qquad \text{30.107}$$

$$v_2 = v_1 \left(\frac{p_1}{p_2}\right)^{\frac{1}{k}} \qquad \text{30.108}$$

$$= v_1 \left(\frac{T_1}{T_2}\right)^{\frac{1}{k-1}} \qquad \text{30.109}$$

$$q = 0 \qquad \text{30.110}$$

$$u_2 - u_1 = -w \qquad \text{30.111}$$

$$= c_v(T_2 - T_1) \qquad \text{30.112}$$

$$= \frac{c_v(p_2 v_2 - p_1 v_1)}{R} \qquad \text{30.113}$$

$$= \frac{p_2 v_2 - p_1 v_1}{k - 1} \qquad \text{30.114}$$

$$w = u_1 - u_2 \qquad \text{30.115}$$

$$= c_v(T_1 - T_2) \qquad \text{30.116}$$

$$= \frac{p_1 v_1 - p_2 v_2}{k - 1} \qquad \text{30.117}$$

$$= \frac{p_1 v_1}{k - 1}\left(1 - \left(\frac{p_2}{p_1}\right)^{\frac{k-1}{k}}\right) \qquad \text{30.118}$$

$$s_2 - s_1 = 0 \qquad \text{30.119}$$

$$h_2 - h_1 = c_p(T_2 - T_1) \qquad \text{30.120}$$

$$= \frac{k(p_2 v_2 - p_1 v_1)}{k - 1} \qquad \text{30.121}$$

Polytropic, Closed Systems

For $n = 1$, use constant temperature equations. For $n = k$, use adiabatic equations. For $n = 0$, use constant pressure equations. For $n = \infty$, use constant volume equations.

$$p_2 = p_1 \left(\frac{v_1}{v_2}\right)^n \qquad \text{30.122}$$

$$= p_1 \left(\frac{T_2}{T_1}\right)^{\frac{n}{n-1}} \qquad \text{30.123}$$

$$T_2 = T_1 \left(\frac{v_1}{v_2}\right)^{n-1} \qquad \text{30.124}$$

$$= T_1 \left(\frac{p_2}{p_1}\right)^{\frac{n-1}{n}} \qquad \text{30.125}$$

$$v_2 = v_1 \left(\frac{p_1}{p_2}\right)^{\frac{1}{n}} \qquad \text{30.126}$$

$$= v_1 \left(\frac{T_1}{T_2}\right)^{\frac{1}{n-1}} \qquad \text{30.127}$$

$$q = \frac{c_v(n - k)(T_2 - T_1)}{n - 1} \qquad \text{30.128}$$

$$u_2 - u_1 = c_v(T_2 - T_1) \qquad \text{30.129}$$

$$= q - w \qquad \text{30.130}$$

$$w = \frac{R(T_1 - T_2)}{n - 1} \qquad \text{30.131}$$

$$= \frac{p_1 v_1 - p_2 v_2}{n - 1} \qquad \text{30.132}$$

$$= \frac{p_1 v_1}{n - 1}\left(1 - \left(\frac{p_2}{p_1}\right)^{\frac{n-1}{n}}\right) \qquad \text{30.133}$$

$$s_2 - s_1 = \frac{c_v(n - k)}{n - 1}\ln\left(\frac{T_2}{T_1}\right) \qquad \text{30.134}$$

$$h_2 - h_1 = c_p(T_2 - T_1) \qquad \text{30.135}$$

$$= \frac{n(p_2 v_2 - p_1 v_1)}{n - 1} \qquad \text{30.136}$$

Isentropic, Steady-Flow Systems

p_2, v_2, and T_2 are the same as for isentropic, closed systems.

$$q = 0 \qquad \text{30.137}$$

$$w = h_1 - h_2 \qquad \text{30.138}$$

$$= \frac{k p_1 v_1}{k - 1}\left(1 - \left(\frac{p_2}{p_1}\right)^{\frac{k-1}{k}}\right)$$

$$= c_p T_1 \left(1 - \left(\frac{p_2}{p_1}\right)^{\frac{k-1}{k}}\right) \qquad \text{30.139}$$

$$u_2 - u_1 = c_v(T_2 - T_1) \qquad \text{30.140}$$

$$h_2 - h_1 = -w \qquad \text{30.141}$$

$$= c_p(T_2 - T_1) \qquad \text{30.142}$$

$$= \frac{k(p_2 v_2 - p_1 v_1)}{k - 1} \qquad \text{30.143}$$

$$s_2 - s_1 = 0 \qquad \text{30.144}$$

Ventilation

Polytropic, Steady-Flow Systems

p_2, v_2, and T_2 are the same as for polytropic, closed systems. $w \neq \Delta h$ because the process is not adiabatic.

$$q = \frac{c_v(n-k)(T_2 - T_1)}{n-1} \qquad 30.145$$

$$w = h_1 - h_2 - |q_{loss}| \qquad 30.146$$

$$= \frac{nR(T_1 - T_2)}{n-1} \qquad 30.147$$

$$u_2 - u_1 = c_v(T_2 - T_1) \qquad 30.148$$

$$h_2 - h_1 = -w + |q_{loss}| \qquad 30.149$$

$$= c_p(T_2 - T_1) \qquad 30.150$$

$$= \frac{n(p_2 v_2 - p_1 v_1)}{n-1} \qquad 30.151$$

$$s_2 - s_1 = \frac{c_v(n-k)}{n-1} \ln\left(\frac{T_2}{T_1}\right) \qquad 30.152$$

Throttling, Steady-Flow Systems

$$p_1 v_1 = p_2 v_2 \qquad 30.153$$

$$p_2 < p_1 \qquad 30.154$$

$$v_2 > v_1 \qquad 30.155$$

$$T_2 = T_1 \quad \text{[ideal gas]} \qquad 30.156$$

$$q = 0 \qquad 30.157$$

$$w = 0 \qquad 30.158$$

$$u_2 - u_1 = 0 \qquad 30.159$$

$$h_2 - h_1 = 0 \qquad 30.160$$

$$s_2 - s_1 = R \ln\left(\frac{p_1}{p_2}\right) \qquad 30.161$$

$$= R \ln\left(\frac{v_2}{v_1}\right) \qquad 30.162$$

Example 30.3

2.0 lbm (0.9 kg) of hydrogen are cooled from 760°F to 660°F (400°C to 350°C) in a constant volume process. The specific heat at constant volume, c_v, is 2.435 Btu/lbm-°R (10.2 kJ/kg·K). How much heat is removed?

SI Solution

Equation 30.75 is on a per unit mass basis. The total heat transfer for a mass m is

$$Q = mc_v(T_2 - T_1)$$
$$= (0.9 \text{ kg})\left(10.2 \ \frac{\text{kJ}}{\text{kg·K}}\right)(350°\text{C} - 400°\text{C})$$
$$= -459 \text{ kJ}$$

The minus sign is consistent with the convention that a heat loss is negative.

Customary U.S. Solution

The total heat transfer for a mass m is

$$Q = mc_v(T_2 - T_1)$$
$$= (2.0 \text{ lbm})\left(2.435 \ \frac{\text{Btu}}{\text{lbm-°R}}\right)(660°\text{F} - 760°\text{F})$$
$$= -487 \text{ Btu}$$

Example 30.4

4 lbmol (1.8 kmol) of air initially at 1 atm and 530°R (295K) are compressed isothermally to 8 atm. How much total heat is removed during the compression?

SI Solution

Equation 30.99 applies to constant temperature compression. R^* is used in place of R to convert the calculations to a molar basis.

$$Q = nR^* T \ln\left(\frac{p_1}{p_2}\right)$$
$$= (1.8 \text{ kmol})\left(8.314 \ \frac{\text{kJ}}{\text{kmol·K}}\right)(295\text{K}) \ln\left(\frac{1 \text{ atm}}{8 \text{ atm}}\right)$$
$$= -9180 \text{ kJ}$$

Customary U.S. Solution

Writing Eq. 30.99 on a molar basis,

$$Q = nR^* T \ln\left(\frac{p_1}{p_2}\right)$$
$$= (4.0 \text{ lbmol})\left(1545 \ \frac{\text{ft-lbf}}{\text{lbmol-°R}}\right)(530°\text{R}) \ln\left(\frac{1 \text{ atm}}{8 \text{ atm}}\right)$$
$$= -6.81 \times 10^6 \text{ ft-lbf}$$

14. PROPERTY CHANGES IN INCOMPRESSIBLE FLUIDS AND SOLIDS

In order to simplify the solution to practical problems, enthalpy, entropy, internal energy, and specific volume in liquids and solids are often considered to be functions of temperature only. The effect of pressure is disregarded.

There are times, however, when it is necessary to evaluate the property changes in a solid or liquid system, no matter how small they may be. Changes for liquids can be evaluated by using compressed liquid tables. If certain assumptions are made, some property changes can also be calculated without knowing the initial and final values. The main assumptions are incompressibility and constant specific heats.

$$c_p = c_v = c \qquad 30.163$$

$$dv = 0 \qquad 30.164$$

$$du = c\,dT \qquad 30.165$$

$$ds = \frac{du}{T} \qquad 30.166$$

$$dh = c\,dT + v\,dp \qquad 30.167$$

If the specific heat is constant, Eq. 30.168 through Eq. 30.171 are valid.

$$v_2 - v_1 = 0 \qquad 30.168$$

$$u_2 - u_1 = c(T_2 - T_1) \qquad 30.169$$

$$s_2 - s_1 = c\ln\left(\frac{T_2}{T_1}\right) \qquad 30.170$$

$$h_2 - h_1 = c(T_2 - T_1) + v(p_2 - p_1) \qquad 30.171$$

15. HEAT RESERVOIRS

It is convenient to show a source of energy as an infinite constant-temperature *reservoir*. Figure 30.4 illustrates a source of energy (known as a *high-temperature reservoir* or *source reservoir*). By convention, the reservoir temperature is designated T_H, and the heat transfer from it is Q_H. The energy derived from such a theoretical source might actually be supplied by combustion, electrical heating, or nuclear reaction.

Figure 30.4 Energy Reservoirs

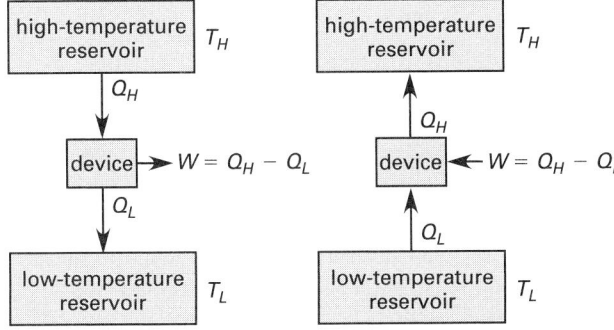

(a) power generation   (b) refrigeration effect

Similarly, energy is released to a *low-temperature reservoir* known as a *sink reservoir* or *energy sink*. The most common practical sink is the local environment. T_L and

Q_L are used to represent the reservoir temperature and energy absorbed. It is common to refer to Q_L as the rejected energy or energy released to the environment.

16. CYCLES

Although heat can be extracted and work performed in a single process, a cycle is necessary to obtain work in a useful quantity and duration. A *cycle* is a series of processes that eventually brings the system back to its original condition. Most cycles are continually repeated.

A cycle is completely defined by the working substance, the high- and low-temperature reservoirs, the means of doing work on the system, and the means of removing energy from the system.[9] (See Fig. 30.5.)

Figure 30.5 Net Work and Net Heat

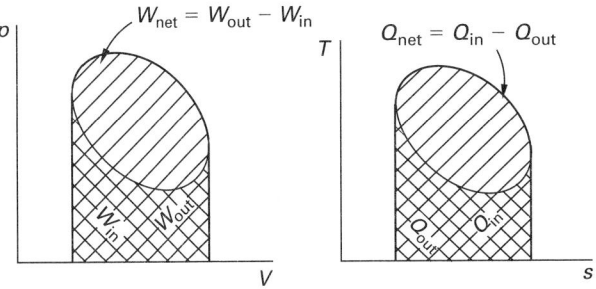

A cycle will appear as a closed curve when plotted on p-V and T-s diagrams. The area within the p-V or T-s curve represents the net work or net heat, respectively.

17. THERMAL EFFICIENCY

The *thermal efficiency* of a *power cycle* is defined as the ratio of useful work output to the supplied input energy.[10] W in Eq. 30.172 is the *net work*, since some of the gross output work may be used to run certain parts of the cycle. For example, a small amount of turbine output power may run boiler feed pumps.

$$\eta_{\text{th}} = \frac{\text{net work output}}{\text{energy input}} = \frac{W_{\text{net}}}{Q_{\text{in}}}$$

$$= \frac{W_{\text{out}} - W_{\text{in}}}{Q_{\text{in}}} \qquad 30.172$$

The first law can be written as

$$Q_{\text{in}} = Q_{\text{out}} + W_{\text{net}} \qquad 30.173$$

---

[9]The *Carnot cycle* depends only on the source and sink temperatures, not on the working fluid. However, most practical cycles depend on the working fluid.
[10]The effectiveness of refrigeration and compression cycles is measured by other parameters.

Equation 30.172 and Eq. 30.173 can be combined to define the thermal efficiency in terms of heat variables alone.

$$\eta_{\text{th}} = \frac{Q_{\text{in}} - Q_{\text{out}}}{Q_{\text{in}}} \qquad \textit{30.174}$$

Equation 30.174 shows that obtaining the maximum efficiency requires minimizing the Q_{out} term. The most efficient power cycle possible is the *Carnot cycle*.

18. SECOND LAW OF THERMODYNAMICS

The *second law of thermodynamics* can be stated in several ways. Equation 30.175 is the mathematical relation defining the second law. The equality holds for reversible processes; the inequality holds for irreversible processes.

$$\Delta s \geq \int_{T_1}^{T_2} \frac{dq}{T} \qquad \textit{30.175}$$

Equation 30.175 effectively states that net entropy must always increase in practical (irreversible) cyclical processes.

> A natural process that starts in one equilibrium state and ends in another will go in the direction that causes the entropy of the system and the environment to increase.

The *Kelvin-Planck statement* of the second law effectively says that it is impossible to build a cyclical engine that will have a thermal efficiency of 100%.

> It is impossible to operate an engine operating in a cycle that will have no other effect than to extract heat from a reservoir and turn it into an equivalent amount of work.

This formulation is not a contradiction of the first law of thermodynamics. The first law does not preclude the possibility of converting heat entirely into work—it only denies the possibility of creating or destroying energy. The second law says that if some heat is converted entirely into work, some other energy must be rejected to a low-temperature sink (i.e., lost to the surroundings).

Figure 30.6 illustrates violations of the second law.

19. AVAILABILITY

The maximum possible work that can be obtained from a cycle is known as the *availability*. Availability is independent of the device but is dependent on the temperature of the local environment. Both the first and second law must be applied to determine availability.

In Fig. 30.7, the *heat engine* is surrounded by an environment at absolute temperature T_L. The net heat available for conversion to useful work is $Q = Q_H - Q_L$.

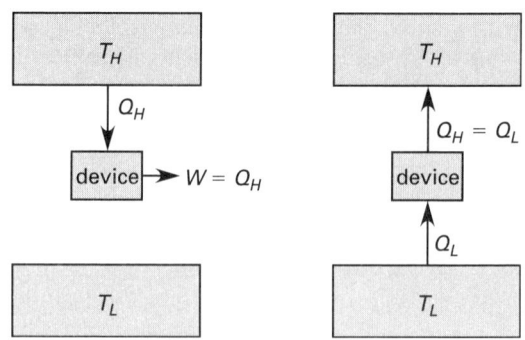

Figure 30.6 *Second Law Violations*

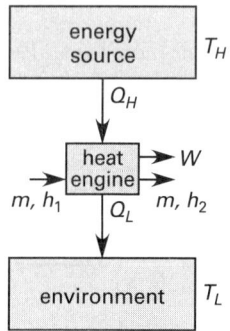

Figure 30.7 *Typical Heat Engine*

Work, W, is performed at a constant rate on the environment by the engine. This is the traditional development of the availability concept. Actually, the symbols $\dot{Q}$, $\dot{m}$, and P (power) should be used to represent the time rate of heat, mass, and work. Assuming steady flow, and neglecting kinetic and potential energies, the first law can be written as

$$mh_1 + Q = mh_2 + W \qquad \textit{30.176}$$

Entropy is not a part of the first law. The leaving entropy, however, can be calculated from the entering entropy and the entropy production.

$$ms_2 = ms_1 + \frac{Q}{T_L} \qquad \textit{30.177}$$

Since Eq. 30.176 and Eq. 30.177 both contain Q, they can be combined.

$$W = m(h_1 - T_L s_1 - h_2 + T_L s_2) \qquad \textit{30.178}$$

This equation can be simplified by introducing the *steady-flow availability function*, Φ. The maximum work output (availability) is

$$W_{\text{max}} = m(\Phi_1 - \Phi_2) \qquad \textit{30.179}$$

$$\Phi = h - T_L s \qquad \textit{30.180}$$

If the equality in Eq. 30.179 holds, both the process within the control volume and the energy transfers between the system and the environment must be reversible. Maximum work output, therefore, will be obtained in a reversible process. The difference between the maximum and the actual work output is known as the *process irreversibility*.

Example 30.5

What is the maximum useful work that can be produced per pound (per kilogram) of steam that enters a steady flow system at 800 psia (5.0 MPa) saturated and leaves in equilibrium with the environment at 70°F and 14.7 psia (20°C and 100 kPa)?

SI Solution

From the saturated steam table, $h_1 = 2794.2$ kJ/kg, and $s_1 = 5.9737$ kJ/kg·K.

The final properties are obtained from the saturated steam table for water at 20°C. $h_2 = 83.91$ kJ/kg, and $s_2 = 0.2965$ kJ/kg·K.

The availability is calculated from Eq. 30.178 using $T_L = 20°C + 273° = 293\text{K}$.

$$W_{\max} = m\left(h_1 - h_2 - T_L(s_1 - s_2)\right)$$

$$= (1 \text{ kg})\left(\begin{array}{c} 2794.2\ \dfrac{\text{kJ}}{\text{kg}} - 83.91\ \dfrac{\text{kJ}}{\text{kg}} \\ -(293\text{K})\left(\begin{array}{c} 5.9737\ \dfrac{\text{kJ}}{\text{kg·K}} \\ -0.2965\ \dfrac{\text{kJ}}{\text{kg·K}} \end{array}\right) \end{array}\right)$$

$$= 1047 \text{ kJ}$$

Customary U.S. Solution

From the saturated steam table, $h_1 = 1199.3$ Btu/lbm, and $s_1 = 1.4162$ Btu/lbm-°R.

The final properties are obtained from the saturated steam table for water at 70°F. $h_2 = 38.08$ Btu/lbm, and $s_2 = 0.007459$ Btu/lbm-°R.

The availability is calculated from Eq. 30.178 using $T_L = 70°F + 460° = 530°R$.

$$W_{\max} = m\left(h_1 - h_2 - T_L(s_1 - s_2)\right)$$

$$= (1 \text{ lbm})\left(\begin{array}{c} 1199.3\ \dfrac{\text{Btu}}{\text{lbm}} - 38.08\ \dfrac{\text{Btu}}{\text{lbm}} \\ -(530°\text{R})\left(\begin{array}{c} 1.4162\ \dfrac{\text{Btu}}{\text{lbm-°R}} \\ -0.07459\ \dfrac{\text{Btu}}{\text{lbm-°R}} \end{array}\right) \end{array}\right)$$

$$= 450.2 \text{ Btu}$$

Ventilation

31 HVAC: Psychrometrics

Nomenclature

ADP	apparatus dew point	°F	°C
B	volumetric fraction	–	–
BF	bypass factor	–	–
C	cycles of concentration	ppm	mg/L
CF	contact factor	–	–
c_p	specific heat	Btu/lbm-°F	kJ/kg·°C
G	gravimetric fraction	–	–
h	enthalpy	Btu/lbm	kJ/kg
m	mass	lbm	kg
n	number of moles	–	–
p	pressure	lbf/ft²	kPa
PF	performance factor	–	–
q	heat	Btu/lbm	J/kg
Q	volumetric flow rate	gal/min	L/s
R	specific gas constant	ft-lbf/lbm-°R	kJ/kg·K
RF	rating factor	–	–
SHR	sensible heat ratio	–	–
T	temperature	°F	°C
TDS	total dissolved solids	ppm	mg/L
TU	tower units	–	–
V	volume	ft³	m³
x	mole fraction	–	–

Symbols

η	efficiency	–	–
μ	degree of saturation	–	–
ρ	mass density	lbm/ft³	kg/m³
υ	specific volume	ft³/lbm	m³/kg
ϕ	relative humidity	–	–
ω	humidity ratio	lbm/lbm	kg/kg

Subscripts

a	dry air
db	dry-bulb
dp	dew-point
fg	vaporization
l	latent
s	sensible
sat	saturation
t	total
unsat	unsaturated
v	vapor
w	water
wb	wet-bulb

1. INTRODUCTION TO PSYCHROMETRICS

Atmospheric air contains small amounts of moisture and can be considered to be a mixture of two ideal gases—dry air and water vapor. All of the thermodynamic rules relating to the behavior of nonreacting gas mixtures apply to atmospheric air. From Dalton's law, for example, the total atmospheric pressure is the sum of the dry air partial pressure and the water vapor pressure.[1]

$$p = p_a + p_w \qquad 31.1$$

The study of the properties and behavior of atmospheric air is known as *psychrometrics*. Properties of atmospheric air are seldom evaluated, however, from theoretical thermodynamic principles. Rather, specialized techniques and charts have been developed for that purpose.

2. PROPERTIES OF ATMOSPHERIC AIR

At first, psychrometrics seems complicated by three different definitions of temperature. These three terms are not interchangeable.

- *dry-bulb temperature, T_{db}*: This is the equilibrium temperature that a regular thermometer measures if exposed to atmospheric air.

---

[1]Equation 31.1 points out a problem in semantics. The term *air* means *dry air*. The term *atmosphere* refers to the combination of dry air and water vapor. It is common to refer to the atmosphere as *moist air*.

- *wet-bulb temperature*, T_{wb}: This is the temperature of air that has gone through an adiabatic saturation process. (See Sec. 31.13.)

- *dew-point temperature*, T_{dp}: This is the dry-bulb temperature at which water starts to condense out when moist air is cooled in a constant pressure process.

For every temperature, there is a unique vapor pressure, p_{sat}, which represents the maximum pressure the water vapor can exert. The actual vapor pressure, p_w, can be less than or equal to, but not greater than, the saturation value. The saturation pressure is found from steam tables as the pressure corresponding to the dry-bulb temperature of the atmospheric air.

$$p_w \leq p_{sat} \qquad 31.2$$

If the vapor pressure equals the saturation pressure, the air is said to be saturated.[2] *Saturated air* is a mixture of dry air and saturated water vapor. When the air is saturated, all three temperatures are equal.

$$T_{db} = T_{wb} = T_{dp}\Big|_{sat} \qquad 31.3$$

Unsaturated air is a mixture of dry air and superheated water vapor.[3] When the air is unsaturated, the dew-point temperature will be less than the wet-bulb temperature. The *wet-bulb depression* is the difference between the dry-bulb and wet-bulb temperatures.

$$T_{dp} < T_{wb} < T_{db}\Big|_{unsat} \qquad 31.4$$

The amount of water vapor in atmospheric air is specified by three different parameters. The *humidity ratio*, ω (also known as the *specific humidity*), is the mass ratio of water vapor to dry air. If both masses are expressed in pounds (kilograms), the units of humidity ratio are lbm/lbm (kg/kg). However, since there is so little water vapor, the water vapor mass is often reported in *grains* of water. (There are 7000 grains per pound.) Accordingly, the humidity ratio will have the units of grains per pound.

$$\omega = \frac{m_w}{m_a} \qquad 31.5$$

Since $m = \rho V$, and since $V_w = V_a$, the humidity ratio can be written as

$$\omega = \frac{\rho_w}{\rho_a} \qquad 31.6$$

---

[2]Actually, the water vapor is saturated, not the air. However, this particular inconsistency in terms is characteristic of psychrometrics.
[3]As strange as it sounds, atmospheric water vapor is almost always superheated. This can be shown by drawing an isotherm passing through the vapor dome on a p-V diagram. The only place where the water vapor pressure is less than the saturation pressure is in the superheated region.

From the equation of state for an ideal gas, $m = pV/RT$. Since $V_w = V_a$ and $T_w = T_a$, the humidity ratio can be written in one additional form.

$$\omega = \frac{R_a p_w}{R_w p_a} = \frac{53.35 p_w}{85.78 p_a} = 0.622\left(\frac{p_w}{p_a}\right) \qquad 31.7$$

The *degree of saturation*, μ (also known as the *saturation ratio* and the *percentage humidity*), is the ratio of the actual humidity ratio to the saturated humidity ratio at the same temperature and pressure.

$$\mu = \frac{\omega}{\omega_{sat}} \qquad 31.8$$

A third index of moisture content is the *relative humidity*—the partial pressure of the water vapor divided by the saturation pressure.

$$\phi = \frac{p_w}{p_{sat}} \qquad 31.9$$

From the equation of state for an ideal gas, $\rho = p/RT$, so the relative humidity can be written as

$$\phi = \frac{\rho_w}{\rho_{sat}} \qquad 31.10$$

Combining the definitions of specific and relative humidities,

$$\phi = 1.608\omega\left(\frac{p_a}{p_{sat}}\right) \qquad 31.11$$

3. VAPOR PRESSURE

There are at least six ways of determining the partial pressure, p_w, of the water vapor in the air. The first method, derived from Eq. 31.9, is to multiply the relative humidity, ϕ, by the water's saturation pressure. The saturation pressure, in turn, is obtained from steam tables as the pressure corresponding to the air's dry-bulb temperature.

$$p_w = \phi p_{sat,db} \qquad 31.12$$

A more direct method is to read the saturation pressure (from the steam tables) corresponding to the air's dew-point temperature.

$$p_w = p_{sat,dp} \qquad 31.13$$

The third method can be used if water's mole (volumetric) fraction is known.

$$p_w = x_w p_t = B_w p_t \qquad 31.14$$

The fourth method is to calculate the actual vapor pressure from the empirical *Carrier equation*, valid for customary U.S. units only.[4]

$$p_w = p_{sat,wb} - \frac{(p_t - p_{sat,wb})(T_{db} - T_{wb})}{2830 - 1.44\,T_{wb}}$$

[U.S. only] *31.15*

The fifth method is based on the humidity ratio.

$$p_w = \frac{p_t \omega}{0.622 + \omega}$$

31.16

The sixth (and easiest) method is to read the water vapor pressure from a psychrometric chart. Some, but not all, psychrometric charts have water vapor scales.

Example 31.1

Use the methods described in the previous section to determine the partial pressure of water vapor in standard atmospheric air at 60°F (16°C) dry-bulb and 50% relative humidity.

SI Solution

method 1: From the steam tables, the saturation pressure corresponding to 16°C is 0.01819 bars. The partial pressure of the vapor is

$$p_w = \phi p_{sat}$$

$$= (0.50)(0.01819 \text{ bars})\left(100\,\frac{kPa}{bar}\right)$$

$$= 0.910 \text{ kPa}$$

method 2: The dew-point temperature (reading straight across on the psychrometric chart) is approximately 5°C. The saturation pressure from the steam table corresponding to 5°C is approximately 0.0087 bars (0.87 kPa).

method 3: The humidity ratio is 0.0056 kg/kg. From Eq. 31.16,

$$p_w = \frac{p_t \omega}{0.622 + \omega}$$

$$= \frac{(101.3 \text{ kPa})\left(0.0056\,\frac{kg}{kg}\right)}{0.622 + 0.0056\,\frac{kg}{kg}}$$

$$= 0.904 \text{ kPa}$$

---

[4]Equation 31.15 uses updated constants and is more accurate than the equation originally published by Carrier.

Customary U.S. Solution

method 1: From the steam tables, the saturation pressure corresponding to 60°F is 0.2564 lbf/in². The partial pressure of the vapor is

$$p_w = \phi p_{sat} = (0.50)\left(0.2564\,\frac{lbf}{in^2}\right)$$

$$= 0.128 \text{ lbf/in}^2$$

method 2: The dew-point temperature (reading straight across on the psychrometric chart) is approximately 41°F. The saturation pressure from the steam table corresponding to 41°F is approximately 0.127 lbf/in².

method 3: Use the Carrier equation. The wet-bulb temperature of the air is approximately 50°F. From the steam tables, the saturation pressure corresponding to that temperature is 0.1780 lbf/in².

$$p_w = p_{sat,wb} - \frac{(p_t - p_{sat,wb})(T_{db} - T_{wb})}{2830 - 1.44\,T_{wb}}$$

$$= 0.1780\,\frac{lbf}{in^2} - \frac{\left(14.7\,\frac{lbf}{in^2} - 0.1780\,\frac{lbf}{in^2}\right)}{2830 - (1.44)(50°F)}$$

$$= 0.125 \text{ lbf/in}^2$$

4. ENERGY CONTENT OF AIR

Since moist air is a mixture of dry air and water vapor, its total enthalpy, h (i.e., energy content), takes both components into consideration. Total enthalpy is conveniently shown on the diagonal scales of the psychrometric chart, but it can also be calculated. As Eq. 31.18 indicates, the reference temperature (i.e., the temperature that corresponds to a zero enthalpy) for the enthalpy of dry air is 0°F (0°C). Steam properties correspond to a low-pressure superheated vapor at room temperature.

$$h_t = h_a + \omega h_w$$

31.17

$$h_a = c_{p,air}\,T \approx \left(1.005\,\frac{kJ}{kg\cdot°C}\right)T_{°C}$$

[SI] *31.18(a)*

$$h_a = c_{p,air}\,T \approx \left(0.240\,\frac{Btu}{lbm\text{-}°F}\right)T_{°F}$$

[U.S.] *31.18(b)*

$$h_w = c_{p,water\,vapor}\,T + h_{fg}$$

$$\approx \left(1.805\,\frac{kJ}{kg\cdot°C}\right)T_{°C} + 2501\,\frac{kJ}{kg}$$

[SI] *31.19(a)*

$$h_w = c_{p,water\,vapor}\,T + h_{fg}$$

$$\approx \left(0.444\,\frac{Btu}{lbm\text{-}°F}\right)T_{°F} + 1061\,\frac{Btu}{lbm}$$

[U.S.] *31.19(b)*

5. THE PSYCHROMETRIC CHART

It is possible to develop mathematical relationships for enthalpy and specific volume (the two most useful thermodynamic properties) for atmospheric air. However, these relationships are almost never used. Rather, psychrometric properties can be read directly from *psychrometric charts* ("psych charts," as they are usually referred to), as illustrated in App. 31.A and App. 31.B. There are different psychrometric charts for low, medium, and high temperature ranges, as well as charts for different atmospheric pressures (i.e., elevations).

The usage of several scales varies somewhat from chart to chart. In particular, the use of the enthalpy scale depends on the chart used. Furthermore, not all psychrometric charts contain all scales.

A psychrometric chart is easy to use, despite the multiplicity of scales. The thermodynamic state (i.e., the position on the chart) is defined by specifying the values of any two parameters on intersecting scales (e.g., dry-bulb and wet-bulb temperature, or dry-bulb temperature and relative humidity). Once the state point has been located on the chart, all other properties can be read directly.

6. ENTHALPY CORRECTIONS

Some psychrometric charts have separate lines or scales for wet-bulb temperature and enthalpy. However, the deviation between lines of constant wet-bulb temperature and lines of constant enthalpy is small. Therefore, other psychrometric charts use only one set of diagonal lines for both scales. The error introduced is small—seldom greater than 0.1–0.2 Btu/lbm (0.23–0.46 kJ/kg). When extreme precision is needed, correction factors from the psychrometric chart can be used.

Example 31.2

Air at 50°F (10°C) dry bulb has a humidity ratio of 0.006 lbm/lbm (0.006 kg/kg). (a) Use the psychrometric chart to determine the enthalpy of the air. (b) Calculate the enthalpy of the air directly. (c) How much heat is needed to heat one unit mass of the air from 50°F to 140°F (10°C to 60°C) without changing the moisture content?

SI Solution

(a) Use the moisture content and dry-bulb temperature scales to locate the point corresponding to the original conditions. From the psychrometric chart, the enthalpy is approximately 25 kJ/kg.

(b) Use Eq. 31.17 and Eq. 31.18.

$$h_a = c_{p,\text{air}} T \approx \left(1.005 \ \frac{\text{kJ}}{\text{kg}\cdot{}^\circ\text{C}}\right) T_{{}^\circ\text{C}}$$

$$= \left(1.005 \ \frac{\text{kJ}}{\text{kg}\cdot{}^\circ\text{C}}\right)(10{}^\circ\text{C})$$

$$= 10 \ \text{kJ/kg}$$

$$h_w = c_{p,\text{water vapor}} T + h_{fg}$$

$$\approx \left(1.805 \ \frac{\text{kJ}}{\text{kg}\cdot{}^\circ\text{C}}\right) T_{{}^\circ\text{C}} + 2501 \ \frac{\text{kJ}}{\text{kg}}$$

$$= \left(1.805 \ \frac{\text{kJ}}{\text{kg}\cdot{}^\circ\text{C}}\right)(10{}^\circ\text{C}) + 2501 \ \frac{\text{kJ}}{\text{kg}}$$

$$= 2519 \ \text{kJ/kg}$$

$$h_t = h_a + \omega h_w$$

$$= 10 \ \frac{\text{kJ}}{\text{kg}} + \left(0.006 \ \frac{\text{kg}}{\text{kg}}\right)\left(2519 \ \frac{\text{kJ}}{\text{kg}}\right)$$

$$= 25.1 \ \text{kJ/kg}$$

(c) The psychrometric chart does not go up to 60°C. Therefore, the energy difference must be calculated mathematically. Although the initial enthalpy could be subtracted from the calculated final enthalpy, it is equivalent merely to calculate the difference based on the variable terms.

$$q = h_{t,2} - h_{t,1} = (c_{p,\text{air}} + \omega c_{p,\text{water vapor}})(T_2 - T_1)$$

$$= \left(\left(1.005 \ \frac{\text{kJ}}{\text{kg}\cdot{}^\circ\text{C}}\right) + \left(0.006 \ \frac{\text{kg}}{\text{kg}}\right)\left(1.805 \ \frac{\text{kJ}}{\text{kg}\cdot{}^\circ\text{C}}\right)\right)$$

$$\times (60{}^\circ\text{C} - 10{}^\circ\text{C})$$

$$= 50.8 \ \text{kJ/kg}$$

Customary U.S. Solution

(a) Use the moisture content and dry-bulb temperature scales to locate the point corresponding to the original conditions. From the psychrometric chart, the enthalpy is approximately 18.5 Btu/lbm.

(b) Use Eq. 31.17 and Eq. 31.18.

$$h_a = c_{p,\text{air}} T \approx \left(0.240 \ \frac{\text{Btu}}{\text{lbm-}{}^\circ\text{F}}\right) T_{{}^\circ\text{F}}$$

$$= \left(0.240 \ \frac{\text{Btu}}{\text{lbm-}{}^\circ\text{F}}\right)(50{}^\circ\text{F})$$

$$= 12 \ \text{Btu/lbm}$$

$$h_w = c_{p,\text{water vapor}} T + h_{fg}$$

$$\approx \left(0.444 \; \frac{\text{Btu}}{\text{lbm-}^\circ\text{F}}\right) T_{^\circ\text{F}} + 1065 \; \frac{\text{Btu}}{\text{lbm}}$$

$$= \left(0.444 \; \frac{\text{Btu}}{\text{lbm-}^\circ\text{F}}\right)(50^\circ\text{F}) + 1065 \; \frac{\text{Btu}}{\text{lbm}}$$

$$= 1087.2 \; \text{Btu/lbm}$$

$$h_t = h_a + \omega h_w$$

$$= 12 \; \frac{\text{Btu}}{\text{lbm}} + \left(0.006 \; \frac{\text{lbm}}{\text{lbm}}\right)\left(1087.2 \; \frac{\text{Btu}}{\text{lbm}}\right)$$

$$= 18.5 \; \text{Btu/lbm}$$

(c) Psychrometric charts for room temperature do not go up to 140°F. (Appendix 31.D could be used.) Therefore, the energy difference must be calculated mathematically. Although the initial enthalpy could be subtracted from the calculated final enthalpy, it is equivalent merely to calculate the difference based on the variable terms.

$$q = h_{t,2} - h_{t,1}$$

$$= (c_{p,\text{air}} + \omega c_{p,\text{water vapor}})(T_2 - T_1)$$

$$= \left(\begin{array}{c} 0.240 \; \dfrac{\text{Btu}}{\text{lbm-}^\circ\text{F}} + \left(0.006 \; \dfrac{\text{lbm}}{\text{lbm}}\right) \\ \times \left(0.444 \; \dfrac{\text{Btu}}{\text{lbm-}^\circ\text{F}}\right) \end{array} \right)$$

$$\times (140^\circ\text{F} - 50^\circ\text{F})$$

$$= 21.84 \; \text{Btu/lbm}$$

7. BASIS OF PROPERTIES

Several of the properties read from the psychrometric chart (specific volume, enthalpy, etc.) are given "per pound of dry air." This basis does not mean that the water vapor's contribution is absent. For example, if the enthalpy of atmospheric air is 28.0 Btu per pound of dry air, the energy content of the water vapor has been included. However, to get the energy of a mass of moist air, the enthalpy of 28 Btu/lbm would be multiplied by the mass of the dry air (m_a) only, not by the combined air and water masses.

$$h_t = m_a h_{\text{chart}} \qquad \textit{31.20}$$

Example 31.3

During the summer, air in a room reaches 75°F and 50% relative humidity. Find the air's (a) wet-bulb temperature, (b) humidity ratio, (c) enthalpy, (d) specific volume, (e) dew-point temperature, (f) actual vapor pressure, and (g) degree of saturation.

Solution

Locate the point where the 75°F vertical line intersects the curved 50% humidity line. Read all other values directly from the chart.

(a) Follow the diagonal line up to the left until it intersects the wet-bulb temperature scale. Read $T_{\text{wb}} = 62.6^\circ\text{F}$.

(b) Follow the horizontal line to the right until it intersects the humidity ratio scale. Read $\omega = 64.8$ gr (0.0093 lbm) of moisture per pound of dry air.

(c) Finding the enthalpy is different on different charts. Some charts use the same diagonal lines for wet-bulb temperature and humidity. Corrections are required in such cases. Other charts employ two alignment scales to use in conjunction with a straightedge. Read 28.1 Btu per pound of dry air.

(d) Interpolate between diagonal specific volume lines. Read $v = 13.68$ cubic feet per pound of dry air.

(e) Follow the horizontal line to the left until it intersects the dew-point scale. Read $T_{\text{dp}} = 55.1^\circ\text{F}$.

(f) From the steam tables, the saturation pressure corresponding to a dry-bulb temperature of 75°F is approximately 0.43 psia. From Eq. 31.9, the water vapor pressure is

$$p_w = \phi p_{\text{sat}} = (0.50)(0.43 \; \text{psia}) = 0.215 \; \text{psia}$$

(g) The humidity ratio at 75°F saturated is 131.5 gr (0.0188 lbm) per pound. From Eq. 31.8, the degree of saturation is

$$\mu = \frac{\omega}{\omega_{\text{sat}}} = \frac{64.8 \; \dfrac{\text{gr}}{\text{lbm}}}{131.5 \; \dfrac{\text{gr}}{\text{lbm}}} = 0.49$$

8. LEVER RULE

With few exceptions (e.g., relative humidity and enthalpy correction), the scales on a psychrometric chart are linear. Because they are linear, any one property can be used as the basis for interpolation or extrapolation for another property on an intersecting linear scale. This applies regardless of orientation of the scales. The scales do not have to be orthogonal.

Furthermore, since psychrometric properties are extensive properties (i.e., they depend on the quantity of air present), the mass of air can be used as the basis for interpolation or extrapolation. This principle, known as the *lever rule* or *inverse lever rule*, is used when determining the properties of a mixture. (See Sec. 31.9.)

Ventilation

9. ADIABATIC MIXING OF TWO AIR STREAMS

Figure 31.1 shows the mixing of two moist air streams. The state of the mixture can be determined if the flow rates and psychrometric properties of the two component streams are known.

Figure 31.1 *Mixing of Two Air Streams*

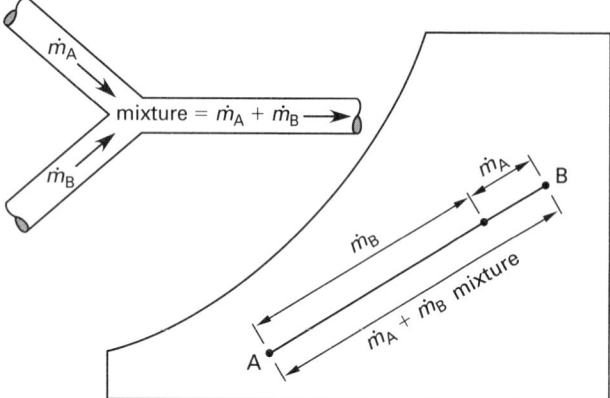

The two input states are located on the psychrometric chart and a straight line is drawn between them. The state of the mixture air will be on the straight line. The lever rule based on air masses is used to locate the mixture point. (Since the water vapor adds little to the mixture mass, the ratio of moist air masses can be approximated by the ratio of dry air masses, which, in turn, can be approximated by the ratio of air flow volumes.)

The lever rule can be used to find the mixture properties algebraically. Density changes can generally be disregarded, allowing volumetric flow rates to be used in place of mass flow rates. For the dry-bulb temperature (or any other property with a linear scale), the mixture temperature is

$$T_{\text{mixture}} = T_A + \left(\frac{\dot{m}_B}{\dot{m}_A + \dot{m}_B}\right)(T_B - T_A)$$

$$\approx T_A + \left(\frac{\dot{V}_B}{\dot{V}_A + \dot{V}_B}\right)(T_B - T_A) \qquad 31.21$$

Example 31.4

5000 ft³/min (2.36 m³/s) of air at 40°F (4°C) dry-bulb and 35°F (2°C) wet-bulb are mixed with 15,000 ft³/min (7.08 m³/s) of air at 75°F (24°C) dry-bulb and 50% relative humidity. Find the mixture dry-bulb temperature.

SI Solution

An approximate mixture temperature can be found by disregarding the change in density and taking a volumetrically weighted average. (The psychrometric chart

can be used to determine a more precise value mixture temperature. The more precise approach is used in the customary U.S. solution.)

$$T_{\text{mixture}} \approx \frac{\dot{V}_A T_A + \dot{V}_B T_B}{\dot{V}_A + \dot{V}_B}$$

$$= \frac{\left(2.36 \ \frac{\text{m}^3}{\text{s}}\right)(4°C) + \left(7.08 \ \frac{\text{m}^3}{\text{s}}\right)(24°C)}{2.36 \ \frac{\text{m}^3}{\text{s}} + 7.08 \ \frac{\text{m}^3}{\text{s}}}$$

$$= 19°C$$

Customary U.S. Solution

Locate the two points on the psychrometric chart, and draw a line between them. Estimate the specific volumes.

$$v_A = 12.65 \ \text{ft}^3/\text{lbm}$$

$$v_B = 13.68 \ \text{ft}^3/\text{lbm}$$

Calculate the dry air masses.

$$\dot{m}_A = \frac{\dot{V}_A}{v_A} = \frac{5000 \ \frac{\text{ft}^3}{\text{min}}}{12.65 \ \frac{\text{ft}^3}{\text{lbm}}}$$

$$= 395 \ \text{lbm/min}$$

$$\dot{m}_B = \frac{\dot{V}_B}{v_B} = \frac{15{,}000 \ \frac{\text{ft}^3}{\text{min}}}{13.68 \ \frac{\text{ft}^3}{\text{lbm}}}$$

$$= 1096 \ \text{lbm/min}$$

Use Eq. 31.21.

$$T_{\text{mixture}} = T_A + \left(\frac{\dot{m}_B}{\dot{m}_A + \dot{m}_B}\right)(T_B - T_A)$$

$$= 40°F + \left(\frac{1096 \ \frac{\text{lbm}}{\text{min}}}{1096 \ \frac{\text{lbm}}{\text{min}} + 395 \ \frac{\text{lbm}}{\text{min}}}\right)$$

$$\times (75°F - 40°F)$$

$$= 65.7°F$$

10. AIR CONDITIONING PROCESSES

The psychrometric chart is particularly useful in analyzing air conditioning processes because the paths of many processes are straight lines. Sensible heating and cooling processes, for example, follow horizontal straight lines. Adiabatic saturation processes follow lines of constant enthalpy (essentially parallel to lines of constant wet-bulb temperature). The paths of pure humidification and dehumidification follow vertical paths. Figure 31.2 summarizes the directions of these paths.

Figure 31.2 *Common Psychrometric Processes*

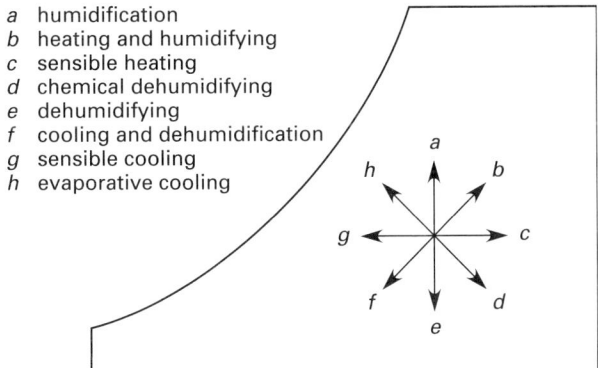

    a   humidification
    b   heating and humidifying
    c   sensible heating
    d   chemical dehumidifying
    e   dehumidifying
    f   cooling and dehumidification
    g   sensible cooling
    h   evaporative cooling

11. SENSIBLE HEAT RATIO

In general, the slope of any process line on the psychrometric chart is determined from the *sensible heat ratio*, also known as the *sensible heat factor*, SHF, and *sensible-total ratio*, S/T, scale on the chart. In an air conditioning process, the sensible heat ratio, SHR, is the ratio of sensible heat added (or removed) to total heat added (or removed). (The use of such scales varies from chart to chart. The process slope is determined from the sensible heat factor protractor and then translated (i.e., moved) to the appropriate point on the chart.)

$$\text{SHR} = \frac{q_s}{q_t} = \frac{q_s}{q_s + q_l} \qquad 31.22$$

The sensible heat ratio is always the slope of the line representing the change from the beginning point to the ending point on the psychrometric chart. Different designations are given to the sensible heat ratio, however, depending on where the changes occur.

If the sensible and latent energies change as the air passes through an occupied room, the term *room sensible heat ratio*, RSHR, is used. If the changes occur as the air passes through an air conditioning coil (apparatus), the term *coil* (or *apparatus*) *sensible heat ratio* is used, CSHR. Since the air conditioning apparatus usually removes heat and moisture from both the conditioned room and from outside makeup air, the term *grand sensible heat ratio*, GSHR, can be used in place of the coil sensible heat ratio. The *effective sensible heat ratio*, ESHR, is the slope of the line between the apparatus dew point on the saturation line and the design conditions of the conditioned space.

The sensible heat ratio is a psychrometric slope; it is not a geometric slope.

Example 31.5

During the summer, air from a conditioner enters an occupied space at 55°F (13°C) dry-bulb and 30% relative humidity. The ratio of sensible to total loads in the space is 0.45:1. The humidity ratio of the air leaving the room is 60 gr/lbm (8.6 g/kg). What is the dry-bulb temperature of the leaving air?

SI Solution

The sensible heat ratio is 0.45. Use the psychrometric chart (see App. 31.B) to determine the slope corresponding to this ratio. Draw a temporary line from the center of the protractor to the 0.45 mark on the sensible heat factor (inside) scale.

Locate 13°C dry-bulb and 30% relative humidity on the psychrometric chart. Draw a line through this point parallel to the temporary line, which is drawn with a slope of 0.45. The intersection of this line and the horizontal line corresponding to 8.6 g/kg determines the condition of the leaving air. The dry-bulb temperature is approximately 25.2°C.

Customary U.S. Solution

The sensible heat ratio is 0.45. Use the psychrometric chart (see App. 31.A) to determine the slope corresponding to this ratio. Draw a temporary line from the center of the protractor to 0.45 on the sensible heat factor (inside) scale.

Locate 55°F dry-bulb and 30% relative humidity on the psychrometric chart. Draw a line through this point parallel to the temporary line, which is drawn with a slope of 0.45. The intersection of this line and the horizontal line corresponding to 60 gr/lbm determines the condition of the leaving air. The dry-bulb temperature is approximately 76°F.

12. STRAIGHT HUMIDIFICATION

Straight (pure) *humidification* increases the water content of the air without changing the dry-bulb temperature. This is represented by a vertical condition line on the psychrometric chart. The *humidification load* is the mass of water added to the air per unit time (usually per hour).

13. BYPASS FACTOR AND COIL EFFICIENCY

Conditioning of air is accomplished by passing it through cooling or heating coils. Ideally, all of the air will come into contact with the coil for a long enough time and will leave at the coil temperature. In reality, this does not occur, and the air does not reach the coil temperature. The *bypass factor* can be thought of as the percentage of the air that is not cooled (or heated) by the coil. Under this interpretation, the remaining air (which is cooled or heated by the coil) is assumed to reach the coil temperature. The bypass factor expressed in decimal form is

$$BF = \frac{T_{\text{db,out}} - T_{\text{coil}}}{T_{\text{db,in}} - T_{\text{coil}}} \qquad 31.23$$

Bypass factors depend largely on the type of coil used. Bypass factors for large commercial units (such as those used in department stores) are small—around 10%. For small residential units, they are approximately 35%.

The *coil efficiency* is the complement of the bypass factor.

$$\eta_{\text{coil}} = 1.0 - BF \qquad 31.24$$

14. SENSIBLE COOLING AND HEATING

There is no change in the dew point or moisture content of the air with *sensible heating* and *cooling*. Since the moisture content is constant, these processes are represented by horizontal condition lines on the psychrometric chart (moving right for heating and left for cooling). (See Fig. 31.3.)

Figure 31.3 Sensible Cooling

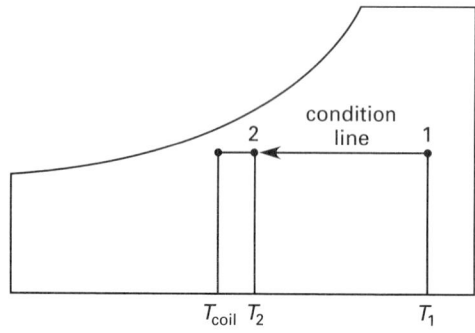

The energy change during the process can be calculated from enthalpies read directly from the psychrometric chart or approximated from the dry-bulb temperatures. In Eq. 31.25, $c_{p,\text{air}}$ is usually taken as 0.240 Btu/lbm-°F (1.005 kJ/kg·°C), and $c_{p,\text{moisture}}$ is taken as approximately 0.444 Btu/lbm-°F (1.805 kJ/kg·°C).

$$q = m_a(h_2 - h_1)$$
$$= m_a(c_{p,\text{air}} + \omega c_{p,\text{moisture}})(T_2 - T_1) \qquad 31.25$$

15. COOLING WITH COIL DEHUMIDIFICATION

If the cooling coil's temperature is below the air's dew point (as is usually the case), moisture will condense on the coil. The *effective coil temperature* in this instance is referred to as the *apparatus dew point*, ADP (also known as the *coil apparatus dew point*), and is determined from the intersection of the condition line (i.e., *coil load line*)

and the curved saturation line on the psychrometric chart. The apparatus dew point is the temperature to which the air would be cooled if 100% of it contacted the coil. (The term "apparatus dew point" is generally only used with cooling-dehumidification processes.)

The mass of condensing water will be

$$m_w = m_a(\omega_1 - \omega_2) \qquad 31.26$$

The total energy removed from the air includes both sensible and latent components. The latent heat is calculated from the heat of vaporization evaluated at the pressure of the water vapor.

$$q_t = q_s + q_l = m_a(h_1 - h_2) \qquad 31.27$$
$$q_l = m_a(\omega_1 - \omega_2)h_{fg} \qquad 31.28$$

Referring to Fig. 31.4, it is convenient to think of air experiencing sensible cooling from point 1 to point 3, after which the air follows the saturation line down from point 3 to point 4 (the apparatus dew point). Water condenses out between points 3 and 4. For convenience, the condition line is drawn as a straight line between points 1 and 4. The slope of the ADP-2-1 line corresponds to the sensible heat ratio. (See Sec. 31.11.) Since some of the air does not contact the coil at all, the final condition of the air will actually be at point 2 on the condition line.

Figure 31.4 Cooling and Dehumidification

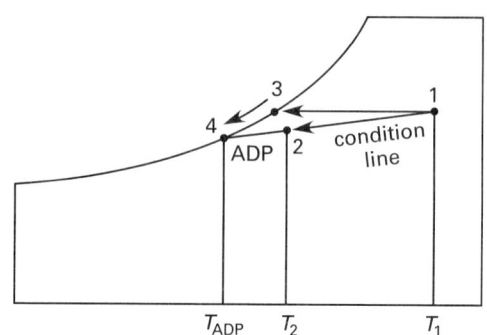

In practice, point 1 is usually known and either point 2 or point 4 are unknown. If point 2 is known, point 4 (the apparatus dew point) can be found graphically by extending the condition line over to the saturation line. (In some cases, the sensible heat ratio must be used to locate the apparatus dew point.) If point 4 is known, point 2 can be found from the bypass factor. The *contact factor*, CF, is essentially a dehumidification efficiency, calculated as the complement of the bypass factor.

$$CF = 1 - BF = 1 - \frac{T_{2,\text{db}} - ADP}{T_{1,\text{db}} - ADP} \qquad 31.29$$

Water condenses out over the entire temperature range from point 3 to point 4. The temperature of the water being removed is assumed to be the dew-point temperature at point 2.

Example 31.6

A coil has a bypass factor of 20% and an apparatus dew point of 55°F (13°C). Air enters the coil at 85°F (29°C) dry-bulb and 69°F (21°C) wet-bulb. What are the (a) latent heat loss, (b) sensible heat loss, and (c) sensible heat ratio?

SI Solution

(a) Locate the point corresponding to the entering air on the psychrometric chart. The enthalpy and humidity ratio are approximately

$$h_1 = 60.4 \text{ kJ/kg}$$
$$\omega_1 = 0.0123 \text{ kg/kg}$$

Use Eq. 31.29 to calculate the dry-bulb temperature of the air leaving the coil.

$$\begin{aligned} T_{2,\text{db}} &= \text{ADP} + \text{BF}(T_{1,\text{db}} - \text{ADP}) \\ &= 13°\text{C} + (0.20)(29°\text{C} - 13°\text{C}) \\ &= 16.2°\text{C} \end{aligned}$$

Draw a condition line between the entering air and the apparatus dew point on the psychrometric chart. Locate the point corresponding to 16.2°C dry-bulb on the condition line. The leaving enthalpy and humidity ratio are approximately

$$h_2 = 40.8 \text{ kJ/kg}$$
$$\omega_2 = 0.0100 \text{ kg/kg}$$

The total energy loss per kilogram is

$$\begin{aligned} q_t = h_1 - h_2 &= 60.4 \ \frac{\text{kJ}}{\text{kg}} - 40.8 \ \frac{\text{kJ}}{\text{kg}} \\ &= 19.6 \text{ kJ/kg of dry air} \end{aligned}$$

Since the partial pressure of the water vapor is unknown, estimate $h_{fg} \approx 2501 \text{ kJ/kg}$.

From Eq. 31.28, on a kilogram basis,

$$\begin{aligned} \frac{q_l}{m_a} &= (\omega_1 - \omega_2)h_{fg} \\ &= \left(0.0123 \ \frac{\text{kg}}{\text{kg}} - 0.0100 \ \frac{\text{kg}}{\text{kg}}\right)\left(2501 \ \frac{\text{kJ}}{\text{kg}}\right) \\ &= 5.75 \text{ kJ/kg of dry air} \end{aligned}$$

(b) The sensible heat loss is

$$\begin{aligned} q_s = q_t - q_l &= 19.6 \ \frac{\text{kJ}}{\text{kg}} - 5.75 \ \frac{\text{kJ}}{\text{kg}} \\ &= 13.9 \text{ kJ/kg of dry air} \end{aligned}$$

(c) The sensible heat ratio is

$$\begin{aligned} \text{SHR} = \frac{q_s}{q_t} &= \frac{13.9 \ \dfrac{\text{kJ}}{\text{kg}}}{19.6 \ \dfrac{\text{kJ}}{\text{kg}}} \\ &= 0.71 \end{aligned}$$

Customary U.S. Solution

(a) Locate the point corresponding to the entering air on the psychrometric chart. The enthalpy and humidity ratio are approximately

$$h_1 = 33.1 \text{ Btu/lbm}$$
$$\omega_1 = 0.0116 \text{ lbm/lbm}$$

Use Eq. 31.29 to calculate the dry-bulb temperature of the air leaving the coil.

$$\begin{aligned} T_{2,\text{db}} &= \text{ADP} + \text{BF}(T_{1,\text{db}} - \text{ADP}) \\ &= 55°\text{F} + (0.20)(85°\text{F} - 55°\text{F}) \\ &= 61°\text{F} \end{aligned}$$

Draw a condition line between the entering air and apparatus dew point on the psychrometric chart. Locate the point corresponding to 61°F dry-bulb on the condition line. The leaving enthalpy and humidity ratio are approximately

$$h_2 = 25.1 \text{ Btu/lbm}$$
$$\omega_2 = 0.0097 \text{ lbm/lbm}$$

The total energy loss per pound is

$$\begin{aligned} q_t = h_1 - h_2 &= 33.1 \ \frac{\text{Btu}}{\text{lbm}} - 25.1 \ \frac{\text{Btu}}{\text{lbm}} \\ &= 8.0 \text{ Btu/lbm of dry air} \end{aligned}$$

Since the partial pressure of the water vapor is not known, estimate $h_{fg} \approx 1060 \text{ Btu/lbm}$.

From Eq. 31.28, on a pound basis,

$$\begin{aligned} \frac{q_l}{m_a} &= (\omega_1 - \omega_2)h_{fg} \\ &= \left(0.0116 \ \frac{\text{lbm}}{\text{lbm}} - 0.0097 \ \frac{\text{lbm}}{\text{lbm}}\right)\left(1060 \ \frac{\text{Btu}}{\text{lbm}}\right) \\ &= 2.01 \text{ Btu/lbm of dry air} \end{aligned}$$

(b) The sensible heat loss is

$$\begin{aligned} q_s = q_t - q_l &= 8.0 \ \frac{\text{Btu}}{\text{lbm}} - 2.0 \ \frac{\text{Btu}}{\text{lbm}} \\ &= 6.0 \text{ Btu/lbm of dry air} \end{aligned}$$

Ventilation

(c) The sensible heat ratio is

$$\mathrm{SHR} = \frac{q_s}{q_t} = \frac{6.0 \, \dfrac{\mathrm{Btu}}{\mathrm{lbm}}}{8.0 \, \dfrac{\mathrm{Btu}}{\mathrm{lbm}}}$$

$$= 0.75$$

16. ADIABATIC SATURATION PROCESSES

To measure the wet-bulb temperature, air must experience an *adiabatic saturation process*, also known as *evaporative cooling*. Adiabatic saturation processes occur in cooling towers, air washers, and evaporative coolers ("swamp coolers"). To become saturated, the air must pick up the maximum amount of moisture it can hold at that temperature. This moisture comes from the vaporization of liquid water. For the process to be adiabatic, there can be no external source of energy to vaporize the liquid water needed to saturate the air.

At first analysis, the terms adiabatic and saturation seem contradictory. Adiabatic saturation is possible, however, if the latent heat of vaporization comes from the air itself. If the air gives up sensible heat, that energy can be used to vaporize liquid water. Of course, the air temperature decreases when sensible heat is given up. That is the reason that the wet-bulb temperature is generally less than the dry-bulb temperature. Only when the air is saturated will the two temperatures be equal.

An adiabatic saturation process can be produced with a *sling psychrometer*, which is essentially a regular thermometer with its bulb wrapped in wet cotton or gauze. Rapidly twirling the thermometer through the air at the end of a cord will cause the water in the gauze to evaporate. The latent heat needed to vaporize the water will come from the sensible heat of the air, and the thermometer will measure the wet-bulb temperature.

Since the increase in the water vapor's latent heat content equals the decrease in the air's sensible heat, the total enthalpies before and after adiabatic saturation are the same. Therefore, an adiabatic saturation process follows a line of constant enthalpy on the psychrometric chart. These lines are, for approximation purposes, parallel to lines of constant wet-bulb temperature.

The bypass factor concept is not used with adiabatic saturation processes. Instead, the *saturation efficiency* (*humidification efficiency*) is used. The saturation efficiency of large commercial air washers is typically 90% to 95%. The wet bulb temperature does not change during the saturation process. .

$$\eta_{\mathrm{sat}} = \frac{T_{\mathrm{db,air,in}} - T_{\mathrm{db,air,out}}}{T_{\mathrm{db,air,in}} - T_w} \qquad \textit{31.30}$$

17. AIR WASHERS

An *air washer* is a device that passes air through a dense spray of recirculating water. The water is used to change the properties of the air. Air washers are used in air purifying and cleaning processes (i.e., removal of solids, liquids, gases, vapors, and odors), as well as for evaporative cooling and dew-point control.[5]

The difference between a spray humidifier and spray dehumidifier is the temperature of the spray water. In an *adiabatic air washer*, the spray water is recirculated without being heated or cooled. After equilibrium is reached, the water temperature will be equal to the air's entering wet-bulb temperature. The air will be cooled and humidified, leaving partially or completely saturated at its entering wet-bulb temperature. However, if the spray water is chilled, the air will be cooled and dehumidified. And, if the spray water is heated, the air will be humidified and (possibly) heated.

An air washer's *saturation efficiency*, typically 90% to 95%, is measured by the drop in dry-bulb temperature relative to the entering wet-bulb depression.

$$\eta_{\mathrm{sat}} = \frac{T_{\mathrm{in,db}} - T_{\mathrm{out,db}}}{T_{\mathrm{in,db}} - T_{\mathrm{in,wb}}} \qquad \textit{31.31}$$

Air velocity through washers is approximately 500 ft/min (2.6 m/s). Velocities outside the range of 300 ft/min to 750 ft/min (1.5 m/s to 3.8 m/s) are probably faulty. The water pressure is typically 20 psig to 40 psig (140 kPa to 280 kPa). The spray quantity per bank of nozzles is in the range of 1.5 gal/min to 5 gal/min per 1000 ft^3 (3.3 L/s to 11 L/s per 1000 m^3) of air. Screens, louvers, and mist eliminator plates will generate a static pressure drop of approximately 0.2 in wg to 0.5 in wg (50 kPa to 125 kPa) at 500 ft/min (2.6 m/s). Other operating parameters used to describe air washer performance include air mass flow rate per unit area (lbm/hr-ft^2 or kg/m^2·s), air and liquid heat transfer coefficients per volume of chamber (Btu/hr-°F-ft^3 or kW/°C·m^3), and the *spray ratio* (the mass of water sprayed to the mass of air passing through the washer per unit time).

18. COOLING WITH HUMIDIFICATION

When air passes through a water spray (as in an *air washer*), an *adiabatic saturation process* known as *evaporative cooling* occurs.[6] (See Sec. 31.16.) The air leaves with a lower temperature and a higher moisture content. This is represented on the psychrometric chart by

---

[5]Air washers are generally not used for removing carbonaceous or greasy particles.

[6]An *air washer* is basically a *spray chamber* through which air passes. When supplied with chilled water from a refrigeration source, the air washer can cool, dehumidify, or humidify the air. Air washers can be used without refrigeration to cool and humidify the air through an evaporative cooling process.

a condition line parallel to the lines of constant enthalpy (essentially constant wet-bulb temperature).

Adiabatic saturation is a constant-enthalpy process, since any evaporation of the water requires heat to be drawn from the air. Since the removed heat goes into the remaining water, the water temperature increases. When the water spray is continuously recirculated, the water temperature gradually increases to the wet-bulb temperature of the incoming air. The minimum leaving air temperature will be the water temperature (i.e., the wet-bulb temperature of the incoming air).

During steady-state operation, the temperature of the water spray will normally be stable at the air's wet-bulb temperature. However, the water temperature can also be artificially maintained by refrigeration at less than the wet-bulb temperature (but more than the dew-point temperature). Line 1–3 in Fig. 31.5 illustrates such a process.

Figure 31.5 *Cooling with Humidification (adiabatic saturation)*

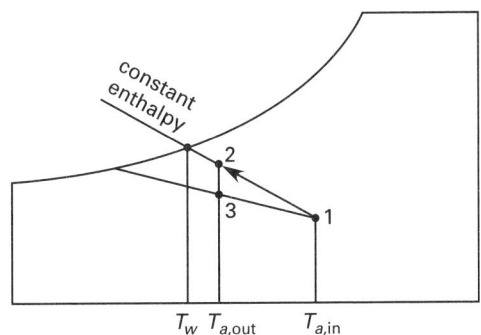

To prevent ice buildup, the cooled air temperature should be kept from dropping below the freezing point of water. The entering wet-bulb temperature should be kept above 35°F (1.7°C).

Example 31.7

Air at 90°F (32°C) dry-bulb and 65°F (18°C) wet-bulb enters an evaporative cooler. The air leaves at 90% relative humidity. The continuously recirculated spray water is stable at 65°F (18°C). What is the dry-bulb temperature of the leaving air?

Solution

Since the spray water is the same temperature as the wet-bulb temperature of the entering air, the cooler has reached its steady-state operating conditions. Locate the entering point on the psychrometric chart and draw a line of constant enthalpy (or constant 65°F (18°C) wet-bulb temperature) up to the 90% relative humidity curve. Read the dry-bulb temperature as approximately 67°F (19°C).

19. COOLING WITH SPRAY DEHUMIDIFICATION

If air passes through a water spray whose temperature is less than the entering air's wet-bulb temperature, both the dry-bulb and wet-bulb temperatures will decrease.[7] If the leaving water temperature is below the entering air's dew point, dehumidification will occur. As with any evaporative cooling, the air will give up thermal energy to the water. The final water temperature will depend on the thermal energy pickup and water flow rate. All air temperatures decrease, and some moisture condenses. The *performance factor* is defined as

$$PF = 1 - \frac{T_{air,wb,out} - T_{w,out}}{T_{air,wb,in} - T_{w,in}} \qquad 31.32$$

20. HEATING WITH HUMIDIFICATION

If air is humidified by injecting steam (*steam humidification*) or by passing the air through a hot water spray, the dry-bulb temperature and enthalpy of the air will increase.[8] The final air enthalpy and/or the required steam enthalpy can be determined from a conservation of energy equation. In Eq. 31.33, the mass of the air used is the dry air mass, which does not change. h_a, though expressed per pound of dry air, includes the energy of all vaporized water.

$$m_a h_{a,in} + m_w h_w = m_a h_{a,out} \qquad 31.33$$

From a conservation of mass for the water,

$$m_a \omega_{in} + m_w = m_a \omega_{out} \qquad 31.34$$

Figure 31.6 illustrates that the condition line will be above the line of constant enthalpy that radiates from the point corresponding to the incoming air. However, even though heat is added to the water, the air temperature can either decrease (as in the 1–2 process shown) or increase (as in the 1–3 process shown).

Figure 31.6 *Heating with Steam Humidification*

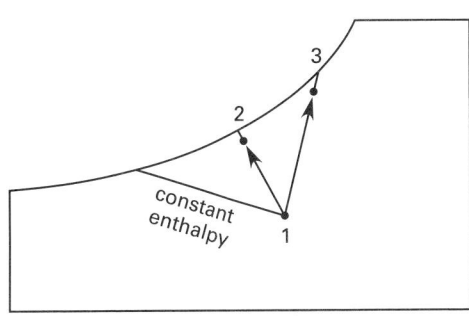

[7]This can unintentionally occur during the start-up of an air washer used for humidification, or the water can be kept intentionally chilled.
[8]When a spray of hot water is used, the water must be continually heated. Unlike a cold water spray, a natural equilibrium water temperature is not achieved.

21. HEATING AND DEHUMIDIFICATION

Air passing through a solid or liquid *adsorbent bed*, such as silica gel or activated alumina, will decrease in humidity. This is sometimes referred to as *chemical dehumidification*, *chemical dehydration*, or "*absorbent*" *dehumidification*.[9] If only latent heat was involved, this process would be the reverse of an adiabatic saturation process. However, as moisture is removed, exothermic chemical energy is generated in addition to the heat of vaporization liberated. Since thermal energy is generated, this is not an adiabatic process. (See Fig. 31.7.)

Figure 31.7 *Heating and Dehumidification*

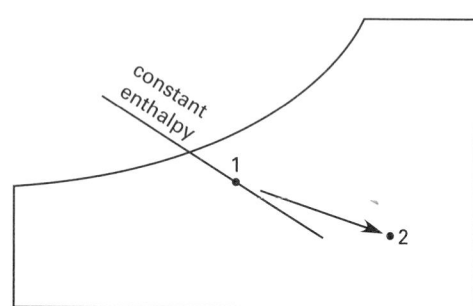

22. WET COOLING TOWERS

Conventional *wet cooling towers* cool warm water by exposing it to colder air.[10] They are usually used to provide cold water to power plant and large refrigeration condensers. The air is used to change the properties of the water, which leaves cooler. As it leaves, the saturated (or nearly saturated) warmed air takes sensible and latent heat from the water.

Cooling towers are generally counterflow, crossflow, or a combination. Though natural-draft and atmospheric towers exist, limited space usually requires that cooling towers operate with mechanical draft. Fans are located at the base of *forced draft towers* and blow air into the water cascading down. With *induced mechanical draft*, fans are located at the top of the tower, drawing air upward. Some portion of the exhaust air might reenter the cooling tower. This is known as *recycle air* (*recirculation air*). Recycle air decreases the efficiency of the tower.

During countercurrent operation, warm water is introduced at the top of the tower and is distributed by troughs or spray nozzles. The water passes over

staggered slats or interior *fill* (also known as *packing*).[11] Air flows upward, contacting the water on its downward path. A portion of the water evaporates, cooling the remainder of the water. The water temperature cannot decrease below the wet-bulb temperature. The actual final water temperature depends on a number of factors, including the state of the incoming air, the heat load, and the design (and efficiency) of the cooling tower. (See Fig. 31.8.)

Figure 31.8 *Counterflow Wet Cooling Tower*

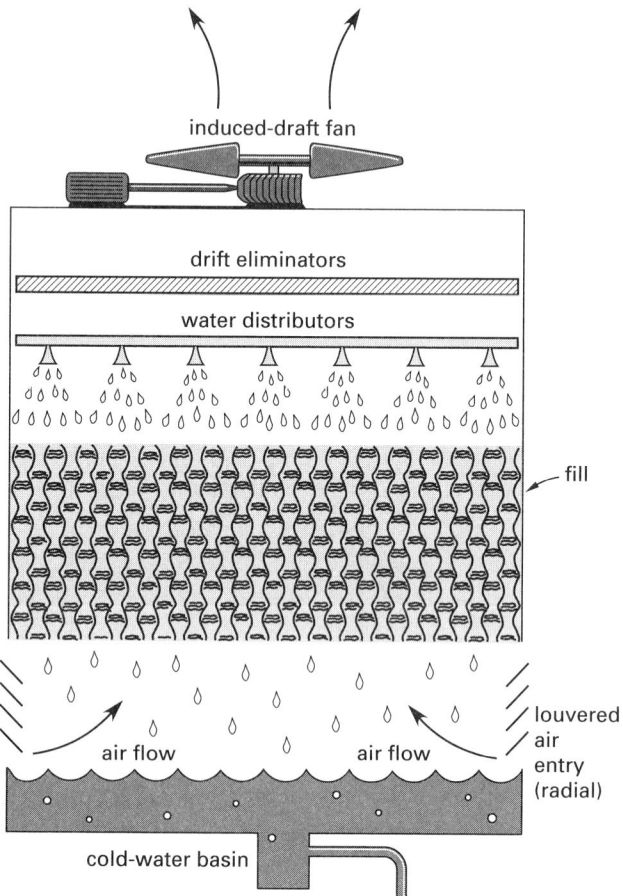

There are several environmental issues associated with wet cooling towers. Makeup water, though relatively little is needed, may be difficult to obtain. Moist plume discharges cause shadowing of adjacent areas and fogging and icing on nearby highways. Disposal of blowdown wastewater is also problematic.

Equation 31.35 is a per-unit energy balance that can be used to evaluate cooling tower performance. Each term, including the circulating water flow rate, is per unit

---

[9]The correct term for a substance that collects water on its surface is an *adsorbent*. By virtue of their great porosities, adsorbent particles have large surface areas. The attractive forces on the surfaces of these solids cause a thin layer of condensed water to form. Adsorbents are reactivated by heating.

[10]Though larger in size, a cooling tower is similar in operation to an air washer. In fact, an air washer can be used to cool water. Since air washer operation is not countercurrent, however, larger air flows are required to obtain the same cooling effect.

[11]Modern *filled towers* use corrugated *cellular fill* to maximize the air-water contact area. Standard polyvinyl chloride (PVC) fill is useful up to about 125°F (52°C). From 125°F to 140°F (52°C to 60°C), chlorinated PVC fill is recommended. Polypropylene fill should be used above 140°F (60°C). "Fill-less" towers, where the sprayed water merely falls through oncoming air, are used in some industries (food, steel, and paper processes) where a high-product carryover can lead to coating or buildup on the fill material.

mass (e.g., pound or kilogram) of dry air. Since the energy contribution of the makeup water is small, that term can be omitted for a first approximation.

$$m_{w,\text{in}} h_{w,\text{in}} + h_{a,\text{in}} + (\omega_{a,\text{out}} - \omega_{a,\text{in}}) h_{\text{makeup}}$$
$$= m_{w,\text{out}} h_{w,\text{out}} + h_{a,\text{out}}$$
$$= (m_{w,\text{in}} + \omega_{a,\text{in}} - \omega_{a,\text{out}}) h_{w,\text{out}} + h_{a,\text{out}} \qquad 31.35$$

If operation is at standard pressure, a psychrometric chart can be used to obtain the air enthalpies. For operation at different altitudes (i.e., different atmospheric pressures), the mathematical psychrometric relationships in Sec. 31.4 can be used to calculate the enthalpy. From Eq. 31.16, the humidity ratio is

$$\omega = \frac{0.622 p_{\text{water vapor}}}{p_{\text{total}} - p_{\text{water vapor}}} \qquad 31.36$$

When a cooling tower is used to provide cold water for the condenser of a refrigeration system, the water circulation will be approximately 3 gal/min per ton (0.19 L/s) of refrigeration. Approximately 2 gal/min to 4 gal/min of water are distributed per square foot (1.4 L/s to 2.7 L/s per square meter) of tower, and the air velocity should be approximately 700 ft/min (3.6 m/s) through the net free area. Coolants for condensers in reciprocating refrigeration systems usually call for an 85°F to 90°F (29°C to 32°C) water temperature. (This corresponds to a condensing temperature of approximately 100°F to 110°F (38°C to 43°C).) Various valves, mixing, louvers, and dampers are used to maintain a constant output water temperature.

The lowest temperature to which water can be cooled by purely evaporative means is the wet-bulb temperature of the entering air. The *cooling efficiency*, η_w, is based on the water temperature. The *water range* (*cooling range* or *range*) is defined as the actual difference between the entering and leaving water temperatures. (For water-cooled refrigeration condensers, this is equal to the water's temperature increase in the condenser.) The *approach* is defined as the difference between the leaving water temperature and the entering air wet-bulb temperatures.[12] Cooling efficiency is typically 50% to 70%.[13] Natural draft towers can cool the water to within 10°F to 12°F (5.5°C to 6.7°C) of the wet-bulb temperature. Forced draft towers can cool the water to within 5°F to 6°F (2.8°C to 3.3°C).

$$\eta_w = \frac{\text{range}}{\text{approach} + \text{range}} = \frac{T_{w,\text{in}} - T_{w,\text{out}}}{T_{w,\text{in}} - T_{\text{air,wb,in}}} \qquad 31.37$$

As Eq. 31.37 indicates, the actual wet-bulb temperature of the cooling air is particularly important in determining cooling tower performance. The higher the wet-bulb temperature, the lower the efficiency. (This is because when the denominator in Eq. 31.37 decreases, the numerator decreases even more.) In rating their cooling towers, most manufacturers have adopted the practice of using wet-bulb temperatures that will be exceeded only 2.5% of the time or less.

Performance of a cooling tower also depends on the relative humidity of the air. High relative humidities decrease the water evaporation rate, decreasing the efficiency.

The *heat load* (*tower load* or *cooling duty*) is calculated from the range and the water mass flow rate.

$$q = m_w c_p (T_{w,\text{in}} - T_{w,\text{out}}) = m_w (h_{w,\text{in}} - h_{w,\text{out}}) \qquad 31.38$$

Cooling towers are sometimes rated in tower units, which are essentially proportional to the tower cost. The number of *tower units*, TU, is equal to a rating factor multiplied by the flow rate. *Rating factors* define the relative difficulty in cooling, essentially the relative amount of contact area or fill volume required. Manufacturers provide charts showing the relationship between rating factor, approach, range, and wet-bulb temperature.

$$\text{TU} = \text{RF} \times Q_{\text{gpm}} \qquad 31.39$$

23. COOLING TOWER BLOWDOWN

Water losses occur from evaporation, windage, and blowdown. *Evaporation loss* can be calculated from the humidity ratio increase and is approximately 0.1% per °F (0.18% per °C) decrease in water temperature.[14] *Windage loss*, also known as *drift*, is water lost in small droplets and carried away by the air flow. Windage loss is typically in the 0.1% to 0.3% range for mechanical draft towers. Since windage droplets are a mechanical mixture (not a thermodynamic solution of two gases), they are not adequately accounted for by the humidity ratio.

Makeup water must be provided to replace all water losses. As more and more water enters the system, *total dissolved solids*, TDS (e.g., chlorides), will build up over time. Water can be treated to prevent deposit, and a portion of the water can be periodically or continuously bled off. *Cycles of concentration* (*ratio of concentration*), C, is the ratio of total dissolved solids in the

---

[12]Thus, approach for a cooling tower is analogous to the terminal temperature difference in the surface condenser.
[13]The term "thermal efficiency" is sometimes used here inappropriately.

[14]This value is approximate and is reported in various ways. Some authorities state "0.1% per degree Fahrenheit"; others say "1% per 10°F"; and yet others, "1% per 10°F to 13°F."

recirculating water to the total dissolved solids in the makeup water.[15]

$$C = \frac{(\text{TDS})_{\text{recirculating}}}{(\text{TDS})_{\text{makeup}}}$$

$$= \frac{m_{\text{evaporation}} + m_{\text{blowdown}} + m_{\text{windage}}}{m_{\text{blowdown}} + m_{\text{windage}}} \qquad 31.40$$

Though windage removes some of the solids, most must be removed by bleeding some of the water off. This is known as *blowdown* or *bleed-off*. If the maximum cycles of concentration are known, the blowdown is

$$m_{\text{blowdown}} = \frac{m_{\text{evaporation}} + (1 - C_{\text{max}})m_{\text{windage}}}{C_{\text{max}} - 1} \qquad 31.41$$

Additives should be used to prevent specific problems encountered, such as scale buildup, corrosion, biological growth, foaming, and discoloration.

24. DRY COOLING TOWERS

Dry cooling is used when environmental protection and water conservation are issues. It is used primarily by nonutility generators (e.g., waste-to-energy and cogeneration plants).

There are two types of dry cooling towers. Both use finned-tube heat exchangers. In a *direct-condensing tower*, steam travels through large-diameter "trunks" to a crossflow heat exchanger where it is condensed and cooled by the cooler air.[16] In an *indirect-condensing dry cooling tower*, steam is condensed by cold water jets (surface or jet condenser) and is subsequently cooled by air. The hot condensate is then pumped to crossflow heat exchangers where it is sensibly cooled (no condensation) by the air. Air flow may be mechanical or natural draft. Most U.S. installations are direct-condensing. Worldwide, natural-draft indirect systems are more predominant, particularly for power plants with capacities in excess of 100 MW. (See Fig. 31.9.)

Figure 31.9 *Dry Cooling Towers*

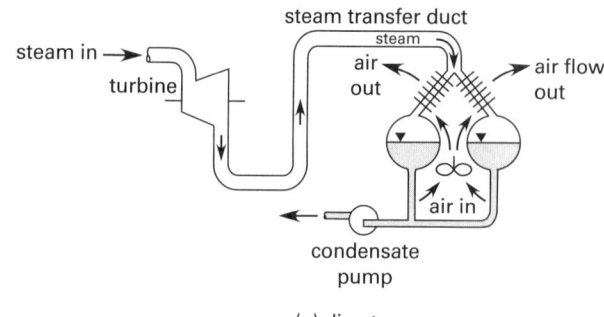

(a) direct

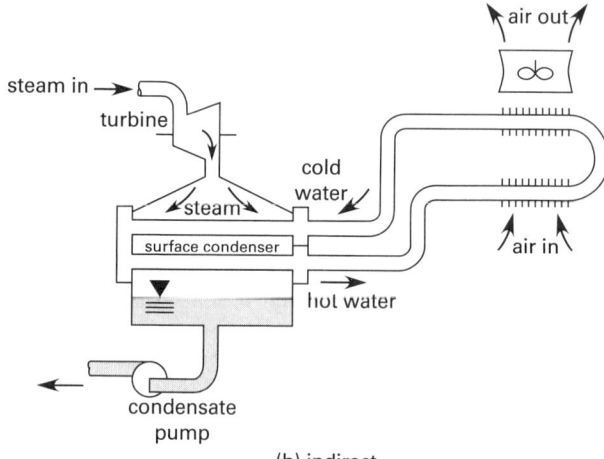

(b) indirect

---

[15]Multiply grains/gallon (gr/gal) by 17.1 to obtain parts per million (ppm) or milligrams per liter (mg/L).
[16]The term "direct contact" does not mean that the air and steam are combined in a single vessel.

32 HVAC: Ventilation and Humidification Requirements

Nomenclature

A	area	ft^2	m^2
ACH	number of air changes per hour	$1/hr$	$1/h$
B	crack coefficient	$ft^3/ft\cdot min$	$m^3/m\cdot min$
c_p	specific heat	$Btu/lbm\text{-}°F$	$kJ/kg\cdot°C$
C	concentration	lbm/ft^3	kg/m^3
C_d	coefficient of discharge	–	–
C_p	coefficient of pressure	–	–
D	occupancy diversity	–	–
E	efficiency	–	–
g	gravitational acceleration, 32.2 (9.81)	ft/sec^2	m/s^2
g_c	gravitational constant, 32.2	$ft\text{-}lbm/lbf\text{-}sec^2$	n.a.
h	elevation (height)	ft	m
h	enthalpy	Btu/lbm	kJ/kg
K	mixing factor	–	–
L	length	ft	m
$\dot{m}$	mass flow rate	lbm/min	kg/min
MW	molecular weight	$lbm/lbmol$	$kg/kmol$
n	exponent	–	–
p	pressure	lbf/in^2	Pa
P	population	persons	persons
$\dot{q}$	heat transfer rate	Btu/min	W
R	contaminant generation rate	lbm/min	kg/min
R_a	rate per unit area	$ft^3/min\text{-}ft^2$	$L/s\cdot m^2$
R_p	rate per person	$ft^3/min\text{-}person$	$L/s\cdot person$
s	sheltering coefficient	–	–
SG	specific gravity	–	–
t	time	hr	h
T	temperature	$°F$ or $°R$	$°C$ or K
TLV	threshold limit value	various	various
v	velocity	mi/hr	km/h
V	volume	ft^3	m^3
$\dot{V}$	flow rate	ft^3/min	m^3/min
y	elevation	ft	m

Symbols

η	efficiency	–	–
ρ	density	lbm/ft^3	kg/m^3
ω	humidity ratio	lbm/lbm	kg/kg

Subscripts

a	area
abs	absolute
ADP	apparatus dew point
bz	breathing zone
fg	vaporization
id	indoor design
in	entering the room
l	latent
NPL	neutral pressure level
o	neutral or reference
ot	outdoor total
ou	outdoor uncorrected
oz	outdoor zone
p	people or primary
pz	primary zonal
s	sensible or system
v	ventilation
w	wind
z	zone

1. VENTILATION

Ventilation primarily refers to air that is necessary to satisfy the needs of occupants.[1] The term may mean the air that is introduced into an occupied space, or it may refer to the new air that is deliberately drawn in from the outside and mixed with return air. Ventilation, however, does not normally include unintentional infiltration through cracks and openings.

---

[1]The term *process air* is the most common designation given to "ventilation" needed for manufacturing processes.

Ventilation air is provided to the occupied space primarily to remove heat and moisture generated in the space. Heat and moisture can both be generated metabolically as well as by equipment and processes. To a lesser extent, ventilation is also used to remove odors, provide oxygen, prevent carbon dioxide buildup, and remove noxious fumes. Generally, however, all of the other needs will be met if removal of body heat is accomplished.

2. VENTILATION STANDARDS

Few ventilation codes have the force of law, but there are several recommended standards. The U.S. Occupational Safety and Health Act (OSHA) contains a handful of mandatory standards, including 29 CFR 1910.146, dealing with minimum oxygen percentage in confined spaces, and 29 CFR 1910.94, dealing primarily with local exhaust systems. Almost all cities, counties, states, and municipalities have building codes, a few of which have their own ventilation requirements. Most building codes now incorporate provisions of the *International Mechanical Code* (IMC), published by the International Code Council (ICC). The American National Standards Institute (ANSI) has joined forces with the American Society of Heating, Refrigerating and Air-Conditioning Engineers (ASHRAE) in publishing indoor air quality standards, but it has several specialized standards of its own, primarily ANSI Z9. ANSI Z9 is published in conjunction with the American Industrial Hygiene Association (AIHA) and covers, for example, ventilation of laboratories, paint spray booths, and grinding stations. The U.S. National Institute for Occupational Safety and Health (NIOSH) has a few standards covering foundry ventilation, recirculation, and push-pull hoods. The National Fire Protection Association (NFPA) has standards that cover some specialized ventilation requirements such as NFPA 45, which covers lab fume hoods. The ASTM International ASTM D6245 describes how carbon dioxide can be used as an index of ventilation rate and objectionable body odor. The Air Movement and Control Association (AMCA) and the Sheet Metal and Air Conditioning Contractors' National Association (SMACNA) cover ventilation peripherally in their fan and duct publications, such as AMCA 201, but they do not recommend ventilation rates.

Perhaps the most authoritative and comprehensive ventilation requirements are published by American Conference of Governmental Industrial Hygienists (ACGIH) and ASHRAE. The ACGIH Industrial Ventilation Committee publishes *Industrial Ventilation: A Manual of Recommended Practice for Design*, which is used throughout the world. ASHRAE publishes, in conjunction with ANSI, the most complete guidance, in particular, *Ventilation for Acceptable Indoor Air Quality* (ASHRAE Standard 62.1) and *Ventilation and Acceptable Indoor Air Quality in Low-Rise Residential Buildings* (ASHRAE Standard 62.2). ASHRAE Standard 62.1 sets voluntary minimum standards for new and substantially renovated commercial buildings. ASHRAE Standard 62.1 is a consensus standard that

has been incorporated into NFPA 5000 and LEED green building qualifications. ASHRAE Standard 62.2 specifies voluntary minimum standards for single-family houses and multifamily dwellings three stories or less in height. The *ASHRAE Handbook: Fundamentals* volume also provides useful design guidance.

Compliance with *Energy Standard for Buildings Except Low-Rise Residential Buildings* (ASHRAE Standard 90.1), along with state and federal regulations intended to minimize environmental impact and energy loss (e.g., California's *Energy Efficiency Standards for Residential and Nonresidential Buildings*, California Energy Code, Title 24, Part 6 of the *California Code of Regulations*), may also be required.

Minimum ventilation requirements are given by local building codes, local ordinances, health regulations, and construction specifications. A common minimum design standard (as specified by ASHRAE Standard 62.1) is 20 ft^3/min (0.57 m^3/min; 9.4 L/s) of new, outside air per person. In interior areas that permit heavy smoking (e.g., casinos and smoking lounges), 30 ft^3/min to 60 ft^3/min (0.84 m^3/min to 1.68 m^3/min; 14.1 L/s to 28.2 L/s) per person is required. Some nonsmoking areas may require more than 20 ft^3/min (0.57 m^3/min; 9.4 L/s) anyway to avoid the "sick building syndrome" (as when formaldehyde-emitting furniture and building materials are present).

Some ventilation requirements are specified by the number of air changes (i.e., room volumes or "cubical contents" without allowance for room contents) required per hour, ACH. This is known as the *air change method*. A typical minimum value for toilet rooms, for example, is four air changes per hour. For other uses (automotive, boiler rooms, engine rooms, etc.) the number of air changes can be significantly higher (e.g., 25 to 100 per hour).

3. ASHRAE VENTILATION RATE: SINGLE ZONE

ASHRAE Standard 62.1 prescribes two methods for determining the amount of outdoor ventilation air: a *ventilation rate procedure* (VRP) and an *indoor air quality procedure* (IAQP). Since it is not practical to monitor all air contaminants in all locations, and since some contaminants (e.g., mold and fungi spores) cannot be monitored in real time, using the IAQP is associated with significant risk. Due to a multiplicity of air contaminants that (a) are not monitored, (b) are not detected, and (c) do not even have definite limits, and since the straightforward VRP is so much simpler, the rate-based methodology is preferred.

Modern rate-based ventilation standards, including ASHRAE Standard 62.1, specify the required *breathing zone outdoor air* (i.e., the outdoor ventilation air in the breathing zone), $\dot{V}_{bz}$, as a function of both zone occupancy, P_z, and zone floor area, A_z. The first term accounts for contaminants produced by occupants,

Table 32.1 *Representative Minimum Ventilation Rates in Breathing Zone*[a,b]

occupancy category	people outdoor air rate, R_p		area outdoor air rate, R_a		default values			
					occupant density, people per 1000 ft^2 (100 m^2)	combined outdoor air rate		
	cfm/person	L/s·person	cfm/ft^2	L/s·m^2		cfm/ person	L/s·person	cfm/ft^2
educational:								
classrooms (age 9 and up)	10	5	0.12	0.6	35	13	6.7	0.46
science laboratories	10	5	0.18	0.9	25	17	8.6	0.43
general:								
conference and meeting	5	2.5	0.06	0.3	50	6	3.1	0.30
hotel:								
bedrooms and living rooms	5	2.5	0.06	0.3	10	11	5.5	0.11
office building:								
main lobbies	5	2.5	0.06	0.3	10	11	5.5	0.11
offices	5	2.5	0.06	0.3	5	17	8.5	0.09
miscellaneous:								
bank vaults	5	2.5	0.06	0.3	5	17	8.5	0.09
public assembly:								
auditorium seating areas	5	2.5	0.06	0.3	150	5	2.7	0.75
retail:								
sales areas	7.5	3.8	0.12	0.6	15	16	7.8	0.24
sports and entertainment:								
spectator areas	7.5	3.8	0.06	0.3	150	8	4.0	1.2

(Multiply cfm by 0.02832 to obtain m^3/min.)
(Multiply cfm by 0.4719 to obtain L/s.)
(Multiply cfm/ft^2 by 5.08 to obtain L/s·m^2.)
(Multiply people/1000 ft^2 by 0.929 to obtain people/100 m^2.)
(Multiply cfm/ft^2 by 0.3048 to obtain m^3/min·m^2.)
[a]This table applies to environmental tobacco smoke (ETS)-free areas only. Refer to ASHRAE Standard 62.1 Sec. 5.17 for requirements for buildings containing ETS areas and ETS-free areas.
[b]Rates are based on an air density of 0.075 lbm/ft^3 (1.2 kg/m^3).

From ASHRAE Standard 62.1, Table 6.1, copyright © 2010, by American Society of Heating, Refrigerating and Air-Conditioning Engineers, Inc. Reproduced with permission.

while the second term accounts for contaminants produced by the building. The maximum number of occupants expected in the zone during typical usage is normally used rather than a value based on building code classification occupancy densities. However, different short-term time-averaging methods prescribed by ASHRAE Standard 62.1 may also be used if the zone population fluctuates.[2] ASHRAE Standard 62.1 requires that the rate specified by Eq. 32.1 be maintained during operation under all load conditions.

$$\dot{V}_{bz} = R_p P_z + R_a A_z \qquad 32.1$$

Table 32.1 contains representative values of R_p and R_a, although local codes, federal regulations, and contract requirements take precedence. The default columns are used only if the actual occupant density is unknown.

The outdoor ventilation rate specified by Eq. 32.1 is affected by the *distribution effectiveness*, E_z, as specified in Table 32.2. The *zone outdoor airflow*, $\dot{V}_{oz}$, at the diffusers is given by Eq. 32.2. For single-zone systems, this is also the system total outdoor air requirement, $\dot{V}_{ot}$, as shown in Eq. 32.3.

$$\dot{V}_{oz} = \frac{\dot{V}_{bz}}{E_z} \qquad 32.2$$

$$\dot{V}_{oz} = \dot{V}_{ot} \quad \text{[single-zone system]} \qquad 32.3$$

---

[2]Outdoor airflow rates can be reduced dynamically in the critical zones that have variable occupancy. Changes in outdoor air demand (i.e., changes in occupancy) can be detected several ways, including measurement of carbon dioxide, CO_2.

Table 32.2 *Zone Air Distribution Effectiveness*

air distribution configuration	effectiveness, E_z
ceiling supply of cool air	1.0
ceiling supply of warm air with floor return	1.0
ceiling supply of warm air 15°F (8°C) or more above space temperature, with ceiling return	0.8
ceiling supply of warm air less than 15°F (8°C) above space temperature, with ceiling return, provided that the 150 fpm (0.8 m/s) supply air jet reaches to within 4.5 ft (1.4 m) of floor level	1.0
ceiling supply of warm air less than 15°F (8°C) above space temperature, with ceiling return, with supply jet air velocity less than 150 fpm (0.8 m/s)	0.8
floor supply of cool air with ceiling return, provided that the 150 fpm (0.8 m/s) supply air jet reaches 4.5 ft (1.4 m) or more above the floor level[*]	1.0
floor supply of cool air with ceiling return, provided the low-velocity displacement ventilation achieves unidirectional flow and thermal stratification	1.2
floor supply of warm air with floor return	1.0
floor supply of warm air with ceiling return	0.7

[*]This describes most underfloor air distribution systems.

From ASHRAE Standard 62.1, Table 6-2, copyright © 2010, by American Society of Heating, Refrigerating and Air-Conditioning Engineers, Inc. Reproduced with permission.

Example 32.1

A single-zone high school classroom has a floor area of 1600 ft^2 and seats 35 students. Cool air is supplied from ceiling diffusers. How much outdoor ventilation air is required per ASHRAE Standard 62.1?

Solution

From Table 32.1 (educational: classroom (age 9 and up)), $R_p = 10$ cfm/person, and $R_a = 0.12$ cfm/ft^2. From Eq. 32.1,

$$
\begin{aligned}
\dot{V}_{bz} &= R_p P_z + R_a A_z \\
&= \left(10 \ \frac{\text{ft}^3}{\text{min-person}} \right)(35 \text{ people}) \\
&\quad + \left(0.12 \ \frac{\text{ft}^3}{\text{min-ft}^2} \right)(1600 \text{ ft}^2) \\
&= 542 \ \text{ft}^3/\text{min}
\end{aligned}
$$

Since cool air is supplied from the ceiling, from Table 32.2, $E_z = 1.0$. The total outdoor air to the zone is

$$
\dot{V}_{oz} = \frac{\dot{V}_{bz}}{E_z} = \frac{542 \ \dfrac{\text{ft}^3}{\text{min}}}{1.0} = 542 \ \text{ft}^3/\text{min}
$$

4. ASHRAE VENTILATION RATE: MULTIZONE

There are two types of multiple-zone (multizone) systems drawing outside air: 100% outside air (OA) systems and recirculating air systems. For 100% outside air systems, the system total outside air requirement is the sum of all of the zonal requirements.

$$
\dot{V}_{ot} = \sum \dot{V}_{oz} \quad \text{[multiple 100\% OA zones]} \qquad \textit{32.4}
$$

For recirculating systems with outside air intakes, a correction is made for *occupancy diversity* (also known as *occupant diversity* or *population diversity*), D, which is the ratio of the *system population* (the maximum number of simultaneous occupants in the space served by the system), P_s, to the sum of the zonal peak occupancies.

$$
D = \frac{P_s}{\sum P_z} \qquad \textit{32.5}
$$

Only the population outdoor air component is affected by the diversity term. The uncorrected system outdoor air requirement is corrected for diversity but not for distribution effectiveness.

$$
\dot{V}_{ou} = D \sum R_p P_z + \sum R_a A_z \qquad \textit{32.6}
$$

Since multizone recirculating systems are not as efficient as 100% OA systems, the system outdoor rate is determined from the *system ventilation efficiency*, E_v. The system ventilation efficiency, in turn, depends on the maximum primary outdoor air fraction. ASHRAE Standard 62.1 gives two methods for determining this efficiency: the default method using Table 32.3 (ASHRAE Table 6-3), and the more accurate and more involved calculated method using ASHRAE App. A. These methods produce significantly different results, but either may be used.

The system ventilation efficiency depends on the maximum *primary outdoor air fraction* evaluated over all of the zones. A zone's primary outdoor air fraction, Z_p, is the fraction of total air (including the outdoor and recirculated airflows) from the air handler, known as the *zonal primary airflow*, $\dot{V}_{pz}$, that is outdoor air.

$$
Z_p = \frac{\dot{V}_{oz}}{\dot{V}_{pz}} \qquad \textit{32.7}
$$

Table 32.3 System Ventilation Efficiency[*]

maximum Z_p	E_v
≤ 0.15	1.0
≤ 0.25	0.9
≤ 0.35	0.8
≤ 0.45	0.7
≤ 0.55	0.6
> 0.55	Use ASHRAE Standard 62.1 App. A method.

[*]Interpolation may be used between tabulated values.

From ASHRAE Standard 62.1, Table 6-3, copyright © 2010, by American Society of Heating, Refrigerating and Air-Conditioning Engineers, Inc. Reproduced with permission.

Table 32.3 can be used with the maximum value of Z_p to determine the system ventilation efficiency, E_v.

$$\dot{V}_{ot} = \frac{\dot{V}_{ou}}{E_v} \quad \left[\begin{array}{l} \text{multiple zones} \\ \text{with recirculation} \end{array} \right] \qquad 32.8$$

5. SPECIAL ASHRAE VENTILATION REQUIREMENTS

As specified in ASHRAE Standard 62.1, some conditions trigger special ventilation requirements.[3]

- Natural ventilation may be relied on when certain requirements are met.

- Outdoor air drawn from National Ambient Air Quality Standards (NAAQS) *nonattainment areas* must be treated to reduce particulate matter, ozone, carbon dioxide, sulfur oxides, nitrogen dioxide, and/or lead to specified levels. Specifically, coils and other devices with wetted surfaces must have MERV 6 filters upstream if the outdoor air does not meet NAAQS for PM-10 particulates. An ozone air cleaner (minimum 40% efficiency) is generally required if the average ozone concentration exceeds 0.107 ppm.

- Ventilation in areas exposed to *environmental tobacco smoke* (ETS) (i.e., in smoking areas) requires the use of methods in ASHRAE Standard 62.1 Sec. 5.17. Smoke-free areas must be maintained at higher static pressures relative to adjacent ETS areas.

- Variable air volume (VAV) systems with fixed outside air dampers must comply at the minimum supply airflow.

- Residential spaces in buildings over three stories have special requirements.

ASHRAE Standard 90.1 Sec. 6.4.3.9 (*Ventilation Controls for High-Occupancy Areas*) specifies that *demand controlled ventilation* (DCV) is required for spaces larger than 500 ft^2 (47 m^2) and with a design occupancy for ventilation of greater than 40 people per 1000 ft^2 (100 m^2) of floor area and served by systems with one or more of the following: (a) an air-side economizer, (b) automatic modulating control of the outdoor air damper, or (c) a design outdoor airflow greater than 3000 cfm.

6. INFILTRATION

Infiltration (also known as *accidental infiltration*) refers to the air that unintentionally enters an occupied space through cracks around doors and windows and through openings in a building. Accidental infiltration may be as high as 0.4 to 1.0 ACH.

With the *crack length method*, the amount of infiltration, $\dot{V}$, is determined from the *crack coefficient*, B, and the crack length, L. Values of the crack coefficient vary greatly and depend on the type of window or door, wind velocity, orientation, and degree of closure.[4] Alternatively, the infiltration may be found from the plane area. (This method is more common when determining infiltration through entire walls.) As with the crack length method, the crack area coefficient B' depends on many factors.

$$\dot{V} = BL = B'A \qquad 32.9$$

More sophisticated correlations recognize the dependence on the difference in outside and inside pressures. Values of B'' and n must be known or assumed and must be consistent with the units of pressure.

$$\dot{V} = B''A(\Delta p)^n \qquad 32.10$$

Since air entering through cracks on the windward side must leave through cracks on the leeward side, only half of the total crack length is used when all four sides of a building are exposed to wind. However, the amount of crack length used also depends on the building orientation. When only one wall is exposed to wind, that wall's total crack length is used. With two exposed walls, the wall with the larger crack length is used. When three walls are exposed, only two walls contribute to crack length. The crack length used should never be less than half of the total crack length.

The air change method can also be used for infiltration. Infiltration into modern (tight) residential construction may be as low as 0.2 air changes per hour, while older residences in good condition may experience ten times as much. In the past, a rule of thumb used (to size furnaces) in the absence of other information was that infiltration into residences with windows on one, two, or three sides would be one, one and one-half, or two air changes per

---

[3]An interesting situation is a repair garage. For a repair garage, $R_p = 0$, and $R_a = 0$. Therefore, $\dot{V}_{bz} = 0$, and $\dot{V}_{oz} = 0$. Outside air is not required in a repair garage, although removal of exhaust is. Make-up air would most likely, but not necessarily, be outside air.

[4]Typical values of the crack length coefficient are given in most HVAC books. Manufacturers' literature should be used for specific name-brand windows and doors.

hour, respectively. Experience is needed to modify these values for use with modern, energy-efficient construction.

7. INFILTRATION IN TALL BUILDINGS

For tall buildings (i.e., those over 100 ft (30 m) in height), the pressure difference, Δp, in Eq. 32.11 is a combination of the wind velocity pressure and the *stack effect* (*chimney effect*). The stack effect is particularly important during the winter. The combined infiltration due to static pressure (including wind velocity pressure) and the stack effect is proportional to the square root of the sum of the heads acting on the building.

$$\Delta p = \sqrt{p_w^2 + \Delta p_{\text{stack effect}}^2} \qquad 32.11$$

If the infiltrations due to wind alone and stack effect alone are known, Eq. 32.12 is equivalent to taking the square root of the sum of the heads.

$$\dot{V} = \sqrt{\dot{V}_w^2 + \dot{V}_{\text{stack effect}}^2} \qquad 32.12$$

In addition to traditional HVAC methods based on crack length, infiltration from wind may be calculated from extensions of traditional fluid principles. Specifically, if the opening area or effective leakage area, A, and *entrance* or *discharge coefficient*, C_d, are known, the infiltration is

$$\dot{V}_w = C_d A \sqrt{\frac{2p_w}{\rho}} \qquad \text{[SI]} \quad 32.13(a)$$

$$\dot{V}_w = C_d A \sqrt{\frac{2g_c p_w}{\rho}} \qquad \text{[U.S.]} \quad 32.13(b)$$

The *wind pressure*, p_w, is based on the theoretical *velocity pressure* (*stagnation pressure*) modified by a *wind surface pressure coefficient*, C_p, which is a function of wind direction, building orientation, and vertical location. Ideally, $C_p = 1.0$ for wind perpendicular to a surface, but rarely does C_p exceed 0.9 in practice, and values between 0.5 and 0.9 are typical. In cases where the wind flow is affected by adjacent structures or vegetation, a *shelter factor* (*sheltering coefficient*), s, may be incorporated into the calculation of wind pressure. For an unsheltered building projecting vertically above level surroundings, $s = 1$.

$$p_w = \frac{C_p s^2 \rho v^2}{2} \qquad \text{[SI]} \quad 32.14(a)$$

$$p_w = \frac{C_p s^2 \rho v^2}{2g_c} \qquad \text{[U.S.]} \quad 32.14(b)$$

Several different methods (theoretical, heuristic, and code-based) can be used to determine the stack effect. The theoretical value is given by Eq. 32.15, where h_{NPL}

is the elevation of the *neutral pressure level* and all temperatures are absolute.

$$\Delta p_{\text{stack effect}} = (\rho_{\text{outside}} - \rho_{\text{inside}})g(h - h_{\text{NPL}})$$
$$= \frac{\rho_{\text{inside}}g(h - h_{\text{NPL}})(T_{\text{abs,outside}} - T_{\text{abs,inside}})}{T_{\text{abs,inside}}}$$
$$\text{[SI]} \quad 32.15(a)$$

$$\Delta p_{\text{stack effect}} = \frac{(\rho_{\text{outside}} - \rho_{\text{inside}})g(h - h_{\text{NPL}})}{g_c}$$
$$= \frac{\rho_{\text{inside}}g(h - h_{\text{NPL}})(T_{\text{abs,outside}} - T_{\text{abs,inside}})}{g_c T_{\text{abs,inside}}}$$
$$\text{[U.S.]} \quad 32.15(b)$$

Some methods determine infiltration by using an *effective wind velocity* (*equivalent wind velocity*), $v_{\text{effective}}$, to combine the effects of wind and stack effect. For example, a table of crack coefficients usually requires knowing the wind velocity. For short buildings, the actual wind velocity, v_o, at the opening (window, door, opening, crack, etc.) elevation is used. For tall buildings, v_o at the opening elevation is modified for stack effect. For rough estimates of infiltration made by such methods, Eq. 32.16 gives an effective wind velocity at a location y above or below a building's midheight, the assumed location of the neutral pressure level. v_o is the wind velocity that would be used if the stack effect was neglected. y is positive above the midheight and negative below the midheight. This correctly reflects infiltration into the building due to both wind and stack effect below midheight, but infiltration due to wind balanced against exfiltration due to stack effect above midheight.

$$v_{\text{effective,mph}} = \sqrt{v_{o,\text{mph}}^2 - 1.75y_{\text{ft}}} \quad \text{[U.S. only]} \quad 32.16$$

8. INDOOR DESIGN CONDITION

The *indoor design condition* refers to the thermodynamic state of the air that is removed from an occupied space. The inside design temperature, T_{id}, represents the maximum dry-bulb air temperature—a not-to-be-exceeded limit—that the space will reach.

HVAC books contain tables of recommended inside design conditions and charts of comfort ranges that can be used to select suitable combinations of temperature and humidity. Within a *comfort range*, choice of the actual inside design condition is subjective and requires modification based on experience for the needs of the particular industries, the season, and the levels of physical exertion.

Most people feel comfortable when the dry-bulb temperature is kept between 74°F and 77°F (23.3°C and 25°C) and the relative humidity is 30% to 35% (in the winter) or 45% to 50% (in the summer). 75°F (23.9°C) dry-bulb and 50% relative humidity is often selected as an inside design condition for initial studies. This temperature is

Ventilation

for the breathing line, 3 ft to 5 ft (0.9 m to 1.5 m) above the floor.[5]

Attention also needs to be given to the *temperature swing*, the difference between the thermostat's on and off settings. For commercial applications, the swing during the summer should be approximately 2°F to 4°F (1.1°C to 2.2°C) above the indoor design (i.e., "off") setting, and the swing during the winter should be approximately 4°F (2.2°C) below the indoor design (i.e., "off") setting.

9. HUMIDIFICATION

Ventilation provides humidification to the occupied space, particularly during the winter. Air should not be completely dry when it enters an occupied space. Air that is too dry will cause discomfort and susceptibility to respiratory ailments. Also, some pathogenic bacteria that survive in low- and high-humidity air will die very quickly in air with midrange humidities.[6]

Some manufacturing and materials handling processes require specific humidity for efficient operation. *Hygroscopic materials*, such as wood, paper, textiles, leather, and many food and chemical products, readily absorb moisture. A constant humidity level is required to obtain consistent manufacturing conditions with such products.

Dry air prevents static electricity from dissipating (into the air). Therefore, dry air can cause intermittent electrical/electronic failures; affect the handling of static-prone materials such as paper, films, and plastics; and ignite potentially explosive atmospheres of dust and gases.

Humidification can be provided by evaporating water in the occupied space (the *evaporative pan method*) or by injecting water (the *water spray method*) or steam into the duct flow. Most commercial humidification is accomplished by placing one or more steam manifolds in the air distribution duct. *Booster humidification* (*spot humidification*) from a separate, independent source is required when a higher humidity is needed in a limited area within a larger controlled space. Steam flow is controlled by humidistats placed downstream of the steam manifold.[7]

10. OXYGEN NEEDS

Providing oxygen is not an issue in traditional buildings. Infiltration alone provides the oxygen needed. However, in closed and confined environments such as mines, tunnels, manholes, and closed tanks, forced ventilation and/or oxygen masks are necessary.

Air is approximately 20.9% oxygen by volume, independent of altitude. For confined spaces, OSHA (29 CFR 1910.134) specifies a minimum oxygen content of 19.5%, which is adequate for elevations below 3000 ft (914 m). NIOSH and ACGIH specifications differ slightly. Concentrations less than 19.5% are known as *oxygen-deficient atmospheres*. Reaction to oxygen deficiency varies with individuals, but in general, significant impairments to work rate, perception, concentration, and judgment can be expected with lower values. Some individuals may experience coronary, pulmonary, and circulatory problems. Concentrations below 12% pose immediate danger to life. OSHA defines an *oxygen-enriched atmosphere* as one with an oxygen concentration greater than 23.5% (29 CFR 1910.146(b)). (OSHA 1915.12(a)(2) pertaining to shipyard operations specifies concentrations above 22% as being enriched.) Oxygen-enriched environments pose extreme fire and explosion hazards, especially if combustible material is present.

11. CARBON DIOXIDE BUILDUP

Diluting exhaled carbon dioxide, like providing oxygen, is only an issue in completely closed environments. Infiltration alone provides the dilution needed. Healthy individuals can usually tolerate a concentration of 0.5% (by volume), though the air will be noticeably stale.[8] Equation 32.17 is used for finding the approximate time (in hours) for a 3% buildup of carbon dioxide in a closed area.[9] The carbon dioxide concentration should not exceed 5% under any circumstances.

$$t_{\mathrm{h}} = \frac{1.4 V_{\mathrm{room,m^3}}}{\text{no. of occupants}} \quad \text{[SI]} \quad 32.17(a)$$

$$t_{\mathrm{hr}} = \frac{0.04 V_{\mathrm{room,ft^3}}}{\text{no. of occupants}} \quad \text{[U.S.]} \quad 32.17(b)$$

12. ODOR REMOVAL

The airflow required through a room to remove body odors depends on the room size and level of activity. Body odors become more pronounced when the relative humidity is above 55%. Except for very cramped areas, common ventilation standards are normally sufficient for odor removal. Good practice requires approximately one-third of the air to be new air.

---

[5]The temperature variation with height above the floor is approximately 0.75°F/ft (1.4°C/m).

[6]In particular, airborne type 1 pneumococcus, group C staphylococcus, and staphylococcus are quickly killed in relative humidities of 45% to 55%. Other viruses, including measles, influenza, and encephalomyelitis, survive longer in very dry air than in midrange relative humidities.

[7]Steam flow is turned off when the humidity reaches the high-limit humidistat setting, typically 90% relative humidity. This prevents oversaturation of the air when there is a failure in the air conditioning system or controlling humidistat.

---

[8]Some submarines have operated at 1% by volume carbon dioxide.

[9]This equation assumes an initial (atmospheric) carbon dioxide content of 0.03% and carbon dioxide production of 0.011 ft³/min (0.00031 m³/min; 5.2 mL/s) per person.

13. SENSIBLE AND LATENT HEAT

Metabolic heat contains both sensible and latent components. *Sensible heat* is pure thermal energy that increases the air's dry-bulb temperature. *Latent heat* is moisture that increases air's humidity ratio. Table 32.4 gives the approximate amounts of metabolic heat in a 75°F (23.9°C) environment. The "adjusted" column refers to a normal mix of men, women, and children. For design purposes, the heat gain for an adult female is approximately 85% of the adult male rate; the heat gain for a child is 75% of the adult male rate.

Table 32.4 Approximate Heat Generation by Occupants (Btu/hr)

activity	total adult males	total adjusted	sensible* adjusted	latent* adjusted
seated, at rest, theater, classroom	390	330	225	105
moderately active office work	475	450	250	200
standing, light work, slowly walking	550	450	250	200
moderate dancing	900	850	305	545
walking 3 mph, moderately heavy work	1000	1000	375	625
heavy work	1500	1450	580	870

(Multiply Btu/hr by 0.293 to obtain watts.)
*The sensible-latent splits given are for a 75°F (23.9°C) environment. For an 80°F (26.7°C) environment, total heat remains the same, but sensible heat decreases approximately 20% and latent heat increases accordingly.

14. VENTILATION FOR HEAT REMOVAL

Ventilation requirements can be calculated from sensible heat and/or moisture (i.e., latent heat) generation rates. In Eq. 32.18, T_{in} is the dry-bulb temperature of the air entering the room.

$$\dot{q}_s = \dot{m}c_p(T_{id} - T_{in})$$
$$= \dot{V}\rho c_p(T_{id} - T_{in}) \qquad 32.18$$

Equation 32.18 can be written in terms of the number of air changes per hour, ACH, and the temperature of the ventilation air, T_{out}.

$$\dot{q} = \frac{\rho c_p V_{room}(ACH)(T_{in} - T_{out})}{60 \frac{min}{hr}} \qquad 32.19$$

In ventilation work, volumetric flow rates are traditionally given in ft³/min (cfm), m³/min, or L/s. The constant 1.08 (0.02) in Eq. 32.20 is the product of an air density of 0.075 lbm/ft³ (1.2 kg/m³), a specific heat of 0.24 Btu/lbm-°F (1.0 kJ/kg·°C), and the factor 60 min/hr (60 s/min).[10]

$$\dot{V}_{m^3/min} = \frac{\dot{q}_{s,kW}}{\left(0.02 \frac{kJ\cdot min}{m^3\cdot s\cdot °C}\right)(T_{id,°C} - T_{in,°C})}$$
$$\text{[SI]} \quad 32.20(a)$$

$$\dot{V}_{ft^3/min} = \frac{\dot{q}_{s,Btu/hr}}{\left(1.08 \frac{Btu\text{-}min}{ft^3\text{-}hr\text{-}°F}\right)(T_{id,°F} - T_{in,°F})}$$
$$\text{[U.S.]} \quad 32.20(b)$$

$$\dot{V}_{L/s} = \frac{\dot{q}_{s,Btu/hr}}{\left(2.28 \frac{Btu\text{-}sec}{L\text{-}hr\text{-}°F}\right)(T_{id,°F} - T_{in,°F})}$$
$$\text{[mixed units]} \quad 32.20(c)$$

$$\dot{V}_{L/s} = \frac{\dot{q}_{s,W}}{\left(1.20 \frac{W\cdot s}{L\cdot °C}\right)(T_{id,°C} - T_{in,°C})}$$
$$\text{[SI]} \quad 32.20(d)$$

The sensible heat loads will usually be more significant than the latent load, and ventilation will be determined solely on that basis. When large moisture sources are present, however, the latent loads may control.

$$\dot{q}_l = \dot{m}_{water}h_{fg} = \dot{m}_{air}\Delta\omega h_{fg}$$
$$= \dot{V}\rho\Delta\omega h_{fg} \qquad 32.21$$

The constant 4775 (49.36) in Eq. 32.22 is the product of the air density of 0.075 lbm/ft³ (1.2 kg/m³), a latent heat of vaporization at the approximate partial pressure of the water vapor in the air of 1061 Btu/lbm (2468 kJ/kg), and the factor 60 min/hr (60 s/min).[11]

$$\dot{V}_{m^3/min} = \frac{\dot{q}_{l,kW}}{\left(49.36 \frac{kJ\cdot min}{m^3\cdot s}\right)\Delta\omega_{kg/kg}} \quad \text{[SI]} \quad 32.22(a)$$

$$\dot{V}_{ft^3/min} = \frac{\dot{q}_{l,Btu/hr}}{\left(4775 \frac{Btu\text{-}min}{ft^3\text{-}hr}\right)\Delta\omega_{lbm/lbm}} \quad \text{[U.S.]} \quad 32.22(b)$$

15. VENTILATION FOR MOLD CONTROL

Humidity as low as 70%, even without condensing infiltration, can provide sufficient moisture for mold and fungi growth in as little as six hours. Moisture management should be specifically considered in ventilation

---

[10]c_p = 0.24 Btu/lbm-°F (1.0 kJ/kg·°C) is applicable to dry air. For air with normal amounts of water vapor, the specific heat is closer to c_p = 0.244 Btu/lbm-°F (1.02 kJ/kg·°C).

[11]There is some variation in these constants depending on what heat of vaporization is used. For example, some sources use 1076 Btu/lbm (2503 kJ/kg), in which case the constant is 4840 Btu-min/ft³-hr (50.06 kJ·min/m³·s).

design, and it should take precedence over energy management. Best practice requires that (a) infiltration of unfiltered and unconditioned humid air be prevented; (b) negative interior pressures be avoided and net positive pressure with respect to outdoors be maintained (in the absence of wind and stack effects) while dehumidification is occurring; (c) building and system design, operation, and maintenance provide for dehumidification (drying) of surfaces and materials prone to moisture accumulation under normal operating conditions; and (d) the HVAC should specifically monitor and control humidity.

With only a few exceptions, ASHRAE Standard 62.1 Sec 5.9.1 limits maximum humidity in occupied spaces to 65% during periods of peak outdoor dew point.[12] Without dehumidification equipment, meeting this limit may be difficult with high outdoor air latent loading and low inside space sensible heat ratio. Meeting acceptable humidity ratios is facilitated in buildings maintained at small positive pressures.

Even if the interior relative humidity is maintained below 65%, as prescribed by ASHRAE Standard 62.1, local areas with higher humidity can exist. Areas with higher spot humidity include carpet over concrete, window sills, under sinks, under outside sliding doors, near defective roof installations, and, ironically, near dehumidification equipment. Although not required by ASHRAE, relative humidities less than 35% may be required to avoid condensation on cold surfaces during the winter. Alternatively, relative humidity should be kept above 30% to prevent generation of static electricity.

16. FUME EXHAUST HOODS

Air hoods (also known as *fume hoods* or *exhaust hoods*) are used to provide localized protection from and removal of hazardous vapors, dusts, and biological materials. Table 32.5 lists representative *control velocities* needed for air to capture and entrain materials generated by various processes. When there is sufficient air in the work room to create the face velocity, a separate blower bringing air from the outside may be used, in which case, the term *auxiliary-air hood* is used. Hoods may be of the nonenclosure or closure varieties. *Nonenclosure air hoods* for nontoxic materials may be simple canopies over open tanks (for vapors that rise) or periphery slots (for contents that do not rise). Flanges around the exterior of canopies extending over the edges of the tank increase the collection efficiency and reduce the required airflow, but nonenclosure hoods are particularly inefficient at best.

With a conventional *enclosure air hood*, air is drawn through the front opening into the *fume chamber* and across the work surface, entraining the captured

---

[12]Exceptions where humidity may exceed 65% include kitchens, hot tub rooms, refrigerated areas including ice rinks, shower rooms, spas, and pools.

Table 32.5 *Minimum Control Velocities for Enclosure Hoods*

process	minimum control velocity	
	ft/min	m/min
evaporation from open tanks	50–100	15–30
paint spraying, welding, plating	100–200	36–60
stone cutting, mixing, conveying	200–500	60–150
grinding, crushing	500–2000	150–600

(Multiply ft/min by 0.3048 to obtain m/min.)

material. In *nonbypass air hoods*, the resulting high-velocity air jet sweeping over the work surface and noise is often disconcerting, leading to a reluctance by users to close the sash fully. *Bypass air hoods* address this issue. (See Fig. 32.1.) In air hoods with bypass, air enters the fume chamber through the bypass opening when the sash is closed. When the sash is open, air enters the fume chamber through the sash opening. The bypass area is typically 20–30% of the all-open sash area, and the resulting terminal face velocities with a fully closed sash are in the 300–500 ft/min (90–150 m/s) range, significantly higher than required for most processes. Therefore, most users prefer a combination of bypass and partially open sash.

Figure 32.1 *Conventional Cabinet Bypass Air Hood*

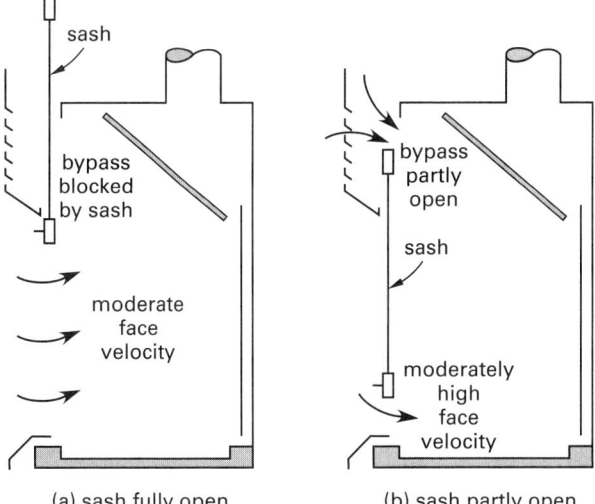

(a) sash fully open   (b) sash partly open

With *auxiliary-air hood units* (also known as *makeup air fume hoods*), outside air is drawn in by a supply blower and discharges downward through a face plenum along the top width of the bypass hood. Typically, 50–70% of the discharged air can be supplied by the auxiliary blower. Unfortunately, even when adjusted properly, auxiliary air hoods provide relatively poor containment and result in significantly higher worker exposure compared to conventional (non-auxiliary air) hoods. The air curtain created may even pull vapors out of the hood interior. Therefore, most authorities, including *Laboratory Ventilation* (ANSI/AIHA Standard Z9.5), recommend against auxiliary air hoods.

Some contaminants are released with almost no velocity of their own. This is the case with products of natural evaporation. Other contaminants (paint booth overspray, dust from grinding wheels, etc.) are released at high velocity. With distance, the velocity dissipates and reaches zero at the *null point*. Although capture is easiest at the null point, it is difficult in most situations to determine the distance to the null point. Even when the distance to the null point is known, the directionality may vary. Therefore, a high air intake velocity is needed to capture the moving contaminants near the point of generation.

Air velocity decreases with increasing distance from the source, varying almost inversely with the square of the distance. Therefore, the hood opening should be as close as possible to the contaminant source.

Since heated air rises, different design principles are needed for high-temperature (e.g., molten metal) processes. The heated air mixes with the surrounding air, and a larger volume of diluted air must be captured.

17. DILUTION VENTILATION

Dilution ventilation (*toxicity dilution*) refers to the dilution of contaminated air in order to reduce its health or explosion hazard. Dilution is less effective than outright removal of hazards by exhaust ventilation. Dilution is generally applicable to organic liquids and solvents whose toxicities are low, when the workers are not too close to the source, and when vapor generation is fairly uniform.[13] The required volume of dilution air must not be so great as to make air velocities unreasonable.

Dilution is achieved by providing enough air to reduce a vapor's concentration to an acceptable level. Various designations are given to acceptable levels, although the *threshold limit value* (TLV) in parts per million (ppm) by volume (mg/m^3 in SI) and *lower explosive limit* (LEL) in parts per hundred (pph) by volume (mg/m^3 in SI) are the most common.[14] The TLV is assumed to be the concentration that workers may be continuously exposed to during a certain period. TLVs are subject to ever-changing legislation and ongoing research.[15]

Three types of TLVs are used. The *time-weighted average* (TLV-TWA) is the time-weighted average concentration that workers may be exposed to for eight hours per day, day after day, without experiencing adverse effects. The *short-term exposure limit* (TLV-STEL) is the time-weighted average concentration that workers may be exposed to for fifteen minutes, up to four times per eight-hour period.[16] The *ceiling value* (TLV-C) is the concentration that should not be exceeded, even instantaneously. Depending on the substance, one, two, or all three of these limits may be applicable.[17]

Assuming that a contaminant is uniformly distributed throughout the plant air, at equilibrium the contaminant generation rate, R, is equal to the ventilated removal rate.

$$R = C\dot{V} \qquad\qquad 32.23$$

To account for irregular vapor evolution, inefficient ventilation, and toxicity, an empirical multiplicative "effectiveness of mixing factor" (the *K-factor*) between 3 and 10 is used.

$$KR = C\dot{V} \qquad\qquad 32.24$$

The maximum concentration, $C_{\max}$, is usually the threshold limit value or the *permissible exposure limit* (PEL), both with a safety factor. The required airflow is

$$\dot{V}_{L/s} = \frac{(4.02 \times 10^8)KR_{kg/min}}{(MW)(C_{\max})}$$

$$= \frac{(4.02 \times 10^8)K(SG)R_{L/min}}{(MW)(C_{\max})} \quad \text{[SI]} \quad 32.25(a)$$

$$\dot{V}_{ft^3/min} = \frac{(3.86 \times 10^8)KR_{lbm/min}}{(MW)(C_{\max})}$$

$$= \frac{(4.03 \times 10^8)K(SG)R_{pints/min}}{(MW)(C_{\max})} \quad \text{[U.S.]} \quad 32.25(b)$$

When two or more hazardous substances that have similar toxicologic effects are simultaneously present (i.e., act on the same organ or metabolic process), the combined effect should be considered. If the air is to be breathed, the airflow rates for each substance must be calculated and the separate flow rates summed. The mixture threshold value is exceeded when

$$\left(\frac{C}{TLV}\right)_1 + \left(\frac{C}{TLV}\right)_2 + \cdots > 1 \qquad 32.26$$

---

[13]Dilution ventilation is not recommended for carbon tetrachloride, chloroform, and gasoline, among others. Dusts are seldom removed successfully by dilution.

[14]Another unit used for dust concentrations in respirable air is *millions of particles per cubic foot* (mppcf) determined by *midget impinger* techniques. The conversion between mppcf and other units is not exact, depending primarily on the particle size and density. However, equivalences of 5.6 mppcf and 6.4 mppcf to 1.0 mg/m^3 are quoted. In the absence of any other information, an average value of 6 mppcf is recommended.

[15]In the United States, TLVs are updated annually in *Industrial Ventilation: A Manual of Recommended Practice*, published by the American Conference of Governmental Industrial Hygienists.

[16]Other restrictions may apply to TLV-STEL. For example, there may be a sixty-minute waiting period between successive exposures at this level.

[17]For example, irritant gases may be controlled only by the TLV-C value.

The additive nature implied by Eq. 32.26 is assumed unless the two substances are known to act independently instead of additively. In that case, the threshold limit is exceeded only when the ratio C/TLV for at least one component in the mixture exceeds unity. The highest ventilation rate calculated for each component independently is the design ventilation rate.

18. RECIRCULATION OF CLEANED AIR

The volume of ventilation air required will be reduced if some of the contaminated air can be cleaned and returned. Dust and particulate matter in air can be removed by two types of air cleaners. *Air filters* are applicable when the concentration is between 0.5 grains and 50 grains per 1000 cubic feet (1.1 milligrams per cubic meter to 110 milligrams per cubic meter). *Dust collectors* are used at the higher concentrations normally found in manufacturing processes. Equilibrium will be achieved when the contaminant generation rate equals the rate at which the air filter removes particles from the air.

$$R = \eta_{\text{filter}} C \dot{V} \qquad \textit{32.27}$$

19. CLEAN ROOMS

Clean rooms are defined by the number of particles (pollen, skin flakes, etc.) above a given size (usually 0.5 microns) in a cubic foot of air. Most semiconductor clean rooms are Class-100 or better, meaning that there will be no more than one hundred 0.5 micron-sized particles per cubic foot.

Clean rooms technology generally relies on high-efficiency, prefilters, either *high-efficiency particulate arresting* (HEPAs) or *ultra-low penetration air* (ULPA) filters in the supply, positive room pressure, fast air movement, and floor grates (i.e., downflow air movement). A positive pressure of approximately 0.1 in of water (25 Pa) is typical. Twenty air changes per hour is a typical minimum, while airflow velocities of 75 ft/min to 100 ft/min (0.38 m/s to 0.5 m/s) are used in the best clean rooms.

High-efficiency (60% to 90%) prefilters reduce the load on HEPAs. Usually, HEPAs (99.97% efficiency at the 0.3 micron level) are suitable for Class-100 clean rooms, while ULPAs are needed for Class-10 or better clean rooms. Large centralized equipment may be used, or modular *air handling units* (AHUs) drawing air from the main general supply may be used for individual clean rooms. Adjustable-frequency drives can be used to change the airflow in order to reduce energy usage or change the cleanliness. Stainless steel is the preferred material for ducts and hoods, as it does not have the flaking problem associated with galvanized metals.

Ventilation requirements are similar to regular designs. Makeup air should be 25% of the total airflow and not less than 20 ft^3/min (0.57 m^3/min; 9.4 L/s) per person. Clean rooms are normally maintained with a positive pressure relative to the surroundings.

20. VENTILATION AND PRESSURIZATION IN LABORATORIES AND CLEAN ROOMS

Clean rooms should be maintained at positive pressure with respect to the surrounding areas. However, OSHA, NFPA, and ANSI require most laboratories to be maintained at negative pressure. Recirculation of air from laboratories is strongly discouraged, if not prohibited.

21. CLOSED RECIRCULATING ATMOSPHERES

The closed recirculating atmosphere in submarines and spacecraft presents unique challenges. There are four primary requirements for closed recirculation of atmosphere within closed environments such as submarines and spacecraft: oxygen replacement, carbon dioxide removal, moisture removal, and, in some cases, heat removal. Water vapor is removed in a dehumidification process. Replacement oxygen is added from tanks, electrolysis of water, or oxygen generators. Oxygen can be released continuously by a monitoring system that senses the percentage of oxygen in the air, or it can be released periodically in bulk.

Exhaled air is 4–5% by volume carbon dioxide. The carbon dioxide content is chemically reduced to a normal atmospheric concentration, approximately 0.04% by volume, in a *scrubber*.[18,19] Scrubbers use chemical aqueous absorbents (e.g., soda lime, consisting of mostly calcium hydroxide ($Ca(OH)_2$) with small amounts of sodium hydroxide (NaOH) and/or potassium hydroxide (KOH)) to remove carbon dioxide. The absorbent can be rejuvenated by heating.

---

[18]Similar but separate scrubbing operations are required to remove other contaminants, such as carbon monoxide, hydrogen, and refrigerants in the closed system.
[19]Atmospheric air is 0.038% (380 ppm) carbon dioxide by volume.

33
HVAC: Heating Load

Nomenclature

A	area	ft^2	m^2
C	thermal conductance	Btu/hr-ft^2-°F	W/m^2·°C
CDD	cooling degree-days	°F-day	°C·d
HDD	heating degree-days	°F-day	°C·d
DD	degree-days	°F-day	°C·d
E	effective emissivity	–	–
F	slab edge coefficient	Btu/hr-ft-°F	W/m·°C
h	enthalpy	Btu/lbm	kJ/kg
h	surface heat transfer coefficient	Btu/hr-ft^2-°F	W/m^2·°C
HV	heating value	various	various
k	efficiency factor	–	–
k	empirical Hitchin's exponent	1/°F	1/°C
k	thermal conductivity	Btu-ft/hr-ft^2-°F	W·m/m^2·°C
L	length	ft	m
$\dot{m}$	mass flow rate	lbm/hr	kg/h
M	masonry M factor	–	–
N	number of days in heating season	–	–
p	perimeter length	ft	m
P	power	hp	W
$\dot{q}$	heat transfer rate	Btu/hr	W
R	total thermal resistance	hr-ft^2-°F/Btu	m^2·°C/W
SF	service factor	–	–
T	temperature	°F	°C
U	overall coefficient of heat transfer	Btu/hr-ft^2-°F	W/m^2·°C
$\dot{V}$	volumetric flow rate	ft^3/min	m^3/s

Symbols

ϵ	emissivity	–	–
η	efficiency	–	–
ρ	density	lbm/ft^3	kg/m^3

Subscripts

a	dry air
b	base
fg	vaporization
i	inside design
o	outside design

1. INTRODUCTION

A building's *heating load* is the maximum heat loss (typically expressed in Btu/hr or kW) during the heating season.[1] The *maximum heating load* occurs when the outside temperature is the lowest. The maximum heating load corresponds to the minimum furnace size, even though the lowest temperature occurs only a few times each year. The *average heating load* can be derived from the maximum heating load and is used to determine the annual fuel requirements.

Heating load consists of heat to make up for transmission and infiltration losses. Determining transmission losses is essentially a heat transfer problem. *Transmission loss* is heat lost through the walls, roof, and floor. *Infiltration loss* is heat required to warm ventilation and infiltration air. Though no "credit" for solar heat gain is taken in heating load calculations, reliable sources of internal heating are considered.[2] Modifications for thermal inertia due to high-mass walls and ceilings are generally not made in calculations of heating load. When thermal inertia is considered, the approach taken is simplistic. (See Sec. 33.11.)

Calculation of the heating load is greatly simplified by having access to tabulations of data.[3] Data on climatological conditions are essential, and data on building materials and construction will greatly simplify the task of calculating heat transfer coefficients. Data of this nature is available in a variety of formats. Heat transmission data are available for specific materials as well as for composite walls of specific construction. Both types of data are useful.

---

[1] A *therm* per hour is 100,000 Btu/hr.
[2] The sky is assumed to be overcast during the heating season, so solar heat gain is minimal. Reliable sources of internal heating include permanently mounted equipment and lights.
[3] It is essential that engineers working in this area obtain their own compilations of this type of data.

Ventilation

Calculations of heating load are based on many assumptions. Because of the intrinsic unreliability of some of the data, an *exposure allowance* of up to 15% may be added to the calculated ideal heating load. This helps to account for unexpected heat losses and severe climatic conditions.

2. INSIDE DESIGN CONDITIONS

For the purposes of initial heating load calculations for residences and office spaces, the *inside design temperature* is generally taken as 70°F to 72°F (21.1°C to 22.2°C).

For industrial spaces, such as factories and warehouses, the inside design temperature is lower: 60°F to 65°F (15.6°C to 18.3°C). Humidity is typically 30% to 35% relative humidity, but is generally not considered except in certain manufacturing industries (e.g., textiles and printing) where moisture content is critical.

3. OUTSIDE DESIGN CONDITIONS

The outside temperature and average wind speed (for infiltration) are needed to determine heating load. For estimates of annual heating costs, information on the winter degree days is needed. (See Sec. 33.13.) These values are almost always obtained from tables of climatological data.

Design conditions are probabilistic in nature. There is always some probability that a temperature will be exceeded. Depending on the nature of the facility, it may be desirable to select an outside design temperature that will be exceeded (for example) 5 days out of 100 days. Many tables give design temperatures for 1%, 2.5%, and 5% exceedance probabilities.

4. ADJACENT SPACE CONDITIONS

Since the conductive heat transfer through shared walls depends on the temperatures on both faces, determining the heating load for single rooms and separately heated offices also requires knowing the temperatures in adjacent spaces. For residential calculations, this may require knowing the temperature in attics, large closets, and basements. For attics ventilated by large open louvers, the approximate attic temperature is the average of the inside and outside design temperatures.

5. WALLS AND CEILINGS

Each material used in constructing a wall, ceiling, and so on, contributes resistance to heat flow. This resistance can be specified in a variety of ways. *Total resistance, R* (with units of hr-ft^2-°F/Btu or m^2·°C/W), is the total resistance to heat flow through all of the material. *Unit resistance* (with units of hr-ft^2-°F/Btu-ft or m^2·°C/W·cm) is the

resistance per unit thickness of the material.[4] The total resistance is the product of the resistivity and the material thickness. *Conductance, C,* and *conductivity, k,* are the reciprocals of total resistance and unit resistance, respectively.

$$R = \frac{L}{k} = \frac{1}{C} \qquad 33.1$$

The heat transfer through walls, doors, windows, and ceilings is calculated from the traditional heat transfer equation, Eq. 33.2. The *overall coefficient of heat transfer, U,* can be calculated for each transmission path from the conductivities and resistances of the individual components in that path, or it can be obtained from tabulations of typical wall/ceiling construction. Table 33.1 contains typical values.

$$\dot{q} = UA(T_i - T_o) \qquad 33.2$$

$$U = \frac{1}{\sum R_i} = \frac{1}{\dfrac{1}{h_i} + \sum \dfrac{L}{k} + \sum \dfrac{1}{C} + \dfrac{1}{h_o}} \qquad 33.3$$

Unlike most heat transfer problems, little effort is expended in calculating surface heat transfer (film) coefficients from theoretical correlations. Tables of typical values are used. (See Table 33.2.) When tabulations of overall coefficients of heat transfer are used, it is important to know if an outside film coefficient has been included. If it has, the table's assumption about outside wind speed must be known and compared with actual wind conditions. Multiplicative corrections for other wind speeds generally accompany such tables.

6. AIR SPACES

The thermal conductance of an air space, C, used in Eq. 33.3 depends on the space thickness, emissivities of both sides, orientation, mean temperature, and temperature differential. Typical values for a 50°F (27.8°C) temperature differential are given in Table 33.3.[5] The effective space emissivity, E, used in the table is a function of the surface emissivity, ϵ, and is defined by Eq. 33.4. Surface emissivities range from approximately 0.05 for bright aluminum foil through 0.25 for bright galvanized steel to 0.90 for typical building materials (wood, sheetrock, masonry, etc.).

$$E = \frac{1}{\dfrac{1}{\epsilon_1} + \dfrac{1}{\epsilon_2} - 1} \qquad 33.4$$

---

[4]Unit resistance and conductivity are combined with a length when calculating thermal resistance. There are several sets of units in use for conductivity, depending on how the length is measured. For lengths in feet, Btu/hr-ft-°F and Btu-ft/hr-ft^2-°F are the same. For lengths in inches (centimeters), the units Btu-in/hr-ft^2-°F (W·cm/h·m^2·°C) must be used.

[5]Data in Table 33.3 depend on many factors and are merely representative. Values can vary by 20% or more.

Table 33.1 *Typical Overall Coefficients of Heat Transfer (with film coefficients) (Btu/hr-ft²-°F)*

exterior walls	average U-factors, insulation			
	none	1 in	2 in	4 in
• standard 2 by 4 construction sheathed in wood or insulating board, covered with wood siding, shingles, brick, and sheetrocked or plastered	0.22	0.12	0.09	0.07
• 12 in concrete blocks	0.49	0.14	—	—
• 8 in poured concrete walls	0.70	0.16	—	—

interior ceilings	average U-factors, insulation between rafters					
	none	1 in	2 in	3 in	4 in	6 in
• ceiling applied directly to wood rafters with wood sheathing covered by asphalt or cedar shingles	0.64	0.19	0.12	0.09	0.077	0.05
	average U-factors, insulated between joists as shown					
	none	1 in	2 in	3 in	4 in	6 in
• horizontal ceiling under a pitched roof with no flooring on the ceiling; rafters covered by wood sheathing and asphalt or cedar shingles	0.32	0.15	0.10	0.09	0.077	0.05

There is no heat loss in ceilings between heated floors. For ceilings of rooms over insulated crawl spaces below, use ½ of the U-value shown. For ceilings of rooms over vented or unheated crawl spaces, use the indicated U-value. For ceilings of rooms over unheated basements, use ⅓ of the indicated U-value.

concrete slab on grade floors	average U-factors, per linear foot of slab edge, Btu/hr-ft-°F insulation*		
	none	1 in × 12 in	1 in × 24 in
	0.81	0.46	0.21

windows	average U-factors
• single glazing	1.31
• with storm window	0.45
• thermopane	0.61

doors	
• ¾ in wood	0.69
• 1⅝ in wood	0.46
• 1⅝ in wood with storm door	0.32

(Multiply in by 2.54 to obtain cm.)
(Multiply Btu/hr-ft²-°F by 5.68 to obtain W/m²·°C.)
*1 in of foam board insulation extending 12 in or 24 in vertically (exterior insulation) or horizontally (interior insulation)

7. GROUND SLABS

Heat is lost from ground slabs both through the face and from the exposed edges. Since the ground temperature is usually higher than the winter air temperature, a lower temperature difference (typically 5°F (2.8°C)) should be used to find the heat loss through the face.

This loss is generally small, and an overall coefficient of 0.05 Btu/hr-ft²-°F (0.3 W/m²·°C) is typical. The loss is essentially constant throughout the year since the soil temperature under a building does not vary appreciably. The radial loss from the edges can be found from the *slab edge coefficient*, F, and from the perimeter of the exposed edge. Thickness of the slab is disregarded. Coefficients for concrete slabs on grade depend on the construction method, the amount of insulation, and weather conditions. The coefficient varies from

Table 33.2 *Representative Surface Heat Transfer (Film) Coefficients for Air (nonreflecting surfaces)*

surface orientation, heat flow direction	air speed*	Btu/hr-ft²-°F	W/m²·°C
horizontal, heat flow up	0	1.63	9.26
vertical, heat flow horizontal	0	1.46	8.29
	7½ mph	4.00	22.7
	15 mph	6.00	34.1
horizontal, heat flow down	0	1.08	6.13
	7½ mph	4.00	22.7
	15 mph	6.00	34.1

(Multiply mph by 1.61 to obtain kph.)
(Multiply Btu/hr-ft²-°F by 5.68 to obtain W/m²·°C.)
*Use 0 indoors. Use 7½ mph (12 kph) in summer. Use 15 mph (24 kph) in winter.

Compiled from a variety of sources.

approximately 0.5 Btu/hr-ft-°F (0.9 W/m·°C) for slabs with insulated edges in warm climates to approximately 2.7 Btu/hr-ft-°F (4.9 W/m·°C) for slabs with no insulation located in cold climates.[6]

$$\dot{q} = pF(T_i - T_o) \qquad 33.5$$

8. VENTILATION AND INFILTRATION AIR

Outside ventilation air must be warmed before being introduced into the occupied space. All infiltrated air (from window sashes, door jams, and other cracks) will also go through the air conditioner, so its sensible heat load is combined with the ventilation air deliberately drawn in. The amount of sensible heat required is covered in Chap. 31.

Based on dry or low-humidity air, the sensible heating load for infiltration and ventilation air whose temperature is increased from T_1 to T_2 is[7]

$$\dot{q}_{kW} = \left(72 \frac{kW \cdot min}{m^3 \cdot °C}\right) \dot{V}_{m^3/min}(T_{2,°C} - T_{1,°C})$$
$$= \left(1.2 \frac{kW \cdot s}{m^3 \cdot °C}\right) \dot{V}_{m^3/s}(T_{2,°C} - T_{1,°C}) \quad [SI] \quad 33.6(a)$$

$$\dot{q}_W = \left(1.2 \frac{W \cdot s}{L \cdot °C}\right) \dot{V}_{L/s}(T_{2,°C} - T_{1,°C})$$
$$= \left(1200 \frac{W \cdot s}{m^3 \cdot °C}\right) \dot{V}_{m^3/s}(T_{2,°C} - T_{1,°C}) \quad [SI] \quad 33.6(b)$$

$$\dot{q}_{Btu/hr} = \left(1.08 \frac{Btu \cdot min}{hr \cdot ft^3 \cdot °F}\right) \dot{V}_{cfm}(T_{2,°F} - T_{1,°F})$$
$$= \left(0.018 \frac{Btu}{ft^3 \cdot °F}\right) \dot{V}_{cfh}(T_{2,°F} - T_{1,°F})$$
$$[U.S.] \quad 33.6(c)$$

---

[6]The slab edge coefficient may be given in Btu/hr-°F (W/°C) per unit distance.
[7]The constant 1.08 in Eq. 33.6(c) is reported by other authorities as 1.085 and 1.1 depending on the degree of precision intended.

Ventilation

Table 33.3 *Representative Thermal Conductances of Planar Air Spaces* (Btu/hr-ft^2-$\circ$F (W/m^2.$\circ$C))*

air space orientation	heat flow direction	space thickness (in (mm))	effective space emissivity, E			
			0.05	0.20	0.50	0.82
horizontal	up	0.75–4.0	0.41	0.55	0.82	1.11
		(19–102)	(2.3)	(3.1)	(4.7)	(6.30)
horizontal	down	0.75	0.28	0.42	0.69	0.98
		(19)	(1.6)	(2.4)	(3.9)	(5.6)
		1.5	0.18	0.31	0.58	0.87
		(38)	(1.0)	(1.8)	(3.3)	(4.9)
		4.0	0.11	0.25	0.52	0.81
		(102)	(0.62)	(1.4)	(3.0)	(4.6)
vertical	horizontal	0.75–4.0	0.28	0.42	0.69	0.99
		(19–102)	(1.6)	(2.4)	(3.9)	(5.6)

(Multiply in by 25.4 to obtain mm.)
(Multiply Btu/hr-ft^2-$\circ$F by 5.68 to obtain W/m^2.$\circ$C.)
*For a 50$\circ$F (27.8$\circ$C) temperature differential.

Compiled from a variety of sources.

If the outside design temperature is 32$\circ$F (0$\circ$C) or cooler, there will be little or no moisture in the incoming air. However, for air with higher temperatures, heat is required to warm any moisture that enters with outside air.

Although the sensible heating of the dry air and accompanying moisture can be calculated separately and added, it is more expedient to use enthalpy values read from the psychrometric chart.

$$\dot{q}_{kW} = \dot{m}_{a,kg/s}(h_i - h_o)_{kJ/kg}$$
$$= \dot{V}_{m^3/s}\rho_{kg/m^3}(h_i - h_o)_{kJ/kg}$$
$$\approx \left(1.2\ \frac{kg}{m^3}\right)\dot{V}_{m^3/s}(h_i - h_o)_{kJ/kg}$$
$$\approx \left(0.0012\ \frac{kg}{L}\right)\dot{V}_{L/s}(h_i - h_o)_{kJ/kg} \quad \text{[SI]} \quad \textbf{33.7(a)}$$

$$\dot{q}_{Btu/hr} = \dot{m}_{a,lbm/hr}(h_i - h_o)_{Btu/lbm}$$
$$= \left(60\ \frac{min}{hr}\right)\dot{V}_{ft^3/min}\rho_{lbm/ft^3}(h_i - h_o)_{Btu/lbm}$$
$$\approx \left(4.5\ \frac{lbm\text{-}min}{ft^3\text{-}hr}\right)\dot{V}_{ft^3/min}(h_i - h_o)_{Btu/lbm}$$
$$\text{[U.S.]} \quad \textbf{33.7(b)}$$

Special accounting for infiltrated air is necessary when the conditioned space supplies combustion air. This normally is the case for residential fireplaces and space heaters. Each volume of gaseous fuel gas burned uses approximately nine volumes of heated air. Unless the face of the fireplace is blocked off, heated air continues to be lost up the flue, even after combustion stops.

9. HUMIDIFICATION

Outside air entering at very low temperatures may need to be humidified. The latent heat required to add moisture to the outside air is covered in Chap. 31.

10. INTERNAL HEAT SOURCES

Internal heat sources are heat sources within the conditioned space.[8] Heat sources may introduce sensible and latent loads. Two general methods are used to estimate internal sources. The load can be based on the equipment nameplate rating. Alternatively, tables of typical values can be used. Such tables are helpful in estimating the fractions of each load that are sensible and latent.[9] The sensible heat supplied by permanent machinery and lighting should be subtracted from the heating load.

Some equipment is rated in horsepower; other equipment is rated by wattage.[10] The *service factor*, SF, used in Eq. 33.8 is the fraction of the rated power being developed. Service factors greater than 1.00 are possible in overload conditions.

---

[8]Residential heating sources such as toasters and coffee brewers are sometimes referred to as *domestic heat sources*.

[9]In the absence of tabular data, simple, logical rules of thumb can be used. For example, when cooking under a hood, all moisture is captured, the latent load is zero, and the sensible load transferred due to radiation is taken as some assumed fraction of the nameplate rating. If not cooking under a hood, two-thirds of the load is sensible and one-third is latent. However, tables are usually essential when determining the heat gain for esoteric sources such as doughnut machines, deep-fry kettles, and waffle irons for ice cream sandwiches.

[10]The *watt density* (strictly, the *linear watt density*) for baseboard heaters, heat tracing, and similar devices is the amount of heating generated per foot (meter). It is not based on area.

$$\dot{q}_{kW} = \left(0.7457 \ \frac{kW}{hp}\right)(SF)\left(\frac{P_{hp}}{\eta}\right)$$

$$= (SF)\left(\frac{P_{kW}}{\eta}\right) \qquad \text{[SI]} \quad \textit{33.8(a)}$$

$$\dot{q}_{Btu/hr} = \left(2545 \ \frac{Btu}{hp\text{-}hr}\right)(SF)\left(\frac{P_{hp}}{\eta}\right)$$

$$= \left(3413 \ \frac{Btu}{kW\text{-}hr}\right)(SF)\left(\frac{P_{kW}}{\eta}\right) \qquad \text{[U.S.]} \quad \textit{33.8(b)}$$

The equipment efficiency, η, is included in the denominator in Eq. 33.8. This is appropriate when the motor and equipment being driven are both in the conditioned space. If the motor is not actually in the space but the driven equipment is, then the efficiency should be omitted.[11] (An example of this configuration would be an in-duct fan that is driven by a motor located outside of the duct.) Typical motor efficiencies are 80% for 1 hp motors and 90% for 10 hp and higher motors.

For fluorescent lights, the rated wattage should be increased by 20% to 25% to account for wound (transformer type) ballast heating. (No increase is used with incandescent lights or fluorescent lights with power-saving, electronic ballasts.) However, it may be inappropriate to assume that all heat generated enters the conditioned space. If the air space above the lights in a dropped ceiling is not directly conditioned, then only some fraction (e.g., 60%) of the heat enters the occupied space.

Unless the room is reasonably occupied on a permanent basis, the heating load should not be reduced by the metabolic heat of the occupants.

Some buildings have internal sources generating large amounts of energy. Under no circumstance, however, should the theoretical heating load be reduced by these internal heat gains to a point where the inside temperature would be 40°F (4.4°C) or lower in the absence of these internal sources.

11. THERMAL INERTIA

Buildings with massive masonry (including concrete) walls are thermally more stable than those with thin walls. However, the effect of *thermal lag* (*thermal inertia* or *thermal flywheel effect*) is difficult to incorporate in most studies. A simplified approach known as the

M-factor method has been developed. This method correlates a factor, M, with the degree-days and mass per unit wall area (lbm/ft^2 or kg/m^2). (See Fig. 33.1.) The M-factor modifies the overall coefficient of heat transfer, U. The actual heat transfer is

$$\dot{q} = MUA(T_i - T_o) \qquad \textit{33.9}$$

As written, Eq. 33.9 is appropriate for analyzing the heat loss through a wall. For design, particularly when wall construction with a maximum heat loss coefficient, U, is specified by the building code, a wall may be designed to have an instantaneous heat loss coefficient of U/M and still be up to code.

Figure 33.1 *M-Factor*

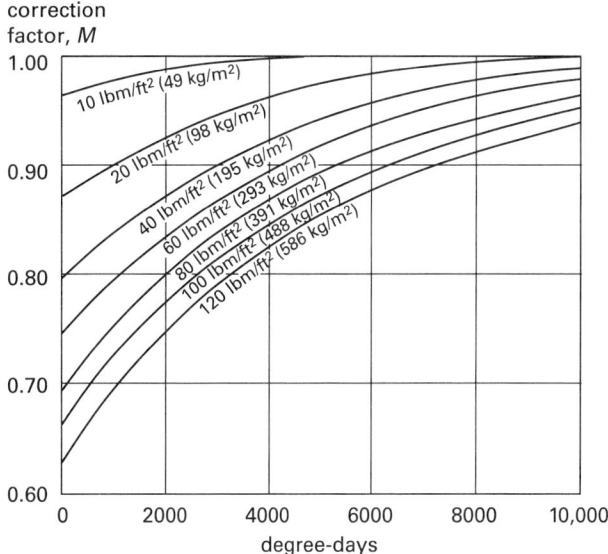

Source: "Mass Masonry Energy," Masonry Industry Committee, ca. 1979.

12. FURNACE SIZING

A furnace must be capable of keeping a building warm on the coldest days of the heating season. Therefore, a furnace should be sized based on the coldest temperature reasonably expected (i.e., on the outside design temperature). However, not all days in the heating season will have temperatures that cold, and the entire capacity will rarely be utilized.

Pickup load is the furnace capacity needed to bring a cold building up to the inside design temperature in a reasonable time. Since the outside design temperature provides excess capacity most of the year, a pickup allowance may not be necessary. However, churches,

---

[11]There is a third, less likely alternative. The conditioned space may contain only the motor, with the driven machinery being someplace else. In that case, only the energy lost in the energy conversion appears in the load. The nameplate power is multiplied by $(1-\eta)/\eta$.

auditoriums, and office buildings needing to be "ready" at specific times need special attention.[12] If a building is heated during the day only, a 10% increase in the rated furnace size is sometimes added as the pickup load to allow for starting up. If a building is left unheated for extended periods, the increase should be 25% or higher.

Some furnaces have nameplate ratings for both fuel input and output heating. The furnace efficiency is incorporated in a rating for "output Btus." Output ratings apply to newly installed furnaces. Output can be expected to decrease with time.

13. DEGREE-DAYS AND KELVIN-DAYS

The *degree-day (kelvin-day)* concept can be used to estimate heating costs during the heating season.[13] The assumption is made that the heating system operates when degree-days are being accumulated, and there is an indoor *base temperature*, T_b, that triggers the heating system to turn on. In the past, the base temperature was typically 65°F in the United States (19°C in Europe), although other temperatures such as 50°F and 15.5°C are used. The base temperature affects the estimate materially, so it should not be selected casually. The base temperature depends on the temperature that a building is heated to and the sources of internal heat gain. For example, if a building was going to be heated to 19°C, and if the average internal heat gain from people, lights, and equipment in a building was estimated to increase the temperature by 3.5°C, it would be appropriate to use 19°C − 3.5°C = 15.5°C as the base temperature.

The heating season is assumed to last for N days, different for each geographical location, and determined from meteorological data. Each day whose 24-hour average temperature, $\overline{T}$, is less than the base temperature, T_b, will accumulate $T_b - \overline{T}$ degree-days (kelvin-days), DD. The sum of these degree-day terms over the entire heating season of N days is the total *heating degree-days (heating kelvin-days)*, HDD, also known as *winter degree-days*, for that location.[14]

$$\text{HDD} = \sum_{i=1}^{N}(T_b - \overline{T}_i) \qquad 33.10$$

Ultimately, use of the degree-day concept comes down to the meteorological data that are available. Degree-day information is not always available, although

average monthly temperature usually is available from meteorological data. *Hitchin's formula* (1983) can be used to estimate degree-days from the average weekly or monthly temperature. The average number of degree-days per day during the heating period is calculated from Eq. 33.11. The empirical exponent k has a value of 0.39 1/°F (0.71 1/°C).

$$\frac{\text{HDD}}{N} = \frac{T_b - \overline{T}}{1 - e^{-k(T_b - \overline{T})}} \qquad 33.11$$

The degree-day concept is intended to help in estimating heating cost during the heating season. There may be times during some days when the heating system does not operate, and for that reason, the degree-day accumulation does not represent a complete record of the exterior temperatures encountered. Only the cold parts of the heating season are represented. Degree-days cannot be used to determine the average temperature during the entire heating season. However, Hitchin's formula can be extended to the entire heating season and used to calculate the average outside temperature. For regions where heating is continuous throughout the heating season (as represented by large differences between the base and average temperatures), this is equivalent to Eq. 33.12.

$$\frac{\text{HDD}}{N} = T_b - \overline{T} \qquad 33.12$$

14. FUEL CONSUMPTION

The design heating load used to size the furnace is not the average heating load over the heating season. However, when combined with degree-days, the design heating load can be used with smaller, simple structures to calculate the total fuel consumption and heating costs during the entire heating season. (This method assumes that there will be no heating when the outside temperature is equal to the base temperature or higher. This assumption is appropriate, particularly in residential buildings, even though the interior temperature is maintained at 68°F to 72°F (20°C to 22.2°C) because of the existence of internal heat sources.) The units of fuel consumption in Eq. 33.13 depend on the units of the heating value. Equation 33.13 does not include factors for operating pumps, fans, stokers, and other devices, nor are costs of maintenance, tank insurance, and so on, included.[15] The

---

[12]There is not much information on this subject. The increase for pickup load is largely a matter of judgment.

[13]Although the degree-day concept continues to appear in ASHRAE publications, ASHRAE stopped promoting the method in the 1980s in favor of more accurate surface-by-surface heat loss calculations.

[14]A similar concept for *cooling degree-days (summer degree-days)*, CDD, exists for use in estimating average cooling costs during the cooling season.

---

[15]Equation 33.13 appears to imply that the lower T_o is, the lower the fuel consumption will be. This is obviously untrue. The temperature difference in the denominator actually cancels the same temperature difference used to calculate the heat transfer terms in $\dot{q}$. Thus, $\dot{q}$ is put on a per-degree basis. The average temperature difference used in the calculation of degree days converts the per-degree heat loss to an average heat loss.

minimum efficiency, η, of gas- and oil-fired residential furnaces is approximately 70% to 75%.

fuel consumption (in units/heating season)

$$= \frac{\left(86\,400\,\frac{\text{s}}{\text{d}}\right)\dot{q}_{\text{kW}}(\text{HDD})}{(T_i - T_o)(\text{HV}_{\text{kJ/unit}})\eta_{\text{furnace}}}$$

[SI]   *33.13(a)*

fuel consumption (in units/heating season)

$$= \frac{\left(24\,\frac{\text{hr}}{\text{day}}\right)\dot{q}_{\text{Btu/hr}}(\text{HDD})}{(T_i - T_o)(\text{HV}_{\text{Btu/unit}})\eta_{\text{furnace}}}$$

[U.S.]   *33.13(b)*

Equation 33.13 has traditionally been used to estimate fuel consumption. In recent years, various improvements to the "model" have been made. Specifically, Eq. 33.13 is multiplied by an empirical correction, C_D, to correct for the difference between calculated values and actual performance.[16] Values of C_D range from about 0.60 to 0.87 and are correlated with the number of degree-days. Also, the furnace efficiency, η, is replaced with an efficiency factor, k, that includes the effects of rated full-load efficiency, part-load performance, over-sizing, and energy conservation devices. A value of 1.0 should be used for k for electric heating. Values of 0.55 and 0.65 are appropriate for older and energy-efficient houses, respectively.

15. CONSERVATION THROUGH THERMOSTAT SETBACK

A variety of *conservation methods* are used to reduce energy consumption during the heating season. These include installing insulation, weather stripping, and reducing the thermostat setting for all or part of the day. An extreme case of thermostat setback occurs when the heating system is turned off entirely at night.

If a building is to be maintained at two different temperatures during different parts of the day, the average heating can be found from the duration-weighted average of the two inside temperatures. Alternatively, separate calculations can be made for the periods of different temperatures.

16. FREEZE-UP OF HEATING AND PREHEATING COILS

Coil freeze-up (*freezing of the coils*) has two different causes. In the winter, cold make-up (ventilation) air from outside may cause water to freeze inside the heating coils. In the summer, coils with effective surface temperatures of 32°F (0°C) or less can cause moisture (humidity) in the air to freeze on the outside of the cooling (evaporator) coils.

Steam is usually the source of heating in coils because the heat of vaporization is so much larger than the sensible energy available from hot water alone. Steam in the coil condenses as it releases the heat of vaporization. To prevent freezing inside the coil, the condensate must drain from the coil rapidly, before the cold outside air can reduce the water's temperature to freezing. A rapid-drain plumbing design includes a vacuum breaker, large trap, long drip line, drain line below the supply line, adequate drain line slope, and clean strainers, among other features. Regular maintenance is required to ensure everything works as designed.

In the winter, when the outside air temperature is below freezing, coil freeze-up can be avoided by ensuring cold make-up air is thoroughly mixed with warm return air before the mixture enters the coils.[17] In the event that the mixture is still too cold, freeze-resistant heating coils with adequate piping should be used. In rare cases, some form of electric air preheating can be provided prior to the steam coils.

---

[16]It has been shown that Eq. 33.13 overestimates the fuel requirement in most cases. C_D simply reduces the estimate.

[17]Mixing louvers can fail. So, proper coil protection should always be provided.

34 HVAC: Cooling Load

Nomenclature

A	area	ft^2	m^2
ACH50	air changes per hour at 50 Pa	1/hr	1/h
ACHnat	air changes per hour at 4 Pa	1/hr	1/h
BF	bypass factor	–	–
C	leakage curve coefficient	–	–
C_d	coefficient of discharge	–	–
CDD	cooling degree days	°F-day	°C·d
CFM50	airflow rate at 50 Pa	ft^3/min	n.a.
CLF	cooling load factor	–	–
CLTD	cooling load temperature difference	°F	°C
COP	coefficient of performance	–	–
EER	energy-efficiency ratio	Btu/W-hr	n.a.
ELA	effective leakage area	in^2	m^2
g_c	gravitational constant, 32.2	lbm-ft/ lbf-sec^2	n.a.
GSHR	grand sensible heat ratio	–	–
h	enthalpy	Btu/lbm	kJ/kg
h	height	in	m
I	current	A	A
LBL	energy climate factor	–	–
LF	latent factor	–	–
n	leakage curve exponent	–	–
NL	normalized leakage	1/hr	1/h
Δp	pressurization	Pa	Pa
P	power	hp	W
$\dot{q}$	heat transfer rate	Btu/hr	W
Q	airflow	ft^3/min	m^3/h
Q50	air permeability	$\text{ft}^3/\text{ft}^2\text{-min}$	$\text{m}^3/\text{m}^2\cdot\text{h}$
RSHR	room sensible heat ratio	–	–
S	building envelope surface area	ft^2	m^2
SC	shading coefficient	–	–
SCL	solar cooling load factor	Btu/hr-ft^2	W/m^2
SHR	sensible heat ratio	–	–
SEER	seasonal energy-efficiency ratio	Btu/W-hr	n.a.
t	time (duration)	hr	h
T	temperature	°F	°C
T^*	mixture temperature	°F	°C
U	overall coefficient of heat transfer	Btu/ hr-ft^2-°F	W/m$^2\cdot$°C
V	building volume	ft^3	m^3
V	voltage	V	V
$\dot{V}$	volumetric flow rate	ft^3/min	L/s

Symbols

η	efficiency	–	–
ρ	density	lbm/ft^3	kg/m^3
v	specific volume	ft^3/lbm	m^3/kg

Subscripts

c	cooling
co	output of air conditioner
i	indoor design
in	in (entering the space)
l	latent
m	mean
o	outdoor design
ref	reference
s	sensible
t	total
te	total equivalent

1. INTRODUCTION

The procedure for finding the *cooling load* (also referred to as the *air conditioning load*) is similar in some respects to the procedure for finding the heating load.[1] The aspects of determining inside and outside design conditions, heat transfer from adjacent spaces, ventilation air requirements, and internal heat gains are the same as for heating load calculations and are not covered in this chapter.

However, the calculation of cooling load is complicated considerably by the thermal lag of the exterior surfaces (i.e., walls and roof). Depending on construction, the

---

[1]The *refrigeration load* or *coil loading* is the cooling load expressed in appropriate units (e.g., tons of refrigeration).

solar energy absorbed by exterior surfaces can take hours to appear as an interior cooling load.[2] Further complicating the determination of cooling load are the direct transmission of solar energy through windows and the facts that the delay is different for each surface, the solar energy absorbed changes with time of day, and instantaneous heat gain into the room contributes to instantaneous and delayed cooling loads.

It is important to distinguish between three terms. The *instantaneous heat absorption* is the solar energy that is absorbed at a particular moment. The *instantaneous heat gain* is the energy that enters the conditioned space at that moment. Due to solar lag, the heat gain is a complex combination of heat absorptions from previous hours. The *instantaneous cooling load* is a portion (i.e., is essentially the convective portion) of the instantaneous heat gain.

2. SOURCE OF COOLING

Once the cooling load is determined, the source and size of the cooling unit must be considered. Cooling normally comes from liquefied refrigerant passing through a cooling coil. Some or all of the airflow passes across the cooling coil. The refrigerant is continuously vaporized in the coil as part of a complete refrigeration cycle. Alternatively, cold water may be used in the coil to cool the airflow. In such cases, the water is cooled in a *chiller* running its own refrigeration cycle.

An *economizer* is an electromechanical system that changes a portion of the cooling process in order to decrease cost, usually by taking advantage of cold ambient air. A *water-side economizer* substitutes natural cooling from a cooling tower for the chiller's more expensive refrigeration cycle when the ambient air temperature drops below the desired coil temperature. An *air-side economizer* increases the amount of outside air that is brought into a space when the ambient air characteristics (temperature, humidity, or enthalpy) are better than the return airflow. The outside air may still be conditioned by passing through coils, but less change will be required. Use of cold ambient air by either type of economizer is known as *free cooling*.

3. REFRIGERATION SYSTEM SIZE AND RATINGS

Refrigeration equipment for HVAC use is typically rated in tons of cooling, where a *ton of cooling* is equal to 12,000 Btu/hr (3517 W). Prior to the emphasis on energy conservation and detailed energy analyses, a common rule of thumb, particularly in residential and small commercial installations, was to size refrigeration systems at one ton for every 400 ft[2] (37 m[2]) of conditioned space. For modern homes meeting minimum insulation, window, and sealing code requirements, the

rule of thumb has increased to one ton for every 600 ft[2], 800 ft[2], or 1000 ft[2] (56 m[2], 74 m[2], or 93 m[2]), although oversizing is still likely without a detailed analysis.

The theoretical power dissipated by a refrigeration cycle's single-phase compressor motor is calculated from the motor's actual measured voltage and current draw or from the cooling load. Various efficiencies must be considered.

$$P_{compressor,watts} = I_{amps} V_{volts} = \frac{\dot{q}_{c,Btu/hr}}{3.412 \eta_{electrical} \eta_{compressor}}$$

34.1

Similarly, the theoretical motor horsepower is calculated as

$$P_{compressor,hp} = \frac{\dot{q}_{c,Btu/hr}}{2544 \eta_{electrical} \eta_{compressor}}$$

34.2

In reality, the efficiencies depend on the refrigerant and its condition. Compressor manufacturers provide charts showing horsepower versus cooling load at various suction pressures for various refrigerants. A general rule for air conditioning is 1 hp per 10,000 Btu/hr (2.9 kW), although for colder temperatures (e.g., as required by freezers), 1 hp for each 3000–5000 Btu/hr (0.9–1.5 kW) may be required. Power requirements are also affected by ambient conditions and by the amount of free cooling used.

Because of consumer difficulties in evaluating the various efficiencies and other details affecting operating costs of unitary systems, the *energy-efficiency ratio*, EER, is used as a simple, comparable measure to describe the efficiency of cooling systems (see Eq. 34.3).[3] The total input power, P_{total}, includes the power required to run the compressor as well as fans, controls, and all other parts of the air conditioning system. EER is typically evaluated at 50% relative humidity with a 95°F (35°C) outside temperature and an 80°F (27°C) inside temperature (return air). EER is equivalent to 3.412 times the coefficient of performance, COP, as shown.

$$EER = \frac{\dot{q}_{c,Btu/hr}}{P_{total,watts}} = 3.412(COP)$$

34.3

The *seasonal energy-efficiency ratio*, SEER, of the Air-Conditioning, Heating, and Refrigeration Institute (AHRI) has the same definition as the EER, but is averaged over a range of outside temperatures using a standardized cooling season. This is unlike the EER, which is evaluated at a specific operating point.[4] Although subject to local and future legislation as well

---

[2]That is, the *instantaneous heat gain* is the heat that enters the conditioned space. Due to thermal lag, it is not the same as the *instantaneous heat absorption* by the building at that same moment. This terminology is not rigidly adhered to.

[3]"Other details" affecting energy usage include the local ambient air conditions, ducting and sealing, type of unit (unitary or split), and the differences between air-cooled and water-cooled units.
[4]ANSI/AHRI 210/240: *Standard for Performance Rating of Unitary Air-Conditioning & Air-Source Heat Pump Equipment*, AHRI.

as exceptions by type of unit, by U.S. law, all newly manufactured and installed air conditioning equipment must be at least 13 SEER. "High efficiency" units are 14 SEER and higher. Commercially available units up to approximately 20 SEER are available, but most newly installed equipment is 16 SEER or lower. Equation 34.4 gives an approximate relationship between EER and SEER.

$$EER = 1.12(SEER) - 0.02(SEER)^2 \qquad 34.4$$

4. ENERGY STAR CONSTRUCTION

Energy Star is a joint program of the U.S. Environmental Protection Agency (EPA) and the U.S. Department of Energy (DOE) that promotes and rewards energy-efficient construction and remodeling, primarily for residential houses. For construction to be Energy Star certified, it must have energy-efficient features that enable the structure to reduce total energy consumption by 15–20% (depending on climate) compared to a structure built according to the local energy code. Such features typically include high-performance windows, increased insulation levels, high-efficiency heating, cooling and water heating equipment, fluorescent lighting, Energy Star appliances, duct sealing, and air sealing of the building envelope.

Certification is performed by a Home Energy Rating System (HERS) rater. The rater determines the structure's characteristics via computer modeling, inspects the insulation and sealing of the structure, and tests the structure's duct and building envelope tightness.

5. INSIDE AND OUTSIDE DESIGN CONDITIONS

It is impossible to maintain the air passing through a conditioned space at a particular temperature. Cool air enters a space, and warm air leaves. The *inside design temperature*, T_i, is understood to be the temperature of the air removed from the conditioned space. Indoor design temperatures for summer use are generally a few degrees warmer than winter design temperatures—approximately 75°F (23.9°C).

6. INSTANTANEOUS COOLING LOAD FROM WALLS AND ROOFS

Three general methods are used to determine instantaneous cooling load: (1) total equivalent temperature difference method, (2) transfer function method, and (3) cooling load factor and temperature difference methods.

The total equivalent temperature difference (TETD) method determines the instantaneous heat gain. The *total equivalent temperature difference*, ΔT_{te}, depends on the type of construction, geographical location, time of day, and wall orientation. It is read from extensive

tabulations. The instantaneous heat gain consists of stored radiant and convective portions. In the TETD/TA (time averaging) method, weighting factors are used to average the radiant portions from current and previous hours. The sum of the convective portions and the weighted average of the series of radiant portions are taken as the cooling load. Computer analysis and considerable judgment are required to use this method.

$$\dot{q}_{\text{heat gain}} = UA\Delta T_{te} \qquad 34.5$$

The *transfer function method* is similar to the total equivalent temperature difference method. A series of weighting factors, known as *room transfer functions*, is applied to cooling load values from the current and previous hours. The transfer functions are related to spatial geometry, configuration, mass, and other characteristics.

The only modern method of calculating the cooling load suitable for quick (manual) analysis is the *cooling load temperature difference method*, CLTD, using the related *solar cooling load factor*, SCL, and *cooling load factor*, CLF, covered in subsequent sections. For exterior surfaces, the cooling load is calculated by Eq. 34.6. Tables of CLTD are needed. Values depend on time of year, location and orientation, type, configuration, and orientation of the surface, as well as other factors. Using the CLTD/SCL/CLF method, the instantaneous cooling load for conduction through opaque walls and roofs is

$$\dot{q}_c = UA(\text{CLTD}_{\text{corrected}}) \qquad 34.6$$

The base conditions used to calculate the values of CLTD are

- a clear sky on July 21
- exposed, sunlit, flat roofs
- walls at 40°N latitude based on roof and wall construction and orientation
- an indoor temperature of 78°F
- an outdoor maximum temperature of 95°F with a mean temperature of 85°F
- a daily temperature range of 21°F

These base conditions generally don't coincide precisely with actual conditions during the study period, so CLTD is corrected according to Eq. 34.7. T_i is the indoor design temperature, and T_m is the mean outdoor temperature.

$$\text{CLTD}_{\text{corrected}} = \text{CLTD}_{\text{table}} + (78°F - T_i)$$
$$+ (T_m - 85°F) \qquad 34.7$$

$$T_m = T_{\text{outdoor,max}} - \tfrac{1}{2}(\text{daily range}) \qquad 34.8$$

Ventilation

7. INSTANTANEOUS COOLING LOAD FROM WINDOWS

Using the CLTD/SCL/CLF method, the cooling load due to solar energy received through windows is calculated in two parts.[5] The first is an immediate conductive part; the second is a radiant part. Appropriate tables are needed to evaluate the *shading coefficient*, SC, and the SCL for the radiant portions.

$$\dot{q}_c = \dot{q}_{\text{conductive}} + \dot{q}_{\text{radiant}}$$
$$= UA(\text{CLTD}_{\text{corrected}}) + A(\text{SC})(\text{SCL}) \qquad 34.9$$

8. COOLING LOAD FROM INTERNAL HEAT SOURCES

Latent loads (including metabolic latent loads) are considered instantaneous cooling loads. Only a portion, given by the CLF, of the sensible heat sources show up as instantaneous cooling load. CLF is a function of time and depends on zone type, occupancy period, interior and exterior shading, and other factors. Although tables are usually necessary to evaluate CLF, there are some cases where CLF is assumed to be 1.0. These include when the cooling system is shut down during the night, when there is a high occupant density (as in theaters and auditoriums), and when lights and other sources are operated for 24 hours a day.

$$\dot{q}_{c,\text{internal sources}} = \dot{q}_l + \dot{q}_s(\text{CLF}) \qquad 34.10$$

9. VENTILATION AND INFILTRATION

Since all air passes through the air conditioner, sensible and latent loads from ventilation and infiltration air are instantaneous cooling loads. *Building air leakage*, Q, is a function of the pressure differential, Δp, between the inside and outside of a building. The *building leakage curve* is defined by a coefficient, C, and an exponent, n, both obtained from a curve fit correlation of at least 12 test points between 15–20 Pa and 60–75 Pa. In U.S. models, the values of C and n are used to determine leakage in ft³/min even though Δp is in pascals.

$$Q_{\text{ft}^3/\text{min}} = C\Delta p_{\text{Pa}}^n \qquad 34.11$$

Air leaks can be identified and leakage can be quantified and minimized by pressure testing.[6] It is common to base leakage testing on ACH50 (also known as ACH-50, ACH_{50}, or n50), a fan-induced pressurization at 50 Pa (0.00725 lbf/in²). ACH50 is used to determine the *air change rate*, the number of air changes per hour. The pressurization is accomplished by temporarily inserting a *blower door*, a frame with a built-in fan, pressure and airflow rate measurements, and other instrumentation,

into an exterior door opening. The instrumentation in the blower door gives a direct reading of CFM50, the leakage in cubic feet per minute, which can also be calculated from ACH50 as shown in Eq. 34.12. Once the pressurization level is reached, leaks are subsequently located by a smoke puffer or an infrared camera.

$$\text{CFM50} = \frac{(\text{ACH50})\, V_{\text{structure,ft}^3}}{60\,\dfrac{\text{min}}{\text{hr}}} \qquad 34.12$$

CFM50 can be used to determine the *effective leakage area*, ELA (or EfLA). ELA was defined by the Lawrence Berkeley Laboratory in its infiltration model as the area of a nozzle-shaped orifice (with rounded edges, similar to the inlet of a blower door fan) that leaks the same amount of air that the building leaks at a pressure difference of 4 Pa.[7] The coefficient of discharge, C_d, for a smooth, rounded orifice is 0.97–0.98, essentially 1.00. The coefficient C and exponent n are the same as in Eq. 34.11. The approximate density, ρ, of dry air is 0.075 lbm/ft³ (1.2 kg/m³).

$$\text{ELA}_{\text{in}^2} = \frac{Q\sqrt{\dfrac{\rho}{2g_c\Delta p}}}{C_d} \quad \text{[consistent units]}$$

$$= \frac{C\Delta p_{\text{Pa}}^n \left(12\,\dfrac{\text{in}}{\text{ft}}\right)^2 \times \sqrt{\dfrac{\rho_{\text{lbm/ft}^3}}{(2)\left(32.2\,\dfrac{\text{lbm-ft}}{\text{lbf-sec}^2}\right)\Delta p_{\text{Pa}}\left(0.02089\,\dfrac{\frac{\text{lbf}}{\text{ft}^2}}{\text{Pa}}\right)}}}{C_d\left(60\,\dfrac{\text{sec}}{\text{min}}\right)}$$

$$\approx \frac{\text{CFM50}}{18}$$
$$\qquad 34.13$$

Specific leakage is the ELA reported per unit floor area or per unit building envelope area.[8] The *air permeability* is the leakage rate per unit building envelope area. In Eq. 34.14, the above-grade surface area, S, includes floor, ceiling, wall, and window areas. Various units are used to report air permeability. When reported in units of m³/m²·h, the designation "Q50" is often used. When reported in ft³/ft²-min, the designation "MLR," the *Minneapolis leakage ratio*, may be used.

$$\text{Q50} = \frac{(\text{ACH50})\, V_{\text{structure}}}{S} \qquad 34.14$$

---

[5]The term *fenestration* refers to windows or other openings transparent to solar radiation.

[6]The greatest accuracy is achieved when results of pressurization and depressurization (i.e., negative pressure) tests are averaged.

[7]The *equivalent leakage area*, EqLA, is defined by the Canadian National Research Council as the area of a sharp-edged orifice (a sharp, round hole cut into a thin plate with $C_d = 0.61$) that would leak the same amount of air as the building does at a pressurization of 10 Pa. The 10 Pa EqLA is approximately two times the 4 Pa ELA.

[8]The building envelope is also referred to as the *building fabric*.

In order to compare buildings with different floor areas and heights, ASHRAE defines the *normalized leakage*, NL, as a measure of the tightness of a building envelope relative to the building size and number of stories.[9] The "normal" reference condition is a new, 1100 ft^2 (100 m^2), single-story, non-energy efficient, slab-on-grade, non-low income house. The *reference height*, h_{ref}, in Eq. 34.15 is 98 in (2.5 m).

$$\text{NL} = 1000\left(\frac{\text{ELA}_{\text{in}^2}}{A_{\text{floor,in}^2}}\right)\left(\frac{h_{\text{building,in}}}{h_{\text{ref,in}}}\right)^{0.3} \qquad 34.15$$

Pressurization to 50 Pa corresponds to a 20 mph (32 km/h) wind impacting all sides of a structure. Since infiltration from all exterior surfaces would not occur naturally, an estimate of the natural infiltration rate, ACHnat (also known as ACHn or ENIR), is estimated from the *LBL factor* (also known as the *N factor* or the *energy climate factor*), which is dependent on the climate region, the number of stories of the structure, and sheltering from wind.[10] LBL factors range from 4 to 40, with typical values ranging from 10 to 20. Energy Star has established a natural infiltration target threshold of 0.35 air changes per hour, but well-sealed structures can achieve ACHnat values much lower than this. As calculated, ACHnat is only a rough estimate, and true values can range from 50% lower to 100% higher. Table 34.1 uses ACHnat and other factors to categorize the air-tightness of moderately sized houses.

$$\text{ACHnat} = \frac{\text{ACH50}}{\text{LBL}} \qquad 34.16$$

Table 34.1 *Approximate Air-Tightness of Moderately Sized Houses*

	tight	moderate	leaky
CFM50	< 1500	1500–4000	> 4000
ACH50	< 5	5–10	> 10
ACHnat	< 0.35	0.35–1	> 1

The annual cost of the additional cooling load due to air leakage with a SEER-rated appliance can be estimated from Eq. 34.17.

$$\text{annual cooling cost} = \frac{\dot{q}_{c,\text{Btu/hr}}\, t_{\text{hr/yr}}(\text{cost})_{\text{\$/kW-hr}}}{\left(1000\,\frac{\text{W}}{\text{kW}}\right)(\text{SEER})}$$
$$\approx \frac{0.26(\text{CDD})(\text{CFM50})(\text{cost})_{\text{\$/kW-hr}}}{(\text{LBL})(\text{SEER})} \qquad 34.17$$

10. HEAT GAIN TO AIR CONDITIONING DUCTS

The calculation of heat absorbed by air conditioning ducts that pass through unconditioned spaces is not sophisticated. The heat transfer is generally estimated from tables or figures of standard configurations (e.g., the heat loss per fixed length of duct per 10 degrees of temperature difference). Extrapolation is used for other duct lengths and temperature differences.

The logarithmic mean temperature difference (generally used when the temperature difference varies along the length) is seldom used in the HVAC industry.[11] Rather, the heat transfer is based on the temperature difference between the environment and the midlength temperature of the duct. If the temperature of the duct at its mid-length is not known, one or more iterations will be needed to calculate the temperature drop and heat transfer.

11. DEGREE-DAYS

Some sources present tables of *cooling degree-days*, CDD (*summer degree-days*).[12] Data in these tables are usually related to a base temperature of 65°F (18.3°C), but they may be related to a 70°F (21.1°C) base. Cooling is considered to occur only when the temperature is higher than 65°F (18.3°C).

12. SEASONAL COOLING ENERGY

The approximate total energy used during the cooling season can be calculated from the cooling degree days. SEER is the *seasonal energy-efficiency ratio*, which incorporates various equipment and process efficiencies as well as the conversion from Btus to kilowatt-hours. SEER for electrically driven refrigeration is typically in the range of 10 Btu/W-hr to 12 Btu/W-hr (2.9 W/W to 3.5 W/W). The seasonal cooling cost is determined from the cost per kilowatt-hour.

$$\text{energy}_{\text{kW·h/season}}$$
$$= \frac{\left(24\,\frac{\text{h}}{\text{d}}\right)\dot{q}_{\text{design cooling,W}}(\text{CDD})}{\left(1000\,\frac{\text{W}}{\text{kW}}\right)(T_o - T_i)(\text{SEER}_{\text{W/W}})}$$
$$\text{[SI]} \quad 34.18(a)$$

$$\text{energy}_{\text{kW-hr/season}}$$
$$= \frac{\left(24\,\frac{\text{hr}}{\text{day}}\right)\dot{q}_{\text{design cooling,Btu/hr}}(\text{CDD})}{\left(1000\,\frac{\text{W}}{\text{kW}}\right)(T_o - T_i)(\text{SEER}_{\text{Btu/W-hr}})}$$
$$\text{[U.S.]} \quad 34.18(b)$$

---

[9]ASHRAE Standard 119: *Air Leakage Performance for Detached Single-Family Residential Buildings*, ASHRAE.
[10]The LBL factor is named after Lawrence Berkeley Labs, where the correlations between a structure's characteristics and leakiness were first evaluated in the 1980s.

[11]This is probably because the accuracy of other data does not warrant a high level of sophistication.
[12]Tables of cooling degree-days are far less common than tables of heating degree-days.

13. LATENT AND SENSIBLE LOADS

Latent loads increase the cooling load. The main sources of residential latent loads are infiltration, perspiration and exhalation by occupants, cooking, laundry, showering, and bathing. The often quoted rule of thumb is that residential latent load is 30% of the total load, although the actual latent load varies widely depending on infiltration rate, climate, and occupancy.

The *sensible heat ratio*, SHR (also known as the *sensible heat factor*, SHF), is the sensible load divided by the total load (including the latent load). The *latent factor*, LF, is the reciprocal of the SHR. Most air conditioning equipment is designed to operate at a sensible heat ratio in the range of 0.70–0.75. According to ASHRAE, a latent factor of 1.3 or a sensible heat ratio of 0.77 matches the performance of typical residential vapor compression cooling systems.

$$\text{SHR} = \frac{1}{\text{LF}} = \frac{\dot{q}_{\text{sensible}}}{\dot{q}_{\text{sensible}} + \dot{q}_{\text{latent}}} = \frac{c_p \Delta T}{\Delta h} \qquad 34.19$$

14. RECIRCULATING AIR BYPASS

Some of the return air may be bypassed around the air conditioner through a bypass channel in the air handling unit. Figure 34.1 illustrates such a *recirculating air bypass* configuration.[13]

Figure 34.1 *Recirculating Air Bypass*

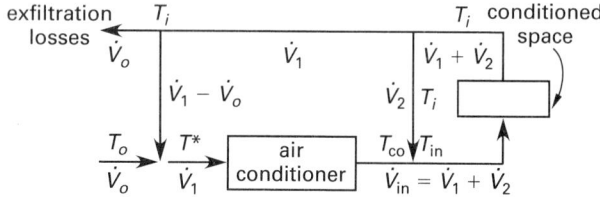

The inside design temperature, T_i, and the sensible and latent loads are generally known, as are the outside design conditions. The temperature, T_{in}, and flow rate of the air entering the conditioned space are generally not known. The following procedure can be used to determine these unknowns.[14] This procedure can also be used when there is no bypass (i.e., straight recirculation) by setting $\dot{V}_2 = 0$. The procedure is presented in

---

[13]The separate bypass duct shown in Fig. 34.1 does not actually exist. Bypassed air actually flows unchanged through a separate channel in the air conditioner.

There is a variation of this configuration in which the outdoor air is mixed with the return air before the bypass takeoff. The primary difference between the variations is when the $\dot{V}_o$ term is used.

[14]Slight modifications of the procedure may be necessary, depending on what is known.

Care must be taken in distinguishing between subscripts "i," "in," and "1."

terms of volumetric flows, making use of flow rates that are typically known at each branch. Using volumetric flow rates is sufficiently accurate for situations that do not involve large temperature differences (i.e., $T_i - T_o$ is not too large). In situations where this assumption is inappropriate, Eq. 34.20, Eq. 34.22, and others should be reformulated in terms of mass flow rates.

step 1: Locate the indoor, i, and outdoor, o, design conditions on the psychrometric chart. Read h_i, h_o, and v_o.

step 2: Draw a line between the indoor and outdoor points. This line represents all possible ratios of mixing indoor and outdoor air. The ratio of outdoor ventilation air, $\dot{V}_o$, to the fraction of conditioned recirculating air, $\dot{V}_1$, determines the actual mixture point, *. For an initial estimate, assume that the densities of the two air streams are the same. Then, the air masses are proportional to the air volumes. Calculate the temperature T^* from Eq. 34.20.[15] (See Fig. 34.2.)

$$\frac{T^* - T_i}{T_o - T_i} = \frac{\dot{V}_o}{\dot{V}_1} \qquad 34.20$$

Figure 34.2 *Adiabatic Mixing of Inside and Outside Air*

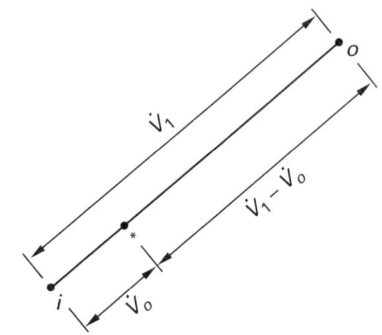

step 3: If the sensible and latent loads in the conditioned space are known, calculate the *room sensible heat ratio*, RSHR. The latent and sensible loads from outside ventilation air are not included.

$$\text{RSHR} = \frac{\dot{q}_s}{\dot{q}_s + \dot{q}_l} = \frac{c_p \Delta T}{\Delta h} \qquad 34.21$$

step 4: Draw a line with the slope RSHR (based on the psychrometric chart's sensible heat ratio scale) through point "i." Since the air conditioner must bring the air through a process that takes

---

[15]Though dry-bulb temperatures are commonly used in Eq. 34.20, they need not be. Since all of the temperature scales are linear on the psychrometric chart, wet-bulb and dew-point temperatures could be used.

nonbypassed air from condition "i" to condition "co," the "co" point must be somewhere along this line.

Alternatively, if the condition of the air leaving the air conditioner is known, draw a line from that point ("co" for conditioner output) to point "i." (See Fig. 34.3.)

Figure 34.3 *Adiabatic Mixing of Conditioned and Bypass Air*

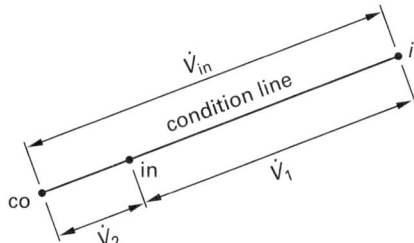

step 5: The condition of the air entering the conditioned space lies along the line drawn in step 4. The line represents all possible ratios of mixing conditioned and bypassed air. The amounts of conditioned air, $\dot{V}_1$, and bypassed air, $\dot{V}_2$, determine the mixture point "in." For an initial estimate, assume that the densities of the two air streams are the same. Calculate the temperature $T_{\rm in}$. The ratio of air flows determines the *system bypass factor*.

$$\mathrm{BF}_{\rm system} = \frac{T_{\rm in} - T_{\rm co}}{T_i - T_{\rm co}}$$

$$= \frac{\dot{V}_2}{\dot{V}_1 + \dot{V}_2} \qquad 34.22$$

step 6: If neither $T_{\rm in}$ nor $\dot{V}_{\rm in}$ is known, $T_{\rm in}$ should be chosen such that it is 15°F to 20°F (8°C to 11°C) less than T_i. The temperature of the air entering the conditioned space and the flow rate through the space are related; one determines the other. The larger the temperature difference $T_i - T_{\rm in}$ (representing the temperature rise as the air flows through the conditioned space), the lower the flow rate. However, very large temperature differences require extremely efficient mixing within the space, and very low $T_{\rm in}$ temperatures are uncomfortable for occupants near the discharge registers. Therefore, the temperature difference should not exceed 15°F to 20°F (8°C to 11°C).

step 7: If $\dot{V}_{\rm in}$ is known, calculate $T_{\rm in}$ from the sensible heating relationship, Eq. 34.23. (Also, see Eq. 34.20.)

$$\dot{V}_{\rm in,L/s} = \dot{V}_1 + \dot{V}_2$$

$$= \frac{\dot{q}_{s,\rm W}}{\left(1.20\ \dfrac{\rm J}{\rm L\cdot °C}\right)(T_i - T_{\rm in})} \qquad \text{[SI]} \quad 34.23(a)$$

$$\dot{V}_{\rm in,ft^3/min} = \dot{V}_1 + \dot{V}_2$$

$$= \frac{\dot{q}_{s,\rm Btu/hr}}{\left(1.08\ \dfrac{\rm Btu\text{-}min}{\rm hr\text{-}ft^3\text{-}°F}\right)(T_i - T_{\rm in})}$$

$$\text{[U.S.]} \quad 34.23(b)$$

step 8: Locate point "in" corresponding to $T_{\rm in}$ on the RSHR condition line from step 4.

step 9: Knowing $T_{\rm in}$ establishes the ratio of $\dot{V}_1$ and $\dot{V}_2$ in Eq. 34.22. Calculate $\dot{V}_1$ and $\dot{V}_2$.

$$\dot{V}_2 = (\mathrm{BF}_{\rm system})\dot{V}_{\rm in}$$

$$= (\mathrm{BF}_{\rm system})(\dot{V}_1 + \dot{V}_2) \qquad 34.24$$

$$\dot{V}_1 = \dot{V}_{\rm in} - \dot{V}_2 \qquad 34.25$$

step 10: Draw a line through the mixture point * and the conditioner output point "co." This line represents the process occurring in the air conditioner. Heat from outside air and from within the space are both removed by the air conditioner. Therefore, the slope of this line is the *grand sensible heat ratio*, GSHR, also known as the *coil sensible heat ratio* and *grand sensible heat factor*. If this slope is known in advance, it can be used (with the sensible heat ratio scale on the psychrometric chart) to draw a line through either * or "co," thereby establishing point "co" or *, respectively. If it is not known in advance, it can be determined from the sensible heat ratio scale.

$$\mathrm{GSHR} = \frac{\dot{q}_{s,\rm room} + \dot{q}_{s,\rm outside\ air}}{\dot{q}_{t,\rm room} + \dot{q}_{t,\rm outside\ air}} \qquad 34.26$$

The air will be cooled and dehumidified as it passes through the coil. (See Fig. 34.4.) The coil *apparatus dew point* (ADP) is determined by extending the line containing points "co" and * to the saturation line. The ADP should be greater than 32°F (0°C) so that moisture does not freeze on the coil.

step 11: The required air conditioning capacity is

$$\dot{q}_t = \dot{q}_{t,\rm room} + \dot{q}_{t,\rm outside\ air}$$

$$= \dot{q}_{s,\rm room} + \dot{q}_{l,\rm room} + \frac{(h_o - h_i)\dot{V}_o}{v_o} \qquad 34.27$$

Equation 34.28(a) expresses the air conditioning capacity in traditional HVAC units. The constant 4.5 lbm-min/ft³-hr is the product of air density (0.075 lbm/ft³) and 60 min/hr. The constant 1.2 kg/L is the product of air density (1.2 kg/m^3) and conversions 1000 W/kW and 0.001 m^3/L.

$$\dot{q}_{t,\mathrm{W}} = \dot{q}_{t,\mathrm{room}} + \dot{q}_{l,\mathrm{room}}$$
$$+ \left(1.2\ \frac{\mathrm{kg}}{\mathrm{L}}\right)(h_o - h_i)\,\dot{V}_{o,\mathrm{L/s}} \qquad \text{[SI]} \quad \textbf{\textit{34.28(a)}}$$

$$\dot{q}_{t,\mathrm{Btu/hr}} = \dot{q}_{s,\mathrm{room}} + \dot{q}_{l,\mathrm{room}}$$
$$+ \left(4.5\ \frac{\mathrm{lbm\text{-}min}}{\mathrm{ft}^3\text{-}\mathrm{hr}}\right)(h_o - h_i)\,\dot{V}_{o,\mathrm{ft}^3/\mathrm{min}}$$
$$\text{[U.S.]} \quad \textbf{\textit{34.28(b)}}$$

Figure 34.4 *Total Heat Removal Process*

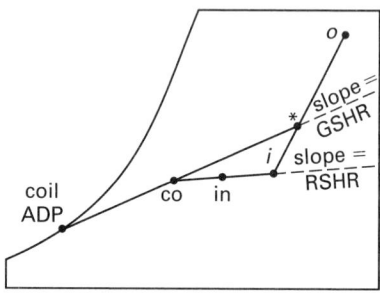

Example 34.1

The inside design condition for a conditioned space with partial recirculation is 75°F dry-bulb (23.9°C) and 62.5°F (16.9°C) wet-bulb. The outside air is at 94°F dry-bulb (34.4°C) and 78°F (25.6°C) wet-bulb. The sensible space load is 160,320 Btu/hr (46.5 kW). The latent load from occupants and infiltration, but excluding intentional ventilation, is 19,210 Btu/hr (5.6 kW). A total of 1275 ft^3/min (600 L/s) of outside air is required. The air temperature increases 19°F (10.6°C) as it passes through the conditioned space. Air leaves the conditioner saturated.

Find the (a) coil apparatus dew point, (b) volume of air passing through the space, (c) wet-bulb temperature of the air entering the space, (d) system bypass ratio, and (e) dry-bulb temperature of the air entering the conditioner.

SI Solution

(a) The room sensible heat ratio, RSHR, is

$$\mathrm{RSHR} = \frac{\dot{q}_s}{\dot{q}_s + \dot{q}_l}$$
$$= \frac{46.5\ \mathrm{kW}}{46.5\ \mathrm{kW} + 5.6\ \mathrm{kW}}$$
$$= 0.89$$

Locate the indoor, i, and outdoor, o, design conditions on the psychrometric chart. Draw a line with the slope 0.89 through point "i." Extend the line to the left to 11.9°C on the saturation line. This is T_{co}. Since the air leaves the conditioner saturated, the coil apparatus dew point coincides with the conditioner output.

(b) The dry-bulb temperature of the air as it enters the conditioned space is

$$T_{\mathrm{in}} = T_i - 10.6°\mathrm{C} = 23.9°\mathrm{C} - 10.6°\mathrm{C} = 13.3°\mathrm{C}$$

Calculate the air flow through the space.

$$\dot{V}_{\mathrm{in,L/s}} = \frac{\dot{q}_{s,\mathrm{W}}}{\left(1.20\ \dfrac{\mathrm{J}}{\mathrm{L\cdot°C}}\right)(T_i - T_{\mathrm{in}})}$$
$$= \frac{(46.5\ \mathrm{kW})\left(1000\ \dfrac{\mathrm{W}}{\mathrm{kW}}\right)}{\left(1.20\ \dfrac{\mathrm{J}}{\mathrm{L\cdot°C}}\right)(23.9°\mathrm{C} - 13.3°\mathrm{C})}$$
$$= 3656\ \mathrm{L/s}$$

(c) Locate point "in" corresponding to T_{in} on the RSHR condition line from step 4. Read the wet-bulb temperature at this point as 12.5°C.

(d) Calculate the system bypass ratio from Eq. 34.22.

$$\mathrm{BF}_{\mathrm{system}} = \frac{T_{\mathrm{in}} - T_{co}}{T_i - T_{co}} = \frac{13.3°\mathrm{C} - 11.9°\mathrm{C}}{23.9°\mathrm{C} - 11.9°\mathrm{C}}$$
$$= 0.117$$

(e) The flow rates are

$$\dot{V}_2 = (\mathrm{BF}_{\mathrm{system}})\,\dot{V}_{\mathrm{in}} = (0.117)\left(3656\ \frac{\mathrm{L}}{\mathrm{s}}\right) = 428\ \mathrm{L/s}$$
$$\dot{V}_1 = 3656\ \frac{\mathrm{L}}{\mathrm{s}} - 428\ \frac{\mathrm{L}}{\mathrm{s}} = 3228\ \mathrm{L/s}$$

Use Eq. 34.20 to locate the point corresponding to the air entering the conditioner.

$$\frac{\dot{V}_o}{\dot{V}_1} = \frac{600\ \dfrac{\mathrm{L}}{\mathrm{s}}}{3228\ \dfrac{\mathrm{L}}{\mathrm{s}}} = 0.186$$
$$\frac{T^* - T_i}{T_o - T_i} = 0.186$$
$$T^* = T_i + (0.186)(T_o - T_i)$$
$$= 23.9°\mathrm{C} + (0.186)(34.4°\mathrm{C} - 23.9°\mathrm{C})$$
$$= 25.9°\mathrm{C}$$

Customary U.S. Solution

(a) The room sensible heat ratio, RSHR, is

$$\text{RSHR} = \frac{\dot{q}_s}{\dot{q}_s + \dot{q}_l} = \frac{160{,}320 \, \frac{\text{Btu}}{\text{hr}}}{160{,}320 \, \frac{\text{Btu}}{\text{hr}} + 19{,}210 \, \frac{\text{Btu}}{\text{hr}}}$$

$$= 0.89$$

Locate the indoor, i, and outdoor, o, design conditions on the psychrometric chart. Draw a line with the slope 0.89 through point "i." Extend the line to the left to 53.4°F on the saturation line. This is T_{co}. Since the air leaves the conditioner saturated, the coil apparatus dew point coincides with the conditioner output.

(b) The dry-bulb temperature of the air as it enters the conditioned space is

$$T_{\text{in}} = T_i - 19°\text{F} = 75°\text{F} - 19°\text{F}$$

$$= 56°\text{F}$$

Calculate the air flow through the space.

$$\dot{V}_{\text{in},\text{ft}^3/\text{min}} = \frac{\dot{q}_{s,\text{Btu/hr}}}{\left(1.08 \, \frac{\text{Btu-min}}{\text{hr-ft}^3\text{-}°\text{F}}\right)(T_i - T_{\text{in}})}$$

$$= \frac{160{,}320 \, \frac{\text{Btu}}{\text{hr}}}{\left(1.08 \, \frac{\text{Btu-min}}{\text{hr-ft}^3\text{-}°\text{F}}\right)(75°\text{F} - 56°\text{F})}$$

$$= 7813 \, \text{ft}^3/\text{min}$$

(c) Locate point "in" corresponding to T_{in} on the RSHR condition line from step 4. Read the wet-bulb temperature at this point as 54.6°F.

(d) Calculate the system bypass ratio. From Eq. 34.22,

$$\text{BF}_{\text{system}} = \frac{T_{\text{in}} - T_{\text{co}}}{T_i - T_{\text{co}}} = \frac{56°\text{F} - 53.4°\text{F}}{75°\text{F} - 53.4°\text{F}}$$

$$= 0.120$$

(e) The flow rates are

$$\dot{V}_2 = (\text{BF}_{\text{system}}) \dot{V}_{\text{in}} = (0.120)\left(7813 \, \frac{\text{ft}^3}{\text{min}}\right)$$

$$= 938 \, \text{ft}^3/\text{min}$$

$$\dot{V}_1 = 7813 \, \frac{\text{ft}^3}{\text{min}} - 938 \, \frac{\text{ft}^3}{\text{min}}$$

$$= 6875 \, \text{ft}^3/\text{min}$$

Use Eq. 34.20 to locate the point corresponding to the air entering the conditioner.

$$\frac{\dot{V}_o}{\dot{V}_1} = \frac{1275 \, \frac{\text{ft}^3}{\text{min}}}{6875 \, \frac{\text{ft}^3}{\text{min}}} = 0.185$$

$$\frac{T^* - T_i}{T_o - T_i} = 0.185$$

$$T^* = T_i + (0.185)(T_o - T_i)$$

$$= 75°\text{F} + (0.185)(94°\text{F} - 75°\text{F})$$

$$= 78.5°\text{F}$$

15. FREEZE-UP OF COOLING COILS

Coil freeze-up (freezing of the coils) has two different causes. In the winter, cold make-up (ventilation) air from outside may cause water to freeze inside the heating coils. In the summer, coils with effective surface temperatures of 32°F (0°C) or less can cause moisture (humidity) in the air to freeze on the outside of the cooling (evaporator) coils.

In the summer, water (humidity) condensing on the outside of a cooling coil is normally removed by a built-in pan and drain. However, frost (rather than liquid condensate) will form when the dew point temperature is below freezing. Coil freeze-up can be prevented by designing the system so that the coil temperature is above freezing. If the extension of the line containing points "co" and * does not intersect the saturation line above 32°F (0°C) (or, if the line does not intersect the saturation line at all), heating may be required.

Having an above-freezing coil apparatus dew point may not be economical or practical, and with preexisting systems, it may not be possible at all. Freeze-up can be functionally prevented by ensuring that (a) the cooling load is high, and (b) the refrigerant flow rate is full. Low cooling load is usually attributable to low airflow or low entering temperature. Low airflow can be caused by a plugged coil, fouled air filter, low blower speed, broken fan belt, failed blower motor, closed distribution register, or undersized ductwork. Low entering air temperatures can be caused by setting the thermostat too low during the noncooling part of the daily schedule or by cold outside (make-up) air. Low refrigerant flow rates are usually attributable to refrigerant leaks and can be detected by pressure gauges, but low flow rates can also be caused by restricted liquid or suction filter-driers and faulty metering devices.

Ventilation

16. REHEATING COOLED AIR

If the sensible heat load from the occupied region decreases, or if the air leaving the air conditioner ("co") is too cold for any reason (not necessarily because of below-freezing temperatures), the air can be reheated in a *reheat coil*. Reheating can occur within the air conditioner, in an intermediate distribution box, or in an air terminal device. As the moisture content does not change, sensible reheating is represented on the psychrometric chart by a horizontal line from the point representing the conditioner output ("co") toward the right. If the air leaves the conditioner saturated, the line will start at the equipment's apparatus dew point on the psychrometric chart's saturation curve. Since the energy of heating must be removed by the coil when the air returns through it, the cooling load (refrigeration load) will be increased by the reheat, as shown in Fig. 34.5. The additional cooling load can be calculated from either the enthalpy difference or (since the specific heat and humidity ratio are known) the dry-bulb temperature difference.

Figure 34.5 Reheat

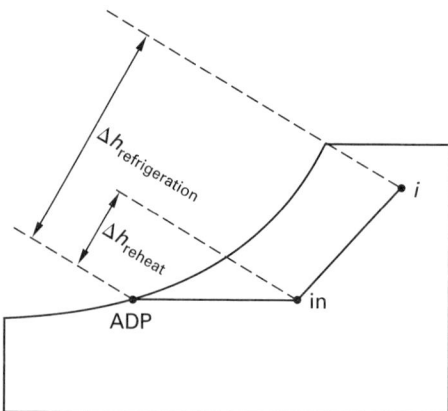

35 Air Conditioning Equipment and Controls

Nomenclature

R	range	°F	°C
T	temperature	°F	°C

1. TYPES OF AIR CONDITIONING SYSTEMS

Depending on the medium delivered to the conditioned space, air conditioning systems are categorized as all-air, air-and-water, all-water, and unitary. *All-air systems* maintain the temperature by distributing only air, and most systems rely on internal loads for heating, sending only cold air to the space. Most central units are *single-duct*, which means that the cooling and heating coils are in series. In *dual-duct* units, the heating and cooling coils are in parallel ducts.

In *air-and-water systems*, air and water are both distributed to the conditioned space. In *all-water systems*, the cooling and heating effects are provided solely by cooled and/or heated water pumped to the conditioned space. With *unitary equipment*, the fan, condenser, and cooling and heating coils are combined in a stand-alone unit for window and through-the-wall installation.

2. CONTROL EQUIPMENT

Control equipment in HVAC systems consists of sensors, actuators, motors, relays, and controllers. *Sensors (transducers)* are used to monitor temperature (*thermostats* or just "*stats*"), enthalpy, and humidity (*humidistats* or *hygrostats*). In complex delivery systems, pressure may also be monitored (i.e., by *pressurestats*). Thermostats are designated as "room," "insertion," or "immersion" according to their placement (i.e., in the occupied space, in a duct, or in a water/steam manifold, respectively).

The signal from a sensor is received by the *controller*, which energizes or deenergizes the appropriate equipment. Since most control systems operate at low voltages (e.g., 24 V AC), a *relay* must be used when the equipment operates at a higher voltage. Relays allow low-voltage thermostats and controllers to control high-voltage, high-drain devices. For example, a room thermostat might energize a relay that, in turn, would provide power to a line-voltage fan motor. Alternatively, the thermostat could activate a pneumatic relay (*electro-pneumatic switch*).

Actuators (also referred to as *operators* and *motors*) provide the force to open and close valves and dampers. The control signal can be electrical, electronic, or pneumatic. Actuators are designated as *normally open* (NO) or *normally closed* (NC) depending on their position when deenergized. Most actuators act relatively slowly. *Solenoids*, however, act quickly in response to signals.

A *pneumatic actuator* is essentially a piston/cylinder arrangement or diaphragm/bellows. With pneumatic actuators, a separate compressed-air system is required to supply the force for changing damper settings. With a typical 18 psig (124 kPa) source, the pneumatic signal will be approximately 3 psig to 15 psig (21 kPa to 100 kPa). Pneumatic actuator performance is essentially linear. The actuator position is proportional to the air pressure.

3. HVAC PROCESS CONTROL

Control of basic commercial HVAC systems has traditionally meant controlling either the amount of bypass air or the amount of reheat. These two constant volume methods are known as face-and-bypass damper control and reheat control, respectively. *Face-and-bypass damper control* is normally used to control only the dry-bulb temperature. Because of the possibility of bringing in too much moisture (an uncontrolled variable), this method is generally not used with a high percentage of outside air unless the outside air can be dehumidified. *Reheat control* is needed when both room temperature and humidity control are needed. Once the proper humidity level is achieved, reheat ensures the proper room temperature.

A third method, *air volume control*, relies on variations in flow rate through the conditioned space. Only one parameter (i.e., dry-bulb temperature) can be adequately controlled in this manner. Volume control is more applicable in the largest systems where the additional cost and complexity can be economically justified. Advances in noise control, monitoring of other comfort parameters, and ability to provide sufficient outside ventilation when volume is low may overcome the criticisms this method has received.

Ventilation

4. TYPICAL CONTROLS INTEGRATION

There are numerous variations in equipment layout, mixing sequence, and control methodology. Figure 35.1 schematically illustrates a central station *air handling unit* (AHU) in a multizone system with reheat control. Return air enters from the top; ventilation air enters from the left. Conditioned air is discharged into the distribution ductwork. The cooling effect may be either provided by the evaporator coils of a vapor-compression refrigeration cycle or from liquid chiller cooling coils. The heating effect (within the individual zone ducts) may be provided by hot water or steam coils or from the condenser section of a vapor-compression heat pump. The air flow through various components is controlled by remotely actuated dampers.

Until the fan motor starts, all the dampers are in their deenergized positions. The bypass damper is normally open. The outside air damper is normally closed. (Interlocking the outside air damper to the fan motor prevents induction of cold air by the stack effect and potential coil freeze-up whenever the fan is not running.)

The control sequence begins when the fan motor starts. The fan voltage energizes a relay and/or electric-pneumatic valve (EP), which provides air to the controllers. When the fan starts, damper motor (or damper actuator) DM1 opens the outside damper to a predetermined minimum position, permitting outside air to enter.

Damper motor DM2 is controlled by two sensors: the return air humidistat (humidity controller, HC) and the supply air temperature controller (TC), also known as a *mixing thermostat*. The duplex pressure selector (DS) selects the higher of the two pressure signals from either the HC or TC sensors and positions damper motor DM2 appropriately.[1]

The cooling coil both cools and dehumidifies the air stream. If the latent loads are low, the air is merely cooled to TC's set point temperature. If latent loads are high, more air is passed through the cooling coil to remove the moisture. Reheating is used to prevent overcooling of the space. Reheating is controlled by room thermostats (TR1, TR2, and TR3). When a TR set point is reached, the corresponding steam or hot water valve (V) is opened.

Temperature controller TC also acts as a high-limit controller, preventing the supply air control from increasing above what is required for adequate zone cooling.

When the space has low latent loads, humidification is required. Figure 35.1 does not show the humidification system and controlling humidistats. Small humidification increases can be obtained by increasing the amount

of bypass air. Steam or hot water injection is needed for large humidity changes.

Also not shown is an outside enthalpy controller. This controller compares the heat content of the return air with that of the outside air. When the refrigeration load can be reduced, the enthalpy controller overrides the temperature controller and increases the outdoor air damper opening. In smaller installations, an *outside air thermostat* can work almost as well.

5. FREEZE PROTECTION

When the outdoor air is at subfreezing temperature, water in preheat, reheat, and chilled water coils can freeze whether or not the coils are in operation. Freezing is caused by direct contact with or incomplete mixing (*stratification*) of the cold outside air with return air, although reduced warm air flows due to clogged filters can also be a contributing factor. In some systems, freeze-up can be prevented by using antifreeze or by draining the coils when they are not in use.

It is appropriate to protect the coils with thermostats (*freeze stats*). For example, the face dampers upstream of the coils can be closed down when the plenum temperature drops to approximately 35°F (2°C) or when the temperature of the incoming heated water in the heating coils (as determined by an *immersion thermostat*) drops below 120°F to 150°F (50°C to 65°C). Furthermore, when the fan is not running, the outside air dampers should be closed and minimum heat should continue to be provided to the heating coils.

6. FEEDBACK AND CONTROL

Together, sensors and their controllers constitute a traditional analog feedback loop and control system. (See Fig. 35.2.) Using temperature as the controlled variable, the *set point*, $T_{\text{set point}}$, is the temperature that the conditioned space would like to maintain. The *control point* is the actual temperature in the room. The *offset* is the difference between the set and control points. The time required for the room temperature to become established at the set point is known as the *settling time*.

$$T_{\text{offset}} = T_{\text{set point}} - T_{\text{control point}} \qquad 35.1$$

There are several basic control methods. The most simple is *two-position control*. The controlled device (e.g., a valve or damper) is either fully on or fully off. Because of thermal inertia and other delays, the temperature will continue to increase for a short time after the heat is turned off. This is known as *temperature overshoot*. Similarly, the temperature will continue to drop for a short time after the heat is turned on. The room temperature oscillates around the set point. The range of temperatures experienced by the room is the *operating differential*, while the difference between the on and off set point temperatures is the *control differential*.

---

[1]Motor DM2 controls two dampers to vary the air passing through the bypass and coil. The bypass damper is normally open; the coil damper is normally closed. In some systems, the coil damper is a *face damper*. The face damper is installed immediately before the face of the cooling coil, hence its name.

Figure 35.1 *Bypass Air Handling Unit (reheat control configuration)*

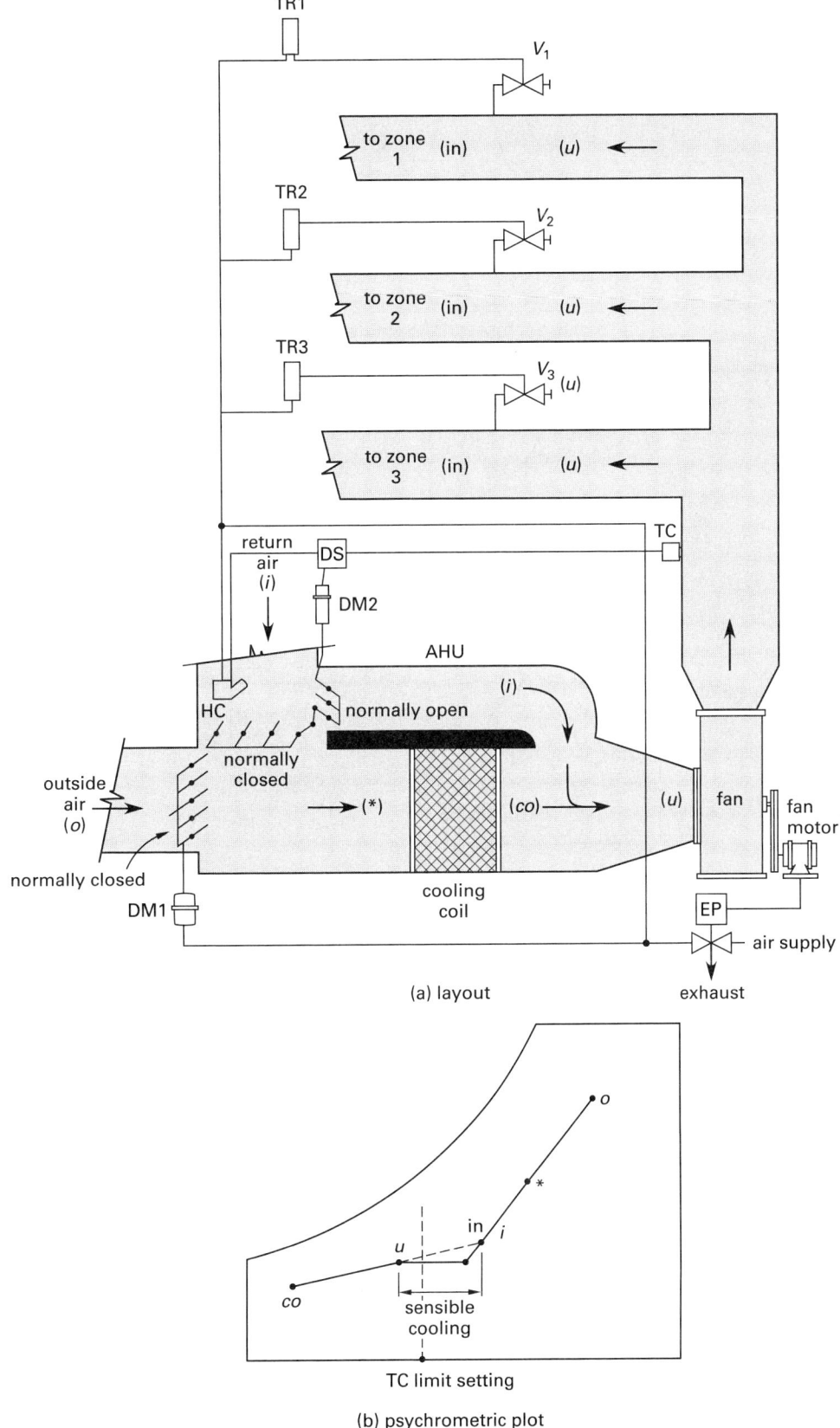

(a) layout

(b) psychrometric plot

Ventilation

Figure 35.2 *Response of Temperature Control Methods*

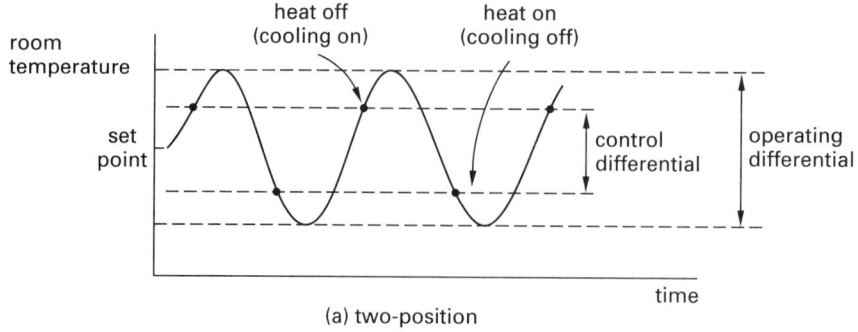

(a) two-position

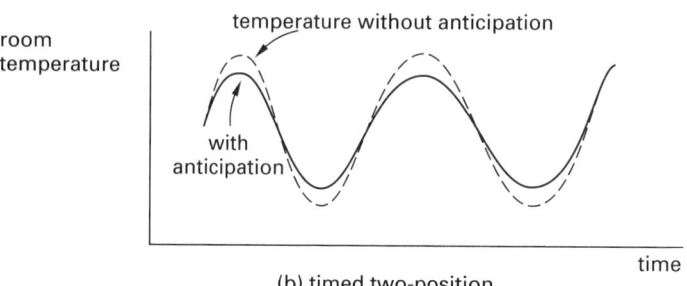

(b) timed two-position

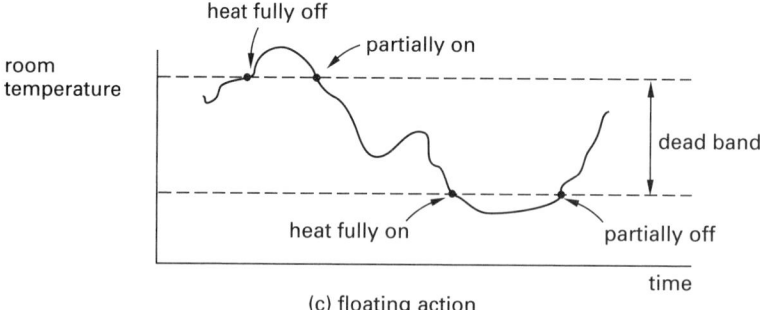

(c) floating action

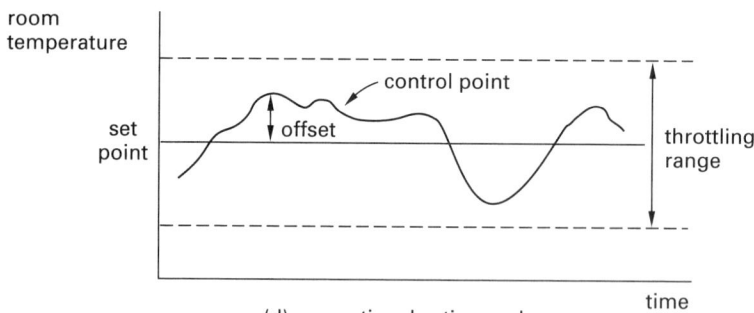

(d) proportional action and
proportion with reset

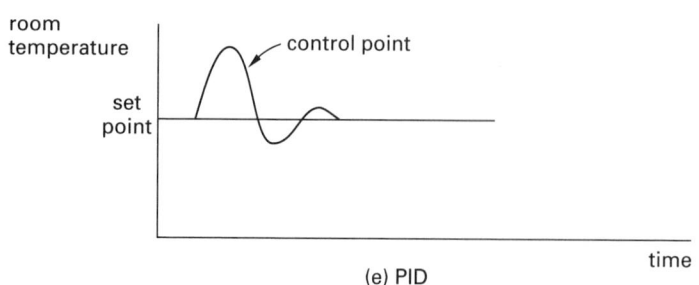

(e) PID

Ventilation

For *timed two-position control* (*anticipation control*), a small heater is built into the thermostat. While the room is being heated, the thermostat is also being heated, and this turns the thermostat off sooner than otherwise. The overshoot is reduced considerably.

With *floating action control*, the controlled device has three positions: fully on, fully off, and a fixed intermediate position. The controller has three corresponding signals: fully on, fully off, and a *neutral position* signal that is generated while the temperature is within a *dead band* range. Within the dead band, there is an intermediate amount of air heating (or cooling). If the amount of heating (cooling) corresponds to the heat loss (gain), the temperature remains within the dead band. Otherwise, the controller will generate a fully on or fully off signal.

With *proportional action*, the position (e.g., percentage opening) of the damper or valve is proportional to the offset. The *throttling range*, R_{throttle}, is the temperature range over which the damper or valve changes from fully closed to fully open. The throttling range should coincide with the normal range of temperatures encountered. (When the temperature is outside of the throttling range, the system is "out of control.") Since the damper or valve should be 50% open at the set point, within the limits of 0 to 1.00, the fraction open is

$$\text{fraction open} = 0.50 + \frac{T_{\text{offset}}}{R_{\text{throttle}}} \qquad 35.2$$

Proportional action does not provide extra heating (cooling) to compensate for changes; it tends to maintain the existing control point. With proportional action, the settling time is very long.

Proportional action with automatic reset, also known as *proportional plus integral control* (PI), attempts to return the room temperature back to the set point. In effect, the controller "overreacts" and the signal is more than proportional to the offset.

With *proportional plus integral plus derivative control* (PID), the control action responds to three different parameters: (1) the magnitude of the offset, (2) the duration of the offset, and (3) the rate at which the temperature is changing. These three aspects correspond to the terms "proportional," "integral," and "derivative," respectively, in the name. Because of the complexity of the algorithm, and since an accurate time base is needed, PID control is implemented through digital control.

7. DIGITAL CONTROL

Direct digital control (DDC) is an alternative to traditional analog control. Devices in analog and digital control systems are analogous. (See Fig. 35.3.) Digital sensors replace analog sensors one for one and are located in the same locations. Digital controllers replace analog controllers. Digital actuators replace analog (electric and pneumatic) actuators. Digital devices are connected by simple

Figure 35.3 *Basic Hook-Up Diagrams*

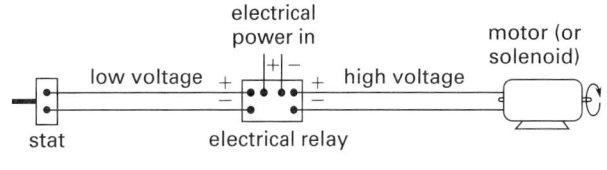

(a) analog all-electric

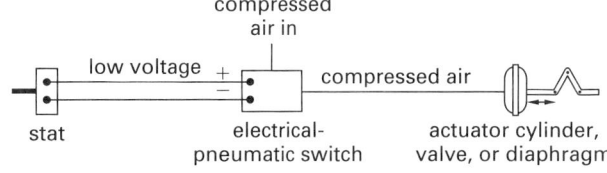

(b) analog pneumatic

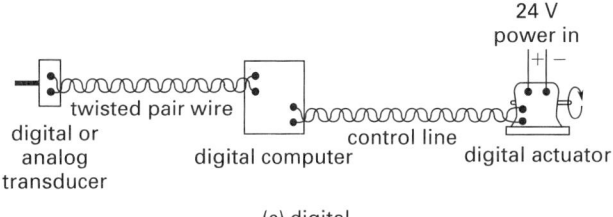

(c) digital

"twisted pair" control wiring. Each device has its own address, and the digital signal generated by the controller includes the device address.

Digital controllers are essentially *local control computers* (LCC) running algorithms preprogrammed by the manufacturer. PI and (in some cases) PID control algorithms are easily implemented. Changes to the dead band, proportional band, set points, low- and high-limits, lockouts, and so on, can be programmed for all the controlled devices after installation.

When installing a digital controller in a system already equipped with analog (pneumatic or electric) control devices, a *digital-to-analog* (*digital-to-proportional*) *staging module* is needed. The staging module is essentially an electropneumatic switch that translates digital signals into signals compatible with the pre-existing devices.

8. RELAY LOGIC DIAGRAMS

Controllers used for HVAC systems receive analog and digital inputs from temperature, humidity, and pressure sensors, as well as from other devices. Controller outputs can be analog (on/off for relays or infinitely variable for dampers, valves, and actuators) or digital to control and communicate with digital devices. Analog and digital relays can, themselves, be used to implement simple control sequences. Contacts in a relay can be normally open (NO) or normally closed (NC), and relays may have several sets of contacts in any combination of NO and NC. Since relay contacts are either

Ventilation

off or on (open or closed), *relay logic* is binary or *Boolean*. For example, two basic relays in parallel constitute a simple OR gate, while two relays in series constitute a simple AND gate. Figure 35.4 contains the symbols of devices that primarily behave in a Boolean manner or appear in HVAC circuit diagrams.

The symbols in Fig. 35.4 are combined into *relay logic diagrams*, also known as *ladder logic diagrams*, *line diagrams*, and *elementary diagrams*. Each row in the diagram is known as a *rung*. In such a diagram, the voltage sources (120 V, 24 V, etc.) are shown as a *supply rail (supply bus)*, typically labeled "L1," and a *ground*

Figure 35.4 *NEMA Boolean Control Element Symbols**

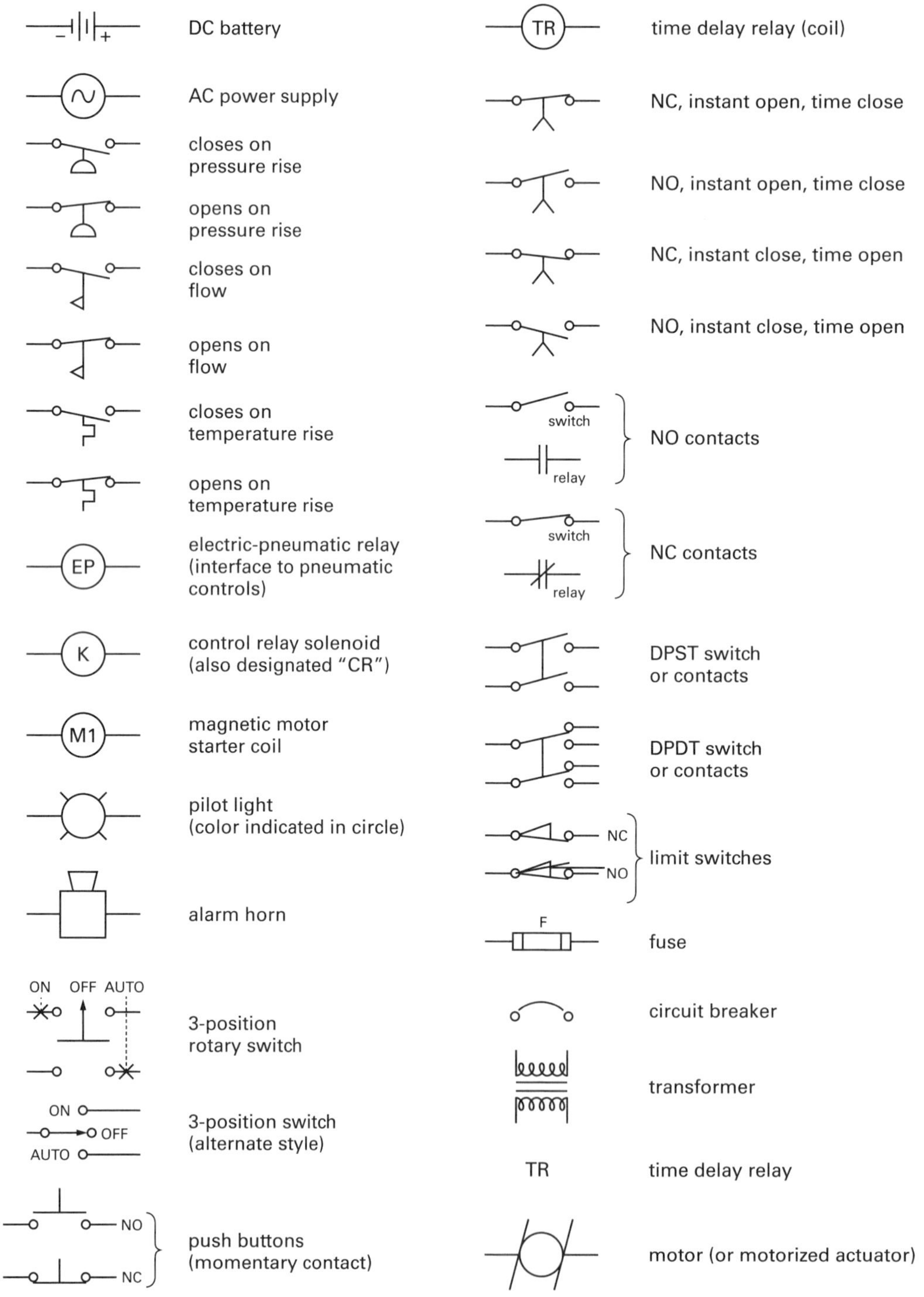

*Slight differences exist between NEMA and IEC symbols.

rail (*ground bus*) labeled "L2." Relay logic diagrams are drawn according to the following rules.

1. A rung is numbered on its left-hand side. Optional comments and descriptions of a rung's function appear on the right-hand side.

2. Control devices (inputs, such as switches) appear to the left of a rung, while controlled devices (outputs, such as motors) appear to the right of a rung. Controlled devices cannot appear between control devices.

3. Control devices can be connected in series and are shown on the same rung, while control devices connected in parallel are shown on a rung and its sub-rungs. The same control device (i.e., a switch) may appear on multiple rungs.

4. Output devices cannot be placed in series. Each output device is shown on its own rung or sub-rung. For example, a switch that controls a motor and a pilot light will be represented by either a rung with a sub-rung, or by two rungs.

5. Rungs can be electrically connected (with a *tie line* or *tie bus*) vertically, but control and controlled devices cannot be connected between rungs in the tie line.

6. When analyzing functionality, the control sequence (or, alternatively, the current) always moves from left-to-right (or, occasionally, vertically). The control sequence never moves right-to-left, not even for a single sub-rung or component.

7. A relay's coil and its contacts are shown on separate rungs. (For example, a switch may energize a relay coil that controls a fan. The input switch and output coil are shown on a single rung. The relay contacts and the output fan motor would be shown on a separate rung.) The numbers of the rungs containing a relay's contacts are listed as a comment for the rung that contains the relay's coil.

Example 35.1

Describe the function of the relay logic diagram shown.

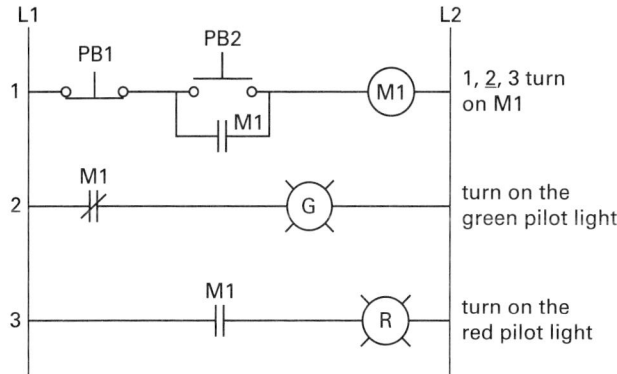

Solution

The appropriate voltage is applied across rails L1 and L2. There are three rungs, labeled 1, 2, and 3. The controlled device on rung 1 is a relay coil. The controlled device on rung 2 is a green pilot light. The controlled device on rung 3 is a red pilot light. Both pilot lights are controlled by relay M1 contacts. The physical circuit contains two momentary push-buttons, PB1 and PB2, a relay M1, and green and red pilot lights. The relay has two NO (normally open) contacts and one NC (normally closed) contact. The numbers to the right of the first rung show that coil M1 controls contacts on rungs 1, 2, and 3. PB1 is normally closed, and pushing it disconnects the rung from power; PB1 probably acts as an "off" or interrupt switch. Pushing PB2 energizes the (circled) M1 coil. The energized coil closes the M1 contacts on sub-rung 1, maintaining continuous power to relay coil (circled) M1. When relay coil M1 is energized, the NC M1 contacts on rung 2 open, extinguishing the green (safe) light. When relay coil M1 is energized, the NO contacts on rung 3 close, illuminating the red (running) light.

36

Fans, Ductwork, and Terminal Devices

Nomenclature[1]

A	area	ft^2	m^2
ACFM	actual flow rate	ft^3/min	L/s
AHP	air horsepower	hp	n.a.
AkW	air kilowatts	n.a.	kW
BHP	brake horsepower	hp	n.a.
BkW	brake kilowatts	n.a.	kW
c_p	specific heat at constant pressure	Btu/lbm-°F	kJ/kg·K
C	coefficient	–	–
C_L	leakage class	cfm/100 ft^2	n.a.
d	diameter	in	mm
D	diameter	ft	m
E	energy	ft-lbf	J
EHP	electrical power	hp	n.a.
F	leakage factor	cfm/100 ft^2	n.a.
FHP	friction horsepower	hp	n.a.
FP	friction pressure	in wg	Pa
g	acceleration of gravity, 32.2 (9.81)	ft/sec^2	m/s^2
h	head or height	ft	m
H	duct height	ft	mm
ICFM	inlet flow rate	cfm	L/s
k	Atkinson friction factor	–	–
K	factor	–	–
L	length	ft	m
L	sound pressure level	dB	dB
L	space characteristic length	ft	mm
$\dot{m}$	mass flow rate	lbm/sec	kg/s
ME	mechanical efficiency	–	–
n	fan speed	rev/min	rev/min
N	exponent	–	–
p	pressure	in wg	Pa
P	power	hp	kW
Q	volumetric flow rate	ft^3/min	L/s
r	radius	ft	m
R	aspect ratio	–	–
R	Atkinson resistance	atkinsons	gauls
R	regain coefficient	–	–
S	perimeter	ft	mm
SCFM	standard flow rate	ft^3/min	L/s
SE	static efficiency	–	–
SP	static pressure	in wg	Pa
SR	static regain	in wg	Pa
t	thickness	in	mm

[1]There is only marginal consistency in the symbols used by this industry. For example, the symbol for total pressure can be p_t, p_T, TP, P_t, T_p, h_t, and many other variations. The symbol for fitting loss coefficient is almost universally K in the industry; ASHRAE uses C.

Ventilation

T	absolute temperature	°R	K
TP	total pressure	in wg	Pa
v	velocity	ft/min	m/s
VP	velocity pressure	in wg	Pa
W	duct width	ft	mm
X_{50}	throw to 50 fpm	ft	mm
z	elevation above sea level	ft	m

Symbols

γ	specific weight	lbf/ft^3	N/m^3
η	efficiency	–	–
ρ	density	lbm/ft^3	kg/m^3
υ	specific volume	ft^3/lbm	m^3/kg
ϕ	relative humidity	%	%

Subscripts

br	branch
d	density or discharge
down	downstream
e	entry or equivalent
f	friction
k	kinetic
m	motor
s	specific
sat	saturation
std	standard
up	upstream
v	velocity

1. STANDARD AND ACTUAL FLOW RATES

Airflow through fans is typically measured in units of cubic feet per minute, ft^3/min (L/s). When the flow is at the *standard conditions* of 70°F (21°C) and 14.7 psia (101 kPa), the airflow is designated as SCFM (*standard cubic feet per minute*). As described in Sec. 36.3, the density, ρ, of standard air is 0.075 lbm/ft^3 (1.2 kg/m^3). The specific volume, υ, of standard air is the reciprocal of the density of standard air, or 13.33 lbm/ft^3 (0.8333 m^3/kg). Airflow at any other condition is designated as ACFM (*actual cubic feet per minute*). The two quantities are related by the density factor, K_d.[2] In Eq. 36.2, absolute temperature must be used.[3] Table 36.1 gives the ratio of p_{actual} to p_{std}.

$$\text{SCFM} = \frac{\text{ACFM}}{K_d} \qquad 36.1$$

$$K_d = \frac{\rho_{std}}{\rho_{actual}} = \left(\frac{p_{std}}{p_{actual}}\right)\left(\frac{T_{actual}}{T_{std}}\right) \qquad 36.2$$

---

[2]Some sources use an *air density ratio* that is the reciprocal of the density factor, K_d, defined by Eq. 36.2. In some confusing cases, the same name (i.e., density factor) is used with the reciprocal value.

[3]The temperature correction should be based on the temperature and pressure of the air through the duct system. Though atmospheric pressure and temperature both decrease with higher altitudes, air entering any occupied space will generally be heated to normal temperatures. Therefore, the temperature correction will not generally be used unless the duct system carries air for process heating or cooling.

Table 36.1 *Pressure at Altitudes*

altitude (ft (m))	ratio of p_{actual}/p_{std}
sea level (0)	1.00
1000 (305)	0.965
2000 (610)	0.930
3000 (915)	0.896
4000 (1220)	0.864
5000 (1525)	0.832
6000 (1830)	0.801
7000 (2135)	0.772

(Multiply ft by 0.3048 to obtain m.)

Standard air is implicitly dry air. A correction for water vapor can be made if the relative humidity, ϕ, is known. The saturation pressure, p_{sat}, is read from a saturated steam table for the dry bulb temperature of the air.

$$K_d = \left(\frac{p_{std}}{p_{actual} - p_{vapor}}\right)\left(\frac{T_{actual}}{T_{std}}\right)$$
$$= \left(\frac{p_{std}}{p_{actual} - \phi p_{sat}}\right)\left(\frac{T_{actual}}{T_{std}}\right) \qquad 36.3$$

Other air flow designations are ICFM (inlet cubic feet per minute), SDCFM (standard dry cubic feet per minute, the time rate of DSCF, dry standard cubic feet), MSCFD (thousand standard cubic feet per day), and MMSCFD (million standard cubic feet per day). The term ICFM is not normally used in duct design. It is used by compressor manufacturers and suppliers to specify conditions before and after filters, boosters, and other equipment. If the conditions before and after the equipment are the same, then ICFM and ACFM will be identical. Otherwise, Eq. 36.4 can be used.

$$\text{ACFM} = \text{ICFM}\left(\frac{p_{before}}{p_{after}}\right)\left(\frac{T_{after}}{T_{before}}\right) \qquad 36.4$$

Example 36.1

A manufacturing application in Denver, Colorado ($p_{actual} = 12.2$ psia, $T_{actual} = 60°F$, relative humidity of 75%) requires 100 SCFM of compressed air at 125 psig. What is the ICFM in Denver?

Solution

Use Eq. 36.3 to calculate the density factor, K_d. From App. 29.A, the saturation pressure at 60°F is 0.2564 lbf/in^2.

$$K_d = \left(\frac{p_{std}}{p_{actual} - \phi p_{sat}}\right)\left(\frac{T_{actual}}{T_{std}}\right)$$
$$= \left(\frac{14.7 \, \dfrac{\text{lbf}}{\text{in}^2}}{12.2 \, \dfrac{\text{lbf}}{\text{in}^2} - (0.75)\left(0.2564 \, \dfrac{\text{lbf}}{\text{in}^2}\right)}\right)\left(\frac{60°F + 460°}{70°F + 460°}\right)$$
$$= 1.20$$

From Eq. 36.1, the inlet flow rate is

$$\text{ICFM} = \text{ACFM} = K_d(\text{SCFM}) = (1.20)\left(100\ \frac{\text{ft}^3}{\text{min}}\right)$$

$$= 120\ \text{ft}^3/\text{min}$$

2. STATIC PRESSURE

The force of moving or stationary air perpendicular to a duct wall is known as the *static pressure*, SP. Static pressure can be measured in the field by a static tube or static tap. It is usually reported in inches of water, abbreviated "in wg" (or "in. w.g.") for "inches of water gage," or in pascals. The pressure, height of a fluid column, and specific weight of the fluid are related by Eq. 36.5. The specific weight, γ, of water is approximately 0.0361 lbf/in^3 (9810 N/m^3).

$$h = \frac{p}{\gamma} \qquad 36.5$$

$$\text{SP}_{\text{in wg}} = \frac{p_{\text{psig}}}{0.0361\ \frac{\text{lbf}}{\text{in}^3}} \qquad 36.6$$

3. VELOCITY PRESSURE

The *velocity pressure*, VP, is the kinetic energy of the air expressed in inches of water. Velocity pressure is measured with a pitot tube, hand-held velometer, hot wire or rotating vane anemometer, or calibrated orifice or nozzle. The velocity head of a mass of moving air is

$$h_v = \frac{v^2}{2g} \qquad 36.7$$

Air velocities are typically measured in ft/min (fpm) in the United States and in m/s in countries that use SI units. (The designation "LFM," for "linear feet per minute," is occasionally encountered.) As expressed, Eq. 36.7 calculates the kinetic energy as the height of an air column, not a height of a water column. The specific weights of air and water are approximately 0.075 lbf/ft^3 (density of 1.2 kg/m^3) and 62.4 lbf/ft^3 (density of 1000 kg/m^3), respectively. Therefore, the velocity pressure in inches of water is[4]

$$\text{VP}_{\text{in wg}} = \left(\frac{\left(\frac{v}{60\ \frac{\text{sec}}{\text{min}}}\right)^2\left(12\ \frac{\text{in}}{\text{ft}}\right)}{(2)\left(32.2\ \frac{\text{ft}}{\text{sec}^2}\right)}\right)\left(\frac{0.075\ \frac{\text{lbf}}{\text{ft}^3}}{62.4\ \frac{\text{lbf}}{\text{ft}^3}}\right)$$

$$\approx \left(\frac{v_{\text{ft/min}}}{4005}\right)^2 \quad \text{[standard conditions]} \qquad 36.8$$

---

[4]The constant 4005 is reported as 4004 and 4004.4 in some references. 4005 is the most common.

In SI units, with pressure in pascals, the velocity pressure is

$$\text{VP}_{\text{Pa}} = \frac{\rho v_{\text{m/s}}^2}{2} = \frac{\left(1.2\ \frac{\text{kg}}{\text{m}^3}\right)v_{\text{m/s}}^2}{2}$$

$$= 0.6v_{\text{m/s}}^2 \quad \text{[standard conditions]} \qquad 36.9$$

4. TOTAL PRESSURE

The *total pressure*, TP, is the sum of the velocity and static pressures. Contributions from potential energy are insignificant in virtually all duct design problems.

The total pressure decreases in the direction of flow. (However, the static pressure can increase with diameter increases.)

$$\text{TP} = \text{SP} + \text{VP} \qquad 36.10$$

The change in total pressure is the algebraic sum of the changes in static and velocity pressures. In straight ducts with no branches or diameter changes, the change in total pressure is the same as the friction loss. ΔTP is positive when total pressure decreases.

$$\Delta\text{TP} = \text{TP}_1 - \text{TP}_2 = \Delta\text{SP} + \Delta\text{VP}$$

$$= \text{SP}_1 - \text{SP}_2 + \text{VP}_1 - \text{VP}_2 \qquad 36.11$$

5. AIR HANDLERS

Air handling unit (*AHU* or *air handler*) is the term used to describe a unit that combines a fan with other process equipment, such as heating coils, cooling/dehumidification coils, and filters, as well as various dampers and bypass paths. An air handler that conditions outside air only, without receiving any recirculated air, is known as a *make-up air unit* (MAU). An air handler designed for exterior use is known as a *packaged unit* (PU) or *rooftop unit* (RTU). Smaller units containing an air filter, coil, and blower are known as *terminal units*, *blower-coil units*, or *fan-coil units*.

6. VARIABLE AIR VOLUME BOXES

Historically, most HVAC systems were constant air volume (CAV) systems that varied the temperature of the delivered air in order to maintain space conditions. In its simplest form, a *variable air volume* (VAV) *terminal box* is a unit that varies the amount of air entering a zone while keeping the temperature of the delivered air constant. A VAV box typically contains a motorized damper controlled by a controller that receives signals from sensors. Zone air temperature is the primary input variable affecting flow quantity, although velocity in the supply duct may also be measured. The controller may be pneumatic, single loop digital, or microprocessor. In addition to increased reliability and accuracy, microprocessor controllers accommodate daily schedules, automatic adjustment of hot and cold set points, multiple after-hours (unoccupied) setbacks, outdoor air ventilation control, and on-demand operation. The controller's *deadband* (*deadzone* or *neutral zone*) is the temperature range over which the controller does not generate any control signal. For a VAV with both heating and cooling, a deadband separates the ranges of temperatures over which heating and cooling coils are turned on.

There are many types of VAV units, including single- and multizone; variable and constant volume; with and without fan power; constant- and variable-speed fans; with and without reheat, and hybrid systems; and with and without void air induction, among others. Single-zone, VAV-with-reheat (VRH) and parallel fan-powered VAV (FPV, or *fan-coil unit* (FCU)) units are most common. A multizone (MZ) system consists of a unit with multiple output ducts, each duct serving a separate zone with its own sensors. Conventional MZ systems perform both heating and cooling functions. Selection (sizing and design) of VAV boxes is complex, as input and output duct quantity and sizes, pressure drop across the box, noise generation, heating and cooling capacity, and installation size are all related. Selection is further complicated by consideration of life-cycle costs that include energy usage.

In a true VRH installation, the central air handler's heating and cooling coils remove sensible and latent heat from all zones served, but the air supplied to the VAV box is heated only to 55°F. Any additional heating required by the zone is done by the reheat coil in the VAV terminal box. In a cooling mode, the VAV damper will be all or mostly open, supplying as much 55°F air as is required to cool the zone. Since VAV boxes vary the volume of supplied air instead of the supply air temperature, operation in cooling mode is energy efficient. In the heating mode, or in the cooling mode after the zone has been adequately cooled, electric or hydronic heating elements turn on, and the VAV damper closes substantially, supplying only enough air to meet outside air ventilation requirements. Air typically enters the VAV terminal box at 55°F (13°C) and leaves (enters the zone) after its temperature is increased approximately 20°F (11°C), to 75°F (24°C), by the reheat coil.[5] The heating coil should be sized to satisfy the conditioned zone's heating load as well as to heat the supply air from 55°F (13°C) to the room supply temperature of 75°F (24°C).

VAV controllers have minimum and maximum volumetric setpoints. Regardless of the type of VAV, the maximum is typically the airflow required to provide the design cooling. For a VRH unit, the minimum airflow is the largest of (1) outside air ventilation requirements, (2) the airflow that meets the design heating load at a reasonable temperature (e.g., below

90°F (32°C)), and (3) the airflow required to prevent dumping and poor distribution. For an FPV unit, typically only the ventilation requirement is important because the parallel fan operation ensures high supply rates. Both California Title 24 and ASHRAE Standard 90.1 limit the minimum airflow to the largest of (1) 30% of the maximum air flow, (2) the minimum required for ventilation, (3) 0.4 cfm/ft^2 (2 L/s per m^2), and (4) 300 cfm (142 L/s).

VAV boxes in the duct system's index run will affect the system pressure. Since the inlet velocity will normally be greater than the outlet velocity, there will be static regain, so the static pressure drop across a VAV box will normally be less than the total pressure drop (the sum of static and velocity pressure changes). Total pressure drop is the true measure of fan power required. Drops in total pressure should be used to evaluate and select VAV boxes, since the fan has to supply both static and velocity pressures.

7. AXIAL FANS

Axial-flow fans are essentially propellers mounted with small tip clearances in ducts. They develop static pressure by changing the airflow velocity. Axial flow fans are usually used when it is necessary to move large quantities of air (i.e., greater than 500,000 ft^3/min; 235 000 L/s) against low static pressures (i.e., less than 12 in of water, 3 kPa), although the pressures and flow rates are much lower at most installations. An axial-flow fan may be followed by a *diffuser* (i.e., an *evase*) to convert some of the kinetic energy to static pressure.

Compared with centrifugal fans, axial flow fans are more compact and less expensive. However, they run faster than centrifugals, draw more power, are less efficient, and are noisier. Axial flow fans are capable of higher velocities than centrifugal fans. In addition, overloading is less likely due to the flatter power curve. (See Fig. 36.1(a).) Fan noise is lowest at maximum efficiencies.

Axial fans can be further categorized into propeller, tubeaxial, and vaneaxial varieties. *Propeller fans* (such as the popular ceiling-mounted fans) are usually used only for exhaust and make-up air duty, where the system static pressure is not more than $^1/_2$ in wg (125 Pa). Since they generally don't have housings, they are not capable of generating static pressures in excess of about 1 in wg (250 Pa). Though they are light and inexpensive, they are the least efficient (efficiency of about 50%) and the most noisy axial fans.

Tubeaxial fans, also known as *duct fans*, generally move air against less than 3 in of water (750 Pa). They have four to eight blades, and the clearance between the blade tips and surrounding duct is small. Fan efficiency is approximately 75–80%. Tubeaxials can be recognized by their hub diameters, which are less than 50% of the tip-to-tip diameters.

---

[5]To allow for early morning warm-up, the reheat coil can be oversized slightly to allow for an 80°F (27°C) discharge. Maximum air supply temperature is approximately 90°F (32°C).

Figure 36.1 *Typical Fan Curves*

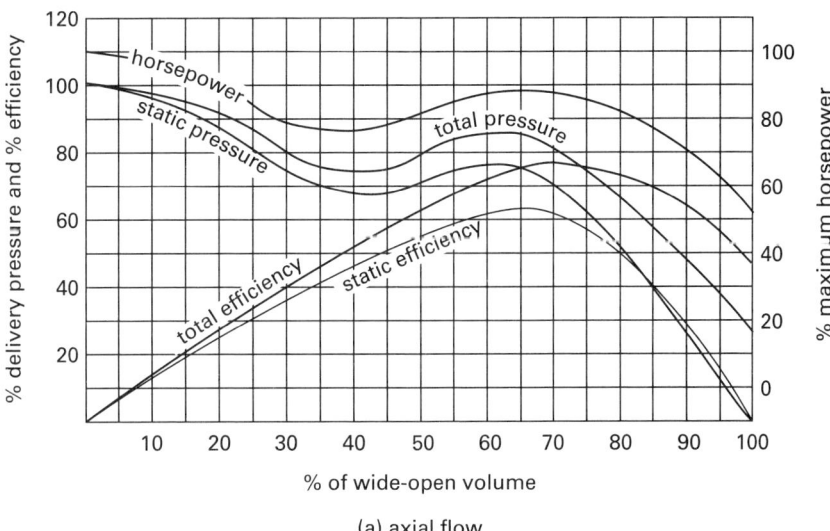

(a) axial flow

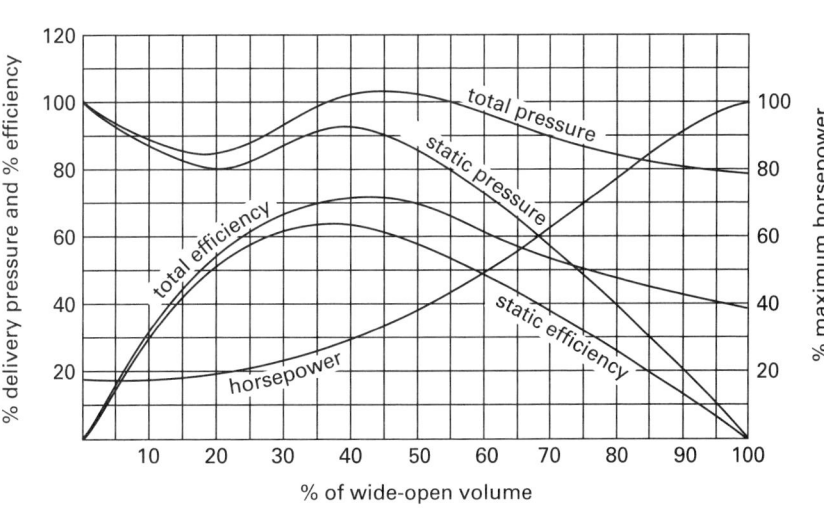

(b) forward-curved centrifugal

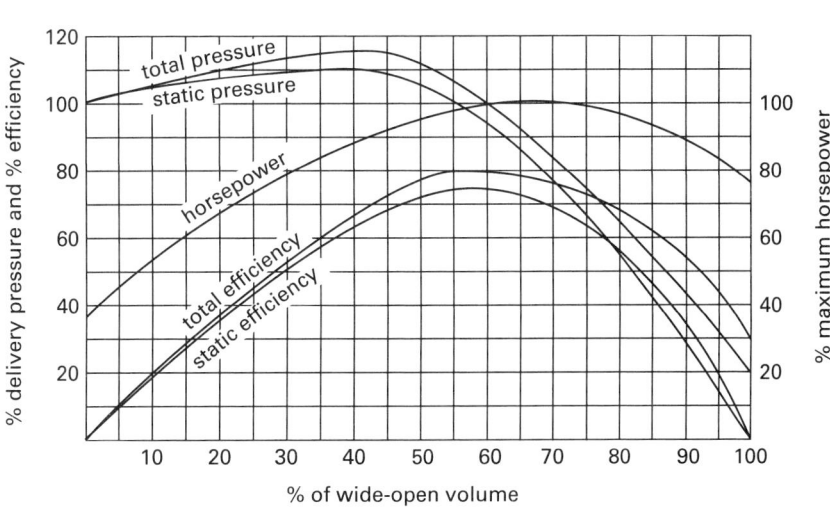

(c) backward-curved centrifugal

Vaneaxial fans can be distinguished from tubeaxial fans by their hub diameters, which are greater than 50% of the tip-to-tip diameters. Furthermore, the fan assembly will usually have vanes downstream from the fan to straighten the airflow and recover the rotational kinetic energy that would otherwise be lost. Vaneaxials typically have as many as 24 blades, and the blades may have cross sections similar to airfoils. Because they recover the rotational energy, vaneaxials are capable of moving air against pressures of up to 12 in of water (3.8 kPa). Their efficiencies are typically 85–90%.

8. CENTRIFUGAL FANS

Centrifugal fans are used in installations moving less than 1×10^6 ft³/min (470 000 L/s) and pressures less than 60 in of water (15 kPa). Like centrifugal pumps, they develop static pressure by imparting a centrifugal force on the rotating air. Depending on the blade curvature, kinetic energy can be made greater (forward-curved blades) or less (backward-curved blades) than the tangential velocity of the impeller blades.

Forward-curved centrifugals (also called *squirrel cage fans*) are the most widely used centrifugals for general ventilation and packaged units. They operate at relatively low speeds, about half that of backward-curved fans. This makes them useful in high-temperature applications where stress due to rotation is a factor. Compared with backward-curved centrifugals, forward-curved blade fans have a greater capacity (due to their higher velocities) but require larger scrolls. However, since the fan blades are "cupped," they cannot be used when the air contains particles or contaminants. Efficiencies are the lowest of all centrifugals—70–75%.

Motors driving centrifugal fans with forward-curved blades can be overloaded if the duct losses are not calculated correctly. The power drawn increases rapidly with increases in the delivery rate. The motors are usually sized with some safety factor to compensate for the possibility that the actual system pressure will be less than the design pressure. For forward-curved blades, the maximum efficiency occurs near the point of maximum static pressure. Since their tip speeds are low, these fans are quiet. The fan noise is lowest at maximum pressure.

Radial fans (also called *straight-blade fans*, *paddle wheel fans*, and *shaving wheel fans*) have blades that are neither forward- nor backward-inclined. Radial fans are the workhorses of most industrial exhaust applications and can be used in material-handling and conveying systems where large amounts of bulk material pass through them. Such fans are low-volume, high-pressure (up to 60 in wg; 15 kPa), high-noise, high-temperature, and low-efficiency (65–70%) units. *Radial tip fans* constitute a subcategory of radial fans. Their performance characteristics are between those of forward-curved and conventional radial fans.

Backward-curved centrifugals are quiet, medium- to high-volume and pressure, and high-efficiency units. They can

be used in most applications with clean air below 1000°F (540°C) and up to about 40 in wg (10 kPa). They are available in three styles: flat, curved, and airfoil. Airfoil fans have the highest efficiency (up to 90%), while the other types have efficiencies between 80% and 90%. Because of these high efficiencies, power savings easily compensate for higher installation or replacement costs.

Motor overloading is less likely with backward-curved blades than with forward-curved blades, and, therefore, may be referred to as *non-overloading fans*. These fans are normally equipped with motors sized to the peak power requirement so that the motors will not overload at any other operating condition.

Such fans operate over a great range of flows without encountering unstable air. Though their efficiencies are greater, they are noisier than forward-curved fans. The fan noise is lowest at the highest efficiencies. For the same operating speed, backward-curved blade fans develop more pressure than forward-curved fans.

9. FAN SPECIFIC SPEED

The fan specific speed is calculated from Eq. 36.12. Specific speed ranges will be 10,000 to 20,000 (110 to 220) for radial centrifugals; 12,000 to 50,000 (130 to 550) for centrifugals; 40,000 to 170,000 (440 to 1900) for vaneaxials; 100,000 to 200,000 (1100 to 2200) for tubeaxials; and 120,000 to 300,000 (1300 to 3300) for propeller fans.

$$n_s = \frac{n_{\rm rpm}\sqrt{Q_{\rm L/s}}}{{\rm SP}_{\rm Pa}^{0.75}} \qquad \text{[SI]} \quad 36.12(a)$$

$$n_s = \frac{n_{\rm rpm}\sqrt{Q_{\rm ft^3/min}}}{{\rm SP}_{\rm in\,wg}^{0.75}} \qquad \text{[U.S.]} \quad 36.12(b)$$

10. FAN POWER

The *air horsepower* (*blower horsepower*), AHP, or *air kilowatts*, AkW, is the power required to move the air.

$$\text{AkW} = \frac{Q_{\rm L/s}(\text{TP}_{\rm Pa})}{10^6} \qquad \text{[SI]} \quad 36.13(a)$$

$$\text{AHP} = \frac{Q_{\rm ft^3/min}(\text{TP}_{\rm in\,wg})}{6356} \qquad \text{[U.S.]} \quad 36.13(b)$$

The actual power delivered to a fan from its motor is the *brake horsepower*, BHP, or *brake kilowatts*, BkW. Centrifugal fan efficiencies are in the range of 50–65%, although values as high as 80% are possible. The mechanical efficiency, ME, of a fan is normally read from the *total efficiency* fan curves once the operating point is known, but can be calculated from Eq. 36.14.

$$\text{ME} = \frac{\text{AHP}}{\text{BHP}} = \frac{\text{AkW}}{\text{BkW}} \qquad 36.14$$

The electrical power delivered to the motor will include the friction, windage, and other electrical losses in the motor. The electrical power is

$$\text{EHP} = \frac{\text{BHP}}{\eta_m} = \frac{\text{AHP}}{\eta_m(\text{ME})} \qquad 36.15$$

The *static efficiency*, SE, of a fan is defined as

$$\text{SE} = (\text{ME})\left(\frac{\text{SP}}{\text{TP}}\right) \qquad 36.16$$

The primary advantage of using a variable speed fan motor is the ability to accommodate minor changes in flow rates and resistance (e.g., changes in the system friction, as when filter resistance increases over time.) Variable speed fan drives are often variable frequency drives (VFD).

Example 36.2

A fan moves 27,000 ft³/min (12 700 L/s) of air at 1800 ft/min (9.2 m/s) against a static pressure of 2 in wg (500 Pa). The electrical motor driving the fan delivers 13.12 hp (9.77 kW) to it. What are the (a) mechanical efficiency and (b) static efficiency?

SI Solution

(a) From Eq. 36.9, the velocity pressure is

$$\text{VP}_{\text{Pa}} = 0.6\text{v}_{\text{m/s}}^2 = (0.6)\left(9.2 \ \frac{\text{m}}{\text{s}}\right)^2 = 51 \ \text{Pa}$$

From Eq. 36.10, the total pressure is

$$\text{TP} = \text{SP} + \text{VP} = 500 \ \text{Pa} + 51 \ \text{Pa} = 551 \ \text{Pa}$$

From Eq. 36.13, the air kilowatts are

$$\begin{aligned} \text{AkW} &= \frac{Q_{\text{L/s}}(\text{TP}_{\text{Pa}})}{10^6} = \frac{\left(12\,700 \ \frac{\text{L}}{\text{s}}\right)(551 \ \text{Pa})}{10^6} \\ &= 7 \ \text{kW} \end{aligned}$$

From Eq. 36.14, the fan efficiency is

$$\text{ME} = \frac{\text{AHP}}{\text{BHP}} = \frac{7 \ \text{kW}}{9.77 \ \text{kW}} = 0.72 \quad (72\%)$$

(b) From Eq. 36.16, the static efficiency is

$$\begin{aligned} \text{SE} &= (\text{ME})\left(\frac{\text{SP}}{\text{TP}}\right) = (0.72)\left(\frac{500 \ \text{Pa}}{551 \ \text{Pa}}\right) \\ &= 0.65 \quad (65\%) \end{aligned}$$

Customary U.S. Solution

(a) From Eq. 36.8, the velocity pressure is

$$\text{VP}_{\text{in wg}} = \frac{\text{v}_{\text{ft/min}}^2}{4005} = \left(\frac{1800 \ \frac{\text{ft}}{\text{min}}}{4005}\right)^2 = 0.2 \ \text{in wg}$$

From Eq. 36.10, the total pressure is

$$\begin{aligned} \text{TP} &= \text{SP} + \text{VP} = 2.0 \ \text{in wg} + 0.2 \ \text{in wg} \\ &= 2.2 \ \text{in wg} \end{aligned}$$

From Eq. 36.13, the air horsepower is

$$\begin{aligned} \text{AHP} &= \frac{Q_{\text{ft}^3/\text{min}}(\text{TP}_{\text{in wg}})}{6356} \\ &= \frac{\left(27{,}000 \ \frac{\text{ft}^3}{\text{min}}\right)(2.2 \ \text{in wg})}{6356} \\ &= 9.35 \ \text{hp} \end{aligned}$$

From Eq. 36.14, the fan efficiency is

$$\text{ME} = \frac{\text{AHP}}{\text{BHP}} = \frac{9.35 \ \text{hp}}{13.12 \ \text{hp}} = 0.71 \quad (71\%)$$

(b) From Eq. 36.16, the static efficiency is

$$\begin{aligned} \text{SE} &= (\text{ME})\left(\frac{\text{SP}}{\text{TP}}\right) = (0.71)\left(\frac{2 \ \text{in wg}}{2.2 \ \text{in wg}}\right) \\ &= 0.65 \quad (65\%) \end{aligned}$$

11. VARIABLE FLOW RATES

Ventilation and air conditioning rates usually vary with time. It is essential in modern, large systems to be able to vary the flow rate, as large amounts of energy are saved. *Variable flow rates* (i.e., *capacity control, flow rate modulation*) can be achieved through use of system dampers, fan speed control, variable blade pitch, and inlet vanes.

System dampers downstream of the fan are rarely used for capacity control. They increase friction loss, are noisy, and are nonlinear in their response. Speed control of the fan through fluid or magnetic coupling has low noise levels, but a high initial cost. For that reason, it also is seldom used.

With *blade pitch control* (*controllable pitch*), all blades are linked and simultaneously controlled while the fan is operating. Changing the blade angle of attack is efficient, quiet, and linear in response.

Inlet vanes are commonly used with centrifugal and in-line fans. Inlet vanes pre-spin and throttle the air prior to its entry into the wheel. Inlet vanes are relatively inefficient, noisy, and nonlinear in response.

Ventilation

Though a direct drive with flexible coupling can be used when flow is steady, most fans are run by v-belts.[6] In some applications requiring variable volume, either *variable-pitch pulleys* or variable (multispeed) motors can be used. However, most modern designs use pulse-width modulation (PWM) (also known as *pulse duration modulation*) with variable frequency drives (VFDs) to achieve energy-saving and efficient control of supply and exhaust fans, pumps, and other HVAC system components.

12. TEMPERATURE INCREASE ACROSS THE FAN

The difference between the brake horsepower and air horsepower represents the power lost in the fan. This *friction horsepower*, FHP, heats the air passing through the fan. If the fan motor is also in the airstream, it will also contribute to the heating effect to the extent that the motor's efficiency is not 100%.[7]

$$FHP = BHP - AHP = BHP(1 - ME) \quad\quad 36.17$$

The temperature increase across the fan is given by Eq. 36.18. Consistent units must be used.

$$\Delta T = \frac{FHP}{\dot{m}c_p} \quad\quad 36.18$$

If traditional units are used and the air is essentially at standard conditions, Eq. 36.19 and Eq. 36.20 give the relationship between temperature change and the sensible heating or cooling effect.

$$\Delta T_{\circ F} = \frac{3160(\text{heating or cooling effect in kW})}{Q_{ft^3/min}}$$

$$= \frac{0.926\left(\text{heating or cooling effect in } \frac{Btu}{hr}\right)}{Q_{ft^3/min}}$$

$$= \frac{2356(\text{heating or cooling effect in hp})}{Q_{ft^3/min}}$$

$$\quad\quad 36.19$$

$$\Delta T_{\circ C} = \frac{829(\text{heating or cooling effect in kW})}{Q_{L/s}}$$

$$= \frac{1760(\text{heating or cooling effect in kW})}{Q_{ft^3/min}}$$

$$\quad\quad 36.20$$

---

[6]When selecting v-belts for fans, use a load factor of 1.4 (i.e., a *power rating* of 1.4 times the motor power).
[7]If the electrical input power and air horsepower are known, it isn't necessary to determine where the friction losses occur. The bearings, pulleys, and belts may all contribute to friction. However, the heating depends only on the difference between the input power and the power contributing to pressure and velocity.

13. FAN CURVES

The operational parameters of fans are usually presented graphically by fan manufacturers. Total pressure, power, and efficiency are typically plotted on *fan characteristic curves*. Figure 36.1 contains typical curves for the three main types of fans. The dip in total pressure for axial flow and forward-curved centrifugal fans is characteristic.

14. MULTIRATING TABLES

Some manufacturers provide *fan rating tables* similar to Table 36.2. These tables, known as *multirating tables*, give the fan curve data in tabular, rather than in graphical, format. The highest mechanical efficiency for each pressure range will be in the middle third of the flow rate (Q) range. Manufacturers often indicate (by underlining or shading) points of operation that are within 2% of the peak efficiency. If the peak efficiency point is not indicated, the actual efficiency can be calculated for each point using Eq. 36.14. Otherwise, selections can be limited to the middle third of the column.

Table 36.2 Typical Fan Rating Table (portion)

Q (ft^3/min)	SP = 1 in wg		SP = 2 in wg	
	n (rev/min)	P (BHP)	n (rev/min)	P (BHP)
5000	440	1.20	617	2.67
10,000	492	2.18	626	4.20
15,000	600	4.06	706	6.45
20,000	816	9.59	830	10.83

15. SYSTEM CURVE

The ductwork's *index run* (or, *critical path*) has the highest overall pressure drop and determines the total pressure, and therefore, the fan power required. The index run is generally the longest run in the system. Only equipment, such as VAV boxes, in the index run affects the design pressure. *Resistance pressure*, *friction pressure*, and *external static pressure* are terms that are used to designate the minimum total system pressure that the fan must provide.

As with liquid flow in pipes, the pressure loss due to friction of air flowing in ducts varies with the square of the velocity. And, since $Q = Av$, the pressure loss varies with the square of Q. The graph of the friction loss versus the flow rate is the *system characteristic curve* (*system curve*), as shown in Fig. 36.2.

If one point on the curve is known, the remainder of the curve can be found or generated from Eq. 36.21. Static head is assumed to be insignificant, and due to the low density of air, this is almost always true.

$$\frac{p_2}{p_1} = \left(\frac{Q_2}{Q_1}\right)^2 \quad\quad 36.21$$

Figure 36.2 Typical System Curve

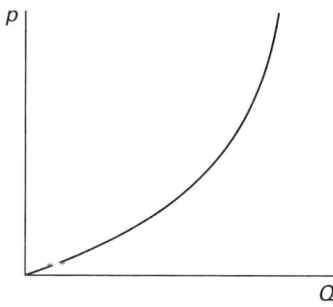

Figure 36.3 Operating Point and Unstable Region

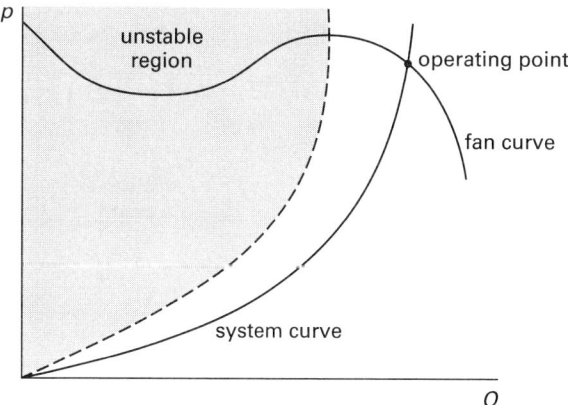

16. SYSTEM EFFECT

Almost all fans are rated under ideal laboratory conditions. Not only is the air at standard conditions, but also many of the physical features that would normally cause turbulence (bearings, diffusers, plenums, duct corners, etc.) are not present when the fan is tested. For that reason, rated performance is rarely achieved in practice.

Most fans are tested without being attached to a duct system. Merely connecting a duct system to a fan will produce a degradation in fan performance from rated values. This degradation, known as the *system effect*, is in addition to the duct friction and other losses. The *system effect factor* is the additional pressure (in inches of water) that must be added to the calculated duct friction. The system effect factor depends on the flow rate (velocity) through the fan and the type of fan. For that reason, it must be based on information provided by the fan manufacturer.

17. OPERATING POINT

The intersection of the fan and system curves defines the *operating point* (*point of operation*).[8] If a fan is to be chosen by plotting the system curve on various fan curves, the following guidelines should be observed.

- To minimize the required motor power, the operating point should be as close as possible to the peak efficiency.

- For fans whose pressure characteristics have a dip (e.g., forward-curved centrifugals and axial flow fans), the operating point should be to the right of the peak fan pressure. This will avoid the noise and uneven motor loading that accompany pressure and volume fluctuations. (See Fig. 36.3.) The unstable

region is shown in Fig. 36.3 as being to the left of the peak. Technically, the region of instability includes a horizontal band containing all system pressures for which there are two or three different airflows. Depending on transient factors, the airflow could fluctuate between the corresponding rates while satisfying the requirement that duct friction equals the supply pressure. This is known as airflow *surging*. The curves of axial fans and those with forward curved airfoils and backward curved blades have regions of instability and are more prone to surging. Surging is more difficult with steep fan curves.

- A fan with a steep pressure curve should be chosen to avoid large variations in flow rate with changes in duct friction.

System configurations (length of runs, bends, equipment, etc.) are usually less flexible than fans when it comes to changing performance. If the operating point for a specific fan does not provide the required airflow (efficiency, power, etc.), there are several different steps that can be taken.

- Use a different fan.

- Change the fan speed.

- Change the fan size.

- Use two fans in parallel. The combined flow at a particular pressure will be the sum of the individual fan flows corresponding to that pressure.

- Use two fans in series. The combined pressure at a particular flow rate will be the sum of the individual fan pressures corresponding to that flow rate.

Example 36.3

The pressure loss due to friction in a system is 1.5 in wg (375 Pa) when the flow rate is 3500 ft^3/min (1650 L/s). Velocity head and outlet pressure are negligible. What will be the flow rate if a fan with the characteristics shown is used?

---

[8]The *rating point* (*point of rating*) is the one single point on the fan curve that corresponds to the stated (often the rated) performance. The *duty point* (*point of duty*) is one single point on the system curve where a fan is to operate.

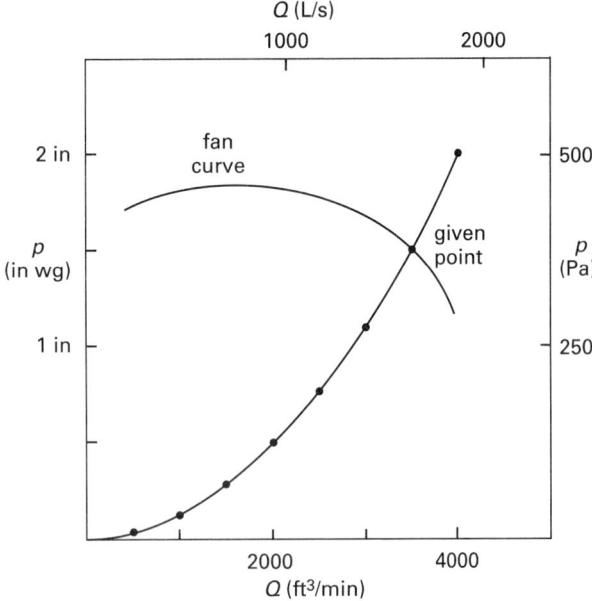

SI Solution

The fan curve is given. One point on the system curve is known. Use Eq. 36.21 to derive the remainder of the system curve. For 1500 L/s, the pressure drop would be

$$p_2 = p_1 \left(\frac{Q_2}{Q_1}\right)^2 = (375 \text{ Pa}) \left(\frac{1500 \frac{\text{L}}{\text{s}}}{1650 \frac{\text{L}}{\text{s}}}\right)^2 = 310 \text{ Pa}$$

The remainder of the system curve can be determined in the same manner. The fan and system curves intersect at approximately 1550 L/s.

Q (L/s)	p (Pa)
500	35
750	80
1000	140
1250	215
1500	310
1650	375
1750	420

Customary U.S. Solution

The fan curve is given. One point on the system curve is known. Use Eq. 36.21 to derive the remainder of the system curve. For 3000 ft³/min, the pressure drop would be

$$p_2 = p_1 \left(\frac{Q_2}{Q_1}\right)^2 = (1.5 \text{ in wg}) \left(\frac{3000 \frac{\text{ft}^3}{\text{min}}}{3500 \frac{\text{ft}^3}{\text{min}}}\right)^2$$

$$= 1.1 \text{ in wg}$$

The remainder of the system curve can be determined in the same manner. The fan and system curves intersect at approximately 3300 ft³/min.

Q (ft³/min)	p (in wg)
500	0.03
1000	0.12
1500	0.28
2000	0.49
2500	0.77
3000	1.1
3500	1.5
4000	2.0

18. AFFINITY LAWS

Within reasonable limits, the speed of v-belt driven fans can be easily changed by changing pulleys. The following *affinity laws* (*fan laws*) can be used to predict performance of a particular fan at different speeds. These fan laws assume the fan size, fan efficiency, and air density are the same.[9,10]

$$\frac{Q_2}{Q_1} = \frac{n_2}{n_1} \qquad\qquad 36.22$$

$$\frac{p_2}{p_1} = \left(\frac{n_2}{n_1}\right)^2 \begin{bmatrix} \text{static, velocity, or} \\ \text{total pressure} \end{bmatrix} \qquad 36.23$$

$$\frac{\text{AHP}_2}{\text{AHP}_1} = \left(\frac{n_2}{n_1}\right)^3 \qquad\qquad 36.24$$

Since the efficiency at the two speeds is assumed to be the same, Eq. 36.24 can be rewritten in terms of power drawn.

$$\frac{\text{BHP}_2}{\text{BHP}_1} = \left(\frac{n_2}{n_1}\right)^3 \qquad\qquad 36.25$$

A change in the fan speed cannot be used to put the operating point into a stable region. Changing the fan speed does not change the relative position of the operating point with respect to the dip present in some fan curves. The locus of peak points follows the Q^2 rule also. So, if an operating point is to the left of the peak point (i.e., is in an unstable region) at one fan speed, the new operating point will be to the left of the peak point on the new fan curve corresponding to the new speed.

19. FAN SIMILARITY

The performance of one fan can be used to predict the performance of a *dynamically similar* (*homologous*) fan.

---

[9]These fan laws are simplifications of the similarity laws presented in the next section. The similarity laws must be used if the density changes.

[10]For any given efficiency, the locus of equal-efficiency points on the pressure-capacity (*p-Q*) diagram is a parabola starting at the origin and crossing the different fan curves corresponding to different speeds. The intersection points of the fan curves and the parabolic equal-efficiency curve are known as *corresponding points*. Theoretically, the fan laws can only be used at these points.

This can be done by using Eq. 36.26 through Eq. 36.30. Pressures may be static, velocity, or total.

$$\frac{Q_A}{Q_B} = \left(\frac{D_A}{D_B}\right)^3 \left(\frac{n_A}{n_B}\right)$$
$$= \left(\frac{D_A}{D_B}\right)^2 \sqrt{\frac{p_A}{p_B}}\sqrt{\frac{\gamma_B}{\gamma_A}} \qquad 36.26$$

$$\frac{p_A}{p_B} = \left(\frac{D_A}{D_B}\right)^2 \left(\frac{n_A}{n_B}\right)^2 \left(\frac{\gamma_A}{\gamma_B}\right) \qquad 36.27$$

$$\frac{AHP_A}{AHP_B} = \left(\frac{D_A}{D_B}\right)^5 \left(\frac{n_A}{n_B}\right)^3 \left(\frac{\gamma_A}{\gamma_B}\right)$$
$$= \left(\frac{D_A}{D_B}\right)^2 \left(\frac{p_A}{p_B}\right)^{3/2} \sqrt{\frac{\gamma_B}{\gamma_A}}$$
$$= \left(\frac{Q_A}{Q_B}\right)^2 \left(\frac{p_A}{p_B}\right) \qquad 36.28$$

$$\frac{n_A}{n_B} = \left(\frac{D_B}{D_A}\right)\sqrt{\frac{p_A}{p_B}}\sqrt{\frac{\gamma_B}{\gamma_A}}$$
$$= \sqrt{\frac{Q_B}{Q_A}}\left(\frac{p_A}{p_B}\right)^{3/4}\left(\frac{\gamma_B}{\gamma_A}\right)^{3/4} \qquad 36.29$$

$$\frac{D_A}{D_B} = \sqrt{\frac{Q_A}{Q_B}}\left(\frac{p_B}{p_A}\right)^{1/4}\left(\frac{\gamma_A}{\gamma_B}\right)^{1/4} \qquad 36.30$$

Similarity laws may be used to predict the performance of a larger fan from a smaller fan's performance, since the efficiency of the larger fan can be expected to exceed that of the smaller fan. Larger fans should not be used to predict the performance of smaller fans. Even extrapolations to larger fans should be viewed cautiously when there is a significant decrease in air density or when the ratio of the larger-to-smaller fan diameters, the ratio of speed, or the product of the diameter and speed ratios exceed 3.0.

Example 36.4

A fan turning at 800 rev/min develops 6.2 hp (4.6 kW) against a static pressure of 2.25 in wg (560 Pa). (a) If the fan is driven at 1400 rev/min, what will be the power developed? (b) If duct length is increased such that the static pressure loss is 2.85 in wg (713 Pa) and the fan efficiency remains the same, what speed will be required?

SI Solution

(a) Use Eq. 36.24.

$$AkW_2 = AkW_1\left(\frac{n_2}{n_1}\right)^3 = (4.6\text{ kW})\left(\frac{1400\frac{rev}{min}}{800\frac{rev}{min}}\right)^3$$
$$= 24.7\text{ kW}$$

(It is unlikely that the same motor will be able to provide this increased power.)

(b) Use Eq. 36.23.

$$n_2 = n_1\sqrt{\frac{p_2}{p_1}} = \left(800\ \frac{rev}{min}\right)\sqrt{\frac{713\text{ Pa}}{560\text{ Pa}}}$$
$$= 903\text{ rev/min}$$

Customary U.S. Solution

(a) Use Eq. 36.24.

$$AHP_2 = AHP_1\left(\frac{n_2}{n_1}\right)^3 = (6.2\text{ hp})\left(\frac{1400\frac{rev}{min}}{800\frac{rev}{min}}\right)^3$$
$$= 33.2\text{ hp}$$

(It is unlikely that the same motor will be able to provide this increased power.)

(b) Use Eq. 36.23.

$$n_2 = n_1\sqrt{\frac{p_2}{p_1}} = \left(800\ \frac{rev}{min}\right)\sqrt{\frac{2.85\text{ in wg}}{2.25\text{ in wg}}}$$
$$= 900\text{ rev/min}$$

20. OPERATION AT NONSTANDARD CONDITIONS

Fan tests used to develop curves and rating tables are based on air at standard conditions—70°F and 14.7 psia (21°C and 101 kPa). Small variations in density due to normal temperature and humidity fluctuations can be disregarded. However, if the system operates at extremely elevated temperatures or reduced atmospheric pressures, corrections will be necessary.

Fans are constant-volume devices. They deliver the same volume of air (and at the same duct speed) regardless of temperature, pressure, and humidity ratio. Therefore, the actual flow rate, ACFM, should be used to select a fan from rating tables. The speed can be read directly from the fan table. The standard (i.e., table) values of power and pressure (static, velocity, and total) should be modified by the *density factor*, K_d (see Eq. 36.2), with the density.[11]

$$K_d = \frac{SP_{table}}{SP_{actual}} = \frac{VP_{table}}{VP_{actual}} = \frac{TP_{table}}{TP_{actual}}$$
$$= \frac{FP_{table}}{FP_{actual}} = \frac{BHP_{table}}{BHP_{actual}} \qquad 36.31$$

[11]There is also a correction for viscosity. However, the viscosity change is so insignificant that it is disregarded.

Example 36.5

A fan is chosen to move 18,000 SCFM (8500 L/s) of air against a static pressure of 0.85 in wg (210 Pa). The fan draws 4.2 hp (3.1 kW) when moving standard air in that configuration. The fan is used in a nonstandard environment to provide 150°F (66°C) air for drying. What will be the (a) required power, and (b) friction loss?

SI Solution

(a) Equation 36.2 gives the density factor. The pressure is unchanged.

$$K_d = \frac{T_{actual}}{T_{std}} = \frac{66°C + 273°}{21°C + 273°} = 1.15$$

From Eq. 36.31,

$$AkW_{actual} = \frac{AkW_{std}}{K_d} = \frac{3.1 \text{ kW}}{1.15} = 2.7 \text{ kW}$$

(b) The friction loss is

$$p_{f,actual} = \frac{p_{f,std}}{K_d} = \frac{210 \text{ Pa}}{1.15} = 183 \text{ Pa}$$

Customary U.S. Solution

(a) Equation 36.2 gives the density factor. The pressure is unchanged.

$$K_d = \frac{T_{actual}}{T_{std}} = \frac{150°F + 460°}{70°F + 460°} = 1.15$$

From Eq. 36.31,

$$BHP_{actual} = \frac{BHP_{std}}{K_d} = \frac{4.2 \text{ hp}}{1.15} = 3.7 \text{ hp}$$

(b) The friction loss is

$$p_{f,actual} = \frac{p_{f,std}}{K_d} = \frac{0.85 \text{ in wg}}{1.15} = 0.74 \text{ in wg}$$

Example 36.6

70°F (21°C) air in a duct located at an altitude of 5000 ft (1525 m) moves at 1500 ft/min (7.7 m/s). The actual flow rate is 39,000 ft³/min (18 300 L/s). The duct resistance at that altitude is 1.5 in wg (375 Pa). (a) What duct resistance should be used with fan rating tables? (b) At standard conditions, what input power to the fan is required if the fan efficiency is 75%?

SI Solution

(a) From Table 36.1, the atmospheric pressure ratio at 1525 m altitude is 0.832. The pressure is

$$p_{actual} = (0.832)(101.3 \text{ kPa}) = 84.3 \text{ kPa}$$

From Eq. 36.2, the density factor is

$$K_d = \frac{p_{std}}{p_{actual}} = \frac{101.3 \text{ kPa}}{84.3 \text{ kPa}} = 1.2$$

From Eq. 36.31, the standard friction pressure loss is

$$FP_{table} = K_d(FP_{actual}) = (1.2)(375 \text{ Pa}) = 450 \text{ Pa}$$

(b) Since the volumetric flow rate does not change, the duct speed is also unchanged. The fan supplies both velocity and static pressure. From Eq. 36.9, the velocity pressure is

$$VP = (0.832)\left(\frac{1.2 \frac{\text{kg}}{\text{m}^3}}{2}\right) v_{m/s}^2$$

$$= (0.832)\left(\frac{1.2 \frac{\text{kg}}{\text{m}^3}}{2}\right)\left(7.7 \frac{\text{m}}{\text{s}}\right)^2$$

$$= 30 \text{ Pa}$$

From Eq. 36.10, the total pressure energy supplied by the fan is

$$TP = SP + VP = 375 \text{ Pa} + 30 \text{ Pa} = 405 \text{ Pa}$$

From Eq. 36.13, the original power drawn is

$$AkW = \frac{Q_{L/s}(TP_{Pa})}{10^6} = \frac{\left(18\,300 \frac{\text{L}}{\text{s}}\right)(405 \text{ Pa})}{10^6}$$

$$= 7.41 \text{ kW}$$

From Eq. 36.14 and Eq. 36.31, the standardized brake kilowatts are

$$BkW_{table} = \frac{K_d(AkW_{actual})}{ME} = \frac{(1.2)(7.41 \text{ kW})}{0.75}$$

$$= 11.9 \text{ kW}$$

Customary U.S. Solution

(a) From Table 36.1, the atmospheric pressure ratio at 5000 ft altitude is 0.832. The pressure is

$$p_{actual} = (0.832)(14.7 \text{ psia}) = 12.2 \text{ psia}$$

From Eq. 36.2, the density factor is

$$K_d = \frac{p_{std}}{p_{actual}} = \frac{14.7 \text{ psia}}{12.2 \text{ psia}} = 1.2$$

From Eq. 36.31, the standard friction pressure loss is

$$FP_{table} = K_d(FP_{actual}) = (1.2)(1.5 \text{ in wg})$$

$$= 1.8 \text{ in wg}$$

(b) Since the volumetric flow rate does not change, the duct speed is also unchanged. The fan supplies both velocity and static pressure. From Eq. 36.8, the velocity pressure is

$$VP = (0.832)\left(\frac{v}{4005}\right)^2 = (0.832)\left(\frac{1500\ \frac{ft}{min}}{4005}\right)^2$$

$$= 0.12\ \text{in wg}$$

From Eq. 36.10, the total pressure energy supplied by the fan is

$$TP = SP + VP = 1.5\ \text{in wg} + 0.12\ \text{in wg}$$

$$= 1.62\ \text{in wg}$$

From Eq. 36.13, the original power drawn is

$$AHP = \frac{Q(TP)}{6356} = \frac{\left(39{,}000\ \frac{ft^3}{min}\right)(1.62\ \text{in wg})}{6356}$$

$$= 9.94\ \text{hp}$$

From Eq. 36.14 and Eq. 36.31, the standardized brake horsepower is

$$BHP_{table} = \frac{K_d(AHP_{actual})}{ME} = \frac{(1.2)(9.94\ \text{hp})}{0.75}$$

$$= 15.9\ \text{hp}$$

21. DUCT SYSTEMS

There are two primary types of air ducting designs: trunk and radial. The most common duct design is the *trunk system*, also known as an *extended plenum system*. A large main supply duct (trunk duct) connects to and extends the air handler plenum. Smaller branch ducts, known as *runout ducts*, deliver air from the trunk to the individual outlets. Particularly in residential applications, the trunk is usually rectangular, while the branch ducts are usually round. In a *reducing trunk system*, the trunk is proportionately reduced after each branch take-off. Because of additional design and construction costs, reducing trunk systems are generally used only in commercial and high-end residential applications.

Radial duct systems are used less often than trunk systems and are typically used where the air handling equipment may be centrally located and where it is not necessary to conceal ductwork. All of the radial branch ducts connect directly to the equipment plenum. *Gravity duct systems* are essentially radial systems that circulate heated air through ductwork by natural convection, without fan assistance. They are suitable only for small residences, and they are typically associated with coal- and wood-burning furnaces.

In *dual duct systems*, heated and cooled air flowing from hot and cold "decks," respectively, are both available to the zone. The mixture of the two flows is determined by sensors in the conditioned zone, and the two air flows are combined in a *mixing plenum* through use of a *mixing damper*. The entire apparatus is usually combined into a single *dual duct VAV box*. Numerous variations exist. The system can be constant- or variable-volume; a single fan can be used, or each deck can have its own fan; the mixing plenum can have its own heating and cooling capabilities.

22. SHEET METAL DUCT

Most commercial and residential ductwork, whether rectangular or round, is manufactured from plain and galvanized sheet steel, stainless steel, and aluminum with folded seams. Ducts may subsequently be painted or powder coated or wrapped with fiberglass insulation. Table 36.3 lists the sheet metal gauges used in ducts.

Table 36.3 Thickness of Sheet Metal Used in Ducts (in (mm))

gauge	plain mild steel	galvanized steel	stainless steel	aluminum
16	0.0598	0.0635	0.0625	0.0159
	(1.52)	(1.62)	(1.59)	(0.404)
18	0.0478	0.0516	0.0500	0.0201
	(1.21)	(1.31)	(1.27)	(0.511)
20	0.0359	0.0396	0.0375	0.0253
	(0.91)	(1.01)	(0.953)	(0.643)
22	0.0299	0.0336	0.0312	0.0319
	(0.76)	(0.85)	(0.792)	(0.810)
24	0.0239	0.0276	0.0250	0.0403
	(0.61)	(0.70)	(0.635)	(1.02)
26	0.0179	0.0217	0.0187	0.0508
	(0.45)	(0.55)	(0.475)	(1.29)

(Multiply in by 25.4 to obtain mm.)

23. SPIRAL DUCT

Spiral duct (with a spiral seam, as distinguished from traditional round duct with a longitudinal seam) is preferred for high velocity systems, when visual aesthetics are important, and where space is available. It can be manufactured from any sheet metal. Double-wall and oval varieties are available. Spiral duct can be painted, powder coated, and/or insulated just like rectangular ducting. *Polyvinyl coated spiral duct* (PCD) is available for underground ducts and fume exhaust systems. PCD combines the strength of steel and the chemical inertness of plastic; it is lightweight, weather resistant, and corrosion resistant. Although seams for spiral duct can be welded, a folded lock seam is adequate for normal ductwork. The seam is external to the duct, presenting a smooth surface to the airflow. Standard spiral duct is uncorrugated; corrugations add mechanical strength and are used with underground ducts. Purchase and installation costs for spiral duct are comparable or lower than for traditional rectangular sheet metal ducts. Flow resistance (friction) is generally lower. Spiral duct generally has less leakage, reduced noise, greater mechanical strength, greater bursting (seam failure) resistance to positive pressures, and greater collapse resistance to negative pressures.

Ventilation

There is no significant difference between spiral and traditional ducts in duct layout and design methodology, although accurate determination of system friction may require use of manufacturer's charts and tables. Traditional duct has a specific roughness of 0.0005 ft (0.15 mm). Spiral duct is usually put into the ASHRAE category of "medium smooth" and has a specific roughness of 0.0003 ft (0.09 mm). Corrugated spiral duct is categorized as "medium rough" and has a specific roughness of 0.0024 ft (0.74 mm). Corrugations increase the duct resistance by 10–30%. Spiral ducts have less friction pressure loss, FP, than traditional ducts, so standard friction charts predict pressure losses for spiral ducts that are slightly (i.e., 5–10%) higher than actual. The error is conservative and well within the acceptable range considering all other inaccuracies and assumptions.

Burst and collapse pressures of ductwork depend somewhat on manufacturing methods, and so, are based on a combination of theoretical methods and testing. Pressures are typically correlated to the wall thickness-diameter ratio, t/D, as illustrated in Fig. 36.4.

Figure 36.4 *Typical Collapse Pressure of Steel Spiral Ducts*

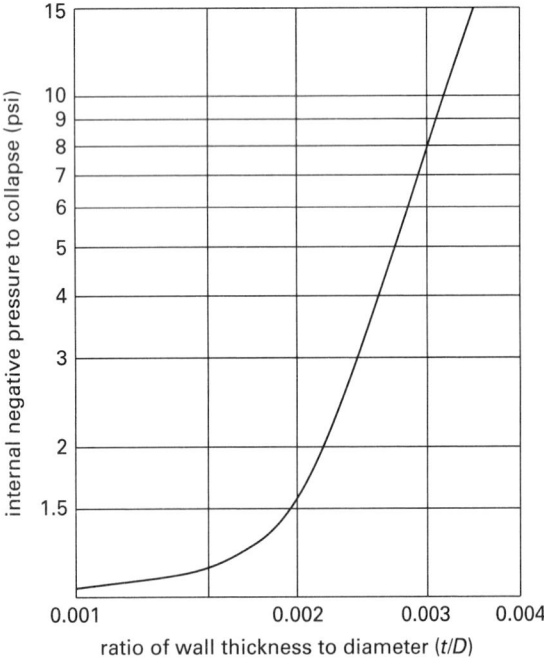

From *A Complete Line of High Pressure Ductwork*, Vol. 5, copyright © 2006, by Spiral Manufacturing Co., Inc. Reproduced with permission.

24. FIBERGLASS DUCT

Fiberglass (also known as *fiber-reinforced plastic, fiber-reinforced polymer* (FRP) and *fibrous*) *duct*, of rigid and flexible varieties, is generally more expensive than standard metal, but may be easier to install without sheet metal training. Rectangular ducts can be manufactured using FRP *ductboard*. FRP duct has comparable friction

but is lighter and stronger than light gauge steel. It generally has better acoustical qualities and is corrosion resistant. FRP duct may be the best material in corrosive environments and where the duct must resist positive and negative pressure extremes. Since FRP duct cannot be grounded, it can accumulate static electricity and become an ignition source in dusty environments. Therefore, FRP duct should not be used in dust collection systems.

25. FLEXIBLE DUCT

Flexible duct (*flex*) is typically manufactured by wrapping a plastic sheet over a metal wire coil. Flex is commonly used for connecting supply/return grills to trunk and branch lines. Due to the significantly increased friction (e.g., three times as much as smooth metal ductwork), runs of flex line are kept as short as possible, generally less than 15 ft (5 m). Since flex does not tolerate large negative pressures, its use in return air systems is not favored.[12]

26. FRICTION LOSSES IN ROUND DUCTS

Friction loss (*friction pressure*, FP) can be calculated from the standard Moody equation. Equation 36.32 expresses the Moody equation in typical air-moving units for standard air (0.075 lbm/ft³) through average, clean, round galvanized duct (specific roughness of 0.0005 ft) having a typical number of connections, joints, and slip couplings.[13,14]

$$FP_{in\,wg,100\,ft} = \frac{2.74}{D_{in}^{1.22}}\left(\frac{v_{fpm}}{1000}\right)^{1.9} = \frac{0.109Q_{cfm}^{1.9}}{D_{in}^{5.02}} \qquad 36.32$$

However, Eq. 36.32 is almost never used in the HVAC industry. Rather, friction losses are typically determined from graphs. Figure 36.5 and Fig. 36.6 assume clean, commercial-quality round ducts with a normal number of joints and conditions close to standard air. For smooth ducts with no joints, the friction loss is 60–95% of the value determined from Fig. 36.5 and Fig. 36.6.[15]

Duct flow areas are calculated from their nominal dimensions. Any size duct can be manufactured. However, there are standard sizes of premanufactured round duct, and these sizes should be chosen to minimize cost.

[12]UL181 testing specification requires functionality at a negative pressure of 0.03 lbf/in² (200 Pa), so flexible duct can tolerate moderate negative pressures.
[13]Specific roughness is the reciprocal of the number of duct diameters required to cause a static pressure loss of one velocity pressure. Typical values for galvanized duct are 0.0003 ft to 0.0005 ft (0.09 mm to 0.15 mm).
[14]Some authorities, such as Carrier, report the 1.9 exponent as 1.82.
[15]For corrugated ductwork, the friction loss is approximately twice that shown in Fig. 36.5 and Fig. 36.6. However, the analysis is not precise enough to make most corrections, including those for a different number of joints or duct material. Corrections should be limited to extreme cases when the ductwork deviates significantly from commercial (e.g., airflow through brickwork or corrugated duct).

Figure 36.5 *Standard Friction Loss in Standard Duct* (inches of water per 100 ft of duct; ±5% for temperatures of 40° F to 100° F, elevations to 1500 ft, and duct pressures of −20 in wg to +20 in wg) (Recommended operating points shown as shaded region.)*

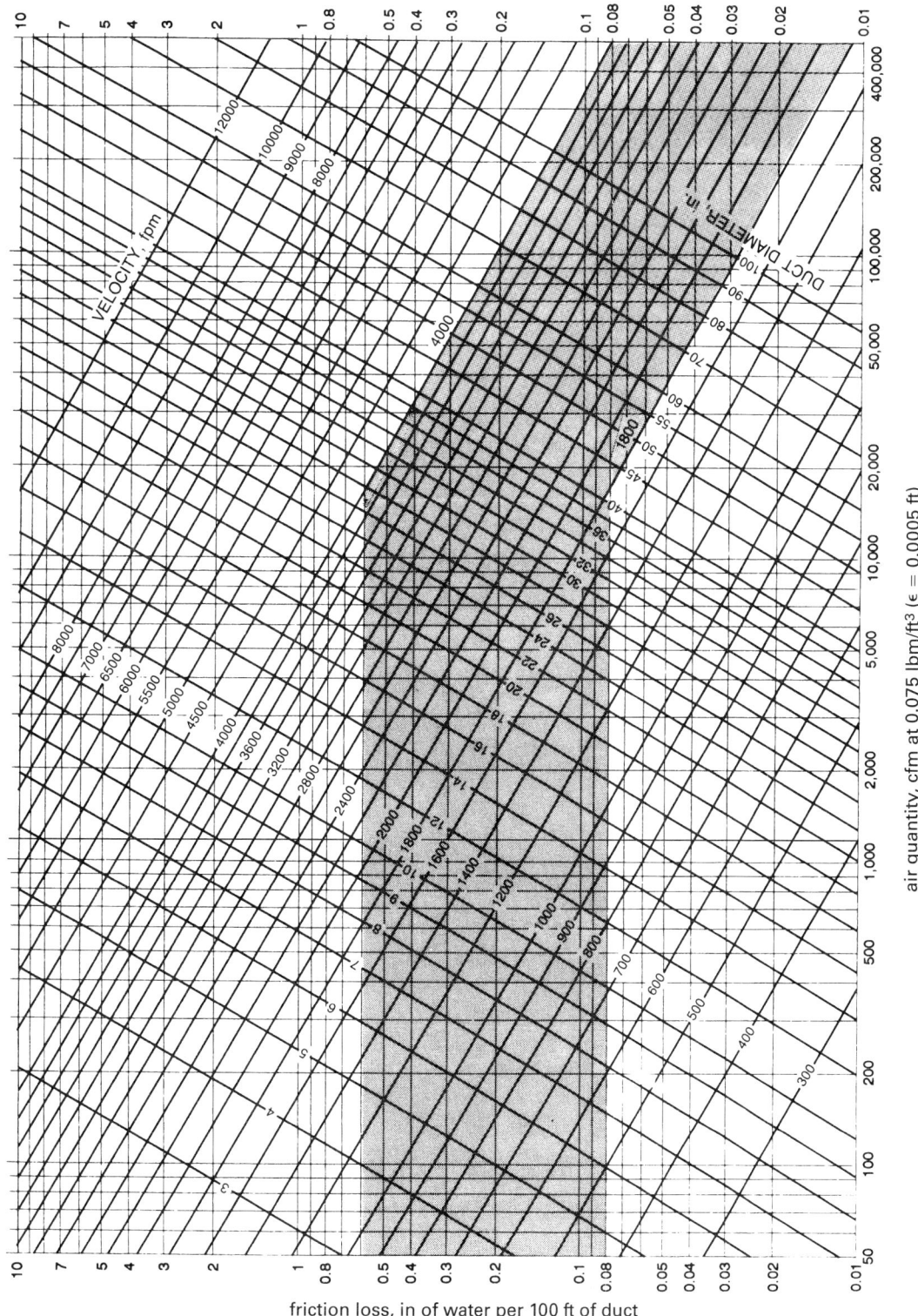

*Clean, round galvanized duct with a specific roughness of 0.0005 ft (0.15 mm) and approximately 25 beaded slip-couplings (joints) per 100 ft (30 m). Can also be used for smooth commercial spiral duct with about 10 joints per 100 ft (30 m).

From *2009 ASHRAE Handbook: Fundamentals, Inch-Pound Edition*, copyright © 2009, by American Society of Heating, Refrigerating, and Air-Conditioning Engineers, Inc. Reproduced with permission.

Figure 36.6 *Standard Friction Loss in Standard Duct* (pascals per meter of duct; ±5% for temperatures of 5°C to 35°C, elevations to 500 m, and duct pressures of −5 kPa to +5 kPa) (Recommended operating points shown as shaded region.)*

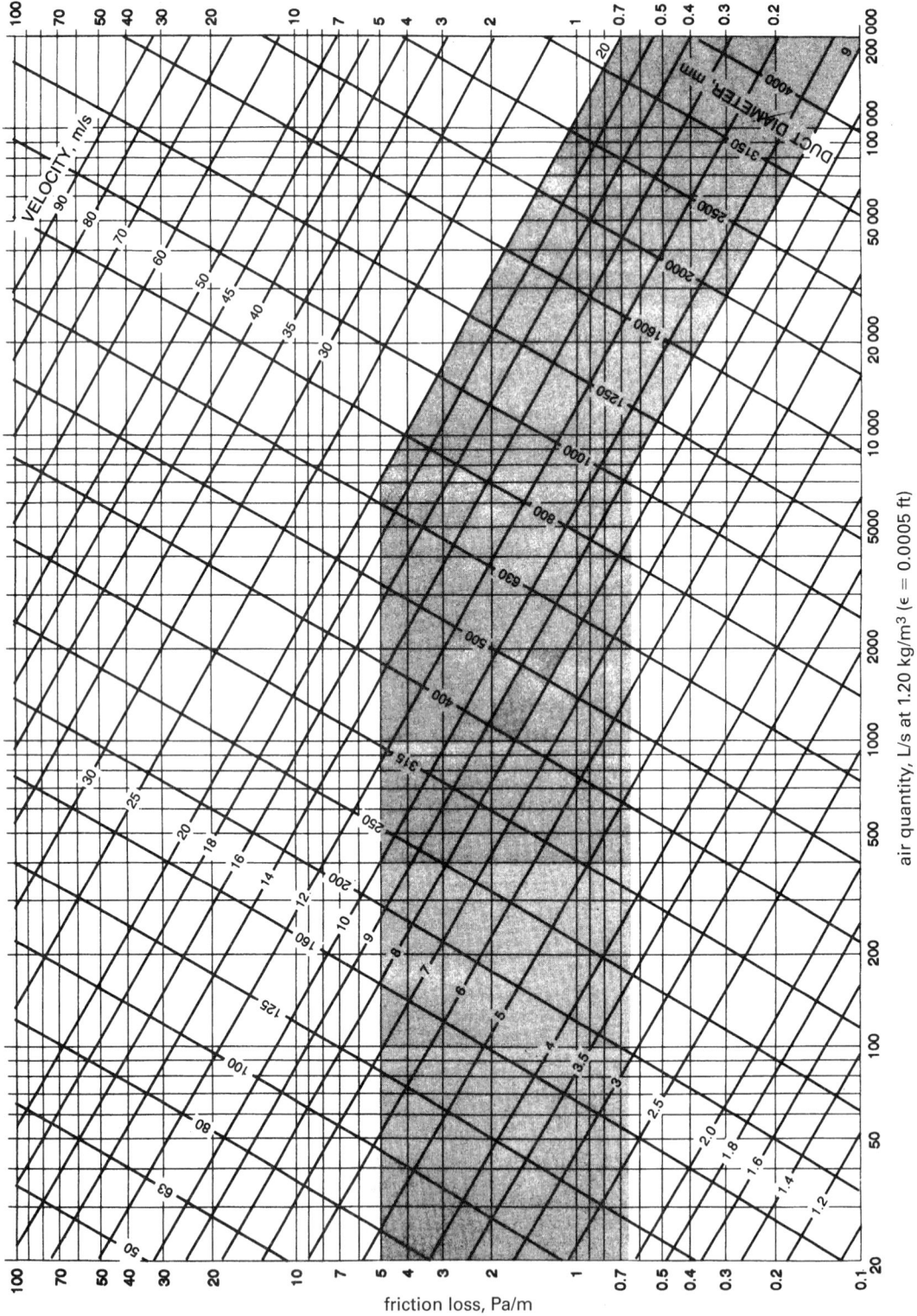

*Clean, round galvanized duct with a specific roughness of 0.0005 ft (0.15 mm) and approximately 1 beaded slip-coupling (joint) per meter. Can also be used for smooth commercial spiral duct with about 1 joint per 3 m.

From *2009 ASHRAE Handbook: Fundamentals, Inch-Pound Edition,* copyright © 2009, by American Society of Heating, Refrigerating, and Air-Conditioning Engineers, Inc. Reproduced with permission.

Generally, commercial duct manufacturers produce every whole-inch size up to at least 20 in (510 mm) in diameter. After that, ducts are available in 2 in (50 mm) increments. Odd-number sizes may be available with premium pricing.

Example 36.7

2000 ft³/min (1000 L/s) of air flows in a 13 in (315 mm) diameter duct. What are the (a) velocity and (b) friction loss per 100 ft (per meter) of duct?

SI Solution

(a) Use Fig. 36.6. Locate the intersection of the 1000 L/s and 315 mm lines. The velocity is approximately 13 m/s.

(b) Move horizontally to the left and read from the vertical scale. The friction loss is approximately 6 Pa/m of duct.

Customary U.S. Solution

(a) Use Fig. 36.5. Locate the intersection of the 2000 ft³/min and 13 in lines. The velocity is approximately 2200 ft/min. (This answer could also be calculated from $v = Q/A$.)

(b) Drop straight down to the horizontal scale. The friction loss is approximately 0.5 in wg per 100 ft of duct.

Example 36.8

2000 ft³/min (1000 L/s) of standard air moves through a duct with a velocity of 1600 ft/min (8 m/s). (a) What size duct is required? (b) What is the friction loss?

SI Solution

(a) Locate the intersection of 1000 L/s and 8 m/s. The required duct diameter is 400 mm.

(b) Move horizontally to the left and read from the vertical scale. The friction loss is approximately 1.7 Pa/m.

Customary U.S. Solution

(a) Locate the intersection of 2000 ft³/min and 1600 ft/min. The required duct diameter is approximately 15 in.

(b) Drop straight down to the horizontal scale. The friction loss is approximately 0.23 in wg per 100 ft of duct.

27. PRESSURE DROP FOR NONSTANDARD CONDITIONS

Common pressure drop equations and friction charts assume standard air with a density of 0.075 lbm/ft³, which corresponds to air at sea level at about 65°F and 40% relative humidity, or air at 70°F and 0% relative humidity. The density of air depends on the temperature, T, and elevation above sea level, z. For nonstandard temperatures and elevations, the pressure drop is given by Eq. 36.33.

$$\mathrm{FP}_{actual} = K_{elevation} K_{temperature} (\mathrm{FP})_{std} \qquad 36.33$$

$$K_{elevation} = (1 - 6.8754 \times 10^{-6} z_{ft})^{4.73} \qquad 36.34$$

$$K_{temperature} = \left(\frac{530°\mathrm{R}}{T_{°\mathrm{F}} + 460°} \right)^{0.825} \qquad 36.35$$

28. RECTANGULAR DUCTS

Duct systems are initially designed for round ducts. Then, conversions to rectangular ducts are made as required. A round duct with diameter D can be converted to a rectangular duct with equal friction per unit length if the desired aspect ratio is known. The *aspect ratio*, R, of a rectangular duct should be kept below 8 for ease of manufacture.

$$\text{short side} = \frac{D(1 + R)^{1/4}}{1.3 R^{5/8}} \qquad 36.36$$

$$R = \frac{\text{long side}}{\text{short side}} \qquad 36.37$$

The *equivalent diameter* of a rectangular duct with an aspect ratio less than 8 is given by the *Huebscher equation*, Eq. 36.38. The round duct will have the same friction and capacity as the rectangular duct. Figure 36.5 and Fig. 36.6 can be used with D_e and the actual flow rate to find the friction loss.[16] The velocity indicated by the chart will be incorrect but can be calculated as Q/A.

$$D_e = \frac{1.3(\text{short side} \times \text{long side})^{5/8}}{(\text{short side} + \text{long side})^{1/4}} \qquad 36.38$$

29. FRICTION LOSSES IN FITTINGS

Friction losses (*dynamic losses*) due to bends, fittings, enlargements, contractions, and obstructions are calculated in the same ways as for liquid friction losses—either by loss coefficient or equivalent length methods.[17,18]

With the *loss coefficient method*, the friction loss is calculated as a multiple of the velocity pressure. Though reported values vary widely, typical values of the loss coefficient, K, for common features are given in Table 36.4.[19] Loss coefficients are usually based on the upstream velocity pressure. However, there are some cases (i.e., where plenum

---

[16]This is the formula used by ASHRAE and most other authorities for the *equivalent* diameter. Some sources merely equate formulas for round and rectangular areas and use the *hydraulic* diameter, which is $\sqrt{(4 \times \text{short side} \times \text{long side})/\pi}$. When rounded to the nearest whole duct size, the difference is often insignificant. Although the two ducts may have the same cross-sectional area, they will not have the same capacity or friction.

[17]Unlike losses for liquid flows, however, fitting losses for duct systems are significant. They are not "minor" losses.

[18]Any fitting or feature that causes a static pressure loss of 0.75 in wg (200 Pa) or higher is a potential source of unwanted noise.

[19]Static pressure losses due to equipment (e.g., filters, coils, and heat exchangers) are determined from manufacturers' literature.

Table 36.4 Typical Fitting Loss Coefficients[a,b,g]

feature			K
abrupt expansion	$v_{down}/v_{up} = A_{up}/A_{down} =$	0	1.00
from A_{up} to A_{down}		0.2	0.64
(referred to v_{up})		0.4	0.36
$K_{up} = (1-(A_{up}/A_{down}))^2$		0.6	0.16
		0.8	0.04
abrupt contraction	$v_{up}/v_{down} = A_{down}/A_{up} =$	0.20	0.32
from A_{up} to A_{down}		0.25	0.30
(referred to v_{down})		0.40	0.25
		0.50	0.20
		0.60	0.16
		0.75	0.10
		0.80	0.06
round pipe of diameter D_1	$D_1/D_2 =$	0.10	0.20
across (through) duct of		0.25	0.55
diameter D_2		0.50	2.00
tapered reducing section[c]	taper angle[d] $=$	20°	0.012
		30°	0.020
		40°	0.032
		45°	0.040
		60°	0.070
bell-mouthed entrance[e]			0.040
90° round elbows[f]	$r/D =$	0.00	1.20
continuous die-stamped		(miter)	
(bend radius, r)		0.50	0.83
		0.75	0.46
		1.00	0.31
		1.25	0.27
		1.50	0.22
		1.75	0.20
		2.00	0.19
		2.25	0.18
		2.50	0.17
		2.75	0.16
		3.00	0.15

30°, 45°, 60° continuous, die-stamped elbows
multiply 90° loss coefficients by 0.33 (30°), 0.50 (45°), and 0.67 (60°)

90° mitered and gored elbows (round)
straight miter: 1.2
straight miter with turning vanes: 0.5

	$r/D = 0.75$	$r/D = 1.0$	$r/D = 1.5$	$r/D = 2.0$
3 piece	0.54	0.42	0.34	0.33
4 piece	0.50	0.37	0.27	0.24
5 piece	0.46	0.33	0.24	0.19

30°, 45°, 60° gored elbows
multiply 90° loss coefficients by 0.45 (30°), 0.60 (45°), and 0.78 (60°)

[a]Subscripts "up" and "down" refer to upstream and downstream, respectively.
[b]In multiples of velocity pressure.
[c]The total energy loss is small for all taper angles. The advantage of a very long taper is insignificant.
[d]The *taper angle* is the angle one side makes with the straight wall. The *included angle* refers to twice the taper angle.
[e]When stationary air is drawn into a bell-mouthed opening, the fan must supply the velocity pressure (1.0) as well as overcome the entrance friction (0.04). Because of that, some sources report this value as 1.04.
[f]Also, see Table 36.5.
[g]Specific roughness is $\epsilon = 0.0005$ ft; $f \approx 0.0185$.

air with negligible velocity enters a duct) where the downstream velocity is used by convention. The coefficient should always be used with the velocity at the point corresponding to the coefficient's subscript.

$$FP_{Pa} = K(VP_{Pa}) = 0.6Kv_{m/s}^2 \qquad \text{[SI]} \qquad 36.39(a)$$

$$FP_{in\,wg} = K(VP_{in\,wg}) = K\left(\frac{v_{ft/min}}{4005}\right)^2 \qquad \text{[U.S.]} \qquad 36.39(b)$$

In a straight duct of constant diameter, the velocity pressure is unchanged. The change in static pressure is the friction loss. Since there is no change in the velocity pressure, the friction loss will produce an equivalent decrease in total pressure. In other words, the friction loss is the change in total pressure. The same loss coefficient is used for calculating the change in static pressure and the change in total pressure.

$$FP = TP_1 - TP_2 = SP_1 - SP_2$$
$$= K(VP)_{up} \quad \begin{bmatrix} \text{constant area duct} \\ \text{with no branches} \end{bmatrix} \qquad 36.40$$

Loss coefficients are *zero-length losses*. That is, they only include the dynamic effects. If a large fitting has a specific length, that length must be included in the run-of-duct length when calculating the duct friction.

The fitting loss can be assumed to be the result of an equivalent length of duct. These lengths are given in multiples of duct diameter in Table 36.5. This is the *equivalent length method*.

Table 36.5 Typical Equivalent Lengths[a,b]

		L_e
90° continuous, round elbows	$r/D = 0.00$ (miter)	$65D$
of bend radius r and	0.50	$45D$
diameter D	0.75	$23D$
	1.00	$17D$
	1.25	$15D$
	1.50	$12D$
	1.75	$11D$
	2.00	$10D$
	2.25	$9.7D$
	2.50	$9.2D$
	2.75	$8.6D$
	3.00	$8.1D$

30°, 45°, 60° continuous elbows
multiply 90° equivalent lengths by 0.33 (30°), 0.50 (45°), and 0.67 (60°)

90° mitered and gored elbows (round)
straight miter: $65D$
straight miter with turning vanes: $27D$

	$r/D = 0.75$	$r/D = 1.0$	$r/D = 1.5$	$r/D = 2.0$
3 piece	$29D$	$23D$	$18D$	$18D$
4 piece	$27D$	$20D$	$15D$	$13D$
5 piece	$25D$	$18D$	$13D$	$10D$

30°, 45°, 60° gored elbows
multiply 90° equivalent lengths by 0.45 (30°), 0.60 (45°), and 0.78 (60°)

[a]In terms of inside duct diameter, D.
[b]Specific roughness is $\epsilon = 0.0005$ ft; $f \approx 0.0185$.

Ventilation

Duct elbows can be die-formed (i.e., stamped) or gored. Gore elbows typically have three or five gores per 90° of bend (e.g., a "3-gore, 90° elbow"), although 2- and 4-piece elbows are available. A *radius-to-diameter ratio* (i.e., the ratio of *centerline radius* (CLR) and diameter) of 1.5 is typical, although values of 2.0 and 2.5 are also used. Data on friction losses in bends are usually dependent on the radius-diameter ratio. The *throat radius* (i.e., the radius to the inside of the bend) is not commonly used to categorize friction losses.

The equivalent length of a smooth-radius rectangular duct elbow depends on aspect ratio and radius. For a duct with a width W, height H, and bend radius r, the equivalent length of a 90° elbow can be estimated from Eq. 36.41.

$$L_e = W\left(0.33\,\frac{r}{W}\right)^{-2.13(H/W)^{0.126}} \quad \begin{bmatrix} \text{90° smooth} \\ \text{rectangular elbow} \end{bmatrix}$$
$$36.41$$

30. COEFFICIENT OF ENTRY

The *coefficient of entry*, C_e, is the ratio of the actual to ideal velocities as stationary air is drawn into an inlet.[20] The coefficient of entry is not the same as the loss coefficient, K_e, which is the entrance friction expressed as a multiple of the velocity pressure. Typical values of the coefficient of entry are given in Table 36.6. Equation 36.44 is the relationship between the coefficient of entry and the loss coefficient.

$$C_e = \sqrt{\frac{\mathrm{VP}_{\mathrm{duct}}}{\mathrm{SP}_{\mathrm{duct}}}} \qquad 36.42$$

$$Q = \mathrm{v}_{\mathrm{actual}}A = C_e \mathrm{v}_{\mathrm{ideal}}A$$
$$= 4005\,C_e A\sqrt{\mathrm{SP}_{\mathrm{duct}}} \qquad 36.43$$

$$K_e = \frac{1 - C_e^2}{C_e^2} \qquad 36.44$$

Table 36.6 *Typical Coefficients of Entry*

entrance	C_e
plain opening (round, rectangular, square)	0.72
flanged openings (round, rectangular, square)	0.82
bell-mouthed	0.98
tapered square[*]	0.93
conical[*]	0.96

[*]For included taper angle of 30° to 60°. Values of other angles are close.

---

[20]This is analogous to the coefficient of velocity, C_v, used in liquid flow measurement devices.

31. STATIC REGAIN

Disregarding friction loss, energy is constant along the run of a duct. If the velocity pressure decreases due to an increase in duct area or a branch takeoff, the static pressure will increase. This increase is known as the *static regain*, SR.[21]

Ideally, the regain would be exactly equal to the decrease in velocity pressure. Actually, 10% to 25% of the energy is lost due to friction, turbulence, and other factors. The portion of the theoretical regain that is realized is given by the *static regain coefficient*, R. R has typical values of 0.75 to 0.90 for well-designed ducts without reducing sections. (When $\mathrm{v}_{\mathrm{down}} > \mathrm{v}_{\mathrm{up}}$, the regain will be a static pressure loss. Use $R = 1.1$ in that case.)

$$R = \frac{\mathrm{SR}_{\mathrm{actual}}}{\mathrm{SR}_{\mathrm{ideal}}} = \frac{\mathrm{SR}_{\mathrm{actual,Pa}}}{0.6(\mathrm{v}_{\mathrm{up}}^2 - \mathrm{v}_{\mathrm{down}}^2)} \qquad \text{[SI]} \quad 36.45(a)$$

$$R = \frac{\mathrm{SR}_{\mathrm{actual}}}{\mathrm{SR}_{\mathrm{ideal}}} = \frac{\mathrm{SR}_{\mathrm{actual,in\,wg}}}{\dfrac{\mathrm{v}_{\mathrm{up}}^2 - \mathrm{v}_{\mathrm{down}}^2}{(4005)^2}} \qquad \text{[U.S.]} \quad 36.45(b)$$

32. DIVIDED-FLOW FITTINGS

Figure 36.7 illustrates typical commercial *divided-flow* (i.e., *branch takeoff*) *fittings* and the terminology that describes them. Air enters upstream, from the left. After the air reduction at the branch takeoff, the downstream velocity will (generally) be less than the upstream velocity. The change in static pressure due to the change in velocity is

$$\mathrm{SP}_{\mathrm{up}} - \mathrm{SP}_{\mathrm{down}} = (\mathrm{TP}_{\mathrm{up}} - \mathrm{TP}_{\mathrm{down}}) - (\mathrm{VP}_{\mathrm{up}} - \mathrm{VP}_{\mathrm{down}})$$
$$= R(\mathrm{VP}_{\mathrm{up}} - \mathrm{VP}_{\mathrm{down}}) \quad \begin{bmatrix} \text{positive for} \\ \text{decreases} \end{bmatrix}$$
$$36.46$$

The total pressure change from upstream to downstream is

$$\mathrm{TP}_{\mathrm{up}} - \mathrm{TP}_{\mathrm{down}} = (\mathrm{SP}_{\mathrm{up}} - \mathrm{SP}_{\mathrm{down}}) + (\mathrm{VP}_{\mathrm{up}} - \mathrm{VP}_{\mathrm{down}})$$
$$= (R + 1)(\mathrm{VP}_{\mathrm{up}} - \mathrm{VP}_{\mathrm{down}}) \quad \begin{bmatrix} \text{positive for} \\ \text{decreases} \end{bmatrix}$$
$$36.47$$

The total pressure change from upstream through the branch is

$$\mathrm{TP}_{\mathrm{up}} - \mathrm{TP}_{\mathrm{br}} = (\mathrm{SP}_{\mathrm{up}} - \mathrm{SP}_{\mathrm{br}}) + (\mathrm{VP}_{\mathrm{up}} - \mathrm{VP}_{\mathrm{br}})$$
$$= K_{\mathrm{br}}(\mathrm{VP}_{\mathrm{up}}) \qquad 36.48$$

---

[21]If a downstream or branch velocity is low enough, it is possible for the regain to actually exceed the dynamic losses due to fitting turbulence. This may hide the true inefficiency of the fitting.

Figure 36.7 *Types of Commercial Divided-Flow Fittings*

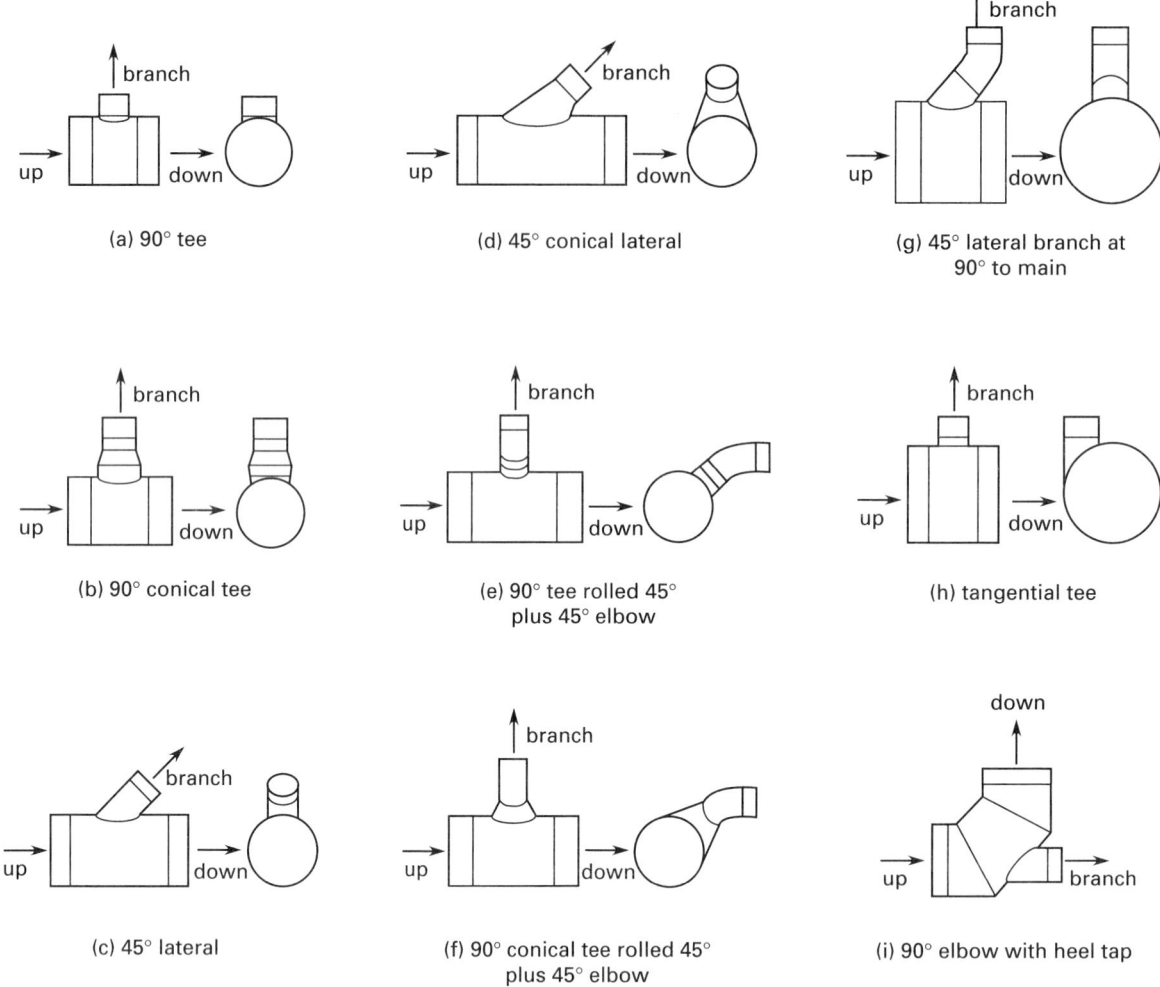

(a) 90° tee

(d) 45° conical lateral

(g) 45° lateral branch at 90° to main

(b) 90° conical tee

(e) 90° tee rolled 45° plus 45° elbow

(h) tangential tee

(c) 45° lateral

(f) 90° conical tee rolled 45° plus 45° elbow

(i) 90° elbow with heel tap

The change in static pressure from upstream through the branch is

$$SP_{up} - SP_{br} = (TP_{up} - TP_{br}) + (VP_{up} - VP_{br})$$

$$= K_{br}(VP_{up}) - (VP_{up} - VP_{br})$$

$$= (K_{br} - 1)(VP_{up}) + VP_{br} \quad \begin{bmatrix} \text{positive for} \\ \text{decreases} \end{bmatrix}$$

$$36.49$$

Manufacturers of commercial fittings provide graphs of the loss coefficient as a function of the ratio of branch-to-upstream velocities.[22] Typical values of the *branch loss coefficient*, K_{br}, are given in Table 36.7.

Example 36.9

The velocity in a main duct before a branch is 3200 ft/min (16 m/s). The velocity in the branch duct is 2560 ft/min (12.8 m/s). What is the change in static pressure from the main duct through the branch?

---
[22]Some manufacturers also provide direct-reading charts that give the friction loss directly in terms of in wg.

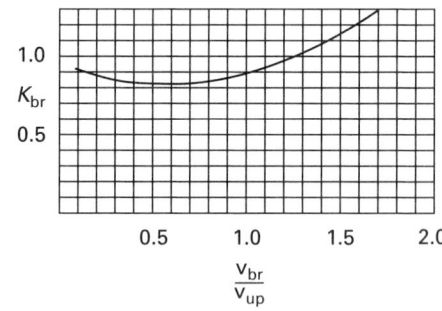

SI Solution

The ratio of the branch-to-upstream velocity is

$$\frac{v_{br}}{v_{up}} = \frac{12.8 \ \frac{m}{s}}{16 \ \frac{m}{s}} = 0.8$$

From the graph, the loss coefficient is $K_{br} = 0.85$.

From Eq. 36.9, the upstream and branch velocity pressures are

$$\text{VP}_{\text{up}} = 0.6v^2_{\text{m/s}} = (0.6)\left(16 \ \frac{\text{m}}{\text{s}}\right)^2 = 154 \ \text{Pa}$$

$$\text{VP}_{\text{br}} = (0.6)\left(12.8 \ \frac{\text{m}}{\text{s}}\right)^2 = 98 \ \text{Pa}$$

From Eq. 36.49, the change in static pressure is

$$\text{SP}_{\text{up}} - \text{SP}_{\text{br}} = (K_{\text{br}} - 1)(\text{VP}_{\text{up}}) + \text{VP}_{\text{br}}$$
$$= (0.85 - 1)(154 \ \text{Pa}) + 98 \ \text{Pa}$$
$$= 75 \ \text{Pa} \quad [\text{decrease}]$$

Customary U.S. Solution

The ratio of the branch-to-upstream velocity is

$$\frac{v_{\text{br}}}{v_{\text{up}}} = \frac{2560 \ \dfrac{\text{ft}}{\text{min}}}{3200 \ \dfrac{\text{ft}}{\text{min}}} = 0.8$$

From the graph, the loss coefficient is $K_{\text{br}} = 0.85$. From Eq. 36.8, the upstream and branch velocity pressures are

$$\text{VP}_{\text{up}} = \left(\frac{v_{\text{up}}}{4005}\right)^2 = \left(\frac{3200 \ \dfrac{\text{ft}}{\text{min}}}{4005}\right)^2 = 0.64 \ \text{in wg}$$

$$\text{VP}_{\text{br}} = \left(\frac{2560 \ \dfrac{\text{ft}}{\text{min}}}{4005}\right)^2 = 0.41 \ \text{in wg}$$

From Eq. 36.49, the change in static pressure is

$$\text{SP}_{\text{up}} - \text{SP}_{\text{br}} = (K_{\text{br}} - 1)(\text{VP}_{\text{up}}) + \text{VP}_{\text{br}}$$
$$= (0.85 - 1)(0.64 \ \text{in wg}) + 0.41 \ \text{in wg}$$
$$= 0.31 \ \text{in wg} \quad [\text{decrease}]$$

33. DUCT DESIGN PRINCIPLES

Duct systems are categorized as low-velocity (up to 2000 ft/min to 2500 ft/min or 10.2 m/s to 12.8 m/s) and high-velocity (above 2500 ft/min or 12.8 m/s). *Low-velocity systems*, also known as *conventional systems*, are usually designed with the velocity-reduction and equal-friction methods. *High-velocity systems* are able to take advantage of benefits associated with the static regain method.

Duct systems are categorized according to the static pressure at the fan: low-pressure (0–2 in wg; 0–500 Pa), medium-pressure (2–6 in wg; 500–1500 Pa), and high-pressure (6–10 in wg; 1500–2500 Pa). Residential and commercial ducts are typically designed such that the pressure drop at the fan is between 0.08 in wg and 0.15 in wg.

Table 36.7 *Typical Branch Loss Coefficient (K_{br}) Values**

ratio of $v_{\text{br}}/v_{\text{up}}$	angle of takeoff		
	90°	60°	45°
0.5	1.1	0.8	0.5
1.0	1.5	0.8	0.5
1.5	2.2	1.1	0.9
2.0	3.0	2.9	2.8
2.5	4.3	3.3	3.2
3.0	5.6	5.2	4.9

*Round ducts only.

Supply duct systems take air from the fan and bring it to the ventilated space. *Exhaust duct systems* (*return air systems*) carry air from the ventilated space back to the fan.

Compared with conventional systems, high-pressure and high-velocity systems require less space, cost less for ductwork, and provide better control of the conditioned space. However, they are noisier, require a more precise duct design, and require larger (more expensive) fans.

The following general recommendations apply to all duct designs and design methods.

- Make duct routes as direct as possible.

- Avoid sudden changes in direction and diameter.

- Use radius-to-diameter ratios of 1.5 or higher.

- Eliminate obstructions in and through the ducts.

- Use radiused elbows whenever possible, and when not, use turning vanes.

- Make rectangular ducts as square as possible. Avoid aspect ratios greater than 8:1, and use 4:1 or less whenever space permits.

- Use smooth metal construction whenever possible.

- Maintain an incremental size difference of at least 2 in in adjacent duct sections.

- Include a small volume allowance above the sum of all the outlet volumes to account for leakage.[23]

- Size the fan with excess capacity to compensate for inaccuracies in the design.

- Install balancing dampers in all branches, even when the static regain method is used for the design. In order to minimize noise, install dampers as close as possible to the main duct.

---

[23]Some sources say to include up to 10% excess air to account for leaks. While this may sound nominal, the fan laws show that increasing the fan speed 10% to obtain the extra flow will increase the horsepower 30%. It is unlikely that a motor would be able to provide 30% more power. Therefore, more reliance should be placed on tight ductwork than on excess air.

- Use the lowest possible duct velocities in order to minimize fan power and noise.

- In practice, to prevent undersizing supply ducts in residential applications, supply-side designs should be based on no greater than 0.1 in wg per 100 ft; and, a value of 0.06 in wg is more appropriate. For return lines, values in the range of 0.04–0.05 in wg are appropriate.

- Conventional rule-of-thumb "wisdom" specifies the gross area of return grilles as 1 ft^2 (144 in^2) per ton of refrigeration. However, this generally results in the average system return being undersized by 30% or more. A better rule of thumb is to have 1 in^2 of gross grille area for every 2 ft^3/min of air flow.

34. ECONOMICAL DUCT DESIGN

All other factors being equal, economical duct design is achieved by using standard, factory-manufactured round duct, keeping runs straight, minimizing the aspect ratio of rectangular duct, minimizing the total amount of sheet metal (i.e., minimizing the total mass) used, and by maintaining trunk size until a reduction of 2 in (51 mm) or more is warranted (this is known as the *2-inch rule*).

35. LEAKAGE

Ducts are not intended to be leak-free. However, the volumetric leakage should be less than 1% for well-sealed ducts and 2–5% for unsealed ducts.[24] Ducts are not pressure vessels and are not intended to be tested by sealing and pressurization. (The term "airtight" should be avoided.) Leakage should be tested volumetrically with the air in motion.

Leakage can be classified and quantified by a duct leakage class. The *leakage class*, C_L, is defined as actual leakage in cubic feet per minute per 100 square feet of duct area, a quantity known as the *leakage factor*, F, when the gage pressure, p, within the duct is 1 in wg. An exponent, N, is used to correlate the leakage class and leakage factor. N depends on turbulence within the duct, but it has a reliable average value of 0.65. Duct leakage is essentially independent of duct velocity. Equation 36.50 is valid for both positive and negative pressures.

$$Q_{\text{leakage}} = C_L p^N \qquad 36.50$$

For convenience, the leakage class of ducts constructed by skilled, trained technicians can be predicted by the SMACNA *seal class*.[25] Seal class A, applicable to ducts with pressurizations 4 in wg and higher, requires all transverse joints, longitudinal seams, and duct wall

penetrations to be sealed. For seal class A, the leakage class, C_L, can be estimated as 3 cfm/100 ft^2 for round metal ducts and as 6 cfm/100 ft^2 for rectangular metal ducts. Seal class B, applicable to pressurizations of 3–4 in wg, requires sealing of transverse joints and longitudinal seams. The approximate leakage class is 6 cfm/100 ft^2 for round metal ducts and 12 cfm/100 ft^2 for rectangular metal ducts. Seal class C, applicable to pressurizations of 2 in wg and less, requires sealing only of transverse joints. The approximate leakage class is 12 cfm/100 ft^2 for round metal ducts and 24 cfm/100 ft^2 for rectangular metal ducts. Unsealed ductwork can be expected to exhibit a leakage class of 24 cfm/100 ft^2 for round metal ducts and 48 cfm/100 ft^2 for rectangular metal ducts.

Actual construction and sealing can be used to predict the seal class when ducts are manufactured customarily. Since Eq. 36.50 was developed from measurements of ducts constructed with normal and customary quality, it should only be used to predict leakage from ducts whose construction is customary for the intended pressurization range. It should not be used when construction is inconsistent with intended use. For example, a duct expected to operate at a pressure less than 2 in wg would not normally be constructed with a class A seal class, and Eq. 36.50 cannot be expected to predict leakage accurately in that instance.

Ductwork carries flows of 2–5 cfm/ft^2 (cfm per square foot of duct area). Systems with a lot of ductwork and small air flows are nearer to the lower end, while systems with minimal ductwork and large air flows are nearer to the upper end. The leakage as a percentage of the supplied air flow is

$$\text{leakage}_{\% \text{ of supply}} = \frac{Q_{\text{leakage,cfm}}}{Q_{\text{supply,cfm}}} \times 100\%$$

$$= \frac{Q_{\text{leakage,cfm/100 ft}^2}}{Q_{\text{supply,cfm/ft}^2}} \qquad 36.51$$

Example 36.10

Air flows through a rectangular duct (seal class C) at the rate of 3 cfm/ft^2. The gage pressure in the duct is 1.7 in wg. What is the leakage as a percentage of the supply rate?

Solution

Rectangular duct with seal class C can be expected to have a leakage class, C_L, of 24 cfm/100 ft^2.

From Eq. 36.50,

$$Q_{\text{leakage}} = C_L p^N = \left(24 \, \frac{\text{cfm}}{100 \, \text{ft}^2}\right)(1.7 \text{ in wg})^{0.65}$$

$$= 33.88 \text{ cfm/100 ft}^2$$

---

[24]ASHRAE and SMACNA recommendations.
[25]*Duct Construction Standards—Metal and Flexible*, SMACNA, 1985 ed.

From Eq. 36.51, combining the area and flow rate terms,

$$\text{leakage}_{\%\,\text{of supply}} = \frac{Q_{\text{leakage,cfm/100 ft}^2}}{Q_{\text{supply,cfm/ft}^2}} = \frac{33.88\,\dfrac{\text{cfm}}{100\ \text{ft}^2}}{3\,\dfrac{\text{cfm}}{\text{ft}^2}}$$

$$= 11.29\%$$

36. COLLAPSE OF DUCTS

Under certain conditions, ducts may collapse inward. This may happen in medium- and high-velocity systems when a fire damper or blast gate suddenly closes, but can also occur in long, large-diameter air return systems. The negative pressure created between a closed damper and the retreating mass of air may collapse the duct. The negative pressure required to collapse a duct depends on the duct construction and must be specified by the duct manufacturer.

To prevent collapse in air return systems, increased metal gage and/or angle rings may be used. To prevent collapse due to sudden closures, a negative-pressure relief valve can be installed immediately downstream of each fire damper.

37. FILTERS

Duct filters are used to remove dust, pollen, spores, bacteria, and other particles. In residential and light commercial applications, traditional pleated furnace-type filters are used, along with fiberglass and foam media filters. Most commercial/industrial filter units contain two or more stages of successively finer filtration, starting with a pleated pre-filter. *Hogs hair* filters (made from latex-coated organic fibers) are washable and reusable and can be used as pre-filters. Commercial and industrial environments may require box and bag cloth filters. In more demanding environments, such as clean rooms and hospitals, multi-stage, high-efficiency particulate air (HEPA) and electrostatic filters can be used. Filters with layers of granulated activated carbon (GAC) are somewhat useful in removing gases, VOCs, and odors.

Filtration efficiency is measured by the percentage of particles removed. For most filters, the removal efficiency varies nonlinearly with the particle size. The *arrestance* is the percentage of macroscopic particles (lint, hair, dust, etc.) removed. Arrestance for most filter types is usually well above 80%. However, most filters have lower *filtration efficiencies* with smaller particles. The 0.3 micron particle size is an industry standard comparison point. For example, when new, a typical, high-quality pleated furnace filter for residential use has a removal efficiency of approximately 80% at 10 microns, 40% at 1 micron, and less than 10% at 0.3 micron. Electrostatic filters have efficiencies approaching 95%. HEPA filters have efficiencies of 99.97% at the 0.3 micron level, and some manufacturers claim efficiencies better than "5 nines" (i.e., 99.999%) for multi-stage units. Efficiencies of all filters decrease as the filter is used and the filter becomes clogged.

Filters are rated by their *minimum efficiency reporting value* (MERV), a standard used to categorize the overall efficiency of the filter. MERV ratings range from 1 to 16, with the more efficient filters receiving the higher values. Typical residential pleated filters have poor performance below 10 microns and have MERV ratings of 1 to 4. High-quality filters with MERV ratios 5 to 8 can remove particles as small as 3 microns. Filters with 9 to 12 MERV ratings are used in commercial and industrial applications and will stop particles in the 1 to 3 micron range. The most efficient filters have MERV ratings of 13 to 16 and will stop particles as small as 0.3 microns. These filters are used in hospitals and clean rooms.

Pressure drops in most pleated filters are less than 0.5 in wg, and for residential and light industrial applications are generally 0.2–0.3 in wg. Manufacturer's data must be used for accurate assessments. Filters with MERV ratings greater than 13 generally have high pressure drops. Because of this, they may be installed in parallel (not inline) with the return duct and filter only a portion of the air at a time.

38. DAMPERS

Since friction loss is proportional to the distance from the fan to the outlet, duct runs to outlets near the fan will have lower friction losses than the main duct run. If not constrained, most of the airflow will escape out of the lower-friction runs. Limited pressure balancing can be achieved with *jumper ducts*, also known as *crossover ducts*, which are ducts (without equipment) that run between zones and terminate at simple grilles in order to equalize pressure in the zones. Jumper ducts are generally only used in residential construction where duct runs are short.

Duct systems are not self-equalizing or self-balancing. Even when exquisitely designed, installed ducts rarely perform as designed, and each run must be adjusted individually after installation. Balancing adjusts the flow rate in each duct to match the design value for the corresponding zone. Dampers are used to balance airflow in ducts and to regulate the quantity of outside make-up air drawn in. Dampers can be motorized, but they are usually operated manually. A significant characteristic (i.e., disadvantage) of manual dampers is that they are generally left forever in their originally installed position. Pressure loss through dampers can be substantial, even when fully open. A typical pressure-loss coefficient, K, for a fully open damper is 0.52.

Dampers are included in all runs to keep the pressure drop the same in all duct runs, even those that are short. *Balancing* is the act of closing down the dampers to equalize the friction losses. It is a good idea to install dampers even when sophisticated design methods are used. Dampers can be manually operated (for balancing or other occasional use), motorized (for zone control and variable volume), or gravity operated.

Figure 36.8 illustrates three general types of dampers. *Volume dampers* should be used only in branch ducts when a splitter damper cannot be used.[26] *Splitter dampers* should be used at the junction of the main and branch ducts. *Automatic dampers* are usually chosen with parallel blades for applications with two distinct positions. Dampers with opposed blades are chosen when airflow is to be controlled over a wide range. *Gravity dampers* are self-closing and are intended to prevent backflow. *Fire dampers* and *smoke dampers* close automatically to prevent the spread of smoke throughout the system.

Figure 36.8 *Types of Dampers*

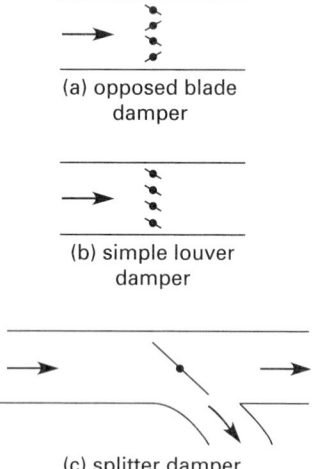

(a) opposed blade
damper

(b) simple louver
damper

(c) splitter damper

Balancing/volume-adjusting dampers should be installed close to the main supply, as far away as possible from the outlets. However, outlets can be designed to act as dampers. Such decorative *grilles* may be *fixed pattern dampers* (i.e., *perforated plate* or *fixed-bar grilles*) or *adjustable bar grilles* with a manual control lever. The disadvantage of combining terminal air distribution with dampering is that the noise created by the friction is projected directly into the room. In order to limit noise, the maximum flow velocities and flow rates are specified with standard commercial grilles by the manufacturers.

39. VELOCITY-REDUCTION METHOD

In the *velocity-reduction method*, the fan discharge velocity is selected by judgment. Arbitrary reductions in velocity are made down the run, usually at branch takeoffs. This method requires expertise on the part of the designer. It is used primarily for estimating simple layouts. The following steps constitute the velocity-reduction method.

[26]In the ventilation industry, dampers used only to adjust volume are also known as *blast gates*.

step 1: Select the velocity leaving the fan from judgment. As the duct branches off, use judgment to select a reduced velocity for each branch. Table 36.8 lists typical maximum values of duct velocities for conventional low-velocity systems.

Table 36.8 *Typical Maximum Duct Velocities (in ft/min)*

application	large supply ducts	small supply ducts	return ducts
residences	800	600	600
apartments/hotel bedrooms	1500	1100	1000
theaters	1600	1200	1200
deluxe offices		1100	800
average offices		1300	1000
general offices	2200	1400	1200
restaurants	1800	1400	1200
small shops		1500	1200
department stores			
lower floors	2100	1600	1200
upper floors	1800	1400	1200

(Multiply ft/min by 0.00508 to obtain m/s.)

step 2: Determine the airflow requirement, Q, for each outlet.

step 3: Calculate the duct size from $A = Q/\mathrm{v}$.

step 4: By inspection, find the highest-resistance (i.e., longest) duct run. Calculate the static pressure drop in the longest run.

step 5: Specify dampers in all of the runs for balancing.

40. EQUAL-FRICTION METHOD

The *equal-friction method* is applicable to simple low-velocity systems. The method gets its name from the procedure that arbitrarily keeps the friction loss per unit length the same in all duct runs. Velocity pressure and regain are disregarded. The system will require extensive dampering, as no attempt is made to equalize pressure drops in the branches.

A common assumption in preliminary design studies is a friction loss of 0.08 in wg per 100 ft (0.65 Pa/m) for virtually all situations except offices (0.10 in wg per 100 ft; 0.82 Pa/m) and industrial uses (0.15 in wg per 100 ft; 1.2 Pa/m). This value is referred to as the *design pressure drop* (dpd). These values ensure duct velocities are low enough to avoid excessive noise, and they represent a good compromise between duct and fan installation and operating costs.

step 1: Select the main duct velocity from Table 36.8, contract specifications, or judgment.

step 2: From the velocity and flow rate in the main duct, find the friction loss per unit length from Fig. 36.5 or Fig. 36.6.

step 3: After each branch, reduce the main duct flow rate by the branch flow. Find the new velocity and duct size to keep the same friction loss per unit length. (This means that all points will be along a vertical line on Fig. 36.5 or Fig. 36.6.)

step 4: Determine the static pressure drop in the highest-resistance duct. The fan must supply this static pressure drop plus the desired outlet pressure.

step 5: Compare the actual system pressure with the design pressure, if known. If they are significantly different, repeat all the steps with a different main duct velocity.

step 6: Size branch runs the same way—keeping the same friction loss per unit length. Use dampers to equalize the pressure drops.

Example 36.11

A theater duct system is shown. All bends have a radius-to-diameter ratio of 1.5. The branch takeoff between sections A and B has a branch loss coefficient of $K_{br} = 1.5$. The design pressure at each outlet is 0.15 in wg (38 Pa). Disregard the divided flow fitting loss. Use the equal-friction method to size the system.

SI Solution

step 1: From Table 36.8, choose the main duct velocity (section A) as 1600 ft/min. From the table footnote, the SI velocity is

$$v_{main} = \left(1600\ \frac{ft}{min}\right)\left(0.00508\ \frac{m \cdot min}{s \cdot ft}\right) = 8.1\ m/s$$

step 2: The total airflow from the fan is 2000 L/s. From Fig. 36.6, the main duct diameter (section A) is approximately 560 mm. (This may not correspond to a standard duct size.) The friction loss is 1.2 Pa/m.

step 3: After the first takeoff, the flow rate in section B is

$$2000\ \frac{L}{s} - 600\ \frac{L}{s} = 1400\ L/s$$

From Fig. 36.6 for 1400 L/s and 1.2 Pa/m, the diameter is 490 mm, and the velocity is 7.4 m/s.

step 4: By inspection, the longest run is ABC. From Table 36.5, the equivalent length of each bend is 12D.

$$L_{e,bend} = 12D = \frac{(12)(560\ mm)}{1000\ \frac{mm}{m}} = 6.72\ m\quad(7\ m)$$

The equivalent length of the entire run is

6 m	(from fan to first bend)
7 m	(equivalent length of first bend)
9 m	(first bend to second bend)
7 m	(equivalent length of second bend)
3 m	(jog between sections A and B)
6 m	(section B)
12 m	(section C)
total: 50 m	

The straight-through friction loss in the longest run is

$$(50\ m)\left(1.2\ \frac{Pa}{m}\right) = 60\ Pa$$

Use Eq. 36.48 to find the friction loss in the branch takeoff between sections A and B.

$$TP_A - TP_B = K_{br}(VP_{up}) = K_{br}(0.6)v_{up}^2$$
$$= (1.5)(0.6)\left(8.1\ \frac{m}{s}\right)^2$$
$$= 59\ Pa$$

The fan must be able to supply a static pressure of

$$SP_{fan} = 60\ Pa + 59\ Pa + 38\ Pa = 157\ Pa$$

The total pressure supplied by the fan is

$$TP_{fan} = SP_{fan} + 0.6v_{m/s}^2$$
$$= 157\ Pa + (0.6)\left(8.1\ \frac{m}{s}\right)^2$$
$$= 196\ Pa$$

Customary U.S. Solution

step 1: From Table 36.8, choose the main duct velocity (section A) as 1600 ft/min.

step 2: The total airflow from the fan is 4000 ft³/min. From Fig. 36.5, the main duct diameter (section A) is 21 in. The friction loss is 0.15 in wg per 100 ft.

step 3: After the first takeoff, the flow rate in section B is

$$4000\ \frac{ft^3}{min} - 1200\ \frac{ft^3}{min} = 2800\ ft^3/min$$

From Fig. 36.5 for 2800 ft³/min and 0.15 in wg per 100 ft, the diameter is 18 in, and the velocity is 1500 ft/min. Similarly, the diameters at sections C and D are 11.5 in (say 12 in) and 13 in, respectively. The velocity at section C is 1000 ft/min.

step 4: By inspection, the longest run is ABC. From Table 36.5, the equivalent length of each bend is 12D.

$$L_{e,\text{bend}} = 12D = \frac{(12)(21 \text{ in})}{12 \frac{\text{in}}{\text{ft}}} = 21 \text{ ft}$$

The equivalent length of the entire run is

20 ft	(from fan to first bend)
21 ft	(equivalent length of first bend)
30 ft	(first bend to second bend)
21 ft	(equivalent length of second bend)
10 ft	(jog between sections A and B)
20 ft	(section B)
40 ft	(section C)

total: 162 ft

The straight-through friction loss in the longest run is

$$\left(\frac{162 \text{ ft}}{100 \text{ ft}}\right)(0.15 \text{ in wg per } 100 \text{ ft}) = 0.24 \text{ in wg}$$

Use Eq. 36.48 to find the friction loss in the branch takeoff between sections A and B.

$$TP_A - TP_B = K_{\text{br}}(VP_{\text{up}})$$
$$= (1.5)\left(\frac{1600 \frac{\text{ft}}{\text{min}}}{4005}\right)^2$$
$$= 0.24 \text{ in wg}$$

The fan must be able to supply a static pressure of

$$SP_{\text{fan}} = 0.24 \text{ in wg} + 0.24 \text{ in wg} + 0.15 \text{ in wg}$$
$$= 0.63 \text{ in wg}$$

The total pressure supplied by the fan is

$$TP_{\text{fan}} = SP_{\text{fan}} + VP_{\text{up}}$$
$$= 0.63 \text{ in wg} + \left(\frac{1600 \frac{\text{ft}}{\text{min}}}{4005}\right)^2$$
$$= 0.79 \text{ in wg}$$

41. COMBINATION METHOD

A *combination method* is sometimes used. The main duct is sized by the equal-friction method. The branch runs are sized so as to dissipate the remaining friction

(compared with the main duct run) and to (theoretically) eliminate the need for dampers.

The desired outlet pressure is subtracted from the pressure at the main duct branch takeoff to get the pressure that must be dissipated in the branch run. This pressure is divided by the estimated equivalent length to find the pressure drop per unit length. Figure 36.5 and Fig. 36.6 can be used to find the duct size and velocity.[27]

Example 36.12

An air supply system consists of a long run and two branches. The total friction loss in the long run is 0.15 in wg (38 Pa). The longest duct was sized with the equal-friction method using a pressure drop of 0.2 in wg per 100 ft (1.6 Pa/m). The equivalent length of the branch takeoff at A is 12 ft (3.6 m). The equivalent length of the elbow in duct A is 18 ft (5.4 m). Rather than use a damper in duct A to equal the pressure drop, duct A will be sized small enough to equalize the losses through increased velocity. Use the combination method to size duct A.

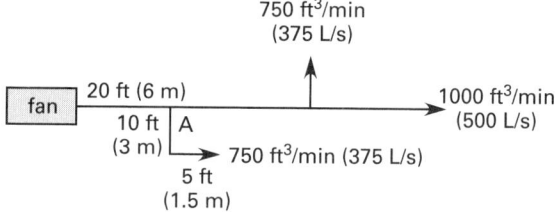

SI Solution

Subtracting the pressure drop from the fan to the branch takeoff, the pressure left to be dissipated in duct A is

$$38 \text{ Pa} - (6 \text{ m})\left(1.6 \frac{\text{Pa}}{\text{m}}\right) = 28 \text{ Pa}$$

The required loss per meter in duct A is

$$\frac{28 \text{ Pa}}{3 \text{ m} + 1.5 \text{ m} + 3.6 \text{ m} + 5.4 \text{ m}} = 2.07 \text{ Pa/m}$$

Use Fig. 36.6. With 375 L/s and 2.07 Pa/m, the velocity is approximately 6.9 m/s, and the diameter is approximately 260 mm.

Customary U.S. Solution

Subtracting the pressure drop from the fan to the branch takeoff, the pressure left to be dissipated in duct A is

$$0.15 \text{ in wg} - \left(\frac{20 \text{ ft}}{100 \text{ ft}}\right)(0.2 \text{ in wg per } 100 \text{ ft})$$
$$= 0.11 \text{ in wg}$$

---
[27]To limit noise, very high velocities should be avoided.

The required loss per 100 ft in duct A is

$$\frac{(0.11 \text{ in wg})\left(100 \; \dfrac{\text{ft}}{100 \text{ ft}}\right)}{10 \text{ ft} + 5 \text{ ft} + 12 \text{ ft} + 18 \text{ ft}} = 0.24 \text{ in wg per 100 ft}$$

Use Fig. 36.5. With 750 ft^3/min and 0.24 in wg, the velocity is 1240 ft/min, and the diameter is 10 in.

42. STATIC REGAIN METHOD

In the *static regain method*, the diameter of each successive branch is reduced in order to increase the static pressure at the branch entrance back to the fan discharge pressure. The reduction is such that the static regain offsets the friction loss in the succeeding section. (This method can also be used to size branch ducts as long as the duct sizes are reasonable.) In Eq. 36.52, point A is before the branch takeoff, point B is immediately after the branch takeoff, and point C is just prior to the next branch takeoff. (See Fig. 36.9.)

$$SP_A - SP_B = FP_{B-C} \qquad 36.52$$

Figure 36.9 Duct Runs for Static Regain

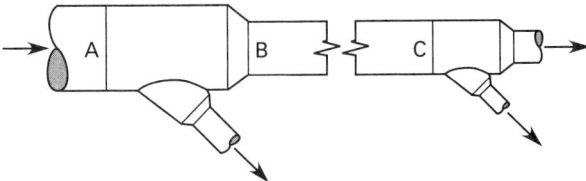

The analytical relationship between the velocities in the sections is given by Eq. 36.53. C is equal to 0.0832 for a static regain coefficient, R, of 0.75. Values of C for other values of R can be calculated from Eq. 36.54.[28]

$$v_A^2 - v_B^2 = Cv_B^{2.43}\left(\frac{L_{BC}}{Q_B^{0.61}}\right) \quad \text{[U.S. only]} \qquad 36.53$$

$$C = \frac{6.256 \times 10^{-2}}{R} \quad \text{[U.S. only]} \qquad 36.54$$

The velocity v_B appears on both sides, making Eq. 36.53 difficult to use. For that reason, duct diameters are often chosen by trial and error. However, graphical aids (e.g., Fig. 36.10) can be used to determine the unknown velocity without extensive trial and error iterations. To use Fig. 36.10, the quantity $L_{AB}/Q_B^{0.61}$ is calculated. The intersection of the L/Q line and the v_A line defines v_B. In most cases, it is assumed that the regain will equal the friction loss in the following section. In that case, v_B is read directly from the horizontal scale. However, Fig. 36.10 can also be used to determine a velocity that will increase or decrease the static pressure by some given amount. If a loss in static pressure is required, move to the right of the intersection point until the vertical separation between the two curves equals the desired loss. (Use the vertical scale on the right edge to

[28]Equation 36.53 is based on a friction factor of 0.0270.

determine the vertical separation.) Then, drop down and read the velocity from the horizontal scale. A gain is handled similarly—by moving to the left.

The following steps constitute a simplified static regain method. In practice, the prediction of the regain coefficient, R, is quite difficult, rendering this method generally unusable.

step 1: Use Table 36.8 to choose a velocity in the main run.

step 2: Size the main run using $A = Q/\text{v}$.

step 3: Find the equivalent length of the main duct from the fan to the first branch takeoff. Assume any unknown bend radii.

step 4: Use Fig. 36.5 or Fig. 36.6 to find the friction loss, FP_{main} in the main run up to the branch takeoff.

step 5: Determine the fan pressure. Assuming that all subsequent friction after the first branch takeoff will be recovered with static regain, the static pressure supplied by the fan will be

$$SP_{\text{fan}} = FP_{\text{main}} + \text{grille discharge pressure} \qquad 36.55$$

Equation 36.55 assumes that the fan discharge velocity and velocity in the main duct run are the same. If the velocities are different, Eq. 36.45 is used to determine a static regain that reduces the pressure supplied by the fan.

step 6: Calculate the flow rate in the duct after the branch takeoff.

$$Q = Q_{\text{main}} - \text{branch flow} \qquad 36.56$$

step 7: Knowing the flow rate and length of the next section, determine the velocity in that section from Fig. 36.10. For other values of R, or when a regain chart is not available, the duct size must be found by trial and error. The duct size is varied until the friction loss equals the regain.

step 8: Solve for the duct size from $A = Q/\text{v}$.

Example 36.13

The fan in the duct system shown moves a total of 1500 ft^3/min. The velocity of the air in the fan is 1700 ft/min. Bends have equivalent lengths of 15 ft. The required outlet grille pressure is 0.25 in wg. Use the static regain method with a static regain coefficient of 0.75 to size the main duct run fan-A-F.

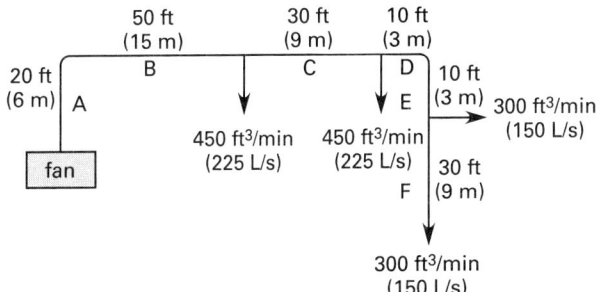

Ventilation

Figure 36.10 *Static Regain Chart (R = 0.75)*

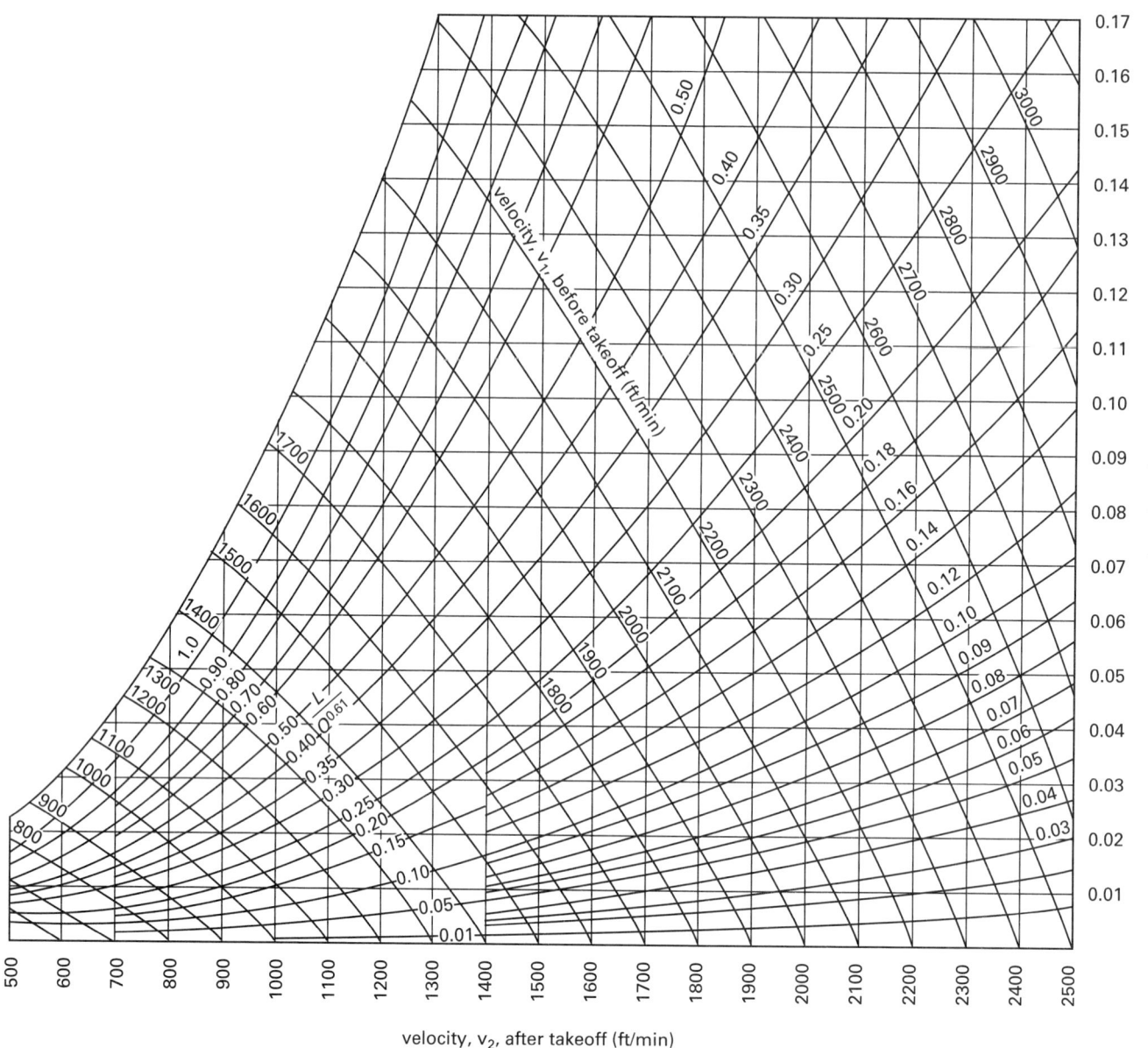

velocity, v_2, after takeoff (ft/min)

net gain or loss (in of water)

Ventilation

Solution

step 1: Choose 1500 ft/min as the main duct velocity.

step 2: The area and diameter of the main duct are

$$A = \frac{Q}{v} = \frac{1500 \ \dfrac{\text{ft}^3}{\text{min}}}{1500 \ \dfrac{\text{ft}^3}{\text{min}}} = 1 \ \text{ft}^2$$

$$D = \sqrt{\frac{4A}{\pi}} = \sqrt{\frac{(4)(1 \ \text{ft}^2)}{\pi}} \left(12 \ \frac{\text{in}}{\text{ft}}\right)$$
$$= 13.5 \ \text{in} \quad (14 \ \text{in})$$

step 3: The equivalent length of the main duct from the fan to the first takeoff and bend is

$$L = 20 \ \text{ft} + 15 \ \text{ft} + 50 \ \text{ft} = 85 \ \text{ft}$$

step 4: From Fig. 36.5, the friction loss in the main run up to the branch takeoff is approximately 0.20 in wg per 100 ft. The actual friction loss is

$$\text{FP}_{\text{main}} = (0.20 \ \text{in wg per 100 ft}) \left(\frac{85 \ \text{ft}}{100 \ \text{ft}}\right)$$
$$= 0.17 \ \text{in wg}$$

step 5: Since the main duct velocity is lower than the fan discharge velocity, there will be static regain from the fan. From Eq. 36.45,

$$SR_{fan} = \frac{R(v_{fan}^2 - v_{main}^2)}{(4005)^2}$$

$$= (0.75)$$

$$\times \left(\frac{\left(1700 \frac{ft}{min}\right)^2 - \left(1500 \frac{ft}{min}\right)^2}{(4005)^2} \right)$$

$$= 0.03 \text{ in wg}$$

$$SP_{fan} = FP_{main} + \text{grille pressure} - SR_{fan}$$

$$= 0.17 \text{ in wg} + 0.25 \text{ in wg} - 0.03 \text{ in wg}$$

$$= 0.39 \text{ in wg}$$

step 6: The flow rates, equivalent lengths, and $L/Q^{0.61}$ ratios for each section are

section	L	Q	$\frac{L}{Q^{0.61}}$
C	30	1050	0.43
D to E	10 + 15 + 10 = 35	600	0.71
F	30	300	0.92

step 7: From Fig. 36.10, the velocities are

section	v (ft/min)
C	1130
D to E	800
F	560

step 8: Solve for the duct size from $A = Q/v$.

$$D = \sqrt{\frac{4A}{\pi}} = \sqrt{\frac{4Q}{\pi v}}$$

$$D_C = \sqrt{\frac{(4)\left(1050 \frac{ft^3}{min}\right)}{\pi\left(1130 \frac{ft}{min}\right)}}\left(12 \frac{in}{ft}\right)$$

$$= 13.1 \text{ in} \quad (13 \text{ in})$$

$$D_{D/E} = \sqrt{\frac{(4)\left(600 \frac{ft^3}{min}\right)}{\pi\left(800 \frac{ft}{min}\right)}}\left(12 \frac{in}{ft}\right)$$

$$= 11.7 \text{ in} \quad (12 \text{ in})$$

$$D_F = \sqrt{\frac{(4)\left(300 \frac{ft^3}{min}\right)}{\pi\left(560 \frac{ft}{min}\right)}}\left(12 \frac{in}{ft}\right)$$

$$= 9.91 \text{ in} \quad (10 \text{ in})$$

43. TOTAL PRESSURE DESIGN METHOD

The previous duct design methods focus on static pressure. The static regain method actually compensates for the inefficiency of a fitting by increasing subsequent duct sizes. None of the duct design methods mentioned attempts to minimize the friction loss. Nevertheless, energy is lost due to friction and turbulence at fittings.

The true measure of the loss in a fitting is represented by the change in total pressure it causes. Unlike static pressure, total pressure along a duct run will always decrease, never increase. Features (i.e., fittings) where the total pressure drops by a significant amount represent inefficient features (i.e., the "wrong" fitting for that location). These should be replaced with fittings with lower loss coefficients. This is the basic premise of the *total pressure design method.*

44. AIR DISTRIBUTION

An *outlet* is a supply opening through which air enters the ventilated space. An *inlet* is a return opening through which air is removed from the ventilated space. In residential construction, outlets are usually placed in floors under windows, and inlets are placed in the ceiling or on walls near the ceilings. In commercial construction, locations are determined by numerous factors and can be anywhere. The terms *grille*, *register*, and *diffuser* are used to describe coverings for the openings, and the terms are used somewhat interchangeably. A *grille* is a decorative covering for an opening. For example, perforated plate grilles are used to cover inlets to return air ducts. A *filter grille* accommodates a furnace filter behind its face. A *diffuser* is a grille with fixed or moveable louvers that guides the supply or return air. The number of louvers is given in bars per unit length. In residential applications, units with 2 bars to the inch are used for heating, while units with 3 bars to the inch are for mixed use. Diffusers can be 1-, 2-, 3-, and 4-way, referring to the number of orthogonal directions the air is directed by louvers. A *register* is a grille with an internal damper. Registers may have louvers.

Incorrect location of registers results in drafts, hot and cold spots, and noise. Locating a register requires knowledge of register performance regarding throw, spread, drop, and terminal velocity for the given airflow and velocity. The *Coanda effect* (*ceiling effect*) causes air to adhere to the ceiling after discharge from a wall register at the ceiling level. The suction effect is proportional to the square of the discharge velocity. With improper designs, the terminal velocity is too low, eventually decreasing to a point (about 4.5 ft/sec (1.5 m/s)) where the ceiling effect suction is inadequate, and the discharged air drops downward, a characteristic known as *dumping*. Occupants find dumping to be uncomfortable, as it places them in drafts. Once dumping begins, the discharge velocity must be increased to 30–40% above the original velocity in order to reattach the airflow to the ceiling.

ASHRAE has two suggestions: For systems in the cooling mode, diffuser selection should be based on the ratio of the diffuser's throw to the length of the zone being supplied. For systems in the heating mode, the diffuser to room temperature difference (DT) should not exceed 15°F to avoid excessive temperature stratification.

For systems in the cooling mode, ASHRAE has developed the *air diffusion performance index* (ADPI) to categorize occupant thermal comfort in sedentary environments with ceilings of at least 8 ft. The ADPI is a single-digit index derived from temperatures and velocities at specific locations (prescribed by the ASHRAE test method) around the outlet diffuser. The ADPI essentially represents the percentage of occupants that would feel comfortable. In general, velocities experienced by occupants should be less than 70 fpm (50 fpm ideally), and the ADPI should predict greater than 70%, and preferably 80–90%, of occupant acceptance. ASHRAE suggests combinations of velocity and temperatures that accomplish these goals.[29] Using manufacturer's data, diffusers should be selected that satisfy the suggested X_{50}/L_{throw} ratios. X_{50} is the manufacturer's reported throw to 50 fpm. L is the *space characteristic length*. This is usually the distance from the outlet to the wall or mid-plane between outlets. The desired throw value can be determined by multiplying the desired throw ratio by the characteristic length. The throw ratio is based on a 9 ft ceiling height. The throw can be increased or decreased by the same amount that the ceiling height exceeds or is less than 9 ft.

For an outlet (grille, register, etc.) to work properly, the air must have a minimum static pressure (typically 0.1–0.3 in wg; 25–75 Pa) at the grille outlet. This *grille pressure* (*terminal pressure*) is added to the static and velocity pressures of the air when the fan is sized.

The *gross area* or *core area* of the grille is its total cross-sectional area. The net opening left when the gross area is reduced by the area of the louvers or dividers is the *free area* or *daylight area*, also known as the *effective area*. The outlet velocity can be found from the core velocity and outlet's coefficient of discharge, C_d, which is typically between 0.7 and 0.9.

$$\text{v}_{\text{outlet}} = \frac{\text{v}_{\text{core}} A_{\text{core}}}{C_d A_{\text{free}}} \qquad 36.57$$

The *throw* (also known as the *blow*) is the distance from the outlet to the *distribution point*. (See Fig. 36.11.) When an outlet has been properly selected, the average *terminal velocity* at the distribution point should be approximately 50 ft/min (0.25 m/s) for sedentary occupants up to 75 ft/min (0.38 m/s) for slightly active occupants.[30,31] Therefore, the throw is roughly the distance from the outlet where the average air velocity is 50 ft/min (0.25 m/s). As it emerges from the outlet, duct air will entrain room air. The increase in airflow width is known as the *rise*, and the absolute width of the airflow is the *spread*. (Even straight outlets have airflows that diverge with a total included angle of up to 20°.)

Figure 36.11 *Air Distribution Terminology*

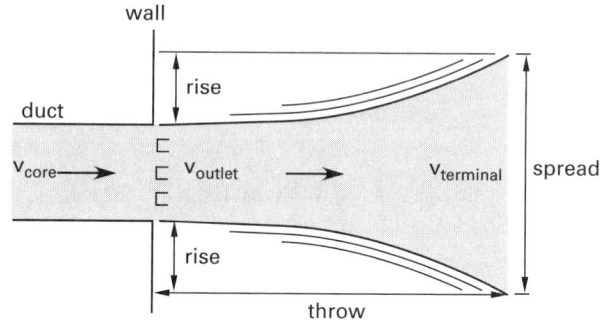

The conservation of momentum law predicts the amount of entrained air. The velocity of the room air is initially zero. The *induction ratio*, IR, is the ratio of combined to outlet air masses.

$$m_{\text{outlet}} + \text{v}_{\text{outlet}} = (m_{\text{outlet}} + m_{\text{entrained}})\text{v}_{\text{combined}} \qquad 36.58$$

$$\text{IR} = \frac{\text{v}_{\text{outlet}}}{\text{v}_{\text{combined}}} = \frac{m_{\text{outlet}} + m_{\text{entrained}}}{m_{\text{outlet}}} \qquad 36.59$$

The centerline velocity can be predicted for a distance x from the outlet. In Eq. 36.60, K is the *outlet constant* supplied by the outlet manufacturer, and R_{fa} is the ratio of free area to core (gross) area. The outlet velocity should be 300–500 ft/min (1.5–2.5 m/s) for ultra-quiet areas, 500–750 ft/min (2.5–3.8 m/s) for residences, theaters, and libraries, and 600–1000 ft/min (3.0–5.1 m/s) for offices and service areas. In noisy industrial areas, velocities as high as 2000 ft/min (10.2 m/s) may be tolerable.

$$\text{v}_{\text{centerline at distance } x} = \frac{K Q_{\text{outlet}}}{x\sqrt{\dfrac{C_d A_{\text{core}} R_{fa}}{A}}}$$

$$= \frac{K Q_{\text{outlet}}}{x\sqrt{C_d A_{\text{free}}}} \qquad 36.60$$

The centerline velocity of the air emerging from a duct is approximately twice that of the average velocity across the duct face. Since the throw is roughly the distance at

---

[29]*ASHRAE Handbook: Fundamentals, Inch-Pound Edition*, American Society of Heating, Refrigerating and Air-Conditioning Engineers, Inc., Atlanta, GA, 2009.

[30]For industrial work, the velocity may be as high as 300 ft/min (1.5 m/s).
[31]Some sources say the minimum air movement should be 20 ft/min (0.1 m/s) or above.

which the average distribution velocity is 50 ft/min (0.25 m/s), the throw is

$$\text{throw} = \frac{K Q_{\text{outlet}}}{\left(100 \ \frac{\text{ft}}{\text{min}}\right) \sqrt{\frac{C_d A_{\text{core}} A_{\text{free}}}{A_{\text{gross}}}}} \quad \begin{bmatrix} \text{consistent} \\ \text{units} \end{bmatrix}$$

$$36.61$$

Generally, the throw should be 75% of the distance from the outlet face to the opposing normal surface. For example, for a ceiling-mounted outlet and a 12 ft (3.6 m) ceiling height, the throw would be approximately 9 ft (2.7 m). The throw should be increased 25% to 50% when the air is released along a wall or near the ceiling in order to compensate for the friction between the air and that surface.

45. EXHAUST DUCT SYSTEMS

Exhaust (return air) duct systems are designed somewhat differently from supply systems.[32] Low-velocity designs often use the equal-friction method. Since the duct operates under a negative pressure, collapse is always a consideration. The negative suction rating of the fan should also not be exceeded.

With a single-fan system, in order for air to enter the return ducts through return grilles and then exhaust through relief dampers, the return air must enter the return air duct above atmospheric pressure. This requires the building to be continuously over-pressurized, causing doors to blow open and other problems. In order to avoid these over-pressurization problems, a *return air fan* after the return grilles and before the relief dampers is used. Air enters the return ductwork near atmospheric pressure, and the return air fan raises the static pressure in the duct to above atmospheric as required for the relief dampers. Since no additional friction sources (as compared to the single-fan system) are added, this does not increase the total fan power required by the system, although it does increase the total initial installation cost.

Converging-flow fittings behave differently than do divided-flow fittings. Figure 36.12 shows a typical converging-flow fitting. The branch loss coefficient (branch-to-downstream) for the fitting is defined by Eq. 36.62.

$$K_{\text{br}} = \frac{\text{TP}_{\text{br}} - \text{TP}_{\text{down}}}{\text{VP}_{\text{down}}} \qquad 36.62$$

The *main loss coefficient* (upstream-to-downstream) is

$$K_{\text{main}} = \frac{\text{TP}_{\text{up}} - \text{TP}_{\text{down}}}{\text{VP}_{\text{down}}} \qquad 36.63$$

---

[32]Some buildings (e.g., those with high-velocity supply systems and where space is limited) don't have return air systems. Even when there is a return air system, some rooms (e.g., those generating odors) may not have return air inlets.

Figure 36.12 Converging-Flow Fitting

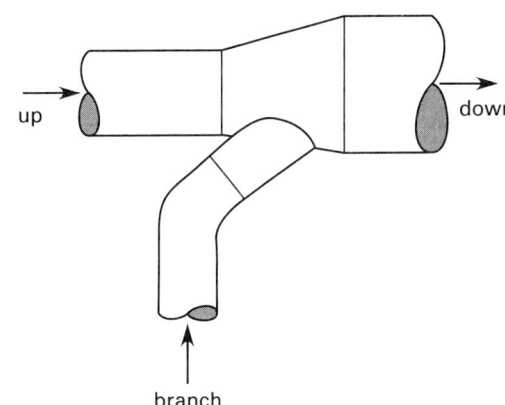

The change in static pressure from the branch to downstream is

$$\text{SP}_{\text{br}} - \text{SP}_{\text{down}} = (1 + K_{\text{br}}) \text{VP}_{\text{down}} - \text{VP}_{\text{br}} \qquad 36.64$$

The change in static pressure from upstream to downstream is

$$\text{SP}_{\text{up}} - \text{SP}_{\text{down}} = (1 + K_{\text{main}}) \text{VP}_{\text{down}} - \text{VP}_{\text{up}} \qquad 36.65$$

46. DUCT SYSTEM NOISE

Fan noise and *duct noise* have many sources.

1. *Vortex shedding* describes air separation from the blade surface and trailing edge. The resulting noise is broadband (i.e., containing a wide range of frequencies). Noise from vortex shedding is minimized by good blade profile design, use of proper pitch angle, and the presence of notched or serrated trailing blade edges.

2. Broadband noise can also be caused by turbulence in the air stream caused by inlet and outlet disturbances, sharp edges, and bends. High-impedance systems (i.e., those with high back-pressure) are noisier than low-impedance systems.

3. Fan speed is a major factor in fan noise. The variation in sound level, L (measured in decibels), with rotational speed, n, is predicted by Eq. 36.66.[33]

$$L_2 = L_1 - 50 \log_{10} \frac{n_1}{n_2} \qquad 36.66$$

4. A fan is generally quieter when operated near its peak efficiency. Noise will vary as the system load varies and the operating point shifts.

5. Substantial noise can be generated by structural vibration related to unbalance, bearings, rotor to stator eccentricity, and motor mounting.

---

[33]A 3 dB change is barely noticeable; a 5 dB change is clearly noticeable; a 10 dB change is twice (or half) as loud.

Ventilation

47. DUST COLLECTION SYSTEMS

Design of dust collection systems draws on concepts similar to the design of duct systems for conditioned air, with notable differences. Dust collection systems are primarily return-air systems, and the fan is often incorporated into a cyclone type *dust collector*.[34] *Two-stage dust collectors* combine a cyclone with a fine-dust filter. Each dust type should be collected separately. To minimize explosion hazards, wood dust, metal dust, and fumes should be separated. Solvent fumes must be collected using non-ferrous ducts (e.g., aluminum) and explosion-proof blowers.

Spiral duct is preferred for dust collection because of its smoothness. 22 gauge metal duct is the most common; 18 gauge is used for heavy duty systems where high collapsing strength and abrasion resistance are required. 24 gauge and 20 gauge can also be used. Branch ducts serve individual dust-generating equipment; branch ducts join a common main duct feeding the dust collector. Branches should enter the main duct horizontally to prevent dust in the mains from falling back into the branches. Ducts should be equipped with access doors for duct cleaning and blockage clearing.

Dust-generating equipment attached to the system is categorized as primary or secondary. All primary machines operate simultaneously and are served by direct branch connections to the main duct. Secondary machines operate sporadically and are isolated from the main duct by sliding-blade blast gates. Equipment requiring the highest airflow should be placed closest to the dust collector. Branch diameters can be determined from the sizes of factory-installed equipment collars or from $Q = Av$ if the airflows are known. Main duct diameters increase as new primary branches enter the main duct in order to accommodate the airflow of all branches while maintaining the main duct velocity. Duct velocities are high in order to entrain dust in the airflow. Duct noise on the suction side can be minimized by using heavier gauge duct, proper hanging, and exterior insulation. Duct silencers (i.e., mufflers) should only be placed in the discharge side. Table 36.9 lists typical duct speeds.

Table 36.9 Typical Duct Velocities for Dust Collection Systems (fpm (m/s))

type of dust	branch duct velocity	main duct velocity
metalworking	4500 (23)	4000 (20)
woodworking	4000 (20)	3500 (18)
other light dust	4000 (20)	3500 (18)

(Multiply ft/min by 0.00508 to obtain m/s.)

The system friction pressure loss is determined using standard methods (e.g., loss coefficients, equivalent lengths, and standard duct friction loss charts). Friction in the ducts, silencers, and up-blast stack caps on the discharge side of the dust collection should be added to the friction from the suction index run ducts, along with losses from entrances, capture nozzles, fittings, filters, and flexible duct. Rules of thumb and assumptions may be needed in the absence of manufacturer's data. Convenient assumptions include (1) 1 in wg entrance loss, (2) 2 in wg filter loss, and (3) flex hose loss equal to three times smooth duct loss.

48. ATKINSON RESISTANCE

In the mining industry, duct resistance is characterized by specialized units and unique assumptions. Some parts of a mine are ventilated using the mine passageway as the duct. Air may flow in manufactured ducts into a mine, but exhaust air may simply escape through the mine tunnel. The pressure drop through a mine tunnel is known as the *Atkinson resistance*.[35] The friction and pressure drop are related to the resistance, R, measured in atkinsons. One *atkinson* is the resistance that results in a pressure drop of 1 lbf/ft^2 with a flow rate of 1000 ft^3/sec at a standard air density, ρ_{std}, of 0.075 lbm/ft^3. Similarly, for use with SI units, one *gaul* is the resistance that results in a pressure drop of 1 Pa with a flow rate of 1 m^3/s at a standard air density of 1.2 kg/m^3.[36,37]

The Atkinson resistance is

$$\Delta p_{Pa} = \left(\frac{\rho_{actual}}{\rho_{std}}\right) R_{gauls} Q^2_{m^3/s} \qquad \text{[SI]} \quad 36.67(a)$$

$$\Delta p_{lbf/ft^2} = \left(\frac{\rho_{actual}}{\rho_{std}}\right) R_{atkinsons} \left(\frac{Q_{cfm}}{1000}\right)^2 \qquad \text{[U.S.]} \quad 36.67(b)$$

The Atkinson resistance, R, can be calculated from the *Atkinson friction factor*, k, and the duct perimeter, S.

$$R = \frac{kLS}{A^3} \qquad 36.68$$

Equation 36.67 is noteworthy because of its derivation. As Eq. 36.68 shows, all constants and physical characteristics, including the flow area, perimeter length, mine length, and friction factor, have been incorporated into the resistance term. While mines in England may all have been similarly constructed near the earth's surface during Atkinson's time, conditions in modern mines are considerably different. In addition to variability in mine length and cross section, the density of air miles below the surface increases greatly, increasing the required fan power. Still, Atkinson resistance remains in widespread use in the modern mining industry, although resistance is often quoted by manufacturers of flexible ventilation ducts in terms of atkinsons or gauls per unit length (e.g., atkinsons/1000 yd or gauls/100 m).

---

[34]Cyclone design is covered in Chap. 42.

[35]Atkinson resistance is named after J. J. Atkinson who investigated mine ventilation in 1862.

[36]One of the reasons that the Atkinson method remains in use today is the relative ease of making onsite measurements. Only a pressure gage and anemometer are required.

[37]Multiply gauls by 16.747 to obtain atkinsons.

Topic V: Air Resources #2: Combustion and Air Quality

Combustion

37 Fuels and Combustion

Nomenclature

A	area	ft^2	m^2
B	volumetric fraction	–	–
°Be	Baumé specific gravity	degree	degree
c_p	specific heat at constant pressure	Btu/lbm-°R	kJ/kg·°C
C_c	Cunningham slip factor	–	–
C_v	coefficient of velocity	–	
d	hydraulic diameter	ft	m
D	draft	in wg	kPa
F	factor	–	–
g	acceleration of gravity, 32.2 (9.81)	ft/sec^2	m/s^2
g_c	gravitational constant, 32.2	$lbm\text{-}ft/lbf\text{-}sec^2$	n.a.
G	gravimetric fraction	–	–
h	enthalpy	Btu/lbm	kJ/kg
H	stack height	ft	m
HHV	higher heating value	Btu/lbm	kJ/kg
HV	heating value	Btu/lbm	kJ/kg
J	gravimetric air-fuel ratio	lbm/lbm	kg/kg
K	friction coefficient	–	–
K	volumetric air-fuel ratio	ft^3/ft^3	m^3/m^3
L	length	ft	m
LHV	lower heating value	Btu/lbm	kJ/kg
m	mass	lbm	kg
M	moisture fraction	–	–
MON	motor octane number	–	–
ON	octane number	–	–
p	pressure	lbf/ft^2	kPa
PN	performance number	–	–
q	heat loss	Btu/lbm	kJ/kg
Q	volumetric flow rate	ft^3/sec	m^3/s
r	radius	ft	m
$R_{a/f}$	air/fuel ratio	lbm/lbm	kg/kg
RON	research octane number	–	–
SG	specific gravity	–	–
t	time	sec	s
T	temperature	°R	K
v	velocity	ft/sec	m/s
V	volume	ft^3	m^3

Symbols

γ	specific weight	ft^3/lbm	m^3/kg
η	efficiency	–	–
λ	mean free path length	ft	m
μ	absolute viscosity	$lbf\text{-}sec/ft^2$	Pa·s
ρ	density	lbm/ft^3	kg/m^3
ω	angular velocity	rad/sec	rad/s
ω	humidity ratio	lbm/lbm	kg/kg

Subscripts

a/f	air/fuel
C	carbon
e	equivalent
f	liquid (fluid)
fg	vaporization

Combustion

g	gas (vapor)
H	hydrogen
i	initial or inner
max	maximum
o	outer
O	oxygen
p	constant pressure or particle
r	residence
s	solid
SE	stack effect
Stk	Stokes
v	velocity
w	water

1. HYDROCARBONS

With the exception of sulfur and related compounds, most fuels are hydrocarbons. Hydrocarbons are further categorized into subfamilies such as *alkynes* (C_nH_{2n-2}, such as acetylene C_2H_2), *alkenes* (C_nH_{2n}, such as ethylene C_2H_4), and *alkanes* (C_nH_{2n+2}, such as octane C_8H_{18}). The alkynes and alkenes are referred to as *unsaturated hydrocarbons*, while the alkanes are referred to as *saturated hydrocarbons*. The alkanes are also known as the *paraffin series* and *methane series*. The alkenes are subdivided into the chain-structured *olefin series* and the ring-structured *naphthalene series*. *Aromatic hydrocarbons* (C_nH_{2n-6}, such as benzene C_6H_6) constitute another subfamily. Names for common hydrocarbon compounds are listed in App. 37.A.

2. CRACKING OF HYDROCARBONS

Cracking is the process of splitting hydrocarbon molecules into smaller molecules. For example, alkane molecules crack into a smaller member of the alkane subfamily and a member of the alkene subfamily. Cracking is used to obtain lighter hydrocarbons (such as those in gasoline) from heavy hydrocarbons (e.g., crude oil).

Cracking can proceed under the influence of high temperatures (*thermal cracking*) or catalysts (*catalytic cracking* or "cat cracking"). Since (from Le Châtelier's principle) cracking at high pressure favors recombination, catalytic cracking is performed at pressures near atmospheric. Catalytic cracking produces gasolines with better antiknock properties than does thermal cracking.

3. FUEL ANALYSIS

Fuel analyses are reported as either percentages by weight (for liquid and solid fuels) or percentages by volume (for gaseous fuels). Percentages by weight are known as *gravimetric analyses*, while percentages by volume are known as *volumetric analyses*. An *ultimate analysis* is a type of gravimetric analysis in which the constituents are reported by atomic species rather than by compound. In an ultimate analysis, combined hydrogen from moisture in the fuel is added to hydrogen from the combustive compounds. (See Sec. 37.6.)

A *proximate analysis* (not "approximate") gives the gravimetric fraction of moisture, volatile matter, fixed carbon, and ash. Sulfur may be combined with the ash or may be specified separately.

A *combustible analysis* considers only the combustible components, disregarding moisture and ash.

Compositions of gaseous fuels are typically given as volumetric fractions of each component. For gas A in a mixture, its *volumetric fraction*, B_A, *mole fraction*, x_A, and *partial pressure ratio*, p_A/p_t, are all the same. This means the same value can be used with Dalton's and Henry's laws.

A volumetric fraction can be converted to a gravimetric fraction by multiplying by the molecular weight and then dividing by the sum of the products of all the volumetric fractions and molecular weights.

4. WEIGHTING OF THERMODYNAMIC PROPERTIES

Many gaseous fuels (and all gaseous combustion products) are mixtures of different compounds. Some thermodynamic properties of mixtures are gravimetrically weighted, while others are volumetrically weighted. Specific heat, specific gas constant, enthalpy, internal energy, and entropy are gravimetrically weighted. For gases, molecular weight, density, and all molar properties are volumetrically weighted.[1]

When a compound experiences a large temperature change, the thermodynamic properties should be evaluated at the average temperature. Table 37.1 can be used to find the specific heat of gases at various temperatures.

5. STANDARD CONDITIONS

Many gaseous fuel properties are specified per unit volume. The phrases "standard cubic foot" (SCF) and "normal cubic meters" (nm or Nm^3) are incorporated into the designations of volumes and flow rates. These volumes are "standard" or "normal" because they are based on standard conditions. Unfortunately, there are numerous definitions of *standard temperature and pressure (standard conditions, STP)*. For nearly a century, the natural gas and oil industries in North America and the Organization of the Petroleum Exporting Countries (OPEC) defined standard conditions as 14.73 psia and 60°F (15.56°C; 288.71K). The *international standard metric conditions* for natural gas and similar fuels are 288.15K (59.00°F; 15.00°C) and 101.325 kPa. Some gas flows related to environmental engineering are based on

---

[1]For gases, molar properties include molar specific heats, enthalpy per mole, and internal energy per mole.

Table 37.1 *Approximate Specific Heats (at Constant Pressure) of Gases (c_p in Btu/lbm-°R; at 1 atm)*

	temperature (°R)							
gas	500	1000	1500	2000	2500	3000	4000	5000
air	0.240	0.249	0.264	0.277	0.286	0.294	0.302	–
carbon dioxide	0.196	0.251	0.282	0.302	0.314	0.322	0.332	0.339
carbon monoxide	0.248	0.257	0.274	0.288	0.298	0.304	0.312	0.316
hydrogen	3.39	3.47	3.52	3.63	3.77	3.91	4.14	4.30
nitrogen	0.248	0.255	0.270	0.284	0.294	0.301	0.310	0.315
oxygen	0.218	0.236	0.253	0.264	0.271	0.276	0.286	0.294
sulfur dioxide	0.15	0.16	0.18	0.19	0.20	0.21	0.23	–
water vapor	0.444	0.475	0.519	0.566	0.609	0.645	0.696	0.729

(Multiply Btu/lbm-°R by 4.187 to obtain kJ/kg·K.)

standard conditions of either 15°C or 20°C and 101.325 kPa. These are different from the standard conditions used for scientific work, historically 32.00°F and 14.696 psia (0°C and 101.325 kPa) but now transitioning to 0°C and 100.00 kPa, and the various temperatures, usually 20°C or 25°C, used to tabulate standard enthalpies of formation and reaction of components. Clearly, it is essential to know the "standard" conditions onto which any gaseous fuel flows are referred.

Some combustion equipment (e.g., particulate collectors) operates within a narrow range of temperatures and pressures. These conditions are referred to as *normal temperature and pressure* (NTP).

6. MOISTURE

If an ultimate analysis of a solid or liquid fuel is given, all of the oxygen is assumed to be in the form of free water.[2] The amount of hydrogen combined as free water is assumed to be one-eighth of the oxygen weight.[3] All remaining hydrogen, known as the *available hydrogen*, is assumed to be combustible. (Also, see Sec. 37.35.)

$$G_{\text{H,combined}} = \frac{G_O}{8} \qquad 37.1$$

$$G_{\text{H,available}} = G_{\text{H,total}} - \frac{G_O}{8} \qquad 37.2$$

Moisture in fuel is undesirable because it increases fuel weight (transportation costs) and decreases available combustion heat.[4] For coal, the *"bed" moisture level* refers to the moisture level when the coal is mined. The terms *dry* and *as fired* are often used in commercial coal specifications. The "as fired" condition corresponds to a specific moisture content when placed in the furnace. The "as fired" heating value should be used, since the moisture actually decreases the useful combustion

energy. The approximate relationship between the two heating values is given by Eq. 37.3, where M is the moisture content from a proximate analysis.[5]

$$\text{HV}_{\text{as fired}} = \text{HV}_{\text{dry}}(1 - M) \qquad 37.3$$

7. ASH AND MINERAL MATTER

Mineral matter is the noncombustible material in a fuel. *Ash* is the residue remaining after combustion. Ash may contain some combustible carbon as well as the original mineral matter. The two terms ("mineral matter" and "ash") are often used interchangeably when reporting fuel analyses.

Ash may also be categorized according to where it is recovered. Dry and wet *bottom ashes* are recovered from *ash pits*. However, as little as 10% of the total ash content may be recovered in the ash pit. *Fly ash* is carried out of the boiler by the flue gas. Fly ash can be deposited on walls and heat transfer surfaces. It will be discharged from the stack if not captured. *Economizer ash* and *air heater ash* are recovered from the devices the ashes are named after.

The finely powdered ash that covers combustion grates protects them from high temperatures.[6] If the ash has a low (i.e., below 2200°F; 1200°C) fusion temperature (melting point), it may form *clinkers* in the furnace and/or *slag* in other high-temperature areas. In extreme cases, it can adhere to the surfaces. Ashes with high melting (fusion) temperatures (i.e., above 2600°F; 1430°C) are known as *refractory ashes*. The T_{250} *temperature* is used as an index of slagging tendencies of an ash. This is the temperature at which the slag becomes molten with a viscosity of 250 poise. Slagging will be experienced when the T_{250} temperature is exceeded.

---

[2]This assumes that none of the oxygen is in the form of carbonates.
[3]The value of 1/8 follows directly from the combustion reaction of hydrogen and oxygen.
[4]A moisture content up to 5% is reported to be beneficial in some mechanically fired boilers. The moisture content contributes to lower temperatures, protecting grates from slag formation and sintering.

[5]Equation 37.3 corrects for the portion of the as-fired coal that isn't combustible, but it neglects other moisture-related losses. Section 37.35 and Sec. 37.37 contain more rigorous discussions.
[6]Some boiler manufacturers rely on the thermal protection the ash provides. For example, coal burned in cyclone boilers should have a minimum ash content of 7% to cover and protect the cyclone barrel tubes. Boiler wear and ash carryover will increase with lower ash contents.

The actual melting point depends on the ash composition. Ash is primarily a mixture of silica (SiO_2), alumina (Al_2O_3), and ferric oxide (Fe_2O_3).[7] The relative proportions of each will determine the melting point, with lower melting points resulting from high amounts of ferric oxide and calcium oxide. The melting points of pure alumina and pure silica are in the 2700–2800°F (1480–1540°C) range.

Coal ash is either of a bituminous type or lignite type. Bituminous-type ash (from midwestern and eastern coals) contains more ferric oxide than lime and magnesia. Lignite-type ash (from western coals) contains more lime and magnesia than ferric oxide.

8. SULFUR

Several forms of sulfur are present in coal and fuel oils. *Pyritic sulfur* (FeS_2) is the primary form. *Organic sulfur* is combined with hydrogen and carbon in other compounds. *Sulfate sulfur* is iron sulfate and gypsum ($CaSO_4 \cdot 2H_2O$). Sulfur in elemental, organic, and pyritic forms oxidizes to sulfur dioxide. *Sulfur trioxide* can be formed under certain conditions. Sulfur trioxide combines with water to form sulfuric acid and is a major source of boiler/stack corrosion and acid rain.

$$SO_3 + H_2O \rightarrow H_2SO_4$$

9. WOOD

Wood is not an industrial fuel, though it may be used in small quantities in developing countries. Most woods have higher heating values around 8300 Btu/lbm (19 MJ/kg), with specific values depending on the species and moisture content.

10. WASTE FUELS

Waste fuels are increasingly being used as fuels in industrial boilers and furnaces. Such fuels include digester and landfill gases, waste process gases, flammable waste liquids, and volatile organic compounds (VOCs) such as benzene, toluene, xylene, ethanol, and methane. Other waste fuels include oil shale, tar sands, green wood, seed and rice hulls, biomass refuse, peat, tire shreddings, and shingle/roofing waste.

The term *refuse-derived fuels* (RDF) is used to describe fuel produced from municipal waste. After separation (removal of glass, plastics, metals, corrugated cardboard, etc.), the waste is merely pushed into the combustion chamber. If the waste is to be burned elsewhere, it is compressed and baled.

The heating value of RDF depends on the moisture content and fraction of combustible material. For RDFs derived from typical municipal wastes, the heating value ranges from 3000 Btu/lbm to 6000 Btu/lbm (7 MJ/kg to 14 MJ/kg). Higher ranges (such as 7500–8500 Btu/lbm (17.5–19.8 MJ/kg)) can be obtained by careful selection of ingredients. Pelletized RDF (containing some coal and a limestone binder) with heating values around 8000 Btu/lbm (18.6 MJ/kg) can be used as a supplemental fuel in coal-fired units.

Scrap tires are an attractive fuel source due to their high heating values, which range from 12,000 Btu/lbm to 16,000 Btu/lbm (28 MJ/kg to 37 MJ/kg). To be compatible with existing coal-loading equipment, tires are chipped or shredded to 1 in (25 mm) size. Tires in this form are known as *tire derived fuel* (TDF). Metal (from tire reinforcement) may or may not be present.

TDF has been shown to be capable of supplying up to 90% of a steam-generating plant's total Btu input without any deterioration in particulate emissions, pollutants, and stack opacity. In fact, compared with some low-quality coals (e.g., lignite), TDF is far superior: about 2.5 times the heating value and about 2.5 times less sulfur per Btu.

11. INCINERATION

Many toxic wastes are incinerated rather than "burned." Incineration and combustion are not the same. *Incineration* is the term used to describe a disposal process that uses combustion to render wastes ineffective (nonharmful, nontoxic, etc.). Wastes and combustible fuel are combined in a furnace, and the heat of combustion destroys the waste.[8] Wastes may themselves be combustible, though they may not be self-sustaining if the moisture content is too high.

Wastes destined for industrial incineration are usually classified by composition and heating value. The Incinerator Institute of America published broad classifications to categorize incinerator waste in 1968. Although the compositions of many wastes, particularly municipal, have changed since then, these categories remain in use. For example, biological and pathological waste is referred to as "type 4 waste." Incinerated wastes are categorized into seven types. Type 0 is *trash* (highly combustible paper and wood, with 10% or less moisture); type 1 is *rubbish* (combustible waste with up to 25% moisture); type 2 is *refuse* (a mixture of rubbish and garbage, with up to 50% moisture); type 3 is *garbage* (residential waste with up to 70% moisture); type 4 is animal solids and pathological wastes (85% moisture); type 5 is industrial process wastes in gaseous, liquid, and semiliquid form; and type 6 is industrial process wastes in solid and semisolid form requiring incineration in hearth, retort, or grate burning equipment.

---

[7]Calcium oxide (CaO), magnesium oxide ("magnesia," MgO), titanium oxide ("titania," TiO_2), ferrous oxide (FeO), and alkalies (Na_2O and K_2O) may be present in smaller amounts.

[8]Rotary kilns can accept waste in many forms. They are "workhorse" incinerators.

Combustion

12. COAL

Coal consists of volatile matter, fixed carbon, moisture, noncombustible mineral matter ("ash"), and sulfur. *Volatile matter* is driven off as a vapor when the coal is heated, and it is directly responsible for flame size. *Fixed carbon* is the combustible portion of the solid remaining after the volatile matter is driven off. Moisture is present in the coal as free water and (for some mineral compounds) as water of hydration. Sulfur, an undesirable component, contributes to heat content.

Coals are categorized into anthracitic, bituminous, and lignitic types. *Anthracite coal* is clean, dense, and hard. It is comparatively difficult to ignite but burns uniformly and smokelessly with a short flame. *Bituminous coal* varies in composition, but generally has a higher volatile content than anthracite, starts easily, and burns freely with a long flame. Smoke and soot are possible if bituminous coal is improperly fired. *Lignite coal* is a coal of woody structure, is very high in moisture, and has a low heating value. It normally ignites slowly due to its moisture, breaks apart when burning, and burns with little smoke or soot.

Coal is burned efficiently in a particular furnace only if it is uniform in size. Screen sizes are used to grade coal, but descriptive terms can also be used.[9] *Run-of-mine coal*, ROM, is coal as mined. *Lump coal* is in the 1–6 in (25–150 mm) range. *Nut coal* is smaller, followed by even smaller *pea coal screenings*, and *fines* (dust).

13. LOW-SULFUR COAL

Switching to low-sulfur coal is a way to meet sulfur emission standards. Western and eastern low-sulfur coals have different properties.[10,11] Eastern low-sulfur coals are generally low-impact coals (that is, few changes need to be made to the power plant when switching to them). Western coals are generally high-impact coals. Properties of typical high- and low-sulfur fuels are shown in Table 37.2.

The lower sulfur content results in less boiler corrosion. However, of all the coal variables, the different ash characteristics are the most significant with regard to the steam generator components. The slagging and fouling tendencies are prime concerns.

14. CLEAN COAL TECHNOLOGIES

A lot of effort has been put into developing technologies that will reduce acid rain, pollution and air toxics (NOx

---

[9]The problem with descriptive terms is that one company's "pea coal" may be as small as 1/4 in (6 mm), while another's may start at 1/2 in (13 mm).

[10]In the United States, low-sulfur coals predominately come from the western United States ("western subbituminous"), although some come from the east ("eastern bituminous").

[11]Some parameters dependent on coal type are coal preparation, firing rate, ash volume and handling, slagging, corrosion rates, dust collection and suppression, and fire and explosion prevention.

Table 37.2 *Typical Properties of High- and Low-Sulfur Coals*

property	high-sulfur	low-sulfur eastern	western
higher heating value,			
Btu/lbm	10,500	13,400	8000
(MJ/kg)	(24.4)	(31.2)	(18.6)
moisture content, %	11.7	6.9	30.4
ash content, %	11.8	4.5	6.4
sulfur content, %	3.2	0.7	0.5
slag melting			
temperature, °F	2400	2900	2900
(°C)	(1320)	(1590)	(1590)

(Multiply Btu/lbm by 2.326 to obtain kJ/kg.)
*All properties are "as received."

and SO_2), and carbon dioxide emissions. These technologies are loosely labeled as *clean coal technologies* (CCTs). Whether or not these technologies can be retrofitted into an existing plant or designed into a new plant depends on the economics of the process.

With *coal cleaning*, coal is ground to ultrafine sizes to remove sulfur and ash-bearing minerals.[12] However, finely ground coal creates problems in handling, storage, and dust production. The risk of fire and explosion increases. Different approaches to reducing the problems associated with transporting and storing finely ground coal include the use of dust suppression chemicals, pelletizing, transportation of coal in liquid slurry form, and pelletizing followed by reslurrying. Some of these technologies may not be suitable for retrofit into existing installations.

With *coal upgrading*, moisture is thermally removed from low-rank coal (e.g., lignite or subbituminous coal). With some technologies, sulfur and ash are also removed when the coal is upgraded.

Reduction in sulfur dioxide emissions is the goal of SO_2 *control* technologies. These technologies include conventional use of lime and limestone in *flue gas desulfurization* (FGD) systems, *furnace sorbent-injection* (FSI), and *duct sorbent-injection*. *Advanced scrubbing* is included in FGD technologies.

Redesigned burners and injectors and adjustment of the flame zone are typical types of *NOx control*. Use of secondary air, injection of ammonia or urea, and selective catalytic reduction (SCR) are also effective in NOx reduction.

Fluidized-bed combustion (FBC) reduces NOx emissions by reducing combustion temperatures to around 1500°F (815°C). FBC is also effective in removing up to 90% of the SO_2. *Atmospheric FBC* operates at atmospheric pressure, but higher thermal efficiencies are achieved in *pressurized FBC* units operating at pressures up to 10 atm.

---

[12]80% or more of *micronized coal* is 44 microns or less in size.

Combustion

Integrated gasification/combined cycle (IGCC) processes are able to remove 99% of all sulfur while reducing NOx to well below current emission standards. *Synthetic gas* (*syngas*) is derived from coal. Syngas has a lower heating value than natural gas, but it can be used to drive gas turbines in combined cycles or as a reactant in the production of other liquid fuels.

15. COKE

Coke, typically used in blast furnaces, is produced by heating coal in the absence of oxygen. The heavy hydrocarbons crack (i.e., the hydrogen is driven off), leaving only a carbonaceous residue containing ash and sulfur. Coke burns smokelessly. *Breeze* is coke smaller than 5/8 in (16 mm). It is not suitable for use in blast furnaces, but steam boilers can be adapted to use it. *Char* is produced from coal in a 900°F (500°C) carbonization process. The volatile matter is removed, but there is little cracking. The process is used to solidify tars, bitumens, and some gases.

16. LIQUID FUELS

Liquid fuels are lighter hydrocarbon products refined from crude petroleum oil. They include liquefied petroleum gas (LPG), gasoline, kerosene, jet fuel, diesel fuels, and heating oils. JP-4 ("jet propellant") is a 50-50 blend of kerosene and gasoline. JP-8 and Jet A are kerosene-like fuels for aircraft gas turbine engines. Important characteristics of a liquid fuel are its composition, ignition temperature, flash point,[13] viscosity, and heating value.

17. FUEL OILS

In the United States, fuel oils are categorized into grades 1 through 6 according to their viscosities.[14] Viscosity is the major factor in determining firing rate and the need for preheating for pumping or atomizing prior to burning. Grades 1 and 2 can be easily pumped at ambient temperatures. In the United States, the heaviest fuel oil used is grade 6, also known as *Bunker C oil*.[15,16] Fuel oils are also classified according to their viscosities as *distillate oils* (lighter) and *residual fuel oils* (heavier).

Like coal, fuel oils contain sulfur and ash that may cause pollution, slagging on the hot end of the boiler, and corrosion in the cold end. Table 37.3 lists typical properties of common commercial fuels, while Table 37.4 lists typical properties of fuel oils.

---

[13]This is different from the *flash point* that is the temperature at which fuel oils generate enough vapor to sustain ignition in the presence of spark or flame.

[14]Grade 3 became obsolete in 1948. Grade 5 is also subdivided into light and heavy categories.

[15]120°F (48°C) is the optimum temperature for pumping no. 6 fuel oil. At that temperature, no. 6 oil has a viscosity of approximately 3000 SSU. Further heating is necessary to lower the viscosity to 150–350 SSU for atomizing.

[16]To avoid *coking* of oil, heating coils in contact with oil should not be hotter than 240°F (116°C).

18. GASOLINE

Gasoline is not a pure compound. It is a mixture of various hydrocarbons blended to give a desired flammability, volatility, heating value, and octane rating. There is an infinite number of blends that can be used to produce gasoline.

Gasoline's heating value depends only slightly on composition. Within a variation of $1^{1}/_{2}$%, the heating value can be taken as 20,200 Btu/lbm (47.0 MJ/kg) for regular gasoline and as 20,300 Btu/lbm (47.2 MJ/kg) for high-octane aviation fuel.

Since gasoline is a mixture of hydrocarbons, different fractions will evaporate at different temperatures. The *volatility* is the percentage of the fuel that evaporates by a given temperature. Typical volatility specifications call for 10% or greater at 167°F (75°C), 50% at 221°F (105°C), and 90% at 275°F (135°C). Low volatility causes difficulty in starting and poor engine performance at low temperatures.

The *octane number* (ON) is a measure of *knock resistance*. It is based on comparison, performed in a standardized one-cylinder engine, with the burning of isooctane and *n*-heptane. *n*-heptane, C_7H_{16}, is rated zero and produces violent knocking. Isooctane, C_8H_{18}, is rated 100 and produces relatively knock-free operation. The percentage blend by volume of these fuels that matches the performance of the gasoline is the octane rating. The *research octane number* (RON) is a measure of the fuel's antiknock characteristics while idling; the *motor octane number* (MON) applies to high-speed, high-acceleration operations. The octane rating reported for commercial gasoline is an average of the two.

Gasolines with octanes greater than 100 (including aviation gasoline) are rated by their performance number. The *performance number* (PN) of gasoline containing antiknock compounds (e.g., tetraethyl lead, TEL, used in aviation gasoline) is related to the octane number.

$$ON = 100 + \frac{PN - 100}{3} \qquad 37.4$$

19. OXYGENATED GASOLINE

In parts of the United States, gasoline is "oxygenated" during the cold winter months. This has led to use of the term "winterized gasoline." The addition of *oxygenates* raises the combustion temperature, reducing carbon monoxide and unburned hydrocarbons.[17] Common oxygenates used in *reformulated gasoline* (RFG) include

---

[17]Oxygenation may not be successful in reducing carbon dioxide. Since the heating value of the oxygenates is lower, fuel consumption of oxygenated fuels is higher. On a per-gallon (per-liter) basis, oxygenation reduces carbon dioxide. On a per-mile (per-kilometer) basis, however, oxygenation appears to increase carbon dioxide. In any case, claims of CO_2 reduction are highly controversial, as the CO_2 footprint required to plant, harvest, dispose of decaying roots, stalks, and leaves (i.e., silage), and refine alcohol is generally ignored.

Table 37.3 Typical Properties of Common Commercial Fuels

	butane	no. 1 diesel	no. 2 diesel	ethanol	gasoline	JP-4	methanol	propane
chemical formula	C_4H_{10}	–	–	C_2H_5OH	–	–	CH_3OH	C_3H_8
molecular weight	58.12	≈ 170	≈ 184	46.07	≈ 126		32.04	44.09
heating value								
higher, Btu/lbm	21,240	19,240	19,110	12,800	20,260		9838	21,646
lower, Btu/lbm	19,620	18,250	18,000	11,500	18,900	18,400	8639	19,916
lower, Btu/gal	102,400	133,332	138,110	76,152	116,485	123,400	60,050	81,855
latent heat of								
vaporization,		115	105	361	142		511 (20°C)	147
Btu/lbm								
specific gravity*	2.01	0.876	0.920	0.794	0.68–0.74	0.8017	0.793	1.55

(Multiply Btu/lbm by 2.326 to obtain kJ/kg.)
(Multiply Btu/gal by 0.2786 to obtain MJ/m³.)
*Specific gravities of propane and butane are with respect to air.

Table 37.4 Typical Properties of Fuel Oils[a]

grade	specific gravity	heating value (MBtu/gal)[b]	heating value (GJ/m³)
1	0.805	134	37.3
2	0.850	139	38.6
4	0.903	145	40.4
5	0.933	148	41.2
6	0.965	151	41.9

(Multiply MBtu/gal by 0.2786 to obtain GJ/m³.)
[a]Actual values will vary depending on composition.
[b]One MBtu equals one thousand Btus.

methyl tertiary-butyl ether (MTBE) and ethanol. Methanol, ethyl tertiary-butyl ether (ETBE), tertiary-amyl methyl ether (TAME), and tertiary-amyl ethyl ether (TAEE), may also be used. (See Table 37.5.) Oxygenates are added to bring the minimum oxygen level to 2–3% by weight.[18]

20. DIESEL FUEL

Properties and specifications for various grades of diesel fuel oil are similar to specifications for fuel oils. Grade 1-D ("D" for diesel) is a light distillate oil for high-speed engines in service requiring frequent speed and load changes. Grade 2-D is a distillate of lower volatility for engines in industrial and heavy mobile service. Grade 4-D is for use in medium speed engines under sustained loads.

Diesel oils are specified by a *cetane number*, which is a measure of the ignition quality (ignition delay) of a fuel. A cetane number of approximately 30 is required for satisfactory operation of low-speed diesel engines. High-speed engines, such as those used in cars, require a cetane number of 45 or more. Like the octane number for gasoline, the cetane number is determined by comparison with standard fuels. Cetane, $C_{16}H_{34}$, has a

cetane number of 100. *n*-methyl-naphthalene, $C_{11}H_{10}$, has a cetane number of zero. The cetane number can be increased by use of such additives as amyl nitrate, ethyl nitrate, and ether.

A diesel fuel's *pour point* number refers to its viscosity. A fuel with a pour point of 10°F (–12°C) will flow freely above that temperature. A fuel with a high pour point will thicken in cold temperatures.

The *cloud point* refers to the temperature at which wax crystals cloud the fuel at lower temperatures. The cloud point should be 20°F (–7°C) or higher. Below that temperature, the engine will not run well.

21. ALCOHOL

Both methanol and ethanol can be used in internal combustion engines. *Methanol (methyl alcohol)* is produced from natural gas and coal, although it can also be produced from wood and organic debris. *Ethanol (ethyl alcohol, grain alcohol)* is distilled from grain, sugarcane, potatoes, and other agricultural products containing various amounts of sugars, starches, and cellulose.

Although methanol generally works as well as ethanol, only ethanol can be produced in large quantities from inexpensive agricultural products and by-products.

Alcohol is water soluble. The concentration of alcohol is measured by its *proof*, where 200 proof is pure alcohol. (180 proof is 90% alcohol and 10% water.)

Gasohol is a mixture of approximately 90% gasoline and 10% alcohol (generally ethanol).[19] Alcohol's heating value is less than gasoline's, so fuel consumption (per distance traveled) is higher with gasohol. Also, since alcohol absorbs moisture more readily than gasoline, corrosion of fuel tanks becomes problematic. In some engines, significantly higher percentages of alcohol may require such modifications as including larger carburetor jets, timing advances, heaters for preheating fuel in cold weather, tank lining to prevent rusting, and alcohol-resistant gaskets.

[18]Other restrictions on gasoline during the winter months intended to reduce pollution may include maximum percentages of benzene and total aromatics and limits on Reid vapor pressure, as well as specifications covering volatile organic compounds, nitric oxide (NOx), and toxins.

[19]In fact, oxygenated gasoline may use more than 10% alcohol.

Table 37.5 *Typical Properties of Common Oxygenates*

	ethanol	MTBE[a]	TAME	ETBE	TAEE
specific gravity	0.794	0.744	0.740	0.770	0.791
octane[b]	115	110	112	105	100
heating value, MBtu/gal[c]	76.2	93.6			
Reid vapor pressure, psig[d]	18	8	15–4	3–4	2
percent oxygen by weight	34.73	18.15	15.66	15.66	13.8
volumetric percent needed to achieve gasoline					
2.7% oxygen by weight	7.8	15.1	17.2	17.2	19.4
2.0% oxygen by weight	5.6	11.0	12.4	12.7	13.0

(Multiply MBtu/gal by 0.2786 to obtain MJ/m^3.)

[a]MTBE is water soluble and does not degrade. It imparts a foul taste to water and is a possible carcinogen. It has been legislatively banned in most states, including California.
[b]Octane is equal to $(1/2)(\text{MON} + \text{RON})$.
[c]One MBtu equals one thousand Btus.
[d]The Reid vapor pressure is the vapor pressure when heated to 100°F (38°C). This may also be referred to as the "blending vapor pressure."

Mixtures of gasoline and alcohol can be designated by the first letter and the fraction of the alcohol. E10 is a mixture of 10% ethanol and 90% gasoline. M85 is a blend of 85% methanol and 15% gasoline.

Alcohol is a poor substitute for diesel fuel because alcohol's cetane number is low—from −20 to +8. Straight injection of alcohol results in poor performance and heavy knocking.

22. GASEOUS FUELS

Various gaseous fuels are used as energy sources, but most applications are limited to natural gas and *liquefied petroleum gases* (LPGs) (i.e., propane, butane, and mixtures of the two).[20,21] *Natural gas* is a mixture of methane (between 55% and 95%), higher hydrocarbons (primarily ethane), and other noncombustible gases. Typical heating values for natural gas range from 950 Btu/ft^3 to 1100 Btu/ft^3 (35 MJ/m^3 to 41 MJ/m^3).

The production of *synthetic gas* (*syngas*) through coal gasification may be applicable to large power generating plants. The cost of gasification, though justifiable to reduce sulfur and other pollutants, is too high for syngas to become a widespread substitute for natural gas.

23. IGNITION TEMPERATURE

The *ignition temperature* (*autoignition temperature*) is the minimum temperature at which combustion can be sustained. It is the temperature at which more heat is generated by the combustion reaction than is lost to the surroundings, after which combustion becomes self-sustaining. For coal, the minimum ignition temperature

varies from around 800°F (425°C) for bituminous varieties to 900–1100°F (480–590°C) for anthracite. For sulfur and charcoal, the ignition temperatures are approximately 470°F (240°C) and 650°F (340°C), respectively.

For gaseous fuels, the ignition temperature depends on the air/fuel ratio, temperature, pressure, and length of time the source of heat is applied. Ignition can be instantaneous or delayed, depending on the temperature. Generalizations can be made for any gas, but the generalized temperatures will be meaningless without specifying all of these factors.

24. ATMOSPHERIC AIR

It is important to make a distinction between "air" and "oxygen." Atmospheric air is a mixture of oxygen, nitrogen, and small amounts of carbon dioxide, water vapor, argon, and other inert ("rare") gases. For the purpose of combustion calculations, all constituents except oxygen are grouped with nitrogen. (See Table 37.6.) It is necessary to supply 4.32 (i.e., 1/0.2315) masses of air to obtain one mass of oxygen. Similarly, it is necessary to supply 4.773 volumes of air to obtain one volume of oxygen. The average molecular weight of air is 28.97, and its specific gas constant is 53.35 ft-lbf/lbm-°R (287.03 J/kg·K).

Table 37.6 *Composition of Dry Air*[a]

component	percent by weight	percent by volume
oxygen	23.15	20.95
nitrogen/inerts	76.85	79.05
ratio of nitrogen to oxygen	3.320	3.773[b]
ratio of air to oxygen	4.320	4.773

[a]Inert gases and CO$_2$ are included as N$_2$.
[b]The value is also reported by various sources as 3.76, 3.78, and 3.784.

[20]A number of *manufactured gases* are of practical (and historical) interest in specific industries, including *coke-oven gas, blast-furnace gas, water gas, producer gas,* and *town gas.* However, these gases are not now in widespread use.
[21]At atmospheric pressure, propane boils at −44°F (−42°C), while butane boils at 31°F (−0.5°C).

25. COMBUSTION REACTIONS

A limited number of elements appear in combustion reactions. Carbon, hydrogen, sulfur, hydrocarbons, and oxygen are the reactants. Carbon dioxide and water vapor are the main products, with carbon monoxide, sulfur dioxide, and sulfur trioxide occurring in lesser amounts. Nitrogen and excess oxygen emerge hotter but unchanged from the stack.

Combustion reactions occur according to the normal chemical reaction principles. Balancing combustion reactions is usually easiest if carbon is balanced first, followed by hydrogen and then by oxygen. When a gaseous fuel has several combustible gases, the volumetric fuel composition can be used as coefficients in the chemical equation.

Table 37.7 lists ideal combustion reactions. These reactions do not include any nitrogen or water vapor that are present in the combustion air.

Table 37.7 Ideal Combustion Reactions

fuel	formula	reaction equation (excluding nitrogen)
carbon (to CO)	C	$2C + O_2 \rightarrow 2CO$
carbon (to CO_2)	C	$C + O_2 \rightarrow CO_2$
sulfur (to SO_2)	S	$S + O_2 \rightarrow SO_2$
sulfur (to SO_3)	S	$2S + 3O_2 \rightarrow 2SO_3$
carbon monoxide	CO	$2CO + O_2 \rightarrow 2CO_2$
methane	CH_4	$CH_4 + 2O_2$ $\rightarrow CO_2 + 2H_2O$
acetylene	C_2H_2	$2C_2H_2 + 5O_2$ $\rightarrow 4CO_2 + 2H_2O$
ethylene	C_2H_4	$C_2H_4 + 3O_2$ $\rightarrow 2CO_2 + 2H_2O$
ethane	C_2H_6	$2C_2H_6 + 7O_2$ $\rightarrow 4CO_2 + 6H_2O$
hydrogen	H_2	$2H_2 + O_2 \rightarrow 2H_2O$
hydrogen sulfide	H_2S	$2H_2S + 3O_2$ $\rightarrow 2H_2O + 2SO_2$
propane	C_3H_8	$C_3H_8 + 5O_2$ $\rightarrow 3CO_2 + 4H_2O$
n-butane	C_4H_{10}	$2C_4H_{10} + 13O_2$ $\rightarrow 8CO_2 + 10H_2O$
octane	C_8H_{18}	$2C_8H_{18} + 25O_2$ $\rightarrow 16CO_2 + 18H_2O$
olefin series	C_nH_{2n}	$2C_nH_{2n} + 3nO_2$ $\rightarrow 2nCO_2 + 2nH_2O$
paraffin series	C_nH_{2n+2}	$2C_nH_{2n+2} + (3n+1)O_2$ $\rightarrow 2nCO_2$ $+ (2n+2)H_2O$

(Multiply oxygen volumes by 3.773 to get nitrogen volumes.)

Example 37.1

A gaseous fuel is 20% hydrogen and 80% methane by volume. What stoichiometric volume of oxygen is required to burn 120 volumes of fuel at the same conditions?

Solution

Write the unbalanced combustion reaction.

$$H_2 + CH_4 + O_2 \rightarrow CO_2 + H_2O$$

Use the volumetric analysis as coefficients of the fuel.

$$0.2H_2 + 0.8CH_4 + O_2 \rightarrow CO_2 + H_2O$$

Balance the carbons.

$$0.2H_2 + 0.8CH_4 + O_2 \rightarrow 0.8CO_2 + H_2O$$

Balance the hydrogens.

$$0.2H_2 + 0.8CH_4 + O_2 \rightarrow 0.8CO_2 + 1.8H_2O$$

Balance the oxygens.

$$0.2H_2 + 0.8CH_4 + 1.7O_2 \rightarrow 0.8CO_2 + 1.8H_2O$$

For gaseous components, the coefficients correspond to the volumes. Since one $(0.2 + 0.8)$ volume of fuel requires 1.7 volumes of oxygen, the required oxygen is

$$(1.7)(120 \text{ volumes of fuel}) = 204 \text{ volumes of oxygen}$$

26. STOICHIOMETRIC REACTIONS

Stoichiometric quantities (*ideal quantities*) are the exact quantities of reactants that are needed to complete a combustion reaction without any reactants left over. Table 37.7 contains some of the more common chemical reactions. Stoichiometric volumes and masses can always be determined from the balanced chemical reaction equation. Table 37.8 can be used to quickly determine stoichiometric amounts for some fuels.

27. STOICHIOMETRIC AIR

Stoichiometric air (*ideal air*) is the air necessary to provide the exact amount of oxygen for complete combustion of a fuel. Stoichiometric air includes atmospheric nitrogen. For each volume of oxygen, 3.773 volumes of nitrogen pass unchanged through the reaction.[22]

Stoichiometric air can be stated in units of mass (pounds or kilograms of air) for solid and liquid fuels, and in units of volume (cubic feet or cubic meters of air) for gaseous fuels. When stated in terms of mass, the

---

[22]The only major change in the nitrogen gas is its increase in temperature. Dissociation of nitrogen and formation of nitrogen compounds can occur but are essentially insignificant.

Combustion

Table 37.8 Consolidated Combustion Data[a,b,c,d]

fuel	for 1 mole of fuel — air O₂	N₂	other products CO₂	H₂O	SO₂	for 1 ft³ of fuel[e] — air O₂	N₂	other products CO₂	H₂O	for 1 lbm of fuel — air O₂	N₂	other products CO₂	H₂O	SO₂	units of fuel
C carbon	1.0	3.773	1.0							0.0833	0.3143	0.0833			moles
	379.5	1432	379.5							31.63	119.3	31.63			ft³
	32.0	106	44.0							2.667	8.883	3.667			lbm
H₂ hydrogen	0.5	1.887		1.0		0.001317	0.004969		0.002635	0.248	0.9357		0.496		moles
	189.8	716.1		379.5		0.5	1.887		1.0	94.12	355.1		188.25		ft³
	16.0	53.0		18.0		0.04216	0.1397		0.04747	7.936	26.29		8.936		lbm
S sulfur	1.0	3.773			1.0					0.03119	0.1177			0.03119	moles
	379.5	1432			379.5					11.84	44.67			11.84	ft³
	32.0	106.0			64.06					0.998	3.306			1.998	lbm
CO carbon monoxide	0.5	1.887	1.0			0.001317	0.004969	0.002635		0.01785	0.06735	0.03570			moles
	189.8	716.1	379.5			0.5	1.887	1.0		6.774	25.56	13.55			ft³
	16.0	53.0	44.01			0.04216	0.1397	0.1160		0.5712	1.892	1.572			lbm
CH₄ methane	2.0	7.546	1.0	2.0		0.00527	0.01988	0.002635	0.00527	0.1247	0.4705	0.06233	0.1247		moles
	759	2864	379.5	758		2.0	7.546	1.0	2.0	47.31	178.5	23.66	47.31		ft³
	64.0	212.0	44.01	36.03		0.1686	0.5586	0.1160	0.0949	3.989	13.21	2.743	2.246		lbm
C₂H₂ acetylene	2.5	9.433	2.0	1.0		0.006588	0.02486	0.00527	0.002635	0.09601	0.3622	0.07681	0.03841		moles
	948.8	3580	758	379.5		2.5	9.443	2.0	1.0	36.44	137.5	29.15	14.57		ft³
	80.0	265.0	88.02	18.02		0.2108	0.6983	0.2319	0.04747	3.072	10.18	3.380	0.6919		lbm
C₂H₄ ethylene	3.0	11.32	2.0	2.0		0.007905	0.02983	0.00527	0.00527	0.1069	0.4033	0.07129	0.07129		moles
	1139	4297	758	758		3.0	11.32	2.0	2.0	40.58	153.1	27.05	27.05		ft³
	96.0	318.0	88.02	36.03		0.2530	0.8380	0.2319	0.0949	3.422	11.34	3.137	1.284		lbm
C₂H₆ ethane	3.5	13.21	2.0	3.0		0.009223	0.03480	0.00527	0.007905	0.1164	0.4392	0.06651	0.09977		moles
	1328	5010	758	1139		3.5	13.21	2.0	3.0	44.17	166.7	25.24	37.86		ft³
	112.0	371.0	88.02	54.05		0.2951	0.9776	0.2319	0.1424	3.724	12.34	2.927	1.797		lbm

(Multiply lbm/ft³ by 0.06243 to obtain kg/m³.)

[a]Rounding of molecular weights and air composition may introduce slight inconsistencies in the table values. This table is based on atomic weights with at least four significant digits, a ratio of 3.773 volumes of nitrogen per volume of oxygen, and 379.5 ft³ per mole at 1 atm and 60°F.

[b]Volumes per unit mass are at 1 atm and 60°F (16°C). To obtain volumes at other temperatures, multiply by (T_F + 460°)/520° or (T_C + 273°)/289°.

[c]The volume of water applies only when the combustion products are at such high temperatures that all of the water is in vapor form.

[d]This table can be used to directly determine some SI ratios. For kg/kg ratios, the values are the same as lbm/lbm. For L/L or m³/m³, use ft³/ft³. For mol/mol, use mole/mole. For mixed units (e.g., ft³/lbm), conversions are required.

[e]Sulfur is not used in gaseous form.

Combustion

stoichiometric ratio of air to fuel masses is known as the ideal *air/fuel ratio*, $R_{a/f}$.

$$R_{a/f,\text{ideal}} = \frac{m_{\text{air,ideal}}}{m_{\text{fuel}}} \qquad 37.5$$

The ideal air/fuel ratio can be determined from the combustion reaction equation. It can also be determined by adding the oxygen and nitrogen amounts listed in Table 37.8.

For fuels whose ultimate analysis is known, the approximate stoichiometric air (oxygen and nitrogen) requirement in pounds of air per pound of fuel (kilograms of air per kilogram of fuel) can be quickly calculated by using Eq. 37.6.[23] All oxygen in the fuel is assumed to be free moisture. All of the reported oxygen is assumed to be locked up in the form of water. Any free oxygen (i.e., oxygen dissolved in liquid fuels) is subtracted from the oxygen requirements.

$$R_{a/f,\text{ideal}} = 34.5\left(\frac{G_C}{3} + G_H - \frac{G_O}{8} + \frac{G_S}{8}\right)$$

$$[\text{solid and liquid fuels}] \qquad 37.6$$

For fuels consisting of a mixture of gases, Eq. 37.7 and the constant J_i from Table 37.9 can be used to quickly determine the stoichiometric air requirements.

$$R_{a/f,\text{ideal}} = \sum G_i J_i \quad [\text{gaseous fuels}] \qquad 37.7$$

Table 37.9 *Approximate Air/Fuel Ratio Coefficients for Components of Natural Gas**

fuel component	J (gravimetric)	K (volumetric)
acetylene, C_2H_2	13.25	11.945
butane, C_4H_{10}	15.43	31.06
carbon monoxide, CO	2.463	2.389
ethane, C_2H_6	16.06	16.723
ethylene, C_2H_4	14.76	14.33
hydrogen, H_2	34.23	2.389
hydrogen sulfide, H_2S	6.074	7.167
methane, CH_4	17.20	9.556
oxygen, O_2	−4.320	−4.773
propane, C_3H_8	15.65	23.89

*Rounding of molecular weights and air composition may introduce slight inconsistencies in the table values. This table is based on atomic weights with at least four significant digits and a ratio of 3.773 volumes of nitrogen per volume of oxygen.

[0]This is a "compromise" equation. Variations in the atomic weights will affect the coefficients slightly. The coefficient 34.5 is reported as 34.43 in some older books. 34.5 is the exact value needed for carbon and hydrogen, which constitute the bulk of the fuel. 34.43 is the correct value for sulfur, but the error is small and is disregarded in this equation.

For fuels consisting of a mixture of gases, the air/fuel ratio can also be expressed in volumes of air per volume of fuel using values of K_i from Table 37.9.

$$\text{volumetric air/fuel ratio} = \sum B_i K_i \quad [\text{gaseous fuels}]$$
$$37.8$$

Example 37.2

Use Table 37.8 to determine the theoretical volume of 90°F (32°C) air required to burn 1 volume of 60°F (16°C) carbon monoxide to carbon dioxide.

Solution

From Table 37.8, 0.5 volumes of oxygen are required to burn 1 volume of carbon monoxide to carbon dioxide. 1.887 volumes of nitrogen accompany the oxygen. The total amount of air at the temperature of the fuel is $0.5 + 1.887 = 2.387$ volumes.

This volume will expand at the higher temperature. The volume at the higher temperature is

$$V_2 = \frac{T_2 V_1}{T_1}$$
$$= \frac{(90°F + 460°)(2.387 \text{ volumes})}{60°F + 460°}$$
$$= 2.53 \text{ volumes}$$

Example 37.3

How much air is required for the ideal combustion of (a) coal with an ultimate analysis of 93.5% carbon, 2.6% hydrogen, 2.3% oxygen, 0.9% nitrogen, and 0.7% sulfur; (b) fuel oil with a gravimetric analysis of 84% carbon, 15.3% hydrogen, 0.4% nitrogen, and 0.3% sulfur; and (c) natural gas with a volumetric analysis of 86.92% methane, 7.95% ethane, 2.81% nitrogen, 2.16% propane, and 0.16% butane?

Solution

(a) Use Eq. 37.6.

$$R_{a/f,\text{ideal}} = 34.5\left(\frac{G_C}{3} + G_H - \frac{G_O}{8} + \frac{G_S}{8}\right)$$
$$= (34.5)\left(\frac{0.935}{3} + 0.026 - \frac{0.023}{8} + \frac{0.007}{8}\right)$$
$$= 11.58 \text{ lbm/lbm (kg/kg)}$$

(b) Use Eq. 37.6.

$$R_{a/f,\text{ideal}} = 34.5\left(\frac{G_C}{3} + G_H + \frac{G_S}{8}\right)$$
$$= (34.5)\left(\frac{0.84}{3} + 0.153 + \frac{0.003}{8}\right)$$
$$= 14.95 \text{ lbm/lbm (kg/kg)}$$

(c) Use Eq. 37.8 and the coefficients from Table 37.9.

$$\frac{\text{volumetric}}{\text{air/fuel ratio}} = \sum B_i K_i$$

$$= (0.8692)\left(9.556 \; \frac{\text{ft}^3}{\text{ft}^3}\right)$$

$$+ (0.0795)\left(16.723 \; \frac{\text{ft}^3}{\text{ft}^3}\right)$$

$$+ (0.0216)\left(23.89 \; \frac{\text{ft}^3}{\text{ft}^3}\right)$$

$$+ (0.0016)\left(31.06 \; \frac{\text{ft}^3}{\text{ft}^3}\right)$$

$$= 10.20 \; \text{ft}^3/\text{ft}^3 \; (\text{m}^3/\text{m}^3)$$

28. INCOMPLETE COMBUSTION

Incomplete combustion occurs when there is insufficient oxygen to burn all of the hydrogen, carbon, and sulfur in the fuel. Without enough available oxygen, carbon burns to carbon monoxide.[24] Carbon monoxide in the flue gas indicates incomplete and inefficient combustion. Incomplete combustion is caused by cold furnaces, low combustion temperatures, poor air supply, smothering from improperly vented stacks, and insufficient mixing of air and fuel.

29. SMOKE

The amount of smoke can be used as an indicator of combustion completeness. Smoky combustion may indicate improper air/fuel ratio, insufficient draft, leaks, insufficient preheat, or misadjustment of the fuel system.

Smoke measurements are made in a variety of ways, with the standards depending on the equipment used. Photoelectric sensors in the stack are used to continuously monitor smoke. The *smoke spot number* (SSN) and ASTM smoke scale are used with continuous stack monitors. For coal-fired furnaces, the maximum desirable smoke number is SSN 4. For grade 2 fuel oil, the SSN should be less than 1; for grade 4, SSN 4; for grades 5L, 5H, and low-sulfur residual fuels, SSN 3; for grade 6, SSN 4.

The *Ringelmann scale* is a subjective method in which the smoke density is visually compared to five standardized white-black grids. Ringelmann chart no. 0 is solid white; chart no. 5 is solid black. Ringelmann chart no. 1, which is 20% black, is the preferred (and required) operating point for most power plants.

30. FLUE GAS ANALYSIS

Combustion products that pass through a furnace's exhaust system are known as *flue gases* (*stack gases*).

Flue gases are almost all nitrogen.[25] Nitrogen oxides are not present in large enough amounts to be included separately in combustion reactions.

The actual composition of flue gases can be obtained in a number of ways, including by modern electronic detectors, less expensive "length-of-stain" detectors, and direct sampling with an Orsat-type apparatus.

The antiquated *Orsat apparatus* determines the volumetric percentages of CO_2, CO, O_2, and N_2 in a flue gas. The sampled flue gas passes through a series of chemical compounds. The first compound absorbs only CO_2, the next only O_2, and the third only CO. The unabsorbed gas is assumed to be N_2 and is found by subtracting the volumetric percentages of all other components from 100%. An Orsat analysis is a dry analysis; the percentage of water vapor is not usually determined. A wet volumetric analysis (needed to compute the dew-point temperature) can be derived if the volume of water vapor is added to the Orsat volumes.

The Orsat procedure is now rarely used, although the term "Orsat" may be generally used to refer to any flue gas analyzer. Modern electronic analyzers can determine free oxygen (and other gases) independently of the other gases.

31. ACTUAL AND EXCESS AIR

Complete combustion occurs when all of the fuel is burned. Usually, *excess air* is required to achieve complete combustion. Excess air is expressed as a percentage of the theoretical air requirements. Different fuel types burn more efficiently with different amounts of excess air. Coal-fired boilers need approximately 30–35% excess air, oil-based units need about 15%, and natural gas burners need about 10%.

The actual air/fuel ratio for dry, solid fuels with no unburned carbon can be estimated from the volumetric flue gas analysis and the gravimetric fractions of carbon and sulfur in the fuel.

$$R_{a/f,\text{actual}} = \frac{m_{\text{air,actual}}}{m_{\text{fuel}}} = \frac{3.04 B_{N_2}\left(G_C + \dfrac{G_S}{1.833}\right)}{B_{CO_2} + B_{CO}} \qquad 37.9$$

Too much free oxygen or too little carbon dioxide in the flue gas are indicative of excess air. Because the relationship between oxygen and excess air is relatively insensitive to fuel composition, oxygen measurements are replacing standard carbon dioxide measurements in determining combustion efficiency.[26] The relationship between excess air and the volumetric fraction of oxygen in the flue gas is given in Table 37.10.

---

[24]Toxic alcohols, ketones, and aldehydes may also be formed during incomplete combustion.

[25]This assumption is helpful in making quick determinations of the thermodynamic properties of flue gases.
[26]The relationship between excess air required and CO_2 is much more dependent on fuel type (e.g., liquid) and furnace design.

Table 37.10 *Approximate Volumetric Percentage of Oxygen in Stack Gas*

fuel*	excess air							
	0%	1%	5%	10%	20%	50%	100%	200%
fuel oils,								
no. 2–6	0	0.22	1.06	2.02	3.69	7.29	10.8	14.2
natural gas	0	0.25	1.18	2.23	4.04	7.83	11.4	14.7
propane	0	0.23	1.08	2.06	3.75	7.38	10.9	14.3

*Values for coal are only marginally lower than the values for fuel oils.

Reducing the air/fuel ratio will have several outcomes. (a) The furnace temperature will increase due to a reduction in cooling air. (b) The flue gas will decrease in quantity. (c) The heat loss will decrease. (d) The furnace efficiency will increase. (e) Pollutants will (usually) decrease.

With a properly adjusted furnace and good mixing, the flue gas will contain no carbon monoxide, and the amount of carbon dioxide will be maximized. The stoichiometric amount of carbon dioxide in the flue gas is known as the *ultimate CO_2* and is the theoretical maximum level of carbon dioxide. The air/fuel mixture should be adjusted until the maximum level of carbon dioxide is attained.

Example 37.4

Propane (C_3H_8) is burned completely with 20% excess air. What is the volumetric fraction of carbon dioxide in the flue gas?

Solution

From Table 37.7, the balanced chemical reaction equation is

$$C_3H_8 + 5O_2 \rightarrow 3CO_2 + 4H_2O$$

With 20% excess air, the oxygen volume is $(1.2)(5) = 6$.

$$C_3H_8 + 6O_2 \rightarrow 3CO_2 + 4H_2O + O_2$$

From Table 37.6, there are 3.773 volumes of nitrogen for every volume of oxygen.

$$(3.773)(6) = 22.6$$

$$C_3H_8 + 6O_2 + 22.6N_2 \rightarrow 3CO_2 + 4H_2O + O_2 + 22.6N_2$$

For gases, the coefficients can be interpreted as volumes. The volumetric fraction of carbon dioxide is

$$B_{CO_2} = \frac{3}{3 + 4 + 1 + 22.6}$$
$$= 0.0980 \quad (9.8\%)$$

32. CALCULATIONS BASED ON FLUE GAS ANALYSIS

Equation 37.10 gives the approximate percentage (by volume) of actual excess air.

$$\frac{\text{actual excess air}}{\text{\% by volume}} = \frac{(B_{O_2} - 0.5B_{CO})}{0.264B_{N_2} - B_{O_2} + 0.5B_{CO}} \times 100\%$$

37.10

The ultimate CO_2 (i.e., the maximum theoretical carbon dioxide) can be determined from Eq. 37.11.

$$\frac{\text{ultimate } CO_2}{\text{\% by volume}} = \frac{B_{CO_2,\text{actual}}}{1 - 4.773B_{O_2,\text{actual}}} \times 100\% \qquad 37.11$$

The mass ratio of dry flue gases to solid fuel is given by Eq. 37.12.

$$\frac{\text{mass of flue gas}}{\text{mass of solid fuel}} = \frac{\left(11B_{CO_2} + 8B_{O_2} + 7(B_{CO} + B_{N_2})\right) \times \left(G_C + \dfrac{G_S}{1.833}\right)}{3(B_{CO_2} + B_{CO})}$$

37.12

Example 37.5

A sulfur-free coal has a proximate analysis of 75% carbon. The volumetric analysis of the flue gas is 80.2% nitrogen, 12.6% carbon dioxide, 6.2% oxygen, and 1.0% carbon monoxide. Calculate the (a) actual air/fuel ratio, (b) percentage excess air, (c) ultimate carbon dioxide, and (d) mass of flue gas per mass of fuel.

Solution

(a) Use Eq. 37.9.

$$R_{a/f,\text{actual}} = \frac{3.04B_{N_2}G_C}{B_{CO_2} + B_{CO}}$$
$$= \frac{(3.04)(0.802)(0.75)}{0.126 + 0.01}$$
$$= 13.4 \text{ lbm air/lbm fuel (kg air/kg fuel)}$$

(b) Use Eq. 37.10.

$$\frac{\text{actual}}{\text{excess air}} = \frac{B_{O_2} - 0.5B_{CO}}{0.264B_{N_2} - B_{O_2} + 0.5B_{CO}} \times 100\%$$
$$= \frac{0.062 - (0.5)(0.01)}{(0.264)(0.802) - 0.062 + (0.5)(0.01)} \times 100\%$$
$$= 36.8\% \text{ by volume}$$

(c) Use Eq. 37.11.

$$\text{ultimate } CO_2 = \frac{B_{CO_2,\text{actual}}}{1 - 4.773 B_{O_2,\text{actual}}} \times 100\%$$

$$= \frac{0.126}{1 - (4.773)(0.062)} \times 100\%$$

$$= 17.9\% \text{ by volume}$$

(d) Use Eq. 37.12.

$$\frac{\text{mass of flue gas}}{\text{mass of solid fuel}} = \frac{\left(11 B_{CO_2} + 8 B_{O_2} + 7(B_{CO} + B_{N_2})\right) \times \left(G_C + \dfrac{G_S}{1.833}\right)}{3(B_{CO_2} + B_{CO})}$$

$$= \frac{\left(\begin{array}{c}(11)(0.126) + (8)(0.062) \\ + (7)(0.01 + 0.802)\end{array}\right)(0.75)}{(3)(0.126 + 0.01)}$$

$$= 13.9 \text{ lbm flue gas/lbm fuel}$$

$$(\text{kg flue gas/kg fuel})$$

33. TEMPERATURE OF FLUE GAS

The temperature of the gas at the furnace outlet—before the gas reaches any other equipment—should be approximately 550°F (300°C). Overly low temperatures mean there is too much excess air. Overly high temperatures—above 750°F (400°C)—mean that heat is being wasted to the atmosphere and indicate other problems (ineffective heat transfer surfaces, overfiring, defective combustion chamber, etc.).

The *net stack temperature* is the difference between the stack and local environment temperatures. The net stack temperature should be as low as possible without causing corrosion of the low end.

34. DEW POINT OF FLUE GAS MOISTURE

The *dew point* is the temperature at which the water vapor in the flue gas begins to condense in a constant pressure process. To avoid condensation and corrosion in the stack, the temperature of the flue gases must be above the dew point. When there is no sulfur in the fuel, the dew point is typically around 100°F (40°C). The presence of sulfur in virtually any quantity increases the actual dew point to approximately 300°F (150°C).[27]

Dalton's law predicts the dew point of moisture in the flue gas. The partial pressure of the water vapor depends on the mole fraction (i.e., the volumetric fraction) of water vapor. The higher the water vapor pressure, the higher the dew point. Air entering a furnace can also contribute to moisture in the flue gas. This moisture should be added to the water vapor from combustion when calculating the mole fraction.

$$\begin{array}{c}\text{water vapor} \\ \text{partial pressure}\end{array} = \begin{array}{c}(\text{water vapor mole fraction}) \\ \times (\text{flue gas pressure})\end{array}$$

$$37.13$$

Once the water vapor's partial pressure is known, the dew point can be found from steam tables as the saturation temperature corresponding to the partial pressure.

35. HEAT OF COMBUSTION

The *heating value* of a fuel can be determined experimentally in a *bomb calorimeter*, or it can be estimated from the fuel's chemical analysis. The *higher heating value* (HHV), or *gross heating value*, of a fuel includes the heat of vaporization (condensation) of the water vapor formed from the combustion of hydrogen in the fuel. The *lower heating value* (LHV), or *net heating value*, assumes that all the products of combustion remain gaseous. The LHV is generally the value to use in calculations of thermal energy generated, since the heat of vaporization is not recovered within the furnace.

Traditionally, heating values have been reported on an HHV basis for coal-fired systems but on an LHV basis for natural gas-fired combustion turbines. There is an 11% difference between HHV and LHV thermal efficiencies for gas-fired systems and a 4% difference for coal-fired systems, approximately.

The HHV can be calculated from the LHV if the enthalpy of vaporization, h_{fg}, is known at the pressure of the water vapor.[28] In Eq. 37.14, m_{water} is the mass of water produced per unit (lbm, mole, m³, etc.) of fuel.

$$HHV = LHV + m_{\text{water}} h_{fg} \qquad 37.14$$

As presented in Sec. 37.6, only the hydrogen that is not locked up with oxygen in the form of water is combustible. This is known as the *available hydrogen*. The correct percentage of combustible hydrogen, $G_{H,\text{available}}$, is calculated from the hydrogen and oxygen fraction. Equation 37.15 (same as Eq. 37.2) assumes that all of the oxygen is present in the form of water.

$$G_{H,\text{available}} = G_{H,\text{total}} - \frac{G_O}{8} \qquad 37.15$$

Dulong's formula calculates the higher heating value of coals and coke with a 2–3% accuracy range for moisture

---

[27]The theoretical dew point is even higher—up to 350–400°F (175–200°C). For complex reasons, the theoretical value is not attained.

[28]For the purpose of initial studies, the heat of vaporization is usually assumed to be 1040 Btu/lbm (2.42 MJ/kg). This corresponds to a partial pressure of approximately 1 psia (7 kPa) and a dew point of 100°F (40°C).

contents below approximately 10%.[29] The gravimetric or volumetric analysis percentages for each combustible element (including sulfur) are multiplied by the heating value per unit (mass or volume) from App. 37.A and summed.

$$\text{HHV}_{\text{MJ/kg}} = 32.78 G_C + 141.8\left(G_H - \frac{G_O}{8}\right)$$
$$+ 9.264 G_S$$
$$\text{[SI]} \quad 37.16(a)$$

$$\text{HHV}_{\text{Btu/lbm}} = 14{,}093 G_C + 60{,}958\left(G_H - \frac{G_O}{8}\right)$$
$$+ 3983 G_S$$
$$\text{[U.S.]} \quad 37.16(b)$$

The higher heating value of gasoline can be approximated from the Baumé specific gravity, °Be.

$$\text{HHV}_{\text{gasoline,MJ/kg}} = 42.61 + 0.093(°\text{Be} - 10)$$
$$\text{[SI]} \quad 37.17(a)$$

$$\text{HHV}_{\text{gasoline,Btu/lbm}} = 18{,}320 + 40(°\text{Be} - 10)$$
$$\text{[U.S.]} \quad 37.17(b)$$

The heating value of petroleum oils (including diesel fuel) can also be approximately determined from the oil's specific gravity. The values derived by using Eq. 37.18 may not exactly agree with values for specific oils because the equation does not account for refining methods and sulfur content. Equation 37.18 was originally intended for combustion at constant volume, as in a gasoline engine. However, variations in heating values for different oils are very small; therefore, Eq. 37.18 is widely used as an approximation for all types of combustion, including constant pressure combustion in industrial boilers.

$$\text{HHV}_{\text{fuel oil,MJ/kg}} = 51.92 - 8.792(\text{SG})^2 \quad \text{[SI]} \quad 37.18(a)$$

$$\text{HHV}_{\text{fuel oil,Btu/lbm}} = 22{,}320 - 3780(\text{SG})^2 \quad \text{[U.S.]} \quad 37.18(b)$$

Example 37.6

A coal has an ultimate analysis of 93.9% carbon, 2.1% hydrogen, 2.3% oxygen, 0.3% nitrogen, and 1.4% ash. What are its (a) higher and (b) lower heating values in Btu/lbm?

---

[29]The coefficients in Eq. 37.16 are slightly different from the coefficients originally proposed by Dulong. Equation 37.16 reflects currently accepted heating values that were unavailable when Dulong developed his formula. Equation 37.16 makes these assumptions: (1) None of the oxygen is in carbonate form. (2) There is no free oxygen. (3) The hydrogen and carbon are not combined as hydrocarbons. (4) Carbon is amorphous, not graphitic. (5) Sulfur is not in sulfate form. (6) Sulfur burns to sulfur dioxide.

Solution

(a) The noncombustible ash, oxygen, and nitrogen do not contribute to heating value. Some of the hydrogen is in the form of water. From Eq. 37.2, the available hydrogen fraction is

$$G_{H,\text{available}} = G_{H,\text{total}} - \frac{G_O}{8}$$
$$= 2.1\% - \frac{2.3\%}{8}$$
$$= 1.8\%$$

From App. 37.A, the higher heating values of carbon and hydrogen are 14,093 Btu/lbm and 60,958 Btu/lbm, respectively. From Eq. 37.16, the total heating value per pound of coal is

$$\text{HHV} = 14{,}093 G_C + 60{,}958 G_{H,\text{available}}$$
$$= \left(14{,}093 \frac{\text{Btu}}{\text{lbm}}\right)(0.939)$$
$$+ \left(60{,}958 \frac{\text{Btu}}{\text{lbm}}\right)(0.018)$$
$$= 14{,}331 \text{ Btu/lbm}$$

(b) All of the combustible hydrogen forms water vapor. The mass of water produced is equal to the hydrogen mass plus eight times as much oxygen.

$$m_w = m_{H,\text{available}} + 8m_O$$
$$= 0.018 + (8)(0.018)$$
$$= 0.162 \text{ lbm water/lbm coal}$$

Assume that the partial pressure of the water vapor is approximately 1 psia (7 kPa). Then, from steam tables, the heat of condensation will be 1040 Btu/lbm. From Eq. 37.14, the lower heating value is approximately

$$\text{LHV} = \text{HHV} - m_w h_{fg}$$
$$= 14{,}331 \frac{\text{Btu}}{\text{lbm}} - \left(0.162 \frac{\text{lbm}}{\text{lbm}}\right)\left(1040 \frac{\text{Btu}}{\text{lbm}}\right)$$
$$= 14{,}163 \text{ Btu/lbm}$$

Alternatively, the lower heating value of the coal can be calculated from Eq. 37.16 by substituting the lower (net) hydrogen heating value from App. 37.A, 51,623 Btu/lbm, for the gross heating value of 60,598 Btu/lbm. This yields 14,167 Btu/lbm.

36. MAXIMUM THEORETICAL COMBUSTION (FLAME) TEMPERATURE

It can be assumed that the maximum theoretical increase in flue gas temperature will occur if all of the combustion energy is absorbed adiabatically by the smallest possible quantity of combustion products. This provides a method of estimating the *maximum theoretical combustion temperature*, also sometimes called the *maximum flame temperature* or *adiabatic flame temperature*.[30]

In Eq. 37.19, the mass of the products is the sum of the fuel, oxygen, and nitrogen masses for stoichiometric combustion. The mean specific heat is a gravimetrically weighted average of the values of c_p for all combustion gases. (Since nitrogen comprises the majority of the combustion gases, the mixture's specific heat will be approximately that of nitrogen.) The heat of combustion can be found either from the lower heating value, LHV, or from a difference in air enthalpies across the furnace.

$$T_{max} = T_i + \frac{LHV}{m_{products}c_{p,mean}}$$ *37.19*

Due to thermal losses, incomplete combustion, and excess air, actual flame temperatures are always lower than the theoretical temperature. Most fuels produce flame temperatures in the range of 3350–3800°F (1850–2100°C).

37. COMBUSTION LOSSES

A portion of the combustion energy is lost in heating the dry flue gases, dfg.[31] This is known as *dry flue gas loss*. In Eq. 37.20, $m_{flue\,gas}$ is the mass of dry flue gas per unit mass of fuel. It can be estimated from Eq. 37.12. Although the full temperature difference is used, the specific heat should be evaluated at the average temperature of the flue gas. For quick estimates, the dry flue gas can be assumed to be pure nitrogen.

$$q_1 = m_{flue\,gas}c_p(T_{flue\,gas} - T_{incoming\,air})$$ *37.20*

Heat is lost in vaporizing the water formed during the combustion of hydrogen. In Eq. 37.21, m_{vapor} is the mass of vapor per pound of fuel. G_H is the gravimetric fraction of hydrogen in the fuel. The coefficient 8.94 is essentially $8 + 1 = 9$ and converts the gravimetric mass of hydrogen to gravimetric mass of water formed. h_g is the enthalpy of superheated steam at the flue gas temperature and the partial pressure of the water vapor. h_f is the enthalpy of saturated liquid at the air's entrance temperature.

$$q_2 = m_{vapor}(h_g - h_f) = 8.94G_H(h_g - h_f)$$ *37.21*

Heat is lost when it is absorbed by moisture originally in the combustion air (and by free moisture in the fuel, if any). In Eq. 37.22, $m_{combustion\,air}$ is the mass of combustion air per pound of fuel. ω is the humidity ratio. h'_g is the enthalpy of superheated steam at the air's entrance temperature and partial pressure of the water vapor.

$$q_3 = m_{atmospheric\,water\,vapor}(h_g - h'_g)$$
$$= \omega m_{combustion\,air}(h_g - h'_g)$$ *37.22*

When carbon monoxide appears in the flue gas, potential energy is lost in incomplete combustion. The higher heating value of carbon monoxide in combustion to CO_2 is 4347 Btu/lbm (9.72 MJ/kg).[32]

$$q_4 = \frac{2.334HHV_{CO}G_C B_{CO}}{B_{CO_2} + B_{CO}}$$ *37.23*

For solid fuels, energy is lost in unburned carbon in the ash. (Some carbon may be carried away in the flue gas, as well.) This is known as *combustible loss* or *unburned fuel loss*. In Eq. 37.24, m_{ash} is the mass of ash produced per pound of fuel consumed. $G_{C,ash}$ is the gravimetric fraction of carbon in the ash. The heating value of carbon is 14,093 Btu/lbm (32.8 MJ/kg).

$$q_5 = HHV_C m_{ash} G_{C,ash}$$ *37.24*

Energy is also lost through radiation from the exterior boiler surfaces. This can be calculated if enough information is known. The *radiation loss* is fairly insensitive to different firing rates, and once calculated it can be considered constant for different conditions.

Other conditions where energy can be lost include air leaks, poor pulverizer operation, excessive blowdown, steam leaks, missing or loose insulation, and excessive soot-blower operation. Losses due to these sources must be evaluated on a case-by-case basis. The term *manufacturer's margin* is used to describe an accumulation of various unaccounted-for losses, which can include incomplete combustion to CO, energy loss in ash, instrument errors, energy carried away by atomizing steam, sulfation and calcination reactions in fluidized bed combustion boilers, and loss due to periodic blowdown. It can be 0.25–1.5% of all energy inputs, depending on the type of combuster/boiler.

---

[30]Flame temperature is limited by the dissociation of common reaction products (CO_2, N_2, etc.). At high enough temperatures (3400–3800°F; 1880–2090°C), the endothermic dissociation process reabsorbs combustion heat and the temperature stops increasing. The temperature at which this occurs is known as the *dissociation temperature* (*maximum flame temperature*). This definition of flame temperature is not a function of heating values and flow rates.

[31]The abbreviation "dfg" for dry flue gas is peculiar to the combustion industry. It may not be recognized outside of that field.

[32]2.334 is the ratio of the molecular weight of carbon monoxide (28.01) to carbon (12), which is necessary to convert the higher heating value of carbon monoxide from a mass of CO to a mass of C. The product $2.334HHV_{CO}$ is often stated as 10,160 Btu/lbm (23.63 MJ/kg), although the actual calculated value is somewhat less. Other values such as 10,150 Btu/lbm (23.61 MJ/kg) and 10,190 Btu/lbm (23.70 MJ/kg) are encountered.

38. COMBUSTION EFFICIENCY

The *combustion efficiency* (also referred to as *boiler efficiency*, *furnace efficiency*, and *thermal efficiency*) is the overall thermal efficiency of the combustion reaction. Furnace/boilers for all fuels (e.g., coal, oil, and gas) with air heaters and economizers have 75–85% efficiency ranges, with all modern installation trending to the higher end of the range.

In Eq. 37.25, m_{steam} is the mass of steam produced per pound of fuel burned. The useful heat may also be determined from the boiler rating. One *boiler horsepower* equals approximately 33,475 Btu/hr (9.8106 kW).[33]

$$\eta = \frac{\text{useful heat extracted}}{\text{heating value}}$$
$$= \frac{m_{steam}(h_{steam} - h_{feedwater})}{\text{HHV}} \qquad 37.25$$

Calculating the efficiency by subtracting all known losses is known as the *loss method*. Minor sources of thermal energy, such as the entering air and feedwater, are essentially disregarded.

$$\eta = \frac{\text{HHV} - q_1 - q_2 - q_3 - q_4 - q_5 - \text{radiation}}{\text{HHV}}$$
$$= \frac{\text{LHV} - q_1 - q_4 - q_5 - \text{radiation}}{\text{HHV}} \qquad 37.26$$

Combustion efficiency can be improved by decreasing either the temperature or the volume of the flue gas or both. Since the latent heat of moisture is a loss, and since the amount of moisture generated corresponds to the hydrogen content of the fuel, a minimum efficiency loss due to moisture formation cannot be eliminated. This minimum loss is approximately 13% for natural gas, 8% for oil, and 6% for coal.

39. DRAFT

The amount of air that flows through a furnace is determined by the furnace draft. *Draft* is the difference in static pressures that causes the flue gases to flow.[34] It is usually stated in inches of water (kPa). *Natural draft* (ND) furnaces rely on the *stack effect* (*chimney draft*) to draw off combustion gases. Air flows through the furnace and out the stack due to the pressure differential caused by reduced densities in the stack.[35]

Forced draft (FD) fans located before the furnace are used to supply air for burning. Combustion occurs under pressure, hence the descriptive term "pressure-fired unit" for such fans. FD fans are run at relatively high speeds (1200 rpm to 1800 rpm) with direct-drive motors. Two or more fans are used in parallel to provide for efficient operation at low furnace demand.

FD fans create a positive pressure (e.g., 2 in wg to 10 in wg; 0.5 kPa to 2.5 kPa). This pressure is reduced to a very small negative pressure after passing through the air heater, ducts, and windbox system. The negative pressure in the furnace serves to keep combustion gases from leaking out into the furnace and boiler areas. The pressure continues to drop as it passes through the boiler, economizer, air heater, and pollution-control equipment.

Whereas FD fans force air through the system, *induced draft* (ID) fans are used to draw combustion products through the furnace bed, stack, and pollution control system by injecting air into the stack after combustion. The term "suction units" is used with ID fans.

ID fans are located after dust collectors and precipitators (often at the base of the stack) in order to reduce the abrasive effects of fly ash. They are run at slower speeds than forced draft fans in order to reduce the abrasive effects even further. Unlike FD fans, ID fans are usually very large and powerful because they have to handle all of the combustion gases, not just combustion air.

Modern welded stacks are essentially "air-tight" and can operate at pressures above atmospheric, often eliminating the need for ID fans.

Pure FD and ID systems are rarely used. Modern stack systems operate in a condition of *balanced draft*. Balanced draft is the term used when the static pressure is equal to atmospheric pressure. This requires the use of both ID and FD fans. In order to keep combustion products inside the combustion chamber and stack system, balanced draft systems may actually operate with a slight negative pressure.

The *draft loss* is the static pressure drop due to friction through the boiler and stack. The *available draft* is the difference between the theoretical draft (stack effect) and the draft loss. The available draft is zero in a balanced system.

$$D_{available} = D_{theoretical} - D_{friction} \qquad 37.27$$

The *net rating* or *fan boost* of the fan is the total pressure (the difference of draft losses and draft gains) supplied by the fan at maximum operating conditions. The fan power for ID and FD fans is calculated, as with any fan application, from the net rating and the flow rate. It is customary in sizing ID and FD draft fans to include increases of approximately 15% for flow rate, 30% for pressure, and 20°C (10°C) for temperature.

---

[33]Boiler horsepower is sometimes equated with a gross heating rate of 44,633 Btu/hr (13.08 kW). However, this is the total incoming heating value assuming a standardized 75% combustion efficiency.
[34]The British term "draught" is synonymous with "draft."
[35]Chimneys that rely on natural draft are sometimes referred to as *gravity chimneys*.

40. STACK EFFECT

Stack effect (*chimney action* or *natural draft*) is a pressure difference caused by the difference in atmospheric air and flue gas densities. It can be relied on to draw combustion products at least partially up through the stack. The stack effect is determined with no flow of flue gas. With flow, some of the stack effect is converted to velocity head, and the remainder is used to overcome friction.

Generally, the higher the chimney, the greater the stack effect. However, the stack effect in modern plants is greatly reduced by the friction and cooling effects of economizers, air heaters, and precipitators. Fans supply the needed pressure differential. Therefore, the primary function of modern stacks is to carry the combustion products a sufficient distance upward to dilute the combustion products, not to generate draft. Modern stacks for coal-fired power plants are seldom built shorter than 200 ft (60 m) in order to meet their dispersion requirements, and most are higher than 500 ft (150 m).

The theoretical stack effect, $D_{\text{theoretical}}$, is calculated from the densities of the flue gas and atmospheric air. In Eq. 37.28, H_{stack} is the total height of the stack. The average temperature is used to determine the average density in Eq. 37.28.[36]

$$D_{\text{theoretical}} = H_{\text{stack}}(\gamma_{\text{air}} - \gamma_{\text{flue gas,ave}}) \qquad 37.28$$

If a few assumptions are made, the stack effect can be calculated from the average temperature of the flue gas (i.e., the temperature at the average stack elevation).[37] The average flue gas temperature is the temperature halfway up the stack. H_{stack} is the total height of the stack.

$$D_{\text{theoretical}} = \left(\frac{p_{\text{air}}H_{\text{stack}}g}{R_{\text{air}}}\right)\left(\frac{1}{T_{\text{air}}} - \frac{1}{T_{\text{flue gas,ave}}}\right)$$
$$\text{[SI]} \quad 37.29(a)$$

$$D_{\text{theoretical}} = \left(\frac{p_{\text{air}}H_{\text{stack}}g}{R_{\text{air}}g_c}\right)\left(\frac{1}{T_{\text{air}}} - \frac{1}{T_{\text{flue gas,ave}}}\right)$$
$$\text{[U.S.]} \quad 37.29(b)$$

---

[36]The actual stack effect will be less than the value calculated with Eq. 37.28. For realistic problems, the achievable stack effect probably should be considered to be 80% of the ideal.

[37]One assumption is that the pressure in the stack is equal to atmospheric pressure. Then, the density difference will be due to only the temperature difference. Also, the flue gas is assumed to be air. This makes it unnecessary to know the flue gas composition, molecular weight, and so on.

The *stack effect head*, h_{SE}, is used to calculate the stack velocity. The coefficient of velocity, C_v, is approximately 0.30 to 0.50.

$$h_{\text{SE}} = \frac{D_{\text{theoretical}}}{\gamma_{\text{flue gas,ave}}} \qquad 37.30$$

$$v = C_v v_{\text{ideal}} = C_v\sqrt{2gh_{\text{SE}}} \qquad 37.31$$

The required stack area is

$$A_{\text{stack}} = \frac{\dot{Q}_{\text{flue gas}}}{v} \qquad 37.32$$

Example 37.7

The air surrounding an 80 ft (24 m) tall vertical stack is at 70°F (21°C) and one standard atmospheric pressure. The temperature of the flue gas at the average stack elevation is 400°F (200°C). What is the theoretical natural draft?

SI Solution

The absolute temperatures are

$$T_{\text{air}} = 21°C + 273° = 294\text{K}$$
$$T_{\text{flue gas}} = 200°C + 273° = 473\text{K}$$

Calculate the stack effect from Eq. 37.29.

$$
\begin{aligned}
D_{\text{theoretical}} &= \left(\frac{p_{\text{air}}H_{\text{stack}}g}{R_{\text{air}}}\right)\left(\frac{1}{T_{\text{air}}} - \frac{1}{T_{\text{flue gas}}}\right) \\
&= \left(\frac{(101.3\text{ kPa})(24\text{ m})\left(9.81\frac{\text{m}}{\text{s}^2}\right)}{287\frac{\text{J}}{\text{kg·K}}}\right) \\
&\quad \times \left(\frac{1}{294\text{K}} - \frac{1}{473\text{K}}\right) \\
&= 0.107\text{ kPa}
\end{aligned}
$$

Customary U.S. Solution

The absolute temperatures are

$$T_{\text{air}} = 70°F + 460° = 530°R$$
$$T_{\text{flue gas}} = 400°F + 460° = 860°R$$

Calculate the stack effect from Eq. 37.29.

$$D_{\text{theoretical}} = \left(\frac{p_{\text{air}}H_{\text{stack}}g}{R_{\text{air}}g_c}\right)\left(\frac{1}{T_{\text{air}}} - \frac{1}{T_{\text{flue gas}}}\right)$$

$$= \left(\frac{\left(14.7\,\frac{\text{lbf}}{\text{in}^2}\right)\left(12\,\frac{\text{in}}{\text{ft}}\right)^2(80\,\text{ft})\left(32.2\,\frac{\text{ft}}{\text{sec}^2}\right)}{\left(53.3\,\frac{\text{ft-lbf}}{\text{lbm-}^\circ\text{R}}\right)\left(32.2\,\frac{\text{lbm-ft}}{\text{lbf-sec}^2}\right)}\right)$$

$$\times\left(\frac{1}{530^\circ\text{R}} - \frac{1}{860^\circ\text{R}}\right)$$

$$= 2.3\,\text{lbf/ft}^2$$

Convert this to inches of water.

$$D_{\text{theoretical}} = \frac{p}{\gamma} = \frac{\left(2.3\,\frac{\text{lbf}}{\text{ft}^2}\right)\left(12\,\frac{\text{in}}{\text{ft}}\right)}{62.4\,\frac{\text{lbf}}{\text{ft}^3}}$$

$$= 0.44\,\text{in wg}$$

41. STACK FRICTION

Flue gas velocities of 15 ft/sec to 40 ft/sec (4.5 m/s to 12 m/s) are typical. As with most turbulent fluids, the friction pressure drop, D_{friction}, through stack "piping" is proportional to the velocity head.[38]

$$D_{\text{friction}} = \frac{K\text{v}^2}{2g} \qquad \textit{37.33}$$

Values of the friction coefficient, K, for chimney fittings are similar to those used in ventilating ductwork. Values for specific pieces of equipment (burners, pressure regulators, vents, etc.) must be provided by the manufacturers.

The approximate friction loss coefficient for a circular section of ductwork or chimney of length L and diameter d is given by Eq. 37.34. Equation 37.34 assumes a Darcy friction factor of 0.0233. For other friction factors, the constant terms can be scaled up or down proportionally. For noncircular ducts, the hydraulic diameter, d, can be replaced by four times the hydraulic radius (i.e., four times the duct area divided by the duct perimeter).

$$D_{\text{friction,kPa}} = \frac{0.119 L_{\text{m}}\rho_{\text{kg/m}^3}Q_{\text{L/s}}^2}{d_{\text{cm}}^5} \qquad [\text{SI}] \quad \textit{37.34}(a)$$

$$D_{\text{friction,in wg}} = \frac{28 L_{\text{ft}}\gamma_{\text{lbf/ft}^3}Q_{\text{cfs}}^2}{d_{\text{in}}^5} \qquad [\text{U.S.}] \quad \textit{37.34}(b)$$

---

[38]The term *piping* is used to mean the stack passages.

42. EMISSIONS SAMPLING

There are numerous sampling procedures, as most pollutants have their own specific characteristics, nuances, and idiosyncracies.

Sampling of emissions in flue gases can either be continuous or by spot sampling using wet chemistry or whole air methods. With *wet chemistry sampling methods*, a sample is collected in a container, solid adsorbent, or liquid-impinger train. The sample is then moved from the field to the laboratory for analysis.

With *whole air sampling*, a volume of flue gas is collected in a bag-like container (e.g., a *Tedlar* bag), and the container is transported to the laboratory for analysis by gas chromatography (GC). Flame ionization detection (GC-FID) and a mass spectrophotometer (GC-MS) are both used. For some pollutants, residence time in the container is critical. Volatile organic compounds (VOCs) with short half-lives can be captured and tested in a sampling train (VOST), which allows for longer sample-holding times.

Particulates are detected and measured by EPA method 5, which, though complex, has become a de facto standard, or EPA method 17 (in-stack filtration). With method 5, the sampling device includes a series of absorbers connected in tandem. The absorption train is followed by a gas-drying tube and a vacuum pump. Several samples are usually taken. The amount of particulates is determined from the filter's weight increase. Particles in the sampling path are washed out, dried, and included in the weight. Stack gas moisture content is determined from the impinger train. Oxygen, carbon dioxide, and (by difference) nitrogen volumes are determined by analyzing a sample collected separately in a sample bag.

Tests of gas streams should be unbiased, which requires sampling to be isokinetic. The gas velocity and ratio of dry gas mass to pollutant mass are not disturbed by *isokinetic testing*.[39,40]

The number of samples taken depends on the proximity of the sampling point to upstream and downstream disturbances. The number of samples can be minimized by locating the sampling point such that flow is undisturbed by upstream and downstream changes of direction, changes in diameter, and equipment in the duct (i.e., after a "good straight run"). Sampling should be at least eight equivalent diameters downstream and two equivalent diameters upstream from disturbances in

---

[39]With *super-isokinetic* sampling, too little pollution is indicated. This can occur if the vacuum draws a sample such that sampling velocity is higher than the bulk flow velocity. Since particles are subject to inertial forces and gases are not, the particle count will be essentially unaffected, but the metered volume will be higher, resulting in a lower apparent concentration. Conversely, if the sample is drawn too slowly (i.e., *sub-isokinetic* sampling), the metered volume would be lower, and the apparent pollution concentration will be too high.

[40]Solid and liquid substances in flue gases require isokinetic sampling. Gaseous substances do not.

flow that would cause the flow pattern to be unsymmetrical (i.e., would cause swirling). For rectangular ducts, the equivalent diameter is

$$d_e = \frac{(2)(\text{height})(\text{width})}{\text{height} + \text{width}} \qquad 37.35$$

Since a pollutant may not be dispersed evenly across the duct, the average concentration must be determined by taking multiple samples across the flow cross section. This process is known as *traversing* the area, and the sampling points are known as *traverse points*. For circular ducts, the cross section is divided into annular areas according to a standardized procedure. A minimum of six measurements should be taken along each perpendicular diameter. 24 is the usual maximum number of traverse points. (Figure 37.1 shows the locations for a 10-point traverse.) For rectangular ducts, the cross section is divided into a minimum of nine equal rectangular areas of the same shape and with an aspect ratio between 1:1 and 1:2. The traverse points are located at the centroids of these areas.

Figure 37.1 *Traverse Sampling Points*

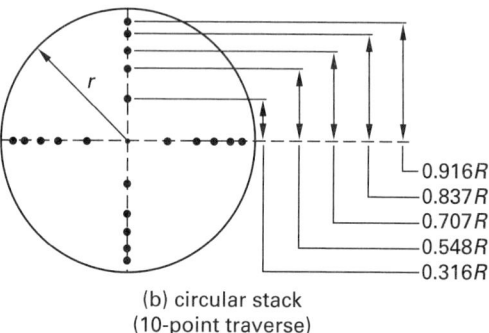

(a) rectangular stack
(measure at center of at least 9 equal areas)

0.916R
0.837R
0.707R
0.548R
0.316R

(b) circular stack
(10-point traverse)

43. EMISSIONS MONITORING SYSTEMS

Continuous emissions monitoring systems (CEMs) are used in large installations, though smaller installations may also be subject to CEM requirements. CEM not only ensures compliance with regulatory limits, but also permits the installation to participate in any allowance trading components of the regulations. CEMs are meant as compliance monitors, not merely operational indicators. The name "continuous emissions monitor system" has come to mean combined equipment for determining and recording SO_2 and NOx concentrations, volumetric flow, opacity, *diluent gases* (O_2 or CO_2), total hydrocarbons, and HCl.

CEMs consist of components that (1) obtain the sample, (2) analyze the sample, and (3) acquire and record analyses in real time.[41] With *straight extractive* CEMs, a sample is drawn from the stack by a vacuum pump and sent down a heated line to an analyzer at the base of the stack. This method must keep the stack gas heated above its 250°F to 300°F (120°C to 150°C) dew point to keep the moisture from condensing. Since samples are drawn by vacuum, leaks in the tube can lead to errors.

With *dilution extraction* CEMs, the gas is diluted with clean air or other inert carrier gas and passed to the analyzer located at a lower level. Dilution ratios are generally high—between 50:1 and 300:1, although they may be as low as approximately 12:1 at the outlet of an FGD unit. Since the dew point is lowered to below −10°F (−23°C) by the dilution, samples passed to the analyzer do not need to be heated. Samples are passed under positive pressure, ensuring that outside air and other gases cannot enter the tube. Dilution extraction analyses are on a wet basis.

With in situ CEM methods, direct measurements are taken by advanced electro-optical techniques. All instrumentation is located on the stack or duct. Data is sent by wire to remote monitors and the *data acquisition/reporting system* (DAS). The term "in situ" also implies that the flue gas analysis will be "wet" (i.e., the volumetric fraction will include water vapor).

There are three common methods of determining volumetric flow rates: (1) thermal sensing using a hotwire anemometer or thermal dispersion, (2) differential pressure using a single-point pitot tube or multipoint annubar, and (3) acoustic velocimetry using ultrasonic transducers.

With the advent of mass emission measurement requirements, pollutant (e.g., SO_2 and CO_2) concentration monitors and velocity monitors are being combined.

---

[41]The term "continuous" is somewhat of a misnomer. Though sampling is "automatic and frequent," data are rarely collected more frequently than once every 15 minutes.

38 Air Quality

Nomenclature

a_c	centrifugal acceleration	m/s^2
A	area	m^2
A_x	cross-sectional area	m^2
c	specific heat	$kJ/kg \cdot ^\circ C$
C	Cunningham correction factor	–
C_D	drag coefficient	–
C_e	gas concentration dissolved in liquid	kg/L
C_f	gas concentration exiting adsorber	kg/m^3
C_p	particulate concentration	kg/m^3
$C_{x,0}$	concentration along plume centerline at distance x	kg/m^3
$C_{x,y,0}$	concentration along plume centerline at distance x and at crosswind distance y	kg/m^3
d	diameter	m
D	cylinder diameter	m
D_e	outlet diameter	m
DRE	destruction and removal efficiency	%
E	efficiency	%
E_f	electric field	V/m
E_f	fractional target efficiency	–
f	friction factor	–
f	wettability coefficient	–
f_v	volumetric fraction	–
g	acceleration of gravity, 9.81	m/s^2
h	stack height	m
h_{fg}	latent heat of vaporization	kJ/kg
H	effective stack height	m
i_o	permittivity constant, 8.85×10^{-12}	$C^2/N \cdot m^2$
k	pressure drop coefficient	–
K	constant	–
K_1	fabric resistance	$kPa \cdot s/m$
K_2	cake resistance	$kPa \cdot s \cdot m/kg$
K_F	Freundlich equation intercept	–
K_H	Henry's constant	kPa

K_p	venturi coefficient	–
L	packing media bed depth	m
L_1	cylinder length	m
L_2	cone length	m
m	mass	kg
$\dot{m}$	mass flow rate	kg/s
MW	molecular weight	kg/mol
n	number of moles	–
$1/n$	Freundlich equation slope	–
p	pressure	kPa
p_g	gas partial pressure	kPa
p_v	vapor pressure	kPa
P_c	corrected particulate matter concentration	kg/m^3
P_m	measured particulate matter concentration	kg/m^3
PSI	pollution standard index	–
q_a	heat rate	kJ/s
Q_g	gas flow rate	m^3/s
Q_g	pollutant emission rate	kg/s
Q_l	cooling liquid flow rate	L/s
r	radius	m
r_a	average radius at particle location	m
r_e	radius at D_e	m
R	specific gas constant	$kJ/kg \cdot K$
R^*	universal gas constant	$kJ/kmol \cdot K$
Re	Reynolds number	–
SCA	specific collection area	$m^2 \cdot s/m^3$
S_w	chemical solubility in water	kg/L
t	time	s
T	temperature	K
U	overall heat transfer coefficient	$kJ/^\circ C \cdot m^2 \cdot h$
v	velocity	m/s
v_t	terminal settling velocity of the 100% removed particle	m/s
V	volume	m^3
V_d	voltage drop between electrodes	V
w	drift velocity	m/s
W_{in}	mass feed rate of one POHC in the waste stream feeding the incinerator	kg/s
W_{out}	mass emission rate of the same POHC in the exhaust emissions prior to emission	kg/s
x	downwind distance	m
x	mole fraction	–
x_o	fraction removed corresponding to v_t	–
X	mass of chemical removed	kg
y	crosswind distance	m

Combustion

Y	oxygen concentration measured in stack gas	%
z	elevation	m
Z_e	distance separating discharge electrode and collector	m
Z_o	scrubber contact zone length	m

Symbols

α	packing media bed porosity	–
β	volume shape factor	–
Γ	lapse rate	°C/m
μ	average windspeed through the plume	m/s
μ_g	gas absolute viscosity	kg/m·s
ρ	density	kg/m³
σ_y	Gaussian dispersion coefficient for horizontal plume concentration at downwind distance x	m
σ_z	Gaussian dispersion coefficient for vertical plume concentration at downwind distance x	m

Subscripts

a	VOC
d	droplet
f	filtration
g	gas
i	increment
l	liquid
m	mean
p	particle or particle removed to 100%
pi	incremental particle removed to 100%
s	settling
t	terminal

1. STATUTES AND REGULATIONS

Clean Air Act and Amendments

Air quality issues have been addressed by federal legislation since 1955, and many updated statutes and amendments have followed. The statute providing regulatory authority for national air quality issues is the Clean Air Act (CAA) and amendments. Federal air pollution regulations address air quality issues through a variety of enforcement and incentive mechanisms.

The National Ambient Air Quality Standards (NAAQS) attempt to implement a uniform national air pollution control program. For each pollutant included in the NAAQS, primary and secondary standards are defined. *Primary standards* are intended to protect human health, and *secondary standards* are intended to protect public welfare. The NAAQS are applied to a limited number of substances, known as the *criteria pollutants*. The NAAQS and the criteria pollutants to which they apply are presented in Table 38.1.

To ensure application of the NAAQS in each of the 50 states, the United States Environmental Protection Agency (EPA) has required the states to prepare and submit State Implementation Plans (SIPs). Each SIP must address a control strategy for each of the criteria pollutants included in the NAAQS. If a state is unable or unwilling to submit an SIP acceptable to the EPA, the EPA has the authority to develop an SIP for the state and then require the state to enforce it.

Prevention of significant deterioration (PSD) is a program that is designed to protect air quality in clean-air areas that already meet the NAAQS. To apply the PSD principle, the EPA has developed a regional classification system. Class I regions, generally including national monuments and parks and national wilderness areas, allow very little air quality deterioration and, thereby, limit economic development. Class II regions allow moderate air quality deterioration, and class III regions allow deterioration to the secondary NAAQS.

Because of geographic conditions, population density, climate, and other factors, some areas in the United States are unable to meet the NAAQS. These areas are designated as *nonattainment areas* (NA). Any new or modified sources in nonattainment areas are required to control emissions at the *lowest achievable emission rate* (LAER). Recognizing the economic problems posed by the CAA in nonattainment areas, the EPA has allowed some flexibility, through *emissions trading*, in how industry goes about meeting the compliance requirements. Emissions trading can occur by the following methods.

- *Emission Reduction Credit:* By applying a higher level of treatment than required by regulation, businesses earn credits that can be redeemed through other CAA programs.

- *Offset:* The offset allows industrial expansion by letting businesses compensate for new emissions by offsetting them with credits acquired by other businesses already existing in the region.

- *Bubble:* All activities in a single plant or among a group of proximate industries can emit at various rates as long as the resulting total emission does not exceed the allowable emission for an individual source.

- *Netting:* Businesses can expand without acquiring a new permit as long as the net increase in emissions is not significant. Consequently, by applying improved control technology, businesses can earn credits that can be subsequently applied to plant expansion with no net increase in emissions.

New source performance standards (NSPS) are applied to specific source categories (e.g., asphalt concrete plants, incinerators) to prevent the emergence of other air pollution problems from new construction. The NSPS are intended to reflect the best emission control measures achievable at reasonable cost, and are to be applied during initial construction of new industrial facilities.

Combustion

Table 38.1 National Ambient Air Quality Standards[a]

pollutant	time basis	primary standard[b]	secondary standard
carbon monoxide (CO)	8 h	9 ppm (10 mg/m^3)	none
	1 h	35 ppm (40 mg/m^3)	none
nitrogen dioxide (NO$_2$)	1 h	100 ppb	none
	mean annual	53 ppb (100 μg/m^3)	same
ground-level ozone (O$_3$)	8 h	0.075 ppm (147 μg/m^3)	same
sulfur dioxide (SO$_2$)	3 h	none	0.50 ppm (1300 μg/m^3)
	1 h	0.075 ppm	none
PM10	24 h	150 μg/m^3	same
PM2.5	annual	12 μg/m^3	15 μg/m^3
	24 h	35 μg/m^3	same
lead (Pb)	3 mo	0.15 μg/m^3	same
hydrocarbons[c]	–	–	–

[a] Typical values. The NAAQS are subject to ongoing legislation, and current values may be revised from those presented here.
[b] Equivalents in parentheses may or may not be part of the NAAQS.
[c] Hydrocarbon standards have been revoked by the EPA since hydrocarbons, as precursors of ozone, are adequately regulated by ozone standards.

Source: National Ambient Air Quality Standards (NAAQS), U.S. Environmental Protection Agency, epa.gov.

National Emission Standards for Hazardous Air Pollutants (NESHAPs) focus on hazardous pollutants not included as criteria pollutants. *Hazardous pollutants* are compounds that pose particular, usually localized, hazards to the exposed population. On a localized level, their impact is more severe than that of the criteria pollutants. Compounds identified as hazardous pollutants are limited to asbestos, inorganic arsenic, benzene, mercury, beryllium, radionuclides, and vinyl chloride. The list is short because onerous demands are placed on the EPA by the CAA for listed pollutants and because a very high level of abatement, based on risk to the exposed population, is required once a pollutant is listed. The 1990 CAA amendments listed 187 other *hazardous air pollutants* (HAPs) whose abatement is based on the *maximum achievable control technology* (MACT), a less demanding criterion than risk. Many states have adopted their own regulations to designate and control other hazardous air pollutants.

The 1990 CAA amendments placed limits on allowable emissions of sulfur oxides (SOx) and nitrogen oxides (NOx)—two categories of compounds associated with acid rain. These limits require acid deposition emission controls, which are implemented through a system of emission allowances applied to fossil fuel-fired boilers in the 48 contiguous states. Control options may include such things as use of low-sulfur fuels, flue-gas desulfurization, and replacing older facilities.

The 1977 CAA amendments authorized the EPA to regulate any substance or practice with reasonable potential to damage the stratosphere, especially with regard to ozone. In response to this regulatory authority, the EPA has closely regulated the production and sale of chlorinated fluorocarbons (CFCs) through out-and-out bans on their production and use, taxes on emissions from their use, and a permitting program.

Resource Conservation and Recovery Act (RCRA)

Regulations promulgated under the Resource Conservation and Recovery Act (RCRA) impose restrictions on emissions from hazardous waste incinerators. The regulations define *principal organic hazardous constituents* (POHCs) present in the waste feed and establish performance standards for their destruction and removal in the incineration process.

For most wastes, the performance standard for POHCs is defined by a *destruction and removal efficiency* (DRE) of 99.99% (referred to as "four nines") for each POHC in the waste feed. However, for selected wastes, primarily those associated with dioxins and furans, a DRE of 99.9999% ("six nines") is required. The DRE includes the destruction efficiency of the incinerator as well as the removal efficiency of any air pollution control equipment. The DRE is calculated from Eq. 38.1.

$$\text{DRE} = \frac{\dot{m}_{in} - \dot{m}_{out}}{\dot{m}_{in}} \times 100\% \qquad 38.1$$

The RCRA regulations also apply to hydrochloric acid (HCl) and particulate emissions. The HCl emissions are based on a mass emission rate of 1.8 kg/h, and the particulate emissions corrected for oxygen (21% oxygen by volume in the combustion air) are calculated from Eq. 38.2.

$$P_c = P_m\left(\frac{14}{21 - Y}\right) \qquad 38.2$$

Example 38.1

The POHC feed rate to an incinerator is 2.73 kg/h. Stack emissions monitoring shows the POHC at 0.011 kg/h, particulate at 176 mg/m^3, and O$_2$ at 5%. (a) What is the DRE for the POHC? (b) What is the corrected particulate matter concentration?

Solution

(a) Use Eq. 38.1.

$$\text{DRE} = \frac{\dot{m}_{\text{in}} - \dot{m}_{\text{out}}}{\dot{m}_{\text{in}}} \times 100\%$$

$$= \frac{2.73\ \frac{\text{kg}}{\text{h}} - 0.011\ \frac{\text{kg}}{\text{h}}}{2.73\ \frac{\text{kg}}{\text{h}}} \times 100\%$$

$$= 99.60\%$$

(b) Use Eq. 38.2.

$$P_c = P_m\left(\frac{14}{21-Y}\right) = \left(176\ \frac{\text{mg}}{\text{m}^3}\right)\left(\frac{14}{21-5}\right)$$

$$= 154\ \text{mg/m}^3$$

2. POLLUTANT CLASSIFICATION

Introduction

Air pollutants may be classified as primary and secondary. *Primary pollutants* are those that exist in the air in the same form in which they were emitted. An example of a primary pollutant is carbon monoxide emitted directly from gasoline-powered automobiles. *Secondary pollutants* are those that are formed in the air from other emitted compounds. Ground-level ozone formed from the nitrogen dioxide photolytic cycle is an example of a common secondary pollutant. Primary and secondary pollutants are not the same as, and should not be confused with, the primary and secondary NAAQS. The major source of air pollutants in the United States and most other industrialized countries is fossil fuel combustion from both stationary and mobile sources. To a much lesser degree, pollutants may be generated from wind erosion, plant and animal bio-effluents, vapors from architectural coatings and other hydrocarbon-based materials (e.g., new asphalt pavements), industrial manufacturing, and a variety of other sources.

Particulate Matter

Suspended *particulate matter* (PM) is produced in a variety of forms. These include *dust* from manufacturing, agricultural, and construction activities, *fumes* produced by chemical processes, *mists* formed by condensed water vapor, *smoke* generated by incomplete combustion from transportation and industrial activities, and *sprays* formed by atomization of liquids. Natural wind erosion and forest fires also produce dust and smoke. The health risks associated with particulate matter exposure are exacerbated by the adsorption from the air of toxic organics onto particle surfaces.

Particulate matter may be liquid or solid with a broad size range, from 0.005 μm to 100 μm. The *particulate size* is taken as the aerodynamic diameter of an

equivalent perfect sphere having the same settling velocity. Particulate matter in the size range from 0.1 μm to 10 μm presents the greatest potential impact on air quality. Particulate matter smaller than 0.1 μm tends to coagulate into larger particles, and particles larger than 10 μm settle relatively rapidly. Of particular concern is particulate matter with aerodynamic diameters less than 10 μm (PM10) and less than 2.5 μm (PM2.5) for which NAAQS are included. Particles making up PM2.5 are called *respirable particulate matter* and present the greatest potential risk to human health because they are able to reach and remain in the lung. The sources, forms, and sizes of particulate matter are summarized in Table 38.2.

Table 38.2 Particulate Matter Types

type	source	form	size
dust	emissions from handling and mechanical activities (e.g., grain elevators, sandblasting, construction)	solid	1 μm to 10 mm
fume	condensed vapors from chemical reactions (e.g., calcination, distillation, sublimation reactions)	solid	0.03 μm to 0.3 μm
mist	condensed vapors from chemical reactions (e.g., sulfuric acid from sulfur trioxide)	liquid	0.5 μm to 3 μm
smoke	particulate emissions from combustion activities (e.g., forest fires, coal-fired power plants)	solid	0.05 μm to 1 μm
spray	atomized liquids (e.g., sprayed coatings, pesticide application)	liquid	10 μm to 10 mm

Adapted from: Vesilind, P.A., J.J. Pierce, and R.F. Weiner, 1994, *Environmental Engineering*, 3rd ed., Butterworth-Heinemann, Newton, MA.

The *settling velocity* for a particle of particulate matter can be calculated from *Stokes' law*.

$$v_s = \frac{d_p^2(\rho_p - \rho_g)g}{18\mu_g} \quad \begin{bmatrix}\text{consistent}\\\text{units}\end{bmatrix} \quad 38.3$$

The same parameters can also be used to calculate the *settling Reynolds number*.

$$\text{Re} = \frac{v_s d_p \rho_g}{\mu_g} \quad 38.4$$

Equation 38.3 is valid for particle diameters between about 0.1 μm and 100 μm and for settling velocities less than about 1.0 m/s, both of which are well within the range of interest in air pollution applications.

Particulate lead pollution is most pronounced in areas where automobile traffic has historically been high. Although the use of unleaded gasoline has greatly reduced airborne lead pollution from automobile emissions, residual contamination persists in soils along roadways, in parks, and in other areas where soil is exposed. This contaminated soil can act as an ongoing source of airborne lead when disturbed by vehicle traffic, construction activities, natural wind action, and other events.

An important source of airborne lead not associated with leaded gasoline use is chipping and flaking lead-based paint. When lead-based paint, used either indoors or outdoors, is sanded or when it chips or flakes, the lead-containing particles become airborne and can be inhaled or contribute to soil contamination. Lead is no longer used in paints intended for indoor applications in the United States, but it is still used in some exterior paints.

Exposure to lead-based paint presents a significant health hazard to children and is associated with behavioral changes, learning disabilities, and other health problems. Lead is included in the NAAQS.

Asbestos is a fibrous silicate material that is inert, strong, and incombustible. As *asbestos-containing materials* (ACM) age or are damaged, asbestos fibers are released into the air. Although the greatest potential exposure to asbestos occurs indoors where asbestos-containing materials were used in construction, soil containing mineral asbestos has contributed to significant asbestos exposures in some areas of the United States.

The most substantial asbestos exposure has historically been occupational. Asbestos poses health risks in relatively localized areas from industrial and construction-related activities. Asbestos is included in the national emission standards for hazardous air pollutants (NESHAP).

Example 38.2

What is the settling velocity for particles with an aerodynamic diameter of 10 μm and a density of 700 kg/m³ in an atmosphere at 20°C and 1 atm?

Solution

From App. 38.C, air viscosity is $\mu_g = 1.82 \times 10^{-5}$ kg/m·s and air density is $\rho_g = 1.21$ kg/m³ at 20°C.

Use Eq. 38.3.

$$v_s = \frac{d_p^2(\rho_p - \rho_g)g}{18\mu_g}$$

$$= -\frac{(10 \ \mu m)^2 \left(10^{-6} \ \frac{m}{\mu m}\right)^2}{\times \left(700 \ \frac{kg}{m^3} - 1.21 \ \frac{kg}{m^3}\right)\left(9.81 \ \frac{m}{s^2}\right)}{(18)\left(1.82 \times 10^{-5} \ \frac{kg}{m \cdot s}\right)}$$

$$= 2.09 \times 10^{-3} \ m/s$$

Gaseous Pollutants

The principal gaseous pollutants of concern include oxides of nitrogen, sulfur, and carbon; ozone; and hydrocarbons. Most gaseous pollutants are associated with combustion processes and the characteristics of the fuels burned. Potential emissions from combustion increase as combustion efficiency decreases, as available oxygen decreases, and as the presence of compounds other than carbon and hydrogen increases. The sources of gaseous pollutants and their fates in the atmosphere are summarized in Table 38.3.

Table 38.3 Gaseous Emissions Sources

gas	source	fate
methane and other organic gases and vapors	combustion, volatilization	reaction with ·OH
carbon monoxide and carbon dioxide	combustion	oxidation
nitrous oxide and nitrogen dioxide	combustion	oxidation, precipitation, dry deposition
sulfur dioxide and sulfur trioxide	combustion	oxidation, precipitation, dry deposition

The two nitrogen oxides (NOx) that are significant air pollutants are nitric oxide (NO) and nitrogen dioxide (NO₂). Most nitrogen oxide emissions occur from fuel release and fossil fuel combustion as relatively benign NO. However, NO can rapidly oxidize to NO₂, which is associated with respiratory ailments and which reacts with hydrocarbons in the presence of sunlight to form photochemical oxidants. Photochemical oxidants are a significant constituent of smog. Nitrogen dioxide also reacts with the hydroxyl radical (·OH) in the atmosphere to form nitric acid (HNO₃), a contributor to acid rain.

Most sulfur emissions result from the combustion of fossil fuels, especially coal. When burned, the fuels release the sulfur primarily as sulfur dioxide (SO_2) with much smaller amounts of sulfur trioxide (SO_3). These are known as *sulfur oxides* (SOx). Sulfur oxide aerosols contribute to particulate matter concentrations in the air and are a constituent of acid rain. Sulfur oxides also can significantly impact long-range visibility.

Combustion of fossil fuels under ideal conditions produces carbon dioxide (CO_2) and water. However, where conditions reduce combustion efficiency, *carbon monoxide* (CO) is formed instead of CO_2. Although CO_2 is associated with global warming, it typically is not included as an air pollutant. Carbon monoxide, however, represents as much as half the total mass of air pollutants emitted in the United States, with most of these emissions coming from motor vehicles. Carbon monoxide does not seem to present any particular environmental hazards, but it does present a potential serious threat to human health by displacing oxygen (O_2) in the blood.

Ozone (O_3) forms in the stratosphere and at ground level. Stratospheric ozone is maintained through naturally occurring processes and acts as a filter to ultraviolet light radiation from the sun. Stratospheric ozone is not an air pollutant. Ground-level ozone, on the other hand, is a secondary pollutant formed through reactions with nitrogen oxides and hydrocarbons in the presence of sunlight, making it a *photochemical oxidant*. Among the photochemical oxidants, ozone is the most common. As a secondary pollutant, emissions limitations are not applicable to ozone. Instead, emission controls for those compounds, such as nitrogen oxides and hydrocarbons, that are precursors to ozone formation are required. Ground-level ozone is a major contributor to photochemical smog and is an irritant to the mucosa and lungs.

Hydrocarbons are included as air pollutants because of their role in the formation of photochemical oxidants, particularly ground-level ozone. The hydrocarbons involved in air pollution are primarily alkanes, comprising the methane series. The methane series includes methane (CH_4), ethane (C_2H_6), propane (C_3H_8), butane (C_4H_{10}), and so on. Alkanes have the general formula C_nH_{2n+2}. If an alkane loses one hydrogen atom, it becomes a radical or alkyl (i.e., $\cdot CH_3$) that is able to readily react with other compounds. When oxidized, alkanes form compounds of the carbonyl group (*aldehydes*), where an oxygen atom is double-bonded to a terminal carbon atom. Formaldehyde (CH_2O) and acrolein (CH_2CHCOH) are common aldehyde constituents of photochemical smog that are associated with eye irritation.

Hydrocarbons include other aliphatic chemicals, in addition to alkanes, and some aromatic hydrocarbons. Chlorinated solvents such as carbon tetrachloride, tetrachloroethene, 1,1,1-trichloroethane, trichloroethene, and their degradation products; and benzene, toluene, ethylbenzene, and xylene are among the most common of these and comprise those compounds typically called *volatile organic compounds* (VOCs). Occurrence of VOCs in the air can be a major contributor to formation of photochemical smog.

Structural formulas for a few representative aliphatic and aromatic hydrocarbons are shown in Fig. 38.1.

Hazardous Air Pollutants

The hazardous air pollutants identified under NESHAP as HAPs comprise approximately 200 chemicals and related groups of compounds, and have been identified as potentially significant sources of health risk to humans. These pollutants occur from a variety of sources and include asbestos; compounds of antimony, arsenic, beryllium, lead, manganese, mercury, and nickel; benzene and benzene derivatives such as toluene, xylene, and phenols; halogen gases; halogenated alkanes; chloroform and related compounds; and radionuclides.

Allergens and Microorganisms

Allergens and microorganisms comprise *biological air pollutants* and include such things as algae, animal dander and hair, bacteria, excreta from animals and insects, pollens, mold spores, viruses, and other organic structures, living or dead. Exposure to biological air pollutants can be very dramatic, as in the outbreak of Legionnaires' disease at the American Legion convention in Philadelphia in 1976, or, more commonly, in the form of the symptoms of allergies and other common illnesses.

3. POLLUTANT AND GAS CHEMISTRY

Gas Laws and Chemistry

Boyle's law states that the volume of a gas varies inversely with its pressure at constant temperature. For any given gas, the product of the pressure and the volume will be constant at constant temperature. The mathematical representation of Boyle's law is

$$pV = \text{constant} \qquad 38.5$$

Boyle's law is most useful when expressed in the following form, which is used to find the volume of a gas at a new pressure given a current pressure and volume at constant temperature.

$$p_1 V_1 = p_2 V_2 \qquad 38.6$$

Charles' law states that the volume of a gas at constant pressure varies in direct proportion to the absolute temperature. The ratio of the volume occupied by a gas and the absolute temperature remains constant for any gas. Mathematically, Charles' law is

$$\frac{V}{T} = \text{constant} \qquad 38.7$$

Figure 38.1 *Representative Organic Chemicals Important to Air Pollution*

Aliphatics – straight or branched chain
Alkanes (methane series)
all C-C bonds are single

```
    H
    |
H – C – H
    |
    H
```
methane (CH_4)

```
  H   H
  |   |
H–C– C–H
  |   |
  H   H
```
ethane (C_2H_6)

Alkynes or *acetylenes*
at least one C-C bond is triple

$$H-C \equiv C-H$$

ethyl acetylene (C_2H_2)

Alkenes or *olefins* (ethene series)
at least one C-C bond is double

```
  H   H
  |   |
  C = C
  |   |
  H   H
```
ethene (C_2H_4)

```
  H   H   H
  |   |   |
  C = C - C–H
  |       |
  H       H
```
propene (C_3H_6)

Aldehydes
oxidation products of primary alcohols
contain carbonyl group

```
    O
    ||
   –C–
```

```
      H
      |
  C – C=O
```
methanal or formaldehyde (CH_2O)

Aromatics – contain the 6-carbon
benzene ring

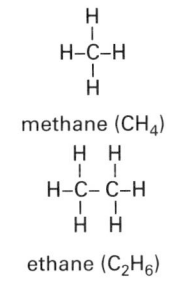

benzene (C_6H_6)

toluene (C_7H_9)

Halogenated aliphatics and aromatics –
(includes VOCs and CFCs)

```
 Cl  Cl
  |   |
  C = C
  |   |
  H   Cl
```
trichloroethene ($CHClCCl_2$)

```
 Cl  Cl
  |   |
Cl–C– C–F
  |   |
  F   F
```
CFC 113 (1,1,2-trichloro-1,2,2-trifluoroethane)

```
    H
    |
H–C–Cl
    |
    H
```
chloroform (CH_3Cl)

dioxins (2,3,7,8-TCDD)

Charles' law is usually expressed as a ratio of volume and temperature under two different conditions as follows.

$$\frac{V_1}{T_1} = \frac{V_2}{T_2} \qquad 38.8$$

Dalton's law of partial pressures states that in a mixture of gases, such as air, each gas exerts pressure independently of the others. The resulting "partial pressure" of each gas is proportional to the amount (percent by volume or mole fraction) of that gas in the mixture.

Each gas exerts pressure equal to that which it would exert if it were the sole occupant of the volume available to the mixture. For example, in an atmosphere of

80% nitrogen and 20% oxygen at 1.0 atm pressure, the pressure exerted by nitrogen is 0.8 atm and by oxygen is 0.2 atm.

Henry's law states that the amount of any gas that will dissolve in a given volume of a liquid at constant temperature is directly proportional to the pressure that the gas exerts above the liquid. Henry's law is commonly expressed in terms of *Henry's constant*, K_H, as

$$K_H = \frac{p_g}{C_e} \qquad 38.9$$

The concentration of the gas dissolved in the liquid, C_e, can be taken as either the solubility of the chemical in water at a known temperature, S_w, in units of mg/L, or as the mole fraction of the gas, x. p_g is the gas partial pressure and is usually taken as the vapor pressure, p_v, of the chemical at a known temperature.

Values for Henry's law constants for common gases are presented in App. 38.A and App. 38.B.

The *universal gas law* is a combination of Boyle's law and Charles' law and defines a relationship among pressure, volume, and absolute temperature for gases.

$$pV = nR^*T \qquad 38.10$$

Table 38.4 presents the value and units for R^* corresponding to the units selected for pressure, volume, and temperature.

Table 38.4 *Universal Gas Constant, R^**

numerical value	units
8.314	J/mol·K
8314	J/kmol·K
8.314×10^{-3}	kJ/mol·K
1.99	cal/mol·K
1.99×10^{-3}	kcal/mol·K
0.082	atm·L/mol·K
8.2×10^{-5}	atm·m³/mol·K

The universal gas law can be written in a variety of forms, depending on the application. Alternative forms include the following.

$$pV = mRT \qquad 38.11$$

$$p(\text{MW}) = \rho_g R^*T \qquad 38.12$$

$$\frac{p_1 V_1}{T_1} = \frac{p_2 V_2}{T_2} \qquad 38.13$$

The universal gas law can also be used to calculate the *molar volume*, V/n, of a gas.

$$\frac{V}{n} = \frac{R^*T}{p} \qquad 38.14$$

The molecular weight of gas mixtures is a mole- or volumetric-weighted average.

average molecular weight
$$= f_{v1}(\text{MW})_1 + f_{v2}(\text{MW})_2$$
$$+ f_{v3}(\text{MW})_3 + \cdots + f_{vn}(\text{MW})_n \qquad 38.15$$

The *specific volume* of a gas is the inverse of its density. The density can be calculated using Eq. 38.12.

Example 38.3

A 2.0 m³ closed tank is filled halfway with water. The tank contains a mixture of nitrogen and oxygen gases at 20°C and 5 atm. The respective proportions of each gas in the mixture are 72% and 28% by volume. What is the concentration in mg/L of the oxygen in the water?

Solution

From Dalton's law, the partial pressure of O_2 in the tank is $(0.28)(5 \text{ atm}) = 1.4$ atm.

From App. 38.B, K_H for O_2 at 20°C is 40 100 atm.

Use Eq. 38.9.

$$C_e = \frac{p_g}{K_H} = \frac{1.4 \text{ atm}}{40\,100 \text{ atm}}$$
$$= 3.5 \times 10^{-5}$$

C_e can be interpreted as the mole fraction of gas in the liquid.

$$x = 3.5 \times 10^{-5} = \frac{n_g}{n_g + n_l}$$

For 1 L of water,

$$n_l = \frac{\rho}{(\text{MW})_{\text{H}_2\text{O}}} = \frac{1000 \frac{\text{g}}{\text{L}}}{18 \frac{\text{g}}{\text{mol}}}$$
$$= 56 \text{ mol/L}$$

$$3.5 \times 10^{-5} = \frac{n_g}{n_g + 56 \frac{\text{mol}}{\text{L}}}$$
$$n_g = 0.0020 \text{ mol/L}$$

C_e can also be expressed in mass units.

$$C_e = n_g(\text{MW})_{\text{O}_2} = \left(0.0020 \frac{\text{mol}}{\text{L}}\right)\left(32 \frac{\text{g}}{\text{mol}}\right)\left(10^3 \frac{\text{mg}}{\text{g}}\right)$$
$$= 64 \text{ mg/L}$$

Example 38.4

Volatile organic compound (VOC) vapors at 1.3 atm and 28°C are vented from a storage tank into the building where the tank is located. The building volume is

$35\,000$ m^3, and the allowable average VOC concentration in the building is 800 $\mu g/m^3$. The temperature and pressure in the building are $18°C$ and 1 atm, respectively. The molecular weight of the VOC is 165 g/mol. What volume of VOC can be vented from the tank without exceeding the allowable VOC concentration?

Solution

The maximum number of moles in the VOC is

$$n = \frac{\left(800\ \frac{\mu g}{m^3}\right)(35\,000\ m^3)}{\left(165\ \frac{g}{mol}\right)\left(10^6\ \frac{\mu g}{g}\right)} = 0.17\ mol$$

Use Eq. 38.10. The VOC volume allowed in the building is V_2.

$$V_2 = \frac{nR^*T_2}{p_2}$$

$$= \frac{(0.17\ mol)\left(8.2 \times 10^{-5}\ \frac{atm \cdot m^3}{mol \cdot K}\right)(18°C + 273°)}{1\ atm}$$

$$= 4.1 \times 10^{-3}\ m^3 \quad (4.1\ L)$$

Use Eq. 38.13. The VOC volume vented from the tank is V_1.

$$V_1 = \frac{V_2 p_2 T_1}{p_1 T_2}$$

$$= \frac{(4.1 \times 10^{-3}\ m^3)(1\ atm)(28°C + 273°)}{(1.3\ atm)(18°C + 273°)}$$

$$= 3.3 \times 10^{-3}\ m^3 \quad (3.3\ L)$$

Example 38.5

By volume, air is a mixture of N_2 at 78%, O_2 at 20.9%, Ar at 0.93%, CO_2 at 0.032%, and many other gases at lesser concentrations (see App. 38.D). What is the average molecular weight of air?

Solution

Use Eq. 38.15.

gas	volumetric gas fraction	MW (g/mol)	fractional MW (g/mol)
N_2	0.78	28	21.8
O_2	0.209	32	6.7
Ar	0.0093	40	0.37
CO_2	0.00032	44	0.01
			28.88 g/mol

The apparent molecular weight of air is 28.88 g/mol.

Transformations and Fate

Photochemical smog is produced from a variety of complex physical and chemical reactions involving nitrogen oxides, carbon monoxide, hydrocarbons, sunlight, and many other factors under favorable conditions. The primary components of photochemical smog are ozone and oxidized hydrocarbons, particularly aldehydes. Also present in photochemical smog are organic nitrates.

Nitric oxide (NO) reacts with oxygen in the air to form nitrogen dioxide (NO_2). Under favorable conditions and in the presence of sunlight, the NO_2 will photolyze back to NO and release a free oxygen atom (O). The atomic oxygen reacts with an oxygen molecule (O_2) to form ozone (O_3). Finally, the O_3 reacts with NO to form NO_2 and O_2. The cycle is shown by the following reactions.

$$2NO + O_2 \rightarrow 2NO_2 \qquad \textit{38.16}$$

$$NO_2 + sunlight \rightarrow NO + O \qquad \textit{38.17}$$

$$O + O_2 + N_2 \rightarrow O_3 + N_2 \qquad \textit{38.18}$$

$$O_3 + NO \rightarrow NO_2 + O_2 \qquad \textit{38.19}$$

This sequence represents the normal nitrogen photolytic cycle and does not contribute to ground-level ozone accumulation as long as available NO and NO_2 remain balanced. However, if conditions develop that increase the availability of NO_2 over the NO, more ozone will be produced (see Eq. 38.17 and Eq. 38.18) than is destroyed (see Eq. 38.19), and ground-level ozone concentrations will increase.

The presence of hydrocarbons in the air is an important condition affecting increased NO_2 production. The following reactions show one of many ways in which NO_2 production is increased by the presence of hydrocarbons.

$$RH + \cdot OH \rightarrow \cdot R + H_2O \qquad \textit{38.20}$$

$$\cdot R + O_2 \rightarrow \cdot RO_2 \qquad \textit{38.21}$$

$$\cdot RO_2 + NO \rightarrow \cdot RO + NO_2 \qquad \textit{38.22}$$

$$\cdot RO + O_2 \rightarrow \cdot HO_2 + RCHO \qquad \textit{38.23}$$

$$\cdot HO_2 + NO \rightarrow NO_2 + \cdot OH \qquad \textit{38.24}$$

The RH in Eq. 38.20 is a hydrocarbon, $\cdot$R represents a hydrocarbon radical, and RCHO in Eq. 38.23 represents an aldehyde of a different hydrocarbon. The hydroxyl radical ($\cdot$OH) also plays an important role in the reaction and is available from reaction with atomic oxygen and water, as well as being a product of Eq. 38.24. Not only does ground-level ozone increase through the production of excess NO_2, but other constituents of photochemical smog, such as aldehydes, are also produced. Figure 38.2 illustrates these reactions.

Nitrogen and sulfur oxides are subject to transformations associated with *acid formation*. One of the primary mechanisms for NO_2 removal from the atmosphere is its photochemical conversion to nitric acid (HNO_3).

Combustion

Figure 38.2 *Nitrogen Photolytic Cycle and Ground Level Ozone Accumulation*

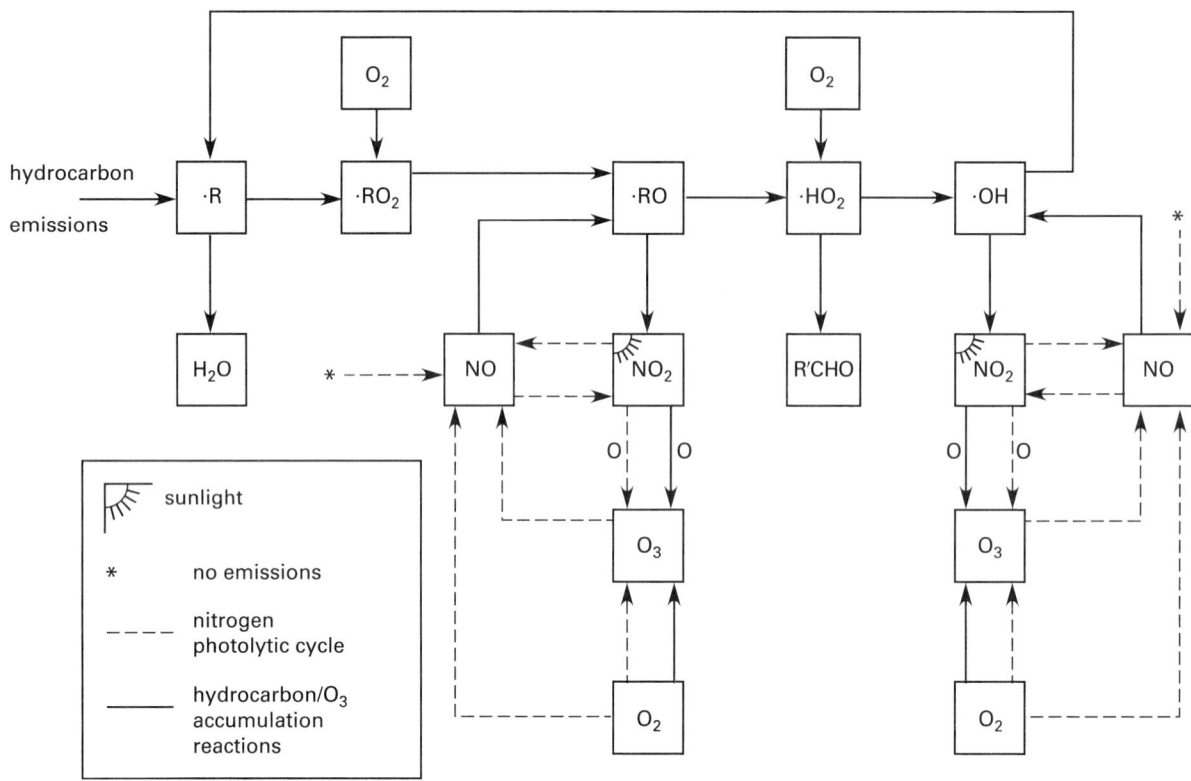

Adapted from Masters, G.M., *Introduction to Environmental Engineering and Science*, 2nd ed., © Prentice-Hall, Inc., Upper Saddle River, NJ.

The nitric acid is subsequently removed as aerosols by dry- and wet-deposition processes. In dry deposition, the HNO_3 aerosols settle out onto surfaces. In wet deposition, the aerosols either form rain droplets or are washed out by falling rain. The reaction, involving the hydroxyl radical, is as follows.

$$NO_2 + \cdot OH \rightarrow HNO_3 \qquad 38.25$$

Sulfur dioxide reacts with oxygen to form sulfur trioxide (SO_3), which subsequently reacts with water to form *sulfuric acid* (H_2SO_4). The reaction from SO_2 to H_2SO_4 is relatively slow, requiring one or more days. Since sulfuric acid dissociates in solution to sulfate (SO_4^{-2}), the resulting aerosols are known as *sulfate aerosols*. The sulfate aerosols, as well as unreacted SO_3, are removed by dry and wet deposition. Sulfur dioxide may also react directly with water in the atmosphere to form *sulfurous acid* (H_2SO_3). The reaction of SO_3 with water is rapid and SO_3 condenses at the relatively low temperature of 22°C. Simplified reactions for the formation of sulfuric acid and sulfurous acid from sulfur oxides are as follows.

For sulfuric acid formation,

$$2SO_2 + O_2 \rightarrow 2SO_3 \qquad 38.26$$

$$SO_3 + H_2O \rightarrow H_2SO_4 \qquad 38.27$$

For sulfurous acid formation,

$$SO_2 + H_2O \rightarrow H_2SO_3 \qquad 38.28$$

Example 38.6

Assume RH in Eq. 38.20 is methane (CH_4). (a) What are Eq. 38.20 through Eq. 38.24 with the hydrocarbon, hydrocarbon radical, and aldehyde substituted for RH, ·R, and RCHO? (b) What are the chemical names of the radical and aldehyde?

Solution

(a) For Eq. 38.20 through Eq. 38.24,

$$CH_4 + \cdot OH \rightarrow \cdot CH_3 + H_2O$$

$$\cdot CH_3 + O_2 \rightarrow \cdot CH_3O_2$$

$$\cdot CH_3O_2 + NO \rightarrow \cdot CH_3O + NO_2$$

$$\cdot CH_3O + O_2 \rightarrow \cdot HO_2 + HCHO$$

$$\cdot HO_2 + NO \rightarrow NO_2 + \cdot OH$$

(b) The hydrocarbon radical, R, is methyl, and the aldehyde, RCHO, is formaldehyde (also called methylaldehyde and methanal).

Example 38.7

During a rainstorm, 0.2 m^3 of rainwater scrubs 1 m^3 of air with an SO$_2$ concentration of 0.10 ppmv. The air temperature is 20°C, and the pressure is 1 atm. What is the concentration of sulfite (SO$_3^{-2}$) in the rainwater?

Solution

From Eq. 38.28,

$$SO_2 + H_2O \rightarrow H_2SO_3$$

One mole of SO$_2$ scrubbed will yield one mole of SO$_3^{-2}$ in the rainwater.

$$MW_{SO_2} = 32 \ \frac{g}{mol} + (2)\left(16 \ \frac{g}{mol}\right) = 64 \ g/mol$$

$$MW_{SO_3} = 32 \ \frac{g}{mol} + (3)\left(16 \ \frac{g}{mol}\right) = 80 \ g/mol$$

Use Eq. 38.12.

$$\rho_g = \frac{p(MW)}{R^*T} = \frac{(1 \ atm)\left(64 \ \frac{g}{mol}\right)}{\left(8.2 \times 10^{-5} \ \frac{atm \cdot m^3}{mol \cdot K}\right)(20°C + 273°)}$$

$$= 2664 \ g/m^3$$

$$mass \ SO_2 = \left(2664 \ \frac{g}{m^3}\right)\left(\frac{0.1 \ m^3 \ SO_2}{10^6 \ m^3 \ air}\right)$$

$$= 2.7 \times 10^{-4} \ g/m^3 \ air$$

The SO$_2$ concentration in rainwater is

$$\left(2.7 \times 10^{-4} \ \frac{g}{m^3 \ air}\right)\left(\frac{1 \ m^3 \ air}{0.2 \ m^3 \ rainwater}\right)$$

$$= 1.3 \times 10^{-3} \ g/m^3 \ rainwater$$

The SO$_3$ concentration in rainwater is

$$\left(1.3 \times 10^{-3} \ \frac{g}{m^3 \ rainwater}\right)\left(\frac{1 \ m^3}{1000 \ L}\right)\left(\frac{1 \ mol \ SO_3^{-2}}{1 \ mol \ SO_2}\right)$$

$$\times \left(80 \ \frac{g}{mol \ SO_3^{-2}}\right)\left(10^6 \ \frac{\mu g}{g}\right)\left(\frac{1 \ mol \ SO_2}{64 \ g}\right)$$

$$= 1.6 \ \mu g/L$$

4. AIR QUALITY MONITORING

Introduction

Air quality monitoring involves three different categories of measurements: emissions, ambient air, and meteorological. Emissions measurements apply to both stationary and mobile sources and are taken by sampling or monitoring at the point where the emission leaves the source. Ambient air is measured to provide a "control" against which to compare emissions monitoring results and to assess pollutant levels. Meteorological measurements include wind speed and direction, air temperature profiles, and other parameters necessary to evaluate pollutant dispersion and fate.

Monitoring may collect short-duration, single-measurement grab samples or longer-term, continuous samples. The type of sample is dictated by the pollutant being measured, the pollutant source, and the objective of the sampling event.

Particulate Matter

Particulate monitors are of three general types: filtering devices, impactor devices, and photometric devices.

The most common filtering device is the *high-volume* or *hi-vol sampler*. The sampler, typically operated over a period of 24 h or less, uses a blower to draw a large volume of air across a filter material. For gross measurements of total particulate, a single filter may be used. If it is necessary to classify particulate by particle size, stacked filters with openings selected to provide the desired particle size distribution are used. A hi-vol sampler is illustrated in Fig. 38.3.

Figure 38.3 *Hi-Vol Particulate Sampler*

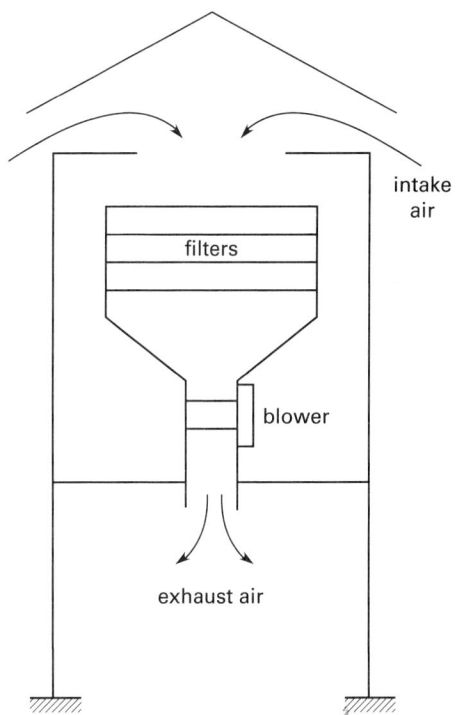

Impactor (impingement) samplers rely on velocity, air flow geometry, and individual particle mass to sample and characterize particulate matter. The sampler works in a cascading fashion, with successively larger particles being collected on a membrane or film in successive stages. The air flow through each stage occurs at successively higher velocities so the higher-mass particles are unable to follow the gas flow-path and, therefore, impinge on the membrane. Impactor samplers are used for sampling particulate matter with diameters below 10 μm and are especially effective for particles with diameters less than 2.5 μm. An impactor sampler is illustrated in Fig. 38.4.

Figure 38.4 *Impactor Particulate Sampler*

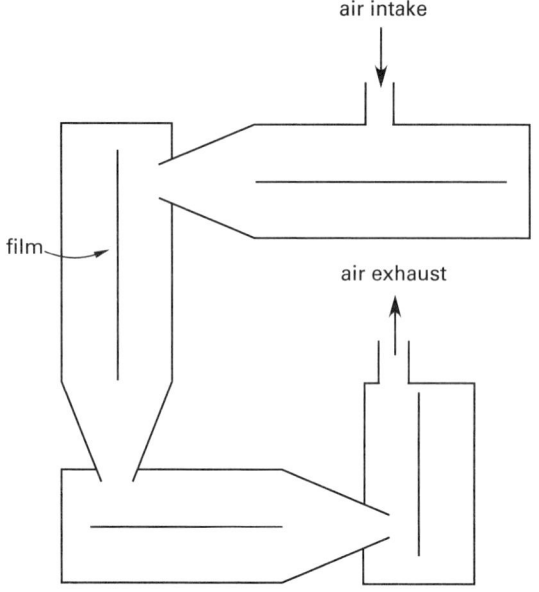

Photometric devices provide an indirect measurement of particles by correlating particulate concentration with light intensity. The air stream is passed between a light beam and detector. As concentrations of particulate increase, the light becomes more scattered, and the light intensity measured by the detector is less. Photometric devices offer the advantage of having a continuous real-time read-out, but they do not differentiate among particle sizes.

Example 38.8

Particulate sampling produced the following results for a representative 24 h period.

particle size collected, μm	less than 10	less than 2.5
clean filter mass, g	8.4910	8.7391
filter mass at 24 h, g	8.7818	8.9461
initial air flow, m³/min	1.5	1.5
final air flow, m³/min	1.46	1.47

What are the PM10 and PM2.5 concentrations?

Solution

The average air flow for PM10 is

$$\frac{1.5\ \frac{m^3}{min} + 1.46\ \frac{m^3}{min}}{2} = 1.48\ m^3/min$$

$$PM10 = \frac{(8.7818\ g - 8.4910\ g)\left(10^6\ \frac{\mu g}{g}\right)}{\left(1.48\ \frac{m^3}{min}\right)\left(24\ \frac{h}{d}\right)\left(60\ \frac{min}{h}\right)}$$

$$= 136\ \mu g/m^3$$

The average air flow for PM2.5 is

$$\frac{1.5\ \frac{m^3}{min} + 1.47\ \frac{m^3}{min}}{2} = 1.485\ m^3/min$$

$$PM2.5 = \frac{(8.9461\ g - 8.7391\ g)\left(10^6\ \frac{\mu g}{g}\right)}{\left(1.485\ \frac{m^3}{min}\right)\left(24\ \frac{h}{d}\right)\left(60\ \frac{min}{h}\right)}$$

$$= 97\ \mu g/m^3$$

Gases

Unlike particulate matter, gases do not typically require long collection periods, being well-suited to detection by monitoring instruments that can provide continuous-flow measurements with instantaneous results. Devices employed for gas monitoring are typically specific to only one gas, so multiple instruments may be required to monitor gas mixtures. Monitoring of gaseous emissions employs wet chemistry techniques; visible, infra-red, and ultraviolet light analyzers; and electrochemical analyzers. Alternatively, samples can be collected in appropriate containers and analyzed using gas chromatography. Gas chromatography is useful when there are several compounds with similar chemical structures such as N_2, O_2, CO_2, CO, and CH_4, or such as VOCs.

Smoke

Smoke presents an observable, inexpensive, and unobtrusive opportunity to assess emissions from stacks. Regulatory agency staff, plant operations personnel, and other interested persons can become certified to provide quantitative measurements of smoke opacity based solely on observation.

Observations of qualified persons are compared to opacity standards represented by the *Ringelmann scale*. The Ringelmann scale uses a percent opacity from completely transparent (Ringelmann No. 0) to completely opaque (Ringelmann No. 5) and is applied to black or dark smoke. For white smoke, opacity is expressed as "percent opacity" rather than by Ringelmann number. Each Ringelmann number corresponds to 20% opacity, so 5% opacity, the accuracy with which opacity observations are typically reported, would represent $^1/_4$ Ringelmann number. The Ringelmann scale is summarized in Fig. 38.5.

Air Quality Index

The *air quality index* (AQI) is a scale used to quantify and report daily air quality to the public. The index was devised by the EPA to replace the formerly used *pollution standard index* (PSI). The AQI considers five of the six criteria pollutants identified by the NAAQS: ground-level ozone, particulates (PM2.5 and PM10), carbon monoxide, sulfur dioxide, and nitrogen dioxide. Lead is not included in the AQI.

Figure 38.5 *Ringelmann Scale For Smoke Opacity*

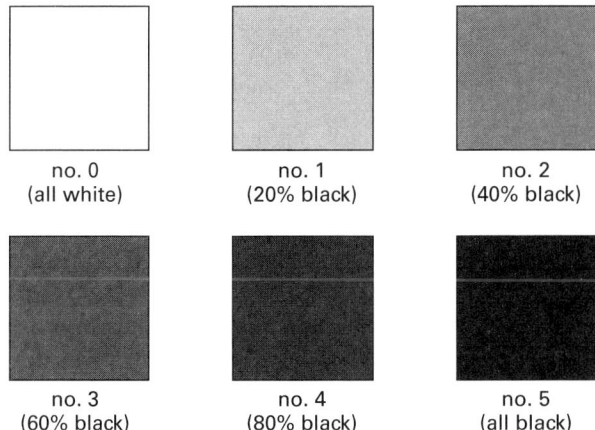

no. 0
(all white)

no. 1
(20% black)

no. 2
(40% black)

no. 3
(60% black)

no. 4
(80% black)

no. 5
(all black)

Data from these pollutants are condensed into a single number that is scaled down to a qualitative descriptor of air quality as shown in Table 38.5. As the AQI increases, public health officials may issue public health advisories and warnings and may impose restrictions on industrial and transportation activities that contribute to air pollution.

Table 38.5 *Air Quality Index (AQI) Categories*

AQI value	air quality descriptor	color
0–50	good	green
51–100	moderate	yellow
101–150	unhealthy for sensitive groups	orange
151–200	unhealthy	red
201–300	very unhealthy	purple
301–500	hazardous	maroon

Source: *Technical Assistance Document for the Reporting of Daily Air Quality—the Air Quality Index (AQI)*, U.S. Environmental Protection Agency, December 2013.

Each AQI category designates upper and lower limits for the concentration of each pollutant. These concentrates are referred to as the *breakpoints*, and are the maximum and minimum acceptable concentrations in each category. (See Table 38.6.)

The AQI is the highest of the subindices, I_p, calculated for each pollutant from the following procedure.

step 1: Truncate the highest pollutant concentration, C_p, observed during the reporting period as follows.

ozone—truncate to 3 decimal places
PM2.5—truncate to 1 decimal place
PM10—truncate to an integer
CO—truncate to 1 decimal place
SO_2—truncate to an integer
NO_2—truncate to an integer

step 2: Using Table 38.6, find the high and low breakpoints, BP, and the high and low indices, I, that contain the truncated observed concentration, C_p.

step 3: Calculate the subindex from Eq. 38.29.

$$I_p = I_{low} + \left(\frac{C_p - BP_{low}}{BP_{high} - BP_{low}}\right)(I_{high} - I_{low}) \qquad 38.29$$

step 4: Round the subindex to the nearest integer.

Example 38.9

The maximum concentrations of air pollutants measured in a metropolitan area during a single day are as follows.

pollutant	duration (h)	concentration
O_3	1	0.14 ppm
CO	8	8 ppm
PM10	24	190 $\mu g/m^3$
SO_2	1	5 ppm
NO_2	1	0.4 ppm

What is the day's AQI value and descriptor?

Solution

The concentrations for SO_2 and NO_2 are reported in ppm. In order to calculate their subindices, those concentrations must be converted to ppb.

$$C_{p,SO_2,ppb} = C_{p,SO_2,ppm}\left(1000\ \frac{ppb}{ppm}\right)$$
$$= (0.07\ ppm)\left(1000\ \frac{ppb}{ppm}\right)$$
$$= 70\ ppb$$
$$C_{p,NO_2,ppb} = C_{p,NO_2,ppm}\left(1000\ \frac{ppb}{ppm}\right)$$
$$= (0.4\ ppm)\left(1000\ \frac{ppb}{ppm}\right)$$
$$= 400\ ppb$$

The 1-hour O_3 concentration is 0.14 ppm. From Table 38.6, this value is within the C_p range of 0.125–0.164 ppm, corresponding to the subindex range of 101–150. From Eq. 38.29, the O_3 subindex is

$$I_{p,O_3} = I_{low} + \left(\frac{C_p - BP_{low}}{BP_{high} - BP_{low}}\right)(I_{high} - I_{low})$$
$$= 101 + \left(\frac{0.14 - 0.125}{0.164 - 0.125}\right)(150 - 101)$$
$$= 120$$

Similarly,

$$I_{p,CO} = 51 + \left(\frac{8 - 4.4}{9.4 - 4.4}\right)(100 - 51) = 86$$
$$I_{p,PM10} = 101 + \left(\frac{190 - 155}{254 - 155}\right)(150 - 101) = 123$$
$$I_{p,SO_2} = 51 + \left(\frac{70 - 36}{75 - 36}\right)(100 - 51) = 94$$
$$I_{p,NO_2} = 151 + \left(\frac{400 - 361}{649 - 361}\right)(200 - 151) = 158$$

Combustion

Table 38.6 *Representative Air Quality Index (AQI) Health Impact Breakpoints*

AQI category	subindex breakpoints, BP						
	1 h O_3 (ppm)[a]	8 h O_3 (ppm)	24 h PM10 ($\mu g/m^3$)	24 h PM2.5 ($\mu g/m^3$)	8 h CO (ppm)	1 h SO_2 (ppb)	1 h NO_2 (ppb)
0–50 (good)	–	0–0.059	0–54	0–12.0	0–4.4	0–35	0–53
51–100 (moderate)	–	0.060–0.075	55–154	12.1–35.4	4.5–9.4	36–75	54–100
101–150 (unhealthy for sensitive groups)	0.125–0.164	0.076–0.095	155–254	35.5–55.4	9.5–12.4	76–185	101–360
151–200 (unhealthy)	0.165–0.204	0.096–0.115	255–354	55.5–150.4[b]	12.5–15.4	186–304[c]	361–649
201–300 (very unhealthy)	0.205–0.404	0.116–0.374	355–424	150.5–250.4[b]	15.5–30.4	305–604[c]	650–1249
301–500 (hazardous)	0.405–0.604	–[d]	425–604	250.5–500.4[b]	30.5–50.4	605–1004[c]	1250–2049

[a]Usually, areas are required to report 8 h ozone values instead of 1 h ozone values. However, some areas are required to calculate 1 h values as well. These areas report the higher of the two values.
[b]Values may change for different significant harm levels (SHLs).
[c]AQIs ≥ 185 use 24 h values instead of 1 h values.
[d]AQIs ≥ 0.374 use 1 h values for ozone instead of 8 h values.

Adapted from: *Technical Assistance Document for the Reporting of Daily Air Quality—the Air Quality Index (AQI)* (EPA-454/B-13-001), Table 2, U.S. Environmental Protection Agency, December 2013.

The highest subindex is 158 for NO_2, so 158 is the AQI. From Table 38.5, this corresponds to the category "unhealthy."

5. METEOROLOGY

Introduction

Although the earth's atmosphere is approximately 100 mi deep, the region extending from ground surface to about 12 mi, called the *troposphere*, is where air pollution problems are encountered. Weather that affects daily life activities also occurs within the troposphere. The dispersion of air pollution in the atmosphere and subsequent human exposure is largely a function of meteorological (weather) conditions.

Wind

The earth's rotation and solar radiation induce global wind patterns. Local factors such as topography, land mass, diurnal cycles, seasonal temperature, cloud conditions, and pressure contribute to local wind patterns.

Local wind is measured as a horizontal velocity from the originating direction. Hence, a north wind originates from the north and is blowing to the south. Localized wind patterns may be represented by a *wind rose*. A variety of configurations are employed for wind roses, but they all show wind direction, magnitude, and frequency. A typical wind rose is illustrated in Fig. 38.6.

Example 38.10

Using the wind rose in Fig. 38.6, what percent of the time do winds blow from the southwest, and what is their maximum velocity?

Solution

The wind blows from the southwest about 19% of the time with a maximum velocity of between 16 mph and 30 mph.

Lapse Rate and Stability

The *lapse rate*, Γ, is the rate of air temperature change with elevation and is used to assess the stability of atmospheric conditions. The lapse rate will determine the configuration of an air pollutant plume and the dispersion pattern of the pollutants. The *dry adiabatic lapse rate* is the temperature change as a parcel of dry air rises or falls without exchanging heat with its

Figure 38.6 *Typical Wind Rose*

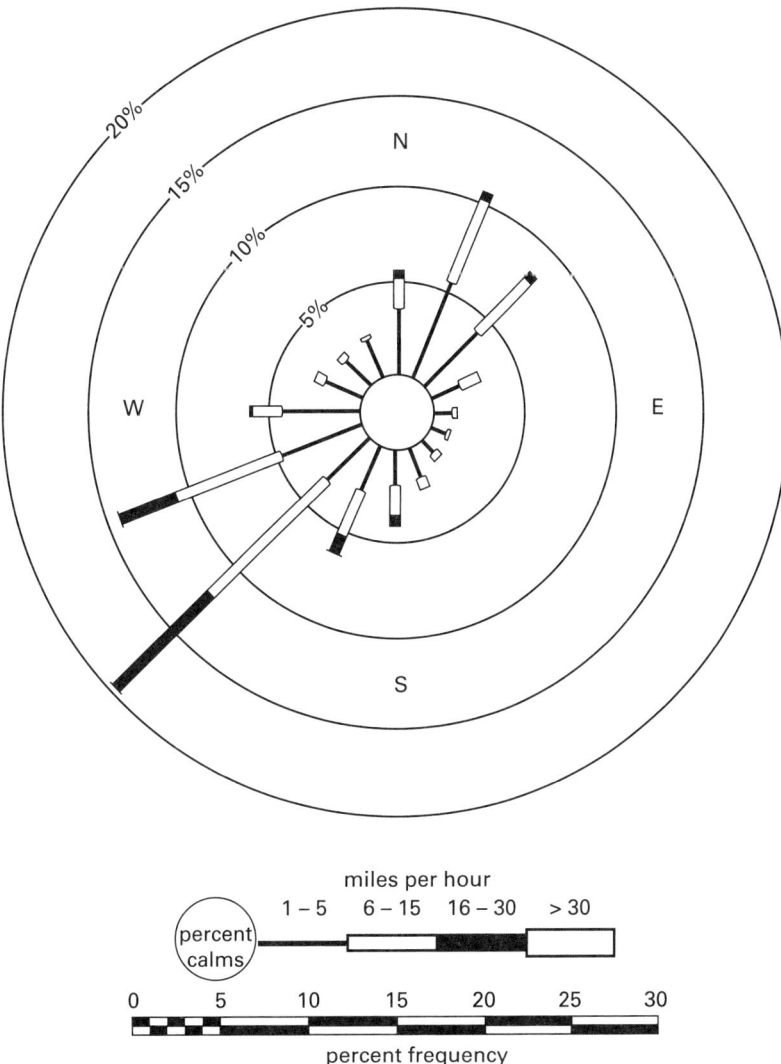

Source: *Meteorology and Atomic Energy* © 1955, U.S. Dept. of Commerce Weather Bureau, U.S. Atomic Energy Commission, Washington, DC.

surroundings. The *wet adiabatic lapse rate*, where moisture exits in the atmosphere, is slightly less than the dry adiabatic lapse rate. However, for most practical purposes, they are approximately equal, and $-0.0098°C/m$ can be used for either.

As an air pollutant is emitted from a stack, its temperature will likely be greater than the temperature of the ambient air. As the pollutant cools, the actual lapse rate of the surrounding air (the *ambient lapse rate*) will define how the pollutant is dispersed. The four general conditions illustrated in Fig. 38.7 and described as follows can result.

- *Superadiabatic lapse rate* occurs when the surrounding air cools faster than the plume as elevation increases. This represents an unstable condition that will allow pollutants to readily disperse.

- *Adiabatic lapse rate* occurs when the surrounding air and the plume cool at the same rate. When this occurs, there will be no tendency for the plume to rise or fall due to temperature differences.

- *Subadiabatic lapse rate* occurs when the surrounding air cools slower than the plume as elevation increases. This represents a stable condition that interferes with the dispersion of the plume.

- *Inversion* occurs when the lapse rate is inverted so that the ambient air temperature is cooler near the ground than at elevation. This represents a stable condition.

Lapse rates can vary at different elevations to produce a variety of stability conditions and plume configurations as shown in Fig. 38.8.

Figure 38.7 *Lapse Rates*

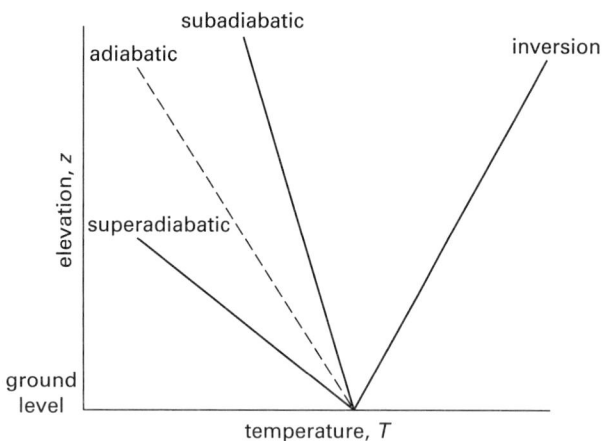

Example 38.11

An 80 m stack emits a plume at 32°C. The air temperature at ground level is 23°C, and the ambient lapse rate up to an elevation of 300 m is +0.0081 °C/m. Above 300 m, the lapse rate is −0.0053 °C/m. How high will the plume rise?

Solution

The plume cools at the dry adiabatic lapse rate of 0.0098 °C/m as it rises and will stop rising when the plume and air temperatures are equal.

The plume temperature at 300 m is

$$32°C + \left(-0.0098 \; \frac{°C}{m}\right)(300 \; m - 80 \; m) = 29.8°C$$

The air warms at the ambient lapse rate of +0.0081 °C/m up to an elevation of 300 m.

The air temperature at 300 m is

$$23°C + \left(+0.0081 \; \frac{°C}{m}\right)(300 \; m) = 25.4°C$$

The air cools above 300 m at the ambient lapse rate of −0.0053 °C/m.

$$29.8°C + \left(-0.0098 \; \frac{°C}{m}\right)z = 25.4°C + \left(-0.0053 \; \frac{°C}{m}\right)z$$

z is the distance above 300 m where the air and plume temperatures are equal.

$$z = \frac{29.8°C - 25.4°C}{\left(-0.0053 \; \frac{°C}{m}\right) - \left(-0.0098 \; \frac{°C}{m}\right)} = 978 \; m$$

The plume will rise to 300 m + 978 m = 1278 m.

Maximum Mixing Depth

As air pollutants are emitted into the atmosphere, their dispersion is aided by mixing through convection and turbulence. The greater the volume of air that is available for mixing, the more effective dispersion will be in reducing the impact of the emission. The dry adiabatic lapse rate and the ambient lapse rate are used to define the height to which a plume will rise in the atmosphere. The two lapse rates are used to calculate the elevation at which the temperatures of the ambient air and the emitted plume are equal. This elevation defines the height of the *mixing zone*, commonly referred to as the maximum mixing depth. Figure 38.9 illustrates the *maximum mixing depth* under different conditions of atmospheric stability.

Example 38.12

Using a graphical solution, what is the maximum mixing depth for the conditions described in Ex. 38.11?

Solution

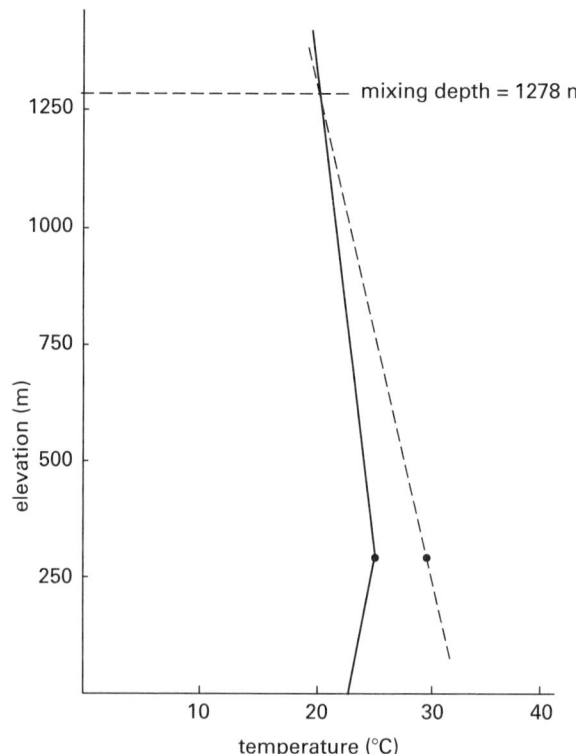

The maximum mixing depth is 1278 m.

Dispersion

Dispersion involves the horizontal and vertical spreading of a pollutant plume as it moves away from the source. This spreading is most commonly

Figure 38.8 *Representative Plume Stability Configurations
(adiabatic lapse rate shown as dashed line for comparison)*

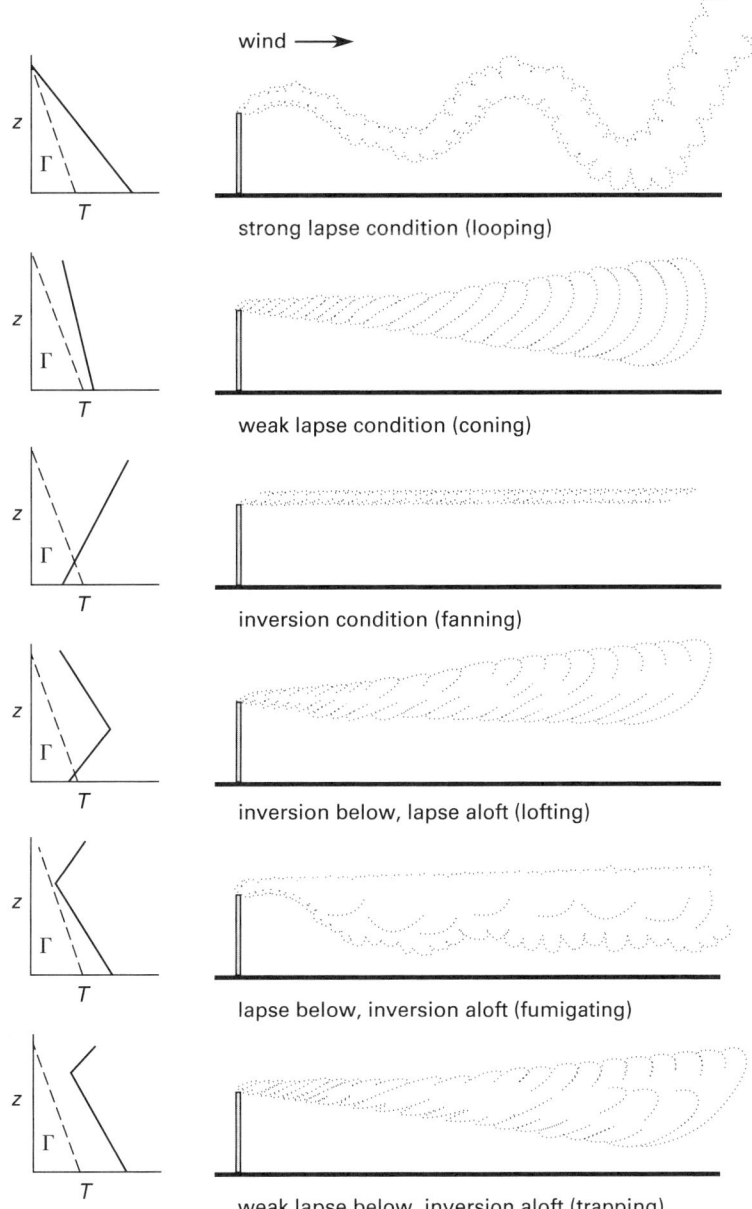

strong lapse condition (looping)

weak lapse condition (coning)

inversion condition (fanning)

inversion below, lapse aloft (lofting)

lapse below, inversion aloft (fumigating)

weak lapse below, inversion aloft (trapping)

Source: *Meteorology and Atomic Energy* © 1955, U.S. Dept. of Commerce Weather Bureau, U.S. Atomic Energy Commission, Washington, DC.

Figure 38.9 *Representative Maximum Mixing Depths*

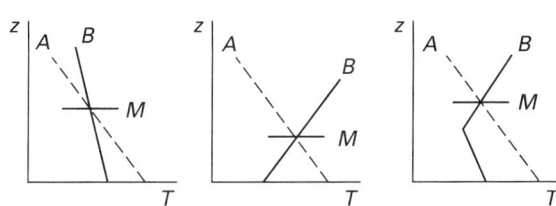

A = adiabatic lapse rate
B = ambient lapse rate
M = maximum mixing depth

described using the *Gaussian plume* model. The physical features of the model are shown in Fig. 38.10 using a three-dimensional x-y-z coordinate system.

The model can estimate the downwind configuration of the plume and hence concentrations of emitted pollutants at specific locations of interest. Selected applications of the model are presented as follows.

- Assuming the ground is a perfect reflection, the ground-level concentration at some x-y coordinate location downwind of the source and originating

Combustion

Figure 38.10 Plume Coordinates For Gaussian Dispersion Model

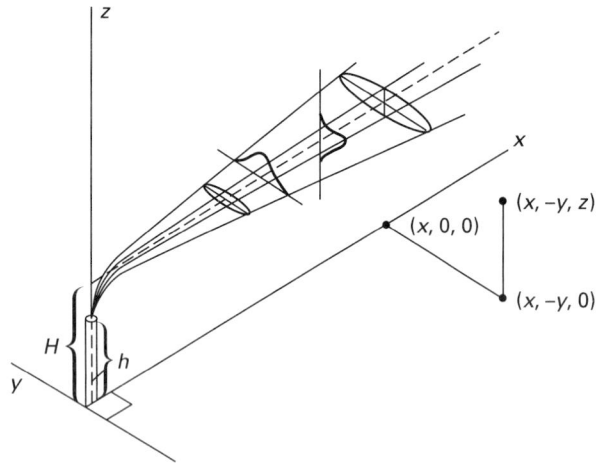

from a stack with an effective height H, with no reflections from the ground, is

$$C_{x,y,0} = \frac{Q_g e^{(-0.5)\left(\frac{y}{\sigma_y}\right)^2} e^{(-0.5)\left(\frac{H}{\sigma_z}\right)^2}}{\pi \mu \sigma_z \sigma_y} \qquad 38.30$$

The *effective stack height*, H, is equal to the stack height, h, plus an incremental height, Δh, to correct for the initial rise of the plume as it leaves the stack. However, the incremental stack height is commonly ignored, and the stack height and effective stack height are assumed to be equal. The effective stack height can be calculated from empirical relationships.

- The ground-level concentration at some distance along the plume centerline and originating from a stack with an effective height H, with no reflections from the ground, is

$$C_{x,0} = \frac{Q_g e^{(-0.5)\left(\frac{H}{\sigma_z}\right)^2}}{\pi \mu \sigma_z \sigma_y} \qquad 38.31$$

- The ground-level concentration at some distance along the plume centerline and originating from a ground-level source, with no reflections from the ground, is

$$C_{x,0} = \frac{Q_g}{\pi \mu \sigma_z \sigma_y} \qquad 38.32$$

- The location of the maximum ground-level concentration along the plume centerline and originating from a stack with an effective height H, with no reflections from the ground, is

$$\sigma_z = 0.707H \qquad 38.33$$

The *Gaussian dispersion coefficients*, σ_y and σ_z, are taken from Fig. 38.11 and Fig. 38.12 for the atmospheric stability conditions defined in Table 38.7.

Table 38.7 Atmospheric Stability Categories

surface wind speed at 10 m (m/s)	day incoming solar radiation			night cloud cover	
	strong	moderate	slight	mostly overcast	mostly clear
class[*]	1	2	3	4	5
< 2	A	A-B	B	E	F
2–3	A-B	B	C	E	F
3–5	B	B-C	C	D	E
5–6	C	C-D	D	D	D
>6	C	D	D	D	D

[*]The neutral class, D, should be assumed for overcast conditions during day or night. Class A is the most unstable, and class F is the most stable. Class B is moderately unstable, and class E is slightly stable.

Source: Wark, K. and C.F. Warner, *Air Pollution*, 2nd ed., © 1981, Harper & Row, NY.

Example 38.13

An air pollutant is emitted at 3.6 kg/s through a stack with an effective height of 225 m. The wind speed at 10 m above grade is 4 m/s, and atmospheric stability conditions are class B. (a) What is the maximum ground level concentration location along the plume centerline? (b) What is the maximum ground-level concentration?

Solution

(a) From Fig. 38.12 with $\sigma_z = 159$ m and a curve for stability class B, $x = 1500$ m.

Use Eq. 38.31 to find the maximum ground-level concentration along the plume centerline.

From Fig. 38.11 with $x = 1500$ m and a curve for stability class B, $\sigma_y = 200$ m.

$$C_{x,0} = \frac{Q_g e^{(-0.5)\left(\frac{H}{\sigma_z}\right)^2}}{\pi \mu \sigma_z \sigma_y}$$

$$C_{1700,0} = \frac{\left(3.6 \ \frac{\text{kg}}{\text{s}}\right)\left(10^3 \ \frac{\text{g}}{\text{kg}}\right) \times e^{(-0.5)\left(\frac{225 \ \text{m}}{159 \ \text{m}}\right)^2}}{\pi \left(4 \ \frac{\text{m}}{\text{s}}\right)(159 \ \text{m})(200 \ \text{m})}$$

$$= 0.0033 \ \text{g/m}^3$$

$$\left(0.0033 \ \frac{\text{g}}{\text{m}^3}\right)\left(10^6 \ \frac{\mu\text{g}}{\text{g}}\right) = 3300 \ \mu\text{g/m}^3$$

(b) Use Eq. 38.33 to find the location of maximum ground-level concentration.

$$\sigma_z = 0.707H$$

$$= (0.707)(225 \ \text{m})$$

$$= 159 \ \text{m}$$

Figure 38.11 *Gaussian Dispersion Coefficient, σ_y, for Vertical Plume Concentration at Downwind Distance, x*
(see Table 38.7 for definition of curves)

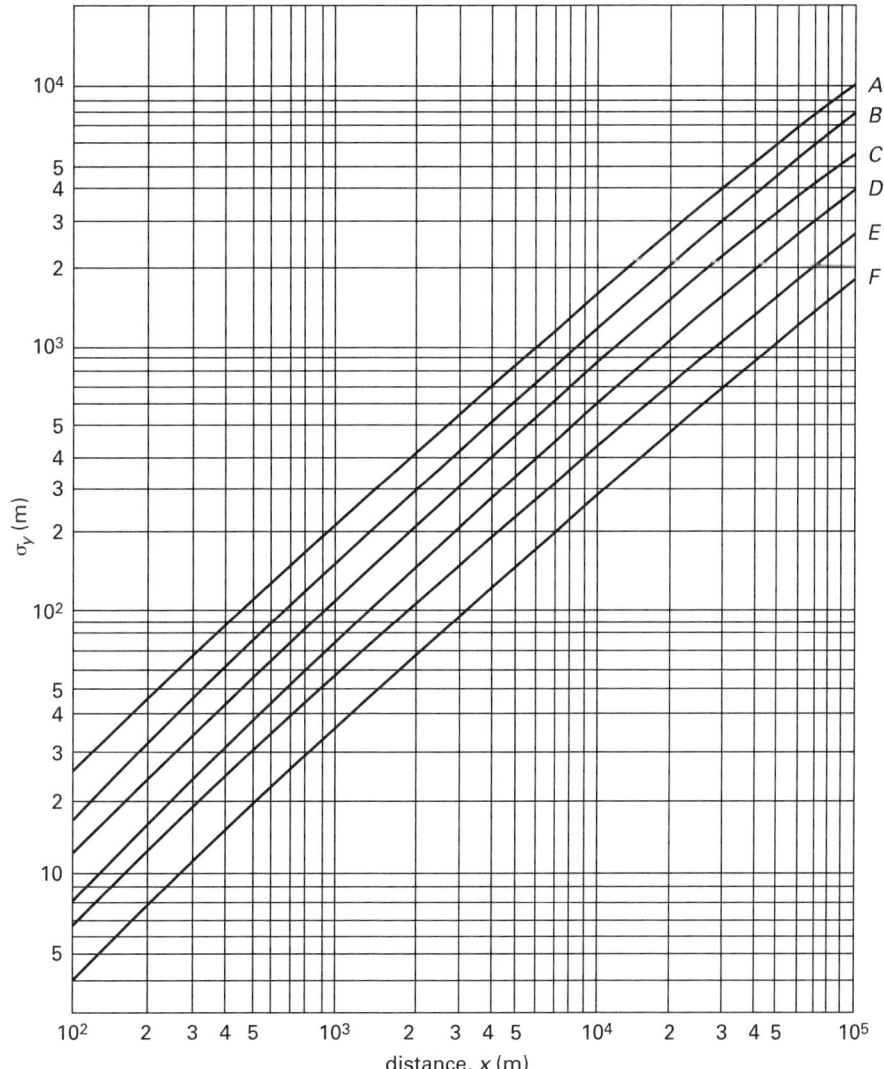

Source: Wark, K. and C.F. Warner, *Air Pollution*, 2nd ed., © 1981, Harper & Row, NY.

6. STATIONARY SOURCE CONTROL

Baghouses and Filters

Baghouses consist of banks of fabric filters, usually bags, typically suspended from overhead supports. Contaminated air flows into the bottom of the baghouse and must pass through the filter fabric to escape as cleaned air through the top. The particulate is trapped at the fabric surface and accumulates there until limiting pressure drops of between 1.5 kPa and 2.0 kPa are measured. The bags are then cleaned by a mechanical shaker, by reverse air, or by pulse-jetting. The particulate freed from the filters during cleaning falls to the bottom of the baghouse where it is collected as solid waste. Shaker and pulse jet baghouses are illustrated in Fig. 38.13.

Removal efficiencies of baghouses are in excess of 99% and are often as high as 99.99% (weight basis). In some cases, liquid pollutants may condense on smaller-diameter (less than 10 μm) particles at temperatures below 150°C. The condensed liquids are then removed with the particles.

Efficiency and operating conditions depend on the particulate characteristics and the filter fabric selected. Filters are manufactured from a variety of natural and synthetic fabrics and may be felted or woven. Important factors influencing bag selection include temperature and corrosivity of the gas stream, particulate abrasiveness and hygroscopicity (water affinity), particle size and concentration, desired filtering velocities, and others. Although tube configurations are most common, other configurations include

Figure 38.12 *Gaussian Dispersion Coefficient, σ_z, for Horizontal Plume Concentration at Downwind Distance, x (see Table 38.7 for definition of curves)*

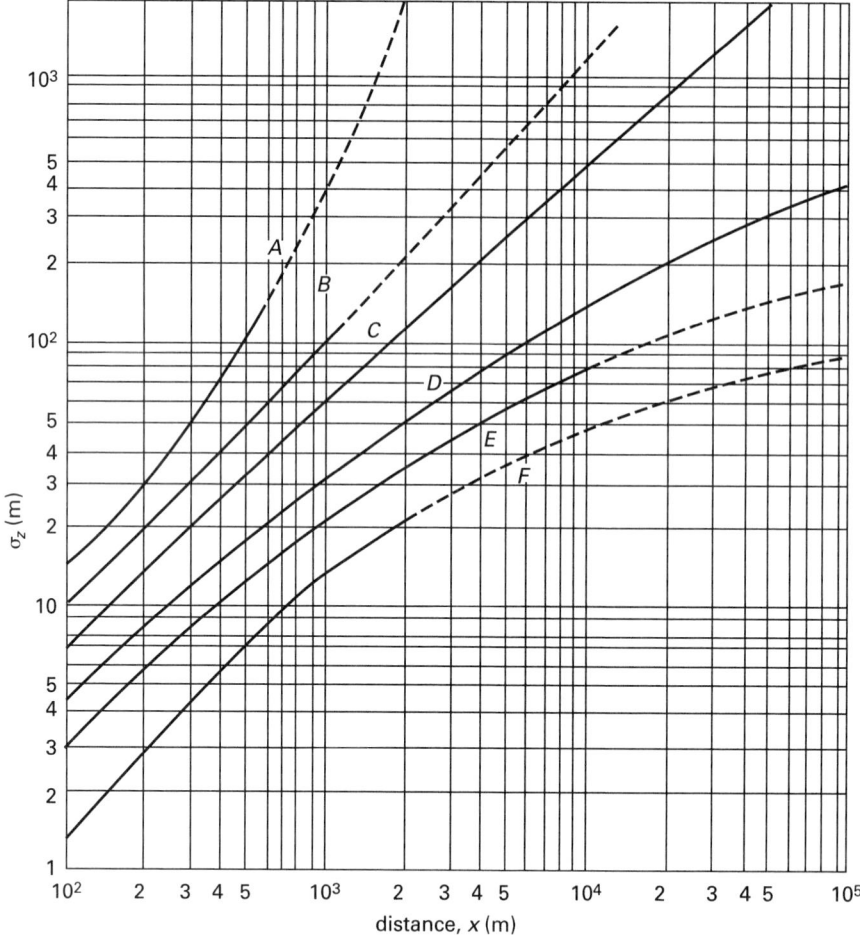

Source: Wark, K. and C.F. Warner, *Air Pollution*, 2nd ed., © 1981, Harper & Row, NY.

envelopes (i.e., flat bags) and pleated cartridges. A typical bag diameter is 20 cm, with lengths varying from 1.8 m to 9 m.

Baghouses are characterized by their air-to-cloth ratios and pressure drops. The *air-to-cloth ratio*, designated A/C ratio, is alternatively known as *filter ratio*, *superficial face velocity*, and *filtering velocity*. Typical A/C ratios as a function of cleaning method are shown in Table 38.8.

The A/C ratio is used to calculate the required filter fabric area. For this calculation, the A/C ratio is called the filtering velocity, and the following equation is applied.

$$A = \frac{Q_g}{v_f} \qquad 38.34$$

Table 38.8 *Air-to-Cloth (A/C) Ratios*

cleaning method	A/C $(\mathrm{m^3/m^2 \cdot s})$
shaking	0.01–0.03
reverse air	0.005–0.02
pulsed jet	0.02–0.08

Adapted from: Noll, K.E., *Fundamentals of Air Quality Systems: Design of Air Pollution Control Devices*, © 1999, American Academy of Environmental Engineers, Annapolis, MD.

The minimum cleaning frequency for a desired pressure drop or the pressure drop for a specified operating period can be calculated using Eq. 38.35 in the appropriate form. K_1 and K_2 are the *fabric resistance* and *cake resistance*, respectively.

$$-\Delta p = K_1 v_f + K_2 C_p v_f^2 t \qquad 38.35$$

Figure 38.13 *Types of Baghouses*

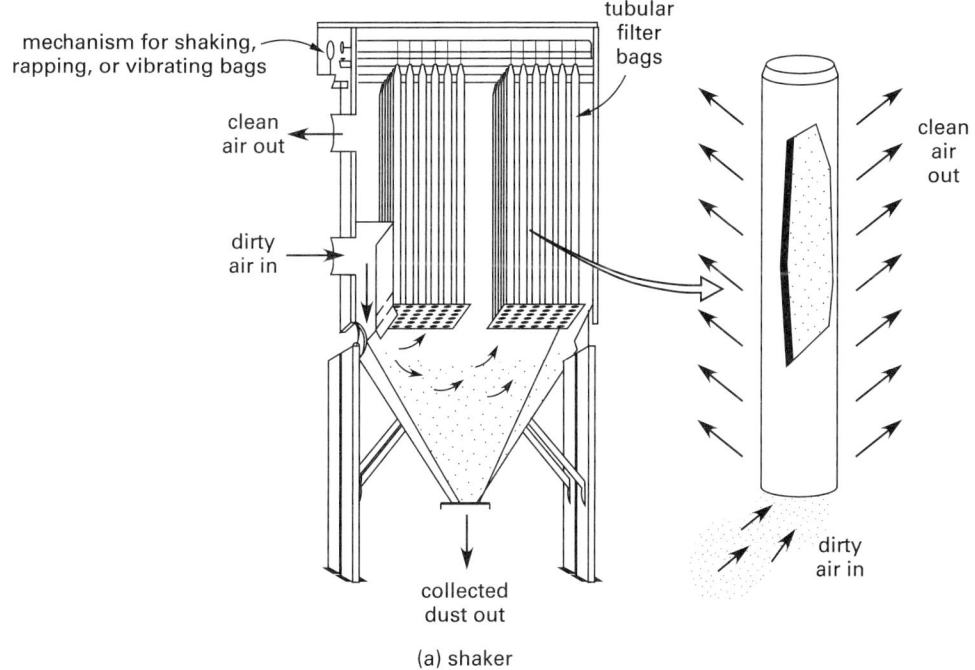

(a) shaker

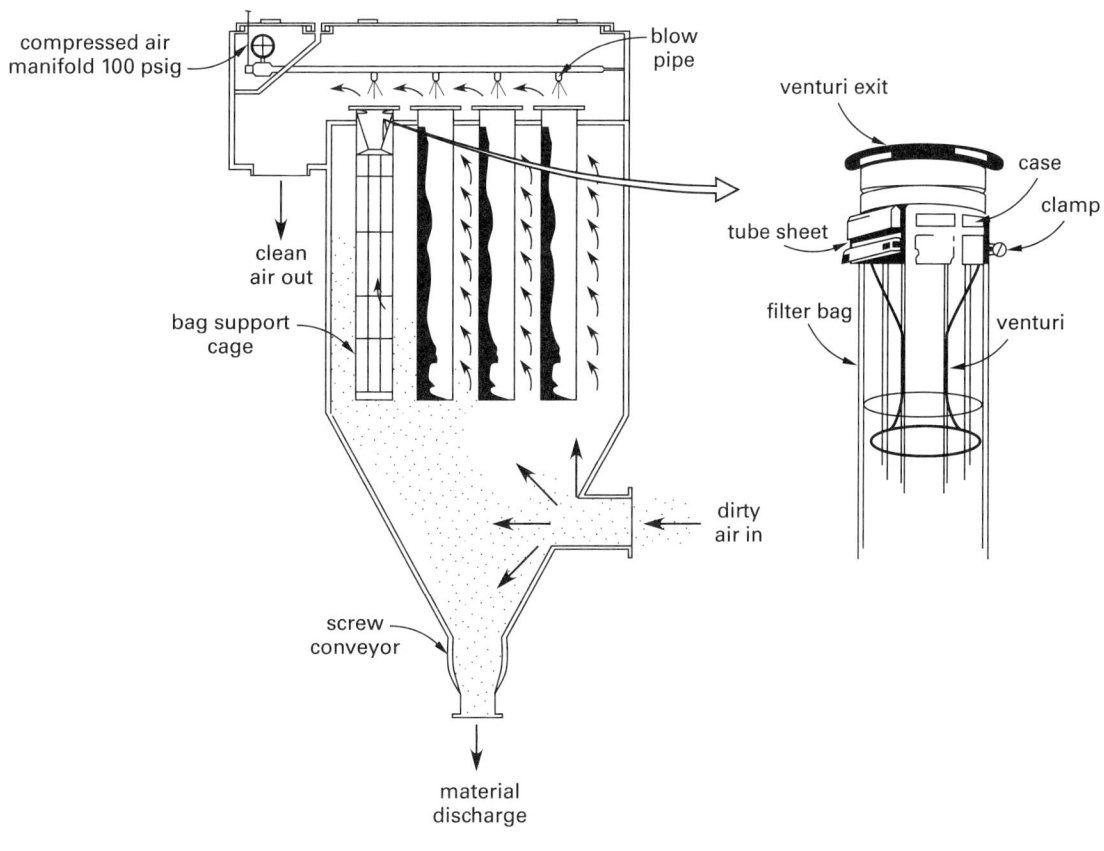

(b) pulse jet

Environmental Engineering: A Design Approach by Sincero and Sincero, © 1995. Reprinted by permission of Prentice-Hall, Inc., Upper Saddle River, NJ.

Example 38.14

A pulsed-jet baghouse is described by the following.

air flow rate	8 m³/s
particulate concentration	45 g/m³
A/C ratio	0.025 m³/m²·s
fabric resistance	0.2 kPa·s/cm
cake resistance	1.8 kPa·s·cm/g
allowable pressure drop	0.65 kPa

What are the (a) required total fabric area, and (b) minimum cleaning frequency?

Solution

(a) Use Eq. 38.34.

The total fabric area is

$$A = \frac{Q_g}{v_f} = \frac{8 \ \dfrac{m^3}{s}}{0.025 \ \dfrac{m^3}{m^2 \cdot s}}$$

$$= 320 \ m^2$$

(b) Use Eq. 38.35.

$$-\Delta p = K_1 v_f + K_2 C_p v_f^2 t$$

$$0.65 \ kPa = \frac{\left(0.2 \ \dfrac{kPa \cdot s}{cm}\right)\left(0.025 \ \dfrac{m}{s}\right)}{\dfrac{1 \ m}{100 \ cm}}$$

$$+ \frac{\left(1.8 \ \dfrac{kPa \cdot s \cdot cm}{g}\right)\left(45 \ \dfrac{g}{m^3}\right)\left(0.025 \ \dfrac{m}{s}\right)^2 t}{100 \ \dfrac{cm}{m}}$$

$$t = 296 \ s \quad [\text{clean every 4.9 min}]$$

Cyclones

Cyclone separators are used to remove particulate matter having aerodynamic diameters greater than about 15 μm. Separation occurs as centrifugal force, generated by accelerating the particulate-laden air in the cyclone cone, forcing particles to impinge on the cyclone wall and slough off into a collection hopper. The clean air is exhausted out the top of the cyclone.

Cyclones are classified as high-throughput, conventional, or high-efficiency. *High-throughput cyclones* are designed to handle high gas volumes where lower efficiencies are acceptable. *High-efficiency cyclones* handle lower flow volumes but provide higher efficiencies. *Conventional cyclones* fall between high-throughput and high-efficiency cyclones. Standard cyclone design dimensions, illustrated in Fig. 38.14, are summarized in Table 38.9.

Figure 38.14 *Standard Cyclone Dimensions*

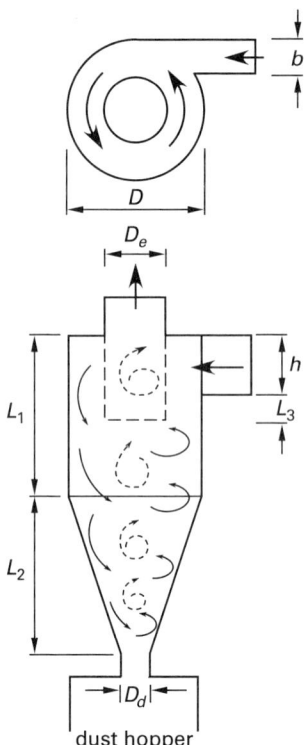

Cyclone design involves selecting dimensions to provide the desired efficiency as a function of particle physical characteristics (i.e., particle size distribution and density) and terminal settling velocity. There are several different methods used to design and analyze cyclone collectors. All methods yield results that are in the same order of magnitude, but large variations in predicted performance are part of cyclone analysis.

The centrifugal acceleration is

$$a_c = \frac{\left(\dfrac{4Q_g}{(D-D_e)(2L_1+L_2)}\right)^2}{r_a} \quad \text{38.36}$$

$$r_a = r_e + 0.5(r - r_e)$$

$$= 0.5D_e + 0.25(D - D_e) \quad \text{38.37}$$

The diameter of the 100% removed particles is

$$d_p = 4\sqrt{\frac{Q_g \mu_g}{a_c \rho_p (2L_1 + L_2)(D + D_e)}} \quad \text{38.38}$$

The terminal settling velocity of the 100% removed particles is

$$v_t = \frac{Q_g}{0.5\pi(L_1 + 0.5L_2)(D + D_e)} \qquad 38.39$$

The *overall removal efficiency* of the cyclone uses the following equations. β is the *volume shape factor*.

$$E = 1 - x_o + \frac{\sum(\Delta x)v_i}{v_t} \qquad 38.40$$

$$v_i = \sqrt{\frac{4a_c\rho_p d_i}{3C_D\rho_g}} \qquad 38.41$$

$$C_D = \frac{24}{Re} \qquad 38.42$$

$$Re = \frac{d_i v_i \rho_g}{\mu_g} \qquad 38.43$$

$$d_i = 1.24\beta^{1/3} d_{pi} \qquad 38.44$$

Example 38.15

A 0.8 m diameter conventional cyclone receives a contaminated gas stream at 6 m³/s and 200°C. The particulates in the gas stream have a density of 1600 kg/m³. What particle diameter will the cyclone be able to completely remove?

Solution

Use Eq. 38.36 and Eq. 38.37, and refer to Fig. 38.14 and Table 38.9.

$$D_e = 0.5D = (0.5)(0.8 \text{ m}) = 0.4 \text{ m}$$

$$L_1 = 2D = (2)(0.8 \text{ m}) = 1.6 \text{ m}$$

$$L_2 = 2D = (2)(0.8 \text{ m}) = 1.6 \text{ m}$$

$$r_a = r_e + 0.5(r - r_e)$$
$$= 0.5D_e + 0.25(D - D_e)$$
$$= (0.5)(0.4 \text{ m}) + (0.25)(0.8 \text{ m} - 0.4 \text{ m})$$
$$= 0.3 \text{ m}$$

$$a_c = \frac{\left(\dfrac{4Q_g}{(D - D_e)(2L_1 + L_2)}\right)^2}{r_a}$$

$$= \frac{\left(\dfrac{(4)\left(6 \dfrac{\text{m}^3}{\text{s}}\right)}{(0.8 \text{ m} - 0.4 \text{ m})\big((2)(1.6 \text{ m}) + 1.6 \text{ m}\big)}\right)^2}{0.3 \text{ m}}$$

$$= 521 \text{ m/s}^2$$

From App. 38.C, $\mu_g = 2.6 \times 10^{-5}$ kg/m·s for air at 200°C.

Use Eq. 38.38.

$$d_p = 4\sqrt{\frac{Q_g\mu_g}{a_c\rho_p(2L_1 + L_2)(D + D_e)}}$$

$$= 4\sqrt{\frac{\left(6 \dfrac{\text{m}^3}{\text{s}}\right)\left(2.6 \times 10^{-5} \dfrac{\text{kg}}{\text{m·s}}\right)}{\begin{array}{c}\left(521 \dfrac{\text{m}}{\text{s}^2}\right)\left(1600 \dfrac{\text{kg}}{\text{m}^3}\right) \\ \times \big((2)(1.6 \text{ m}) + 1.6 \text{ m}\big)(0.8 \text{ m} + 0.4 \text{ m})\end{array}}}$$

$$= 2.3 \times 10^{-5} \text{ m}$$

$$d_p = (2.3 \times 10^{-5} \text{ m})\left(10^6 \frac{\mu\text{m}}{\text{m}}\right) = 23 \ \mu\text{m}$$

Table 38.9 *Standard Cyclone Design Dimensions*

dimension	symbol*	conventional	high throughput	high efficiency
cylinder diameter	D	D	D	D
cylinder length	L_1	$2D$	$1.5D$	$1.5D$
cone length	L_2	$2D$	$2.5D$	$2.5D$
total length	$L_1 + L_2$	$4D$	$4D$	$4D$
outlet length	$h + L_3$	$0.675D$	$0.875D$	$0.5D$
inlet height	h	$0.5D$	$0.75D$	$0.5D$
inlet width	b	$0.25D$	$0.375D$	$0.2D$
outlet diameter	D_e	$0.5D$	$0.75D$	$0.5D$
dust exit diameter	D_d	$0.25D$	$0.375D$	$0.375D$

*corresponding to Fig. 38.14

Environmental Engineering: A Design Approach by Sincero and Sincero, © 1995. Reprinted by permission of Prentice-Hall, Inc., Upper Saddle River, NJ.

Combustion

Example 38.16

The grain size distribution for the particulate being removed by the cyclone described in Ex. 38.15 is as follows.

particle size (μm)	mass %
0–15	12
15–25	64
25–35	24

Assume a volume shape factor of 0.9 and a particulate density of 1600 kg/m³. What is the overall removal efficiency of the cyclone?

Solution

Use Eq. 38.39 and the results from Ex. 38.15.

$$v_t = \frac{Q_g}{0.5\pi(L_1 + 0.5L_2)(D + D_e)}$$

$$= \frac{6 \ \frac{\text{m}^3}{\text{s}}}{0.5\pi(1.6 \text{ m} + (0.5)(1.6 \text{ m}))(0.8 \text{ m} + 0.4 \text{ m})}$$

$$= 1.3 \text{ m/s}$$

Use Eq. 38.44.

$$d_i = 1.24\beta^{1/3}d_{\text{pi}} = (1.24)(0.9)^{1/3}d_{\text{pi}} = 1.2d_{\text{pi}}$$

From App. 38.C, $\rho_g = 0.748$ kg/m³ for air at 200°C.

Use Eq. 38.41.

$$v_i = \sqrt{\frac{4a_c\rho_p d_i}{3C_D\rho_g}}$$

$$= \sqrt{\frac{(4)\left(521 \ \frac{\text{m}}{\text{s}^2}\right)\left(1600 \ \frac{\text{kg}}{\text{m}^3}\right)(1.2)d_{\text{pi}}}{3C_D\left(0.748 \ \frac{\text{kg}}{\text{m}^3}\right)}}$$

$$= 1335\sqrt{\frac{d_{\text{pi}}}{C_D}}$$

Use Eq. 38.43.

$$\text{Re} = \frac{d_i v_i \rho_g}{\mu_g} = \frac{v_i(1.2)d_{\text{pi}}\left(0.748 \ \frac{\text{kg}}{\text{m}^3}\right)}{2.6 \times 10^{-5} \ \frac{\text{kg}}{\text{m}\cdot\text{s}}}$$

$$= (3.5 \times 10^4)v_i d_{\text{pi}}$$

Use Eq. 38.42.

$$C_D = \frac{24}{\text{Re}} = \frac{24}{(3.5 \times 10^4)v_i d_{\text{pi}}} = \frac{7.0 \times 10^{-4}}{v_i d_{\text{pi}}}$$

Substituting the result from Eq. 38.42 into the result from Eq. 38.41 gives

$$v_i = 1335\sqrt{\frac{d_{\text{pi}}^2 v_i}{7.0 \times 10^{-4}}} = (2.5 \times 10^9)d_{\text{pi}}^2$$

particle size (μm)	mass percentage	cumulative mass fraction	d_{pi} (10^{-6} m)	v_i (m/s)
0–15	12	0.12	7.5	0.14
15–25	64	0.76	20	1.0
25–35	24	1.0	30	2.3
	100%			

In the above table, d_{pi} is taken as the average particle diameter within each particle size range (i.e., for particle size range 15 μm through 25 μm, $d_{\text{pi}} = (15 \ \mu\text{m} + 25 \ \mu\text{m})/2 = 20 \ \mu\text{m}$).

Cumulative mass fraction and incremental settling velocity can be roughly plotted.

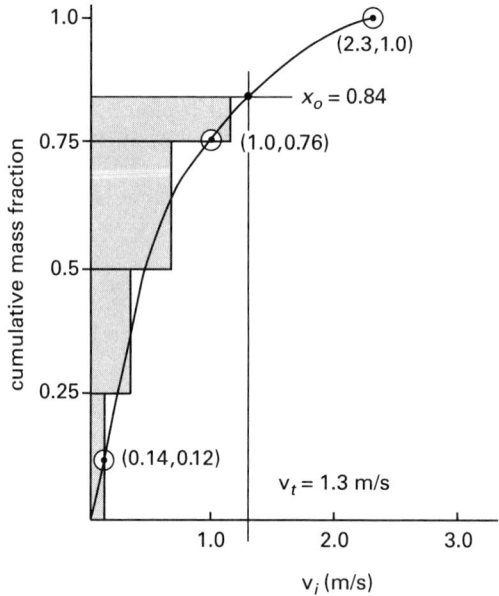

The incremental mass fraction, Δx, and the corresponding incremental settling velocity, v_i, resulting from the integration of the plot above are presented as follows.

Δx	v_i (m/s)	$\Delta x v_i$ (m/s)
0.25	0.15	0.04
0.25	0.36	0.09
0.25	0.69	0.17
0.09	1.15	0.10
		0.40

Use Eq. 38.40.

$$E = 1 - x_o + \frac{\sum \Delta x v_i}{v_t} = 1 - 0.84 + \frac{0.40 \, \frac{m}{s}}{1.3 \, \frac{m}{s}}$$

$$= 0.467 \quad (47\%)$$

Electrostatic Precipitators

Electrostatic precipitators (ESPs) are used to remove particulate matter from gas streams. ESPs are preferred over many other particulate removal processes because they are economical to operate, provide high efficiencies (near 99%), are dependable and predictable, and do not produce a moisture plume. However, ESPs generally cannot be used with moist flows, mists, or sticky or hygroscopic particles, and they must be heated during start-up and shut-down to avoid corrosion from acid gas condensation. Although humid gas streams can be treated with ESPs, performance is inhibited since water droplets can insulate particles and reduce their resistivities.

In operation, the gas passes through a charging field created by high-voltage electrode wires called *discharge electrodes*. The negatively charged particles are attracted to positively charged collection plates known as *collectors*. The speed at which the particles migrate toward the collectors is known as the *drift velocity* or the *electric wind velocity*. Periodically, rappers vibrate the collectors and dislodge the particles, which drop into hoppers. Figure 38.15 illustrates several features of ESPs.

In addition to affecting drift velocity, the *specific collection area* (SCA) affects *removal (collection) efficiency*. The higher the SCA, the higher the removal efficiency, with 180 $m^2 \cdot s/m^3$ defining the upper limit. Pressure drops across ESPs are low, typically on the order of 0.13 kPa. The drift velocity, w, is calculated from Eq. 38.45 and Eq. 38.46, where i_o is the permittivity constant, 8.85×10^{-12} $C^2/N \cdot m^2$.

$$w = \frac{8 i_o E_f^2 d_p}{24 \mu_g} \qquad \textit{38.45}$$

$$E_f = \frac{V_d}{Z_e} \qquad \textit{38.46}$$

E_f is the average electric field and is expressed in units of newtons per coulomb (N/C), equivalent to volts/meter (V/m), or the applied voltage divided by the distance separating the discharge electrodes from the collectors. The separation distance is one-half the distance between collectors. The collector area, A, and the theoretical efficiency, E, can be calculated using Eq. 38.47, and the SCA can be calculated using Eq. 38.48.

$$A = \frac{\left(-\ln\left(1 - \frac{E}{100\%}\right)\right) Q_g}{w} \qquad \textit{38.47}$$

$$\text{SCA} = \frac{A}{Q_g} \qquad \textit{38.48}$$

Use of ESPs is most efficient when particles have resistivities in the 10^{10} $\Omega \cdot cm$ to 10^{14} $\Omega \cdot cm$ range. If the particle resistivity is too low, particles may release their charge upon contacting the collector and become reentrained in the gas. If the resistivity is too high, the particles may be difficult to dislodge and may insulate the collectors. Very small particles (0.1 μm to 1.0 μm) reduce the efficiency of the charging field, and particles less than 2 μm to 3 μm are susceptible to reentrainment. Typical design parameter values are presented in Table 38.10.

Table 38.10 Electrostatic Precipitator Design Parameters

parameter	symbol	typical value
efficiency	E	90–98%
gas velocity	v_g	0.6–1.2 m/s
maximum gas temperature	T	700°C
drift velocity	w	0.03–0.21 m/s
residence time	t	2–10 s
pressure loss	Δp	0.025–0.125 kPa
collector spacing	$2Z_e$	30–41 cm
collector height	h	9–15 m
collector length:height	$l{:}h$	1:1–1:2
maximum specific collection area	SCA	180 $m^2 \cdot s/m^3$
applied voltage	V_d	30–100 kV

Example 38.17

An ESP will treat a 260°C gas stream containing particulates with an average particle diameter of 4 μm and concentration of 6 g/m^3. The average electric field for the ESP is 350 000 N/C. What is the SCA required to reduce the particulate concentration to 0.08 g/m^3?

Solution

From App. 38.C, $\mu_g = 2.78 \times 10^{-5}$ kg/m·s for air at 260°C.

Use Eq. 38.45.

$$w = \frac{8 i_o E_f^2 d_p}{24 \mu_g}$$

$$= \frac{(8)\left(8.85 \times 10^{-12} \, \frac{C^2}{N \cdot m^2}\right)\left(350\,000 \, \frac{N}{C}\right)^2 (4 \, \mu m)}{(24)\left(2.78 \times 10^{-5} \, \frac{kg}{m \cdot s}\right)\left(10^6 \, \frac{\mu m}{m}\right)}$$

$$= 0.052 \text{ m/s}$$

Combustion

Figure 38.15 *Electrostatic Precipitator*

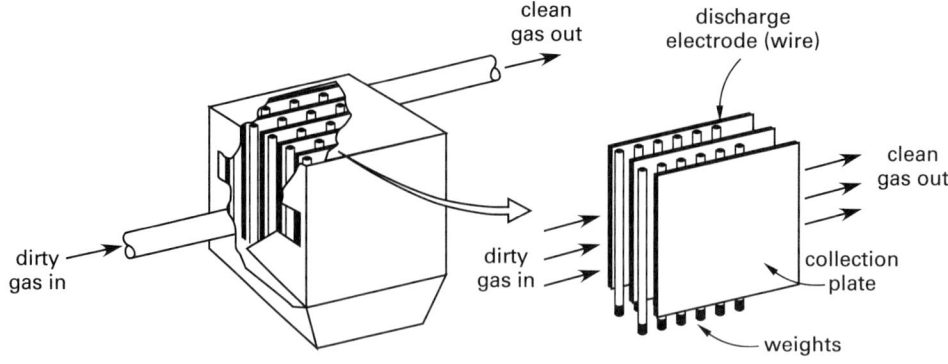

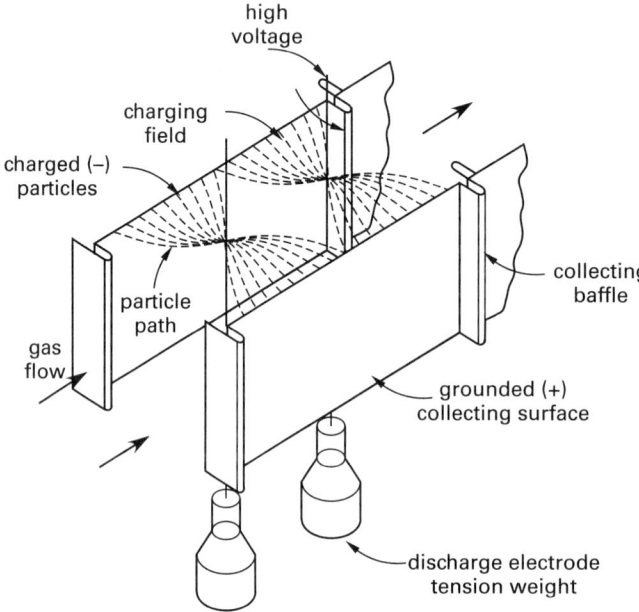

Environmental Engineering: A Design Approach by Sincero and Sincero, © 1995. Reprinted by permission of Prentice-Hall, Inc., Upper Saddle River, NJ.

The required efficiency is

$$E = \frac{6 \ \frac{g}{m^3} - 0.08 \ \frac{g}{m^3}}{6 \ \frac{g}{m^3}} = 0.987 \quad (98.7\%)$$

Combine Eq. 38.47 and Eq. 38.48.

$$SCA = \frac{-\ln\left(1 - \frac{E}{100\%}\right)}{w} = \frac{-\ln\left(1 - \frac{98.7\%}{100\%}\right)}{0.052 \ \frac{m}{s}}$$

$$= 83 \ s/m \quad (83 \ m^2 \cdot s/m^3)$$

Scrubbers

Scrubbers for particulate control are constructed in a variety of different configurations including upward countercurrent and crosscurrent flow and incorporating various methods to contact the particulate matter with the water. Regardless of the configuration, the objective of the scrubber is to entrain the particulate matter in water droplets. The water subsequently flows from the bottom of the scrubber, the particulate is allowed to settle, and clarified water is recirculated. Compared to dry-particulate removal processes, scrubbers provide the advantages of reduced explosion risk and quenching of hot gases, but they present the disadvantage of having to settle the particulate and manage a wet sludge.

Scrubbers may be classified by the method used to contact the particles with the water droplets. General classifications are spray towers, cyclones, and venturi scrubbers. Several typical scrubber configurations are shown in Fig. 38.16, and design parameters for these are summarized in Table 38.11. The *cut diameter* in Table 38.11 represents the particle size removed at 50% efficiency.

Because particulates are removed by entrainment in water droplets, the most important factors affecting scrubber efficiency are water droplet diameter and quantity. Efficiency increases with decreasing water droplet diameter and with increasing droplet quantity. Using the Calvert model, scrubber efficiency can be calculated using Eq. 38.49.

$$E = (100\%)(1 - e^{-k(-\Delta p)}) \qquad 38.49$$

For countercurrent, vertical flow scrubbers, $k(-\Delta p)$ in Eq. 38.49 is determined by Eq. 38.50.

$$k(-\Delta p) = \left(\frac{3Z_o E_f \mathrm{v}_d}{2 d_d (\mathrm{v}_d - \mathrm{v}_g)}\right)\left(\frac{Q_l}{Q_g}\right) \qquad 38.50$$

For crosscurrent flow scrubbers, $k(-\Delta p)$ in Eq. 38.49 is determined by Eq. 38.51.

$$k(-\Delta p) = \left(\frac{3Z_o E_f}{2 d_d}\right)\left(\frac{Q_l}{Q_g}\right) \qquad 38.51$$

The *fractional target efficiency*, E_f, in Eq. 38.50 and Eq. 38.51 is calculated from Eq. 38.52.

$$E_f = \left(\frac{\dfrac{C\rho_p d_p^2 \mathrm{v}_p}{9\mu_g d_d}}{\dfrac{C\rho_p d_p^2 \mathrm{v}_p}{9\mu_g d_d} + 0.7}\right)^2 \qquad 38.52$$

For venturi scrubbers, $k(-\Delta p)$ in Eq. 38.49 is calculated by Eq. 38.53.

$$k(-\Delta p) = \left(\frac{\mathrm{v}_g \rho_l d_d}{55\mu_g}\right)\left(\frac{Q_l}{Q_g}\right)$$

$$\times \left(-0.7 - K_p f + 1.4\ln\left(\frac{K_p f + 0.7}{0.7}\right)\right.$$
$$\left. + \frac{0.49}{0.7 + K_p f}\right)$$

$$\times \left(\frac{1}{K_p}\right) \qquad 38.53$$

In Eq. 38.53, f is 0.25 for hydrophobic particles and 0.50 for hydrophilic particles. K_p is calculated by Eq. 38.54.

$$K_p = \frac{C\rho_p d_p^2 \mathrm{v}_p}{9\mu_g d_d} \qquad 38.54$$

Values for C, the *Cunningham correction factor*, in Eq. 38.52 and Eq. 38.54 corresponding to liquid droplet diameters are provided in Table 38.12.

The scrubber cross-sectional area can be calculated from the gas flow rate and velocity by Eq. 38.55.

$$A_x = \frac{Q_g}{\mathrm{v}_g} \qquad 38.55$$

Example 38.18

A crosscurrent flow scrubber is used to remove particulates from a gas stream. The gas and scrubber characteristics are as follows.

gas temperature	60°C
gas velocity	32 cm/s
liquid-gas flow ratio	2×10^{-4}
scrubber contact length	4 m
liquid droplet diameter	0.025 cm
particle diameter	12 μm
particle density	2 g/cm³

What is the scrubber efficiency?

Solution

Use Eq. 38.52 and Eq. 38.51 with $\mu_g = 1.97 \times 10^{-5}$ kg/m·s at 60°C from App. 38.C and $C = 1.0$ from Table 38.12. Assume particle and gas velocities are equal.

To simplify, convert all units to meters and kilograms.

$$d_p = 12 \ \mu\mathrm{m} = 1.2 \times 10^{-5} \ \mathrm{m}$$
$$d_d = 0.025 \ \mathrm{cm} = 2.5 \times 10^{-4} \ \mathrm{m}$$
$$\rho_p = 2 \ \mathrm{g/cm^3} = 2000 \ \mathrm{kg/m^3}$$
$$\mathrm{v}_p = 32 \ \mathrm{cm/s} = 0.32 \ \mathrm{m/s}$$

$$E_f = \left(\frac{\dfrac{C\rho_p d_p^2 \mathrm{v}_p}{9\mu_g d_d}}{\dfrac{C\rho_p d_p^2 \mathrm{v}_p}{9\mu_g d_d} + 0.7}\right)^2$$

$$= \left(\frac{\dfrac{(1.0)\left(2000 \ \frac{\mathrm{kg}}{\mathrm{m^3}}\right)(1.2\times10^{-5} \ \mathrm{m})^2\left(0.32 \ \frac{\mathrm{m}}{\mathrm{s}}\right)}{(9)\left(1.97\times10^{-5} \ \frac{\mathrm{kg}}{\mathrm{m\cdot s}}\right)(2.5\times10^{-4} \ \mathrm{m})}}{\dfrac{(1.0)\left(2000 \ \frac{\mathrm{kg}}{\mathrm{m^3}}\right)(1.2\times10^{-5} \ \mathrm{m})^2\left(0.32 \ \frac{\mathrm{m}}{\mathrm{s}}\right)}{(9)\left(1.97\times10^{-5} \ \frac{\mathrm{kg}}{\mathrm{m\cdot s}}\right)(2.5\times10^{-4} \ \mathrm{m})} + 0.7}\right)^2$$

$$= 0.56$$

$$k(-\Delta p) = \left(\frac{3Z_o E_f}{2 d_d}\right)\left(\frac{Q_l}{Q_g}\right)$$

$$= \frac{(3)(4 \ \mathrm{m})(0.56)(2\times10^{-4})}{(2)(2.5\times10^{-4} \ \mathrm{m})}$$

$$= 2.69$$

Figure 38.16 *Types of Scrubbers*

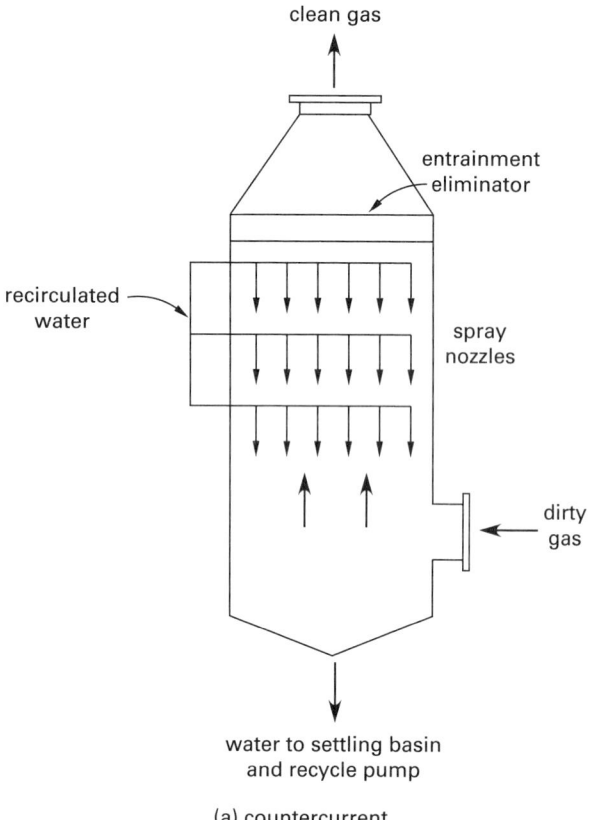

(a) countercurrent

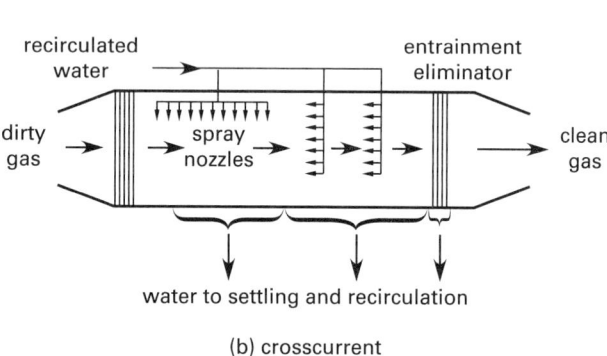

(b) crosscurrent

(c) venturi

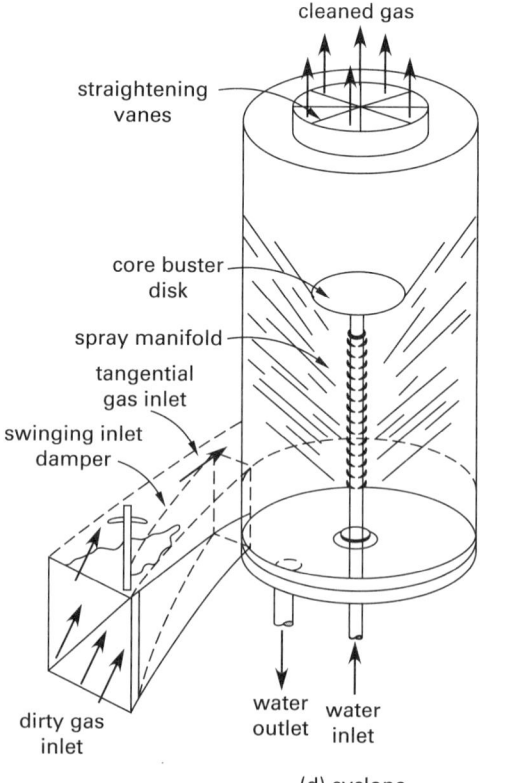

(d) cyclone

Environmental Engineering: A Design Approach, by Sincero and Sincero, © 1995. Reprinted by permission of Prentice-Hall, Inc., Upper Saddle River, NJ.

Table 38.11 Particulate Scrubber Design Parameters

scrubber type	pressure drop (kPa)	Q_l/Q_g (m³/m³)	liquid inlet pressure (kPa)	particle cut diameter (µm)
baffle spray	0.25–0.75	$1 \times 10^{-4} - 2 \times 10^{-4}$	<100	10
spray tower	0.12–0.75	$7 \times 10^{-4} - 3 \times 10^{-3}$	65–2800	2–8
cyclone	0.35–2.5	$3 \times 10^{-4} - 2 \times 10^{-3}$	275–2800	2–3
venturi ejector	0.12–1.3	$7 \times 10^{-3} - 2 \times 10^{-2}$	100–825	1
venturi	1.3–25	$4 \times 10^{-4} - 3 \times 10^{-3}$	<5–100	0.2

Adapted from: Noll, K.E., *Fundamentals of Air Quality Systems: Design of Air Pollution Control Devices*, © 1999, American Academy of Environmental Engineers, Annapolis, MD.

Table 38.12 Cunningham Correction Factor

particle diameter (µm)	correction factor, C^*
10	1.0
2.0	1.1
1.0	1.2
0.50	1.3
0.05	5.0
0.01	23.0

*at 1 atm and 25°C

Source: *Environmental Engineering: A Design Approach* by Sincero and Sincero, © 1995 Prentice-Hall, Inc., Upper Saddle River, NJ.

Use Eq. 38.49.

$$E = (100\%)(1 - e^{-k(-\Delta p)})$$
$$= (100\%)(1 - e^{-2.69})$$
$$= 93\%$$

Absorption

Absorption processes are selected to remove gaseous air pollutants by dissolution into a liquid solvent such as water or a caustic or acid solution. Absorption is used to remove sulfur dioxide, hydrogen sulfide, chlorine, ammonia, nitrogen oxides, and hydrocarbons. Absorption air pollution control processes include various types of packed and spray towers.

In a *packed tower*, clean liquid flows from top to bottom over the tower packing media. The media typically consist of plastic or ceramic shapes designed to maximize surface area for gas-liquid contact. The contaminated gas flows countercurrently, from bottom to top, or crosscurrently, from side to side. The liquid may be recirculated, and a demister may be used to dry the effluent gas stream. Head loss through packed towers varies from 0.25 kPa to 2.0 kPa, and liquid-gas ratios are typically 1.0–3.0 L liquid/m³ gas. Contaminant removal efficiencies exceed 90%. A typical packed tower is shown in Fig. 38.17.

Packed tower design follows principles of fluid mechanics and equilibrium chemistry. The head loss through a

Figure 38.17 Packed Tower

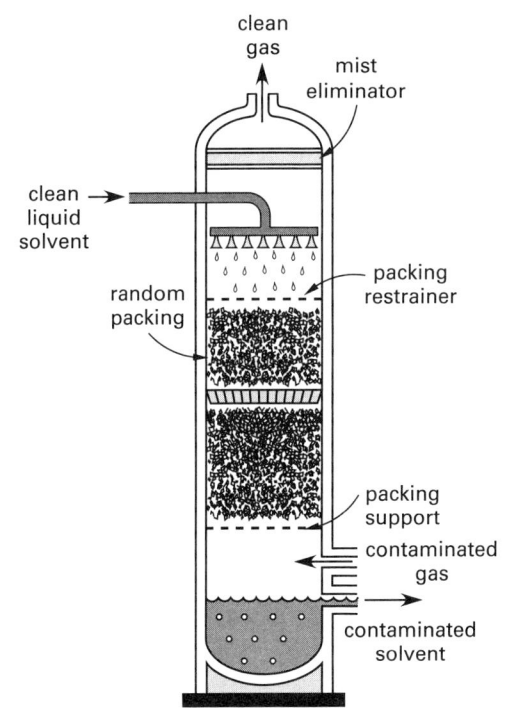

packed tower with uniform media can be calculated using the following equations.

$$-\Delta p = \frac{f(1-\alpha)Lv_g^2}{\alpha^3 d_p g} \qquad 38.56$$

$$f = \frac{150(1-\alpha)}{\text{Re}} + 1.75 \qquad 38.57$$

Reynolds number, Re, is calculated using Eq. 38.4.

In *spray towers*, the liquid and gas usually flow countercurrently or crosscurrently. The gas moves through a liquid spray that is carried downward by gravity. As with packed towers, the liquid may be recirculated and a demister may be used to dry the effluent gas stream. Pressure drops through spray towers are somewhat lower than through packed towers, varying from 0.25 kPa to 0.5 kPa. With values between 3 L and 14 L liquid/m³ gas, liquid-gas ratios are much higher than in packed towers. Contaminant removal efficiencies are between 50% and 75%, with efficiencies near 90% reported for flue gas desulfurization. A typical spray tower is shown in Fig. 38.16(a) and Fig. 38.16(b).

Spray tower design is similar to packed tower design. However, because water use is much higher and surface area for liquid-gas contact is much lower than in packed towers, spray towers are typically problematic to maintain and operate.

Example 38.19

A countercurrent packed tower uses 0.005 M sodium hydroxide (NaOH) to remove hydrochloric acid (HCl) from a gas stream at 34°C and 1 atm. The HCl concentration in the gas stream is 260 ppmv. What is the liquid-gas ratio based on the stoichiometric HCl requirement?

Solution

Use Eq. 38.10 to find the volume of 1 mole of HCl.

$$V = \frac{nR^*T}{p}$$

$$= \frac{(1 \text{ mol})\left(8.2 \times 10^{-5} \frac{\text{atm·m}^3}{\text{mol·K}}\right)(34°C + 273°)}{1 \text{ atm}}$$

$$= 0.025 \text{ m}^3$$

mole weight HCl $= 1 \frac{g}{mol} + 35 \frac{g}{mol} = 36$ g/mol

The HCl concentration in the gas stream is

$$\frac{(260 \text{ m}^3)\left(36 \frac{g}{mol}\right)}{(10^6 \text{ m}^3)\left(0.025 \frac{m^3}{mol}\right)} = 0.37 \text{ g/m}^3$$

The stoichiometric reaction equation is

$$HCl + NaOH \rightarrow Cl^- + Na^+ + H_2O$$

1 mole HCl requires 1 mole NaOH.

$$\frac{\left(0.37 \frac{g \text{ HCl}}{m^3 \text{ gas}}\right)\left(\frac{1 \text{ L NaOH solution}}{0.005 \text{ mol NaOH}}\right)}{\left(36 \frac{g \text{ HCl}}{mol \text{ HCl}}\right)\left(\frac{1 \text{ mol HCl}}{1 \text{ mol NaOH}}\right)}$$

$$= 2.1 \text{ L NaOH solution/m}^3 \text{ gas}$$

Example 38.20

A packed tower consists of a 1.5 m bed of 10 mm ceramic media. The gas velocity through the bed is 1 m/s, and the bed porosity is 0.45. The gas temperature is 39°C, and the pressure in the bed is 1 atm. What is the head loss through the bed?

Solution

From App. 38.C, $\rho_g = 1.14$ kg/m^3 and $\mu_g = 1.94 \times 10^{-5}$ kg/m·s at 39°C.

Use Eq. 38.4.

$$Re = \frac{v_s d_p \rho_g}{\mu_g}$$

$$= \frac{\left(1.0 \frac{m}{s}\right)(10 \text{ mm})\left(1.14 \frac{kg}{m^3}\right)}{\left(1.94 \times 10^{-5} \frac{kg}{m·s}\right)\left(1000 \frac{mm}{m}\right)}$$

$$= 588$$

Use Eq. 38.57.

$$f = \frac{150(1-\alpha)}{Re} + 1.75$$

$$= \frac{(150)(1-0.45)}{588} + 1.75$$

$$= 1.89$$

Use Eq. 38.56 and $\rho_l = 1000$ kg/m^3.

$$-\Delta p = \frac{f(1-\alpha)Lv_g^2}{\alpha^3 d_p g}$$

$$= \frac{(1.89)(1-0.45)(1.5 \text{ m})\left(1.0 \frac{m}{s}\right)^2\left(1000 \frac{mm}{m}\right)}{(10 \text{ mm})\left(9.81 \frac{m}{s^2}\right)(0.45)^3}$$

$$= 174 \text{ m of air}$$

$$-\Delta p = \frac{(174 \text{ m of air})\left(1.14 \frac{kg}{m^3}\right)\left(100 \frac{cm}{m}\right)}{1000 \frac{kg}{m^3}}$$

$$= 19.8 \text{ cm of H}_2O$$

$$-\Delta p = (19.8 \text{ cm of H}_2O)\left(\frac{0.249 \text{ kPa}}{2.54 \text{ cm of H}_2O}\right)$$

$$= 1.94 \text{ kPa}$$

Adsorption

Adsorption occurs when the target contaminant becomes attached to the surface of the media. Typically, the media is *activated carbon* and is available in either granular (GAC) or powdered (PAC) form. The activated carbon is produced by starved air combustion of coal, coconut shells, wood, and other organic materials. Adsorption is effective in removing heavy metals such as lead and mercury, and in removing aromatic and aliphatic organic chemicals such as VOCs and dioxins. Adsorption processes may be employed to recover gases for reuse or recycling, or for destruction. Very high removal efficiencies, exceeding 99%, are possible.

The activated carbon process provides essentially constant removal until *breakthrough* occurs. At breakthrough, the effluent concentration begins to increase and continues to do so until the carbon is saturated and no pollutant removal occurs. Because optimum operation uses carbon to saturation, adsorption processes are frequently operated in a *two-in-series lead-follow mode* as shown in Fig. 38.18. The carbon is regenerated or reactivated sometime during the interval between breakthrough and saturation. *Regeneration* is associated with contaminant recovery and is typically a chemical process. *Reactivation* is associated with contaminant destruction and is a thermal process.

Activated carbon adsorption process design is based on the X/M ratio determined from an adsorption isotherm specific to the target contaminant. The X/M ratio represents the mass of chemical adsorbed per unit mass of carbon and is influenced by the gas temperature and moisture content. Many adsorption isotherm relationships have been developed, but one of the most common is the *Freundlich equation.*

$$\frac{X}{M} = K_F C_f^{1/n} \qquad 38.58$$

Head loss through the carbon bed is calculated using Eq. 38.56 and Eq. 38.57.

Example 38.21

A gas containing a VOC at 90 $\mu g/m^3$ is vented to a GAC adsorber at 1.3 m^3/s. The GAC is effective in removing 99.9% of the VOC, adsorption equipment is available with 75 kg GAC capacity, and reactivation should occur at no less than 28 day intervals. The isotherm values for the VOC at the temperature and moisture conditions of the vented gas are

$$K_F = 210$$
$$1/n = 1.6$$
$$X/M \text{ units} = \text{mg/g}$$

What is the daily mass of GAC required to remove the VOC from the gas stream?

Solution

The final concentration is

$$C_f = \left(90 \ \frac{\mu g}{m^3}\right)(1 - 0.999) = 0.09 \ \mu g/m^3$$

Figure 38.18 *Two-in-Series Activated Carbon Adsorption*

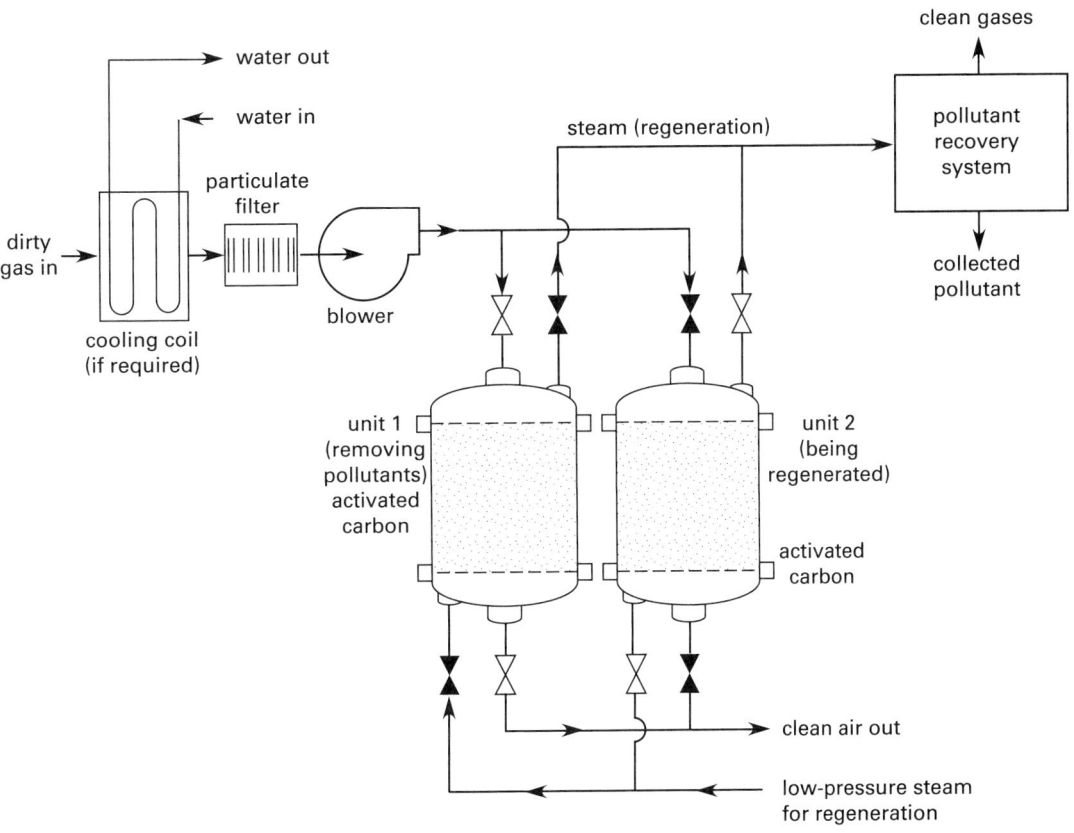

Environmental Engineering: A Design Approach by Sincero and Sincero, © 1995. Prentice-Hall, Inc., Upper SaddleRiver, NJ.

Use Eq. 38.58.

$$\frac{X}{M} = K_F C_f^{1/n}$$

$$= (210)\left(0.09 \ \frac{\mu g}{m^3}\right)^{1.6}$$

$$= 4.5 \text{ mg VOC/g GAC}$$

$$\dot{m} = \frac{\Delta C Q_g}{X/M}$$

$$= \frac{\left(90 \ \frac{\mu g}{m^3} - 0.09 \ \frac{\mu g}{m^3}\right)\left(1.3 \ \frac{m^3}{s}\right)\left(86\,400 \ \frac{s}{d}\right)}{\left(4.5 \ \frac{\text{mg VOC}}{\text{g GAC}}\right)\left(1000 \ \frac{\mu g}{mg}\right)\left(1000 \ \frac{g}{kg}\right)}$$

$$= 2.2 \text{ kg GAC/d}$$

The maximum time between regenerations is

$$\frac{75 \text{ kg}}{2.2 \ \frac{\text{kg GAC}}{d}} = 34 \text{ d} \qquad \begin{bmatrix} \text{acceptable since} \\ 34 \text{ d} > 28 \text{ d} \end{bmatrix}$$

Incineration

Incineration is applied to air pollution control to destroy combustible air pollutants such as gases, vapors, and odors where the composition of these substances is limited to carbon, hydrogen, and oxygen atoms. When used for this purpose, incineration may be referred to as *afterburning*. Unlike incineration for solid waste disposal, power generation, or manufacturing uses (which are typically sources of air pollutants), incineration for air pollution control should not require subsequent emission control processes. Incineration may be divided into three classifications: direct flame, thermal, and catalytic.

Direct flame incineration is typically applied where the gas stream has sufficient energy value to maintain combustion without the need to provide a supplemental fuel source. Gases with heating values above about 3200 kJ/m^3 typically satisfy this criterion. Gases with high energy values can result in temperatures exceeding $1300°C$, a condition that may produce nitrogen oxides if combustion occurs with excess air. A common application of direct flame incineration is the *flare*. Flares are used to burn landfill and digester gases as well as waste refining gases.

Flares will burn without smoke when the hydrogen-carbon ratio exceeds $1/3$ on a weight basis. This can be illustrated by comparing methane (CH_4) with an H:C ratio of $1/3$ and acetylene (C_2H_2) with an H:C ratio of $1/12$. Methane burns without smoke while acetylene burns with soot. Gases with an H:C ratio less than $1/3$ can be made to burn without smoke by injecting steam into the flame zone to provide turbulence and enhance mixing between the gas and the air. Steam injection also makes for a shorter flame and hence improves stability under windy conditions.

Thermal incineration is applied to gas streams with low energy values that require supplemental fuel to sustain combustion. These gases may have heating values between about 40 kJ/m^3 and 750 kJ/m^3. Efficiency is based on the three Ts: time, temperature, and turbulence. Normal residence times are between 0.2 s and 0.8 s, and typical temperatures range from $500°C$ to $850°C$. If the incinerator is deficient in any of the three Ts, carbon monoxide and intermediate oxidation products may be formed. However, if operated properly, hydrocarbon destruction efficiencies can approach 100%. Figure 38.19 presents a schematic of a thermal incinerator.

Figure 38.19 Thermal Incinerator

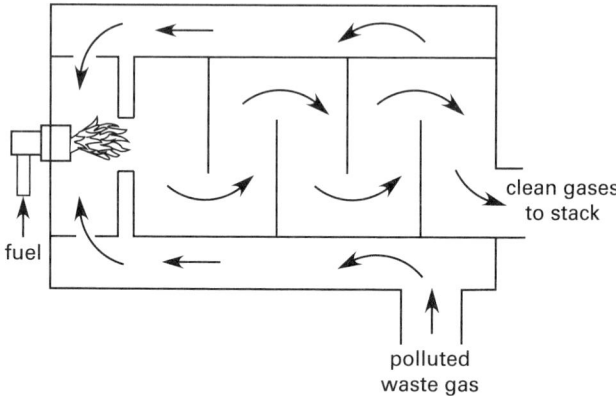

Catalytic incinerators rely on a catalyst material to produce oxidation at the desired temperature. Catalytic incineration can be applied to gas streams similar to those suited for thermal incineration, but because of the catalyst, residence times and temperatures can be greatly reduced. Residence times may be as much as 50 times shorter than those required for thermal incineration, and temperatures can be as much as $250°C$ to $300°C$ lower while achieving similar destruction efficiencies.

Catalytic materials include metals such as platinum, molybdenum, cobalt, and copper formed into a variety of shapes supported on inert *carriers* constructed of alumina, porcelain, and other materials. *Fouling* or *poisoning* of the catalyst is the most common operating problem. Catalytic reactivity is influenced by molecular structure in the following order for hydrocarbons: aromatics; branched alkanes; normal alkanes; alkenes; alkynes. A catalytic incinerator is illustrated in Fig. 38.20.

Example 38.22

Four hydrocarbons with their chemical formulas are listed as follows.

benzene	C_6H_6
cyclopropane	C_3H_6
ethene	C_2H_4
propane	C_3H_8

Will any of the listed compounds burn without smoke?

Solution

MW carbon = 12 g/mol

MW hydrogen = 1 g/mol

compound	carbon moles	hydrogen moles	carbon mass (g)	hydrogen mass (g)	H:C
benzene	6	6	72	6	1/12
cyclopropane	3	6	36	6	1/6
ethene	2	4	24	4	1/6
propane	3	8	36	8	1/4.5

None of the compounds will burn without smoke since none have an H:C ratio greater than $1/3$.

Figure 38.20 *Catalytic Incinerator*

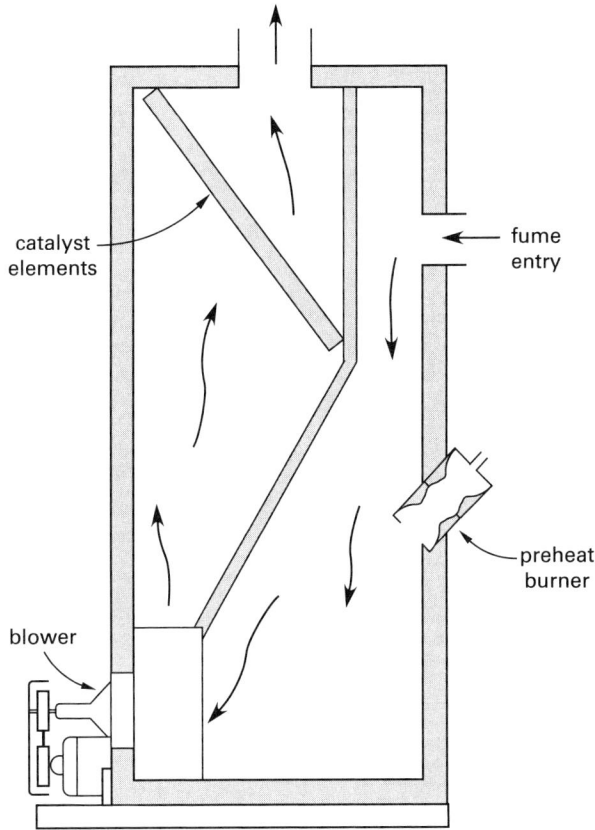

Condensers

Condensers are used for pretreatment before other emission control processes to remove VOCs when they occur as high-concentration vapors. Condensers can reduce the operating costs of subsequent emission control processes and provide a potential economic benefit from the recovery of chemical vapors. The utility and efficiency of condensers improve as VOC concentrations increase. Recovery efficiencies over 95% are possible for vapor streams containing VOCs at over 10 000 ppmv, but they fall to as low as 50% when vapor concentrations are near 500 ppmv.

The two different types of condensers are contact and noncontact. Because they allow the VOCs and the condensing liquid to become mixed and thereby create disposal problems, *contact condensers* are not as widely used as *noncontact condensers* for air pollution control. The most common noncontact condensers are shell-and-tube *surface condensers*. A noncontact surface condenser is illustrated in Fig. 38.21.

Figure 38.21 *Surface Condenser*

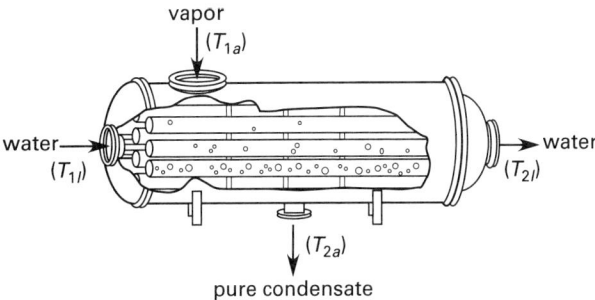

pure condensate

Noncontact surface condensers are designed to promote heat transfer between the VOC and the cooling liquid. The required cooling liquid flow rate, Q_l, can be calculated using Eq. 38.59 with Eq. 38.60 to estimate heat loss, q_a, from the VOC during condensation.

$$Q_l = \frac{q_a}{c_l(T_{2l} - T_{1l})\rho_l} \qquad 38.59$$

$$q_a = \dot{m}c_a(T_{1a} - T_{2a}) + \dot{m}h_{fg} \quad \left[\begin{array}{c}\text{no}\\\text{subcooling}\end{array}\right] \qquad 38.60$$

The *log mean temperature difference*, ΔT_m, between the vapor and the cooling liquid can be determined from Eq. 38.61.

$$\Delta T_m = \frac{(T_{1a} - T_{2l}) - (T_{2a} - T_{1l})}{\ln\left(\dfrac{T_{1a} - T_{2l}}{T_{2a} - T_{1l}}\right)} \qquad 38.61$$

With the heat loss and temperature difference defined, the surface area of the condenser tubes is found by Eq. 38.62.

$$A_s = \frac{q_a}{U\Delta T_m} \qquad 38.62$$

Example 38.23

A surface condenser is selected to remove trichloroethene (TCE) vapor from a gas stream. The mass flow rate of the TCE is 1200 kg/h, and a maximum outlet temperature of 42°C is required. The condensing liquid is water with 16°C inlet and 28°C outlet temperatures. For TCE, the latent heat of vaporization is 239 kJ/kg, the specific heat is 0.96 kJ/kg·°C, and the boiling point is 87°C at the gas stream pressure. TCE enters the surface condenser saturated.

(a) What is the log mean temperature difference?
(b) What is the required cooling water flow rate?

Combustion

Solution

(a) Use Eq. 38.61. Since the TCE is saturated, $T_{1a} = 87°C$.

$$\Delta T_m = \frac{(T_{1a} - T_{2l}) - (T_{2a} - T_{1l})}{\ln\left(\dfrac{T_{1a} - T_{2l}}{T_{2a} - T_{1l}}\right)}$$

$$= \frac{(87°C - 28°C) - (42°C - 16°C)}{\ln\left(\dfrac{87°C - 28°C}{42°C - 16°C}\right)}$$

$$= 40°C$$

(b) Use Eq. 38.60.

$$q_a = \dot{m}c_a(T_{1a} - T_{2a}) + \dot{m}h_{fg}$$

$$= \left(1200\ \frac{kg}{h}\right)\left(0.96\ \frac{kJ}{kg·°C}\right)$$

$$\times (87°C - 42°C)\left(24\ \frac{h}{d}\right)$$

$$+ \left(1200\ \frac{kg}{h}\right)\left(239\ \frac{kJ}{kg}\right)\left(24\ \frac{h}{d}\right)$$

$$= 8.1 \times 10^6\ kJ/d$$

Use Eq. 38.59. For water, take the specific heat as 4.186 kJ/kg·°C and density as 1 kg/L.

$$Q_l = \frac{q_a}{c_l(T_{2l} - T_{1l})\rho_l}$$

$$= \left(\frac{8.1 \times 10^6\ \dfrac{kJ}{d}}{\left(4.186\ \dfrac{kJ}{kg·°C}\right)(28°C - 16°C)\left(1\ \dfrac{kg}{L}\right) \times \left(24\ \dfrac{h}{d}\right)\left(60\ \dfrac{min}{h}\right)}\right)$$

$$= 112\ L/min$$

7. MOBILE SOURCE CONTROL

Introduction

The genesis of mobile source emission regulation occurred in California in 1959 when the state adopted standards to control hydrocarbons and carbon monoxide. Since then, standards have been expanded to include other pollutants as technology has developed to facilitate their control.

Mobile sources of air pollution include cars, trucks, buses, trains, planes, and so on. In the United States, there are nearly 200 million such mobile sources, compared to over 30,000 stationary sources. Mobile sources present unique problems for air pollution control and policy. They emit many of the same pollutants as stationary sources, especially hydrocarbons, carbon monoxide, and nitrogen dioxide. Also, gasoline is a significant source of benzene emissions.

Institutional Issues

One major solution to mobile source air pollution is to reduce the number of miles driven. This could be accomplished by increased mass transit access, increased bicycle use, improved pedestrian access, and so forth. However, the demand is not for additional metro lines, bike paths, and sidewalks, but for more roads and parking.

The use of the automobile is subsidized by taxes for building roads, and parking is provided free of cost to the user by those who develop commercial property, city governments, and others. Programs to facilitate bicycle use are not provided, except in a very few places. Also, it is hard to use a bicycle for transportation when people live, work, and shop in places that are widely separated. The subsidized use of the automobile has allowed people to disperse homes far and wide. It is no longer feasible to provide alternatives to automobile transportation in the United States because the dispersed population has developed a lifestyle dependent on the automobile.

One problem with automobile use is that the associated costs are not internalized. Possible ways to internalize these costs include the following.

- Higher fuel taxes: These would be potentially very high and carry the possible disadvantage of drivers in lower pollution areas subsidizing those in higher pollution areas.

- Restricted access periods: During rush hour, automobile traffic to higher pollution areas excludes delivery trucks and is limited to commuters and residents.

- Toll roads: Toll fees fluctuate for various use periods and are higher for roads accessing higher pollution areas. Toll roads that cover the actual cost of building and maintaining the road are also possible deterrents to heavy vehicle use.

- Reserved lanes for buses and carpoolers: The traffic in these lanes is lighter, which makes traveling faster. Faster travel reduces the time vehicles spend on the road and, therefore, reduces emissions.

- Parking costs: Charge for the actual cost of providing parking.

Emissions Regulation Strategies

It is difficult to find effective control strategies because of the number of mobile sources and because those involved are individuals, not corporations. This problem is overcome by regulating mobile source emissions through activities in two categories: regulation at the point of manufacture and regulation at the point of use. A combination of these two options has been applied in the United States in the form of a preconsumer certification and enforcement program.

The certification program incorporates testing of new automobiles under controlled conditions at the manufacturers' facilities. Automobiles in each engine class are tested and can only be sold if they are able to demonstrate compliance with the certification criteria. These tests are conducted using prototype vehicles, before new engines are placed into full production.

The enforcement program applies to automobiles in production. The EPA tests a representative number of vehicles as they come off the assembly line to ensure that they comply with the performance standards established during the certification program. Only 60% of the vehicles need to pass the EPA test for the certification to remain valid. Regulations also include requirements for the manufacturer to warrant emissions control equipment, creating an incentive for the equipment to be reliable.

In some locales where mobile sources contribute significantly to nonattainment, additional restrictions may be placed on the automobile manufacturer and on the user. Stricter California emissions standards, for example, are applied to new vehicles. Also, some states require owners to have their vehicles inspected for compliance with emissions restrictions and to make repairs where noncompliance occurs.

Some locales, most notably California, have implemented regulatory and policy measures to promote the availability and sale of *low emission vehicles* (LEVs) and *zero emission vehicles* (ZEVs). In urban areas with chronic air pollution problems, LEVs and ZEVs are particularly attractive as a mechanism to improve air quality.

As mobile source operating conditions change, so do emissions. Therefore, in an effort to define a representative range of conditions for applying EPA performance standards, a standard driving cycle test is used. The *Federal Test Procedure* (FTP, also known as the *EPA75 test* and *CVS-75 test*) is a constant volume sampler test conducted while a test vehicle is running on a dynamometer. It simulates cold start, idle, stop-go traffic, highway, and many other conditions modeled on a 7.5 mi course and taking 22.8 minutes. There are other federal tests, including the *Urban Dynamometer Driving Schedule* (UDDS, also known as the *LA4 test* and the "*City Test*"), which is part of the FTP and simulates light vehicle city driving, and the *Highway Fuel Economy Driving Schedule* (HWFET), which simulates highway driving conditions under 60 mph. Individual states (e.g., California and New York) and cities may have their own specialized tests.

The EPA also employs the FTP for evaluating the city part of fuel economy ratings. The highway part is determined by a simulated 10 mi, nonstop, 48 mph test. A harmonic average of the two tests, 55% for city and 45% for highway, is used to determine the average fuel economy. The average fuel economy determined in this way represents the *Corporate Average Fuel Economy* (CAFE).

Example 38.24

What would be the EPA average fuel economy for a car rated at 18 mpg city and 23 mpg highway?

Solution

CAFE applies 55% to city mileage and 45% for highway mileage. Assume 1 mi of driving. The average fuel economy is

$$\frac{1 \text{ mi}}{\dfrac{(1 \text{ mi})(0.55)}{18 \dfrac{\text{mi}}{\text{gal}}} + \dfrac{(1 \text{ mi})(0.45)}{23 \dfrac{\text{mi}}{\text{gal}}}} = 19.95 \text{ mpg}$$

Emissions Controls

A schematic of an automobile engine without air pollution control devices is presented in Fig. 38.22(a) and identifies four emission sources: (1) volatilization of hydrocarbons to the atmosphere from the carburetor or throttle body, (2) volatilization of hydrocarbons to the atmosphere from a vented gasoline tank, (3) emission of noncombusted and partially oxidized hydrocarbons from the crankcase, and (4) emission of criteria pollutants from the exhaust.

Each of the four emission sources is controlled in new and well-maintained older cars and, to varying degrees, in other types of mobile sources. The emission control devices applied to the four sources are represented by the schematic in Fig. 38.22(b) and include the following.

- a vent from the air cleaner to capture volatilized hydrocarbons. The vented hydrocarbons are adsorbed on activated carbon contained in a canister typically located in the engine compartment.

- a vent from the gasoline tank to the activated carbon canister described above to capture volatilized hydrocarbons

- a vent from the crankcase to the activated carbon canister to capture noncombusted and partially oxidized hydrocarbons from the crankcase

- a catalytic converter on the exhaust pipe to thermally oxidize criteria pollutants, primarily hydrocarbons, nitrogen oxides, and carbon monoxide

Pollution control equipment performance deteriorates with age and requires maintenance and a well-tuned engine to perform properly. Therefore, one way to remove pollutants contributed by mobile sources is to get rid of old cars, reduce the sale of used cars, and increase the sale of new cars.

Diesel engines require different approaches to air pollution control. Diesel engines operate with very lean fuel mixtures (rich in oxygen) and at relatively high temperatures. This results in inherently low hydrocarbon and carbon monoxide emissions, but creates higher nitrogen oxide emissions than do gasoline engines. The nitrogen oxides are less efficiently removed by catalytic converters on diesel exhaust systems because of the rich oxygen mixture. Furthermore, obstacles exist for controlling the carbonaceous soot emissions (designated as a probable human carcinogen) which are associated with diesel exhaust.

Figure 38.22 Internal Combustion Engine

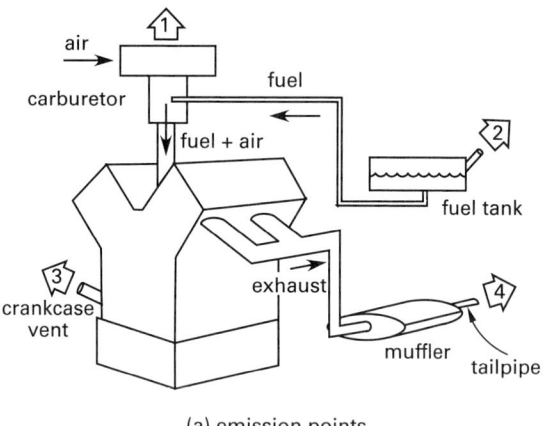

(a) emission points

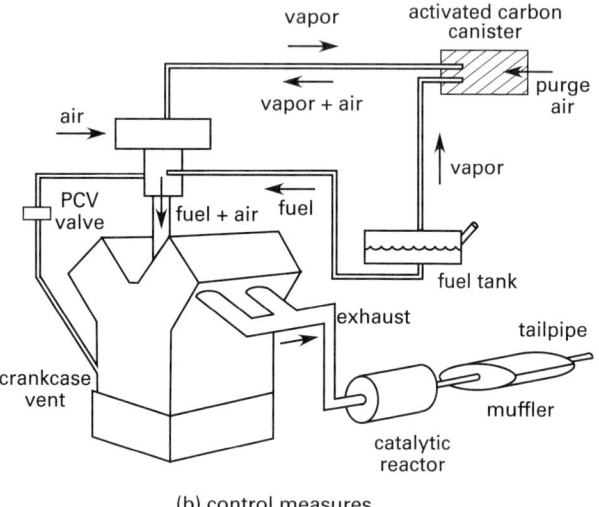

(b) control measures

8. INDOOR AIR

Introduction

Pollution of indoor air is a concern for at least three reasons.

- The average person in the United States spends about 87% of his/her time indoors.

- Multiple air pollution sources emitting a variety of pollutants are located indoors.

- Interior spaces tend to concentrate pollutants since dispersion into the atmosphere is limited by the finite volume of the room.

Table 38.13 lists the major indoor air pollutants and their sources. Of these, environmental tobacco smoke (ETS) and radon are the most significant because of their relative toxicity, frequency of occurrence, and potential to exist at elevated concentrations. Of somewhat lesser significance are formaldehyde and asbestos, followed by the other listed pollutants.

Table 38.13 Indoor Air Pollutant Sources and Concerns

source	pollutants of concern
combustion	criteria pollutants
tobacco smoke	environmental tobacco smoke (ETS)
woodburning stoves	carbon monoxide
kerosene heaters	nitrogen dioxide
gas stoves	hydrocarbons
automobiles in attached garages	
building materials and furnishings	formaldehyde
particleboard	VOCs
paints	hydrocarbons
adhesives	
caulks	
carpeting	
consumer products	VOCs
pesticides	hydrocarbons
cleaning materials	particulates
waxes, polishes	
cosmetics, personal products	
hobby materials	
people and pets	particulate bioeffluents
outdoor air	criteria pollutants

Environmental Tobacco Smoke

Tobacco smoke may be characterized as either mainstream or sidestream. *Mainstream* tobacco smoke is that taken directly into the lungs by the smoker. *Sidestream* smoke is what is commonly referred to as secondhand tobacco smoke. Sidestream smoke and exhaled mainstream smoke are the sources of ETS. Indoor levels of the criteria pollutants CO and PM2.5 attributable to tobacco smoke may be as much as one to two orders of magnitude greater than in similar spaces where smoking does not occur. Environmental tobacco smoke is also the source of a variety of other chemicals including formaldehyde and other aldehydes, acetone and other ketones, benzene, phenol, and other aromatics, organic acids, amines, methyl chloride, and nicotine, and many more.

The size of particulate matter associated with ETS is from about 0.1 μm to 0.5 μm, making it particularly prone to deposition in the lungs. Radon exposure is increased by ETS since it provides a mechanism for transport into the lungs.

Radon

Radon, a noble gas, results from the radioactive decay of radium which is found naturally in soils in many regions of the United States. The half-life of radon is a relatively short 3.8 days. However, the decay products, or progeny, of radon have much shorter half-lives of no more than several minutes. Although radon is often described as an alpha source, Fig. 38.23 shows that three forms of radiation are emitted during the decay lifetime.

Figure 38.23 *Radon Decay Products*

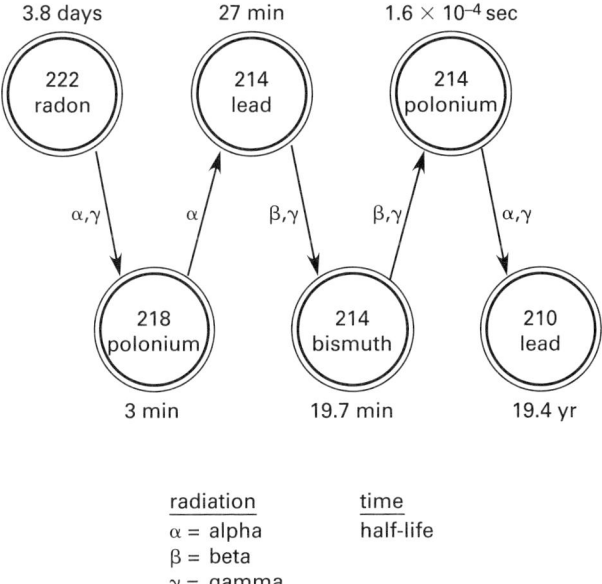

```
radiation            time
α = alpha            half-life
β = beta
γ = gamma
```

The radon decay products are electrically charged and, consequently, adsorb to airborne particulates which can be inhaled and deposited in the lungs. The primary risk from radon exposure is lung cancer.

The primary source of radon in indoor air is soil, contributing up to 90% of all exposures. Masonry and other building products account for some of the remaining 10% and are not the primary source as is sometimes reported. The radon gas enters buildings through cracks and other breaches in foundations, floors, and other parts of the structure proximate to the soil. Negative pressures caused by temperature gradients between indoor and outdoor air, wind, and fluctuating barometric pressure draw the radon into buildings. If buildings are drafty, a chimney effect may act to distribute radon gas in upper floors as well.

Radon concentrations are measured in pCi/L. The EPA guideline for action to mitigate indoor radon exposure is 4 pCi/L. Test kits to measure radon concentrations in indoor air are easy to use, inexpensive (typically costing less than $25), and available from retail outlets and online.

Radon mitigation techniques are implemented with the purpose of reducing indoor air concentrations. Examples of mitigation techniques include the following.

- improving ventilation in basement and crawlspace areas by passive measures such as simply opening windows or installing vent openings, or by more active measures

- sealing cracks and other breaches and creating physical barriers to prevent entry

- sealing drafty windows and doors to reduce air currents that draw radon from lower to upper floors

- constructing subslab and crawlspace vents to create pressure barriers that prevent entry and exhaust the gas to the outside atmosphere

Formaldehyde

The sources of formaldehyde in indoor air are mostly associated with building materials, especially plywood, particle board, and fiberboard, which contain urea formaldehyde resins. When exposed to moisture from humid air or other sources, the urea formaldehyde hydrolyzes to release formaldehyde vapor to the indoor air. Other, far less significant sources are furnishings and fabrics such as carpets, draperies, and clothing.

Symptoms of low level exposure to formaldehyde include headaches, sinus congestion, depression, and other ailments. The National Institute for Occupational Safety and Health (NIOSH) has set a recommended action level for formaldehyde exposure at 0.1 ppm.

Asbestos

Asbestos has been used in many building materials such as floor tiles, textured ceilings, heating pipe insulation, and structural fire protection insulation. When any of these asbestos-containing materials (ACMs) are disturbed or deteriorate, asbestos fibers may become airborne and eventually find their way into the lungs. Consequently, mitigation measures may be directed at simply protecting the ACM from activities that would disturb it. However, public pressure may require removal of asbestos as the only acceptable mitigation measure.

Nonoccupational exposure to asbestos has been linked to lung cancer and *mesothelioma. Asbestosis* is typically associated with occupational exposure. One outcome of the potential health risks associated with exposure to asbestos in indoor air was the Asbestos Hazard Emergency Response Act (AHERA) of 1986, specifically passed by the United States Congress to address asbestos in schools.

9. GLOBAL ISSUES

Introduction

Pollution becomes a global issue when it crosses national boundaries. Ready examples are rivers that flow through more than one country, ocean outfalls near border cities, and air pollution carried across borders. The reason global pollution issues are usually politically important is that pollution generated somewhere else, often by a society with different social or political objectives, is deposited at a location far removed from the source.

Prominent global air pollution issues include acid rain, global warming associated with greenhouse gases, and ozone depletion. None of these are universally accepted

Combustion

by experts as significant global problems, although, in the case of acid rain, regional impacts are well documented.

Acid Rain

Acid rain results from deposition of nitric acid and sulfuric acid formed when nitrogen and sulfur oxides mix with atmospheric moisture. Once formed, the acids may fall to earth in rainfall or they may dry in the atmosphere and ultimately be deposited in a dry form. The acids depress the pH of rainwater and of surface water, soils, and vegetation surfaces.

Most rainfall worldwide has a natural background pH of around 5.0, which is fairly acidic. Some evidence exists to suggest that this pH has been influenced by air emissions since the early 1900s. Areas in the United States with the most significantly acidified surface waters are somewhat localized in New York and in Florida. From 14% to 23% of lakes in these regions are acidified. The close proximity of many countries of western and eastern Europe has created some significant impacts from emissions originating in less regulated countries.

One problem has been the application of stack height criteria as a means of emission control. The stack height concept is to place the pollutants high enough for them to be dispersed in the atmosphere over a wide area away from the source. This works well for local air quality control, but is not very effective for regional or global pollution control. Nevertheless, it has been a dominant air pollution control strategy in the United States and elsewhere in the world.

Example 38.25

Assume rainwater with a natural pH of 5.0 reacts with SO_3 in the air to form H_2SO_4 at 320 μg/L. What is the pH of the rainwater after reacting with SO_3?

Solution

From Eq. 38.27 and dissociation of H_2SO_4,

$$SO_3 + H_2O \rightarrow H_2SO_4 \rightarrow 2H^+ + SO_4^{-2}$$

1 mol of H_2SO_4 will yield 2 mol of H^+ in the rainwater. The molecular weight of H_2SO_4 is 98 g/mol.

At pH 5.0, the H^+ concentration of natural rainwater is 10^{-5} mol/L.

$$H^+ \text{ added} = \frac{\left(2 \; \frac{\text{mol } H^+}{\text{mol } H_2SO_4}\right)\left(320 \; \frac{\mu g}{L}\right)}{\left(98 \; \frac{g}{\text{mol } H_2SO_4}\right)\left(10^6 \; \frac{\mu\text{mol}}{\text{mol}}\right)}$$
$$= 6.5 \times 10^{-6} \text{ mol/L}$$

The final H^+ concentration of the rainwater is

$$10^{-5} \; \frac{\text{mol}}{L} + 6.5 \times 10^{-6} \; \frac{\text{mol}}{L} = 1.65 \times 10^{-5} \text{ mol/L}$$

The final pH of the rainwater is

$$pH = -\log[H^+] = -\log\left(1.65 \times 10^{-5} \; \frac{\text{mol}}{L}\right)$$
$$= 4.78$$

Ozone Depletion

The problem with chlorinated fluorocarbons (CFCs) in the upper atmosphere is their photolytic decomposition, which releases free chlorine atoms. The free chlorine then reacts with ozone to form oxygen. If the cycle repeats, a net decrease in ozone can potentially occur. A simplified reaction sequence is shown below where the CFC is represented by Freon-12 (CF_2Cl_2).

$$CF_2Cl_2 + \text{light} \rightarrow CF_2 + 2Cl \qquad 38.63$$

$$Cl + O_3 \rightarrow ClO + O_2 \qquad 38.64$$

$$ClO + O \rightarrow Cl + O_2 \qquad 38.65$$

Fear of damage to the stratospheric ozone layer has produced remarkable alliances among many nations. The use of CFCs is a suspected significant contributor to destruction of the ozone layer. Although the evidence is not conclusive, the potential consequences of taking no action until unquestioned evidence can be obtained have been used to justify action to control CFCs worldwide.

Current international law closely regulates the production and sale of CFCs through out-and-out bans on their use and production, taxes on emissions from their use, and a permit program. A fund was also established to assist poorer countries in implementing alternatives to ozone-depleting chemicals. Research to develop alternatives to CFCs is also funded, primarily by the industries that stand to benefit.

Global Warming

The *greenhouse gases* (GHG) include carbon dioxide (CO_2), methane (CH_4), nitrous oxide (N_2O), oxygen (O_2), ozone (O_3), and water vapor (H_2O). These gases, most notably carbon dioxide, trap heat from the earth's surface and the atmosphere. Not being able to radiate into space, the trapped heat creates increased ambient temperatures. Global temperature increases of 1°C to 2°C would be very significant, causing dramatic changes in global climate and a rise in the sea level.

Not all nations view global warming with the same level of concern since its impacts are not uniform. Whereas an increased temperature in middle latitude countries like the United States may turn agricultural land to desert, countries like Russia and Canada would find vast areas of previously marginally useable land to be much more hospitable.

Although several measures have been proposed to address global warming, the two primary areas of focus are reducing dependence on fossil fuels (reducing combustion sources and CO_2 production) and preventing deforestation combined with restoring deforested areas (increasing CO_2 use by increasing vegetation). These are identified as prevention and mitigation strategies and exist in conflict with economic goals, at least in the short term.

Proposals directed at reducing sources that emit CO_2 (such as combustion associated with mobile sources of air pollution) include increasing the tax on fuels.

Automobile fuel taxes in the United States have been somewhat less than \$0.50 per gallon, while in other developed countries, fuel taxes may exceed something equivalent to \$3.00 per gallon. Consequently, no economic incentive is evident to reduce fuel consumption, and any consequences of current activities pertaining to global warming are in the distant future. The delay is compounded by a belief held by many that global warming is not a significant threat in the first place.

Combustion

Topic VI: Solid and Hazardous Waste

Chapter

Solid Waste

39
Municipal Solid Waste

Nomenclature

a	distance	ft	m
A	area	ft^2	m^2
b	distance	ft	m
B	volumetric fraction	–	–
C	concentration	mg/L	mg/L
CF	compaction factor	–	–
D	depth of flow	ft	m
g	acceleration of gravity, 32.2 (9.81)	ft/sec^2	m/s^2
g_c	gravitational constant, 32.2	$lbm\text{-}ft/lbf\text{-}sec^2$	n.a.
G	MSW generation rate, per capita	lbm/day	kg/d
H	total hydraulic head	ft	m
HRR	heat release rate	$Btu/hr\text{-}ft^2$	W/m^2
HV	heating value	Btu/lbm	J/kg
i	hydraulic gradient	ft/ft	m/m
K	hydraulic conductivity	ft/sec	m/s
K_s	solubility	ft^3/ft^3	L/L
L	spacing or distance	ft	m
LF	loading factor	–	–
m	mass	lbm	kg
MW	molecular weight	lbm/lbmol	g/mol
N	population size	–	–
O	overlap	–	–
p	pressure	lbf/ft^2	Pa
Q	flow rate	ft^3/sec	m^3/s
Q	recharge rate	$ft^3/ft^2\text{-}sec$	$m^3/m^2\cdot s$
R	radius of influence	ft	m
R^*	universal gas constant, 1545.35 (8314.47)	ft-lbf/ lbmol-°R	J/kmol·K
SFC	specific feed characteristics	Btu/lbm	J/kg
T	absolute temperature	°R	K
V	volume	ft^3	m^3
x	mass fraction	–	–

Symbols

γ	specific weight	lbf/ft^3	n.a.
ρ	density	lbm/ft^3	kg/m^3

Subscripts

c	compacted
i	gas component i
K	Kelvin
L	leachate
MSW	municipal solid waste
o	original
s	solubility
t	total
w	water

1. MUNICIPAL SOLID WASTE

Municipal solid waste (MSW, known for many years as *"garbage"*) consists of the solid material discarded by a community, including excess food, containers and packaging, residential garden wastes, other household discards, and light industrial debris. Hazardous wastes, including paints, insecticides, lead car batteries, used crankcase oil, dead animals, raw animal wastes, and radioactive substances present special disposal problems and are not included in MSW.

MSW is generated at the average rate of approximately 5–8 lbm/capita-day (2.3–3.6 kg/capita·d). A value of 5 lbm/capita-day (2.3 kg/capita·d) is commonly used in design studies.

Although each community's MSW has its own characteristics, Table 39.1 gives average composition in the United States. The moisture content of MSW is approximately 20%.

Solid Waste

Table 39.1 Typical Characteristics of MSW[a,b]

component	percentage by weight
paper, cardboard	35.6
yard waste	20.1
food waste	8.9
metal	8.9
glass	8.4
plastics	7.3
wood	4.1
rubber, leather	2.8
cloth	2.0
other (ceramic, stone, dirt)	1.9

[a]United States national average
[b]dry composition

2. LANDFILLS

Municipal solid waste has traditionally been disposed of in *municipal solid waste landfills* (MSWLs, for many years referred to as "*dumps*"). Other wastes, including incineration ash and water treatment sludge, may be included with MSW in some landfills.

The fees charged to deposit waste are known as *tipping fees*. Tipping fees are generally higher in the eastern United States (where landfill sites are becoming scarcer). Tipping fees for resource recovery plants are approximately double those of traditional landfills.

Many early landfills were not designed with liners, not even simple compacted-soil bottom layers. Design was often by rules of thumb, such as the bottom of the

landfill had to be at least 5 ft (1.5 m) above maximum elevation of groundwater and 5 ft (1.5 m) above bedrock. Percolation through the soil was believed sufficient to make the leachate bacteriologically safe. However, inorganic pollutants can be conveyed great distances, and the underlying soil itself can no longer be used as a barrier to pollution and to perform filtering and cleansing functions.

Landfills without bottom liners are known as *natural attenuation* (NA) *landfills*, since the native soil is used to reduce the concentration of leachate components. Landfills lined with clay (pure or bentonite-amended soil) or synthetic liners are known as *containment landfills*.

Landfills can be incrementally filled in a number of ways, as shown in Fig. 39.1. With the *area method*, solid waste is spread and compacted on flat ground before being covered. With the *trench method*, solid waste is placed in a trench and covered with the trench soil. In the *progressive method* (also known as the *slope/ramp method*), cover soil is taken from the front (toe) of the working face.

Landvaults are essentially above-ground piles of MSW placed on flat terrain. This design, with appropriate covers, liners, and instrumentation, has been successfully used with a variety of industrial wastes, including incineration ash and wastewater sludge. However, except when piggybacking a new landfill on top of a closed landfill, it is not widely used with MSW.

Waste is placed in layers, typically 2–3 ft thick (0.6–0.9 m), and is compacted before soil is added as a cover. Two to five passes by a tracked bulldozer are

Figure 39.1 Landfill Creation Methods

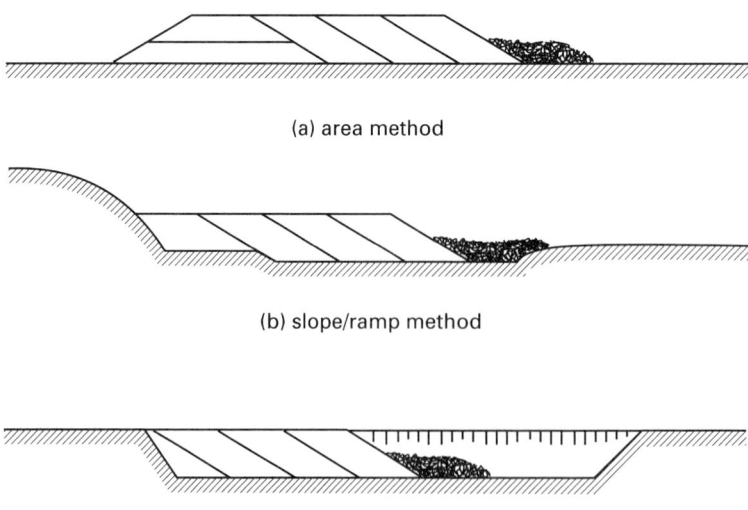

(a) area method

(b) slope/ramp method

(c) trench method

sufficient to compact the MSW to 800–1500 lbm/yd³ (470–890 kg/m³). (A density of 1000 lbm/yd³ or 590 kg/m³ is used in design studies.) The *lift* is the height of the covered layer, as shown in Fig. 39.2. When the landfill layer has reached full height, the ratio of solid waste volume to soil cover volume will be approximately between 4:1 and 3:1.

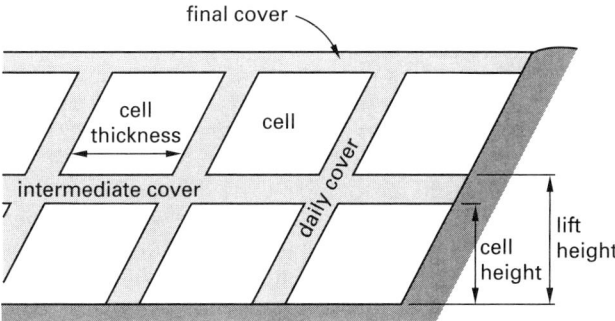

Figure 39.2 Landfill Cells

The *cell height* is typically taken as 8 ft (2.4 m) for design studies, although it can actually be much higher. The height should be chosen to minimize the cover material requirement consistent with the regulatory requirements. Cell slopes will be less than 40°, and typically 20–30°.

Each *cell* is covered by a soil layer 0.5–1.0 ft (0.15–0.3 m) in thickness. The final top and side covers should be at least 2 ft (0.6 m) thick. Daily, intermediate, and final soil covers are essential to proper landfill operation. The daily and intermediate covers prevent fly emergence, discourage rodents, reduce odors, and reduce windblown litter. In the event of ignition, the soil layers form fire stops within the landfill.

The final soil cover (cap) is intended to reduce or eliminate entering moisture. The soil cover also contains and channels landfill gas and provides a pleasing appearance and location for growing vegetation.

3. LANDFILL CAPACITY

The *compaction factor*, CF, is a multiplicative factor used to calculate the compacted volume for a particular type of waste and compaction method. The compacted volume of MSW or components within the waste is

$$V_c = (\text{CF})V_o \qquad 39.1$$

The daily increase in landfill volume is predicted by using Eq. 39.2. N is the population size, G is the per-capita MSW generation rate per day, and γ is the landfill overall specific weight calculated as a weighted average of soil and compacted MSW specific weights. Soils have specific weights between 70 lbf/ft³ and 130 lbf/ft³ (densities between 1100 kg/m³ and 2100 kg/m³). A value of 100 lbf/ft³ (1600 kg/m³) is commonly used in design studies.

$$\Delta V = \frac{NG(\text{LF})}{\gamma} = \frac{NG(\text{LF})}{\rho g} \qquad \text{[SI]} \quad 39.2(a)$$

$$\Delta V = \frac{NG(\text{LF})}{\gamma} = \frac{NG(\text{LF})g_c}{\rho g} \qquad \text{[U.S.]} \quad 39.2(b)$$

The *loading factor*, LF, is 1.25 for a 4:1 volumetric ratio and is calculated from Eq. 39.3 for other ratios.

$$\text{LF} = \frac{V_{\text{MSW}} + V_{\text{cover soil}}}{V_{\text{MSW}}} \qquad 39.3$$

Most large-scale sanitary landfills do not apply daily cover to deposited waste. Time, cost, and reduced capacity are typically cited as the reasons that the landfill is not covered with soil. In the absence of such cover, the loading factor has a value of 1.0.

4. SUBTITLE D LANDFILLS

Since 1992, new and expanded municipal landfills in the United States have had to satisfy strict design regulations and are designated "*Subtitle D landfills*," referring to Subtitle D of the Resource Conservation and Recovery Act (RCRA). The regulations apply to any landfill designed to hold municipal solid waste, biosolids, and ash from MSW incineration.

Construction of landfills is prohibited near sensitive areas such as airports, floodplains, wetlands, earthquake zones, and geologically unstable terrain. Air quality control methods are required to control emission of dust, odors (from hydrogen sulfide and volatile organic vapors), and landfill gas. Runoff from storms must be controlled.

Subtitle D landfills must have *double-liner systems*, as shown in Fig. 39.3, and which are defined as "two or more liners and a leachate collection system above and between such liners." While states can specify greater protection, the minimum (basic) bottom layer requirements are a 30-mil flexible PVC membrane liner (FML) and at least 2 ft (0.6 m) and up to 5 ft (1.5 m) of compacted soil with a maximum hydraulic conductivity of 1.2×10^{-5} ft/hr (1×10^{-7} cm/s). If the membrane is high-density polyethylene (HDPE), the minimum thickness is 60 mils. 30-mil PVC costs less than 60-mil HDPE, but the 60-mil product offers superior protection.

The preferred double liner consists of two FMLs separated by a drainage layer (approximately 2 ft (0.6 m) thick) of sand, gravel, or *drainage netting* and placed on low-permeability soil. This is the standard design for *Subtitle C* hazardous waste landfills. The advantage of selecting the preferred design for municipal landfills is realized in the permitting process.

The minimum thickness is 20 mils for the top FML, and the maximum hydraulic conductivity of the cover soil is 1.8×10^{-5} ft/hr (1.5×10^{-7} cm/s).

A series of wells and other instrumentation are required to detect high hydraulic heads and the accumulation of heavy metals and volatile organic compounds (VOCs) in the leachate.

5. CLAY LINERS

Liners constructed of well-compacted bentonite clays fail by fissuring (as during freeze-thaw cycles), dessication (i.e., drying out), chemical interaction, and general disruption. Permeability increases dramatically with organic fluids (hydrocarbons) and acidic or caustic leachates. Clays need to be saturated and swollen to retain their high impermeabilities. Permeabilities measured in the lab as 1.2×10^{-5} ft/hr (10^{-7} cm/s) can actually be 1.2×10^{-3} ft/hr (10^{-5} cm/s) in the field.

Although a 0.5–1.0 ft (0.15–0.30 m) thickness could theoretically contain the leachate, the probability of success is not high when typical exposure issues are considered. Clay liners with thicknesses of 4–5 ft (1.2–1.5 m) will still provide adequate protection, even when the top half has degraded.

6. FLEXIBLE MEMBRANE LINERS

Synthetic *flexible membrane liners* (FMLs, the term used by the United States EPA), also known as *synthetic membranes, synthetic membrane liners* (SMLs), *geosynthetics*, and *geomembranes*, are highly attractive because of their negligible permeabilities, and good chemical resistance. The term *geocomposite liner* is used when referring to a combination of FML and one or more compacted clay layers.

FML materials include polyvinyl chloride (PVC); chlorinated polyethylene (CPE); low-density polyethylene (LDPE); linear, low-density polyethylene (LLDPE); very low density polyethylene (VLDPE); high-density polyethylene (HDPE); chlorosulfonated polyethylene, also known as Hypalon® (CSM or CSPE); ethylene propylene dienemonomer (EP or EPDM); polychloroprene, also known as neoprene (CR); isobutylene isoprene, also known as butyl (IIR); and oil-resistant polyvinyl chloride (ORPVC). LDPE covers polyethylene with densities in the range of about 56.2–58.3 lbm/ft³ (0.900–0.935 g/cm³), while HDPE covers the range of about 58.3–60.5 lbm/ft³ (0.935–0.970 g/cm³).

HDPE is preferred for municipal landfills and hazardous waste sites because of its high tensile strength, high toughness (resistance to tear and puncture), and ease of seaming. HDPE has captured the majority of the FML market.

Some FMLs can be manufactured with an internal nylon or dacron mesh, known as a *scrim*, or a spread coating on one side. The resulting membrane is referred to as a *reinforced* or *supported membrane*. An "R" is added to the name to indicate the reinforcement (e.g., "HDPE-R"). Membranes without internal scrims are *unreinforced* or *unsupported membranes*.

FMLs can be embossed or textured to increase the friction between the FML and any soil layer adjacent to it.

FMLs are sealed to the original grade and restrained against movement by deep *anchor trenches* around the periphery of the landfill. (See Fig. 39.3.)

Figure 39.3 *Subtitle D Liner and Cover Detail*

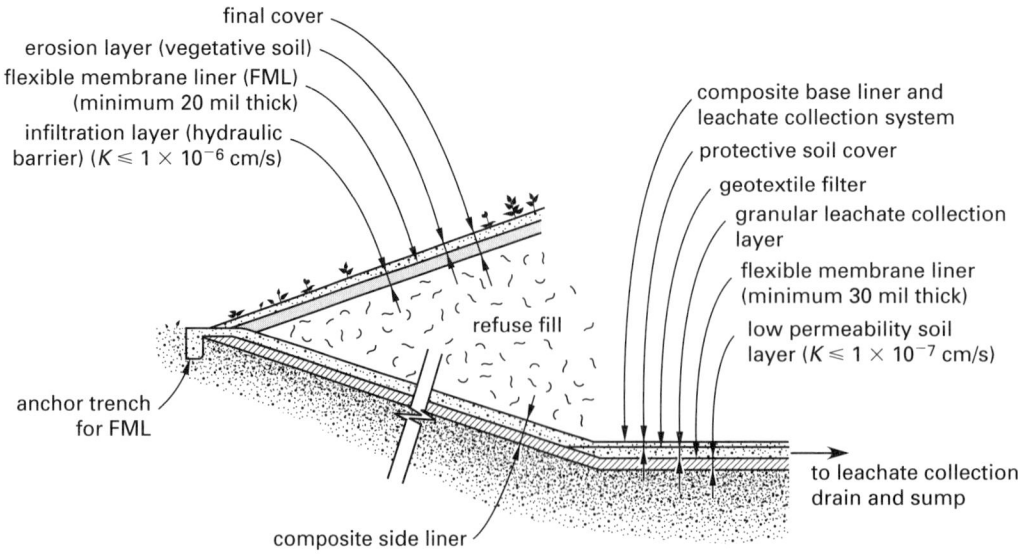

final cover
erosion layer (vegetative soil)
flexible membrane liner (FML)
(minimum 20 mil thick)
infiltration layer (hydraulic barrier) ($K \leq 1 \times 10^{-6}$ cm/s)

composite base liner and leachate collection system
protective soil cover
geotextile filter
granular leachate collection layer
flexible membrane liner (minimum 30 mil thick)
low permeability soil layer ($K \leq 1 \times 10^{-7}$ cm/s)

refuse fill

anchor trench for FML

to leachate collection drain and sump

composite side liner

FMLs fail primarily by puncturing, tearing, and long-term exposure to ultraviolet (UV) rays. Freezing temperatures may also weaken them. FMLs are protected from UV radiation by a soil backfill with a minimum thickness of approximately 6 in (15 cm). The cover soil should be placed promptly.

7. LANDFILL CAPS

The landfill *cap* (*top cap*) is the *final cover* placed over the landfill. Under Subtitle D regulations, caps cannot be any more permeable than the bottom liners. Therefore, only preexisting landfills can be capped with clay.

Caps fail by desiccation, penetration by roots, rodent and other animal activity, freezing and thawing cycles, erosion, settling, and breakdown of synthetic liners from sunlight exposure. Failure can also occur when the cover collapses into the disposal site due to voids opening below or excess loading (rainwater ponding) above.

Synthetic FML caps offer several advantages over clay caps—primarily lower permeability, better gas containment, faster installation, and easier testing and inspection.

In the past, some landfills had slopes as steep as 0.7:1 (horizontal:vertical). Preferred modern landfill design incorporates maximum slopes of 3:1 and occasionally as steep as 2:1 (horizontal:vertical), with horizontal tiers (benches) every so often. This is steep enough to cause concern about cap slope instability. Although clay caps can be constructed this steep, synthetic membranes have lower *interfacial friction angles* and are too smooth for use on steep slopes. (See Table 39.2.) Slippage between different layers in the cap (e.g., between the FML and the soil cover) is known as *veneer failure*. Some relief is possible with *textured synthetics*.

A common cap system is a textured synthetic with a nonwoven geotextile. Vertical cap collapse can be prevented by incorporating a high-strength plastic *geogrid*.

Table 39.2 Typical Interfacial Friction Angles

interface	typical friction angle
nonwoven geotextile/smooth FML	10–13°
compacted sand/smooth FML	15–25°
compacted clay/smooth FML	5–20°
nonwoven geotextile/textured FML	25–35°
compacted sand/textured FML	20–35°
compacted clay/textured FML	15–32°

8. LANDFILL SITING

Sanitary landfills should be designed for a five-year minimum life. Sanitary landfill sites are selected on the basis of many relevant factors, including (a) economics and the availability of inexpensive land;

(b) location, including ease of access, transport distance, acceptance by the population, and aesthetics; (c) availability of cover soil, if required; (d) wind direction and speed, including odor, dust, and erosion considerations; (e) flat topography; (f) dry climate and low infiltration rates; (g) location (elevation) of the water table; (h) low risk of aquifer contamination; (i) types and permeabilities of underlying strata; (j) avoidance of winter freezing; (k) future growth and capacity requirements; and (l) ultimate use.

Suitable landfill sites are becoming difficult to find, and once identified, they are subjected to a rigorous permitting process. They are also objected to by residents near the site and MSW transport corridor. This is referred to as the *NIMBY syndrome* (not in my backyard). People agree that landfills are necessary, but they do not want to live near them.

9. VECTORS

In landfill parlance, a *vector* is an insect or arthropod, rodent, or other animal of public health significance capable of causing human discomfort or injury, or capable of harboring or transmitting the causative agent of human disease. Vectors include flies, domestic rats, field rodents, mosquitoes, wasps, and cockroaches. Standards for vector control should state the *vector thresholds* (e.g., "six or more flies per square yard").

10. ULTIMATE LANDFILL DISPOSITION

When filled to final capacity and closed, the covered landfill can be used for grassed green areas, shallow-rooted agricultural areas, and recreational areas (soccer fields, golf courses, etc.). Light construction uses are also possible, although there may be problems with gas accumulation, corrosion of pipes and foundation piles, and settlement.

Settlement will generally be uneven. Settlement in areas with high rainfall can be up to 20% of the overall landfill height. In dry areas, the settlement may be much less, 2–3%. Little information on bearing capacity of landfills is available. Some studies have suggested capacities of 500–800 lbf/ft^2 (24–38 kPa).

11. LANDFILL GAS

Once covered, the organic material within a landfill cell provides an ideal site for decomposition. Aerobic conditions prevail for approximately the first month. Gaseous products of the initial aerobic decomposition include CO_2 and H_2O. The following are typical reactions involving carbohydrates and stearic acid.

$$C_6H_{12}O_6 + 6O_2 \rightarrow 6CO_2 + 6H_2O \qquad 39.4$$

$$C_{18}H_{36}O_2 + 26O_2 \rightarrow 18CO_2 + 18H_2O \qquad 39.5$$

After the first month or so, the decomposing waste will have exhausted the oxygen in the cell. Digestion continues anaerobically, producing a low-heating value *landfill gas* (LFG) consisting of approximately 50% methane (CH_4), 50% carbon dioxide, and trace amounts of other gases (e.g., CO, N_2, and H_2S). Decomposition occurs at temperatures of 100–120°F (40–50°C), but may increase to up to 160°F (70°C).

$$C_6H_{12}O_6 \rightarrow 3CO_2 + 3CH_4 \qquad 39.6$$

$$C_{18}H_{36}O_2 + 8H_2O \rightarrow 5CO_2 + 13CH_4 \qquad 39.7$$

LFG is essentially saturated with water vapor when formed. However, if the gas collection pipes are vertical or sloped, some of the water vapor will condense on the pipes and drain back into the landfill. Most of the remaining moisture is removed in condensate traps located along the gas collection system line.

From Dalton's law, the total gas pressure within a landfill is the sum of the partial pressures of the component gases. Partial pressure is volumetrically weighted and can be found from the volumetric fraction, B, of the gas.

$$p_t = p_{CH_4} + p_{CO_2} + p_{H_2O} + p_{N_2} + p_{other} \qquad 39.8$$

$$p_i = B_i p_t \qquad 39.9$$

If uncontrolled, LFG will migrate to the surface. If the LFG accumulates, an explosion hazard results, since methane is highly explosive in concentrations of 5–15% by volume. Methane will pass through the explosive concentration as it is being diluted. Other environmental problems, including objectionable odors, can also occur. Therefore, various methods are used to prevent gas from escaping or spreading laterally.

Landfill gases can be collected by either passive or active methods. *Passive collection* uses the pressure of the gas itself as the driving force. Passive collection is applicable for sites that generate low volumes of gas and where offsite migration of gas is not expected. This generally applies to small municipal landfills—those with volumes less than approximately 50,000 yd³ (40 000 m³). Common passive control methods include (a) isolated gas (pressure relief) vents in the landfill cover, with or without a common flare; (b) perimeter interceptor gravel trenches; (c) perimeter barrier trenches (e.g., slurry walls); (d) impermeable barriers within the landfill; and (e) sorptive barriers in the landfill.

Active gas collection draws the landfill gas out with a vacuum from *extraction wells*. Various types of horizontal and vertical wells can be used through the landfill to extract gas in vertical movement, while perimeter facilities are used to extract gases in lateral movement. Vertical wells are usually perforated PVC pipes packed in sleeves of gravel and bentonite clay. PVC pipe is resistant to the chlorine and sulfur compounds present in the landfill.

Determining the location of isolated vents is essentially heuristic—one vent per 10,000 yd³ (7500 m³) of landfill

is probably sufficient. Regardless, extraction wells should be spaced with overlapping zones of influence. If the *radius of influence*, R, of each well and the desired fractional overlap, O, are known, the spacing between wells is given by Eq. 39.10. For example, if the extraction wells are placed on a square grid with spacing of $L = 1.4R$, the overlap will be 60%. For a 100% overlap, the spacing would equal the radius of influence.

$$\frac{L}{R} = 2 - O \qquad 39.10$$

In many locations, the LFG is incinerated in flares. However, emissions from flaring are problematic. Alternatives to flaring include using the LFG to produce hot water or steam for heating or electricity generation. During the 1980s, reciprocating engines and combustion turbines powered by LFG were tried. However, such engines generated relatively high emissions of their own due to impurities and composition variations in the fuel. True Rankine-cycle power plants (generally without reheat) avoid this problem, since boilers are less sensitive to impurities.

One problem with using LFG commercially is that LFG is withdrawn from landfills at less than atmospheric pressure. Conventional furnace burners need approximately 5 psig at the boiler front. Low-pressure burners that require 2 psig are available, but they are expensive. Therefore, some of the plant power must be used to pressurize the LFG in blowers.

Although production is limited, LFG is produced for a long period after a landfill site is closed. Production slowly drops 3–5% annually to approximately 30% of its original value after about 20–25 years, which is considered to be the economic life of a gas-reclamation system. The theoretical ultimate production of LFG has been estimated by other researchers as 15,000 ft³ per ton (0.45 m³/kg) of solid waste, with an estimated volumetric gas composition of 54% methane and 46% carbon dioxide. (See Table 39.3.) However, unfavorable and non-ideal conditions in the landfill often reduce this yield to approximately 1000–3000 ft³/ton (0.03–0.09 m³/kg) or even lower.

Table 39.3 Properties of Methane and Carbon Dioxide

	CH_4	CO_2
color	none	none
odor	none	none
density, at STP		
(g/L)	0.717	1.977
(lbm/ft³)	0.0447	0.123
specific gravity, at STP, ref. air	0.554	1.529
solubility (760 mm Hg, 20°C), volumes in one volume of water	0.33	0.88
solubility, qualitative	slight	moderate

(Multiply lbm/ft³ by 16.02 to obtain g/L.)

12. LANDFILL LEACHATE

Leachates are liquid wastes containing dissolved and finely suspended solid matter and microbial waste produced in landfills. Leachate becomes more concentrated as the landfill ages. Leachate forms from liquids brought into the landfill, water run-on, and precipitation. Leachate in a natural attenuation landfill will contaminate the surrounding soil and groundwater. In a lined containment landfill, leachate will percolate downward through the refuse and collect at the first landfill liner. Leachate must be removed to reduce hydraulic head on the liner and to reduce unacceptable concentrations of hazardous substances.

When a layer of liquid sludge is disposed of in a landfill, the consolidation of the sludge by higher layers will cause the water to be released. This released water is known as *pore-squeeze liquid.*

Some water will be absorbed by the MSW and will not percolate down. The quantity of water that can be held against the pull of gravity is referred to as the *field capacity*, FC. The potential quantity of leachate is the amount of moisture within the landfill in excess of the FC.

In general, the amount of leachate produced is directly related to the amount of external water entering the landfill. Theoretically, the leachate generation rate can be determined by writing a water mass balance on the landfill. This can be done on a preclosure and a post-closure basis. Typical units for all the terms are units of length (e.g., "1.2 in of rain") or mass per unit volume (e.g., "a field capacity of 4 lbm/yd^3").

The preclosure leachate generation rate is

preclosure leachate generation
    = moisture released by incoming waste,
          including pore-squeezed liquid
      + precipitation
      − moisture lost due to evaporation
      − field capacity   *39.11*

The postclosure water balance is

postclosure leachate generation
    = precipitation
      − surface runoff
      − evapotranspiration
      − moisture lost in formation
          of landfill gas and other
          chemical compounds
      − water vapor removed along
          with landfill gas
      − change in soil moisture storage   *39.12*

13. LEACHATE MIGRATION FROM LANDFILLS

From Darcy's law, migration of leachate contaminants that have passed through liners into aquifers or the groundwater table is proportional to the hydraulic conductivity, K, and the hydraulic gradient, i. Hydraulic conductivities of clay liners are 1.2×10^{-7} ft/hr to 1.2×10^{-5} ft/hr (10^{-9} cm/s to 10^{-7} cm/s). However, the properties of clay liners can change considerably over time due to interactions with materials in the landfill. If the clay dries out (desiccates), it will be much more permeable. For synthetic FMLs, hydraulic conductivities are 1.2×10^{-10} ft/hr to 1.2×10^{-7} ft/hr (10^{-12} cm/s to 10^{-9} cm/s). The average permeability of high-density polyethylene is approximately 1.2×10^{-11} ft/hr (1×10^{-13} cm/s).

$$Q = KiA \qquad\qquad 39.13$$

$$i = \frac{dH}{dL} \qquad\qquad 39.14$$

14. GROUNDWATER DEWATERING

It may be possible to prevent or reduce contaminant migration by reducing the elevation of the groundwater table (GWT). This is accomplished by dewatering the soil with relief-type *extraction drains* (*relief drains*). The *ellipse equation*, also known as the *Donnan formula* and the *Colding equation*, used for calculating pipe spacing, L, in draining agricultural fields, can be used to determine the spacing of groundwater dewatering systems. In Eq. 39.15, K is the hydraulic conductivity with units of length/time, a is the distance between the pipe and the impervious layer barrier (a is zero if the pipe is installed on the barrier), b is the maximum allowable table height above the barrier, and Q is the *recharge rate*, also known as the *drainage coefficient*, with dimensions of length/time. The units of K and Q must be on the same time basis.

$$L = 2\sqrt{\left(\frac{K}{Q}\right)(b^2 - a^2)} = 2\sqrt{\frac{b^2 - a^2}{i}} \qquad 39.15$$

Equation 39.15 is often used because of its simplicity, but the accuracy is only approximately ±20%. Therefore, the calculated spacing should be decreased by 10–20%.

For pipes above the impervious stratum, as illustrated in Fig. 39.4, the total discharge per unit length of each pipe (in ft^3/ft-sec or m^3/m·s) is given by Eq. 39.16. H is the maximum height of the water table above the pipe invert elevation. D is the average depth of flow.

$$Q_{\text{unit length}} = \frac{2\pi KHD}{L} \qquad\qquad 39.16$$

Solid Waste

$$D = a + \frac{H}{2} \qquad \text{39.17}$$

For pipes on the impervious stratum, the total discharge per unit length from the ends of each pipe (in ft³/ft-sec or m³/m·s) is

$$Q_{\text{unit length}} = \frac{4KH^2}{L} \qquad \text{39.18}$$

Equation 39.19 gives the total discharge per pipe. If the pipe drains from both ends, the discharge per end would be half of that amount.

$$Q_{\text{pipe}} = L Q_{\text{unit length}} \qquad \text{39.19}$$

Figure 39.4 Geometry for Groundwater Dewatering Systems

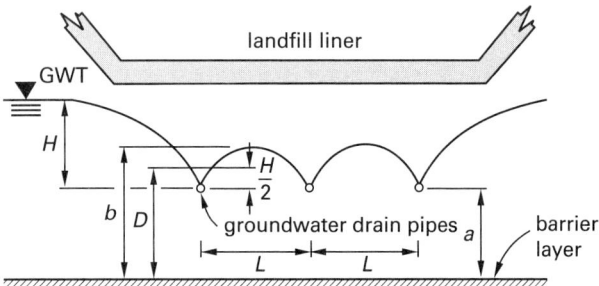

(a) pipes above impervious stratum

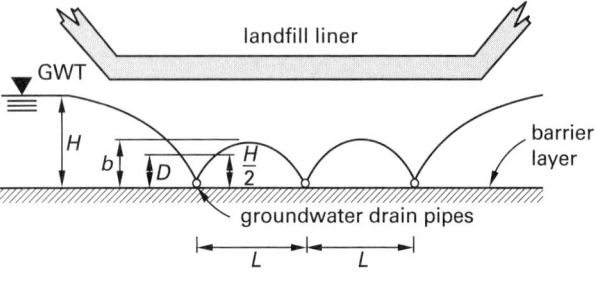

(b) pipes on impervious stratum

Example 39.1

Subsurface leachate migration from a landfill is to be mitigated by maintaining the water table that surrounds the landfill site lower than the natural level. The surrounding area consists of 15 ft of saturated, homogeneous soil over an impervious rock layer. The water table, originally at the surface, has been lowered to a depth of 9 ft. Fully pervious parallel collector drains at a depth of 12 ft are present. The hydraulic conductivity of the soil is 0.23 ft/hr. The natural water table will recharge the site in 30 days if drainage stops. What collector drain separation is required?

Solution

Referring to Fig. 39.4(a),

$$a = 15 \text{ ft} - 12 \text{ ft} = 3 \text{ ft}$$

$$b = 15 \text{ ft} - 9 \text{ ft} = 6 \text{ ft}$$

$$K = \left(0.23 \ \frac{\text{ft}}{\text{hr}}\right)\left(24 \ \frac{\text{hr}}{\text{day}}\right) = 5.52 \text{ ft/day}$$

To maintain the lowered water table level, the drainage rate must equal the recharge rate.

$$Q = \frac{9 \text{ ft}}{30 \text{ days}} = 0.3 \text{ ft/day}$$

Use Eq. 39.15 to find the drain spacing.

$$L = 2\sqrt{\left(\frac{K}{Q}\right)(b^2 - a^2)}$$

$$= 2\sqrt{\left(\frac{5.52 \ \frac{\text{ft}}{\text{day}}}{0.3 \ \frac{\text{ft}}{\text{day}}}\right)\left((6 \text{ ft})^2 - (3 \text{ ft})^2\right)}$$

$$= 44.57 \text{ ft} \quad (45 \text{ ft})$$

15. LEACHATE RECOVERY SYSTEMS

At least two distinct leachate recovery systems are required in landfills: one within the landfill to limit the hydraulic head of leachate that has reached the top liner, and another to catch the leachate that has passed through the top liner and drainage layer and has reached the bottom liner.

By removing leachate at the first liner, the hydrostatic pressure on the liner is reduced, minimizing the pressure gradient and hydraulic movement through the liner. A pump is used to raise the collected leachate to the surface once a predetermined level has been reached. Tracer compounds (e.g., lithium compounds or radioactive hydrogen) can be buried with the wastes to signal migration and leakage.

Leachate collection and recovery systems fail because of clogged drainage layers and pipe, crushed collection pipes due to waste load, pump failures, and faulty design.

16. LEACHATE TREATMENT

Leachate is essentially a very strong municipal wastewater, and it tends to become more concentrated as the landfill ages. Leachate from landfills contains extremely high concentrations of compounds and cannot be discharged directly into rivers or other water sources.

Leachate is treated with biological (i.e., trickling filter and activated sludge) and physical/chemical processes very similar to those used in wastewater treatment plants. For large landfills, these treatment facilities are located on the landfill site. A typical large landfill treatment facility would include an equalization tank, a primary clarifier, a first-stage activated sludge aerator and clarifier, a second-stage activated sludge aerator and clarifier, and a rapid sand filter. Additional equipment for sludge dewatering and digestion would also be required. Liquid effluent would be discharged to the municipal wastewater treatment plant.

17. LANDFILL MONITORING

Monitoring is conducted at sanitary landfills to ensure that no contaminants are released from the landfill. Monitoring is conducted in the vadose zone for gases and liquid, in the groundwater for leachate movements, and in the air for air quality monitoring.

Landfills use *monitoring wells* located outside of the covered cells and extending past the bottom of the disposal site into the aquifer. In general, individual monitoring wells should extend into each stratum upstream and downstream (based on the hydraulic gradient) of the landfill.

18. CARBON DIOXIDE IN LEACHATE

Carbon dioxide is formed during both aerobic and anaerobic decomposition. Carbon dioxide combines with water to produce *carbonic acid*, H_2CO_3. In the absence of other mitigating compounds, this will produce a slightly acidic leachate.

The concentration of carbon dioxide (in mg/L) in the leachate (assumed to be water) can be calculated by using the *absorption coefficient (solubility)*, K_s, as given in Table 39.4. The absorption coefficient for carbon dioxide at 32°F (0°C) and 1 atm is approximately $K_s = 0.88$ L/L.

$$\rho_{g/L} = \frac{p(MW)}{R^* T}$$

$$= \frac{B_{CO_2} p_{t,atm} \left(44 \frac{g}{mol}\right)}{\left(0.08206 \frac{atm \cdot L}{mol \cdot K}\right) T_K} \qquad 39.20$$

$$T_K = T_{°C} + 273° \qquad 39.21$$

At a typical internal landfill temperature of 100°F (38°C), Eq. 39.20 reduces to

$$\rho_{g/L} = 1.72 B_{CO_2} \qquad 39.22$$

The concentration of carbon dioxide in the leachate is

$$C_{CO_2,mg/L} = \rho_{g/L} K_{s,L/L} \left(1000 \frac{mg}{g}\right) \qquad 39.23$$

Table 39.4 *Typical Absorption Coefficients (L/L at 0°C and 1 atm)*

element	K_s (L/L)
hydrogen	0.017
nitrogen	0.015
oxygen	0.028
carbon monoxide	0.025
methane	0.33
carbon dioxide	0.88

Values of solubility for carbon dioxide (or any gas) in leachate depend on the leachate composition as well as temperature and are difficult to find. Values of solubilities of gases in water, with typical units of mass of gas per mass of pure water, are commonly available, however. Solubility values with units of L/L (i.e., liters of gas per liter of water) will usually have to be calculated from published values with units of g/kg (i.e., grams of gas per kilograms of water). The total pressure within the landfill is essentially one atmosphere, and pressure of each gas is volumetrically weighted. Reasonable assumptions about the volumetric fraction (or, partial pressure ratio), B, of carbon dioxide must be made. Carbon dioxide constitutes approximately 0.033–0.039% by volume of atmospheric air, although values within a landfill will be significantly higher. (At maturity, landfill gas is 40–60% carbon dioxide, 45–60% methane, 2–5% nitrogen, and less than 1% oxygen by volume.) Solubility values are also dependent on temperature within the landfill, and solubility decreases with an increase in temperature, as Fig. 39.5 shows. Considering the heat of decomposition, an assumed landfill temperature of 100°F (38°C) is reasonable.

Figure 39.5 *Solubility of Carbon Dioxide in Water*

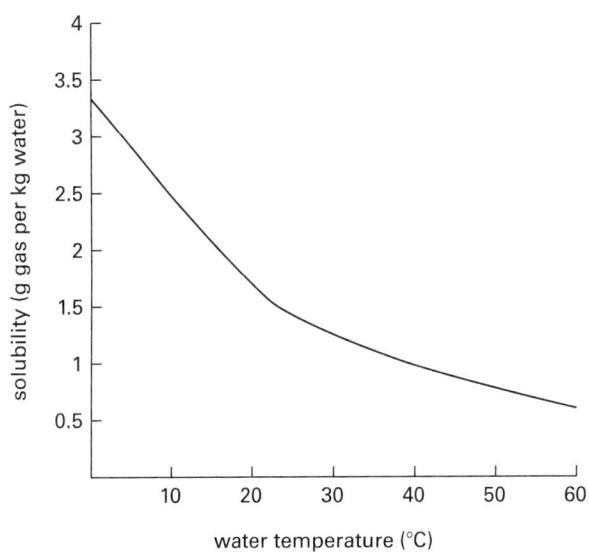

19. INCINERATION OF MUNICIPAL SOLID WASTE

Incineration of MSW results in a 90% reduction in waste disposal volume and a mass reduction of 75%.

In *no-boiler incinerators*, MSW is efficiently burned without steam generation. However, incineration at large installations is often accompanied by recycling and steam and/or electrical power generation. Incinerator/generator facilities with separation capability are known as *resource-recovery plants*. Facilities with boilers are referred to as *waste-to-steam plants*. Facilities with boilers and electrical generators are referred to as *waste-to-energy* (WTE) *facilities*. (See Fig. 39.6.) In WTE plants, the combustion heat is used to generate steam and electrical power.

Mass burning is the incineration of unprocessed MSW, typically on stoker grates or in rotary and waterwall combustors to generate steam. With mass burning, MSW is unloaded into a pit and then moved by crane to the furnace conveying mechanism. Approximately 27% of the MSW remains as ash, which consists of glass, sand, stones, aluminum, and other noncombustible materials. It and the fly ash are usually collected and disposed of in municipal landfills. (The U.S. Environmental Protection Agency (EPA) has ruled that ash from the incineration of MSW is not a hazardous waste. However, this is hotly contested and is subject to ongoing evaluation.) Capacities of typical mass burn units vary from less than 400 tons/day (360 Mg/d) to a high of 3000 tons/day (2700 Mg/d), with the majority of units processing 1000–2000 tons/day (910–1800 Mg/d).

MSW, as collected, has a heating value of approximately 4500 Btu/lbm (10 MJ/kg), though higher values have been reported. For 1000 tons/day (1000 Mg/d) of MSW incinerated, the yields are approximately 150,000–200,000 lbm/hr (75–100 Mg/h) of 650–750 psig (4.5–5.2 MPa) steam at 700–750°F (370–400°C) and 25–30 MW (27.6–33.1 MW) of gross electrical power. Approximately 10% of the electrical power is used internally, and units generating much less than approximately 10 MW (gross) may use all of their generated electrical power internally.

The *burning rate* is approximately 40–60 lbm/hr-ft^2 (200–300 kg/h·m^2) and is the fueling rate divided by the total effective grate area. Maximum heat release rates are approximately 300,000 Btu/hr-ft^2 (940 kW/m^2). The *heat release rate*, HRR, is defined as

$$\text{HRR} = \frac{(\text{fueling rate})(\text{HV})}{\text{total effective grate area}} \qquad 39.24$$

20. REFUSE-DERIVED FUEL

Refuse-derived fuel (RDF) is derived from MSW. (See Table 39.5.) First-generation RDF plants use "crunch and burn" technology. In these plants, the MSW is shredded after ferrous metals are removed magnetically. First-generation RDF plants suffer from the same problems that have plagued early mass burn units, including ash with excessive quantities of noncombustible materials such as glass, grit and sand, and aluminum.

Figure 39.6 *Typical Mass-Burn Waste-to-Energy Plant*

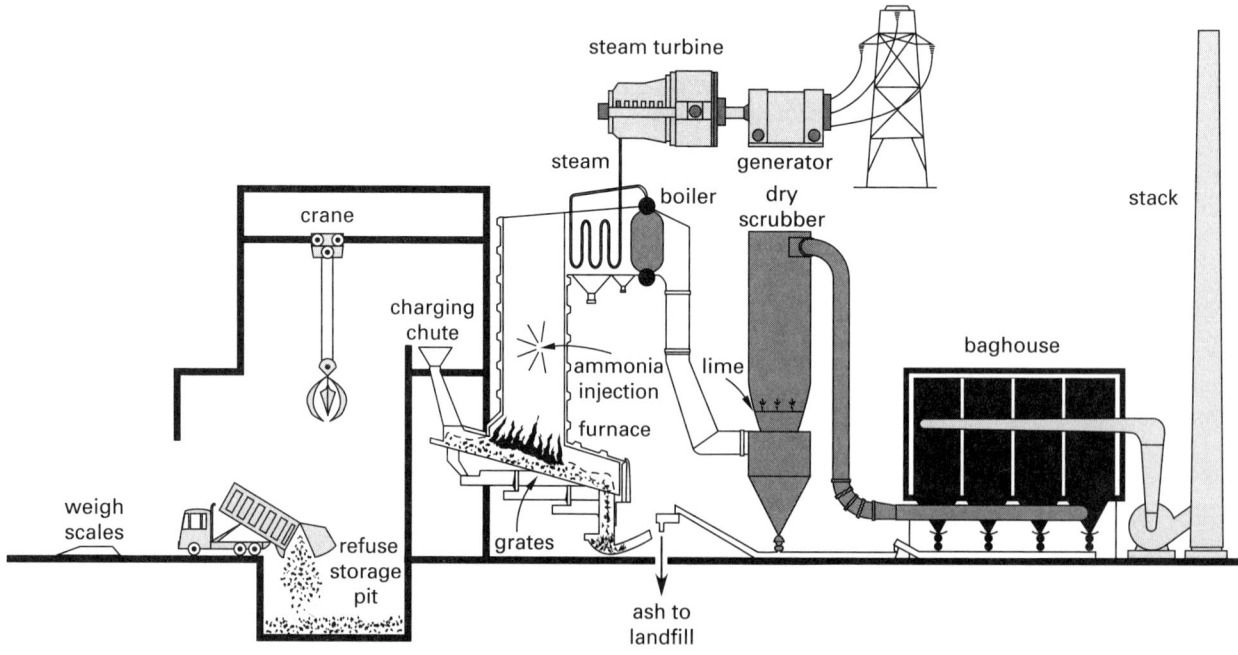

Table 39.5 *Typical Ultimate Analyses of MSW and RDF*

element	percentage by weight	
	MSW	RDF
carbon	26.65	31.00
water	25.30	27.14
ash	23.65	13.63
oxygen	19.61	22.72
hydrogen	3.61	4.17
chlorine	0.55	0.66
nitrogen	0.46	0.49
sulfur	0.17	0.19
total	100.00	100.00

Second-generation plants incorporate screens and air classifiers to reduce noncombustible materials and to increase the recovery of some materials. *Material recovery facilities* (MRFs) specialize in sorting out recyclables from MSW. The ash content is reduced, and the energy content of the RDF is increased. The MSW is converted into RDF pellets between $2^1/_2$ in and 6 in (6.4 cm and 15 cm) in size and is introduced through feed ports above a traveling grate. Some of fuel is burned in suspension, with the rest burned on the grate. Grate speed is varied so that the fuel is completely incinerated by the time it reaches the ash rejection ports at the front of the burner.

In large (2000–3000 tons/day (1800–2700 Mg/d)), third-generation RDF plants, illustrated in Fig. 39.7, more than 95% of the original MSW combustibles are retained while reducing *mass yield* (i.e., the ratio of RDF mass to MSW mass) to below 85%.

RDF has a heating value of approximately 5500–5900 Btu/lbm (12–14 MJ/kg). The moisture and ash contents of RDF are approximately 24% and 12%, respectively.

The performance of a typical third-generation facility burning RDF is similar to a mass-burn unit. For each 1000 tons/day (1000 Mg/d) of MSW collected, approximately 150,000–250,000 lbm/hr (75–125 Mg/h) of 750–850 psig (5.2–5.9 MPa) steam at 750–825°F (400–440°C) can be generated, resulting in approximately 30–40 MW (33.1–44.1 MW) of electrical power.

Figure 39.7 *Typical RDF Processing*

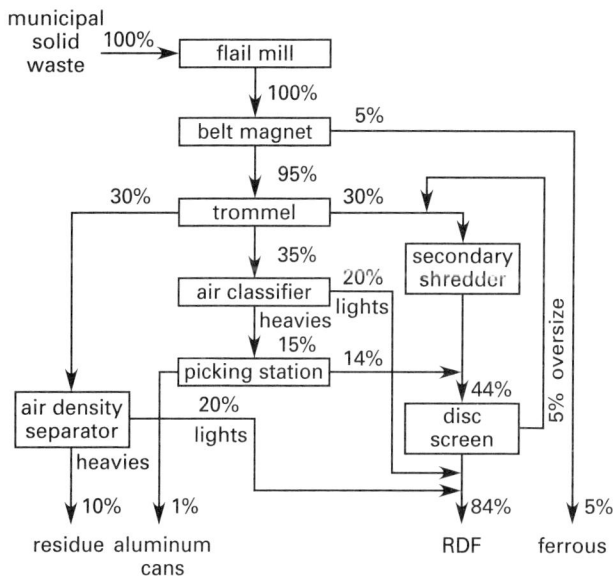

Natural gas is introduced at startup to bring the furnace up to the required 1800°F (1000°C) operating temperature. Natural gas can also be used upon demand, as when a load of particularly wet RDF lowers the furnace temperature.

Coal, oil, or natural gas can be used as a back-up fuel if RDF is unavailable or as an intended part of a *co-firing* design. Co-firing installations are relatively rare, and many have discontinued burning RDF due to poor economic performance. Typical problems with co-firing are (a) upper furnace wall slagging, (b) decreased efficiencies in electrostatic precipitators, (c) increased boiler tube corrosion, (d) excessive amounts of bottoms ash, and (e) difficulties in receiving and handling RDF.

RDF is a low-sulfur fuel, but like MSW, RDF is high in ash and chlorine. Relative to coal, RDF produces less SO_2 but more hydrogen chloride. Bottom ash can also contain trace organics and lead, cadmium, and other heavy metals.

Solid Waste

40 Environmental Pollutants

Nomenclature

a	interfacial area per volume of packing	1/ft	1/m
A	area	ft^2	m^2
b	max table height above barrier	ft	m
B	volumetric fraction	–	–
C	concentration	various	various
D	flow depth	ft	m
E	mass emission rate	lbm/MMBtu	kg/MJ
G	MSW generation rate per capita	lbm/day	kg/d
H	cyclone inlet height	ft	m
i	hydraulic gradient	ft/ft	m/m
K	hydraulic conductivity	ft/hr	cm/s
L	length	ft	m
LF	loading factor	–	–
$\dot{m}$	mass flow rate	lbm/hr	kg/h
MW	molecular weight	lbm/lbmol	kg/kmol
N	population size	–	–
p	vapor pressure	lbf/in^2	kPa
q	heat generate rate or recharge rate	MMBtu/hr	kW
Q	total discharge per unit length of pipe	ft^3/ft-sec	m^3/m·s
Q	volumetric flow rate or recharge rate	ft^3/sec	m^3/s
T	temperature	°R	K
V	volume	ft^3	m^3

Symbols

γ	specific weight	lbf/ft^3	–

Subscripts

dp	dew point
s	sulfur
t	total
w	water

PART 1: GENERAL CONCEPTS

1. POLLUTION PREVENTION AND CONTROL

Pollution prevention (sometimes referred to as "P2") can be achieved in a number of ways. Table 40.1 lists some examples in order of desirability. The most desirable method is reduction at the source. Reduction is accomplished through process modifications, raw material quantity reduction, substitution of materials, improvements in material quality, and increased efficiencies. However, traditional *end-of-pipe* treatment and disposal processes after the pollution is generated remain the main focus. *Pollution control* is the limiting of pollutants in a planned and systematic manner.

Table 40.1 *Pollution Prevention Hierarchy**

source reduction
recycling
waste separation and concentration
waste exchange
energy/material recovery
waste treatment
disposal

*from most to least desirable

Pollution control, hazardous waste, and other environmental regulations vary from nation to nation and are constantly changing. Therefore, regulation-specific issues, including timetables, permit application processes, enforcement, and penalties for violations, are either omitted from this chapter or discussed in general terms. For a similar reason, few limits on pollutant emission rates and concentrations are given in this chapter.[1] Those that are given should be considered merely representative and typical of the general range of values.

PART 2: TYPES AND SOURCES OF POLLUTION

2. THE ENVIRONMENT

Specific regulations often deal with parts of the *environment*, such as the atmosphere (i.e., the "air"), oceans and other surface water, subsurface water, and the soil. *Nonattainment areas* are geographical areas identified by regulation that do not meet *national ambient air quality standards* (NAAQS). Nonattainment is usually the result of geography, concentrations of industrial facilities, and excessive vehicular travel, and can apply to any of the regulated substances (e.g., ozone, oxides of sulfur and nitrogen, and heavy metals).

3. POLLUTANTS

A *pollutant* is a material or substance that is accidentally or intentionally introduced to the environment in a quantity that exceeds what occurs naturally. Not all pollutants are toxic or hazardous, but the issue is moot when regulations limiting permissible concentrations are

specific. As defined by regulations in the United States, *hazardous air pollutants* (HAPs), also known as *air toxics*, consist of trace metals (e.g., lead, beryllium, mercury, cadmium, nickel, and arsenic) and other substances for a total of approximately 200 "listed" substances.[2]

Another categorization defined by regulation in the United States separates pollutants into *criteria pollutants* (e.g., sulfur dioxide, nitrogen oxides, carbon monoxide, volatile organic compounds, particulate matter, and lead) and *noncriteria pollutants* (e.g., fluorides, sulfuric acid mist, reduced sulfur compounds, vinyl chloride, asbestos, beryllium, mercury, and other heavy metals).

4. WASTES

Process wastes are generated during manufacturing. *Intrinsic wastes* are part of the design of the product and manufacturing process. Examples of intrinsic wastes are impurities in the reactants and raw materials, by-products, residues, and spent materials. Reduction of intrinsic wastes usually means redesigning the product or manufacturing process. *Extrinsic wastes*, on the other hand, are usually reduced by administrative controls, maintenance, training, or recycling. Examples of extrinsic wastes are fugitive leaks and discharges during material handling, testing, or process shutdown.

Solid wastes are garbage, refuse, biosolids, and containerized solid, liquid, and gaseous wastes.[3] *Hazardous waste* is defined as solid waste, alone or in combination with other solids, that because of its quantity, concentration, or physical, chemical, or infectious characteristics may either (a) cause an increase in mortality or serious (irreversible or incapacitating) illness, or (b) pose a present or future hazard to health or the environment when improperly treated, stored, transported, disposed of, or managed.

A substance is *reactive* if it reacts violently with water; if its pH is less than 2 or greater than 12.5, it is *corrosive*; if its flash point is less than 140°F (60°C), it is *ignitable*. The *toxicity characteristic leaching procedure* (TCLP)[4] test is used to determine if the substance is *toxic*. The TCLP tests for the presence of eight metals (e.g., chromium, lead, and mercury) and 25 organic compounds (e.g., benzene and chlorinated hydrocarbons).

---

[1]Another factor complicating the publication of specific regulations is that the maximum permitted concentrations and emissions depend on the size and nature of the source.

[2]The most dangerous air toxics include asbestos, benzene, cadmium, carbon tetrachloride, chlorinated dioxins and dibenzofurans, chromium, ethylene dibromide, ethylene dichloride, ethylene oxide and methylene chloride, radionuclides, vinyl chloride, and emissions from coke ovens. Most of these substances are carcinogenic.

[3]The term *biosolids* is replacing the term *sludge* when it refers to organic waste produced from biological wastewater treatment processes. Sludge from industrial processes and flue gas cleanup (FGC) devices retains its name.

[4]Probably no subject in this book has more acronyms than environmental engineering. All the acronyms used in this chapter are in actual use; none was invented for the benefit of the chapter.

Hazardous wastes can be categorized by the nature of their sources. *F-wastes* (e.g., spent solvents and distillation residues) originate from nonspecific sources; *K-wastes* (e.g., separator sludge from petroleum refining) are generated from industry-specific sources; *P-wastes* (e.g., hydrogen cyanide) are acutely hazardous discarded commercial products, off-specification products, and spill residues; *U-wastes* (e.g., benzene and hydrogen sulfide) are other discarded commercial products, off-specification products, and spill residues.

Two notable "rules" pertain to hazardous wastes. The *mixture rule* states that any solid waste mixed with hazardous waste becomes hazardous. The *derived from rule* states that any waste derived from the treatment of a hazardous waste (e.g., ash from the incineration of hazardous waste) remains a hazardous waste.[5]

5. POLLUTION SOURCES

A *pollution source* is any facility that produces pollution. A *generator* is any facility that generates hazardous waste. The term "major" (i.e., a *major source*) is defined differently for each class of nonattainment areas.[6]

The combustion of fossil fuels (e.g., coal, fuel oil, and natural gas) to produce steam in electrical generating plants is the most significant source of airborne pollution. For this reason, this industry is among the most highly regulated.

Specific regulations pertain to generators of hazardous waste. Generators of more than certain amounts (e.g., 100 kg/month) must be registered (i.e., with the Environmental Protection Agency (EPA)). Restrictions are placed on generators in the areas of storage, personnel training, shipping, treatment, and disposal.

6. ENVIRONMENTAL IMPACT REPORTS

New installations and large-scale construction projects must be assessed for potential environmental damage before being approved by state and local building officials. The assessment is documented in an *environmental impact report* (EIR). The EIR evaluates all of the potential ways that a project could affect the ecological balance and environment. A report that alleges the absence of any environmental impacts is known as a *negative declaration* ("negative dec").

The following issues are typically addressed in an EIR.

1. the nature of the proposed project

2. the nature of the project area, including distinguishing natural and human-made characteristics

3. current and proposed percentage uses by zoning: residential, commercial, industrial, public, and planned development

4. current and proposed percentage uses by application: built up, landscaped, agricultural, paved streets and highways, paved parking, surface and aerial utilities, railroad, and vacant and unimproved

5. the nature and degree that the earth will be altered: (a) changes in topology from excavation and earth movement to and from the site, (b) changes in slope, (c) changes in chemical composition of the soil, and (d) changes in structural capacity of the soil as a result of compaction, tilling, shoring, or changes in moisture content

6. the nature of known geologic hazards such as earthquake faults and soil instability (subsidence, landslides, and severe erosion)

7. increases in dust, smoke, or air pollutants generated by hauling during construction

8. changes in path, direction, and capacity of natural draining or tendency to flood

9. changes in erosion rates on and off the site

10. the extent to which the project will affect the potential use, extraction, or conservation of the earth's resources, such as crops, minerals, and groundwater

11. after construction is complete, the nature and extent to which the completed project's sewage, drainage, airborne particulate matter, and solid waste will affect the quality and characteristics of the soil, water, and air in the immediate project area and in the surrounding community

12. the quantity and source of fresh water consumed as a result of the project

13. effects on the plant and animal life currently on site

14. effects on any unique (i.e., not found anywhere else in the city, county, state, or nation) natural or human-made features

15. effects on any historically significant or archeological site

16. changes to the view of the site from points around the project

17. changes affecting wilderness use, open space, landscaping, recreational use, and other aesthetic considerations

18. effects on the health and safety of the people in the project area

---

[5]Hazardous waste should not be incinerated with nonhazardous waste, as all of the ash would be considered hazardous by these rules.
[6]A nonattainment area is classified as marginal, moderate, serious, severe, or extreme based on the average pollution (e.g., ozone) level measured in the area.

$$p_s = B_s p_t$$

$$= \frac{(100 \text{ ppm})(30.2 \text{ in Hg})}{\left(29.92 \, \dfrac{\text{in Hg}}{\text{atm}}\right)(10^6 \text{ ppm})}$$

$$= 1.01 \times 10^{-4} \text{ atm}$$

Use Eq. 40.1.

$$\frac{1000}{T_{\text{dp},K}} = 1.7842 + 0.0269 \log_{10}(0.0807 \text{ atm})$$

$$- 0.1029 \log_{10}(1.01 \times 10^{-4} \text{ atm})$$

$$+ 0.0329 \log_{10}(0.0807 \text{ atm})$$

$$\times \log_{10}(1.01 \times 10^{-4} \text{ atm})$$

$$= 2.3097$$

$$T_{\text{dp}} = \frac{1000\text{K}}{2.3097} = 433\text{K} \quad (160°\text{C}, \ 320°\text{F})$$

9. ACID RAIN

Acid rain consists of weak solutions of sulfuric, hydrochloric, and to a lesser extent, nitric acids. These acids are formed when emissions of sulfur oxides (SOx), hydrogen chloride (HCl), and nitrogen oxides (NOx) return to the ground in rain, fog, or snow, or as dry particles and gases. Acid rain affects lakes and streams, damages buildings and monuments, contributes to reduced visibility, and affects certain forest tree species. Acid rain may also represent a health hazard. (See also Sec. 40.8 and Sec. 40.47.)

10. ALLERGENS AND MICROORGANISMS

Allergens such as molds, viruses, bacteria, animal droppings, mites, cockroaches, and pollen can cause allergic reactions in humans. One form of bacteria, *legionella*, causes the potentially fatal *Legionnaires' disease*. Inside buildings, allergens and microorganisms become particularly concentrated in standing water, carpets, HVAC filters and humidifier pads, and in locations where birds and rodents have taken up residence. Legionella bacteria can also be spread by aerosol mists generated by cooling towers and evaporative condensers if the bacteria are present in the recirculating water.

Some buildings cause large numbers of people to simultaneously become sick, particularly after a major renovation or change has been made. This is known as *sick building syndrome* or *building-related illness*. This phenomenon can be averted by using building materials that do not release vapor over time (e.g., as does plywood impregnated with formaldehyde) or harbor other irritants. Carpets can accumulate dusts. Repainting, wallpapering, and installing new flooring can release new airborne chemicals. Areas must be flushed with fresh air until all noticeable effects have been eliminated.

Care must also be taken to ensure that filters in the HVAC system do not harbor microorganisms and are not contaminated by bird or rodent droppings. Air intakes must not be located near areas of chemical storage or parking garages.

11. ASBESTOS

Asbestos is a fibrous silicate mineral material that is inert, strong, and incombustible. Once released into the air, its fibers are light enough to stay airborne for a long time.

Asbestos has typically been used in woven and compressed forms in furnace insulation, gaskets, pipe coverings, boards, roofing felt, and shingles, and has been used as a filler and reinforcement in paint, asphalt, cement, and plastic. Asbestos is no longer banned outright in industrial products. However, regulations, well-publicized health risks associated with cancer and *asbestosis*, and potential liabilities have driven producers to investigate alternatives.

No single product has emerged as a suitable replacement for all asbestos applications. (Almost all replacements are more costly.) *Fiberglass* is an insulator with superior tensile strength but low heat resistance. Fiberglass has a melting temperature of approximately 1000°F (538°C). However, fiberglass treated with hydrochloric acid to leach out most of the silica (SiO_2) can withstand 2000–3500°F (1090–1930°C).

In typical static sealing applications, *aramid fibers* (known by the trade names Kevlar™ and Twaron™) are particularly useful up to approximately 800°F (427°C). However, aramid fibers cannot withstand the caustic, high-temperature environment encountered in curing concrete.

12. BOTTOM ASH

Ash is the residue left after combustion of a fossil fuel. *Bottom ash* (*bottoms ash* or *bottoms*) is the ash that is removed from the combustor after a fuel is burned. (The rest of the ash is fly ash.) The ash falls through the combustion grates into quenching troughs below. It may be continuously removed by *submerged scraper conveyors* (SSCs), screw-type devices, or ram dischargers. The ash can be dewatered to approximately 15% moisture content by compression or by being drawn up a dewatering slope. Bottom ash is combined with conditioned fly ash on the way to the ash storage bunker.[11] Most combined ash is eventually landfilled.

13. DISPOSAL OF ASH

Combined ash (bottom ash and fly ash) is usually landfilled. Other occasional uses for high-quality combustion

---

[11]Approximately half of the electrical generating plants in the United States use wet fly ash handling.

Solid Waste

ash (not ash from the incineration of municipal solid waste) include roadbed subgrades, road surfaces ("asphalt"), and building blocks.

14. CARBON DIOXIDE

Carbon dioxide, though an environmental issue, is not a hazardous material and is not regulated as a pollutant.[12] Carbon dioxide is not an environmental or human toxin, is not flammable, and is not explosive. Skin contact with solid or liquid carbon dioxide presents a freezing hazard. Other than the remote potential for causing frostbite, carbon dioxide has no long-term health effects. Its major hazard is that of asphyxiation, by excluding oxygen from the lungs.

As with other products of combustion, the fraction of carbon dioxide in a flue gas can be arbitrarily reduced by the introduction of dilution air into the flue stream. Therefore, carbon dioxide is reported on a standardized basis—typically as a dry volumetric fraction at some percentage (e.g., 3%) of oxygen.[13] The standardized value can be calculated from stoichiometric relationships. Since there is essentially no carbon dioxide in air (approximately 0.03% by volume), carbon dioxide in a flue gas has a unique source—carbon in the fuel. Knowing the theoretical flue gas composition is sufficient to calculate the standardized value. The analysis is independent of stack temperature.

For a standardized value of 3% oxygen, Table 40.2 gives the dry carbon dioxide volumetric fraction directly, based on the volumetric ratio of hydrogen to carbon in the fuel. For any other percentage of oxygen, the following procedure can be used.

step 1: Obtain the volumetric fuel composition. Gaseous fuel compositions are normally reported on a volumetric basis. Solid and liquid fuels are reported on a weight (gravimetric) basis. Convert weight basis analyses to a volumetric basis by dividing the gravimetric percentages by their respective atomic (or, molecular) weights. Combustion products are gaseous. (For gases, molar and volumetric ratios are the same.)

step 2: Write and balance the stoichiometric combustion equation using the volumetric fractions as coefficients for the fuel elements. Disregard trace emissions (NO, SO_2, CO, etc.) that contribute less than 1% to the flue gas volume, and disregard oxygen contributed by the fuel. Include a variable amount of excess air. Include nitrogen for the combustion and excess air at the ratio of 3.773 volumes of nitrogen for each volume of oxygen.

step 3: Divide the oxygen volume (i.e., the balanced reaction coefficient) by the sum of all flue gas volumes, excluding water vapor, and set this ratio equal to the standardized volumetric oxygen fraction. Solve for excess air.

step 4: Divide the carbon dioxide volume by the sum of all flue gas volumes, excluding water vapor. (The fraction can be multiplied by 10^6 to obtain the volumetric fraction in ppm, though this is seldom done for carbon dioxide.)

The carbon dioxide concentration (mass emission per dry standard volume), C, can be calculated from the carbon dioxide's molecular weight (MW = 44) and Eq. 40.17.

Table 40.2 *Theoretical CO_2 Fraction at 3% Oxygen (dry volumetric basis)*

H/C ratio	CO_2	H/C ratio	CO_2
0	0.18	2.1	0.12723
0.1	0.17651	2.2	0.12548
0.2	0.17816	2.3	0.12378
0.3	0.16993	2.4	0.12212
0.4	0.16682	2.5	0.12050
0.5	0.16382	2.6	0.11893
0.6	0.16093	2.7	0.11740
0.7	0.15814	2.8	0.11590
0.8	0.15544	2.9	0.11445
0.9	0.15283	3.0	0.11303
1.0	0.15031	3.1	0.11165
1.1	0.14787	3.2	0.11029
1.2	0.14551	3.3	0.10898
1.3	0.14323	3.4	0.10768
1.4	0.14101	3.5	0.10643
1.5	0.13886	3.6	0.10520
1.6	0.13678	3.7	0.10400
1.7	0.13476	3.8	0.10283
1.8	0.13279	3.9	0.10168
1.9	0.13089	4.0	0.10056
2.0	0.12903		

Example 40.2

Coal has the following gravimetric composition: carbon, 86.5%; hydrogen, 11.75%; nitrogen (N_2), 0.39%; sulfur, 0.40%; ash, 0.01%; andoxygen (O_2), 0.96%. Determine the theoretical carbon dioxide concentration in parts per million on a dry volumetric (ppmvd) basis at 3% oxygen by volume.

Solution

step 1: Disregarding the combustion of sulfur to SO_2 and other elements present in small quantities, flue gases will be products of carbon and hydrogen combustion and the excess air. The atomic weight of carbon is 12 lbm/lbmol. Since the hydrogen is present in elemental form, not as H_2 gas, the atomic weight is 1 lbm/lbmol. Consider 100 lbm of fuel. The carbon content will be

[12]Industrial exposure is regulated by the U.S. Occupational Safety and Health Administration (OSHA).
[13]The phrases "at 3% O_2" and "at 3% excess air" are not equivalent. The former means that oxygen comprises 3% of the gaseous reaction products by volume. The latter means that 3% more air (3% more oxygen) is provided in reactants than is needed.

PPI • **www.ppi2pass.com**

$(0.865)(100 \text{ lbm}) = 86.5 \text{ lbm}$. The volumetric ratios (number of moles) are

$$\text{C:} \quad \frac{86.5 \text{ lbm}}{12 \frac{\text{lbm}}{\text{lbmol}}} = 7.208 \text{ lbmol}$$

$$\text{H:} \quad \frac{11.75 \text{ lbm}}{1 \frac{\text{lbm}}{\text{lbmol}}} = 11.75 \text{ lbmol}$$

step 2: The unbalanced combustion reaction is

$$7.208\text{C} + 11.75\text{H} + n_1\text{O}_2 + 3.773n_1\text{N}_2$$
$$\rightarrow n_2\text{CO}_2 + n_3\text{H}_2\text{O} + 3.773n_2\text{N}_2$$

The balanced combustion reaction is

$$7.208\text{C} + 11.75\text{H} + 10.146\text{O}_2 + 38.281\text{N}_2$$
$$\rightarrow 7.208\text{CO}_2 + 5.875\text{H}_2\text{O} + 38.281\text{N}_2$$

Let e represent the excess air fraction. All of the excess oxygen and nitrogen will appear in the flue gas.

$$7.208\text{C} + 11.75\text{H} + (1 + e)10.146\text{O}_2$$
$$+ (1 + e)38.281\text{N}_2$$
$$\rightarrow 7.208\text{CO}_2 + 5.875\text{H}_2\text{O}$$
$$+ (1 + e)38.281\text{N}_2 + 10.146e\text{O}_2$$

step 3: At 3% O_2,

$$0.03 = \frac{10.146e}{7.208 + (1 + e)(38.281) + 10.146e}$$

$$e \approx 0.157 \quad (15.7\% \text{ excess air})$$

The balanced combustion reaction at 3% oxygen is

$$7.208\text{C} + 11.75\text{H} + 11.739\text{O}_2 + 44.291\text{N}_2$$
$$\rightarrow 7.208\text{CO}_2 + 5.875\text{H}_2\text{O} + 44.291\text{N}_2 + 1.593\text{O}_2$$

step 4: The theoretical carbon dioxide fraction, on a dry volumetric basis at 3% oxygen, is

$$\frac{7.208}{7.208 + 44.291 + 1.593} = 0.1358 \quad (13.58\%)$$

(This is the same value as obtained from Table 40.2 for a volumetric fuel ratio of $\text{H}/\text{C} = 11.75/7.208 = 1.63$.)

Example 40.3

When the fuel described in Ex. 40.2 is burned, the carbon dioxide in the stack gas is measured on a wet, volumetric basis to be 10.4%. What is this value corrected to 3% O_2?

Solution

Since the theoretical carbon dioxide volumetric fraction can be calculated from the fuel composition, the actual carbon dioxide fraction in the stack is irrelevant. (The amount of excess air could be found from this value, however.) The theoretical carbon dioxide fraction, on a dry volumetric basis at 3% oxygen, is still 13.55%.

15. CARBON MONOXIDE

Carbon monoxide, CO, is formed during incomplete combustion of carbon in fuels. This is usually the result of an oxygen deficiency at lower temperatures. Carbon monoxide displaces oxygen in the bloodstream, so it represents an asphyxiation hazard. Carbon monoxide does not contribute to smog.

The generation of carbon monoxide can be minimized by furnace monitoring and control. For industrial sources, the American Boiler Manufacturers Association (ABMA) recommends limiting carbon monoxide to 400 ppm (corrected to 3% O_2) in oil- and gas-fired industrial boilers. This value can usually be met with reasonable ease. Local ordinances may be more limiting, however.

Most carbon monoxide released in highly populated areas comes from vehicles. Vehicular traffic may cause the CO concentration to exceed regulatory limits. For this reason, *oxygenated fuels* are required to be sold in those areas during certain parts of the year. Oxygenated gasoline has a minimum oxygen content of approximately 2.0%. Oxygen is increased in gasoline with additives such as ethanol (ethyl alcohol). Methyl tertiary butyl ether (MTBE) continues to be used as an oxygenate outside of the United States.

Minimization of carbon monoxide is compromised by efforts to minimize nitrogen oxides. Control of these pollutants is inversely related.

16. CHLOROFLUOROCARBONS

Most atmospheric oxygen is in the form of two-atom molecules, O_2. However, there is a thin layer in the stratosphere about 12 miles up where *ozone* molecules, O_3, are found in large quantities. Ozone filters out ultraviolet radiation that damages crops and causes skin cancer.

Chlorofluorocarbons (i.e., chlorinated fluorocarbons, such as Freon™) contribute to the deterioration of the earth's ozone layer. Ozone in the atmosphere is depleted in a complex process involving pollutants, wind patterns, and atmospheric ice. As chlorofluorocarbon molecules rise through the atmosphere, solar energy breaks the chlorine free. The chlorine molecules attach themselves to ozone molecules, and the new structure eventually decomposes into chlorine oxide and normal oxygen, O_2. The depletion process is particularly pronounced in the Antarctic because that continent's dry, cold air is filled with ice crystals on whose surfaces the chlorine and ozone can combine. Also, the prevailing winter wind isolates and concentrates the chlorofluorocarbons.

Solid Waste

The depletion is not limited to the Antarctic, but occurs as well throughout the northern hemisphere, including virtually all of the United States.

The 1987 Montreal Protocol (conference) resulted in an international agreement to phase out worldwide production of chemical compounds that have ozone-depletion characteristics. Since 2000 (the peak of Antarctic ozone depletion), concentrations of atmospheric chlorine have decreased, and ozone layer recovery is increasing.

Over the years, the provisions of the 1987 Montreal Protocol have been modified numerous times. Substance lists have been amended, and action deadlines have been extended. In many cases, though production may have ceased in a particular country, significant recycling and stockpiling keeps chemicals in use. Voluntary compliance by some nations, particularly developing countries, is spotty or nonexistent. In the United States, provisions have been incorporated into the Clean Air Act (Title VI) and other legislation, but such provisions are subject to constant amendment.

Special allowances are made for aviation safety, national security, and fire suppression and explosion prevention if safe or effective substitutes are not available for those purposes. Excise taxes are used as interim disincentives for those who produce the compounds. Large reserves and recycling, however, probably ensure that chlorofluorocarbons and halons will be in use for many years after the deadlines have passed.

Possible replacements for chlorofluorocarbons (CFCs) include hydrochlorofluorocarbons (HCFCs) and hydrofluorocarbons (HFCs), both of which are environmentally more benign than CFCs, and blends of HCFCs and HFCs. (See Table 40.3.) The additional hydrogen atoms in the molecules make them less stable, allowing nearly all chlorine to dissipate in the lower atmosphere before reaching the ozone layer. The lifetime of HCFC molecules is 2 years to 25 years, compared with 100 years or longer for CFCs. The net result is that HCFCs have only 2–10% of the ozone-depletion ability of CFCs. HFCs have no chlorine and thus cannot deplete the ozone layer.

Most chemicals intended to replace CFCs still have chlorine, but at reduced levels. Additional studies are determining if HCFCs and HFCs accumulate in the atmosphere, how they decompose, and whether any by-products could damage the environment.

17. COOLING TOWER BLOWDOWN

State-of-the art reuse programs in *cooling towers* (CTs) may recirculate water 15 to 20 times before it is removed through blowdown. Pollutants such as metals, herbicides, and pesticides originally in the makeup water are concentrated to five or six times their incoming concentrations. Most CTs are constructed with copper alloy condenser tubes, so the recirculated water becomes contaminated with copper ions as well. CT water also usually contains chlorine compounds or other biocides

Table 40.3 *Typical Replacement Compounds for Chlorofluorocarbons*

designation	applications
HCFC 22	low- and medium-temperature refrigerant; blowing agent; propellant
HCFC 123	replacement for CFC-11; industrial chillers and applications where potential for exposure is low; somewhat toxic; blowing agent; replacement for perchloroethylene (dry cleaning fluid)
HCFC 124	industrial chillers; blowing agent
HFC 134a	replacement for CFC-12; medium-temperature refrigeration systems; centrifugal and reciprocating chillers; propellant
HCFC 141b	replacement for CFC-11 as a blowing agent; solvent
HCFC 142b	replacement for CFC-12 as a blowing agent; propellant
IPC (isopropyl chloride)	replacement for CFC-11 as a blowing agent

added to inhibit biofouling. Ideally, no water should leave the plant (i.e., a *zero-discharge facility*). If discharged, CT blowdown must be treated prior to disposal.

18. CONDENSER COOLING WATER

Approximately half of the electrical generation plants in the United States use *once-through* (OT) *cooling water*. The discharged cooling water may be a chemical or thermal pollutant. Since fouling in the main steam condenser significantly reduces performance, water can be treated by the intermittent addition of chlorine, chlorine dioxide, bromine, or ozone, and these chemicals may be present in residual form. *Total residual chlorine* (TRC) is regulated. Methods of chlorine control include *targeted chlorination* (the frequent application of small amounts of chlorine where needed) and *dechlorination*.

19. DIOXINS

Dioxins are a family of chlorinated dibenzo-*p*-dioxins (CDDs). The term *dioxin*, however, is commonly used to refer to the specific congener 2,3,7,8-tetrachlorodibenzo-*p*-dioxin (TCDD). Primary dioxin sources include herbicides containing 2,4-D, 2,4,5-trichlorophenol, and hexachlorophene. Other potential sources include incinerated municipal and industrial waste, leaded gasoline exhaust, chlorinated chemical wastes, incinerated polychlorinated biphenyls (PCBs), and any combustion in the presence of chlorine.

The exact mechanism of dioxin formation during incineration is complex but probably requires free chlorine (in the form of HCl vapor), heavy metal concentrations (often found in the ash), and a critical temperature window of 570–840°F (300–450°C). Dioxins in incinerators probably form near waste heat boilers, which operate in this temperature range.

Dioxin destruction is difficult because it is a large organic molecule with a high boiling point. Most destruction methods rely on high temperature since temperatures of 1550°F (850°C) denature the dioxins. Other methods include physical immobilization (i.e., vitrification), dehalogenation, oxidation, and catalytic cracking using catalysts such as platinum.

Dioxins liberated during the combustion of municipal solid waste (MSW) and refuse-derived fuel (RDF) can be controlled by the proper design and operation of the furnace combustion system. Once formed, they can be removed by end-of-pipe processes, including activated charcoal (AC) injection. Success has also been reported using the vanadium oxide catalyst used for NOx removal, as well as manganese oxide and tungsten oxide.

20. DUST, GENERAL

Dust or *fugitive dust* is any solid particulate matter (PM) that becomes airborne, with the exception of PM emitted from the exhaust stack of a combustion process. Nonhazardous fugitive dusts are commonly generated when a material (e.g., coal) is unloaded from trucks and railcars. Dusts are also generated by manufacturing, construction, earth-moving, sand blasting, demolition, and vehicle movement.

Dusts pose three types of hazards. (a) Inhalation of airborne dust or vapors, particularly those that carry hazardous compounds, is the major concern. Even without toxic compounds, odors can be objectionable. Dusts are easily observed and can cover cars and other objects left outside. (b) Dusts can transport hazardous materials, contaminating the environment far from the original source. (c) In closed environments, even nontoxic dusts can represent an explosion hazard.

Dusts are categorized by size according to how deep they can penetrate into the respiratory system. Particle size is based on *aerodynamic equivalent diameter* (AED), the diameter of a sphere with the density of water (62.4 lbm/ft³ (1000 kg/m³)) that would have the same settling velocity as the particle. The distribution of dust sizes is divided into three fractions, also known as *conventions*. The *inhalable fraction* (*inhalable convention*) ($< 100 \ \mu m$ AED; $d_{50} = 100 \ \mu m$) can be breathed into the nose and mouth; the *thoracic fraction* ($< 25 \ \mu m$ AED; $d_{50} = 10 \ \mu m$) can enter the larger lung airways; and the *respirable fraction* ($< 10 \ \mu m$ AED; $d_{50} = 4 \ \mu m$) can penetrate beyond terminal bronchioles into the gas exchange regions. (See Fig. 40.1.)

Dust emission reduction from *spot sources* (e.g., manufacturing processes such as grinders) is accomplished by

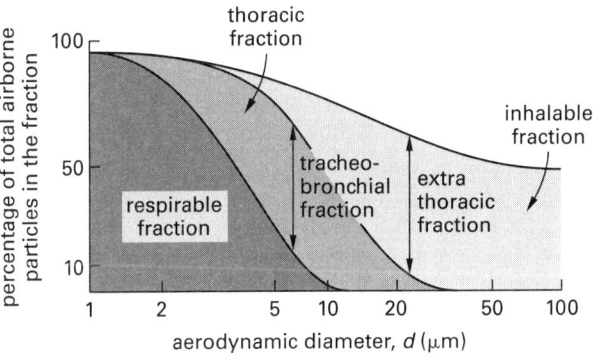

Figure 40.1 Airborne Particle Fractions

inertial separators such as cyclone separators. Potential dust sources (e.g., truck loads and loose piles) can be covered, and dust generation can be reduced by spraying water mixed with other compounds.

There are three mechanisms of dust control by spraying. (a) In *particle capture* (as occurs in a spray curtain at a railcar unloading station), suspended particles are knocked down, wetted, and captured by liquid droplets. (b) In *bulk agglomeration* (as when a material being carried on a screw conveyor is sprayed), the moisture keeps the dust with the material being transported. (c) Spraying roads and coal piles to inhibit wind-blown dust is an example of *surface stabilization*.

Wetting agents are *surfactant* formulations added to water to improve water's ability to wet and agglomerate fine particles.[14] The resulting solution can be applied as liquid spray or as a foam.[15] *Humectant binders* (e.g., magnesium chloride and calcium chloride) and adhesive binders (e.g., waste oil) may also be added to the mixture to make the dust adhere to the contact surface if other water-based methods are ineffective.[16]

Surface stabilization of materials stored outside and exposed to wind, rain, freeze-thaw cycles, and ultraviolet radiation is enhanced by the addition of *crusting agents*.

21. DUST, COAL

Clean coals, western low-sulfur coal, eastern low-sulfur coal, eastern high-sulfur coal, low-rank lignite coal, and varieties in between have their own peculiar handling characteristics.[17] *Dry ultra-fine coal* (DUC) and coal slurries have their own special needs.

---

[14]A *surface-acting agent* (*surfactant*) is a soluble compound that reduces a liquid's surface tension or reduces the interfacial tension between a liquid and a solid.

[15]Collapsible aqueous foam is an increasingly popular means of reducing the potential for explosions in secondary coal crushers.

[16]A *humectant* is a substance that absorbs or retains moisture.

[17]A valuable resource for this subject is NFPA 850, *Recommended Practice for Fire Protection for Electrical Generating Plants and High-Voltage Direct Current Converter Stations*, National Fire Protection Association, Quincy, MA.

Solid Waste

Western coals have a lower sulfur content, but because they are easily fractured, they generate more dust. Western coals also pose higher fire and explosion hazards than eastern coals. Water misting or foam must be applied to coal cars, storage piles, and conveyer transfer points. Adequate ventilation in storage silos and bunkers is also required, with the added benefit of reducing methane accumulation.

DUC is as fine as talcum (10 μm with less than 10% moisture).[18] It must be containerized for transport, because traditional railcars are not sufficiently airtight. Also, since the minimum oxygen content for combustion of 14–15% is satisfied by atmospheric air, DUC should be maintained in a pressurized, oxygen-depleted environment until used. Pneumatic flow systems are used to transport DUC through the combustion plant.

22. FUGITIVE EMISSIONS

Equipment leaks from plant equipment are called *fugitive emissions* (FEs). Leaks from pump and valve seals are common sources, though compressors, pressure-relief devices, connectors, sampling systems, closed vents, and storage vessels are also potential sources. FEs are reduced administratively by *leak detection and repair programs* (LDARs).

Some common causes of fugitive emissions include (a) equipment in poor condition, (b) off-design pump operation, (c) inadequate seal characteristics, and (d) inadequate boiling-point margin in the seal chamber. Other pump/shaft/seal problems that can increase emissions include improper seal-face material, excessive seal-face loading and seal-face deflections, and improper pressure balance ratio.

Inadequate *boiling-point margin* (BPM) results in poor seal performance and face damage. BPM is the difference between the seal chamber temperature and the initial boiling point temperature of the pumped product at the seal chamber pressure. When seals operate close to the boiling point, seal-generated heat can cause the pumped product between the seal faces to flash and the seal to run dry. A minimum BPM of 15°F (9°C) or a 25 psig (172 kPa) pressure margin is recommended to avoid flashing. Even greater margins will result in longer seal life and reduced emissions.

23. GASOLINE-RELATED POLLUTION

Gasoline-related pollution is primarily in the form of unburned hydrocarbons, nitrogen oxides, lead, and carbon monoxide. (The requirement for lead-free automobile gasoline has severely curtailed gasoline-related lead pollution. Low-lead aviation fuel remains in use.) Though a small reduction in gasoline-related pollution can be achieved by blending detergents into gasoline,

reformulation is required for significant improvements. Table 40.4 lists typical characteristics for traditional and reformulated gasolines.

The potential for smog formation can be reduced by reformulating gasoline and/or reducing the summer *Reid vapor pressure* (RVP) of the blend. Regulatory requirements to reduce the RVP can be met by using less butane in the gasoline. However, reformulating to remove pentanes is required when the RVP is required to achieve 7 psig (48 kPa) or below. Sulfur content can be reduced by pretreating the refinery feed in hydrodesulfurization (HDS) units. Heavier gasoline components must be removed to lower the 50% and 90% *distillation temperatures* (T50 and T90).

Table 40.4 Typical Gasoline Characteristics

fuel parameter	traditional	reformulated
sulfur	150 ppmw[*]	40 ppmw
benzene	2% by vol	1% by vol
olefins	9.9% by vol	4% by vol
aromatic hydrocarbons	32% by vol	25% by vol
oxygen	0	1.8 to 2.2% by wt
T90	330°F (166°C)	300°F (149°C)
T50	220°F (104°C)	210°F (99°C)
RVP	8.5 psig (59 kPa)	7 psig (48 kPa)

(Multiply psi by 6.89 to obtain kPa.)
[*]ppmw stands for parts per million by weight (same as ppmm, parts per million by mass).

24. GLOBAL WARMING

The *global warming* theory is that increased levels of atmospheric carbon dioxide, CO_2, from the combustion of carbon-rich fossil fuels and other *greenhouse gases* (e.g., water vapor, methane, nitrous oxide, and chlorofluorocarbons) trap an increasing amount of solar radiation in a *greenhouse effect*, gradually increasing the earth's temperature.

Recent studies have shown that atmospheric carbon dioxide has increased at the rate of about 0.4% per year since 1958. (For example, in one year carbon dioxide might increase from 390 ppmv to 392 ppmv. By comparison, oxygen is approximately 209,500 ppmv.) There has also been a 100% increase in atmospheric methane since the beginning of the industrial revolution. However, increases in carbon dioxide have not correlated conclusively with surface temperature change. According to most researchers, there has been a global temperature increase of approximately 1.6°F (0.91°C) during the past 50 years.[19] However, the 2014 NOAA/NASA *GISS Surface Temperature Analysis* data noted that the 5-year mean surface temperature has not changed in more than ten years. Other studies have shown that temperatures since 2005 are not well-predicted by climate models, which produced higher results.

---

[18]A μm is a *micrometer*, 10^{-6} m, and is commonly referred to as a *micron*.

[19]Other researchers detect no discernible upward trend.

In addition to a temperature increase, other evidence of the global warming theory are record-breaking hot summers, widespread aberrations in the traditional seasonal weather patterns (e.g., hurricane-like storms in England), an increase in the length of the arctic melt season, melting of glaciers and the ice caps, and an average 7 in (18 cm) rise in sea level over the last century.[20] It has been predicted that there will be a temperature increase of 2–11.5°F (1.1–6.4°C) by the year 2100.

Although global warming is generally accepted, its anthropogenic (human-made) causes are not. The global warming theory is disputed by some scientists and has not been proven to be an absolute truth. Arguments against the theory center around the fact that manufactured carbon dioxide is a small fraction of what is naturally released (e.g., by wetlands, in rain forest fires, and during volcanic eruptions). It is argued that, in the face of such massive contributors, and since the earth's temperature has remained essentially constant for millenia, the earth already possesses some built-in mechanism, currently not perfectly understood, that reduces the earth's temperature swings.

Most major power generation industries have adopted goals of reducing carbon dioxide emissions. However, efforts to reduce carbon dioxide emissions by converting fossil fuel from one form to another are questioned by many engineers. Natural gas produces the least amount of carbon dioxide of any fossil fuel. Therefore, conversion of coal to a gas or liquid fuel in order to lower the carbon-to-hydrogen ratio would appear to lessen carbon dioxide emissions at the point of final use. However, the conversion processes consume energy derived from carbon-containing fuel. This additional consumption, taken over all sites, results in a net increase in carbon dioxide emission of 10–200%, depending on the process.

The use of ethanol as an alternative for gasoline is also problematic. Manufacturing processes that produce ethanol give off (at least) twice as much carbon dioxide as the gasoline being replaced produces during combustion.

Most synthetic fuels are intrinsically less efficient (based on their actual heating values compared with those theoretically obtainable from the fuels' components in elemental form). This results in an increase in the amount of fuel consumed. Thus, fossil fuels should be used primarily in their raw forms until cleaner sources of energy are available.

25. LANDFILL GAS

Most closed landfills are covered by a thick soil layer. As the covered refuse decays beneath the soil, the natural anaerobic biological reaction generates a low-Btu *landfill gas* (LFG) consisting of approximately 50% methane, carbon dioxide, and trace amounts of other gases. If uncontrolled, LFG migrates to the surface. If the LFG accumulates, an explosion hazard results. If it escapes, other environmental problems (including objectionable odors) occur. Therefore, synthetic and compacted clay liners and clay trenches are used to prevent gas from spreading laterally. Wells and collection pipes are used to collect and incinerate LFG in flares. However, emissions from flaring are also problematic.

Alternatives to flaring include using the LFG to produce hot water or steam for heating or electricity generation. During the 1980s, reciprocating engines and combustion turbines powered by LFG were tried. However, such engines generated relatively high emissions of their own due to impurities and composition variations in the fuel. True Rankine-cycle power plants (generally without reheat) avoid this problem, since boilers are less sensitive to impurities.

One problem with using LFG commercially is that LFG is withdrawn from landfills at less than atmospheric pressure. Conventional furnace burners need approximately 5 psig at the boiler front. Low-pressure burners requiring 2 psig are available, though expensive. Therefore, some of the plant power must be used to pressurize the LFG.

Although production is limited, LFG is produced for a long period after a landfill site is closed. Production slowly drops 3% to 5% annually to approximately 30% of its original value after about 20 to 25 years, which is considered to be the economical life of a gas-reclamation system. Approximately 40 ft³ of gas will be produced per cubic yard (1.5 m³ of gas per cubic meter) of landfill.

26. LANDFILL LEACHATE

Leachates are liquid wastes containing dissolved and finely suspended solid matter and microbial waste produced in landfills. Leachate becomes more concentrated as the landfill ages. Leachate forms from liquids brought into the landfill, water run-on, and precipitation. Leachate in a natural attenuation landfill will contaminate the surrounding soil and groundwater. In a lined containment landfill, leachate will percolate downward through the refuse and collect at the first landfill liner. Leachate must be removed to reduce hydraulic head on the liner and to reduce unacceptable concentrations of hazardous substances.

When a layer of liquid sludge is disposed of in a landfill, the consolidation of the sludge by higher layers will cause the water to be released. This released water is known as *pore-squeeze liquid*.

Some water will be absorbed by the MSW and will not percolate down. The quantity of water that can be held against the pull of gravity is referred to as the *field capacity*, FC. The potential quantity of leachate is the amount of moisture within the landfill in excess of the FC.

---

[20]The rise in the level of the oceans is generally accepted though the causes are disputed. Satellite altimetry indicates a current rise rate of 12 in (30 cm) per century.

Solid Waste

In general, the amount of leachate produced is directly related to the amount of external water entering the landfill. Theoretically, the leachate generation rate can be determined by writing a water mass balance on the landfill. This can be done on a preclosure and a post-closure basis. Typical units for all the terms are units of length (e.g., "1.2 in of rain") or mass per unit volume (e.g., "a field capacity of 4 lbm/yd^3").

The preclosure leachate generation rate is

preclosure leachate generation
  = moisture released by incoming waste,
      including pore-squeezed liquid
  + precipitation
  − moisture lost due to evaporation
  − field capacity   40.2

The postclosure water balance is

postclosure leachate generation
  = precipitation
    − surface runoff
    − evapotranspiration
    − moisture lost in formation
        of landfill gas and other
        chemical compounds
    − water vapor removed along
        with landfill gas
    − change in soil moisture storage   40.3

27. LEACHATE MIGRATION FROM LANDFILLS

From Darcy's law, migration of leachate contaminants that have passed through liners into aquifers or the groundwater table is proportional to the hydraulic conductivity, K, and the hydraulic gradient, i. Hydraulic conductivities of clay liners are 1.2×10^{-7} ft/hr to 1.2×10^{-5} ft/hr (10^{-9} cm/s to 10^{-7} cm/s). However, the properties of clay liners can change considerably over time due to interactions with materials in the landfill. If the clay dries out (desiccates), it will be much more permeable. For synthetic FMLs, hydraulic conductivities are 1.2×10^{-10} ft/hr to 1.2×10^{-7} ft/hr (10^{-12} cm/s to 10^{-9} cm/s). The average permeability of high-density polyethylene is approximately 1.2×10^{-11} ft/hr (1×10^{-13} cm/s).

$$Q = KiA \qquad 40.4$$

$$i = \frac{dH}{dL} \qquad 40.5$$

28. GROUNDWATER DEWATERING

It may be possible to prevent or reduce contaminant migration by reducing the elevation of the groundwater table (GWT). This is accomplished by dewatering the soil with relief-type *extraction drains* (*relief drains*). The *ellipse equation*, also known as the *Donnan formula* and the *Colding equation*, used for calculating pipe spacing, L, in draining agricultural fields, can be used to determine the spacing of groundwater dewatering systems. In Eq. 40.6, K is the hydraulic conductivity with units of length/time, a is the distance between the pipe and the impervious layer barrier (a is zero if the pipe is installed on the barrier), b is the maximum allowable table height above the barrier, and Q is the *recharge rate*, also known as the *drainage coefficient*, with dimensions of length/time. The units of K and Q must be on the same time basis.

$$L = 2\sqrt{\left(\frac{K}{Q}\right)(b^2 - a^2)} = 2\sqrt{\frac{b^2 - a^2}{i}} \qquad 40.6$$

Equation 40.6 is often used because of its simplicity, but the accuracy is only approximately ±20%. Therefore, the calculated spacing should be decreased by 10–20%.

For pipes above the impervious stratum, as illustrated in Fig. 40.2, the total discharge per unit length of each pipe (in ft^3/ft-sec or m^3/m·s) is given by Eq. 40.7. H is the maximum height of the water table above the pipe invert elevation. D is the average depth of flow.

$$Q_{\text{unit length}} = \frac{2\pi KHD}{L} \qquad 40.7$$

$$D = a + \frac{H}{2} \qquad 40.8$$

For pipes on the impervious stratum, the total discharge per unit length from the ends of each pipe (in ft^3/ft-sec or m^3/m·s) is

$$Q_{\text{unit length}} = \frac{4KH^2}{L} \qquad 40.9$$

Equation 40.10 gives the total discharge per pipe. If the pipe drains from both ends, the discharge per end would be half of that amount.

$$Q_{\text{pipe}} = LQ_{\text{unit length}} \qquad 40.10$$

Example 40.4

Subsurface leachate migration from a landfill is to be mitigated by maintaining the water table that surrounds the landfill site lower than the natural level. The surrounding area consists of 15 ft of saturated, homogeneous soil over an impervious rock layer. The water table, originally at the surface, has been lowered to a depth of 9 ft. Fully pervious parallel collector drains at a depth of 12 ft are present. The hydraulic conductivity of the soil is 0.23 ft/hr. The natural water table will recharge the site in 30 days if drainage stops. What collector drain separation is required?

Solution

Referring to Fig. 40.2(a),

$$a = 15 \text{ ft} - 12 \text{ ft} = 3 \text{ ft}$$

$$b = 15 \text{ ft} - 9 \text{ ft} = 6 \text{ ft}$$

$$K = \left(0.23 \ \frac{\text{ft}}{\text{hr}}\right)\left(24 \ \frac{\text{hr}}{\text{day}}\right) = 5.52 \text{ ft/day}$$

To maintain the lowered water table level, the drainage rate must equal the recharge rate.

$$Q = \frac{9 \text{ ft}}{30 \text{ days}} = 0.3 \text{ ft/day}$$

Use Eq. 40.6 to find the drain spacing.

$$L = 2\sqrt{\left(\frac{K}{Q}\right)(b^2 - a^2)}$$

$$= 2\sqrt{\left(\frac{5.52 \ \dfrac{\text{ft}}{\text{day}}}{0.3 \ \dfrac{\text{ft}}{\text{day}}}\right)\left((6 \text{ ft})^2 - (3 \text{ ft})^2\right)}$$

$$= 44.57 \text{ ft} \quad (45 \text{ ft})$$

29. LEACHATE RECOVERY SYSTEMS

At least two distinct leachate recovery systems are required in landfills: one within the landfill to limit the hydraulic head of leachate that has reached the top liner, and another to catch the leachate that has passed through the top liner and drainage layer and has reached the bottom liner.

By removing leachate at the first liner, the hydrostatic pressure on the liner is reduced, minimizing the pressure gradient and hydraulic movement through the liner. A pump is used to raise the collected leachate to the surface once a predetermined level has been reached. Tracer compounds (e.g., lithium compounds or radioactive hydrogen) can be buried with the wastes to signal migration and leakage.

Leachate collection and recovery systems fail because of clogged drainage layers and pipe, crushed collection pipes due to waste load, pump failures, and faulty design.

Figure 40.2 Geometry for Groundwater Dewatering Systems

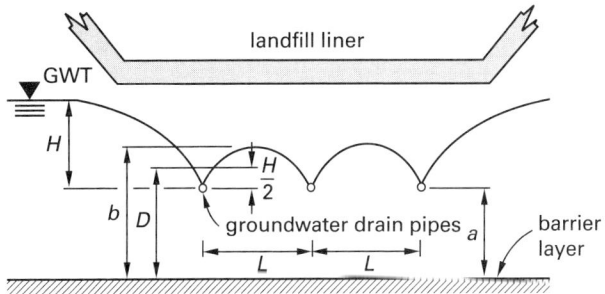

(a) pipes above impervious stratum

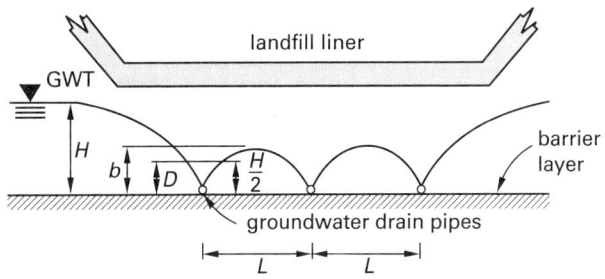

(b) pipes on impervious stratum

30. LEACHATE TREATMENT

Leachate is essentially a very strong municipal wastewater, and it tends to become more concentrated as the landfill ages. Leachate from landfills contains extremely high concentrations of compounds and cannot be discharged directly into rivers or other water sources.

Leachate is treated with biological (i.e., trickling filter and activated sludge) and physical/chemical processes very similar to those used in wastewater treatment plants. For large landfills, these treatment facilities are located on the landfill site. A typical large landfill treatment facility would include an equalization tank, a primary clarifier, a first-stage activated sludge aerator and clarifier, a second-stage activated sludge aerator and clarifier, and a rapid sand filter. Additional equipment for sludge dewatering and digestion would also be required. Liquid effluent would be discharged to the municipal wastewater treatment plant.

31. LEAD

Lead, even in low concentrations, is toxic. Inhaled lead accumulates in the blood, bones, and vital organs. It can produce stomach cramps, fatigue, aches, and nausea. It causes irreparable damage to the brain, particularly in young and unborn children, and high blood pressure in adults. At high concentrations, lead damages the nervous system and can be fatal.

Solid Waste

Lead was outlawed in paint in the late 1970s.[21] Lead has also been removed from most gasoline blends. Lead continues to be used in large quantities in automobile batteries and plating and metal-finishing operations. However, these manufacturing operations are tightly regulated. Lead enters the atmosphere during the combustion of fossil fuels and the smelting of sulfide ore. Lead enters lakes and streams primarily from acid mine drainage.

For lead in industrial and municipal wastewater, current remediation methods include pH adjustment with lime or alkali hydroxides, coagulation-sedimentation, reverse osmosis, and zeolite ion exchange.

32. MUNICIPAL SOLID WASTE

Municipal solid waste (MSW, previously known as *garbage*) has traditionally been disposed of in *municipal solid waste landfills* (MSWLs, previously referred to as *dumps*).[22] Waste is placed in layers, typically 2 ft to 3 ft thick (0.6 m to 0.9 m), and is compacted to the cell height before soil is added as a cover. Two to five passes by a tracked bulldozer are sufficient to compact the MSW to 800 lbm/yd^3 to 1500 lbm/yd^3 (470 kg/m^3 to 890 kg/m^3). (A density of 1000 lbm/ft^3 or 590 kg/m^3 is used in design studies.) The *lift* is the height of the covered layer, as shown in Fig. 40.3. When the landfill layer is full, the ratio of solid waste volume to soil cover volume will be approximately between 4:1 and 3:1.

Figure 40.3 Landfill Cells

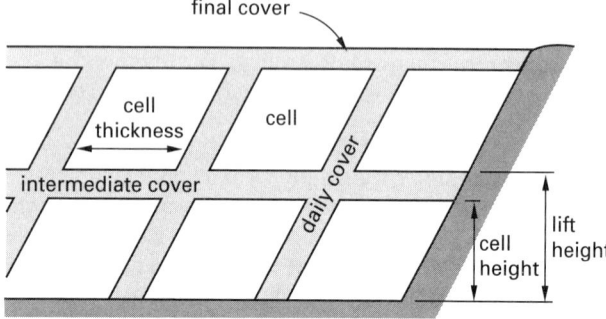

The *cell height* is typically taken as 8 ft (2.4 m) for design studies, although it can actually be much higher. The height should be chosen to minimize the cover material requirement consistent with the regulatory requirements. Cell slopes will be less than 40°, and typically less than 30°.

MSW is generated in the United States at the average rate of about 5 lbm/capita-day to 8 lbm/capita-day

(2.3 kg/capita-day to 3.6 kg/capita-day).[23] (Characteristics of MSW in the United States are given in Table 40.5.) The daily increase in landfill volume is predicted by Eq. 40.11.[24] N is the population size, G is the per-capita MSW generation rate, and γ is the landfill overall specific weight, calculated as a weighted average of soil and compacted MSW specific weights.[25]

$$\Delta V_{\text{day}} = \frac{NG(\text{LF})}{\gamma} \qquad 40.11$$

Table 40.5 Typical Nationwide Characteristics of MSW

component	percentage by weight
paper	40
yard waste	18
glass	8
plastic	7
steel	7
food waste	6
aluminum	2
other	12

The *loading factor*, LF, is 1.25 for a 4:1 volumetric ratio and is calculated from for other ratios.

$$\text{LF} = \frac{V_{\text{MSW}} + V_{\text{cover soil}}}{V_{\text{MSW}}} \qquad 40.12$$

Suitable (economical and safe) landfill sites are becoming difficult to find, and once identified, are objected to by residents near the site and in the MSW transport corridor. This is referred to as the *NIMBY* (not in my backyard) *syndrome*.[26]

33. NITROGEN OXIDES

Nitrogen oxides (NOx) are one of the primary causes of smog formation. NOx from the combustion of coal is primarily *nitric oxide* (NO) with small quantities of *nitrogen dioxide* (NO$_2$).[27] NO$_2$ can be a primary or a secondary pollutant. Although some NO$_2$ is emitted directly from combustion sources, it is also produced from the oxidation of nitric oxide, NO, in the atmosphere.

NOx is produced in two ways: (a) *thermal NOx* produced at high temperatures from free nitrogen and

[21]Workers can be exposed to lead if they strip away lead-based paint, demolish old buildings, or weld or cut metals that have been coated with lead-based paint.
[22]In 1988, 73% of the nation's MSW was landfilled, 14% was incinerated, and 13% was recycled. In 2010, 54% was landfilled, 12% was incinerated with energy recovery, and 34% was recycled.

[23]In 2010, the EPA reported that the generation rate was 4.43 lbm/capita-day. 5 lbm/capita-day is a conservative estimate commonly used in design studies.
[24]Multiply cubic yards by 27/43,560 to get acre-feet (ac-ft).
[25]Soils have specific weights between 70 lbf/ft^3 and 130 lbf/ft^3 (densities between 1100 kg/w^3 and 2100 kg/m^3). A value of 100 lbf/ft^3 (1600 kg/m^3) is commonly used in design studies.
[26]In fact, NIMBY applies to landfills, incinerators, and any commercial or processing plant dealing with hazardous wastes.
[27]Other oxides are produced in insignificant amounts. These include *nitrous oxide* (N$_2$O), N$_2$O$_4$, N$_2$O$_5$, and NO$_3$, all of which are eventually oxidized to (and reported as) NO$_2$.

oxygen, and (b) *fuel NOx* (or *fuel-bound NOx*) formed from the decomposition/combustion of fuel-bound nitrogen.[28] When natural gas and light distillate oil are burned, almost all of the NOx produced is thermal. Residual fuel oil, however, can have a nitrogen content as high as 0.3%, and 50–60% of this can be converted to NOx. Coal has an even higher nitrogen content.[29]

Thermal NOx is usually produced in small (but significant) quantities when excess oxygen is present at the highest temperature point in a furnace, such that nitrogen (N_2) can dissociate.[30] Dissociation of N_2 and O_2 is negligible, and little or no thermal NOx is produced below approximately 3000°F (1650°C).

Formation of thermal NOx is approximately exponential with temperature. For this reason, many NOx-reduction techniques attempt to reduce the *peak flame temperature* (PFT). NOx formation can also be reduced by injecting urea or ammonia reagents directly into the furnace.[31] The relationship between NOx production and excess oxygen is complex, but NOx production appears to vary directly with the square root of the oxygen concentration.

In existing plants, retrofit NOx-reduction techniques include fuel-rich combustion (i.e., staged air burners), flue gas recirculation, changing to a low-nitrogen fuel, reduced air preheat, installing low-NOx burners, and using overfire air.[32] Low-NOx burners using controlled flow/split flame or internal fuel staging technology are essentially direct replacement (i.e., "plug in") units, differing from the original burners primarily in nozzle configuration. However, some fuel supply and air modifications may also be needed. Use of overfire air requires windbox modifications and separate ducts.

Lime spray-dryer scrubbers of the type found in electrical generating plants do not remove all of the NOx. Further reduction requires that the remaining NOx be destroyed. Reburn, selective catalytic reduction (SCR), and selective noncatalytic methods are required.

Scrubbing, incineration, and other end-of-pipe methods typically have been used to reduce NOx emissions from stationary sources such as gas turbine-generators, although these methods are unwieldy for the smallest units. Water/steam injection, catalytic combustion, and *selective catalytic reduction* (SCR) are particularly suited for gas-turbine and combined-cycle installations. SCR is also effective in NOx reduction in all heater and boiler applications.

The volumetric fraction (in ppm) of NOx in the flue gas is calculated from the molecular weight and mass flow rates. The molecular weight of NOx is assumed to be 46.[33] The ratio of $\dot{m}_{NOx}/\dot{m}_{flue\,gas}$, percentage of water vapor, and flue gas molecular weight are derived from flue gas analysis.

$$B_{NOx} = \left(\frac{\dot{m}_{NOx}}{\dot{m}_{flue\,gas}}\right)(10^6)\left(\frac{MW_{flue\,gas}}{MW_{NOx}}\right)$$
$$\times \left(\frac{100\%}{100\% - \%H_2O}\right) \qquad 40.13$$

Since the apparent concentration of NOx could be arbitrarily decreased without reducing the NOx production rate by simply diluting the combustion gas with excess air, it is common to correct NOx readings to 3% O_2 by volume, dry basis. (This is indicated by the units *ppmvd*.) Standardizing is accomplished by multiplying the measured NOx reading (in ppmvd) by the O_2 correction factor from Eq. 40.14. (Corrections to 7%, 12%, and 15% oxygen content are also used. For a correction to any other percentage of O_2, replace the 3% in Eq. 40.14 with the new value.)

$$O_2 \text{ correction factor} = \frac{21\% - 3\%}{21\% - \%O_2}$$
$$= \frac{18\%}{21\% - \%O_2} \qquad 40.14$$

If the flue gas analysis is on a wet basis, Eq. 40.15 can be used to calculate the multiplicative factor.

$$O_2 \text{ correction factor} = \frac{21\% - 3\%}{21\% - \left(\frac{100\%}{100\% - \%H_2O}\right)(\%O_2)}$$
$$= \frac{18\%}{21\% - \left(\frac{100\%}{100\% - \%H_2O}\right)(\%O_2)} \qquad 40.15$$

Some regulations specify NOx limitations in terms of pounds per hour or in terms of pounds of NOx per million Btus (MMBtu) of gross heat released. The mass emission rate, E, in lbm/MMBtu, is calculated from the

[28]Some engineers further divide the production of fuel-bound NOx into low- and high-temperature processes and declare a third fuel-related NOx-production method known as *prompt-NOx*. Prompt-NOx is the generation of the first 15–20 ppm of NOx from partial combustion of the fuel at lower temperatures.
[29]In order of increasing fuel-bound NOx production potential, common boiler fuels are: methanol, ethanol, natural gas, propane, butane, ultralow-nitrogen fuel oil, fuel oil No. 2, fuel oil No. 6, and coal.
[30]The *mean residence time* at the high temperature points of the combustion gases is also an important parameter. At temperatures below 2500°F (1370°C), several minutes of exposure may be required to generate any significant quantities of NOx. At temperatures above 3000°F (1650°C), dissociation can occur in less than a second. At the highest temperatures—3600°F (1980°C) and above—dissociation takes less than a tenth of a second.
[31]*Urea* (NH_2CONH_2), also known as *carbamide urea*, is a water-soluble organic compound prepared from ammonia. Urea has significant biological and industrial usefulness.
[32]Air is injected into the furnace at high velocity over the combustion bed to create turbulence and to provide oxygen.

[33]46 is the molecular weight of NO_2. The predominant oxide in the flue gas is NO, and NO_2 may be only 10% to 15% of the total NOx. Furthermore, NO is measured, not NO_2. However, NO has a short half-life and is quickly oxidized to NO_2 in the atmosphere.

concentration in pounds per dry, standard cubic foot (dscf).

$$E_{lbm/MMBtu} = \frac{C_{lbm/dscf} V_{NOx,dscf/hr}}{q_{MMBtu/hr}}$$

$$= \frac{C_{lbm/dscf}(100) V_{th,CO_2}}{C_{m,CO_2,\%}(HHV_{MMBtu/lbm})} \qquad 40.16$$

The relationship between the NOx concentrations, C, expressed in pounds per dry standard volume and ppm can be calculated from Eq. 40.17.[34] Conversions between ppm and pounds are made assuming NOx has a molecular weight of 46.

$$C_{g/m^3} = (4.15 \times 10^{-5}) C_{ppm}(MW) \qquad [SI] \quad 40.17(a)$$

$$C_{lbm/ft^3} = (2.59 \times 10^{-9}) C_{ppm}(MW) \qquad [U.S.] \quad 40.17(b)$$

The theoretical volume of carbon dioxide, V_{th,CO_2}, produced can be determined stoichiometrically. However, Table 40.6 can be used for quick estimates with an accuracy of approximately ±5.9%.

Table 40.6 Approximate CO_2 Production for Various Fuels (with 0% excess air)

fuel	standard* ft³/10⁶ Btu	standard* m³/10⁶ cal
coal, anthracite	1980	0.222
coal, bituminous	1810	0.203
coal, lignite	1810	0.203
gas, butane	1260	0.412
gas, natural	1040	0.117
gas, propane	1200	0.135
oil	1430	0.161

*Standard conditions are 70°F (21°C) and 1 atm pressure.

Equation 40.16 is based on the theoretical volume of carbon dioxide gas produced per pound of fuel. Other approximations and correlations are based on the total dry volume of the flue gas in standard cubic feet per million Btu at 3% oxygen, $V_{t,dry}$. For quick estimates on furnaces burning natural gas, propane, and butane, V_t is approximately 10,130 ft³/MMBtu; for fuel oil, V_t is approximately 10,680 ft³/MMBtu.

$$E_{lbm/MMBtu} = (C_{ppmv\,at\,3\%\,O_2})(V_{t,dry})$$
$$\times \left(\frac{46 \frac{lbm}{lbmol}}{\left(379.3 \frac{ft^3}{lbmol}\right)(10^6)}\right) \qquad 40.18$$

As with all pollutants, the maximum allowable concentration or discharge of NOx is subject to continuous review and revision. Actual limits may depend on the

[34]The constants in Eq. 40.17 are the same and can be used for any gas.

type of geographical location, type of fuel, size of the facility, and so on. For steam/electrical (gas turbine) plants, the general target is 25 ppm to 40 ppm. The lower values apply to combustion of natural gas, and the higher values apply to combustion of distillate oil. Even lower values (down to 9 ppm to 10 ppm) are imposed in some areas.

Example 40.5

A combustion turbine produces 25 lbm/hr of NOx while generating 550,000 lbm/hr of exhaust gases. The exhaust gas contains 10% water vapor and 11% oxygen by volume. Assume the molecular weight of NOx is 46 lbm/lbmol. What is the NOx concentration in ppm, dry volume basis, corrected to 3% oxygen?

Solution

Some of the gas analysis is missing, so the flue gas is assumed to be mostly nitrogen (the largest component of air) with a molecular weight of 28 lbm/lbmol.

The uncorrected, dry NOx volumetric fraction is

$$B_{NOx} = \left(\frac{\dot{m}_{NOx}}{\dot{m}_{flue\,gas}}\right)$$
$$\times \left(\frac{MW_{flue\,gas}}{MW_{NOx}}\right)(10^6)\left(\frac{100\%}{100\% - \%H_2O}\right)$$
$$= \left(\frac{25 \frac{lbm}{hr}}{550,000 \frac{lbm}{hr}}\right)\left(\frac{28 \frac{lbm}{lbmol}}{46 \frac{lbm}{lbmol}}\right)$$
$$\times (10^6)\left(\frac{100\%}{100\% - 10\%}\right)$$
$$= 30.7 \text{ ppmvd} \quad [uncorrected]$$

Equation 40.14 corrects the value to 3% oxygen.

$$O_2 \text{ correction factor} = \frac{21\% - 3\%}{21\% - \%O_2}$$
$$= \frac{18\%}{21\% - 11\%}$$
$$= 1.8$$

The corrected value is

$$C = (1.8)(30.7 \text{ ppmvd}) = 55.3 \text{ ppmvd}$$

34. ODORS

Odors of unregulated substances can be eliminated at their source, contained by sealing and covering, diluted to unnoticeable levels with clean air, or removed by simple water washing, chemical scrubbing (using acid, alkali, or sodium hypochlorite), bioremediation, and activated carbon adsorption.

35. OIL SPILLS IN NAVIGABLE WATERS

Intentional and accidental releases of oil in navigable waters are prohibited. Deleterious effects of such spills include large-scale biological (i.e., sea life and wildlife) damage and destruction of scenic and recreational sites. Long-term toxicity can be harmful for microorganisms that normally live in the sediment.

36. OZONE

Ground-level ozone is a secondary pollutant. Ozone is not usually emitted directly, but is formed from hydrocarbons and nitrogen oxides (NO and NO_2) in the presence of sunlight. *Oxidants* are by-products of reactions between combustion products. Nitrogen oxides react with other organic substances (e.g., hydrocarbons) to form the oxidants ozone and peroxyacyl nitrates (PAN) in complex *photochemical reactions*. Ozone and PAN are usually considered to be the major components of smog.[35] (See also Sec. 40.44.)

37. PARTICULATE MATTER

Particulate matter (PM), also known as *aerosols*, is defined as all particles that are emitted by a combustion source. Particulate matter with aerodynamic diameters of less than or equal to a nominal 10 μm is known as *PM-10*. Particulate matter is generally inorganic in nature. It can be categorized into metals (or heavy metals), acids, bases, salts, and nonmetallic inorganics.

Metallic inorganic PM from incinerators is controlled with baghouses or electrostatic precipitators (ESPs), while nonmetallic inorganics are removed by scrubbing (wet absorption). Flue gas PM, such as fly ash and lime particles from desulfurization processes, can be removed by fabric baghouses and electrostatic precipitators. These processes must be used with other processes that remove NOx and SO_2.

High temperatures cause the average flue gas particle to decrease in size toward or below 10 μm. With incineration, emission of trace metals into the atmosphere increases significantly. Because of this, incinerators should not operate above 1650°F (900°C).[36]

38. PCBs

Polychlorinated biphenyls (PCBs) are organic compounds (i.e., *chlorinated organics*) manufactured in oily liquid and solid forms through the late 1970s, and subsequently prohibited. PCBs are carcinogenic and can cause skin lesions and reproductive problems. PCBs

build up, rather than dissipate, in the food chain, accumulating in fatty tissues. Most PCBs were used as dielectric and insulating liquids in large electrical transformers and capacitors, and in ballasts for fluorescent lights (which contain capacitors). PCBs were also used as heat transfer and hydraulic fluids, as dye carriers in carbonless copy paper, and as plasticizers in paints, adhesives, and caulking compounds.

Incineration of PCB liquids and PCB-contaminated materials (usually soil) has long been used as an effective mediation technique. Removal and landfilling of contaminated soil is expensive and regulated, but may be appropriate for quick clean-ups. In addition to incineration and other thermal destruction processes, methods used to routinely remediate PCB-laden soils include biodegradation, ex situ thermal desorption and soil washing, in situ vaporization by heating or steam injection and subject vacuum extraction, and stabilization to prevent leaching. The application of quicklime (CaO) or high-calcium fly ash is now known to be ineffective.

Specialized PCB processes targeted at cleaning PCB from spent oil are also available. Final PCB concentrations are below detectable levels, and the cleaned oil can be recycled or used as fuel.

39. PESTICIDES

The term *traditional organochlorine pesticide* refers to a narrow group of persistent pesticides, including DDT, the *"drins"* (aldrin, endrin, and dieldrin), chlordane, endosulfan, hexachlorobenzene, lindane, mirex, and toxaphene. Traditional organochlorine insecticides used extensively between the 1950s and early 1970s have been widely banned because of their environmental persistence. There are, however, notable exceptions, and environmental levels of traditional organochlorine pesticides (especially DDT) are not necessarily declining throughout the world, especially in developing countries and countries with malaria. DDT, with its half-life of up to 60 years, does not always remain in the country where it is used. The semi-volatile nature of the chemicals means that at high temperatures they will tend to evaporate from the land, only to condense in cooler air. This *global distillation* is thought to be responsible for levels of organochlorines increasing in the Arctic.

The term *chlorinated pesticides* refers to a much wider group of insecticides, fungicides, and herbicides that contain organically bound chlorine. A major difference between traditional organochlorine and other chlorinated pesticides is that the former have been perceived to have high persistence and build up in the food chain, and the latter do not. However, even some chlorinated pesticides are persistent in the environment. There is little information available concerning the overall environmental impact of chlorinated pesticides. Far more studies exist concerning the effects of traditional organochlorines in the public domain than on chlorinated pesticides in general. Pesticides that have, for example, active organophosphate (OP), carbamate, or triazine

---

[35]The term *oxidant* sometimes is used to mean the original emission products of NOx and hydrocarbons.
[36]Incineration of biosolids (sewage sludge) concentrates trace metals into the combustion ash, which is subsequently landfilled. Due to the high initial investment required, difficulty in using combustion energy, and increasingly stringent air quality and other environmental regulations, biosolid incineration is not widespread.

Solid Waste

parts of their molecules are chlorinated pesticides and may pose long-lasting environmental dangers.

In the United States, about 30–40% of pesticides are chlorinated. All the top five pesticides are chlorinated. Worldwide, half of the ten top-selling herbicides are chlorinated (alachlor, metolachlor, 2,4-D, cyanazine, and atrazine). Four of the top ten insecticides are chlorinated (chlorpyrifos, fenvalerate, endosulfan, and cypermethrin). Four (propiconazole, chlorothalonil, prochloraz, and triadimenol) of the ten most popular fungicides are chlorinated.

40. PLASTICS

Plastics, of which there are six main chemical polymers as given in Table 40.7, generally do not degrade once disposed of and are considered a disposal issue. Disposal is not a problem per se, however, since plastics are lightweight, inert, and do not harm the environment when discarded.

Table 40.7 Polymers

polymer	plastic ID number	common use
polyethylene terephthalate (PET)	1	clear beverage containers
high-density polyethylene (HDPE)	2	detergent; milk bottles; oil containers; toys
polyvinyl chloride (PVC)	3	clear bottles
low-density polyethylene (LDPE)	4	grocery bags; food wrap
polypropylene (PP)	5	labels; bottles; housewares
polystyrene (PS)	6	styrofoam cups; "clam shell" food containers

A distinction is made between *biodegrading* and *recycling*. Most plastics, such as the polyethylene bags used to protect pressed shirts from the dry cleaners and to mail some magazines, are not biodegradable but are recyclable. Also, all plastics can be burned for their fuel value.

The collection and sorting problems often render low-volume plastic recycling efforts uneconomical. Complicating the drive toward recycling is the fact that many of the six different types cannot easily be distinguished visually, and they cannot be recycled successfully when intermixed. Also, some plastic products consist of layers of different polymers that cannot be separated mechanically. Some plastic products are marked with a plastic type identification number.

Sorting in low-volume applications is performed visually and manually. Commercial high-volume methods include hydrocycloning, flotation with flocculation (for all polymers), X-ray fluorescence (primarily for PVC detection), and near-infrared (NIR) spectroscopy (primarily for separating PVC, PET, PP, PE, and PS). Mass NIR spectroscopy is also promising, but has yet to be commercialized.

Unsorted plastics can be melted and reformed into some low-value products. This operation is known as *downcycling*, since each successful cycle further degrades the material. This method is suitable only for a small fraction of the overall recyclable plastic.

Other operations that can reuse the compounds found in plastic products include hydrogenation, pyrolysis, and gasification. *Hydrogenation* is the conversion of mixed plastic scrap to "syncrude" (synthetic crude oil), in a high-temperature (i.e., 750–880°F (400–470°C)), high-pressure (i.e., 2200–4400 psig (15–30 MPa)), hydrogen-rich atmosphere. Since the end product is a crude oil substitute, hydrogenation operations must be integrated into refinery or petrochemical operations.

Gasification and pyrolysis are stand-alone operations that do not require integration with a refinery. *Pyrolysis* takes place in a fluidized bed between 750°F and 1475°F (400°C and 800°C). Cracked polymer gas or other inert gas fluidizes the sand bed, which promotes good mixing and heat transfer, resulting in liquid and gaseous petroleum products.

Gasification operates at higher temperatures, 1650–3600°F (900–2000°C), and lower pressures, around 870 psig (6 MPa). The waste stream is pyrolyzed at lower temperatures before being processed by the gasifier. The gas can be used on-site to generate steam. Gasification has the added advantage of being able to treat the entire municipal solid waste stream, avoiding the need for sorting plastics.

Biodegradable plastics have focused on polymers that are derived from agricultural sources (e.g., corn, potato, tapioca, and soybean starches), rather than from petroleum. *Bioplastics* have various degrees of *biodegradation* (i.e., breaking down into carbon dioxide, water, and biomass), *disintegration* (losing their shapes and identities and becoming invisible in the compost without needing to be screened out), and *eco-toxicity* (containing no toxic material that prevents plant growth in the compost). A bioplastic that satisfies all three characteristics is a *compostable plastic*. A *biodegradable plastic* will eventually be acted on by naturally occurring bacteria or fungi, but the time required is indeterminate. Also, biodegradable plastics may leave some toxic components. A *degradable plastic* will experience a significant change in its chemical structure and properties under specific environmental conditions, although it may not be affected by bacteria (i.e., be biodegradable) or satisfy any of the requirements for a compostable plastic.

Some engineers point out that biodegrading is not even a desirable characteristic for plastics and that being nonbiodegradable is not harmful. Biodegrading converts materials (such as the paper bags often preferred over plastic bags) to water and carbon dioxide, contributing to the greenhouse effect without even receiving the energy benefit of incineration. Biodegrading of most substances also results in gases and leachates that can be more harmful to the environment than the original substance. In a landfill, biodegrading serves no useful purpose, since the space occupied by the degraded plastic does not create additional useful space (volume).

41. RADON

Radon gas is a radioactive gas produced from the natural decay of radium within the rocks beneath a building. Radon accumulates in unventilated areas (e.g., basements), in stagnant water, and in air pockets formed when the ground settles beneath building slabs. Radon also can be brought into the home by radon-saturated well water used in baths and showers. The EPA's action level of 4 pCi/L for radon in air is contested by many as being too high.

Radon mitigation methods include (a) pressurizing to prevent the infiltration of radon, (b) installing depressurization systems to intercept radon below grade and vent it safely, (c) removing radon-producing soil, and (d) abandoning radon-producing sites.

42. RAINWATER RUNOFF

Rainwater percolating through piles of coal, fly ash, mine tailings, and other stored substances can absorb toxic compounds and eventually make its way into the earth, possibly contaminating soil and underground aquifers.

43. RAINWATER RUNOFF FROM HIGHWAYS

The *first flush* of a storm is generally considered to be the first half-inch of storm runoff or the runoff from the first 15 min of the storm. Along highways and other paved transportation corridors, the first flush contains potent pollutants such as petroleum products, asbestos fibers from brake pads, tire rubber, and fine metal dust from wearing parts. Under the National Pollutant Discharge Elimination System (NPDES), stormwater runoff in newly developed watersheds must be cleaned before it reaches existing drainage facilities, and runoff must be maintained at or below the present undeveloped runoff rate.

A good stormwater system design generally contains two separate basins or a single basin with two discrete compartments. The function of one compartment is water quality control, and the function of the other is peak runoff control.

The *water quality compartment* (WQC) should normally have sufficient volume and discharge rates to provide a minimum of one hour of detention time for 90–100% of the first flush volume. "Treatment" in the WQC consists of sedimentation of suspended solids and evaporation of volatiles. A removal goal of 75% of the suspended solids is reasonable in all but the most environmentally sensitive areas.

In environmentally sensitive areas, a filter berm of sand, a sand chamber, or a sand filter bed can further clarify the discharge from the WQC.

After the WQC becomes full from the first flush, subsequent runoff will be diverted to the *peak-discharge compartment* (PDC). This is done by designing a junction structure with an inlet for the incoming runoff and separate outlets from each compartment. If the elevation of the WQC outlet is lower than the inlet to the PDF, the first flush will be retained in the WQC.

Sediment from the WQC chamber and any filters should be removed every one to three years, or as required. The sediment must be properly handled, as it may be considered to be hazardous waste under the EPA's "mixture" and "derived-from" rules and its "contained-in" policy.

When designing chambers and filters, sizing should accommodate the first flush of a 100-year storm. The top of the berm between the WQC and PDC should include a minimum of 1 ft (0.3 m) of freeboard. In all cases, an emergency overflow weir should allow a storm greater than the design storm to discharge into the PDC or receiving water course.

Minimum chamber and berm width is approximately 8 ft (2.4 m). Optimum water depth in each chamber is approximately 2–5 ft (0.6–1.5 m). Each compartment can be sized by calculating the divided flows and staging each compartment. The outlet from the WQC should be sufficient to empty the compartment in approximately 24–28 hours after a 25-year storm.

A *filter berm* is essentially a sand layer between the WQC and the receiving chamber that filters water as it flows between the two compartments. The filter should be constructed as a layer of sand placed on geosynthetic fabric, protected with another sheet of geosynthetic fabric, and covered with coarse gravel, another geosynthetic cover, and finally a layer of medium stone. The sand, gravel, and stone layers should all have a minimum thickness of 1 ft (0.3 m). The rate of permeability is controlled by the sand size and front-of-fill material. Permeability calculations should assume that 50% of the filter fabric is clogged.

A *filter chamber* consists of a concrete structure with a removable filter pack. The filter pack consists of geosynthetic fabric wrapped around a plastic frame (core) that can be removed for backwashing and maintenance. The filter is supported on a metal screen mounted in the

Solid Waste

concrete chamber. The outlet of the chamber should be located at least 1 ft (0.3 m) behind the filter pack. The opening's size will determine the discharge rate through the filter, which should be designed as less than 2 ft/sec (60 cm/s) assuming that 50% of the filter area is clogged.

A *sand filter bed* is similar to the filter beds used for tertiary sewage treatment. The sand filter consists of a series of 4 in (100 mm) perforated PVC pipes in a gravel bed. The gravel bed is covered by geotextile fabric and 8 in (200 mm) of fine-to-medium sand. The perforated underdrains lead to the outlet channel or chamber.

44. SMOG

Photochemical smog (usually, just *smog*) consists of ground-level ozone and peroxyacyl nitrates (PAN). Smog is produced by the sunlight-induced reaction of ozone *precursors*, primarily nitrogen dioxide (NOx), hydrocarbons, and volatile organic compounds (VOCs). NOx and hydrocarbons are emitted by combustion sources such as automobiles, refineries, and industrial boilers. VOCs are emitted by manufacturing processes, dry cleaners, gasoline stations, print shops, painting operations, and municipal wastewater treatment plants. (See also Sec. 40.36.)

45. SMOKE

Smoke results from incomplete combustion and indicates unacceptable combustion conditions. In addition to being a nuisance problem, smoke contributes to air pollution and reduced visibility. Smoke generation can be minimized by proper furnace monitoring and control.

Opacity can be measured by a variety of informal and formal methods, including transmissometers mounted on the stack. The sum of the *opacity* (the fraction of light blocked) and the *transmittance* (the fraction of light transmitted) is 1.0.

Optical density is calculated from Eq. 40.19. The *smoke spot number* (SSN) can also be used to quantify smoke levels.

$$\text{optical density} = \log_{10} \frac{1}{1 - \text{opacity}} \qquad 40.19$$

Visible moisture plumes with opacities of 40% are common at large steam generators even when there are no unburned hydrocarbons emitted. High-sulfur fuels and the presence of ammonium chloride (a by-product of some ammonia-injection processes) seem to increase formation of visible plumes. Moisture plumes from saturated gas streams can be avoided by reheating prior to discharge to the atmosphere.

46. SPILLS, HAZARDOUS

Contamination by a hazardous material can occur accidentally (e.g., a spill) or intentionally (e.g., a previously used chemical-holding lagoon). Soil that has been contaminated with hazardous materials from spills or leaks from *underground storage tanks* (commonly known as *UST wastes*) is itself a hazardous waste.

The type of waste determines what laws are applicable, what permits are required, and what remediation methods are used. With contaminated soil, spilled substances can be (a) solid and nonhazardous or (b) nonhazardous liquid petroleum products (e.g., "UST nonhazardous"), and Resource Conservation and Recovery Act- (RCRA-) listed (c) hazardous substances, and (d) toxic substances.

Cleaning up a hazardous waste requires removing the waste from whatever air, soil, and water (lakes, rivers, and oceans) have been contaminated. The term *remediation* is used to mean the corrective steps taken to return the environment to its original condition. *Stabilization* refers to the act of reducing the waste concentrations to lower levels so that the waste can be transported, stored, or landfilled.

Remediation methods are classified as available or innovative. *Available methods* can be implemented immediately without being further tested. *Innovative methods* are new, unproven methods in various stages of study.

The remediation method used depends on the waste type. The two most common available methods are incineration and landfilling after stabilization.[37] Incineration can occur in rotary kilns, injection incinerators, infrared incineration, and fluidized-bed combustors. Landfilling requires the contaminated soil to be stabilized chemically or by other means prior to disposal. Technologies for general VOC-contaminated soil include vacuum extraction, bioremediation, thermal desorption, and soil washing.

47. SULFUR OXIDES

Sulfur oxides (SOx), consisting of *sulfur dioxide* (SO_2) and *sulfur trioxide* (SO_3), are the primary cause of acid rain. *Sulfurous acid* (H_2SO_3) and sulfuric acid (H_2SO_4) are produced when oxides of sulfur react with moisture in the flue gas. Both of these acids are corrosive.

$$SO_2 + H_2O \rightarrow H_2SO_3 \qquad 40.20$$

$$SO_3 + H_2O \rightarrow H_2SO_4 \qquad 40.21$$

Fuel switching (*coal substitution/blending* (CS/B)) is the burning of low-sulfur fuel. (Low-sulfur fuels are approximately 0.25–0.65% sulfur by weight, compared to high-sulfur coals with 2.4–3.5% sulfur.) However, unlike nitrogen oxides, which can be prevented during

---

[37]Other technologies include in situ and ex situ bioremediation, chemical treatment, in situ flushing, in situ vitrification, soil vapor extraction, soil washing, solvent extraction, and thermal desorption.

combustion, formation of sulfur oxides cannot be avoided when low-cost, high-sulfur fuels are burned.

Some air quality regulations regarding SOx production may be met by a combination of options. These options include fuel switching, flue gas scrubbing, derating, and allowance trading. The most economical blend of these options will vary from location to location.

In addition to fuel switching, available technology options for retrofitting existing coal-fired plants include wet scrubbing, dry scrubbing, sorbent injection, repowering with clean coal technology (CCT), and co-firing with natural gas.

Sulfur dioxide (like carbon dioxide and nitrogen oxides) emissions must be reported on a standardized basis. Equation 40.22 can be used to calculate the volumetric, dry basis concentration from a wet stack gas sample.

$$\frac{SO_{2,dry,\,at\,3\%\,O_2}}{SO_{2,wet,\,at\,stack\,O_2}} = \frac{CO_{2,th,\,dry,at\,3\%\,O_2}}{CO_{2,wet,\,at\,stack\,O_2}} \qquad 40.22$$

The sulfur dioxide concentration (mass per dry standard volume, C) can be calculated from sulfur dioxide's molecular weight (MW = 64.07) and Eq. 40.17.[38]

Example 40.6

A wet stack gas analysis shows 230 ppm of SO_2 and 10.5% CO_2. Based on stoichiometric combustion, the theoretical carbon dioxide percentage at 3% oxygen should have been 13.4%. What is the SO_2 concentration in ppm on a dry basis, standardized to 3% oxygen?

Solution

Use Eq. 40.22.

$$SO_{2,dry,at\,3\%\,O_2} = SO_{2,wet,at\,stack\,O_2}$$
$$\times \frac{CO_{2,th,dry,at\,3\%\,O_2}}{CO_{2,wet,at\,stack\,O_2}}$$
$$= \frac{(230\ ppm)(0.134)}{0.105}$$
$$= 294\ ppm$$

48. TIRES

Discarded tires are more than a disposal problem. At 15,000 Btu/lbm (35 MJ/kg), their energy content is nearly 80% that of crude oil. Discarded tires are wasted energy resources. While tires can be incinerated, other processes can be used to gasify them to produce clean synthetic gas (i.e., *syngas*). The low-sulfur, hydrogen-rich syngas can subsequently be burned in combined-cycle plants or used as a feedstock for ammonia and methanol production, or the hydrogen can be recovered for separate use. Tires can also be converted in a non-chemical process into a strong asphalt-rubber pavement.

---

[38]See Ftn. 33.

49. TRIHALOMETHANES

Trihalomethanes (THMs) are halogenated disinfection by-products (DBPs). These organic chemicals are produced during the disinfection of water and, therefore, constitute a drinking water quality problem. The chemically active elements of chlorine, iodine, and bromine react with various *organic precursors* to produce THMs.

Chlorine in various forms is widely used to disinfect water. Bromine can be present in gaseous chlorine as an impurity. Bromine also results from reacting chlorine with the bromide present in high-salinity water. Iodine is seldom used in disinfection. Therefore, only four THMs are found in significant quantities: trichloromethane, also known as chloroform ($CHCl_3$); tribromomethane, also known as bromoform ($CHBr_3$); bromodichloromethane ($CHBrCl_2$); and dibromochloromethane ($CHBr_2Cl$).

The *organic precursors* that react with chlorine to produce THMs are primarily naturally occurring (e.g., humic and fulvic acids from decaying vegetation). The precursors are not themselves harmful, but the THMs produced from them are presumed to be carcinogenic.[39]

Two main options exist for reducing THMs: formation avoidance and removal after formation. The first option includes (1) using ozone, chlorine dioxide, or potassium permanganate (60% to 90% THM formation reduction), (2) dechlorination after chlorination to prevent reaction of chlorine, and (3) adding ammonia to water prior to discharge to enhance chloramineformation, since chloramines suppress THM formation. The second option includes (1) using activated carbon to remove THMs (25% to 60% THM reduction), (2) using activated carbon, adsorbents, or zeolite filters to remove precursors, (3) moving the chlorination point to the end of the treatment process so that most precursors are removed in earlier processes prior to disinfection (70% to 75% THM reduction), (4) optimizing coagulation and settling water treatment processes to improve precursor removal, and (5) selecting water with fewer precursors.

Follow-up studies are necessary when changing to alternative disinfectants. An alternative disinfectant and its by-products should be evaluated to determine disinfecting power, residual power, toxicity, and other health effects. Costs of operation will increase when alternative disinfectants are used, although moving the point of application may not result in any significant increase in operating costs after a modest capital expenditure is made.

Ozone, frequently considered as an alternative disinfectant, is less expensive than chlorine dioxide, but it costs more than using chloramines or changing the point of chlorine application. An emerging consensus holds that ozonation alone is relatively ineffective (5% to 20% reduction at typical ozone doses) in controlling THM.

---

[39]Tests have shown that chloroform in high doses is carcinogenic to rats. Other THMs are considered carcinogenic by association.

Also, ozonation converts the precursors to *biodegradable organic matter* (BOM). Unless a biological process subsequently removes the BOM, aftergrowths in the distribution system may become a problem. Finally, ozonation does not leave any residual disinfectant in the water for use throughout the distribution system. Therefore, emphasis is on ozonation in combination with biological treatment to destroy precursors, followed by chlorine application for residual formation.

50. VOLATILE INORGANIC COMPOUNDS

Volatile inorganic compounds (VICs) include H_2S, NOx (except N_2O), SO_2, HCl, NH_3, and many other less common compounds.

51. VOLATILE ORGANIC COMPOUNDS

Volatile organic compounds (VOCs) (e.g., benzene, chloroform, formaldehyde, methylene chloride, napthalene, phenol, toluene, and trichloroethylene) are highly soluble in water. VOCs that have leaked from storage tanks or have been discharged often end up in groundwater and drinking supplies.

There are a large number of methods for removing VOCs, including incineration, chemical scrubbing with oxidants, water washing, air or steam stripping, activated charcoal adsorption processes, SCR (selective catalytic reduction), and bioremediation. Incineration of VOCs is fast and 99%+ effective, but incineration requires large amounts of fuel and produces NOx. Using SCR with heat recovery after incineration reduces the energy input but adds to the expense.

52. WATER VAPOR

Emitted water vapor is not generally considered to be a pollutant.

41 Disposition of Hazardous Materials

1. GENERAL STORAGE

Storage of *hazardous materials* (*hazmats*) is often governed by local building codes in addition to state and federal regulations. Types of construction, maximum floor areas, and building layout may all be restricted.[1]

Good engineering judgment is called for in areas not specifically governed by the building code. Engineering consideration will need to be given to the following aspects of storage facility design: (1) spill containment provisions, (2) chemical resistance of construction and storage materials, (3) likelihood of and resistance to explosions, (4) exiting, (5) ventilation, (6) electrical design, (7) storage method, (8) personnel emergency equipment, (9) security, and (10) spill cleanup provisions.

2. STORAGE TANKS

Underground storage tanks (USTs) have traditionally been used to store bulk chemicals and petroleum products. Fire and explosion risks are low with USTs, but subsurface pollution is common since inspection is limited. Since 1988, the U.S. Environmental Protection Agency (EPA) has required USTs to have secondary containment, corrosion protection, and leak detection. UST operators also must carry insurance in an amount sufficient to clean up a tank failure.

Because of the cost of complying with UST legislation, above-ground storage tanks (ASTs) are becoming more popular. AST strengths and weaknesses are the reverse of USTs: ASTs reduce pollution caused by leaks, but the expected damage due to fire and explosion is greatly increased. Because of this, some local ordinances prohibit all ASTs for petroleum products.[2]

The following factors should be considered when deciding between USTs and ASTs: (1) space available, (2) zoning ordinances, (3) secondary containment, (4) leak-detection equipment, (5) operating limitations, and (6) economics.

Most ASTs are constructed of carbon or stainless steel. These provide better structural integrity and fire resistance than fiberglass-reinforced plastic and other composite tanks. Tanks can be either field-erected or factory-fabricated (capacities greater than approximately 50,000 gal (190 kL)). Factory-fabricated ASTs are usually designed according to UL-142 (Underwriters Laboratories *Standard for Safety*), which dictates steel type, wall thickness, and characteristics of compartments, bulkheads, and fittings. Most ASTs are not pressurized, but those that are must be designed in accordance with the ASME *Boiler and Pressure Vessel Code*, Section VIII.

NFPA 30 (*Flammable and Combustible Liquids Code*, National Fire Protection Association, Quincy, MA) specifies the minimum separation distances between ASTs, other tanks, structures, and public right-of-ways. The separation is a function of tank type, size, and contents. NFPA 30 also specifies installation, spill control, venting, and testing.

ASTs must be double-walled, concrete-encased, or contained in a dike or vault to prevent leaks and spills, and they must meet fire codes. Dikes should have a capacity in excess (e.g., 110% to 125%) of the tank volume. ASTs (as do USTs) must be equipped with overfill prevention systems. Piping should be above-ground wherever possible. Reasonable protection against vandalism and hunters' bullets is also necessary.[3]

Though they are a good idea, leak-detection systems are not typically required for ASTs. Methodology for leak detection is evolving, but currently includes vacuum or pressure monitoring, electronic gauging, and optical and sniffing sensors. Double-walled tanks may also be fitted with sensors within the interstitial space.

Operationally, ASTs present special problems. In hot weather, volatile substances vaporize and represent an additional leak hazard. In cold weather, viscous contents may need to be heated (often by steam tracing).

ASTs are not necessarily less expensive than USTs, but they are generally thought to be so. Additional hidden costs of regulatory compliance, secondary containment, fire protection, and land acquisition must also be considered.

Solid Waste

---

[1]For example, flammable materials stored in rack systems are typically limited to heights of 25 ft (8.3 m).
[2]The American Society of Petroleum Operations Engineers (ASPOE) policy statement states, "Above-ground storage of liquid hydrocarbon motor fuels is inherently less safe than underground storage. Above-ground storage of Class 1 liquids (gasoline) should be prohibited at facilities open to the public."

---

[3]Approximately 20% of all spills from ASTs result from vandalism.

3. DISPOSITION OF HAZARDOUS WASTES

When a hazardous waste is disposed of, it must be taken to a registered *treatment, storage, or disposal facility* (TSDF). The EPA's *land ban* specifically prohibits the disposal of hazardous wastes on land prior to treatment. Incineration at sea is also prohibited. Wastes must be treated to specific maximum concentration limits by specific technology prior to disposal in landfills.

Once treated to specific regulated concentrations, hazardous waste residues can be disposed of by incineration, by landfilling, or, less frequently, by deep-well injection. All disposal facilities must meet detailed design and operational standards.

42 Environmental Remediation

Nomenclature

a	interfacial area per volume of packing	1/ft	1/m
A	area	ft^2	m^2
B	cyclone inlet width	ft	m
c	scaling constant	–	–
C	concentration	lbm/ft^3	kg/m^3
C_c	Cunningham slip factor	–	–
CCR	corona current ratio	A/ft^2	A/m^2
CPR	corona power ratio	$W\text{-}min/ft^3$	$W{\cdot}h/m^3$
d	diameter	ft	m
D_e	cyclone exit diameter	ft	m
E	electric field strength	V/m	V/m
g	gravitational acceleration, 32.2	ft/sec^2	m/s^2
g_c	gravitational constant, 32.2	$lbm\text{-}ft/lbf\text{-}sec^2$	n.a.
h	height	ft	m
H	cyclone inlet height	ft	m
H	Henry's law constant	atm	atm
HHV	higher heating value	MMBtu/lbm	kJ/kg
HTU	height of a transfer unit	ft	m
I	current	A	A
k	reaction rate constant	1/sec	1/s
K	ESP constant	–	–
K	factor	various	various
K	ratio of area to volume	1/ft	1/m
L	length	ft	m
L	liquid loading rate	$ft^3/min\text{-}ft^2$	$m^3/s{\cdot}m^2$
L/G	liquid-gas ratio	gal-min/ 1000 ft^3	$L{\cdot}h/m^3$
m	mass	lbm	kg
$\dot{m}$	mass flow rate	lbm/hr	kg/h
N	number (quantity)	–	–
NTU	number of transfer units	–	–
p	pressure	lbf/ft^2	Pa
p	vapor pressure	lbf/in^2	kPa
P	power	ft-lbf/sec	W
Pt	penetration	–	–
q	charge	C	C
Q	flow rate	ft^3/sec	m^3/s
r	cyclone radius	ft	m
R	specific gas constant	ft-lbf/lbm-°R	J/kg·K
R	stripping factor	–	–

Solid Waste

s	distance	ft	m
S	drag	lbf-min/ft^3	Pa·min/m
S	separation factor	–	–
SCA	specific collection area	ft^2-min/ 1000 ft^3	m^2·h/ 1000 m^3
SFC	specific feed characteristic	Btu/lbm	J/kg
t	time	sec	s
T	temperature	°R	K
v	velocity	ft/sec	m/s
V	voltage	V	V
w	drift velocity	ft/sec	m/s
W	areal dust density	lbm/ft^2	kg/m^2
x	fraction by weight	–	–
x	humidity ratio	lbm/lbm	kg/kg
y	exponent	–	–
z	packing height	ft	m

Symbols

α	coefficient of linear thermal expansion	ft/ft-°F	m/m·°C
ϵ	dielectric constant	–	–
ϵ_0	permittivity of free space, 8.854×10^{-12}	F/m	F/m
η	efficiency	%	%
μ	absolute viscosity	lbf-sec/ft^2	Pa·s
ρ	density	lbm/ft^3	kg/m^3

Subscripts

a	air
A	areal (per unit area)
d	droplet
e	electron or effective
f	filtering
g	gas
G	gas
L	liquid
o	reference
p	particle
r	residence
sat	saturation
t	total
w	water

1. INTRODUCTION

This chapter discusses (in alphabetical order) the methods and equipment that can be used to reduce or eliminate pollution. Legislation often requires use of the *best available control technology* (BACT—also known as the *best available technology*—BAT) and the *maximum achievable control technology* (MACT), *lowest achievable emission rate* (LAER), and *reasonably available control technology* (RACT) in the design of pollution-prevention systems.

2. ABSORPTION, GAS (GENERAL)

Gas *absorption processes* remove a gas (the *target substance*) from a gas stream by dissolving it in a liquid solvent.[1] Absorption can be used for flue gas cleanup (FGC) to remove sulfur dioxide, hydrogen sulfide, hydrogen chloride, chlorine, ammonia, nitrogen oxides, and light hydrocarbons. Gas absorption equipment includes packed towers, spray towers and chambers, and venturi absorbers.[2]

3. ABSORPTION, GAS (SPRAY TOWERS)

In a general spray tower or spray chamber (i.e., a *wet scrubber*), liquid and gas flow countercurrently or cross-currently. The gas moves through a liquid spray that is carried downward by gravity. A mist eliminator removes entrained liquid from the gas flow. (See Fig. 42.1.) The liquid can be recirculated. The removal efficiency is moderate.

Spray towers are characterized by low pressure drops—typically 1 in wg to 2 in wg[3] (0.25 kPa to 0.5 kPa)—and their liquid-to-gas ratios—typically 20 gal/1000 ft^3 to 100 gal/1000 ft^3 (3 L/m^3 to 14 L/m^3). For self-contained units, fan power is also low—approximately 3×10^{-4} kW/ft^3 (0.01 kW/m^3) of gas moved.

Flooding of spray towers is an operational difficulty where the liquid spray is carried up the column by the gas stream. Flooding occurs when the gas stream velocity approaches the *flooding velocity*. To prevent this, the tower diameter is chosen as approximately 50% to 75% of the flooding velocity.

Flue gas desulfurization (FGD) wet scrubbers can remove approximately 90% to 95% of the sulfur with efficiencies of 98% claimed by some installations. Stack effluent leaves at approximately 150°F (65°C) and is saturated (or nearly saturated) with moisture. Dense steam plumes may be present unless the scrubbed stream is reheated.

---

[1]Scrubbing, gas absorption, and stripping are distinguished by their target substances, the carrier flow phase, and the directions of flow. *Scrubbing* is the removal of particulate matter from a gas flow by exposing the flow to a liquid or slurry spray. *Gas absorption* is a countercurrent operation for the removal of a target gas from a gas mixture by exposing the mixture to a liquid bath or spray. *Stripping*, also known as *gas desorption* (also a countercurrent operation), is the removal of a dissolved gas or other volatile component from liquid by exposing the liquid to air or steam. Stripping is the reverse operation of gas absorption.

Packed towers can be used for both gas absorption and stripping processes, and the processes look similar. The fundamental difference is that in gas absorption processes, the target substance (i.e., the substance to be removed) is in the gas flow moving up the tower, and in stripping processes, the target substance is in the liquid flow moving down the tower.

[2]Spray and packed towers, though they are capable, are not generally used for desulfurization of furnace or incineration combustion gases, as scrubbers are better suited for this task.

[3]"in wg" and "iwg" are abbreviations for "inches water gage," also referred to as iwc for "inches water column" and "inches of water."

Collected sludge waste and fly ash are removed within the scrubber by purely inertial means. After being dewatered, the sludge is landfilled.[4]

Figure 42.1 Spray Tower Absorber

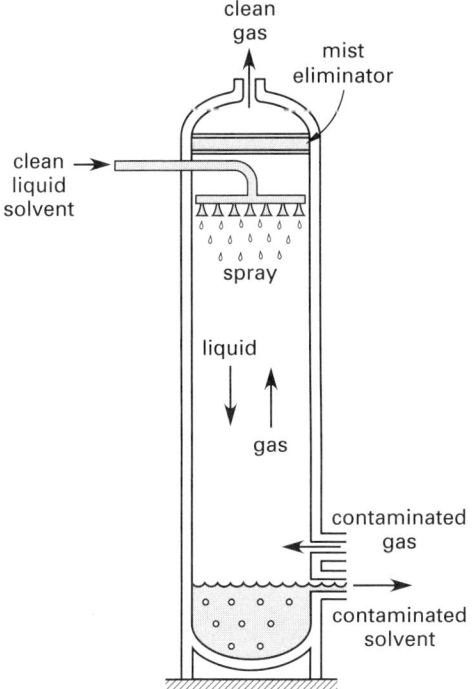

Spray towers, though simple in concept and requiring little energy to operate, are not simple to operate and maintain. Other disadvantages of spray scrubbers include high water usage, generation of wastewater or sludge, and low efficiency at removing particles smaller than $5\,\mu$m. Relative to dry scrubbing, the production of wet sludge and the requirement for a sludge-handling system is the major disadvantage of wet scrubbing.

4. ABSORPTION, GAS (IN PACKED TOWERS)

In a packed tower, clean liquid flows from top to bottom over the tower packing media, usually consisting of synthetic engineered shapes designed to maximize liquid-surface contact area. (See Fig. 42.2.) The contaminated gas flows countercurrently, from bottom to top, although crosscurrent designs also exist. As in spray towers, the liquid can be recirculated, and a mist eliminator is used.

Pressure drops are in the 1 in wg to 8 in wg (0.25 kPa to 2.0 kPa) range. Typical liquid-to-gas ratios are 10 gal/1000 ft^3 to 20 gal/1000 ft^3 (1 L/m^3 to 3 L/m^3). Although the pressure drop is greater than in spray towers, the removal efficiency is much higher.

Figure 42.2 Packed Bed Spray Tower

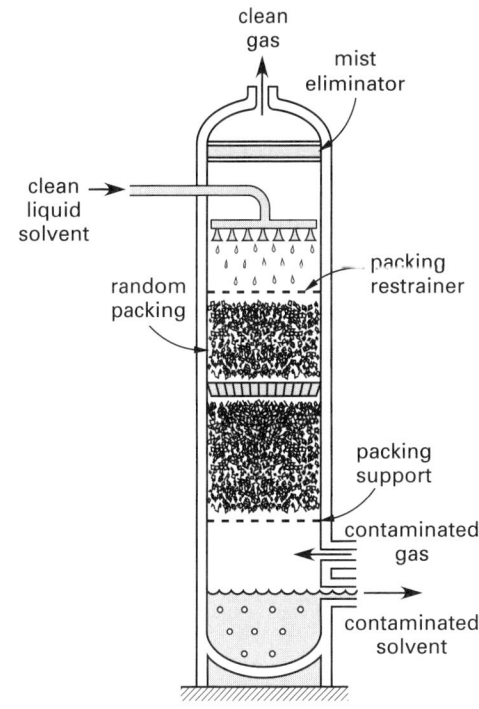

5. ADSORPTION (ACTIVATED CARBON)

Granular activated carbon (GAC), also known as *activated carbon* and *activated charcoal* (AC), processes are effective in removing a number of compounds, including volatile organic compounds (VOCs), heavy metals (e.g., lead and mercury), and dioxins.[5] AC is an effective adsorbent[6] for both air and water streams, and a VOC removal efficiency of 99% can be achieved. AC can be manufactured from almost any carbonaceous raw material, but wood, coal, and coconut shell are widely used.

Pollution control processes use AC in both solvent capture-recovery (i.e., recycling) and capture-destruction processes. AC is available in powder and granular form. Granules are preferred for use in recovery systems since the AC can be regenerated when *breakthrough*, also known as *breakpoint*, occurs. This is when the AC has become saturated with the solvent and traces of the solvent begin to appear in the exit air. Until breakthrough, removal efficiency is essentially constant. The *retentivity* of the AC is the ratio of adsorbed solvent mass to carbon mass.

---

[4]A typical 500 MW plant burning high-sulfur fuel can produce as much as 10^7 ft^3 (300,000 m^3) of dewatered sludge per year.

[5]The term *activated* refers to the high-temperature removal of tarry substances from the interior of the carbon granule, leaving a highly porous structure.

[6]An *adsorbent* is a substance with high surface area per unit weight, an intricate pore structure, and a hydrophobic surface. An *adsorbent material* traps substances in fluid (liquid and gaseous) form on its exposed surfaces. In addition to activated carbon, other common industrial adsorbents include alumina, bauxite, bone char, Fuller's earth, magnesia, and silica gel.

6. ADSORPTION (SOLVENT RECOVERY)

AC for solvent recovery is used in a cyclic process where it is alternately exposed to the target substance and then regenerated by the removal of the target substance. For solvent recovery to be effective, the VOC inlet concentration should be at least 700 ppm. Regeneration of the AC is accomplished by heating, usually by passing low-pressure (e.g., 5 psig) steam over the AC to raise the temperature above the solvent-capture temperature. Figure 42.3 shows a stripping-AC process with recovery.

Figure 42.3 *Stripping-AC Process with Recovery*

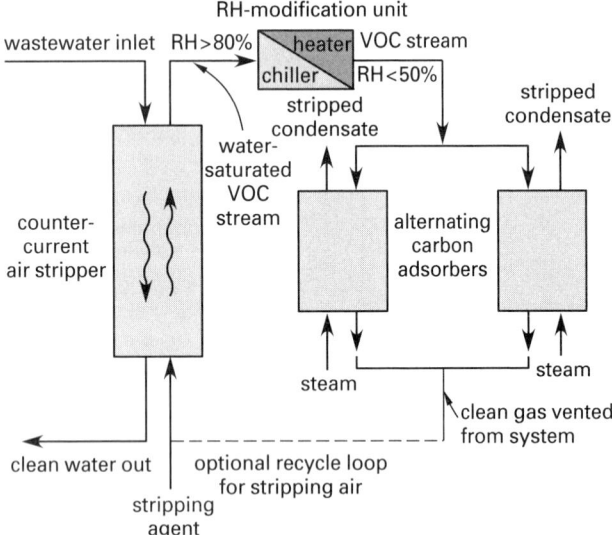

There are three main processes for solvent recovery. In a traditional *open-loop recovery system*, a countercurrent air stripper separates VOC from the incoming stream. The resulting VOC air/vapor stream passes through an AC bed. Periodically, the air/vapor stream is switched to an alternate bed, and the first bed is regenerated by passing steam through it. The VOC is recovered from the steam. The VOC-free air stream is freely discharged. Operation of a *closed-loop recovery system* is the same as an open loop system, except that the VOC-free air stream is returned to the air stripper.

For VOC-laden gas flows from 5000 SCFM to 100,000 SCFM (2.3 kL/s to 47 kL/s), a third type of solvent recovery system involves a *rotor concentrator*. VOCs are continuously adsorbed onto a multilayer, corrugated wheel whose honeycomb structure has been coated with powdered AC. The wheel area is divided into three zones: one for adsorption, one for desorption (i.e., regeneration), and one for cooling. Each of the zones is isolated from the other by tight sealing. The wheel rotates at low speed, continuously exposing new portions of the streams to each of the three zones. Though the equipment is expensive, operational efficiencies are high—95% to 98%.

7. ADSORPTION, HAZARDOUS WASTE

AC is particularly attractive for flue gas cleanup (FGC) at installations that burn spent oil, electrical cable, biosolids (i.e., sewage sludge), waste solvents, or tires as supplemental fuels. As with liquids, flue gases can be cleaned by passing them through fixed AC. Heavy metal dioxins are quickly adsorbed by the AC—in the first 8 in (20 cm) or so. AC injection (direct or spray) can remove 60% to 90% of the target substance present.

The spent AC creates its own waste disposal problem. For some substances (e.g., dioxins), AC can be incinerated. Heavy metals in AC must be removed by a wash process. Another process in use is vitrifying the AC and fly ash into a glassy, unleachable substance. *Vitrification* is a high-temperature process that turns incinerator ash into a safe, glass-like material. In some processes, heavy-metal salts are recovered separately. No gases and no hazardous wastes are formed.

8. ADVANCED FLUE GAS CLEANUP

Most air pollution control systems reduce NOx in the burner and remove SOx in the stack. *Advanced flue gas cleanup* (AFGC) methods combine processes to remove both NOx and SOx in the stack. Particulate matter is removed by an electrostatic precipitator or baghouse as is typical. Promising AFGC methods include wet scrubbing with metal chelates such as ferrous ethylenediaminetetraacetate (Fe(II)-EDTA) (NOx/SOx removal efficiencies of 60%/90%), adding sodium hydroxide injection to dry scrubbing operations (35% NOx removal), in-duct sorbent injection of urea (80% NOx removal), in-duct sorbent injection of sodium bicarbonate (35% NOx removal), and the NOXSO process using a fluidized bed of sodium-impregnated alumina sorbent at approximately 250°F (120°C) (70% to 90%/90% NOx/SOx removal).

9. ADVANCED OXIDATION

The term *advanced oxidation* refers to the use of ozone, hydrogen peroxide, ultraviolet radiation, and other exotic methods that produce free hydroxyl radicals (OH^-). (See also Sec. 42.38.) Table 42.1 shows the relative oxidation powers of common oxidants.

10. BAGHOUSES

Baghouses have a reputation for excellent particulate removal, down to 0.005 grains/ft^3 (0.18 grains/m^3) (dry), with particulate emissions of 0.01 grains/ft^3 (0.35 grains/m^3) being routine. Removal efficiencies are in excess of 99% and are often as high as 99.99% (weight basis). Fabric filters have a high efficiency for removing particular matter less than 10 μm in size. Because of this, baghouses are effective at collecting air toxics, which preferentially condense on these particles at the baghouse operating temperature—less than

Table 42.1 *Relative Oxidation Powers of Common Oxidants*

oxidant	oxidation power (relative to chlorine)
fluorine	2.25
hydroxyl radical (OH)	2.05
ozone (O_3)	1.52
permanganate radical (MnO_4)	1.23
chlorine dioxide (ClO_2)	1.10
hypochlorous acid (HClO)	1.10
chlorine	1.00
bromine	0.80
iodine	0.40
oxygen	0.29

Figure 42.4 *Typical Baghouse*

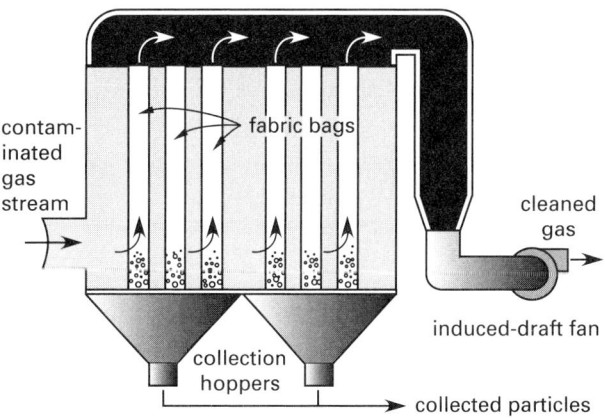

$300°F$ ($150°C$). Gas temperatures can be up to $500°F$ ($260°C$) with short periods up to about $550°F$ ($288°C$), depending on the filter fabric. Most of the required operational power is to compensate for the system pressure drops due to bags, cake, and ducting.

Baghouse filter fabric has micro-sized holes but may be felted or woven, with the material depending on the nature of the flue gas and particulates. Woven fabrics are better suited for lower filtering velocities in the 1 ft/min to 2 ft/min (0.3 m/min to 0.7 m/min) range, while felted fabrics are better at 5 ft/min (1.7 m/min). The fabric is often used in tube configuration, but envelopes (i.e., flat bags) and pleated cartridges are also available.

The particulate matter collects on the outside of the bag, forming a cake-like coating. If lime has been introduced into the stream in a previous step, some of the cake will consist of unreacted lime. As the gases pass through this cake, additional neutralizing takes place.

When the pressure drop across the filter reaches a preset limit (usually 6 in wg to 8 in wg (1.5 kPa to 2 kPa)), the cake is dislodged by mechanical shaking, reverse-air cleaning, or pulse-jetting. The dislodged fly ash cake falls into collection hoppers below the bags. Fly ash is transported by pressure or vacuum conveying systems to the conditioning system (consisting of surge bins, rotary feeders, and pug mills). Figure 42.4 shows a typical baghouse.

Baghouses are characterized by their air-to-cloth ratios and pressure drop. The *air-to-cloth ratio*, also known as *filter ratio*, *superficial face velocity*, and *filtering velocity*, is the ratio of the air volumetric flow rate in ft^3/min (m^3/s) to the exposed surface area in ft^2 (m^2). After canceling, the units are ft/min (m/s), hence the name *filtering velocity*. The higher the ratio, the smaller the baghouse and the higher the pressure drop. Shaker and reverse-air baghouses with woven fabrics have typical air-to-cloth ratios ranging from 1.0 ft/min to 6.0 ft/min (0.056 m/s to 0.34 m/s). Pulse-jet collectors with felted fabrics have higher ratios, ranging from 3.5 ft/min to 5.0 ft/min (0.0175 m/s to 0.025 m/s).

Advantages of baghouses include high efficiency and performance that is essentially independent of flow rate, particle size, and particle (electrical) resistivity. Also, baghouses produce the lowest opacity (generally less than 10, which is virtually invisible). Disadvantages include clogging, difficult cleaning, and bag breakage.

In a baghouse with N compartments, each with multiple bags, the *gross filtering velocity*, *gross air-to-cloth ratio*, and *gross filtering area*, $v_{f,N}$, refer to having all N compartments operating simultaneously. (See Eq. 42.1.) The *design filtering velocity*, *net filtering velocity*, *net air-to-cloth ratio*, and *net filtering area*, $v_{f,N-1}$, refer to one compartment taken offline for cleaning, meaning $N-1$ compartments are operating simultaneously. (See Eq. 42.2.) *Net net* refers to having $N-2$ compartments operating simultaneously.

$$v_{f,N} = \frac{Q_t}{A_{N\,\text{compartments}}} = \frac{Q_t}{NA_{\text{one compartment}}} \qquad 42.1$$

$$v_{f,N-1} = \frac{Q_t}{(N-1)A_{\text{one compartment}}} \qquad 42.2$$

The number of bags is

$$n_{\text{bags}} = \frac{A_N}{A_{\text{bag}}} = \frac{A_N}{\pi dh} \qquad 42.3$$

The *areal dust density*, W, is the mass of dust cake on the filter per unit area. The areal dust density can be calculated from the incoming particle concentration (*dust loading*, *fabric loading*), C, filtering velocity, and collection efficiency. Typical units are lbm/ft^2 (kg/m^2).

$$W = \eta C v_f t \approx C v_f t \qquad 42.4$$

In the *filter drag model*, the pressure drop through the baghouse is calculated from the *filter drag* (*filter resistance*), S, in units of in wg·min/ft (Pa·min/m), which depends on the permeabilities of the cloth, K_1, K_0, or K_e, and the particle cake, K_2 or K_s. (K_1 is also known as the *flow resistance* of the clean fabric; K_2 is the *specific resistance* of the cake.) Since Eq. 42.6 represents a

Solid Waste

straight line, K_e and K_s can be determined by plotting S versus W for a few test points. Maximum pressure drop is typically in the 5–20 in wg (1.2–5.0 kPa) range.

$$\Delta p = v_{\text{filtering}} S \qquad 42.5(a)$$

$$\Delta p_{\text{in wg}} = v_{f,\text{ft/min}} S_{\text{in wg-min/ft}} \qquad 42.5(b)$$

$$S = K_e + K_s W \qquad 42.6$$

Baghouses commonly have multiple compartments, and sufficient capacity must remain when one compartment is taken offline for cleaning. If there are N compartments, the time for N cycles of filtration (known as the *filtration time*) will be[7]

$$t_{\text{filtration}} = N(t_{\text{run}} + t_{\text{cleaning}}) - t_{\text{cleaning}} \qquad 42.7$$

Filtering velocity and pressure drop depend on the number of compartments that are online (i.e., they increase when one of the compartments is taken offline). The maximum pressure drop, Δp_m, occurs several times during the filtration time. The maximum pressure drop can be calculated from the maximum filter drag, which in turn, depends on the maximum areal density. The *actual filtering velocity*, $v_{f,\max\Delta p}$, is the velocity at the time the maximum pressure drop occurs.[8] c in Eq. 42.11 is a scaling constant that depends on the number of compartments. (See Table 42.2.)

$$\Delta p_{\max} = S_{\max\Delta p} v_{f,\max\Delta p} \qquad 42.8$$

$$S_{\max\Delta p} = K_e + K_s W_{\max\Delta p} \qquad 42.9$$

$$W_{\max\Delta p} = (N-1)C(v_{f,N}t_{\text{run}} + v_{f,N-1}t_{\text{cleaning}}) \qquad 42.10$$

$$v_{f,\max\Delta p} = cv_{f,N-1} \qquad 42.11$$

Table 42.2 *Values for Scaling Constant, c, Based on Number of Compartments, N*

number of compartments, N	c
3	0.87
4	0.80
5	0.76
7	0.71
10	0.67
12	0.65
15	0.64
20	0.62

The motor for the fan (blower) is sized based on the maximum pressure loss.

$$P = Q_t \Delta p_{\max} \qquad 42.12(a)$$

$$P_{\text{hp}} = \frac{Q_{t,\text{ft}^3/\text{min}} \Delta p_{\max,\text{in wg}}}{6356\eta_{\text{motor}}} \quad \text{[U.S. only]} \qquad 42.12(b)$$

There is no single formula derived from basic principles that predicts baghouse collection efficiency, η. Baghouse designs are normally based on experience, not on fractional efficiency curves. The instantaneous collection efficiency (as well as the pressure drop) of a baghouse increases with time as pores fill with particles. The empirical rate constant, k, is derived from a curve fit of actual performance.

$$\eta = 1 - e^{-kt} \qquad 42.13$$

11. BIOREMEDIATION

Bioremediation encompasses the methods of biofiltration, bioreaction, bioreclamation, activated sludge, trickle filtration, fixed-film biological treatment, landfilling, and injection wells (for in situ treatment of soils and groundwater). Bioremediation relies on microorganisms in a moist, oxygen-rich environment to oxidize solid, liquid, or gaseous organic compounds, producing carbon dioxide and water. Bioremediation is effective for removing volatile organic compounds (VOCs) and easy-to-degrade organic compounds such as BTEXs (benzene, toluene, ethylbenzene, and xylene).[9] Wood-preserving wastes, such as creosote and other polynuclear aromatic hydrocarbons (PAHs), can also be treated.

Bioremediation can be carried out in open tanks, packed columns, beds of porous synthetic materials, composting piles, or soil. The effectiveness of bioremediation depends on the nature of the process, the time and physical space available, and the degradability of the substance. Table 42.3 categorizes gases according to their general degradabilities.

Though slow, limited by microorganisms with the specific affinity for the chemicals present, and susceptible to compounds that are toxic to the microorganisms, bioremediation has the advantage of destroying substances rather than merely concentrating them. Bioremediation is less effective when a variety of different compounds are present simultaneously.

12. BIOFILTRATION

The term *biofiltration* refers to the use of composting and soil beds. A *biofilter* is a bed of soil or compost through which runs a distribution system of perforated pipe. Contaminated air or liquid flows through the pipes and into the bed. Volatile organic compounds (VOCs) are oxidized to CO_2 by microorganisms. Volatile inorganic compounds (VICs) are oxidized to acids (e.g., HNO_3 and H_2SO_4) and salts.

---

[7]By convention, the filtration time includes only $N-1$ cleanings.
[8]The name "actual filtering velocity" is definitely confusing.
[9]BTEXs are common ingredients in gasoline.

Table 42.3 *Degradability of Volatile Organic and Inorganic Gases*

rapidly degradable VOCs	rapidly reactive VICs	slowly degradable VOCs	very slowly degradable VOCs
alcohols	H_2S	hydrocarbons[a]	halogenated hydrocarbons[b]
aldehydes	NOx (but not N_2O)	phenols	polyaromatic hydrocarbons
ketones		methylene chloride	
ethers	SO_2		
esters	HCl		CS_2
organic acids	NH_3		
amines	PH_3		
thiols	SiH_4		
other molecules containing O, N, or S functional groups	HF		

[a]Aliphatics degrade faster than aromatics, such as xylene, toluene, benzene, and styrene.
[b]These include trichloroethylene, trichloroethane, carbon tetrachloride, and pentachlorophenol.

Biofilters require no fuel or chemicals when processing VOCs, and the operational lifetime is essentially infinite. For VICs, the lifetime depends on the soil's capacity to neutralize acids. Though reaction times are long and absorption capacities are low (hence the large areas required), the oxidation continually regenerates (rather than depletes) the treatment capacity. Once operational, biofiltration is probably the least expensive method of eliminating VOCs and VICs.

The *removal efficiency* of a biofilter is given by Eq. 42.14. k is an empirical *reaction rate constant*, and t is the *bed residence time* of the carrier fluid (water or air) in the bed. The reaction rate constant depends on the temperature but is primarily a function of the biodegradability of the target substance. Biofiltration typically removes 80% to 99% of volatile organic and inorganic compounds (VOCs and VICs).

$$\eta = 1 - \frac{C_{\text{out}}}{C_{\text{in}}} = 1 - e^{-kt} \qquad 42.14$$

13. BIOREACTION

Bioreactors (*reactor tanks*) are open or closed tanks containing dozens or hundreds of slowly rotating disks covered with a biological film of microorganisms (i.e., colonies). Closed tanks can be used to maintain anaerobic or other design atmosphere conditions. For example, methanotrophic bacteria are useful in breaking down chlorinated hydrocarbons (e.g., trichlorethylene (TCE), dichloroethylene (DCE), and vinyl chloride (VC)) that would otherwise be considered nonbiodegradable. Methanotrophic bacteria derive their food from methane gas that is added to the bioreactor. An enzyme, known as MMO, secreted by the bacteria breaks down the chlorinated hydrocarbons.

14. BIOVENTING

Bioventing is the treatment of contaminated soil in a large plastic-covered tank. Clean air, water, and nutrients are continuously supplied to the tank while off-gases are suctioned off. The off-gas is cleaned with activated carbon (AC) adsorption or with thermal or catalytic oxidation prior to discharge. Bioventing has been used successfully to remove volatile hydrocarbon compounds (e.g., gasoline and BTEX compounds) from soil.

15. COAL CONDITIONING

Coal intended for electrical generating plants can be modified into "self-scrubbing" coal by conditioning prior to combustion. Conditioning consists of physical separation by size, by cleaning (e.g., cycloning to remove noncombustible material, including up to 90% of the pyritic sulfur), and by the optional addition of limestone and other additives to capture SO_2 during combustion. Small-sized material is pelletized to reduce loss of fines and particulate emissions (dust).

16. CYCLONE SEPARATORS

Inertial separators of the *double-vortex, single cyclone* variety are suitable for collecting medium- and large-sized (i.e., greater than 5 μm) particles from spot sources. During operation, particulate matter in the incoming gas stream spirals downward at the outside and upward at the inside.[10] The particles, because of their greater mass, move toward the outside wall, where they drop into a collection bin. Figure 42.5 shows a double-vortex, single cyclone separator.

Figure 42.5 *Double-Vortex, Single Cyclone*

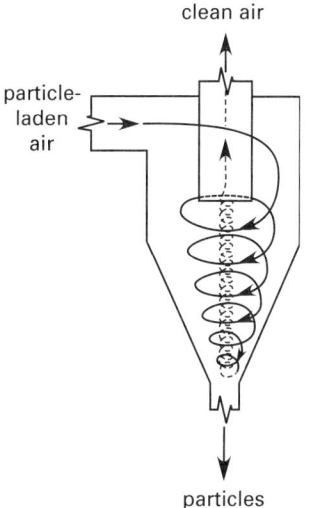

clean air

particle-laden air

particles

---

[10]The number of gas revolutions varies approximately between 0.5 and 10.0, with averages of 1.5 revolutions for simple cyclones and 5 revolutions for high-efficiency cyclones.

Solid Waste

The *cut size* is the diameter of particles collected with a 50% efficiency. Separation efficiency varies directly with (1) particle diameter, (2) particle density, (3) inlet velocity, (4) cyclone body length (ratio of cyclone body diameter to outlet diameter), (5) smoothness of inner wall, (6) number of gas revolutions, and (7) amount of particles in the flow. Collection efficiency decreases with increases in (1) gas viscosity, (2) gas density, (3) cyclone diameter, (4) gas outlet diameter, (5) gas inlet duct width, and (6) inlet area.

Collection efficiencies are not particularly high, and dusts ($5\,\mu\mathrm{m}$ to $10\,\mu\mathrm{m}$) are too fine for most cyclones. For geometrically similar cyclones, the collection efficiency varies directly with the dimensionless *separation factor, S.*

$$S = \frac{\mathrm{v}_{\mathrm{inlet}}^2}{rg} \qquad\qquad 42.15$$

Pressure drop, h, in feet (meters) of air across a cyclone can be roughly (i.e., with an accuracy of only approximately $\pm30\%$) estimated by Eq. 42.16. H and B are the height and width of the rectangular cyclone inlet duct, respectively; D_e is the gas exit duct diameter; and K is an empirical constant that varies from approximately 7.5 to 18.4.

$$h = \frac{KBH\mathrm{v}_{\mathrm{inlet}}^2}{2gD_e^2} \qquad\qquad 42.16$$

17. DECHLORINATION

Industrial and municipal wastewaters containing excessive amounts of *total residual chlorine* (TRC) must be dechlorinated prior to discharge. Sulfur dioxide, sodium metabisulfate, and sulfite salts are effective dechlorinating agents. For sulfur dioxide, the dose is approximately 0.9 lbm/lbm (0.9 kg/kg) of chlorine to be removed. Reaction is essentially instantaneous, being completed within 10 pipe diameters at turbulent flow. For open channels, a submerged weir may be necessary to obtain the necessary turbulence. TRC is reduced to less than detectable levels.

18. ELECTROSTATIC PRECIPITATORS

Electrostatic precipitators (ESPs) are used to remove particulate matter from gas streams. Collection efficiency for particulate matter is usually in the 95% to 99% range. ESPs are preferred over scrubbers because they are more economical to operate, dependable, and predictable and because they don't produce a moisture plume.

ESPs used on steam generator/electrical utility units treat gas that is approximately 280°F to 300°F (140°C to 150°C) with moisture being superheated.[11] This high temperature enhances buoyancy and plume dissipation. However, ESPs generally cannot be used with moist flows, mists, or sticky or hygroscopic particles. Scrubbers should be used in those cases. Relatively humid flows can be treated, although entrained water droplets can insulate particles, lowering their resistivities. Table 42.4 gives typical design parameters for ESPs.

Table 42.4 *Typical Electrostatic Precipitator Design Parameters*

parameter	typical range	
	U.S.	SI
efficiency	90% to 98%	
gas velocity	2 ft/sec to 4 ft/sec	0.6 m/s to 1.2 m/s
gas temperature		
standard	$\leq700°$F	$\leq370°$C
high-temperature	$\leq1000°$F	$\leq540°$C
special	$\leq1300°$F	$\leq700°$C
drift velocity	0.1 ft/sec to 0.7 ft/sec	0.03 m/s to 0.21 m/s
treatment/residence time	2 sec to 10 sec	
draft pressure loss	0.1 iwg to 0.5 iwg	0.025 kPa to 0.125 kPa
plate spacing	12 in to 16 in	30 cm to 41 cm
plate height	30 ft to 50 ft	9 m to 15 m
plate length/height ratio	1.0 to 2.0	
applied voltage	30 kV to 75 kV	

In operation, the gas passes over negatively charged tungsten *corona wires* or grids. Particles are attracted to positively charged collection plates.[12] The speed at which the particles move toward the plate is known as the *drift velocity, w.* Drift velocity is approximately 0.20 ft/sec to 0.30 ft/sec (0.06 m/s to 0.09 m/s), with 0.25 ft/sec (0.075 m/s) being a reasonable design value. Periodically, *rappers* vibrate the collection plates and dislodge the particles, which drop into collection hoppers.

Advantages of ESPs include high reliability, low maintenance, low power requirements, and low pressure drop. Disadvantages are sensitivity to particle size and resistivity, and the need to heat ESPs during start-up and shutdown to avoid corrosion from acid gas condensation.

For a rectangular electrostatic precipitator (ESP) with interior dimensions $W \times H \times L$, the flow rate is

$$Q = \mathrm{v}_g WH \qquad\qquad 42.17$$

---

[11]The flue gas, at approximately 1400°F (760°C), is cooled to this temperature range in a conditioning tower. Injected water increases the moisture content of the flue gas to approximately 25% by volume.
[12]The *tubular ESP*, used for collecting moist or sticky particles, is a variation on this design.

Solid Waste

The *residence time (exposure time)*, t_r, is

$$t_r = \frac{L}{v_g} \qquad 42.18$$

Collection efficiency is affected by particle *resistivity*, which is divided into three categories: low (less than 1×10^8 $\Omega\cdot$cm); medium, moderate, or normal (1×10^8 $\Omega\cdot$cm to 2×10^{11} $\Omega\cdot$cm); and high (more than 2×10^{11} $\Omega\cdot$cm). Particles in the medium resistivity range are collected most easily. Particles with low resistivity are easily charged, but upon contact with the charged plate, they rapidly lose their charges and are re-entrained into the gas flow. Particles with high resistivity coat the collection plate with a layer of insulation, causing *flashovers* and *back corona*, a localized electrical discharge. High resistivity particles can be brought into the moderate range by *particle conditioning*. When the gas temperature is below 350°F (177°C), adding moisture to the gas stream, reducing the temperature, or adding SO_3 and ammonia will reduce resistivity into the moderate range. When the gas temperature is above 350°F (177°C), increasing the temperature lowers particle resistivity.

Collection efficiency is also affected by the particle size. The theoretical *drift velocity (migration velocity or precipitation rate)* is proportional to the particle diameter. The theoretical drift velocity, w, of a particle in an electric field can be calculated from basic principles. In Eq. 42.19, q is the charge on the particle, which can be many times the charge on an electron, q_e (1.602×10^{-19} C); E is the electric field strength in V/m; ϵ is the particle's *dielectric constant (relative permittivity)*, approximately 1.5–2.6 for fly ash;[13] and ϵ_0 is the *permittivity of free space*, equal to 8.854×10^{-12} F/m (same as C/V, $C^2/N\cdot m^2$, and $A^2\cdot s^4/kg\cdot m^3$). C_c is the *Cunningham slip factor (Cunningham correction factor)*, which accounts for the reduction in drag on particles less than 1 μm in size when gas molecules slip past them. C_c approaches 1.0 for particles with $d_p > 5$ μm. λ is the *mean free path length* of the gas. For air at 68°F (20°C) and 1 atm, λ is approximately 0.0665 μm. (See Eq. 42.20.)

$$w \approx \frac{qEC_c}{3\pi\mu_g d_p} = \frac{Nq_eEC_c}{3\pi\mu_g d_p} \approx \left.\frac{\epsilon_0 E^2 d_p C_c}{(\epsilon+2)\mu_g}\right|_{sat} \qquad 42.19$$

$$C_c \approx 1 + \frac{\lambda}{d_{p,\text{Stk}}}\left(2.514 + 0.80\exp\left(\frac{-0.55d_{p,\text{Stk}}}{\lambda}\right)\right) \qquad 42.20$$

A particle will become saturated if it remains in a strong electric field sufficiently long. The *equilibrium charge* is found from the number of electron charges on a saturated spherical particle. The charging electric field is not necessarily the same as the drift electric field.

$$N_{\text{sat,spherical}} = \frac{3\epsilon_0\pi E_{\text{charging}}d_p^2}{(\epsilon+2)q_e} \qquad 42.21$$

In practice, the theoretical drift velocity is seldom used. Rather, an *effective drift velocity*, w_e, is used. Although it shares the "velocity" name and units, the effective drift velocity is not an actual velocity, but is an empirical design parameter derived from pilot tests of collection efficiency.

The *Deutsch-Anderson equation* predicts the single-particle fractional *collection efficiency (removal efficiency)*, η, of an electrostatic precipitator with total collection area, A, of all plates. The exponent, y, is 1 for fly ash and for anything else in the absence of specific knowledge otherwise. *Penetration*, Pt, is the fraction of particles that pass through the ESP uncollected. Penetration is the complement of collection efficiency.

$$\eta = \frac{C_{\text{in}}-C_{\text{out}}}{C_{\text{in}}} = \frac{\dot{m}_{\text{in}}-\dot{m}_{\text{out}}}{\dot{m}_{\text{in}}} = 1-\exp\left(\frac{-Aw_e}{Q}\right)^y \qquad 42.22$$

$$\text{Pt} = 1-\eta \qquad 42.23$$

The effect of flow rate on the collection efficiency is predicted by Eq. 42.24.

$$\frac{Q_1}{Q_2} = \ln(\eta_1-\eta_2) \qquad 42.24$$

Collection efficiency can also be expressed as a function of the *residence time*. K is the ratio of collection plate area to internal volume.

$$\eta = 1-\exp(-Kw_e t_r) = 1-\exp\left(-\left(\frac{A}{WHL}\right)w_e t_r\right) \qquad 42.25$$

One of the most important factors affecting the collection efficiency is the *specific collection area*, SCA, which is the ratio of the total collection surface area to the gas flow rate into the collector, usually reported in $ft^2/1000$ ft^3·min ($m^2/1000$ m^3·h). The higher the SCA, the higher the collection efficiency will be.

$$\text{SCA} = \frac{A}{Q} \qquad 42.26$$

---

[13]The dielectric constant of fly ash depends greatly on the temperature and amount of unburned carbon.

Using units common to the industry, the instantaneous collection efficiency of an ESP collector is

$$\eta = 1 - \exp(-0.06 w_{e,\text{ft/sec}} \text{SCA}_{\text{ft}^2/1000\,\text{ACFM}}) \qquad 42.27$$

The *corona power ratio*, CPR, is the power consumed by the corona in developing the electric field divided by the airflow. Generally, collection efficiency increases with increased CPR. The *corona current ratio*, CCR, is the current per unit plate area. The *power density*, P_A, is the power per unit plate area.

$$\text{CPR}_{\text{W-min/ft}^3} = \frac{P_{c,\text{W}}}{Q_{\text{ft}^3/\text{min}}} = \frac{I_{c,\text{A}} V_{\text{V}}}{Q_{\text{ft}^3/\text{min}}} \qquad 42.28$$

$$\text{CCR}_{\mu\text{A/ft}^2} = \frac{I_{c,\mu\text{A}}}{A_{\text{ft}^2}} \qquad 42.29$$

$$P_{A,\text{W/ft}^2} = \frac{P}{A} \qquad 42.30$$

Typical values for ESP design parameters are given in Table 42.5.

Table 42.5 Typical ESP Design Parameters

design parameter	typical value
particle effective drift velocity	0.05–1.0 ft/sec (0.015–0.3 m/s)
gas velocity	2–5 ft/sec typical; 15 ft/sec max (0.6–1.5 m/s typical; 4.5 m/s max)
number of stages	1–7
total specific collection area	100–1000 ft^2/1000 ACFM (5.5–55 m^2/1000 m^3·h)
pressure drop	> 0.5 in wg (0.13 kPa)

19. FLUE GAS RECIRCULATION

NOx emissions can be reduced when thermal dissociation is the primary NOx source (as it is when low-nitrogen fuels such as natural gas are burned) by recirculating a portion (15% to 25%) of the flue gas back into the furnace. This process is known as *flue gas recirculation* (FGR). The recirculated gas absorbs heat energy from the flame and lowers the peak temperature. Thermal NOx formation can be reduced by up to 50%. The recirculated gas should not be more than 600°F (315°C).

20. FLUIDIZED-BED COMBUSTORS

Fluidized-bed combustors (FBCs) are increasingly being used in steam/electric generation systems and for destruction of hazardous wastes. A *bubbling bed FBC*, as shown in Fig. 42.6, consists of four major components: (1) a windbox (plenum) that receives the fluidizing/combustion air, (2) an air distribution plate that transmits the air at 10 ft/sec to 30 ft/sec (3 m/s to 9 m/s) from the windbox to the bed and prevents the

Figure 42.6 Fluidized-Bed Combustor

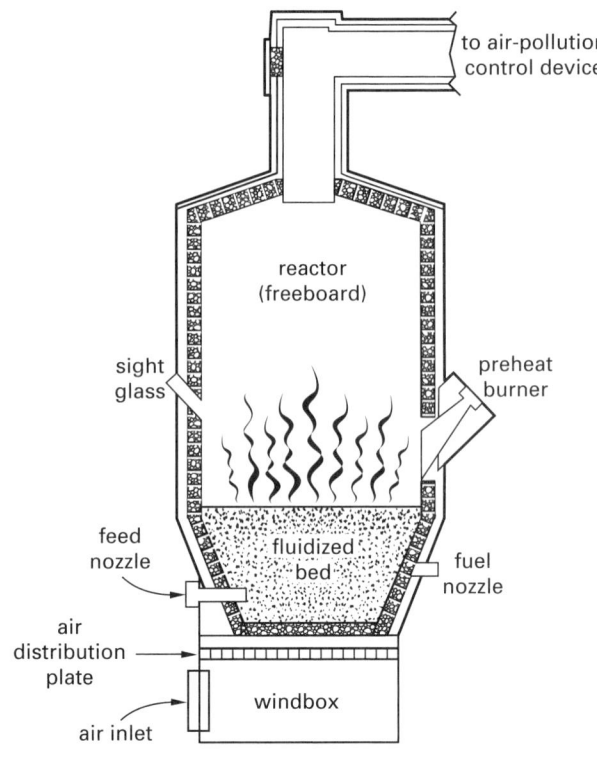

bed material from sifting into the windbox, (3) the fluid bed of inert material (usually sand or product ash), and (4) the freeboard area above the bed.

During the operation of a bubbling bed fluidized-bed combustor, the inert bed is levitated by the upcoming air, taking on many characteristics of a fluid (hence, the name FBC). The bed "boils" vigorously, offering an extremely large heat-transfer area, resulting in the thorough mixing of feed combustibles and air. Combustible material usually represents less than 1% of the bed mass, so the rest of the bed acts as a large thermal flywheel.

Combustion of volatile materials is completed in the freeboard area. Ash is generally reduced to small size so that it exits with the flue gas. In some cases (e.g., deliberate pelletization or wastes with high ash content), ash can accumulate in the bed. The fluid nature of the bed allows the ash to float on its surface, where it is removed through an overflow drain.

Most FBC systems use forced air with a single blower at the front end. If there are significant losses due to heat recovery or pollution control systems, an exhaust fan may also be used. In that case, the *null (balanced draft) point* should be in the freeboard area.

Large variations in the composition of the flue gases (known as *puffing*) is minimized by the long residence time and the large heat reservoir of the bed. Air pollution control equipment common to most boilers and incinerators is used with fluidized-bed combustors. Either wet scrubbers or baghouses can be used.

The temperature of the bed can be as high as approximately 1900°F (1040°C), though in most applications, temperatures this high are neither required nor desirable.[14] Most systems operate in the 1400°F to 1650°F (760°C to 900°C) range.

Three main options exist for reducing temperatures with overautogeneous fuels:[15] (1) Water can be injected into the bed. This has the disadvantage of reducing downstream heat recovery. (2) Excess air can be injected. This requires the entire system to be sized for the excess air, increasing its cost. (3) Heat-exchange coils can be placed within the bed itself, in which case, the name *fluidized-bed boiler* is applicable.

Air can be preheated to approximately 1000°F to 1250°F (540°C to 680°C) for use with subautogenous fuels, or auxiliary fuel can be used.

Contempory FBC boilers for steam/electricity generation are typically of the *circulating fluidized-bed boiler design*. An important aspect of FBC boiler operation is the in-bed gas desulfurization and dechlorination that occurs when limestone and other solid reagents are injected into the combustion area. In addition, circulating fluidized-bed boilers have very low NOx emissions (i.e., less than 200 ppm).

21. INJECTION WELLS

Properly treated and stabilized liquid and low-viscosity wastes can be injected under high pressure into appropriate strata 2000 ft to 6000 ft (600 m to 1800 m) below the surface. The wastes displace natural fluids, and the injection well is capped to maintain the pressure. Injection wells fail primarily by waste plumes through fractures, cracks, fault slips, and seepage around the well casing. Figure 42.7 shows a typical injection well installation.

22. INCINERATORS, FLUIDIZED-BED COMBUSTION

In an FBC, combustion is efficient and excess air required is low (e.g., 25% to 50%). Destruction is essentially complete due to the long residence time (5 sec to 8 sec for gases, and even longer for solids), high turbulence, and exposure to oxygen. Combustion control is simple, and combustion is often maintained within a 15°F (8°C) band. Both overautogeneous and subautogenous feeds can be handled. Due to the thermal mass of the bed, the FBC temperature drops slowly—at the rate of about 10°F/hr (5°C/h) after shutdown. Start-up is fast and operation can be intermittent.

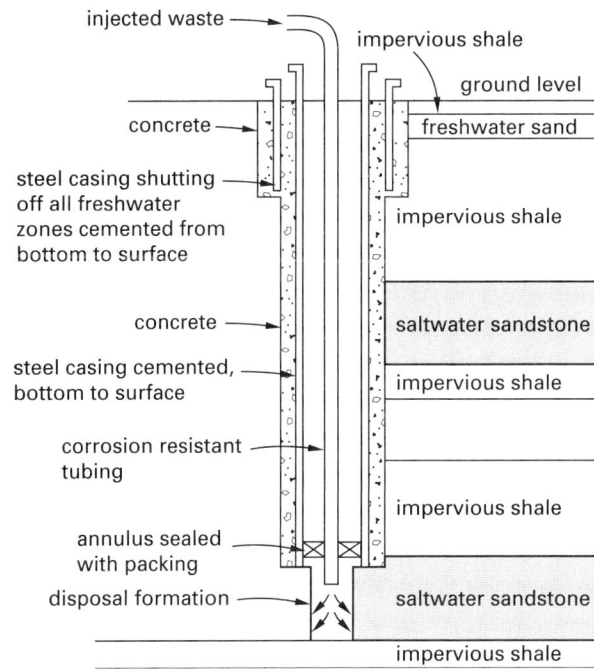

Figure 42.7 *Typical Injection Well Installation*

Relative to other hazardous waste disposal methods, NOx and metal emissions are low, and the organic content of the ash is very low (e.g., below 1%). Because of the long residence time, FBC systems do not usually require afterburners or secondary combustion chambers (SCCs). Due to the turbulent combustion process, the combustion temperature can be 200°F to 300°F (110°C to 170°C) lower than in rotary kilns. These factors translate into fuel savings, fewer or no NOx emissions, and lower metal emissions.

Most limitations of FBC systems relate to the feed. The feed must be of a form (small size and roughly regular in shape) that can be fluidized. This makes FBCs ideal for non-atomizable liquids, slurries, sludges, tars, and granular solids. It is more appropriate (and economical) to destroy atomizable liquid wastes in a boiler or special injection furnace and large bulky wastes in a rotary kiln or incinerator.

FBCs are usually inapplicable when the feed material or its ash melts below the bed temperature.[16] If the feed melts, it will agglomerate and defluidize the bed. When it is necessary to burn a feed with low melting temperature, two methods can be used. The bed can be operated as a *chemically active bed*, where operation is close to but below the ash melting point of 1350°F to 1450°F (730°C to 790°C). A small amount of controlled melting is permitted to occur, and the agglomerated ash is removed.

When a higher temperature is desirable in order to destroy hazardous materials, the melting point of feed

---

[14]It is a common misconception that extremely high temperatures are required to destroy hazardous waste.

[15]A *subautogenous waste* has a heating value too low to sustain combustion and requires a supplemental fuel. Conversely, an *overautogeneous waste* has a heating value in excess of what is required to sustain combustion and requires temperature control.

[16]The melting temperature of a eutectic mixture of two components may be lower than the individual melting temperatures.

Solid Waste

can be increased by injecting low-cost additives (e.g., calcium hydroxide or kaolin clay). These combine with salts of alkali metals to form refractory-like materials with melting points in the 1950°F to 2350°F (1070°C to 1290°C) range.

The design of a fluidizing-bed incinerator starts with a heat balance. Energy enters the FBC during combustion of the feed and auxiliary fuel and from sensible heat contributions of the air and fuel. Some of the heat is recovered and reused. The remainder of the heat is lost through sensible heating of the combustion products, water vapor, excess air, and ash, and through radiation from the combustor vessel and associated equipment. Since these items may not all be known in advance, the following assumptions are reasonable.

- Radiation losses are approximately 5%.

- Sensible heat losses from the ash are minimal, particularly if the feed contains significant amounts of moisture.

- Excess air will be approximately 40%.

- Combustion temperature will be approximately 1400°F (760°C).

- If air is preheated, the preheat temperature will be approximately 1000°F (540°C).

The fuel's *specific feed characteristic* (SFC) is the ratio of the higher (gross) heating value to the moisture content.

$$SFC = \frac{\text{total heat of combustion}}{\text{mass of water}} = \frac{m_{\text{solid}}\text{HHV}}{m_w} \qquad 42.31$$

Typical values of the specific feed characteristic range from a low of approximately 1000 Btu/lbm (2300 kJ/kg) for wastewater treatment plant biosolids, to more than 60,000 Btu/lbm (140 MJ/kg) for barks, sawdust, and RDF.[17] Making the previous assumptions and using the SFC as an indicator, the following generalizations can be made.

- Fuels with SFCs that are less than 2600 Btu/lbm (6060 kJ/kg) require a 1000°F (540°C) hot windbox, and are autogenous at that value and subautogenous below that value. The SFC drops to 2400 Btu/lbm (5600 kJ/kg) with air preheated to 1200°F (650°C). Below the subautogenous SFC, water evaporation is the controlling design factor, and auxiliary fuel is required.

- Fuels with an SFC of 4000 Btu/lbm (9300 kJ/kg) are autogenous with a cold windbox and are overautogenous above that value. Combustion is the controlling design factor for overautogenous SFCs.

---

[17]Although the units are the same, the specific feed characteristic is not the same as the heating value.

Example 42.1

A wastewater treatment sludge is dewatered to 75% water by weight. The solids have a heating value of 6500 Btu/lbm (15 000 kJ/kg). The dewatered sludge enters a fluidized-bed combustor with a 1000°F (540°C) windbox at the rate of 15,000 lbm/hr (1.9 kg/s). (a) What is the specific feed characteristic? (b) Approximately what energy (in Btu/hr) must the auxiliary fuel supply?

SI Solution

(a) The dewatered sludge consists of 25% combustible solids and 75% moisture. The total mass of sludge required to contain 1.0 kg water is

$$m_{\text{sludge}} = \frac{m_w}{x_w} = \frac{1 \text{ kg}}{0.75} = 1.333 \text{ kg}$$

The mass of combustible solids per kilogram of water is

$$m_{\text{solids}} = 0.25 m_{\text{sludge}} = (0.25)(1.333 \text{ kg})$$
$$= 0.3333 \text{ kg}$$

From Eq. 42.31, the specific feed characteristic is

$$SFC = \frac{\text{total heat of combustion}}{m_w} = \frac{m_{\text{solid}}\text{HHV}}{m_w}$$

$$= \frac{(0.3333 \text{ kg})\left(15\,000 \,\frac{\text{kJ}}{\text{kg}}\right)\left(1000 \,\frac{\text{J}}{\text{kJ}}\right)}{1 \text{ kg}}$$

$$= 5.0 \times 10^6 \text{ J/kg}$$

(b) Making the listed assumptions, operation with a 540°C hot windbox requires an SFC of approximately 6060 kJ/kg. Therefore, the energy per kilogram of water in the fuel that an auxiliary fuel must provide is

$$(6060 \text{ kJ})\left(1000 \,\frac{\text{J}}{\text{kJ}}\right) - 5 \times 10^6 \text{ J} = 1.06 \times 10^6 \text{ J}$$

The energy supplied by the auxiliary fuel is

$$\dot{m}_{\text{fuel}} x_w \times \text{SFC deficit}$$
$$= \left(1.9 \,\frac{\text{kg}}{\text{s}}\right)(0.75)\left(1.06 \times 10^6 \,\frac{\text{J}}{\text{kg}}\right)$$
$$= 1.51 \times 10^6 \text{ J/s} \quad (1.5 \text{ MW})$$

Customary U.S. Solution

(a) The dewatered sludge consists of 25% combustible solids and 75% moisture. The total mass of sludge required to contain 1.0 lbm water is

$$m_{\text{sludge}} = \frac{m_w}{x_w} = \frac{1 \text{ lbm}}{0.75} = 1.333 \text{ lbm}$$

Solid Waste

The mass of combustible solids per pound of water is

$$m_{\text{solids}} = 0.25 m_{\text{sludge}} = (0.25)(1.333 \text{ lbm})$$
$$= 0.3333 \text{ lbm}$$

From Eq. 42.31, the specific feed characteristic is

$$\text{SFC} = \frac{\text{total heat of combustion}}{m_w} = \frac{m_{\text{solid}}\text{HHV}}{m_w}$$
$$= \frac{(0.3333 \text{ lbm})\left(6500 \dfrac{\text{Btu}}{\text{lbm}}\right)}{1 \text{ lbm}}$$
$$= 2166 \text{ Btu/lbm}$$

(b) Making the listed assumptions, operation with a 1000°F hot windbox requires an autogenous SFC of approximately 2600 Btu/lbm. Therefore, the energy per pound of water in the fuel that an auxiliary fuel must provide is

$$2600 \text{ Btu} - 2166 \text{ Btu} = 434 \text{ Btu}$$

The energy supplied by the auxiliary fuel is

$$\dot{m}_{\text{fuel}} x_w \times \text{SFC deficit}$$
$$= \left(15{,}000 \frac{\text{lbm}}{\text{hr}}\right)(0.75)\left(434 \frac{\text{Btu}}{\text{lbm}}\right)$$
$$= 4.88 \times 10^6 \text{ Btu/hr}$$

23. INCINERATION, GENERAL

Most rotary kiln and liquid injection incinerators have *primary* and *secondary combustion chambers* (SCCs). Kiln temperatures are approximately 1200°F to 1400°F (650°C to 760°C) for soil incineration and up to 1700°F (930°C) for other waste types.[18] SCC temperatures are higher—1800°F (980°C) for most hazardous wastes and 2200°F (1200°C) for liquid PCBs. The waste heat may be recovered in a boiler, but the combustion gas must be cooled prior to further processing. *Thermal ballast* can be accomplished by injecting large amounts of excess air (typical when liquid fuels are burned) or by quenching with a water spray (typical in rotary kilns).

SCCs are necessary to destroy toxics in the off-gases. SCCs are vertical units with high-swirl, vortex-type burners. These produce high *destruction removal efficiencies* (DREs) with low retention times (e.g., 0.5 sec) and moderate-to-high temperatures, even for chlorinated compounds. When soil with fine clay is incinerated, fines can build up in the SCC, causing slagging and other problems. A refractory-lined cyclone located

after the primary combustion chamber can be used to reduce particle carryover to the SCC.

Prior to full operation, incinerators must be tested in a trial burn with a *principal organic hazardous constituent* (POHC) that is in or has been added to the waste. The POHC must be destroyed with a DRE of at least 99.99% by weight.

For nontoxic organics, a DRE of 95% is a common requirement. Hazardous wastes require a DRE of 99.99%. Certain hazardous wastes, including PCBs and dioxins, require a 99.9999% DRE. This is known as the *six nines rule*.

Emission limitations of some pollutants depend on the incoming concentration and the height of the stack.[19] Thus, in certain circumstances, raising the stack is the most effective method of being in compliance. This is considered justified on the basis that ground-level concentrations will be lower with higher stacks.

Rules of thumb regarding incinerator performance are:

- Stoichiometric combustion requires approximately 725 lbm (330 kg) of air for each million Btu of fuel or waste burned.

- 100% excess air is required.

- Stack gas dew point is approximately 180°F (80°C).

- Water-spray-quenched flue gas is approximately 40% moisture by weight.

Table 42.6 gives performance parameters typical of incinerators.

Table 42.6 Representative Incinerator Performance

	type of incinerator/use			
			soil incinerators	
	rotary kiln	liquid injection	hazard-ous	nonhaz-ardous
waste heating value				
(Btu/lbm)	15,000	20,000	0	0
(MJ/kg)	35	46	0	0
kiln temp				
(°F)	1700	–	1650	850
(°C)	925	–	900	450
SCC temp				
(°F)	1800–2200	2000–2200	1800	1400
(°C)	980–1200	1100–1200	980	750
SCC mean residence time (sec)	2	2	2	1
O_2 in stack gas (%)	10%	12%	9%	6%

---

[18]Temperatures higher than 1400°F (760°C) may cause incinerated soil to vitrify and clog the incinerator.

[19]Some of the metallic pollutants treated this way include antimony, arsenic, barium, beryllium, cadmium, chromium, lead, mercury, silver, and thallium.

Solid Waste

Example 42.2

What is the mass (in lbm/hr) of water required to spray-quench combustion gases from the incineration of hazardous waste with a heating rate of 50 MBtu/hr? No additional fuel is added to the incinerator.

Solution

Use the rules of thumb. Assume 100% excess air is required. The total dry air required is

$$m_a = \frac{(2)\left(50 \times 10^6 \; \dfrac{\text{Btu}}{\text{hr}}\right)(725 \; \text{lbm})}{10^6 \; \dfrac{\text{Btu}}{\text{hr}}}$$

$$= 72{,}500 \; \text{lbm/hr} \quad [\text{dry}]$$

Since the quenched combustion gas is 40% water by weight, it is 60% dry air by weight. The total mass of wet combustion gas produced per hour is

$$m_t = \frac{m_a}{x_a} = \frac{72{,}500 \; \dfrac{\text{lbm}}{\text{hr}}}{0.6}$$

$$= 1.21 \times 10^5 \; \text{lbm/hr}$$

The required mass of quenching water is

$$m_w = x_w m_t = (0.4)\left(1.21 \times 10^5 \; \frac{\text{lbm}}{\text{hr}}\right)$$

$$= 4.84 \times 10^4 \; \text{lbm/hr}$$

24. INCINERATION, HAZARDOUS WASTES

Most hazardous waste incinerators use rotary kilns. Waste in solid and paste form enters a rotating drum where it is burned at 1850°F to 2200°F (1000°C to 1200°C). (See Table 42.6 for other representative performance characteristics.) Slag is removed at the bottom, and toxic gases exit to a tall, vortex secondary combustion chamber (SCC). Gases remain in the SCC for 2 sec to 4 sec where they are completely burned at approximately 1850°F (1000°C). Liquid wastes are introduced into and destroyed by the SCC as well. Heat from off-gases may be recovered in a boiler. Typical flue gas cleaning processes include electrostatic precipitation, two-stage scrubbing, and NOx removal.

Common problems with hazardous waste incinerators include (1) inadequate combustion efficiency (easily caused by air leakage in the drum and uneven fuel loading) resulting in incomplete combustion of the primary organic hazardous component (POHC), emission of CO, NOx, and *products of incomplete combustion* (also known as *partially incinerated compounds* or PICS) and metals, (2) meeting low dioxin limits, and (3) minimizing the toxicity of slag and fly ash.

These problems are addressed by (1) reducing air leaks in the drum and (2) introducing air to the SCC through multiple sets of ports at specific levels. Gas is burned in the SCC in substoichiometric conditions, with the vortex ensuring adequate mixing to obtain complete combustion. Dioxin formation can be reduced by eliminating the waste-heat recovery process, since the lower temperatures present near waste-heat boilers are ideal for dioxin formation. Once formed, dioxin is removed by traditional end-of-pipe methods. Figure 42.8 shows a large-scale hazardous waste incinerator.

25. INCINERATION, INFRARED

Infrared incineration (II) is effective for reducing dioxins to undetectable levels. The basic II system consists of a waste feed conveyor, an electrical-heated primary chamber, a gas-fired afterburner, and a typical flue gas cleanup (FGC) system (i.e., scrubber, electrostatic precipitator, and/or baghouse). Electrical heating elements heat organic wastes to their combustion temperatures. Off-gas is burned in a secondary combustion chamber (SCC).

26. INCINERATION, LIQUIDS

Liquid-injection incinerators can be used for atomizable liquids. Such incinerators have burners that fire directly into a refractory-lined chamber. If the liquid waste contains salts or metals, a downfired liquid-injection incinerator is used with a submerged quench to capture the molten material. A typical flue gas cleanup (FGC) system (i.e., scrubber, electrostatic precipitator, and/or baghouse) completes the system.

Incineration of organic liquid wastes usually requires little external fuel, since the wastes are overautogenous and have good heating values. The heating value is approximately 20,000 Btu/lbm (47 MJ/kg) for solvents and approximately 8000 Btu/lbm to 18,000 Btu/lbm (19 MJ/kg to 42 MJ/kg) for chlorinated compounds.

27. INCINERATION, OXYGEN-ENRICHED

Oxygen-enriched incineration is intended primarily for dioxin removal and is operationally similar to that of a rotary kiln. However, the burner includes oxidant jets. The jets aspirate furnace gases to provide more oxygen for combustion. Apparent advantages are low NOx production and increased incinerator feed rates.

28. INCINERATION, PLASMA

A wide variety of solid, liquid, and gaseous wastes can be treated in a *plasma incinerator*. Wastes are heated by an electric arc to higher than 5000°F (2760°C), dissociating them into component atoms. Upon cooling, atoms recombine into hydrogen gas, nitrogen gas, carbon monoxide, hydrochloric acid, and particulate carbon. The ash cools to a nonleachable, vitrified matrix. Off-gases pass through a normal train of cyclone, baghouse, and scrubbing operations. The process has a very high DRE. However, energy requirements are high.

Figure 42.8 Large-Scale Hazardous Waste Incinerator

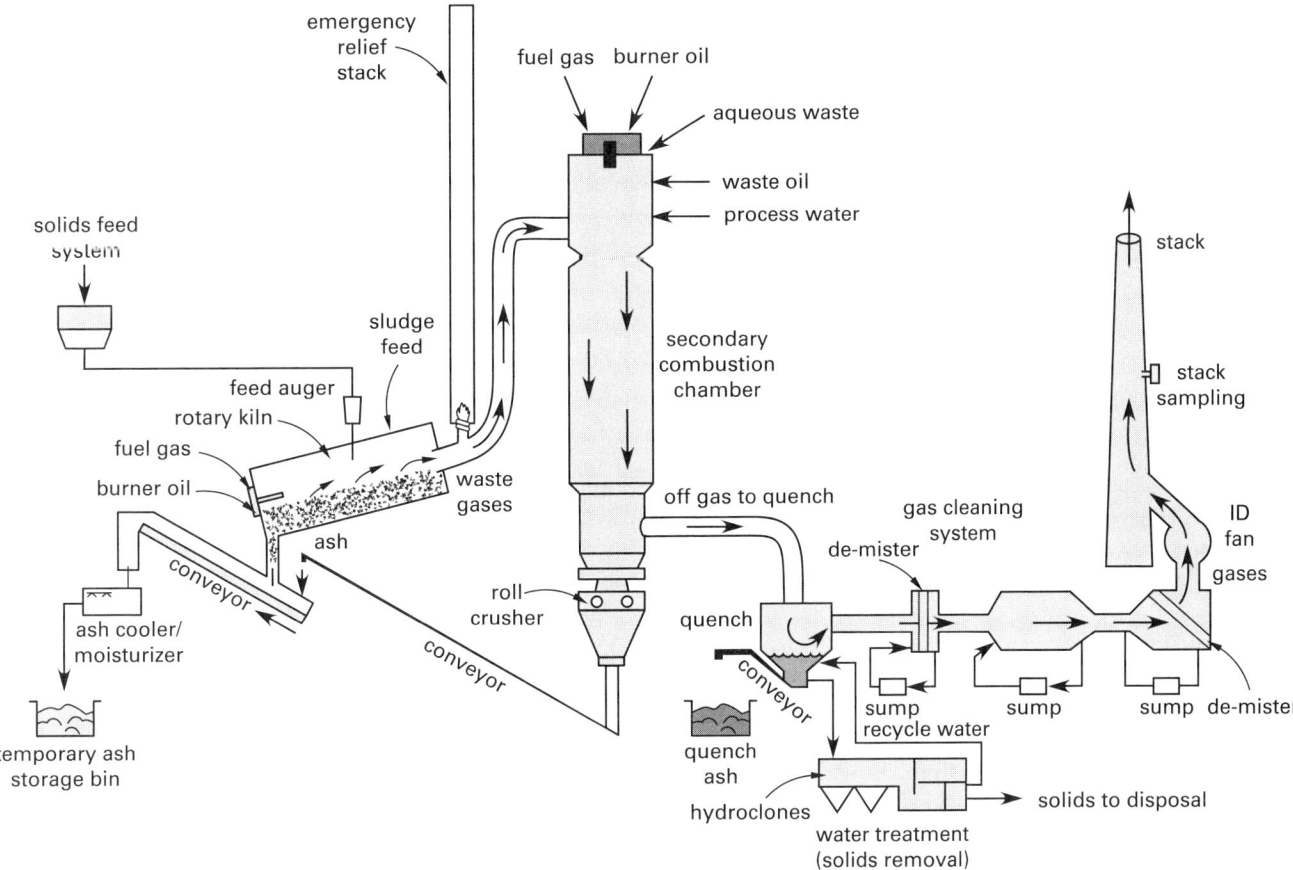

29. INCINERATION, SOIL

Incineration can completely decontaminate soil. Incinerators are often thought of as being fixed facilities, as are cement kilns and special-use (e.g., Superfund) incinerators. However, mobile incinerators can be brought to sites when the soil quantities are large (e.g., 2000 tons to 100,000 tons (1800 Mg to 90,000 Mg)) and enough setup space is available.[20] (See also Sec. 42.30.)

30. INCINERATION, SOLIDS AND SLUDGES

Rotary kilns and fluidized-bed incinerators are commonly used to incinerate solids and sludges. The feed system depends on the waste's physical characteristics. Ram feeders are used for boxed or drummed solids. Bulk solids are fed via chutes or screw feeders. Sludges are fed via lances or by premixing with solids.

The constant rotation of the shell moves wastes through rotary kilns and promotes incineration. External fuel is required in rotary kilns if the heating value of the waste is below 1200 Btu/lbm (2800 kJ/mg) (i.e., is subautogenous). Additional fuel is required in the secondary chamber, as well.

Fluidized-bed incinerators work best when the waste is consistent in size and texture. An important benefit is the ability to introduce limestone and other solid reagents to the bed in order to remove HCl and SO_2.

31. INCINERATION, VAPORS

Vapor incinerators, also known as *afterburners* and *flares*, convert combustible materials (gases, vapors, and particulate matter) in the stack gas to carbon dioxide and water. Afterburners can be either direct-flame or catalytic in operation.

32. LOW EXCESS-AIR BURNERS

NOx formation in gas-fired boilers can be reduced by maintaining excess air below 5%.[21] *Low excess-air burners* use a forced-draft and self-recirculating combustion chamber configuration to approximate multistaged combustion.

33. LOW-NOX BURNERS

Low (or *ultralow*) *NOx burners* (LNB) in gas-fired applications use a combination of staged-fuel burning and

---

[20]Modified asphalt batch processing plants can be used to incinerate soils contaminated with low-heating value, nonchlorinated hydrocarbons.

[21]Reducing excess air from 30% to 10%, for example, can reduce NOx emissions by 30%.

Solid Waste

internal flue gas recirculation (FGR). Recirculation within a burner is induced by either the pressure of the fuel gas or other agents (e.g., medium-pressure steam or compressed air).

34. MECHANICAL SEALS

Fugitive emissions from pumps and other rotating equipment can be reduced or eliminated using current technology by the proper selection and installation of mechanical seals. The three major classes of mechanical seals are single seals, tandem seals (dual seals placed next to each other), and double seals (dual seals mounted back to back or face to face).

The most economical and reliable sealing device is a *single seal*. It has a minimum number of parts and requires no support devices. However, since the pumped product is usually the lubricant for the seal face, small amounts of the product escape into the environment. Emissions are generally below 1000 ppmv, and are often below 100 ppmv.

Tandem seals consist of two seal assemblies separated by a buffer fluid at a pressure lower than the seal-chamber liquid. The primary inner seal operates at full pump pressure. The outer seal operates in the buffer fluid. Tandem seals can achieve zero emissions when used with a vapor recovery system, provided that the pumped product's specific gravity is less than that of the buffer fluid and the product is immiscible with the buffer fluid.

If the buffer fluid used in a tandem dual seal is a controlled substance, emissions from the outer seal must also meet the emission limits. Seals using glycol as the buffer fluid typically achieve zero emissions. Seals using diesel oil or kerosene have emissions in the 25 ppmv to 100 ppmv range.

Double seals are recommended for hazardous fluids with specific gravities less than 0.4 and are a good choice when a vapor recovery system is not available. (Liquids with low specific gravities do not lubricate the seal well.) Double seals consist of two seal assemblies connected by a common collar; they operate with a barrier fluid kept at a pressure higher than that of the pumped product. Double seals can reduce the emission rate to zero.

Figure 42.9 shows the relationship between typical emissions and specific gravity for the different seal types.

35. MULTIPLE PIPE CONTAINMENT

The U.S. Environmental Protection Agency (EPA) requires *multiple pipe containment* (MPC) in the storage and transmission of hazardous fluids.[22] MPC systems consist of a carrier pipe or bundle of pipes, a

Figure 42.9 *Representative Emissions to Atmosphere for Mechanical Seals* (1 cm from source)*

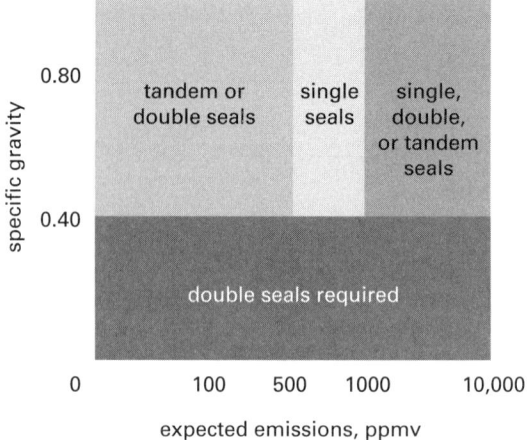

*maximum seal size, 6 in (153 mm); maximum pressure, 600 psig (40 bar); maximum speed, 3600 rpm.

common casing, and a leak detector consisting of redundant instrumentation with automatic shutdown features. Casings are generally not pressurized. If one of the pipes within the bundle develops a leak, the fluid will remain in the casing and the detector will signal an alarm and automatically shut off the liquid flow.

Some of the many factors that go into the design of an MPC system are (1) the number of pipes to be grouped together; (2) the weight and strength of the pipes, carrier, and supports; (3) the compatibility of pipe materials and fluids; (4) differential expansion of the container and carrier pipes; and (5) a method of leak detection.

Carrier pipes within the casing should be supported and separated from each other by internal supports (i.e., perforated baffles within the casing). Carrier pipes pass through oval-shaped holes in the internal supports. Holes are oval and oversized to allow flexing of the carrier pipes due to expansion. Other holes through the internal supports allow venting and draining of the casing in the event of a leak. If multiple detectors are used, isolation baffles can be included to separate the casing into smaller sections.

When a pipe bundle changes elevation as well as direction, the carrier pipes may change their positions relative to one another. A pipe on the outside (i.e., at 3 o'clock) may end up being the lower pipe (i.e., at 6 o'clock).[23] This change in relative positioning is known as *rotation*. Rotation complicates the accommodation of pipes that must enter and exit the bundle at specific locations.

Rotation occurs when a change in direction is located at the top or bottom of an elevation change. Therefore, unless rotation can be tolerated, changes in direction

---

[22]In addition to multiple pipe containment, other factors that contribute to reduction of fugitive emissions are (1) the use of tongue-in-groove flanges, and (2) the reduction of as many nozzles as possible from storage tanks and reactor vessels.

[23]It is important to consistently "face" the same way (e.g., in the direction of flow).

should not be combined with changes in elevation. In Fig. 42.10(a), pipe A remains to the left of pipe B. In Fig. 42.10(b), pipe A starts out to the left of pipe B and ends up above pipe B.

Another problem associated with MPC systems is expansion of the containment casing and carrier pipes caused by differences in ambient and fluid temperatures and the use of different pipe materials. For pipes containing direct changes, there must be space at each elbow to permit the pipe expansion. If both ends of a pipe are fixed, expansion and contraction will cause the pipes to flex unless Z-bends or loops are included.

Figure 42.10 Rotation in Pipe Bundles

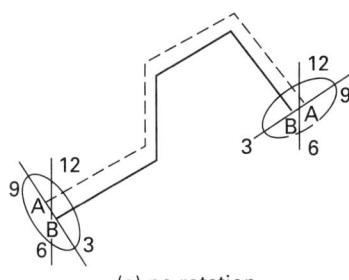

(a) no rotation

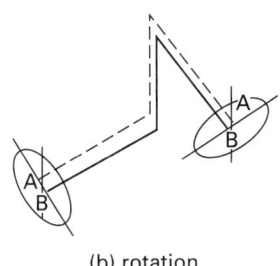

(b) rotation

Differential expansion must be calculated from the largest expected temperature difference using Eq. 42.32. α is the coefficient of thermal expansion found from Table 42.7.

$$\Delta L = \alpha L \Delta T \qquad 42.32$$

Computer programs are commonly used to design pipe networks. These programs are helpful in locating guides and anchors and in performing stress analyses on the final design. Drains, purge points, baffles, and joints are generally chosen by experience.

36. OZONATION

Ozonation is one of several advanced oxidation methods capable of reducing pollutants from water. Ozone is a powerful oxidant produced by electric discharge through liquid oxygen. Ozone is routinely used in the water treatment industry for disinfection. Table 42.8 shows reactions used for oxidation of several common industrial wastewater pollutants.

Table 42.7 Approximate Coefficients of Thermal Expansion of Piping Materials (at 70°F (21°C))

pipe material	U.S. (ft/ft-°F)	SI (m/m·°C)
carbon steel	6.1×10^{-6}	1.1×10^{-5}
chlorinated polyvinyl chloride (CPVC)	3.8×10^{-5}	6.8×10^{-5}
copper	9.5×10^{-6}	1.7×10^{-5}
fiberglass-reinforced polyethylene (FRP)	8.5×10^{-6}	1.5×10^{-5}
polyethylene (PE)	8.3×10^{-5}	1.5×10^{-4}
polyvinyl chloride (PVC)	3.0×10^{-5}	5.4×10^{-5}
stainless steel	9.1×10^{-6}	1.6×10^{-5}

Table 42.8 Oxidation of Industrial Wastewater Pollutants*

	Cl_2	ClO_2	$KMnO_4$	O_3	H_2O_2	OH^-
amines	C	P	P	P	P	C
ammonia	C	N	P	N	N	N
bacteria	C	C	C	C	P	C
carbohydrates	P	P	P	P	N	C
chlorinated solvents	P	P	P	P	N	C
phenols	P	C	C	C	P	C
sulfides	C	C	C	C	C	C

*C = complete reaction; P = partially effective; N = not effective

37. SCRUBBING, GENERAL

Scrubbing is the act of removing particulate matter from a gaseous stream, although substances in gaseous and liquid forms may also be removed.[24]

38. SCRUBBING, CHEMICAL

Chemical scrubbing using oxidizing compounds such as chlorine, ozone, hypochlorite, or permanganate rapidly destroys volatile organic compounds (VOCs). Efficiencies are typically 95% for highly reactive substances, but are lower for hydrocarbons and substances with low reactivities.

39. SCRUBBING, DRY

Dry scrubbing, also known as *dry absorption* and *semi-dry scrubbing*, is one form of sorbent injection. It is commonly used to remove SO_2 from flue gas. In operation, flue gases pass through a scrubbing chamber where

[24]As it relates to air pollution control, the term *scrubbing* is loosely used. The term may be used in reference to any process that removes any substance from flue gas. In particular, any process that removes SO_2 by passing flue gas through lime or limestone is referred to as *scrubbing*.

a slurry of lime and water is sprayed through them.[25] The slurry is produced by high-speed rotary atomizers. The sorbent (a reagent such as lime, limestone, hydrated lime, sodium bicarbonate, or trona, also called sodium sesqui-carbonate) is injected into the flue gas in either dry or slurry form. The flue gas heat drives off the water. In either case, a dry powder is carried through the flue gas system. An electrostatic precipitator or baghouse captures the fly ash and calcium sulfate particulates.

Dry scrubbers do not saturate the gas stream, even when some moisture is used as a carrier. The moisture that is added is evaporated and does not condense, so unlike wet scrubbers, there is no visible moisture plume.

Dry scrubbing has an SO_2 removal efficiency of approximately 50% to 75%, with some installations reporting 90% efficiencies. NOx removal is not usually intended and is essentially zero. Though the removal efficiency is lower than with wet scrubbing, an advantage of dry scrubbing is that the waste is dry, requiring no sludge-handling equipment. The lower efficiency may be sufficient in older power plants with less stringent regulations. (See also Sec. 42.44.)

Particulate removal efficiencies of 90% can be achieved. Since wet scrubbers normally have a lower particulate removal efficiency than baghouses, they are usually combined with electrostatic precipitators.

40. SCRUBBING, VENTURI

Venturi scrubbers (also known as *atomizing scrubbers*) are effective with particle sizes greater than 0.2 μm. In a fixed-throat venturi scrubber, the gas stream enters a converging section and is accelerated toward the throat. In the throat section, the high velocity gas stream strikes liquid streams that are injected at right angles to the gas flow, shattering the liquid into small droplets. The particles impact the slower moving high-mass droplets and are collected in the droplets. Some particles are also collected in the diverging section where the gas stream slows and the droplets agglomerate. The droplet-entrained particles are collected in the flooded elbow and cyclonic separator. (See Fig. 42.11.)

The pressure drop between the entrance of the converging section and the exit of the diverging section is proportional to the square of the gas velocity at the throat. Q_w/Q_g is known as the *liquid-gas ratio*, L/G. Typical operating ranges are: Δp, less than 60 in wg maximum, 5–25 in wg typical (less than 15 kPa maximum, 1.2–6.2 typical); $v_{g,\text{throat}}$, 18–45 ft/sec (60–150 m/s); Q_w/Q_g,

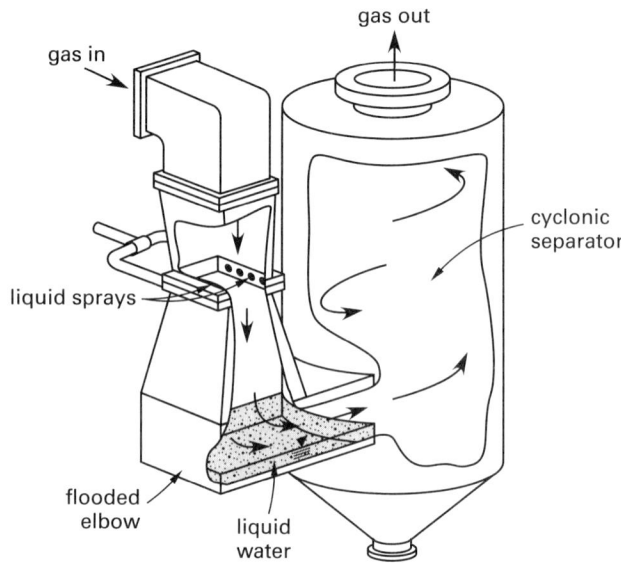

Figure 42.11 Wet Venturi Scrubber

10–30 gal/1000 ACFM (80–240 L·h/1000 m³). The high pressure drop results in a high operating cost.

$$\Delta p \propto \frac{Q_w v_{g,\text{throat}}^2}{Q_g} \qquad 42.33$$

The gas is considered incompressible and in a saturated condition at the throat. The throat velocity can be calculated from basic principles.

$$v_{g,\text{throat}} = \frac{Q_{\text{sat}}}{A_{\text{throat}}} = K_{\text{Calvert}}\sqrt{\frac{\Delta p}{\rho_{\text{sat}}}} \qquad 42.34$$

$$A_{\text{throat}} = \frac{A_{\text{inlet}} v_{g,\text{inlet}}}{v_{g,\text{throat}}} \qquad 42.35$$

$$\rho_{\text{sat}} = \rho_{\text{dry air}} + \rho_w = \frac{p_{\text{air}}}{R_{\text{air}} T} + \frac{p_w}{R_w T}$$

$$= \rho_{\text{dry air}}\left(\frac{1+x}{1+1.609x}\right) \qquad 42.36$$

For v in ft/sec, Δp in in wg, and ρ in lbm/ft³, *Calvert's constant* is

$$K_{\text{Calvert}} = \sqrt{\frac{1850}{\dfrac{L}{G}}} \qquad 42.37$$

The derivation of an explicit equation for venturi scrubber efficiency is dependent on the assumptions made about the extent of liquid atomization, droplet sizes, acceleration and velocity of the droplets, coefficient of drag, and location of impaction. Accordingly, there are numerous models of varying complexities, none of which are generically applicable or particularly accurate. For example, the common *Calvert model* assumes an

---

[25]Limestone, sodium carbonate, and magnesium oxide can also be used instead of lime. Because they cost less, lime and limestone are the most common.

infinitely long throat section where all droplets travel at the throat velocity. The Calvert model incorporates an *impaction parameter*, $K_{\text{impaction}}$, defined as the ratio of particle stopping distance, $s_{p,\text{stopping}}$, to droplet diameter, d_d. The predicted efficiency is

$$\eta = \left(\frac{K_{\text{impaction}}}{K_{\text{impaction}}+0.35}\right)^2 \qquad 42.38$$

$$K_{\text{impaction}} = \frac{s_{p,\text{stopping}}}{d_d} = \frac{C_c \rho_s d_p^2 |v_g \quad v_d|}{18\mu_g g_c d_d} \qquad 42.39$$

Equation 42.40 is a generic equation for calculating collection efficiency for many collection system types, including scrubbers. K' is based on observation and is unique to the characteristics of the process and equipment. It is determined from pilot tests or actual performance measurements. L is a characteristic length, such as the height of a tower or length or width of a chamber, and may or may not be incorporated into K'.

$$\eta = 1 - \exp\left(\frac{K'L}{d_p}\right) \qquad 42.40$$

Collection efficiency may be expressed in terms of penetration. *Penetration*, Pt, is the fraction of particles that pass through the scrubber uncollected. Penetration is the complement of collection efficiency.

$$\text{Pt} = 1 - \eta \qquad 42.41$$

In order to compare relative (not absolute) changes in collection operations, penetration can be expressed in terms of the *number of transfer units*, NTU.

$$\text{NTU} = -\ln(\text{Pt}) \qquad 42.42$$

41. SELECTIVE CATALYTIC REDUCTION

Selective catalytic reduction (SCR) uses ammonia in combination with a catalyst to reduce nitrogen oxides (NOx) to nitrogen gas and water.[26] Vaporized ammonia is injected into the flue gas; the mixture passes through a catalytic reaction bed approximately 0.5 sec to 1.0 sec later. The reaction bed can be constructed from honeycomb plates or parallel-ridged plates. Alternatively, the reaction bed may consist of a packed bed of rings, pellets, or other shapes. The catalyst lowers the NOx decomposition activation energy. NOx and NH_3 combine on the catalyst's surface, producing nitrogen and water.

One mole of ammonia is required for every mole of NOx removed. However, in order to maximize the reduction

efficiency, approximately 5 ppm to 10 ppm of unreacted ammonia is left behind.[27] The chemical reaction is

$$O_2 + 4NO + 4NH_3 \longrightarrow 4N_2 + 6H_2O$$

The optimum temperature for SCR is 600°F to 700°F (315°C to 370°C). Gas velocities are typically around 20 ft/sec (6 m/s). The pressure drop is 3 in wg to 4 in wg (0.75 kPa to 1 kPa). NOx removal efficiency is typically 90% but can range from 70% to 95% depending on the application.[28] SCR produces no liquid waste.

Several conditions can cause deactivation of the catalyst. *Poisoning* occurs when trace quantities of specific materials (e.g., arsenic, lead, phosphorous, other heavy metals, silicon, and sulfur from SO_2) react with the catalyst and lower its activity. Poisoning by SO_2 can be reduced by keeping the flue gas temperature above 608°F (320°C) and/or using poison-resisting compounds. *Masking* (or *plugging*) occurs when the catalytic surface becomes covered with fine particle dust, unburned solids, or ammonium salts. Internal cleaning devices remove surface contaminants such as ash deposits and are used to increase the activity of the equipment.

42. SELECTIVE NONCATALYTIC REDUCTION

Selective noncatalytic reduction (SNCR) involves injecting ammonia or urea into the upper parts of the combustion chamber (or into a thermally favorable location downstream of the combustion chamber) to reduce NOx.[29] SNCR is effective when the oxygen content is low (e.g., 1%) and the combustion temperature is controlled. If the temperature is too high, the NH_3 will react more with oxygen than with NOx, forming even more NOx. If the temperature is too low, the reactions slow and unreacted ammonia enters the flue gas. The optimal temperature range is 1600°F to 1750°F (870°C to 950°C) for NH_3 injection and up to 1900°F (1040°C) for urea. (In general, the temperature ranges for NH_3 and urea are similar. However, various hydrocarbon

---

[26]Vanadium oxide, titanium, and platinum are metallic catalysts; zeolites and ceramic catalysts are also used.

[27]Leftover ammonia in the flue gas is referred to as *ammonia slip* and is usually measured in ppm. 50 ppm to 100 ppm would be considered an excessive ammonia slip. Since NOx and NH_3 react on a 1:1 molar basis, slip is easily calculated. In the following equations, the units can be either molar or volumetric (ppmvd, lbm/hr, mol/hr, scfm, etc.), but they must be consistent.

$$NH_{3,\text{slip}} = NH_{3,\text{feed}} - NH_{3,\text{reacted}}$$
$$= NH_{3,\text{feed}} - (NOx_{\text{in}} - NOx_{\text{out}})$$
$$= NH_{3,\text{feed}} - (NOx_{\text{in}})(\text{removal efficiency})$$

[28]Removal efficiencies of 95% to 99% are possible when SCR is used for VOC removal.

[29]*Urea*, which decomposes to NH_3 and carbon dioxide inside the combustion chamber, is safer and easier to handle.

additives can be used with urea to lower the temperature range.) The reactions are

$$6NO + 4NH_3 \longrightarrow 5N_2 + 6H_2O$$
$$6NO_2 + 8NH_3 \longrightarrow N_2 + 12H_2O$$

The NOx reduction efficiency is approximately 20% to 50% with an ammonia slip of 20 ppm to 30 ppm. The actual efficiency is highly dependent on the injection geometries and interrelations between ammonia slip, ash, and sulfur.

Since the NOx reduction efficiency is relatively low, the SNCR process cannot usually satisfy NOx regulations by itself. The use of urea must be balanced against the potential for ammonia slip and the conversion of NO to nitrous oxide. These and other problems relating to formation of ammonium salts have kept SNCR from gaining widespread popularity in small installations.

43. SOIL WASHING

Soil washing is effective in removing heavy metals, wood preserving wastes (PAHs), and BTEX compounds from contaminated soil. Soil washing is a two-step process. In the first step, soil is mixed with water to dissolve the contaminants. Additives are used as required to improve solubility. In the second step, additional water is used to flush the soil and to separate the fine soil from coarser particles. (Semi-volatile materials concentrate in the fines.) Metals are extracted by adding chelating agents to the wash water.[30] The contaminated wash water is subsequently treated.

44. SORBENT INJECTION

Sorbent injection (FSI) involves injecting a limestone slurry directly into the upper parts of the combustion chamber (or into a thermally favorable location downstream) to reduce SOx. Heat calcines the limestone into reactive lime. Fast drying prevents wet particles from building up in the duct. Lime particles are captured in a scrubber with or without an electrostatic precipitator (ESP). With only an ESP, the SOx removal efficiency is approximately 50%.

45. SPARGING

Sparging is the process of using air injection wells to bubble air through groundwater. The air pushes volatile contaminants into the overlying soil above the aquifer where they can be captured by vacuum extraction.

---

[30]A *chelate* is a ring-like molecular structure formed by unshared electrons of neighboring atoms. A *chelating agent* (*chelant*) is an organic compound in which atoms form bonds with metals in solution. By combining with metal ions, chelates control the damaging effects of trace metal contamination. Ethylenediaminetetraacetic acid (EDTA) types are the leading industrial chelants.

46. STAGED COMBUSTION

Staged combustion methods are primarily used with gas-fired burners to reduce formation of nitrogen oxides (NOx). Both staged-air burner and staged-fuel burner systems are used. *Staged-air burner systems* reduce NOx production 20% to 35% by admitting the combustion air through primary and secondary paths around each fuel nozzle. The fuel burns partially in the fuel-rich zone. Fuel-borne nitrogen is converted to compounds that are subsequently oxidized to nitrogen gas. Secondary air completes combustion and controls the flame size and shape. Combustion temperature is lowered by recirculation of combustion products within the burner. Staged burners have few disadvantages, the main one being longer flames.

In *staged-fuel burner systems*, a portion of the fuel gas is initially burned in a fuel-lean (air-rich) combustion. The peak flame temperature is reduced, with a corresponding reduction in thermal NOx production. The remainder of the fuel is injected through secondary nozzles. Combustion gases from the first stage dilute the combustion in the second stage, reducing peak temperature and oxygen content. Reductions in NOx formation of 50% to 60% are possible. Flame length is less than with staged-air burners, and the required excess air is lower.

47. STRIPPING, AIR

Air strippers are primarily used to remove volatile organic compounds (VOCs) or other target substances from water. In operation, contaminated water enters a stripping tower at the top, and fresh air enters at the bottom. The effectiveness of the process depends on the volatility of the compound, its temperature and concentration, and the liquid-air contact area. However, removal efficiencies of 80% to 90% (and above) are common for VOCs. Figure 42.12 shows a typical air stripping operation.

There are three types of stripping towers—*packed towers* filled with synthetic *packing media* (polypropylene balls, rings, saddles, etc.) like those shown in Fig. 42.13, *spray towers*, and (less frequently for VOC removal) *tray towers* with horizontal trays arranged in layers vertically. *Redistribution rings* (*wall wipers*) prevent channeling down the inside of the tower. (*Channeling* is the flow of liquid through a few narrow paths rather than an over-the-bed packing.) The stripping air is generated by a small blower at the column base. A mist eliminator at the top eliminates entrained water from the air.

As the contaminated water passes over packing media in a packed tower, the target substance leaves the liquid and enters the air where the concentration is lower. The mole fraction of the target substance in the water, x, decreases; the mole fraction of the target substance in the air, y, increases.

Solid Waste

Figure 42.12 *Schematic of Air Stripping Operation*

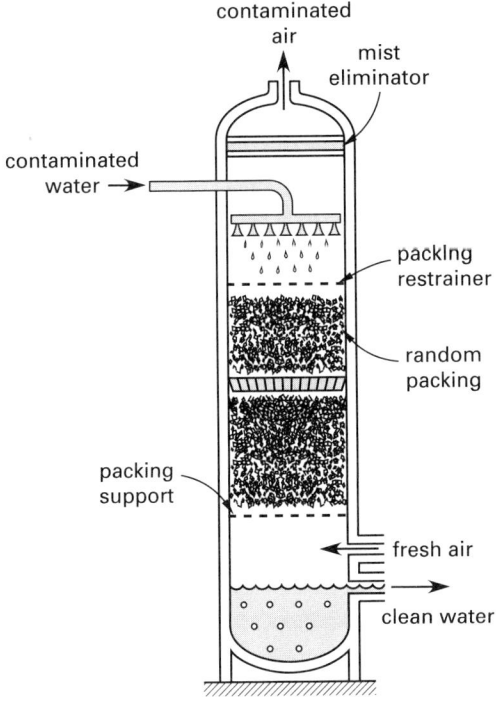

Henry's law, as it applies to water treatment, states that at equilibrium, the vapor pressure of target substance A is directly proportional to the target substance's mole fraction, x_A. (The maximum mole fraction is the substance's *solubility*.) Table 42.9 lists representative values of Henry's law constant, H.[31] When multiple compounds are present in the water, the *key component* is the one that is the most difficult to remove (i.e., the component with the highest concentration and lowest Henry's constant).

$$p_A = H_A x_A \qquad 42.43$$

Table 42.9 *Typical Henry's Law Constants for Selected Volatile Organic Compounds (at low pressures and 25°C)*

VOC	Henry's law constant, H (atm)[*]
1,1,2,2-tetrachloroethane	24.02
1,1,2-trichloroethane	47.0
propylene dichloride	156.8
methylene chloride	177.4
chloroform	188.5
1,1,1-trichloroethane	273.56
1,2-dichloroethene	295.8
1,1-dichloroethane	303.0
hexachloroethane	547.7
hexachlorobutadiene	572.7
trichloroethylene	651.0
1,1-dichloroethene	834.03
perchloroethane	1596.0
carbon tetrachloride	1679.17

(Multiply atm by 101.3 to obtain kPa.)

[*]Henry's law indicates that the units of H are atmospheres per mole fraction. Mole fraction is dimensionless and does not appear in the units for H.

Figure 42.13 *Packing Media Types*

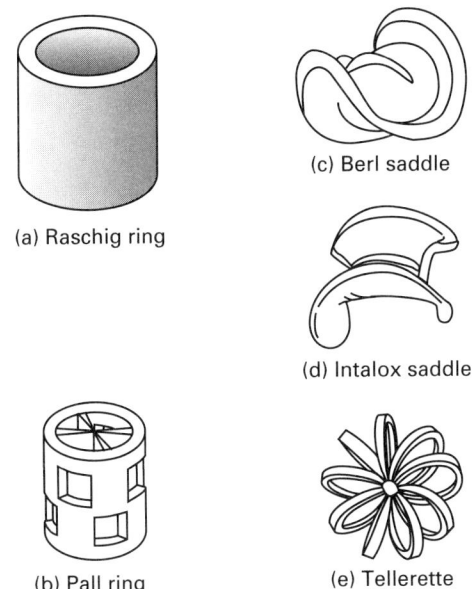

(a) Raschig ring

(c) Berl saddle

(d) Intalox saddle

(b) Pall ring

(e) Tellerette

The liquid *mass-transfer coefficient*, $K_L a$, is the product of the *coefficient of liquid mass transfer*, K, and the *interfacial area* per volume of packing, a. The mass transfer coefficient is a measure of the efficiency of the air stripper as a whole. The higher the mass transfer coefficient, the higher the efficiency. The $K_L a$ value is largely a function of the size and shape of the packing material, and it is usually obtained from the packing manufacturer or theoretical correlations. For some types of packing, the gas-phase resistance is greater than the liquid-phase resistance, and the gas mass-transfer coefficient, $K_G a$, may be given.

The *stripping factor*, R (the reciprocal of the *absorption factor*), is given by Eq. 42.44. Units for the gas and liquid flow rates, G and L, depend on the correlations used to solve the stripping equations. The air flow rate is

The discharged air, known as *off-gas*, is discharged to a process that destroys or recovers the target substance. (Since the quantities are small, recovery is rarer.) Destruction of the target substance can be accomplished by flaring, carbon absorption, and incineration. Since flaring is dangerous and carbon absorption creates a secondary waste if the AC is not regenerated, incineration is often preferred.

---

[31]Henry's law is fairly accurate for most gases when the partial pressure is less than 1 atm. If the partial pressure is greater than 1 atm, Henry's law constants will vary with partial pressure.

Solid Waste

limited by the acceptable pressure drop through the tower.

$$R = \frac{HG}{L} \qquad \textit{42.44}$$

The *transfer unit method* is a convenient way of designing and analyzing the performance of stripping towers. A *transfer unit* is a measure of the difficulty of the mass-transfer operation and depends on the solubility and concentrations. The overall number of transfer units is expressed as NTU_{OG} or NTU_{OL} (alternatively, N_{OG} or N_{OL}), depending on whether the gas or liquid resistance dominates.[32] The NTU value depends on the incoming and desired concentrations and the material flow rates.[33]

The *height of the packing*, z, is the effective height of the tower. It is calculated from the NTUs and the *height of a transfer unit*, HTU, using Eq. 42.45. The HTU value depends on the packing media.

$$z = (\text{HTU}_{OG})(\text{NTU}_{OG}) = (\text{HTU}_{OL})(\text{NTU}_{OL}) \qquad \textit{42.45}$$

10% of additional height should be added to the theoretical value as a safety factor. Packing heights greater than 30 ft to 40 ft (9.1 m to 12.2 m) are not recommended since the packing might be crushed, and greater heights produce little or no increase in removal efficiency.

The blower power depends on the packing shape and size and the air-to-water ratio. Typical pressure drops are 0.5 in wg to 1.0 in wg per vertical foot (0.38 kPa/m to 0.75 kPa/m) of packing. To minimize the blower power requirement, the ratio of tower diameter to gross packing dimension should be between 8 and 15. Blower fan power increases with increasing air-to-water ratios.

A significant problem for air strippers removing contaminants from water is fouling of the packing media through biological growth and solids (e.g., iron complexes) deposition. Biological growth can be inhibited by continuous or batch use of a *biocide* that does not interfere with the off-gas system.[34] Another problem is *flooding*, which occurs when excess liquid flow rates impede the flow of the air.

---

[32]The gas film resistance controls when the solubility of the target substance in the liquid is high. Conversely, when the solubility is low, the liquid film resistance controls. In flue gas cleanup (FGC) work, the gas film resistance usually controls.

[33]Stripping is one of many mass-transfer operations studied by chemical engineers. Determining the number of transfer units required and the height of a single transfer unit are typical chemical engineering calculations.

[34]A *biocide* is a chemical that kills living things, particularly microorganisms. Biocide categories include chlorinated isocyanurates (used in swimming pool disinfectants and dishwashing detergents), sodium bromide, inorganics (used in wood treatments), quaternaries ("quats") used in hard-surface cleaners and sanitizers, and iodophors (used in human-skin disinfectants). By comparison, *biostats* do not kill microorganisms already present, but they retard further growth of microorganisms from the moment they are incorporated. Biostats are organic acids and salts (e.g., sodium and potassium benzoate, sorbic acid, and potassium sorbate used "to preserve freshness" in foods).

48. STRIPPING, STEAM

Steam stripping is more effective than air stripping for removing semi- and non-volatile compounds, such as diesel fuel, oil, and other organic compounds with boiling points up to approximately 400°F (200°C). Operation of a steam stripper is similar to that of an air stripper, except that steam is used in place of the air. Steam strippers can be operated at or below atmospheric pressure. Higher vacuums will remove greater amounts of the compound.

49. THERMAL DESORPTION

Thermal desorption is primarily used to remove volatile organic compounds (VOCs) from contaminated soil. The soil is heated directly or indirectly to approximately 1000°F (540°C) to evaporate the volatiles. This method differs from incineration in that the released gases are not burned but are captured in a subsequent process step (e.g., activated carbon filtration).

50. VACUUM EXTRACTION

Vacuum extraction is used to remove many types of volatile organic compounds (VOCs) from soil. The VOCs are pulled from the soil through a well dug in the contaminated area. Air is withdrawn from the well, vacuuming volatile substances with it.

51. VAPOR CONDENSING

Some vapors can be removed simply by cooling them to below their dew points. Traditional contact (open) and surface (closed) condensers can be used.

52. VITRIFICATION

Vitrification melts and forms slag and ash wastes into glass-like pellets. Heavy metals and toxic compounds cannot leach out, and the pellets can be disposed of in hazardous waste landfills. Vitrification can occur in the incineration furnace or in a stand-alone process. Stand-alone vitrification occurs in an electrically heated vessel where the temperature is maintained at 2200°F to 2370°F (1200°C to 1300°C) for up to 20 hr or so. Since the electric heating is nonturbulent, flue gas cleaning systems are not needed.

53. WASTEWATER TREATMENT PROCESSES

Many industrial processes use water for cooling, rinsing, or mixing. Such water must be treated prior to being discharged to *publicly owned treatment works* (POTWs).[35] Table 42.10 lists some of the polluting characteristics of industrial wastewaters.

---

[35]Rainwater runoff from some industrial plants can also be a hazardous waste.

Table 42.10 Types of Pollution from Industrial Wastewater

ammonia
biochemical oxygen demand (BOD)
carbon, total organic (TOC)
chemical oxygen demand (COD)
chloride
flow rate
metals, soluble
metals, nonsoluble
nitrate
nitrite
organic compounds, acid-extractable
organic compounds, base/neutral-extractable
organic compounds, volatile (VOC)
pH
phosphorus
sodium
solids, total suspended (TSS)
sulfate
surfactants
temperature
whole-effluent toxicity (LC_{50})

Most large-volume industrial wastewaters go through the following processes: (1) flow equalization, (2) neutralization, (3) oil and grease removal, (4) suspended solids removal, (5) metals removal, and (6) VOC removal. These processes are similar, in many cases, to processes with the same names used to treat municipal wastewater in a POTW.

Flow equalization reduces the chance of under- and overloading a treatment process. Equalization of both hydraulic and chemical loading is required. *Hydraulic loading* is usually equalized by storing wastewater during high flow periods and discharging it during periods of low flow. *Chemical loading* is equalized by use of an equalization basin with mechanical mixing or air agitation. Mixing with air has the added advantages of oxidizing reducing compounds, stripping away volatiles, and eliminating odors.

There are two reasons for water *neutralization*. First, the pH of water is regulated and must be between 6.0 and 9.0 when discharged. Second, water that is too acidic or alkaline may not be properly processed by subsequent, particularly biological, processes. The optimum pH for biological processes is between 6.5 and 7.5. The effectiveness of the processes is greatly reduced with pHs below 4.0 and above 9.5.

Acidic water is neutralized by adding lime (oxides and hydroxides of calcium and magnesium), limestone, or some other caustic solution. Alkaline waters are treated with sulfuric or hydrochloric acid or carbon dioxide gas.[36] Large volumes of oil and grease float to the surface and are skimmed off. Smaller volumes may require a dissolved-air process. Removing emulsified oil is more complex and may require use of chemical coagulants.

The method used to remove suspended solids depends on the solid size. Most industrial plants do not require strainers, bar screens, or fine screens to remove large solids (i.e., those larger than 1 in (25 mm)). *Grit* (i.e., sand and gravel) is removed in grit chambers by simple sedimentation. Fine solids are categorized into *settleable solids* (diameters more than 1 μm) and *colloids* (diameters between 0.001 μm and 1 μm). Settleable solids and colloids are removed in a sedimentation tank, with chemical coagulants or dissolved-air flotation being used to assist colloidal particles in settling out.

Most metal (e.g., lead) removal occurs by precipitating its hydroxide, although precipitation as carbonates or sulfides is also used.[37] A caustic substance (e.g., lime) is added to the water to raise the pH below the solubility limit of the metal ion. When they come out of solution, the metallic compounds are *flocculated* into larger flakes that ultimately settle out. Floc is mechanically removed as inorganic heavy-metal sludge, which has a 96% to 99% water content by weight. The sludge is dewatered in a drying bed, vacuum filter, or filter press to 65% to 85% water. Depending on its composition, the sludge can be landfilled or treated as a hazardous waste by incineration or other methods.

VOCs such as benzene and toluene are removed by air stripping or adsorption in activated charcoal (AC) towers. Nonvolatile organic compounds (NVOCs) are removed by biological processes such as lagooning, trickle filtration, and activated sludge.

Although destruction of biological health hazards is not always needed for industrial wastewater, chemical oxidants (see Table 42.1) are still needed to destroy odors, control bacterial growth downstream, and eliminate sulfur compounds.[38] Chemical oxidants can also reduce heavy metals (e.g., iron, manganese, silver, and lead) that were not removed in previous operations.

---

[36]In water, carbon dioxide gas forms *carbonic acid.*

[37]Ion exchange, activated charcoal, and reverse osmosis methods can be used but may be more expensive.

[38]Meat packing and dairy (e.g., cheese) plants are examples of industrial processes that require chlorination to destroy bacteria in wastewater.

Solid Waste

Topic VII: Environmental Health, Safety, and Welfare

Health, Safety, Welfare

43 Organic Chemistry

1. INTRODUCTION TO ORGANIC CHEMISTRY

Organic chemistry deals with the formation and reaction of compounds of carbon, many of which are produced by living organisms. Organic compounds typically have one or more of the following characteristics.

- They are insoluble in water.[1]

- They are soluble in concentrated acids.

- They are relatively non-ionizing.

- They are unstable at high temperatures.

The method of naming organic compounds was standardized in 1930 at the International Union Chemistry meeting in Belgium. Names conforming to the established guidelines are known as *IUC names* or *IUPAC names.*[2]

2. FUNCTIONAL GROUPS

Certain combinations (groups) of atoms occur repeatedly in organic compounds and remain intact during reactions. Such combinations are called *functional groups* or *moieties.*[3] For example, the radical OH^- is known as a *hydroxyl group.* Table 43.1 contains some common functional groups. In this table and others similar to it, the symbols R and R' usually denote an attached hydrogen atom or other hydrocarbon chain of any length. They may also denote some other group of atoms.

---

[1]This is especially true for hydrocarbons. However, many organic compounds containing oxygen are water soluble. The sugar family is an example of water-soluble compounds.
[2]IUPAC stands for *International Union of Pure and Applied Chemistry.*
[3]Although "moiety" is often used synonymously with "functional group," there is a subtle difference. When a functional group combines into a compound, its moiety may gain/lose an atom from/to its combinant. Therefore, moieties differ from their original functional groups by one or more atoms.

Table 43.1 *Selected Functional Groups*

name	standard symbol	formula	number of single bonding sites
aldehyde		CHO	1
alkyl	[R]	C_nH_{2n+1}	1
alkoxy	[RO]	$C_nH_{2n+1}O$	1
amine (amino, $n = 2$)		NH_n	$3 - n$ $[n = 0,1,2]$
aryl (benzene ring)	[Ar]	C_6H_5	1
carbinol		COH	3
carbonyl (keto)	[CO]	CO	2
carboxyl		COOH	1
ester		COO	1
ether		O	2
halogen (halide)	[X]	Cl, Br, I, or F	1
hydroxyl		OH	1
nitrile		CN	1
nitro		NO_2	1

3. FAMILIES OF ORGANIC COMPOUNDS

For convenience, organic compounds are categorized into families, or *chemical classes.* "R" is an abbreviation for the word "radical," though it is unrelated to ionic radicals. Compounds within each family have similar structures, being based on similar combinations of groups. For example, all compounds in the alcohol family have the structure [R]-OH, where [R] is any alkyl group and -OH is the hydroxyl group.

Families of compounds can be further subdivided into subfamilies. For example, the hydrocarbons are classified into alkanes (single carbon-carbon bond), alkenes (double carbon-carbon bond), and alkynes (triple carbon-carbon bond).[4]

Table 43.2 contains some common organic families, and Table 43.3 gives synthesis routes for various classes of organic compounds.

4. SYMBOLIC REPRESENTATION

The nature and structure of organic groups and families cannot be explained fully without showing the types of

---

[4]Hydrocarbons with two double carbon-carbon bonds are known as *dienes.*

bonds between the elements. Figure 43.1 illustrates the symbolic representation of some of the functional groups and families, and Fig. 43.2 illustrates reactions between organic compounds.

5. FORMATION OF ORGANIC COMPOUNDS

There are usually many ways of producing an organic compound. The types of reactions contained in this

section deal only with the interactions between the organic families. The following processes are referred to.

- *oxidation:* replacement of a hydrogen atom with a hydroxyl group

- *reduction:* replacement of a hydroxyl group with a hydrogen atom

- *hydrolysis:* addition of one or more water molecules

- *dehydration:* removal of one or more water molecules

Table 43.2 Families (Chemical Classes) of Organic Compounds

family (chemical class)	structure[a]	example
organic acids		
carboxylic acids	[R]-COOH	acetic acid ((CH_3)COOH)
fatty acids	[Ar]-COOH	benzoic acid (C_6H_5COOH)
alcohols		
aliphatic	[R]-OH	methanol (CH_3OH)
aromatic	[Ar]-[R]-OH	benzyl alcohol ($C_6H_5CH_2OH$)
aldehydes	[R]-CHO	formaldehyde (HCHO)
alkyl halides		
(haloalkanes)	[R]-[X]	chloromethane (CH_3Cl)
amides	[R]-CO-NH_n	β-methylbutyramide ($C_4H_9CONH_2$)
amines	$[R]_{3-n}$-NH_n	methylamine (CH_3NH_2)
	$[Ar]_{3-n}$-NH_n	aniline ($C_6H_5NH_2$)
primary amines	$n = 2$	
secondary amines	$n = 1$	
tertiary amines	$n = 0$	
amino acids	CH-[R]-(NH_2)COOH	glycine ($CH_2(NH_2)COOH$)
anhydrides	[R]-CO-O-CO-[R']	acetic anhydride ($(CH_3CO)_2O$)
arenes (aromatics)	ArH = C_nH_{2n-6}	benzene (C_6H_6)
aryl halides	[AR]-[X]	fluorobenzene (C_6H_5F)
carbohydrates	$C_x(H_2O)_y$	dextrose ($C_6H_{12}O_6$)
sugars		
polysaccharides		
esters	[R]-COO-[R']	methyl acetate (CH_3COOCH_3)
ethers	[R]-O-[R]	diethyl ether ($C_2H_5OC_2H_5$)
	[Ar]-O-[R]	methyl phenyl ether ($CH_3OC_6H_5$)
	[Ar]-O-[Ar]	diphenyl ether ($C_6H_5OC_6H_5$)
glycols	$C_nH_{2n}(OH)_2$	ethylene glycol ($C_2H_4(OH)_2$)
hydrocarbons		
alkanes (single bonds)[b]	RH = C_nH_{2n+2}	octane (C_8H_{18})
saturated hydrocarbons		
cycloalkanes (cycloparaffins)	C_nH_{2n}	cyclohexane (C_6H_{12})
alkenes (double bonds between two carbons)[c]	C_nH_{2n}	ethylene (C_2H_4)
unsaturated hydrocarbons		
cycloalkenes	C_nH_{2n-2}	cyclohexene (C_6H_{10})
alkynes (triple bonds between two carbons)	C_nH_{2n-2}	acetylene (C_2H_2)
unsaturated hydrocarbons		
ketones	[R]-[CO]-[R]	acetone (($CH_3)_2CO$)
nitriles	[R]-CN	acetonitrile (CH_3CN)
phenols	[Ar]-OH	phenol (C_6H_5OH)

[a]See Table 43.1 for definitions of [R], [Ar], [X], and [CO].
[b]Alkanes are also known as the *paraffin series* and *methane series*.
[c]Alkenes are also known as the *olefin series*.

Table 43.3 *Synthesis Routes for Various Classes of Organic Compounds*

organic acids
    oxidation of primary alcohols
    oxidation of ketones
    oxidation of aldehydes
    hydrolysis of esters
alcohols
    oxidation of hydrocarbons
    reduction of aldehydes
    reduction of organic acids
    hydrolysis of esters
    hydrolysis of alkyl halides
    hydrolysis of alkenes (aromatic hydrocarbons)
aldehydes
    oxidation of primary and tertiary alcohols
    oxidation of esters
    reduction of organic acids
amides
    replacement of hydroxyl group in an acid with an amino
       group
anhydrides
    dehydration of organic acids (withdrawal of one water
       molecule from two acid molecules)
carbohydrates
    oxidation of alcohols
esters
    reaction of acids with alcohols (*ester alcohols*)[*]
    reaction of acids with phenols (*ester phenols*)
    dehydration of alcohols
    dehydration of organic acids
ethers
    dehydration of alcohol
hydrocarbons
    alkanes: reduction of alcohols and organic acids
                 hydrogenation of alkenes
    alkenes: dehydration of alcohols
                 dehydrogenation of alkanes
ketones
    oxidation of secondary and tertiary alcohols
    reduction of organic acids
phenols
    hydrolysis of aryl halides

[*]The reaction of an organic acid with an alcohol is called *esterification*.

Figure 43.1 *Representation of Functional Groups and Families*

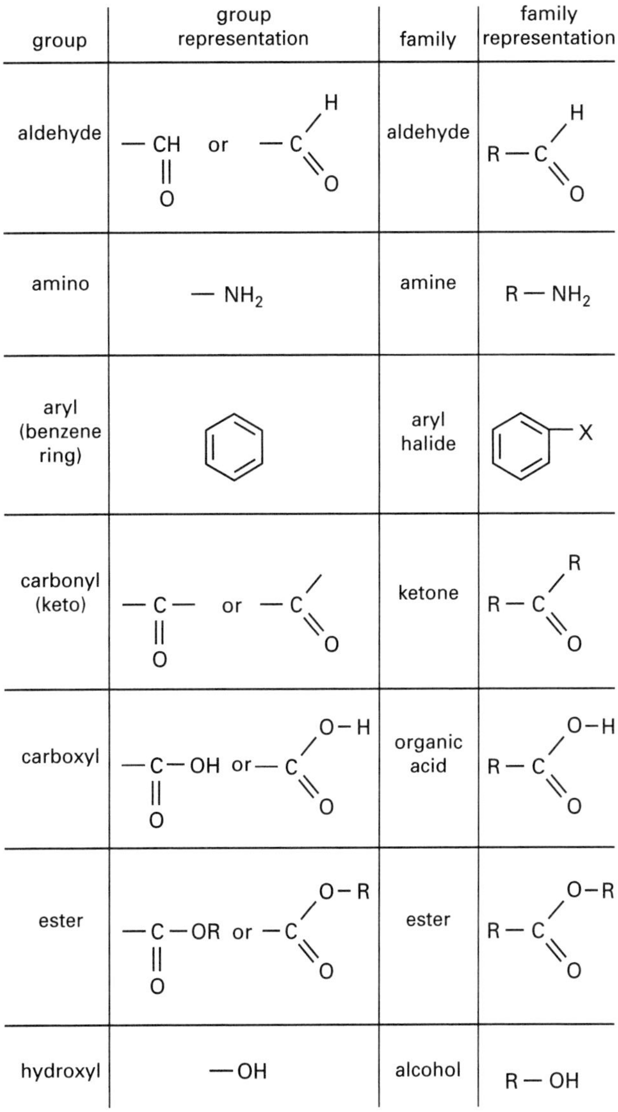

group	group representation	family	family representation
aldehyde	$-CH$ or $-C$ with $\parallel O$ / $=O$ and $-H$	aldehyde	$R-C$ with $-H$ and $=O$
amino	$-NH_2$	amine	$R-NH_2$
aryl (benzene ring)	(benzene ring)	aryl halide	(benzene ring)—X
carbonyl (keto)	$-C-$ with $\parallel O$ or $-C$ with $=O$	ketone	$R-C$ with R and $=O$
carboxyl	$-C-OH$ with $\parallel O$ or $-C$ with $O-H$ and $=O$	organic acid	$R-C$ with $O-H$ and $=O$
ester	$-C-OR$ with $\parallel O$ or $-C$ with $O-R$ and $=O$	ester	$R-C$ with $O-R$ and $=O$
hydroxyl	$-OH$	alcohol	$R-OH$

Figure 43.2 *Reactions Between Organic Compounds*

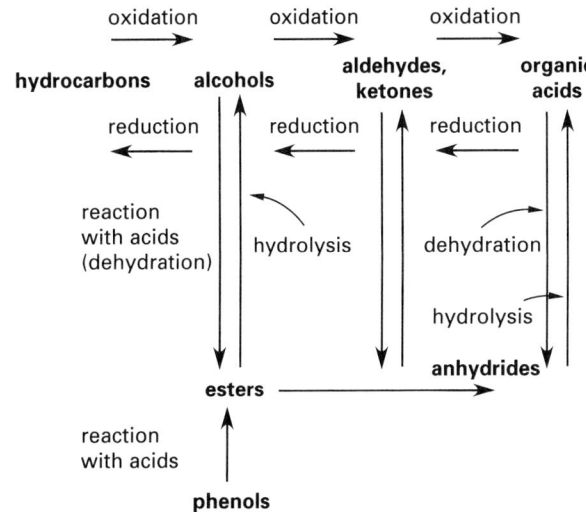

Health, Safety, Welfare

44 Biology and Bacteriology

1. MICROORGANISMS

Microorganisms occur in untreated water and represent a potential human health risk if they enter the water supply. Microorganisms include viruses, bacteria, fungi, algae, protozoa, worms, rotifers, and crustaceans.

Microorganisms are organized into three broad groups based upon their structural and functional differences. The groups are called *kingdoms*. The three kingdoms are animals (rotifers and crustaceans), plants (mosses and ferns), and *Protista* (bacteria, algae, fungi, and protozoa). Bacteria and protozoa of the kingdom *Protista* constitute the major groups of microorganisms in the biological systems that are used in secondary treatment of wastewater.

2. PATHOGENS

Many infectious diseases in humans or animals are caused by organisms categorized as *pathogens*. Pathogens are found in fecal wastes that are transmitted and transferred through the handling of wastewater. Pathogens will proliferate in areas where sanitary disposal of feces is not adequately practiced and where contamination of water supply from infected individuals is not properly controlled. The wastes may also be improperly discharged into surface waters, making the water *non-potable* (unfit for drinking). Certain shellfish can become toxic when they concentrate pathogenic organisms in

their tissues, increasing the toxic levels much higher than the levels in the surrounding waters.

Organisms that are considered to be pathogens include bacteria, protozoa, viruses, and helminths (worms). Table 44.1 lists potential waterborne diseases, the causative organisms, and the typical infection sources.

Not all microorganisms are considered pathogens. Some microorganisms are exploited for their usefulness in wastewater processing. Most wastewater engineering (and an increasing portion of environmental engineering) involves designing processes and operating facilities that utilize microorganisms to destroy organic and inorganic substances.

3. MICROBE CATEGORIZATION

Carbon is the basic building block for cell synthesis, and it is prevalent in large quantities in wastewater. Wastewater treatment converts carbon into microorganisms that are subsequently removed from the water by settling. Therefore, the growth of organisms that use organic material as energy is encouraged.

If a microorganism uses organic material as a carbon supply, it is *heterotrophic*. *Autotrophs* require only carbon dioxide to supply their carbon needs. Organisms that rely only on the sun for energy are called *phototrophs*. *Chemotrophs* extract energy from organic or inorganic oxidation/reduction (redox) reactions. *Organotrophs* use organic materials, while *lithotrophs* oxidize inorganic compounds.

Most microorganisms in wastewater treatment processes are bacteria. Conditions in the treatment plant are readjusted so that chemoheterotrophs predominate.

Each species of bacteria reproduces most efficiently within a limited range of temperatures. Four temperature ranges are used to classify bacteria. Those bacteria that grow best below 68°F (20°C) are called *psychrophiles* (i.e., are *psychrophilic*, also known as *cryophilic*). *Mesophiles* (i.e., *mesophilic bacteria*) grow best at temperatures in a range starting around 68°F (20°C) and ending around 113°F (45°C). Between 113°F (45°C) and 140°F (60°C), the *thermophiles* (*thermophilic bacteria*) grow best. Above 140°F (60°C), *stenothermophiles* grow best.

Because most reactions proceed slowly at these temperatures, cells use *enzymes* to speed up the reactions and control the rate of growth. Enzymes are proteins,

Health, Safety, Welfare

Table 44.1 *Potential Pathogens*

name of organism	major disease	source
Bacteria		
Salmonella typhi	typhoid fever	human feces
Salmonella paratyphi	paratyphoid fever	human feces
other Salmonella	salmonellosis	human/animal feces
Shigella	bacillary dysentery	human feces
Vibrio cholerae	cholera	human feces
Enteropathogenic coli	gastroenteritis	human feces
Yersinia enterocolitica	gastroenteritis	human/animal feces
Campylobacter jejuni	gastroenteritis	human/animal feces
Legionella pneumophila	acute respiratory illness	thermally enriched waters
Mycobacterium	tuberculosis	human respiratory exudates
other Mycobacteria	pulmonary illness	soil and water
opportunistic bacteria	variable	natural waters
Enteric Viruses/Enteroviruses		
Polioviruses	poliomyelitis	human feces
Coxsackieviruses A	aseptic meningitis	human feces
Coxsackieviruses B	aseptic meningitis	human feces
Echoviruses	aseptic meningitis	human feces
other Enteroviruses	encephalitis	human feces
Reoviruses	upper respiratory and gastrointestinal illness	human/animal feces
Rotaviruses	gastroenteritis	human feces
Adenoviruses	upper respiratory and gastrointestinal illness	human feces
Hepatitis A virus	infectious hepatitis	human feces
Norwalk and related gastrointestinal viruses	gastroenteritis	human feces
Fungi		
Aspergillus	ear, sinus, lung, and skin infections	airborne spores
Candida	yeast infections	various
Protozoa		
Acanthamoeba castellani	amoebic meningoencephalitis	soil and water
Balantidium coli	balantidosis (dysentery)	human feces
Cryptosporidium[*]	cryptosporidiosis	human/animal feces
Entamoeba histolytica	amoebic dysentery	human feces
Giardia lamblia	giardiasis (gastroenteritis)	human/animal feces
Naegleria fowleri	amoebic meningoencephalitis	soil and water
Algae (blue-green)		
Anabaena flos-aquae	gastroenteritis (possible)	natural waters
Microcystis aeruginosa	gastroenteritis (possible)	natural waters
Alphanizomenon flos-aquae	gastroenteritis (possible)	natural waters
Schizothrix calciola	gastroenteritis (possible)	natural waters
Helminths (intestinal parasites/worms)		
Ascaris lumbricoides (roundworm)	digestive disturbances	ingested worm eggs
E. vericularis (pinworm)	any part of the body	ingested worm eggs
Hookworm	pneumonia, anemia	ingested worm eggs
Threadworm	abdominal pain, nausea, weight loss	ingested worm eggs
T. trichiuro (whipworm)	trichinosis	ingested worm eggs
Tapeworm	digestive disturbances	ingested worm eggs

[*]Disinfectants have little effect on *Cryptosporidia*. Most large systems now use filtration, the most effective treatment to date against *Cryptosporidia*.

ranging from simple structures to complex conjugates, and are specialized for the reactions they catalyze.

The temperature ranges are qualitative and somewhat subjective. The growth range of facultative thermophiles extends from the thermophilic range into the mesophilic range. Bacteria will grow in a wide range of temperatures and will survive at a very large range of temperatures. *E. coli*, for example, is classified as a mesophile. It grows best at temperatures between 68°F (20°C) and 122°F (50°C) but can continue to reproduce at temperatures down to 32°F (0°C).

Nonphotosynthetic bacteria are classified into two heterotrophic and autotrophic groups depending on their sources of nutrients and energy. *Heterotrophs* use organic matter as both an energy source and a carbon source for synthesis. Heterotrophs are further subdivided into groups depending on their behavior toward free oxygen: aerobes, anaerobes, and facultative bacteria. *Obligate aerobes* require free dissolved oxygen while they decompose organic matter to gain energy for growth and reproduction. *Obligate anaerobes* oxidize organics in the complete absence of dissolved oxygen by using the oxygen bound in other compounds, such as nitrate and sulfate. *Facultative bacteria* comprise a group that uses free dissolved oxygen when available but that can also behave anaerobically in the absence of free dissolved oxygen (i.e., *anoxic conditions*). Under anoxic conditions, a group of facultative anaerobes, called *denitrifiers*, utilizes nitrites and nitrates instead of oxygen. Nitrate nitrogen is converted to nitrogen gas in the absence of oxygen. This process is called *anoxic denitrification*.

Autotrophic bacteria (*autotrophs*) oxidize inorganic compounds for energy, use free oxygen, and use carbon dioxide as a carbon source. Significant members of this group are the *Leptothrix* and *Crenothrix* families of *iron bacteria*. These have the ability to oxidize soluble ferrous iron into insoluble ferric iron. Because soluble iron is often found in well waters and iron pipe, these bacteria deserve some attention. They thrive in water pipes where dissolved iron is available as an energy source and bicarbonates are available as a carbon source. As the colonies die and decompose, they release foul tastes and odors and have the potential to cause staining of porcelain or fabrics.

4. VIRUSES

Viruses are parasitic organisms that pass through filters that retain bacteria, can only be seen with an electron microscope, and grow and reproduce only inside living cells, but they can survive outside the host. They are not cells, but particles composed of a protein sheath surrounding a nucleic-acid core. Most viruses of interest in water supply range in size from 10 nm to 25 nm.

The *viron particles* invade living cells, and the viral genetic material redirects cell activities toward production of new viral particles. A large number of viruses are released to infect other cells when the infected cell dies. Viruses are host-specific, attacking only one type of organism.

There are more than 100 types of human enteric viruses. Those of interest in drinking water are *Hepatitis A*, *Norwalk*-type viruses, *Rotaviruses*, *Adenoviruses*, *Enteroviruses*, and *Reoviruses*.

5. BACTERIA

Bacteria are microscopic organisms (*microorganisms*) having round, rodlike, spiral or filamentous single-celled or noncellular bodies. Bacteria are *prokaryotes* (i.e., they lack nuclei structures). They are often aggregated into colonies. Bacteria use soluble food and reproduce through binary fission. Most bacteria are not pathogenic to humans, but they do play a significant role in the decomposition of organic material and can have an impact on the aesthetic quality of water.

6. FUNGI

Fungi are aerobic, multicellular, nonphotosynthetic, heterotrophic, eukaryotic protists. Most fungi are saprophytes that degrade dead organic matter. Fungi grow in low-moisture areas, and they are tolerant of low-pH environments. Fungi release carbon dioxide and nitrogen during the breakdown of organic material.

Fungi are obligate aerobes that reproduce by a variety of methods including fission, budding, and spore formation. They form normal cell material with one-half the nitrogen required by bacteria. In nitrogen-deficient wastewater, they may replace bacteria as the dominant species.

7. ALGAE

Algae are autotrophic, photosynthetic organisms (*photoautotrophs*) and may be either unicellular or multicellular. They take on the color of the pigment that is the catalyst for photosynthesis. In addition to chlorophyll (green), different algae have different pigments, such as carotenes (orange), phycocyanin (blue), phycoerythrin (red), fucoxanthin (brown), and xanthophylls (yellow).

Algae derive carbon from carbon dioxide and bicarbonates in water. The energy required for cell synthesis is obtained through photosynthesis. Algae utilize oxygen for respiration in the absence of light. Algae and bacteria have a symbiotic relationship in aquatic systems, with the algae producing oxygen used by the bacterial population.

In the presence of sunlight, the photosynthetic production of oxygen is greater than the amount used in respiration. At night algae use up oxygen in respiration. If the daylight hours exceed the night hours by a reasonable amount, there is a net production of oxygen.

Health, Safety, Welfare

Excessive algal growth (*algal blooms*) can result in supersaturated oxygen conditions in the daytime and anaerobic conditions at night.

Some algae cause tastes and odors in natural water. While they are not generally considered pathogenic to humans, algae do cause turbidity, and turbidity provides a residence for microorganisms that are pathogenic.

8. PROTOZOA

Protozoa are single-celled animals that reproduce by *binary fission* (dividing in two). Most are aerobic chemoheterotrophs (i.e., *facultative heterotrophs*). Protozoa have complex digestive systems and use solid organic matter as food, including algae and bacteria. Therefore, they are desirable in wastewater effluent because they act as polishers in consuming the bacteria.

Flagellated protozoa are the smallest protozoans. The *flagella* (long hairlike strands) provide motility by a whiplike action. Amoeba move and take in food through the action of a mobile protoplasm. Free-swimming protozoa have *cilia*, small hairlike features, used for propulsion and gathering in organic matter.

9. WORMS AND ROTIFERS

Rotifers are aerobic, multicellular chemoheterotrophs. The rotifer derives its name from the apparent rotating motion of two sets of cilia on its head. The cilia provide mobility and a mechanism for catching food. Rotifers consume bacteria and small particles of organic matter.

Many *worms* are aquatic parasites. *Flatworms* of the class *Trematoda* are known as *flukes*, and the *Cestoda* are tapeworms. *Nemotodes* of public health concern are *Trichinella*, which causes trichinosis; *Necator*, which causes pneumonia; *Ascaris*, which is the common roundworm that takes up residence in the human intestine (ascariasis); and *Filaria*, which causes filariasis.

10. MOLLUSKS

Mollusks, such as mussels and clams, are characterized by a shell structure. They are aerobic chemoheterotrophs that feed on bacteria and algae. They are a source of food for fish and are not found in wastewater treatment systems to any extent, except in underloaded lagoons. Their presence is indicative of a high level of dissolved oxygen and a very low level of organic matter.

Macrofouling is a term referring to infestation of water inlets and outlets by clams and mussels. *Zebra mussels*, accidentally introduced into the United States in 1986, are particularly troublesome for several reasons. First, young mussels are microscopic, easily passing through intake screens. Second, they attach to anything, even other mussels, producing thick colonies. Third, adult mussels quickly sense some biocides, most notably those that are halogen-based (including chlorine), quickly closing and remaining closed for days or weeks.

The use of biocides in the control of zebra mussels is controversial. Chlorination is a successful treatment that is recommended with some caution since it results in increased toxicity, affecting other species and THM production. Nevertheless, an ongoing biocide program aimed at pre-adult mussels, combined with slippery polymer-based and copper-nickle alloy surface coatings, is probably the best approach at prevention. Once a pipe is colonized, mechanical removal by scraping or water blasting is the only practical option.

11. INDICATOR ORGANISMS

The techniques for comprehensive bacteriological examination for pathogens are complex and time-consuming. Isolating and identifying specific pathogenic microorganisms is a difficult and lengthy task. Many of these organisms require sophisticated tests that take several days to produce results. Because of these difficulties, and also because the number of pathogens relative to other microorganisms in water can be very small, *indicator organisms* are used as a measure of the quality of the water. The primary function of an indicator organism is to provide evidence of recent fecal contamination from warm-blooded animals.

The characteristics of a good indicator organism are: (a) The indicator is always present when the pathogenic organism of concern is present. It is absent in clean, uncontaminated water. (b) The indicator is present in fecal material in large numbers. (c) The indicator responds to natural environmental conditions and to treatment processes in a manner similar to the pathogens of interest. (d) The indicator is easy to isolate, identify, and enumerate. (e) The ratio of indicator to pathogen should be high. (f) The indicator and pathogen should come from the same source (e.g., gastrointestinal tract).

While there are several microorganisms that meet these criteria, *total coliform* and *fecal coliform* are the indicators generally used. *Total coliform* refers to the group of aerobic and facultatively anaerobic, gram-negative, non-spore-forming, rod-shaped bacteria that ferment lactose with gas formation within 48 hr at 95°F (35°C). This encompasses a variety of organisms, mostly of intestinal origin, including *Escherichia coli* (*E. coli*), the most numerous facultative bacterium in the feces of warm-blooded animals. Unfortunately, this group also includes *Enterobacter*, *Klebsiella*, and *Citrobacter*, which are present in wastewater but can be derived from other environmental sources such as soil and plant materials.

Fecal coliforms are a subgroup of the total coliforms that come from the intestines of warm-blooded animals. They are measured by running the standard total coliform fermentation test at an elevated temperature of 112°F (44.5°C), providing a means to distinguish false positives in the total coliform test.

Results of fermentation tests are reported as a *most probable number* (MPN) *index.* This is an index of the number of coliform bacteria that, more probably than any other number, would give the results shown by the laboratory examination; it is not an actual enumeration.

12. METABOLISM/METABOLIC PROCESSES

Metabolism is a term given to describe all chemical activities performed by a cell. The cell uses *adenosine triphosphate* (ATP) as the principle energy currency in all processes. Those processes that allow the bacterium to synthesize new cells from the energy stored within its body are said to be *anabolic*. All biochemical processes in which cells convert substrate into useful energy and waste products are said to be *catabolic*.

13. DECOMPOSITION OF WASTE

Decomposition of waste involves oxidation/reduction reactions and is classified as aerobic or anaerobic. The type of electron acceptor available for catabolism determines the type of decomposition used by a mixed culture of microorganisms. Each type of decomposition has its own peculiar characteristics that affect its use in waste treatment.

14. AEROBIC DECOMPOSITION

Molecular oxygen (O_2) must be present as the terminal electron acceptor in order for decomposition to proceed by aerobic oxidation. As in natural water bodies, the dissolved oxygen content is measured. When oxygen is present, it is the only terminal electron acceptor used. Hence, the chemical end products of decomposition are primarily carbon dioxide, water, and new cell material

as demonstrated by Table 44.2. Odoriferous, gaseous end products are kept to a minimum. In healthy natural water systems, aerobic decomposition is the principal means of self-purification.

A wider spectrum of organic material can be oxidized by aerobic decomposition than by any other type of decomposition. Because of the large amount of energy released in aerobic oxidation, most aerobic organisms are capable of high growth rates. Consequently, there is a relatively large production of new cells in comparison with the other oxidation systems. This means that more biological sludge is generated in aerobic oxidation than in the other oxidation systems.

Aerobic decomposition is the preferred method for large quantities of dilute ($BOD_5 < 500$ mg/L) wastewater because decomposition is rapid and efficient and has a low odor potential. For high-strength wastewater ($BOD_5 > 1000$ mg/L), aerobic decomposition is not suitable because of the difficulty in supplying enough oxygen and because of the large amount of biological sludge that is produced.

15. ANOXIC DECOMPOSITION

Some microorganisms will use nitrates in the absence of oxygen needed to oxidize carbon. The end products from such *denitrification* are nitrogen gas, carbon dioxide, water, and new cell material. The amount of energy made available to the cell during denitrification is about the same as that made available during aerobic decomposition. The production of cells, although not as high as in aerobic decomposition, is relatively high.

Denitrification is of importance in wastewater treatment when nitrogen must be removed. In such cases, a special treatment step is added to the conventional

Table 44.2 *Waste Decomposition End Products*

substrates	representative end products		
	aerobic decomposition	anoxic decomposition	anaerobic decomposition
proteins and other organic nitrogen compounds	amino acids ammonia → nitrites → nitrates alcohols organic acids $\Big\} \to CO_2 + H_2O$	amino acids nitrates → nitrites → N_2 alcohols organic acids $\Big\} \to CO_2 + H_2O$	amino acids ammonia hydrogen sulfide methane carbon dioxide alcohols organic acids
carbohydrates	alcohols fatty acids $\Big\} \to CO_2 + H_2O$	alcohols fatty acids $\Big\} \to CO_2 + H_2O$	carbon dioxide alcohols fatty acids
fats and related substances	fatty acids + glycerol alcohols lower fatty acids $\Big\} \to CO_2 + H_2O$	fatty acids + glycerol alcohols lower fatty acids $\Big\} \to CO_2 + H_2O$	fatty acids + glycerol carbon dioxide alcohols lower fatty acids

Health, Safety, Welfare

process for removal of carbonaceous material. One other important aspect of denitrification is in final clarification of the treated wastewater. If the final clarifier becomes anoxic, the formation of nitrogen gas will cause large masses of sludge to float to the surface and escape from the treatment plant into the receiving water. Thus, it is necessary to ensure that anoxic conditions do not develop in the final clarifier.

16. ANAEROBIC DECOMPOSITION

In order to achieve anaerobic decomposition, molecular oxygen and nitrate must not be present as terminal electron acceptors. Sulfate, carbon dioxide, and organic compounds that can be reduced serve as terminal electron acceptors. The reduction of sulfate results in the production of hydrogen sulfide (H_2S) and a group of equally odoriferous organic sulfur compounds called *mercaptans*.

The anaerobic decomposition of organic matter, also known as *fermentation*, generally is considered to be a two-step process. In the first step, complex organic compounds are fermented to low molecular weight *fatty acids* (*volatile acids*). In the second step, the organic acids are converted to methane. Carbon dioxide serves as the electron acceptor.

Anaerobic decomposition yields carbon dioxide, methane, and water as the major end products. Additional end products include ammonia, hydrogen sulfide, and mercaptans. As a consequence of these last three compounds, anaerobic decomposition is characterized by a malodorous stench.

Because only small amounts of energy are released during anaerobic oxidation, the amount of cell production is low. Thus, sludge production is correspondingly low. Wastewater treatment based on anaerobic decomposition is used to stabilize sludges produced during aerobic and anoxic decomposition.

Direct anaerobic decomposition of wastewater generally is not feasible for dilute waste. The optimum growth temperature for the anaerobic bacteria is at the upper end of the mesophilic range. Thus, to get reasonable biodegradation, the temperature of the culture must first be elevated. For dilute wastewater, this is not practical. For concentrated wastes ($BOD_5 >$ 1000 mg/L), anaerobic digestion is quite appropriate.

17. FACTORS AFFECTING DISEASE TRANSMISSION

Waterborne disease transmission is influenced by the latency, persistence, and quantity (dose) of the pathogens. *Latency* is the period of time between excretion of a pathogen and its becoming infectious to a new host. *Persistence* is the length of time that a pathogen remains viable in the environment outside a human host. The *infective dose* is the number of organisms that must be ingested to result in disease.

18. AIDS AND HIV

Acquired immunodeficiency syndrome (AIDS) is caused by the *human immunodeficiency virus* (HIV). HIV is present in virtually all body excretions of infected persons, and therefore, is present in wastewater. Present research, statistics, and available information indicates that the risk of contracting AIDS through contact with wastewater or working at a wastewater plant is nil. This is based on the following facts.

- HIV is relatively weak and does not remain viable for long periods of time in harsh environments (e.g., wastewater).

- HIV is quickly inactivated by alcohol, chlorine, and exposure to air. The chlorine concentration present in water used to flush the toilet would probably deactivate HIV.

- HIV that survives disinfection would be too dilute to be infectious.

- HIV replicates in white blood cells, not in the human intestinal tract, and, therefore, would not reproduce in wastewater.

- There is no evidence that HIV can be transmitted through water, air, food, or contact. HIV must enter the bloodstream directly, through a wound. It cannot enter the body through unbroken skin or through respiration.

- HIV is less infectious than the hepatitis virus.

- There are no reported AIDS cases that have been linked to occupational exposure in wastewater collection and treatment.

45 Toxicology

Nomenclature

AT	averaging time	yr
BCF	bioconcentration factor (mg/kg) in tissue/(mg/L) in water	L/kg
BW	body mass (weight)	kg
C	concentration	ppm, mg/m^3
CDI	chronic daily intake	mg/kg·d
CPF	carcinogen potency factor (same as cancer slope factor)	(mg/kg·d)$^{-1}$
CR	contact rate	d^{-1}
CSF	cancer slope factor (same as carcinogen potency factor)	(mg/kg·d)$^{-1}$
E	time-weighted average exposure	ppm, mg/m^3
E_m	equivalent exposure of mixture	–
EC	excess cancers	–
ED	exposure duration	yr
EED	estimated exposure dose	mg/kg·d
EF	exposure factor	–
EP	exposed population	–
HI	hazard index	–
k	decay constant	d^{-1}
L	exposure limit of particular contaminants	mg/m^3
LC$_{50}$	lethal concentration	mg/m^3
LD$_{50}$	lethal dose	mg/kg
LOAEL	lowest observed adverse effect level	mg/kg·d
LOEL	lowest observed effect level	mg/kg·d
MF	modifying factor	–
NOAEL	no observed adverse effect level	mg/kg·d
NOEL	no observed effect level	mg/kg·d
R	risk; probability of excess cancer	–
RfD	reference dose	mg/kg·d
SF	slope factor (same as carcinogen potency factor)	(mg/kg·d)$^{-1}$
T	time of exposure	s
UF	uncertainty factor	–

Subscripts

m	mixture
o	initial condition
org	organism
t	time
w	water (or other medium)
x	toxicant

1. INTRODUCTION

Sixteenth century physician Paracelsus is known as the father of toxicology. He articulated the following four toxicological principles while studying the illnesses and diseases of workers exposed to toxic substances.

- Experiments are needed to identify and understand responses to chemicals.

- The curative and toxic properties of chemicals are different.

- The healing and harmful effects of chemicals can be differentiated based on dose.

- The properties and effects of specific chemicals can be identified.

From these principles grew the science of toxicology, which is presently defined as the study of adverse effects of chemicals on living organisms. Over the nearly five centuries since Paracelsus suggested his principles of toxicology, the effects of ionizing radiation, biological agents, and naturally occurring toxins have been added to the science and study of toxicology.

This chapter examines the pathways of human exposure to chemicals and other toxicants, the effects on workers of exposure to toxicants, dose-response relationships, methods for determining safe human doses, and standards for worker protection.

2. EXPOSURE PATHWAYS

Before a toxicant can do damage to a "target" tissue or organ, it must first enter the body. "Target" means the tissue or organ where the toxicant exerts its effects. There are three major exposure pathways: dermal absorption, inhalation, and ingestion (see Fig. 45.1). The eyes are an additional exposure pathway because they are particularly vulnerable to damage in the workplace.

Dermal Absorption

A common exposure pathway for toxicants is through the skin. The skin is composed of the epidermis, the dermis, and the subcutaneous layer. The *epidermis* is the upper layer, which is composed of several layers of flattened and scale-like cells. These cells do not contain blood vessels; they obtain their nutrients from the underlying dermis. The cells of the epidermis migrate to the surface, die, and leave behind a protein called *keratin*.

Figure 45.1 *Exposure Routes for Chemical Agents*

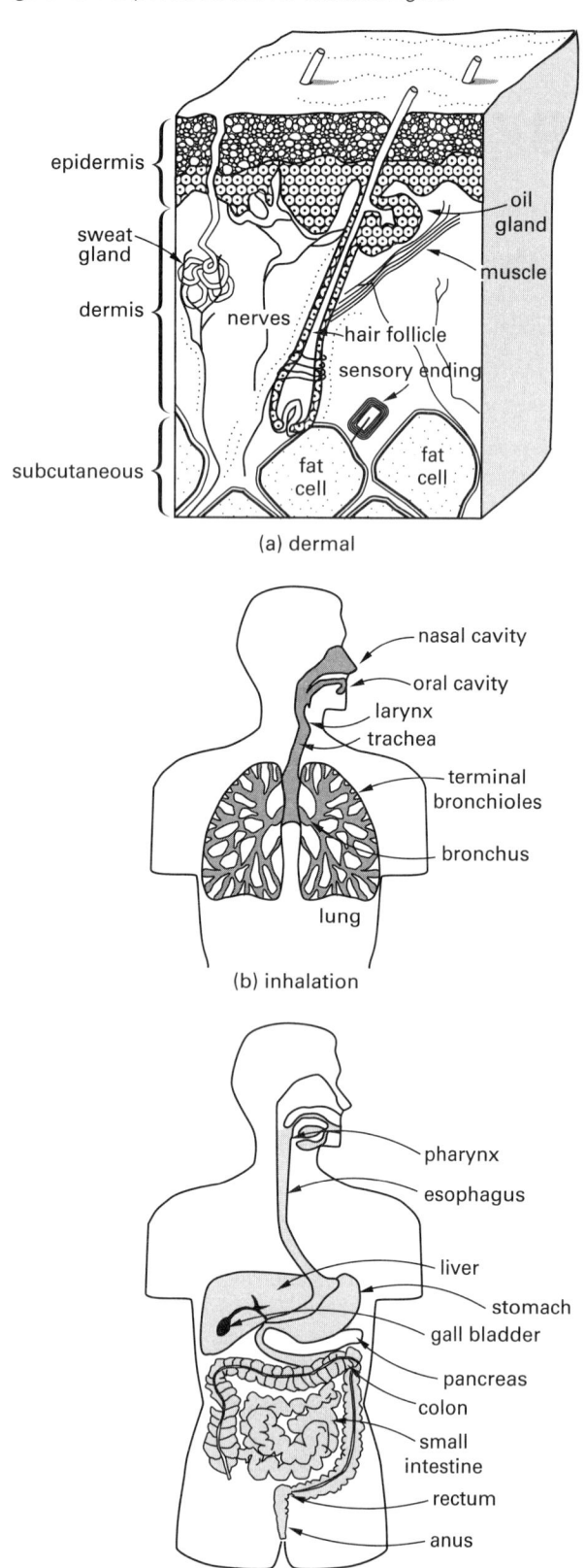

(a) dermal

(b) inhalation

(c) ingestion

Keratin is the most insoluble of all proteins, and, together with the scale-like cells, provides extreme resistance to substances and environmental conditions. Beneath the epidermis lies the *dermis*, which contains blood vessels, connective tissue, hair follicles, sweat glands, and other glands. The dermis supplies the nutrients for itself and for the epidermis. The innermost layer is called the *subcutaneous fatty tissue*, which provides a cushion for the skin and connection to the underlying tissue.

The condition of the skin, and the chemical nature of any toxic substance it contacts, affect whether and at what rate the skin absorbs that substance. The epidermis is impermeable to many gases, water, and chemicals. However, if the epidermis is damaged by cuts and abrasions, or is broken down by repeated exposure to soaps, detergents, or organic solvents, toxic substances can readily penetrate and enter the bloodstream. Chemical burns, such as those from acids, can also destroy the protection afforded by the epidermis, allowing toxicants to enter the bloodstream. Inorganic chemicals (and organic chemicals that are dissolved in water) are not readily absorbed through healthy skin. However, many organic solvents are lipid- (fat-) soluble and can easily penetrate skin cells and enter the body. After a toxicant has penetrated the skin and entered the bloodstream, the blood can transport it to target organs in the body.

Inhalation

Because the lungs are intended to transfer oxygen and carbon dioxide to and from the blood, they are a "super-highway" for toxicants to enter the body through the same process of molecular absorption. The respiratory tract consists of the nasal cavity, pharynx, larynx, trachea, primary bronchi, bronchioles, and alveoli. The first line of defense for the lungs consists of the mucus membranes found in the nasal cavity, the pharynx, the larynx, the trachea, and the bronchi. Particles are trapped in the mucus linings of the respiratory tract and are moved by the action of small hairs, called *cilia*, to the throat and then into the digestive system or are expectorated through the nose or mouth. Molecular transfer of oxygen and carbon dioxide between the bloodstream and the lungs occurs in the millions of air sacs, known as *alveoli*. Toxicants that reach the alveoli can be transferred to the blood through the respiratory system.

Toxicants that reach the alveoli will not be transferred to the blood at the same rate; various factors can increase or decrease the transfer rate of one toxicant relative to another. As in all gas transfer, the rate of diffusion depends on the ratio of the concentration of the toxicant in the air to its concentration in the blood. When the concentration of the toxicant is the same in the air and the blood, equilibrium exists and diffusion ceases. The solubility of a toxicant in blood will affect its absorption in the bloodstream. Toxicants with a high solubility in blood, such as chloroform, will be rapidly absorbed into the bloodstream, while other less-soluble toxicants will diffuse more slowly. Large molecule toxicants will transfer more slowly across the cell membrane in the alveoli than toxicants of smaller-size molecules. Both the respiration rate and the duration of exposure will affect the mass of toxicant transferred to the bloodstream over a given period. Finally, lungs that have been damaged through fibrosis or another type of disease will not be able to function as efficiently as those in a healthy individual, which will affect the transfer of a toxicant to the bloodstream.

Ingestion

Toxicants can enter the bloodstream through the digestive tract, which consists of the mouth, pharynx, esophagus, stomach, small intestine, large intestine, and rectum. The small intestine is the principal site in the digestive tract where the beneficial nutrients from food, and toxics from contaminated substances, are absorbed. Within the small intestine, millions of *villi* (projections) provide a huge surface area to absorb substances into the bloodstream. The large intestine is small, with much less surface area, and completes the digestion process by absorbing primarily water and inorganic salts.

Toxic substances will be absorbed in the intestines at various rates depending on the specific toxicant, its molecular size, and its degree of lipid (fat) solubility. Small molecular size and high lipid solubility facilitate diffusion of toxicants in the digestive tract.

The Eye

The eye is a pathway for toxicants to enter the body, through exposure to splashed chemicals, mists, particulates, and other workplace hazards. Transparent tissue in the front of the eye is known as the *cornea* and is the most likely eye tissue to come in contact with toxic substances. Toxicants can damage the cornea beyond its ability to repair itself by replacement of the outer cells from lower, basal cells. Under the cornea are the *anterior chamber*, the *lens*, and the *posterior chamber*. The transparency of the cornea and the lens is critical for the eye to function properly. Permanent damage can occur from scar tissue and death of the cells of the cornea.

Site of Effect

The exposure pathways described are routes of entry of toxicants into the body. However, the effects of such exposure may be either localized or systemic. If *localized*, the immediate area of entry is affected. If *systemic*, the toxicant's effects will occur at other sites in the body and may affect an organ or the central nervous system, causing damage to one or more functional systems. An example of a localized effect would be acid burns on the skin at the site of the exposure. An example of a systemic effect would be a toxicant that enters the bloodstream and is transported to a target organ where its effects occur.

For systemic effects to occur, the rate of accumulation of a toxicant must exceed the body's ability to *excrete* (eliminate) it or to *biotransform* it (transform it to less

harmful substances). A toxicant can be eliminated from the body through the *kidneys*, which are the primary organs for eliminating toxicants from the body. The kidneys biotransform a toxicant into a water-soluble form and then eliminate it though the urine. The *liver* is also an important organ for eliminating a toxicant from the body, first by biotransformation, then by excretion into the bile where it is eliminated through the small intestine as feces.

A toxicant may also be stored in tissues for long periods before an effect occurs. Toxic substances stored in tissue (primarily in fat but also in bones, the liver, and the kidneys) may exert no effect for many years, or at all within the affected person's life. DDT, a pesticide, for instance, can be stored in body fat for many years and not exert any adverse effect on the body.

When toxic substances cannot be eliminated fast enough to keep up with the exposure, or when they cannot be stored without adverse effect in fatty tissue, then the target organ is overwhelmed and systemic effects occur. The liver, kidneys, and central nervous system are target organs that are commonly affected systemically.

3. EFFECTS OF EXPOSURE TO TOXICANTS

After toxicants are absorbed into the body through one or more of the pathways, a wide variety of effects on the human body is possible. When the toxic agents concentrate in target tissue or organs, the agents may interfere with the normal functioning of enzymes and cells or may cause genetic mutations.

Pulmonary Toxicity

Pulmonary toxicity refers to adverse effects on the respiratory system from toxic agents.

- Various toxicants, particulates, and gases and vapors can irritate or damage the nasal passages and the nerve cells responsible for smell. Excessive mucus flow, constriction of airflow, and a decrease in the effectiveness of the mucus system to clear toxicants may occur.

- *Nasal cancer* may result from prolonged exposure to chromate, nickel, isopropyl alcohol, hexavalent chromium, formaldehyde, nitrosamines, acrolein, actaldehyde, and vinyl chloride.

- Nitrogen dioxide, sulfur dioxide, ozone, and cigarette smoke may damage the cilia, which move particulates to the mouth for removal from the body. These toxicants, plus ammonia, may also lead to the proliferation of cells and the development of cancer.

- Chlorine, hexavalent chromium, ammonia, sulfur dioxide, and cigarette smoke may cause excessive mucus secretion, known as *bronchitis*.

- Ammonia, chlorine, hydrogen fluoride, nickel, and oxides of nitrogen, ozone, and perchloroethylene can cause *pulmonary edema*, the excessive accumulation of fluid in the alveoli of the lungs. This can interfere with transfer of oxygen and carbon dioxide and may develop into pneumonia.

- Silica, coal dust, asbestos, beryllium, cadmium, and the pesticide paraquat can result in an increased amount of connective tissue around the alveoli. This is known as *fibrosis*. Asbestos fibers can also cause lung cancer.

- Crystalline silica can cause increased deposition of connective tissue around the alveoli, known as *silicosis*.

- Cigarette smoke, ozone, nitrogen dioxide, and cadmium can cause enlargement and joining of the alveoli, causing the lungs to lose the ability to expand and contract, known as *emphysema*.

- Benzo(a)pyrene, nickel, chromium, arsenic, beryllium, and cigarette smoke can cause lung cancer.

Cardiotoxicity

Cardiotoxicity refers to the effects of toxic agents on the heart. The heart pumps blood through the aorta to the arteries and capillaries where oxygen and nutrients are delivered to tissues. Waste products are pumped from the capillaries to the lungs, kidney, and liver for removal from the body. The heart rate is controlled by the autonomic nervous system, and contractions are caused by electrochemical activity in the heart muscle.

- Toxicants can interfere with the normal contractions of the heart, which decreases its ability to transport oxygen and nutrients. The heart rate may be changed, and the strength of contractions may be diminished.

- Certain metals can affect the contractions of the heart and can interfere with cell metabolism.

- Organic solvents can alter the heart rate and the strength of contractions. Solvents act by suppressing the central nervous system, which affects the portion of the brain responsible for regulating the heart rate.

- Carbon monoxide can result in a decrease in the oxygen supply, causing improper functioning of the nervous system controlling the heart rate.

Hematoxicity

Hematoxicity refers to damage to the body's blood supply, which includes red blood cells, white blood cells, platelets, and plasma. The *red blood cells* transport oxygen to the body's cells and carbon dioxide to the lungs. *White blood cells* perform a variety of functions associated with the immune system. *Platelets* are important in blood clotting. *Plasma* is the noncellular portion of blood and contains proteins, nutrients, gases, and waste products.

- Benzene, lead, methylene chloride, nitrobenzene, naphthalene, and insecticides are capable of red blood cell destruction and can cause a decrease in the oxygen-carrying capacity of the blood. The resulting anemia can affect normal nerve cell functioning and control of the heart rate, and can cause shortness of breath, pale skin, and fatigue.

- Carbon tetrachloride, pesticides, benzene, and ionizing radiation can affect the ability of the bone marrow to produce red blood cells.

- Mercury, cadmium, and other toxicants can affect the ability of the kidneys to stimulate the bone marrow to produce more red blood cells when needed to counteract low oxygen levels in the blood.

- Some chemicals, including carbon monoxide, can interfere with the blood's capacity to carry oxygen, resulting in lowered blood pressure, dizziness, fainting, increased heart rate, muscular weakness, nausea, and, after prolonged exposure, death.

- Hydrogen cyanide and hydrogen sulfide can stimulate cells in the aorta, causing increased heart and respiratory rate. At high concentrations, death can result from respiratory failure.

- Benzene, carbon tetrachloride, and trinitrotoluene can suppress stem cell production and the production of white blood cells. This can affect the clotting mechanism and the immune system.

- Benzene can cause high levels of white blood cells, a condition known as *leukemia*.

Hepatoxicity

Hepatoxicity refers to adverse effects on the liver that impede its ability to function properly. The liver converts carbohydrates, fats, and proteins to maintain the proper levels of glucose in the blood and converts excess protein and carbohydrates to fat. It also converts excess amino acids to ammonia and urea, which are removed in the kidneys. The liver also provides storage of vitamins and beneficial metals, as well as carbohydrates, fats, and proteins. Red blood cells that have degenerated are removed by the liver. Substances needed for other metabolic processes are provided by the liver. Finally, the liver detoxifies metabolically produced substances and toxicants that enter the body.

- Hexavalent chromium and arsenic cause cell damage in the liver.

- Carbon tetrachloride and alcohol can cause damage and death of liver cells, a condition known as *cirrhosis of the liver*.

- Chemicals or viruses can cause inflammation of the liver, known as *hepatitis*. Cell death and enlargement of the liver can occur.

Nephrotoxicity

Nephrotoxicity refers to adverse effects on the kidneys. The kidneys excrete ammonia as urea to rid the body of metabolic wastes. They maintain blood pH by exchanging hydrogen ions for sodium ions, and maintain the ion and water balance by excreting excess ions or water as needed. They also secrete hormones needed to regulate blood pressure. Like the liver, the kidneys function to detoxify substances.

- Heavy metals—primarily lead, mercury, and cadmium—cause impaired cell function and cell death. These metals can be stored in the kidneys, interfering with the functioning of enzymes in the kidneys.

- Chloroform and other organic substances can cause cell dysfunction, cell death, and cancer.

- Ethylene glycol can cause renal failure from obstruction of the normal flow of liquid through the kidneys.

Neurotoxicity

Neurotoxicity refers to toxic effects on the nervous system, which consists of the central nervous system (CNS) and the peripheral nervous system (PNS). The *central nervous system* includes the brain and the spinal cord, while the *peripheral nervous system* includes the remaining nerves, which are distinguished as sensory and motor nerves.

The basic functional unit of the nervous system is the *neuron*, which has the job of transmitting stimuli to and from the CNS. *Axons* are fiber-like groups of cells wrapped in a membrane sheath that carry the impulses within the neuron. The nerves function through electrochemical changes that occur in response to *neurotransmitters*, which are chemical substances released to initiate a response in a nerve, muscle, or gland. The exchange of sodium and potassium ions across a nerve membrane produces a voltage potential that stimulates the receptor. Enzymes cause the neurotransmitter to be rapidly broken down so the receptor will not be constantly stimulated.

Neurotoxic effects fall into two basic types: *destruction* of nerve cells and *interference* with neurotransmission.

- Methyl-mercury causes neurotoxicity in both high and low doses. In adults, damage from methyl-mercury includes loss of muscle control and coordination, a condition known as *ataxia*. In children, it can cause mental retardation and paralysis.

- Aluminum has been associated with Alzheimer's disease, but its effects on nerve cell functioning are not clear.

- Manganese can result in damage to the CNS. Its effects include irritability, difficulty in walking, and speech abnormalities. Prolonged exposure to manganese can cause symptoms similar to Parkinson's disease, including tremors and spastic muscle contractions.

- Methanol can cause nerve damage leading to blindness and death.

- Carbon monoxide and hydrogen cyanide can disrupt nerve cell metabolism and lead to death.

- Acrylamide, carbon disulfide, hexacarbons, and organophosphate pesticides can cause loss of peripheral sensations and can impair skeletal muscle functioning.

- Hexachlorophene is an antibacterial agent that interferes with neurotransmission and can cause blurred vision, muscle weakness, confusion, seizures, coma, and death.

- Lead affects the motor nerves.

- DDT interferes with neurotransmission, which can lead to tremors, irritability, hypersensitivity, dizziness, and convulsions.

- Methylene chloride, carbon tetrachloride, and chloroform depress the CNS and can lead to dizziness, loss of muscle control, unconsciousness, and respiratory or cardiac arrest.

- Organophosphate pesticides and carbamate insecticides can cause muscle twitching, weakness, paralysis, and respiratory arrest.

Immunotoxicity

Immunotoxicity refers to toxic effects on the immune system, which includes the lymph system, blood cells, and antibodies in the blood.

The *lymph system* includes lymph fluid from the blood that passes through lymph nodules where microorganisms and cellular wastes are filtered out. The *spleen* also serves to destroy microorganisms as blood passes through it, and the *thymus* produces cells that protect the body from microorganisms. The cells of the immune system are produced in the *bone marrow*, and have the capabilities to destroy bacteria- and virus-infected cells. These cells are known as *natural killer cells*. *Antibodies* that are specific for antigens are also produced in the cells of the immune system.

- Halogenated aromatic hydrocarbons (HAHs) include polychlorinated biphenyls (PCBs), polybrominated byphenyls (PBBs), polychlorinated dibenzo-*p*-dioxins (TCDDs), and dibenzofurans. HAHs can cause atrophy of the thymus, decreased antibody production, and tumor promotion. TCDDs can affect stem cell division in the bone marrow.

- Polycyclic aromatic hydrocarbons (PAHs) are carcinogens due (possibly) to their ability to suppress the body's immune system. Examples of PAHs are benzo(a)pyrene (BaP) and 7,12-dimethylbenz(a)anthracene (DMBA).

- Heavy metals (e.g., lead, arsenic, mercury, and cadmium) can suppress the immune system. Lead, arsenic, and cadmium cause decreased levels of antibodies, resulting in susceptibility to infection.

Mercury can decrease antibody response and cause an autoimmune response whereby the immune system attacks normal tissue, primarily in the kidneys, which is a major storage site for mercury.

- Organophosphate pesticides can decrease production of antibodies and killer cells, leading to greater susceptibility to infection.

- Some organic solvents affect antibody production and the formation of natural killer cells.

Reproductive Toxicity

The effect of toxicants on the male or female reproductive system is known as *reproductive toxicity*. For the male reproductive system, toxicants primarily affect the division of sperm cells and the development of healthy sperm. For the female reproductive system, toxicants can affect the endocrine system, the brain, and the reproductive tract.

The male reproductive system can be affected by the following toxicants.

- the fungicide 1,2-dibromo-3-chloropropane
- the fumigant ethylene dibromide
- glycol ethers used in paint
- lead
- ethanol
- acetaldehyde
- phthalate esters
- methylene chloride
- the soil fumigant dibromochloropropane
- cadmium
- pesticides (e.g., DDT, DDE, methoxychlor, toxaphene, chlordecone, and endosulfan)

The female reproductive system can be affected by the following toxicants.

- heavy metals (e.g., inorganic mercury, lead, manganese, tin, and cadmium)
- styrene and polystyrene
- polycyclic aromatic hydrocarbons
- cigarette smoke
- pesticides

Toxic Effects on the Eye

There are a wide variety of toxic substances that can cause damage to the eye through contact with the cornea. Acids and bases can change the pH of the eye tissue and cause the cells to break down or disintegrate through coagulation of the proteins in the eye. Lime, metals, solvents, arsenicals, organic mercury, organic

toxic chemicals, methanol, smog, dust, and allergens can cause damage to the eye by causing interference with the neurological functions, irritation or cloudiness of the cornea, damage to blood vessels, scratches on the cornea, formation of cataracts in the lens, and decreased vision, among other effects. Caustics, hydroxides of light metals, can cause greater damage to the eye than acids because acids precipitate a protein barrier that prevents further penetration into the tissue. Caustics, however, will continue to penetrate the eye until they are removed.

4. DOSE-RESPONSE RELATIONSHIPS

Dose-response relationships can be used to relate the response of an organism to increasing dose levels of toxicants. They show the level of death or other effect at specific doses of a toxic substance.

Studies involving animals are known as *toxicological studies*, whereas studies involving humans are called *epidemiological studies*. The results from both types of studies are sometimes used together to establish dose-response relationships and to establish safe human doses.

Dose-Response Curves

An objective of toxicity tests is to establish the *dose-response curve*, as illustrated in Fig. 45.2.

Figure 45.2 Dose-Response Relationships

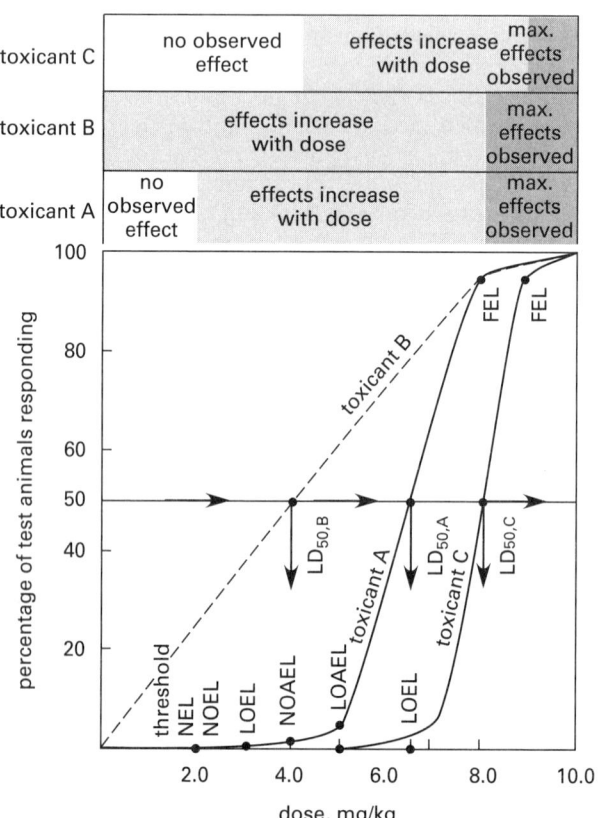

Several important features of a dose-response curve are illustrated and described as follows for toxicant A of Fig. 45.2.

- *Response:* The ordinate is typically the percent of the test animals that respond at the given dose.

- *Dose:* The abscissa is typically the dose applied to the test animal, such as milligram of toxicant per kilogram of body mass. Other measures of dose are possible.

- *No Observed Effect:* The range of the curve below which no effect is observed is the range of no observed effect. The upper end of the range is known as the *threshold,* the *no effect level* (NEL), or the *no observed effect level* (NOEL). The terms all mean the dose below which no measurable effect can be observed.

- *Lowest Observed Effect:* The dose where minor effects first can be measured, but the effects are not directly related to the response being measured, is known as the *lowest observed effect level* (LOEL).

- *No Observed Adverse Effect:* The dose where effects related to the response being measured first can be measured is known as the *no observed adverse effect level* (NOAEL). However, at this level, the effects observed at the higher doses are not observed. The terms all mean the dose below which no measurable effect can be observed.

- *Lowest Observed Adverse Effect:* The dose where effects related to the response being measured first can be measured, and are the same effects as the effects observed at the higher doses, is known as the *lowest observed adverse effect level* (LOAEL).

- *Frank Effect:* The *frank effect level* (FEL) dose marks the point where maximum effects are observed with little increase in effect for increasing dose.

For some toxicants, primarily those believed to be carcinogens, there is no apparent threshold. Any dose is considered to have an effect even though such effect may be unmeasurable at low doses. Such toxicants have no safe exposure level. This is illustrated as toxicant B in Fig. 45.2. Lead is an example of a toxicant with no threshold dose.

Toxicants A and C in Fig. 45.2 have similarly shaped dose-response curves, which means that they show the same relative effects at both high and low doses (i.e., toxicant A is more toxic than toxicant C at both high and low doses). If the dose-response curves for toxicants A and C were to cross, then one toxicant would be more toxic at low doses, and the other would be more toxic at high doses. Toxicant A is more toxic than toxicant C because a lower dose of toxicant A results in a pronounced response.

Acute and Chronic Toxicity Tests

To establish a dose-response relationship, toxicity tests are necessary. Both acute and chronic toxicity tests can be used for this purpose.

Acute toxicity tests are conducted for a relatively short duration, typically 24 hours to 14 days, with death of the test animal typically being the "response." Test animals typically include rats, mice, rabbits, and guinea pigs. The acute test must include a *control group* that is subjected to identical environmental conditions as the test animals except that it is not exposed to the toxicant. The response of the control group enables the responses at low toxicant doses to be differentiated from environmental or other factors that may affect the test animals.

The results of acute toxicity tests are typically expressed as the *lethal dose* or *lethal concentration*—the concentration of toxicant at which a specified percentage of the test animals died. The lethal dose is expressed as the mass of toxicant per unit mass of test animal. Thus, LD_{50} means the dose in milligrams of toxicant per kilogram of body mass at which 50% of the test animals died.

For acute tests involving inhalation as the exposure pathway, the concentration (in parts per million) of the toxicant in air is used. If the toxicant is in particulate form, the concentration in milligrams of toxic particles per cubic meter of air is used. For example, LC_{50} means the concentration of the toxicant in air at which 50% of the test animals died.

The lethal concentration is associated with the inhalation pathway, while the lethal dose may be associated with either the ingestion or skin absorption pathway. However, this convention is not always followed, and the LD_{50} sometimes refers to the inhalation pathway. (Usually the pathway is noted with the lethal dose.)

Chronic toxicity tests can also be performed to determine the long-term dose-response relationship of a toxicant. Chronic tests may be conducted for 30 days or more, for years, or for the life of the test animal. Although death is often the end point in acute toxicity tests, in chronic toxicity tests the administered doses are selected such that most of the animals survive for the duration of the study. Chronic tests monitor diet, perform urinalysis, observe changes in blood composition and chemistry, and perform gross and microscopic examination of major tissues and organs.

5. SAFE HUMAN DOSE

EPA Methods

Several approaches exist for selecting a safe human dose from the data obtained from toxicological and epidemiological studies. The approach most likely to be encountered by the environmental engineer is the one recommended by the U.S. Environmental Protection Agency (EPA).

First, the differences between carcinogens and noncarcinogens (referred to by the EPA as *systemic toxicants*) are compared. Chemicals are classified as carcinogens if they can, or are believed to, produce tumors after exposure. *Carcinogens* generally do not exhibit thresholds of response at low doses of exposure, and any exposure is assumed to have an associated risk. The EPA model used to evaluate carcinogenic risk assumes no threshold and a linear response to any amount of exposure. *Noncarcinogens* (i.e., systemic toxicants) are chemicals that do not produce tumors (or gene mutations) but instead adversely interfere with the functions of enzymes in the body, thereby causing abnormal metabolic responses. Noncarcinogens have a dose threshold below which no adverse health response can be measured. Some substances can produce both carcinogenic and noncarcinogenic responses.

Noncarcinogens

The threshold below which adverse health effects in humans are not measurable or observable is defined by the EPA as the *reference dose* (RfD). The reference dose is used by the EPA and often results in more restrictive intake values than the values published by other organizations. The reference dose is the safe daily intake that is believed not to cause adverse health effects. The reference dose relates to the ingestion and dermal contact pathways and is route specific. For gases and vapors (exposure by the inhalation pathway), the threshold may be defined as the *reference concentration* (RfC). Sometimes the term "reference dose" is also used for exposure by the inhalation pathway. Note also that prior to 1993, the reference dose was referred to as the *acceptable daily intake* (ADI). The terms mean the same thing, but to distinguish between various aspects of risk assessment and management, the EPA adopted less value-laden terminology in 1993—hence the current use of RfD.

The reference dose is set for the most sensitive response when more than one response, or endpoint, is measured. For example, a chemical may affect the central nervous system, the cardiovascular system, or other particular organs. The most sensitive of these responses would be the basis for selecting the reference dose.

The reference dose (or reference concentration) is determined by dividing the NOAEL by the *uncertainty* and *modifying factors* that represent uncertainty of the procedure (UF and MF, respectively). The factors used by the EPA and the World Health Organization are given in Table 45.1.

$$\text{RfD} = \frac{\text{NOAEL}}{(\text{UF})(\text{MF})} \qquad 45.1$$

After an RfD has been identified for one or more toxicants, the *hazard ratio* can be determined to assess whether exposures indicate an unacceptable hazard. The hazard ratio (HR) is the *estimated exposure dose* (EED) divided by the reference dose for each of the toxicants from all routes of exposure. If the sum of the ratios exceeds 1.0, the risk is unacceptable. The hazard ratio should be considered a preliminary assessment.

$$\text{HR} = \frac{\text{EED}}{\text{RfD}} \qquad 45.2$$

Example 45.1

A chronic oral toxicity study was conducted to determine the effects of copper cyanide (CuCN) on rats. The quality of the data is judged by the toxicologist to be "average." The following results were obtained from the study.

NOEL	adverse effect observed at all doses
LOEL	adverse effect observed at all doses
NOAEL	adverse effect observed at all doses
LOAEL	25 mg/kg·d
FEL	100 mg/kg·d

In the absence of other, more reliable published information, what should be used as the reference dose (RfD) for human exposure from drinking water containing copper cyanide (CuCN)?

Solution

From Table 45.1, uncertainty factors 10A, 10H, and 10L apply. Therefore,

$$\text{UF} = (10)(10)(10) = 1000$$

Select an average modifying factor of MF = 5.

From Eq. 45.1,

$$\text{RfD} = \frac{\text{NOAEL}}{(\text{UF})(\text{MF})} = \frac{25 \frac{\text{mg}}{\text{kg·d}}}{(1000)(5)}$$
$$= 5 \times 10^{-3} \text{ mg/kg·d} \quad (5 \ \mu\text{g/kg·d})$$

Example 45.2

A community water supply well was found to be contaminated with copper cyanide (CuCN), methanol (CH$_3$OH), and potassium cyanide (KCN). The concentrations in the water and the EPA reference doses are as follows.

toxicant	exposure (μg/L)	RfD (oral) (μg/kg·d)
CuCN	40	5
CH$_3$OH	1000	500
KCN	600	50

Is a 70 kg person who drinks 2 L of water daily exceeding the safe human dose for these noncarcinogens?

Solution

Determine the estimated exposure dose for the toxicants.

$$\text{EED}_{\text{CuCN}} = \frac{E_{\text{CuCN}}(\text{CR})}{m} = \frac{\left(40 \frac{\mu\text{g}}{\text{L}}\right)\left(2 \frac{\text{L}}{\text{d}}\right)}{70 \text{ kg}}$$
$$= 1.14 \ \mu\text{g/kg·d}$$

Table 45.1 *Typical EPA Factors Used to Calculate Reference Dose*[*]

designation	description	uncertainty factor (UF)
10A, interspecies variability	used when extrapolating from results of long-term studies on experimental animals when results from human exposure studies are inadequate or unavailable	10
10H, intraspecies variability	used when extrapolating from valid experimental results in studies using prolonged exposure to average healthy humans	10
10L	used when deriving a reference dose (RfD) from LOAEL values rather than NOAEL values	10
10S	used when extrapolating from subchronic results on experimental animals when long-term human results are inadequate or absent	10

[*]A professional judgment modifying factor (MF)—greater than zero and less than or equal to ten—is used to reflect the scientific uncertainties associated with the study not mentioned explicitly in this table.

From Eq. 45.2,

$$HR_{CuCN} = \frac{EED_{CuCN}}{RfD_{CuCN}} = \frac{1.14 \frac{\mu g}{kg \cdot d}}{5 \frac{\mu g}{kg \cdot d}}$$

$$= 0.229$$

Similarly, for the other toxicants,

toxicant	exposure, E ($\mu g/L$)	estimated exposure dose, EED ($\mu g/kg \cdot d$)	RfD ($\mu g/kg \cdot d$)	hazard ratio, HR
CuCN	40	1.14	5	0.229
CH$_3$OH	1000	28.57	500	0.057
KCN	600	17.14	50	0.343
			total	0.629

Because the sum of the hazard ratios is less than 1.0, the safe human dose is not exceeded.

Carcinogens

The distinguishing feature of cancer is the uncontrolled growth of cells into masses of tissue called *tumors*. Tumors may be *benign*, in which the mass of cells remains localized, or *malignant*, in which the tumors spread through the bloodstream to other sites within the body. This latter process is known as *metastasis* and determines whether the disease is characterized as cancer. The term *neoplasm* (new and abnormal tissue) is also used to describe tumors.

Cancer occurs in three stages: initiation, promotion, and progression. During the *initiation* stage, a cell mutates and the DNA is not repaired by the body's normal DNA repair mechanisms. During the *promotion* stage, the mutated cells increase in number and undergo differentiation to create new genes. During *progression*, the cancer cells invade adjacent tissue and move through the bloodstream to other sites in the body. It is believed that continued exposure to the agent that initiated genetic mutation is necessary for progression to continue. Many mutations are believed to be required for the progression of cancer cells to occur at remote sites in the body.

Direct Human Exposure

The EPA's classification system for carcinogenicity is based on a consensus of expert opinion called *weight of evidence*.

The EPA maintains a database of toxicological information known as the Integrated Risk Information System (IRIS). The IRIS data include chemical names, chemical abstract service registry numbers (CASRN), reference doses for systemic toxicants, carcinogen potency factors (CPF) for carcinogens, and the carcinogenicity group classification, which is shown in Table 45.2.

Table 45.2 *EPA Carcinogenicity Classification System*

group	description
A	human carcinogen
B1 or B2	probable human carcinogen B1 indicates that human data are available. B2 indicates sufficient evidence in animals and inadequate or no evidence in humans.
C	possible human carcinogen
D	not classifiable as to human carcinogenicity
E	evidence of noncarcinogenicity for humans

The dose-response for carcinogens differs substantially from that of noncarcinogens. For carcinogens it is believed that any dose can cause a response (mutation of DNA).

Since there are no levels (i.e., no thresholds) of carcinogens that could be considered safe for continued human exposure, a judgment must be made as to the acceptable level of exposure, which is typically chosen to be an excess lifetime cancer risk of 1×10^{-6} (0.0001%). *Excess lifetime cancer risk* refers to the incidence of cancers developed in the exposed animals minus the incidence in the unexposed control animals. For whole populations exposed to carcinogens, the number of total excess cancers, EC, is the product of the probability of excess cancer, R, and the total exposed population, EP.

$$\text{EC} = (\text{EP})R \qquad 45.3$$

Under the EPA approach, the *carcinogen potency factor* (CPF) is the slope of the dose-response curve at very low exposures. The CPF is also called the *potency factor* or *cancer slope factor* (CSF) and has units of $(\text{mg/kg·d})^{-1}$. The CPF is pathway (route) specific. The CPF is obtained by extrapolation from the high doses typically used in toxicological studies. (See Table 45.3.)

Table 45.3 *EPA Standard Values for Intake Calculations*

parameter	standard value
average body weight, adult (15–44 years)[a]	71.8 kg[b]
average body weight, child (excluding infants)[a]	
1–3 years	13 kg[b]
1–14 years	25 kg[b]
amount of water ingestion, adult	2 L/d
amount of water ingestion, child	1 L/d
amount of air breathed, adult	20 m³/d
amount of air breathed, child	5 m³/d
amount of fish consumed, adult	6.5 g/d
exposure duration, lifetime	70 yr

[a]male and female
[b]Standard weights vary with age, gender, date of study, and source.

The CPF is the probability of risk produced by lifetime exposure to 1.0 mg/kg·d of the known or potential human carcinogen. The slope factor can be multiplied by the long-term daily intake (*chronic daily intake*, CDI) to obtain the lifetime probability of risk, R, for daily doses other than 1.0 mg/kg·d. For less than lifetime exposure, the *exposure duration* must be used to calculate the total intake, which must be divided by the *averaging duration* of 70 years for carcinogens. For noncarcinogens, the averaging duration is the same as the exposure duration (see Eq. 11.94).

$$R = (\text{SF})(\text{CDI}) \qquad 45.4$$

Example 45.3

A village with a stable population of 50,000 has a water supply that has been contaminated with benzene (C_6H_6) from a leaking underground storage tank. The leak occurred during the 20 yr before it was removed.

The estimated average concentration of benzene during this period of leaking was 50 μg/L. It is expected that it will take another 10 yr before the benzene will be below detectable levels. The average estimated concentration of benzene during this period is 20 μg/L. The slope factor for benzene by the oral route is 2.0×10^{-2} $(\text{mg/kg·d})^{-1}$. Assume a 70 yr lifespan, 70 kg adults, 10 kg children, 2 L/d adult water consumption, 1 L/d child water consumption, and 10% children in the village. If the acceptable risks of additional cancer deaths due to benzene in the water are 1 adult and 0.5 child, can the water supply be used for the next 10 yr, or should the village abandon the supply?

Solution

step 1: Find the chronic daily intake (CDI) for adults and children for each time period.

Given a total lifespan (AT) of 70 yr and an exposure factor (EF) of 1, use Table 45.3 and Eq. 11.94.

$$\text{CDI} = \frac{C(\text{CR})(\text{EF})(\text{ED})}{(\text{BW})(\text{AT})}$$

	BW (kg)	CR (L/d)	ED (yr)	C (mg/L)	CDI (mg/kg·d)
adults	70	2	10	20×10^{-3}	8.16×10^{-5}
			20	50×10^{-3}	4.08×10^{-4}
children	10	1	10	20×10^{-3}	2.86×10^{-4}
			20	50×10^{-3}	1.43×10^{-3}

step 2: Find the probable risk of additional cancers, R, for adults and children for each time period.

Given a slope factor for orally ingested benzene of 2.0×10^{-2} $(\text{mg/kg·d})^{-1}$, use Eq. 45.4.

$$R = (\text{SF})(\text{CDI})$$

Equation 45.4 yields the following probabilities of excess cancer risk.

	ED (yr)	CDI (mg/kg·d)	R
adults	10	8.16×10^{-5}	1.63×10^{-6}
	20	4.08×10^{-4}	8.16×10^{-6}
children	10	2.86×10^{-4}	5.71×10^{-6}
	20	1.43×10^{-3}	2.86×10^{-5}

step 3: Find the total excess cancers for adults and children for each time period.

From Eq. 45.3,

$$\text{EC} = (\text{EP})R$$

Health, Safety, Welfare

	EP	ED (yr)	R	EC
adults	45,000	10	1.63×10^{-6}	0.0735
		20	8.16×10^{-6}	0.367
children	5000	10	5.71×10^{-6}	0.0286
		20	2.86×10^{-5}	0.143

The total adult excess cancer risk is

$$\mathrm{EC}_{a,\text{total}} = \mathrm{EC}_{a,10} + \mathrm{EC}_{a,20}$$
$$= 0.0735 + 0.367$$
$$= 0.441$$

Similarly, the total children excess cancer risk is

$$\mathrm{EC}_{c,\text{total}} = \mathrm{EC}_{c,10} + \mathrm{EC}_{c,20}$$
$$= 0.0286 + 0.143$$
$$= 0.171$$

Because the total number of excess cancers for adults and children is less than 1.0 and 0.5, respectively, the water supply can be used.

Bioconcentration Factors

Besides setting factors for direct human exposure to toxicants through water ingestion, inhalation, and skin contact, the EPA also has developed *bioconcentration factors* (BCF) (also referred to as *steady-state BCF*) so that the human intake from consumption of fish and other foods can be determined. Bioconcentration factors have been developed for many toxicants and provide a relationship between the toxicant concentration in the tissue of the organism (e.g., fish) and the concentration in the medium (e.g., water). The concentration in the organism equals the product of the BCF and the concentration in the medium. Not all chemicals or other substances will bioaccumulate, and the BCF pertains to a specific organism, such as fish.

$$C_{\text{org}} = (\mathrm{BCF})C_w \qquad 45.5$$

Selected bioconcentration factors (BCF) for selected chemicals in fish are given in Table 45.4. The substances are arranged in descending order of BCFs to illustrate the substances that have a high potential to bioaccumulate in fish. These substances are of great importance when the oral pathway is present in a particular situation.

The BCF factors can be applied to determine the total dose to humans who ingest fish from water contaminated with toxicants that bioaccumulate. This dose would be added to the dose received from drinking the contaminated water.

Table 45.4 Typical Bioconcentration Factors for Fish[*]

substance	BCF (L/kg)
polychlorinated biphenyls	100 000
4,4′ DDT	54 000
DDE	51 000
heptachlor	15 700
chlordane	14 000
toxaphene	13 100
mercury	5500
2,3,7,8 tetrachlorodibenzo-p-dioxin (TCDD)	5000
dieldrin	4760
copper	200
cadmium	81
lead	49
zinc	47
arsenic	44
tetrachloroethylene	31
aldrin	28
carbon tetrachloride	19
chromium	16
chlorobenzene	10
benzene	5.2
chloroform	3.75
vinyl chloride	1.17
antimony	1

[*]For illustrative purposes only. Subject to change without notice. Local regulations may be more restrictive than federal.

Example 45.4

The town of Central has a water supply from the Middle River that was discovered to be contaminated with 0.03 µg/L of heptachlor (carcinogen class B2) for 5 yr. People in the town have continually enjoyed the large trout from the Middle River and consume twice the EPA standard factor for fish consumption. The slope factor for heptachlor is 4.5 $(\mathrm{mg/kg \cdot d})^{-1}$. What is the risk of excess cancer over the lifetime of an adult if the heptachlor contamination is removed this year?

Solution

Determine the standard factors. From Table 45.3 and Table 45.4,

$$\text{fish consumption} = \left(6.5 \, \frac{\mathrm{g}}{\mathrm{d}}\right)(2) = 13 \text{ g/d}$$
$$\mathrm{CR} = 2 \text{ L/d}$$
$$\mathrm{ED} = 5 \text{ yr}$$
$$\mathrm{BW} = 70 \text{ kg}$$
$$\mathrm{BCF} = 15\,700 \text{ L/kg}$$
$$C_{\text{heptachlor}} = 0.03 \text{ µg/L}$$
$$\mathrm{SF}_{\text{heptachlor}} = 4.5 \, (\mathrm{mg/kg \cdot d})^{-1}$$

Calculate the CDI for heptachlor in water.

From Eq. 11.94,

$$\mathrm{CDI}_w = \frac{C_{\mathrm{heptachlor}}(\mathrm{CR})(\mathrm{EF})(\mathrm{ED})}{(\mathrm{BW})(\mathrm{AT})}$$

$$= \frac{\left(0.03 \times 10^{-6}\ \frac{\mathrm{g}}{\mathrm{L}}\right)\left(2\ \frac{\mathrm{L}}{\mathrm{d}}\right)(1)(5\ \mathrm{yr})\left(10^3\ \frac{\mathrm{mg}}{\mathrm{g}}\right)}{(70\ \mathrm{kg})(70\ \mathrm{yr})}$$

$$= 6.12 \times 10^{-8}\ \mathrm{mg/kg{\cdot}d}$$

Calculate the concentration of heptachlor in fish and determine its CDI.

From Eq. 45.5,

$$C_{\mathrm{heptachlor/fish}} = (\mathrm{BCF})C_{\mathrm{heptachlor}}$$

$$= \left(15\,700\ \frac{\mathrm{L}}{\mathrm{kg}}\right)\left(0.03 \times 10^{-6}\ \frac{\mathrm{g}}{\mathrm{L}}\right)\left(10^3\ \frac{\mathrm{mg}}{\mathrm{g}}\right)$$

$$= 0.471\ \mathrm{mg/kg}$$

From Eq. 11.94,

$$\mathrm{CDI}_{\mathrm{fish}} = \frac{\left(0.471\ \frac{\mathrm{mg}}{\mathrm{kg}}\right)\left(13\ \frac{\mathrm{g}}{\mathrm{d}}\right)(1)(5\ \mathrm{yr})}{(70\ \mathrm{kg})(70\ \mathrm{yr})\left(10^3\ \frac{\mathrm{g}}{\mathrm{kg}}\right)}$$

$$= 6.25 \times 10^{-6}\ \mathrm{mg/kg{\cdot}d}$$

The total CDI is

$$\mathrm{CDI}_{\mathrm{total}} = \mathrm{CDI}_w + \mathrm{CDI}_{\mathrm{fish}}$$

$$= 6.12 \times 10^{-8}\ \frac{\mathrm{mg}}{\mathrm{kg{\cdot}d}} + 6.25 \times 10^{-6}\ \frac{\mathrm{mg}}{\mathrm{kg{\cdot}d}}$$

$$= 6.31 \times 10^{-6}\ \mathrm{mg/kg{\cdot}d}$$

From Eq. 45.4, the excess cancer risk is

$$R = (\mathrm{SF}_{\mathrm{heptachlor}})(\mathrm{CDI}_{\mathrm{total}})$$

$$= \left(4.5\ \frac{\mathrm{kg{\cdot}d}}{\mathrm{mg}}\right)\left(6.31 \times 10^{-6}\ \frac{\mathrm{mg}}{\mathrm{kg{\cdot}d}}\right)$$

$$= 28.4 \times 10^{-6}$$

The total risk of excess cancer is 28.4×10^{-6}.

ACGIH Methods

The American Conference of Governmental Industrial Hygienists (ACGIH) uses methods for determining the safe human dose that are somewhat different from the EPA methods previously described.

Threshold Limit Values

The ACGIH method uses predetermined *threshold limit values* (TLV) for both noncarcinogens and carcinogens. The TLVs are the concentrations in air that workers could be repeatedly exposed to on a daily basis without adverse health effects. The term TLV-TWA means the maximum time-weighted average concentration that all workers may be exposed to during an 8 hour day and 40 hour week. The TLV-TWA is for the inhalation route of exposure.

ACGIH also determines *short-term exposure limits* (TLV-STEL) for airborne toxicants, which are the recommended concentrations workers may be exposed to for short periods during the workday without suffering certain adverse health effects (e.g., irritation, chronic tissue damage, and narcosis). The TLV-STEL is the TWA concentration in air that should not be exceeded for more than 15 minutes of the workday. The TLV-STEL should not occur more than four times daily, and there should be at least 60 minutes between successive STEL exposures. In such cases, the excursions may exceed three times the TLV-TWA for no more than a total of 30 minutes during the workday, but shall not exceed five times the TLV-TWA under any circumstances. In all cases, the 8 hour TLV-TWA may not be exceeded. Short-term exposure limits have not been established by ACGIH for some toxicants.

ACGIH also publishes *ceiling threshold limit values* (TLV-C) that should not be exceeded at any time during the workday. If instantaneous sampling is infeasible, the sampling period for the TLV-C can be up to 15 minutes in duration. Also, the TLV-TWA should not be exceeded.

For mixtures of substances, the *equivalent exposure* over 8 hours is the sum of the individual exposures.

$$E = \frac{1}{8}\sum_{i=1}^{n} C_i T_i \qquad 45.6$$

The *hazard ratio* is the concentration of the contaminant divided by the exposure limit of the contaminant. For mixtures of substances, the total hazard ratio is the sum of the individual hazard ratios and must not exceed unity. This is known as the *law of additive effects*. For this law to apply, the effects from the individual substances in the mixture must act on the same organ. If the effects do not act on the same organ, then each of the individual hazard ratios must not exceed unity. The equivalent exposure of a mixture of gases is

$$E_m = \sum_{i=1}^{n} \frac{C_i}{L_i} \qquad 45.7$$

The TLV values published by ACGIH also include a "skin" designation (i.e., classification) for toxicants for which the dermal exposure route should also be considered. This notation means that ACGIH recognizes that the potential exposure through the skin, particularly the mucus membranes, of gases and vapors and the direct contact of the skin to liquids may be significant for the chemical. ACGIH then recommends biological monitoring to determine the relative exposure by the dermal route compared to the total dose.

Example 45.5

A worker is exposed to contaminants according to the following schedule.

substance	duration and concentration (ppm)				peak
	2 h	1 h	1 h	4 h	
acetone	1200	900	800	900	–
n-butyl alcohol	150	100	50	50	–
toluene	200	250	250	100	450, 10 min

The TWA-TLVs for these substances are as follows.

substance	TWA-TLV (ppm)	ceiling (ppm)	allowable peak (ppm)	duration of peak (min)
acetone	1000	–	–	–
n-butyl alcohol	100	–	–	–
toluene	200	300	500	10

Assume that the target organs are the same for acetone and n-butyl alcohol but different for toluene. Has the worker been exposed in excess of the allowable limitations?

Solution

Calculate the TWA of each contaminant.

Use Eq. 45.6.

$$E = \tfrac{1}{8}\sum_{i=1}^{n} C_i T_i$$

For acetone,

$$E_{\text{acetone}} = \left(\frac{1}{8\ \text{h}}\right)\left(\begin{array}{l}(1200\ \text{ppm})(2\ \text{h}) + (900\ \text{ppm})(1\ \text{h}) \\ +(800\ \text{ppm})(1\ \text{h}) + (900\ \text{ppm})(4\ \text{h})\end{array}\right)$$
$$= 962.5\ \text{ppm}$$

Similarly, for n-butyl alcohol,

$$E_{n\text{-butyl}} = \left(\frac{1}{8\ \text{h}}\right)\left(\begin{array}{l}(150\ \text{ppm})(2\ \text{h}) + (100\ \text{ppm})(1\ \text{h}) \\ +(50\ \text{ppm})(1\ \text{h}) + (50\ \text{ppm})(4\ \text{h})\end{array}\right)$$
$$= 81.3\ \text{ppm}$$

For toluene,

$$E_{\text{toluene}} = \left(\frac{1}{8\ \text{h}}\right)\left(\begin{array}{l}(200\ \text{ppm})(2\ \text{h}) + (250\ \text{ppm})(1\ \text{h}) \\ +(250\ \text{ppm})(1\ \text{h}) + (100\ \text{ppm})(4\ \text{h})\end{array}\right)$$
$$= 162.5\ \text{ppm}$$

Check the mixture, using the law of additive effects for the components that affect the same target organs.

Use Eq. 45.7.

$$E_m = \sum_{i=1}^{n} \frac{C_i}{L_i} = \frac{C_{\text{acetone}}}{L_{\text{acetone}}} + \frac{C_{n\text{-butyl}}}{L_{n\text{-butyl}}} + \frac{C_{\text{toluene}}}{L_{\text{toluene}}}$$
$$= \frac{962.5\ \text{ppm}}{1000\ \text{ppm}} + \frac{81.3\ \text{ppm}}{100\ \text{ppm}}$$
$$= 0.963 + 0.813 = 1.775 > 1.0$$

Check the individual components.

$$\text{acetone} = 0.963 < 1.0$$
$$n\text{-butyl alcohol} = 0.813 < 1.0$$
$$\text{toluene} = \frac{162.5\ \text{ppm}}{200\ \text{ppm}} = 0.813 < 1.0$$

Check the ceiling and peak.

Toluene is the only contaminant with exposure limitations for a ceiling and a peak.

$$\text{maximum} = 250\ \text{ppm} < 300\ \text{ppm ceiling}$$
$$\text{peak} = 450\ \text{ppm},\ 10\ \text{min} < 500\ \text{ppm},\ 10\ \text{min}$$

The worker is not in compliance due to the mixture of acetone and n-butyl alcohol on a TWA basis.

Biological Exposure Indices

Biological exposure indices (BEIs) are published by ACGIH for biological monitoring. BEIs are reference values for evaluation of the total exposure of a worker to chemicals in the workplace. The BEIs are the levels of contaminants that would be observed in specimens collected from a healthy worker who was exposed to chemicals to the same extent as a worker with inhalation exposure to the TLV. The BEIs are based on a normal workday of 8 hr and a workweek of 5 days. ACGIH does not provide safe exposure values for dermal or ingestion pathways except as referenced in the BEIs.

The BEIs require collection of urine, exhaled air, and blood specimens from exposed workers. The assistance of medical personnel is required to collect the specimens, and a qualified industrial hygienist is needed to collect data and interpret the results.

This medical approach provides the total body burden by accounting for substances that are absorbed through the skin, the pulmonary system, and the gastrointestinal tract. The urine analysis accounts for the level of metabolites of toxic agents in the urine. The blood analysis measures the concentration of the toxic agent present in the blood. Analysis of exhaled air is used to determine the rate at which fat-soluble gases and vapors that are not metabolized are cleared from the body.

Carcinogenicity Classification

ACGIH and other organizations do not use the same classification system as the EPA system described previously. ACGIH's carcinogenicity categories are given in Table 45.5.

Table 45.5 ACGIH Carcinogenicity Categories

category	description
A1	confirmed human carcinogen (sufficient weight of evidence from epidemiologic studies)
A2	suspected human carcinogen (limited evidence from human data but sufficient evidence from animal data)
A3	confirmed animal carcinogen with unknown relevance to humans (limited animal data not confirmed by human data)
A4	not classifiable as a human carcinogen (lack of sufficient human and animal data)
A5	not suspected as a human carcinogen (sufficient human or animal data)

ACGIH recommendations are contained in *Threshold Limit Values for Chemical Substances and Physical Agents, Biological Exposure Indices (TLVs and BEIs)*, published annually by the American Conference of Governmental Industrial Hygienists. This publication contains threshold limit values for chemical substances and physical agents in the workplace, along with biological exposure indices. Samples of the data provided in this publication for benzene are given in Table 45.6. A qualified industrial hygienist should be consulted regarding the interpretation and application of TLVs.

Table 45.6 Sample TLV Data Provided by ACGIH[*]

TLV section	Benzene [71-43-2]
TWA	0.5 ppm/mg/m^3
STEL/C	2.5 ppm/mg/m^3
notations	skin; A1: BEI
molecular weight	78.11
TLV basis— critical effect(s)	leukemia
BEI section	
determinant	s-phenylmercapturic acid in urine
sampling time	end of shift
BEI	25 μg/g creatine
notation	B
notation section	
A	refers to Appendix A—carcinogenicity
BEI	identifies substances for which there are BEIs (see BEI section)
B	refers to Appendix B—substances of variable composition

[*]Adapted from *2012 TLVs and BEIs*, ACGIH, Cincinnati, OH

NIOSH Methods

The National Institute for Occupational Safety and Health (NIOSH) was established by the Occupational Safety and Health Act of 1970. NIOSH is part of the Centers for Disease Control and Prevention (CDC) and is the only federal institute responsible for conducting research and making recommendations for the prevention of work-related illnesses and injuries. The Institute's responsibilities include

- investigating hazardous working conditions as requested by employers or workers
- evaluating hazards ranging from chemicals to machinery
- creating and disseminating methods for preventing disease, injury, and disability
- conducting research and providing recommendations for protecting workers
- providing education and training to persons preparing for or actively working in the field of occupational safety and health

NIOSH maintains and publishes important databases and references on chemical hazards and worker protection, including the following.

- Immediately Dangerous to Life and Health (IDLH) Concentrations
- International Chemical Safety Cards
- *NIOSH Manual of Analytical Methods* (NMAM)
- *NIOSH Pocket Guide to Chemical Hazards* (NPG)
- Recommendations for Chemical Protective Clothing
- Specific Medical Tests Published for OSHA Regulated Substances
- Toxicologic Review of Selected Chemicals
- Certified Equipment List

The information given in the IDLH concentration database applies to exposure to airborne contaminants when that exposure is likely either to cause death or immediate or delayed permanent adverse health effects or to prevent escape from such an environment. The purpose of establishing an IDLH concentration is to ensure that workers can escape from a given contaminated environment in the event of failure of the respiratory protection equipment.

The NIOSH chemical *safety cards* database contains chemical information integrated from various national and international organizations. A sample of an International Chemical Safety Card is given in Fig. 45.3.

The *NIOSH Manual of Analytical Methods* (NMAM) is a collection of methods for sampling and analysis of contaminants in workplace air and in the blood and urine of workers who are occupationally exposed. These methods have been developed specifically to have

Figure 45.3 Sample International Chemical Safety Card

International Chemical Safety Cards

ICSC: 0015

BENZENE

CAS # 71-43-2
RTECS # CY1400000
ICSC # 0015
UN # 1114
EC # 601-020-00-8

BENZENE
Cyclohexatriene
Benzol
C_6H_6
Molecular mass: 78.1

TYPES OF HAZARD/ EXPOSURE	ACUTE HAZARDS/ SYMPTOMS	PREVENTION	FIRST AID/ FIRE FIGHTING
FIRE	Highly flammable.	NO open flames, NO sparks, and NO smoking.	Powder, AFFF, foam, carbon dioxide.
EXPLOSION	Vapour/air mixtures are explosive. Risk of fire and explosion: see chemical dangers.	Closed system, ventilation, explosion-proof electrical equipment and lighting. Do NOT use compressed air for filling, discharging, or handling. Use non-sparking handtools.	In case of fire: keep drums, etc., cool by spraying with water.
EXPOSURE		AVOID ALL CONTACT!	
• INHALATION	Dizziness. Drowsiness. Headache. Nausea. Shortness of breath. Convulsions. Unconsciousness.	Ventilation, local exhaust, or breathing protection.	Fresh air, rest. Refer for medical attention.
• SKIN	MAY BE ABSORBED! Dry skin (further see Inhalation).	Protective gloves. Protective clothing.	Remove contaminated clothes. Rinse skin with plenty of water or shower. Refer for medical attention.
• EYES		face shield, or eye protection in combination with breathing protection.	First rinse with plenty of water for several minutes (remove contact lenses if easily possible), then take to a doctor.
• INGESTION	Abdominal pain. Sore throat. Vomiting (further see Inhalation).	Do not eat, drink, or smoke during work.	Rinse mouth. Do NOT induce vomiting. Refer for medical attention.

SPILLAGE DISPOSAL	STORAGE	PACKAGING & LABELLING
Collect leaking and spilled liquid in sealable containers as far as possible. Absorb remaining liquid in sand or inert absorbent and remove to safe place. Do NOT wash away into sewer (extra personal protection: complete protective clothing including self-contained breathing apparatus).	Fireproof. Separated from food and feedstuffs.	Do not transport with food and feedstuffs. F symbol T symbol R: 45-11-48/23/24/25 S: 53-45 UN Hazard Class: 3 UN Packing Group: II

SEE IMPORTANT INFORMATION ON BACK

ICSC: 0015

Prepared in the context of cooperation between the International Programme on Chemical Safety & the Commission of the European Communities © IPCS CEC 1993 No modifications to the International version have been made except to add the OSHA PELs, NIOSH RELs and IDLH values

---

PHYSICAL STATE; APPEARANCE:
COLOURLESS LIQUID, WITH CHARACTERISTIC ODOUR.

PHYSICAL DANGERS:
The vapour is heavier than air and may travel along the ground; distant ignition possible.

CHEMICAL DANGERS:
Reacts violently with oxidants and halogens causing fire and explosion hazard.

OCCUPATIONAL EXPOSURE LIMITS (OELs):
TLV: 10 ppm; 32 mg/m³ (as TWA) A2 (ACGIH 1991-1992).
OSHA PEL: 1910.1028 TWA 1 ppm ST 5 ppm See Appendix F
NIOSH REL: Ca TWA 0.1 ppm ST 1 ppm See Appendix A
NIOSH IDLH: Potential occupational carcinogen 500 ppm

ROUTES OF EXPOSURE:
The substance can be absorbed into the body by inhalation and through the skin.

INHALATION RISK:
A harmful contamination of the air can be reached rather quickly on evaporation of this substance at 20°C; on spraying or dispersion, however, much faster.

EFFECTS OF SHORT-TERM EXPOSURE:
The substance irritates the skin and the respiratory tract. Swallowing the liquid may cause aspiration into the lungs with the risk of chemical pneumonitis. The substance may cause effects on the central nervous system. Exposure far above the occupational exposure limit may result in unconsciousness.

EFFECTS OF LONG-TERM OR REPEATED EXPOSURE:
The liquid defats the skin. The substance may have effects on the blood forming organs, liver and immune system. This substance is carcinogenic to humans.

PHYSICAL PROPERTIES
Boiling point: 80°C
Melting point: 6°C
Relative density (water = 1): 0.9
Solubility in water, g/100 ml at 25°C: 0.18
Vapour pressure, kPa at 20°C: 10
Relative vapour density (air = 1): 2.7
Relative density of the vapour/air-mixture at 20°C (air = 1): 1.2
Flash point: -11°C (c.c.)°C
Auto-ignition temperature: about 500°C
Explosive limits, vol% in air: 1.2-8.0
Octanol/water partition coefficient as log Pow: 2.13

ENVIRONMENTAL DATA

NOTES
Use of alcoholic beverages enhances the harmful effect. Depending on the degree of exposure, periodic medical examination is indicated. The odour warning when the exposure limit value is exceeded is insufficient.
Transport Emergency Card: TEC (R)-7
NFPA Code: H2; F3; R0;

ADDITIONAL INFORMATION

ICSC: 0015   BENZENE

IMPORTANT LEGAL NOTICE: Neither NIOSH, the CEC or the IPCS nor any person acting on behalf of NIOSH, the CEC or the IPCS is responsible for the use which might be made of this information. This card contains the collective views of the IPCS Peer Review Committee and may not reflect in all cases all the detailed requirements included in national legislation on the subject. The user should verify compliance of the cards with the relevant legislation in the country of use. The only modifications made to produce the U.S. version is inclusion of the OSHA PELs, NIOSH RELs and IDLH values.

© IPCS, CEC 1993

Source: International Chemical Safety Cards (ICSC), Centers for Disease Control and Prevention and National Institute for Occupational Safety and Health.

Health, Safety, Welfare

adequate sensitivity to detect the lowest concentrations as regulated by OSHA and recommended by NIOSH, and to have sufficient range to measure concentrations exceeding safe levels of exposure. NMAM also includes chapters on quality assurance, strategies of sampling airborne substances, method development, and discussions of some portable direct-reading instrumentation.

Probably the most important NIOSH toxicological publication is the *NIOSH Pocket Guide to Chemical Hazards*. The *Pocket Guide* is intended as a source of general industrial hygiene information for workers, employers, and occupational health professionals. The *Pocket Guide* presents key information and data in abbreviated tabular form for 677 chemicals or substance groupings (e.g., manganese compounds, tellurium compounds, inorganic tin compounds) that are found in the work environment. The industrial hygiene information in the *Pocket Guide* is intended to help users recognize and control occupational chemical hazards. The chemicals or substances include all substances for which the National Institute for Occupational Safety and Health (NIOSH) has recommended exposure limits (RELs) and those with permissible exposure limits (PELs) as found in the Occupational Safety and Health Administration (OSHA) *General Industry Air Contaminants Standard* (29 CFR 1910.1000).

A sample of the information is given in Fig. 45.4. The codes and abbreviations used in the *Pocket Guide* are given in App. 45.A.

The NIOSH recommended exposure limits (RELs) given in the *Pocket Guide* are time-weighted average (TWA) concentrations for up to a 10 hr workday during a 40 hr workweek. A short-term exposure limit (STEL) is a 15 min TWA exposure that should not be exceeded at any time during a workday. A ceiling REL should not be exceeded at any time. The "skin" designation means there is a potential for dermal absorption, so skin exposure should be prevented as necessary through the use of good work practices and gloves, coveralls, goggles, and other appropriate equipment.

The NIOSH database "Recommendations for Chemical Protective Clothing, A Companion to the NIOSH Pocket Guide to Chemical Hazards" provides recommendations for *chemical protective clothing* (CPC) to prevent direct skin contact and contamination. The database also includes recommendations to prevent physical injury to the unprotected skin from thermal hazards, such as rapidly evaporating liquified gases.

6. LEGAL (OSHA) STANDARDS FOR WORKER PROTECTION

While the EPA provides exposure limitations and risk factors for environmental cleanup projects, ACGIH provides recommendations to industrial hygienists about workplace exposure, and NIOSH provides research and recommendations for workplace exposure limits (recommended exposure limits), the *Occupational Safety and Health Administration* (OSHA) sets the legally enforceable workplace exposure limits. OSHA standards are given in the 29 Code Federal Regulations (CFR). Although OSHA deals with many other aspects of workplace safety, of key importance relative to toxicology are the regulations on air contaminants given in 29 CFR 1910.00, Subpart Z, "Toxic and Hazardous Substances."

Subpart Z includes three tables, Z-1, Z-2, and Z-3, as given in App. 45.B. The current standards should always be reviewed when performing any type of air contaminant evaluation because changes to the exposure limits occur from time to time.

Table Z-1 gives *permissible exposure limits* (PELs) for air contaminants based on an 8 hr TWA and identifies substances for which protection from skin exposure is required. Substances with ceiling limits are also identified. Table Z-2 gives limits for air contaminants on an 8 hr TWA basis with acceptable ceiling concentrations and acceptable maximum peak concentrations and durations for exposure allowed above the ceiling concentrations. Table Z-3 gives exposure limits for mineral dusts.

The calculation procedure for use of the PELs is the same as for ACGIH TLVs. The procedure for mixtures is also the same.

Figure 45.4 *Sample of NIOSH Pocket Guide to Chemical Hazards*

NIOSH Pocket Guide to Chemical Hazards

Benzene	CAS 71-43-2
C_6H_6	RTECS CY1400000
Synonyms & Trade Names Benzol, Phenyl hydride	**DOT ID & Guide** 1114 130

Exposure Limits	NIOSH REL: Ca TWA 0.1 ppm ST 1 ppm See Appendix A
	OSHA PEL: [1910.1028] TWA 1 ppm ST 5 ppm See Appendix F

IDLH Ca [500 ppm] See: 71432	**Conversion** 1 ppm = 3.19 mg/m^3

Physical Description
Colorless to light-yellow liquid with an aromatic odor. [Note: A solid below 42°F.]

MW: 78.1	BP: 176°F	FRZ: 42°F	Sol: 0.07%
VP: 75 mmHg	IP: 9.24 eV		Sp.Gr: 0.88
Fl.P: 12°F	UEL: 7.8%	LEL: 1.2%	

Class IB Flammable Liquid: Fl.P. below 73°F and BP at or above 100°F.

Incompatibilities & Reactivities
Strong oxidizers, many fluorides & perchlorates, nitric acid

Measurement Method
Charcoal tube; CS_2; Gas chromatography/Flame ionization detection; IV [#1500, Hydrocarbons] [Also #3700, #1501] See: NMAM INDEX

Personal Protection & Sanitation Skin: Prevent skin contact Eyes: Prevent eye contact Wash skin: When contaminated Remove: When wet (flammable) Change: N.R. Provide: Eyewash, Quick drench	**First Aid** (See procedures) Eye: Irrigate immediately Skin: Soap wash immediately Breathing: Respiratory support Swallow: Medical attention immediately

Respirator Recommendations NIOSH
At concentrations above the NIOSH REL, or where there is no REL, at any detectable concentration: (APF = 10,000) Any self-contained breathing apparatus that has a full facepiece and is operated in a pressure-demand or other positive-pressure mode/(APF = 10,000) Any supplied-air respirator that has a full facepiece and is operated in a pressure-demand or other positive-pressure mode in combination with an auxiliary self-contained positive-pressure breathing apparatus
Escape: (APF = 50) Any air-purifying, full-facepiece respirator (gas mask) with a chin-style, front- or back-mounted organic vapor canister/Any appropriate escape-type, self-contained breathing apparatus

Exposure Routes inhalation, skin absorption, ingestion, skin and/or eye contact

Symptoms irritation eyes, skin, nose, respiratory system; giddiness; headache, nausea, staggered gait; fatigue, anorexia, lassitude (weakness, exhaustion); dermatitis; bone marrow depressant/depression; [Potential occupational carcinogen]

Target Organs Eyes, skin, respiratory system, blood, central nervous system, bone marrow

Cancer Site [leukemia]

See also: INTRODUCTION   See ICSC CARD: 0015   See MEDICAL TESTS: 0022

Health, Safety, Welfare

Source: *NIOSH Pocket Guide to Chemical Hazards*, DHHS (NIOSH) Publication No. 2005-149, National Institute for Occupational Safety and Health.

46 Industrial Hygiene

Nomenclature

a	speed of sound	m/s
A	activity metabolism	W
A	activity (disintegration rate)	Bq
AM	asymmetry multiplier for lifting	–
B	basal metabolism	W
C	concentration	ppm, mg/m^3
C	constant for calculating sound intensity	–
C	time of noise exposure at specified level	s
CL	ceiling heat limit	°C
CM	coupling multiplier for lifting	–
DM	distance multiplier for lifting	–
E	noise exposure	–
ECT	equivalent chill temperature	°C
f	frequency of sound	Hz
FM	frequency multiplier for lifting	–
HM	horizontal multiplier for lifting	–
I	sound intensity	W/m^2
IL	insertion loss	dB
k	ratio of specific heats	–
L	sound pressure or sound power level	dB
LC	load constant for lifting	kg
m	mass	kg
MW	molecular weight	g/mol
n	number of moles	–
p	pressure or partial pressure	Pa
p	total number of observations	–
P	posture metabolism	W
PEL	permissible exposure limit	mg/m^3
Q	heat flow	W
r	distance from sound source	m
R	specific gas constant	kJ/kg·K
R^*	universal gas constant, 8.314	kJ/kmol·K
RAL	recommended heat alert limit	°C
REL	recommended heat exposure limit	°C
RWL	recommended weight limit	kg
t	rest time, percent of period	%
t	time	s
T	temperature	°C, K
TLV	threshold limit value	–
V	velocity metabolism	W
V	volume	L, m^3
VM	vertical multiplier for lifting	–
W	sound power	W
WBGT	wet-bulb globe temperature	°C
x	mole fraction	–
x_{rms}	root-mean-square value of n observations	–

Symbols

λ	decay constant for radionuclides	s^{-1}
λ	wavelength	m
ρ	density	kg/m^3

Subscripts

0	initial condition or reference
C	convection
db	dry-bulb
E	evaporation
g	globe
in	indoor
m	mixed
max	maximum
M	metabolic
nwb	natural wet-bulb
R	radiation
rest	resting
rms	root-mean-square
rms-ref	reference rms
S	storage
W	sound power

1. INTRODUCTION

Industrial hygiene is the art and science of identifying, evaluating, and controlling environmental factors (including stress) that may cause sickness, health impairment, or discomfort among workers or citizens of the community. Industrial hygiene involves the recognition of health hazards associated with work operations and processes, evaluation and measurement of the magnitudes of hazards, and determining applicable control methods. Occupational health hazards specifically involve illness or impairment for which a worker may be compensated under a worker protection program.

Health, Safety, Welfare

The fundamental law governing worker protection in the United States is the 1970 federal Occupational Safety and Health Act. It requires employers to provide a workplace that is free from hazards by complying with specified safety and health standards. Employees must also comply with standards that apply to their own conduct. The federal regulatory agency responsible for administering the Occupational Safety and Health Act is the Occupational Safety and Health Administration (OSHA). OSHA sets standards, investigates violations of the standards, performs inspections of plants and other facilities, investigates complaints, and takes enforcement action against violators. OSHA also funds state programs, which are permitted if they are at least as stringent as the federal program.

The 1970 act also established the National Institute for Occupational Safety and Health (NIOSH). NIOSH is responsible for safety and health research and makes recommendations for regulations. The recommendations are known as *recommended exposure limits* (RELs). Among other activities, NIOSH also publishes health and safety criteria and health hazard alerts, and is responsible for testing and certifying respiratory protective equipment.

2. HAZARD IDENTIFICATION

Overview of Hazards

There are four basic types of hazards with which industrial hygiene is concerned: chemical hazards, physical hazards, ergonomic hazards, and biological hazards. *Chemical hazards* result from airborne chemicals such as gases, vapors, and particulates in harmful concentrations. Besides inhalation, chemical hazards may affect workers by absorption through the skin. *Physical hazards* include radiation, noise, vibration, and excessive heat or cold. *Ergonomic hazards* include work procedures and arrangements that require motions that result in biomechanical stress and injury. *Biological hazards* include exposure to biological organisms that may lead to illness. Any of these types of hazards may occur at high intensity, resulting in acute or immediate effects, or they may occur at low intensity, resulting in long-term or chronic effects. Industrial hygiene is focused on control or elimination of these hazards in the workplace.

In respect to chemical hazards, the terms "toxicity" and "hazard" are not synonymous. *Toxicity* is the capacity of the chemical to produce harm when it has reached a sufficient concentration at a particular site in the body. *Hazard* refers to the probability that this concentration will occur.

Hazard Communication

Two important preventative measures required by OSHA are *material safety data sheets* (MSDSs) and labeling of containers of hazardous materials. A third important OSHA requirement is that all covered employers must

provide the necessary information and training to affected workers. The OSHA Hazard Communication Standard is given in Title 29 of the *Code of Federal Regulations* (CFR) Part 1910.1200. Other OSHA requirements are also given in Title 29.

MSDSs

An MSDS provides key information about a chemical or substance so that users or emergency responders can determine safe use procedures and necessary emergency response actions. An MSDS provides information on the identification of the material and its manufacturer, identification of hazardous components and their characteristics, physical and chemical characteristics of the ingredients, fire and explosion hazard data, reactivity data, health hazard data, precautions for safe handling and use, and recommended control measures for use of the material. The information on MSDSs is essential for dealing with hazardous chemicals and should be complete; however, it may be necessary to contact the manufacturer if information is lacking.

Container Labeling

Labels are required on hazardous material containers. Labels should provide essential information for the safe use and storage of hazardous materials. Failure to provide adequate labeling of hazardous material containers is a common violation of OSHA standards.

Worker Information and Training

The OSHA standard requires that employers provide workers with information about the potential health hazards from exposure to hazardous chemicals that they use in the workplace. It also requires employers to provide adequate training to workers on how to safely handle and use hazardous materials. The standard requires training in use of the hazardous materials in normal operations and actions to take during emergencies.

3. GASES, VAPORS, AND SOLVENTS

Exposure Factors for Gases and Vapors

The most frequently encountered hazard in the workplace is exposure to gases and vapors from solvents and chemicals. Several factors define the exposure potential for gases and vapors. The most important are how a material is used and what engineering or personal protective controls exist. If the inhalation route of entry is controlled, dermal contact may still be a major route of exposure.

Vapor pressure of a substance is related to temperature. Vapor pressure affects the concentration of the substance in vapor form above the liquid and is dependent upon the temperature and the properties of the substance. Processes that operate at lower temperatures are inherently less hazardous than processes that operate at higher temperatures because fewer chemicals have boiling points and vapor pressures in the operating temperature range.

The *concentration* of a chemical in a liquid (a solution) affects the potential hazard because the vapor pressure varies directly with the mole fraction (concentration) of the solute according to Raoult's law. Thus, solutions with a lower concentration will have lower potentials for volatilization of the hazardous fraction.

Reactivity affects the hazard potential because the products may be volatile or nonvolatile depending on the properties of the combining substances.

Two terms are used to quantify the concentration of a gas in air that a worker can be safely exposed to. These are *threshold limit value* (TLV) and *permissible exposure limit* (PEL). The TLV is a concentration in air that nearly all workers can be exposed to daily without adverse effects. The PEL is a regulatory exposure limit for workers; OSHA publishes PELs as standards. A PEL may be more restrictive than the corresponding TLV. TLVs are updated annually by the American Conference of Governmental Industrial Hygienists (ACGIH). PELs are updated less frequently. Both should be reviewed when evaluating potential hazards from gases and vapors.

Further information on TLVs and PELs is given in Chap. 45.

Solvents

Solvents are widely used throughout industry for many purposes, and their safe use is an important industrial hygiene concern. It is essential that accurate MSDS information be provided to employees on the physical properties and the toxicological effects of exposure to solvents.

Solvent vapors can be inhaled or may be absorbed through the skin. Vapors that are inhaled are absorbed through the lungs into the bloodstream and are then deposited in tissue with a high content of fat and lipids. The central nervous system, liver, and bone marrow are the key organs of the body that are affected. The toxicity of a solvent is greatly affected by the conditions of use.

Further information on toxicity of chemicals is given in Chap. 45.

Water as a Solvent

Aqueous solutions have low vapor pressures at ambient temperatures, so the potential hazard through inhalation is low. Salts in an aqueous solution are essentially not volatile.

Organic Solvents

Organic solvents, and all organic compounds in general, can be classified by their molecular structure. The classification is by the number of carbon atoms in the basic skeletal chain, bonding, and molecular arrangement. Organic solvents can be classified as aliphatic, cyclic, aromatic, halogenated hydrocarbon, ketone, ester, alcohol, and ether. Each class has characteristic molecular structures, properties, and health effects. Since the names of organic solvents can be similar or misleading, the environmental engineer should always refer to the label, the MSDS, or a laboratory before assessing the hazard content of an organic solvent.

Gases and Flammable or Combustible Liquids

Hazardous gases fall into four main types: cryogenic liquids, simple asphyxiants, chemical asphyxiants, and all other gases whose hazards depend on their properties.

Cryogenic liquids can vaporize rapidly, producing a cold gas that is more dense than air and displacing oxygen in confined spaces. After reaching thermal equilibrium, gas from a cryogenic liquid may spread out in the available space. Gas from liquid nitrogen, for example, can cause oxygen to condense out of the atmosphere, creating an explosion hazard.

Simple asphyxiants, which include helium, neon, nitrogen, hydrogen, and methane, can dilute or displace oxygen. *Chemical asphyxiants*, which include carbon monoxide, hydrogen cyanide, and hydrogen sulfide, can pass into blood cells and tissue and interfere with blood-carrying oxygen.

In addition to the hazards from gases and vapors, some liquids pose hazards because their vapors are flammable or combustible when mixed with air in certain percentages. The term *flammable* refers to the ability of an ignition source to propagate a flame throughout the vapor-air mixture and have a closed-cup flash point below 37.8°C (100°F) and a vapor pressure not exceeding 272 atm at 37.8°C (100°F). The phrase *closed-cup flash point* refers to a method of testing for flash points of liquids. The term *combustible* refers to liquids with flash points above 37.8°C (100°F).

For each airborne flammable substance, there are minimum and maximum concentrations in air between which flame propagation will occur. The lower concentration in air is known as the *lower flammable limit* (LFL) or *lower explosive limit* (LEL). The upper limit is known as the *upper flammable limit* (UFL) or *upper explosive limit* (UEL). Below the LEL, there is not enough fuel to propagate a flame. Above the UEL, there is not enough air to propagate a flame. The lower the LEL, the greater the hazard from a flammable liquid. For many common liquids and gases, the LEL is a few percent, and the UEL is 6–12%. If a concentration in air is less than the PEL or the TLV, the concentration will be less than the LEL. The occupational safety requirements for handling and using flammable and combustible liquids are given in Subpart H of 29 CFR 1910.106.

See Table 46.1 for a representative listing of combustible materials and their LELs and UELs.

Table 46.1 Representative Hazardous Concentrations in Air*

	combustibles						toxics				
material	TLV/TWA (ppm)	LFL (%/vol)	UFL (%/vol)	IDLH (ppm)	specific gravity (air = 1.0)	material	TLV/TWA (ppm)	IDLH (ppm)	LFL (ppm)	LFL (%/vol)	specific gravity (air = 1.0)
Acetone	750	2.5	12.8	2500	2.0	Acetone	750	2500	25 000	2.5	2.0
Acetylene	-A-	2.5	100.0	-A-	0.9	Ammonia	25	300	160 000	16.0	0.6
Ammonia	25	15.0	28.0	300	0.6	Benzene	1.0	-C-	12 000	1.2	2.6
Benzene	1.0	1.2	7.8	500	2.6	Butane	800	-U-	16 000	1.6	2.0
Butane	800	1.6	8.4	-U-	2.0	n-Butyl Acetate	150	1700	17 000	1.7	4.0
n-Butyl Acetate	150	1.7	7.6	1700	4.0	Carbon Dioxide	5000	40 000	N/C	N/C	1.5
Diborane	0.1	0.8	88.0	15	1.0	Carbon Monoxide	25	1200	125 000	12.5	1.0
Ethane	-A-	3.0	12.5	-A-	1.0	Chlorine	0.5	10	N/C	N/C	2.5
Ethanol	1000	3.3	19.0	-U-	1.6	Ethylene Oxide	1	-C-	30 000	3.0	1.5
Ethyl Acetate	400	2.0	11.5	2000	3.0	Ethyl Ether	400	19 000	19 000	1.9	2.6
Ethyl Ether	400	1.9	36.0	1900	2.6	Gasoline	300	-U-	14 000	1.4	3–4.0
Ethylene Oxide	1	3.0	100.0	-C-	1.5	Heptane	400	750	10 500	1.05	3.5
Gasoline	300	1.4	7.6	-U-	3–4.0	Hexane	50	1100	11 000	1.0	3.0
Heptane	400	1.05	6.7	750	3.5	Hydrogen Cyanide	10	50	56 000	5.6	0.9
Hexane	50	1.1	7.5	1100	3.0	Hydrogen Sulfide	10	100	40 000	4.0	1.2
Hydrogen	-A-	4.0	75.0	-A-	0.1	Isopropyl Alcohol	400	2000	20 000	2.0	2.1
Isopropyl Alcohol	400	2.0	12.0	2000	2.1	Methyl Acetate	200	3100	31 000	3.1	2.6
Methane	-A-	5.0	15.0	-A-	0.6	Methanol	200	6000	60 000	6.0	1.1
Methanol	200	6.0	36.0	6000	1.1	Methyl Chloride	50	2000	81 000	8.1	1.8
Methyl Ethyl Ketone	200	1.4	11.4	3000	2.5	Methyl Ethyl Ketone	200	3000	14 000	1.4	2.5
Pentane	600	1.5	7.8	15 000	2.5	Methyl Methacrylate	100	1000	17 000	1.7	3.5
Propane	1000	2.1	9.5	2100	1.6	Nitric Oxide	25	100	N/C	N/C	1.0
Propylene Oxide	20	2.3	36.0	400	2.0	Nitrogen Dioxide	3	20	N/C	N/C	1.6
Styrene	50	0.9	6.8	700	3.6	Pentane	600	15 000	15 000	1.5	2.5
Toluene	50	1.1	7.1	500	3.1	n-Propyl Acetate	200	1700	17 000	1.7	3.5
Turpentine	100	0.8	-U-	800	4.7	Styrene	50	700	9000	0.9	3.6
Vinyl Acetate	10	2.6	13.4	-U-	3.0	Sulfur Dioxide	2	100	N/C	N/C	2.2
Vinyl Chloride	1.0	3.6	33.0	-C-	2.2	1,1,1-Trichloroethane	350	700	75 000	7.5	4.6
Xylene	100	0.9	6.7	900	3.7	Toluene	50	500	11 000	1.1	3.2
						Trichloroethylene	50	1000	80 000	8.0	4.5
						Turpentine	100	800	8000	0.8	4.7
						Vinyl Chloride	1.0	-C-	36 000	3.6	2.2
						Xylene	100	900	9000	0.9	3.7

Key: A, asphyxiant; C, carcinogen; U, data not available; N/C, noncombustible
*subject to change without notice

Evaluation and Control of Hazards

The toxicological effects from aqueous solutions include dermatitis, throat irritation, and bronchitis. The hazard is less with aqueous solutions because their vapor pressures are usually low. On the other hand, organic compounds represent a greater hazard because their vapor pressures are usually higher and their toxicities greater. The effects from organic solvents include central nervous system disorders, narcosis, and death.

Vapor-Hazard Ratio

One indicator of hazards from vapors and gases from solvents is the *vapor-hazard ratio*, which is the equilibrium vapor pressure in ppm at 25°C (77°F) divided by the TLV in ppm. The higher the ratio, the greater the hazard. The vapor-hazard ratio accounts for the volatility of a solvent as well as its toxicity. To assess the overall hazard, the vapor-hazard ratio should be evaluated in conjunction with the TLV, ignition temperature, flash point, toxicological information, and degree of exposure. The vapor-hazard ratio accounts for the volatility of a solvent as well as its toxicity.

The best control method is not to use a solvent that is hazardous. Sometimes a process can be redesigned to eliminate the use of a solvent. The following evaluation steps are recommended.

- Use water or an aqueous solution when possible.
- Use a *safety solvent* if it is not possible to use water. Safety solvents have inhibitors and high flash points.
- Use a different process when possible to avoid use of a hazardous solvent.
- Provide a properly designed ventilation system if toxic solvents must be used.
- Never use highly toxic or highly flammable solvents (benzene, carbon tetrachloride, gasoline).

Ventilation

The most effective way to prevent inhalation of vapors from solvents is to provide closed systems or adequate local exhaust ventilation. If limitations exist on the use of closed systems or local exhaust ventilation, then workers should be provided with personal protective equipment.

Personal Protective Equipment

Respirators provide emergency and backup protection but are unreliable as a primary source of protection from hazardous vapors. Face masks can leak or become contaminated around the edges, reduce the efficiency of the worker, and increase the lack of oxygen in oxygen-deficient areas. Other drawbacks are the need to have the respirator properly fitted to the worker and the need for the worker to be trained in its proper use. Additionally, the worker may feel a false sense of security while wearing a respirator.

Besides inhalation, dermal contact is an important concern when working with hazardous solvents. Mechanical equipment should be provided to keep the worker isolated from contact with the solvent. However, since some contact may occur even with mechanical equipment in use, protective clothing should be provided. Protective clothing includes aprons, face shields, goggles, and gloves. The manufacturer's recommendations should be followed for use of all protective clothing and equipment.

One common problem with protective clothing is incorrect selection or misuse of gloves. The time for particular solvents to penetrate gloves that are commonly thought of as "protective" is surprisingly short. Both the permeability and the abrasion resistance of gloves must be considered in their selection and use. For example, methyl chloride will permeate a neoprene glove in less than 15 minutes. The manufacturer should provide the *breakthrough time* and the *permeation rates* for the glove being evaluated. The breakthrough time and permeation rate are dependent on the specific chemical and the composition and thickness of the glove.

Protective eyewear should be provided where the risk of splashing of chemicals is present. Of course, mechanical equipment, barriers, guards, and other engineering measures should be provided as the first line of defense. For chemical splash protection, unvented chemical goggles, indirect-vented chemical goggles, or indirect-vented eyecup goggles should be used. A face shield may also be needed. Direct-vented goggles and normal eyeglasses should not be used, and contact lenses should not be worn.

For more information on personal protection, see Chap. 47.

Example 46.1

A rubber safety glove has an 8 h breakthrough time for ethanol based on a steady-state permeation rate of 12 μg/cm^2·min. The glove has been in continuous use for two 8 h shifts by the same worker who has 300 cm^2 of skin per hand. How many grams of ethanol have reached both hands of the worker at the end of the second shift?

Solution

The exposure duration is

$$\text{total time} - \text{breakthrough time} = 16\text{ h} - 8\text{ h}$$
$$= 8\text{ h}$$

The total exposure is

$$\left(\begin{array}{c}\text{no. of}\\\text{hands}\end{array}\right)\left(\frac{\text{skin area}}{\text{hand}}\right)\left(\begin{array}{c}\text{steady-state}\\\text{permeation rate}\end{array}\right)\left(\begin{array}{c}\text{exposure}\\\text{duration}\end{array}\right)$$

$$= (2)(300\text{ cm}^2)\left(12\ \frac{\mu\text{g}}{\text{cm}^2\cdot\text{min}}\right)$$
$$\times (8\text{ h})\left(60\ \frac{\text{min}}{\text{h}}\right)\left(\frac{1\text{ g}}{10^6\ \mu\text{g}}\right)$$
$$= 3.456\text{ g}$$

The ethanol exposure at the end of the second shift is 3.456 g.

4. PARTICULATES

Particulates include dusts, fumes, fibers, and mists. *Dusts* have a wide range of sizes and usually result from a mechanical process such as grinding. *Fumes* are extremely small particles, less than 1 μm in diameter, and result from combustion and other processes. *Fibers* are thin and long particulates, with asbestos being a prime example. *Mists* are suspended liquids that float in air, such as from the atomization of cutting oil. All of these types of particulate can pose an inhalation hazard if they reach the lungs.

With one known exception, particles larger than approximately 5 μm cannot reach the alveoli or inner recesses of the lungs before being trapped and expelled from the body through the digestive system or from the mouth and nose. Protection is afforded by the presence of mucus and cilia in the nasal passages, throat, larynx, trachea, and bronchi. The exception is asbestos fibers, which can reach the alveoli even though fibers may be larger. Particles smaller than 5 μm are considered respirable dusts and pose an exposure hazard when present in the breathing zone.

The body's reaction to particulates depends primarily on the type of particulate. Lung diseases result from the accumulation of particulates in the lungs, which restrict the ability of the lungs to transfer oxygen and carbon dioxide. This can cause the heart to overwork, leading to damage. Lung disease includes *fibrosis* (scar tissue formation), *bronchitis* (inflammation of the bronchi and an overproduction of mucus), *asthma* (constriction of the bronchial tubes), and cancer. Systemic reactions occur when the blood absorbs inorganic toxic particulates such as lead, mercury, and organic compounds. *Metal fume fever* results from breathing fine fumes of zinc, copper, and other metals. Allergic reactions occur from inhalation or dermal contact with organic particulates such as flour, grains, and chemicals. Bacterial and fungal infections result from the inhalation of particulates containing live organisms such as anthrax from wool particulates. Irritation of the nose and throat results from exposure to acid, alkali, or other irritating dusts or mists. Damage to internal tissues can result from inhalation of radioactive particulates.

There are four factors that affect the health risk from exposure to particulates: the types of particulate, the length of exposure, the concentration of particulates in the breathing zone, and the size of particulates in the breathing zone.

The type of particulate can determine the type of health effect that may result from the exposure. Both organic and inorganic dusts can produce allergic effects, dermatitis, and systemic toxic effects. Particulates that contain free silica can produce pneumoconiosis from chronic exposure. *Pneumoconiosis* is lung disease caused by fibrosis from exposure to both organic and inorganic particulates. Other particulates can cause systemic toxicity to the kidneys, blood, and central nervous system. Asbestos fibers can cause lung scarring and cancer.

The critical duration of exposure varies with the type of particulate. Metal fumes can cause metal fume fever with just a few hours of exposure. Toxic metal particulates can cause toxic effects from exposure of a few days to several months. Pneumoconioses may take several years to become disabling.

The concentration of particulates in the breathing zone is the primary factor in determining the health risk from particulates. The American Conference of Governmental Industrial Hygienists (ACGIH) has established TLVs that should not be exceeded. OSHA establishes PELs for safe exposure levels in the workplace.

The fourth exposure factor is the size of the particulates. Particles larger than 5 μm will normally be filtered out through the upper respiratory system before reaching the alveoli of the lungs.

Silica

Silica (SiO_2) has several associated health hazards. The crystalline form of free silica (quartz) deposited in the lungs causes the growth of fibrous tissue around the deposit. The fibrous tissue reduces the amount of normal lung tissue, thereby reducing the ability of the lungs to transfer oxygen. When the heart tries to pump more blood to compensate, heart strain and permanent damage or death may result. This condition is known as *silicosis*. Mycobacterial infection occurs in about 25% of silicosis cases.

Silica-containing dust is generated from a variety of occupations, including rock mining, sandblasting, stone cutting, and foundry work. Nonspecific defense mechanisms, such as mucus in the nasal passages, entrap and remove some of the particles. However, the particles from 0.5 μm to 3 μm in diameter may reach the alveoli where they can cause damage. Smokers exposed to silica dust have a significantly increased chance of developing lung cancer.

The OSHA PEL for dusts containing crystalline silica (quartz) is based on a formula that uses the percent of silica present in the sample. The PEL for crystalline silica can be found from 29 CFR 1910.1000, Table Z-3, as follows.

$$\mathrm{PEL}_{mg/m^3} = \frac{10}{\%SiO_2 + 2} \qquad 46.1$$

The constant 2 is designed to limit the PEL to 5 mg/m^3 when the percent of free silica ($\%SiO_2$) is low (less than 1%). The time-weighted average (TWA) of the TLV for free silica is 0.1 mg/m^3. This is consistent with the PEL formula for 100% free silica, which yields 0.1 mg/m^3.

The TLVs for other forms of free silica are given in Table 46.2.

Table 46.2 *Typical TLVs for Various Forms of Free Silica*

form of free silica	TLV (mg/m^3)
quartz	0.1
cristobalite (diatomite)	0.05
tridymite	0.05
fused silica dust	0.1
tripoli and silica flour	0.1

Source: 29 CFR 1910.1000

Example 46.2

Workers are crushing and grading rock for a drainfield in a subsurface disposal system. Samples of the air in the breathing zone of the workers show the following.

time	dust concentration (mg/m^3)	percent SiO$_2$
0800	1.5	5.8
1000	1.1	6.3
1300	2.1	4.2
1500	1.6	3.9
1700	1.8	4.1

Has worker exposure exceeded the OSHA PEL for quartz?

Solution

For mixtures containing free silica, use Eq. 46.1.

$$\text{PEL} = \frac{10 \ \dfrac{\text{mg}}{\text{m}^3}}{\%\text{SiO}_2 + 2}$$

percent SiO$_2$	PEL (mg/m^3)
5.8	1.28
6.3	1.20
4.2	1.61
3.9	1.69
4.1	1.64

The readings at 0800, 1300, and 1700 hours exceeded the PEL.

Asbestos

Asbestos is generically described as a naturally occurring, fibrous, hydrated mineral silicate. Inhalation of short asbestos fibers can cause *asbestosis*, a kind of pneumoconiosis, as a nonmalignant scarring of the lungs. *Bronchogenic carcinoma* is a malignancy (cancer) of the lining of the lung's air passages. *Mesothelioma* is a diffuse malignancy of the lining of the chest cavity or the lining of the abdomen.

Asbestos mining, construction activities, and working in shipyards are possible exposure activities that may involve the environmental engineer. The asbestos must be in a form that allows it to be airborne in order for inhalation to occur. Moreover, inhalation does not necessarily mean that lung damage will occur.

The onset of illness seems to be correlated with length and diameter of inhaled asbestos fibers. Fibers 2 μm in length cause asbestosis. Mesothelioma is associated with fibers 5 μm long. Fibers longer than 10 μm produce lung cancer. Fiber diameters greater than 3 μm are more likely to cause asbestosis or lung cancer, while fibers 3 μm or less in diameter are associated with mesothelioma.

There is a considerable latency period for asbestos-related illness. It may be 20 years after exposure to respirable asbestos before lung cancer is clinically evident, and it may be 30 years before peak incidence is noted.

OSHA has established comprehensive regulations relative to exposure to asbestos. The key documents are

- 29 CFR 1910.1001—regulations for general industry
- 29 CFR 1926.1101—regulations for the construction industry
- 29 CFR 1915.1001—regulations for the shipyard industry
- 40 CFR 763—Asbestos-Containing Materials in Schools Rule

The regulations should be reviewed for the requirements associated with the particular workplace activity (e.g., construction—29 CFR 1926.1101). The principal requirement is that no worker can be exposed to airborne asbestos fibers in excess of the PEL.

1. 0.1 fiber/cm^3 of air, 8 h TWA (PEL)
2. 1.0 fiber/cm^3 of air, 30 min short-term exposure limit (STEL)

The OSHA regulations for protection from exposure to asbestos are extensive. They require an employer to perform a negative exposure assessment in many cases. Monitoring must be performed by a competent person who is capable of identifying asbestos hazards and selecting control strategies and who has the authority to make corrective changes. The regulations also specify when medical surveillance is required, when personal protection must be provided, and the engineering controls and work practices that must be implemented.

Lead

The body does not use lead for any metabolic purpose, so any exposure to lead is undesirable. Lead dust and fumes can pose a severe hazard. Acute large doses of lead can cause systemic poisoning or seizures. Chronic exposure can damage the blood-forming bone marrow and the urinary, reproductive, and nervous systems. Lead is probably a human carcinogen, although whether it is causative or facilitative is subject to research.

OSHA has developed several regulations pertaining to exposure to lead in the workplace.

- 29 CFR 1910.1025—general exposure to lead in industry
- 29 CFR 1926.62—lead in construction standard

The general industry exposure standard for lead is a PEL of 50 μg/m^3, 8 h TWA (29 CFR 1910.1025(c)(1)).

Beryllium

Inhalation of metallic beryllium, beryllium oxide, or soluble beryllium compounds can lead to *chronic beryllium disease* (*berylliosis*). Ingestion and dermal contact do not pose a documented hazard, so maintaining beryllium dusts and fumes below the TLV in the breathing zone is a critical protection measure. As with asbestosis, it may be 20 years before the effects of exposure to beryllium dust are detectable.

Chronic beryllium disease is characterized by granulomas on the lungs, skin, and other organs. The disease can result in lung and heart dysfunction and enlargement of certain organs. Beryllium has been classified as a suspected human carcinogen.

The OSHA standard for exposure to beryllium is $2.0 \ \mu g/m^3$, 8 h TWA with a ceiling concentration of $5.0 \ \mu g/m^3$. The standard allows a 30 min exposure of up to $25 \ \mu g/m^3$ during an 8 h shift.

Coal Dust

Coal dust can cause chronic bronchitis, silicosis, and *coal worker's pneumoconiosis*, also known as *black lung disease*. In coal mines, the Mine Safety and Health Administration (MSHA) regulates exposure to coal dust. The MSHA standard is $2 \ mg/m^3$, 8 h TWA, of respirable coal dust. In other settings, OSHA regulates exposure to coal dust. The OSHA PEL is $2.4 \ mg/m^3$ if less than 5% free silica is present; otherwise, the limit is the same as for crystalline silica.

Welding Fumes

Exposure to welding fumes can cause a disease known as *metal fume fever*. This disease results from inhalation of extremely fine oxide particles that have been freshly formed as fume. Zinc oxide fume is the most common source, but magnesium oxide, copper oxide, and other metallic oxides can also cause metal fume fever. Metal fume fever is of short duration, with symptoms including fever and shaking chills appearing 4–12 hours after exposure.

Since a wide variety of welding processes and alloys may be involved, the nature of the process and the system employed need to be analyzed with the fumes in order to determine the hazard and controls. The specific TLVs should be evaluated for the specific constituent involved. Fumes may come from welding aluminum, titanium, steel, ferrous alloys, and stainless steel. Fumes may contain hexavalent chromium, ozone, carbon monoxide, iron, manganese, silicon, nickel, and fluoride.

Radioactive Dusts

Radioactive dusts can cause toxicity in addition to the effects from ionizing radiation. Inhalation of radioactive dust can result in deposition of the radionuclide in the body, which may enter the bloodstream and affect individual organs.

Control measures should be instituted to prevent workers from inhaling radioactive dust, either by restricting access or by providing appropriate personal protection such as respirators. Engineering controls to capture radioactive dust are an absolute necessity to minimize worker exposure.

Biological Particulates

A wide variety of biological organisms can be inhaled as particulates causing respiratory diseases and allergies. Examples include dust that contains anthrax spores from the wool or bones of infected animals, and fungi spores from grain and other agricultural produce.

Control of Particulates

Ventilation

Ventilation is the most effective method for controlling particulates. Enclosed processes should be used wherever possible. Equipment can be enclosed so that only the feed and discharge openings are open. With adequate pressure, enclosed equipment can be nearly as effective as closed processes. Large automated equipment can sometimes be placed in separate enclosures. Workers would have to wear personal protection to enter the enclosures. Local exhaust ventilation with hooded enclosures can be very effective at controlling particulate emissions into general work areas. Where complete enclosure and local exhaust methods are not sufficient, *dilution ventilation* will be necessary to control particulates in the work area. In some instances, the work process can be changed from a dry to a wet process to reduce particulate generation.

Personal Protection

Respirators are an effective means of controlling worker exposure to particulates that remain in the work area after engineering controls have been applied or when access to dusty areas is intermittent. Respirators may also be used to provide additional protection or comfort to workers in areas where local or general ventilation is effective. The NIOSH guidelines for selection of respirators should be followed to ensure that the respirator will be effective at removing the specific particulate to which the workers are exposed. The OSHA respiratory protection regulations in 29 CFR 1910.134 should also be followed.

Besides respirators, workers may also need protective clothing such as suits, eye protection, gloves, hard hats, boots, and so on. 29 CFR 1910.132 provides OSHA general requirements for protective equipment.

5. SOUND AND NOISE

Characteristics of Sound

Sound is pressure variation in air, water, or some other medium that the human ear can detect. *Noise* is unwanted, unpleasant, or painful sound. The *frequency* of sound is the number of pressure variations per second, measured in cycles per second, or hertz (Hz). The frequency range of human audible sound is approximately 20–20 000 Hz.

Sound produces a sensory response in the brain called *hearing*, but excessive pressure can cause pain in the ears. Sound is produced by vibration of a source that causes longitudinal vibration of the medium, which is usually air, but could also be water or solids. Sound waves are elastic waves that can occur in media with both elasticity and mass.

Another characteristic of sound is the *wavelength*, denoted as λ, which is the distance between two analogous and successive points of the sound wave. Wavelength is important because sound will bend around objects with dimensions less than the sound wavelength but will be reflected or scattered by objects larger than the sound wavelength. Thus, barriers will be effective for a short wavelength sound.

Velocity is the rate at which analogous successive sound pressure points pass a point. Velocity is called the *speed of sound* and is equal to the product of the wavelength and the frequency.

$$a = f\lambda \qquad 46.2$$

The speed of sound is dependent upon the medium, as illustrated in Table 46.3.

Table 46.3 *Speed of Sound in Various Media*

medium	speed of sound (m/s)	condition
air	330	1 atm, 0°C
water	1490	1 atm, 20°C
aluminum	4990	1 atm
steel	5150	1 atm

The speed of sound in an ideal gas can be determined from the following formula.

$$a = \sqrt{\frac{kR^*T}{\text{MW}}} \qquad 46.3$$

For air, k is approximately 1.40.

The characteristics of a sound wave are illustrated in Fig. 46.1.

Figure 46.1 *Characteristics of a Sound Wave*

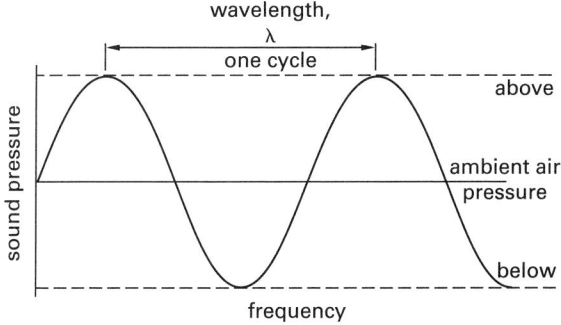

Sound Pressure

Sound pressure measures the intensity of sound and is the variation in atmospheric pressure caused by the disturbance of the air by a vibrating object. (See Fig. 46.1.) Positive pressures are known as *compressions*, and negative pressures are *rarefactions*. Because the sound wave produces both positive and negative pressure, the mean value would be zero, which is not a meaningful measurement. Therefore, the root-mean-square (rms) pressure is used.

$$p_{\text{rms}} = \sqrt{\frac{\sum p_i^2}{n}} \qquad 46.4$$

The rms pressure eliminates the negative pressures by taking the square root of the average of the sum of the squares of the sound pressure over each instant of time. Sound pressure is typically measured in micropascals (μPa) (10^{-6} N/m^2). Because humans can perceive absolute sound pressure in a very wide range of 20–200 million μPa (which is also the range of the *threshold of hearing* to the *threshold of pain*), it is more convenient to use a relative measure called the *bel*. The *decibel* (0.1 bel), abbreviated dB, is the common measure and is the minimum sound difference humans can perceive.

Sound pressure level, L_p, is measured in decibels relative to a reference level, p_0, of 20 μPa, the threshold of hearing at a reference frequency of 1000 Hz, as follows.

$$L_p = 10 \log\left(\frac{p}{p_0}\right)^2$$
$$= 20 \log\left(\frac{p}{p_0}\right) \qquad 46.5$$

Sound Power

Sound power, W, is the absorbed or transmitted sound energy per unit time (in watts). The sound power level, L_W, is measured in decibels relative to a reference level, W_0, of 10^{-12} W.

$$L_W = 10 \log\left(\frac{W}{W_0}\right) \qquad 46.6$$

Sound Intensity

Sound intensity is an areal function of the sound power of a source.

$$I = \frac{W}{4\pi r^2} \qquad 46.7$$

This relationship can be visualized as the sound power on a unit area on the surface of a sphere of radius r from the source (a point source of sound). (The area of the surface of the sphere is $4\pi r^2$.) Sound intensity cannot be measured directly.

Combining Sound Sources

The combined sound pressure level or sound power level from two or more sound sources can be determined by

$$L = 10 \log \sum 10^{L_i/10} \qquad 46.8$$

Example 46.3

What is the combined sound pressure from two grinders with sound pressure levels of 93 dB and 98 dB?

Solution

Use Eq. 46.8.

$$L_{W,\text{total}} = 10 \log \sum_{i=1}^{2} 10^{L_i/10}$$

$$L_{W,1} = 93 \text{ dB}$$

$$L_{W,2} = 98 \text{ dB}$$

$$L_{W,\text{total}} = 10 \log\left(10^{93 \text{ dB}/10} + 10^{98 \text{ dB}/10}\right) = 99.2 \text{ dB}$$

Loudness

Loudness is primarily determined by sound pressure but is also affected by frequency because the human ear is more sensitive to high-frequency sounds than low-frequency sounds. The upper sound limit that humans can perceive is between 16 000 Hz and 20 000 Hz. Because loudness is related to both sound pressure and frequency, a sound pressure-weighting scheme has been developed. This scheme uses sound level meters that have weighting characteristics, called weighting networks A, B, and C. The A network approximates equal loudness curves at low sound pressure levels. The B network approximates medium sound pressure levels, and the C network approximates high levels. The A network is widely used because it relates well to industrial noises and disturbances to the community. OSHA has adopted the A network as the preferred unit of measurement. Measurements using the A-weighted scale are designated dBA. *Slow response* refers to the operating mode of the instrument, in which the indicator will respond more slowly for easier reading.

Noise

Noise is vibration conducted through solids, liquids, or gases. Noise can cause psychological and physiological damage and can interfere with workers' communication, thereby affecting safety. Exposure to excessive noise for a sufficient time can result in hearing loss. Threshold limit value (TLV) criteria have been developed to protect against hearing loss in the speech-frequency range. In 29 CFR 1910.95, OSHA has established the permissible exposure levels (PELs) and duration of exposure to noise to prevent harm to workers. Generally, the TLVs for noise are more restrictive than the PELs. Employers are required by OSHA to implement a hearing conservation program when sound levels exceed 85 dBA on an

8 h time-weighted average (TWA) basis. The permissible noise exposures are given in Table 46.4.

When the daily noise exposure is composed of two or more periods of noise exposure at different levels, their combined effect should be considered, rather than the individual effect of each. If the sum of the fractions $C_1/t_1 + C_2/t_2 \cdots C_n/t_n$ exceeds unity, then the mixed exposure should be considered to exceed the limit value. C_n indicates the total time of exposure at a specified noise level, and t_n indicates the total time of exposure permitted at that level.

Table 46.4 Permissible Noise Exposures

duration per day (h)	sound level (dBA slow response)[*]
8	90
6	92
4	95
3	97
2	100
1.5	102
1	105
0.5	110
0.25 or less	115

[*]"Slow response" refers to the metering circuit being set on slow mode to reduce rapid, hard-to-read needle excursions. Compliance with OSHA regulations requires measurement set on slow response mode.

Source: 29 CFR 1910.95, Table G-16

Hearing Loss

Hearing loss can be caused by sudden intense noise over only a few exposures. This type of loss is known as *acoustic trauma*. Hearing loss can also be caused by exposure over a long duration (months or years) to hazardous noise levels. This type is known as *noise-induced hearing loss*. The permanence and nature of the injury depends on the type of hearing loss.

The main risk factors associated with hearing loss are the intensity of the noise (sound pressure level), the type of noise (frequency), daily exposure time (hours per day), and the total work duration (years of exposure). These are known as *noise exposure factors*. Generally, exposure to sound levels above 115 dBA is considered hazardous, and exposure to levels below 70–75 dBA is considered safe from risk of permanent hearing loss. Also, noise with predominant frequencies above 500 Hz is considered to have a greater potential to cause hearing loss than lower-frequency sounds.

Sound Measurements

Sound measurements are typically made with a sound level meter, which can be set to the A, B, or C network. Meters typically have a range of 40–140 dB (referenced to 20 μPa) and can be set for slow or fast response. In fast mode, the meter responds quickly to changing noise levels. In slow mode, the meter will be easier to read in a factory setting and is the mode required by OSHA for compliance with its requirements.

Other instruments used in noise measurements are octave-band analyzers and noise dosimeters. *Octave-band analyzers* can determine the frequency distribution of sound pressure. *Noise dosimeters* record the total noise energy to which a worker has been exposed during a shift.

Classes of Noise Exposure

Noise exposure can be classified as continuous noise, intermittent noise, and impact noise. *Continuous noise* is broadband noise of a nearly constant sound pressure level and frequency to which a worker is exposed 8 hours daily and 40 hours weekly. *Intermittent noise* involves exposure to a specific broadband sound pressure level several times a day. *Impact noise* is a sharp burst of short duration sound.

OSHA has established permissible noise exposures for a specific duration each workday, known as *permissible exposure levels* (PELs). The PELs are based on continuous 8 hour exposure at a sound pressure level of 90 dBA, which is established as 100%. For other exposure durations, OSHA has established relationships between the sound level and the exposure time. Every 5 dBA increase in noise cuts the allowable exposure time in half. Sound pressure levels below 90 dBA are not considered hazardous and do not have to be determined.

When workers are exposed to different noise levels during the day, the mixed exposure must be calculated as follows.

$$E_m = \sum_{i=1}^{n} \frac{C_i}{t_i} \qquad 46.9$$

If E_m equals or exceeds 1, the mixed exposure exceeds the OSHA standard.

For intermittent noise, the time characteristics of the noise must also be determined. Both short-term and long-term exposure must be measured. A dosimeter is typically used for intermittent noises.

For impact noises, workers should not be exposed to peaks of more than 140 dBA under any circumstances. The threshold limit value for impulse noise should not exceed the values provided in Table 46.5.

Table 46.5 *Typical Threshold Limit Values for Impact Noise*

peak sound level (dB)	maximum number of daily impacts
140	100
130	1000
120	10 000

Example 46.4

A machine shop worker is exposed to the following noise levels during the workday. Is the exposure excessive?

$$80 \text{ dBA for } 2.5 \text{ h}$$
$$92 \text{ dBA for } 2.0 \text{ h}$$
$$95 \text{ dBA for } 3.0 \text{ h}$$
$$115 \text{ dBA for } 0.5 \text{ h}$$

Solution

There is no exposure limit for noise of 85 dBA or lower. From Table 46.4, determine the permissible noise exposure time at each noise level. The noise dose exposure for the worker is given by Eq. 46.9.

$$\begin{aligned} E_m &= \sum_{i=1}^{n} \frac{C_i}{t_i} \\ &= 0 + \frac{2.0 \text{ h}}{6 \text{ h}} + \frac{3.0 \text{ h}}{4 \text{ h}} + \frac{0.5 \text{ h}}{0.25 \text{ h}} \\ &= 3.1 \end{aligned}$$

Because the E_m of 3.1 is greater than 1, the exposure is excessive.

Noise Control

A *hearing conservation program* should include noise measurements, noise control measures, hearing protection, audiometric testing of workers, and information and training programs. Employees are required to properly use the protective equipment provided by employers.

After the noise exposure is compared with acceptable noise levels, the degree of noise reduction needed can be determined. Noise reduction measures can comprise the following three basic methods applied in order.

1. changing the process or equipment
2. limiting the exposure
3. using hearing protection

Noise reduction involves looking at the source of noise generation, the path of noise from the source to the worker, and protection of the worker. Reducing noise generation may require modifying or replacing equipment or reengineering the manufacturing process. The noise path can be altered by enclosing the noise source, increasing the distance from the worker to the source, or placing sound barriers or sound absorbing materials along the noise path. Worker protection can include placing an enclosure around the worker and/or providing ear protection.

The decrease in sound pressure level or sound power level due to insertion of an enclosure or wall between a noise source and a receiver (the *insertion loss*) can be found

Health, Safety, Welfare

from the following equation. The decrease in sound pressure level is the same as the decrease in sound power level.

$$
\begin{aligned}
\text{IL} &= L_{p,1} - L_{p,2} = L_{W,1} - L_{W,2} \\
&= 10 \log \left(\frac{L_{W,1}}{L_{W,2}} \right) \\
&= 20 \log \left(\frac{p_1}{p_2} \right) \qquad \text{46.10}
\end{aligned}
$$

Noise control measures can include engineering controls, administrative controls, and personal hearing protection. Engineering controls are the most reliable and effective approach to noise control. Engineering controls involve modifying a machine to operate with lower noise generation, for example, by reducing vibrating components or substituting quieter components or functions. Engineering controls also include reducing noise along the noise path, for example, through the use of sound barriers. Noise generation should be considered early in the machine or manufacturing design process and should be a part of all procurement decisions.

Administrative controls include changing the exposure of workers to high noise levels by modifying work schedules or locations so as to reduce workers' exposure times. Administrative controls include any administrative decision that limits a worker's exposure to noise.

Personal hearing protection is the final noise control measure, to be implemented only after engineering controls are implemented. Protective devices do not reduce the noise hazard and may not be totally effective, so engineering controls are preferred over hearing protection. Protective devices include helmets, earplugs, canal caps, and earmuffs. Earplugs may be used with helmets to increase the level of noise reduction.

An important characteristic of personal hearing protection is the *noise reduction rating* (NRR). The NRR is established by the Environmental Protection Agency (EPA) and must be printed on the package of a device. The NRR can be used to determine whether a device provides sufficient hearing protection.

Audiometry

Audiometry is the measurement of hearing acuity. It is used to assess a worker's hearing ability by measuring the individual's threshold sound pressure level at various frequencies (250–6000 Hz). The threshold audiogram can be used to create a baseline of hearing ability and to determine changes over time and identify changes resulting from noise control measures. Baseline and annual hearing tests are required where workers are exposed to more than a *time-weighted average* (TWA) over 8 hours of 85 dBA. The average change from the baseline is used to measure the degree of hearing impairment.

6. RADIATION

Radiation can be either nonionizing or ionizing. *Nonionizing radiation* includes electric fields, magnetic fields, electromagnetic radiation, radio frequency and microwave radiation, and optical radiation and lasers.

Dealing with *ionizing radiation* from nuclear sources requires special skills and knowledge. A specialist in health physics should be consulted whenever ionizing radiation is encountered.

Nuclear Radiation

Nuclear radiation is a term that applies to all forms of radiation energy that originate in the nuclei of radioactive atoms. Nuclear radiation includes alpha particles, beta particles, neutrons, X-rays, and gamma rays. The common property of all nuclear radiation is an ability to be absorbed by and transfer energy to the absorbing body.

The preferred unit of ionizing radiation, given in the National Council on Radiation Protection's (NCRP's) *Recommended Limitations for Exposure to Ionizing Radiation*, is the mSv. Sv is the symbol for sievert, which is the SI unit of absorbed dose times the *weighting factor* (previously known as the *quality factor*) of the radiation as compared to gamma radiation. The absorbed dose is measured in grays (Gy). The gray is equal to 1 J of absorbed energy per kilogram of matter. A summary of ionizing radiation units is given in Table 46.6.

Table 46.6 Units for Measuring Ionizing Radiation

property	SI
energy absorbed	gray (Gy) 1 J/kg 1 Gy = 100 rad (obsolete)
biological effect	sievert (Sv) Gy × quality factor 1 Sv = 100 rem (obsolete)

Alpha Particles

Alpha particles consist of two protons and two neutrons, with an atomic mass of four. Alpha particles combine with electrons from the absorbed material and become helium atoms. Alpha particles have a positive charge of two units and react electrically with human tissue. Because of their large mass, they can travel only about 10 cm in air and are stopped by the outer layer of the skin. Alpha-emitters are considered to be only internal radiation hazards, which requires alpha particles to be ingested by eating or breathing. They affect the bones, kidney, liver, lungs, and spleen.

Beta Particles

Beta particles are electrically charged particles ejected from the nuclei of radioactive atoms during disintegration. They have a negative charge of one unit and the mass of an electron. High-energy beta particles can penetrate in human tissue to a depth of 20 mm to 130 mm and travel up to 9 m in air. Skin burns can result from an extremely high dose of low-energy beta radiation, and some high-energy beta sources can penetrate deep into the body, but beta-emitters are primarily internal radiation hazards, which would require them to be ingested. Beta particles are more hazardous than alpha particles because they can penetrate deeper into tissue. High-energy beta radiation can produce a secondary radiation called *bremsstrahlung*. These are X-rays produced when electrons (i.e., beta particles) pass near the nuclei of other atoms. Bremsstrahlung radiation is proportional to the energy of the beta particle and the atomic number of the adjacent nucleus. Materials with low atomic numbers (e.g., plexiglass) are preferred shielding materials.

Neutrons

Neutron particles have no electrical charge and are released upon disintegration of certain radioactive materials. Their range in air and in human tissue depends on their kinetic energy, but the average depth of penetration in human tissue is 60 mm. Neutrons lose velocity when they are absorbed or deflected by the nuclei with which they collide. However, the nuclei are left with higher energy that is later released as protons, gamma rays, beta particles, or alpha particles. It is these secondary emissions from neutrons that produce damage in tissue.

X-Rays

X-rays are produced by electron bombardment of target materials and are highly penetrating electromagnetic radiation. X-rays have a valuable scientific and commercial use in producing shadow pictures of objects. The energy of an X-ray is inversely proportional to its wavelength. X-rays of short wavelength are called *hard*, and they can penetrate several centimeters of steel. Long wavelength X-rays are called *soft*, and they are less penetrating. The power of X-rays and gamma rays to penetrate matter is called *quality*. *Intensity* is the energy flux density.

Gamma Rays

Gamma rays, or gamma radiation, are a class of electromagnetic photons (radiation) emitted from the nuclei of radioactive atoms. They are highly penetrating and are an external radiation hazard. Gamma rays are emitted spontaneously from radioactive materials, and the energy emitted is specific to the radionuclide. Gamma rays present an internal exposure problem because of their deep penetrating ability.

Radioactive Decay

The activity, A, of a radioactive substance is its *decay rate*. The SI unit of activity is the *becquerel* (Bq), equivalent to $1/s$ (i.e., one disintegration per second).

The non-SI Curie (Ci) unit, equivalent to 3.7×10^{10} disintegrations per second, was previously used and may still be encountered, particularly in medical research.

Radioactive decay is measured in terms of *half-life*, the time to lose half of the activity of the original material. Decay activity can be calculated as follows.

$$A = A_o e^{-\lambda t} \qquad 46.11$$

The half-life can be calculated from the decay constant, $t_{1/2} - 0.693/\lambda$.

Example 46.5

(a) How long will it take a 1000 becquerel (Bq) radionuclide to decay to 300 Bq if the decay constant is 0.090 d^{-1}? (b) What is the half-life?

Solution

(a) Use Eq. 46.11.

$$e^{-\lambda t} = \frac{A}{A_o}$$

$$t = \frac{\ln \dfrac{A}{A_o}}{-\lambda} = \frac{\ln \dfrac{300 \text{ Bq}}{1000 \text{ Bq}}}{-0.090 \text{ d}^{-1}}$$

$$= 13.4 \text{ d}$$

(b) Find the half-life.

$$t_{1/2} = \frac{0.693}{\lambda} = \frac{0.693}{0.090 \text{ d}^{-1}}$$

$$= 7.7 \text{ d}$$

Example 46.6

What radioactivity would remain from 2 Bq cobalt-60, which has a half-life of 5.24 yr, after a 20 yr period?

Solution

Use Eq. 46.11.

$$\lambda = \frac{0.693}{t_{1/2}} = \frac{0.693}{5.24 \text{ yr}} = 0.132 \text{ yr}^{-1}$$

$$A = A_o e^{-\lambda t} = (2 \text{ Bq}) e^{-(0.132 \text{ yr}^{-1})(20 \text{ yr})}$$

$$= 0.143 \text{ Bq}$$

Radiation Effects on Humans

Ionizing radiation transfers energy to human tissue when it passes through the body. *Dose* refers to the amount of radiation that a body absorbs when exposed to ionizing radiation. The effects on the body from external radiation are quite different from the effects from internal radiation. Internal radiation is spread throughout the body to tissues and organs according to the chemical properties of the radiation. The effects of internal radiation depend on the energy and the residence time within

Health, Safety, Welfare

the body. The principal effect of radiation on the body is destruction of or damage to cells. Damage may affect reproduction of cells or cause mutation of cells.

The effects of ionizing radiation on individuals include skin, lung, and other cancers; bone damage; cataracts; and a shortening of life. Effects on the population as a whole include possible damage to human reproductive elements, thereby affecting the genes of future generations.

Permissible Exposure Levels

To protect humans from radiation effects, maximum permissible levels of exposure have been developed based on the maximum radiation dose that can be received with little risk of later development of adverse effects. The National Council on Radiation Protection and Measurement (NCRP) has published maximum permissible levels of external and internal radiation. The Nuclear Regulatory Commission (NRC) has established permissible doses, levels, and concentrations in 10 CFR 20. The International Commission on Radiological Protection (ICRP) has also published guides for maximum external exposure. For occupational exposure, the NCRP recommendations are given in Table 46.7.

Table 46.7 NCRP-Recommended Limitations for Exposure to Ionizing Radiation

occupational exposure	mSv
effective dose limit—annual	50
effective dose limit—cumulative	$10 \times$ age
equivalent dose annual limits for tissues and organs—lens of eye	150
equivalent dose annual limits for tissues and organs—skin, hands, and feet	500

Source: Adapted from NCRP Report No. 116, *Limitation of Exposure to Ionizing Radiation*, 1993.

Safety Factors

An environmental engineer should be aware of the basic safety factors for limiting dose. These factors are time, distance, and shielding.

The dose received is directly related to the time exposed, so reducing the time of exposure will reduce the dose. An individual's time of exposure can also be limited by spreading the exposure time among more workers.

Distance is another safety factor that can be changed to reduce the dose. The intensity of external radiation decreases as the inverse of the square of the distance. By increasing the distance to a source from 2 m to 20 m, for example, the exposure would be reduced to 1% (2 m/20 m)².

Shielding involves placing a mass of material between a source and workers. The objective is to use a high-density material that will act as a barrier to X-ray and

gamma-ray radiation. Lead and concrete are often used, with lead being the more effective material because of its greater density. For neutrons, different material is needed than for X-rays and gamma rays because neutrons produce secondary radiation from collisions with nuclei. Neutron shielding requires a light nucleus material. Typically water or graphite is used.

The shielding properties of materials are often compared using the *half-value thickness*, which is the thickness of the material required to reduce the radiation to half of the incident value. The half-value properties vary with the radiation source.

An environmental engineer should keep the following suggestions in mind when radiation is encountered on a project site.

- Obtain the advice of a qualified health physicist or industrial hygienist when dealing with radioactive materials.
- Treat radioactive materials as hazardous to human health.
- Be aware that the properties and hazards associated with internal and external radiation are different.
- Ensure that only properly calibrated instruments are used by trained personnel to monitor radiation.
- Apply the safety factors of time, distance, and shielding to keep exposure at safe levels.
- Ensure that exposure guidelines are achieved and that workers receive the minimum exposure possible in any situation.

7. HEAT AND COLD STRESS

Thermal Stress

Heat and cold, or *thermal*, stress involves three zones of consideration relative to industrial hygiene. In the middle is the *comfort zone*, where workers feel comfortable in the work environment. On either side of the comfort zone is a *discomfort zone*, where workers feel uncomfortable with the heat or cold, but a health risk is not present. Outside of each discomfort zone is a *health risk zone*, where there is a significant risk of health disorders due to heat or cold. Industrial hygiene is primarily concerned with controlling worker exposure in the health risk zone.

The analysis of thermal stress involves taking a *heat balance* of the human body with the objective of determining whether the net heat storage is positive, negative, or zero. A simplified form of the heat balance is

$$Q_S = Q_M + Q_R + Q_C + Q_E \qquad 46.12$$

If the storage, Q_S, is zero, heat gain is balanced by heat loss, and the body is in equilibrium. If Q_S is positive, the body is gaining heat; and if Q_S is negative, the body is losing heat.

The heat balance is affected by environmental and climatic conditions, work demands, and clothing. The metabolic rate, Q_M, is more significant for heat stress than for cold stress when compared with radiation and convection. The metabolic rate can affect heat gain by one to two orders of magnitude compared to radiation and convection, but it affects heat loss to about the same extent as radiation and convection.

Clothing affects the thermal balance through insulation, permeability, and ventilation. *Insulation* provides resistance to heat flow by radiation, convection, and conduction. *Permeability* affects the movement of water vapor and the amount of evaporative cooling. *Ventilation* influences evaporative and convective cooling.

Heat Stress

Heat stress can increase body temperature, heart rate, and sweating, which together constitute *heat strain.* As body temperature increases, blood circulation transports heat from the body core to the skin, where the blood is cooled and returned to the core. At the skin surface, heat is lost by convection and radiation at a rate that depends on the difference in temperature between the skin and the environment. Heat is also lost by evaporation of sweat, when the body attempts to increase sweating to the point where the heat storage rate is zero. Heart rate increases because more blood must be pumped between the body core and the skin to increase the heat loss. Sweat rate (and the total volume of sweat) increases to increase evaporative cooling.

The most serious heat disorder is *heatstroke*, because it involves a high risk of death or permanent damage. Fortunately, heatstroke is rare. Of lesser severity, *heat exhaustion* is the most commonly observed heat disorder for which treatment is sought. *Dehydration* is usually not noticed or reported, but without restoration of water loss, dehydration leads to heat exhaustion. The symptoms of these key heat stress disorders are as follows.

- heatstroke: chills, restlessness, irritability
- heat exhaustion: fatigue, weakness, blurred vision, dizziness, headache
- dehydration: no early symptoms, fatigue or weakness, headache, dry mouth

Appropriate first aid and medical attention should be sought when any heat stress disorder is recognized.

Evaluation of Heat Stress

NIOSH has established a threshold limit value (TLV) for prolonged exposure of 38°C for the acceptable body core temperature. (The *core temperature* is commonly assumed to equal the *oral temperature* plus 0.5°C.) Also, the core temperature can be 39°C for short periods with adequate recovery so long as the time-weighted average (TWA) is no more than 38°C. Moreover, the

average heart rate over a day should not exceed 110 beats per minute (bpm). "Prolonged exposure" is assumed to be an 8 h work shift with breaks every 2 h.

NIOSH has established additional heat stress TLVs that can be used to ensure that the core temperature will not exceed safe levels. The TLVs are a function of the metabolic rate and an index of environmental heat. Three equations can be used to determine the ceiling limit, the recommended exposure limit, and the recommended alert limit.

The *ceiling limit* (CL) is given by

$$\text{CL} = 70.0°\text{C} - 13.5 \log Q_{M,\text{watts}} \qquad 46.13$$

NIOSH recommends that the CL not be exceeded for more than 15 min during the workday.

The NIOSH *recommended exposure limit* (REL) for acclimatized workers is given by

$$\text{REL} = 56.7°\text{C} - 11.5 \log Q_{M,\text{watts}} \qquad 46.14$$

The REL is based on summer-weight work clothes and is based on 8 h workdays with breaks every 2 h for acclimatized workers.

The *recommended alert level* (RAL) for unacclimatized workers is given by

$$\text{RAL} = 60.0°\text{C} - 14.1 \log Q_{M,\text{watts}} \qquad 46.15$$

Heat stress on unacclimatized workers should not be allowed to exceed the RAL.

To use the formulas for CL, REL, or RAL, the time-weighted average of the metabolic rate must be estimated. The metabolic rate for a given task can be estimated from Table 46.8.

Using the metabolic rates for a given task and the following equation, the time-weighted average of the metabolism can be determined for an 8 h workday. B, P, A, and V are determined from Table 46.8. A time-weighted average of the metabolic rate (TWA-M) is then calculated from the duration of the metabolism for each task.

$$Q_{M,\text{task}} = B + P + A + V \qquad 46.16$$

After the REL or RAL is determined, the *wet-bulb globe temperature* (WBGT) must be determined for the working conditions. The WBGT affects heat removal through perspiration and is determined from Eq. 46.17 or Eq. 46.18 for indoor and outdoor (direct sunlight) exposure.

$$\text{WBGT}_{\text{in}} = 0.77 T_{\text{nwb}} + 0.3 T_g \qquad 46.17$$

$$\text{WBGT}_{\text{out}} = 0.7 T_{\text{nwb}} + 0.2 T_g + 0.1 T_{\text{db}} \qquad 46.18$$

Health, Safety, Welfare

Table 46.8 Metabolic Rates for a Task

task	average metabolic rate (W)
basal metabolism, B	70
posture metabolism, P	
sitting	20
standing	40
walking	170
activity metabolism, A	
light hand	30
heavy hand	65
light one arm	70
heavy one arm	120
light both arms	105
heavy both arms	175
light whole body	245
moderate whole body	350
heavy whole body	490
very heavy whole body	630
climbing metabolism, V	$56V_v$*

*V_v = rate of vertical ascent in m/min

Source: Adapted from American Conference of Government Industrial Hygienists, *1999 TLVs and BEIs, Threshold Limit Values for Chemical Substances and Physical Agents, Biological Exposure Indices*, 1999.

For varying conditions, the TWA-WBGT must be determined. Also, the TWA-WBGT must be adjusted for other than summer work clothes by adding an adjustment factor to the calculated WBGT. The adjustment factors are given in Table 46.9.

Table 46.9 WBGT Clothing Adjustment Factors

clothing type	adjustment factor (°C)
summer work clothes	0
coveralls	+2
winter uniform	+4
vapor transmitting suit	+6
vapor barrier suit	+8
full-body encapsulating unit	+11

Source: Adapted from American Conference of Government Industrial Hygienists, *1999 TLVs and BEIs, Threshold Limit Values for Chemical Substances and Physical Agents, Biological Exposure Indices*, 1999.

Another aspect of heat stress evaluation is determining how long a worker can safely work in a particular environment. One approach to this problem is to establish conditions for recovery, such as sitting in a cool area in work clothes, and then to determine the environmental heat stress factors associated with the recovery period. The duration of the recovery period can then be changed until the REL or RAL is no longer exceeded. This analysis can also be performed for a 1 h period for a single set of activities and environmental conditions in order to get a ratio of recovery to work time for those conditions.

Other methods to estimate time-limited exposure to heat stress involve comparison of the TWA-WBGT with exposure limit charts to get the allowable exposure time, graphical solutions, and sophisticated heat balance analyses. Additionally, the International Organization for Standardization (ISO) has published a method for determining the allowable exposure time based on a standard perspiration rate.

Besides evaluating the exposure to heat stress through formulas based on environmental conditions, certain physiological functions can be measured directly to confirm that heat stress is within allowable limits. These functions are core temperature, heart rate, and sweating.

The instantaneous core temperature should never exceed 39°C, and the TWA core temperature should not exceed 38°C. The average heart rate should not exceed 110 beats per minute (bpm). The recovery heart rate after 1 min should be less than 110 bpm. Alternatively, the heart rate at 3 min should be less than 90 bpm, or the 1 min rate minus the 3 min rate should be at least 10 bpm.

Perspiration mass is the initial body mass plus the mass of food or drink minus the mass of excretions minus the final body mass. The perspiration volume in liters is the perspiration mass in kg × 1 L/kg. Perspiration volume should not exceed 5 L, and perspiration rate should not exceed 1 L/h (measured over a 2 h to 4 h period).

Example 46.7

A crew at a toxic spill site along a railroad in Arizona must conduct the cleanup during daylight hours and within one 8 h workday. The anticipated heat exposure levels for the crew are as follows.

activity	duration (h)	clothing
1 sitting, writing	1	coveralls
2 walking	2	coveralls
3 carrying heavy tools	2	coveralls
4 digging soil by hand (moderate, whole body)	1	vapor transmitting
5 filling sample containers (standing, both arms, light)	1	vapor transmitting
6 operating excavating equipment (sitting, both arms, light)	1	encapsulated

All workers are acclimatized to the local environment, and nominal rest breaks are provided in a cool trailer. It is a sunny day, and the environmental heat factors are

$$T_{db} = 22°C$$

$$T_g = 25°C$$

$$T_{nwb} = 18°C$$

$$\text{wind speed} = 2 \text{ m/s}$$

Will the crew exceed the NIOSH recommended exposure level (REL) for this work?

Health, Safety, Welfare

Solution

Calculate the TWA-WBGT that will be experienced by the crew while working in direct sunlight.

From Eq. 46.18,

$$\text{WBGT} = 0.7\,T_{\text{nwb}} + 0.2\,T_g + 0.1\,T_{\text{db}}$$
$$= (0.7)(18°\text{C}) + (0.2)(25°\text{C}) + (0.1)(22°\text{C})$$
$$= 19.8°\text{C}$$

activity	adjustment factor (°C)	duration (h)	TWA adjustment factor (°C)
1	+2	1	2
2	+2	2	4
3	+2	2	4
4	+6	1	6
5	+6	1	6
6	+11	1	11
total		8	33
TWA			4.1

$$\text{TWA-WBGT} = 19.8°\text{C} + 4.1°\text{C}$$
$$= 23.9°\text{C}$$

Calculate the TWA metabolic rate (TWA-M) of the crew.

Use Eq. 46.16.

$$Q_{M,\text{task}} = B + P + A + V$$

From Table 46.8,

activity	B	P	A	V	$Q_{M,\text{task}}$	duration	TWA-M
1	70	20	30	0	120	1	120
2	70	170	0	0	240	2	480
3	70	170	65	0	305	2	610
4	70	40	350	0	460	1	460
5	70	40	105	0	215	1	215
6	70	20	105	0	195	1	195
total						8	2080
TWA							260

Calculate the REL using Eq. 46.14.

$$\text{REL} = 56.7°\text{C} - 11.5\log Q_M$$
$$= 56.7°\text{C} - 11.5\log 260$$
$$= 28.9°\text{C}$$

Since the TWA-WBGT of 23.9°C is less than the REL of 28.9°C, the crew will not exceed the recommended exposure limit.

Control of Heat Stress

Controls that are applicable to any heat stress situation are known as *general controls*. General controls include worker training, heat stress hygiene, and medical monitoring.

Worker training will help the worker understand heat stress and recognize heat stress disorders. The training can also show the worker first aid for heat stress and appropriate hygiene to prevent heat stress problems.

Heat stress hygiene comprises actions taken by workers to prevent heat stress problems. These include fluid replacement, self-determination of heat stress effects, and acclimation to heat stress exposure. Lifestyle, general health, and diet also affect a worker's response to heat stress.

Medical monitoring by a qualified physician of individuals who are at extraordinary risk of heat-related disorders is also a general control measure.

Specific controls are controls that are put in place for a particular job. They include engineering controls, administrative controls, and personal protection. *Engineering controls* include changing the physical work demands to reduce the metabolic heat gain, reducing external heat gain from the air or surfaces, and enhancing external heat loss by increasing sweat evaporation and decreasing air temperature. *Administrative controls* include scheduling the work to allow worker acclimatization to occur, leveling work activity to reduce peak metabolic activity, and sharing or scheduling work so the heat exposure of individual workers is reduced. *Personal protection* includes using systems to circulate air or water through tubes or channels around the body, wearing ice garments, and wearing reflective clothing.

Cold Stress

The body reacts to cold stress by reducing blood circulation to the skin to insulate itself. The body also shivers to increase metabolism. These mechanisms are ineffective against long-term extreme cold stress, so humans react by increasing clothing for more insulation, increasing body activity to increase metabolic heat gain, and finding a warmer location.

There are two main hazards from cold stress: hypothermia and tissue damage. *Hypothermia* depresses the central nervous system, causing sluggishness and slurred speech, and progresses to disorientation and unconsciousness. To avoid hypothermia, the minimum core body temperature must be above 36°C for prolonged exposure and above 35°C for occasional exposure of short duration. Fatigue can increase the risk of severe hypothermia. Tissue damage (e.g., *frostbite* or *frostnip*) can occur when contact is made with objects or air that cause the skin temperature to fall below −1°C (the temperature required for skin to freeze).

First aid should be applied and medical attention should be sought whenever cold stress problems are recognized.

The *equivalent chill temperature* (ECT), also known as the *wind chill index*, is calculated by determining the heat loss under combinations of air temperature and wind speed and then computing an equivalent air temperature with no air motion giving the same rate of heat loss. When workplace temperatures fall below 16°C, monitoring should be performed. Below −1°C, the dry-bulb temperature and air speed should be measured and recorded every 4 h. When the air speed exceeds 2 m/s, the ECT should be determined. When the ECT falls below −7°C, the ECT should be recorded.

Worker training, cold stress hygiene, and medical surveillance can control cold stress. Engineering controls, administrative controls, and personal protection measures can also be used to control cold stress.

8. ERGONOMICS

Ergonomics is the study of human characteristics to determine how a work environment should be designed to make work activities safe and efficient. It includes both physiological and psychological effects on the worker, as well as health, safety, and productivity aspects.

An environmental engineer should have a general knowledge of ergonomics because he or she may assume a position of responsibility for a project, a program, or a corporation that includes responsibility for the health and safety of employees or contractors. On large projects or programs, an environmental engineer will probably be assisted by an industrial hygienist or a health and safety specialist who will have specific responsibility to identify and report ergonomic hazards along with other hazards on the project or in the workplace. On small projects, however, an environmental engineer may not have such expertise readily available and needs to be able to identify situations where expert advice is needed. This section describes situations or conditions that imply a need for further ergonomic evaluation by an industrial hygienist. It also focuses on projects (such as hazardous waste site remediation or treatment plant operation) rather than on traditional manufacturing processes.

Work-Rest Cycles

An environmental engineer should be aware of the need for workers to be given rest periods appropriate for the metabolic energy used on the tasks. Excessively heavy work should be broken by frequent short rest periods to reduce cumulative fatigue. The percentage of time a worker should rest can be estimated by the following equation.

$$t_{\text{rest}} = \frac{Q_{M,\text{max}} - Q_M}{Q_{M,\text{rest}} - Q_M} \times 100\% \qquad 46.19$$

The rate of metabolic energy expended in typical activities for a 70 kg male worker is given in Table 46.10.

Table 46.10 Typical Total Metabolic Energy Expenditure Rate of Various Activities

activity	typical total metabolic energy expenditure rate (W)
resting, prone	105
resting, seated	115
standing, at ease	130
light bench work	120
medium bench work	190
driving	200
walking, casual	230
walking, hard	410
pushing wheelbarrow	410
shoveling	450
rock drilling	500
climbing stairs	700
heavy digging	800

Source: Adapted from Plog, Barbara A., Jill Niland, and Patricia J. Quinlan, *Fundamentals of Industrial Hygiene*, 4th ed., National Safety Council, Itasca, Illinois, 1996.

The maximum energy a worker can produce is approximately 1200 W, about half of which goes to maintain the body and the other half to actual work. The heaviest work a fit young man can sustain is approximately 580 W, but the general population can sustain only about 465–495 W. Women, on average, can sustain about 80% of these values.

The procedure for estimating rest time is illustrated in Ex. 46.8.

Besides metabolic energy, heart rate can be used to evaluate work in terms of light, medium, heavy, very heavy, and extremely heavy work, as shown in Table 46.11.

Table 46.11 Metabolic Classification of Work

class of work	total energy expenditure (W)	heart rate (beats/min)
light	175	90 or less
medium	350	100
heavy	525	120
very heavy	700	140
extremely heavy	1050	160 or more

Source: Adapted from Plog, Barbara A., Jill Niland, and Patricia J. Quinlan, *Fundamentals of Industrial Hygiene*, 4th ed., National Safety Council, Itasca, Illinois, 1996.

The direct correlation between heart rate and metabolic energy allows work to be evaluated through the relatively simple measurement of pulse. During light, medium, and heavy work, the metabolic and other physiological functions of the body can attain a steady-state condition for the duration of the work. During very heavy work when the heart rate is 140 bpm or more, the oxygen deficit and the buildup of metabolic by-products increase, which requires the worker to take rest periods or to cease work. At 160 bpm, frequent rest periods are required; this level of activity usually cannot be sustained for a full workday by even the most capable of workers.

Example 46.8

Workers at a hazardous waste remediation site must dig hardpan from around several tanks in an area not accessible to excavating equipment. The male workers are in poor physical condition, so the maximum metabolic energy the project manager believes is appropriate for sustained work on this project is 300 W. She estimates that the workers will need to spend 2 h pushing a wheelbarrow, 2 h shoveling, and 4 h digging with pickaxes to complete the project. The workers can sit when they rest. How much time should the project manager allocate to rest?

Solution

Calculate the TWA of the metabolic energy for the project.

activity	metabolic energy, Q_M (W)	duration (h)	TWA-M
pushing wheelbarrow	410	2	820
shoveling	450	2	900
digging	800	4	3200
total		8	4920
TWA-M			615

The TWA metabolic energy, Q_M, to be sustained for the duration of the project is 615 W. From Table 46.10, resting metabolism while sitting is 115 W.

Calculate the resting time required. Use Eq. 46.19.

$$t_{\rm rest} = \frac{Q_{M,\max} - Q_M}{Q_{M,\rm rest} - Q_M} \times 100\% = \frac{300\ {\rm W} - 615\ {\rm W}}{115\ {\rm W} - 615\ {\rm W}} \times 100\%$$
$$= 63\%$$

The workers should be given 38 min rest out of every hour (63%).

Manual Handling of Loads

On many projects, loads must be handled manually. Improper handling is the most common cause of injury and of the most severe injuries in the workplace.

Heavy loads can strain the body, particularly the lower back. Even light or small objects can cause risk of injury to the body if they are handled in a way that requires strain-inducing stretching, reaching, or lifting.

Guidelines for Manual Material Handling

Safe manual material handling depends both on the design of the task and on the employee following safe practices. Table 46.12 provides guidelines for both the task design and the work practices.

Table 46.12 *Guidelines for Task Design and Work Practices for Manual Material Handling*

Task Design Guidelines
- Move materials predominately horizontally by pushing and pulling. Avoid lifting, lowering, and bending.
- Lift and lower materials between knuckle height and shoulder height.
- Lift and lower close to and in front of the body. Avoid bending forward or twisting the body.
- Make the material light, compact, and safe to grasp whenever possible. Provide good handles.
- Make sure the material does not have sharp edges, corners, or pinch points.
- If material is in bins, make sure the material can be easily removed.

Work Practices Guidelines
- Train workers in safe and efficient procedures.
- Select workers capable of performing the work.
- Encourage workers to be in good physical shape.
- Place material conveniently within reach.
- Use handling aids where appropriate.
- Keep the workspace clear.
- Get a good grip on the load.
- Test the weight before trying to move the load.
- Get the load close to the body.
- Place the feet close to the load.
- Stand in a stable position with the feet pointing in direction of movement.
- Lift mostly by straightening the legs.
- Do not twist the back or bend sideways.
- Get help when needed.
- Do not lift or lower with the arms extended.
- Do not continue heaving when the load is too heavy.

Load Limits for Manual Material Handling

NIOSH has established a recommended weight limit (RWL) for manually lifting and lowering loads (*NIOSH Guide to Manual Lifting*). The maximum load under the most favorable circumstances is 51 lbm (23 kg). This limit is the maximum load that 90% of American workers, male or female, physically fit, and accustomed to physical labor, can lift and lower without injury. The maximum load of 51 lbm (23 kg) is reduced when the most favorable conditions are not achieved.

NIOSH provides an equation to reduce the maximum load based on six less-than-favorable conditions, as follows.

$$RWL = (LC)(HM)(VM)(DM)(AM)(FM)(CM)$$

46.20

These multipliers can be determined as described below.

$$HM = \frac{25}{H} \qquad \text{[SI]} \qquad 46.21(a)$$

$$HM = \frac{10}{H} \qquad \text{[U.S.]} \qquad 46.21(b)$$

H is the horizontal distance between the object and the ankles at the start and end of the lift, in cm (must be between 25 cm and 63 cm).

$$VM = 1 - 0.003|V - 75| \qquad \text{[SI]} \qquad 46.22(a)$$

$$VM = 1 - 0.0075|V - 30| \qquad \text{[U.S.]} \qquad 46.22(b)$$

V is the height of the hands above the floor at the start and end of the lift, in cm (must be between 25 cm and 175 cm).

$$DM = 0.82 + \frac{4.5}{D} \qquad \text{[SI]} \qquad 46.23(a)$$

$$DM = 0.82 + \frac{1.8}{D} \qquad \text{[U.S.]} \qquad 46.23(b)$$

D is the vertical travel distance from start to end of lift, in cm.

$$AM = 1 - 0.0032A \qquad 46.24$$

A is the angular displacement of the load that forces the worker to twist the body from start to end points of the lift projected on the floor, in degrees.

The frequency multiplier, FM, is specified in Table 5 of the *NIOSH Guide to Manual Lifting*. Some illustrative frequency multipliers are in Table 46.13.

Table 46.13 *Illustrative Frequency Multipliers (V < 30 cm)*

	frequency of lift		
	0.2/min	1/min	5/min
up to 1 h	1.0	0.94	0.80
up to 2 h	0.95	0.88	0.60
up to 8 h	0.85	0.75	0.35

The coupling multiplier, CM, is given in a table in the *NIOSH Guide to Manual Lifting*. CM varies from 1.0 for good couplings to 0.90 for poor couplings.

Use of the NIOSH equation for calculating the recommended weight limit is illustrated in Ex. 46.9.

Example 46.9

Workers in a laboratory are engaged in a project to analyze the contaminants in soil collected from a construction site. Each worker at the lab will need to lift the soil samples from the floor to the workbench. Once picked up, the samples are not placed directly on the table in front, but may be put off to the side. The average angular offset from a straight-ahead ending

location is 20°. The work to be performed by each worker is as follows.

lot	quantity	duration	mass
A	12	1 h	12 kg
B	60	1 h	18 kg
C	300	6 h	8 kg

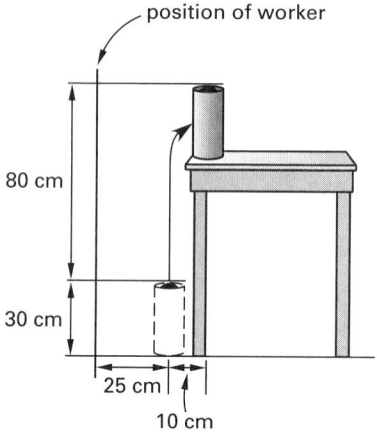

Assume good coupling due to easy-to-use handles. Will any of the work exceed the NIOSH recommended weight limit?

Solution

Use Eq. 46.20.

$$RWL = (LC)(HM)(VM)(DM)(AM)(FM)(CM)$$
$$LC = 23 \text{ kg}$$
$$HM = \frac{25}{H}$$

At the start of the lift, HM = 25/25 = 1.0. At the end of the lift, HM = 25/35 = 0.71.

$$VM = 1 - 0.003|V - 75|$$

At the start of the lift, $V = 30$ cm.

$$VM = 1 - 0.003|30 - 75| = 0.865 \quad \text{[start]}$$

At the end of the lift, $V = 110$ cm.

$$VM = 1 - 0.003|110 - 75| = 0.895 \quad \text{[end]}$$
$$DM = 0.82 + \frac{4.5}{D}$$
$$D = 110 \text{ cm} - 30 \text{ cm} = 80 \text{ cm}$$
$$DM = 0.82 + \frac{4.5}{80} = 0.876 \quad (0.88)$$
$$AM = 1 - 0.0032A$$
$$A = 20°$$
$$AM = 1 - (0.0032)(20) = 0.936 \quad (0.94)$$

For FM,

lot	quantity	duration	frequency	FM
A	12	1 h	0.2/min	1.0
B	60	1 h	1/min	0.94
C	300	6 h	0.83/min	0.77

CM = 1.0 (good handles)

Calculate RWL for the start and end of the lift for each lot.

lot	LC	HM	VM	DM	AM	FM	CM	RWL
A, start	23	1.00	0.87	0.88	0.94	1.00	1.0	16.6
A, end	23	0.71	0.90	0.88	0.94	1.00	1.0	12.2
B, start	23	1.00	0.87	0.88	0.94	0.94	1.0	15.6
B, end	23	0.71	0.90	0.88	0.94	0.94	1.0	11.4
C, start	23	1.00	0.87	0.88	0.94	0.75	1.0	12.7
C, end	23	0.71	0.90	0.88	0.94	0.75	1.0	9.3

The RWL is exceeded for Lot B (18 kg v. RWL of 11.4 kg).

Cumulative Trauma Disorders

Cumulative trauma disorders (CTDs) can occur in almost any work situation. CTDs result from repeated stresses that are not excessive individually but, over time, cause disorders, injuries, and the inability to perform a job. High repetitiveness, or continuous use of the same body part, results in fatigue followed by cumulative muscle strain. These cumulative injuries are usually incurred by tendons, tendon sheaths, and soft tissue. Moreover, the cumulative injuries can result in damage to nerves and restricted blood flow. CTDs are common in the hand, wrist, forearm, shoulder, neck, and back. Bone and the spinal vertebrae may also be damaged.

The manifestations of CTDs on soft tissues include stretched and strained muscles, rough or torn tendons, inflammation of tendon sheaths, irritation and inflammation of bursa, and stretched (sprained) ligaments. Nerves can be affected by pressure from tendons or other soft tissue, resulting in loss of muscle control, numbness, tingling, or pain, and loss of response of nerves that control automatic functions such as body temperature and sweating. Blood vessels may be compressed, resulting in restricted blood flow and impaired control of tissues (muscles) dependent on that blood supply. Vibration, such as from operating vibrating tools, can cause the arteries in the fingers and hands to close down, resulting in numbness, tingling, and eventually loss of sensation and control.

Industrial hygienists have defined *high repetitiveness* as a cycle time of less than 30 s, or more than 50% of a cycle time spent performing the same fundamental motion. If the work activity requires the muscles to remain contracted at about 15–20% of their maximum capability, circulation can be restricted, which also contributes to CTDs. Also, severe deviation of the wrists, forearms, and other body parts can contribute to CTDs.

9. TYPES OF CTDS

This section describes the characteristics of common CTDs and the activities that pose a risk for CTDs. Note that environmental engineers should obtain professional help from a qualified industrial hygienist whenever a potential risk of CTDs is recognized in the workplace. The guidelines described in this section should be considered warning signs for practices that have CTD risk, but should not be used for definitive evaluation.

Carpal Tunnel Syndrome

Carpal tunnel syndrome (CTS) is the best known CTD. The American Industrial Hygiene Association has described CTS as an occupational illness of the hand and arm system. CTS is recognized as a disabling condition of the hand that can be caused or aggravated by certain work activities. CTS results from rapid, repetitious finger and wrist movements.

The wrist has a "tunnel" created by the carpal bones on the outer side and ligaments, which are firmly attached to the bones, across the inner side. In the carpal tunnel, which is roughly oval in shape, are tendons and tendon sheaths of the fingers, several nerves, and arteries. If the wrist is bent up or down or flexed from side to side, the space in the carpal tunnel is reduced. Swelling of the tendons or tendon sheaths can place pressure on the nerves, blood vessels, and tendons. Activities that can lead to this disorder include grinding, sanding, hammering, keyboarding, and assembly work.

Cubital Tunnel Syndrome

This disorder occurs from compression of the nerve in the forearm below the elbow and results in tingling, numbness, or pain in the fingers. Leaning over a workbench and resting the forearm on a hard surface or edge typically causes this disorder.

Epicondylitis

This disorder is also known as "tennis elbow" and "golfer's elbow." It results from irritation of the tendons of the elbow. It is caused by forceful wrist extensions, repeated straightening and bending of the elbow, and impacting throwing motions. Activities that can lead to this disorder include hammering, turning screws, and small-parts assembly.

Ganglionitis

Ganglionitis is a swelling of a tendon sheath in the wrist. Activities that can lead to this disorder include grinding, sanding, sawing, cutting, and using pliers and screwdrivers.

Neck Tension Syndrome

This disorder is characterized by an irritation of the muscles of the neck. It commonly occurs after repeated or sustained overhead work. Activities that can lead to this disorder include belt conveyor assembly, keyboarding, small-parts assembly, packing, or carrying loads on the shoulder.

Pronator Syndrome

This disorder compresses a nerve in the forearm. It results from rapid and forceful strenuous flexing of the elbow and wrist. Activities that can lead to this disorder include buffing, grinding, polishing, and sanding.

Tendonitis

This is an inflammation of a tendon where its surface becomes thickened, bumpy, and irregular. Tendon fibers may become frayed or torn. This disorder can result from repetitious, forceful movements, contact with hard surfaces, and vibrations. Activities that can lead to this disorder include punch press operation, assembly work, packaging, and using pliers.

Shoulder tendonitis is irritation and swelling of the tendon or bursa of the shoulder. It is caused by continuous elevation of the arm. Activities that can lead to this disorder include overhead assembly, overhead welding, overhead painting, belt conveyor assembly work, reaching, lifting, and carrying a load on the shoulder.

Tenosynovitis

This disorder is characterized by swelling of tendon sheaths and irritation of the tendon. It is known as *DeQuervain's syndrome* when it affects the thumb. Activities that can lead to this disorder include grinding, polishing, sanding, sawing, cutting, and using screwdrivers.

Trigger finger is a special case of tenosynovitis that results in the tendon of the trigger finger becoming nearly locked so that its forced movement is jerky. It comes from using hand tools with sharp edges pressing into the tissue of the finger or where the tip of the finger is flexed but the middle part is straight.

Thoracic Outlet Syndrome

This disorder is characterized by reduced blood flow to and from the arm due to compression of nerves and blood vessels between the collarbone and the ribs. It results in a numbing of the arm and constrains muscular activities. The disorder results from hyperextension of the arm and flexing of the shoulder by elevating the arm and carrying loads. Activities that can lead to this disorder include grinding, polishing, sanding, overhead assembly, overhead welding, overhead painting, material handling, and carrying heavy loads with arms extended.

Ulnar Artery Aneurysm

This disorder is characterized by a weakening of an artery in the wrist, causing an expansion that presses on the nerve. This often occurs from pounding or pushing with the heel of the hand, as in assembly work.

Ulnar Nerve Entrapment

This disorder involves pressure on a nerve in the wrist. It occurs from prolonged flexing of the wrist and repeated pressure on the palm. Activities that can lead to this disorder include carpentry, brick laying, and using pliers and hammers.

White Finger

This disorder is also known as "dead finger," *Raynaud's syndrome*, or *vibration syndrome*. In this disorder, the finger turns cold and numb, tingles, and loses sensation and control. The cause is insufficient blood supply, which causes the finger to turn white. It results from closure of the arteries due to vibrations. Gripping vibrating tools, especially in the cold, is a common cause. Activities that can lead to this disorder include operating a chain saw, jackhammering, and using vibrating tools.

10. CTD PREVENTION

The following are general guidelines for the prevention of CTDs.

- Recognize early signs of CTDs.
- Analyze suspicious (forceful, repetitive, and long duration) jobs for the risk of CTDs.
- Seek competent advice and review of work practices by a qualified industrial hygienist.
- Redesign the workstation or the work task to avoid activities that can lead to CTDs.
- Use the proper tools for the task.
- Ensure that workers are trained in proper work habits.
- Provide adequate rest breaks.
- Rotate the work or the worker to change work patterns.
- Provide relief workers when needed.

11. BIOLOGICAL HAZARDS

Since at least the mid-nineteenth century, sanitary engineers and public health engineers have dealt with biological organisms in many different situations. More recently, environmental engineers, who embrace a broader range of concerns and responsibilities for the natural and human environment than did their predecessors, need to be aware of biological hazards that may be encountered both directly with and incidentally to projects. This section describes biological hazards that an environmental engineer needs to be familiar with in order to recognize when to obtain the advice of an industrial hygienist or biosafety specialist.

Biological Agents

Approximately 200 biological agents are known to produce infectious, allergenic, toxic, and carcinogenic reactions in workers. These agents and their reactions are as follows.

- Microorganisms (viruses, bacteria, fungi) and the toxins they produce cause infection and allergic reactions.

- Arthropod (crustaceans), arachnid (spiders, scorpions, mites, and ticks), and insect bites and stings cause skin inflammation, systemic intoxication, transmission of infectious agents, and allergic reactions.

- Allergens and toxins from plants cause dermatitis from skin contact, rhinitis (inflammation of the nasal mucus membranes), and asthma from inhalation.

- Protein allergens (urine, feces, hair, saliva, and dander) from vertebrate animals cause allergic reactions.

Also posing potential biohazards are lower plants other than fungi (e.g., lichens, liverworts, and ferns) and invertebrate animals other than arthropods (e.g., parasites, flatworms, and roundworms).

Microorganisms may be divided into prokaryotes and eukaryotes. *Prokaryotes* are organisms having DNA that is not physically separated from its cytoplasm (cell plasma that does not include the nucleus). They are small, simple, one-celled structures, less than 5 μm in diameter, with a primitive nuclear area consisting of one chromosome. Reproduction is normally by binary fission in which the parent cell divides into two daughter cells. All bacteria, both single-celled and multicellular, are prokaryotes, as are blue-green algae.

Eukaryotes are organisms having a nucleus that is separated from the cytoplasm by a membrane. Eukaryotes are larger cells (greater than 20 μm) than prokaryotes, with a more complex structure, and each cell contains a distinct membrane-bound nucleus with many chromosomes. They may be single-celled or multicellular, reproduction may be asexual or sexual, and complex life cycles may exist. This class of microorganisms includes fungi, algae (except blue-green), and protozoa.

Since prokaryotes and eukaryotes have all of the enzymes and biological elements to produce metabolic energy, they are considered organisms.

In contrast, a *virus* does not contain all of the elements needed to reproduce or sustain itself and must depend on its host for these functions. Viruses are nucleic acid molecules enclosed in a protein coat. A virus is inert outside of a host cell and must invade the host cell and use its enzymes and other elements for the virus's own reproduction. Viruses can infect very small organisms such as bacteria, as well as humans and animals. Viruses are 20 μm to 300 μm in diameter.

Smaller than the viruses by an order of magnitude are *prions*, small proteinaceous infectious particles. Prions have properties similar to viruses and cause degenerative diseases in humans and animals.

Infection

The invasion of the body by pathogenic microorganisms and the reaction of the body to them and to the toxins they produce is called an *infection*. Infection may be *endogenous* where microorganisms that are normally present in the body (*indigenous*) at a particular site (such as *E. coli* in the intestinal tract) reach another site (such as the urinary tract), causing infection there.

Infections from microorganisms not normally found on the body are called *exogenous* infections. The mechanisms of entry include inhalation, indirect or direct contact, penetration (e.g., mosquito bites), and ingestion. Sometimes infectious agents can be transmitted from coworkers by inhalation, such as with measles or tuberculosis. Workers who do not show signs of the disease but are able to transmit the infection are known as *carriers*. The effect of the infection is dependent upon the virulence of the infectious agent, the route of infection, and the relative immunity of the worker.

The most common routes of exposure to infectious agents are through cuts, punctures and bites (insect and animal), abrasions of the skin, inhalation of aerosols generated by accidents or work practices, contact between mucous membranes or contaminated material, and ingestion. In laboratory and medical settings, transmission of blood-borne pathogens can occur through handling of blood products and human tissue.

Biohazardous Workplaces and Activities

Although environmental engineers have long been concerned with waterborne diseases and their prevention in the design and operation of water supply and wastewater systems, pathogens may also be encountered in the workplace through air or direct contact.

Microbiology and Public Health Laboratories

Workers in laboratories handling infectious agents have a potential risk of infection from accidental exposure or work practices. The majority of laboratory-acquired infections occur in research laboratories. Research workers often handle concentrated preparations of infectious agents, and some test procedures require complex manipulation of the samples. Also, these workers often handle infected animals that can expose workers to infection from bites or scratches. Laboratory-acquired infections have included brucellosis, Q fever, hepatitis, typhoid fever, tularemia, tuberculosis, human immunodeficiency virus (HIV), hantavirus (Korean hemorrhagic fever), herpes B virus, and Ebola-related filovirus.

Environmental engineers should use special care when dealing with projects involving laboratories that handle infectious microorganisms. Such situations require the support of a qualified industrial hygienist, and in some cases a biosafety hygienist.

Health, Safety, Welfare

Health Care Facilities

Health care facilities—such as hospitals, medical offices, blood banks, and outpatient clinics—present numerous opportunities for exposure to a wide variety of hazardous and toxic substances, as well as to infectious agents. Hospitals usually have a doctor and an infectious disease control nurse or microbiologist to manage and oversee infection control functions. An environmental engineer may become involved with projects involving ventilation systems, containment systems, and handling of infectious wastes at health care facilities. Infectious wastes include blood, blood products, and tissues, and can cause infection from blood-borne pathogens.

Biotechnology Facilities

Biotechnology is one of the newest technologies and involves a much greater scope and complexity than the historical use of microorganisms in the chemical and pharmaceutical industries. This technology now deals with DNA manipulation and the development of products for medicine, industry, and agriculture. The microorganisms used by the biotechnology industry often are genetically engineered plant and animal cells. Allergies can be a major health issue.

Animal Facilities

Workers exposed to animals are at risk for animal-related allergies and infectious agents. Occupations include agricultural workers, veterinarians, workers in zoos and museums, taxidermists, and workers in animal-product processing plants.

Zoonotic diseases (diseases that affect both humans and animals) are the most common diseases reported by laboratory workers. Infection can occur from bites and scratches, puncture with contaminated needles, aerosols from animal respiration or excretions, and contact with infected tissue. Work-acquired infections from nonhuman primates are common.

Some of the diseases of concern in animal facilities include Q fever, hantavirus, Ebola and Marburg viruses, and simian immunodeficiency viruses. An environmental engineer could encounter potential infectious agents while dealing with solid and liquid wastes from animal processing facilities and animal agricultural activities.

Agriculture

Agricultural workers are exposed to infectious microorganisms through inhalation of aerosols, contact with broken skin or mucus membranes, and inoculation from injuries. Farmers and horticultural workers may be exposed to fungal diseases. Food and grain handlers may be exposed to parasitic diseases. Workers who process animal products may acquire bacterial skin diseases such as anthrax from contaminated hides, tularemia from skinning infected animals, and erysipelas from contaminated fish, shellfish, meat, or poultry. Infected turkeys, geese, and ducks can expose poultry workers to *psittacosis*, a bacterial infection. Workers handling grain may be exposed to *mycotoxins* from fungi and *endotoxins* from bacteria.

Environmental engineers or project workers may potentially be exposed to any number of these diseases while working on projects involving the treatment and disposal of solid and liquid wastes, such as through waste treatment processes and land application of wastewater. Landfills and composting facilities may also handle wastes containing contaminated agricultural and animal wastes or by-products.

Utility Workers

Workers maintaining water systems may be exposed to *Legionella pneumophila* (Legionnaires' disease). Sewage collection and treatment workers may be exposed to enteric bacteria, hepatitis A virus, infectious bacteria, parasitic protozoa (*Giardia*), and allergenic fungi.

Solid waste handling and disposal facility workers may be exposed to blood-borne pathogens from infectious wastes.

Wood-Processing Facilities

Wood-processing workers may be exposed to bacterial endotoxins and allergenic fungi. Environmental engineers may be involved with disposal of wastewater or solid wastes that contain these microorganisms.

Mining

Miners may be exposed to zoonotic bacteria, mycobacteria, fungi, and untreated runoff water and wastewater.

Forestry

Forestry workers may be exposed to zoonotic diseases (rabies virus, Russian spring fever virus, Rocky Mountain spotted fever, Lyme disease, and tularemia) transmitted by ticks and fungi. Environmental engineers may be involved with forestry activities associated with wetlands protection, land application of wastewater, landfills, and other projects where such infectious agents are present.

Blood-Borne Pathogens

Environmental engineers should have a general knowledge of blood-borne pathogens and be able to recognize when expert advice should be obtained on a project involving exposure to infectious wastes.

The risk from hepatitis B and human immunodeficiency virus (HIV) in health care and laboratory situations led OSHA to publish standards for occupational exposure to blood-borne pathogens. Some blood-borne pathogens are summarized as follows.

Health, Safety, Welfare

Human Immunodeficiency Virus (HIV)

HIV is the blood-borne virus that causes acquired immunodeficiency syndrome (AIDS). Contact with infected blood or other body fluids can transmit HIV. Transmission may occur from unprotected sexual intercourse, sharing of infected needles, accidental puncture wounds from contaminated needles or sharp objects, or transfusion with contaminated blood.

Symptoms of HIV include swelling of lymph nodes, pneumonia, intermittent fever, intestinal infections, weight loss, and tuberculosis. Death typically occurs from severe infection causing respiratory failure due to pneumonia.

Hepatitis

The hepatitis virus affects the liver. Symptoms of infection include jaundice, cirrhosis and liver failure, and liver cancer.

Hepatitis A can be contracted through contaminated food or water or by direct contact with body fluids such as blood or saliva. Hepatitis B, known as *serum hepatitis*, may be transmitted through contact with infected blood or other body fluids and through blood transfusions. Hepatitis B is the most significant occupational infector of health care and laboratory workers. Hepatitis C is similar to hepatitis B, but can also be transmitted by shared needles, accidental puncture wounds, blood transfusions, and unprotected sex.

Hepatitis D requires one of the other hepatitis viruses to replicate. Individuals with chronic hepatitis D often develop cirrhosis of the liver. Chronic hepatitis may be present in carriers.

Syphilis

The bacterium responsible for the transmission of syphilis is *Treponema pallidum* (subspecies *pallidum*). Syphilis is most often transmitted through sexual contact, but may also be transmitted from mother to fetus. Symptoms begin with an open sore in the genital region, followed by a rash on the palms and soles of the feet, fever, headache, and loss of appetite. The final stage produces degeneration of the aorta of the heart and affects the central nervous system, producing psychosis. There may be years between stages of the disease.

Toxoplasmosis

Toxoplasmosis is caused by a parasitic organism called *Toxoplasma gondii*, which may be transmitted by ingestion of contaminated meat, across the placenta, and through blood transfusions and organ transplants. The organism can be shed in the fecal matter of infected pets. Infection of pregnant women is of great concern with this disease, which can include blindness, mental retardation, jaundice, anemia, and central nervous system disorders.

Rocky Mountain Spotted Fever

Ticks infected with the pathogen *Rickettsia rickettsii* pass this disease from pets and other animals to humans. Symptoms and effects include headache, rash, fever, chills, nausea, vomiting, cardiac arrhythmia, and kidney dysfunction. Death may occur from renal failure and shock.

Bacteremia

Bacteremia is the presence of bacteria in the bloodstream, whether associated with active disease or not. The bacteria may be introduced through contamination during transfusions, through surgical procedures that allow bacteria to be released from sequestered sites, and from routine medical procedures. Fever is the most common symptom, and severe cases can lead to death.

Bacteria- and Virus-Derived Toxins

Some toxins are derived from bacteria and viruses. The effects of these toxins vary from mild illness to debilitating illness or death.

Botulism

The organism *Clostridium botulinum* produces the toxin that is responsible for botulism. There are four types of botulism: food-borne, infant, adult enteric (intestinal), and wound. *Food botulism* is associated with poorly preserved foods and is the most widely recognized form. *Infant botulism* can occur in the second month after birth when the bacteria colonize the intestinal tract and produce the toxin. *Adult enteric botulism* is similar to infant botulism. *Wound botulism* occurs when the spores enter a wound through contaminated soil or needles. The toxin is absorbed in the bloodstream and blocks the release of a neurotransmitter. Severe cases can result in respiratory paralysis and death.

Lyme Disease

Lyme disease is transmitted to humans through bites of ticks infected with *Borrelia burgdorferi*. Lyme disease follows three stages of development. Stage I involves transmission of the organism throughout the body through the bloodstream. Symptoms include fever, chills, fatigue, headache, pneumonia, hepatitis, and meningitis. Stages II and III result in neurological symptoms, meningitis, and encephalitis, and they produce an irregular heartbeat. Stage III results in arthritis of the large joints.

Tetanus

Tetanus occurs from infection by the bacterium *Clostridium tetani*, which produces two exotoxins, tetanolysin and tetanospasmin. These toxins affect the red blood cells and the nervous system. Symptoms include muscle spasms and a constant state of muscle contraction. Routine immunizations prevent the disease.

Health, Safety, Welfare

Toxic Shock Syndrome

Toxic shock syndrome (TSS) is caused by the bacterium *Staphylococcus aureus*, which produces a pyrogenic toxin. Symptoms include sore throat, fever, fatigue, headache, vomiting, diarrhea, low blood pressure, and rash. Shock or kidney failure can occur, which can be lethal. TSS may occur in children and men as well as in menstruating females.

Ebola (African Hemorrhagic Fever)

The Ebola and Marburg viruses produce an acute hemorrhagic fever in humans. Symptoms include headache, progressive fever, sore throat, and diarrhea. Later, hepatitis and bleeding from the mucus membranes and internal organs occur. Blood does not coagulate and shock may develop; death usually occurs by the ninth day.

Hantavirus

The hantavirus is found in rodents and shrews of the southwest and is spread by contact with their excreta. Symptoms during the early phase of the disease include chills, fever, headache, backache, weakness, and general malaise. Later, gastrointestinal bleeding and hypotension occur, followed by reduced and then increased amounts of urine being produced. Death may result without proper treatment.

Tuberculosis

Tuberculosis (TB) is a bacterial disease from *Mycobacterium tuberculosis*. Humans are the primary source of infection. TB affects a third of the world's population outside the United States. A drug-resistant strain is a serious problem worldwide, including in the United States. The risk of contracting active TB is increased among HIV-infected individuals. The usual route of transmission is by inhalation of infectious droplets suspended in air from coughing, sneezing, singing, or talking by an infected person. Symptoms include fatigue, fever, weight loss, hoarseness, cough, and blood in sputum.

Legionnaires' Disease

Legionnaires' disease (legionellosis) is a type of pneumonia caused by inhaling the bacteria *Legionella pneumophilia*. Symptoms include fever, cough, headache, muscle aches, and abdominal pain. People usually recover in a few weeks and suffer no long-term consequences. *Legionellae* are common in nature and are associated with heat-transfer systems, warm-temperature water, and stagnant water. Sources of exposure include sprays from cooling towers or evaporative condensers and fine mists from showers and humidifiers. Proper design and operation of ventilation, humidification, and water-cooled heat-transfer equipment and other water systems equipment can reduce the risk. Good system maintenance includes regular cleaning and disinfection.

47

Health and Safety

Nomenclature

C	concentration	ppm, kg/m^3
CV	coefficient of variation including sampling error	–
LCL	lower confidence limit	–
m	mass	kg
MW	molecular weight	g/mol
n	number of moles	–
p	pressure	atm
PEL	permissible exposure limit	ppm, mg/m^3
Q	flow rate	m^3/s
R^*	universal gas constant	L·atm/mol·K
s	sample standard deviation	–
SAE	sampling and analytical errors	–
t	time	s
T	temperature	K
TLV	threshhold limit value	ppm, mg/m^3
UCL	upper confidence limit	–
V	volume	L, m^3
x^*	solution phase mole fraction	–

Symbols

ρ	density	mg/m^3

Subscripts

c	calibration conditions
d	dilution
f	final condition
F	field
i	initial condition
PVP	partial vapor
S	standard
TVP	total vapor
v	variation
VP	vapor
w	water

1. INTRODUCTION

The environmental engineer will most likely encounter health and safety concerns while working with remediation of hazardous waste sites. Although health and safety are always a concern in the workplace, especially around industrial process equipment or construction, remediation of hazardous waste sites involves the purposeful exposure of workers to known and unknown hazardous substances. This chapter will focus primarily on health and safety related to work at hazardous waste sites.

Hazardous waste site remediation work can expose workers to many types of physical, chemical, and biological hazards. Such hazards include

- heat and cold stress
- noise
- ionizing radiation
- cumulative trauma disorders
- biological hazards
- electrical hazards
- physical safety hazards
- fire and explosion
- oxygen deficiency
- chemical exposure

Heat and cold stress, noise, ionizing radiation, cumulative trauma disorders, and biological hazards are covered in Chap. 46. The environmental engineer can usually defer electrical and physical safety hazards to general safety personnel, so these hazards will not be covered in this chapter. Explosion hazards are covered in Chap. 51. The principal focus of this chapter is on chemical exposure; fire and explosion from gases and vapors along with oxygen deficiency are also covered briefly.

This chapter describes chemical hazard assessment, health and safety plans for hazardous chemicals, air monitoring for chemical gases and vapors, and personal protective equipment.

2. HAZARD ASSESSMENT

From an OSHA (legal) standpoint, a *health hazard* is a chemical, mixture of chemicals, or pathogen for which there is statistically significant evidence based on at least one study conducted in accordance with established scientific principles that acute or chronic health effects may occur in exposed employees. Health hazards include chemicals that are carcinogens, toxic or highly toxic agents, reproductive toxins, irritants, corrosives, sensitizers, hepatotoxins (affecting the liver), nephrotoxins (affecting the kidneys), neurotoxins (affecting the nervous system), and agents that act on the hematopoietic system (the production of blood cells), along with agents that damage the lungs, skin, eyes, or mucous membranes.

One of the first tasks associated with remediation at a hazardous waste site is assessment of the hazards that must be dealt with so that the necessary precautions (such as the use of personal protective equipment) can be taken. In addition, an assessment must be made to identify conditions that may be immediately dangerous to life or health (IDLH). Chemicals pose three primary types of hazards: exposure to chemicals, fire or explosion, and oxygen deficiency.

Chemical Exposure

Chemical exposure can be acute or chronic. *Acute exposure* is exposure to a high dose over a short period of time (hours or a few days) with symptoms appearing during or within a short time (a few days to two weeks) after exposure. *Chronic exposure* is exposure to a low dose over a long period (months or years) with the onset of symptoms not observable for years or even decades after exposure. Because a toxic chemical may be colorless and odorless (or may desensitize a person's sense of smell (e.g., hydrogen sulfide)), hazard assessment should not depend on the ability of the worker to sense the chemical or feel discomfort.

The primary exposure route for chemicals is inhalation. Toxic chemicals can readily enter the bloodstream through the lungs, a fact that makes this exposure route extremely significant to workers at hazardous waste remediation sites. Another important exposure route is through contact with the skin (dermal) or the eyes. Chemicals can directly injure the skin or the eyes or can pass through the skin or eyes into the bloodstream to attack target organs or tissues anywhere in the body. Exposure via ingestion, although unlikely, can result from chewing gum, smoking, drinking, or eating at a hazardous waste site. Exposure through injection can occur through puncture wounds sustained while working with materials at the site.

Chemical hazard assessment involves determining the concentrations of chemicals present and their associated health hazards. There are four primary sources of information about chemical hazards, which are described in detail in Chap. 45. These sources are

- *Integrated Risk Information System* (IRIS): an EPA database
- *Threshold Limit Values for Chemical Substances and Physical Agents, Biological Exposure Indices*: published annually by the American Conference of Governmental Industrial Hygienists (ACGIH)

- *NIOSH Pocket Guide to Chemical Hazards*: published by the National Institute for Occupational Safety and Health (NIOSH)
- 29 CFR 1910.1000, Subpart Z, *Toxic and Hazardous Substances*: workplace exposure limits established by the Occupational Safety and Health Administration

Because workers performing remediation at a hazardous waste site will need to meet the requirements of the Occupational Safety and Health Act, the environmental engineer should first review the workplace exposure limits of 29 CFR 1910.1000, Subpart Z, *Toxic and Hazardous Substances*, in assessing the chemical exposure hazard. These exposure limits establish the legal air contaminant levels to which workers can safely be exposed over an 8 h workday without personal protection.

The other three sources of data on air contaminants should also be reviewed because they may be more restrictive. The OSHA standards may be more or less restrictive than the NIOSH-, ACGIH-, or EPA-recommended air contaminant levels, so the environmental engineer should understand the differences in order to be able to make a conservative (protective) decision on acceptable exposure levels.

Fire or Explosion

The causes of fires and explosions at hazardous waste sites include chemical reactions, ignition of flammable or explosive chemicals, ignition of materials due to oxygen enrichment, disturbance of shock-sensitive materials, and the sudden release of pressure of contained chemicals. Common causes of fire or explosion at hazardous waste sites are from handling containers or mixing chemicals, or exposing chemicals (usually as vapors) to sparks from equipment.

Oxygen Deficiency

The term *oxygen deficiency* refers to a concentration of oxygen below which atmosphere-supplying respiratory protection must be provided. This condition exists in atmospheres where the percentage of oxygen by volume is less than 19.5%.

Oxygen deficiency can result from the displacement of normal air, which contains 21% oxygen by volume at sea level, with other gases present at the hazardous waste site. Oxygen deficiency is an especially important hazard in enclosed spaces, such as tanks or buildings, or in low-lying areas, such as in ravines, ditches, or depressions at landfills. Although the main physiological effects of oxygen deprivation become apparent at 16% oxygen, in order to account for differences in individuals and measurement errors, air with 19.5% or less oxygen is considered the level at which personal protection is required. Physiological symptoms of oxygen deficiency include impaired attention and judgment and increased breathing and heart rate.

Example 47.1

Workers are remediating an old warehouse that contains a variety of chemicals. The warehouse is made of concrete and has little ventilation. Approximately 7 kg of a combustible carbon material was accidentally burned in the warehouse a few days ago. Assume that 60% of the carbon formed carbon dioxide and 40% formed carbon monoxide. The warehouse contains a maximum of 2 kmol (i.e., 2000 mol or 2 kg·mol) of oxygen in air. Determine whether the air in the warehouse is oxygen deficient.

Solution

Air is approximately 21% O_2 and 79% N_2 by volume. The combustion occurs as follows.

$$C + O_2 \longrightarrow CO_2$$

$$2C + O_2 \longrightarrow 2CO$$

For CO_2, the number of moles of O_2 consumed per mole of carbon is

$$(0.6)\left(\frac{7 \text{ kg C}}{12 \frac{\text{kg C}}{\text{kmol C}}}\right)\left(1 \frac{\text{kmol O}_2}{\text{kmol C}}\right)$$
$$= 0.35 \text{ kmol O}_2 \text{ consumed}$$

For CO, the number of moles of O_2 consumed per mole of carbon is

$$(0.4)\left(\frac{7 \text{ kg C}}{12 \frac{\text{kg}}{\text{kmol C}}}\right)\left(\frac{1 \text{ kmol O}_2}{2 \text{ kmol C}}\right)$$
$$= 0.12 \text{ kmol O}_2 \text{ consumed}$$

The total O_2 consumed is

$$0.35 \text{ kmol} + 0.12 \text{ kmol} = 0.47 \text{ kmol O}_2$$

Calculate the concentration of each gas component in the air.

$$(0.35 \text{ kmol O}_2)\left(1 \frac{\text{kmol CO}_2}{\text{kmol O}_2}\right) = 0.35 \text{ kmol CO}_2$$

$$(0.12 \text{ kmol O}_2)\left(2 \frac{\text{kmol CO}}{\text{kmol O}_2}\right) = 0.23 \text{ kmol CO}$$

$$2 \text{ kmol O}_2 - 0.47 \text{ kmol O}_2 \text{ consumed}$$
$$= 1.53 \text{ kmol O}_2 \text{ remaining}$$

$$(2 \text{ kmol O}_2)\left(\frac{0.79 \text{ kmol N}_2}{0.21 \text{ kmol O}_2}\right) = 7.52 \text{ kmol N}_2$$

The total moles of gases is

$$0.35 \text{ kmol CO}_2 + 0.23 \text{ kmol CO}$$
$$+ 1.53 \text{ kmol O}_2 + 7.52 \text{ kmol N}_2$$
$$= 9.64 \text{ kmol}$$

The mole fraction of O_2 is

$$\frac{1.53 \text{ kmol O}_2}{9.64 \text{ kmol total}} = 0.159 \quad (15.9\%)$$

Because the mole fraction of O_2 is less than 19.5%, the atmosphere in the warehouse is oxygen deficient.

Identification of IDLH Conditions

The phrase "immediately dangerous to life or health" (IDLH) refers to an atmospheric concentration of any toxic, corrosive, or asphyxiant substance that poses an immediate threat to life or would interfere with an individual's ability to escape from a dangerous atmosphere. It is extremely important that conditions that may be immediately dangerous to life or health are systematically identified in the initial hazard assessment. There are several indicators of conditions that may be IDLH.

Tanks that must be entered by remediation personnel are prime candidates for IDLH conditions because vapors may be present that are toxic or explosive or that create an oxygen-deficient environment. Even tanks that are open at the top can present an IDLH hazard because some gases are heavier than air and can accumulate in the bottom of the tank. Buildings and trenches that must be entered present similar types of IDLH hazards.

Flammable or explosive conditions are also IDLH hazards. The presence of gas-generating materials, such as from incompatible mixtures or as suggested by bulging drums, should be regarded as potential for flammable or explosive conditions. Instrument readings can also indicate the presence of such conditions.

The presence of extremely toxic materials, such as cyanide or phosgene, also presents potential IDLH conditions. Depending on concentration, many other chemicals can present IDLH hazards. The National Institute for Occupational Safety and Health (NIOSH) *Pocket Guide to Chemical Hazards* gives the IDLH concentration for many chemicals and also gives flammable and explosive concentrations (see Chap. 45). Appendix 47.A gives typical IDLH values for many chemicals.

Visible vapor clouds should raise concerns about IDLH conditions, as should the presence of dead vegetation or dead animals.

Site Control Zones

Hazardous waste remediation and emergency response sites are typically divided into zones for control of personnel entry and personal protective equipment requirements, as well as for designation of the activities that

Health, Safety, Welfare

can be conducted in each zone. The EPA nomenclature for the various zones and perimeters is shown in Fig. 47.1 along with other common terms for the same items.

Figure 47.1 *Site Control Zone Nomenclature*

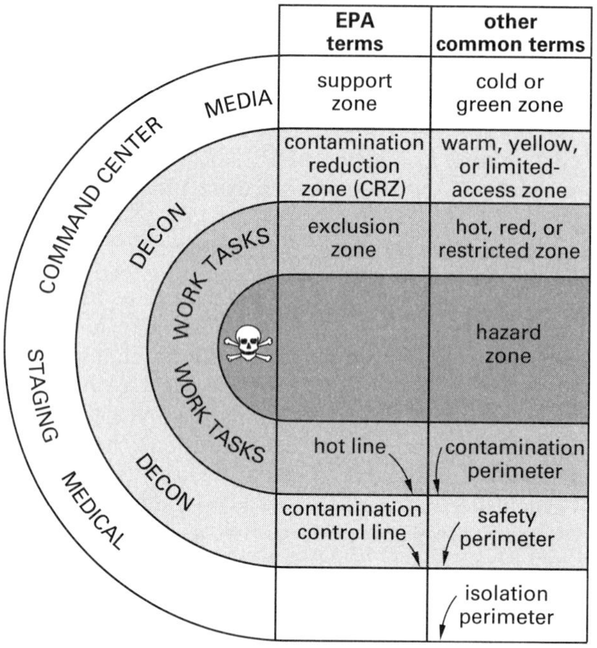

EPA terms	other common terms
support zone	cold or green zone
contamination reduction zone (CRZ)	warm, yellow, or limited-access zone
exclusion zone	hot, red, or restricted zone
	hazard zone
hot line	contamination perimeter
contamination control line	safety perimeter
	isolation perimeter

3. HEALTH AND SAFETY PLANS

Often, an environmental engineer will be in charge of remediation at a hazardous waste site and will be assisted by a health and safety officer. The health and safety officer will normally prepare a health and safety *plan* and be responsible for conducting the health and safety *program*. (That is, the "program" is the implementation of the plan.) However, the environmental engineer needs to understand the elements comprising the health and safety plan and program so he or she can ensure that work tasks are conducted accordingly. The environmental engineer must also be able to identify changing conditions that require revisions or exceptions to a plan or program.

Employers are required to develop and implement a written health and safety plan for their employees involved in hazardous waste operations. The program must be designed to identify, evaluate, and control health and safety hazards, and provide for emergency response for hazardous waste operations. The written health and safety plan must be made available to any contractor or subcontractor who will be involved with the hazardous waste operation; to employees or union representatives; to OSHA personnel; and to personnel of federal, state, or local agencies with regulatory authority over the site.

The health and safety plan must be in writing and address the health and safety hazards of each phase of site

operation and include the requirements and procedures for employee protection. The plan must include at least the following.

- a health and safety risk or hazard analysis for each site task and operation found in the work plan
- employee training assignments
- personal protective equipment to be used by workers for each of the site tasks and operations being conducted
- medical surveillance requirements
- frequency and types of air monitoring, personnel monitoring, and environmental sampling techniques and instrumentation to be used, including methods of maintenance and calibration of monitoring and sampling equipment
- site control measures
- decontamination procedures
- an emergency response plan for safe and effective responses to emergencies, including the necessary personal protective equipment (PPE) and other equipment
- confined space entry procedures
- a spill-containment program
- requirements for pre-entry briefings

The site-specific health and safety plan must provide for pre-entry briefings prior to initiating any site activity, and whenever needed to ensure that workers understand the plan and that the plan is being followed.

The health and safety supervisor is responsible for conducting inspections to determine the effectiveness of the site health and safety plan and to ensure that any deficiencies in the plan are corrected.

4. AIR MONITORING

Ambient air monitoring should include monitoring for toxic substances, combustible gases, inorganic and organic particulates and vapors, oxygen deficiency, and specific chemicals suspected to be present. Air monitoring is necessary both to assess air contaminant hazards at the site for which personal protection may be required and to monitor ongoing exposure to air contaminants so the medical monitoring requirements can be determined. Air monitoring will provide information for preparing a health and safety plan that will define the location where specific levels of personal protection are needed and the degree of hazards present.

Direct-Reading Instruments

Direct-reading air-monitoring instruments are commonly used at hazardous waste remediation sites to determine the level of personal protection needed, identify the types of instruments needed to further characterize the air contaminants, and provide information needed to prepare the health and safety plan. Also, these types of instruments can detect conditions that are immediately dangerous to life and health (IDLH).

Direct-reading instruments are designed for, and are limited to, measuring high concentrations of air contaminants. They are unreliable at detecting low concentrations. Also, though designed to detect air contaminants relative to specific chemicals, these instruments also detect other chemicals, thereby giving false readings.

OSHA has established a database of chemical sampling information, examples of which are given in App. 47.B. The database provides general information about chemicals, including exposure limits, health factors, and monitoring requirements and methods. Similar information can also be obtained from the NIOSH *Manual of Analytical Methods* and from the OSHA *Analytical Methods*.

Limit of Detection

Published sampling procedures give minimum sampling requirements. Small samples may not contain enough of the contaminant to produce a signal distinguishable from the instrument's background noise.

Typically, the *signal-to-noise ratio* ranges from 3:1 to 5:1. The chemical mass that produces a 3:1 signal-to-noise ratio is known as the *limit of detection* (LOD). The LOD is defined as the lowest concentration that can be determined to be statistically different from a blank. For example, if the LOD of a contaminant is 20 ng, the laboratory would report a finding of less than 20 ng as ND or "not detected." Although the actual mass and concentration of the contaminant cannot be determined from such a result, it is possible to use the LOD mass and the sample volume to compute an estimated concentration, which can be useful in the absence of better data. If the LOD concentration is less than 25% of the threshold limit value (TLV) at a specified mass (and preferably less than 10% of the TLV), the exposure can be considered acceptable. If the LOD concentration is more than 25% of the TLV, it would be good practice to resurvey the exposure area.

Limit of Quantification

The *limit of quantification* (LOQ) is the minimum mass of a contaminant above which the precision of a reported result is better than a specified level. *Precision* is the reproducibility of replicate analyses of a sample with the same mass or concentration. Precision refers to the size of deviations from the mean observation. Numerically, precision is the *coefficient of variation* (CV), which is calculated by dividing the estimated standard deviation by the mean of a series of measurements. The ratio of the standard deviation to the mean is a decimal, and when multiplied by 100 can be reported as a percent precision.

Normal and Standard Conditions

After air monitoring is performed, the results are compared with one or more criteria. In industrial hygiene air monitoring, the criteria are based on a defined set of conditions called *normal* conditions, or *normal temperature and pressure* (NTP). This set of conditions differs from the set of conditions used in basic science and engineering called *standard* conditions, or *standard temperature and pressure* (STP). NTP may be called "standard" conditions in the context of industrial hygiene or air monitoring, but note the difference between NTP conditions and STP conditions listed in Table 47.1.

Table 47.1 *Properties of Atmospheric Air*

parameter	NTP	STP
temperature	25°C = 298.15K	0°C = 273.15K
pressure	1 atm, 760 mm Hg, 101 325 Pa	1 atm, 760 mm Hg, 101 325 Pa
dry air molecular weight	28.96 g/mol	28.96 g/mol

Often it is necessary to convert the volume of gases and vapors (or air) from field conditions to normal conditions and vice versa. The conversions are based on the ideal gas law. The ideal gas equation can be written at two different environmental conditions, standard condition (S) and field condition (F).

$$V_F = V_S \left(\frac{p_S}{p_F}\right)\left(\frac{T_F}{T_S}\right) \qquad 47.1$$

$$V_S = V_F \left(\frac{p_F}{p_S}\right)\left(\frac{T_S}{T_F}\right) \qquad 47.2$$

Concentration Conversions

For typical industrial hygiene air monitoring, gases and vapors are presumed to behave as ideal gases. The *ideal gas equation* can be used to calculate the density of a presumed ideal gas.

$$pV = nR^*T \qquad 47.3$$

The *number of moles* is the mass divided by the molecular weight.

$$n = \frac{m}{\text{MW}} \qquad 47.4$$

Density is the mass of the gas or vapor divided by its volume.

$$\rho = \frac{m}{V} \qquad 47.5$$

Concentration is the mass of a substance A divided by the volume of substance B.

$$C_{m/V,A} = \frac{m_A}{V_B} \qquad 47.6$$

Concentration can also be expressed volumetrically as the volume of substance A to 10^6 volumes of substance A plus B, or parts per million (ppm or ppmv).

$$C_{\text{ppm,A}} = \left(\frac{V_A}{V_A + V_B}\right)(10^6) \qquad 47.7$$

However, in virtually all industrial hygiene or air sampling of gas and vapor monitoring, the volume of substance A (the gas or vapor contaminant) is small compared to the volume of substance B (typically air), so Eq. 47.7 can be simplified to eliminate the volume of substance A in the denominator. This assumption results in an error of only about 1% at 10 000 ppm and much less at the lower concentrations typically encountered. The simplified equation is

$$C_{\text{ppm,A}} \approx \left(\frac{V_A}{V_B}\right)(10^6) \qquad 47.8$$

The density can be found by combining Eq. 47.3 through Eq. 47.5.

$$\rho_A = \frac{m_A}{V_A} = \frac{p(\text{MW}_A)}{R^*T} \qquad 47.9$$

Equation 47.8 and Eq. 47.9 can be combined to provide a way to convert between concentrations as ppm and mass per volume.

$$C_{\text{ppmv,A}} = \left(\frac{C_{\text{kg/m}^3,A}\,R^*T}{(\text{MW}_A)p}\right)(10^6) \qquad 47.10$$

$$C_{\text{kg/m}^3} = \left(\frac{C_{\text{ppmv,A}}(\text{MW}_A)p}{R^*T}\right)(10^{-6}) \qquad 47.11$$

In the case of air contaminants, a volumetric concentration expressed in ppmv may be converted to a mass-per-unit volume concentration (e.g., mg/m^3), which is still considered to be a volumetric concentration. In some cases, concentrations calculated as ratios of two masses will also be reported in parts per million. These gravimetric ratios will be distinguished from volumetric ratios by the designation "ppmw" (i.e., parts per million by weight).

Example 47.2

Workers at a hazardous waste site remediation project are expected to encounter toluene ($C_6H_5CH_3$) vapors, so the project manager decides to perform a preliminary survey to collect and analyze air samples to determine the level of personal protection needed. The lab provides the following data on toluene from NIOSH Method No. 1500.

conditions	NTP
ppm conversion	1 ppm = 3.77 mg/m^3 at NTP
range studied	548–2190 mg/m^3 or
	1.13–4.51 mg/sample
ACGIH TLV-TWA	188 mg/m^3
estimated LOD	53 μg
air sampling flow rate	≤ 0.2 L/min
air sampling media	100 mg/50 mg charcoal
molecular weight	92 g/mol

A preliminary sample analyzed by the lab resulted in "not detected" (ND) as the mass of toluene measured. The volume collected was 1.8 L, and the field temperature and pressure during sample collection were 32°C and 720 mm Hg, respectively.

(a) Was the TLV-TWA exceeded in the preliminary sample?

(b) If the concentration of toluene is 30 ppm at field conditions, what sample air pump flow rate at field conditions is needed to collect the minimum sample mass in 1 h? Is this flow rate acceptable?

(c) What is the maximum sample volume that should be collected at field conditions if the concentration of toluene is twice the TLV?

Solution

(a) Correct the sample volume to NTP.

At NTP, $T_S = 25°C + 273° = 298$K.

$$p_S = (760 \text{ mm Hg})\left(\frac{1 \text{ atm}}{760 \text{ mm Hg}}\right) = 1 \text{ atm}$$

At field conditions, $T_F = 32°C + 273° = 305$K.

$$p_F = (720 \text{ mm Hg})\left(\frac{1 \text{ atm}}{760 \text{ mm Hg}}\right) = 0.947 \text{ atm}$$

$$V_F = 1.8 \text{ L}$$

Use Eq. 47.2. At NTP,

$$V_S = V_F\left(\frac{p_F}{p_S}\right)\left(\frac{T_S}{T_F}\right) = (1.8 \text{ L})\left(\frac{720 \text{ mm Hg}}{760 \text{ mm Hg}}\right)\left(\frac{298\text{K}}{305\text{K}}\right)$$
$$= 1.67 \text{ L}$$

The maximum value of the reported concentration of toluene is ND, which is 53 μg. Calculate the LOD concentration at NTP.

From Eq. 47.6,

$$C_{\text{LOD}} = \frac{m}{V_S}$$
$$= \left(\frac{53 \text{ } \mu\text{g}}{1.67 \text{ L}}\right)\left(\frac{1 \text{ g}}{10^6 \text{ } \mu\text{g}}\right)\left(10^3 \frac{\text{mg}}{\text{g}}\right)\left(10^3 \frac{\text{L}}{\text{m}^3}\right)$$
$$= 31.8 \text{ mg/m}^3$$

The percent of TLV is

$$\frac{31.8 \frac{\text{mg}}{\text{m}^3}}{188 \frac{\text{mg}}{\text{m}^3}} = 0.17 \quad (17\%)$$

The TLV is not exceeded. The preliminary sample result was 17% of the TLV at NTP. The preliminary result could be considered acceptable exposure, but additional monitoring could also be appropriate.

(b) From Eq. 47.9,

$$\rho_{\text{toluene}} = \frac{m_{\text{toluene}}}{V_{\text{toluene}}} = \frac{p_F(\text{MW}_{\text{toluene}})}{R^*T}$$

$$= \frac{(0.947\text{ atm})\left(92\ \frac{\text{g}}{\text{mol}}\right)}{\left(0.08205\ \frac{\text{L·atm}}{\text{mol·K}}\right)(305\text{K})}$$

$$= 3.48\text{ g/L}$$

Find the mass concentration of toluene at field conditions.

From Eq. 47.8,

$$C_{\text{ppmv,A}} \approx \left(\frac{V_A}{V_B}\right)(10^6)$$

$$\frac{V_A}{V_B} \approx \frac{C_{\text{ppmv,A}}}{10^6}$$

$$\frac{V_{\text{toluene}}}{V_{\text{air}}} \approx \frac{C_{\text{ppmv,toluene}}}{10^6} = \frac{30}{10^6}$$

Use Eq. 47.6.

$$C_{m/V,A} = \frac{m_A}{V_B} = \left(\frac{m_A}{V_A}\right)\left(\frac{V_A}{V_B}\right)$$

$$C_{m/V,\text{toluene}} = \left(\frac{m_{\text{toluene}}}{V_{\text{toluene}}}\right)\left(\frac{V_{\text{toluene}}}{V_{\text{air}}}\right)$$

$$= \left(3.48\ \frac{\text{g}}{\text{L}}\right)\left(\frac{30}{10^6}\right)$$

$$\times \left(10^3\ \frac{\text{mg}}{\text{g}}\right)\left(10^3\ \frac{\text{L}}{\text{m}^3}\right)$$

$$= 104\text{ mg/m}^3 \quad [\text{field conditions}]$$

Find the flow rate at field conditions to collect the minimum mass.

$$m_{\text{sample}} = Q_F C_F t = 1.13\text{ mg}$$

$$Q_F = \frac{1.13\text{ mg}}{C_F t}$$

$$= \frac{(1.13\text{ mg})\left(10^3\ \frac{\text{L}}{\text{m}^3}\right)}{\left(104\ \frac{\text{mg}}{\text{m}^3}\right)(60\text{ min})}$$

$$= 0.180\text{ L/min} \quad [\text{field conditions}]$$

Convert the maximum allowable flow rate at NTP to field conditions and compare with the calculated flow rate needed.

From Eq. 47.1 and NIOSH Method No. 1500,

$$Q_F = Q_S\left(\frac{p_S}{p_F}\right)\left(\frac{T_F}{T_S}\right)$$

$$= \left(0.2\ \frac{\text{L}}{\text{min}}\right)\left(\frac{760\text{ mm Hg}}{720\text{ mm Hg}}\right)\left(\frac{305\text{K}}{298\text{K}}\right)$$

$$= 0.216\text{ L/min}$$

Since the 0.180 L/min calculated is less than the 0.216 L/min allowable (field), the flow rate is acceptable.

(c) Find the mass concentration of toluene at field conditions. Use the ACGIH TLV-TWA.

$$C_{m/V} = 2 \times \text{TLV} = (2)\left(188\ \frac{\text{mg}}{\text{m}^3}\right)$$

$$= 376\text{ mg/m}^3 \quad [\text{at NTP}]$$

$$C_{m/V} = \left(376\ \frac{\text{mg}}{\text{m}^3}\right)\left(\frac{720\text{ mm Hg}}{760\text{ mm Hg}}\right)\left(\frac{298\text{K}}{305\text{K}}\right)$$

$$= 348\text{ mg/m}^3 \quad [\text{field conditions}]$$

Find the volume to collect the mass required. From NIOSH Method No. 1500, the maximum mass per sample is $m_{\text{max}} = 4.51$ mg.

From Eq. 47.6,

$$V = \frac{m}{C_{m/V}} = \frac{(4.51\text{ mg})\left(10^3\ \frac{\text{L}}{\text{m}^3}\right)}{348\ \frac{\text{mg}}{\text{m}^3}} = 13.0\text{ L}$$

Dynamic Mixing of Gases and Vapors (Dilution)

When performing air monitoring, it is sometimes necessary to prepare continuously flowing mixtures of dilute concentrations of a contaminant in air. Usually a contaminant in high concentrations is diluted to low concentrations after mixing with a flow of clean air. In such situations, the phrase "clean air" means that the contaminant is absent. The principle of the conservation of mass and the continuity equation are used to determine the mixture volumes.

When a contaminant is diluted by mixing with air, the volume increases and the concentration decreases; however, the total mass of contaminant remains constant. The mass of the contaminant entering the mixing point equals the mass of the contaminant leaving the mixing point (conservation of mass). Also, total flow in a system is equal to the sum of the individual flows (continuity equation). A dynamic mixing system is illustrated in Fig. 47.2.

Figure 47.2 Dynamic Mixing

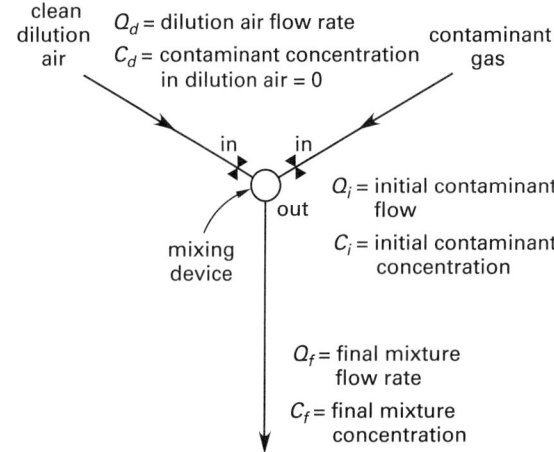

The conservation of mass equation is as follows.

$$Q_d C_d + Q_i C_i = Q_f C_f \qquad 47.12$$

Because clean air is used for dilution, $C_d = 0$. Therefore,

$$Q_f C_f = Q_i C_i \qquad 47.13$$

The continuity equation is as follows.

$$Q_f = Q_i + Q_d \qquad 47.14$$

Example 47.3

A continuous flow of a mixture of nitrogen dioxide (NO_2) and clean air is to be prepared for equipment testing. The desired final mixed concentration is 100 ppm NO_2 with a flow rate of 25 cm^3/min at the field conditions. The system is to operate at 710 mm Hg and 35°C. The NO_2 is available from a cylinder producing a concentration of 945 mg/m^3 at 25°C and 760 mm Hg. What dilution and NO_2 flow rates are required?

Solution

At "field" conditions,

$$T_F = 35°C + 273° = 308K$$

$$p_F = (710 \text{ mm Hg})\left(\frac{1 \text{ atm}}{760 \text{ mm Hg}}\right) = 0.934 \text{ atm}$$

Correct the NO_2 concentration to field conditions. From Eq. 47.1,

$$V_F = V_S \left(\frac{p_S}{p_F}\right)\left(\frac{T_F}{T_S}\right)$$

$$= (1 \text{ m}^3)\left(\frac{760 \text{ mm Hg}}{710 \text{ mm Hg}}\right)\left(\frac{308K}{25°C + 273°}\right)$$

$$= 1.11 \text{ m}^3$$

$$C_{m/V_F,NO_2} = C_{m/V_S,NO_2}\left(\frac{V_S}{V_F}\right)$$

$$= \left(945 \frac{\text{mg}}{\text{m}^3}\right)\left(\frac{1 \text{ m}^3}{1.11 \text{ m}^3}\right)$$

$$= 854 \text{ mg/m}^3$$

Convert the NO_2 concentration to ppm at field conditions. The molecular weight of NO_2 is

$$MW_{NO_2} = (1)\left(14 \frac{\text{g}}{\text{mol}}\right) + (2)\left(16 \frac{\text{g}}{\text{mol}}\right)$$

$$= 46 \text{ g/mol}$$

From Eq. 47.10,

$$C_{\text{ppm,NO}_2} = \left(\frac{C_{m/V,\text{NO}_2} R^* T}{(MW_{NO_2})p}\right)(10^6)$$

$$= \frac{\left(854 \frac{\text{mg}}{\text{m}^3}\right)\left(0.08205 \frac{\text{L·atm}}{\text{mol·K}}\right)}{\left(46 \frac{\text{g}}{\text{mol}}\right)(0.934 \text{ atm})}$$

$$\times \left(10^3 \frac{\text{L}}{\text{m}^3}\right)\left(10^3 \frac{\text{mg}}{\text{g}}\right)$$

$$= 500 \text{ ppm} \quad \text{[field conditions]}$$

From the conservation of mass equation (see Eq. 47.13),

$$Q_i = \frac{Q_f C_f}{C_i} = \frac{\left(25 \frac{\text{cm}^3}{\text{min}}\right)(100 \text{ ppm})}{500 \text{ ppm}}$$

$$= 5 \text{ cm}^3/\text{min} \quad \text{[field conditions]}$$

From the continuity equation (see Eq. 47.14),

$$Q_d = Q_f - Q_i = 25 \frac{\text{cm}^3}{\text{min}} - 5 \frac{\text{cm}^3}{\text{min}}$$

$$= 20 \text{ cm}^3/\text{min} \quad \text{[field conditions]}$$

Determining Concentrations Using Partial Pressures

Raoult's law is used to determine the concentration of the components of a mixture in a closed space. The law applies to the components of a mixture containing two or more volatile components where the mixed vapor phase is in equilibrium with the solution. The solution mole fraction, x^*, is used in these calculations, rather than the vapor phase mole fractions.

$$p_{PVP,i} = x_i^* p_{VP,i} \qquad 47.15$$

$$p_{TVP,\text{mixture}} = \sum x_i^* p_{VP,i} \qquad 47.16$$

The concentration of any gaseous component can be found from the ratio of the partial pressure to its mole fraction in the entire gaseous mixture. This relationship provides a way to determine concentration (volume-based) of an individual component in a gaseous mixture. p_{total} is the atmospheric pressure, which includes the pressures of the air as well as all vapors.

$$C_{\text{ppm},i} = \left(\frac{p_{PVP,i}}{p_{\text{total}}}\right)(10^6)$$

$$= \left(\frac{x_i p_{VP,i}}{p_{\text{total}}}\right)(10^6) \qquad 47.17$$

Example 47.4

At 20°C, the vapor pressures of pure ethanol (CH_3CH_2OH) and pure water (H_2O) are 44 mm Hg and 17.54 mm Hg, respectively. 90-proof whiskey contains 45% ethanol and 55% water by mass. What would the partial vapor pressures of ethanol and water be above a glass containing 400 g of such whiskey in a closed room at 20°C and 750 mm Hg?

Solution

The molecular weights of ethanol and water, respectively, are

$$MW_{ethanol} = (6)\left(1 \ \frac{g}{mol}\right) + (2)\left(12 \ \frac{g}{mol}\right)$$
$$+ (1)\left(16 \ \frac{g}{mol}\right)$$
$$= 46 \ g/mol$$

$$MW_{water} = (2)\left(1 \ \frac{g}{mol}\right) + (1)\left(16 \ \frac{g}{mol}\right)$$
$$= 18 \ g/mol$$

The masses of ethanol and water, respectively, are

$$m_{ethanol} = (0.45)(400 \ g) = 180 \ g$$
$$m_{water} = (0.55)(400 \ g) = 220 \ g$$

Use Eq. 47.4 to determine the number of moles of ethanol and water.

$$n_{ethanol} = \frac{m_{ethanol}}{MW_{ethanol}} = \frac{180 \ g}{46 \ \frac{g}{mol}} = 3.91 \ mol$$

$$n_{water} = \frac{m_{water}}{MW_{water}} = \frac{220 \ g}{18 \ \frac{g}{mol}} = 12.2 \ mol$$

The total number of moles is

$$n_{total} = n_{ethanol} + n_{water}$$
$$= 3.91 \ mol + 12.2 \ mol$$
$$= 16.1 \ mol$$

The mole fraction of ethanol is

$$x^*_{ethanol} = \frac{n_{ethanol}}{n_{total}} = \frac{3.91 \ mol}{16.1 \ mol} = 0.24$$

The mole fraction of water is

$$x^*_{water} = \frac{n_{water}}{n_{total}} = \frac{12.2 \ mol}{16.1 \ mol} = 0.76$$

The partial vapor pressure can be determined from Eq. 47.15.

$$p_{PVP,ethanol} = x^*_{ethanol} p_{VP,ethanol}$$
$$= (0.24)(44 \ mm \ Hg)$$
$$= 10.7 \ mm \ Hg$$

$$p_{PVP,water} = x^*_{water} p_{VP,water}$$
$$= (0.76)(17.54 \ mm \ Hg)$$
$$= 13.3 \ mm \ Hg$$

Use Eq. 47.17 to solve for the ethanol and water concentrations.

$$C_{ppm,ethanol} = \left(\frac{p_{PVP,ethanol}}{p_{total}}\right)(10^6)$$
$$= \left(\frac{10.7 \ mm \ Hg}{750 \ mm \ Hg}\right)(10^6)$$
$$= 14\,200 \ ppm$$

$$C_{ppm,water} = \left(\frac{p_{PVP,water}}{p_{total}}\right)(10^6)$$
$$= \left(\frac{13.3 \ mm \ Hg}{750 \ mm \ Hg}\right)(10^6)$$
$$= 17\,700 \ ppm$$

Exposure Assessment

The purpose of air sampling and analysis is to determine worker exposure to airborne contaminants in relation to standards and guides developed by regulatory and technical service organizations. The time-weighted average calculation method is the primary method used in assessing worker exposure.

Time-Weighted Average Exposure Method

The *time-weighted average* (TWA) method apportions the *measured actual concentrations* (MACs) from air sampling to the times over which the exposures occurred. This method averages the high and low exposures experienced based on the associated time period of each exposure and gives a more accurate representative measure of exposure than averaging measurements alone. The TWA is calculated by multiplying each concentration by the exposure time interval and dividing the sum by the total time period.

$$TWA = \frac{\sum C_i t_i}{\sum t_i} \qquad 47.18$$

The *averaging time (averaging period)* is either 8 h or 10 h, depending on the exposure standard that is being used for the exposure assessment. The standards in use in the United States are given in Table 47.2.

The OSHA and NIOSH standards typically specify allowable air sampling flow rates for the various chemicals.

Health, Safety, Welfare

Table 47.2 Exposure Assessment Standards

organization	standard	averaging period
American Conference of Governmental Industrial Hygienists (ACGIH)	threshold limit value—time-weighted average (TLV-TWA)	480 min (8 h) per day, 40 h per week
American Conference of Governmental Industrial Hygienists (ACGIH)	threshold limit value—short-term exposure limit (TLV-STEL)	15 min TWA; TLV-TWA must not be exceeded for the day
American Conference of Governmental Industrial Hygienists (ACGIH)	threshold limit value—ceiling (TLV-C)	instantaneous reading; use 15 min TWA if no instantaneous reading available
Occupational Safety and Health Administration (OSHA)	permissible exposure limit—time-weighted average (PEL-TWA)	480 min (8 h) per day, 40 h per week
Occupational Safety and Health Administration (OSHA)	permissible exposure limit—short-term limit (PEL-STEL)	15 min TWA unless another exposure period is given; PEL-TWA must not be exceeded
Occupational Safety and Health Administration (OSHA)	permissible exposure limit—ceiling (PEL-C)	instantaneous reading; use 15 min TWA if no instantaneous reading available
National Institute of Occupational Safety and Health (NIOSH)	recommended exposure limit—time-weighted average (REL-TWA)	up to 10 h per day, 40 h per week
National Institute of Occupational Safety and Health (NIOSH)	recommended exposure limit—short-term exposure limit (REL-STEL)	15 min TWA
National Institute of Occupational Safety and Health (NIOSH)	recommended exposure limit—ceiling (REL-C)	instantaneous reading
American Industrial Hygiene Association (AIHA)	workplace environmental exposure level—time-weighted average (WEEL-TWA)	up to 8 h per day, 40 h per week
American Industrial Hygiene Association (AIHA)	workplace environmental exposure level—short-term time-weighted average (WEEL-short term TWA)	averaging time is specified for each chemical
American Industrial Hygiene Association (AIHA)	workplace environmental exposure level—ceiling (WEEL-C)	instantaneous reading

The flow rates are established to prevent loss of sample mass due to breakthrough or collection of a sample mass lower than the limit of detection (LOD) for the method. These standards also indicate total volumes that should be collected during the sampling period.

Example 47.5

A worker is exposed to toluene (C_7H_8) during an 8 h shift as follows. The measured actual concentration (MAC) data are reported at NTP.

duration (h)	measured actual concentration (MAC) (mg/m^3)
1	300
3	400
3	200
1	0

Compare the TWA exposure to the TLV-TWA of 188 mg/m^3 and the PEL-TWA of 200 ppm.

Solution

At NTP,

$$T = 25°C + 273° = 298K$$

$$p = 1 \text{ atm}$$

Find the TWA of the MAC values reported.

Use Eq. 47.18.

$$\text{TWA} = \frac{\sum C_{m/V,i}t_i}{\sum t_i}$$

$$= \frac{\left(300 \frac{\text{mg}}{\text{m}^3}\right)(1 \text{ h}) + \left(400 \frac{\text{mg}}{\text{m}^3}\right)(3 \text{ h}) + \left(200 \frac{\text{mg}}{\text{m}^3}\right)(3 \text{ h}) + \left(0 \frac{\text{mg}}{\text{m}^3}\right)(1 \text{ h})}{1 \text{ h} + 3 \text{ h} + 3 \text{ h} + 1 \text{ h}}$$

$$= 262.5 \text{ mg/m}^3$$

Convert the PEL-TWA to mass concentration. The molecular weight of toluene is

$$\text{MW}_{\text{toluene}} = (8)\left(1 \frac{\text{g}}{\text{mol}}\right) + (7)\left(12 \frac{\text{g}}{\text{mol}}\right) = 92 \text{ g/mol}$$

From Eq. 47.11,

$$C_{m/V,\text{toluene}} = \left(\frac{C_{\text{ppm,toluene}}\text{MW}_{\text{toluene}}p}{R^*T}\right)(10^{-6})$$

$$= \left(\frac{(200 \text{ ppm})\left(92 \frac{\text{g}}{\text{mol}}\right)(1 \text{ atm}) \times \left(10^3 \frac{\text{mg}}{\text{g}}\right)\left(10^3 \frac{\text{L}}{\text{m}^3}\right)}{\left(0.08205 \frac{\text{L·atm}}{\text{mol·K}}\right)(298K)}\right)(10^{-6})$$

$$= 753 \text{ mg/m}^3$$

The MAC-TWA for toluene (262.5 mg/m^3) exceeds the TLV-TWA (188 mg/m^3) but not the PEL-TWA (753 mg/m^3).

Short-Term Exposure Method

When measuring short-term and ceiling exposures, it is sometimes not possible to get an instantaneous reading to compare to ceiling levels. In these cases, the analyst must estimate the MAC peak from TWA 15 min measurements. The ceiling can be estimated by assuming that the MAC rises from the pre-short term value to a peak and then declines to the post-short term value in the form of a triangle. The duration of the peak can be assumed to be 5 min (or some other rational duration) out of the 15 min short-term sampling period. The peak concentration can then be estimated by assuming that

the area of the triangle equals the short-term MAC times 15 min. This procedure is illustrated in the following example.

Example 47.6

Workers are exposed to chemical A (MW$_A$ = 78 g/mol) for a 60 min period as follows.

period	sample no.	MAC
pre-peak: 30 min	1	100 mg/m^3
peak: 15 min	2	160 mg/m^3
post-peak: 30 min	3	120 mg/m^3

The short-term exposure limits for chemical A are as follows.

TLV-STEL	TLV-C	PEL-STEL	PEL-C
50 ppm	100 ppm	239.3 mg/m^3	478.6 mg/m^3

Have the short-term exposure limits been exceeded?

Solution

Assume the peak occurs for 5 min out of the 15 min short-term exposure period. Draw a diagram of exposure and estimate the ceiling exposure.

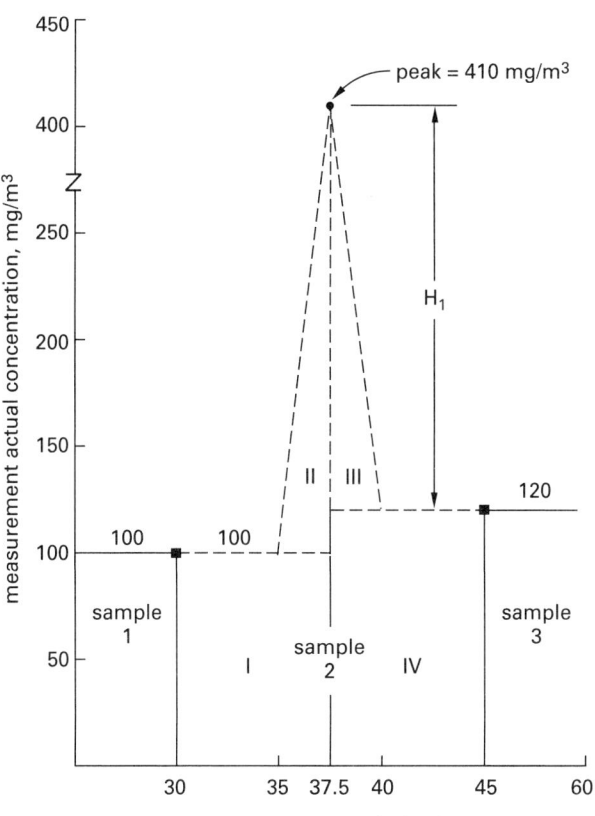

The average exposure during the short-term exposure period is 160 mg/m³. The concentration multiplied by time for the short-term period is

$$Ct = \left(160 \ \frac{mg}{m^3}\right)(15 \ min) = 2400 \ mg{\cdot}min/m^3$$

Assume the peak occurs as shown in the illustration over a 5 min period. The sum of areas I through IV must equal the Ct of 2400 mg·min/m³.

$$A_I = \left(100 \ \frac{mg}{m^3}\right)(7.5 \ min) = 750 \ mg{\cdot}min/m^3$$

$$A_{II} = \left(\frac{1}{2}\right)\left(H_1 + 20 \ \frac{mg}{m^3}\right)(2.5 \ min)$$

$$A_{III} = \left(\frac{1}{2}\right)(H_1)(2.5 \ min)$$

$$A_{IV} = \left(120 \ \frac{mg}{m^3}\right)(7.5 \ min) = 900 \ mg{\cdot}min/m^3$$

$$Ct = A_I + A_{II} + A_{III} + A_{IV}$$

$$2400 \ \frac{mg{\cdot}min}{m^3} = 750 \ \frac{mg{\cdot}min}{m^3} + 25 \ \frac{mg{\cdot}min}{m^3}$$
$$+ \left(\frac{2.5 \ min}{2}\right)H_1 + \left(\frac{2.5 \ min}{2}\right)H_1$$
$$+ 900 \ \frac{mg{\cdot}min}{m^3}$$

$$H_1 = 290 \ mg/m^3$$

The peak is

$$120 \ \frac{mg}{m^3} + 290 \ \frac{mg}{m^3} = 410 \ mg/m^3$$

At NTP,

$$T = 25°C + 273° = 298K$$

$$p = (760 \ mm \ Hg)\left(\frac{1 \ atm}{760 \ mm \ Hg}\right) = 1 \ atm$$

Use Eq. 47.10.

$$C_{ppm,A} = \left(\frac{C_{m/V,A}R^*T}{(MW_A)p}\right)(10^6)$$

$$= \left(\frac{\left(410 \ \frac{mg}{m^3}\right)\left(0.08205 \ \frac{L{\cdot}atm}{mol{\cdot}K}\right)(298K)}{\left(78 \ \frac{g}{mol}\right)(1 \ atm)\left(10^3 \ \frac{L}{m^3}\right)\left(10^3 \ \frac{mg}{g}\right)}\right)(10^6)$$

$$= 129 \ ppm$$

Similarly, a concentration of 160 mg/m³ is 50.2 ppm. Compare the results.

	TLV-STEL (ppm)	TLV-C (ppm)	PEL-STEL (mg/m³)	PEL-C (mg/m³)
standard	50	100	239.3	478.6
MAC/ estimated	50.2	129	160	410
compliance	no	no	yes	yes

Upper and Lower Confidence Limits

It is sometimes necessary to know how close a calculated or measured result is to the standard when *sampling and analytical errors* (SAEs) are considered. It is assumed that the measured value may be higher or lower than the true value by plus or minus the SAE. Usually a 95% confidence level is used in calculating the SAE. The SAE is calculated as follows.

$$SAE = (1.645)(CV_{total}) \quad\quad 47.19$$

The constant 1.645 relates to the 95% confidence level for a normal distribution. With a 99% confidence level, this factor would be 2.329.

The coefficient of variation including sampling and analytical errors, CV_{total}, is found from the following.

$$CV_{total} = \frac{s}{x} \quad\quad 47.20$$

The CV and the SAE are used to calculate the upper and lower confidence limits, UCL and LCL, as follows.

$$UCL \ (95\%) = \frac{\overline{x}}{PEL} + SAE \quad\quad 47.21$$

$$LCL \ (95\%) = \frac{\overline{x}}{PEL} - SAE \quad\quad 47.22$$

The term "$\overline{x}/PEL$" is often referred to as the *standardized concentration*. When the LCL is less than or equal to 1.0, occupants are being underexposed, and OSHA regulations are not being violated. When the LCL is greater than 1.0, an OSHA violation exists. If the LCL is less than or equal to 1.0 and the UCL is greater than 1.0, a possible overexposure may have occurred.

Example 47.7

Workers were exposed to solvent X49 for an 8 h period. The laboratory reported a TWA concentration of 90 ppm of X49 in samples collected for analysis. The PEL for X49 is 102 ppm. The total coefficient of variation including sampling and analytical errors, CV_{total}, is 30%. Was the PEL exceeded?

Solution

Find the SAE. From Eq. 47.19,

$$SAE = (1.645)(CV_{total}) = (1.645)(0.30)$$
$$= 0.49$$

Find the standardized concentration.

$$\frac{\overline{x}}{PEL} = \frac{90 \ ppm}{102 \ ppm} = 0.88$$

Find the UCL and LCL. From Eq. 47.21,

$$\text{UCL} \ (95\%) = \frac{\bar{x}}{\text{PEL}} + \text{SAE} = 0.88 + 0.49$$
$$= 1.38$$

From Eq. 47.22,

$$\text{LCL} \ (95\%) = \frac{\bar{x}}{\text{PEL}} - \text{SAE} = 0.88 - 0.49$$
$$= 0.39$$

Although the 90 ppm did not exceed the PEL of 102 ppm, the 95% UCL did exceed 1.0, so it's 95% certain that a violation has occurred.

5. PERSONAL PROTECTIVE EQUIPMENT

OSHA Requirements

The Occupational Safety and Health Administration has promulgated regulations covering *personal protective equipment* (PPE) for worker respiratory protection as well as regulations for other workplace hazards. The regulations include standards and guidelines for respiratory protection and are found in 29 CFR 1910, Subpart I, *Personal Protective Equipment*. The following standards are relevant.

- 1910.132—General Requirements

- 1910.134—Respiratory Protection

- 1910.134, Appendix A—Fit Testing Procedures (Mandatory)

- 1910.134, Appendix B-1—User Seal Check Procedures (Mandatory)

- 1910.134, Appendix B-2—Respiratory Cleaning Procedures (Mandatory)

- 1910.134, Appendix C—OSHA Respirator Medical Evaluation Questionnaire (Mandatory)

- 1910.134, Appendix D—Information for Employees Using Respirators When Not Required Under Standard (Mandatory)

- 1910 Subpart I, Appendix B—Nonmandatory Compliance Guidelines for Hazard Assessment and Personal Protective Equipment Selection

The first four OSHA standards for respiratory personal protection are partially given in App. 47.C. These standards describe the general requirements for personal protective equipment in the workplace, the standards for the use of respiratory protective equipment, fit testing procedures for respirators, and procedures for the user to check the respirator seal.

The standards for respiratory protection are legal requirements that should be implemented at hazardous waste remediation sites as well as at other workplace locations. Guidelines for application of the OSHA standards for personal protection are given in App. 47.D.

EPA Levels of Protection

The levels of protection developed by the EPA for work at hazardous waste sites are described in Table 47.3. These levels are also accepted by OSHA for hazardous waste site entry with the additional requirement that each PPE ensemble must be tailored to the specific situation in order to provide the most appropriate level of protection.

Table 47.3 *EPA Levels of Protection*

aspect	level A	level B
required	vapor protective suit that meets NFPA 1991; pressure-demand, full-face SCBA; inner chemical-resistant gloves; chemical-resistant safety boots; two-way radio communication	liquid splash-protective suit that meets NFPA 1992; pressure-demand, full-facepiece SCBA; inner chemical-resistant gloves; chemical-resistant safety boots; two-way radio communications; hard hat
optional	cooling system; outer gloves; hard hat	cooling system; outer gloves
protection provided	highest available level of respiratory, skin, and eye protection from solid, liquid, and gaseous chemicals	provides same level of respiratory protection as level A, but less skin protection; liquid splash protection but no protection against chemical vapors or gases
used when	chemicals have been identified and have high level of hazards to respiratory system, skin, and eyes; substances are present with known or suspected skin toxicity or carcinogenicity; operations must be conducted in confined or poorly ventilated areas	chemicals have been identified but do not require a high level of skin protection; initial site surveys are required until higher levels of hazards are identified; primary hazards associated with site entry are from liquid and not vapor contact
limitations	protective clothing must resist permeation by the chemical or mixtures present; ensemble items must allow integration without loss of performance	protective clothing items must resist penetration by the chemicals or mixtures present; ensemble items must allow integration without loss of performance

aspect	level C	level D
required	support function protective garment that meets NFPA 1993; full-facepiece, air-purifying, canister-equipped respirator; chemical-resistant gloves and safety boots; two-way communications system; hard hat	coveralls; safety boots/shoes; safety glasses or chemical splash goggles
optional	face-shield; escape SCBA	gloves; escape SCBA; face-shield
protection provided	same level of skin protection as level B, but a lower level of respiratory protection; liquid splash protection but no protection against chemical vapors or gases	no respiratory protection; minimal skin protection
used when	contact with site chemicals will not affect the skin; air contaminants have been identified and concentrations measured; a canister is available that can remove the contaminant; the site and its hazards have been completely characterized	atmosphere contains no known hazard; work functions preclude splashes, immersion, potential for inhalation, or direct contact with hazardous chemicals
limitations	protective clothing items must resist penetration by the chemical or mixtures present; chemical airborne concentration must be less than IDLH levels; atmosphere must contain at least 19.5% oxygen; not acceptable for chemical emergency response	this level should not be worn in the hot zone; the atmosphere must contain at least 19.5% oxygen; not acceptable for chemical emergency response

Health, Safety, Welfare

48 Biological Effects of Radiation

Nomenclature

A	activity	Bq
A_s	surface area	m^2
d	distance from a radiation source	m
D_T	absorbed dose	Gy
DF	dose factor	$(Sv/yr)/(Bq/cm^2)$
f	fluence	neutrons/cm^2
H_D	deep-dose equivalent	Sv
$H_{E,50}$	committed effective dose equivalent	Sv
H_L	lens dose equivalent	Sv
H_S	shallow-dose equivalent	Sv
H_T	dose equivalent	Sv
$H_{T,50}$	committed dose equivalent	Sv
I	radiation source intensity	C/kg
q	radioactivity in an organ or the body	Bq
t	time	s
$t_{1/2}$	radioisotope half-life	s
TEDE	total effective dose equivalent	rem, Sv
w_R	weighting factor	Sv/Gy or rem/rad
w_T	tissue weighting factor	–
X	dose equivalent rate	Sv/s

Symbols

Γ	gamma constant	(mSv/h)/MBq
λ	decay constant of a radioisotope	s^{-1}

Subscripts

0	initial amount or time
b	beta
bio	biological
eff	effective
g	gamma
rad	radiological
t	time
T	tissue or total body

1. INTRODUCTION

Radiation is classified as either ionizing or non-ionizing. *Ionizing radiation* can create ions through atomic interactions with matter. The rate and magnitude of ion creation in tissue determines the biological effect on it. Ionizing radiation includes particle radiation (e.g., alpha and beta particles) and high-energy electromagnetic radiation (e.g., X-rays and gamma rays).

Nonionizing radiation includes electromagnetic (wave) radiation of relatively low energy such as radio waves, television signals, and microwaves, as well as infrared, visible, and ultraviolet light. This type of radiation can cause tissue damage through heating and induction, but not through ionization. Nonionizing radiation is not addressed in this chapter.

Quantum theory treats particles as possessing some wave properties (interactions following probability patterns) and waves as possessing some properties associated with particles (energy deposited in discrete "packets" or quanta).

Particle radiation includes charged particles such as protons, beta particles (negatively charged electrons, or negatrons (β^-), and positively charged electrons, or positrons (β^+)), and alpha particles (helium-4 atoms stripped of electrons (He^{2+} or α^{2+})). Neutrons, which are themselves chargeless, produce ions (charged particles) through atomic collisions.

All *electromagnetic waves* travel at the speed of light. These waves, also known as *photons*, can be classified by their energies, frequencies, or wavelengths. The energy is proportional to the frequency and inversely proportional to the wavelength. In order of increasing energy, electromagnetic waves include radio waves, television signals, microwaves, infrared light, visible light, ultraviolet light, and X-rays/gamma rays. Gamma rays originate in the nucleus of atoms during atomic changes, while X-rays originate in the electron shells surrounding the nucleus of atoms. Only the origins of gamma rays and X-rays distinguish them when they have equal energies. Photon interactions with tissue also can cause ions to form in the body.

Isotopes

An isotope of an element contains the same number of protons as the element but a different number of neutrons. The notation for element X that has atomic number Z and mass number A is: $_Z^A X$ or X-A. For example, nickel-63, Ni-63, $^{63}_{28}$Ni, or ^{63}Ni are recognized as being equivalent. The atomic number, Z, is determined by the number of protons in the nucleus. Hydrogen has an atomic number of 1; all atoms with only one proton in their nuclei are hydrogen. However, hydrogen can have zero neutrons (H-1, $^1_1$H), one neutron (H-2, $^2_1$H, deuterium), or two neutrons (H-3, $^3_1$H, tritium).

Not all isotopes of an element are radioactive. Some are stable and do not decay. For example, hydrogen-3 is radioactive, but hydrogen-2 and hydrogen-1 are not radioactive. All elements having an atomic number greater than 80 have radioactive isotopes, and all isotopes of elements with an atomic number above 83 are radioactive.

Until a radioisotope undergoes a decay event, no radiation is emitted. An isotope may emit one or more different types of ionizing radiation, depending on the mode of radioactive decay.

2. RADIATION QUANTITIES AND UNITS

Activity

Activity is the rate of "disintegration" ("transformation" or "decay") of radioactive atoms. The preferred unit of activity is the becquerel (Bq), equivalent to 1/s (i.e., one disintegration per second). The non-SI curie (Ci) unit, equivalent to 3.7×10^{10} disintegrations per second, may still be encountered. Activity is not necessarily proportional to physical size or any other physical dimension. A radioactive source the size of a thin wire can contain more activity than a 55 gallon drum full of a radioactive liquid.

Because activity follows an exponential decay pattern, the percentage of atoms undergoing decay in any given period of time is constant, within reasonable probabilistic limits. The time it takes for some mass or number of atoms of a radioisotope to decay to one-half of its initial value is known as the *half-life*, $t_{1/2}$. (See Table 48.1.)

$$A_t = A_0 e^{-\lambda t} \qquad 48.1$$

The *decay constant*, λ, is

$$\lambda = \frac{\ln 2}{t_{1/2}} \approx \frac{0.693}{t_{1/2}} \qquad 48.2$$

Example 48.1

What is the activity of an I-131 ($t_{1/2} = 8.04$ days) source now if the activity was 3.7 TBq 11 days ago?

Solution

From Eq. 48.2,

$$\lambda = \frac{\ln 2}{t_{1/2}} = \frac{\ln 2}{8.04 \text{ d}}$$
$$= 0.0862 \text{ d}^{-1}$$

From Eq. 48.1,

$$A_t = A_0 e^{-\lambda t}$$
$$= (3.7 \text{ TBq}) e^{-(0.0862 \text{ d}^{-1})(11 \text{ d})}$$
$$= 1.43 \text{ TBq}$$

Table 48.1 *Approximate Half-Lives of Selected Radioisotopes*

radioisotope	half-life, $t_{1/2}$
Au-198	2.696 d
C-11	20.38 min
Co-57	270.9 d
Cr-51	27.704 d
Cs-137	30.0 yr
Cu-64	12.701 h
Fe-55	2.7 yr
H-3	12.35 yr
I-125	60.14 d
I-131	8.04 d
K-40	1.277×10^9 yr
O-15	122.24 s
P-32	14.29 d
Pd-109	13.427 h
Po-210	138.38 d
Rb-86	18.66 d
Rn-222	3.8235 d
S-35	87.44 d
Th-230	7.7×10^4 yr
U-235	703.8×10^6 yr

Source: Kocher, D.C., *Radioactive Decay Data Tables*, DOE-TICS-11026; USDOE; Washington, DC, 1981.

Exposure is a measure of X-ray and gamma radiation based on its ability to ionize the molecules in air. The term "exposure" does not apply to any absorbing material except air. In the past, the *roentgen* (R) was used to measure the exposure. One roentgen produces 1 statcoulomb of charge in 1 cm^3 of air at standard temperature and pressure (STP).

The roentgen is no longer used in any of the regulations relating to occupational or public exposure to radiation. It remains useful in several other fields, such as medical diagnosis and therapy, radiation sterilization, and shielding design. The SI unit of exposure is the coulomb per kilogram (C/kg). A roentgen is equivalent to a dose of 2.58×10^{-4} C/kg.

Biological damage caused by radiation passing through a volume of a target is roughly proportional to the amount of radiation energy absorbed by the target. *Absorbed dose*, D_T, is the total energy deposited by ionizing radiation per unit mass of irradiated material. In the case of radiation's biological effects, the absorbing material of interest is human tissue.

The units of absorbed dose are the rad and the gray. The *rad* is an absorbed dose of 100 ergs/gram or

0.01 J/kg (0.01 Gy). The *gray* (Gy) is the SI unit for an absorbed dose of 1 J/kg (100 rad).

Dose Equivalent and Weighting Factor

Even when depositing the same amount of energy per mass (i.e., the same absorbed dose), different types of radiation cause different amounts of biological damage in living tissue. To reflect this difference, a quantity called the *dose equivalent*, H_T, is used to indicate biological effects on a specific tissue type.

To quantify the sensitivity of a biological tissue to different types of radiation (or energies in the case of neutrons), a *weighting factor*, w_R, is used. The weighting factor reflects the ability of a particular type of radiation to cause biological damage, relative to gamma radiation.

The rem and sievert (Sv) are used to measure dose equivalent in radiation protection applications. Given the relationship 1 Gy = 100 rad, the relationship between rem and Sv is 1 Sv = 100 rem. The dose equivalent, H_T, is equal to the absorbed tissue dose, D_T, multiplied by the weighting factor, w_R.

$$H_T = D_T w_R \qquad 48.3$$

Consistent units should be maintained in Eq. 48.3. If the absorbed dose is given in rads, the dose equivalent is in rems, while if the absorbed dose is given in grays, the dose equivalent is in sieverts.

As seen in Table 48.2, the weighting factor for beta particles is 1 Sv/Gy (or rem/rad), and the weighting factor for alpha particles is 20 Sv/Gy (or rem/rad). Since the weighting factor for alphas is higher than for betas, alphas do more biological damage than betas for the same amount of energy deposited per unit mass of tissue (i.e., for the same dose).

Example 48.2

A person is exposed to 5 rad of alpha radiation, 1 rad of beta radiation, and 1 rad of neutrons of unknown energy. What is the dose equivalent for this person?

Solution

From Eq. 48.3 and Table 48.2,

$$H_T = \sum D_T w_R$$
$$= (5 \text{ rad})\left(20 \ \frac{\text{rem}}{\text{rad}}\right) + (1 \text{ rad})\left(1 \ \frac{\text{rem}}{\text{rad}}\right)$$
$$+ (1 \text{ rad})\left(10 \ \frac{\text{rem}}{\text{rad}}\right)$$
$$= 111 \text{ rem} \quad (1.11 \text{ Sv})$$

3. BIOLOGICAL EFFECTS

Background Radiation

Background radiation includes natural radiation sources and human-made radiation sources. *Natural radiation sources* include cosmic sources and *naturally occurring radioactive material* (NORM). Approximately 340 nuclides occur in nature, and 70 of these, such as radon, are radioactive. *Human-made radiation sources* include fallout from nuclear explosive devices, some consumer products, and nuclear accidents such as the one at Chernobyl.

Together, natural and human-made background radiation expose each person to an average of 360 mrem (3.6 mSv) each year.

Acute Versus Chronic Effects

The duration of a radiation exposure is an important factor in determining how the human body reacts. Exposure durations are categorized as acute or chronic. *Acute exposure* is exposure to a large amount of radiation in a short period of time. In an occupational setting, the exposure period is typically less than eight hours. Table 48.3 gives a summary of the clinical effects of acute ionizing radiation doses. *Chronic exposure* is long-term, low-level exposure. Background radiation is an example of chronic radiation exposure.

Stochastic Versus Nonstochastic Effects

The term *stochastic* means involving chance or probability. Radiation increases the probability of stochastic events, such as cancer, occurring.

Stochastic effects are biological effects that occur randomly and for which the probability of the effect occurring, not its severity, is assumed to be a linear function of dose without threshold. Hereditary effects and cancer incidence are examples of stochastic effects.

Nonstochastic (also called *deterministic*) *effects* are biological effects for which the severity varies with the dose and a threshold is believed to exist. Skin reddening and radiation-induced cataract formation are examples of nonstochastic effects.

Somatic, Genetic, and Fetal Effects

Exposure to ionizing radiation can cause three types of biological effects: somatic effects, genetic effects, and fetal effects.

Somatic effects are experienced directly by the irradiated individual and are either prompt or delayed, depending upon the period of time before the effects manifest in the exposed individual. These effects include damage to body tissues and organs that can impair their ability to function normally. The symptoms exhibited during the course of the Chernobyl accident were prompt somatic effects. Delayed somatic effects can be long-term, 20 years to 30 years. The main delayed somatic effect of ionizing radiation is an increase in the probability of developing different types of cancers. For example, without any artificial exposure, of 10,000 people, 2500 may exhibit some form of cancer during their lifetime. If 10,000 people are each irradiated with

Health, Safety, Welfare

Table 48.2 Weighting Factors and Absorbed Dose Equivalencies

type of radiation	weighting factor, w_R (Sv/Gy or rem/rad)	absorbed dose equal to a unit dose equivalent[*] (Gy/Sv or rad/rem)
X-, gamma, or beta radiation	1	1
alpha particles, multiple-charged particles, fission fragments, and heavy particles of unknown charge	20	0.05
neutrons of unknown energy	10	0.1
high-energy protons	10	0.1

[*]The absorbed dose in rad equal to 1 rem, or the absorbed dose in Gy equal to 1 Sv.

Source: Table 1004(b).1, 10 CFR 20.1004, *Units of radiation dose*, U.S. Nuclear Regulatory Commission, Washington, DC.

1 rem of whole-body radiation, it is estimated that the radiation may cause three additional cases of cancer in the group. Of 10,000 people, 1640 may succumb to some form of cancer. If 10,000 people are each irradiated with 1 rem of whole-body radiation, it is estimated that the radiation may cause one additional cancer death in the group.

Genetic effects may be passed on to future generations. Inherited characteristics in human reproduction are controlled by genes in the reproductive cells. Radiation can alter genes and produce mutations that could eventually result in anomalies in offspring. Because these mutations are generally recessive, several generations may pass before the effects become apparent. The current incidence of all types of genetic disorders and traits that cause some type of serious handicap at some time during an individual's lifetime is about 1000 incidents per 10,000 live births. If each 30 year generation receives 1 rem of whole-body irradiation, the radiation may cause an additional 5 to 75 genetic disorders in the first generation, or 60 to 1000 disorders at genetic equilibrium about four generations later.

Fetal effects are those effects that result from the exposure of a fetus or embryo to penetrating radiation. A number of studies have indicated that embryos or fetuses are more sensitive to radiation than adults, particularly during the first three months after conception, when a woman may not know that she is pregnant. The main concerns during this period of the pregnancy may be developmental abnormalities during the growth of the fetus. As the pregnancy progresses, the sensitivity of the fetus to radiation decreases. The main concern later in the pregnancy is an increase in the risk of leukemia in the first 10 years of the child's life. The current incidence of all types of fetal effects is about 700 incidents per 10,000 live births. This includes effects due to measles, alcohol, and drugs. If 10,000 embryos were to receive 1 rem of whole-body irradiation before birth, it is estimated that the radiation may affect 10 additional children.

4. RADIATION DOSE CALCULATIONS

Radiation exposure can come from sources inside as well as outside of the body. To quantify both internal and external exposures, the *total effective dose equivalent*

(TEDE) is used. The TEDE is the sum of the deep-dose equivalent, H_D, for external exposures and the committed effective dose equivalent, $H_{E,50}$, for internal exposures over the next 50 years (assumed remaining lifespan).

$$\frac{\text{total radiation}}{\text{exposure}} = \frac{\text{external}}{\text{exposure}} + \frac{\text{internal}}{\text{exposure}}$$

$$\text{TEDE} = H_D + H_{E,50} \qquad 48.4$$

The *external dose* is the portion of the dose equivalent received from radiation sources outside the whole body. For radiation protection purposes, "whole body" means head, trunk (including male gonads), arms above the elbow, and legs above the knee.

The *deep-dose equivalent*, H_D, is the external whole-body exposure: the portion of the dose equivalent at a tissue depth of 1 cm (1 g/cm^2).

The *shallow-dose equivalent*, H_S, is the external exposure of the skin or an extremity and is taken as the dose equivalent at a tissue depth of 0.007 cm (0.007 g/cm^2) averaged over an area of 1 cm^2.

The *lens dose equivalent* or *eye dose equivalent*, H_L, is the external exposure of the lens of the eye, taken as the dose equivalent at a tissue depth of 0.3 cm (0.3 g/cm^2).

Beta Doses

For beta emitters, the dose equivalent is a function of the activity of the radioactive source, the distance from the source, the energy of the beta particles, and the number of beta particles per decay (the "yield" or "abundance"). The energy of the beta particles and the number of beta particles per decay are incorporated into a unit called the *electron dose rate factor*, DF. (See Table 48.4.)

$$H_T = X_b t = \text{DF}\left(\frac{A}{A_s}\right)t \qquad 48.5$$

Example 48.3

What is the dose equivalent rate at a depth of 0.007 cm for a 3.7×10^7 Bq paste of phosphorus-32 uniformly deposited on 1 in^2 of skin?

Table 48.3 *Summary of Clinical Effects of Acute Ionizing Radiation Doses*

	dose equivalent (rem)					
	0–100	100–200	200–600	600–1000	1000–5000	over 5000
incidence of vomiting	none	100 rem=5% 200 rem=50%	300 rem=100%	100%		
time of onset	–	3 hr	2 hr	1 hr	30 min	
principal affected organs	none	hematopoietic tissue			gastrointestinal tract	central nervous system
characteristic signs	none	moderate leukopenia	severe leukopenia; purpura; hemorrhage; infection; epilation above 300 rem		diarrhea; fever; disturbance of electrolyte balance	convulsions; tremor; ataxia; lethargy
critical period postexposure	–	–	4 to 6 wk		5 to 14 days	1 to 48 hr
prognosis	excellent	excellent	good	guarded	hopeless	
convalescent period	none	several weeks	1 to 12 months	long	–	–
incidence of death	none	none	0 to 80%	80% to 100%	90% to 100%	
death occurs within	–	–	2 months		2 weeks	2 days
cause of death	–	–	hemorrhage; infection		circulatory collapse	respiratory failure; brain edema

Glasstone, S., *The Effects of Nuclear Weapons*, USAEC, Washington, DC, 1962.

Solution

$$A_s = (1 \text{ in}^2)\left(2.54 \frac{\text{cm}}{\text{in}}\right)^2 = 6.45 \text{ cm}^2$$

From Table 48.4 for P-32 at 0.007 cm,

$$\text{DF} = 2.1 \times 10^{-2} \text{ Sv·cm}^2/\text{Bq·yr}$$

Use Eq. 48.5.

$$X_b = \text{DF}\left(\frac{A}{A_s}\right)$$
$$= \left(2.1 \times 10^{-2} \frac{\text{Sv·cm}^2}{\text{Bq·yr}}\right)\left(\frac{3.7 \times 10^7 \text{ Bq}}{6.45 \text{ cm}^2}\right)$$
$$= 1.2 \times 10^5 \text{ Sv/yr}$$

Gamma Doses

For photon emitters, the dose equivalent is a function of the activity of the radioactive source, the distance from the source, the energy of the photons, and the number of photons per decay. The energy of the photons and the number of photons per decay are incorporated into a unit called the *specific gamma ray dose constant*, Γ. (See Table 48.5.)

$$H_T = X_g t = \Gamma A t \quad [\text{at 100 cm}] \qquad 48.6$$

Example 48.4

What is the whole-body dose equivalent rate 100 cm from an unshielded 1.59×10^5 MBq cobalt-57 source?

Solution

From Table 48.5, for Co-57,

$$\Gamma = 4.087 \times 10^{-5} \text{ (mSv/h)/MBq} \quad [\text{at 100 cm}]$$

Use Eq. 48.6.

$$X_g = \Gamma A = \left(4.087 \times 10^{-5} \; \frac{\text{mSv/h}}{\text{MBq}}\right)(1.59 \times 10^5 \; \text{MBq})$$

$$= 6.5 \; \text{mSv/h} \quad [\text{at } 100 \text{ cm}]$$

Neutron Doses

For neutrons, the *fluence, f,* can be used to calculate the neutron dose equivalent. 1 rem (0.01 Sv) of neutron radiation of unknown energies may be assumed to be caused by a total fluence of 25 million neutrons per square centimeter hitting the body. If the approximate energy distribution of the neutrons is known, Table 48.6 can be used to calculate the tissue dose equivalent.

$$H_T = D_T w_R = X_n t$$

$$= \frac{f}{\text{fluence per unit dose}} \qquad 48.7$$

equivalent from Table 48.6

Example 48.5

What is the dose equivalent, H_T, in rems from an unshielded 5 MeV neutron source with a fluence, *f*, of 10^9 neutrons/cm^2?

Solution

From Table 48.6, for 5 MeV neutrons,

$$\frac{f}{H_T} = 23 \times 10^6 \; \text{neutrons/cm}^2\text{·rem}$$

$$H_T = \frac{f}{\text{fluence per unit}}$$

$$\text{dose equivalent}$$

$$= \frac{10^9 \; \dfrac{\text{neutrons}}{\text{cm}^2}}{23 \times 10^6 \; \dfrac{\text{neutrons}}{\text{cm}^2\text{·rem}}}$$

$$= 43 \; \text{rem} \quad (0.43 \; \text{Sv})$$

5. INTERNAL EXPOSURES

The *internal* (or *committed*) *dose* is that portion of the dose equivalent received from radioactive material taken into the body. The longer a radioactive material stays in the body, the larger the radiation dose received.

The *committed dose equivalent, $H_{T,50}$,* is the dose equivalent to organs or tissues that will be received from an intake of radioactive material by an individual during the 50 year period following the intake.

The *committed effective dose equivalent, $H_{E,50}$,* is the sum of the products of (1) the weighting factors applicable to each of the body organs or tissues that are irradiated, and (2) the $H_{T,50}$ to these organs or tissues.

The *tissue weighting factor, w_T,* for an organ or tissue is the proportion of the risk of stochastic effects resulting from irradiation of that organ or tissue to the total risk of stochastic effects when the whole body is irradiated uniformly. (See Table 48.7.)

$$H_{E,50} = \sum w_T H_{T,50} \qquad 48.8$$

The human body cannot distinguish between stable isotopes and radioactive isotopes of the same element. This means that radioactive isotopes inside the body will experience the same absorption and removal rates, and occupy the same deposition sites, that stable isotopes would. Clearance of internal radioisotopes involves two processes: (1) radioactive decay and (2) normal biological removal. In performing clearance calculations, both radiological and biological clearances are assumed to follow exponential laws.

The *radiological clearance rate* is

$$q_{\text{rad},t} = q_0 e^{-\lambda_{\text{rad}} t} \qquad 48.9$$

The *biological clearance rate* is

$$q_{\text{bio},t} = q_0 e^{-\lambda_{\text{bio}} t} \qquad 48.10$$

The *effective clearance rate* due to the combined effects of biological and physical clearance is

$$q_t = q_0 e^{-\lambda_{\text{rad}} t} e^{-\lambda_{\text{bio}} t}$$

$$= q_0 e^{-(\lambda_{\text{rad}} + \lambda_{\text{bio}})t}$$

$$= q_0 e^{-\lambda_{\text{eff}} t} \qquad 48.11$$

The *effective half-life, $t_{1/2,\text{eff}}$,* is the time it takes for a radioactive material in the body to decrease from the original activity to one-half of that activity, considering both the biological and physical clearance.

$$t_{1/2,\text{eff}} = \frac{\ln 2}{\lambda_{\text{eff}}}$$

$$= \frac{0.693}{\lambda_{\text{eff}}} \qquad 48.12$$

The effective half-life is related to the biological and radiological half-lives by

$$t_{1/2,\text{eff}} = \frac{t_{1/2,\text{bio}} \, t_{1/2,\text{rad}}}{t_{1/2,\text{bio}} + t_{1/2,\text{rad}}} \qquad 48.13$$

Table 48.4 Electron Dose Rate Factors in Skin from Selected Radionuclides Deposited on the Body Surface

	dose rate factors (Sv/yr per Bq/cm^2) versus depth in tissue[a,b]			
nuclide	0.004 cm	0.008 cm	0.040 cm	0.007 cm
^{3}H	0.0	0.0	0.0	0.0
^{14}C	7.9×10^{-3}	2.1×10^{-3}	0.0	2.9×10^{-3}
^{32}P	2.4×10^{-2}	2.0×10^{-2}	1.1×10^{-2}	2.1×10^{-2}
^{51}Cr	0.0	0.0	0.0	0.0
^{54}Mn	0.0	0.0	0.0	0.0
^{55}Fe	0.0	0.0	0.0	0.0
^{58}Co	3.6×10^{-3}	2.6×10^{-3}	4.4×10^{-4}	2.8×10^{-3}
^{60}Co	1.6×10^{-2}	8.7×10^{-3}	2.5×10^{-4}	9.9×10^{-3}
^{86}Rb	2.3×10^{-2}	1.9×10^{-2}	1.0×10^{-2}	2.0×10^{-2}
^{89}Sr	2.3×10^{-3}	1.9×10^{-2}	9.8×10^{-3}	2.0×10^{-2}
^{90}Sr	2.1×10^{-2}	1.5×10^{-2}	3.4×10^{-3}	1.6×10^{-2}
^{90}Y	2.4×10^{-2}	2.0×10^{-2}	1.2×10^{-2}	2.1×10^{-2}
^{91}Y	2.3×10^{-2}	1.9×10^{-2}	9.9×10^{-3}	2.0×10^{-2}
^{95}Zr	1.7×10^{-2}	1.0×10^{-2}	7.4×10^{-4}	1.2×10^{-2}
^{95}Nb	6.4×10^{-3}	1.7×10^{-3}	1.8×10^{-5}	2.3×10^{-3}
^{99}Mo	2.3×10^{-2}	1.8×10^{-2}	7.1×10^{-3}	1.9×10^{-2}
^{99m}Tc	2.9×10^{-3}	1.8×10^{-3}	0.0	2.1×10^{-3}
^{103}Ru	1.1×10^{-2}	4.8×10^{-3}	2.1×10^{-4}	5.8×10^{-3}
^{106}Ru	0.0	0.0	0.0	0.0
^{105}Rh	1.8×10^{-2}	1.2×10^{-2}	2.1×10^{-3}	1.3×10^{-2}
^{127}Sb	2.2×10^{-2}	1.7×10^{-2}	5.9×10^{-3}	1.8×10^{-2}
^{127}Te	2.1×10^{-2}	1.5×10^{-2}	4.0×10^{-3}	1.6×10^{-2}
^{127m}Te	1.6×10^{-2}	2.5×10^{-3}	9.4×10^{-5}	4.7×10^{-3}
^{129}Te	2.3×10^{-2}	1.9×10^{-2}	9.1×10^{-3}	2.0×10^{-2}
^{129m}Te	2.3×10^{-2}	1.1×10^{-2}	3.5×10^{-3}	1.3×10^{-2}
^{131}Te	2.8×10^{-2}	2.2×10^{-2}	1.0×10^{-2}	2.3×10^{-2}
^{132}Te	1.3×10^{-2}	5.9×10^{-3}	4.7×10^{-5}	7.0×10^{-3}
^{131}I	2.1×10^{-2}	1.4×10^{-2}	3.0×10^{-3}	1.5×10^{-2}
^{132}I	2.3×10^{-2}	1.8×10^{-2}	8.2×10^{-3}	1.9×10^{-2}
^{135}I	2.2×10^{-2}	1.7×10^{-2}	6.5×10^{-3}	1.8×10^{-2}
^{134}Cs	1.6×10^{-2}	1.1×10^{-2}	2.7×10^{-3}	1.2×10^{-2}
^{136}Cs	2.0×10^{-2}	1.2×10^{-2}	5.9×10^{-4}	1.3×10^{-2}
^{137}Cs	2.0×10^{-2}	1.3×10^{-2}	2.3×10^{-3}	1.4×10^{-2}
^{137m}Ba	2.4×10^{-3}	2.0×10^{-3}	1.2×10^{-3}	2.1×10^{-3}
^{140}Ba	2.2×10^{-2}	1.6×10^{-2}	5.0×10^{-3}	1.7×10^{-2}
^{140}La	2.4×10^{-2}	1.9×10^{-2}	9.2×10^{-3}	2.0×10^{-2}
^{141}Ce	2.5×10^{-2}	1.5×10^{-2}	1.6×10^{-3}	1.7×10^{-2}
^{143}Ce	2.4×10^{-2}	1.8×10^{-2}	7.7×10^{-3}	1.9×10^{-2}
^{144}Ce	1.5×10^{-2}	7.6×10^{-3}	1.7×10^{-4}	8.9×10^{-3}
^{143}Pr	2.2×10^{-2}	1.7×10^{-2}	6.2×10^{-3}	1.8×10^{-2}
^{147}Nd	2.3×10^{-2}	1.5×10^{-2}	4.2×10^{-3}	1.7×10^{-2}
^{239}Np	3.6×10^{-2}	2.0×10^{-2}	1.2×10^{-3}	2.3×10^{-2}
^{238}Pu	0.0	0.0	0.0	0.0
^{239}Pu	3.8×10^{-6}	0.0	0.0	0.0
^{241}Pu	0.0	0.0	0.0	0.0
^{241}Am	4.8×10^{-4}	1.1×10^{-5}	0.0	2.2×10^{-5}
^{242}Cm	0.0	0.0	0.0	0.0
^{244}Cm	0.0	0.0	0.0	0.0

[a]The first three depths are values recommended by Whitton (1973) for various parts of the body surface; the last depth is the average value over the body surface recommended in ICRP Publication 26 1977.
[b]Depths in g/cm^2 are equivalent to depths in cm for water.

Source: Kocher, D.C. and Eckerman, K.F., "Electron Dose-Rate Conversion Factors for External Exposure of the Skin for Uniformly Deposited Activity on the Body Surface," Health Physics, 53:135-141; 1987.

Health, Safety, Welfare

Table 48.5 Specific Gamma Ray Dose Constants (whole body; point sources at 100 cm distance)[*]

radioisotope	Γ $\left(\dfrac{\text{mSv/h}}{\text{MBq}}\right)$
Be-7	9.292×10^{-6}
C-11	1.908×10^{-4}
O-15	1.911×10^{-4}
F-18	1.851×10^{-4}
Na-22	3.590×10^{-4}
K-40	2.197×10^{-5}
Cr-51	6.320×10^{-6}
Fe-59	1.787×10^{-4}
Co-57	4.087×10^{-5}
Co-60	3.697×10^{-4}
Rb-86	1.458×10^{-5}
Pd-109	1.290×10^{-7}
I-125	7.432×10^{-5}
I-131	7.640×10^{-5}
Cs-137	1.017×10^{-4}
Au-198	7.881×10^{-5}
Po-210	1.422×10^{-9}
Rn-222	7.280×10^{-8}
Th-230	1.861×10^{-5}
U-235	9.159×10^{-5}

[*]Use the inverse-square law to calculate dose equivalents at other distances.

Adapted from Unger, L.M. and Trubey, D.K., *Specific Gamma-Ray Dose Constants for Nuclides Important to Dosimetry and Radiological Assessment*. ORNL/RSIC-45; Oak Ridge National Laboratory; Oak Ridge, TN; 1981, as reprinted 1982.

Example 48.6

What is the effective half-life of Mn-54 ($t_{1/2,\text{rad}} = 312.5$ days), which is part of a molecule that is cleared from the body with $t_{1/2,\text{bio}}$ of 25 days?

Solution

From Eq. 48.13,

$$t_{1/2,\text{eff}} = \frac{t_{1/2,\text{bio}} \, t_{1/2,\text{rad}}}{t_{1/2,\text{bio}} + t_{1/2,\text{rad}}} = \frac{(25 \text{ days})(312.5 \text{ days})}{25 \text{ days} + 312.5 \text{ days}}$$
$$= 23 \text{ days}$$

Calculating Internal Dose

The *annual limit on intake* (ALI) and the *derived air concentration* (DAC) can be used to calculate internal doses for intakes of radioactive material.

ALI is the derived limit for the amount of radioactive material that can be taken into the body of an adult worker by inhalation or ingestion in a year. ALI is the smaller value of intake of a given radionuclide in a year that would result in either a committed effective dose equivalent of 5 rems (0.05 Sv) or a committed dose equivalent of 50 rems (0.5 Sv) in any individual organ or tissue.

DAC is the concentration of a given radionuclide in air that, if breathed for a working year of 2000 hours under conditions of light work (i.e., inhalation rate 1.2 m³ of air per hour), would result in an intake of one ALI. The DAC relates to one of two modes of exposure: either external submersion or the internal committed dose equivalents resulting from inhalation of radioactive materials. DACs based upon submersion are for immersion in a semi-infinite cloud of uniform concentration and apply to each radionuclide separately.

A "DAC-hour" is the product of the actual concentration of radioactive material in air (expressed as a fraction or multiple of the DAC for each radionuclide) and the time of exposure for that radionuclide, in hours. Two thousand DAC-hours equal one ALI and produce a committed effective dose equivalent of 5 rems (0.05 Sv).

The DAC is a limit intended to control chronic occupational exposures. The relationship between the DAC and the ALI is given by

$$\text{DAC} = \frac{\text{ALI}}{(2000 \text{ h})\left(60 \, \dfrac{\text{min}}{\text{h}}\right)\left(2 \times 10^4 \, \dfrac{\text{mL}}{\text{min}}\right)}$$
$$= \frac{\text{ALI}}{2.4 \times 10^9 \text{ mL}} \qquad \text{48.14}$$

2×10^4 mL/min is the standard breathing rate under conditions of light work.

Inhalation Class

Class (*lung class* or *inhalation class*) as used in Table 48.8 is a classification scheme for inhaled material categorized according to the rate of clearance from the pulmonary region of the lung. Materials are classified as D, W, or Y as follows.

> class D (days) = less than 10 days
>
> class W (weeks) = from 10 to 100 days
>
> class Y (years) = greater than 100 days

Occupational Values

The ALIs in Table 48.8, part 1 are the annual intakes of a given radionuclide that would result in either

- a committed effective dose equivalent, $H_{E,50}$, of 5 rem (stochastic ALI)

- a committed dose equivalent, $H_{T,50}$, of 50 rem to an organ or tissue (nonstochastic ALI)

When the ALI is defined by the stochastic dose limit, it is the only ALI value listed in Table 48.8. When the committed dose equivalent, $H_{T,50}$, of a particular organ would reach 50 rem before the committed effective dose equivalent, $H_{E,50}$, would reach 5 rem, the ALI is limited by the nonstochastic dose limit to an organ. In that case, the organ or tissue to which the nonstochastic (committed dose equivalent, $H_{T,50} \leq 50$ rem) limit applies is listed, and the ALI for the stochastic (committed effective dose equivalent, $H_{E,50} \leq 5$ rem) limit is shown in parentheses.

Table 48.6 Neutron Weighting Factors and Absorbed Dose Equivalencies (whole body)

neutron energy (MeV)	weighting factor[a], (w_R)	fluence per unit dose equivalent[b].f/H_T (neutrons/cm^2·rem)
2.5×10^{-8} (thermal)	2	980×10^6
1×10^{-7}	2	980×10^6
1×10^{-6}	2	810×10^6
1×10^{-5}	2	810×10^6
1×10^{-4}	2	840×10^6
1×10^{-3}	2	980×10^6
1×10^{-2}	2.5	1010×10^6
1×10^{-1}	7.5	170×10^6
5×10^{-1}	11	39×10^6
1	11	27×10^6
2.5	9	29×10^6
5	8	23×10^6
7	7	24×10^6
10	6.5	24×10^6
14	7.5	17×10^6
20	8	16×10^6
40	7	14×10^6
60	5.5	16×10^6
1×10^2	4	20×10^6
2×10^2	3.5	19×10^6
3×10^2	3.5	16×10^6
4×10^2	3.5	14×10^6

[a] Value of weighting factor, w_R, at the point where the dose equivalent is maximum in a 30 cm diameter cylinder tissue-equivalent phantom
[b] Monoenergetic neutrons incident normally on a 30 cm diameter cylinder tissue-equivalent phantom

Source: Table 1004(b).2, 10 CFR 20.1004, *Units of Radiation Dose*, U.S. Nuclear Regulatory Commission, Washington, DC.

Table 48.7 Organ Dose Tissue Weighting Factors

organ or tissue	w_T
gonads	0.25
breast	0.15
red bone marrow	0.12
lung	0.12
thyroid	0.03
bone surface	0.03
remainder[a]	0.30
whole body[b]	1.00

[a] 0.30 results from 0.06 for each of 5 "remainder" organs (excluding the skin and the lens of the eye) that receive the highest doses.
[b] For the purpose of weighting the external whole-body dose (for adding it to the internal dose), a single weighting factor, $w_T = 1.0$, has been specified.

Source: 10 CFR 20.1003, *Definitions*, U.S. Nuclear Regulatory Commission, Washington, DC.

Effluents, Air, and Water

The concentration values given in columns 1 and 2 of part 2 of Table 48.8 are equivalent to the radionuclide concentrations that, if inhaled or ingested continuously over a year, would produce a total effective dose equivalent, $H_{E,tot}$, of 0.05 rem (50 mrem or 0.5 mSv).

Sewer Disposal

The sewer disposal concentrations were calculated by taking the most restrictive occupational stochastic oral ingestion ALI and dividing by (a) 7.3×10^5 mL (the standard per-capita annual water intake by a "reference man") and (b) a factor of 10. The factor of 10 reduces the 5 rem produced by intake of one ALI to the 0.5 rem committed effective dose equivalent, $H_{E,50}$, regulatory limit for a member of the public.

Example 48.7

A worker inhaled 8 μCi of iodine-131. Calculate (a) the committed dose equivalent, $H_{T,50}$, to the thyroid and (b) the committed effective dose equivalent, $H_{E,50}$.

Solution
(a) From Table 48.8, part 1, column 2, for I-131, the thyroid gland has a nonstochastic ALI of 5×10^1 μCi. A nonstochastic ALI produces 50 rem committed dose equivalent, $H_{T,50}$, in the specified organ (thyroid).

$$50 \text{ rem} = 5 \times 10^1 \ \mu\text{Ci}$$

For the thyroid, the committed dose equivalent produced by 8 μCi is

$$H_{T,50} = 8 \ \mu\text{Ci}$$

Therefore,

$$\frac{H_{T,50}}{8 \ \mu\text{Ci}} = \frac{50 \text{ rem}}{5 \times 10^1 \ \mu\text{Ci}}$$

$$H_{T,50} = \frac{(8 \ \mu\text{Ci})(50 \text{ rem})}{5 \times 10^1 \ \mu\text{Ci}} = 8 \text{ rem} \quad (0.08 \text{ Sv})$$

(b) From Table 48.8, part 1, column 2, for I-131, the stochastic ALI is 2×10^2 μCi. This would produce a 5 rem committed effective dose equivalent, $H_{E,50}$.

$$5 \text{ rem} = 2 \times 10^2 \ \mu\text{Ci}$$

The committed effective dose equivalent produced by 8 μCi is

$$H_{E,50} = 8 \ \mu\text{Ci}$$

Therefore,

$$\frac{H_{E,50}}{8 \ \mu\text{Ci}} = \frac{5 \text{ rem}}{2 \times 10^2 \ \mu\text{Ci}}$$

$$H_{E,50} = \frac{(8 \ \mu\text{Ci})(5 \text{ rem})}{2 \times 10^2 \ \mu\text{Ci}} = 0.2 \text{ rem} \quad (0.002 \text{ Sv})$$

Example 48.8

What is the total effective dose equivalent, $H_{E,tot}$, to a person who inhaled iodine-131 at an average concentration of 9×10^{-12} μCi/mL for one year?

Health, Safety, Welfare

Table 48.8 *Representative Annual Limits on Intake (ALIs) and Derived Air Concentrations (DACs) of Radionuclides for Occupational Exposure, Effluent Concentrations, and Concentrations for Release to Sewerage*

atomic no.	radionuclide	class	part 1 occupational values			part 2 effluent concentrations		part 3 releases to sewers
			oral ingestion ALI (μCi)	inhalation		col. 1 air (μCi/mL)	col. 2 water (μCi/mL)	monthly average concentration (μCi/mL)
				ALI (μCi)	DAC (μCi/mL)			
1	hydrogen-3	water, DAC includes absorption	8×10^4	8×10^4	2×10^{-5}	1×10^{-7}	1×10^{-3}	1×10^{-2}
6	carbon-11	monoxide	–	1×10^6	5×10^{-4}	2×10^{-6}	–	–
		dioxide	–	6×10^5	3×10^{-4}	9×10^{-7}	–	–
		compounds	4×10^5	4×10^5	2×10^{-4}	6×10^{-7}	6×10^{-3}	6×10^{-2}
6	carbon-14	monoxide	–	2×10^6	7×10^{-4}	2×10^{-6}	–	–
		dioxide	–	2×10^5	9×10^{-5}	3×10^{-7}	–	–
		compounds	2×10^3	2×10^3	1×10^{-6}	3×10^{-9}	3×10^{-5}	3×10^{-4}
15	phosphorus-32	D, all compounds except phosphates given for W	6×10^2	9×10^2	4×10^{-7}	1×10^{-9}	9×10^{-6}	9×10^{-5}
		W, phosphates of Zn^{2+}, S^{3+}, Mg^{2+}, Fe^{3+}, Bi^{3+}, and lanthanides	–	4×10^2	2×10^{-7}	5×10^{-10}	–	–
28	nickel-63	D, see Ni-56	9×10^3	2×10^3	7×10^{-7}	2×10^{-9}	1×10^{-4}	1×10^{-3}
		W, see Ni-56	–	3×10^3	1×10^{-6}	4×10^{-9}	–	–
		vapor	–	8×10^2	3×10^{-7}	1×10^{-9}	–	–
20	calcium-45	W, all compounds	2×10^3	8×10^2	4×10^{-7}	1×10^{-9}	2×10^{-5}	2×10^{-4}
53	iodine-125	D, all compounds	4×10^1 thyroid	6×10^1 thyroid	3×10^{-8}	–	–	–
			(1×10^2)	(2×10^2)	–	3×10^{-10}	2×10^{-6}	2×10^{-5}
53	iodine-131	D, all compounds	3×10^1 thyroid	5×10^1 thyroid	2×10^{-8}	–	–	–
			(9×10^1)	(2×10^2)	–	2×10^{-10}	1×10^{-6}	1×10^{-5}
95	americium-241	W, all compounds	8×10^{-1} bone surface	6×10^{-3} bone surface	3×10^{-12}	–	–	–
			(1×10^0)	(1×10^{-2})	–	2×10^{-14}	2×10^{-8}	2×10^{-7}

Source: Appendix B to 10 CFR 20, Index of Radioisotopes, U.S. Nuclear Regulatory Commission, Washington, DC.

Solution

From Table 48.8, part 2, column 1, for I-131, 2×10^{-10} μCi/mL produces a 0.05 rem total effective dose equivalent, $H_{E,\text{tot}}$.

$$0.05 \text{ rem} = 2 \times 10^{-10} \text{ } \mu\text{Ci/mL}$$

The total effective dose equivalent produced by 9×10^{-12} μCi/mL is

$$H_{E,\text{tot}} = 9 \times 10^{-12} \text{ } \mu\text{Ci/mL}$$

Therefore,

$$\frac{H_{E,\text{tot}}}{9 \times 10^{-12} \text{ } \frac{\mu\text{Ci}}{\text{mL}}} = \frac{0.05 \text{ rem}}{2 \times 10^{-10} \text{ } \frac{\mu\text{Ci}}{\text{mL}}}$$

$$H_{E,\text{tot}} = \frac{\left(9 \times 10^{-12} \text{ } \frac{\mu\text{Ci}}{\text{mL}}\right)(0.05 \text{ rem})}{2 \times 10^{-10} \text{ } \frac{\mu\text{Ci}}{\text{mL}}}$$

$$= 2.3 \times 10^{-3} \text{ rem} \quad (2.3 \times 10^{-5} \text{ Sv})$$

Example 48.9

What is the committed effective dose equivalent, $H_{E,50}$, to a person whose drinking water was supplied by a facility containing Ni-63 at an average monthly concentration of 5×10^{-5} μCi/mL?

Solution

From Table 48.8, part 3 for Ni-63, 1×10^{-3} μCi/mL produces 0.5 rem committed effective dose equivalent, $H_{E,50}$, so

$$0.5 \text{ rem} = 1 \times 10^{-3} \ \mu\text{Ci/mL}$$

The committed effective dose equivalent produced by 5×10^{-5} μCi/mL is

$$H_{E,50} = 5 \times 10^{-5} \ \mu\text{Ci/mL}$$

Therefore,

$$\frac{H_{E,50}}{5 \times 10^{-5} \ \frac{\mu\text{Ci}}{\text{mL}}} = \frac{0.5 \text{ rem}}{1 \times 10^{-3} \ \frac{\mu\text{Ci}}{\text{mL}}}$$

$$H_{E,50} = \frac{\left(5 \times 10^{-5} \ \frac{\mu\text{Ci}}{\text{mL}}\right)(0.5 \text{ rem})}{1 \times 10^{-3} \ \frac{\mu\text{Ci}}{\text{mL}}}$$

$$= 2.5 \times 10^{-2} \text{ rem} \quad (2.5 \times 10^{-4} \text{ Sv})$$

6. CONTROLLING EXTERNAL RADIATION EXPOSURE

External radiation exposure is a problem primarily with mid- to high-energy beta emitters and gamma and X-ray emitters. The three most effective ways to reduce the exposure from external radiation hazards are (1) reduced exposure time, (2) increased distance, and (3) thicker shielding.

Time

The exposure received is directly proportional to the amount of time spent near a radiation source. That is, if the time spent near a radiation source is doubled, the radiation dose is doubled.

Distance

The greater the distance from a radiation source, the less radiation received. For sources that can be treated as a *point source* (a source that is a distance at least three times the longest dimension of the source), the radiation dose received is inversely proportional to the square of the distance of separation. This is referred to as the *inverse-square law*. The intensity of a radiation field decreases with the square of the distance from the source, so if the distance from a source is doubled, the

intensity of the radiation field will decrease to one-fourth the initial exposure.

$$I_1 d_1^2 = I_2 d_2^2 \qquad 48.15$$

Not all sources follow the inverse-square law.

- *Point sources:* Radiation intensity decreases with distance approximately as $1/d^2$ (doubling the distance quarters the dose rate).

- *Line sources:* Radiation intensity decreases with distance approximately as $1/d$ (doubling the distance cuts the radiation intensity to about $1/2$).

- *Disk and cylindrical sources:* Radiation intensity decreases with distance between $1/d$ and $1/d^2$ (doubling the distance reduces the radiation intensity to between $1/2$ and $1/4$).

Example 48.10

For the scenario in Ex. 48.4, what is the dose equivalent rate 20 cm from the same source?

Solution

From Eq. 48.15,

$$I_2 = I_1 \left(\frac{d_1}{d_2}\right)^2 = \left(6.5 \ \frac{\text{mSv}}{\text{h}}\right)\left(\frac{1 \text{ m}}{0.2 \text{ m}}\right)^2$$

$$= 1.6 \times 10^2 \text{ mSv/h}$$

Shielding

Any material can act as a radiation shield, and, in general, the denser the material, the better shield it makes. In most cases, the closer the shielding to the source, the better economically, since less shielding material will be required. Selecting the proper shielding material depends on the type of radiation involved.

- Proper shielding for low-energy gamma emitters, such as I-125, requires only thin sheets of lead foil.

- For medium-energy gamma emitters, such as Co-57, about $1/4$ in (6.4 mm) of lead is needed.

- High-energy gamma emitters such as cobalt-60, sodium-22, manganese-54, chromium-51, and iodine-131 require lead bricks to effectively attenuate the gamma radiation.

- Proper shielding of high-energy beta emitters can be achieved with about 1 cm of plexiglass.

- Shielding is not required for low-energy beta emitters, such as sulfur-35 or carbon-14, as these beta particles have very limited ranges in air.

- Shielding is not required for alpha particles, though alpha emitters, such as americium-241, usually also emit gamma radiation.

More specific information on shielding is contained in Chap. 49.

Health, Safety, Welfare

7. RADIATION EXPOSURE REGULATIONS

Part 20 of Title 10 of the Code of Federal Regulations (10 CFR 20) contains the U.S. Nuclear Regulatory Commission's radiation exposure limits.

Occupational Dose to Adults

The occupational dose to adults (10 CFR 20.1201) is the more limiting of

- a total effective dose equivalent, $H_{E,tot}$, of 5 rems (0.05 Sv) in one year

- the sum of the deep-dose equivalent, H_D, and the committed dose equivalent, $H_{T,50}$, to any individual organ or tissue, other than the lens of the eye, equal to 50 rems (0.5 Sv) in one year

The annual limits to the lens of the eye, to the skin, and to the *extremities* (e.g., hand, elbow, arm below the elbow, foot, knee, and leg below the knee) are

- a lens dose equivalent, H_L, of 15 rems (0.15 Sv) in one year

- a shallow-dose equivalent, H_S, of 50 rems (0.50 Sv) to the skin or to any extremity in one year

Occupational Dose to Minors

The annual occupational dose limit for minors is 10% of the annual dose limits for adult workers (10 CFR 20.1207).

Embryo/Fetus

The term *embryo/fetus* means the developing human organism from conception until the time of birth. A *declared pregnant woman* is a woman who has voluntarily informed her employer, in writing, of her pregnancy and the estimated date of conception. The total effective dose equivalent, $H_{E,tot}$, limit for the embryo/fetus during the entire pregnancy, due to the occupational exposure of a declared pregnant woman, is 0.5 rem (5 mSv) (10 CFR 20.1208). Additionally, substantial variation from a uniform monthly exposure rate to a declared pregnant woman must be avoided. The dose equivalent to the embryo/fetus is the sum of

- the deep-dose equivalent to the declared pregnant woman

- the dose equivalent to the embryo/fetus resulting from radionuclides in the embryo/fetus and radionuclides in the declared pregnant woman

Individual Members of the Public

Public dose means the dose received by a member of the public from exposure to radiation or radioactive material, not including occupational dose or doses received from background radiation, from any medical exposure the individual has received, from exposure to individuals administered radioactive material, from voluntary participation in medical research programs, or from disposal of radioactive material into sanitary sewage. The total effective dose equivalent, H_E, limit for individual members of the public is 0.1 rem (1 mSv) in a year (10 CFR 20.1301).

As Low As Reasonably Achievable (ALARA) Regulation

As low as reasonably achievable (ALARA) means making every reasonable effort to maintain exposures to radiation as far below the dose limits as is practical consistent with the purpose for which the activity is undertaken. The USNRC's regulations in 10 CFR 20 require that employees use, to the extent that is practical, procedures and engineering controls based upon sound radiation protection principles to limit occupational doses (and doses to members of the public) to those that are as low as is reasonably achievable. In other words, it is possible to violate the ALARA regulation even if no exposure exceeds the regulatory limits if a reasonable way to reduce radiation exposure is not taken advantage of. The ALARA concept must be applied both to individual doses and to collective doses. *Collective dose* is the sum of all doses received by persons and is reported in person·rem or person·Sv.

Example 48.11

Ten people were exposed to a 2 R/hr X-ray field for three hours. What is the collective dose equivalent?

Solution

$$(10 \text{ persons})\left(2 \ \frac{R}{h}\right)(3 \text{ h}) = 60 \text{ person·R}$$

For radiation protection purposes, 1 R ≈ 1 rem.

$$60 \text{ person·R} \approx 60 \text{ person·rem} = 0.60 \text{ person·Sv}$$

The average dose equivalent would be

$$\frac{60 \text{ person·rem}}{10 \text{ persons}} = 6.0 \text{ rem/person}$$
$$= 0.060 \text{ person·Sv}$$

49 Shielding

Nomenclature

A	atomic mass	amu
B	buildup factor	–
c	speed of light	m/s
d	distance	m
E	energy	J
F	force	N
h	Planck's constant, 6.63×10^{-34}	J·s
I	radiation intensity	various
k_0	electrostatic constant of proportionality, 9×10^9	N·m^2/C^2
k_1	low energy α-particle coefficient in air, 0.56	cm/MeV
k_2	high energy α-particle coefficient in air, 1.24	cm/MeV
k_3	high energy α-particle standardization factor in air, 2.62	cm
k_4	general α-particle coefficient, 0.56×10^{-3}	g/cm^3
k_5	low energy β-particle coefficient, 0.412	g/cm^2
k_6	low energy β-particle exponential standardization factor, 1.265	–
k_7	low energy β-particle exponential coefficient, 0.0954	–
k_8	high energy β-particle coefficient, 0.530	g/cm^2·MeV
k_9	high energy β-particle standardization factor, 0.106	g/cm^2
m	mass	kg
N	atomic (nucleic) density	cm^{-3}
N_A	Avogadro's number, 6.02×10^{23}	mol^{-1}
q	electrical charge	C
R	range	cm
v	velocity	cm/s
x	shield or absorber thickness	cm

Symbols

λ	relaxation length (shielding)	cm
λ	wavelength (electromagnetic energy)	cm
μ	mass attenuation coefficient	cm^2/g
ν	frequency	Hz
ρ	density	g/cm^3
σ	microscopic cross section	cm^2
Σ	macroscopic cross section	cm^{-1}

Subscripts

0	initial
a	absorption
ave	average
max	maximum
s	scattering
t	total

1. RADIATION ENERGY

A particle's kinetic energy can be calculated from its mass and velocity.

$$E = \tfrac{1}{2}mv^2 \qquad \textbf{49.1}$$

For electromagnetic radiation (photons), energy can be calculated from Planck's constant, h, and the frequency, ν. The speed of light, c, in free space is approximately 3×10^8 m/s.

$$E = h\nu \qquad \textbf{49.2}$$

$$c = \nu\lambda \qquad \textbf{49.3}$$

Substituting for ν, the energy equation becomes

$$E = \frac{hc}{\lambda} \qquad \textbf{49.4}$$

The goal of radiation shielding is to capture the radiation's energy in the shield. How much of the radiation's energy is captured depends on (1) the number of particles or photons, (2) the interaction rate with the shield's atoms, and (3) the interaction mechanism.

Example 49.1

Calculate the kinetic energy of a 140 g baseball traveling at 10 m/s.

Solution

From Eq. 49.1,

$$E = \tfrac{1}{2}mv^2 = \frac{\left(\tfrac{1}{2}\right)(140 \text{ g})\left(10 \ \frac{\text{m}}{\text{s}}\right)^2}{1000 \ \frac{\text{g}}{\text{kg}}}$$

$$= 7.0 \text{ J}$$

Health, Safety, Welfare

Example 49.2

Calculate the energy of green visible light at a wavelength of 500 nm.

Solution

From Eq. 49.4,

$$E = \frac{hc}{\lambda} = \frac{(6.63 \times 10^{-34} \text{ J·s})\left(3 \times 10^8 \, \frac{\text{m}}{\text{s}}\right)}{500 \times 10^{-9} \text{ m}}$$
$$= 3.98 \times 10^{-19} \text{ J}$$

Radiation Decay Products

During radioactive decay, an atom emits energy. Since energy and mass are equivalent ($E = mc^2$), an atom can emit energy in the form of charged particles (alpha and beta particles), uncharged particles (neutrons), or electromagnetic energy "photons" (gamma rays, X-rays, and visible light). (See Table 49.1.)

Table 49.1 *Properties of Radioactive Decay Products*

product	symbol	mass (amu)	charge
alpha	α or $^4_2\text{He}^{+2}$	4.0026	$+2$ (3.2×10^{-19} C)
beta	β or β^-	0.000548	-1 (-1.6×10^{-19} C)
gamma	γ	0	0
neutron	n	1.008665	0

Range

The total path length a charged particle travels before it is stopped is called its *range*. Theoretically, the range may be determined by numerical integration of an appropriate stopping power formula that describes the energy lost by the charged particle per unit path length through a given substance. Often, though, it is easier to use empirically derived equations to determine charged particle range. For heavier particles (e.g., alpha particles), range-energy relations expressed as power laws—in which the charged particle's range, R, is proportional to its energy, E, raised to a given power—provide a reasonably accurate measure of charged particle range. It should be borne in mind, however, that because energy loss events are statistical phenomena, discrepancies in range measurement can occur. Two particles with identical initial energies may travel different distances in the same medium before being stopped. For this reason, range should be interpreted as an average value.

Shielding Charged Particles

Charged particles interact and lose energy in a predictable way per unit distance of travel. If a material greater in thickness than their range is placed in their path, all of the charged particles will interact, give up their energy, and be successfully shielded. Charged particles interact with the charged particles in the shield's atoms and experience electrostatic forces. *Coulomb's law* states

that two charged particles will exert a force on each other, the magnitude of which is proportional to the product of their charges and inversely proportional to the square of their separation distance. k_0 has a value of approximately 9×10^9 N·m²/C².

$$F = \frac{k_0 q_1 q_2}{d^2} \qquad \text{49.5}$$

Ionization occurs when a charged particle interacts electrostatically with a surrounding atom's orbital electron, exerting enough force to liberate the electron from its atomic orbit. This ionizes the atom, creating an ion pair consisting of the liberated electron (the negative ion) and the ionized atom (the positive ion).

Excitation occurs when a charged particle interacts with a surrounding atom's orbital electron through electrostatic forces that are not sufficient to ionize the orbital electron, but that cause the orbital electron to "jump" to a higher energy level orbital around the atom.

Example 49.3

Calculate the electrostatic force between two beta particles separated by 2.0×10^{-15} m.

Solution

From Table 49.1, for beta particles,

$$q = -1.6 \times 10^{-19} \text{ C}$$

From Eq. 49.5,

$$F = \frac{k_0 q_1 q_2}{d^2}$$
$$= \frac{\left(9 \times 10^9 \, \frac{\text{N·m}^2}{\text{C}^2}\right)(-1.6 \times 10^{-19} \text{ C})(-1.6 \times 10^{-19} \text{ C})}{(2.0 \times 10^{-15} \text{ m})^2}$$
$$= 57.6 \text{ N}$$

Shielding Uncharged Particles

Unlike alpha and beta particles, neutrons and photons are not electrically charged. They interact and lose their energy through collisions and direction changes in a probabilistic way per unit distance of travel. It is generally possible only to reduce the intensity of the uncharged particle radiation, rather than to completely eliminate it. Complete shielding of neutrons and photons theoretically requires an infinitely thick shield.

2. ALPHA PARTICLE SHIELDING

Alpha particles are "bare" helium atoms that are missing the two orbital electrons. Alpha particles are the least penetrating of the radiations. In air, even the most energetic alpha particles travel only several centimeters. In tissue, alpha particles can only travel on the order of microns.

To determine the required shield thickness for alpha particles, first find their range in air using Eq. 49.6 or Eq. 49.7. Then use Eq. 49.8 to find the range in the shielding material.

For alpha particles in air with energies less than 4 MeV (at 0°C and 1 atm pressure),

$$R_{\alpha,\text{air}} = k_1 E_\alpha \quad [E_\alpha \le 4 \text{ MeV}] \qquad \textbf{49.6}$$

For alpha particles in air with energies between 4 MeV and 8 MeV (at 0°C and 1 atm pressure),

$$R_{\alpha,\text{air}} = k_2 E_\alpha - k_3 \quad [4 \text{ MeV} < E_\alpha < 8 \text{ MeV}] \qquad \textbf{49.7}$$

For alpha particles in materials other than air,

$$R_{\alpha,\text{shield}} = \frac{k_4 (A_{\text{shield}})^{1/3} R_{\alpha,\text{air}}}{\rho_{\text{shield}}} \qquad \textbf{49.8}$$

Example 49.4

What thickness of copper ($\rho = 8.9$ g/cm^3, $A_{\text{Cu}} = 63.5$) is needed to shield 3.5 MeV alpha particles?

Solution

From Eq. 49.6,

$$R_{\alpha,\text{air}} = k_1 E_\alpha = \left(0.56 \; \frac{\text{cm}}{\text{MeV}}\right)(3.5 \text{ MeV})$$
$$= 1.96 \text{ cm}$$

From Eq. 49.8,

$$R_{\alpha,\text{Cu}} = \frac{k_4 (A_{\text{Cu}})^{1/3} R_{\alpha,\text{air}}}{\rho_{\text{Cu}}}$$
$$= \frac{\left(0.56 \times 10^{-3} \; \frac{\text{g}}{\text{cm}^3}\right)(63.5)^{1/3}(1.96 \text{ cm})}{8.9 \; \frac{\text{g}}{\text{cm}^3}}$$
$$= 4.9 \times 10^{-4} \text{ cm}$$

Example 49.5

What thickness of aluminum ($\rho_{\text{Al}} = 2.7$ g/cm^3, $A_{\text{Al}} = 2.7$) is needed to shield the alpha particles from americium-237 ($E_\alpha = 6.0$ MeV)?

Solution

From Eq. 49.7,

$$R_{\alpha,\text{air}} = k_2 E_\alpha - k_3$$
$$= \left(1.24 \; \frac{\text{cm}}{\text{MeV}}\right)(6.0 \text{ MeV}) - 2.62 \text{ cm}$$
$$= 4.82 \text{ cm}$$

From Eq. 49.8,

$$R_{\alpha,\text{Al}} = \frac{k_4 (A_{\text{Al}})^{1/3} R_{\alpha,\text{air}}}{\rho_{\text{Al}}}$$
$$= \frac{\left(0.56 \times 10^{-3} \; \frac{\text{g}}{\text{cm}^3}\right)(27)^{1/3}(4.82 \text{ cm})}{2.7 \; \frac{\text{g}}{\text{cm}^3}}$$
$$= 3.0 \times 10^{-3} \text{ cm}$$

3. BETA PARTICLE SHIELDING

Beta particles are more penetrating than alpha particles because of their lower masses and electrical charges.

A beta particle's mass is 7300 times less than the mass of an alpha particle. For an alpha particle and a beta particle with the same kinetic energy, the beta particle travels 85 times faster than the alpha particle. This allows less time for the beta particle to be affected by electrostatic forces.

Since beta particles have half the electrical charge of alpha particles, Coulomb forces are also halved. This allows beta particles to travel much farther than alpha particles before losing their kinetic energy to Coulomb interactions.

Unlike alpha particles, which are all emitted with the same energy (i.e., alpha particle decay is *monoenergetic*), beta particles are emitted in a spectrum of energies from very low energies up to a maximum energy, $E_{\beta,\text{max}}$. (See Table 49.2.) The average beta energy, $E_{\beta,\text{ave}}$, is roughly equal to one-third of $E_{\beta,\text{max}}$. Since the maximum-energy beta particles are the most penetrating, shielding must be designed to stop them.

Table 49.2 *Energy for Selected Radionuclides*

radioisotope	average beta energy, $E_{\beta,\text{ave}}$ (MeV)	maximum beta energy, $E_{\beta,\text{max}}$ (MeV)
argon-41	1.077	2.492
calcium-47	0.8168	1.988
carbon-14	0.04947	0.1565
cobalt-60	0.09579	0.3179
hydrogen-3	0.005685	0.01860
lead-210	0.01613	0.06300
molybdenum-99	0.4427	1.214
phosphorus-32	0.6949	1.710
plutonium-241	0.005230	0.02081
potassium-40	0.5085	1.312
strontium-89	0.5830	1.491
strontium-90	0.1958	0.5460
yttrium-90	0.9348	2.284

Source: Kocher, D.C., *Radioactive Decay Data Tables*, DOE-TIC-11026, U.S. Department of Energy, Washington, DC, 1981.

Health, Safety, Welfare

To determine the required shield thickness for beta particles, use Eq. 49.9 or Eq. 49.10.

$$R_\beta = \frac{k_5 E_{\beta,\max}^{k_6 - k_7 (\ln E_{\beta,\max})}}{\rho}$$

$$[0.01 \text{ MeV} \leq E_{\beta,\max} \leq 2.5 \text{ MeV}] \quad 49.9$$

$$R_\beta = \frac{k_8 E_{\beta,\max} - k_9}{\rho}$$

$$[E_{\beta,\max} > 2.5 \text{ MeV}] \quad 49.10$$

Example 49.6

What thickness of platinum ($\rho_{Pt} = 21.4$ g/cm^3) is needed to shield beta particles from phosphorus-32 ($E_{\beta,\max} = 1.710$ MeV)?

Solution

From Eq. 49.9,

$$R_{\beta,Pt} = \frac{k_5 E_{\beta,\max}^{k_6 - k_7 (\ln E_{\beta,\max})}}{\rho_{Pt}}$$

$$= \frac{\left(0.412 \ \frac{\text{g}}{\text{cm}^2}\right) \times (1.710 \text{ MeV})^{(1.265 - (0.0954)(\ln 1.710 \text{ MeV}))}}{21.4 \ \frac{\text{g}}{\text{cm}^3}}$$

$$= 3.69 \times 10^{-2} \text{ cm}$$

Example 49.7

How far will the most energetic beta particles from argon-41 ($E_{\beta,\max} = 2.492$ MeV) travel in water ($\rho_{H_2O} = 1$ g/cm^3)?

Solution

From Eq. 49.10,

$$R_{\beta,H_2O} = \frac{k_8 E_{\beta,\max} - k_9}{\rho_{H_2O}}$$

$$= \frac{\left(0.530 \ \frac{\text{g}}{\text{cm}^2 \cdot \text{MeV}}\right)(2.492 \text{ MeV}) - 0.106 \ \frac{\text{g}}{\text{cm}^2}}{1 \ \frac{\text{g}}{\text{cm}^3}}$$

$$= 1.21 \text{ cm}$$

Beta Particle Energy Loss

Ionization

Ionization occurs when a beta particle collides with a shield atom and adds enough energy to liberate one of the electrons in the shield atom. This ionizes the atom

and creates an ion pair consisting of the liberated electron as the negative ion and the ionized atom as the positive ion.

Excitation

During excitation, a beta particle's collision with a shield atom does not add enough energy to remove the orbital electron, but it does add enough to cause the electron to "jump" to a higher energy level.

Bremsstrahlung

In this interaction, the added energy is not strong enough to ionize or "excite" the orbital electron. Rather, when beta particles interact with shield electrons, they produce *bremsstrahlung* (German for "braking radiation") X-ray radiation as they slow down or change direction (similar to heat generated in brake pads when braking).

Thus, by shielding beta particles, a gamma radiation hazard may be created. Since the intensity of the bremsstrahlung is proportional to the number of beta particles, their incident energy, and the effective (average) atomic number, Z, of the target material, beta shields should be designed from materials with low atomic numbers. Since plastics have low effective atomic numbers due to the carbon and hydrogen content (in addition to being inexpensive and easy to get and form), they are useful for beta shielding.

4. GAMMA RAY AND X-RAY SHIELDING

Photons are bundles ("packets") of energy that travel at the speed of light. In many interactions, photons act like particles. Because photons are massless and carry no electrical charge, they interact differently than charged particles. Charged particles interact and give up their kinetic energy continuously along their path, while photons interact only occasionally. Photons interact with shield atoms through the photoelectric effect, Compton scattering, pair production, and photodisintegration.

The combined probabilities of each type of interaction give a total probability of a photon's interaction as it travels through the shield. This combined probability is expressed in terms of a *mass attenuation coefficient*, μ (in cm^2/g). (See Table 49.3.) For a relatively thin shield, the intensity of the gamma rays that can penetrate a distance x is

$$I_x = I_0 e^{-\mu \rho x} \quad [\text{thin shield}] \quad 49.11$$

Health, Safety, Welfare

Table 49.3 Mass Attenuation Coefficients, μ, for Gamma Radiation (cm^2/g)

material	Gamma Ray Energy, MeV																	
	0.1	0.15	0.2	0.3	0.4	0.5	0.6	0.8	1.0	1.25	1.5	2	3	4	5	6	8	10
H	0.295	0.265	0.243	0.212	0.189	0.173	0.160	0.140	0.126	0.113	0.103	0.0876	0.0691	0.0579	0.0502	0.0446	0.0371	0.0321
Be	0.132	0.119	0.109	0.0945	0.0847	0.0773	0.0715	0.0628	0.0565	0.0504	0.0459	0.0394	0.0313	0.0266	0.0234	0.0211	0.0180	0.0161
C	0.149	0.134	0.122	0.106	0.0953	0.0870	0.0805	0.0707	0.0636	0.0568	0.0518	0.0444	0.0356	0.0304	0.0270	0.0245	0.0213	0.0194
N	0.150	0.134	0.123	0.106	0.0955	0.0869	0.0805	0.0707	0.0636	0.0568	0.0517	0.0445	0.0357	0.0306	0.0273	0.0249	0.0218	0.0200
O	0.151	0.134	0.123	0.107	0.0953	0.0870	0.0806	0.0708	0.0636	0.0568	0.0518	0.0445	0.0359	0.0309	0.0276	0.0254	0.0224	0.0206
Na	0.151	0.130	0.118	0.102	0.0912	0.0833	0.0770	0.0676	0.0608	0.0546	0.0496	0.0427	0.0348	0.0303	0.0274	0.0254	0.0229	0.0215
Mg	0.160	0.135	0.122	0.106	0.0944	0.0860	0.0795	0.0699	0.0627	0.0560	0.0512	0.0442	0.0360	0.0315	0.0286	0.0266	0.0242	0.0228
Al	0.161	0.134	0.120	0.103	0.0922	0.0840	0.0777	0.0683	0.0614	0.0548	0.0500	0.0432	0.0353	0.0310	0.0282	0.0264	0.0241	0.0229
Si	0.172	0.139	0.125	0.107	0.0954	0.0869	0.0802	0.0706	0.0635	0.0567	0.0517	0.0447	0.0367	0.0323	0.0293	0.0277	0.0254	0.0243
P	0.174	0.137	0.122	0.104	0.0928	0.0846	0.0780	0.0685	0.0617	0.0551	0.0502	0.0436	0.0358	0.0316	0.0290	0.0273	0.0252	0.0242
S	0.188	0.144	0.127	0.108	0.0958	0.0874	0.0806	0.0707	0.0635	0.0568	0.0519	0.0448	0.0371	0.0328	0.0302	0.0284	0.0266	0.0255
Ar	0.188	0.135	0.117	0.0977	0.0867	0.0790	0.0730	0.0638	0.0573	0.0512	0.0468	0.0407	0.0338	0.0301	0.0279	0.0266	0.0248	0.0241
K	0.215	0.149	0.127	0.106	0.0938	0.0852	0.0786	0.0689	0.0618	0.0552	0.0505	0.0438	0.0365	0.0327	0.0305	0.0289	0.0274	0.0267
Ca	0.238	0.158	0.132	0.109	0.0965	0.0876	0.0809	0.0708	0.0634	0.0566	0.0518	0.0451	0.0376	0.0338	0.0316	0.0302	0.0285	0.0280
Fe	0.344	0.183	0.138	0.106	0.0919	0.0828	0.0762	0.0664	0.0595	0.0531	0.0485	0.0424	0.0361	0.0330	0.0313	0.0304	0.0295	0.0294
Cu	0.427	0.206	0.147	0.108	0.0916	0.0820	0.0751	0.0654	0.0585	0.0521	0.0476	0.0418	0.0357	0.0330	0.0316	0.0309	0.0303	0.0305
Mo	1.03	0.389	0.225	0.130	0.0998	0.0851	0.0761	0.0648	0.0575	0.0510	0.0467	0.0414	0.0365	0.0349	0.0344	0.0344	0.0349	0.0359
Sn	1.58	0.563	0.303	0.153	0.109	0.0886	0.0776	0.0647	0.0568	0.0501	0.0459	0.0408	0.0367	0.0355	0.0355	0.0358	0.0368	0.0383
I	1.83	0.648	0.339	0.165	0.114	0.0913	0.0792	0.0653	0.0571	0.0502	0.0460	0.0409	0.0370	0.0360	0.0361	0.0365	0.0377	0.0394
W	4.21	1.44	0.708	0.293	0.174	0.125	0.101	0.0763	0.0640	0.0544	0.0492	0.0437	0.0405	0.0402	0.0409	0.0418	0.0438	0.0465
Pt	4.75	1.64	0.795	0.324	0.191	0.135	0.107	0.0800	0.0659	0.0554	0.0501	0.0445	0.0414	0.0411	0.0418	0.0427	0.0448	0.0477
Tl	5.16	1.80	0.866	0.346	0.204	0.143	0.112	0.0824	0.0675	0.0563	0.0508	0.0452	0.0420	0.0416	0.0423	0.0433	0.0454	0.0484
Pb	5.29	1.84	0.896	0.356	0.208	0.145	0.114	0.0836	0.0684	0.0569	0.0512	0.0457	0.0421	0.0420	0.0426	0.0536	0.0459	0.0489
U	10.60	2.42	1.17	0.452	0.259	0.176	0.136	0.0952	0.0757	0.0615	0.0548	0.0484	0.0445	0.0440	0.0446	0.0455	0.0479	0.0511
air	0.151	0.134	0.123	0.106	0.0953	0.0868	0.0804	0.0706	0.0655	0.0567	0.0517	0.0445	0.0357	0.0307	0.0274	0.0250	0.0220	0.0202
H$_2$O	0.167	0.149	0.136	0.118	0.106	0.0966	0.0896	0.0786	0.0706	0.0630	0.0575	0.0493	0.0396	0.0339	0.0301	0.0275	0.0240	0.0219
concrete	0.169	0.139	0.124	0.107	0.0954	0.0870	0.0804	0.0706	0.0635	0.0567	0.0517	0.0445	0.0363	0.0317	0.0287	0.0268	0.0243	0.0229
tissue	0.163	0.144	0.132	0.115	0.100	0.0936	0.0867	0.0761	0.0683	0.0600	0.0556	0.0478	0.0384	0.0329	0.0292	0.0267	0.0233	0.0212

Health, Safety, Welfare

Example 49.8

Given a 500 keV gamma ray source producing a 3.0 mR/h exposure rate, what will be the exposure rate behind a 12.5 mm thin aluminum shield ($\rho_{Al} = 2.7$ g/cm^3)?

Solution

From Table 49.3, for 500 keV (0.5 MeV) gamma radiation in aluminum, $\mu = 0.0840$ cm^2/g.

$$x = \frac{12.5 \text{ mm}}{10 \frac{\text{mm}}{\text{cm}}} = 1.25 \text{ cm}$$

From Eq. 49.11,

$$I_x = I_0 e^{-\mu\rho x} = \left(3.0 \ \frac{\text{mR}}{\text{h}}\right) e^{-\left(0.0840 \frac{\text{cm}^2}{\text{g}}\right)\left(2.7 \frac{\text{g}}{\text{cm}^3}\right)(1.25 \text{ cm})}$$

$$= 2.26 \text{ mR/h}$$

Example 49.9

What thickness of lead ($\rho_{Pb} = 11.35$ g/cm^3) is needed to reduce the exposure rate of 1 MeV gamma radiation from 5 R/h to 0.5 mR/h, assuming thin shielding?

Solution

From Table 49.3, for 1 MeV gamma radiation in lead, $\mu = 0.0684$ cm^2/g.

From Eq. 49.11,

$$I_x = I_0 e^{-\mu\rho x}$$

Solving for x yields

$$x = \frac{-\ln\left(\frac{I_x}{I_0}\right)}{\mu\rho} = \frac{-\ln\left(\frac{0.5 \frac{\text{mR}}{\text{h}}}{\left(5 \frac{\text{R}}{\text{h}}\right)\left(1000 \frac{\text{mR}}{\text{R}}\right)}\right)}{\left(0.0684 \frac{\text{cm}^2}{\text{g}}\right)\left(11.35 \frac{\text{g}}{\text{cm}^3}\right)}$$

$$= 11.9 \text{ cm}$$

Buildup Factor, *B*

If the gamma source is strong and the shield is thick, Compton scattering and pair production may not remove all the photon energy from the beam. Residual, lower-energy photons may still pass through the shield. In the case of a thick shield, these photons can interact a second time and scatter in different directions to produce a higher exposure rate outside the shield than that produced by the original gamma rays. The thicker and taller the shield, the greater will be the "buildup" of these scattered photons.

This situation is accounted for with the introduction of a "buildup factor," *B*, in the photon shielding equation (see Eq. 49.11). As shown in Table 49.4, the buildup factor depends on the atomic number of the shield material, gamma ray energy, and size and shape of the shield. For buildup in a thick shield,

$$I_x = BI_0 e^{-\mu\rho x} \quad \text{[thick shield]} \qquad 49.12$$

Example 49.10

Given a 3 MeV gamma ray source creating an unshielded 25 R/h exposure rate, what will be the exposure rate behind a 2 ft thick concrete ($\rho = 2.2$ g/cm^3) shield?

Solution

$$x = (2 \text{ ft})\left(30.48 \ \frac{\text{cm}}{\text{ft}}\right)$$

$$= 60.96 \text{ cm}$$

From Table 49.3, for 3 MeV gamma radiation in concrete, $\mu = 0.0363$ cm^2/g.

$$\mu\rho x = \left(0.0363 \ \frac{\text{cm}^2}{\text{g}}\right)\left(2.2 \ \frac{\text{g}}{\text{cm}^3}\right)(60.96 \text{ cm})$$

$$= 4.87$$

From Table 49.4, $B \approx 4.3$.

From Eq. 49.12,

$$I_x = BI_0 e^{-\mu\rho x}$$

$$= (4.3)\left(25 \ \frac{\text{R}}{\text{h}}\right) e^{-4.87}$$

$$= 0.825 \text{ R/h}$$

5. NEUTRON SHIELDING

Because neutrons have no electrical charge, they do not experience the electrostatic forces that produce energy loss for charged particles. Rather, neutrons interact with the shield nuclei.

A neutron may experience either an *absorption reaction* where it totally disappears or is replaced by one or more secondary radiations (usually heavy charged particles), or a *scattering reaction* where the energy or direction of the neutron is changed. The relative probabilities of neutron interactions heavily depend on the neutron's kinetic energy. (See Table 49.5.)

Neutron absorption can induce target materials to become radioactive and produce gamma rays. Absorption reactions are as follows.

- Charged-particle reactions: The neutron disappears into the nucleus of a target atom, and an alpha particle or proton is ejected. The notation for these reactions is (n, α) and (n, p), respectively.

- Fission: The neutron collides with a nucleus and causes the nucleus to split apart, releasing energy and more neutrons. This is the principal source of nuclear energy in nuclear power plants.

- Neutron-producing reactions: Neutrons are liberated from the target atom's nucleus without causing fission. With energetic neutrons, multiple neutrons may be liberated—for example, (n, 2n) and (n, 3n) reactions.

Table 49.4 *Dose Buildup Factor, B, for a Point Gamma or X-Ray Isotropic Source*

shield	energy (MeV)	$\mu\rho x$ 1	2	4	7	10	15	20
water	0.5	2.52	5.14	14.3	38.8	77.6	178	334
	1.0	2.13	3.71	7.68	16.2	27.1	50.4	82.2
	2.0	1.83	2.77	4.88	8.46	12.4	19.5	27.7
	3.0	1.69	2.42	3.91	6.23	8.63	12.8	17.0
	4.0	1.58	2.17	3.34	5.13	6.94	9.97	12.9
	6.0	1.46	1.91	2.76	3.99	5.18	7.09	8.85
	8.0	1.38	1.74	2.40	3.34	4.25	5.66	6.95
	10.0	1.33	1.63	2.19	2.97	3.72	4.90	5.98
concrete	0.5	2.18	3.67	7.73	16.6	29.1	58.1	98.3
	1.0	1.95	3.10	5.98	11.7	18.7	33.1	50.6
	2.0	1.76	2.52	4.38	7.65	11.4	18.2	25.7
	3.0	1.60	2.24	3.65	6.00	8.58	13.3	18.4
	4.0	1.49	2.01	3.13	4.96	7.00	10.7	14.7
	6.0	1.38	1.79	2.64	4.10	5.76	8.93	12.6
	8.0	1.31	1.62	2.30	3.47	4.83	7.53	10.8
	10.0	1.24	1.49	2.04	3.00	4.16	6.59	9.86
iron	0.5	1.98	3.09	5.98	11.7	19.2	35.4	55.6
	1.0	1.87	2.89	5.39	10.2	16.2	28.3	42.7
	2.0	1.76	2.43	4.13	7.25	10.9	17.6	25.1
	3.0	1.55	2.15	3.51	5.85	8.51	13.5	19.1
	4.0	1.45	1.94	3.03	4.91	7.11	11.2	16.0
	6.0	1.34	1.72	2.58	4.14	6.02	9.89	14.7
	8.0	1.27	1.56	2.23	3.49	5.07	8.50	13.0
	10.0	1.20	1.42	1.95	2.99	4.35	7.54	12.4
lead	0.5	1.24	1.42	1.69	2.00	2.27	2.65	−2.73
	1.0	1.37	1.69	2.26	3.02	3.74	4.81	5.86
	2.0	1.39	1.76	2.51	3.66	4.84	6.87	9.00
	3.0	1.34	1.68	2.43	2.75	5.30	8.44	12.3
	4.0	1.27	1.56	2.25	3.61	5.44	9.80	16.3
	6.0	1.18	1.40	1.97	3.34	5.69	13.8	32.7
	8.0	1.14	1.30	1.74	2.89	5.07	14.1	44.6
	10.0	1.11	1.23	1.58	2.52	4.34	12.5	39.2

Source: *Radiological Health Handbook*, pp. 145–146, U.S. Department of Health, Education and Welfare, January 1970.

- Radiative capture: The neutron is captured by the nucleus, and one or more gamma rays are emitted. The radiative capture reaction (n, γ) is the most probable and is, therefore, important in shielding neutrons.

Scattering reactions are as follows.

- Elastic scattering: A neutron strikes a nucleus, the neutron reappears, and the nucleus is left in the ground state. The notation for this is (n, n).

- Inelastic scattering: This reaction is identical to elastic scattering except the nucleus is left in an excited state. The excited nucleus decays by the emission of gamma radiation. The notation is (n, n′).

The relative probability that a neutron will be removed by mechanism x is defined as σ_x/σ_t, where σ_x is the *cross section* (also known as *microscopic cross section*) appropriate for mechanism x. The *total cross section* is

$$\sigma_t = \sigma_{\text{scatter}} + \sigma_{\text{capture}} + \sigma_{\text{activation}}$$
$$+ \sigma_{\text{absorption}} + \sigma_{\text{fission}} + \cdots \qquad 49.13$$

The total cross section reflects the diminution of a neutron being removed from a beam by any mechanism by a single shield atom. The cross sections are a function of the kinetic energy of the neutrons for any given material and are reported in *barns*, a unit of area equal to 10^{-24} cm^2. Cross-sectional data for naturally occurring elements is provided in App. 49.A.

Health, Safety, Welfare

Table 49.5 *Neutron Classification by Kinetic Energy*

term	derivation	range of energies
thermal	neutrons in thermal equilibrium with surroundings	most probable energy at 20°C, 0.025 eV Maxwellian distribution of 20°C extends to about 0.01 eV
epithermal	neutrons with energy greater than thermal	energies above about 0.2 eV
cadmium	neutrons that are strongly absorbed by cadmium	energies below about 0.4 eV
epicadmium	neutrons that are not strongly absorbed by cadmium	energies above about 0.6 eV
slow	neutrons with energy slightly greater than thermal	usually means less than 1–10 eV occasionally means less than 1 keV
resonant	in pile neutron physics, usually refers to neutrons that are strongly captured in the resonance of U-238	a broad band of energies from about 1 eV to about 300 eV
intermediate	neutrons between slow and fast	from a few hundred eV to about 0.5 MeV
fast		neutrons of energy greater than about 0.5 MeV
ultra fast (relativistic)		neutrons of energy greater than about 20 MeV
pile	neutrons of all energies present in nuclear reactors	about 0.001 eV to approximately 15 MeV
fission	neutrons formed directly in fission	about 100 keV to approximately 15 MeV most probable energy 0.8 MeV average energy 2.0 MeV

The *macroscopic cross section*, Σ, reflects the diminution of a neutron being removed per cubic centimeter of a shield material with N atomic nuclei per cubic centimeter.

$$\Sigma = N\sigma \qquad 49.14$$

The number of neutrons penetrating a shield declines exponentially with the shield's thickness, x, atomic density, N, and energy-dependent total neutron cross section, σ_t.

$$N = \frac{N_A \rho_{\text{shield}}}{A_{\text{shield}}} \qquad 49.15$$

$$I_x = I_0 e^{-N\sigma x} = I_0 e^{-\Sigma x} \qquad 49.16$$

Example 49.11

A source of thermal neutrons is generating 3.5×10^9 neutrons per second. If the source is shielded with 2 cm of lead ($\rho_{\text{Pb}} = 11.35$ g/cm^3, $A_{\text{Pb}} = 207$ g/mol), what will be the rate of neutrons passing through the lead shield?

Solution

Calculate the number of lead nuclei per cubic centimeter.

From Eq. 49.15,

$$N = \frac{N_A \rho_{\text{Pb}}}{A_{\text{Pb}}}$$

$$= \frac{\left(6.02 \times 10^{23} \, \frac{\text{nuclei}}{\text{mol}}\right)\left(11.35 \, \frac{\text{g}}{\text{cm}^3}\right)}{207 \, \frac{\text{g}}{\text{mol}}}$$

$$= 3.30 \times 10^{22} \text{ nuclei/cm}^3$$

Determine the (total) neutron cross section per lead nucleus. From App. 49.A,

$$\sigma_t = 11.2 \text{ barns} \quad (11.2 \times 10^{-24} \text{ cm}^2)$$

From Eq. 49.16,

$$I_x = I_0 e^{-N\sigma x}$$

$$= \left(3.5 \times 10^9 \, \frac{\text{neutrons}}{\text{s}}\right)$$

$$\times e^{-\left(3.30 \times 10^{22} \, \frac{\text{nuclei}}{\text{cm}^3}\right)\left(11.2 \times 10^{-24} \, \frac{\text{cm}^2}{\text{nuclei}}\right)(2 \text{ cm})}$$

$$= 1.67 \times 10^9 \text{ neutrons/s}$$

Example 49.12

What will be the rate of neutrons passing through the shield in Ex. 49.11 if a 2 cm water shield ($\rho_{H_2O} = 1$ g/cm^3, $A_{H_2O} = 18$ g/mol) is used instead of the lead?

Solution

Calculate the number of water molecules per cubic centimeter.

From Eq. 49.15,

$$N = \frac{N_A \rho_{H_2O}}{A_{H_2O}} = \frac{\left(6.02 \times 10^{23} \ \frac{\text{molecules}}{\text{mol}}\right)\left(1 \ \frac{\text{g}}{\text{cm}^3}\right)}{18 \ \frac{\text{g}}{\text{mol}}}$$

$$= 3.34 \times 10^{22} \ \text{molecules/cm}^3$$

Determine the (total) neutron cross section per water molecule.

From App. 49.A,

$$\sigma_t = 103 \ \text{barns} \quad (103 \times 10^{-24} \ \text{cm}^2)$$

From Eq. 49.16,

$$I_x = I_0 e^{-N\sigma x}$$

$$= \left(3.5 \times 10^9 \ \frac{\text{neutrons}}{\text{s}}\right)$$

$$\times e^{-\left(3.34\times10^{22} \ \frac{\text{molecules}}{\text{cm}^2}\right)\left(103\times10^{-24} \ \frac{\text{cm}^2}{\text{molecules}}\right)(2 \text{ cm})}$$

$$= 3.6 \times 10^6 \ \text{neutrons/s}$$

6. RELAXATION LENGTH

The *relaxation length* (or *relaxation distance*), λ, is the thickness of a shield that will attenuate a narrow beam of gammas or neutrons to 36.8% (100%/e) of its original intensity. For gamma radiation, it is equivalent to a shield thickness of $1/\mu$, and for neutrons, it is equivalent to a shield thickness of $1/N\sigma$ (same as $1/\Sigma$). Exact equivalence occurs only for collimated beams of mono-energetic radiation in thin shields. For a specified radiation of a given energy, the relaxation lengths (or their reciprocals) provide a rough means for comparing the shielding effectiveness of different materials. Relaxation lengths may be used in preliminary shielding calculations, but the results are only approximate.

$$\lambda = \frac{1}{\mu\rho_{\text{shield}}} \quad \text{[gamma radiation]} \qquad 49.17$$

$$\lambda = \frac{1}{N\sigma_t} = \frac{1}{\Sigma_t} \quad \text{[neutrons]} \qquad 49.18$$

Example 49.13

What is the absorption relaxation length for thermal neutrons in Pu-239 ($A_{Pu} = 239$ g/mol; $\rho_{Pu} = 19.74$ g/cm^3)?

Solution

Calculate the number of plutonium nuclei per cubic centimeter.

From Eq. 49.15,

$$N = \frac{N_A \rho_{Pu}}{A_{Pu}} = \frac{\left(6.02 \times 10^{23} \ \frac{\text{nuclei}}{\text{mol}}\right)\left(19.74 \ \frac{\text{g}}{\text{cm}^3}\right)}{239 \ \frac{\text{g}}{\text{mol}}}$$

$$= 4.97 \times 10^{22} \ \text{nuclei/cm}^3$$

Determine the (absorbed) neutron cross section per plutonium nucleus.

From App. 49.A,

$$\sigma_a = 1026 \ \text{barns} \quad (1.026 \times 10^{-21} \ \text{cm}^2)$$

The relaxation length for neutrons is given by Eq. 49.18.

$$\lambda_{Pu} = \frac{1}{N\sigma_a}$$

$$= \frac{1}{\left(4.97 \times 10^{22} \ \frac{\text{nuclei}}{\text{cm}^3}\right)\left(1.026 \times 10^{-21} \ \frac{\text{cm}^2}{\text{nuclei}}\right)}$$

$$= 1.96 \times 10^{-2} \ \text{cm}$$

Health, Safety, Welfare

50

Illumination and Sound

Nomenclature

a	speed of sound	ft/sec	m/s
A	area	ft^2	m^2
B	bending stiffness per unit width	ft-lbf	N·m
c	speed of light	ft/sec	m/s
C	exposure duration	hr	h
CU	coefficient of utilization	–	–
D	dose	–	–
E	illuminance	lm/ft^2	lm/m^2
E	modulus of elasticity	lbf/ft^2	Pa
f	frequency	Hz	Hz
g	gravitational acceleration, 32.2 (9.81)	ft/sec^2	m/s^2
g_c	gravitational constant, 32.2	lbm-ft/lbf-sec^2	n.a.
h	height	ft	m
I	light intensity	lm/sr	lm/sr
I	sound intensity	dB	dB
IL	insertion loss	dB	dB
k	ratio of specific heats	–	–
L	length	ft	m
L	sound level	dB	dB
LLF	light loss factor	–	–
m	mass	lbm	kg
M	brightness	lm/ft^2	lm/m^2
MW	molecular weight	lbm/lbmol	kg/kmol
N	number	–	–
NR	noise reduction	dB	dB
NRR	noise reduction rating	dB	dB
p	pressure	lbf/ft^2	Pa
Q	directivity	–	–
Q	quantity of light	lm	lm
r	distance	ft	m
R	acoustic resistance	lbf-sec/ft^5	N·s/m^5
R	room constant	ft^2	m^2
R	specific gas constant, air, 53.35 (287.03)	ft-lbf/lbm-°R	J/kg·K
R^*	universal gas constant, 1545.35 (8314.47)	ft-lbf/lbmol-°R	J/kmol·K
S	surface absorption	sabins	sabins
t	reference duration	hr	h
t	thickness	ft	m
t	time	sec	s
T	period	sec	s
T	temperature	°R	K
TL	transmission loss	dB	dB
TWA	time weighted average	dB	dB
v	velocity	ft/sec	m/s
W	power	W	W
X	acoustic reactance	lbf-sec/ft^5	N·s/m^5
z	specific acoustic impedance	lbf-sec/ft^3	N·s/m^3
Z	acoustic impedance	lbf-sec/ft^5	N·s/m^5
Z_0	specific acoustic impedance of the medium	lbf-sec/ft^3	N·s/m^3

Health, Safety, Welfare

Symbols

α	sound absorption coefficient	–	–
η	efficacy	lm/W	lm/W
η	efficiency	–	–
λ	wavelength	ft	m
ρ	density	lbm/ft^3	kg/m^3
τ	transmission coefficient	–	–
θ	angle	–	–
ν	Poisson's ratio	–	–
Φ	luminous flux	lm	lm
ω	solid angle	sr	sr

Subscripts

l	luminous
o	observer
p	pressure
ref	reference
s	source
S	surface per unit area
t	total
W	power

1. LUMINOUS FLUX

The amount of visible light emitted from a source is the *luminous flux*, Φ, with units of lumens (lm).[1] Not all of the energy emitted will be in the visible region, though. Therefore, the total source power is not a good indicator of the visible lighting effect. The ratio of luminous flux to total radiant energy (also known as the *quantity of light*), Q, is the *luminous efficiency*, η_l.

$$\eta_l = \frac{\Phi}{Q} \qquad 50.1$$

Efficiency of a lamp is the ratio of light output in lumens to the lamp's input in lumens. *Efficacy* is the ratio of the lamp's output in lumens to its input power in watts. (See Table 50.1.)

$$\eta = \frac{\Phi}{W} \qquad 50.2$$

The *luminous emittance* (also known as *brightness*), M, of a source is the luminous flux per unit area of the source. (Compare this with *illumination* of a receiver covered in Sec. 50.3, which is the flux per unit of receiving area.) In the United States, the unit of brightness is the lm/ft^2 (foot-lambert).[2]

$$M = \frac{\Phi}{A_{\text{source}}} \qquad 50.3$$

---

[1]A lumen is approximately equal to 1.47×10^{-3} W of visible light with a wavelength of 5.55×10^{-7} m.
[2]The term *lambert* in the United States is a source of confusion. The old lambert unit is the same as lm/cm^2. A foot-lambert is the same as lm/ft^2.

Table 50.1 *Typical Values of Luminous Efficacy*

source	lm/W
candle	0.1
25 W tungsten lamp	10
100 W tungsten lamp	16
1000 W tungsten lamp	22
40 W fluorescent lamp	80
400 W mercury fluorescent lamp	58
1000 W carbon arc	60
100 W mercury arc	35
400 W metal halide lamp	85
1000 W metal halide lamp	100
400 W high-pressure sodium lamp	125
1000 W high-pressure sodium lamp	130
180 W low-pressure sodium lamp	180

Example 50.1

A 100 W lamp has a luminous flux of 4400 lm. What is the lamp's luminous efficacy?

Solution

From Eq. 50.2,

$$\eta = \frac{\Phi}{W} = \frac{4400 \text{ lm}}{100 \text{ W}} = 44 \text{ lm/W}$$

Example 50.2

A tungsten filament emitting 1600 lm has a surface area of 0.35 in^2 (2.26×10^{-4} m^2). What is its brightness?

SI Solution

$$M = \frac{\Phi}{A_{\text{surface}}} = \frac{1600 \text{ lm}}{2.26 \times 10^{-4} \text{ m}^2} = 7.08 \times 10^6 \text{ lm/m}^2$$

Customary U.S. Solution

$$M = \frac{\Phi}{A_{\text{surface}}} = \frac{(1600 \text{ lm})\left(12 \frac{\text{in}}{\text{ft}}\right)^2}{0.35 \text{ in}^2} = 6.58 \times 10^5 \text{ lm/ft}^2$$

2. LUMINOUS INTENSITY

The *luminous intensity*, I, of a source is a measure of flux per unit solid angle. Its units are lumens per steradian (lm/sr), referred to by the synonymous names of *candle*, *candela*, and *candlepower*.[3] Equation 50.4 relates flux from an omnidirectional point source to intensity measured on a spherical receptor of radius r.

$$I = \frac{\Phi}{\omega} = \frac{\Phi r^2}{A} = \frac{\Phi_t r^2}{4\pi r^2} = \frac{\Phi_t}{4\pi} \qquad 50.4$$

---

[3]Intensity is not a measure of power, as the term *candlepower* may imply.

Example 50.3

Two lumens pass through a circular hole (diameter = 0.5 m) in a screen located 6.0 m from an omnidirectional source. (a) What is the luminous intensity of the source? (b) What luminous flux is emitted by the source?

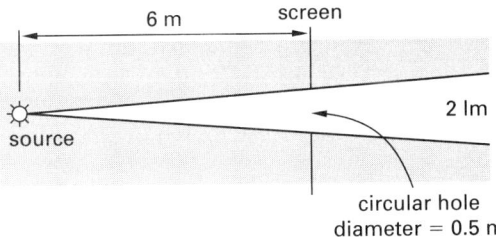

Solution

(a) The solid angle subtended by the hole is

$$\omega = \frac{A}{r^2} = \frac{\left(\frac{\pi}{4}\right)(0.5 \text{ m})^2}{(6 \text{ m})^2} = 0.005454 \text{ sr}$$

From Eq. 50.4, the intensity is

$$I = \frac{\Phi}{\omega} = \frac{2 \text{ lm}}{0.005454 \text{ sr}} = 366.7 \text{ lm/sr}$$

(b) From Eq. 50.4, the luminous flux is

$$\Phi_t = 4\pi I = (4\pi \text{ sr})\left(366.7 \frac{\text{lm}}{\text{sr}}\right) = 4608 \text{ lm}$$

3. ILLUMINANCE

The *illuminance* (*illumination*), E, is a measure of luminous flux per incident area. Typical units are lux (lm/m^2) and foot-candles (lm/ft^2). (See Table 50.2.) The name *irradiance* is used when the units are W/m^2.

One foot-candle is the illumination from a one *candle-power* (cp) source at a distance of one foot.

$$E = \frac{\Phi}{A_{\text{receptor}}} \quad\quad 50.5$$

For an omnidirectional source and a spherical receptor of radius r,

$$E = \frac{\Phi_t}{4\pi r^2} \quad\quad 50.6$$

Equation 50.6 shows that illumination follows the *distance squared law*.

$$E_1 r_1^2 = E_2 r_2^2 \quad\quad 50.7$$

Table 50.2 Recommended Illuminance

location	lm/ft^2 (fc)	lux
roadway	1–2	10–20
living room	5–15	50–150
library (reading)	30–70	300–700
evening sports	30–100	300–1000
factory (assembly)	100–200	1000–2000
office	100–200	1000–2000
factory (fine assembly)	500–1000	5000–10 000
hospital operating room	2000–2500	20 000–25 000

(Multiply foot-candles by 10.764 to obtain lux.)
(Multiply foot-candles by 0.10764 to obtain hectolux.)

Example 50.4

A lamp radiating hemispherically and rated at 2000 lm is positioned 20 ft (6 m) above the ground. What is the illumination on a walkway directly below the lamp?

SI Solution

From Eq. 50.5,

$$E = \frac{\Phi}{A} = \frac{\Phi}{\frac{1}{2}A_{\text{sphere}}} = \frac{2000 \text{ lm}}{\left(\frac{1}{2}\right)(4\pi)(6 \text{ m})^2}$$

$$= 8.84 \text{ lux}$$

Customary U.S. Solution

From Eq. 50.5,

$$E = \frac{\Phi}{A} = \frac{\Phi}{\frac{1}{2}A_{\text{sphere}}} = \frac{2000 \text{ lm}}{\left(\frac{1}{2}\right)(4\pi)(20 \text{ ft})^2}$$

$$= 0.796 \text{ lm/ft}^2 \text{ (fc)}$$

4. LUMINANCE

Illuminance of an object is the quantity of light from a source reaching the object. The amount of light reflected from the object's surface is referred to as the *luminance* or *brightness* of the object. Luminance depends on the *reflectance* of the surface and is measured in essentially the same units as illuminance.[4] For example, if an object has a reflectance of 40% and is illuminated at 150 fc, its brightness is $(0.40)(150 \text{ fc}) = 60 \text{ fL}$.

5. LIGHTING DESIGN

Design of lighting systems has many elements and is primarily performed by architects. A common requirement is to determine the number of lamps (luminaires) required. The *lumen method* (*total flux method*) can be used in rectangular rooms with a gridded lamp arrangement and where uniform illumination is needed. In Eq. 50.8, A is the area at the working plane level, CU is the *coefficient of utilization* (*utilization factor*), an

---

[4]A *lambert* is equal to 929 foot-lamberts. A *millilambert* is equal to 0.929 foot-lamberts. Other units include the *stilb* (candle/cm^2) and *nit* (candle/m^2).

allowance for the light distribution of the lamps and the room surfaces, and LLF is the *light loss factor (maintenance factor)*, an allowance for reduced light output due to deterioration and dirt. The coefficient of utilization depends on wall color, lamp placement, aspect ratio of the room and lamp spacing, and other factors. In the absence of other information, the light loss factor is arbitrarily set to 0.80.

$$N = \frac{EA}{\Phi(\text{CU})(\text{LLF})} \qquad 50.8$$

6. INTERACTION OF LIGHT WITH MATTER

Light travels through a vacuum as an electromagnetic wave. When the light makes contact with matter, some of the wave energy is absorbed by the matter, causing electrons to jump into higher energy states. (A polished metal surface, for example, will absorb only about 10% of the incident energy, reflecting the remaining 90% away.) Some of this absorbed energy is re-emitted when the electrons drop back to a lower energy level. Generally, the re-emitted light will not be at its original wavelength.

If the reflecting surface is smooth, the *reflection angle* for most of the light will be the same as the *incident angle*, and the light is said to be *regularly reflected* (i.e., the case of *specular reflection*). If the surface is rough, however, the light will be scattered and reflected randomly (the case of *diffuse reflection*). (See Fig. 50.1.)

Figure 50.1 *Specular and Diffuse Reflection*

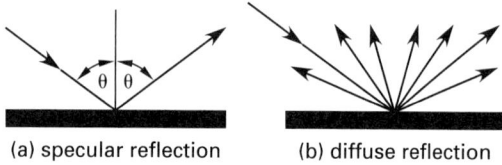

   (a) specular reflection   (b) diffuse reflection

The energy that is absorbed is said to be *refracted*. In the case of an *opaque material*, the refracted energy is absorbed within a very thin layer and converted to heat. Light is able to pass through a *transparent material* without being absorbed. Light is partially absorbed in a *translucent material*.

7. SOUND AND NOISE

Sound is made up of the pressure waves and vibrations that are received from a vibrating source. *Noise* is unwanted sound. Noise can either be continuous or impulsive. For the purpose of classification, *impact noise (impulse noise)* is noise whose peak levels have a duration less than one second and occur at intervals greater than one second.

Background noise is the "normal" noise in the local environment. Background noise consists of environmental noises that are not components of the noise to be

measured or studied. It should be measured separately and subtracted from any noise measurements made while equipment or other sources are operating.

A *pure tone* is one that consists essentially of a single frequency. Relatively pure tones can be generated at simple harmonic frequencies in rotating equipment. For example, a five-bladed fan turning at 50 rev/sec would generate harmonics with a fundamental frequency of 250 Hz. Pure tones and harmonics are of interest because they are difficult to attenuate. However, most noise is white noise, not pure tones. *White noise* is relatively evenly distributed over the frequency spectrum.

Structure-borne noise (also known as *mechanical noise*) is noise that is generated by mechanical elements in moving equipment. Panels, covers, housings, and other membrane-like elements are frequent sources of structure-borne noise. These elements produce noise at their own natural frequencies, not just at the cycling frequency of the equipment.

8. PHYSIOLOGICAL BASIS OF SOUND

Longitudinal sound waves in the 20 Hz to 20,000 Hz range can be heard by most people. The average human ear is most sensitive to frequencies around 3000 Hz. Even in this range, however, the intensity must be great enough for a sound to be detected. *Intensity* is the amount of energy passing through a unit area each second. For example, the *threshold of audibility* at 3000 Hz is approximately 10^{-12} W/m². If the sound is too intense (i.e., the intensity is above the *threshold of pain*, approximately 10 W/m²), the sound will be experienced painfully. The thresholds of audibility and pain are both somewhat frequency dependent.

Loudness is a qualitative sensation that can be quantified approximately by comparing the sound intensity to a reference intensity (usually the assumed threshold of audibility, 10^{-12} W/m²). (See Table 50.3.) Loudness is perceived by most individuals approximately logarithmically and is calculated in that manner.

$$\text{loudness} = 10\log\frac{I}{I_0} = 10\log\frac{I}{10^{-12}\,\frac{\text{W}}{\text{m}^2}} \qquad 50.9$$

Table 50.3 *Approximate Loudness of Various Sources*

source	loudness (dB)	perception
jet engine, thunder	120	painful
jackhammer	110	deafening
sheet metal shop	90	very loud
street noise	70	loud
office noise, normal speech	50	moderate
quiet conversation	30	quiet
whisper	20	faint
anechoic room	10	very faint
none	0	silence

Health, Safety, Welfare

9. LONGITUDINAL SOUND WAVES

Sound, like light and electromagnetic radiation, is a wave phenomenon. However, sound has two distinct differences from electromagnetic waves. First, sound is transmitted as *longitudinal waves*, also known as *compression waves*. Such waves are alternating compressions and expansions of the medium through which they travel. In comparison, electromagnetic waves are *transverse waves*. Second, longitudinal waves require a medium through which to travel—they cannot travel through a vacuum, as can electromagnetic waves.

For ease of visualization, longitudinal waves may be represented by transverse waves. The *wavelength* is the distance between successive compressions (or expansions), as shown in Fig. 50.2.

Figure 50.2 *Longitudinal Waves*

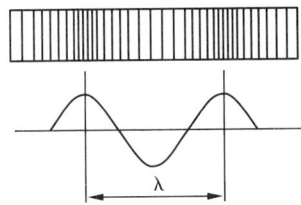

The usual relationships among wavelength, propagation velocity, period, and frequency are valid.

$$a = f\lambda \qquad \textit{50.10}$$

$$T = \frac{1}{f} \qquad \textit{50.11}$$

10. PROPAGATION VELOCITY OF LONGITUDINAL SOUND WAVES

The *propagation velocity* of longitudinal waves, commonly known as the *speed of sound* and *sonic velocity*, depends on the compressibility of the supporting medium. For solids, the compressibility is accounted for by the elastic modulus, E. For liquids, it is customary to refer to the elastic modulus as the *bulk modulus*.

$$a = \sqrt{\frac{E}{\rho}} \qquad \text{[SI]} \quad \textit{50.12(a)}$$

$$a = \sqrt{\frac{Eg_c}{\rho}} \qquad \text{[U.S.]} \quad \textit{50.12(b)}$$

For ideal gases, the propagation velocity is calculated from Eq. 50.13.

$$a = \sqrt{kRT} = \sqrt{\frac{kR^*T}{MW}} \qquad \text{[SI]} \quad \textit{50.13(a)}$$

$$a = \sqrt{kg_cRT} = \sqrt{\frac{kg_cR^*T}{MW}} \qquad \text{[U.S.]} \quad \textit{50.13(b)}$$

Table 50.4 lists approximate values for the speed of sound in various media.

Table 50.4 *Approximate Speeds of Sound (at one atmospheric pressure)*

material	speed of sound (ft/sec)	speed of sound (m/s)
air	1130 at 70°F	330 at 0°C
aluminum	16,400	4990
carbon dioxide	870 at 70°F	260 at 0°C
hydrogen	3310 at 70°F	1260 at 0°C
steel	16,900	5150
water	4880 at 70°F	1490 at 20°C

(Multiply ft/sec by 0.3048 to obtain m/s.)

11. DOPPLER EFFECT

If the distance between a sound source and a listener is changing, the frequency heard will differ from the frequency emitted. If the separation distance is decreasing, the frequency will be shifted higher; if the separation distance is increasing, the frequency will be shifted lower. This shifting is known as the *Doppler effect*. The ratio of observed frequency, f', to emitted frequency, f, depends on the local speed of sound, a, and the absolute velocities of the source and observer, v_s and v_o. In Eq. 50.14, v_s is positive if the source moves away from the observer and is negative otherwise; v_o is positive if the observer moves toward the source and negative otherwise.

$$\frac{f'}{f} = \frac{a + v_o}{a + v_s} \qquad \textit{50.14}$$

Example 50.5

On a particular day, the local speed of sound is 1130 ft/sec (344 m/s). What frequency is heard by a stationary pedestrian when a 1200 Hz siren on an emergency vehicle moves away at 45 mph (72 kph)?

SI Solution

The observer's velocity is zero. The source's velocity is positive because the vehicle is moving away. From Eq. 50.14, the frequency heard is

$$f' = f\left(\frac{a + v_o}{a + v_s}\right)$$

$$= (1200 \text{ Hz}) \left(\frac{344 \, \frac{\text{m}}{\text{s}} + 0 \, \frac{\text{m}}{\text{s}}}{344 \, \frac{\text{m}}{\text{s}} + \frac{\left(72 \, \frac{\text{km}}{\text{h}}\right)\left(1000 \, \frac{\text{m}}{\text{km}}\right)}{3600 \, \frac{\text{s}}{\text{h}}}} \right)$$

$$= 1134 \text{ Hz}$$

Customary U.S. Solution

The observer's velocity is zero. The source velocity is positive because the vehicle is moving away. Convert the source velocity from miles per hour to feet per second.

$$v_s = \frac{\left(45 \; \frac{\text{mi}}{\text{hr}}\right)\left(5280 \; \frac{\text{ft}}{\text{mi}}\right)}{3600 \; \frac{\text{sec}}{\text{hr}}} = 66 \; \text{ft/sec}$$

From Eq. 50.14, the frequency heard is

$$f' = f\left(\frac{a + v_o}{a + v_s}\right) = (1200 \; \text{Hz})\left(\frac{1130 \; \frac{\text{ft}}{\text{sec}} + 0 \; \frac{\text{ft}}{\text{sec}}}{1130 \; \frac{\text{ft}}{\text{sec}} + 66 \; \frac{\text{ft}}{\text{sec}}}\right)$$

$$= 1134 \; \text{Hz}$$

12. ACOUSTIC IMPEDANCE

The *acoustic impedance (sound impedance)*, Z, is a function of the pressure, p, wave velocity, v, and the surface area, A, through which the sound travels. Common units are lbf-sec/ft^5 and N·s/m^5, equivalent to rayls/m^2. The quantity vA is known as the *volume velocity*.

$$Z = \frac{p}{vA} \qquad 50.15$$

The *specific acoustic impedance*, z, is

$$z = \frac{p}{v} = ZA \qquad 50.16$$

The *specific acoustic impedance (characteristic acoustic impedance) of the medium* (e.g., air, water, steel), Z_0, is given by Eq. 50.17. Common units are lbf-sec/ft^3 and rayls, equivalent to N·s/m^3 and Pa·s/m (also known as an *acoustic ohm*). For 68°F (20°C) air at 1 atm, the specific acoustic impedance is between 410 rayls and 420 rayls.

$$Z_0 = \rho a \qquad \text{[SI]} \quad 50.17(a)$$

$$Z_0 = \frac{\rho a}{g_c} \qquad \text{[U.S.]} \quad 50.17(b)$$

Specific acoustic impedance is the vector resultant of two orthogonal vectors. The real, resistive part, R, represents energy lost due to friction and viscous effects when compression sound waves pass through the medium. The imaginary, reactive part, X, represents energy temporarily stored in and subsequently released from the compressed medium.

$$\mathbf{Z_0} = \mathbf{R} + i\mathbf{X} \qquad 50.18$$

13. SOUND METER MEASUREMENTS

Rather than attempt to describe sound in terms of subjective loudness, simple *sound meters (noise meters)*

measure noise using various response curves. (A *response curve* is basically a set of weighting values that can be applied to each frequency detected.) Meter response curves have been standardized worldwide, and most industrial-quality meters respond accurately to three curves, designated as the A-, B-, and C-scales. The slow-response A-scale approximates human hearing the most accurately, so it is used the most frequently.[5,6]

When sound measurements are reported, a sound "level" always means a logarithmic value expressed in decibels. Sound level units are reported along with the scale used. For example, "15 dBA" (also written as "dB(A)" and "dB-A") indicates that the A-scale was used.

Some meters often also provide a fourth, unweighted reading using the basic frequency response of the microphone and meter circuitry. This reading is different for different meters. The terms *unweighted response*, *flat response*, *20 kHz response*, and *linear response* all refer to this case.

Peak levels from impulse sources are measured with special equipment (i.e., an impact meter or oscilloscope) that captures and holds the maximum instantaneous noise level.

14. SOUND FIELDS

Sound that is received directly from a source without any significant reflection (including that from the floor or ground) is known as *free-field* sound. Similarly, free-field measurements are those that do not include any contribution from reflections. Measurements made near walls or other objects where reflected sound is significant are called *reverberant field* measurements. Most in-room measurements are reverberant field measurements. In a reverberant field, sound pressure levels do not fall off with increases in distance from the source. Sound levels are essentially the same anywhere in a reverberant field.

Measurements made close to a noise source are called *near-field measurements*. They can be in error due to directionality effects and reflections from the source itself. The near-field effects will be significant if the measurement is taken closer than one wavelength (based on the lowest frequency of interest) from the source. Measurements made at large distances from a noise source are called *far-field measurements*. Far-field measurements can be in error due to background noise. In a true far-field environment, the sound pressure level will decay 6 dB for each doubling of the distance from the source.

Anechoic rooms (free-field throughout) and *reverberant rooms* (reverberant field throughout) represent extremes in measurement environments. Such rooms are routinely used for measuring directionality effects and absorption material performance.

---

[5]The A-curve discriminates against lower frequencies, the C-curve is nearly flat, and the B-curve falls in between.
[6]Refer to ANSI S1.4 for sound level meters.

15. SOUND PRESSURE LEVEL

Sound propagates as pressure fluctuates. The actual pressure at a particular point is the *sound pressure level*, L_p. (The symbol SPL is also used in some literature.) Sound pressure level is the quantity that is actually measured in sound meters.

Since the human ear responds to variations in air pressure over a range of more than a million to one, a logarithmic power ratio scale is used to measure sound in decibels. The minimum perceptible pressure amplitude (i.e., the approximate *threshold of hearing*) has been standardized as 20 μPa, and this value is used as a reference pressure.[7] The acoustic energy is proportional to the square of the sound pressure. For an rms (root-mean-squared or "effective") sound pressure, p, in pascals, the sound pressure level is

$$L_p = 10 \log \left(\frac{p}{p_{\text{ref}}}\right)^2 = 20 \log \frac{p}{p_{\text{ref}}}$$
$$= 20 \log \frac{p}{20 \times 10^{-6} \text{ Pa}} \qquad \textit{50.19}$$

Usually, the sound pressure level is sampled at several single frequencies over the audio spectrum. These samples are referred to as "narrow band" or "octave band" measurements. The individual measurements can be combined into a *broad band* value by weighting each by standardized values.[8]

16. SOUND POWER LEVEL

The *sound power level* is used to measure the total acoustic power emitted by a source of sound. Sound power is, therefore, independent of the environment. Sound power is not measured by sound meters; it must be calculated. A standard reference level of 1×10^{-12} W is used.[9] The literature uses L_W and PWL as symbols for sound power level. For a source of sound emitting W watts, the sound power level in decibels is

$$L_W = 10 \log \frac{W}{W_{\text{ref}}} = 10 \log \frac{W}{1 \times 10^{-12} \text{ W}} \qquad \textit{50.20}$$

17. DIRECTIVITY

The *directivity*, Q (also known as the *directivity factor*), affects the sound power that appears as sound pressure in a near-field measurement. (Directivity is not an issue in far-field measurements.) Directivity depends on the orientation of the source and the physical location of the source in relationship to walls and other solid surfaces. Directivity is 1 for an isotropic source radiating into free

space, 2 for a source on the surface of an infinite flat plane, 4 for a source at the intersection of two perpendicular planes, and 8 for a source in the corner of three mutually perpendicular reflecting planes.

18. CONVERTING SOUND PRESSURE INTO SOUND POWER

Sound pressure level and sound power level are both reported in decibels. However, there is no simple conversion between them because sound pressure level is affected by directivity and separation distance from the source, while sound power level is not. However, the two scales have been established such that a decibel change in one quantity will produce the same decibel change in the other quantity. Therefore, a decibel decrease or increase in sound power will produce the same decibel decrease or increase in sound pressure.

The sound pressure at a distance r from a continuous noise source with known sound power, directivity, and room constant (which is covered in Sec. 50.27), in a confined space large enough to have both direct and reverberant sound, is

$$L_{p,\text{dBA}} = 0.2 + L_W + 10 \log \left(\frac{Q}{4\pi r^2} + \frac{4}{R}\right) \quad \text{[SI]} \quad \textit{50.21(a)}$$

$$L_{p,\text{dBA}} = 10.5 + L_W + 10 \log \left(\frac{Q}{4\pi r^2} + \frac{4}{R}\right) \quad \text{[U.S.]} \quad \textit{50.21(b)}$$

Example 50.6

The combined sound pressure level from four identical machines at a point equally distant from each is 100 dBA. Three machines are shut down. What is the new sound pressure level?

Solution

When three of the four machines are shut down, the total radiated sound power will be 25% of the original condition. The reduction in sound power level will be

$$L_{W,1} - L_{W,2} = 10 \log \frac{W_1}{W_2} = 10 \log 4$$
$$= 6.02 \text{ dB} \quad \text{[use 6 dB]}$$

The change in sound pressure level is the same as the change in sound power level. Therefore, the new sound pressure level will be

$$L_p = 100 \text{ dBA} - 6 \text{ dB} = 94 \text{ dBA}$$

19. FREQUENCY CONTENT OF NOISE

Detailed knowledge of the frequency content of noise is required for most noise control problems. *Frequency content* is described by dividing the sound spectrum into a series of frequency ranges called *bands*. Frequency

---

[7]At one time, this was thought to be the threshold of hearing for an average young person at 1000 Hz.

[8]Weighting and combining are done automatically by the sound meter.

[9]Many years ago, a reference of 10^{-13} W was used. There may still be some references to this value in older literature.

Health, Safety, Welfare

bands are established by dividing the frequency range of interest into bands of either equal bandwidths or equal percentages.[10] The sound level in each frequency band can be reported or plotted against the midpoint of the frequency band.

20. OCTAVE BAND ANALYSIS

Percentage bandwidths and center frequencies have been standardized worldwide. The most common (and widest) standard bandwidth is 50%, which means that the ratio between the frequencies at the two ends of the band is 1:2. For example, the band whose center frequency is 1000 Hz spans the range from 707 Hz to 1414 Hz. Since musical pitch frequency ratios of 1:2 are called *octaves*, the name *octave band* has been adopted.

Center frequencies for octave bands have been standardized at 63 Hz, 125 Hz, 250 Hz, 500 Hz, 1000 Hz, 2000 Hz, 4000 Hz, and 8000 Hz.[11,12] These bands are numbered 1, 2, 3, 4, 5, 6, 7, and 8, respectively. Octave bands can also be subdivided for better frequency detail. The most common subdivision is the *one-third octave band*.

21. COMBINING MULTIPLE SOURCES

Multiple sound sources combine to produce more sound. Sound pressures from multiple correlated sources combine linearly, while sound pressures combine as sums of squares. However, multiple sound pressure levels or sound power levels expressed in decibels cannot be directly added. Two 90 dB sources do not produce a 180 dB source. For multiple sound sources, the combined power is given by Eq. 50.22. This is known as an *unweighted sum* because the individual readings are used without modification.

$$L_p = 10 \log \sum 10^{L_i/10} \qquad \textit{50.22}$$

Using Eq. 50.22 to add or subtract the sound from two or more sources is not the same as determining the *change* in sound level. As described in Sec. 50.30, Sec. 50.31, and Sec. 50.32, the relationship between the old, new, and change in sound levels is that of a simple linear combination (i.e., addition or subtraction).

---

[10]Equal bandwidths are generally used only when the frequency range of interest is small. There is little standardization of narrow bandwidths, primarily because narrow band analysis is used more for detailed investigation of noise sources than for reporting absolute levels.
[11]A now-obsolete series of octave bands was used until the late 1950s.
[12]Octave bands centered at 31.5 Hz and 16,000 Hz are also occasionally encountered. However, these two end bands are usually omitted, as they are at the extreme limits of most peoples' hearing abilities. For example, the A-weighted correction for 31.5 Hz is −39.2 dB. This is such a large reduction that the contribution from this band is negligible except for all but the most powerful or pure-tone sources.

Equation 50.22 illustrates an important noise reduction principle: The noisiest source must be identified and treated before significant overall noise reduction can be achieved. Sources with sound levels more than a few decibels lower than the noisiest source make a minor contribution to the total sound level.

Example 50.7

What is the unweighted combined sound pressure from two machines, one with a sound pressure of 89 dB and the other with a sound pressure of 94 dB?

Solution

Use Eq. 50.22.

$$\begin{aligned} L_p &= 10 \log \sum 10^{L_i/10} \\ &= 10 \log \left(10^{89 \text{ dB}/10} + 10^{94 \text{ dB}/10} \right) \\ &= 95.2 \text{ dB} \end{aligned}$$

22. BROAD BAND MEASUREMENTS

Octave band measurements contain enough information about the frequency content to permit calculation of the equivalent A-weighted level. The correction from Table 50.5 is made to each octave band measurement, and the corresponding levels are added using Eq. 50.22.

Table 50.5 A-Weighting Corrections from Octave Band Analysis

center frequency of band (Hz)	correction to be added (dB)[*]
31.5	−39.2
63	−26.2
125	−16.1
250	−8.6
500	−3.2
1000	0.0
2000	+1.2
4000	+1.0
8000	−1.1

[*]referenced to 20 μPa

Example 50.8

The octave band measurements of a sound source are: 63 Hz, 60 dB; 125 Hz, 72 dB; 250 Hz, 85 dB; 500 Hz, 90 dB; 1000 Hz, 90 dB; 2000 Hz, 80 dB; 4000 Hz, 83 dB; and 8000 Hz, 70 dB. What is the A-weighted sound level?

Solution

Add the corrections from Table 50.5 to the measurements.

Health, Safety, Welfare

frequency (Hz)	measurement (dB)	correction (dB)	corrected value (dB)
63	60	−26.2	33.8
125	72	−16.1	55.9
250	85	−8.6	76.4
500	90	−3.2	86.8
1000	90	0	90
2000	80	+1.2	81.2
4000	83	+1.0	84
8000	70	−1.1	68.9

Use Eq. 50.22.

$$L_p = 10 \log \sum 10^{L_i/10}$$

$$= 10 \log \left(\begin{array}{c} 10^{33.8 \text{ dBA}/10} + 10^{55.9 \text{ dBA}/10} + 10^{76.4 \text{ dBA}/10} \\ + 10^{86.8 \text{ dBA}/10} + 10^{90 \text{ dBA}/10} + 10^{81.2 \text{ dBA}/10} \\ + 10^{84 \text{ dBA}/10} + 10^{68.9 \text{ dBA}/10} \end{array} \right)$$

$$= 92.8 \text{ dBA}$$

23. LOUDNESS

Human hearing is not equally sensitive to pressure at all frequencies. In addition, some types of noise are more annoying than others. Subjective (i.e., perceived) noise level is called *loudness.*

There are two loudness scales: the *phon scale* and the *sone scale.* One *phon* is numerically equal to the sound pressure level in decibels at a frequency of 1000 Hz. At 1000 Hz, the values of loudness and L_p are the same.

The common loudness scale used by fan and duct manufacturers is the sone scale. By definition, one *sone* is the loudness of a 1000 Hz sound with a sound pressure of 0.02 μbar.[13] Sones are calculated from octave band sound pressure level measurements made through frequency filters in much the same way as are dBA measurements.[14]

Sone values are easier to use than decibel values because the sone scale is linear, not logarithmic. A doubling of sones is perceived by most people to be a doubling of the loudness.

24. NOISE CRITERION CURVES

One way of specifying limits on loudness is to require all octave band measurements to be below a certain value. However, since the human ear is not as sensitive to low frequencies as it is to high frequencies, the limiting value will depend on the frequency. *Noise criterion* (NC) *curves* (also known as *noise criteria curves*) are curves of acceptable sound pressure level plotted against frequency. There are several such curves, and these are

designated NC-15, NC-20, and so on. The NC curve used depends on the type of location. NC-35 is commonly specified for such locations as apartment buildings, hotels, executive offices, and laboratories. NC-45 is commonly specified for washrooms, kitchens, lobbies, banking areas, and open offices.

The method of determining the NC level or rating actually achieved consists of plotting the measured octave band sound pressure levels (the *noise spectrum*) on a set of NC curves. (See Fig. 50.3.) The NC rating is the lowest NC curve that is not exceeded by any measurement in the spectrum.

Figure 50.3 NC Rating Curves

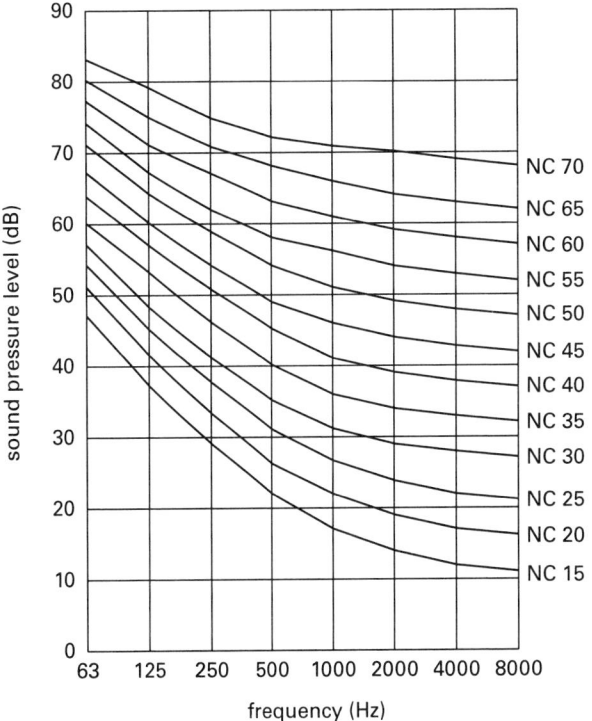

25. ACOUSTIC ABSORBER MATERIALS

Acoustic absorbers are materials that prevent reflection of incident sound. Fibrous glass, draperies, and open cell foams are acoustic absorbers. Absorbers generally are used in combination with barriers and dampers since absorbers do not block noise or reduce vibration when used alone. Absorbers work best at higher frequencies unless they are very thick or are tuned to low frequencies.

26. SOUND ABSORPTION COEFFICIENT

The *sound absorption coefficient,* α, is the ratio of sound energy absorbed by the surface compared to the sound energy striking the surface. Hard surfaces absorb essentially no sound energy and have sound absorption coefficients close to zero. Soft surfaces absorb most of the

---

[13]This is approximately the loudness of a quiet refrigerator in a quiet kitchen.
[14]However, sone values are combined and manipulated differently than decibel (dBA) values.

sound energy striking them. A value of 1.0 indicates a perfect absorber.

Since the sound absorption coefficient varies with frequency, the *noise reduction coefficient* rating (NRC) is used to simplify comparisons of materials. The NRC is the arithmetic average of the four absorption coefficients measured at frequencies of 250 Hz, 500 Hz, 1000 Hz, and 2000 Hz. Coefficients of common building materials and furnishings are listed in App. 50.A.

Absorptive products seldom can reduce ambient noise by more than 6 dB to 8 dB. They reduce noise reflection, but they do not stop sound transmission.

27. ROOM CONSTANT

If a room contains absorbing materials with known absorption coefficients, the average absorption coefficient can be calculated. For this method of determining the average absorption coefficient to be valid, the absorption materials must be well distributed in the room so that the value applies to the majority of locations. In Eq. 50.23, the product αA is known as the *surface absorption* or *sabin area*, S, with units of sabins.[15] One *sabin* (also known as an *open window unit*, OWU) is one unit area of perfectly absorbing surface. Generally, the surface absorption should be 20% to 50% of the room area. Otherwise, the room will be reverberatory.

$$S = \alpha A \qquad 50.23$$

$$\overline{\alpha} = \frac{\sum A_i \alpha_i}{\sum A_i} = \frac{\sum S_i}{\sum A_i} \qquad 50.24$$

The acoustic absorption characteristics of the room can be described by its *room constant*, R. A room with perfectly absorbing surfaces (i.e., an anechoic room) has an infinite room constant. A room with perfectly reflecting surfaces (i.e., a reverberant room) has a room constant of zero.

$$R = \frac{\overline{\alpha} A}{1 - \overline{\alpha}} = \frac{S}{1 - \overline{\alpha}} \qquad 50.25$$

28. ACOUSTIC BARRIER MATERIALS

Acoustic barriers block sound. The best performing *barrier materials* have high mass per unit area. Due to their low cost, gypsum board, plywood, and masonry are often used for this purpose.

29. SOUND TRANSMISSION CLASS

The *sound transmission class* (STC) is a single-number rating that is based on a match of the barrier's transmission loss curve to one of several standard ASTM

[15]Derived from the early 1900s pioneer in room acoustics, Wallace Sabine, the spelling *sabine* is also encountered.

contours. STC is particularly useful in selecting noise barriers for speech isolation. However, STC ratings are not accurate enough for most industrial noise control work because they are functions of the material used, not the design or actual attenuation obtained, and the frequency of the noise to be attenuated is not considered. Typical STC values are given in Table 50.6.

Table 50.6 *Typical Values of Sound Transmission Class (STC)*

barrier construction	STC (dB)
22-gauge steel	25
$^3/_8$ in plasterboard	26
1 lbm barrier or curtain	26
solid wood door	27
2 in fiberglass curtain with 1 lbm barrier	29
$^5/_8$ in gypsum wallboard	30
$^3/_{16}$ in steel	31
two sheets of $^1/_2$ in drywall on wood studs, uninsulated	33
2 in metal panel	35
two sheets of $^1/_2$ in drywall on wood studs, with fiberglass batt insulation	36–39
double $^1/_2$ in drywall on both sides, wood studs, with fiberglass batt insulation	45
$^1/_2$ in drywall on 6 in lightweight concrete block, painted both sides	46
4 in metal panel	41
concrete block, unpainted	44
12 in concrete	53
$^1/_2$ in drywall on 8 in dense concrete block, painted both sides	54
double $^1/_2$ in drywall on both sides, staggered wood studs, with fiberglass batt insulation	55
double $^1/_2$ in drywall on both sides, wood studs, resilient channels on one side, with fiberglass batt insulation	59

A sound barrier or partition is given an STC rating based on measured transmission losses at 16 different one-third octave frequencies between 125 Hz and 4000 Hz, which is consistent with the frequency range of speech. The STC rating does not assess sound transfer at lower or higher frequencies such as might be generated by mechanical equipment or musical instruments. Penetrations and openings through the wall, as well as *flanking paths* around the wall, can degrade the isolation performance of a barrier. Improvement in sound isolation is perceived as almost imperceptible at an STC rating of 1 dB, just barely perceptible at 3 dB, clearly perceptible at 5 dB, and half as loud at 10 dB.

The STC of a partition is determined by comparing the transmission loss spectrum with a standard reference contour. This contour consists of three segments with different vertical increments: 125–400 Hz (15 dB increase), 400–1250 Hz (5 dB increase), and 1250–4000 Hz (0 dB increase), as shown in Fig. 50.4. The STC rating is found by shifting the standard contour vertically until most of the measured values fall below the STC curve and two conditions are met: (1) the sum of all the deficiencies (i.e., measured values above the standard contour) do not exceed 32 dB, and (2) the maximum deficiency at any frequency does not exceed 8 dB. The contour is shifted vertically in 1 dB steps, and when the conforming position is found, the STC rating is read as the transmission loss corresponding to the reference contour at 500 Hz. For the transmission loss spectrum shown in Fig. 50.4, the STC rating is 31 dB.

Figure 50.4 *STC Standard Contour and Sample Rating*

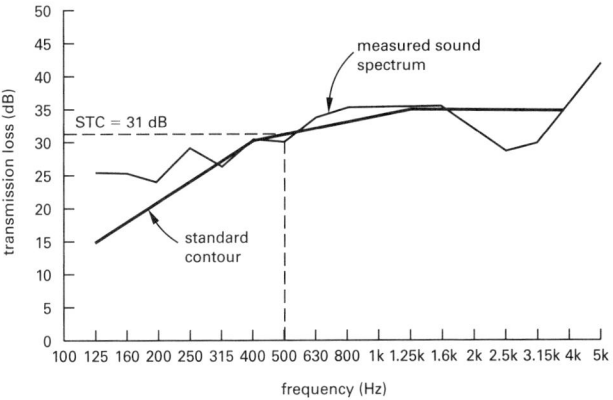

30. NOISE REDUCTION

Received sound can be reduced by modifying either the sound source, the transmission path, or the area occupied by the source. The process of reducing the noise is known as *noise reduction, sound abatement*, and *attenuation*. The amount of noise reduction, NR, is a decrease in sound pressure level due to such modifications.

$$\text{NR} = L_{p,1} - L_{p,2} = L_{W,1} - L_{W,2} \qquad 50.26$$

When adding noise reduction material to a room, the new surface absorption (sabin area) should be 3 to 10 times the original surface absorption. In a room with well-distributed absorption surfaces and at large distances from a single noise source, or for the case of many small noise sources distributed in the room, the noise reduction is

$$\text{NR} = 10 \log \frac{\sum S_1}{\sum S_2} = 10 \log \frac{R_1(1 - \overline{\alpha}_1)}{R_2(1 - \overline{\alpha}_2)} \qquad 50.27$$

31. INSERTION LOSS

Insertion loss, IL, is the term used to describe the amount of noise reduction based on sound pressure levels at a particular place after the noise source has been isolated. The source is usually isolated by covering it with an enclosure or by constructing a wall between

it and occupants. (The wall or enclosure has been "inserted" into the sound transmission path.)

$$\text{IL} = L_{p,1} - L_{p,2} = 20 \log \frac{p_1}{p_2} \qquad 50.28$$

Insertion loss takes into account structure-borne sound and flanking and diffracted sound, which can bend around the edges of a barrier. Including these factors produces a better estimate of the noise reduction than transmission loss.

32. TRANSMISSION LOSS

The *transmission loss*, TL, (also known as the *sound reduction index*, SRI) is based on sound power levels and is the actual portion of the sound energy lost when sound passes through a wall or barrier. The fraction of the incident energy transmitted through a barrier is designated as the *transmission coefficient*, τ. Appendix 50.B lists the transmission loss of several common materials used for noise control.

$$\text{TL} = L_{W,1} - L_{W,2} = 10 \log \frac{W_1}{W_2} = 10 \log \frac{1}{\tau} \qquad 50.29$$

Unlike insertion loss, which is determined from measurements of actual sound pressure levels, with and without a sound barrier, transmission loss is solely related to a material's ability to reduce the flow of sound energy passing through it. Basically, transmission loss is a function of the materials used, while insertion loss is affected by materials, installation, and leaks. According to the *mass law* (i.e., a doubling of the mass of the wall will result in a 6 dB reduction in transmitted sound), transmission loss is greatly dependent on the mass per unit area, ρ_S, (also known as *surface density*) of the barrier. Transmission loss can be predicted at normal incidence for a simple, homogeneous panel (e.g., a concrete block wall or single layer of sheetrock on vertical wood studs) at a specific frequency, f, from the mathematical expression of the mass law, Eq. 50.30. Predicting transmission loss of walls receiving frequencies near their resonant frequencies, at other than normal incidence, and of more complex walls, including double panels with airspace and/or insulation between, requires more sophisticated expressions.

$$\text{TL}_{\text{dB}} = 20 \log(f_{\text{Hz}}\rho_{S,\text{panel,kg/m}^2}) - 47 \qquad [\text{SI}] \quad 50.30(a)$$

$$\text{TL}_{\text{dB}} = 20 \log(4.88 f_{\text{Hz}}\rho_{S,\text{panel,lbm/ft}^2}) - 47 \qquad [\text{U.S.}] \quad 50.30(b)$$

33. RESONANT FREQUENCY OF A PANEL

Most panels have fairly low resonant frequencies. Transmission loss is at a minimum for sound at the resonant frequency. The *fundamental resonant frequency* (*natural frequency, first harmonic frequency*), f_0, of a simple infinite panel (plate) of material with fixed edges, with a density ρ and a Poisson's ratio ν, is given by Eq. 50.31.

The second harmonic frequency is found by multiplying the resonant frequency by 2, and so on.

$$f_{0,\text{Hz}} = 2\pi a_{\text{panel,m/s}}^2 \sqrt{\frac{12\rho_{S,\text{panel,kg/m}^2}(1-\nu_{\text{panel}}^2)}{E_{\text{panel,Pa}}t_m^3}}$$

$$= 2\pi \sqrt{\frac{12E_{\text{panel,Pa}}}{\rho_{S,\text{panel,kg/m}^2}t_m^3(1-\nu_{\text{panel}}^2)}}$$

$$\text{[infinite single panel]} \qquad 50.31$$

$$a_{\text{panel,m/s}} = \sqrt{\frac{E_{\text{panel,Pa}}}{\rho_{S,\text{panel,kg/m}^2}(1-\nu_{\text{panel}}^2)}} \qquad 50.32$$

The resonant frequency of a simple finite panel with fixed edges of length L, height h, and thickness t can be estimated from Eq. 50.33. B is the *bending stiffness per unit width*, with units of ft-lbf or N·m.

$$f_{0,\text{Hz}} = \frac{\pi}{2}\sqrt{\frac{B_{\text{N·m}}}{\rho_{S,\text{panel,kg/m}^2}}}\left(\frac{1}{L_m^2}+\frac{1}{h_m^2}\right)$$

$$= 0.45 a_{\text{panel,m/s}} t_m\left(\frac{1}{L_m^2}+\frac{1}{h_m^2}\right)$$

$$\text{[finite single panel]} \qquad 50.33$$

$$B = \frac{E_{\text{panel,Pa}}t_{\text{panel,m}}^3}{12(1-\nu_{\text{panel}}^2)} \quad \text{[homogeneous panel]} \qquad 50.34$$

The resonant frequency of an infinite double panel with an air cavity of thickness t (e.g., a double-glazed window) depends on the two surface densities and the cavity thickness.

$$f_{0,\text{Hz}} = 2\pi a_{\text{air,m/s}}\sqrt{\left(\frac{\rho_{\text{air,kg/m}^3}}{t_{\text{air,m}}}\right)\times\left(\frac{1}{\rho_{S,\text{panel 1,kg/m}^2}}+\frac{1}{\rho_{S,\text{panel 2,kg/m}^2}}\right)}$$

$$50.35$$

34. ATTENUATION

Attenuation is the amount by which sound is decreased as it travels from the source to the receiver. This general term is used interchangeably with noise reduction, insertion loss, and transmission loss. An *attenuator* is any device (e.g., a muffler) that can be inserted into the sound path to reduce the noise received.

35. TIME WEIGHTED AVERAGE

In the United States, the parameter used in the Occupational Safety and Health Act (OSHA) regulations to assess a worker's exposure to noise is the *time weighted average*, TWA. TWA represents a worker's noise exposure normalized to an eight-hour day, taking into account the average levels of noise and the time spent in each area. Determining the TWA starts with calculating the *noise dose*, D. C_i is the time in hours spent at each noise exposure level, and t_i is the *reference duration* (*permitted duration*) in hours calculated from the sound pressure level in dB for each exposure level. Since they are different representations of the same level, the *OSHA action levels* are based on either TWA or dose percent. These action levels are 85 dB (or 50% dose) and 90 dB (or 100% dose). Equation 50.36 is used when the noise level varies during the measurement period. Preventative steps must be taken if the daily noise dose is more than 1.0.

$$D_\% = \sum \frac{C_i}{t_i}\times 100\% \qquad 50.36$$

$$t_i = \frac{8\text{ hr}}{2^{(L_i-90)/5}} \qquad 50.37$$

$$\text{TWA} = 90 + 16.61\log_{10}\frac{D_\%}{100\%} \qquad 50.38$$

Unlike a traditional sound meter, which measures the instantaneous sound level, a *noise dosimeter* (*dose badge*) worn by an employee monitors and integrates the sound level over the entire work shift in order to determine the dose, D, time weighted average, TWA, average sound level, L_p, and other parameters directly.

36. ALLOWABLE EXPOSURE LIMITS

OSHA sets maximum limits on daily sound exposure. (See Table 50.7.) The "all-day" eight-hour noise level limit in the United States is 90 dBA. This is higher than the maximum level permitted in other countries.[16] In the United States, employees may not be exposed to steady sound levels above 115 dBA, regardless of the duration. Impact sound levels are limited to 140 dBA.

In the United States, hearing protection, educational programs, periodic examinations, and other actions are required for workers whose eight-hour exposure is more than 85 dBA or whose noise dose exceeds 50% of the action levels.

37. HEARING PROTECTIVE DEVICES

For continuous (as opposed to impulsive noise), when engineering or administrative controls are ineffective or not feasible, *hearing protective devices* (HPDs) (such as earmuffs and earplugs) can be used as a last resort. HPDs are labeled with a *noise reduction rating* (NRR)

---

[16]The 2003 *European Union Noise Directive* (2003/10/EC) became effective in 2006. It sets the maximum "all day" eight-hour noise level exposure limit in EU countries at 87 dBA. The first action level, to train personnel and make hearing protection available, occurs at 80 dBA. The second action level, to implement a noise reduction program and require usage of hearing protection, occurs at 85 dBA.

Table 50.7 *Typical Permissible Noise Exposure Levels[a,b]*

sound level (dBA)	exposure (hours per day)
90	8
92	6
95	4
97	3
100	2
102	$1^1/_2$
105	1
110	$^1/_2$
115	$^1/_4$ or less

[a]without hearing protection
[b]Federal Register, Title 29, Sec. 1910.95, Table G-16

which is the manufacturer's claim of how much noise reduction (in dB) a hearing protective device provides. The NRR is essentially an average insertion loss. Theoretically, the entire NRR would be subtracted from the environmental noise level to obtain the noise level at the ear. However, since HPDs generally provide much less protection than their labels claim, OSHA (drawing on NIOSH standards) recommends in its field manual that inspectors calculate a more realistic measurement of effectiveness.

Since sound level meters have multiple scales, the calculation method depends on the measurement scale used to measure the environmental noise. For formable earplugs, when noise is measured in dBA (i.e., on the A-weighted scale), the OSHA formula subtracts 7 dB from the NRR and divides the result by 2. (Values for earmuffs and nonformable earplugs are calculated differently.)

$$\Delta L_{\mathrm{dBA}} = \frac{\mathrm{NRR} - 7}{2} \quad \text{[formable earplugs; A-scale]}$$

$$50.39$$

When noise is measured in dBC (i.e., on the C-weighted scale), the OSHA formula subtracts 7 dB from the NRR.

$$\Delta L_{\mathrm{dBC}} = \mathrm{NRR} - 7 \quad \text{[formable earplugs; C-scale]}$$

$$50.40$$

For environments dominated by frequencies below 500 Hz, the C-weighted noise level should be used. With impulsive noise (e.g., gunfire), the NRR may not be an accurate indicator of hearing protection.

Example 50.9

A construction worker leaves an 86 dB environment after six hours and works in a 92 dB environment for three hours more. What are the (a) dose and (b) TWA?

Solution

Use Eq. 50.37.

$$t_1 = \frac{8 \text{ hr}}{2^{(L_1 - 90)/5}} = \frac{8 \text{ hr}}{2^{(86 \text{ dB} - 90 \text{ dB})/5}} = 13.92 \text{ hr}$$

$$t_2 = \frac{8 \text{ hr}}{2^{(L_2 - 90)/5}} = \frac{8 \text{ hr}}{2^{(92 \text{ dB} - 90 \text{ dB})/5}} = 6.06 \text{ hr}$$

From Eq. 50.36, the noise dose is

$$D_\% = \sum \frac{C_i}{t_i} \times 100\% = \left(\frac{6 \text{ hr}}{13.92 \text{ hr}} + \frac{3 \text{ hr}}{6.06 \text{ hr}} \right) \times 100\%$$
$$= 92.6\%$$

From Eq. 50.38, the time weighted average is

$$\mathrm{TWA} = 90 + 16.61 \log_{10} \frac{D_\%}{100\%}$$
$$= 90 \text{ dB} + 16.61 \log_{10} \frac{92.6\%}{100\%}$$
$$= 89.46 \text{ dB}$$

51 Emergency Management

Nomenclature

ABS	absorption rate	–
AT	averaging time	d
B	volumetric fraction	–
BW	body mass (weight)	kg
c	specific heat	J/kg·K
C	concentration	mg/m^3
CDI	chronic daily intake	mg/kg·d
CR	contact rate	m^3/d
ED	exposure duration	yr
EF	exposure frequency	d/yr
HR	hazard ratio	–
I	intake	mg/kg·d
k	first order reaction rate constant	d^{-1}
k	specific heat ratio	–
LFL	lower flammability limit	–
m	mass	kg
p	pressure, absolute or gage	Pa
PF	potency factor (carcinogens), same as slope factor, SF	(mg/kg·d)$^{-1}$
Q	volumetric flow rate	L/d
R	risk of cancer	–
R^*	universal gas constant	kJ/kmol·K
RfD	reference dose (noncarcinogens)	mg/kg·d
RR	retention rate	–
SF	slope factor (carcinogens), same as potency factor, PF	(mg/kg·d)$^{-1}$
T	temperature	K, °C
UFL	upper flammability limit	–
V	volume	m^3, L
W	energy release from adiabatic compression of gas or vapor	kJ
x	mole fraction	–

Subscripts

0	initial condition
fp	flash point
l	liquid
mix	mixture
p	constant pressure
t	at time t
v	constant volume, vapor, or in vapor phase
vp	vapor pressure

1. TYPES OF EMERGENCIES

Incident Types

There are seven major types of incidents that may require emergency response.

- highway accidents
- railway accidents
- ship and barge accidents
- aircraft accidents
- plant accidents
- pipeline accidents
- container failures

Highway Accidents

Highway accidents are by far the most common type of emergency incident to which an environmental engineer may offer expertise. Hazardous materials are carried by trucks every day over highways and streets. These materials are transported through residential as well as commercial and rural areas, and they often pose a risk to humans and the natural environment. A highway accident may involve injuries and multiple vehicles, so the additional complication of dealing with hazardous materials presents a challenging problem for emergency response personnel. The effects resulting from spills of toxic materials into drains or waterways or the release of toxic gases into the air may be catastrophic. Highway accidents may also result in long-term chronic effects from contamination of soil and water that cannot be cleaned up completely after the accident.

The cause of most highway accidents is human error. However, accidents also occur due to weather, mechanical failure, roadway obstructions, and roadbed defects.

Railway Accidents

Railway accidents are less common than highway accidents, but the potential effects are greater due to the larger volumes of hazardous materials that may be involved in the incident. For example, a typical railroad tank car has a capacity of approximately 130 m^3, while a typical highway tanker has a capacity of approximately 30 m^3 to 40 m^3. Moreover, rail accidents may occur at locations with difficult access compared to highway accidents.

Although track deterioration and failure are common causes of rail accidents, other causes include collision with other trains or vehicles at crossings, terrorist acts, and signal or mechanical failures.

Ship and Barge Accidents

Ship and barge accidents can have catastrophic effects on waterways. Containment and clean-up of ship and barge accidents are very difficult and frequently are insufficient to prevent both short-term catastrophic effects and long-term chronic effects on aquatic life, wildlife, and humans. Releases to waterways can result in an accumulation of pollutants in sediments that is difficult to remediate and may result in decades of chronic effects.

Aircraft Accidents

Fortunately, aircraft accidents are rare compared to other transportation accidents, and they seldom involve large quantities of hazardous cargoes. Aircraft accidents may occur anywhere and can involve releases of petroleum products from the aircraft (as well as release of cargo).

Plant Accidents

Accidents at plants or facilities are not uncommon. They often involve serious health and safety threats to humans as well as to the natural environment. Plant accidents are usually associated with manufacturing operations but may also occur from the use, transportation, or storage of both hazardous and nonhazardous materials. Causes of plant accidents include equipment failure, human error (such as accidental mixing of reactive products), product or tank overflow, physical damage to containers, and exposure to fire, water, or heat.

Besides the concern for the safety of employees at a plant during an accident, there is also concern about hazardous material or pollutant releases to the natural environment (air and water) and effects on humans in the vicinity of the plant site (so-called "downwinders"). In many industrial plants, discharges to the water or emissions to the air are a routine part of operation. During an incident, these routine pathways to the environment may become routes for hazardous materials or additional pollutants to escape.

Pipeline Accidents

Pipeline accidents are not common due to (1) the care taken to design pipelines with generous safety factors, and (2) the fact that pipelines are, for the most part, buried and are not exposed to potential hazards. However, pipeline accidents can be quite dramatic and are often catastrophic. Pipelines carry a wide variety of products such as natural gas, water, petroleum, and wastewater. Some gases or liquids transported by pipelines pose more serious potential hazards than other materials. While accidents involving natural gas pipelines may involve explosions and fire, which are often a serious threat to human safety, the effects on the natural environment may be minimal. On the other hand, failures of wastewater pipelines may not pose a serious threat to the safety of human life, but could have a potentially catastrophic effect on aquatic life.

Container Failures

Container failures can occur in conjunction with highway, railroad, and plant accidents. They can also occur independently of a transportation accident, such as at a storage facility.

Although mishandling of a container is a common cause of failure, other causes include thermal, mechanical, and chemical failures. *Thermal failures* occur when the container is subjected to heat or flame. In a fire, a pressure relief valve may vent or blow out due to pressure buildup, causing a "blowtorch" effect. (See Fig. 51.1.) *Mechanical failures* occur due to overfilling, a defect in the container or pressure relief valve, a gasket or connection failure, or damage from other events. *Chemical failures* result when products cause corrosion.

Figure 51.1 Pressure Venting

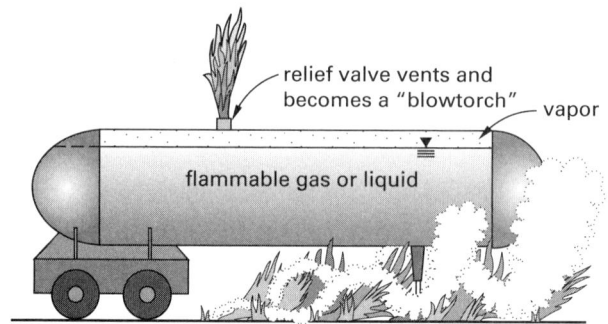

Emergency Types

Flammable and Combustible Liquid Emergencies

The majority of U.S. hazardous material responses involve flammable and combustible liquids. Flammable and combustible liquids are handled by truck, train, barge, and pipeline and are of great concern to incident responders. (The principal distinction between flammable and combustible liquids is that *flammable liquids* give off vapors at temperatures below $37.8°C$ that may be easily ignited. *Combustible liquids* also give off vapors at temperatures below $37.8°C$, but not enough to present an ignition hazard. However, combustible liquids produce large amounts of heat when they are ignited.)

The U.S. Department of Transportation (USDOT) defines a flammable liquid as "a liquid having a flash point of not more than $60.5°C$, or any material in a liquid phase with a flash point at or above $37.8°C$ that is intentionally heated and offered for transportation or is transported at its flash point in a bulk packaging. . ." (49 CFR 173.120(a)). The USDOT definition excludes liquefied compressed gases, compressed gases in solution, cryogenic liquids, and mixtures with flash points above 60.5°C.

USDOT defines a combustible liquid as "any liquid that does not meet the definition of any other hazard class. . . and (1) has a flash point above 60.5°C and below 93°C, or (2) is a flammable liquid with a flash point at or above 37.8°C" (49 CFR 173.120(b)).

Gas Emergencies

Emergencies involving gases are one of the most common types emergency responders must handle. Gases are widely used throughout the nation and are shipped in several forms: compressed, liquefied, and cryogenic. USDOT has established specific definitions for gases shipped on the nation's transportation network.

Flammable gas is defined as "any material which is a gas at 20°C or less and 101.3 kPa of pressure which (1) is ignitable at 101.3 kPa when in a mixture of 13% or less by volume with air, or (2) has a flammable range (at 101.3 kPa) with air of at least 12% regardless of the lower limit" (49 CFR 173.115(a)). Flammable gases are assigned USDOT Hazard Class 2.1.

Compressed gas is defined as "any material or mixture which (1) exerts on the packaging a gage pressure of 200 kPa or greater at 20°C, and (2) does not meet the definition of a flammable gas or a gas poisonous by inhalation. This category includes compressed gas, liquefied gas, pressurized cryogenic gas, and compressed gas in solution" (49 CFR 173.115(b)). Compressed gases are assigned USDOT Hazard Class 2.2.

Gas that is *poisonous* by inhalation is defined as "a material which is a gas at 20°C or less and a pressure of 101.3 kPa and which (1) is known to be so toxic to humans as to pose a hazard to health during transportation, or (2) in the absence of adequate data on human toxicity, is presumed to be toxic to humans based on tests conducted on laboratory animals" (49 CFR 173.115 (c)). Gases that are poisonous by inhalation are assigned USDOT Hazard Class 2.3.

Liquefied compressed gas is defined as "a gas which, when packaged under pressure for transport, is partially liquid at temperatures above −50°C" (49 CFR 173.115(e)).

A *cryogenic liquid* is defined as "a refrigerated liquefied gas having a boiling point colder than −90°C at 101.3 kPa" (49 CFR 173.115(g)).

2. ROLE OF THE ENVIRONMENTAL ENGINEER

Potential Roles

The environmental engineer may have a wide variety of roles relative to emergency management. An environmental engineer employed by a public agency may function as a key member of a team having direct responsibility for response and mitigation or cleanup of an emergency, such as an oil spill from a ship or tanker collision or grounding. An environmental engineer employed by a public agency may also have a more limited role in the emergency and may function as an advisor to utility services, facilities, or environmental regulatory agencies that may be affected by the emergency.

An environmental engineer may also be employed by an emergency response contractor or as a consultant to a potentially affected service provider, such as a water supply utility. The environmental engineer may be asked to evaluate the risk to human health and the environment from the release of hazardous substances and a wide variety of other pollutants in order to advise emergency response personnel and to identify mitigation or remedial measures. Unless prepared by appropriate training and experience, the environmental engineer should not function as a hazmat responder or as a member of an emergency service team (Emergency Medical Services (EMS) providers, police officers, and firefighters). However, the environmental engineer may be called upon to provide advice to hazardous waste responders about potential effects on, or protection of, the natural environment, human health, or water supplies.

Working with Hazardous Materials Responders

There are typically four categories of hazardous materials responders.

- first responders
- technicians
- specialists
- incident (on-scene) commander

The *first responders* may include police, firefighters, public works personnel, industrial spill teams, and EMS personnel. The first responders initially recognize the problem, identify the hazards involved, and notify more qualified responders. They also work to contain spills and try to minimize harm.

Technicians may include hazardous materials response teams, industrial crews, and emergency response teams. Their role is to control the spill and stop the release.

Specialists may include teams with special skills and expertise, state and local regulators, and industrial experts. The specialists provide support to the first responders through consultation and advice about the effects of courses of action, possible mitigation measures, and technical characteristics of the hazardous materials. The environmental engineer may be considered a specialist and may be called upon to fulfill this role.

The *incident commander* is also known as the *on-scene commander*. Incident commanders may be fire chiefs, battalion fire chiefs, sheriffs and deputies, state troopers, or emergency response team supervisors. The incident commander is responsible for the overall safety of personnel on the scene and for directing the proper containment, cleanup, mitigation, and disposal of the hazardous materials.

3. LEGAL REQUIREMENTS

As hazardous waste remediation and emergency response operations have matured over the last two or three decades, so have the legal regulations and requirements. The Occupational Safety and Health Administration (OSHA) has developed explicit standards that apply to workers involved with emergency response and hazardous waste site operations. These standards provide legal obligations incumbent on employers to protect the health and safety of workers who perform such activities. The standards are given in 29 CFR 1910.119 along with four appendices (1910.119 Appendices A through D). (The mandatory OSHA standards of App. A are given in App. 51.A. The standards are subject to change, so the most current version should be used when dealing with an actual cleanup or emergency response incident.)

While 29 CFR 1910.120 contains provisions for worker safety during hazardous clean-up, emergency response, and corrective actions, 29 CFR 1910.119 contains requirements for preventing releases of toxic, reactive, flammable, and explosive chemicals. This section and its four appendices are applicable, among other criteria, to processes with accumulations of chemicals at or above the specified *threshold quantities* (TQs) listed in 19 CFR 1910.119 Appendix A (and presented in App. 51.A). Appendix 51.A, which is from 29 CFR 1910.119, does not relate to the protection of emergency response and clean-up personnel.

4. PROPERTIES OF GASES AND LIQUIDS RELEVANT TO EMERGENCY RESPONSE

Vapor Pressure

When liquids evaporate into a confined space, such as in the top of a tank or railroad tank car, the *vapor pressure* increases until the maximum vapor pressure is reached for the temperature of the space. Exposure of a tank to fire can increase the vapor pressure above atmospheric pressure, at which point the liquid will boil. The greater the mass of vapor and the higher the vapor pressure, the greater the potential for ignition or explosion. Pressure relief valves, designed to release some of the accumulated vapors, may be damaged in an emergency incident and not function properly. A relief valve may also function as a blowtorch if the escaping vapors are ignited.

Raoult's law states that the vapor pressure of a solution at a particular temperature is equal to the mole fraction of the solvent (in the liquid phase) multiplied by the vapor pressure of the pure solvent at the same temperature. For an ideal vapor (gas), the *partial pressure* *ratio, mole fraction,* and *volumetric fraction* are equivalent for a given temperature and pressure. As a result, Eq. 51.1 can be used to determine flammable limits.

$$x_v = \frac{p_{\text{vp,pure}} x_l}{p} \qquad 51.1$$

Vapor Specific Gravity

Vapor specific gravity (sometimes incorrectly referred to as "vapor density") is the specific gravity of a gas or vapor referred to the density of air. (The specific gravity of air is 1.0.) A gas or vapor with a vapor specific gravity less than 1.0 is lighter than air and will rise upward to the top of a confining space, such as a tank or building. A gas or vapor with a vapor specific gravity greater than 1.0 is heavier than air and will sink to the bottom of a confining space. Table 51.1 gives approximate specific gravities of some common vapors. Emergency responders must consider the vapor specific gravity of the gas or vapor involved in an incident.

Table 51.1 Approximate Specific Gravities of Vapors

gas	vapor specific gravity	conditions (at 1 atm, 101.325 kPa)
acetylene	0.906	0°C
air	1.000	21°C
ammonia (anhydrous)	0.597	0°C
carbon dioxide	1.522	21°C
chlorine	2.482	0°C
ethylene	0.978	0°C
hydrogen	0.0695	0°C
hydrogen sulfide	1.2	15°C
methane	0.5549	16°C
nitrogen	0.967	21°C
oxygen	1.105	21°C
sulfur dioxide	2.262	21°C
vinyl chloride	2.15	15°C

Source: Onguard, Inc. 1996 (Appendix D)

Liquid Specific Gravity

The *specific gravity* of a liquid is the ratio of the density of the liquid to the density of pure water. Since most flammable liquids have specific gravities less than 1.0, they will normally accumulate and float on the surface of a body of water.

Boiling Point

The *boiling point* of a liquid is the temperature at which a liquid rapidly becomes a vapor. Below its boiling point, a liquid changes to a vapor slowly through *evaporation*. A liquid with a low boiling point is more dangerous because it will produce vapors sooner than a liquid with a high boiling point.

Expansion Ratio

The *expansion ratio* is defined as the ratio of the volume of gas produced when liquid is vaporized to the original volume of liquid. Examples of expansion ratios for several liquids are given in Table 51.2.

Table 51.2 Approximate Expansion Ratios

liquid	boiling point (°C)	expansion ratio[*]
helium	−269	745
methane	−161	625
nitrogen	−196	696
oxygen	−183	860

[*]volume expansion ratio from liquid at boiling point and 1 atm to gas at 21°C and 1 atm

Critical Point

The *critical point* refers to the condition at which a substance can exist in both gas and liquid states simultaneously. When a liquid is at its *critical temperature*, further heating will cause part or all of it to vaporize instantly. *Critical pressure* is the pressure required to liquefy a gas. These concepts are important in emergency response situations because liquefied gas above its critical temperature will instantly convert to a gas, which may cause a container to fail catastrophically.

Flash Point

The *flash point* is defined as the minimum temperature at which a liquid will ignite if an ignition source is present. Below the flash point, a liquid will not produce enough vapor for combustion (i.e., the mixture will be too lean).

Fire Point

The *fire point* is the minimum temperature at which a liquid will produce enough vapor for sustained combustion. This differs from the flash point, at which a vapor flash can occur but combustion will not be sustained. The fire point is usually only one or two degrees Celsius above the flash point.

Flammable Limits of Gases and Vapors

The concentrations of a substance in air below which there is insufficient fuel to burn or explode (i.e., the mixture is too lean) are known as the *lower flammable limit* (LFL) and *lower explosion limit* (LEL), respectively. The concentrations above which there is too much fuel to burn or explode (i.e., the mixture is too rich) are known as the *upper flammable limit* (UFL) and *upper explosion limit* (UEL), respectively. The LFL and the UFL are typically expressed at normal temperature and pressure (NTP) of 25°C and 1 atm, but this should be verified when using data from a reference.

The *flammable range* refers to the range of volumetric percentages of a gas or vapor in air that will form a flammable mixture. Gases and vapors with low LFLs are inherently more dangerous than those with higher LFLs. Vapors with a wide flammable range also are of greater concern than those with a narrow flammable range. Increased pressure or temperature will lower the LFL and raise the UFL, thereby widening the flammability range. Decreased pressure or temperature will narrow the range between the LFL and the UFL. Flammability limits of some typical flammable products are shown in Fig. 51.2.

Figure 51.2 Flammable Limits

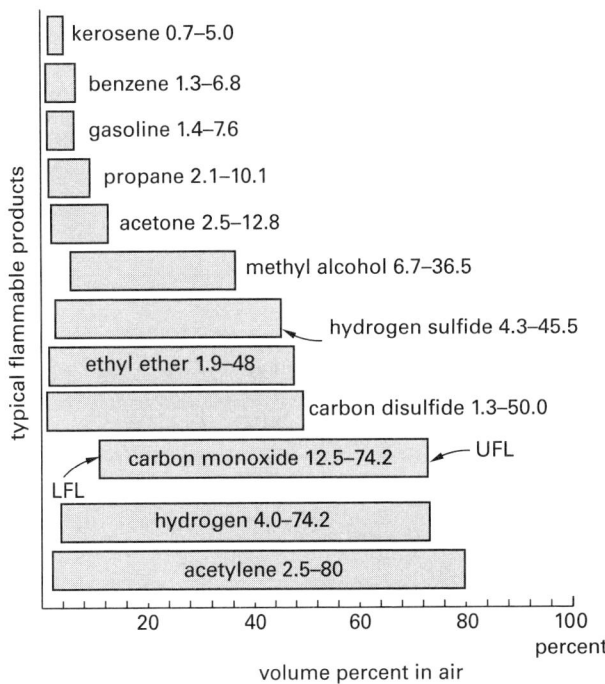

The LFL of a flammable gas can be estimated by dividing the vapor pressure of the gas at its flash point by the pressure at one atmosphere (760 mm Hg, 101.325 kPa, or 14.7 psia).

$$\text{LFL} = \frac{p_{\text{vp,fp}}}{p_{\text{atmospheric}}} \qquad 51.2$$

For a flammable gas, the flash point may be in the cryogenic range. In that case, a gas is classified as flammable if the LFL is less than or equal to 13% by volume at NTP or if the flammability range is more than 12% at NTP. Note that most flammable gases are lighter than air, but flammable vapors are often heavier than air.

The LFL and UFL of mixtures of vapors can be determined from Eq. 51.3 and Eq. 51.4.

$$\text{LFL}_{\text{mix}} = \frac{1}{\sum \dfrac{x_i}{\text{LFL}_i}} \qquad 51.3$$

$$\text{UFL}_{\text{mix}} = \frac{1}{\sum \dfrac{x_i}{\text{UFL}_i}} \qquad 51.4$$

Health, Safety, Welfare

Example 51.1

Workers are clearing a large warehouse containing drums and cylinders that are believed to contain flammable or explosive liquids and gases. While they are working in one of the rooms, the control valve on a cylinder marked "Warning—Hydrogen Gas—Contents Under Pressure" is broken. The room is 10 m by 20 m by 10 m high. The estimated pressure in the cylinder is 20 MPa (gage), and the gas has an estimated volume of 3000 L. Is the resulting mixture flammable?

Solution

Find the volume of hydrogen (H_2) in the room. Assume the H_2 mixes thoroughly in the room almost instantaneously and the cylinder is at the same temperature as the air in the room. Assume the pressure in the room remains at 1 atm.

The volume of the cylinder is

$$V_1 = (3000 \text{ L})\left(\frac{1 \text{ m}^3}{10^3 \text{ L}}\right)$$
$$= 3 \text{ m}^3$$
$$p_1 = (20 \text{ MPa})\left(1000 \frac{\text{kPa}}{\text{MPa}}\right)$$
$$+ (1 \text{ atm})\left(\frac{101.325 \text{ kPa}}{1 \text{ atm}}\right)$$
$$= 20\,101.325 \text{ kPa}$$
$$p_2 = (1 \text{ atm})\left(\frac{101.325 \text{ kPa}}{1 \text{ atm}}\right)$$
$$= 101.325 \text{ kPa}$$

From Boyle's law,

$$p_1 V_1 = p_2 V_2$$
$$V_2 = \frac{p_1 V_1}{p_2} = \frac{(20\,101.325 \text{ kPa})(3 \text{ m}^3)}{101.325 \text{ kPa}}$$
$$= 595.15 \text{ m}^3$$

Find the volume of air in the room.

$$V_{air} = (10 \text{ m})(20 \text{ m})(10 \text{ m})$$
$$= 2000 \text{ m}^3$$

Find the percent H_2 by volume.

$$B_{H_2} = \frac{V_2}{V_{total}} = \frac{595.15 \text{ m}^3}{2000 \text{ m}^3 + 595.15 \text{ m}^3}$$
$$= 0.23 \quad (23\%)$$

Determine the flammability limits of hydrogen gas from Fig. 51.2.

$$LFL = 4.0\%$$
$$UFL = 74.2\%$$

The room contents are flammable because the concentration of H_2 is between the LFL and the UFL.

Example 51.2

A mixture contains three gases.

gas	actual volume	LFL % volume	UFL % volume
benzene	1	1.3	6.8
pentane	4	1.4	7.8
ethyl alcohol	2	3.3	19.0

What are the lower and upper flammability limits of the mixture?

Solution

Determine the mole fraction of each gas. Note that for a given temperature and pressure, the mole fraction of an ideal gas is equal to its volumetric fraction.

For benzene,

$$x_{v,benzene} = B_{benzene} = \frac{1}{1+4+2} = 0.143$$

For pentane,

$$x_{v,pentane} = B_{pentane} = \frac{4}{1+4+2} = 0.571$$

For ethyl alcohol,

$$x_{v,ethyl\ alcohol} = B_{ethyl\ alcohol} = \frac{2}{1+4+2} = 0.286$$

From Eq. 51.3,

$$LFL_{mix} = \frac{1}{\sum \frac{x_i}{LFL_i}}$$
$$= \frac{1}{\frac{0.143}{1.3\%} + \frac{0.571}{1.4\%} + \frac{0.286}{3.3\%}}$$
$$= 1.65\%$$

From Eq. 51.4,

$$UFL_{mix} = \frac{1}{\sum \frac{x_i}{UFL_i}}$$
$$= \frac{1}{\frac{0.143}{6.8\%} + \frac{0.571}{7.8\%} + \frac{0.286}{19.0\%}}$$
$$= 9.15\%$$

Autoignition

The *autoignition* temperature is the minimum temperature at which a substance (liquid, gas, or vapor) will spontaneously initiate self-sustained combustion without the presence of an ignition source. Autoignition can result from an adiabatic (without heat transfer) compression of a gas. In emergency response situations, such compression may occur from damage to or collapse of tanks, tanker trucks, or structures. The gas temperature resulting from adiabatic compression can be found from Eq. 51.5.

$$\frac{T_2}{T_1} = \left(\frac{p_2}{p_1}\right)^{(k-1)/k} \quad \text{[adiabatic]} \qquad 51.5$$

k is the ratio of specific heats.

$$k = \frac{c_p}{c_v} \qquad 51.6$$

For air at room temperature and atmospheric pressure, c_p is approximately 1005 J/kg·K, and c_v is approximately 718 J/kg·K. k is approximately 1.4.

Example 51.3

A large tank contains ethane gas (C_2H_6, $c_p = 1616$ J/kg·K, $c_v = 1340$ J/kg·K) at 2 atm and 100°C. The autoignition temperature of ethane is 472°C. During an emergency incident, the gas in the tank mixes with air and enters its flammable range. If the tank is compressed due to structural failure until the internal pressure in the tank rises to 10 MPa, will an explosion occur?

Solution

The ratio of specific heats is

$$k = \frac{c_p}{c_v} = \frac{1616 \ \dfrac{J}{kg \cdot K}}{1340 \ \dfrac{J}{kg \cdot K}}$$

$$= 1.206$$

$$T_1 = 100°C + 273°$$

$$= 373K$$

$$p_1 = (2 \text{ atm})\left(\frac{101.325 \text{ kPa}}{1 \text{ atm}}\right)$$

$$= 202.65 \text{ kPa}$$

$$p_2 = (10 \text{ MPa})\left(\frac{10^3 \text{ kPa}}{1 \text{ MPa}}\right)$$

$$= 10 \times 10^3 \text{ kPa}$$

Use Eq. 51.5.

$$T_2 = T_1\left(\frac{p_2}{p_1}\right)^{(k-1)/k}$$

$$= (373K)\left(\frac{10 \times 10^3 \text{ kPa}}{202.65 \text{ kPa}}\right)^{(1.206-1)/1.206}$$

$$= 726K - 273°$$

$$= 453°C$$

Since $T_2 = 453°C$ is less than the autoignition temperature of ethane (472°C), combustion and explosion will not occur.

Explosive Energy Release

Pressurized containers are often encountered in emergency response situations. A sudden, catastrophic release of compressed gas can release tremendous amounts of energy, causing destruction of property and injury or death to emergency responders or others in the vicinity. The energy released from an adiabatic failure of a pressurized tank or cylinder can be calculated from the following equation.

$$W = \left(\frac{p_1 V_1}{k - 1}\right)\left(1 - \left(\frac{p_2}{p_1}\right)^{(k-1)/k}\right) \qquad 51.7$$

Example 51.4

Workers at a remediation site uncovered an old tank of compressed nitrogen gas (N_2) with an indicated volume of 20 m³ and pressure of 1500 kPa (absolute). For nitrogen, $c_p = 1043$ J/kg·K, and $c_v = 746$ J/kg·K. If the tank fails catastrophically and its contents are released under adiabatic conditions, what is the energy release in equivalent mass of TNT? Assume TNT releases 4.7 MJ/kg TNT.

Solution

$$p_2 = (1 \text{ atm})\left(\frac{101.325 \text{ kPa}}{1 \text{ atm}}\right)$$

$$= 101.325 \text{ kPa}$$

Use Eq. 51.6.

$$k = \frac{c_p}{c_v} = \frac{1043 \ \dfrac{J}{kg \cdot K}}{746 \ \dfrac{J}{kg \cdot K}}$$

$$= 1.398$$

$$V_1 = 20 \text{ m}^3$$

Use Eq. 51.7.

$$W = \left(\frac{p_1 V_1}{k - 1}\right)\left(1 - \left(\frac{p_2}{p_1}\right)^{(k-1)/k}\right)$$

$$= \left(\frac{(1500 \text{ kPa})(20 \text{ m}^3)}{1.398 - 1}\right)$$

$$\times \left(1 - \left(\frac{101.325 \text{ kPa}}{1500 \text{ kPa}}\right)^{(1.398-1)/1.398}\right)$$

$$= 40.4 \times 10^3 \text{ kJ} \quad (40.4 \text{ MJ})$$

The TNT equivalent is

$$m = \frac{40.4 \text{ MJ}}{4.7 \ \dfrac{\text{MJ}}{\text{kg}}} = 8.60 \text{ kg}$$

Health, Safety, Welfare

Solubility

The solubility of liquids and gases is an important consideration in emergency response situations. Gases that are soluble in water may produce a product that has dangerous properties or is a hazard or pollutant. An example is the solution of hydrogen chloride gas in water—hydrochloric acid, HCl.

BLEVEs

A *BLEVE* is a *boiling liquid expanding vapor explosion.* A BLEVE can occur when a liquid (including water) in a closed container (i.e., tank, drum) is heated well above its boiling point at normal atmospheric pressure. A BLEVE explosion can cause catastrophic loss of life and property.

5. ASSESSMENT OF CHRONIC EFFECTS FROM EMERGENCY INCIDENTS

Releases to Air

Releases to air during emergency incidents are usually rapid and may result in acute effects on hazardous materials responders and the community. Such releases include smoke and vapors containing toxic pollutants, vapor clouds, and toxic or hazardous gases.

The intake of a chemical from long-term exposure to inhalation of vapors can be calculated from Eq. 51.8.

$$I = \frac{C(\text{CR})(\text{EF})(\text{ED})(\text{RR})(\text{ABS})}{(\text{BW})(\text{AT})} \quad \text{51.8}$$

In evaluating chronic effects, the *exposure frequency*, EF, is the number of days per year that the affected population is exposed to vapors from the incident, and the *averaging time*, AT, is the expected time the affected population would be exposed (for noncarcinogens). For carcinogens, the averaging time, AT, is 70 yr. The same procedure is applied to chronic exposure over several years as for acute exposure.

Refer to Chap. 45 and Chap. 11 for a more detailed description of the methods for exposure assessment and risk analysis.

Releases to Water

Water quality emergencies involve release of petroleum products or toxic or hazardous chemicals to the natural environment, such as to drainage ways or directly to waterbodies. The concern of the environmental engineer may be the potential effects on human health from contamination of drinking water supplies or the potential effects from pollutants toxic to aquatic life.

The chronic effects from drinking water sources can be evaluated by determining the instream concentration of the pollutant at the point of use. One approach is to ensure that the instream concentration of the pollutant is at an acceptable level prior to treatment. This approach is appropriate when it is risky to rely upon treatment processes for long-term protection of human health after an emergency response incident. Decay of the pollutant may be accounted for as a part of the mitigation approach. Dilution can also be used, at least in part, to reduce the pollutant concentration at the point of use. The safest approach is to ensure that discharges leaving the control of the responsible party at the incident site are within acceptable limits so that any downstream exposure will not cause unsafe chronic exposure.

The evaluation procedure for chronic effects starts with determining the *chronic daily intake* (CDI) of the pollutant through the oral route from drinking water and through other applicable routes, such as vapors released from the water. The intake can then be compared to the *reference dose* (RfD) for noncarcinogens by computing the *hazard ratio* (HR), CDI/RfD. If the hazard ratio exceeds 1.0, the exposure is unacceptable, and further cleanup efforts will be needed. For carcinogens, the *risk* (R) can be determined by the product of the CDI and the *slope factor* (SF) or the *potency factor* (PF). The SF and the PF both refer to the risk associated with 1 mg/kg·d daily intake. Judgment will be involved in determining an *acceptable risk* for the specific conditions. Acceptable risk is usually set at 1×10^{-4} to 1×10^{-6}, which is a range of risk levels typically used at USEPA cleanup projects.

Refer to Chap. 45 and Chap. 11 for more detailed descriptions of the methods for exposure assessment and risk analysis.

Topic VIII: Systems, Management, and Professional

Systems, Mgmt., Professional

52 Electrical Systems and Equipment

Nomenclature

a	turns ratio	–	–
A	area	ft^2	m^2
B	susceptance	S	S
C	capacitance	F	F
d	diameter	ft	m
E	energy usage (demand)	kW-hr	kW·h
f	linear frequency	Hz	Hz
G	conductance	S	S
i	varying current	A	A
I	constant current	A	A
l	length	ft	m
L	inductance	H	H
n	speed	rev/min	rev/min
N	number of turns	–	–
p	number of poles	–	–
pf	power factor	–	–
P	power	W	W
Q	charge	C	C
Q	reactive power	VAR	VAR
R	resistance	Ω	Ω
s	slip	–	–
sf	service factor	–	–
S	apparent power	VA	VA
SR	speed regulation	–	–
t	time	sec	s
T	period	sec	s
T	torque	ft-lbf	N·m
v	varying voltage	V	V
V	constant voltage	V	V
VR	voltage regulation	–	–
X	reactance	Ω	Ω
Y	admittance	S	S
Z	impedance	Ω	Ω

Symbols

η	efficiency	–	–
θ	phase angle	rad	rad
ρ	resistivity	Ω-ft	Ω·m
σ	conductivity	$1/\Omega$-ft	$1/\Omega$·m
ϕ	angle	rad	rad
ω	angular frequency	rad/sec	rad/s

Subscripts

a	armature
ave	average
C	capacitive
Cu	copper
e	equivalent

eff   effective
f field
l line
L inductive
m maximum
p phase or primary
r rated
R resistive
s secondary
t total

1. ELECTRIC CHARGE

The charge on an electron is one *electrostatic unit* (esu). Since an esu is very small, electrostatic charges are more conveniently measured in *coulombs* (C). One coulomb is approximately 6.242×10^{18} esu. Another unit of charge, the *faraday*, is sometimes encountered in the description of ionic bonding and plating processes. One faraday is equal to one mole of electrons, approximately 96,485 C (96,485 A·s).

2. CURRENT

Current, I, is the movement of electrons. By historical convention, current moves in a direction opposite to the flow of electrons (i.e., current flows from the positive terminal to the negative terminal). Current is measured in *amperes* (A) and is the time rate change in charge.

$$I = \frac{dQ}{dt} \qquad \text{[DC circuits]} \qquad 52.1(a)$$

$$i(t) = \frac{dQ}{dt} \qquad \text{[AC circuits]} \qquad 52.1(b)$$

3. VOLTAGE SOURCES

A net *voltage, V*, causes electrons to move, hence the common synonym *electromotive force* (emf).[1] With *direct-current* (DC) voltage sources, the voltage may vary in amplitude, but not in polarity. In simple problems where a battery serves as the voltage source, the magnitude is also constant.

With *alternating-current* (AC) voltage sources, the magnitude and polarity both vary with time. Due to the method of generating electrical energy, AC voltages are typically sinusoidal.

4. RESISTIVITY AND RESISTANCE

Resistance is the property of a *resistor* or resistive circuit to impede current flow.[2] A circuit with zero resistance is a *short circuit*, whereas an *open circuit* has infinite resistance. Adjustable resistors are known as *potentiometers, rheostats*, and *variable resistors*.

---

[1]The symbol E (derived from the name electromotive force) has been commonly used to represent voltage induced by electromagnetic induction.
[2]*Resistance* is not the same as *inductance* (see Sec. 52.20), which is the property of a device to impede a *change* in current flow.

Resistance of a circuit element depends on the *resistivity*, ρ (in Ω-in or Ω·cm), of the material and the geometry of the current path through the element. The area, A, of circular conductors is often measured in *circular mils*, abbreviated *cmils*, the area of a 0.001 in diameter circle.

$$R = \frac{\rho l}{A} \qquad 52.2$$

$$A_{\text{cmils}} = \left(\frac{d_{\text{in}}}{0.001}\right)^2 \qquad 52.3$$

$$A_{\text{in}^2} = (7.854 \times 10^{-7}) A_{\text{cmils}} \qquad 52.4$$

Resistivity depends on temperature. For most conductors, resistivity increases linearly with temperature. The resistivity of standard *IACS (International Annealed Copper Standard)* copper wire at 20°C is approximately

$$\rho_{\text{IACS Cu,20°C}} = 1.7241 \times 10^{-6} \ \Omega\text{·cm}$$
$$= 1.7241 \times 10^{-8} \ \text{m/S}$$
$$= 0.67879 \times 10^{-6} \ \Omega\text{-in}$$
$$= 10.371 \ \Omega\text{-cmil/ft} \qquad 52.5$$

Example 52.1

What is the resistance of a parallelepiped (1 cm × 1 cm × 1 m long) of IACS copper if current flows between the two smaller faces?

Solution

From Eq. 52.2 and Eq. 52.5, the resistance is

$$R = \frac{\rho l}{A} = \frac{(1.7241 \times 10^{-6} \ \Omega\text{·cm})(1 \ \text{m})\left(100 \ \frac{\text{cm}}{\text{m}}\right)}{(1 \ \text{cm})(1 \ \text{cm})}$$
$$= 1.724 \times 10^{-4} \ \Omega$$

5. CONDUCTIVITY AND CONDUCTANCE

The reciprocals of resistivity and resistance are *conductivity*, σ, and *conductance*, G, respectively. The unit of conductance is the *siemens*, S.[3]

$$\sigma = \frac{1}{\rho} \qquad 52.6$$

$$G = \frac{1}{R} \qquad 52.7$$

Percent conductivity is the ratio of a substance's conductivity to the conductivity of standard IACS copper. (See Sec. 52.4.)

$$\% \text{ conductivity} = \frac{\sigma}{\sigma_{\text{Cu}}} \times 100\%$$
$$= \frac{\rho_{\text{Cu}}}{\rho} \times 100\% \qquad 52.8$$

---

[3]The siemens is the inverse of an ohm and is the same as the obsolete unit, the *mho*.

6. OHM'S LAW

The *voltage drop*, also known as the *IR drop*, across a circuit or circuit element with resistance, R, is given by *Ohm's law*.[4]

$$V = IR \quad \text{[DC circuits]} \qquad 52.9$$

$$v(t) = i(t)R \quad \text{[AC circuits]} \qquad 52.10$$

7. POWER IN DC CIRCUITS

The *power*, P (in watts), dissipated across two terminals with resistance, R, and voltage drop, V, is given by *Joule's law*, Eq. 52.11.

$$P = IV = I^2 R = \frac{V^2}{R}$$

$$= V^2 G \quad \text{[DC circuits]} \qquad 52.11$$

$$P(t) = i(t)v(t) = i(t)^2 R = \frac{v(t)^2}{R}$$

$$= v(t)^2 G \quad \text{[AC circuits]} \qquad 52.12$$

8. ELECTRICAL CIRCUIT SYMBOLS

Figure 52.1 illustrates symbols typically used to diagram electrical circuits in this book.

Figure 52.1 *Symbols for Electrical Circuit Elements*

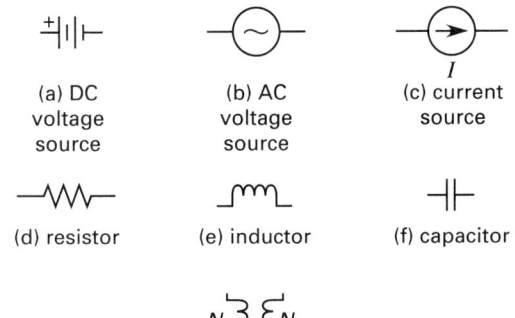

(a) DC voltage source

(b) AC voltage source

(c) current source

(d) resistor

(e) inductor

(f) capacitor

(g) transformer

9. RESISTORS IN COMBINATION

Resistors in series are added to obtain the total (equivalent) resistance of a circuit.

$$R_e = R_1 + R_2 + R_3 + \cdots \quad \text{[series]} \qquad 52.13$$

Resistors in parallel are combined by adding their reciprocals. This is a direct result of the fact that conductances in parallel add.

$$G_e = G_1 + G_2 + G_3 + \cdots \quad \text{[parallel]} \qquad 52.14$$

---

[4]This book uses the convention that uppercase letters represent fixed, maximum, or effective values, and lowercase letters represent values that change with time.

$$\frac{1}{R_e} = \frac{1}{R_1} + \frac{1}{R_2} + \frac{1}{R_3} + \cdots \quad \text{[parallel]} \qquad 52.15$$

For two resistors in parallel, the equivalent resistance is

$$R_e = \frac{R_1 R_2}{R_1 + R_2} \quad \text{[two parallel resistors]} \qquad 52.16$$

10. SIMPLE SERIES CIRCUITS

Figure 52.2 illustrates a simple series DC circuit and its equivalent circuit.

- The current is the same through all circuit elements.

$$I = I_{R_1} = I_{R_2} = I_{R_3} \qquad 52.17$$

- The equivalent resistance is the sum of the individual resistances.

$$R_e = R_1 + R_2 + R_3 \qquad 52.18$$

- The equivalent applied voltage is the sum of all voltage sources (polarities considered).

$$V_e = \pm V_1 \pm V_2 \qquad 52.19$$

- The sum of all of the voltage drops across the components in the circuit (a *loop*) is equal to the equivalent applied voltage. This fact is known as *Kirchhoff's voltage law*.

$$V_e = \sum IR_j = I \sum R_j = IR_e \qquad 52.20$$

Figure 52.2 *Simple Series DC Circuit and Its Equivalent*

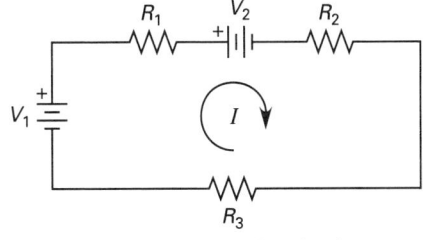

(a) original series circuit

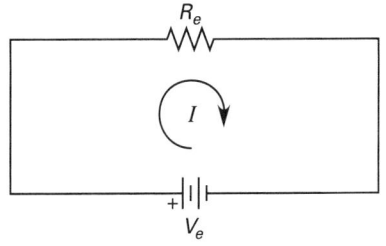

(b) equivalent circuit

11. SIMPLE PARALLEL CIRCUITS

Figure 52.3 illustrates a simple parallel DC circuit with one voltage source and its equivalent circuit.

- The voltage drop is the same across all legs.

$$V = V_{R_1} = V_{R_2} = V_{R_3}$$
$$= I_1 R_1 = I_2 R_2 = I_3 R_3 \qquad 52.21$$

- The reciprocal of the equivalent resistance is the sum of the reciprocals of the individual resistances.

$$\frac{1}{R_e} = \frac{1}{R_1} + \frac{1}{R_2} + \frac{1}{R_3} \qquad 52.22$$
$$G_e = G_1 + G_2 + G_3 \qquad 52.23$$

- The sum of all of the leg currents is equal to the total current. This fact is an extension of *Kirchhoff's current law:* The current flowing out of a connection (*node*) is equal to the current flowing into it.

$$I = I_1 + I_2 + I_3 = \frac{V}{R_1} + \frac{V}{R_2} + \frac{V}{R_3}$$
$$= V(G_1 + G_2 + G_3) \qquad 52.24$$

Figure 52.3 Simple Parallel DC Circuit and Its Equivalent

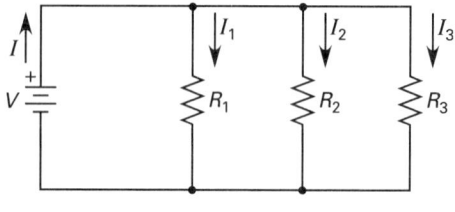

(a) original parallel circuit

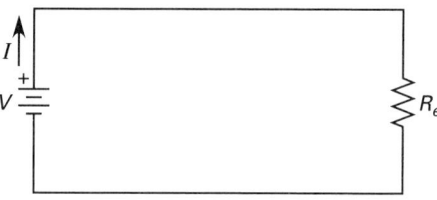

(b) equivalent circuit

12. VOLTAGE AND CURRENT DIVIDERS

Figure 52.4(a) illustrates a *voltage divider circuit*. The voltage across resistor 2 is

$$V_2 = V\left(\frac{R_2}{R_1 + R_2}\right) \qquad 52.25$$

Figure 52.4(b) illustrates a *current divider circuit*. The current through resistor 2 is

$$I_2 = I\left(\frac{R_1}{R_1 + R_2}\right) \qquad 52.26$$

Figure 52.4 Divider Circuits

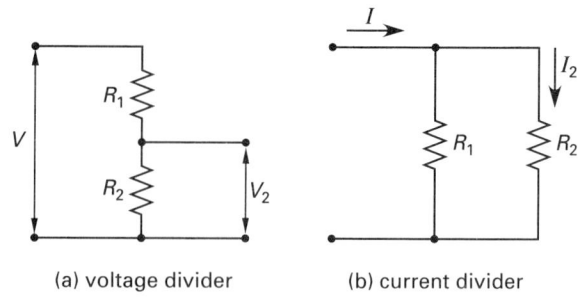

(a) voltage divider   (b) current divider

13. AC VOLTAGE SOURCES

The term *alternating current* (AC) almost always means that the current is produced from the application of a voltage with sinusoidal waveform.[5] Sinusoidal variables can be specified without loss of generality as either sines or cosines. If a sine waveform is used, Eq. 52.27 gives the instantaneous voltage as a function of time. V_m is the *maximum value*, also known as the *amplitude*, of the sinusoid. If $v(t)$ is not zero at $t = 0$, a *phase angle*, θ, must be included.

$$v(t) = V_m \sin(\omega t + \theta) \quad \text{[trigonometric form]} \qquad 52.27$$

Figure 52.5 illustrates the *period*, T, of the waveform. The *frequency*, f (also known as *linear frequency*), of the sinusoid is the reciprocal of the period and is expressed in hertz (Hz).[6] *Angular frequency*, ω, in radians per second (rad/s), can also be specified.

$$f = \frac{1}{T} = \frac{\omega}{2\pi} \qquad 52.28$$

$$\omega = 2\pi f = \frac{2\pi}{T} \qquad 52.29$$

Figure 52.5 Sinusoidal Waveform

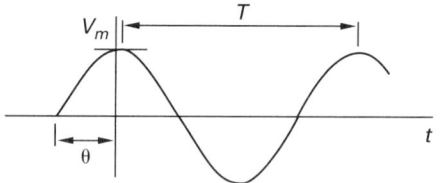

14. MAXIMUM, EFFECTIVE, AND AVERAGE VALUES

The *maximum value* (see Fig. 52.5), V_m, of a sinusoidal voltage is usually not specified in commercial and residential power systems. The *effective value*, also known as the *root-mean-square* (*rms*) *value*, is usually specified

[5]Other alternating waveforms commonly encountered in commercial applications are the square, triangular, and sawtooth waveforms.
[6]In the United States, the standard frequency is 60 Hz. In Japan, the British Isles and Commonwealth countries, continental Europe, and some Mediterranean, Near Eastern, African, and South American countries, the standard is 50 Hz.

Systems, Mgmt., Professional

when referring to single- and three-phase voltages.[7] A DC current equal in magnitude to the effective value of a sinusoidal AC current produces the same heating effect as the sinusoid. The scale reading of a typical AC current meter is proportional to the effective current.

$$V_{\text{eff}} = \frac{V_m}{\sqrt{2}} \approx 0.707\,V_m \qquad 52.30$$

The *average value* of a symmetrical sinusoidal waveform is zero. However, the average value of a rectified sinusoid (or the average value of a sinusoid taken over half of the cycle) is $V_{\text{ave}} = 2\,V_m/\pi$. A DC current equal to the average value of a *rectified AC current* has the same electrolytic action (e.g., capacitor charging, plating, and ion formation) as the rectified sinusoid.[8]

15. IMPEDANCE

Simple alternating current circuits can be composed of three different types of passive circuit components—resistors, inductors, and capacitors. Each type of component affects both the magnitude of the current flowing as well as the phase angle (see Sec. 52.23) of the current. For both individual components and combinations of components, these two effects are quantified by the *impedance*, $\mathbf{Z}$. Impedance is a complex quantity with a magnitude (in ohms) and an associated *impedance angle*, ϕ. It is usually written in *phasor (polar) form* (e.g., $\mathbf{Z} \equiv Z\angle\phi$).

Multiple impedances in a circuit combine in the same manner as resistances: Impedances in series add; reciprocals of impedances in parallel add. However, the addition must use complex (i.e., vector) arithmetic.

16. REACTANCE

Impedance, like any complex quantity, can also be written in rectangular form. In this case, impedance is written as the complex sum of the resistive, R, and reactive, X, components, both having units of ohms. The resistive and reactive components combine trigonometrically in the impedance triangle. (See Fig. 52.6.) The reactive component, X, is known as the *reactance*.

$$\mathbf{Z} \equiv R \pm j\mathbf{X} \qquad 52.31$$

$$R = Z\cos\phi \quad \text{[resistive part]} \qquad 52.32$$

$$X = Z\sin\phi \quad \text{[reactive part]} \qquad 52.33$$

Figure 52.6 *Impedance Triangle of a Complex Circuit*

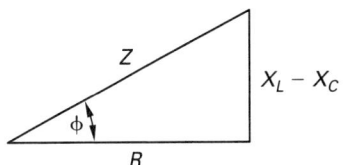

17. ADMITTANCE, CONDUCTANCE, AND SUSCEPTANCE

The reciprocal of impedance is the complex quantity *admittance*, $\mathbf{Y}$. Admittance can be used to analyze parallel circuits, since admittances of parallel circuit elements add together.

$$\mathbf{Y} = \frac{1}{\mathbf{Z}} \equiv \frac{1}{Z}\angle-\phi \qquad 52.34$$

The reciprocal of the resistive part of the impedance is the *conductance*, G, with units of siemens, (S). The reciprocal of the reactive part of impedance is the *susceptance*, B.

$$G = \frac{1}{R} \qquad 52.35$$

$$B = \frac{1}{X} \qquad 52.36$$

18. RESISTORS IN AC CIRCUITS

An ideal resistor, with resistance R, has no inductance or capacitance. The magnitude of the impedance is equal to the resistance, R (in ohms, Ω). The impedance angle is zero. Therefore, current and voltage are in phase in a purely resistive circuit.

$$\mathbf{Z}_R \equiv R\angle 0° \equiv R + j0 \qquad 52.37$$

19. CAPACITORS

A *capacitor* stores electrical charge. The charge on a capacitor is proportional to its *capacitance*, C (in farads, F), and voltage.[9]

$$Q = CV \qquad 52.38$$

An ideal capacitor has no resistance or inductance. The magnitude of the impedance is the *capacitive reactance*, X_C, in ohms. The impedance angle is $-\pi/2$ ($-90°$). Therefore, current leads the voltage by 90° in a purely capacitive circuit.

$$\mathbf{Z}_C \equiv X_C\angle-90° \equiv 0 - jX_C \qquad 52.39$$

$$X_C = \frac{1}{\omega C} = \frac{1}{2\pi f C} \qquad 52.40$$

Systems, Mgmt., Professional

20. INDUCTORS

An ideal *inductor*, with an *inductance*, L (in henrys, H), has no resistance or capacitance. The magnitude of the impedance is the *inductive reactance*, X_L, in ohms. The impedance angle is $\pi/2$ (90°). Therefore, current lags the voltage by 90° in a purely inductive circuit.[10]

$$\mathbf{Z}_L \equiv X_L \angle 90° \equiv 0 + jX_L \qquad 52.41$$

$$X_L = \omega L = 2\pi f L \qquad 52.42$$

21. TRANSFORMERS

Transformers are used to change voltages, isolate circuits, and match impedances. Transformers usually consist of two coils of wire wound on magnetically permeable cores. One coil, designated as the *primary coil*, serves as the input; the other coil, the *secondary coil*, is the output. The primary current produces a magnetic flux in the core; the magnetic flux, in turn, induces a voltage in the secondary coil. In an *ideal transformer* (*lossless transformer* or *100% efficient transformer*), the coils have no electrical resistance, and all magnetic flux lines pass through both coils.

The ratio of the numbers, N, of primary to secondary coil windings is the *turns ratio* (*ratio of transformation*), a. If the turns ratio is greater than 1.0, the transformer decreases voltage and is a *step-down transformer*. If the turns ratio is less than 1.0, the transformer increases voltage and is a *step-up transformer*.

$$a = \frac{N_p}{N_s} \qquad 52.43$$

In an ideal transformer, the power transferred from the primary side equals the power received by the secondary side.

$$P = I_p V_p = I_s V_s \qquad 52.44$$

$$a = \frac{V_p}{V_s} = \frac{I_s}{I_p} = \sqrt{\frac{Z_p}{Z_s}} \qquad 52.45$$

22. OHM'S LAW FOR AC CIRCUITS

Ohm's law (see Sec. 52.6) can be written in phasor (polar) form. Voltage and current can be represented by their maximum, effective (rms), or average values. However, both must be represented in the same manner.

$$\mathbf{V} = \mathbf{IZ} \qquad 52.46$$

$$V = IZ \quad \text{[magnitudes only]} \qquad 52.47$$

$$\phi_V = \phi_I + \phi_Z \quad \text{[angles only]} \qquad 52.48$$

23. AC CURRENT AND PHASE ANGLE

The current and current phase angle of a circuit are determined by using Ohm's law in phasor form. The current *phase angle*, ϕ_I, is the angular difference between when the current and voltage waveforms peak.

$$\mathbf{I} = \frac{\mathbf{V}}{\mathbf{Z}} \qquad 52.49$$

$$I = \frac{V}{Z} \quad \text{[magnitudes only]} \qquad 52.50$$

$$\phi_I = \phi_V - \phi_Z \quad \text{[angles only]} \qquad 52.51$$

Example 52.2

An inventor's black box is connected across standard household voltage (110 V rms). The current drawn is 1.7 A with a lagging phase angle of 14° with respect to the voltage. What is the impedance of the black box?

Solution

From Eq. 52.49,

$$\mathbf{Z} = \frac{\mathbf{V}}{\mathbf{I}} = \frac{110 \text{ V} \angle 0°}{1.7 \text{ A} \angle 14°}$$
$$= 64.71 \ \Omega \angle -14°$$

24. POWER IN AC CIRCUITS

In a purely resistive circuit, all of the current drawn contributes to dissipated energy. The flow of current causes resistors to increase in temperature, and heat is transferred to the environment.

In a typical AC circuit containing inductors and capacitors as well as resistors, some of the current drawn does not cause heating.[11] Rather, the current charges capacitors and creates magnetic fields in inductors. Since the voltage alternates in polarity, capacitors alternately charge and discharge. Energy is repeatedly drawn and returned by capacitors. Similarly, energy is repeatedly drawn and returned by inductors as their magnetic fields form and collapse.

Current through resistors in AC circuits causes heating, just as in DC circuits. The power dissipated is represented by the *real power* vector, **P**. Current through capacitors and inductors contributes to reactive power, represented by the *reactive power* vector, **Q**. Reactive power does not contribute to heating. For convenience, both real and reactive power are considered to be complex (vector) quantities, with magnitudes and associated angles. Real and reactive power combine as vectors into the *complex power* vector, **S**, as shown in Fig. 52.7. The angle, ϕ, is known as the overall *impedance angle*.

---

[10]Use the memory aid "ELI the ICE man" to remember that current (I) comes after voltage (electromagnetic force, E) in inductive (L) circuits. In capacitive (C) circuits, current leads voltage.

[11]The notable exception is a *resonant circuit* in which the inductive and capacitive reactances are equal. In that case, the circuit is purely resistive in nature.

Figure 52.7 *Complex Power Triangle (lagging)*

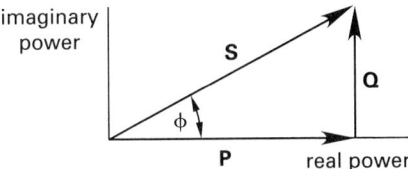

The magnitude of real power is known as the *average power*, P, and is measured in watts. The magnitude of the reactive power vector is known as *reactive power*, Q, and is measured in VARs (volt-amps-reactive). The magnitude of the complex power vector is the *apparent power*, S, measured in VAs (volt-amps). The apparent power is easily calculated from measurements of the line current and line voltage.

$$S = I_{\text{eff}} V_{\text{eff}} = \tfrac{1}{2} I_m V_m \qquad 52.52$$

25. POWER FACTOR

The complex power triangle shown in Fig. 52.7 is congruent to the impedance triangle (see Fig. 52.6), and the *power angle*, ϕ, is identical to the overall impedance angle.

$$S^2 = P^2 + Q^2 \qquad 52.53$$

$$P = S \cos \phi \qquad 52.54$$

$$Q = S \sin \phi \qquad 52.55$$

$$\phi = \arctan \frac{Q}{P} \qquad 52.56$$

The *power factor*, pf (also occasionally referred to as the *phase factor*), for a sinusoidal voltage is $\cos \phi$. By convention, the power factor is usually given in percent rather than by its equivalent decimal value. For a purely resistive circuit, pf = 100%; for a purely reactive circuit, pf = 0%.

$$\text{pf} = \frac{P}{S} \qquad 52.57$$

Since the cosine is positive for both positive and negative angles, the terms "leading" and "lagging" are used to describe the nature of the circuit. In a circuit with a *leading power factor* (i.e., in a *leading circuit*), the load is primarily capacitive in nature. In a circuit with a *lagging power factor* (i.e., in a *lagging circuit*), the load is primarily inductive in nature.

26. COST OF ELECTRICAL ENERGY

Except for large industrial users, electrical meters at service locations usually measure and record real power only. Electrical utilities charge on the basis of the total energy used. Energy usage, commonly referred to as the *usage* or *demand*, is measured in kilowatt-hours, abbreviated kWh or kW-hr.

$$\text{cost} = (\text{cost per kW-hr}) \times E_{\text{kW-hr}} \qquad 52.58$$

The cost per kW-hr may not be a single value but may be tiered so that cost varies with cumulative usage. The lowest rate is the *baseline rate*.[12] To encourage conservation, the incremental cost of energy increases with increases in monthly usage.[13] To encourage cutbacks during the day, the cost may also increase during periods of peak demand.[14] The increase in cost for usage during peak demand may be accomplished by varying the rate, additively, or by use of a *peak demand multiplier*. There may also be different rates for summer and winter usage.

Although only real power is dissipated, reactive power contributes to total current. (Reactive power results from the current drawn in supplying the magnetization energy in motors and charges on capacitors.) The distribution system (wires, transformers, etc.) must be sized to carry the total current, not just the current supplying the heating effect. When real power alone is measured at the service location, the power factor is routinely monitored and its effect is built into the charge per kW-hr. This has the equivalent effect of charging the user for apparent power usage.

Example 52.3

A small office normally uses 700 kW-hr per month of electrical energy. The company adds a 1.5 kW heater (to be used at the rate of 1000 kW-hr/month) and a 5 hp motor with a mechanical efficiency of 90% (to be used at a rate of 240 hr/mo). What is the incremental cost of adding the heater and motor? The tiered rate structure is

electrical usage (kW-hr)	rate ($/kW-hr)
less than 350	0.1255
350–999	0.1427
1000–3999	0.1693

---

[12] There might also be a *lifeline rate* for low-income individuals with very low usage.
[13] There are different rate structures for different categories of users. While increased use within certain categories of users (e.g., residential) results in a higher cost per kW-hr, larger users in another category may pay substantially less per kW-hr due to their volume "buying power."
[14] The day may be divided into *peak periods, partial-peak periods*, and *off-peak periods*.

Solution

Motors are rated by their real power output, which is less than their real power demand. The incremental electrical usage per month is

$$1000 \text{ kW-hr} + \frac{(5 \text{ hp})\left(0.7457 \frac{\text{kW}}{\text{hp}}\right)(240 \text{ hr})}{0.90}$$
$$= 1994 \text{ kW-hr}$$

The cumulative monthly electrical usage is

$$700 \text{ kW-hr} + 1994 \text{ kW-hr} = 2694 \text{ kW-hr}$$

The company was originally in the second tier. The new usage will be billed partially at the second and third tier rates.

The incremental cost is

$$(999 \text{ kW-hr} - 700 \text{ kW-hr})\left(0.1427 \frac{\$}{\text{kW-hr}}\right)$$
$$+ (2694 \text{ kW-hr} - 999 \text{ kW-hr})\left(0.1693 \frac{\$}{\text{kW-hr}}\right)$$
$$= \$329.63$$

27. CIRCUIT BREAKER SIZING

The purpose of a circuit breaker ("breaker") is to protect the wiring insulation from overheating. Breakers operate as heat-activated switches. Overloads, short circuits, and faulty equipment result in large current draws. Extended currents in excess of a breaker's rating will cause it to trip.

A breaker should be sized based on the average expected electrical load for that circuit. *Dedicated circuits* are easily sized based on the amperage of the equipment served, taken either from the nameplate current or calculated from the nameplate power as $I = P/V$. For general purpose circuits, reasonable judgment is required to estimate how many of the outlets will be in use simultaneously, and what equipment will be plugged in. Although circuit breakers are designed to operate at their rated loads for at least three hours, breakers should be sized for about 1.25 times the expected maximum load, not to exceed the capacity of the wire.

28. NATIONAL ELECTRICAL CODE

While a circuit breaker can be simplistically sized by calculating the maximum current from the known loads, unless the practitioner is a licensed electrician or electrical power engineer, a little knowledge of circuit breaker operation and the *National Electrical Code* (NEC) can greatly complicate the analysis. For this reason, most engineers should stop after calculating the maximum current, and they should leave the actual device selection to another trade. Selecting protection for standby generator circuits is particularly complex.

In the United States, the NEC specifies circuit breaker size as a function of the wire size. The wire is measured in American wire gauge and is abbreviated AWG or just GA. (Numerical wire size gauge decreases as the physical size increases.) Typical wire sizes and maximum currents (*ampacity*) for copper wire are 14 AWG (used for lighting and light-duty circuits up to 15 A), 12 AWG (for power outlets, kitchen microwaves, and lighting circuits up to 20 A), 10 AWG (30 A), 8 AWG (40 A), 6 AWG (50 A or 60 A), 4 AWG (75 A), and 3 AWG (100 A). For aluminum wires, the wire size should be one gauge larger. Common household circuit breaker sizes are 15 A, 20 A, and 30 A, although specialty sizes are also available in 5 A increments. Industrial sizes are readily available through 800 A, and larger breakers can handle thousands of amps.

The NEC refers to circuit breakers as *overcurrent protection devices* (OCPD). The NEC permits a circuit breaker to run essentially continuously at 100% of its *rated current*, I_r. This ability is intrinsic to the circuit breaker design, as Fig. 52.8 shows. When the current is greater than the rated current, the circuit breaker will open (i.e., "clear the circuit") if the exposure duration is between the minimum and maximum curve times. The range of times between vertically adjacent points on the two curves represents the maximum delay for the breaker for that current, although exact values should not be expected. The lower minimum curve anticipates tripping due to thermal action; the upper maximum curve anticipates tripping due to magnetic action (i.e., short circuits).

Figure 52.8 shows that this particular curve allows circuit breakers to pass currents as large as 10 times their rated values, albeit for no more than a second, and twice their rated currents for up to 10 seconds. This overcurrent must be considered when selecting breakers with rated currents significantly in excess of the calculated maximum circuit current.

Allowing a circuit breaker to run continuously at its rated value is based on ideal testing conditions, such as mounting in free air, no enclosure, and constant 104°F (40°C) air temperature. For typical installations with heating from adjacent breakers in an enclosure, the NEC requires continuously operated breakers to be derated to 80% of rated capacities. This derating should be applied with continuous operation, which NEC Art. 100 defines as operation that exceeds three hours in duration. Interior office lighting would be considered continuous, while electrical kitchen appliances normally would not be. Most electrical heating for vessels, pipes, and deicing is considered continuous.

Figure 52.8 Typical Thermal-Magnetic Breaker Trip Curve*

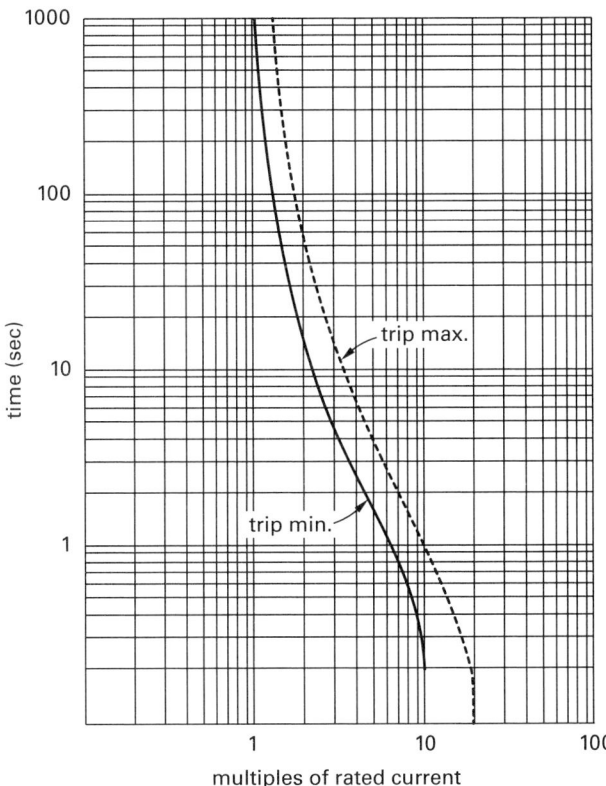

multiples of rated current

---

*Typical curve with Underwriters Laboratories (UL) specifications for use in the United States, 104°F (40°C).
International Electrotechnical Commission trip curves (IEC 947-2) for other countries are similar in shape but differ in specific values from UL curves.

NEC Sec. 210.19(A)(1) and Sec. 210.20(A) specify the required OCPD (i.e., breaker) size as Eq. 52.59, which is sometimes referred to as the *80% rule*, since the reciprocal of 0.80 is 1.25.

$$\text{required rated capacity} = \begin{array}{l} 100\% \text{ of noncontinuous load} \\ + 125\% \text{ of continuous load} \end{array}$$

52.59

Circuit breakers are subject to heating from adjacent breakers in an enclosure regardless of whether their loads are continuous or noncontinuous. This does not mean that all circuit breakers in an enclosure must be derated. *Standard circuit breakers*, also known as *80% rated circuit breakers* and *non-100% rated circuit breakers*, have been tested and certified to function properly with noncontinuous loads in enclosures, so Eq. 52.59 can be used with confidence. Unfortunately, there is also a class of *100% rated circuit breakers* (including solid state devices that do not generate thermal energy), which have been tested and certified to function properly with continuous loads within enclosures. The 125% multiplier does not need to be applied to 100% rated circuit breakers. Clearly, the specification of actual OCPDs should be left to the experts.

Example 52.4

A 120 V (rms) circuit supplies 6 kW of office lighting and 125 A to one phase of a three-phase elevator induction motor. 175 A, 200 A, 225 A, and 250 A standard breakers are available. Which breaker should be specified?

Solution

The elevator motor is a noncontinuous load. The office lighting draws a continuous current of

$$I = \frac{P}{V} = \frac{(6 \text{ kW})\left(1000 \ \frac{\text{W}}{\text{kW}}\right)}{120 \text{ V}} = 50 \text{ A}$$

Since a standard (i.e., 80% rated) circuit breaker is called for, use Eq. 52.59.

$$\begin{aligned} \text{required rated capacity} &= \begin{array}{l} 100\% \text{ of noncontinuous load} \\ + 125\% \text{ of continuous load} \end{array} \\ &= (1.0)(125 \text{ A}) + (1.25)(50 \text{ A}) \\ &= 187.5 \text{ A} \end{aligned}$$

The 200 A circuit breaker should be specified.

29. POWER FACTOR CORRECTION

Inasmuch as apparent power is paid for but only real power is used to drive motors or provide light and heating, it may be possible to reduce electrical utility charges by reducing the power angle without changing the real power. This strategy, known as *power factor correction*, is routinely accomplished by changing the circuit reactance in order to reduce the reactive power. The change in reactive power needed to change the power angle from ϕ_1 to ϕ_2 is

$$\Delta Q = P(\tan\phi_1 - \tan\phi_2) \qquad \textit{52.60}$$

When a circuit is capacitive (i.e., leading), induction motors (see Sec. 52.41) can be connected across the line to improve the power factor. In the more common situation, when a circuit is inductive (i.e., lagging), capacitors or synchronous capacitors (see Sec. 52.43) can be added across the line. The size (in farads) of capacitor required is

$$C = \frac{\Delta Q}{\pi f V_m^2} \quad [V_m \text{ maximum}] \qquad \textit{52.61}$$

$$C = \frac{\Delta Q}{2\pi f V_{\text{eff}}^2} \quad [V_{\text{eff}} \text{ effective}] \qquad \textit{52.62}$$

Capacitors for power factor correction are generally rated in kVA. Equation 52.60 can be used to find that rating.

Example 52.5

A 60 Hz, 5 hp induction motor draws 53 A (rms) at 117 V (rms) with a 78.5% electrical-to-mechanical energy conversion efficiency. What capacitance should be connected across the line to increase the power factor to 92%?

Solution

The apparent power is found from the observed voltage and the current. Use Eq. 52.52.

$$S = IV = \frac{(53 \text{ A})(117 \text{ V})}{1000 \ \frac{\text{VA}}{\text{kVA}}}$$

$$= 6.201 \text{ kVA}$$

The real power drawn from the line is calculated from the real work done by the motor. Use Eq. 52.72.

$$P_{\text{electrical}} = \frac{P_{\text{out}}}{\eta} = \frac{(5 \text{ hp})\left(0.7457 \ \frac{\text{kW}}{\text{hp}}\right)}{0.785}$$

$$= 4.750 \text{ kW}$$

The reactive power and power angle are calculated from the real and apparent powers. From Eq. 52.53,

$$Q_1 = \sqrt{S^2 - P^2} = \sqrt{(6.201 \text{ kVA})^2 - (4.750 \text{ kW})^2}$$

$$= 3.986 \text{ kVAR}$$

From Eq. 52.54,

$$\phi_1 = \arccos \frac{P}{S} = \arccos \frac{4.750 \text{ kW}}{6.201 \text{ kVA}} = 40.00°$$

The desired power factor angle is

$$\phi_2 = \arccos 0.92 = 23.07°$$

The reactive power after the capacitor is installed is given by Eq. 52.56.

$$Q_2 = P \tan \phi_2 = (4.750 \text{ kW})(\tan 23.07°)$$

$$= 2.023 \text{ kVAR}$$

The required capacitance is found from Eq. 52.62.

$$C = \frac{\Delta Q}{2\pi f V_{\text{eff}}^2}$$

$$= \frac{(3.986 \text{ kVAR} - 2.023 \text{ kVAR})\left(1000 \ \frac{\text{VAR}}{\text{kVAR}}\right)}{(2\pi)(60 \text{ Hz})(117 \text{ V})^2}$$

$$= 3.8 \times 10^{-4} \text{ F} \quad (380 \ \mu\text{F})$$

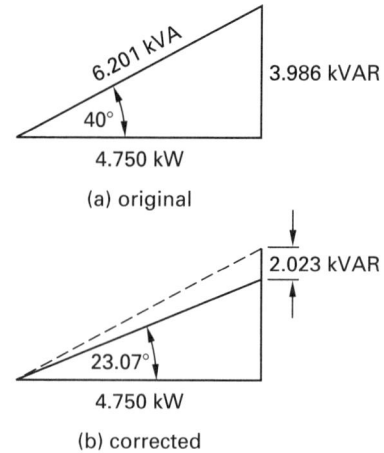

(a) original

(b) corrected

30. THREE-PHASE ELECTRICITY

Smaller electric loads, such as household loads, are served by single-phase power. The power company delivers a sinusoidal voltage of fixed frequency and amplitude connected between two wires—a phase wire and a neutral wire. Large electric loads, large buildings, and industrial plants are served by three-phase power. Three voltage signals are connected between three phase wires and a single neutral wire. The phases have equal frequency and amplitude, but they are out of phase by 120° (electrical) with each other. Such *three-phase systems* use smaller conductors to distribute electricity.[15] For the same delivered power, three-phase distribution systems have lower losses and are more efficient.

Three-phase motors provide a more uniform torque than do single-phase motors whose torque production pulsates.[16] Three-phase induction motors require no additional starting windings or associated switches. When rectified, three-phase voltage has a smoother waveform. (See Fig. 52.9.)

Figure 52.9 Three-Phase Voltage

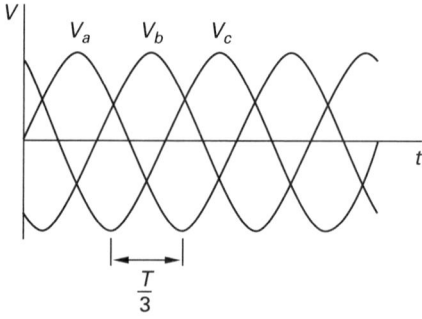

---

[15]The uppercase Greek letter phi is often used as an abbreviation for the word "phase." For example, "3Φ" would be interpreted as "three-phase."

[16]Single-phase motors require auxiliary windings for starting, since one phase alone cannot get the magnetic field rotating.

31. THREE-PHASE LOADS

Three impedances are required to fully load a three-phase voltage source. In three-phase motors and other devices, these impedances are three separate motor windings. Similarly, three-phase transformers have three separate sets of primary and secondary windings.

The impedances in a three-phase system are said to be *balanced loads* when they are identical in magnitude and angle. The voltages, line currents, and real, apparent, and reactive powers are all identical in a balanced system. Also, the power factor is the same for each phase. Therefore, only one phase of a balanced system needs to be analyzed (i.e., a *one-line analysis*).

32. LINE AND PHASE VALUES

The *line current*, I_l, is the current carried by the distribution lines (wires). The *phase current*, I_p, is the current flowing through each of the three separate loads (i.e., the phase) in the motor or device. Line and phase currents are both vector quantities.

Depending on how the motor or device is internally wired, the line and phase currents may or may not be the same. Figure 52.10 illustrates delta- and wye-connected loads. For balanced *wye-connected loads*, the line and phase currents are the same. For balanced *delta-connected loads*, they are not.

$$I_p = I_l \quad \text{[wye]} \qquad 52.63$$

$$I_p = \frac{I_l}{\sqrt{3}} \quad \text{[delta]} \qquad 52.64$$

Similarly, the *line voltage*, V_l (same as *line-to-line voltage*, commonly referred to as the *terminal voltage*), and *phase voltage*, V_p, may not be the same. With balanced delta-connected loads, the full line voltage appears across each phase. With balanced wye-connected loads, the line voltage appears across two loads.

$$V_p = \frac{V_l}{\sqrt{3}} \quad \text{[wye]} \qquad 52.65$$

$$V_p = V_l \quad \text{[delta]} \qquad 52.66$$

33. INPUT POWER IN THREE-PHASE SYSTEMS

Each impedance in a balanced system dissipates the same real *phase power*, P_p. The power dissipated in a balanced three-phase system is three times the phase power and is calculated in the same manner for both delta- and wye-connected loads.

$$P_t = 3P_p = 3 V_p I_p \cos \phi$$
$$= \sqrt{3} V_l I_l \cos \phi \qquad 52.67$$

Figure 52.10 *Wye- and Delta-Connected Loads*

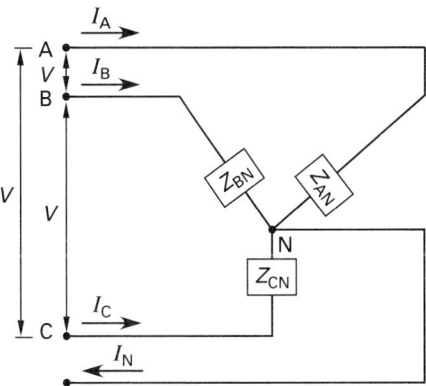

(a) wye-connected loads

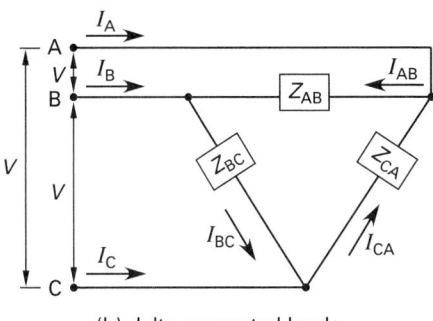

(b) delta-connected loads

The real power component is sometimes referred to as "power in kW." Apparent power is sometimes referred to as "kVA value" or "power in kVA."

$$S_t = 3S_p = 3 V_p I_p$$
$$= \sqrt{3} V_l I_l \qquad 52.68$$

34. ROTATING MACHINES

Rotating machines are categorized as AC and DC machines. Both categories include machines that use electrical power (i.e., motors) and those that generate electrical power (alternators and generators).[17] Machines can be constructed in either single-phase or polyphase configurations, although single-phase machines may be inferior in terms of economics and efficiency. (See Sec. 52.30.)

Large AC motors are almost always three-phase. However, since the phases are balanced, it is necessary to analyze one phase only of the motor. Torque and power are divided evenly among the three phases.

35. REGULATION

Rotating machines (motors and alternators), as well as power supplies, are characterized by changes in voltage

---

[17]An *alternator* produces AC potential. A *generator* produces DC potential.

and speed under load. *Voltage regulation*, VR, is defined as

$$VR = \frac{\text{no-load voltage} - \text{full-load voltage}}{\text{full-load voltage}} \times 100\%$$

52.69

Speed regulation, SR, is defined as

$$SR = \frac{\text{no-load speed} - \text{full-load speed}}{\text{full-load speed}} \times 100\% \quad \textit{52.70}$$

36. TORQUE AND POWER

For rotating machines, torque and power are basic operational parameters. It takes mechanical power to turn an alternator or generator. A motor converts electrical power into mechanical power. In SI units, power is given in watts (W) and kilowatts (kW). One horsepower (hp) is equivalent to 0.7457 kilowatts. The relationship between torque and power is

$$T_{\text{N·m}} = \frac{9549 P_{\text{kW}}}{n_{\text{rpm}}} \qquad \text{[SI]} \quad \textit{52.71(a)}$$

$$T_{\text{ft-lbf}} = \frac{5252 P_{\text{horsepower}}}{n_{\text{rpm}}} \qquad \text{[U.S.]} \quad \textit{52.71(b)}$$

There are many important torque parameters for motors. (See Fig. 52.11.) The *starting torque* (also known as *static torque*, *break-away torque*, and *locked-rotor torque*) is the turning effort exerted by the motor when starting from rest. *Pull-up torque* (*acceleration torque*) is the minimum torque developed during the period of acceleration from rest to full speed. *Pull-in torque* (as developed in synchronous motors) is the maximum torque that brings the motor back to synchronous speed (see Sec. 52.41). *Nominal pull-in torque* is the torque that is developed at 95% of synchronous speed.

The *full-load torque* (*steady-state torque*) occurs at the rated speed and horsepower. Full-load torque is supplied to the load on a continuous basis. The full-load torque establishes the temperature increase that the motor must be able to withstand without deterioration. The *rated torque* is developed at rated speed and rated horsepower. The maximum torque that the motor can develop at its synchronous speed is the *pull-out torque*. *Breakdown torque* is the maximum torque that the motor can develop without stalling (i.e., without coming rapidly to a complete stop).

Motors are rated according to their output power (*rated power* or *brake power*). A 5 hp motor will normally deliver a full five horsepower when running at its rated speed. While the rated power is not affected by the motor's energy conversion efficiency, η, the electrical power input to the motor is.

$$P_{\text{electrical}} = \frac{P_{\text{rated}}}{\eta} \qquad \textit{52.72}$$

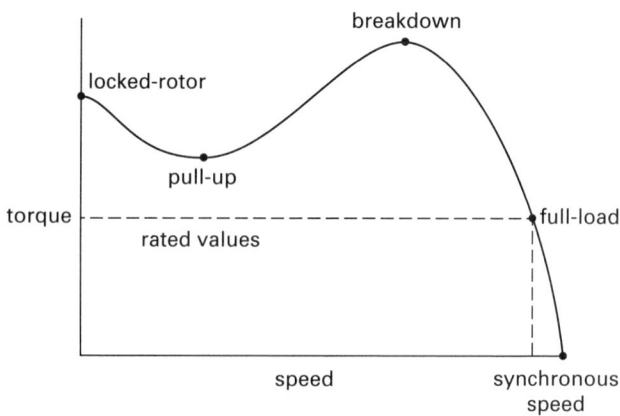

Figure 52.11 *Induction Motor Torque-Speed Characteristic (typical of design B frames)*

Table 52.1 lists standard motor sizes by rated horsepower.[18] The rated horsepower should be greater than the calculated brake power requirements. Since the rated power is the power actually produced, motors are not selected on the basis of their efficiency or electrical power input. The smaller motors listed in Table 52.1 are generally single-phase motors. The larger motors listed are three-phase motors.

Table 52.1 *Typical Standard Motor Sizes (horsepower)**

$\frac{1}{8}$	$\frac{1}{6}$	$\frac{1}{4}$	$\frac{1}{3}$	$\frac{1}{2}$	$\frac{3}{4}$
1	$1\frac{1}{2}$	2	3	5	$7\frac{1}{2}$
10	15	20	25	30	40
50	60	75	100	125	150
200	250				

(Multiply hp by 0.7457 to obtain kW.)
*1/8 hp and 1/6 hp motors are less common.

37. NEMA MOTOR CLASSIFICATIONS

Motors are classified by their NEMA design type—A, B, C, D, and F.[19,20] These designs are collectively referred to as *NEMA frame motors*. All NEMA motors of a given frame size are interchangeable as to bolt holes, shaft diameter, height, length, and various other dimensions.

- Frame design A: three-phase, squirrel-cage motor with high locked-rotor (starting) current but also higher breakdown torques; capable of handling intermittent overloads without stalling; 1–3% slip at full load.

- Frame design B: three-phase, squirrel-cage motor capable of withstanding full-voltage starting; locked-rotor and breakdown torques suitable for

[18]For economics, standard motor sizes should be specified.
[19]NEMA stands for the National Electrical Manufacturers Association.
[20]There is no type E.

most applications; 1–3% slip at full load; often designated for "normal" usage; the most common design.[21]

- Frame design C: three-phase, squirrel-cage motor capable of withstanding full-voltage starting; high locked-rotor torque for special applications (e.g., conveyors and compressors); 1–3% slip at full load.

- Frame design D: three-phase, squirrel-cage motor capable of withstanding full-voltage starting; develops 275% locked rotor torque for special applications; high breakdown torque; depending on design, 5–13% slip at full load; commonly known as a *high-slip motor*; used for cranes, hoists, oil well pump jacks, and valve actuators.

- Frame design F: three-phase, squirrel-cage motor with low starting current and low starting torque; used to meet applicable regulations regarding starting current; limited availability.

38. NAMEPLATE VALUES

A *nameplate* is permanently affixed to each motor's housing or frame. This nameplate is embossed or engraved with the *rated values* of the motor (*nameplate values* or *full-load values*). Nameplate information may include some or all of the following: voltage,[22] frequency,[23] number of phases, rated power (output power), running speed, duty cycle (see Sec. 52.40), locked-rotor and breakdown torques, starting current, current drawn at rated load (in kVA/hp), ambient temperature, temperature rise at rated load, temperature rise at service factor (see Sec. 52.39), insulation rating, power factor, efficiency, frame size, and enclosure type.

It is important to recognize that the rated power of the motor is the actual output power (i.e., power delivered to the load), not the input power. Only the electrical power input is affected by the motor efficiency. (See Eq. 52.72.)

Nameplates are also provided on transformers. Transformer nameplate information includes the two winding voltages (either of which can be the primary winding), frequency, and kVA rating. Apparent power in kVA, not real power, is used in the rating because heating is proportional to the square of the supply current. As with motors, continuous operation at the rated values will not result in excessive heat build-up.

39. SERVICE FACTOR

The horsepower and torque ratings listed on the nameplate of an AC motor can be maintained on a continuous basis without overheating. However, motors can be operated at slightly higher loads without exceeding a safe temperature rise.[24] The ratio of the safe to rated loads (horsepower or torque) is the *service factor*, sf, usually expressed as a decimal.

$$\text{sf} = \frac{\text{maximum safe load}}{\text{nameplate load}} \qquad 52.73$$

Service factors vary from 1.15 to 1.4, with the lower values being applicable to the larger, more efficient motors. Typical values of service factor are 1.4 (up to $1/8$ hp motors), 1.35 ($1/6$–$1/3$ hp motors), 1.25 ($1/2$–1 hp motors), and 1.15 (1 or $1\frac{1}{2}$ to 200 hp).

When running above the rated load, the motor speed, temperature, power factor, full-load current, and efficiency will differ from the nameplate values. However, the locked-rotor and breakdown torques will remain the same, as will the starting current.

Active current is proportional to torque, and hence, is proportional to the horsepower developed (see Eq. 52.75). The active current (line or phase) drawn is also proportional to the service factor. The current drawn per phase is given by Eq. 52.75.[25]

$$I_{\text{active}} = I_l(\text{pf}) \qquad 52.74$$

$$I_{\text{actual},p} = \frac{(\text{sf})P_{\text{rated},p}}{V_p\eta(\text{pf})}$$

$$= \frac{P_{\text{actual},p}}{V_p\eta(\text{pf})} \quad [\text{per phase; } P \text{ in watts}] \qquad 52.75$$

Equation 52.75 can also be used when a motor is developing less than its rated power. In this case, the service factor can be considered as the fraction of the rated power being developed.

Example 52.6

What is the approximate phase current drawn by a 98% efficient, three-phase, 75 hp, 230 V (rms) motor that is running at 88% power factor and at its rated load?

---

[21]Design B is estimated to be used in 90% of all applications.

[22]Standard NEMA nameplate voltages (effective) are 200 V, 230 V, 460 V, and 575 V. The NEMA motor voltage rating assumes that there will be a voltage drop from the network to the motor terminals. A 200 V motor is appropriate for a 208 V network; a 230 V motor on a 240 V network; and, so on. NEMA motors are capable of operating in a range of only ±10% of their rated voltages. Thus, 230 V rated motors should not be used on 208 V systems.

[23]While some 60 Hz motors (notably those intended for 230 V operation) can be used at 50 Hz, most others (e.g., those intended for 200 V operation) are generally not suitable for use at 50 Hz.

---

[24]Higher temperatures have a deteriorating effect on the winding insulation. A general rule of thumb is that a motor loses two or three hours of useful life for each hour run at the service factor load.

[25]It is important to recognize the difference between the rated (i.e., nameplate) power and the actual power developed. The actual power should not be combined with the service factor, since the actual power developed is the product of the rated power and the service factor.

Solution

The service factor is 1.00 because the motor is running at its rated load. From Eq. 52.75, the approximate phase current is

$$I_p = \frac{(\text{sf})P_p}{V_p \eta (\text{pf})}$$

$$= \frac{(1.00)\left(\dfrac{75 \text{ hp}}{3 \text{ phases}}\right)\left(0.7457 \dfrac{\text{kW}}{\text{hp}}\right)\left(1000 \dfrac{\text{W}}{\text{kW}}\right)}{(230 \text{ V})(0.98)(0.88)}$$

$$= 93.99 \text{ A} \quad [\text{per phase}]$$

40. DUTY CYCLE

Motors are categorized according to their *duty cycle: continuous duty* (24 hr/day); *intermittent* or *short-time duty* (15 min to 30 min); and *special duty* (application specific). Duty cycle is the amount of time a motor can be operated out of every hour without overheating the winding insulation. The idle time is required to allow the motor to cool.

41. INDUCTION MOTORS

The three-phase induction motor is by far the most frequently used motor in industry. In an induction motor, the magnetic field rotates at the synchronous speed. The *synchronous speed* can be calculated from the number of poles and frequency. The frequency, f, is either 60 Hz (in the United States) or 50 Hz (in Europe and other locations). The number of poles, p, must be even.[26] The most common motors have 2, 4, or 6 poles.

$$n_{\text{synchronous}} = \frac{120f}{p} \quad [\text{rpm}] \qquad 52.76$$

Due to friction and other factors, rotors (and hence, the motor shafts) in induction motors run slightly slower than their synchronous speeds. The percentage difference is known as the *slip, s*. Slip is seldom greater than 10%, and it is usually much less than that. 4% is a typical value.

$$s = \frac{n_{\text{synchronous}} - n_{\text{actual}}}{n_{\text{synchronous}}} \qquad 52.77$$

The rotor's actual speed is[27]

$$n_{\text{actual}} = (1-s)n_{\text{synchronous}} \qquad 52.78$$

Induction motors are usually specified in terms of the *kVA ratings*. The kVA rating is not the same as the motor power in kilowatts, although one can be calculated from the other if the motor's power factor is known. The power factor generally varies from 0.8 to 0.9 depending on the motor size.

$$\text{kVA rating} = \frac{P_{\text{kW}}}{\text{pf}} \qquad 52.79$$

$$P_{\text{kW}} = 0.7457 P_{\text{mechanical,hp}} \qquad 52.80$$

Induction motors can differ in the manner in which their rotors are constructed. A *wound rotor* is similar to an armature winding in a dynamo. Wound rotors have high-torque and soft-starting capabilities. There are no wire windings at all in a *squirrel-cage* rotor. Most motors use squirrel-cage rotors. Typical torque-speed characteristics of a design B induction motor are shown in Fig. 52.11.

Example 52.7

A pump is driven by a three-phase induction motor running at its rated values. The motor's nameplate lists the following rated values: 50 hp, 440 V, 92% lagging power factor, 90% efficiency, 60 Hz, and 4 poles. The motor's windings are delta-connected. The pump efficiency is 80%. When running under the pump load, the slip is 4%. What are the (a) total torque developed, (b) torque developed per phase, and (c) line current?

Solution

(a) The synchronous speed is given by Eq. 52.76.

$$n_{\text{synchronous}} = \frac{120f}{p} = \frac{(120)(60 \text{ Hz})}{4}$$

$$= 1800 \text{ rpm}$$

From Eq. 52.78, the rotor speed is

$$n_{\text{actual}} = (1-s)n_{\text{synchronous}} = (1 - 0.04)(1800 \text{ rpm})$$

$$= 1728 \text{ rpm}$$

Since the motor is running at its rated values, the motor delivers 50 hp to the pump. The pump's efficiency is

---

[26]There are various forms of Eq. 52.76. As written, the speed is given in rpm, and the number of poles, p, is twice the number of *pole pairs*. When the synchronous speed is specified as f/p, it is understood that the speed is in revolutions per second (rps) and p is the number of pole pairs.

[27]Some motors (i.e., *integral gear motors*) are manufactured with integral speed reducers. Common standard output speeds are 37 rpm, 45 rpm, 56 rpm, 68 rpm, 84 rpm, 100 rpm, 125 rpm, 155 rpm, 180 rpm, 230 rpm, 280 rpm, 350 rpm, 420 rpm, 520 rpm, and 640 rpm. While the integral gear motor is more compact, lower in initial cost, and easier to install than a separate motor with belt drive, coupling, and guard, the separate motor and reducer combination may nevertheless be preferred for its flexibility, especially in replacing and maintaining the motor.

Systems, Mgmt., Professional

irrelevant. From Eq. 52.71, the total torque developed by all three phases is

$$T_t = \frac{5252 P_{\text{horsepower}}}{n_{\text{rpm}}} = \frac{\left(5252 \frac{\text{ft-lbf}}{\text{hp-min}}\right)(50 \text{ hp})}{1728 \text{ rpm}}$$
$$= 152.0 \text{ ft-lbf} \quad [\text{total}]$$

(b) The torque developed per phase is one-third of the total torque developed.

$$T_p = \frac{T_t}{3} = \frac{152.0 \text{ ft-lbf}}{3 \text{ phases}}$$
$$= 50.67 \text{ ft-lbf} \quad [\text{per phase}]$$

(c) The total electrical input power is given by Eq. 52.72.

$$P_{\text{electrical}} = \frac{P_{\text{rated}}}{\eta} = \frac{(50 \text{ hp})\left(0.7457 \frac{\text{kW}}{\text{hp}}\right)}{0.90}$$
$$= 41.43 \text{ kW} \quad [\text{total}]$$

Since the power factor is less than 1.0, more current is being drawn than is being converted into useful work. From Eq. 52.79, the apparent power in kVA per phase is

$$S_{\text{kVA}} = \frac{P_{\text{kW}}}{\text{pf}} = \frac{41.43 \text{ kW}}{(3)(0.92)}$$
$$= 15.01 \text{ kVA} \quad [\text{per phase}]$$

The phase current is given by Eq. 52.52.

$$I = \frac{S}{V} = \frac{(15.01 \text{ kVA})\left(1000 \frac{\text{VA}}{\text{kVA}}\right)}{440 \text{ V}}$$
$$= 34.11 \text{ A}$$

Since the motor's windings are delta-connected across the three lines, the line current is found from Eq. 52.64.

$$I_l = \sqrt{3}I_p = (\sqrt{3})(34.11 \text{ A})$$
$$= 59.08 \text{ A}$$

42. INDUCTION MOTOR PERFORMANCE

The following rules of thumb can be used for initial estimates of induction motor performance.

- At 1800 rpm, a motor will develop a torque of 3 ft-lbf/hp.
- At 1200 rpm, a motor will develop a torque of 4.5 ft-lbf/hp.
- At 550 V, a three-phase motor will draw 1 A/hp.
- At 440 V, a three-phase motor will draw 1.25 A/hp.

- At 220 V, a three-phase motor will draw 2.5 A/hp.
- At 220 V, a single-phase motor will draw 5 A/hp.
- At 110 V, a single-phase motor will draw 10 A/hp.

43. SYNCHRONOUS MOTORS

Synchronous motors are essentially dynamo alternators operating in reverse. The stator field frequency is fixed, so regardless of load, the motor runs only at a single speed—the synchronous speed given by Eq. 52.76. Stalling occurs when the motor's counter-torque is exceeded. For some equipment that must be driven at constant speed, such as large air or gas compressors, the additional complexity of synchronous motors is justified.

Power factor can be adjusted manually by varying the field current. With *normal excitation* field current, the power factor is 1.0. With *over-excitation*, the power factor is leading, and the field current is greater than normal. With *under-excitation*, the power factor is lagging, and the field current is less than normal.

Since a synchronous motor can be adjusted to draw leading current, it can be used for power factor correction. A synchronous motor used purely for power factor correction is referred to as a *synchronous capacitor* or *synchronous condenser*. A power factor of 80% is often specified or used with synchronous capacitors.

44. DC MACHINES

DC motors and generators can be wired in one of three ways: series, shunt, and compound. Operational characteristics are listed in Table 52.2. Equation 52.71 can be used to calculate torque and power.

45. CHOICE OF MOTOR TYPES

Squirrel-cage induction motors are commonly chosen because of their simple construction, low maintenance, and excellent efficiencies.[28] A wound-rotor induction motor should be used only if it is necessary to achieve a low starting kVA, controllable kVA, controllable torque, or variable speed.[29]

While induction motors are commonly used, synchronous motors are suitable for many applications normally handled by a NEMA design B squirrel-cage motor. They have adjustable power factors and higher efficiency. Their initial cost may also be less.

Selecting a motor type is greatly dependent on the power, torque, and speed requirements of the rotating load. Table 52.3, Fig. 52.12, App. 52.A, and App. 52.B can be used as starting points in the selection process.

---

[28]The larger the motor and the higher the speed, the higher the efficiency. Large 3600 rpm induction motors have excellent performance.
[29]A constant-speed motor with a slip coupling could also be used.

Table 52.2 *Operational Characteristics of DC and AC Machines*

	motors — shunt	motors — series	motors — compound	generators — shunt	generators — series	generators — compound
equivalent circuit	(circuit diagram)	(circuit diagram)	(circuit diagram)	(circuit diagram)	(circuit diagram)	(circuit diagram)
line voltage, V	V	V	V	V	V	V
line current, I_l	I_l	$I_l = I_a$	I_l	$I_l = I_a - I_f$	$I_l = I_a$	$I_l = I_a - I_f$
field current, I_f	$I_f = \dfrac{V}{R_f}$	$I_f = I_a$	$I_f = \dfrac{V}{R_{f1}}$	$I_f = \dfrac{V}{R_f}$	$I_f = I_a$	$I_f = \dfrac{V}{R_{f1}}$
armature current, I_a	$I_a = I_l - I_f$	$I_a = I_l$	$I_a = I_l - I_f$	$I_a = I_l + I_f$	$I_a = I_l$	$I_a = I_l + I_f$
armature circuit loss, V_a	$V_a = I_a R_a$	$V_a = I_a(R_a + R_f)$	$V_a = I_a(R_a + R_{f2})$	$V_a = I_a R_a$	$V_a = I_a(R_a + R_f)$	$V_a = I_a(R_a + R_{f2})$
counter emf, E_S	$E_S = V - V_a$ $= V - I_a R_a$	$E_S = V - V_a$ $= V - I_a(R_a + R_f)$	$E_S = V - I_a$ $\times (R_a + R_{f2})$	$E_S = V + V_a$ $= V + I_a R_a$	$E_S = V + V_a$ $= V + I_a(R_a + R_f)$	$E_S = V + V_a$ $= V + I_a(R_a + R_{f2})$
power in kW or hp [DC machines]	$P = VI_l$	$P = VI_l$	$P = VI_l$	$hp = \dfrac{2\pi nT}{33,000}$	$hp = \dfrac{2\pi nT}{33,000}$	$hp = \dfrac{2\pi nT}{33,000}$
power in kVA [AC machines]	$P = VI_l \cos\phi$	$P = VI_l \cos\phi$	$P = VI_l \cos\phi$	$hp = \dfrac{2\pi nT}{33,000}$	$hp = \dfrac{2\pi nT}{33,000}$	$hp = \dfrac{2\pi nT}{33,000}$
power out in hp or kW [DC machines]	$hp = \dfrac{2\pi nT}{33,000}$	$hp = \dfrac{2\pi nT}{33,000}$	$hp = \dfrac{2\pi nT}{33,000}$	$P = VI_l$	$P = VI_l$	$P = VI_l$
power out in kVA [AC machines]	$hp = \dfrac{2\pi nT}{33,000}$	$hp = \dfrac{2\pi nT}{33,000}$	$hp = \dfrac{2\pi nT}{33,000}$	$P = VI_l \cos\phi$	$P = VI_l \cos\phi$	$P = VI_l \cos\phi$

Table 52.3 Recommended Motor Voltage and Power Ranges

voltage	horsepower
direct current	
115	0–30 (max)
230	0–200 (max)
550 or 600	$\frac{1}{2}$ and upward
alternating current, one-phase	
110, 115, or 120	0–1$\frac{1}{2}$
220, 230, or 240	0–10
440 or 550	5–10*
alternating current, two- and three-phase	
110, 115, or 120	0–15
208, 220, 230, or 240	0–200
440 or 550	0–500
2200 or 2300	40 and upward
4000	75 and upward
6600	400 and upward

(Multiply hp by 0.7457 to obtain kW.)
*not recommended

Figure 52.12 Motor Rating According to Speed (general guidelines)

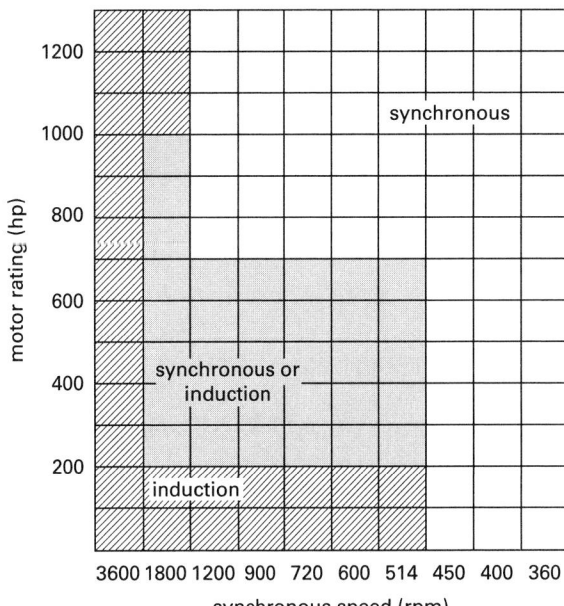

Adapted from *Mechanical Engineering, Design Manual NAVFAC DM-3*, Department of the Navy, © 1972.

46. LOSSES IN ROTATING MACHINES

Losses in rotating machines are typically divided into the following categories: armature copper losses, field copper losses, mechanical losses (including friction and windage), core losses (including hysteresis losses, eddy current losses, and brush resistance losses), and stray losses.

Copper losses (also known as I^2R *losses, impedance losses, heating losses,* and *real losses*) are real power losses due to wire and winding resistance heating.

$$P_{\text{Cu}} = \sum I^2 R \qquad 52.81$$

Core losses (also known as *iron losses*) are constant losses that are independent of the load and, for that reason, are also known as *open-circuit* and *no-load losses.*

Mechanical losses (also known as *rotational losses*) include brush and bearing friction and *windage* (air friction). (Windage is a no-load loss but is not an electrical core loss.) Mechanical losses are determined by measuring the power input at the rated speed and with no load.

Stray losses are due to nonuniform current distribution in the conductors. Stray losses are approximately 1% for DC machines and zero for AC machines.

47. EFFICIENCY OF ROTATING MACHINES

Only real power is used to compute the efficiency of a rotating machine. This efficiency is sometimes referred to as *overall efficiency* and *commercial efficiency.*

$$\begin{aligned} \eta &= \frac{\text{output power}}{\text{input power}} \\ &= \frac{\text{output power}}{\text{output power} + \text{power losses}} \\ &= \frac{\text{input power} - \text{power losses}}{\text{input power}} \qquad 52.82 \end{aligned}$$

Example 52.8

A DC shunt motor draws 40 A at 112 V when fully loaded. When running without a load at the same speed, it draws only 3 A at 106 V. The field resistance is 100 Ω, and the armature resistance is 0.125 Ω. (a) What is the efficiency of the motor? (b) What power (in hp) does the motor deliver at full load?

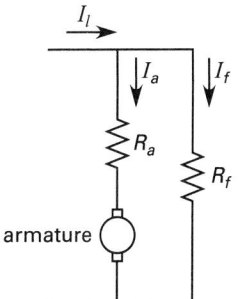

Systems, Mgmt., Professional

Solution

(a) The field current is given by Eq. 52.9.

$$I_f = \frac{V}{R_f} = \frac{112 \text{ V}}{100 \text{ } \Omega}$$
$$= 1.12 \text{ A}$$

The total full-load line current is known to be 40 A. The full-load armature current is

$$I_a = I_l - I_f = 40 \text{ A} - 1.12 \text{ A}$$
$$= 38.88 \text{ A}$$

The field copper loss is given by Eq. 52.11.

$$P_{\text{Cu},f} = I_f^2 R_f = (1.12 \text{ A})^2 (100 \text{ } \Omega)$$
$$= 125.4 \text{ W}$$

The armature copper loss is given by Eq. 52.11.

$$P_{\text{Cu},a} = I_a^2 R_a = (38.88 \text{ A})^2 (0.125 \text{ } \Omega)$$
$$= 189.0 \text{ W}$$

The total copper loss is

$$P_{\text{Cu},t} = P_{\text{Cu},f} + P_{\text{Cu},a} = 125.4 \text{ W} + 189.0 \text{ W}$$
$$= 314.4 \text{ W}$$

Stray power is determined from the no-load conditions. At no load, the field current is given by Eq. 52.9.

$$I_f = \frac{V}{R_f} = \frac{106 \text{ V}}{100 \text{ } \Omega}$$
$$= 1.06 \text{ A}$$

At no load, the armature current is

$$I_a = I_l - I_f = 3 \text{ A} - 1.06 \text{ A}$$
$$= 1.94 \text{ A}$$

The stray power loss is

$$P_{\text{stray}} = V_l I_a - I_a^2 R_a$$
$$= (106 \text{ V})(1.94 \text{ A}) - (1.94 \text{ A})^2 (0.125 \text{ } \Omega)$$
$$= 205.2 \text{ W}$$

The stray power loss is assumed to be independent of the load. The total losses at full load are

$$P_{\text{loss}} = P_{\text{Cu},t} + P_{\text{stray}} = 314.4 \text{ W} + 205.2 \text{ W}$$
$$= 519.6 \text{ W}$$

The power input to the motor when fully loaded is

$$P_{\text{input}} = I_l V_l = (40 \text{ A})(112 \text{ V}) = 4480 \text{ W}$$

The efficiency is

$$\eta = \frac{\text{input power} - \text{power losses}}{\text{input power}}$$
$$= \frac{4480 \text{ W} - 519.6 \text{ W}}{4480 \text{ W}}$$
$$= 0.884 \quad (88.4\%)$$

(b) The real power (in hp) delivered at full load is

$$P_{\text{real}} = \text{input power} - \text{power losses}$$
$$= \frac{4480 \text{ W} - 519.6 \text{ W}}{\left(1000 \text{ } \frac{\text{W}}{\text{kW}}\right)\left(0.7457 \text{ } \frac{\text{kW}}{\text{hp}}\right)}$$
$$= 5.3 \text{ hp}$$

48. HIGH-EFFICIENCY MOTORS AND DRIVES

A premium, energy-efficient motor will have approximately 50% of the losses of a conventional motor. Due to the relatively high overall efficiency enjoyed by all motors, however, this translates into only a 5% increase in overall efficiency.

High-efficiency motors are often combined with *variable-frequency drives* (VFDs) to achieve continuous-variable speed control.[30] VFDs can substantially reduce the power drawn by the process. For example, with a motor-driven pump or fan, the motor can be slowed down instead of closing a valve or damper when the flow requirements decrease. Since the required motor horsepower varies with the cube root of the speed, for processes that do not always run at full-flow (fluid pumping, metering, flow control, etc.), the energy savings can be substantial.

A VFD uses an electronic controller to produce a variable-frequency signal that is not an ideal sine wave. This results in additional heating (e.g., an increase of 20–40%) from copper and core losses in the motor. In pumping applications, though, the process load drops off faster than additional heat is produced. Thus, the process power savings dominate.

Most low- and medium-power motors implement VFD with DC voltage intermediate circuits, a technique known as *voltage-source inversion*. Voltage-source inversion is further subdivided into *pulse-width modulation* (PWM) and *pulse-amplitude modulation* (PAM).

Under VFD control, the motor torque is approximately proportional to the applied voltage and drawn current but inversely proportional to the applied frequency.

$$T \propto \frac{VI}{f} \qquad \qquad 52.83$$

---

[30]Prior to VFDs, there were two common ways to change the speed of an induction motor: (a) increasing the slip, and (b) changing the number of pole-pairs. Increasing the slip was accomplished by undermagnetizing the motor so that it received less input voltage than it was built for. Dropping or adding the number of active poles resulted in the speed changing by a factor of 2 (or ¹/₂).

49. ELECTRICAL SAFETY

Electrical safety hazards include shock, arc flash, explosion, and fire. Electrical safety programs must comply with state and federal requirements and are tailored to the specific needs of the workplace. Safety measures include overcurrent protection, such as circuit breakers, grounding, flame-resistant clothing, and insulated tools.

Electrical shock, when current runs through or across the body, is a hazard that can occur in nearly all industries and workplaces. Shock hazard is a function of current, commonly measured in milliamps. Table 52.4 lists levels of current and their effects on a human body.

Table 52.4 Effects of Current on Humans

current level	probable effect on human body
1 mA	Perception level. Slight tingling sensation. Still dangerous under certain conditions.
5 mA	Slight shock felt; not painful, but disturbing. Average individual can let go. However, strong involuntary reactions to shocks in this range may lead to injuries.
6–16 mA	Painful shock, begin to lose muscular control. Commonly referred to as the freezing current or "let-go" range.
17–99 mA	Extreme pain, respiratory arrest, severe muscular contractions. Individual cannot let go. Death is possible.
100–2000 mA	Ventricular fibrillation (uneven, uncoordinated pumping of the heart). Muscular contraction and nerve damage begins to occur. Death is likely.
> 2000 mA	Cardiac arrest, internal organ damage, and severe burns. Death is probable.

53 Instrumentation and Digital Numbering Systems

Nomenclature

A	area	in^2	m^2
b	base length	in	m
BC	bridge constant	–	–
c	distance from neutral axis to extreme fiber	in	m
C	concentration	various	various
d	diameter	in	m
E	modulus of elasticity	lbf/in^2	Pa
F	force	lbf	N
G	shear modulus	lbf/in^2	Pa
GF	gage factor	–	–
h	height	in	m
I	current	A	A
I	moment of inertia	in^4	m^4
J	polar moment of inertia	in^4	m^4
k	constant	various	various
k	deflection constant	lbf/in	N/m
K	factor	–	–
L	shaft length	in	m
M	moment	in-lbf	N·m
n	number	–	–
n	rotational speed	rev/min	rev/min
p	pressure	lbf/in^2	Pa
P	permeability	various	various
P	power	hp	kW
Q	statical moment	in^3	m^3
r	radius	in	m
R	resistance	Ω	Ω
t	thickness	in	m
T	temperature	°R	K
T	torque	in-lbf	N·m
V	voltage	V	V
VR	voltage ratio	–	–
y	deflection	in	m

Symbols

α	temperature coefficient	1/°R	1/K
β	RTD temperature coefficient	$1/°R^2$	$1/K^2$
β	thermistor constant	°R	K
γ	shear strain	–	–
ϵ	strain	–	–
η	efficiency	–	–
θ	angle of twist	deg	deg
ν	Poisson's ratio	–	–
ρ	resistivity	Ω-in	Ω·cm
σ	stress	lbf/in^2	Pa
τ	shear stress	lbf/in^2	Pa
ϕ	angle of twist	rad	rad

Subscripts

b	battery
g	gage
o	original
r	ratio
ref	reference
t	transverse or total
T	at temperature T
x	in x-direction
y	in y-direction

Systems, Mgmt., Professional

1. ACCURACY

A measurement is said to be *accurate* if it is substantially unaffected by (i.e., is insensitive to) all variation outside of the measurer's control.

For example, suppose a rifle is aimed at a point on a distant target and several shots are fired. The target point represents the "true value" of a measurement—the value that should be obtained. The impact points represent the measured values—what is obtained. The distance from the centroid of the points of impact to the target point is a measure of the alignment accuracy between the barrel and the sights. This difference between the true and measured values is known as the measurement *bias*.

2. PRECISION

Precision is not synonymous with *accuracy*. Precision is concerned with the repeatability of the measured results. If a measurement is repeated with identical results, the experiment is said to be precise. The average distance of each impact from the centroid of the impact group is a measure of precision. Therefore, it is possible to take highly precise measurements and still be inaccurate (i.e., have a large bias).

Most measurement techniques (e.g., taking multiple measurements and refining the measurement methods or procedures) that are intended to improve accuracy actually increase the precision.

Sometimes the term *reliability* is used with regard to the precision of a measurement. A *reliable measurement* is the same as a *precise estimate*.

3. STABILITY

Stability and *insensitivity* are synonymous terms. (Conversely, *instability* and *sensitivity* are synonymous.) A stable measurement is insensitive to minor changes in the measurement process.

Example 53.1

At 65°F (18°C), the centroid of an impact group on a rifle target is 2.1 in (5.3 cm) from the sight-in point. At 80°F (27°C), the distance is 2.3 in (5.8 cm). What is the sensitivity to temperature?

SI Solution

$$\text{sensitivity to temperature} = \frac{\Delta \text{ measurement}}{\Delta \text{ temperature}}$$
$$= \frac{5.8 \text{ cm} - 5.3 \text{ cm}}{27°C - 18°C}$$
$$= 0.0556 \text{ cm/}°C$$

Customary U.S. Solution

$$\text{sensitivity to temperature} = \frac{\Delta \text{ measurement}}{\Delta \text{ temperature}}$$
$$= \frac{2.3 \text{ in} - 2.1 \text{ in}}{80°F - 65°F}$$
$$= 0.0133 \text{ in/}°F$$

4. CALIBRATION

Calibration is used to determine or verify the scale of the measurement device. In order to calibrate a measurement device, one or more known values of the quantity to be measured (temperature, force, torque, etc.) are applied to the device and the behavior of the device is noted. (If the measurement device is linear, it may be adequate to use just a single calibration value. This is known as *single-point calibration*.)

Once a measurement device has been calibrated, the calibration signal should be reapplied as often as necessary to prove the reliability of the measurements. In some electronic measurement equipment, the calibration signal is applied continuously.

5. ERROR TYPES

Measurement errors can be categorized as *systematic* (*fixed*) *errors*, *random* (*accidental*) *errors*, *illegitimate errors*, and *chaotic errors*.

Systematic errors, such as improper calibration, use of the wrong scale, and incorrect (though consistent) technique, are essentially constant or similar in nature over time. *Loading errors* are systematic errors and occur when the act of measuring alters the true value.[1] Some *human errors*, if present in each repetition of the measurement, are also systematic. Systematic errors can be reduced or eliminated by refinement of the experimental method.

Random errors are caused by random and irregular influences generally outside the control of the measurer. Such errors are introduced by fluctuations in the environment, changes in the experimental method, and variations in materials and equipment operation. Since the occurrence of these errors is irregular, their effects can be reduced or eliminated by multiple repetitions of the experiment.

There is no reason to expect or tolerate *illegitimate errors* (e.g., errors in computations and other blunders). These are essentially mistakes that can be avoided through proper care and attention.

Chaotic errors, such as resonance, vibration, or experimental "noise," essentially mask or entirely invalidate the experimental results. Unlike the random errors

---

[1]For example, inserting an air probe into a duct will change the flow pattern and velocity around the probe.

previously mentioned, chaotic disturbances can be sufficiently large to reduce the experimental results to meaninglessness.[2] Chaotic errors must be eliminated.

6. ERROR MAGNITUDES

If a single measurement is taken of some quantity whose true value is known, the *error* is simply the difference between the true and measured values. However, the true value is never known in an experiment, and measurements are usually taken several times, not just once. Therefore, many conventions exist for estimating the unknown error.

When most experimental quantities are measured, the measurements tend to cluster around some "average value." The measurements will be distributed according to some distribution, such as linear, normal, Poisson, and so on. The measurements can be graphed in a *histogram* and the distribution inferred. Usually the data will be normally distributed.[3]

Certain error terms used with normally distributed data have been standardized. These are listed in Table 53.1.

Table 53.1 *Normal Distribution Error Terms*

term	number of standard deviations	percent certainty	approximate odds of being incorrect
probable error	0.6745	50	1 in 2
mean deviation	0.6745	50	1 in 2
standard deviation	1.000	68.3	1 in 3
one-sigma error	1.000	68.3	1 in 3
90% error	1.6449	90	1 in 10
two-sigma error	2.000	95	1 in 20
three-sigma error	3.000	99.7	1 in 370
maximum error[*]	3.29	99.9+	1 in 1000

[*]The true maximum error is theoretically infinite.

7. POTENTIOMETERS

A *potentiometer* (*potentiometer transducer, variable resistor*) is a resistor with a sliding third contact. It converts linear or rotary motion into a variable resistance (voltage).[4] It consists of a resistance oriented in

---

[2]Much has been written about *chaos theory*. This theory holds that, for many processes, the ending state is dependent on imperceptible differences in the starting state. Future weather conditions and the landing orientation of a finely balanced spinning top are often used as examples of states that are greatly affected by their starting conditions.

[3]The results of all numerical experiments are not automatically normally distributed. The throw of a die (one of two dice) is linearly distributed. Emissive power of a heated radiator is skewed with respect to wavelength. However, the means of sets of experimental data generally will be normally distributed, even if the raw measurements are not.

[4]There is a voltage-balancing device that shares the name *potentiometer* (*potentiometer circuit*). An unknown voltage source can be measured by adjusting a calibrated voltage until a null reading is obtained on a voltage meter. The applications are sufficiently different that no confusion occurs when the "pot is adjusted."

---

a linear or angular manner and a variable-position contact point known as the *tap*. A voltage is applied across the entire resistance, causing current to flow through the resistance. The voltage at the tap will vary with tap position. (See Fig. 53.1.)

Figure 53.1 *Potentiometer Circuit Diagram*

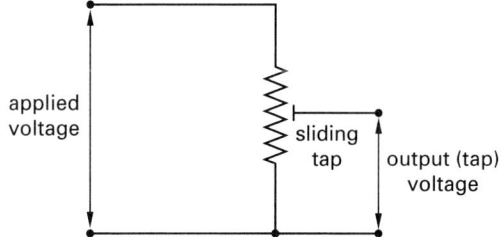

8. TRANSDUCERS

Physical quantities are often measured with transducers (*detector-transducers*). A *transducer* converts one variable into another. For example, a Bourdon tube pressure gauge converts pressure to angular displacement; a strain gauge converts stress to resistance change. Transducers are primarily mechanical in nature (e.g., pitot tube, spring devices, Bourdon tube pressure gauge) or electrical in nature (e.g., thermocouple, strain gauge, moving-core transformer).

9. SENSORS

While the term "transducer" is commonly used for devices that respond to mechanical input (force, pressure, torque, etc.), the term *sensor* is commonly applied to devices that respond to varying chemical conditions.[5] For example, an electrochemical sensor might respond to a specific gas, compound, or ion (known as a *target substance* or *species*). Two types of electrochemical sensors are in use today: potentiometric and amperometric.

Potentiometric sensors generate a measurable voltage at their terminals. In electrochemical sensors taking advantage of half-cell reactions at electrodes, the generated voltage is proportional to the absolute temperature, T, and is inversely proportional to the number of electrons, n, taking part in the chemical reaction at the half-cell. In Eq. 53.1, p_1 is the partial pressure of the target substance at the measurement electrode; p_2 is the partial pressure of the target substance at the reference electrode.

$$V \propto \left(\frac{T_{\text{absolute}}}{n} \right) \ln \frac{p_1}{p_2} \qquad 53.1$$

Amperometric sensors (also known as *voltammetric sensors*) generate a measurable current at their terminals. In the conventional electrochemical sensors known

---

[5]The categorization is common but not universal. The terms "transducer," "sensor," "sending unit," and "pickup" are often used loosely.

as *diffusion-controlled cells*, a high-conductivity acid or alkaline liquid electrolyte is used with a gas-permeable membrane that transmits ions from the outside to the inside of the sensor. A reference voltage is applied to two terminals within the electrolyte, and the current generated at a (third) sensing electrode is measured.

The maximum current generated is known as the *limiting current*. Current is proportional to the concentration, C, of the target substance; the permeability, P; the exposed sensor (membrane) area, A; and the number of electrons transferred per molecule detected, n. The current is inversely proportional to the membrane thickness, t.

$$I \propto \frac{nPCA}{t} \qquad 53.2$$

10. VARIABLE-INDUCTANCE TRANSDUCERS

Inductive transducers contain a wire coil and a moving permeable *core*.[6] As the core moves, the flux linkage through the coil changes. The change in inductance affects the overall impedance of the detector circuit.

The *differential transformer* or *linear variable differential transformer* (LVDT) is an important type of *variable-inductance transducer*. (See Fig. 53.2.) It converts linear motion into a change in voltage. The transformer is supplied with a low AC voltage. When the core is centered between the two secondary windings, the LVDT is said to be in its *null position*.

Figure 53.2 *Linear Variable Differential Transformer Schematic and Performance Characteristic*

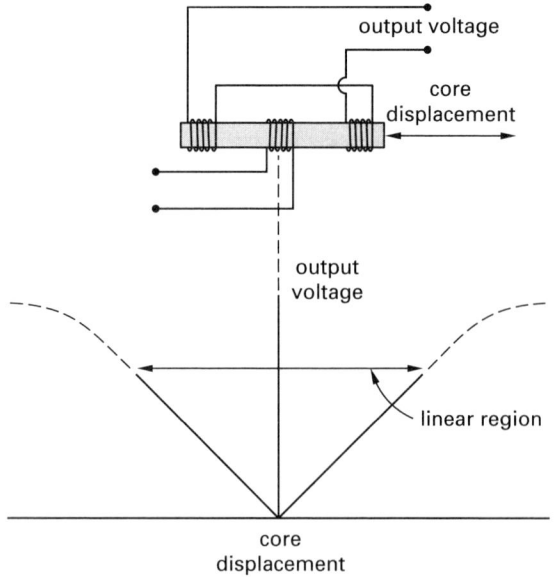

---

[6]The term *core* is used even if the cross sections of the coil and core are not circular.

Movement of the core changes the magnetic flux linkage between the primary and secondary windings. Over a reasonable displacement range, the output voltage is proportional to the displacement of the core from the null position, hence the "linear" designation. The voltage changes phase (by 180°) as the core passes through the null position.

Sensitivity of an LVDT is measured in mV/in (mV/cm) of core movement. The sensitivity and output voltage depend on the frequency of the applied voltage (i.e., the *carrier frequency*) and are directly proportional to the magnitude of the applied voltage.

11. VARIABLE-RELUCTANCE TRANSDUCERS

A *variable-reluctance transducer* (*pickup*) is essentially a permanent magnet and a coil in the vicinity of the process being monitored.[7] There are no moving parts in this type of transducer. However, some of the magnet's magnetic flux passes through the surroundings, and the presence or absence of the process changes the coil voltage. Two typical applications of variable-reluctance pickups are measuring liquid levels and determining the rotational speed of a gear.

12. VARIABLE-CAPACITANCE TRANSDUCERS

In *variable-capacitance transducers*, the capacitance of a device can be modified by changing the plate separation, plate area, or dielectric constant of the medium separating the plates.

13. OTHER ELECTRICAL TRANSDUCERS

The *piezoelectric effect* is the name given to the generation of an electrical voltage when placed under stress.[8] *Piezoelectric transducers* generate a small voltage when stressed (strained). Since voltage is developed during the application of changing strain but not while strain is constant, piezoelectric transducers are limited to dynamic applications. Piezoelectric transducers may suffer from low voltage output, instability, and limited ranges in operating temperature and humidity.

The *photoelectric effect* is the generation of an electrical voltage when a material is exposed to light.[9] Devices using this effect are known as *photocells, photovoltaic*

---

[7]*Reluctance* depends on the area, length, and permeability of the medium through which the magnetic flux passes.
[8]Quartz, table sugar, potassium sodium tartarate (Rochelle salt), and barium titanate are examples of piezoelectric materials. Quartz is commonly used to provide a stable frequency in electronic oscillators. Barium titanate is used in some ultrasonic cleaners and sonar-like equipment.
[9]While the *photogenerative* (*photovoltaic*) definition is the most common definition, the term "photoelectric" can also be used with *photoconductive devices* (those whose resistance changes with light) and *photoemissive devices* (those that emit light when a voltage is applied).

cells, photosensors, or *light-sensitive detectors,* depending on the applications. The sensitivity can be to radiation outside of the visible spectrum. Photoelectric detectors can be made that respond to infrared and ultraviolet radiation. The magnitude of the voltage (or of the current in an attached circuit) will depend on the amount of illumination. If the cell is reverse-biased by an external battery, its operation is similar to a constant-current source.[10] (See Fig. 53.3.)

Figure 53.3 *Photovoltaic Device Characteristic Curves*

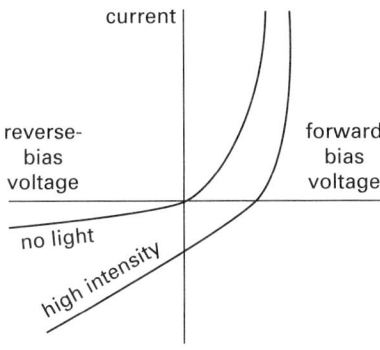

14. PHOTOSENSITIVE CONDUCTORS

Cadmium sulfide and *cadmium selenide* are two compounds that decrease in resistance when exposed to light. Cadmium sulfide is most sensitive to light in the 5000–6000 Å (0.5–0.6 μm) range, while cadmium selenide shows peak sensitivities in the 7000–8000 Å (0.7–0.8 μm) range. These compounds are used in *photosensitive conductors.* Due to hysteresis, photosensitive conductors do not react instantaneously to changes in intensity.[11] High-speed operation requires high light intensities and careful design.

15. RESISTANCE TEMPERATURE DETECTORS

Resistance temperature detectors (RTDs), also known as *resistance thermometers,* make use of changes in their resistance to determine changes in temperature. A fine wire is wrapped around a form and protected with glass or a ceramic coating. Nickel and copper are commonly used for industrial RTDs. Type D 100 Ω and type M 1000 Ω platinum are used when precision resistance thermometry is required. RTDs are connected through

resistance bridges (see Sec. 53.19) to compensate for lead resistance. RTDs are categorized by their precision (tolerance). Class A RTDs are precise to within ±0.06% at 0°C, while class B RTDs are precise to within ±0.12% at 0°C. Tolerance varies with temperature, hence the need to specify a reference temperature.

Resistance in most conductors increases with temperature. The resistance at a given temperature can be calculated from the *coefficients of thermal resistance,* α and β.[12] Positive values of α are known as *positive temperature coefficients* (PTCs); negative values are known as *negative temperature coefficients* (NTCs). Although PTC platinum RTDs are most prevalent, RTDs can be constructed using materials with either positive or negative characteristics. The variation of resistance with temperature is nonlinear, though β is small and is often insignificant over short temperature ranges. Therefore, a linear relationship is often assumed and only α is used. In Eq. 53.3, R_{ref} is the resistance at the reference temperature, T_{ref}, usually 100 Ω at 32°F (0°C).

$$R_T \approx R_{\mathrm{ref}}\left(1 + \alpha\Delta T + \beta\Delta T^2\right) \qquad 53.3$$

$$\Delta T = T - T_{\mathrm{ref}} \qquad 53.4$$

In commercial RTDs, α is referred to by the literal term *alpha-value.* There are two applicable alpha values for platinum, depending on the purity. Commercial platinum RTDs produced in the United States generally have alpha values of 0.00391 1/°C, while RTDs produced in Europe and other countries generally have alpha values of 0.00385 1/°C.

Equation 53.3 and values of α and β are of academic interest. In practice, an equation similar in appearance using the actual temperature, T (not ΔT), is used. Equation 53.5 and Eq. 53.6 constitute a very accurate correlation of resistance and temperature for platinum. The correlation is known as the *Callendar-Van Dusen equation.* R_0 is the resistance of the RTD at 0°C, typically 100 Ω or 1000 Ω for commercial platinum RTDs, which are also referred to as *platinum resistance thermometers* (PRTs).

$$R_T = R_0\left(1 + AT + BT^2 + CT^3(T - 100°C)\right)$$
$$[-200°C < T < 0°C] \qquad 53.5$$

$$R_T = R_0\left(1 + AT + BT^2\right) \quad [0°C < T < 850°C] \qquad 53.6$$

Values of A, B, and C can be calculated from the α and β values, if needed. For platinum RTDs with $\alpha = 0.00385$ 1/°C, these values are

- $A = 3.9083 \times 10^{-3}$ 1/°C
- $B = -5.775 \times 10^{-7}$ 1/°C^2
- $C = -4.183 \times 10^{-12}$ 1/°C^4

---

[10]A semiconductor device is *reverse-biased* when a negative battery terminal is connected to a p-type semiconductor material in the device, or when a positive battery terminal is connected to an n-type semiconductor material.

[11]*Hysteresis* is the tendency for the transducer to continue to respond (i.e., indicate) when the load is removed. Alternatively, hysteresis is the difference in transducer outputs when a specific load is approached from above and from below. Hysteresis is usually expressed in percent of the full-load reading during any single calibration cycle.

---

[12]Higher-order terms (third, fourth, etc.) are used when extreme accuracy is required.

Table 53.2 gives resistances for standard Pt100 RTDs.

Table 53.2 *Resistance of Standard Pt100 RTDs**

T ($°C$)	R (Ω)	T ($°C$)	R (Ω)	T ($°C$)	R (Ω)
−50	80.31	10	103.90	70	127.07
−45	82.29	15	105.85	75	128.98
−40	84.27	20	107.79	80	130.89
−35	86.25	25	109.73	85	132.80
−30	88.22	30	111.67	90	134.70
−25	90.19	35	113.61	95	136.60
−20	92.16	40	115.54	100	138.50
−15	94.12	45	117.47	105	140.39
−10	96.09	50	119.40	110	142.29
−5	98.04	55	121.32	150	157.31
0	100.00	60	123.24	200	175.84
5	101.95	65	125.16		

(Multiply tabulated resistances by 10 for Pt1000 RTDs.)

*Per *DIN International Electrotechnical Commission* (*IEC*) *Standard* 751

16. THERMISTORS

Thermistors are temperature-sensitive semiconductors constructed from oxides of manganese, nickel, and cobalt, and from sulfides of iron, aluminum, and copper. Thermistor materials are encapsulated in glass or ceramic materials to prevent penetration of moisture. Unlike RTDs, the resistance of thermistors decreases as the temperature increases.

Thermistor temperature-resistance characteristics are exponential. Depending on the brand, material, and construction, β typically varies between 3400K and 3900K.

$$R = R_o e^k \qquad 53.7$$

$$k = \beta\left(\frac{1}{T} - \frac{1}{T_o}\right) \quad [T \text{ in K}] \qquad 53.8$$

Thermistors can be connected to measurement circuits with copper wire and soldered connections. Compensation of lead wire effects is not required because resistance of thermistors is very large, far greater than the resistance of the leads. Since the negative temperature characteristic makes it difficult to design customized detection circuits, some thermistor and instrumentation standardization has occurred. The most common thermistors have resistances of 2252 Ω at 77°F (25°C), and most instrumentation is compatible with them. Other standardized resistances are 3000 Ω, 5000 Ω, 10,000 Ω, and 30,000 Ω at 77°F (25°C). (See Table 53.3.)

Thermistors typically are less precise and more unpredictable than metallic resistors. Since resistance varies exponentially, most thermistors are suitable for use only up to approximately 550°F (290°C).

Table 53.3 *Typical Resistivities and Coefficients of Thermal Resistance[a]*

conductor	resistivity[b] ($\Omega\cdot$cm)	α[c] (1/$°C$)
alumel[d]	28.1×10^{-6}	0.0024 @ 212°F (100°C)
aluminum	2.82×10^{-6}	0.0039 @ 68°F (20°C)
		0.0040 @ 70°F (21°C)
brass	7×10^{-6}	0.002 @ 68°F (20°C)
constantan[e,f]	49×10^{-6}	0.00001 @ 68°F (20°C)
chromel[g]	0.706×10^{-6}	0.00032 @ 68°F (20°C)
copper, annealed	1.724×10^{-6}	0.0043 @ 32°F (0°C)
		0.0039 @ 70°F (21°C)
		0.0037 @ 100°F (38°C)
		0.0031 @ 200°F (93°C)
gold	2.44×10^{-6}	0.0034 @ 68°F (20°C)
iron (99.98% pure)	10×10^{-6}	0.005 @ 68°F (20°C)
isoelastic[h]	112×10^{-6}	0.00047
lead	22×10^{-6}	0.0039
magnesium	4.6×10^{-6}	0.004 @ 68°F (20°C)
manganin[i]	44×10^{-6}	0.0000 @ 68°F (20°C)
monel[j]	42×10^{-6}	0.002 @ 68°F (20°C)
nichrome[k]	100×10^{-6}	0.0004 @ 68°F (20°C)
nickel	7.8×10^{-6}	0.006 @ 68°F (20°C)
platinum	10×10^{-6}	0.0039 @ 32°F (0°C)
		0.0036 @ 70°F (21°C)
platinum-iridium[l]	24×10^{-6}	0.0013
platinum-rhodium[m]	18×10^{-6}	0.0017 @ 212°F (100°C)
silver	1.59×10^{-6}	0.004 @ 68°F (20°C)
tin	11.5×10^{-6}	0.0042 @ 68°F (20°C)
tungsten (drawn)	5.8×10^{-6}	0.0045 @ 70°F (21°C)

(Multiply 1/°C by 5/9 to obtain 1/°F.)
(Multiply ppm/°F by 1.8×10^{-6} to obtain 1/°C.)
[a]Compiled from various sources. Data is not to be taken too literally, as values depend on composition and cold working.
[b]At 20°C (68°F)
[c]Values vary with temperature. Common values given when no temperature is specified.
[d]Trade name for 94% Ni, 2.5% Mn, 2% Al, 1% Si, 0.5% Fe (TM of Hoskins Manufacturing Co.)
[e]60% Cu, 40% Ni, also known by trade names *Advance*, *Eureka*, and *Ideal*.
[f]Constantan is also the name given to the composition 55% Cu and 45% Ni, an alloy with slightly different properties.
[g]Trade name for 90% Ni, 10% Cr (TM of Hoskins Manufacturing Co.)
[h]36% Ni, 8% Cr, 0.5% Mo, remainder Fe
[i]9% to 18% Mn, 11% to 4% Ni, remainder Cu
[j]33% Cu, 67% Ni
[k]75% Ni, 12% Fe, 11% Cr, 2% Mn
[l]95% Pt, 5% Ir
[m]90% Pt, 10% Rh

17. THERMOCOUPLES

A *thermocouple* consists of two wires of dissimilar metals joined at both ends.[13] One set of ends, typically called a *junction*, is kept at a known *reference temperature* while the other junction is exposed to the

---

[13]The joint may be made by simply twisting the ends together. However, to achieve a higher mechanical strength and a better electrical connection, the ends should be soldered, brazed, or welded.

unknown temperature.[14] (See Fig. 53.4.) In a laboratory, the reference junction is often maintained at the *ice point*, 32°F (0°C), in an ice/water bath for convenience in later analysis. In commercial applications, the reference temperature can be any value, with appropriate compensation being made.

Figure 53.4 *Thermocouple Circuits*

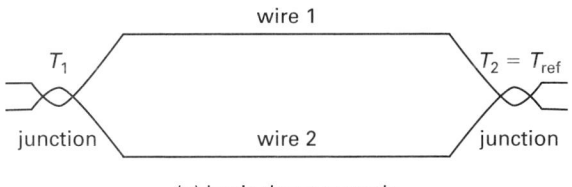

(a) basic thermocouple

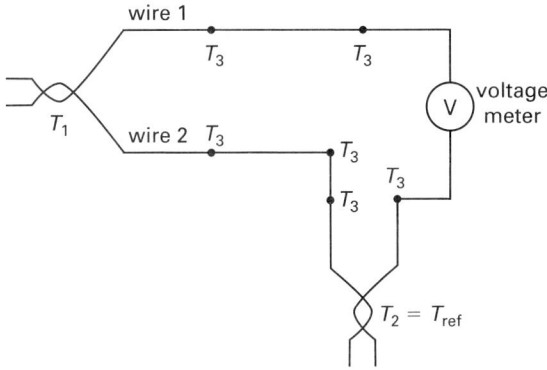

(b) thermocouple in measurement circuit

Thermocouple materials, standard ANSI designations, and approximate useful temperature ranges are given in Table 53.4.[15] The "functional" temperature range can be much larger than the useful range. The most significant factors limiting the useful temperature range, sometimes referred to as the *error limits range*, are linearity, the rate at which the material will erode due to oxidation at higher temperatures, irreversible magnetic effects above magnetic critical points, and longer stabilization periods at higher temperatures.

A voltage is generated when the temperatures of the two junctions are different. This phenomenon is known as the *Seebeck effect*.[16] Referring to the polarities of the

---

[14]The *ice point* is the temperature at which liquid water and ice are in equilibrium. Other standardized temperature references are: *oxygen point*, −297.346°F (90.19K); *steam point*, 212.0°F (373.16K); *sulfur point*, 832.28°F (717.76K); *silver point*, 1761.4°F (1233.96K); and *gold point*, 1945.4°F (1336.16K).

[15]It is not uncommon to list the two thermocouple materials with "vs." (as in *versus*). For example, a copper/constantan thermocouple might be designated as "copper vs. constantan."

[16]The inverse of the Seebeck effect, that current flowing through a junction of dissimilar metals will cause either heating or cooling, is the *Peltier effect*, though the term is generally used in regard to cooling applications. An extension of the Peltier effect, known as the *Thompson effect*, is that heat will be carried along the conductor. Both the Peltier and Thompson effects occur simultaneously with the Seebeck effect. However, the Peltier and Thompson effects are so minuscule that they can be disregarded.

Table 53.4 *Typical Temperature Ranges of Thermocouple Materials*[a]

materials	ANSI designation	useful range (°F (°C))
copper-constantan	T	−300 to 700 (−180 to 370)
chromel-constantan	E	32 to 1600 (0 to 870)
iron-constantan[b]	J	32 to 1400 (0 to 760)
chromel-alumel	K	32 to 2300 (0 to 1260)
platinum-10% rhodium	S	32 to 2700 (0 to 1480)
platinum-13% rhodium	R	32 to 2700 (0 to 1480)
Pt-6% Rh-Pt-30% Rh	B	1600 to 3100 (870 to 1700)
tungsten-Tu-25% rhenium	–	to 4200[c] (2320)
Tu-5% rhenium-Tu-26% rhenium	–	to 4200[c] (2320)
Tu-3% rhenium-Tu-25% rhenium	–	to 4200[c] (2320)
iridium-rhodium	–	to 3500[c] (1930)
nichrome-constantan	–	to 1600[c] (870)
nichrome-alumel	–	to 2200[c] (1200)

[a]Actual values will depend on wire gauge, atmosphere (oxidizing or reducing), use (continuous or intermittent), and manufacturer.
[b]Nonoxidizing atmospheres only.
[c]Approximate usable temperature range. Error limit range is less.

voltage generated, one metal is known as the *positive element* while the other is the *negative element*. The generated voltage is small, and thermocouples are calibrated in $\mu V/°F$ or $\mu V/°C$. An amplifier may be required to provide usable signal levels, although thermocouples can be connected in series (a *thermopile*) to increase the value.[17] The accuracy (referred to as the *calibration*) of thermocouples is approximately $1/2$–$3/4\%$, though manufacturers produce thermocouples with various guaranteed accuracies.

The voltage generated by a thermocouple is given by Eq. 53.9. Since the *thermoelectric constant*, k_T, varies with temperature, thermocouple problems can be solved with published tables of total generated voltage versus temperature. (See App. 53.A.)

$$V = k_T(T - T_{\text{ref}}) \qquad 53.9$$

Generation of thermocouple voltage in a measurement circuit is governed by three laws. The *law of homogeneous circuits* states that the temperature distribution along one or both of the thermocouple leads is irrelevant. Only the junction temperatures contribute to the generated voltage.

The *law of intermediate metals* states that an intermediate length of wire placed within one leg or at the junction of the thermocouple circuit will not affect the voltage generated as long as the two new junctions are

---

[17]There is no special name for a combination of thermocouples connected in parallel.

at the same temperature. This law permits the use of a measuring device, soldered connections, and extension leads.

The *law of intermediate temperatures* states that if a thermocouple generates voltage V_1 when its junctions are at T_1 and T_2, and it generates voltage V_2 when its junctions are at T_2 and T_3, then it will generate voltage $V_1 + V_2$ when its junctions are at T_1 and T_3.

Example 53.2

A type-K (chromel-alumel) thermocouple produces a voltage of 10.79 mV. The "cold" junction is kept at 32°F (0°C) by an ice bath. What is the temperature of the hot junction?

Solution

Since the cold junction temperature corresponds to the reference temperature, the hot junction temperature is read directly from App. 53.A as 510°F.

Example 53.3

A type-K (chromel-alumel) thermocouple produces a voltage of 10.87 mV. The "cold" junction is at 70°F. What is the temperature of the hot junction?

Solution

Use the law of intermediate temperatures. From App. 53.A, the thermoelectric constant for 70°F is 0.84 mV. If the cold junction had been at 32°C, the generated voltage would have been higher. The corrected reading is

$$10.87 \text{ mV} + 0.84 \text{ mV} = 11.71 \text{ mV}$$

The temperature corresponding to this voltage is 550°F.

Example 53.4

A type-T (copper-constantan) thermocouple is connected directly to a voltage meter. The temperature of the meter's screw-terminals is measured by a nearby thermometer as 70°F. The thermocouple generates 5.262 mV. What is the temperature of its hot junction?

Solution

There are two connections at the meter. However, both connections are at the same temperature, so the meter can be considered to be a length of different wire, and the law of intermediate metals applies. Since the meter connections are not at 32°F, the law of intermediate temperatures applies. The corrected voltage is

$$5.262 \text{ mV} + 0.832 \text{ mV} = 6.094 \text{ mV}$$

From App. 53.A, 6.094 mV corresponds to 280°F.

18. STRAIN GAUGES

A *bonded strain gauge* is a metallic resistance device that is cemented to the surface of the unstressed member.[18] The gauge consists of a metallic conductor (known as the *grid*) on a backing (known as the *substrate*).[19] The substrate and grid experience the same strain as the surface of the member. The resistance of the gauge changes as the member is stressed due to changes in conductor cross section and intrinsic changes in resistivity with strain. Temperature effects must be compensated for by the circuitry or by using a second unstrained gauge as part of the bridge measurement system. (See Sec. 53.19.)

When simultaneous strain measurements in two or more directions are needed, it is convenient to use a commercial *rosette strain gauge*. A rosette consists of two or more *grids* properly oriented for application as a single unit. (See Fig. 53.5.)

Figure 53.5 *Strain Gauge*

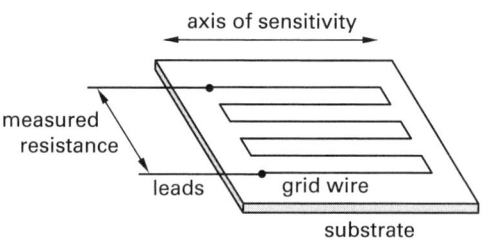

(a) folded-wire strain gauge

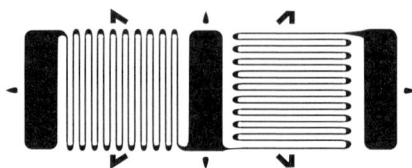

(b) commercial two-element rosette

The *gage factor* (*strain sensitivity factor*), GF, is the ratio of the fractional change in resistance to the fractional change in length (strain) along the detecting axis of the gauge. The higher the gage factor, the greater the sensitivity of the gauge. The gage factor is a function of the gauge material. It can be calculated from the grid material's properties and configuration. From a practical

---

[18]A *bonded strain gauge* is constructed by bonding the conductor to the surface of the member. An *unbonded strain gauge* is constructed by wrapping the conductor tightly around the member or between two points on the member.

Strain gauges on rotating shafts are usually connected through *slip rings* to the measurement circuitry.

[19]The grids of strain gauges were originally of the folded-wire variety. For example, nichrome wire with a total resistance under 1000 Ω was commonly used. Modern strain gauges are generally of the foil type manufactured by printed circuit techniques. Semiconductor gauges are also used when extreme sensitivity (i.e., gage factors in excess of 100) is required. However, semiconductor gauges are extremely temperature-sensitive.

standpoint, however, the gage factor and gage resistance are provided by the gauge manufacturer. Only the change in resistance is measured. (See Table 53.5.)

$$\text{GF} = 1 + 2\nu + \frac{\dfrac{\Delta\rho}{\rho_o}}{\dfrac{\Delta L}{L_o}} = \frac{\dfrac{\Delta R_g}{R_g}}{\dfrac{\Delta L}{L_o}} = \frac{\dfrac{\Delta R_g}{R_g}}{\epsilon} \qquad 53.10$$

Table 53.5 *Approximate Gage Factors*[a]

material	GF
constantan	2.0
iron, soft	4.2
isoelastic	3.5
manganin	0.47
monel	1.9
nichrome	2.0
nickel	−12[b]
platinum	4.8
platinum-iridium	5.1

[a]Other properties of strain gauge materials are listed in Table 53.3.
[b]Value depends on amount of preprocessing and cold working.

Constantan and isoelastic wires, along with metal foil with gage factors of approximately 2 and initial resistances of less than 1000 Ω (typically 120 Ω, 350 Ω, 600 Ω, and 700 Ω) are commonly used. In practice, the gage factor and initial gage resistance, R_g, are specified by the manufacturer of the gauge. Once the strain sensitivity factor is known, the strain, ϵ, can be determined from the change in resistance. Strain is often reported in units of μin/in (μm/m) and is given the name *microstrain*.

$$\epsilon = \frac{\Delta R_g}{(\text{GF})R_g} \qquad 53.11$$

Theoretically, a strain gauge should not respond to strain in its transverse direction. However, the turn-around end-loops are also made of strain-sensitive material, and the end-loop material contributes to a nonzero sensitivity to strain in the transverse direction. Equation 53.12 defines the *transverse sensitivity factor*, K_t, which is of academic interest in most problems. The transverse sensitivity factor is seldom greater than 2%.

$$K_t = \frac{(\text{GF})_{\text{transverse}}}{(\text{GF})_{\text{longitudinal}}} \qquad 53.12$$

Example 53.5

A strain gauge with a nominal resistance of 120 Ω and gage factor of 2.0 is used to measure a strain of 1 μin/in. What is the change in resistance?

Solution

From Eq. 53.11,

$$\Delta R_g = (\text{GF})R_g\epsilon = (2.0)(120\ \Omega)\left(1 \times 10^{-6}\ \frac{\text{in}}{\text{in}}\right)$$

$$= 2.4 \times 10^{-4}\ \Omega$$

19. WHEATSTONE BRIDGES

The *Wheatstone bridge*, shown in Fig. 53.6, is one type of *resistance bridge*.[20] The bridge can be used to determine the unknown resistance of a resistance transducer (e.g., thermistor or resistance-type strain gauge), say R_1 in Fig. 53.6. The potentiometer is adjusted (i.e., the bridge is "balanced") until no current flows through the meter or until there is no voltage across the meter (hence the name *null indicator*).[21,22] When the bridge is balanced and no current flows through the meter leg, Eq. 53.13 through Eq. 53.16 are applicable.

$$I_2 = I_4 \quad \text{[balanced]} \qquad 53.13$$

$$I_1 = I_3 \quad \text{[balanced]} \qquad 53.14$$

$$V_1 + V_3 = V_2 + V_4 \quad \text{[balanced]} \qquad 53.15$$

$$\frac{R_1}{R_2} = \frac{R_3}{R_4} \quad \text{[balanced]} \qquad 53.16$$

Since any one of the four resistances can be the unknown, up to three of the remaining resistances can be fixed or adjustable, and the battery and meter can be connected to either of two diagonal corners, it is sometimes confusing to apply Eq. 53.16 literally. However, the following bridge law statement can be used to help formulate the proper relationship: *When a series Wheatstone bridge is null-balanced, the ratio of resistance of any two adjacent arms equals the ratio of resistance of the remaining two arms, taken in the same sense.* In this statement, "taken in the same sense" means that both ratios must be formed reading either left to right, right to left, top to bottom, or bottom to top.

Figure 53.6 *Wheatstone Bridge*

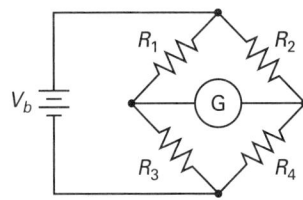

20. STRAIN GAUGE DETECTION CIRCUITS

The resistance of a strain gauge can be measured by placing the gauge in either a ballast circuit or bridge circuit. A *ballast circuit* consists of a voltage source, V_b, of less than 10 V (typical); a current-limiting ballast

---

[20]Other types of resistance bridges are the *differential series balance bridge*, *shunt balance bridge*, and *differential shunt balance bridge*. These differ in the manner in which the adjustable resistor is incorporated into the circuit.
[21]This gives rise to the alternate names of *zero-indicating bridge* and *null-indicating bridge*.
[22]The unknown resistance can also be determined from the amount of voltage unbalance shown by the meter reading, in which case, the bridge is known as a *deflection bridge* rather than a null-indicating bridge. Deflection bridges are described in Sec. 53.20.

resistance, R_b; and the strain gauge of known resistance, R_g, in series. (See Fig. 53.7.) This is essentially a voltage-divider circuit. The change in voltage, ΔV_g, across the strain gauge is measured. The strain, ϵ, can be determined from Eq. 53.17.

$$\Delta V_g = \frac{(\text{GF})\epsilon V_b R_b R_g}{(R_b + R_g)^2} \qquad 53.17$$

Figure 53.7 Ballast Circuit

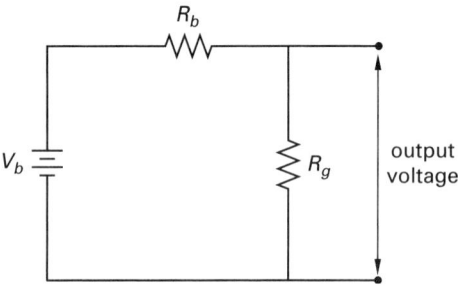

Ballast circuits do not provide temperature compensation, nor is their sensitivity adequate for measuring static strain. Ballast circuits, where used, are often limited to measurement of transient strains. A bridge detection circuit overcomes these limitations.

Figure 53.8 illustrates how a strain gauge can be used with a resistance bridge. Gauge 1 measures the strain, while *dummy gauge* 2 provides temperature compensation.[23] The meter voltage is a function of the input (battery) voltage and the resistors. (As with bridge circuits, the input voltage is typically less than 10 V.) The variable resistance is used for balancing the bridge prior to the strain. When the bridge is balanced, V_{meter} is zero.

When the gauge is strained, the bridge becomes unbalanced. Assuming the bridge is initially balanced, the voltage at the meter (known as the *voltage deflection* from the null condition) will be[24]

$$V_{\text{meter}} = V_b\left(\frac{R_1}{R_1 + R_3} - \frac{R_2}{R_2 + R_4}\right) \quad \left[\text{1/4-bridge}\right]$$
$$53.18$$

For a single strain gauge in a resistance bridge and neglecting lead resistance, the voltage deflection is related to the strain by Eq. 53.19.

$$V_{\text{meter}} = \frac{(\text{GF})\epsilon V_b}{4 + 2(\text{GF})\epsilon}$$
$$\approx \tfrac{1}{4}(\text{GF})\epsilon V_b \quad \left[\text{1/4-bridge}\right] \qquad 53.19$$

---

[23]This is a "quarter-bridge" or "1/4-bridge" configuration, as described in Sec. 53.23. The strain gauge used for temperature compensation is not active.
[24]Equation 53.18 applies to the unstrained condition as well. However, if the gauge is unstrained and the bridge is balanced, the resistance term in parentheses is zero.

Figure 53.8 Strain Gauge in Resistance Bridge

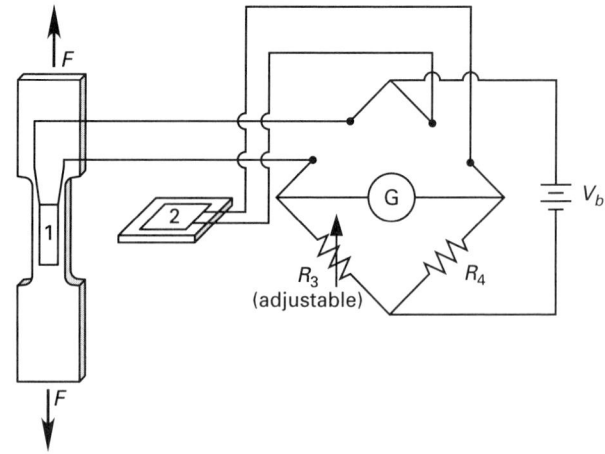

21. STRAIN GAUGE IN UNBALANCED RESISTANCE BRIDGE

A resistance bridge does not need to be balanced prior to use as long as an accurate digital voltmeter is used in the detection circuit. The voltage ratio difference, ΔVR, is defined as the fractional change in the output voltage from the unstrained to the strained condition.

$$\Delta\text{VR} = \left(\frac{V_{\text{meter}}}{V_b}\right)_{\text{strained}} - \left(\frac{V_{\text{meter}}}{V_b}\right)_{\text{unstrained}} \qquad 53.20$$

If the only resistance change between the strained and unstrained conditions is in the strain gauge and lead resistance is disregarded, the fractional change in gage resistance for a single strain gauge in a resistance bridge is

$$\frac{\Delta R_g}{R_g} = \frac{-4\Delta(\text{VR})}{1 + 2\Delta(\text{VR})} \quad \left[\text{1/4-bridge}\right] \qquad 53.21$$

Since the fractional change in gage resistance also occurs in the definition of the gage factor (see Eq. 53.11), the strain is

$$\epsilon = \frac{-4\Delta\text{VR}}{(\text{GF})(1 + 2\Delta\text{VR})} \quad \left[\text{1/4-bridge}\right] \qquad 53.22$$

22. BRIDGE CONSTANT

The voltage deflection can be doubled (or quadrupled) by using two (or four) strain gauges in the bridge circuit. The larger voltage deflection is more easily detected, resulting in more accurate measurements.

Use of multiple strain gauges is generally limited to configurations where symmetrical strain is available on the member. For example, a beam in bending experiences the same strain on the top and bottom faces. Therefore, if the temperature-compensation strain gauge shown in Fig. 53.8 is bonded to the bottom of the beam, the resistance change would double.

The *bridge constant* (BC) is the ratio of the actual voltage deflection to the voltage deflection from a single gauge. Depending on the number and orientation of the gauges used, bridge constants of 1.0, 1.3, 2.0, 2.6, and 4.0 may be encountered (for materials with a Poisson's ratio of 0.3).

Figure 53.9 illustrates how (up to) four strain gauges can be connected in a Wheatstone bridge circuit. The total strain indicated will be the algebraic sum of the four strains detected. For example, if all four strains are equal in magnitude, ϵ_1 and ϵ_4 are tensile (i.e., positive), and ϵ_2 and ϵ_3 are compressive (i.e., negative), then the bridge constant would be 4.

$$\epsilon_t = \epsilon_1 - \epsilon_2 - \epsilon_3 + \epsilon_4 \qquad 53.23$$

Figure 53.9 *Wheatstone Bridge Strain Gauge Circuit**

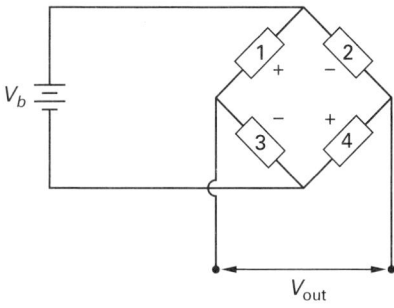

*The 45° orientations shown are figurative. Actual gauge orientation can be in any direction.

23. STRESS MEASUREMENTS IN KNOWN DIRECTIONS

Strain gauges are the most frequently used method of determining the stress in a member. Stress can be calculated from strain, or the measurement circuitry can be calibrated to give the stress directly.

For stress in only one direction (i.e., the *uniaxial stress* case), such as a simple bar in tension, only one strain gauge is required. The stress can be calculated from *Hooke's law*.

$$\sigma = E\epsilon \qquad 53.24$$

When a surface, such as that of a pressure vessel, experiences simultaneous stresses in two directions (the *biaxial stress* case), the strain in one direction affects the strain in the other direction.[25] Therefore, two strain gauges are required, even if the stress in only one direction is needed. The strains actually measured by the gauges are known as the *net strains*.

$$\epsilon_x = \frac{\sigma_x - \nu\sigma_y}{E} \qquad 53.25$$

$$\epsilon_y = \frac{\sigma_y - \nu\sigma_x}{E} \qquad 53.26$$

[25]Thin-wall pressure vessel theory shows that the circumferential (hoop) stress is twice the longitudinal stress. However, the ratio of circumferential to longitudinal strains is closer to 4:1 than to 2:1.

The stresses are determined by solving Eq. 53.25 and Eq. 53.26 simultaneously.

$$\sigma_x = \frac{E(\epsilon_x + \nu\epsilon_y)}{1 - \nu^2} \qquad 53.27$$

$$\sigma_y = \frac{E(\epsilon_y + \nu\epsilon_x)}{1 - \nu^2} \qquad 53.28$$

Figure 53.9 shows how four strain gauges can be interconnected in a bridge circuit. Figure 53.10 shows how (up to) four strain gauges would be physically oriented on a test specimen to measure different types of stress. Not all four gauges are needed in all cases. If four gauges are used, the arrangement is said to be a *full bridge*. If only one or two gauges are used, the terms *quarter-bridge* (¹/₄-bridge) and *half-bridge* (¹/₂-bridge), respectively, apply.

In the case of up to four gauges applied to detect bending strain (see Fig. 53.10(a)), the bridge constant (BC) can be 1.0 (one gauge in position 1), 2.0 (two gauges in positions 1 and 2), or 4.0 (all four gauges). The relationships between the stress, strain, and applied force are

$$\sigma = E\epsilon = \frac{E\epsilon_t}{BC} \qquad 53.29$$

$$\sigma = \frac{Mc}{I} = \frac{Mh}{2I} \qquad 53.30$$

$$I = \frac{bh^3}{12} \quad \text{[rectangular section]} \qquad 53.31$$

For axial strain (see Fig. 53.10(b)) and a material with a Poisson's ratio of 0.3, the bridge constant can be 1.0 (one gauge in position 1), 1.3 (two gauges in positions 1 and 2), 2.0 (two gauges in positions 1 and 3), or 2.6 (all four gauges).

$$\sigma = E\epsilon = \frac{E\epsilon_t}{BC} \qquad 53.32$$

$$\sigma = \frac{F}{A} \qquad 53.33$$

$$A = bh \quad \text{[rectangular section]} \qquad 53.34$$

For shear strain (see Fig. 53.10(c)) and a material with a Poisson's ratio of 0.3, the bridge constant can be 2.0 (two gauges in positions 1 and 2) or 4.0 (all four gauges). The shear strain is twice the axial strain at 45°.

$$\tau = G\gamma = 2G\epsilon = \frac{2G\epsilon_t}{BC} \qquad 53.35$$

$$\tau_{max} = \frac{FQ_{max}}{bI} \qquad 53.36$$

$$\gamma = 2\epsilon \quad \text{[at 45°]} \qquad 53.37$$

$$Q_{max} = \frac{bh^2}{8} \quad \text{[rectangular section]} \qquad 53.38$$

$$G = \frac{E}{2(1 + \nu)} \qquad 53.39$$

For torsional strain (see Fig. 53.10(d)) and a material with a Poisson's ratio of 0.3, the bridge constant can be 2.0 (two gauges in positions 1 and 2) or 4.0 (all four gauges). The shear strain is twice the axial strain at 45°.

$$\tau = G\gamma = 2G\epsilon = \frac{2G\epsilon_t}{\text{BC}} \qquad 53.40$$

$$\tau = \frac{Tr}{J} = \frac{Td}{2J} \qquad 53.41$$

$$\gamma = 2\epsilon \quad \text{[at 45°]} \qquad 53.42$$

$$J = \frac{\pi d^4}{32} \quad \text{[solid circular section]} \qquad 53.43$$

$$\phi = \frac{TL}{JG} \qquad 53.44$$

$$G = \frac{E}{2(1+\nu)} \qquad 53.45$$

Figure 53.10 Orientation of Strain Gauges

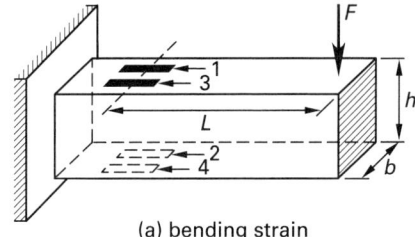

(a) bending strain

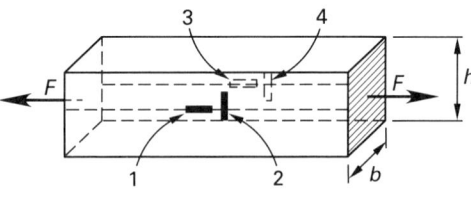

(b) axial strain

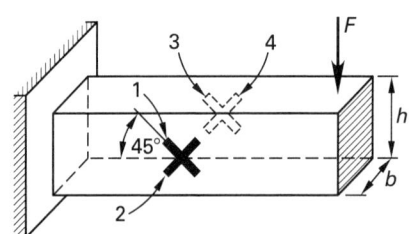

(c) shear strain

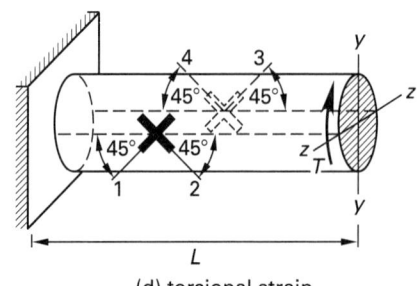

(d) torsional strain

24. STRESS MEASUREMENTS IN UNKNOWN DIRECTIONS

In order to calculate the maximum stresses (i.e., the principal stresses) on the surface shown in Fig. 53.10, the gauges would be oriented in the known directions of the principal stresses.

In most cases, however, the directions of the principal stresses are not known. Therefore, rosettes of at least three gauges are used to obtain information in a third direction. Rosettes of three gauges (*rectangular* and *equiangular (delta) rosettes*) are used for this purpose. *T-delta rosettes* include a fourth strain gauge to refine and validate the results of the three primary gauges. Table 53.6 can be used for calculating principal stresses.

25. LOAD CELLS

Load cells are used to measure forces. A load cell is a transducer that converts a tensile or compressive force into an electrical signal. Though the details of the load cell will vary with the application, the basic elements are (a) a member that is strained by the force and (b) a strain detection system (e.g., strain gauge). The force is calculated from the observed deflection, y. In Eq. 53.46, the spring constant, k, is known as the load cell's *deflection constant*.

$$F = ky \qquad 53.46$$

Because of their low cost and simple construction, *bending beam load cells* are the most common variety of load cells. Two strain gauges, one on the top and the other mounted on the bottom of a cantilever bar, are used. *Shear beam load cells* (which detect force by measuring the shear stress) can be used where the shear does not vary considerably with location, as in the web of an I-beam cross section.[26] The common S-shaped load cell constructed from a machined steel block can be instrumented as either a bending beam or shear beam load cell.

Load cell applications are categorized into grades or accuracy classes, with class III (500 to 10,000 scale divisions) being the most common. Commercial load cells meet standardized limits on errors due to temperature, nonlinearity, and hysteresis. The *temperature effect on output* (TEO) is typically stated in percentage change per 100°F (55.5°C) change in temperature.

---

[26]While shear in a rectangular beam varies parabolically with distance from the neutral axis, shear in the web of an I-beam is essentially constant at F/A. The flanges carry very little of the shear load.

Other advantages of the shear beam load cell include protection from the load and environment, high side load rejection, lower creep, faster RTZ (return to zero) after load removal, and higher tolerance of vibration, dynamic forces, and noise.

Table 53.6 Stress-Strain Relationships for Strain Gauge Rosettes[a]

type of rosette	rectangular	equiangular (delta)	T-delta
principal strains, ϵ_p, ϵ_q	$\dfrac{1}{2}\left(\epsilon_a+\epsilon_c \pm \sqrt{2(\epsilon_a-\epsilon_b)^2+2(\epsilon_b-\epsilon_c)^2}\right)$	$\dfrac{1}{3}\left(\epsilon_a+\epsilon_b+\epsilon_c \pm \sqrt{2(\epsilon_a-\epsilon_b)^2+2(\epsilon_b-\epsilon_c)^2+2(\epsilon_c-\epsilon_a)^2}\right)$	$\dfrac{1}{2}\left(\epsilon_a+\epsilon_d \pm \sqrt{(\epsilon_a-\epsilon_d)^2+\frac{4}{3}(\epsilon_b-\epsilon_c)^2}\right)$
principal stresses, σ_1, σ_2	$\dfrac{E}{2}\left(\dfrac{\epsilon_a+\epsilon_c}{1-\nu} \pm \dfrac{1}{1+\nu} \times \sqrt{2(\epsilon_a-\epsilon_b)^2+2(\epsilon_b-\epsilon_c)^2}\right)$	$\dfrac{E}{3}\left(\dfrac{\epsilon_a+\epsilon_b+\epsilon_c}{1-\nu} \pm \dfrac{1}{1+\nu} \times \sqrt{2(\epsilon_a-\epsilon_b)^2+2(\epsilon_b-\epsilon_c)^2+2(\epsilon_c-\epsilon_a)^2}\right)$	$\dfrac{E}{2}\left(\dfrac{\epsilon_a+\epsilon_d}{1-\nu} \pm \dfrac{1}{1+\nu} \times \sqrt{(\epsilon_a-\epsilon_d)^2+\frac{4}{3}(\epsilon_b-\epsilon_c)^2}\right)$
maximum shear, τ_{max}	$\dfrac{E}{2(1+\nu)} \times \sqrt{2(\epsilon_a-\epsilon_b)^2+2(\epsilon_b-\epsilon_c)^2}$	$\dfrac{E}{3(1+\nu)} \times \sqrt{2(\epsilon_a-\epsilon_b)^2+2(\epsilon_b-\epsilon_c)^2+2(\epsilon_c-\epsilon_a)^2}$	$\dfrac{E}{2(1+\nu)} \times \sqrt{(\epsilon_a-\epsilon_d)^2+\frac{4}{3}(\epsilon_b-\epsilon_c)^2}$
$\tan 2\theta$[b]	$\dfrac{2\epsilon_b-\epsilon_a-\epsilon_c}{\epsilon_a-\epsilon_c}$	$\dfrac{\sqrt{3}(\epsilon_c-\epsilon_b)}{2\epsilon_a-\epsilon_b-\epsilon_c}$	$\dfrac{2}{\sqrt{3}}\left(\dfrac{\epsilon_c-\epsilon_b}{\epsilon_a-\epsilon_d}\right)$
$0 < \theta < +90°$	$\epsilon_b > \dfrac{\epsilon_a+\epsilon_c}{2}$	$\epsilon_c > \epsilon_b$	$\epsilon_c > \epsilon_b$

[a]θ is measured in the counterclockwise direction from the a-axis of the rosette to the axis of the algebraically larger stress.
[b]θ is the angle from a-axis to axis of maximum normal stress.

Nonlinearity errors are reduced in proportion to the load cell's derating (i.e., using the load cell to measure forces less than its rated force). For example, a 2:1 derating will reduce the nonlinearity errors by 50%. Hysteresis is not normally reduced by derating.

The overall error of force measurement can be reduced by a factor of $1/\sqrt{n}$ (where n is the number of load cells that share the load equally) by using more than one load cell. Conversely, the applied force can vary by $\sqrt{n}$ times the known accuracy of a single load cell without decreasing the error.

26. DYNAMOMETERS

Torque from large motors and engines is measured by a *dynamometer. Absorption dynamometers* (e.g., the simple *friction brake, Prony brake, water brake,* and *fan brake*) dissipate energy as the torque is measured. Opposing torque in pumps and compressors must be supplied by a *driving dynamometer*, which has its own power input. *Transmission dynamometers* (e.g., *torque meters, torsion dynamometers*) use strain gauges to sense torque. They do not absorb or provide energy.

Using a brake dynamometer involves measuring a force, a moment arm, and the angular speed of rotation. The familiar torque-power-speed relationships are used with absorption dynamometers.

$$T = Fr \qquad \qquad 53.47$$

$$P_{\text{ft-lbf/min}} = 2\pi T_{\text{ft-lbf}} n_{\text{rpm}} \qquad 53.48$$

$$P_{\text{kW}} = \frac{T_{\text{N·m}} n_{\text{rpm}}}{9549} \qquad \text{[SI]} \quad 53.49(a)$$

$$P_{\text{hp}} = \frac{2\pi F_{\text{lbf}} r_{\text{ft}} n_{\text{rpm}}}{33{,}000}$$
$$= \frac{2\pi T_{\text{ft-lbf}} n_{\text{rpm}}}{33{,}000} \qquad \text{[U.S]} \quad 53.49(b)$$

Some brakes and dynamometers are constructed with a "standard" brake arm whose length is 5.252 ft. In that case, the horsepower calculation conveniently reduces to

$$P_{\text{hp}} = \frac{F_{\text{lbf}} n_{\text{rpm}}}{1000} \quad \text{[standard brake arm]} \qquad 53.50$$

If an absorption dynamometer uses a DC generator to dissipate energy, the generated voltage (V in volts) and line current (I in amps) are used to determine the power. Equation 53.49 and Eq. 53.50 are used to determine the torque.

$$P_{\text{hp}} = \frac{IV}{\eta\left(1000\,\frac{\text{W}}{\text{kW}}\right)\left(0.7457\,\frac{\text{W}}{\text{hp}}\right)} \quad \text{[absorption]} \qquad 53.51$$

Systems, Mgmt., Professional

For a driving dynamometer using a DC motor,

$$P_{\mathrm{hp}} = \frac{IV\eta}{\left(1000 \ \frac{\mathrm{W}}{\mathrm{kW}}\right)\left(0.7457 \ \frac{\mathrm{W}}{\mathrm{hp}}\right)} \quad \text{[driving]} \qquad 53.52$$

Torque can be measured directly by a *torque meter* mounted to the power shaft. Either the angle of twist, ϕ, or the shear strain, τ/G, is measured. The torque in a solid shaft of diameter, d, and length, L, is

$$T = \frac{JG\phi}{L} = \left(\frac{\pi}{32}\right)d^4\left(\frac{G\phi}{L}\right)$$
$$= \frac{\pi}{16} \ d^3\tau \quad \text{[solid round]} \qquad 53.53$$

27. INDICATOR DIAGRAMS

Indicator diagrams are plots of pressure versus volume and are encountered in the testing of reciprocating engines. In the past, indicator diagrams were actually drawn on an *indicator card* wrapped around a drum through a mechanical linkage of arms and springs. This method is mechanically complex and is not suitable for rotational speeds above 2000 rpm. Modern records of pressure and volume are produced by signals from electronic transducers recorded in real time by computers.

Analysis of indicator diagrams produced by mechanical devices requires knowing the spring constant, also known as the *spring scale*. The *mean effective pressure* (MEP) is calculated by dividing the area of the diagram by the width of the plot and then multiplying by the spring constant.

28. POSITIONAL NUMBERING SYSTEMS

A *base-b number*, N_b, is made up of individual *digits*. In a *positional numbering system*, the position of a digit in the number determines that digit's contribution to the total value of the number. Specifically, the position of the digit determines the power to which the *base* (also known as the *radix*), b, is raised. For *decimal numbers*, the radix is 10, hence the description *base-10 numbers*.

$$(a_n a_{n-1} \cdots a_2 a_1 a_0)_b = a_n b^n + a_{n-1} b^{n-1} + \cdots$$
$$+ a_2 b^2 + a_1 b + a_0 \qquad 53.54$$

The leftmost digit, a_n, contributes the greatest to the number's magnitude and is known as the *most significant digit* (MSD). The rightmost digit, a_0, contributes the least and is known as the *least significant digit* (LSD).

29. CONVERTING BASE-b NUMBERS TO BASE-10

Equation 53.54 converts base-b numbers to base-10 numbers. The calculation of the right-hand side of

Eq. 53.54 is performed in the base-10 arithmetic and is known as the *expansion method*.[27]

Converting base-b numbers (i.e., decimals) to base-10 is similar to converting whole numbers and is accomplished by Eq. 53.55.

$$(0.a_1 a_2 \cdots a_m)_b = a_1 b^{-1} + a_2 b^{-2} + \cdots$$
$$+ a_m b^{-m} \qquad 53.55$$

30. CONVERTING BASE-10 NUMBERS TO BASE-b

The *remainder method* is used to convert base-10 numbers to base-b numbers. This method consists of successive divisions by the base, b, until the quotient is zero. The base-b number is found by taking the remainders in the reverse order from which they were found. This method is illustrated in Ex. 53.6 and Ex. 53.8.

Converting a base-10 fraction to base-b requires multiplication of the base-10 fraction and subsequent fractional parts by the base. The base-b fraction is formed from the integer parts of the products taken in the same order in which they were determined. This is illustrated in Ex. 53.7(d).

31. BINARY NUMBER SYSTEM

There are only two *binary digits* (*bits*) in the *binary number system*: zero and one.[28] Thus, all binary numbers consist of strings of bits (i.e., zeros and ones). The leftmost bit is known as the *most significant bit* (MSB), and the rightmost bit is the *least significant bit* (LSB).

As with digits from other numbering systems, bits can be added, subtracted, multiplied, and divided, although only digits 0 and 1 are allowed in the results. The rules of bit addition are

$$0 + 0 = 0$$
$$0 + 1 = 1$$
$$1 + 0 = 1$$
$$1 + 1 = 0 \text{ carry } 1$$

Example 53.6

(a) Convert $(1011)_2$ to base-10. (b) Convert $(75)_{10}$ to base-2.

Solution

(a) Using Eq. 53.54 with $b = 2$,

$$(1)(2)^3 + (0)(2)^2 + (1)(2)^1 + 1 = 11$$

---

[27]Equation 53.54 works with any base number. The *double-dabble* (*double and add*) *method* is a specialized method of converting from base-2 to base-10 numbers.

[28]Alternatively, the binary states may be called *true* and *false*, *on* and *off*, *high* and *low*, or *positive* and *negative*.

(b) Use the remainder method. (See Sec. 53.30.)

$$75 \div 2 = 37 \text{ remainder } 1$$
$$37 \div 2 = 18 \text{ remainder } 1$$
$$18 \div 2 = 9 \text{ remainder } 0$$
$$9 \div 2 = 4 \text{ remainder } 1$$
$$4 \div 2 = 2 \text{ remainder } 0$$
$$2 \div 2 = 1 \text{ remainder } 0$$
$$1 \div 2 = 0 \text{ remainder } 1$$

The binary representation of $(75)_{10}$ is $(1001011)_2$.

32. OCTAL NUMBER SYSTEM

The *octal (base-8) system* is one of the alternatives to working with long binary numbers. Only the digits 0 through 7 are used. The rules for addition in the octal system are the same as for the decimal system except that the digits 8 and 9 do not exist. For example,

$$7 + 1 = 6 + 2 = 5 + 3 = (10)_8$$
$$7 + 2 = 6 + 3 = 5 + 4 = (11)_8$$
$$7 + 3 = 6 + 4 = 5 + 5 = (12)_8$$

Example 53.7

Perform the following operations.

(a) $(2)_8 + (5)_8$

(b) $(7)_8 + (6)_8$

(c) Convert $(75)_{10}$ to base-8.

(d) Convert $(0.14)_{10}$ to base-8.

(e) Convert $(13)_8$ to base-10.

(f) Convert $(27.52)_8$ to base-10.

Solution

(a) The sum of 2 and 5 in base-10 is 7, which is less than 8 and, therefore, is a valid number in the octal system. The answer is $(7)_8$.

(b) The sum of 7 and 6 in base-10 is 13, which is greater than 8 (and, therefore, needs to be converted). Using the remainder method (see Sec. 53.30),

$$13 \div 8 = 1 \text{ remainder } 5$$
$$1 \div 8 = 0 \text{ remainder } 1$$

The answer is $(15)_8$.

(c) Use the remainder method. (See Sec. 53.30.)

$$75 \div 8 = 9 \text{ remainder } 3$$
$$9 \div 8 = 1 \text{ remainder } 1$$
$$1 \div 8 = 0 \text{ remainder } 1$$

The answer is $(113)_8$.

(d) Refer to Sec. 53.30.

$$0.14 \times 8 = 1.12$$
$$0.12 \times 8 = 0.96$$
$$0.96 \times 8 = 7.68$$
$$0.68 \times 8 = 5.44$$
$$0.44 \times 8 = \text{etc.}$$

The answer, $(0.1075\ldots)_8$, is constructed from the integer parts of the products.

(e) Use Eq. 53.54.

$$(1)(8) + 3 = (11)_{10}$$

(f) Use Eq. 53.54 and Eq. 53.55.

$$(2)(8)^1 + (7)(8)^0$$
$$+ (5)(8)^{-1} + (2)(8)^{-2} = 16 + 7 + \frac{5}{8} + \frac{2}{64}$$
$$= (23.656)_{10}$$

33. HEXADECIMAL NUMBER SYSTEM

The *hexadecimal (base-16) system* is a shorthand method of representing the value of four binary digits at a time.[29] Since 16 distinctly different characters are needed, the capital letters A through F are used to represent the decimal numbers 10 through 15. The progression of hexadecimal numbers is illustrated in Table 53.7.

Table 53.7 Binary, Octal, Decimal, and Hexadecimal Equivalents

binary	octal	decimal	hexadecimal
0	0	0	0
1	1	1	1
10	2	2	2
11	3	3	3
100	4	4	4
101	5	5	5
110	6	6	6
111	7	7	7
1000	10	8	8
1001	11	9	9
1010	12	10	A
1011	13	11	B
1100	14	12	C
1101	15	13	D
1110	16	14	E
1111	17	15	F
10000	20	16	10

Example 53.8

(a) Convert $(4D3)_{16}$ to base-10. (b) Convert $(1475)_{10}$ to base-16. (c) Convert $(0.8)_{10}$ to base-16.

---

[29]The term *hex number* is often heard.

Solution

(a) The hexadecimal number D is 13 in base-10. Using Eq. 53.54,

$$(4)(16)^2 + (13)(16)^1 + 3 = (1235)_{10}$$

(b) Use the remainder method. (See Sec. 53.30.)

$$1475 \div 16 = 92 \text{ remainder } 3$$
$$92 \div 16 = 5 \text{ remainder } 12$$
$$5 \div 16 = 0 \text{ remainder } 5$$

Since $(12)_{10}$ is $(C)_{16}$, (or hex C), the answer is $(5C3)_{16}$.

(c) Refer to Sec. 53.30.

$$0.8 \times 16 = 12.8$$
$$0.8 \times 16 = 12.8$$
$$0.8 \times 16 = \text{etc.}$$

Since $(12)_{10} = (C)_{16}$, the answer is $(0.CCCCC\ldots)_{16}$.

34. CONVERSIONS AMONG BINARY, OCTAL, AND HEXADECIMAL NUMBERS

The octal system is closely related to the binary system since $(2)^3 = 8$. Conversion from a binary to an octal number is accomplished directly by starting at the LSB (right-hand bit) and grouping the bits in threes. Each group of three bits corresponds to an octal digit. Similarly, each digit in an octal number generates three bits in the equivalent binary number.

Conversion from a binary to a hexadecimal number starts by grouping the bits (starting at the LSB) into fours. Each group of four bits corresponds to a hexadecimal digit. Similarly, each digit in a hexadecimal number generates four bits in the equivalent binary number.

Conversion between octal and hexadecimal numbers is easiest when the number is first converted to a binary number.

Example 53.9

(a) Convert $(5431)_8$ to base-2. (b) Convert $(1001011)_2$ to base-8. (c) Convert $(1011111101111001)_2$ to base-16.

Solution

(a) Convert each octal digit to binary digits.

$$(5)_8 = (101)_2$$
$$(4)_8 = (100)_2$$
$$(3)_8 = (011)_2$$
$$(1)_8 = (001)_2$$

The answer is $(101100011001)_2$.

(b) Group the bits into threes starting at the LSB.

$$1 \quad 001 \quad 011$$

Convert these groups into their octal equivalents.

$$(1)_2 = (1)_8$$
$$(001)_2 = (1)_8$$
$$(011)_2 = (3)_8$$

The answer is $(113)_8$.

(c) Group the bits into fours starting at the LSB.

$$1011 \quad 1111 \quad 0111 \quad 1001$$

Convert these groups into hexadecimal equivalents.

$$(1011)_2 = (B)_{16}$$
$$(1111)_2 = (F)_{16}$$
$$(0111)_2 = (7)_{16}$$
$$(1001)_2 = (9)_{16}$$

The answer is $(BF79)_{16}$.

35. COMPLEMENT OF A NUMBER

The *complement*, N^*, of a number, N, depends on the machine (computer, calculator, etc.) being used. Assuming that the machine has a maximum number, n, of digits per integer number stored, the b's and $(b-1)$'s complements are

$$N_b^* = b^n - N \qquad 53.56$$

$$N_{b-1}^* = N_b^* - 1 \qquad 53.57$$

For a machine that works in base-10 arithmetic, the *tens* and *nines complements* are

$$N_{10}^* = 10^n - N \qquad 53.58$$

$$N_9^* = N_{10}^* - 1 \qquad 53.59$$

For a machine that works in base-2 arithmetic, the *twos* and *ones complements* are

$$N_2^* = 2^n - N \qquad 53.60$$

$$N_1^* = N_2^* - 1 \qquad 53.61$$

The binary ones complement is easily found by switching all of the ones and zeros to zeros and ones, respectively.

36. APPLICATION OF COMPLEMENTS TO COMPUTER ARITHMETIC

Equation 53.62 and Eq. 53.63 are the practical applications of complements to computer arithmetic.

$$(N^*)^* = N \qquad 53.62$$

$$M - N = M + N^* \qquad 53.63$$

The binary ones complement can be combined with a technique known as *end-around carry* to perform subtraction. End-around carry is the addition of the *overflow bit* to the sum of N and its ones complement.

Example 53.10

(a) Simulate the operation of a base-10 machine with a capacity of four digits per number and calculate the difference $(18)_{10} - (6)_{10}$ with tens complements.

(b) Simulate the operation of a base-2 machine with a capacity of five digits per number and calculate the difference $(01101)_2 - (01010)_2$ with twos complements.

(c) Solve part (b) with a ones complement and end-around carry.

Solution

(a) The tens complement of 6 is

$$(6)_{10}^* = (10)^4 - 6 = 10,000 - 6 = 9994$$

Using Eq. 53.63,

$$18 - 6 = 18 + (6)_{10}^* = 18 + 9994 = 10,012$$

However, the machine has a maximum capacity of four digits. Therefore, the leading 1 is dropped, leaving 0012 as the answer.

(b) The twos complement of $(01010)_2$ is

$$\begin{aligned} N_2^* &= (2)^5 - N = (32)_{10} - N \\ &= (100000)_2 - (01010)_2 \\ &= (10110)_2 \end{aligned}$$

From Eq. 53.63,

$$\begin{aligned} (01101)_2 - (01010)_2 &= (01101)_2 + (10110)_2 \\ &= (100011)_2 \end{aligned}$$

Since the machine has a capacity of only five bits, the leftmost bit is dropped, leaving $(00011)_2$ as the difference.

(c) The ones complement is found by reversing all the digits.

$$(01010)_1^* = (10101)_2$$

Adding the ones complement,

$$(01101)_2 + (10101)_2 = (100010)_2$$

The leading bit is the overflow bit, which is removed and added to give the difference.

$$(00010)_2 + (1)_2 = (00011)_2$$

37. COMPUTER REPRESENTATION OF NEGATIVE NUMBERS

On paper, a minus sign indicates a negative number. This representation is not possible in a machine. Hence, one of the n digits, usually the MSB, is reserved for sign representation. (This reduces the machine's capacity to represent numbers to $n - 1$ bits per number.) It is arbitrary whether the sign bit is 1 or 0 for negative numbers as long as the MSB is different for positive and negative numbers.

The ones complement is ideal for forming a negative number since it automatically reverses the MSB. For example, $(00011)_2$ is a five-bit representation of decimal 3. The ones complement is $(11100)_2$, which is recognized as a negative number because the MSB is 1.

Example 53.11

Simulate the operation of a six-digit binary machine that uses ones complements for negative numbers.

(a) What is the machine representation of $(-27)_{10}$? (b) What is the decimal equivalent of the twos complement of $(-27)_{10}$? (c) What is the decimal equivalent of the ones complement of $(-27)_{10}$?

Solution

(a) $(27)_{10} = (011011)_2$. The negative of this number is the same as the ones complement: $(100100)_2$.

(b) The twos complement is one more than the ones complement. (See Eq. 53.60 and Eq. 53.61.) Therefore, the twos complement is

$$(100100)_2 + 1 = (100101)_2$$

This represents $(-26)_{10}$.

(c) From Eq. 53.62, the complement of a complement of a number is the original number. Therefore, the decimal equivalent is -27.

54 Engineering Economic Analysis

Nomenclature

A	annual amount	\$
B	present worth of all benefits	\$
BV_j	book value at end of the jth year	\$
C	cost or present worth of all costs	\$
d	declining balance depreciation rate	decimal
D	demand	various
D	depreciation	\$
DR	present worth of after-tax depreciation recovery	\$
e	constant inflation rate	decimal
$\mathcal{E}$	expected value	various
E_0	initial amount of an exponentially growing cash flow	\$
EAA	equivalent annual amount	\$
$EUAC$	equivalent uniform annual cost	\$
f	federal income tax rate	decimal
F	forecasted quantity	various
F	future worth	\$
g	exponential growth rate	decimal
G	uniform gradient amount	\$
i	effective interest rate	decimal
$i\%$	effective interest rate	%
i'	effective interest rate corrected for inflation	decimal
j	number of years	—

Systems, Mgmt., Professional

k	number of compounding periods per year	–
m	an integer	–
n	number of compounding periods or years in life of asset	–
p	probability	decimal
P	present worth	$
r	nominal rate per year (rate per annum)	decimal per unit time
ROI	return on investment	$
ROR	rate of return	decimal per unit time
s	state income tax rate	decimal
S_n	expected salvage value in year n	$
t	composite tax rate	decimal
t	time	years (typical)
T	a quantity equal to $\frac{1}{2}n(n+1)$	–
TC	tax credit	$
z	a quantity equal to $\dfrac{1+i}{1-d}$	decimal

Symbols

| α | smoothing coefficient for forecasts | – |
| ϕ | effective rate per period | decimal |

Subscripts

0	initial
j	at time j
n	at time n
t	at time t

1. INTRODUCTION

In its simplest form, an *engineering economic analysis* is a study of the desirability of making an investment.[1] The decision-making principles in this chapter can be applied by individuals as well as by companies. The nature of the spending opportunity or industry is not important. Farming equipment, personal investments, and multimillion dollar factory improvements can all be evaluated using the same principles.

Similarly, the applicable principles are insensitive to the monetary units. Although *dollars* are used in this chapter, it is equally convenient to use pounds, yen, or euros.

Finally, this chapter may give the impression that investment alternatives must be evaluated on a year-by-year basis. Actually, the *effective period* can be defined as a day, month, century, or any other convenient period of time.

2. MULTIPLICITY OF SOLUTION METHODS

Most economic conclusions can be reached in more than one manner. There are usually several different

analyses that will eventually result in identical answers.[2] Other than the pursuit of elegant solutions in a timely manner, there is no reason to favor one procedural method over another.[3]

3. PRECISION AND SIGNIFICANT DIGITS

The full potential of electronic calculators will never be realized in engineering economic analyses. Considering that calculations are based on estimates of far-future cash flows and that unrealistic assumptions (no inflation, identical cost structures of replacement assets, etc.) are routinely made, it makes little sense to carry cents along in calculations.

The calculations in this chapter have been designed to illustrate and review the principles presented. Because of this, greater precision than is normally necessary in everyday problems may be used. Though used, such precision is not warranted.

Unless there is some compelling reason to strive for greater precision, the following rules are presented for use in reporting final answers to engineering economic analysis problems.

- Omit fractional parts of the dollar (i.e., cents).

- Report and record a number to a maximum of four significant digits unless the first digit of that number is 1, in which case, a maximum of five significant digits should be written. For example,

$49	not	$49.43
$93,450	not	$93,453
$1,289,700	not	$1,289,673

4. NONQUANTIFIABLE FACTORS

An engineering economic analysis is a quantitative analysis. Some factors cannot be introduced as numbers into the calculations. Such factors are known as *nonquantitative factors*, *judgment factors*, and *irreducible factors*. Typical nonquantifiable factors are

- preferences
- political ramifications
- urgency
- goodwill
- prestige
- utility
- corporate strategy
- environmental effects

---

[1]This subject is also known as *engineering economics* and *engineering economy*. There is very little, if any, true economics in this subject.

[2]Because of round-off errors, particularly when factors are taken from tables, these different calculations will produce slightly different numerical results (e.g., $49.49 versus $49.50). However, this type of divergence is well known and accepted in engineering economic analysis.

[3]This does not imply that approximate methods, simplifications, and rules of thumb are acceptable.

- health and safety rules
- reliability
- political risks

Since these factors are not included in the calculations, the policy is to disregard the issues entirely. Of course, the factors should be discussed in a final report. The factors are particularly useful in breaking ties between competing alternatives that are economically equivalent.

5. YEAR-END AND OTHER CONVENTIONS

Except in short-term transactions, it is simpler to assume that all receipts and disbursements (cash flows) take place at the end of the year in which they occur.[4] This is known as the *year-end convention*. The exceptions to the year-end convention are initial project cost (purchase cost), trade-in allowance, and other cash flows that are associated with the inception of the project at $t = 0$.

On the surface, such a convention appears grossly inappropriate since repair expenses, interest payments, corporate taxes, and so on seldom coincide with the end of a year. However, the convention greatly simplifies engineering economic analysis problems, and it is justifiable on the basis that the increased precision associated with a more rigorous analysis is not warranted (due to the numerous other simplifying assumptions and estimates initially made in the problem).

There are various established procedures, known as *rules* or *conventions*, imposed by the Internal Revenue Service on U.S. taxpayers. An example is the *half-year rule*, which permits only half of the first-year depreciation to be taken in the first year of an asset's life when certain methods of depreciation are used. These rules are subject to constantly changing legislation and are not covered in this book. The implementation of such rules is outside the scope of engineering practice and is best left to accounting professionals.

6. CASH FLOW DIAGRAMS

Although they are not always necessary in simple problems (and they are often unwieldy in very complex problems), *cash flow diagrams* can be drawn to help visualize and simplify problems having diverse receipts and disbursements.

The following conventions are used to standardize cash flow diagrams.

- The horizontal (time) axis is marked off in equal increments, one per period, up to the duration (or *horizon*) of the project.

- Two or more transfers in the same period are placed end to end, and these may be combined.

- Expenses incurred before $t = 0$ are called *sunk costs*. Sunk costs are not relevant to the problem unless they have tax consequences in an after-tax analysis.

- *Receipts* are represented by arrows directed upward. *Disbursements* are represented by arrows directed downward. The arrow length is proportional to the magnitude of the cash flow.

Example 54.1

A mechanical device will cost $20,000 when purchased. Maintenance will cost $1000 each year. The device will generate revenues of $5000 each year for five years, after which the salvage value is expected to be $7000. Draw and simplify the cash flow diagram.

Solution

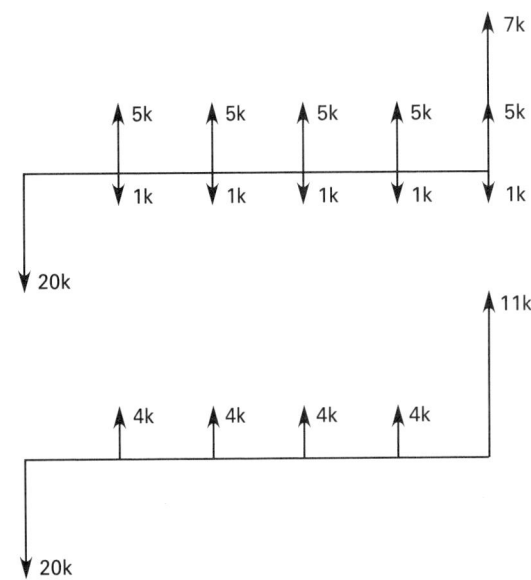

7. TYPES OF CASH FLOWS

To evaluate a real-world project, it is necessary to present the project's cash flows in terms of standard cash flows that can be handled by engineering economic analysis techniques. The standard cash flows are single payment cash flow, uniform series cash flow, gradient series cash flow, and the infrequently encountered exponential gradient series cash flow. (See Fig. 54.1.)

A *single payment cash flow* can occur at the beginning of the time line (designated as $t = 0$), at the end of the time line (designated as $t = n$), or at any time in between.

The *uniform series cash flow* consists of a series of equal transactions starting at $t = 1$ and ending at $t = n$.

---

[4]A *short-term transaction* typically has a lifetime of five years or less and has payments or compounding that are more frequent than once per year.

Figure 54.1 *Standard Cash Flows*

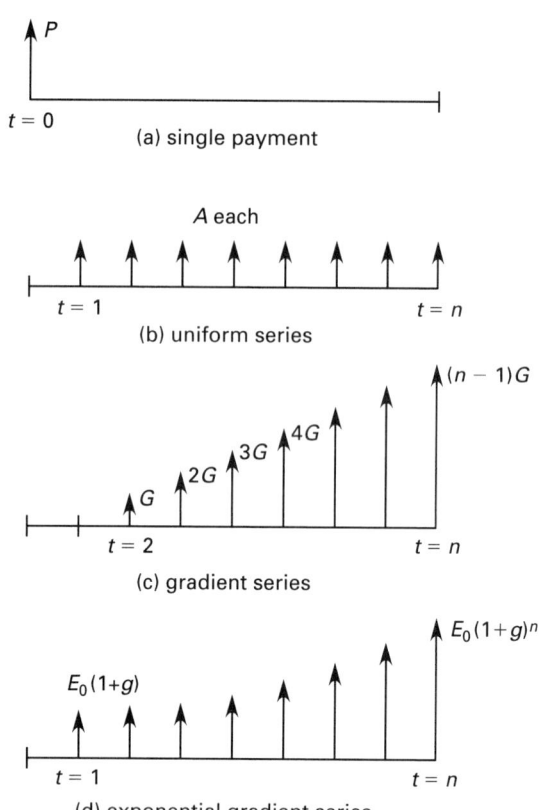

(a) single payment

(b) uniform series

(c) gradient series

(d) exponential gradient series

The symbol A is typically given to the magnitude of each individual cash flow.[5]

The *gradient series cash flow* starts with a cash flow (typically given the symbol G) at $t = 2$ and increases by G each year until $t = n$, at which time the final cash flow is $(n - 1)G$.

An *exponential gradient series cash flow* is based on a phantom value (typically given the symbol E_0) at $t = 0$ and grows or decays exponentially according to the following relationship.[6]

$$\text{amount at time } t = E_t = E_0(1 + g)^t$$
$$[t = 1, 2, 3, \ldots, n] \qquad 54.1$$

In Eq. 54.1, g is the *exponential growth rate*, which can be either positive or negative. Exponential gradient cash

flows are rarely seen in economic justification projects assigned to engineers.[7]

8. TYPICAL PROBLEM TYPES

There is a wide variety of problem types that, collectively, are considered to be engineering economic analysis problems.

By far, the majority of engineering economic analysis problems are *alternative comparisons*. In these problems, two or more mutually exclusive investments compete for limited funds. A variation of this is a *replacement/retirement analysis*, which is repeated each year to determine if an existing asset should be replaced. Finding the percentage return on an investment is a *rate of return problem*, one of the alternative comparison solution methods.

Investigating interest and principal amounts in loan payments is a *loan repayment problem*. An *economic life analysis* will determine when an asset should be retired. In addition, there are miscellaneous problems involving economic order quantity, learning curves, break-even points, product costs, and so on.

9. IMPLICIT ASSUMPTIONS

Several assumptions are implicitly made when solving engineering economic analysis problems. Some of these assumptions are made with the knowledge that they are or will be poor approximations of what really will happen. The assumptions are made, regardless, for the benefit of obtaining a solution.

The most common assumptions are the following.

- The year-end convention is applicable.

- There is no inflation now, nor will there be any during the lifetime of the project.

- Unless otherwise specifically called for, a before-tax analysis is needed.

- The effective interest rate used in the problem will be constant during the lifetime of the project.

- Nonquantifiable factors can be disregarded.

- Funds invested in a project are available and are not urgently needed elsewhere.

- Excess funds continue to earn interest at the effective rate used in the analysis.

This last assumption, like most of the assumptions listed, is almost never specifically mentioned in the body of a solution. However, it is a key assumption when comparing two alternatives that have different initial costs.

---

[5]The cash flows do not begin at $t = 0$. This is an important concept with all of the series cash flows. This convention has been established to accommodate the timing of annual maintenance (and similar) cash flows for which the year-end convention is applicable.

[6]By convention, for an exponential cash flow series: The first cash flow, E_0, is at $t = 1$, as in the uniform annual series. However, the first cash flow is $E_0(1 + g)$. The cash flow of E_0 at $t = 0$ is absent (i.e., is a *phantom cash flow*).

[7]For one of the few discussions on exponential cash flow, see *Capital Budgeting*, Robert V. Oakford, The Ronald Press Company, New York, 1970.

For example, suppose two investments, one costing $10,000 and the other costing $8000, are to be compared at 10%. It is obvious that $10,000 in funds is available, otherwise the costlier investment would not be under consideration. If the smaller investment is chosen, what is done with the remaining $2000? The last assumption yields the answer: The $2000 is "put to work" in some investment earning (in this case) 10%.

10. EQUIVALENCE

Industrial decision makers using engineering economic analysis are concerned with the magnitude and timing of a project's cash flow as well as with the total profitability of that project. In this situation, a method is required to compare projects involving receipts and disbursements occurring at different times.

By way of illustration, consider $100 placed in a bank account that pays 5% effective annual interest at the end of each year. After the first year, the account will have grown to $105. After the second year, the account will have grown to $110.25.

Assume that you will have no need for money during the next two years, and any money received will immediately go into your 5% bank account. Then, which of the following options would be more desirable?

option A: $100 now

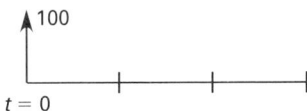

option B: $105 to be delivered in one year

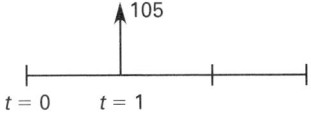

option C: $110.25 to be delivered in two years

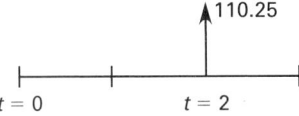

As illustrated, none of the options is superior under the assumptions given. If the first option is chosen, you will immediately place $100 into a 5% account, and in two years the account will have grown to $110.25. In fact, the account will contain $110.25 at the end of two years regardless of the option chosen. Therefore, these alternatives are said to be *equivalent*.

Equivalence may or may not be the case, depending on the interest rate, so an alternative that is acceptable to one decision maker may be unacceptable to another.

The interest rate that is used in actual calculations is known as the *effective interest rate*.[8] If compounding is once a year, it is known as the *effective annual interest rate*. However, effective quarterly, monthly, daily, and so on, interest rates are also used.

The fact that $100 today grows to $105 in one year (at 5% annual interest) is an example of what is known as the *time value of money* principle. This principle simply articulates what is obvious: Funds placed in a secure investment will increase to an equivalent future amount. The procedure for determining the present investment from the equivalent future amount is known as *discounting*.

11. SINGLE-PAYMENT EQUIVALENCE

The equivalence of any present amount, P, at $t = 0$, to any future amount, F, at $t = n$, is called the *future worth* and can be calculated from Eq. 54.2.

$$F = P(1 + i)^n \qquad 54.2$$

The factor $(1 + i)^n$ is known as the single payment *compound amount factor* and has been tabulated in App. 54.B for various combinations of i and n.

Similarly, the equivalence of any future amount to any present amount is called the *present worth* and can be calculated from Eq. 54.3.

$$P = F(1 + i)^{-n} = \frac{F}{(1 + i)^n} \qquad 54.3$$

The factor $(1 + i)^{-n}$ is known as the *single payment present worth factor*.[9]

The interest rate used in Eq. 54.2 and Eq. 54.3 must be the effective rate per period. Also, the basis of the rate (annually, monthly, etc.) must agree with the type of period used to count n. Therefore, it would be incorrect to use an effective annual interest rate if n was the number of compounding periods in months.

Example 54.2

How much should you put into a 10% (effective annual rate) savings account in order to have $10,000 in five years?

---

[8]The adjective *effective* distinguishes this interest rate from other interest rates (e.g., nominal interest rates) that are not meant to be used directly in calculating equivalent amounts.
[9]The *present worth* is also called the *present value* and *net present value*. These terms are used interchangeably and no significance should be attached to the terms *value*, *worth*, and *net*.

Solution

This problem could also be stated: What is the equivalent present worth of $10,000 five years from now if money is worth 10% per year?

$$P = F(1 + i)^{-n} = (\$10{,}000)(1 + 0.10)^{-5}$$

$$= \$6209$$

The factor 0.6209 would usually be obtained from the tables.

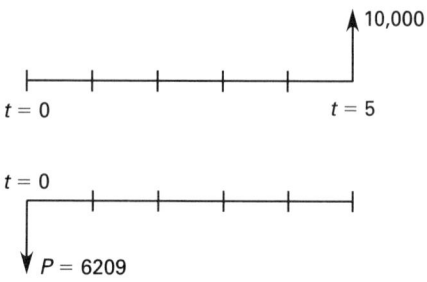

12. STANDARD CASH FLOW FACTORS AND SYMBOLS

Equation 54.2 and Eq. 54.3 may give the impression that solving engineering economic analysis problems involves a lot of calculator use, and, in particular, a lot of exponentiation. Such calculations may be necessary from time to time, but most problems are simplified by the use of tabulated values of the factors.

Rather than actually writing the formula for the compound amount factor (which converts a present amount to a future amount), it is common convention to substitute the standard functional notation of $(F/P, i\%, n)$. Therefore, the future value in n periods of a present amount would be symbolically written as

$$F = P(F/P, i\%, n) \qquad 54.4$$

Similarly, the present worth factor has a functional notation of $(P/F, i\%, n)$. The present worth of a future amount n periods hence would be symbolically written as

$$P = F(P/F, i\%, n) \qquad 54.5$$

Values of these *cash flow (discounting) factors* are tabulated in App. 54.B. There is often initial confusion about whether the (F/P) or (P/F) column should be used in a particular problem. There are several ways of remembering what the functional notations mean.

One method of remembering which factor should be used is to think of the factors as conditional probabilities. The conditional probability of event **A** given that event **B** has occurred is written as $p\{\mathbf{A}|\mathbf{B}\}$, where the given event comes after the vertical bar. In the standard notational form of discounting factors, the given amount is similarly placed after the slash. What you

want comes before the slash. (F/P) would be a factor to find F given P.

Another method of remembering the notation is to interpret the factors algebraically. The (F/P) factor could be thought of as the fraction F/P. Algebraically, Eq. 54.4 would be

$$F = P\left(\frac{F}{P}\right) \qquad 54.6$$

This algebraic approach is actually more than an interpretation. The numerical values of the discounting factors are consistent with this algebraic manipulation. The (F/A) factor could be calculated as $(F/P) \times (P/A)$. This consistent relationship can be used to calculate other factors that might be occasionally needed, such as (F/G) or (G/P). For instance, the annual cash flow that would be equivalent to a uniform gradient may be found from

$$A = G(P/G, i\%, n)(A/P, i\%, n) \qquad 54.7$$

Formulas for the compounding and discounting factors are contained in Table 54.1. Normally, it will not be necessary to calculate factors from the formulas. For solving most problems, App. 54.B is adequate.

Example 54.3

What factor will convert a gradient cash flow ending at $t = 8$ to a future value at $t = 8$? (That is, what is the $(F/G, i\%, n)$ factor?) The effective annual interest rate is 10%.

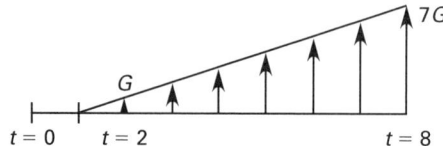

Solution

method 1:

From Table 54.1, the $(F/G, i\%, n)$ factor is

$$(F/G, 10\%, 8) = \frac{(1 + i)^n - 1}{i^2} - \frac{n}{i}$$

$$= \frac{(1 + 0.10)^8 - 1}{(0.10)^2} - \frac{8}{0.10}$$

$$= 34.3589$$

method 2:

The tabulated values of (P/G) and (F/P) in App. 54.B can be used to calculate the factor.

$$(F/G, 10\%, 8) = (P/G, 10\%, 8)(F/P, 10\%, 8)$$

$$= (16.0287)(2.1436)$$

$$= 34.3591$$

Table 54.1 *Discount Factors for Discrete Compounding*

factor name	converts	symbol	formula
single payment compound amount	P to F	$(F/P, i\%, n)$	$(1 + i)^n$
single payment present worth	F to P	$(P/F, i\%, n)$	$(1 + i)^{-n}$
uniform series sinking fund	F to A	$(A/F, i\%, n)$	$\dfrac{i}{(1 + i)^n - 1}$
capital recovery	P to A	$(A/P, i\%, n)$	$\dfrac{i(1 + i)^n}{(1 + i)^n - 1}$
uniform series compound amount	A to F	$(F/A, i\%, n)$	$\dfrac{(1 + i)^n - 1}{i}$
uniform series present worth	A to P	$(P/A, i\%, n)$	$\dfrac{(1 + i)^n - 1}{i(1 + i)^n}$
uniform gradient present worth	G to P	$(P/G, i\%, n)$	$\dfrac{(1 + i)^n - 1}{i^2(1 + i)^n} - \dfrac{n}{i(1 + i)^n}$
uniform gradient future worth	G to F	$(F/G, i\%, n)$	$\dfrac{(1 + i)^n - 1}{i^2} - \dfrac{n}{i}$
uniform gradient uniform series	G to A	$(A/G, i\%, n)$	$\dfrac{1}{i} - \dfrac{n}{(1 + i)^n - 1}$

The (F/G) factor could also have been calculated as the product of the (A/G) and (F/A) factors.

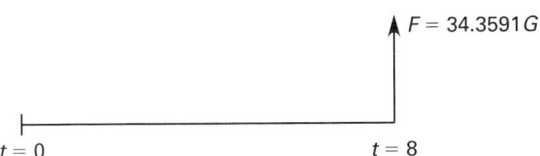

13. CALCULATING UNIFORM SERIES EQUIVALENCE

A cash flow that repeats each year for n years without change in amount is known as an *annual amount* and is given the symbol A. As an example, a piece of equipment may require annual maintenance, and the maintenance cost will be an annual amount. Although the equivalent value for each of the n annual amounts could be calculated and then summed, it is more expedient to use one of the uniform series factors. For example, it is possible to convert from an annual amount to a future amount by use of the (F/A) factor.

$$F = A(F/A, i\%, n) \qquad 54.8$$

A *sinking fund* is a fund or account into which annual deposits of A are made in order to accumulate F at $t = n$ in the future. Since the annual deposit is calculated as $A = F(A/F, i\%, n)$, the (A/F) factor is known as the *sinking fund factor*. An *annuity* is a series of equal payments, A, made over a period of time.[10] Usually, it is necessary to "buy into" an investment (a bond, an insurance policy, etc.) in order to ensure the annuity. In the simplest case of an annuity that starts at the end of the first year and continues for n years, the purchase price, P, is

$$P = A(P/A, i\%, n) \qquad 54.9$$

The present worth of an *infinite (perpetual) series* of annual amounts is known as a *capitalized cost*. There is no $(P/A, i\%, \infty)$ factor in the tables, but the capitalized cost can be calculated simply as

$$P = \frac{A}{i} \quad [i \text{ in decimal form}] \qquad 54.10$$

Alternatives with different lives will generally be compared by way of *equivalent uniform annual cost* (EUAC). An EUAC is the annual amount that is equivalent to all of the cash flows in the alternative. The EUAC differs in sign from all of the other cash flows. Costs and expenses expressed as EUACs, which

---

[10]An annuity may also consist of a lump sum payment made at some future time. However, this interpretation is not considered in this chapter.

Systems, Mgmt., Professional

would normally be considered negative, are actually positive. The term *cost* in the designation EUAC serves to make clear the meaning of a positive number.

Example 54.4

Maintenance costs for a machine are \$250 each year. What is the present worth of these maintenance costs over a 12-year period if the interest rate is 8%?

Solution

Using Eq. 54.9,

$$P = A(P/A, 8\%, 12) = (-\$250)(7.5361)$$
$$= -\$1884$$

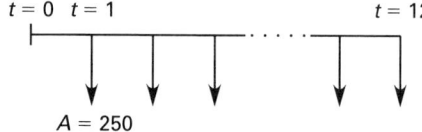

$$A = 250$$

$$P = 1884$$

14. FINDING PAST VALUES

From time to time, it will be necessary to determine an amount in the past equivalent to some current (or future) amount. For example, you might have to calculate the original investment made 15 years ago given a current annuity payment.

Such problems are solved by placing the $t = 0$ point at the time of the original investment, and then calculating the past amount as a P value. For example, the original investment, P, can be extracted from the annuity, A, by using the standard cash flow factors.

$$P = A(P/A, i\%, n) \qquad 54.11$$

The choice of $t = 0$ is flexible. As a general rule, the $t = 0$ point should be selected for convenience in solving a problem.

Example 54.5

You currently pay \$250 per month to lease your office phone equipment. You have three years (36 months) left on the five-year (60-month) lease. What would have been an equivalent purchase price two years ago? The effective interest rate per month is 1%.

Solution

The solution of this example is not affected by the fact that investigation is being performed in the middle of the horizon. This is a simple calculation of present worth.

$$P = A(P/A, 1\%, 60)$$
$$= (-\$250)(44.9550)$$
$$= -\$11,239$$

15. TIMES TO DOUBLE AND TRIPLE AN INVESTMENT

If an investment doubles in value (in n compounding periods and with $i\%$ effective interest), the ratio of current value to past investment will be 2.

$$F/P = (1+i)^n = 2 \qquad 54.12$$

Similarly, the ratio of current value to past investment will be 3 if an investment triples in value. This can be written as

$$F/P = (1+i)^n = 3 \qquad 54.13$$

It is a simple matter to extract the number of periods, n, from Eq. 54.12 and Eq. 54.13 to determine the *doubling time* and *tripling time*, respectively. For example, the doubling time is

$$n = \frac{\log 2}{\log(1+i)} \qquad 54.14$$

When a quick estimate of the doubling time is needed, the *rule of 72* can be used. The doubling time is approximately $72/i$.

The tripling time is

$$n = \frac{\log 3}{\log(1+i)} \qquad 54.15$$

Equation 54.14 and Eq. 54.15 form the basis of Table 54.2.

16. VARIED AND NONSTANDARD CASH FLOWS

Gradient Cash Flow

A common situation involves a uniformly increasing cash flow. If the cash flow has the proper form, its present worth can be determined by using the *uniform gradient factor*, $(P/G, i\%, n)$. The uniform gradient factor finds the present worth of a uniformly increasing cash flow that starts in year two (not in year one).

There are three common difficulties associated with the form of the uniform gradient. The first difficulty is that the initial cash flow occurs at $t = 2$. This convention recognizes that annual costs, if they increase uniformly, begin with some value at $t = 1$ (due to the year-end convention) but do not begin to increase until $t = 2$.

Table 54.2 Doubling and Tripling Times for Various Interest Rates

interest rate (%)	doubling time (periods)	tripling time (periods)
1	69.7	110.4
2	35.0	55.5
3	23.4	37.2
4	17.7	28.0
5	14.2	22.5
6	11.9	18.9
7	10.2	16.2
8	9.01	14.3
9	8.04	12.7
10	7.27	11.5
11	6.64	10.5
12	6.12	9.69
13	5.67	8.99
14	5.29	8.38
15	4.96	7.86
16	4.67	7.40
17	4.41	7.00
18	4.19	6.64
19	3.98	6.32
20	3.80	6.03

The tabulated values of (P/G) have been calculated to find the present worth of only the increasing part of the annual expense. The present worth of the base expense incurred at $t = 1$ must be found separately with the (P/A) factor.

The second difficulty is that, even though the factor $(P/G, i\%, n)$ is used, there are only $n - 1$ actual cash flows. It is clear that n must be interpreted as the *period number* in which the last gradient cash flow occurs, not the number of gradient cash flows.

The sign convention used with gradient cash flows may seem confusing. If an expense increases each year (as in Ex. 54.6), the gradient will be negative, since it is an expense. If a revenue increases each year, the gradient will be positive. In most cases, the sign of the gradient depends on whether the cash flow is an expense or a revenue.[11] Figure 54.2 illustrates positive and negative gradient cash flows.

Example 54.6

Maintenance on an old machine is $100 this year but is expected to increase by $25 each year thereafter. What is the present worth of five years of the costs of maintenance? Use an interest rate of 10%.

Figure 54.2 Positive and Negative Gradient Cash Flows

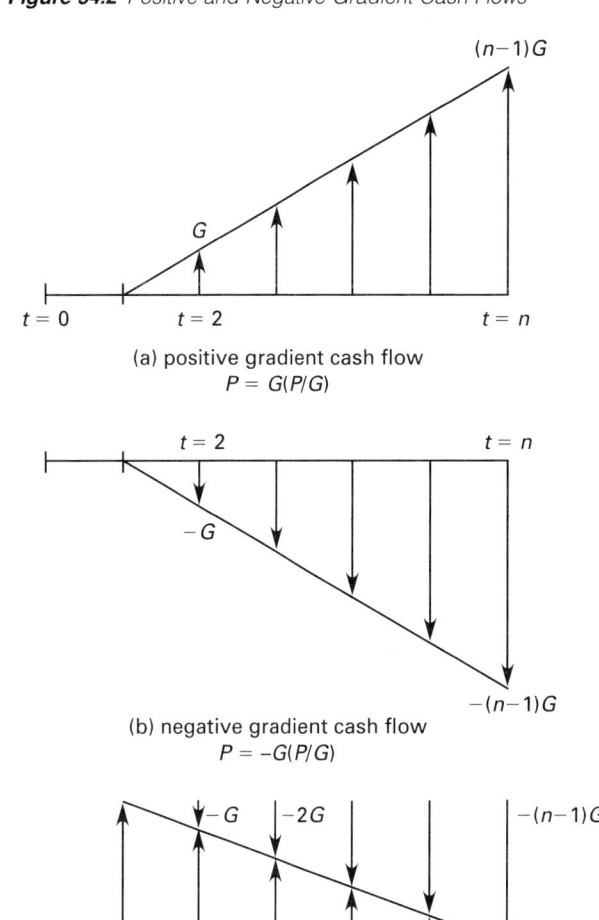

(a) positive gradient cash flow
$P = G(P/G)$

(b) negative gradient cash flow
$P = -G(P/G)$

(c) decreasing revenue incorporating a negative gradient
$P = A(P/A) - G(P/G)$

Solution

In this problem, the cash flow must be broken down into parts. (The five-year gradient factor is used even though there are only four nonzero gradient cash flows.)

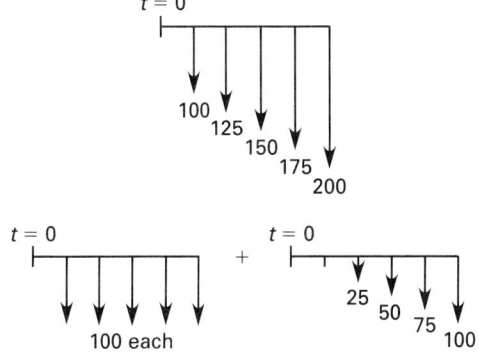

---

[11]This is not a universal rule. It is possible to have a uniformly decreasing revenue as in Fig. 54.2(c). In this case, the gradient would be negative.

Therefore, the present worth is

$$P = A(P/A, 10\%, 5) + G(P/G, 10\%, 5)$$
$$= (-\$100)(3.7908) - (\$25)(6.8618)$$
$$= -\$551$$

Stepped Cash Flows

Stepped cash flows are easily handled by the technique of *superposition of cash flows*. This technique is illustrated by Ex. 54.7.

Example 54.7

An investment costing $1000 returns $100 for the first five years and $200 for the following five years. How would the present worth of this investment be calculated?

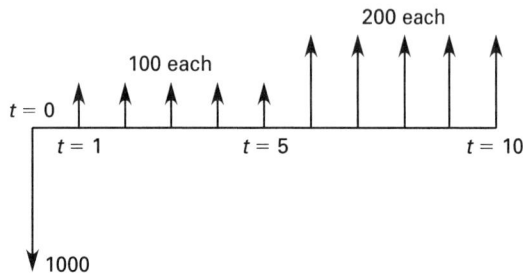

Solution

Using the principle of superposition, the revenue cash flow can be thought of as $200 each year from $t = 1$ to $t = 10$, with a negative revenue of $100 from $t = 1$ to $t = 5$. Superimposed, these two cash flows make up the actual performance cash flow.

$$P = -\$1000 + (\$200)(P/A, i\%, 10) - (\$100)(P/A, i\%, 5)$$

Missing and Extra Parts of Standard Cash Flows

A missing or extra part of a standard cash flow can also be handled by superposition. For example, suppose an annual expense is incurred each year for ten years, except in the ninth year, as is illustrated in Fig. 54.3. The present worth could be calculated as a subtractive process.

$$P = A(P/A, i\%, 10) - A(P/F, i\%, 9) \qquad 54.16$$

Figure 54.3 *Cash Flow with a Missing Part*

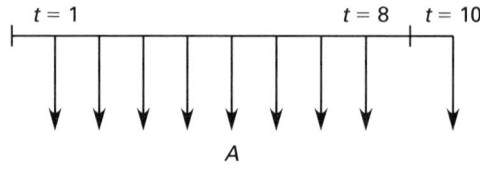

Alternatively, the present worth could be calculated as an additive process.

$$P = A(P/A, i\%, 8) + A(P/F, i\%, 10) \qquad 54.17$$

Delayed and Premature Cash Flows

There are cases when a cash flow matches a standard cash flow exactly, except that the cash flow is delayed or starts sooner than it should. Often, such cash flows can be handled with superposition. At other times, it may be more convenient to shift the time axis. This shift is known as the *projection method*. Example 54.8 illustrates the projection method.

Example 54.8

An expense of $75 is incurred starting at $t = 3$ and continues until $t = 9$. There are no expenses or receipts until $t = 3$. Use the projection method to determine the present worth of this stream of expenses.

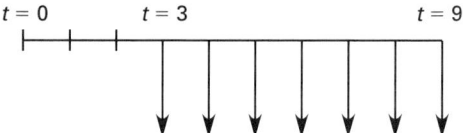

Solution

Determine a cash flow at $t = 2$ that is equivalent to the entire expense stream. If $t = 0$ was where $t = 2$ actually is, the present worth of the expense stream would be

$$P' = (-\$75)(P/A, i\%, 7)$$

P' is a cash flow at $t = 2$. It is now simple to find the present worth (at $t = 0$) of this future amount.

$$P = P'(P/F, i\%, 2) = (-\$75)(P/A, i\%, 7)(P/F, i\%, 2)$$

Cash Flows at Beginnings of Years: The Christmas Club Problem

This type of problem is characterized by a stream of equal payments (or expenses) starting at $t = 0$ and ending at $t = n - 1$, as shown in Fig. 54.4. It differs from the standard annual cash flow in the existence of a cash flow at $t = 0$ and the absence of a cash flow at $t = n$. This problem gets its name from the service provided by some savings institutions whereby money is automatically deposited each week or month (starting immediately, when the savings plan is opened) in order to accumulate money to purchase Christmas presents at the end of the year.

It may seem that the present worth of the savings stream can be determined by directly applying the (P/A) factor. However, this is not the case, since the

Figure 54.4 *Cash Flows at Beginnings of Years*

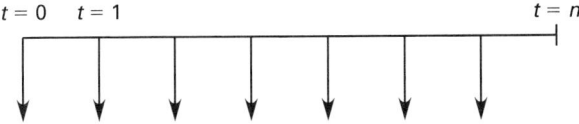

Christmas Club cash flow and the standard annual cash flow differ. The Christmas Club problem is easily handled by superposition, as illustrated by Ex. 54.9.

Example 54.9

How much can you expect to accumulate by $t = 10$ for a child's college education if you deposit \$300 at the beginning of each year for a total of ten payments?

Solution

The first payment is made at $t=0$, and there is no payment at $t=10$. The future worth of the first payment is calculated with the (F/P) factor. The absence of the payment at $t=10$ is handled by superposition. This "correction" is not multiplied by a factor.

$$F = (\$300)(F/P, i\%, 10) + (\$300)(F/A, i\%, 10) - \$300$$
$$= (\$300)(F/A, i\%, 11) - \$300$$

17. THE MEANING OF PRESENT WORTH AND *i*

If \$100 is invested in a 5% bank account (using annual compounding), you can remove \$105 one year from now; if this investment is made, you will receive a *return on investment* (ROI) of \$5. The cash flow diagram, as shown in Fig. 54.5, and the present worth of the two transactions are as follows.

$$P = -\$100 + (\$105)(P/F, 5\%, 1)$$
$$= -\$100 + (\$105)(0.9524)$$
$$= 0$$

Figure 54.5 *Cash Flow Diagram*

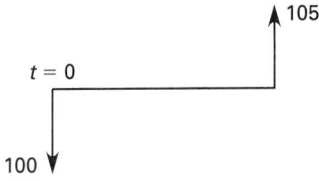

The present worth is zero even though you will receive a \$5 return on your investment.

However, if you are offered \$120 for the use of \$100 over a one-year period, the cash flow diagram, as shown in Fig. 54.6, and present worth (at 5%) would be

$$P = -\$100 + (\$120)(P/F, 5\%, 1)$$
$$= -\$100 + (\$120)(0.9524)$$
$$= \$14.29$$

Figure 54.6 *Cash Flow Diagram*

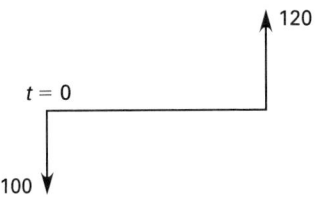

Therefore, the present worth of an alternative is seen to be equal to the equivalent value at $t = 0$ of the increase in return above that which you would be able to earn in an investment offering $i\%$ per period. In the previous case, \$14.29 is the present worth of (\$20 − \$5), the difference in the two ROIs.

The present worth is also the amount that you would have to be given to dissuade you from making an investment, since placing the initial investment amount along with the present worth into a bank account earning $i\%$ will yield the same eventual return on investment. Relating this to the previous paragraphs, you could be dissuaded from investing \$100 in an alternative that would return \$120 in one year by a $t = 0$ payment of \$14.29. Clearly, (\$100 + \$14.29) invested at $t = 0$ will also yield \$120 in one year at 5%.

Income-producing alternatives with negative present worths are undesirable, and alternatives with positive present worths are desirable because they increase the average earning power of invested capital. (In some cases, such as municipal and public works projects, the present worths of all alternatives are negative, in which case, the least negative alternative is best.)

The selection of the interest rate is difficult in engineering economics problems. Usually, it is taken as the average rate of return that an individual or business organization has realized in past investments. Alternatively, the interest rate may be associated with a particular level of risk. Usually, i for individuals is the interest rate that can be earned in relatively *risk-free investments*.

18. SIMPLE AND COMPOUND INTEREST

If \$100 is invested at 5%, it will grow to \$105 in one year. During the second year, 5% interest continues to be accrued, but on \$105, not on \$100. This is the principle of *compound interest*: The interest accrues interest.[12]

If only the original principal accrues interest, the interest is said to be *simple interest*. Simple interest is rarely encountered in long-term engineering economic analyses, but the concept may be incorporated into short-term transactions.

19. EXTRACTING THE INTEREST RATE: RATE OF RETURN

An intuitive definition of the *rate of return* (ROR) is the effective annual interest rate at which an investment accrues income. That is, the rate of return of a project is the interest rate that would yield identical profits if all money were invested at that rate. Although this definition is correct, it does not provide a method of determining the rate of return.

It was previously seen that the present worth of a $100 investment invested at 5% is zero when $i = 5\%$ is used to determine equivalence. Therefore, a working definition of rate of return would be the effective annual interest rate that makes the present worth of the investment zero. Alternatively, rate of return could be defined as the effective annual interest rate that will discount all cash flows to a total present worth equal to the required initial investment.

It is tempting, but impractical, to determine a rate of return analytically. It is simply too difficult to extract the interest rate from the equivalence equation. For example, consider a $100 investment that pays back $75 at the end of each of the first two years. The present worth equivalence equation (set equal to zero in order to determine the rate of return) is

$$P = 0 = -\$100 + (\$75)(1+i)^{-1} + (\$75)(1+i)^{-2}$$

54.18

Solving Eq. 54.18 requires finding the roots of a quadratic equation. In general, for an investment or project spanning n years, the roots of an nth-order polynomial would have to be found. It should be obvious that an analytical solution would be essentially impossible for more complex cash flows. (The rate of return in this example is 31.87%.)

If the rate of return is needed, it can be found from a trial-and-error solution. To find the rate of return of an investment, proceed as follows.

step 1: Set up the problem as if to calculate the present worth.

step 2: Arbitrarily select a reasonable value for i. Calculate the present worth.

step 3: Choose another value of i (not too close to the original value), and again solve for the present worth.

step 4: Interpolate or extrapolate the value of i that gives a zero present worth.

step 5: For increased accuracy, repeat steps 2 and 3 with two more values that straddle the value found in step 4.

A common, although incorrect, method of calculating the rate of return involves dividing the annual receipts or returns by the initial investment. (See Sec. 54.56.) However, this technique ignores such items as salvage, depreciation, taxes, and the time value of money. This technique also is inadequate when the annual returns vary.

It is possible that more than one interest rate will satisfy the zero present worth criteria. This confusing situation occurs whenever there is more than one change in sign in the investment's cash flow.[13] Table 54.3 indicates the numbers of possible interest rates as a function of the number of sign reversals in the investment's cash flow.

Table 54.3 Multiplicity of Rates of Return

number of sign reversals	number of distinct rates of return
0	0
1	0 or 1
2	0, 1, or 2
3	0, 1, 2, or 3
4	0, 1, 2, 3, or 4
m	$0, 1, 2, 3, \ldots, m-1, m$

Difficulties associated with interpreting the meaning of multiple rates of return can be handled with the concepts of external investment and external rate of return. An *external investment* is an investment that is distinct from the investment being evaluated (which becomes known as the internal investment). The *external rate of return*, which is the rate of return earned by the external investment, does not need to be the same as the rate earned by the internal investment.

Generally, the multiple rates of return indicate that the analysis must proceed as though money will be invested outside of the project. The mechanics of how this is done are not covered here.

---

[12]This assumes, of course, that the interest remains in the account. If the interest is removed and spent, only the remaining funds accumulate interest.

[13]There will always be at least one change of sign in the cash flow of a legitimate investment. (This excludes municipal and other tax-supported functions.) At $t = 0$, an investment is made (a negative cash flow). Hopefully, the investment will begin to return money (a positive cash flow) at $t = 1$ or shortly thereafter. Although it is possible to conceive of an investment in which all of the cash flows were negative, such an investment would probably be classified as a *hobby*.

Example 54.10

What is the rate of return on invested capital if $1000 is invested now with $500 being returned in year 4 and $1000 being returned in year 8?

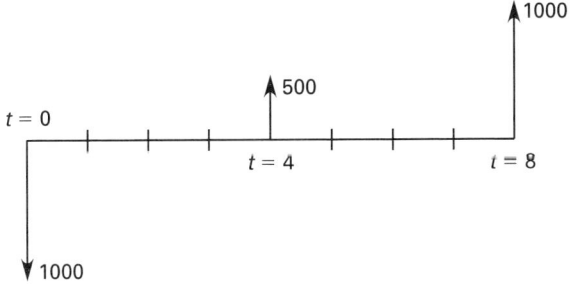

Solution

First, set up the problem as a present worth calculation.

Try $i = 5\%$.

$$P = -\$1000 + (\$500)(P/F, 5\%, 4)$$
$$+ (\$1000)(P/F, 5\%, 8)$$
$$= -\$1000 + (\$500)(0.8227) + (\$1000)(0.6768)$$
$$= \$88$$

Next, try a larger value of i to reduce the present worth. If $i = 10\%$,

$$P = -\$1000 + (\$500)(P/F, 10\%, 4)$$
$$+ (\$1000)(P/F, 10\%, 8)$$
$$= -\$1000 + (\$500)(0.6830) + (\$1000)(0.4665)$$
$$= -\$192$$

Using simple interpolation, the rate of return is

$$\text{ROR} = 5\% + \left(\frac{\$88}{\$88 + \$192}\right)(10\% - 5\%)$$
$$= 6.57\%$$

A second iteration between 6% and 7% yields 6.39%.

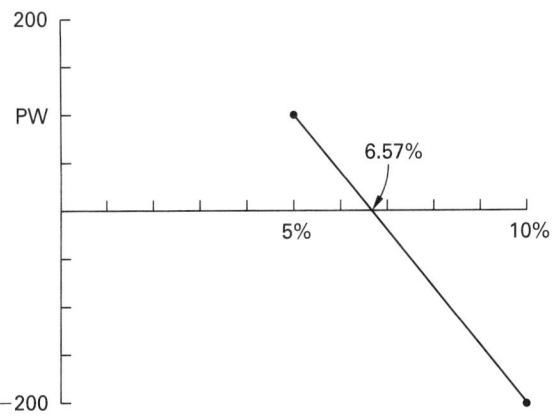

Example 54.11

A biomedical company is developing a new drug. A venture capital firm gives the company $25 million initially and $55 million more at the end of the first year. The drug patent will be sold at the end of year 5 to the highest bidder, and the biomedical company will receive $80 million. (The venture capital firm will receive everything in excess of $80 million.) The firm invests unused money in short-term commercial paper earning 10% effective interest per year through its bank. In the meantime, the biomedical company incurs development expenses of $50 million annually for the first three years. The drug is to be evaluated by a government agency and there will be neither expenses nor revenues during the fourth year. What is the biomedical company's rate of return on this investment?

Solution

Normally, the rate of return is determined by setting up a present worth problem and varying the interest rate until the present worth is zero. Writing the cash flows, though, shows that there are two reversals of sign: one at $t = 2$ (positive to negative) and the other at $t = 5$ (negative to positive). Therefore, there could be two interest rates that produce a zero present worth. (In fact, there actually are two interest rates: 10.7% and 41.4%.)

time	cash flow (millions)
0	+25
1	+55 − 50 = +5
2	−50
3	−50
4	0
5	+80

However, this problem can be reduced to one with only one sign reversal in the cash flow series. The initial $25 million is invested in commercial paper (an *external investment* having nothing to do with the drug development process) during the first year at 10%. The accumulation of interest and principal after one year is

$$(25)(1 + 0.10) = 27.5$$

This 27.5 is combined with the 5 (the money remaining after all expenses are paid at $t = 1$) and invested externally, again at 10%. The accumulation of interest and principal after one year (i.e., at $t = 2$) is

$$(27.5 + 5)(1 + 0.10) = 35.75$$

This 35.75 is combined with the development cost paid at $t = 2$.

The cash flow for the development project (the internal investment) is

time	cash flow (millions)
0	0
1	0
2	$35.75 - 50 = -14.25$
3	-50
4	0
5	$+80$

There is only one sign reversal in the cash flow series. The *internal rate of return* on this development project is found by the traditional method to be 10.3%. This is different from the rate the company can earn from investing externally in commercial paper.

20. RATE OF RETURN VERSUS RETURN ON INVESTMENT

Rate of return (ROR) is an effective annual interest rate, typically stated in percent per year. *Return on investment* (ROI) is a dollar amount. *Rate of return* and *return on investment* are not synonymous.

Return on investment can be calculated in two different ways. The accounting method is to subtract the total of all investment costs from the total of all net profits (i.e., revenues less expenses). The time value of money is not considered.

In engineering economic analysis, the return on investment is calculated from equivalent values. Specifically, the present worth (at $t = 0$) of all investment costs is subtracted from the future worth (at $t = n$) of all net profits.

When there are only two cash flows, a single investment amount and a single payback, the two definitions of return on investment yield the same numerical value. When there are more than two cash flows, the returns on investment will be different depending on which definition is used.

21. MINIMUM ATTRACTIVE RATE OF RETURN

A company may not know what effective interest rate, i, to use in engineering economic analysis. In such a case, the company can establish a minimum level of economic performance that it would like to realize on all investments. This criterion is known as the *minimum attractive rate of return* (MARR). Unlike the effective interest rate, i, the minimum attractive rate of return is not used in numerical calculations.[14] It is used only in comparisons with the rate of return.

Once a rate of return for an investment is known, it can be compared to the minimum attractive rate of return. To be a viable alternative, the rate of return must be greater than the minimum attractive rate of return.

The advantage of using comparisons to the minimum attractive rate of return is that an effective interest rate, i, never needs to be known. The minimum attractive rate of return becomes the correct interest rate for use in present worth and equivalent uniform annual cost calculations.

22. TYPICAL ALTERNATIVE-COMPARISON PROBLEM FORMATS

With the exception of some investment and rate of return problems, the typical problem involving engineering economics will have the following characteristics.

- An interest rate will be given.
- Two or more alternatives will be competing for funding.
- Each alternative will have its own cash flows.
- It will be necessary to select the best alternative.

23. DURATIONS OF INVESTMENTS

Because they are handled differently, short-term investments and short-lived assets need to be distinguished from investments and assets that constitute an infinitely lived project. Short-term investments are easily identified: a drill press that is needed for three years or a temporary factory building that is being constructed to last five years.

Investments with perpetual cash flows are also (usually) easily identified: maintenance on a large flood control dam and revenues from a long-span toll bridge. Furthermore, some items with finite lives can expect renewal on a repeated basis.[15] For example, a major freeway with a pavement life of 20 years is unlikely to be abandoned; it will be resurfaced or replaced every 20 years.

Actually, if an investment's finite lifespan is long enough, it can be considered an infinite investment because money 50 or more years from now has little impact on current decisions. The $(P/F, 10\%, 50)$ factor, for example, is 0.0085. Therefore, one dollar at $t = 50$ has an equivalent present worth of less than one penny. Since these far-future cash flows are eclipsed by present cash flows, long-term investments can be considered finite or infinite without significant impact on the calculations.

24. CHOICE OF ALTERNATIVES: COMPARING ONE ALTERNATIVE WITH ANOTHER ALTERNATIVE

Several methods exist for selecting a superior alternative from among a group of proposals. Each method has its own merits and applications.

[14]Not everyone adheres to this rule. Some people use "minimum attractive rate of return" and "effective interest rate" interchangeably.

[15]The term *renewal* can be interpreted to mean replacement or repair.

ENGINEERING ECONOMIC ANALYSIS **54-15**

Present Worth Method

When two or more alternatives are capable of performing the same functions, the superior alternative will have the largest present worth. The *present worth method* is restricted to evaluating alternatives that are mutually exclusive and that have the same lives. This method is suitable for ranking the desirability of alternatives.

Example 54.12

Investment A costs $10,000 today and pays back $11,500 two years from now. Investment B costs $8000 today and pays back $4500 each year for two years. If an interest rate of 5% is used, which alternative is superior?

Solution

$$P(A) = -\$10,000 + (\$11,500)(P/F, 5\%, 2)$$
$$= -\$10,000 + (\$11,500)(0.9070)$$
$$= \$431$$
$$P(B) = -\$8000 + (\$4500)(P/A, 5\%, 2)$$
$$= -\$8000 + (\$4500)(1.8594)$$
$$= \$367$$

Alternative A is superior and should be chosen.

Capitalized Cost Method

The present worth of a project with an infinite life is known as the *capitalized cost* or *life-cycle cost*. Capitalized cost is the amount of money at $t = 0$ needed to perpetually support the project on the earned interest only. Capitalized cost is a positive number when expenses exceed income.

In comparing two alternatives, each of which is infinitely lived, the superior alternative will have the lowest capitalized cost.

Normally, it would be difficult to work with an infinite stream of cash flows since most economics tables do not list factors for periods in excess of 100 years. However, the (A/P) discounting factor approaches the interest rate as n becomes large. Since the (P/A) and (A/P) factors are reciprocals of each other, it is possible to divide an infinite series of equal cash flows by the interest rate in order to calculate the present worth of the infinite series. This is the basis of Eq. 54.19.

$$\text{capitalized cost} = \text{initial cost} + \frac{\text{annual costs}}{i} \quad \textit{54.19}$$

Equation 54.19 can be used when the annual costs are equal in every year. If the operating and maintenance costs occur irregularly instead of annually, or if the costs vary from year to year, it will be necessary to somehow determine a cash flow of equal annual amounts (EAA) that is equivalent to the stream of original costs.

The equal annual amount may be calculated in the usual manner by first finding the present worth of all the actual costs and then multiplying the present worth by the interest rate (the (A/P) factor for an infinite series). However, it is not even necessary to convert the present worth to an equal annual amount since Eq. 54.20 will convert the equal amount back to the present worth.

$$\text{capitalized cost} = \text{initial cost} + \frac{\text{EAA}}{i}$$
$$= \text{initial cost} + \frac{\text{present worth}}{\text{of all expenses}}$$

54.20

Example 54.13

What is the capitalized cost of a public works project that will cost $25,000,000 now and will require $2,000,000 in maintenance annually? The effective annual interest rate is 12%.

Solution

Worked in millions of dollars, from Eq. 54.19, the capitalized cost is

$$\text{capitalized cost} = 25 + (2)(P/A, 12\%, \infty)$$
$$= 25 + \frac{2}{0.12}$$
$$= 41.67$$

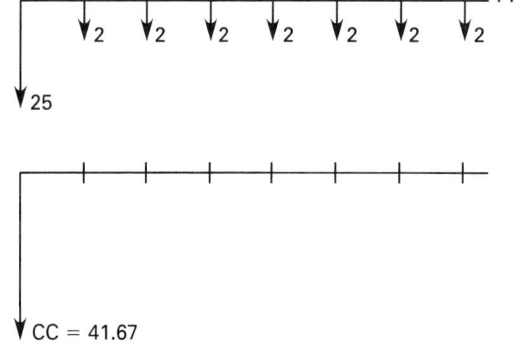

Annual Cost Method

Alternatives that accomplish the same purpose but that have unequal lives must be compared by the *annual cost method*.[16] The annual cost method assumes that each alternative will be replaced by an identical twin at the end of its useful life (infinite renewal). This method, which may also be used to rank alternatives according

---

[16]Of course, the annual cost method can be used to determine the superiority of assets with identical lives as well.

Systems, Mgmt., Professional

to their desirability, is also called the *annual return method* or *capital recovery method*.

Restrictions are that the alternatives must be mutually exclusive and repeatedly renewed up to the duration of the longest-lived alternative. The calculated annual cost is known as the *equivalent uniform annual cost* (EUAC) or just *equivalent annual cost*. Cost is a positive number when expenses exceed income.

Example 54.14

Which of the following alternatives is superior over a 30-year period if the interest rate is 7%?

	alternative A	alternative B
type	brick	wood
life	30 years	10 years
initial cost	$1800	$450
maintenance	$5/year	$20/year

Solution

$$\text{EUAC(A)} = (\$1800)(A/P, 7\%, 30) + \$5$$
$$= (\$1800)(0.0806) + \$5$$
$$= \$150$$
$$\text{EUAC(B)} = (\$450)(A/P, 7\%, 10) + \$20$$
$$= (\$450)(0.1424) + \$20$$
$$= \$84$$

Alternative B is superior since its annual cost of operation is the lowest. It is assumed that three wood facilities, each with a life of 10 years and a cost of $450, will be built to span the 30-year period.

25. CHOICE OF ALTERNATIVES: COMPARING AN ALTERNATIVE WITH A STANDARD

With specific economic performance criteria, it is possible to qualify an investment as acceptable or unacceptable without having to compare it with another investment. Two such performance criteria are the benefit-cost ratio and the minimum attractive rate of return.

Benefit-Cost Ratio Method

The *benefit-cost ratio* method is often used in municipal project evaluations where benefits and costs accrue to different segments of the community. With this method, the present worth of all benefits (irrespective of the beneficiaries) is divided by the present worth of all costs. The project is considered acceptable if the ratio equals or exceeds 1.0, that is, if $B/C \geq 1.0$.

When the benefit-cost ratio method is used, disbursements by the initiators or sponsors are *costs*. Disbursements by the users of the project are known as *disbenefits*. It is often difficult to determine whether a cash flow is a cost or a disbenefit (whether to place it in the numerator or denominator of the benefit-cost ratio calculation).

Regardless of where the cash flow is placed, an acceptable project will always have a benefit-cost ratio greater than or equal to 1.0, although the actual numerical result will depend on the placement. For this reason, the benefit-cost ratio method should not be used to rank competing projects.

The benefit-cost ratio method of comparing alternatives is used extensively in transportation engineering where the ratio is often (but not necessarily) written in terms of annual benefits and annual costs instead of present worths. Another characteristic of highway benefit-cost ratios is that the route (road, highway, etc.) is usually already in place and that various alternative upgrades are being considered. There will be existing benefits and costs associated with the current route. Therefore, the *change* (usually an increase) in benefits and costs is used to calculate the benefit-cost ratio.[17]

$$B/C = \frac{\Delta \text{ user benefits}}{\Delta \text{ investment cost} + \Delta \text{ maintenance} - \Delta \text{ residual value}}$$

54.21

The change in *residual value* (*terminal value*) appears in the denominator as a negative item. An increase in the residual value would decrease the denominator.

Example 54.15

By building a bridge over a ravine, a state department of transportation can shorten the time it takes to drive through a mountainous area. Estimates of costs and benefits (due to decreased travel time, fewer accidents, reduced gas usage, etc.) have been prepared. Should the bridge be built? Use the benefit-cost ratio method of comparison.

	millions
initial cost	40
capitalized cost of perpetual annual maintenance	12
capitalized value of annual user benefits	49
residual value	0

---

[17]This discussion of highway benefit-cost ratios is not meant to imply that everyone agrees with Eq. 54.21. In *Economic Analysis for Highways* (International Textbook Company, Scranton, PA, 1969), author Robley Winfrey took a strong stand on one aspect of the benefits versus disbenefits issue: highway maintenance. According to Winfrey, regular highway maintenance costs should be placed in the numerator as a subtraction from the user benefits. Some have called this mandate the *Winfrey method*.

Solution

If Eq. 54.21 is used, the benefit-cost ratio is

$$B/C = \frac{49}{40 + 12 - 0} = 0.942$$

Since the benefit-cost ratio is less than 1.00, the bridge should not be built.

If the maintenance costs are placed in the numerator (per Ftn. 17), the benefit-cost ratio value will be different, but the conclusion will not change.

$$B/C_{\text{alternate method}} = \frac{49 - 12}{40} = 0.925$$

Rate of Return Method

The minimum attractive rate of return (MARR) has already been introduced as a standard of performance against which an investment's actual *rate of return* (ROR) is compared. If the rate of return is equal to or exceeds the minimum attractive rate of return, the investment is qualified. This is the basis for the *rate of return method* of alternative selection.

Finding the rate of return can be a long, iterative process. Usually, the actual numerical value of rate of return is not needed; it is sufficient to know whether or not the rate of return exceeds the minimum attractive rate of return. This *comparative analysis* can be accomplished without calculating the rate of return simply by finding the present worth of the investment using the minimum attractive rate of return as the effective interest rate (i.e., $i = $ MARR). If the present worth is zero or positive, the investment is qualified. If the present worth is negative, the rate of return is less than the minimum attractive rate of return.

26. RANKING MUTUALLY EXCLUSIVE MULTIPLE PROJECTS

Ranking of multiple investment alternatives is required when there is sufficient funding for more than one investment. Since the best investments should be selected first, it is necessary to place all investments into an ordered list.

Ranking is relatively easy if the present worths, future worths, capitalized costs, or equivalent uniform annual costs have been calculated for all the investments. The highest ranked investment will be the one with the largest present or future worth, or the smallest capitalized or annual cost. Present worth, future worth, capitalized cost, and equivalent uniform annual cost can all be used to rank multiple investment alternatives.

However, neither rates of return nor benefit-cost ratios should be used to rank multiple investment alternatives. Specifically, if two alternatives both have rates of return exceeding the minimum acceptable rate of return, it is not sufficient to select the alternative with the highest rate of return.

An *incremental analysis*, also known as a *rate of return on added investment study*, should be performed if rate of return is used to select between investments. An incremental analysis starts by ranking the alternatives in order of increasing initial investment. Then, the cash flows for the investment with the lower initial cost are subtracted from the cash flows for the higher-priced alternative on a year-by-year basis. This produces, in effect, a third alternative representing the costs and benefits of the added investment. The added expense of the higher-priced investment is not warranted unless the rate of return of this third alternative exceeds the minimum attractive rate of return as well. The choice criterion is to select the alternative with the higher initial investment if the incremental rate of return exceeds the minimum attractive rate of return.

An incremental analysis is also required if ranking is to be done by the benefit-cost ratio method. The incremental analysis is accomplished by calculating the ratio of differences in benefits to differences in costs for each possible pair of alternatives. If the ratio exceeds 1.0, alternative 2 is superior to alternative 1. Otherwise, alternative 1 is superior.[18]

$$\frac{B_2 - B_1}{C_2 - C_1} \geq 1 \quad \text{[alternative 2 superior]} \qquad 54.22$$

27. ALTERNATIVES WITH DIFFERENT LIVES

Comparison of two alternatives is relatively simple when both alternatives have the same life. For example, a problem might be stated: "Which would you rather have: car A with a life of three years, or car B with a life of five years?"

However, care must be taken to understand what is going on when the two alternatives have different lives. If car A has a life of three years and car B has a life of five years, what happens at $t = 3$ if the five-year car is chosen? If a car is needed for five years, what happens at $t = 3$ if the three-year car is chosen?

In this type of situation, it is necessary to distinguish between the length of the need (the *analysis horizon*) and the lives of the alternatives or assets intended to meet that need. The lives do not have to be the same as the horizon.

Finite Horizon with Incomplete Asset Lives

If an asset with a five-year life is chosen for a three-year need, the disposition of the asset at $t = 3$ must be known in order to evaluate the alternative. If the asset is sold at $t = 3$, the salvage value is entered into the analysis (at $t = 3$) and the alternative is evaluated as a three-year investment. The fact that the asset is sold

---

[18]It goes without saying that the benefit-cost ratios for all investment alternatives by themselves must also be equal to or greater than 1.0.

when it has some useful life remaining does not affect the analysis horizon.

Similarly, if a three-year asset is chosen for a five-year need, something about how the need is satisfied during the last two years must be known. Perhaps a rental asset will be used. Or, perhaps the function will be "farmed out" to an outside firm. In any case, the costs of satisfying the need during the last two years enter the analysis, and the alternative is evaluated as a five-year investment.

If both alternatives are "converted" to the same life, any of the alternative selection criteria (present worth method, annual cost method, etc.) can be used to determine which alternative is superior.

Finite Horizon with Integer Multiple Asset Lives

It is common to have a long-term horizon (need) that must be met with short-lived assets. In special instances, the horizon will be an integer number of asset lives. For example, a company may be making a 12-year transportation plan and may be evaluating two cars: one with a three-year life, and another with a four-year life.

In this example, four of the first car or three of the second car are needed to reach the end of the 12-year horizon.

If the horizon is an integer number of asset lives, any of the alternative selection criteria can be used to determine which is superior. If the present worth method is used, all alternatives must be evaluated over the entire horizon. (In this example, the present worth of 12 years of car purchases and use must be determined for both alternatives.)

If the equivalent uniform annual cost method is used, it may be possible to base the calculation of annual cost on one lifespan of each alternative only. It may not be necessary to incorporate all of the cash flows into the analysis. (In the running example, the annual cost over three years would be determined for the first car; the annual cost over four years would be determined for the second car.) This simplification is justified if the subsequent asset replacements (renewals) have the same cost and cash flow structure as the original asset. This assumption is typically made implicitly when the annual cost method of comparison is used.

Infinite Horizon

If the need horizon is infinite, it is not necessary to impose the restriction that asset lives of alternatives be integer multiples of the horizon. The superior alternative will be replaced (renewed) whenever it is necessary to do so, forever.

Infinite horizon problems are almost always solved with either the annual cost or capitalized cost method. It is common to (implicitly) assume that the cost and cash flow structure of the asset replacements (renewals) are the same as the original asset.

28. OPPORTUNITY COSTS

An *opportunity cost* is an imaginary cost representing what will not be received if a particular strategy is rejected. It is what you will lose if you do or do not do something. As an example, consider a growing company with an existing operational computer system. If the company trades in its existing computer as part of an upgrade plan, it will receive a *trade-in allowance*. (In other problems, a *salvage value* may be involved.)

If one of the alternatives being evaluated is not to upgrade the computer system at all, the trade-in allowance (or, salvage value in other problems) will not be realized. The amount of the trade-in allowance is an opportunity cost that must be included in the problem analysis.

Similarly, if one of the alternatives being evaluated is to wait one year before upgrading the computer, the *difference in trade-in allowances* is an opportunity cost that must be included in the problem analysis.

29. REPLACEMENT STUDIES

An investigation into the retirement of an existing process or piece of equipment is known as a *replacement study*. Replacement studies are similar in most respects to other alternative comparison problems: An interest rate is given, two alternatives exist, and one of the previously mentioned methods of comparing alternatives is used to choose the superior alternative. Usually, the annual cost method is used on a year-by-year basis.

In replacement studies, the existing process or piece of equipment is known as the *defender*. The new process or piece of equipment being considered for purchase is known as the *challenger*.

30. TREATMENT OF SALVAGE VALUE IN REPLACEMENT STUDIES

Since most defenders still have some market value when they are retired, the problem of what to do with the salvage arises. It seems logical to use the salvage value of the defender to reduce the initial purchase cost of the challenger. This is consistent with what would actually happen if the defender were to be retired.

By convention, however, the defender's salvage value is subtracted from the defender's present value. This does not seem logical, but it is done to keep all costs and benefits related to the defender with the defender. In this case, the salvage value is treated as an opportunity cost that would be incurred if the defender is not retired.

If the defender and the challenger have the same lives and a present worth study is used to choose the superior alternative, the placement of the salvage value will have no effect on the net difference between present worths for the challenger and defender. Although the values of the two present worths will be different depending on the placement, the difference in present worths will be the same.

If the defender and the challenger have different lives, an annual cost comparison must be made. Since the salvage value would be "spread over" a different number of years depending on its placement, it is important to abide by the conventions listed in this section.

There are a number of ways to handle salvage value in retirement studies. The best way is to calculate the cost of keeping the defender one more year. In addition to the usual operating and maintenance costs, that cost includes an opportunity interest cost incurred by not selling the defender, and also a drop in the salvage value if the defender is kept for one additional year. Specifically,

$$
\begin{aligned}
\text{EUAC (defender)} = {} & \text{next year's maintenance costs} \\
& + i(\text{current salvage value}) \\
& + \text{current salvage} \\
& - \text{next year's salvage}
\end{aligned}
$$

$$54.23$$

It is important in retirement studies not to double count the salvage value. That is, it would be incorrect to add the salvage value to the defender and at the same time subtract it from the challenger.

Equation 54.23 contains the difference in salvage value between two consecutive years. This calculation shows that the defender/challenger decision must be made on a year-by-year basis. One application of Eq. 54.23 will not usually answer the question of whether the defender should remain in service indefinitely. The calculation must be repeatedly made as long as there is a drop in salvage value from one year to the next.

31. ECONOMIC LIFE: RETIREMENT AT MINIMUM COST

As an asset grows older, its operating and maintenance costs typically increase. Eventually, the cost to keep the asset in operation becomes prohibitive, and the asset is retired or replaced. However, it is not always obvious when an asset should be retired or replaced.

As the asset's maintenance cost is increasing each year, the amortized cost of its initial purchase is decreasing. It is the sum of these two costs that should be evaluated to determine the point at which the asset should be retired or replaced. Since an asset's initial purchase price is likely to be high, the amortized cost will be the controlling factor in those years when the maintenance costs are low. Therefore, the EUAC of the asset will decrease in the initial part of its life.

However, as the asset grows older, the change in its amortized cost decreases while maintenance cost increases. Eventually, the sum of the two costs reaches a minimum and then starts to increase. The age of the asset at the minimum cost point is known as the *economic life* of the asset. The economic life generally is less than the length of need and the technological lifetime of the asset, as shown in Fig. 54.7.

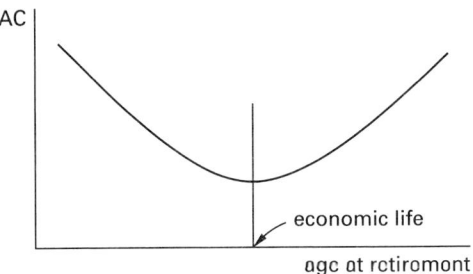

Figure 54.7 *EUAC Versus Age at Retirement*

The determination of an asset's economic life is illustrated by Ex. 54.16.

Example 54.16

Buses in a municipal transit system have the characteristics listed. In order to minimize its annual operating expenses, when should the city replace its buses if money can be borrowed at 8%?

initial cost of bus: $120,000

year	maintenance cost	salvage value
1	35,000	60,000
2	38,000	55,000
3	43,000	45,000
4	50,000	25,000
5	65,000	15,000

Solution

The annual maintenance is different each year. Each maintenance cost must be spread over the life of the bus. This is done by first finding the present worth and then amortizing the maintenance costs. If a bus is kept for one year and then sold, the annual cost will be

$$
\begin{aligned}
\text{EUAC(1)} = {} & (\$120{,}000)(A/P, 8\%, 1) \\
& + (\$35{,}000)(A/F, 8\%, 1) \\
& - (\$60{,}000)(A/F, 8\%, 1) \\
= {} & (\$120{,}000)(1.0800) + (\$35{,}000)(1.000) \\
& - (\$60{,}000)(1.000) \\
= {} & \$104{,}600
\end{aligned}
$$

If a bus is kept for two years and then sold, the annual cost will be

$$
\begin{aligned}
\text{EUAC(2)} = {} & \big(\$120{,}000 + (\$35{,}000)(P/F, 8\%, 1)\big) \\
& \times (A/P, 8\%, 2) \\
& + (\$38{,}000 - \$55{,}000)(A/F, 8\%, 2) \\
= {} & \big(\$120{,}000 + (\$35{,}000)(0.9259)\big)(0.5608) \\
& + (\$38{,}000 - \$55{,}000)(0.4808) \\
= {} & \$77{,}296
\end{aligned}
$$

If a bus is kept for three years and then sold, the annual cost will be

$$
\begin{aligned}
\text{EUAC}(3) = & \big(\$120{,}000 + (\$35{,}000)(P/F, 8\%, 1) \\
& + (\$38{,}000)(P/F, 8\%, 2)\big)(A/P, 8\%, 3) \\
& + (\$43{,}000 - \$45{,}000)(A/F, 8\%, 3) \\
= & \big(\$120{,}000 + (\$35{,}000)(0.9259) \\
& + (\$38{,}000)(0.8573)\big)(0.3880) \\
& - (\$2000)(0.3080) \\
= & \$71{,}158
\end{aligned}
$$

This process is continued until the annual cost begins to increase. In this example, EUAC(4) is $71,700. Therefore, the buses should be retired after three years.

32. LIFE-CYCLE COST

The *life-cycle cost* of an alternative is the equivalent value (at $t = 0$) of the alternative's cash flow over the alternative's lifespan. Since the present worth is evaluated using an effective interest rate of i (which would be the interest rate used for all engineering economic analyses), the life-cycle cost is the same as the alternative's present worth. If the alternative has an infinite horizon, the life-cycle cost and capitalized cost will be identical.

33. CAPITALIZED ASSETS VERSUS EXPENSES

High expenses reduce profit, which in turn reduces income tax. It seems logical to label each and every expenditure, even an asset purchase, as an expense. As an alternative to this *expensing the asset*, it may be decided to capitalize the asset. *Capitalizing the asset* means that the cost of the asset is divided into equal or unequal parts, and only one of these parts is taken as an expense each year. Expensing is clearly the more desirable alternative, since the after-tax profit is increased early in the asset's life.

There are long-standing accounting conventions as to what can be expensed and what must be capitalized.[19] Some companies capitalize everything—regardless of cost—with expected lifetimes greater than one year. Most companies, however, expense items whose purchase costs are below a cutoff value. A cutoff value in the range of $250–500, depending on the size of the company, is chosen as the maximum purchase cost of an expensed asset. Assets costing more than this are capitalized.

It is not necessary for a large corporation to keep track of every lamp, desk, and chair for which the purchase price is greater than the cutoff value. Such assets, all of which

have the same lives and have been purchased in the same year, can be placed into groups or *asset classes*. A group cost, equal to the sum total of the purchase costs of all items in the group, is capitalized as though the group was an identifiable and distinct asset itself.

34. PURPOSE OF DEPRECIATION

Depreciation is an *artificial expense* that spreads the purchase price of an asset or other property over a number of years.[20] Depreciating an asset is an example of capitalization, as previously defined. The inclusion of depreciation in engineering economic analysis problems will increase the after-tax present worth (profitability) of an asset. The larger the depreciation, the greater will be the profitability. Therefore, individuals and companies eligible to utilize depreciation want to maximize and accelerate the depreciation available to them.

Although the entire property purchase price is eventually recognized as an expense, the net recovery from the expense stream never equals the original cost of the asset. That is, depreciation cannot realistically be thought of as a fund (an annuity or sinking fund) that accumulates capital to purchase a replacement at the end of the asset's life. The primary reason for this is that the depreciation expense is reduced significantly by the impact of income taxes, as will be seen in later sections.

35. DEPRECIATION BASIS OF AN ASSET

The *depreciation basis* of an asset is the part of the asset's purchase price that is spread over the *depreciation period*, also known as the *service life*.[21] Usually, the depreciation basis and the purchase price are not the same.

A common depreciation basis is the difference between the purchase price and the expected salvage value at the end of the depreciation period. That is,

$$
\text{depreciation basis} = C - S_n \qquad \textit{54.24}
$$

There are several methods of calculating the year-by-year depreciation of an asset. Equation 54.24 is not

---

[19]For example, purchased vehicles must be capitalized; payments for leased vehicles can be expensed. Repainting a building with paint that will last five years is an expense, but the replacement cost of a leaking roof must be capitalized.

[20]In the United States, the tax regulations of Internal Revenue Service (IRS) allow depreciation on almost all forms of *business property* except land. The following types of property are distinguished: *real* (e.g., buildings used for business), *residential* (e.g., buildings used as rental property), and *personal* (e.g., equipment used for business). Personal property does *not* include items for personal use (such as a personal residence), despite its name. *Tangible personal property* is distinguished from *intangible property* (goodwill, copyrights, patents, trademarks, franchises, agreements not to compete, etc.).

[21]The *depreciation period* is selected to be as short as possible within recognized limits. This depreciation will not normally coincide with the *economic life* or *useful life* of an asset. For example, a car may be capitalized over a depreciation period of three years. It may become uneconomical to maintain and use at the end of an economic life of nine years. However, the car may be capable of operation over a useful life of 25 years.

universally compatible with all depreciation methods. Some methods do not consider the salvage value. This is known as an *unadjusted basis*. When the depreciation method is known, the depreciation basis can be rigorously defined.[22]

36. DEPRECIATION METHODS

Generally, tax regulations do not allow the cost of an asset to be treated as a deductible expense in the year of purchase. Rather, portions of the depreciation basis must be allocated to each of the n years of the asset's depreciation period. The amount that is allocated each year is called the *depreciation*.

Various methods exist for calculating an asset's depreciation each year.[23] Although the depreciation calculations may be considered independently (for the purpose of determining book value or as an academic exercise), it is important to recognize that depreciation has no effect on engineering economic analyses unless income taxes are also considered.

Straight Line Method

With the *straight line method*, depreciation is the same each year. The depreciation basis $(C - S_n)$ is allocated uniformly to all of the n years in the depreciation period. Each year, the depreciation will be

$$D = \frac{C - S_n}{n} \qquad 54.25$$

Constant Percentage Method

The *constant percentage method*[24] is similar to the straight line method in that the depreciation is the same each year. If the fraction of the basis used as depreciation is $1/n$, there is no difference between the constant percentage and straight line methods. The two methods differ only in what information is available. (With the straight line method, the life is known. With the constant percentage method, the depreciation fraction is known.)

Each year, the depreciation will be

$$D = (\text{depreciation fraction})(\text{depreciation basis})$$
$$= (\text{depreciation fraction})(C - S_n) \qquad 54.26$$

Sum-of-the-Years' Digits Method

In *sum-of-the-years' digits* (SOYD) depreciation, the digits from 1 to n inclusive are summed. The total, T, can also be calculated from

$$T = \tfrac{1}{2}n(n + 1) \qquad 54.27$$

The depreciation in year j can be found from Eq. 54.28. Notice that the depreciation in year j, D_j, decreases by a constant amount each year.

$$D_j = \frac{(C - S_n)(n - j + 1)}{T} \qquad 54.28$$

Double Declining Balance Method[25]

Double declining balance[26] (DDB) depreciation is independent of salvage value. Furthermore, the book value never stops decreasing, although the depreciation decreases in magnitude. Usually, any book value in excess of the salvage value is written off in the last year of the asset's depreciation period. Unlike any of the other depreciation methods, double declining balance depends on accumulated depreciation.

$$D_{\text{first year}} = \frac{2C}{n} \qquad 54.29$$

$$D_j = \frac{2\left(C - \sum_{m=1}^{j-1} D_m\right)}{n} \qquad 54.30$$

Calculating the depreciation in the middle of an asset's life appears particularly difficult with double declining balance, since all previous years' depreciation amounts seem to be required. It appears that the depreciation in the sixth year, for example, cannot be calculated unless the values of depreciation for the first five years are calculated. However, this is not true.

---

[22]For example, with the Accelerated Cost Recovery System (ACRS) the *depreciation basis* is the total purchase cost, regardless of the expected salvage value. With declining balance methods, the depreciation basis is the purchase cost less any previously taken depreciation.
[23]This discussion gives the impression that any form of depreciation may be chosen regardless of the nature and circumstances of the purchase. In reality, the IRS tax regulations place restrictions on the higher-rate (accelerated) methods, such as declining balance and sum-of-the-years' digits methods. Furthermore, the *Economic Recovery Act of 1981* and the *Tax Reform Act of 1986* substantially changed the laws relating to personal and corporate income taxes.
[24]The *constant percentage method* should not be confused with the declining balance method, which used to be known as the *fixed percentage on diminishing balance method*.

[25]In the past, the *declining balance method* has also been known as the *fixed percentage of book value* and *fixed percentage on diminishing balance method*.
[26]Double declining balance depreciation is a particular form of *declining balance depreciation*, as defined by the IRS tax regulations. Declining balance depreciation includes 125% declining balance and 150% declining balance depreciations that can be calculated by substituting 1.25 and 1.50, respectively, for the 2 in Eq. 54.29.

Systems, Mgmt., Professional

Depreciation in the middle of an asset's life can be found from the following equations. (d is known as the *depreciation rate*.)

$$d = \frac{2}{n} \qquad 54.31$$

$$D_j = dC(1-d)^{j-1} \qquad 54.32$$

Statutory Depreciation Systems

In the United States, property placed into service in 1981 and thereafter must use the *Accelerated Cost Recovery System* (ACRS), and after 1986, the *Modified Accelerated Cost Recovery System* (MACRS) or other statutory method. Other methods (straight line, declining balance, etc.) cannot be used except in special cases.

Property placed into service in 1980 or before must continue to be depreciated according to the method originally chosen (e.g., straight line, declining balance, or sum-of-the-years' digits). ACRS and MACRS cannot be used.

Under ACRS and MACRS, the cost recovery amount in the jth year of an asset's cost recovery period is calculated by multiplying the initial cost by a factor.

$$D_j = C \times \text{factor} \qquad 54.33$$

The initial cost used is not reduced by the asset's salvage value for ACRS and MACRS calculations. The factor used depends on the asset's cost recovery period. (See Table 54.4.) Such factors are subject to continuing legislation changes. Current tax publications should be consulted before using this method.

Table 54.4 *Representative MACRS Depreciation Factors**

depreciation rate for recovery period, n

year, j	3 years	5 years	7 years	10 years
1	33.33%	20.00%	14.29%	10.00%
2	44.45%	32.00%	24.49%	18.00%
3	14.81%	19.20%	17.49%	14.40%
4	7.41%	11.52%	12.49%	11.52%
5		11.52%	8.93%	9.22%
6		5.76%	8.92%	7.37%
7			8.93%	6.55%
8			4.46%	6.55%
9				6.56%
10				6.55%
11				3.28%

*Values are for the "half-year" convention. This table gives typical values only. Since these factors are subject to continuing revision, they should not be used without consulting an accounting professional.

Production or Service Output Method

If an asset has been purchased for a specific task and that task is associated with a specific lifetime amount of output or production, the depreciation may be calculated by the fraction of total production produced during the year. Under the *units of production* method, the depreciation is not expected to be the same each year.

$$D_j = (C - S_n)\left(\frac{\text{actual output in year } j}{\text{estimated lifetime output}}\right) \qquad 54.34$$

Sinking Fund Method

The *sinking fund method* is seldom used in industry because the initial depreciation is low. The formula for sinking fund depreciation (which increases each year) is

$$D_j = (C - S_n)(A/F, i\%, n)(F/P, i\%, j-1) \qquad 54.35$$

Example 54.17

An asset is purchased for $9000. Its estimated economic life is ten years, after which it will be sold for $200. Find the depreciation in the first three years using straight line, double declining balance, and sum-of-the-years' digits depreciation methods.

Solution

SL: $\quad D = \dfrac{\$9000 - \$200}{10} = \$880$ each year

DDB: $\quad D_1 = \dfrac{(2)(\$9000)}{10} = \$1800$ in year 1

$\quad D_2 = \dfrac{(2)(\$9000 - \$1800)}{10} = \$1440$ in year 2

$\quad D_3 = \dfrac{(2)(\$9000 - \$3240)}{10} = \$1152$ in year 3

SOYD: $\quad T = \left(\frac{1}{2}\right)(10)(11) = 55$

$\quad D_1 = \left(\frac{10}{55}\right)(\$9000 - \$200) = \$1600$ in year 1

$\quad D_2 = \left(\frac{9}{55}\right)(\$8800) = \$1440$ in year 2

$\quad D_3 = \left(\frac{8}{55}\right)(\$8800) = \$1280$ in year 3

37. ACCELERATED DEPRECIATION METHODS

An *accelerated depreciation method* is one that calculates a depreciation amount greater than a straight line amount. Double declining balance and sum-of-the-years' digits methods are accelerated methods. The ACRS and MACRS methods are explicitly accelerated methods. Straight line and sinking fund methods are not accelerated methods.

Use of an accelerated depreciation method may result in unexpected tax consequences when the depreciated asset or property is disposed of. Professional tax advice should be obtained in this area.

38. BONUS DEPRECIATION

Bonus depreciation is a special, one-time, depreciation authorized by legislation for specific types of equipment, to be taken in the first year, in addition to the standard depreciation normally available for the equipment. Bonus depreciation is usually enacted in order to stimulate investment in or economic recovery of specific industries (e.g., aircraft manufacturing).

39. BOOK VALUE

The difference between original purchase price and accumulated depreciation is known as *book value*.[27] At the end of each year, the book value (which is initially equal to the purchase price) is reduced by the depreciation in that year.

It is important to properly synchronize depreciation calculations. It is difficult to answer the question, "What is the book value in the fifth year?" unless the timing of the book value change is mutually agreed upon. It is better to be specific about an inquiry by identifying when the book value change occurs. For example, the following question is unambiguous: "What is the book value at the end of year 5, after subtracting depreciation in the fifth year?" or "What is the book value after five years?"

Unfortunately, this type of care is seldom taken in book value inquiries, and it is up to the respondent to exercise reasonable care in distinguishing between beginning-of-year book value and end-of-year book value. To be consistent, the book value equations in this chapter have been written in such a way that the year subscript, j, has the same meaning in book value and depreciation calculations. That is, BV_5 means the book value at the end of the fifth year, after five years of depreciation, including D_5, have been subtracted from the original purchase price.

There can be a great difference between the book value of an asset and the *market value* of that asset. There is no legal requirement for the two values to coincide, and no intent for book value to be a reasonable measure of market value.[28] Therefore, it is apparent that book value is merely an accounting convention with little practical use. Even when a depreciated asset is disposed of, the book value is used to determine the consequences of disposal, not the price the asset should bring at sale.

The calculation of book value is relatively easy, even for the case of the declining balance depreciation method.

For the straight line depreciation method, the book value at the end of the jth year, after the jth depreciation deduction has been made, is

$$\text{BV}_j = C - \frac{j(C - S_n)}{n} = C - jD \qquad 54.36$$

For the sum-of-the-years' digits method, the book value is

$$\text{BV}_j = (C - S_n)\left(1 - \frac{j(2n + 1 - j)}{n(n + 1)}\right) + S_n \qquad 54.37$$

For the declining balance method, including double declining balance (see Ftn. 26), the book value is

$$\text{BV}_j = C(1 - d)^j \qquad 54.38$$

For the sinking fund method, the book value is calculated directly as

$$\text{BV}_j = C - (C - S_n)(A/F, i\%, n)(F/A, i\%, j) \qquad 54.39$$

Of course, the book value at the end of year j can always be calculated for any method by successive subtractions (i.e., subtraction of the accumulated depreciation), as Eq. 54.40 illustrates.

$$\text{BV}_j = C - \sum_{m=1}^{j} D_m \qquad 54.40$$

Figure 54.8 illustrates the book value of a hypothetical asset depreciated using several depreciation methods. Notice that the double declining balance method initially produces the fastest write-off, while the sinking fund method produces the slowest write-off. Also, the book value does not automatically equal the salvage value at the end of an asset's depreciation period with the double declining balance method.[29]

Example 54.18

For the asset described in Ex. 54.17, calculate the book value at the end of the first three years if sum-of-the-years' digits depreciation is used. The book value at the beginning of year 1 is $9000.

---

[27]The balance sheet of a corporation usually has two asset accounts: the *equipment account* and the *accumulated depreciation account*. There is no book value account on this financial statement, other than the implicit value obtained from subtracting the accumulated depreciation account from the equipment account. The book values of various assets, as well as their original purchase cost, date of purchase, salvage value, and so on, and accumulated depreciation appear on detail sheets or other peripheral records for each asset.

[28]Common examples of assets with great divergences of book and market values are buildings (rental houses, apartment complexes, factories, etc.) and company luxury automobiles (Porsches, Mercedes, etc.) during periods of inflation. Book values decrease, but actual values increase.

[29]This means that the straight line method of depreciation may result in a lower book value at some point in the depreciation period than if double declining balance is used. A *cut-over* from double declining balance to straight line may be permitted in certain cases. Finding the *cut-over point*, however, is usually done by comparing book values determined by both methods. The analytical method is complicated.

Solution

From Eq. 54.40,

$$BV_1 = \$9000 - \$1600 = \$7400$$
$$BV_2 = \$7400 - \$1440 = \$5960$$
$$BV_3 = \$5960 - \$1280 = \$4680$$

Figure 54.8 *Book Value with Different Depreciation Methods*

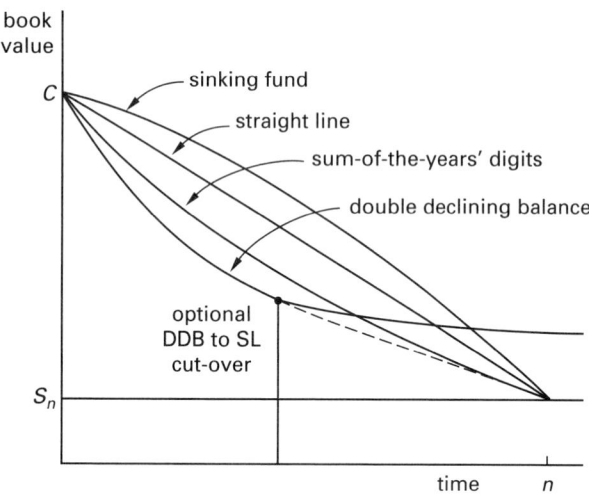

40. AMORTIZATION

Amortization and depreciation are similar in that they both divide up the cost basis or value of an asset. In fact, in certain cases, the term "amortization" may be used in place of the term "depreciation." However, depreciation is a specific form of amortization.

Amortization spreads the cost basis or value of an asset over some base. The base can be time, units of production, number of customers, and so on. The asset can be tangible (e.g., a delivery truck or building) or intangible (e.g., goodwill or a patent).

If the asset is tangible, if the base is time, and if the length of time is consistent with accounting standards and taxation guidelines, then the term "depreciation" is appropriate. However, if the asset is intangible, if the base is some other variable, or if some length of time other than the customary period is used, then the term "amortization" is more appropriate.[30]

Example 54.19

A company purchases complete and exclusive patent rights to an invention for $1,200,000. It is estimated that once commercially produced, the invention will have a

---

[30]From time to time, the U.S. Congress has allowed certain types of facilities (e.g., emergency, grain storage, and pollution control) to be written off more rapidly than would otherwise be permitted in order to encourage investment in such facilities. The term "amortization" has been used with such write-off periods.

specific but limited market of 1200 units. For the purpose of allocating the patent right cost to production cost, what is the amortization rate in dollars per unit?

Solution

The patent should be amortized at the rate of

$$\frac{\$1,200,000}{1200 \text{ units}} = \$1000 \text{ per unit}$$

41. DEPLETION

Depletion is another artificial deductible operating expense, designed to compensate mining organizations for decreasing mineral reserves. Since original and remaining quantities of minerals are seldom known accurately, the *depletion allowance* is calculated as a fixed percentage of the organization's gross income. These percentages are usually in the 10–20% range and apply to such mineral deposits as oil, natural gas, coal, uranium, and most metal ores.

42. BASIC INCOME TAX CONSIDERATIONS

The issue of income taxes is often overlooked in academic engineering economic analysis exercises. Such a position is justifiable when an organization (e.g., a non-profit school, a church, or the government) pays no income taxes. However, if an individual or organization is subject to income taxes, the income taxes must be included in an economic analysis of investment alternatives.

Assume that an organization pays a fraction, f, of its profits to the federal government as income taxes. If the organization also pays a fraction s of its profits as state income taxes and if state taxes paid are recognized by the federal government as tax-deductible expenses, then the composite tax rate is

$$t = s + f - sf \qquad \textit{54.41}$$

The basic principles used to incorporate taxation into engineering economic analyses are the following.

- Initial purchase expenditures are unaffected by income taxes.

- Salvage revenues are unaffected by income taxes.

- Deductible expenses, such as operating costs, maintenance costs, and interest payments, are reduced by the fraction t (i.e., multiplied by the quantity $(1 - t)$).

- Revenues are reduced by the fraction t (i.e., multiplied by the quantity $(1 - t)$).

- Since tax regulations allow the depreciation in any year to be handled as if it were an actual operating expense, and since operating expenses are deductible from the income base prior to taxation, the after-tax profits will be increased. If D is the depreciation, the net result to the after-tax cash flow will be the addition of tD. Depreciation is multiplied by t and added to the appropriate year's cash flow, increasing that year's present worth.

For simplicity, most engineering economics practice problems involving income taxes specify a single income tax rate. In practice, however, federal and most state tax rates depend on the income level. Each range of incomes and its associated tax rate are known as *income bracket* and *tax bracket*, respectively. For example, the state income tax rate might be 4% for incomes up to and including $30,000, and 5% for incomes above $30,000. The income tax for a taxpaying entity with an income of $50,000 would have to be calculated in two parts.

$$\text{tax} = (0.04)(\$30,000) + (0.05)(\$50,000 - \$30,000)$$
$$= \$2200$$

Income taxes and depreciation have no bearing on municipal or governmental projects since municipalities, states, and the U.S. government pay no taxes.

Example 54.20

A corporation that pays 53% of its profit in income taxes invests $10,000 in an asset that will produce a $3000 annual revenue for eight years. If the annual expenses are $700, salvage after eight years is $500, and 9% interest is used, what is the after-tax present worth? Disregard depreciation.

Solution

$$P = -\$10,000 + (\$3000)(P/A, 9\%, 8)(1 - 0.53)$$
$$- (\$700)(P/A, 9\%, 8)(1 - 0.53)$$
$$+ (\$500)(P/F, 9\%, 8)$$
$$= -\$10,000 + (\$3000)(5.5348)(0.47)$$
$$- (\$700)(5.5348)(0.47) + (\$500)(0.5019)$$
$$= -\$3766$$

43. TAXATION AT THE TIMES OF ASSET PURCHASE AND SALE

There are numerous rules and conventions that governmental tax codes and the accounting profession impose on organizations. Engineers are not normally expected to be aware of most of the rules and conventions, but occasionally it may be necessary to incorporate their effects into an engineering economic analysis.

Tax Credit

A *tax credit* (also known as an *investment tax credit* or *investment credit*) is a one-time credit against income taxes.[31] Therefore, it is added to the after-tax present worth as a last step in an engineering economic analysis. Such tax credits may be allowed by the government from time to time for equipment purchases, employment of various classes of workers, rehabilitation of historic landmarks, and so on.

A tax credit is usually calculated as a fraction of the initial purchase price or cost of an asset or activity.

$$TC = \text{fraction} \times \text{initial cost} \qquad 54.42$$

When the tax credit is applicable, the fraction used is subject to legislation. A professional tax expert or accountant should be consulted prior to applying the tax credit concept to engineering economic analysis problems.

Since the investment tax credit reduces the buyer's tax liability, a tax credit should be included only in after-tax engineering economic analyses. The credit is assumed to be received at the end of the year.

Gain or Loss on the Sale of a Depreciated Asset

If an asset that has been depreciated over a number of prior years is sold for more than its current book value, the difference between the book value and selling price is taxable income in the year of the sale. Alternatively, if the asset is sold for less than its current book value, the difference between the selling price and book value is an expense in the year of the sale.

Example 54.21

One year, a company makes a $5000 investment in a historic building. The investment is not depreciable, but it does qualify for a one-time 20% tax credit. In that same year, revenue is $45,000 and expenses (exclusive of the $5000 investment) are $25,000. The company pays a total of 53% in income taxes. What is the after-tax present worth of this year's activities if the company's interest rate for investment is 10%?

Solution

The tax credit is

$$TC = (0.20)(\$5000) = \$1000$$

---

[31]Strictly, *tax credit* is the more general term, and applies to a credit for doing anything creditable. An *investment tax credit* requires an investment in something (usually real property or equipment).

This tax credit is assumed to be received at the end of the year. The after-tax present worth is

$$P = -\$5000 + (\$45,000 - \$25,000)(1 - 0.53)$$
$$\times (P/F, 10\%, 1) + (\$1000)(P/F, 10\%, 1)$$
$$= -\$5000 + (\$20,000)(0.47)(0.9091)$$
$$+ (\$1000)(0.9091)$$
$$= \$4455$$

44. DEPRECIATION RECOVERY

The economic effect of depreciation is to reduce the income tax in year j by tD_j. The present worth of the asset is also affected: The present worth is increased by $tD_j(P/F, i\%, j)$. The after-tax present worth of all depreciation effects over the depreciation period of the asset is called the *depreciation recovery* (DR).[32]

$$DR = t\sum_{j=1}^{n} D_j(P/F, i\%, j) \qquad 54.43$$

There are multiple ways depreciation can be calculated, as summarized in Table 54.5. *Straight line depreciation recovery* from an asset is easily calculated, since the depreciation is the same each year. Assuming the asset has a constant depreciation of D and depreciation period of n years, the depreciation recovery is

$$DR = tD(P/A, i\%, n) \qquad 54.44$$

$$D = \frac{C - S_n}{n} \qquad 54.45$$

Sum-of-the-years' digits depreciation recovery is also relatively easily calculated, since the depreciation decreases uniformly each year.

$$DR = \left(\frac{t(C - S_n)}{T}\right)\left(n(P/A, i\%, n) - (P/G, i\%, n)\right)$$
$$54.46$$

Finding *declining balance depreciation recovery* is more involved. There are three difficulties. The first (the apparent need to calculate all previous depreciations in order to determine the subsequent depreciation) has already been addressed by Eq. 54.32.

The second difficulty is that there is no way to ensure (that is, to force) the book value to be S_n at $t = n$. Therefore, it is common to write off the remaining book value (down to S_n) at $t = n$ in one lump sum. This assumes $BV_n \geq S_n$.

The third difficulty is that of finding the present worth of an *exponentially decreasing cash flow*. Although the

proof is omitted here, such exponential cash flows can be handled with the *exponential gradient factor*, (P/EG).[33]

$$(P/EG, z - 1, n) = \frac{z^n - 1}{z^n(z - 1)} \qquad 54.47$$

$$z = \frac{1 + i}{1 - d} \qquad 54.48$$

Then, as long as $BV_n > S_n$, the declining balance depreciation recovery is

$$DR = tC\left(\frac{d}{1 - d}\right)(P/EG, z - 1, n) \qquad 54.49$$

Example 54.22

For the asset described in Ex. 54.17, calculate the after-tax depreciation recovery with straight line and sum-of-the-years' digits depreciation methods. Use 6% interest with 48% income taxes.

Solution

SL:
$$DR = (0.48)(\$880)(P/A, 6\%, 10)$$
$$= (0.48)(\$880)(7.3601)$$
$$= \$3109$$

SOYD: The depreciation series can be thought of as a constant \$1600 term with a negative \$160 gradient.

$$DR = (0.48)(\$1600)(P/A, 6\%, 10)$$
$$- (0.48)(\$160)(P/G, 6\%, 10)$$
$$= (0.48)(\$1600)(7.3601)$$
$$- (0.48)(\$160)(29.6023)$$
$$= \$3379$$

The ten-year (P/G) factor is used even though there are only nine years in which the gradient reduces the initial \$1600 amount.

Example 54.23

What is the after-tax present worth of the asset described in Ex. 54.20 if straight line, sum-of-the-years' digits, and double declining balance depreciation methods are used?

Solution

Using SL, the depreciation recovery is

$$DR = (0.53)\left(\frac{\$10,000 - \$500}{8}\right)(P/A, 9\%, 8)$$
$$= (0.53)\left(\frac{\$9500}{8}\right)(5.5348)$$
$$= \$3483$$

---

[32]Since the depreciation benefit is reduced by taxation, depreciation cannot be thought of as an annuity to fund a replacement asset.

[33]The (P/A) columns in App. 54.B can be used for (P/EG) as long as the interest rate is assumed to be $z - 1$.

Table 54.5 Depreciation Calculation Summary

method	depreciation basis	depreciation in year j, D_j	book value after jth depreciation, BV_j	present worth of after-tax depreciation recovery, DR	supplementary formulas
straight line (SL)	$C - S_n$	$\dfrac{C - S_n}{n}$ (constant)	$C - jD$	$tD(P/A, i\%, n)$	
constant percentage	$C - S_n$	fraction $\times (C - S_n)$ (constant)	$C - jD$	$tD(P/A, i\%, n)$	
sum-of-the-years' digits (SOYD)	$C - S_n$	$\dfrac{(C - S_n) \times (n - j + 1)}{T}$	$(C - S_n) \times \left(1 - \dfrac{j(2n + 1 - j)}{n(n + 1)}\right) + S_n$	$\dfrac{t(C - S_n)}{T} \times \big(n(P/A, i\%, n) - (P/G, i\%, n)\big)$	$T = \tfrac{1}{2}n(n + 1)$
double declining balance (DDB)	C	$dC(1 - d)^{j-1}$	$C(1 - d)^j$	$tC\left(\dfrac{d}{1 - d}\right) \times (P/EG, z - 1, n)$	$d = \dfrac{2}{n}; \; z = \dfrac{1 + i}{1 - d}$ $(P/EG, z - 1, n) = \dfrac{z^n - 1}{z^n(z - 1)}$
sinking fund (SF)	$C - S_n$	$(C - S_n) \times (A/F, i\%, n) \times (F/P, i\%, j - 1)$	$C - (C - S_n) \times (A/F, i\%, n) \times (F/A, i\%, j)$	$\dfrac{t(C - S_n)(A/F, i\%, n)}{1 + i}$	
accelerated cost recovery system (ACRS/ MACRS)	C	$C \times$ factor	$C - \displaystyle\sum_{m=1}^{j} D_m$	$t \displaystyle\sum_{j=1}^{n} D_j(P/F, i\%, j)$	
units of production or service output	$C - S_n$	$(C - S_n) \times \left(\dfrac{\text{actual output in year } j}{\text{lifetime output}}\right)$	$C - \displaystyle\sum_{m=1}^{j} D_m$	$t \displaystyle\sum_{j=1}^{n} D_j(P/F, i\%, j)$	

Using SOYD, the depreciation recovery is calculated as follows.

$$T = \left(\tfrac{1}{2}\right)(8)(9) = 36$$

$$\text{depreciation base} = \$10{,}000 - \$500 = \$9500$$

$$D_1 = \left(\tfrac{8}{36}\right)(\$9500) = \$2111$$

$$G = \left(\tfrac{1}{36}\right)(\$9500) = \$264$$

$$\text{DR} = (0.53)\Big(($2111)(P/A, 9\%, 8)$$

$$- (\$264)(P/G, 9\%, 8)\Big)$$

$$= (0.53)\Big((\$2111)(5.5348)$$

$$- (\$264)(16.8877)\Big)$$

$$= \$3830$$

Using DDB, the depreciation recovery is calculated as follows.[34]

$$d = \frac{2}{8} = 0.25$$

$$z = \frac{1 + 0.09}{1 - 0.25} = 1.4533$$

$$(P/EG, z - 1, n) = \frac{(1.4533)^8 - 1}{(1.4533)^8(0.4533)} = 2.095$$

---

[34]This method should start by checking that the book value at the end of the depreciation period is greater than the salvage value. In this example, such is the case. However, the step is not shown.

ON

54-28 ENVIRONMENTAL ENGINEERING REFERENCE MANUAL

From Eq. 54.49,

$$DR = (0.53)\left(\frac{(0.25)(\$10{,}000)}{0.75}\right)(2.095)$$

$$= \$3701$$

The after-tax present worth, neglecting depreciation, was previously found to be $-\$3766$.

The after-tax present worths, including depreciation recovery, are

SL: $P = -\$3766 + \$3483 = -\$283$

SOYD: $P = -\$3766 + \$3830 = \$64$

DDB: $P = -\$3766 + \$3701 = -\$65$

45. OTHER INTEREST RATES

The *effective interest rate per period*, i (also called *yield* by banks), is the only interest rate that should be used in equivalence equations. The interest rates at the top of the factor tables in App. 54.B are implicitly all effective interest rates. Usually, the period will be one year, hence the name *effective annual interest rate*. However, there are other interest rates in use as well.

The term *nominal interest rate*, r (*rate per annum*), is encountered when compounding is more than once per year. The nominal rate does not include the effect of compounding and is not the same as the effective rate. And, since the effective interest rate can be calculated from the nominal rate only if the number of compounding periods per year is known, nominal rates cannot be compared unless the method of compounding is specified. The only practical use for a nominal rate per year is for calculating the effective rate per period.

46. RATE AND PERIOD CHANGES

If there are k compounding periods during the year (two for semiannual compounding, four for quarterly compounding, twelve for monthly compounding, etc.) and the nominal rate is r, the *effective rate per compounding period* is

$$\phi = \frac{r}{k} \qquad 54.50$$

The effective annual rate, i, can be calculated from the effective rate per period, ϕ, by using Eq. 54.51.

$$i = (1+\phi)^k - 1$$

$$= \left(1+\frac{r}{k}\right)^k - 1 \qquad 54.51$$

Sometimes, only the effective rate per period (e.g., per month) is known. However, that will be a simple

problem since compounding for n periods at an effective rate per period is not affected by the definition or length of the period.

The following rules may be used to determine which interest rate is given.

- Unless specifically qualified, the interest rate given is an annual rate.

- If the compounding is annual, the rate given is the effective rate. If compounding is other than annual, the rate given is the nominal rate.

The effective annual interest rate determined on a *daily compounding basis* will not be significantly different than if *continuous compounding* is assumed.[35] In the case of continuous (or daily) compounding, the discounting factors can be calculated directly from the nominal interest rate and number of years, without having to find the effective interest rate per period. Table 54.6 can be used to determine the discount factors for continuous compounding.

Table 54.6 Discount Factors for Continuous Compounding (n is the number of years)

symbol	formula
$(F/P, r\%, n)$	e^{rn}
$(P/F, r\%, n)$	e^{-rn}
$(A/F, r\%, n)$	$\dfrac{e^r - 1}{e^{rn} - 1}$
$(F/A, r\%, n)$	$\dfrac{e^{rn} - 1}{e^r - 1}$
$(A/P, r\%, n)$	$\dfrac{e^r - 1}{1 - e^{-rn}}$
$(P/A, r\%, n)$	$\dfrac{1 - e^{-rn}}{e^r - 1}$

Example 54.24

A savings and loan offers a nominal rate of 5.25% compounded daily over 365 days in a year. What is the effective annual rate?

Solution

method 1: Use Eq. 54.51.

$$r = 0.0525, \quad k = 365$$

$$i = \left(1+\frac{r}{k}\right)^k - 1 = \left(1+\frac{0.0525}{365}\right)^{365} - 1 = 0.0539$$

---

[35]The number of *banking days in a year* (250, 360, etc.) must be specifically known.

PPI • www.ppi2pass.com

method 2: Assume daily compounding is the same as continuous compounding.

$$i = (F/P, r\%, 1) - 1$$
$$= e^{0.0525} - 1$$
$$= 0.0539$$

Example 54.25

A real estate investment trust pays \$7,000,000 for an apartment complex with 100 units. The trust expects to sell the complex in ten years for \$15,000,000. In the meantime, it expects to receive an average rent of \$900 per month from each apartment. Operating expenses are expected to be \$200 per month per occupied apartment. A 95% occupancy rate is predicted. In similar investments, the trust has realized a 15% effective annual return on its investment. Compare to those past investments the expected present worth of this investment when calculated assuming (a) annual compounding (i.e., the year-end convention), and (b) monthly compounding. Disregard taxes, depreciation, and all other factors.

Solution

(a) The net annual income will be

$$(0.95)(100 \text{ units})\left(\frac{\$900}{\text{unit-mo}} - \frac{\$200}{\text{unit-mo}}\right)\left(12 \frac{\text{mo}}{\text{yr}}\right)$$
$$= \$798,000/\text{yr}$$

The present worth of ten years of operation is

$$P = -\$7,000,000 + (\$798,000)(P/A, 15\%, 10)$$
$$+ (\$15,000,000)(P/F, 15\%, 10)$$
$$= -\$7,000,000 + (\$798,000)(5.0188)$$
$$+ (\$15,000,000)(0.2472)$$
$$= \$713,000$$

(b) The net monthly income is

$$(0.95)(100 \text{ units})\left(\frac{\$900}{\text{unit-mo}} - \frac{\$200}{\text{unit-mo}}\right)$$
$$= \$66,500/\text{mo}$$

Equation 54.51 is used to calculate the effective monthly rate, ϕ, from the effective annual rate, $i = 15\%$, and the number of compounding periods per year, $k = 12$.

$$\phi = (1 + i)^{1/k} - 1$$
$$= (1 + 0.15)^{1/12} - 1$$
$$= 0.011715 \quad (1.1715\%)$$

The number of compounding periods in ten years is

$$n = (10 \text{ yr})\left(12 \frac{\text{mo}}{\text{yr}}\right) = 120 \text{ mo}$$

The present worth of 120 months of operation is

$$P = -\$7,000,000 + (\$66,500)(P/A, 1.1715\%, 120)$$
$$+ (\$15,000,000)(P/F, 1.1715\%, 120)$$

Since table values for 1.1715% discounting factors are not available, the factors are calculated from Table 54.1.

$$(P/A, 1.1715\%, 120) = \frac{(1+i)^n - 1}{i(1+i)^n}$$
$$= \frac{(1 + 0.011715)^{120} - 1}{(0.011715)(1 + 0.011715)^{120}}$$
$$= 64.261$$

$$(P/F, 1.1715\%, 120) = (1+i)^{-n} = (1 + 0.011715)^{-120}$$
$$= 0.2472$$

The present worth over 120 monthly compounding periods is

$$P = -\$7,000,000 + (\$66,500)(64.261)$$
$$+ (\$15,000,000)(0.2472)$$
$$= \$981,357$$

47. BONDS

A *bond* is a method of long-term financing commonly used by governments, states, municipalities, and very large corporations.[36] The bond represents a contract to pay the bondholder specific amounts of money at specific times. The holder purchases the bond in exchange for specific payments of interest and principal. Typical municipal bonds call for quarterly or semiannual interest payments and a payment of the *face value of the bond* on the *date of maturity* (end of the bond period).[37] Due to the practice of discounting in the bond market, a bond's face value and its purchase price generally will not coincide.

In the past, a bondholder had to submit a coupon or ticket in order to receive an interim interest payment. This has given rise to the term *coupon rate*, which is the nominal annual interest rate on which the interest payments are made. Coupon books are seldom used with modern bonds, but the term survives. The coupon rate determines the magnitude of the semiannual (or otherwise) interest payments during the life of the bond. The

---

[36]In the past, 30-year bonds were typical. Shorter term 10-year, 15-year, 20-year, and 25-year bonds are also commonly issued.
[37]A *fully amortized bond* pays back interest and principal throughout the life of the bond. There is no balloon payment.

bondholder's own effective interest rate should be used for economic decisions about the bond.

Actual *bond yield* is the bondholder's actual rate of return of the bond, considering the purchase price, interest payments, and face value payment (or, value realized if the bond is sold before it matures). By convention, bond yield is calculated as a nominal rate (rate per annum), not an effective rate per year. The bond yield should be determined by finding the effective rate of return per payment period (e.g., per semiannual interest payment) as a conventional rate of return problem. Then, the nominal rate can be found by multiplying the effective rate per period by the number of payments per year, as in Eq. 54.51.

Example 54.26

What is the maximum amount an investor should pay for a 25-year bond with a $20,000 face value and 8% coupon rate (interest only paid semiannually)? The bond will be kept to maturity. The investor's effective annual interest rate for economic decisions is 10%.

Solution

For this problem, take the compounding period to be six months. Then, there are 50 compounding periods. Since 8% is a nominal rate, the effective bond rate per period is calculated from Eq. 54.50 as

$$\phi_{bond} = \frac{r}{k} = \frac{8\%}{2} = 4\%$$

The bond payment received semiannually is

$$(0.04)(\$20{,}000) = \$800$$

10% is the investor's effective rate per year, so Eq. 54.51 is again used to calculate the effective analysis rate per period.

$$0.10 = (1+\phi)^2 - 1$$
$$\phi = 0.04881 \quad (4.88\%)$$

The maximum amount that the investor should be willing to pay is the present worth of the investment.

$$P = (\$800)(P/A, 4.88\%, 50)$$
$$+ (\$20{,}000)(P/F, 4.88\%, 50)$$

Table 54.1 can be used to calculate the following factors.

$$(P/A, 4.88\%, 50) = \frac{(1+i)^n - 1}{i(1+i)^n}$$
$$= \frac{(1+0.0488)^{50} - 1}{(0.0488)(1.0488)^{50}}$$
$$= 18.600$$

$$(P/F, 4.88\%, 50) = \frac{1}{(1+i)^n}$$
$$= \frac{1}{(1+0.0488)^{50}}$$
$$= 0.09233$$

Then, the present worth is

$$P = (\$800)(18.600) + (\$20{,}000)(0.09233)$$
$$= \$16{,}727$$

48. PROBABILISTIC PROBLEMS

If an alternative's cash flows are specified by an implicit or explicit probability distribution rather than being known exactly, the problem is *probabilistic*.

Probabilistic problems typically possess the following characteristics.

- There is a chance of loss that must be minimized (or, rarely, a chance of gain that must be maximized) by selection of one of the alternatives.

- There are multiple alternatives. Each alternative offers a different degree of protection from the loss. Usually, the alternatives with the greatest protection will be the most expensive.

- The magnitude of loss or gain is independent of the alternative selected.

Probabilistic problems are typically solved using annual costs and expected values. An *expected value* is similar to an *average value* since it is calculated as the mean of the given probability distribution. If cost 1 has a probability of occurrence, p_1, cost 2 has a probability of occurrence, p_2, and so on, the expected value is

$$\mathcal{E}\{\text{cost}\} = p_1(\text{cost } 1) + p_2(\text{cost } 2) + \cdots \qquad 54.52$$

Example 54.27

Flood damage in any year is given according to the following table. What is the present worth of flood damage for a ten-year period? Use 6% as the effective annual interest rate.

damage	probability
0	0.75
$10,000	0.20
$20,000	0.04
$30,000	0.01

Solution

The expected value of flood damage in any given year is

$$\mathcal{E}\{\text{damage}\} = (0)(0.75) + (\$10{,}000)(0.20)$$
$$+ (\$20{,}000)(0.04) + (\$30{,}000)(0.01)$$
$$= \$3100$$

The present worth of ten years of expected flood damage is

$$\text{present worth} = (\$3100)(P/A, 6\%, 10)$$
$$= (\$3100)(7.3601)$$
$$= \$22,816$$

Example 54.28

A dam is being considered on a river that periodically overflows and causes \$600,000 damage. The damage is essentially the same each time the river causes flooding. The project horizon is 40 years. A 10% interest rate is being used.

Three different designs are available, each with different costs and storage capacities.

design alternative	cost	maximum capacity
A	\$500,000	1 unit
B	\$625,000	1.5 units
C	\$900,000	2.0 units

The National Weather Service has provided a statistical analysis of annual rainfall runoff from the watershed draining into the river.

units annual rainfall	probability
0	0.10
0.1–0.5	0.60
0.6–1.0	0.15
1.1–1.5	0.10
1.6–2.0	0.04
2.1 or more	0.01

Which design alternative would you choose assuming the dam is essentially empty at the start of each rainfall season?

Solution

The sum of the construction cost and the expected damage should be minimized. If alternative A is chosen, it will have a capacity of 1 unit. Its capacity will be exceeded (causing \$600,000 damage) when the annual rainfall exceeds 1 unit. Therefore, the expected value of the annual cost of alternative A is

$$\mathcal{E}\{\text{EUAC(A)}\} = (\$500,000)(A/P, 10\%, 40)$$
$$+ (\$600,000)(0.10 + 0.04 + 0.01)$$
$$= (\$500,000)(0.1023) + (\$600,000)(0.15)$$
$$= \$141,150$$

Similarly,

$$\mathcal{E}\{\text{EUAC(B)}\} = (\$625,000)(A/P, 10\%, 40)$$
$$+ (\$600,000)(0.04 + 0.01)$$
$$= (\$625,000)(0.1023) + (\$600,000)(0.05)$$
$$= \$93,938$$

$$\mathcal{E}\{\text{EUAC(C)}\} = (\$900,000)(A/P, 10\%, 40)$$
$$+ (\$600,000)(0.01)$$
$$= (\$900,000)(0.1023) + (\$600,000)(0.01)$$
$$= \$98,070$$

Alternative B should be chosen.

49. WEIGHTED COSTS

The reliability of preliminary cost estimates can be increased by considering as much historical data as possible. For example, the cost of finishing a concrete slab when the contractor has not yet been selected should be estimated from actual recent costs from as many local jobs as is practical. Most jobs are not directly comparable, however, because they differ in size or in some other characteristic. A *weighted cost* (*weighted average cost*) is a cost that has been averaged over some rational basis. The weighted cost is calculated by weighting the individual cost elements, C_i, by their respective weights, w_i. Respective weights are usually *relative fractions* (*relative importance*) based on other characteristics, such as length, area, number of units, frequency of occurrence, points, etc.

$$C_{\text{weighted}} = w_1 C_1 + w_2 C_2 + \cdots + w_N C_N \qquad 54.53$$

$$w_j = \frac{A_j}{\sum_{i=1}^{N} A_i} \qquad \text{[weighting by area]} \qquad 54.54$$

It is implicit in calculating weighted average costs that the costs with the largest weights are more important to the calculation. For example, costs from a job of 10,000 ft^2 are considered to be twice as reliable (important, relevant, etc.) as costs from a 5000 ft^2 job. This assumption must be carefully considered.

Determining a weighted average from Eq. 54.53 disregards the fixed and variable natures of costs. In effect, fixed costs are allocated over entire jobs, increasing the apparent variable costs. For that reason, it is important to include in the calculation only costs of similar cases. This requires common sense segregation of the initial cost data. For example, it is probably not appropriate to include very large (e.g., supermarket and mall) concrete finishing job costs in the calculation of a weighted average cost used to estimate residential slab finishing costs.

Example 54.29

Calculate a weighted average cost per unit length of concrete wall to be used for estimating future job costs. Data from three similar walls are available.

wall 1: length, 50 ft; actual cost, \$9000
wall 2: length, 90 ft; actual cost, \$15,000
wall 3: length, 65 ft; actual cost, \$11,000

Solution

Consider wall 1. The cost per foot of wall is $9000/50 ft. According to Eq. 54.54, accuracy of cost data is proportional to wall length (i.e., cost per foot from a wall of length $2X$ is twice as reliable as cost per foot from a wall of length X). This cost per foot value is given a weight of 50 ft/(50 ft + 90 ft + 65 ft). So, the weighted cost per foot for wall 1 is

$$ w_1 C_{ft,1} = \frac{(50 \text{ ft}) \left(\dfrac{\$9000}{50 \text{ ft}} \right)}{50 \text{ ft} + 90 \text{ ft} + 65 \text{ ft}} = \frac{\$9000}{50 \text{ ft} + 90 \text{ ft} + 65 \text{ ft}} $$

Similarly, the weighted cost per foot for wall 2 is $w_2 C_{ft,2} = \$15,000/(50 \text{ ft} + 90 \text{ ft} + 65 \text{ ft})$, and the weighted cost per foot for wall 3 is $w_3 C_{ft,3} = \$11,000/(50 \text{ ft} + 90 \text{ ft} + 65 \text{ ft})$.

From Eq. 54.53, the total weighted cost per foot is

$$ C_{ft} = w_1 C_{ft,1} + w_2 C_{ft,2} + w_3 C_{ft,3} $$
$$ = \frac{\$9000 + \$15,000 + \$11,000}{50 \text{ ft} + 90 \text{ ft} + 65 \text{ ft}} $$
$$ = \$170.73 \text{ per foot} $$

50. FIXED AND VARIABLE COSTS

The distinction between fixed and variable costs depends on how these costs vary when an independent variable changes. For example, factory or machine production is frequently the independent variable. However, it could just as easily be vehicle miles driven, hours of operation, or quantity (mass, volume, etc.). Examples of fixed and variable costs are given in Table 54.7.

If a cost is a function of the independent variable, the cost is said to be a *variable cost*. The change in cost per unit variable change (i.e., what is usually called the *slope*) is known as the *incremental cost*. Material and labor costs are examples of variable costs. They increase in proportion to the number of product units manufactured.

If a cost is not a function of the independent variable, the cost is said to be a *fixed cost*. Rent and lease payments are typical fixed costs. These costs will be incurred regardless of production levels.

Some costs have both fixed and variable components, as Fig. 54.9 illustrates. The fixed portion can be determined by calculating the cost at zero production.

An additional category of cost is the *semivariable cost*. This type of cost increases stepwise. Semivariable cost structures are typical of situations where *excess capacity* exists. For example, supervisory cost is a stepwise function of the number of production shifts. Also, labor cost for truck drivers is a stepwise function of weight

Table 54.7 *Summary of Fixed and Variable Costs*

fixed costs
    rent
    property taxes
    interest on loans
    insurance
    janitorial service expense
    tooling expense
    setup, cleanup, and tear-down expenses
    depreciation expense
    marketing and selling costs
    cost of utilities
    general burden and overhead expense
variable costs
    direct material costs
    direct labor costs
    cost of miscellaneous supplies
    payroll benefit costs
    income taxes
    supervision costs

Figure 54.9 *Fixed and Variable Costs*

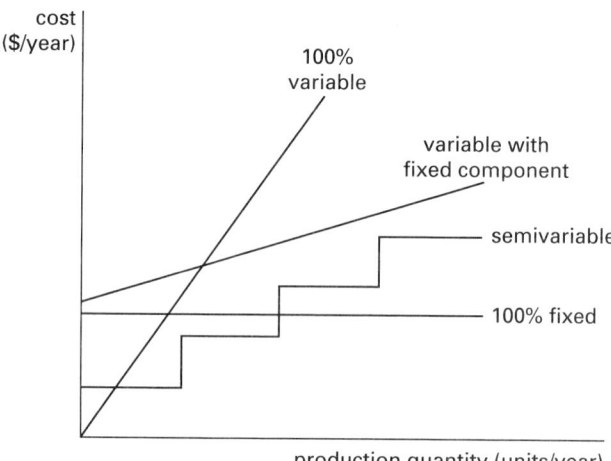

(volume) transported. As long as a truck has room left (i.e., excess capacity), no additional driver is needed. As soon as the truck is filled, labor cost will increase.

51. ACCOUNTING COSTS AND EXPENSE TERMS

The accounting profession has developed special terms for certain groups of costs. When annual costs are incurred due to the functioning of a piece of equipment, they are known as *operating and maintenance* (O&M) *costs*. The annual costs associated with operating a business (other than the costs directly attributable to production) are known as *general, selling, and administrative* (GS&A) *expenses*.

Direct labor costs are costs incurred in the factory, such as assembly, machining, and painting labor costs. *Direct material costs* are the costs of all materials that go into production.[38] Typically, both direct labor and direct material costs are given on a per-unit or per-item basis. The sum of the direct labor and direct material costs is known as the *prime cost*.

There are certain additional expenses incurred in the factory, such as the costs of factory supervision, stock-picking, quality control, factory utilities, and miscellaneous supplies (cleaning fluids, assembly lubricants, routing tags, etc.) that are not incorporated into the final product. Such costs are known as *indirect manufacturing expenses* (IME) or *indirect material and labor costs*.[39] The sum of the per-unit indirect manufacturing expense and prime cost is known as the *factory cost*.

Research and development (R&D) *costs* and *administrative expenses* are added to the factory cost to give the *manufacturing cost* of the product.

Additional costs are incurred in marketing the product. Such costs are known as *selling expenses* or *marketing expenses*. The sum of the selling expenses and manufacturing cost is the *total cost* of the product. Figure 54.10 illustrates these terms.[40]

Figure 54.10 Costs and Expenses Combined

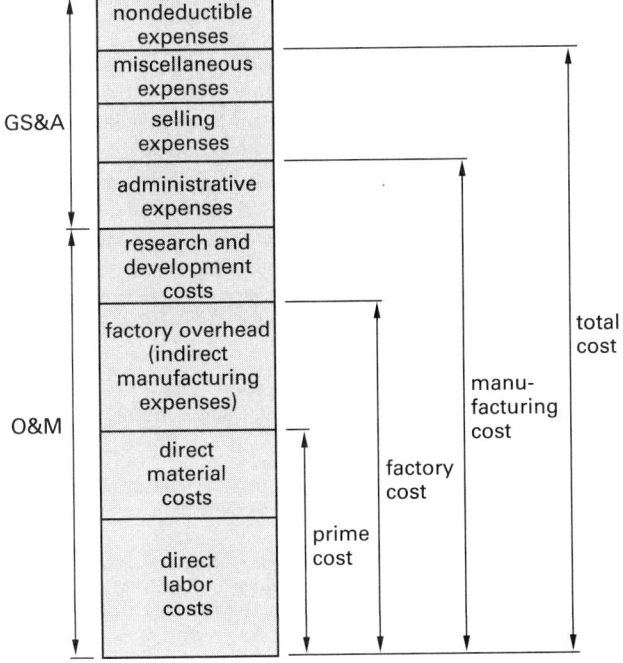

The distinctions among the various forms of cost (particularly with overhead costs) are not standardized. Each company must develop a classification system to deal with the various cost factors in a consistent manner. There are also other terms in use (e.g., *raw*

materials, *operating supplies, general plant overhead*), but these terms must be interpreted within the framework of each company's classification system. Table 54.8 is typical of such classification systems.

52. ACCOUNTING PRINCIPLES

Basic Bookkeeping

An accounting or *bookkeeping system* is used to record historical financial transactions. The resultant records are used for product costing, satisfaction of statutory requirements, reporting of profit for income tax purposes, and general company management.

Bookkeeping consists of two main steps: recording the transactions, followed by categorization of the transactions.[41] The transactions (receipts and disbursements) are recorded in a *journal* (*book of original entry*) to complete the first step. Such a journal is organized in a simple chronological and sequential manner. The transactions are then categorized (into interest income, advertising expense, etc.) and posted (i.e., entered or written) into the appropriate *ledger account*.[42]

The ledger accounts together constitute the *general ledger* or *ledger*. All ledger accounts can be classified into one of three types: *asset accounts, liability accounts*, and *owners' equity accounts*. Strictly speaking, income and expense accounts, kept in a separate journal, are included within the classification of owners' equity accounts.

Together, the journal and ledger are known simply as "the books" of the company, regardless of whether bound volumes of pages are actually involved.

Balancing the Books

In a business environment, *balancing the books* means more than reconciling the checkbook and bank statements. All accounting entries must be posted in such a way as to maintain the equality of the *basic accounting equation*,

$$\text{assets} = \text{liability} + \text{owner's equity} \qquad 54.55$$

In a *double-entry bookkeeping system*, the equality is maintained within the ledger system by entering each transaction into two balancing ledger accounts. For example, paying a utility bill would decrease the cash account (an asset account) and decrease the utility expense account (a liability account) by the same amount.

---

[38]There may be problems with pricing the material when it is purchased from an outside vendor and the stock on hand derives from several shipments purchased at different prices.
[39]The *indirect material and labor costs* usually exclude costs incurred in the office area.
[40]*Total cost* does not include income taxes.

[41]These two steps are not to be confused with the *double-entry bookkeeping method*.
[42]The two-step process is more typical of a *manual bookkeeping system* than a computerized *general ledger system*. However, even most computerized systems produce reports in journal entry order, as well as account summaries.

Table 54.8 *Typical Classification of Expenses*

direct labor expenses
  machining and forming
  assembly
  finishing
  inspection
  testing
direct material expenses
  items purchased from other vendors
  manufactured assemblies
factory overhead expenses (indirect manufacturing expenses)
  supervision
  benefits
    pension
    medical insurance
    vacations
  wages overhead
    unemployment compensation taxes
    social security taxes
    disability taxes
  stock-picking
  quality control and inspection
  expediting
  rework
  maintenance
  miscellaneous supplies
    routing tags
    assembly lubricants
    cleaning fluids
    wiping cloths
    janitorial supplies
  packaging (materials and labor)
  factory utilities
  laboratory
  depreciation on factory equipment
research and development expenses
  engineering (labor)
  patents
  testing
  prototypes (material and labor)
  drafting
  O&M of R&D facility
administrative expenses
  corporate officers
  accounting
  secretarial/clerical/reception
  security (protection)
  medical (nurse)
  employment (personnel)
  reproduction
  data processing
  production control
  depreciation on nonfactory equipment
  office supplies
  office utilities
  O&M of offices
selling expenses
  marketing (labor)
  advertising
  transportation (if not paid by customer)
  outside sales force (labor and expenses)
  demonstration units
  commissions
  technical service and support
  order processing
  branch office expenses
miscellaneous expenses
  insurance
  property taxes
  interest on loans
nondeductible expenses
  federal income taxes
  fines and penalties

Transactions are either *debits* or *credits*, depending on their sign. Increases in asset accounts are debits; decreases are credits. For liability and equity accounts, the opposite is true: Increases are credits, and decreases are debits.[43]

Cash and Accrual Systems[44]

The simplest form of bookkeeping is based on the *cash system*. The only transactions that are entered into the journal are those that represent cash receipts and disbursements. In effect, a checkbook register or bank deposit book could serve as the journal.

During a given period (e.g., month or quarter), expense liabilities may be incurred even though the payments for those expenses have not been made. For example, an invoice (bill) may have been received but not paid. Under the *accrual system*, the obligation is posted into the appropriate expense account before it is paid.[45] Analogous to expenses, under the accrual system, income will be claimed before payment is received. Specifically, a sales transaction can be recorded as income when the customer's order is received, when the outgoing invoice is generated, or when the merchandise is shipped.

Financial Statements

Each period, two types of corporate financial statements are typically generated: the *balance sheet* and *profit and loss* (P&L) *statement*.[46] The profit and loss statement, also known as a *statement of income and retained earnings*, is a summary of sources of *income* or *revenue* (interest, sales, fees charged, etc.) and *expenses* (utilities, advertising, repairs, etc.) for the period. The expenses are subtracted from the revenues to give a *net income* (generally, before taxes).[47] Figure 54.11 illustrates a simplified profit and loss statement.

The *balance sheet* presents the *basic accounting equation* in tabular form. The balance sheet lists the major

---

[43]There is a difference in sign between asset and liability accounts. An increase in an expense account is actually a decrease. The accounting profession, apparently, is comfortable with the common confusion that exists between debits and credits.

[44]There is also a distinction made between cash flows that are known and those that are expected. It is a *standard accounting principle* to record losses in full, at the time they are recognized, even before their occurrence. In the construction industry, for example, losses are recognized in full and projected to the end of a project as soon as they are foreseeable. Profits, on the other hand, are recognized only as they are realized (typically, as a percentage of project completion). The difference between cash and accrual systems is a matter of *bookkeeping*. The difference between loss and profit recognition is a matter of *accounting convention*. Engineers seldom need to be concerned with the accounting tradition.

[45]The expense for an item or service might be accrued even *before* the invoice is received. It might be recorded when the purchase order for the item or service is generated, or when the item or service is received.

[46]Other types of financial statements (*statements of changes in financial position, cost of sales statements*, inventory and asset reports, etc.) also will be generated, depending on the needs of the company.

[47]Financial statements also can be prepared with percentages (of total assets and net revenue) instead of dollars, in which case they are known as *common size financial statements*.

Figure 54.11 *Simplified Profit and Loss Statement*

```
revenue
   interest           2000
   sales           237,000
   returns         (23,000)
   net revenue                216,000
expenses
   salaries        149,000
   utilities          6000
   advertising      28,000
   insurance         4000
   supplies          1000
   net expenses               188,000
period net income                       28,000
beginning retained earnings             63,000
net year-to-date earnings                          91,000
```

categories of assets and outstanding liabilities. The difference between asset values and liabilities is the *equity*, as defined in Eq. 54.55. This equity represents what would be left over after satisfying all debts by liquidating the company.

Figure 54.12 is a simplified balance sheet.

Figure 54.12 *Simplified Balance Sheet*

```
                   ASSETS
current assets
   cash                   14,000
   accounts receivable    36,000
   notes receivable       20,000
   inventory              89,000
   prepaid expenses        3000
   total current assets           162,000
plant, property, and equipment
   land and buildings    217,000
   motor vehicles         31,000
   equipment              94,000
   accumulated
     depreciation        (52,000)
   total fixed assets             290,000
total assets                               452,000
        LIABILITIES AND OWNERS' EQUITY
current liabilities
   accounts payable       66,000
   accrued income taxes   17,000
   accrued expenses        8000
   total current liabilities       91,000
long-term debt
   notes payable         117,000
   mortgage               23,000
   total long-term debt           140,000
owners' and stockholders' equity
   stock                 130,000
   retained earnings      91,000
   total owners' equity           221,000
total liabilities and owners' equity       452,000
```

There are several terms that appear regularly on balance sheets.

- *current assets:* cash and other assets that can be converted quickly into cash, such as accounts receivable, notes receivable, and merchandise (inventory). Also known as *liquid assets.*

- *fixed assets:* relatively permanent assets used in the operation of the business and relatively difficult to convert into cash. Examples are land, buildings, and equipment. Also known as *nonliquid assets.*

- *current liabilities:* liabilities due within a short period of time (e.g., within one year) and typically paid out of current assets. Examples are accounts payable, notes payable, and other accrued liabilities.

- *long-term liabilities:* obligations that are not totally payable within a short period of time (e.g., within one year).

Analysis of Financial Statements

Financial statements are evaluated by management, lenders, stockholders, potential investors, and many other groups for the purpose of determining the *health of the company.* The health can be measured in terms of *liquidity* (ability to convert assets to cash quickly), *solvency* (ability to meet debts as they become due), and *relative risk* (of which one measure is *leverage*—the portion of total capital contributed by owners).

The analysis of financial statements involves several common ratios, usually expressed as percentages. The following are some frequently encountered ratios.

- *current ratio:* an index of short-term paying ability.

$$\text{current ratio} = \frac{\text{current assets}}{\text{current liabilities}} \quad 54.56$$

- *quick* (or *acid-test*) *ratio:* a more stringent measure of short-term debt-paying ability. The *quick assets* are defined to be current assets minus inventories and prepaid expenses.

$$\text{quick ratio} = \frac{\text{quick assets}}{\text{current liabilities}} \quad 54.57$$

- *receivable turnover:* a measure of the average speed with which accounts receivable are collected.

$$\text{receivable turnover} = \frac{\text{net credit sales}}{\text{average net receivables}} \quad 54.58$$

- *average age of receivables:* number of days, on the average, in which receivables are collected.

$$\text{average age of receivables} = \frac{365}{\text{receivable turnover}} \quad 54.59$$

- *inventory turnover:* a measure of the speed with which inventory is sold, on the average.

$$\text{inventory turnover} = \frac{\text{cost of goods sold}}{\text{average cost of inventory on hand}}$$
$$54.60$$

- *days supply of inventory on hand:* number of days, on the average, that the current inventory would last.

$$\text{days supply of inventory on hand} = \frac{365}{\text{inventory turnover}}$$
$$54.61$$

- *book value per share of common stock:* number of dollars represented by the balance sheet owners' equity for each share of common stock outstanding.

$$\text{book value per share of common stock}$$
$$= \frac{\text{common shareholders' equity}}{\text{number of outstanding shares}}$$
$$54.62$$

- *gross margin:* gross profit as a percentage of sales. (Gross profit is sales less cost of goods sold.)

$$\text{gross margin} = \frac{\text{gross profit}}{\text{net sales}}$$
$$54.63$$

- *profit margin ratio:* percentage of each dollar of sales that is net income.

$$\text{profit margin} = \frac{\text{net income before taxes}}{\text{net sales}}$$
$$54.64$$

- *return on investment ratio:* shows the percent return on owners' investment.

$$\text{return on investment} = \frac{\text{net income}}{\text{owners' equity}}$$
$$54.65$$

- *price-earnings ratio:* indication of relationship between earnings and market price per share of common stock, useful in comparisons between alternative investments.

$$\text{price-earnings} = \frac{\text{market price per share}}{\text{earnings per share}}$$
$$54.66$$

53. COST ACCOUNTING

Cost accounting is the system that determines the cost of manufactured products. Cost accounting is called *job cost accounting* if costs are accumulated by part number or contract. It is called *process cost accounting* if costs are accumulated by departments or manufacturing processes.

Cost accounting is dependent on historical and recorded data. The unit product cost is determined from actual expenses and numbers of units produced. Allowances (i.e., budgets) for future costs are based on these historical figures. Any deviation from historical figures is called a *variance*. Where adequate records are available,

variances can be divided into *labor variance* and *material variance*.

When determining a unit product cost, the direct material and direct labor costs are generally clear-cut and easily determined. Furthermore, these costs are 100% variable costs. However, the indirect cost per unit of product is not as easily determined. Indirect costs (*burden, overhead,* etc.) can be fixed or semivariable costs. The amount of indirect cost allocated to a unit will depend on the unknown future overhead expense as well as the unknown future production (*vehicle size*).

A typical method of allocating indirect costs to a product is as follows.

step 1: Estimate the total expected indirect (and overhead) costs for the upcoming year.

step 2: Determine the most appropriate vehicle (basis) for allocating the overhead to production. Usually, this vehicle is either the number of units expected to be produced or the number of direct hours expected to be worked in the upcoming year.

step 3: Estimate the quantity or size of the overhead vehicle.

step 4: Divide expected overhead costs by the expected overhead vehicle to obtain the unit overhead.

step 5: Regardless of the true size of the overhead vehicle during the upcoming year, one unit of overhead cost is allocated per unit of overhead vehicle.

Once the prime cost has been determined and the indirect cost calculated based on projections, the two are combined into a *standard factory cost* or *standard cost*, which remains in effect until the next budgeting period (usually a year).

During the subsequent manufacturing year, the standard cost of a product is not generally changed merely because it is found that an error in projected indirect costs or production quantity (vehicle size) has been made. The allocation of indirect costs to a product is assumed to be independent of errors in forecasts. Rather, the difference between the expected and actual expenses, known as the *burden (overhead) variance*, experienced during the year is posted to one or more *variance accounts*.

Burden (overhead) variance is caused by errors in forecasting both the actual indirect expense for the upcoming year and the overhead vehicle size. In the former case, the variance is called *burden budget variance*; in the latter, it is called *burden capacity variance*.

Example 54.30

A company expects to produce 8000 items in the coming year. The current material cost is $4.54 each. Sixteen minutes of direct labor are required per unit. Workers are paid $7.50 per hour. 2133 direct labor hours are forecasted for the product. Miscellaneous overhead costs are estimated at $45,000.

Find the per-unit (a) expected direct material cost, (b) direct labor cost, (c) prime cost, (d) burden as a function of production and direct labor, and (e) total cost.

Solution

(a) The direct material cost was given as $4.54.

(b) The direct labor cost is

$$\left(\frac{16\ \text{min}}{60\ \frac{\text{min}}{\text{hr}}}\right)\left(\frac{\$7.50}{\text{hr}}\right) = \$2.00$$

(c) The prime cost is

$$\$4.54 + \$2.00 = \$6.54$$

(d) If the burden vehicle is production, the burden rate is $45,000/8000 = $5.63 per item.

If the burden vehicle is direct labor hours, the burden rate is $45,000/2133 = $21.10 per hour.

(e) If the burden vehicle is production, the total cost is

$$\$4.54 + \$2.00 + \$5.63 = \$12.17$$

If the burden vehicle is direct labor hours, the total cost is

$$\$4.54 + \$2.00 + \left(\frac{16\ \text{min}}{60\ \frac{\text{min}}{\text{hr}}}\right)\left(\frac{\$21.10}{\text{hr}}\right) = \$12.17$$

Example 54.31

The actual performance of the company in Ex. 54.30 is given by the following figures.

$$\text{actual production: } 7560$$

$$\text{actual overhead costs: } \$47,000$$

What are the burden budget variance and the burden capacity variance?

Solution

The burden capacity variance is

$$\$45,000 - (7560)(\$5.63) = \$2437$$

The burden budget variance is

$$\$47,000 - \$45,000 = \$2000$$

The overall burden variance is

$$\$47,000 - (7560)(\$5.63) = \$4437$$

The sum of the burden capacity and burden budget variances should equal the overall burden variance.

$$\$2437 + \$2000 = \$4437$$

54. COST OF GOODS SOLD

Cost of goods sold (COGS) is an accounting term that represents an inventory account adjustment.[48] Cost of goods sold is the difference between the starting and ending inventory valuations. That is,

$$\text{COGS} = \text{starting inventory valuation}$$
$$- \text{ending inventory valuation} \quad \textit{54.67}$$

Cost of goods sold is subtracted from *gross profit* to determine the *net profit* of a company. Despite the fact that cost of goods sold can be a significant element in the profit equation, the inventory adjustment may not be made each accounting period (e.g., each month) due to the difficulty in obtaining an accurate inventory valuation.

With a *perpetual inventory system*, a company automatically maintains up-to-date inventory records, either through an efficient stocking and stock-releasing system or through a *point of sale* (POS) *system* integrated with the inventory records. If a company only counts its inventory (i.e., takes a *physical inventory*) at regular intervals (e.g., once a year), it is said to be operating on a *periodic inventory system*.

Inventory accounting is a source of many difficulties. The inventory value is calculated by multiplying the quantity on hand by the standard cost. In the case of completed items actually assembled or manufactured at the company, this standard cost usually is the manufacturing cost, although factory cost also can be used. In the case of purchased items, the standard cost will be the cost per item charged by the supplying vendor. In some cases, delivery and transportation costs will be included in this standard cost.

It is not unusual for the elements in an item's inventory to come from more than one vendor, or from one vendor in more than one order. Inventory valuation is more difficult if the price paid is different for these different purchases. There are four methods of determining the cost of elements in inventory. Any of these methods can be used (if applicable), but the method must be used consistently from year to year. The four methods are as follows.

- *specific identification method:* Each element can be uniquely associated with a cost. Inventory elements with serial numbers fit into this costing scheme. Stock, production, and sales records must include the serial number.

---

[48]The cost of goods sold inventory adjustment is posted to the COGS *expense account.*

- *average cost method:* The standard cost of an item is the average of (recent or all) purchase costs for that item.

- *first-in, first-out* (FIFO) *method:* This method keeps track of how many of each item were purchased each time and the number remaining out of each purchase, as well as the price paid at each purchase. The inventory system assumes that the oldest elements are issued first.[49] Inventory value is a weighted average dependent on the number of elements from each purchase remaining. Items issued no longer contribute to the inventory value.

- *last-in, first-out* (LIFO) *method:* This method keeps track of how many of each item were purchased each time and the number remaining out of each purchase, as well as the price paid at each purchase.[50] The inventory value is a weighted average dependent on the number of elements from each purchase remaining. Items issued no longer contribute to the inventory value.

55. BREAK-EVEN ANALYSIS

Special Nomenclature

a	*incremental cost* to produce one additional item (also called *marginal cost* or *differential cost*)
C	total cost
f	fixed cost that does not vary with production
p	*incremental value* (price)
Q	quantity sold
Q^*	quantity at break-even point
R	total revenue

Break-even analysis is a method of determining when the value of one alternative becomes equal to the value of another. A common application is that of determining when costs exactly equal revenue. If the manufactured quantity is less than the break-even quantity, a loss is incurred. If the manufactured quantity is greater than the break-even quantity, a profit is made. (See Fig. 54.13.)

Assuming no change in the inventory, the *break-even point* can be found by setting costs equal to revenue ($C = R$).

$$C = f + aQ \qquad 54.68$$

$$R = pQ \qquad 54.69$$

$$Q^* = \frac{f}{p - a} \qquad 54.70$$

---

[49]If all elements in an item's inventory are identical, and if all shipments of that item are agglomerated, there will be no way to guarantee that the oldest element in inventory is issued first. But, unless *spoilage* is a problem, it really does not matter.

[50]See previous footnote.

Figure 54.13 *Break-Even Quality*

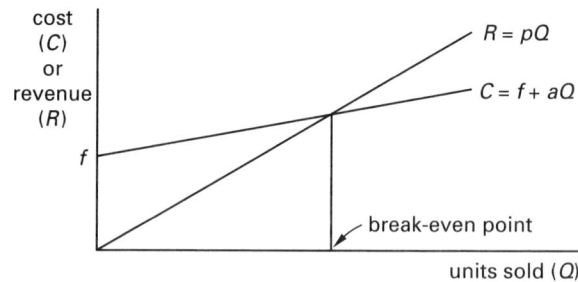

An alternative form of the break-even problem is to find the number of units per period for which two alternatives have the same total costs. Fixed costs are to be spread over a period longer than one year using the equivalent uniform annual cost (EUAC) concept. One of the alternatives will have a lower cost if production is less than the break-even point. The other will have a lower cost for production greater than the break-even point.

Example 54.32

Two plans are available for a company to obtain automobiles for its sales representatives. How many miles must the cars be driven each year for the two plans to have the same costs? Use an interest rate of 10%. (Use the year-end convention for all costs.)

plan 1: Lease the cars and pay $0.15 per mile.

plan 2: Purchase the cars for $5000. Each car has an economic life of three years, after which it can be sold for $1200. Gas and oil cost $0.04 per mile. Insurance is $500 per year.

Solution

Let x be the number of miles driven per year. Then, the EUAC for both alternatives is

$$
\begin{aligned}
\text{EUAC(A)} &= 0.15x \\
\text{EUAC(B)} &= 0.04x + \$500 + (\$5000)(A/P, 10\%, 3) \\
&\quad - (\$1200)(A/F, 10\%, 3) \\
&= 0.04x + \$500 + (\$5000)(0.4021) \\
&\quad - (\$1200)(0.3021) \\
&= 0.04x + 2148
\end{aligned}
$$

Setting EUAC(A) and EUAC(B) equal and solving for x yields 19,527 miles per year as the break-even point.

56. PAY-BACK PERIOD

The *pay-back period* is defined as the length of time, usually in years, for the cumulative net annual profit to equal the initial investment. It is tempting to introduce equivalence into pay-back period calculations, but by convention, this is generally not done.[51]

$$\text{pay-back period} = \frac{\text{initial investment}}{\text{net annual profit}} \qquad 54.71$$

Example 54.33

A ski resort installs two new ski lifts at a total cost of $1,800,000. The resort expects the annual gross revenue to increase by $500,000 while it incurs an annual expense of $50,000 for lift operation and maintenance. What is the pay-back period?

Solution

From Eq. 54.71,

$$\text{pay-back period} = \frac{\$1,800,000}{\dfrac{\$500,000}{\text{yr}} - \dfrac{\$50,000}{\text{yr}}} = 4 \text{ years}$$

57. MANAGEMENT GOALS

Depending on many factors (market position, age of the company, age of the industry, perceived marketing and sales windows, etc.), a company may select one of many production and marketing strategic goals. Three such strategic goals are

- maximization of product demand

- minimization of cost

- maximization of profit

Such goals require knowledge of how the dependent variable (e.g., demand quantity or quantity sold) varies as a function of the independent variable (e.g., price). Unfortunately, these three goals are not usually satisfied simultaneously. For example, minimization of product cost may require a large production run to realize economies of scale, while the actual demand is too small to take advantage of such economies of scale.

If sufficient data are available to plot the independent and dependent variables, it may be possible to optimize the dependent variable graphically. (See Fig. 54.14.) Of course, if the relationship between independent and dependent variables is known algebraically, the dependent variable can be optimized by taking derivatives or by use of other numerical methods.

---

[51]Equivalence (i.e., interest and compounding) generally is not considered when calculating the "pay-back period." However, if it is desirable to include equivalence, then the term *pay-back period* should not be used. Other terms, such as *cost recovery period* or *life of an equivalent investment*, should be used. Unfortunately, this convention is not always followed in practice.

Figure 54.14 *Graphs of Management Goal Functions*

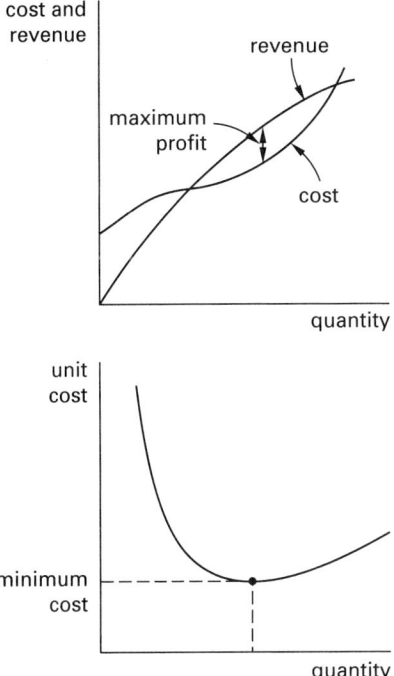

58. INFLATION

It is important to perform economic studies in terms of *constant value dollars*. One method of converting all cash flows to constant value dollars is to divide the flows by some annual *economic indicator* or price index.

If indicators are not available, cash flows can be adjusted by assuming that inflation is constant at a decimal rate, e, per year. Then, all cash flows can be converted to $t = 0$ dollars by dividing by $(1 + e)^n$, where n is the year of the cash flow.

An alternative is to replace the effective annual interest rate, i, with a value corrected for inflation. This corrected value, i', is

$$i' = i + e + ie \qquad 54.72$$

This method has the advantage of simplifying the calculations. However, precalculated factors are not available for the non-integer values of i'. Therefore, Table 54.1 must be used to calculate the factors.

Example 54.34

What is the uninflated present worth of a $2000 future value in two years if the average inflation rate is 6% and i is 10%?

Solution

$$P = \frac{F}{(1+i)^n (1+e)^n}$$

$$= \frac{\$2000}{(1+0.10)^2 (1+0.06)^2}$$

$$= \$1471$$

Example 54.35

Repeat Ex. 54.34 using Eq. 54.72.

Solution

$$i' = i + e + ie$$

$$= 0.10 + 0.06 + (0.10)(0.06) = 0.166$$

$$P = \frac{F}{(1+i')^2} = \frac{\$2000}{(1+0.166)^2} = \$1471$$

59. CONSUMER LOANS

Special Nomenclature

BAL_j	balance after the jth payment
j	payment or period number
LV	principal total value loaned (cost minus down payment)
N	total number of payments to pay off the loan
PI_j	jth interest payment
PP_j	jth principal payment
PT_j	jth total payment
ϕ	effective rate per period (r/k)

Many different arrangements can be made between a borrower and a lender. With the advent of creative financing concepts, it often seems that there are as many variations of loans as there are loans made. Nevertheless, there are several traditional types of transactions. Real estate or investment texts, or a financial consultant, should be consulted for more complex problems.

Simple Interest

Interest due does not compound with a *simple interest loan*. The interest due is merely proportional to the length of time that the principal is outstanding. Because of this, simple interest loans are seldom made for long periods (e.g., more than one year). (For loans less than one year, it is commonly assumed that a year consists of 12 months of 30 days each.)

Example 54.36

A $12,000 simple interest loan is taken out at 16% per annum interest rate. The loan matures in two years with no intermediate payments. How much will be due at the end of the second year?

Solution

The interest each year is

$$PI = (0.16)(\$12{,}000) = \$1920$$

The total amount due in two years is

$$PT = \$12{,}000 + (2)(\$1920) = \$15{,}840$$

Example 54.37

$4000 is borrowed for 75 days at 16% per annum simple interest. There are 360 banking days per year. How much will be due at the end of 75 days?

Solution

$$\text{amount due} = \$4000 + (0.16)\left(\frac{75 \text{ days}}{360 \frac{\text{days}}{\text{bank yr}}}\right)(\$4000)$$

$$= \$4133$$

Loans with Constant Amount Paid Toward Principal

With this loan type, the payment is not the same each period. The amount paid toward the principal is constant, but the interest varies from period to period. (See Fig. 54.15.) The equations that govern this type of loan are

$$BAL_j = LV - j(PP) \qquad 54.73$$

$$PI_j = \phi(BAL)_{j-1} \qquad 54.74$$

$$PT_j = PP + PI_j \qquad 54.75$$

$$PP = \frac{LV}{N} \qquad 54.76$$

$$N = \frac{LV}{PP} \qquad 54.77$$

$$LV = (PP + PI_1)(P/A, \phi, N)$$
$$\quad - PI_N (P/G, \phi, N) \qquad 54.78$$

$$1 = \left(\frac{1}{N} + \phi\right)(P/A, \phi, N)$$
$$\quad - \left(\frac{\phi}{N}\right)(P/G, \phi, N) \qquad 54.79$$

Systems, Mgmt., Professional

Figure 54.15 Loan with Constant Amount Paid Toward Principal

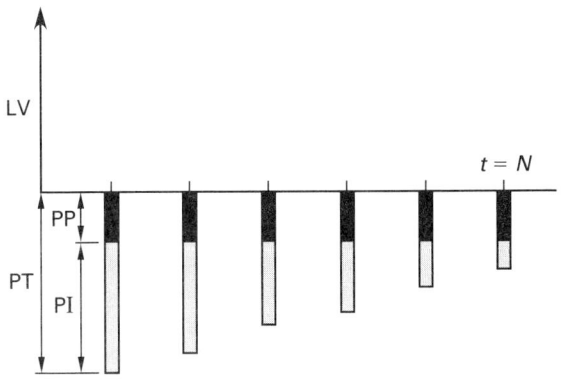

Example 54.38

A \$12,000 six-year loan is taken from a bank that charges 15% effective annual interest. Payments toward the principal are uniform, and repayments are made at the end of each year. Tabulate the interest, total payments, and the balance remaining after each payment is made.

Solution

The amount of each principal payment is

$$PP = \frac{LV}{N} = \frac{\$12,000}{6} = \$2000$$

At the end of the first year (before the first payment is made), the principal balance is \$12,000 (i.e., $BAL_0 = \$12,000$). From Eq. 54.74, the interest payment is

$$PI_1 = \phi(BAL)_0 = (0.15)(\$12,000) = \$1800$$

The total first payment is

$$PT_1 = PP + PI = \$2000 + \$1800$$
$$= \$3800$$

The following table is similarly constructed.

j	BAL_j	PP_j	PI_j	PT_j
	(in dollars)			
0	12,000	–	–	–
1	10,000	2000	1800	3800
2	8000	2000	1500	3500
3	6000	2000	1200	3200
4	4000	2000	900	2900
5	2000	2000	600	2600
6	0	2000	300	2300

Direct Reduction Loans

This is the typical "interest paid on unpaid balance" loan. The amount of the periodic payment is constant, but the amounts paid toward the principal and interest both vary. (See Fig. 54.16.)

$$BAL_{j-1} = PT\left(\frac{1-(1+\phi)^{j-1-N}}{\phi}\right) \qquad 54.80$$

$$PI_j = \phi(BAL)_{j-1} \qquad 54.81$$

$$PP_j = PT - PI_j \qquad 54.82$$

$$BAL_j = BAL_{j-1} - PP_j \qquad 54.83$$

$$N = \frac{-\ln\left(1 - \frac{\phi(LV)}{PT}\right)}{\ln(1+\phi)} \qquad 54.84$$

Figure 54.16 Direct Reduction Loan

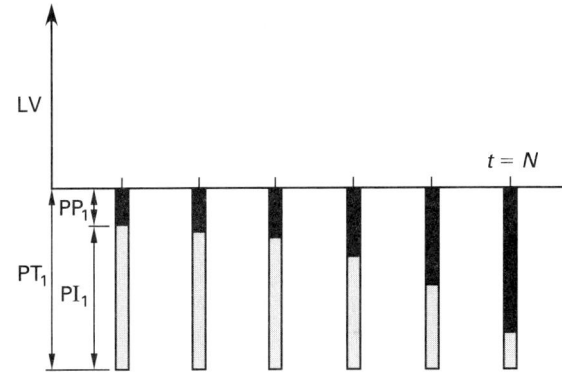

Equation 54.84 calculates the number of payments necessary to pay off a loan. This equation can be solved with effort for the total periodic payment (PT) or the initial value of the loan (LV). It is easier, however, to use the $(A/P, I\%, n)$ factor to find the payment and loan value.

$$PT = LV(A/P, \phi\%, N) \qquad 54.85$$

If the loan is repaid in yearly installments, then i is the effective annual rate. If the loan is paid off monthly, then i should be replaced by the effective rate per month (ϕ from Eq. 54.51). For monthly payments, N is the number of months in the loan period.

Example 54.39

A \$45,000 loan is financed at 9.25% per annum. The monthly payment is \$385. What are the amounts paid toward interest and principal in the 14th period? What is the remaining principal balance after the 14th payment has been made?

Solution

The effective rate per month is

$$\phi = \frac{r}{k} = \frac{0.0925}{12}$$
$$= 0.0077083 \ldots \quad [\text{use } 0.007708]$$

From Eq. 54.84,

$$N = \frac{-\ln\left(1 - \frac{\phi(\text{LV})}{\text{PT}}\right)}{\ln(1 + \phi)}$$

$$= \frac{-\ln\left(1 - \frac{(0.007708)(45{,}000)}{385}\right)}{\ln(1 + 0.007708)}$$

$$= 301$$

From Eq. 54.80,

$$\text{BAL}_{14-1} = \text{PT}\left(\frac{1 - (1 + \phi)^{14-1-N}}{\phi}\right)$$

$$= (\$385)\left(\frac{1 - (1 + 0.007708)^{14-1-301}}{0.007708}\right)$$

$$= \$44{,}476.39$$

From Eq. 54.81,

$$\text{PI}_{14} = \phi(\text{BAL})_{14-1}$$
$$= (0.007708)(\$44{,}476.39)$$
$$= \$342.82$$

From Eq. 54.82,

$$\text{PP}_{14} = \text{PT} - \text{PI}_{14} = \$385 - \$342.82 = \$42.18$$

Therefore, using Eq. 54.83, the remaining principal balance is

$$\text{BAL}_{14} = \text{BAL}_{14-1} - \text{PP}_{14}$$
$$= \$44{,}476.39 - \$42.18$$
$$= \$44{,}434.21$$

Direct Reduction Loans with Balloon Payments

This type of loan has a constant periodic payment, but the duration of the loan is insufficient to completely pay back the principal (i.e., the loan is not fully amortized). Therefore, all remaining unpaid principal must be paid back in a lump sum when the loan matures. This large

payment is known as a *balloon payment*.[52] (See Fig. 54.17.)

Equation 54.80 through Eq. 54.84 also can be used with this type of loan. The remaining balance after the last payment is the balloon payment. This balloon payment must be repaid along with the last regular payment calculated.

Figure 54.17 *Direct Reduction Loan with Balloon Payment*

60. FORECASTING

There are many types of forecasting models, although most are variations of the basic types.[53] All models produce a *forecast*, F_{t+1}, of some quantity (*demand* is used in this section) in the next period based on actual measurements, D_j, in current and prior periods. All of the models also try to provide *smoothing* (or *damping*) of extreme data points.

Forecasts by Moving Averages

The method of *moving average forecasting* weights all previous demand data points equally and provides some smoothing of extreme data points. The amount of smoothing increases as the number of data points, n, increases.

$$F_{t+1} = \frac{1}{n} \sum_{m=t+1-n}^{t} D_m \qquad \textit{54.86}$$

Forecasts by Exponentially Weighted Averages

With *exponentially weighted forecasts*, the more current (most recent) data points receive more weight. This method uses a *weighting factor*, α, also known as a *smoothing coefficient*, which typically varies between 0.01 and 0.30. An initial forecast is needed to start the method. Forecasts immediately following are sensitive

---

[52]The term *balloon payment* may include the final interest payment as well. Generally, the problem statement will indicate whether the balloon payment is inclusive or exclusive of the regular payment made at the end of the loan period.

[53]For example, forecasting models that take into consideration steady (linear), cyclical, annual, and seasonal trends are typically variations of the exponentially weighted model. A truly different forecasting tool, however, is *Monte Carlo simulation*.

to the accuracy of this first forecast. It is common to choose $F_0 = D_1$ to get started.

$$F_{t+1} = \alpha D_t + (1 - \alpha) F_t \qquad \textbf{54.87}$$

61. LEARNING CURVES

Special Nomenclature

b	learning curve constant
n	total number of items produced
R	decimal learning curve rate (2^{-b})
T_1	time or cost for the first item
T_n	time or cost for the nth item

The more products that are made, the more efficient the operation becomes due to experience gained. Therefore, direct labor costs decrease.[54] Usually, a *learning curve* is specified by the decrease in cost each time the cumulative quantity produced doubles. If there is a 20% decrease per doubling, the curve is said to be an 80% learning curve (i.e., the *learning curve rate*, R, is 80%).

Then, the time to produce the nth item is

$$T_n = T_1 n^{-b} \qquad \textbf{54.88}$$

The total time to produce units from quantity n_1 to n_2 inclusive is approximately given by Eq. 54.89. T_1 is a constant, the time for item 1, and does not correspond to n unless $n_1 = 1$.

$$\int_{n_1}^{n_2} T_n\, dn \approx \left(\frac{T_1}{1-b}\right)\left(\left(n_2 + \tfrac{1}{2}\right)^{1-b} - \left(n_1 - \tfrac{1}{2}\right)^{1-b}\right)$$

$$\textbf{54.89}$$

The *average time per unit* over the production from n_1 to n_2 is the above total time from Eq. 54.89 divided by the quantity produced, $(n_2 - n_1 + 1)$.

$$T_{\text{ave}} = \frac{\displaystyle\int_{n_1}^{n_2} T_n\, dn}{n_2 - n_1 + 1} \qquad \textbf{54.90}$$

Table 54.9 lists representative values of the *learning curve constant*, b. For learning curve rates not listed in the table, Eq. 54.91 can be used to find b.

$$b = \frac{-\log_{10} R}{\log_{10}(2)} = \frac{-\log_{10} R}{0.301} \qquad \textbf{54.91}$$

---

[54]Learning curve reductions apply only to direct labor costs. They are not applied to indirect labor or direct material costs.

Table 54.9 *Learning Curve Constants*

learning curve rate, R	b
0.70 (70%)	0.515
0.75 (75%)	0.415
0.80 (80%)	0.322
0.85 (85%)	0.234
0.90 (90%)	0.152
0.95 (95%)	0.074

Example 54.40

A 70% learning curve is used with an item whose first production time is 1.47 hr. (a) How long will it take to produce the 11th item? (b) How long will it take to produce the 11th through 27th items?

Solution

(a) From Eq. 54.88,

$$T_{11} = T_1 n^{-b} = (1.47 \text{ hr})(11)^{-0.515}$$

$$= 0.428 \text{ hr}$$

(b) The time to produce the 11th item through 27th item is given by Eq. 54.89.

$$T \approx \left(\frac{T_{11}}{1-b}\right)\left(\left(n_{27} + \tfrac{1}{2}\right)^{1-b} - \left(n_{11} - \tfrac{1}{2}\right)^{1-b}\right)$$

$$T \approx \left(\frac{1.47 \text{ hr}}{1 - 0.515}\right)\left((27.5)^{1-0.515} - (10.5)^{1-0.515}\right)$$

$$= 5.643 \text{ hr}$$

62. ECONOMIC ORDER QUANTITY

Special Nomenclature

a	constant depletion rate (items/unit time)
h	inventory storage cost ($/item-unit time)
H	total inventory storage cost between orders ($)
K	fixed cost of placing an order ($)
Q	order quantity (original quantity on hand)
t^*	time at depletion

The *economic order quantity* (EOQ) is the order quantity that minimizes the inventory costs per unit time. Although there are many different EOQ models, the simplest is based on the following assumptions.

- Reordering is instantaneous. The time between order placement and receipt is zero, as shown in Fig. 54.18.

- Shortages are not allowed.

- Demand for the inventory item is deterministic (i.e., is not a random variable).

- Demand is constant with respect to time.

- An order is placed when the inventory is zero.

Figure 54.18 *Inventory with Instantaneous Reorder*

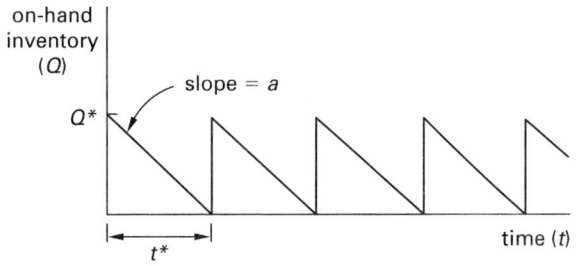

If the original quantity on hand is Q, the stock will be depleted at

$$t^* = \frac{Q}{a} \qquad \qquad 54.92$$

The total inventory storage cost between t_0 and t^* is

$$H = \tfrac{1}{2}Qht^* = \frac{Q^2 h}{2a} \qquad \qquad 54.93$$

The total inventory and ordering cost per unit time is

$$C_t = \frac{aK}{Q} + \frac{hQ}{2} \qquad \qquad 54.94$$

C_t can be minimized with respect to Q. The economic order quantity and time between orders are

$$Q^* = \sqrt{\frac{2aK}{h}} \qquad \qquad 54.95$$

$$t^* = \frac{Q^*}{a} \qquad \qquad 54.96$$

63. SENSITIVITY ANALYSIS

Data analysis and forecasts in economic studies require estimates of costs that will occur in the future. There are always uncertainties about these costs. However, these uncertainties are insufficient reason not to make the best possible estimates of the costs. Nevertheless, a decision between alternatives often can be made more confidently if it is known whether or not the conclusion is sensitive to moderate changes in data forecasts. Sensitivity analysis provides this extra dimension to an economic analysis.

The sensitivity of a decision is determined by inserting a range of estimates for critical cash flows and other parameters. If radical changes can be made to a cash flow without changing the decision, the decision is said to be *insensitive* to uncertainties regarding that cash flow. However, if a small change in the estimate of a cash flow will alter the decision, that decision is said to be very *sensitive* to changes in the estimate. If the decision is sensitive only for a limited range of cash flow

values, the term *variable sensitivity* is used. Figure 54.19 illustrates these terms.

Figure 54.19 *Types of Sensitivity*

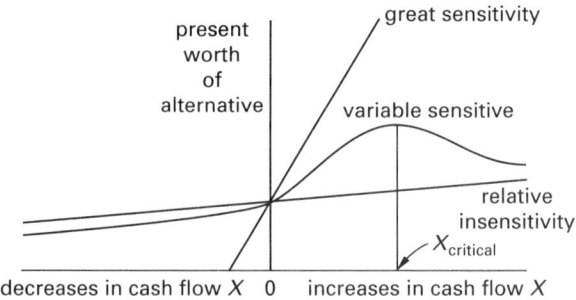

An established semantic tradition distinguishes between risk analysis and uncertainty analysis. *Risk analysis* addresses variables that have a known or estimated probability distribution. In this regard, statistics and probability theory can be used to determine the probability of a cash flow varying between given limits. On the other hand, *uncertainty analysis* is concerned with situations in which there is not enough information to determine the probability or frequency distribution for the variables involved.

As a first step, sensitivity analysis should be applied one at a time to the dominant factors. Dominant cost factors are those that have the most significant impact on the present value of the alternative.[55] If warranted, additional investigation can be used to determine the sensitivity to several cash flows varying simultaneously. Significant judgment is needed, however, to successfully determine the proper combinations of cash flows to vary. It is common to plot the dependency of the present value on the cash flow being varied in a two-dimensional graph. Simple linear interpolation is used (within reason) to determine the critical value of the cash flow being varied.

64. VALUE ENGINEERING

The *value* of an investment is defined as the ratio of its return (performance or utility) to its cost (effort or investment). The basic object of *value engineering* (VE, also referred to as *value analysis*) is to obtain the maximum per-unit value.[56]

Value engineering concepts often are used to reduce the cost of mass-produced manufactured products. This is done by eliminating unnecessary, redundant, or superfluous features, by redesigning the product for a less

---

[55]In particular, engineering economic analysis problems are sensitive to the choice of effective interest rate, i, and to accuracy in cash flows at or near the beginning of the horizon. The problems will be less sensitive to accuracy in far-future cash flows, such as salvage value and subsequent generation replacement costs.
[56]Value analysis, the methodology that has become today's value engineering, was developed in the early 1950s by Lawrence D. Miles, an analyst at General Electric.

expensive manufacturing method, and by including features for easier assembly without sacrificing utility and function.[57] However, the concepts are equally applicable to one-time investments, such as buildings, chemical processing plants, and space vehicles. In particular, value engineering has become an important element in all federally funded work.[58]

Typical examples of large-scale value engineering work are using stock-sized bearings and motors (instead of custom manufactured units), replacing rectangular concrete columns with round columns (which are easier to form), and substituting custom buildings with prefabricated structures.

Value engineering is usually a team effort. And, while the original designers may be on the team, usually outside consultants are utilized. The cost of value engineering is usually returned many times over through reduced construction and life-cycle costs.

---

[57]Some people say that value engineering is "the act of going over the plans and taking out everything that is interesting."

[58]U.S. Government Office of Management and Budget Circular A-131 outlines value engineering for federally funded construction projects.

55 Project Management, Budgeting, and Scheduling

Nomenclature

D	duration
EF	earliest finish
ES	earliest start
LF	latest finish
LS	latest start
t	time
z	standard normal variable

Symbols

μ	mean
σ	standard deviation

1. PROJECT MANAGEMENT

Project management is the coordination of the entire process of completing a job, from its inception to final move-in and post-occupancy follow-up. In many cases, project management is the responsibility of one person. Large projects can be managed with *partnering*. With this method, the various stakeholders of a project, such as the architect, owner, contractor, engineer, vendors, and others are brought into the decision making process. Partnering can produce much closer communication on a project and shared responsibilities. However, the day-to-day management of a project may be difficult with so many people involved. A clear line of communications and delegation of responsibility should be established and agreed to before the project begins.

Many project managers follow the procedures outlined in *A Guide to the Project Management Body of Knowledge* (PMBOK Guide), published by the Project Management Institute. The PMBOK Guide is an internationally recognized standard (IEEE Std 1490) that defines the fundamentals of project management as they apply to a wide range of projects, including construction, engineering,

software, and many other industries. The PMBOK Guide is process-based, meaning it describes projects as being the outcome of multiple processes. Processes overlap and interact throughout the various phases of a project. Each process occurs within one of five *process groups*, which are related as shown in Fig. 55.1. The five PMBOK process groups are (1) initiating, (2) planning, (3) controlling and monitoring, (4) executing, and (5) closing.

Figure 55.1 PMBOK Process Groups

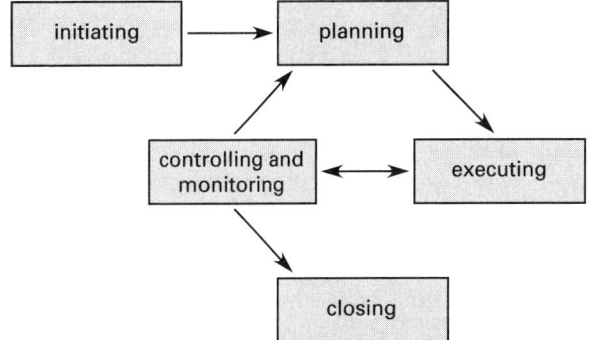

The PMBOK Guide identifies nine project management knowledge areas that are typical of nearly all projects. The nine knowledge areas and their respective processes are summarized as follows.

1. *Integration:* develop project charter, project scope statement, and management plan; direct and manage execution; monitor and control work; integrate change control; close project

2. *Scope:* plan, define, create work breakdown structure (WBS), verify, and control scope

3. *Time:* define, sequence, estimate resources, and estimate duration of activities; develop and control schedule

4. *Cost:* estimate, budget, and control costs

5. *Quality:* plan; perform quality assurance and quality control

6. *Human Resources:* plan; acquire, develop, and manage project team

7. *Communications:* plan; distribute information; report on performance; manage stakeholders

8. *Risk:* identify risks; plan risk management and response; perform qualitative and quantitative risk analysis; monitor and control

9. *Procurement:* plan purchases, acquisitions, and contracts; request seller responses; select sellers; administer and close contracts

Each of the processes also falls into one of the five basic process groups, creating a matrix so that every process is related to one knowledge area and one process group. Additionally, processes are described in terms of inputs (documents, plans, designs, etc.), tools and techniques (mechanisms applied to inputs), and outputs (documents, products, etc.).

Establishing a budget (knowledge area 4) and scheduling design and construction (knowledge area 3) are two of the most important parts of project management because they influence many of the design decisions to follow and can determine whether a project is even feasible.

2. BUDGETING

Budgets may be established in several ways. For speculative or for-profit projects, the owner or developer works out a pro forma statement listing the expected income of the project and the expected costs to build it. An estimated selling price of the developed project or rent per square foot is calculated and balanced against all the various costs, one of which is the construction price. In order to make the project economically feasible, there will be a limit on the building costs. This becomes the budget within which the work must be completed.

Budgets for municipal and other public projects are often established through public funding or legislation. In these cases, the construction budget is often fixed without the architect's or engineer's involvement, and the project must be designed and built for the fixed amount. Unfortunately, when public officials estimate the cost to build a project, they sometimes neglect to include all aspects of development, such as professional fees, furnishings, and other items.

Budgets may also be based on the proposed project specifics. This is the most realistic and accurate way to establish a preliminary budget because it is based on the actual characteristics of the project. However, this method assumes funding will be available for the project as designed.

There are four basic variables in developing any construction budget: quantity, quality, available funds, and time. There is always a balance among these four variables, and changing one or more affects the others. For instance, if an owner needs a certain area (quantity), needs the project built at a certain time, and has a fixed amount of money to spend, then the quality of construction will have to be adjusted to meet the other constraints. If time, quality, and the total budget are fixed, then the area must be adjusted. For more

information, see the "Time-Cost Trade-Off" section in this chapter. In some cases, *value engineering* can be performed during which individual systems and materials are reviewed to see if the same function can be accomplished in a less expensive way.

Fee Projections

A *fee projection* is one of the earliest and most important tasks that a project manager must complete. A fee projection takes the total fee the designer will receive for the project and allocates it to the schedule and staff members who will work on the project, after deducting amounts for profit, overhead, and other expenses that will not be used for professional time.

Ideally, fee projections should be developed from a careful projection of the scope of work, its associated costs (direct personnel expenses, indirect expenses, and overhead), consultant fees, reimbursable expenses, and profit desired. These should be determined as a basis for setting the final fee agreement with the client. If this is done correctly, there should be enough money to complete the project within the allotted time.

There are many methods for estimating and allocating fees. Figure 55.2 shows a simple manual form that combines time scheduling with fee projections. In this example, the total working fee, that is, the fee available to pay people to do the job after subtracting for profit, consultants, and other expenses, is listed in the upper right corner of the chart. The various phases or work tasks needed to complete the job are listed in the left-hand column, and the time periods (most commonly in weeks) are listed across the top of the chart.

The project manager estimates the percentage of the total amount of work or fee that he or she thinks each phase will require. This estimate is based on experience and any common rules of thumb the design or construction office may use. The percentages are placed in the third column on the right and multiplied by the total working fee to get the allotted fee for each phase (the figure in the second column on the right). This allotted fee is then divided among the number of time periods in the schedule and placed in the individual columns under each time period.

If phases or tasks overlap (as they do in Fig. 55.2), sum the fees in each period and place this total dollar amount at the bottom of the chart. The total dollar amount can then be divided by an average billing rate for the people working on the project to determine an approximate budgeted number of hours that the office can afford to spend on the project each week and still make a profit. Of course, if the number of weekly hours exceeds about 40, then more than one person will be needed to do the work.

By monitoring time sheets, the project manager can compare the actual hours (or fees) expended against the budgeted time (or fees) and take corrective action if actual time exceeds budgeted time.

Figure 55.2 *Fee Projection Chart*

Project: Mini-mall		Project No.: 9274								Date: 10/14/2012		
Completed by: JBL		Project Manager: JBL								Total Fee: $26,400		

Phase or Task	Period	1	2	3	4	5	6	7	8	9	% of total fee	fee allocation by phase or task	person-hrs est.
	Date	11/16-22	11/23	11/30	12/7	12/14	12/21	12/28	1/4	1/11			
SD—design		1320	1320								10	2640	
SD—presentation			1320								5	1320	
DD—arch. work				1980	1980						15	3960	
DD—consultant coord.				530	790						5	1320	
DD—approvals					1320						5	1320	
CD—plans/elevs.						1056	1056	1056	1056	1056	20	5280	
CD—details								2640	2640		20	5280	
CD—consultant coord.						440		440	440		5	1320	
CD—specs.									1320	1320	10	2640	
CD—material sel.						660	660				5	1320	

budgeted fees/period	1320	2640	2510	4090	2156	1716	4136	5456	2376	100%	$26,400	
person-weeks or hours	53 / 1.3	106 / 2.6	100 / 2.5	164 / 4	108 / 2.7	86 / 2.2	207 / 5	273 / 6.8	119 / 3			
staff assigned	JLK	JLK AST JBC	JLK AST EMW-(1/2)	JLK AST JBC EMW	JLK AST EMW	JLK AST	JLK AST EMW ⟶	JLK SBS → BFD	JLK AST EMW			
actual fees expended												

Quality planning involves determining with the client what the expectations are concerning design, cost, and other aspects of the project. Quality does not simply mean high-cost finishes, but rather the requirements of the client based on his or her needs. These needs should be clearly defined in the programming phase of a project and written down and approved by the client before design work begins.

Cost Estimating

Estimators compile and analyze data on all of the factors that can influence costs, such as materials, labor, location, duration of the project, and special machinery requirements. The methods for estimating costs can differ greatly by industry. On a construction project, for example, the estimating process begins with the decision to submit a bid. After reviewing various preliminary drawings and specifications, the estimator visits the site of the proposed project. The estimator needs to gather information on access to the site; the availability of electricity, water, and other services; and surface topography and drainage. The estimator usually records this information in a signed report that is included in the final project estimate.

After the site visit, the estimator determines the quantity of materials and labor the firm will need to furnish. This process, called the quantity survey or "takeoff," involves completing standard estimating forms, filling in dimensions, numbers of units, and other information. Table 55.1 is a small part of a larger takeoff report illustrating the degree of detail needed to estimate the project cost. A cost estimator working for a general contractor, for example, uses a construction project's plans and specifications to estimate the materials dimensions and count the quantities of all items associated with the project.

Though the quantity takeoff process can be done manually using a printout, a red pen, and a clicker, it can also be done with a digitizer that enables the user to take measurements from paper bid documents, or with an integrated *takeoff viewer* program that interprets electronic bid documents. In any case, the objective is to generate a set of takeoff elements (counts, measurements, and other conditions that affect cost) that is used to establish cost estimates. Table 55.1 is an example of a typical quantity take-off report for lumber needed for a construction project.

Although subcontractors estimate their costs as part of their own bidding process, the general contractor's cost estimator often analyzes bids made by subcontractors. Also during the takeoff process, the estimator must make decisions concerning equipment needs, the sequence of operations, the size of the crew required, and physical constraints at the site. Allowances for wasted materials, inclement weather, shipping delays,

Table 55.1 *Partial Lumber and Hardware Take-Off Report*

size	description	usage	pieces	total length
foundation framing				
2×4	DF STD/BTR	stud	38	8
2×6	DF #2/BTR	stud	107	8
2×6	PTDF	mudsill	–	RL
2×6	DF #2/BTR	bracing	–	RL
2×4	DF STD/BTR	blocking	–	RL
$11^{7}/_{8}''$	TJI/250	floor joist	11	18
$11^{7}/_{8}''$	TJI/250	floor joist	12	10
2×4	DF STD/BTR	plate	–	RL
2×6	DF #2/BTR	plate	–	RL
4×4	PTDF	posts	7	4
4×6	PTDF	posts	23	4
6×6	PTDF	posts	1	4
2×12	DF #2/BTR	rim	–	RL
$^{15}/_{32}''$	CDX plywood	subfloor	12	4×8
pre-cut doors				
$3^{1}/_{2}'' \times 7^{1}/_{4}''$	LVL	header	1	100''
$3^{1}/_{2}'' \times 7^{1}/_{4}''$	LVL	header	1	130''
$3^{1}/_{2}'' \times 7^{1}/_{4}''$	LVL	header	10	24''
exterior sheathing and shearwall				
$^{1}/_{2}''$	CDX plywood	shear wall	100	4×8
$^{1}/_{2}''$	CDX plywood	exterior sheathing	89	4×8
roof sheathing				
$^{1}/_{2}''$	CDX plywood	roof sheathing	67	4×8
building A & B hardware				
Simpson HD64	6 pieces each			
Simpson HD22	23 pieces each			

and other factors that may increase costs also must be incorporated in the estimate. After completing the quantity surveys, the estimator prepares a cost summary for the entire project, including the costs of labor, equipment, materials, subcontracts, overhead, taxes, insurance, markup, and any other costs that may affect the project. The chief estimator then prepares the bid proposal for submission to the owner. Construction cost estimators also may be employed by the project's architect or owner to estimate costs or to track actual costs relative to bid specifications as the project develops.

Estimators often specialize in large construction companies employing more than one estimator. For example, one may estimate only electrical work and another may concentrate on excavation, concrete, and forms.

Computers play an integral role in cost estimation because estimating often involves numerous mathematical calculations requiring access to various historical databases. For example, to undertake a parametric analysis (a process used to estimate costs per unit based on square footage or other specific requirements of a

project), cost estimators use a computer database containing information on the costs and conditions of many other similar projects. Although computers cannot be used for the entire estimating process, they can relieve estimators of much of the drudgery associated with routine, repetitive, and time-consuming calculations.

Cost Influences

There are many variables that affect project cost. Construction cost is only one part of the total project development budget. Other factors include such things as site acquisition, site development, fees, and financing. Table 55.2 lists most of the items commonly found in a project budget and a typical range of values based on construction cost. Not all of these are part of every development, but they illustrate the things that must be considered.

Building cost is the money required to construct the building, including structure, exterior cladding, finishes, and electrical and mechanical systems. *Site development*

Table 55.2 *Project Budget Line Items*

	line item		example
A	site acquisition		$1,100,000
B	building costs	area times cost per ft^2	(assume) $6,800,000
C	site development	10% to 20% of B	(15%) $1,020,000
D	total construction cost	B + C	$7,820,000
E	movable equipment	5% to 10% of B	(5%) $340,000
F	furnishings		$200,000
G	total construction and furnishings	D + E + F	$8,360,000
H	professional services	5% to 10% of D	(7%) $547,400
I	inspection and testing		$15,000
J	escalation estimate	2% to 20% of G per year	(10%) $836,000
K	contingency	5% to 10% of G	(8%) $668,800
L	financing costs		$250,000
M	moving expenses		(assume) $90,000
N	total project budget	G + H through M	$11,867,200

costs are usually a separate item. They include such things as parking, drives, fences, landscaping, exterior lighting, and sprinkler systems. If the development is large and affects the surrounding area, a developer may be required to upgrade roads, extend utility lines, and do other major off-site work as a condition of getting approval from public agencies.

Movable equipment and furnishings include furniture, accessories, window coverings, and major equipment necessary to put the facility into operation. These are often listed as separate line items because the funding for them may come out of a separate budget and because they may be supplied under separate contracts.

Professional services are architectural and engineering fees as well as costs for such things as topographic surveys, soil tests, special consultants, appraisals and legal fees, and the like. Inspection and testing involve money required for special on-site, full-time inspection (if required), and testing of such things as concrete, steel, window walls, and roofing.

Because construction takes a great deal of time, a factor for inflation should be included. Generally, the present budget estimate is escalated to a time in the future at the expected midpoint of construction. Although it is impossible to predict the future, by using past cost indexes and inflation rates and applying an estimate to the expected condition of the construction, the architect can usually make an educated guess.

A *contingency cost* should also be added to account for unforeseen changes by the client and other conditions that add to the cost. For an early project budget, the percentage of the contingency should be higher than contingencies applied to later budgets, because there are more unknowns. Normally, 5% to 10% should be included.

Financing includes not only the long-term interest paid on permanent financing but also the immediate costs of

loan origination fees, construction loan interest, and other administrative costs. On long-term loans, the cost of financing can easily exceed all of the original building and development costs. In many cases, long-term interest, called *debt service*, is not included in the project budget because it is an ongoing cost to the owner, as are maintenance costs.

Finally, many clients include moving costs in the development budget. For large companies and other types of clients, the money required to physically relocate, including changing stationery, installing telephones, and the like, can be a substantial amount.

Methods of Budgeting

The costs described in the previous section and shown in Table 55.2 represent a type of budget done during programming or even prior to programming to test the feasibility of a project. The numbers are preliminary, often based on sketchy information. For example, the building cost may simply be an estimated cost per square foot multiplied by the number of gross square feet needed. The square footage cost may be derived from similar buildings in the area, from experience, or from commercially available cost books.

Budgeting, however, is an ongoing activity. At each stage of the design process, there should be a revised budget reflecting the decisions made to that time. As shown in the example, pre-design budgets are usually based only on area, but other units can also be used. For example, many companies have rules of thumb for making estimates based on cost per hospital bed, cost per student, cost per hotel room, or similar functional units.

After the pre-programming budget, the architect usually begins to concentrate on the building and site development costs. At this stage an average cost per square foot may still be used, or the building may be divided into several functional parts and different square footage prices may be assigned to each part. A school, for example, may be classified into classroom space, laboratory space, shop space, office space, and gymnasium space, each having a different cost per square foot. This type of division can be developed concurrently with the programming of the space requirements.

During schematic design, when more is known about the space requirements and general configuration of the building and site, *system budgeting* is based on major subsystems. Historical cost information on each type of subsystem can be applied to the design. At this point it is easier to see where the money is being used in the building. Design decisions can then be based on studies of alternative systems. A typical subsystem budget is shown in Table 55.3.

Table 55.3 *System Cost Budget of Office Buildings*

| | average cost | |
subsystem	($/ft^2)	(% of total)
foundations	3.96	5.2
floors on grade	3.08	4.0
superstructure	16.51	21.7
roofing	0.18	0.2
exterior walls	9.63	12.6
partitions	5.19	6.8
wall finishes	3.70	4.8
floor finishes	3.78	5.0
ceiling finishes	2.79	3.7
conveying systems	6.45	8.5
specialties	0.70	0.9
fixed equipment	2.74	3.6
HVAC	9.21	12.1
plumbing	3.61	4.6
electrical	4.68	6.1
	76.21	100.0

Values for low-, average-, and high-quality construction for different building types can be obtained from cost databases and published estimating manuals and applied to the structure being budgeted. The dollar amounts included in system cost budgets usually include markup for contractor's overhead and profit and other construction administrative costs.

During the later stages of schematic design and early stages of construction documents, more detailed estimates are made. The procedure most often used is the *parameter method*, which involves an expanded itemization of construction quantities and assignment of unit costs to these quantities. For example, instead of using one number for floor finishes, the cost is broken down into carpeting, vinyl tile, wood strip flooring, unfinished concrete, and so forth. Using an estimated cost per square foot, the cost of each type of flooring can be estimated based on the area. With *parametric budgeting*, it is possible to evaluate the cost implications of each building component and to make decisions concerning both quantity and quality in order to meet the original budget estimate. If floor finishes are over

budget, the architect and the client can review the parameter estimate and decide, for example, that some wood flooring must be replaced with less expensive carpeting. Similar decisions can be made concerning any of the parameters in the budget.

Another way to compare and evaluate alternative construction components is with *matrix costing*. With this technique, a matrix is drawn showing, along one side, the various alternatives and, along the other side, the individual elements that combine to produce the total cost of the alternatives. For example, in evaluating alternatives for workstations, all of the factors that would comprise the final cost could be compared. These factors might include the cost of custom-built versus pre-manufactured workstations, task lighting that could be planned with custom-built units versus higher-wattage ambient lighting, and so on.

Parameter line items are based on commonly used units that relate to the construction element under study. For instance, a gypsum board partition would have an assigned cost per square foot of complete partition of a particular construction type rather than separate costs for metal studs, gypsum board, screws, and finishing. There would be different costs for single-layer gypsum board partitions, 1-hr rated walls, 2-hr rated walls, and other partition types.

Overhead and Profit

Two additional components of construction cost are the contractor's overhead and profit. Overhead can be further divided into general overhead and project overhead. *General overhead* is the cost to run a contracting business, and involves office rent, secretarial help, heat, and other recurring costs. *Project overhead* is the money it takes to complete a particular job, not including labor, materials, or equipment. Temporary offices, project telephones, sanitary facilities, trash removal, insurance, permits, and temporary utilities are examples of project overhead. The total overhead costs, including both general and project expenses, can range from about 10% to 20% of the total costs for labor, materials, and equipment.

Profit is the last item a contractor adds onto an estimate and is listed as a percentage of the total of labor, materials, equipment, and overhead. This is one of the most highly variable parts of a budget. Profit depends on the type of project, its size, the amount of risk involved, how much money the contractor wants to make, the general market conditions, and, of course, whether or not the job is being bid.

During extremely difficult economic conditions, a contractor may cut the profit margin to almost nothing simply to get the job and keep his or her workforce employed. If the contract is being negotiated with only one contractor, the profit percentage will be much higher. In most cases, however, profit will range from 5% to 20% of the total cost of the job. Overall, overhead and profit can total about 15% to 40% of construction cost.

Cost Information

One of the most difficult aspects of developing project budgets is obtaining current, reliable prices for the kinds of construction units being used. There is no shortage of commercially produced cost books that are published yearly. These books list costs in different ways; some are very detailed, giving the cost for labor and materials for individual construction items, while others list parameter costs and subsystem costs. The detailed price listings are of little use to architects because they are too specific and make comparison of alternate systems difficult.

There are also computerized cost estimating services that only require the architect or engineer to provide general information about the project, location, size, major materials, and so forth. The computer service then applies its current price database to the information and produces a cost budget. Many architects and engineers also work closely with general contractors to develop a realistic budget.

Commercially available cost information, however, is the average of many past construction projects from around the country. Local variations and particular conditions may affect the value of their use on a specific project.

Two conditions that must be accounted for in developing any project budget are geographical location and inflation. These variables can be adjusted by using cost indexes that are published in a variety of sources, including the major architectural and construction trade magazines. Using a base year as index 1000, for example, for selected cities around the country, new indexes are developed each year that reflect the increase in costs (both material and labor) that year.

The indexes can be used to apply costs from one part of the country to another and to escalate past costs to the expected midpoint of construction of the project being budgeted.

Example 55.1

The cost index in your city is 1257 and the cost index for another city in which you are designing a building is 1308. If the expected construction cost is $1,250,000 based on prices for your city, what will be the expected cost in the other region?

Solution

$$\frac{\text{cost A}}{\text{index A}} = \frac{\text{cost B}}{\text{index B}}$$

$$\begin{aligned}\text{cost in} \atop \text{other region} &= \left(\text{index in} \atop \text{other region}\right)\left(\frac{\text{your city cost}}{\text{your city index}}\right)\\ &= (1308)\left(\frac{\$1,250,000}{1257}\right)\\ &= \$1,300,716\end{aligned}$$

Life-Cycle Cost Analysis

Life-cycle cost analysis (LCC) is a method for determining the total cost of a building or building component or system over a specific period of time. It takes into account the initial cost of the element or system under consideration as well as the cost of financing, operation, maintenance, and disposal. Any residual value of the components is subtracted from the other costs. The costs are estimated over a length of time called the *study period*. The duration of the study period varies with the needs of the client and the useful life of the material or system. For example, investors in a building may be interested in comparing various alternate materials over the expected investment time frame, while a city government may be interested in a longer time frame representing the expected life of the building. All future costs are discounted back to a common time, usually the base date, to account for the time value of money. The *discount rate* is used to convert future costs to their equivalent present values.

Using life-cycle cost analysis allows two or more alternatives to be evaluated and their total costs to be compared. This is especially useful when evaluating energy conservation measures where one design alternative may have a higher initial cost than another, but a lower overall cost because of energy savings. Some of the specific costs involved in an LCC of a building element include the following.

- initial costs, which include the cost of acquiring and installing the element

- operational costs for electricity, water, and other utilities

- maintenance costs for the element over the length of the study period, including any repair costs

- replacement costs, if any, during the length of the study period

- finance costs required during the length of the study period

- taxes, if any, for initial costs and operating costs

The *residual value* is the remaining value of the element at the end of the study period based on resale value, salvage value, value in place, or scrap value. All of the costs listed are estimated, discounted to their present value, and added together. Any residual value is discounted to its present value and then subtracted from the total to get the final life-cycle cost of the element.

A life-cycle cost analysis is not the same as a *life-cycle assessment* (LCA). An LCA analyzes the environmental impact of a product or building system over the entire life of the product or system.

3. SCHEDULING

There are two primary elements that affect a project's schedule: design sequencing and construction sequencing. The architect has control over design and the production of contract documents, while the contractor has control over construction. The entire project should be scheduled for the best course of action to meet the client's goals. For example, if the client must move by a certain date and normal design and construction sequences make this impossible, the engineer, architect, or contractor may recommend a fast-track schedule or some other approach to meet the deadline.

Design Sequencing

Prior to creating a design, the architect needs to gather information about a client's specific goals and objectives, as well as analyzing any additional factors that may influence a project's design. (This gathering of information is known as *programming*.)

Once this preliminary information has been gathered, the design process may begin. The design process normally consists of several clearly defined phases, each of which must be substantially finished and approved by the client before the next phase may begin. These phases are outlined in the "Owner-Architect Agreement," which is published by the American Institute of Architects (AIA), as well as in other AIA documents. The architectural profession commonly refers to the phases as follows.

1. *Schematic Design Phase:* develops the general layout of the project through schematic design drawings, along with any preliminary alternate studies for materials and building systems

2. *Design Development Phase:* refines and further develops any decisions made during the schematic design phase; preliminary or outline specifications are written and a detailed project budget is created

3. *Construction Documents Phase:* final working drawings, as well as the project manual and any bidding or contract documents, are solidified

4. *Bidding or Negotiation Phase:* bids from several contractors are obtained and analyzed; negotiations with a contractor begin and a contractor is selected

5. *Construction Phase:* see the following section, "Construction Sequencing"

The time required for each of these five phases is highly variable and depends on the following factors.

- *Size and complexity of the project:* A 500,000 ft^2 (46 450 m^2) hospital will take much longer to design than a 30,000 ft^2 (2787 m^2) office building.

- *Number of people working on the project:* Although adding more people to the job can shorten the schedule, there is a point of diminishing returns. Having too many people only creates a management and coordination problem, and for some phases only a few people are needed, even for very large jobs.

- *Abilities and design methodology of the project team:* Younger, less-experienced designers will usually need more time to do the same amount of work than would more senior staff members.

- *Type of client, client decision-making, and approval processes of the client:* Large corporations or public agencies are likely to have a multilayered decision-making and approval process. Getting necessary information or approval for one phase from a large client may take weeks or even months, while a small, single-authority client might make the same decision in a matter of days.

The construction schedule may be established by the contractor or construction manager, or it may be estimated by the architect during the programming phase so that the client has some idea of the total time required from project conception to move-in.

Many variables can affect construction time. Most can be controlled in one way or another, but others, like weather, are independent of anyone's control. Beyond the obvious variables of size and complexity, the following is a partial list of some of the more common variables.

- management ability of the contractor to coordinate the work of direct employees with that of any subcontractors

- material delivery times

- quality and completeness of the architect's drawings and specifications

- weather

- labor availability and labor disputes

- new construction or remodeling (remodeling generally takes more time and coordination than for new buildings of equal areas)

- site conditions (construction sites or those with subsurface problems usually take more time to build on)

- characteristics of the architect (some professionals are more diligent than others in performing their duties during construction)

- lender approvals

- agency and governmental approvals

Construction Sequencing

Construction sequencing involves creating and following a work schedule that balances the timing and sequencing of land disturbance activities (e.g., earthwork) and the installation of *erosion and sedimentation control* (ESC) measures. The objective of construction sequencing is to reduce on-site erosion and off-site sedimentation that might affect the water quality of nearby water bodies.

The project manager should confirm that the general construction schedule and the construction sequencing schedule are compatible. Key construction activities and associated ESC measures are listed in Table 55.4.

Table 55.4 Construction Activities and ESC Measures

construction activity	ESC measures
designate site access	Stabilize exposed areas with gravel and/or temporary vegetation. Immediately apply stabilization to areas exposed throughout site development.
protect runoff outlets and conveyance systems	Install principle sediment traps, fences, and basins prior to grading; stabilize stream banks and install storm drains, channels, etc.
land clearing	Mark trees and buffer areas for preservation.
site grading	Install additional ESC measures as needed during grading.
site stabilization	Install temporary and permanent seeding, mulching, sodding, riprap, etc.
building construction and utilities install-ation	Install additional ESC measures as needed during construction.
landscaping and final site stabilization[*]	Remove all temporary control measures; install topsoil, trees and shrubs, permanent seeding, mulching, sodding, riprap, etc.; stabilize all open areas, including borrow and spoil areas.

[*]This is the last construction phase.

Time-Cost Trade-Off

A project's completion time and its cost are intricately related. Though some costs are not directly related to the time a project takes, many costs are. This is the essence of the time-cost trade-off: The cost increases as the project time is decreased and vice versa. A project manager's roles include understanding the time-cost relationship, optimizing a project's pace for minimal cost, and predicting the impact of a schedule change on project cost.

The costs associated with a project can be classified as direct costs or indirect costs. The project cost is the sum of the direct and indirect costs.

Direct costs, also known as *variable costs*, *operating costs*, *prime costs*, and *on costs*, are costs that vary directly with the level of output (e.g., labor, fuel, power, and the cost of raw material). Generally, direct costs

increase as a project's completion time is decreased, since more resources need to be allocated to increase the pace.

Indirect costs, also known as *fixed costs*, are costs that are not directly related to a particular function or product. Indirect costs include taxes, administration, personnel, and security costs. Such costs tend to be relatively steady over the life of the project and decrease as the project duration decreases.

The time required to complete a project is determined by the critical path, so to compress (or "crash") a project schedule (accelerate the project activities in order to complete the project sooner), a project manager must focus on critical path activities.

A procedure for determining the optimal project time, or time-cost-trade-off, is to determine the normal completion time and direct cost for each critical path activity and compare it to its respective "crash time" and direct cost. The *crash time* is the shortest time in which an activity can be completed. If a new critical path emerges, consider this in subsequent time reductions. In this way, one can step through the critical path activities and calculate the total direct project cost versus the project time. (To minimize the cost, those activities that are not on the critical path can be extended without increasing the project completion time.) The indirect, direct, and total project costs can then be calculated for different project durations. The optimal duration is the one with the lowest cost. This model assumes that the normal cost for an activity is lower than the crash cost, the time and cost are linearly related, and the resources needed to shorten an activity are available. If these assumptions are not true, then the model would need to be adapted. Other cost considerations include incentive payments, marketing initiatives, and the like.

Fast Tracking

Besides efficient scheduling, construction time can be compressed with *fast-track scheduling*. This method overlaps the design and construction phases of a project. Ordering of long-lead materials and equipment can occur, and work on the site and foundations can begin before all the details of the building are completely worked out. With fast-track scheduling, separate contracts are established so that each major system can be bid and awarded by itself to avoid delaying other construction.

Although the fast-track method requires close coordination between the architect, contractor, subcontractors, owner, and others, it makes it possible to construct a high-quality building in 10% to 30% less time than with a conventional construction contract.

Schedule Management

Several methods are used to schedule and monitor projects. The most common and easiest is the *bar chart* or

Gantt chart, such as Fig. 55.3. The various activities of the schedule are listed along the vertical axis. Each activity is given a starting and finishing date, and overlaps are indicated by drawing the bars for each activity so that they overlap. Bar charts are simple to make and understand and are suitable for small to midsize projects. However, they cannot show all the sequences and dependencies of one activity on another.

Critical path techniques are used to graphically represent the multiple relationships between stages in a complicated project. The graphical network shows the *precedence relationships* between the various activities. The graphical network can be used to control and monitor the progress, cost, and resources of a project. A critical path technique will also identify the most critical activities in the project.

Critical path techniques use *directed graphs* to represent a project. These graphs are made up of *arcs* (arrows) and *nodes* (junctions). The placement of the arcs and nodes completely specifies the precedences of the project. Durations and precedences are usually given in a *precedence table* (matrix).

4. RESOURCE LEVELING

Resource leveling is used to address *overallocation* (i.e., situations that demand more resources than are available). Usually, people and equipment are the limited resources, although project funding may also be limited if it becomes available in stages. Two common ways are used to level resources: (1) Tasks can be delayed (either by postponing their start dates or extending their completion dates) until resources become available. (2) Tasks can be split so that the parts are completed when planned and the remainders are completed when resources becomes available. The methods used depend on the limitations of the project, including budget, resource availability, finish date, and the amount of flexibility available for scheduling tasks. If resource leveling is used with tasks on a project's critical path, the project's completion date will inevitably be extended.

Example 55.2

Because of a shortage of reusable forms, a geotechnical contractor decides to build a 250 ft long concrete wall in 25 ft segments. Each segment requires the following crews and times.

- set forms: 3 carpenters; 2 days
- pour concrete: 3 carpenters; 1 day
- strip forms: 4 carpenters; 1 day

The contractor has budgeted for only three carpenters. Compared to the ideal schedule, how many additional days will it take to construct the wall if the number of carpenters is leveled to three?

Figure 55.3 *Gantt Chart*

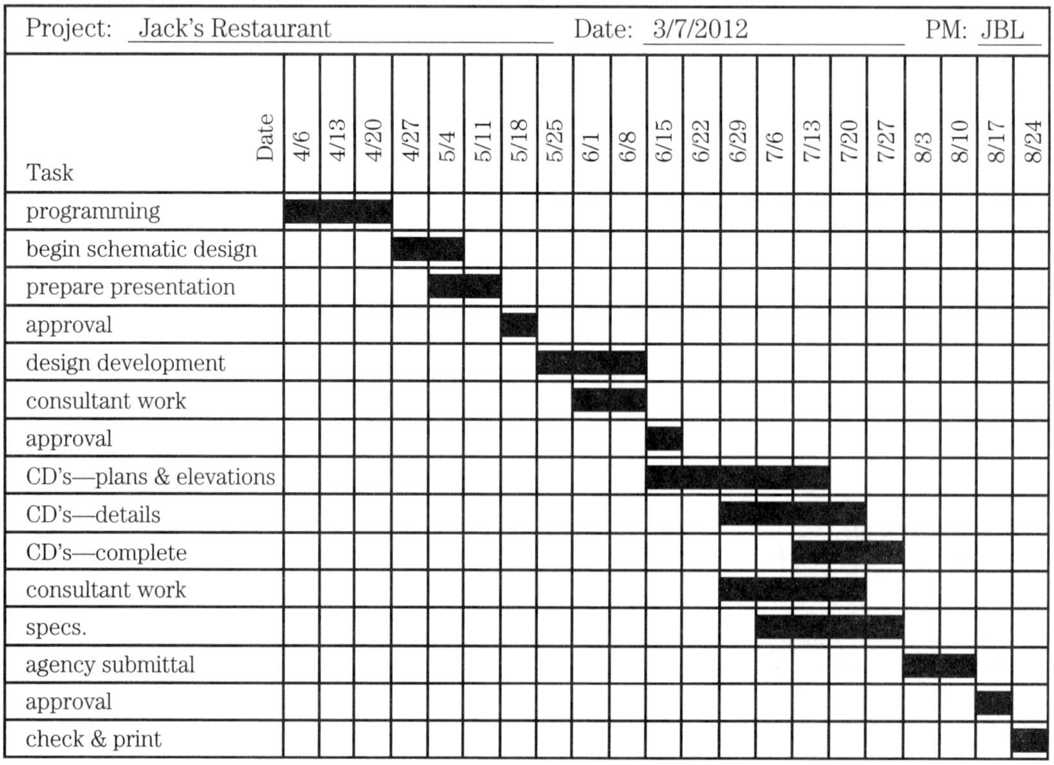

| Project: Jack's Restaurant | Date: 3/7/2012 | PM: JBL |

Task	Date	4/6	4/13	4/20	4/27	5/4	5/11	5/18	5/25	6/1	6/8	6/15	6/22	6/29	7/6	7/13	7/20	7/27	8/3	8/10	8/17	8/24
programming		■	■																			
begin schematic design					■																	
prepare presentation						■																
approval							■															
design development									■													
consultant work										■												
approval											■											
CD's—plans & elevations												■	■									
CD's—details														■	■							
CD's—complete																■						
consultant work														■								
specs.																■	■					
agency submittal																		■				
approval																				■		
check & print																						■

Systems, Mgmt., Professional

Solution

Having only three carpenters affects the form stripping operation, but it does not affect the form setting and concrete pouring operations. If the contractor had four carpenters, stripping the forms from each 25 ft segment would take 1 day; stripping forms for the entire 250 ft wall would take (10 segments)(1 day/segment) = 10 days.

With only three carpenters, form stripping productivity is reduced to $^3/_4$ of the four-carpenter rate. Each form stripping operation takes 4/3 days, and the entire wall takes (10 segments)(4/3 days/segment) = 13.3 days. The increase in time is 13.3 days − 10 days = 3.3 days.

5. ACTIVITY-ON-NODE NETWORKS

The *critical path method* (CPM) is one of several critical path techniques that uses a directed graph to describe the precedence of project activities. CPM requires that all activity durations be specified by single values. That is, CPM is a *deterministic method* that does not intrinsically support activity durations that are distributed as random variables.

Another characteristic of CPM is that each activity (task) is traditionally represented by a node (junction), hence the name *activity-on-node network*. Each node is typically drawn on the graph as a square box and labeled with a capital letter, although these are not absolute or universally observed conventions. Each activity can be thought of as a continuum of work, each with its own implicit "start" and "finish" *events*. For example, the activity "grub building site" starts with the event of a bulldozer arriving at the native site, followed by several days of bulldozing, and ending with the bulldozer leaving the cleaned site. An activity, including its start and finish events, occurs completely within its box (node).

Each activity in a CPM diagram is connected by arcs (connecting arrows, lines, etc.) The arcs merely show precedence and dependencies. Events are not represented on the graph, other than, perhaps, as the heads of tails of the arcs. Nothing happens along the arcs, and the arcs have zero durations. Because of this, arcs are not labeled. (See Fig. 55.4.)

Figure 55.4 *Activity-on-Node Network*

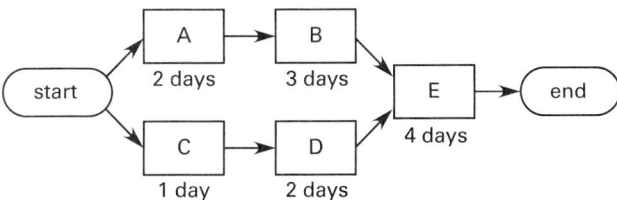

For convenience, when a project starts or ends with multiple simultaneous activities, *dummy nodes* with zero durations may be used to specify the start and/or finish of the entire project. Dummy nodes do not add time to the project. They are included for convenience.

Since all CPM arcs have zero durations, the arcs connecting dummy nodes are not different from arcs connecting other activities.

A CPM graph depicts the activities required to complete a project and the sequence in which the activities must be completed. No activity can begin until all of the activities with arcs leading into it have been completed. The duration and various dates are associated with each node, and these dates can be written on the CPM graph for convenience. These dates include the earliest possible start date, ES; the latest start date, LS; the earliest finish date, EF; and, the latest finish date, LF. The ES, EF, LS, and LF dates are calculated from the durations and the activity interdependencies.

After the ES, EF, LS, and LF dates have been determined for each activity, they can be used to identify the critical path. The *critical path* is the sequence of activities (the *critical activities*) that must all be started and finished exactly on time in order to not delay the project. Delaying the starting time of any activity on the critical path, or increasing its duration, will delay the entire project. The critical path is the longest path through the network. If it is desired that the project be completed sooner than expected, then one or more activities in the critical path must be shortened. The critical path is generally identified on the network with heavier (thicker) arcs.

Activities not on the critical path are known as *noncritical activities*. Other paths through the network will require less time than the critical path, and hence, will have inherent delays. The noncritical activities can begin or finish earlier or later (within limits) without affecting the overall schedule. The amount of time that an activity can be delayed without affecting the overall schedule is known as the *float* (*float time* or *slack time*). Float can be calculated in two ways, with identical results. The first way uses Eq. 55.1. The second way is described in the following section, "Solving a CPM Problem."

$$\text{float} = \text{LS} - \text{ES} = \text{LF} - \text{EF} \qquad 55.1$$

Float is zero along the critical path. This fact can be used to identify the activities along the critical path from their respective ES, EF, LS, and LF dates.

It is essential to maintain a distinction between time, date, length, and duration. Like the timeline for engineering economics problems, all projects start at time = 0, but this corresponds to starting on day = 1. That is, work starts on the first day. Work ends on the last day of the project, but this end rarely corresponds to midnight. So, if a project has a critical path length (duration-to-completion) of 15 days and starts on (at the beginning of) May 1, it will finish on (at the end of) May 15, not May 16.

6. SOLVING A CPM PROBLEM

As previously described, the solution to a critical path method problem reveals the earliest and latest times that an activity can be started and finished, and it also identifies the critical path and generates the float for each activity.

As an alternative to using Eq. 55.1, the following procedure may be used to solve a CPM problem. To facilitate the solution, each node should be replaced by a square that has been quartered. The compartments have the meanings indicated by the key.

ES	EF
LS	LF

key

ES: Earliest Start

EF: Earliest Finish

LS: Latest Start

LF: Latest Finish

step 1: Place the project start time or date in the **ES** and **EF** positions of the start activity. The start time is zero for relative calculations.

step 2: Consider any unmarked activity, all of whose predecessors have been marked in the **EF** and **ES** positions. (Go to step 4 if there are none.) Mark in its **ES** position the largest number marked in the **EF** position of those predecessors.

step 3: Add the activity time to the **ES** time and write this in the **EF** box. Go to step 2.

step 4: Place the value of the latest finish date in the **LS** and **LF** boxes of the finish mode.

step 5: Consider unmarked predecessors whose successors have all been marked. Their **LF** is the smallest **LS** of the successors. Go to step 7 if there are no unmarked predecessors.

step 6: The **LS** for the new node is **LF** minus its activity time. Go to step 5.

step 7: The float for each node is **LS − ES** and **LF − EF**.

step 8: The critical path encompasses nodes for which the float equals **LS − ES** from the start node. There may be more than one critical path.

Example 55.3

Using the precedence table given, construct the precedence matrix and draw an activity-on-node network.

activity	duration (days)	predecessors
A, start	0	–
B	7	A
C	6	A
D	3	B
E	9	B, C
F	1	D, E
G	4	C
H, finish	0	F, G

Solution

The precedence matrix is

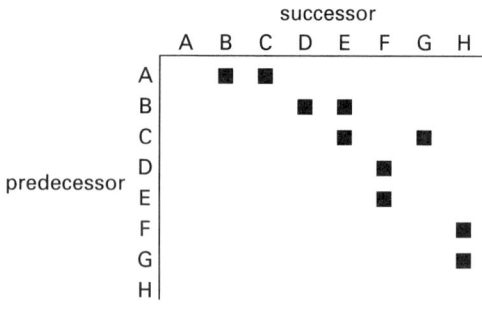

The activity-on-node network is

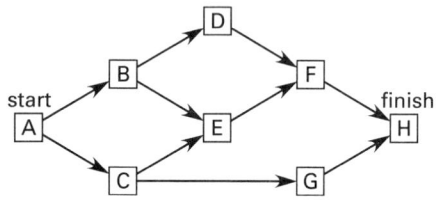

Example 55.4

Complete the network for the previous example and find the critical path. Assume the desired completion duration is in 19 days.

Solution

The critical path is shown with darker lines.

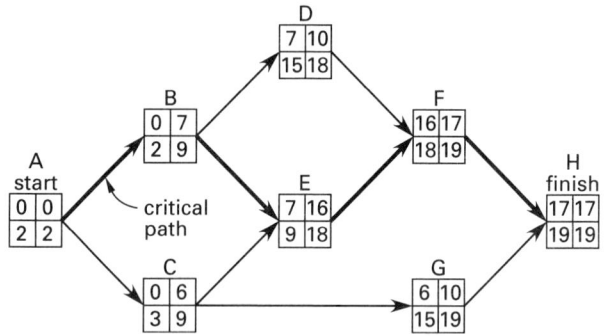

7. ACTIVITY-ON-ARC NETWORKS

Another variety of deterministic critical path techniques represents project activities as arcs. With *activity-on-arc networks* (also known as *activity-on-branch networks*), the continuum of work occupies the arcs, while the nodes represent instantaneous starting and ending events. The arcs have durations, while the nodes do not. Nothing happens on a node, as it represents an instant in time only. As shown in Fig. 55.5, each node is typically drawn on the graph as a circle and labeled with a number, although this format is not universally adhered to. Since the activities are described by the same precedence table as would be used with an activity-on-node graph, the arcs are labeled with the activity capital

Figure 55.5 *Activity-on-Arc Network*

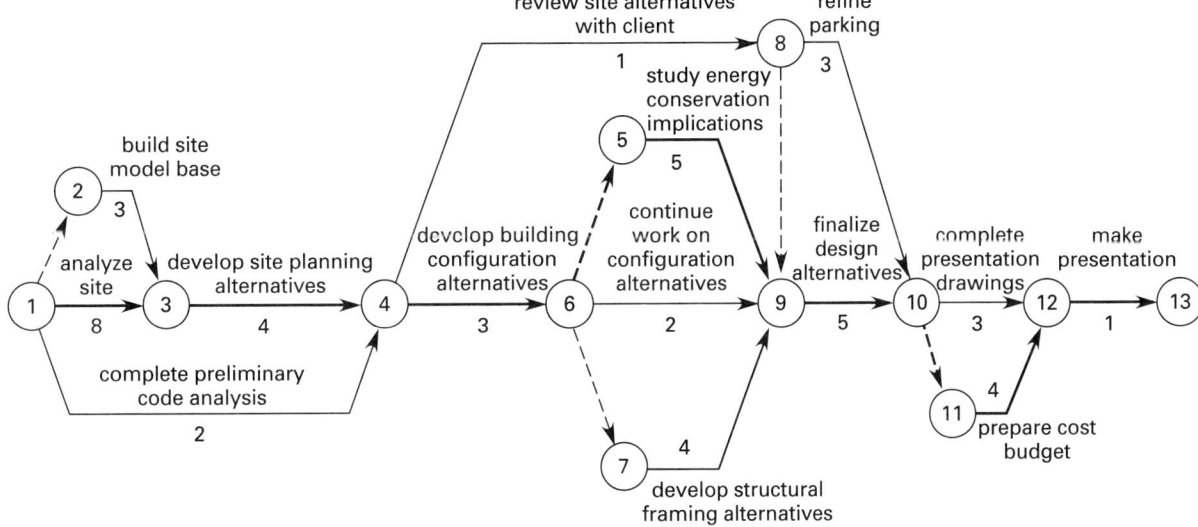

critical time path: 30 days

(Numbers in circles are beginning and ending points; numbers between circles indicate days.)

letter identifiers or the activity descriptions. The duration of the activity may be written adjacent to its arc. (This is the reason for identifying activities with letters, so that any number appearing with an arc can be interpreted as a duration.)

Although the concepts of ES, EF, LS, and LF dates, critical path, and float are equally applicable to activity-on-arc and activity-on-node networks, the two methods cannot be combined within a single project graph. Calculations for activity-on-arc networks may seem less intuitive, and there are other possible complications.

The activity-on-arc method is complicated by the frequent requirement for *dummy activities* and nodes to maintain precedence. Consider the following part of a precedence table.

activity	predecessors
L	–
M	–
N	L, M
P	M

Activity P depends on the completion of only M. Figure 55.6(a) is an activity-on-arc representation of this precedence. However, N depends on the completion of both L and M. It would be incorrect to draw the network as Fig. 55.6(b) since the activity N appears twice. To represent the project, the dummy activity X must be used, as shown in Fig. 55.6(c).

If two activities have the same starting and ending events, a *dummy node* is required to give one activity a uniquely identifiable completion event. This is illustrated in Fig. 55.7(b).

Figure 55.6 *Activity-on-Arc Network with Predecessors*

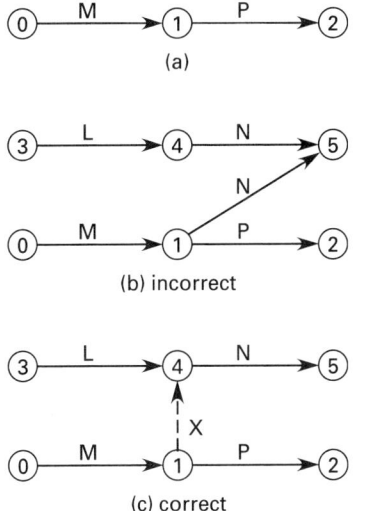

The solution method for an activity-on-arc problem is essentially the same as for the activity-on-node problem, requiring forward and reverse passes to determine earliest and latest dates.

Figure 55.7 *Use of a Dummy Node*

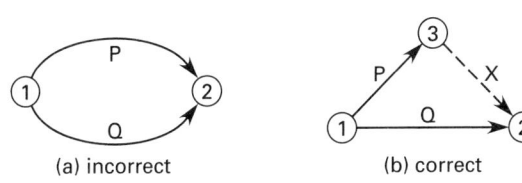

Example 55.5

Represent the project in Ex. 55.3 as an activity-on-arc network.

Solution

event	event description
0	start project
1	finish B, start D
2	finish C, start G
3	finish B and C, start E
4	finish D and E, start F
5	finish F and G

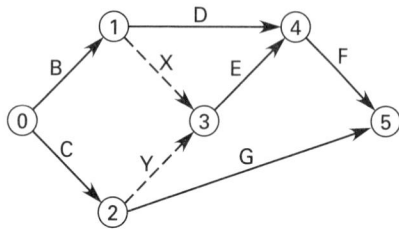

8. STOCHASTIC CRITICAL PATH MODELS

Stochastic models differ from deterministic models only in the way in which the activity durations are found. Whereas durations are known explicitly for the deterministic model, the time for a stochastic activity is distributed as a random variable.

This stochastic nature complicates the problem greatly since the actual distribution is often unknown. Such problems are solved as a deterministic model using the mean of an assumed duration distribution as the activity duration.

The most common stochastic critical path model is PERT, which stands for *program evaluation and review technique*. PERT diagrams have traditionally been drawn as activity-on-arc networks. In PERT, all duration variables are assumed to come from a *beta distribution*, with mean and standard deviation given by Eq. 55.2 and Eq. 55.3, respectively.

$$\mu = \tfrac{1}{6}(t_{\min} + 4t_{\text{most likely}} + t_{\max}) \qquad 55.2$$

$$\sigma = \tfrac{1}{6}(t_{\max} - t_{\min}) \qquad 55.3$$

The project *completion time* for large projects is assumed to be normally distributed with mean equal to the critical path length and with overall variance equal to the sum of the variances along the critical path.

The probability that a project duration will exceed some length, D, can be found from Eq. 55.4. z is the standard normal variable.

$$p\{\text{duration} > D\} = p\{t > z\} \qquad 55.4$$

$$z = \frac{D - \mu_{\text{critical path}}}{\sigma_{\text{critical path}}} \qquad 55.5$$

Example 55.6

The mean times and variances for activities along a PERT critical path are given. What is the probability that the project's completion time will be (a) less than 14 days, (b) more than 14 days, (c) more than 23 days, and (d) between 14 and 23 days?

activity mean time (days)	activity standard deviation (days)
9	1.3
4	0.5
7	2.6

Solution

The most likely completion time is the sum of the mean activity times.

$$\mu_{\text{critical path}} = 9 \text{ days} + 4 \text{ days} + 7 \text{ days}$$
$$= 20 \text{ days}$$

The variance of the project's completion times is the sum of the variances along the critical path. Variance, σ^2, is the square of the standard deviation, σ.

$$\sigma^2_{\text{critical path}} = (1.3 \text{ days})^2 + (0.5 \text{ days})^2$$
$$+ (2.6 \text{ days})^2$$
$$= 8.7 \text{ days}^2$$
$$\sigma_{\text{critical path}} = \sqrt{8.7 \text{ days}^2}$$
$$= 2.95 \text{ days} \quad [\text{use 3 days}]$$

The standard normal variable corresponding to 14 days is given by Eq. 55.5.

$$z = \frac{D - \mu_{\text{critical path}}}{\sigma_{\text{critical path}}} = \frac{14 \text{ days} - 20 \text{ days}}{3 \text{ days}}$$
$$= -2.0$$

From App. 11.A, the area under the standard normal curve is 0.4772 for $0.0 < z < -2.0$. Since the normal curve is symmetrical, the negative sign is irrelevant in determining the area.

The standard normal variable corresponding to 23 days is given by Eq. 55.5.

$$z = \frac{D - \mu_{\text{critical path}}}{\sigma_{\text{critical path}}} = \frac{23 \text{ days} - 20 \text{ days}}{3 \text{ days}}$$
$$= 1.0$$

The area under the standard normal curve is 0.3413 for $0.0 < z < 1.0$.

(a) $p\{\text{duration} < 14\} = p\{z < -2.0\} = 0.5 - 0.4772$
$= 0.0228 \quad (2.28\%)$

(b) $p\{\text{duration} > 14\} = p\{z > -2.0\} = 0.4772 + 0.5$
$= 0.9772 \quad (97.72\%)$

(c) $p\{\text{duration} > 23\} = p\{z > 1.0\} = 0.5 - 0.3413$
$= 0.1587 \quad (15.87\%)$

(d) $p\{14 < \text{duration} < 23\} = p\{-2.0 < z < 1.0\}$
$= 0.4772 + 0.3413$
$= 0.8185 \quad (81.85\%)$

9. MONITORING

Monitoring is keeping track of the progress of the job to see if the planned aspects of time, fee, and quality are being accomplished. The original fee projections can be monitored by comparing weekly time sheets with the original estimate. This can be done manually or with project management software. A manual method is shown in Fig. 55.8, which uses the same example project estimated in Fig. 55.2.

In Fig. 55.8, the budgeted weekly costs are placed in the table under the appropriate time-period column and phase-of-work row. The actual costs expended are written next to them. At the bottom of the chart, a cumulative graph is plotted that shows the actual money expended against the budgeted fees. The cumulative ratio of percentage completion to cost can also be plotted.

Monitoring quality is more difficult. At regular times during a project, the project manager, designers, and office principals should review the progress of the job to determine if the original project goals are being met and if the job is being produced according to the client's and design firm's expectations. The work in progress can also be reviewed to see whether it is technically correct and if all the contractual obligations are being met.

10. COORDINATING

During the project, the project manager must constantly coordinate the various people involved: the architect's staff, the consultants, the client, the building code officials, firm management, and, of course, the construction contractors. This may be done on a weekly, or even daily, basis to make sure the schedule is being maintained and the necessary work is getting done.

The coordination can be done by using checklists, holding weekly project meetings to discuss issues and assign

Figure 55.8 *Project Monitoring Chart*

Project: Mini-mall

Phase/People/Departments	fee dollars	1	2	3	4	5	6	7	8	9	10	11	12	total
schematic design	budgeted	1320	2640											
	actual	2000	2900											
design development	budgeted			2510	4090									
	actual			3200										
construction docs.	budgeted					2156	1716	4136	5456	2376				
	actual													
	budgeted													
	actual													
	budgeted													
	actual													
	budgeted													
	actual													
	budgeted													
	actual													
total (cumulative)	budgeted	1320	3960	6470	10,560	12,716	14,432	18,568	24,024	26,400				
	actual	2000	4900	8100										

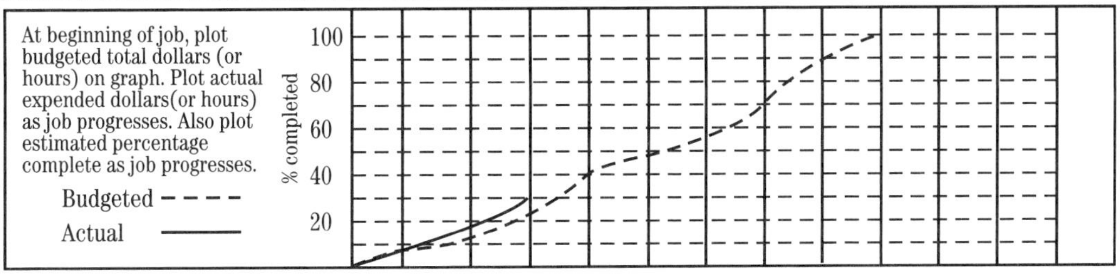

work, and exchanging drawings or project files among the consultants.

11. DOCUMENTATION

Everything that is done on a project must be documented in writing. This documentation provides a record in case legal problems develop and serves as a project history to use for future jobs. Documentation is also a vital part of communication. An email or written memo is more accurate, communicates more clearly, and is more difficult to forget than a simple phone call, for example.

Most design firms have standard forms or project management software for documents such as transmittals, job observation reports, time sheets, and the like. Such software makes it easy to record the necessary information. In addition, all meetings should be documented with meeting notes. Phone call logs (listing date, time, participants, and discussion topics), emails, personal daily logs, and formal communications like letters and memos should also be generated and preserved to serve as documentation.

Two types of documents, *change orders* due to unexpected conditions or changes to the plans after bidding, and *as-built construction documents* to record what was actually installed (as opposed to what was shown in the original construction documents) are particularly important.

12. EARNED VALUE METHOD

The *earned value method* (EVM), also known as *earned value management*, is a project management technique that correlates actual *project value* (PV) with *earned value* (EV).[1] *Value* is generally defined in dollars, but it may also be defined in hours. Close monitoring of earned value makes it possible to forecast cost and schedule overruns early in a project. In its simplest form, the method monitors the project plan, actual work performed, expenses, and cumulative value to see if the project is on track. Earned value shows how much of the budget and time should have been spent, with regard to the amount of work actually completed. The earned value method differs from typical budget versus expenses models by requiring the cost of work in progress to be quantified. Because of its complexity, the method is best implemented in its entirety in very large projects.

A *work breakdown structure* (WBS) is at the core of the method. A WBS is a hierarchical structure used to organize tasks for reporting schedules and tracking costs. For monitoring, the WBS is subsequently broken down into manageable *work packages*—small sets of activities at lower and lowest levels of the WBS that collectively constitute the overall project scope. Each work package has a relatively short completion time and can be divided into a series of milestones whose status can be objectively measured. Each work package has start and finish dates and a budget value.

The earned value method uses specific terminology for otherwise common project management and accounting principles. There are three primary measures of project performance: BCWS, ACWP, and BCWP. The *budgeted cost of work scheduled* (BCWS) is a spending plan for the project as a function of schedule and performance. For any specified time period, the *cumulative planned expenditures* is the total amount budgeted for the project up to that point. With EVM, the spending plan serves as a performance baseline for making predictions about cost and schedule variance and estimates of completion.

The *actual cost of work performed* (ACWP) is the actual spending as a function of time or performance. It is the cumulative actual expenditures on the project viewed at regular intervals within the project duration. The *budgeted cost of work performed* (BCWP) is the actual earned value based on the technical accomplishment. BCWP is the cumulative budgeted value of the work actually completed. It may be calculated as the sum of the values budgeted for the work packages actually completed, or it may be calculated by multiplying the fraction of work completed by the planned cost of the project.

The primary measures are used to derive secondary measures that give a different view of the project's current and future health. Figure 55.9 illustrates how some of these measures can be presented graphically.

Cost variance (CV) is the difference between the planned and actual costs of the work completed.

$$CV = BCWP - ACWP \qquad 55.6$$

Schedule variance (SV) is the difference between the value of work accomplished for a given period and the value of the work planned. It measures how much a project is ahead of or behind schedule. Schedule variance is measured in value (e.g., dollars), not time.

$$SV = BCWP - BCWS \qquad 55.7$$

Cost performance index (CPI) is a cost efficiency factor representing the relationship between the actual cost expended and the earned value. A CPI ≥ 1 suggests efficiency.

$$CPI = \frac{BCWP}{ACWP} \qquad 55.8$$

---

[1]The earned value method may also be referred to as *performance measurement, management by objectives, budgeted cost of work performed,* and *cost/schedule control systems.*

Figure 55.9 *Earned Value Management Measures*

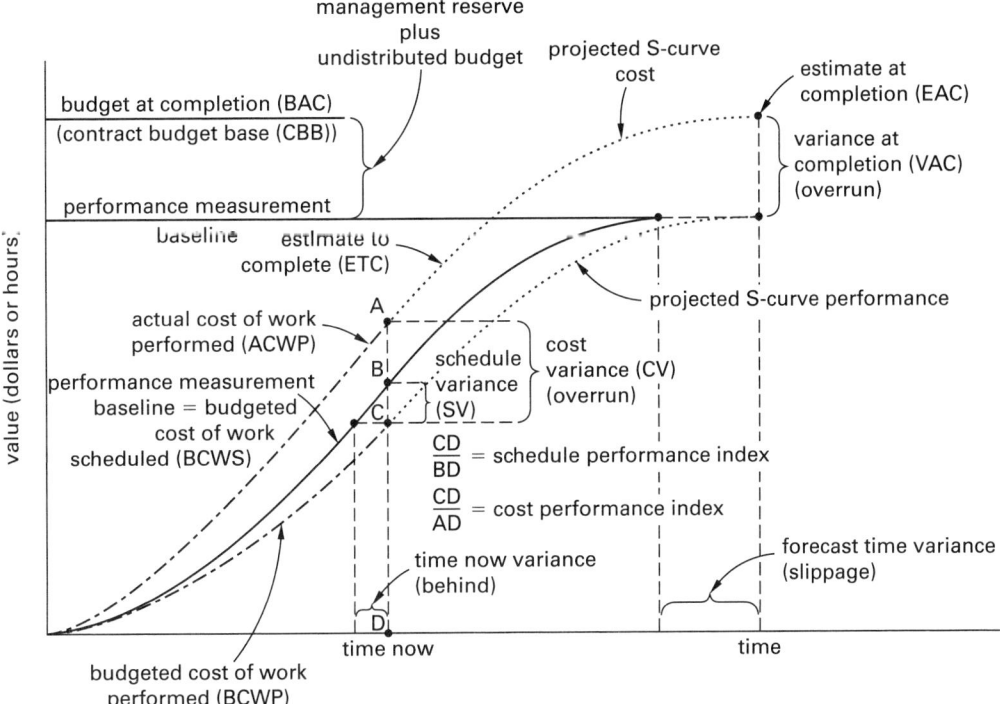

Schedule performance index (SPI) is a measure of schedule effectiveness, as determined from the earned value and the initial planned schedule. SPI ≥ 1 suggests work is ahead of schedule.

$$\text{SPI} = \frac{\text{BCWP}}{\text{BCWS}} \qquad 55.9$$

Budget at completion (BAC), also known as *performance measurement baseline*, is the sum total of the time-phased budget.

Estimate to complete (ETC) is a calculated value representing the cost of work required to complete remaining project tasks.

$$\text{ETC} = \text{BAC} - \text{BCWP} \qquad 55.10$$

Estimate at completion (EAC) is a calculated value that represents the projected total final cost of work when completed.

$$\text{EAC} = \text{ACWP} + \text{ETC} \qquad 55.11$$

Example 55.7

A project was baselined at $10,000 per week for 8 weeks. The project was supposed to be completed at 8 weeks but has already taken 11 weeks. One additional week is estimated to complete the project. Determine the (a) budgeted cost of work scheduled; (b) actual cost of work performed; (c) budget at completion; (d) percentage of project planned; (e) estimate to complete; (f) estimate at completion; (g) actual percentage of project completed; (h) budgeted cost of work performed; (i) variance at completion; (j) schedule variance; (k) schedule performance index; (l) cost variance; and (m) cost performance index.

Solution

(a) The budgeted cost of work scheduled, BCWS, or the project value (planned value), PV, is

$$\text{BCWS} = \text{PV} = (8 \text{ wk})\left(10,000 \ \frac{\$}{\text{wk}}\right) = \$80,000$$

(b) The actual cost of work performed, ACWP, or the actual cost, AC, is

$$\text{ACWP} = \text{AC} = (11 \text{ wk})\left(10,000 \ \frac{\$}{\text{wk}}\right) = \$110,000$$

(c) The budget at completion, BAC, is

$$\text{BAC} = (8 \text{ wk})\left(10,000 \ \frac{\$}{\text{wk}}\right) = \$80,000$$

(d) The percentage of project planned is 100%, since the project should have been completed by now.

$$\% \text{ completed, planned} = \frac{\text{PV}}{\text{BAC}} = \frac{\$80,000}{\$80,000} = 1 \quad (100\%)$$

Systems, Mgmt., Professional

(e) The estimate to complete, ETC, is

$$ \text{ETC} = (1 \text{ wk})\left(10{,}000 \ \frac{\$}{\text{wk}}\right) = \$10{,}000 $$

(f) The estimate at completion, EAC, is

$$ \text{EAC} = \text{AC} + \text{ETC} = \$110{,}000 + \$10{,}000 = \$120{,}000 $$

(g) The actual percentage of project completed is

$$ \%\text{ completed, actual} = \frac{\text{AC}}{\text{EAC}} = \frac{\$110{,}000}{\$120{,}000} $$
$$ = 0.917 \quad (91.7\%) $$

(h) The budgeted cost of work performed, BCWP, or earned value, EV, is

$$ \text{BCWP} = \text{EV} = (\text{fraction completed})(\text{BCWS}) $$
$$ = (0.917)(\$80{,}000) $$
$$ = \$73{,}400 $$

(i) The variance at completion, VAC, is

$$ \text{VAC} = \text{BAC} - \text{EAC} $$
$$ = \$80{,}000 - \$120{,}000 $$
$$ = -\$40{,}000 $$

The VAC is negative, indicating a cost overrun.

(j) The schedule variance, SV, is

$$ \text{SV} = \text{EV} - \text{PV} = \$73{,}400 - \$80{,}000 = -\$6600 $$

(k) The schedule performance index, SPI, is

$$ \text{SPI} = \frac{\text{BCWP}}{\text{BCWS}} = \frac{\$73{,}400}{\$80{,}000} = 0.917 $$

(l) The cost variance, CV, is

$$ \text{CV} = \text{EV} - \text{AC} = \$73{,}400 - \$110{,}000 = -\$36{,}600 $$

The CV is negative, indicating a cost overrun.

(m) The cost performance index, CPI, is

$$ \text{CPI} = \frac{\text{BCWP}}{\text{ACWP}} = \frac{\$73{,}400}{\$110{,}000} = 0.667 $$

The CPI is less than 1.0, which indicates that the project is over budget.

56 Professional Services, Contracts, and Engineering Law[1]

1. FORMS OF COMPANY OWNERSHIP

There are three basic forms of company ownership in the United States: (a) sole proprietorship, (b) partnership, and (c) corporation.[2] Each of these forms of ownership has advantages and disadvantages.

2. SOLE PROPRIETORSHIPS

A *sole proprietorship* (*single proprietorship*) is the easiest form of ownership to establish. Other than the necessary licenses, filings, and notices (which apply to all forms of ownership), no legal formalities are required to start business operations. A sole proprietor (the owner) has virtually total control of the business and makes all supervisory and management decisions.

Legally, there is no distinction between the sole proprietor and the sole proprietorship (the business). This is the greatest disadvantage of this form of business. The owner is solely responsible for the operation of the business, even if the owner hires others for assistance. The owner assumes personal, legal, and financial liability for all acts and debts of the company. If the company debts remain unpaid, or in the event there is a legal judgment against the company, the owner's personal assets (home, car, savings, etc.) can be seized or attached.

Another disadvantage of the sole proprietorship is the lack of significant organizational structure. In times of business crisis or trouble, there may be no one to share the responsibility or to help make decisions. When the owner is sick or dies, there may be no way to continue the business.

There is also no distinction between the incomes of the business and the owner. Therefore, the business income is taxed at the owner's income tax rate. Depending on the owner's financial position, the success of the business, and the tax structure, this can be an advantage or a disadvantage.[3]

3. PARTNERSHIPS

A *partnership* (also known as a *general partnership*) is ownership by two or more persons known as *general partners*. Legally, this form is very similar to a sole proprietorship, and the two forms of business have many of the same advantages and disadvantages. For example, with the exception of an optional *partnership agreement*, there are a minimum of formalities to setting up business. The partners make all business and management decisions themselves according to an agreed-upon process. The business income is split among the partners and taxed at the partners' individual tax rates.[4] Continuity of the business is still a problem since most partnerships are automatically dissolved upon the withdrawal or death of one of the partners.[5]

One advantage of a partnership over a sole proprietorship is the increase in available funding. Not only do

---

[1]This chapter is not intended to be a substitute for professional advice. Law is not always black and white. For every rule there are exceptions. For every legal principle, there are variations. For every type of injury, there are numerous legal precedents. This chapter covers the superficial basics of a small subset of U.S. law affecting engineers.

[2]The discussion of forms of company ownership in Sec. 56.2, Sec. 56.3, and Sec. 56.4 applies equally to service-oriented companies (e.g., consulting engineering firms) and product-oriented companies.

[3]To use a simplistic example, if the corporate tax rates are higher than the individual tax rates, it would be *financially* better to be a sole proprietor because the company income would be taxed at a lower rate.

[4]The percentage split is specified in the partnership agreement.

[5]Some or all of the remaining partners may want to form a new partnership, but this is not always possible.

more partners bring in more start-up capital, but the resource pool may make business credit easier to obtain. Also, the partners bring a diversity of skills and talents.

Unless the partnership agreement states otherwise, each partner can individually obligate (i.e., *bind*) the partnership without the consent of the other partners. Similarly, each partner has personal responsibility and liability for the acts and debts of the partnership company, just as sole proprietors do. In fact, each partner assumes the *sole* responsibility, not just a proportionate share. If one or more partners are unable to pay, the remaining partners shoulder the entire debt. The possibility of one partner having to pay for the actions of another partner must be considered when choosing this form of business ownership.

A *limited partnership* differs from a general partnership in that one (or more) of the partners is silent. The *limited partners* make a financial contribution to the business and receive a share of the profit but do not participate in the management and cannot bind the partnership. While *general partners* have unlimited personal liabilities, limited partners are generally liable only to the extent of their investment.[6] A written partnership agreement is required, and the agreement must be filed with the proper authorities.

4. CORPORATIONS

A corporation is a legal entity (i.e., a legal person) distinct from the founders and owners. The separation of ownership and management makes the corporation a fundamentally different business form than a sole proprietorship or partnership, with very different advantages and disadvantages.

A corporation becomes legally distinct from its founders upon formation and proper registration. Ownership of the corporation is through shares of stock, distributed to the founders and investors according to some agreed-upon investment and distribution rule. The founders and investors become the stockholders (i.e., owners) of the corporation. A *closely held* (*private*) corporation is one in which all stock is owned by a family or small group of co-investors. A *public corporation* is one whose stock is available for the public-at-large to purchase.

There is no mandatory connection between ownership and management functions. The decision-making power is vested in the executive officers and a *board of directors* that governs by majority vote. The stockholders elect the board of directors which, in turn, hires the executive officers, management, and other employees. Employees of the corporation may or may not be stockholders.

Disadvantages (at least for a person or persons who could form a partnership or sole proprietorship) include

the higher corporate tax rate, difficulty and complexity of formation (some states require a minimum number of persons on the board of directors), and additional legal and accounting paperwork.

However, since a corporation is distinctly separate from its founders and investors, those individuals are not liable for the acts and debts of the corporation. Debts are paid from the corporate assets. Income to the corporation is not taxable income to the owners. (Only the salaries, if any, paid to the employees by the corporation are taxable to the employees.) Even if the corporation were to go bankrupt, the assets of the owners would not ordinarily be subject to seizure or attachment.

A corporation offers the best guarantee of continuity of operation in the event of the death, incapacitation, or retirement of the founders since, as a legal entity, it is distinct from the founders and owners.

5. LIMITED LIABILITY ENTITIES

A variety of other legal entities have been established that blur the lines between the three traditional forms of business (i.e., proprietorship, partnership, and corporation). The *limited liability partnership*, LLP, extends a measure of corporate-like protection to professionals while permitting partnership-like personal participation in management decisions. Since LLPs are formed and operated under state laws, actual details vary from state to state. However, most LLPs allow the members to participate in management decisions, while not being responsible for the misdeeds of other partners. As in a corporation, the debts of the LLP do not become debts of the members.[7] The *double taxation* characteristic of traditional corporations is avoided, as profits to the LLP flow through to the members.

For engineers and architects (as well as doctors, lawyers, and accountants), the *professional corporation*, PC, offers protection from the actions (e.g., malpractice) of other professionals within a shared environment, such as a design firm. While a PC does not shield the individual from responsibility for personal negligence or malpractice, it does permit the professional to be associated with a larger entity, such as a partnership of other PCs, without accepting responsibility for the actions of the other members. In that sense, the protection is similar to that of an LLP. Unlike a traditional corporation, a PC may have a board of directors consisting of only a single individual, the professional.

The *limited liability company*,[8] LLC, also combines advantages from partnerships and corporations. In an LLC, the members are shielded from debts of the LLC

---

[6]That is, if the partnership fails or is liquidated to pay debts, the limited partners lose no more than their initial investments.

[7]Depending on the state, the shield may be complete or limited. It is common that the protection only applies to negligence-related claims, as opposed to intentional tort claims, contract-related obligations, and day-to-day operating expenses such as rent, utilities, and employees.

[8]LLC does not mean *limited liability corporation*. LLCs are not corporations.

while enjoying the pass-through of all profits.[9] Like a partnership or shareholder, a member's obligation is limited to the *membership interest* in (i.e., contribution to) the LLC. LLCs are directed and controlled by one or more managers who may also be members. A variation of the LLC specifically for design, medical, and other professionals is the *professional limited liability company*, PLLC.

The traditional corporation, as described in Sec. 56.4, is referred to as a *Subchapter C corporation* or "*C corp.*"[10] A variant is the *S corporation* ("*S corp*") which combines characteristics of the C corporation with pass-through for taxation. S corporations can be limited or treated differently than C corporations by state and federal law.

6. PIERCING THE CORPORATE VEIL

An individual operating as a corporation, LLP, LLC, or PC entity may lose all protection if his or her actions are fraudulent, or if the court decides the business is an "alter ego" of the individual. Basically, this requires the business to be run as a business. Business and personal assets cannot be intermingled, and business decisions must be made and documented in a business-like manner. If operated fraudulently or loosely, a court may assign liability directly to an individual, an action known as *piercing the corporate veil*.

7. AGENCY

In some contracts, decision-making authority and right of action are transferred from one party (the owner, or *principal*) who would normally have that authority to another person (the *agent*). For example, in construction contracts, the engineer may be the agent of the owner for certain transactions. Agents are limited in what they can do by the scope of the agency agreement. Within that scope, however, an agent acts on behalf of the principal, and the principal is liable for the acts of the agent and is bound by contracts made in the principal's name by the agent.

Agents are required to execute their work with care, skill, and diligence. Specifically, agents have *fiduciary responsibility* toward their principal, meaning that agent must be honest and loyal. Agents are liable for damages resulting from a lack of diligence, loyalty, and/or honesty. If the agents misrepresented their skills when obtaining the agency, they can be liable for breach of contract or fraud.

8. GENERAL CONTRACTS

A *contract* is a legally binding agreement or promise to exchange goods or services.[11] A written contract is merely a documentation of the agreement. Some agreements must be in writing, but most agreements for engineering services can be verbal, particularly if the parties to the agreement know each other well.[12] Written contract documents do not need to contain intimidating legal language, but all agreements must satisfy three basic requirements to be enforceable (binding).

- There must be a clear, specific, and definite *offer* with no room for ambiguity or misunderstanding.
- There must be some form of conditional future *consideration* (i.e., payment).[13]
- There must be an *acceptance* of the offer.

There are other conditions that the agreement must meet to be enforceable. These conditions are not normally part of the explicit agreement but represent the conditions under which the agreement was made.

- The agreement must be *voluntary* for all parties.
- All parties must have *legal capacity* (i.e., be mentally competent, of legal age, not under coercion, and uninfluenced by drugs).
- The purpose of the agreement must be *legal*.

For small projects, a simple *letter of agreement* on one party's stationery may suffice. For larger, complex projects, a more formal document may be required. Some clients prefer to use a *purchase order*, which can function as a contract if all basic requirements are met.

Regardless of the format of the written document—letter of agreement, purchase order, or standard form—a contract should include the following features.[14]

- introduction, preamble, or preface indicating the purpose of the contract
- name, address, and business forms of both contracting parties
- signature date of the agreement

---

[11]Not all agreements are legally binding (i.e., enforceable). Two parties may agree on something, but unless the agreement meets all of the requirements and conditions of a contract, the parties cannot hold each other to the agreement.

[12]All states have a *statute of frauds* that, among other things, specifies what types of contracts must be in writing to be enforceable. These include contracts for the sale of land, contracts requiring more than one year for performance, contracts for the sale of goods over $500 in value, contracts to satisfy the debts of another, and marriage contracts. Contracts to provide engineering services do not fall under the statute of frauds.

[13]Actions taken or payments made prior to the agreement are irrelevant. Also, it does not matter to the courts whether the exchange is based on equal value or not.

[14]*Construction contracts* are unique unto themselves. Items that might also be included as part of the *contract documents* are the agreement form, the general conditions, drawings, specifications, and addenda.

---

[9]LLCs enjoy *check the box taxation*, which means they can elect to be taxed as sole proprietorships, partnerships, or corporations.

[10]The reference is to subchapter C of the Internal Revenue Code.

- effective date of the agreement (if different from the signature date)
- duties and obligations of both parties
- deadlines and required service dates
- fee amount
- fee schedule and payment terms
- agreement expiration date
- standard boilerplate clauses
- signatures of parties or their agents
- declaration of authority of the signatories to bind the contracting parties
- supporting documents

9. STANDARD BOILERPLATE CLAUSES

It is common for full-length contract documents to include important *boilerplate clauses*. These clauses have specific wordings that should not normally be changed, hence the name "boilerplate." Some of the most common boilerplate clauses are paraphrased here.

- Delays and inadequate performance due to war, strikes, and acts of God and nature are forgiven (*force majeure*).
- The contract document is the complete agreement, superseding all prior verbal and written agreements.
- The contract can be modified or canceled only in writing.
- Parts of the contract that are determined to be void or unenforceable will not affect the enforceability of the remainder of the contract (*severability*). Alternatively, parts of the contract that are determined to be void or unenforceable will be rewritten to accomplish their intended purpose without affecting the remainder of the contract.
- None (or one, or both) of the parties can (or cannot) assign its (or their) rights and responsibilities under the contract (*assignment*).
- All notices provided for in the agreement must be in writing and sent to the address in the agreement.
- Time is of the essence.[15]
- The subject headings of the agreement paragraphs are for convenience only and do not control the meaning of the paragraphs.
- The laws of the state in which the contract is signed must be used to interpret and govern the contract.
- Disagreements shall be arbitrated according to the rules of the American Arbitration Association.

---
[15]Without this clause in writing, damages for delay cannot be claimed.

- Any lawsuits related to the contract must be filed in the county and state in which the contract is signed.
- Obligations under the agreement are unique, and in the event of a breach, the defaulting party waives the defense that the loss can be adequately compensated by monetary damages (*specific performance*).
- In the event of a lawsuit, the prevailing party is entitled to an award of reasonable attorneys' and court fees.[16]
- Consequential damages are not recoverable in a lawsuit.

10. SUBCONTRACTS

When a party to a contract engages a third party to perform the work in the original contract, the contract with the third party is known as a *subcontract*. Whether or not responsibilities can be subcontracted under the original contract depends on the content of the *assignment clause* in the original contract.

11. PARTIES TO A CONSTRUCTION CONTRACT

A specific set of terms has developed for referring to parties in consulting and construction contracts. The *owner* of a construction project is the person, partnership, or corporation that actually owns the land, assumes the financial risk, and ends up with the completed project. The *developer* contracts with the architect and/or engineer for the design and with the contractors for the construction of the project. In some cases, the owner and developer are the same, in which case the term *owner-developer* can be used.

The *architect* designs the project according to established codes and guidelines but leaves most stress and capacity calculations to the *engineer*.[17] Depending on the construction contract, the engineer may work for the architect, or vice versa, or both may work for the developer.

Once there are approved plans, the developer hires *contractors* to do the construction. Usually, the entire construction project is awarded to a *general contractor*. Due to the nature of the construction industry, separate *subcontracts* are used for different tasks (electrical, plumbing, mechanical, framing, fire sprinkler installation, finishing, etc.). The general contractor who hires all of these different *subcontractors* is known as the *prime contractor* (or *prime*). (The subcontractors can also work directly for the owner-developer, although this is less

---
[16]Without this clause in writing, attorneys' fees and court costs are rarely recoverable.
[17]On simple small projects, such as wood-framed residential units, the design may be developed by a *building designer*. The legal capacities of building designers vary from state to state.

common.) The prime contractor is responsible for all acts of the subcontractors and is liable for any damage suffered by the owner-developer due to those acts.

Construction is managed by an agent of the owner-developer known as the *construction manager*, who may be the engineer, the architect, or someone else.

12. STANDARD CONTRACTS FOR DESIGN PROFESSIONALS

Several professional organizations have produced standard agreement forms and other standard documents for design professionals.[18] Among other standard forms, notices, and agreements, the following standard contracts are available.[19]

- standard contract between engineer and client
- standard contract between engineer and architect
- standard contract between engineer and contractor
- standard contract between owner and construction manager

Besides completeness, the major advantage of a standard contract is that the meanings of the clauses are well established, not only among the design professionals and their clients but also in the courts. The clauses in these contracts have already been litigated many times. Where a clause has been found to be unclear or ambiguous, it has been rewritten to accomplish its intended purpose.

13. CONSULTING FEE STRUCTURE

Compensation for consulting engineering services can incorporate one or more of the following concepts.

---

[18]There are two main sources of standardized construction and design agreements: EJCDC and AIA. Consensus documents, known as *ConsensusDOCS*, for every conceivable situation have been developed by the *Engineers Joint Contracts Documents Committee*, EJCDC. EJCDC includes the American Society of Civil Engineers (ASCE), the American Council of Engineering Companies (ACEC), National Society of Professional Engineers' (NSPE's) Professional Engineers in Private Practice Division, Associated General Contractors of America (AGC), and more than fifteen other participating professional engineering design, construction, owner, legal, and risk management organizations, including the Associated Builders and Contractors; American Subcontractors Association; Construction Users Roundtable; National Roofing Contractors Association; Mechanical Contractors Association of America; and National Plumbing, Heating-Cooling Contractors Association. The American Institute of Architects, AIA, has developed its own standardized agreements in a less collaborative manner. Though popular with architects, AIA provisions are considered less favorable to engineers, contractors, and subcontractors who believe the AIA documents assign too much authority to architects, too much risk and liability to contractors, and too little flexibility in how construction disputes are addressed and resolved.

[19]The Construction Specifications Institute (CSI) has produced standard specifications for materials. The standards have been organized according to a UNIFORMAT structure consistent with ASTM Standard E1557.

- *lump-sum fee:* This is a predetermined fee agreed upon by client and engineer. This payment can be used for small projects where the scope of work is clearly defined.

- *cost plus fixed fee:* All costs (labor, material, travel, etc.) incurred by the engineer are paid by the client. The client also pays a predetermined fee as profit. This method has an advantage when the scope of services cannot be determined accurately in advance. Detailed records must be kept by the engineer in order to allocate costs among different clients.

- *per diem fee:* The engineer is paid a specific sum for each day spent on the job. Usually, certain direct expenses (e.g., travel and reproduction) are billed in addition to the per diem rate.

- *salary plus:* The client pays for the employees on an engineer's payroll (the salary) plus an additional percentage to cover indirect overhead and profit plus certain direct expenses.

- *retainer:* This is a minimum amount paid by the client, usually in total and in advance, for a normal amount of work expected during an agreed-upon period. None of the retainer is returned, regardless of how little work the engineer performs. The engineer can be paid for additional work beyond what is normal, however. Some direct costs, such as travel and reproduction expenses, may be billed directly to the client.

- *percentage of construction cost:* This method, which is widely used in construction design contracts, pays the architect and/or the engineer a percentage of the final total cost of the project. Costs of land, financing, and legal fees are generally not included in the construction cost, and other costs (plan revisions, project management labor, value engineering, etc.) are billed separately.

14. MECHANIC'S LIENS

For various reasons, providers and material, labor, and design services to construction sites may not be promptly paid or even paid at all. Such providers have, of course, the right to file a lawsuit demanding payment, but due to the nature of the construction industry, such relief may be insufficient or untimely. Therefore, such providers have the right to file a *mechanic's lien* (also known as a *construction lien, materialman's lien, supplier's lien,* or *laborer's lien*) against the property. Although there are strict requirements for deadlines, filing, and notices, the procedure for obtaining (and removing) such a lien is simple. The lien establishes the supplier's security interest in the property. Although the details depend on the state, essentially the property owner is prevented from transferring title (i.e., selling) the property until the lien has been removed by the supplier. The act of filing a lawsuit to obtain payment is known as "perfecting the lien." Liens are perfected by forcing a judicial foreclosure sale. The court orders the property sold, and the proceeds are used to pay off any lien-holders.

15. DISCHARGE OF A CONTRACT

A contract is normally discharged when all parties have satisfied their obligations. However, a contract can also be terminated for the following reasons:

- mutual agreement of all parties to the contract

- impossibility of performance (e.g., death of a party to the contract)

- illegality of the contract

- material breach by one or more parties to the contract

- fraud on the part of one or more parties

- failure (i.e., loss or destruction) of consideration (e.g., the burning of a building one party expected to own or occupy upon satisfaction of the obligations)

Some contracts may be dissolved by actions of the court (e.g., bankruptcy), passage of new laws and public acts, or a declaration of war.

Extreme difficulty (including economic hardship) in satisfying the contract does not discharge it, even if it becomes more costly or less profitable than originally anticipated.

16. BREACH OF CONTRACT, NEGLIGENCE, MISREPRESENTATION, AND FRAUD

A *breach of contract* occurs when one of the parties fails to satisfy all of its obligations under a contract. The breach can be *willful* (as in a contractor walking off a construction job) or *unintentional* (as in providing less than adequate quality work or materials). A *material breach* is defined as nonperformance that results in the injured party receiving something substantially less than or different from what the contract intended.

Normally, the only redress that an *injured party* has through the courts in the event of a breach of contract is to force the breaching party to provide *specific performance*—that is, to satisfy all remaining contract provisions and to pay for any damage caused. Normally, *punitive damages* (to punish the breaching party) are unavailable.

Negligence is an action, willful or unwillful, taken without proper care or consideration for safety, resulting in damages to property or injury to persons. "Proper care" is a subjective term, but in general it is the diligence that would be exercised by a reasonably prudent

person.[20] Damages sustained by a negligent act are recoverable in a tort action. (See Sec. 56.17.) If the plaintiff is partially at fault (as in the case of *comparative negligence*), the defendant will be liable only for the portion of the damage caused by the defendant.

Punitive damages are available, however, if the breaching party was fraudulent in obtaining the contract. In addition, the injured party has the right to void (nullify) the contract entirely. A *fraudulent act* is basically a special case of *misrepresentation* (i.e., an intentionally false statement known to be false at the time it is made). Misrepresentation that does not result in a contract is a tort. When a contract is involved, misrepresentation can be a breach of that contract (i.e., *fraud*).

Unfortunately, it is extremely difficult to prove *compensatory fraud* (i.e., fraud for which damages are available). Proving fraud requires showing *beyond a reasonable doubt* (a) a reckless or intentional misstatement of a material fact, (b) an intention to deceive, (c) it resulted in misleading the innocent party to contract, and (d) it was to the innocent party's detriment.

For example, if an engineer claims to have experience in designing steel buildings but actually has none, the court might consider the misrepresentation a fraudulent action. If, however, the engineer has some experience, but an insufficient amount to do an adequate job, the engineer probably will not be considered to have acted fraudulently.

17. TORTS

A *tort* is a civil wrong committed by one person causing damage to another person or person's property, emotional well-being, or reputation.[21] It is a breach of the rights of an individual to be secure in person or property. In order to correct the wrong, a civil lawsuit (*tort action* or *civil complaint*) is brought by the alleged injured party (the *plaintiff*) against the *defendant*. To be a valid *tort action* (i.e., lawsuit), there must have been injury (i.e., damage). Generally, there will be no contract between the two parties, so the tort action cannot claim a breach of contract.[22]

Tort law is concerned with compensation for the injury, not punishment. Therefore, tort awards usually consist

---

[20]Negligence of a design professional (e.g., an engineer or architect) is the absence of a *standard of care* (i.e., customary and normal care and attention) that would have been provided by other engineers. It is highly subjective.

[21]The difference between a *civil tort* (*lawsuit*) and a *criminal lawsuit* is the alleged injured party. A *crime* is a wrong against society. A criminal lawsuit is brought by the state against a defendant.

[22]It is possible for an injury to be both a breach of contract and a tort. Suppose an owner has an agreement with a contractor to construct a building, and the contract requires the contractor to comply with all state and federal safety regulations. If the owner is subsequently injured on a stairway because there was no guardrail, the injury could be recoverable both as a tort and as a breach of contract. If a third party unrelated to the contract was injured, however, that party could recover only through a tort action.

guidelines is to guide the conduct and decision making of engineers. Most codes are primarily educational. Nevertheless, from time to time they have been used by the societies and regulatory agencies as the basis for disciplinary actions.

Fundamental to ethical codes is the requirement that engineers render faithful, honest, professional service. In providing such service, engineers must represent the interests of their employers or clients and, at the same time, protect public health, safety, and welfare.

There is an important distinction between what is legal and what is ethical. Many actions that are legal can be violations of codes of ethical or professional behavior.[3] For example, an engineer's contract with a client may give the engineer the right to assign the engineer's responsibilities, but doing so without informing the client would be unethical.

Ethical guidelines can be categorized on the basis of who is affected by the engineer's actions—the client, vendors and suppliers, other engineers, or the public at large.[4]

3. ETHICAL PRIORITIES

There are frequently conflicting demands on engineers. While it is impossible to use a single decision-making process to solve every ethical dilemma, it is clear that ethical considerations will force engineers to subjugate their own self-interests. Specifically, the ethics of engineers dealing with others need to be considered in the following order, from highest to lowest priority.

- society and the public
- the law
- the engineering profession
- the engineer's client
- the engineer's firm
- other involved engineers
- the engineer personally

4. DEALING WITH CLIENTS AND EMPLOYERS

The most common ethical guidelines affecting engineers' interactions with their employer (the *client*) can be summarized as follows.[5]

- An engineer should not accept assignments for which he/she does not have the skill, knowledge, or time.
- An engineer must recognize his/her own limitations, and use associates and other experts when the design requirements exceed his/her ability.
- The client's interests must be protected. The extent of this protection exceeds normal business relationships and transcends the legal requirements of the engineer-client contract.
- An engineer must not be bound by what the client wants in instances where such desires would be unsuccessful, dishonest, unethical, unhealthy, or unsafe.
- Confidential client information remains the property of the client and must be kept confidential.
- An engineer must avoid conflicts of interest and should inform the client of any business connections or interests that might influence his/her judgment. An engineer should also avoid the *appearance* of a conflict of interest when such an appearance would be detrimental to the profession, the client, or the engineer.
- The engineer's sole source of income for a particular project should be the fee paid by the client. An engineer should not accept compensation in any form from more than one party for the same services.
- If the client rejects the engineer's recommendations, the engineer should fully explain the consequences to the client.
- An engineer must freely and openly admit to the client any errors made.

All courts of law have required an engineer to perform in a manner consistent with normal professional standards. This is not the same as saying an engineer's work must be error-free. If an engineer completes a design, has the design and calculations checked by another competent engineer, and an error is subsequently shown to have been made, the engineer may be held responsible, but the engineer will probably not be considered negligent.

---

[3]Whether the guidelines emphasize ethical behavior or professional conduct is a matter of wording. The intention is the same: to provide guidelines that transcend the requirements of the law.
[4]Some authorities also include ethical guidelines for dealing with the employees of an engineer. However, these guidelines are no different for an engineering employer than they are for a supermarket, automobile assembly line, or airline employer. Ethics is not a unique issue when it comes to employees.

[5]These general guidelines contain references to contractors, plans, specifications, and contract documents. This language is common, though not unique, to the situation of an engineer supplying design services to an owner-developer or architect. However, most of the ethical guidelines are general enough to apply to engineers in industry as well.

57 Engineering Ethics

1. CREEDS, CODES, CANONS, STATUTES, AND RULES

It is generally conceded that an individual acting on his or her own cannot be counted on to always act in a proper and moral manner. Creeds, statutes, rules, and codes all attempt to complete the guidance needed for an engineer to do "...the correct thing."

A *creed* is a statement or oath, often religious in nature, taken or assented to by an individual in ceremonies. For example, the *Engineers' Creed* adopted by the National Society of Professional Engineers (NSPE) in 1954 is[1]

> As a professional engineer, I dedicate my professional knowledge and skill to the advancement and betterment of human welfare.
>
> I pledge...
>
> ...to give the utmost of performance;
>
> ...to participate in none but honest enterprise;
>
> ...to live and work according to the laws of man and the highest standards of professional conduct;
>
> ...to place service before profit, the honor and standing of the profession before personal advantage, and the public welfare above all other considerations.
>
> In humility and with need for Divine Guidance, I make this pledge.

A *code* is a system of nonstatutory, nonmandatory canons of personal conduct. A *canon* is a fundamental belief that usually encompasses several rules. For example, the code of ethics of the American Society of Civil Engineers (ASCE) contains seven canons.

1. Engineers shall hold paramount the safety, health, and welfare of the public and shall strive to comply with the principles of sustainable development in the performance of their professional duties.

2. Engineers shall perform services only in areas of their competence.

3. Engineers shall issue public statements only in an objective and truthful manner.

4. Engineers shall act in professional matters for each employer or client as faithful agents or trustees and shall avoid conflicts of interest.

5. Engineers shall build their professional reputation on the merit of their service and shall not compete unfairly with others.

6. Engineers shall act in such a manner as to uphold and enhance the honor, integrity, and dignity of the engineering profession, and shall act with zero tolerance for bribery, fraud, and corruption.

7. Engineers shall continue their professional development throughout their careers and shall provide opportunities for the professional development of those engineers under their supervision.

A *rule* is a guide (principle, standard, or norm) for conduct and action in a certain situation. A *statutory rule* is enacted by the legislative branch of a state or federal government and carries the weight of law. Some U.S. engineering registration boards have statutory *rules of professional conduct.*

2. PURPOSE OF A CODE OF ETHICS

Many different sets of *codes of ethics* (*canons of ethics, rules of professional conduct,* etc.) have been produced by various engineering societies, registration boards, and other organizations.[2] The purpose of these ethical

---

[1]The *Faith of an Engineer* adopted by the Accreditation Board for Engineering and Technology (ABET), formerly the Engineer's Council for Professional Development (ECPD), is a similar but more detailed creed.

[2]All of the major engineering technical and professional societies in the United States (ASCE, IEEE, ASME, AIChE, NSPE, etc.) and throughout the world have adopted codes of ethics. Most U.S. societies have endorsed the *Code of Ethics of Engineers* developed by the Accreditation Board for Engineering and Technology (ABET). The National Council of Examiners for Engineering and Surveying (NCEES) has developed its *Model Rules of Professional Conduct* as a guide for state registration boards in developing guidelines for the professional engineers in those states.

example of the defendant (i.e., to deter others from doing the same thing).

- *Consequential damages* provide compensation for indirect losses incurred by the injured party but not directly related to the contract.

21. INSURANCE

Most design firms and many independent design professionals carry *errors and omissions insurance* to protect them from claims due to their mistakes. Such policies are costly, and for that reason, some professionals choose to "go bare."[27] Policies protect against inadvertent mistakes only, not against willful, knowing, or conscious efforts to defraud or deceive.

---

[27]Going bare appears foolish at first glance, but there is a perverted logic behind the strategy. One-person consulting firms (and perhaps, firms that are not profitable) are "judgment-proof." Without insurance or other assets, these firms would be unable to pay any large judgments against them. When damage victims (and their lawyers) find this out in advance, they know that judgments will be uncollectable. So, often the lawsuit never makes its way to trial.

of general, compensatory, and special damages and rarely include punitive and exemplary damages. (See Sec. 56.20 for definitions of these damages.)

18. STRICT LIABILITY IN TORT

Strict liability in tort means that the injured party wins if the injury can be proven. It is not necessary to prove negligence, breach of explicit or implicit warranty, or the existence of a contract (*privity of contract*). Strict liability in tort is most commonly encountered in product liability cases. A defect in a product, regardless of how the defect got there, is sufficient to create strict liability in tort.

Case law surrounding defective products has developed and refined the following requirements for winning a strict liability in tort case. The following points must be proved.

- The product was defective in manufacture, design, labeling, and so on.
- The product was defective when used.
- The defect rendered the product unreasonably dangerous.
- The defect caused the injury.
- The specific use of the product that caused the damage was reasonably foreseeable.

19. MANUFACTURING AND DESIGN LIABILITY

Case law makes a distinction between *design professionals* (architects, structural engineers, building designers, etc.) and manufacturers of consumer products. Design professionals are generally consultants whose primary product is a design service sold to sophisticated clients. Consumer product manufacturers produce specific product lines sold through wholesalers and retailers to the unsophisticated public.

The law treats design professionals favorably. Such professionals are expected to meet a *standard of care* and skill that can be measured by comparison with the conduct of other professionals. However, professionals are not expected to be infallible. In the absence of a contract provision to the contrary, design professionals are not held to be guarantors of their work in the strict sense of legal liability. Damages incurred due to design errors are recoverable through tort actions, but proving a breach of contract requires showing negligence (i.e., not meeting the standard of care).

On the other hand, the law is much stricter with consumer product manufacturers, and perfection is (essentially) expected of them. They are held to the standard of strict liability in tort without regard to negligence.

A manufacturer is held liable for all phases of the design and manufacturing of a product being marketed to the public.[23]

Prior to 1916, the court's position toward product defects was exemplified by the expression *caveat emptor* ("let the buyer beware").[24] Subsequent court rulings have clarified that "...a manufacturer is strictly liable in tort when an article [it] places on the market, knowing that it will be used without inspection, proves to have a defect that causes injury to a human being."[25]

Although all defectively designed products can be traced back to a design engineer or team, only the manufacturing company is usually held liable for injury caused by the product. This is more a matter of economics than justice. The company has liability insurance; the product design engineer (who is merely an employee of the company) probably does not. Unless the product design or manufacturing process is intentionally defective, or unless the defect is known in advance and covered up, the product design engineer will rarely be punished by the courts.[26]

20. DAMAGES

An injured party can sue for *damages* as well as for specific performance. Damages are the award made by the court for losses incurred by the injured party.

- *General* or *compensatory damages* are awarded to make up for the injury that was sustained.
- *Special damages* are awarded for the direct financial loss due to the breach of contract.
- *Nominal damages* are awarded when responsibility has been established but the injury is so slight as to be inconsequential.
- *Liquidated damages* are amounts that are specified in the contract document itself for nonperformance.
- *Punitive* or *exemplary damages* are awarded, usually in tort and fraud cases, to punish and make an

---

[23]The reason for this is that the public is not considered to be as sophisticated as a client who contracts with a design professional for building plans.
[24]1916, *MacPherson v. Buick.* MacPherson bought a Buick from a car dealer. The car had a defective wheel, and there was evidence that reasonable inspection would have uncovered the defect. MacPherson was injured when the wheel broke and the car collapsed, and he sued Buick. Buick defended itself under the ancient *prerequisite of privity* (i.e., the requirement of a face-to-face contractual relationship in order for liability to exist), since the dealer, not Buick, had sold the car to MacPherson, and no contract between Buick and MacPherson existed. The judge disagreed, thus establishing the concept of *third party liability* (i.e., manufacturers are responsible to consumers even though consumers do not buy directly from manufacturers).
[25]1963, *Greenman v. Yuba Power Products.* Greenman purchased and was injured by an electric power tool.
[26]The engineer can expect to be discharged from the company. However, for strategic reasons, this discharge probably will not occur until after the company loses the case.

5. DEALING WITH SUPPLIERS

Engineers routinely deal with manufacturers, contractors, and vendors (*suppliers*). In this regard, engineers have great responsibility and influence. Such a relationship requires that engineers deal justly with both clients and suppliers.

An engineer will often have an interest in maintaining good relationships with suppliers since this often leads to future work. Nevertheless, relationships with suppliers must remain highly ethical. Suppliers should not be encouraged to feel that they have any special favors coming to them because of a long-standing relationship with the engineer.

The ethical responsibilities relating to suppliers are listed as follows.

- An engineer must not accept or solicit gifts or other valuable considerations from a supplier during, prior to, or after any job. An engineer should not accept discounts, allowances, commissions, or any other indirect compensation from suppliers, contractors, or other engineers in connection with any work or recommendations.

- An engineer must enforce the plans and specifications (i.e., the *contract documents*), but must also interpret the contract documents fairly.

- Plans and specifications developed by an engineer on behalf of the client must be complete, definite, and specific.

- Suppliers should not be required to spend time or furnish materials that are not called for in the plans and contract documents.

- An engineer should not unduly delay the performance of suppliers.

6. DEALING WITH OTHER ENGINEERS

Engineers should try to protect the engineering profession as a whole, to strengthen it, and to enhance its public stature. The following ethical guidelines apply.

- An engineer should not attempt to maliciously injure the professional reputation, business practice, or employment position of another engineer. However, if there is proof that another engineer has acted unethically or illegally, the engineer should advise the proper authority.

- An engineer should not review someone else's work while the other engineer is still employed unless the other engineer is made aware of the review.

- An engineer should not try to replace another engineer once the other engineer has received employment.

- An engineer should not use the advantages of a salaried position to compete unfairly (i.e., moonlight) with other engineers who have to charge more for the same consulting services.

- Subject to legal and proprietary restraints, an engineer should freely report, publish, and distribute information that would be useful to other engineers.

7. DEALING WITH (AND AFFECTING) THE PUBLIC

In regard to the social consequences of engineering, the relationship between an engineer and the public is essentially straightforward. Responsibilities to the public demand that the engineer place service to humankind above personal gain. Furthermore, proper ethical behavior requires that an engineer avoid association with projects that are contrary to public health and welfare or that are of questionable legal character.

- An engineer must consider the safety, health, and welfare of the public in all work performed.

- An engineer must uphold the honor and dignity of his/her profession by refraining from self-laudatory advertising, by explaining (when required) his/her work to the public, and by expressing opinions only in areas of knowledge.

- When an engineer issues a public statement, he/she must clearly indicate if the statement is being made on anyone's behalf (i.e., if anyone is benefitting from his/her position).

- An engineer must keep his/her skills at a state-of-the-art level.

- An engineer should develop public knowledge and appreciation of the engineering profession and its achievements.

- An engineer must notify the proper authorities when decisions adversely affecting public safety and welfare are made.[6]

8. COMPETITIVE BIDDING

The ethical guidelines for dealing with other engineers presented here and in more detailed codes of ethics no longer include a prohibition on *competitive bidding*. Until 1971, most codes of ethics for engineers considered competitive bidding detrimental to public welfare, since cost-cutting normally results in a lower quality design.

However, in a 1971 case against NSPE that went all the way to the U.S. Supreme Court, the prohibition against competitive bidding was determined to be a violation of the Sherman Antitrust Act (i.e., it was an unreasonable restraint of trade).

---

[6]This practice has come to be known as *whistle-blowing*.

The opinion of the Supreme Court does not *require* competitive bidding—it merely forbids a prohibition against competitive bidding in NSPE's code of ethics. The following points must be considered.

- Engineers and design firms may individually continue to refuse to bid competitively on engineering services.

- Clients are not required to seek competitive bids for design services.

- Federal, state, and local statutes governing the procedures for procuring engineering design services, even those statutes that prohibit competitive bidding, are not affected.

- Any prohibitions against competitive bidding in individual state engineering registration laws remain unaffected.

- Engineers and their societies may actively and aggressively lobby for legislation that would prohibit competitive bidding for design services by public agencies.

9. MODERN ETHICAL ISSUES

With few exceptions, ethical concepts have changed little for engineers since they were first developed by the fledgling engineering societies. Ethical rules covering such issues as service-before-self, conflicts of interest, bribery and kickbacks, moonlighting and unfair competition, respectable dealings with engineering competitors, and respect for the profession and its image may have seen wording changes, but their basic concepts have not changed. Even protection for whistle-blowers can be traced back to the early 1900s, although environmental whistle-blowing came into its own in the 1970s.

Modern engineers, however, have to deal with newer concepts such as environmental protection, sustainable (green) design, offshoring, energy efficiency and dependency, cultural diversity and tolerance, nutrition and safety of foodstuffs and drugs, national sufficiency, and issues affecting national security and personal privacy and security. Most engineers have little experience or training in these concepts, and for some issues, neither regulations nor technology exist to help engineers arrive at definitive decisions.

Environmental protection, from an ethical standpoint, places burdens on engineers that go far beyond adhering to EPA and U.S. Army Corps of Engineers regulations. Issues include waste, pollution, loss of biodiversity, introduction of invasive species, release of genetically modified organisms, genetically modified foodstuffs, and release of toxic substances.

In 2006, the American Society of Civil Engineers amended its first canon to include mention of *sustainable development*, which is subsequently defined as follows: "Sustainable development is the process of applying natural, human, and economic resources to enhance the safety, welfare, and quality of life for all of society while maintaining the availability of the remaining natural resources."

Offshoring broadly means the replacement of a service originally performed within an organization by a service located in a foreign country. Offshoring applies to designing facilities and processes sited in foreign countries, as well as the more common issue of sending jobs to foreign countries. Closer to home, the engineering professional itself has struggled with respecting and accepting offshore engineering services. The very cost-cutting moves that engineers have pursued in service to their employers have hurt engineers and their profession and has given rise to a form of nationalism founded on resentment.

58 Engineering Licensing in the United States

1. ABOUT LICENSING

Engineering licensing (also known as *engineering registration*) in the United States is an examination process by which a state's *board of engineering licensing* (typically referred to as the "engineers' board" or "board of registration") determines and certifies that an engineer has achieved a minimum level of competence.[1] This process is intended to protect the public by preventing unqualified individuals from offering engineering services.

Most engineers in the United States do not need to be licensed.[2] In particular, most engineers who work for companies that design and manufacture products are exempt from the licensing requirement. This is known as the *industrial exemption*, something that is built into the laws of most states.[3]

Nevertheless, there are many good reasons to become a licensed engineer. For example, an engineer cannot offer consulting engineering services in any state unless he or she is licensed in that state. Even within a product-oriented corporation, an engineer may find that employment, advancement, and managerial positions are limited to licensed engineers.

Once an engineer has met the licensing requirements, he or she will be allowed to use the titles *Professional Engineer* (PE), *Registered Engineer* (RE), and *Consulting Engineer* (CE) as permitted by the state.

Although the licensing process is similar in each of the 50 states, each has its own licensing law. Unless an engineer offers consulting engineering services in more than one state, however, there is no need to be licensed in the other states.

2. THE U.S. LICENSING PROCEDURE

The licensing procedure is similar in all states. A candidate will take two written examinations. The full process requires the candidate to complete two applications, one for each of the two examinations. The first examination is the *Fundamentals of Engineering* (FE) *examination*, formerly known (and still commonly referred to) as the *Engineer-In-Training* (EIT) *examination*.[4] This examination covers basic subjects from all of the engineering courses required by an undergraduate program.

The second examination is the *Professional Engineering* (PE) *examination*, also known as the *Principles and Practice* (P&P) *of Engineering examination*. This examination covers only the subjects specific to an engineering discipline (e.g., civil, mechanical, electrical, and others).

The actual details of licensing qualifications, experience requirements, minimum education levels, fees, and examination schedules vary from state to state. Each state's licensing board has more information. Contact information (websites, telephone numbers, email addresses, etc.) for all U.S. state and territorial boards of registration is provided at **ppi2pass.com/stateboards**.

3. NATIONAL COUNCIL OF EXAMINERS FOR ENGINEERING AND SURVEYING

The *National Council of Examiners for Engineering and Surveying* (NCEES) in Clemson, South Carolina, writes, distributes, and scores the national FE and PE examinations.[5] The individual states purchase the examinations from NCEES and administer them in a uniform, controlled environment dictated by NCEES.

---

[1]Licensing of engineers is not unique to the United States. However, the practice of requiring a degreed engineer to take an examination is not common in other countries. Licensing in many countries requires a degree and may also require experience, references, and demonstrated knowledge of ethics and law, but no technical examination.

[2]Less than one-third of the degreed engineers in the United States are licensed.

[3]Only one or two states have abolished the industrial exemption. There has always been a lot of "talk" among engineers about abolishing it, but there has been little success in actually trying to do so. One of the reasons is that manufacturers' lobbies are very strong.

[4]The terms *engineering intern* (EI) and *intern engineer* (IE) have also been used in the past to designate the status of an engineer who has passed the first exam. These uses are rarer but may still be encountered in some states.

[5]National Council of Examiners for Engineering and Surveying, P.O. Box 1686, Clemson, SC 29633, (800) 250-3196, ncees.org.

4. UNIFORM EXAMINATIONS

Although each state has its own licensing law and is, theoretically, free to administer its own exams, none does so for the major disciplines. All states have chosen to use the NCEES exams. The exams from all the states are sent to NCEES to be graded. Each state adopts the cut-off passing scores recommended by NCEES. These practices have led to the term *uniform examination*.

5. RECIPROCITY AMONG STATES

With minor exceptions, having a license from one state will not permit an engineer to practice in another state. A professional engineering license from each state is required to work in that state. Most engineers do not work across state lines or in multiple states, but some do. Luckily, it is not too difficult to get a license from every state worked in once a license from one of them is obtained.

All states use the NCEES examinations. If an engineer takes and passes the FE or PE examination in one state, the certificate or license will be honored by all of the other states. Upon proper application, payment of fees, and proof of engineering license, a license by the new state will be issued. Although there may be other special requirements imposed by a state, it will not be necessary to retake the FE or PE examinations.[6] The issuance of an engineering license based on another state's licensing is known as *reciprocity* or *comity*.

6. APPLYING FOR THE FE EXAMINATION

The FE exam is administered at approved Pearson VUE testing centers. Registration is open year-round and can be completed online through an NCEES account.[7] Registration fees may be paid online. Once a candidate receives notification of eligibility from NCEES, the exam can be scheduled through an NCEES account. A letter from Pearson VUE confirming the exam location and date will be sent via email.

Whether or not applying for and taking the exam is the same as applying for a state's FE certificate depends on the state. In most cases, a candidate might take the exam without a state board ever knowing about it. In fact, as part of the NCEES online exam application process, candidates have to agree to the following statement:

> Passage of the FE exam alone does not ensure certification as an engineer intern or engineer-in-training in any U.S. state or territory. To obtain certification, you must file an application with an engineering licensing board and meet that board's requirements.

[6]For example, California requires all civil engineering applicants to pass special examinations in seismic design and surveying in addition to their regular eight-hour PE exams. Licensed engineers from other states only have to pass these two special exams. They do not need to retake the PE exam.
[7]PPI is not associated with NCEES.

After a state's education requirements have been met and a candidate is ready to obtain an FE (EIT, IE, etc.) certificate, he or she must apply and pay an additional fee to that state. In some cases, passing an additional nontechnical exam related to professional practice in that state may be required. Actual procedures will vary from state to state.

Approximately 7–10 days after the exam, a candidate will receive an email notification that the exam results are ready for viewing through the NCEES account. This notification may include information about additional state-specific steps in the process.

All states make special accommodations for candidates who are physically challenged or who have religious or other special needs. Needs must be communicated to Pearson VUE well in advance of the examination day.

7. APPLYING FOR THE PE EXAMINATION

As with the FE exam, applying for the PE exam includes the steps of registering with NCEES, registering with a state, paying fees to that state, paying fees to NCEES, sending documentation to that state, and notifying that state and NCEES of what has been done. The sequence of these steps varies from state to state.[8]

In some states, the first thing a candidate does is reserve a seat for the upcoming exam through an NCEES account. In other states, the first thing a candidate does is to register with his or her state. In some states, registering with NCEES is the very last thing done, and a candidate may not even be able to notify NCEES until the entire application has been approved by his or her state. Some states use a third-party testing organization with whom a candidate must register. In some states, all fees are paid directly to the state. In other states, all fees are paid to NCEES. In some states, fees are paid to both. In some states, NCEES forms are included with the state application; in other states, state forms are included with the NCEES application. Both the states and NCEES have their own application deadlines, and a state's deadlines will not necessarily coincide with NCEES deadlines.

If a candidate has special physical needs or requirements, these must be coordinated directly with the state. The method and deadline for requesting special accommodations depends on the state. A candidate should keep copies of all forms and correspondence with NCEES and the state.

All states require an application containing a detailed work history. The application form may be available online to print out, or the state may have it mailed. All states require submitted evidence of qualifying engineering experience in the form of written corroborative statements (typically referred to as "professional

[8]Each state board of registration (licensure board) is the most definitive source of information about the application process. A summary of each state's application process is also listed on the NCEES website, ncees.org.

references") provided and signed by supervising engineers, usually other professional engineers. The nature and format of these applications and corroborative statements depend on the state. An application is not complete until all professional references have been received by the state. The state will notify a candidate if one or more references have not been received in a timely manner.

Approximately 2–3 weeks before the exam, NCEES will email the candidate a link to the Exam Authorization Admittance notice, which should be printed out. This document must be presented along with a valid, government-issued photo ID in order to enter the exam site.

The examination results will be released by NCEES to each state 8–10 weeks after the exam. Soon thereafter, states will notify candidates via mail, email, or web access, or alternatively, NCEES may notify candidates via email that results are accessible from an NCEES account. If a state-specific examination was taken, but was not written or graded by NCEES, the state will notify a candidate separately of those results.

8. EXAMINATION DATES

The FE examination is administered eight months out of the year: January, February, April, May, July, August, October, and November. There are multiple testing dates within each of those months. No exams are administered in March, June, September, or December.

The PE examination is administered twice a year (usually in mid-April and late October), on the same weekends in all states. For a current exam schedule, check **ppi2pass.com/CE**. Click on the Exam FAQs link.

9. FE EXAMINATION FORMAT

The FE exam is a computer-based test that contains 110 multiple-choice questions given over two consecutive sessions (sections, parts, etc.). Each session contains approximately 55 multiple-choice questions that are grouped together by knowledge area (subject, topic, etc.). The subjects are not explicitly labeled, and the beginning and ending of the subjects are not noted. No subject spans the two exam sessions. That is, if a subject appears in the first session of the exam, it will not appear in the second.

The exam is six hours long and includes an 8-minute tutorial, a 25-minute break, and a brief survey at the conclusion of the exam. Questions from all undergraduate technical courses appear in the examination, including mathematics, physics, and engineering.

The FE exam is essentially entirely in SI units. There may be a few non-SI problems and some opportunities to choose between identical SI and non-SI problems.

The FE exam is a limited-reference exam. NCEES provides its own reference handbook for use in the examination. Personal books or notes may not be used, though candidates may bring an approved calculator. A reusable, erasable notepad and compatible writing instrument to use for scratchwork are provided during the exam.

10. PE EXAMINATION FORMAT

The NCEES PE examination consists of two four-hour sessions, separated by a one-hour lunch period. It covers only subjects in your major field of study (e.g., civil engineering). There are several variations in PE exam format, but the most common type has 40 questions in each session.

All questions are *objectively scored*. This means that each question is machine graded. There is no penalty for guessing or wrong answers, but there is also no partial credit for correct methods and assumptions.

Unlike the FE exam where SI units are used extensively, virtually all of the questions on the PE exam are in customary U.S. units. There are a few exceptions (e.g., environmental topics), but most problems use pounds, feet, seconds, gallons, and British thermal units.

The PE examination is open book. While some states have a few restrictions, generally a candidate may bring in an unlimited number of personal references. Pencils and erasers are provided, and calculators must be selected from a list of calculators approved by NCEES for exam use.

Systems, Mgmt., Professional

Topic IX: Support Material

Appendices

Support
Material

APPENDIX 1.A
Conversion Factors

multiply	by	to obtain	multiply	by	to obtain
acres	0.40468	hectares	feet	30.48	centimeters
	43,560.0	square feet		0.3048	meters
	1.5625×10^{-3}	square miles		1.645×10^{-4}	miles (nautical)
ampere-hours	3600.0	coulombs		1.894×10^{-4}	miles (statute)
angstrom units	3.937×10^{-9}	inches	feet/min	0.5080	centimeters/sec
	1×10^{-4}	microns	feet/sec	0.592	knots
astronomical units	1.496×10^{8}	kilometers		0.6818	miles/hr
atmospheres	76.0	centimeters of mercury	foot-pounds	1.285×10^{-3}	Btu
atomic mass unit	9.3149×10^{8}	electron-volts		5.051×10^{-7}	horsepower-hours
	1.4924×10^{-10}	joules		3.766×10^{-7}	kilowatt-hours
	1.6605×10^{-27}	kilograms	foot-pound/sec	4.6272	Btu/hr
BeV (also GeV)	1×10^{9}	electron-volts		1.818×10^{-3}	horsepower
Btu	3.93×10^{-4}	horsepower-hours		1.356×10^{-3}	kilowatts
	778.2	foot-pounds	furlongs	660.0	feet
	1055.1	joules		0.125	miles (statute)
	2.931×10^{-4}	kilowatt hours	gallons	0.1337	cubic feet
	1.0×10^{-5}	therms		3.785	liters
Btu/hr	0.2161	foot-pounds/sec	gallons H_2O	8.3453	pounds H_2O
	3.929×10^{-4}	horsepower	gallons/min	8.0208	cubic feet/hr
	0.2931	watts		0.002228	cubic feet/sec
bushels	2150.4	cubic inches	GeV (also BeV)	1×10^{9}	electron-volts
calories, gram (mean)	3.9683×10^{-3}	Btu (mean)	grams	1×10^{-3}	kilograms
centares	1.0	square meters		3.527×10^{-2}	ounces (avoirdupois)
centimeters	1×10^{-5}	kilometers		3.215×10^{-2}	ounces (troy)
	1×10^{-2}	meters		2.205×10^{-3}	pounds
	10.0	millimeters	hectares	2.471	acres
	3.281×10^{-2}	feet		1.076×10^{5}	square feet
	0.3937	inches	horsepower	2545.0	Btu/hr
chains	792.0	inches		42.44	Btu/min
coulombs	1.036×10^{-5}	faradays		550	foot-pounds/sec
cubic centimeters	0.06102	cubic inches		0.7457	kilowatts
	2.113×10^{-3}	pints (U.S. liquid)		745.7	watts
cubic feet	0.02832	cubic meters	horsepower-hours	2545.0	Btu
	7.4805	gallons		1.976×10^{-6}	foot-pounds
cubic feet/min	62.43	pounds H_2O/min		0.7457	kilowatt-hours
cubic feet/sec	448.831	gallons/min	hours	4.167×10^{-2}	days
	0.64632	millions of gallons per day		5.952×10^{-3}	weeks
cubits	18.0	inches	inches	2.540	centimeters
days	86,400.0	seconds		1.578×10^{-5}	miles
degrees (angle)	1.745×10^{-2}	radians	inches, H_2O	5.199	pounds force/ft^2
degrees/sec	0.1667	revolutions/min		0.0361	psi
dynes	1×10^{-5}	newtons		0.0735	inches, mercury
electron-volts	1.0735×10^{-9}	atomic mass units	inches, mercury	70.7	pounds force/ft^2
	1×10^{-9}	BeV (also GeV)		0.491	pounds force/in^2
	1.60218×10^{-19}	joules		13.60	inches, H_2O
	1.78266×10^{-36}	kilograms	joules	6.705×10^{9}	atomic mass units
	1×10^{-6}	MeV		9.478×10^{-4}	Btu
faradays/sec	96,485	amperes		1×10^{7}	ergs
fathoms	6.0	feet		6.2415×10^{18}	electron-volts
				1.1127×10^{-17}	kilograms

(continued)

APPENDIX 1.A *(continued)*
Conversion Factors

multiply	by	to obtain	multiply	by	to obtain
kilograms	6.0221×10^{26}	atomic mass units	pascal-sec	1000	centipoise
	5.6096×10^{35}	electron-volts		10	poise
	8.9875×10^{16}	joules		0.02089	pound force-sec/ft^2
	2.205	pounds		0.6720	pound mass/ft-sec
kilometers	3281.0	feet		0.02089	slug/ft-sec
	1000.0	meters	pints (liquid)	473.2	cubic centimeters
	0.6214	miles		28.87	cubic inches
kilometers/hr	0.5396	knots		0.125	gallons
kilowatts	3412.9	Btu/hr		0.5	quarts (liquid)
	737.6	foot-pounds/sec	poise	0.002089	pound-sec/ft^2
	1.341	horsepower	pounds	0.4536	kilograms
kilowatt-hours	3413.0	Btu		16.0	ounces
knots	6076.0	feet/hr		14.5833	ounces (troy)
	1.0	nautical miles/hr		1.21528	pounds (troy)
	1.151	statute miles/hr	pounds/ft^2	0.006944	pounds/in^2
light years	5.9×10^{12}	miles	pounds/in^2	2.308	feet, H$_2$O
links (surveyor)	7.92	inches		27.7	inches, H$_2$O
liters	1000.0	cubic centimeters		2.037	inches, mercury
	61.02	cubic inches		144	pounds/ft^2
	0.2642	gallons (U.S. liquid)	quarts (dry)	67.20	cubic inches
	1000.0	milliliters	quarts (liquid)	57.75	cubic inches
	2.113	pints		0.25	gallons
MeV	1×10^6	electron-volts		0.9463	liters
meters	100.0	centimeters	radians	57.30	degrees
	3.281	feet		3438.0	minutes
	1×10^{-3}	kilometers	revolutions	360.0	degrees
	5.396×10^{-4}	miles (nautical)	revolutions/min	6.0	degrees/sec
	6.214×10^{-4}	miles (statute)	rods	16.5	feet
	1000.0	millimeters		5.029	meters
microns	1×10^{-6}	meters	rods (surveyor)	5.5	yards
miles (nautical)	6076	feet	seconds	1.667×10^{-2}	minutes
	1.853	kilometers	square meters/sec	1×10^6	centistokes
	1.1516	miles (statute)		10.76	square feet/sec
miles (statute)	5280.0	feet		1×10^4	stokes
	1.609	kilometers	slugs	32.174	pounds mass
	0.8684	miles (nautical)	stokes	0.0010764	square feet/sec
miles/hr	88.0	feet/min	tons (long)	1016.0	kilograms
milligrams/liter	1.0	parts/million		2240.0	pounds
milliliters	1×10^{-3}	liters		1.120	tons (short)
millimeters	3.937×10^{-2}	inches	tons (short)	907.1848	kilograms
newtons	1×10^5	dynes		2000.0	pounds
ounces	28.349527	grams		0.89287	tons (long)
	6.25×10^{-2}	pounds	watts	3.4129	Btu/hr
ounces (troy)	1.09714	ounces (avoirdupois)		1.341×10^{-3}	horsepower
parsecs	3.086×10^{13}	kilometers	yards	0.9144	meters
	1.9×10^{13}	miles		4.934×10^{-4}	miles (nautical)
				5.682×10^{-4}	miles (statute)

APPENDIX 1.B
Common SI Unit Conversion Factors

1 acre = 4046.86 m (handwritten)

multiply	by	to obtain
AREA		
circular mil	506.7	square micrometer
square foot	0.0929	square meter
square kilometer	0.3861	square mile
square meter	10.764	square foot
	1.196	square yard
square micrometer	0.001974	circular mil
square mile	2.590	square kilometer
square yard	0.8361	square meter
ENERGY		
Btu (international)	1.0551	kilojoule
erg	0.1	microjoule
foot-pound	1.3558	joule
horsepower-hour	2.6485	megajoule
joule	0.7376	foot-pound
	0.10197	meter·kilogram force
kilogram·calorie (international)	4.1868	kilojoule
kilojoule	0.9478	Btu
	0.2388	kilogram·calorie
kilowatt·hour	3.6	megajoule
megajoule	0.3725	horsepower-hour
	0.2778	kilowatt-hour
	0.009478	therm
meter·kilogram force	9.8067	joule
microjoule	10.0	erg
therm	105.506	megajoule
FORCE		
dyne	10.0	micronewton
kilogram force	9.8067	newton
kip	4448.2	newton
micronewton	0.1	dyne
newton	0.10197	kilogram force
	0.0002248	kip
	3.597	ounce force
	0.2248	pound force
ounce force	0.2780	newton
pound force	4.4482	newton
HEAT		
Btu/ft^2-hr	3.1546	$watt/m^2$
Btu/hr-ft^2-°F	5.6783	$watt/m^2$·°C
Btu/ft^3	0.0373	$megajoule/m^3$
Btu/ft^3-°F	0.06707	$megajoule/m^3$·°C
Btu/hr	0.2931	watt
Btu/lbm	2326	joule/kg
Btu/lbm-°F	4186.8	joule/kg·°C
Btu-inch/hr-ft^2-°F	0.1442	watt/meter·°C
joule/kg	0.000430	Btu/lbm
joule/kg·°C	0.0002388	Btu/lbm-°F
$megajoule/m^3$	26.839	Btu/ft^3
$megajoule/m^3$·°C	14.911	Btu/ft^3-°F
watt	3.4121	Btu/hr
watt/m·°C	6.933	Btu-in/hr-ft^2-°F

(continued)

APPENDIX 1.B *(continued)*
Common SI Unit Conversion Factors

multiply	by	to obtain
HEAT *(continued)*		
watt/m^2	0.3170	Btu/ft^2-hr
watt/m^2·°C	0.1761	Btu/hr-ft^2-°F
LENGTH		
angstrom	0.1	nanometer
foot	0.3048	meter
inch	25.4	millimeter
kilometer	0.6214	mile
	0.540	mile (nautical)
meter	3.2808	foot
	1.0936	yard
micrometer	1.0	micron
micron	1.0	micrometer
mil	0.0254	millimeter
mile	1.6093	kilometer
mile (nautical)	1.852	kilometer
millimeter	0.0394	inch
	39.370	mil
nanometer	10.0	angstrom
yard	0.9144	meter
MASS		
grain	64.799	milligram
gram	0.0353	ounce (avoirdupois)
	0.03215	ounce (troy)
kilogram	2.2046	pound mass
	0.068522	slug
	0.0009842	ton (long—2240 lbm)
	0.001102	ton (short—2000 lbm)
milligram	0.0154	grain
ounce (avoirdupois)	28.350	gram
ounce (troy)	31.1035	gram
pound mass	0.4536	kilogram
slug	14.5939	kilogram
ton (long—2240 lbm)	1016.047	kilogram
ton (short—2000 lbm)	907.185	kilogram
PRESSURE		
bar	100.0	kilopascal
inch, H$_2$O (20°C)	0.2486	kilopascal
inch, Hg (20°C)	3.3741	kilopascal
kilogram force/cm^2	98.067	kilopascal
kilopascal	0.01	bar
	4.0219	inch, H$_2$O (20°C)
	0.2964	inch, Hg (20°C)
	0.0102	kilogram force/cm^2
	7.528	millimeter, Hg (20°C)
	0.1450	pound force/in^2
	0.009869	standard atmosphere (760 torr)
	7.5006	torr
millimeter, Hg (20°C)	0.13332	kilopascal
pound force/in^2	6.8948	kilopascal
standard atmosphere (760 torr)	101.325	kilopascal
torr	0.13332	kilopascal

(continued)

APPENDIX 1.B *(continued)*
Common SI Unit Conversion Factors

multiply	by	to obtain
POWER		
Btu (international)/hr	0.2931	watt
foot-pound/sec	1.3558	watt
horsepower	0.7457	kilowatt
kilowatt	1.341	horsepower
	0.2843	ton of refrigeration
meter·kilogram force/second	9.8067	watt
ton of refrigeration	3.517	kilowatt
watt	3.4122	Btu (international)/hr
	0.7376	foot-pound/sec
	0.10197	meter·kilogram force/second
TEMPERATURE		
Celsius	$\frac{9}{5}°C + 32°$	Fahrenheit
Fahrenheit	$\frac{5}{9}(°F - 32°)$	Celsius
Kelvin	$\frac{9}{5}$	Rankine
Rankine	$\frac{5}{9}$	Kelvin
TORQUE		
gram force·centimeter	0.098067	millinewton·meter
kilogram force·meter	9.8067	newton·meter
millinewton	10.197	gram force·centimeter
newton·meter	0.10197	kilogram force·meter
	0.7376	foot-pound
	8.8495	inch-pound
foot-pound	1.3558	newton·meter
inch-pound	0.1130	newton·meter
VELOCITY		
feet/sec	0.3048	meters/second
kilometers/hr	0.6214	miles/hr
meters/second	3.2808	feet/sec
	2.2369	miles/hr
miles/hr	1.60934	kilometers/hr
	0.44704	meters/second
VISCOSITY		
centipoise	0.001	pascal·seconds
centistoke	1×10^{-6}	meter^2/second
pascal·second	1000	centipoise
meter^2/second	1×10^{6}	centistoke
VOLUME (capacity)		
cubic centimeter	0.06102	cubic inch
cubic foot	28.3168	liter
cubic inch	16.3871	cubic centimeter
cubic meter	1.308	cubic yard
cubic yard	0.7646	cubic meter
gallon (U.S. fluid)	3.785	liter
liter	0.2642	gallon (U.S. fluid)
	2.113	pint (U.S. fluid)
	1.0567	quart (U.S. fluid)
	0.03531	cubic foot
milliliter	0.0338	ounce (U.S. fluid)
ounce (U.S. fluid)	29.574	milliliter

(handwritten: 32°F = 491.67°R)

(continued)

APPENDIX 1.B *(continued)*
Common SI Unit Conversion Factors

multiply	by	to obtain
VOLUME (capacity) *(continued)*		
pint (U.S. fluid)	0.4732	liter
quart (U.S. fluid)	0.9464	liter
VOLUME FLOW (gas-air)		
cubic meter/sec	2119	standard cubic foot/min
liter/sec	2.119	standard cubic foot/min
microliter/sec	0.000127	standard cubic foot/hr
milliliter/sec	0.002119	standard cubic foot/min
	0.127133	standard cubic foot/hr
standard cubic foot/min	0.0004719	cubic meter/sec
	0.4719	liter/sec
	471.947	milliliter/sec
standard cubic foot/hr	7866	microliter/sec
	7.8658	milliliter/sec
VOLUME FLOW (liquid)		
gallon/hr (U.S. fluid)	0.001052	liter/sec
gallon/min (U.S. fluid)	0.06309	liter/sec
liter/sec	951.02	gallon/hr (U.S. fluid)
	15.850	gallon/min (U.S. fluid)

APPENDIX 2.A
Centroids and Area Moments of Inertia for Basic Shapes

shape		centroidal location x_c	centroidal location y_c	area, A	area moment of inertia (rectangular and polar), I, J	radius of gyration, r
rectangle		$\dfrac{b}{2}$	$\dfrac{h}{2}$	bh	$I_x = \dfrac{bh^3}{3}$ $I_{cx} = \dfrac{bh^3}{12}$ $J_c = \left(\dfrac{1}{12}\right)bh(b^2 + h^2)^*$ (see note below)	$r_x = \dfrac{h}{\sqrt{3}}$ $r_{cx} = \dfrac{h}{2\sqrt{3}}$
triangular area		$\dfrac{2b}{3}$	$\dfrac{h}{3}$	$\dfrac{bh}{2}$	$I_x = \dfrac{bh^3}{12}$ $I_{cx} = \dfrac{bh^3}{36}$	$r_x = \dfrac{h}{\sqrt{6}}$ $r_{cx} = \dfrac{h}{3\sqrt{2}}$
trapezoid		$\dfrac{2tz + t^2 + zb + tb + b^2}{3b + 3t}$	$h\left(\dfrac{b + 2t}{3b + 3t}\right)$	$\dfrac{(b + t)h}{2}$	$I_x = \dfrac{(b + 3t)h^3}{12}$ $I_{cx} = \dfrac{(b^2 + 4bt + t^2)h^3}{36(b + t)}$	$r_x = \left(\dfrac{h}{\sqrt{6}}\right)\sqrt{\dfrac{b + 3t}{b + t}}$ $r_{cx} = \dfrac{h\sqrt{2(b^2 + 4bt + t^2)}}{6(b + t)}$
circle		0	0	πr^2	$I_x = I_y = \dfrac{\pi r^4}{4}$ $J_c = \dfrac{\pi r^4}{2}$	$r_x = \dfrac{r}{2}$
quarter-circular area		$\dfrac{4r}{3\pi}$	$\dfrac{4r}{3\pi}$	$\dfrac{\pi r^2}{4}$	$I_x = I_y = \dfrac{\pi r^4}{16}$ $J_o = \dfrac{\pi r^4}{8}$	
semicircular area		0	$\dfrac{4r}{3\pi}$	$\dfrac{\pi r^2}{2}$	$I_x = I_y = \dfrac{\pi r^4}{8}$ $I_{cx} = 0.1098r^4$ $J_o = \dfrac{\pi r^4}{4}$ $J_c = 0.5025r^4$	$r_x = \dfrac{r}{2}$ $r_{cx} = 0.264r$
quarter-elliptical area		$\dfrac{4a}{3\pi}$	$\dfrac{4b}{3\pi}$	$\dfrac{\pi ab}{4}$	$I_x = \dfrac{\pi ab^3}{8}$ $I_y = \dfrac{\pi a^3 b}{8}$ $J_o = \dfrac{\pi ab(a^2 + b^2)}{8}$	
semielliptical area		0	$\dfrac{4b}{3\pi}$	$\dfrac{\pi ab}{2}$		
semiparabolic area		$\dfrac{3a}{8}$	$\dfrac{3h}{5}$	$\dfrac{2ah}{3}$		
parabolic area		0	$\dfrac{3h}{5}$	$\dfrac{4ah}{3}$	$I_x = \dfrac{4ah^3}{7}$ $I_y = \dfrac{4ha^3}{15}$ $I_{cx} = \dfrac{16ah^3}{175}$	$r_x = h\sqrt{\dfrac{3}{7}}$ $r_y = \dfrac{a}{\sqrt{5}}$
parabolic spandrel		$\dfrac{3a}{4}$	$\dfrac{3h}{10}$	$\dfrac{ah}{3}$	$I_x = \dfrac{ah^3}{21}$ $I_y = \dfrac{3ha^3}{15}$	
general spandrel		$\left(\dfrac{n + 1}{n + 2}\right)a$	$\left(\dfrac{n + 1}{4n + 2}\right)h$	$\dfrac{ah}{n + 1}$	(note to accompany rectangular area above) *Theoretical definition based on $J = I_x + I_y$. However, in torsion, not all parts of the shape are effective. Effective values will be lower.	
circular sector [α in radians]		$\dfrac{2r\sin\alpha}{3\alpha}$	0	αr^2	$J = C\left(\dfrac{b^2 + h^2}{b^3 h^3}\right)$	

For the circular sector entry, the following subtable appears:

b/h	C
1	3.56
2	3.50
4	3.34
8	3.21

APPENDIX 7.A
Mensuration of Two-Dimensional Areas

Nomenclature
A total surface area
b base
c chord length
d distance
h height
L length
p perimeter
r radius
s side (edge) length, arc length
θ vertex angle, in radians
ϕ central angle, in radians

Circular Sector

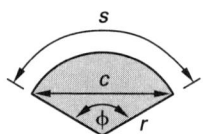

$$A = \tfrac{1}{2}\phi r^2 = \tfrac{1}{2}sr$$

$$\phi = \frac{s}{r}$$

$$s = r\phi$$

$$c = 2r\sin\left(\frac{\phi}{2}\right)$$

Triangle

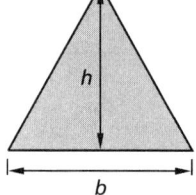 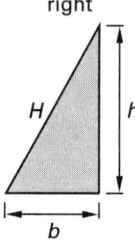 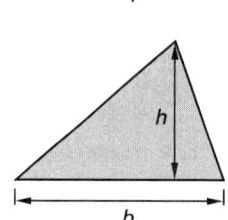

equilateral   right   oblique

$$A = \tfrac{1}{2}bh = \frac{\sqrt{3}}{4}b^2 \qquad A = \tfrac{1}{2}bh \qquad A = \tfrac{1}{2}bh$$

$$h = \frac{\sqrt{3}}{2}b \qquad\qquad H^2 = b^2 + h^2$$

Parabola

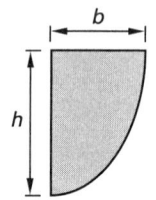

$$A = \tfrac{2}{3}bh$$

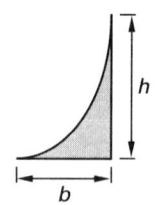

$$A = \tfrac{1}{3}bh$$

Circle

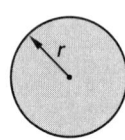

$$p = 2\pi r$$

$$A = \pi r^2 = \frac{p^2}{4\pi}$$

Circular Segment

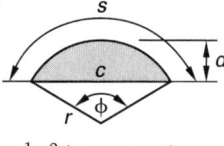

$$A = \tfrac{1}{2}r^2(\phi - \sin\phi)$$

$$\phi = \frac{s}{r} = 2\left(\arccos\frac{r-d}{r}\right)$$

$$c = 2r\sin\left(\frac{\phi}{2}\right)$$

Ellipse

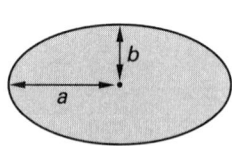

$$A = \pi ab$$

$$p \approx 2\pi\sqrt{\tfrac{1}{2}(a^2 + b^2)} \qquad \begin{bmatrix} \text{Euler's} \\ \text{upper bound} \end{bmatrix}$$

(continued)

APPENDIX 7.A *(continued)*
Mensuration of Two-Dimensional Areas

Trapezoid

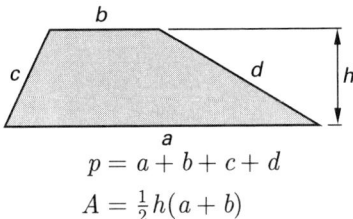

$$p = a + b + c + d$$
$$A = \tfrac{1}{2}h(a + b)$$

If $c = d$, the trapezoid is isosceles.

Parallelogram

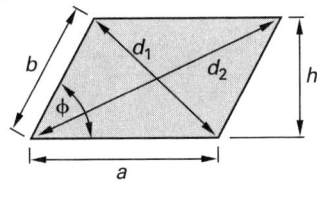

$$p = 2(a + b)$$
$$d_1 = \sqrt{a^2 + b^2 - 2ab\cos\phi}$$
$$d_2 = \sqrt{a^2 + b^2 + 2ab\cos\phi}$$
$$d_1^2 + d_2^2 = 2(a^2 + b^2)$$
$$A = ah = ab\sin\phi$$

If $a = b$, the parallelogram is a rhombus.

Regular Polygon (*n* equal sides)

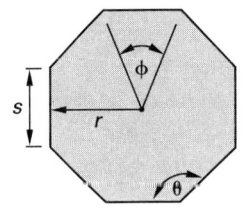

$$\phi = \frac{2\pi}{n}$$
$$\theta = \frac{\pi(n - 2)}{n} = \pi - \phi$$
$$p = ns$$
$$s = 2r\tan\frac{\theta}{2}$$
$$A = \tfrac{1}{2}nsr$$

regular polygons							
sides	name	area (A) when diameter of inscribed circle = 1	area (A) when side = 1	radius (r) of circumscribed circle when side = 1	length (L) of side when radius (r) of circumscribed circle = 1	length (L) of side when perpendicular to circle = 1	perpendicular (p) to center when side = 1
3	triangle	1.299	0.433	0.577	1.732	3.464	0.289
4	square	1.000	1.000	0.707	1.414	2.000	0.500
5	pentagon	0.908	1.720	0.851	1.176	1.453	0.688
6	hexagon	0.866	2.598	1.000	1.000	1.155	0.866
7	heptagon	0.843	3.634	1.152	0.868	0.963	1.038
8	octagon	0.828	4.828	1.307	0.765	0.828	1.207
9	nonagon	0.819	6.182	1.462	0.684	0.728	1.374
10	decagon	0.812	7.694	1.618	0.618	0.650	1.539
11	undecagon	0.807	9.366	1.775	0.563	0.587	1.703
12	dodecagon	0.804	11.196	1.932	0.518	0.536	1.866

APPENDIX 7.B
Mensuration of Three-Dimensional Volumes

Nomenclature

A	surface area
b	base
h	height
r	radius
R	radius
s	side (edge) length
V	internal volume

Sphere

$$V = \tfrac{4}{3}\pi r^3 = \tfrac{4}{3}\pi \left(\frac{d}{2}\right)^3 = \tfrac{1}{6}\pi d^3$$

$$A = 4\pi r^2$$

Right Circular Cone (excluding base area)

$$V = \tfrac{1}{3}\pi r^2 h = \tfrac{1}{3}\pi \left(\frac{d}{2}\right)^2 h = \tfrac{1}{12}\pi d^2 h$$

$$A = \pi r \sqrt{r^2 + h^2}$$

Right Circular Cylinder (excluding end areas)

$$V = \pi r^2 h$$

$$A = 2\pi r h$$

Spherical Segment (spherical cap)

Surface area of a spherical segment of radius r cut out by an angle θ_0 rotated from the center about a radius, r, is

$$A = 2\pi r^2 (1 - \cos \theta_0)$$

$$\omega = \frac{A}{r^2} = 2\pi (1 - \cos \theta_0)$$

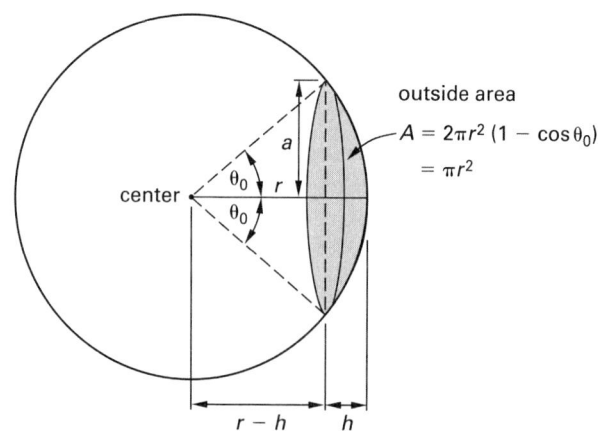

outside area

$A = 2\pi r^2 \,(1 - \cos\theta_0)$

$= \pi r^2$

center

θ_0 r
θ_0

$r - h$ h

$$V_{\text{cap}} = \tfrac{1}{6}\pi h (3a^2 + h^2)$$

$$= \tfrac{1}{3}\pi h^2 (3r - h)$$

$$a = \sqrt{h(2r - h)}$$

Paraboloid of Revolution

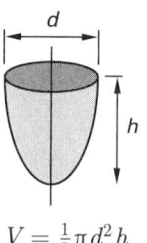

$$V = \tfrac{1}{8}\pi d^2 h$$

Torus

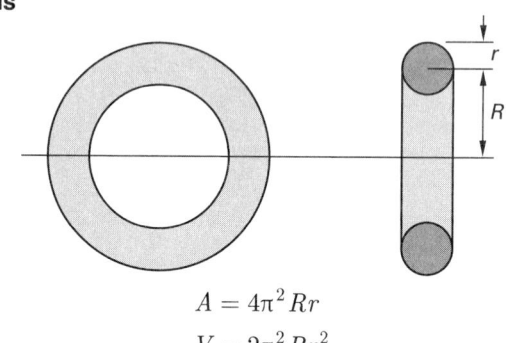

$$A = 4\pi^2 R r$$

$$V = 2\pi^2 R r^2$$

Regular Polyhedra (identical faces)

name	number of faces	form of faces	total surface area	volume
tetrahedron	4	equilateral triangle	$1.7321\,s^2$	$0.1179\,s^3$
cube	6	square	$6.0000\,s^2$	$1.0000\,s^3$
octahedron	8	equilateral triangle	$3.4641\,s^2$	$0.4714\,s^3$
dodecahedron	12	regular pentagon	$20.6457\,s^2$	$7.6631\,s^3$
icosahedron	20	equilateral triangle	$8.6603\,s^2$	$2.1817\,s^3$

The radius of a sphere inscribed within a regular polyhedron is

$$r = \frac{3\,V_{\text{polyhedron}}}{A_{\text{polyhedron}}}$$

APPENDIX 9.A
Abbreviated Table of Indefinite Integrals
(In each case, add a constant of integration. All angles are measured in radians.)

General Formulas

1. $\int dx = x$

2. $\int c\,dx = c \int dx$

3. $\int (dx + dy) = \int dx + \int dy$

4. $\int u\,dv = uv - \int v\,du$ $\begin{bmatrix} \text{integration by parts; } u \text{ and } v \text{ are} \\ \text{functions of the same variable} \end{bmatrix}$

Algebraic Forms

5. $\int x^n\,dx = \dfrac{x^{n+1}}{n+1}$ $\quad [n \neq -1]$

6. $\int x^{-1}\,dx = \int \dfrac{dx}{x} = \ln|x|$

7. $\int (ax+b)^n\,dx = \dfrac{(ax+b)^{n+1}}{a(n+1)}$ $\quad [n \neq -1]$

8. $\int \dfrac{dx}{ax+b} = \dfrac{1}{a}\ln(ax+b)$

9. $\int \dfrac{x\,dx}{ax+b} = \dfrac{1}{a^2}\left(ax+b - b\ln(ax+b)\right)$

10. $\int \dfrac{x\,dx}{(ax+b)^2} = \dfrac{1}{a^2}\left(\dfrac{b}{ax+b} + \ln(ax+b)\right)$

11. $\int \dfrac{dx}{x(ax+b)} = \dfrac{1}{b}\ln\left(\dfrac{x}{ax+b}\right)$

12. $\int \dfrac{dx}{x(ax+b)^2} = \dfrac{1}{b(ax+b)} + \dfrac{1}{b^2}\ln\left(\dfrac{x}{ax+b}\right)$

13. $\int \dfrac{dx}{x^2+a^2} = \dfrac{1}{a}\tan^{-1}\left(\dfrac{x}{a}\right)$

14. $\int \dfrac{dx}{a^2-x^2} = \dfrac{1}{a}\tanh^{-1}\left(\dfrac{x}{a}\right)$

15. $\int \dfrac{x\,dx}{ax^2+b} = \dfrac{1}{2a}\ln(ax^2+b)$

16. $\int \dfrac{dx}{x(ax^n+b)} = \dfrac{1}{bn}\ln\left(\dfrac{x^n}{ax^n+b}\right)$

17. $\int \dfrac{dx}{ax^2+bx+c} = \dfrac{1}{\sqrt{b^2-4ac}}\ln\left(\dfrac{2ax+b-\sqrt{b^2-4ac}}{2ax+b+\sqrt{b^2-4ac}}\right)$ $\quad [b^2 > 4ac]$

18. $\int \dfrac{dx}{ax^2+bx+c} = \dfrac{2}{\sqrt{4ac-b^2}}\tan^{-1}\left(\dfrac{2ax+b}{\sqrt{4ac-b^2}}\right)$ $\quad [b^2 < 4ac]$

19. $\int \sqrt{a^2-x^2}\,dx = \dfrac{x}{2}\sqrt{a^2-x^2} + \dfrac{a^2}{2}\sin^{-1}\left(\dfrac{x}{a}\right)$

20. $\int x\sqrt{a^2-x^2}\,dx = -\dfrac{1}{3}\left(a^2-x^2\right)^{3/2}$

21. $\int \dfrac{dx}{\sqrt{a^2-x^2}} = \sin^{-1}\left(\dfrac{x}{a}\right)$

22. $\int \dfrac{x\,dx}{\sqrt{a^2-x^2}} = -\sqrt{a^2-x^2}$

APPENDIX 10.A
Laplace Transforms

$f(t)$	$\mathcal{L}\big(f(t)\big)$	$f(t)$	$\mathcal{L}\big(f(t)\big)$
$\delta(t)$ [unit impulse at $t=0$]	1	$1 - \cos at$	$\dfrac{a^2}{s(s^2 + a^2)}$
$\delta(t - c)$ [unit impulse at $t=c$]	e^{-cs}	$\cosh at$	$\dfrac{s}{s^2 - a^2}$
1 or u_0 [unit step at $t=0$]	$\dfrac{1}{s}$	$t \cos at$	$\dfrac{s^2 - a^2}{(s^2 + a^2)^2}$
u_c [unit step at $t=c$]	$\dfrac{e^{-cs}}{s}$	t^n [n is a positive integer]	$\dfrac{n!}{s^{n+1}}$
t [unit ramp at $t=0$]	$\dfrac{1}{s^2}$	e^{at}	$\dfrac{1}{s - a}$
rectangular pulse, magnitude M, duration a	$\left(\dfrac{M}{s}\right)(1 - e^{-as})$	$e^{at}\sin bt$	$\dfrac{b}{(s - a)^2 + b^2}$
triangular pulse, magnitude M, duration $2a$	$\left(\dfrac{M}{as^2}\right)(1 - e^{-as})^2$	$e^{at}\cos bt$	$\dfrac{s - a}{(s - a)^2 + b^2}$
sawtooth pulse, magnitude M, duration a	$\left(\dfrac{M}{as^2}\right)\big(1 - (as + 1)e^{-as}\big)$	$e^{at}t^n$ [n is a positive integer]	$\dfrac{n!}{(s - a)^{n+1}}$
sinusoidal pulse, magnitude M, duration π/a	$\left(\dfrac{Ma}{s^2 + a^2}\right)(1 + e^{-\pi s/a})$	$1 - e^{-at}$	$\dfrac{a}{s(s + a)}$
$\dfrac{t^{n-1}}{(n - 1)!}$	$\dfrac{1}{s^n}$	$e^{-at} + at - 1$	$\dfrac{a^2}{s^2(s + a)}$
$\sin at$	$\dfrac{a}{s^2 + a^2}$	$\dfrac{e^{-at} - e^{-bt}}{b - a}$	$\dfrac{1}{(s + a)(s + b)}$
$at - \sin at$	$\dfrac{a^3}{s^2(s^2 + a^2)}$	$\dfrac{(c - a)e^{-at} - (c - b)e^{-bt}}{b - a}$	$\dfrac{s + c}{(s + a)(s + b)}$
$\sinh at$	$\dfrac{a}{s^2 - a^2}$	$\dfrac{1}{ab} + \dfrac{be^{-at} - ae^{-bt}}{ab(a - b)}$	$\dfrac{1}{s(s + a)(s + b)}$
$t \sin at$	$\dfrac{2as}{(s^2 + a^2)^2}$	$t \sinh at$	$\dfrac{2as}{(s^2 - a^2)^2}$
$\cos at$	$\dfrac{s}{s^2 + a^2}$	$t \cosh at$	$\dfrac{s^2 + a^2}{(s^2 - a^2)^2}$

APPENDIX 11.A
Areas Under the Standard Normal Curve
(0 to z)

Areas Under the Standard Normal Curve
(0 to z)

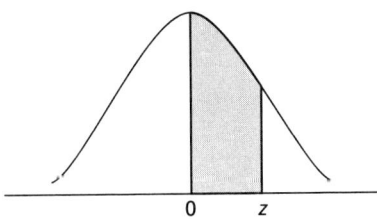

z	0	1	2	3	4	5	6	7	8	9
0.0	0.0000	0.0040	0.0080	0.0120	0.0160	0.0199	0.0239	0.0279	0.0319	0.0359
0.1	0.0398	0.0438	0.0478	0.0517	0.0557	0.0596	0.0636	0.0675	0.0714	0.0754
0.2	0.0793	0.0832	0.0871	0.0910	0.0948	0.0987	0.1026	0.1064	0.1103	0.1141
0.3	0.1179	0.1217	0.1255	0.1293	0.1331	0.1368	0.1406	0.1443	0.1480	0.1517
0.4	0.1554	0.1591	0.1628	0.1664	0.1700	0.1736	0.1772	0.1808	0.1844	0.1879
0.5	0.1915	0.1950	0.1985	0.2019	0.2054	0.2088	0.2123	0.2157	0.2190	0.2224
0.6	0.2258	0.2291	0.2324	0.2357	0.2389	0.2422	0.2454	0.2486	0.2518	0.2549
0.7	0.2580	0.2612	0.2642	0.2673	0.2704	0.2734	0.2764	0.2794	0.2823	0.2852
0.8	0.2881	0.2910	0.2939	0.2967	0.2996	0.3023	0.3051	0.3078	0.3106	0.3133
0.9	0.3159	0.3186	0.3212	0.3238	0.3264	0.3289	0.3315	0.3340	0.3365	0.3389
1.0	0.3413	0.3438	0.3461	0.3485	0.3508	0.3531	0.3554	0.3577	0.3599	0.3621
1.1	0.3643	0.3665	0.3686	0.3708	0.3729	0.3749	0.3770	0.3790	0.3810	0.3830
1.2	0.3849	0.3869	0.3888	0.3907	0.3925	0.3944	0.3962	0.3980	0.3997	0.4015
1.3	0.4032	0.4049	0.4066	0.4082	0.4099	0.4115	0.4131	0.4147	0.4162	0.4177
1.4	0.4192	0.4207	0.4222	0.4236	0.4251	0.4265	0.4279	0.4292	0.4306	0.4319
1.5	0.4332	0.4345	0.4357	0.4370	0.4382	0.4394	0.4406	0.4418	0.4429	0.4441
1.6	0.4452	0.4463	0.4474	0.4484	0.4495	0.4505	0.4515	0.4525	0.4535	0.4545
1.7	0.4554	0.4564	0.4573	0.4582	0.4591	0.4599	0.4608	0.4616	0.4625	0.4633
1.8	0.4641	0.4649	0.4656	0.4664	0.4671	0.4678	0.4686	0.4693	0.4699	0.4706
1.9	0.4713	0.4719	0.4726	0.4732	0.4738	0.4744	0.4750	0.4756	0.4761	0.4767
2.0	0.4772	0.4778	0.4783	0.4788	0.4793	0.4798	0.4803	0.4808	0.4812	0.4817
2.1	0.4821	0.4826	0.4830	0.4834	0.4838	0.4842	0.4846	0.4850	0.4854	0.4857
2.2	0.4861	0.4864	0.4868	0.4871	0.4875	0.4878	0.4881	0.4884	0.4887	0.4890
2.3	0.4893	0.4896	0.4898	0.4901	0.4904	0.4906	0.4909	0.4911	0.4913	0.4916
2.4	0.4918	0.4920	0.4922	0.4925	0.4927	0.4929	0.4931	0.4932	0.4934	0.4936
2.5	0.4938	0.4940	0.4941	0.4943	0.4945	0.4946	0.4948	0.4949	0.4951	0.4952
2.6	0.4953	0.4955	0.4956	0.4957	0.4959	0.4960	0.4961	0.4962	0.4963	0.4964
2.7	0.4965	0.4966	0.4967	0.4968	0.4969	0.4970	0.4971	0.4972	0.4973	0.4974
2.8	0.4974	0.4975	0.4976	0.4977	0.4977	0.4978	0.4979	0.4979	0.4980	0.4981
2.9	0.4981	0.4982	0.4982	0.4983	0.4984	0.4984	0.4985	0.4985	0.4986	0.4986
3.0	0.4987	0.4987	0.4987	0.4988	0.4988	0.4989	0.4989	0.4989	0.4990	0.4990
3.1	0.4990	0.4991	0.4991	0.4991	0.4992	0.4992	0.4992	0.4992	0.4993	0.4993
3.2	0.4993	0.4993	0.4994	0.4994	0.4994	0.4994	0.4994	0.4995	0.4995	0.4995
3.3	0.4995	0.4995	0.4996	0.4996	0.4996	0.4996	0.4996	0.4996	0.4996	0.4997
3.4	0.4997	0.4997	0.4997	0.4997	0.4997	0.4997	0.4997	0.4997	0.4997	0.4998
3.5	0.4998	0.4998	0.4998	0.4998	0.4998	0.4998	0.4998	0.4998	0.4998	0.4998
3.6	0.4998	0.4998	0.4999	0.4999	0.4999	0.4999	0.4999	0.4999	0.4999	0.4999
3.7	0.4999	0.4999	0.4999	0.4999	0.4999	0.4999	0.4999	0.4999	0.4999	0.4999
3.8	0.4999	0.4999	0.4999	0.4999	0.4999	0.4999	0.4999	0.4999	0.4999	0.4999
3.9	0.5000	0.5000	0.5000	0.5000	0.5000	0.5000	0.5000	0.5000	0.5000	0.5000

APPENDIX 11.B
Chi-Squared Distribution

degrees of freedom	probability of exceeding the critical value, α				
	0.10	0.05	0.025	0.01	0.001
1	2.706	3.841	5.024	6.635	10.828
2	4.605	5.991	7.378	9.210	13.816
3	6.251	7.815	9.348	11.345	16.266
4	7.779	9.488	11.143	13.277	18.467
5	9.236	11.070	12.833	15.086	20.515
6	10.645	12.592	14.449	16.812	22.458
7	12.017	14.067	16.013	18.475	24.322
8	13.362	15.507	17.535	20.090	26.125
9	14.684	16.919	19.023	21.666	27.877
10	15.987	18.307	20.483	23.209	29.588
11	17.275	19.675	21.920	24.725	31.264
12	18.549	21.026	23.337	26.217	32.910
13	19.812	22.362	24.736	27.688	34.528
14	21.064	23.685	26.119	29.141	36.123
15	22.307	24.996	27.488	30.578	37.697
16	23.542	26.296	28.845	32.000	39.252
17	24.769	27.587	30.191	33.409	40.790
18	25.989	28.869	31.526	34.805	42.312
19	27.204	30.144	32.852	36.191	43.820
20	28.412	31.410	34.170	37.566	45.315
21	29.615	32.671	35.479	38.932	46.797
22	30.813	33.924	36.781	40.289	48.268

APPENDIX 11.C
t-Distribution
(values of *t* for degrees of freedom (df); $1 - \alpha$ confidence level)

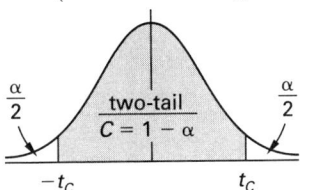

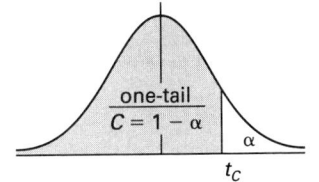

	two-tail, t_C								
	$t_{99\%}$	$t_{98\%}$	$t_{95\%}$	$t_{90\%}$	$t_{80\%}$	$t_{60\%}$	$t_{50\%}$	$t_{40\%}$	$t_{20\%}$	$t_{10\%}$
ν (df)	one-tail, t_C								
	$t_{99.5\%}$	$t_{99\%}$	$t_{97.5\%}$	$t_{95\%}$	$t_{90\%}$	$t_{80\%}$	$t_{75\%}$	$t_{70\%}$	$t_{60\%}$	$t_{55\%}$
1	63.66	31.82	12.71	6.31	3.08	1.376	1.000	0.727	0.325	0.158
2	9.92	6.96	4.30	2.92	1.89	1.061	0.816	0.617	0.289	0.142
3	5.84	4.54	3.18	2.35	1.64	0.978	0.765	0.584	0.277	0.137
4	4.60	3.75	2.78	2.13	1.53	0.941	0.741	0.569	0.271	0.134
5	4.03	3.36	2.57	2.02	1.48	0.920	0.727	0.559	0.267	0.132
6	3.71	3.14	2.45	1.94	1.44	0.906	0.718	0.553	0.265	0.131
7	3.50	3.00	2.36	1.90	1.42	0.896	0.711	0.549	0.263	0.130
8	3.36	2.90	2.31	1.86	1.40	0.889	0.706	0.546	0.262	0.130
9	3.25	2.82	2.26	1.83	1.38	0.883	0.703	0.543	0.261	0.129
10	3.17	2.76	2.23	1.81	1.37	0.879	0.700	0.542	0.260	0.129
11	3.11	2.72	2.20	1.80	1.36	0.876	0.697	0.540	0.260	0.129
12	3.06	2.68	2.18	1.78	1.36	0.873	0.695	0.539	0.259	0.128
13	3.01	2.65	2.16	1.77	1.35	0.870	0.694	0.538	0.259	0.128
14	2.98	2.62	2.14	1.76	1.34	0.868	0.692	0.537	0.258	0.128
15	2.95	2.60	2.13	1.75	1.34	0.866	0.691	0.536	0.258	0.128
16	2.92	2.58	2.12	1.75	1.34	0.865	0.690	0.535	0.258	0.128
17	2.90	2.57	2.11	1.74	1.33	0.863	0.689	0.534	0.257	0.128
18	2.88	2.55	2.10	1.73	1.33	0.862	0.688	0.534	0.257	0.127
19	2.86	2.54	2.09	1.73	1.33	0.861	0.688	0.533	0.257	0.127
20	2.84	2.53	2.09	1.72	1.32	0.860	0.687	0.533	0.257	0.127
21	2.83	2.52	2.08	1.72	1.32	0.859	0.686	0.532	0.257	0.127
22	2.82	2.51	2.07	1.72	1.32	0.858	0.686	0.532	0.256	0.127
23	2.81	2.50	2.07	1.71	1.32	0.858	0.685	0.532	0.256	0.127
24	2.80	2.49	2.06	1.71	1.32	0.857	0.685	0.531	0.256	0.127
25	2.79	2.48	2.06	1.71	1.32	0.856	0.684	0.531	0.256	0.127
26	2.78	2.48	2.06	1.71	1.32	0.856	0.684	0.531	0.256	0.127
27	2.77	2.47	2.05	1.70	1.31	0.855	0.684	0.531	0.256	0.127
28	2.76	2.47	2.05	1.70	1.31	0.855	0.683	0.530	0.256	0.127
29	2.76	2.46	2.04	1.70	1.31	0.854	0.683	0.530	0.256	0.127
30	2.75	2.46	2.04	1.70	1.31	0.854	0.683	0.530	0.256	0.127
40	2.70	2.42	2.02	1.68	1.30	0.851	0.681	0.529	0.255	0.126
60	2.66	2.39	2.00	1.67	1.30	0.848	0.679	0.527	0.254	0.126
120	2.62	2.36	1.98	1.66	1.29	0.845	0.677	0.526	0.254	0.126
∞	2.58	2.33	1.96	1.645	1.28	0.842	0.674	0.524	0.253	0.126

APPENDIX 11.D
Values of the Error Function and Complementary Error Function
for positive values of x

x	erf(x)	erfc(x)	x	erf(x)	erfc(x)	x	erf(x)	erfc(x)	x	erf(x)	erfc(x)	x	erf(x)	erfc(x)
0	0.0000	1.0000												
0.01	0.0113	0.9887	0.51	0.5292	0.4708	1.01	0.8468	0.1532	1.51	0.9673	0.0327	2.01	0.9955	0.0045
0.02	0.0226	0.9774	0.52	0.5379	0.4621	1.02	0.8508	0.1492	1.52	0.9684	0.0316	2.02	0.9957	0.0043
0.03	0.0338	0.9662	0.53	0.5465	0.4535	1.03	0.8548	0.1452	1.53	0.9695	0.0305	2.03	0.9959	0.0041
0.04	0.0451	0.9549	0.54	0.5549	0.4451	1.04	0.8586	0.1414	1.54	0.9706	0.0294	2.04	0.9961	0.0039
0.05	0.0564	0.9436	0.55	0.5633	0.4367	1.05	0.8624	0.1376	1.55	0.9716	0.0284	2.05	0.9963	0.0037
0.06	0.0676	0.9324	0.56	0.5716	0.4284	1.06	0.8661	0.1339	1.56	0.9726	0.0274	2.06	0.9964	0.0036
0.07	0.0789	0.9211	0.57	0.5798	0.4202	1.07	0.8698	0.1302	1.57	0.9736	0.0264	2.07	0.9966	0.0034
0.08	0.0901	0.9099	0.58	0.5879	0.4121	1.08	0.8733	0.1267	1.58	0.9745	0.0255	2.08	0.9967	0.0033
0.09	0.1013	0.8987	0.59	0.5959	0.4041	1.09	0.8768	0.1232	1.59	0.9755	0.0245	2.09	0.9969	0.0031
0.1	0.1125	0.8875	0.6	0.6039	0.3961	1.1	0.8802	0.1198	1.6	0.9763	0.0237	2.1	0.9970	0.0030
0.11	0.1236	0.8764	0.61	0.6117	0.3883	1.11	0.8835	0.1165	1.61	0.9772	0.0228	2.11	0.9972	0.0028
0.12	0.1348	0.8652	0.62	0.6194	0.3806	1.12	0.8868	0.1132	1.62	0.9780	0.0220	2.12	0.9973	0.0027
0.13	0.1459	0.8541	0.63	0.6270	0.3730	1.13	0.8900	0.1100	1.63	0.9788	0.0212	2.13	0.9974	0.0026
0.14	0.1569	0.8431	0.64	0.6346	0.3654	1.14	0.8931	0.1069	1.64	0.9796	0.0204	2.14	0.9975	0.0025
0.15	0.1680	0.8320	0.65	0.6420	0.3580	1.15	0.8961	0.1039	1.65	0.9804	0.0196	2.15	0.9976	0.0024
0.16	0.1790	0.8210	0.66	0.6494	0.3506	1.16	0.8991	0.1009	1.66	0.9811	0.0189	2.16	0.9977	0.0023
0.17	0.1900	0.8100	0.67	0.6566	0.3434	1.17	0.9020	0.0980	1.67	0.9818	0.0182	2.17	0.9979	0.0021
0.18	0.2009	0.7991	0.68	0.6638	0.3362	1.18	0.9048	0.0952	1.68	0.9825	0.0175	2.18	0.9980	0.0020
0.19	0.2118	0.7882	0.69	0.6708	0.3292	1.19	0.9076	0.0924	1.69	0.9832	0.0168	2.19	0.9980	0.0020
0.2	0.2227	0.7773	0.7	0.6778	0.3222	1.2	0.9103	0.0897	1.7	0.9838	0.0162	2.2	0.9981	0.0019
0.21	0.2335	0.7665	0.71	0.6847	0.3153	1.21	0.9130	0.0870	1.71	0.9844	0.0156	2.21	0.9982	0.0018
0.22	0.2443	0.7557	0.72	0.6914	0.3086	1.22	0.9155	0.0845	1.72	0.9850	0.0150	2.22	0.9983	0.0017
0.23	0.2550	0.7450	0.73	0.6981	0.3019	1.23	0.9181	0.0819	1.73	0.9856	0.0144	2.23	0.9984	0.0016
0.24	0.2657	0.7343	0.74	0.7047	0.2953	1.24	0.9205	0.0795	1.74	0.9861	0.0139	2.24	0.9985	0.0015
0.25	0.2763	0.7237	0.75	0.7112	0.2888	1.25	0.9229	0.0771	1.75	0.9867	0.0133	2.25	0.9985	0.0015
0.26	0.2869	0.7131	0.76	0.7175	0.2825	1.26	0.9252	0.0748	1.76	0.9872	0.0128	2.26	0.9986	0.0014
0.27	0.2974	0.7026	0.77	0.7238	0.2762	1.27	0.9275	0.0725	1.77	0.9877	0.0123	2.27	0.9987	0.0013
0.28	0.3079	0.6921	0.78	0.7300	0.2700	1.28	0.9297	0.0703	1.78	0.9882	0.0118	2.28	0.9987	0.0013
0.29	0.3183	0.6817	0.79	0.7361	0.2639	1.29	0.9319	0.0681	1.79	0.9886	0.0114	2.29	0.9988	0.0012
0.3	0.3286	0.6714	0.8	0.7421	0.2579	1.3	0.9340	0.0660	1.8	0.9891	0.0109	2.3	0.9989	0.0011
0.31	0.3389	0.6611	0.81	0.7480	0.2520	1.31	0.9361	0.0639	1.81	0.9895	0.0105	2.31	0.9989	0.0011
0.32	0.3491	0.6509	0.82	0.7538	0.2462	1.32	0.9381	0.0619	1.82	0.9899	0.0101	2.32	0.9990	0.0010
0.33	0.3593	0.6407	0.83	0.7595	0.2405	1.33	0.9400	0.0600	1.83	0.9903	0.0097	2.33	0.9990	0.0010
0.34	0.3694	0.6306	0.84	0.7651	0.2349	1.34	0.9419	0.0581	1.84	0.9907	0.0093	2.34	0.9991	0.0009
0.35	0.3794	0.6206	0.85	0.7707	0.2293	1.35	0.9438	0.0562	1.85	0.9911	0.0089	2.35	0.9991	0.0009
0.36	0.3893	0.6107	0.86	0.7761	0.2239	1.36	0.9456	0.0544	1.86	0.9915	0.0085	2.36	0.9992	0.0008
0.37	0.3992	0.6008	0.87	0.7814	0.2186	1.37	0.9473	0.0527	1.87	0.9918	0.0082	2.37	0.9992	0.0008
0.38	0.4090	0.5910	0.88	0.7867	0.2133	1.38	0.9490	0.0510	1.88	0.9922	0.0078	2.38	0.9992	0.0008
0.39	0.4187	0.5813	0.89	0.7918	0.2082	1.39	0.9507	0.0493	1.89	0.9925	0.0075	2.39	0.9993	0.0007
0.4	0.4284	0.5716	0.9	0.7969	0.2031	1.4	0.9523	0.0477	1.9	0.9928	0.0072	2.4	0.9993	0.0007
0.41	0.4380	0.5620	0.91	0.8019	0.1981	1.41	0.9539	0.0461	1.91	0.9931	0.0069	2.41	0.9993	0.0007
0.42	0.4475	0.5525	0.92	0.8068	0.1932	1.42	0.9554	0.0446	1.92	0.9934	0.0066	2.42	0.9994	0.0006
0.43	0.4569	0.5431	0.93	0.8116	0.1884	1.43	0.9569	0.0431	1.93	0.9937	0.0063	2.43	0.9994	0.0006
0.44	0.4662	0.5338	0.94	0.8163	0.1837	1.44	0.9583	0.0417	1.94	0.9939	0.0061	2.44	0.9994	0.0006
0.45	0.4755	0.5245	0.95	0.8209	0.1791	1.45	0.9597	0.0403	1.95	0.9942	0.0058	2.45	0.9995	0.0005
0.46	0.4847	0.5153	0.96	0.8254	0.1746	1.46	0.9611	0.0389	1.96	0.9944	0.0056	2.46	0.9995	0.0005
0.47	0.4937	0.5063	0.97	0.8299	0.1701	1.47	0.9624	0.0376	1.97	0.9947	0.0053	2.47	0.9995	0.0005
0.48	0.5027	0.4973	0.98	0.8342	0.1658	1.48	0.9637	0.0363	1.98	0.9949	0.0051	2.48	0.9995	0.0005
0.49	0.5117	0.4883	0.99	0.8385	0.1615	1.49	0.9649	0.0351	1.99	0.9951	0.0049	2.49	0.9996	0.0004
0.5	0.5205	0.4795	1	0.8427	0.1573	1.5	0.9661	0.0339	2	0.9953	0.0047	2.5	0.9996	0.0004

APPENDIX 14.A
Properties of Water at Atmospheric Pressure
(customary U.S. units)

temperature (°F)	density (lbm/ft^3)	absolute viscosity (lbf-sec/ft^2)	kinematic viscosity (ft^2/sec)	surface tension (lbf/ft)	vapor pressure head^{a,b,c} (ft)	bulk modulus (lbf/in^2)
32	62.42	3.746×10^{-5}	1.931×10^{-5}	0.518×10^{-2}	0.20	293×10^3
40	62.43	3.229×10^{-5}	1.664×10^{-5}	0.514×10^{-2}	0.28	294×10^3
50	62.41	2.735×10^{-5}	1.410×10^{-5}	0.509×10^{-2}	0.41	305×10^3
60	62.37	2.359×10^{-5}	1.217×10^{-5}	0.504×10^{-2}	0.59	311×10^3
70	62.30	2.050×10^{-5}	1.059×10^{-5}	0.500×10^{-2}	0.84	320×10^3
80	62.22	1.799×10^{-5}	0.930×10^{-5}	0.492×10^{-2}	1.17	322×10^3
90	62.11	1.595×10^{-5}	0.826×10^{-5}	0.486×10^{-2}	1.62	323×10^3
100	62.00	1.424×10^{-5}	0.739×10^{-5}	0.480×10^{-2}	2.21	327×10^3
110	61.86	1.284×10^{-5}	0.667×10^{-5}	0.473×10^{-2}	2.97	331×10^3
120	61.71	1.168×10^{-5}	0.609×10^{-5}	0.465×10^{-2}	3.96	333×10^3
130	61.55	1.069×10^{-5}	0.558×10^{-5}	0.460×10^{-2}	5.21	334×10^3
140	61.38	0.981×10^{-5}	0.514×10^{-5}	0.454×10^{-2}	6.78	330×10^3
150	61.20	0.905×10^{-5}	0.476×10^{-5}	0.447×10^{-2}	8.76	328×10^3
160	61.00	0.838×10^{-5}	0.442×10^{-5}	0.441×10^{-2}	11.21	326×10^3
170	60.80	0.780×10^{-5}	0.413×10^{-5}	0.433×10^{-2}	14.20	322×10^3
180	60.58	0.726×10^{-5}	0.385×10^{-5}	0.426×10^{-2}	17.87	313×10^3
190	60.36	0.678×10^{-5}	0.362×10^{-5}	0.419×10^{-2}	22.29	313×10^3
200	60.12	0.637×10^{-5}	0.341×10^{-5}	0.412×10^{-2}	27.61	308×10^3
212	59.83	0.593×10^{-5}	0.319×10^{-5}	0.404×10^{-2}	35.38	300×10^3

^{a}Based on actual densities, not on standard "cold, clear water."
^{b}Can also be calculated from steam tables as $(p_{\text{saturation}})(12 \text{ in/ft})^2(v_f)(g_c/g)$.
^{c}Multiply the vapor pressure head by the specific weight and divide by $(12 \text{ in/ft})^2$ to obtain lbf/in^2.

APPENDIX 14.B
Properties of Water at Atmospheric Pressure
(SI units)

temperature (°C)	density (kg/m^3)	absolute viscosity (Pa·s)	kinematic viscosity (m^2/s)	vapor pressure (kPa)	bulk modulus (kPa)
0	999.87	1.7921×10^{-3}	1.792×10^{-6}	0.611	204×10^4
4	1000.00	1.5674×10^{-3}	1.567×10^{-6}	0.813	206×10^4
10	999.73	1.3077×10^{-3}	1.371×10^{-6}	1.228	211×10^4
20	998.23	1.0050×10^{-3}	1.007×10^{-6}	2.338	220×10^4
25	997.08	0.8937×10^{-3}	8.963×10^{-7}	3.168	221×10^4
30	995.68	0.8007×10^{-3}	8.042×10^{-7}	4.242	223×10^4
40	992.25	0.6560×10^{-3}	6.611×10^{-7}	7.375	227×10^4
50	988.07	0.5494×10^{-3}	5.560×10^{-7}	12.333	230×10^4
60	983.24	0.4688×10^{-3}	4.768×10^{-7}	19.92	228×10^4
70	977.81	0.4061×10^{-3}	4.153×10^{-7}	31.16	225×10^4
80	971.83	0.3565×10^{-3}	3.668×10^{-7}	47.34	221×10^4
90	965.34	0.3165×10^{-3}	3.279×10^{-7}	70.10	216×10^4
100	958.38	0.2838×10^{-3}	2.961×10^{-7}	101.325	207×10^4

Compiled from various sources.

APPENDIX 14.C
Viscosity of Water in Other Units
(customary U.S. units)

temperature	absolute viscosity	kinematic viscosity	
(°F)	(cP)	(cSt)	(SSU)
32	1.79	1.79	33.0
50	1.31	1.31	31.6
60	1.12	1.12	31.2
70	0.98	0.98	30.9
80	0.86	0.86	30.6
85	0.81	0.81	30.4
100	0.68	0.69	30.2
120	0.56	0.57	30.0
140	0.47	0.48	29.7
160	0.40	0.41	29.6
180	0.35	0.36	29.5
212	0.28	0.29	29.3

Reprinted with permission from *Hydraulic Handbook*, 10th ed., by Colt Industries, Fairbanks Morse Pump Division, Kansas City, Kansas, 1977.

APPENDIX 14.D
Properties of Air at Atmospheric Pressure
(customary U.S. units)

absolute temperature,[a] T_{abs} (°R)	temperature,[a] T (°F)	density,[b] ρ (lbm/ft³)	absolute (dynamic) viscosity,[c] μ (lbf-sec/ft²)	kinematic viscosity,[d] ν (ft²/sec)
300	−160	0.1322	2.378×10^{-7}	5.786×10^{-5}
350	−110	0.1133	2.722×10^{-7}	7.728×10^{-5}
400	−60	0.0992	3.047×10^{-7}	9.886×10^{-5}
450	−10	0.0881	3.355×10^{-7}	12.24×10^{-5}
460	0	0.0863	3.412×10^{-7}	12.72×10^{-5}
470	10	0.0845	3.471×10^{-7}	13.22×10^{-5}
480	20	0.0827	3.530×10^{-7}	13.73×10^{-5}
490	30	0.0810	3.588×10^{-7}	14.25×10^{-5}
492	32	0.0807	3.599×10^{-7}	14.35×10^{-5}
495	35	0.0802	3.616×10^{-7}	14.51×10^{-5}
500	40	0.0794	3.645×10^{-7}	14.77×10^{-5}
505	45	0.0786	3.674×10^{-7}	15.04×10^{-5}
510	50	0.0778	3.702×10^{-7}	15.30×10^{-5}
515	55	0.0771	3.730×10^{-7}	15.57×10^{-5}
520	60	0.0763	3.758×10^{-7}	15.84×10^{-5}
525	65	0.0756	3.786×10^{-7}	16.11×10^{-5}
528	68	0.0752	3.803×10^{-7}	16.28×10^{-5}
530	70	0.0749	3.814×10^{-7}	16.39×10^{-5}
535	75	0.0742	3.842×10^{-7}	16.66×10^{-5}
540	80	0.0735	3.869×10^{-7}	16.94×10^{-5}
545	85	0.0728	3.897×10^{-7}	17.21×10^{-5}
550	90	0.0722	3.924×10^{-7}	17.49×10^{-5}
555	95	0.0715	3.951×10^{-7}	17.78×10^{-5}
560	100	0.0709	3.978×10^{-7}	18.06×10^{-5}
570	110	0.0696	4.032×10^{-7}	18.63×10^{-5}
580	120	0.0684	4.085×10^{-7}	19.21×10^{-5}
590	130	0.0673	4.138×10^{-7}	19.79×10^{-5}
600	140	0.0661	4.190×10^{-7}	20.38×10^{-5}
610	150	0.0651	4.242×10^{-7}	20.98×10^{-5}
650	190	0.0610	4.448×10^{-7}	23.45×10^{-5}
660	200	0.0601	4.496×10^{-7}	24.06×10^{-5}
700	240	0.0567	4.693×10^{-7}	26.65×10^{-5}
710	250	0.0559	4.740×10^{-7}	27.28×10^{-5}
750	290	0.0529	4.930×10^{-7}	29.99×10^{-5}
760	300	0.0522	4.975×10^{-7}	30.65×10^{-5}
800	340	0.0496	5.159×10^{-7}	33.47×10^{-5}
850	390	0.0467	5.380×10^{-7}	37.09×10^{-5}
900	440	0.0441	5.594×10^{-7}	40.84×10^{-5}
950	490	0.0418	5.802×10^{-7}	44.71×10^{-5}
1000	540	0.0397	6.004×10^{-7}	48.70×10^{-5}
1050	590	0.0378	6.201×10^{-7}	52.81×10^{-5}
1100	640	0.0361	6.393×10^{-7}	57.04×10^{-5}
1150	690	0.0345	6.580×10^{-7}	61.37×10^{-5}
1200	740	0.0331	6.762×10^{-7}	65.82×10^{-5}
1250	790	0.0317	6.941×10^{-7}	70.37×10^{-5}
1300	840	0.0305	7.115×10^{-7}	75.03×10^{-5}
1350	890	0.0294	7.286×10^{-7}	79.78×10^{-5}
1400	940	0.0283	7.454×10^{-7}	84.64×10^{-5}
1450	990	0.0274	7.618×10^{-7}	89.60×10^{-5}
1500	1040	0.0264	7.779×10^{-7}	94.65×10^{-5}
1550	1090	0.0256	7.937×10^{-7}	99.79×10^{-5}
1600	1140	0.0248	8.093×10^{-7}	105.0×10^{-5}
1650	1190	0.0240	8.246×10^{-7}	110.4×10^{-5}
1700	1240	0.0233	8.396×10^{-7}	115.8×10^{-5}
1800	1340	0.0220	8.689×10^{-7}	126.9×10^{-5}
1900	1440	0.0209	8.974×10^{-7}	138.3×10^{-5}
2000	1540	0.0198	9.250×10^{-7}	150.1×10^{-5}
2100	1340	0.0189	9.519×10^{-7}	162.1×10^{-5}
2200	1740	0.0180	9.781×10^{-7}	174.5×10^{-5}
2300	1840	0.0172	10.04×10^{-7}	187.2×10^{-5}
2400	1940	0.0165	10.29×10^{-7}	200.2×10^{-5}
2500	2040	0.0159	10.53×10^{-7}	213.5×10^{-5}
2600	2140	0.0153	10.77×10^{-7}	227.1×10^{-5}
2800	2340	0.0142	11.23×10^{-7}	255.1×10^{-5}
3000	2540	0.0132	11.68×10^{-7}	284.1×10^{-5}
3200	2740	0.0124	12.11×10^{-7}	314.2×10^{-5}
3400	2940	0.0117	12.52×10^{-7}	345.3×10^{-5}

[a]Temperatures are rounded to the nearest whole degree, but all significant digits were used in calculations.
[b]Density is calculated from the ideal gas law using $p = 14.696$ lbf/in² and $R_{air} = 53.35$ ft-lbf/lbm-°R.
[c]Absolute viscosity is calculated from Sutherland's formula using $C_{Sutherland} = 109.1$°C and $\mu_0 = 0.14592$ kg/m·s, and is subsequently converted to customary U.S. units. Error is ~0 at 530°R. Error is expected to be less than 0.7% at 300°R and 0.2% at 3400°R.
[d]Kinematic viscosity is calculated as $\nu = \mu g_c / \rho$, with $g_c = 32.1742$ lbm-ft/lbf-sec².

APPENDIX 14.E
Properties of Air at Atmospheric Pressure
(SI units)

absolute temperature,[a] T_{abs} (K)	temperature,[a] T (°C)	density,[b] ρ (kg/m³)	absolute (dynamic) viscosity,[c] μ (kg/m·s or Pa·s)	kinematic viscosity,[d] ν (m²/s)
175	−98	2.015	1.190×10^{-5}	0.5905×10^{-5}
200	−73	1.764	1.336×10^{-5}	0.7576×10^{-5}
225	−48	1.568	1.475×10^{-5}	0.9408×10^{-5}
250	−23	1.411	1.607×10^{-5}	1.139×10^{-5}
273	0	1.292	1.723×10^{-5}	1.334×10^{-5}
275	2	1.283	1.733×10^{-5}	1.351×10^{-5}
283	10	1.247	1.772×10^{-5}	1.422×10^{-5}
293	20	1.204	1.821×10^{-5}	1.512×10^{-5}
300	27	1.176	1.854×10^{-5}	1.577×10^{-5}
303	30	1.164	1.868×10^{-5}	1.605×10^{-5}
313	40	1.127	1.915×10^{-5}	1.699×10^{-5}
323	50	1.092	1.961×10^{-5}	1.795×10^{-5}
325	52	1.086	1.970×10^{-5}	1.815×10^{-5}
333	60	1.060	2.006×10^{-5}	1.894×10^{-5}
343	70	1.029	2.051×10^{-5}	1.994×10^{-5}
350	77	1.008	2.082×10^{-5}	2.065×10^{-5}
353	80	0.9995	2.095×10^{-5}	2.096×10^{-5}
363	90	0.9720	2.138×10^{-5}	2.200×10^{-5}
373	100	0.9459	2.181×10^{-5}	2.306×10^{-5}
375	102	0.9409	2.190×10^{-5}	2.327×10^{-5}
400	127	0.8821	2.294×10^{-5}	2.600×10^{-5}
423	150	0.8342	2.386×10^{-5}	2.861×10^{-5}
450	177	0.7841	2.492×10^{-5}	3.178×10^{-5}
473	200	0.7460	2.579×10^{-5}	3.457×10^{-5}
500	227	0.7057	2.679×10^{-5}	3.796×10^{-5}
523	250	0.6747	2.762×10^{-5}	4.093×10^{-5}
550	277	0.6416	2.856×10^{-5}	4.452×10^{-5}
573	300	0.6159	2.935×10^{-5}	4.765×10^{-5}
600	327	0.5881	3.025×10^{-5}	5.143×10^{-5}
623	350	0.5664	3.100×10^{-5}	5.473×10^{-5}
650	377	0.5429	3.186×10^{-5}	5.868×10^{-5}
673	400	0.5244	3.258×10^{-5}	6.213×10^{-5}
700	427	0.5041	3.341×10^{-5}	6.626×10^{-5}
723	450	0.4881	3.410×10^{-5}	6.985×10^{-5}
750	477	0.4705	3.489×10^{-5}	7.415×10^{-5}
773	500	0.4565	3.556×10^{-5}	7.788×10^{-5}
800	527	0.4411	3.632×10^{-5}	8.234×10^{-5}
850	577	0.4152	3.771×10^{-5}	9.082×10^{-5}
900	627	0.3921	3.905×10^{-5}	9.958×10^{-5}
950	677	0.3715	4.035×10^{-5}	10.86×10^{-5}
1000	727	0.3529	4.161×10^{-5}	11.79×10^{-5}
1050	777	0.3361	4.284×10^{-5}	12.74×10^{-5}
1100	827	0.3208	4.403×10^{-5}	13.72×10^{-5}
1150	877	0.3069	4.520×10^{-5}	14.73×10^{-5}
1200	927	0.2941	4.634×10^{-5}	15.76×10^{-5}
1250	977	0.2823	4.745×10^{-5}	16.81×10^{-5}
1300	1027	0.2715	4.854×10^{-5}	17.88×10^{-5}
1350	1077	0.2614	4.961×10^{-5}	18.98×10^{-5}
1400	1127	0.2521	5.065×10^{-5}	20.09×10^{-5}
1500	1227	0.2353	5.269×10^{-5}	22.39×10^{-5}
1600	1327	0.2206	5.464×10^{-5}	24.77×10^{-5}
1700	1427	0.2076	5.654×10^{-5}	27.23×10^{-5}
1800	1527	0.1961	5.837×10^{-5}	29.77×10^{-5}
1900	1627	0.1858	6.015×10^{-5}	32.38×10^{-5}

[a]Temperatures are rounded to the nearest whole degree, but all significant digits were used in calculations.

[b]Density is calculated from the ideal gas law using $p = 101.325$ kPa; and $R_{air} = 287.058$ J/kg·K.

[c]Absolute viscosity is calculated from Sutherland's formula using $C_{Sutherland} = 109.1$°C and $\mu_0 = 0.14592$ kg/m·s. Error is ∼0 at 293K. Error is expected to be less than 0.7% at 175K and 0.2% at 1900K.

[d]Kinematic viscosity is calculated as $\nu = \mu/\rho$.

APPENDIX 14.F
Properties of Common Liquids

liquid	temp (°F)	specific gravity*	absolute viscosity (cP)	kinematic viscosity		
				centistokes	SSU	ft²/sec
acetone	68	0.792		0.41		
alcohol, ethyl (ethanol)	68	0.789		1.52	31.7	1.65×10^{-5}
(C_2H_5OH)	104	0.772		1.2	31.5	
alcohol, methyl (methanol)	68	0.79				
(CH_3OH)	59			0.74		
	0	0.810				
ammonia	0	0.662		0.30		
benzene	60	0.885				
	32	0.899				
butane	−50			0.52		
	30			0.35		
	60	0.584				
carbon tetrachloride	68	1.595				
castor oil	68	0.96				1110×10^{-5}
	104	0.95		259–325	1200–1500	
	130			98–130	450–600	
ethylene glycol	0 (−18°C)	1.16	310	267	1210	2.88×10^{-3}
(mono, MEG)	40 (4.4°C)	1.145	48	42	195	4.51×10^{-4}
	68 (20°C)	1.115	17	15	77	1.64×10^{-5}
	140 (60°C)	1.107	5.2	4.7	41	5.06×10^{-5}
	200 (93°C)	1.084	1.4	1.3	32	1.39×10^{-5}
Freon-11	70	1.49		21.1	0.21	
Freon-12	70	1.33		21.1	0.27	
fuel oils, no. 1 to no. 6	60	0.82–0.95				
fuel oil no. 1	70			2.39–4.28	34–40	
	100			−2.69	32–35	
fuel oil no. 2	70			3.0–7.4	36–50	
	100			2.11–4.28	33–40	
fuel oil no. 3	70			2.69–5.84	35–45	
	100			2.06–3.97	32.8–39	
fuel oil no. 5A	70			7.4–26.4	50–125	
	100			4.91–13.7	42–72	
fuel oil no. 5B	70			26.4–	125–	
	100			13.6–67.1	72–310	
fuel oil no. 6	122			97.4–660	450–3000	
	160			37.5–172	175–780	
gasoline	60	0.728				0.73×10^{-5}
	80	0.719				0.66×10^{-5}
	100	0.710				0.60×10^{-5}
glycerine	68	1.261				
kerosene	60	0.78–0.82				
	68			2.17	35	
jet fuel (JPI, 3, 4, 5, 6)	−30			7.9	52	
	60	0.78–0.82				
mercury	70	13.55		21.1	0.118	
	100	13.55		37.8	0.11	
oil, SAE 5 to 150	60	0.88–0.94				
SAE-5W	0			1295 max	6000 max	
SAE-10W	0			1295–2590	6000–12,000	
SAE-20W	0			2590–10,350	12,000–48,000	
SAE-20	210			5.7–9.6	45–58	
SAE-30	210			9.6–12.9	58–70	
SAE-40	210			12.9–16.8	70–85	
SAE-50	210			16.8–22.7	85–110	
propylene glycol	0 (−18°C)	1.063	~1600	~1500	~6900	$\sim 1.63 \times 10^{-2}$
	25 (−4°C)	1.054	~380	~360	~1600	$\sim 3.89 \times 10^{-3}$
	50 (10°C)	1.044	~130	~125	~570	$\sim 1.35 \times 10^{-3}$
	77 (25°C)	1.032	49	47	220	5.11×10^{-4}
	140 (60°C)	1.006	8.4	8.3	53	8.99×10^{-5}
	200 (93°C)	0.985	2.7	2.7	35	2.95×10^{-5}
saltwater (5%)	39	1.037				
	68			1.097	31.1	
saltwater (25%)	39	1.196				
	60	1.19		2.4	34	
seawater	59	1.025				

*Measured with respect to 60°F water.

APPENDIX 14.G
Viscosities Versus Temperature of
Hydrocarbons, Lubricants, Fuels, and Heat Transfer Fluids

kinematic viscosity (cSt)

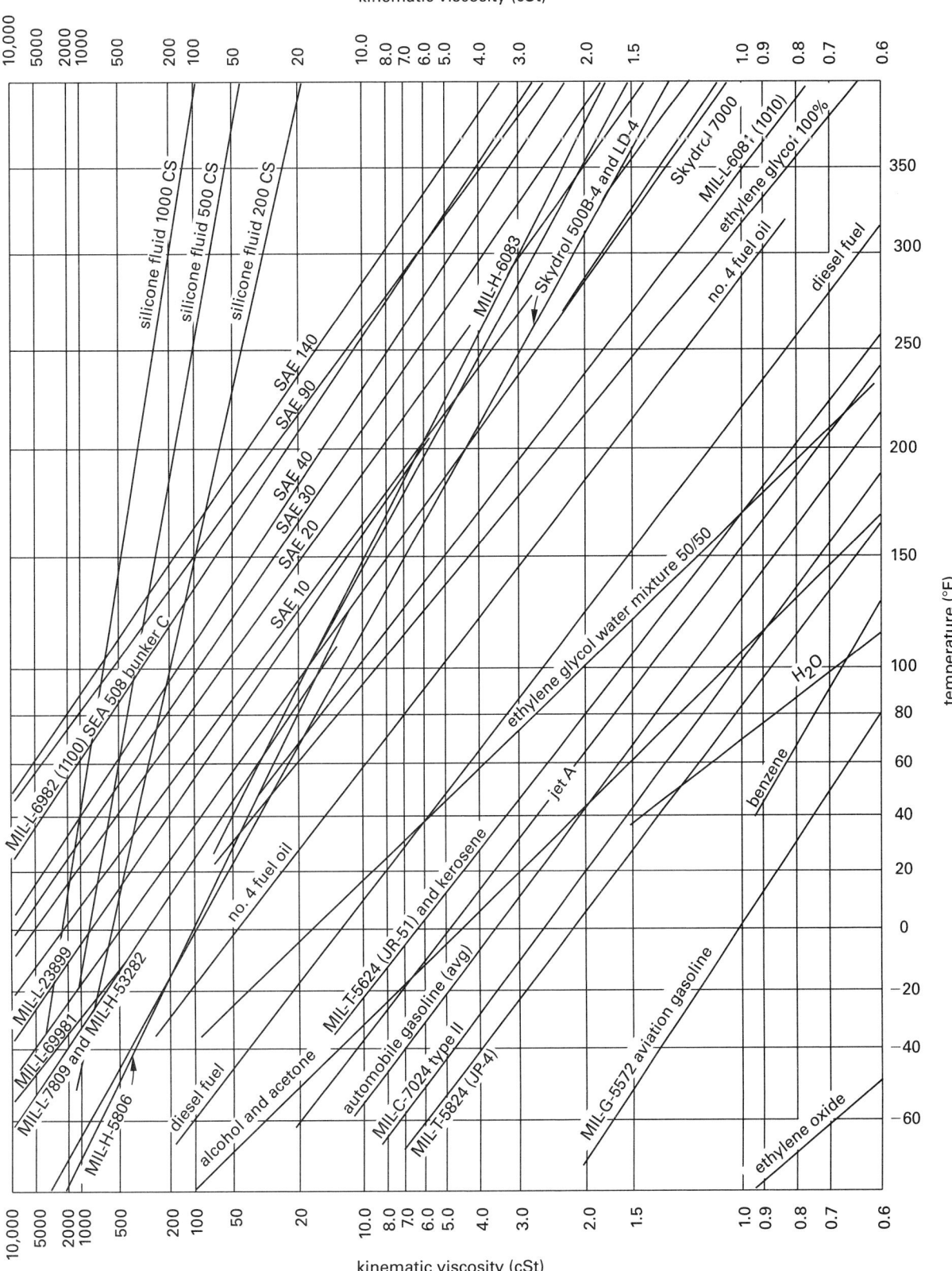

kinematic viscosity (cSt)

APPENDIX 14.H
Properties of Uncommon Fluids

Typical fluid viscosities are listed in the following table. The values given for thixotropic fluids are effective viscosities at normal pumping shear rates. Effective viscosity can vary greatly with changes in solids content, concentration, and so on.

Fluid Type: N—Newtonian   T—Thixotropic

fluid	specific gravity	viscosity cP	fluid type	fluid	specific gravity	viscosity cP	fluid type
adhesives				meat products			
"box" adhesives	1	3000	T	animal fat, melted	0.9	43 at 110°F	N
PVA	1.3	100	T	ground beef fat	0.9	11,000 at 60°F	T
rubber and solvents	1.0	15,000	N	meat emulsion	1.0	22,000 at 40°F	T
bakery				pet food	1.0	11,000 at 40°F	T
batter	1	2200	T	pork fat slurry	1.0	650 at 40°F	T
butter, melted	0.98	18 at 140°F	N	paint			
egg, whole	0.5	60 at 50°F	N	auto paint, metalic		200	T
emulsifier		20	T	solvents	0.8–0.9	0.5–10	N
frosting	1	10,000	T	titanium dioxide slurry		10,000	T
lecithin		3250 at 125°F	T	varnish	1.06	140 at 100°F	
77% sweetened condensed milk	1.3	10,000 at 77°F	N	turpentine	0.86	2 at 60°F	
				paper and textile			
yeast slurry, 15%	1	180	T	paper coating, 35%		400	
beer, wine				sulfide, 6%		1600	
beer	1.0	1.1 at 40°F	N	black liquor	1.3	1100 at 122°F	
brewers concentrated yeast, 80% solids		16,000 at 40°F	T	black liquor tar		2000 at 300°F	
				black liquor soap		7000 at 122°F	
wine	1.0			petroleum and			
confectionary				petroleum products			
caramel	1.2	400 at 140°F		asphalt, unblended	1.3	500–2500	
chocolate	1.1	17,000 at 120°F	T	gasoline	0.7	0.8 at 60°F	N
fudge, hot	1.1	36,000	T	kerosene	0.8	3 at 68°F	N
toffee	1.2	87,000	T	fuel oil no. 6	0.9	660 at 122°F	N
cosmetics, soaps				auto lube oil SAE 40	0.9	200 at 100°F	N
face cream		10,000	T	auto trans oil SAE 90	0.9	320 at 100°F	N
gel, hair	1.4	5000	T	propane	0.46	0.2 at 100°F	N
shampoo		5000	T	tars	1.2	wide range	
toothpaste		20,000	T	pharmaceuticals			
hand cleaner		2000	T	caster oil	0.96	350	N
dairy				cough syrup	1.0	190	N
cottage cheese	1.08	225	T	"stomach" remedy		1500	T
cream	1.02	20 at 40°F	N	pill pastes		5000	T
milk	1.03	1.2 at 60°F	N	plastic, resins			
cheese, process		30,000 at 160°F	T	butadiene	0.94	0.17 at 40°F	
yogurt		1100	T	polyester resin (typ)	1.4	3000	
detergents				PVA resin (typ)	1.3	65,000	T
detergent concentrate		10	N	starches, gums			
dyes and inks				corn starch sol 22°B	1.18	32	T
ink, printers	1–1.38	10,000	T	corn starch sol 25°B	1.21	300	T
dye	1.1	10	N	sugar syrups molasses			
gum		5000	T	corn syrup 41°Be	1.39	15,000 at 60°F	N
fats and oils				corn syrup 45°Be	1.45	12,000 at 130°F	N
corn oils	0.92	30	N	glucose	1.42	10,000 at 100°F	
lard	0.96	60 at 100°F	N	molasses			
linseed oil	0.93	30 at 100°F	N	A (light)	1.4–1.47	280–5000 at 100°F	
peanut oil	0.92	42 at 100°F	N				
soybean oil	0.95	36 at 100°F	N	B (medium, dark)	1.43–1.48	1400–13,000 at 100°F	
vegetable oil	0.92	3 at 300°F	N				
foods, miscellaneous				C (blackstrap)	1.46–1.49	2600–5500 at 100°F	
black bean paste		10,000	T				
cream style corn		130 at 190°F	T	sugar syrups			
catsup	1.11	560 at 145°F	T	60 brix	1.29	75 at 60°F	N
pablum		4500	T	68 brix	1.34	360 at 60°F	N
pear pulp		4000 at 160°F	T	78 brix	1.39	4000 at 60°F	N
potato, mashed	1.0	20,000	T	water and waste			
potato skins and caustic		20,000 at 100°F	T	treatment			
prune juice	1.0	60 at 120°F	T	clarified sewage sludge	1.1	2000 range	
orange juice concentrate	1.1	5000 at 38°F	T				
tapioca pudding	0.7	1000 at 235°F	T				
mayonnaise	1.0	5000 at 75°F	T				
tomato paste, 33%	1.14	7000	T				
honey	1.5	1500 at 100°F	T				

Used with permission from *Waukesha Pump Engineering Manual*, © 2002 by Waukesha Cherry-Burrell. A United Dominion Company.

APPENDIX 14.I
Vapor Pressure of Various Hydrocarbons and Water
(Cox Chart)

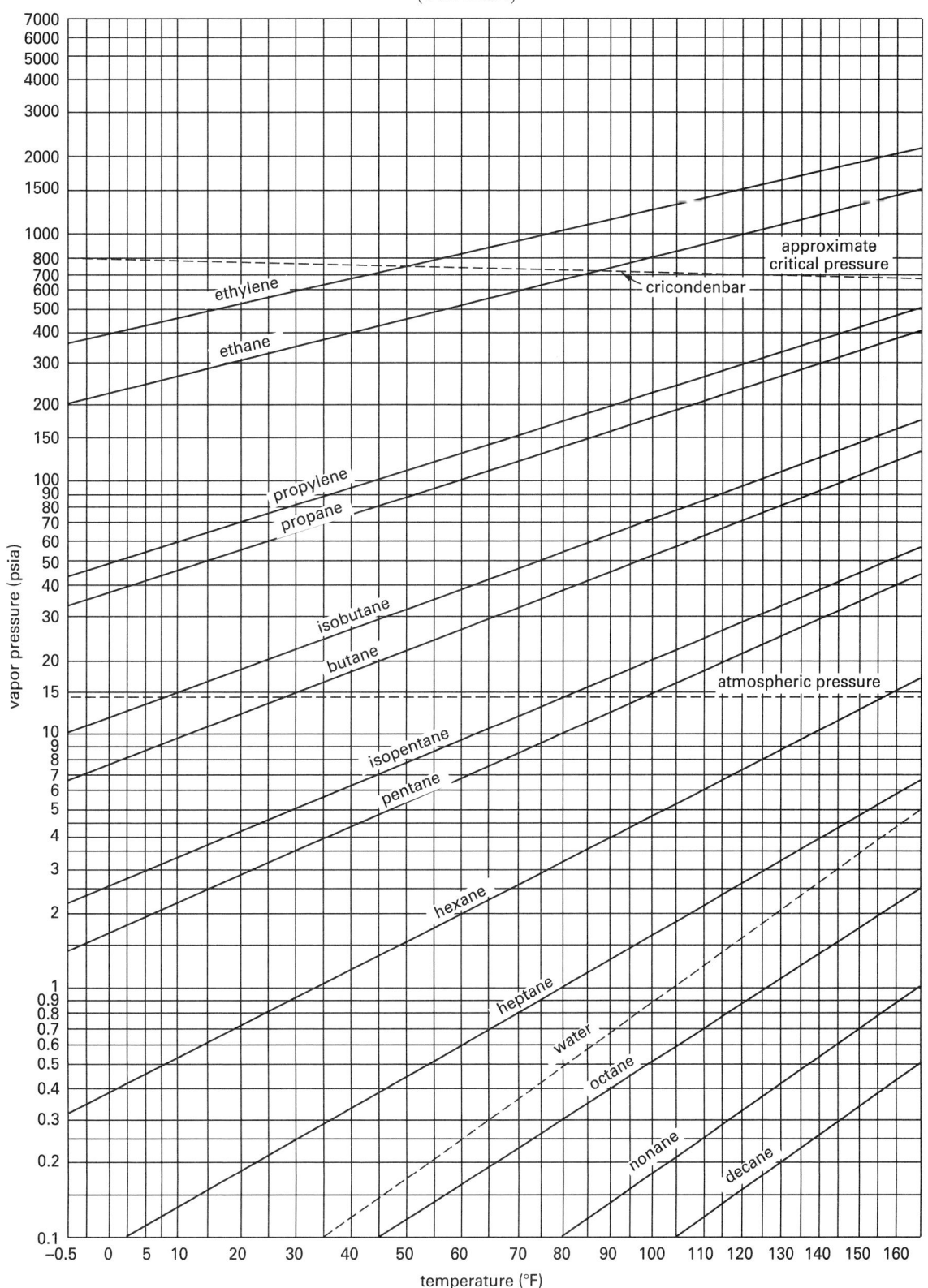

Used with permission from *Hydraulic Handbook*, 10th ed., by Colt Industries, Fairbanks Morse Pump Division, Kansas City, Kansas, 1977.

APPENDIX 14.J
Specific Gravity of Hydrocarbons

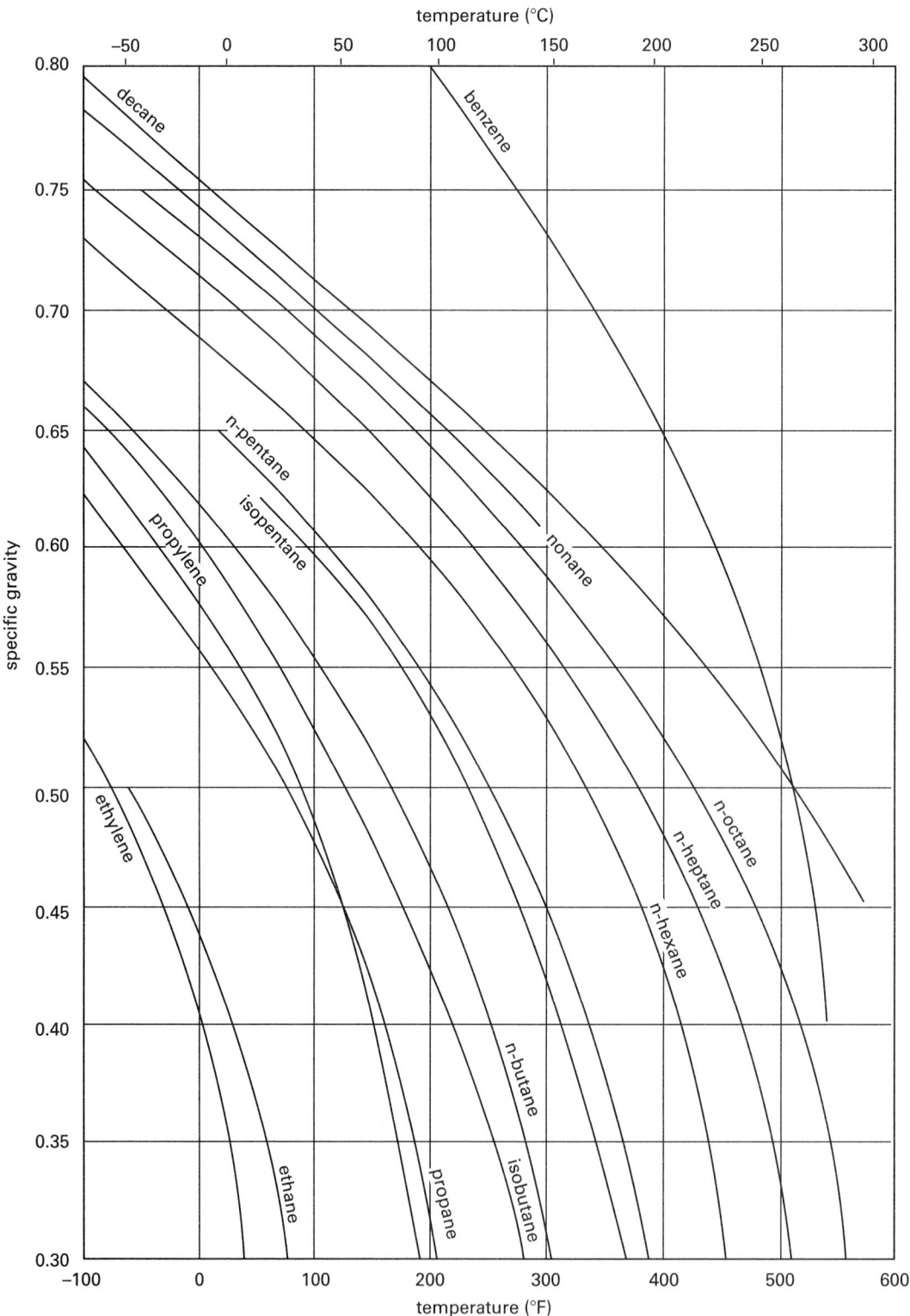

Reproduced with permission of Ingersoll-Dresser Pump Company from *Cameron Hydraulic Data: A handy reference on the subject of hydraulics and steam*, 18th edition, edited by C. C. Heald, Ingersoll-Dresser Pumps, © 1995.

APPENDIX 14.K
Viscosity Conversion Chart
(Approximate Conversions for Newtonian Fluids)

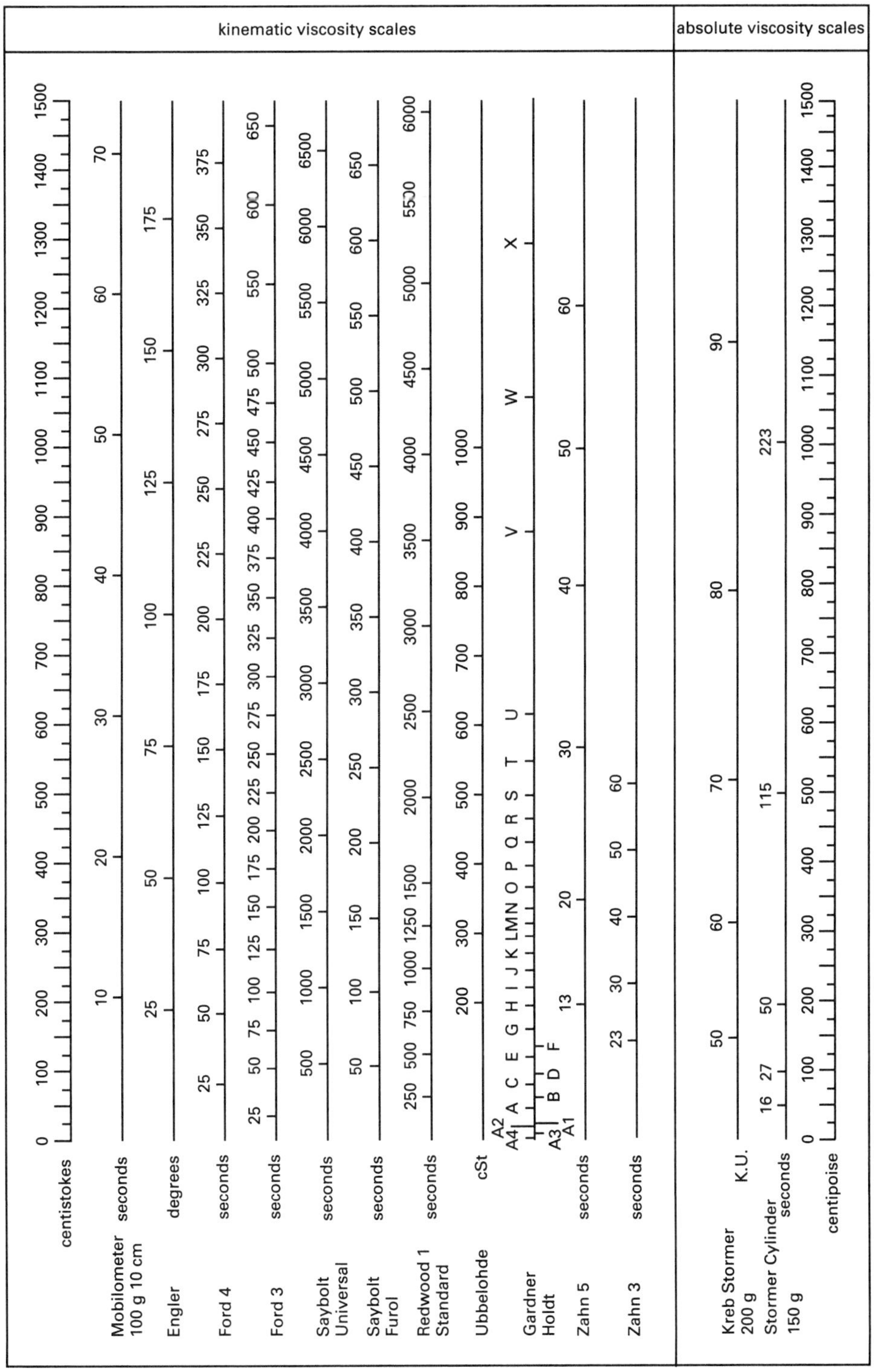

Support Material

kinematic viscosity scales	absolute viscosity scales

centistokes

Mobilometer 100 g 10 cm — seconds

Engler — degrees

Ford 4 — seconds

Ford 3 — seconds

Saybolt Universal — seconds

Saybolt Furol — seconds

Redwood 1 Standard — seconds

Ubbelohde — cSt

Gardner Holdt

Zahn 5 — seconds

Zahn 3 — seconds

Kreb Stormer 200 g — K.U.

Stormer Cylinder 150 g — seconds

centipoise

(Multiply centistokes by 1.0764×10^{-5} to get ft^2/sec.)
(Multiply centistokes by 1.000×10^{-6} to get m^2/s.)

Support
Material

APPENDIX 14.L
Viscosity Index Chart: 0–100 VI

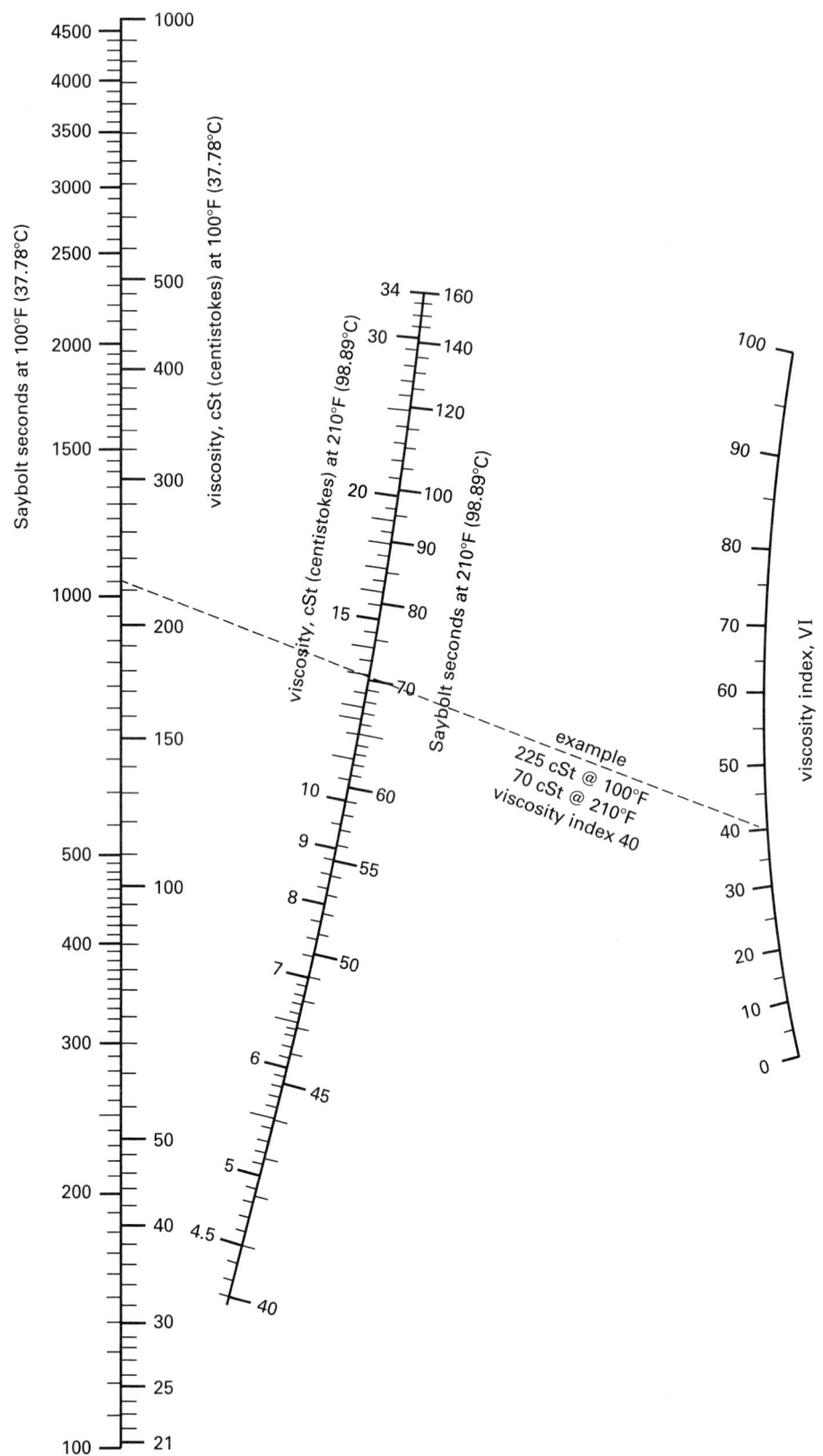

Based on correlations presented in ASTM D2270.

APPENDIX 16.A
Area, Wetted Perimeter, and Hydraulic Radius of Partially Filled Circular Pipes

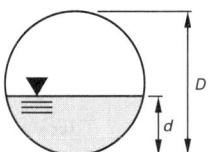

$\frac{d}{D}$	$\frac{area}{D^2}$	$\frac{\text{wetted perimeter}}{D}$	$\frac{r_h}{D}$	$\frac{d}{D}$	$\frac{area}{D^2}$	$\frac{\text{wetted perimeter}}{D}$	$\frac{r_h}{D}$
0.01	0.0013	0.2003	0.0066	0.51	0.4027	1.5908	0.2531
0.02	0.0037	0.2838	0.0132	0.52	0.4127	1.6108	0.2561
0.03	0.0069	0.3482	0.0197	0.53	0.4227	1.6308	0.2591
0.04	0.0105	0.4027	0.0262	0.54	0.4327	1.6509	0.2620
0.05	0.0147	0.4510	0.0326	0.55	0.4426	1.6710	0.2649
0.06	0.0192	0.4949	0.0389	0.56	0.4526	1.6911	0.2676
0.07	0.0242	0.5355	0.0451	0.57	0.4625	1.7113	0.2703
0.08	0.0294	0.5735	0.0513	0.58	0.4723	1.7315	0.2728
0.09	0.0350	0.6094	0.0574	0.59	0.4822	1.7518	0.2753
0.10	0.0409	0.6435	0.0635	0.60	0.4920	1.7722	0.2776
0.11	0.0470	0.6761	0.0695	0.61	0.5018	1.7926	0.2797
0.12	0.0534	0.7075	0.0754	0.62	0.5115	1.8132	0.2818
0.13	0.0600	0.7377	0.0813	0.63	0.5212	1.8338	0.2839
0.14	0.0688	0.7670	0.0871	0.64	0.5308	1.8546	0.2860
0.15	0.0739	0.7954	0.0929	0.65	0.5404	1.8755	0.2881
0.16	0.0811	0.8230	0.0986	0.66	0.5499	1.8965	0.2899
0.17	0.0885	0.8500	0.1042	0.67	0.5594	1.9177	0.2917
0.18	0.0961	0.8763	0.1097	0.68	0.5687	1.9391	0.2935
0.19	0.1039	0.9020	0.1152	0.69	0.5780	1.9606	0.2950
0.20	0.1118	0.9273	0.1206	0.70	0.5872	1.9823	0.2962
0.21	0.1199	0.9521	0.1259	0.71	0.5964	2.0042	0.2973
0.22	0.1281	0.9764	0.1312	0.72	0.6054	2.0264	0.2984
0.23	0.1365	1.0003	0.1364	0.73	0.6143	2.0488	0.2995
0.24	0.1449	1.0239	0.1416	0.74	0.6231	2.0714	0.3006
0.25	0.1535	1.0472	0.1466	0.75	0.6318	2.0944	0.3017
0.26	0.1623	1.0701	0.1516	0.76	0.6404	2.1176	0.3025
0.27	0.1711	1.0928	0.1566	0.77	0.6489	2.1412	0.3032
0.28	0.1800	1.1152	0.1614	0.78	0.6573	2.1652	0.3037
0.29	0.1890	1.1373	0.1662	0.79	0.6655	2.1895	0.3040
0.30	0.1982	1.1593	0.1709	0.80	0.6736	2.2143	0.3042
0.31	0.2074	1.1810	0.1755	0.81	0.6815	2.2395	0.3044
0.32	0.2167	1.2025	0.1801	0.82	0.6893	2.2653	0.3043
0.33	0.2260	1.2239	0.1848	0.83	0.6969	2.2916	0.3041
0.34	0.2355	1.2451	0.1891	0.84	0.7043	2.3186	0.3038
0.35	0.2450	1.2661	0.1935	0.85	0.7115	2.3462	0.3033
0.36	0.2546	1.2870	0.1978	0.86	0.7186	2.3746	0.3026
0.37	0.2642	1.3078	0.2020	0.87	0.7254	2.4038	0.3017
0.38	0.2739	1.3284	0.2061	0.88	0.7320	2.4341	0.3008
0.39	0.2836	1.3490	0.2102	0.89	0.7384	2.4655	0.2995
0.40	0.2934	1.3694	0.2142	0.90	0.7445	2.4981	0.2980
0.41	0.3032	1.3898	0.2181	0.91	0.7504	2.5322	0.2963
0.42	0.3130	1.4101	0.2220	0.92	0.7560	2.5681	0.2944
0.43	0.3229	1.4303	0.2257	0.93	0.7612	2.6061	0.2922
0.44	0.3328	1.4505	0.2294	0.94	0.7662	2.6467	0.2896
0.45	0.3428	1.4706	0.2331	0.95	0.7707	2.6906	0.2864
0.46	0.3527	1.4907	0.2366	0.96	0.7749	2.7389	0.2830
0.47	0.3627	1.5108	0.2400	0.97	0.7785	2.7934	0.2787
0.48	0.3727	1.5308	0.2434	0.98	0.7816	2.8578	0.2735
0.49	0.3827	1.5508	0.2467	0.99	0.7841	2.9412	0.2665
0.50	0.3927	1.5708	0.2500	1.00	0.7854	3.1416	0.2500

APPENDIX 16.B
Dimensions of Welded and Seamless Steel Pipe[a,b]
(selected sizes)[c]
(customary U.S. units)

nominal diameter (in)	schedule	outside diameter (in)	wall thickness (in)	internal diameter (in)	internal area (in^2)	internal diameter (ft)	internal area (ft^2)
$\frac{1}{8}$	40 (S)	0.405	0.068	0.269	0.0568	0.0224	0.00039
	80 (X)		0.095	0.215	0.0363	0.0179	0.00025
$\frac{1}{4}$	40 (S)	0.540	0.088	0.364	0.1041	0.0303	0.00072
	80 (X)		0.119	0.302	0.0716	0.0252	0.00050
$\frac{3}{8}$	40 (S)	0.675	0.091	0.493	0.1909	0.0411	0.00133
	80 (X)		0.126	0.423	0.1405	0.0353	0.00098
$\frac{1}{2}$	40 (S)	0.840	0.109	0.622	0.3039	0.0518	0.00211
	80 (X)		0.147	0.546	0.2341	0.0455	0.00163
	160		0.187	0.466	0.1706	0.0388	0.00118
	(XX)		0.294	0.252	0.0499	0.0210	0.00035
$\frac{3}{4}$	40 (S)	1.050	0.113	0.824	0.5333	0.0687	0.00370
	80 (X)		0.154	0.742	0.4324	0.0618	0.00300
	160		0.218	0.614	0.2961	0.0512	0.00206
	(XX)		0.308	0.434	0.1479	0.0362	0.00103
1	40 (S)	1.315	0.133	1.049	0.8643	0.0874	0.00600
	80 (X)		0.179	0.957	0.7193	0.0798	0.00500
	160		0.250	0.815	0.5217	0.0679	0.00362
	(XX)		0.358	0.599	0.2818	0.0499	0.00196
$1\frac{1}{4}$	40 (S)	1.660	0.140	1.380	1.496	0.1150	0.01039
	80 (X)		0.191	1.278	1.283	0.1065	0.00890
	160		0.250	1.160	1.057	0.0967	0.00734
	(XX)		0.382	0.896	0.6305	0.0747	0.00438
$1\frac{1}{2}$	40 (S)	1.900	0.145	1.610	2.036	0.1342	0.01414
	80 (X)		0.200	1.500	1.767	0.1250	0.01227
	160		0.281	1.338	1.406	0.1115	0.00976
	(XX)		0.400	1.100	0.9503	0.0917	0.00660
2	40 (S)	2.375	0.154	2.067	3.356	0.1723	0.02330
	80 (X)		0.218	1.939	2.953	0.1616	0.02051
	160		0.343	1.689	2.240	0.1408	0.01556
	(XX)		0.436	1.503	1.774	0.1253	0.01232
$2\frac{1}{2}$	40 (S)	2.875	0.203	2.469	4.788	0.2058	0.03325
	80 (X)		0.276	2.323	4.238	0.1936	0.02943
	160		0.375	2.125	3.547	0.1771	0.02463
	(XX)		0.552	1.771	2.464	0.1476	0.01711
3	40 (S)	3.500	0.216	3.068	7.393	0.2557	0.05134
	80 (X)		0.300	2.900	6.605	0.2417	0.04587
	160		0.437	2.626	5.416	0.2188	0.03761
	(XX)		0.600	2.300	4.155	0.1917	0.02885

(continued)

APPENDIX 16.B *(continued)*
Dimensions of Welded and Seamless Steel Pipe[a,b]
(selected sizes)[c]
(customary U.S. units)

nominal diameter (in)	schedule	outside diameter (in)	wall thickness (in)	internal diameter (in)	internal area (in²)	internal diameter (ft)	internal area (ft²)
$3\frac{1}{2}$	40 (S)	4.000	0.226	3.548	9.887	0.2957	0.06866
	80 (X)		0.318	3.364	8.888	0.2803	0.06172
	(XX)		0.636	2.728	5.845	0.2273	0.04059
4	40 (S)	4.500	0.237	4.026	12.73	0.3355	0.08841
	80 (X)		0.337	3.826	11.50	0.3188	0.07984
	120		0.437	3.626	10.33	0.3022	0.07171
	160		0.531	3.438	9.283	0.2865	0.06447
	(XX)		0.674	3.152	7.803	0.2627	0.05419
5	40 (S)	5.563	0.258	5.047	20.01	0.4206	0.1389
	80 (X)		0.375	4.813	18.19	0.4011	0.1263
	120		0.500	4.563	16.35	0.3803	0.1136
	160		0.625	4.313	14.61	0.3594	0.1015
	(XX)		0.750	4.063	12.97	0.3386	0.09004
6	40 (S)	6.625	0.280	6.065	28.89	0.5054	0.2006
	80 (X)		0.432	5.761	26.07	0.4801	0.1810
	120		0.562	5.501	23.77	0.4584	0.1650
	160		0.718	5.189	21.15	0.4324	0.1469
	(XX)		0.864	4.897	18.83	0.4081	0.1308
8	20	8.625	0.250	8.125	51.85	0.6771	0.3601
	30		0.277	8.071	51.16	0.6726	0.3553
	40 (S)		0.322	7.981	50.03	0.6651	0.3474
	60		0.406	7.813	47.94	0.6511	0.3329
	80 (X)		0.500	7.625	45.66	0.6354	0.3171
	100		0.593	7.439	43.46	0.6199	0.3018
	120		0.718	7.189	40.59	0.5990	0.2819
	140		0.812	7.001	38.50	0.5834	0.2673
	(XX)		0.875	6.875	37.12	0.5729	0.2578
	160		0.906	6.813	36.46	0.5678	0.2532
10	20	10.75	0.250	10.250	82.52	0.85417	0.5730
	30		0.307	10.136	80.69	0.84467	0.5604
	40 (S)		0.365	10.020	78.85	0.83500	0.5476
	60 (X)		0.500	9.750	74.66	0.8125	0.5185
	80		0.593	9.564	71.84	0.7970	0.4989
	100		0.718	9.314	68.13	0.7762	0.4732
	120		0.843	9.064	64.53	0.7553	0.4481
	140 (XX)		1.000	8.750	60.13	0.7292	0.4176
	160		1.125	8.500	56.75	0.7083	0.3941
12	20	12.75	0.250	12.250	117.86	1.0208	0.8185
	30		0.330	12.090	114.80	1.0075	0.7972
	(S)		0.375	12.000	113.10	1.0000	0.7854
	40		0.406	11.938	111.93	0.99483	0.7773
	(X)		0.500	11.750	108.43	0.97917	0.7530
	60		0.562	11.626	106.16	0.96883	0.7372
	80		0.687	11.376	101.64	0.94800	0.7058
	100		0.843	11.064	96.14	0.92200	0.6677

(continued)

APPENDIX 16.B *(continued)*
Dimensions of Welded and Seamless Steel Pipe[a,b]
(selected sizes)[c]
(customary U.S. units)

nominal diameter (in)	schedule	outside diameter (in)	wall thickness (in)	internal diameter (in)	internal area (in^2)	internal diameter (ft)	internal area (ft^2)
12	120 (XX)		1.000	10.750	90.76	0.89583	0.6303
(continued)	140		1.125	10.500	86.59	0.87500	0.6013
	160		1.312	10.126	80.53	0.84383	0.5592
14 O.D.	10	14.00	0.250	13.500	143.14	1.1250	0.9940
	20		0.312	13.376	140.52	1.1147	0.9758
	30 (S)		0.375	13.250	137.89	1.1042	0.9575
	40		0.437	13.126	135.32	1.0938	0.9397
	(X)		0.500	13.000	132.67	1.0833	0.9213
	60		0.593	12.814	128.96	1.0679	0.8956
	80		0.750	12.500	122.72	1.0417	0.8522
	100		0.937	12.126	115.48	1.0105	0.8020
	120		1.093	11.814	109.62	0.98450	0.7612
	140		1.250	11.500	103.87	0.95833	0.7213
	160		1.406	11.188	98.31	0.93233	0.6827
16 O.D.	10	16.00	0.250	15.500	188.69	1.2917	1.3104
	20		0.312	15.376	185.69	1.2813	1.2895
	30 (S)		0.375	15.250	182.65	1.2708	1.2684
	40 (X)		0.500	15.000	176.72	1.2500	1.2272
	60		0.656	14.688	169.44	1.2240	1.1767
	80		0.843	14.314	160.92	1.1928	1.1175
	100		1.031	13.938	152.58	1.1615	1.0596
	120		1.218	13.564	144.50	1.1303	1.0035
	140		1.437	13.126	135.32	1.0938	0.9397
	160		1.593	12.814	128.96	1.0678	0.8956
18 O.D.	10	18.00	0.250	17.500	240.53	1.4583	1.6703
	20		0.312	17.376	237.13	1.4480	1.6467
	(S)		0.375	17.250	233.71	1.4375	1.6230
	30		0.437	17.126	230.36	1.4272	1.5997
	(X)		0.500	17.000	226.98	1.4167	1.5762
	40		0.562	16.876	223.68	1.4063	1.5533
	60		0.750	16.500	213.83	1.3750	1.4849
	80		0.937	16.126	204.24	1.3438	1.4183
	100		1.156	15.688	193.30	1.3073	1.3423
	120		1.375	15.250	182.65	1.2708	1.2684
	140		1.562	14.876	173.81	1.2397	1.2070
	160		1.781	14.438	163.72	1.2032	1.1370
20 O.D.	10	20.00	0.250	19.500	298.65	1.6250	2.0739
	20 (S)		0.375	19.250	291.04	1.6042	2.0211
	30 (X)		0.500	19.000	283.53	1.5833	1.9689
	40		0.593	18.814	278.00	1.5678	1.9306
	60		0.812	18.376	265.21	1.5313	1.8417
	80		1.031	17.938	252.72	1.4948	1.7550
	100		1.281	17.438	238.83	1.4532	1.6585
	120		1.500	17.000	226.98	1.4167	1.5762
	140		1.750	16.500	213.83	1.3750	1.4849
	160		1.968	16.064	202.67	1.3387	1.4075

(continued)

APPENDIX 16.B *(continued)*
Dimensions of Welded and Seamless Steel Pipe[a,b]
(selected sizes)[c]
(customary U.S. units)

nominal diameter (in)	schedule	outside diameter (in)	wall thickness (in)	internal diameter (in)	internal area (in^2)	internal diameter (ft)	internal area (ft^2)
24 O.D.	10	24.00	0.250	23.500	433.74	1.9583	3.0121
	20 (S)		0.375	23.250	424.56	1.9375	2.9483
	(X)		0.500	23.000	415.48	1.9167	2.8852
	30		0.562	22.876	411.01	1.9063	2.8542
	40		0.687	22.626	402.07	1.8855	2.7922
	60		0.968	22.060	382.20	1.8383	2.6542
	80		1.218	21.564	365.21	1.7970	2.5362
	100		1.531	20.938	344.32	1.7448	2.3911
	120		1.812	20.376	326.92	1.6980	2.2645
	140		2.062	19.876	310.28	1.6563	2.1547
	160		2.343	19.310	292.87	1.6092	2.0337
30 O.D.	10	30.00	0.312	29.376	677.76	2.4480	4.7067
	(S)		0.375	29.250	671.62	2.4375	4.6640
	20 (X)		0.500	29.000	660.52	2.4167	4.5869
	30		0.625	28.750	649.18	2.3958	4.5082

(Multiply in by 25.4 to obtain mm.)
(Multiply in^2 by 645 to obtain mm^2.)
[a]Designations are per ANSI B36.10.
[b]The "S" wall thickness was formerly designated as "standard weight." Standard weight and schedule-40 are the same for all diameters through 10 in. For diameters between 12 in and 24 in, standard weight pipe has a wall thickness of 0.375 in. The "X" (or "XS") wall thickness was formerly designated as "extra strong." Extra strong weight and schedule-80 are the same for all diameters through 8 in. For diameters between 10 in and 24 in, extra strong weight pipe has a wall thickness of 0.50 in. The "XX" (or "XXS") wall thickness was formerly designed as "double extra strong." Double extra strong weight pipe does not have a corresponding schedule number.
[c]Pipe sizes and weights in most common usage are listed. Other weights and sizes exist.

APPENDIX 16.C
Dimensions of Welded and Seamless Steel Pipe
(SI units)

nominal diameter (in)	(mm)	outside diameter (mm)	schedule		wall thickness (mm)	internal diameter (mm)	internal area (cm²)
$\frac{1}{8}$	6	10.3	10S		1.245	7.811	0.479
			(S)	40	1.727	6.846	0.368
			(X)	80	2.413	5.474	0.235
$\frac{1}{4}$	8	13.7	10S		1.651	10.398	0.846
			(S)	40	2.235	9.23	0.669
			(X)	80	3.023	7.654	0.460
$\frac{3}{8}$	10	17.145	10S		1.651	13.843	1.505
			(S)	40	2.311	12.523	1.232
			(X)	80	3.2	10.745	0.907
$\frac{1}{2}$	15	21.336	5S		1.651	18.034	2.554
			10S		2.108	17.12	2.302
			(S)	40	2.769	15.798	1.960
			(X)	80	3.734	13.868	1.510
				160	4.75	11.836	1.100
			(XX)		7.468	6.4	0.322
$\frac{3}{4}$	20	26.67	5S		1.651	23.368	4.289
			10S		2.108	22.454	3.960
			(S)	40	2.87	20.93	3.441
			(X)	80	3.912	18.846	2.790
				160	5.537	15.596	1.910
			(XX)		7.823	11.024	0.954
1	25	33.401	5S		1.651	30.099	7.115
			10S		2.769	27.863	6.097
			(S)	40	3.378	26.645	5.576
			(X)	80	4.547	24.307	4.640
				160	6.35	20.701	3.366
			(XX)		9.093	15.215	1.818
$1\frac{1}{4}$	32	42.164	5S		1.651	38.862	11.862
			10S		2.769	36.626	10.563
			(S)	40	3.556	35.052	9.650
			(X)	80	4.851	32.462	8.276
				160	6.35	29.464	6.818
			(XX)		9.703	22.758	4.068
$1\frac{1}{2}$	40	48.26	5S		1.651	44.958	15.875
			10S		2.769	42.722	14.335
			(S)	40	3.683	40.894	13.134
			(X)	80	5.08	38.1	11.401
				160	7.137	33.986	9.072
			(XX)		10.16	27.94	6.131
					13.335	21.59	3.661
					15.875	16.51	2.141

(continued)

APPENDIX 16.C *(continued)*
Dimensions of Welded and Seamless Steel Pipe
(SI units)

nominal diameter (in)	(mm)	outside diameter (mm)	schedule		wall thickness (mm)	internal diameter (mm)	internal area (cm²)
2	50	60.325	5S		1.651	57.023	25.538
			10S		2.769	54.787	23.575
			(S)	40	3.912	52.501	21.648
			(X)	80	5.537	49.251	19.051
				160	8.712	42.901	14.455
			(XX)		11.074	38.177	11.447
					14.275	31.775	7.930
					17.45	25.425	5.077
2½	65	73.025	5S		2.108	68.809	37.186
			10S		3.048	66.929	35.182
			(S)	40	5.156	62.713	30.889
			(X)	80	7.01	59.005	27.344
				160	9.525	53.975	22.881
			(XX)		14.021	44.983	15.892
					17.145	38.735	11.784
					20.32	32.385	8.237
3	80	88.9	5S		2.108	84.684	56.324
			10S		3.048	82.804	53.851
			(S)	40	5.486	77.928	47.696
			(X)	80	7.62	73.66	42.614
				160	11.1	66.7	34.942
			(XX)		15.24	58.42	26.805
					18.415	52.07	21.294
					21.59	45.72	16.417
3½	90	101.6	5S		2.108	97.384	74.485
			10S	40	3.048	95.504	71.636
			(S)	80	5.74	90.12	63.787
			(X)		8.077	85.446	57.342
			(XX)		16.154	69.292	37.710
4	100	114.3	5S		2.108	110.084	95.179
			10S		3.048	108.204	91.955
					4.775	104.75	86.179
			(S)	40	6.02	102.26	82.130
			(X)	80	8.56	97.18	74.173
				120	11.1	92.1	66.621
					12.7	88.9	62.072
				160	13.487	87.326	59.893
			(XX)		17.12	80.06	50.341
					20.32	73.66	42.614
					23.495	67.31	35.584
5	125	141.3	5S		2.769	135.762	144.76
			10S		3.404	134.492	142.06
			(S)	40	6.553	128.194	129.07
			(X)	80	9.525	122.25	117.38
				120	12.7	115.9	105.50
				160	15.875	109.55	94.254
			(XX)		19.05	103.2	83.647
					22.225	96.85	73.670
					25.4	90.5	64.326

(continued)

APPENDIX 16.C *(continued)*
Dimensions of Welded and Seamless Steel Pipe
(SI units)

nominal diameter (in)	(mm)	outside diameter (mm)	schedule		wall thickness (mm)	internal diameter (mm)	internal area (cm²)
			5S		2.769	162.737	208.00
			10S		3.404	161.467	204.77
					5.563	157.149	193.96
			(S)	40	7.112	154.051	186.39
			(X)	80	10.973	146.329	168.17
6	150	168.275		120	14.275	139.725	153.33
				160	18.237	131.801	136.44
			(XX)		21.946	124.383	121.51
					25.4	117.475	108.39
					28.575	111.125	96.987
			5S		2.769	213.537	358.13
			10S		3.759	211.557	351.52
					5.563	207.949	339.63
				20	6.35	206.375	334.51
				30	7.036	205.003	330.07
			(S)	40	8.179	202.717	322.75
				60	10.312	198.451	309.31
8	200	219.075	(X)	80	12.7	193.675	294.60
				100	15.062	188.951	280.41
				120	18.237	182.601	261.88
				140	20.625	177.825	248.36
				160	23.012	173.051	235.20
					25.4	168.275	222.40
					28.575	161.925	205.93

(Multiply in by 25.4 to obtain mm.)
(Multiply lbf/in² by 6.895 to obtain kPa.)

APPENDIX 16.D
Dimensions of Rigid PVC and CPVC Pipe
(customary U.S. units)

nominal diameter (in)	outside diameter (in)	schedule-40 ASTM D1785			schedule-80 ASTM D1785			class 200 ASTM D2241		
		internal diameter (in)	wall thickness (in)	pressure rating* (lbf/in²)	internal diameter (in)	wall thickness (in)	pressure rating* (lbf/in²)	internal diameter (in)	wall thickness (in)	pressure rating* (lbf/in²)
$\frac{1}{8}$	0.405	0.249	0.068	810	0.195	0.095	1230	–	–	–
$\frac{1}{4}$	0.540	0.344	0.088	780	0.282	0.119	1130	–	–	–
$\frac{3}{8}$	0.675	0.473	0.091	620	0.403	0.126	920	–	–	–
$\frac{1}{2}$	0.840	0.622	0.109	600	0.546	0.147	850	0.716	0.062	200
$\frac{3}{4}$	1.050	0.824	0.113	480	0.742	0.154	690	0.930	0.060	200
1	1.315	1.049	0.133	450	0.957	0.179	630	1.189	0.063	200
$1\frac{1}{4}$	1.660	1.380	0.140	370	1.278	0.191	520	1.502	0.079	200
$1\frac{1}{2}$	1.900	1.610	0.145	330	1.500	0.200	470	1.720	0.090	200
2	2.375	2.067	0.154	280	1.939	0.218	400	2.149	0.113	200
$2\frac{1}{2}$	2.875	2.469	0.203	300	2.323	0.276	420	2.601	0.137	200
3	3.500	3.068	0.216	260	2.900	0.300	370	3.166	0.167	200
4	4.500	4.026	0.237	220	3.826	0.337	320	4.072	0.214	200
5	5.563	5.047	0.258	190	4.768	0.375	290	–	–	–
6	6.625	6.065	0.280	180	5.761	0.432	280	5.993	0.316	200
8	8.625	7.961	0.332	160	7.565	0.500	250	7.740	0.410	200
10	10.750	9.976	0.365	140	9.492	0.593	230	9.650	0.511	200
12	12.750	11.890	0.406	130	11.294	0.687	230	11.450	0.606	200
14	14.000	13.073	0.447	130	12.410	0.750	220	–	–	–
16	16.000	14.940	0.500	130	14.213	0.843	220	–	–	–
18	18.000	16.809	0.562	130	16.014	0.937	220	–	–	–
20	20.000	18.743	0.5937	130	17.814	1.031	220	–	–	–
24	24.000	22.554	0.687	120	21.418	1.218	210	–	–	–

(Multiply in by 25.4 to obtain mm.)
(Multiply lbf/in² by 6.895 to obtain kPa.)
*Pressure ratings are for a pipe temperature of 68°F (20°C) and are subject to the following temperature derating factors. Operation above 140°F (60°C) is not permitted.

pipe operating temperature °F (°C)	73 (23)	80 (27)	90 (32)	100 (38)	110 (43)	120 (49)	130 (54)	140 (60)
derating factor	1.0	0.88	0.75	0.62	0.51	0.40	0.31	0.22

APPENDIX 16.E
Dimensions of Large Diameter, Nonpressure, PVC Sewer and Water Pipe
(customary U.S. units)

nominal size (in)	nominal size (mm)	designations and dimensional controls	minimum wall thickness (in)	outside diameter (inside diameter for profile wall pipes) (in)
ASTM F679 – Gravity Sewer Pipe (solid wall)				
18		PS-46, T-2	0.499	
		PS-46, T-1	0.536	
		PS-115, T-2	0.671	18.701
		PS-115, T-1	0.720	
21		PS-46, T-2	0.588	
		PS-46, T-1	0.632	
		PS-115, T-2	0.791	22.047
		PS-115, T-1	0.849	
24		PS-46, T-2	0.661	
		PS-46, T-1	0.709	
		PS-115, T-2	0.889	24.803
		PS-115, T-1	0.954	
27		PS-46, T-2	0.745	
		PS-46, T-1	0.745	
		PS-115, T-2	1.002	27.953
		PS-115, T-1	1.075	
30		PS-46, T-2	0.853	
		PS-46, T-1	0.914	
		PS-115, T-2	1.148	32.00
		PS-115, T-1	1.231	
36		PS-46, T-2	1.021	
		PS-46, T-1	1.094	
		PS-115, T-2	1.373	38.30
		PS-115, T-1	1.473	
42		PS-46, T-2	1.186	
		PS-46, T-1	1.271	
		PS-115, T-2	1.596	44.50
		PS-115, T-1	1.712	
48		PS-46, T-2	1.354	
		PS-46, T-1	1.451	
		PS-115, T-2	1.822	50.80
		PS-115, T-1	1.954	
ASTM F758 – PVC Highway Underdrain Pipe (perforated)				
4			0.120	4.215
6			0.180	6.275
8			0.240	8.40

ASTM F789 – PVC Sewage Pipe (solid wall)

available in sizes 4–18 in (obsolete)

available in two stiffnesses and three material grades:
PS-46; T-1, T-2, T-3
PS-115; T-1, T-2, T-3

(continued)

APPENDIX 16.E *(continued)*
Dimensions of Large Diameter, Nonpressure, PVC Sewer and Water Pipe
(customary U.S. units)

nominal size (in)	(mm)	designations and dimensional controls	minimum wall thickness (in)	outside diameter (inside diameter for profile wall pipes) (in)

ASTM F794 – Open Profile Wall Pipe

available in sizes 4–18 in   PS-46   internal diameter controlled

AWWA C900 – Water and Wastewater Pressure Pipe (solid wall)

nominal size	designations	min wall	OD
4	DR-25 (PC-100,		4.80
6	PR-165, CL100);		6.90
8	DR-18 (PC-150,	$t_{min} = D_o/DR$	9.05
10	PR-235, CL150);		11.10
12	DR-14 (PC-200, PC-305, CL200)		13.20

AWWA C905 – Water and Wastewater Pipe (solid wall)

nominal size	designations	min wall	OD
14	DR-51 (PR-80);		15.30
16	DR-41 (PR-100);		17.40
18	DR-32.5 (PR-125);		19.50
20	DR-25 (PR-165);	$t_{min} = D_o/DR$	21.60
24	DR-21 (PR-200);		25.8
30	DR-18 (PR-235);		32.0
36	DR-14 (PR-305)		38.3

ASTM F949 – Open Profile Wall Pipe

available in sizes 4–36 in   PS-46   internal diameter controlled

ASTM D1785
See App. 16.D.

ASTM F1803 – Closed Profile Wall "Truss" Pipe (DWCP)

available in sizes 18–60 in   PS-46   internal diameter controlled

(continued)

APPENDIX 16.E *(continued)*
Dimensions of Large Diameter, Nonpressure, PVC Sewer and Water Pipe
(customary U.S. units)

nominal size (in)	(mm)	designations and dimensional controls	minimum wall thickness (in)	outside diameter (inside diameter for profile wall pipes) (in)

ASTM D2241 – PVC Sewer Pipe (solid wall)

nominal size (in)	designations and dimensional controls	minimum wall thickness (in)	outside diameter (in)
1			1.315
$1\frac{1}{4}$			1.660
$1\frac{1}{2}$			1.900
2			2.375
$2\frac{1}{2}$	SDR-41		2.875
3	SDR-32.5		3.500
	SDR-26	$t_{\min} = D_o/\text{SDR}$	
$3\frac{1}{2}$	SDR-21		4.000
4	SDR-17		4.500
	SDR-13.5		
6			6.625
8			8.625
10			10.750
12			12.750

ASTM D2729 – PVC Drainage Pipe (solid wall; perforated)

nominal size (in)	minimum wall thickness (in)	outside diameter (in)
3	0.070	3.250
4	0.075	4.215
6	0.100	6.275

ASTM D3034 – PVC Gravity Sewer Pipe and Perforated Drain Pipe (solid wall)

(in)	(mm)	designations and dimensional controls	minimum wall thickness (in)	outside diameter (in)
4	100	SDR-35 & PS-46 (regular);		4.215
5	135	SDR-26 & PS-115 (HW);		5.640
6	150	SDR-23.5 & PS-135		6.275
8	200	SDR-41 & PS-28;	$t_{\min} = D_o/\text{SDR}$	8.400
10	250	SDR-35 & PS-46 (regular);		10.50
12	300	SDR-26 & PS-115		12.50
15	375	(HW)		15.30

abbreviations

CL – pressure class (same as PC)
DR – dimensional ratio (constant ratio of outside diameter/wall thickness)
DWCP – double wall corrugated pipe
HW – heavy weight
PC – pressure class (same as CL)
PS – pipe stiffness (resistance to vertical diametral deflection due to compression from burial soil and surface loading, in lbf/in^2; calculated uncorrected as vertical load in lbf per inch of pipe length divided by deflection in inches)
SDR – standard dimension ratio (constant ratio of outside diameter/wall thickness)

(Multiply in by 25.4 to obtain mm.)
(Multiply lbf/in^2 by 6.895 to obtain kPa.)

APPENDIX 16.F
Dimensions and Weights of Concrete Sewer Pipe
(customary U.S. units)

ASTM C14 – Nonreinforced Round Sewer and Culvert Pipe for Nonpressure (Gravity Flow) Applications

nominal size and internal diameter		minimum wall thickness and required ASTM C14 $D_{0.01}$-load rating[a] (in and lbf/ft)		
(in)	(mm)	class 1[b]	class 2[b]	class 3[b]
4	100	0.625 (1500)	0.75 (2000)	0.875 (2400)
6	150	0.625 (1500)	0.75 (2000)	1.0 (2400)
8	200	0.75 (1500)	0.875 (2000)	1.125 (2400)
10	250	0.875 (1600)	1.0 (2000)	1.25 (2400)
12	300	1.0 (1800)	1.375 (2250)	1.75 (2500)
15	375	1.25 (2000)	1.625 (2500)	1.875 (2900)
18	450	1.5 (2200)	2.0 (3000)	2.25 (3300)
21	525	1.75 (2400)	2.25 (3300)	2.75 (3850)
24	600	2.125 (2600)	3.0 (3600)	3.375 (4400)
27	675	3.25 (2800)	3.75 (3950)	3.75 (4600)
30	750	3.5 (3000)	4.25 (4300)	4.25 (4750)
33	825	3.75 (3150)	4.5 (4400)	4.5 (4850)
36	900	4.0 (3300)	4.75 (4500)	4.75 (5000)

(Multiply in by 25.4 to obtain mm.)
(Multiply lbf/ft by 14.594 to obtain kN/m.)
[a]For C14 pipe, D-loads are stated in lbf/ft (pound per foot length of pipe).
[b]As defined by the strength (D-load) requirements of ASTM C14.

ASTM C76 – Reinforced Round Concrete Culvert, Storm Drain, and Sewer Pipe for Nonpressure (Gravity Flow) Applications

nominal size and internal diameter		minimum wall thickness and maximum ASTM C76 $D_{0.01}$-load rating (in and lbf/ft^2)		
(in)	(mm)	wall A[a]	wall B[a]	wall C[a]
8[b]	200		2.0	
10[b]	250		2.0	
12	300	1.75 (2000)	2.0 (3000)	2.75 (3000)
15	375	1.875 (2000)	2.25 (3000)	3.0 (3000)
18	450	2.0 (2000)	2.5 (3000)	3.25 (3000)
21	525	2.25 (2000)	2.75 (3000)	3.50 (3000)
24	600	2.5 (2000)	3.0 (3000)	3.75 (3000)
27	675	2.625 (2000)	3.25 (3000)	4.0 (3000)
30	750	2.75 (2000)	3.5 (3000)	4.25 (3000)
33	825	2.875 (1350)	3.75 (3000)	4.5 (3000)
36	900	3.0 (1350)	4.0 (3000)	4.75 (3000)
42	1050	3.5 (1350)	4.5 (3000)	5.25 (3000)
48	1200	4.0 (1350)	5.0 (3000)	5.75 (3000)
54	1350	4.5 (1350)	5.5 (2000)	6.25 (3000)
60	1500	5.0 (1350)	6.0 (2000)	6.75 (3000)
66	1650	5.5 (1350)	6.5 (2000)	7.25 (3000)
72	1800	6.0 (1350)	7.0 (2000)	7.75 (3000)
78	1950	6.5 (1350)	7.5 (1350)	8.25 (2000)
84	2100	7.0 (1350)	8.0 (1350)	8.75 (1350)
90	2250	7.5 (1350)	8.5 (1350)	9.25 (1350)
96	2400	8.0 (1350)	9.0 (1350)	9.75 (1350)
102	2550	8.5 (1350)	9.5 (1350)	10.25 (1350)
108	2700	9.0 (1350)	10.0 (1350)	10.75 (1350)

(Multiply in by 25.4 to obtain mm.)
(Multiply lbf/ft^2 by 0.04788 to obtain kN/m^2 (kPa).)
[a]wall A thickness in inches = diameter in feet; wall B thickness in inches = diameter in feet + 1 in; wall C thickness in inches = diameter in feet + 1.75 in
[b]Although not specifically called out in ASTM C76, 8 in and 10 in diameter circular concrete pipes are routinely manufactured.

(continued)

APPENDIX 16.F *(continued)*
Dimensions and Weights of Concrete Sewer Pipe
(customary U.S. units)

ASTM C76 – Large Sizes of Pipe for Nonpressure (Gravity Flow) Applications

nominal size and internal diameter		minimum wall thickness (in)
(in)	(mm)	wall A[*]
114	2850	9.5
120	3000	10.0
126	3150	10.5
132	3300	11.0
138	3450	11.5
144	3600	12.0
150	3750	12.5
156	3900	13.0
162	4050	13.5
168	4200	14.0
174	4350	14.5
180	4500	15.0

(Multiply in by 25.4 to obtain mm.)
[*]wall A thickness in inches = diameter in feet

ASTM C361 – RCPP: Reinforced Concrete Low Head Pressure Pipe

"Low head" means 125 ft (54 psi; 375 kPa) or less. Standard pipe with 12 in (300 mm) to 108 in (2700 mm) diameters are available. ASTM C361 limits tensile stress (and, therefore, strain) and flexural deformation. Any reinforcement design that meets these limitations is permitted. Internal and external dimensions typically correspond to C76 (wall B) pipe, but wall C may also be used.

ASTM C655 – Reinforced Concrete D-Load Culvert, Storm Drain, and Sewer Pipe

Pipes smaller than 72 in (1800 mm) are often specified by D-load. Pipe manufactured to satisfy ASTM C655 will support a specified concentrated vertical load (applied in three-point loading) in pounds per ft of length per ft of diameter. Some standard C76 pipe meets this standard. In order to use C76 (and other) pipes, D-loads are mapped onto C76 pipe classes. In some cases, special pipe designs are required. Pipe selection by D-load is accomplished by first determining the size of pipe required to carry the flow, then choosing the appropriate class of pipe according to the required D-load.

$D_{0.01}$-load range	C76 pipe class
1–800	class 1 (I)
801–1000	class 2 (II)
1001–1350	class 3 (III)
1351–2000	class 4 (IV)
2000–3000	class 5 (V)
>3000	no equivalent class pipe

(Multiply in by 25.4 to obtain mm.)
(Multiply lbf/ft^2 by 0.04788 to obtain kN/m^2 (kPa).)

APPENDIX 16.G
Dimensions of Cast-Iron and Ductile Iron Pipe
Standard Pressure Classes
(customary U.S. units)

ANSI/AWWA C106 Gray Cast-Iron Pipe

The ANSI/AWWA C106 standard is obsolete. The outside diameters of 4–48 in (100–1200 mm) diameter gray cast-iron pipes are the same as for C150 pipes. Ductile iron pipe is interchangeable with respect to joining diameters, accessories, and fittings. However, inside diameters of cast-iron pipe are substantially greater than C150 pipe. Outside diameters and thicknesses of AWWA 1908 standard cast-iron pipe are substantially different from C150 values.

ANSI/AWWA C150/A21.50 and ANSI/AWWA C151/A21.51

		calculated minimum wall thickness[a], t (in)						
		pressure class and head (lbf/in^2 and ft)					casting	
nominal size (in)	outside diameter, D_o (in)	150 (346)	200 (462)	250 (577)	300	350	tolerance (in)	inside diameter
3	3.96					0.25[b]	0.05[c]	
4	4.80					0.25[b]	0.05[c]	
6	6.90					0.25[b]	0.05[c]	
8	9.05					0.25[b]	0.05[c]	
10	11.10					0.26	0.06	
12	13.20					0.28	0.06	
14	15.30			0.28	0.30	0.31	0.07	
16	17.40			0.30	0.32	0.34	0.07	
18	19.50			0.31	0.34	0.36	0.07	$D_i = D_o - 2t$
20	21.60			0.33	0.36	0.38	0.07	
24	25.80		0.33	0.37	0.40	0.43	0.07	
30	32.00	0.34	0.38	0.42	0.45	0.49	0.07	
36	38.30	0.38	0.42	0.47	0.51	0.56	0.07	
42	44.50	0.41	0.47	0.52	0.57	0.63	0.07	
48	50.80	0.46	0.52	0.58	0.64	0.70	0.08	
54	57.56	0.51	0.58	0.65	0.72	0.79	0.09	
60	61.61	0.54	0.61	0.68	0.76	0.83	0.09	
64	65.67	0.56	0.64	0.72	0.80	0.87	0.09	

(Multiply in by 25.4 to obtain mm.)
(Multiply lbf/in^2 by 6.895 to obtain kPa.)
[a]Per ANSI/AWWA C150/A21.50, the tabulated minimum wall thicknesses include a 0.08 in (2 mm) service allowance and the appropriate casting tolerance. Listed thicknesses are adequate for the rated water working pressure plus a surge allowance of 100 lbf/in^2 (690 kPa). Values are based on a yield strength of 42,000 lbf/in^2 (290 MPa), the sum of the working pressure and 100 lbf/in^2 (690 kPa) surge allowance, and a safety factor of 2.0.
[b]Pressure class is defined as the rated gage water pressure of the pipe in lbf/in^2.
[c]Limited by manufacturing. Calculated required thickness is less.

APPENDIX 16.H
Dimensions of Ductile Iron Pipe
Special Pressure Classes
(customary U.S. units)

ANSI/AWWA C150/A21.50 and ANSI/AWWA C151/A21.51

nominal size[a] (in)	outside diameter (in)	minimum wall thickness (in) special pressure class[b]						
		50	51	52	53	54	55	56
4	4.80	–	0.26	0.29	0.32	0.35	0.38	0.41
6	6.90	0.25	0.28	0.31	0.34	0.37	0.40	0.43
8	9.05	0.27	0.30	0.33	0.36	0.39	0.42	0.45
10	11.10	0.29	0.32	0.35	0.38	0.41	0.44	0.47
12	13.20	0.31	0.34	0.37	0.40	0.43	0.46	0.49
14	15.30	0.33	0.36	0.39	0.42	0.45	0.48	0.51
16	17.40	0.34	0.37	0.40	0.43	0.46	0.49	0.52
18	19.50	0.35	0.38	0.41	0.44	0.47	0.50	0.53
20	21.60	0.36	0.39	0.42	0.45	0.48	0.51	0.54
24	25.80	0.38	0.41	0.44	0.47	0.50	0.53	0.56
30	32.00	0.39	0.43	0.47	0.51	0.55	0.59	0.63
36	38.30	0.43	0.48	0.53	0.58	0.63	0.68	0.73
42	44.50	0.47	0.53	0.59	0.65	0.71	0.77	0.83
48	50.80	0.51	0.58	0.65	0.72	0.79	0.86	0.93
54	57.56	0.57	0.65	0.73	0.81	0.89	0.97	1.05

(Multiply in by 25.4 to obtain mm.)
(Multiply lbf/in^2 by 6.895 to obtain kPa.)
[a]Formerly designated "standard thickness classes." These special pressure classes are as shown in AWWA C150 and C151. Special classes are most appropriate for some threaded, grooved, or ball and socket pipes, or for extraordinary design conditions. They are generally less available than standard pressure class pipe.
[b]60 in (1500 mm) and 64 in (1600 mm) pipe sizes are not available in special pressure classes.

ASTM A746 Cement-Lined Gravity Sewer Pipe
The thicknesses for cement-lined pipe are the same as those in AWWA C150 and C151.

ASTM A746 Flexible Lining Gravity Sewer Pipe
The thicknesses for flexible lining pipe are the same as those in AWWA C150 and C151.

APPENDIX 16.I
Standard ASME/ANSI Z32.2.3 Piping Symbols*

	flanged	screwed	bell and spigot	welded	soldered
joint					
elbow—90°					
elbow—45°					
elbow—turned up					
elbow—turned down					
elbow—long radius					
reducing elbow					
tee					
tee—outlet up					
tee—outlet down					
side outlet tee—outlet up					
cross					
reducer—concentric					
reducer—eccentric					
lateral					
gate valve					
globe valve					
check valve					
stop cock					
safety valve					
expansion joint					
union					
sleeve					
bushing					

*Similar to MIL-STD-17B.

APPENDIX 16.J
Dimensions of Copper Water Tubing
(customary U.S. units)

classification	nominal tube size (in)	outside diameter (in)	wall thickness (in)	inside diameter (in)	transverse area (in²)	safe working pressure (psi)
hard	$1/4$	$3/8$	0.025	0.325	0.083	1000
	$3/8$	$1/2$	0.025	0.450	0.159	1000
	$1/2$	$5/8$	0.028	0.569	0.254	890
	$3/4$	$7/8$	0.032	0.811	0.516	710
	1	$1 1/8$	0.035	1.055	0.874	600
	$1 1/4$	$1 3/8$	0.042	1.291	1.309	590
type "M" 250 psi working pressure	$1 1/2$	$1 5/8$	0.049	1.527	1.831	580
	2	$2 1/8$	0.058	2.009	3.17	520
	$2 1/2$	$2 5/8$	0.065	2.495	4.89	470
	3	$3 1/8$	0.072	2.981	6.98	440
	$3 1/2$	$3 5/8$	0.083	3.459	9.40	430
	4	$4 1/8$	0.095	3.935	12.16	430
	5	$5 1/8$	0.109	4.907	18.91	400
	6	$6 1/8$	0.122	5.881	27.16	375
	8	$8 1/8$	0.170	7.785	47.6	375
hard	$3/8$	$1/2$	0.035	0.430	0.146	1000
	$1/2$	$5/8$	0.040	0.545	0.233	1000
	$3/4$	$7/8$	0.045	0.785	0.484	1000
	1	$1 1/8$	0.050	1.025	0.825	880
	$1 1/4$	$1 3/8$	0.055	1.265	1.256	780
	$1 1/2$	$1 5/8$	0.060	1.505	1.78	720
type "L" 250 psi working pressure	2	$2 1/8$	0.070	1.985	3.094	640
	$2 1/2$	$2 5/8$	0.080	2.465	4.77	580
	3	$3 1/8$	0.090	2.945	6.812	550
	$3 1/2$	$3 5/8$	0.100	3.425	9.213	530
	4	$4 1/8$	0.110	3.905	11.97	510
	5	$5 1/8$	0.125	4.875	18.67	460
	6	$6 1/8$	0.140	5.845	26.83	430
hard	$1/4$	$3/8$	0.032	0.311	0.076	1000
	$3/8$	$1/2$	0.049	0.402	0.127	1000
	$1/2$	$5/8$	0.049	0.527	0.218	1000
	$3/4$	$7/8$	0.065	0.745	0.436	1000
	1	$1 1/8$	0.065	0.995	0.778	780
	$1 1/4$	$1 3/8$	0.065	1.245	1.217	630
type "K" 400 psi working pressure	$1 1/2$	$1 5/8$	0.072	1.481	1.722	580
	2	$2 1/8$	0.083	1.959	3.014	510
	$2 1/2$	$2 5/8$	0.095	2.435	4.656	470
	3	$3 1/8$	0.109	2.907	6.637	450
	$3 1/2$	$3 5/8$	0.120	3.385	8.999	430
	4	$4 1/8$	0.134	3.857	11.68	420
	5	$5 1/8$	0.160	4.805	18.13	400
	6	$6 1/8$	0.192	5.741	25.88	400
soft	$1/4$	$3/8$	0.032	0.311	0.076	1000
	$3/8$	$1/2$	0.049	0.402	0.127	1000
	$1/2$	$5/8$	0.049	0.527	0.218	1000
	$3/4$	$7/8$	0.065	0.745	0.436	1000
	1	$1 1/8$	0.065	0.995	0.778	780
	$1 1/4$	$1 3/8$	0.065	1.245	1.217	630
type "K" 250 psi working pressure	$1 1/2$	$1 5/8$	0.072	1.481	1.722	580
	2	$2 1/8$	0.083	1.959	3.014	510
	$2 1/2$	$2 5/8$	0.095	2.435	4.656	470
	3	$3 1/8$	0.109	2.907	6.637	450
	$1 1/2$	$2 5/8$	0.120	3.385	8.999	430
	4	$4 1/8$	0.134	3.857	11.68	420
	5	$5 2/8$	0.160	4.805	18.13	400
	6	$6 1/8$	0.192	5.741	25.88	400

(Multiply in by 25.4 to obtain mm.)
(Multiply in² by 645 to obtain mm².)

APPENDIX 16.K
Dimensions of Brass and Copper Tubing
(customary U.S. units)

regular

pipe size (in)	nominal dimensions (in)			cross-sectional area of bore (in²)	lbm/ft	
	O.D.	I.D.	wall		red brass	copper
1/8	0.405	0.281	0.062	0.062	0.253	0.259
1/4	0.540	0.376	0.082	0.110	0.447	0.457
3/8	0.675	0.495	0.090	0.192	0.627	0.641
1/2	0.840	0.626	0.107	0.307	0.934	0.955
3/4	1.050	0.822	0.114	0.531	1.270	1.300
1	1.315	1.063	0.126	0.887	1.780	1.820
1 1/4	1.660	1.368	0.146	1.470	2.630	2.690
1 1/2	1.900	1.600	0.150	2.010	3.130	3.200
2	2.375	2.063	0.156	3.340	4.120	4.220
2 1/2	2.875	2.501	0.187	4.910	5.990	6.120
3	3.500	3.062	0.219	7.370	8.560	8.750
3 1/2	4.000	3.500	0.250	9.620	11.200	11.400
4	4.500	4.000	0.250	12.600	12.700	12.900
5	5.562	5.062	0.250	20.100	15.800	16.200
6	6.625	6.125	0.250	29.500	19.000	19.400
8	8.625	8.001	0.312	50.300	30.900	31.600
10	10.750	10.020	0.365	78.800	45.200	46.200
12	12.750	12.000	0.375	113.000	55.300	56.500

extra strong

pipe size (in)	nominal dimensions (in)			cross-sectional area of bore (in²)	lbm/ft	
	O.D.	I.D.	wall		red brass	copper
1/8	0.405	0.205	0.100	0.033	0.363	0.371
1/4	0.540	0.294	0.123	0.068	0.611	0.625
3/8	0.675	0.421	0.127	0.139	0.829	0.847
1/2	0.840	0.542	0.149	0.231	1.230	1.250
3/4	1.050	0.736	0.157	0.425	1.670	1.710
1	1.315	0.951	0.182	0.710	2.460	2.510
1 1/4	1.660	1.272	0.194	1.270	3.390	3.460
1 1/2	1.990	1.494	0.203	1.750	4.100	4.190
2	2.375	1.933	0.221	2.94	5.670	5.800
2 1/2	2.875	2.315	0.280	4.21	8.660	8.850
3	3.500	2.892	0.304	6.57	11.600	11.800
3 1/2	4.000	3.358	0.321	8.86	14.100	14.400
4	4.500	3.818	0.341	11.50	16.900	17.300
5	5.562	4.812	0.375	18.20	23.200	23.700
6	6.625	5.751	0.437	26.00	32.200	32.900
8	8.625	7.625	0.500	45.70	48.400	49.500
10	10.750	9.750	0.500	74.70	61.100	62.400

(Multiply in by 25.4 to obtain mm.)
(Multiply in² by 645 to obtain mm².)

APPENDIX 16.L
Dimensions of Seamless Steel Boiler (BWG) Tubing[a,b,c]
(customary U.S. units)

O.D. (in)	BWG	wall thickness (in)	O.D. (in)	BWG	wall thickness (in)
1	13	0.095	3	12	0.109
	12	0.109		11	0.120
	11	0.120		10	0.134
	10	0.134		9	0.148
1 1/4	13	0.095	3 1/4	11	0.120
	12	0.109		10	0.134
	11	0.120		9	0.148
	10	0.134		8	0.165
1 1/2	13	0.095	3 1/2	11	0.120
	12	0.109		10	0.134
	11	0.120		9	0.148
	10	0.134		8	0.165
1 3/4	13	0.095	4	10	0.134
	12	0.109		9	0.148
	11	0.120		8	0.165
	10	0.134		7	0.180
2	13	0.095	4 1/2	10	0.134
	12	0.109		9	0.148
	11	0.120		8	0.165
	10	0.134		7	0.180
2 1/4	13	0.095	5	9	0.148
	12	0.109		8	0.165
	11	0.120		7	0.180
	10	0.134		6	0.203
2 1/2	12	0.109	5 1/2	9	0.148
	11	0.120		8	0.165
	10	0.134		7	0.180
	9	0.148		6	0.203
2 3/4	12	0.109	6	7	0.180
	11	0.120		6	0.203
	10	0.134		5	0.220
	9	0.148		4	0.238

(Multiply in by 25.4 to obtain mm.)
(Multiply in² by 645 to obtain mm².)
[a]Abstracted from information provided by United States Steel Corporation.
[b]Values in this table are not to be used for tubes in condensers and heat-exchangers unless those tubes are specified by BWG.
[c]Birmingham wire gauge, commonly used for ferrous tubing, is identical with Stubs iron-wire gauge.

APPENDIX 17.A
Specific Roughness and Hazen-Williams Constants for Various Water Pipe Materials[a]

type of pipe or surface	ϵ (ft) range	ϵ (ft) design	C range	C clean	C design[b]
steel					
welded and seamless	0.0001–0.0003	0.0002	150–80	140	100
interior riveted, no projecting rivets				139	100
projecting girth rivets				130	100
projecting girth and horizontal rivets				115	100
vitrified, spiral-riveted, flow with lap				110	100
vitrified, spiral-riveted, flow against lap				100	90
corrugated			80–40	80	60
mineral					
concrete	0.001–0.01	0.004	150–60	120	100
cement-asbestos			160–140	150	140
vitrified clays					110
brick sewer					100
iron					
cast, plain	0.0004–0.002	0.0008	150–80	130	100
cast, tar (asphalt) coated	0.0002–0.0006	0.0004	145–50	130	100
cast, cement lined		0.00001		150	140
cast, bituminous lined		0.00001	160–130	148	140
cast, centrifugally spun	0.00001	0.00001			
ductile iron	0.0004–0.002	0.0008	100–150	150	140
cement lined		0.00001	120–150	150	140
asphalt coated	0.0002–0.0006	0.0004	145–50	130	160
galvanized, plain	0.0002–0.0008	0.0005			
wrought, plain	0.0001–0.0003	0.0002	150–80	130	100
miscellaneous					
aluminum, irrigation pipe			135–100	135	130
copper and brass	0.000005	0.000005	150–120	140	130
wood stave	0.0006–0.003	0.002	145–110	120	110
transite	0.000008	0.000008			
lead, tin, glass		0.000005	150–120	140	130
plastic (PVC, ABS, and HDPE)		0.000005	150–120	155	150
fiberglass	0.000017	0.000017	160–150	155	150

(Multiply ft by 0.3048 to obtain m.)
[a]C values for sludge pipes are 20% to 40% less than the corresponding water pipe values.
[b]The following guidelines are provided for selecting Hazen-Williams coefficients for cast-iron pipes of different ages. Values for welded steel pipe are similar to those of cast-iron pipe 5 years older. New pipe, all sizes: $C = 130$. 5 year-old pipe: $C = 120$ ($d < 24$ in); $C = 115$ ($d \geq 24$ in). 10 year-old pipe: $C = 105$ ($d = 4$ in); $C = 110$ ($d = 12$ in); $C = 85$ ($d \geq 30$ in). 40 year-old pipe: $C = 65$ ($d = 4$ in); $C = 80$ ($d = 16$ in).

APPENDIX 17.B
Darcy Friction Factors (turbulent flow)

relative roughness, ϵ/D

Reynolds no.	0.00000	0.000001	0.0000015	0.00001	0.00002	0.00004	0.00005	0.00006	0.00008
2×10^3	0.0495	0.0495	0.0495	0.0495	0.0495	0.0495	0.0495	0.0495	0.0495
2.5×10^3	0.0461	0.0461	0.0461	0.0461	0.0461	0.0461	0.0461	0.0461	0.0461
3×10^3	0.0435	0.0435	0.0435	0.0435	0.0435	0.0436	0.0436	0.0436	0.0436
4×10^3	0.0399	0.0399	0.0399	0.0399	0.0399	0.0399	0.0400	0.0400	0.0400
5×10^3	0.0374	0.0374	0.0374	0.0374	0.0374	0.0374	0.0374	0.0375	0.0375
6×10^3	0.0355	0.0355	0.0355	0.0355	0.0355	0.0356	0.0356	0.0356	0.0356
7×10^3	0.0340	0.0340	0.0340	0.0340	0.0340	0.0341	0.0341	0.0341	0.0341
8×10^3	0.0328	0.0328	0.0328	0.0328	0.0328	0.0328	0.0329	0.0329	0.0329
9×10^3	0.0318	0.0318	0.0318	0.0318	0.0318	0.0318	0.0318	0.0319	0.0319
1×10^4	0.0309	0.0309	0.0309	0.0309	0.0309	0.0309	0.0310	0.0310	0.0310
1.5×10^4	0.0278	0.0278	0.0278	0.0278	0.0278	0.0279	0.0279	0.0279	0.0280
2×10^4	0.0259	0.0259	0.0259	0.0259	0.0259	0.0260	0.0260	0.0260	0.0261
2.5×10^4	0.0245	0.0245	0.0245	0.0245	0.0246	0.0246	0.0246	0.0247	0.0247
3×10^4	0.0235	0.0235	0.0235	0.0235	0.0235	0.0236	0.0236	0.0236	0.0237
4×10^4	0.0220	0.0220	0.0220	0.0220	0.0220	0.0221	0.0221	0.0222	0.0222
5×10^4	0.0209	0.0209	0.0209	0.0209	0.0210	0.0210	0.0211	0.0211	0.0212
6×10^4	0.0201	0.0201	0.0201	0.0201	0.0201	0.0202	0.0203	0.0203	0.0204
7×10^4	0.0194	0.0194	0.0194	0.0194	0.0195	0.0196	0.0196	0.0197	0.0197
8×10^4	0.0189	0.0189	0.0189	0.0189	0.0190	0.0190	0.0191	0.0191	0.0192
9×10^4	0.0184	0.0184	0.0184	0.0184	0.0185	0.0186	0.0186	0.0187	0.0188
1×10^5	0.0180	0.0180	0.0180	0.0180	0.0181	0.0182	0.0183	0.0183	0.0184
1.5×10^5	0.0166	0.0166	0.0166	0.0166	0.0167	0.0168	0.0169	0.0170	0.0171
2×10^5	0.0156	0.0156	0.0156	0.0157	0.0158	0.0160	0.0160	0.0161	0.0163
2.5×10^5	0.0150	0.0150	0.0150	0.0151	0.0152	0.0153	0.0154	0.0155	0.0157
3×10^5	0.0145	0.0145	0.0145	0.0146	0.0147	0.0149	0.0150	0.0151	0.0153
4×10^5	0.0137	0.0137	0.0137	0.0138	0.0140	0.0142	0.0143	0.0144	0.0146
5×10^5	0.0132	0.0132	0.0132	0.0133	0.0134	0.0137	0.0138	0.0140	0.0142
6×10^5	0.0127	0.0128	0.0128	0.0129	0.0131	0.0133	0.0135	0.0136	0.0139
7×10^5	0.0124	0.0124	0.0124	0.0126	0.0127	0.0131	0.0132	0.0134	0.0136
8×10^5	0.0121	0.0121	0.0121	0.0123	0.0125	0.0128	0.0130	0.0131	0.0134
9×10^5	0.0119	0.0119	0.0119	0.0121	0.0123	0.0126	0.0128	0.0130	0.0133
1×10^6	0.0116	0.0117	0.0117	0.0119	0.0121	0.0125	0.0126	0.0128	0.0131
1.5×10^6	0.0109	0.0109	0.0109	0.0112	0.0114	0.0119	0.0121	0.0123	0.0127
2×10^6	0.0104	0.0104	0.0104	0.0107	0.0110	0.0116	0.0118	0.0120	0.0124
2.5×10^6	0.0100	0.0100	0.0101	0.0104	0.0108	0.0113	0.0116	0.0118	0.0123
3×10^6	0.0097	0.0098	0.0098	0.0102	0.0105	0.0112	0.0115	0.0117	0.0122
4×10^6	0.0093	0.0094	0.0094	0.0098	0.0103	0.0110	0.0113	0.0115	0.0120
5×10^6	0.0090	0.0091	0.0091	0.0096	0.0101	0.0108	0.0111	0.0114	0.0119
6×10^6	0.0087	0.0088	0.0089	0.0094	0.0099	0.0107	0.0110	0.0113	0.0118
7×10^6	0.0085	0.0086	0.0087	0.0093	0.0098	0.0106	0.0110	0.0113	0.0118
8×10^6	0.0084	0.0085	0.0085	0.0092	0.0097	0.0106	0.0109	0.0112	0.0118
9×10^6	0.0082	0.0083	0.0084	0.0091	0.0097	0.0105	0.0109	0.0112	0.0117
1×10^7	0.0081	0.0082	0.0083	0.0090	0.0096	0.0105	0.0109	0.0112	0.0117
1.5×10^7	0.0076	0.0078	0.0079	0.0087	0.0094	0.0104	0.0108	0.0111	0.0116
2×10^7	0.0073	0.0075	0.0076	0.0086	0.0093	0.0103	0.0107	0.0110	0.0116
2.5×10^7	0.0071	0.0073	0.0074	0.0085	0.0093	0.0103	0.0107	0.0110	0.0116
3×10^7	0.0069	0.0072	0.0073	0.0084	0.0092	0.0103	0.0107	0.0110	0.0116
4×10^7	0.0067	0.0070	0.0071	0.0084	0.0092	0.0102	0.0106	0.0110	0.0115
5×10^7	0.0065	0.0068	0.0070	0.0083	0.0092	0.0102	0.0106	0.0110	0.0115

(continued)

APPENDIX 17.B *(continued)*
Darcy Friction Factors (turbulent flow)

relative roughness, ϵ/D

Reynolds no.	0.0001	0.00015	0.00020	0.00025	0.00030	0.00035	0.0004	0.0006	0.0008
2×10^3	0.0495	0.0496	0.0496	0.0496	0.0497	0.0497	0.0498	0.0499	0.0501
2.5×10^3	0.0461	0.0462	0.0462	0.0463	0.0463	0.0463	0.0464	0.0466	0.0467
3×10^3	0.0436	0.0437	0.0437	0.0437	0.0438	0.0438	0.0439	0.0441	0.0442
4×10^3	0.0400	0.0401	0.0401	0.0402	0.0402	0.0403	0.0403	0.0405	0.0407
5×10^3	0.0375	0.0376	0.0376	0.0377	0.0377	0.0378	0.0378	0.0381	0.0383
6×10^3	0.0356	0.0357	0.0357	0.0358	0.0359	0.0359	0.0300	0.0302	0.0365
7×10^3	0.0341	0.0342	0.0343	0.0343	0.0344	0.0345	0.0345	0.0348	0.0350
8×10^3	0.0329	0.0330	0.0331	0.0331	0.0332	0.0333	0.0333	0.0336	0.0339
9×10^3	0.0319	0.0320	0.0321	0.0321	0.0322	0.0323	0.0323	0.0326	0.0329
1×10^4	0.0310	0.0311	0.0312	0.0313	0.0313	0.0314	0.0315	0.0318	0.0321
1.5×10^4	0.0280	0.0281	0.0282	0.0283	0.0284	0.0285	0.0285	0.0289	0.0293
2×10^4	0.0261	0.0262	0.0263	0.0264	0.0265	0.0266	0.0267	0.0272	0.0276
2.5×10^4	0.0248	0.0249	0.0250	0.0251	0.0252	0.0254	0.0255	0.0259	0.0264
3×10^4	0.0238	0.0239	0.0240	0.0241	0.0243	0.0244	0.0245	0.0250	0.0255
4×10^4	0.0223	0.0224	0.0226	0.0227	0.0229	0.0230	0.0232	0.0237	0.0243
5×10^4	0.0212	0.0214	0.0216	0.0218	0.0219	0.0221	0.0223	0.0229	0.0235
6×10^4	0.0205	0.0207	0.0208	0.0210	0.0212	0.0214	0.0216	0.0222	0.0229
7×10^4	0.0198	0.0200	0.0202	0.0204	0.0206	0.0208	0.0210	0.0217	0.0224
8×10^4	0.0193	0.0195	0.0198	0.0200	0.0202	0.0204	0.0206	0.0213	0.0220
9×10^4	0.0189	0.0191	0.0194	0.0196	0.0198	0.0200	0.0202	0.0210	0.0217
1×10^5	0.0185	0.0188	0.0190	0.0192	0.0195	0.0197	0.0199	0.0207	0.0215
1.5×10^5	0.0172	0.0175	0.0178	0.0181	0.0184	0.0186	0.0189	0.0198	0.0207
2×10^5	0.0164	0.0168	0.0171	0.0174	0.0177	0.0180	0.0183	0.0193	0.0202
2.5×10^5	0.0158	0.0162	0.0166	0.0170	0.0173	0.0176	0.0179	0.0190	0.0199
3×10^5	0.0154	0.0159	0.0163	0.0166	0.0170	0.0173	0.0176	0.0188	0.0197
4×10^5	0.0148	0.0153	0.0158	0.0162	0.0166	0.0169	0.0172	0.0184	0.0195
5×10^5	0.0144	0.0150	0.0154	0.0159	0.0163	0.0167	0.0170	0.0183	0.0193
6×10^5	0.0141	0.0147	0.0152	0.0157	0.0161	0.0165	0.0168	0.0181	0.0192
7×10^5	0.0139	0.0145	0.0150	0.0155	0.0159	0.0163	0.0167	0.0180	0.0191
8×10^5	0.0137	0.0143	0.0149	0.0154	0.0158	0.0162	0.0166	0.0180	0.0191
9×10^5	0.0136	0.0142	0.0148	0.0153	0.0157	0.0162	0.0165	0.0179	0.0190
1×10^6	0.0134	0.0141	0.0147	0.0152	0.0157	0.0161	0.0165	0.0178	0.0190
1.5×10^6	0.0130	0.0138	0.0144	0.0149	0.0154	0.0159	0.0163	0.0177	0.0189
2×10^6	0.0128	0.0136	0.0142	0.0148	0.0153	0.0158	0.0162	0.0176	0.0188
2.5×10^6	0.0127	0.0135	0.0141	0.0147	0.0152	0.0157	0.0161	0.0176	0.0188
3×10^6	0.0126	0.0134	0.0141	0.0147	0.0152	0.0157	0.0161	0.0176	0.0187
4×10^6	0.0124	0.0133	0.0140	0.0146	0.0151	0.0156	0.0161	0.0175	0.0187
5×10^6	0.0123	0.0132	0.0139	0.0146	0.0151	0.0156	0.0160	0.0175	0.0187
6×10^6	0.0123	0.0132	0.0139	0.0145	0.0151	0.0156	0.0160	0.0175	0.0187
7×10^6	0.0122	0.0132	0.0139	0.0145	0.0151	0.0155	0.0160	0.0175	0.0187
8×10^6	0.0122	0.0131	0.0139	0.0145	0.0150	0.0155	0.0160	0.0175	0.0187
9×10^6	0.0122	0.0131	0.0139	0.0145	0.0150	0.0155	0.0160	0.0175	0.0187
1×10^7	0.0122	0.0131	0.0138	0.0145	0.0150	0.0155	0.0160	0.0175	0.0186
1.5×10^7	0.0121	0.0131	0.0138	0.0144	0.0150	0.0155	0.0159	0.0174	0.0186
2×10^7	0.0121	0.0130	0.0138	0.0144	0.0150	0.0155	0.0159	0.0174	0.0186
2.5×10^7	0.0121	0.0130	0.0138	0.0144	0.0150	0.0155	0.0159	0.0174	0.0186
3×10^7	0.0120	0.0130	0.0138	0.0144	0.0150	0.0155	0.0159	0.0174	0.0186
4×10^7	0.0120	0.0130	0.0138	0.0144	0.0150	0.0155	0.0159	0.0174	0.0186
5×10^7	0.0120	0.0130	0.0138	0.0144	0.0150	0.0155	0.0159	0.0174	0.0186

Support
Material

(continued)

APPENDIX 17.B *(continued)*
Darcy Friction Factors (turbulent flow)

Reynolds no.	relative roughness, ϵ/D								
	0.001	0.0015	0.002	0.0025	0.003	0.0035	0.004	0.006	0.008
2×10^3	0.0502	0.0506	0.0510	0.0513	0.0517	0.0521	0.0525	0.0539	0.0554
2.5×10^3	0.0469	0.0473	0.0477	0.0481	0.0485	0.0489	0.0493	0.0509	0.0524
3×10^3	0.0444	0.0449	0.0453	0.0457	0.0462	0.0466	0.0470	0.0487	0.0503
4×10^3	0.0409	0.0414	0.0419	0.0424	0.0429	0.0433	0.0438	0.0456	0.0474
5×10^3	0.0385	0.0390	0.0396	0.0401	0.0406	0.0411	0.0416	0.0436	0.0455
6×10^3	0.0367	0.0373	0.0378	0.0384	0.0390	0.0395	0.0400	0.0421	0.0441
7×10^3	0.0353	0.0359	0.0365	0.0371	0.0377	0.0383	0.0388	0.0410	0.0430
8×10^3	0.0341	0.0348	0.0354	0.0361	0.0367	0.0373	0.0379	0.0401	0.0422
9×10^3	0.0332	0.0339	0.0345	0.0352	0.0358	0.0365	0.0371	0.0394	0.0416
1×10^4	0.0324	0.0331	0.0338	0.0345	0.0351	0.0358	0.0364	0.0388	0.0410
1.5×10^4	0.0296	0.0305	0.0313	0.0320	0.0328	0.0335	0.0342	0.0369	0.0393
2×10^4	0.0279	0.0289	0.0298	0.0306	0.0315	0.0323	0.0330	0.0358	0.0384
2.5×10^4	0.0268	0.0278	0.0288	0.0297	0.0306	0.0314	0.0322	0.0352	0.0378
3×10^4	0.0260	0.0271	0.0281	0.0291	0.0300	0.0308	0.0317	0.0347	0.0374
4×10^4	0.0248	0.0260	0.0271	0.0282	0.0291	0.0301	0.0309	0.0341	0.0369
5×10^4	0.0240	0.0253	0.0265	0.0276	0.0286	0.0296	0.0305	0.0337	0.0365
6×10^4	0.0235	0.0248	0.0261	0.0272	0.0283	0.0292	0.0302	0.0335	0.0363
7×10^4	0.0230	0.0245	0.0257	0.0269	0.0280	0.0290	0.0299	0.0333	0.0362
8×10^4	0.0227	0.0242	0.0255	0.0267	0.0278	0.0288	0.0298	0.0331	0.0361
9×10^4	0.0224	0.0239	0.0253	0.0265	0.0276	0.0286	0.0296	0.0330	0.0360
1×10^5	0.0222	0.0237	0.0251	0.0263	0.0275	0.0285	0.0295	0.0329	0.0359
1.5×10^5	0.0214	0.0231	0.0246	0.0259	0.0271	0.0281	0.0292	0.0327	0.0357
2×10^5	0.0210	0.0228	0.0243	0.0256	0.0268	0.0279	0.0290	0.0325	0.0355
2.5×10^5	0.0208	0.0226	0.0241	0.0255	0.0267	0.0278	0.0289	0.0325	0.0355
3×10^5	0.0206	0.0225	0.0240	0.0254	0.0266	0.0277	0.0288	0.0324	0.0354
4×10^5	0.0204	0.0223	0.0239	0.0253	0.0265	0.0276	0.0287	0.0323	0.0354
5×10^5	0.0202	0.0222	0.0238	0.0252	0.0264	0.0276	0.0286	0.0323	0.0353
6×10^5	0.0201	0.0221	0.0237	0.0251	0.0264	0.0275	0.0286	0.0323	0.0353
7×10^5	0.0201	0.0221	0.0237	0.0251	0.0264	0.0275	0.0286	0.0322	0.0353
8×10^5	0.0200	0.0220	0.0237	0.0251	0.0263	0.0275	0.0286	0.0322	0.0353
9×10^5	0.0200	0.0220	0.0236	0.0251	0.0263	0.0275	0.0285	0.0322	0.0353
1×10^6	0.0199	0.0220	0.0236	0.0250	0.0263	0.0275	0.0285	0.0322	0.0353
1.5×10^6	0.0198	0.0219	0.0235	0.0250	0.0263	0.0274	0.0285	0.0322	0.0352
2×10^6	0.0198	0.0218	0.0235	0.0250	0.0262	0.0274	0.0285	0.0322	0.0352
2.5×10^6	0.0198	0.0218	0.0235	0.0249	0.0262	0.0274	0.0285	0.0322	0.0352
3×10^6	0.0197	0.0218	0.0235	0.0249	0.0262	0.0274	0.0285	0.0321	0.0352
4×10^6	0.0197	0.0218	0.0235	0.0249	0.0262	0.0274	0.0284	0.0321	0.0352
5×10^6	0.0197	0.0218	0.0235	0.0249	0.0262	0.0274	0.0284	0.0321	0.0352
6×10^6	0.0197	0.0218	0.0235	0.0249	0.0262	0.0274	0.0284	0.0321	0.0352
7×10^6	0.0197	0.0218	0.0234	0.0249	0.0262	0.0274	0.0284	0.0321	0.0352
8×10^6	0.0197	0.0218	0.0234	0.0249	0.0262	0.0274	0.0284	0.0321	0.0352
9×10^6	0.0197	0.0218	0.0234	0.0249	0.0262	0.0274	0.0284	0.0321	0.0352
1×10^7	0.0197	0.0218	0.0234	0.0249	0.0262	0.0273	0.0284	0.0321	0.0352
1.5×10^7	0.0197	0.0217	0.0234	0.0249	0.0262	0.0273	0.0284	0.0321	0.0352
2×10^7	0.0197	0.0217	0.0234	0.0249	0.0262	0.0273	0.0284	0.0321	0.0352
2.5×10^7	0.0196	0.0217	0.0234	0.0249	0.0262	0.0273	0.0284	0.0321	0.0352
3×10^7	0.0196	0.0217	0.0234	0.0249	0.0262	0.0273	0.0284	0.0321	0.0352
4×10^7	0.0196	0.0217	0.0234	0.0249	0.0262	0.0273	0.0284	0.0321	0.0352
5×10^7	0.0196	0.0217	0.0234	0.0249	0.0262	0.0273	0.0284	0.0321	0.0352

(continued)

APPENDIX 17.B *(continued)*
Darcy Friction Factors (turbulent flow)

relative roughness, ϵ/D

Reynolds no.	0.01	0.015	0.02	0.025	0.03	0.035	0.04	0.045	0.05
2×10^3	0.0568	0.0602	0.0635	0.0668	0.0699	0.0730	0.0760	0.0790	0.0819
2.5×10^3	0.0539	0.0576	0.0610	0.0644	0.0677	0.0709	0.0740	0.0770	0.0800
3×10^3	0.0519	0.0557	0.0593	0.0628	0.0661	0.0694	0.0725	0.0756	0.0787
4×10^3	0.0491	0.0531	0.0570	0.0606	0.0641	0.0674	0.0707	0.0739	0.0770
5×10^3	0.0473	0.0515	0.0555	0.0592	0.0628	0.0662	0.0696	0.0728	0.0759
6×10^3	0.0460	0.0504	0.0544	0.0583	0.0619	0.0654	0.0688	0.0721	0.0752
7×10^3	0.0450	0.0495	0.0537	0.0576	0.0613	0.0648	0.0682	0.0715	0.0747
8×10^3	0.0442	0.0489	0.0531	0.0571	0.0608	0.0644	0.0678	0.0711	0.0743
9×10^3	0.0436	0.0484	0.0526	0.0566	0.0604	0.0640	0.0675	0.0708	0.0740
1×10^4	0.0431	0.0479	0.0523	0.0563	0.0601	0.0637	0.0672	0.0705	0.0738
1.5×10^4	0.0415	0.0466	0.0511	0.0553	0.0592	0.0628	0.0664	0.0698	0.0731
2×10^4	0.0407	0.0459	0.0505	0.0547	0.0587	0.0624	0.0660	0.0694	0.0727
2.5×10^4	0.0402	0.0455	0.0502	0.0544	0.0584	0.0621	0.0657	0.0691	0.0725
3×10^4	0.0398	0.0452	0.0499	0.0542	0.0582	0.0619	0.0655	0.0690	0.0723
4×10^4	0.0394	0.0448	0.0496	0.0539	0.0579	0.0617	0.0653	0.0688	0.0721
5×10^4	0.0391	0.0446	0.0494	0.0538	0.0578	0.0616	0.0652	0.0687	0.0720
6×10^4	0.0389	0.0445	0.0493	0.0536	0.0577	0.0615	0.0651	0.0686	0.0719
7×10^4	0.0388	0.0443	0.0492	0.0536	0.0576	0.0614	0.0650	0.0685	0.0719
8×10^4	0.0387	0.0443	0.0491	0.0535	0.0576	0.0614	0.0650	0.0685	0.0718
9×10^4	0.0386	0.0442	0.0491	0.0535	0.0575	0.0613	0.0650	0.0684	0.0718
1×10^5	0.0385	0.0442	0.0490	0.0534	0.0575	0.0613	0.0649	0.0684	0.0718
1.5×10^5	0.0383	0.0440	0.0489	0.0533	0.0574	0.0612	0.0648	0.0683	0.0717
2×10^5	0.0382	0.0439	0.0488	0.0532	0.0573	0.0612	0.0648	0.0683	0.0717
2.5×10^5	0.0381	0.0439	0.0488	0.0532	0.0573	0.0611	0.0648	0.0683	0.0716
3×10^5	0.0381	0.0438	0.0488	0.0532	0.0573	0.0611	0.0648	0.0683	0.0716
4×10^5	0.0381	0.0438	0.0487	0.0532	0.0573	0.0611	0.0647	0.0682	0.0716
5×10^5	0.0380	0.0438	0.0487	0.0531	0.0572	0.0611	0.0647	0.0682	0.0716
6×10^5	0.0380	0.0438	0.0487	0.0531	0.0572	0.0611	0.0647	0.0682	0.0716
7×10^5	0.0380	0.0438	0.0487	0.0531	0.0572	0.0611	0.0647	0.0682	0.0716
8×10^5	0.0380	0.0437	0.0487	0.0531	0.0572	0.0611	0.0647	0.0682	0.0716
9×10^5	0.0380	0.0437	0.0487	0.0531	0.0572	0.0610	0.0647	0.0682	0.0716
1×10^6	0.0380	0.0437	0.0487	0.0531	0.0572	0.0610	0.0647	0.0682	0.0716
1.5×10^6	0.0379	0.0437	0.0487	0.0531	0.0572	0.0610	0.0647	0.0682	0.0716
2×10^6	0.0379	0.0437	0.0487	0.0531	0.0572	0.0610	0.0647	0.0682	0.0716
2.5×10^6	0.0379	0.0437	0.0487	0.0531	0.0572	0.0610	0.0647	0.0682	0.0716
3×10^6	0.0379	0.0437	0.0487	0.0531	0.0572	0.0610	0.0647	0.0682	0.0716
4×10^6	0.0379	0.0437	0.0486	0.0531	0.0572	0.0610	0.0647	0.0682	0.0716
5×10^6	0.0379	0.0437	0.0486	0.0531	0.0572	0.0610	0.0647	0.0682	0.0716
6×10^6	0.0379	0.0437	0.0486	0.0531	0.0572	0.0610	0.0647	0.0682	0.0716
7×10^6	0.0379	0.0437	0.0486	0.0531	0.0572	0.0610	0.0647	0.0682	0.0716
8×10^6	0.0379	0.0437	0.0486	0.0531	0.0572	0.0610	0.0647	0.0682	0.0716
9×10^6	0.0379	0.0437	0.0486	0.0531	0.0572	0.0610	0.0647	0.0682	0.0716
1×10^7	0.0379	0.0437	0.0486	0.0531	0.0572	0.0610	0.0647	0.0682	0.0716
1.5×10^7	0.0379	0.0437	0.0486	0.0531	0.0572	0.0610	0.0647	0.0682	0.0716
2×10^7	0.0379	0.0437	0.0486	0.0531	0.0572	0.0610	0.0647	0.0682	0.0716
2.5×10^7	0.0379	0.0437	0.0486	0.0531	0.0572	0.0610	0.0647	0.0682	0.0716
3×10^7	0.0379	0.0437	0.0486	0.0531	0.0572	0.0610	0.0647	0.0682	0.0716
4×10^7	0.0379	0.0437	0.0486	0.0531	0.0572	0.0610	0.0647	0.0682	0.0716
5×10^7	0.0379	0.0437	0.0486	0.0531	0.0572	0.0610	0.0647	0.0682	0.0716

APPENDIX 17.C
Water Pressure Drop in Schedule-40 Steel Pipe

	pressure drop per 1000 ft of schedule-40 steel pipe (lbf/in²)																	
discharge (gal/min)	velocity (ft/sec)	pressure drop	velocity (ft/sec)	pressure drop	velocity (ft/sec)	pressure drop	velocity (ft/sec)	pressure drop	velocity (ft/sec)	pressure drop	velocity (ft/sec)	pressure drop	velocity (ft/sec)	pressure drop	velocity (ft/sec)	pressure drop	velocity (ft/sec)	pressure drop
	1 in																	
1	0.37	0.49	1¼ in															
2	0.74	1.70	0.43	0.45	1½ in													
3	1.12	3.53	0.64	0.94	0.47	0.44												
4	1.49	5.94	0.86	1.55	0.63	0.74												
5	1.86	9.02	1.07	2.36	0.79	1.12	2 in											
6	2.24	12.25	1.28	3.30	0.95	1.53	0.57	0.46										
8	2.98	21.1	1.72	5.52	1.26	2.63	0.76	0.75	2½ in									
10	3.72	30.8	2.14	8.34	1.57	3.86	0.96	1.14	0.67	0.48								
15	5.60	64.6	3.21	17.6	2.36	8.13	1.43	2.33	1.00	0.99	3 in							
20	7.44	110.5	4.29	29.1	3.15	13.5	1.91	3.86	1.34	1.64	0.87	0.59	3½ in					
25			5.36	43.7	3.94	20.2	2.39	5.81	1.68	2.48	1.08	0.67	0.81	0.42				
30			6.43	62.9	4.72	29.1	2.87	8.04	2.01	3.43	1.30	1.21	0.97	0.60	4 in			
35			7.51	82.5	5.51	38.2	3.35	10.95	2.35	4.49	1.52	1.58	1.14	0.79	0.88	0.42		
40					6.30	47.8	3.82	13.7	2.68	5.88	1.74	2.06	1.30	1.00	1.01	0.53		
45					7.08	60.6	4.30	17.4	3.00	7.14	1.95	2.51	1.46	1.21	1.13	0.67		
50					7.87	74.7	4.78	20.6	3.35	8.82	2.17	3.10	1.62	1.44	1.26	0.80		
60							5.74	29.6	4.02	12.2	2.60	4.29	1.95	2.07	1.51	1.10	5 in	
70							6.69	38.6	4.69	15.3	3.04	5.84	2.27	2.71	1.76	1.50	1.12	0.48
80							7.65	50.3	5.37	21.7	3.48	7.62	2.59	3.53	2.01	1.87	1.28	0.63
90	6 in						8.60	63.6	6.04	26.1	3.91	9.22	2.92	4.46	2.26	2.37	1.44	0.80
100	1.11	0.39					9.56	75.1	6.71	32.3	4.34	11.4	3.24	5.27	2.52	2.81	1.60	0.95
125	1.39	0.56							8.38	48.2	5.42	17.1	4.05	7.86	3.15	4.38	2.00	1.48
150	1.67	0.78							10.06	60.4	6.51	23.5	4.86	11.3	3.78	6.02	2.41	2.04
175	1.94	1.06							11.73	90.0	7.59	32.0	5.67	14.7	4.41	8.20	2.81	2.78
200	2.22	1.32	8 in								8.68	39.7	6.48	19.2	5.04	10.2	3.21	3.46
225	2.50	1.66	1.44	0.44							9.77	50.2	7.29	23.1	5.67	12.9	3.61	4.37
250	2.78	2.05	1.60	0.55							10.85	61.9	8.10	28.5	6.30	15.9	4.01	5.14
275	3.06	2.36	1.76	0.63							11.94	75.0	8.91	34.4	6.93	18.3	4.41	6.22
300	3.33	2.80	1.92	0.75							13.02	84.7	9.72	40.9	7.56	21.8	4.81	7.41
325	3.61	3.29	2.08	0.88									10.53	45.5	8.18	25.5	5.21	8.25
350	3.89	3.62	2.24	0.97									11.35	52.7	8.82	29.7	5.61	9.57
375	4.16	4.16	2.40	1.11									12.17	60.7	9.45	32.3	6.01	11.0
400	4.44	4.72	2.56	1.27									12.97	68.9	10.08	36.7	6.41	12.5
425	4.72	5.34	2.72	1.43									13.78	77.8	10.70	41.5	6.82	14.1
450	5.00	5.96	2.88	1.60	10 in								14.59	87.3	11.33	46.5	7.22	15.0
475	5.27	6.66	3.04	1.69	1.93	0.30									11.96	51.7	7.62	16.7
500	5.55	7.39	3.20	1.87	2.04	0.63									12.59	57.3	8.02	18.5
550	6.11	8.94	3.53	2.26	2.24	0.70									13.84	69.3	8.82	22.4
600	6.66	10.6	3.85	2.70	2.44	0.86									15.10	82.5	9.62	26.7
650	7.21	11.8	4.17	3.16	2.65	1.01	12 in										10.42	31.3
700	7.77	13.7	4.49	3.69	2.85	1.18	2.01	0.48									11.22	36.3
750	8.32	15.7	4.81	4.21	3.05	1.35	2.15	0.55									12.02	41.6
800	8.88	17.8	5.13	4.79	3.26	1.54	2.29	0.62	14 in								12.82	44.7
850	9.44	20.2	5.45	5.11	3.46	1.74	2.44	0.70	2.02	0.43							13.62	50.5
900	10.00	22.6	5.77	5.73	3.66	1.94	2.58	0.79	2.14	0.48							14.42	56.6
950	10.55	23.7	6.09	6.38	3.87	2.23	2.72	0.88	2.25	0.53							15.22	63.1
1000	11.10	26.3	6.41	7.08	4.07	2.40	2.87	0.98	2.38	0.59							16.02	70.0
1100	12.22	31.8	7.05	8.56	4.48	2.74	3.16	1.18	2.61	0.68	16 in						17.63	84.6
1200	13.32	37.8	7.69	10.2	4.88	3.27	3.45	1.40	2.85	0.81	2.18	0.40						
1300	14.43	44.4	8.33	11.3	5.29	3.86	3.73	1.56	3.09	0.95	2.36	0.47						
1400	15.54	51.5	8.97	13.0	5.70	4.44	4.02	1.80	3.32	1.10	2.54	0.54						
1500	16.65	55.5	9.62	15.0	6.10	5.11	4.30	2.07	3.55	1.19	2.73	0.62						
1600	17.76	63.1	10.26	17.0	6.51	5.46	4.59	2.36	3.80	1.35	2.91	0.71	18 in					
1800	19.98	79.8	11.54	21.6	7.32	6.91	5.16	2.98	4.27	1.71	3.27	0.85	2.58	0.48				
2000	22.20	98.5	12.83	25.0	8.13	8.54	5.73	3.47	4.74	2.11	3.63	1.05	2.88	0.56				
2500			16.03	39.0	10.18	12.5	7.17	5.41	5.92	3.09	4.54	1.63	3.59	0.88	20 in			
3000			19.24	52.4	12.21	18.0	8.60	7.31	7.12	4.45	5.45	2.21	4.31	1.27	3.45	0.73		
3500			22.43	71.4	14.25	22.9	10.03	9.95	8.32	6.18	6.35	3.00	5.03	1.52	4.03	0.94	24 in	
4000			25.65	93.3	16.28	29.9	11.48	13.0	9.49	7.92	7.25	3.92	5.74	2.12	4.61	1.22	3.19	0.51
4500					18.31	37.8	12.90	15.4	10.67	9.36	8.17	4.97	6.47	2.50	5.19	1.55	3.59	0.60
5000					20.35	46.7	14.34	18.9	11.84	11.6	9.08	5.72	7.17	3.08	5.76	1.78	3.99	0.74
6000					24.42	67.2	17.21	27.3	14.32	15.4	10.88	8.24	8.62	4.45	6.92	2.57	4.80	1.00
7000					28.50	85.1	20.08	37.2	16.60	21.0	12.69	12.2	10.04	6.06	8.06	3.50	5.68	1.36
8000							22.95	45.1	18.98	27.4	14.52	13.6	11.48	7.34	9.23	4.57	6.38	1.78
9000							25.80	57.0	21.35	34.7	16.32	17.2	12.92	9.20	10.37	5.36	7.19	2.25
10,000							28.63	70.4	23.75	42.9	18.16	21.2	14.37	11.5	11.53	6.63	7.96	2.78
12,000							34.38	93.6	28.50	61.8	21.80	30.9	17.23	16.5	13.83	9.54	9.57	3.71
14,000									33.20	84.0	25.42	41.6	20.10	20.7	16.14	12.0	11.18	5.05
16,000											29.05	54.4	22.96	27.1	18.43	15.7	12.77	6.60

(Multiply gal/min by 0.0631 to obtain L/s.) (Multiply in by 25.4 to obtain mm.)
(Multiply ft/sec by 0.3048 to obtain m/s.) (Multiply lbf/in²-1000 ft by 2.3 to obtain kPa/100 m.)

Reproduced with permission from *Design of Fluid Systems Hook-Ups*, published by Spirax Sarco, Inc., © 1997.

APPENDIX 17.D
Equivalent Length of Straight Pipe for Various (Generic) Fittings
(in feet, turbulent flow only, for any fluid)

fittings			1/4	3/8	1/2	3/4	1	1¼	1½	2	2½	3	4	5	6	8	10	12	14	16	18	20	24
regular 90° ell	screwed	steel	2.3	3.1	3.6	4.4	5.2	6.6	7.4	8.5	9.3	11.0	13.0										
		cast iron										9.0	11.0										
	flanged	steel			0.92	1.2	1.6	2.1	2.4	3.1	3.6	4.4	5.9	7.3	8.9	12.0	14.0	17.0	18.0	21.0	23.0	25.0	30.0
		cast iron										3.6	4.8		7.2	9.8	12.0	15.0	17.0	19.0	22.0	24.0	28.0
long radius 90° ell	screwed	steel	1.5	2.0	2.2	2.3	2.7	3.2	3.4	3.6	3.6	4.0	4.6										
		cast iron										3.3	3.7										
	flanged	steel			1.1	1.3	1.6	2.0	2.3	2.7	2.9	3.4	4.2	5.0	5.7	7.0	8.0	9.0	9.4	10.0	11.0	12.0	14.0
		cast iron										2.8	3.4		4.7	5.7	6.8	7.8	8.6	9.6	11.0	11.0	13.0
regular 45° ell	screwed	steel	0.34	0.52	0.71	0.92	1.3	1.7	2.1	2.7	3.2	4.0	5.5										
		cast iron										3.3	4.5										
	flanged	steel			0.45	0.59	0.81	1.1	1.3	1.7	2.0	2.6	3.5	4.5	5.6	7.7	9.0	11.0	13.0	15.0	16.0	18.0	22.0
		cast iron										2.1	2.9		4.5	6.3	8.1	9.7	12.0	13.0	15.0	17.0	20.0
tee-line flow	screwed	steel	0.79	1.2	1.7	2.4	3.2	4.6	5.6	7.7	9.3	12.0	17.0										
		cast iron										9.9	14.0										
	flanged	steel			0.69	0.82	1.0	1.3	1.5	1.8	1.9	2.2	2.8	3.3	3.8	4.7	5.2	6.0	6.4	7.2	7.6	8.2	9.6
		cast iron										1.9	2.2		3.1	3.9	4.6	5.2	5.9	6.5	7.2	7.7	8.8
tee-branch flow	screwed	steel	2.4	3.5	4.2	5.3	6.6	8.7	9.9	12.0	13.0	17.0	21.0										
		cast iron										14.0	17.0										
	flanged	steel			2.0	2.6	3.3	4.4	5.2	6.6	7.5	9.4	12.0	15.0	18.0	24.0	30.0	34.0	37.0	43.0	47.0	52.0	62.0
		cast iron										7.7	10.0		15.0	20.0	25.0	30.0	35.0	39.0	44.0	49.0	57.0
180° return bend	screwed	steel	2.3	3.1	3.6	4.4	5.2	6.6	7.4	8.5	9.3	11.0	13.0										
		cast iron										9.0	11.0										
	regular flanged	steel			0.92	1.2	1.6	2.1	2.4	3.1	3.6	4.4	5.9	7.3	8.9	12.0	14.0	17.0	18.0	21.0	23.0	25.0	30.0
		cast iron										3.6	4.8		7.2	9.8	12.0	15.0	17.0	19.0	22.0	24.0	28.0
	long radius flanged	steel			1.1	1.3	1.6	2.0	2.3	2.7	2.9	3.4	4.2	5.0	5.7	7.0	8.0	9.0	9.4	10.0	11.0	12.0	14.0
		cast iron										2.8	3.4		4.7	5.7	6.8	7.8	8.6	9.6	11.0	11.0	13.0
globe valve	screwed	steel	21.0	22.0	22.0	24.0	29.0	37.0	42.0	54.0	62.0	79.0	110.0										
		cast iron										65.0	86.0										
	flanged	steel			38.0	40.0	45.0	54.0	59.0	70.0	77.0	94.0	120.0	150.0	190.0	260.0	310.0	390.0					
		cast iron										77.0	99.0		150.0	210.0	270.0	330.0					
gate valve	screwed	steel	0.32	0.45	0.56	0.67	0.84	1.1	1.2	1.5	1.7	1.9	2.5										
		cast iron										1.6	2.0										
	flanged	steel								2.6	2.7	2.8	2.9	3.1	3.2	3.2	3.2	3.2	3.2	3.2	3.2	3.2	3.2
		cast iron										2.3	2.4		2.6	2.7	2.8	2.9	2.9	3.0	3.0	3.0	3.0
angle valve	screwed	steel	12.8	15.0	15.0	15.0	17.0	18.0	18.0	18.0	18.0	18.0	18.0										
		cast iron										15.0	15.0										
	flanged	steel			15.0	15.0	17.0	18.0	18.0	21.0	22.0	28.0	38.0	50.0	63.0	90.0	120.0	140.0	160.0	190.0	210.0	240.0	300.0
		cast iron										23.0	31.0		52.0	74.0	98.0	120.0	150.0	170.0	200.0	230.0	280.0
swing check valve	screwed	steel	7.2	7.3	8.0	8.8	11.0	13.0	15.0	19.0	22.0	27.0	38.0										
		cast iron										22.0	31.0										
	flanged	steel			3.8	5.3	7.2	10.0	12.0	17.0	21.0	27.0	38.0	50.0	63.0	90.0	120.0	140.0					
		cast iron										22.0	31.0		52.0	74.0	98.0	120.0					
coupling or union	screwed	steel	0.14	0.18	0.21	0.24	0.29	0.36	0.39	0.45	0.47	0.53	0.65										
		cast iron										0.44	0.52										
inlet	bell mouth inlet	steel	0.04	0.07	0.10	0.13	0.18	0.26	0.31	0.43	0.52	0.67	0.95	1.3	1.6	2.3	2.9	3.5	4.0	4.7	5.3	6.1	7.6
		cast iron										0.55	0.77		1.3	1.9	2.4	3.0	3.6	4.3	5.0	5.7	7.0
	square mouth inlet	steel	0.44	0.68	0.96	1.3	1.8	2.6	3.1	4.3	5.2	6.7	9.5	13.0	16.0	23.0	29.0	35.0	40.0	47.0	53.0	61.0	76.0
		cast iron										5.5	7.7		13.0	19.0	24.0	30.0	36.0	43.0	50.0	57.0	70.0
	re-entrant pipe	steel	0.88	1.4	1.9	2.6	3.6	5.1	6.2	8.5	10.0	13.0	19.0	25.0	32.0	45.0	58.0	70.0	80.0	95.0	110.0	120.0	150.0
		cast iron										11.0	15.0		26.0	37.0	49.0	61.0	73.0	86.0	100.0	110.0	140.0

pipe size (in)

(Multiply in by 25.4 to obtain mm.)
(Multiply ft by 0.3048 to obtain m.)

APPENDIX 17.E
Hazen-Williams Nomograph ($C = 100$)

Quantity (i.e., flow rate) and velocity are proportional to the C-value. For values of C other than 100, the quantity and velocity must be converted according to $\dot{V}_{actual} = \dot{V}_{chart} C_{actual}/100$. When quantity is the unknown, use the chart with known values of diameter, slope, or velocity to find $\dot{V}_{chart}$, and then convert to $\dot{V}_{actual}$. When velocity is the unknown, use the chart with the known values of diameter, slope, or quantity to find $\dot{V}_{chart}$, then convert to $\dot{V}_{actual}$. If $\dot{V}_{actual}$ is known, it must be converted to $\dot{V}_{chart}$ before this nomograph can be used. In that case, the diameter, loss, and quantity are as read from this chart.

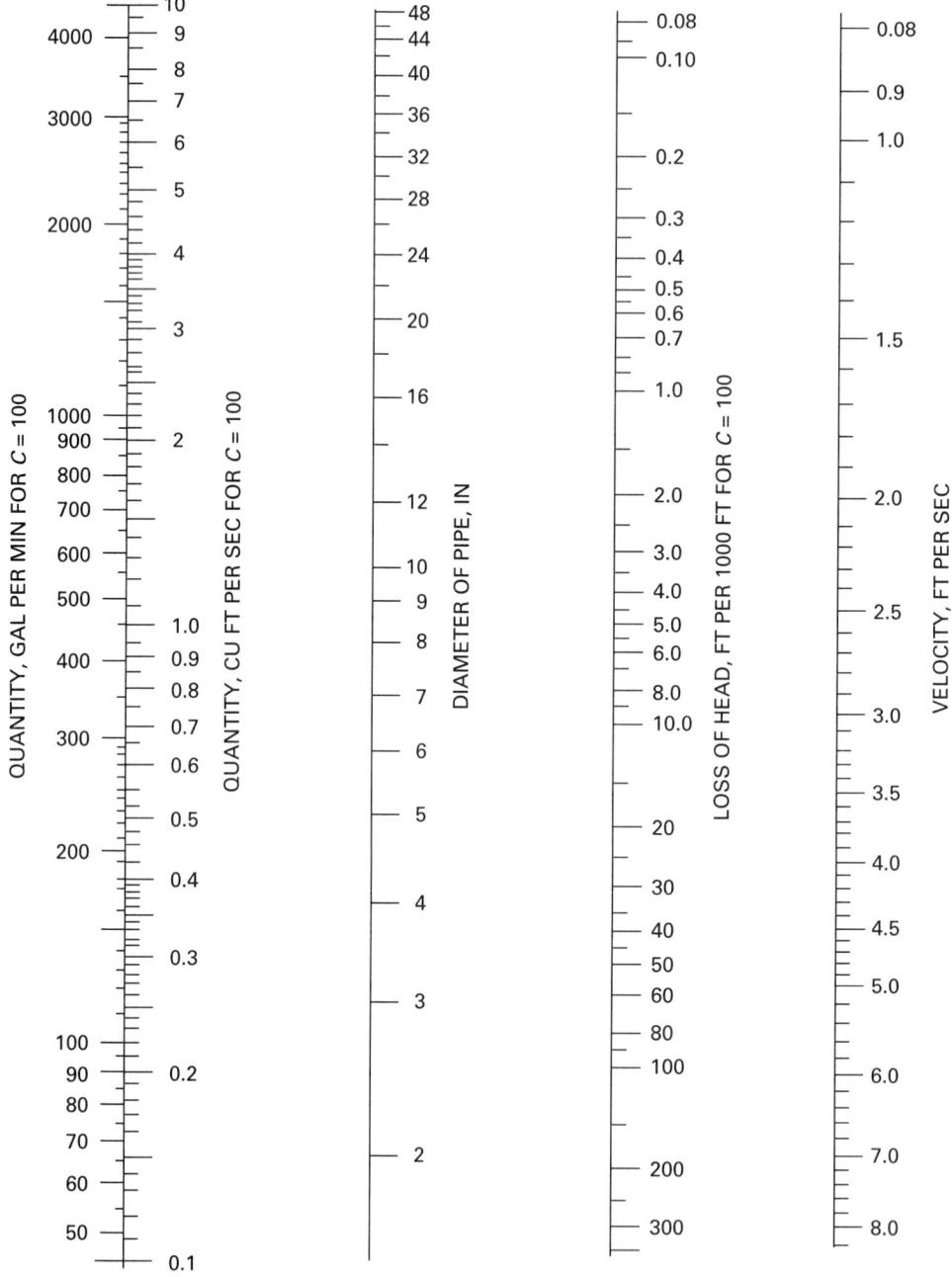

(Multiply gal/min by 0.0631 to obtain L/s.)
(Multiply ft³/sec by 28.3 to obtain L/s.)
(Multiply in by 25.4 to obtain mm.)
(Multiply ft/1000 ft by 0.1 to obtain m/100 m.)
(Multiply ft/sec by 0.3048 to obtain m/s.)

APPENDIX 17.F
Corrugated Metal Pipe

Corrugated metal pipe (CMP, also known as *corrugated steel pipe*) is frequently used for culverts. Pipe is made from corrugated sheets of galvanized steel that are rolled and riveted together along a longitudinal seam. Aluminized steel may also be used in certain ranges of soil pH. Standard round pipe diameters range from 8 in to 96 in (200 mm to 2450 mm). Metric dimensions of standard diameters are usually rounded to the nearest 25 mm or 50 mm (e.g., a 42 in culvert would be specified as a 1050 mm culvert, not 1067 mm).

Larger and noncircular culverts can be created out of curved steel plate. Standard section lengths are 10 ft to 20 ft (3 m to 6 m). Though most corrugations are transverse (i.e., annular), helical corrugations are also used. Metal gages of 8, 10, 12, 14, and 16 are commonly used, depending on the depth of burial.

The most common corrugated steel pipe has transverse corrugations that are $1/2$ in (13 mm) deep and $2^2/3$ in (68 mm) from crest to crest. These are referred to as "$2^1/2$ inch" or "68×13" corrugations. For larger culverts, corrugations with a 2 in (25mm) depth and 3, 5, and 6 in (76, 125, or 152 mm) pitches are used. Plate-based products using 6 in by 2 in (152 mm by 51 mm) corrugations are known as *structural plate corrugated steel pipe* (SPCSP) and *multiplate* after the trade-named product "Multi-Plate™."

The flow area for circular culverts is based on the nominal culvert diameter, regardless of the gage of the plate metal used to construct the pipe. Flow area is calculated to (at most) three significant digits.

A Hazen-Williams coefficient, C, of 60 is typically used with all sizes of corrugated pipe. Values of C and Manning's constant, n, for corrugated pipe are generally not affected by age. *Design Charts for Open Channel Flow* (U.S. Department of Transportation, 1979) recommends a Manning constant of $n = 0.024$ for all cases. The U.S. Department of the Interior recommends the following values. For standard ($2^2/3$ in by $1/2$ in or 68 mm by 13 mm) corrugated pipe with the diameters given: 12 in (457 mm), 0.027; 24 in (610 mm), 0.025; 36 in to 48 in (914 mm to 1219 mm), 0.024; 60 in to 84 in (1524 mm to 2134 mm), 0.023; 96 in (2438 mm), 0.022. For (6 in by 2 in or 152 mm by 51 mm) multiplate construction with the diameters given: 5 ft to 6 ft (1.5 m to 1.8 m), 0.034; 7 ft to 8 ft (2.1 m to 2.4 m), 0.033; 9 ft to 11 ft (2.7 m to 3.3 m), 0.032; 12 ft to 13 ft (3.6 m to 3.9 m), 0.031; 14 ft to 15 ft (4.2 m to 4.5 m), 0.030; 16 ft to 18 ft (4.8 m to 5.4 m), 0.029; 19 ft to 20 ft (5.8 m to 6.0 m), 0.028; 21 ft to 22 ft (6.3 m to 6.6 m), 0.027.

If the inside of the corrugated pipe has been asphalted completely smooth 360° circumferentially, Manning's n ranges from 0.009 to 0.011. For culverts with 40% asphalted inverts, $n = 0.019$. For other percentages of paved invert, the resulting value is proportional to the percentage and the values normally corresponding to that diameter pipe. For field-bolted corrugated metal pipe arches, $n = 0.025$.

It is also possible to calculate the Darcy friction loss if the corrugation depth, 0.5 in (13 mm) for standard corrugations and 2.0 in (51 mm) for multiplate, is taken as the specific roughness.

APPENDIX 18.A
International Standard Atmosphere

customary U.S. units				SI units		
altitude (ft)	temperature (°R)	pressure (psia)		altitude (m)	temperature (K)	pressure (bar)
0	518.7	14.696		0	288.15	1.01325
1000	515.1	14.175				
2000	511.6	13.664		500	284.9	0.9546
3000	508.0	13.168		1000	281.7	0.8988
4000	504.4	12.692		1500	278.4	0.8456
				2000	275.2	0.7950
5000	500.9	12.225		2500	271.9	0.7469
6000	497.3	11.778				
7000	493.7	11.341		3000	268.7	0.7012
8000	490.2	10.914		3500	265.4	0.6578
9000	486.6	10.501		4000	262.2	0.6166
				4500	258.9	0.5775
10,000	483.0	10.108		5000	255.7	0.5405
11,000	479.5	9.720				
12,000	475.9	9.347		5500	252.4	0.5054
13,000	472.3	8.983		6000	249.2	0.4722
14,000	468.8	8.630		6500	245.9	0.4408
				7000	242.7	0.4111
15,000	465.2	8.291		7500	239.5	0.3830
16,000	461.6	7.962				
17,000	458.1	7.642		8000	236.2	0.3565
18,000	454.5	7.338		8500	233.0	0.3315
19,000	450.9	7.038		9000	229.7	0.3080
				9500	226.5	0.2858
20,000	447.4	6.753		10 000	223.3	0.2650
21,000	443.8	6.473				
22,000	440.2	6.203		10 500	220.0	0.2454
23,000	436.7	5.943		11 000	216.8	0.2270
24,000	433.1	5.693		11 500	216.7	0.2098
				12 000	216.7	0.1940
25,000	429.5	5.452		12 500	216.7	0.1793
26,000	426.0	5.216				
27,000	422.4	4.990		13 000	216.7	0.1658
28,000	418.8	4.774		13 500	216.7	0.1533
29,000	415.3	4.563		14 000	216.7	0.1417
				14 500	216.7	0.1310
30,000	411.7	4.362		15 000	216.7	0.1211
31,000	408.1	4.165				
32,000	404.6	3.978		15 500	216.7	0.1120
33,000	401.0	3.797		16 000	216.7	0.1035
34,000	397.5	3.625		16 500	216.7	0.09572
				17 000	216.7	0.08850
35,000	393.9	3.458		17 500	216.7	0.08182
36,000	392.7	3.296				
37,000	392.7	3.143		18 000	216.7	0.07565
38,000	392.7	2.996		18 500	216.7	0.06995
39,000	392.7	2.854		19 000	216.7	0.06467
				19 500	216.7	0.05980
40,000	392.7	2.721		20 000	216.7	0.05529
41,000	392.7	2.593				
42,000	392.7	2.475		22 000	218.6	0.04047
43,000	392.7	2.358		24 000	220.6	0.02972
44,000	392.7	2.250		26 000	222.5	0.02188
				28 000	224.5	0.01616
45,000	392.7	2.141		30 000	226.5	0.01197
46,000	392.7	2.043				
47,000	392.7	1.950		32 000	228.5	0.00889
48,000	392.7	1.857				
49,000	392.7	1.768				
50,000	392.7	1.690				
51,000	392.7	1.611				
52,000	392.7	1.532				
53,000	392.7	1.464				
54,000	392.7	1.395				
55,000	392.7	1.331				
56,000	392.7	1.267				
57,000	392.7	1.208				
58,000	392.7	1.154				
59,000	392.7	1.100				
60,000	392.7	1.046				
61,000	392.7	0.997				
62,000	392.7	0.953				
63,000	392.7	0.909				
64,000	392.7	0.864				
65,000	392.7	0.825				

troposphere
tropopause
stratosphere
(to approximately
160,000 ft)

troposphere
tropopause
stratosphere
(to approximately
50 000 m)

APPENDIX 19.A
Manning's Roughness Coefficient[a,b]
(design use)

channel material	n
plastic (PVC and ABS)	0.009
clean, uncoated cast iron	0.013–0.015
clean, coated cast iron	0.012–0.014
dirty, tuberculated cast iron	0.015–0.035
riveted steel	0.015–0.017
lock-bar and welded steel pipe	0.012–0.013
galvanized iron	0.015–0.017
brass and glass	0.009–0.013
wood stave	
small diameter	0.011–0.012
large diameter	0.012–0.013
concrete	
average value used	0.013
typical commercial, ball and spigot	
rubber gasketed end connections	
– full (pressurized and wet)	0.010
– partially full	0.0085
with rough joints	0.016–0.017
dry mix, rough forms	0.015–0.016
wet mix, steel forms	0.012–0.014
very smooth, finished	0.011–0.012
vitrified sewer	0.013–0.015
common-clay drainage tile	0.012–0.014
asbestos	0.011
planed timber (flume)	0.012 (0.010–0.014)
canvas	0.012
unplaned timber (flume)	0.013 (0.011–0.015)
brick	0.016
rubble masonry	0.017
smooth earth	0.018
firm gravel	0.023
corrugated metal pipe (CMP)	0.022–0.033
natural channels, good condition	0.025
rip rap	0.035
natural channels with stones and weeds	0.035
very poor natural channels	0.060

[a]compiled from various sources
[b]Values outside these ranges have been observed, but these values are typical.

APPENDIX 19.B
Manning Equation Nomograph

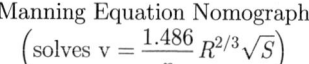

$$\left(\text{solves } v = \frac{1.486}{n} R^{2/3}\sqrt{S}\right)$$

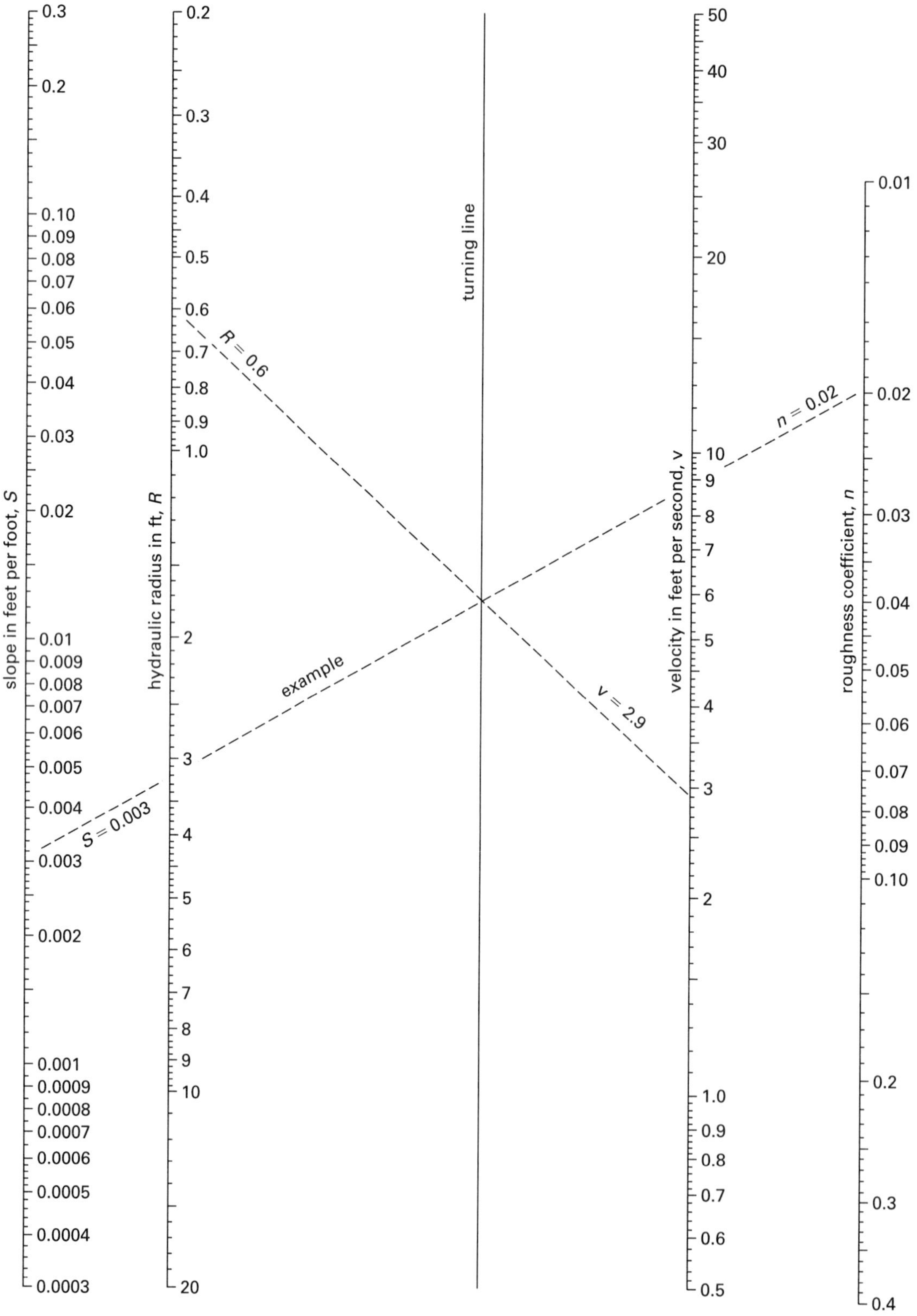

APPENDIX 19.C
Circular Channel Ratios[*]

Experiments have shown that n varies slightly with depth. This figure gives velocity and flow rate ratios for varying n (solid line) and constant n (broken line) assumptions.

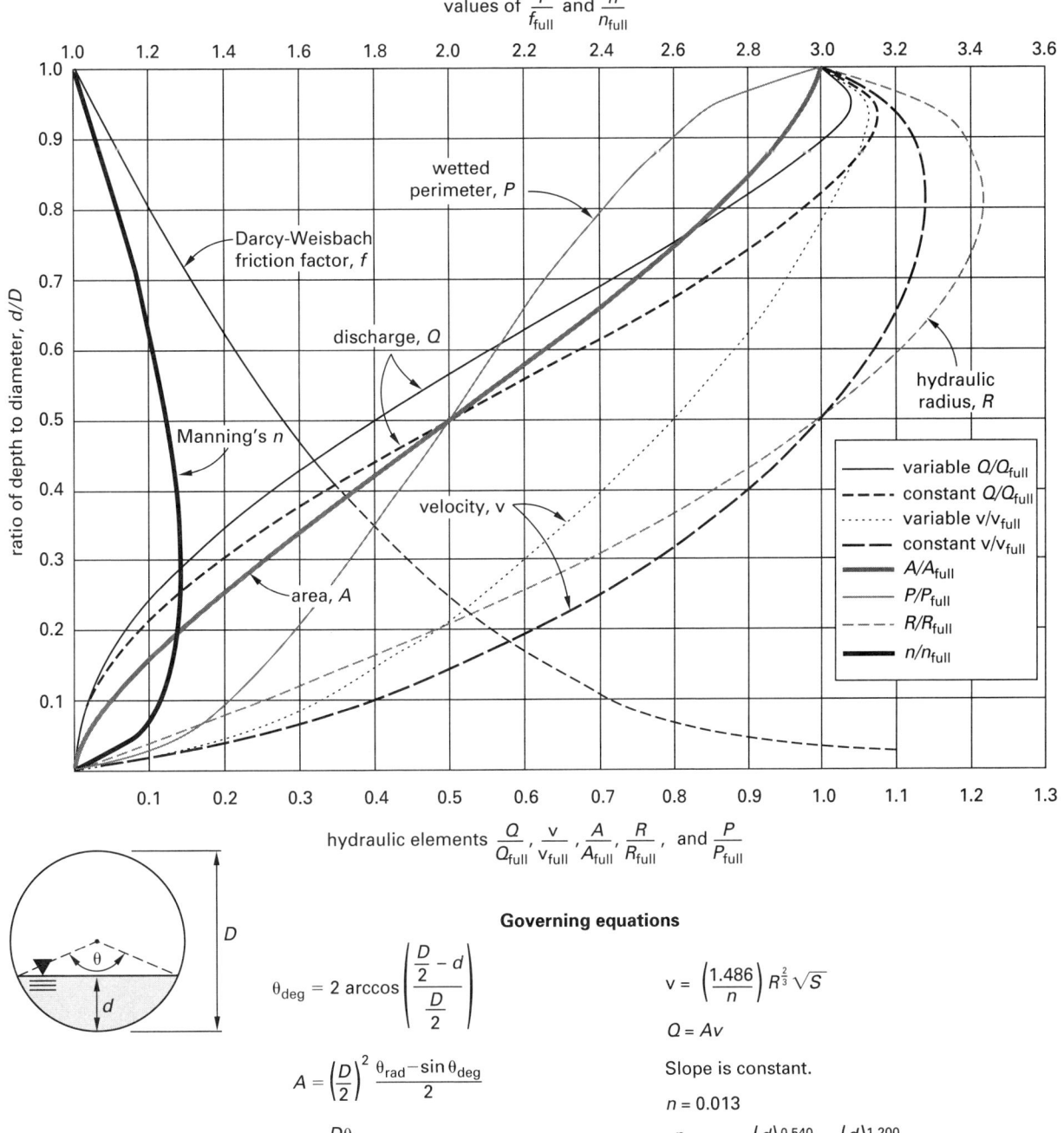

Governing equations

$$\theta_{deg} = 2 \arccos\left(\frac{\frac{D}{2} - d}{\frac{D}{2}}\right)$$

$$A = \left(\frac{D}{2}\right)^2 \frac{\theta_{rad} - \sin\theta_{deg}}{2}$$

$$P = \frac{D\theta_{rad}}{2}$$

$$R = \frac{A}{P}$$

$$v = \left(\frac{1.486}{n}\right) R^{\frac{2}{3}} \sqrt{S}$$

$$Q = Av$$

Slope is constant.

$$n = 0.013$$

$$\frac{n}{n_{full}} = 1 + \left(\frac{d}{D}\right)^{0.540} - \left(\frac{d}{D}\right)^{1.200}$$

[*]For $n = 0.013$.

Adapted from *Design and Construction of Sanitary and Storm Sewers*, p. 87, ASCE, 1969, as originally presented in "Design of Sewers to Facilitate Flow", Camp, T. R., *Sewage Works Journal*, 18, 3 (1946).

APPENDIX 19.D
Critical Depths in Circular Channels

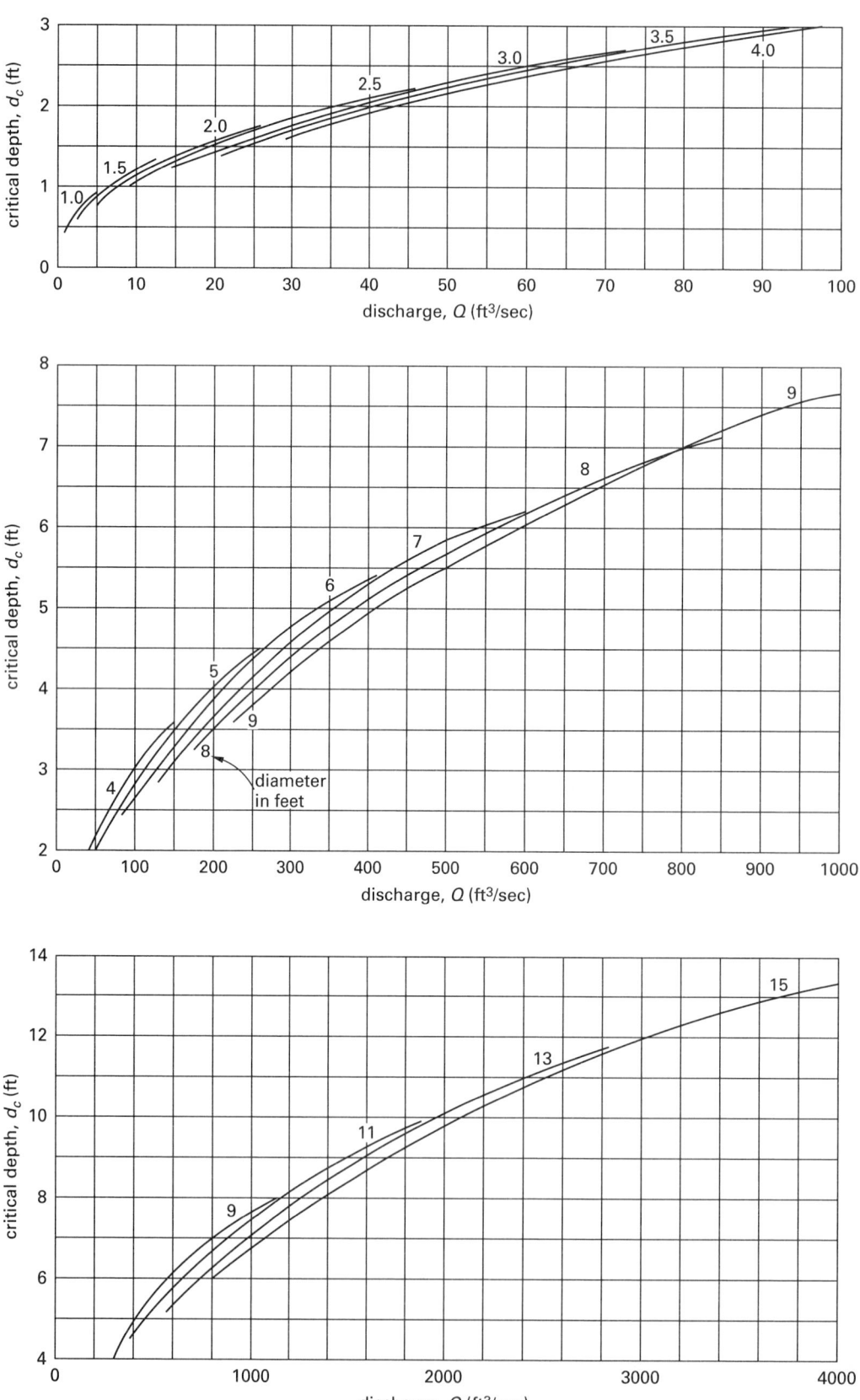

APPENDIX 19.E
Conveyance Factor, K
Symmetrical Rectangular,[a] Trapezoidal, and V-Notch[b] Open Channels
(use for determining Q or b when d is known)
(customary U.S. units[c,d])

$$K \text{ in } Q = K\left(\frac{1}{n}\right) d^{8/3}\sqrt{S}$$

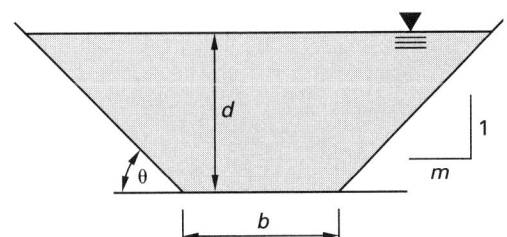

					m and θ					
	0.0	0.25	0.5	0.75	1.0	1.5	2.0	2.5	3.0	4.0
$x = d/b$	90°	76.0°	63.4°	53.1°	45.0°	33.7°	26.6°	21.8°	18.4°	14.0°
0.01	146.7	147.2	147.6	148.0	148.3	148.8	149.2	149.5	149.9	150.5
0.02	72.4	72.9	73.4	73.7	74.0	74.5	74.9	75.3	75.6	76.3
0.03	47.6	48.2	48.6	49.0	49.8	49.8	50.2	50.6	50.9	51.6
0.04	35.3	35.8	36.3	36.6	36.9	37.4	37.8	38.2	38.6	39.3
0.05	27.9	28.4	28.9	29.2	29.5	30.0	30.5	30.9	31.2	32.0
0.06	23.0	23.5	23.9	24.3	24.6	25.1	25.5	26.0	26.3	27.1
0.07	19.5	20.0	20.4	20.8	21.1	21.6	22.0	22.4	22.8	23.6
0.08	16.8	17.3	17.8	18.1	18.4	18.9	19.4	19.8	20.2	21.0
0.09	14.8	15.3	15.7	16.1	16.4	16.9	17.4	17.8	18.2	19.0
0.10	13.2	13.7	14.1	14.4	14.8	15.3	15.7	16.2	16.6	17.4
0.11	11.83	12.33	12.76	13.11	13.42	13.9	14.4	14.9	15.3	16.1
0.12	10.73	11.23	11.65	12.00	12.31	12.8	13.3	13.8	14.2	15.0
0.13	9.80	10.29	10.71	11.06	11.37	11.9	12.4	12.8	13.3	14.1
0.14	9.00	9.49	9.91	10.26	10.57	11.1	11.6	12.0	12.5	13.4
0.15	8.32	8.80	9.22	9.67	9.88	10.4	10.9	11.4	11.8	12.7
0.16	7.72	8.20	8.61	8.96	9.27	9.81	10.29	10.75	11.2	12.1
0.17	7.19	7.67	8.08	8.43	8.74	9.28	9.77	10.23	10.68	11.6
0.18	6.73	7.20	7.61	7.96	8.27	8.81	9.30	9.76	10.21	11.1
0.19	6.31	6.78	7.19	7.54	7.85	8.39	8.88	9.34	9.80	10.7
0.20	5.94	6.40	6.81	7.16	7.47	8.01	8.50	8.97	9.43	10.3
0.22	5.30	5.76	6.16	6.51	6.82	7.36	7.86	8.33	8.79	9.70
0.24	4.77	5.22	5.62	5.96	6.27	6.82	7.32	7.79	8.26	9.18
0.26	4.32	4.77	5.16	5.51	5.82	6.37	6.87	7.35	7.81	8.74
0.28	3.95	4.38	4.77	5.12	5.48	5.98	6.48	6.96	7.43	8.36
0.30	3.62	4.05	4.44	4.78	5.09	5.64	6.15	6.63	7.10	8.04
0.32	3.34	3.77	4.15	4.49	4.80	5.35	5.86	6.34	6.82	7.75
0.34	3.09	3.51	3.89	4.23	4.54	5.10	5.60	6.09	6.56	7.50
0.36	2.88	3.29	3.67	4.01	4.31	4.87	5.38	5.86	6.34	7.28
0.38	2.68	3.09	3.47	3.81	4.11	4.67	5.17	5.66	6.14	7.09
0.40	2.51	2.92	3.29	3.62	3.93	4.48	4.99	5.48	5.96	6.91
0.42	2.36	2.76	3.13	3.46	3.77	4.32	4.83	5.32	5.80	6.75
0.44	2.22	2.61	2.98	3.31	3.62	4.17	4.68	5.17	5.66	6.60
0.46	2.09	2.48	2.85	3.18	3.48	4.04	4.55	5.04	5.52	6.47
0.48	1.98	2.36	2.72	3.06	3.36	3.91	4.43	4.92	5.40	6.35
0.50	1.87	2.26	2.61	2.94	3.25	3.80	4.31	4.81	5.29	6.24
0.55	1.65	2.02	2.37	2.70	3.00	3.55	4.07	4.56	5.05	6.00
0.60	1.46	1.83	2.17	2.50	2.80	3.35	3.86	4.36	4.84	5.80
0.70	1.18	1.53	1.87	2.19	2.48	3.03	3.55	4.04	4.53	5.49

(continued)

APPENDIX 19.E *(continued)*
Conveyance Factor, K
Symmetrical Rectangular,[a] Trapezoidal, and V-Notch[b] Open Channels
(use for determining Q or b when d is known)
(customary U.S. units[c,d])

$$K \text{ in } Q = K\left(\frac{1}{n}\right)d^{8/3}\sqrt{S}$$

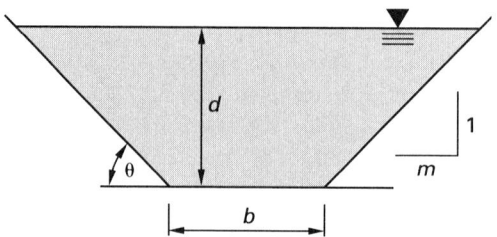

	m and θ									
	0.0	0.25	0.5	0.75	1.0	1.5	2.0	2.5	3.0	4.0
$x = d/b$	90°	76.0°	63.4°	53.1°	45.0°	33.7°	26.6°	21.8°	18.4°	14.0°
0.80	0.982	1.31	1.64	1.95	2.25	2.80	3.31	3.81	4.30	5.26
0.90	0.831	1.15	1.47	1.78	2.07	2.62	3.13	3.63	4.12	5.08
1.00	0.714	1.02	1.33	1.64	1.93	2.47	2.99	3.48	3.97	4.93
1.20	0.548	0.836	1.14	1.43	1.72	2.26	2.77	3.27	3.76	4.72
1.40	0.436	0.708	0.998	1.29	1.57	2.11	2.62	3.12	3.60	4.57
1.60	0.357	0.616	0.897	1.18	1.46	2.00	2.51	3.00	3.49	4.45
1.80	0.298	0.546	0.820	1.10	1.38	1.91	2.42	2.91	3.40	4.36
2.00	0.254	0.491	0.760	1.04	1.31	1.84	2.35	2.84	3.33	4.29
2.25	0.212	0.439	0.700	0.973	1.24	1.77	2.28	2.77	3.26	4.22
∞	0.00	0.091	0.274	0.499	0.743	1.24	1.74	2.23	2.71	3.67

[a]For rectangular channels, use the 0.0 (90°, vertical sides) column.
[b]For V-notch triangular channels, use the $d/b = \infty$ row.
[c]Q = flow rate, ft³/sec; d = depth of flow, ft; b = bottom width of channel, ft; S = geometric slope, ft/ft; n = Manning's roughness constant.
[d]For SI units (i.e., Q in m³/s and d and b in m), divide each table value by 1.486.

APPENDIX 19.F
Conveyance Factor, K'
Symmetrical Rectangular,[a] Trapezoidal Open Channels
(use for determining Q or d when b is known)
(customary U.S. units[b,c])

$$K' \text{ in } Q = K'\left(\frac{1}{n}\right)b^{8/3}\sqrt{S}$$

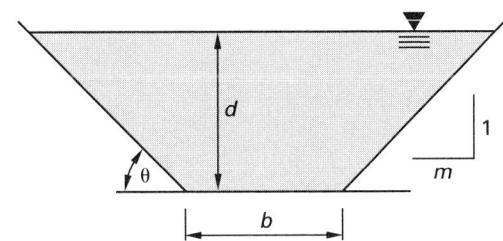

	m and θ									
	0.0	0.25	0.5	0.75	1.0	1.5	2.0	2.5	3.0	4.0
$x = d/b$	90°	76.0°	63.4°	53.1°	45.0°	33.7°	26.6°	21.8°	18.4°	14.0°
0.01	0.00068	0.00068	0.00069	0.00069	0.00069	0.00069	0.00069	0.00069	0.00070	0.00070
0.02	0.00213	0.00215	0.00216	0.00217	0.00218	0.00220	0.00221	0.00222	0.00223	0.00225
0.03	0.00414	0.00419	0.00423	0.00426	0.00428	0.00433	0.00436	0.00439	0.00443	0.00449
0.04	0.00660	0.00670	0.00679	0.00685	0.00691	0.00700	0.00708	0.00716	0.00723	0.00736
0.05	0.00946	0.00964	0.00979	0.00991	0.01002	0.01019	0.01033	0.01047	0.01060	0.01086
0.06	0.0127	0.0130	0.0132	0.0134	0.0136	0.0138	0.0141	0.0148	0.0145	0.0150
0.07	0.0162	0.0166	0.0170	0.0173	0.0175	0.0180	0.0183	0.0187	0.0190	0.0197
0.08	0.0200	0.0206	0.0211	0.0215	0.0219	0.0225	0.0231	0.0236	0.0240	0.0250
0.09	0.0241	0.0249	0.0256	0.0262	0.0267	0.0275	0.0282	0.0289	0.0296	0.0310
0.10	0.0284	0.0294	0.0304	0.0311	0.0318	0.0329	0.0339	0.0348	0.0358	0.0376
0.11	0.0329	0.0343	0.0354	0.0364	0.0373	0.0387	0.0400	0.0413	0.0424	0.0448
0.12	0.0376	0.0393	0.0408	0.0420	0.0431	0.0450	0.0466	0.0482	0.0497	0.0527
0.13	0.0425	0.0446	0.0464	0.0480	0.0493	0.0516	0.0537	0.0556	0.0575	0.0613
0.14	0.0476	0.0502	0.0524	0.0542	0.0559	0.0587	0.0612	0.0636	0.0659	0.0706
0.15	0.0528	0.0559	0.0585	0.0608	0.0627	0.0662	0.0692	0.0721	0.0749	0.0805
0.16	0.0582	0.0619	0.0650	0.0676	0.0700	0.0740	0.0777	0.0811	0.0845	0.0912
0.17	0.0638	0.0680	0.0716	0.0748	0.0775	0.0823	0.0866	0.0907	0.0947	0.1026
0.18	0.0695	0.0744	0.0786	0.0822	0.0854	0.0910	0.0960	0.1008	0.1055	0.1148
0.19	0.0753	0.0809	0.0857	0.0899	0.0936	0.1001	0.1059	0.1115	0.1169	0.1277
0.20	0.0812	0.0876	0.0931	0.0979	0.1021	0.1096	0.1163	0.1227	0.1290	0.1414
0.22	0.0934	0.1015	0.109	0.115	0.120	0.130	0.139	0.147	0.155	0.171
0.24	0.1061	0.1161	0.125	0.133	0.140	0.152	0.163	0.173	0.184	0.204
0.26	0.119	0.131	0.142	0.152	0.160	0.175	0.189	0.202	0.215	0.241
0.28	0.132	0.147	0.160	0.172	0.182	0.201	0.217	0.234	0.249	0.281
0.30	0.146	0.163	0.179	0.193	0.205	0.228	0.248	0.267	0.287	0.324
0.32	0.160	0.180	0.199	0.215	0.230	0.256	0.281	0.304	0.327	0.371
0.34	0.174	0.198	0.219	0.238	0.256	0.287	0.316	0.343	0.370	0.423
0.36	0.189	0.216	0.241	0.263	0.283	0.319	0.353	0.385	0.416	0.478
0.38	0.203	0.234	0.263	0.288	0.312	0.353	0.392	0.429	0.465	0.537
0.40	0.218	0.253	0.286	0.315	0.341	0.389	0.434	0.476	0.518	0.600
0.42	0.233	0.273	0.309	0.342	0.373	0.427	0.478	0.526	0.574	0.668
0.44	0.248	0.293	0.334	0.371	0.405	0.467	0.525	0.580	0.633	0.740
0.46	0.264	0.313	0.359	0.401	0.439	0.509	0.574	0.636	0.696	0.816
0.48	0.279	0.334	0.385	0.432	0.474	0.553	0.625	0.695	0.763	0.897
0.50	0.295	0.355	0.412	0.463	0.511	0.598	0.679	0.757	0.833	0.983
0.55	0.335	0.410	0.482	0.548	0.609	0.722	0.826	0.926	1.025	1.22
0.60	0.375	0.468	0.557	0.640	0.717	0.858	0.990	1.117	1.24	1.49
0.70	0.457	0.592	0.722	0.844	0.959	1.17	1.37	1.56	1.75	2.12

(continued)

APPENDIX 19.F *(continued)*
Conveyance Factor, K'
Symmetrical Rectangular,[a] Trapezoidal Open Channels
(use for determining Q or d when b is known)
(customary U.S. units[b,c])

$$K' \text{ in } Q = K'\left(\frac{1}{n}\right)b^{8/3}\sqrt{S}$$

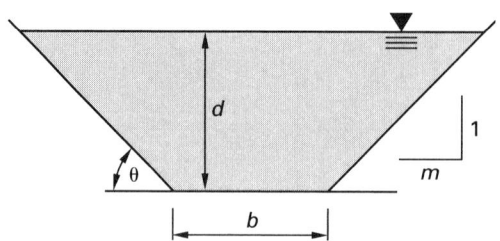

					m and θ					
	0.0	0.25	0.5	0.75	1.0	1.5	2.0	2.5	3.0	4.0
$x = d/b$	90°	76.0°	63.4°	53.1°	45.0°	33.7°	26.6°	21.8°	18.4°	14.0°
0.80	0.542	0.725	0.906	1.078	1.24	1.54	1.83	2.10	2.37	2.90
0.90	0.628	0.869	1.11	1.34	1.56	1.98	2.36	2.74	3.11	3.83
1.00	0.714	1.022	1.33	1.64	1.93	2.47	2.99	3.48	3.97	4.93
1.20	0.891	1.36	1.85	2.33	2.79	3.67	4.51	5.32	6.11	7.67
1.40	1.07	1.74	2.45	3.16	3.85	5.17	6.42	7.64	8.84	11.2
1.60	1.25	2.16	3.14	4.14	5.12	6.99	8.78	10.52	12.2	15.6
1.80	1.43	2.62	3.93	5.28	6.60	9.15	11.6	14.0	16.3	20.9
2.00	1.61	3.12	4.82	6.58	8.32	11.7	14.9	18.1	21.2	27.3
2.25	1.84	3.81	6.09	8.46	10.8	15.4	19.8	24.1	28.4	36.7

[a]For rectangular channels, use the 0.0 (90°, vertical sides) column.
[b]Q = flow rate, ft³/sec; d = depth of flow, ft; b = bottom width of channel, ft; S = geometric slope, ft/ft; n = Manning's roughness constant.
[c]For SI units (i.e., Q in m³/s, and d and b in m), divide each table value by 1.486.

APPENDIX 20.A
Rational Method Runoff *C*-Coefficients

categorized by surface

forested	0.059–0.2
asphalt	0.7–0.95
brick	0.7–0.85
concrete	0.8–0.95
shingle roof	0.75–0.95
lawns, well-drained (sandy soil)	
up to 2% slope	0.05–0.1
2% to 7% slope	0.10–0.15
over 7% slope	0.15–0.2
lawns, poor drainage (clay soil)	
up to 2% slope	0.13–0.17
2% to 7% slope	0.18–0.22
over 7% slope	0.25–0.35
driveways, walkways	0.75–0.85

categorized by use

farmland	0.05–0.3
pasture	0.05–0.3
unimproved	0.1–0.3
parks	0.1–0.25
cemeteries	0.1–0.25
railroad yards	0.2–0.35
playgrounds (except asphalt or concrete)	0.2–0.35
business districts	
neighborhood	0.5–0.7
city (downtown)	0.7–0.95
residential	
single family	0.3–0.5
multiplexes, detached	0.4–0.6
multiplexes, attached	0.6–0.75
suburban	0.25–0.4
apartments, condominiums	0.5–0.7
industrial	
light	0.5–0.8
heavy	0.6–0.9

APPENDIX 20.B
Random Numbers[*]

78466	83326	96589	88727	72655	49682	82338	28583	01522	11248
78722	47603	03477	29528	63956	01255	29840	32370	18032	82051
06401	87397	72898	32441	88861	71803	55626	77847	29925	76106
04754	14489	39420	94211	58042	43184	60977	74801	05931	73822
97118	06774	87743	60156	38037	16201	35137	54513	68023	34380
71923	49313	59713	95710	05975	64982	79253	93876	33707	84956
78870	77328	09637	67080	49168	75290	50175	34312	82593	76606
61208	17172	33187	92523	69895	28284	77956	45877	08044	58292
05033	24214	74232	33769	06304	54676	70026	41957	40112	66451
95983	13391	30369	51035	17042	11729	88647	70541	36026	23113
19946	55448	75049	24541	43007	11975	31797	05373	45893	25665
03580	67206	09635	84610	62611	86724	77411	99415	58901	86160
56823	49819	20283	22272	00114	92007	24369	00543	05417	92251
87633	31761	99865	31488	49947	06060	32083	47944	00449	06550
95152	10133	52693	22480	50336	49502	06296	76414	18358	05313
05639	24175	79438	92151	57602	03590	25465	54780	79098	73594
65927	55525	67270	22907	55097	63177	34119	94216	84861	10457
59005	29000	38395	80367	34112	41866	30170	84658	84441	03926
06626	42682	91522	45955	23263	09764	26824	82936	16813	13878
11306	02732	34189	04228	58541	72573	89071	58066	67159	29633
45143	56545	94617	42752	31209	14380	81477	36952	44934	97435
97612	87175	22613	84175	96413	83336	12408	89318	41713	90669
97035	62442	06940	45719	39918	60274	54353	54497	29789	82928
62498	00257	19179	06313	07900	46733	21413	63627	48734	92174
80306	19257	18690	54653	07263	19894	89909	76415	57246	02621
84114	84884	50129	68942	93264	72344	98794	16791	83861	32007
58437	88807	92141	88677	02864	02052	62843	21692	21373	29408
15702	53457	54258	47485	23399	71692	56806	70801	41548	94809
59966	41287	87001	26462	94000	28457	09469	80416	05897	87970
43641	05920	81346	02507	25349	93370	02064	62719	45740	62080
25501	50113	44600	87433	00683	79107	22315	42162	25516	98434
98294	08491	25251	26737	00071	45090	68628	64390	42684	94956
52582	89985	37863	60788	27412	47502	71577	13542	31077	13353
26510	83622	12546	00489	89304	15550	09482	07504	64588	92562
24755	71543	31667	83624	27085	65905	32386	30775	19689	41437
38399	88796	58856	18220	51056	04976	54062	49109	95563	48244
18889	87814	52232	58244	95206	05947	26622	01381	28744	38374
51774	89694	02654	63161	54622	31113	51160	29015	64730	07750
88375	37710	61619	69820	13131	90406	45206	06386	06398	68652
10416	70345	93307	87360	53452	61179	46845	91521	32430	74795

[*]To use, enter the table randomly and arbitrarily select any direction (i.e., up, down, to the right, left, or diagonally).

APPENDIX 22.A
Atomic Numbers and Weights of the Elements
(referred to carbon-12)

name	symbol	atomic number	atomic weight	name	symbol	atomic number	atomic weight
actinium	Ac	89	–	meitnerium	Mt	109	–
aluminum	Al	13	26.9815	mendelevium	Md	101	–
americium	Am	95	–	mercury	Hg	80	200.59
antimony	Sb	51	121.760	molybdenum	Mo	42	95.96
argon	Ar	18	39.948	neodymium	Nd	60	144.242
arsenic	As	33	74.9216	neon	Ne	10	20.1797
astatine	At	85	–	neptunium	Np	93	237.048
barium	Ba	56	137.327	nickel	Ni	28	58.693
berkelium	Bk	97	–	niobium	Nb	41	92.906
beryllium	Be	4	9.0122	nitrogen	N	7	14.0067
bismuth	Bi	83	208.980	nobelium	No	102	–
bohrium	Bh	107	–	osmium	Os	76	190.23
boron	B	5	10.811	oxygen	O	8	15.9994
bromine	Br	35	79.904	palladium	Pd	46	106.42
cadmium	Cd	48	112.411	phosphorus	P	15	30.9738
calcium	Ca	20	40.078	platinum	Pt	78	195.084
californium	Cf	98	–	plutonium	Pu	94	–
carbon	C	6	12.0107	polonium	Po	84	–
cerium	Ce	58	140.116	potassium	K	19	39.0983
cesium	Cs	55	132.9054	praseodymium	Pr	59	140.9077
chlorine	Cl	17	35.453	promethium	Pm	61	–
chromium	Cr	24	51.996	protactinium	Pa	91	231.0359
cobalt	Co	27	58.9332	radium	Ra	88	–
copernicium	Cn	112	–	radon	Rn	86	226.025
copper	Cu	29	63.546	rhenium	Re	75	186.207
curium	Cm	96	–	rhodium	Rh	45	102.9055
darmstadtium	Ds	110	–	roentgenium	Rg	111	–
dubnium	Db	105	–	rubidium	Rb	37	85.4678
dysprosium	Dy	66	162.50	ruthenium	Ru	44	101.07
einsteinium	Es	99	–	rutherfordium	Rf	104	–
erbium	Er	68	167.259	samarium	Sm	62	150.36
europium	Eu	63	151.964	scandium	Sc	21	44.956
fermium	Fm	100	–	seaborgium	Sg	106	–
fluorine	F	9	18.9984	selenium	Se	34	78.96
francium	Fr	87	–	silicon	Si	14	28.0855
gadolinium	Gd	64	157.25	silver	Ag	47	107.868
gallium	Ga	31	69.723	sodium	Na	11	22.9898
germanium	Ge	32	72.64	strontium	Sr	38	87.62
gold	Au	79	196.9666	sulfur	S	16	32.065
hafnium	Hf	72	178.49	tantalum	Ta	73	180.94788
hassium	Hs	108	–	technetium	Tc	43	–
helium	He	2	4.0026	tellurium	Te	52	127.60
holmium	Ho	67	164.930	terbium	Tb	65	158.925
hydrogen	H	1	1.00794	thallium	Tl	81	204.383
indium	In	49	114.818	thorium	Th	90	232.038
iodine	I	53	126.90447	thulium	Tm	69	168.934
iridium	Ir	77	192.217	tin	Sn	50	118.710
iron	Fe	26	55.845	titanium	Ti	22	47.867
krypton	Kr	36	83.798	tungsten	W	74	183.84
lanthanum	La	57	138.9055	uranium	U	92	238.0289
lawrencium	Lr	103	–	vanadium	V	23	50.942
lead	Pb	82	207.2	xenon	Xe	54	131.293
lithium	Li	3	6.941	ytterbium	Yb	70	173.054
lutetium	Lu	71	174.9668	yttrium	Y	39	88.906
magnesium	Mg	12	24.305	zinc	Zn	30	65.38
manganese	Mn	25	54.9380	zirconium	Zr	40	91.224

APPENDIX 22.B
Periodic Table of the Elements (referred to carbon-12)

The Periodic Table of Elements (Long Form)

The number of electrons in filled shells is shown in the column at the extreme left; the remaining electrons for each element are shown immediately below the symbol for each element. Atomic numbers are enclosed in brackets. Atomic weights (rounded, based on carbon-12) are shown above the symbols. Atomic weight values in parentheses are those of the isotopes of longest half-life for certain radioactive elements whose atomic weights cannot be precisely quoted without knowledge of origin of the element.

APPENDIX 22.C
Water Chemistry CaCO$_3$ Equivalents

cations	formula	ionic weight	equivalent weight	substance to CaCO$_3$ factor
aluminum	Al^{+3}	27.0	9.0	5.56
ammonium	NH$_4^+$	18.0	18.0	2.78
calcium	Ca^{+2}	40.1	20.0	2.50
cupric copper	Cu^{+2}	63.6	31.8	1.57
cuprous copper	Cu^{+3}	63.6	21.2	2.36
ferric iron	Fe^{+3}	55.8	18.6	2.69
ferrous iron	Fe^{+2}	55.8	27.9	1.79
hydrogen	H$^+$	1.0	1.0	50.00
manganese	Mn^{+2}	54.9	27.5	1.82
magnesium	Mg^{+2}	24.3	12.2	4.10
potassium	K$^+$	39.1	39.1	1.28
sodium	Na$^+$	23.0	23.0	2.18

anions	formula	ionic weight	equivalent weight	substance to CaCO$_3$ factor
bicarbonate	HCO$_3^-$	61.0	61.0	0.82
carbonate	CO$_3^{-2}$	60.0	30.0	1.67
chloride	Cl$^-$	35.5	35.5	1.41
fluoride	F$^-$	19.0	19.0	2.66
hydroxide	OH$^-$	17.0	17.0	2.94
nitrate	NO$_3^-$	62.0	62.0	0.81
phosphate (tribasic)	PO$_4^{-3}$	95.0	31.7	1.58
phosphate (dibasic)	HPO$_4^{-2}$	96.0	48.0	1.04
phosphate (monobasic)	H$_2$PO$_4^-$	97.0	97.0	0.52
sulfate	SO$_4^{-2}$	96.1	48.0	1.04
sulfite	SO$_3^{-2}$	80.1	40.0	1.25

(handwritten: CO$_2$ above formula column; 2.27 above substance to CaCO$_3$ factor column)

compounds	formula	molecular weight	equivalent weight	substance to CaCO$_3$ factor
aluminum hydroxide	Al(OH)$_3$	78.0	26.0	1.92
aluminum sulfate	Al$_2$(SO$_4$)$_3$	342.1	57.0	0.88
aluminum sulfate	Al$_2$(SO$_4$)$_3$·18H$_2$O	666.1	111.0	0.45
alumina	Al$_2$O$_3$	102.0	17.0	2.94
sodium aluminate	Na$_2$Al$_2$O$_4$	164.0	27.3	1.83
calcium bicarbonate	Ca(HCO$_3$)$_2$	162.1	81.1	0.62
calcium carbonate	CaCO$_3$	100.1	50.1	1.00
calcium chloride	CaCl$_2$	111.0	55.5	0.90
calcium hydroxide (pure)	Ca(OH)$_2$	74.1	37.1	1.35
calcium hydroxide (90%)	Ca(OH)$_2$	–	41.1	1.22
calcium oxide (lime)	CaO	56.1	28.0	1.79
calcium sulfate (anhydrous)	CaSO$_4$	136.2	68.1	0.74
calcium sulfate (gypsum)	CaSO$_4$·2H$_2$O	172.2	86.1	0.58
calcium phosphate	Ca$_3$(PO$_4$)$_2$	310.3	51.7	0.97
disodium phosphate	Na$_2$HPO$_4$·12H$_2$O	358.2	119.4	0.42
disodium phosphate (anhydrous)	Na$_2$HPO$_4$	142.0	47.3	1.06
ferric oxide	Fe$_2$O$_3$	159.6	26.6	1.88
iron oxide (magnetic)	Fe$_3$O$_4$	321.4	–	–
ferrous sulfate (copperas)	FeSO$_4$·7H$_2$O	278.0	139.0	0.36

(continued)

APPENDIX 22.C *(continued)*
Water Chemistry CaCO$_3$ Equivalents

compounds	formula	molecular weight	equivalent weight	substance to CaCO$_3$ factor
magnesium oxide	MgO	40.3	20.2	2.48
magnesium bicarbonate	Mg(HCO$_3$)$_2$	146.3	73.2	0.68
magnesium carbonate	MgCO$_3$	84.3	42.2	1.19
magnesium chloride	MgCl$_2$	95.2	47.6	1.05
magnesium hydroxide	Mg(OH)$_2$	58.3	29.2	1.71
magnesium phosphate	Mg$_3$(PO$_4$)$_2$	263.0	43.8	1.14
magnesium sulfate	MgSO$_4$	120.4	60.2	0.83
monosodium phosphate	NaH$_2$PO$_4$·H$_2$O	138.1	46.0	1.09
monosodium phosphate (anhydrous)	NaH$_2$PO$_4$	120.1	40.0	1.25
metaphosphate	NaPO$_3$	102.0	34.0	1.47
silica	SiO$_2$	60.1	30.0	1.67
sodium bicarbonate	NaHCO$_3$	84.0	84.0	0.60
sodium carbonate	Na$_2$CO$_3$	106.0	53.0	0.94
sodium chloride	NaCl	58.5	58.5	0.85
sodium hydroxide	NaOH	40.0	40.0	1.25
sodium nitrate	NaNO$_3$	85.0	85.0	0.59
sodium sulfate	Na$_2$SO$_4$	142.0	71.0	0.70
sodium sulfite	Na$_2$SO$_3$	126.1	63.0	0.79
tetrasodium EDTA	(CH$_2$)$_2$N$_2$(CH$_2$COONa)$_4$	380.2	95.1	0.53
trisodium phosphate	Na$_3$PO$_4$·12H$_2$O	380.2	126.7	0.40
trisodium phosphate (anhydrous)	Na$_3$PO$_4$	164.0	54.7	0.91
trisodium NTA	(CH$_2$)$_3$N(COONa)$_3$	257.1	85.7	0.58

gases	formula	molecular weight	equivalent weight	substance to CaCO$_3$ factor
ammonia	NH$_3$	17	17	2.94
carbon dioxide	CO$_2$	44	22	2.27
hydrogen	H$_2$	2	1	50.00
hydrogen sulfide	H$_2$S	34	17	2.94
oxygen	O$_2$	32	8	6.25

acids	formula	molecular weight	equivalent weight	substance to CaCO$_3$ factor
carbonic	H$_2$CO$_3$	62.0	31.0	1.61
hydrochloric	HCl	36.5	36.5	1.37
phosphoric	H$_3$PO$_4$	98.0	32.7	1.53
sulfuric	H$_2$SO$_4$	98.1	49.1	1.02

(Multiply the concentration (in mg/L) of the substance by the corresponding factors to obtain the equivalent concentration in mg/L as CaCO$_3$. For example, 70 mg/L of Mg^{++} would be (70 mg/L)(4.1) = 287 mg/L as CaCO$_3$.)

APPENDIX 22.D
Saturation Concentrations of Dissolved Oxygen in Water[*]

temperature (°C)	chloride concentration in water (mg/L)			difference per 100 mg chloride	vapor pressure (mm Hg)
	0	5000	10,000		
	dissolved oxygen (mg/L)				
0	14.62	13.79	12.97	0.017	5
1	14.23	13.41	12.61	0.016	5
2	13.84	13.05	12.28	0.015	5
3	13.48	12.72	11.98	0.015	6
4	13.13	12.41	11.69	0.014	6
5	12.80	12.09	11.39	0.014	7
6	12.48	11.79	11.12	0.014	7
7	12.17	11.51	10.85	0.013	8
8	11.87	11.24	10.61	0.013	8
9	11.59	10.97	10.36	0.012	9
10	11.33	10.73	10.13	0.012	9
11	11.08	10.49	9.92	0.011	10
12	10.83	10.28	9.72	0.011	11
13	10.60	10.05	9.52	0.011	11
14	10.37	9.85	9.32	0.010	12
15	10.15	9.65	9.14	0.010	13
16	9.95	9.46	8.96	0.010	14
17	9.74	9.26	8.78	0.010	15
18	9.54	9.07	8.62	0.009	16
19	9.35	8.89	8.45	0.009	17
20	9.17	8.73	8.30	0.009	18
21	8.99	8.57	8.14	0.009	19
22	8.83	8.42	7.99	0.008	20
23	8.68	8.27	7.85	0.008	21
24	8.53	8.12	7.71	0.008	22
25	8.38	7.96	7.56	0.008	24
26	8.22	7.81	7.42	0.008	25
27	8.07	7.67	7.28	0.008	27
28	7.92	7.53	7.14	0.008	28
29	7.77	7.39	7.00	0.008	30
30	7.63	7.25	6.86	0.008	32

[*]For saturation at barometric pressures other than 760 mm Hg (29.92 in Hg), C'_s is related to the corresponding tabulated value, C_s, by the following equation.

$$C'_s = C_s \left(\frac{P - p}{760 - p} \right)$$

C'_s = solubility at barometric pressure P and given temperature, mg/L
C_s = saturation solubility at given temperature from appendix, mg/L
P = barometric pressure, mm Hg
p = pressure of saturated water vapor at temperature of the water selected from appendix, mm Hg

APPENDIX 22.E
Names and Formulas of Important Chemicals

common name	chemical name	chemical formula
acetone	acetone	$(CH_3)_2CO$
acetylene	acetylene	C_2H_2
ammonia	ammonia	NH_3
ammonium	ammonium hydroxide	NH_4OH
aniline	aniline	$C_6H_5NH_2$
bauxite	hydrated aluminum oxide	$Al_2O_3 \cdot 2H_2O$
bleach	calcium hypochlorite	$Ca(ClO)_2$
borax	sodium tetraborate	$Na_2B_4O_7 \cdot 10H_2O$
carbide	calcium carbide	CaC_2
carbolic acid	phenol	C_6H_5OH
carbon dioxide	carbon dioxide	CO_2
carborundum	silicon carbide	SiC
caustic potash	potassium hydroxide	KOH
caustic soda/lye	sodium hydroxide	$NaOH$
chalk	calcium carbonate	$CaCO_3$
cinnabar	mercuric sulfide	HgS
ether	diethyl ether	$(C_2H_5)_2O$
formic acid	methanoic acid	$HCOOH$
Glauber's salt	decahydrated sodium sulfate	$Na_2SO_4 \cdot 10H_2O$
glycerine	glycerine	$C_3H_5(OH)_3$
grain alcohol	ethanol	C_2H_5OH
graphite	crystalline carbon	C
gypsum	calcium sulfate	$CaSO_4 \cdot 2H_2O$
halite	sodium chloride	$NaCl$
iron chloride	ferrous chloride	$FeCl_2 \cdot 4H_2O$
laughing gas	nitrous oxide	N_2O
limestone	calcium carbonate	$CaCO_3$
magnesia	magnesium oxide	MgO
marsh gas	methane	CH_4
muriate of potash	potassium chloride	KCl
muriatic acid	hydrochloric acid	HCl
niter	sodium nitrate	$NaNO_3$
niter cake	sodium bisulfate	$NaHSO_4$
oleum	fuming sulfuric acid	SO_3 in H_2SO_4
potash	potassium carbonate	K_2CO_3
prussic acid	hydrogen cyanide	HCN
pyrites	ferrous sulfide	FeS
pyrolusite	manganese dioxide	MnO_2
quicklime	calcium oxide	CaO
sal soda	decahydrated sodium carbonate	$NaCO_3 \cdot 10H_2O$
salammoniac	ammonium chloride	NH_4Cl
sand or silica	silicon dioxide	SiO_2
salt cake	sodium sulfate (crude)	Na_2SO_4
slaked lime	calcium hydroxide	$Ca(OH)_2$
soda ash	sodium carbonate	Na_2CO_3
soot	amorphous carbon	C
stannous chloride	stannous chloride	$SnCl_2 \cdot 2H_2O$
superphosphate	monohydrated primary calcium phosphate	$Ca(H_2PO_4)_2 \cdot H_2O$
table salt	sodium chloride	$NaCl$
table sugar	sucrose	$C_{12}H_{22}O_{11}$
trilene	trichloroethylene	C_2HCl_3
urea	urea	$CO(NH_2)_2$
vinegar (acetic acid)	ethanoic acid	CH_3COOH
washing soda	decahydrated sodium carbonate	$Na_2CO_3 \cdot 10H_2O$
wood alcohol	methanol	CH_3OH
zinc blende	zinc sulfide	ZnS

APPENDIX 22.F
Approximate Solubility Product Constants at 25°C

substance	formula	K_{sp}
aluminum hydroxide	$Al(OH)_3$	1.3×10^{-33}
aluminum phosphate	$AlPO_4$	6.3×10^{-19}
barium carbonate	$BaCO_3$	5.1×10^{-9}
barium chromate	$BaCrO_4$	1.2×10^{-10}
barium fluoride	BaF_2	1.0×10^{-6}
barium hydroxide	$Ba(OH)_2$	5×10^{-3}
barium sulfate	$BaSO_4$	1.1×10^{-10}
barium sulfite	$BaSO_3$	8×10^{-7}
barium thiosulfate	BaS_2O_3	1.6×10^{-6}
bismuthyl chloride	$BiOCl$	1.8×10^{-31}
bismuthyl hydroxide	$BiOOH$	4×10^{-10}
cadmium carbonate	$CdCO_3$	5.2×10^{-12}
cadmium hydroxide	$Cd(OH)_2$	2.5×10^{-14}
cadmium oxalate	CdC_2O_4	1.5×10^{-8}
cadmium sulfide[a]	CdS	8×10^{-28}
calcium carbonate[b]	$CaCO_3$	2.8×10^{-9}
calcium chromate	$CaCrO_4$	7.1×10^{-4}
calcium fluoride	CaF_2	5.3×10^{-9}
calcium hydrogen phosphate	$CaHPO_4$	1×10^{-7}
calcium hydroxide	$Ca(OH)_2$	5.5×10^{-6}
calcium oxalate	CaC_2O_4	2.7×10^{-9}
calcium phosphate	$Ca_3(PO_4)_2$	2.0×10^{-29}
calcium sulfate	$CaSO_4$	9.1×10^{-6}
calcium sulfite	$CaSO_3$	6.8×10^{-8}
chromium (II) hydroxide	$Cr(OH)_2$	2×10^{-16}
chromium (III) hydroxide	$Cr(OH)_3$	6.3×10^{-31}
cobalt (II) carbonate	$CoCO_3$	1.4×10^{-13}
cobalt (II) hydroxide	$Co(OH)_2$	1.6×10^{-15}
cobalt (III) hydroxide	$Co(OH)_3$	1.6×10^{-44}
cobalt (II) sulfide[a]	CoS	4×10^{-21}
copper (I) chloride	$CuCl$	1.2×10^{-6}
copper (I) cyanide	$CuCN$	3.2×10^{-20}
copper (I) iodide	CuI	1.1×10^{-12}
copper (II) arsenate	$Cu_3(AsO_4)_2$	7.6×10^{-36}
copper (II) carbonate	$CuCO_3$	1.4×10^{-10}
copper (II) chromate	$CuCrO_4$	3.6×10^{-6}
copper (II) ferrocyanide	$Cu[Fe(CN)_6]$	1.3×10^{-16}
copper (II) hydroxide	$Cu(OH)_2$	2.2×10^{-20}
copper (II) sulfide[a]	CuS	6×10^{-37}
iron (II) carbonate	$FeCO_3$	3.2×10^{-11}
iron (II) hydroxide	$Fe(OH)_2$	8.0×10^{-16}
iron (II) sulfide[a]	FeS	6×10^{-19}
iron (III) arsenate	$FeAsO_4$	5.7×10^{-21}
iron (III) ferrocyanide	$Fe_4[Fe(CN)_6]_3$	3.3×10^{-41}
iron (III) hydroxide	$Fe(OH)_3$	4×10^{-38}
iron (III) phosphate	$FePO_4$	1.3×10^{-22}

(continued)

APPENDIX 22.F *(continued)*
Approximate Solubility Product Constants at 25°C

substance	formula	K_{sp}
lead (II) arsenate	$Pb_3(AsO_4)_2$	4×10^{-36}
lead (II) azide	$Pb(N_3)_2$	2.5×10^{-9}
lead (II) bromide	$PbBr_2$	4.0×10^{-5}
lead (II) carbonate	$PbCO_3$	7.4×10^{-14}
lead (II) chloride	$PbCl_2$	1.6×10^{-5}
lead (II) chromate	$PbCrO_4$	2.8×10^{-13}
lead (II) fluoride	PbF_2	2.7×10^{-8}
lead (II) hydroxide	$Pb(OH)_2$	1.2×10^{-15}
lead (II) iodide	PbI_2	7.1×10^{-9}
lead (II) sulfate	$PbSO_4$	1.6×10^{-8}
lead (II) sulfide[a]	PbS	3×10^{-28}
lithium carbonate	Li_2CO_3	2.5×10^{-2}
lithium fluoride	LiF	3.8×10^{-3}
lithium phosphate	Li_3PO_4	3.2×10^{-9}
magnesium ammonium phosphate	$MgNH_4PO_4$	2.5×10^{-13}
magnesium arsenate	$Mg_3(AsO_4)_2$	2×10^{-20}
magnesium carbonate	$MgCO_3$	3.5×10^{-8}
magnesium fluoride	MgF_2	3.7×10^{-8}
magnesium hydroxide	$Mg(OH)_2$	1.8×10^{-11}
magnesium oxalate	MgC_2O_4	8.5×10^{-5}
magnesium phosphate	$Mg_3(PO_4)_2$	1×10^{-25}
manganese (II) carbonate	$MnCO_3$	1.8×10^{-11}
manganese (II) hydroxide	$Mn(OH)_2$	1.9×10^{-13}
manganese (II) sulfide[a]	MnS	3×10^{-14}
mercury (I) bromide	Hg_2Br_2	5.6×10^{-23}
mercury (I) chloride	Hg_2Cl_2	1.3×10^{-18}
mercury (I) iodide	Hg_2I_2	4.5×10^{-29}
mercury (II) sulfide[a]	HgS	2×10^{-53}
nickel (II) carbonate	$NiCO_3$	6.6×10^{-9}
nickel (II) hydroxide	$Ni(OH)_2$	2.0×10^{-15}
nickel (II) sulfide[a]	NiS	3×10^{-19}
scandium fluoride	ScF_3	4.2×10^{-18}
scandium hydroxide	$Sc(OH)_3$	8.0×10^{-31}
silver acetate	$AgC_2H_3O_2$	2.0×10^{-3}
silver arsenate	Ag_3AsO_4	1.0×10^{-22}
silver azide	AgN_3	2.8×10^{-9}
silver bromide	$AgBr$	5.0×10^{-13}
silver chloride	$AgCl$	1.8×10^{-10}
silver chromate	Ag_2CrO_4	1.1×10^{-12}
silver cyanide	$AgCN$	1.2×10^{-16}
silver iodate	$AgIO_3$	3.0×10^{-8}
silver iodide	AgI	8.5×10^{-17}
silver nitrite	$AgNO_2$	6.0×10^{-4}
silver sulfate	Ag_2SO_4	1.4×10^{-5}
silver sulfide[a]	Ag_2S	6×10^{-51}
silver sulfite	Ag_2SO_3	1.5×10^{-14}
silver thiocyanate	$AgSCN$	1.0×10^{-12}

(continued)

APPENDIX 22.F *(continued)*
Approximate Solubility Product Constants at 25°C

substance	formula	K_{sp}
strontium carbonate	$SrCO_3$	1.1×10^{-10}
strontium chromate	$SrCrO_4$	2.2×10^{-5}
strontium fluoride	SrF_2	2.5×10^{-9}
strontium sulfate	$SrSO_4$	3.2×10^{-7}
thallium (I) bromide	$TlBr$	3.4×10^{-6}
thallium (I) chloride	$TlCl$	1.7×10^{-4}
thallium (I) iodide	TlI	6.5×10^{-8}
thallium (III) hydroxide	$Tl(OH)_3$	6.3×10^{-46}
tin (II) hydroxide	$Sn(OH)_2$	1.4×10^{-28}
tin (II) sulfide[a]	SnS	1×10^{-26}
zinc carbonate	$ZnCO_3$	1.4×10^{-11}
zinc hydroxide	$Zn(OH)_2$	1.2×10^{-17}
zinc oxalate	ZnC_2O_4	2.7×10^{-8}
zinc phosphate	$Zn_3(PO_4)_2$	9.0×10^{-33}
zinc sulfide[a]	ZnS	2×10^{-25}

[a]Sulfide equilibrium of the type:
$MS(s) + H_2O(l) \rightleftharpoons M^{2+}(aq) + HS^-(aq) + OH^-(aq)$
[b]Solubility product depends on mineral form.

APPENDIX 22.G
Dissociation Constants of Acids at 25°C

acid		K_a
acetic	K_1	1.8×10^{-5}
arsenic	K_1	5.6×10^{-3}
	K_2	1.2×10^{-7}
	K_3	3.2×10^{-12}
arsenious	K_1	1.4×10^{-9}
benzoic	K_1	6.3×10^{-5}
boric	K_1	5.9×10^{-10}
carbonic	$K_1{}^*$	4.5×10^{-7}
	K_2	5.6×10^{-11}
chloroacetic	K_1	1.4×10^{-3}
chromic	K_2	3.2×10^{-7}
citric	K_1	7.4×10^{-4}
	K_2	1.7×10^{-5}
	K_3	3.9×10^{-7}
ethylenedinitrilotetracetic	K_1	1.0×10^{-2}
	K_2	2.1×10^{-3}
	K_3	6.9×10^{-7}
	K_4	7.4×10^{-11}
formic	K_1	1.8×10^{-4}
hydrocyanic	K_1	4.9×10^{-10}
hydrofluoric	K_1	6.8×10^{-4}
hydrogen sulfide	K_1	1.0×10^{-8}
	K_2	1.2×10^{-14}
hypochlorous	K_1	2.8×10^{-8}
iodic	K_1	1.8×10^{-1}
nitrous	K_1	4.5×10^{-4}
oxalic	K_1	5.4×10^{-2}
	K_2	5.1×10^{-5}
phenol	K_1	1.1×10^{-10}
phosphoric (ortho)	K_1	7.1×10^{-3}
	K_2	6.3×10^{-8}
	K_3	4.4×10^{-13}
o-phthalic	K_1	1.1×10^{-3}
	K_2	3.9×10^{-6}
salicylic	K_1	1.0×10^{-3}
	K_2	4.0×10^{-14}
sulfamic	K_1	1.0×10^{-1}
sulfuric	K_1	1.1×10^{-2}
sulfurous	K_1	1.7×10^{-2}
	K_2	6.3×10^{-8}
tartaric	K_1	9.2×10^{-4}
	K_2	4.3×10^{-5}
thiocyanic	K_1	1.4×10^{-1}

*apparent constant based on $C_{H_2CO_3} = [CO_2] + [H_2CO_3]$

APPENDIX 22.H
Dissociation Constants of Bases at 25°C

base		K_b
2-amino-2-(hydroxymethyl)-1,3-propanediol	K_1	1.2×10^{-6}
ammonia	K_1	1.8×10^{-5}
aniline	K_1	4.2×10^{-10}
diethylamine	K_1	1.3×10^{-3}
hexamethylenetetramine	K_1	1.0×10^{-9}
hydrazine	K_1	9.8×10^{-7}
hydroxylamine	K_1	9.6×10^{-9}
lead hydroxide	K_1	1.2×10^{-4}
piperidine	K_1	1.3×10^{-3}
pyridine	K_1	1.5×10^{-9}
silver hydroxide	K_1	6.0×10^{-5}

APPENDIX 23.A
National Primary Drinking Water Standards
Code of Federal Regulations (CFR), Title 40, Ch. I, Part 141, July 2013

microorganisms	MCLG^a (mg/L)^b	MCL or TT^a (mg/L)^b	potential health effects from ingestion of water	sources of contaminant in drinking water
Cryptosporidium	0	TT^c	gastrointestinal illness (e.g., diarrhea, vomiting, cramps)	human and animal fecal waste
Giardia lamblia	0	TT^c	gastrointestinal illness (e.g., diarrhea, vomiting, cramps)	human and animal fecal waste
heterotrophic plate count (HPC)	n/a	TT^c	HPC has no health effects; it is an analytic method used to measure the variety of bacteria that are common in water. The lower the concentration of bacteria in drinking water, the better maintained the water is.	HPC measures a range of bacteria that are naturally present in the environment.
Legionella	0	TT^c	Legionnaires' disease, a type of pneumonia	found naturally in water; multiplies in heating systems
total coliforms (including fecal coliform and *E. coli*)	0	5.0%^{d}	Not a health threat in itself; it is used to indicate whether other potentially harmful bacteria may be present.^{e}	Coliforms are naturally present in the environment as well as in feces; fecal coliforms and *E. coli* only come from human and animal fecal waste.
turbidity	n/a	TT^c	Turbidity is a measure of the cloudiness of water. It is used to indicate water quality and filtration effectiveness (e.g., whether disease-causing organisms are present). Higher turbidity levels are often associated with higher levels of disease-causing microorganisms such as viruses, parasites, and some bacteria. These organisms can cause symptoms such as nausea, cramps, diarrhea, and associated headaches.	soil runoff
viruses (enteric)	0	TT^c	gastrointestinal illness (e.g., diarrhea, vomiting, cramps)	human and animal fecal waste

disinfection products	MCLG^a (mg/L)^b	MCL or TT^a (mg/L)^b	potential health effects from ingestion of water	sources of contaminant in drinking water
bromate	0	0.010	increased risk of cancer	byproduct of drinking-water disinfection
chlorite	0.8	1.0	anemia in infants and young children; nervous system effects	byproduct of drinking-water disinfection
haloacetic acids (HAA5)	n/a^f	0.060	increased risk of cancer	byproduct of drinking-water disinfection
total trihalomethanes (TTHMs)	none^g n/a^f	0.10 0.080	liver, kidney, or central nervous system problems; increased risk of cancer	byproduct of drinking-water disinfection

(continued)

APPENDIX 23.A (continued)
National Primary Drinking Water Standards
Code of Federal Regulations (CFR), Title 40, Ch. I, Part 141, July 2013

disinfection products	$MCLG^a$ $(mg/L)^b$	MCL or TT^a $(mg/L)^b$	potential health effects from ingestion of water	sources of contaminant in drinking water
chloramines (as Cl_2)	4^a	4.0^a	eye/nose irritation, stomach discomfort, anemia	water additive used to control microbes
chlorine (as Cl_2)	4^a	4.0^a	eye/nose irritation, stomach discomfort	water additive used to control microbes
chlorine dioxide (as ClO_2)	0.8^a	0.8^a	anemia in infants and young children, nervous system effects	water additive used to control microbes

inorganic chemicals	$MCLG^a$ $(mg/L)^b$	MCL or TT^a $(mg/L)^b$	potential health effects from ingestion of water	sources of contaminant in drinking water
antimony	0.006	0.006	increase in blood cholesterol; decrease in blood sugar	discharge from petroleum refineries; fire retardants; ceramics; electronics; solder
arsenic	0^g	0.010 as of January 23, 2003	skin damage or problems with circulatory systems; may increase cancer risk	erosion of natural deposits; runoff from orchards; runoff from glass and electronics production wastes
asbestos (fiber > 10 micrometers)	7 million fibers per liter	7 MFL	increased risk of developing benign intestinal polyps	decay of asbestos cement in water mains; erosion of natural deposits
barium	2	2	increase in blood pressure	discharge of drilling wastes; discharge from metal refineries; erosion of natural deposits
beryllium	0.004	0.004	intestinal lesions	discharge from metal refineries and coal-burning factories; discharge from electrical, aerospace, and defense industries
cadmium	0.005	0.005	kidney damage	corrosion of galvanized pipes; erosion of natural deposits; discharge from metal refineries; runoff from waste batteries and paints
chromium (total)	0.1	0.1	allergic dermatitis	discharge from steel and pulp mills; erosion of natural deposits
copper	1.3	TT^h, action level = 1.3	short-term exposure: gastrointestinal distress long-term exposure: liver or kidney damage People with Wilson's disease should consult their personal doctor if the amount of copper in their water exceeds the action level.	corrosion of household plumbing systems; erosion of natural deposits
cyanide (as free cyanide)	0.2	0.2	nerve damage or thyroid problems	discharge from steel/metal factories; discharge from plastic and fertilizer factories
fluoride	4.0	4.0	bone disease (pain and tenderness of the bones); children may get mottled teeth	water additive that promotes strong teeth; erosion of natural deposits; discharge from fertilizer and aluminum factories

(continued)

APPENDIX 23.A *(continued)*
National Primary Drinking Water Standards
Code of Federal Regulations (CFR), Title 40, Ch. I, Part 141, July 2013

inorganic chemicals	MCLG^a (mg/L)^b	MCL or TT^a (mg/L)^b	potential health effects from ingestion of water	sources of contaminant in drinking water
lead	0	TT^h, action level = 0.015	infants and children: delays in physical or mental development; children could show slight deficits in attention span and learning disabilities	corrosion of household plumbing systems; erosion of natural deposits
			adults: kidney problems, high blood pressure	
mercury (inorganic)	0.002	0.002	kidney damage	erosion of natural deposits; discharge from refineries and factories; runoff from landfills and croplands
nitrate (measured as nitrogen)	10	10	Infants below the age of six months who drink water containing nitrate in excess of the MCL could become seriously ill and, if untreated, may die. Symptoms include shortness of breath and blue baby syndrome.	runoff from fertilizer use; leaching from septic tanks/sewage; erosion of natural deposits
nitrite (measured as nitrogen)	1	1	Infants below the age of six months who drink water containing nitrite in excess of the MCL could become seriously ill and, if untreated, may die. Symptoms include shortness of breath and blue baby syndrome.	runoff from fertilizer use; leaching from septic tanks/sewage; erosion of natural deposits
selenium	0.05	0.05	hair and fingernail loss; numbness in fingers or toes; circulatory problems	discharge from petroleum refineries; erosion of natural deposits; discharge from mines
thalium	0.0005	0.002	hair loss; changes in blood; kidney, intestine, or liver problems	leaching from ore-processing sites; discharge from electronics, glass, and drug factories

organic chemicals	MCLG^a (mg/L)^b	MCL or TT^a (mg/L)^b	potential health effects from ingestion of water	sources of contaminant in drinking water
acrylamide	0	TT^i	nervous system or blood problems; increased risk of cancer	added to water during sewage/wastewater treatment
alachlor	0	0.002	eye, liver, kidney, or spleen problems; anemia; increased risk of cancer	runoff from herbicide used on row crops
atrazine	0.003	0.003	cardiovascular system or reproductive problems	runoff from herbicide used on row crops
benzene	0	0.005	anemia; decrease in blood platelets; increased risk of cancer	discharge from factories; leaching from gas storage tanks and landfills
benzo(a)pyrene (PAHs)	0	0.0002	reproductive difficulties; increased risk of cancer	leaching from linings of water storage tanks and distribution lines
carbofuran	0.04	0.04	problems with blood, nervous system, or reproductive system	leaching of soil fumigant used on rice and alfalfa
carbon tetrachloride	0	0.005	liver problems; increased risk of cancer	discharge from chemical plants and other industrial activities

(continued)

APPENDIX 23.A *(continued)*
National Primary Drinking Water Standards
Code of Federal Regulations (CFR), Title 40, Ch. I, Part 141, July 2013

organic chemicals	MCLG[a] (mg/L)[b]	MCL or TT[a] (mg/L)[b]	potential health effects from ingestion of water	sources of contaminant in drinking water
chlordane	0	0.002	liver or nervous system problems; increased risk of cancer	residue of banned termiticide
chlorobenzene	0.1	0.1	liver or kidney problems	discharge from chemical and agricultural chemical factories
2,4-D	0.07	0.7	kidney, liver, or adrenal gland problems	runoff from herbicide used on row crops
dalapon	0.2	0.2	minor kidney changes	runoff from herbicide used on rights of way
1,2-dibromo-3-chloropropane (DBCP)	0	0.0002	reproductive difficulties; increased risk of cancer	runoff/leaching from soil fumigant used on soybeans, cotton, pineapples, and orchards
o-dichlorobenzene	0.6	0.6	liver, kidney, or circulatory system problems	discharge from industrial chemical factories
p-dichlorobenzene	0.075	0.075	anemia; liver, kidney, or spleen damage; changes in blood	discharge from industrial chemical factories
1,2-dichloroethane	0	0.005	increased risk of cancer	discharge from industrial chemical factories
1,1-dichloroethylene	0.007	0.007	liver problems	discharge from industrial chemical factories
cis-1,2-dichloro-ethylene	0.07	0.07	liver problems	discharge from industrial chemical factories
trans-1,2-dichloro-ethylene	0.1	0.1	liver problems	discharge from industrial chemical factories
dichloromethane	0	0.005	liver problems; increased risk of cancer	discharge from industrial chemical factories
1,2-dichloropropane	0	0.005	increased risk of cancer	discharge from industrial chemical factories
di(2-ethylhexyl) adipate	0.4	0.4	general toxic effects or reproductive difficulties	discharge from chemical factories
di(2-ethylhexyl) phthalate	0	0.006	reproductive difficulties; liver problems; increased risk of cancer	discharge from rubber and chemical factories
dinoseb	0.007	0.007	reproductive difficulties	runoff from herbicide used on soybeans and vegetables
dioxin (2,3,7,8-TCDD)	0	0.00000003	reproductive difficulties; increased risk of cancer	emissions from waste incineration and other combustion; discharge from chemical factories
diquat	0.02	0.02	cataracts	runoff from herbicide use
endothall	0.1	0.1	stomach and intestinal problems	runoff from herbicide use
endrin	0.002	0.002	liver problems	residue of banned insecticide
epichlorohydrin	0	TT[i]	increased cancer risk; over a long period of time, stomach problems	discharge from industrial chemical factories; an impurity of some water treatment chemicals
ethylbenzene	0.7	0.7	liver or kidney problems	discharge from petroleum refineries
ethylene dibromide	0	0.00005	problems with liver, stomach, reproductive system, or kidneys; increased risk of cancer	discharge from petroleum refineries

(continued)

APPENDIX 23.A (continued)
National Primary Drinking Water Standards
Code of Federal Regulations (CFR), Title 40, Ch. I, Part 141, July 2013

organic chemicals	MCLG[a] (mg/L)[b]	MCL or TT[a] (mg/L)[b]	potential health effects from ingestion of water	sources of contaminant in drinking water
glyphosate	0.7	0.7	kidney problems; reproductive difficulties	runoff from herbicide use
heptachlor	0	0.0004	liver damage; increased risk of cancer	residue of banned termiticide
heptachlor epoxide	0	0.0002	liver damage; increased risk of cancer	breakdown of heptachlor
hexachlorobenzene	0	0.001	liver or kidney problems; reproductive difficulties; increased risk of cancer	discharge from metal refineries and agricultural chemical factories
hexachlorocyclo-pentadiene	0.05	0.05	kidney or stomach problems	discharge from chemical factories
lindane	0.0002	0.0002	liver or kidney problems	runoff/leaching from insecticide used on cattle, lumber, and gardens
methoxychlor	0.04	0.04	reproductive difficulties	runoff/leaching from insecticide used on fruits, vegetables, alfalfa, and livestock
oxamyl (vydate)	0.2	0.2	slight nervous system effects	runoff/leaching from insecticide used on apples, potatoes, and tomatoes
polychlorinated biphenyls (PCBs)	0	0.0005	skin changes; thymus gland problems; immune deficiencies; reproductive or nervous system difficulties; increased risk of cancer	runoff from landfills; discharge of waste chemicals
pentachlorophenol	0	0.001	liver or kidney problems; increased cancer risk	discharge from wood preserving factories
picloram	0.5	0.5	liver problems	herbicide runoff
simazine	0.004	0.004	problems with blood	herbicide runoff
styrene	0.1	0.1	liver, kidney, or circulatory system problems	discharge from rubber and plastic factories; leaching from landfills
tetrachloroethylene	0	0.005	liver problems; increased risk of cancer	discharge from factories and dry cleaners
toluene	1	1	nervous system, kidney, or liver problems	discharge from petroleum factories
toxaphene	0	0.003	kidney, liver, or thyroid problems; increased risk of cancer	runoff/leaching from insecticide used on cotton and cattle
2,4,5-TP (silvex)	0.05	0.05	liver problems	residue of banned herbicide
1,2,4-trichlorobenzene	0.07	0.07	changes in adrenal glands	discharge from textile finishing factories
1,1,1-trichloroethane	0.20	0.2	liver, nervous system, or circulatory problems	discharge from metal degreasing sites and other factories
1,1,2-trichloroethane	0.003	0.005	liver, kidney, or immune system problems	discharge from industrial chemical factories
trichloroethylene	0	0.005	liver problems; increased risk of cancer	discharge from metal degreasing sites and other factories
vinyl chloride	0	0.002	increased risk of cancer	leaching from PVC pipes; discharge from plastic factories
xylenes (total)	10	10	nervous system damage	discharge from petroleum factories; discharge from chemical factories

(continued)

APPENDIX 23.A *(continued)*
National Primary Drinking Water Standards
Code of Federal Regulations (CFR), Title 40, Ch. I, Part 141, July 2013

radionuclides	MCLG^a (mg/L)^b	MCL or TT^a (mg/L)^b	potential health effects from ingestion of water	sources of contaminant in drinking water
alpha particles	none^g	15 pCi/L	increased risk of cancer	erosion of natural deposits of certain minerals that are radioactive and may emit a form of radiation known as alpha radiation
beta particles emitters	none^g	4 mrem/yr	increased risk of cancer	decay of natural and artificial deposits of certain minerals that are radioactive and may emit forms of radiation known as photons and beta radiation
radium 226 and radium 228 (combined)	none^g	5 pCi/L	increased risk of cancer	erosion of natural deposits
uranium	0	30 μg/L as of December 8, 2003	increased risk of cancer; kidney toxicity	erosion of natural deposits

^{a}Definitions:
Maximum Contaminant Level (MCL)—The highest level of a contaminant that is allowed in drinking water. MCLs are set as close to MCLGs as feasible using the best available treatment technology and taking cost into consideration. MCLs are enforceable standards.
Maximum Contaminant Level Goal (MCLG)—The level of a contaminant in drinking water below which there is no known or expected risk to health. MCLGs allow for a margin of safety and are non-enforceable public health goals.
Maximum Residual Disinfectant Level (MRDL)—The highest level of a disinfectant allowed in drinking water. There is convincing evidence that addition of a disinfectant is necessary for control of microbial contaminants.
Maximum Residual Disinfectant Level Goal (MRDLG)—The level of a drinking water disinfectant below which there is no known or expected risk to health. MRDLGs do not reflect the benefits of the use of disinfectants to control microbial contaminants.
Treatment Technique (TT)—A required process intended to reduce the level of a contaminant in drinking water.
^{b}Units are in milligrams per liter (mg/L) unless otherwise noted. Milligrams per liter are equivalent to parts per million.
^{c}The EPA's surface water treatment rules require systems using surface water or ground water under the direct influence of surface water to (1) disinfect their water, and (2) filter their water or meet criteria for avoiding filtration so that the following contaminants are controlled at the following levels.
• cryptosporidium (as of January 1, 2002, for systems serving > 10,000 and January 14, 2005, for systems serving < 10,000): 99% removal
• *Giardia lamblia*: 99.9% removal/inactivation
• Viruses: 99.99% removal/inactivation
• *Legionella*: No limit, but the EPA believes that if *Giardia* and viruses are removed/inactivated, *Legionella* will also be controlled.
• Turbidity: At no time can turbidity (cloudiness of water) go above 5 nephelolometric turbidity units (NTU); systems that filter must ensure that the turbidity go no higher than 1 NTU (0.5 NTU for conventional or direct filtration) in at least 95% of the daily samples in any month. As of January 1, 2002, turbidity may never exceed 1 NTU, and must not exceed 0.3 NTU in 95% of daily samples in any month.
• heterotrophic plate count (HPC): No more than 500 bacterial colonies per milliliter.
• Long Term 1 Enhanced Surface Water Treatment (as of January 14, 2005): Surface water systems or ground water under direct influence (GWUDI) systems serving fewer than 10,000 people must comply with the applicable Long Term 1 Enhanced Surface Water Treatment Rule provisions (e.g., turbidity standards, individual filter monitoring, cryptosporidium removal requirements, updated watershed control requirements for unfiltered systems).
• Filter Backwash Recycling: The Filter Backwash Recycling Rule requires systems that recycle to return specific recycle flows through all processes of the systems' existing conventional or direct filtration system or at an alternate location approved by the state.
^{d}More than 5.0% of samples are total coliform-positive in a month. (For water systems that collect fewer than 40 routine samples per month, no more than one sample can be total coliform-positive per month.) Every sample that has total coliform must be analyzed for either fecal coliforms or *E. coli*: If two consecutive samples are TC-positive, and one is also positive for *E. coli* or fecal coliforms, the system has an acute MCL violation.
^{e}Fecal coliform and *E. coli* are bacteria whose presence indicates that the water may be contaminated with human or animal wastes. Disease-causing microbes (pathogens) in these wastes can cause short-term effects, such as diarrhea, cramps, nausea, headaches, or other symptoms. These pathogens may pose a special health risk for infants, young children, some of the elderly, and people with severely compromised immune systems.
^{f}Although there is no collective MCLG for this contaminant group, there are individual MCLGs for some of the individual contaminants.
• Haloacetic acids: dichloroacetic acid (0); trichloroacetic acid (0.3 mg/L). Monochloroacetic acid, bromoacetic acid, and dibromoacetic acid are regulated with this group but have no MCLGs.
• Trihalomethanes: bromodichloromethane (0); bromoform (0); dibromochloromethane (0.06 mg/L). Chloroform is regulated with this group but has no MCLG.

(continued)

APPENDIX 23.A *(continued)*
National Primary Drinking Water Standards
Code of Federal Regulations (CFR), Title 40, Ch. I, Part 141, July 2013

[g]MCLGs were not established before the 1986 Amendments to the Safe Drinking Water Act. Therefore, there is no MCLG for this contaminant.
[h]Lead and copper are regulated by a treatment technique that requires systems to control the corrosiveness of their water. If more than 10% of tap water samples exceed the action level, water systems must take additional steps. For copper, the action level is 1.3 mg/L, and for lead it is 0.015 mg/L.
[i]Each water system agency must certify, in writing, to the state (using third party or manufacturers' certification) that, when acrylamide and epichlorohydrin are used in drinking water systems, the combination (or product) of dose and monomer level does not exceed the levels specified, as follows.
- acrylamide = 0.05% dosed at 1 mg/L (or equivalent)
- epichlorohydrin = 0.01% dosed at 20 mg/L (or equivalent)

APPENDIX 24.A
Properties of Chemicals Used in Water Treatment

chemical name	formula	use	molecular weight	equivalent weight
activated carbon	C	taste and odor control	12.0	–
aluminum hydroxide	$Al(OH)_3$	–	78.0	26.0
aluminum sulfate (filter alum)	$Al_2(SO_4)_3 \cdot 14.3H_2O$	coagulation	600	100
ammonia	NH_3	chloramine disinfection	17.0	–
ammonium fluosilicate	$(NH_4)_2SiF_6$	fluoridation	178	89.0
ammonium sulfate	$(NH_4)_2SO_4$	coagulation	132	66.1
calcium bicarbonate*	$Ca(HCO_3)_2$	–	162	81.0
calcium carbonate	$CaCO_3$	corrosion control	100	50.0
calcium fluoride	CaF_2	fluoridation	78.1	39.0
calcium hydroxide	$Ca(OH)_2$	softening	74.1	37.0
calcium hypochlorite	$Ca(ClO)_2 \cdot 2H_2O$	disinfection	179	–
calcium oxide (lime)	CaO	softening	56.1	28.0
carbon dioxide	CO_2	recarbonation	44.0	22.0
chlorine	Cl_2	disinfection	71.0	–
chlorine dioxide	ClO_2	taste and odor control	67.0	–
copper sulfate	$CuSO_4$	algae control	160	79.8
ferric chloride	$FeCl_3$	coagulation	162	54.1
ferric hydroxide	$Fe(OH)_3$	arsenic removal	107	35.6
ferric sulfate	$Fe_2(SO_4)_3$	coagulation	400	66.7
ferrous sulfate (copperas)	$FeSO_4 \cdot 7H_2O$	coagulation	278	139
fluosilicic acid	H_2SiF_6	fluoridation	144	72.0
hydrochloric acid	HCl	pH adjustment	36.5	36.5
magnesium hydroxide	$Mg(OH)_2$	defluoridation	58.3	29.2
oxygen	O_2	aeration	32.0	16.0
ozone	O_3	disinfection	48.0	16.0
potassium permanganate	$KMnO_4$	oxidation	158	158
sodium aluminate	$NaAlO_2$	coagulation	82.0	82.0
sodium bicarbonate (baking soda)	$NaHCO_3$	alkalinity adjustment	84.0	84.0
sodium carbonate (soda ash)	Na_2CO_3	softening	106	53.0
sodium chloride (common salt)	$NaCl$	ion-exchange regeneration	58.4	58.4
sodium diphosphate	Na_2HPO_4	corrosion control	142	71.0
sodium fluoride	NaF	fluoridation	42.0	42.0
sodium fluosilicate	Na_2SiF_6	fluoridation	188	99.0
sodium hexametaphosphate	$(NaPO_3)_n$	corrosion control	–	–
sodium hydroxide	$NaOH$	pH adjustment	40.0	40.0
sodium hypochlorite	$NaClO$	disinfection	74.0	–
sodium phosphate	NaH_2PO_4	corrosion control	120	120
sodium silicate	Na_2OSiO_2	corrosion control	184	92.0
sodium thiosulfate	$Na_2S_2O_3$	dechlorination	158	79.0
sodium tripolyphosphate	$Na_5P_3O_{10}$	corrosion control	368	–
sulfur dioxide	SO_2	dechlorination	64.1	–
sulfuric acid	H_2SO_4	pH adjustment	98.1	49.0
trisodium phosphate	Na_3PO_4	corrosion control	118	–
water	H_2O	–	18.0	–

*Exists only in aqueous form as a mixture of ions.

APPENDIX 26.A
Selected *Ten States' Standards**

[11.243] **Hydraulic Load:** Use 100 gpcd (0.38 m^3/d) for new systems in undeveloped areas unless other information is available.

[42.3] **Pumps:** At least two pumps are required. Both pumps should have the same capacity if only two pumps are used. This capacity must exceed the total design flow. If three or more pumps are used, the capacities may vary, but capacity (peak hourly flow) pumping must be possible with one pump out of service.

[42.7] **Pump Well Ventilation:** Provide 30 complete air changes per hour for both wet and dry wells using intermittent ventilation. For continuous ventilation, the requirement is reduced to 12 (for wet wells) and 6 (for dry wells) air changes per hour. In general, ventilation air should be forced in, as opposed to air being extracted and replaced by infiltration.

[61.12] **Racks and Bar Screens:** All racks and screens shall have openings less than 1.75 in (45 mm) wide. The smallest opening for manually cleaned screens is 1 in (2.54 cm). The smallest opening for mechanically cleaned screens may be smaller. Flow velocity must be 1.25–3.0 ft/sec (38–91 cm/s).

[62.2–62.3] **Grinders and Shredders:** Comminutors are required if there is no screening. Gravel traps or grit-removal equipment should precede comminutors.

[63.3–63.4] **Grit Chambers:** Grit chambers are required when combined storm and sanitary sewers are used. A minimum of two grit chambers in parallel should be used, with a provision for bypassing. For channel-type grit chambers, the optimum velocity is 1 ft/sec (30 cm/s) throughout. The detention time for channel-type grit chambers is dependent on the particle sizes to be removed.

[71–72] **Settling Tanks:** Multiple units are desirable, and multiple units must be provided if the average flow exceeds 100,000 gal/day (379 m^3/d). For primary settling, the sidewater depth should be 10 ft (3.0 m) or greater. For tanks not receiving activated sludge, the design average overflow rate is 1000 gal/day-ft^2 (41 m^3/d·m^2); the maximum peak overflow rate is 1500–2000 gal/day-ft^2 (61–81 m^3/d·m^2). The maximum weir loading for flows greater than 1 MGD (3785 m^3/d) is 30,000 gal/day-ft (375 m^3/d·m). The basin size shall also be calculated based on the average design flow rate and a maximum settling rate of 1000 gal/day-ft^2 (41 m^3/d·m^2). The larger of the two sizes shall be used. If the flow rate is less than 1 MGD (3785 m^3/d), the maximum weir loading is reduced to 20,000 gal/day-ft (250 m^3/d·m).

For settling tanks following trickling filters and rotating biological contactors, the peak settling rate is 1500 gal/day-ft^2 (61 m^3/d·m^2).

For settling tanks following activated sludge processes, the maximum hydraulic loadings are: 1200 gal/day-ft^2 (49 m^3/d·m^2) for conventional, step, complete mix, and contact units; 1000 gal/day-ft^2 (41 m^3/d·m^2) for extended aeration units; and 800 gal/day-ft^2 (33 m^3/d·m^2) for separate two-stage nitrification units.

[84] **Anaerobic Digesters:** Multiple units are required. Minimum sidewater depth is 20 ft (6.1 m). For completely mixed digesters, maximum loading of volatile solids is 80 lbm/day-1000 ft^3 (1.3 kg/d·m^3) of digester volume. For moderately mixed digesters, the limit is 40 lbm/day-1000 ft^3 (0.65 kg/d·m^3).

[88.22] **Sludge Drying Beds:** For design purposes, the maximum sludge depth is 8 in (200 mm).

[91.3] **Trickling Filters:** All media should have a minimum depth of 6 ft (1.8 m). Rock media depth should not exceed 10 ft (3 m). Manufactured media depth should not exceed the manufacturer's recommendations. The rock media should be 1–4.5 in (2.5–11.4 cm) in size. Freeboard of 4 ft (1.2 ft) or more is required for manufactured media. The drain should slope at 1% or more, and the average drain velocity should be 2 ft/sec (0.6 m/s).

[92.3] **Activated Sludge Processes:** The maximum BOD loading shall be 40 lbm/day-1000 ft^3 (0.64 kg/d·m^3) for conventional, step, and complete-mix units; 50 lbm/day-1000 ft^3 (0.8 kg/d·m^3) for contact stabilization units; and 15 lbm/day-1000 ft^3 (0.24 kg/d·m^3) for extended aeration units and oxidation ditch designs.

Aeration tank depths should be 10–30 ft (3–9 m). Freeboard should be 18 in (460 mm) or more. At least two aeration tanks shall be used. The dissolved oxygen content should not be allowed to drop below 2 mg/L at any time. The aeration rate should be 1500 ft^3 of oxygen per lbm of BOD$_5$ (94 m^3/kg). For extended aeration, the rate should be 2050 ft^3 of oxygen per lbm of BOD$_5$ (128 m^3/kg).

(continued)

APPENDIX 26.A *(continued)*
Selected *Ten States' Standards**

[93] **Wastewater Treatment Ponds:** (Applicable to controlled-discharge, flow-through (facultative), and aerated pond systems.) Pond bottoms must be at least 4 ft (1.2 m) above the highest water table elevation. Pond primary cells are designed based on average BOD_5 loads of 15–35 lbm/ac-day (17–40 kg/ha·d) at the mean operating depth. Detention time for controlled-discharge ponds shall be at least 180 days between a depth of 2 ft (0.61 m) and the maximum operating depth. Detention time for flow-through (facultative) ponds should be 90–120 days, modified for cold weather. Detention time for aerated ponds may be estimated from the percentage, E, of BOD_5 to be removed: $t_{days} = E/2.3k_1(100\% - E)$, where k_1 is the base-10 reaction coefficient, assumed to be 0.12/d at 68°F (20°C) and 0.06/d at 34°F (1°C). Although two cells can be used in very small systems, a minimum of three cells is normally required. Maximum cell size is 40 ac (16 ha). Ponds may be round, square, or rectangular (with length no greater than three times the width). Islands and sharp corners are not permitted. Dikes shall be at least 8 ft (2.4 m) wide at the top. Inner and outer dike slopes cannot be steeper than 1:3 (*V:H*). Inner slopes cannot be flatter than 1:4 (*V:H*). At least 3 ft (0.91 m) freeboard is required unless the pond system is very small, in which case 2 ft (0.6 m) is acceptable. Pond depth shall never be less than 2 ft (0.6 m). Depth shall be: (a) for controlled discharged and flow-through (facultative) ponds, a maximum of 6 ft (1.8 m) for primary cells; greater for subsequent cells with mixing and aeration as required; and (b) for aerated ponds, a design depth of 10–15 ft (3–4.5 m).

[102.4] **Chlorination:** Minimum contact time is 15 min at peak flow.

*Numbers in square brackets refer to sections in *Recommended Standards for Sewage Works*, 2004 ed., Great Lakes-Upper Mississippi River Board of State Sanitary Engineers, published by Health Education Service, NY, from which these guidelines were extracted. Refer to the original document for complete standards.

APPENDIX 29.A
Properties of Saturated Steam by Temperature
(customary U.S. units)

temp. (°F)	absolute pressure (psia)	specific volume (ft³/lbm)		internal energy (Btu/lbm)		enthalpy (Btu/lbm)			entropy (Btu/lbm-°R)		temp. (°F)
		sat. liquid, v_f	sat. vapor, v_g	sat. liquid, u_f	sat. vapor, u_g	sat. liquid, h_f	evap., h_{fg}	sat. vapor, h_g	sat. liquid, s_f	sat. vapor, s_g	
32	0.0886	0.01602	3302	−0.02	1021.0	−0.02	1075.2	1075.2	−0.0004	2.1868	32
34	0.0961	0.01602	3059	2.00	1021.7	2.00	1074.1	1076.1	0.00405	2.1797	34
36	0.1040	0.01602	2836	4.01	1022.3	4.01	1072.9	1076.9	0.00812	2.1727	36
38	0.1126	0.01602	2632	6.02	1023.0	6.02	1071.8	1077.8	0.01217	2.1658	38
40	0.1217	0.01602	2443	8.03	1023.7	8.03	1070.7	1078.7	0.01620	2.1589	40
42	0.1316	0.01602	2270	10.04	1024.3	10.04	1069.5	1079.6	0.02022	2.1522	42
44	0.1421	0.01602	2111	12.05	1025.0	12.05	1068.4	1080.4	0.02421	2.1454	44
46	0.1533	0.01602	1964	14.06	1025.6	14.06	1067.3	1081.3	0.02819	2.1388	46
48	0.1653	0.01602	1828	16.06	1026.3	16.06	1066.1	1082.2	0.03215	2.1322	48
50	0.1781	0.01602	1703	18.07	1026.9	18.07	1065.0	1083.1	0.03609	2.1257	50
52	0.1918	0.01603	1587	20.07	1027.6	20.07	1063.8	1083.9	0.04001	2.1192	52
54	0.2065	0.01603	1481	22.07	1028.2	22.07	1062.7	1084.8	0.04392	2.1128	54
56	0.2221	0.01603	1382	24.07	1028.9	24.08	1061.6	1085.7	0.04781	2.1065	56
58	0.2387	0.01603	1291	26.08	1029.6	26.08	1060.5	1086.6	0.05168	2.1002	58
60	0.2564	0.01604	1206	28.08	1030.2	28.08	1059.3	1087.4	0.05554	2.0940	60
62	0.2752	0.01604	1128	30.08	1030.9	30.08	1058.2	1088.3	0.05938	2.0879	62
64	0.2953	0.01604	1055	32.08	1031.5	32.08	1057.1	1089.2	0.06321	2.0818	64
66	0.3166	0.01604	987.7	34.08	1032.2	34.08	1055.9	1090.0	0.06702	2.0758	66
68	0.3393	0.01605	925.2	36.08	1032.8	36.08	1054.8	1090.9	0.07081	2.0698	68
70	0.3634	0.01605	867.1	38.07	1033.5	38.08	1053.7	1091.8	0.07459	2.0639	70
72	0.3889	0.01606	813.2	40.07	1034.1	40.07	1052.5	1092.6	0.07836	2.0581	72
74	0.4160	0.01606	763.0	42.07	1034.8	42.07	1051.4	1093.5	0.08211	2.0523	74
76	0.4448	0.01606	716.3	44.07	1035.4	44.07	1050.3	1094.4	0.08585	2.0466	76
78	0.4752	0.01607	672.9	46.07	1036.1	46.07	1049.1	1095.2	0.08957	2.0409	78
80	0.5075	0.01607	632.4	48.06	1036.7	48.07	1048.0	1096.1	0.09328	2.0353	80
82	0.5416	0.01608	594.7	50.06	1037.4	50.06	1046.9	1097.0	0.09697	2.0297	82
84	0.5778	0.01608	559.5	52.06	1038.0	52.06	1045.7	1097.8	0.1007	2.0242	84
86	0.6160	0.01609	526.7	54.05	1038.7	54.06	1044.6	1098.7	0.1043	2.0187	86
88	0.6564	0.01610	496.0	56.05	1039.3	56.05	1043.4	1099.5	0.1080	2.0133	88
90	0.6990	0.01610	467.4	58.05	1039.9	58.05	1042.4	1100.4	0.1116	2.0079	90
92	0.7441	0.01611	440.7	60.04	1040.6	60.05	1041.3	1101.3	0.1152	2.0026	92
94	0.7917	0.01611	415.6	62.04	1041.2	62.04	1040.1	1102.1	0.1189	1.9974	94
96	0.8418	0.01612	392.3	64.04	1041.9	64.04	1039.0	1103.0	0.1225	1.9922	96
98	0.8947	0.01613	370.3	66.03	1042.5	66.04	1037.8	1103.8	0.1260	1.9870	98
100	0.9505	0.01613	349.8	68.03	1043.2	68.03	1036.7	1104.7	0.1296	1.9819	100
105	1.103	0.01615	304.0	73.02	1044.8	73.03	1033.8	1106.8	0.1385	1.9693	105
110	1.277	0.01617	265.0	78.01	1046.4	78.02	1031.0	1109.0	0.1473	1.9570	110
115	1.473	0.01619	231.6	83.01	1048.0	83.01	1028.1	1111.1	0.1560	1.9450	115
120	1.695	0.01621	203.0	88.00	1049.6	88.00	1025.2	1113.2	0.1647	1.9333	120
125	1.945	0.01623	178.3	92.99	1051.1	93.00	1022.3	1115.3	0.1732	1.9218	125

(continued)

APPENDIX 29.A *(continued)*
Properties of Saturated Steam by Temperature
(customary U.S. units)

temp. (°F)	absolute pressure (psia)	specific volume (ft³/lbm)		internal energy (Btu/lbm)		enthalpy (Btu/lbm)			entropy (Btu/lbm-°R)		temp. (°F)
		sat. liquid, v_f	sat. vapor, v_g	sat. liquid, u_f	sat. vapor, u_g	sat. liquid, h_f	evap., h_{fg}	sat. vapor, h_g	sat. liquid, s_f	sat. vapor, s_g	
130	2.226	0.01625	157.1	97.99	1052.7	97.99	1019.4	1117.4	0.1818	1.9106	130
135	2.541	0.01627	138.7	102.98	1054.3	102.99	1016.5	1119.5	0.1902	1.8996	135
140	2.893	0.01629	122.8	107.98	1055.8	107.99	1013.6	1121.6	0.1986	1.8888	140
145	3.286	0.01632	109.0	112.98	1057.4	112.99	1010.7	1123.7	0.2069	1.8783	145
150	3.723	0.01634	96.93	117.98	1059.0	117.99	1007.7	1125.7	0.2151	1.8680	150
155	4.209	0.01637	86.40	122.98	1060.5	122.99	1004.8	1127.8	0.2233	1.8580	155
160	4.747	0.01639	77.18	127.98	1062.0	128.00	1001.8	1129.8	0.2314	1.8481	160
165	5.343	0.01642	69.09	132.99	1063.6	133.01	998.9	1131.9	0.2394	1.8384	165
170	6.000	0.01645	61.98	138.00	1065.1	138.01	995.9	1133.9	0.2474	1.8290	170
175	6.724	0.01648	55.71	143.01	1066.6	143.03	992.9	1135.9	0.2553	1.8197	175
180	7.520	0.01651	50.17	148.02	1068.1	148.04	989.9	1137.9	0.2632	1.8106	180
185	8.393	0.01654	45.27	153.03	1069.6	153.06	986.8	1139.9	0.2710	1.8017	185
190	9.350	0.01657	40.92	158.05	1071.0	158.08	983.7	1141.8	0.2788	1.7930	190
195	10.396	0.01660	37.05	163.07	1072.5	163.10	980.7	1143.8	0.2865	1.7844	195
200	11.538	0.01663	33.61	168.09	1074.0	168.13	977.6	1145.7	0.2941	1.7760	200
205	12.782	0.01667	30.54	173.12	1075.4	173.16	974.4	1147.6	0.3017	1.7678	205
210	14.14	0.01670	27.79	178.2	1076.8	178.2	971.3	1149.5	0.3092	1.7597	210
212	14.71	0.01672	26.78	180.2	1077.4	180.2	970.1	1150.3	0.3122	1.7565	212
220	17.20	0.01677	23.13	188.2	1079.6	188.3	965.0	1153.3	0.3242	1.7440	220
230	20.80	0.01685	19.37	198.3	1082.4	198.4	958.5	1156.9	0.3389	1.7288	230
240	24.99	0.01692	16.31	208.4	1085.1	208.5	952.0	1160.5	0.3534	1.7141	240
250	29.84	0.01700	13.82	218.5	1087.7	218.6	945.4	1164.0	0.3678	1.7000	250
260	35.45	0.01708	11.76	228.7	1090.3	228.8	938.6	1167.4	0.3820	1.6863	260
270	41.88	0.01717	10.06	238.9	1092.8	239.0	931.7	1170.7	0.3960	1.6730	270
280	49.22	0.01726	8.64	249.0	1095.2	249.2	924.7	1173.9	0.4099	1.6601	280
290	57.57	0.01735	7.46	259.3	1097.5	259.5	917.6	1177.0	0.4236	1.6476	290
300	67.03	0.01745	6.466	269.5	1099.8	269.7	910.3	1180.0	0.4372	1.6354	300
310	77.69	0.01755	5.626	279.8	1101.9	280.1	902.8	1182.8	0.4507	1.6236	310
320	89.67	0.01765	4.914	290.1	1103.9	290.4	895.1	1185.5	0.4640	1.6120	320
330	103.07	0.01776	4.308	300.5	1105.9	300.8	887.2	1188.0	0.4772	1.6007	330
340	118.02	0.01787	3.788	310.9	1107.7	311.2	879.3	1190.5	0.4903	1.5897	340
350	134.63	0.01799	3.343	321.3	1109.4	321.7	871.0	1192.7	0.5032	1.5789	350
360	153.03	0.01811	2.958	331.8	1111.0	332.3	862.5	1194.8	0.5161	1.5684	360
370	173.36	0.01823	2.625	342.3	1112.5	342.9	853.8	1196.7	0.5289	1.5580	370
380	195.74	0.01836	2.336	352.9	1113.9	353.5	845.0	1198.5	0.5415	1.5478	380
390	220.3	0.01850	2.084	363.5	1115.1	364.3	835.8	1200.1	0.5541	1.5378	390
400	247.3	0.01864	1.864	374.2	1116.2	375.1	826.4	1201.4	0.5667	1.5279	400
410	276.7	0.01879	1.671	385.0	1117.1	385.9	816.7	1202.6	0.5791	1.5182	410
420	308.8	0.01894	1.501	395.8	1117.8	396.9	806.8	1203.6	0.5915	1.5085	420
430	343.6	0.01910	1.351	406.7	1118.4	407.9	796.4	1204.3	0.6038	1.4990	430

(continued)

APPENDIX 29.A *(continued)*
Properties of Saturated Steam by Temperature
(customary U.S. units)

temp. (°F)	absolute pressure (psia)	specific volume (ft³/lbm)		internal energy (Btu/lbm)		enthalpy (Btu/lbm)			entropy (Btu/lbm-°R)		temp. (°F)
		sat. liquid, v_f	sat. vapor, v_g	sat. liquid, u_f	sat. vapor, u_g	sat. liquid, h_f	evap., h_{fg}	sat. vapor, h_g	sat. liquid, s_f	sat. vapor, s_g	
440	381.5	0.01926	1.218	417.6	1118.9	419.0	785.8	1204.8	0.6161	1.4895	440
450	422.5	0.01944	1.0999	428.7	1119.1	430.2	774.9	1205.1	0.6283	1.4802	450
460	466.8	0.01962	0.9951	439.8	1119.2	441.5	763.6	1205.1	0.6405	1.4708	460
470	514.5	0.01981	0.9015	451.1	1119.0	452.9	752.0	1204.9	0.6527	1.4615	470
480	566.0	0.02001	0.8179	462.4	1118.7	464.5	739.8	1204.3	0.6648	1.4522	480
490	621.2	0.02022	0.7429	473.8	1118.1	476.1	727.4	1203.5	0.6769	1.4428	490
500	680.6	0.02044	0.6756	485.4	1117.2	487.9	714.4	1202.3	0.6891	1.4335	500
510	744.1	0.02068	0.6149	497.1	1116.2	499.9	700.9	1200.8	0.7012	1.4240	510
520	812.1	0.02092	0.5601	508.9	1114.8	512.0	686.9	1198.9	0.7134	1.4146	520
530	884.7	0.02119	0.5105	520.8	1113.1	524.3	672.4	1196.7	0.7256	1.4050	530
540	962.2	0.02146	0.4655	533.0	1111.1	536.8	657.2	1194.0	0.7378	1.3952	540
550	1044.8	0.02176	0.4247	545.3	1108.8	549.5	641.4	1190.9	0.7501	1.3853	550
560	1132.7	0.02208	0.3874	557.8	1106.0	562.4	624.8	1187.2	0.7625	1.3753	560
570	1226.2	0.02242	0.3534	570.5	1102.8	575.6	607.4	1183.0	0.7750	1.3649	570
580	1325.5	0.02279	0.3223	583.5	1099.2	589.1	589.3	1178.3	0.7876	1.3543	580
590	1430.8	0.02319	0.2937	596.7	1095.0	602.8	570.0	1172.8	0.8004	1.3434	590
600	1542.5	0.02363	0.2674	610.3	1090.3	617.0	549.6	1166.6	0.8133	1.3320	600
610	1660.9	0.02411	0.2431	624.2	1084.8	631.6	528.0	1159.6	0.8266	1.3201	610
620	1786.2	0.02465	0.2206	638.5	1078.6	646.7	504.8	1151.5	0.8401	1.3077	620
630	1918.9	0.02525	0.1997	653.4	1071.5	662.4	480.1	1142.4	0.8539	1.2945	630
640	2059.2	0.02593	0.1802	668.9	1063.2	678.7	453.1	1131.8	0.8683	1.2803	640
650	2207.8	0.02672	0.1618	685.1	1053.6	696.0	423.7	1119.7	0.8833	1.2651	650
660	2364.9	0.02766	0.1444	702.4	1042.2	714.5	390.8	1105.3	0.8991	1.2482	660
670	2531.2	0.02883	0.1277	721.1	1028.4	734.6	353.6	1088.2	0.9163	1.2292	670
680	2707.3	0.03036	0.1113	742.2	1011.1	757.4	309.4	1066.8	0.9355	1.2070	680
690	2894	0.03258	0.0946	767.1	987.7	784.5	253.8	1038.3	0.9582	1.1790	690
700	3093	0.03665	0.0748	801.5	948.2	822.5	168.5	991.0	0.9900	1.1353	700
705.103	3200.11	0.04975	0.04975	866.6	866.6	896.1	0	896.1	1.0526	1.0526	705.103

Values in this table were calculated from *NIST Standard Reference Database 10*, "NIST/ASME Steam Properties," Ver. 2.11, National Institute of Standards and Technology, U.S. Department of Commerce, Gaithersburg, MD, 1997, which has been licensed to PPI.

APPENDIX 29.B
Properties of Saturated Steam by Pressure
(customary U.S. units)

absolute press (psia)	temp. (°F)	specific volume (ft³/lbm)		internal energy (Btu/lbm)		enthalpy (Btu/lbm)			entropy (Btu/lbm-°R)			absolute press (psia)
		sat. liquid, v_f	sat. vapor, v_g	sat. liquid, u_f	sat. vapor, u_g	sat. liquid, h_f	evap., h_{fg}	sat. vapor, h_g	sat. liquid, s_f	evap., s_{fg}	sat. vapor, s_g	
0.4	72.83	0.01606	791.8	40.91	1034.4	40.91	1052.1	1093.0	0.0799	1.9758	2.0557	0.4
0.6	85.18	0.01609	539.9	53.23	1038.4	53.24	1045.1	1098.3	0.1028	1.9182	2.0210	0.6
0.8	94.34	0.01611	411.6	62.38	1041.3	62.38	1039.9	1102.3	0.1195	1.8770	1.9965	0.8
1.0	101.69	0.01614	333.5	69.72	1043.7	69.72	1035.7	1105.4	0.1326	1.8450	1.9776	1.0
1.2	107.87	0.01616	280.9	75.88	1045.7	75.89	1032.2	1108.1	0.1435	1.8187	1.9622	1.2
1.5	115.64	0.01619	227.7	83.64	1048.2	83.65	1027.8	1111.4	0.1571	1.7864	1.9435	1.5
2.0	126.03	0.01623	173.7	94.02	1051.5	94.02	1021.8	1115.8	0.1750	1.7445	1.9195	2.0
3.0	141.42	0.01630	118.7	109.4	1056.3	109.4	1012.8	1122.2	0.2009	1.6849	1.8858	3.0
4.0	152.91	0.01636	90.63	120.9	1059.9	120.9	1006.0	1126.9	0.2199	1.6423	1.8621	4.0
5.0	162.18	0.01641	73.52	130.2	1062.7	130.2	1000.5	1130.7	0.2349	1.6089	1.8438	5.0
6.0	170.00	0.01645	61.98	138.0	1065.1	138.0	995.9	1133.9	0.2474	1.5816	1.8290	6.0
7.0	176.79	0.01649	53.65	144.8	1067.1	144.8	991.8	1136.6	0.2581	1.5583	1.8164	7.0
8.0	182.81	0.01652	47.34	150.8	1068.9	150.9	988.1	1139.0	0.2676	1.5380	1.8056	8.0
9.0	188.22	0.01656	42.40	156.3	1070.5	156.3	984.8	1141.1	0.2760	1.5201	1.7961	9.0
10	193.16	0.01659	38.42	161.2	1072.0	161.3	981.9	1143.1	0.2836	1.5040	1.7876	10
14.696	211.95	0.01671	26.80	180.1	1077.4	180.2	970.1	1150.3	0.3122	1.4445	1.7566	14.696
15	212.99	0.01672	26.29	181.2	1077.7	181.2	969.5	1150.7	0.3137	1.4412	1.7549	15
20	227.92	0.01683	20.09	196.2	1081.8	196.3	959.9	1156.2	0.3358	1.3961	1.7319	20
25	240.03	0.01692	16.31	208.4	1085.1	208.5	952.0	1160.5	0.3535	1.3606	1.7141	25
30	250.30	0.01700	13.75	218.8	1087.8	218.9	945.2	1164.1	0.3682	1.3313	1.6995	30
35	259.25	0.01708	11.90	227.9	1090.1	228.0	939.2	1167.2	0.3809	1.3064	1.6873	35
40	267.22	0.01715	10.50	236.0	1092.1	236.1	933.7	1169.8	0.3921	1.2845	1.6766	40
45	274.41	0.01721	9.402	243.3	1093.9	243.5	928.7	1172.2	0.4022	1.2650	1.6672	45
50	280.99	0.01727	8.517	250.1	1095.4	250.2	924.0	1174.2	0.4113	1.2475	1.6588	50
55	287.05	0.01732	7.788	256.2	1096.8	256.4	919.7	1176.1	0.4196	1.2316	1.6512	55
60	292.68	0.01738	7.176	262.0	1098.1	262.2	915.6	1177.8	0.4273	1.2170	1.6443	60
65	297.95	0.01743	6.656	267.4	1099.3	267.6	911.8	1179.4	0.4344	1.2035	1.6379	65
70	302.91	0.01748	6.207	272.5	1100.4	272.7	908.1	1180.8	0.4411	1.1908	1.6319	70
75	307.58	0.01752	5.816	277.3	1101.4	277.6	904.6	1182.1	0.4474	1.1790	1.6264	75
80	312.02	0.01757	5.473	281.9	1102.3	282.1	901.2	1183.3	0.4534	1.1679	1.6212	80
85	316.24	0.01761	5.169	286.2	1103.2	286.5	898.0	1184.5	0.4590	1.1573	1.6163	85
90	320.26	0.01765	4.897	290.4	1104.0	290.7	894.9	1185.6	0.4643	1.1474	1.6117	90
95	324.11	0.01770	4.653	294.4	1104.8	294.7	891.9	1186.6	0.4694	1.1379	1.6073	95
100	327.81	0.01774	4.433	298.2	1105.5	298.5	889.0	1187.5	0.4743	1.1289	1.6032	100
110	334.77	0.01781	4.050	305.4	1106.8	305.8	883.4	1189.2	0.4834	1.1120	1.5954	110
120	341.25	0.01789	3.729	312.2	1107.9	312.6	878.2	1190.7	0.4919	1.0965	1.5884	120
130	347.32	0.01796	3.456	318.5	1109.0	318.9	873.2	1192.1	0.4998	1.0821	1.5818	130
140	353.03	0.01802	3.220	324.5	1109.9	324.9	868.5	1193.4	0.5071	1.0686	1.5757	140
150	358.42	0.01809	3.015	330.1	1110.8	330.6	863.9	1194.5	0.5141	1.0559	1.5700	150
160	363.54	0.01815	2.835	335.5	1111.6	336.0	859.5	1195.5	0.5206	1.0441	1.5647	160
170	368.41	0.01821	2.675	340.6	1112.3	341.2	855.2	1196.4	0.5268	1.0328	1.5596	170
180	373.07	0.01827	2.532	345.5	1112.9	346.1	851.2	1197.3	0.5328	1.0222	1.5549	180
190	377.52	0.01833	2.404	350.3	1113.5	350.9	847.2	1198.1	0.5384	1.0119	1.5503	190
200	381.80	0.01839	2.288	354.8	1114.1	355.5	843.3	1198.8	0.5438	1.0022	1.5460	200
250	400.97	0.01865	1.844	375.2	1116.2	376.1	825.5	1201.6	0.5679	0.9591	1.5270	250

(continued)

APPENDIX 29.B *(continued)*
Properties of Saturated Steam by Pressure
(customary U.S. units)

absolute press (psia)	temp. (°F)	specific volume (ft³/lbm)		internal energy (Btu/lbm)		enthalpy (Btu/lbm)			entropy (Btu/lbm-°R)			absolute press (psia)
		sat. liquid, v_f	sat. vapor, v_g	sat. liquid, u_f	sat. vapor, u_g	sat. liquid, h_f	evap., h_{fg}	sat. vapor, h_g	sat. liquid, s_f	evap., s_{fg}	sat. vapor, s_g	
300	417.35	0.01890	1.544	392.9	1117.7	394.0	809.4	1203.3	0.5882	0.9229	1.5111	300
350	431.74	0.01913	1.326	408.6	1118.5	409.8	794.6	1204.4	0.6059	0.8915	1.4974	350
400	444.62	0.01934	1.162	422.7	1119.0	424.2	780.9	1205.0	0.6217	0.8635	1.4852	400
450	456.31	0.01955	1.032	435.7	1119.2	437.3	767.8	1205.1	0.6360	0.8382	1.4742	450
500	467.04	0.01975	0.928	447.7	1119.1	449.5	755.5	1205.0	0.6491	0.8152	1.4642	500
550	476.98	0.01995	0.842	458.9	1118.8	461.0	743.5	1204.5	0.6611	0.7939	1.4550	550
600	486.24	0.02014	0.770	469.5	1118.3	471.7	732.1	1203.8	0.6724	0.7739	1.4463	600
700	503.13	0.02051	0.656	489.0	1116.9	491.7	710.2	1201.9	0.6929	0.7376	1.4305	700
800	518.27	0.02088	0.569	506.8	1115.0	509.9	689.4	1199.3	0.7113	0.7050	1.4162	800
900	532.02	0.02124	0.501	523.3	1112.7	526.8	669.4	1196.2	0.7280	0.6750	1.4030	900
1000	544.65	0.02160	0.446	538.7	1110.1	542.7	649.9	1192.6	0.7435	0.6472	1.3907	1000
1100	556.35	0.02196	0.401	553.2	1107.1	557.7	631.0	1188.6	0.7580	0.6211	1.3790	1100
1200	567.26	0.02233	0.362	567.0	1103.8	571.9	612.3	1184.2	0.7715	0.5963	1.3678	1200
1300	577.49	0.02270	0.330	580.2	1100.2	585.6	593.9	1179.5	0.7844	0.5726	1.3570	1300
1400	587.14	0.02307	0.302	592.9	1096.3	598.9	575.5	1174.4	0.7967	0.5498	1.3465	1400
1500	596.26	0.02346	0.277	605.2	1092.1	611.7	557.3	1169.0	0.8085	0.5278	1.3363	1500
1600	604.93	0.02386	0.255	617.1	1087.7	624.1	539.1	1163.2	0.8198	0.5064	1.3262	1600
1700	613.18	0.02428	0.236	628.7	1082.9	636.3	520.8	1157.1	0.8308	0.4854	1.3162	1700
1800	621.07	0.02471	0.218	640.1	1077.9	648.3	502.3	1150.6	0.8415	0.4648	1.3063	1800
1900	628.61	0.02516	0.203	651.3	1072.5	660.1	483.6	1143.7	0.8520	0.4443	1.2963	1900
2000	635.85	0.02564	0.188	662.4	1066.8	671.8	464.6	1136.4	0.8623	0.4240	1.2863	2000
2250	652.74	0.02696	0.157	689.7	1050.6	701.0	415.1	1116.0	0.8875	0.3731	1.2606	2250
2500	668.17	0.02859	0.131	717.6	1031.1	730.8	360.8	1091.6	0.9130	0.3199	1.2329	2500
2750	682.34	0.03080	0.108	747.6	1006.3	763.2	297.8	1061.0	0.9404	0.2607	1.2011	2750
3000	695.41	0.03434	0.085	783.4	969.9	802.5	214.4	1016.9	0.9733	0.1856	1.1589	3000
3200.11	705.1028	0.04975	0.04975	866.6	866.6	896.1	0	896.1	1.0526	0	1.0526	3200.11

Values in this table were calculated from *NIST Standard Reference Database 10*, "NIST/ASME Steam Properties," Ver. 2.11, National Institute of Standards and Technology, U.S. Department of Commerce, Gaithersburg, MD, 1997, which has been licensed to PPI.

APPENDIX 29.C
Properties of Superheated Steam
(customary U.S. units)
specific volume, v, in ft³/lbm; enthalpy, h, in Btu/lbm; entropy, s, in Btu/lbm-°R

absolute pressure (psia) (sat. temp., °F)		temperature (°F)								
		200	300	400	500	600	700	800	900	1000
1.0	v	392.5	452.3	511.9	571.5	631.1	690.7	750.3	809.9	869.5
(101.69)	h	1150.1	1105.7	1241.8	1288.6	1336.2	1384.6	1433.9	1484.1	1535.1
	s	2.0510	2.1153	2.1723	2.2238	2.2710	2.3146	2.3554	2.3937	2.4299
5.0	v	78.15	90.25	102.25	114.21	126.15	138.09	150.02	161.94	173.86
(162.18)	h	1148.5	1194.8	1241.3	1288.2	1335.9	1384.4	1433.7	1483.9	1535.0
	s	1.8716	1.9370	1.9944	2.0461	2.0934	2.1371	2.1779	2.2162	2.2525
10.0	v	38.85	44.99	51.04	57.04	63.03	69.01	74.98	80.95	86.91
(193.16)	h	1146.4	1193.8	1240.6	1287.8	1335.6	1384.2	1433.5	1483.8	1534.9
	s	1.7926	1.8595	1.9174	1.9693	2.0167	2.0605	2.1014	2.1397	2.1760
14.696	v	....	30.53	34.67	38.77	42.86	46.93	51.00	55.07	59.13
(211.95)	h	....	1192.7	1240.0	1287.4	1335.3	1383.9	1433.3	1483.6	1534.8
	s	....	1.8160	1.8744	1.9266	1.9741	2.0179	2.0588	2.0972	2.1335
20.0	v	....	22.36	25.43	28.46	31.47	34.47	37.46	40.45	43.44
(227.92)	h	....	1191.6	1239.3	1286.9	1334.9	1383.6	1433.1	1483.4	1534.6
	s	....	1.7808	1.8398	1.8922	1.9399	1.9838	2.0247	2.0631	2.0994
60.0	v	....	7.260	8.355	9.400	10.426	11.440	12.448	13.453	14.454
(292.68)	h	....	1181.9	1233.7	1283.1	1332.2	1381.6	1431.5	1482.1	1533.5
	s	....	1.6496	1.7138	1.7682	1.8168	1.8613	1.9026	1.9413	1.9778
100.0	v	....	....	4.936	5.588	6.217	6.834	7.446	8.053	8.658
(327.81)	h	....	....	1227.7	1279.3	1329.4	1379.5	1429.8	1480.7	1532.4
	s	....	....	1.6521	1.7089	1.7586	1.8037	1.8453	1.8842	1.9209
150.0	v	....	....	3.222	3.680	4.112	4.531	4.944	5.353	5.759
(358.42)	h	....	....	1219.7	1274.3	1325.9	1376.8	1427.7	1479.0	1531.0
	s	....	....	1.6001	1.6602	1.7114	1.7573	1.7994	1.8386	1.8755
200.0	v	....	....	2.362	2.725	3.059	3.380	3.693	4.003	4.310
(381.80)	h	....	....	1210.9	1269.0	1322.3	1374.1	1425.6	1477.3	1529.6
	s	....	....	1.5602	1.6243	1.6771	1.7238	1.7664	1.8060	1.8430
250.0	v	....	....	....	2.151	2.426	2.688	2.943	3.193	3.440
(400.97)	h	....	....	....	1263.6	1318.6	1371.3	1423.5	1475.6	1528.1
	s	....	....	....	1.5953	1.6499	1.6975	1.7406	1.7804	1.8177
300.0	v	....	....	....	1.767	2.005	2.227	2.442	2.653	2.861
(417.35)	h	....	....	....	1257.9	1314.8	1368.6	1421.3	1473.9	1526.7
	s	....	....	....	1.5706	1.6271	1.6756	1.7192	1.7594	1.7969
400.0	v	....	....	....	1.285	1.477	1.651	1.817	1.978	2.136
(444.62)	h	....	....	....	1245.6	1306.9	1362.9	1416.9	1470.4	1523.9
	s	....	....	....	1.5288	1.5897	1.6402	1.6849	1.7257	1.7637

(continued)

APPENDIX 29.C *(continued)*
Properties of Superheated Steam
(customary U.S. units)
specific volume, v, in ft^3/lbm; enthalpy, h, in Btu/lbm; entropy, s, in Btu/lbm-°R

absolute pressure (psia) (sat. temp., °F)		temperature (°F)									
		500	600	700	800	900	1000	1100	1200	1400	1600
450.0	v	1.123	1.300	1.458	1.608	1.753	1.894	2.034	2.172	2.445	
(456.31)	h	1238.9	1302.8	1360.0	1414.7	1468.6	1522.4	1576.5	1631.0	1742.0	
	s	1.5103	1.5737	1.6253	1.6706	1.7118	1.7499	1.7858	1.8196	1.8828	
500.0	v	0.993	1.159	1.304	1.441	1.573	1.701	1.827	1.952	2.199	
(467.04)	h	1231.9	1298.6	1357.0	1412.5	1466.9	1521.0	1575.3	1630.0	1741.2	
	s	1.4928	1.5591	1.6118	1.6576	1.6992	1.7376	1.7736	1.8076	1.8708	
600.0	v	0.795	0.946	1.073	1.190	1.302	1.411	1.518	1.623	1.830	
(486.24)	h	1216.5	1289.9	1351.0	1408.0	1463.3	1518.1	1572.8	1627.9	1739.7	
	s	1.4596	1.5326	1.5877	1.6348	1.6771	1.7160	1.7523	1.7865	1.8501	
700.0	v	....	0.793	0.908	1.011	1.109	1.204	1.296	1.387	1.566	
(503.13)	h	....	1280.7	1344.8	1403.4	1459.7	1515.1	1570.4	1625.9	1738.2	
	s	....	1.5087	1.5666	1.6151	1.6581	1.6975	1.7341	1.7686	1.8324	
800.0	v	....	0.678	0.783	0.877	0.964	1.048	1.130	1.211	1.368	
(518.27)	h	....	1270.9	1338.4	1398.7	1456.0	1512.2	1568.0	1623.8	1736.7	
	s	....	1.4867	1.5476	1.5975	1.6414	1.6812	1.7182	1.7529	1.8171	
900.0	v	....	0.588	0.686	0.772	0.852	0.928	1.001	1.073	1.214	
(532.02)	h	....	1260.4	1331.8	1393.9	1452.3	1509.2	1565.5	1621.8	1735.2	
	s	....	1.4658	1.5302	1.5816	1.6263	1.6667	1.7040	1.7389	1.8034	
1000	v	....	0.514	0.608	0.688	0.761	0.831	0.898	0.963	1.091	1.216
(544.65)	h	....	1249.3	1325.0	1389.0	1448.6	1506.2	1563.0	1619.7	1733.6	1849.6
	s	....	1.4457	1.5140	1.5671	1.6126	1.6535	1.6912	1.7264	1.7912	1.8504
1200	v	....	0.402	0.491	0.562	0.626	0.686	0.743	0.798	0.906	1.012
(567.26)	h	....	1224.2	1310.6	1379.0	1441.0	1500.1	1558.1	1615.5	1730.6	1847.3
	s	....	1.4061	1.4842	1.5408	1.5882	1.6302	1.6686	1.7043	1.7698	1.8294
1400	v	....	0.318	0.406	0.472	0.529	0.582	0.632	0.681	0.774	0.866
(587.14)	h	....	1193.8	1295.1	1368.5	1433.1	1494.0	1553.0	1611.3	1727.5	1845.0
	s	....	1.3649	1.4566	1.5174	1.5668	1.6100	1.6491	1.6853	1.7515	1.8115
1600	v	....	....	0.342	0.404	0.456	0.504	0.549	0.592	0.676	0.756
(604.93)	h	....	....	1278.3	1357.6	1425.1	1487.7	1547.9	1607.1	1724.5	1842.7
	s	....	....	1.4302	1.4959	1.5475	1.5920	1.6319	1.6686	1.7354	1.7958
1800	v	....	....	0.291	0.350	0.399	0.443	0.484	0.524	0.599	0.671
(621.07)	h	....	....	1260.0	1346.2	1416.9	1481.3	1542.8	1602.8	1721.4	1840.3
	s	....	....	1.4044	1.4759	1.5299	1.5756	1.6164	1.6537	1.7211	1.7819
2000	v	....	....	0.249	0.308	0.354	0.395	0.433	0.469	0.537	0.603
(635.85)	h	....	....	1239.7	1334.3	1408.5	1474.8	1537.6	1598.5	1718.3	1838.0
	s	....	....	1.3783	1.4568	1.5135	1.5606	1.6022	1.6400	1.7082	1.7693

Values in this table were calculated from *NIST Standard Reference Database 10*, "NIST/ASME Steam Properties," Ver. 2.11, National Institute of Standards and Technology, U.S. Department of Commerce, Gaithersburg, MD, 1997, which has been licensed to PPI.

APPENDIX 29.D
Properties of Compressed Water
(customary U.S. units)

T (°F)	p (psia)	ρ (lbm/ft³)	v (ft³/lbm)	x	h (Btu/lbm)	s (Btu/lbm-°R)	u (Btu/lbm)
32	200	62.46	0.01601	subcooled	0.5852	−0.00001534	−0.007334
100	200	62.03	0.01612	subcooled	68.56	0.1295	67.96
200	200	60.16	0.01662	subcooled	168.6	0.2939	167.9
300	200	57.34	0.01744	subcooled	270	0.437	269.3
381.8	200	54.39	0.01839	0	355.5	0.5438	354.8
381.8	200	0.437	2.288	1	1199	1.546	1114
32	400	62.5	0.016	subcooled	1.187	0.00000448	0.003032
100	400	62.07	0.01611	subcooled	69.09	0.1294	67.89
200	400	60.2	0.01661	subcooled	169	0.2936	167.8
300	400	57.39	0.01742	subcooled	270.3	0.4366	269.1
400	400	53.7	0.01862	subcooled	375.2	0.5662	373.8
444.6	400	51.7	0.01934	0	424.2	0.6217	422.7
444.6	400	0.8609	1.162	1	1205	1.485	1119
32	600	62.55	0.01599	subcooled	1.788	0.00002258	0.01296
100	600	62.1	0.0161	subcooled	69.61	0.1292	67.83
200	600	60.24	0.0166	subcooled	169.5	0.2934	167.6
300	600	57.44	0.01741	subcooled	270.7	0.4362	268.8
400	600	53.77	0.0186	subcooled	375.4	0.5657	373.4
486.2	600	49.65	0.02014	0	471.7	0.6724	469.5
486.2	600	1.298	0.7702	1	1204	1.446	1118
32	800	62.59	0.01598	subcooled	2.388	0.00003896	0.02244
100	800	62.14	0.01609	subcooled	70.14	0.1291	67.76
200	800	60.28	0.01659	subcooled	169.9	0.2931	167.5
300	800	57.49	0.01739	subcooled	271.1	0.4359	268.5
400	800	53.84	0.01857	subcooled	375.7	0.5652	372.9
500	800	48.99	0.02041	subcooled	487.9	0.6885	484.9
518.3	800	47.9	0.02088	0	509.9	0.7112	506.8
518.3	800	1.757	0.5692	1	1199	1.416	1115
32	1000	62.63	0.01597	subcooled	2.986	0.00005365	0.0315
100	1000	62.18	0.01608	subcooled	70.67	0.129	67.69
200	1000	60.31	0.01658	subcooled	170.4	0.2929	167.3
300	1000	57.54	0.01738	subcooled	271.5	0.4355	268.2
400	1000	53.9	0.01855	subcooled	375.9	0.5646	372.5
500	1000	49.1	0.02037	subcooled	487.8	0.6876	484
544.6	1000	46.3	0.0216	0	542.7	0.7435	538.7
544.6	1000	2.242	0.446	1	1193	1.391	1110
32	1500	62.74	0.01594	subcooled	4.477	0.00008309	0.05229
100	1500	62.27	0.01606	subcooled	71.98	0.1287	67.53
200	1500	60.41	0.01655	subcooled	171.5	0.2923	166.9
300	1500	57.65	0.01734	subcooled	272.4	0.4346	267.6
400	1500	54.06	0.0185	subcooled	376.5	0.5633	371.4
500	1500	49.36	0.02026	subcooled	487.6	0.6855	482
596.3	1500	42.62	0.02346	0	611.7	0.8085	605.2
596.3	1500	3.61	0.277	1	1169	1.336	1092

(continued)

APPENDIX 29.D *(continued)*
Properties of Compressed Water
(customary U.S. units)

T (°F)	p (psia)	ρ (lbm/ft^3)	v (ft^3/lbm)	x	h (Btu/lbm)	s (Btu/lbm-°R)	u (Btu/lbm)
32	2000	62.85	0.01591	subcooled	5.959	0.0001023	0.0705
100	2000	62.36	0.01603	subcooled	73.3	0.1284	67.36
200	2000	60.51	0.01653	subcooled	172.7	0.2917	166.5
300	2000	57.77	0.01731	subcooled	273.3	0.4338	266.9
400	2000	54.22	0.01844	subcooled	377.1	0.5621	370.3
500	2000	49.62	0.02015	subcooled	487.5	0.6835	480.1
600	2000	42.89	0.02332	subcooled	614.4	0.809	605.8
635.8	2000	39.01	0.02564	0	671.8	0.8623	662.3
635.8	2000	5.316	0.1881	1	1136	1.286	1067
32	3000	63.06	0.01586	subcooled	8.903	0.0001113	0.09948
100	3000	62.55	0.01599	subcooled	75.91	0.1278	67.04
200	3000	60.7	0.01648	subcooled	174.9	0.2905	165.8
300	3000	58	0.01724	subcooled	275.2	0.4321	265.7
400	3000	54.53	0.01834	subcooled	378.4	0.5596	368.2
500	3000	50.1	0.01996	subcooled	487.5	0.6796	476.4
600	3000	43.94	0.02276	subcooled	610.1	0.8009	597.4
695.4	3000	29.12	0.03434	0	802.5	0.9733	783.4
695.4	3000	11.81	0.08466	1	1017	1.159	969.9
32	4000	63.26	0.01581	subcooled	11.82	0.00008277	0.119
100	4000	62.73	0.01594	subcooled	78.52	0.1271	66.72
200	4000	60.88	0.01642	subcooled	177.2	0.2894	165.1
300	4000	58.22	0.01718	subcooled	277.1	0.4304	264.4
400	4000	54.83	0.01824	subcooled	379.7	0.5572	366.2
500	4000	50.55	0.01978	subcooled	487.7	0.676	473.1
600	4000	44.81	0.02231	subcooled	607	0.794	590.5
700	4000	34.83	0.02871	subcooled	763.6	0.9347	742.3

Values in this table were calculated from *NIST Standard Reference Database 10*, "NIST/ASME Steam Properties," Ver. 2.11, National Institute of Standards and Technology, U.S. Department of Commerce, Gaithersburg, MD, 1997, which has been licensed to PPI.

APPENDIX 29.E
Enthalpy-Entropy (Mollier) Diagram for Steam
(customary U.S. units)

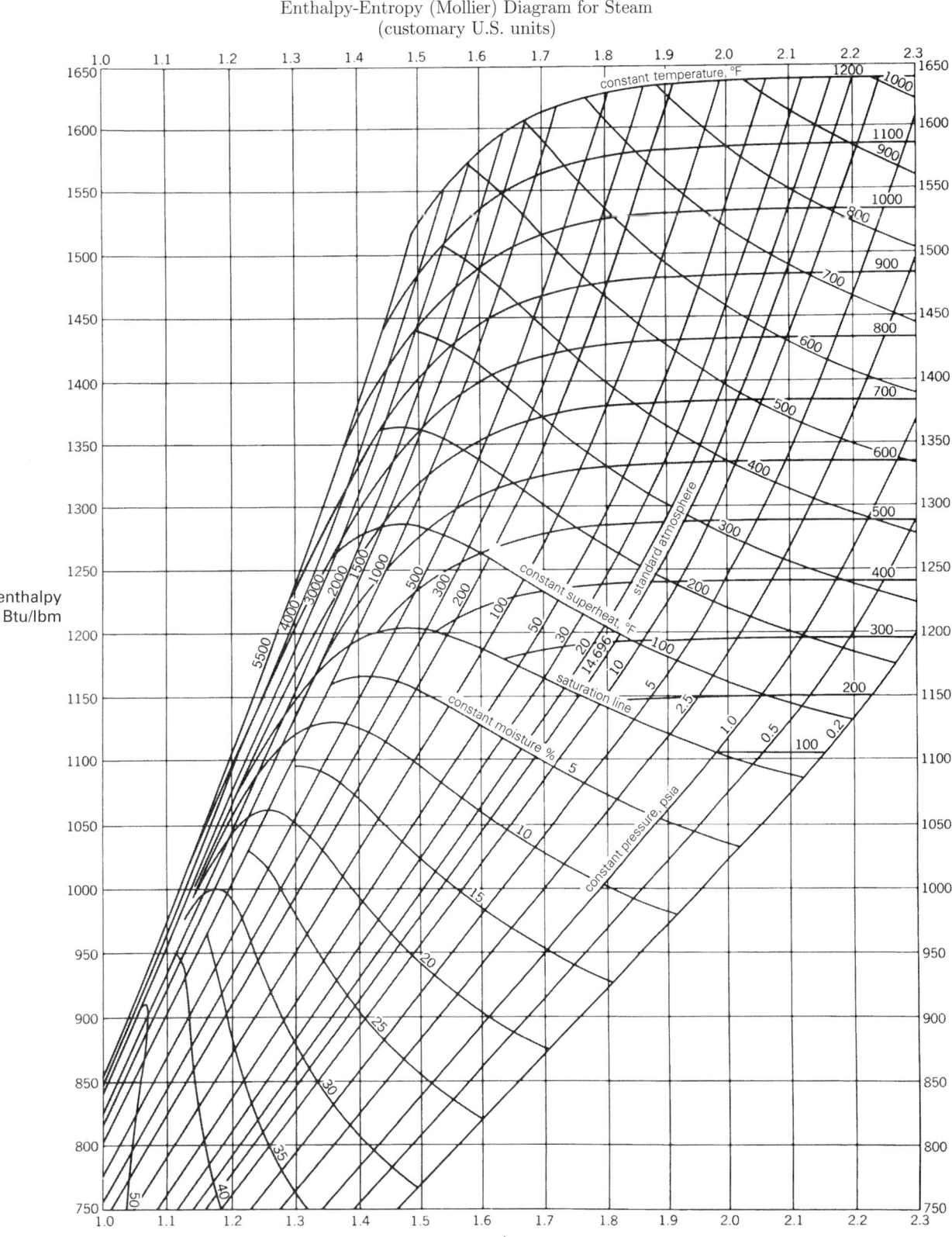

Support Material

APPENDIX 29.F
Properties of Low-Pressure Air
(customary U.S. units)

T in °R; h and u in Btu/lbm; ϕ in Btu/lbm-°R

T	h	p_r	u	v_r	ϕ
200	47.67	0.04320	33.96	1714.9	0.36303
220	52.46	0.06026	37.38	1352.5	0.38584
240	57.25	0.08165	40.80	1088.8	0.40666
260	62.03	0.10797	44.21	892.0	0.42582
280	66.82	0.13986	47.63	741.6	0.44356
300	71.61	0.17795	51.04	624.5	0.46007
320	76.40	0.22290	54.46	531.8	0.47550
340	81.18	0.27545	57.87	457.2	0.49002
360	85.97	0.3363	61.29	396.6	0.50369
380	90.75	0.4061	64.70	346.6	0.51663
400	95.53	0.4858	68.11	305.0	0.52890
420	100.32	0.5760	71.52	270.1	0.54058
440	105.11	0.6776	74.93	240.6	0.55172
460	109.90	0.7913	78.36	215.33	0.56235
480	114.69	0.9182	81.77	193.65	0.57255
500	119.48	1.0590	85.20	174.90	0.58233
520	124.27	1.2147	88.62	158.58	0.59172
537	128.34	1.3593	91.53	146.34	0.59945
540	129.06	1.3860	92.04	144.32	0.60078
560	133.86	1.5742	95.47	131.78	0.60950
580	138.66	1.7800	98.90	120.70	0.61793
600	143.47	2.005	102.34	110.88	0.62607
620	148.28	2.249	105.78	102.12	0.63395
640	153.09	2.514	109.21	94.30	0.64159
660	157.92	2.801	112.67	87.27	0.64902
680	162.74	3.111	116.12	80.96	0.65621
700	167.56	3.446	119.58	75.25	0.66321
720	172.39	3.806	123.04	70.07	0.67002
740	177.23	4.193	126.51	65.38	0.67665
760	182.08	4.607	129.99	61.10	0.68312
780	186.94	5.051	133.47	57.20	0.68942
800	191.81	5.526	136.97	53.63	0.69558
820	196.69	6.033	140.47	50.35	0.70160
840	201.56	6.573	143.98	47.34	0.70747
860	206.46	7.149	147.50	44.57	0.71323
880	211.35	7.761	151.02	42.01	0.71886
900	216.26	8.411	154.57	39.64	0.72438
920	221.18	9.102	158.12	37.44	0.72979
940	226.11	9.834	161.68	35.41	0.73509
960	231.06	10.61	165.26	33.52	0.74030
980	236.02	11.43	168.83	31.76	0.74540
1000	240.98	12.30	172.43	30.12	0.75042
1040	250.95	14.18	179.66	27.17	0.76019
1080	260.97	16.28	186.93	24.58	0.76964
1120	271.03	18.60	194.25	22.30	0.77880

(continued)

APPENDIX 29.F (continued)
Properties of Low-Pressure Air
(customary U.S. units)

T in °R; h and u in Btu/lbm; ϕ in Btu/lbm-°R

T	h	p_r	u	v_r	ϕ
1160	281.14	21.18	201.63	20.29	0.78767
1200	291.30	24.01	209.05	18.51	0.79628
1240	301.52	27.13	216.53	16.93	0.80466
1280	311.79	30.55	224.05	15.52	0.81280
1320	322.11	34.31	231.63	14.25	0.82075
1360	332.48	38.41	239.25	13.12	0.82848
1400	342.90	42.88	246.93	12.10	0.83604
1440	353.37	47.75	254.66	11.17	0.84341
1480	363.89	53.04	262.44	10.34	0.85062
1520	374.47	58.78	270.26	9.578	0.85767
1560	385.08	65.00	278.13	8.890	0.86456
1600	395.74	71.73	286.06	8.263	0.87130
1650	409.13	80.89	296.03	7.556	0.87954
1700	422.59	90.95	306.06	6.924	0.88758
1750	436.12	101.98	316.16	6.357	0.89542
1800	449.71	114.0	326.32	5.847	0.90308
1850	463.37	127.2	336.55	5.388	0.91056
1900	477.09	141.5	346.85	4.974	0.91788
1950	490.88	157.1	357.20	4.598	0.92504
2000	504.71	174.0	367.61	4.258	0.93205
2050	518.61	192.3	378.08	3.949	0.93891
2100	532.55	212.1	388.60	3.667	0.94564
2150	546.54	233.5	399.17	3.410	0.95222
2200	560.59	256.6	409.78	3.176	0.95868
2250	574.69	281.4	420.46	2.961	0.96501
2300	588.82	308.1	431.16	2.765	0.97123
2350	603.00	336.8	441.91	2.585	0.97732
2400	617.22	367.6	452.70	2.419	0.98331
2450	631.48	400.5	463.54	2.266	0.98919
2500	645.78	435.7	474.40	2.125	0.99497
2550	660.12	473.3	485.31	1.996	1.00064
2600	674.49	513.5	496.26	1.876	1.00623
2650	688.90	556.3	507.25	1.765	1.01172
2700	703.35	601.9	518.26	1.662	1.01712
2750	717.83	650.4	529.31	1.566	1.02244
2800	732.33	702.0	540.40	1.478	1.02767
2850	746.88	756.7	551.52	1.395	1.03282
2900	761.45	814.8	562.66	1.318	1.03788
2950	776.05	876.4	573.84	1.247	1.04288
3000	790.68	941.4	585.04	1.180	1.04779
3050	805.34	1011	596.28	1.118	1.05264
3100	820.03	1083	607.53	1.060	1.05741
3150	834.75	1161	618.82	1.006	1.06212
3200	849.48	1242	630.12	0.9546	1.06676

(continued)

APPENDIX 29.F *(continued)*
Properties of Low-Pressure Air
(customary U.S. units)

T in °R; h and u in Btu/lbm; ϕ in Btu/lbm-°R

T	h	p_r	u	v_r	ϕ
3250	864.24	1328	641.46	0.9069	1.07134
3300	879.02	1418	652.81	0.8621	1.07585
3350	893.83	1513	664.20	0.8202	1.08031
3400	908.66	1613	675.60	0.7807	1.08470
3450	923.52	1719	687.04	0.7436	1.08904
3500	938.40	1829	698.48	0.7087	1.09332
3550	953.30	1946	709.95	0.6759	1.09755
3600	968.21	2068	721.44	0.6449	1.10172
3650	983.15	2196	732.95	0.6157	1.10584
3700	998.11	2330	744.48	0.5882	1.10991
3750	1013.1	2471	756.04	0.5621	1.11393
3800	1028.1	2618	767.60	0.5376	1.11791
3850	1043.1	2773	779.19	0.5143	1.12183
3900	1058.1	2934	790.80	0.4923	1.12571
3950	1073.2	3103	802.43	0.4715	1.12955
4000	1088.3	3280	814.06	0.4518	1.13334
4050	1103.4	3464	825.72	0.4331	1.13709
4100	1118.5	3656	837.40	0.4154	1.14079
4150	1133.6	3858	849.09	0.3985	1.14446
4200	1148.7	4067	860.81	0.3826	1.14809
4300	1179.0	4513	884.28	0.3529	1.15522
4400	1209.4	4997	907.81	0.3262	1.16221
4500	1239.9	5521	931.39	0.3019	1.16905
4600	1270.4	6089	955.04	0.2799	1.17575
4700	1300.9	6701	978.73	0.2598	1.18232
4800	1331.5	7362	1002.5	0.2415	1.18876
4900	1362.2	8073	1026.3	0.2248	1.19508
5000	1392.9	8837	1050.1	0.2096	1.20129
5100	1423.6	9658	1074.0	0.1956	1.20738
5200	1454.4	10539	1098.0	0.1828	1.21336
5300	1485.3	11481	1122.0	0.1710	1.21923

Gas Tables: Thermodynamic Properties of Air, Products of Combustion and Component Gases, Compressible Flow Functions, 2nd Edition, Joseph H. Keenan, Jing Chao, and Joseph Kaye, copyright © 1980. Reproduced with permission of John Wiley & Sons, Inc.

APPENDIX 29.G
Properties of Saturated Refrigerant-12 (R-12) by Temperature
(customary U.S. units)

temp. (°F)	pressure (psia)	specific volume (ft³/lbm)		enthalpy (Btu/lbm)			entropy (Btu/lbm-°R)			temp. (°F)
		sat. liquid	sat. vapor	sat. liquid	evap.	sat. vapor	sat. liquid	evap.	sat. vapor	
T	p	v_f	v_g	h_f	h_{fg}	h_g	s_f	s_{fg}	s_g	T
−60	5.37	0.01036	6.516	−4.20	75.33	71.13	−0.0102	0.1681	0.1783	−60
−50	7.13	0.01047	5.012	−2.11	74.42	72.31	−0.0050	0.1717	0.1767	−50
−40	9.32	0.0106	3.911	0.00	73.50	73.50	0.00000	0.17517	0.17517	−40
−30	12.02	0.0107	3.088	2.03	72.67	74.70	0.00471	0.16916	0.17387	−30
−20	15.28	0.0108	2.474	4.07	71.80	75.87	0.00940	0.16335	0.17275	−20
−10	19.20	0.0109	2.003	6.14	70.91	77.05	0.01403	0.15772	0.17175	−10
0	23.87	0.0110	1.637	8.25	69.96	78.21	0.01869	0.15222	0.17091	0
5	26.51	0.0111	1.485	9.32	69.47	78.79	0.02097	1.14955	0.17052	5
10	29.35	0.0112	1.351	10.39	68.97	79.36	0.02328	0.14687	0.17015	10
20	35.75	0.0113	1.121	12.55	67.94	80.49	0.02783	0.14166	0.16949	20
30	43.16	0.0115	0.939	14.76	66.85	81.61	0.03233	0.13654	0.16887	30
40	51.68	0.0116	0.792	17.00	65.71	82.71	0.03680	0.13153	0.16833	40
50	61.39	0.0118	0.673	19.27	64.51	83.78	0.04126	0.12659	0.16785	50
60	72.41	0.0119	0.575	21.57	63.25	84.82	0.04568	0.12173	0.16741	60
70	84.82	0.0121	0.493	23.90	61.92	85.82	0.05009	0.11692	0.16701	70
80	98.76	0.0123	0.425	26.28	60.52	86.80	0.05446	0.11215	0.16662	80
86	107.9	0.0124	0.389	27.72	59.65	87.37	0.05708	0.10932	0.16640	86
90	114.3	0.0125	0.368	28.70	59.04	87.74	0.05882	0.10742	0.16624	90
100	131.6	0.0127	0.319	31.16	57.46	88.62	0.06316	0.10268	0.16584	100
110	150.7	0.0129	0.277	33.65	55.78	89.43	0.06749	0.09793	0.16542	110
120	171.8	0.0132	0.240	36.16	53.99	90.15	0.07180	0.09315	0.16495	120
233	596.9	0.02870	0.02870	78.86	0	78.86	0.1359	0	0.1359	233

Reproduced with permission from the DuPont Company.

APPENDIX 29.H
Properties of Saturated Refrigerant-12 (R-12) by Pressure
(customary U.S. units)

pressure (psia)	temp. (°F)	specific volume (ft³/lbm)		enthalpy (Btu/lbm)			entropy (Btu/lbm-°R)			pressure (psia)
		sat. liquid	sat. vapor	sat. liquid	evap.	sat. vapor	sat. liquid	evap.	sat. vapor	
p	T	v_f	v_g	h_f	h_{fg}	h_g	s_f	s_{fg}	s_g	p
5	−62.5	0.01034	6.953	−4.73	75.56	70.83	−0.0115	0.1943	0.1788	5
10	−37.3	0.0106	3.662	0.54	73.28	73.82	0.00127	0.17360	0.17487	10
15	−20.8	0.0108	2.518	3.91	71.87	75.78	0.00902	0.16381	0.17283	15
20	−8.2	0.0109	1.925	6.53	70.74	77.27	0.01488	0.15672	0.17160	20
30	11.1	0.0112	1.324	10.62	68.86	79.48	0.02410	0.14597	0.17007	30
40	25.9	0.0114	1.009	13.86	67.30	81.16	0.03049	0.13865	0.16914	40
50	38.3	0.0116	0.817	16.58	65.94	82.52	0.03597	0.13244	0.16841	50
60	48.7	0.0117	0.688	18.96	64.69	83.65	0.04065	0.12726	0.16791	60
80	66.3	0.0120	0.521	23.01	62.44	85.45	0.04844	0.11872	0.16716	80
100	80.9	0.0123	0.419	26.49	60.40	86.89	0.05483	0.11176	0.16659	100
120	93.4	0.0126	0.419	29.53	58.52	88.05	0.06030	0.10580	0.16610	120
140	104.5	0.0128	0.298	32.28	56.71	88.99	0.06513	0.10053	0.16566	140
160	114.5	0.0130	0.260	34.78	54.99	89.77	0.06958	0.09564	0.16522	160
180	123.7	0.0133	0.228	37.07	53.31	90.38	0.07337	0.09139	0.16476	180
200	132.1	0.0135	0.202	39.21	51.65	90.86	0.07694	0.08730	0.16424	200
220	139.9	0.0138	0.181	41.22	50.28	91.50	0.08021	0.08354	0.16375	220
596	233.6	0.02870	0.02870	78.86	0	78.86	0.1359	0	0.1359	596

Reproduced with permission from the DuPont Company.

APPENDIX 29.I
Properties of Superheated Refrigerant-12 (R-12)
(customary U.S. units)

specific volume, v, in ft^3/lbm; enthalpy, h, in Btu/lbm; entropy, s, in Btu/lbm-°R

pressure (psia) (sat. temp.)		temperature (°F)											
		−40	−20	0	20	40	60	80	100	150	200	250	300
5 (−62.5)	v	7.363	7.726	8.088	8.450	8.812	9.173	9.533	9.893	10.79	11.69	......	......
	h	73.72	76.36	79.05	81.78	84.56	87.41	90.30	93.25	100.84	108.75	......	......
	s	0.1859	0.1920	0.1979	0.2038	0.2095	0.2150	0.2205	0.2258	0.2388	0.2518	......	......
10 (−37.3)	v	......	3.821	4.006	4.189	4.371	4.556	4.740	4.923	5.379	5.831	6.281	......
	h	......	76.11	78.81	81.56	84.35	87.19	90.11	93.05	100.66	108.63	116.88	......
	s	......	0.1801	0.1861	0.1919	0.1977	0.2033	0.2087	0.2141	0.2271	0.2396	0.2517	......
15 (−20.8)	v	......	2.521	2.646	2.771	2.895	3.019	3.143	3.266	3.571	3.877	4.191	......
	h	......	75.89	78.59	81.37	84.18	87.03	89.94	92.91	100.53	108.49	116.78	......
	s	......	0.17307	0.17913	0.18499	0.19074	0.19635	0.20185	0.20723	0.22028	0.23282	0.24491	......
20 (−8.2)	v	......	......	1.965	2.060	2.155	2.250	2.343	2.437	2.669	2.901	3.130	......
	h	......	......	78.39	81.14	83.97	86.85	89.78	92.75	100.40	108.38	116.67	......
	s	......	......	0.17407	0.17996	0.18573	0.19138	0.19688	0.20229	0.21537	0.22794	0.24005	......
25 (2.2)	v	......	......	......	1.712	1.793	1.873	1.952	2.031	2.227	2.422	2.615	......
	h	......	......	......	80.95	83.78	86.67	89.61	92.56	100.26	108.26	116.56	......
	s	......	......	......	0.17637	0.18216	0.18783	0.19336	0.19748	0.21190	0.22450	0.23665	......
30 (11.1)	v	......	......	......	1.364	1.430	1.495	1.560	1.624	1.784	1.943	2.099	......
	h	......	......	......	80.75	83.59	86.49	89.43	92.42	100.12	108.13	116.45	......
	s	......	......	......	0.17278	0.17859	0.18429	0.18983	0.19527	0.20843	0.22105	0.23325	......
35 (18.9)	v	......	......	......	1.109	1.237	1.295	1.352	1.409	1.550	1.689	1.827	......
	h	......	......	......	80.49	83.40	86.30	89.26	92.26	99.98	108.01	116.33	......
	s	......	......	......	0.16963	0.17591	0.18162	0.18719	0.19266	0.20584	0.21849	0.23069	......
40 (25.9)	v	......	......	......	......	1.044	1.095	1.144	1.194	1.315	1.435	1.554	......
	h	......	......	......	......	83.20	86.11	89.09	92.09	99.83	107.88	116.21	......
	s	......	......	......	......	0.17322	0.17896	0.18455	0.19004	0.20325	0.21592	0.22813	......
50 (38.3)	v	......	......	......	......	0.821	0.863	0.904	0.944	1.044	1.142	1.239	1.332
	h	......	......	......	......	82.76	85.72	88.72	91.75	99.54	107.62	116.00	124.69
	s	......	......	......	......	0.16895	0.17475	0.19040	0.18591	0.19923	0.21196	0.22419	0.23600
60 (48.7)	v	......	......	......	......	......	0.708	0.743	0.778	0.863	0.946	1.028	1.108
	h	......	......	......	......	......	85.33	88.35	91.41	99.24	107.36	115.54	124.29
	s	......	......	......	......	......	0.17120	0.17689	0.18246	0.19585	0.20865	0.22094	0.23280
70 (57.9)	v	......	......	......	......	......	0.553	0.642	0.673	0.750	0.824	0.896	0.967
	h	......	......	......	......	......	84.94	87.96	91.05	98.94	107.10	115.54	124.29
	s	......	......	......	......	......	0.16765	0.17399	0.17961	0.19310	0.20597	0.21830	0.23020
80 (66.3)	v	......	......	......	......	......	......	0.540	0.568	0.636	0.701	0.764	0.826
	h	......	......	......	......	......	......	87.56	90.68	98.64	106.84	115.30	124.08
	s	......	......	......	......	......	......	0.17108	0.17675	0.19035	0.20328	0.21566	0.22760

(continued)

APPENDIX 29.I *(continued)*
Properties of Superheated Refrigerant-12 (R-12)
(customary U.S. units)

specific volume, v, in ft^3/lbm; enthalpy, h, in Btu/lbm; entropy, s, in Btu/lbm-°R

pressure (psia) (sat. temp.)		temperature (°F)											
		−40	−20	0	20	40	60	80	100	150	200	250	300
90 (73.6)	v	......	......	......	......	......	......	......	0.505	0.568	0.627	0.685	0.742
	h	......	......	......	......	......	......	......	90.31	98.32	106.56	115.07	123.88
	s	......	......	......	......	......	......	......	0.17443	0.18813	0.20111	0.21356	0.22554
100 (80.9)	v	......	......	......	......	......	......	......	0.442	0.499	0.553	0.606	0.657
	h	......	......	......	......	......	......	......	89.93	97.99	106.29	114.84	123.67
	s	......	......	......	......	......	......	......	0.17210	0.18590	0.19894	0.21145	0.22347
120 (93.4)	v	......	......	......	......	......	......	......	0.357	0.407	0.454	0.500	0.543
	h	......	......	......	......	......	......	......	89.13	97.30	105.70	114.35	123.25
	s	......	......	......	......	......	......	......	0.16803	0.18207	0.19529	0.20792	0.22000
140 (104.5)	v	......	......	......	......	......	......	......	......	0.341	0.383	0.423	0.462
	h	......	......	......	......	......	......	......	......	96.65	105.14	113.85	122.85
	s	......	......	......	......	......	......	......	......	0.17868	0.19205	0.20479	0.21701
160 (114.5)	v	......	......	......	......	......	......	......	......	0.318	0.335	0.372	0.408
	h	......	......	......	......	......	......	......	......	95.82	104.50	113.33	122.39
	s	......	......	......	......	......	......	......	......	0.17561	0.18927	0.20213	0.21444
180 (123.7)	v	......	......	......	......	......	......	......	......	0.294	0.287	0.321	0.353
	h	......	......	......	......	......	......	......	......	94.99	103.85	112.81	121.92
	s	......	......	......	......	......	......	......	......	0.17254	0.18648	0.19947	0.21187
200 (132.1)	v	......	......	......	......	......	......	......	......	0.241	0.255	0.288	0.317
	h	......	......	......	......	......	......	......	......	94.16	103.12	112.20	121.42
	s	......	......	......	......	......	......	......	......	0.16970	0.18395	0.19717	0.20970
220 (139.9)	v	......	......	......	......	......	......	......	......	0.188	0.232	0.254	0.282
	h	......	......	......	......	......	......	......	......	93.32	102.39	111.59	120.91
	s	......	......	......	......	......	......	......	......	0.16685	0.18142	0.19387	0.20753

Reproduced with permission from the DuPont Company.

APPENDIX 29.J
Pressure-Enthalpy Diagram for Refrigerant-12 (R-12)
(customary U.S. units)

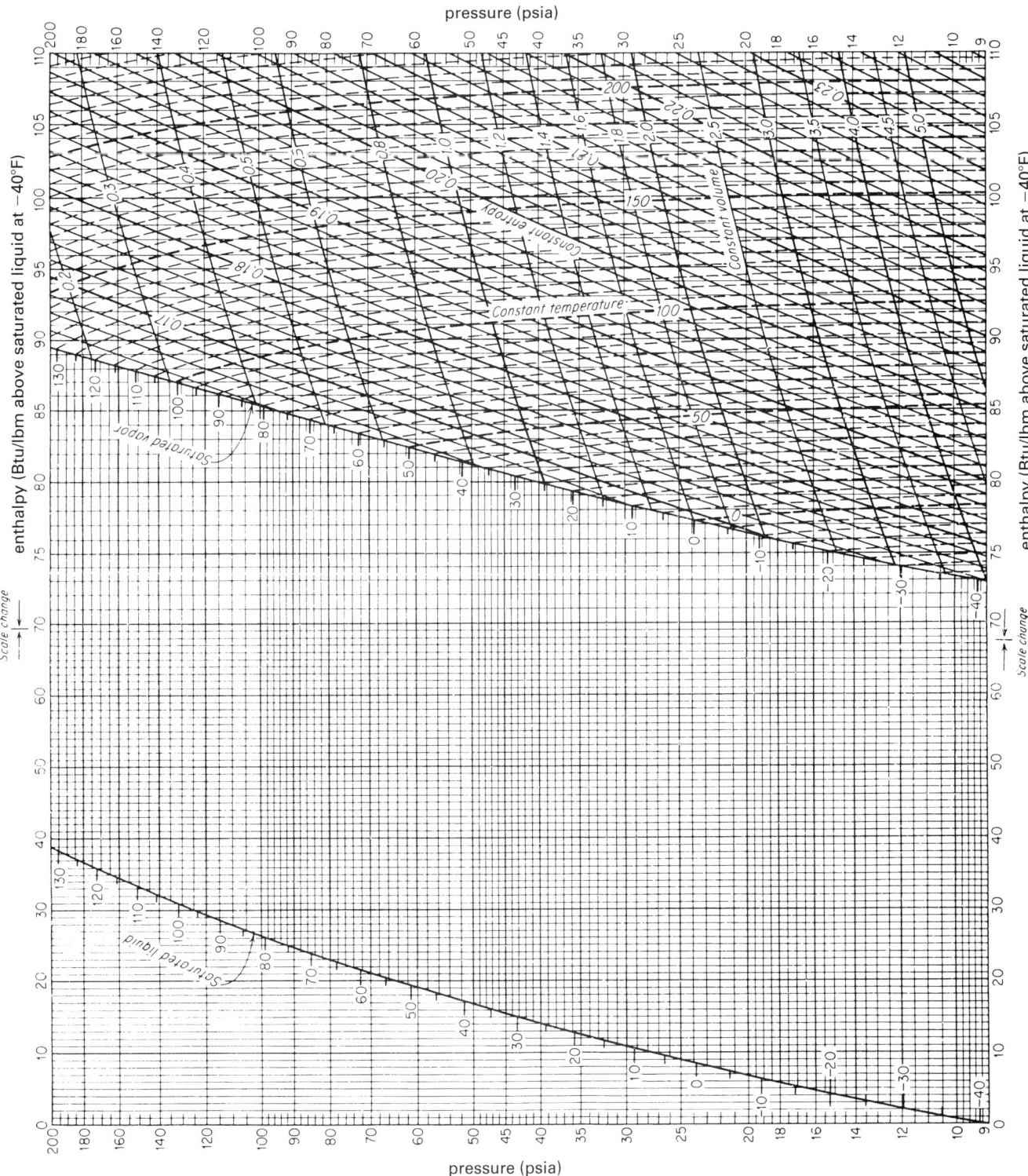

Support Material

APPENDIX 29.K
Properties of Saturated Refrigerant-22 (R-22) by Temperature
(customary U.S. units)

temp. (°F)	saturation pressure (psia)	saturation pressure (psig)	volume (ft³/lbm) v_g	density (lbm/ft³) ρ_f	enthalpy (Btu/lbm) h_f	enthalpy (Btu/lbm) h_g	entropy (Btu/lbm-°R) s_f	entropy (Btu/lbm-°R) s_g
−130	0.68858	28.519[a]	59.170	96.313	−24.388	89.888	−0.065456	0.28118
−120	1.0725	27.738[a]	39.078	95.416	−21.538	91.049	−0.056942	0.27452
−110	1.6199	26.623[a]	26.578	94.509	−18.738	92.211	−0.048818	0.26848
−100	2.3802	25.075[a]	18.558	93.590	−15.980	93.371	−0.041046	0.26298
−90	3.4111	22.976[a]	13.268	92.660	−13.259	94.527	−0.033590	0.25798
−80	4.7793	20.191[a]	9.6902	91.717	−10.570	95.676	−0.026418	0.25342
−75	5.6131	18.493[a]	8.3419	91.241	−9.2346	96.247	−0.022929	0.25128
−70	6.5603	16.564[a]	7.2139	90.761	−7.9050	96.815	−0.019500	0.24924
−65	7.6317	14.383[a]	6.2655	90.278	−6.5802	97.380	−0.016128	0.24728
−60	8.8386	11.926[a]	5.4641	89.791	−5.2593	97.942	−0.012808	0.24541
−50	11.707	6.0851[a]	4.2039	88.807	−2.6263	99.055	−0.006316	0.24189
−48	12.361	4.7548[a]	3.9962	88.608	−2.1007	99.275	−0.005039	0.24122
−46	13.042	3.3666[a]	3.8007	88.408	−1.5753	99.495	−0.003770	0.24056
−44	13.754	1.9186[a]	3.6168	88.208	−1.0501	99.714	−0.002507	0.23991
−42	14.495	0.4090[a]	3.4437	88.007	−0.5250	99.932	−0.001250	0.23927
−41.47	14.696	0.0	3.3997	87.954	−0.3865	99.990	−0.000920	0.23910
−40	15.268	0.5717	3.2805	87.806	0.0	100.15	0.0	0.23864
−38	16.072	1.3763	3.1267	87.604	0.5250	100.37	0.001244	0.23802
−36	16.910	2.2138	2.9816	87.401	1.0500	100.58	0.002482	0.23741
−34	17.781	3.0852	2.8446	87.197	1.5751	100.80	0.003714	0.23681
−32	18.687	3.9914	2.7152	86.993	2.1003	101.01	0.004940	0.23622
−30	19.629	4.9333	2.5930	86.788	2.6257	101.22	0.006161	0.23564
−28	20.608	5.9119	2.4774	86.582	3.1512	101.44	0.007377	0.23506
−26	21.624	6.9283	2.3680	86.375	3.6771	101.65	0.008587	0.23450
−24	22.679	7.9832	2.2645	86.168	4.2032	101.86	0.009792	0.23394
−22	23.774	9.0778	2.1664	85.960	4.7297	102.07	0.010993	0.23340
−20	24.909	10.213	2.0735	85.751	5.2566	102.28	0.012188	0.23285
−18	26.086	11.390	1.9854	85.542	5.7840	102.48	0.013379	0.23232
−16	27.306	12.610	1.9018	85.331	6.3119	102.69	0.014566	0.23180
−14	28.569	13.873	1.8225	85.120	6.8403	102.90	0.015748	0.23128
−12	29.877	15.181	1.7472	84.908	7.3693	103.10	0.016926	0.23077
−10	31.231	16.535	1.6757	84.695	7.8989	103.30	0.018100	0.23027
−8	32.632	17.936	1.6077	84.481	8.4292	103.51	0.019270	0.22977
−6	34.081	19.385	1.5430	84.266	8.9603	103.71	0.020436	0.22928
−4	35.579	20.883	1.4815	84.051	9.4921	103.91	0.021598	0.22880
−2	37.127	22.431	1.4230	83.834	10.025	104.10	0.022757	0.22832
0	38.726	24.030	1.3672	83.617	10.558	104.30	0.023912	0.022785
2	40.378	25.682	1.3141	83.399	11.093	104.50	0.025064	0.22738
4	42.083	27.387	1.2635	83.179	11.628	104.69	0.026213	0.22693
6	43.843	29.147	1.2152	82.959	12.164	104.89	0.027359	0.22647
8	45.658	30.962	1.1692	82.738	12.702	105.08	0.028502	0.22602

(continued)

APPENDIX 29.K *(continued)*
Properties of Saturated Refrigerant-22 (R-22) by Temperature
(customary U.S. units)

temp. (°F)	saturation pressure (psia)	saturation pressure (psig)	volume (ft³/lbm) v_g	density (lbm/ft³) ρ_f	enthalpy (Btu/lbm) h_f	enthalpy (Btu/lbm) h_g	entropy (Btu/lbm-°R) s_f	entropy (Btu/lbm-°R) s_g
10	47.530	32.834	1.1253	82.516	13.240	105.27	0.029642	0.22558
12	49.461	34.765	1.0833	82.292	13.779	105.46	0.030779	0.22515
14	51.450	36.754	1.0433	82.008	14.320	105.64	0.031013	0.22471
16	53.501	38.805	1.0050	81.843	14.862	105.83	0.033045	0.22429
18	55.612	40.916	0.96841	81.616	15.405	106.02	0.034175	0.22387
20	57.786	43.090	0.93343	81.389	15.950	106.20	0.035302	0.22345
25	63.505	48.809	0.85246	80.815	17.317	106.65	0.038110	0.22243
30	69.641	54.945	0.77984	80.234	18.693	107.09	0.040905	0.22143
35	76.215	61.519	0.71454	79.645	20.078	107.52	0.043689	0.22046
40	83.246	68.550	0.65571	79.049	21.474	107.94	0.046464	0.21951
45	90.754	76.058	0.60258	78.443	22.880	108.35	0.049229	0.21858
50	98.758	84.062	0.55451	77.829	24.298	108.74	0.051987	0.21767
55	107.28	92.583	0.51093	77.206	25.728	109.12	0.054739	0.21677
60	116.34	101.64	0.47134	76.572	27.170	109.49	0.057486	0.21589
65	125.95	111.26	0.43531	75.928	28.626	109.84	0.060228	0.21502
70	136.15	121.45	0.40245	75.273	30.095	110.18	0.062968	0.21416
80	158.36	143.66	0.34497	73.926	33.077	110.80	0.068441	0.21246
90	183.14	168.44	0.29668	72.525	36.121	111.35	0.073911	0.21077
100	210.67	195.97	0.25582	71.061	39.233	111.81	0.079400	0.20907
110	241.13	226.44	0.22102	69.524	42.422	112.17	0.084906	0.20734
120	274.73	260.03	0.19118	67.901	45.694	112.42	0.090444	0.20554
130	311.66	296.96	0.16542	66.174	49.064	112.52	0.096033	0.20365
140	352.14	337.45	0.14300	64.319	52.550	112.47	0.10170	0.20161
150	396.42	381.72	0.12334	62.301	56.177	112.20	0.10749	0.19938
160	444.75	430.06	0.10590	60.068	59.989	111.67	0.11345	0.19684
170	497.46	482.76	0.090228	57.532	64.055	110.76	0.11970	0.19386
180	554.89	540.19	0.075819	54.533	68.504	109.30	0.12640	0.19018
190	617.52	602.82	0.061991	50.703	73.617	106.88	0.13399	0.18518
200	686.02	671.32	0.046923	44.671	80.406	101.99	0.14394	0.17666
205.07[b]	723.4	708.7	0.03123	32.03	91.58	91.58	0.1605	0.1605

[a] in Hg vacuum
[b] critical point

From *2009 ASHRAE Handbook—Fundamentals*, copyright © 2009, by American Society of Heating, Refrigerating, and Air-Conditioning Engineers, Inc. Reproduced with permission.

APPENDIX 29.L
Pressure-Enthalpy Diagram for Refrigerant-22 (R-22)
(customary U.S. units)

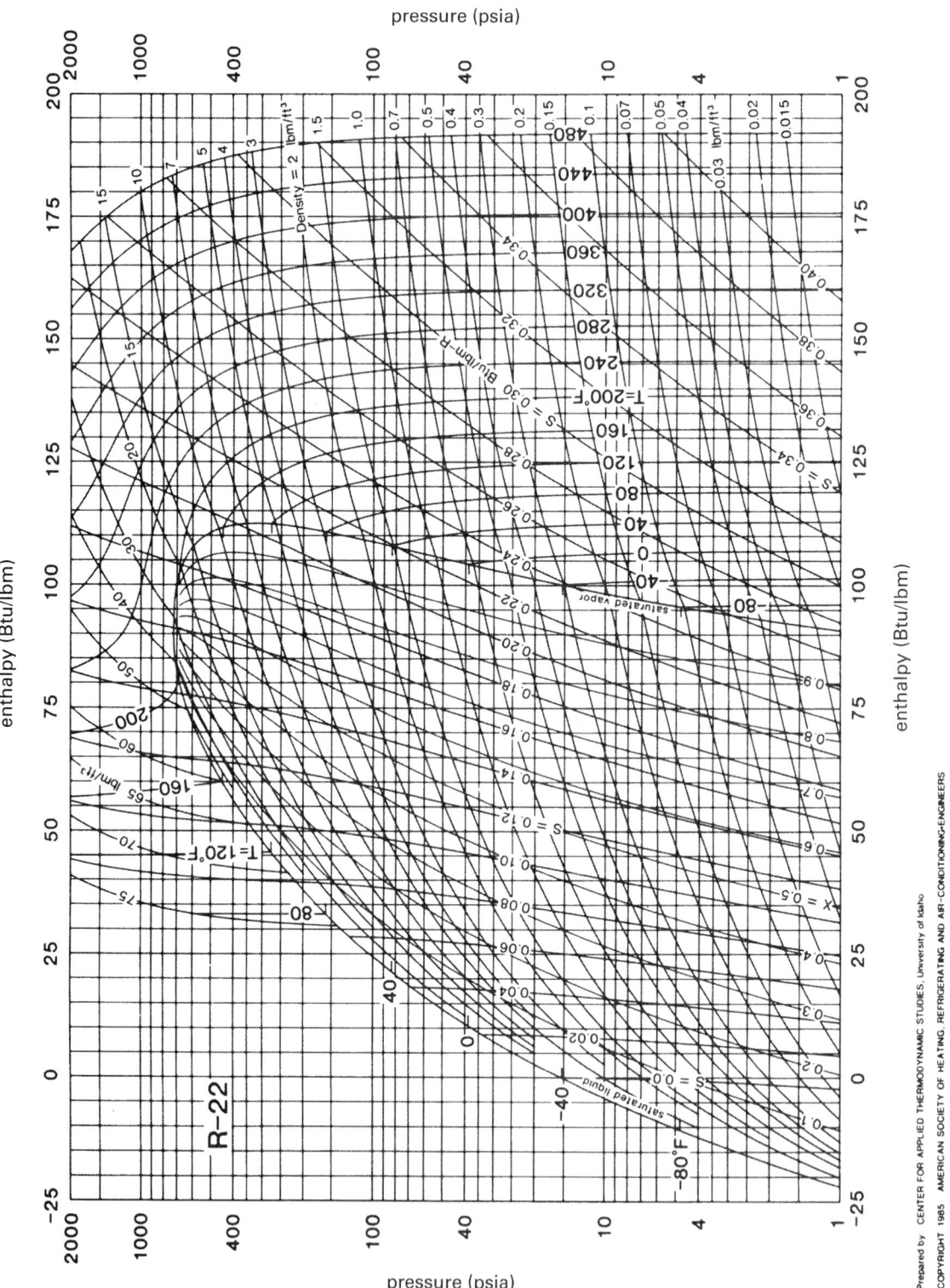

From 2009 *ASHRAE Handbook—Fundamentals*, copyright © 2009, by American Society of Heating, Refrigerating, and Air-Conditioning Engineers, Inc. Reproduced with permission.

APPENDIX 29.M
Pressure-Enthalpy Diagram for Refrigerant HFC-134a
(customary U.S. units)

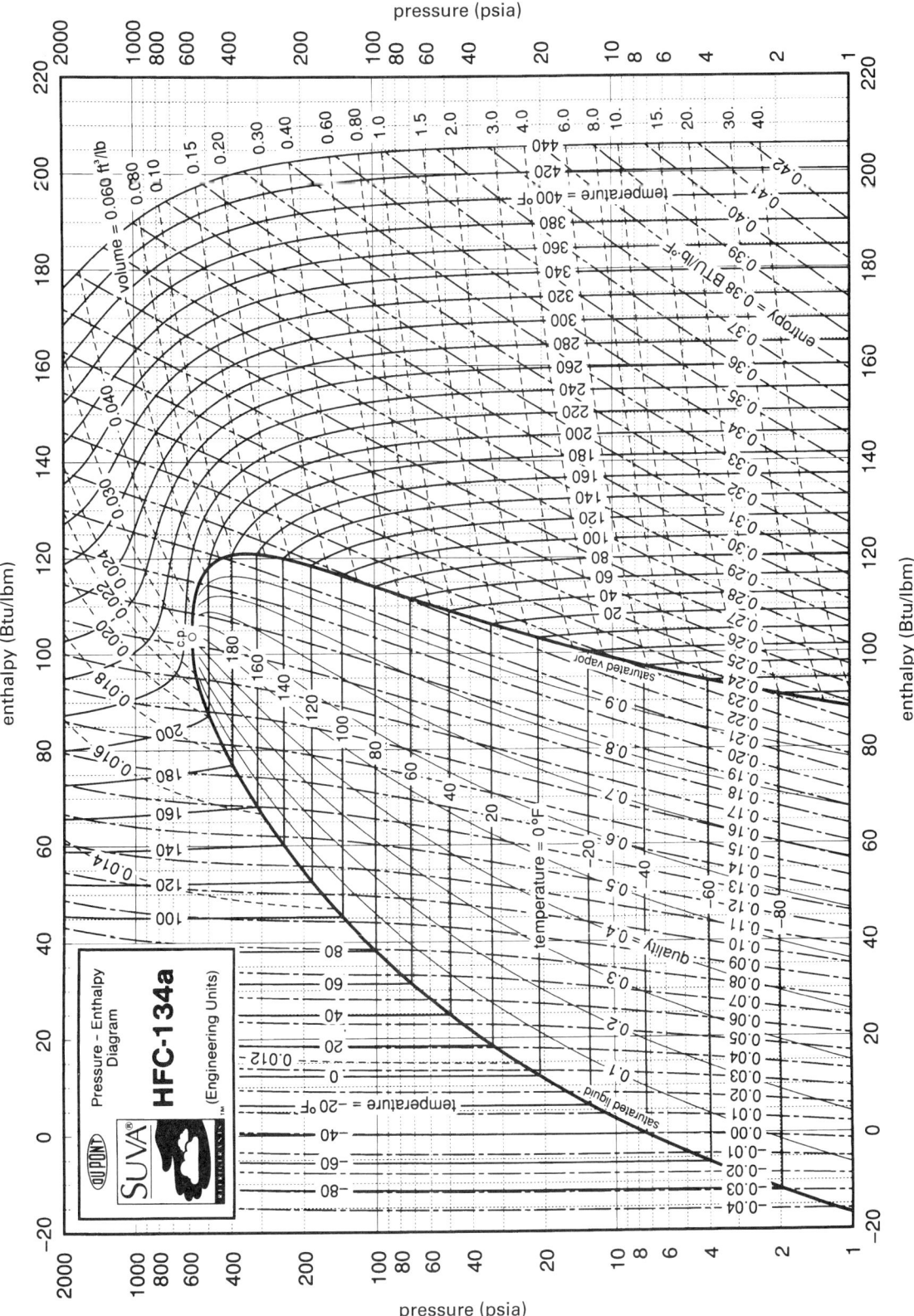

Reproduced by permission of the DuPont Company.

APPENDIX 29.N
Properties of Saturated Steam by Temperature
(SI units)

temp. (°C)	absolute pressure (bars)	specific volume (cm³/g)		internal energy (kJ/kg)		enthalpy (kJ/kg)			entropy (kJ/kg·K)		temp. (°C)
		sat. liquid, v_f	sat. vapor, v_g	sat. liquid, u_f	sat. vapor, u_g	sat. liquid, h_f	evap., h_{fg}	sat. vapor, h_g	sat. liquid, s_f	sat. vapor, s_g	
0.01	0.006117	1.0002	205991	0.00	2374.9	0.0006	2500.9	2500.9	0.0000	9.1555	0.01
4	0.00814	1.0001	157116	16.81	2380.4	16.81	2491.4	2508.2	0.0611	9.0505	4
5	0.00873	1.0001	147011	21.02	2381.8	21.02	2489.1	2510.1	0.0763	9.0248	5
6	0.00935	1.0001	137633	25.22	2383.2	25.22	2486.7	2511.9	0.0913	8.9993	6
8	0.01073	1.0002	120829	33.63	2385.9	33.63	2482.0	2515.6	0.1213	8.9491	8
10	0.01228	1.0003	106303	42.02	2388.6	42.02	2477.2	2519.2	0.1511	8.8998	10
11	0.01313	1.0004	99787	46.22	2390.0	46.22	2474.8	2521.0	0.1659	8.8754	11
12	0.01403	1.0005	93719	50.41	2391.4	50.41	2472.5	2522.9	0.1806	8.8513	12
13	0.01498	1.0007	88064	54.60	2392.8	54.60	2470.1	2524.7	0.1953	8.8274	13
14	0.01599	1.0008	82793	58.79	2394.1	58.79	2467.7	2526.5	0.2099	8.8037	14
15	0.01706	1.0009	77875	62.98	2395.5	62.98	2465.3	2528.3	0.2245	8.7803	15
16	0.01819	1.0011	73286	67.17	2396.9	67.17	2463.0	2530.2	0.2390	8.7570	16
17	0.01938	1.0013	69001	71.36	2398.2	71.36	2460.6	2532.0	0.2534	8.7339	17
18	0.02065	1.0014	64998	75.54	2399.6	75.54	2458.3	2533.8	0.2678	8.7111	18
19	0.02198	1.0016	61256	79.73	2401.0	79.73	2455.9	2535.6	0.2822	8.6884	19
20	0.02339	1.0018	57757	83.91	2402.3	83.91	2453.5	2537.4	0.2965	8.6660	20
21	0.02488	1.0021	54483	88.10	2403.7	88.10	2451.2	2539.3	0.3107	8.6437	21
22	0.02645	1.0023	51418	92.28	2405.0	92.28	2448.8	2541.1	0.3249	8.6217	22
23	0.02811	1.0025	48548	96.46	2406.4	96.47	2446.4	2542.9	0.3391	8.5998	23
24	0.02986	1.0028	45858	100.64	2407.8	100.65	2444.1	2544.7	0.3532	8.5781	24
25	0.03170	1.0030	43337	104.83	2409.1	104.83	2441.7	2546.5	0.3672	8.5566	25
26	0.03364	1.0033	40973	109.01	2410.5	109.01	2439.3	2548.3	0.3812	8.5353	26
27	0.03568	1.0035	38754	113.19	2411.8	113.19	2436.9	2550.1	0.3952	8.5142	27
28	0.03783	1.0038	36672	117.37	2413.2	117.37	2434.5	2551.9	0.4091	8.4933	28
29	0.04009	1.0041	34716	121.55	2414.6	121.55	2432.2	2553.7	0.4229	8.4725	29
30	0.04247	1.0044	32878	125.73	2415.9	125.73	2429.8	2555.5	0.4368	8.4520	30
31	0.04497	1.0047	31151	129.91	2417.3	129.91	2427.4	2557.3	0.4505	8.4316	31
32	0.04760	1.0050	29526	134.09	2418.6	134.09	2425.1	2559.2	0.4642	8.4113	32
33	0.05035	1.0054	27998	138.27	2420.0	138.27	2422.7	2561.0	0.4779	8.3913	33
34	0.05325	1.0057	26560	142.45	2421.3	142.45	2420.4	2562.8	0.4916	8.3714	34
35	0.05629	1.0060	25205	146.63	2422.7	146.63	2417.9	2564.5	0.5051	8.3517	35
36	0.05948	1.0064	23929	150.81	2424.0	150.81	2415.5	2566.3	0.5187	8.3321	36
38	0.06633	1.0071	21593	159.17	2426.7	159.17	2410.7	2569.9	0.5456	8.2935	38
40	0.07385	1.0079	19515	167.53	2429.4	167.53	2406.0	2573.5	0.5724	8.2555	40
45	0.09595	1.0099	15252	188.43	2436.1	188.43	2394.0	2582.4	0.6386	8.1633	45
50	0.1235	1.0121	12027	209.33	2442.7	209.34	2382.0	2591.3	0.7038	8.0748	50
55	0.1576	1.0146	9564	230.24	2449.3	230.26	2369.8	2600.1	0.7680	7.9898	55
60	0.1995	1.0171	7667	251.16	2455.9	251.18	2357.6	2608.8	0.8313	7.9081	60
65	0.2504	1.0199	6194	272.09	2462.4	272.12	2345.4	2617.5	0.8937	7.8296	65
70	0.3120	1.0228	5040	293.03	2468.9	293.07	2333.0	2626.1	0.9551	7.7540	70

(continued)

APPENDIX 29.N (continued)
Properties of Saturated Steam by Temperature
(SI units)

temp. (°C)	absolute pressure (bars)	specific volume (cm³/g)		internal energy (kJ/kg)		enthalpy (kJ/kg)			entropy (kJ/kg·K)		temp. (°C)
		sat. liquid, v_f	sat. vapor, v_g	sat. liquid, u_f	sat. vapor, u_g	sat. liquid, h_f	evap., h_{fg}	sat. vapor, h_g	sat. liquid, s_f	sat. vapor, s_g	
75	0.3860	1.0258	4129	313.99	2475.2	314.03	2320.6	2634.6	1.0158	7.6812	75
80	0.4741	1.0291	3405	331.06	2481.6	335.01	2308.0	2643.0	1.0756	7.6111	80
85	0.5787	1.0324	2826	355.95	2487.8	356.01	2295.3	2651.3	1.1346	7.5434	85
90	0.7018	1.0360	2359	376.97	2494.0	377.04	2282.5	2659.5	1.1929	7.4781	90
95	0.8461	1.0396	1981	398.00	2500.0	398.09	2269.5	2667.6	1.2504	7.4151	95
100	1.014	1.0435	1672	419.06	2506.0	419.17	2256.4	2675.6	1.3072	7.3541	100
105	1.209	1.0474	1418	440.15	2511.9	440.27	2243.1	2683.4	1.3633	7.2952	105
110	1.434	1.0516	1209	461.26	2517.7	461.42	2229.6	2691.1	1.4188	7.2381	110
115	1.692	1.0559	1036	482.42	2523.3	482.59	2216.0	2698.6	1.4737	7.1828	115
120	1.987	1.0603	891.2	503.60	2528.9	503.81	2202.1	2705.9	1.5279	7.1291	120
125	2.322	1.0649	770.0	524.83	2534.3	525.07	2188.0	2713.1	1.5816	7.0770	125
130	2.703	1.0697	668.0	546.10	2539.5	546.38	2173.7	2720.1	1.6346	7.0264	130
135	3.132	1.0746	581.7	567.41	2544.7	567.74	2159.1	2726.9	1.6872	6.9772	135
140	3.615	1.0798	508.5	588.77	2549.6	589.16	2144.3	2733.4	1.7392	6.9293	140
145	4.157	1.0850	446.0	610.19	2554.4	610.64	2129.2	2739.8	1.7907	6.8826	145
150	4.762	1.0905	392.5	631.66	2559.1	632.18	2113.8	2745.9	1.8418	6.8371	150
155	5.435	1.0962	346.5	653.19	2563.5	653.79	2098.0	2751.8	1.8924	6.7926	155
160	6.182	1.1020	306.8	674.79	2567.8	675.47	2082.0	2757.4	1.9426	6.7491	160
165	7.009	1.1080	272.4	696.46	2571.9	697.24	2065.6	2762.8	1.9923	6.7066	165
170	7.922	1.1143	242.6	718.20	2575.7	719.08	2048.8	2767.9	2.0417	6.6650	170
175	8.926	1.1207	216.6	740.02	2579.4	741.02	2031.7	2772.7	2.0906	6.6241	175
180	10.03	1.1274	193.8	761.92	2582.8	763.05	2014.2	2777.2	2.1392	6.5840	180
185	11.23	1.1343	173.9	783.91	2586.0	785.19	1996.2	2781.4	2.1875	6.5447	185
190	12.55	1.1415	156.4	806.00	2589.0	807.43	1977.9	2785.3	2.2355	6.5059	190
195	13.99	1.1489	140.9	828.18	2591.7	829.79	1959.0	2788.8	2.2832	6.4678	195
200	15.55	1.1565	127.2	850.47	2594.2	852.27	1939.7	2792.0	2.3305	6.4302	200
205	17.24	1.1645	115.1	872.87	2596.4	874.88	1920.0	2794.8	2.3777	6.3930	205
210	19.08	1.1727	104.3	895.39	2598.3	897.63	1899.6	2797.3	2.4245	6.3563	210
215	21.06	1.1813	94.68	918.04	2599.9	920.53	1878.8	2799.3	2.4712	6.3200	215
220	23.20	1.1902	86.09	940.82	2601.3	943.58	1857.4	2801.0	2.5177	6.2840	220
225	25.50	1.1994	78.40	963.74	2602.2	966.80	1835.4	2802.2	2.5640	6.2483	225
230	27.97	1.2090	71.50	986.81	2602.9	990.19	1812.7	2802.9	2.6101	6.2128	230
235	30.63	1.2190	65.30	1010.0	2603.2	1013.8	1789.4	2803.2	2.6561	6.1775	235
240	33.47	1.2295	59.71	1033.4	2603.1	1037.6	1765.4	2803.0	2.7020	6.1423	240
245	36.51	1.2403	54.65	1057.0	2602.7	1061.6	1740.7	2802.2	2.7478	6.1072	245
250	39.76	1.2517	50.08	1080.8	2601.8	1085.8	1715.2	2800.9	2.7935	6.0721	250
255	43.23	1.2636	45.94	1104.8	2600.5	1110.2	1688.8	2799.1	2.8392	6.0369	255
260	46.92	1.2761	42.17	1129.0	2598.7	1135.0	1661.6	2796.6	2.8849	6.0016	260
265	50.85	1.2892	38.75	1153.4	2596.5	1160.0	1633.5	2793.5	2.9307	5.9661	265
270	55.03	1.3030	35.62	1178.1	2593.7	1185.3	1604.4	2789.7	2.9765	5.9304	270

(continued)

APPENDIX 29.N *(continued)*
Properties of Saturated Steam by Temperature
(SI units)

temp. (°C)	absolute pressure (bars)	specific volume (cm³/g)		internal energy (kJ/kg)		enthalpy (kJ/kg)			entropy (kJ/kg·K)		temp. (°C)
		sat. liquid, v_f	sat. vapor, v_g	sat. liquid, u_f	sat. vapor, u_g	sat. liquid, h_f	evap., h_{fg}	sat. vapor, h_g	sat. liquid, s_f	sat. vapor, s_g	
275	59.46	1.3175	32.77	1203.1	2590.3	1210.9	1574.3	2785.2	3.0224	5.8944	275
280	64.17	1.3328	30.15	1228.3	2586.4	1236.9	1543.0	2779.9	3.0685	5.8579	280
285	69.15	1.3491	27.76	1253.9	2581.8	1263.3	1510.5	2773.7	3.1147	5.8209	285
290	74.42	1.3663	25.56	1279.9	2576.5	1290.0	1476.7	2766.7	3.1612	5.7834	290
295	79.99	1.3846	23.53	1306.2	2570.5	1317.3	1441.4	2758.7	3.2080	5.7451	295
300	85.88	1.4042	21.66	1332.9	2563.6	1345.0	1404.6	2749.6	3.2552	5.7059	300
305	92.09	1.4252	19.93	1360.2	2555.9	1373.3	1366.1	2739.4	3.3028	5.6657	305
310	98.65	1.4479	18.34	1387.9	2547.1	1402.2	1325.7	2728.0	3.3510	5.6244	310
315	105.6	1.4724	16.85	1416.3	2537.2	1431.8	1283.2	2715.1	3.3998	5.5816	315
320	112.8	1.4990	15.47	1445.3	2526.0	1462.2	1238.4	2700.6	3.4494	5.5372	320
325	120.5	1.5283	14.18	1475.1	2513.4	1493.5	1190.8	2684.3	3.5000	5.4908	325
330	128.6	1.5606	12.98	1505.8	2499.2	1525.9	1140.2	2666.0	3.5518	5.4422	330
335	137.1	1.5967	11.85	1537.6	2483.0	1559.5	1085.9	2645.4	3.6050	5.3906	335
340	146.0	1.6376	10.78	1570.6	2464.4	1594.5	1027.3	2621.9	3.6601	5.3356	340
345	155.4	1.6846	9.769	1605.3	2443.1	1631.5	963.4	2594.9	3.7176	5.2762	345
350	165.3	1.7400	8.802	1642.1	2418.1	1670.9	892.8	2563.6	3.7784	5.2110	350
355	175.7	1.8079	7.868	1682.0	2388.4	1713.7	812.9	2526.7	3.8439	5.1380	355
360	186.7	1.8954	6.949	1726.3	2351.8	1761.7	719.8	2481.5	3.9167	5.0536	360
365	198.2	2.0172	6.012	1777.8	2303.8	1817.8	605.2	2423.0	4.0014	4.9497	365
370	210.4	2.2152	4.954	1844.1	2230.3	1890.7	443.8	2334.5	4.1112	4.8012	370
374	220.64000	3.1056	3.1056	2015.70	2015.70	2084.30	0	2084.3	4.4070	4.4070	373.9

(Multiply MPa by 10 to obtain bars.)

Values in this table were calculated from *NIST Standard Reference Database 10*, "NIST/ASME Steam Properties," Ver. 2.11, National Institute of Standards and Technology, U.S. Department of Commerce, Gaithersburg, MD, 1997, which has been licensed to PPI.

APPENDIX 29.O
Properties of Saturated Steam by Pressure
(SI units)

absolute pressure (bars)	temp. (°C)	specific volume (cm³/g)		internal energy (kJ/kg)		enthalpy (kJ/kg)			entropy (kJ/kg·K)		absolute pressure (bars)
		sat. liquid, v_f	sat. vapor, v_g	sat. liquid, u_f	sat. vapor, u_g	sat. liquid, h_f	evap., h_{fg}	sat. vapor, h_g	sat. liquid, s_f	sat. vapor, s_g	
0.04	28.96	1.0041	34791	121.38	2414.5	121.39	2432.3	2553.7	0.4224	8.4734	0.04
0.06	36.16	1.0065	23733	151.47	2424.2	151.48	2415.2	2566.6	0.5208	8.3290	0.06
0.08	41.51	1.0085	18099	173.83	2431.4	173.84	2402.4	2576.2	0.5925	8.2273	0.08
0.10	45.81	1.0103	14670	191.80	2437.2	191.81	2392.1	2583.9	0.6492	8.1488	0.10
0.20	60.06	1.0172	7648	251.40	2456.0	251.42	2357.5	2608.9	0.8320	7.9072	0.20
0.30	69.10	1.0222	5228	289.24	2467.7	289.27	2335.3	2624.5	0.9441	7.7675	0.30
0.40	75.86	1.0264	3993	317.58	2476.3	317.62	2318.4	2636.1	1.0261	7.6690	0.40
0.50	81.32	1.0299	3240	340.49	2483.2	340.54	2304.7	2645.2	1.0912	7.5930	0.50
0.60	85.93	1.0331	2732	359.84	2489.0	359.91	2292.9	2652.9	1.1454	7.5311	0.60
0.70	89.93	1.0359	2365	376.68	2493.9	376.75	2282.7	2659.4	1.1921	7.4790	0.70
0.80	93.49	1.0385	2087	391.63	2498.2	391.71	2273.5	2665.2	1.2330	7.4339	0.80
0.90	96.69	1.0409	1869	405.10	2502.1	405.20	2265.1	2670.3	1.2696	7.3943	0.90
1.00	99.61	1.0432	1694	417.40	2505.6	417.50	2257.4	2674.9	1.3028	7.3588	1.00
1.01325	99.97	1.0434	1673	418.95	2506.0	419.06	2256.5	2675.5	1.3069	7.3544	1.01325
1.50	111.3	1.0527	1159	466.97	2519.2	467.13	2226.0	2693.1	1.4337	7.2230	1.50
2.00	120.2	1.0605	885.7	504.49	2529.1	504.70	2201.5	2706.2	1.5302	7.1269	2.00
2.50	127.4	1.0672	718.7	535.08	2536.8	535.35	2181.1	2716.5	1.6072	7.0524	2.50
3.00	133.5	1.0732	605.8	561.10	2543.2	561.43	2163.5	2724.9	1.6717	6.9916	3.00
3.50	138.9	1.0786	524.2	583.88	2548.5	584.26	2147.7	2732.0	1.7274	6.9401	3.50
4.00	143.6	1.0836	462.4	604.22	2553.1	604.65	2133.4	2738.1	1.7765	6.8955	4.00
4.50	147.9	1.0882	413.9	622.65	2557.1	623.14	2120.2	2743.4	1.8205	6.8560	4.50
5.00	151.8	1.0926	374.8	639.54	2560.7	640.09	2108.0	2748.1	1.8604	6.8207	5.00
6.00	158.8	1.1006	315.6	669.72	2566.8	670.38	2085.8	2756.1	1.9308	6.7592	6.00
7.00	164.9	1.1080	272.8	696.23	2571.8	697.00	2065.8	2762.8	1.9918	6.7071	7.00
8.00	170.4	1.1148	240.3	719.97	2576.0	720.86	2047.4	2768.3	2.0457	6.6616	8.00
9.00	175.4	1.1212	214.9	741.55	2579.6	742.56	2030.5	2773.0	2.0940	6.6213	9.00
10.0	179.9	1.1272	194.4	761.39	2582.7	762.52	2014.6	2777.1	2.1381	6.5850	10.0
15.0	198.3	1.1539	131.7	842.83	2593.4	844.56	1946.4	2791.0	2.3143	6.4430	15.0
20.0	212.4	1.1767	99.59	906.14	2599.1	908.50	1889.8	2798.3	2.4468	6.3390	20.0
25.0	224.0	1.1974	79.95	958.91	2602.1	961.91	1840.0	2801.9	2.5543	6.2558	25.0
30.0	233.9	1.2167	66.66	1004.7	2603.2	1008.3	1794.9	2803.2	2.6455	6.1856	30.0
35.0	242.6	1.2350	57.06	1045.5	2602.9	1049.8	1752.8	2802.6	2.7254	6.1243	35.0
40.0	250.4	1.2526	49.78	1082.5	2601.7	1087.5	1713.3	2800.8	2.7968	6.0696	40.0
45.0	257.4	1.2696	44.06	1116.5	2599.7	1122.3	1675.7	2797.9	2.8615	6.0197	45.0
50.0	263.9	1.2864	39.45	1148.2	2597.0	1154.6	1639.6	2794.2	2.9210	5.9737	50.0
55.0	270.0	1.3029	35.64	1177.9	2593.7	1185.1	1604.6	2789.7	2.9762	5.9307	55.0
60.0	275.6	1.3193	32.45	1206.0	2589.9	1213.9	1570.7	2784.6	3.0278	5.8901	60.0
65.0	280.9	1.3356	29.73	1232.7	2585.7	1241.4	1537.5	2778.9	3.0764	5.8516	65.0
70.0	285.8	1.3519	27.38	1258.2	2581.0	1267.7	1504.9	2772.6	3.1224	5.8148	70.0
75.0	290.5	1.3682	25.33	1282.7	2575.9	1292.9	1473.0	2765.9	3.1662	5.7793	75.0

(continued)

APPENDIX 29.O *(continued)*
Properties of Saturated Steam by Pressure
(SI units)

absolute pressure (bars)	temp. (°C)	specific volume (cm³/g)		internal energy (kJ/kg)		enthalpy (kJ/kg)			entropy (kJ/kg·K)		absolute pressure (bars)
		sat. liquid, v_f	sat. vapor, v_g	sat. liquid, u_f	sat. vapor, u_g	sat. liquid, h_f	evap., h_{fg}	sat. vapor, h_g	sat. liquid, s_f	sat. vapor, s_g	
80.0	295.0	1.3847	23.53	1306.2	2570.5	1317.3	1441.4	2758.7	3.2081	5.7450	80.0
85.0	299.3	1.4013	21.92	1329.0	2564.7	1340.9	1410.1	2751.0	3.2483	5.7117	85.0
90.0	303.3	1.4181	20.49	1351.1	2558.5	1363.9	1379.0	2742.9	3.2870	5.6791	90.0
95.0	307.2	1.4352	19.20	1372.6	2552.0	1386.2	1348.2	2734.4	3.3244	5.6473	95.0
100	311.0	1.4526	18.03	1393.5	2545.2	1408.1	1317.4	2725.5	3.3606	5.6160	100
100	311.0	1.4526	18.03	1393.5	2545.2	1408.1	1317.4	2725.5	3.3606	5.6160	100
110	318.1	1.4885	15.99	1434.1	2530.5	1450.4	1255.9	2706.3	3.4303	5.5545	110
120	324.7	1.5263	14.26	1473.1	2514.3	1491.5	1193.9	2685.4	3.4967	5.4939	120
130	330.9	1.5665	12.78	1511.1	2496.5	1531.5	1131.2	2662.7	3.5608	5.4336	130
140	336.7	1.6097	11.49	1548.4	2477.1	1571.0	1066.9	2637.9	3.6232	5.3727	140
150	342.2	1.6570	10.34	1585.3	2455.6	1610.2	1000.5	2610.7	3.6846	5.3106	150
160	347.4	1.7094	9.309	1622.3	2431.8	1649.7	931.1	2580.8	3.7457	5.2463	160
170	352.3	1.7693	8.371	1659.9	2405.2	1690.0	857.5	2547.5	3.8077	5.1787	170
180	357.0	1.8398	7.502	1699.0	2374.8	1732.1	777.7	2509.8	3.8718	5.1061	180
190	361.5	1.9268	6.677	1740.5	2339.1	1777.2	688.8	2466.0	3.9401	5.0256	190
200	365.7	2.0400	5.865	1786.4	2295.0	1827.2	585.1	2412.3	4.0156	4.9314	200
210	369.8	2.2055	4.996	1841.2	2233.7	1887.6	451.0	2338.6	4.1064	4.8079	210
220.64	373.95	3.1056	3.1056	2015.7	2015.7	2084.3	0	2084.3	4.4070	4.4070	220.64

(Multiply MPa by 10 to obtain bars.)

Values in this table were calculated from *NIST Standard Reference Database 10*, "NIST/ASME Steam Properties," Ver. 2.11, National Institute of Standards and Technology, U.S. Department of Commerce, Gaithersburg, MD, 1997, which has been licensed to PPI.

APPENDIX 29.P
Properties of Superheated Steam
(SI units)
specific volume, v, in m³/kg; enthalpy, h, in kJ/kg; entropy, s, in kJ/kg·K

absolute pressure (kPa) (sat. temp. °C)		temperature (°C)							
		100	150	200	250	300	360	420	500
10	v	17.196	19.513	21.826	24.136	26.446	29.216	31.986	35.680
(45.81)	h	2687.5	2783.0	2879.6	2977.4	3076.7	3197.9	3321.4	3489.7
	s	8.4489	8.6892	8.9049	9.1015	9.2827	9.4837	9.6700	9.8998
50	v	3.419	3.890	4.356	4.821	5.284	5.830	6.394	7.134
(81.32)	h	2682.4	2780.2	2877.8	2976.1	3075.8	3197.2	3320.8	3489.3
	s	7.6953	7.9413	8.1592	8.3568	8.5386	8.7401	8.9266	9.1566
75	v	2.270	2.588	2.900	3.211	3.521	3.891	4.262	4.755
(91.76)	h	2679.2	2778.4	2876.6	2975.3	3075.1	3196.7	3320.4	3489.0
	s	7.5011	7.7509	7.9702	8.1685	8.3507	8.5524	8.7391	8.9692
100	v	1.6959	1.9367	2.172	2.406	2.639	2.917	3.195	3.566
(99.61)	h	2675.8	2776.6	2875.5	2974.5	3074.5	3196.3	3320.1	3488.7
	s	7.3610	7.6148	7.8356	8.0346	8.2172	8.4191	8.6059	8.8361
150	v	......	1.2855	1.4445	1.6013	1.7571	1.9433	2.129	2.376
(111.35)	h	......	2772.9	2873.1	2972.9	3073.3	3195.3	3319.4	3488.2
	s	......	7.4208	7.6447	7.8451	8.0284	8.2309	8.4180	8.6485
400	v	......	0.4709	0.5343	0.5952	0.6549	0.7257	0.7961	0.8894
(143.61)	h	......	2752.8	2860.9	2964.5	3067.1	3190.7	3315.8	3485.5
	s	......	6.9306	7.1723	7.3804	7.5677	7.7728	7.9615	8.1933
700	v	......	......	0.3000	0.3364	0.3714	0.4126	0.4533	0.5070
(164.95)	h	......	......	2845.3	2954.0	3059.4	3185.1	3311.5	3482.3
	s	......	......	6.8884	7.1070	7.2995	7.5080	7.6986	7.9319
1000	v	......	......	0.2060	0.2328	0.2580	0.2874	0.3162	0.3541
(179.88)	h	......	......	2828.3	2943.1	3051.6	3179.4	3307.1	3479.1
	s	......	......	6.6955	6.9265	7.1246	7.3367	7.5294	7.7641
1500	v	......	......	0.13245	0.15201	0.16971	0.18990	0.2095	0.2352
(198.29)	h	......	......	2796.0	2923.9	3038.2	3169.8	3299.8	3473.7
	s	......	......	6.4536	6.7111	6.9198	7.1382	7.3343	7.5718
2000	v	......	......	......	0.11150	0.12551	0.14115	0.15617	0.17568
(212.38)	h	......	......	......	2903.2	3024.2	3159.9	3292.3	3468.2
	s	......	......	......	6.5475	6.7684	6.9937	7.1935	7.4337
2500	v	......	......	......	0.08705	0.09894	0.11188	0.12416	0.13999
(223.95)	h	......	......	......	2880.9	3009.6	3149.8	3284.8	3462.7
	s	......	......	......	6.4107	6.6459	6.8788	7.0824	7.3254
3000	v	......	......	......	0.07063	0.08118	0.09236	0.10281	0.11620
(233.85)	h	......	......	......	2856.5	2994.3	3139.5	3277.1	3457.2
	s	......	......	......	6.2893	6.5412	6.7823	6.9900	7.2359

(Multiply kPa by 0.01 to obtain bars.)

Values in this table were calculated from *NIST Standard Reference Database 10*, "NIST/ASME Steam Properties," Ver. 2.11, National Institute of Standards and Technology, U.S. Department of Commerce, Gaithersburg, MD, 1997, which has been licensed to PPI.

APPENDIX 29.Q
Properties of Compressed Water
(SI units)

T (°C)	p (bars)	ρ (kg/cm³)	v (cm³/kg)	x	h (kJ/kg)	s (kJ/kg·K)	u (kJ/kg)
0	25	0.0010011	998.94	subcooled	2.5	0.00000380	0.0026204
25	25	0.00099813	1001.9	subcooled	107.14	0.36658	104.63
50	25	0.00098908	1011	subcooled	211.49	0.70266	208.96
75	25	0.00097591	1024.7	subcooled	316.02	1.0142	313.45
100	25	0.00095947	1042.2	subcooled	420.97	1.3053	418.36
125	25	0.00094018	1063.6	subcooled	526.64	1.5794	523.98
150	25	0.00091815	1089.1	subcooled	633.43	1.8395	630.71
175	25	0.00089332	1119.4	subcooled	741.87	2.0885	739.07
200	25	0.00086538	1155.6	subcooled	852.65	2.329	849.76
223.95	25	0.00083512	1197.4	0	961.91	2.5543	958.91
223.95	25	0.000012508	79949	1	2801.9	6.2558	2602.1
0	50	0.0010023	997.68	subcooled	5.0325	0.0001383	0.044068
25	50	0.00099925	1000.8	subcooled	109.45	0.36592	104.44
50	50	0.00099016	1009.9	subcooled	213.64	0.7015	208.59
75	50	0.00097701	1023.5	subcooled	318.03	1.0127	312.92
100	50	0.00096063	1041	subcooled	422.85	1.3034	417.64
125	50	0.00094144	1062.2	subcooled	528.37	1.5771	523.06
150	50	0.00091956	1087.5	subcooled	634.98	1.8368	629.55
175	50	0.00089493	1117.4	subcooled	743.19	2.0852	737.61
200	50	0.00086726	1153.1	subcooled	853.68	2.3251	847.91
225	50	0.00083599	1196.2	subcooled	967.38	2.5592	961.4
250	50	0.00080009	1249.9	subcooled	1085.7	2.791	1079.5
263.94	50	0.00077737	1286.4	0	1154.6	2.921	1148.2
263.94	50	0.000025351	39446	1	2794.2	5.9737	2597
0	75	0.0010036	996.44	subcooled	7.5555	0.0002494	0.082204
25	75	0.0010004	999.64	subcooled	111.75	0.36526	104.25
50	75	0.00099124	1008.8	subcooled	215.79	0.70035	208.22
75	75	0.0009781	1022.4	subcooled	320.05	1.0111	312.38
100	75	0.00096179	1039.7	subcooled	424.73	1.3015	416.93
125	75	0.00094269	1060.8	subcooled	530.11	1.5748	522.15
150	75	0.00092095	1085.8	subcooled	636.54	1.8341	628.4
175	75	0.00089651	1115.4	subcooled	744.54	2.082	736.17
200	75	0.00086911	1150.6	subcooled	854.73	2.3212	846.1
225	75	0.00083824	1193	subcooled	968.01	2.5545	959.06
250	75	0.00080293	1245.4	subcooled	1085.7	2.7851	1076.4
275	75	0.0007614	1313.4	subcooled	1210.2	3.0174	1200.4
290.54	75	0.00073088	1368.2	0	1292.9	3.1662	1282.7
290.54	75	0.000039479	25330	1	2765.9	5.7793	2575.9
0	100	0.0010048	995.2	subcooled	10.069	0.00033757	0.1171
25	100	0.0010015	998.54	subcooled	114.05	0.3646	104.06
50	100	0.00099231	1007.8	subcooled	217.94	0.6992	207.86
75	100	0.00097919	1021.3	subcooled	322.07	1.0096	311.85
100	100	0.00096293	1038.5	subcooled	426.62	1.2996	416.23
125	100	0.00094393	1059.4	subcooled	531.84	1.5725	521.25
150	100	0.00092232	1084.2	subcooled	638.11	1.8313	627.27
175	100	0.00089807	1113.5	subcooled	745.89	2.0788	734.75
200	100	0.00087094	1148.2	subcooled	855.8	2.3174	844.31
225	100	0.00084044	1189.9	subcooled	968.68	2.5499	956.78
250	100	0.0008057	1241.2	subcooled	1085.8	2.7792	1073.4
275	100	0.00076513	1307	subcooled	1209.3	3.0097	1196.2
300	100	0.00071529	1398	subcooled	1343.3	3.2488	1329.4

(continued)

APPENDIX 29.Q *(continued)*
Properties of Compressed Water
(SI units)

T (°C)	p (bars)	ρ (kg/cm³)	ν (cm³/kg)	x	h (kJ/kg)	s (kJ/kg·K)	μ (kJ/kg)
0	150	0.0010073	992.76	subcooled	15.069	0.00044686	0.17746
25	150	0.0010037	996.35	subcooled	118.63	0.36325	103.69
50	150	0.00099443	1005.6	subcooled	222.23	0.6969	207.15
75	150	0.00098135	1019	subcooled	326.1	1.0065	310.81
100	150	0.0009652	1036.1	subcooled	430.39	1.2958	414.85
125	150	0.00094638	1056.7	subcooled	535.33	1.568	519.48
150	150	0.00092503	1081	subcooled	641.27	1.826	625.05
175	150	0.00090114	1109.7	subcooled	748.63	2.0725	731.98
200	150	0.0008745	1143.5	subcooled	857.99	2.31	840.84
225	150	0.00084471	1183.8	subcooled	970.12	2.5409	952.36
250	150	0.00081103	1233	subcooled	1086.1	2.768	1067.6
275	150	0.00077216	1295.1	subcooled	1207.8	2.9951	1188.3
300	150	0.00072555	1378.3	subcooled	1338.3	3.2279	1317.6
0	200	0.0010097	990.36	subcooled	20.033	0.00046962	0.22569
25	200	0.0010058	994.19	subcooled	123.2	0.36187	103.32
50	200	0.00099653	1003.5	subcooled	226.51	0.69461	206.44
75	200	0.00098348	1016.8	subcooled	330.13	1.0035	309.79
100	200	0.00096744	1033.7	subcooled	434.17	1.292	413.5
125	200	0.00094879	1054	subcooled	538.84	1.5635	517.76
150	200	0.00092769	1077.9	subcooled	644.45	1.8208	622.89
175	200	0.00090414	1106	subcooled	751.42	2.0664	729.3
200	200	0.00087797	1139	subcooled	860.27	2.3027	837.49
225	200	0.00084882	1178.1	subcooled	971.69	2.5322	948.13
250	200	0.00081609	1225.4	subcooled	1086.7	2.7573	1062.2
275	200	0.00077871	1284.2	subcooled	1206.7	2.9814	1181
300	200	0.00073471	1361.1	subcooled	1334.4	3.2091	1307.1
0	250	0.0010122	988	subcooled	24.962	0.00040919	0.26234
25	250	0.001008	992.07	subcooled	127.75	0.36047	102.95
50	250	0.00099861	1001.4	subcooled	230.79	0.69233	205.75
75	250	0.00098559	1014.6	subcooled	334.16	1.0004	308.79
100	250	0.00096965	1031.3	subcooled	437.95	1.2883	412.17
125	250	0.00095116	1051.3	subcooled	542.35	1.5591	516.07
150	250	0.0009303	1074.9	subcooled	647.66	1.8156	620.78
175	250	0.00090707	1102.5	subcooled	754.25	2.0604	726.69
200	250	0.00088133	1134.6	subcooled	862.61	2.2956	834.24
225	250	0.00085279	1172.6	subcooled	973.38	2.5237	944.06
250	250	0.00082092	1218.1	subcooled	1087.4	2.7471	1057
275	250	0.00078485	1274.1	subcooled	1206	2.9685	1174.2
300	250	0.00074302	1345.9	subcooled	1331.3	3.1919	1297.6
0	300	0.0010145	985.67	subcooled	29.858	0.00026879	0.28791
25	300	0.0010101	989.98	subcooled	132.28	0.35905	102.58
50	300	0.0010007	999.33	subcooled	235.05	0.69005	205.07
75	300	0.00098767	1012.5	subcooled	338.19	0.99746	307.81
100	300	0.00097182	1029	subcooled	441.74	1.2847	410.87
125	300	0.0009535	1048.8	subcooled	545.88	1.5548	514.42
150	300	0.00093286	1072	subcooled	650.89	1.8106	618.73
175	300	0.00090994	1099	subcooled	757.11	2.0545	724.15
200	300	0.00088462	1130.4	subcooled	865.02	2.2888	831.1
225	300	0.00085664	1167.4	subcooled	975.17	2.5156	940.15
250	300	0.00082556	1211.3	subcooled	1088.4	2.7373	1052
275	300	0.00079064	1264.8	subcooled	1205.7	2.9563	1167.7
300	300	0.00075066	1332.2	subcooled	1328.9	3.176	1288.9

(Multiply MPa by 10 to obtain bars.)

Values in this table were calculated from *NIST Standard Reference Database 10*, "NIST/ASME Steam Properties," Ver. 2.11, National Institute of Standards and Technology, U.S. Department of Commerce, Gaithersburg, MD, 1997, which has been licensed to PPI.

APPENDIX 29.R
Enthalpy-Entropy (Mollier) Diagram for Steam
(SI units)

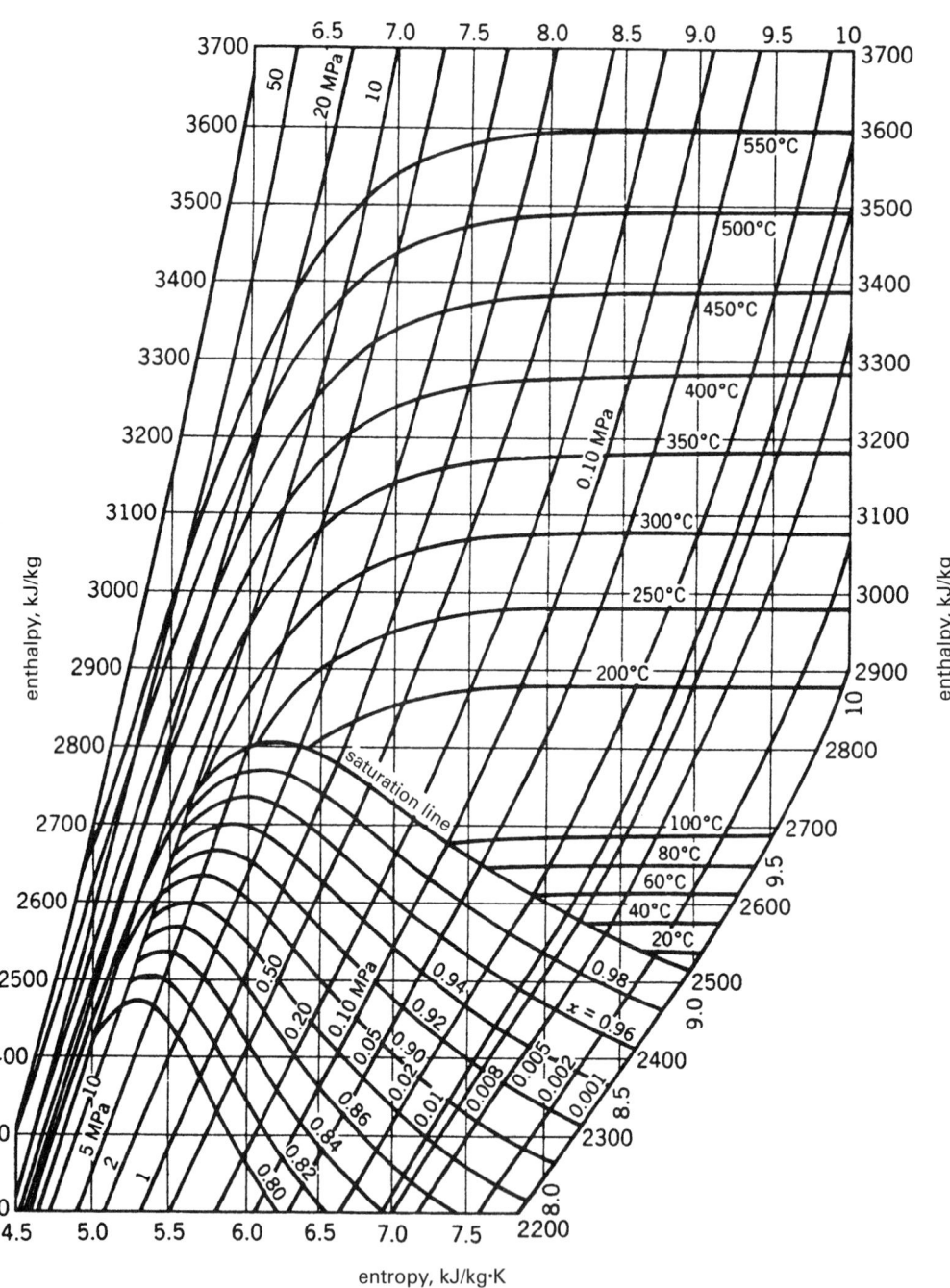

(Multiply MPa by 10 to obtain bars.)

Reproduced from *Steam Tables: Thermodynamic Properties of Water Including Vapor, Liquid, and Solid Phase* (SI version), by John H. Keenan, Frederick G. Keyes, Philip G. Hill, and Joan G. Moore, copyright © 1978. Reproduced with permission of John Wiley & Sons, Inc.

APPENDIX 29.S
Properties of Low-Pressure Air
(SI units)

T in K; h and u in kJ/kg; φ in kJ/kg·K

T	h	p_r	u	v_r	φ
200	199.97	0.3363	142.56	1707	1.29559
210	209.97	0.3987	149.69	1512	1.34444
220	219.97	0.4690	156.82	1346	1.39105
230	230.02	0.5477	164.00	1205	1.43557
240	240.02	0.6355	171.13	1084	1.47824
250	250.05	0.7329	178.28	979	1.51917
260	260.09	0.8405	185.45	887.8	1.55848
270	270.11	0.9590	192.60	808.0	1.59634
280	280.13	1.0889	199.75	738.0	1.63279
285	285.14	1.1584	203.33	706.1	1.65055
290	290.16	1.2311	206.91	676.1	1.66802
295	295.17	1.3068	210.49	647.9	1.68515
300	300.19	1.3860	214.07	621.2	1.70203
305	305.22	1.4686	217.67	596.0	1.71865
310	310.24	1.5546	221.25	572.3	1.73498
315	315.27	1.6442	224.85	549.8	1.75106
320	320.29	1.7375	228.42	528.6	1.76690
325	325.31	1.8345	232.02	508.4	1.78249
330	330.34	1.9352	235.61	489.4	1.79783
340	340.42	2.149	242.82	454.1	1.82790
350	350.49	2.379	250.02	422.2	1.85708
360	360.58	2.626	257.24	393.4	1.88543
370	370.67	2.892	264.46	367.2	1.91313
380	380.77	3.176	271.69	343.4	1.94001
390	390.88	3.481	278.93	321.5	1.96633
400	400.98	3.806	286.16	301.6	1.99194
410	411.12	4.153	293.43	283.3	2.01699
420	421.26	4.522	300.69	266.6	2.04142
430	431.43	4.915	307.99	251.1	2.06533
440	441.61	5.332	315.30	236.8	2.08870
450	451.80	5.775	322.62	223.6	2.11161
460	462.02	6.245	329.97	211.4	2.13407
470	472.24	6.742	337.32	200.1	2.15604
480	482.49	7.268	344.70	189.5	2.17760
490	492.74	7.824	352.08	179.7	2.19876
500	503.02	8.411	359.49	170.6	2.21952
510	513.32	9.031	366.92	162.1	2.23993
520	523.63	9.684	374.36	154.1	2.25997
530	533.98	10.37	381.84	146.7	2.27967
540	544.35	11.10	389.34	139.7	2.29906
550	554.74	11.86	396.86	133.1	2.31809
560	565.17	12.66	404.42	127.0	2.33685
570	575.59	13.50	411.97	121.2	2.35531
580	586.04	14.38	419.55	115.7	2.37348
590	596.52	15.31	427.15	110.6	2.39140

(continued)

APPENDIX 29.S *(continued)*
Properties of Low-Pressure Air
(SI units)

T in K; h and u in kJ/kg; ϕ in kJ/kg·K

T	h	p_r	u	v_r	ϕ
600	607.02	16.28	434.78	105.8	2.40902
610	617.53	17.30	442.42	101.2	2.42644
620	628.07	18.36	450.09	96.92	2.44356
630	638.63	19.84	457.78	92.84	2.46048
640	649.22	20.64	465.50	88.99	2.47716
650	659.84	21.86	473.25	85.34	2.49364
660	670.47	23.13	481.01	81.89	2.50985
670	681.14	24.46	488.81	78.61	2.52589
680	691.82	25.85	496.62	75.50	2.54175
690	702.52	27.29	504.45	72.56	2.55731
700	713.27	28.80	512.33	69.76	2.57277
710	724.04	30.38	520.23	67.07	2.58810
720	734.82	32.02	528.14	64.53	2.60319
730	745.62	33.72	536.07	62.13	2.61803
740	756.44	35.50	544.02	59.82	2.63280
750	767.29	37.35	551.99	57.63	2.64737
760	778.18	39.27	560.01	55.54	2.66176
770	789.11	41.31	568.07	53.39	2.67595
780	800.03	43.35	576.12	51.64	2.69013
790	810.99	45.55	584.21	49.86	2.70400
800	821.95	47.75	592.30	48.08	2.71787
820	843.98	52.59	608.59	44.84	2.74504
840	866.08	57.60	624.95	41.85	2.77170
860	888.27	63.09	641.40	39.12	2.79783
880	910.56	68.98	657.95	36.61	2.82344
900	932.93	75.29	674.58	34.31	2.84856
920	955.38	82.05	691.28	32.18	2.87324
940	977.92	89.28	708.08	30.22	2.89748
960	1000.55	97.00	725.02	28.40	2.92128
980	1023.25	105.2	741.98	26.73	2.94468
1000	1046.04	114.0	758.94	25.17	2.96770
1020	1068.89	123.4	776.10	23.72	2.99034
1040	1091.85	133.3	793.36	22.39	3.01260
1060	1114.86	143.9	810.62	21.14	3.03449
1080	1137.89	155.2	827.88	19.98	3.05608
1100	1161.07	167.1	845.33	18.896	3.07732
1120	1184.28	179.7	862.79	17.886	3.09825
1140	1207.57	193.1	880.35	16.946	3.11883
1160	1230.92	207.2	897.91	16.064	3.13916
1180	1254.34	222.2	915.57	15.241	3.15916
1200	1277.79	238.0	933.33	14.470	3.17888
1220	1301.31	254.7	951.09	13.747	3.19834
1240	1324.93	272.3	968.95	13.069	3.21751
1260	1348.55	290.8	986.90	12.435	3.23638
1280	1372.24	310.4	1004.76	11.835	3.25510

(continued)

APPENDIX 29.S *(continued)*
Properties of Low-Pressure Air
(SI units)

T in *K*; *h* and *u* in kJ/kg; ϕ in kJ/kg·K

T	h	p_r	u	v_r	ϕ
1300	1395.97	330.9	1022.82	11.275	3.27345
1320	1419.76	352.5	1040.88	10.747	3.29160
1340	1443.60	375.3	1058.94	10.247	3.30959
1360	1467.49	399.1	1077.10	9.780	3.32724
1380	1491.44	424.2	1095.26	9.337	3.34474
1400	1515.42	450.5	1113.52	8.919	3.36200
1420	1539.44	478.0	1131.77	8.526	3.37901
1440	1563.51	506.9	1150.13	8.153	3.39586
1460	1587.63	537.1	1168.49	7.801	3.41247
1480	1611.79	568.8	1186.95	7.468	3.42892
1500	1635.97	601.9	1205.41	7.152	3.44516
1520	1660.23	636.5	1223.87	6.854	3.46120
1540	1684.51	672.8	1242.43	6.569	3.47712
1560	1708.82	710.5	1260.99	6.301	3.49276
1580	1733.17	750.0	1279.65	6.046	3.50829
1600	1757.57	791.2	1298.30	5.804	3.52364
1620	1782.00	834.1	1316.96	5.574	3.53879
1640	1806.46	878.9	1335.72	5.355	3.55381
1660	1830.96	925.6	1354.48	5.147	3.56867
1680	1855.50	974.2	1373.24	4.949	3.58335
1700	1880.1	1025	1392.7	4.761	3.5979
1750	1941.6	1161	1439.8	4.328	3.6336
1800	2003.3	1310	1487.2	3.944	3.6684
1850	2065.3	1475	1534.9	3.601	3.7023
1900	2127.4	1655	1582.6	3.295	3.7354
1950	2189.7	1852	1630.6	3.022	3.7677
2000	2252.1	2068	1678.7	2.776	3.7994
2050	2314.6	2303	1726.8	2.555	3.8303
2100	2377.4	2559	1775.3	2.356	3.8605
2150	2440.3	2837	1823.8	2.175	3.8901
2200	2503.2	3138	1872.4	2.012	3.9191
2250	2566.4	3464	1921.3	1.864	3.9474

APPENDIX 29.T
Properties of Saturated Refrigerant-12 (R-12) by Temperature
(SI units)

temp. (°C)	absolute pressure (MPa)	volume vapor (m³/kg)	density liquid (kg/m³)	enthalpy liquid (kJ/kg)	enthalpy vapor (kJ/kg)	entropy liquid (kJ/kg·K)	entropy vapor (kJ/kg·K)
−100	0.001174	10.122	1678.0	112.69	306.46	0.60441	1.7235
−95	0.001851	6.6005	1665.0	116.92	308.67	0.62845	1.7048
−90	0.002836	4.4264	1651.9	121.14	310.90	0.65182	1.6879
−85	0.004230	3.0449	1638.7	125.36	313.16	0.67457	1.6727
−80	0.006160	2.1438	1625.5	129.59	315.44	0.69673	1.6589
−75	0.008774	1.5416	1612.1	133.81	317.74	0.71835	1.6465
−70	0.012246	1.1301	1598.7	138.06	320.05	0.73948	1.6353
−65	0.016776	0.84332	1585.2	142.32	322.38	0.76015	1.6252
−60	0.022591	0.63956	1571.5	146.58	324.71	0.78040	1.6161
−55	0.029944	0.49230	1557.8	150.87	327.05	0.80025	1.6079
−50	0.039115	0.38415	1543.9	155.18	329.40	0.81974	1.6005
−45	0.050408	0.30355	1529.9	159.51	331.74	0.83890	1.5938
−40	0.064152	0.24264	1515.7	163.86	334.09	0.85775	1.5879
−35	0.080701	0.19603	1501.4	168.25	336.43	0.87632	1.5825
−30	0.10043	0.15993	1486.9	172.67	338.76	0.89462	1.5777
−29.79	0.101325	0.15861	1486.3	172.85	338.86	0.89538	1.5775
−25	0.12373	0.13166	1472.3	177.12	341.08	0.91269	1.5734
−20	0.15101	0.10929	1457.4	181.61	343.39	0.93053	1.5696
−15	0.18272	0.09142	1442.4	186.14	345.69	0.94817	1.5662
−10	0.21928	0.07702	1427.1	190.72	347.96	0.96561	1.5632
−5	0.26117	0.06531	1411.5	195.33	350.22	0.98289	1.5605
0	0.30885	0.05571	1395.6	200.00	352.44	1.0000	1.5581
2	0.32966	0.05236	1389.2	201.88	353.32	1.0068	1.5572
4	0.35150	0.04925	1382.7	203.77	354.20	1.0136	1.5564
6	0.37441	0.04637	1376.2	205.66	355.07	1.0203	1.5556
8	0.39842	0.04369	1369.6	207.57	355.93	1.0271	1.5548
10	0.42356	0.04119	1363.0	209.48	356.79	1.0338	1.5541
12	0.44986	0.03887	1356.2	211.40	357.65	1.0405	1.5533
14	0.47737	0.03670	1349.5	213.33	358.49	1.0472	1.5527
16	0.50610	0.03468	1342.6	215.27	359.33	1.0538	1.5520
18	0.53610	0.03279	1335.7	217.22	360.16	1.0605	1.5514
20	0.56740	0.03102	1328.7	219.18	360.98	1.0671	1.5508
22	0.60003	0.02937	1321.6	221.14	361.80	1.0737	1.5502
24	0.63403	0.02782	1314.5	223.12	362.60	1.0803	1.5497
26	0.66943	0.02637	1307.3	225.11	363.40	1.0868	1.5491
28	0.70626	0.02500	1299.9	227.10	364.19	1.0934	1.5486
30	0.74457	0.02372	1292.5	229.11	364.96	1.0999	1.5481
32	0.78439	0.02252	1285.0	231.12	365.73	1.1064	1.5476
34	0.82574	0.02138	1277.4	233.15	366.48	1.1130	1.5471
36	0.86868	0.02032	1269.7	235.18	367.22	1.1195	1.5466
38	0.91324	0.01931	1261.9	237.23	367.95	1.1259	1.5461

(continued)

APPENDIX 29.T *(continued)*
Properties of Saturated Refrigerant-12 (R-12) by Temperature
(SI units)

temp. (°C)	absolute pressure (MPa)	volume vapor (m³/kg)	density liquid (kg/m³)	enthalpy liquid (kJ/kg)	enthalpy vapor (kJ/kg)	entropy liquid (kJ/kg·K)	entropy vapor (kJ/kg·K)
40	0.95944	0.01836	1253.9	239.29	368.67	1.1324	1.5456
42	1.0073	0.01746	1245.9	241.36	369.37	1.1389	1.5451
44	1.0570	0.01662	1237.7	243.44	370.06	1.1453	1.5446
46	1.1084	0.01581	1299.3	245.54	370.73	1.1518	1.5441
48	0.1616	0.01506	1220.9	247.64	371.38	1.1582	1.5435
50	1.2167	0.01434	1212.2	249.76	372.02	1.1647	1.5430
52	1.2736	0.01366	1203.5	251.90	372.64	1.1711	1.5425
54	1.3325	0.01301	1194.5	254.04	373.24	1.1776	1.5419
56	1.3934	0.01239	1185.4	256.21	373.82	1.1840	1.5413
58	1.4562	0.01181	1176.1	258.38	374.38	1.1904	1.5407
60	1.5212	0.01126	1166.6	260.58	374.91	1.1969	1.5401
62	1.5883	0.01073	1156.9	262.79	375.42	1.2033	1.5394
64	1.6575	0.01023	1146.9	265.02	375.90	1.2098	1.5387
66	1.7289	0.009746	1136.7	267.27	376.36	1.2162	1.5379
68	1.8026	0.009289	1126.3	269.54	376.78	1.2227	1.5371
70	1.8786	0.008852	1115.6	271.83	377.17	1.2292	1.5362
72	1.9570	0.008434	1104.6	274.15	377.53	1.2357	1.5353
74	2.0378	0.008034	1093.3	276.49	377.85	1.2423	1.5343
76	2.1210	0.007651	1081.6	278.86	378.13	1.2489	1.5332
78	2.2069	0.007283	1069.6	281.27	378.36	1.2555	1.5320
80	2.2953	0.006931	1057.2	283.70	378.54	1.2622	1.5308
85	2.5282	0.006106	1024.1	289.97	378.75	1.2792	1.5271
90	2.7790	0.005350	987.60	296.56	378.52	1.2968	1.5225
95	3.0490	0.004648	946.44	303.58	377.67	1.3152	1.5165
100	3.3399	0.003980	898.55	311.26	375.88	1.3351	1.5083
105	3.6538	0.003311	839.10	320.08	372.41	1.3576	1.4960
110	3.9943	0.002517	746.58	331.91	364.02	1.3875	1.4713
*111.80	4.125	0.00179	558	348.4	348.4	1.430	1.430

*critical point

From *2009 ASHRAE Handbook—Fundamentals, SI Edition*, copyright © 2009, by American Society of Heating, Refrigerating, and Air-Conditioning Engineers, Inc. Reproduced with permission.

APPENDIX 29.U
Pressure-Enthalpy Diagram for Refrigerant-12 (R-12)
(SI units)

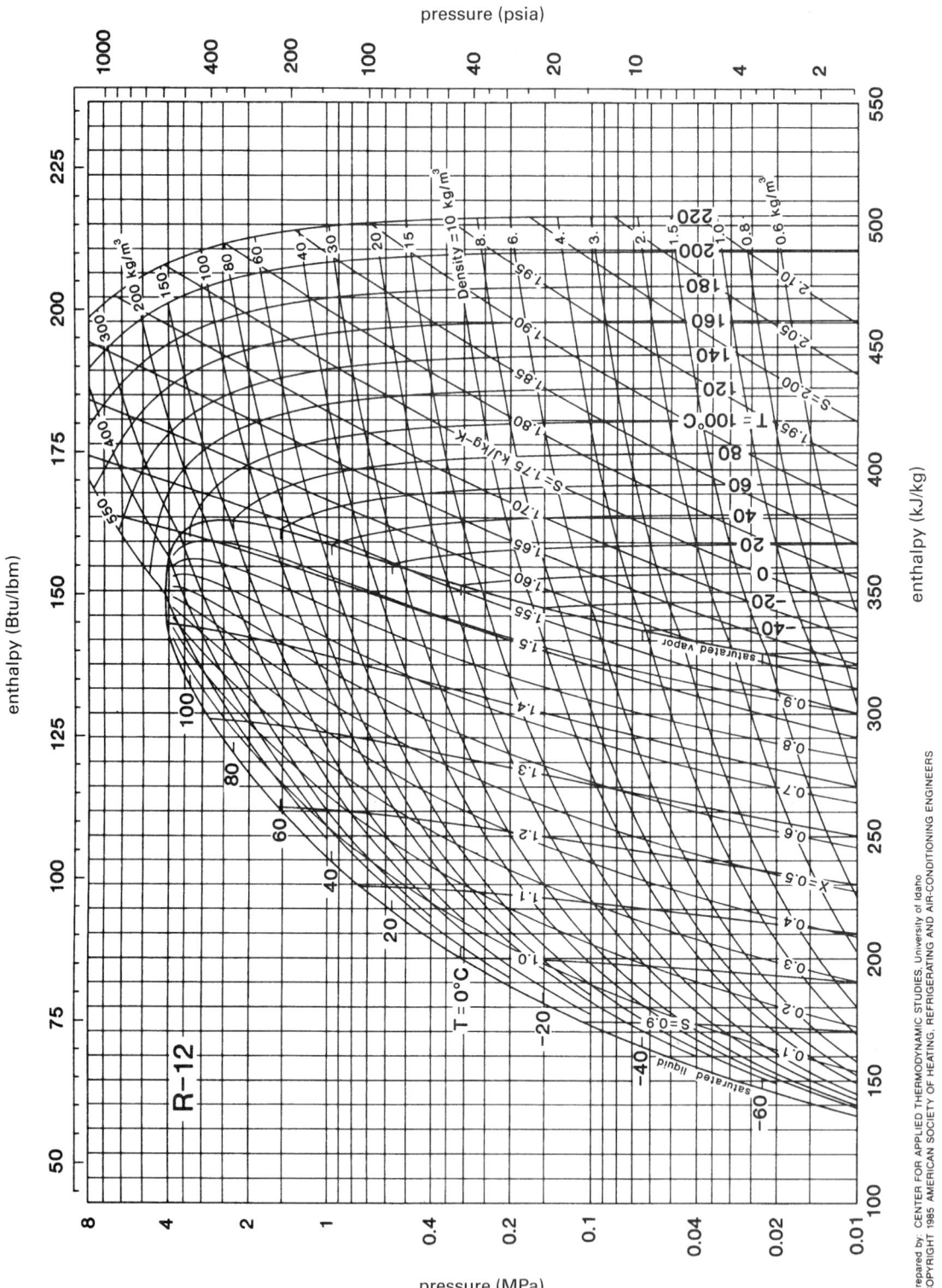

From *2009 ASHRAE Handbook—Fundamentals, SI Edition*, copyright © 2009, by American Society of Heating, Refrigerating, and Air-Conditioning Engineers, Inc. Reproduced with permission.

APPENDIX 29.V
Properties of Saturated Refrigerant-22 (R-22) by Temperature
(SI units)

temp. (°C)	absolute pressure (MPa)	volume vapor (m³/kg)	density liquid (kg/m³)	enthalpy liquid (kJ/kg)	vapor (kJ/kg)	entropy liquid (kJ/kg·K)	vapor (kJ/kg·K)
−90	0.004748	3.6939	1542.8	98.575	364.20	0.55032	2.006
−85	0.007084	2.5394	1529.9	104.54	366.63	0.58246	1.9754
−80	0.010308	1.7883	1516.8	110.42	369.06	0.61326	1.9524
−75	0.014662	1.2870	1503.6	116.21	371.49	0.64284	1.9312
−70	0.020424	0.94477	1490.3	121.92	373.91	0.67132	1.9117
−65	0.027914	0.70609	1476.7	127.58	376.32	0.69879	1.8938
−60	0.037491	0.53641	1463.1	133.18	378.72	0.72535	1.8773
−55	0.049556	0.41362	1449.2	138.74	381.10	0.75109	1.8621
−50	0.064549	0.32330	1435.2	144.27	383.45	0.77610	1.8479
−45	0.082947	0.25586	1421.0	149.77	385.77	0.80044	1.8348
−40.82	0.101325	0.21223	1408.9	154.37	387.69	0.82034	1.8246
−40	0.10527	0.20480	1406.5	155.26	388.06	0.82419	1.8227
−38	0.11542	0.18790	1400.7	157.46	388.96	0.83354	1.8180
−36	0.12632	0.17268	1394.8	159.66	389.86	0.84281	1.8135
−34	0.13801	0.15894	1388.9	161.86	390.75	0.85200	1.8091
−32	0.15053	0.14651	1382.9	164.06	391.64	0.86113	1.8049
−30	0.16391	0.13524	1376.9	166.26	392.52	0.87018	1.8007
−28	0.17821	0.12502	1370.9	168.46	393.39	0.87917	1.7967
−26	0.19346	0.11573	1364.8	170.67	394.25	0.88810	1.7927
−24	0.20969	0.10726	1358.7	172.89	395.10	0.89697	1.7889
−22	0.22696	0.09954	1352.6	175.10	395.95	0.90579	1.7851
−20	0.24531	0.09249	1346.4	177.33	396.79	0.91455	1.7815
−18	0.26477	0.08603	1340.1	179.56	397.62	0.92327	1.7779
−16	0.28540	0.08012	1333.8	181.79	398.43	0.93194	1.7744
−14	0.30724	0.07470	1327.5	184.04	399.24	0.94057	1.7710
−12	0.33034	0.06971	1321.1	186.29	400.04	0.94916	1.7677
−10	0.35474	0.06513	1314.6	188.55	400.83	0.95771	1.7644
−8	0.38049	0.06090	1308.1	190.82	401.61	0.96623	1.7612
−6	0.40763	0.05701	1301.5	193.10	402.37	0.97471	1.7581
−4	0.43622	0.05341	1294.9	195.39	403.12	0.98317	1.7550
−2	0.46630	0.05008	1288.2	197.69	403.87	0.99160	1.7520
0	0.49792	0.04700	1281.5	200.00	404.59	1.0000	1.7490
2	0.53113	0.04415	1274.7	202.32	405.31	1.0084	1.7461
4	0.56599	0.04150	1267.8	204.66	406.01	1.0167	1.7432
6	0.60254	0.03904	1260.8	207.01	406.70	1.0251	1.7404
8	0.64083	0.03675	1253.8	209.37	407.37	1.0334	1.7376
10	0.68091	0.03462	1246.7	211.74	408.03	1.0417	1.7349
12	0.72285	0.03263	1239.5	214.13	408.67	1.0500	1.7322
14	0.76668	0.03078	1232.3	216.54	409.29	1.0583	1.7295
16	0.81246	0.02905	1224.9	218.96	409.90	1.0665	1.7269
18	0.86025	0.02743	1217.5	221.40	410.49	1.0748	1.7243

(continued)

APPENDIX 29.V *(continued)*
Properties of Saturated Refrigerant-22 (R-22) by Temperature
(SI units)

temp. (°C)	absolute pressure (MPa)	volume vapor (m³/kg)	density liquid (kg/m³)	enthalpy liquid (kJ/kg)	enthalpy vapor (kJ/kg)	entropy liquid (kJ/kg·K)	entropy vapor (kJ/kg·K)
20	0.91009	0.02592	1210.0	223.85	411.06	1.0831	1.7217
22	0.96205	0.02451	1202.4	226.32	411.61	1.0913	1.7191
24	1.0162	0.02318	1194.6	228.80	412.14	1.0996	1.7165
26	1.0725	0.02193	1186.8	231.31	412.65	1.1078	1.7140
28	1.1312	0.02076	1178.9	233.83	413.13	1.1160	1.7114
30	1.1921	0.01967	1170.8	236.38	413.60	1.1243	1.7089
32	1.2555	0.01863	1162.6	238.94	414.03	1.1325	1.7063
34	1.3213	0.01766	1154.3	241.52	414.45	1.1408	1.7038
36	1.3896	0.01674	1145.9	244.13	414.83	1.1490	1.7012
38	1.4605	0.01588	1137.3	246.75	415.19	1.1573	1.6987
40	1.5340	0.01506	1128.6	249.40	415.52	1.1656	1.6961
42	1.6102	0.01429	1119.7	252.07	415.82	1.1739	1.6934
44	1.6892	0.01356	1110.6	254.77	416.08	1.1822	1.6908
46	1.7710	0.01287	1101.4	257.49	416.31	1.1905	1.6881
48	1.8556	0.01221	1091.9	260.24	416.50	1.1989	1.6854
50	1.9432	0.01159	1082.3	263.02	416.65	1.2072	1.6826
52	2.0339	0.01101	1072.4	265.83	416.75	1.2156	1.6798
54	2.1276	0.01045	1062.3	268.67	416.81	1.2241	1.6769
56	2.2244	0.009915	1051.9	271.55	416.83	1.2326	1.6739
58	2.3245	0.009409	1041.3	274.46	416.79	1.2411	1.6709
60	2.4279	0.008927	1030.3	277.41	416.69	1.2497	1.6677
65	2.7015	0.007816	1001.3	284.98	416.16	1.2714	1.6594
70	2.9975	0.006819	969.68	292.88	416.14	1.2937	1.6500
75	3.3175	0.005917	934.38	301.22	413.46	1.3169	1.6393
80	3.6633	0.005086	893.89	310.18	410.88	1.3414	1.6265
85	4.0370	0.004301	845.17	320.13	406.90	1.3681	1.6104
90	4.4413	0.003517	780.60	331.96	400.28	1.3996	1.5877
95	4.8808	0.002547	660.94	350.67	384.95	1.4490	1.5421
*96.15	4.988	0.00195	513	368.1	368.1	1.496	1.496

*critical point

From *2009 ASHRAE Handbook—Fundamentals, SI Edition,* copyright © 2009, by American Society of Heating, Refrigerating, and Air-Conditioning Engineers, Inc. Reproduced with permission.

APPENDIX 29.W
Pressure-Enthalpy Diagram for Refrigerant-22 (R-22)
(SI units)

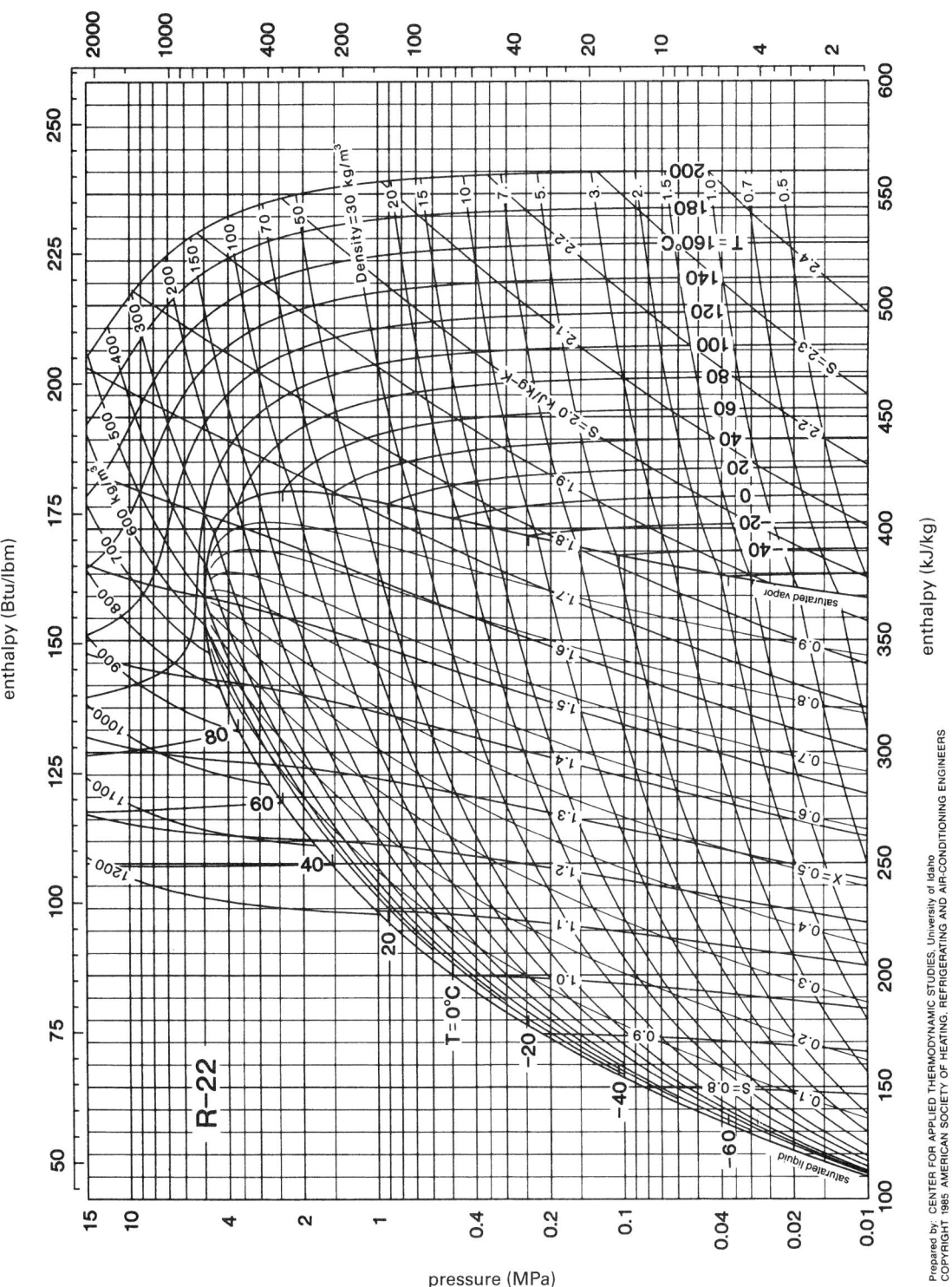

From *2009 ASHRAE Handbook—Fundamentals, SI Edition*, copyright © 2009, by American Society of Heating, Refrigerating, and Air-Conditioning Engineers, Inc. Reproduced with permission.

APPENDIX 29.X
Pressure-Enthalpy Diagram for Refrigerant HFC-134a
(SI units)

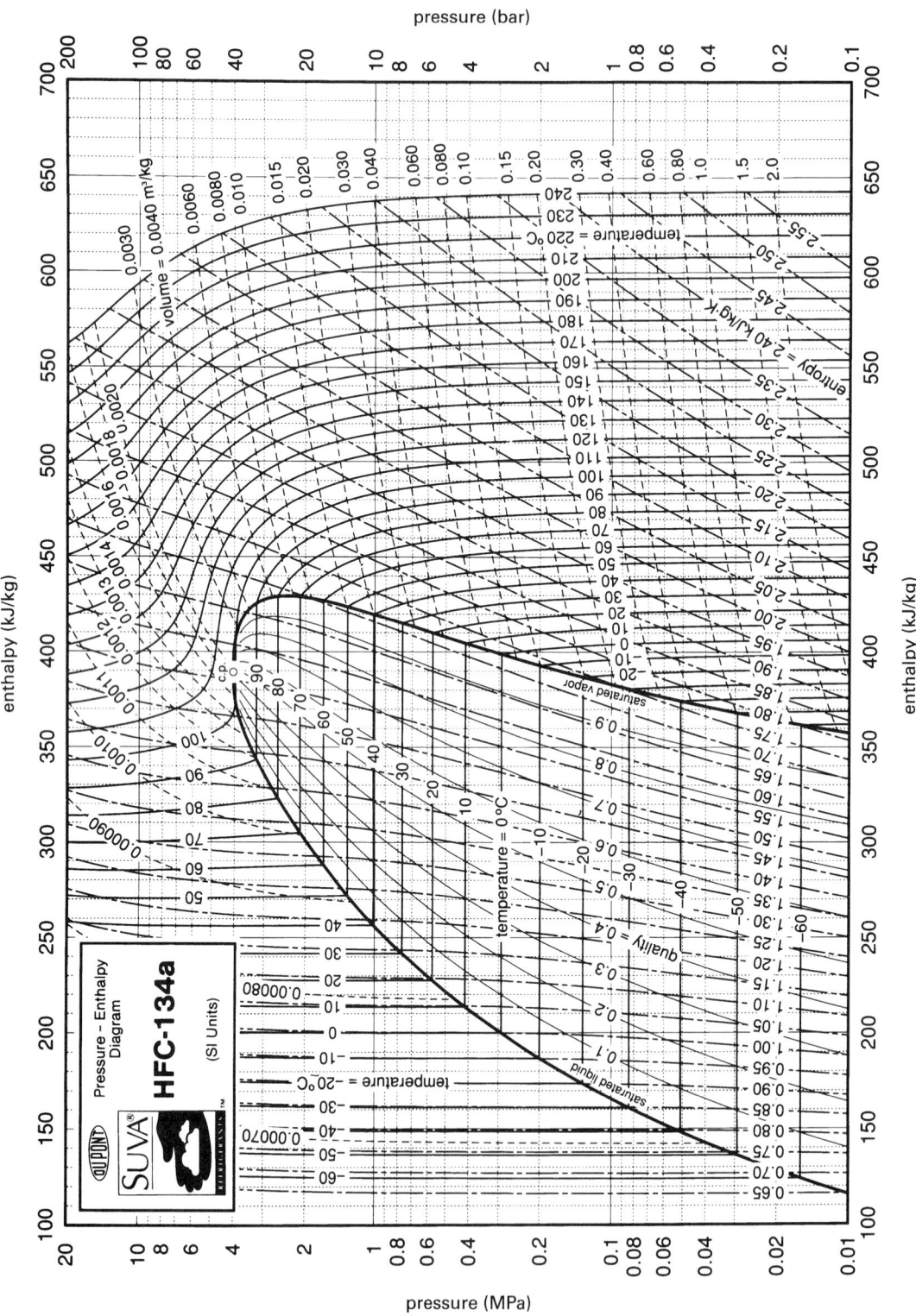

APPENDIX 29.Y
Thermal Properties of Selected Solids
(customary U.S. units)

material	ρ (lbm/ft^3) (68°F) (20°C)	c_p (Btu/lbm-°F) (68°F) (20°C)	α (ft^2/hr) (68°F) (20°C)	k (Btu/hr-ft-°F) (68°F) (20°C)	(212°F) (100°C)	(572°F) (300°C)
metals						
aluminum	168.6	0.224	3.55	132	132	133
copper	555	0.092	3.98	223	219	213
gold	1206	0.031	4.52	169	170	172
iron	492	0.122	0.83	42.3	39.0	31.6
lead	708	0.030	0.80	20.3	19.3	17.2
magnesium	109	0.248	3.68	99.5	96.8	91.4
nickel	556	0.111	0.87	53.7	47.7	36.9
platinum	1340	0.032	0.09	40.5	41.9	43.5
silver	656	0.057	6.42	240	237	209
tin	450	0.051	1.57	36	36	
tungsten	1206	0.032	2.44	94	87	77
uranium α	1167	0.027	0.53	16.9	17.2	19.6
zinc	446	0.094	1.55	65	63	58
alloys						
aluminum 2024	173	0.23	1.76	70.2		
brass (70% Cu, 30% Zn)	532	0.091	1.27	61.8	73.9	85.3
constantan (60% Cu, 40% Ni)	557	0.098	0.24	13.1	15.4	
iron, cast	455	0.100	0.65	29.6	26.8	
nichrome V	530	0.106	0.12	7.06	7.99	9.94
stainless steel	488	0.110	0.17	9.4	10.0	13
steel, mild (1% C)	488	0.113	0.45	24.8	24.8	22.9
nonmetals						
asbestos	36	0.25		0.092	0.11	0.125
brick (fire clay)	144	0.22			0.65	
brick (masonry)	106	0.20		0.38		
brick (chrome)	188	0.20			0.67	
concrete	150	0.21		0.70		
corkboard	10	0.4		0.025		
diatomaceous earth, powdered	14	0.2		0.03		
glass, window	170	0.2		0.45		
glass, Pyrex™	140	0.2		0.63	0.67	0.84
Kaolin firebrick	19					0.052
85% magnesia	17			0.038	0.041	
sandy loam, 4% H$_2$O	104	~0.4		0.54		
sandy loam, 10% H$_2$O	121			1.08		
rock wool	~10	0.2		0.023	0.033	
wood, oak, $\perp$ to grain	51	0.57		0.12		
wood, oak, || to grain	51	0.57		0.23		

(Multiply lbm/ft^3 by 16.018 to obtain kg/m^3.)
(Multiply Btu/lbm-°F by 4186.8 to obtain J/kg·K.)
(Multiply ft^2/hr by 0.0929 to obtain m^2/h.)
(Multiply Btu/hr-ft-°F by 1.7307 to obtain W/m·K.)

Engineering Heat Transfer, James R. Welty, copyright © 1974. Reproduced with permission of John Wiley & Sons, Inc.

Support Material

APPENDIX 29.Z
Generalized Compressibility Charts

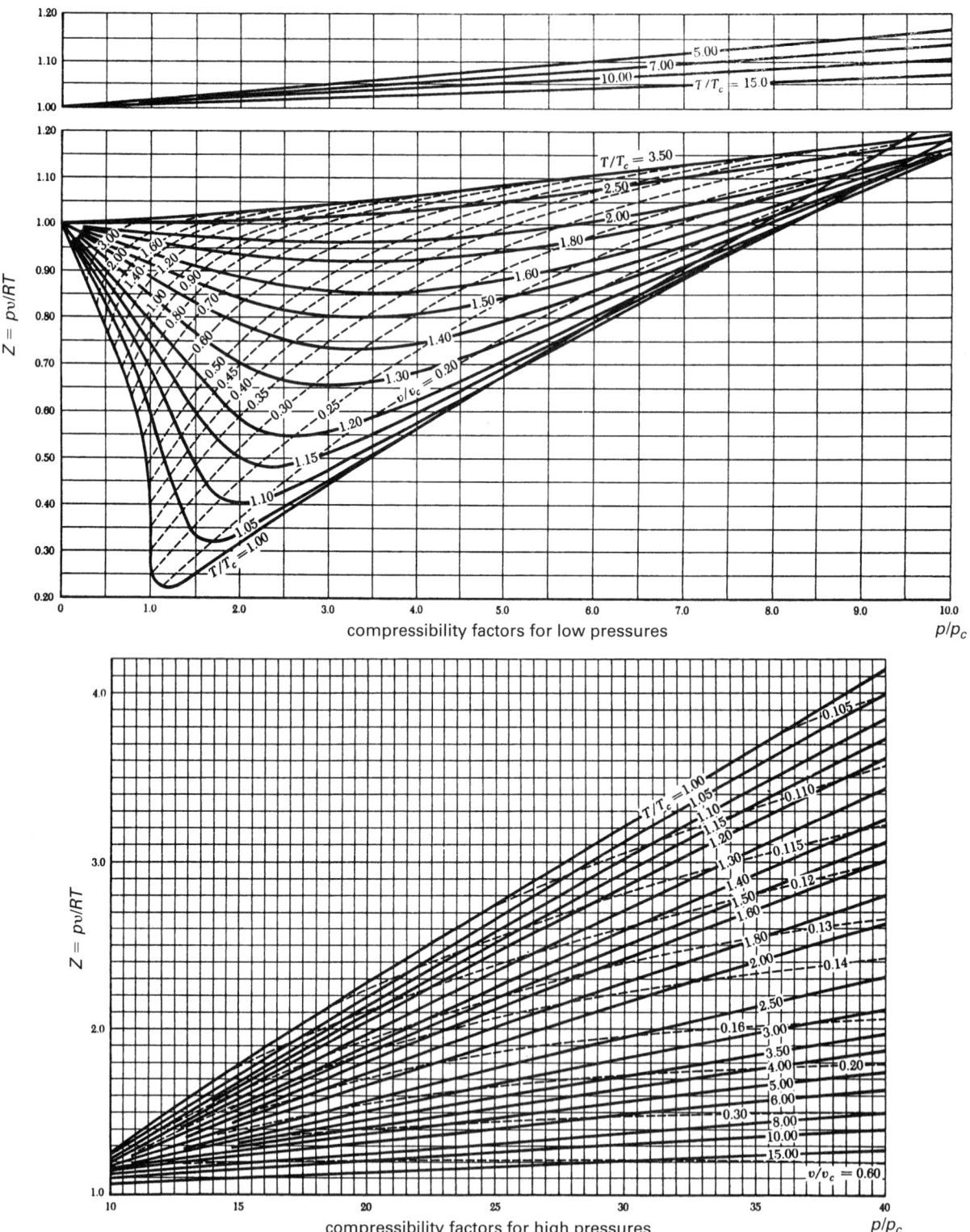

compressibility factors for low pressures

compressibility factors for high pressures

Reprinted with permission from "Generalized PVT Properties of Gases," *ASME Transactions*, Vol. 76, pp. 1057-1066, Edward F. Obert and L.C. Nelson, American Society of Mechanical Engineers, 1954.

APPENDIX 31.A
ASHRAE Psychrometric Chart No. 1, Normal Temperature—Sea Level (32–120°F)
(customary U.S. units)

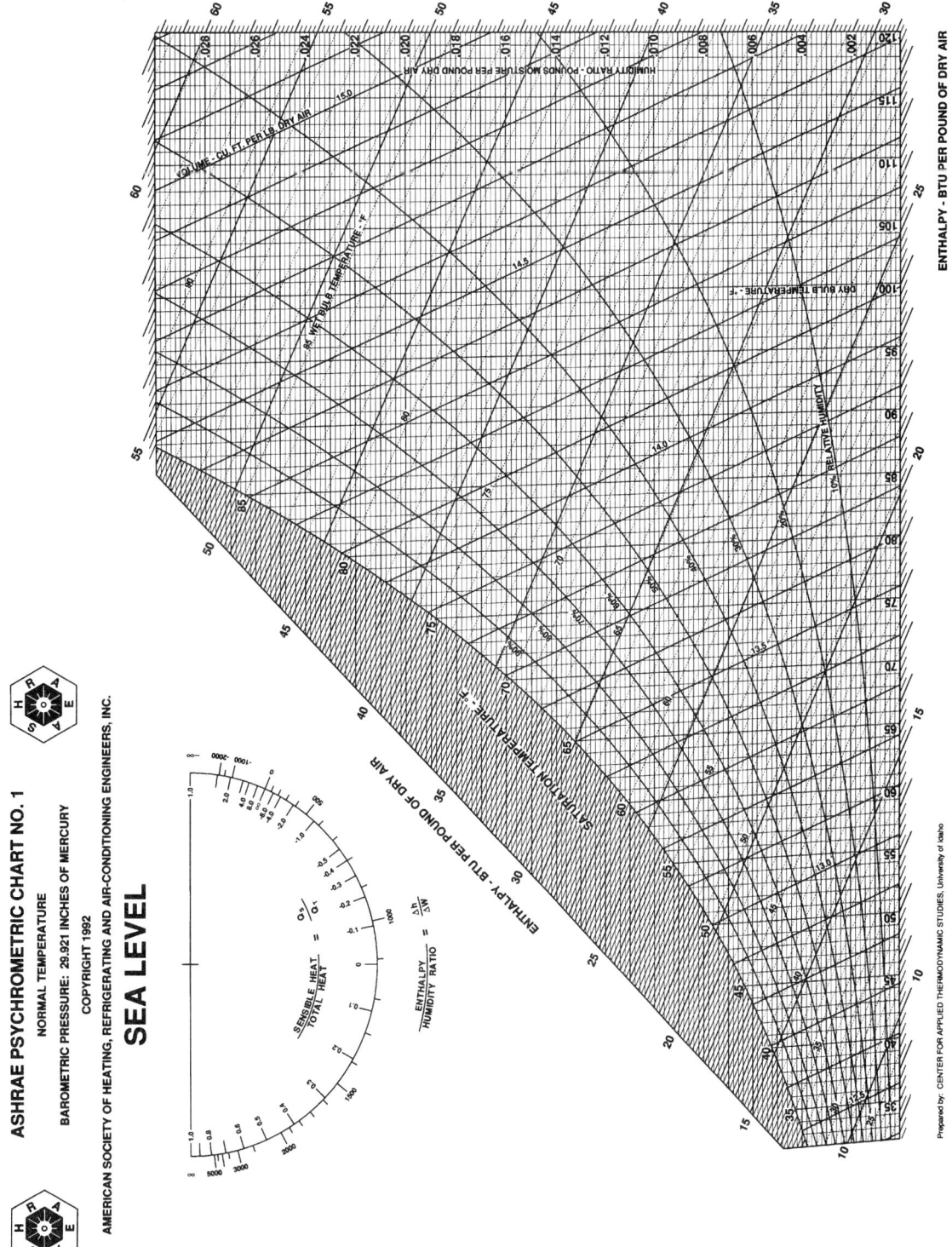

APPENDIX 31.B
ASHRAE Psychrometric Chart No. 1, Normal Temperature—Sea Level (0–50°C)
(SI units)

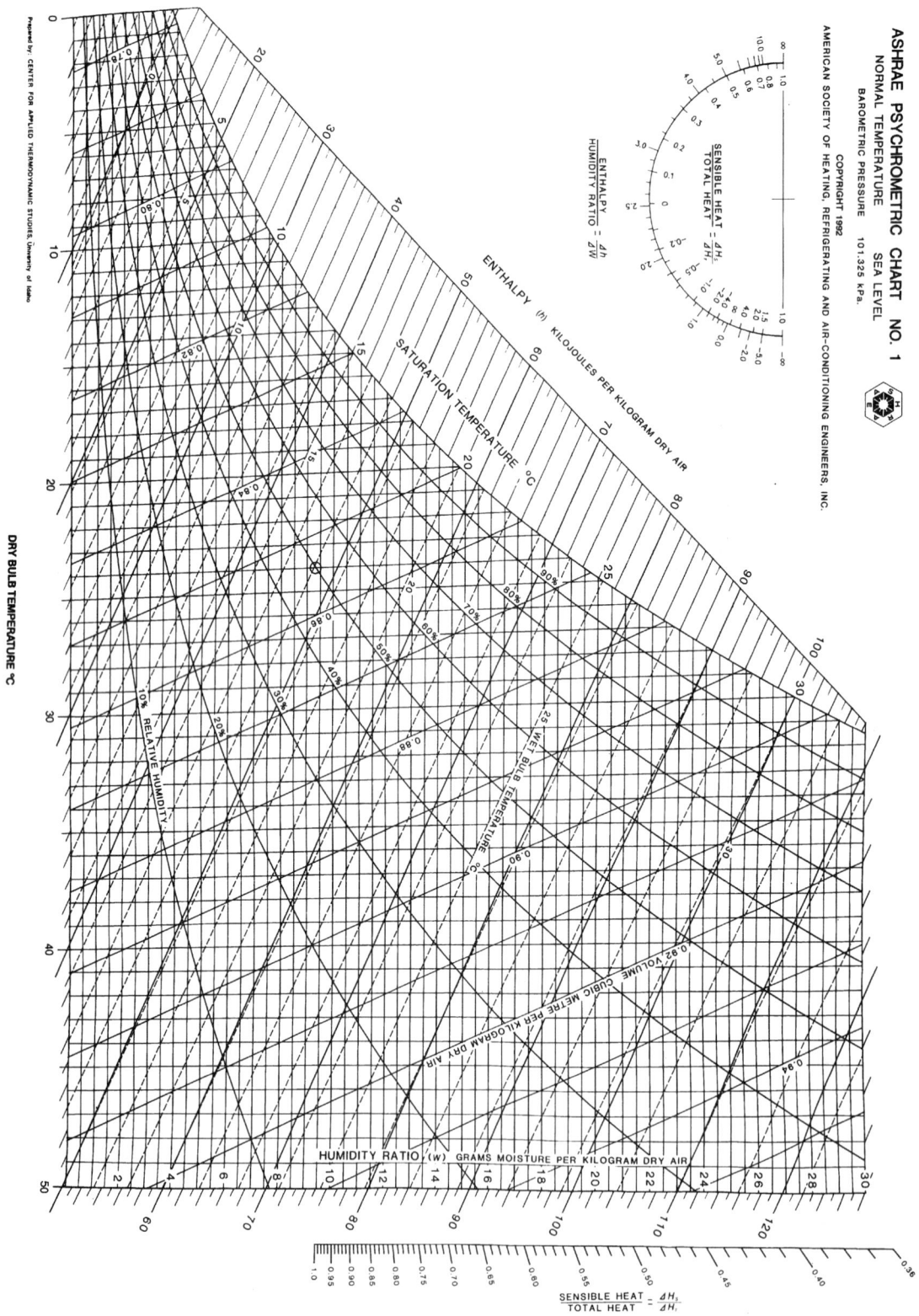

APPENDIX 31.C
ASHRAE Psychrometric Chart No. 2, Low Temperature—Sea Level (–40–50°F)
(customary U.S. units)

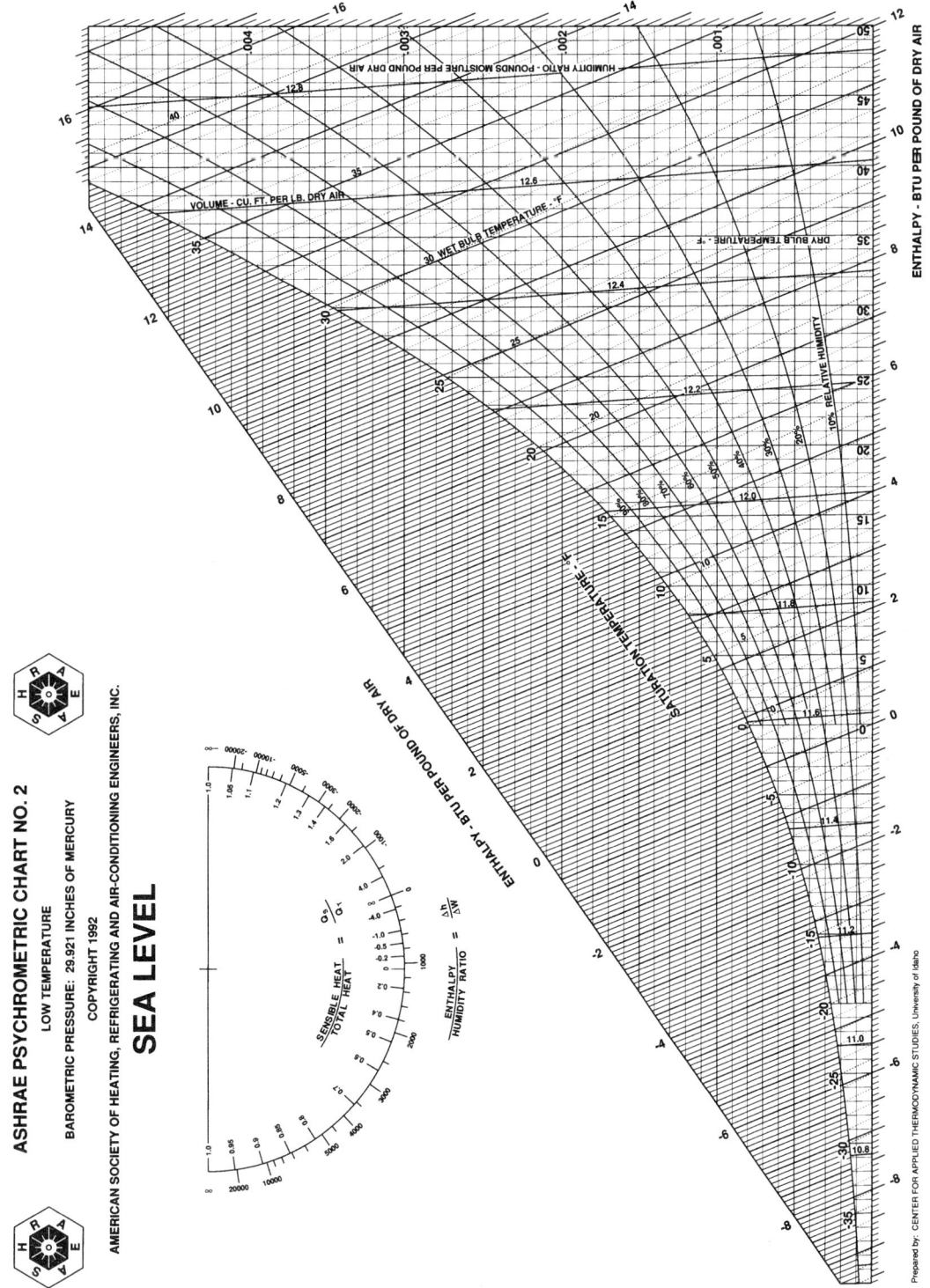

APPENDIX 31.D
ASHRAE Psychrometric Chart No. 3, High Temperature—Sea Level (60–250°F)
(customary U.S. units)

ASHRAE PSYCHROMETRIC CHART NO. 3

HIGH TEMPERATURE

BAROMETRIC PRESSURE: 29.921 INCHES OF MERCURY

COPYRIGHT 1992

AMERICAN SOCIETY OF HEATING, REFRIGERATING AND AIR-CONDITIONING ENGINEERS, INC.

SEA LEVEL

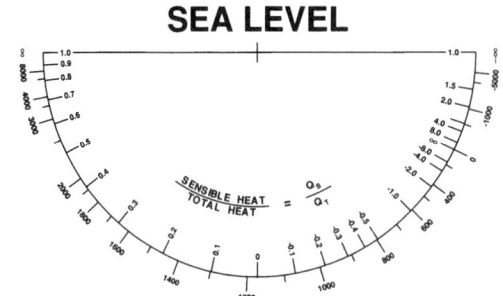

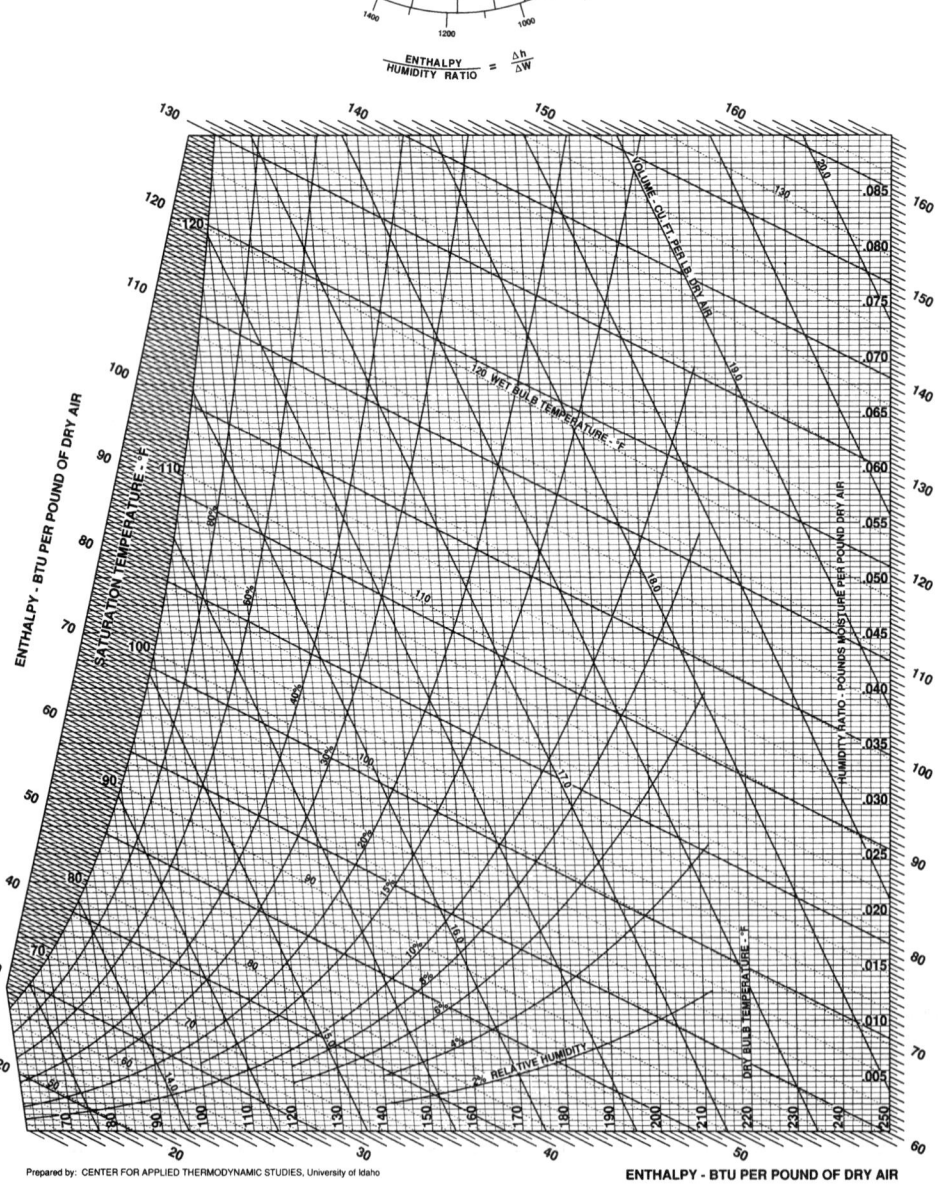

Prepared by: CENTER FOR APPLIED THERMODYNAMIC STUDIES, University of Idaho

ENTHALPY - BTU PER POUND OF DRY AIR

APPENDIX 33.A
Representative Insulating Properties of Building Materials[a]

usage	description	density (lbm/ft^3)	density (lbm/ft^2)	resistance per inch $\left(\dfrac{\text{hr-ft}^2\text{-}°\text{F}}{\text{Btu-in}}\right)$	resistance for thickness listed $\left(\dfrac{\text{hr-ft}^2\text{-}°\text{F}}{\text{Btu}}\right)$
building board	asbestos-cement, $1/8$ in	120	1.25	0.25	0.03
	gypsum or plaster board, $1/2$ in	50	5.00		0.45
	gypsum or plaster board, $5/8$ in	50	6.25		0.56
	plywood, $1/2$ in	34	1.42	1.25	0.63
	wood fiber board	26		2.38	
	fir/pine sheathing, 0.72 in	32	2.08		0.98
flooring	asphalt tile, $1/8$ in	120	1.25		0.04
	carpet and fiber pad				2.08
	carpet and rubber pad				1.23
	ceramic tile, 1 in				0.08
	cork tile, $1/8$ in	25	0.26	2.22	0.28
	linoleum, $1/8$ in	80	0.83		0.08
	plywood subfloor, $5/8$ in	34	1.77		0.78
	rubber/plastic tile, $1/8$ in	110	1.15		0.02
	hardwood, $3/4$ in	45	2.18		0.68
insulation (blanket/batt)	fibrous mineral wool, including fiberglass	0.8–2.0		3.85	
	wood fiber	3.2–3.6		4.00	
insulation (board/slab)	glass fiber	9.5		4.00	
	acoustical tile, $1/2$ in	22.4	0.93		1.19
	acoustical tile, $3/4$ in	22.4	1.4		1.78
	cellular glass	9.0		2.5	
	foamed plastic	1.62		3.45	
insulation (loose)	paper/pulp products	2.5–3.5		3.57	
	mineral wool (glass, slag, or rock)	2.0–5.0		3.33	
	expanded vermiculite	7.0		2.08	
roof insulation	all types for use above deck	15.6	1.3 per inch thickness	2.78	
masonry materials (concrete)	cement mortar	116		0.20	
	lightweight aggregates, depending on density	120		0.19	
		80		0.40	
		40		0.86	
		20		1.43	
	oven-dried sand, gravel, or stone aggregate	140		0.11	
	stucco	116		0.20	
plastering materials	cement (sand), $1/2$ in	116	4.8	0.20	0.10
	cement (sand), $3/4$ in	116	7.2	0.20	0.15
	gypsum plaster, lightweight aggregate, $1/2$ in	45	1.88		0.32
	lightweight aggregate on metal lath	45		0.59	
	sand aggregate, $1/2$ in	105	4.4	0.18	0.09
	sand aggregate on metal lath, $3/4$ in	105	6.6		0.13
roofing	asbestos-cement shingle	120			0.21
	asphalt roll roofing	70			0.15
	asphalt shingles	70			0.44
	built-up roofing, $3/8$ in	70	2.2		0.33
	wood shingles	40			0.94

(continued)

APPENDIX 33.A *(continued)*
Representative Insulating Properties of Building Materials[a]

usage	description	density (lbm/ft³)	density (lbm/ft²)	resistance per inch $\left(\frac{\text{hr-ft}^2\text{-}°F}{\text{Btu-in}}\right)$	resistance for thickness listed $\left(\frac{\text{hr-ft}^2\text{-}°F}{\text{Btu}}\right)$
siding material (on flat face)	16 in shingles				
	with $7^1/_2$ in exposure				0.87
	with 12 in exposure				1.19
	siding				
	asbestos-cement, $1/_4$ in lap				0.21
	asphalt roll siding				0.15
	wood, bevel ($1/_2$ in × 8 in lap)				0.81
	wood, bevel ($3/_4$ in × 10 in lap)				1.05
	structural glass				0.10
woods	hardwoods	45		0.91	
	softwoods	32		1.25	
masonry	brick, common, 4 in	120	40	0.20	0.80
	brick, face, 4 in	130	43	0.11	0.44
	hollow clay tile, 4 in	48			1.11
	stone (lime or sand)	150		0.08	
	concrete blocks (3 oval core)				
	with sand/gravel aggregate,				
	3 in	76	19		0.40
	6 in	64	32		0.91
	12 in	63	63		1.28
	lightweight aggregate,				
	3 in	60	15		0.56
	6 in	46	23		1.50
	12 in	43	43		1.89
poured concrete	8 in thick	30	20		10.0[b]
		80	53		4.00[b]
		140	93		1.72[b]
	12 in thick	30	30		14.28[b]
		80	80		5.55[b]
		140	140		2.17[b]
miscellaneous	adobe 8 in		26		2.94[b]
	glass (single sheet, $1/_4$ in, vertical)				0.88[b]
	glass (double, vertical)				
	$1/_4$ in air space				1.63[b]
	$1/_2$ in air space				1.82[b]
	glass (single, horizontal)				summer: 1.16[b]
					winter: 0.71[b]
	glass (double, horizontal) $1/_4$ in air space				summer: 2.00[b]
					winter: 1.43[b]
	door, 1 in nominal, wood				1.45[b]
	$1^1/_2$ in nominal, wood				1.92[b]
	2 in nominal, wood				2.18[b]
	door, glass (herculite), $3/_4$ in thick				0.95[b]

(Multiply lbm/ft³ by 16.0 to obtain kg/m³.)
(Multiply lbm/ft² by 4.88 to obtain kg/m².)
(Multiply hr-ft²-°F/Btu-in by 6.93 to obtain m·°C/W.)
(Multiply hr-ft²-°F/Btu by 0.176 to obtain m²·°C/W.)
[a]Compiled from a variety of sources.
[b]Includes appropriate inside and outside film coefficients.

APPENDIX 37.A
Heats of Combustion for Common Compounds[a]

| | | | | heat of combustion | | | |
| | | | | Btu/ft^3 | | Btu/lbm | |
substance	formula	molecular weight	specific volume (ft^3/lbm)	gross (high)	net (low)	gross (high)	net (low)
carbon	C	12.01				14,093	14,093
carbon dioxide	CO_2	44.01	8.548				
carbon monoxide	CO	28.01	13.506	322	322	4347	4347
hydrogen	H_2	2.016	187.723	325	275	60,958	51,623
nitrogen	N_2	28.016	13.443				
oxygen	O_2	32.000	11.819				
paraffin series (alkanes)							
methane	CH_4	16.041	23.565	1013	913	23,879	21,520
ethane	C_2H_6	30.067	12.455	1792	1641	22,320	20,432
propane	C_3H_8	44.092	8.365	2590	2385	21,661	19,944
n-butane	C_4H_{10}	58.118	6.321	3370	3113	21,308	19,680
isobutane	C_4H_{10}	58.118	6.321	3363	3105	21,257	19,629
n-pentane	C_5H_{12}	72.144	5.252	4016	3709	21,091	19,517
isopentane	C_5H_{12}	72.144	5.252	4008	3716	21,052	19,478
neopentane	C_5H_{12}	72.144	5.252	3993	3693	20,970	19,396
n-hexane	C_6H_{14}	86.169	4.398	4762	4412	20,940	19,403
olefin series (alkenes and alkynes)							
ethylene	C_2H_4	28.051	13.412	1614	1513	21,644	20,295
propylene	C_3H_6	42.077	9.007	2336	2186	21,041	19,691
n-butene	C_4H_8	56.102	6.756	3084	2885	20,840	19,496
isobutene	C_4H_8	56.102	6.756	3068	2869	20,730	19,382
n-pentene	C_5H_{10}	70.128	5.400	3836	3586	20,712	19,363
aromatic series							
benzene	C_6H_6	78.107	4.852	3751	3601	18,210	17,480
toluene	C_7H_8	92.132	4.113	4484	4284	18,440	17,620
xylene	C_8H_{10}	106.158	3.567	5230	4980	18,650	17,760
miscellaneous fuels							
acetylene	C_2H_2	26.036	14.344	1499	1448	21,500	20,776
air		28.967	13.063				
ammonia	NH_3	17.031	21.914	441	365	9668	8001
digester gas[b]	–	25.8	18.3	658	593	15,521	13,988
ethyl alcohol	C_2H_5OH	46.067	8.221	1600	1451	13,161	11,929
hydrogen sulfide	H_2S	34.076	10.979	647	596	7100	6545
iso-octane	C_8H_{18}	114.2	0.0232[c]	106	98.9	20,590	19,160
methyl alcohol	CH_3OH	32.041	11.820	868	768	10,259	9078
naphthalene	$C_{10}H_8$	128.162	2.955	5854	5654	17,298	16,708
sulfur	S	32.06				3983	3983
sulfur dioxide	SO_2	64.06	5.770				
water vapor	H_2O	18.016	21.017				

(Multiply Btu/lbm by 2.326 to obtain kJ/kg.)
(Multiply Btu/ft^3 by 37.25 to obtain kJ/m^3.)
[a]Gas volumes listed are at 60°F (16°C) and 1 atm.
[b]Digester gas from wastewater treatment plants is approximately 65% methane and 35% carbon dioxide by volume. Use composite properties of these two gases.
[c]liquid form; stoichiometric mixture

APPENDIX 37.B
Approximate Properties of Selected Gases

gas	symbol	temp °F	MW	customary U.S. units			SI units			k
				R ft-lbf / lbm-°R	c_p Btu / lbm-°R	c_v Btu / lbm-°R	R J / kg·K	c_p J / kg·K	c_v J / kg·K	
acetylene	C_2H_2	68	26.038	59.35	0.350	0.274	319.32	1465	1146	1.279
air		100	28.967	53.35	0.240	0.171	287.03	1005	718	1.400
ammonia	NH_3	68	17.032	90.73	0.523	0.406	488.16	2190	1702	1.287
argon	Ar	68	39.944	38.69	0.124	0.074	208.15	519	311	1.669
n-butane	C_4H_{10}	68	58.124	26.59	0.395	0.361	143.04	1654	1511	1.095
carbon dioxide	CO_2	100	44.011	35.11	0.207	0.162	188.92	867	678	1.279
carbon monoxide	CO	100	28.011	55.17	0.249	0.178	296.82	1043	746	1.398
chlorine	Cl_2	100	70.910	21.79	0.115	0.087	117.25	481	364	1.322
ethane	C_2H_6	68	30.070	51.39	0.386	0.320	276.50	1616	1340	1.206
ethylene	C_2H_4	68	28.054	55.08	0.400	0.329	296.37	1675	1378	1.215
Freon (R-12)[*]	CCl_2F_2	200	120.925	12.78	0.159	0.143	68.76	666	597	1.115
helium	He	100	4.003	386.04	1.240	0.744	2077.03	5192	3115	1.667
hydrogen	H_2	100	2.016	766.53	3.420	2.435	4124.18	14 319	10 195	1.405
hydrogen sulfide	H_2S	68	34.082	45.34	0.243	0.185	243.95	1017	773	1.315
krypton	Kr		83.800	18.44	0.059	0.035	99.22	247	148	1.671
methane	CH_4	68	16.043	96.32	0.593	0.469	518.25	2483	1965	1.264
neon	Ne	68	20.183	76.57	0.248	0.150	411.94	1038	626	1.658
nitrogen	N_2	100	28.016	55.16	0.249	0.178	296.77	1043	746	1.398
nitric oxide	NO	68	30.008	51.50	0.231	0.165	277.07	967	690	1.402
nitrous oxide	NO_2	68	44.01	35.11	0.221	0.176	188.92	925	736	1.257
octane vapor	C_8H_{18}		114.232	13.53	0.407	0.390	72.78	1704	1631	1.045
oxygen	O_2	100	32.000	48.29	0.220	0.158	259.82	921	661	1.393
propane	C_3H_8	68	44.097	35.04	0.393	0.348	188.55	1645	1457	1.129
sulfur dioxide	SO_2	100	64.066	24.12	0.149	0.118	129.78	624	494	1.263
water vapor[*]	H_2	212	18.016	85.78	0.445	0.335	461.50	1863	1402	1.329
xenon	Xe		131.300	11.77	0.038	0.023	63.32	159	96	1.661

(Multiply Btu/lbm-°F by 4186.8 to obtain J/kg·K.)
(Multiply ft-lbf/lbm-°R by 5.3803 to obtain J/kg·K.)
[*]Values for steam and Freon are approximate and should be used only for low pressures and high temperatures.

APPENDIX 38.A
Approximate Henry's Law Constants for Common Gases and VOCs[a]

gas	K_H $(atm)^b$	ΔH (cal/mol)	k
ammonia (NH_3)	0.76	3750	6.31
benzene (C_6H_6)	240	3680	8.68
carbon dioxide (CO_2)	151	2070	6.73
chlorine (Cl_2)	585	1740	5.75
chlorine dioxide (ClO_2)	54	2930	6.76
chloroform ($CHCl_3$)	170	4000	9.10
hydrogen sulfide (H_2S)	515	1850	5.88
methane (CH_4)	38 000	1540	7.22
nitrogen (N_2)	86 000	1120	6.85
oxygen (O_2)	43 000	1450	7.11
ozone (O_3)	5000	2520	8.05
sulfur dioxide (SO_2)	38	2400	5.68
trichloroethene (C_2HCl_3)	550	3410	8.59
vinyl chloride (C_2H_2Cl)	355 000	–	–

(Multiply atm by 1.8×10^{-5} to get atm·m^3/mol.)

(Multiply atm by 1.8×10^{-2} to get atm·L/mol.)

(Multiply atm by 1.8×10^3 to get Pa·L/mol.)

(Multiply atm by $0.22/(30°C + 273°)$ to get a unitless number.)

[a]Values are for 20°C. K_H at other temperatures is calculated from $\log K_H = \dfrac{-\Delta H}{R^* T} + k$.

For methane at 30°C,

$$\log K_H = \frac{-1540 \ \dfrac{cal}{mol}}{\left(1.986 \ \dfrac{cal}{mol \cdot K}\right)(30°C + 273°)} + 7.22 = 4.6608$$

$K_H = 45\,800$ atm

[b]equivalent to atmospheres per mole fraction

Adapted from Yen, T.F., *Environmental Chemistry*, Volume 4B, © 1999 Prentice Hall, Upper Saddle River, NJ.

APPENDIX 38.B
Approximate Henry's Law Constants for Air and Common Gases

K_H (atm*)

T (°C)	air	CH_4	CO	CO_2	H_2S	N_2	O_2
0	43 200	22 400	35 200	728	268	52 900	25 500
10	54 900	29 700	44 200	1040	367	66 800	32 700
20	66 400	37 600	53 600	1420	483	80 400	40 100
30	77 100	44 900	62 000	1860	609	92 400	47 500
40	87 000	52 000	69 600	2330	745	104 000	53 500
50	94 600	57 700	76 100	2830	884	113 000	58 800
60	101 000	62 600	82 100	3410	1030	120 000	62 900

(Multiply atm by 1.8×10^{-5} to get atm·m^3/mol.)
(Multiply atm by 1.8×10^{-2} to get atm·L/mol.)
(Multiply atm by 1.8×10^{3} to get Pa·L/mol.)
(Multiply atm by $0.22/(30°C + 273°)$ to get a unitless number.)
*equivalent to atmospheres per mole fraction

Adapted from Metcalf and Eddy, Inc., *Wastewater Engineering: Treatment, Disposal, and Reuse,* © 1991 McGraw-Hill, NY.

APPENDIX 38.C
Air Viscosity and Density

temperature ($^\circ$C)	absolute viscosity (kg/m·s)	density[*] (kg/m^3)
0	1.69×10^{-5}	1.30
10	1.75×10^{-5}	1.25
20	1.82×10^{-5}	1.21
30	1.90×10^{-5}	1.17
40	1.94×10^{-5}	1.13
50	1.95×10^{-5}	1.09
60	1.97×10^{-5}	1.06
70	2.06×10^{-5}	1.03
80	2.10×10^{-5}	1.00
90	2.13×10^{-5}	0.974
100	2.17×10^{-5}	0.948
150	2.39×10^{-5}	0.836
200	2.60×10^{-5}	0.748
250	2.75×10^{-5}	0.676
300	2.82×10^{-5}	0.617
350	3.05×10^{-5}	0.568
400	3.23×10^{-5}	0.525
450	3.39×10^{-5}	0.489
500	3.56×10^{-5}	0.458
600	3.86×10^{-5}	0.405
700	4.14×10^{-5}	0.363
800	4.44×10^{-5}	0.330
900	4.72×10^{-5}	0.301
1000	5.05×10^{-5}	0.278

[*]Calculated at 1 atm pressure. Density at other temperatures and pressures can be calculated using an approximate molecular weight for air of 29 g/mol.

$$\rho_g = \frac{(\text{MW})p}{R^*T} = \frac{\left(29\,\frac{\text{g}}{\text{mol}}\right)\left(\frac{1\,\text{kg}}{1000\,\text{g}}\right)p}{\left(8.2 \times 10^{-5}\,\frac{\text{atm·m}^3}{\text{mol·K}}\right)(^\circ\text{C} + 273^\circ)}$$

APPENDIX 38.D
Normal Chemical Composition of Dry Air

component	concentration (ppmv)	concentration (%)
nitrogen	780 900	78.09
oxygen	209 400	20.94
argon	9300	0.93
carbon dioxide	315	0.0315
neon	18	–
helium	5.2	–
methane	1.0–1.2	–
krypton	1.0	–
nitrous oxide	0.5	–
hydrogen	0.5	–
xenon	0.008	–
nitrogen dioxide	0.02	–
ozone	0.01–0.04	–

APPENDIX 45.A
NIOSH Pocket Guide to Chemical Hazards

Representative information shown only. This is a public document and can be downloaded in its entirety at ppi2pass.com/ENVRM/originalsources.html.

NIOSH POCKET GUIDE TO CHEMICAL HAZARDS

Introduction

Chemical and Physical Properties

The following abbreviations are used for the chemical and physical properties given for each substance. "NA" indicates that a property is not applicable, and a question mark (?) indicates that it is unknown.

MW   Molecular weight

BP   Boiling point at 1 atmosphere, °F

Sol   Solubility in water at 68°F (unless a different temperature is noted), % by weight (i.e., g/100 ml)

Fl.P   Flash point (i.e., the temperature at which the liquid phase gives off enough vapor to flash when exposed to an external ignition source), closed cup (unless annotated "(oc)" for open cup), °F

IP   Ionization potential, eV (electron volts) [Ionization potentials are given as a guideline for the selection of photoionization detector lamps used in some direct-reading instruments.]

VP   Vapor pressure at 68°F (unless a different temperature is noted), mm Hg; "approx" indicates approximately

MLT   Melting point for solids, °F

FRZ   Freezing point for liquids and gases, °F

UEL   Upper explosive (flammable) limit in air, % by volume (at room temperature unless otherwise noted)

LEL   Lower explosive (flammable) limit in air, % by volume (at room temperature unless otherwise noted)

MEC   Minimum explosive concentration, g/m^3 (when available)

Sp.Gr   Specific gravity at 68°F (unless a different temperature is noted) referenced to water at 39.2°F (4°C)

RGasD   Relative density of gases referenced to air = 1 (indicates how many times a gas is heavier than air at the same temperature)

When available, the flammability/combustibility of a substance is listed at the bottom of the chemical and physical properties section. The following OSHA criteria (29 CFR 1910.106) were used to classify flammable or combustible liquids:

Class IA flammable liquid
     Fl.P. below 73°F and BP below 100°F.
Class IB flammable liquid
     Fl.P. below 73°F and BP at or above 100°F.
Class IC flammable liquid
     Fl.P. at or above 73°F and below 100°F.
Class II combustible liquid
     Fl.P. at or above 100°F and below 140°F.
Class IIIA combustible liquid
     Fl.P. at or above 140°F and below 200°F.
Class IIIB combustible liquid
     Fl.P. at or above 200°F.

Personal Protection and Sanitation Recommendations

This section presents a summary of recommended practices for each substance. These recommendations supplement general work practices (e.g., no eating, drinking, or smoking where chemicals are used) and should be followed if additional controls are needed after using all feasible process, equipment, and task controls. Each category is described as follows:

SKIN: Recommends the need for personal protective clothing.

EYES: Recommends the need for eye protection.

WASH SKIN: Recommends when workers should wash the spilled chemical from the body in addition to normal washing (e.g., before eating).

REMOVE: Advises workers when to remove clothing that has accidentally become wet or significantly contaminated.

CHANGE: Recommends whether the routine changing of clothing is needed.

PROVIDE: Recommends the need for eyewash fountains and/or quick drench facilities.

APPENDIX 45.B
OSHA Air Contaminant Regulations
(Standards—29 CFR 1910.1000)

> **Representative page shown only. This is a public document and can be downloaded in its entirety at**
> **ppi2pass.com/ENVRM/originalsources.html.**

Regulations (Standards - 29 CFR) - Table of Contents

- **Part Number:** 1910
- **Part Title:** Occupational Safety and Health Standards
- **Subpart:** Z
- **Subpart Title:** Toxic and Hazardous Substances
- **Standard Number:** 1910.1000
- **Title:** Air contaminants.

An employee's exposure to any substance listed in Tables Z-1, Z-2, or Z-3 of this section shall be limited in accordance with the requirements of the following paragraphs of this section.

1910.1000(a)

Table Z-1 --.

1910.1000(a)(1)

Substances with limits preceded by "C" - Ceiling Values. An employee's exposure to any substance in Table Z-1, the exposure limit of which is preceded by a "C", shall at no time exceed the exposure limit given for that substance. If instantaneous monitoring is not feasible, then the ceiling shall be assessed as a 15-minute time weighted average exposure which shall not be exceeded at any time during the working day.

1910.1000(a)(2)

Other substances -- 8-hour Time Weighted Averages. An employee's exposure to any substance in Table Z-1, the exposure limit of which is not preceded by a "C", shall not exceed the 8-hour Time Weighted Average given for that substance any 8-hour work shift of a 40-hour work week.

1910.1000(b)

Table Z-2. An employee's exposure to any substance listed in Table Z-2 shall not exceed the exposure limits specified as follows:

1910.1000(b)(1)

8-hour time weighted averages. An employee's exposure to any substance listed in Table Z-2, in any 8-hour work shift of a 40-hour work week, shall not exceed the 8-hour time weighted average limit given for that substance in Table Z-2.

1910.1000(b)(2)

APPENDIX 47.A
NIOSH IDLH Values

Representative page shown only. This is a public document and can be downloaded in its entirety at
ppi2pass.com/ENVRM/originalsources.html.

NIOSH CHEMICAL LISTING OF REVISED IDLH VALUES
(AS OF 3/1/95)

Source: National Institute Occupational Safety and Health,
Documentation for Immediately Dangerous to Safety and Health Concentrations, 1995

Warning: obtain the most current IDLH values before placing personnel at risk

SUBSTANCE	IDLH VALUE
Acetaldehyde	2,000 ppm
Acetic acid	50 ppm
Acetic anhydride	200 ppm
Acetone	2,500 ppm [LEL]
Acetonitrile	500 ppm
Acetylene tetrabromide	8 ppm
Acrolein	2 ppm
Acrylamide	60 mg/m^3
Acrylonitrile	85 ppm
Aldrin	25 mg/m^3
Allyl alcohol	20 ppm
Allyl chloride	250 ppm
Allyl glycidyl ether	50 ppm
2 Aminopyridine	5 ppm
Ammonia	300 ppm
Ammonium sulfamate	1,500 mg/m^3
n-Amyl acetate	1,000 ppm
sec-Amyl acetate	1,000 ppm
Aniline	100 ppm
o-Anisidine	50 mg/m^3
p-Anisidine	50 mg/m^3
Antimony compounds (as Sb)	50 mg Sb/m^3
ANTU	100 mg/m^3
Arsenic (inorganic compounds, as As)	5 mg As/m^3
Arsine	3 ppm
Azinphosmethyl	10 mg/m^3
Barium (soluble compounds, as Ba)	50 mg Ba/m^3
Benzene	500 ppm
Benzoyl peroxide	1,500 mg/m^3
Benzyl chloride	10 ppm
Beryllium compounds (as Be)	4 mg Be/m^3
Boron oxide	2,000 mg/m^3
Boron trifluoride	25 ppm
Bromine	3 ppm
Bromoform	850 ppm
1,3-Butadiene	2,000 ppm [LEL]
2-Butanone	3,000 ppm

APPENDIX 47.B
Sampling For Air Contaminants
(OSHA Technical Manual, Sec. II, Chap. 1)

OSHA Technical Manual

SECTION II: CHAPTER 1

PERSONAL SAMPLING FOR AIR CONTAMINANTS

Contents:

I.   INTRODUCTION

This chapter provides basic information related to sampling air contaminants. Other reference resources are OSHA's Chemical Sampling Information (CSI) file and the OSHA Field Operations Manual (FOM). Sampling and analytical methods that have been validated by either OSHA or the National Institute for Occupational Safety and Health (NIOSH) should be used whenever possible. Sometimes the Salt Lake Technical Center (SLTC) will approve the use of procedures developed by other organizations. Only procedures approved by the SLTC should be used. The use of sampling methods not approved by the SLTC may require resampling with an approved sampling procedure. The SLTC is aware that unique sampling situations will arise during some inspections and it is essential that OSHA Compliance Safety and Health Officers (CSHOs) contact, and work closely with, the SLTC whenever questions arise.

APPENDIX 47.C
Selected OSHA Standards for Respiratory Personal Protective Equipment
(Standards—29 CFR 1910.132)

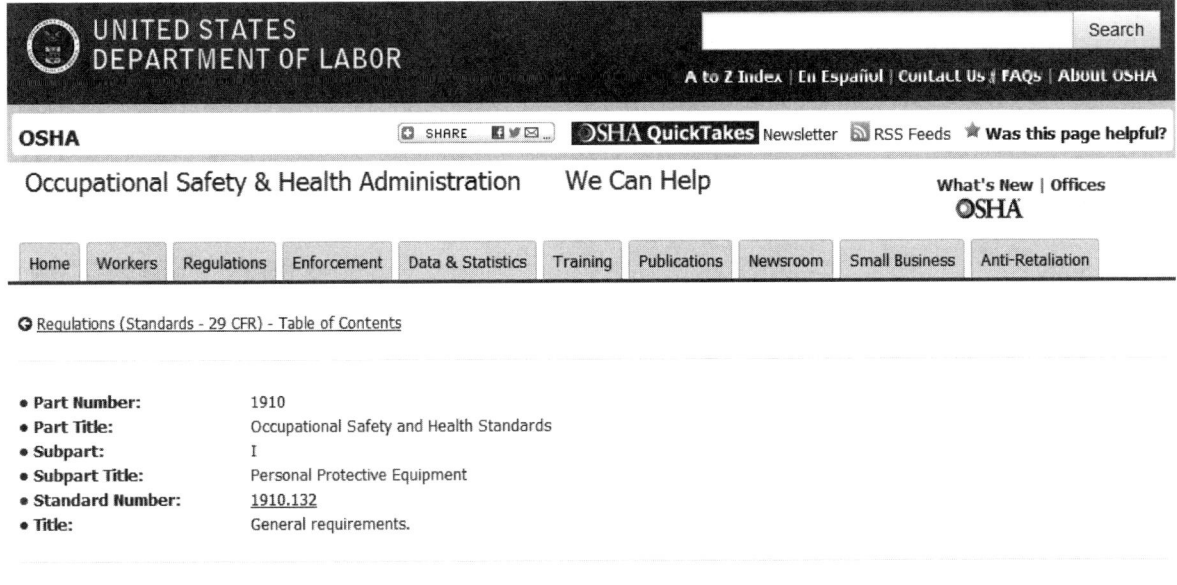

Regulations (Standards - 29 CFR) - Table of Contents

- **Part Number:**   1910
- **Part Title:**   Occupational Safety and Health Standards
- **Subpart:**   I
- **Subpart Title:**   Personal Protective Equipment
- **Standard Number:**   1910.132
- **Title:**   General requirements.

1910.132(a)

Application. Protective equipment, including personal protective equipment for eyes, face, head, and extremities, protective clothing, respiratory devices, and protective shields and barriers, shall be provided, used, and maintained in a sanitary and reliable condition wherever it is necessary by reason of hazards of processes or environment, chemical hazards, radiological hazards, or mechanical irritants encountered in a manner capable of causing injury or impairment in the function of any part of the body through absorption, inhalation or physical contact.

1910.132(b)

Employee-owned equipment. Where employees provide their own protective equipment, the employer shall be responsible to assure its adequacy, including proper maintenance, and sanitation of such equipment.

1910.132(c)

Design. All personal protective equipment shall be of safe design and construction for the work to be performed.

1910.132(d)

Hazard assessment and equipment selection.

1910.132(d)(1)

The employer shall assess the workplace to determine if hazards are present, or are likely to be present, which necessitate the use of personal protective equipment (PPE). If such hazards are present, or likely to be present, the employer shall:

1910.132(d)(1)(i)

Select, and have each affected employee use, the types of PPE that will protect the affected employee from the hazards identified in the hazard assessment;

Support Material

APPENDIX 47.D
Selected Extracts from OSHA Technical Manual

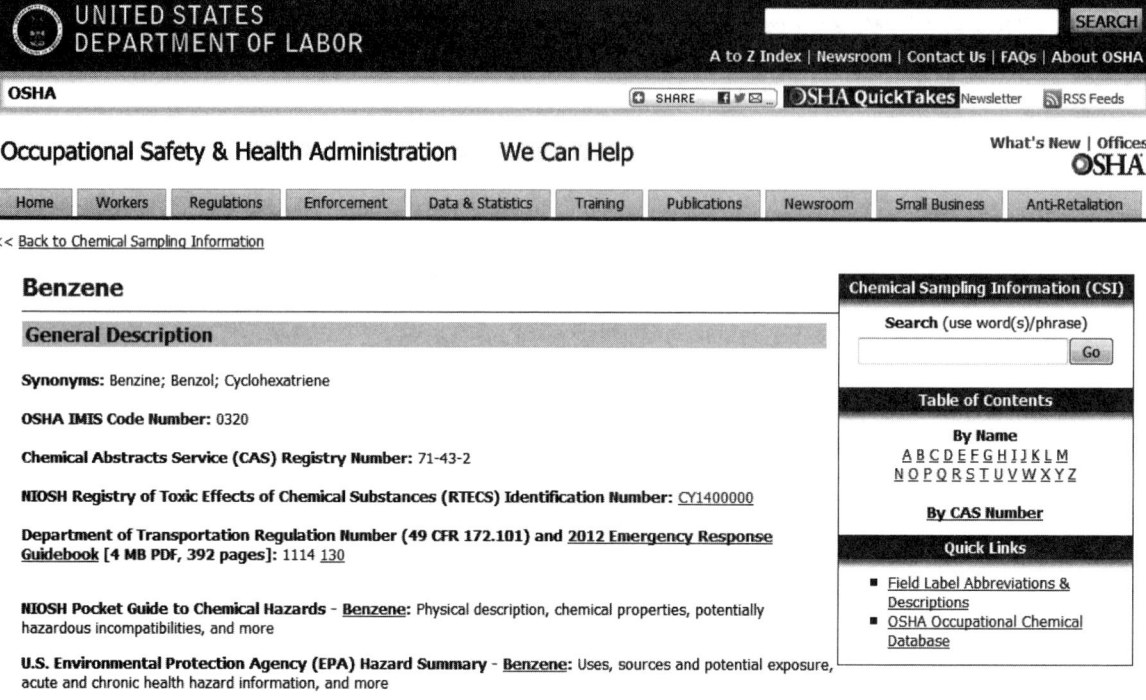

Benzene

General Description

Synonyms: Benzine; Benzol; Cyclohexatriene

OSHA IMIS Code Number: 0320

Chemical Abstracts Service (CAS) Registry Number: 71-43-2

NIOSH Registry of Toxic Effects of Chemical Substances (RTECS) Identification Number: CY1400000

Department of Transportation Regulation Number (49 CFR 172.101) and 2012 Emergency Response Guidebook [4 MB PDF, 392 pages]: 1114 130

NIOSH Pocket Guide to Chemical Hazards - Benzene: Physical description, chemical properties, potentially hazardous incompatibilities, and more

U.S. Environmental Protection Agency (EPA) Hazard Summary - Benzene: Uses, sources and potential exposure, acute and chronic health hazard information, and more

Chemical Sampling Information (CSI)

Search (use word(s)/phrase)

[Go]

Table of Contents

By Name
A B C D E F G H I J K L M
N O P Q R S T U V W X Y Z

By CAS Number

Quick Links

- Field Label Abbreviations & Descriptions
- OSHA Occupational Chemical Database

Exposure Limits and Health Effects

Exposure Limit	Limit Values	HE Codes	Health Factors and Target Organs
OSHA Permissible Exposure Limit (PEL) - General Industry See 29 CFR 1910.1028	1 ppm TWA 5 ppm STEL	HE1	Leukemia
		HE7	Central nervous system excitation followed by central nervous system depression
		HE8	Loss of consciousness, respiratory paralysis, death (very high concentrations)
		HE12	Nonmalignant blood disorders (bleeding, anemia, aplastic anemia, thrombocytopenia, leukopenia)
		HE14	Eye, nose, and respiratory irritation
OSHA PEL - Sectors Excluded from General Industry See 1910.1000 Table Z-2 (See also Z37.40-1969) Note: These values apply to the industry segments exempt from the 1 ppm 8-hour TWA and 5 ppm STEL of	10 ppm TWA 25 ppm Ceiling 50 ppm Maximum peak	HE12	Blood disorders (anemia, leukopenia, aplastic anemia)
		HE14	Eye, nose, and respiratory irritation

APPENDIX 47.E
OSHA Process Hazard Analysis Regulations
(Standards—29 CFR 1910.119(e))

(e)
Process hazard analysis.

(e)(1)
The employer shall perform an initial process hazard analysis (hazard evaluation) on processes covered by this standard. The process hazard analysis shall be appropriate to the complexity of the process and shall identify, evaluate, and control the hazards involved in the process. Employers shall determine and document the priority order for conducting process hazard analyses based on a rationale which includes such considerations as extent of the process hazards, number of potentially affected employees, age of the process, and operating history of the process. The process hazard analysis shall be conducted as soon as possible, but not later than the following schedule: [schedule omitted]

(e)(2)
The employer shall use one or more of the following methodologies that are appropriate to determine and evaluate the hazards of the process being analyzed.

(e)(2)(i)
What-If;

... *1910.119(e)(2)(ii)*

(e)(2)(ii)
Checklist;

(e)(2)(iii)
What-If/Checklist;

(e)(2)(iv)
Hazard and Operability Study (HAZOP);

(e)(2)(v)
Failure Mode and Effects Analysis (FMEA);

(e)(2)(vi)
Fault Tree Analysis; or

(e)(2)(vii)
An appropriate equivalent methodology.

(e)(3)
The process hazard analysis shall address;

(e)(3)(i)
The hazards of the process;

(e)(3)(ii)
The identification of any previous incident which had a likely potential for catastrophic consequences in the workplace;

(e)(3)(iii)
Engineering and administrative controls applicable to the hazards and their interrelationships such as appropriate application of detection methodologies to provide early warning of releases. (Acceptable detection methods might include process monitoring and control instrumentation with alarms, and detection hardware such as hydrocarbon sensors.);

... *1910.119(e)(3)(iv)*

(e)(3)(iv)
Consequences of failure of engineering and administrative controls;

(e)(3)(v)
Facility siting;

(e)(3)(vi)
Human factors; and

(e)(3)(vii)
A qualitative evaluation of a range of the possible safety and health effects of failure of controls on employees in the workplace.

(e)(4)
The process hazard analysis shall be performed by a team with expertise in engineering and process operations, and the team shall include at least one employee who has experience and knowledge specific to the process being evaluated. Also, one member of the team must be knowledgeable in the specific process hazard analysis methodology being used.

(e)(5)
The employer shall establish a system to promptly address the team's findings and recommendations; assure that the recommendations are resolved in a timely manner and that the resolution is documented; document what actions are to be taken; complete actions as soon as possible; develop a written schedule of when these actions are to be completed; communicate the actions to operating, maintenance, and other employees whose work assignments are in the process and who may be affected by the recommendations or actions.

Source: U.S. Department of Labor

APPENDIX 48.A
ALIs and DACs of Radionuclides for Occupational Exposure
(Standards—10 CFR, Part 20, App. B)

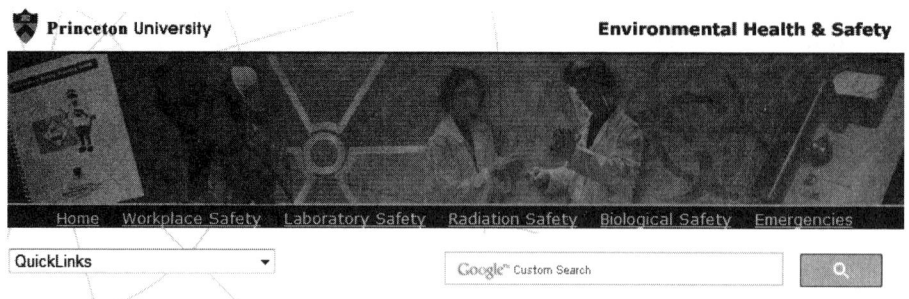

Princeton University   **Environmental Health & Safety**

Home | Workplace Safety | Laboratory Safety | Radiation Safety | Biological Safety | Emergencies

QuickLinks

Google Custom Search

Radiation Safety Guide

APPENDIX B

Appendix B is an abridged version of the Appendix B to "10 CFR Part 20" For the Table 1, 2 or 3 values for any radioisotopes which are not listed in this Appendix, contact the Office of Environmental Health and Safety.

Table 1: **ANNUAL LIMITS ON INTAKE (ALIs) AND DERIVED AIR CONCENTRATIONS (DACs) OF RADIONUCLIDES FOR OCCUPATIONAL EXPOSURE**

For each radionuclide, Table 1 indicates the chemical form which is to be used for selecting the appropriate ALI or DAC value. The ALIs and DACs for inhalation are given for an aerosol with an activity median aerodynamic diameter (AMAD) of 1 mm and for three classes (D, W, Y) of radioactive material, which refer to their retention (approximately days, weeks or years) in the pulmonary region of the lung. This classification applies to a range of clearance half-times of less than 10 days for D, for W from 10 to 100 days, and for Y greater than 100 days. The class (D, W, or Y) given in the column headed "Class" applies only to the inhalation ALIs and DACs given in Table 1, columns 2 and 3.

Table 2: **EFFLUENT CONCENTRATIONS**
Table 2 provides concentrations limits for airborne and liquid effluents released to the general environment.

Table 3: **CONCENTRATIONS FOR RELEASE TO SEWERAGE**
Table 3 provides concentration limits for discharges to sanitary sewer systems.

Page B.2, which comes directly from Appendix B to "10 CFR Part 20", provides additional information about the derivation and use of Tables 1, 2, and 3.

Atomic No.	Radionuclide	Class	Table 1 Occupational Values			Table 2 Effluent Concentrations		Table 3 Release to Sewers
			Col 1	Col 2	Col 3	Col 1	Col 2	
			Oral Ingestion ALI (μCi)	Inhalation ALI (μCi)	DAC (μCi/ml)	Air (μCi/ml)	Water (μCi/ml)	Monthly Average concentration (μCi/ml)
1	Hydrogen-3	Water, DAC includes skin absorption	8E+4	8E+4	2E-5	1E-7	1E-3	1E-2
		Gas (HT or T2) Submersion[1]: Use above values as HT and T2 oxidize in air and in the body to HTO.						
4	Beryllium-7	W, all compounds except those given for Y	4E+4	2E+4	9E-6	3E-8	6E-4	6E-3
		Y, oxides, halides, and nitrates	--	2E+4	8E-6	3E-8	--	--

APPENDIX 49.A
Cross Sections for Naturally Occurring Elements[*]
2200 m/s (Thermal) Neutrons

atomic no.	element or compound	atomic or mol. wt.	density (g/cm^3)	microscopic cross section (barns)		
				σ_a	σ_s	σ_t
1	H	1.008	8.9[a]	0.33	38	38
	H$_2$O	18.016	1	0.66	103	103
	D$_2$O	20.030	1.10	0.001	13.6	13.6
2	He	4.003	17.8[a]	0.007	0.8	0.807
3	Li	6.940	0.534	71	1.4	72.4
4	Be	9.013	1.85	0.010	7.0	7.01
	BeO	25.02	3.025	0.010	6.8	6.8
5	B	10.82	2.45	755	4	759
6	C	12.011	1.60	0.004	4.8	4.80
7	N	14.008	0.0013	1.88	10	11.9
8	O	16.000	0.0014	20[a]	4.2	4.2
9	F	19.00	0.0017	0.001	3.9	3.90
10	Ne	20.183	0.0009	<2.8	2.4	5.2
11	Na	22.991	0.971	0.525	4	4.53
12	Mg	24.32	1.74	0.069	3.6	3.67
13	Al	26.98	2.699	0.241	1.4	1.64
14	Si	28.09	2.42	0.16	1.7	1.86
15	P	30.975	1.82	0.20	5	5.20
16	S	32.066	2.07	0.52	1.1	1.62
17	Cl	35.457	0.0032	33.8	16	49.8
18	Ar	39.944	0.0018	0.66	1.5	2.16
19	K	39.100	0.87	2.07	1.5	3.57
20	Ca	40.08	1.55	0.44	3.0	3.44
21	Sc	44.96	2.5	24	24	48
22	Ti	47.90	4.5	5.8	4	9.8
23	V	50.95	5.96	5	5	10.0
24	Cr	52.01	7.1	3.1	3	6.1
25	Mn	54.94	7.2	13.2	2.3	15.5
26	Fe	55.85	7.86	2.62	11	13.6
27	Co	58.94	8.9	38	7	45
28	Ni	58.71	8.90	4.6	17.5	22.1
29	Cu	63.54	8.94	3.85	7.2	11.05
30	Zn	65.38	7.14	1.10	3.6	4.70
31	Ga	69.72	5.91	2.80	4	6.80
32	Ge	72.60	5.36	2.45	3	5.45
33	As	74.91	5.73	4.3	6	10.3
34	Se	78.96	4.8	12.3	11	23.3
35	Br	79.916	3.12	6.7	6	12.7
36	Kr	83.80	0.0037	31	7.2	38.2
37	Rb	85.48	1.53	0.73	12	12.7
38	Sr	87.63	2.54	1.21	10	11.2
39	Yt	88.92	5.51	1.31	3	4.3
40	Zr	91.22	6.4	0.185	8	8.2
41	Nb	92.91	8.4	1.16	5	6.16
42	Mo	95.95	10.2	2.70	7	9.70
43	Tc	98	—	22	—	—
44	Ru	101.1	12.2	2.56	6	8.56
45	Rh	102.91	12.5	149	5	154
46	Pd	106.4	12.16	8	3.6	11.6
47	Ag	107.88	10.5	63	6	69
48	Cd	112.41	8.65	2450	7	2457

(continued)

APPENDIX 49.A *(continued)*
Cross Sections for Naturally Occurring Elements[*]
2200 m/s (Thermal) Neutrons

atomic no.	element or compound	atomic or mol. wt.	density (g/cm^3)	microscopic cross section (barns)		
				σ_a	σ_s	σ_t
49	In	114.82	7.28	191	2.2	193
50	Sn	118.70	6.5	0.625	4	4.6
51	Sb	121.76	6.69	5.7	4.3	10.0
52	Te	127.61	6.24	4.7	5	9.7
53	I	126.91	4.93	7.0	3.6	10.6
54	Xe	131.30	0.0059	35	4.3	39.3
55	Cs	132.91	1.873	28	20	48
56	Ba	137.36	3.5	1.2	8	9.2
57	La	138.92	6.19	8.9	15	24
58	Ce	140.13	6.78	0.73	9	9.7
59	Pr	140.92	6.78	11.3	4	15.3
60	Nd	144.27	6.95	46	16	62
61	Pm	145	–	60	–	–
62	Sm	150.35	7.7	5600	5	5605
	Sm$_2$O$_3$	348.70	7.43	16 500	22.6	16 500
63	Eu	152	5.22	4300	8	4308
	Eu$_2$O$_3$	352.00	7.42	8740	30.2	8770
64	Gd	157.26	7.95	46 000	–	–
65	Tb	158.93	8.33	46	–	–
66	Dy	162.51	8.56	950	100	1050
	Dy$_2$O$_3$	372.92	7.81	2200	214	2414
67	Ho	164.94	8.76	65	–	–
68	Er	167.27	9.16	173	15	188
69	Tm	168.94	9.35	127	7	134
70	Yb	173.04	7.01	37	12	49
71	Lu	174.99	9.74	112	–	–
72	Hf	178.5	13.3	105	8	113
73	Ta	180.95	16.6	21	5	26
74	W	183.86	19.3	19.2	5	24.2
75	Re	186.22	20.53	86	14	100
76	Os	190.2	22.48	15.3	11	26.3
77	Ir	192.2	22.42	440	–	–
78	Pt	195.09	21.37	8.8	10	18.8
79	Au	197	19.32	98.8	9.3	107.3
80	Hg	200.61	13.55	380	20	400
81	Tl	204.39	11.85	3.4	14	17.4
82	Pb	207.21	11.35	0.170	11	11.2
83	Bi	209	9.747	0.034	9	9
84	Po	210	9.24	–	–	–
85	At	211	–	–	–	–
86	Rn	222	0.0097	0.7	–	–
87	Fr	223	–	–	–	–
88	Ra	226.05	5	20	–	–
89	Ac	227	–	510	–	–
90	Th	232.05	11.3	7.56	12.6	20.2
91	Pa	231	15.4	200	–	–
92	U	238.07	18.9	7.68	8.3	16.0
	UO$_2$	270.07	10	7.6	16.7	24.3
93	Np	237	–	170	–	–
94	Pu	239	19.74	1026	9.6	1036
95	Am	242	–	8.000	–	–

(continued)

APPENDIX 49.A *(continued)*
Cross Sections for Naturally Occurring Elements[*]
2200 m/s (Thermal) Neutrons

atomic no.	element or compound	atomic or mol. wt.	density (g/cm^3)	microscopic cross section (barns)		
				σ_a	σ_s	σ_t
96	Cm	245	–	–	–	–
97	Bk	249	–	500	–	–
98	Cf	249	–	900	–	–
99	E	253	–	160	–	–
100	Fm	256	–	–	–	–
101	Mv	260	–	–	–	–

[*]Value has been multiplied by 10^5.

Source: "Reactor Physics Constants," 2nd ed., Argonne National Library, ANL-5800, U.S. Atomic Energy Commission, July 1963. See this source for a complete list of references.

APPENDIX 50.A
Noise Reduction and Absorption Coefficients

material	noise absorption coefficients						NRC
	125 Hz	250 Hz	500 Hz	1000 Hz	2000 Hz	4000 Hz	
brick, unglazed	0.03	0.03	0.03	0.04	0.05	0.07	0.04
brick, unglazed, painted	0.01	0.01	0.02	0.02	0.02	0.03	0.02
carpet							
$^1/_8$ in pile height	0.05	0.05	0.10	0.20	0.30	0.40	0.16
$^1/_4$ in pile height	0.05	0.10	0.15	0.30	0.50	0.55	0.26
$^3/_{16}$ in combined pile and foam	0.05	0.10	0.10	0.30	0.40	0.50	0.23
$^5/_{16}$ in combined pile and foam	0.05	0.15	0.30	0.40	0.50	0.60	0.34
concrete block, painted	0.10	0.05	0.06	0.07	0.09	0.08	0.07
fabrics							
light velour, 10 oz per sq yd, hung straight, in contact with wall	0.03	0.04	0.11	0.17	0.24	0.35	0.14
medium velour, 14 oz per sq yd, draped to half area	0.07	0.31	0.49	0.75	0.70	0.60	0.56
heavy velour, 18 oz per sq yd, draped to half area	0.14	0.35	0.55	0.72	0.70	0.65	0.62
floors							
concrete or terrazzo	0.01	0.01	0.01	0.02	0.02	0.02	0.02
linoleum, asphalt, rubber or cork tile on concrete	0.02	0.03	0.03	0.03	0.03	0.02	0.03
wood	0.15	0.11	0.10	0.07	0.06	0.07	0.09
wood parquet in asphalt on concrete	0.04	0.04	0.07	0.06	0.06	0.07	0.06
glass							
$^1/_4$ in, sealed, large panes	0.05	0.03	0.02	0.02	0.03	0.02	0.03
24 oz, operable windows (in closed condition)	0.10	0.05	0.04	0.03	0.03	0.03	0.04
gypsum board, $^1/_2$ in nailed to 2 × 4's 16 in o.c., painted	0.10	0.08	0.05	0.03	0.03	0.03	0.05
marble or glazed tile	0.01	0.01	0.01	0.01	0.02	0.02	0.01
plaster, gypsum or lime,							
rough finish or lath	0.02	0.03	0.04	0.05	0.04	0.03	0.04
same, with smooth finish	0.02	0.02	0.03	0.04	0.04	0.03	0.03
hardwood plywood paneling $^1/_4$ in thick, wood frame	0.58	0.22	0.07	0.04	0.03	0.07	0.09
water surface, as in a swimming pool	0.01	0.01	0.01	0.01	0.02	0.03	0.01
wood roof decking, tongue-and-groove cedar	0.24	0.19	0.14	0.08	0.13	0.10	0.14
air, sabins per 1000 cubic ft at 50% RH				0.9	2.3	7.2	
audience, seated, depending on spacing and upholstery of seats[*]	2.5–4.0	3.5–5.0	4.0–5.5	4.5–6.5	5.0–7.0	4.5–7.0	
seats, heavily upholstered with fabric[*]	1.5–3.5	3.5–4.5	4.0–5.0	4.0–5.5	3.5–5.5	3.5–4.5	
seats, heavily upholstered with leather, plastic, etc.[*]	2.5–3.5	3.0–4.5	3.0–4.0	2.0–4.0	1.5–3.5	1.0–3.0	
seats, lightly upholstered with leather, plastic, etc.[*]				1.5–2.0			
seats, wood veneer, no upholstery[*]	0.15	0.20	0.25	0.30	0.50	0.50	

[*]Values given are in sabins per person or unit of seating at the indicated frequency.

Derived from *Performance Data for Acoustical Materials Bulletin*, Acoustical and Board Products Association, 1975.

APPENDIX 50.B
Transmission Loss Through Common Materials
(decibels)

material	density per unit area (lbm/in^2)	125 Hz	250 Hz	500 Hz	1000 Hz	2000 Hz	4000 Hz	8000 Hz
lead								
$^1/_{32}$ in thick	2	22	24	29	33	40	43	49
$^1/_{64}$ in thick	1	19	20	24	27	33	39	43
plywood								
$^3/_4$ in thick	2	24	22	27	28	25	27	35
$^1/_4$ in thick	0.7	17	15	20	24	28	27	25
lead vinyl	0.5	11	12	15	20	26	32	37
lead vinyl	1.0	15	17	21	28	33	37	43
steel								
18-gauge	2.0	15	19	31	32	35	48	53
16-gauge	2.5	21	30	34	37	40	47	52
sheet metal (viscoelastic laminate core)	2	15	25	28	32	39	42	47
plexiglass								
$^1/_4$ in thick	1.45	16	17	22	28	33	35	35
$^1/_2$ in thick	2.9	21	23	26	32	32	37	37
1 in thick	5.8	25	28	32	32	34	46	46
glass								
$^1/_8$ in thick	1.5	11	17	23	25	26	27	28
$^1/_4$ in thick	3	17	23	25	27	28	29	30
double glass								
$^1/_4 \times ^1/_2 \times ^1/_4$ in		23	24	24	27	28	30	36
$^1/_4 \times 6 \times ^1/_4$ in		25	28	31	37	40	43	47
$^5/_8$ in gypsum								
on 2×2 in stud		23	28	33	43	50	49	50
on staggered stud		26	35	42	52	57	55	57
concrete, 4 in thick	48	29	35	37	43	44	50	55
concrete block, 6 in	36	33	34	35	38	46	52	65
panels of 16 gauge steel, 4 in absorbent, 20 gauge steel		25	35	43	48	52	55	56

(Multiply lbm/ft^2 by 4.882 to obtain kg/m^2.)

Reprinted from *Industrial Noise Control Manual*, HEW Publication NIOSH 75-183, 1975.

APPENDIX 51.A
Highly Hazardous Chemicals, Toxics and Reactives
(Standards—29 CFR 1910.119, App. A)

chemical name	CAS[a]	TQ[b]
Acetaldehyde	75-07-0	2500
Acrolein (2-Propenal)	107-02-8	150
Acrylyl Chloride	814-68-6	250
Allyl Chloride	107-05-1	1000
Allylamine	107-11-9	1000
Alkylaluminums	Varies	5000
Ammonia, Anhydrous	7664-41-7	10000
Ammonia solutions (greater than 44% ammonia by weight)	7664-41-7	15000
Ammonium Perchlorate	7790-98-9	7500
Ammonium Permanganate	7787-36-2	7500
Arsine (also called Arsenic Hydride)	7784-42-1	100
Bis(Chloromethyl) Ether	542-88-1	100
Boron Trichloride	10294-34-5	2500
Boron Trifluoride	7637-07-2	250
Bromine	7726-95-6	1500
Bromine Chloride	13863-41-7	1500
Bromine Pentafluoride	7789-30-2	2500
Bromine Trifluoride	7787-71-5	15000
3-Bromopropyne (also called Propargyl Bromide)	106-96-7	100
Butyl Hydroperoxide (Tertiary)	75-91-2	5000
Butyl Perbenzoate (Tertiary)	614-45-9	7500
Carbonyl Chloride (see Phosgene)	75-44-5	100
Carbonyl Fluoride	353-50-4	2500
Cellulose Nitrate (concentration greater than 12.6% nitrogen)	9004-70-0	2500
Chlorine	7782-50-5	1500
Chlorine Dioxide	10049-04-4	1000
Chlorine Pentafluoride	13637-63-3	1000
Chlorine Trifluoride	7790-91-2	1000
Chlorodiethylaluminum (also called Diethylaluminum Chloride)	96-10-6	5000
1-Chloro-2,4-Dinitrobenzene	97-00-7	5000
Chloromethyl Methyl Ether	107-30-2	500
Chloropicrin	76-06-2	500
Chloropicrin and Methyl Bromide mixture	None	1500
Chloropicrin and Methyl Chloride mixture	None	1500
Commune Hydroperoxide	80-15-9	5000
Cyanogen	460-19-5	2500
Cyanogen Chloride	506-77-4	500
Cyanuric Fluoride	675-14-9	100
Diacetyl Peroxide (concentration greater than 70%)	110-22-5	5000
Diazomethane	334-88-3	500
Dibenzoyl Peroxide	94-36-0	7500
Diborane	19287-45-7	100
Dibutyl Peroxide (Tertiary)	110-05-4	5000
Dichloro Acetylene	7572-29-4	250
Dichlorosilane	4109-96-0	2500
Diethylzinc	557-20-0	10000
Diisopropyl Peroxydicarbonate	105-64-6	7500

(continued)

APPENDIX 51.A *(continued)*
Highly Hazardous Chemicals, Toxics and Reactives
(Standards—29 CFR 1910.119, App. A)

chemical name	CAS^a	TQ^b
Dilauroyl Peroxide	105-74-8	7500
Dimethyldichlorosilane	75-78-5	1000
Dimethylhydrazine, 1,1-	57-14-7	1000
Dimethylamine, Anhydrous	124-40-3	2500
2,4-Dinitroaniline	97-02-9	5000
Ethyl Methyl Ketone Peroxide (also Methyl Ethyl Ketone Peroxide; concentration greater than 60%)	1338-23-4	5000
Ethyl Nitrite	109-95-5	5000
Ethylamine	75-04-7	7500
Ethylene Fluorohydrin	371-62-0	100
Ethylene Oxide	75-21-8	5000
Ethyleneimine	151-56-4	1000
Fluorine	7782-41-4	1000
Formaldehyde (Formalin)	50-00-0	1000
Furan	110-00-9	500
Hexafluoroacetone	684-16-2	5000
Hydrochloric Acid, Anhydrous	7647-01-0	5000
Hydrofluoric Acid, Anhydrous	7664-39-3	1000
Hydrogen Bromide	10035-10-6	5000
Hydrogen Chloride	7647-01-0	5000
Hydrogen Cyanide, Anhydrous	74-90-8	1000
Hydrogen Fluoride	7664-39-3	1000
Hydrogen Peroxide (52% by weight or greater)	7722-84-1	7500
Hydrogen Selenide	7783-07-5	150
Hydrogen Sulfide	7783-06-4	1500
Hydroxylamine	7803-49-8	2500
Iron, Pentacarbonyl	13463-40-6	250
Isopropylamine	75-31-0	5000
Ketene	463-51-4	100
Methacrylaldehyde	78-85-3	1000
Methacryloyl Chloride	920-46-7	150
Methacryloyloxyethyl Isocyanate	30674-80-7	100
Methyl Acrylonitrile	126-98-7	250
Methylamine, Anhydrous	74-89-5	1000
Methyl Bromide	74-83-9	2500
Methyl Chloride	74-87-3	15000
Methyl Chloroformate	79-22-1	500
Methyl Ethyl Ketone Peroxide (concentration greater than 60%)	1338-23-4	5000
Methyl Fluoroacetate	453-18-9	100
Methyl Fluorosulfate	421-20-5	100
Methyl Hydrazine	60-34-4	100
Methyl Iodide	74-88-4	7500
Methyl Isocyanate	624-83-9	250
Methyl Mercaptan	74-93-1	5000
Methyl Vinyl Ketone	79-84-4	100
Methyltrichlorosilane	75-79-6	500
Nickel Carbonly (Nickel Tetracarbonyl)	13463-39-3	150
Nitric Acid (94.5% by weight or greater)	7697-37-2	500
Nitric Oxide	10102-43-9	250

(continued)

APPENDIX 51.A *(continued)*
Highly Hazardous Chemicals, Toxics and Reactives
(Standards—29 CFR 1910.119, App. A)

chemical name	CAS[a]	TQ[b]
Nitroaniline (para Nitroaniline)	100-01-6	5000
Nitromethane	75-52-5	2500
Nitrogen Dioxide	10102-44-0	250
Nitrogen Oxides (NO; NO(2); N2O4; N2O3)	10102-44-0	250
Nitrogen Tetroxide (also called Nitrogen Peroxide)	10544-72-6	250
Nitrogen Trifluoride	7783-54-2	5000
Nitrogen Trioxide	10544-73-7	250
Oleum (65% to 80% by weight; also called Fuming Sulfuric Acid)	8014-94-7	1000
Osmium Tetroxide	20816-12-0	100
Oxygen Difluoride (Fluorine Monoxide)	7783-41-7	100
Ozone	10028-15-6	100
Pentaborane	19624-22-7	100
Peracetic Acid (concentration greater than 60% Acetic Acid; also called Peroxyacetic Acid)	79-21-0	1000
Perchloric Acid (concentration greater than 60% by weight)	7601-90-3	5000
Perchloromethyl Mercaptan	594-42-3	150
Perchloryl Fluoride	7616-94-6	5000
Peroxyacetic Acid (concentration greater than 60% Acetic Acid; also called Peracetic Acid)	79-21-0	1000
Phosgene (also called Carbonyl Chloride)	75-44-5	100
Phosphine (Hydrogen Phosphide)	7803-51-2	100
Phosphorus Oxychloride (also called Phosphoryl Chloride)	10025-87-3	1000
Phosphorus Trichloride	7719-12-2	1000
Phosphoryl Chloride (also called Phosphorus Oxychloride)	10025-87-3	1000
Propargyl Bromide	106-96-7	100
Propyl Nitrate	627-3-4	2500
Sarin	107-44-8	100
Selenium Hexafluoride	7783-79-1	1000
Stibine (Antimony Hydride)	7803-52-3	500
Sulfur Dioxide (liquid)	7446-09-5	1000
Sulfur Pentafluoride	5714-22-7	250
Sulfur Tetrafluoride	7783-60-0	250
Sulfur Trioxide (also called Sulfuric Anhydride)	7446-11-9	1000
Sulfuric Anhydride (also called Sulfur Trioxide)	7446-11-9	1000
Tellurium Hexafluoride	7783-80-4	250
Tetrafluoroethylene	116-14-3	5000
Tetrafluorohydrazine	10036-47-2	5000
Tetramethyl Lead	75-74-1	1000
Thionyl Chloride	7719-09-7	250
Trichloro (chloromethyl) Silane	1558-25-4	100
Trichloro (dichlorophenyl) Silane	27137-85-5	2500
Trichlorosilane	10025-78-2	5000

(continued)

APPENDIX 51.A *(continued)*
Highly Hazardous Chemicals, Toxics and Reactives
(Standards—29 CFR 1910.119, App. A)

chemical name	CAS[a]	TQ[b]
Trifluorochloroethylene	79-38-9	10000
Trimethyoxysilane	2487-90-3	1500

[a]Chemical Abstract Service Number
[b]Threshold Quantity in Pounds (amount necessary to be covered by this standard).

APPENDIX 52.A
Polyphase Motor Classifications and Characteristics

speed regulations	speed control	starting torque	breakdown torque	application
general-purpose squirrel cage (NEMA design B)				
Drops about 3% for large to 5% for small sizes.	None, except multispeed types, designed for two to four fixed speeds.	100% for large; 275% for 1 hp 4 pole unit.	200% of full load.	Constant-speed service where starting is not excessive. Fans, blowers, rotary compressors, and centrifugal pumps.
high-torque squirrel cage (NEMA design C)				
Drops about 3% for large to 6% for small sizes.	None, except multispeed types, designed for two and four fixed speeds.	250% of full load for high-speed to 200% for low-speed designs.	200% of full load.	Constant-speed where fairly high starting torque is required infrequently with starting current about 550% of full load. Reciprocating pumps and compressors, crushers, etc.
high-slip squirrel cage (NEMA design D)				
Drops about 10% to 15% from no load to full load.	None, except multispeed types, designed for two to four fixed speeds.	225% to 300% full load, depending on speed with rotor resistance.	200%. Will usually not stall until loaded to maximum torque, which occurs at standstill.	Constant-speed and high starting torque, if starting is not too frequent, and for high-peak loads with or without flywheels. Punch presses, shears, elevators, etc.
low-torque squirrel cage (NEMA design F)				
Drops about 3% for large to 5% for small sizes.	None, except multispeed types, designed for two to four fixed speeds.	50% of full load for high-speed to 90% for low-speed designs.	135% to 170% of full load.	Constant-speed service where starting duty is light. Fans, blowers, centrifugal pumps, and similar loads.
wound rotor				
With rotor rings short circuited, drops about 3% for large to 5% for small sizes.	Speed can be reduced to 50% by rotor resistance. Speed varies inversely as load.	Up to 300% depending on external resistance in rotor circuit and how distributed.	300% when rotor slip rings are short circuited.	Where high starting torque with low starting current or where limited speed control is required. Fans, centrifugal and plunger pumps, compressors, conveyors, hoists, cranes, etc.
synchronous				
Constant.	None, except special motors designed for two fixed speeds.	40% for slow to 160% for medium-speed 80% pf. Specials develop higher.	Unity-pf motors 170%, 80%-pf motors 225%. Specials, up to 300%	For constant-speed service, direct connection to slow-speed machines and where pf correction is required.

Adapted from *Mechanical Engineering, Design Manual*, NAVFAC DM-3, Department of the Navy, copyright © 1972.

APPENDIX 52.B
DC and Single-Phase Motor Classifications and Characteristics

speed regulations	speed control	starting torque	breakdown torque	application
series				
Varies inversely as load. Races on light loads and full voltage.	Zero to maximum depending on control and load.	High. Varies as square of voltage. Limited by commutation, heating, and line capacity.	High. Limited by commutation, heating, and line capacity.	Where high starting torque is required and speed can be regulated. Traction, bridges, hoists, gates, car dumpers, car retarders.
shunt				
Drops 3% to 5% from no load to full load.	Any desired range depending on design, type of system.	Good. With constant field, varies directly as voltage applied to armature.	High. Limited by commutation, heating, and line capacity.	Where constant or adjustable speed is required and starting conditions are not severe. Fans, blowers, centrifugal pumps, conveyors, wood and metal-working machines, elevators.
compound				
Drops 7% to 20% from no load to full load depending on amount of compounding.	Any desired range, depending on design, type of control.	Higher than for shunt, depending on amount of compounding.	High. Limited by commutation, heating and line capacity.	Where high starting torque and fairly constant speed is required. Plunger pumps, punch presses, shears, bending rolls, geared elevators, conveyors, hoists.
split-phase				
Drops about 10% from no load to full load.	None.	75% for large to 175% for small sizes.	150% for large to 200% for small sizes.	Constant-speed service where starting is easy. Small fans, centrifugal pumps and light-running machines, where polyphase is not available.
capacitors				
Drops about 5% for large to 10% for small sizes.	None.	150% to 350% of full load depending on design and size.	50% for large to 200% for small sizes.	Constant-speed service for any starting duty and quiet operation, where polyphase current cannot be used.
commutator type				
Drops about 5% for large to 10% for small sizes.	Repulsion induction, none. Brush-shifting types, four to one at full load.	250% for large to 350% for small sizes.	150% for large to 250%.	Constant-speed service for any starting duty where speed control is required and polyphase current cannot be used.

Adapted from *Mechanical Engineering, Design Manual*, NAVFAC DM-3, Department of the Navy, copyright © 1972.

APPENDIX 53.A
Thermoelectric Constants for Thermocouples[*]
(mV, reference 32°F (0°C))

(a) chromel-alumel (type K)

°F	0	10	20	30	40	50	60	70	80	90
−300	−5.51	−5.60			millivolts					
−200	−4.29	−4.44	−4.58	−4.71	−4.84	−4.96	−5.08	−5.20	−5.30	−5.41
−100	−2.65	−2.83	−3.01	−3.19	−3.36	−3.52	−3.69	−3.84	−4.00	−4.15
−0	−0.68	−0.89	−1.10	−1.30	−1.50	−1.70	−1.90	−2.09	−2.28	−2.47
+0	−0.68	−0.49	−0.26	−0.04	0.18	0.40	0.62	0.84	1.06	1.29
100	1.52	1.74	1.97	2.20	2.43	2.66	2.89	3.12	3.36	3.59
200	3.82	4.05	4.28	4.51	4.74	4.97	5.20	5.42	5.65	5.87
300	6.09	6.31	6.53	6.76	6.98	7.20	7.42	7.64	7.87	8.09
400	8.31	8.54	8.76	8.98	9.21	9.43	9.66	9.88	10.11	10.34
500	10.57	10.79	11.02	11.25	11.48	11.71	11.94	12.17	12.40	12.63
600	12.86	13.09	13.32	13.55	13.78	14.02	14.25	14.48	14.71	14.95
700	15.18	15.41	15.65	15.88	16.12	16.35	16.59	16.82	17.06	17.29
800	17.53	17.76	18.00	18.23	18.47	18.70	18.94	19.18	19.41	19.65
900	19.89	20.13	20.36	20.60	20.84	21.07	21.31	21.54	21.78	22.02
1000	22.26	22.49	22.73	22.97	23.20	23.44	23.68	23.91	24.15	24.39
1100	24.63	24.86	25.10	25.34	25.57	25.81	26.05	26.28	26.52	26.75
1200	26.98	27.22	27.45	27.69	27.92	28.15	28.39	28.62	28.86	29.09
1300	29.32	29.56	29.79	30.02	30.25	30.49	30.72	30.95	31.18	31.42
1400	31.65	31.88	32.11	32.34	32.57	32.80	33.02	33.25	33.48	33.71
1500	33.93	34.16	34.39	34.62	34.84	35.07	35.29	35.52	35.75	35.97
1600	36.19	36.42	36.64	36.87	37.09	37.31	37.54	37.76	37.98	38.20
1700	38.43	38.65	38.87	39.09	39.31	39.53	39.75	39.96	40.18	40.40
1800	40.62	40.84	41.05	41.27	41.49	41.70	41.92	42.14	42.35	42.57
1900	42.78	42.99	43.21	43.42	43.63	43.85	44.06	44.27	44.49	44.70
2000	44.91	45.12	45.33	45.54	45.75	45.96	46.17	46.38	46.58	46.79

(b) iron-constantan (type J)

°F	0	10	20	30	40	50	60	70	80	90
−300	−7.52	−7.66			millivolts					
−200	−5.76	−5.96	−6.16	−6.35	−6.53	−6.71	−6.89	−7.06	−7.22	−7.38
−100	−3.49	−3.73	−3.97	−4.21	−4.44	−4.68	−4.90	−5.12	−5.34	−5.55
−0	−0.89	−1.16	−1.43	−1.70	−1.96	−2.22	−2.48	−2.74	−2.99	−3.24
+0	−0.89	−0.61	−0.34	−0.06	0.22	0.50	0.79	1.07	1.36	1.65
100	1.94	2.23	2.52	2.82	3.11	3.41	3.71	4.01	4.31	4.61
200	4.91	5.21	5.51	5.81	6.11	6.42	6.72	7.03	7.33	7.64
300	7.94	8.25	8.56	8.87	9.17	9.48	9.79	10.10	10.41	10.72
400	11.03	11.34	11.65	11.96	12.26	12.57	12.88	13.19	13.50	13.81
500	14.12	14.42	14.73	15.04	15.34	15.65	15.96	16.26	16.57	16.88
600	17.18	17.49	17.80	18.11	18.41	18.72	19.03	19.34	19.64	19.95
700	20.26	20.56	20.87	21.18	21.48	21.79	22.10	22.40	22.71	23.01
800	23.32	23.63	23.93	24.24	24.55	24.85	25.16	25.47	25.78	26.09
900	26.40	26.70	27.02	27.33	27.64	27.95	28.26	28.58	28.89	29.21
1000	29.52	29.84	30.16	30.48	30.80	31.12	31.44	31.76	32.08	32.40
1100	32.72	33.05	33.37	33.70	34.03	34.36	34.68	35.01	35.35	35.68
1200	36.01	36.35	36.69	37.02	37.36	37.71	38.05	38.39	38.74	39.08
1300	39.43	39.78	40.13	40.48	40.83	41.19	41.54	41.90	42.25	42.61

(continued)

APPENDIX 53.A *(continued)*
Thermoelectric Constants for Thermocouples[*]
(mV, reference 32°F (0°C))

(c) copper-constantan (type T)

°F	0	10	20	30	40	50	60	70	80	90
−300	−5.284	−5.379			millivolts					
−200	−4.111	−4.246	−4.377	−4.504	−4.627	−4.747	−4.863	−4.974	−5.081	−5.185
−100	−2.559	−2.730	−2.897	−3.062	−3.223	−3.380	−3.533	−3.684	−3.829	−3.972
−0	−0.670	−0.872	−1.072	−1.270	−1.463	−1.654	−1.842	−2.026	−2.207	−2.385
+0	−0.670	−0.463	−0.254	−0.042	0.171	0.389	0.609	0.832	1.057	1.286
100	1.517	1.751	1.987	2.226	2.467	2.711	2.958	3.207	3.458	3.712
200	3.967	4.225	4.486	4.749	5.014	5.280	5.550	5.821	6.094	6.370
300	6.647	6.926	7.208	7.491	7.776	8.064	8.352	8.642	8.935	9.229
400	9.525	9.823	10.123	10.423	10.726	11.030	11.336	11.643	11.953	12.263
500	12.575	12.888	13.203	13.520	13.838	14.157	14.477	14.799	15.122	15.447

[*]This appendix is included to support general study and noncritical applications. *NBS Circular 561* has been superseded by *NBS Monograph 125, Thermocouple Reference Tables Based on the IPTS-68: Reference Tables in degrees Fahrenheit for Thermoelements versus Platinum,* presenting the same information as British Standards Institution standard *B.S. 4937.* In the spirit of globalization, these have been superseded in other countries by *IEC 584-1 (60584-1) Thermocouples Part 1: Reference Tables,* published in 1995. In addition to being based on modern correlations and the 1990 International Temperature Scale (ITS), the 160-page IEC 60584-1 duplicates all text in English and French, refers only to the Celsius temperature scale, and is available only by purchase or licensing.

... (truncated)

APPENDIX 54.A
Standard Cash Flow Factors

	multiply	by	to obtain	

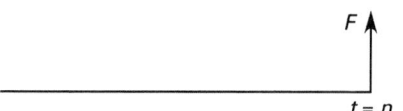

$$P = F(1+i)^{-n}$$
F $\quad (P/F, i\%, n) \quad$ P

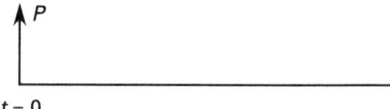

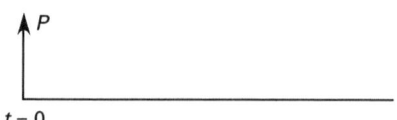

$$F = P(1+i)^n$$
P $\quad (F/P, i\%, n) \quad$ F

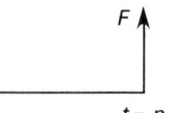

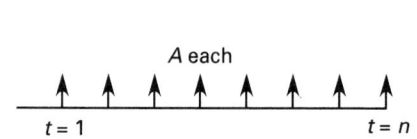

$$P = A\left(\frac{(1+i)^n - 1}{i(1+i)^n}\right)$$
A $\quad (P/A, i\%, n) \quad$ P

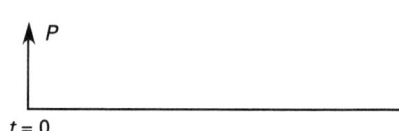

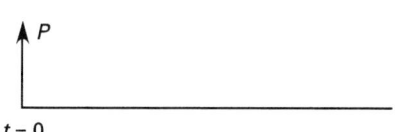

$$A = P\left(\frac{i(1+i)^n}{(1+i)^n - 1}\right)$$
P $\quad (A/P, i\%, n) \quad$ A

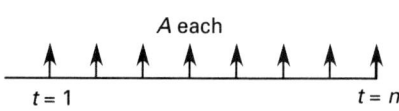

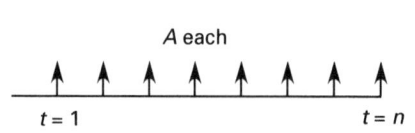

$$F = A\left(\frac{(1+i)^n - 1}{i}\right)$$
A $\quad (F/A, i\%, n) \quad$ F

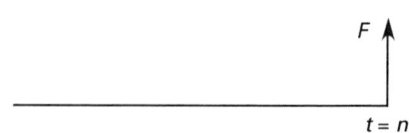

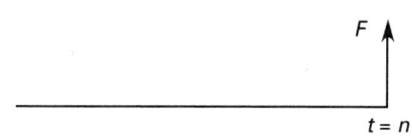

$$A = F\left(\frac{i}{(1+i)^n - 1}\right)$$
F $\quad (A/F, i\%, n) \quad$ A

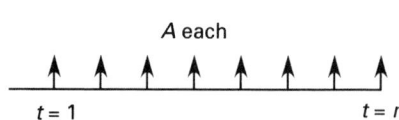

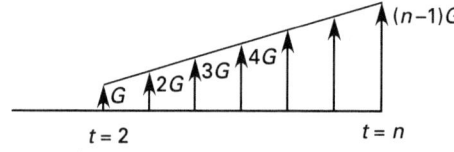

$$P = G\left(\frac{(1+i)^n - 1}{i^2(1+i)^n} - \frac{n}{i(1+i)^n}\right)$$
G $\quad (P/G, i\%, n) \quad$ P

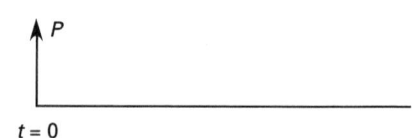

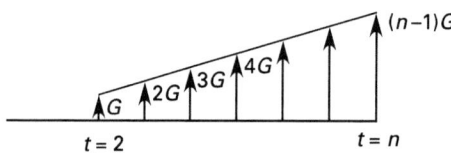

$$A = G\left(\frac{1}{i} - \frac{n}{(1+i)^n - 1}\right)$$
G $\quad (A/G, i\%, n) \quad$ A

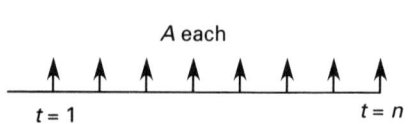

APPENDIX 54.B
Cash Flow Equivalent Factors

$i = 0.50\%$

n	P/F	P/A	P/G	F/P	F/A	A/P	A/F	A/G	n
1	0.9950	0.9950	0.0000	1.0050	1.0000	1.0050	1.0000	0.0000	1
2	0.9901	0.9851	0.9901	1.0100	2.0050	0.5038	0.4988	0.4988	2
3	0.9851	2.9702	2.9604	1.0151	3.0150	0.3367	0.3317	0.9967	3
4	0.9802	3.9505	5.9011	1.0202	4.0301	0.2531	0.2481	1.4938	4
5	0.9754	4.9259	9.8026	1.0253	5.0503	0.2030	0.1980	1.9900	5
6	0.9705	5.8964	14.6552	1.0304	6.0755	0.1696	0.1646	2.4855	6
7	0.9657	6.8621	20.4493	1.0355	7.1059	0.1457	0.1407	2.9801	7
8	0.9609	7.8230	27.1755	1.0407	8.1414	0.1278	0.1228	3.4738	8
9	0.9561	8.7791	34.8244	1.0459	9.1821	0.1139	0.1089	3.9668	9
10	0.9513	9.7304	43.3865	1.0511	10.2280	0.1028	0.0978	4.4589	10
11	0.9466	10.6770	52.8526	1.0564	11.2792	0.0937	0.0887	4.9501	11
12	0.9419	11.6189	63.2136	1.0617	12.3356	0.0861	0.0811	5.4406	12
13	0.9372	12.5562	74.4602	1.0670	13.3972	0.0796	0.0746	5.9302	13
14	0.9326	13.4887	86.5835	1.0723	14.4642	0.0741	0.0691	6.4190	14
15	0.9279	14.4166	99.5743	1.0777	15.5365	0.0694	0.0644	6.9069	15
16	0.9233	15.3399	113.4238	1.0831	16.6142	0.0652	0.0602	7.3940	16
17	0.9187	16.2586	128.1231	1.0885	17.6973	0.0615	0.0565	7.8803	17
18	0.9141	17.1728	143.6634	1.0939	18.7858	0.0582	0.0532	8.3658	18
19	0.9096	18.0824	160.0360	1.0994	19.8797	0.0553	0.0503	8.8504	19
20	0.9051	18.9874	177.2322	1.1049	20.9791	0.0527	0.0477	9.3342	20
21	0.9006	19.8880	195.2434	1.1104	22.0840	0.0503	0.0453	9.8172	21
22	0.8961	20.7841	214.0611	1.1160	23.1944	0.0481	0.0431	10.2993	22
23	0.8916	21.6757	233.6768	1.1216	24.3104	0.0461	0.0411	10.7806	23
24	0.8872	22.5629	254.0820	1.1272	25.4320	0.0443	0.0393	11.2611	24
25	0.8828	23.4456	275.2686	1.1328	26.5591	0.0427	0.0377	11.7407	25
26	0.8784	24.3240	297.2281	1.1385	27.6919	0.0411	0.0361	12.2195	26
27	0.8740	25.1980	319.9523	1.1442	28.8304	0.0397	0.0347	12.6975	27
28	0.8697	26.0677	343.4332	1.1499	29.9745	0.0384	0.0334	13.1747	28
29	0.8653	26.9330	367.6625	1.1556	31.1244	0.0371	0.0321	13.6510	29
30	0.8610	27.7941	392.6324	1.1614	32.2800	0.0360	0.0310	14.1265	30
31	0.8567	28.6508	418.3348	1.1672	33.4414	0.0349	0.0299	14.6012	31
32	0.8525	29.5033	444.7618	1.1730	34.6086	0.0339	0.0289	15.0750	32
33	0.8482	30.3515	471.9055	1.1789	35.7817	0.0329	0.0279	15.5480	33
34	0.8440	31.1955	499.7583	1.1848	36.9606	0.0321	0.0271	16.0202	34
35	0.8398	32.0354	528.3123	1.1907	38.1454	0.0312	0.0262	16.4915	35
36	0.8356	32.8710	557.5598	1.1967	39.3361	0.0304	0.0254	16.9621	36
37	0.8315	33.7025	587.4934	1.2027	40.5328	0.0297	0.0247	17.4317	37
38	0.8274	34.5299	618.1054	1.2087	41.7354	0.0290	0.0240	17.9006	38
39	0.8232	35.3531	649.3883	1.2147	42.9441	0.0283	0.0233	18.3686	39
40	0.8191	36.1722	681.3347	1.2208	44.1588	0.0276	0.0226	18.8359	40
41	0.8151	36.9873	713.9372	1.2269	45.3796	0.0270	0.0220	19.3022	41
42	0.8110	37.7983	747.1886	1.2330	46.6065	0.0265	0.0215	19.7678	42
43	0.8070	38.6053	781.0815	1.2392	47.8396	0.0259	0.0209	20.2325	43
44	0.8030	39.4082	815.6087	1.2454	49.0788	0.0254	0.0204	20.6964	44
45	0.7990	40.2072	850.7631	1.2516	50.3242	0.0249	0.0199	21.1595	45
46	0.7950	41.0022	886.5376	1.2579	51.5758	0.0244	0.0194	21.6217	46
47	0.7910	41.7932	922.9252	1.2642	52.8337	0.0239	0.0189	22.0831	47
48	0.7871	42.5803	959.9188	1.2705	54.0978	0.0235	0.0185	22.5437	48
49	0.7832	43.3635	997.5116	1.2768	55.3683	0.0231	0.0181	23.0035	49
50	0.7793	44.1428	1035.6966	1.2832	56.6452	0.0227	0.0177	23.4624	50
51	0.7754	44.9182	1074.4670	1.2896	57.9284	0.0223	0.0173	23.9205	51
52	0.7716	45.6897	1113.8162	1.2961	59.2180	0.0219	0.0169	24.3778	52
53	0.7677	46.4575	1153.7372	1.3026	60.5141	0.0215	0.0165	24.8343	53
54	0.7639	47.2214	1194.2236	1.3091	61.8167	0.0212	0.0162	25.2899	54
55	0.7601	47.9814	1235.2686	1.3156	63.1258	0.0208	0.0158	25.7447	55
60	0.7414	51.7256	1448.6458	1.3489	69.7700	0.0193	0.0143	28.0064	60
65	0.7231	55.3775	1675.0272	1.3829	76.5821	0.0181	0.0131	30.2475	65
70	0.7053	58.9394	1913.6427	1.4178	83.5661	0.0170	0.0120	32.4680	70
75	0.6879	62.4136	2163.7525	1.4536	90.7265	0.0160	0.0110	34.6679	75
80	0.6710	65.8023	2424.6455	1.4903	98.0677	0.0152	0.0102	36.8474	80
85	0.6545	69.1075	2695.6389	1.5280	105.5943	0.0145	0.0095	39.0065	85
90	0.6383	72.3313	2976.0769	1.5666	113.3109	0.0138	0.0088	41.1451	90
95	0.6226	75.4757	3265.3298	1.6061	121.2224	0.0132	0.0082	43.2633	95
100	0.6073	78.5426	3562.7934	1.6467	129.3337	0.0127	0.0077	45.3613	100

(continued)

APPENDIX 54.B *(continued)*
Cash Flow Equivalent Factors

$i = 0.75\%$

n	P/F	P/A	P/G	F/P	F/A	A/P	A/F	A/G	n
1	0.9926	0.9926	0.0000	1.0075	1.0000	1.0075	1.0000	0.0000	1
2	0.9852	1.9777	0.9852	1.0151	2.0075	0.5056	0.4981	0.4981	2
3	0.9778	2.9556	2.9408	1.0227	3.0226	0.3383	0.3308	0.9950	3
4	0.9706	3.9261	5.8525	1.0303	4.0452	0.2547	0.2472	1.4907	4
5	0.9633	4.8894	9.7058	1.0381	5.0756	0.2045	0.1970	1.9851	5
6	0.9562	5.8456	14.4866	1.0459	6.1136	0.1711	0.1636	2.4782	6
7	0.9490	6.7946	20.1808	1.0537	7.1595	0.1472	0.1397	2.9701	7
8	0.9420	7.7366	26.7747	1.0616	8.2132	0.1293	0.1218	3.4608	8
9	0.9350	8.6716	34.2544	1.0696	9.2748	0.1153	0.1078	3.9502	9
10	0.9280	9.5996	42.6064	1.0776	10.3443	0.1042	0.0967	4.4384	10
11	0.9211	10.5207	51.8174	1.0857	11.4219	0.0951	0.0876	4.9253	11
12	0.9142	11.4349	61.8740	1.0938	12.5076	0.0875	0.0800	5.4110	12
13	0.9074	12.3423	72.7632	1.1020	13.6014	0.0810	0.0735	5.8954	13
14	0.9007	13.2430	84.4720	1.1103	14.7034	0.0755	0.0680	6.3786	14
15	0.8940	14.1370	96.9876	1.1186	15.8137	0.0707	0.0632	6.8606	15
16	0.8873	15.0243	110.2973	1.1270	16.9323	0.0666	0.0591	7.3413	16
17	0.8807	15.9050	124.3887	1.1354	18.0593	0.0629	0.0554	7.8207	17
18	0.8742	16.7792	139.2494	1.1440	19.1947	0.0596	0.0521	8.2989	18
19	0.8676	17.6468	154.8671	1.1525	20.3387	0.0567	0.0492	8.7759	19
20	0.8612	18.5080	171.2297	1.1612	21.4912	0.0540	0.0465	9.2516	20
21	0.8548	19.3628	188.3253	1.1699	22.6524	0.0516	0.0441	9.7261	21
22	0.8484	20.2112	206.1420	1.1787	23.8223	0.0495	0.0420	10.1994	22
23	0.8421	21.0533	224.6682	1.1875	25.0010	0.0475	0.0400	10.6714	23
24	0.8358	21.8891	243.8923	1.1964	26.1885	0.0457	0.0382	11.1422	24
25	0.8296	22.7188	263.8029	1.2054	27.3849	0.0440	0.0365	11.6117	25
26	0.8234	23.5422	284.3888	1.2144	28.5903	0.0425	0.0350	12.0800	26
27	0.8173	24.3595	305.6387	1.2235	29.8047	0.0411	0.0336	12.5470	27
28	0.8112	25.1707	327.5416	1.2327	31.0282	0.0397	0.0322	13.0128	28
29	0.8052	25.9759	350.0867	1.2420	32.2609	0.0385	0.0310	13.4774	29
30	0.7992	26.7751	373.2631	1.2513	33.5029	0.0373	0.0298	13.9407	30
31	0.7932	27.5683	397.0602	1.2607	34.7542	0.0363	0.0288	14.4028	31
32	0.7873	28.3557	421.4675	1.2701	36.0148	0.0353	0.0278	14.8636	32
33	0.7815	29.1371	446.4746	1.2796	37.2849	0.0343	0.0268	15.3232	33
34	0.7757	29.9128	472.0712	1.2892	38.5646	0.0334	0.0259	15.7816	34
35	0.7699	30.6827	498.2471	1.2989	39.8538	0.0326	0.0251	16.2387	35
36	0.7641	31.4468	524.9924	1.3086	41.1527	0.0318	0.0243	16.6946	36
37	0.7585	32.2053	552.2969	1.3185	42.4614	0.0311	0.0236	17.1493	37
38	0.7528	32.9581	580.1511	1.3283	43.7798	0.0303	0.0228	17.6027	38
39	0.7472	33.7053	608.5451	1.3383	45.1082	0.0297	0.0222	18.0549	39
40	0.7416	34.4469	637.4693	1.3483	46.4465	0.0290	0.0215	18.5058	40
41	0.7361	35.1831	666.9144	1.3585	47.7948	0.0284	0.0209	18.9556	41
42	0.7306	35.9137	696.8709	1.3686	49.1533	0.0278	0.0203	19.4040	42
43	0.7252	36.6389	727.3297	1.3789	50.5219	0.0273	0.0198	19.8513	43
44	0.7198	37.3587	758.2815	1.3893	51.9009	0.0268	0.0193	20.2973	44
45	0.7145	38.0732	789.7173	1.3997	53.2901	0.0263	0.0188	20.7421	45
46	0.7091	38.7823	821.6283	1.4102	54.6898	0.0258	0.0183	21.1856	46
47	0.7039	39.4862	854.0056	1.4207	56.1000	0.0253	0.0178	21.6280	47
48	0.6986	40.1848	886.8404	1.4314	57.5207	0.0249	0.0174	22.0691	48
49	0.6934	40.8782	920.1243	1.4421	58.9521	0.0245	0.0170	22.5089	49
50	0.6883	41.5664	953.8486	1.4530	60.3943	0.0241	0.0166	22.9476	50
51	0.6831	42.2496	988.0050	1.4639	61.8472	0.0237	0.0162	23.3850	51
52	0.6780	42.9276	1022.5852	1.4748	63.3111	0.0233	0.0158	23.8211	52
53	0.6730	43.6006	1057.5810	1.4859	64.7859	0.0229	0.0154	24.2561	53
54	0.6680	44.2686	1092.9842	1.4970	66.2718	0.0226	0.0151	24.6898	54
55	0.6630	44.9316	1128.7869	1.5083	67.7688	0.0223	0.0148	25.1223	55
60	0.6387	48.1734	1313.5189	1.5657	75.4241	0.0208	0.0133	27.2665	60
65	0.6153	51.2963	1507.0910	1.6253	83.3709	0.0195	0.0120	29.3801	65
70	0.5927	54.3046	1708.6065	1.6872	91.6201	0.0184	0.0109	31.4634	70
75	0.5710	57.2027	1917.2225	1.7514	100.1833	0.0175	0.0100	33.5163	75
80	0.5500	59.9944	2132.1472	1.8180	109.0725	0.0167	0.0092	35.5391	80
85	0.5299	62.6838	2352.6375	1.8873	118.3001	0.0160	0.0085	37.5318	85
90	0.5104	65.2746	2577.9961	1.9591	127.8790	0.0153	0.0078	39.4946	90
95	0.4917	67.7704	2807.5694	2.0337	137.8225	0.0148	0.0073	41.4277	95
100	0.4737	70.1746	3040.7453	2.1111	148.1445	0.0143	0.0068	43.3311	100

(continued)

APPENDIX 54.B *(continued)*
Cash Flow Equivalent Factors

$$i = 1.00\%$$

n	P/F	P/A	P/G	F/P	F/A	A/P	A/F	A/G	n
1	0.9901	0.9901	0.0000	1.0100	1.0000	1.0100	1.0000	0.0000	1
2	0.9803	1.9704	0.9803	1.0201	2.0100	0.5075	0.4975	0.4975	2
3	0.9706	2.9410	2.9215	1.0303	3.0301	0.3400	0.3300	0.9934	3
4	0.9610	3.9020	5.8044	1.0406	4.0604	0.2563	0.2463	1.4876	4
5	0.9515	4.8534	9.6103	1.0510	5.1010	0.2060	0.1960	1.9801	5
6	0.9420	5.7955	14.3205	1.0615	6.1520	0.1725	0.1625	2.4710	6
7	0.9327	6.7282	19.0168	1.0721	7.2135	0.1480	0.1386	2.9602	7
8	0.9235	7.6517	26.3812	1.0829	8.2857	0.1307	0.1207	3.4478	8
9	0.9143	8.5660	33.6959	1.0937	9.3685	0.1167	0.1067	3.9337	9
10	0.9053	9.4713	41.8435	1.1046	10.4622	0.1056	0.0956	4.4179	10
11	0.8963	10.3676	50.8067	1.1157	11.5668	0.0965	0.0865	4.9005	11
12	0.8874	11.2551	60.5687	1.1268	12.6825	0.0888	0.0788	5.3815	12
13	0.8787	12.1337	71.1126	1.1381	13.8093	0.0824	0.0724	5.8607	13
14	0.8700	13.0037	82.4221	1.1495	14.9474	0.0769	0.0669	6.3384	14
15	0.8613	13.8651	94.4810	1.1610	16.0969	0.0721	0.0621	6.8143	15
16	0.8528	14.7179	107.2734	1.1726	17.2579	0.0679	0.0579	7.2886	16
17	0.8444	15.5623	120.7834	1.1843	18.4304	0.0643	0.0543	7.7613	17
18	0.8360	16.3983	134.9957	1.1961	19.6147	0.0610	0.0510	8.2323	18
19	0.8277	17.2260	149.8950	1.2081	20.8109	0.0581	0.0481	8.7017	19
20	0.8195	18.0456	165.4664	1.2202	22.0190	0.0554	0.0454	9.1694	20
21	0.8114	18.8570	181.6950	1.2324	23.2392	0.0530	0.0430	9.6354	21
22	0.8034	19.6604	198.5663	1.2447	24.4716	0.0509	0.0409	10.0998	22
23	0.7954	20.4558	216.0660	1.2572	25.7163	0.0489	0.0389	10.5626	23
24	0.7876	21.2434	234.1800	1.2697	26.9735	0.0471	0.0371	11.0237	24
25	0.7798	22.0232	252.8945	1.2824	28.2432	0.0454	0.0354	11.4831	25
26	0.7720	22.7952	272.1957	1.2953	29.5256	0.0439	0.0339	11.9409	26
27	0.7644	23.5596	292.0702	1.3082	30.8209	0.0424	0.0324	12.3971	27
28	0.7568	24.3164	312.5047	1.3213	32.1291	0.0411	0.0311	12.8516	28
29	0.7493	25.0658	333.4863	1.3345	33.4504	0.0399	0.0299	13.3044	29
30	0.7419	25.8077	355.0021	1.3478	34.7849	0.0387	0.0287	13.7557	30
31	0.7346	26.5423	377.0394	1.3613	36.1327	0.0377	0.0277	14.2052	31
32	0.7273	27.2696	399.5858	1.3749	37.4941	0.0367	0.0267	14.6532	32
33	0.7201	27.9897	422.6291	1.3887	38.8690	0.0357	0.0257	15.0995	33
34	0.7130	28.7027	446.1572	1.4026	40.2577	0.0348	0.0248	15.5441	34
35	0.7059	29.4086	470.1583	1.4166	41.6603	0.0340	0.0240	15.9871	35
36	0.6989	30.1075	494.6207	1.4308	43.0769	0.0332	0.0232	16.4285	36
37	0.6920	30.7995	519.5329	1.4451	44.5076	0.0325	0.0225	16.8682	37
38	0.6852	31.4847	544.8835	1.4595	45.9527	0.0318	0.0218	17.3063	38
39	0.6784	32.1630	570.6616	1.4741	47.4123	0.0311	0.0211	17.7428	39
40	0.6717	32.8347	596.8561	1.4889	48.8864	0.0305	0.0205	18.1776	40
41	0.6650	33.4997	623.4562	1.5038	50.3752	0.0299	0.0199	18.6108	41
42	0.6584	34.1581	650.4514	1.5188	51.8790	0.0293	0.0193	19.0424	42
43	0.6519	34.8100	677.8312	1.5340	53.3978	0.0287	0.0187	19.4723	43
44	0.6454	35.4555	705.5853	1.5493	54.9318	0.0282	0.0182	19.9006	44
45	0.6391	36.0945	733.7037	1.5648	56.4811	0.0277	0.0177	20.3273	45
46	0.6327	36.7272	762.1765	1.5805	58.0459	0.0272	0.0172	20.7524	46
47	0.6265	37.3537	790.9938	1.5963	59.6263	0.0268	0.0168	21.1758	47
48	0.6203	37.9740	820.1460	1.6122	61.2226	0.0263	0.0163	21.5976	48
49	0.6141	38.5881	849.6237	1.6283	62.8348	0.0259	0.0159	22.0178	49
50	0.6080	39.1961	879.4176	1.6446	64.4632	0.0255	0.0155	22.4363	50
51	0.6020	39.7981	909.5186	1.6611	66.1078	0.0251	0.0151	22.8533	51
52	0.5961	40.3942	939.9175	1.6777	67.7689	0.0248	0.0148	23.2686	52
53	0.5902	40.9844	970.6057	1.6945	69.4466	0.0244	0.0144	23.6823	53
54	0.5843	41.5687	1001.5743	1.7114	71.1410	0.0241	0.0141	24.0945	54
55	0.5785	42.1472	1032.8148	1.7285	72.8525	0.0237	0.0137	24.5049	55
60	0.5504	44.9550	1192.8061	1.8167	81.6697	0.0222	0.0122	26.5333	60
65	0.5237	47.6266	1358.3903	1.9094	90.9366	0.0210	0.0110	28.5217	65
70	0.4983	50.1685	1528.6474	2.0068	100.6763	0.0199	0.0099	30.4703	70
75	0.4741	52.5871	1702.7340	2.1091	110.9128	0.0190	0.0090	32.3793	75
80	0.4511	54.8882	1879.8771	2.2167	121.6715	0.0182	0.0082	34.2492	80
85	0.4292	57.0777	2059.3701	2.3298	132.9790	0.0175	0.0075	36.0801	85
90	0.4084	59.1609	2240.5675	2.4486	144.8633	0.0169	0.0069	37.8724	90
95	0.3886	61.1430	2422.8811	2.5735	157.3538	0.0164	0.0064	39.6265	95
100	0.3697	63.0289	2605.7758	2.7048	170.4814	0.0159	0.0059	41.3426	100

(continued)

APPENDIX 54.B *(continued)*
Cash Flow Equivalent Factors

$i = 1.50\%$

n	P/F	P/A	P/G	F/P	F/A	A/P	A/F	A/G	n
1	0.9852	0.9852	0.0000	1.0150	1.0000	1.0150	1.0000	0.0000	1
2	0.9707	1.9559	0.9707	1.0302	2.0150	0.5113	0.4963	0.4963	2
3	0.9563	2.9122	2.8833	1.0457	3.0452	0.3434	0.3284	0.9901	3
4	0.9422	3.8544	5.7098	1.0614	4.0909	0.2594	0.2444	1.4814	4
5	0.9283	4.7826	9.4229	1.0773	5.1523	0.2091	0.1941	1.9702	5
6	0.9145	5.6972	13.9956	1.0934	6.2296	0.1755	0.1605	2.4566	6
7	0.9010	6.5982	19.4018	1.1098	7.3230	0.1516	0.1366	2.9405	7
8	0.8877	7.4859	25.6157	1.1265	8.4328	0.1336	0.1186	3.4219	8
9	0.8746	8.3605	32.6125	1.1434	9.5593	0.1196	0.1046	3.9008	9
10	0.8617	9.2222	40.3675	1.1605	10.7027	0.1084	0.0934	4.3772	10
11	0.8489	10.0711	48.8568	1.1779	11.8633	0.0993	0.0843	4.8512	11
12	0.8364	10.9075	58.0571	1.1956	13.0412	0.0917	0.0767	5.3227	12
13	0.8240	11.7315	67.9454	1.2136	14.2368	0.0852	0.0702	5.7917	13
14	0.8118	12.5434	78.4994	1.2318	15.4504	0.0797	0.0647	6.2582	14
15	0.7999	13.3432	89.6974	1.2502	16.6821	0.0749	0.0599	6.7223	15
16	0.7880	14.1313	101.5178	1.2690	17.9324	0.0708	0.0558	7.1839	16
17	0.7764	14.9076	113.9400	1.2880	19.2014	0.0671	0.0521	7.6431	17
18	0.7649	15.6726	126.9435	1.3073	20.4894	0.0638	0.0488	8.0997	18
19	0.7536	16.4262	140.5084	1.3270	21.7967	0.0609	0.0459	8.5539	19
20	0.7425	17.1686	154.6154	1.3469	23.1237	0.0582	0.0432	9.0057	20
21	0.7315	17.9001	169.2453	1.3671	24.4705	0.0559	0.0409	9.4550	21
22	0.7207	18.6208	184.3798	1.3876	25.8376	0.0537	0.0387	9.9018	22
23	0.7100	19.3309	200.0006	1.4084	27.2251	0.0517	0.0367	10.3462	23
24	0.6995	20.0304	216.0901	1.4295	28.6335	0.0499	0.0349	10.7881	24
25	0.6892	20.7196	232.6310	1.4509	30.0630	0.0483	0.0333	11.2276	25
26	0.6790	21.3986	249.6065	1.4727	31.5140	0.0467	0.0317	11.6646	26
27	0.6690	22.0676	267.0002	1.4948	32.9867	0.0453	0.0303	12.0992	27
28	0.6591	22.7267	284.7958	1.5172	34.4815	0.0440	0.0290	12.5313	28
29	0.6494	23.3761	302.9779	1.5400	35.9987	0.0428	0.0278	12.9610	29
30	0.6398	24.0158	321.5310	1.5631	37.5387	0.0416	0.0266	13.3883	30
31	0.6303	24.6461	340.4402	1.5865	39.1018	0.0406	0.0256	13.8131	31
32	0.6210	25.2671	359.6910	1.6103	40.6883	0.0396	0.0246	14.2355	32
33	0.6118	25.8790	379.2691	1.6345	42.2986	0.0386	0.0236	14.6555	33
34	0.6028	26.4817	399.1607	1.6590	43.9331	0.0378	0.0228	15.0731	34
35	0.5939	27.0756	419.3521	1.6839	45.5921	0.0369	0.0219	15.4882	35
36	0.5851	27.6607	439.8303	1.7091	47.2760	0.0362	0.0212	15.9009	36
37	0.5764	28.2371	460.5822	1.7348	48.9851	0.0354	0.0204	16.3112	37
38	0.5679	28.8051	481.5954	1.7608	50.7199	0.0347	0.0197	16.7191	38
39	0.5595	29.3646	502.8576	1.7872	52.4807	0.0341	0.0191	17.1246	39
40	0.5513	29.9158	524.3568	1.8140	54.2679	0.0334	0.0184	17.5277	40
41	0.5431	30.4590	546.0814	1.8412	56.0819	0.0328	0.0178	17.9284	41
42	0.5351	30.9941	568.0201	1.8688	57.9231	0.0323	0.0173	18.3267	42
43	0.5272	31.5212	590.1617	1.8969	59.7920	0.0317	0.0167	18.7227	43
44	0.5194	32.0406	612.4955	1.9253	61.6889	0.0312	0.0162	19.1162	44
45	0.5117	32.5523	635.0110	1.9542	63.6142	0.0307	0.0157	19.5074	45
46	0.5042	33.0565	657.6979	1.9835	65.5684	0.0303	0.0153	19.8962	46
47	0.4967	33.5532	680.5462	2.0133	67.5519	0.0298	0.0148	20.2826	47
48	0.4894	34.0426	703.5462	2.0435	69.5652	0.0294	0.0144	20.6667	48
49	0.4821	34.5247	726.6884	2.0741	71.6087	0.0290	0.0140	21.0484	49
50	0.4750	34.9997	749.9636	2.1052	73.6828	0.0286	0.0136	21.4277	50
51	0.4680	35.4677	773.3629	2.1368	75.7881	0.0282	0.0132	21.8047	51
52	0.4611	35.9287	796.8774	2.1689	77.9249	0.0278	0.0128	22.1794	52
53	0.4543	36.3830	820.4986	2.2014	80.0938	0.0275	0.0125	22.5517	53
54	0.4475	36.8305	844.2184	2.2344	82.2952	0.0272	0.0122	22.9217	54
55	0.4409	37.2715	868.0285	2.2679	84.5296	0.0268	0.0118	23.2894	55
60	0.4093	39.3803	988.1674	2.4432	96.2147	0.0254	0.0104	25.0930	60
65	0.3799	41.3378	1109.4752	2.6320	108.8028	0.0242	0.0092	26.8393	65
70	0.3527	43.1549	1231.1658	2.8355	122.3638	0.0232	0.0082	28.5290	70
75	0.3274	44.8416	1352.5600	3.0546	136.9728	0.0223	0.0073	30.1631	75
80	0.3039	46.4073	1473.0741	3.2907	152.7109	0.0215	0.0065	31.7423	80
85	0.2821	47.8607	1592.2095	3.5450	169.6652	0.0209	0.0059	33.2676	85
90	0.2619	49.2099	1709.5439	3.8189	187.9299	0.0203	0.0053	34.7399	90
95	0.2431	50.4622	1824.7224	4.1141	207.6061	0.0198	0.0048	36.1602	95
100	0.2256	51.6247	1937.4506	4.4320	228.8030	0.0194	0.0044	37.5295	100

(continued)

APPENDIX 54.B *(continued)*
Cash Flow Equivalent Factors

$i = 2.00\%$

n	P/F	P/A	P/G	F/P	F/A	A/P	A/F	A/G	n
1	0.9804	0.9804	0.0000	1.0200	1.0000	1.0200	1.0000	0.0000	1
2	0.9612	1.9416	0.9612	1.0404	2.0200	0.5150	0.4950	0.4950	2
3	0.9423	2.8839	2.8458	1.0612	3.0604	0.3468	0.3268	0.9868	3
4	0.9238	3.8077	5.6173	1.0824	4.1216	0.2626	0.2426	1.4752	4
5	0.9057	4.7135	9.2403	1.1041	5.2040	0.2122	0.1922	1.9604	5
6	0.8880	5.6014	13.6801	1.1262	6.3081	0.1785	0.1585	2.4423	6
7	0.8706	6.4720	18.9035	1.1487	7.4343	0.1545	0.1345	2.9208	7
8	0.8535	7.3255	24.8779	1.1717	8.5830	0.1365	0.1165	3.3961	8
9	0.8368	8.1622	31.5720	1.1951	9.7546	0.1225	0.1025	3.8681	9
10	0.8203	8.9826	38.9551	1.2190	10.9497	0.1113	0.0913	4.3367	10
11	0.8043	9.7868	46.9977	1.2434	12.1687	0.1022	0.0822	4.8021	11
12	0.7885	10.5753	55.6712	1.2682	13.4121	0.0946	0.0746	5.2642	12
13	0.7730	11.3484	64.9475	1.2936	14.6803	0.0881	0.0681	5.7231	13
14	0.7579	12.1062	74.7999	1.3195	15.9739	0.0826	0.0626	6.1786	14
15	0.7430	12.8493	85.2021	1.3459	17.2934	0.0778	0.0578	6.6309	15
16	0.7284	13.5777	96.1288	1.3728	18.6393	0.0737	0.0537	7.0799	16
17	0.7142	14.2919	107.5554	1.4002	20.0121	0.0700	0.0500	7.5256	17
18	0.7002	14.9920	119.4581	1.4282	21.4123	0.0667	0.0467	7.9681	18
19	0.6864	15.6785	131.8139	1.4568	22.8406	0.0638	0.0438	8.4073	19
20	0.6730	16.3514	144.6003	1.4859	24.2974	0.0612	0.0412	8.8433	20
21	0.6598	17.0112	157.7959	1.5157	25.7833	0.0588	0.0388	9.2760	21
22	0.6468	17.6580	171.3795	1.5460	27.2990	0.0566	0.0366	9.7055	22
23	0.6342	18.2922	185.3309	1.5769	28.8450	0.0547	0.0347	10.1317	23
24	0.6217	18.9139	199.6305	1.6084	30.4219	0.0529	0.0329	10.5547	24
25	0.6095	19.5235	214.2592	1.6406	32.0303	0.0512	0.0312	10.9745	25
26	0.5976	20.1210	229.1987	1.6734	33.6709	0.0497	0.0297	11.3910	26
27	0.5859	20.7069	244.4311	1.7069	35.3443	0.0483	0.0283	11.8043	27
28	0.5744	21.2813	259.9392	1.7410	37.0512	0.0470	0.0270	12.2145	28
29	0.5631	21.8444	275.7064	1.7758	38.7922	0.0458	0.0258	12.6214	29
30	0.5521	22.3965	291.7164	1.8114	40.5681	0.0446	0.0246	13.0251	30
31	0.5412	22.9377	307.9538	1.8476	42.3794	0.0436	0.0236	13.4257	31
32	0.5306	23.4683	324.4035	1.8845	44.2270	0.0426	0.0226	13.8230	32
33	0.5202	23.9886	341.0508	1.9222	46.1116	0.0417	0.0217	14.2172	33
34	0.5100	24.4986	357.8817	1.9607	48.0338	0.0408	0.0208	14.6083	34
35	0.5000	24.9986	374.8826	1.9999	49.9945	0.0400	0.0200	14.9961	35
36	0.4902	25.4888	392.0405	2.0399	51.9944	0.0392	0.0192	15.3809	36
37	0.4806	25.9695	409.3424	2.0807	54.0343	0.0385	0.0185	15.7625	37
38	0.4712	26.4406	426.7764	2.1223	56.1149	0.0378	0.0178	16.1409	38
39	0.4619	26.9026	444.3304	2.1647	58.2372	0.0372	0.0172	16.5163	39
40	0.4529	27.3555	461.9931	2.2080	60.4020	0.0366	0.0166	16.8885	40
41	0.4440	27.7995	479.7535	2.2522	62.6100	0.0360	0.0160	17.2576	41
42	0.4353	28.2348	497.6010	2.2972	64.8622	0.0354	0.0154	17.6237	42
43	0.4268	28.6616	515.5253	2.3432	67.1595	0.0349	0.0149	17.9866	43
44	0.4184	29.0800	533.5165	2.3901	69.5027	0.0344	0.0144	18.3465	44
45	0.4102	29.4902	551.5652	2.4379	71.8927	0.0339	0.0139	18.7034	45
46	0.4022	29.8923	569.6621	2.4866	74.3306	0.0335	0.0135	19.0571	46
47	0.3943	30.2866	587.7985	2.5363	76.8172	0.0330	0.0130	19.4079	47
48	0.3865	30.6731	605.9657	2.5871	79.3535	0.0326	0.0126	19.7556	48
49	0.3790	31.0521	624.1557	2.6388	81.9406	0.0322	0.0122	20.1003	49
50	0.3715	31.4236	642.3606	2.6916	84.5794	0.0318	0.0118	20.4420	50
51	0.3642	31.7878	660.5727	2.7454	87.2710	0.0315	0.0115	20.7807	51
52	0.3571	32.1449	678.7849	2.8003	90.0164	0.0311	0.0111	21.1164	52
53	0.3501	32.4950	696.9900	2.8563	92.8167	0.0308	0.0108	21.4491	53
54	0.3432	32.8383	715.1815	2.9135	95.6731	0.0305	0.0105	21.7789	54
55	0.3365	33.1748	733.3527	2.9717	98.5865	0.0301	0.0101	22.1057	55
60	0.3048	34.7609	823.6975	3.2810	114.0515	0.0288	0.0088	23.6961	60
65	0.2761	36.1975	912.7085	3.6225	131.1262	0.0276	0.0076	25.2147	65
70	0.2500	37.4986	999.8343	3.9996	149.9779	0.0267	0.0067	26.6632	70
75	0.2265	38.6771	1084.6393	4.4158	170.7918	0.0259	0.0059	28.0434	75
80	0.2051	39.7445	1166.7868	4.8754	193.7720	0.0252	0.0052	29.3572	80
85	0.1858	40.7113	1246.0241	5.3829	219.1439	0.0246	0.0046	30.6064	85
90	0.1683	41.5869	1322.1701	5.9431	247.1567	0.0240	0.0040	31.7929	90
95	0.1524	42.3800	1395.1033	6.5617	278.0850	0.0236	0.0036	32.9189	95
100	0.1380	43.0984	1464.7527	7.2446	312.2323	0.0232	0.0032	33.9863	100

(continued)

APPENDIX 54.B *(continued)*
Cash Flow Equivalent Factors

$$i = 3.00\%$$

n	P/F	P/A	P/G	F/P	F/A	A/P	A/F	A/G	n
1	0.9709	0.9709	0.0000	1.0300	1.0000	1.0300	1.0000	0.0000	1
2	0.9426	1.9135	0.9426	1.0609	2.0300	0.5226	0.4926	0.4926	2
3	0.9151	2.8286	2.7729	1.0927	3.0909	0.3535	0.3235	0.9803	3
4	0.8885	3.7171	5.4383	1.1255	4.1836	0.2690	0.2390	1.4631	4
5	0.8626	4.5797	8.8888	1.1593	5.3091	0.2184	0.1884	1.9409	5
6	0.8375	5.4172	13.0762	1.1941	6.4684	0.1846	0.1546	2.4138	6
7	0.8131	6.2303	17.9547	1.2299	7.6625	0.1605	0.1305	2.8819	7
8	0.7894	7.0197	23.4806	1.2668	8.8923	0.1425	0.1125	3.3450	8
9	0.7664	7.7861	29.6119	1.3048	10.1591	0.1284	0.0984	3.8032	9
10	0.7441	8.5302	36.3088	1.3439	11.4639	0.1172	0.0872	4.2565	10
11	0.7224	9.2526	43.5330	1.3842	12.8078	0.1081	0.0781	4.7049	11
12	0.7014	9.9540	51.2482	1.4258	14.1920	0.1005	0.0705	5.1485	12
13	0.6810	10.6350	59.4196	1.4685	15.6178	0.0940	0.0640	5.5872	13
14	0.6611	11.2961	68.0141	1.5126	17.0863	0.0885	0.0585	6.0210	14
15	0.6419	11.9379	77.0002	1.5580	18.5989	0.0838	0.0538	6.4500	15
16	0.6232	12.5611	86.3477	1.6047	20.1569	0.0796	0.0496	6.8742	16
17	0.6050	13.1661	96.0280	1.6528	21.7616	0.0760	0.0460	7.2936	17
18	0.5874	13.7535	106.0137	1.7024	23.4144	0.0727	0.0427	7.7081	18
19	0.5703	14.3238	116.2788	1.7535	25.1169	0.0698	0.0398	8.1179	19
20	0.5537	14.8775	126.7987	1.8061	26.8704	0.0672	0.0372	8.5229	20
21	0.5375	15.4150	137.5496	1.8603	28.6765	0.0649	0.0349	8.9231	21
22	0.5219	15.9369	148.5094	1.9161	30.5368	0.0627	0.0327	9.3186	22
23	0.5067	16.4436	159.6566	1.9736	32.4529	0.0608	0.0308	9.7093	23
24	0.4919	16.9355	170.9711	2.0328	34.4265	0.0590	0.0290	10.0954	24
25	0.4776	17.4131	182.4336	2.0938	36.4593	0.0574	0.0274	10.4768	25
26	0.4637	17.8768	194.0260	2.1566	38.5530	0.0559	0.0259	10.8535	26
27	0.4502	18.3270	205.7309	2.2213	40.7096	0.0546	0.0246	11.2255	27
28	0.4371	18.7641	217.5320	2.2879	42.9309	0.0533	0.0233	11.5930	28
29	0.4243	19.1885	229.4137	2.3566	45.2189	0.0521	0.0221	11.9558	29
30	0.4120	19.6004	241.3613	2.4273	47.5754	0.0510	0.0210	12.3141	30
31	0.4000	20.0004	253.3609	2.5001	50.0027	0.0500	0.0200	12.6678	31
32	0.3883	20.3888	265.3993	2.5751	52.5028	0.0490	0.0190	13.0169	32
33	0.3770	20.7658	277.4642	2.6523	55.0778	0.0482	0.0182	13.3616	33
34	0.3660	21.1318	289.5437	2.7319	57.7302	0.0473	0.0173	13.7018	34
35	0.3554	21.4872	301.6267	2.8139	60.4621	0.0465	0.0165	14.0375	35
36	0.3450	21.8323	313.7028	2.8983	63.2759	0.0458	0.0158	14.3688	36
37	0.3350	22.1672	325.7622	2.9852	66.1742	0.0451	0.0151	14.6957	37
38	0.3252	22.4925	337.7956	3.0748	69.1594	0.0445	0.0145	15.0182	38
39	0.3158	22.8082	349.7942	3.1670	72.2342	0.0438	0.0138	15.3363	39
40	0.3066	23.1148	361.7499	3.2620	75.4013	0.0433	0.0133	15.6502	40
41	0.2976	23.4124	373.6633	3.3599	78.6633	0.0427	0.0127	15.9597	41
42	0.2890	23.7014	385.5024	3.4607	82.0232	0.0422	0.0122	16.2650	42
43	0.2805	23.9819	397.2852	3.5645	85.4839	0.0417	0.0117	16.5660	43
44	0.2724	24.2543	408.9972	3.6715	89.0484	0.0412	0.0112	16.8629	44
45	0.2644	24.5187	420.6325	3.7816	92.7199	0.0408	0.0108	17.1556	45
46	0.2567	24.7754	432.1856	3.8950	96.5015	0.0404	0.0104	17.4441	46
47	0.2493	25.0247	443.6515	4.0119	100.3965	0.0400	0.0100	17.7285	47
48	0.2420	25.2667	455.0255	4.1323	104.4084	0.0396	0.0096	18.0089	48
49	0.2350	25.5017	466.3031	4.2562	108.5406	0.0392	0.0092	18.2852	49
50	0.2281	25.7298	477.4803	4.3839	112.7969	0.0389	0.0089	18.5575	50
51	0.2215	25.9512	488.5535	4.5154	117.1808	0.0385	0.0085	18.8258	51
52	0.2150	26.1662	499.5191	4.6509	121.6962	0.0382	0.0082	19.0902	52
53	0.2088	26.3750	510.3742	4.7904	126.3471	0.0379	0.0079	19.3507	53
54	0.2027	26.5777	521.1157	4.9341	131.1375	0.0376	0.0076	19.6073	54
55	0.1968	26.7744	531.7411	5.0821	136.0716	0.0373	0.0073	19.8600	55
60	0.1697	27.6756	583.0526	5.8916	163.0534	0.0361	0.0061	21.0674	60
65	0.1464	28.4529	631.2010	6.8300	194.3328	0.0351	0.0051	22.1841	65
70	0.1263	29.1234	676.0869	7.9178	230.5941	0.0343	0.0043	23.2145	70
75	0.1089	29.7018	717.6978	9.1789	272.6309	0.0337	0.0037	24.1634	75
80	0.0940	30.2008	756.0865	10.6409	321.3630	0.0331	0.0031	25.0353	80
85	0.0811	30.6312	791.3529	12.3357	377.8570	0.0326	0.0026	25.8349	85
90	0.0699	31.0024	823.6302	14.3005	443.3489	0.0323	0.0023	26.5667	90
95	0.0603	31.3227	853.0742	16.5782	519.2720	0.0319	0.0019	27.2351	95
100	0.0520	31.5989	879.8540	19.2186	607.2877	0.0316	0.0016	27.8444	100

(continued)

APPENDIX 54.B *(continued)*
Cash Flow Equivalent Factors

$$i = 4.00\%$$

n	P/F	P/A	P/G	F/P	F/A	A/P	A/F	A/G	n
1	0.9615	0.9615	0.0000	1.0400	1.0000	1.0400	1.0000	0.0000	1
2	0.9246	1.8861	0.9246	1.0816	2.0400	0.5302	0.4902	0.4902	2
3	0.8890	2.7751	2.7025	1.1249	3.1216	0.3603	0.3203	0.9739	3
4	0.8548	3.6299	5.2670	1.1699	4.2465	0.2755	0.2355	1.4510	4
5	0.8219	4.4518	8.5547	1.2167	5.4163	0.2246	0.1846	1.9216	5
6	0.7903	5.2421	12.5062	1.2653	6.6330	0.1908	0.1508	2.3857	6
7	0.7599	6.0021	17.0657	1.3150	7.8083	0.1666	0.1266	2.8433	7
8	0.7307	6.7327	22.1806	1.3686	9.2142	0.1485	0.1085	3.2944	8
9	0.7026	7.4353	27.8013	1.4233	10.5828	0.1345	0.0945	3.7391	9
10	0.6756	8.1109	33.8814	1.4802	12.0061	0.1233	0.0833	4.1773	10
11	0.6496	8.7605	40.3772	1.5395	13.4864	0.1141	0.0741	4.6090	11
12	0.6246	9.3851	47.2477	1.6010	15.0258	0.1066	0.0666	5.0343	12
13	0.6006	9.9856	54.4546	1.6651	16.6268	0.1001	0.0601	5.4533	13
14	0.5775	10.5631	61.9618	1.7317	18.2919	0.0947	0.0547	5.8659	14
15	0.5553	11.1184	69.7355	1.8009	20.0236	0.0899	0.0499	6.2721	15
16	0.5339	11.6523	77.7441	1.8730	21.8245	0.0858	0.0458	6.6720	16
17	0.5134	12.1657	85.9581	1.9479	23.6975	0.0822	0.0422	7.0656	17
18	0.4936	12.6593	94.3498	2.0258	25.6454	0.0790	0.0390	7.4530	18
19	0.4746	13.1339	102.8933	2.1068	27.6712	0.0761	0.0361	7.8342	19
20	0.4564	13.5903	111.5647	2.1911	29.7781	0.0736	0.0336	8.2091	20
21	0.4388	14.0292	120.3414	2.2788	31.9692	0.0713	0.0313	8.5779	21
22	0.4220	14.4511	129.2024	2.3699	34.2480	0.0692	0.0292	8.9407	22
23	0.4057	14.8568	138.1284	2.4647	36.6179	0.0673	0.0273	9.2973	23
24	0.3901	15.2470	147.1012	2.5633	39.0826	0.0656	0.0256	9.6479	24
25	0.3751	15.6221	156.1040	2.6658	41.6459	0.0640	0.0240	9.9925	25
26	0.3607	15.9828	165.1212	2.7725	44.3117	0.0626	0.0226	10.3312	26
27	0.3468	16.3296	174.1385	2.8834	47.0842	0.0612	0.0212	10.6640	27
28	0.3335	16.6631	183.1424	2.9987	49.9676	0.0600	0.0200	10.9909	28
29	0.3207	16.9837	192.1206	3.1187	52.9663	0.0589	0.0189	11.3120	29
30	0.3083	17.2920	201.0618	3.2434	56.0849	0.0578	0.0178	11.6274	30
31	0.2965	17.5885	209.9556	3.3731	59.3283	0.0569	0.0169	11.9371	31
32	0.2851	17.8736	218.7924	3.5081	62.7015	0.0559	0.0159	12.2411	32
33	0.2741	18.1476	227.5634	3.6484	66.2095	0.0551	0.0151	12.5396	33
34	0.2636	18.4112	236.2607	3.7943	69.8579	0.0543	0.0143	12.8324	34
35	0.2534	18.6646	244.8768	3.9461	73.6522	0.0536	0.0136	13.1198	35
36	0.2437	18.9083	253.4052	4.1039	77.5983	0.0529	0.0129	13.4018	36
37	0.2343	19.1426	261.8399	4.2681	81.7022	0.0522	0.0122	13.6784	37
38	0.2253	19.3679	270.1754	4.4388	85.9703	0.0516	0.0116	13.9497	38
39	0.2166	19.5845	278.4070	4.6164	90.4091	0.0511	0.0111	14.2157	39
40	0.2083	19.7928	286.5303	4.8010	95.0255	0.0505	0.0105	14.4765	40
41	0.2003	19.9931	294.5414	4.9931	99.8265	0.0500	0.0100	14.7322	41
42	0.1926	20.1856	302.4370	5.1928	104.8196	0.0495	0.0095	14.9828	42
43	0.1852	20.3708	310.2141	5.4005	110.0124	0.0491	0.0091	15.2284	43
44	0.1780	20.5488	317.8700	5.6165	115.4129	0.0487	0.0087	15.4690	44
45	0.1712	20.7200	325.4028	5.8412	121.0294	0.0483	0.0083	15.7047	45
46	0.1646	20.8847	332.8104	6.0748	126.8706	0.0479	0.0079	15.9356	46
47	0.1583	21.0429	340.0914	6.3178	132.9454	0.0475	0.0075	16.1618	47
48	0.1522	21.1951	347.2446	6.5705	139.2632	0.0472	0.0072	16.3832	48
49	0.1463	21.3415	354.2689	6.8333	145.8337	0.0469	0.0069	16.6000	49
50	0.1407	21.4822	361.1638	7.1067	152.6671	0.0466	0.0066	16.8122	50
51	0.1353	21.6175	367.9289	7.3910	159.7738	0.0463	0.0063	17.0200	51
52	0.1301	21.7476	374.5638	7.6866	167.1647	0.0460	0.0060	17.2232	52
53	0.1251	21.8727	381.0686	7.9941	174.8513	0.0457	0.0057	17.4221	53
54	0.1203	21.9930	387.4436	8.3138	182.8454	0.0455	0.0055	17.6167	54
55	0.1157	22.1086	393.6890	8.6464	191.1592	0.0452	0.0052	17.8070	55
60	0.0951	22.6235	422.9966	10.5196	237.9907	0.0442	0.0042	18.6972	60
65	0.0781	23.0467	449.2014	12.7987	294.9684	0.0434	0.0034	19.4909	65
70	0.0642	23.3945	472.4789	15.5716	364.2905	0.0427	0.0027	20.1961	70
75	0.0528	23.6804	493.0408	18.9453	448.6314	0.0422	0.0022	20.8206	75
80	0.0434	23.9154	511.1161	23.0498	551.2450	0.0418	0.0018	21.3718	80
85	0.0357	24.1085	526.9384	28.0436	676.0901	0.0415	0.0015	21.8569	85
90	0.0293	24.2673	540.7369	34.1193	827.9833	0.0412	0.0012	22.2826	90
95	0.0241	24.3978	552.7307	41.5114	1012.7846	0.0410	0.0010	22.6550	95
100	0.0198	24.5050	563.1249	50.5049	1237.6237	0.0408	0.0008	22.9800	100

(continued)

APPENDIX 54.B *(continued)*
Cash Flow Equivalent Factors

$i = 5.00\%$

n	P/F	P/A	P/G	F/P	F/A	A/P	A/F	A/G	n
1	0.9524	0.9524	0.0000	1.0500	1.0000	1.0500	1.0000	0.0000	1
2	0.9070	1.8594	0.9070	1.1025	2.0500	0.5378	0.4878	0.4878	2
3	0.8638	2.7232	2.6347	1.1576	3.1525	0.3672	0.3172	0.9675	3
4	0.8227	3.5460	5.1028	1.2155	4.3101	0.2820	0.2320	1.4391	4
5	0.7835	4.3295	8.2369	1.2763	5.5256	0.2310	0.1810	1.9025	5
6	0.7462	5.0757	11.9680	1.3401	6.8019	0.1970	0.1470	2.3579	6
7	0.7107	5.7864	16.2321	1.4071	8.1420	0.1728	0.1228	2.8052	7
8	0.6768	6.4632	20.9700	1.4775	9.5491	0.1547	0.1047	3.2445	8
9	0.6446	7.1078	26.1268	1.5513	11.0266	0.1407	0.0907	3.6758	9
10	0.6139	7.7217	31.6520	1.6289	12.5779	0.1295	0.0795	4.0991	10
11	0.5847	8.3064	37.4988	1.7103	14.2068	0.1204	0.0704	4.5144	11
12	0.5568	8.8633	43.6241	1.7959	15.9171	0.1128	0.0628	4.9219	12
13	0.5303	9.3936	49.9879	1.8856	17.7130	0.1065	0.0565	5.3215	13
14	0.5051	9.8986	56.5538	1.9799	19.5986	0.1010	0.0510	5.7133	14
15	0.4810	10.3797	63.2880	2.0789	21.5786	0.0963	0.0463	6.0973	15
16	0.4581	10.8378	70.1597	2.1829	23.6575	0.0923	0.0423	6.4736	16
17	0.4363	11.2741	77.1405	2.2920	25.8404	0.0887	0.0387	6.8423	17
18	0.4155	11.6896	84.2043	2.4066	28.1324	0.0855	0.0355	7.2034	18
19	0.3957	12.0853	91.3275	2.5270	30.5390	0.0827	0.0327	7.5569	19
20	0.3769	12.4622	98.4884	2.6533	33.0660	0.0802	0.0302	7.9030	20
21	0.3589	12.8212	105.6673	2.7860	35.7193	0.0780	0.0280	8.2416	21
22	0.3418	13.1630	112.8461	2.9253	38.5052	0.0760	0.0260	8.5730	22
23	0.3256	13.4886	120.0087	3.0715	41.4305	0.0741	0.0241	8.8971	23
24	0.3101	13.7986	127.1402	3.2251	44.5020	0.0725	0.0225	9.2140	24
25	0.2953	14.0939	134.2275	3.3864	47.7271	0.0710	0.0210	9.5238	25
26	0.2812	14.3752	141.2585	3.5557	51.1135	0.0696	0.0196	9.8266	26
27	0.2678	14.6430	148.2226	3.7335	54.6691	0.0683	0.0183	10.1224	27
28	0.2551	14.8981	155.1101	3.9201	58.4026	0.0671	0.0171	10.4114	28
29	0.2429	15.1411	161.9126	4.1161	62.3227	0.0660	0.0160	10.6936	29
30	0.2314	15.3725	168.6226	4.3219	66.4388	0.0651	0.0151	10.9691	30
31	0.2204	15.5928	175.2333	4.5380	70.7608	0.0641	0.0141	11.2381	31
32	0.2099	15.8027	181.7392	4.7649	75.2988	0.0633	0.0133	11.5005	32
33	0.1999	16.0025	188.1351	5.0032	80.0638	0.0625	0.0125	11.7566	33
34	0.1904	16.1929	194.4168	5.2533	85.0670	0.0618	0.0118	12.0063	34
35	0.1813	16.3742	200.5807	5.5160	90.3203	0.0611	0.0111	12.2498	35
36	0.1727	16.5469	206.6237	5.7918	95.8363	0.0604	0.0104	12.4872	36
37	0.1644	16.7113	212.5434	6.0814	101.6281	0.0598	0.0098	12.7186	37
38	0.1566	16.8679	218.3378	6.3855	107.7095	0.0593	0.0093	12.9440	38
39	0.1491	17.0170	224.0054	6.7048	114.0950	0.0588	0.0088	13.1636	39
40	0.1420	17.1591	229.5452	7.0400	120.7998	0.0583	0.0083	13.3775	40
41	0.1353	17.2944	234.9564	7.3920	127.8398	0.0578	0.0078	13.5857	41
42	0.1288	17.4232	240.2389	7.7616	135.2318	0.0574	0.0074	13.7884	42
43	0.1227	17.5459	245.3925	8.1497	142.9933	0.0570	0.0070	13.9857	43
44	0.1169	17.6628	250.4175	8.5572	151.1430	0.0566	0.0066	14.1777	44
45	0.1113	17.7741	255.3145	8.9850	159.7002	0.0563	0.0063	14.3644	45
46	0.1060	17.8801	260.0844	9.4343	168.6852	0.0559	0.0059	14.5461	46
47	0.1009	17.9810	264.7281	9.9060	178.1194	0.0556	0.0056	14.7226	47
48	0.0961	18.0772	269.2467	10.4013	188.0254	0.0553	0.0053	14.8943	48
49	0.0916	18.1687	273.6418	10.9213	198.4267	0.0550	0.0050	15.0611	49
50	0.0872	18.2559	277.9148	11.4674	209.3480	0.0548	0.0048	15.2233	50
51	0.0831	18.3390	282.0673	12.0408	220.8154	0.0545	0.0045	15.3808	51
52	0.0791	18.4181	286.1013	12.6428	232.8562	0.0543	0.0043	15.5337	52
53	0.0753	18.4934	290.0184	13.2749	245.4990	0.0541	0.0041	15.6823	53
54	0.0717	18.5651	293.8208	13.9387	258.7739	0.0539	0.0039	15.8265	54
55	0.0683	18.6335	297.5104	14.6356	272.7126	0.0537	0.0037	15.9664	55
60	0.0535	18.9293	314.3432	18.6792	353.5837	0.0528	0.0028	16.6062	60
65	0.0419	19.1611	328.6910	23.8399	456.7980	0.0522	0.0022	17.1541	65
70	0.0329	19.3427	340.8409	30.4264	588.5285	0.0517	0.0017	17.6212	70
75	0.0258	19.4850	351.0721	38.8327	756.6537	0.0513	0.0013	18.0176	75
80	0.0202	19.5965	359.6460	49.5614	971.2288	0.0510	0.0010	18.3526	80
85	0.0158	19.6838	366.8007	63.2544	1245.0871	0.0508	0.0008	18.6346	85
90	0.0124	19.7523	372.7488	80.7304	1597.6073	0.0506	0.0006	18.8712	90
95	0.0097	19.8059	377.6774	103.0347	2040.6935	0.0505	0.0005	19.0689	95
100	0.0076	19.8479	381.7492	131.5013	2610.0252	0.0504	0.0004	19.2337	100

(continued)

APPENDIX 54.B *(continued)*
Cash Flow Equivalent Factors

$$i = 6.00\%$$

n	P/F	P/A	P/G	F/P	F/A	A/P	A/F	A/G	n
1	0.9434	0.9434	0.0000	1.0600	1.0000	1.0600	1.0000	0.0000	1
2	0.8900	1.8334	0.8900	1.1236	2.0600	0.5454	0.4854	0.4854	2
3	0.8396	2.6730	2.5692	1.1910	3.1836	0.3741	0.3141	0.9612	3
4	0.7921	3.4651	4.9455	1.2625	4.3746	0.2886	0.2286	1.4272	4
5	0.7473	4.2124	7.9345	1.3382	5.6371	0.2374	0.1774	1.8836	5
6	0.7050	4.9173	11.4594	1.4185	6.9753	0.2034	0.1434	2.3304	6
7	0.6651	5.5824	15.4497	1.5036	8.3938	0.1791	0.1191	2.7676	7
8	0.6274	6.2098	19.8416	1.5938	9.8975	0.1610	0.1010	3.1952	8
9	0.5919	6.8017	24.5768	1.6895	11.4913	0.1470	0.0870	3.6133	9
10	0.5584	7.3601	29.6023	1.7908	13.1808	0.1359	0.0759	4.0220	10
11	0.5268	7.8869	34.8702	1.8983	14.9716	0.1268	0.0668	4.4213	11
12	0.4970	8.3838	40.3369	2.0122	16.8699	0.1193	0.0593	4.8113	12
13	0.4688	8.8527	45.9629	2.1329	18.8821	0.1130	0.0530	5.1920	13
14	0.4423	9.2950	51.7128	2.2609	21.0151	0.1076	0.0476	5.5635	14
15	0.4173	9.7122	57.5546	2.3966	23.2760	0.1030	0.0430	5.9260	15
16	0.3936	10.1059	63.4592	2.5404	25.6725	0.0990	0.0390	6.2794	16
17	0.3714	10.4773	69.4011	2.6928	28.2129	0.0954	0.0354	6.6240	17
18	0.3503	10.8276	75.3569	2.8543	30.9057	0.0924	0.0324	6.9597	18
19	0.3305	11.1581	81.3062	3.0256	33.7600	0.0896	0.0296	7.2867	19
20	0.3118	11.4699	87.2304	3.2071	36.7856	0.0872	0.0272	7.6051	20
21	0.2942	11.7641	93.1136	3.3996	39.9927	0.0850	0.0250	7.9151	21
22	0.2775	12.0416	98.9412	3.6035	43.3923	0.0830	0.0230	8.2166	22
23	0.2618	12.3034	104.7007	3.8197	46.9958	0.0813	0.0213	8.5099	23
24	0.2470	12.5504	110.3812	4.0489	50.8156	0.0797	0.0197	8.7951	24
25	0.2330	12.7834	115.9732	4.2919	54.8645	0.0782	0.0182	9.0722	25
26	0.2198	13.0032	121.4684	4.5494	59.1564	0.0769	0.0169	9.3414	26
27	0.2074	13.2105	126.8600	4.8223	63.7058	0.0757	0.0157	9.6029	27
28	0.1956	13.4062	132.1420	5.1117	68.5281	0.0746	0.0146	9.8568	28
29	0.1846	13.5907	137.3096	5.4184	73.6398	0.0736	0.0136	10.1032	29
30	0.1741	13.7648	142.3588	5.7435	79.0582	0.0726	0.0126	10.3422	30
31	0.1643	13.9291	147.2864	6.0881	84.8017	0.0718	0.0118	10.5740	31
32	0.1550	14.0840	152.0901	6.4534	90.8898	0.0710	0.0110	10.7988	32
33	0.1462	14.2302	156.7681	6.8406	97.3432	0.0703	0.0103	11.0166	33
34	0.1379	14.3681	161.3192	7.2510	104.1838	0.0696	0.0096	11.2276	34
35	0.1301	14.4982	165.7427	7.6861	111.4348	0.0690	0.0090	11.4319	35
36	0.1227	14.6210	170.0387	8.1473	119.1209	0.0684	0.0084	11.6298	36
37	0.1158	14.7368	174.2072	8.6361	127.2681	0.0679	0.0079	11.8213	37
38	0.1092	14.8460	178.2490	9.1543	135.9042	0.0674	0.0074	12.0065	38
39	0.1031	14.9491	182.1652	9.7035	145.0585	0.0669	0.0069	12.1857	39
40	0.0972	15.0463	185.9568	10.2857	154.7620	0.0665	0.0065	12.3590	40
41	0.0917	15.1380	189.6256	10.9029	165.0477	0.0661	0.0061	12.5264	41
42	0.0865	15.2245	193.1732	11.5570	175.9505	0.0657	0.0057	12.6883	42
43	0.0816	15.3062	196.6017	12.2505	187.5076	0.0653	0.0053	12.8446	43
44	0.0770	15.3832	199.9130	12.9855	199.7580	0.0650	0.0050	12.9956	44
45	0.0727	15.4558	203.1096	13.7646	212.7435	0.0647	0.0047	13.1413	45
46	0.0685	15.5244	206.1938	14.5905	226.5081	0.0644	0.0044	13.2819	46
47	0.0647	15.5890	209.1681	15.4659	241.0986	0.0641	0.0041	13.4177	47
48	0.0610	15.6500	212.0351	16.3939	256.5645	0.0639	0.0039	13.5485	48
49	0.0575	15.7076	214.7972	17.3775	272.9584	0.0637	0.0037	13.6748	49
50	0.0543	15.7619	217.4574	18.4202	290.3359	0.0634	0.0034	13.7964	50
51	0.0512	15.8131	220.0181	19.5254	308.7561	0.0632	0.0032	13.9137	51
52	0.0483	15.8614	222.4823	20.6969	328.2814	0.0630	0.0030	14.0267	52
53	0.0456	15.9070	224.8525	21.9387	348.9783	0.0629	0.0029	14.1355	53
54	0.0430	15.9500	227.1316	23.2550	370.9170	0.0627	0.0027	14.2402	54
55	0.0406	15.9905	229.3222	24.6503	394.1720	0.0625	0.0025	14.3411	55
60	0.0303	16.1614	239.0428	32.9877	533.1282	0.0619	0.0019	14.7909	60
65	0.0227	16.2891	246.9450	44.1450	719.0829	0.0614	0.0014	15.1601	65
70	0.0169	16.3845	253.3271	59.0759	967.9322	0.0610	0.0010	15.4613	70
75	0.0126	16.4558	258.4527	79.0569	1300.9487	0.0608	0.0008	15.7058	75
80	0.0095	16.5091	262.5493	105.7960	1746.5999	0.0606	0.0006	15.9033	80
85	0.0071	16.5489	265.8096	141.5789	2342.9817	0.0604	0.0004	16.0620	85
90	0.0053	16.5787	268.3946	189.4645	3141.0752	0.0603	0.0003	16.1891	90
95	0.0039	16.6009	270.4375	253.5463	4209.1042	0.0602	0.0002	16.2905	95
100	0.0029	16.6175	272.0471	339.3021	5638.3681	0.0602	0.0002	16.3711	100

(continued)

APPENDIX 54.B (*continued*)
Cash Flow Equivalent Factors

$i = 7.00\%$

n	P/F	P/A	P/G	F/P	F/A	A/P	A/F	A/G	n
1	0.9346	0.9346	0.0000	1.0700	1.0000	1.0700	1.0000	0.0000	1
2	0.8734	1.8080	0.8734	1.1449	2.0700	0.5531	0.4831	0.4831	2
3	0.8163	2.6243	2.5060	1.2250	3.2149	0.3811	0.3111	0.9549	3
4	0.7629	3.3872	4.7947	1.3108	4.4399	0.2952	0.2252	1.4155	4
5	0.7130	4.1002	7.6467	1.4026	5.7507	0.2439	0.1739	1.8650	5
6	0.6663	4.7665	10.9784	1.5007	7.1533	0.2098	0.1398	2.3032	6
7	0.6227	5.3893	14.7149	1.6058	8.6540	0.1856	0.1156	2.7304	7
8	0.5820	5.9713	18.7889	1.7182	10.2598	0.1675	0.0975	3.1465	8
9	0.5439	6.5152	23.1404	1.8385	11.9780	0.1535	0.0835	3.5517	9
10	0.5083	7.0236	27.7156	1.9672	13.8164	0.1424	0.0724	3.9461	10
11	0.4751	7.4987	32.4665	2.1049	15.7836	0.1334	0.0634	4.3296	11
12	0.4440	7.9427	37.3506	2.2522	17.8885	0.1259	0.0559	4.7025	12
13	0.4150	8.3577	42.3302	2.4098	20.1406	0.1197	0.0497	5.0648	13
14	0.3878	8.7455	47.3718	2.5785	22.5505	0.1143	0.0443	5.4167	14
15	0.3624	9.1079	52.4461	2.7590	25.1290	0.1098	0.0398	5.7583	15
16	0.3387	9.4466	57.5271	2.9522	27.8881	0.1059	0.0359	6.0897	16
17	0.3166	9.7632	62.5923	3.1588	30.8402	0.1024	0.0324	6.4110	17
18	0.2959	10.0591	67.6219	3.3799	33.9990	0.0994	0.0294	6.7225	18
19	0.2765	10.3356	72.5991	3.6165	37.3790	0.0968	0.0268	7.0242	19
20	0.2584	10.5940	77.5091	3.8697	40.9955	0.0944	0.0244	7.3163	20
21	0.2415	10.8355	82.3393	4.1406	44.8652	0.0923	0.0223	7.5990	21
22	0.2257	11.0612	87.0793	4.4304	49.0057	0.0904	0.0204	7.8725	22
23	0.2109	11.2722	91.7201	4.7405	53.4361	0.0887	0.0187	8.1369	23
24	0.1971	11.4693	96.2545	5.0724	58.1767	0.0872	0.0172	8.3923	24
25	0.1842	11.6536	100.6765	5.4274	63.2490	0.0858	0.0158	8.6391	25
26	0.1722	11.8258	104.9814	5.8074	68.6765	0.0846	0.0146	8.8773	26
27	0.1609	11.9867	109.1656	6.2139	74.4838	0.0834	0.0134	9.1072	27
28	0.1504	12.1371	113.2264	6.6488	80.6977	0.0824	0.0124	9.3289	28
29	0.1406	12.2777	117.1622	7.1143	87.3465	0.0814	0.0114	9.5427	29
30	0.1314	12.4090	120.9718	7.6123	94.4608	0.0806	0.0106	9.7487	30
31	0.1228	12.5318	124.6550	8.1451	102.0730	0.0798	0.0098	9.9471	31
32	0.1147	12.6466	128.2120	8.7153	110.2182	0.0791	0.0091	10.1381	32
33	0.1072	12.7538	131.6435	9.3253	118.9334	0.0784	0.0084	10.3219	33
34	0.1002	12.8540	134.9507	9.9781	128.2588	0.0778	0.0078	10.4987	34
35	0.0937	12.9477	138.1353	10.6766	138.2369	0.0772	0.0072	10.6687	35
36	0.0875	13.0352	141.1990	11.4239	148.9135	0.0767	0.0067	10.8321	36
37	0.0818	13.1170	144.1441	12.2236	160.3374	0.0762	0.0062	10.9891	37
38	0.0765	13.1935	146.9730	13.0793	172.5610	0.0758	0.0058	11.1398	38
39	0.0715	13.2649	149.6883	13.9948	185.6403	0.0754	0.0054	11.2845	39
40	0.0668	13.3317	152.2928	14.9745	199.6351	0.0750	0.0050	11.4233	40
41	0.0624	13.3941	154.7892	16.0227	214.6096	0.0747	0.0047	11.5565	41
42	0.0583	13.4524	157.1807	17.1443	230.6322	0.0743	0.0043	11.6842	42
43	0.0545	13.5070	159.4702	18.3444	247.7765	0.0740	0.0040	11.8065	43
44	0.0509	13.5579	161.6609	19.6285	266.1209	0.0738	0.0038	11.9237	44
45	0.0476	13.6055	163.7559	21.0025	285.7493	0.0735	0.0035	12.0360	45
46	0.0445	13.6500	165.7584	22.4726	306.7518	0.0733	0.0033	12.1435	46
47	0.0416	13.6916	167.6714	24.0457	329.2244	0.0730	0.0030	12.2463	47
48	0.0389	13.7305	169.4981	25.7289	353.2701	0.0728	0.0028	12.3447	48
49	0.0363	13.7668	171.2417	27.5299	378.9990	0.0726	0.0026	12.4387	49
50	0.0339	13.8007	172.9051	29.4570	406.5289	0.0725	0.0025	12.5287	50
51	0.0317	13.8325	174.4915	31.5190	435.9860	0.0723	0.0023	12.6146	51
52	0.0297	13.8621	176.0037	33.7253	467.5050	0.0721	0.0021	12.6967	52
53	0.0277	13.8898	177.4447	36.0861	501.2303	0.0720	0.0020	12.7751	53
54	0.0259	13.9157	178.8173	38.6122	537.3164	0.0719	0.0019	12.8500	54
55	0.0242	13.9399	180.1243	41.3150	575.9286	0.0717	0.0017	12.9215	55
60	0.0173	14.0392	185.7677	57.9464	813.5204	0.0712	0.0012	13.2321	60
65	0.0123	14.1099	190.1452	81.2729	1146.7552	0.0709	0.0009	13.4760	65
70	0.0088	14.1604	193.5185	113.9894	1614.1342	0.0706	0.0006	13.6662	70
75	0.0063	14.1964	196.1035	159.8760	2269.6574	0.0704	0.0004	13.8136	75
80	0.0045	14.2220	198.0748	224.2344	3189.0627	0.0703	0.0003	13.9273	80
85	0.0032	14.2403	199.5717	314.5003	4478.5761	0.0702	0.0002	14.0146	85
90	0.0023	14.2533	200.7042	441.1030	6287.1854	0.0702	0.0002	14.0812	90
95	0.0016	14.2626	201.5581	618.6697	8823.8535	0.0701	0.0001	14.1319	95
100	0.0012	14.2693	202.2001	867.7163	12381.6618	0.0701	0.0001	14.1703	100

(*continued*)

APPENDIX 54.B *(continued)*
Cash Flow Equivalent Factors

$i = 8.00\%$

n	P/F	P/A	P/G	F/P	F/A	A/P	A/F	A/G	n
1	0.9259	0.9259	0.0000	1.0800	1.0000	1.0800	1.0000	0.0000	1
2	0.8573	1.7833	0.8573	1.1664	2.0800	0.5608	0.4808	0.4808	2
3	0.7938	2.5771	2.4450	1.2597	3.2464	0.3880	0.3080	0.9487	3
4	0.7350	3.3121	4.6501	1.3605	4.5061	0.3019	0.2219	1.4040	4
5	0.6806	3.9927	7.3724	1.4693	5.8666	0.2505	0.1705	1.8465	5
6	0.6302	4.6229	10.5233	1.5869	7.3359	0.2163	0.1363	2.2763	6
7	0.5835	5.2064	14.0242	1.7138	8.9228	0.1921	0.1121	2.6937	7
8	0.5403	5.7466	17.8061	1.8509	10.6366	0.1740	0.0940	3.0985	8
9	0.5002	6.2469	21.8081	1.9990	12.4876	0.1601	0.0801	3.4910	9
10	0.4632	6.7101	25.9768	2.1589	14.4866	0.1490	0.0690	3.8713	10
11	0.4289	7.1390	30.2657	2.3316	16.6455	0.1401	0.0601	4.2395	11
12	0.3971	7.5361	34.6339	2.5182	18.9771	0.1327	0.0527	4.5957	12
13	0.3677	7.9038	39.0463	2.7196	21.4953	0.1265	0.0465	4.9402	13
14	0.3405	8.2442	43.4723	2.9372	24.2149	0.1213	0.0413	5.2731	14
15	0.3152	8.5595	47.8857	3.1722	27.1521	0.1168	0.0368	5.5945	15
16	0.2919	8.8514	52.2640	3.4259	30.3243	0.1130	0.0330	5.9046	16
17	0.2703	9.1216	56.5883	3.7000	33.7502	0.1096	0.0296	6.2037	17
18	0.2502	9.3719	60.8426	3.9960	37.4502	0.1067	0.0267	6.4920	18
19	0.2317	9.6036	65.0134	4.3157	41.4463	0.1041	0.0241	6.7697	19
20	0.2145	9.8181	69.0898	4.6610	45.7620	0.1019	0.0219	7.0369	20
21	0.1987	10.0168	73.0629	5.0338	50.4229	0.0998	0.0198	7.2940	21
22	0.1839	10.2007	76.9257	5.4365	55.4568	0.0980	0.0180	7.5412	22
23	0.1703	10.3711	80.6726	5.8715	60.8933	0.0964	0.0164	7.7786	23
24	0.1577	10.5288	84.2997	6.3412	66.7648	0.0950	0.0150	8.0066	24
25	0.1460	10.6748	87.8041	6.8485	73.1059	0.0937	0.0137	8.2254	25
26	0.1352	10.8100	91.1842	7.3964	79.9544	0.0925	0.0125	8.4352	26
27	0.1252	10.9352	94.4390	7.9881	87.3508	0.0914	0.0114	8.6363	27
28	0.1159	11.0511	97.5687	8.6271	95.3388	0.0905	0.0105	8.8289	28
29	0.1073	11.1584	100.5738	9.3173	103.9659	0.0896	0.0096	9.0133	29
30	0.0994	11.2578	103.4558	10.0627	113.2832	0.0888	0.0088	9.1897	30
31	0.0920	11.3498	106.2163	10.8677	123.3459	0.0881	0.0081	9.3584	31
32	0.0852	11.4350	108.8575	11.7371	134.2135	0.0875	0.0075	9.5197	32
33	0.0789	11.5139	111.3819	12.6760	145.9506	0.0869	0.0069	9.6737	33
34	0.0730	11.5869	113.7924	13.6901	158.6267	0.0863	0.0063	9.8208	34
35	0.0676	11.6546	116.0920	14.7853	172.3168	0.0858	0.0058	9.9611	35
36	0.0626	11.7172	118.2839	15.9682	187.1021	0.0853	0.0053	10.0949	36
37	0.0580	11.7752	120.3713	17.2456	203.0703	0.0849	0.0049	10.2225	37
38	0.0537	11.8289	122.3579	18.6253	220.3159	0.0845	0.0045	10.3440	38
39	0.0497	11.8786	124.2470	20.1153	238.9412	0.0842	0.0042	10.4597	39
40	0.0460	11.9246	126.0422	21.7245	259.0565	0.0839	0.0039	10.5699	40
41	0.0426	11.9672	127.7470	23.4625	280.7810	0.0836	0.0036	10.6747	41
42	0.0395	12.0067	129.3651	25.3395	304.2435	0.0833	0.0033	10.7744	42
43	0.0365	12.0432	130.8998	27.3666	329.5830	0.0830	0.0030	10.8692	43
44	0.0338	12.0771	132.3547	29.5560	356.9496	0.0828	0.0028	10.9592	44
45	0.0313	12.1084	133.7331	31.9204	386.5056	0.0826	0.0026	11.0447	45
46	0.0290	12.1374	135.0384	34.4741	418.4261	0.0824	0.0024	11.1258	46
47	0.0269	12.1643	136.2739	37.2320	452.9002	0.0822	0.0022	11.2028	47
48	0.0249	12.1891	137.4428	40.2106	490.1322	0.0820	0.0020	11.2758	48
49	0.0230	12.2122	138.5480	43.4274	530.3427	0.0819	0.0019	11.3451	49
50	0.0213	12.2335	139.5928	46.9016	573.7702	0.0817	0.0017	11.4107	50
51	0.0197	12.2532	140.5799	50.6537	620.6718	0.0816	0.0016	11.4729	51
52	0.0183	12.2715	141.5121	54.7060	671.3255	0.0815	0.0015	11.5318	52
53	0.0169	12.2884	142.3923	59.0825	726.0316	0.0814	0.0014	11.5875	53
54	0.0157	12.3041	143.2229	63.8091	785.1141	0.0813	0.0013	11.6403	54
55	0.0145	12.3186	144.0065	68.9139	848.9232	0.0812	0.0012	11.6902	55
60	0.0099	12.3766	147.3000	101.2571	1253.2133	0.0808	0.0008	11.9015	60
65	0.0067	12.4160	149.7387	148.7798	1847.2481	0.0805	0.0005	12.0602	65
70	0.0046	12.4428	151.5326	218.6064	2720.0801	0.0804	0.0004	12.1783	70
75	0.0031	12.4611	152.8448	321.2045	4002.5566	0.0802	0.0002	12.2658	75
80	0.0021	12.4735	153.8001	471.9548	5886.9354	0.0802	0.0002	12.3301	80
85	0.0014	12.4820	154.4925	693.4565	8655.7061	0.0801	0.0001	12.3772	85
90	0.0010	12.4877	154.9925	1018.9151	12723.9386	0.0801	0.0001	12.4116	90
95	0.0007	12.4917	155.3524	1497.1205	18701.5069	0.0801	0.0001	12.4365	95
100	0.0005	12.4943	155.6107	2199.7613	27484.5157	0.0800	0.0000	12.4545	100

(continued)

APPENDIX 54.B *(continued)*
Cash Flow Equivalent Factors

$$i = 9.00\%$$

n	P/F	P/A	P/G	F/P	F/A	A/P	A/F	A/G	n
1	0.9174	0.9174	0.0000	1.0900	1.0000	1.0900	1.0000	0.0000	1
2	0.8417	1.7591	0.8417	1.1881	2.0900	0.5685	0.4785	0.4785	2
3	0.7722	2.5313	2.3860	1.2950	3.2781	0.3951	0.3051	0.9426	3
4	0.7084	3.2397	4.5113	1.4116	4.5731	0.3087	0.2187	1.3925	4
5	0.6499	3.8897	7.1110	1.5386	5.9847	0.2571	0.1671	1.8282	5
6	0.5963	4.4859	10.0924	1.6771	7.5233	0.2229	0.1329	2.2498	6
7	0.5470	5.0330	13.3746	1.8280	9.2004	0.1987	0.1087	2.6574	7
8	0.5019	5.5348	16.8877	1.9926	11.0285	0.1807	0.0907	3.0512	8
9	0.4604	5.9952	20.5711	2.1719	13.0210	0.1668	0.0768	3.4312	9
10	0.4224	6.4177	24.3728	2.3674	15.1929	0.1558	0.0658	3.7978	10
11	0.3875	6.8052	28.2481	2.5804	17.5603	0.1469	0.0569	4.1510	11
12	0.3555	7.1607	32.1590	2.8127	20.1407	0.1397	0.0497	4.4910	12
13	0.3262	7.4869	36.0731	3.0658	22.9534	0.1336	0.0436	4.8182	13
14	0.2992	7.7862	39.9633	3.3417	26.0192	0.1284	0.0384	5.1326	14
15	0.2745	8.0607	43.8069	3.6425	29.3609	0.1241	0.0341	5.4346	15
16	0.2519	8.3126	47.5849	3.9703	33.0034	0.1203	0.0303	5.7245	16
17	0.2311	8.5436	51.2821	4.3276	36.9737	0.1170	0.0270	6.0024	17
18	0.2120	8.7556	54.8860	4.7171	41.3013	0.1142	0.0242	6.2687	18
19	0.1945	8.9501	58.3868	5.1417	46.0185	0.1117	0.0217	6.5236	19
20	0.1784	9.1285	61.7770	5.6044	51.1601	0.1095	0.0195	6.7674	20
21	0.1637	9.2922	65.0509	6.1088	56.7645	0.1076	0.0176	7.0006	21
22	0.1502	9.4424	68.2048	6.6586	62.8733	0.1059	0.0159	7.2232	22
23	0.1378	9.5802	71.2359	7.2579	69.5319	0.1044	0.0144	7.4357	23
24	0.1264	9.7066	74.1433	7.9111	76.7898	0.1030	0.0130	7.6384	24
25	0.1160	9.8226	76.9265	8.6231	84.7009	0.1018	0.0118	7.8316	25
26	0.1064	9.9290	79.5863	9.3992	93.3240	0.1007	0.0107	8.0156	26
27	0.0976	10.0266	82.1241	10.2451	102.7231	0.0997	0.0097	8.1906	27
28	0.0895	10.1161	84.5419	11.1671	112.9682	0.0989	0.0089	8.3571	28
29	0.0822	10.1983	86.8422	12.1722	124.1354	0.0981	0.0081	8.5154	29
30	0.0754	10.2737	89.0280	13.2677	136.3076	0.0973	0.0073	8.6657	30
31	0.0691	10.3428	91.1024	14.4618	149.5752	0.0967	0.0067	8.8083	31
32	0.0634	10.4062	93.0690	15.7633	164.0370	0.0961	0.0061	8.9436	32
33	0.0582	10.4644	94.9314	17.1820	179.8003	0.0956	0.0056	9.0718	33
34	0.0534	10.5178	96.6935	18.7284	196.9823	0.0951	0.0051	9.1933	34
35	0.0490	10.5668	98.3590	20.4140	215.7108	0.0946	0.0046	9.3083	35
36	0.0449	10.6118	99.9319	22.2512	236.1247	0.0942	0.0042	9.4171	36
37	0.0412	10.6530	101.4162	24.2538	258.3759	0.0939	0.0039	9.5200	37
38	0.0378	10.6908	102.8158	26.4367	282.6298	0.0935	0.0035	9.6172	38
39	0.0347	10.7255	104.1345	28.8160	309.0665	0.0932	0.0032	9.7090	39
40	0.0318	10.7574	105.3762	31.4094	337.8824	0.0930	0.0030	9.7957	40
41	0.0292	10.7866	106.5445	34.2363	369.2919	0.0927	0.0027	9.8775	41
42	0.0268	10.8134	107.6432	37.3175	403.5281	0.0925	0.0025	9.9546	42
43	0.0246	10.8380	108.6758	40.6761	440.8457	0.0923	0.0023	10.0273	43
44	0.0226	10.8605	109.6456	44.3370	481.5218	0.0921	0.0021	10.0958	44
45	0.0207	10.8812	110.5561	48.3273	525.8587	0.0919	0.0019	10.1603	45
46	0.0190	10.9002	111.4103	52.6767	574.1860	0.0917	0.0017	10.2210	46
47	0.0174	10.9176	112.2115	57.4176	626.8628	0.0916	0.0016	10.2780	47
48	0.0160	10.9336	112.9625	62.5852	684.2804	0.0915	0.0015	10.3317	48
49	0.0147	10.9482	113.6661	68.2179	746.8656	0.0913	0.0013	10.3821	49
50	0.0134	10.9617	114.3251	74.3575	815.0836	0.0912	0.0012	10.4295	50
51	0.0123	10.9740	114.9420	81.0497	889.4411	0.0911	0.0011	10.4740	51
52	0.0113	10.9853	115.5193	88.3442	970.4908	0.0910	0.0010	10.5158	52
53	0.0104	10.9957	116.0593	96.2951	1058.8349	0.0909	0.0009	10.5549	53
54	0.0095	11.0053	116.5642	104.9617	1155.1301	0.0909	0.0009	10.5917	54
55	0.0087	11.0140	117.0362	114.4083	1260.0918	0.0908	0.0008	10.6261	55
60	0.0057	11.0480	118.9683	176.0313	1944.7921	0.0905	0.0005	10.7683	60
65	0.0037	11.0701	120.3344	270.8460	2998.2885	0.0903	0.0003	10.8702	65
70	0.0024	11.0844	121.2942	416.7301	4619.2232	0.0902	0.0002	10.9427	70
75	0.0016	11.0938	121.9646	641.1909	7113.2321	0.0901	0.0001	10.9940	75
80	0.0010	11.0998	122.4306	986.5517	10950.5741	0.0901	0.0001	11.0299	80
85	0.0007	11.1038	122.7533	1517.9320	16854.8003	0.0901	0.0001	11.0551	85
90	0.0004	11.1064	122.9758	2335.5266	25939.1842	0.0900	0.0000	11.0726	90
95	0.0003	11.1080	123.1287	3593.4971	39916.6350	0.0900	0.0000	11.0847	95
100	0.0002	11.1091	123.2335	5529.0408	61422.6755	0.0900	0.0000	11.0930	100

(continued)

APPENDIX 54.B *(continued)*
Cash Flow Equivalent Factors

$$i = 10.00\%$$

n	P/F	P/A	P/G	F/P	F/A	A/P	A/F	A/G	n
1	0.9091	0.9091	0.0000	1.1000	1.0000	1.1000	1.0000	0.0000	1
2	0.8264	1.7355	0.8264	1.2100	2.1000	0.5762	0.4762	0.4762	2
3	0.7513	2.4869	2.3291	1.3310	3.3100	0.4021	0.3021	0.9366	3
4	0.6830	3.1699	4.3781	1.4641	4.6410	0.3155	0.2155	1.3812	4
5	0.6209	3.7908	6.8618	1.6105	6.1051	0.2638	0.1638	1.8101	5
6	0.5645	4.3553	9.6842	1.7716	7.7156	0.2296	0.1296	2.2236	6
7	0.5132	4.8684	12.7631	1.9487	9.4872	0.2054	0.1054	2.6216	7
8	0.4665	5.3349	16.0287	2.1436	11.4359	0.1874	0.0874	3.0045	8
9	0.4241	5.7590	19.4215	2.3579	13.5795	0.1736	0.0736	3.3724	9
10	0.3855	6.1446	22.8913	2.5937	15.9374	0.1627	0.0627	3.7255	10
11	0.3505	6.4951	26.3963	2.8531	18.5312	0.1540	0.0540	4.0641	11
12	0.3186	6.8137	29.9012	3.1384	21.3843	0.1468	0.0468	4.3884	12
13	0.2897	7.1034	33.3772	3.4523	24.5227	0.1408	0.0408	4.6988	13
14	0.2633	7.3667	36.8005	3.7975	27.9750	0.1357	0.0357	4.9955	14
15	0.2394	7.6061	40.1520	4.1772	31.7725	0.1315	0.0315	5.2789	15
16	0.2176	7.8237	43.4164	4.5950	35.9497	0.1278	0.0278	5.5493	16
17	0.1978	8.0216	46.5819	5.0545	40.5447	0.1247	0.0247	5.8071	17
18	0.1799	8.2014	49.6395	5.5599	45.5992	0.1219	0.0219	6.0526	18
19	0.1635	8.3649	52.5827	6.1159	51.1591	0.1195	0.0195	6.2861	19
20	0.1486	8.5136	55.4069	6.7275	57.2750	0.1175	0.0175	6.5081	20
21	0.1351	8.6487	58.1095	7.4002	64.0025	0.1156	0.0156	6.7189	21
22	0.1228	8.7715	60.6893	8.1403	71.4027	0.1140	0.0140	6.9189	22
23	0.1117	8.8832	63.1462	8.9543	79.5430	0.1126	0.0126	7.1085	23
24	0.1015	8.9847	65.4813	9.8497	88.4973	0.1113	0.0113	7.2881	24
25	0.0923	9.0770	67.6964	10.8347	98.3471	0.1102	0.0102	7.4580	25
26	0.0839	9.1609	69.7940	11.9182	109.1818	0.1092	0.0092	7.6186	26
27	0.0763	9.2372	71.7773	13.1100	121.0999	0.1083	0.0083	7.7704	27
28	0.0693	9.3066	73.6495	14.4210	134.2099	0.1075	0.0075	7.9137	28
29	0.0630	9.3696	75.4146	15.8631	148.6309	0.1067	0.0067	8.0489	29
30	0.0573	9.4269	77.0766	17.4494	164.4940	0.1061	0.0061	8.1762	30
31	0.0521	9.4790	78.6395	19.1943	181.9434	0.1055	0.0055	8.2962	31
32	0.0474	9.5264	80.1078	21.1138	201.1378	0.1050	0.0050	8.4091	32
33	0.0431	9.5694	81.4856	23.2252	222.2515	0.1045	0.0045	8.5152	33
34	0.0391	9.6086	82.7773	25.5477	245.4767	0.1041	0.0041	8.6149	34
35	0.0356	9.6442	83.9872	28.1024	271.0244	0.1037	0.0037	8.7086	35
36	0.0323	9.6765	85.1194	30.9127	299.1268	0.1033	0.0033	8.7965	36
37	0.0294	9.7059	86.1781	34.0039	330.0395	0.1030	0.0030	8.8789	37
38	0.0267	9.7327	87.1673	37.4043	364.0434	0.1027	0.0027	8.9562	38
39	0.0243	9.7570	88.0908	41.1448	401.4478	0.1025	0.0025	9.0285	39
40	0.0221	9.7791	88.9525	45.2593	442.5926	0.1023	0.0023	9.0962	40
41	0.0201	9.7991	89.7560	49.7852	487.8518	0.1020	0.0020	9.1596	41
42	0.0183	9.8174	90.5047	54.7637	537.6370	0.1019	0.0019	9.2188	42
43	0.0166	9.8340	91.2019	60.2401	592.4007	0.1017	0.0017	9.2741	43
44	0.0151	9.8491	91.8508	66.2641	652.6408	0.1015	0.0015	9.3258	44
45	0.0137	9.8628	92.4544	72.8905	718.9048	0.1014	0.0014	9.3740	45
46	0.0125	9.8753	93.0157	80.1795	791.7953	0.1013	0.0013	9.4190	46
47	0.0113	9.8866	93.5372	88.1975	871.9749	0.1011	0.0011	9.4610	47
48	0.0103	9.8969	94.0217	97.0172	960.1723	0.1010	0.0010	9.5001	48
49	0.0094	9.9063	94.4715	106.7190	1057.1896	0.1009	0.0009	9.5365	49
50	0.0085	9.9148	94.8889	117.3909	1163.9085	0.1009	0.0009	9.5704	50
51	0.0077	9.9226	95.2761	129.1299	1281.2994	0.1008	0.0008	9.6020	51
52	0.0070	9.9296	95.6351	142.0429	1410.4293	0.1007	0.0007	9.6313	52
53	0.0064	9.9360	95.9679	156.2472	1552.4723	0.1006	0.0006	9.6586	53
54	0.0058	9.9418	96.2763	171.8719	1708.7195	0.1006	0.0006	9.6840	54
55	0.0053	9.9471	96.5619	189.0591	1880.5914	0.1005	0.0005	9.7075	55
60	0.0033	9.9672	97.7010	304.4816	3034.8164	0.1003	0.0003	9.8023	60
65	0.0020	9.9796	98.4705	490.3707	4893.7073	0.1002	0.0002	9.8672	65
70	0.0013	9.9873	98.9870	789.7470	7887.4696	0.1001	0.0001	9.9113	70
75	0.0008	9.9921	99.3317	1271.8954	12708.9537	0.1001	0.0001	9.9410	75
80	0.0005	9.9951	99.5606	2048.4002	20474.0021	0.1000	0.0000	9.9609	80
85	0.0003	9.9970	99.7120	3298.9690	32979.6903	0.1000	0.0000	9.9742	85
90	0.0002	9.9981	99.8118	5313.0226	53120.2261	0.1000	0.0000	9.9831	90
95	0.0001	9.9988	99.8773	8556.6760	85556.7605	0.1000	0.0000	9.9889	95
100	0.0001	9.9993	99.9202	13780.6123	137796.1234	0.1000	0.0000	9.9927	100

(continued)

APPENDIX 54.B *(continued)*
Cash Flow Equivalent Factors

$$i = 12.00\%$$

n	P/F	P/A	P/G	F/P	F/A	A/P	A/F	A/G	n
1	0.8929	0.8929	0.0000	1.1200	1.0000	1.1200	1.0000	0.0000	1
2	0.7972	1.6901	0.7972	1.2544	2.1200	0.5917	0.4717	0.4717	2
3	0.7118	2.4018	2.2208	1.4049	3.3744	0.4163	0.2963	0.9246	3
4	0.6355	3.0373	4.1273	1.5735	4.7793	0.3292	0.2092	1.3589	4
5	0.5674	3.6048	6.3970	1.7623	6.3528	0.2774	0.1574	1.7746	5
6	0.5066	4.1114	8.9302	1.9738	8.1152	0.2432	0.1232	2.1720	6
7	0.4523	4.5638	11.6443	2.2107	10.0890	0.2191	0.0991	2.5515	7
8	0.4039	4.9676	14.4714	2.4760	12.2997	0.2013	0.0813	2.9131	8
9	0.3606	5.3282	17.3563	2.7731	14.7757	0.1877	0.0677	3.2574	9
10	0.3220	5.6502	20.2541	3.1058	17.5487	0.1770	0.0570	3.5847	10
11	0.2875	5.9377	23.1288	3.4785	20.6546	0.1684	0.0484	3.8953	11
12	0.2567	6.1944	25.9523	3.8960	24.1331	0.1614	0.0414	4.1897	12
13	0.2292	6.4235	28.7024	4.3635	28.0291	0.1557	0.0357	4.4683	13
14	0.2046	6.6282	31.3624	4.8871	32.3926	0.1509	0.0309	4.7317	14
15	0.1827	6.8109	33.9202	5.4736	37.2797	0.1468	0.0268	4.9803	15
16	0.1631	6.9740	36.3670	6.1304	42.7533	0.1434	0.0234	5.2147	16
17	0.1456	7.1196	38.6973	6.8660	48.8837	0.1405	0.0205	5.4353	17
18	0.1300	7.2497	40.9080	7.6900	55.7497	0.1379	0.0179	5.6427	18
19	0.1161	7.3658	42.9979	8.6128	63.4397	0.1358	0.0158	6.8375	19
20	0.1037	7.4694	44.9676	9.6463	72.0524	0.1339	0.0139	6.0202	20
21	0.0926	7.5620	46.8188	10.8038	81.6987	0.1322	0.0122	6.1913	21
22	0.0826	7.6446	48.5543	12.1003	92.5026	0.1308	0.0108	6.3514	22
23	0.0738	7.7184	50.1776	13.5523	104.6029	0.1296	0.0096	6.5010	23
24	0.0659	7.7843	51.6929	15.1786	118.1552	0.1285	0.0085	6.6406	24
25	0.0588	7.8431	53.1046	17.0001	133.3339	0.1275	0.0075	6.7708	25
26	0.0525	7.8957	54.4177	19.0401	150.3339	0.1267	0.0067	6.8921	26
27	0.0469	7.9426	55.6369	21.3249	169.3740	0.1259	0.0059	7.0049	27
28	0.0419	7.9844	56.7674	23.8839	190.6989	0.1252	0.0052	7.1098	28
29	0.0374	8.0218	57.8141	26.7499	214.5828	0.1247	0.0047	7.2071	29
30	0.0334	8.0552	58.7821	29.9599	241.3327	0.1241	0.0041	7.2974	30
31	0.0298	8.0850	59.6761	33.5551	271.2926	0.1237	0.0037	7.3811	31
32	0.0266	8.1116	60.5010	37.5817	304.8477	0.1233	0.0033	7.4586	32
33	0.0238	8.1354	61.2612	42.0915	342.4294	0.1229	0.0029	7.5302	33
34	0.0212	8.1566	61.9612	47.1425	384.5210	0.1226	0.0026	7.5965	34
35	0.0189	8.1755	62.6052	52.7996	431.6635	0.1223	0.0023	7.6577	35
36	0.0169	8.1924	63.1970	59.1356	484.4631	0.1221	0.0021	7.7141	36
37	0.0151	8.2075	63.7406	66.2318	543.5987	0.1218	0.0018	7.7661	37
38	0.0135	8.2210	64.2394	74.1797	609.8305	0.1216	0.0016	7.8141	38
39	0.0120	8.2330	64.6967	83.0812	684.0102	0.1215	0.0015	7.8582	39
40	0.0107	8.2438	65.1159	93.0510	767.0914	0.1213	0.0013	7.8988	40
41	0.0096	8.2534	65.4997	104.2171	860.1424	0.1212	0.0012	7.9361	41
42	0.0086	8.2619	65.8509	116.7231	964.3595	0.1210	0.0010	7.9704	42
43	0.0076	8.2696	66.1722	130.7299	1081.0826	0.1209	0.0009	8.0019	43
44	0.0068	8.2764	66.4659	146.4175	1211.8125	0.1208	0.0008	8.0308	44
45	0.0061	8.2825	66.7342	163.9876	1358.2300	0.1207	0.0007	8.0572	45
46	0.0054	8.2880	66.9792	183.6661	1522.2176	0.1207	0.0007	8.0815	46
47	0.0049	8.2928	67.2028	205.7061	1705.8838	0.1206	0.0006	8.1037	47
48	0.0043	8.2972	67.4068	230.3908	1911.5898	0.1205	0.0005	8.1241	48
49	0.0039	8.3010	67.5929	258.0377	2141.9806	0.1205	0.0005	8.1427	49
50	0.0035	8.3045	67.7624	289.0022	2400.0182	0.1204	0.0004	8.1597	50
51	0.0031	8.3076	67.9169	323.6825	2689.0204	0.1204	0.0004	8.1753	51
52	0.0028	8.3103	68.0576	362.5243	3012.7029	0.1203	0.0003	8.1895	52
53	0.0025	8.3128	68.1856	406.0273	3375.2272	0.1203	0.0003	8.2025	53
54	0.0022	8.3150	68.3022	454.7505	3781.2545	0.1203	0.0003	8.2143	54
55	0.0020	8.3170	68.4082	509.3206	4236.0050	0.1202	0.0002	8.2251	55
60	0.0011	8.3240	68.8100	897.5969	7471.6411	0.1201	0.0001	8.2664	60
65	0.0006	8.3281	69.0581	1581.8725	13173.9374	0.1201	0.0001	8.2922	65
70	0.0004	8.3303	69.2103	2787.7998	23223.3319	0.1200	0.0000	8.3082	70
75	0.0002	8.3316	69.3031	4913.0558	40933.7987	0.1200	0.0000	8.3181	75
80	0.0001	8.3324	69.3594	8658.4831	72145.6925	0.1200	0.0000	8.3241	80
85	0.0001	8.3328	69.3935	15259.2057	127151.7140	0.1200	0.0000	8.3278	85
90	0.0000	8.3330	69.4140	26891.9342	224091.1185	0.1200	0.0000	8.3300	90
95	0.0000	8.3332	69.4263	47392.7766	394931.4719	0.1200	0.0000	8.3313	95
100	0.0000	8.3332	69.4336	83522.2657	696010.5477	0.1200	0.0000	8.3321	100

(continued)

APPENDIX 54.B *(continued)*
Cash Flow Equivalent Factors

$i = 15.00\%$

n	P/F	P/A	P/G	F/P	F/A	A/P	A/F	A/G	n
1	0.8696	0.8696	0.0000	1.1500	1.0000	1.1500	1.0000	0.0000	1
2	0.7561	1.6257	0.7561	1.3225	2.1500	0.6151	0.4651	0.4651	2
3	0.6575	2.2832	2.0712	1.5209	3.4725	0.4380	0.2880	0.9071	3
4	0.5718	2.8550	3.7864	1.7490	4.9934	0.3503	0.2003	1.3263	4
5	0.4972	3.3522	5.7751	2.0114	6.7424	0.2983	0.1483	1.7228	5
6	0.4323	3.7845	7.9368	2.3131	8.7537	0.2642	0.1142	2.0972	6
7	0.3759	4.1604	10.1924	2.0000	11.0668	0.2404	0.0904	2.4498	7
8	0.3269	4.4873	12.4807	3.0590	13.7268	0.2229	0.0729	2.7813	8
9	0.2843	4.7716	14.7548	3.5179	16.7858	0.2096	0.0596	3.0922	9
10	0.2472	5.0188	16.9795	4.0456	20.3037	0.1993	0.0493	3.3832	10
11	0.2149	5.2337	19.1289	4.6524	24.3493	0.1911	0.0411	3.6549	11
12	0.1869	5.4206	21.1849	5.3503	29.0017	0.1845	0.0345	3.9082	12
13	0.1625	5.5831	23.1352	6.1528	34.3519	0.1791	0.0291	4.1438	13
14	0.1413	5.7245	24.9725	7.0757	40.5047	0.1747	0.0247	4.3624	14
15	0.1229	5.8474	26.9630	8.1371	47.5804	0.1710	0.0210	4.5650	15
16	0.1069	5.9542	28.2960	9.3576	55.7175	0.1679	0.0179	4.7522	16
17	0.0929	6.0472	29.7828	10.7613	65.0751	0.1654	0.0154	4.9251	17
18	0.0808	6.1280	31.1565	12.3755	75.8364	0.1632	0.0132	5.0843	18
19	0.0703	6.1982	32.4213	14.2318	88.2118	0.1613	0.0113	5.2307	19
20	0.0611	6.2593	33.5822	16.3665	102.4436	0.1598	0.0098	5.3651	20
21	0.0531	6.3125	34.6448	18.8215	118.8101	0.1584	0.0084	5.4883	21
22	0.0462	6.3587	35.6150	21.6447	137.6316	0.1573	0.0073	5.6010	22
23	0.0402	6.3988	36.4988	24.8915	159.2764	0.1563	0.0063	5.7040	23
24	0.0349	6.4338	37.3023	28.6252	184.1678	0.1554	0.0054	5.7979	24
25	0.0304	6.4641	38.0314	32.9190	212.7930	0.1547	0.0047	5.8834	25
26	0.0264	6.4906	38.6918	37.8568	245.7120	0.1541	0.0041	5.9612	26
27	0.0230	6.5135	39.2890	43.5353	283.5688	0.1535	0.0035	6.0319	27
28	0.0200	6.5335	39.8283	50.0656	327.1041	0.1531	0.0031	6.0960	28
29	0.0174	6.5509	40.3146	57.5755	377.1697	0.1527	0.0027	6.1541	29
30	0.0151	6.5660	40.7526	66.2118	434.7451	0.1523	0.0023	6.2066	30
31	0.0131	6.5791	41.1466	76.1435	500.9569	0.1520	0.0020	6.2541	31
32	0.0114	6.5905	41.5006	87.5651	577.1005	0.1517	0.0017	6.2970	32
33	0.0099	6.6005	41.8184	100.6998	664.6655	0.1515	0.0015	6.3357	33
34	0.0086	6.6091	42.1033	115.8048	765.3654	0.1513	0.0013	6.3705	34
35	0.0075	6.6166	42.3586	133.1755	881.1702	0.1511	0.0011	6.4019	35
36	0.0065	6.6231	42.5872	153.1519	1014.3457	0.1510	0.0010	6.4301	36
37	0.0057	6.6288	42.7916	176.1246	1167.4975	0.1509	0.0009	6.4554	37
38	0.0049	6.6338	42.9743	202.5433	1343.6222	0.1507	0.0007	6.4781	38
39	0.0043	6.6380	43.1374	232.9248	1546.1655	0.1506	0.0006	6.4985	39
40	0.0037	6.6418	43.2830	267.8635	1779.0903	0.1506	0.0006	6.5168	40
41	0.0032	6.6450	43.4128	308.0431	2046.9539	0.1505	0.0005	6.5331	41
42	0.0028	6.6478	43.5286	354.2495	2354.9969	0.1504	0.0004	6.5478	42
43	0.0025	6.6503	43.6317	407.3870	2709.2465	0.1504	0.0004	6.5609	43
44	0.0021	6.6524	43.7235	468.4950	3116.6334	0.1503	0.0003	6.5725	44
45	0.0019	6.6543	43.8051	538.7693	3585.1285	0.1503	0.0003	6.5830	45
46	0.0016	6.6559	43.8778	619.5847	4123.8977	0.1502	0.0002	6.5923	46
47	0.0014	6.6573	43.9423	712.5224	4743.4824	0.1502	0.0002	6.6006	47
48	0.0012	6.6585	43.9997	819.4007	5456.0047	0.1502	0.0002	6.6080	48
49	0.0011	6.6596	44.0506	942.3108	6275.4055	0.1502	0.0002	6.6146	49
50	0.0009	6.6605	44.0958	1083.6574	7217.7163	0.1501	0.0001	6.6205	50
51	0.0008	6.6613	44.1360	1246.2061	8301.3737	0.1501	0.0001	6.6257	51
52	0.0007	6.6620	44.1715	1433.1370	9547.5798	0.1501	0.0001	6.6304	52
53	0.0006	6.6626	44.2031	1648.1075	10980.7167	0.1501	0.0001	6.6345	53
54	0.0005	6.6631	44.2311	1895.3236	12628.8243	0.1501	0.0001	6.6382	54
55	0.0005	6.6636	44.2558	2179.6222	14524.1479	0.1501	0.0001	6.6414	55
60	0.0002	6.6651	44.3431	4383.9987	29219.9916	0.1500	0.0000	6.6530	60
65	0.0001	6.6659	44.3903	8817.7874	58778.5826	0.1500	0.0000	6.6593	65
70	0.0001	6.6663	44.4156	17735.7200	118231.4669	0.1500	0.0000	6.6627	70
75	0.0000	6.6665	44.4292	35672.8680	237812.4532	0.1500	0.0000	6.6646	75
80	0.0000	6.6666	44.4364	71750.8794	478332.5293	0.1500	0.0000	6.6656	80
85	0.0000	6.6666	44.4402	144316.6470	962104.3133	0.1500	0.0000	6.6661	85
90	0.0000	6.6666	44.4422	290272.3252	1935142.1680	0.1500	0.0000	6.6664	90
95	0.0000	6.6667	44.4433	583841.3276	3892268.8509	0.1500	0.0000	6.6665	95
100	0.0000	6.6667	44.4438	1174313.4507	7828749.6713	0.1500	0.0000	6.6666	100

(continued)

APPENDIX 54.B *(continued)*
Cash Flow Equivalent Factors

$$i = 20.00\%$$

n	P/F	P/A	P/G	F/P	F/A	A/P	A/F	A/G	n
1	0.8333	0.8333	0.0000	1.2000	1.0000	1.2000	1.0000	0.0000	1
2	0.6944	1.5278	0.6944	1.4400	2.2000	0.6545	0.4545	0.4545	2
3	0.5787	2.1065	1.8519	1.7280	3.6400	0.4747	0.2747	0.8791	3
4	0.4823	2.5887	3.2986	2.0736	5.3680	0.3863	0.1863	1.2742	4
5	0.4019	2.9906	4.9061	2.4883	7.4416	0.3344	0.1344	1.6405	5
6	0.3349	3.3255	6.5806	2.9860	9.9299	0.3007	0.1007	1.9788	6
7	0.2791	3.6046	8.2551	3.5832	12.9159	0.2774	0.0774	2.2902	7
8	0.2326	3.8372	9.8831	4.2998	16.4991	0.2606	0.0606	2.5756	8
9	0.1938	4.0310	11.4335	5.1598	20.7989	0.2481	0.0481	2.8364	9
10	0.1615	4.1925	12.8871	6.1917	25.9587	0.2385	0.0385	3.0739	10
11	0.1346	4.3271	14.2330	7.4301	32.1504	0.2311	0.0311	3.2893	11
12	0.1122	4.4392	15.4667	8.9161	39.5805	0.2253	0.0253	3.4841	12
13	0.0935	4.5327	16.5883	10.6993	48.4966	0.2206	0.0206	3.6597	13
14	0.0779	4.6106	17.6008	12.8392	59.1959	0.2169	0.0169	3.8175	14
15	0.0649	4.6755	18.5095	15.4070	72.0351	0.2139	0.0139	3.9588	15
16	0.0541	4.7296	19.3208	18.4884	87.4421	0.2114	0.0114	4.0851	16
17	0.0451	4.7746	20.0419	22.1861	105.9306	0.2094	0.0094	4.1976	17
18	0.0376	4.8122	20.6805	26.6233	128.1167	0.2078	0.0078	4.2975	18
19	0.0313	4.8435	21.2439	31.9480	154.7400	0.2065	0.0065	4.3861	19
20	0.0261	4.8696	21.7395	38.3376	186.6880	0.2054	0.0054	4.4643	20
21	0.0217	4.8913	22.1742	46.0051	225.0256	0.2044	0.0044	4.5334	21
22	0.0181	4.9094	22.5546	55.2061	271.0307	0.2037	0.0037	4.5941	22
23	0.0151	4.9245	22.8867	66.2474	326.2369	0.2031	0.0031	4.6475	23
24	0.0126	4.9371	23.1760	79.4968	392.4842	0.2025	0.0025	4.6943	24
25	0.0105	4.9476	23.4276	95.3962	471.9811	0.2021	0.0021	4.7352	25
26	0.0087	4.9563	23.6460	114.4755	567.3773	0.2018	0.0018	4.7709	26
27	0.0073	4.9636	23.8353	137.3706	681.8528	0.2015	0.0015	4.8020	27
28	0.0061	4.9697	23.9991	164.8447	819.2233	0.2012	0.0012	4.8291	28
29	0.0051	4.9747	24.1406	197.8136	984.0680	0.2010	0.0010	4.8527	29
30	0.0042	4.9789	24.2628	237.3763	1181.8816	0.2008	0.0008	4.8731	30
31	0.0035	4.9824	24.3681	284.8516	1419.2579	0.2007	0.0007	4.8908	31
32	0.0029	4.9854	24.4588	341.8219	1704.1095	0.2006	0.0006	4.9061	32
33	0.0024	4.9878	24.5368	410.1863	2045.9314	0.2005	0.0005	4.9194	33
34	0.0020	4.9898	24.6038	492.2235	2456.1176	0.2004	0.0004	4.9308	34
35	0.0017	4.9915	24.6614	590.6682	2948.3411	0.2003	0.0003	4.9406	35
36	0.0014	4.9929	24.7108	708.8019	3539.0094	0.2003	0.0003	4.9491	36
37	0.0012	4.9941	24.7531	850.5622	4247.8112	0.2002	0.0002	4.9564	37
38	0.0010	4.9951	24.7894	1020.6747	5098.3735	0.2002	0.0002	4.9627	38
39	0.0008	4.9959	24.8204	1224.8096	6119.0482	0.2002	0.0002	4.9681	39
40	0.0007	4.9966	24.8469	1469.7716	7343.8578	0.2001	0.0001	4.9728	40
41	0.0006	4.9972	24.8696	1763.7259	8813.6294	0.2001	0.0001	4.9767	41
42	0.0005	4.9976	24.8890	2116.4711	10577.3553	0.2001	0.0001	4.9801	42
43	0.0004	4.9980	24.9055	2539.7653	12693.8263	0.2001	0.0001	4.9831	43
44	0.0003	4.9984	24.9196	3047.7183	15233.5916	0.2001	0.0001	4.9856	44
45	0.0003	4.9986	24.9316	3657.2620	18281.3099	0.2001	0.0001	4.9877	45
46	0.0002	4.9989	24.9419	4388.7144	21938.5719	0.2000	0.0000	4.9895	46
47	0.0002	4.9991	24.9506	5266.4573	26327.2863	0.2000	0.0000	4.9911	47
48	0.0002	4.9992	24.9581	6319.7487	31593.7436	0.2000	0.0000	4.9924	48
49	0.0001	4.9993	24.9642	7583.6985	37913.4923	0.2000	0.0000	4.9935	49
50	0.0001	4.9995	24.9698	9100.4382	45497.1908	0.2000	0.0000	4.9945	50
51	0.0001	4.9995	24.9744	10920.5258	54597.6289	0.2000	0.0000	4.9953	51
52	0.0001	4.9996	24.9783	13104.6309	65518.1547	0.2000	0.0000	4.9960	52
53	0.0001	4.9997	24.9816	15725.5571	78622.7856	0.2000	0.0000	4.9966	53
54	0.0001	4.9997	24.9844	18870.6685	94348.3427	0.2000	0.0000	4.9971	54
55	0.0000	4.9998	24.9868	22644.8023	113219.0113	0.2000	0.0000	4.9976	55
60	0.0000	4.9999	24.9942	56347.5144	281732.5718	0.2000	0.0000	4.9989	60
65	0.0000	5.0000	24.9975	140210.6469	701048.2346	0.2000	0.0000	4.9995	65
70	0.0000	5.0000	24.9989	348888.9569	1744439.7847	0.2000	0.0000	4.9998	70
75	0.0000	5.0000	24.9995	868147.3693	4340731.8466	0.2000	0.0000	4.9999	75

(continued)

APPENDIX 54.B (continued)
Cash Flow Equivalent Factors

$i = 25.00\%$

n	P/F	P/A	P/G	F/P	F/A	A/P	A/F	A/G	n
1	0.8000	0.8000	0.0000	1.2500	1.0000	1.2500	1.0000	0.0000	1
2	0.6400	1.4400	0.6400	1.5625	2.2500	0.6944	0.0444	0.4444	2
3	0.5120	1.9520	1.6640	1.9531	3.8125	0.5123	0.2623	0.8525	3
4	0.4096	2.3616	2.8928	2.4414	5.7656	0.4234	0.1734	1.2249	4
5	0.3277	2.6893	4.2035	3.0518	8.2070	0.3718	0.1218	1.5631	5
6	0.2621	2.9514	5.5142	3.8147	11.2588	0.3383	0.0888	1.8683	6
7	0.2007	3.1611	6.7725	4.7684	15.0735	0.3163	0.0663	2.1424	7
8	0.1678	3.3289	7.9469	5.9605	19.8419	0.3004	0.0504	2.3872	8
9	0.1342	3.4631	9.0207	7.4506	25.8023	0.2888	0.0388	2.6048	9
10	0.1074	3.5705	9.9870	9.3132	33.2529	0.2801	0.0301	2.7971	10
11	0.0859	3.6564	10.8460	11.6415	42.5661	0.2735	0.0235	2.9663	11
12	0.0687	3.7251	11.6020	14.5519	54.2077	0.2684	0.0184	3.1145	12
13	0.0550	3.7801	12.2617	18.1899	68.7596	0.2645	0.0145	3.2437	13
14	0.0440	3.8241	12.8334	22.7374	86.9495	0.2615	0.0115	3.3559	14
15	0.0352	3.8593	13.3260	28.4217	109.6868	0.2591	0.0091	3.4530	15
16	0.0281	3.8874	13.7482	35.5271	138.1085	0.2572	0.0072	3.5366	16
17	0.0225	3.9099	14.1085	44.4089	173.6357	0.2558	0.0058	3.6084	17
18	0.0180	3.9279	14.4147	55.5112	218.0446	0.2546	0.0046	3.6698	18
19	0.0144	3.9424	14.6741	69.3889	273.5558	0.2537	0.0037	3.7222	19
20	0.0115	3.9539	14.8932	86.7362	342.9447	0.2529	0.0029	3.7667	20
21	0.0092	3.9631	15.0777	108.4202	429.6809	0.2523	0.0023	3.8045	21
22	0.0074	3.9705	15.2326	135.5253	538.1011	0.2519	0.0019	3.8365	22
23	0.0059	3.9764	15.3625	169.4066	673.6264	0.2515	0.0015	3.8634	23
24	0.0047	3.9811	15.4711	211.7582	843.0329	0.2512	0.0012	3.8861	24
25	0.0038	3.9849	15.5618	264.6978	1054.7912	0.2509	0.0009	3.9052	25
26	0.0030	3.9879	15.6373	330.8722	1319.4890	0.2508	0.0008	3.9212	26
27	0.0024	3.9903	15.7002	413.5903	1650.3612	0.2506	0.0006	3.9346	27
28	0.0019	3.9923	15.7524	516.9879	2063.9515	0.2505	0.0005	3.9457	28
29	0.0015	3.9938	15.7957	646.2349	2580.9394	0.2504	0.0004	3.9551	29
30	0.0012	3.9950	15.8316	807.7936	3227.1743	0.2503	0.0003	3.9628	30
31	0.0010	3.9960	15.8614	1009.7420	4034.9678	0.2502	0.0002	3.9693	31
32	0.0008	3.9968	15.8859	1262.1774	5044.7098	0.2502	0.0002	3.9746	32
33	0.0006	3.9975	15.9062	1577.7218	6306.8872	0.2502	0.0002	3.9791	33
34	0.0005	3.9980	15.9229	1972.1523	7884.6091	0.2501	0.0001	3.9828	34
35	0.0004	3.9984	15.9367	2465.1903	9856.7613	0.2501	0.0001	3.9858	35
36	0.0003	3.9987	15.9481	3081.4879	12321.9516	0.2501	0.0001	3.9883	36
37	0.0003	3.9990	15.9574	3851.8599	15403.4396	0.2501	0.0001	3.9904	37
38	0.0002	3.9992	15.9651	4814.8249	19255.2994	0.2501	0.0001	3.9921	38
39	0.0002	3.9993	15.9714	6018.5311	24070.1243	0.2500	0.0000	3.9935	39
40	0.0001	3.9995	15.9766	7523.1638	30088.6554	0.2500	0.0000	3.9947	40
41	0.0001	3.9996	15.9809	9403.9548	37611.8192	0.2500	0.0000	3.9956	41
42	0.0001	3.9997	15.9843	11754.9435	47015.7740	0.2500	0.0000	3.9964	42
43	0.0001	3.9997	15.9872	14693.6794	58770.7175	0.2500	0.0000	3.9971	43
44	0.0001	3.9998	15.9895	18367.0992	73464.3969	0.2500	0.0000	3.9976	44
45	0.0000	3.9998	15.9915	22958.8740	91831.4962	0.2500	0.0000	3.9980	45
46	0.0000	3.9999	15.9930	28698.5925	114790.3702	0.2500	0.0000	3.9984	46
47	0.0000	3.9999	15.9943	35873.2407	143488.9627	0.2500	0.0000	3.9987	47
48	0.0000	3.9999	15.9954	44841.5509	179362.2034	0.2500	0.0000	3.9989	48
49	0.0000	3.9999	15.9962	56051.9386	224203.7543	0.2500	0.0000	3.9991	49
50	0.0000	3.9999	15.9969	70064.9232	280255.6929	0.2500	0.0000	3.9993	50
51	0.0000	4.0000	15.9975	87581.1540	350320.6161	0.2500	0.0000	3.9994	51
52	0.0000	4.0000	15.9980	109476.4425	437901.7701	0.2500	0.0000	3.9995	52
53	0.0000	4.0000	15.9983	136845.5532	547378.2126	0.2500	0.0000	3.9996	53
54	0.0000	4.0000	15.9986	171056.9414	684223.7658	0.2500	0.0000	3.9997	54
55	0.0000	4.0000	15.9989	213821.1768	855280.7072	0.2500	0.0000	3.9997	55
60	0.0000	4.0000	15.9996	652530.4468	2610117.7872	0.2500	0.0000	3.9999	60

(continued)

APPENDIX 54.B *(continued)*
Cash Flow Equivalent Factors

$i = 30.00\%$

n	P/F	P/A	P/G	F/P	F/A	A/P	A/F	A/G	n
1	0.7692	0.7692	0.0000	1.3000	1.0000	1.3000	1.0000	0.000	1
2	0.5917	1.3609	0.5917	1.6900	2.3000	0.7348	0.4348	0.434	2
3	0.4552	1.8161	1.5020	2.1970	3.9900	0.5506	0.2506	0.827	3
4	0.3501	2.1662	2.5524	2.8561	6.1870	0.4616	0.1616	1.178	4
5	0.2693	2.4356	3.6297	3.7129	9.0431	0.4106	0.1106	1.490	5
6	0.2072	2.6427	4.6656	4.8268	12.7560	0.3784	0.0784	1.765	6
7	0.1594	2.8021	5.6218	6.2749	17.5828	0.3569	0.0569	2.006	7
8	0.1226	2.9247	6.4800	8.1573	23.8577	0.3419	0.0419	2.215	8
9	0.0943	3.0190	7.2343	10.6045	32.0150	0.3312	0.0312	2.396	9
10	0.0725	3.0915	7.8872	13.7858	42.6195	0.3235	0.0235	2.551	10
11	0.0558	3.1473	8.4452	17.9216	56.4053	0.3177	0.0177	2.683	11
12	0.0429	3.1903	8.9173	23.2981	74.3270	0.3135	0.0135	2.795	12
13	0.0330	3.2233	9.3135	30.2875	97.6250	0.3102	0.0102	2.889	13
14	0.0254	3.2487	9.6437	39.3738	127.9125	0.3078	0.0078	2.968	14
15	0.0195	3.2682	9.9172	51.1859	167.2863	0.3060	0.0060	3.034	15
16	0.0150	3.2832	10.1426	66.5417	218.4722	0.3046	0.0046	3.089	16
17	0.0116	3.2948	10.3276	86.5042	285.0139	0.3035	0.0035	3.134	17
18	0.0089	3.3037	10.4788	112.4554	371.5180	0.3027	0.0027	3.171	18
19	0.0068	3.3105	10.6019	146.1920	483.9734	0.3021	0.0021	3.202	19
20	0.0053	3.3158	10.7019	190.0496	630.1655	0.3016	0.0016	3.227	20
21	0.0040	3.3198	10.7828	247.0645	820.2151	0.3012	0.0012	3.248	21
22	0.0031	3.3230	10.8482	321.1839	1067.2796	0.3009	0.0009	3.264	22
23	0.0024	3.3254	10.9009	417.5391	1388.4635	0.3007	0.0007	3.278	23
24	0.0018	3.3272	10.9433	542.8008	1806.0026	0.3006	0.0006	3.289	24
25	0.0014	3.3286	10.9773	705.6410	2348.8033	0.3004	0.0004	3.297	25
26	0.0011	3.3297	11.0045	917.3333	3054.4443	0.3003	0.0003	3.305	26
27	0.0008	3.3305	11.0263	1192.5333	3971.7776	0.3003	0.0003	3.310	27
28	0.0006	3.3312	11.0437	1550.2933	5164.3109	0.3002	0.0002	3.315	28
29	0.0005	3.3317	11.0576	2015.3813	6714.6042	0.3001	0.0001	3.318	29
30	0.0004	3.3321	11.0687	2619.9956	8729.9855	0.3001	0.0001	3.321	30
31	0.0003	3.3324	11.0775	3405.9943	11349.9811	0.3001	0.0001	3.324	31
32	0.0002	3.3326	11.0845	4427.7926	14755.9755	0.3001	0.0001	3.326	32
33	0.0002	3.3328	11.0901	5756.1304	19183.7681	0.3001	0.0001	3.327	33
34	0.0001	3.3329	11.0945	7482.9696	24939.8985	0.3000	0.0000	3.328	34
35	0.0001	3.3330	11.0980	9727.8604	32422.8681	0.3000	0.0000	3.329	35
36	0.0001	3.3331	11.1007	12646.2186	42150.7285	0.3000	0.0000	3.330	36
37	0.0001	3.3331	11.1029	16440.0841	54796.9471	0.3000	0.0000	3.331	37
38	0.0000	3.3332	11.1047	21372.1094	71237.0312	0.3000	0.0000	3.331	38
39	0.0000	3.3332	11.1060	27783.7422	92609.1405	0.3000	0.0000	3.331	39
40	0.0000	3.3332	11.1071	36118.8648	120392.8827	0.3000	0.0000	3.332	40
41	0.0000	3.3333	11.1080	46954.5243	156511.7475	0.3000	0.0000	3.332	41
42	0.0000	3.3333	11.1086	61040.8815	203466.2718	0.3000	0.0000	3.332	42
43	0.0000	3.3333	11.1092	79353.1460	264507.1533	0.3000	0.0000	3.332	43
44	0.0000	3.3333	11.1096	103159.0898	343860.2993	0.3000	0.0000	3.332	44
45	0.0000	3.3333	11.1099	134106.8167	447019.3890	0.3000	0.0000	3.333	45
46	0.0000	3.3333	11.1102	174338.8617	581126.2058	0.3000	0.0000	3.333	46
47	0.0000	3.3333	11.1104	226640.5202	755465.0675	0.3000	0.0000	3.333	47
48	0.0000	3.3333	11.1105	294632.6763	982105.5877	0.3000	0.0000	3.333	48
49	0.0000	3.3333	11.1107	383022.4792	1276738.2640	0.3000	0.0000	3.333	49
50	0.0000	3.3333	11.1108	497929.2230	1659760.7433	0.3000	0.0000	3.333	50

(continued)

APPENDIX 54.B *(continued)*
Cash Flow Equivalent Factors

$i = 40.00\%$

n	P/F	P/A	P/G	F/P	F/A	A/P	A/F	A/G	n
1	0.7143	0.7143	0.0000	1.4000	1.0000	1.4000	1.0000	0.000	1
2	0.5102	1.2245	0.5102	1.9600	2.4000	0.8167	0.4167	0.416	2
3	0.3644	1.5889	1.2391	2.7440	4.3600	0.6294	0.2294	0.779	3
4	0.2603	1.8492	2.0200	3.8416	7.1040	0.5408	0.1408	1.092	4
5	0.1859	2.0352	2.7637	5.3782	10.9456	0.4914	0.0914	1.358	5
6	0.1328	2.1680	3.4278	7.5295	16.3238	0.4613	0.0613	1.581	6
7	0.0949	2.2628	3.9970	10.5114	23.8534	0.4419	0.0419	1.766	7
8	0.0678	2.3306	4.4713	14.7579	34.3947	0.4291	0.0291	1.918	8
9	0.0484	2.3790	4.8585	20.6610	49.1526	0.4203	0.0203	2.042	9
10	0.0346	2.4136	5.1696	28.9255	69.8137	0.4143	0.0143	2.141	10
11	0.0247	2.4383	5.4166	40.4957	98.7391	0.4101	0.0101	2.221	11
12	0.0176	2.4559	5.6106	56.6939	139.2348	0.4072	0.0072	2.284	12
13	0.0126	2.4685	5.7618	79.3715	195.9287	0.4051	0.0051	2.334	13
14	0.0090	2.4775	5.8788	111.1201	275.3002	0.4036	0.0036	2.372	14
15	0.0064	2.4839	5.9688	155.5681	386.4202	0.4026	0.0026	2.403	15
16	0.0046	2.4885	6.0376	217.7953	541.9883	0.4018	0.0018	2.426	16
17	0.0033	2.4918	6.0901	304.9135	759.7837	0.4013	0.0013	2.444	17
18	0.0023	2.4941	6.1299	426.8789	1064.6971	0.4009	0.0009	2.457	18
19	0.0017	2.4958	6.1601	597.6304	1491.5760	0.4007	0.0007	2.468	19
20	0.0012	2.4970	6.1828	836.6826	2089.2064	0.4005	0.0005	2.476	20
21	0.0009	2.4979	6.1998	1171.3556	2925.8889	0.4003	0.0003	2.482	21
22	0.0006	2.4985	6.2127	1639.8978	4097.2445	0.4002	0.0002	2.486	22
23	0.0004	2.4989	6.2222	2295.8569	5737.1423	0.4002	0.0002	2.490	23
24	0.0003	2.4992	6.2294	3214.1997	8032.9993	0.4001	0.0001	2.492	24
25	0.0002	2.4994	6.2347	4499.8796	11247.1990	0.4001	0.0001	2.494	25
26	0.0002	2.4996	6.2387	6299.8314	15747.0785	0.4001	0.0001	2.495	26
27	0.0001	2.4997	6.2416	8819.7640	22046.9099	0.4000	0.0000	2.496	27
28	0.0001	2.4998	6.2454	12347.6696	30866.6739	0.4000	0.0000	2.497	28
29	0.0001	2.4999	6.2454	17286.7374	43214.3435	0.4000	0.0000	2.498	29
30	0.0000	2.4999	6.2466	24201.4324	60501.0809	0.4000	0.0000	2.498	30
31	0.0000	2.4999	6.2475	33882.0053	84702.5132	0.4000	0.0000	2.499	31
32	0.0000	2.4999	6.2482	47434.8074	118584.5185	0.4000	0.0000	2.499	32
33	0.0000	2.5000	6.2487	66408.7304	166019.3260	0.4000	0.0000	2.499	33
34	0.0000	2.5000	6.2490	92972.2225	232428.0563	0.4000	0.0000	2.499	34
35	0.0000	2.5000	6.2493	130161.1116	325400.2789	0.4000	0.0000	2.499	35
36	0.0000	2.5000	6.2495	182225.5562	455561.3904	0.4000	0.0000	2.499	36
37	0.0000	2.5000	6.2496	255115.7786	637786.9466	0.4000	0.0000	2.499	37
38	0.0000	2.5000	6.2497	357162.0901	892902.7252	0.4000	0.0000	2.499	38
39	0.0000	2.5000	6.2498	500026.9261	1250064.8153	0.4000	0.0000	2.499	39
40	0.0000	2.5000	6.2498	700037.6966	1750091.7415	0.4000	0.0000	2.499	40
41	0.0000	2.5000	6.2499	980052.7752	2450129.4381	0.4000	0.0000	2.500	41
42	0.0000	2.5000	6.2499	1372073.8853	3430182.2133	0.4000	0.0000	2.500	42
43	0.0000	2.5000	6.2499	1920903.4394	4802256.0986	0.4000	0.0000	2.500	43
44	0.0000	2.5000	6.2500	2689264.8152	6723159.5381	0.4000	0.0000	2.500	44
45	0.0000	2.5000	6.2500	3764970.7413	9412424.3533	0.4000	0.0000	2.500	45

Glossary

A

AASHTO: American Association of State and Highway Transportation Officials.

Abandonment: The reversion of title to the owner of the underlying fee where an easement for highway purposes is no longer needed.

Absorbed dose: The energy deposited by radiation as it passes through a material.

Absorption: The process by which a liquid is drawn into and tends to fill permeable pores in a porous body. Also, the increase in weight of a porous solid body resulting from the penetration of liquid into its permeable pores.

Accelerated flow: A form of varied flow in which the velocity is increasing and the depth is decreasing.

Acid: Any compound that dissociates in water into H^+ ions. (The combination of H^+ and water, H_3O^+, is known as the hydronium ion.) Acids conduct electricity in aqueous solutions, have a sour taste, turn blue litmus paper red, have a pH between 0 and 7, and neutralize bases, forming salts and water.

Acidic: Term applied to water with a pH less than 5.5.

Active pressure: Pressure causing a wall to move away from the soil.

Adenosine triphosphate (ATP): The macromolecule that functions as an energy carrier in cells. The energy is stored in a high-energy bond between the second and third phosphates.

Adjudication: A court proceeding to determine rights to the use of water on a particular stream or aquifer.

Admixture: Material added to a concrete mixture to increase its workability, strength, or imperviousness, or to lower its freezing point.

Adsorbed water: Water held near the surface of a material by electrochemical forces.

Adsorption edge: The pH range where solute adsorption changes sharply.

Advection: The transport of solutes along stream lines at the average linear seepage flow velocity.

Aeration: Mixing water with air, either by spraying water or diffusing air through water.

Aerobe: A microorganism whose growth requires free oxygen.

Aerobic: Requiring oxygen. Descriptive of a bacterial class that functions in the presence of free dissolved oxygen.

Aggregate, coarse: Aggregate retained on a no. 4 sieve.

Aggregate, fine: Aggregate passing the no. 4 sieve and retained on the no. 200 sieve.

Aggregate, lightweight: Aggregate having a dry density of 70 lbm/ft^3 (32 kg/m^3) or less.

Agonic line: A line with no magnetic declination.

Air change (Air flush): A complete replacement of the air in a room or other closed space.

Algae: Simple photosynthetic plants having neither roots, stems, nor leaves.

Alidade: A tachometric instrument consisting of a telescope similar to a transit, an upright post that supports the standards of the horizontal axis of the telescope, and a straightedge whose edges are essentially in the same direction as the line of sight. An alidade is used in the field in conjunction with a plane table.

Alkaline: Term applied to water with a pH greater than 7.4.

Alkalinity: A measure of the capacity of a water to neutralize acid without significant pH change. It is usually associated with the presence of hydroxyl, carbonate, and/or bicarbonate radicals in the water.

Alluvial deposit: A material deposited within the alluvium.

Alluvium: Sand, silt, clay, gravel, etc., deposited by running water.

Alternate depths: For a particular channel geometry and discharge, two depths at which water flows with the same specific energy in uniform flow. One depth corresponds to subcritical flow; the other corresponds to supercritical flow.

Amictic: Experiencing no overturns or mixing. Typical of polar lakes.

Amphoteric behavior: Ability of an aqueous complex or solid material to have a negative, neutral, or positive charge.

Anabolism: The phase of metabolism involving the formation of organic compounds; usually an energy-utilizing process.

Anabranch: The intertwining channels of a braided stream.

Anaerobe: A microorganism that grows only or best in the absence of free oxygen.

Anaerobic: Not requiring oxygen. Descriptive of a bacterial class that functions in the absence of free dissolved oxygen.

Anion: Negative ion that migrates to the positive electrode (anode) in an electrolytic solution.

Anticlinal spring: A portion of an exposed aquifer (usually on a slope) between two impervious layers.

Apparent specific gravity of asphalt mixture: A ratio of the unit weight of an asphalt mixture (excluding voids permeable to water) to the unit weight of water.

Appurtenance: That which belongs with or is designed to complement something else. For example, a manhole is a sewer appurtenance.

Apron: An underwater "floor" constructed along the channel bottom to prevent scour. Aprons are almost always extensions of spillways and culverts.

Aquiclude: A saturated geologic formation with insufficient porosity to support any significant removal rate or contribute to the overall groundwater regime. In groundwater analysis, an aquiclude is considered to confine an aquifer at its boundaries.

Aquifer: Rock or sediment in a formation, group of formations, or part of a formation that is saturated and sufficiently permeable to transmit economic quantities of water to wells and springs.

Aquifuge: An underground geological formation that has absolutely no porosity or interconnected openings through which water can enter or be removed.

Aquitard: A saturated geologic unit that is permeable enough to contribute to the regional groundwater flow regime, but not permeable enough to supply a water well or other economic use.

Arterial highway: A general term denoting a highway primarily for through traffic, usually on a continuous route.

Artesian formation: An aquifer in which the piezometric height is greater than the aquifer height. In an artesian formation, the aquifer is confined and the water is under hydrostatic pressure.

Artesian spring: Water from an artesian formation that flows to the ground surface naturally, under hydrostatic pressure, due to a crack or other opening in the formation's confining layer.

Asphalt emulsion: A mixture of asphalt cement with water. Asphalt emulsions are produced by adding a small amount of emulsifying soap to asphalt and water. The asphalt sets when the water evaporates.

Atomic mass unit (AMU): One AMU is one twelfth the atomic weight of carbon.

Atomic number: The number of protons in the nucleus of an atom.

Atomic weight: Approximately, the sum of the numbers of protons and neutrons in the nucleus of an atom.

Autotroph: An organism than can synthesize all of its organic components from inorganic sources.

Auxiliary lane: The portion of a roadway adjoining the traveled way for truck climbing, speed change, or other purposes supplementary to through traffic movement.

Avogadro's law: A gram-mole of any substance contains 6.022×10^{23} molecules.

Azimuth: The horizontal angle measured from the plane of the meridian to the vertical plane containing the line. The azimuth gives the direction of the line with respect to the meridian and is usually measured in a clockwise direction with respect to either the north or south meridian.

B

Backward pass: The steps in a critical path analysis in which the latest start times are determined, usually after the earliest finish times have been determined in the forward pass.

Backwater: Water upstream from a dam or other obstruction that is deeper than it would normally be without the obstruction.

Backwater curve: A plot of depth versus location along the channel containing backwater.

Bar: An elongated landform generated by waves and currents, usually running parallel to the shore, composed predominantly of unconsolidated sand, gravel, stones, cobbles, or rubble and with water on two sides.

Base: (a) A layer of selected, processed, or treated aggregate material of planned thickness and quality placed immediately below the pavement and above the subbase or subgrade soil. (b) Any compound that dissociates in water into OH^- ions. Bases conduct electricity in aqueous solutions, have a bitter taste, turn red litmus paper blue, have a pH between 7 and 14, and neutralize acids, forming salts and water.

Base course: The bottom portion of a pavement where the top and bottom portions are not the same composition.

Base flow: Component of stream discharge that comes from groundwater flow. Water infiltrates and moves through the ground very slowly; up to 2 years may elapse between precipitation and discharge.

Batter pile: A pile inclined from the vertical.

Beach: A sloping landform on the shore of larger water bodies, generated by waves and currents and extending from the water to a distinct break in landform or substrate type (e.g., a foredune, cliff, or bank).

Bed: A layer of rock in the earth. Also the bottom of a body of water such as a river, lake, or sea.

Bell: An enlarged section at the base of a pile or pier used as an anchor.

Belt highway: An arterial highway carrying traffic partially or entirely around an urban area.

Bent: A supporting structure (usually of a bridge) consisting of a beam or girder transverse to the supported roadway and that is supported, in turn, by columns at each end, making an inverted "U" shape.

Benthic zone: The bottom zone of a lake, where oxygen levels are low.

Benthos: Organisms (typically anaerobic) occupying the benthic zone.

Bentonite: A volcanic clay that exhibits extremely large volume changes with moisture content changes.

Berm: A shelf, ledge, or pile.

Bifurcation ratio: The average number of streams feeding into the next side (order) waterway. The range is usually 2 to 4.

Binary fission: An asexual reproductive process in which one cell splits into two independent daughter cells.

Bioaccumulation: The process by which chemical substances are ingested and retained by organisms, either from the environment directly or through consumption of food containing the chemicals.

Bioaccumulation factor: *See* Bioconcentration factor.

Bioactivation process: A process using sedimentation, trickling filter, and secondary sedimentation before adding activated sludge. Aeration and final sedimentation are the follow-up processes.

Bioassay: The determination of kinds, quantities, or concentrations, and in some cases, the locations of material in the body, whether by direct measurement (in vivo counting) or by analysis and evaluation of materials excreted or removed (in vitro) from the body.

Bioavailability: A measure of what fraction, how much, or the rate that a substance (ingested, breathed in, dermal contact) is actually absorbed biologically. Bioavailability measurements are typically based on absorption into the blood or liver tissue.

Biochemical oxygen demand (BOD): The quantity of oxygen needed by microorganisms in a body of water to decompose the organic matter present. An index of water pollution.

Bioconcentration: The increase in concentration of a chemical in an organism resulting from tissue absorption (bioaccumulation) levels exceeding the rate of metabolism and excretion (biomagnification).

Bioconcentration factor: The ratio of chemical concentration in an organism to chemical concentration in the surrounding environment.

Biodegradation: The use of microorganisms to degrade contaminants.

Biogas: A mixture of approximately 55% methane and 45% carbon dioxide that results from the digestion of animal dung.

Biomagnification: The cumulative increase in the concentration of a persistent substance in successively higher levels of the food chain.

Biomagnification factor: *See* Bioconcentration factor.

Biomass: Renewable organic plant and animal material such as plant residue, sawdust, tree trimmings, rice straw, poultry litter and other animal wastes, some industrial wastes, and the paper component of municipal solid waste that can be converted to energy.

Biosorption process: A process that mixes raw sewage and sludge that have been pre-aerated in a separate tank.

Biosphere: The part of the world in which life exists.

Biota: All of the species of plants and animals indigenous to an area.

Bleeding: A form of segregation in which some of the water in the mix tends to rise to the surface of freshly placed concrete.

Blind drainage: Geographically large (with respect to the drainage basin) depressions that store water during a storm and therefore stop it from contributing to surface runoff.

Bloom: A phenomenon whereby excessive nutrients within a body of water results in an explosion of plant life, resulting in a depletion of oxygen and fish kill. Usually caused by urban runoff containing fertilizers.

Bluff: A high and steep bank or cliff.

Body burden: The total amount of a particular chemical in the body.

Boulder: Rock fragment larger than 60 cm (24 in) in diameter.

Brackish: Marine and estuarine waters with mixohaline salinity. The term should not be applied to inland waters.

Braided stream: A wide, shallow stream with many anabranches.

Branch sewer: A sewer off the main sewer.

Breaking chain: A technique used when the slope is too steep to permit bringing the full length of the chain or tape to a horizontal position. When breaking chain, the distance is measured in partial tape lengths.

Breakpoint chlorination: Application of chlorine that results in a minimum of chloramine residuals. No significant free chlorine residual is produced unless the breakpoint is reached.

Broad-leaved deciduous: Woody angiosperms (trees or shrubs) with relatively wide, flat leaves that are shed during the cold or dry season (e.g., black ash (*Fraxinus nigra*)).

Broad-leaved evergreen: Woody angiosperms (trees or shrubs) with relatively wide, flat leaves that generally remain green and are usually persistent for a year or more (e.g., red mangrove (*Rhizophora mangle*)).

Bulking: *See* Sludge bulking.

Bulk specific gravity of asphalt mixture: Ratio of the unit weight of an asphalt mixture (including permeable and impermeable voids) to the unit weight of water.

Butte: A hill with steep sides that usually stands away from other hills.

C

Caisson: An air- and watertight chamber used as a foundation and/or used to work or excavate below the water level.

Calcareous: Formed of calcium carbonate or magnesium carbonate by biological deposition or inorganic precipitation in sufficient quantities to effervesce carbon dioxide visibly when treated with cold 0.1N hydrochloric acid. Calcareous sands are usually a mixture of fragments of mollusk shell, echinoderm spines and skeletal material, coral, foraminifera, and algal platelets (e.g., *Halimeda*).

Capillary water: Water just above the water table that is drawn up out of an aquifer due to capillary action of the soil.

Carbonaceous demand: Oxygen demand due to biological activity in a water sample, exclusive of nitrogenous demand.

Carbonate hardness: Hardness associated with the presence of bicarbonate radicals in the water.

Carcinogen: A cancer-causing agent.

Carrier (biological): An individual harboring a disease agent without apparent symptoms.

Cascade impactor: *See* Impactor.

Cased hole: An excavation whose sides are lined or sheeted.

Catabolism: The chemical reactions by which food materials or nutrients are converted into simpler substances for the production of energy and cell materials.

Catena: A group of soils of similar origin occurring in the same general locale but that differ slightly in properties.

Cation: Positive ion that migrates to the negative electrode (cathode) in an electrolytic solution.

Cell wall: The cell structure exterior to the cell membrane of plants, algae, bacteria, and fungi. It gives cells form and shape.

Cement-treated base: A base layer constructed with good-quality, well-graded aggregate mixed with up to 6% cement.

CFR: The Code of (U.S.) Federal Regulations, a compilation of all federal documents that have general applicability and legal effect, as published by the Office of the Federal Register.

Channel: An open conduit, either naturally or artificially created, which periodically or continuously contains moving water, or which forms a connecting link between two bodies of standing water.

Channel bank: The sloping land bordering a channel. The bank has a steeper slope than the bottom of the channel and is usually steeper than the land surrounding the channel.

Channelization: The separation or regulation of conflicting traffic movements into definite paths of travel by use of pavement markings, raised islands, or other means.

Chat: Small pieces of crushed rock and gravel. May be used for paving roads and roofs.

Check: A short section of built-up channel placed in a canal or irrigation ditch and provided with gates or flashboards to control flow or raise upstream level for diversion.

Chelate: A chemical compound into which a metallic ion (usually divalent) is tightly bound.

Chelation: (a) The process in which a compound or organic material attracts, combines with, and removes a metallic ion. (b) The process of removing metallic contaminants by having them combine with special added substances.

Chemical precipitation: Settling out of suspended solids caused by adding coagulating chemicals.

Chemocline: A steep chemical (saline) gradient separating layers in meromictic lakes.

Chloramine: Compounds of chlorine and ammonia (e.g., NH_2Cl, $NHCl_2$, or NCl_3).

Chlorine demand: The difference between applied chlorine and the chlorine residual. Chlorine demand is chlorine that has been reduced in chemical reactions and is no longer available for disinfection.

Circumneutral: Term applied to water with a pH of 5.5 to 7.4.

Class: A major taxonomic subdivision of a phylum. Each class is composed of one or more related orders.

Clean-out: A pipe through which snakes can be pushed to unplug a sewer.

Clearance: Distance between successive vehicles as measured between the vehicles, back bumper to front bumper.

Cobble: Rock fragment 8 cm (3 in) to 25 cm (10 in) in diameter.

Codominant: Two or more species providing about equal areal cover which, in combination, control the environment.

Coliform: Gram-negative, lactose-fermenting rods, including escherichieae coli and similar species that normally inhabit the colon (large intestine). Commonly included in the coliform are Enterobacteria aerogenes, Klebsiella species, and other related bacteria.

Colloid: A fine particle ranging in size from 1 to 500 millimicrons. Colloids cause turbidity because they do not easily settle out.

Combined residuals: Compounds of an additive (such as chlorine) that have combined with something else. Chloramines are examples of combined residuals.

Combined system: A system using a single sewer for disposal of domestic waste and storm water.

Comminutor: A device that cuts solid waste into small pieces.

Compaction: Densification of soil by mechanical means, involving the expulsion of excess air.

Compensation level: In a lake, the depth of the limnetic area where oxygen production from light and photosynthesis are exactly balanced by depletion.

Complete mixing: Mixing accomplished by mechanical means (stirring).

Compound: A homogeneous substance composed of two or more elements that can be decomposed by chemical means only.

Concrete: A mixture of portland cement, fine aggregate, coarse aggregate, and water, with or without admixtures.

Concrete, normal weight: Concrete having a hardened density of approximately 150 lbm/ft^3.

Concrete, plain: Concrete that is not reinforced with steel.

Concrete, structural lightweight: A concrete containing lightweight aggregate.

Condemnation: The process by which property is acquired for public purposes through legal proceedings under power of eminent domain.

Cone of depression: The shape of the water table around a well during and immediately after use. The cone's water surface level differs from the original water table by the well's drawdown.

Confined water: Artesian water overlaid with an impervious layer, usually under pressure.

Confirmed test: A follow-up test used if the presumptive test for coliforms is positive.

Conflagration: Total involvement or engulfment (as in a fire).

Conjugate depths: The depths on either side of a hydraulic jump.

Connate water: Pressurized water (usually, high in mineral content) trapped in the pore spaces of sedimentary rock at the time it was formed.

Consolidation: Densification of soil by mechanical means, involving expulsion of excess water.

Contraction: A decrease in the width or depth of flow caused by the geometry of a weir, orifice, or obstruction.

Control of access: The condition where the right of owners or occupants of abutting land or other persons to access in connection with a highway is fully or partially controlled by public authority.

Critical depth: The depth that minimizes the specific energy of flow.

Critical flow: Flow at the critical depth and velocity. Critical flow minimizes the specific energy and maximizes discharge.

Critical slope: The slope that produces critical flow.

Critical velocity: The velocity that minimizes specific energy. When water is moving at its critical velocity, a disturbance wave cannot move upstream since the wave moves at the critical velocity.

Cuesta: (From the Spanish word for cliff); a hill with a steep slope on one side and a gentle slope on the other.

Cunette: A small channel in the invert of a large combined sewer for dry weather flow.

Curing: The process and procedures used for promoting the hydration of cement. It consists of controlling the temperature and moisture from and into the concrete.

Cyclone impactor: *See* Impactor.

D

Dead load: An inert, inactive load, primarily due to the structure's own weight.

Deciduous stand: A plant community where deciduous trees or shrubs represent more than 50% of the total areal coverage of trees or shrubs.

Decision sight distance: Sight distance allowing for additional decision time in cases of complex conditions.

Delta: A deposit of sand and other sediment, usually triangular in shape. Deltas form at the mouths of rivers where the water flows into the sea.

Deoxygenation: The act of removing dissolved oxygen from water.

Deposition: The laying down of sediment such as sand, soil, clay, or gravel by wind or water. It may later be compacted into hard rock and buried by other sediment.

Depression storage: Initial storage of rain in small surface puddles.

Depth-area-duration analysis: A study made to determine the maximum amounts of rain within a given time period over a given area.

Detrial mineral: Mineral grain resulting from the mechanical disintegration of a parent rock.

Dewatering: Removal of excess moisture from sludge waste.

Digestion: Conversion of sludge solids to gas.

Dilatancy: The tendency of a material to increase in volume when undergoing shear.

Dilution disposal: Relying on a large water volume (lake or stream) to dilute waste to an acceptable concentration.

Dimictic: Experiencing two overturns per year. Dimictic lakes are usually found in temperate climates.

Dimiper lake: The freely circulating surface water with a small but variable temperature gradient.

Dimple spring: A depression in the earth below the water table.

Distribution coefficient: *See* Partition coefficient.

Divided highway: A highway with separated roadbeds for traffic in opposing directions.

Domestic waste: Waste that originates from households.

Dominant: The species controlling the environment.

Dormant season: That portion of the year when frosts occur.

Downpull: A force on a gate, typically less at lower depths than at upper depths due to increased velocity, when the gate is partially open.

Drainage density: The total length of streams in a watershed divided by the drainage area.

Drawdown: The lowering of the water table level of an unconfined aquifer (or of the potentiometric surface of a confined aquifer) by pumping of wells.

Drawdown curve: *See* Cone of depression.

Dredge line: *See* Mud line.

Dry weather flow: *See* Base flow.

Dystrophic: Receiving large amounts of organic matter from surrounding watersheds, particularly humic materials from wetlands that stain the water brown.

Dystrophic lakes have low plankton productivity, except in highly productive littoral zones.

E

Easement: A right to use or control the property of another for designated purposes.

Effective specific gravity of an asphalt mixture: Ratio of the unit weight of an asphalt mixture (excluding voids permeable to asphalt) to the unit weight of water.

Effluent: That which flows out of a process.

Effluent stream: A stream that intersects the water table and receives groundwater. Effluent streams seldom go completely dry during rainless periods.

Element: A pure substance that cannot be decomposed by chemical means.

Elutriation: A counter-current sludge washing process used to remove dissolved salts.

Elutriator: A device that purifies, separates, or washes material passing through it.

Embankment: A raised structure constructed of natural soil from excavation or borrow sources.

Emergent hydrophytes: Erect, rooted, herbaceous angiosperms that may be temporarily or permanently flooded at the base but do not tolerate prolonged inundation of the entire plant (e.g., bulrushes (*Scirpus* spp.), or saltmarsh cordgrass).

Emergent mosses: Mosses occurring in wetlands, but generally not covered by water.

Eminent domain: The power to take private property for public use without the owner's consent upon payment of just compensation.

Emulsion: *See* Asphalt emulsion.

Encroachment: Use of the highway right-of-way for nonhighway structures or other purposes.

Energy gradient: The slope of the specific energy line (i.e., the sum of the potential and velocity heads).

Enteric: Intestinal.

Enzyme: An organic (protein) catalyst that causes changes in other substances without undergoing any alteration itself.

Ephemeral stream: A stream that goes dry during rainless periods.

Epilimnion: (Gr. for "upper lake"); the freely circulating surface water with a small but variable temperature gradient.

Equivalent weight: The amount of substance (in grams) that supplies one mole of reacting units. It is calculated as the molecular weight divided by the change in oxidation number experienced in a chemical reaction. An alternative calculation is the atomic weight of an element divided by its valence or the molecular weight of a radical or compound divided by its valence.

Escarpment: A steep slope or cliff.

Escherichieae coli (E. coli): *See* Coliform.

Estuary: An area where fresh water meets salt water.

Eutrophic: Nutrient-rich; a eutrophic lake typically has a high surface area-to-volume ratio.

Eutrophic lake: Lake that has a high concentration of plant nutrients such as nitrogen and phosphorous.

Eutrophication: The enrichment of water bodies by nutrients (e.g., phosphorus). Eutroficaction of a lake normally contributes to a slow evolution into a bog, marsh, and ultimately, dry land.

Evaporite: Sediment deposited when sea water evaporates. Gypsum, salt, and anhydrite are evaporites.

Evapotranspiration: Evaporation of water from a study area due to all sources including water, soil, snow, ice, vegetation, and transpiration.

Evergreen stand: A plant community where evergreen trees or shrubs represent more than 50% of the total areal coverage of trees and shrubs. The canopy is never without foliage; however, individual trees or shrubs may shed their leaves.

Extreme high water of spring tides: The highest tide occurring during a lunar month, usually near the new or full moon. This is equivalent to extreme higher high water of mixed semidiurnal tides.

Extreme low water of spring tides: The lowest tide occurring during a lunar month, usually near the new or full moon. This is equivalent to extreme lower low water of mixed semidiurnal tides.

F

Facultative: Able to live under different or changing conditions. Descriptive of a bacterial class that functions either in the presence or absence of free dissolved oxygen.

Fecal coliform: Coliform bacterium present in the intestinal tracts and feces of warm-blooded animals.

Fines: Silt- and/or clay-sized particles.

First-stage demand: *See* Carbonaceous demand.

Flat: A level landform composed of unconsolidated sediments—usually mud or sand. Flats may be irregularly shaped or elongated and continuous with the shore, whereas bars are generally elongated, parallel to the shore, and separated from the shore by water.

Flexible pavement: A pavement having sufficiently low bending resistance to maintain intimate contact with the underlying structure, yet having the required stability furnished by aggregate interlock, internal friction, and cohesion to support traffic.

Float: The amount of time that an activity can be delayed without delaying any succeeding activities.

Floating-leaved plant: A rooted, herbaceous hydrophyte with some leaves floating on the water surface (e.g., white water lily (*Nymphaea odorata*), floating-leaved pondweed (*Potamogeton natans*)). Plants such as yellow water lily (*Nuphar luteum*) which sometimes have leaves raised above the surface are considered either floating-leaved plants or emergents, depending on their growth habit at a particular site.

Floating plant: A nonanchored plant that floats freely in the water or on the surface (e.g., water hyacinth (*Eichhornia crassipes*) or common duckweed (*Lemna minor*)).

Floc: Agglomerated colloidal particles.

Floodplain: A flat expanse of land bordering a river.

Flotation: Addition of chemicals and bubbled air to liquid waste in order to get solids to float to the top as scum.

Flow regime: (a) Subcritical, critical, or supercritical. (b) Entrance control or exit control.

Flowing well: A well that flows under hydrostatic pressure to the surface. Also called an Artesian well.

Flume: In general, any open channel for carrying water. More specifically, an open channel constructed above the earth's surface, usually supported on a trestle or on piers.

Force main: A sewer line that is pressured.

Forebay: A reservoir holding water for subsequent use after it has been discharged from a dam.

Forward pass: The steps in a critical path analysis in which the earliest finish times are determined, usually before the latest start times are determined in the backward pass.

Free residuals: Ions or compounds not combined or reduced. The presence of free residuals signifies excess dosage.

Freeboard distance: The vertical distance between the water surface and the crest of a dam or top of a channel side. The distance the water surface can rise before it overflows.

Freehaul: Pertaining to hauling "for free" (i.e., without being able to bill an extra amount over the contract charge).

Freeway: A divided arterial highway with full control of access.

Freeze (in piles): A large increase in the ultimate capacity (and required driving energy) of a pile after it has been driven some distance.

Fresh: Term used to characterize water with salinity less than 0.5% dissolved salts.

Friable: Easily crumbled.

Frontage road: A local street or road auxiliary to and located on the side of an arterial highway for service to abutting property and adjacent areas, and for control of access.

Frost susceptibility: Susceptible to having water continually drawn up from the water table by capillary action, forming ice crystals below the surface (but above the frost line).

Fulvic acid: The alkaline-soluble portion of organic material (i.e., humus) that remains in solution at low pH and is of lower molecular weight. A breakdown product of cellulose from vascular plants.

Fungi: Aerobic, multicellular, nonphotosynthetic heterotrophic, eucaryote protists that degrade dead organic matter, releasing carbon dioxide and nitrogen.

G

Gap: Corresponding time between successive vehicles (back bumper to front bumper) as they pass a point on a roadway.

Gap-graded: A soil with a discontinuous range of soil particle sizes; for example, containing large particles and small particles but no medium-sized particles.

Geobar: A polymeric material in the form of a bar.

Geocell: A three-dimensional, permeable, polymeric (synthetic or natural) honeycomb or web structure, made of alternating strips of geotextiles, geogrids, or geomembranes.

Geocomposite: A manufactured or assembled material using at least one geosynthetic product among the components.

Geofoam: A polymeric material that has been formed by the application of the polymer in semiliquid form through the use of a foaming agent. Results in a lightweight material with high void content.

Geographic Information System (GIS): A digital database containing geographic information.

Geogrid: A planar, polymeric (synthetic or natural) structure consisting of a regular open network of integrally connected tensile elements that may be linked or formed by extrusion, bonding, or interlacing (knitting or lacing).

Geomat: A three-dimensional, permeable, polymeric (synthetic or natural) structure made of bonded filaments, used for soil protection and to bind roots and small plants in erosion control applications.

Geomembrane: A planar, relatively impermeable, polymeric (synthetic or natural) sheet. May be bituminous, elastomeric, or plastomeric.

Geonet: A planar, polymeric structure consisting of a regular dense network whose constituent elements are linked by knots or extrusions and whose openings are much larger than the constituents.

Geopipe: A polymeric pipe.

Geospacer: A three-dimensional polymeric structure with large void spaces.

Geostrip: A polymeric material in the form of a strip, with a width less than approximately 200 mm.

Geosynthetic: A planar, polymeric (synthetic or natural) material.

Geotextile: A planar, permeable, polymeric (synthetic or natural) textile material that may be woven, nonwoven, or knitted.

Glacial till: Soil resulting from a receding glacier, consisting of mixed clay, sand, gravel, and boulders.

GMT: *See* UTC.

Gobar gas: *See* Biogas.

Gore: The area immediately beyond the divergence of two roadways bounded by the edges of those roadways.

Gradient: The energy (head) loss per unit distance. *See also* Slope.

Gravel: Granular material retained on a no. 4 sieve.

Gravitational water: Free water in transit downward through the vadose (unsaturated) zone.

Grillage: A footing or part of a footing consisting of horizontally laid timbers or steel beams.

Groundwater: Loosely, all water that is underground as opposed to on the surface of the ground. Usually refers to water in the saturated zone below the water table.

Growing season: The frost-free period of the year.

Gumbo: Silty soil that becomes soapy, sticky, or waxy when wet.

H

Haline: Term used to characterize water that contains salt in approximately the same percentage as ocean water.

Hard water: Water containing dissolved salts of calcium and magnesium, typically associated with bicarbonates, sulfates, and chlorides.

Hardpan: A shallow layer of earth material that has become relatively hard and impermeable, usually through the decomposition of minerals.

Head (Total hydraulic): the sum of the elevation head, pressure head, and velocity head at a given point in an aquifer.

Headwall: Entrance to a culvert or sluiceway.

Headway: The time between successive vehicles as they pass a common point.

Heat of hydration: The exothermic heat given off by concrete as it cures.

Herbaceous: With the characteristics of an herb; a plant with no persistent woody stem above ground.

Histosols: Organic soils.

Horizon: A layer of soil with different color or composition than the layers above and below it.

HOV: High-occupancy vehicle (e.g., bus).

Humic acid: The alkaline-soluble portion of organic material (i.e., humus) that precipitates from solution at low pH and is of higher molecular weight. A breakdown product of cellulose from vascular plants.

Humus: A grayish-brown sludge consisting of relatively large particle biological debris, such as the material sloughed off from a trickling filter.

Hydration: The chemical reaction between water and cement.

Hydraulic depth: Ratio of area in flow to the width of the channel at the fluid surface.

Hydraulic jump: An abrupt increase in flow depth that occurs when the velocity changes from supercritical to subcritical.

Hydraulic radius: Ratio of area in flow to wetted perimeter.

Hydric soil: Soil that is wet long enough to periodically produce anaerobic conditions, thereby influencing the growth of plants.

Hydrogen ion: The hydrogen atom stripped of its one orbital electron (H^+). It associates with a water molecule to form the hydronium ion (H_3O^+).

Hydrological cycle: The cycle experienced by water in its travel from the ocean, through evaporation and precipitation, percolation, runoff, and return to the ocean.

Hydrometeor: Any form of water falling from the sky.

Hydronium ion: *See* Hydrogen ion.

Hydrophilic: Seeking or liking water.

Hydrophobic: Avoiding or disliking water.

Hydrophyte, hydrophytic: Any plant growing in water or on a substrate that is at least periodically deficient in oxygen as a result of excessive water content.

Hygroscopic: Absorbing moisture from the air.

Hygroscopic water: Moisture tightly adhering in a thin film to soil grains that is not removed by gravity or capillary forces.

Hyperhaline: Term used to characterize waters with salinity greater than 40% due to ocean-derived salts.

Hypersaline: Term used to characterize waters with salinity greater than 40% due to land-derived salts.

Hypolimnion: (Gr. for "lower lake"); the deep, cold layer in a lake, below the epiliminion and metalimnion, cut off from the air above.

I

Igneous rock: Rock that forms when molten rock (magma or lava) cools and hardens.

Impactor: An environmental device that removes and measures micron-sized dusts, particles, and aerosols from an air stream.

Impervious layer: A geologic layer through which no water can pass.

Independent float: The amount of time that an activity can be delayed without affecting the float on any preceding or succeeding activities.

Infiltration: (a) Groundwater that enters sewer pipes through cracks and joints. (b) The movement of water downward from the ground surface through the upper soil.

Influent: Flow entering a process.

Influent stream: A stream above the water table that contributes to groundwater recharge. Influent streams may go dry during the rainless season.

Initial loss: The sum of interception and depression loss, excluding blind drainage.

In situ: "In place"; without removal; in original location.

Interception: The process by which precipitation is captured on the surfaces of vegetation and other impervious surfaces and evaporates before it reaches the land surface.

Interflow: Infiltrated subsurface water that travels to a stream without percolating down to the water level.

Intrusion: An igneous rock formed from magma that pushed its way through other rock layers. Magma often moves through rock fractures, where it cools and hardens.

Inverse condemnation: The legal process that may be initiated by a property owner to compel the payment of fair compensation when the owner's property has been taken or damaged for a public purpose.

Inversion layer: An extremely stable layer in the atmosphere in which temperature increases with elevation and mobility of airborne particles is restricted.

Inverted siphon: A sewer line that drops below the hydraulic grade line.

In vitro: Removed or obtained from an organism.

In vivo: Within an organism.

Ion: An atom that has either lost or gained one or more electrons, becoming an electrically charged particle.

Isogonic line: A line representing the magnetic declination.

Isotopes: Atoms of the same atomic number but having different atomic weights due to a variation in the number of neutrons.

J

Jam density: The density at which vehicles or pedestrians come to a halt.

Juvenile water: Water formed chemically within the earth from magma that has not participated in the hydrologic cycle.

K

Kingdom: A major taxonomic category consisting of several phyla or divisions.

Krause process: Mixing raw sewage, activated sludge, and material from sludge digesters.

L

Lagging: Heavy planking used to construct walls in excavations and braced cuts.

Lamp holes: Sewer inspection holes large enough to lower a lamp into but too small for a person.

Lane occupancy (ratio): The ratio of a lane's occupied time to the total observation time. Typically measured by a lane detector.

Lapse rate, dry: The rate that the atmospheric temperature decreases with altitude for a dry, adiabatic air mass.

Lapse rate, wet: The rate that the atmospheric temperature decreases with altitude for a moist, adiabatic air mass. The exact rate is a function of the moisture content.

Lateral: A sewer line that branches off from another.

Lava: Hot, liquid rock above ground. Also called lava once it has cooled and hardened.

Limnetic: Open water; extending down to the compensation level. Limnetic lake areas are occupied by suspended organisms (plankton) and free swimming fish.

Limnology: The branch of hydrology that pertains to the study of lakes.

Lipids: A group of organic compounds composed of carbon and hydrogen (e.g., fats, phospholipids, waxes, and steroids) that are soluble in a nonpolar, organic liquid (e.g., ether or chloroform); a constituent of living cells.

Lipiphilic: Having an affinity for lipids.

Littoral: Shallow; heavily oxygenated. In littoral lake areas, light penetrates all the way through to the bottom, and the zone is usually occupied by a diversity of rooted plants and animals.

Live load: The weight of all nonpermanent objects in a structure, including people and furniture. Live load does not include seismic or wind loading.

Loess: A deposit of wind-blown silt.

Lysimeter: A container used to observe and measure percolation and mineral leaching losses due to water percolating through the soil in it.

M

Macrophytic algae: Algal plants large enough either as individuals or communities to be readily visible without the aid of optical magnification.

Magma: Hot, liquid rock under the earth's surface.

Main: A large sewer at which all other branches terminate.

Malodorous: Offensive smelling.

Marl: An earthy substance containing 35% to 65% clay and 65% to 35% carbonate formed under marine or freshwater conditions.

Mean high water: The average height of the high water over 19 years.

Mean higher high tide: The average height of the higher of two unequal daily high tides over 19 years.

Mean low water: The average height of the low water over 19 years.

Mean lower low water: The average height of the lower of two unequal daily low tides over 19 years.

Mean tide level: A plane midway between mean high water and mean low water.

Meander corner: A survey point set where boundaries intersect the bank of a navigable stream, wide river, or large lake.

Meandering stream: A stream with large curving changes of direction.

Median: The portion of a divided highway separating traffic traveling in opposite directions.

Median lane: A lane within the median to accommodate left-turning vehicles.

Meridian: A great circle of the earth passing through the poles.

Meromictic lake: A lake with a permanent hypolimnion layer that never mixes with the epilimnion. The hypolimnion layer is perennially stagnant and saline.

Mesa: A flat-topped hill with steep sides.

Mesohaline: Term used to characterize waters with salinity of 5 to 18% due to ocean-derived salts.

Mesophyllic bacteria: Bacteria growing between 10°C and 40°C, with an optimum temperature of 37°C. 40°C is, therefore, the upper limit for most wastewater processes.

Mesophyte, mesophytic: Any plant growing where moisture and aeration conditions lie between extremes. (Plants typically found in habitats with average moisture conditions, not usually dry or wet.

Mesosaline: Term used to characterize waters with salinity of 5 to 18% due to land-derived salts.

Metabolism: All cellular chemical reactions by which energy is provided for vital processes and new cell substances are assimilated.

Metalimnion: A middle portion of a lake, between the epilimnion and hypolimnion, characterized by a steep and rapid decline in temperature (e.g., 1°C for each meter of depth).

Metamorphic rock: Rock that has changed from one form to another by heat or pressure.

Meteoric water: *See* Hydrometeor.

Methylmercury: A form of mercury that is readily absorbed through the gills of fish, resulting in large bioconcentration factors. Methymercury is passed on to organisms higher in the food chain.

Microorganism: A microscopic form of life.

Mineral soil: Soil composed of predominantly mineral rather than organic materials.

Mixohaline: Term used to characterize water with salinity of 0.5% to 30% due to ocean salts. The term is roughly equivalent to the term "brackish."

Mixosaline: Term used to characterize waters with salinity of 0.5% to 30% due to land-derived salts.

Mixture: A heterogeneous physical combination of two or more substances, each retaining its identity and specific properties.

Mohlman index: *See* Sludge volume index.

Mole: A quantity of substance equal to its molecular weight in grams (gmole or gram-mole) or in pounds (pmole or pound-mole).

Molecular weight: The sum of the atomic weights of all atoms in a molecule.

Monomictic: Experiencing one overturn per year. Monomictic lakes are typically very large and/or deep.

Mud: Wet soft earth composed predominantly of clay and silt—fine mineral sediments less than 0.074 mm (0.0029 in) in diameter.

Mud line: The lower surface of an excavation or braced cut.

N

Needle-leaved deciduous: Woody gymnosperms (trees or shrubs) with needle-shaped or scale-like leaves that are shed during the cold or dry season (e.g., bald cypress (*Taxodium distichum*)).

Needle-leaved evergreen: Woody gymnosperms with green, needle-shaped, or scale-like leaves that are retained by plants throughout the year (e.g., black spruce (*Picea mariana*)).

Nephelometric turbidity unit: The unit of measurement for visual turbidity in water and other solutions.

Net rain: That portion of rain that contributes to surface runoff.

Nitrogen fixation: The formation of nitrogen compounds (NH_3, organic nitrogen) from free atmospheric nitrogen (N_2).

Nitrogenous demand: Oxygen demand from nitrogen consuming bacteria.

Node: An activity in a precedence (critical path) diagram.

Nonpathogenic: Not capable of causing disease.

Nonpersistent emergents: Emergent hydrophytes whose leaves and stems break down at the end of the growing season so that most above-ground portions of the plants are easily transported by currents, waves, or ice. The breakdown may result from normal decay or the physical force of strong waves or ice. At certain seasons of the year there are no visible traces of the plants above the surface of the water (e.g., wild rice (*Zizania aquatica*), arrow arum (*Peltandra virginica*)).

Nonpoint source: A pollution source caused by sediment, nutrients, and organic and toxic substances originating from land-use activities, usually carried to lakes and streams by surface runoff from rain or snowmelt. Nonpoint pollutants include fertilizers, herbicides, insecticides, oil and grease, sediment, salt, and bacteria. Nonpoint sources do not generally require NPDES permits.

Nonstriping sight distance: Nonstriping sight distances are in between stopping and passing distances, and they exceed the minimum sight distances required for marking no-passing zones. They provide a practical distance to complete the passing maneuver in a reasonably safe manner, eliminating the need for a no-passing zone pavement marking.

Normal depth: The depth of uniform flow. This is a unique depth of flow for any combination of channel conditions. Normal depth can be determined from the Manning equation.

Normally consolidated soil: Soil that has never been consolidated by a greater stress than presently existing.

NPDES: National Pollutant Discharge Elimination System.

NTU: *See* Nephelometric turbidity unit.

O

Obligate hydrophytes: Species that are found only in wetlands (e.g., cattail (*Typha latifolia*) as opposed to ubiquitous species that grow either in wetlands or on uplands (e.g., red maple (*Acer rubrum*)).

Observation well: A nonpumping well used to observe the elevation of the water table or the potentiometric surface. An observation well is generally of larger diameter than a piezometer well and typically is screened or slotted throughout the thickness of the aquifer.

Odor number: *See* Threshold odor number.

Oligohaline: Term used to characterize water with salinity of 0.5 to 5.0% due to ocean-derived salts.

Oligosaline: Term used to characterize water with salinity of 0.5 to 5.0% due to land-derived salts.

Oligotrophic: Nutrient-poor. Oligotrophic lakes typically have low surface area-to-volume ratios and largely inorganic sediments and are surrounded by nutrient-poor soil.

Order: In taxonomy, a major subdivision of a class. Each order consists of one or more related families.

Organic soil: Soil composed of predominantly organic rather than mineral material. Same as "Histosol."

Orthotropic bridge deck: A bridge deck, usually steel plate covered with a wearing surface, reinforced in one direction with integral cast-in-place concrete ribs. Used to reduce the bridge deck mass.

Orthotropic material: A material with different strengths (stiffnesses) along different axes.

Osmosis: The flow of a solvent through a semipermeable membrane separating two solutions of different concentrations.

Outcrop: A natural exposure of a rock bed at the earth's surface.

Outfall: A pipe that discharges treated wastewater into a lake, stream, or ocean.

Overchute: A flume passing over a canal to carry floodwaters away without contaminating the canal water below. An elevated culvert.

Overhaul: Pertaining to billable hauling (i.e., with being able to bill an extra amount over the contract charge).

Overland flow: Water that travels over the ground surface to a stream.

Overturn: (a) The seasonal (fall) increase in epilimnion depth to include the entire lake depth, generally aided by unstable temperature/density gradients. (b) The seasonal (spring) mixing of lake layers, generally aided by wind.

Oxidation: The loss of electrons in a chemical reaction. Opposite of *reduction.*

Oxidation number: An electrical charge assigned by a set of prescribed rules, used in predicting the formation of compounds in chemical reactions.

P

Pan: A container used to measure surface evaporation rates.

Parkway: An arterial highway for noncommercial traffic, with full or partial control of access, usually located within a park or a ribbon of park-like development.

Partial treatment: Primary treatment only.

Partition coefficient: The ratio of the contaminant concentration in the solid (unabsorbed) phase to the contaminant concentration in the liquid (absorbed) phase when the system is in equilibrium; typically represented as K_d.

Passing sight distance: The length of roadway ahead required to pass another vehicle without meeting an oncoming vehicle.

Passive pressure: A pressure acting to counteract active pressure.

Pathogenic: Capable of causing disease.

Pathway: An environmental route by which chemicals can reach receptors.

Pay as you throw: An administration scheme by which individuals are charged based on the volume of municipal waste discarded.

Pedology: The study of the formation, development, and classification of natural soils.

Penetration treatment: Application of light liquid asphalt to the roadbed material. Used primarily to reduce dust.

Perched spring: A localized saturated area that occurs above an impervious layer.

Percolation: The movement of water through the subsurface soil layers, usually continuing downward to the groundwater table.

Permanent hardness: Hardness that cannot be removed by heating.

Persistent emergent: Emergent hydrophytes that normally remain standing at least until the beginning of the next growing season (e.g., cattails (*Typha* spp.) or bulrushes (*Scirpus* spp.)).

Person-rem: A unit of the amount of total radiation received by a population. It is the product of the average radiation dose in rems times the number of people exposed in the population group.

pH: A measure of a solution's hydrogen ion concentration (acidity).

Photic zone: The upper water layer down to the depth of effective light penetration where photosynthesis balances respiration. This level (the compensation level) usually occurs at the depth of 1% light penetration and forms the lower boundary of the zone of net metabolic production.

Phreatic zone: The layer the water table down to an impervious layer.

Phreatophytes: Plants that send their roots into or below the capillary fringe to access groundwater.

Phytoplankton: Small drifting plants.

Pier shaft: The part of a pier structure that is supported by the pier foundation.

Piezometer: A nonpumping well, generally of small diameter, that is used to measure the elevation of the water table or potentiometric surface. A piezometer generally has a short well screen through which water can enter.

Piezometer nest: A set of two or more piezometers set close to each other but screened to different depths.

Piezometric level: The level to which water will rise in a pipe due to its own pressure.

Pile bent: A supporting substructure of a bridge consisting of a beam or girder transverse to the roadway and that is supported, in turn, by a group of piles.

Pioneer plants: Herbaceous annual and seedling perennial plants that colonize bare areas as a first stage in secondary succession.

Pitot tube traverse: A volume or velocity measurement device that measures the impact energy of an air or liquid flow simultaneously at various locations in the flow area.

Planimeter: A device used to measure the area of a drawn shape.

Plant mix: A paving mixture that is not prepared at the paving site.

Plat: (a) A plan showing a section of land. (b) A small plot of land.

pOH: A measure of a solution's hydroxyl radical concentration (alkalinity).

Point source: A source of pollution that discharges into receiving waters from easily identifiable locations (e.g., a pipe or feedlot). Common point sources are factories and municipal sewage treatment plants. Point sources typically require NPDES permits.

Pollutant: Any solute or cause of change in physical properties that renders water, soil, or air unfit for a given use.

Polyhaline: Term used to characterize water with salinity of 18% to 30% due to ocean salts.

Polymictic: Experiencing numerous or continual overturns. Polymictic lakes, typically in the high mountains of equatorial regions, experience little seasonable temperature change.

Polysaline: Term used to characterize water with salinity of 18% to 30% due to land-derived salts.

Porosity: The ratio of pore volume to total rock, sediment, or formation volume.

Post-chlorination: Addition of chlorine after all other processes have been completed.

Potable: Suitable for human consumption.

Prechlorination: Addition of chlorine prior to sedimentation to help control odors and to aid in grease removal.

Presumptive test: A first-stage test in coliform fermentation. If positive, it is inconclusive without follow-up testing. If negative, it is conclusive.

Prime coat: The initial application of a low-viscosity liquid asphalt to an absorbent surface, preparatory to any subsequent treatment, for the purpose of hardening or toughening the surface and promoting adhesion between it and the superimposed constructed layer.

Probable maximum rainfall: The rainfall corresponding to some given probability (e.g., 1 in 100 years).

Protium: The stable ^{1}H isotope of hydrogen.

Protozoa: Single-celled aquatic animals that reproduce by binary fission. Several classes are known pathogens.

Putrefaction: Anaerobic decomposition of organic matter with accompanying foul odors.

Pycnometer: A closed flask with graduations.

Q

q-curve: A plot of depth of flow versus quantity flowing for a channel with a constant specific energy.

R

Rad: Abbreviation for "radiation absorbed dose." A unit of the amount of energy deposited in or absorbed by a material. A rad is equal to 62.5×10^6 MeV per gram of material.

Radical: A charged group of atoms that act together as a unit in chemical reactions.

Ranger: See Wale.

Rapid flow: Flow at less than critical depth, typically occurring on steep slopes.

Rating curve: A plot of quantity flowing versus depth for a natural watercourse.

Reach: A straight section of a channel, or a section that is uniform in shape, depth, slope, and flow quantity.

Redox reaction: A chemical reaction in which oxidation and reduction occur.

Reduction: The loss of oxygen or the gain of electrons in a chemical reaction. Opposite of *oxidation*.

Refractory: Dissolved organic materials that are biologically resistant and difficult to remove.

Regulator: A weir or device that diverts large volume flows into a special high-capacity sewer.

Rem: Abbreviation for "Roentgen equivalent mammal." A unit of the amount of energy absorbed by human tissue. It is the product of the absorbed dose (rad) times the quality factor.

Residual: A chemical that is left over after some of it has been combined or inactivated.

Resilient modulus: The modulus of elasticity of the soil.

Respiration: Any biochemical process in which energy is released. Respiration may be aerobic (in the presence of oxygen) or anaerobic (in the absence of oxygen).

Restraint: Any limitation (e.g., scarcity of resources, government regulation, or nonnegativity requirement) placed on a variable or combination of variables.

Resurfacing: A supplemental surface or replacement placed on an existing pavement to restore its riding qualities or increase its strength.

Retarded flow: A form of varied flow in which the velocity is decreasing and the depth is increasing.

Retrograde solubility: Solubility that decreases with increasing temperature. Typical of calcite (calcium carbonate, $CaCO_3$) and radon.

Right of access: The right of an abutting land owner for entrance to or exit from a public road.

Rigid pavement: A pavement structure having portland cement concrete as one course.

Rip rap: Pieces of broken stone used as lining to protect the sides of waterways from erosion.

Road mix: A low-quality asphalt surfacing produced from liquid asphalts and used when plant mixes are not available or economically feasible and where volume is low.

Roadbed: That portion of a roadway extending from curb line to curb line or from shoulder line to shoulder line. Divided highways are considered to have two roadbeds.

Roentgen: The amount of energy absorbed in air by the passage of gamma or X-rays. A roentgen is equal to 5.4×10^7 MeV per gram of air or 0.87 rad per gram of air and 0.96 rad per gram of tissue.

S

Safe yield: The maximum rate of water withdrawal that is economically, hydrologically, and ecologically feasible.

Sag pipe: *See* Inverted siphon.

Saline: Dominated by anionic carbonate, chloride, and sulfate ions.

Salinity: The total amount of solid material in grams contained in 1 kg of water when all the carbonate has been converted to oxide, the bromine and iodine replaced by chlorine, and all the organic matter completely oxidized.

Salt: An ionic compound formed by direct union of elements, reactions between acids and bases, reaction of acids and salts, and reactions between different salts.

Sand: Granular material passing through a no. 4 sieve but predominantly retained on a no. 200 sieve.

Sand trap: A section of channel constructed deeper than the rest of the channel to allow sediment to settle out.

Scour: Erosion typically occurring at the exit of an open channel or toe of a spillway.

Scrim: An open-weave, woven or nonwoven, textile product that is encapsulated in a polymer (e.g., polyester) material to provide strength and reinforcement to a watertight membrane.

Seal coat: An asphalt coating, with or without aggregate, applied to the surface of a pavement for the purpose of waterproofing and preserving the surface, altering the surface texture of the pavement, providing delineation, or providing resistance to traffic abrasion.

Second-stage demand: *See* Nitrogenous demand.

Sedimentary rock: Rocks formed from sediment, broken rocks, or organic matter. Sedimentary rocks are formed when wind or water deposits sediment into layers, which are pressed together by more layers of sediment above.

Seed: The activated sludge initially taken from a secondary settling tank and returned to an aeration tank to start the activated sludge process.

Seep: *See* Spring.

Seiche, external: An oscillation of the surface of a landlocked body of water.

Seiche, internal: An alternating pattern in the directions of layers of lake water movement.

Sensitivity: The ratio of a soil's undisturbed strength to its disturbed strength.

Separate system: A system with separate sewers for domestic and storm wastewater.

Septic: Produced by putrefaction.

Settling basin: A large, shallow basin through which water passes at low velocity, where most of the suspended sediment settles out.

Sheeted pit: *See* Cased hole.

Shooting flow: *See* Rapid flow.

Shrub: A woody plant which at maturity is usually less than 6 m (20 ft) tall and generally exhibits several erect, spreading, or prostrate stems and has a bushy appearance (e.g., speckled alder (*Alnus rugosa*) or buttonbush (*Cephalanthus occidentalis*)).

Sight distance: The length of roadway that a driver can see.

Sinkhole: A natural dip or hole in the ground formed when underground salt or other rocks are dissolved by water and the ground above collapses into the empty space.

Sinuosity: The stream length divided by the valley length.

Slickenside: A surface (plane) in stiff clay that is a potential slip plane.

Slope: The tangent of the angle made by the channel bottom. *See also* Gradient.

Sludge: The precipitated solid matter produced by water and sewage treatment.

Sludge bulking: Failure of suspended solids to completely settle out.

Sludge volume index (SVI): The volume of sludge that settles in 30 min out of an original volume of 1000 mL. May be used as a measure of sludge bulking potential.

Sol: A homogenous suspension or dispersion of colloidal matter in a fluid.

Soldier pile: An upright pile used to hold lagging.

Solution: A homogeneous mixture of solute and solvent.

Sorption: A generic term covering the processes of absorption and adsorption.

Sound: A body of water that is usually broad, elongated, and parallel to the shore between the mainland and one or more islands.

Space mean speed: One of the measures of average speed of a number of vehicles over a common (fixed) distance. Determined as the inverse of the average time per unit distance. Usually less than time mean speed.

Spacing: Distance between successive vehicles, measured front bumper to front bumper.

Specific activity: The activity per gram of a radioisotope.

Specific storage: *See* Specific yield.

Specific yield: (a) The ratio of water volume that will drain freely (under gravity) from a sample to the total volume. Specific yield is always less than porosity. (b) The amount of water released from or taken into storage per unit volume of a porous medium per unit change in head.

Split chlorination: Addition of chlorine prior to sedimentation and after final processing.

Spring: A place where water flows or ponds on the surface due to the intersection of an aquifer with the earth surface.

Spring tide: The highest high and lowest low tides during the lunar month.

Stadia method: Obtaining horizontal distances and differences in elevation by indirect geometric methods.

Stage: Elevation of flow surface above a fixed datum.

Standing wave: A stationary wave caused by an obstruction in a water course. The wave cannot move (propagate) because the water is flowing at its critical speed.

Steady flow: Flow in which the flow quantity does not vary with time at any location along the channel.

Stilling basin: An excavated pool downstream from a spillway used to decrease tailwater depth and to produce an energy-dissipating hydraulic jump.

Stoichiometry: The study of how elements combine in fixed proportions to form compounds.

Stone: Rock fragments larger than 25 cm (10 in) but less than 60 cm (24 in).

Stopping sight distance: The distance that allows a driver traveling at the maximum speed to stop before hitting an observed object.

Storage, specific: The amount of water released from or taken into storage per unit volume of a porous medium per unit change in head.

Stratum: Layer.

Stream gaging: A method of determining the velocity in an open channel.

Stream order: An artificial categorization of stream genealogy. Small streams are first order. Second-order streams are fed by first-order streams, third-order streams are fed by second-order streams, and so on.

Stringer: *See* Wale.

Structural section: The planned layers of specific materials, normally consisting of subbase, base, and pavement, placed over the subbase soil.

Subbase: A layer of aggregate placed on the existing soil as a foundation for the base.

Subcritical flow: Flow with depth greater than the critical depth and velocity less than the critical velocity.

Subgrade: The portion of a roadbed surface that has been prepared as specified, upon which a subbase, base, base course, or pavement is to be constructed.

Submain: *See* Branch sewer.

Submergent plant: A vascular or nonvascular hydrophyte, either rooted or nonrooted, that lies entirely beneath the water surface, except for flowering parts in some species (e.g., wild celery (*Vallisneria americana*) or the stoneworts (*Chara* spp.)).

Substrate: A substance acted upon by an organism, chemical, or enzyme. Sometimes used to mean organic material.

Subsurface runoff: *See* Interflow.

Superchlorination: Chlorination past the breakpoint.

Supercritical flow: Flow with depth less than the critical depth and velocity greater than the critical velocity.

Superelevation: Roadway banking on a horizontal curve for the purpose of allowing vehicles to maintain the traveled speed.

Supernatant: The clarified liquid rising to the top of a sludge layer.

Surcharge: An additional loading. (a) In geotechnical work, any force loading added to the in situ soil load. (b) In water resources, any additional pressurization of a fluid in a pipe.

Surcharged sewer: (a) A sewer that is flowing under pressure (e.g., as a force main). (b) A sewer that is supporting an additional loading (e.g., a truck parked above it).

Surface detention: *See* Surface retention.

Surface retention: The part of a storm that does not contribute to runoff. Retention is made up of depression storage, interception, and evaporation.

Surface runoff: Water flow over the surface that reaches a stream after a storm.

Surficial: Pertaining to the surface.

Swale: (a) A low-lying portion of land, below the general elevation of the surroundings. (b) A natural ditch or long, shallow depression through which accumulated water from adjacent watersheds drains to lower areas.

T

Tack coat: The initial application of asphalt material to an existing asphalt or concrete surface to provide a bond between the existing surface and the new material.

Tail race: An open waterway leading water out of a dam spillway and back to a natural channel.

Tailwater: The water into which a spillway or outfall discharges.

Taxonomy: The description, classification, and naming of organisms.

Temporary hardness: Hardness that can be removed by heating.

Terrigenous: Derived from or originating on the land (usually referring to sediments) as opposed to material or sediments produced in the ocean (marine) or as a result of biologic activity (biogenous).

Theodolite: A survey instrument used to measure or lay off both horizontal and vertical angles.

Thermocline: The temperature gradient in the metalimnion.

Thermophilic bacteria: Bacteria that thrive in the 45°C to 75°C range. The optimum temperature is near 55°C.

Thixotropy: A property of a soil that regains its strength over time after being disturbed and weakened.

Threshold odor number: A measure of odor strength, typically the number of successive dilutions required to reduce an odorous liquid to undetectable (by humans) level.

Till: *See* Glacial till.

Time mean speed: One of the measures of average speed of a number of vehicles over a common (fixed) distance. Determined as the average vehicular speed over a distance. Usually greater than space mean speed.

Time of concentration: The time required for water to flow from the most distant point on a runoff area to the measurement or collection point.

TON: *See* Threshold odor number.

Topography: Physical features such as hills, valleys, and plains that shape the surface of the earth.

Total float: The amount of time that an activity in the critical path (e.g., project start) can be delayed without delaying the project completion date.

Township: A square parcel of land 6 mi on each side.

Toxin: A toxic or poisonous substance.

Tranquil flow: Flow at greater than the critical depth.

Transmissivity: The rate at which water moves through a unit width of an aquifer or confining bed under a unit hydraulic gradient. It is a function of properties of the liquid, the porous media, and the thickness of the porous media.

Transpiration: The process by which water vapor escapes from living plants (principally from the leaves) and enters the atmosphere.

Traveled way: The portion of the roadway for the movement of vehicles, exclusive of shoulders and auxiliary lanes.

Tree: A woody plant which at maturity is usually 6 m (20 ft) or more in height and generally has a single trunk, unbranched for 1 m (3 ft) or more above the ground, and a more or less definite crown (e.g., red maple (*Acer rubrum*), northern white cedar (*Thuja occidentalis*)).

Turbidity: (a) Cloudiness in water caused by suspended colloidal material. (b) A measure of the light-transmitting properties of water.

Turbidity unit: *See* Nephelometric turbidity unit.

Turnout: (a) A location alongside a traveled way where vehicles may stop off of the main road surface without impeding following vehicles. (b) A pipe placed through a canal embankment to carry water from the canal for other uses.

U

Uniform flow: Flow that has constant velocity along a streamline. For an open channel, uniform flow implies constant depth, cross-sectional area, and shape along its course.

Unit process: A process used to change the physical, chemical, or biological characteristics of water or wastewater.

Unit stream power: The product of velocity and slope, representing the rate of energy expenditure per unit mass of water.

Uplift: Elevation or raising of part of the earth's surface through forces within the earth.

UTC: Coordinated Universal Time, the international time standard; previously referred to as Greenwich Meridian Time (GMT).

V

Vadose water: All underground water above the water table, including soil water, gravitational water, and capillary water.

Vadose zone: A zone above the water table containing both saturated and empty soil pores.

Valence: The relative combining capacity of an atom or group of atoms compared to that of the standard hydrogen atom. Essentially equivalent to the oxidation number.

Varied flow: Flow with different depths along the water course.

Varve: A layer of different material in the soil; fine layers of alluvium sediment deposited in glacial lakes.

Vitrification: Encapsulation in or conversion to an extremely stable, insoluble, glasslike solid by melting (usually electrically) and cooling. Used to destroy or immobilize hazardous compounds in soils.

Volatile organic compounds (VOCs): A class of toxic chemicals that easily evaporate or mix with the atmosphere and environment.

Volatile solid: Solid material in a water sample or in sludge that can be burned away or vaporized at high temperature.

Volatilization: The driving off or evaporation of a liquid in a solid or one or more phases in a liquid mixture.

W

Wah gas: *See* Biogas.

Wale: A horizontal brace used to hold timbers in place against the sides of an excavation or to transmit the braced loads to the lagging.

Wasteway: A canal or pipe that returns excess irrigation water to the main channel.

Water table: The piezometric surface of an aquifer, defined as the locus of points where the water pressure is equal to the atmospheric pressure.

Waving the rod: A survey technique used when reading a rod for elevation data. By waving the rod (the rod is actually inclined toward the instrument and then brought more vertical), the lowest rod reading will indicate the point at which the rod is most vertical. The reading at that point is then used to determine the difference in elevation.

Wet well: A short-term storage tank from which liquid is pumped.

Wetted perimeter: The length of the channel cross section that has water contact. The air-water interface is not included in the wetted perimeter.

Woody plant: A seed plant (gymnosperm or angiosperm) that develops persistent, hard, fibrous tissues, basically xylem (e.g., trees and shrubs).

X

Xeriscape: Creative landscaping for water and energy efficiency and lower maintenance.

Xerophytes: Drought-resistant plants, typically with root systems well above the water table.

Z

Zone of aeration: *See* Vadose zone.

Zone of saturation: *See* Phreatic zone.

Zoogloea: The gelatinous film (i.e., "slime") of aerobic organisms that covers the exposed surfaces of a biological filter.

Zooplankton: Small, drifting animals capable of independent movement.

Index

Download a printable copy of this index at **ppi2pass.com/envrmindex**.

INDEX - A

INDEX - B

INDEX - C

Comminutor, 26-7
Committed
  dose, 48-6
  dose equivalent, 48-6
  effective dose equivalent, 48-6
Common
  ion effect, 22-12
  logarithm, 3-5
  ratio, 3-11
Communication factor, 24-24
Commutative
  law, addition, 3-3, 4-4
  law, multiplication, 3-3
  oot, law, 11-2
Compaction factor, 39-3
Company health, 54-35
Comparative
  analysis, 54-17
  negligence, 56-6
Comparison
  alternative, 54-4
  test, 3-12
Compartment, water quality, 40-19
Compensatory
  damages, 56-7
  fraud, 56-6
  mitigation, 28-11
Competitive bidding, 57-3
Complaint, 56-6
  civil, 56-6
Complement
  nines, 53-16
  number, 53-16
  ones, 53-16
  set, 11-1
  set, law, 11-2
  tens, 53-16
  twos, 53-16
Complementary
  angle, 6-2
  equation, 10-1
  error function, 11-9, A-16
  probability, 11-4
  solution, 10-3
Complete
  combustion, 37-12
  filtration, 24-2
  isothermal flow equation, 17-10
  -mix aeration, 27-3, 27-6
  mixing model, 24-10
Completion time, 55-14
Complex
  conjugate, 3-8
  matrix, 4-1
  number, 3-1, 3-7
  number operations, 3-8
  plane, 3-7
  power, 52-6
Component
  alias, 9-8 (ftn)
  of a vector, 5-2
Compositing wastewater, 25-15
Composition
  atmospheric air, 27-8 (tbl)
  dry air, 37-8 (tbl)
  percentage, 22-7
Compostable plastic, 40-18
Composting
  in-vessel, 27-20
  sludge, 27-20
  static pile, 27-20
Compound, 22-5
  amount factor, 54-5
  binary, 22-5
  interest, 54-11
  organic, 43-1
  organic family, 43-3
  properties, A-87
  ternary, 22-5
  tertiary, 22-5
  volatile inorganic, 40-22
  volatile organic, 25-14, 38-6, 38-8, 40-22

Compounding
  discrete, 54-7 (tbl)
  period, 54-28
Compressed
  gas, 51-3
  gas, liquefied, 51-3
  liquid, 29-11
  liquid table, using, 29-11
  water, properties, A-97, A-118
Compressibility, 14-1, 14-13, 29-10
  adiabatic, 14-13
  chart, A-132
  chart, generalized, 29-20
  coefficient of, 14-13
  factor, 14-13 (ftn), 29-20 (ftn)
  isentropic, 14-13, 29-10
  isobaric, 29-10
  isothermal, 14-13, 29-10
Compressible, fluid, flow, 17-32
Compression, 46-9
  adiabatic, 15-13, 51-7
  air, 27-9
  isentropic, 27-9
  isothermal, 15-12
  polytropic, 15-13
  wave, 50-5
  wave speed, 17-38
  zone, 27-14
Compressor efficiency, 27-9
Computer
  complement, 53-16, 53-17
  local control (LCC), 35-5
Concave, 7-2
  down curve, 7-2
  up curve, 7-2
Concavity, 7-2
Concentration, 46-3
  cell corrosion, 22-19
  curve, 20-7
  cycle of, 31-13
  derived air, 48-8
  derived air, radionuclides, 48-10 (tbl)
  gas, 47-5
  hazardous, 46-4 (tbl)
  ionic, 22-11
  measured actual, 47-9
  metal, 25-15
  ppm, 47-5
  ppmw, 47-6
  ratio of, 31-13
  reference, 45-8
  saturation, 22-9
  standardized, 47-12
  time of, 20-3
  tower build up, 31-13
Concentrator, rotor, 42-4
Concentric cylinder viscometer, 14-6 (ftn)
Concrete, A-41, A-42
  cylinder pipe, prestressed, 16-10
  density, 15-11 (ftn)
  pipe, 16-9, 16-10, 25-3, A-41, A-42
  pipe, prestressed, 16-10
  pipe, reinforced, 16-10
  water pipe, 24-25
Condensate, polishing, 22-24
Condenser, 38-33
  contact, 38-33
  cooling water, 40-8
  noncontact, 38-33
  surface, 38-33 (fig)
Condensing vapor, 42-22
Condition
  adjacent space, 33-2
  antecedent moisture, 20-17
  antecedent runoff, 20-17
  design, inside, 33-2, 34-3
  design, outside, 33-2, 34-3
  indoor design, 32-6
  initial, 9-4
  international standard metric, 37-2
  natural gas, standard, 37-2
  normal, 47-5
  standard, 14-4, 36-2, 37-2, 47-5

Conditional
  convergence, 3-13
  probability, 11-4
  probability of failure, 11-9
Conditioner, sludge, 22-24
Conditioning, particle, 42-9
Conductance, 33-2, 52-2, 52-5
  thermal, 33-2
Conductivity, 33-2, 52-2
  hydraulic, 21-2
  water, 23-11, 24-19
Conduit, pressure, 16-5
Cone of depression, 21-5
Confidence
  level, 11-15
  level, z-values, 11-16 (tbl)
  limit, 11-15, 47-12
  limit, one-tail, 11-16 (tbl)
  limit, two-tail, 11-16 (tbl)
Confined aquifer, 21-2
Congruency, 7-3
Conic section, 7-8
Conjugate
  axis, 7-11
  complex, 3-8
  depth, 19-23, 19-26
Conscious order, 11-2
ConsensusDOC, 56-5 (ftn)
Consequential damages, 56-8
Conservation
  method, 33-7
  method, heat, 33-7
  of energy, 16-1
  of mass, 47-7
  of momentum, law, 17-33
  through thermostat setback, 33-7
Consideration, 56-3
Consistent
  system, 3-7
  unit system, 1-2
Constant
  acid, 22-15
  acid dissociation, A-78
  base, 22-15
  base dissociation, A-79
  Boltzmann, 29-17
  bridge, 53-10
  Calvert's, 42-18
  coefficient, 10-1, 10-2, 10-3
  decay, 48-2
  deflection, 53-12
  deoxygenation, 25-8
  dielectric, 42-9
  dissociation, 22-15
  equilibrium, 22-14
  Euler's, 9-8
  force, work, 13-2 (fig)
  formation, 22-15
  gamma ray dose, 48-8 (tbl)
  gravitational, 1-2
  growth rate, 3-11
  Hazen-Williams, A-49
  Henry's, 38-8
  Henry's law, 22-9 (tbl), A-141, A-142
  ionization, 22-15 (tbl)
  Joule's, 13-1, 29-6
  learning curve, 54-43 (tbl)
  Manning, 19-4
  Manning roughness, A-59
  matrix, 4-6
  meter, 17-29
  moisture content, 29-10
  of integration, 9-1, 10-1
  outlet, 36-30
  percentage method, 54-21 (ftn)
  Planck's, 49-1
  pressure, closed system, 30-8
  pressure process, 30-2
  reaction rate, 22-13, 42-7
  reaeration, 25-7
  reoxygenation, 25-7
  room, 50-10
  scaling, 42-6 (tbl)
  self-purification, 25-11

building, 55-4
capitalized, 54-7
construction, 55-4
contingency, 55-5
direct, 55-9
effect of inflation, 55-7
effect of location, 55-7
electricity, 18-9
equivalent uniform annual, 54-16
estimating, 55-3
financing, 55-5
fixed, 55-9
indirect, 55-9
information, 55-7
life-cycle, 54-20
movable equipment, 55-5
of goods sold, 54-37
on, 55-9
operating, 55-9
opportunity, 54-18
overhead, 55-6
plus fixed fee, 56-5
prime, 55-9
professional services, 55-5
site development, 55-4
sunk, 54-3
-time trade-off, 55-9
total, 54-33
variable, 55-9
Costing matrix, 55-6
Cotangent, hyperbolic, 6-5
Coulomb, 52-2
Coulomb's law, 49-2
Count, drag, 17-42
Coupon rate, 54-29
Cover
final, 39-5
type, 20-17
Covers function, 6-4
Coversed sine, 6-4
Cox chart, 14-10
CPF, 45-9
CPM, 55-11
CPVC pipe, 16-9, A-37
dimensions, A-37
Crack
coefficient, 32-5
length, 32-5
length method, 32-5
Cracking, 37-2
catalytic, 37-2
hydrocarbon, 37-2
thermal, 37-2
Cramer's rule, 3-7, 4-7
Crash time, 55-9
Credit, 54-34
emission reduction, 38-2
investment tax, 54-25
tax, 54-25
Creed, 57-1
Crest, dam, 15-11
Crevice
corrosion, 22-19
salt, 22-23
Criteria, pollutant, 38-2, 40-2
Critical
activity, 55-11
cavitation number, 18-16
depth, 19-17, A-62
depth, circular channel, A-62
distance, 17-5
flow, 16-7, 19-17
flow, nonrectangular channel, 19-18
flow, occurrences, 19-19
isobar, 29-3
oxygen deficit, 25-11
path, 55-9, 55-11
path, duct, 36-8
path method, 55-11
point, 8-2, 25-10, 25-11, 29-3, 29-8, 51-5
property, 29-8
Reynolds number, 16-7
settling velocity, 24-7
slope, 19-3

speed, shaft, 24-12
stress, 17-12
velocity, 24-6
velocity, sludge, 18-5, 18-6
zone, 16-7
Croplands, Prior Converted, 28-9
Cross
product, triple, 5-5
product, vector, 5-4 (fig)
section, 49-7
section, macroscopic, 49-7, 49-8
section, microscopic, 49-7
section, most efficient, 19-9
section, neutron, A-153
section, total, 49-7
section, trapezoidal, 19-8 (fig)
Crossover duct, 36-23
Crown, 19-27 (ftn)
Crud, 22-24 (ftn), 24-18
Crustacean, 44-1, 44-4
Crusting agent, 40-9
Cryogen, 29-2 (ftn)
Cryogenic
boiling point, 51-3
fluids, 29-2 (ftn)
liquid, 46-3, 51-3
Cryptosporidiosis, 44-2 (tbl)
Cryptosporidium, 44-2 (tbl)
Crystallization, water of, 22-6
CSO, 26-13
CSTR, 27-6
CTD, 46-21
CTS, 46-21
Cubical contents, 32-2
Cubital tunnel syndrome, 46-21
Culvert, 17-19, 19-26
classifications, 19-28
design, 19-31
entrance loss, 17-20
flow classifications, 19-28 (fig)
pipe, 19-26
simple pipe, 17-19 (fig)
Cumulative
frequency table, 11-11
mass fraction curve, 24-6
rainfall curve, 20-2
trauma disorder, 46-21
Cunningham
correction factor, 38-27, 38-29 (tbl), 42-9
slip factor, 42-9
Cup-and-bob viscometer, 14-6 (ftn)
Curb, inlet, 25-5
Curie, 46-13, 48-2
Curl, 8-7
Current
asset, 54-35
circuit breaker, rated, 52-8
circuit breaker, trip, 52-8
clear the, 52-8
divider, 52-4
electrical, 52-2
liability, 54-35
meter, 17-27
protection, circuit breaker, 52-8
ratio, 54-35
Curve, 7-3
area under standard normal, A-13 (tbl)
bell-shaped, 11-7
concave down, 7-2
concave up, 7-2
concentration, 20-7
cumulative rainfall, 20-2
degree, 7-3
dose-response, 45-6
even symmetry, 7-4
fan, 36-8
fan characteristic, 36-8
graph, 7-3
IDF, 20-5
intensity-duration-frequency, 20-5
inversion, 30-3
isenthalpic, 30-3
learning, 54-43
noise criteria, 50-9

number, NRCS, 20-16
number, runoff, 20-18, 20-19
number, SCS, 20-16
performance, 18-16
recession, 20-7
response, 50-6
sag, 25-10
system, 18-16, 36-8
tail, 11-7
Cut
diameter, 38-27
-over point, 54-23 (ftn)
size, 42-8
CVS 7.5 test, 38 35
Cyanazine, 40-18
Cycle, 30-11
Carnot, 30-11, 30-12
hydrologic, 20-1
integrated gasification/combined, 37-6
of concentration, 31-13
ozone, 38-10 (fig)
photolytic, 38-10 (fig)
power, 30-11
Cycloalkene, 43-2 (tbl)
Cycloid, 7-4
Cyclone, 42-7
conventional, 38-22
dimensions, 38-22 (fig)
high-efficiency, 38-22
high-throughput, 38-22
removal efficiency, 38-23
separator, 38-22
single, 42-7
standard, design dimensions, 38-23 (tbl)
Cylinder
-operated pump, 18-2
pipe, prestressed concrete, 16-10
Cylindrical
coordinate system, 7-3 (fig) (tbl)
source, 48-11
Cypermethrin, 40-18
Cypress dome hydroperiod, 28-3 (fig)

D

D'Alembert's paradox, 17-2 (ftn)
DAC
-hour, 48-8
radionuclide, A-152
Daily intake, chronic, 45-10
Dalton's law, 31-1, 37-14, 38-7, 39-6, 40-4
of partial pressure, 29-18
Dam, 15-11 (fig)
flood control, 20-21
force on, 15-11
gravity, 15-11 (ftn)
Damage (see also type)
hydrogen, 22-19
Damages, 56-7
compensatory, 56-7
consequential, 56-8
exemplary, 56-7
punitive, 56-6
special, 56-7
Damaging, flood, 20-6
Damper
automatic, 36-24
control, face-and-bypass, 35-1
duct, 36-23
face, 35-2 (ftn)
fire, 36-24
gravity, 36-24
mixing, 36-13
pulsation (pump), 18-2
smoke, 36-24
splitter, 36-24
volume, 36-24
Darcy, 21-2
equation, 17-6, 17-8
friction factor, 17-5 (ftn), A-50, A-51, A-52, A-53
velocity, 21-4
-Weisbach equation, 17-6, 17-7
Darcy's law, 21-3, 39-7, 40-12

vector, 5-1 (ftn)
water chemistry, A-71, A-72
weight, 22-6
weight, milligram, 22-10, 22-11
wind velocity, 32-6
Equivalents
calcium carbonate, A-71
population, 25-3
velocity, 21-3
Erf, 9-8
Erg, 1-5, A-3
Ergonomic hazard, 46-2
Ergonomics, 46-18
Erodible
channel, 19-26
channel side-slopes, 19-26
channel velocities, 19-26
Erosion corrosion, 22-19
Error, 11-12
false positive, 11-15
function, 9-8, 11-9, A-16
function, complementary, 11-9, A-16
limits range, 53-7
magnitude, 53-3
sampling and analytical, 47-12
standard, 11-13
terms, 53-3
type, 53-2
type I, 11-15
type II, 11-15
Errors and omissions insurance, 56-8
ESP, 38-25 (tbl), 38-26 (fig)
collection efficiency, 42-9
power density, 42-10
removal efficiency, 42-9
Espey
conveyance factors, 20-12
synthetic unit hydrograph, 20-12
Ester, 43-1 (tbl), 43-3 (tbl)
Estimate, precise, 53-2
Estimating
cost, 55-3
parameter method, 55-6
Estimator, 55-3
unbiased, 11-12, 11-13
Estuarine, water, 28-2
Ethanol, 37-7
Ether, 43-1 (tbl), 43-3 (tbl)
Ethical priority, 57-2
Ethics, 57-1
code of, 57-1
Ethyl alcohol, 37-7
ETS smoke, 32-5
Eukaryote, 46-23
Euler's
constant, 9-8
equation, 3-8, 10-5
number, 1-9 (tbl)
Eutrophic, 28-5
Eutrophication, 23-7
Evaporation, 10-10
loss, 31-13
pan, 20-22
reservoir, 20-22
Evaporative
cooling, 31-10
pan humidification, 32-7
pan method, 32-7
Evase, 36-4
Even symmetry, 9-7 (tbl)
curve, 7-4
Event, 11-3, 55-11
dependent, 11-3
independent, 11-3
numerical, 11-4, 11-5
sample space, 11-3
Evidence, weight of, 45-9
Exact, first-order, 10-3
Exam
date, 58-3
engineering, professional, 58-1
Fundamentals of Engineering, 58-1
licensing, 58-1, 58-2

Examination
date, 58-3
licensing, 58-1, 58-2
Exceedance ratio, 11-13
Excess
air, 22-8, 37-12
kurtosis, 11-14
lifetime cancer risk, 45-10
spherical, 6-6
Exchange ion, 22-21
Excitation, 49-2, 49-4
motor, 52-15
Excretion, 45-3
Exemplary
damages, 56-7
Exemption, industrial, 58-1
Exertion, BOD, 25-9
Exfoliation, 22-18
Exhaust
duct system, 36-21, 36-31
hood, 32-9
Exhaustion
heat, 46-15
premature, 26-13
Exit pipe, 17-13
Exogenous infection, 46-23
Exothermic reaction, 22-18
Expansion
by cofactor, 4-3
factor, 17-32 (fig)
method, 53-14
ratio, 51-5
ratio, liquids, 51-5 (tbl)
series, 8-8
thermal coefficient of, 14-13 (ftn)
Expected value, 11-9, 54-30
Expense, 54-32
administrative, 54-33
artificial, 54-20
marketing, 54-33
selling, 54-33
Expensing an asset, 54-20
Experiment
accuracy, 11-12
insensitivity, 11-12
precision, 11-12
reliability, 11-12
stability, 11-12
Explosion
boiling liquid expanding vapor, 51-8
limit, 51-5
Explosive
energy release, 51-7
limit, lower, 32-10, 46-3
limit, upper, 46-3
power, ton of, 1-7
Exponent, 3-5
polytropic, 15-13, 30-2
rules, 3-5
Exponential
decay, 10-9
decreasing cash flow, 54-26
distribution, 11-6, 11-9, 11-10
distribution, negative, 11-9
form, 3-7, 3-8
function, integral, 9-8
gradient factor, 54-26
gradient series cash flow, 54-4
growth, 3-11, 10-9
growth rate, 54-4
reliability, 11-10
weighted forecast, 54-42
Exposed fraction, 15-15
Exposure, 47-1
acute, 11-23, 47-2, 48-3
allowance, 33-2
assessment, 47-9
assessment standards, 47-10 (tbl)
chemical, 47-2
chronic, 11-23, 47-2, 48-3
direct human, carcinogen, 45-9
duration, carcinogen, 45-10
equivalent, 45-12
factor, 24-24

frequency, 51-8
level, short-term, 45-12
limit, radiation, 48-12
limit, recommended, 46-2
limit, short-term, 32-10, 45-12
pathway, 45-1
permissible level, noise, 46-11
point, 11-22
radiation, 48-2, 48-12
time, particle, 42-9
Exsec function, 6-4
Exsecant, 6-4
Extended
aeration, 27-2
Bernoulli equation, 17-15
plenum system, 36-13
Extensive property, 29-4
External
dose, 48-4
investment, 54-13
pressure, 15-13
rate of return, 54-12
static pressure, 36-8
work, 13-2
Extraction, 28-6
dilution, 37-20
drain, 39-7, 40-12
straight, 37-20
vacuum, 42-22
well, gas, 39-6
Extraneous root, 3-4
Extrema point, 8-2
Extreme
fiber, 2-7
point, 8-2
Extremity, 48-12
Extrinsic waste, 40-2
Eye, 45-3, 45-6
anterior chamber, 45-3
dose equivalent, 48-4
posterior chamber, 45-3
Eyewear, protective, 46-5
Eötvös number, 1-9 (tbl), 17-45

F

F-waste, 40-3
FAA formula, 20-4
Fabric
building, 34-4 (ftn)
loading, 42-5
resistance, 38-20
Face
-and-bypass damper control, 35-1
angle, 6-6
damper, 35-2 (ftn)
value, 54-29
velocity, 21-4, 42-5
velocity, superficial, 38-20
Facility
material recovery, 39-11
waste-to-energy, 39-10
Factor (see also type)
absorption, 42-21
bioconcentration, 45-11
buildup, 49-6
bypass, 31-7, 34-7
carcinogen potency, 11-20, 45-9, 45-10
cash flow, 54-6, A-166, A-167
cash flow, discounting, 54-6
coil, 31-7
compaction, 39-3
compressibility, 14-13 (ftn), 29-20 (ftn)
contact, 31-8
conversion, A-1, A-2
conveyance, open channel, A-63, A-64, A-65, A-66
conveyance, trapezoidal channel, A-63, A-64
cooling load, 34-3
Cunningham correction, 38-27, 38-29 (tbl), 42-9
Cunningham slip, 42-9
Darcy friction, 17-5, A-50, A-51, A-52, A-53

Heating
degree-days, 33-6
friction, 17-4 (ftn)
kelvin-days, 33-6
load, 33-1
load, average, 33-1
load, maximum, 33-1
losses, 52-17
sensible, 31-8
value, 37-14
value, gross, 37-14
value, higher, 37-14
value, lower, 37-14
value, methane, 27-17
with dehumidification, 31-12
with humidification, 31-11
Heatstroke, 46-15
Heave, 21-9
Heavy
hydrogen, 22-2
metal, 22-2
metal, in wastewater, 25-14
Heel (verb), 15-17 (ftn)
dam, 15-11
Height
cell, 39-3, 40-14
effective stack, 38-18
of packing, 42-22
of transfer unit, 42-22
piezometric, 21-2
Helix, 7-12
pitch, 7-12
Helmholtz function, 29-9
Helminth, 44-1, 44-2 (tbl)
Hematoxicity, 45-4
Henry's
constant, 38-8
law, 22-9, 38-8, 42-21
law constant, 22-9 (tbl), A-141, A-142
HEPA, 32-11
filter, 36-23
Hepatitis, 44-2 (tbl), 45-4, 46-25
serum, 46-25
Hepatoxicity, 45-4
Herbicide, 40-17
Hess' law, 22-18
Heterotroph, 44-1, 44-3
facultative, 44-4
Hex number, 53-15 (ftn)
Hexachlorobenzene, 40-17
Hexadecimal system, 53-15
HFC-134a pressure-enthalpy
diagram, A-111, A-130
Hi-vol
particulate sampler, 38-11 (fig)
sampler, 38-11 (fig)
Hideout, 22-23
High
-efficiency cyclone, 38-22
-efficiency particle arresting, 32-11
-efficiency particulate air filter, 36-23
-efficiency particulate arresting, 32-11
-lift safety valve, 16-12
purity oxygen aeration, 27-4
-rate aeration, 27-4
-rate filter, 26-8
repetitiveness, stress, 46-21
-temperature reservoir, 30-11
-throughput cyclone, 38-22
-velocity system, 36-21
-volume sampler, 38-11
Higher heating value, 37-14
Highly significant results, 11-15
Highway
accident, 51-1
Fuel Economy Driving Schedule, 38-35
rainwater runoff, 40-19
Hindered settling zone, 27-14
Histogram, 11-11
Hitchin's formula, degree-days, 33-6
HIV, 44-6, 46-25
Hobby, 54-12 (ftn)
HOCl fraction, ionized, 24-20
Hogs hair filters, 36-23

Hole, tap, 15-3
Holomorphic function, 8-1
Homogeneous
circuits, law of, 53-7
differential equation, 10-1, 10-2
linear equation, 3-7
second-order, 10-3
unit system, 1-2 (ftn)
Homologous
fan, 36-10
pump, 18-13
Hood
air, 32-9
air, enclosure, 32-9
air, nonbypass, 32-9
air, nonenclosure, 32-9
conventional cabinet bypass air,
32-9 (fig)
exposure, minimum control velocity for,
32-9 (tbl)
fume exhaust, 32-9
minimum control velocity for exposure,
32-9 (tbl)
Hooke's law, 53-11
Hookworm, 44-2 (tbl)
Horizon, 54-3
analysis, 54-17
Horizontal flow grit chamber, 26-6
Horsepower, 13-6, A-1, A-2
aero, 17-42
air, 36-6
blower, 36-6
boiler, 37-17
brake, 36-6
friction, 36-6
hydraulic, 17-15, 18-8 (tbl)
theoretical, 17-15
water, 17-15
Horton
coefficient, 19-13
-Einstein equation, 19-10
equation, 19-13, 21-9
Hot
deck, 36-13
-wire anemometer, 17-27
Hour, DAC-, 48-8
HPD, 50-12
Huebscher equation, 36-17
Hull, convex, 7-2, 7-3
Human
direct exposure, carcinogen, 45-9
error, 53-2
immunodeficiency virus, 44-6, 46-25
-made radiation, 48-3
Humectant, 40-9 (ftn)
Humidification, 31-7, 32-7, 33-4
booster, 32-7
efficiency, 31-10
load, 31-7
pure, 31-7
spot, 32-7
steam, 31-11
straight, 31-7
Humidistat, 35-1
Humidity
maximum indoor, 32-9
mold, 32-8
percentage, 31-2
ratio, 31-2
relative, 31-2
specific, 31-2
Hundred year
flood, 20-6
storm, 20-6
HVAC
controller, 35-1
process control, 35-1
HWFET, 38-35
Hydrant, fire, 24-24
Hydrated
lime, 24-15
molecule, 22-6
Hydration, 22-10
water of, 22-6

Hydraulic
conductivity, 21-2
depth, 16-6 (ftn), 19-3
detention time, 27-5, 27-8
diameter, 16-6 (ftn), 17-9, 36-17 (ftn)
drop, 16-9 (ftn), 19-20 (ftn), 19-26
element, circular pipes, A-29
elements, A-61
grade line, 16-8, 17-16, A-29
gradient, 21-2
head, 24-13
horsepower, 17-15, 18-8 (tbl)
jack, 15-14
jump, 16-9 (ftn), 19-23
kilowatt, 18-8 (tbl)
loading, 25-3, 26-10, 42-23
mean depth, 19-2 (ftn)
NPSHR, 18-21
power, 18-7
press, 15-14
radius, 16-5, 19-2, A-29
ram, 15-14
retention time, 27-7
Hydraulically
long, 19-27
short, 19-27
Hydraulics, 17-2
Hydrocarbon, 37-2, 38-6, 43-1, 43-3 (tbl)
aromatic, 37-2
pollutant, 38-5
pumping, 18-21
saturated, 37-2, 43-2 (tbl)
specific gravity, A-26
unsaturated, 37-2
vapor pressure, A-25
Hydrochloric acid, 23-9, 38-3, 40-4
Hydrodynamics, 17-2
Hydroelectric generating plant, 18-23
Hydrogen
available, 37-3
damage, 22-19
embrittlement, 22-19
heavy, 22-2
normal, 22-2
sulfide, 25-5, 44-6
Hydrogenation, 40-18
Hydrograph, 20-7
analysis, 20-7
separation, 20-7
Snyder, 20-11
synthesis, 20-11, 20-13
triangular unit, 20-12
unit, 20-8
Hydrologic
cycle, 20-1
soil group, 20-16
Hydrology
changes, 28-2
of lands, 28-2
wetlands, 28-2
Hydrolysis, 43-2
Hydrolyzing metal ion, 24-8
Hydrometer, 14-4
Hydroperiod, 28-2
bottomland forest, 28-3 (fig)
coastal marsh, 28-3 (fig)
cypress dome, 28-3 (fig)
nontidal, 28-3 (tbl)
tidal, 28-3 (tbl)
tropical floodplain, 28-3 (fig)
Hydrophilic, 14-12
Hydrophobic, 14-12
Hydrostatic, 15-4 (ftn)
moment, 15-11
paradox, 15-4
pressure, 15-4
pressure, vertical plane surface, 15-7 (fig)
resultant, 15-6
torque, 15-11
Hydrostatics, 17-2
Hydroxyl, 43-1 (tbl)
Hyetograph, 20-2
Hygiene, industrial, 46-1

Injection
  sorbent, 42-20
  well, 42-11
Injured party, 56-6
Inland wetlands, 28-2
Inlet, 36-29
  control, 19-27
  curb, 25-5
  grate, 25-5
  gutter, 25-5
  pressure, net positive, 18-2 (ftn)
  vane, 36-7
Inline sampling, 22-23
Inner transition element, 22-2
Innovative method, 40-20
Inorganic
  chemical, 25-14
  compound, volatile, 40-22
  salt, removal, in wastewater, 26-3
Insecticide (see also Pesticide), 40-17
Insensitivity, 53-2
  experiment, 11-12
Insertion loss, 46-11, 50-11
Inside
  and outside design conditions, 34-3
  design conditions, 33-2
  design temperature, 33-2, 34-3
Inspection, polynomial, 3-4
Instability, 53-2
Instantaneous
  cooling load, 34-2
  cooling load from walls and roofs, 34-3
  cooling load from windows, 34-3, 34-4
  growth, 3-11
  heat absorption, 34-2
  heat gain, 34-2
  reorder inventory, 54-44 (fig)
Institutional issues, 38-34
Instrument, coefficient of, 17-28
Insulation clothing, 46-15
Insurance, 56-8
  Services Office, 24-23
Intake
  annual limit on, 48-8, 48-10 (tbl)
  annual limit, radionuclides, 48-10 (tbl)
  chronic daily, 51-8
Intangible property, 54-20 (ftn)
Integral, 9-1, 9-2
  convolution, 10-6
  cosine function, 9-8
  definite, 9-1 (ftn), 9-4
  double, 9-3
  elliptic function, 9-8 (ftn)
  exponential function, 9-8
  Fresnel function, 9-8 (ftn)
  function, 9-8
  gamma function, 9-8 (ftn)
  indefinite, 9-1, 9-4, A-11
  of combination of functions, 9-2
  of hyperbolic function, 9-2
  of transcendental function, 9-1
  sine function, 9-8
  table, 9-1 (ftn)
  triple, 9-3
Integrand, 9-1
Integrated
  gasification/combined cycle, 37-6
  Risk Information System, 45-9
Integrating factor, 10-2
Integration, 9-1
  by parts, 9-2
  by separation of terms, 9-3
  constant of, 9-1, 10-1
  method, 2-1, 2-4
  typical controls, 35-2
Intensity
  -duration-frequency curve, 20-5
  luminous, 50-2
  of sound, 50-4
  radiation, 46-13
  rainfall, 20-4
  sound, 46-9
Intensive property, 29-4

Intercept, 7-5
  form, 7-5
Interceptor, 25-3
Interest
  compound, 54-11
  rate, effective, 54-5, 54-28
  rate, effective annual, 54-5
  rate, nominal, 54-28
  simple, 54-11
Interfacial
  area, 42-21
  friction angle, 39-5 (tbl)
Interference nerve, 45-5
Intergranular, 22-19
  attack, 22-19
  corrosion, 22-19
Intermediate
  clarifier, 26-12
  metals, law of, 53-7
  neutron, 49-8
  temperatures, law of, 53-8
Intermittent
  duty, 52-14
  noise, 46-11
  sand filter, 26-12
Intern, engineer, 58-1 (ftn)
Internal
  combustion engine, 38-36 (fig)
  dose, 48-6, 48-8
  energy, 13-4, 29-5
  energy, molar, 29-5
  heat gain, 33-5, 34-1
  heat source, 33-4, 34-4
  rate of return, 54-12
  specific energy, 29-5
  work, 13-2
International standard
  atmosphere, A-58
  metric conditions, 37-2
Interpolating polynomial, Newton's, 12-3
Interpolation, nonlinear, 12-2, 12-3
Intersection
  angle, 7-6, 7-8
  line, 7-6
  set, 11-1
Intertidal substrate, 28-2
Interval
  proof test, 11-9
  recurrence, 20-5
Intestine, 45-3
  small, 45-3
Intrinsic
  permeability, 21-2
  waste, 40-2
Inventory, 54-37
  instantaneous reorder, 54-44 (fig)
  supply, 54-36
  turnover, 54-35, 54-36
  value, 54-37
Inverse
  function, 6-4
  Laplace transform, 10-6
  lever rule, 31-5
  matrix, 4-6
  square law, 48-11
  trigonometric operation, 6-4
Inversion, 38-15
  curve, 30-3 (fig)
  Fourier, 9-6
  point, 30-3
  temperature, 30-3 (tbl)
Invert, 19-27
Inverted bucket steam trap, 16-14
Investment
  external, 54-13
  return on, 54-11, 54-14
  risk-free, 54-11
  tax credit, 54-25
IOC, 25-14
Iodine, 24-21
Ion
  bar chart, 23-1
  common, effect, 22-12
  exchange, 22-21

  exchange regeneration, 24-18
  exchange resin, 22-22
  exchange softening, 24-17
  product, 22-15
Ionic concentration, 22-11
Ionization, 49-2, 49-4
  constant, 22-15
Ionized HOCl fraction, 24-20
Ionizing radiation, 46-12, 48-1
IRIS, 45-9
Iron
  bacteria, 44-3
  ion, in water, 23-6, 24-15
  loss, 52-17
  pipe, A-43
  pipe dimensions, A-43, A-44
  pipe, standard pressure classes, A-43
  removal processes, 24-15
Irradiance, 50-3
Irradiation, 24-21
Irrational number, 3-1
Irreducible factor, 54-2
Irregular
  areas, 7-1 (fig)
  boundary, 7-1
Irreversibility, 30-13
Irreversible
  process, 30-3
  reaction, 22-13
ISA nozzle, 17-32 (ftn)
Isenthalpic
  curve, 30-3
  process, 30-2
Isentropic
  closed system, reversible adiabatic, 30-9
  compressibility, 14-13, 29-10
  compression, 27-9
  efficiency, 27-9
  process, 30-2, 30-9
  steady-flow, 30-9
Isobar, 29-3
  critical, 29-3
Isobaric
  compressibility, 29-10
  process, 30-2
Isochoric process, 30-2
Isohyet, 20-2
Isohyetal method, 20-2
Isokinetic
  sub-, 37-19 (ftn)
  super-, 37-19 (ftn)
  test, 37-19
Isolated system, 30-2
Isometric process, 30-2
Isometry, 7-3
Isopiestic equilibrium, 14-10 (ftn)
Isotherm, 29-12
Isothermal
  compressibility, 14-13, 29-10
  compression, 15-12
  process, 30-2
Isotope, 22-2, 48-1
  effect, 22-2 (ftn)
Iteration, fixed-point, 12-2

J

Jack, hydraulic, 15-14
Jacob's equation, 21-7
Jet
  force on blade, 17-35
  force on plate, 17-34
  propulsion, 17-34
  pump, 18-2
Job cost accounting, 54-36
Joint (see also type)
  efficiency, 16-10
  probability, 11-4
  processing, 25-2
  quality factor, 16-10
Joule, 1-5, A-3
  equivalent, 13-1
  -Kelvin coefficient, 30-3
  -Thomson coefficient, 30-3

crack, 32-5
  equivalent, 17-12, A-55
  equivalent, fittings, A-55
  method, crack, 32-5
  of hydraulic jump, 19-26
  ratio, 17-45
  relaxation, 49-9
Lens, 45-3
  dose equivalent, 48-4
Leptokurtic distribution, 11-14
Letter of agreement, 56-3
Leukemia, 45-4
Level
  confidence, 11-15
  maximum contaminant, 25-15
  neutral pressure, 32-6
  of protection, 47-13, 47-14 (tbl)
  OSHA noise action, 50-12
  peak, 50-6
  permissible exposure, noise, 46-11
  recommended alert, 46-15
  significance, 11-15
  sound power, 50-7
  sound pressure, 46-9, 50-7
Leveling resource, 55-10
Lever
  effectiveness, 15-14
  efficiency, 15-14
  rule, 31-5
  rule, inverse, 31-5
Leverage, 54-35
Lewis number, 1-9 (tbl)
Liability
  account, 54-33
  current, 54-35
  in tort, 56-7
  limited, company, 56-2
Licensing, 58-1
  exam, 58-3
  examination, 58-3
Lien
  construction, 56-5
  laborer's, 56-5
  materialman's, 56-5
  mechanic's, 56-5
  perfecting, 56-5
  supplier's, 56-5
Life
  analysis, economical, 54-4
  -cycle, 55-7
  -cycle assessment, 55-7
  -cycle cost, 54-15, 54-20
  -cycle cost analysis, 55-7
  economic, 54-19, 54-20
  half-, 22-13
  service, 54-20
  useful, 54-20
Lifeline rate, 52-7 (ftn)
Lifetime
  cancer risk, excess, 45-10
  water treatment plant, 24-2
Lift, 17-39, 39-3
  coefficient of, 17-40
  rotating cylinders, 17-41
  valve, 16-12
Lifting, 46-19
  velocity, water, 21-5
Light
  heat gain from, 33-5
  loss factor, 50-4
  metal, 22-4
  quantity of, 50-2
  reflection, 50-4
  refracted, 50-4
  -sensitive detector, 53-4, 53-5
Lignite coal, 37-5
Limb
  falling, 20-7
  rising, 20-7
Lime, 24-9
  hydrated, 24-15
  slaked, 24-15
  -soda ash process, 24-15
  softening, 24-15 (ftn)

Limit, 3-9
  class, 11-11
  confidence, 11-15, 47-12
  explosion, 51-5
  flammable, 51-5 (fig)
  lower, 9-4
  lower explosive, 32-10, 46-3
  lower flammable, 46-3
  of detection, 47-5
  of quantification, 47-5
  radiation dose, 48-12
  recommended exposure, 46-2
  short-term exposure, 32-10, 45-12
  simplifying, 3-9
  threshold value, noise, 46-11
  upper, 9-4
  upper explosive, 46-3
  upper flammable, 46-3
  value, threshold, 32-10, 45-12
Limited
  liability company, 56-2
  liability partnership, 56-2
  partner, 56-2
  partnership, 56-2
Limiting
  current, sensor, 53-4
  reactant, 22-8
  value, 3-9
Lindane, 40-17
Line, 6-1 (ftn), 7-3
  centroid of a, 2-3
  coil load, 31-8
  current, 52-11
  diagram, 35-6
  energy, 16-8 (ftn)
  energy grade, 16-8, 17-15
  equipotential, 17-3, 21-7
  flow, 21-7, 26-2
  hydraulic grade, 16-8, 17-16
  intersection, 7-6
  normal vector, 8-6
  of action, 5-1
  saturated liquid, 29-3
  saturated vapor, 29-3
  source, 48-11
  straight, 7-4 (fig)
  tie, 35-7
Lineal, 1-8
  measurement, 1-8
Linear, 1-8
  algebra, 4-4 (ftn)
  equation, 3-6, 3-7
  equation, differential, 10-1
  equation, graphing, 3-7
  equation, reduction, 3-7
  equation, simultaneous, 3-6, 3-7
  equation, substitution, 3-7
  first-order, 10-2
  frequency, 52-4
  growth rate, 3-11
  momentum, 17-33
  regression, 11-17
  response, 50-6
  second-order, 10-3
  spring, 13-2 (ftn)
  variable differential transformer, 53-4
  watt density, 33-4 (ftn)
Linearity, 10-6
Lined concrete pipe, 16-10
Liner
  clay, 39-4
  double, 39-3
  flexible membrane, 39-3, 39-4
  geocomposite, 39-4
  synthetic membrane, 39-4
Lines, angle between, 6-1
Lining, pipe, 16-11
Liquefied
  compressed gas, 51-3
  petroleum gas, 37-8
Liquid
  aromatic, 14-10 (ftn)
  combustible, 51-2
  compressed, 29-11

cryogenic, 46-3, 51-3
  expansion ratio, 51-5 (tbl)
  flammable, 51-2
  fuel, 37-6
  -gas ratio, venturi scrubber, 42-18
  immiscible, 15-12
  incineration, 42-14
  pore-squeeze, 39-7, 40-11
  saturated, 29-2
  saturated, properties of, 29-13
  specific gravity, 51-4
  subcooled, 29-2, 29-11
  subcooled, properties, 29-12
  table, compressed, 29-11
  thixotropic, 14-7
  -vapor mixture, 29-2
  -vapor mixture, properties of, 29-13
  volatile, 14-10, 22-10 (ftn)
Liquidated damages, 56-7
Liquidity, 54-35
Liquor, mixed, 27-2
List (verb), 15-17 (ftn)
Liter, 1-6
Lithotroph, 44-1
Liver, 45-3
Lloyd-Davies equation, 20-14 (ftn)
LMTD, 38-33
Load
  air conditioning, 34-1
  average heating, 33-1
  cell, 53-12
  coil, 34-1
  cooling, 34-1
  cooling, from internal heat sources, 34-4
  factor, v-belt, 36-8 (ftn)
  heat, 31-13
  heating, 33-1
  humidification, 31-7
  instantaneous cooling, 34-2
  instantaneous cooling, from walls and
    roofs, 34-3
  instantaneous cooling, from windows,
    34-3, 34-4
  latent, 34-6
  manual handling, 46-19
  maximum heating, 33-1
  overhung, 18-11
  pickup, 33-5
  refrigeration, 34-1 (ftn)
  shock, septage, 26-2
  solar cooling, 34-3
  tower, 31-13
Loading
  BOD, 25-9, 26-9
  chemical, 42-23
  coil, 34-1 (ftn)
  dust, 42-5
  error, 53-2
  fabric, 42-5
  factor, 39-3, 40-14
  hydraulic, 25-3, 26-10, 42-23
  organic, 25-3, 26-9
  pump shaft, 18-11
  rate, 24-13
  sewage treatment plant, 25-3
  surface, 24-6, 26-7, 26-8
  volumetric, 27-8
  weir, 24-6, 26-7
LOAEL, 45-7
Loan, 54-40
  repayment, 54-4
Local
  control computer (LCC), 35-5
  loss, 17-12, A-56
Localized, 45-3
Location
  effect on cost, 55-7
  landfill, 39-5
  parameter, 11-8
  water treatment plant, 24-2
Locus of points, 7-2
LOEL, 45-7
Lofting plume, 38-17 (fig)

transformation, 5-3
triangular, 4-2
type, 4-1
unit, 4-2
variable, 4-6
zero, 4-2
Matter
mineral, 37-3
particulate, 38-4
respirable particulate, 38-4
types, particulate, 38-4 (tbl)
volatile, 37-5
Maturation pond, 26-4
Maturity date, 54-29
Maxima point, 8-2
Maximum
achievable control technology, 38-3
contaminant level, 23-6, 25-15
contaminant level goal, 23-6
flame temperature, 37-16 (ftn)
flood, probable, 20-6
fluid velocity, 17-3
heating load, 33-3
mixing depth, 38-16, 38-17 (fig)
point, 8-2
precipitation, probable, 20-6
specific growth rate, 27-7
theoretical combustion temperature,
37-15, 37-16
value, sinusoid, 52-4
velocity in pipe, 17-3
velocity, open channel, 19-26
yield coefficient, 25-6
Maxwell
-Boltzmann distribution, 29-17
relations, 30-6
MCL, 23-6, 25-15
MCLG, 23-6
Mean, 11-12, 42-9
arithmetic, 11-12
cell residence time, 27-5, 27-8
depth, hydraulic, 19-2 (ftn)
effective pressure, 53-14
fourth moment of, 11-14
free path, air, 42-9
free path, gas, 42-9
geometric, 11-12
harmonic, 11-12
residence time, 26-7
speed of a molecule, 29-17
standard error of, 11-14
third moment about, 11-14
time before failure, 11-9 (ftn)
time to failure, 11-9
velocity, open channel, 19-2
Measured actual concentration, 47-9
Measurement
areal, 1-8
board foot, 1-8
flow, 17-26, 24-3
reliable, 53-2
ton, 1-7
Mechanic's lien, 56-5
Mechanical
advantage, 15-14
noise, 50-4
seal, 42-16
similarity, 17-45
Mechanism transport, 11-22
Media
factor, 26-11
packing, 42-20
Median, 11-12
Medium
acoustic impedance, 50-6
screen, wastewater, 26-6
Megagram, 1-6
Meinzer unit, 21-2
Member, set, 11-1
Membership interest, 56-3
Membrane
reinforced, 39-4
semipermeable, 14-10
supported, 39-4

synthetic, 39-4
unreinforced, 39-4
unsupported, 39-4
Meningitis, 44-2 (tbl)
Meningoencephalitis, 44-2 (tbl)
Meniscus, 14-12
Mensuration, 7-1, A-8
of area, A-3, A-5
of volume, A-10
three-dimensional, A-10
two-dimensional, A-8
MEP, 53-14
Mer, 24-8
Mercaptan, 44-6
Mercury density, 14-3
MERV, 36-23
Mesokurtic distribution, 11-14
Mesophile, 44-1
Mesothelioma, 38-37, 46-7
Metabolic
heat, 32-8
oxygen, 32-7
Metabolism, 44-5
Metacenter, 15-17
Metacentric height, 15-17
Metal, 22-2, 22-4
alkali, 22-2
alkaline earth, 22-2
concentration, 25-15
fume fever, 46-6, 46-8
heavy, 22-2, 25-14
heavy, in wastewater, 25-14
light, 22-4
total, in wastewater, 25-14
transition, 22-4 (ftn)
Metallic properties, 22-2
Metalloid, 22-2, 22-4
Metals, properties, A-131
Metastasis, 45-9
Metathesis, 22-7 (ftn)
Meteorology, 38-14
Meter
constant, 17-29
current, 17-27
displacement, 17-26
noise, 50-6
normal cubic, 37-2
obstruction, 17-26
orifice, 17-30 (fig)
pressure, 14-2 (ftn)
sound, 50-6
torque, 53-13
turbine, 17-27
variable-area, 17-27
venturi, 17-29, 17-30 (fig)
Metering pump, 18-3
Methane
gas, 27-16
heating value, 27-17
landfill, 39-6, 40-11
properties, 39-6 (tbl)
series, 37-2
sludge, 27-16
Methanol, 37-7
Methemoglobinemia, 23-8
Method (see also type)
air change, 32-2
annual cost, 54-15
annual return, 54-15, 54-16
bisection, 3-4, 12-1
combination, 36-26
conservation, 33-7
cooling load temperature difference, 34-3
crack length, 32-5
double and add, 53-14 (ftn)
double-dabble, 53-14 (ftn)
equal-friction, 36-24
equivalent length, duct, 36-18
evaporative pan, 32-7
expansion, 53-14
false position, 12-2 (ftn)
Hardy Cross, 17-24
integration, 2-1, 2-4
isohyetal, 20-2

lagging storm, 20-13
loss, 37-17
loss coefficient, 36-17
lumen, 50-3
M-factor, 33-5
Newton's, 12-2
nonsequential drought, 20-20
normal-ratio, 20-3
numerical, 12-1
numerical, polynomial, 3-4
of discs, volume, 9-6
of equivalent lengths, 17-12
of least squares, 11-17
of loss coefficients, 17-12
of shells, volume, 9-6
of undetermined coefficients, 3-6, 10-4
parallelogram, 5-3
polygon, 5-3
projection, 54-10
rational, modified, 20-20
rational, peak runoff, 20-14
regula falsi, 12-2 (ftn)
remainder, 53-14
S-curve, 20-13
secant, 12-2 (ftn)
short-term exposure, 47-11
static regain, 36-27
storage indication, 20-21
straight line, 54-21
Thiessen, 20-2
total flux, 50-3
total pressure, 36-29
transfer function, 34-3
velocity-reduction, 36-24
water spray, 32-7
Winfrey, 54-16 (ftn)
Methyl
alcohol, 37-7
orange alkalinity, 22-21
Metolachlor, 40-18
Metric
system, 1-4
ton, 1-7
mho, 52-2 (ftn)
Microbe, 44-1
growth, 25-6
Micronized coal, 37-5 (ftn)
Microorganism, 44-1
hazard, 46-23
Microscopic cross section, 49-7
neutron, A-153
Microstrain, 53-9
Microstrainer, 24-3, 24-15
Midget impinger, 32-10 (ftn)
Migration
leachate, 39-7, 40-12
velocity, 42-9
Mil, 22-19, A-3, A-4
Milliequivalent per liter chart, 23-1
Milligram
equivalent weight, 22-10, 22-11
per liter, 22-10, 22-11, 22-23 (ftn)
Millilambert, 50-3 (ftn)
Million
of particles per cubic foot, 32-10 (ftn)
part per, 22-10, 22-11
Mine tunnel duct, 36-32
Mineral matter, 37-3
Minima point, 8-2
Minimum
attractive rate of return, 54-14, 54-16
control velocity for exposure hood,
32-9 (tbl)
efficiency reporting value, 36-23
point, 8-2
velocity, open channel, 19-2
Mining duct, 36-32
Minneapolis leakage ratio, 34-4
Minor
axis, 7-11
entrance loss coefficient, 19-28 (tbl)
loss, 17-12, A-56
of entry, 4-2
plant, 25-3

I-30

ENVIRONMENTAL ENGINEERING REFERENCE MANUAL

INDEX - N

Mirex, 40-17
Miscellaneous formulas, angles, 6-4
Misrepresentation, 56-6
Mission time, 11-9
Mist, 38-4, 46-6
Mitigation
    compensatory, 28-11
    wetlands, 28-11
Mixed
    bed unit, naked, 22-24
    -flow pump, 18-4 (ftn)
    -flow reaction turbine, 18-23
    liquor, 27-2
    liquor suspended solids, 27-4
    liquor volatile suspended solids, 27-4
    triple product, 5-4, 5-5
Mixer, 24-10
    flash, 24-10
    paddle, 24-10
    quick, 24-10
    rapid, 24-10
Mixing
    complete, 24-10
    damper, 36-13
    depth, maximum, 38-16, 38-17 (fig)
    gases and vapors, 47-7
    model, complete, 24-10
    model, plug flow, 24-10
    opportunity parameter, 24-11
    plenum, 36-13
    plug flow, 24-10
    problem, 10-8
    rate constant, 24-10
    Reynolds number, 24-11
    thermostat, 35-2
    two air streams, 31-6
    velocity, 24-11
    zone, 38-16
Mixture
    gas, 29-2, 29-18
    liquid-vapor, 29-2
    of liquid-vapor, properties, 29-13
    rule, 40-3
    vapor/gas, 29-2
mks system, 1-4, 1-5
MLSS, 27-4
MLVSS, 27-4
MLθT system, 1-7
MMSCFD, 36-2
Mobile pollution source, 38-34
Mode, 11-12
    free-flow, 19-14
    submerged, 19-14
Model, 17-45
    Bingham-plastic, 17-12
    Calvert, venturi scrubber, 42-18
    distorted, 17-45
    filter drag, 42-5
    power-law, 17-12
    scale, 17-45
Modeling
    stormwater, 20-23
    watershed, 20-23
Modified rational method, 20-20
Modulation
    flow rate, 36-7
    pulse duration, 36-8
    width-pulse, 36-8
Module, digital-to-
    analog staging, 35-5
    proportional staging, 35-5
Modulus, 3-7, 3-8
    bulk, 14-14, 50-5
    bulk, water, 14-14
    of elasticity, steel, 17-39
    point bulk, 14-14
    secant bulk, 14-14
    section, 2-7
    tangent bulk, 14-14
Mohr's circle, 2-8
Moiety, 43-1
Moist air, 31-2
    enthalpy of, 31-3

Moisture
    condition, antecedent, 20-17 (ftn)
    content, 29-9
    content, constant, 29-10
    content, soil, 21-2
    level, bed, 37-3
    management, mold, 32-8
mol, 22-5
Molality, 22-10
Molar
    enthalpy, 29-4
    internal energy, 29-5
    property, 29-4
    specific heat, 13-4, 29-7
    specific volume, 29-5
    volume, 22-6, 38-8
Molarity, 22-10
Mold, humidity, 32-8
Mole, 22-5, 47-5
    fraction, 14-6, 22-10, 22-11, 29-18,
        37-2, 51-4
    percent, 14-6
    -weighted average, 38-8
Molecular weight, 22-6
    average, gas, 38-8
Molecule, 22-1
    hydrated, 22-6
    mean speed, 29-17
    most probable speed, 29-17
    spacing, 14-2
Mollier diagram
    steam, A-99, A-120
    using, 29-10
Moment
    first area, 2-2
    fourth standardized, 11-14
    hydrostatic, 15-11
    of a function, first, 9-6
    of a function, second, 9-6
    of inertia, 2-3, A-7
    of inertia, area, 2-3, A-7
    of inertia, polar, 2-6
    of inertia, principal, 2-8
    overturning, dam, 15-11
    righting, 15-16
    second area, 2-4
    statical, 2-3
Momentum
    angular, 17-33
    fluid, 17-33
    flux, 17-14 (ftn)
    linear, 17-33
Money time value, 54-5
Monitor, well, 21-4, 39-9
Monitoring
    air, 47-4
    air quality, 38-11
    continuous emissions, 37-20
    emissions, 37-20
    landfill, 39-9
    project, 55-15
Monod's equation, 25-6
Monoenergetic, 49-3
Monofill sludge, 27-20
Monomial, 3-3
Monoxide, carbon, 38-6
Monte Carlo simulation, 20-22
Montreal Protocol, 40-8
Moody friction factor chart, 17-6, 17-7 (fig)
Mortality, infant, 11-9
Most
    efficient cross section, 19-9
    probable number index, 44-5
    probable speed, 29-17
    significant bit, 53-14, 53-17
    significant digit, 53-14
Motor, 35-1, 52-12, 52-15, A-163
    electrical, 52-12
    gear, 18-11
    induction, 52-14
    octane number, 37-6
    polyphase, A-162
    service factor, 18-10

size, 18-10
speed control, 52-15
synchronous, 52-15
Movable equipment cost, 55-5
Movement, Brownian, 29-17
Moving, average forecast, 54-42
MPN, 44-5
MSB, 53-14
MSCFD, 36-2
MSD, 53-14
MSDS, 46-2
MSW, 39-1
    typical nationwide characteristics,
        40-14 (tbl)
MTBF, 11-9 (ftn)
MTON, 1-7
MTTF, 11-9
Multi-media filter, 24-13
Multiloop pipe system, 17-24
Multiplate, 16-11
Multiple
    hypergeometric distribution, 11-6
    pipe containment, 42-16
    reservoir, 17-22
    stage pump, 18-4
Multiplication
    matrix, 4-4
    vector, 5-3
Multirating table, 36-8
Multizone ventilation, 32-4
Municipal
    solid waste, 39-1, 39-2, 39-10, 40-14
    solid waste landfill, 39-2, 40-14
    wastewater, 25-2
    water demand, 24-22
Mussel, 44-4
    zebra, 44-4
Mycobacteria, 44-2 (tbl)
Mycobacterium tuberculosis, 46-26
Mycotoxin, 46-24
Mycrocystis aeruginosa, 44-2 (tbl)

N

N factor, 34-5
NAAQS, 38-2, 38-3 (tbl)
Naegleria fowleri, 44-2 (tbl)
Naked mixed bed unit, 22-24
Nameplate
    motor, 52-13
    rating, 18-10
Names and formulas of chemicals, A-74
Nanofiltration, 24-22
Naphthalene series, 37-2
Napierian logarithm, 3-5
Nappe, 19-11
Nasal cancer, 45-3
National
    Ambient Air Quality Standards,
        38-2, 38-3 (tbl)
    Board of Fire Underwriters, 24-24
    Council of Examiners for Engineering
        and Surveying, 58-1
    Electrical Code, 52-8
    Institute for Occupational Safety and
        Health, 45-14, 46-2
    primary drinking water standards,
        23-5, A-80
    Research Council, 26-10
    Society of Professional Engineers,
        57-1, 57-3
Nationwide permit, 28-10
Natural
    attenuation landfill, 39-2
    draft, 37-17, 37-18
    frequency, 9-6
    frequency, panel, 50-11
    gas, 37-11
    gas, standard conditions, 37-2
    killer cells, 45-5
    logarithm, 3-5
    radiation, 48-3
    watercourse, 19-10

PPI • www.ppi2pass.com

Naturally occurring radioactive
  material, 48-3
NC, 35-1
NCRP, 46-12
Near-field measurement, 50-6
NEC, 52-8
  continuous operation, 52-8
Neck tension syndrome, 46-21
Need oxygen, 32-7
Needed fire flow, ISO, 24-23
Needle valve, 16-12
Negative
  declaration, 40-3
  element, 53-7
  exponential distribution, 11-9
  number, computer representation, 53-17
  temperature coefficient, 53-5
Negligence, 56-6
  comparative, 56-6
NEL, 45-7
NEMA
  motor size, 18-10, 52-12
Nematode, 44-4
Neoplasm, 45-9
Nephelometric
  turbidity unit, 23-8
Nephrotoxicity, 45-5
Nernst theorem, 29-6
Nerve entrapment, ulnar, 46-22
Nervous system, 45-5
Net
  air-to-cloth ratio, 42-5
  filtering area, 42-5
  filtering velocity, 42-5
  flow, 20-7, 21-7
  growth rate, 25-7
  heating value, 37-14
  inlet pressure required, 18-14 (ftn)
  net filtering velocity, 42-5
  positive inlet pressure, 18-2 (ftn)
  positive suction head, 18-14
  rain, 20-7 (ftn), 20-17
  rating, 37-17
  stack temperature, 37-14
  strains, 53-11
  work, 30-11
Netting, 38-2
  draining, 39-3
Network
  activity-on-arc, 55-12
  activity-on-branch, 55-12
  activity-on-node, 55-11
  pipe, 17-24
Neurotoxicity, 45-5
Neurotransmitter, 45-5
Neutral
  position, 35-5
  pressure, 21-9
  pressure level, 32-6
  solution, 22-11, 22-12, 23-1
  stress coefficient, 21-9
  zone, 36-4
Neutralization, 22-12, 42-23
Neutralizing amine, 22-23
Neutron, 49-2 (tbl)
  absorption, 49-6
  cadmium, 49-8 (tbl)
  classification, 49-8 (tbl)
  cross section, A-153
  epicadmium, 49-8 (tbl)
  epithermal, 49-8 (tbl)
  fast, 49-8 (tbl)
  fission, 49-6
  intermediate, 49-8 (tbl)
  particle, radiation, 46-13
  radiation, 48-6
  relativistic, 49-8 (tbl)
  resonant, 49-8 (tbl)
  shielding, 49-6
  slow, 49-8 (tbl)
  speeds, 49-8 (tbl)
  thermal, 49-8
  ultra fast, 49-8 (tbl)
  weighting factor, 48-9

New source performance standards, 38-2
Newton, 1-1, 1-5
  form, 12-3 (ftn)
Newton's
  interpolating polynomial, 12-3
  law of cooling, 10-10
  law of viscosity, 14-6, 14-7
  method, 12-2
  notation, 8-1
Newtonian
  fluid, 14-2, 14-7
  non-, viscosity, 17-12
Nikuradse equation, 17-5
NIMBY, 39-5
  syndrome, 40-14
Nines
  complement, 53-16
  four, 38-3
  six, 38-3
NIOSH, 45-14, 46-2
  IDLH values, A-147
Nit, 50-3 (ftn)
Nitrate in
  wastewater, 25-14
  water, 23-8
Nitric oxide, 40-14
Nitrification, 25-5
  /denitrification process, 26-12
Nitrile, 43-1 (tbl), 43-2 (tbl)
Nitro, 43-1 (tbl)
Nitrogen
  dioxide, 40-14
  in wastewater, 25-14
  ion, in water, 23-8
  Kjeldahl, 25-14
  organic, 23-8, 25-14
  oxide, 38-3, 38-5, 40-14
  -oxygen ratio, 37-8
  photolytic cycle, 38-10 (fig)
  total, 23-8, 25-14
Nitrogenous demand, 25-9
Nitrous oxide, 40-14 (ftn)
NO, 35-1
  -boiler incinerator, 39-10
  effect level, 45-7
  observable adverse effect level, 11-20
  observed adverse effect level, 45-7
  observed effect level, 45-7
NOAEL, 11-20, 45-7
Noble gas, 22-4
Node, 7-2, 55-10
  dummy, 55-13
NOEL, 45-7
Noise, 46-8, 46-10, 50-4
  action level, OSHA, 50-12
  background, 50-4
  criteria curve, 50-9
  dose, 50-12
  dosimeter, 46-11, 50-12
  duct, 36-31
  exposure factor, 46-10
  fan, 36-31
  impact, 50-4
  impulse, 50-4
  -induced hearing loss, 46-10
  mechanical, 50-4
  meter, 50-6
  permissible exposure, 46-10 (tbl)
  reduction, 50-11
  reduction coefficient, 50-10
  reduction rating, 46-12, 50-12
  spectrum, 50-9
  structure-borne, 50-4
  white, 50-4
Nominal
  damages, 56-7
  dimension, pipe, 16-10
  interest rate, 54-28
  system voltage, 52-5 (ftn)
Nomograph, Hazen-Williams, A-56
Non
  bypass air hood, 32-9
  enclosure air hood, 32-9

-Newtonian fluid, 14-2
-Newtonian viscosity, 17-12
-overloading fan, 36-6
-SI units, 1-7
Nonattainment area, 38-2, 40-2
  NAAQS, 32-5
Noncarbonate hardness, 22-21, 23-3, 23-4
  removal, 24-15
Noncarcinogen, 45-8
Noncatalytic reduction, selective, 42-19
Noncircular duct, 17-9
Nonclog pump, 18-5
Noncontact condenser, 38-33
Noncriteria pollutant, 40-2
Noncritical activity, 55-11
Nonhomogeneous
  differential equation, 10-1, 10-3
  linear equation, 3-7
Nonionizing radiation, 46-12, 48-1
Nonlinear
  equation, 10-1
  interpolation, 12-2, 12-3
  regression curves, 11-17
Nonmetal, 22-2, 22-4
  properties, A-131
Nonparametric equation, 3-2
Nonphotosynthetic bacteria, 44-3
Nonpotable water, 44-1
Nonquantifiable factor, 54-2
Nonquantitative factor, 54-2
Nonreacting ideal gas mixtures,
    properties of, 29-19
Nonrectangular channel, 19-18
Nonreverse-flow valve, 16-12
Nonsequential drought method, 20-20
Nonsingular matrix, 4-5
Nonstochastic effect, 48-3
Nontidal hydroperiod, 28-3 (tbl)
Nonuniform flow, 19-2, 19-15
Nonvolatile solids, 27-4
Normal
  condition, 47-5
  cubic meter, 37-2
  curve, area under standard, A-13 (tbl)
  depth, 19-6
  depth, rectangular channel, 19-6
  distribution, 11-6, 11-7
  farming, 28-9
  force, on a dam, 15-11
  form, 7-5
  hydrogen, 22-2
  line vector, 8-6
  -ratio method, 20-3
  slope, 19-3
  temperature and pressure, 37-3, 47-5
  vector, 7-6
Normality, 22-10, 22-11
Normalized leakage, 34-5
Normally
  closed (NC), 35-1
  open (NO), 35-1
Northern peatland, 28-2
Notation, Newton's, 8-1
NOx, 38-3, 40-14
  burner, low, 42-15
  burner, ultralow, 42-15
  control, 37-5
  fuel-bound, 40-14, 40-15
Nozzle
  ASME long radius, 17-32 (ftn)
  coefficient, 18-22
  converging-diverging, 17-29
  flow, 17-32
  ISA, 17-32 (ftn)
  loss, turbine, 18-22
NPDES, 26-1, 40-19
NPSE, 57-1
NPSHR, hydrocarbon, 18-21
NRC equation, 26-10
NRCS, 20-3, 20-11
  curve number, 20-16
  dimensionless unit hydrograph, 20-11
  graphical method, 20-17
  lag equation, 20-4

effect on output, 53-12
equivalent chill, 46-18
flame, 37-15
flue gas, 37-14
ignition, 37-8
inside design, 33-2, 34-3
inversion, 30-3
maximum flame, 37-16 (ftn)
maximum theoretical combustion, 37-15, 37-16
net stack, 37-14
overshoot, 35-2
reference, 53-6
rise, 13-4
saturation, 29-4
scale, absolute, 29-4, 29-5
scales, 29-4
surface, 10-10
swing, 32-7
T_{250}, 37-3
variation constant, 25-8
wet-bulb, 31-2
wet-bulb globe, 46-15
Temporary hardness, 22-21
Ten States' Standards, 25-2, 27-5, A-88, A-89
Tendency, central, 11-12
Tendonitis, 46-22
Tennis elbow, 46-21
Tenosynovitis, 46-22
Tens complement, 53-16
Tensiometer, 21-2
Tension, surface, 14-11
Tensor, 5-1
Term
    differential, 9-1
    general, 3-10
    harmonic, 9-6
    in a sequence, 3-10
    settling velocity, 38-22
Terminal
    box, variable air volume, 36-3
    box, VAV, 36-3
    point, 5-1
    pressure, 36-30
    unit, 36-3
    value, 54-16
    velocity, 17-43
    velocity, duct, 36-30
Terms, separation of, 9-3
Ternary compound, 22-5
Tertiary
    compound, 22-5
    pond, 26-4
    treatment, 26-3
Test (see also type)
    BOD, 25-9
    comparison, 3-12
    double-ring infiltration, 21-9
    financial, 54-34, 54-35
    for convergence, 3-12
    hypothesis, 11-16
    isokinetic, 37-19
    power of the, 11-15
    practicable alternatives, 28-11
    proof, 11-9
    ratio, 3-12
    sliding plate viscometer, 14-6
    three-edge bearing, 16-11
Testing, boiler feedwater, 22-23
Tetanus, 46-25
Textured synthetic cap, 39-5
Theis equation, 21-7
Theorem, 10-6
    Bayes', 11-4
    binomial, 3-3
    Buckingham pi-, 1-10
    buoyancy, 15-14
    Cauchy-Schwartz, 5-3
    central limit, 11-14
    de Moivre's, 3-8
    Gibbs, 29-19
    Kutta-Joukowsky, 17-41
    linearity, 10-6

Nernst, 29-6
of calculus, fundamental, 9-4
of Fourier, 9-6
of Pappus, 9-5
of Pappus-Guldinus, 2-3
parallel axis, 2-4
perpendicular axis, 2-6
Pythagorean, 6-2
superposition, 10-6
time-shifting, 10-6
transfer axis, 2-4
Theoretical
    horsepower, 17-15
    yield, 22-8
Theory
    kinetic gas, 29-17
    Prandtl's boundary layer, 17-44
    probability, 11-3
Therm (unit), 13-1, 33-1 (ftn)
Thermal
    ballast, 42-13
    coefficient of expansion, 14-13 (ftn)
    cracking, 37-2
    desorption, 42-22
    efficiency, 30-11, 37-17
    energy, 13-4 (ftn)
    equilibrium, 29-4, 30-5
    flywheel effect, 33-5
    incineration, 38-32
    incinerator, 38-32 (fig)
    inertia, 33-5
    lag, 33-5
    neutron, 49-8 (tbl)
    NOx, 40-14
    properties, A-131
    resistance, 33-2 (ftn)
    resistance, coefficient of, 53-6
    stress, 46-14
    unit, British, 29-5
Thermistor, 53-6
Thermocouple, 53-6
    constants, A-164, A-165
    material, 53-7
    thermoelectric constants, A-164, A-165
Thermodynamic
    basic relations, 30-6
    disc steam trap, 16-14
    state, 29-2, 29-4
    system, 30-1
Thermodynamics, 29-2
    first law of, 30-2 (ftn)
    first law of, closed system, 30-4
    first law of, open system, 30-5
    second law of, 29-7 (ftn), 30-12
    third law of, 29-6
    zeroth law of, 29-4
Thermoelectric constant, 53-7, A-164, A-165
Thermometer resistance, 53-5
Thermophile, 44-1
Thermopile, 53-7
Thermostat, 35-1
    immersion, 35-2
    mixing, 35-2
    outside air, 35-2
    setback, 33-7
    setback, conservation through, 33-7
Thermostatic steam trap, 16-14
Thickening
    batch gravity, 27-14
    dissolved air flotation, 27-14
    gravity, 27-14
    gravity belt, 27-15
    sludge, 27-14
Thickness
    -diameter ratio, 36-14 (fig)
    half-value, 46-14
    shield, 49-2, 49-3, 49-8
Thiem equation, 21-6
Thiessen method, 20-2
Third
    law of thermodynamics, 29-6
    moment about the mean, 11-14

Thixotropic
    fluid, 14-7
    liquid, 14-7
THM, 23-10, 40-21
Thompson effect, 53-7 (ftn)
Thoracic
    fraction, 40-9
    outlet syndrome, 46-22
Threaded fitting, 17-12 (ftn)
Threadworm, 44-2 (tbl)
Three
    -dimensional mensuration, A-10
    -edge bearing test, 16-11
    -phase electricity, 52-10
    -reservoir problem, 17-22
Threshold, 45-7
    limit value, 32-10, 45-12, 46-3
    limit value, ceiling, 45-12
    limit value, noise, 46-11
    odor number, 24-15
    of audibility, 50-4
    of hearing, 46-9, 50-7
    of pain, 46-9, 50-4
    quantities, 51-4
    vector, 39-5
Throat radius, 36-19
Throttling
    process, 30-2, 30-3, 30-10
    range, 35-5
    service valve, 16-12
    steady-flow system, 30-10
Throw, 36-30
    ratio, 36-30
Thymus, 45-5
Ticks, 46-25
Tidal
    freshwater marsh, 28-2
    hydroperiod, 28-3 (tbl)
    salt marsh, 28-2
Tie
    bus, 35-7
    line, 35-7
Tile
    bed, 26-2
    field, 26-2
Time (see also type)
    averaging, 47-9, 51-8
    base, 20-7
    bed residence, 42-7
    breakthrough, glove, 46-5
    constant, 3-11
    -cost trade-off, 55-9
    crash, 55-9
    decoloration, 25-13
    detention, 24-6, 26-7
    doubling, 3-11, 54-8
    e-folding, 3-11
    filtration, 42-6
    float, 55-11
    hydraulic detention, 27-8
    lag, 20-11
    mean cell residence, 27-8
    mean residence, 26-7
    mission, 11-9
    of concentration, 20-3
    retention, 24-6, 26-7
    retention, hydraulic, 27-7
    -series analysis, 9-8 (ftn)
    settling, 35-2
    -shifting theorem, 10-6
    slack, 55-11
    stabilization, 25-13
    to double, 54-8
    to empty tank, 17-19
    to peak, 20-11
    to triple, 54-8
    tripling, 54-8
    value of money, 54-5
    -weighted average, 32-10, 45-12, 46-12, 47-9, 50-12
Timed two-position control, 35-5
Tip speed, 18-4
Tipping fee, 39-2

INDEX - Z

ATOMIC NUMBERS AND WEIGHTS OF THE ELEMENTS
(referred to Carbon-12)

name	symbol	atomic number	atomic weight	name	symbol	atomic number	atomic weight
actinium	Ac	89	–	meitnerium	Mt	109	–
aluminum	Al	13	26.9815	mendelevium	Md	101	–
americium	Am	95	–	mercury	Hg	80	200.59
antimony	Sb	51	121.760	molybdenum	Mo	42	95.96
argon	Ar	18	39.948	neodymium	Nd	60	144.242
arsenic	As	33	74.9216	neon	Ne	10	20.1797
astatine	At	85	–	neptunium	Np	93	237.048
barium	Ba	56	137.327	nickel	Ni	28	58.693
berkelium	Bk	97	–	niobium	Nb	41	92.906
beryllium	Be	4	9.0122	nitrogen	N	7	14.0067
bismuth	Bi	83	208.980	nobelium	No	102	–
bohrium	Bh	107	–	osmium	Os	76	190.23
boron	B	5	10.811	oxygen	O	8	15.9994
bromine	Br	35	79.904	palladium	Pd	46	106.42
cadmium	Cd	48	112.411	phosphorus	P	15	30.9738
calcium	Ca	20	40.078	platinum	Pt	78	195.084
californium	Cf	98	–	plutonium	Pu	94	–
carbon	C	6	12.0107	polonium	Po	84	–
cerium	Ce	58	140.116	potassium	K	19	39.0983
cesium	Cs	55	132.9054	praseodymium	Pr	59	140.9077
chlorine	Cl	17	35.453	promethium	Pm	61	–
chromium	Cr	24	51.996	protactinium	Pa	91	231.0359
cobalt	Co	27	58.9332	radium	Ra	88	–
copernicium	Cn	112	–	radon	Rn	86	226.025
copper	Cu	29	63.546	rhenium	Re	75	186.207
curium	Cm	96	–	rhodium	Rh	45	102.9055
darmstadtium	Ds	110	–	roentgenium	Rg	111	–
dubnium	Db	105	–	rubidium	Rb	37	85.4678
dysprosium	Dy	66	162.50	ruthenium	Ru	44	101.07
einsteinium	Es	99	–	rutherfordium	Rf	104	–
erbium	Er	68	167.259	samarium	Sm	62	150.36
europium	Eu	63	151.964	scandium	Sc	21	44.956
fermium	Fm	100	–	seaborgium	Sg	106	–
fluorine	F	9	18.9984	selenium	Se	34	78.96
francium	Fr	87	–	silicon	Si	14	28.0855
gadolinium	Gd	64	157.25	silver	Ag	47	107.868
gallium	Ga	31	69.723	sodium	Na	11	22.9898
germanium	Ge	32	72.64	strontium	Sr	38	87.62
gold	Au	79	196.9666	sulfur	S	16	32.065
hafnium	Hf	72	178.49	tantalum	Ta	73	180.94788
hassium	Hs	108	–	technetium	Tc	43	–
helium	He	2	4.0026	tellurium	Te	52	127.60
holmium	Ho	67	164.930	terbium	Tb	65	158.925
hydrogen	H	1	1.00794	thallium	Tl	81	204.383
indium	In	49	114.818	thorium	Th	90	232.038
iodine	I	53	126.90447	thulium	Tm	69	168.934
iridium	Ir	77	192.217	tin	Sn	50	118.710
iron	Fe	26	55.845	titanium	Ti	22	47.867
krypton	Kr	36	83.798	tungsten	W	74	183.84
lanthanum	La	57	138.9055	uranium	U	92	238.0289
lawrencium	Lr	103	–	vanadium	V	23	50.942
lead	Pb	82	207.2	xenon	Xe	54	131.293
lithium	Li	3	6.941	ytterbium	Yb	70	173.054
lutetium	Lu	71	174.9668	yttrium	Y	39	88.906
magnesium	Mg	12	24.305	zinc	Zn	30	65.38
manganese	Mn	25	54.9380	zirconium	Zr	40	91.224